The Birds of Africa
Volume IV

The BIRDS of AFRICA
Volume IV

Edited by

STUART KEITH
*Department of Ornithology, American Museum of Natural History,
New York, New York, USA*

EMIL K. URBAN
*Department of Biology, Augusta College,
Augusta, Georgia, USA*

C. HILARY FRY
*Department of Biology, Sultan Qaboos University,
Muscat, Sultanate of Oman*

Colour Plates by Martin Woodcock

Line Drawings by Ian Willis
Acoustic References by Claude Chappuis

ACADEMIC PRESS
Harcourt Brace Jovanovich, Publishers
London · San Diego · New York · Boston · Sydney · Tokyo · Toronto

ACADEMIC PRESS LIMITED
24/28 Oval Road, London NW1 7DX

United States Edition published by
ACADEMIC PRESS, INC.
San Diego, CA 92101

Copyright © 1992 by
ACADEMIC PRESS LIMITED

All rights reserved. No part of this book may be reproduced or transmitted in any form or by any means, electronic or mechanical, including photocopy, recording, or any information retrieval system without permission in writing from the publisher.

British Library Cataloguing in Publication Data
The Birds of Africa.
    Vol. IV, Broadbills to chats.
    1. Africa. Birds
    I. Keith, Stuart, *1931*–   II. Urban, Emil K., *1934*–
    III. Fry, C. Hilary, *1937*–

ISBN 0–12–137304–5

Editorial and production services by Fisher Duncan Ltd, 10 Barley Mow Passage, London W4 4PH

Colour plates printed in the UK by George Over Ltd, Rugby
Printed and bound in the UK by Mackays of Chatham, PLC, Chatham, Kent

ACKNOWLEDGEMENTS

Once again we wish to express our sincere appreciation of the authors. Researching their subjects in field, museum and library, and writing species accounts to exacting standards, are extremely time-consuming. Authors' drafts have been subjected to the seemingly endless process of revision by colleagues, referees and editors. Our authors have borne all this with patience and goodwill for which, and for their diligence, we warmly thank them.

During the preparation of species accounts authors have drawn heavily upon the special knowledge of numerous colleagues, who have given them access to unpublished and pre-published works and field notes, provided measurements and weights, given information on specimens, distribution and habits, copied library materials, and responded to a host of queries. Many colleagues have also given of their valuable time to referee the accounts. For all their help we take great pleasure in thanking: Passeriformes: J. Cracraft, S. L. Olson and R. J. Raikow; Eurylaimidae, A. Brosset, T. Butynski, C. Erard, J. Kalina and E. F. G. Smith; Pittidae, P. R. Fogarty, J. Jones and A. N. B. Masterson; Alaudidae, D. G. Allan, J. P. Angle, J. S. Ash, D. R. Aspinwall, H. Boyer, R. K. Brooke, T. Cassidy, C. Chappuis, M. Charron, P. A. Clancey, C. Clinning, J. F. R. Colebrook-Robjent, P. R. Colston, R. J. Dowsett, F. Dowsett-Lemaire, E. K. Dunn, R. A. Earlé, J. H. Elgood, L. Freysen, J. D. Gerhart, S. M. Goodman, A. Harris, D. L. Harrison, J. A. Harrison, P. A. R. Hockey, P. Hogg, K. Hustler, M. P. S. Irwin, H. D. Jackson, A. C. Kemp, J. Komen, R. Liversidge, I. A. W. Macdonald, G. L. Maclean, A. P. Martin, R. Martin, G. R. McLachlan, S. J. Milton, J. E. Miskell, J. M. Mendelsohn, D. C. Moyer, G. Nikolaus, S. L. Olson, P. Perry, D. Rockingham-Gill, D. Schultz, W. R. Siegfried, J. C. Sinclair, R. Stjernstedt, W. R. Tarboton, C. J. Vernon and D. A. Zimmerman; Hirundinidae, P. D. Alexander-Marrack, D. G. Allan, J. S. Ash, D. R. Aspinwall, P. Becker, L. Birch, A. Brosset, T. Butynski, M. Carswell, P. Christy, P. R. Colston, H. Q. P. Crick, R. J. Dowsett, M. Dyer, R. A. Earlé, J. H. Elgood, C. Erard, J. Farrand Jr, K. Gamble, R. M. Glen, S. M. Goodman, L. Grant, L. G. Grimes, D. N. Johnson, L. B. Kiff, B. Lamarche, A. D. Lewis, M. Louette, R. D. Medland, J. Millward, H. Morand, G. J. Morel, D. C. Moyer, V. Parker, D. J. Pearson, E. H. Penry, R. Penry, D. E. Pomeroy, the late A. Prigogine, J. Schmidt, M. L. Schmidt, W. Serle, N. J. Skinner, S. N. Stuart, W. R. Tarboton, A. K. Turner, S. J. Tyler, L. L. Urban, D. M. Ward, R. T. Wilson and R. Zusi; Motacillidae, J. S. Ash, D. R. Aspinwall, G. C. Backhurst, P. R. Colston, R. J. Dowsett, F. Dowsett-Lemaire, A. J. Harris, M. P. S. Irwin, M. G. Kelsey, G. L. Maclean, R. A. M. McVicker, M. E. D. Nhlane, T. B. Oatley, S. E. Piper, D. K. Read, J. C. Sinclair, M. Thévenot, S. J. Tyler, A. Vale and B. Wood; Campephagidae, J. S. Ash, D. R. Aspinwall, A. Brosset, C. Chappuis, P. R. Colston, W. Gatter, M. Louette, the late A. Prigogine, N. J. Skinner and J.-P. Vande weghe; Pycnonotidae, especially F. Dowsett-Lemaire (referee) and C. Chappuis, also D. Amadon, D. R. Aspinwall, P. Becker, W. P. Cane, P. A. Clancey, P. R. Colston, G. Cowles, T. M. Crowe, W. R. J. Dean, R. W. Dickerman, R. J. Dowsett, W. Gatter, L. B. Kiff, M. LeCroy, R. Liversidge, B. L. Monroe Jr, T. B. Oatley, S. L. Olson, D. J. Pearson, J. Plunkett, the late R. Plunkett, the late A. Prigogine, L. L. Short, C. G. Sibley, S. N. Stuart, P. B. Taylor, F. Vuilleumier, M. Walters, M. W. Woodcock and D. A. Zimmerman; Cinclidae, S. J. Ormerod and S. J. Tyler; Troglodytidae, M. Thévenot; Prunellidae, N. J. Davies; Turdidae, J. S. Ash, G. C. Backhurst, P. Becker, A. Brosset, T. Butynski, T. Cassidy, C. Chappuis, W. R. J. Dean, P. Devillers, F. Dowsett-Lemaire, E. K. Dunn, A. D. Forbes-Watson, K. Gamble, P. Giraudoux, A. Harris, K. Joysey, J. Kalina, M. Louette, D. Merrie, J. E. Miskell, G. J. Morel, G. Nikolaus, T. B. Oatley, D. J. Pearson, P. Steyn, W. R. Tarboton, P. B. Taylor, K. Thangavelu, Y. Thonnerieux, M. A. Traylor, D. A. Turner, E. O. Willis, M. W. Woodcock and D. A. Zimmerman.

During the preparation of their accounts authors have naturally consulted all the available literature, but we wish to single out for special mention the two major works on which they have drawn most heavily: Cramp (1988) for North Africa and Maclean (1985) for southern Africa. Special thanks are due to Dr Linda Birch, Alexander Librarian at the Edward Grey Institute, and to L. R. Macaulay, for extensive help with the literature. Authors also owe a special debt of gratitude to C. Chappuis and F. Dowsett-Lemaire for providing copies of tape recordings, to M. P. S. Irwin for providing copies of his extensive bibliography, to K. Garrett and M. C. Wimer for providing detailed printouts of specimens in the Los Angeles County Museum of Natural History, to L. B. Kiff for providing lists of specimens in the Western Foundation of Vertebrate Zoology, and to EURING for providing records of ringing recoveries. Stuart Keith also wishes to give special thanks to P. B. and C. Taylor for their hospitality and assistance during an extended visit to the Natural History Museum at Tring, to R. A. Ranney for extended hospitality and in particular for providing space in her house for the final co-ordinating of this volume, and to A. D. Forbes-Watson and D. A. Turner for their help in the field in Kenya.

Again we are greatly indebted to the many museums holding African avian collections. For study facilities and specimen loans we thank the Trustees and staff of: The American Museum of Natural History, Department of Ornithology; Augusta College Reese Library (inter-library loan service); Natural History Museum (London), Department of Ornithology (Tring); Cambridge University Museum of Zoology; Carnegie Museum of Natural History; Durban Museum; East London Museum; Field Museum of Natural History, Bird Division; Los Angeles County Museum, Bird Division; Merseyside County Museum, Liverpool; Musée Royal de l'Afrique Centrale,

Tervuren; Museum of Comparative Zoology, Harvard; Museum Alexander Koenig, Bonn; National Museum of Zimbabwe, Bulawayo; National Museums of Kenya, Nairobi; Percy Fitzpatrick Institute of African Ornithology; Philadelphia Academy of Natural Sciences; Smithsonian Institution, US National Museum of Natural History, Department of Ornithology; South African Museum; State Museum (Namibia); and Transvaal Museum, Bird Department. For copious assistance with literature we thank especially the Edward Grey Institute, Oxford University. For the loan of bird sound recordings we are grateful to the British Library of Wildlife Sounds, Cornell University Laboratory of Ornithology, and the Transvaal Museum Bird Department. We owe particular thanks to the following staff members of these institutions: A. Andors, G. Barrowclough, C. Blake, P. J. K. Burton, P. A. Clancey, P. Colston, G. Cowles, T. M. Crowe, J. Fitzpatrick, K. Garrett, F. B. Gill, J. Gulledge, A. Harris, M. P. S. Irwin, H. Adan Isack, H. D. Jackson, A. C. Kemp, R. Kettle, A. Knox, M. LeCroy, P. Lorber, M. Louette, J. Mendelsohn, P. Morgan, J. P. Myers, S. L. Olson, K. C. Parkes, R. A. Paynter, D. Read, H. Schifter, the late R. W. Schreiber, K.-L. Schuchmann, L. L. Short, W. R. Siegfried, J. C. Sinclair, D. Steadman, M. A. Traylor, F. Vuilleumier, M. Walters and D. S. Wood. We would like to single out for special mention P. R. Colston of the Natural History Museum at Tring, who has willingly supplied authors with information on numerous occasions.

We acknowledge with thanks the following sources of material for the preparation of black and white line drawings: R. M. Bloomfield in 'Complete Book of Southern African Birds' (1989) (*Oenanthe pileata*); G. J. Broekhuysen in *Bokmakierie* (1961) (*Hirundo rustica*); M. Brown (*Ammomanes cincturus* and *Cercomela sordida*); J. P. Busby in 'Birds of the Western Palearctic' (1988) (*Alaemon alaudipes, Galerida cristata*); J. P. Chapin in 'Birds of the Belgian Congo' (1953) (*Smithornis rufolateralis*); R. A. Cheke in *Malimbus* (1982) (*Hirundo preussi*); W. R. J. Dean (*Pinarornis plumosus*); A. R. Devez in Brosset and Erard (1986) (*Smithornis rufolateralis*), in *Biol. Gabonica* (1971) (*Andropadus virens, Phyllastrephus icterinus, Criniger calurus, Chlorocichla simplex*), and in *Terre Vie* (1974) (*Bleda eximia, Criniger barbatus*); M. Dyer in *Nyala* (1988) (*Hirundo atrocaerulea*); R. A. Earlé (*Hirundo cucullata*) and in *Navors. Mus. Bloemfontein* (1985) (*Hirundo spilodera*); M. D. England in 'Birds of the Western Palearctic' (1988) (*Cercotrichas galactotes*); H. Eriksen and J. Eriksen (loan of photo library); P. R. Fogarty (*Pitta angolensis*); D. Foster (*Alaemon alaudipes*); C. H. Fry (*Ammomanes spp., Ramphocoris clotbey, Hirundo rustica, H. lucida, Neocossyphus* spp., *Cercomela* spp.); P. J. Ginn in 'Birds of Botswana' (1979) (*Anthus novaeseelandiae*); E. Hosking in 'Birds of the World' (IPC) 6, Part 5 (*Eremophila bilopha*); R. A. Jensen in 'Birdlife of Southern Africa' (1979) (*Smithornis capensis*) and in 'Complete Book of Southern African Birds' (1989) (*Namibornis herero*); J. Jones (*Pitta angolensis*); G. Langsbury in 'Complete Book of Southern African Birds' (1989) (*Tmetothylacus tenellus*); C. Laubscher in 'Complete Book of Southern African Birds' (1989) (*Coracina pectoralis*); R. Liversidge, unpublished PhD thesis (1970) (*Pycnonotus capensis*); A. N. B. Masterson (*Pitta angolensis*); W. G. McIlleron in 'Complete Book of Southern African Birds' (1989) (*Cercotrichas leucophrys*); H. Morand (*Pseudochelidon eurystomina*); 'Newman's Birds of Southern Africa' (1983) (*Cercomela* spp.); D. J. Pearson (*Luscinia* spp., *Phoenicurus* spp.); C. Rose in 'Birds of the Western Palearctic' (1988) (*Cercomela melanura*); J. C. Sinclair: 'Field Guide to Birds of Southern Africa' (1984) (*Oenanthe bifasciata, Cercomela sinuata*); P. Steyn in *Ostrich* (*Pinarornis plumosus*), in 'Birdlife in Southern Africa' (1979) (*Macronyx capensis*) and in 'Complete Book of Southern African Birds' (1989) (*Smithornis capensis*); W. R. Tarboton in Sinclair: 'Field Guide to Birds of Southern Africa' (1984) (*Oenanthe bifasciata*); P. B. Taylor (*Macronyx* spp.); J. Trotignon/Jacana, in 'Birds of the World' (IPC) 6, Part 5 (*Chersophilus duponti*); C. J. Uys in 'Halcyon Days' (1963) (*Galerida magnirostris, Pycnonotus capensis*); J. L. Viljoen in 'Complete Book of Southern African Birds' (1989) (*Campephaga flava*); and D. Waters (*Hypocolius ampelinus, Cercotrichas podobe*).

All colour plates in this volume are the work of Martin Woodcock, also the monochrome paintings illustrating plumage topography and bill shapes and details of head feathering and ornamentation in the Introduction. All black and white line illustrations except as noted below were drawn by Ian Willis. Many drawings are based on sketches he made in the field. Like the authors, the artists have accommodated all referees' and editors' demands, and we are particularly grateful to them for their skill and ready co-operation. The following line drawings in the text are by Martin Woodcock: *Ixonotus guttatus*, and tail patterns of some cuckoo-shrikes *Campephaga*.

Martin Woodcock has made further extensive studies in the field in Africa, and these have been invaluable during the preparation of the colour plates. He is grateful to all those friends and correspondents who have kindly given advice, criticism, help and photographs; in particular to the Editors and authors, and to his wife Barbara for her support both at home and in the field. Many people in Africa have generously offered hospitality and support in many ways, especially A. Boswell, N. and E. Baker, W. and N. Cooper, M. and E. Coverdale, Mrs J. Hartley, K. Howell, and D. and J. Moyer. He is especially indebted to the Trustees and staff of the Natural History Museum, Department of Ornithology at Tring for the loan of specimens and for their help. Ian Willis likewise wishes to thank R. McGowan of the Royal Museums of Scotland for his help in providing specimens.

Again it is our pleasure to thank C. Chappuis for preparing the acoustic references, Lois Urban for indexing the book, our wives Sallyann Keith, Lois Urban and Kathie Fry for their ever-present support and encouragement, and lastly Andrew Richford, a Senior Editor at Academic Press Ltd, who continues to oversee the project with his customary skill, enthusiasm and tact.

May 1992

Stuart Keith
Emil K. Urban
C. Hilary Fry

CONTENTS

ACKNOWLEDGEMENTS v

LIST OF PLATES ix

INTRODUCTION x

ORDER PASSERIFORMES 1

    Eurylaimidae, broadbills (C. H. Fry) 1

    Pittidae, pittas (C. H. Fry) 9

    Alaudidae, larks (W. R. J. Dean, C. H. Fry, S. Keith and P. Lack) 13

    Hirundinidae, swallows and martins (C. H. Fry, E. K. Urban and S. Keith) 125

    Motacillidae, wagtails, pipits and longclaws (C. H. Fry, D. J. Pearson and P. B. Taylor) .. 197

    Campephagidae, cuckoo-shrikes (D. J. Pearson and S. Keith) 263

    Pycnonotidae, bulbuls (S. Keith) 279

    Bombycillidae, waxwings, Grey Hypocolius and silky flycatchers (C. H. Fry) 377

    Cinclidae, dippers (C. H. Fry) 380

    Troglodytidae, wrens (C. H. Fry) 382

    Prunellidae, accentors (C. H. Fry) 384

    Turdidae, thrushes (W. R. J. Dean, C. Erard, C. H. Fry, V. Haas, R. A. C. Jensen, S. Keith, P. Lack, T. B. Oatley, D. J. Pearson, S. N. Stuart and A. Tye) 387

BIBLIOGRAPHY

    1. General and Regional References 557

    2. References for Each Family 561

    3. Acoustic References 587

ERRATA, VOLUMES I, II AND III 590

INDEXES

    1. Scientific Names 591

    2. English Names 604

    3. French Names 608

Authors' Contributions to Shared Families

ALAUDIDAE

Dean and Keith: *Mirafra* (except *M. pulpa*, *M. hypermetra*, *M. ashi*, *M. degodiensis* and *M. poecilosterna*), *Heteromirafra ruddi* and *H. archeri*, *Certhilauda*, *Pinarocorys*, *Chersomanes*, *Ammomanes deserti* and *A. grayi*, *Calandrella cinerea*, *Spizocorys*, *Eremalauda*, *Pseudalaemon*, *Galerida modesta* and *G. magnirostris*, *Eremopterix australis*, *E. leucotis* and *E. verticalis*.
Fry: *Mirafra ashi* and *M. degodiensis*, *Heteromirafra sidamoensis*, *Alaemon*, *Rhamphocoris*, *Melanocorypha*, *Calandrella brachydactyla* and *C. rufescens*, *Chersophilus*, *Galerida cristata* and *G. malabarica*, *Lullula*, *Alauda* and *Eremophila*.
Keith: *Calandrella somalica*.
Lack: *Mirafra pulpa*, *M. hypermetra* and *M. poecilosterna*, *Eremopterix nigriceps*, *E. signata* and *E. leucopareia*.

HIRUNDINIDAE

Fry: *Pseudochelidon*, *Phedina*, *Riparia*, *Pseudhirundo*, *Hirundo daurica*, *H. fuliginosa*, *H. preussi*, *H. rufigula*, *H. spilodera*, *H. perdita*, *H. fuligula*, *H. rupestris*, *H. rustica* and *Delichon*.
Urban and Keith: *Psalidoprocne*, *Hirundo semirufa*, *H. senegalensis*, *H. abyssinica*, *H. cucullata*, *H. atrocaerulea*, *H. nigrorufa*, *H. smithii*, *H. nigrita*, *H. leucosoma*, *H. megaensis*, *H. dimidiata*, *H. aethiopica*, *H. albigularis*, *H. angolensis* and *H. lucida*.

MOTACILLIDAE

Fry: *Motacilla*.
Pearson: *Tmetothylacus*, *Anthus*.
Taylor: *Macronyx*.

TURDIDAE

Dean: *Cichladusa*.
Erard: *Neocossyphus*.
Fry: *Erithacus*, *Cossypha albicapilla*, *Pinarornis*, *Cercotrichas galactotes* and *C. podobe*, *Cercomela*, *Myrmecocichla cinnamomeiventris* and *M. semirufa*.
Keith: *Alethe*.
Keith and Jensen: *Namibornis*.
Lack: *Myrmecocichla* (except *M. cinnamomeivenrtis* and *M. semirufa*).
Lack and Haas: *Myrmecocichla aethiops*.
Oatley, Fry, Keith and Tye: *Pogonocichla*, *Swynnertonia*, *Stiphrornis*, *Sheppardia* (except *S. montana*, *S. lowei*), *Cossyphicula*, *Cossypha* (except *C. isabellae*, *C. archeri*, *C. anomala* and *C. albicapilla*), *Xenocopsychus*, *Cercotrichas* (except *C. galactotes* and *C. podobe*).
Pearson: *Luscinia*, *Irania*, *Phoenicurus*, *Saxicola torquata* and *S. rubetra*.
Stuart: *Sheppardia montana* and *S. lowei*, *Cossypha isabellae*, *C. archeri* and *C. anomala*, *Modulatrix*, *Arcanator*.
Tye: *Saxicola bifasciata*, *Oenanthe*.

LIST OF PLATES

Plate		Facing Page
1	Broadbills and pittas	32
2	Larks (northern Africa)	33
3	Larks (West, central and East Africa)	48
4	Larks (West, central and East Africa)	49
5	Larks (southern Africa)	96
6	Larks (remainder)	97
7	Swallows, martins	112
8	Swallows	113
9	Swallows, saw-wings	160
10	Swallows, martins	161
11	Wagtails	176
12	Pipits, Sharpe's Longclaw	177
13	Pipits	240
14	Longclaws	241
15	Cuckoo-Shrikes	256
16	Waxwing, Hypocolius, Dipper, Wren, accentors, *Pycnonotus* bulbuls	257
17	Greenbuls (*Andropadus*)	304
18	Greenbuls (*Andropadus, Phyllastrephus*)	305
19	Greenbuls (*Phyllastrephus*)	320
20	Greenbuls (*Chlorocichla, Criniger, Bleda*)	321
21	Greenbuls (remainder)	352
22	Thrushes (mainly *Sheppardia*)	353
23	Thrushes (robins, *Alethe*)	368
24	Thrushes (robin-chats)	369
25	Thrushes (robin-chats and others)	400
26	Thrushes (juveniles)	401
27	Thrushes (scrub-robins and others)	416
28	Thrushes (*Phoenicurus, Saxicola*)	417
29	Wheatears	448
30	Wheatears	449
31	Chats (*Cercomela*)	464
32	Chats (*Myrmecocichla*)	465

INTRODUCTION

Some time ago we decided to expand the number of passerine volumes to four. Originally we had thought that the larger number of passerine than non-passerine species (roughly 1200:800) would be offset by the greater amount of knowledge about the larger birds. It turns out that more is known about African passerines than we had supposed. At the same time the work has been gradually expanding in scope with each volume, and species accounts have been getting longer. Rather than retain the abridged style of earlier volumes, we thought it preferable to include as much information as possible while keeping the text readable; to this end we have increased the average length of species accounts from 750 to 1000 words. The Editors wish to thank Academic Press for agreeing to again increase the size of the work (it was originally planned for just four volumes). Each of the four passerine volumes will contain roughly 300 species. This volume begins with the two suboscine families, Eurylaimidae and Pittidae, and ends with the Turdidae; the last family is so large (125 species in Africa) that four genera (*Monticola*, *Zoothera*, *Psophocichla* and *Turdus*) have been taken over to Volume V.

The literature has been reviewed up to August 1991.

The system of rotating editorship, explained in the introduction to Volume III, has been continued. The Editor whose name appears first is, however, rather more than the '*primus inter pares*' described there. To him falls the job of co-ordinating the entire work, tying up all loose ends, and also preparing much of the front and back matter. Incidentally, Hilary Fry should have been credited for accomplishing that thankless task for Volume III.

Style and layout are the same as in the previous two volumes, with the following exceptions: at the beginning of the Voice section, after 'Tape-recorded', individual recordists are now credited with letters rather than numbers, usually the first 3 or 4 letters of their name. In this way we can give a little more credit to those who perform the exacting task of tape-recording birds in the field, rather than designating them by a number. We have also abandoned the arbitrary limit of six entries after 'Tape-recorded'; while continuing to weed out recordings of lesser quality or value, we thought it worthwhile to cite a larger number in cases where a great many quality recordings exist, e.g. White-browed Scrub-Robin *Cercotrichas leucophrys*. The more recordings cited, the more likely the reader is to find one readily available. Further, in the case of African-Palearctic species, a distinction has been made between recordings made in Africa and those made in Europe. When both are available, the recordings made in Africa come first, separated by a dash (–); e.g. under Woodlark *Lullula arborea* the entries are CHA – 62, 73, B, showing that all recordings except those of Claude Chappuis were made in Europe.

The Range and Status sections are becoming increasingly extensive with each volume. This is partly due to the increased word allowance, but more importantly, it reflects greater knowledge of bird distribution, thanks to the wealth of atlases and country and regional checklists that have been published. One has only to compare the maps in this volume with those of Hall and Moreau (1970) to see what has been accomplished in the last 20 years. Several more atlases and checklists are in an advanced stage of preparation. These works are naturally based on published observations and local checklists, but they also draw heavily on unpublished observations of individuals. We take this opportunity to thank all those individuals who, by publishing lists and records and contributing to atlas projects, have made this advance in knowledge possible.

Readers are reminded that in a species account, a reference which does not appear in the family bibliography should be looked for under 'General and Regional References' at the beginning of the bibliographic section. Further, the family bibliography lists significant works consulted during the preparation of the text; not all of these are necessarily cited in the text.

'The Birds of Africa' is not primarily a taxonomic work, but our classification is naturally as up to date as possible. Authors make their own revisions, and frequently have to choose between alternative treatments, especially at the subspecies level. There is not enough space for a full discussion of the reasons for their decisions, but we have added an optional paragraph 'TAXONOMIC NOTE' at the end of the Description section where they can present a brief summary.

Systematic Treatment

Since the publication of Volume III two landmark publications on avian systematics have appeared: Sibley and Ahlquist (1990) and Sibley and Monroe (1990). They will undoubtedly have a profound influence on avian systematics; however, their classifications and even their methodology are still being questioned, see the review by Gill and Sheldon in *Science*, Vol. 252, 1990, pp. 1003–1005. The dust has not yet settled, and until it does we continue to follow the traditional classification of Campbell and Lack (1985) at higher levels. Among the classifications of Sibley and Monroe (1990) that we do not accept here are the treatment of bulbuls Pycnonotidae as a family of warblers in the Superfamily Sylvioidea, and the removal of wagtails, pipits and longclaws to the Family Passeridae (sparrows and allies), as the Subfamily Motacillinae.

We continue to apply the superspecies concept, although its usefulness is somewhat marred by the differing interpretations of it. For an overview of the concept, see the introduction to Volume II. The superspecies represents one further step in the process of differentiation from the megasubspecies of Amadon and Short (1976), which refers to well-differentiated, conspecific populations which 'approach the level of differentiation of species... but show by their interbreeding in contact or, if allopatric, provide some evidence... that they are still conspecific, i.e. have not evolved effective reproductive isolation' (Short 1980). 'Superspecies' is a term used to designate a group of parapatric or essentially allopatric species (allospecies) that have *very* recently evolved from a common ancestor, so recently that they may be considered conspecific by some writers. When parapatric, allospecies sometimes hybridize in the zone of contact, but such hybridization reinforces the reproductive isolating mechanisms. Later, such species become fully reproductively isolated; on establishing contact they may either '(1) remain parapatric, showing reproductive isolation but no ecological compatibility, hence competitive exclusion occurs; (2) overlap somewhat, showing reduced competition; or (3) overlap broadly with minimal interactions' (Short 1980). Such species comprise a 'species-group'. For allopatric populations, the taxonomist must still judge which stage they are in.

Hall and Moreau (1970) emphasized the great value of the superspecies concept but did not apply it in the strict sense intended by Amadon (1966). Their interpretation tends to be broader, and while many of their groupings fall within the definition, others do not. As a result, some of their superspecies are being dismantled, either because allospecies have been found to be broadly sympatric without hybridization (e.g. *Alethe poliocephala/A. poliophrys*) or, in the case of allopatric species, because vocal and behavioural characters strongly suggest lack of close relationship (e.g. *Andropadus montanus/A. masukuensis*). It must be emphasized that the superspecies was never intended by its definers to be simply a convenient place to put species that show some degree of relationship. Some writers are becoming somewhat disenchanted with Hall and Moreau's superspecies (Vande weghe 1988, R. J. Dowsett, M. P. S. Irwin and L. L. Short, pers. comm.). Nevertheless, Pat Hall and the late Reg Moreau would probably be the first to admit that their groupings were tentative, and delighted at the controversies some are causing; that is how science progresses. One must, however, guard against the tendency to 'quote them as if they were holy writ' (M. P. S. Irwin, pers. comm.).

Maps

We have followed all the conventions established in Volume II in preparing and presenting the range maps. A full explanation of these appears in the introduction to that volume, but for convenience we have reproduced the explanatory maps here. Figure 1 shows the shading conventions used to denote the various categories of occurrence. Figure 2 updates the political map featured in Volume I. The inset map, drawn at the same size as the range maps in the text, provides an extra aid for interpreting distributions defined by geographical coordinates. Figure 3 shows the five climatic zones of East Africa referred to under Laying Dates.

References

Amadon, D. (1966). The superspecies concept. *Syst. Zool.* **15**, 245–249.

Amadon, D. and Short, L. L. (1976). Treatment of subspecies approaching species status. *Syst. Zool.* **25**, 161–167.

Campbell, B. and Lack, E. (1985). 'A Dictionary of Birds'. T. and A. D. Poyser, Calton.

Hall, B. P. and Moreau, R. E. (1970). 'An Atlas of Speciation in African Passerine Birds'. British Museum (Natural History), London.

Short, L. L. (1980). Speciation in African woodpeckers. *Proc. IV Pan-Afr. Orn. Congr.* pp. 1–8.

Sibley, C. G. and Ahlquist, J. E. (1990). 'Phylogeny and Classification of Birds. A Study in Molecular Evolution'. Yale University Press, New Haven.

Sibley, C. G. and Monroe, B. L. Jr. (1990). 'Distribution and Taxonomy of Birds of the World'. Yale University Press, New Haven.

Vande weghe, J.-P. (1988). Problems in passerine speciation in Rwanda, Burundi and adjacent areas. *Proc. XIX Int. Orn. Congr.* pp. 2547–2552.

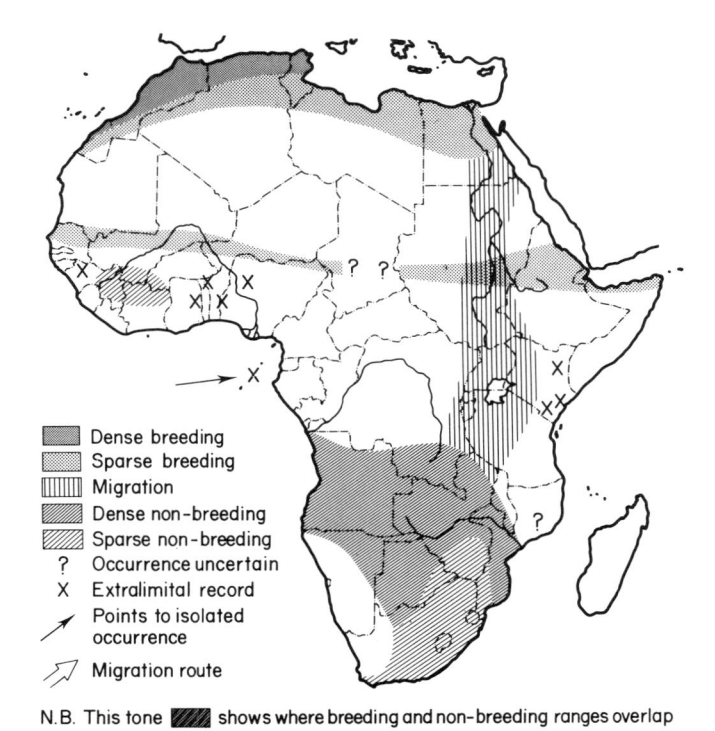

Fig. 1. Shading, symbols and arrows used on maps.

Fig. 2. Updated political map.

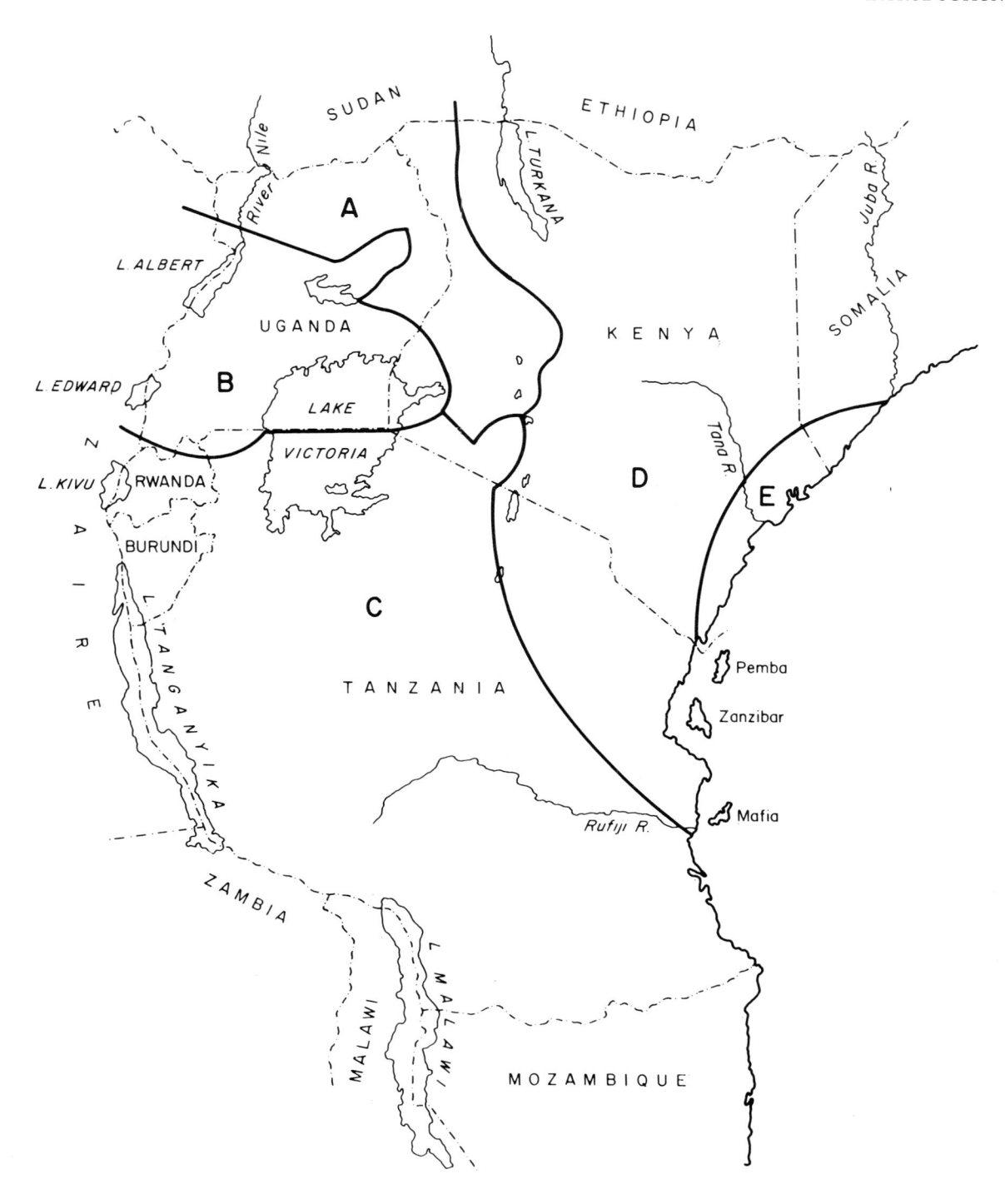

Fig. 3. Climatic zones of East Africa (Uganda, Kenya, Tanzania), referred to under specific Laying Dates (from Brown and Britton 1980 after J. F. Griffiths, Climatic zones of East Africa, *E. Afr. Agric. J.* 1958, 179–185).

Topography of African Passerine Birds I

Topography of African Passerine Birds II

Order PASSERIFORMES

'Perching birds', mostly 'songbirds', much the largest order of birds, embracing approximately 5300 species worldwide, or well over half of all living birds; commonly called passerines for short. Cosmopolitan. Land-birds, only a very few species feeding in (fresh) water; mainly small, a few medium-sized – ravens *Corvus* spp. among the largest. Foot adapted for perching; always 4 toes, the hallux enlarged, directed backwards and not reversible; toes never webbed. In most species, syrinx adapted for singing; syrinx usually tracheo-bronchial with complex and varied musculature, but tracheal in 4 Neotropical families, and simple in a few groups. Wing eutaxic. Primaries 10, the outer 1 reduced or vestigial, or 9 in pipits, swallows, finches, buntings, some larks, and a few New World families. 12 tail-feathers. Aegithognathous palate, 'passerine' tensor propatagialis brevis, bundled spermatozoa with coiled head and large acrosome, and type VII deep plantar tendons. Young nidicolous, hatching blind, naked and helpless. Parental care well developed; but 3 groups brood-parasitic (2 in Africa).

Passeriformes are a monophyletic group (Raikow 1982). Plumages, bill form and life-styles are very varied; 82 families recognized by Campbell and Lack (1985), many biologically distinctive. Evolutionary convergence rife, hence relationships, boundaries and true composition of families often questionable. Morphological and DNA-hybridization studies confirm that pittas, broadbills, tyrant-flycatchers, antbirds, ovenbirds, gnateaters and tapaculos form a natural assemblage (Suborder Tyranni – most families New World). They also suggest that 'true' songbirds of Suborder Passeri (Oscines) are related among themselves in ways radically different from those recognized conventionally (Sibley and Ahlquist 1990).

In the main, classification of Passeriformes and sequence of passerine families adopted here follow those of Mayr and Greenway (1956) and Campbell and Lack (1985). However, subordinal nomenclature, and classification of Suborder Tyranni, follow Sibley *et al.* (1988).

Suborder TYRANNI (Suboscines)

More or less synonymous with Deutero-oscines, Clamatores, Suboscines, Oligomyodi and Tyranniformes: 10–14 families embracing *c.* 1100 species. Syrinx mesomyodian, with not more than 3 pairs of intrinsic muscles (Ames 1971); form of ear ossicle or stapes unique (except in New Zealand wrens: Feduccia 1980). Contains 3 infraorders – 1 large and Neotropical (Tyrannides) and 2 small and Paleotropical (Acanthisittides, New Zealand wrens, and Eurylaimides).

Infraorder EURYLAIMIDES

Pittas Pittidae, broadbills Eurylaimidae, and asities Philepittidae (confined to Madagascar) (Olson 1971, Raikow 1982).

Superfamily EURYLAIMOIDEA: broadbills, asities

Family EURYLAIMIDAE: broadbills

Small to medium birds (weight range 20–160 g) with strikingly-coloured plumage. Bill moderately or extremely wide; outermost primary long, wing rounded; tail short or medium, more or less graduated. Tarsus with scutes anteriorly and small hexagonal scales posteriorly; toes 3 and 4 united basally. Main artery of leg is ischiatic. Non-furcate sterna spinalis. Tendon of flexor hallucis longus connected by vinculum to tendon of flexor digitorum longus. Inhabit humid forests; sedentary, somewhat sluggish; eat plant and some animal material.

EURYLAIMIDAE

W Africa to Philippines. 14 species in 8 genera; in Africa, 4 species in 2 endemic genera. The 2 African genera have generally been allocated to separate subfamilies, but 'recognition of subfamilies within the Eurylaimidae is discouraged' (Olson 1971).

Genus *Pseudocalyptomena* Rothschild

Endemic, monotypic. Differs from *Calyptomena* (3 spp.; SE Asia) in lacking dense feather fan concealing base of bill and nostrils. Bill flat; rictal bristles < 5 mm. Tarsus slender and quite long, covered behind with soft skin with small rounded scales. First joint of toes 3 and 4 united. Outer primaries stiff and straight, with tip of inner web slightly twisted down. Carotids paired.

Plate 1
(Opp. p. 32)

Pseudocalyptomena graueri Rothschild. African Green Broadbill; Grauer's Broadbill. Eurylaime de Grauer.

Pseudocalyptomena graueri Rothschild, 1909. Ibis, p. 690; west of Ruzizi R., north of L. Tanganyika, 2000 m.

Range and Status. Endemic resident. Main population in Itombwe Mts, Zaïre, where known at 5 localities (Ibachilo, Miki, Luvumba, Karungu, Muusi) between 3° 41′S, 28° 31′E and 3° 4′S, 28° 48′E and from 1760 to 2480 m; common around Miki. 80% of *c*. 30 birds collected were at 1940–2390 m. Known also from mountains west of L. Kivu, Zaïre (Mt Kahuzi: Nyawaronga, 2° 0′S, 28° 49′E) and Impenetrable Forest, Uganda (Bwindi, Ruhizha, 2100–2200 m). Threatened by forest clearance around villages; L. Kivu population probably safeguarded in Kahuzi-Biega Nat. Park; other 2 areas need protection (Collar and Stuart 1985).

Description. ADULT ♂: forehead and crown greenish buff, thickly covered with small black streaks; nape, back, wings, rump and tail green; lores black, with very narrow black line over and behind eye; ear-coverts, chin, throat and upper breast pale blue; a few black or dusky spots form irregular moustachial streak; lower breast, belly and flanks green (paler and yellower than upperparts); undertail-coverts blue, or green with blue tips. Blue areas clearly demarcated from green ones. Remiges blackish; underwing and undertail black. Bill black; eye very dark brown, eye-rim black; legs and feet greyish green to blackish. Impenetrable Forest population may be narrower-billed than others (Friedmann and Williams 1968). Sexes alike. SIZE: (3 ♂♂, 2 ♀♀) wing, ♂ 74·2, 75·5, 76, ♀ 75, 76·5; tail, ♂ 39, 39, 39·5, ♀ 39, 39; bill, ♂ 11·8, 14, 15, ♀ 15, 15; bill breadth at gape, ♂ 8·5, 11·5, ♀ 10·6, 10·6; tarsus, ♂ 19, 20, ♀ 20, 20. WEIGHT: ♂ (n = 1) 32·5, ♀ (n = 4) 29·0–32·5 (31·1).

IMMATURE: like adult but slightly duller; undertail-coverts green.
NESTLING: unknown.

Field Characters. Readily identified, even at considerable height in trees, by uniquely green and pale blue plumage: green upperparts, pale blue throat and undertail-coverts, pale green belly. Colours somewhat bee-eater-like, but short-tailed silhouette and stocky, upright stance are more flycatcher-like. Bill, short eye-mask, and underside of tail black.

Pseudocalyptomena graueri

Voice. Not tape-recorded. A very feeble 'tsi-tsi' repeated 3–8 times at rate of *c*. 4 notes per s; sounds like voice of Oriole Finch *Linurgus olivaceus*. Flight call of breeding bird an extremely high-pitched and quite prolonged bell-like ringing (E. F. G. Smith, pers. comm.).

General Habits. Inhabits primary forest, occurring especially in middle storey and canopy, at edge of forest, at 1900–2400 m; also isolated trees in cleared ground. Usually avoids dense interior of forest. Perches mainly at 7–20 m, in dense foliage, sometimes quite immobile, like small barbet. 1 bird climbed along a bough, clung underneath it woodpecker-like, and clung to an upright bough (Chapin 1978) – perhaps feeding; also said to feed by sallying outward from canopy (Prigogine 1971), although these 'sallies' may in fact be display flights (Friedmann 1970). Bird at Ruhizha (Uganda), Nov 1986, gleaned insects actively, constantly fluttering up to pick insect from leaf, then landing briefly again. Another, in *Chrysophyllum gorungosanum* forest there, Dec 1987, flew from dry *Mimulopsis* on log lying on ground, to join mixed foraging party of White-headed Wood-Hoopoes *Phoeniculus bollei*, Black-backed Puffbacks *Dryoscopus cubla* and Masked Apalis *Apalis binotata* 7 m up; later, flock of 4–5 African Green Broadbills foraged actively 3–30 m high in forest, each perching usually for less than 15 s, then flying for only a few m, always calling in flight (T. Butynski and J. Kalina, pers. comm.). In Uganda said to feed more in manner of waxbill *Estrilda* in *Neoboutonia* trees and 2–3 m high in forest understorey. Not particularly shy; sometimes forages near human habitation. Flies up to 30 m between trees; flight direct, not undulating, rather slow, with gliding. Occurs singly, in pairs, or parties of up to 10. Often joins mixed-species bird parties.

Food. Seeds, flowers, buds, fruit, beetles, larvae and snails. Bird picked up 5-cm hairy caterpillar and discarded it after 5 s.

Breeding Habits. For possible display flight, see above.

NEST: only 1 found was spherical, diam. *c.* 200–250, with round entrance *c.* 50 across, made of Spanish-moss-like green lichen, sited *c.* 11 m up in 20 m tree in outermost branches and directly overhanging a stream; in valley floor with open shrubby vegetation and many 20 m high trees (Impenetrable Forest, Uganda: E. F. G. Smith, pers. comm.); ♀ was in nest and ♂ evidently fed her there.

LAYING DATES: Zaïre, 4 adults with somewhat enlarged gonads, late July; an immature bird collected at same time, so breeding season probably long. Uganda (pair at nest, Apr).

References
Aspenlind, L. J. (1935).
Collar, N. J. and Stuart, S. N. (1985).
Friedmann, H. (1970).
Prigogine, A. (1971).

Genus *Smithornis* Bonaparte

Small, sexually dichromatic, streaky brown broadbills with white backs. 4–5 outer primaries straight and stiff, narrow-based, with inner webs down-turned towards tip. Outermost primary covert elongated. Rictal bristles 8–10 mm long. Toes 3 and 4 united. Tongue thick. Short, circular display flight, unique among birds, accompanied by loud wing-rattling noise.

Endemic; 3 species, 2 forming a superspecies (*S. rufolateralis*/*S. capensis*).

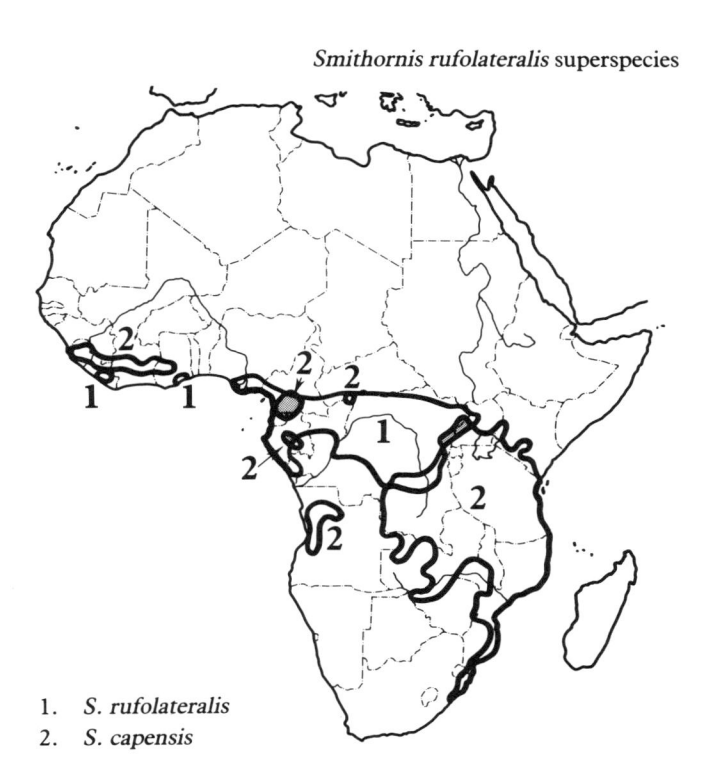

Smithornis rufolateralis superspecies

1. *S. rufolateralis*
2. *S. capensis*

EURYLAIMIDAE

Plate 1
(Opp. p. 32)

Smithornis sharpei Alexander. Grey-headed Broadbill. Eurylaime à tête grise.

Smithornis sharpei Alexander, 1903. Bull. Br. Orn. Club 13, p. 34; Mt St Isabel, 4000 feet, Fernando Po.

Range and Status. Endemic, resident. 2 populations: in Bioko, Cameroon east to Yokadouma and N Gabon (Woleu-N'Tem region; 1 record Makokou); and in E Zaïre from about Bondo Mabe, Ituri, to Kanyaa, Itombwe. Frequent to common.

Description. *S. s. zenkeri* (Reichenow): S Cameroon, N Gabon. ADULT ♂: forehead and crown dark grey, the feathers with blackish centres; forehead suffused with orange at sides and just above nostrils; nape, cheeks and sides of neck dark brownish grey; mantle, back, rump and uppertail-coverts rufescent olive-brown, the mantle and back feathers with silky white bases, rufescent brown tips, and grey-black intervening areas (black and particularly white showing through irregularly); uppertail-coverts rich rufescent brown. Tail-feathers dark brown or blackish brown, with rufescent edges. Lores buff or orange buff; chin and throat white or creamy white in midline with a few diffuse dark grey streaks, and at sides (bordering cheeks) bright orange; breast bright rufous; belly silky white with fine or broad blackish streaks; upper belly, sides of belly and flanks strongly suffused tawny; undertail-coverts pale rufous. Primaries and secondaries blackish brown with rufous-brown edges; scapulars and tertials rufescent brown; upperwing-coverts dark brown with small rufous tips and narrow rufous edges. Underwing-coverts pale rufous; underside of remiges and rectrices brown-grey. Upper mandible black, lower mandible pale greyish white; eye dark brown; legs and feet greenish grey or greyish olive. ADULT ♀: upperparts like ♂ but crown slightly greyer; underparts like ♂ but centre of chin and throat suffused buff and sides of chin and throat paler rufous; breast rufous-buff (rather than brick red), diffusely striped with grey; belly and flanks black-streaked white, washed buff, with only small area in midline free of buff wash. SIZE: (5 ♂♂, 6 ♀♀) wing, ♂ 80–83 (81·6), ♀ 79–84 (81·4); tail, ♂ 54–58 (56·2), ♀ 50–54 (52·0); bill, ♂ 15–16 (15·4), ♀ 14–15 (14·7), breadth at base, ♂♀ 13–14; tarsus, ♂♀ 16–17 (16·5). WEIGHT: unsexed (n = 6) 34–40·3 (37·5) (Bowden 1986).

IMMATURE: differs from adult in having crown olive-brown, and white patch on back vestigial or absent; back feathers lack black marks; upperwing-coverts tipped orange-buff; breast streaked with black.

NESTLING: unknown.

S. s. sharpei Alexander: Bioko. Crown blue-grey (rather than sooty grey of *zenkeri*); black streaking on underparts less heavy.

S. s. eurylaemus Neumann: E Zaïre. ♂ with clearer grey crown and ♀ with sides of breast more olive-brown than in nominate race. Smaller: wing (♂♀) 75–78, bill (♂♀) 12–13.

Field Characters. The largest African broadbill, length 15 cm; a somewhat undistinguished bird with large grey head, buffy lores, white or whitish chin and throat, olive-brown upperparts, bright rufous breast, whitish belly and streaked flanks. Rufous breast distinguishes it from African Broadbill *S. capensis*, and grey head from Rufous-sided Broadbill *S. rufolateralis*. Upper mandible black, lower mandible white (like congeners); bill flat and broad. Back feathers have white bases, not always visible. Juvenile more uniformly olive-brown; breast duller rufous. Mechanical flight 'song', like congeners.

Voice. Tape-recorded (78). Non-vocal sound made in display flight is very loud, rapid, staccato trill of 0.5–0.6 s duration, starting and finishing abruptly. Just perceptibly, trill slows and then accelerates. Like trill of African Broadbill *S. capensis* but shorter and deeper-pitched.

General Habits. Inhabits lower and middle storeys of primary lowland rain forest and heavily wooded mountainsides, from sea level (Cameroon), but mainly at 1000–1500 m on Bioko and in Itombwe. In pairs. Sluggish, sitting quietly for long periods. In evenings and at times during day, makes short display flight in tight circle at perch accompanied by loud, mechanical trill.

Food. Unknown.

Breeding Habits

NEST: 4 described; large pocket-shaped mass of fresh green moss, suspended from twig 1–2 m above forest floor, with lining of dry leaves, fibres and stems. 1 nest built entirely of fine black fungal fibres and dried leaves. Height 22 cm, diam. 9–10 cm, entrance hole 6 × 3·5 cm, on upper side, once with porch.

EGGS: 1–2. Long, pointed ovals; pure white. SIZE: (n = 7) 22–25 × 16–17·5 (23·3 × 16·9).

LAYING DATES: Cameroon, Jan, Mar, and brood-patch Feb; Zaïre, Apr.

Smithornis rufolateralis Gray. Rufous-sided Broadbill; Red-sided Broadbill. Eurylaime à flancs roux. Plate 1

Smithornis rufolateralis Gray, 1864. Proc. Zool. Soc. Lond., p. 143; Gold Coast. (Opp. p. 32)

Forms a superspecies with *S. capensis*.

Range and Status. Endemic resident, lowland forests from Liberia to W Uganda, south to Kananga (Luluaburg) in Zaïre. Frequent to common; probably more widespread than shown on map. Density of 12 pairs per km^2 (M'Passa, Gabon).

Description. *S. r. rufolateralis* Gray: Liberia to about 20°E in Zaïre. ADULT ♂: forehead, crown, nape, hindneck, sides of neck and cheeks black, sometimes with small line of white below eye; mantle black in centre, brown at sides, back and rump black; black feathers of mantle to rump have white bases which sometimes show irregularly through. Rump ochreous dark brown, tail dull black, the tail-feathers narrowly edged brown. Lores buff. Long, soft, black rictal bristles. Chin and throat creamy white, occasionally with irregular dark streaks; sides of throat and breast creamy white, heavily streaked black; large area at sides of breast orange; belly cream-white, heavily streaked black except in midline and around vent. Flanks cream-white heavily streaked black, or uniformly brown where concealed below wing. Remiges dull blackish, edged with brown. Scapulars and tertials dark ochreous brown; lesser coverts dark ochreous brown; remaining upperwing-coverts black or blackish; median and greater primary coverts with triangular or tear-shaped white tips, forming 2 rows of white spots on closed wing. Underwing-coverts creamy white; underside of remiges and rectrices grey. Upper mandible black, lower mandible white or horn, tinged pink; eye dark brown; legs and feet olive green. ADULT ♀: forehead, crown and nape greyish or blackish brown; sides of neck and cheeks warm ochreous brown, mantle, back, rump, uppertail-coverts, scapulars and tertials warm ochreous brown; mantle and back with white feather bases showing through irregularly; mantle feathers white proximally, brown distally, with black intervening mark. Rest of plumage like adult ♂, but sides of breast orange-brown rather than bright orange, greater and median primary coverts brown-black with smaller, occasionally buffier, white tips, and tail browner. SIZE: (41 ♂♂, 13 ♀♀) wing, ♂ 60–72 (66·7), ♀ 60–65 (63·1); tail, ♂ 40–43 (41·8), ♀ 38–40 (39·5); bill, ♂ 13–17 (15·7), ♀ 13–16 (14·9), breadth at base, ♂ 13, ♀ 12; tarsus, unsexed 13–15 (14·0). WEIGHT: (Liberia) ♂ (n = 12) 20·5 ± 2·7, ♀ (n = 6) 10–23·1 (17·7).

IMMATURE: like adult ♀ but somewhat duller.
NESTLING: unknown.

S. r. budongoensis van Someren: central Zaïre to W Uganda. Dark stripes on underside fewer and shorter; crown of ♀ greyer than that of nominate ♀. Smaller. WEIGHT: (Uganda) ♀ (n = 8) 19–29 (20·9).

Field Characters. A flycatcher-like bird of thick undergrowth. ♂ distinguished by: black head and cheeks, sharply demarcated from buff lores and white chin and throat; brown upperparts with back irregularly white (hard to see at rest, can be very conspicuous in flight) and 2 white wing-bars; and white underparts, with dark-streaked breast and flanks and large bright orange patch on side of breast. Bill flat and broad, upper mandible black, lower whitish (like congeners). ♀ similar, duller, less white in back, wing-bars buffy, and head brown where ♂ is black. Commonly makes display flight accompanied by loud, mechanical rattle (see *Voice*); rattle higher-pitched than that of African Broadbill *S. capensis* and much higher-pitched than rattle of Grey-headed Broadbill *S. sharpei*. Flycatching behaviour like wattle-eye *Platysteira*. Orange sides of breast distinguish it from African Broadbill; black (not grey) head and white wing-bars distinguish ♂ from ♂ Grey-headed Broadbill; ♀ told from Grey-headed Broadbill by small size (length 12 cm) and 2 buffy wing-bars. All 3 species are partly sympatric.

Voice. Tape-recorded (78). Voice, used in courtship, is short, weak, high-pitched whistle 'whee, whee . . . .' or 'huiiii'. Much more frequent noise is mechanical flight display (see above), a trill exactly like that of African Broadbill (q.v.) but higher-pitched.

General Habits. Inhabits thick undergrowth in lower storey and at edges of lowland forests from sea level to 1300 m (1500 m on Mt Nyombe, E Zaïre); often near water. Restricted to primary and old secondary forests, so locally range complements that of African Broadbill, q.v. Occurs solitarily or in pairs, ♂ and ♀ foraging some m apart; sits quietly on thin horizontal branch or creeper 3–15 m above ground in thick vegetation with long periods of immobility. Gives rapid chase to passing insect, through open spaces and among foliage, up to 25 m above ground; also picks insect from leaf or trunk in flight. Sometimes, as bird twists and turns in pursuit of prey, wings make toad-like noise of flight display.

6   EURYLAIMIDAE

A

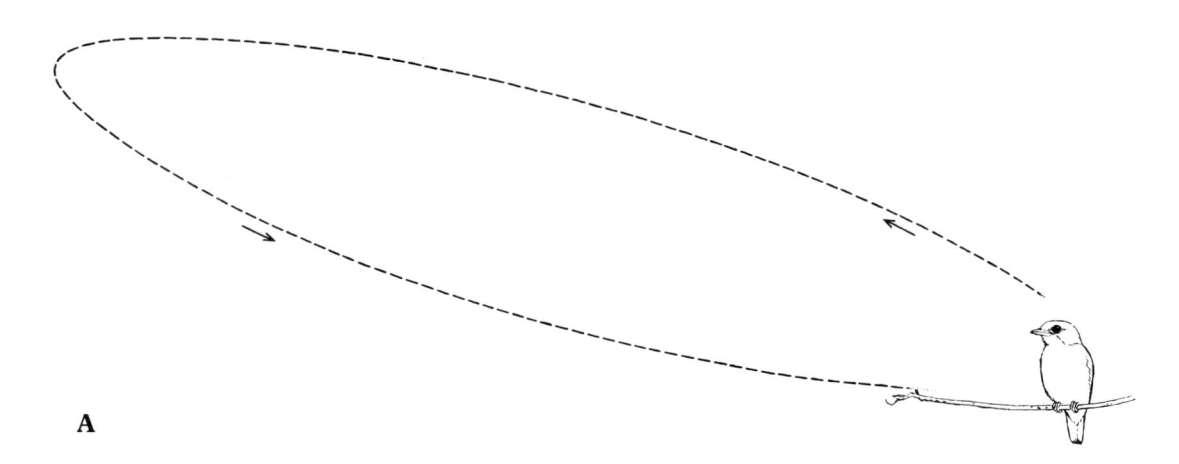

Flight display performed solitarily, by both sexes but mainly by ♂, at all times of day and year, but particularly at dawn and dusk in all but drier months. Frequency of 1 flight per min for $c.$ 1 h, flights performed rather regularly, but 28 flights in 14 min by bird responding to sound of displaying neighbours (flights are also made regularly in response to playback of tape-recording). Flights much less frequent in middle of day. Flight display performed from horizontal branch or creeper, 5–15 m above ground. Bird makes 1–2 abrupt jumps into air (sometimes turning about as it does so), then flies in ellipse of 1 m greatest diam., making burst of sound audible 150 m away, returning to same point on perch. Sound is made by vibration of stiff, twisted outer primaries. During flight, white back feathers puffed out. Plane of ellipse varies from horizontal to inclination of 45° (**A**), the angle of inclination modifying timbre of sound. Successive display flights are made from same perch, bird sometimes giving weak 'huiiii' call between flights; or (usually) from sequence of dozens of perches as bird travels up to 1500 m in 15 min. Displaying bird usually solitary, and flight serves territorial function (Brosset and Erard 1986); display flights also sometimes given during pair-formation and before copulation. Flight frequency increases when rival heard or seen near territory border, but birds do not fight. Bird responding to another at border of territory puffs out plumage, orange patches being conspicuous, wags closed tail slowly up and down through wide arc, trembles wing-tips, gives thin, stifled calls, and makes several abrupt jumps into air.

Food. Arthropods. Small cicadas and small, hairless caterpillars commonest; beetles, grasshoppers, and an earwig, spider and millipede also recorded.

Breeding Habits. Monogamous, territorial. Courting ♂ approaches ♀, giving sweet 'huiiii' calls, his plumage pressed flat and head turned to one side. He then puffs feathers out, showing white back and orange breast-patches, adopting horizontal posture with legs bent and bill pointing up. Wing-tips droop down and are rhythmically, quickly opened and closed, sometimes being

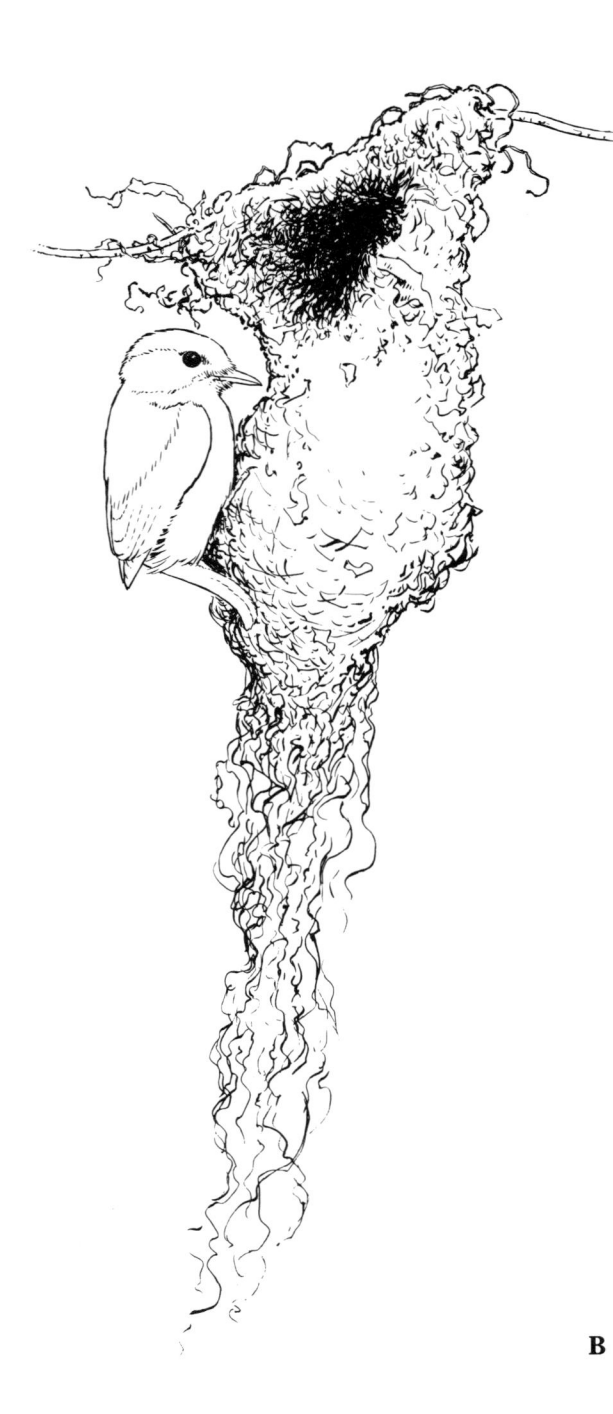

B

flicked up high; closed tail inclined up at 45° and wags in time with calls. He then makes several small, excited jumps up and down in front of ♀ and gives rhythmic calls that run together into a simple song. Performance may alternate with display flights, and functions in pair-formation and as prelude to copulation.

NEST: bag of dead leaves, vegetable fibres, mossy roots and dry twigs, thickly woven together with long black strands of fungus *Marasmius*, with side entrance near top, suspended from thin horizontal branch or creeper by fibres thickly draped over *c*. 12 cm length of branch (**B**). Nest lining of softer dead leaves. A few strands often hang far down below nest. Resembles sunbird's nest, but larger, more globular, lacks porch; height 15 cm excluding 'beard' below; entrance 5 cm diam. and *c*. 3 cm from top.

EGGS: 1–2, av. (n = 9 clutches) 1.8. Rather thin-shelled; glossy (Bannerman 1936) or matt (Brosset and Erard 1986); pure white. SIZE: (n = 3) 22–23 × 15·5–16 (22·4 × 15·7).

LAYING DATES: Cameroon, Jan, Oct; Gabon, Sept–Apr (1 clutch each month but 5 in Feb); Zaïre (Lukolela, Sept; Kamituga, Dec; Ituri, probably Apr–Oct; middle Congo R., probably Aug–Jan); Liberia (gonads enlarged Aug–Apr).

INCUBATION: by ♀ alone.

DEVELOPMENT AND CARE OF YOUNG: young fed in nest by ♀ and ♂. In absence of ♀, ♂ guards young by clinging to nest entrance.

Reference
Brosset, A. and Erard, C. (1986).

Smithornis capensis (Smith). African Broadbill. Eurylaime du Cap.

Platyrhynchus capensis Smith, 1840. Illustr. Zool. South Africa, Aves, pl. 27; N Zululand.

Forms a superspecies with *S. rufolateralis*.

Plate 1
(Opp. p. 32)

Range and Status. Endemic resident. Dense forests (but not primary rain forest) from Sierra Leone to central Ghana; in S Cameroon, Gabon and S Central African Republic; W Angola; and – much more widespread – in SE Africa from NE Zaïre, Uganda and S Kenya to middle Zambezi valley, E Zimbabwe, Mozambique and South Africa (NE Transvaal, and Natal littoral plain south to Port Shepstone). On Zambezi occurs west to Nampini (Zimbabwe) and in Zaïre west to 22°E. Locally frequent to common.

Description. *S. c. capensis* (Smith): Natal. ADULT ♂: forehead and crown sooty black, forehead with buffy feathers immediately above nostrils; nape and sides of neck grey-brown, indistinctly streaked blackish; ear-coverts grey with white feather shafts; mantle and back grey-brown, heavily mottled sooty black and with white feather bases showing through irregularly; rump, uppertail-coverts and tail brown. Lores buff-grey; chin and throat white in midline, heavily but diffusely streaked with blackish brown at edges; breast, flanks and belly silky white, heavily streaked with dark brown, breast sometimes washed palest buff and greyer at sides, centre of belly free of stripes. Undertail-coverts white. Primaries, secondaries and tertials dark brown, scapulars grey-brown or warm brown, each feather darker along midline. Upperwing-coverts dark blackish brown, with warm brown or buff fringes; median coverts with 1–2 mm wide buff tips, forming pale bar in closed wing. Underwing-coverts white; undersides of remiges and rectrices silvery grey. Upper mandible black, lower mandible white or horn; eye dark brown; legs and feet olive- or yellow-green. ADULT ♀: forehead rufous-buff with small dark streaks, crown grey, heavily but diffusely streaked blackish; nape, sides of neck, ear-coverts, mantle, back, rump, uppertail-coverts and scapulars rufescent brown with diffuse blackish shaft streaks. Tail-feathers brown with rufescent brown edges. Lores grey-brown. Underparts like ♂ but washed with buff at sides of chin, throat, breast, and on flanks. Wings like ♂ but tips and edges of feathers buffer. Soft parts like ♂. SIZE: (2 ♂♂) wing, 75, 76; tail, 55·5, 56·5; bill, 18·5, 19; tarsus, 14·5, 16.

IMMATURE: like adult ♀ but forehead and crown dark grey-brown, finely streaked black.

NESTLING: unknown.

S. c. cryptoleucus Clancey: NE Zululand, E Swaziland, NE Transvaal, E Zimbabwe and Mozambique to S Malaŵi and SE Tanzania. ♂ and ♀ with whiter, more sparsely streaked underparts than nominate race. Smaller. SIZE: (5 ♂♂, 4 ♀♀) wing, ♂ 68–71 (69·7), ♀ 68·5–71·5 (70·4); tail, ♂ 50–52 (51·0), ♀ 47–50·5 (49·2); bill, ♂ 17–18 (17·4), ♀ 16·5–18 (17·2), width 12. Rictal bristles up to 6 mm long. WEIGHT: 1 ♂ 26·3.

S. c. suahelicus Grote: littoral districts from Beira, Mozambique, to Shimba Hills, Kenya, inland in Tanzania to Nguru and Uluguru Mts. Mantle and rump more uniform, less streaked, than in above races; ♂ with dark grey ear-coverts; ♂ and ♀ with white underparts washed buff. WEIGHT: 1 juvenile, June, 20.

S. c. conjunctus Clancey: Caprivi Strip and middle Zambezi valley to NW Mozambique. Breast washed yellowish buff; ear-coverts of ♂ buff.

S. c. albigularis Hartert: Angola, Zaïre, N Zambia, W (and central?) Tanzania and N Malaŵi. Upperparts darker than in above races; ear-coverts of ♂ grey, streaked white; crown of ♀ blackish, feathers edged buff; breast of ♂ and ♀ washed yellowish buff.

S. c. medianus Hartert and van Someren: N Tanzania and S Kenya (except coast). ♂ and ♀ like *albigularis* but less heavily streaked above; crown of ♀ black, unstreaked.

S. c. meinertzhageni van Someren: NE Zaïre, Uganda and W Kenya (Elgon to Kavirondo). More heavily streaked than *medianus*, underparts less buffy; crown of ♀ blackish. WEIGHT: 1 ♂ 25, ♀ (n = 19) 20–31·1 (24·8).

S. c. camerunensis Sharpe: Cameroon, Gabon, Central African Republic. Mantle to rump and wings of ♂ and ♀ richer, more rufescent brown than in other races; underparts heavily streaked; breast strongly washed buff. WEIGHT: 1 ♂ 26.

S. c. delacouri Bannerman: Sierra Leone to Ghana. Like *camerunensis* but slightly paler and nape greyer. WEIGHT: 1 ♂ 25·3.

Field Characters. A stolid, flycatcher-like bird of dense forest undergrowth. Plumage rather variable geographically. ♂ has black crown, buffy forehead, grey or buff cheeks; greyish- or rufescent pale or dark brown upperparts with back irregularly white (hard to see at rest, conspicuous in flight), and buffy tips to median wing-coverts usually forming wing-bar; chin to undertail-coverts silky white, more or less heavily streaked black on moustache, breast and flanks; breast washed grey or buff. Bill flat and broad, upper mandible black, lower white (like congeners). ♀ similar, duller, less white in back; crown black or brown. Commonly makes elliptical display flight, lasting 1 s, with loud mechanical rattle. Distinguished from Rufous-sided Broadbill *S. rufolateralis* by white or buffy (not orange-sided) breast, and from Grey-headed Broadbill *S. sharpei* by black (not grey) crown and smaller size.

Voice. Tape-recorded (32, 58, 72, 78, 88, F). Voice, used in courtship, is weak, plaintive whistle 'huiii' or 'twee-oo'; also mewing call (Manson 1983). Much more frequent is mechanical noise made during flight display, a very loud, fast, staccato trill of 0·7 s duration (rather less than duration of flight itself), starting and finishing abruptly, like fast rattle produced by pinging free end of wooden ruler clamped across edge of wooden table. Trill slows and then accelerates: 4–6 notes in 0·25 s followed immediately by 15 notes in 0·5 s – 'ttttt-rrrrrrrrrrrrrrrr'. Pitched lower than trill of Rufous-sided Broadbill and higher than that of Grey-headed Broadbill.

General Habits. Inhabits lower storey of lowland evergreen forest and adjacent dense savanna woodland, especially with stands of bamboo *Oxytenanthera abyssinica*; dense deciduous thickets; riparian fringing forest; forest shrub mosaic; forested hillsides and montane forests; valleys with tree-ferns. In Gabon, strictly in secondary forest, hence range complements that of Rufous-sided Broadbill (strictly in primary forest); plantations, cultivation around villages. Sea level up to 1100 m (Zimbabwe), 1800 m (E Africa) and 2150 m (Zambia/Malaŵi), but commoner below 700 m. Habits, so far as known, do not differ from better-known Rufous-sided Broadbill, q.v. Both sexes perform display flights. Display flight frequency up to 1 every 45 s; elliptical flight diam. *c.* 60–80 cm; puffed-out back feathers very conspicuous in display flight; sound audible at least 60 m away. Keeps mainly between 1 and 6 m above ground in dense cover. Catches insect in sally from perch, like flycatcher *Muscicapa* or wattle-eye *Platysteira*, returning to same perch to eat it, or by searching under leaves and branches like wattle-eye. Occasionally takes prey from ground. Occurs in mixed-species foraging parties.

Food. Insects, including beetles, orthopterans, hemipterans, eggs and caterpillars.

Breeding Habits. Monogamous, territorial. Breeds in dispersed, solitary pairs. Habits not as well known as in Rufous-sided Broadbill q.v., but appear not to differ; however, 2 birds *c.* 45 cm apart on low horizontal branch in forest at Chimonzo (S Mozambique) flicked wings like chat, then swung on perch to hang upside-down under it, where they continued to flick wings and to 'emit their loud klaxon notes' for *c.* 30 s (Lawson 1961).

NEST: bag of green moss, vegetable fibres, dead leaves and a few twigs, thickly woven together with long black strands of fungus *Marasmius* or long dry leaves, which are twined in broad band over supporting thin horizontal branch; with slightly porched entrance high up on side; lined with soft bark, dry stems, leaves and grasses, matted together with spiders' silk. Some fibres hang down below nest. Nest bag *c.* 37 cm long; nest proper 6–7 cm wide and *c.* 9 cm deep (Vincent 1946). Nest placed in deep shade in thick vegetation, 1–2·5 m above ground.

EGGS: 1–3, av. (n = 20 clutches) 2·4. Rather long, pointed ovals; glossy; pure white. SIZE: (n = 11) 19–23·6 × 14·5–16 (21·2 × 15·1).

LAYING DATES: Gabon, Jan; E Africa: Region C, Nov; Region D, Dec; Zaïre (Itombwe), Apr, May, Sept, Oct; Zambia, Oct, Nov, Mar; Malaŵi, Nov, Feb; Zimbabwe, Oct–Feb (Oct 8, Nov 25, Dec 8, Jan 4, Feb 1 clutches). Angola (breeding condition Dec–Mar).

INCUBATION: unknown.

DEVELOPMENT AND CARE OF YOUNG: unknown.

BREEDING SUCCESS/SURVIVAL: probably parasitized by Barred Long-tailed Cuckoo *Cercococcyx montanus* (Dean *et al.* 1974; see also Vol. III, p. 82).

Superfamily PITTOIDEA: pittas

Brilliantly coloured, rather thrush-like ground birds. Stocky, long-legged, short-tailed; weight range 40–220 g. Sexually monochromatic or dichromatic. Large temporal fossae across occiput, nearly meeting in midline; intrinsic syringeal muscles usually absent, but 1 pair in *Pitta angolensis* is mesomyodian; main thigh artery is sciatic; tarsal envelope entire and smooth anteriorly (Sibley 1970). Eat mainly insects. Highly secretive; many species strongly migratory. Systematically enigmatic; possibly closer to Neotropical Tyranni than to Eurylaimoidea. Inhabit forests.

Paleotropics, mainly SE Asia. A single family and genus of 29 species; 2 in Africa, possibly conspecific, and perhaps forming a superspecies with 5 Oriental and Australasian species; of these, *P. brachyura* (Kashmir to S Japan and Borneo) is a particularly close ally.

Family PITTIDAE: pittas

Genus *Pitta* Vieillot

2 African species, forming a superspecies (see above and below).

Pitta reichenowi superspecies

1. *P. reichenowi*
2. *P. angolensis*

Pitta reichenowi **Madarasz. Green-breasted Pitta. Brève à poitrine verte.**

Pitta reichenowi Madarasz, 1901. Orn. Monatsber., p. 133; Middle Congo R. above Stanley Pool.

Forms a superspecies with *P. angolensis*. Sometimes treated as conspecific with it; colour and size intermediates between *P. reichenowi* and *P. angolensis* known in Cameroon (Sakbayeme, Nkolngem, Bafia, Nkom: Louette 1981).

Plate 1
(Opp. p. 32)

Range and Status. Endemic resident. Rain forests of Congo Basin and Uganda. Frequent in S Cameroon from Ebolowa and Sangmelima southwards; common in NE Gabon. Occurs in W, central and (more widely) NE Zaïre; and in Uganda at 1100–1400 m from Budongo, Bugoma and Bwamba forests south to Kigezi and east to Jinja. Shy and retiring; more often trapped than seen; probably more widespread than shown on map, and may be common.

Description. ADULT ♂: centre of forehead and of crown black, nape black; sides of forehead, crown and nape buff, forming a long, broad superciliary stripe from nostril to shoulder. Lores, ear-coverts and sides of neck black. Mantle, back, scapulars and upper rump dark green; lower rump brilliant azure blue; uppertail-coverts the same with violet wash; tail matt black. Chin and throat white or very pale pink; hind edge of throat black, forming neat narrow transverse bar between throat and breast; breast pale green to buffy green, the feathers bluish or greyish green with 1-mm broad buff tips; belly and undertail-coverts bright pink or pink-red, sometimes with greyish feather bases showing through; thighs brown. Primaries black, with grey-buff tips. P6–P8 with white bases, forming patch on outstretched wing c. 12 mm deep. Outer 2 secondaries narrowly bordered white near tip of outer web; inner secondaries edged bluish; tertials dark blue-green; wing-coverts black, lesser coverts with 4–7 mm azure blue tips, median coverts with 2–5 mm azure blue tips, and inner greater coverts with 5–7 mm azure blue tips. Underside of wing glossy black except for greyish tips to outer primaries. Underside of tail black. Bill black with reddish base to lower mandible and with red spot on culmen; tongue and inside of mouth red; eye dark reddish brown; legs, feet and nails pale greyish flesh or greyish white. Sexes alike. SIZE: (10 ♂♂, 4 ♀♀) wing, ♂ 111–122 (118), ♀ 116–124 (118); tail, ♂ 39–47 (42·9), ♀ 40–44 (42·3); bill to skull, ♂ 22–27 (25·6), ♀ 26; tarsus, ♂ 34–40·5 (36·5), ♀ 34–37 (35·5). WEIGHT: unknown.

IMMATURE: like adult but breast dark olive-brown; belly and undertail-coverts pink; bill dull brown-red with blackish band across centre; legs and feet dull brown-red. Some very young birds have azure blue tips to wing-coverts, others lack them.

NESTLING: pre-feathered nestling unknown. Bill of young fledgling has conspicuous orange cutting edges and gape like that of African Pitta *P. angolensis* q.v.

Field Characters. Pittas are robust, short-tailed, long-legged birds of the forest floor. Shy and rarely seen; but shape is unmistakable. This species has green breast, white throat, black head with broad buff superciliary stripe, azure blue rump, scarlet belly, and pale pink legs. Makes 'prrrt' noise (see below). Differs from African Pitta mainly in having green, not buff, breast; almost impossible to tell from it from behind, when flushed.

Voice. Tape-recorded (32, 78). Adult (and nestling) frequently utters short, pure, bell-like whistle, like call of Fire-crested Alethe *Alethe diademata*; duration varies from 0·25 to 0·50 s, repeated a few times at rate of 10 calls in 13–15 s. Pitch 1·9 kHz. Makes short, soft purring sound 'brrrt' or 'prrrt', probably mechanical not vocal, evidently when bird flies from ground up to perch 0·5 m high (e.g. log). Noise is more often heard than bird is seen. Sounds like display flight noise of African Broadbill *Smithornis capensis*, but not so loud or 'wooden'.

Pitta reichenowi

May be repeated 3 times in 30 s. Mournful whistle continuously uttered by full-grown young bird in captivity (Keith and Gunn 1971).

General Habits. Generally prefers mature secondary forest and thickets to primary forest, but in Bwamba Forest (Uganda) occurs in stands of mature ironwood (*Cynometra alexandri*) (S. Keith, pers. comm.). Said by African hunters to be solitary, to hop while feeding on the ground, and occasionally to perch in trees (Chapin 1953).

Food. Invertebrates. Of 7 birds (Zaïre), caterpillars in 5, beetles in 5, beetle larvae in 2, small millipedes in 5, and small centipede, termite, maggot, orthopteran, isopod and snail in one each.

Breeding Habits.
NEST: large, domed, globular, with side entrance; made of large dry leaves with a few twigs, lined with plant stems and petioles, enclosing cup mainly of rootlets. Placed 1·5–2·5 m above ground, on horizontal or gently sloping tree trunk or bough.

EGGS: 1–3, av. (n = 6 clutches) 2·5. Oval, slightly glossy; creamy white, spotted or blotched with dark brown and grey, mainly at large end. SIZE: (n = 14) 24·3–30 × 21·3–22·8.

LAYING DATES: Cameroon, May; Gabon, Jan; Uganda, May; Zaïre (Medje), half-grown young Sept; (Lukolela), enlarged gonads about Oct–Nov.

DEVELOPMENT AND CARE OF YOUNG: plumage and soft parts of half-grown nestling like those of immature. Hind ends of pale superciliary stripes project like horns.

Reference
Chapin, J. P. (1953).

Pitta angolensis Vieillot. African Pitta. Brève d'Angola.

Pitta angolensis Vieillot, 1816. Nouv. Dict. d'Hist. Nat. 4, p. 356; Angola.

Forms a superspecies with *P. reichenowi*.

Range and Status. Endemic resident and intra-African migrant. Most W African populations, from Guinea to S-central Ivory Coast, in S Ghana, and from S Nigeria to Cabinda and NW Angola (forests of Cuanza Norte, Malanje and Lunda) appear to be resident. Some birds occurring in Congo and W Zaïre towns are probably migrants, however. In E Africa breeds (from sea level up to 500 m) from SE Tanzania to E Zimbabwe and 'winters' north to N Zaïre, S Central African Republic (Lobaye, Mbomou), Uganda (north to Budongo) and coastal Kenya (north to Gedi). Vagrant north to Ethiopia (Addis Ababa) and south to South Africa (Port Elizabeth). Seldom seen, so appears rare (e.g. only 1 record W Usambaras, Tanzania: 1979); but the many night migrants killed at lights suggest otherwise.

Description. *P. a. longipennis* (Reichenow): range of species east of 20°E. ADULT ♂: centre of forehead and of crown black, nape black; sides of forehead, crown and nape buff, forming a long, broad superciliary stripe from nostril to shoulder. Lores, ear-coverts and sides of neck black. Mantle, back, scapulars and upper rump yellowish green; lower rump dark violaceous azure blue; tail (including underside) matt black. Chin and throat white, strongly suffused pink; hind edge of throat black, forming neat narrow transverse bar between throat and breast; breast and flanks warm buff; belly and undertail-coverts scarlet (red area not as wide as in *P. reichenowi*); thighs brown. Outer primary blackish with small white mark at base of inner web; next, blackish with pale tip and larger white patch at base of inner web; next 3 primaries with pale greyish white tips to outer webs and large white patch across both webs at basal third of feather; next, blackish with small oval white patch at base of outer web. Secondaries blackish, the outermost 3 with large white patch at tip of outer web; inner secondaries edged bluish; tertials dark blue-green. Wing-coverts black, lesser coverts broadly tipped pale azure blue, median coverts tipped violet-blue, and inner greater coverts with 5–7 mm azure blue tips. Underside of wing glossy black except for greyish tips to outer primaries. Bill black with reddish base to lower mandible and with red spot on culmen; tongue and inside of mouth red; eye dark reddish brown; legs, feet and nails pale greyish flesh or greyish white. Sexes alike. SIZE: (6 ♂♂, 3 ♀♀) wing, ♂ 122–124 (123), ♀ 123–131 (126); tail, ♂ 42–48 (44·0), ♀ 42–47 (46·0); bill, ♂ 21–27 (24·0), ♀ 21–24 (23·0); tarsus, ♂ 36–40 (38·5), ♀ 33–40 (37·0). WEIGHT: ♂ (n = 1, Central African Republic, June) 84, (n = 1, Uganda, Oct) 92.

IMMATURE: like adult but sometimes without any azure tips to wing-coverts; breast and flanks not as clear a buff; belly and undertail-coverts washed with pink. Bill dull brown-red with blackish central portion.

NESTLING: 1, *c.* 7 days old, skin pitch black with wispy black down, black eye and long black quills, except for skin around cloaca which shows orange when cloaca distended, and 1 mm wide plastic-like strip of orange along side of bill from tip to gape below eye, with orange bulge near base of upper mandible and smaller bulge just in front of it at base of lower mandible. Bill orange, with 1·5 mm black band enclosing nostrils and encircling whole bill in middle; inside of mouth immaculate orange (A. Masterson 1987 and pers. comm. 1989).

P. a. pulih Fraser: Guinea to Cameroon (west of line Ebolowa–Yaoundé). Upperparts blue-green (like *P. reichenowi*). Median wing-coverts, rump and uppertail-coverts pale azure blue without violaceous tone. Wing rounder than in *longipennis* (P10 = P4). White patches in remiges smaller. Smaller: wing 102–115; tarsus 33–35. WEIGHT: ♂ (n = 2, Liberia) 45·2, 48·2 (adult birds, even if half the weight of *longipennis*).

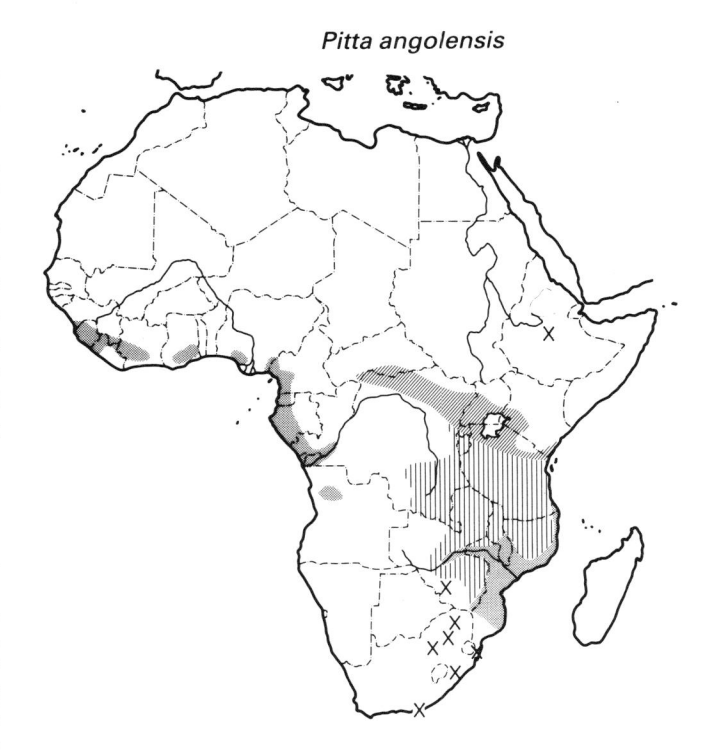

P. a. angolensis Vieillot: Cameroon (east of *pulih*) to NW Angola. Like *pulih* but larger: wing 116–124.

Sometimes treated as conspecific with *P. reichenowi* q.v.

Field Characters. A brightly-coloured, somewhat thrush-like ground bird of deeply-shaded thickets; retiring and seldom seen (particularly on breeding grounds). Extremely similar to Green-breasted Pitta, q.v., but can be distinguished by buff, not green, breast. Wings mainly black, in flight showing panel of white on primaries.

Voice. Tape-recorded (78, 86, F, LEM). A low, abrupt, repeated and remarkably frog-like croak '(g)rk', duration 0·2 s, pitch rising from 0·5 to 1·5–2·0 kHz. Main sound on breeding grounds is short, loud, deep, mellow, liquid ascending trill 'p-r-r-r-p', 'prrrrouuut' or 'ffrrueeep', coinciding with rapid wing beats as bird jumps up *c.* 25 cm into air from low perch. Other calls, of captive birds, are (1) a brief whinnying, (2) a low-pitched, querulous 'skeeow', and (3) a grunt (bird standing almost on tiptoe, fanning out red belly feathers) (Sclater and Moreau 1933). Last call and low croaks given by migrants circling in mist (C. B. Cottrell *in* Benson and Irwin 1964) are probably same as call recorded in (78). Alarm (?) a husky 'hggg'. Fledglings have mewing 'pi'-u'.

General Habits. Breeding habitat is dense deciduous thickets (in leaf), where ground is covered with forbs c. 1 m tall, and dense understorey of tall, closed evergreen forest; 'winters' (e.g. at Gedi, Kenya) in lowland semi-deciduous forest with some clearings, dominated by large *Combretum*, *Gyrocarpus* and *Lecaniodiscus* trees. Disoriented night migrants often occur at lighted windows of country and town houses.

In 'winter', feeds solitarily on ground, keeping to deep shade, in forest home range of 3000–3600 m². Home range patches are exclusive; no territorial behaviour noted. Lone bird feeds by standing motionless on forest floor on 3–5 cm-deep leaf litter, periods of watchful immobility lasting 3–5 min, after which bird hops to new place 1–2 m away. When prey detected, bird hops and lunges towards it, captures it in bill; 10-cm earthworm is mandibulated and partially swallowed several times before being swallowed finally, entire. Foraging behaviour very thrush-like, bird waiting, then moving with long easy hops each of c. 25 cm (producing crunching sound on dry litter); once bird swept litter aside with bill, then paused as if examining cleared spot for prey. If startled, flies to limb c. 3 m above ground and perches motionless, or flies away with low, direct, fast flight. If approached slowly, it takes short flight at distance of c. 7 m and then maintains that distance by hopping (Rathbun 1978). Said also to run, with head down. At rest bobs head and flirts tail. Curious or suspicious bird has vertical stance, stretching to full height (Masterson 1987). In captivity, suspicious bird sinks low and momentarily puffs out breast feathers; it may drop almost on heels, and puff feathers so as to hide feet. At same time head feathers raised, the buff 'horns' standing out. Captive bird also stretches up to full height, with buff 'horns' erect, and breast feathers puffed out with midline feather cleavage from breast to belly (**A**) (A. Masterson, pers. comm.). In aggressive posture crouches low, with bill pointing up and wings extended sideways. Seldom drinks and does not bathe. Preens elaborately; once plunged bill twice per min into fresh earth and resumed preening with earthy bill (Sclater and Moreau 1933).

In W Africa (where poorly known), only evidence of migration is occurrence at lights (Nigeria) and in towns (Congo, W Zaïre). Long-winged E African race is long-distance night migrant (see map); breeds in SE Africa Nov–Feb and 'winters' 2000 km to north, across equator, about Apr–Sept (extreme dates at Gedi, Kenya, where same bird returned to same place in 3 successive 'winters', 22 Apr to 6 Nov). Regular migrant at some localities, e.g. L. Kivu (Zaïre); in 6 years, arrived Amani, Tanzania, between 1 May and 10 June; 40 attracted on misty night to lights at Kilima Tea Factory, 8° 36′S 35° 22′E, Tanzania, 22 Apr–5 May, with 7 on one night (Beakbane and Boswell 1984). Seen every year in late Nov at Mbala (Abercorn), Zambia, where on misty night 1–2 dozen birds occur at Lake View Hotel. Elsewhere in Zambia unknown west of 27°E, and occurs in Feira, where breeds, from late Nov to early Feb (Taylor 1979). In Zimbabwe, main migration crosses northern half of central Mashonaland Plateau from mid-Nov to mid-Dec, when birds occur also in Inyanga/Umtali/Sabi area, probably heading for S Mozambique; return migration, about Apr, much less evident (Irwin 1981). Vagrants south of Limpopo R. all Nov–Dec or Apr–May.

Food. Invertebrates, including ants, termites, beetles, larvae, snails and 10-cm earthworms.

Breeding Habits. Presumably monogamous, pairs widely dispersed. In display, captive ♂ drew itself up until almost on tiptoe, fanned out red belly feathers, and grunted (Sclater and Moreau 1933). Area of c. 28 ha in Zambezi Valley 'jesse' habitat (dense deciduous thicket), where 2 nests found, evidently contained 6 pairs, some only 150 m apart, which called persistently 2 weeks before laying but not thereafter (A. Masterson et al., pers. comm.). Calling was mainly in early morning and evening, with delivery confined to established display posts. Most posts were stout horizontal branches 5–6 m above ground, in relatively open part of canopy, free from overhanging foliage; pair used several posts in radius of 30–40 m situated up to 100 m from nest. Every time bird called 'ffrrueeep' it leapt 30–40 cm up, declining its head slightly and extending wings horizontally, parachuting gently back to perch with a few quick shallow wingbeats. Call and jump repeated at irregular intervals of c. 15 s, for up to 30 min. Between calls and particularly between bouts of calling, bird swayed on hips from horizontal to vertical posture while puffing out breast feathers to expose red belly (A. Masterson et al., pers. comm.). 'Call' may be mechanical (Petit 1899). One bird repeatedly leapt into air and called, to be answered by 2nd bird nearby (Berry and Ansell 1974). 2 birds called to each other 6 times in 1 min (Masterson 1987).

NEST: large, loosely-built, untidy, domed structure of dead leaves and twigs, with side entrance like coucal's

A

nest, but characteristically encased by numerous knitting-needle-sized twigs 2–5 mm thick and 10–20 mm long. Entrance more compactly built, of leaf petioles; cup lined with papery vegetation, skeletonized leaves, fine twigs and tendrils. Size and shape of nests vary with nest site: 2 nests in *Garcinia* saplings and in dense thorny tangle had fringe of sticks projecting out at sides of roof, like brim of upturned hat. Others on horizontal outer branches of *Zizyphus* and *Ximenia* trees had much more extensive bases including platform in front of entrance (**B**) (Masterson 1987 and pers. comm. 1989). Approx. dimensions (2 nests): base 25 × 35 cm, height 18–20 cm, entrance 7–8 cm wide, 6–7 cm high; internally 10 cm wide, 11 cm long, 8 cm high. Placed 2–8 m above ground.

EGGS: 1–3, usually 3. Blunt oval or almost round; creamy white, loosely marked with scattering of irregular small dark blotches, and a few lines and scrawls on underlying grey marks, in wide zone at blunt end or evenly distributed. SIZE: (n = 10) 27·5–29 × 21·7–25 (28·0 × 23·1).

LAYING DATES: Cameroon, Sept, Oct; Angola, Nov; E Africa: Region C, Dec; Zambia, Feb; Malawi, Dec–Jan; Zimbabwe, Nov–Jan; Liberia (nestlings Sept).

DEVELOPMENT AND CARE OF YOUNG: black nestling difficult to see inside nest, except for orange on bill (and cloaca when defaecating). When scared, captive young flattened itself on bottom of nest. When offered food it lay in nest, gave thin, long drawn-out whistle 2–3 times, stretched up for food with open mouth, then subsided quietly into nest. Not vociferous nor demanding; inactive. Faecal sac, ejected out of nest entrance, is as large as nestling's head (Masterson 1987).

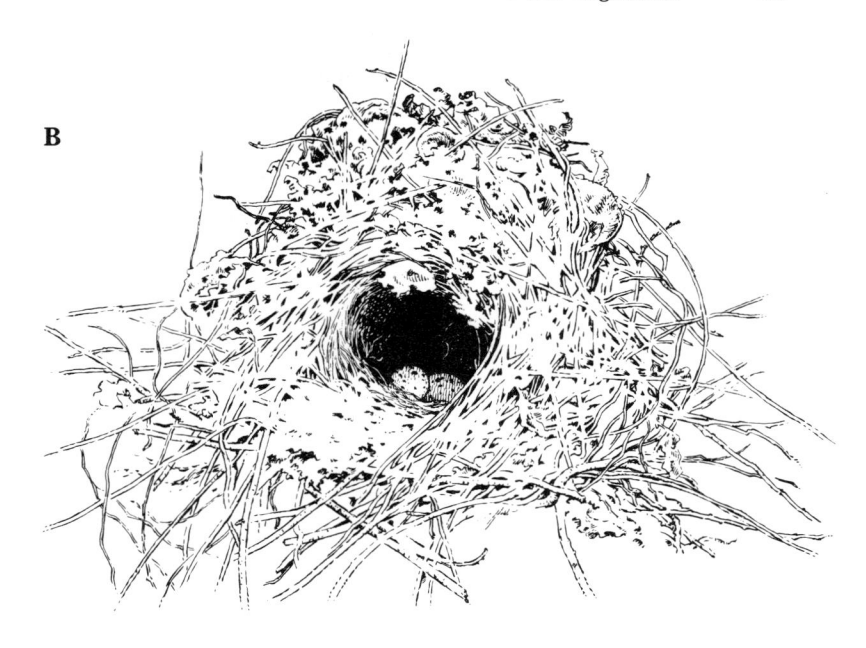

References
Benson, C. W. and Irwin, M. P. S. (1964).
Masterson, A. N. B. (1987).
Rathbun, G. B. (1978).
Sclater, W. L. and Moreau, R. E. (1933).

Suborder PASSERI (Oscines)

All song-birds or perching-birds other than the Tyranni (see above). Syrinx acromyodian: pessulus present (except in Alaudidae) and fused; no internal cartilages; 3–4 pairs of intrinsic muscles (Ames 1971). 10 primaries, or only 9 in Fringillidae, Emberizidae, Thraupidae and a few other families.

Families are defined by plumage and behavioural (rather than anatomical) criteria, and many are so similar that boundaries remain contentious. Moreover, numerous taxa as presently constituted are probably not monophyletic. 51 families recognized by Van Tyne and Berger (1959), 44 by Peters (1960–1979), and 34 by Sibley and Ahlquist (1990). Conventionally, Oscine families have not been grouped into higher taxa within the suborder. However, on the basis of DNA-DNA hybridization, Sibley and Ahlquist (1990) propose 2 parvorders, each of 3 superfamilies, with principal groups as follows: Parvorder Corvida: (a) bowerbirds, (b) honeyeaters, (c) true shrikes, vireos, crows, birds of paradise, cuckoo-shrikes, drongos, and malaconotine shrikes. Parvorder Passerida: (a) waxwings, dippers, thrushes, flycatchers, starlings, and mockingbirds, (b) nuthatches, wrens, tits, swallows, bulbuls, warblers, and babblers, and (c) larks, sunbirds, sparrows, pipits, weavers, estrildines, finches, buntings, parulines, tanagers, and icterines.

Family ALAUDIDAE: larks

Terrestrial songbirds, mainly rather small, nesting on ground and advertising territory with sustained song flights. Plumage usually cryptic, in streaky browns, often matching dominant shade of soil; sexes generally alike. Bill hard, strong; shape varies from long, slender and decurved (*Alaemon*) to short, deep and conical (*Ramphocoris*). Legs rather short to quite long; back of tarsus latiplantar (rounded, with small scutes, unique among Passeriformes); foot strong, some species with long, straight hind claw. Syrinx without ossified pessulus and with only 5 pairs of muscles, uniquely in the Passeriformes. Wing generally quite long and pointed: 10th primary variable, minute and pointed or long and broad. Tail short or medium-length. Bill morphology and condition of 10th primary suggest distant affinity with sparrows, weavers and finches, which is indicated also by biochemical studies.

About 80 species, conventionally placed in *c.* 15 genera, occurring mainly in African and Palearctic grasslands. Only 1 species, *Eremophila alpestris*, extends outside Old World. Generic boundaries and affinities ill-understood. Many genera have, in the recent past, been recognized for species that are here included in *Mirafra*. *Calandrella* sens. lat. probably comprises 3 genera: *Calandrella*, *Spizocorys* and *Eremalauda* (Dean 1989a). Conversely, there is little justification, other than conventional usage, for keeping *Alauda*, *Lullula* and *Galerida* separate (Harrison 1966); moreover, their species grade imperceptibly in morphological and biological characters with *Calandrella* spp. and *Melanocorypha* spp. Pending taxonomic revision of larks, our treatment follows White (1961) as modified by Cramp (1988) and Dean (1989a).

67 species in Africa: 57 breed in subsaharan Africa (52 endemic), 16 breed in Sahara and N Africa (none endemic), and 1 Eurasian species winters in NE Africa.

Genus *Mirafra* Horsfield

Small to large larks, occurring in moist grasslands, open savannas, and arid dwarf shrublands. Plumage varies from grey-brown to rufous-brown. Wings short, often rounded, outer primary about half the length or less of next primary; most species show some red in flight. Bill short and conical to long and decurved, nostrils exposed. Legs slender to robust, hind claw usually short and curved. Nest cup-shaped, with material built up at sides and back to form dome.

25 species worldwide, 21 in Africa, 20 endemic. 3 superspecies in Africa, *M. africana/M. hypermetra*, *M. rufocinnamomea/M. apiata* and *M. gilletti/M. degodiensis*.

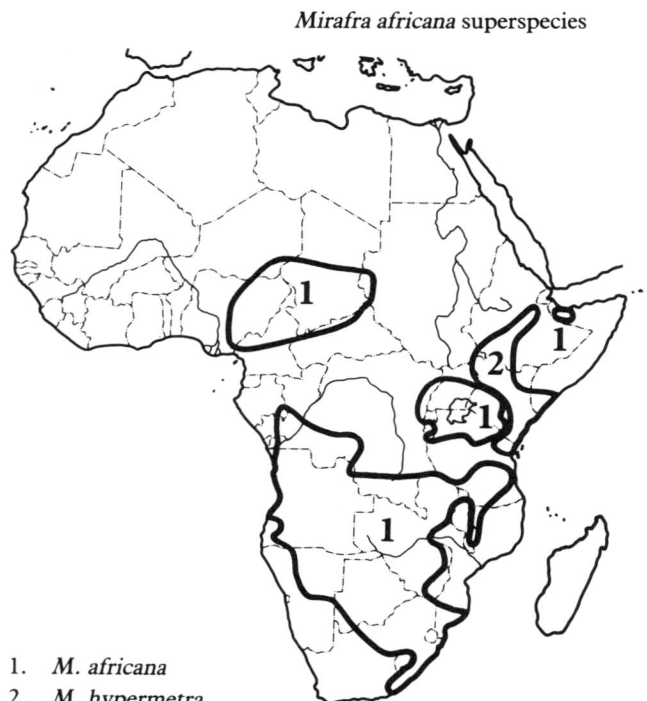

Mirafra africana superspecies

1. *M. africana*
2. *M. hypermetra*

Mirafra rufocinnamomea superspecies

Mirafra gilletti superspecies

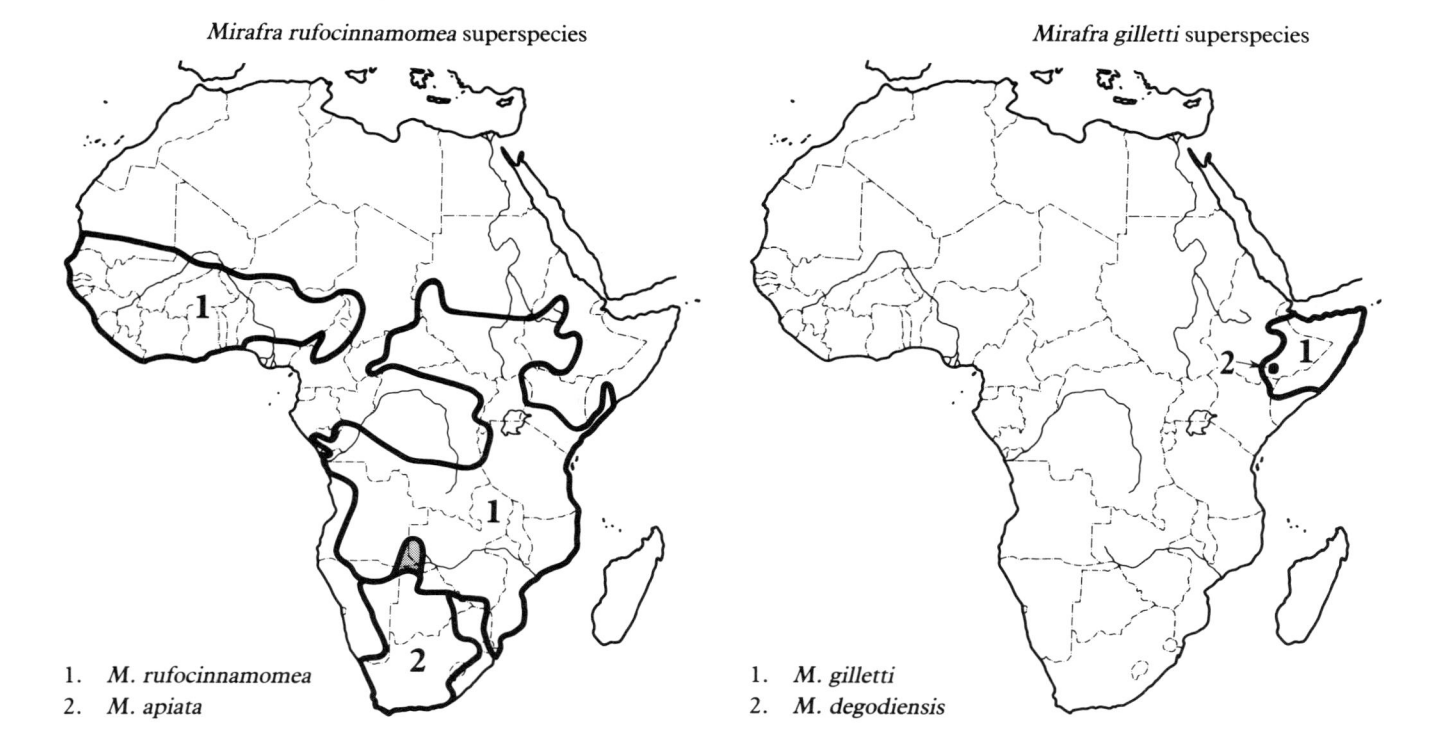

1. *M. rufocinnamomea*
2. *M. apiata*

1. *M. gilletti*
2. *M. degodiensis*

Mirafra cantillans Blyth. Singing Bush-Lark. Alouette chanteuse.

Mirafra cantillans Blyth, 1844. J. Asiat. Soc. Bengal 13, p. 960; India.

Forms a superspecies with Asian *M. javanica*.

Range and Status. Africa, S Arabia, India.

Resident and intra-African migrant. S Mauritania, N Senegambia, Mali (Sahel), Burkina Faso (on dry floodplains); widespread in Niger (Kori Torare, Aouderas, Tessaoua, Niamey, Maradi, Kaadjia, Tillia); N Togo (Tantigou, Mandouri); Nigeria (occasional at Zaria, not uncommon in NE), W-central Chad, Cameroon in extreme N near L. Chad, central and S Sudan, W and NE Ethiopia, NW and S Somalia (Ionte), N Central African Republic, NE Uganda (Karamoja, Teso); W, central and S Kenya from L. Turkana basin and Marsabit to Tsavo Nat. Park, L. Magadi, Narossura; NE Tanzania (lowlands around Mt Kilimanjaro and Mt Meru; Dar-es-Salaam). Common to locally abundant; density in Senegal 27–35 pairs/ha (Mullié and Keith in press).

Description. *M. c. marginata* Hawker (including '*schillingsi*'): S Sudan, NE Ethiopia and Somalia to Kenya, Uganda and Tanzania. ADULT ♂: crown, nape and hindneck streaked dark brown, feathers edged sandy or grey-brown, sides of neck, foreneck and ear-coverts buffy with light brown speckles. Chin and throat whitish, supercilium buff. Mantle dark brown, feathers edged light brown, back dark brown, feathers edged buff and grey-brown. Rump greyish brown, dark feather centres narrower than back, uppertail-coverts grey-brown,

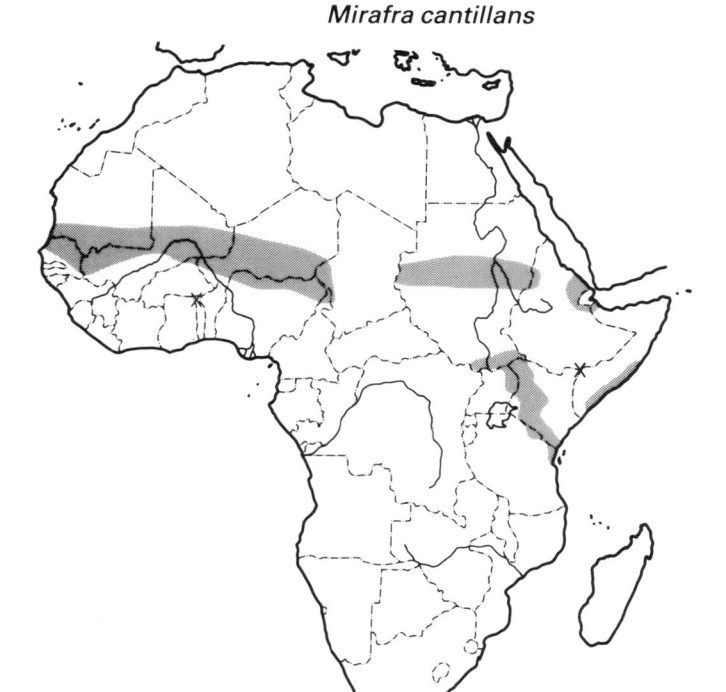

Mirafra cantillans

feathers dark along shafts. Tail dark brown, outer feathers entirely white, adjacent pair white on outer webs only. Breast buff to rufous-buff, streaked lightly with wedge-shaped brown markings, heavier on sides; rest of underparts buff, becoming whitish in centre of belly. Primaries and secondaries dark brown edged reddish buff, scapulars, tertials and upperwing-coverts dark brown edged buff. Underside of primaries and secondaries with buff inner edges, axillaries and underwing-coverts reddish buff. Bill flesh to horn, upper mandible horn-grey; eye brown; legs and feet flesh-white. Sexes alike. SIZE: (11 ♂♂, 7 ♀♀) wing, ♂ 74–80 (76·5), ♀ 67–75 (71·5); tail, ♂ 45–50 (47·8), ♀ 43–46 (44·6); bill, ♂ 13·4–14·8 (14·0), ♀ (n = 6) 12·5–14·1 (13·5); tarsus, ♂ 19–21 (20·1), ♀ 18–21 (19·8). WEIGHT: Senegal, ♂ (n = 24) 18·6 ± 1·5, ♀ (n = 9) 18·8 ± 1·4.

IMMATURE: generally paler, more sandy-buff; upperparts mottled, feathers with broader whitish margins; breast spotting less distinct.

NESTLING: unknown.

M. c. chadensis Alexander: Senegal and Mali to Sudan north of 10°N and W Ethiopia. Paler and more sandy, less heavily streaked above.

Field Characters. A small, short-tailed lark with short, stout bill, streaked upperparts and breast, rufous wing-patch and white outer tail-feathers. Race *marginata* very similar to sympatric White-tailed Bush-Lark *M. albicauda*; upperparts paler, bill lacks pronounced 2-tone effect, either uniformly pale or upper mandible only somewhat darker, but best distinguished by different song and display flight. When flushed, flies away low and hovers, sometimes flying on and hovering again, before dropping into grass.

Voice. Tape-recorded (10, 78, C, F). Song, a series of variable phrases of 9–20 chirps, whistles and buzzing notes interspersed with high piping, usually accelerating, and lasting 3–7 s, although often repeated in an unbroken stream. 'In speed and continuity fully comparable to song of Skylark *Alauda arvensis* ... piping arpeggios resemble song of Woodlark *Lullula arborea*' (North and McChesney 1964). Song the same whether given from perch or in the air (Chappuis 1985).

General Habits. Inhabits low-altitude grassy plains, open grassland with scattered bushes, acacia with short grass cover; also cultivation. In N Senegal lives in semi-arid thornbush savanna of N Sahel, characterized by low trees (57% < 2·5 m high), bushes and annua grasses, e.g. *Aristida* spp. and *Cenchrus biflorus*. On study plot near Richard-Toll, 97% of trees were *Boscia senegalensis*, *Balanites aegyptiaca* and *Acacia* spp. (e.g. *A. senegal*, *A. tortilis*, *A. seyal*) (Mullié and Keith in press). Forages on ground, including burnt ground shortly after fire; once recorded taking flying insect (Lack 1985).

Partly migratory. In W Africa moves north in rains to breed in subdesert in Mauritania, Mali, Nigeria and doubtless elsewhere; apparent new arrivals N Mali (Taberreshat) Sept, singing and displaying (Bates 1934).

Not uncommon migrant along W shores of L. Chad in dry season Sept–Apr; possible local migrant Sudan; some local W–E movement in Somalia Oct–Nov. Partial migrant in E Africa, arriving immediately after or during protracted wet periods; recorded at lights at night in Dec–Jan (Kenya, Tsavo West Nat. Park).

Food. Invertebrates and seeds (mainly grasses). In Senegal, stomach contents of adults (n = 17) were 81% grasshoppers (larvae of *Oedaleus senegalensis*), 6·6% beetles, 5·3% seeds; of juveniles (n = 3) 87% grasshoppers, 2·2% solifugids, 6·5% beetles, 4·3% caterpillars (Mullié and Keith in press).

Breeding Habits. ♂ in display flies around singing, fairly low down, alternately fluttering and gliding.

NEST: cup of fine grass and other plant material, lined with finer grass, placed on ground against tuft; built up to form dome. Nests in Senegal face north, often under grass bent in the same direction by wind.

EGGS: 2–4, mean (11 clutches) 3·6. Ovate; glossy; whitish, heavily but finely spotted with dark brown and underlying greyish brown spots, markings sometimes concentrated on obtuse end. SIZE: (n = 14) 17·5–19·5 × 13·5–15 (18·6 × 14·4).

LAYING DATES: Mauritania, May–Sept, occasionally Oct; Senegambia, June–Sept; Mali, May–Oct; Niger, Aug–Sept; Nigeria, Aug; Sudan, July–Sept; Ethiopia, Apr–June; E Africa: Region D, Apr–June.

BREEDING SUCCESS/SURVIVAL: of 27 eggs in 8 nests, Senegal, 21 hatched and 11 fledged, i.e. breeding success 40%; however, many other nests were less successful (4 destroyed by predators, 2 deserted, no eggs laid in 5), and overall productivity per nest estimated to be between 7 and 13%, i.e. of 100 eggs laid, only 7–13 produced a fledgling (Mullié and Keith in press).

Reference
Mullié, W. M. and Keith, J. O. (in press).

Mirafra passerina Gyldenstolpe. Monotonous Lark. Alouette monotone.

Mirafra passerina Gyldenstolpe, 1926. Ark. Zool. Stockholm 19 (A), (1), p. 24; Mohapoani, South Africa.

Plate 5
(Opp. p. 96)

Range and Status. Endemic resident and nomad. SW Angola (Cainde), S Zambia (Mfubakazi, Livingstone), N Namibia (scattered localities on fringe of N and central Namib, central and S Damaraland, and on NE border), E and central Botswana, Zimbabwe (scattered localities in W Matetsi, Middle Zambesi Valley, NW Matabeleland, SE lowveld to Chipinda Pools, Limpopo R. valley), South Africa in N Cape (Upington, Kimberley District), Transvaal (widely, but sparsely recorded in bushveld in N, central grasslands and E lowveld). Locally common.

Description. ADULT ♂: upperparts brown, feathers with dark centres and pale edges and tips, giving streaked appearance; back somewhat darker than mantle, rump more greyish brown. Tail dark brown, outer feathers entirely white, adjacent pair white on outer webs only. Sides of neck and foreneck brown, lightly speckled with darker brown. Ear-coverts light brown speckled and streaked brown, rather poorly marked pale supercilium. Chin white, throat greyish white, faintly speckled. Breast buff, speckled and streaked with dark brown, rest of underparts white to greyish white. Primaries and secondaries dark brown edged rufous, scapulars, tertials and wing-coverts dark brown, edged buff to rufous-buff. Inner webs of underside of primaries and secondaries edged pale rufous-buff, underwing-coverts and axillaries rufous-buff. Bill blackish horn above, pinkish horn below; eye hazel; legs and feet flesh pink. Sexes alike. SIZE: (7 ♂♂, 6 ♀♀) wing, ♂ 82–87 (85·0), ♀ 77–80 (78·0); tail, ♂ 51–58 (54·0), ♀ 45–50 (48·0); bill, ♂ 14·1–16 (15·0), ♀ 14–15 (14·3); tarsus, ♂ 20–24 (21·7), ♀ 20–22 (21·1). WEIGHT: ♂ (n = 4) 24–27·5 (25·4), 1 ♀ 24, unsexed (n = 5) 21–27 (24·2).

IMMATURE and NESTLING: unknown.

Field Characters. A small, dumpy, short-tailed lark with short, heavy bill, rather poorly-marked white eyebrow that stops short of bill, and white outer tail-feathers; reddish edges to primaries show as patch in flight. Similar to Melodious Lark *M. cheniana* except for white belly (buff in Melodious), and different song and display, but although geographically sympatric, occupies different habitat, so confusion unlikely even when not singing.

Voice. Tape-recorded (88, F, GIB). Song a quick 4–5 note phrase of liquid notes, 'for-syrup-is-sweet', 'toll-it-to-him', 'cool-cooler-chee', 'willow-way-wit', emphasis on the last syllable, repeated monotonously over and over; with pause of *c.* 5 s between phrases.

General Habits. Occurs on bare stony patches in open shrubby woodlands and on stony ridges in *Combretum* woodlands; occupies the ecotone between miombo woodland on Kalahari sand and vegetation along drainage lines, also large clearings in mopane woodland, and sparsely wooded stony plains in acacia savanna; in *Brachystegia boehmii* woodland, inhabits sparse grassland on old lands and firebreaks. Calls from perch on tree or bush; while calling, holds body at *c.* 45°, pushes head forward and puffs out throat.

Nomadic; has no regular migration, but moves into areas in response to rainfall. May suddenly appear in large numbers following rains, breed and move on, perhaps not to return for some years.

Food. Invertebrates and grass seeds.

Breeding Habits. Breeding birds call at night as well as most of day. ♂ has short display flight, rising slowly to *c.* 15 m before gliding down and returning to perch; throughout flight continues to give same monotonous call.

NEST: deep cup of grass, domed over with grass, placed on ground against a tuft.

EGGS: 2–4, mean (13 clutches) 3·0. Ovate; white, heavily mottled or speckled with brown and grey-brown, with underlying slate speckles, markings sometimes concentrated at obtuse end. SIZE: (n = 19) 18·5–21 × 13·5–15·2 (19·4 × 14·3).

LAYING DATES: Zimbabwe, Oct–Mar; South Africa: Transvaal, Nov–Dec; Cape Province, Feb–Mar.

References
Thomson, R. (1983).
Vernon, C. J. (1983).

18　ALAUDIDAE

Plate 3
(Opp. p. 48)

Mirafra albicauda Reichenow. **White-tailed Bush-Lark. Alouette à queue blanche.**

Mirafra albicauda Reichenow, 1891. J. Orn., p. 223; Gonda, Tabora District, Tanganyika Territory.

Range and Status. Endemic resident. W Chad (localized in area of L. Chad), NE Zaïre (locally common), E Sudan (uncommon and local), Uganda (Karamoja, Teso in NE, Butiaba, Rwenzori Nat. Park in W, uncommon), W and central Kenya (in higher rainfall areas from Pokot, Bondo, Karunga Bay in W, to Kitui and Namanga, mainly west of Rift Valley, uncommon), Tanzania (Rukwa valley, common), Tabora, Serengeti Nat. Park, Arusha Nat. Park, Ardai Plains, Engaruka). Generally uncommon and local.

Description. ADULT ♂: upperparts dark grey-brown, feathers with dark, almost black centres and light grey-brown to buff margins, giving a markedly streaked appearance. Nape and uppertail-coverts paler grey, streaked dark grey-brown. Tail dark brown, outer feathers entirely white, adjacent pair white on outer webs only, central pair with narrow pale margins. Supercilium, cheeks and ear-coverts brownish, throat white, breast spotted and streaked with dark brown in a narrow band shoulder to shoulder. Underparts buff. Primaries dark brown with narrow rufous outer edges and broad rufous margins to basal half of inner webs. Secondaries dark brown with pale outer edges and buff inner edges. Upperwing-coverts dark brown with buff margins. Underside of primaries and secondaries with reddish buff inner edges, axillaries and underwing-coverts reddish buff. Bill blackish slate, lower mandible paler; eye brown; legs and feet flesh brown. Sexes alike. SIZE: (5 ♂♂, 3 ♀♀) wing, ♂ 80–85 (82·6), ♀ 75–80 (77·6); tail, ♂ 40–48 (45·4), ♀ 42–44 (43·3); bill, ♂ 15–16 (15·4), ♀ 14–15·5 (14·7); tarsus, ♂ 21–22 (21·6), ♀ 21–22 (21·6). WEIGHT: 1 ♂ 25, ♀ (n = 3) 20–23 (21·3), 1 unsexed juvenile 22.
IMMATURE and NESTLING: unknown.

Field Characters. A small, short-tailed lark with short, stout bill, streaked upperparts and breast, rufous wing-patch and white outer tail-feathers. Very similar to race *marginata* of Singing Bush-Lark *M. cantillans*; upperparts blacker, bill has more 2-tone effect, upper mandible darker than lower, but best distinguished by different song and display flight. Shares habit of hovering before dropping into grass.

Voice. Tape-recorded (F, LEM, McVIC). Song a variable-length refrain of harsh and mellow chirps, interspersed with whistles, uttered on the wing; lacks the trills and the range of Singing Bush-Lark *M. cantillans*.

General Habits. Inhabits open or lightly bushed grassland, favouring higher and wetter short-grass plains in Sudan and Kenya, especially on black cotton soils from 600 to 2000 m. Also common around L. Rukwa, Tanzania, in short grassland dominated by *Diplachne fusca* and *Sporobolus spicatus* close to the shores of the lake, in perimeter grasslands where *Sporobolus marginatus* is abundant, and on sedge *Cyperus longus* and short grass *Cynodon dactylon* 'lawns' in *Echinochloa pyramidalis* flood-plain grassland.

Mirafra albicauda

Shelters from heat in shade of grass tussocks. Metabolic water and invertebrates apparently primary water source for part of year; drinks regularly if surface water available (Vesey-FitzGerald 1957).

Food. Invertebrates (Orthoptera), seeds and green vegetable matter.

Breeding Habits. In display, ♂ soars high in the sky in circular flight, calling continuously; after singing, descends steeply to ground.
NEST: domed or open cup of dry grass, ext. diam. c. 70, placed in saucer-shaped depression in damp ground against a tuft, often quite exposed. In moist sites, mud is incorporated into grassy rim of nest. Nests often sited on flooded black alluvial soils, on patches of bare ground at side of path, in areas trampled bare by game or in bare patches in grassland ecotones.
EGGS: 2 (n = 4). Off-white, freckled all over with brown, sometimes with concentration of small brown blotches at broad end. SIZE: (n = 3) 16·4–19·4 × 12·1–13·1 (17·8 × 12·8).
LAYING DATES: Sudan, May; E Africa: Region C, Apr–May; Region D, Mar.

Reference
Vesey-FitzGerald, D. F. (1957).

Mirafra cheniana Smith. Melodious Lark. Alouette melodieuse.

Mirafra cheniana A. Smith, 1843. Illustr. Zool. South Africa, Aves, pl. 89, fig. 1 (and text); Latakoo, South Africa.

Plate 5
(Opp. p. 96)

Range and Status. Endemic resident. Highly fragmented and patchy range in Zimbabwe (Bulawayo District, Daisyfield), Botswana (Patlana Flats, L. Ngami area), South Africa (Transvaal from northeast of Bronkhorstspruit, Pretoria, west to Krugersdorp, Potchefstroom, Wolmaransstad-Bloemhof, occasional Barberspan; central and E Orange Free State; N Cape Province (Kuruman type locality; no recent records), E Cape (Queenstown, Bedford); Natal interior (Matatiele, Franklin, Bergville, Winterton, Colenso)). Frequent to locally common.

Description. ADULT ♂: upperparts light brown variably streaked dark brown; some birds darker, i.e. more heavily streaked than others; feathers with dark centres and light brown margins varying in width. Rump buff to light brown, feathers with dark centres. Tail dark brown, outer webs of outer feathers white. Lores whitish, ear-coverts brown, chin and throat white, breast buff to light brown, speckled dark brown, rest of underparts buff. Primaries and secondaries dark brown with narrow rufous outer edges and pale rufous inner edges. Tertials dark brown with buff to brown margins, scapulars and upperwing-coverts dark brown with broad light brown to rufous-brown margins. Underwing pale rufous. Bill light brown to horn; eye brown; legs and feet pinkish to flesh. Sexes alike. SIZE: (10 ♂♂, 7 ♀♀) wing, ♂ 69–78 (74·0), ♀ 71–76 (74·0); tail, ♂ 43–48 (45·0), ♀ 40–45 (42·0); bill, ♂ 13–14·8 (13·8), ♀ 13–14·1 (13·7); tarsus, ♂ 18–22 (19·9), ♀ 18–21 (19·5).

IMMATURE: crown mottled, back streaked and mottled, face dark with very thin whitish supercilium; breast streaked. Upper mandible flesh-horn, lower mandible darker.

NESTLING: down on crown, back, wings and thighs. Gape-flanges yellow, inside of mouth orange, with 3 black spots on tongue, 1 apical and 1 on either side of anterior tongue; 1 black spot on tip and 2 dark marks on side of lower mandible.

Field Characters. A small short-tailed lark with heavy conical bill, fairly heavily-streaked upperparts and white outer tail-feathers; red on primaries shows as patch in flight. Similar to Monotonous Lark *M. passerina*, q.v., differing mainly in buff (not white) belly and flanks, different song and display, but confusion unlikely because they live in different habitats.

Voice. Tape-recorded (75, 81, 88, F, WALK). Song a wide variety of chatters, trills, whistles and scratchy notes, broken into a succession of jerky phrases, often given in a continuous stream; includes many imitations of other species. Alarm call, a repeated 'chuk-chuk-chucker-chuk'.

General Habits. Present in climax grasslands dominated by *Themeda triandra*, occasionally in planted *Eragrostis curvula* pastures, at 550–1750 m altitude, with annual rainfall 400–800 mm; localized to optimal habitat, where numbers may vary from year to year. Solitary or in pairs. Calls from perch or, more commonly, in flight.

Food. Invertebrates and seeds.

Breeding Habits. In flight display, ♂ rises to 20 m or more with wings beating rapidly, body feathers puffed out, flies in circle 50 m in diam. for 25 min or more, singing continuously, ends by dropping suddenly into grass.

NEST: deep cup of grass, lined with rootlets and finer material, with grass dome, placed in scrape in ground, usually against tuft of tall grass.

EGGS: 2–4, mean (10 clutches) 2·8. Ovate; white to buff, heavily spotted and speckled with brown and grey-brown, markings sometimes obscuring ground colour completely, sometimes concentrated at obtuse end. SIZE: (n = 20) 17·8–20·9 × 13·8–14·7 (19·3 × 14·2).

LAYING DATES: Zimbabwe, Sept, Nov–Jan; South Africa: Transvaal, Nov–Feb; Orange Free State, Oct–Dec.

20 ALAUDIDAE

Plate 2
(Opp. p. 33)

Mirafra cordofanica Strickland. Kordofan Lark. Alouette du Kordofan.

Mirafra cordofanica Strickland, 1852. Proc. Zool. Soc. Lond. (1850), p. 218; Kordofan, Sudan.

Range and Status. Endemic resident and intra-African migrant. Senegambia, Mauritania (almost throughout), Mali (between 15° and 23°N, uncommon, and local in Tombouctou area, Gao, In Alahy), Niger (uncommon, records from Tahoua, Tazza); east to Sudan (rare, in Darfur and Kordofan).

Description. ADULT ♂: crown, nape, hindneck, sides of neck and foreneck pale rufous-brown or cinnamon-rufous, feathers with buffy white edges. Lores, supercilium and area below eye buffy white, cheeks and ear-coverts brown streaked white, chin and throat white. Mantle like crown, but some feathers with dark shafts, streaks darker towards tip. Back pale rufous-brown, some feathers with buffy white tips and inner edges and subterminal blackish brown inner margins. Rump and uppertail-coverts plain pinkish brown. Tail-feathers dark brown, outer pair mainly white with dark brown basal half of inner web, next pair with white outer web and tip, 2 central pairs pale rufous with dark shafts, central pair with buffy margins. Underparts buffy white, breast lightly washed pale rufous with a few dark brown to rufous wedge-shaped spots. Flight feathers rufous-brown, primaries with outer webs and basal half of inner webs pinkish brown, secondaries similar but margins broader, tips same colour. Scapulars and wing-coverts pinkish brown, tipped and edged buffy white, tertials pinkish brown, inner margins blackish brown, outer margins buff. Underwing pale rufous-buff. Bill pale flesh-white, tip and culmen browner; eye brown; legs and feet flesh pink. Sexes alike. SIZE: (3 ♂♂, 4 ♀♀) wing, ♂ 78–87 (82·3), ♀ 78–84 (80·7); tail, ♂ 51–56 (53·6), ♀ 53–56 (55·0); bill, ♂ 14–14·6 (14·3), ♀ 13·5–15 (14·1); tarsus, ♂ 21·5–23 (22·1), ♀ 22–24 (22·8).
IMMATURE: crown feathers tipped subterminally with dusky black, wing-coverts with subterminal black inner margin and buff outer margin. Underparts white, feathers downy, chest spotted with dull blackish.
NESTLING: unknown.

Field Characters. A medium-small, reddish lark with streaked breast, distinctively tricoloured tail – blackish with broad white outer edges and rufous centre. Sympatric Dunn's Lark *Eremalauda dunni* has paler upperparts, unstreaked white underparts, tail black and rufous only.

Voice. Not tape-recorded. Song said to be sweeter and more musical than that of Singing Bush-Lark *M. cantillans* (Lynes 1924).

General Habits. Present in arid open plains with *Aristida* grasses and scattered bushes on red sands. Seasonal movements in Mauritania and Mali.

Food. Insects and seeds.

Breeding Habits. ♂ sings from perch on low bush or tree in display. Nest and eggs undescribed.
LAYING DATES: Mauritania, May–Aug; Mali, May–July; Niger, Aug; Sudan, May.

Reference
Lynes, H. (1924).

Plate 3
(Opp. p. 48)

Mirafra williamsi Macdonald. Williams's Lark. Alouette de Williams.

Mirafra williamsi Macdonald, 1956. Bull. Br. Orn. Club 76, p. 71; Marsabit, Kenya.

Range and Status. Endemic resident in N Kenya. 2 disjunct populations: (1) Dida Galgalla Desert, Marsabit; (2) between Isiolo and Garba Tula. Rare and local.

Description. ADULT ♂: crown, nape and hindneck brown, feathers with dark centres and paler edges, giving streaked effect; sides of neck and foreneck whitish, streaked and spotted brown, ear-coverts brown; lores, supercilium and line from bill under eye white; chin white with a few brown marks, throat white. Mantle sepia, lightly streaked, feathers with dark centres, paler edges. Back sepia, feathers with distinct buff tips and buff inner edges, rump and uppertail-coverts plain sepia. Tail-feathers dark brown, outer pair white, with dark brown on inner web tapering from full width of inner web at base to a point about two-thirds along feather, next pair with white outer webs and tips, central pair brown with lighter brown edges. Breast light brown, dark brown wedge-shaped spots, lower breast plain brown, remainder of underparts buffy white, flanks with a few dark streaks and light brown wash. Flight feathers brown, primaries with dark rufous edge to outer webs (except on apical one-third), secondaries similar but rufous edge narrower. Scapulars dark brown with dark shafts, buffy tips, tertials dark brown with buff margins, upperwing-coverts brown with blackish shafts and subterminal band, buffy tips. Underwing-coverts and axillaries light rufous-brown. Bill horn-grey, cutting edge and lower mandible pinkish white; eye dark brown; legs and feet pinkish white with yellow tinge. Sexes alike. SIZE: (2 ♂♂) wing 84, 84; tail 49, 56; bill 14·5, 15·7; tarsus 21, 22.

IMMATURE and NESTLING: unknown.

Field Characters. A smallish lark, extremely similar to Singing Bush-Lark *M. cantillans*, with same small rusty wing-patch and white outer tail-feathers; mainly differs in darker, less streaked upperparts. Race *intercedens* of Fawn-coloured Lark *M. africanoides* is about same size but much redder, upperparts heavily streaked, large red wing-patch, much less white in tail.

Voice. Unknown.

General Habits. Frequents arid or semi-arid short grass plains on black lava soils at 600–1350 m, and well-grazed, grassy areas with scattered bushes on sandy soil.

Food. Insects and seeds.

Breeding Habits. Unknown.

Mirafra williamsi

Mirafra pulpa Friedmann. Friedmann's Lark. Alouette de Friedmann.

Plate 3
(Opp. p. 48)

Mirafra pulpa Friedmann, 1930. Occ. Pap. Boston Soc. Nat. Hist. 5, p. 257; Sagan River, southern Shoa, S Abyssinia.

Range and Status. Endemic, E Africa. Status uncertain but apparently a migrant or wanderer. Known from type locality in S Ethiopia, Archer's Post and Kiboko, Ngulia and L. Jipe in Tsavo West Nat. Park, and appears rarely in Tsavo East Nat. Park during rains (Lack 1977). Uncommon in Tsavo area in rainy seasons; rare or vagrant elsewhere. The records from Kamathia and south of Kapedo reported by Pearson *et al.* (East African Bird Reports for 1984 and 1986) and by Lewis and Pomeroy (1989) may refer to the Singing Bush-Lark *M. cantillans* (D. J. Pearson, pers. comm.).

Description. ADULT ♂: upperparts brown with paler edges to feathers, especially on nape, and contrasting darker centres to feathers; slight pale eye-stripe; ear-coverts brown, separated by whitish line from a dark brown conglomeration of spots on side of upper breast; chin and throat white; breast slightly tawny with dark brown spots; belly white with brownish tinge (soil stain?); outermost tail-feather (T6) white, T5 brown with outer web white, inner web of T5, T4 and T3 very dark brown, T2 and T1 paler and slightly rufous with edges paler still. Primaries and secondaries brown with rufous outer edges; tertials reddish brown with pale edges and darker centres; wing-coverts like upperparts but more reddish brown and with very dark subterminal spots. Upper mandible dark horn, lower paler; eye

Mirafra pulpa

brown; legs dark flesh; feet fleshy pink. Sexes alike, so far as known. SIZE: (4 ♂♂, 1 unsexed) wing (n = 4), 81–85 (83·3); tail (n = 4), 54–57 (55·0); bill to feathers (n = 3), 11·5–12 (11·7); bill to skull (n = 2), 14, 14·5; tarsus (n = 5), 20–23 (21·5); hind toe (n = 5), 6–6·5 (6·2); hind claw (n = 5), 5–6 (5·6). WEIGHT: (Kenya, Dec) 2 ♂♂ 21·2, 21·5, 1 unsexed 23·1.
IMMATURE AND NESTLING: unknown.

TAXONOMIC NOTE: authors now generally agree that *M. candida* Friedmann, which was described 3 months after *M. pulpa*, is the same species (Lack 1977).

Field Characters. A small, reddish brown, streaked lark of bush country. Most easily distinguished by characteristic song, but otherwise very similar to several other species (Lack 1977). Singing Bush-Lark *M. cantillans* is smaller and much greyer, has less patterning and no rufous on tertials and wing-coverts. Williams's Lark *M. williamsi* is also greyer, especially on upperparts, has spots on breast extending onto throat, and heavier bill. Somali Short-toed Lark *Calandrella somalica* is generally paler and greyer, has less rufous in wings and only a small amount of white in tail.

Voice. Tape-recorded (GREG, McVIC). Song a characteristic, single long drawn-out 'hoo-ee-oo', with a slight but definite emphasis on middle part. Repeated at intervals of 1–2 s. No other calls known.

General Habits. In Tsavo East Nat. Park (Kenya), prefers rather open bushed grassland with fairly thick grass layer; rarer in areas with denser bushes. It may prefer wetter areas (Hall 1961) and wetter than Tsavo East can provide (Lack 1977). Solitary and rather wary, not allowing approach closer than about 50 m. When disturbed, flies rather slowly and drops abruptly into grass.

Apparently a migrant but movements poorly known. Records in Tsavo East are Dec and Jan, also Mar (1) and Apr (1). The Tsavo West records are from L. Jipe in Aug and specimens at Ngulia Lodge collected at night in Dec, together with many Palearctic and intra-African migrants.

Food. Bird from Tsavo East Nat. Park contained grass seeds, small grasshoppers, beetles (probably weevils), other insect remains and many small grains of quartz (Lack 1977).

Breeding Habits. In possible display flight, ♂ sings while 'floating' around in undulating flight at *c.* 5 m, sometimes up to 10 m. Seldom flies around for more than 20–30 s at a time; also sings from top of small bush. Sings all day, and at night when moon full.

Reference
Lack, P. C. (1977).

Plates 3, 5
(Opp. pp. 48, 96)

Mirafra africana Smith. Rufous-naped Lark. Alouette à nuque rousse.

Mirafra africana A. Smith, 1836. Rep. Exped. Expl. Cent. Afr., p. 47; eastern province of the Colony (= Algoa Bay).

Forms a superspecies with *M. hypermetra*.

Range and Status. Endemic resident. Range discontinuous: SE Guinea, N Liberia and W Ivory Coast (Mt Nimba). Extreme SE Niger (near L. Chad), Nigeria (Jos, Bauchi plateaux, uncommon and very local, Mambilla Plateau, locally common, Gashaka-Gumti Game Reserve, frequent); SE Gabon, Congo, Cameroon (scarce and local, Adamawa Plateau and Montane District (Ngaoundéré, Sabga Pass, Kounden)), W Chad, W Sudan (fairly common), fairly common in NW Somalia, locally common in central and SW Kenya, S and SW Uganda, Rwanda, SW Burundi, NE, SE and S Zaïre, and N, W and E Tanzania. Local in Malaŵi (Mangochi District, Nyika Plateau), widespread in Zambia. Locally common in NE, central, W and SE Angola (on central plateau and from Luanda south along the coast to Leba), Zimbabwe (locally common on central plateau, patchily distributed in NE and SW where locally common to sparse), S Mozambique (locally common south of Save R.), Swaziland (locally common), Botswana (locally common to sparse), N and NE Namibia (south to Okahandja; also in Caprivi Strip). In South Africa common, widespread in Transvaal, Natal and N-central and E Orange Free State; fairly common locally in E Cape.

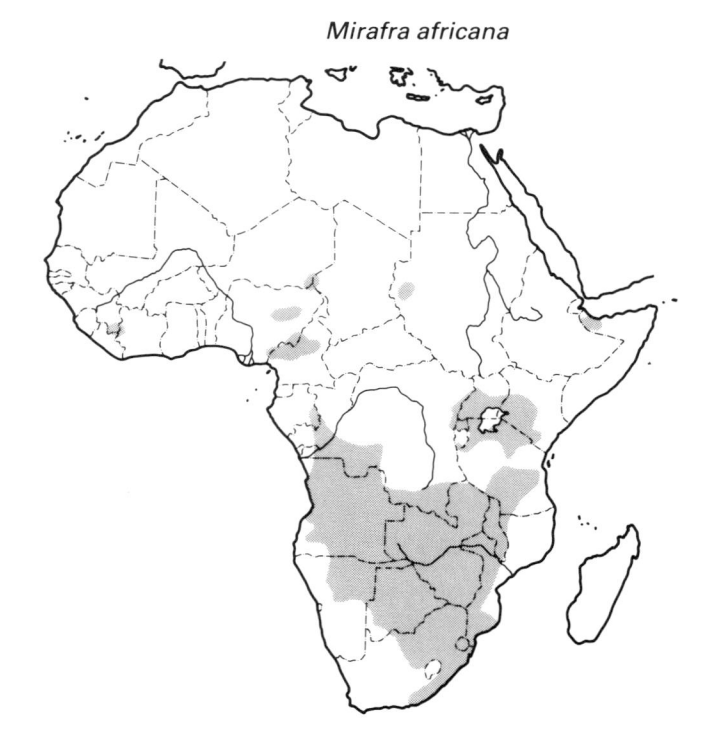

Mirafra africana

Description. *M. a. africana* Smith: South Africa (E Cape, Natal). ADULT ♂: crown, nape and hindneck streaked, feathers with dark centres and lighter margins, those of hindcrown and nape with much rufous, producing distinctive patch. Sides of neck and foreneck buffy speckled brown, lores and supercilium buff, cheeks and ear-coverts brown, chin buff, throat whitish. Mantle rufous-brown, streaked, back darker brown than mantle, feathers with dark centres, buff and rufous-brown margins; rump dark brown, feathers with indistinct dark centres; uppertail-coverts grey-brown with dark centres. Tail-feathers dark brown, outer webs of outer pair buff. Breast pale rufous, spotted darker brown, rest of underparts rufous-buff, flanks a shade darker. Flight feathers dark brown, both webs rufous at base, only outer webs rufous towards tip. Scapulars brown, with blackish centres, buff outer edges and tips, tertials dark brown with narrow dark inner margin and pale buff fringes; upperwing-coverts dark brown edged buff and rufous. Underwing rufous, outer half of flight feathers brown. Bill brown; eye hazel; legs and feet pink. Sexes alike. SIZE: wing, ♂ (n = 26) 89–102 (95·4), ♀ (n = 21) 79–99 (87·5); tail, ♂ (n = 23) 56–65 (59·7), ♀ (n = 20) 47–62 (56·8); bill, ♂ (n = 23) 18·5–23·5 (21·8), ♀ (n = 20) 18–22·1 (19·4); tarsus, ♂ (n = 23) 26·7–32 (29·0), ♀ (n = 21) 25–32 (27·6). WEIGHT: ♂ (n = 25) 41·3–47·5 (44·0), ♀ (n = 12) 33–49 (40·3).

IMMATURE: feathers of crown and back with buff edges.

NESTLING: pale grey to buff down on crown, back, wings and thighs.

M. a. transvaalensis Hartert: SE Botswana, South Africa (Orange Free State, Transvaal, Natal interior), Zimbabwe, Mozambique, Malaŵi, SE Zambia, S Tanzania. Paler above, less heavily streaked than *africana*, more greyish or pale reddish, underparts paler and whiter.

M. a. grisescens Sharpe: NW Zimbabwe (Hwange), N Botswana (Makgadikgadi, L. Ngami), W Zambia. Upperparts paler and more greyish than *transvaalensis*, underparts whiter.

M. a. ghansiensis (Roberts): W Botswana (Ghanzi), Namibia (Rietfontein). Upperparts pale pinkish sandy, lightly streaked.

M. a. pallida Sharpe: Namibia (Windhoek north to Kaokoveld and Ovamboland). Upperparts paler and more yellowish brown than *africana*, lightly to heavily streaked.

M. a. occidentalis Hartlaub: W Angola (Huila north to Kisama, south of Luanda). Upperparts darker and more grey-brown above than *pallida*.

M. a. chapini Grant and Praed: S Zaïre (Marungu, Shaba), NW Zambia (Mwinilunga, Bangweulu). Upperparts dark greyish brown, heavily streaked with blackish, bill heavy.

M. a. gomesi White: NW Zambia (Kabompo), E Angola (Macondo). Upperparts greyer than *chapini*, blackish streaking more sharply defined, upper back more pinkish brown.

M. a. kaballii White: W Zambia (Balovale), NE Angola (Luacana, Kassai). Upperparts light bluish grey and not heavily streaked, lighter and less streaked than *gomesi*, rufous areas pale and pinkish; bill heavy and curved. Intergrades with *malbranti*.

M. a. malbranti Chapin: central (Djambala) and S Zaïre (Petianga, Kasai). Upperparts light reddish brown, lightly streaked.

M. a. nigrescens Reichenow: NE Zambia (Lundazi), S Tanzania (Ukinga, Njombe). Upperparts dark reddish brown with heavy blackish streaking, underparts dark buff with blackish streaks on chest extending over flanks. Bill fine and rather decurved. WEIGHT: ♂ (n = 8) 50–54·5 (51·9), ♀ (n = 2) 44, 45·5.

M. a. isolata Clancey: Malaŵi (Mangochi District, Namweru). Upperparts greyish, underparts tawny.

M. a. nyikae Benson: Nyika Plateau above 2000 m and high-altitude grasslands in E Zambia and Malaŵi. Upperparts dark reddish brown or chestnut, with blackish streaking. Underparts buffy, teardrop markings on throat, upper breast and flanks brown. WEIGHT: ♂ (n = 2) 51, 51·7.

M. a. tropicalis Hartert: N and NW Tanzania (Hanang, Bukoba, Mbulu, Kasulu), S-central Uganda (equator north to Masindi), W Kenya (Kavirondo, Kisumu). Upperparts rich orange-red with blackish streaking, underparts rufous-orange, much more richly coloured than *africana*.

M. a. ruwenzoria Kinnear: SW Uganda (Kigezi, Ankole, Toro) and E Zaïre (Semliki to Rwenzori, Rutshuru, Rwindi Plain). Upperparts greyer and more heavily streaked than *tropicalis*.

M. a. athi Hartert: NE Tanzania (L. Manyara, Arusha), central Kenya (Nairobi, Nakuru). Upperparts greyish to buffy grey with blackish streaking. Some variation in the population: birds from SE are greyest with little or no rufous on crown or wing-coverts; birds from W parts of range are more brownish above; SW birds have crown rufous like *tropicalis*.

M. a. harterti Neumann: E Kenya (Ukamba). Like *athi* but feathers of upperparts with deep orange margins.

M. a. sharpei Elliot: Somalia (Silo Plain, Tuyo Plain, Bankisah). Upperparts bright coppery red with very little streaking.

M. a. kurrae Lynes: Sudan (Kurra, Darfur). Upperparts vinous rufous heavily streaked black; underparts, including throat, deep tawny.

M. a. stresemanni Bannerman: N Cameroon (Ngaoundéré). Crown and upperparts rich rufous with very little black streaking; underparts very dark, rich cinnamon.

M. a. bamendae Serle: W Cameroon. Upperparts and underparts as *stresemanni* but heavily streaked blackish above. Small: wing, ♂ (n = 4) 89.

M. a. batesi Bannerman: Niger, Nigeria. Similar to *kurrae* but upperparts more sandy red.

M. a. henrici Bates: Guinea, Liberia. Crown plain but remainder of upperparts heavily streaked blackish on sandy red.

Field Characters. Common and widespread, the 'basic' lark in many parts of Africa. Often sings from conspicuous perch on top of bush, its short, sweet whistle a familiar open-country sound. Large, stocky and heavy-billed, with short crest giving head peaked appearance; rufous nape distinctive but only visible at close quarters. Large red wing-patch conspicuous in flight. General plumage colour not a useful guide as it varies greatly according to race.

Voice. Tape-recorded (10, 78, 81, 88, F, RS). Song from perch, 2 clear, drawn-out whistles, each one slurred and sounding like 2 notes: 'tseep-tseeoo', 'tsuwee-tsweeoo', 'tiree-tiroo', 'teeoo-teewee', 'treelee-treeloo', 'tooyoo-tooyoo'; repeated endlessly, with 1–4 s (usually 1–2 s) interval between phrases. Wings often flapped during song, producing rattled 'phrrrp'. Flight song rambling whistles, tweets and trills, sometimes with imitations of other species. More rarely heard call from perch is single, rather ventriloquial repeated whistle, sometimes a double note, a higher and a lower, reminiscent of tit *Parus* (R. Stjernstedt, tape). Alarm, 'peewit' or 'tweekiree'.

General Habits. Present in a variety of habitats, including open grassland with bare patches and termitaria or scattered bushes, open scrubby acacia woodland with short, thin grass cover in clearings or between trees, and bare patches in cultivated lands and fallow fields.

Solitary or in pairs. Forages at bases of grass tufts, in patches of ungulate droppings and in bare areas, taking food from soil surface and gleaning from leaves of grasses and forbs. Takes insects in flight, e.g. termite alates.

Food. Invertebrates (adult and larval beetles, larval Lepidoptera, grasshoppers, termites, mantids, larval Diptera, spiders, millipedes, earthworms) and seeds (of grasses).

Breeding Habits. Monogamous, territorial. In display ♂ sings from perch, holding head angled upwards with neck slightly drawn in, feathers of crown erected to form crest, bill wide open. Singing often accompanied by rapid flapping of wings: every 3–5 songs bird flaps wings rapidly and may lift off slightly, to $c.$ 50 mm above perch. Also sings during upward spiral flight; from top of climb planes down on stiffened wings.

NEST: cup of dry grass, int. diam. $c.$ 75, lined with rootlets and finer material, placed in scrape in ground against grass tuft or shrub, back and sides built up to form dome $c.$ 75–90 mm high over nest. Dome may be thin and flimsy or quite substantial.

EGGS: 2–4, mean (74 clutches) 2·4. Ovate; dull white to pale cream, spotted and blotched with grey, greyish brown, chocolate-brown or yellowish brown, over underlying slate or ashy speckles, markings often concentrated at obtuse end. SIZE: (n = 137) 20·2–24·7 × 14·9–18·2 (22·2 × 15·8).

LAYING DATES: E Africa: Region B, Feb, May–June, Nov–Dec; Region C, Jan, Apr–June; Region D, Feb–July, Nov–Dec; Malaŵi, Nov; Zambia, Oct–Jan, Mar; Zimbabwe, July–Apr; South Africa: Transvaal, Oct–Feb, July; Natal, Sept–Dec, Mar; Orange Free State, Oct–Nov, Mar; E Cape, Sept–Dec, Feb.

DEVELOPMENT AND CARE OF YOUNG: young fed by both adults, mainly by ♀; fed more often in morning and late afternoon. Nestling period $c.$ 12 days.

BREEDING SUCCESS/SURVIVAL: of 6 eggs in 2 nests at Nylsvley, Transvaal, 3 young fledged and left nest.

Reference
van Someren, V. G. L. (1956).

Plate 3
(Opp. p. 48)

Mirafra hypermetra (Reichenow). Red-winged Bush-Lark. Alouette polyglotte.

Spilocorydon hypermetrus Reichenow, 1879. Orn. Centralb., p. 155; Kibaradja, Tana River, Kenya.

Forms a superspecies with *M. africana*.

Range and Status. Endemic resident. Extreme SE Sudan, NE Uganda south to Sebei, S Ethiopia with narrow northeasterly extension along Rift Valley nearly to Djibouti, Somalia south of 3°N, E Kenya and lowlands of NE Tanzania east of Mt Kilimanjaro and L. Manyara. One record from Loliondo (Tanzania). Common to abundant.

Description. *M. h. hypermetra* (Reichenow): Somalia to N Tanzania. ADULT ♂: crown greyish brown with heavy dark brown streaks along centre of feathers; a slight crest; nape and hindneck paler; face buff with dark speckling; stripe over eye buffish; chin and throat white, merging to buff on breast which has large, very dark brown spots; on side of throat a dark brown smudge; mantle feathers brown with darker centres, darker subterminal spot and with grey tips; rump similar but greyer; tail mainly brown with pale buff outer edges and tips (especially outermost), and innermost broadly fringed buff distally, rufous proximally; lower breast and belly buff; outer primaries brown with basal half rufous; inner primaries brown with basal three-quarters rufous; secondaries brown with pale edges; tertials dark brown, tinged chestnut, with dark midrib, dark subterminal stripe all round and pale edges; scapulars like mantle with chestnut tinge; all wing-coverts like tertials with rufous and pale stripes around edges; underwing mainly rufous brown. Upper mandible dark horn to sepia, lower paler horn; eye brown; legs and feet flesh to greyish white. Sexes alike. SIZE: (7 ♂♂, 1 ♀) wing, ♂ 112–123 (117), ♀ 103; tail, ♂ (n = 6) 85–95 (89·5), ♀ 77; bill, ♂ 22–24 (23·1), ♀ 22; tarsus, ♂ 33·5–35·5 (34·4), ♀ 31. WEIGHT: (Somalia) 1 ♂ 67·9; (Kenya, Dec) 1 unsexed, 55.

IMMATURE: darker than adult, with buff tips to feathers of upperparts; underparts dirty white with brown spots on breast.

M. h. gallarum (Hartert): E and S Ethiopia. Greyer and less well-marked on back, less buff on belly, more rufous in primaries and wing-coverts, no dark smudge on sides of neck.

M. h. kidepoensis Macdonald: S Sudan and NE Uganda. Upperparts browner and more chestnut than in nominate race; wings rufous, like *gallarum*. Smaller: wing, ♂ (n = 3) 108–111 (110), ♀ 97. WEIGHT: 1 ♂ 51, 1 ♀ 44.

M. h. kathangorensis Cave: SE Sudan. Darker than other races, with pale underwing-coverts and bases to flight feathers.

TAXONOMIC NOTE: sometimes considered a subspecies of Rufous-naped Lark *M. africana* (e.g. White 1961), but there are no signs of intergradation between them in N Kenya where they are parapatric.

Field Characters. A large, bulky, brown, dark-streaked lark with rufous wings, usually seen on top of low bush in grassland. Large bill, slight crest, dark smudges on side of neck and comparatively long tail distinguish it from the similar but smaller Rufous-naped Lark *M. africana*. Perches on bushes more often than latter. Voice and slow flapping flight also distinctive.

Voice. Tape-recorded (C, CHA, LEM, McVIC). Song is several clear, powerful, fluty whistles, in phrases of variable length, which may carry several hundred m. Rarely repeats a phrase more than 3–4 times before switching to new one. Individual has its own phrases but often imitates neighbours and other species, usually in very short bursts. 20 other species were mimicked in 15-min song (in Tsavo West Nat. Park, Kenya: Dowsett-Lemaire and Dowsett 1978); they were of 8 non-passerine and 8 passerine families and included 4 Palearctic migrants and 2 other lark species (all but one present at the time: R. J. Dowsett, pers. comm.). Also has clear, loud, 2-note call, sometimes incorporated into song.

General Habits. Inhabits open plains, usually with quite thick grass and where there are a few bushes or dead sticks, which it uses as song posts. Occurs from sea level to 1350 m. Spends much time on ground, where it runs in short spurts. Flies to bush or termite mound if pursued. Flies strongly but not usually far. When it drops to ground, runs a considerable distance and is difficult to flush again. Shelters in heat of day under bush. Usually takes food from ground; once seen to jump up from ground to catch insect in air.

Largely sedentary. However, 1 flew to lights of Ngulia Lodge, Tsavo West Nat. Park (Kenya) at night in Dec 1976 with many other night-flying migrants (G. C. Backhurst, pers. comm.).

Food. Of 32 items taken from ground, 15 were insects, mainly termites from their galleries on a fallen log, and the rest unidentified (Lack 1985).

Breeding Habits. Monogamous and territorial. Sings from perch on bush, termite mound or dead stick, or in short soaring flight, rising to 4–6 m and hovering with rather slow flaps. Sings all year.

NEST: well built cup of grass and roots, domed, and placed in tuft of grass.

EGGS: 1–4; dull white, well-marked with brown or olive-brown. SIZE: *c*. 23 × 18·5.

LAYING DATES: Somalia, June; E Africa: Kenya (bird with nest material late Jan); Tanzania, Nov.

References
Dowsett-Lemaire, F. and Dowsett, R. J. (1978).
Lack, P. C. (1985).

Mirafra somalica (Witherby). Somali Lark. Alouette de Somalie.

Certhilauda somalica Witherby, 1903. Bull. Br. Orn. Club 14, p. 29; Dibbit, Somaliland.

Plate 3
(Opp. p. 48)

Range and Status. Resident; endemic to Somalia. Fairly common and widespread south of 10° 30′N, along coast to as far as 2°N.

Description. *M. s. somalica* Witherby: N Somalia (Burao, Arori Plain, Gardo). ADULT ♂: crown, nape and hindneck pinky rufous-brown, nape and hindneck streaked white. Sides of neck and foreneck whitish speckled rufous-brown, cheeks and ear-coverts streaked brown and whitish, supercilium white. Mantle and back rufous, feathers with whitish inner edges, rump rufous, feathers tipped white with black subterminal bar; uppertail-coverts with black shafts and subterminal bars, brown centres and buff edges and tips. Tail-feathers dark brown, outer pair with white outer webs, central pair rufous with dark centres and pale margins. Chin and throat white, lower throat speckled brown; breast and remainder of underparts buff, breast streaked rufous-brown. Primaries and secondaries brown, outer webs and basal half of inner webs rufous, secondaries tipped whitish. Scapulars rufous-brown, tipped and edged whitish; tertials rufous, darker along shafts, shading to paler rufous, with black inner margin and white outer margin. Upperwing-coverts rufous, tipped and edged whitish. Underwing-coverts and axillaries pinkish brown; underside of primaries and secondaries with pale rufous margins to inner webs. Bill dark horn, lower mandible paler horn; eye light chocolate brown; legs and feet creamy white, pale creamy brown or yellowish white. Sexes alike. SIZE: (12 ♂♂, 9 ♀♀) wing, ♂ 94–106 (100·6), ♀ 89–108 (97·0); tail, ♂ 63–71 (66·7), ♀ 60–72 (63·1); bill, ♂ 22–27 (26·1), ♀ 22–27 (24·4); tarsus, ♂ 28–34 (31·4), ♀ 30–33 (31·1). WEIGHT: 1 ♂ 43·6, 2 unsexed 46, 50·2.

IMMATURE and NESTLING: unknown.

M. s. rochei Colston: central and S Somalia (Geriban, Garad, Dibbit, Uarsciek, Mallable). Upperparts rich cinnamon-rufous, darker than *somalica*, crown darker chestnut-brown, wings browner, breast and sides of neck with dark cinnamon wash, more heavily streaked than *somalica*. Smaller: (5 ♂♂, 1 ♀) wing ♂ 95–102 (98·4), ♀ 90.

26 ALAUDIDAE

Field Characters. A large, red lark with very long, straight bill, dark tail with rufous centre and white edges to outer feathers. Overlaps in N with race *sharpei* of Rufous-naped Lark *M. africana*, which has almost identical red plumage but shorter, heavier bill, no white in tail. In S might be confused with Red-winged Bush-Lark *M. hypermetra*, which is same size but has much stouter bill, darker upperparts, blackish breast streaks.

Voice. Unknown.

General Habits. Found in open coastal and inland grasslands, often on dry patches of red sand. Generally seen singly or in pairs. Runs fast; flies only short distance.

Food. Unknown.

Breeding Habits. Said to have a short courtship flight (Archer and Godman 1961).
NEST: only 1 described: a scrape in the ground, unlined, but surrounded with circle of plucked grass, placed at base of grass tuft; not domed.
EGGS: only 2 clutches found, 1 of 3, 1 of 4 eggs. Elongate; cream, uniformly and boldly spotted and splashed with red-brown or sepia over underlying ash-grey markings. SIZE: (n = 3) 24 × 15–16 (24·0 × 15·3).
LAYING DATES: Somalia, June, Sept.

References
Archer, G. F. and Godman, E. M. (1961).
Colston, P. R. (1982).

Plate 3
(Opp. p. 48)

Mirafra ashi Colston. Ash's Lark. Alouette d'Ash.

Mirafra ashi Colston, 1982. Bull. Br. Orn. Club 120, pp. 106–114; Uarsciek, Somalia.

Range and Status. Endemic, Somalia, probably resident. Known only from series of 6 taken 13 km north of Uarsciek (= Warsheikh), 80 km northeast of Mogadishu, Somalia, July 1981 (Colston 1982).

Description. ADULT ♂: unknown. ADULT ♀: forehead, crown, nape and mantle greyish brown with strong pink tinge to mantle and centre of crown, dark-streaked, each feather with dark brown centre and pale buff edge and tip. Inconspicuous whitish superciliary stripe. Back, scapulars and wing-coverts variegated pinkish, brown, rufous and pale buff, each feather blackish along shaft grading outwards through rufous to grey-brown, with 2 mm broad buff fringe separated by narrow black subterminal line from brown centre of feather. Rump uniform pale pinkish brown; uppertail-coverts pale brown, each with broad whitish fringes and black subterminal mark. Rectrices dark grey, each with blackish shaft, pale pinkish buff fringe 1 mm wide, and dark line all around feather inside the fringe; outer rectrix variable, either mostly white with black along shaft and with base of inner web dark grey, or dark grey-brown with 2–3 mm wide whiter outer edge and tip. Lores and circle of circumorbital featherlets grey; ear-coverts and sides of neck pale pinkish brown with fine dark streaks. Chin white; throat white with dark grey-brown speckles or short streaks; breast white or off-white, finely streaked with rufous-brown or pink-brown in centre, rather more heavily streaked with dark brown at sides of breast which are strongly suffused with pink-brown. Belly off-white, unstreaked; flanks pale pinkish brown; undertail-coverts pinkish white. Remiges brown, P9–P6 with basal three-quarters of outer webs rufous and P5–P3 with all of broad outer webs rufous except near tip, where blackish. Secondaries mid-brown but innermost ones and tertials slightly rufescent dark grey-brown; each feather with blackish shaft, white or off-white fringe 2 mm broad, and narrow black line between white fringe and brown interior. Underwing-coverts pale pinkish brown. Underside of remiges and rectrices grey. Upper mandible dark grey, lower mandible pale bluish grey; eye dark brown; legs and feet creamy. SIZE: (unsexed, n = 5, 1 ♀) wing, unsexed 80–94 (86·6), ♀ 82; tail, unsexed 50–56 (52·8), ♀ 47 (moulting); bill, unsexed 21–24 (22·9), ♀ 22; tarsus, unsexed 29–30 (29·0), ♀ 29. WEIGHT: unsexed (n = 5) 31·3–41·9 (37·0), 1 ♀ 35·7.

IMMATURE: like adult ♀ but crown more heavily streaked blackish (2 specimens) or more uniform reddish brown (1 specimen). Central 3 pairs of rectrices with broad buff edges (1 specimen). Belly slightly buffier than adult ♀.

Field Characters. No information. From skins, evidently a medium-sized *Mirafra* with greyish, variegated upperparts scalloped with white, and buffy white underparts with lightly and finely streaked breast. Juvenile has white undertail feathers. Larger than Singing Bush-Lark *M. cantillans* but smaller than Rufous-naped Lark *M. africana* or Red-winged Bush-Lark *M. hypermetra*. Of the 3 *Mirafra* spp. which occur near Mogadishu: Collared Lark *M. collaris* distinguished by black throat-collar, sandy upperparts and white-sided blackish tail; Somali Lark *M. somalica* by large size, rufous-brown upperparts, long bill, and white-sided rufous-and-black tail; and Pink-breasted Lark *M. poecilosterna* by plain foreparts (grey crown, pinkish breast) and all-dark tail.

Voice. Unknown.

General Habits. Confined to areas of grazed tufted grass on fixed dunes in grassy maritime plain with fossil coral reef outcrops. Often sings from tops of small stunted bushes (Collar and Stuart 1985). The ♀, in breeding condition, was also in full wing and tail moult.

Food. Unknown.

Breeding Habits. ♀ in full breeding condition (12 mm large ova) July, Somalia. No other information.

References
Collar, N. J. and Stuart, S. N. (1985).
Colston, P. R. (1982).

Mirafra angolensis Bocage. Angola Lark. Alouette d'Angola.

Mirafra angolensis Bocage, 1880. J. Sci. Nat. Lisbōa 8, p. 59; Caconda, Angola.

Plate 5
(Opp. p. 96)

Range and Status. Endemic resident, sedentary. 4 disjunct populations: (1) N Angola (Camabatela, Quiculungo); (2) W Angola (Mombolo, Mt Moco, Huambo, Caconda); (3) extreme E Angola (Luacano), S Zaïre (Dilolo) and NW Zambia (Mwinilunga, Salujinga, Chitunta and Nyambela Plains, Minyanya Plain); and (4) SE Zaïre (Marungu Plateau). Uncommon and local.

Description. *M. a. angolensis* Bocage: W-central and N Angola. ADULT ♂: crown, nape and hindneck reddish brown, centre of feathers streaked dark brown; sides of face and neck buffy brown with dark streaks, chin and throat whitish. Mantle and back reddish brown, broadly streaked dark brown, uppertail-coverts cinnamon or pinkish grey, less streaked, tail dark brown with rufous edging to central feathers, outermost pair white with dark inner edge on outer half, next pair with white outer edge. Underparts buff, brighter on breast which is lightly streaked. Wings brown, mottled and dark streaked on reddish brown on coverts; flight feathers with rufous edges, outer half of secondaries with pattern of narrow dark and broader light brown barring. Bill dark horn, whitish base to lower mandible; eye chestnut; legs and feet pink. Sexes alike; ♀ may be less streaked on chest. SIZE: wing, ♂ (n = 24) 80–90 (85·0), ♀ (n = 20) 76–81 (78·0); tail, ♂ (n = 21) 46–58 (50·0), ♀ (n = 17) 42–50 (46·0); bill, ♂ (n = 24) 17·5–21 (18·9), ♀

(n = 22) 16–20 (17·6); tarsus, ♂ (n = 14) 22·4–26 (24·7), ♀ (n = 12) 22·2–26 (23·7). WEIGHT: ♂ (n = 12) 35–42 (38·8), ♀ (n = 11) 30–36·5 (33·8).

IMMATURE: like adult, but lacks rufous nape. Crown, nape and hindneck finely barred. Lacks longitudinal streaking on feathers of back which have buff edge to rounded tip. Underparts buffy, less brown than adult, chest streaking less sharply defined.

NESTLING: unknown.

M. a. antonii Hall (including '*minyanyae*'): E Angola, S Zaïre, NW Zambia. Darker, dark feather centres broader, particularly on crown, mantle and central rectrices. Uppertail-coverts greyish with blackish markings or barring; more heavily streaked on chest. Outer tail-feathers white with narrow dark inner edge.

M. a. marungensis Hall: SE Zaïre (Marungu Plateau). More richly coloured than *angolensis*, more rufous on nape and lower back, underparts more rufous with heavier streaking on chest.

Field Characters. Similar in size and plumage to Rufous-naped Lark *M. africana* but upperparts and wings considerably darker, rufous wing-patch smaller, more white in tail (outer feathers conspicuous in flight), bill slender, sharply pointed. Smaller Flappet Lark *M. rufocinnamomea* varies in plumage according to race, some greyer, some more rufous, but lacks rufous nape and white in tail. Best told from both by trilling song, very different from mellow whistles of Rufous-naped and wing-clapping of Flappet.

Voice. Tape-recorded (ASP, 86, F). Flight song a series of clear, sweet trills on different pitches, 'z^{i}z^{i}z^{i}z^{i}z^{i}z^{u}z^{u}z^u-z^{u}z^{u}z^{i}z^{i}z^{i}z^{i}z^{i}z^{u}z^{u}z^{u}z^u'; *c.* 9 trills per song, which lasts 4–5 s; trills longer towards beginning. Pitch and tone vary. Another song/call, 'trrp-trrp-t'tiiiii-tuu' given from perch on termite mound.

General Habits. In Zambia occurs mainly on moist grass plains, and on Angolan plateau in moist valleys. Inhabits montane grassland on Mt Moco, Angola and Marungu Plateau, Zaïre. In Zambia may occur in drier grassland with termitaria and low bushes, and on recently burned areas with short green grass; overlaps in habitat with Rufous-naped Lark and Flappet Lark. In area of overlap tends to be in wetter habitat, though moves at times to drier watershed areas occupied by Rufous-naped Lark. Occurs on dry watershed grassland and pastures, typical of Rufous-naped Lark habitat elsewhere, when that species absent.

Food. Insects (beetles) and seeds.

Breeding Habits. In presumed displays, (1) ♂ gives flight song while rising steeply upward to *c.* 25 m, returns to ground in vertical 'parachute' descent with wings held in upward curve; (2) after song from perch (see *Voice*) flies up steeply to *c.* 5 m, then descends quickly with purring wings.

NEST: only 1 recorded, a cup of coarse, dry grass, int. diam. 65, depth 42, rim *c.* 20 thick, with 'chopped' grass rootlets on rim, thinly lined with very fine fresh grass, with slight dome of fresh and dry grass stems; placed in scrape under grass tuft under fleshy leaved shrub (Colebrook-Robjent 1988).

EGGS: 3. Oval; glossy; cream, densely spotted medium-brown over ashy-grey shell markings, forming ill-defined zone at obtuse end. SIZE: (n = 3) 20·5–21 × 14·7–15 (20·7 × 14·8).

LAYING DATES: Zambia, Oct (adults with enlarged gonads Sept–Jan, young bird (Mwinilunga) Oct); Angola (Mt Moco), ♂♂ with enlarged testes, Oct.

References
Aspinwall, D. R. (1979).
Benson, C. W. *et al.* (1971).
Colebrook-Robjent, J. F. R. (1988).

Plates 3, 5
(Opp. pp. 48, 96)

Mirafra rufocinnamomea (Salvadori). Flappet Lark. Alouette bourdonnante.

Megalophoneus rufocinnamomea Salvadori, 1865. Atti. Soc. Ital. Sci. Nat. 8, p. 378; N Abyssinia.

Forms a superspecies with *M. apiata*.

Range and Status. Endemic resident. Mauritania (Sahel), S Senegambia, Guinea, Mali (common south of 15° N), N and coastal Ivory Coast, Burkina Faso, N Ghana, S Niger, Togo, Benin, Nigeria except S (locally common), W Chad, Cameroon (uncommon and local), S Gabon (uncommon and local), Central African Republic, W and S Sudan (uncommon to common), W and S Ethiopia (frequent to common), Somalia (rare, south of 3° N), widespread and common S and SW Kenya and almost throughout Uganda and Tanzania; E Rwanda, E Burundi, widespread Zaïre except forest zone, widespread NE, W and central Angola from NE Lunda west and south to Huila, and on Cabinda coast; widespread and common Zambia, Malaŵi below 1700 m, and Mozambique, widespread but sparse and local in Zimbabwe, frequent to sparse N and NE Botswana and NE Namibia (only in Caprivi and Etosha Pan, sparse and local), South Africa (widespread and common resident in E Transvaal, locally common in central Transvaal bushveld and NE Natal) and Swaziland (locally common). Density varies with habitat; in Tanzania (Serengeti) 6 birds/km^2 in open habitat, 12 birds/km^2 in partly wooded habitat.

Description. *M. r. buckleyi* (Shelley): Senegambia to Ghana, N Nigeria, N Cameroon. ADULT ♂: top of head and upperparts dull rufous, sometimes with greyish cast, streaked blackish (feathers with dark centres), rump less streaked, uppertail-coverts rufous with black shafts. Tail-feathers dark brown, outer pair pale rufous, next pair with outer webs rufous, central pair rufous-brown with blackish shafts. Lores, poorly marked supercilium and eye-ring whitish, ear-coverts brown streaked with buff. Throat and chin white to buff, rest of underparts orange-buff to creamy buff, dark spots on upper breast, reddish spots on lower breast. Primaries brown with variable amount of rufous on outer webs, secondaries brown to rufous-brown with narrow line of blackish brown inside rufous-brown to buff fringes; all flight feathers with rufous panel on inner web. Underwing-coverts and axillaries orange-buff. Bill dark brown, paler on lower mandible; eye dark brown; legs and feet pinkish brown to flesh-brown. Sexes alike. SIZE: (14 ♂♂, 16 ♀♀) wing, ♂ 75–81 (78·5), ♀ 67–80 (75·3); tail, ♂ 48–55 (51·0), ♀ 45–53 (48·9); bill, ♂ 14·8–17·5 (15·9), ♀ 14·1–17 (15·5); tarsus, ♂ 21–25 (23·4), ♀ 22–26 (24·3). WEIGHT: ♂ (n = 15) 21–30 (25·7), ♀ (n = 6) 21–32·4 (26·9).
IMMATURE: generally duller than adult, upperparts more barred, feathers with pale tips.
NESTLING: unknown.
M. r. serlei White: SE Nigeria. Upperparts redder and richer than *buckleyi*, deeper reddish ochre below.
M. r. tigrina Oustalet: W Cameroon to N Zaïre. Upperparts rich vinous pink, less reddish brown than *serlei*, with heavy blackish grey centres to back feathers, almost obscuring red pigments.
M. r. furensis Lynes: Sudan (W Darfur). Upperparts yellowish tawny, lightly marked with black.
M. r. sobatensis Lynes: White Nile–Sobat R. confluence, Sudan. Upperparts black, feathers edged creamy buff, giving markedly scaled appearance.
M. r. rufocinnamomea Salvadori: NW and Central Ethiopia. Upperparts bright rufous-red, heavily marked with black.
M. r. omoensis Neumann: SW Ethiopia (Omo to Madji and Baro R.). More heavily marked with black above than *rufocinnamomea*.
M. r. torrida Shelley: S Ethiopia, SE Sudan, N and central Uganda, central Kenya, Tanzania (to Iringa). Upperparts darker rufous than *rufocinnamomea* and heavily marked with blackish.
M. r. kawirondensis van Someren: W Kenya, W Uganda, E Zaïre. Upperparts blackish, feathers edged grey-brown.
M. r. fischeri Reichenow: S Somalia (Juba R.), E Kenya (from Mombasa), E Tanzania (to Morogoro), S and SW Tanzania (to Kigoma), Malaŵi, N Mozambique, Shaba, N and central Angola, N Zambia. Upperparts dark reddish brown, not heavily patterned on back but variable, some birds more heavily marked than others.
M. r. schoutedeni White: W Zaïre, NE Angola, Central African Republic, Gabon (Oyem). Upperparts rather pale brown, lightly marked.
M. r. lwenarum White: N Zambia (Balovale). Upperparts paler and more pinkish than *fischeri*.
M. r. smithersi White: Zambia (Livingstone, Lusaka), Zimbabwe, NE Botswana, South Africa (N Transvaal). Upperparts pale brick red, back lightly marked.
M. r. mababiensis Roberts: W Zambia to central Botswana, E Namibia (Caprivi). Upperparts greyish.
M. r. pintoi White: South Africa (E Transvaal, NE Natal), Swaziland, S Mozambique. Upperparts dark and heavily marked, paler areas deep vinous brown. Breast darker than other races.

Field Characters. A medium-small lark, in air appearing rather dumpy with narrow, shortish tail. Easily identified in most of Africa as the only lark with wing-clapping display flight. Very similar to Clapper Lark *M. apiata* of southern Africa which it meets but barely overlaps, so usually identifiable by range; for further distinctions see Clapper Lark. Less easy to identify when not displaying or calling since plumage colour very variable throughout its wide range; plumage of local race must be learned before comparing bird with other species. All races have red wing-patch.

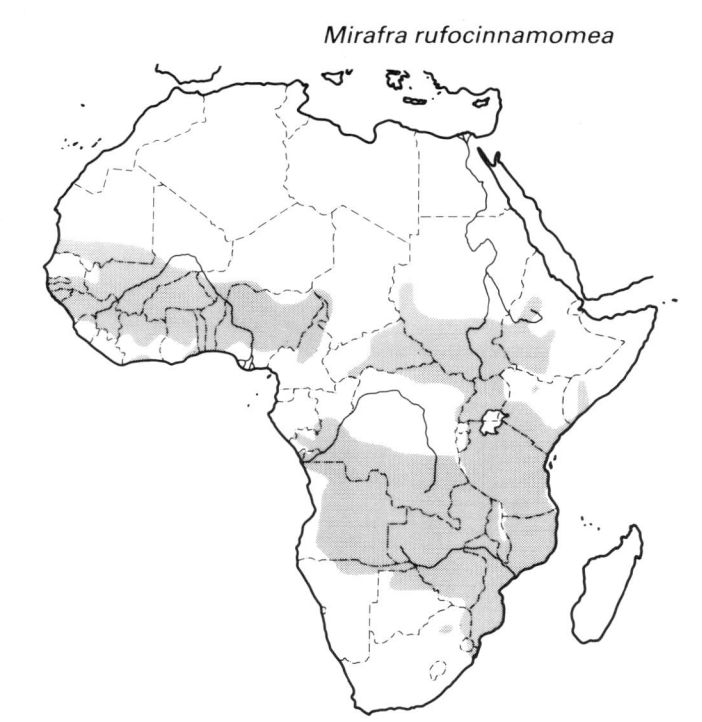

Voice. Tape-recorded (10, 25, 78, 86, C, F). 2 song types: (1) thin, wispy, sunbird-like 'chik-chik-a-wee', 'wichipi-wichipi' or 'wichawo-weeyo-weewi' given during wing-clapping display; (2) 3–4 note clear whistle 'tuee-tui', 'teeyi-tooi', 'chi-chi-weeo-chuwee' given from perch or ground, usually at dawn; uncommon. Much more common than vocalizations, and more important as an auditory signal, are wing-flaps given during display. Wings are flapped in series of bursts, 3–8 flaps per burst, 3–4 bursts in series, terminal bursts 5–21 flaps. Except at very close range, flaps sound dry and muffled, i.e. unlike harsh, strident noise made by wings of broadbills *Smithornis* spp. In typical pattern, bird gives several short flaps followed by a long one, but much variation, and wing-flap dialects vary geographically (Payne 1978).

General Habits. Present in wide range of mainly grassy habitats including grassy clearings and drainage lines in broad-leaved woodland (*Burkea*, *Baikiaea*, *Brachystegia*), and fine-leaved (acacia) savanna woodland, and coastal grasslands. Has marked preference for shrublands in W Africa.

Usually solitary or in pairs. Territories maintained year-round.

Food. Invertebrates (grasshoppers, mantids, larval Lepidoptera, beetles, termites) and grass seeds.

Breeding Habits. Monogamous, territorial. In display flight, ♂ flies up gradually in stages, giving short burst of wing-clapping as it climbs, to height of 50–100 m, then circles repeatedly over an area of *c.* 200 ha. Drops abruptly to ground after display.

NEST: domed cup of dry grass lined with rootlets and finer material, in scrape in ground against tuft of grass.

EGGS: 2–3, mean (25 clutches) 2·2. Ovate; dull white, spotted, speckled or streaked with brown and greyish brown over underlying ashy or slate, markings often concentrated at obtuse end. SIZE: (n = 41) 18·5–22·4 × 13·7–16·1 (20·8 × 14·8).

LAYING DATES: Mauritania, May–Sept, rarely Oct; Nigeria, June–July; Mali, May–Oct; Sudan, July–Aug; E Africa: Region B, June–Aug; Region C, May, Nov–Dec; Region E, July–Aug, Nov–Dec; Zaïre, Nov; Malaŵi, Oct–Nov, Jan–Apr; Zambia, Sept–Feb; Zimbabwe, Oct–Apr; South Africa, Oct–Feb.

Reference
Payne, R. B. (1978).

Plate 5
(Opp. p. 96)

Mirafra apiata (Vieillot). **Clapper Lark. Alouette bateleuse.**

Alauda apiata Vieillot, 1816. Nouv. Dict. d' Hist. Nat. 1, p. 342; Swartland (= Malmesbury), Cape Province.

Forms a superspecies with *M. rufocinnamomea*.

Range and Status. Endemic resident. W Zambia (confined to sparsely grassed semi-arid Kalahari sand plains west of Zambesi R., from Liuwa Plain south to Mashi R.), widespread Botswana (absent from arid SW), N and NW Namibia (local and uncommon); widespread in South Africa: common in W and central highveld, locally in E highveld and escarpment of Transvaal; widespread and common to fairly common in Orange Free State; locally common in E and SW Cape Province; uncommon and local in Karoo and N Cape; locally common in Lesotho.

Description. *M. a. apiata* Vieillot: South Africa (SW Cape Province north to Namaqualand, across Little Karoo to Port Elizabeth and Grahamstown). ADULT ♂: top of head and upperparts variable, feathers rufous with black central and subterminal bars and pale edges, giving mottled and scaled appearance. Hindneck and mantle paler and greyer, less rufous, feathers with greyish margins; rump mainly rufous barred black, uppertail-coverts similar but barring finer, tips pale. Tail-feathers dark grey-brown, outer pair with outer webs and part of inner webs buff, central pair with plain centres, webs barred black, buff and rufous. Sides of neck and foreneck buffy, streaked dark brown, cheeks and ear-coverts mottled buff and brown, chin and throat whitish, throat lightly speckled dark brown. Rest of underparts reddish to rufous-buff, deeper and richer on breast and flanks; breast with dark brown teardrop spots. Flight feathers grey-brown, primaries with buff or rufous outer webs, secondaries with rufous-buff margins, broader on outer webs. Scapulars and wing-coverts rufous-brown with blackish bars and buff and grey margins, tertials smoky brown, barred black along shafts, with blackish inner margins and buff fringes. Underwing-coverts and axillaries pale rufous-brown. Bill brownish horn; eye light brown; legs and feet pinkish. Sexes alike. SIZE: (12 ♂♂, 4 ♀♀) wing, ♂ 78–92 (84·0), ♀ 73–84 (78·0); tail, ♂ 46–58 (53·2), ♀ 50–55 (52·7); bill, ♂ (n = 10) 14·5–18 (16·4), ♀ 14·7–16 (15·3); tarsus, ♂ 21–27 (23·9), ♀ 22·8–26 (23·9). WEIGHT: ♂ (n = 32) 24–44 (29·0), ♀ (n = 10) 23–27 (24·9).

IMMATURE: browner and less clearly barred than adult, crown and back feathers with pale margins, giving scaly appearance.

NESTLING: unknown.

M. a. marjoriae Winterbottom: South Africa south of 34°S, from Cape Peninsula to Knysna. Upperparts more greyish, underparts paler.

M. a. hewitti (Roberts): South Africa (N, central and NE Cape Province, Orange Free State, W, central and E Transvaal), Lesotho, SW Botswana. Upperparts rufous-tawny, blackish barring not as marked as *apiata*, underparts darker rufous-buff. Larger: (7 ♂♂, 3 ♀♀) wing, ♂ 87–95 (89·7), ♀ 81–89 (85·0); bill, ♂ 16·4–17·8 (16·9), ♀ 15·3–18·2 (16·6).

M. a. deserti (Roberts): W and central Botswana, N Namibia. Paler above and below than *hewitti*. Bill stout and strong. Larger: (2 ♂♂, 3 ♀♀) wing, ♂ 93, 95, ♀ 76–83 (79·0); bill, ♂ 18·3, 19, ♀ 16·8–17·1 (16·9).

M. a. reynoldsi Benson and Irwin: N Botswana (Gemsbok Pan, L. Dow), Zambia between Zambesi and Kwando rivers. Upperparts ashy grey, variably tinged with pale sandy yellowish. Underparts variable, generally light creamy white, unevenly tinged with buffy brown, some individuals darker brown than others.

M. a. nata Smithers: NE Botswana (Makgadikgadi Pan). Similar to *reynoldsi*, but mantle paler, greyish to very pale yellowish brown. Rest of upperparts ash-grey. Crown variable, generally reddish.

M. a. jappi Traylor: Zambia (Kalabo, Luanginga R.). Upperparts darker, barring on feathers darker, more brownish. Underparts browner.

Field Characters. A medium-small lark, the only one in its range with wing-clapping display flight. Very similar to Flappet Lark *M. rufocinnamomea*, which it meets but hardly overlaps at northern edge of range, and usually in different habitat. Both birds have very variable plumage, though Clapper usually redder, with outer tail-feathers white, not buff. Best distinguished by display flights and calls, q.v.; both clatter wings but with different rhythm; Clapper gives loud whistle following claps, Flappet does not.

Voice. Tape-recorded (39, 81, 88, F, LEM). Song a clear, drawn-out whistled 'poooeee' or 'pheee-oo', rising in pitch, given in flight as part of display, together with loud wing-claps. Calls include 'peeek', grating 'chrk-chrk' and mewing alarm. Occasionally flies low, giving call which includes imitations of other species' calls (Vernon 1973). Breeding ♂♂ respond to playback of display call with imitations of other species (Cardwell and Dunning 1971).

General Habits. Occurs in grassland, especially tall grass with rocks and boulders, often on slope or hillside. Also inhabits dwarf shrublands in semi-arid areas and interdune valleys with dense cover of long grass in the Kalahari.

Usually solitary or in pairs. Difficult to flush, usually runs away; when flushed may perch on bush or fence post. Forages at bases of tufts and on bare ground. Has been recorded in foraging association with antbear *Orycteropus afer*, accompanying it at dusk and foraging in scrapes it dug (Vernon and Dean 1988).

Food. Ants (Formicidae) and harvester termites (*Hodotermes mossambicus*); also other invertebrates and seeds.

Breeding Habits. Display flight has 2 forms. In commonest version, ♂ cruises in fluttering flight at 20–30 m, then rises with exaggerated wing-flaps; towards top of climb claps wings; at top of climb holds wings stiffly out in steep dihedral and drops in steep dive uttering plaintive rising whistle, 'pheeeooo'. Resumes fluttering flight before repeating climb, clap and whistle. Repeats each sequence at intervals of 15–30 s, whole song-flight may last 10 min or more. Less frequent is similar display from perch on bush or from ground; bird rises from perch to height of *c.* 10 m, claps wings at top of climb, then descends to starting point with steep glide and whistle.

NEST: cup of grass lined with rootlets, domed over from back with grass; dome either thin or so well made that nest appears to be ball-shaped with side entrance; placed in scrape in ground next to tuft of grass or forb.

EGGS: 2–3, mean (26 clutches) 2·5. Ovate; white to creamy white or pale greenish grey, densely freckled or blotched with brown, with underlying lavender and purplish speckles, or blotched with purplish grey, lavender and brown. SIZE: (n = 12) 20·3–24 × 14·3–16 (21·9 × 15·2).

LAYING DATES: South Africa: SW Cape, Aug–Oct; Karoo, Aug; E Cape, Oct; N Cape, Jan; Orange Free State, Sept–Nov, Feb; Transvaal, Oct–Dec, Jan–Feb.

DEVELOPMENT AND CARE OF YOUNG: in distraction and alarm display both parents bounce along in flipping flight, snapping wings at each flip, at *c.* 1 s intervals.

References
Cardwell, P. and Dunning, J. (1971).
Vernon, C. J. (1973).
Vernon, C. J. and Dean, W. R. J. (1988).

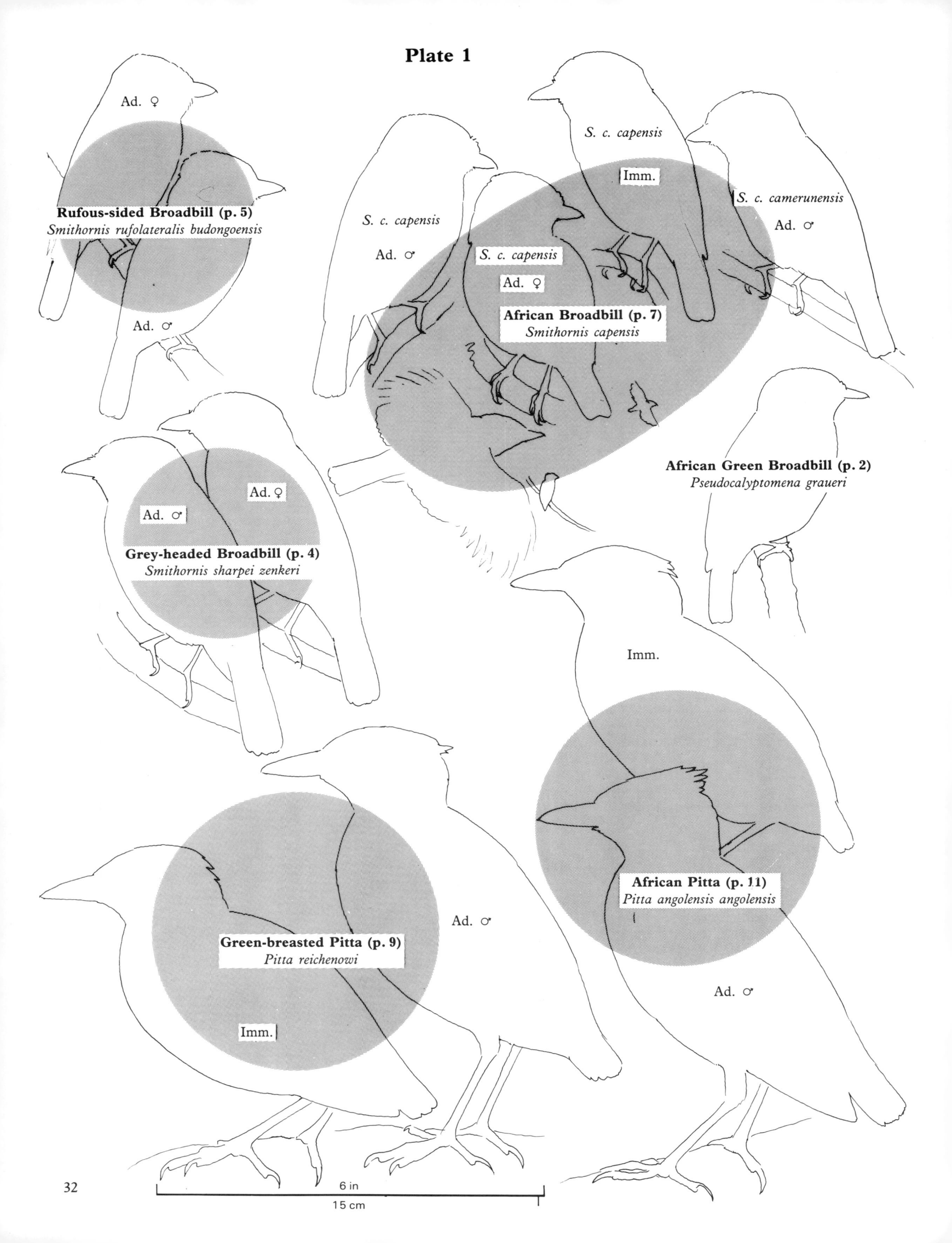

Plate 1

Rufous-sided Broadbill (p. 5)
Smithornis rufolateralis budongoensis
Ad. ♀
Ad. ♂

S. c. capensis
Ad. ♂

S. c. capensis
Imm.

S. c. capensis
Ad. ♀

S. c. camerunensis
Ad. ♂

African Broadbill (p. 7)
Smithornis capensis

Grey-headed Broadbill (p. 4)
Smithornis sharpei zenkeri
Ad. ♂
Ad. ♀

African Green Broadbill (p. 2)
Pseudocalyptomena graueri
Imm.
Ad. ♂

Green-breasted Pitta (p. 9)
Pitta reichenowi
Ad. ♂
Imm.

African Pitta (p. 11)
Pitta angolensis angolensis
Ad. ♂

32

6 in
15 cm

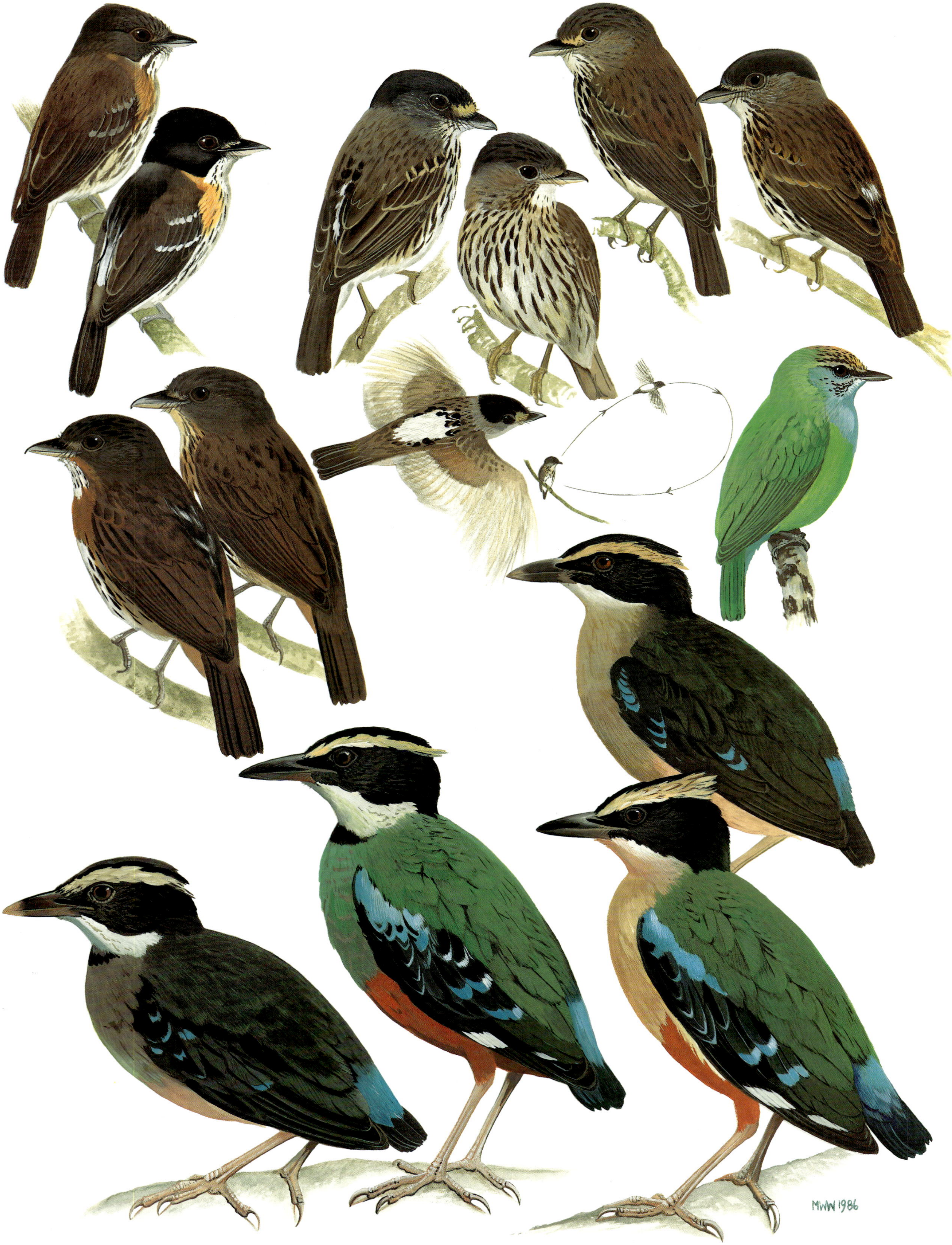
MWW 1986

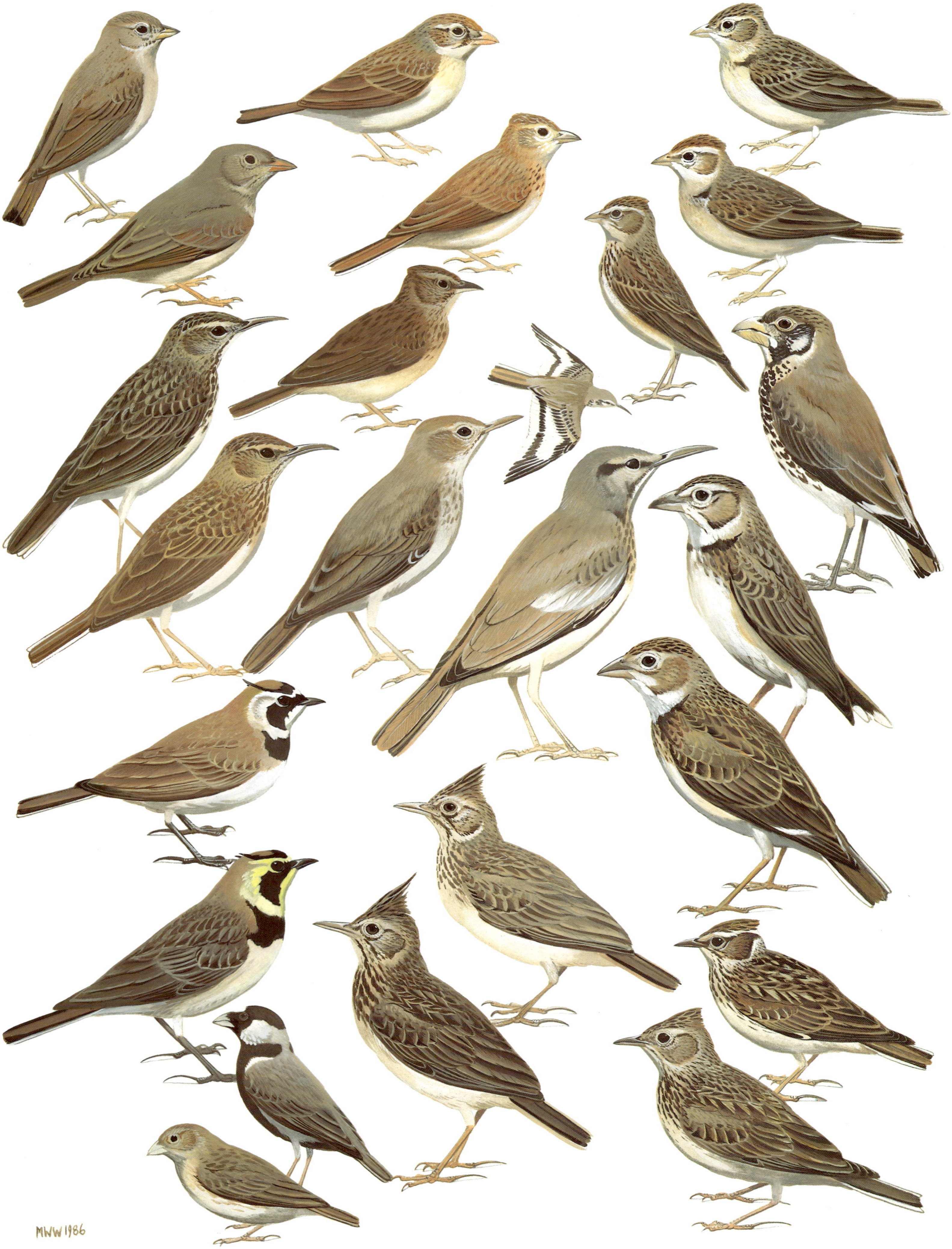
MWW 1986

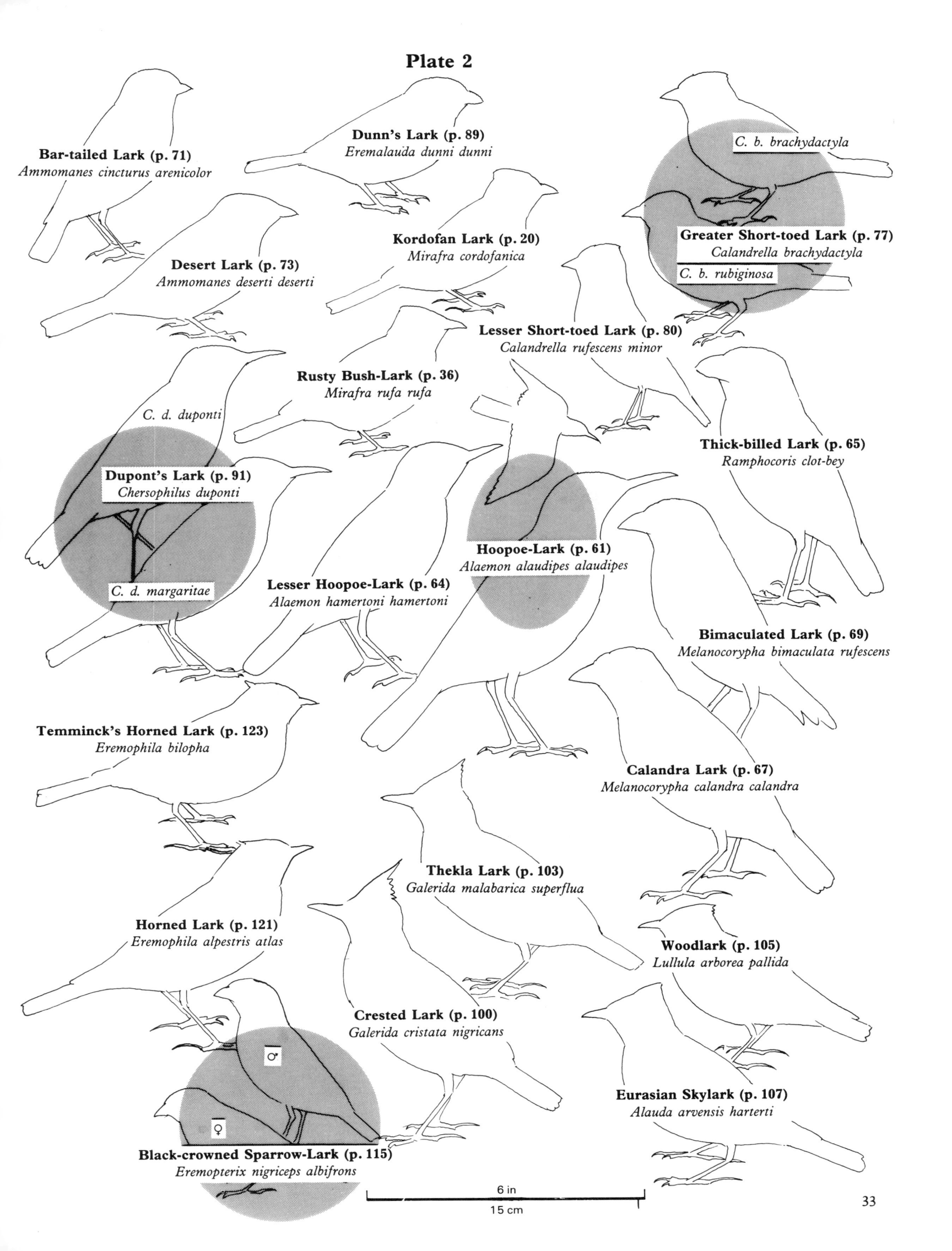

Plate 2

Bar-tailed Lark (p. 71)
Ammomanes cincturus arenicolor

Dunn's Lark (p. 89)
Eremalauda dunni dunni

Desert Lark (p. 73)
Ammomanes deserti deserti

Kordofan Lark (p. 20)
Mirafra cordofanica

C. b. brachydactyla

Greater Short-toed Lark (p. 77)
Calandrella brachydactyla

C. b. rubiginosa

Lesser Short-toed Lark (p. 80)
Calandrella rufescens minor

Rusty Bush-Lark (p. 36)
Mirafra rufa rufa

Thick-billed Lark (p. 65)
Ramphocoris clot-bey

C. d. duponti

Dupont's Lark (p. 91)
Chersophilus duponti

C. d. margaritae

Hoopoe-Lark (p. 61)
Alaemon alaudipes alaudipes

Lesser Hoopoe-Lark (p. 64)
Alaemon hamertoni hamertoni

Bimaculated Lark (p. 69)
Melanocorypha bimaculata rufescens

Temminck's Horned Lark (p. 123)
Eremophila bilopha

Calandra Lark (p. 67)
Melanocorypha calandra calandra

Thekla Lark (p. 103)
Galerida malabarica superflua

Horned Lark (p. 121)
Eremophila alpestris atlas

Woodlark (p. 105)
Lullula arborea pallida

Crested Lark (p. 100)
Galerida cristata nigricans

♂

♀

Eurasian Skylark (p. 107)
Alauda arvensis harterti

Black-crowned Sparrow-Lark (p. 115)
Eremopterix nigriceps albifrons

6 in
15 cm

34 ALAUDIDAE

Plate 3
(Opp. p. 48)

Mirafra collaris **Sharpe. Collared Lark. Alouette à collier.**

Mirafra collaris Sharpe, 1896. Bull. Br. Orn. Club 5, p. 24; Aimola, Kenya.

Range and Status. Endemic resident. SE Ethiopia (local and uncommon), N-central, central and SW Somalia (locally common) south to NE and E Kenya (very local and uncommon: Marsabit, Moyale, Garissa, Hola).

Description. ADULT ♂: crown, mantle, back, wing-coverts and uppertail-coverts rich cinnamon-brown, lightly streaked with white (feathers with one side edged white); sides of neck and collar around hindneck whitish, streaked and speckled black; foreneck whitish, ear-coverts cinnamon with fine pale streaks, chin, throat and supercilium white. Tail-feathers very dark brown, central pair edged rufous. Throat bordered below by broad blackish collar, rest of underparts pale tawny to buff, breast variably blotched and streaked with reddish brown. Flight feathers blackish brown, edged pale buff, secondaries also with buff tips. Scapulars rich cinnamon-brown, edged white, tertials cinnamon-brown, with inner margin of dark brown, outer margin buff. Underwing-coverts and axillaries rufous-buff. Upper mandible dark brown, lower mandible flesh-white; eye brown; legs and feet pale flesh-brown. Sexes alike. SIZE: (5 ♂♂, 3 ♀♀) wing, ♂ 83–89 (86·4), ♀ 76–86 (81·0); tail, ♂ 53–57 (54·6), ♀ 48–53 (50·6); bill, ♂ 14·5–16 (15·6), ♀ 14–16 (15·0); tarsus, ♂ 21–23·5 (22·6), ♀ 20–26 (23·0).
IMMATURE: feathers of crown, upperparts, scapulars, tertials and upperwing-coverts with whitish fringes and subterminal black band, producing scaly appearance. Underparts whiter than adult, especially on belly; breast buff with blackish spots not coalescing to form collar.
NESTLING: unknown.

Field Characters. A medium-small lark with reddish upperparts and distinctive black collar separating white throat from reddish breast. Strikingly red and black in flight, red upperparts contrasting with blackish wings and tail, red underwing-coverts contrasting with black underside of flight feathers.

Voice. Not tape-recorded. Said to be a 'curious, plaintive whistle which gradually mounts the scale, at first increasing and then decreasing in volume as the pitch rises' (Serle 1943).

General Habits. Present in arid or semi-arid tussock grasslands with scattered bushes, trees and termite mounds, and bare patches on red sandy soils, at 100–1350 m (Kenya). Wary and elusive; runs fast if followed; reluctant to fly.

Food. Invertebrates (grasshoppers, lepidopteran larvae) and seeds.

Mirafra collaris

Breeding Habits. In display, ♂ rises steeply, clapping wings below the body, to 10–15 m above ground, then ceases clapping and begins gradual descent, during which it may change direction as it glides down to ground or perch on tree. Wing-beats during descent are slow and regular. May give song during descent, although usually waits until flight completed (Serle 1943).
NEST: domed cup of grass, placed in scrape in ground against tuft of grass.
EGGS: 2–3, mean (4 clutches) 2·7. Dull white, with large and small olive-brown and reddish brown speckles over underlying ashy markings, either uniformly speckled over whole surface or concentrated in zone on obtuse end. SIZE: (n = 11) 19·2–21 × 14–15·2 (19·9 × 14·9).
LAYING DATES: Ethiopia, May.

References
Serle, W. (1943).
von Erlanger, C. F. (1907).

Mirafra africanoides Smith. **Fawn-coloured Lark. Alouette fauve.**

Mirafra africanoides A. Smith, 1836. Rep. Exped. Expl. Cent. Afr., p. 47; eastern province of the Colony and the country towards Latakoo (= Colesberg).

Forms a superspecies with Asian *M. erythroptera* and *M. affinis*.

Range and Status. Resident, sedentary. 2 disjunct groups: (1) local and patchy in S and SE Ethiopia, NW and W Somalia; more widespread Kenya (mainly west of 38°E), NE Uganda (Karamoja), N Tanzania (Serengeti to Dodoma, Tarangire Nat. Park and foothills of N Pare Mts); and (2) SE Angola, Zambia (W of Zambesi, very local), N and E Namibia (N Kaokoveld, Ovamboland and Caprivi Strip south through Etosha and W Damaraland east of the Namib to Kalahari Gemsbok Nat. Park), Botswana (widespread except for Okavango delta; mainly NW, NE, central and S), Zimbabwe (local and patchy on Kalahari and granite sands from Marondera and Beatrice southwest along central watershed to Selukwe and Fairfield; and from Mkien and Chesa Forest area west to Igusu and Nyamandhlovu, W Gwaai to Lupane, Nata R. and Dzivanini Pan north to Tamafupa Pan along W border of Hwange Nat. Park). Mozambique (lowlands from Save R. to Delagoa Bay) and South Africa from N Cape Province (widespread but local north of 31°S and east of 20°E), Transvaal (central bushveld, Waterberg, Pafuri in NE, and extreme SW; sometimes common; centred around patches of Kalahari sands) and Orange Free State (W and N, locally common on Kalahari sands).

Description. *M. a. africanoides* Smith (including '*transvaalensis*', '*gobabisensis*', '*tsumebensis*' and '*quaesita*'): South Africa (Cape Province south of Orange R. to Kimberley), S Botswana (Molopo R.), S and E Namibia (north to Tsumeb and Gobabis). ADULT ♂: crown, nape, hindneck, mantle and back light rufous, streaked dusky brown, feathers with dark centres and light rufous margins. Stripe under eye, lores, supercilium and line around brown ear-coverts whitish; dark line from base of bill to eye; chin and throat white. Rump and uppertail-coverts light rufous, less streaked than back, feathers with darker centres. Tail dark brown, outer feathers with narrow buff outer edge, central pair with rufous margins. Breast white, streaked and spotted reddish brown, remainder of underparts white. Flight feathers brown, outer primaries narrowly edged rufous, inner primaries and secondaries broadly edged rufous. Scapulars brown with rufous-brown margins, tertials brown with broad rufous-buff margins, wing-coverts brown, margins buffy to rufous-brown. Underwing-coverts, axillaries and edges of inner webs of underside of primaries and secondaries pale rufous. Bill light yellowish horn; eye light brown; legs and feet pinkish. Sexes alike. SIZE: (34 ♂♂, 4 ♀♀) wing, ♂ 88–97 (92·5), ♀ 85–89 (87·2); tail, ♂ 53–63 (58·7), ♀ 51–57 (53·7); bill, ♂ 15–17 (16·2), ♀ 15–16 (15·7); tarsus, ♂ 18–24 (21·4), ♀ 19–22 (20·7). WEIGHT: ♂ (n = 34) 21–30 (23·5), ♀ (n = 11) 20·5–27·5 (23·1).

IMMATURE: crown and back feathers tipped buff, appears spotted above rather than streaked. Outer edge of tail more broadly white.

NESTLING: light purplish pink, darker on feather tracts, eyes blackish; long pale grey down on head and back, gape pale yellow. Inside of mouth deep yellow, with variable number of spots, usually 2 black spots on either side of tongue, and 1 black spot at tip of upper and lower mandibles.

M. a. austin-robertsi White: W Zimbabwe (Bulawayo), E Botswana (Mahalapye, Serowe, Tsepe, Digomo di Kae), South Africa (W Transvaal, NE Cape Province). Upperparts dark reddish brown, heavily streaked with blackish brown. Smaller: wing, ♂ (n = 8) 85–93 (89·9), ♀ (n = 2) 84, 87.

M. a. harei Roberts (including '*omaruru*', '*isseli*' and '*rubidior*'): NW Namibia (Windhoek, Okahandja, Omaruru, Kaokoveld). Paler, yellower and less heavily streaked than *africanoides*, crown sometimes very lightly streaked. Underparts white.

M. a. sarwensis (Roberts): NE Namibia (Ovamboland), W Botswana (Kakia, Lehututu, Tsabong, Murwamusa Pan, Lephepe). Paler and more sandy above than *harei*, varying from pinkish sandy to yellowish sandy. Breast reddish buff, plain to lightly streaked, underparts buff.

M. a. makarikari (Roberts) (including '*ovambensis*'): SE Angola (Chimporo), N Namibia (Namutoni, Onguma), N and central Botswana (Nata, Makgadikgadi, Bushman Pits, Maun), W Zimbabwe (Hwange), W Zambia (Sesheke, Nangweshi, Loma, Shangombo). Upperparts whitish grey, or brownish grey, paler than *sarwensis*, rump pinky brown. Breast lightly streaked, underparts grey-white.

M. a. trapnelli White: W Zambia (west of Zambesi R., Balovale), E Angola. Upperparts darker and more greyish brown than *makarikari*, generally more heavily streaked above.

M. a. vincenti (Roberts) (including '*mossambiquensis*'): central Zimbabwe (Marondera, Umvuma, Selukwe), Mozambique (Sul do Save, Panda, Manhica). Upperparts variable, greyish pink to grey, heavily streaked. Underparts buff, breast not well marked. Smaller: wing, ♂ (n = 10) 85–91 (88·0), ♀ (n = 2) 82, 84. WEIGHT: 1 ♀ 21·7.

M. a. intercedens Reichenow: SE and S Ethiopia, SW Somalia, Tanzania (Serengeti, Dodoma), Kenya (Lodwar, Pokot, Marsabit to Loita Plains, Amboseli, Tsavo and Taru), NE

Uganda. Upperparts variable, dark sandy brown to reddish brown, heavily streaked. Breast darker, more streaked. Supercilium and underparts buff.

M. a. macdonaldi White: S Ethiopia (Yaballo, Mega area). Brighter and more vinous red above and less heavily streaked than *intercedens*, crown only lightly streaked or unstreaked.

M. a. alopex Sharpe: SE Ethiopia, NW Somalia (Haud). Upperparts bright coppery red, plain or lightly streaked.

Field Characters. A medium-sized lark, difficult to characterize because it has no distinctive features. Plumage variable throughout wide range and large number of subspecies; important to know colour of local subspecies before comparing with other species. Upperparts rufous to dark or pale brown, outer tail-feathers edged white or buff, reddish wing-patch conspicuous in flight. White eyebrow prominent, white stripe below eye useful but not always easy to see. Bustling song distinctive.

Voice. Tape-recorded (88, F, G1B, McVIC, WALK). Song 4–5 staccato notes followed by a rapid jumble, e.g. 'chip-chip-chip-chip-chiree-chiree-chiree-tswit-tswit-tswit-chweeer', lasting 4–5 s, song phrases 7–10 s apart. Often ends with a harsh, down-slurred 'chweer', but may also trail off into a melodious trill. Alarm, a loud 'peeek'.

General Habits. Occurs in savanna and shrublands on sands; in the Kalahari more common in shrubby areas, especially on *Rhigozum trichotomum* steppes. Also in acacia woodlands on sands and on edges of miombo woodlands on granite sands.

Solitary or in pairs. Runs about on bare patches, but moves to shelter of vegetation if disturbed; often perches on shrub, bush or tree to sing, and sings commonly in flight.

Food. Invertebrates: termites *Hodotermes*, ants, grasshoppers, ant lion larvae (Myrmeleonidae), spiders, other arthropods (Myriapoda); also seeds of grasses (*Enneapogon* and *Schmidtia* spp.) and forbs (*Chenopodium*). Pieces of gravel up to 5 mm in diam. are present in most stomachs.

Breeding Habits. In song-flight lasting several min, ♂ cruises with exaggerated wing-beats at height of 20–30 m.

NEST: domed cup of grass and rootlets, placed in excavation at base of dwarf shrub or grass tuft. Bird first excavates hole for nest cup, then builds roof over nest, extending lower sides of roof forwards to form rim of nest; finally builds and lines cup. In Kalahari entrance usually faces east, southeast or south.

EGGS: 2–4, mean (13 clutches) 2·4. Ovate; white, covered with grey spots and overlying brown spots, spotted and blotched with chocolate-brown and yellow, markings sometimes concentrated at obtuse end. SIZE: (n = 42), 17·2–23·4 × 11·9–15·6 (20·7 × 14·6).

LAYING DATES: E Africa: Region C, Jan; Region D, Apr–May; Zimbabwe, Sept–Feb; South Africa, Sept–Oct, Jan–Mar. In Kalahari, season varies according to rainfall.

INCUBATION: period 12 days.

DEVELOPMENT AND CARE OF YOUNG: young fed by both parents; nestling period 12–14 days; remain dependent on parents after leaving nest. When disturbed at nest with young, both adults give 'peeek' alarm call, and may also flutter about in butterfly-like alarm flight giving alarm call.

BREEDING SUCCESS/SURVIVAL: from 33 eggs laid in 13 nests, 10 fledglings left nest (Maclean 1970). Adults and juveniles preyed on by Gabar Goshawk *Micronisus gabar*.

Reference
Maclean, G. L. (1970).

Plate 2
(Opp. p. 33)

Mirafra rufa Lynes. Rusty Bush-Lark. Alouette rousse.

Mirafra rufa Lynes, 1920. Bull. Br. Orn. Club 41, p. 15; Juga Juga, El Fasher, Sudan.

Range and Status. Endemic, resident. E Mali (common but local in rains), Niger (Tilaberi, Ouallam, Tahoua), N Togo (Tantigou), Chad (fairly common in SE of Ouadi Rimé-Ouadi Achim Faunal Reserve, not recorded north of 15° N), W-central and central Sudan (between 10° and 14°N; fairly common).

Description. *M. r. rufa* Lynes: W Sudan (Darfur), Chad. ADULT ♂: top of head and upperparts variable, from almost plain rufous with narrow blackish shaft streaks to heavily streaked black, rufous and buff, feathers rufous with broad dark centres and buff fringes; uppertail-coverts unmarked (rufous birds) or with dark shafts (streaked birds). Tail-feathers dark brown, outer pair with narrow buff outer edges, central pair rufous with dark shafts. Cheeks and ear-coverts brown with some pale streaks, supercilium, border to lower ear-coverts, chin and throat pale buff, rest of underparts buff, breast variably streaked rufous to dark brown. Flight feathers dark brown, outer edges buff, shading to rufous basally (rufous birds); tertials rufous-brown with darker shafts and narrow buff outer edges; scapulars and wing-coverts rufous-brown with darker shafts. Underwing-coverts and axillaries rufous-buff. Bill blackish, lower mandible whitish below; eye reddish brown; legs and feet greyish brown. Sexes alike. SIZE: (7 ♂♂, 4 ♀♀) wing, ♂ 80–87 (84·0), ♀ 81–86 (83·0); tail, ♂ 53–64 (59·5), ♀ 50–54 (52·7); bill, ♂ (n = 6) 15–16·1 (15·5), ♀ 14·9–16·3 (15·4); tarsus, ♂ 21–23 (21·8), ♀ 22–23 (22·7).

IMMATURE: upperparts paler, more cinnamon-buff, dark feather centres broader, giving spotted appearance, underparts pale buff, upper breast darker buff, breast with obscure brown spots.

NESTLING: unknown.

M. r. lynesi Grant and Mackworth-Praed: central Sudan (Kordofan). Lacks dark streaking on upperparts.

M. r. nigriticola Bates: Mali, Niger. Upperparts darker than *rufa*, less streaked. Underparts deeper buff with breast streaking heavier, more reddish.

Field Characters. A medium-sized, reddish lark with fairly long, dark tail with rufous centre and buffy edges. Other reddish larks in same area have shorter tails; Flappet Lark *M. rufocinnamomea* has darker upperparts, reddish underparts, distinctive flight; Kordofan Lark *M. cordofanica* is paler above and whiter below, with much white in outer tail; Dunn's Lark *Eremalauda dunni* is even paler above, pure white below with no breast streaks, short bill.

Voice. Not tape-recorded. Said to have a 'pleasing' song (Lynes 1924).

General Habits. Present in open savanna on plains and in open *Combretum* woodland on stony ridges. Solitary, in pairs or small family parties.

Food. Insects (termites) and other arthropods; also seeds.

Breeding Habits. ♂ displays, singing, with a prolonged aerial cruise rather high up; concludes flight by closing wings and dropping to ground or perch on tree top.

LAYING DATES: Chad, Sept; Mali, May–July (season varies according to onset of rains); Niger, July; Sudan, July.

References
Bates, G. L. (1934).
Lynes, H. (1924).

Mirafra gilletti Sharpe. Gillett's Lark. Alouette du Gillett.

Mirafra gilletti Sharpe, 1895. Bull. Br. Orn. Club 4, p. 29; Sibbe, Ogaden, Abyssinia.

Forms a superspecies with *M. degodiensis*.

Range and Status. Endemic resident. E and SE Ethiopia (Harar, Dire Ela, Dire Dawa, Awash, Ogaden; common), Somalia north of 2°N (Hargeisa, Dagaho-Meda, Berbera, Webbe-Dawa, Galbeid; fairly common and widespread); extreme NE Kenya (Handotu, Karo-Lola, Sarigo, Gordoba-Dschira, Damaso, Djeroko).

Description. *M. g. gilletti* Sharpe: Ethiopia, NW Somalia. ADULT ♂: crown rufous finely streaked dark brown, nape, hindneck, sides of neck and foreneck brown, feathers with buffy white margins. Ear-coverts rufous, streaked white, lores, face and supercilium whitish, dark line from base of bill to eye, blackish malar streaks. Mantle and back rufous-brown, feathers with dark centres, feathers of lower back with paler

Plate 3
(Opp. p. 48)

inner edge giving white streaked appearance; rump and uppertail-coverts dark smoky brown with pale margins. Tail-feathers dark brown, central pair with narrow rufous-brown margins. Underparts white to pale buff, lower throat to breast heavily streaked brown to orange-brown. Flight feathers dark brown, narrow buff outer edge, pale rufous-buff edge to basal half of inner webs; scapulars with dark centres shading to rufous-brown outer edge, buff inner edge; tertials with dark centres shading to rufous-brown and buff margins. Wing-coverts rufous-brown edged pale buff. Underwing-coverts and axillaries grey-brown, sometimes with rufous tinge, bend of wing barred brown and white. Bill pale brown; eye light hazel; legs and feet pale brown. Sexes alike. SIZE: (12 ♂♂, 8 ♀♀) wing, ♂ 82–90 (85·1), ♀ 80–86 (82·5); tail, ♂ 53–62 (58·5), ♀ 52–59 (56·1); bill, ♂ 16–18·3 (17·0), ♀ (n = 7) 14·5–18 (16·7); tarsus, ♂ 20–23 (21·5), ♀ 20–22 (21·2). WEIGHT: unsexed (n = 7) 20·3–24·8 (22·3).

IMMATURE and NESTLING: unknown.

M. g. arorihensis Erard: Somalia, except NW; possibly NE Kenya. Paler above, generally more reddish, streaking finer and paler than *gilletti*. Breast streaking redder and finer. WEIGHT: 3 ♂♂ 20·1–24·2 (22·0), 1 ♀ 20·9, unsexed (n = 2) 21·8, 26·1.

Field Characters. A medium-sized lark, similar to race *alopex* of Fawn-coloured Lark *M. africanoides* but rather darker and less reddish above, and lacks red wing-patch; underparts whiter, streaks on breast broader. In S may just meet Pink-breasted Lark *M. poecilosterna*, which differs in grey crown, pink face and pinkish underparts, breast streaks suffused.

Voice. Not tape-recorded. Sings 'dsee-dsit' from bush-top perch (J. S. Ash, pers. comm.), also 'sisisidetio' and 'da-di-da-di-da-di-da' (von Erlanger 1907).

General Habits. Occurs in open scrub savanna on sand and on hard stony soils, and in sparse thorny scrub and aloe scrub on edges of rocky outcrops, 1000–1500 m. Solitary or in pairs. ♂♂ sing from tops of trees.

Food. Insects (beetles, lepidopteran larvae, small grasshoppers) and seeds.

Mirafra gilletti

Breeding Habits. In display, ♂ flies high, to at least 100 m, circling over territory in distinctive flapping flight for *c.* 10 min, singing continuously, then plunges bill-first straight to ground in spectacular dive.

NEST: cup of fine stems and rootlets, on stony ground, against tuft of grass or at base of shrub.

EGGS: 3 (4 clutches). Ivory-white, uniformly speckled and boldly blotched and spotted on the obtuse end with rich sepia. SIZE: (n = 6) 19–21 × 14·8–15·2 (19·5 × 15·0).

LAYING DATES: Somalia, Apr–June.

References
Archer, G. F. and Godman, E. M. (1961).
von Erlanger, C. F. (1907).

Plate 3
(Opp. p. 48)

Mirafra degodiensis **Erard. Degodi Lark. Alouette de Degodi.**

Mirafra degodiensis Erard, 1975. L'Oiseau 45, p. 310; Bogo-Mayo, Dolo, Degodi, Sidamo Province, Ethiopia.

Forms a superspecies with *M. gilletti*.

Range and Status. Endemic, S Ethiopia. Known from 2 specimens taken 11 km from 'Bogol-Mayo' (evidently Bogol Manya, 4° 34′N, 41° 29′E: Collar and Stuart 1985) in the direction of Dolo, Ethiopia, close to the Ethiopia–Kenya–Somalia junction, Nov 1971 (Erard 1975b), and from field observations of Ash and Gullick (1990). Parapatric with Gillett's Lark *M. gilletti*, which does not occur in the Degodi area although its range surrounds it. Should be looked for in NE Kenya.

Description. ADULT ♂: differs from *M. gilletti* chiefly in size and proportion (Erard 1975b); plumage very similar, but upperparts slightly paler, greyish with yellowish tinge; black streaks on forehead, crown, nape, mantle, back and scapulars narrower and denser; chin, throat and breast less heavily marked than in *gilletti*, lightly speckled rather than boldly streaked. Bill fawn; eye not described; legs pinkish red. See also Ash and Gullick (1990). Tail, and secondaries, short or very short. Sexes alike. SIZE: 1 ♂, wing 81·5, tail 54, bill 14·5, tarsus 22; 1 ♀, wing 77, tail 51·5, bill 13·5, tarsus 21.

IMMATURE AND NESTLING: unknown.

Field Characters. Ash and Gullick (1990) observed 4 birds at Bogol Manya, presumed to be *M. degodiensis*: a small, slim, dull lark, smaller and duller than *M. gilletti*, unobtrusive and almost fearless. Skins look pipit-like or even *Cisticola*-like: short-tailed and round-winged lark, thin-billed, streaky brown above with closely-striated crown and all-dark tail, and white underparts with dark brown or tawny speckles on throat and breast. Ear-coverts dark, contrasting with pale, narrow superciliary stripe and particularly with pale cream-ochre area immediately behind and below them; whitish hindneck collar.

Voice. Unknown.

General Habits. Specimens collected on the ground in poor, arid savanna with low acacias scattered on otherwise bare soil. Ash and Gullick's birds fed on ground in shade of 3–4 m high acacias with scattered *Commiphora*, in very dry area without any herb layer. They ran fast and when moving across hot sunny patches of ground, or when disturbed, flew reluctantly for short distance just above soil; once alighted on bush.

Food. Stomachs contained small caterpillars and orthopterans.

Breeding Habits. Unknown.

References
Ash, J. S. and Gullick, T. M. (1990).
Collar, N. J. and Stuart, S. N. (1985).
Erard, C. (1975b).

Mirafra poecilosterna (Reichenow). **Pink-breasted Lark. Alouette à poitrine rose.**

Alauda poecilosterna Reichenow, 1879. Orn. Centralb., p. 155; Kibaradja, Tana River, Kenya.

Plate 3
(Opp. p. 48)

Range and Status. Endemic resident, S Ethiopia to N Tanzania and E Uganda to coast. Scattered records west of Rift Valley (Kenya) to Moroto (NE Uganda) and Lokorowa valley (SE Sudan); through S Ethiopia and Somalia south of *c.* 3°N; in semi-arid country south through Kenya below 1800 m, from Rift Valley to coast, especially in Northeast Province and the Tsavo Nat. Parks; reaches lowlands around Mt Kilimanjaro, Arusha Chini, Kahe and Longido in N Tanzania. Common, locally abundant.

Description. ADULT ♂: forehead and crown grey with centre of feathers brown; nape and mantle brown to grey-brown with darker centres to feathers; sides of neck, face and stripe over eye orange-brown; chin white; throat and breast whitish with orange-brown centres to feathers, giving streaky appearance; rump brownish grey merging to grey on uppertail-coverts, which have paler tips; tail uniform greyish brown; belly white, merging to orange-brown on flanks; primaries, secondaries and tertials brown; scapulars and all wing-coverts brown, with paler edges and tips to feathers making paler bar on wing when closed; underwing pale orange-brown. Bill horn with base of lower mandible paler, whole bill greyer and more slate-coloured in northern individuals; eye brown to red-brown; legs and feet flesh to pale brown, sometimes white. All plumage subject to soil stain. Sexes alike. SIZE: wing, ♂ (n = 18) 85–97 (91·4), ♀ (n = 9) 84–93 (87·6); tail, ♂ (n = 17) 64–79 (69·1), ♀ (n = 9) 60–72 (66·2); bill, ♂ (n = 18) 14–16·5 (15·5), ♀ (n = 7) 14·5–16·5 (15·4); tarsus, ♂ (n = 18) 21–25·5 (23·3), ♀ (n = 7) 22–24 (23·1). WEIGHT: (Somalia, Feb) 1 ♂ 26. (Kenya, n = 4) 23·5–26·4 (24·4).

IMMATURE: like adult, but throat and breast rustier, with blackish spots.

NESTLING: undescribed.

Field Characters. A slim lark with thin bill and fairly long tail, easily mistaken for a pipit; indistinct streaks make it appear plain in the field. Pinkish orange-brown face, breast and flanks easily distinguish it from other larks. Usually seen uttering weak, squeaky song from top of tree or bush.

Voice. Tape-recorded (CHA, LEM, McVIC, STJ). Song a rather squeaky trill, with 3–4 fairly slow introductory notes followed by 3–4 rather quicker ones on a slightly descending scale. Always uttered from a perch; given at any season. Call a thin 'tweet', usually uttered several times.

General Habits. Inhabits lightly bushed and wooded savannas, especially on sandy soils in arid and semi-arid areas with little or no grass. Widespread and common throughout Tsavo East Nat. Park, Kenya, except in areas without any bushes and areas of thick scrub with little open ground (Lack 1985).

Mainly solitary or in pairs, appearing to be territorial all year. Commonly perches with rather horizontal stance on topmost twigs of bush or small tree. Feeds on ground, walking or running along between stones or grass tufts.

Food. In Tsavo East Nat. Park, insects, often termites (of 60 food items, 47% insects, rest unidentified: Lack 1985). Large hard seeds and perhaps insects recorded from Somalia (specimen in BMNH).

A

Breeding Habits. Evidently monogamous. Sings only from perches, but once recorded in flight (**A**: I. Willis, pers. comm.).

NEST: raggedly roofed cup of dry grass, in depression in soil under small bush or beside log.

EGGS: 2. Oval; pale or slightly bluish grey, heavily spotted, especially at larger end, with chocolate-brown.

LAYING DATES: E Africa: SE Kenya (adults with nest material Apr, Dec, Jan, and with food Dec), Region D, Mar, June, Dec; Sudan, May. Apparently breeds in rains.

References
Lack, P. C. (1977, 1985).

Plate 5
(Opp. p. 96)

Mirafra sabota Smith. Sabota Lark. Alouette sabota.

Mirafra sabota A. Smith, 1836. Rep. Exped. Expl. Cent. Afr., p. 47; between Latakoo and the Tropic (= Rustenburg).

Range and Status. Endemic resident. Coastal Angola, from Cabinda (local and uncommon) south to Namibia (widespread and common); widespread Botswana, locally common S Zimbabwe, Mozambique south of Save R.; locally common South Africa south to central Natal, S-central Cape Province.

Description. *M. s. sabota* Smith: Zimbabwe (central plateau), Botswana (west to Dikgomo-di-Kae, north to Francistown), South Africa (N Cape Province south to Barkly West, W Orange Free State, Transvaal, Natal). ADULT ♂: top of head and upperparts streaked blackish, crown feathers edged buff, those of nape and hindneck edged grey-brown producing pale collar, those of mantle, back and rump edged rufous; uppertail-coverts with narrower dark centres, appearing less streaked than rest of upperparts. Tail-feathers sepia, outer pair with buff outer webs, central pair with broad rufous-buff margins, all feathers with narrow buff or whitish tips and sometimes narrow buff outer edges. Sides of neck and foreneck buffy, lightly streaked and speckled; lores buff, pronounced white supercilium continuing behind eye towards nape, dark line from bill to eye, narrow black moustachial stripe; chin and throat white, breast light brown speckled and streaked dark brown, rest of underparts buffy, browner on flanks. Flight feathers dark brown, primaries with narrow buff outer edges, secondaries similar, but buff extends around to feather tips. Scapulars dark brown, broadly edged rufous and light brown, tertials dark brown with broad rufous and light brown outer edges, narrow buff inner edges. Wing-coverts dark brown, broadly edged light brown to pale rufous. Underwing-coverts and axillaries light brown. Bill dark horn, shading to flesh at base of lower mandible; eye brown; legs and feet flesh brown. Sexes alike. SIZE: (8 ♂♂, 9 ♀♀) wing, ♂ 82–87 (84·5), ♀ 74–81 (76·7); tail, ♂ 45–51 (48·6), ♀ 41–46 (43·0); bill, ♂ 14–16·6 (15·4), ♀ 13·3–15·5 (14·3); tarsus, ♂ 20–22 (21·2), ♀ 20–22 (21·0). WEIGHT: ♂ (n = 45) 21–30 (25·0), ♀ (n = 16) 21–31 (24·0), unsexed (n = 26) 22–28 (24·5).

IMMATURE: crown and mantle feathers dark brown, tipped light brown, back and rump dark brown, feathers tipped buff. Uppertail-coverts dark brown, broadly tipped light brown. Breast spotted rather than streaked, feathers dark with buff tips.

NESTLING: unknown.

M. s. waibeli Grote (including '*hoeschi*', '*elfriedae*' and '*veseyfitzgeraldi*'): N Namibia (Etosha, Ovamboland), N Botswana (Ghanzi, L. Dow). Paler and more whitish grey than any other race, heavily streaked dark brown above. Bill slender, as in *sabota*.

M. s. sabotoides (Roberts) (including '*fradei*'): W Zimbabwe (Bulawayo, Sentinel Ranch), central and S Botswana (Makgadikgadi Pan, Gemsbok Pan), SE Botswana (Lethlakeng, Molepole, Lephepe), South Africa (NW Transvaal). Close to *herero* in colour but generally more vinaceous, less buffy, feather centres broader and darker. Bill slender, as in *sabota*.

M. s. suffusca Clancey: SE Zimbabwe (Sabi–Lundi confluence), S Mozambique (Porto Henrique, Changalane, Canicado, Maqueze), South Africa (NE Natal, E Transvaal lowveld), Swaziland (Lubuli, Stegi). Darker than *sabota*, streaking of upperparts almost black; crown, mantle and scapulars duller and less rusty brown. Bill slender, as in *sabota*.

M. s. ansorgei Sclater: coastal Angola from Mossamedes to Novo Redondo. Similar to *naevia* but paler. Bill slender, as in *sabota*.

M. s. plebeja Cabanis: Angola (Cabinda coast). Browner above than *ansorgei*. Bill slender, as in *sabota*.

M. s. bradfieldi (Roberts): South Africa (central, E and N Cape Province). Generally paler and more sandy than *sabota*. Bill heaviest of all races.

M. s. herero (Roberts): S and E Namibia (north to Etosha), South Africa (NW Cape Province). Paler and more sandy above than *bradfieldi*; underparts whiter, spotting on breast paler. Bill stout.

M. s. naevia (Strickland) (including '*uis*' and '*erongo*'): NW Namibia (Khorixas, Sesfontein, Orupembe). Paler and more yellowish above than *herero*, dark centres to feathers smaller and feather margins paler. Bill stout, similar to *herero*.

TAXONOMIC NOTE: following Clancey (1966) and Hall and Moreau (1970), we recognize two groups of subspecies: (1) the *naevia* group, and (2) the *sabota* group. The *naevia* group comprises 3 large-billed subspecies (*bradfieldi*, *herero*, *naevia*) of semi-arid and arid regions in SW and W South Africa and W Namibia. Variation is clinal, birds with heavy bills and dark plumage in S, finer bills and paler plumage in N.

The *sabota* group comprises 6 slender-billed subspecies (*sabota*, *waibeli*, *sabotoides*, *suffusca*, *ansorgei*, *plebeja*) of mesic savanna woodlands in E and N South Africa, Swaziland, Zimbabwe, Mozambique, NE Namibia, Botswana and Angola. Variation is clinal, smallest and darkest birds in extreme SE, largest and palest birds in NE Namibia. Birds from coastal Angola resemble pale extremes of *naevia* in plumage, and *sabota* in bill shape.

The 2 groups are largely allopatric, meeting in a narrow contact zone in NE Cape Province; it is possible they represent incipient species.

Field Characters. A medium-sized lark, which lacks the red wing-patch typical of *Mirafra* spp. Very like Fawn-coloured *M. africanoides*, with conspicuous white supercilium stretching round to nape and giving somewhat capped effect, and dark line from bill to eye, but lacks white stripe below eye and instead has dark moustachial stripe lacking in Fawn-coloured. Plumage colour very variable in both species but Sabota is more heavily streaked above and has buff, not white, outer tail-feathers. Bill of Sabota varies in shape but heavier than Fawn-coloured; songs different.

Voice. Tape-recorded (75, 81, 88, F, STJ, WALK). Song variable, a pleasing medley of melodious notes, whistles and canary-like trills, interspersed with some buzzy sounds. Imitates songs, calls and alarm notes of other species (over 60 species identified). Calls include loud repeated 'pip-pip-peeu'; alarm, thin 'si-si-si' and rapid chattering.

General Habits. Inhabits moist to semi-arid fine-leaved savannas on clays and sands, rocky slopes with trees and bushes, acacia drainage lines, mopane scrub and mixed broad-leaved and fine-leaved woodlands on stony soils. In Karoo restricted to areas of open woodland along drainage lines where there is some grass; avoids dwarf shrublands on clays and sands, and dunes. In Namib Desert occurs in areas of low, open shrublands on rocks, mainly above 800 m.

Usually solitary, but density high in optimal habitat: 1 pair/2·4 ha in mesic acacia savanna woodlands at Nylsvley, Transvaal (Tarboton *et al.* 1987).

Forages on open ground or among tufts of grass and at bases of shrubs, pecking food from soil surface or gleaning from low foliage. Does not drink water, metabolizes it from food.

Food. Invertebrates: termites (*Hodotermes*, *Macrotermes*, *Odontotermes*), grasshoppers, beetles, beetle larvae, weevils, ants; also seeds (including grasses *Stipagrostis* spp. and forbs *Cleome* spp.) and green vegetation (grass stems). In Namib, diet consists of 40% insects, 60% seeds (n = 3).

Breeding Habits. In display, ♂ sings from perch on bush or tree, or in soaring flight 30–50 m up.

NEST: cup of dry grass, lined with finer grass and rootlets, placed in scrape in ground against tuft of grass or shrub, or under leaves of aloe. Back and sides of nest often built up and over to form dome, but dome not always present, especially when nest well shaded by vegetation. Scrape either so deep that rim of nest cup is level with ground, or shallower, with rim of nest above ground level.

EGGS: 2–4, mean (43 clutches) 2·3. Ovate; white, spotted and speckled with brown, yellowish brown and greyish brown, often concentrated in zone on obtuse end. SIZE: (n = 49) 18·6–23 × 13·9–15·5 (20·6 × 14·8).

LAYING DATES: Zimbabwe, Oct–Feb; Namibia, Dec–Mar, May; Mozambique, Sept; South Africa: Transvaal, Sept–Feb; Natal, Nov; N Cape, Nov, Jan–Feb.

INCUBATION: incubating bird flushed from nest with eggs flew to top of tree, gave long series of other species' calls, starting with their alarm calls and running without pause into string of other calls.

DEVELOPMENT AND CARE OF YOUNG: both adults feed young.

BREEDING SUCCESS/SURVIVAL: of 13 eggs in 7 nests, 5 young were reared and left nest.

References
Clancey, P. A. (1966).
Hall, B. P. and Moreau, R. E. (1970).
Tarboton, W. R. *et al.* (1987).

Genus *Heteromirafra* Grant

Small larks, inhabiting moist, short grasslands (2 species) and open savanna (1 species) at medium to high altitudes. Plumage brown, streaked, crown dark. Close to *Mirafra* but differs in narrow, short tail, disproportionately large feet and legs, and extremely long, straight hind claw. Bill strong and curved in *ruddi*, finer in *archeri* and *sidamoensis*. Nest like *Mirafra*.

3 species, endemic to Africa, with very restricted ranges, 2 in NE and 1 in S; they form a superspecies.

Heteromirafra ruddi superspecies

1. *H. ruddi*
2. *H. archeri*
3. *H. sidamoensis*

Heteromirafra ruddi (Grant). Rudd's Lark. Alouette de Rudd.

Heteronyx ruddi Grant, 1908. Bull. Br. Orn. Club 21, p. 111; Wakkerstroom, Transvaal.

Forms a superspecies with *H. archeri* and *H. sidamoensis*.

Range and Status. Resident, endemic to South Africa; confined to SE Transvaal (Belfast, Waterval Boven, Ermelo, Amersfoort, Volksrust, Wakkerstroom) and E Orange Free State (Warden, Verkykerskop), with an isolated population in W Natal (Matatiele). Scarce and local; range restricted but no evidence of reduction, although populations have decreased locally as a consequence of land-use practices (Hockey *et al.* 1988).

Description. ADULT ♂: crown mainly black with pale line down centre, feathers with very narrow buff margins; nape paler, feathers with broader buff edges, hindneck light brown, sides of neck brown, speckled darker brown, foreneck pale grey-brown. Cheeks greyish, ear-coverts darker; supercilium, chin and throat white. Mantle pale grey-brown, speckled darker brown, back streaked, feathers black or dark brown with buff margins, rump and uppertail-coverts streaked, feathers black or brown-black with buff and brown margins. Tail thin and short, dark brown with white outer edges to outer feathers, central pair with broad buff-brown margins. Breast rich buff, streaked lightly with brown; rest of underparts pale buff to whitish. Flight feathers dark brown with narrow rufous outer edges; scapulars and upperwing-coverts brown-black with buff or brown edges, tertials brown-black with buff margins. Underwing-coverts and broad margins to inner webs of basal half of underside of flight feathers pale brownish buff, axillaries orange-buff. Bill horn; eye grey-brown; legs and feet flesh pink. Sexes alike. SIZE: (5 ♂♂, 6 ♀♀) wing, ♂ 74–76 (75.0), ♀ 69–74 (71.0); tail, ♂ 37–42 (40.2), ♀ 35–43 (39.0); bill, ♂ 15–16.4 (15.9), ♀ 15–16.8 (15.4); tarsus, ♂ 24–26 (25.2), ♀ 24–26 (25.3); hind claw (17 ♂♀) 12–20.3 (15.0). WEIGHT 2 ♀♀ 26, 27.2.

IMMATURE: feathers of upperparts edged buff, giving a more boldly marked appearance than adults.

NESTLING: flesh-pink with blackish skin around closed eyes and on down tracts. Down present on crown, back, scapulars, wings and rump. Interior of mouth deep yellow with 2 black spots on sides of anterior tongue, and 2 black spots at tip of upper mandible.

Field Characters. A small, big-headed lark, with dark of crown extending down to nape, whitish line down centre of crown, heavy dark mottling on back, tail thin and short. Posture upright, legs relatively long; wings rounded, showing dull reddish brown patch in flight. Only other lark in same habitat is larger Spike-heeled Lark *Chersomanes albofasciata*, which has longer, finer bill, white-tipped tail, and white throat contrasting with darker underparts.

Voice. Tape-recorded (F, ALDG Allan). Aerial song a phrase of 3–8 notes, 'squeaker-wee', 'is-it-wee', 'is-it-wee-prr', 'for-see-is-it-wee', lasting 1 s, repeated at intervals of 5–6 s; at least 8 variations recorded (D. G. Allan, pers. comm.). In different brief song from ground, flies up from grass, whistles plaintively, drops back into grass. Call (alarm?), 5–7 identical notes given from ground near nest.

General Habits. Confined to patches of high rainfall (> 600 mm) grassland at 1700–2200 m, with short, dense grass cover dominated by *Themeda triandra*, *Tristachya leucothrix*, *Trachypogon capensis*, *Heteropogon contortus* and *Eragrostis chalcantha*, only within Acocks's (1975) vegetation types 56 and 57, with a single winter record from type 54. Habitat typically has grass *c.* 0.6 m high, and low mean projected ground cover (Hockey *et al.* 1988), usually on hill tops, plateaux, ridges. Avoids areas where grass is higher than 0.7 m and cover dense. Does not occur in old croplands. Optimal habitat conditions occur under a regime of annual burning and heavy winter grazing (Hockey *et al.* 1988).

Unobtrusive and overlooked unless calling. Usually solitary or in pairs, exceptionally in family party. Forages at bases of grass tufts, does not dig for food.

Food. Insects (beetles) and seeds.

Breeding Habits. Territorial; in breeding season 15 ♂♂ found displaying in area of 4 km². In display flight, bird rises at *c.* 60° angle to 15–30 m, flies round singing in circle 20–50 m in diam. Display flight lasts 1–30 min, average call rate 11 calls/min (n = 30 min). Displaying ♂♂ frequently change from one song to another during display (D. G. Allan, pers. comm.). In pre-copulatory display, ♂ struts in front of ♀ with bill full of nest material.

NEST: domed cup of grass stems, externally 80–100 long, 70–100 wide, 60–75 high, int. diam. 60–70, entrance 40–50 high, 45–55 wide, depth of cup 20–35;

lined with rootlets and dry grass, placed at base of or against grass tuft, usually facing northwest, west or south in SE Transvaal. Built by 1 bird only; all material collected within 10 m of nest site.

EGGS: 2–3, mean (10 clutches) 2·8. Ovate; white to pinky buff, heavily spotted with light brown, darker brown and slate, mainly on broad end. SIZE: (n = 22) 19·9–23·1 × 14·3–16·1 (21·7 × 15·4).

LAYING DATES: W Natal, Oct; SE Transvaal, Jan–Feb; Orange Free State, Dec.

BREEDING SUCCESS/SURVIVAL: 7 of 9 clutches deserted by birds. Of 6 eggs laid in 2 nests, no young reared.

References
Acocks, J. P. H. (1975).
Hockey, P. A. R. et al. (1988).

Plate 3
(Opp. p. 48)

Heteromirafra archeri Clarke. Archer's Lark. Alouette d'Archer.

Heteromirafra archeri S. Clarke, 1920. Bull. Br. Orn. Club 40, p. 64; Jifa, Somaliland.

Forms a superspecies with *H. ruddi* and *H. sidamoensis*.

Range and Status. Resident, endemic to W Somalia. Range very restricted, originally known only from a strip of grassland of *c*. 200 km² at 1500 m from Jifa Medir and Jifa Uri south to Ban Wujaleh, west of Hargeisa. A small population was later found 15–40 km northwest of Buramo, *c*. 100 km northwest of original range. Distribution patchy even within this range. Not recorded in original range since 1922; in recent range in 1955 "very uncommon, few seen" (J. G. Williams, *in* Collar and Stuart 1985).

Description. ADULT ♂: crown feathers with dark centres and buffy edges, pale line down centre of crown, nape and hindneck buffy brown, sides of neck and hindneck buffy, speckled brownish black, face buffy with dark border to back of ear-coverts, chin and throat buffy white. Mantle buff, speckled dark brown, feathers with dark centres and buff edges. Feathers of back, scapulars, upperwing-coverts and tertials with similar pattern—dark brown centre and narrow subterminal band separated by light brown, fringed buff to whitish, fringe broader on inner edge. Rump feathers with dark centres and buff tips, uppertail-coverts buff. Tail dark brown, outer feathers whitish or pale buff, adjacent pair pale buff on outer webs, central pair margined buffy brown. Breast to undertail-coverts buff, richer on breast which is spotted dark brown. Flight feathers dark brown, edged and tipped pale rufous to buff-brown. Underwing-coverts, axillaries and inner margins of underside of flight feathers rufous buff. SIZE: (6 ♂♂, 5 ♀♀) wing, ♂ 75–86 (79·5), ♀ 73–85 (78·5); tail, ♂ 44–56 (48·2), ♀ 43–52 (47·8); bill, ♂ 14–17 (15·6), ♀ 14–17 (15·8); tarsus, ♂ 27–30 (28·1), ♀ 26–28 (26·8), hind claw (9 ♂♀) 13·5–19·5 (15·4).

IMMATURE and NESTLING: unknown.

Field Characters. A small, short-tailed lark with pale outer tail-feathers and rather stout bill. Similar to Singing Bush-Lark *M. cantillans* but pale fringes of back feathers and wing-coverts give distinctive pattern in flight, and wings lack the red patch of Singing Bush-Lark. Extremely secretive.

Voice. Unknown.

General Habits. Restricted to patches of open, short grassland at *c*. 1500 m, and open, short, sparse grassland with rocks and scattered bush at the same altitude. Usually solitary. Shy and wary, creeping away in grass and very reluctant to cross bare ground, flying only when flushed; flight weak and undulating.

Food. Unknown.

Breeding Habits. Display unknown.

NEST: 'funnel-shaped' cup of dry grass, placed on ground in scrape at base of tuft or grass clump.

EGGS: 3 (n = 7). Ovate; white or cream, lightly speckled or spotted, sometimes blotched, with sepia or brown and with violet-grey shell marks. SIZE: (n = 7) 21–23 × 14–16 (21·8 × 14·4).

LAYING DATES: Somalia, June.

References
Archer, G. F and Godman, E. M. (1961).
Collar, N. J. and Stuart, S. N. (1985).

Heteromirafra sidamoensis (Erard). Sidamo Lark. Alouette d'Erard.

Mirafra sidamoensis Erard, 1975. Alauda 43, pp. 115–124; Negelli, Sidamo Province, Ethiopia.

Plate 3
(Opp. p. 48)

Forms a superspecies with *H. archeri* and *H. ruddi*.

Range and Status. Endemic, Ethiopia, probably resident. Known from 2 specimens, 11 km apart, at 1450 m near Negelli (5° 20'N, 39° 35'E, Borana, Sidamo), May 1968 (Erard 1975a) and Apr 1974 (Ash and Olson 1985). Similar birds seen in grassy areas 24 and 48 km southeast and 48 km south of Negelli, Apr 1971 and Apr 1973, probably the same species. Searched for but not found in Negelli grasslands in Oct–Nov 1971, when region was in drought. Not found at either site in 1989 (Ash and Gullick 1989).

Description. ADULT ♂: forehead and crown cream-buff with small blackish streaks, except in midline, forming a conspicuous pale midline stripe on dark-streaked crown; lores pale; long, narrow, creamy superciliary stripe; nape, sides of neck and ear-coverts pale pinkish fawn, very finely streaked blackish; mantle, back, rump, scapulars and wing-coverts rufescent brown, with complicated feather patterns giving mottled, striped and scalloped effect of black, white, cream, buff and (back and rump) deep pinkish fawn (11 feather types illustrated by Erard 1975a). Mantle with long blackish and buff streaks; back and rump feathers mainly with blackish centres, buff or cream tips, and brown or deep pinkish fawn sides; wing-coverts mainly with blackish median stripes flanking diagnostic whitish shaft, cream tips and brown or pinkish fawn sides. Tail with central feathers rufescent brown, outer (T6) deep pinkish fawn, T5 with blackish inner web and deep pinkish fawn outer web, and T2–T4 black-brown. Chin and throat white; breast white in centre, almost unmarked, buffy at sides, with small but well-demarcated blackish streaks; belly and undertail-coverts cream-white; flanks pinkish buff. Remiges rufous-brown; tertials rufous-brown with whitish shaft, blackish brown inner web, and narrow cream border all around; undersides of wings pinkish buff. Bill short, pale horn; eye dark brown; legs and feet pale brown; tarsus robust and hind claw long. ♀ unknown. SIZE: (2 ♂♂) wing 86, 87, tail 55·5, 56·5, bill 14, 15·5, tarsus 29, 29, hind claw 16·1, 16·5. WEIGHT: 1 ♂ 30·2.

IMMATURE and NESTLING: unknown.

Field Characters. A small, short-billed, rather short-tailed lark with robust, pale legs and long hind claw, finely-streaked brown crown with pale median stripe, dark pink-fawn outer rectrices, and richly-patterned rufescent-brown back, with black and buff stripes on mantle and cream scales in scapulars and wing-coverts. Distinct whitish superciliary stripe, pale buffy neck, buff flanks, and white underparts with close, fine and distinct small streaks on sides of breast.

Voice. Unknown.

General Habits. The 2 known birds were in 5–6 km diam. patches of undulating grassland, surrounded by thick *Acacia/Commiphora* bush. The grassland, grazed by domestic animals, gazelles and oryx, consisted of much new grass, extensive areas of tall dead grass, and open stands of low whistling thorn *Acacia drepanolobium*. Specimen collected on ground in Apr 1974 was thought to be one of several which had been 'flappeting' in air (Ash and Olson 1985). Very shy.

Food. Unknown.

Breeding Habits. Bird collected May 1968 was in breeding condition (Erard 1975a).

References
Ash, J. S. and Olson, S. L. (1985).
Collar, N. J. and Stuart, S. N. (1985).
Erard, C. (1975a).

46 ALAUDIDAE

Genus *Certhilauda* Swainson

Medium to large larks of open savannas, semi-arid and arid dwarf shrublands and sparse grasslands on sands. Plumage varies from greyish brown to rufous-brown above, usually streaked, and whitish below. Wings and tail long, legs strong, hind claw usually long and straight. Bill long and often decurved, nostrils exposed. Nest cup-shaped, sometimes domed, but not consistently, even within species.

Endemic to Africa; 5 species, 2 superspecies (*C. curvirostris/C. chuana* and *C. albescens/C. erythrochlamys*).

Certhilauda curvirostris superspecies

1 *C. curvirostris*
2 *C. chuana*

Certhilauda albescens superspecies

1. *C. albescens*
2. *C. erythrochlamys*

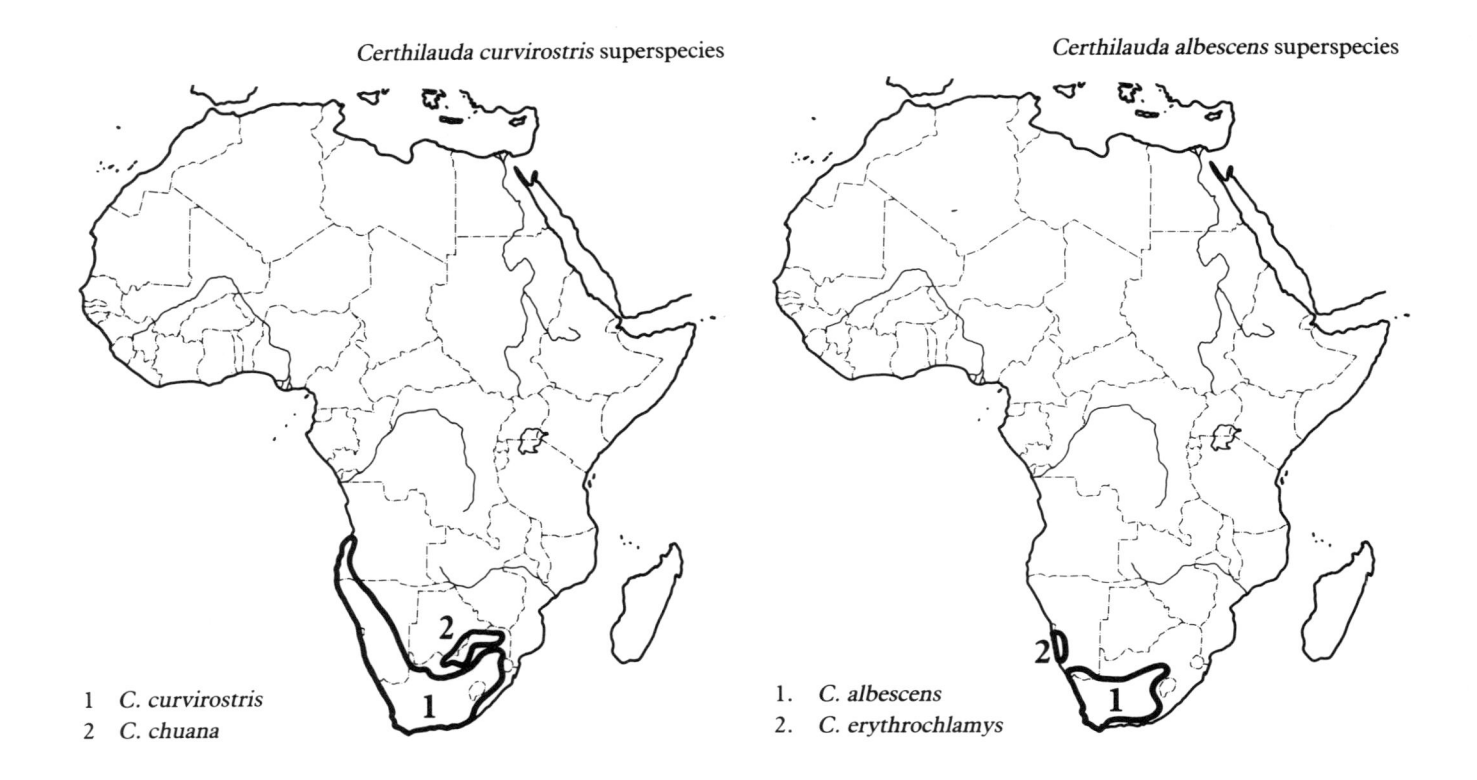

Plate 5
(Opp. p. 96)

Certhilauda curvirostris (Hermann). Long-billed Lark. Alouette à long bec.

Alauda curvirostris Hermann, 1783. Tabl. Affin. Anim., p. 216; Cape of Good Hope.

Forms a superspecies with *C. chuana*.

Range and Status. Resident, endemic. SE Angola north to Benguela; W and S Namibia (locally common); South Africa: locally common in S Transvaal, NW and W Natal and Orange Free State, widespread in Cape Province, particularly in Karoo; lowlands of Lesotho (not common).

Description. *C. c. curvirostris* (Hermann): South Africa (SW Cape Province from Cape Agulhas to Olifants R.). ADULT ♂: crown feathers dark brown edged lighter brown, narrowly tipped white, nape and hindneck same but white tips broader; sides of neck and foreneck whitish, streaked and speckled dark brown. Cheeks whitish, speckled dark brown, ear-coverts brown with narrow pale streaks, chin white, throat white, lightly speckled dark brown, lores and supercilium white; dark line from bill to eye. Mantle brown, feathers with dark centres, brown edges; back brown, streaked, feathers with dark centres, paler edges, narrow white tip. Rump and uppertail-coverts brown with dark shaft streaks, tipped whitish. Tail-feathers dark brown, central pair with rufous-brown margins, remainder with narrow rufous-brown outer edges. Throat to undertail-coverts streaked dark brown on grey-white, heaviest on breast. Flight feathers dark brown, narrowly edged pale rufous; scapulars dark brown with dark centres, brown edges, variably tipped white; tertials dark brown, outer edges variably whitish to rufous-buff, rufous-buff on inner edges. Wing-coverts dark brown, variably tipped and edged whitish to grey-brown. Underwing pale rufous-buff, streaked lightly on underwing-coverts. Bill blackish horn; eye brown; legs and feet pinkish-brown. Sexes alike. SIZE: (5 ♂♂, 2 ♀♀) wing, ♂ 103–111 (108), ♀ 96, 105; tail, ♂ 64–80 (73·0), ♀ 65, 66; bill, ♂ 30–37 (33·4), ♀ 27, 28; tarsus, ♂ 28–31 (30·2), ♀ 26, 30.

IMMATURE: mottled above, feathers of upperparts with buffy tips; speckled below, less streaked than adult.

NESTLING: no data on skin colour or down. Gape pale yellow, mouth spots black, 2 laterally on back part of anterior tongue, 1 black spot at tip of upper and lower mandibles.

Certhilauda curvirostris

Most populations vary in shade of reddish on upperparts and degree of streaking below (see Clancey 1957). Bill size varies throughout the species; longest bills in SW and W coast, smallest in NE. Birds may be divided into 3 broad groupings: (1) large-sized, long-billed, pale brown or grey birds with heavily streaked underparts occurring on aeolian sands of W and S coasts; (2) smaller birds with smaller bills, rufous above and less streaked, occurring on reddish sands and clays of W Angola, Namibia, N Cape and Karoo; (3) small-billed birds, buffy below, streaking confined to the breast, occurring on clays in E Cape, Natal and Transvaal.

C. c. brevirostris Roberts: Cape Agulhas to Knysna. Upperparts grey-brown, bill markedly smaller: 1 ♂ 22·5, 1 ♀ 20·5.

C. c. falcirostris Reichenow: SW coast from Olifants R. to Holgat R. Upperparts paler and greyer than *curvirostris*. Bill long and curved: ♂ (n = 10) 32 –38 (34·9), ♀ (n = 7) 27–29 (27·9).

C. c. subcoronata Smith: NE Karoo in De Aar, Carnarvon, Hanover districts. Upperparts much darker and redder, hindneck greyer, underside greyish white. Bill: ♂ (n = 11) 26·2–32 (27·6), ♀ (n = 5) 21–23·8 (22·5).

C. c. gilli Roberts: central and S Karoo in Beaufort West, Touws Rivier districts. Darker and more heavily streaked than *subcoronata*, but bill similar in size.

C. c. bradshawi (Sharpe): NW Cape Province (Little Namaqualand on lower Orange R., Pofadder, Upington districts); S Namibia (north to Tiraz Mts). Upperparts lighter red and less heavily streaked than *subcoronata*. Bill: ♂ (n = 6), 24–27 (25·9), ♀ (n = 4), 20–22 (21·5). WEIGHT: 1 ♂ 31, unsexed (n = 4) 32–53 (44·2).

C. c. kaokensis Bradfield: N Namibia (Naukluft Mts north to Kaokoveld). Upperparts paler and less streaked than *bradshawi*. Bill: 1 ♂ 26. WEIGHT: 1 unsexed 45.

C. c. benguelensis (Sharpe): NW Namibia, SW Angola. Upperparts paler and greyer than *kaokensis*. Bill: 1 ♂ 24, 1 ♀ 21.

C. c. semitorquata Smith (including '*daviesi*', '*algida*'): E Cape (Grahamstown), Transkei, W Natal (Matatiele), Orange Free State. Upperparts brighter and paler red and less streaked than *subcoronata*; underparts buffy. Bill small: ♂ (n = 5) 24·4–26·4 (25·5), ♀ (n = 3) 20–20·8 (20·3). WEIGHT: 1 ♂ 45.

C. c. infelix White (replaces *C. semitorquata transvaalensis*, preoccupied): N Cape Province, Transvaal, N Natal. Intergrades with *semitorquata*. Upperparts paler and redder, less tinged olive and less heavily streaked than *semitorquata*; darker buff below. Bill: ♂ (n = 11) 23·7–26 (24·7), ♀ (n = 6) 19·5–21·2 (20·5).

Field Characters. A large, variable lark; upperparts grey-brown to reddish with some streaking, underparts whitish to rufous-buff with streaking variable, from fine streaks on breast to heavy streaks on most of underparts. Distinctive long bill also variable, longest and most decurved in west, less so in east and northeast. Also has conspicuous white supercilium, longish tail, crouched posture and distinctive display flight with long whistled call. Only similar lark in same habitat, smaller Spikeheeled Lark *Chersomanes albofasciata*, has short, white-tipped tail, upright stance, very different voice.

Voice. Tape-recorded (88, F, GIB, LEM). Advertising call or song, a long, high-pitched, descending whistle with a short, soft introductory note, 'ti-PEEEEEOO'. Also, lower-pitched, melodious whistles with a burry, rolling quality, variable and with no set rhythm, strung together into a quite different type of song.

General Habits. Present in montane grasslands, dwarf Karoo shrublands, thinly vegetated coastal dunes, sparsely vegetated rocky ridges and stony hills and mountains.

Usually solitary or in pairs. When foraging searches bases of plants, turns over stones and knocks down termite surface workings to eat termite workers, digs in soft soil for larvae and pupae, and takes fruit from dwarf shrubs. Reputed to be a pest in croplands, digging out planted grains.

Sedentary, apparently holding territories throughout the year.

Food. Insects (termites, including *Microhodotermes viator*, flies, orthopterans, beetles, ants *Tetramorium* and *Anoplolepis*), other arthropods, including spiders and solifuges, seeds of forbs (*Sesamum triphyllum*, *Osteospermum*) and shrubs (*Hermannia*), fruits (*Lycium*) and small corms of Iridaceae. All stomachs examined contained gravel up to 3 mm diam.

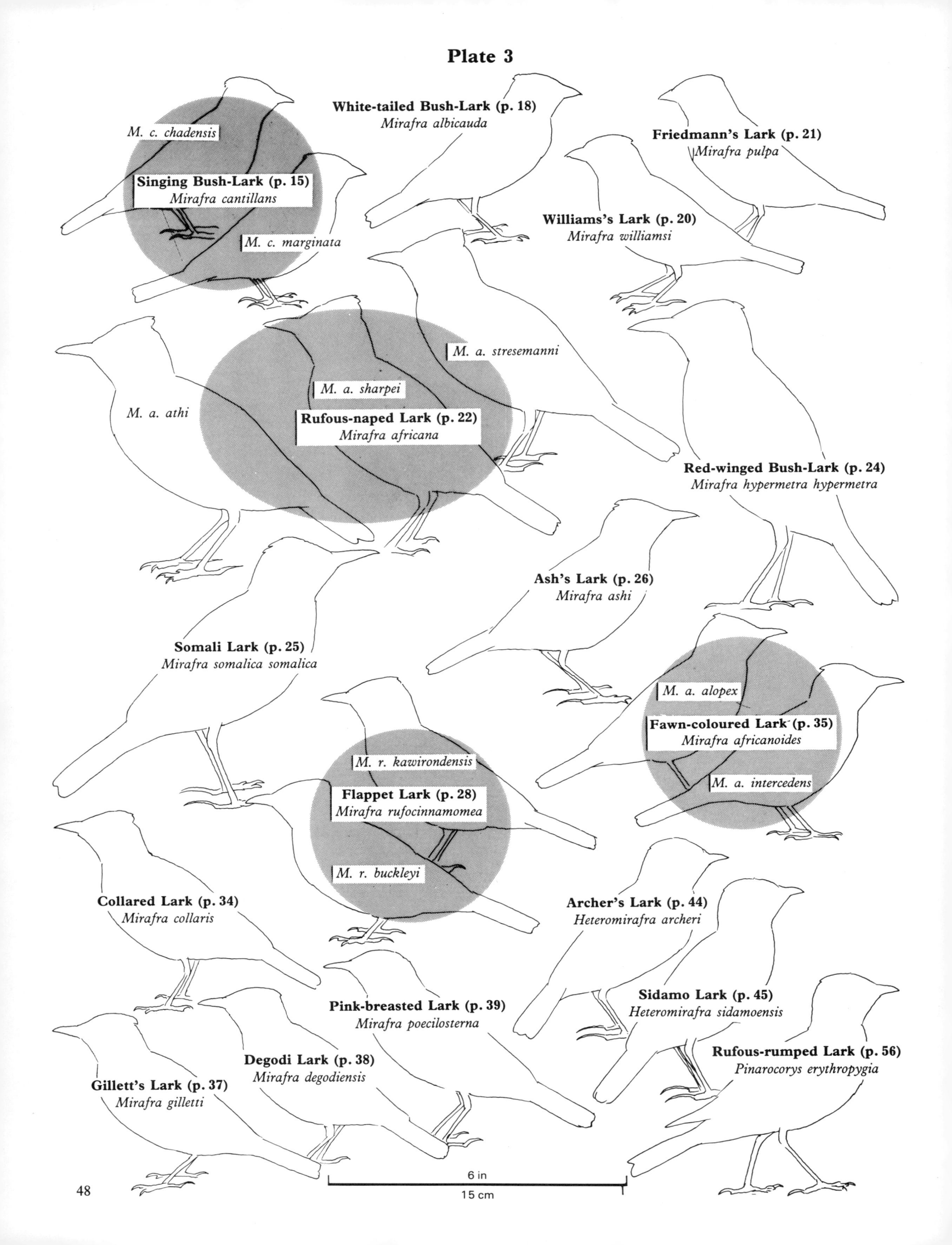

Plate 3

White-tailed Bush-Lark (p. 18)
Mirafra albicauda

Friedmann's Lark (p. 21)
Mirafra pulpa

M. c. chadensis

Singing Bush-Lark (p. 15)
Mirafra cantillans

M. c. marginata

Williams's Lark (p. 20)
Mirafra williamsi

M. a. stresemanni

M. a. sharpei

M. a. athi

Rufous-naped Lark (p. 22)
Mirafra africana

Red-winged Bush-Lark (p. 24)
Mirafra hypermetra hypermetra

Ash's Lark (p. 26)
Mirafra ashi

Somali Lark (p. 25)
Mirafra somalica somalica

M. a. alopex

Fawn-coloured Lark (p. 35)
Mirafra africanoides

M. a. intercedens

M. r. kawirondensis

Flappet Lark (p. 28)
Mirafra rufocinnamomea

M. r. buckleyi

Collared Lark (p. 34)
Mirafra collaris

Archer's Lark (p. 44)
Heteromirafra archeri

Sidamo Lark (p. 45)
Heteromirafra sidamoensis

Pink-breasted Lark (p. 39)
Mirafra poecilosterna

Rufous-rumped Lark (p. 56)
Pinarocorys erythropygia

Degodi Lark (p. 38)
Mirafra degodiensis

Gillett's Lark (p. 37)
Mirafra gilletti

6 in
15 cm

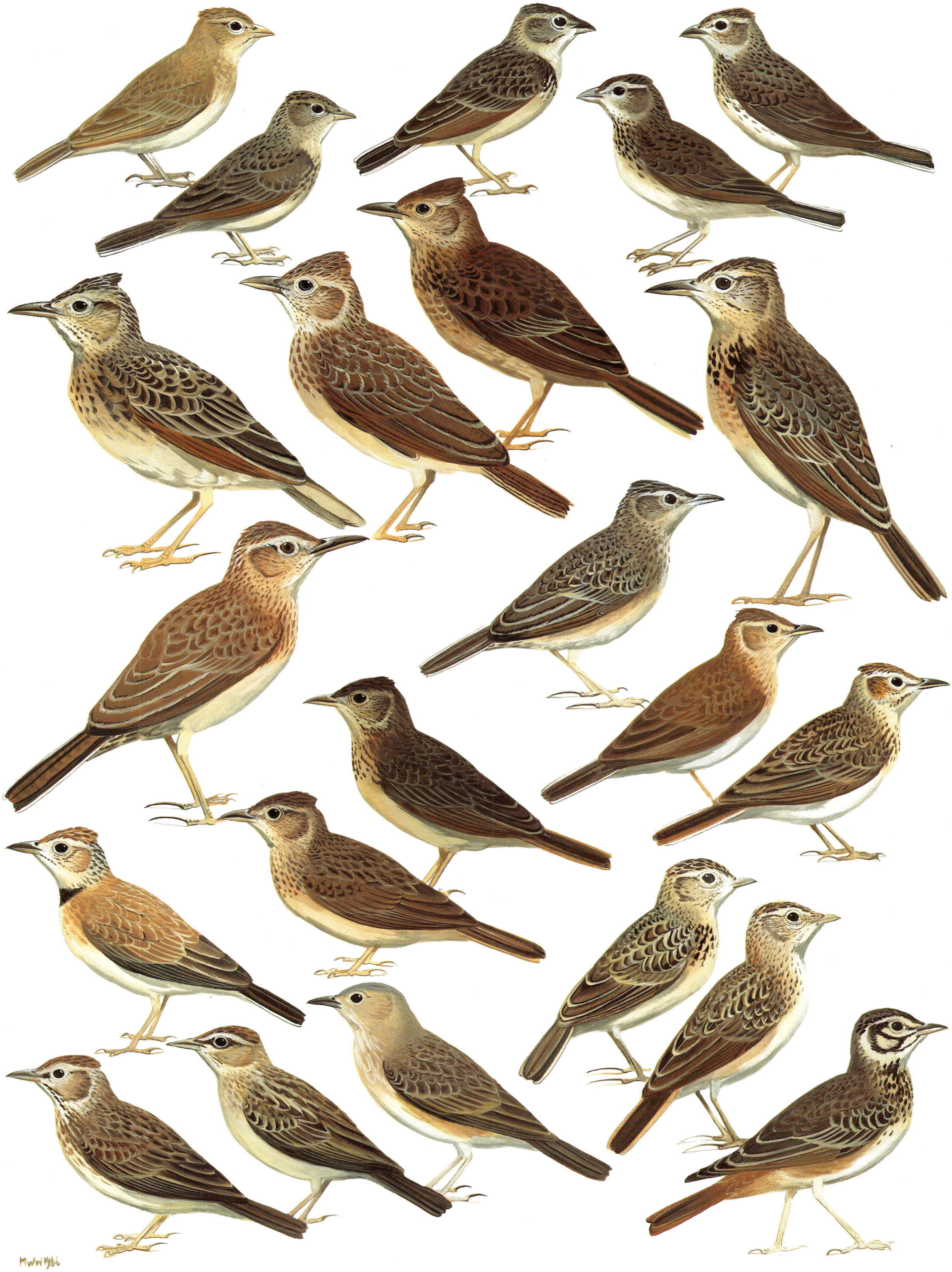
MWW 1986

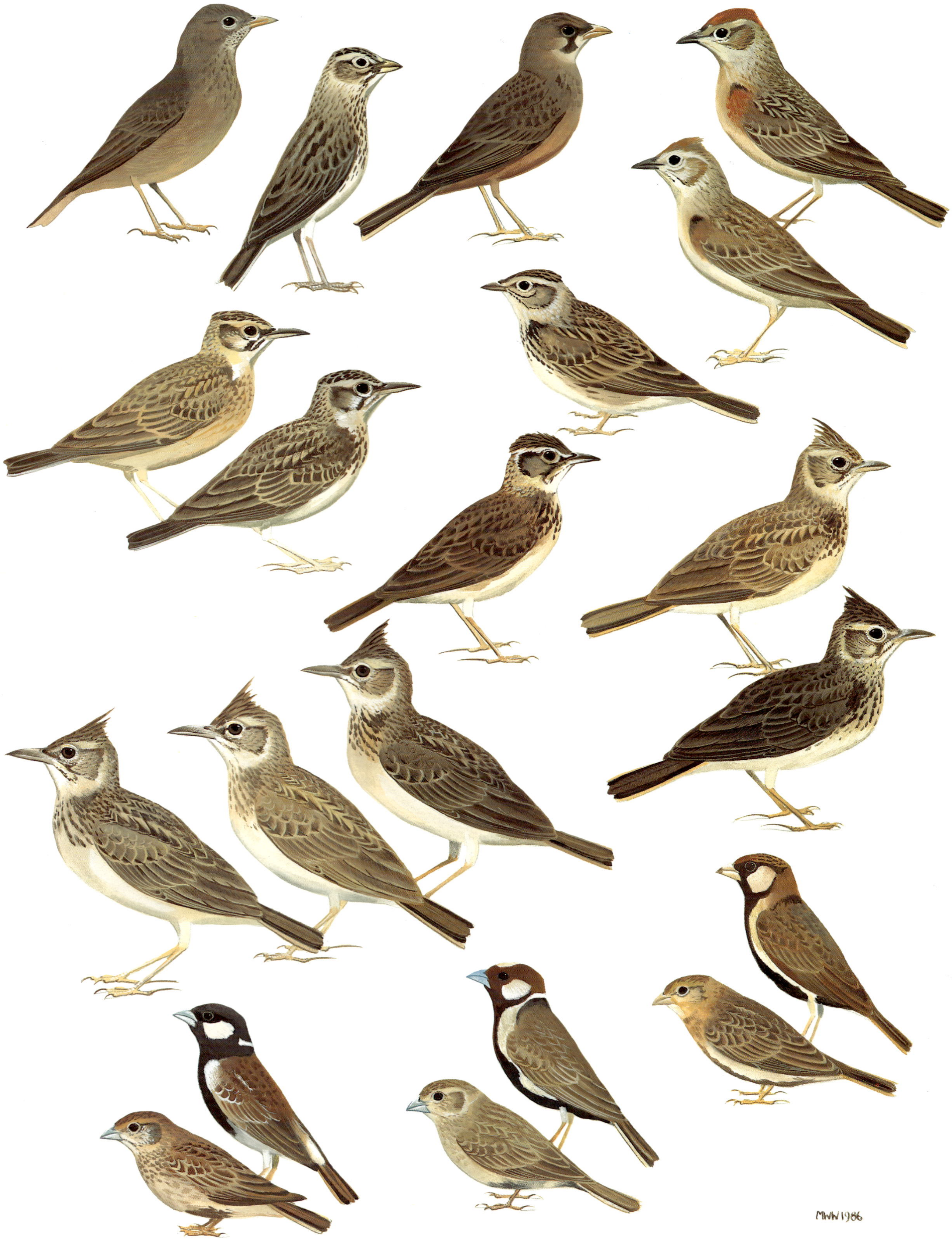
MWW 1986

Plate 4

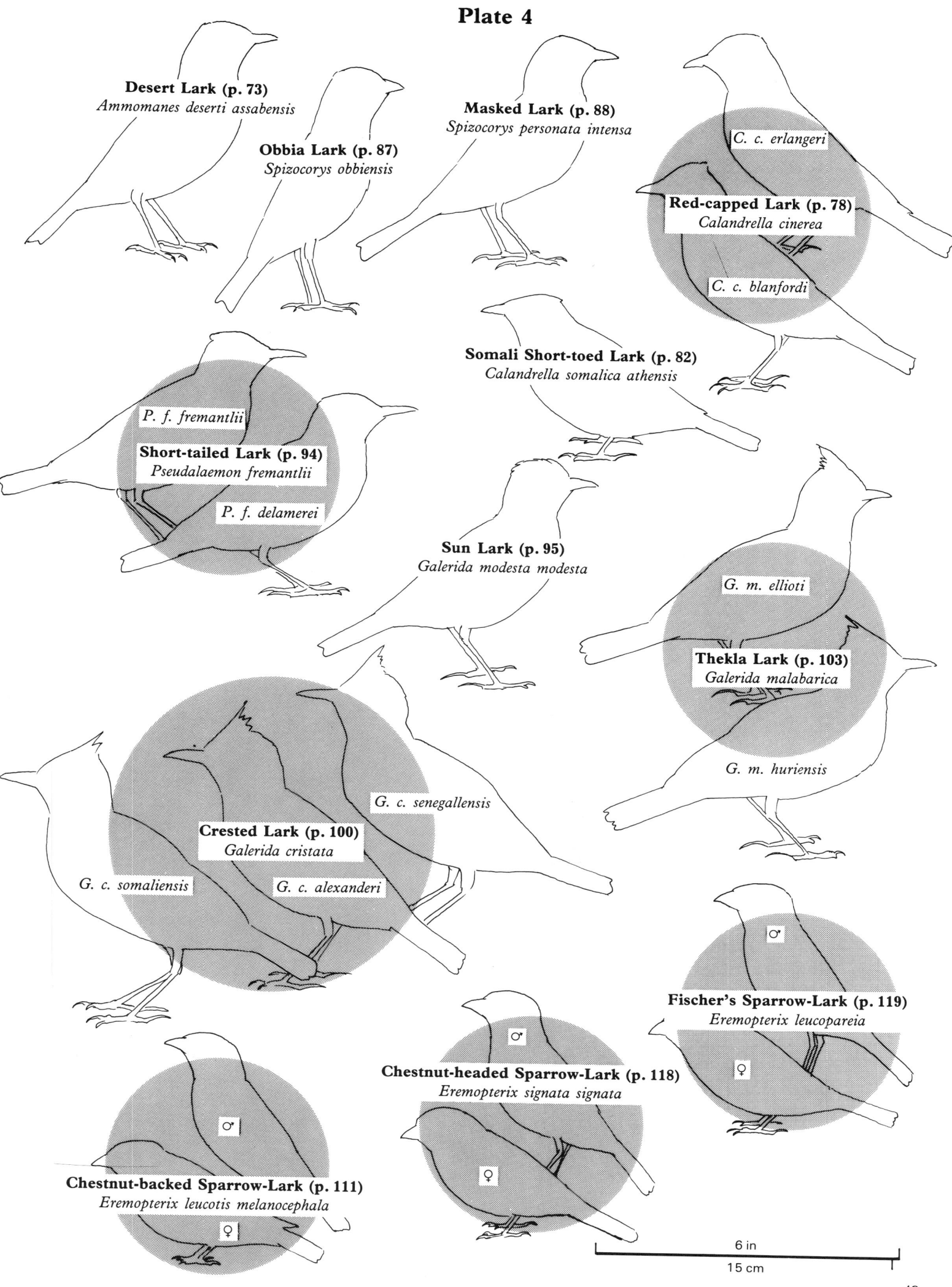

Desert Lark (p. 73)
Ammomanes deserti assabensis

Obbia Lark (p. 87)
Spizocorys obbiensis

Masked Lark (p. 88)
Spizocorys personata intensa

Red-capped Lark (p. 78)
Calandrella cinerea
C. c. erlangeri
C. c. blanfordi

Somali Short-toed Lark (p. 82)
Calandrella somalica athensis

Short-tailed Lark (p. 94)
Pseudalaemon fremantlii
P. f. fremantlii
P. f. delamerei

Sun Lark (p. 95)
Galerida modesta modesta

Thekla Lark (p. 103)
Galerida malabarica
G. m. ellioti
G. m. huriensis

Crested Lark (p. 100)
Galerida cristata
G. c. senegallensis
G. c. somaliensis
G. c. alexanderi

Chestnut-backed Sparrow-Lark (p. 111)
Eremopterix leucotis melanocephala

Chestnut-headed Sparrow-Lark (p. 118)
Eremopterix signata signata

Fischer's Sparrow-Lark (p. 119)
Eremopterix leucopareia

6 in
15 cm

49

Breeding Habits. Territorial. In display, ♂ utters 'pheeeuw' repeatedly from prominent perch and in flight: rises vertically to 10–15 m, closes wings at or just before top of climb, and calls 'pheeeuw', then drops, opening wings just before landing on bush or ground.

NEST: cup of grass, sometimes domed, lined with finer material, placed on ground against grass tuft or stone, or between dwarf shrubs.

EGGS: 2–3, mean (26 clutches) 2·6. Ovate; white to pinkish white, speckled and spotted with yellowish brown, darker brown and grey. SIZE: (n = 18) 21·8–24·9 × 15·5–17·5 (23·4 × 16·5).

LAYING DATES: Namibia, Aug; South Africa: Transvaal, Nov; Natal, Oct–Nov; Orange Free State, Sept–Dec; E Cape, Sept, Nov–Dec; SW Cape, Aug–Oct.

DEVELOPMENT AND CARE OF YOUNG: both adults feed young, landing on ground at least 5 m from nest and walking to it, then walking similar distance away after feeding young before taking wing; ♂ may perform full display while carrying food to nest. Adult runs from the nest if disturbed, does not fly.

Reference
Clancey, P. A. (1957).

Plate 5
(Opp. p. 96)

Certhilauda chuana (Smith). **Short-clawed Lark. Alouette à ongles courts.**

Alauda chuana A. Smith, 1836. Rep. Exped. Expl. Cent. Afr., p. 46; north of Latakoo.

Forms a superspecies with *C. curvirostris*.

Range and Status. Endemic resident. Botswana (Lethlakeng, Molepole), SW and N-central Transvaal (Schweizer-Reneke, Wolmaransstad, Pietersburg, Bandolierskop), W Orange Free State (widespread but scarce). Locally common N Transvaal, status unclear elsewhere.

Description. ADULT ♂: top of head and upperparts heavily streaked blackish, feathers of crown edged brown to rufous, those of hindneck edged buff, producing pale collar, those of back edged rufous; rump and uppertail-coverts rufous with blackish shaft streaks. Tail-feathers dark brown, central pair with rufous margins, outer pair broadly edged buff, rest narrowly edged buff. Lores and supercilium whitish, dark line from bill to eye, ear-coverts brown with narrow black rim, sides of neck brown, foreneck whitish; chin and throat white to pale buff, rest of underparts buff, breast reddish buff streaked dark brown (more heavily on sides than in centre). Primaries and secondaries dark brown with narrow buff outer edges. Scapulars and wing-coverts blackish with broad rufous-brown outer edges, paler, rufous-buff inner edges; tertials dark brown, broadly edged rufous-buff. Underwing-coverts and axillaries rufous-buff, under primary coverts barred blackish along bend of wing. Bill blackish or brownish horn, flesh at base; eye light brown; legs and feet flesh-brown. Sexes alike. SIZE: (10 ♂♂, 2 ♀♀) wing, ♂ 100–107 (102), ♀ 91, 91; tail, ♂ 64–72 (69·2), ♀ 61, 62; bill, ♂ (n = 9) 19–23 (20·9), ♀ 18·6, 19·1; tarsus, ♂ 27–30 (28·8), ♀ 24, 29. WEIGHT: 1 ♂ 44·1.

IMMATURE and NESTLING: unknown.

Field Characters. A large lark with long bill, pronounced white eye-stripe and rufous rump showing in flight; upperparts tawny, boldly streaked black. Very similar to Long-billed Lark *C. curvirostris* but ranges do not overlap. In hand, hind claw curved and short (long and straight in Long-billed). Same size as Rufous-naped Lark *Mirafra africana* but different shape, less chunky, and long bill and lack of crest give head slimmer look; also lacks red in wings, and voice different.

Voice. Tape-recorded (88, F, GIB, HUS) and extensively analysed (Hustler 1980). Song, a variable clear, shrill 2–3 note whistle, 'pree-prree-prree', 'pip-peeu-peeu', 'peeu-weeu', 'wip-peeee-prrr', 'pip-peeet-pree' and other variations (Maclean 1985). Alarm call a soft 'kwert-kwert-kwert'; warning call a harsh 'crare-crare-crare' or 'krerr-krerr-krerr'.

General Habits. Occurs in open acacia savanna woodlands, or grassy plains with scattered acacia trees, particularly where grass has been heavily grazed.

Usually solitary or in pairs. Forages by walking through grass and gleaning invertebrates from leaves and stems; dislodges prey by shaking stems.

Food. Invertebrates (grasshoppers, weevils, lepidopteran larvae, termites *Hodotermes* and *Macrotermes*, ants *Anoplolepis*). Young fed on insect larvae.

Breeding Habits. ♂ displays in fluttering flight with wings clapping; straight and on same plane, low down over dense 1 m tall grass.

NEST: only 1 recorded, a cup of grass stems and roots in scrape in ground at base of geophyte *Hypoxis rigidula*, in well-grazed grassland. 2 young *c*. 10 days old in nest (Hustler 1985).
EGGS: unknown.
LAYING DATES: South Africa (Transvaal), Mar.

References
Hustler, C. W. (1980, 1985).

Certhilauda albescens (Lafresnaye). Karoo Lark. Alouette du Karoo.

Alauda albescens Lafresnaye, 1839. Rev. Zool., p. 259; Blouberg, Cape Town.

Forms a superspecies with *C. erythrochlamys*.

Range and Status. Endemic resident. Extreme S Namibia (north to Witsputs) and South Africa (Cape Province, from Namaqualand (common) south to SW Cape and Worcester Karoo, east through Little and Great Karoo to W Orange Free State). Frequent to locally common.

Description. *C. a. albescens* (Lafresnaye): South Africa (SW Cape from Cape Town to Berg R.). ADULT ♂: crown, nape and hindneck smoky brown, feathers darker in centre, edged and tipped buffy, sides of neck brown, foreneck whitish with dark markings; lores, supercilium, stripe under eye and eye-ring white; dark line from bill to ear-coverts, ear-coverts brown, thin black malar and moustachial stripes. Chin and throat white with a few dark marks. Mantle grey-brown, feathers with dark centres, back smoky brown, tinged pink, feathers with broad blackish centres, rump and uppertail-coverts pinkish brown, uppertail-coverts with dark centres. Tail-feathers brownish black, central pair broadly margined smoky brown, outer pair with very narrow pale outer edges. Breast to undertail-coverts white, breast and upper belly with broad blackish streaks, flanks with narrower streaks. Primaries and secondaries dark brown with narrow buff outer edges; tertials and scapulars with dark centres, broad buff outer edges, narrow buff inner edges and tips. Upperwing-coverts dark brown with buff margins. Axillaries and underwing-coverts grey-brown. Bill blackish horn; eye brown; legs and feet grey-brown. Sexes alike. SIZE: (16 ♂♂, 16 ♀♀) wing, ♂ 85–92 (88·3), ♀ 81–90 (84·6); tail, ♂ 57–65 (61·0), ♀ 55–63 (57·8); bill ♂ 17–20·5 (19·1), ♀ (n = 15) 17–19·1 (17·9); tarsus, ♂ 22–26 (23·5), ♀ 21–25 (23·2). WEIGHT: (3 ♂♂) 31–35 (32·6), unsexed (n = 4) 27–31 (28·7)

IMMATURE: feathers of crown, nape, back and uppertail-coverts with buff tips, giving a spotted appearance. Breast spotted and streaked, less heavily marked than adult.
NESTLING: skin pink, blackish purple on feather tracts and over closed eyes. Down long and grey-white on crown, back and wings. Inside of mouth yellow, with 1 black tongue spot on either side of back of tongue, and 1 black spot at tip of upper and lower mandibles.

C. a. codea (Smith) (including 'saldanhae' and 'patae'): South African coast from Saldanha Bay to Port Nolloth. Upperparts washed with pale vinous pink, back less boldly streaked than *albescens*. Streaks on breast and flanks extend to belly.
C. a. guttata (Lafresnaye): Cape Province from Clanwilliam to Little Namaqualand and S Richtersveld. Upperparts deep vinous red, upperparts, breast and flanks heavily streaked dark brown.

Plate 6
(Opp. p. 97)

C. a. karruensis (Roberts) (including '*calviniensis*'): Cape Province from Calvinia and Williston east to De Aar and Murraysburg, south to Oudtshoorn. Darker and duller than *guttata*, less deep vinous red on upperparts.

C. a. cavei Macdonald: S Namibia (from Witsputs south to Orange R.), South Africa (Cape Province from lower Orange R. and S Great Namaqualand to Holgat, Little Namaqualand). Upperparts rich cinnamon-brown, less streaked on back and breast than *codea*.

Field Characters. A medium-sized lark with a relatively slender bill, bold face pattern, streaked upperparts, heavily streaked underparts, shortish, dark tail and rather heavy flight. Upperparts brown in south, redder in north. Closely related Dune Lark *C. erythrochlamys* does not overlap in range, has paler, unstreaked sandy rufous back, thin, pale rufous streaks on breast; larger Red Lark *C. burra* has thick bill, brighter, unstreaked upperparts, black wings.

Voice. Tape-recorded (88, F, GIB). Song repeated, stereo-typed phrase of 5 rather buzzy notes 'tip-tip-tip-zree-trr', 'chip-chip-chip-cheepy-rrr' or 'chip-chip-chip-chizzee' (Maclean 1985), 'chawi-cheewi-cheekrrrrr' (Gillard tape); also a shorter 'eezer-eeit', given in flight or from perch on bush or shrub. On ground utters a low-pitched chittering call.

General Habits. Occurs in coastal dunes, dune troughs with scattered rocks and sparse scrub, dwarf shrublands on sands and clays and sparse grasslands on sands; occupies sparsely vegetated old fallow fields, but avoids land being farmed.

Solitary or in pairs. Forages on ground, walking about or occasionally running; searches bases of grass tufts and shrubs; occasionally gleans food from leaves or takes fruits. Digs in sand of dry river beds for seed. Takes termites from fresh diggings of aardvark *Orycteropus afer*, going down into excavations.

Food. Invertebrates: termites (*Hodotermes mossambicus*, *Microhodotermes viator*), lacewings (Nemopteridae), mantids, grasshoppers, weevils, sapsuckers and leafhoppers (Homoptera: Cicadellidae), beetles (Coleoptera: Salpingidae, Cassidae, Carabidae), coccid beetles (Lacciferidae), ants (*Messor, Pheidole, Tetramorium, Crematogaster, Acantholepis*), snails and spiders. Also seeds of grasses (*Centropodia glauca, Eragrostis, Stipagrostis, Cladoraphis spinosa*), shrubs (*Protasparagus, Aridaria, Sphalmanthus, Psilocaulon, Hermannia, Lotononis, Requienia sphaerosperma, Plexipus*), annual forbs (*Polygonum, Chenopodium, Atriplex, Amaranthus, Limeum, Galenia, Tetragonia echinata, Talinum, Indigofera, Tribulus, Dimorphotheca*) and juicy fruits (*Lycium*). Leaves and flower parts also eaten. Stomachs contain pieces of gravel up to 5 mm diam. Nestling fed insects, including termites.

Breeding Habits. Monogamous, territorial. In display, ♂ calls from perch on bush or in flight; flies on rapidly beating wings, often almost hovering, or cruises at 15–25 m in wavy flight, with slow wing-beats and depressed tail.

NEST: cup of dry grass, ext. diam. *c.* 100, depth of cup *c.* 30, lined with finer material including awns of grasses *Stipagrostis* spp. and woolly seeds of *Eriocephalus* spp.; in scrape in ground at base of grass tuft or shrub; material frequently built up and over back and sides to form dome over nest; sometimes no dome, only a slightly higher rim at back if sheltering plant is dense. Built by ♀, accompanied by ♂.

EGGS: 2–4, mean (33 clutches) 2·6. Ovate; white, spotted or blotched with grey-brown, brown, reddish brown and grey, or speckled with brown and violet-grey; markings sometimes concentrated at obtuse end. SIZE: (n = 10) 21·5–24 × 14·6–17 (22·4 × 15·9).

LAYING DATES: South Africa: SW Cape, Aug–Nov; Karoo, July–Nov, Feb; Orange Free State, Sept.

DEVELOPMENT AND CARE OF YOUNG: young fed by both adults; leave nest before they can fly.

Reference
Vincent, A. W. (1946).

Plate 6
(Opp. p. 97)

Certhilauda erythrochlamys (Strickland). Dune Lark. Alouette à dos roux.

Alauda erythrochlamys Strickland, 1852. Jardine's Contr. Ornith., p. 151; Damaraland.

Forms a superspecies with *C. albescens*.

Range and Status. Resident, endemic to Namibia; confined to W Namib from *c.* 30 km north of Aus north to Rooibank, Walvis Bay. Locally common.

Description. ADULT ♂: crown, ear-coverts, sides of neck, back, rump, uppertail-coverts, scapulars and upperwing-coverts plain sandy rufous; wing-coverts with pale outer edges.

Mantle usually like crown and back but variable, may be paler or more rufous. Tail-feathers dark brown, outer pair with whitish to rufous-buff outer edges, central pair like upperparts but with dark shafts, next pair (T2) grey-brown edged rufous. Dark line from bill to ear-coverts; lores, supercilium, stripe under eye, eye-ring, chin and throat white, rest of underparts whitish to buff, breast lightly streaked rufous-brown. Flight feathers brown, primaries narrowly, secondaries more broadly edged rufous on outer webs, tertials rufous with narrow buff margins. Underwing-coverts and axillaries pale rufous-buff. Bill brownish horn; eye brown; legs and feet pinkish brown. Sexes alike. SIZE: (6 ♂♂, 6 ♀♀) wing, ♂ 90–95 (91·5), ♀ 81–91 (86·0); tail, ♂ 64–66 (64·7), ♀ 59–64 (60·6); bill, ♂ 18·8–20·8 (19·5), ♀ 17–19 (17·9); tarsus, ♂ 27–29 (27·8), ♀ 24–27 (25·3). WEIGHT: ♂ (n = 8) 26·3–33·1 (29·4), ♀ (n = 10) 24·8–28·9 (26·8).

IMMATURE: feathers of upperparts edged buff, giving spotted appearance.

NESTLING: flesh-pink, with long grey-white down on crown, back, scapulars and wings. Gape yellow, inside of mouth deep yellow, with 3 black spots on tip and sides of anterior tongue, 1 black spot on tip of upper and lower mandibles. Tongue spots variable, sometimes absent.

TAXONOMIC NOTE: treated as a subspecies of *C. albescens* by Macdonald (1953), Lawson (1961) and White (1961), and regarded by Meinertzhagen (1951) as not systematically close to *albescens*. Treated by Clancey (1980) as a full species, and recent studies have shown it differs from *albescens* in call and biology.

Field Characters. A medium-sized, pale rufous lark confined to the sand dunes of W Namib and so occurring with very few other species. Same shape, proportions and slender bill as closely related Karoo Lark *C. albescens*; latter has darker reddish, streaked upperparts, heavily streaked underparts, and lives in different habitat, although their ranges come close.

Voice. Tape-recorded (F, BOY). Flight song similar to that of Karoo Lark *C. albescens*, but with longer introductory notes: 'tip-tip-tip-tip-tip-zree-trr'. Introductory notes sharp, last part of song trilled. Also gives a short song 'tew-tew-tew-TEW-tew-tew' from perch early in breeding season; as season advances, this is replaced by 'chee-chee-ch-ee-chirr-chiree-chi-ch-rrr-chrrr', which may be interspersed with mellow, bubbling notes. Contact call a buzzy 'zeet-zeet'; also a buzzing threat call.

General Habits. Occurs only on thinly vegetated dunes and sandy inter-dune flats of the Namib dune system. Absent from gravel plains and mobile dunes.

Sometimes solitary, or may remain in pairs both during breeding season and throughout most of year; at times in small groups. Mean group size (n = 59) 2·2. Territories vary in size from 2 to 4 ha at Gobabeb, Namib, and all activities of resident pair or group are restricted to confines of territory during breeding season.

Runs quickly across bare sand slopes between vegetation patches and forages rapidly at bases of grasses and forbs before moving on to next patch. Takes insects and jumps up at *Trianthema hereroensis* and *Stipagrostis sabulicola* foliage to glean food items from leaves. Ants (*Camponotus*) are pounded several times before being eaten. Leaps or flies up to take aerial prey. Forages for seeds by digging in soft sand ('bill-cratering': Cox 1983), leaving distinctive small, conical holes. Metabolic water and invertebrates are apparently primary water source; has not been recorded drinking. Roosts at night and during midday break in grass clumps, where midday sand surface temperatures are 15·1–21° C cooler than on exposed sand (Safriel 1990); roosts may also serve as lookout points for predators. Fearless of man, but frequently inspects sky.

Food. Seeds 32% (*Stipagrostis sabulicola*, *Trianthema hereroensis*, other grasses and forbs), invertebrates 68% (ants *Camponotus detritus*, termites *Hodotermes mossambicus*, locusts, beetles, caterpillars). One record of white lady spider (*Leucorchestris*). Nestlings fed exclusively invertebrates (moths, small locusts, larvae).

Breeding Habits. Monogamous, territorial. ♂ advertises territory by singing from elevated perch. Also sings during aerial display while flying 10–20 m above ground.

NEST: cup of coarse material obtained mostly from the grass *Stipagrostis sabulicola*; ext. diam. 110–145, int. diam. 65–100, depth of cup 60–85; lined with finer material (wind-blown plant material, feathers, hair and reptile skin), often plastered together with webs from nests of burrowing spiders; usually domed over but not always, dome supported by vegetation at back of nest; placed in depression dug in sand at base of plant. Built by ♀ only, in 7–9 days.

EGGS: 1–2, mean (11 clutches) 1·9. Ovate; white, densely spotted with dark red-brown and some pale purple-grey, markings often concentrated at obtuse end. SIZE: (n = 16) 20·1–23·7 × 14·3–17·5 (22·0 × 15·9).

ALAUDIDAE

LAYING DATES: Namibia, Jan–Apr, June, Aug–Oct.

INCUBATION: begins when clutch complete; by ♀ only; period (n = 3) 13–14 days.

DEVELOPMENT AND CARE OF YOUNG: feathers in sheaths begin to emerge on day 4, sheaths rupture on day 7, young fully feathered but with traces of down still present at 11–12 days. All eggs hatch on same day; young flatten themselves in nest, head lowered to hide yellow gape. Brooded by female up to day 4, possibly longer during inclement weather. Both adults feed nestlings; 2 small nestlings fed 7 times in 30 min, 2 larger (8-days-old) fed 7 times in 90 min; often bring more than 1 food item per visit. Nestling period (n = 2) 12–14 days. Young leave nest before they are able to fly and hide in vegetation for first few days.

BREEDING SUCCESS/SURVIVAL: of 18 eggs laid in 10 nests, 11 eggs hatched and 5 young reared (fledged and out of nest but still dependent on parents). Overall breeding success 0·38 fledged young/nest attempt at Gobabeb (Boyer 1988).

References
Boyer, H. J. (1988).
Cox, G. W. (1983).
Lawson, W. J. (1961).
Willoughby, E. J. (1971).

Plate 6
(Opp. p. 97)

Certhilauda burra (Bangs). Red Lark. Alouette ferruginea.

Ammomanes burra Bangs, 1930. Bull. Mus. Comp. Zool. 70, p. 368; South Africa (= Great Namaqualand). *Nom. nov* for *Alauda ferruginea* Lafresnaye, 1839, pre-occupied.

Range and Status. Resident, endemic to Cape Province, South Africa: Koa R. valley from east of Steinkopf east to Aggeneys and south to the Kliprand area; Pofadder, Kenhardt, Van Wyksvlei area (on sand dunes from east of Verneukpan to east of Fortuinkolksepan), Brandvlei south and east to isolated sand dunes in the Brospan area. Local but not uncommon.

Description. ADULT ♂: crown, nape, back, scapulars, median and lesser coverts, rump and uppertail-coverts unstreaked rufous, mantle rufous-brown, a shade darker than back and lightly streaked darker brown. Some variation in colour of back, some birds distinctly brownish with heavier streaking. Tail-feathers blackish, T2 and T6 edged rufous outwardly, central pair rufous with black shaft streaks. Facial area boldly patterned, with dark mark from base of bill to ear-coverts; white lores, supercilium and line below eye, brown ear-coverts, black malar and moustachial streaks. Chin and throat greyish white, some dark spots on chin. Rest of underparts off-white, breast heavily marked with dark brown wedge-shaped spots. Flight feathers brown, tertials brown with rufous margins becoming broader inwardly, greater coverts brown with rufous-buff margins. Axillaries and underwing-coverts light brown with pale fringes. Bill blackish horn, paler at base; eye brown; legs and feet greyish brown. Sexes alike but ♀ often appears paler in the field. SIZE: (7 ♂♂, 7 ♀♀) wing, ♂ 100–110 (105), ♀ 85–103 (94·4); tail, ♂ 75–88 (78·9), ♀ 66–77 (70·7); bill, ♂ 16·5–18 (17·6), ♀ 15–17·2 (16·3); tarsus, ♂ 26–28 (27·2), ♀ 23–28 (25·5). WEIGHT: 7 ♂♂ 38·5–43 (40·5), 4 ♀♀ 32–38 (34·5).

IMMATURE: upperparts redder and brighter rufous-brown than adult; feathers of crown and back tipped whitish, giving spotted appearance, tertials with broader whitish margins. Central pair of tail-feathers plain rufous.

NESTLING: skin pink, blackish purple on feather tracts and over closed eyes, with long, grey-white down on crown, back and wings. Inside of mouth yellow, with 2 inconspicuous black spots, 1 either side about half-way along tongue, and 1 black spot at tips of upper and lower mandibles.

TAXONOMIC NOTE: some variation is evident in general colour and in amount of streaking on back. Brownish birds were named *harei* (Roberts 1937) but rufous and brown birds occur together and interbreed (Myburgh and Steyn 1989). A second subspecies, *aridula*, was named on the basis of heavier streaking on the breast, but this difference is not consistent. We follow Meinertzhagen (1951), White (1961) and Winterbottom (1967) in not accepting *harei* or *aridula* as valid subspecies.

The Red Lark has had a tangled taxonomic history (Brooke 1984). Its generic allocation has shifted between *Ammomanes*, *Pseudammomanes*, *Certhilauda* and *Mirafra*, and it has been considered a subspecies of both *C. albescens* and *C. erythrochlamys*.

Field Characters. A large red to brownish lark with unstreaked to lightly streaked upperparts, heavily streaked breast, strongly patterned face and short, thick bill; wings and tail black and red in flight; stance upright. Richer and deeper red above than Dune Lark *C. erythrochlamys*, which is more sandy, with finely streaked breast, longer, thinner bill, no black in wings. Karoo Lark *C. albescens* has similarly streaked breast but upperparts duller and paler, with dark streaks, and has blackish, finer bill. Does not overlap in habitat with either Dune or Karoo Larks. Has a very characteristic habit of hovering above shrubland or dune grasses before dropping on to the sand.

Voice. Tape-recorded (F, GIB). Song, very like Karoo Lark *C. albescens* but rattle at end of call longer; 'chee-chee-chrrrrrr-chee' or 'chwee-chwee-tirrrrr-chee' repeated over and over, given from prominent perch or in flight. Several calls given from ground; a very fast 'teeteetrrr-chee', 'tee-tee-tirrrr-chee', a metallic clinking and a low, repeated rattle in alarm. Gives a short 'chrrk' or 'chuk' when flushed.

General Habits. Found mainly on red sand dunes and sandy plains where large-seeded grasses (*Brachiaria glomerata*, *Stipagrostis ciliata* and *S. brevifolia*) dominate the vegetation, particularly where there are scattered hair-trees *Parkinsonia africana* to provide perch sites. Perennial vegetation cover in optimal habitat ranges from 5 to 25%, with low-growing forbs and annual plants increasing cover to *c.* 40% after rains. Also present in dwarf shrubland on shale soils where large-seeded grasses present. Requires multi-layered vegetation, with scattered emergent bushes to provide perches and shade, and perennial grasses to provide nest-sites. Avoids well-grassed dune troughs.

Runs about on ground and forages at bases of plants; food items pecked from soil surface, fruit taken directly from bushes, caterpillars gleaned from leaves. Seldom digs for food. Frequently perches on bushes and trees; shelters in shade of bushes during heat of day. Solitary or in pairs.

Sedentary.

Food. Invertebrates (termites, lepidopteran larvae, weevils, beetles and larvae, grasshoppers, mantids and ants), seeds (grasses *Brachiaria glomerata*, *Stipagrostis ciliata*, forbs *Arctotis* sp., *Gisekia*, *Hermannia*, *Limeum africanum*, *L. myosotis*, *Polygonum* sp., *Lotononis* sp., *Tribulus terrestris*, *Dimorphotheca*) and juicy fruits (*Lycium* spp.). Prefers large, smooth seeds to small, fluffy or hairy seeds (Dean *et al.* 1991).

Breeding Habits. Monogamous, territorial. In display, ♂ rises from perch to 15–20 m, fluttering and hovering against wind, tail usually fanned and depressed, circles with slow wing-beats, calling, descends in dipping dive to land on perch or ground. May continue to call from perch with tail fanned. Flight and perch song identical.

NEST: cup of coarse grass, placed in scrape between 2 tufts or against a tuft; lined with strips of old weathered grass and awns of *Stipagrostis* spp., 1 with pieces of sloughed reptile skin; cup *c.* 25 deep, int. diam. 65, ext. diam. 75 × 90; roofed over with old greyish grass leaves, on foundation of *Stipagrostis amabilis* stems; entrance to nest (from cup to lower edge of dome) 75–90 long, roofing extended well back from cup; 2 nests faced south.

EGGS: 2–3 (3 clutches). Ovate; white, speckled with sepia and brown over underlying grey-brown speckles, with a few larger sepia blotches on obtuse end where markings denser, or finely speckled dark grey-brown over entire shell. SIZE: (n = 3) 22·7–23·2 × 16·5–16·8 (23·0 × 16·6).

LAYING DATES: South Africa: Aggeneys, Jan, Aug, Oct; Brandvlei, Mar, Oct (adults with enlarged gonads May, following rains); Pofadder, Apr, May.

DEVELOPMENT AND CARE OF YOUNG: both adults feed young; feeding rate (n = 1) 06·00–08·00 h, 8 items/h; adults approach nest circumspectly; both fly to vicinity of nest, land *c.* 20 m away, walk to nest; one feeds young, and when it leaves nest, the other approaches and feeds. When young small, ♂ feeds them first, ♀ may remain to brood them.

References
Dean, W. R. J. (1989b).
Dean, W. R. J. *et al.* (1991).
Myburgh, N. and Steyn, P. (1989).

ALAUDIDAE

Genus *Pinarocorys* Shelley

Large, dark larks of open savanna, close to *Mirafra* but with dark streaks on underparts, bold facial pattern, strong bill, and wing-flicking behaviour while foraging. Wing long and pointed, outer primary much reduced; flight undulating. Intra-African migrants.

Endemic to Africa. 2 species, 1 in northern and 1 in southern savannas, forming a superspecies.

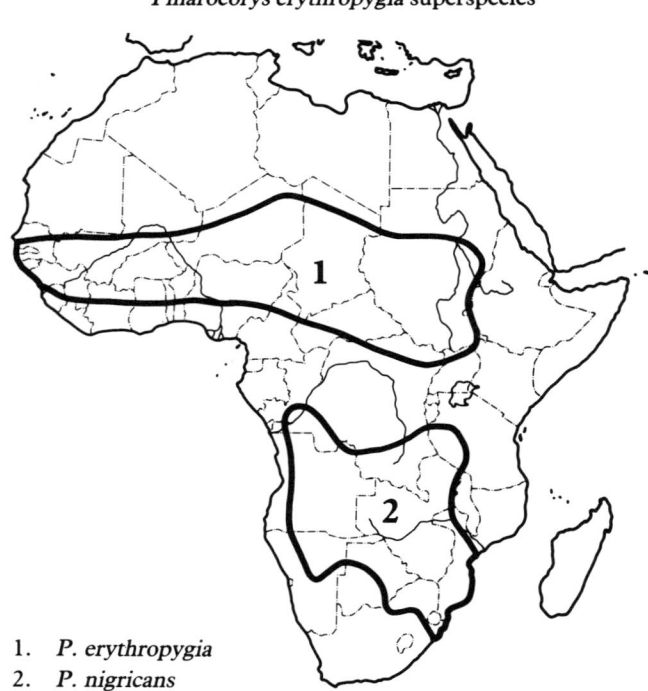

Pinarocorys erythropygia superspecies

1. *P. erythropygia*
2. *P. nigricans*

Plate 3
(Opp. p. 48)

Pinarocorys erythropygia (Strickland). Rufous-rumped Lark. Alouette à queue rousse.

Alauda erythropygia Strickland, 1852. Proc. Zool. Soc. Lond. (1850), p. 219; Kordofan, Sudan.

Forms a superspecies with *P. nigricans*.

Range and Status. Endemic intra-African migrant, breeding from Ivory Coast through N Ghana, N Togo, Nigeria, Cameroon, S Chad, Central African Republic, S Sudan and N Uganda, and migrating in non-breeding season to S Senegal, Gambia, Sierra Leone, the Sahel in N Niger, N Chad and N Sudan. Not uncommon but local, even where regular; vagrant over much of range in N. Locally common on burnt ground.

Description. ADULT ♂: crown and nape dark brown. Sides of neck, foreneck, face and supercilium whitish, lores dark brown, cheeks and ear-coverts dark brown with a whitish spot. Chin and throat whitish. Mantle and back dark brown, rump and uppertail-coverts rufous, tail dark brown with rufous base, all except central feathers edged rufous. Underparts white, breast heavily spotted and streaked dark brown. Flanks and undertail-coverts russet. Primaries dark brown, outwardly edged rufous and pale rufous on inner edges. Secondaries tipped pale rufous. Wing-coverts tipped buff. Bill horn above, paler below; eye dark brown; legs and feet whitish. ADULT ♀: similar to ♂ but primaries with broader rufous edging on outer webs, giving wing more rufous appearance in the field. SIZE: (15 ♂♂, 10 ♀♀) wing, ♂ 100–113 (108), ♀ 99–108 (103); tail, ♂ 64–75 (69·0), ♀ 61–72 (69·0); bill, ♂ 17–20·4 (18·3), ♀ 16·8–19 (17·6); tarsus, ♂ 22·5–26 (24·5), ♀ 22·4–26 (24·3). WEIGHT: 1 ♂ 30·1.

IMMATURE: paler brown above, feathers tipped cinnamon,

broadest on secondaries and greater coverts. Breast buffy, streaked lighter brown.

NESTLING: skin grey, covered with smoke-grey down, paler at tips, on crown, back, wings and thighs. Bill greyish, flanges cream. Inside of mouth orange; 2 black spots at base of tongue and 1 at tip (Serle 1957).

Pinarocorys erythropygia

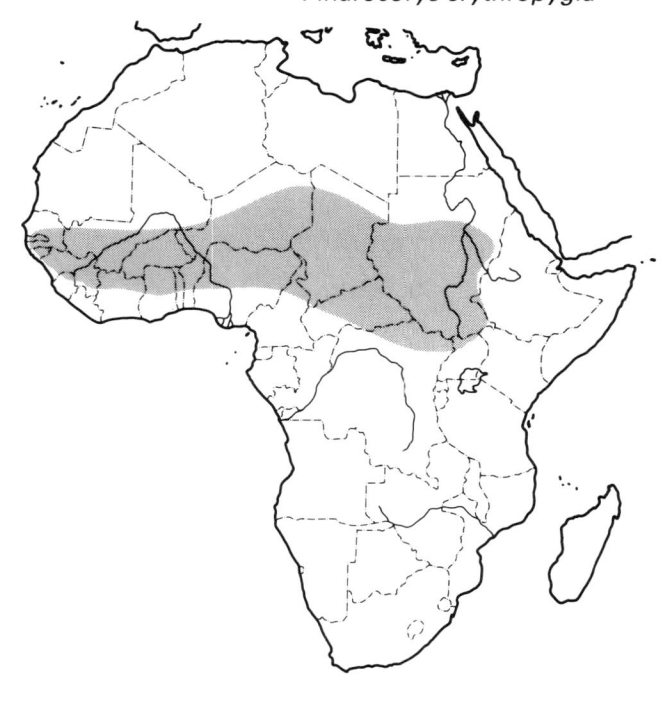

Field Characters. A large, dark brown lark with conspicuous rufous rump and tail edges and heavily spotted breast; cannot be confused with any other lark within its range; closely resembles allopatric Dusky Lark *P. nigricans*, which lacks rufous on rump and tail. Dangles feet in flight, a very good field character; often perches on trees when flushed.

Voice. Tape-recorded (78, F, MEES, STJ). Flight song a series of down-slurred, clear, loud, far-carrying whistles, 'piiuup' or 'pyooo', often interspersed with some buzzy trills, 'zeezeezeezeep' or 'brrrreep'; trill sometimes runs into whistle, 'trrriiiuuup'; also a double note rather like tit *Parus*, 'tit-teeyu, tit-teeyu'.

General Habits. Breeds in open broad-leafed savanna woodlands, grasslands intersecting forest patches, drainage lines and cultivated fields. Post-breeding habitat is arid, open savanna woodlands. Attracted to burnt areas, where small groups often seen foraging on recently burnt ground. Forages on bare ground, among rocks, in cattle pastures; runs about.

Migrates north in Mar–Apr to the Sahel, returning south Oct–Nov. Post-breeding flocks migrate rapidly north and return south equally rapidly (Elgood *et al.* 1973); seldom observed on passage between breeding and non-breeding areas.

Food. Grasshoppers and other invertebrates.

Breeding Habits. ♂ displays in soaring song-flight high in air for 1–2 min; lands on tree after descent.

NEST: only 1 found, natural hole in side of low bank in bare, recently burnt ground; nest hollow lined with fibre, plant down and charred grass stems (Serle 1957).

EGGS: 2 in the only described nest. Ovate; smooth and glossy; stone-grey, thickly and evenly speckled and spotted with ashy shell marks and pale olive-brown markings. SIZE: 23·6 × 16·7.

LAYING DATES: Nigeria, Jan–Mar; E Africa: Region A, Feb.

Reference
Serle, W. (1957).

Pinarocorys nigricans (Sundevall). Dusky Lark. Alouette brune.

Alauda nigricans Sundevall, 1850. Oefv. K. Sv. Vet.-Akad. Förh. 7(4), p. 99; Aapies R., Pretoria, South Africa.

Plate 6
(Opp. p. 97)

Forms a superspecies with *P. erythropygia*.

Range and Status. Endemic intra-African migrant, breeding in central Angola, W and S Zaïre, N Zambia and W Tanzania, migrating in non-breeding season south to N Namibia, N Botswana, Zimbabwe, Mozambique and N and E South Africa. 1 record Malaŵi (Mangochi District). Sporadic and irregular over most of non-breeding range, frequent but local in breeding areas.

Description. *P. n. nigricans* (Sundevall): Zaïre (Shaba), N Zambia, W Tanzania. ADULT ♂: upperparts dark brown, feathers unmarked or variably edged buff or buffy white. Supercilium and area around eyes buffy white, lores dark brown, cheeks and ear-coverts dark brown with a central whitish spot. Tail-feathers dark brown, narrowly edged paler brown. Underparts dull white. Sides of throat, breast and flanks with broad, dark brown (almost black) spots and streaks, belly and undertail-coverts unmarked. Wings quite variable; flight feathers dark brown, edged buffy to buffy white, tertials may be edged rufous; bases of inner webs of primaries buff. Bill dark horn, shading to yellowish at base of lower mandible; eye brown; legs and feet whitish. ADULT ♀: similar, but lighter brown above, facial pattern less bold; breast pale buff to buff, streaking paler, more brown than black. SIZE: wing, ♂ (n = 46) 115–123 (119), ♀ (n = 60) 105–118 (111); tail,

♂ (n = 46) 70–80 (76·0), ♀ (n = 60) 68–80 (74·0); bill, ♂ (n = 43) 16·5–20·4 (18·3), ♀ (n = 60) 16–20 (17·7); tarsus, ♂ (n = 45) 24–27 (26·0), ♀ (n = 60) 22–27 (25·0). WEIGHT: ♂ (n = 7) 30–46 (39·7), ♀ (n = 8) 31–41 (36·6), unsexed (n = 2) 34, 36·5.

IMMATURE: similar to adult, but feathers of crown and back edged buffy brown, giving a scaly appearance. Buff brown edging on wing and tail-feathers more pronounced, streaking on breast less dark than adult; breast more buffy.

NESTLING: unknown.

P. n. occidentis Clancey: Angola, Zaïre (Kwango, Kasai). Streaking on breast of ♂ less bold. Appears paler in the field.

Field Characters. A large, dark lark that forages on bare or thinly covered ground in distinctive manner: takes a few steps, stops, flicks wings, takes a few more steps, stops, flicks wings. Cannot be confused with any other lark within its range; has a superficial resemblance to Groundscraper Thrush *Psophocichla litsitsirupa*, from which it differs in darker back, shorter and stouter bill, shorter and paler legs, shape and gait. Flight undulating. Perches frequently on trees and bushes.

Voice. Tape-recorded (86, F, 296). Aerial song reported (Benson 1956, Dean 1974): a repeated 'zhreep' or 'drree-yup'. Also a soft 'wek-wek-wek' when foraging, probably a contact call.

General Habits. In north of range, occurs in open miombo woodlands, particularly where understorey recently burnt, in open *Uapaca* savanna along drainage lines, and on recently burnt grassland on vertisols and drier uplands. Nesting habitat is uneven, broken ground with clods and tufts of grass and termitaria in open *Uapaca* woodland on edge of drainage line, recently burnt out. In south occurs in short grass on edge of *Combretum* woodlands, in scrub mopane woodland, in heavily grazed cattle paddocks and often on gravel roads through broad-leafed and fine-leafed savannas.

Walks about restlessly while foraging, picking food items from among stones, cow dung and bases of grass tufts. Forages near grazing ungulates, and is attracted to burnt areas after fires. Often observed in small groups or loose flocks in south of range, and migrates both north and south in small parties.

After breeding follows regular migration routes south through Zimbabwe and Mozambique in late Oct and Nov to reach Transvaal and NE Natal in Dec; returns to breeding areas Apr–June.

Food. Insects, including grasshoppers, beetles (Carabidae), ants (*Anoplolepis*, *Camponotus* spp.), termites (*Odontotermes*, *Macrotermes* spp., both workers and alates); also seeds.

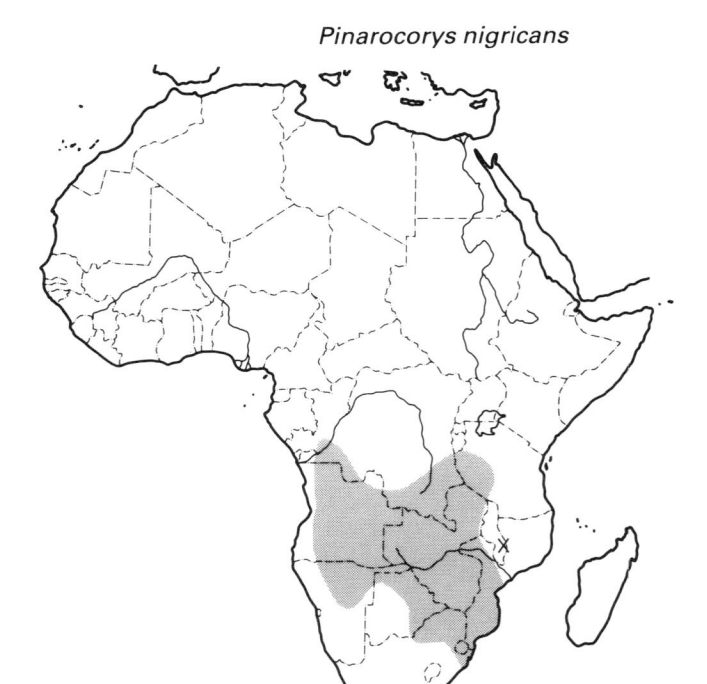

Pinarocorys nigricans

Breeding Habits. In display, ♂ rises in slow spiral flight, climbing to *c*. 30 m and then circling for 1–2 min with alternate flapping and gliding; returns in series of glides to perch on top of tree. Gives 'zhreep' call throughout display. Also displays with an aerial song and fluttering wings (Benson 1956). In suitable breeding habitat a number of birds may display simultaneously.

NEST: a cavity, probably scraped out by bird, under clod or tuft, creating a partly domed nest. Nest hollow *c*. 100 diam., entrance *c*. 60 deep; hollow lined with coarse grass stems and finer material. The 2 nests found both faced south; 1 had pieces of bark scattered about the entrance (Dean 1974).

EGGS: of the 2 known nests, 1 had 2 eggs, the other 2 young. Ovate; dull white with zone of sepia and dark brown freckles around broad end, scattered speckles over rest of egg. SIZE: (n = 1) 24·8 × 17·7.

LAYING DATES: Angola, Aug–Sept; Zambia, Aug–Oct.

DEVELOPMENT AND CARE OF YOUNG: at 1 nest both adults fed young (Dean 1974).

References
Benson, C. W. (1956).
Dean, W. R. J. (1974).

Genus *Chersomanes* Cabanis

A small lark with wide range of habitats: grassland, semi-desert, desert edge. Tail short, white-tipped; hind claw long and straight; bill slender, curved. Plumage colouration varies locally according to rainfall, soil and vegetation. Nest a simple cup, without dome.

Monotypic; endemic to Africa.

Chersomanes albofasciata (Lafresnaye). Spike-heeled Lark. Alouette éperonée.

Plate 6 (Opp. p. 97)

Certhilauda albofasciata Lafresnaye, 1836. Mag. Zool., pl. 58, p. 3; Cape of Good Hope (= Deelfontein).

Range and Status. Endemic resident. An isolated population in Tanzania (north of Arusha); central and SW Angola (sparse and local on central plateau, common in arid SW), south to Namibia (widespread and common), Botswana (widespread, but absent from NE) and South Africa (widespread, often common on Transvaal highveld, locally on edge of bushveld and escarpment, isolated records from Pietersburg plateau; local in Natal, widespread and common to fairly common in Orange Free State, widespread in Cape Province, particularly common in the Karoo). Reported occurrence near Amboseli, Kenya, is questionable (Turner 1985).

Description. *C. a. albofasciata* (Lafresnaye) (including '*baddeleyi*'): South Africa (N, E and central Cape Province, Orange Free State), Transkei, S Botswana (Kanye). ADULT ♂: crown to hindneck rufous-brown streaked blackish, streaking on nape finer; sides of neck brown, speckled darker brown, foreneck whitish, speckled brown. Lores, supercilium and short stripe from bill to just below eye rich buff; thin malar line black, ear-coverts pale rufous-brown, chin and throat white. Mantle rufous-brown streaked darker, back feathers with broader dark centres and buffy white tips, rump rufous-brown, indistinctly streaked, uppertail-coverts rufous-brown. Tail dark brown, broadly tipped white on all but central feathers. Underparts cinnamon to pale rufous-brown with variable amount of dark streaking on breast, almost absent in some birds. Flight feathers brown with narrow buff outer edges, scapulars and upperwing-coverts dark brown with buff edges, tertials dark reddish brown with rufous-brown inner margins and buff outer margins. Underwing-coverts and axillaries brown. Bill blackish horn, paler at base; eye brown; legs and feet pinkish brown. Sexes alike. SIZE: (20 ♂♂, 25 ♀♀) wing, ♂ 80–95 (88·0), ♀ 76–86 (80·0); tail, ♂ 43–55 (47·0), ♀ 39–46 (43·0); bill, ♂ 21·9–25 (23·0), ♀ 17–21·8 (18·5); tarsus, ♂ 25–30 (27·0), ♀ 24–28 (25·0). WEIGHT: ♂ (n = 14) 22·5–34 (26·7), ♀ (n = 7) 19·5–27 (22·7), unsexed (n = 8) 22–33 (28·2).

IMMATURE: crown dark brown, feathers tipped white, hindneck and mantle feathers tipped buff, back dark brown, feathers tipped white, face feathers tipped white, breast indistinctly spotted brown, feathers tipped white. Primaries and secondaries tipped buffy, tertials with whitish margins, wing-coverts with broad white tips.

NESTLING: above deep, slaty purple, with buff down above eyes, on either side of back and on flanks; eyes closed, covered by prominent blackish purple skin; below purplish pink; gape pale creamy yellow, shading to pinkish yellow bill with blackish slate tip; bill short and stout with strongly curved culmen; legs and feet brownish pink.

C. a. garrula Smith (including '*bushmanensis*', '*meinertzhageni*'): South Africa (W and NW Cape Province). Upperparts dark, heavily marked with blackish, feathers with broad dark centres, narrow reddish brown edges. Underparts fawn-brown.

C. a. macdonaldi Winterbottom: South Africa (S and E Karoo). Upperparts similar to *garrula* but greyer, underparts paler, greyer.

C. a. alticola Roberts (including '*subpallida*'): South Africa (S and central Transvaal). Upperparts darker than *albofasciata*, underparts light fawn-brown.

C. a. arenaria Reichenow: S Namibia (north to Windhoek), South Africa (N Cape south to Van Wyksvlei, east to Britstown). Upperparts paler and more yellowish red, dark feather centres narrower, underparts paler than *alticola*.

C. a. kalahariae (Ogilvie-Grant) (including '*bathoeni*'): S and W Botswana (Rietfontein, Kakia), South Africa (N Cape at Stella, Vryburg). Upperparts pale yellowish sandy, underparts paler than *arenaria*.

C. a. erikssoni (Hartert): N Namibia (Ovamboland, Outjo). Upperparts greyer and more heavily streaked than *kalahariae*, uppertail-coverts reddish.

C. a. barlowi White: E Botswana (L. Dow). Upperparts greyer and less streaked than *erikssoni*, underparts paler, belly whiter, chest finely streaked brown.

C. a. boweni (de Schauensee): NW Namibia (Swakopmund, Usakos, Karibib, north to Cunene). Upperparts pale pinkish sandy, with little or no streaking. Underparts paler than *erikssoni*.

C. a. obscurata (Hartert): SW, central and E Angola. Feathers of crown and back with broad blackish centres, reddish margins, feathers of mantle with whitish margins, underparts deep rufous.

C. a. beesleyi Benson: Tanzania. Upperparts similar to *obscurata* but less dark, more blackish brown than black, similar whitish margins on mantle and wing-coverts, feathers of nape and crown with whitish rather than reddish margins. Underparts paler russet than *obscurata*, breast with more pronounced dusky streaking.

Field Characters. A medium-sized lark whose long bill, short tail and upright stance give it a characteristic shape. Lower mandible straight, but decurved upper mandible makes bill appear curved. White tip to blackish tail conspicuous in flight. Plumage colour very variable but underparts typically rufous contrasting with white throat; white fringes to back feathers give scaly appearance. Often found in small groups.

Voice. Tape-recorded (88, F, GIB, LUT). Song a brief, rapid trill with a hollow, muffled quality, 'pirri-pirri-pirri-pirri-pirri . . .', lasting only *c.* 1·5 s; starts loud and trails off at the end; given in flight just after take-off and repeated several times. Alarm, harsh 'sheee'. Other vocalizations reported include 'jumbled "chip-kwip-kwip-kwip, ti-ti-ti-ti-ti-" (probably song) and "trrr-trrr-trrr" in flight display' (Maclean 1985).

General Habits. Present in well-grazed high rainfall grasslands, sparse well-drained grassland with bare patches, succulent and non-succulent dwarf shrublands in the Karoo, desert margins on red and white sands and on gravel plains, and on lawns and sports fields with new-grown grass.

Occurs in pairs or groups of up to 10 birds, usually 3–5. When disturbed, rises with 'bouncy' flight to settle a few m further away. Runs about when foraging, digging in soft soil at bases of grasses and forbs; in soft sand may dig down to 40 mm. Buried prey items are uncovered with rapid sideways jerks of the bill. Also takes food items from soil or gravel surface, and leaps up to glean food from dwarf shrubs. Catches flying prey such as termite alates. In desert and on desert margins tends to forage around rodent burrows. ♂♂ and ♀♀ may use different foraging sub-niches. Avoids heat by perching above soil surface or by going into rodent burrows.

Food. Insects: ants, termites (*Hodotermes*, *Microhodotermes*), beetles (Carabidae, Curculionidae, Tenebrionidae, Chrysomelidae); also other invertebrates (solifugids) and seeds (grasses *Stipagrostis* sp., forbs *Monsonia umbellata*). In Namib average annual food intake is 84% invertebrates, 16% seeds. ♂♂ eat more tenebrionid beetles, ♀♀ more ants and harvester termites.

Breeding Habits. Territorial. Has been recorded breeding cooperatively with 1 helper feeding young (Steyn 1988). In display, ♂ flies up to 2 m above ground and sails down with backward stretched wings, calling 'trrr trrr trrr'.

NEST: cup of dry grass and rootlets; int. diam. 64, depth 31 (n = 28); placed in scrape in sandy soil at base of grass tuft or shrub; usually on foundation of stones, earth clods or sticks; top of cup usually just above surrounding substrate.

EGGS: 2–3, mean (99 clutches) 2·4. Ovate; white or greyish white, speckled and spotted with brown over slate or underlying grey, sometimes blotched with rusty brown and yellow, markings often concentrated on obtuse end. SIZE: (n = 138) 19·2–22·5 × 13·7–16·3 (20·9 × 14·8).

LAYING DATES: E Africa: Region D, Mar–Apr, Nov; Angola, Dec; Namibia, Feb, Apr, Aug, Oct; South Africa: Transvaal, July–Nov; Natal, Sept–Oct; Orange Free State, Aug–Nov; N Cape, June–Apr; Karoo, Aug, Oct–Dec; E Cape, Sept–Oct, Dec, Mar. In Kalahari, breeding commences with rainfall.

INCUBATION: period 12 days.

DEVELOPMENT AND CARE OF YOUNG: young fed by both adults. Food-begging is released by characteristic feeding call, similar to flight call but quieter, given by adult as it comes to nest. After feeding young, adult waits for it to defaecate and then removes faecal sac. Young led from nest by parent when *c.* 10 days old, before they can fly; still dependent on parents after leaving nest.

BREEDING SUCCESS/SURVIVAL: of 69 eggs laid in 28 nests, 21 young fledged and left nest. Adults preyed on by Red-necked Falcon *Falco chicquera*.

References
Maclean, G. L. (1970).
Steyn, P. (1988).

Genus *Alaemon* Keyserling and Blasius

1 small and 1 large species of long-legged, sand-desert-dwelling larks with long or very long, narrow bills. Bill of small species, *A. hamertoni*, long, narrow and straight, with perforate nostrils placed 2 mm in front of feather line; bill of large species, *A. alaudipes*, the longest of all larks, narrow and strongly decurved, with perforate nostrils placed immediately in front of feather line but not concealed. Plumage very lax and soft; plain, sandy or pink-brown, whitish below with soft dapples on breast. Primaries and inner secondaries of *A. alaudipes* largely white. P10 25–30 mm long. Hind toe nail of *A. alaudipes* short and curved (7–8 mm) and of a *A. hamertoni* short and straight (6–10 mm).

1 species endemic to Africa (*A. hamertoni*); the other ranges from Cape Verde Is through Sahara to NW India. The 2 are parapatric in Somalia, each with habitat specializations near parapatric border which do not maintain elsewhere. We follow Hall and Moreau (1970) in treating them as a superspecies, but with considerable reservations on account of the differences in size and plumage. Knowledge of the breeding biology of *A. hamertoni* will help to resolve the question.

Alaemon alaudipes superspecies

1. *A. alaudipes*
2. *A. hamertoni*

Alaemon alaudipes (Desfontaines). Hoopoe-Lark. Sirli du desert.

Plate 2
(Opp. p. 33)

Upupa alaudipes Desfontaines, 1789. Mem. Acad. Sci. Paris (1787), p. 504; between Gafsa and Tozeur, Tunis.

Forms a superspecies with *A. hamertoni*.

Range and Status. Resident N Africa, Cape Verde Is, Sinai and Syria to Oman, Baluchistan and NW India. Rare visitor Malta (26 records), accidental Italy, Greece, Lebanon.

In NW Africa (Morocco, Algeria, Tunisia) absent north of Atlas Mts and $29\frac{1}{2}°$N, but may breed near L. Boughzoul in N Algeria; widely distributed, frequent and locally common in S Morocco and Algeria; in Tunisia widespread but sparse in central areas, 33–35°N, north to Gafsa, Mezzouna and Gabès, commoner southwest of Chott Djerid; in Libya frequent in littoral and sublittoral zone, frequent or common south in Fezzan but status in S Hammada el Homra and Oubari sand sea uncertain; in Egypt locally common around delta, on Mediterranean and Red Sea sand dunes and in 7+ inland oases. Probably breeds on Banc d'Arguin (Mauritania) and occurs inland of it and in north, west of 10°W; occurs in Aouker region in SE Mauritania. 4 records, NW Senegambia. In W Mali occurs from 17°N (Azzawadi, Awana) northward. Widespread in and around central Saharan mountains: Hoggar (Algeria), Aïr (Niger) and Tibesti (Chad), and widespread in Niger except south (Giraudoux et al. 1988). Extended range in late 1970s in Wadi Rimé–Wadi Achim Reserve, E central Chad, owing to creeping desertification (Newby 1980). Common resident in N Sudan north of 14°N and along Red Sea coast and on offshore islands including Dahlac Archipelago (Nikolaus 1987), and on Ethiopian coast. Common on N Djibouti coast (17 in 65 km between Ras Siyan and Obock; Welch and Welch 1986). Common resident in coastal N Somalia, east to 49°E.

Alaemon alaudipes

Description. *A. a. alaudipes* (Desfontaines): African range of species except Red Sea and Somalia coasts. ADULT ♂: RUFOUS MORPH: forehead creamy at sides, grey in centre; crown grey, nape and sides of neck grey with strong buff wash; well-marked long creamy superciliary stripe. Mantle, back and scapulars plain, pale sandy or buffy brown with greyish wash; rump and uppertail-coverts the same but greyer. Central tail-feather warm sandy or buffy brown, turning dark grey-brown at centre, with dark brown shaft; outer tail-feather with outer web white or creamy and inner web blackish brown; T2–T5 blackish brown, narrowly fringed whitish buff all round. Lores and upper half of ear-coverts dark grey, forming narrow dark line through eye; area under eye white; cheeks creamy, washed with brown at rear; moustachial region dark grey forming an irregular, rather ill-defined blackish stripe. Sides of neck uniform buff like back, or softly blackish forming continuation of blackish moustachial stripe. Chin and throat snowy white; breast creamy with soft, partly-concealed small black or dark grey spots; flanks and undertail-coverts creamy white. Outspread wing boldly blackish and creamy; P9–P5 dark brown, P7–P5 with conspicuous white patch on basal quarter of outer web, P4–P1 with similar white base and also with increasingly large white tip; secondaries creamy white with dark brown band across centre of feathers; innermost secondaries with white part of outer web strongly suffused buff; tertials plain buffy, same shade as back; all upperwing-coverts (except marginal, median and lesser coverts) dark greyish brown, broadly fringed buff. Axillaries and underwing-coverts silky white. Underside of primaries and rectrices blackish, but outer rectrix and inner secondaries mainly white. GREY MORPH (rare in Africa): like rufous morph but upperparts uniform grey (same shade as nape of rufous morph). Upper mandible dark brown or green-grey, lower similar but paler; eye dark brown; legs and feet grey-white. ADULT ♀: like ♂ rufous morph but grey parts buffier. SIZE: wing, ♂ (n = 20) 123–134 (128), ♀ (n = 13) 111–120 (115); tail, ♂ (n = 23) 79–92 (84·4), ♀ (n = 12) 72–80 (76·4); bill to skull, ♂ (n = 26) 29·5–34·6 (32·4), ♀ (n = 14) 27·4–30·3 (29·0), bill to feathers av. 5 shorter; tarsus, ♂ (n = 28) 32–36 (34·0), ♀ (n = 15) 30–33 (31·0). WEIGHT: (Sahara) ♂ (n = 10) 39–47, ♀ (n = 6) 30–39.
IMMATURE: paler, yellower and slightly more variegated than adult. Forehead, crown, nape, mantle, scapulars, back feathers and wing-coverts pink-brown with 1–3 mm broad yellowish buffy tips and fringes, and wavy dark brown transverse subterminal bar; all upperparts appear yellowish brown or pinkish brown with small dark wavy bars. Chin and belly white, breast yellowish or greyish or pale pinkish brown, unspotted or with soft dark grey dapples.
NESTLING: on hatching, upperside (or at least crown, scapulars and flanks) covered with long, rather thick, silky white or pale yellowish down. Mouth yellowish orange. On leaving nest entire underparts are very fluffy.
A. a. desertorum Stanley: coasts from Port Sudan (Sudan) to N Somalia (and Aden). Upperparts greyer, especially nape, mantle, scapulars and upperwing-coverts.

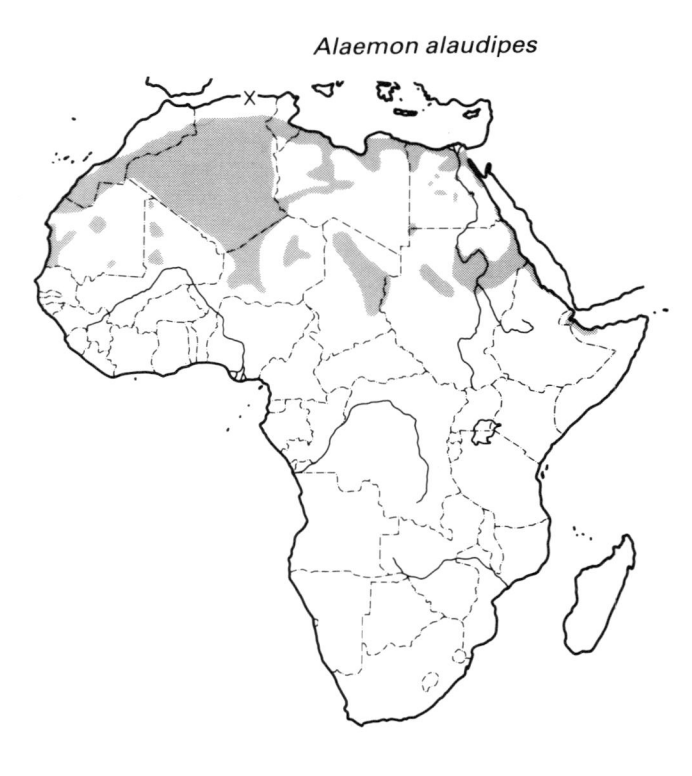

Field Characters. A medium-sized (♀) or large (♂), slender, long-bodied and long-legged lark (**A**) of sandy desert and coastal dunes. Adult unmistakable in flight by reason of dramatic pied appearance of both surfaces of wings (**B**). Flight floppy. Bill long, thin and strongly decurved. Plumage mainly dun brown above, with grey nape, long white supercilium, blackish line through eye and another (variable) above white moustache; underparts white with dense black speckling on upper breast.

A

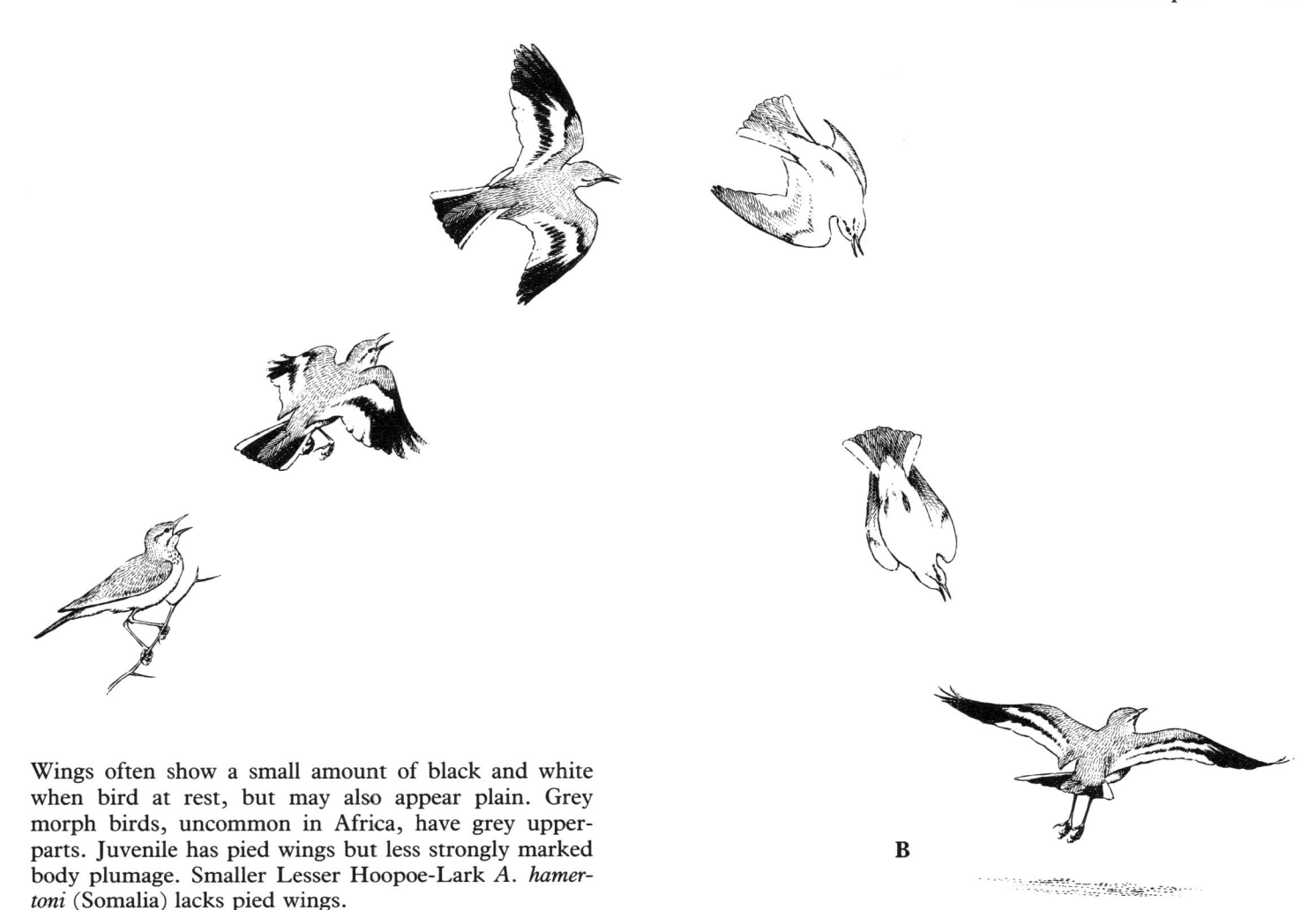

Wings often show a small amount of black and white when bird at rest, but may also appear plain. Grey morph birds, uncommon in Africa, have grey upperparts. Juvenile has pied wings but less strongly marked body plumage. Smaller Lesser Hoopoe-Lark *A. hamertoni* (Somalia) lacks pied wings.

Voice. Tape-recorded (63, 73, 78, HAZ, HOL, ROC). Song a simple, plaintive piping, quite melodious, rather uniform, in 3 parts: a series of fluty whistles, a brief trill, and a long series of fluty whistles. Whole song lasts 12 s, and is on one pitch except for higher then falling whistles immediately after trill. Song repeated after as little as 5 s. Song is somewhat ventriloquial, and at distance can sound mechanical. Contact call a buzzing 'zeee' or soft 'jinzing', like Trumpeter Finch *Rhodopechys githaginea*; threat a harsh 'shweee'; alarm, 'too' (Cramp 1988).

General Habits. Inhabits flat or undulating sand with well-spaced clumps or tussocks of low, xerophytic, halophytic or succulent vegetation, or sometimes with practically no vegetation at all. Prefers soft, dry substrates, including coastal sand dunes, sand desert with or without gravel, open wadis, edges of dried-out pasture, abandoned cultivation, *Stipa* steppe, dry salt marsh, and flat areas of powdered coral (Dahlac Archipelago: Clapham 1964). Avoids wetted shorelines, hammada, erg, much-dissected ground, and plains with many trees.

Solitary or in pairs; seldom > 3 together. The most cursorial of songbirds, sometimes courser-like (*Cursorius* spp.). Walks and runs with long stride; exposed and feathered part of tibia nearly as long as tarsus (**A**). Often runs for several hundred m; tame and approachable (particularly when living near human habitation), and keeps distance by running rather than flying. Commonly flies *c.* 100 m when foraging undisturbed, when flight loose, free, undulating, often curving. Forages by walking slowly, suddenly stopping to dig rapidly with bill or to peck at insect on ground. Scrambles over prostrate shrub, seizing insects from it. Digs up to 5 cm into sand and hardened ground, never failing to find insect (Meinertzhagen 1951). Picks at carcasses; forages over rubbish tips; hawks airborne moths, butterflies and dragonflies; drops snails onto stones from height of 6–23 m, and beats them against stone to break shell (Hegazi 1981). Occasionally squats, and crouches on sand (Walker 1981).

Resident, often evidently quite sedentary; but many Nov–Feb records from coastal plains of W Morocco, Algeria and Libya, and the many records from Malta (mainly Aug–Dec) suggest regular dispersal amounting to partial migration (Ledant *et al.* 1981).

Food. Insects, some snails and plant material, once a gecko excavated from sand. Locusts, grasshoppers, beetles and larvae, ants, sand-flies, termites, mantises; *Eremina* snails (in Egypt in Feb–June, 6–11 birds ate

1090 snails or 6·5 kg of flesh in 2·4 ha area: Hegazi 1981). Also spiders, hard seeds and bits of grass. 8 birds (Saudi Arabia) all contained ant-lions (Cramp 1988). Once a butterfly *Vanessa cardui* (Smith 1965).

Breeding Habits. Solitary, territorial breeder. Territory size evidently varies from c. 2 to 250 ha; 1·25 birds per km^2 in N Sahara, up to 5 per km^2 in Saudi Arabia (Cramp 1988). Song display, from dawn onwards throughout breeding season, Mar–May (rarely sings Jan–Sept), by ♂ (♀ occasionally sings) is spectacular. ♂ flies up from knoll or bush, with wings fluttering and tail spread, rising vertically for 4 (sometimes 10) m, rolls over and nose-dives back to same perch with wings closed until last instant (**B**, p. 63). Bird usually sings 1st and 2nd parts of song (see *Voice*) before flying up, and 3rd part during song-flight.

NEST: a flattish structure made of large twigs threaded into bush, with small unlined cup of finer twigs, grass, rags, paper and palm fibres (Gallagher and Rogers 1978). Typically 300–600 mm above ground in top centre of bush; also sited in tussock of *Panicum* grass and occasionally on ground in shelter of bush. In Somalia nests cemented to top of bush by mud (Archer and Godman 1961). Int. diam. 70–75, depth 35–40.

EGGS: 2–4; N Africa mean 2·88. Subelliptical; glossy; white or pale buff, with reddish brown spots and blotches on blue-grey undermarkings. SIZE: (n = 25, N Africa) 22·3–27·9 × 16–19 (24·3 × 17·1). WEIGHT: c. 3·65.

LAYING DATES: N Africa, late Feb to May; Libya, Apr; Egypt, Apr; Chad, Aug; Somalia, Apr.

INCUBATION: by ♀, beginning with last or penultimate egg; but both sexes also said to incubate, in 40 min spells (Currie 1965). Period: c. 2 weeks.

DEVELOPMENT AND CARE OF YOUNG: young stay in nest for 12–13 days, flying several days later, and remain with parents for at least 1 month (Cramp 1988).

Reference
Cramp, S. (1988).

Plate 2
(Opp. p. 33)

Alaemon hamertoni **Witherby. Lesser Hoopoe-Lark. Sirli de Witherby.**

Alaemon hamertoni Witherby, 1905. Ibis, p. 513; Obbia.

Forms a superspecies with *A. alaudipes*.

Range and Status. Endemic resident, Somalia. Common on tussocky grass plains away from coast in north, north of 6°N, and extends down east coast to 2°N.

Description. *A. h. hamertoni* Witherby: Somalia south of 7°N. ADULT ♂: forehead, crown, nape, mantle, sides of neck, back, wing-coverts and rump plain grey-brown or sandy brown, darkest on mantle and palest on rump; uppertail-coverts plain pale buffy brown. Rectrices mid-brown to dark brown, central one somewhat greyer, all narrowly fringed pale grey or buffy on outer edges. Short, pale superciliary stripe. Lores grey, line or area behind eye white. Ear-coverts plain grey or sandy brown. Chin and throat white, throat faintly mottled with grey; breast creamy white heavily mottled with grey and with a few dark brownish streaks; belly, flanks and undertail-coverts creamy white or pale pinkish buff. Wings greyish brown, the outer edges of remiges and coverts narrowly edged buff, and primary coverts broadly tipped whitish. Axillaries and under-wing-coverts silky white. Bill horn brown; eye dark brown; legs and feet whitish. Sexes alike. SIZE: (6 ♂♂, 2 ♀♀) wing, ♂ 103–107 (104), ♀ 90, 93; tail, ♂ 67·5–74·5 (70·7), ♀ 61, 62.7; bill to skull, ♂ 19–21·5 (20·2), ♀ 17, 19; tarsus, ♂ 28–31·5 (30·5), ♀ 22, 22·5. WEIGHT: unknown.

IMMATURE and NESTLING: unknown.

A. h. altera Witherby: N Somalia from c. 47° to 49°E. Entire upperparts warm, sandy brown, less grey than in nominate form (the same shade as *A. a. alaudipes*, whilst *A. h. hamertoni* is same shade as *A. a. desertorum*). Smaller: wing, ♂ (n = 4) 98–106 (102), ♀ (n = 2) 89, 92.

A. h. tertia Clarke: N Somalia west of c. 47°E. Like *altera* but upperparts more rufescent, especially from nape to back.

Race of population in NE Somalia (east of c. 49°E, north of 7°N) unknown.

Alaemon hamertoni

Field Characters. A poorly known lark with no distinguishing features; a very plain bird, unmarked uniform grey-brown or tawny-brown above, white or pale buffy below, with greyish white line through eye and contiguous area below eye, obscure grey mottling on throat, rather heavier mottling on breast, very thin bill, and rather long pale legs. Very pipit-like; stands upright and runs swiftly, but more uniform above and below than any pipit, or any lark except *Ammomanes* spp. and ♀ Black-crowned Sparrow-Lark *Eremopterix nigriceps*, which are all thick-billed. Best distinguished from Plain-backed Pipit *Anthus leucophrys* by lack of typical pipit scalloped median upperwing-coverts.

Voice. Unknown.

General Habits. Inhabits open grass plains, preferably where grass is tussocky, far from any stunted scrub. Solitary or in pairs, never in flocks. Runs at great speed, rather like a courser, winding its way between patches of long grass and running when alarmed far beyond, onto open sand. Wary.

Food. Unknown.

Breeding Habits. 2 nests found.
NEST: built of root-like, brown fibrous grass stems, sunk well into ground; one nest in open sandy soil under cover of grass stems, the other in open red sandy soil well away from tall grasses but at foot of some short stems.
EGGS: 1 clutch of 2, 1 of 3 eggs. Elongated; ivory-white, well covered, particularly at broad end, with reddish sepia spots, streaks and splashes on ash-grey undermarkings; or minutely speckled all over, most densely at broad end, with umber-brown.

Reference
Archer, G. and Godman, E. M. (1961).

Genus *Ramphocoris* Bonaparte

A single species of desert-dwelling lark in NW Africa and N Arabia with massive bill, the upper and lower mandible contra-curved so that closed bill commonly has small aperture (**A**). Bill up to 15 mm deep at base, white or yellowish with dark tip. Nostrils covered by feathers. Face and breast blackish. Tail mainly white with dark tip and wheatear-like pattern. Secondaries and inner primaries broadly white-tipped, very like some species of *Melanocorypha* with which genus, in our opinion, *Ramphocoris* is closely allied (and not with *Eremopterix*, *cf.* Heim de Balsac and Mayaud 1962, Voous 1977).

Ramphocoris clot-bey (Bonaparte). Thick-billed Lark. Alouette de Clotbey.

Melanocorypha clot-Bey Bonaparte, 1850. Consp. Av. 1, p. 242; Egyptian Desert.

Plate 2
(Opp. p. 33)

Range and Status. NW Africa, central Jordan, N Saudi Arabia, dispersing to Kuwait and central Arabia.

Widespread, frequent to sparse resident from NW Mauritania (23½°N, 15°W; 26°N, 12°W) and S Morocco to interior of N Algeria, central Tunisia and NW Libya. Nomadic outside breeding season, not a true migrant, but may move partially to coastal plains in winter; in Errachidia region (Morocco) found much more commonly in spring than in winter; flocks of 20–200 near Aïn Beni Mathar, Boumalne du Dadès and Ouarzazarte, Morocco. In Tunisia a scarce breeder, flocks dispersing north of breeding area. In Libya extends east to *c.* 16°E, south to *c.* 31°N; 2 old records from El Mechili, Cyrenaica.

Ramphocoris clot-bey

Description. ADULT ♂: forehead and forecrown pinkish brown with minute dark grey spots; crown pinkish grey with narrow dark grey-brown streaks; nape, mantle, back, rump, scapulars and sides of neck uniform pinkish grey or pinkish brown; uppertail-coverts creamy white, irregularly suffused with buff. Central tail-feather with distal half dark brown and proximal half pale pink-brown or white; outer tail-feather (T6) white with long, dark brown, oval mark on inner web at tip and with outer web suffused with pinkish; intervening tail-feathers (T2–T5) have increasingly large amount of dark brown at proximal end of both webs, so that fanned tail shows white with triangle of dark brown pointing forward in centre; inner webs of T2–T6 white, forming white tip to spread tail. Underside of closed tail white with inner webs blackish distally. Lores blackish, mixture of black and white feathers around eye, mainly black featherlets above and white featherlets below eye; moustachial area, cheeks and upper throat mainly black, the feathers minutely fringed buff; small white patch at base of lower mandible, separated by black from another small white patch below cheek. Chin white; upper throat variable; all black, or black with white scales and other marks; lower throat mainly white in midline, spotted with black at sides; breast white or creamy or very pale buff, heavily marked with black spots; belly off-white or creamy, the lower breast and forebelly very heavily marked in midline with large black spots or blotches; sides of breast and flanks grey or pink-grey, with half-concealed dark brown or blackish spots and streaks; undertail-coverts white. Outer 2–3 primaries grey-brown; remaining primaries and secondaries dark blackish brown with narrow white tips to P9–P5 and broad white tips to inner primaries and all except innermost secondaries, forming white trailing edge to wing 15 mm deep; tertials blackish brown with broad buff fringes; upperwing-coverts strongly variegated, mostly dark brown with 1–3 mm wide whitish or buffy or sandy borders, and many coverts with the outer web grey; median and lesser coverts blackish brown with broad whitish or buffy fringe; outer web of P9 whitish. Underwing-coverts dusky or dark brown; underside of remiges grey. Bill yellowish horn with blackish tip; eye brown; legs and feet dull bluish-grey or pale straw. ADULT ♀: like ♂ but black patch from bill to ear-coverts is browner and less well demarcated, spots on breast and forebelly are dark grey (not black) and narrower, and axillaries and underwing-coverts are greyish or brownish black. Bill yellower. SIZE: wing, ♂ (n = 26) 125–134 (130), ♀ (n = 12) 119–125 (122); tail, ♂ (n = 10) 57–64 (59·4), ♀ (n = 12) 53–58 (56·0); bill to skull, ♂ (n = 10) 17·8–22 (20·0), ♀ (n = 15) 18·2–20·9 (19·3), bill to feathers av. 2·5 shorter; tarsus, ♂ (n = 10) 22–25 (23·0), ♀ (n = 15) 21–24 (23·0). WEIGHT: (Algeria, Nov–Dec) 2 ♂♂ 52, 55, 2 ♀♀ 45, 45.

IMMATURE: upperparts, wings and central rectrix cinnamon-pink, with grey-brown centres to tertial feathers, greater coverts and greater primary coverts. Sides of head and neck cinnamon-pink, with paler eye-ring, whitish spot below eye, and grey mottling on lores and cheeks. Breast and flanks cinnamon-pink, breast dappled with grey at sides; belly creamy and undertail-coverts white. Underwing dark grey with cinnamon-pink fringes (Cramp 1988).

NESTLING: plumage dense, soft, uniform very pale pink-brown, darkest on head and back, palest on underparts and thighs. At age when wing still <50 mm, only variegation in plumage is blackish concealed bases to growing primaries. Pre-feathered nestling not known.

Field Characters. A medium-sized or large, heavy-headed NW African lark with huge, arched, yellow beak, black cheeks with distinct white patches, heavily black-spotted breast, plain pinkish brown back, and pied wings and tail. In flight looks rather long-winged and short-tailed; wing-coverts pink-brown, remiges black, trailing edge of wing broadly white; tail white with wheatear-like black end. Adult unlikely to be confused with any other lark in range; the most similar species are the Calandra Lark *Melanocorypha calandra* and Bimaculated Lark *M. bimaculata*, which are readily distinguished by their pale superciliary stripes and blackish patches on sides of breast. Juvenile Thick-billed Lark is plain cinnamon-pink, with creamy throat and belly, looking somewhat like a huge ♀ sparrow-lark *Eremopterix* or giant ♀ Trumpeter Finch *Rhodopechys githaginea*; best feature is wings, blackish and creamy (in flight) where adult is black and white.

Voice. Tape-recorded (60, CHA). Song variously described as a frenzied little jingle of sweet notes, a sweet, rather quiet, tinkling warble, and varied phrases of liquid notes with quality of Corn Bunting *Emberiza calandra*, each phrase separated by several s (Cramp 1988). Flight call: 'prit', 'quip', 'wick-wick', 'coo-ee' or 'co-ep'. Alarm a long plaintive whistle 'tsee', like Curlew *Numenius arquata*. Vocal range wide, including a quiet 'sreee', 'wheet-wheet-wheet', and a hard spitting noise (Smith 1965).

General Habits. Inhabits stony, sparsely-vegetated sub-desert plains and open hammada, stony clay flats with quite thick growth of succulents such as *Aizoon*, wadis, rocky plateaux and stony slopes. Sometimes feeds in newly-turned ploughed fields and short cereal crops (Smith 1965).

Singly, in pairs or scattered parties of 6–12; flocks of 10 or 20 to 200 in winter; once, numerous flocks of 4–50,

Morocco, mid–Feb, totalling tens of thousands in area of 2000 km² (Brosset 1961). Approachable; runs rather than flying from danger. One once flattened itself on ground in alarm, remaining motionless, flying only when approached to within 1 m; another retreated with fast, swerving and twisting flight (Cramp 1988). Flies regularly to drink, taking only 1 drop of water at a time; capable of going for long periods without water. Flight powerful; escape-flight short and low, ending abruptly. Stance on ground upright; runs from danger, and walks and hops when foraging. Feeds by reaching up to plant-head and pulling it to ground, and uses bill as clippers to cut green shoots (Cramp 1988). Hard seeds swallowed whole, with much grit. In hand, bite surprisingly weak.

Food. Mainly green plant material, also seeds and insects (ants, a locust, a large beetle; once, numerous painted lady *Vanessa cardui* caterpillars recorded). Once a small lizard (Algeria). Pieces of *Plantago ovata* and *Euphorbia kahirensis* fruits (Jordan).

Breeding Habits. Solitary, territorial nester. Song-flight poorly known, but ♂ flies high to sing, descending either parachute-like or in zigzags.

NEST: shallow depression in soil, lined with vegetation, often with rim of pebbles at open side; sited under bush, by stone or (uncommonly) in open.
EGGS: 2–6, av. 3·51. Subelliptical; glossy; white, creamy or pink, finely speckled with red-brown or chestnut on underlying greyish mottling. SIZE: (n = 40, NW Africa) 23·6–29 × 16·9–19 (25·5 × 18·3). WEIGHT: *c*. 4·4.
LAYING DATES: NW Africa, Mar–May, exceptionally Jan–Feb.
INCUBATION: by ♀; unknown whether by ♂ also. Nest sited so that it (and incubating bird) usually in shade of bush or stone at midday.
DEVELOPMENT AND CARE OF YOUNG: young fed by both parents, largely on locusts, before and after fledging.

References
Cramp, S. (1988).
Heim de Balsac, H. and Mayaud, N. (1962).

Genus *Melanocorypha* Boie

Large, robust larks, closely related with *Alauda*, distributed from Mediterranean to central Asia. Bill robust and finch-like, rather short in 5 species, long in *M. maxima*. Nostrils concealed below fringe of stiff feathers. Legs robust, not long; hind claw elongated, slightly decurved or (usually) straight. At least 1 species strongly dimorphic in plumage, i.e. *M. yeltoniensis*, ♂♂ of which are all black. ♀♀ of this species, and the remaining ones, have similar plumages characterized by large, paired, black breast marks and by white trailing edges to the wings, barely developed in *M. bimaculata*, but forming a large white patch in *M. mongolica* and *M. leucoptera*. 2 species have dark rufous crowns and upperwing-coverts. *M. mongolica* and *M. leucoptera* have deep rufous rumps. Juveniles of *M. maxima* blackish, like adult *M. yeltoniensis*, the pale parts of the plumage having strong pale yellow suffusion.

6 species, 2 in Africa, 1 breeding (*M. calandra*), 1 a non-breeding visitor (*M. bimaculata*). They are very closely allied and would comprise a superspecies were there not considerable areas of sympatry in Turkey and Iran. They are incompletely segregated altitudinally, with *M. bimaculata* higher.

Melanocorypha calandra (Linnaeus). Calandra Lark. Alouette calandra.

Alauda calandra Linnaeus, 1766. Syst. Nat. (12th ed.), p. 288; Pyrenees.

Plate 2
(Opp. p. 33)

Range and Status. N Africa, Iberia, S Europe, SW USSR, Turkey, Jordan, N Iran east to L. Balkhash. Resident; some migration in Asia.

Nomadic resident, N Morocco, N Algeria and N Tunisia, in cultivated plains and plateaux up to 1250 m, common in Oranie and the Constantinois (Algeria), with flocks of 50–100 June–Mar (Le Berre and Rostan 1976); also common in coastal plains. Not proved to breed in NW Libya, where scarce on Tripoli coast in Mar–Apr and Sept–Oct, but breeds in NE in coastal plain from Gheminez to Martuba, commonly in Barce cornlands. Winter visitor all along Egyptian Mediterranean coast.

Description. *M. c. calandra* (Linnaeus): Mediterranean basin east to E Turkey, NW Iran, Transcaucasia and Urals. ADULT ♂: forehead, crown, nape, mantle, sides of neck and cheeks grey-brown, with evenly distributed small dark brown streaks; back, rump, uppertail-coverts, scapulars and upperwing-coverts noticeably darker brown than head – brown or grey-brown or faintly rufescent brown, mottled with dark brown. Central tail-feather brown; T6 wholly white; T5 dark brown with outer web narrowly edged white and both webs with broad white tip; T4 dark brown with 2 mm deep white tip to inner web; T3 and T2 dark brown with narrow whitish tip to inner web. Lores whitish; narrow whitish or pale buff superciliary stripe from base of bill to well behind eye; ear-coverts plain brown, unstreaked, surrounded by lighter brown (i.e. above by hind end of whitish superciliary stripe, behind by grey-brown sides of neck, and in front by contrasting whitish sides of breast above black breast mark). Chin white; throat white, sometimes with diffuse pale or mid-brown spots; large brownish black mark on side of upper breast; breast pale buff, finely spotted with small or minute brownish black spots in centre and with larger brown or rufescent brown diffuse spots behind black mark at sides; belly white; flanks buffy; undertail-coverts white. P9–P5 wholly dark brown; remaining inner primaries dark brown with broad buffy tips; all of secondaries dark brown with white tips forming 6 mm wide trailing edge to wing; tertials all dark brown. Axillaries and underwing-coverts buffy brown. Upper mandible dark horn or blackish brown, lower mandible pale horn, yellower at gape; eye light to dark brown; legs and feet bright pink-brown, flesh-brown or yellowish brown. Sexes alike. SIZE: wing, ♂ (n = 63) 126–141 (132), ♀ (n = 31) 115–122 (119); tail, ♂ (n = 36) 59–68 (63·8), ♀ (n = 25) 52–62 (56·5); bill to skull, ♂ (n = 37) 17·7–23 (20·8), ♀ (n = 24) 15·9–20·2 (18·1), bill to feathers av. 3·7 shorter; tarsus, ♂ (n = 39) 27–30 (29·0), ♀ (n = 26) 25·5–28 (27·0). Wing evidently shorter in Tunisia (♂, n = 15, av. 133·6) than in Algerian high plateaux (♂, n = 6, av. 136·5). WEIGHT: (Turkey, June) ♂ (n = 8) 57–73 (64·4).

IMMATURE: upperparts somewhat paler than adult, each feather with white tip, dark brown subterminal mark, and pale rufous fringe, giving white-spotted, scaly effect. Underparts like adult but with pale brownish smudge in place of adult's black breast mark.

NESTLING: unknown.

(*M. c. psammochroa* Hartert: N Iraq, Iran to Kazakhstan. Paler; spots on breast and behind black patch absent. Has occurred in Egypt (Sinai) but not yet certainly found in Egypt west of Sinai, although likely to occur there.)

Field Characters. A large N African lark, readily distinguished from all other larks except Bimaculated Lark *M. bimaculata* and Greater Short-toed Lark *Calandrella brachydactyla* by conspicuous black patch on side of upper breast. Very similar to Bimaculated Lark (winters in NE Africa) but has white trailing edge in wing and white sides to tail (Bimaculated Lark has white tip to tail). Much larger than Greater Short-toed Lark, with large, robust bill and white superciliary stripe, and told from it by white trailing edge of wing. Thick-billed Lark *Ramphocoris clot-bey* looks quite different on ground (yellow bill, black sides of head, blackish breast) but can resemble Calandra Lark in flight (underside of wings black with white trailing edge, white sides of tail); however Thick-billed Lark differs in having plain pink-brown back, and black-tipped tail.

Voice. Tape-recorded (HOL, JOHN – 62, 73, BERG). Song like that of Eurasian Skylark *Alauda arvensis*, q.v., but louder, more complex, richer, and with grating sounds constantly interpolated; a continuous flow of short phrases and single notes characterized by repeated inclusion of flight-call 'khitra-a'a'a'a' (like song of Corn Bunting *Emberiza calandra*). Strongly imitative; mimics many other birds including Greater Short-toed Lark, Barn Swallow *Hirundo rustica*, European Goldfinch *Carduelis carduelis*, Brown Linnet *C. cannabina*, Wood Warbler *Phylloscopus sibilatrix*, Carrion Crow *Corvus corone* and Green Woodpecker *Picus viridis* (Cramp 1988). Flight call a characteristic 'kleetra'.

General Habits. Inhabits agricultural lowlands, stubble, pasture, cultivated plains, coasts, grassland and plateaux in Atlas Mts up to 1250 m; avoids bare steppe and desert.

Solitary or in pairs and family parties in breeding season, otherwise strongly gregarious. Wintering flocks typically of c. 50 birds; outside Africa migrant flocks sometimes of up to 1000 birds. Often associates with other larks, and with Corn Buntings. Comes to water in dense, noisy flocks, circling first then drinking hastily; birds pack closely when drinking; visits watering place twice daily (Morocco: Brosset 1961).

Forages by walking and running, pecking at and digging into ground. Shy, hard to approach. Digs pits for dust-bathing (Cramp 1988). Loafs on telegraph wires and bushes (Morocco: Meinertzhagen 1951). ♂ sings from ground or bush and in song-flight, from well before dawn until dusk. Sings mainly Apr–May (Morocco).

Resident in NW Africa but flocks evidently wander widely, e.g. to Laghouat, Algeria (south of Atlas Mts), Oct. Occurs Tripoli (Libya), usually singly, Mar–Apr and Sept–Oct; 2 in Libyan desert, May. Winters Egypt Oct–Mar.

Food. Insects, some quite large, and seeds. For list, see Cramp (1988). NW Africa records include small seeds, wheat, barley and grit.

Breeding Habits. Not well known, in or out of Africa. Solitary, territorial breeder, monogamous or (Cyprus) perhaps polyandrous (Took 1972). In song-flight circles at *c.* 10 m high, singing, then ascends very high (higher than Eurasian Skylark), circling with wings fully extended and with deep, slow beats; descent may be vertical and silent, with gentle glide before landing. Also has low form of song-flight, circling with deep, even beats alternating with gliding; in Tunisia low song-flights commoner than high ones, which are performed mainly at dawn in Mar (Zedlitz 1909). Pair-bonding involves aerial chases; after ♂'s low song-flight ♀ said to take off and fly in gliding curves, ♂ follows calling, with tail fanned, looking larger than usual (Tunisia: Zedlitz 1909, Cramp 1988).

NEST: shallow depression in ground; foundation of grass leaves and stems, lined with softer plant matter; sited under tussock.

EGGS: 3–6, av. of 88 clutches, Algeria, 4·2. Subelliptical to oval; glossy; whitish, yellowish or greenish, heavily spotted and sometimes blotched with dark brown or red-brown and pale purple, mainly at broad end. SIZE: (n = 150) 22–27·2 × 16·2–19·5 (24·5 × 18·0). WEIGHT: *c.* 4·1.

LAYING DATES: Algeria, Apr–June.

INCUBATION: by ♀; probably also by ♂ (which has brood-patches). Period: 16 days.

DEVELOPMENT AND CARE OF YOUNG: young cared for and fed by both parents; spend *c.* 10 days in nest, leaving before they can fly.

References
Cramp, S. (1988).
Zedlitz, O. G. (1909).

Melanocorypha bimaculata (Ménétries). **Bimaculated Lark. Alouette monticole.**

Alauda bimaculata Ménétries, 1832. Cat. Raisonné, p. 37; mountains near Talysh (Transcaucasia).

Plate 2
(Opp. p. 33)

Range and Status. Breeds Turkey, Lebanon, NW Jordan, SW Syria, NW Iraq, Transcaucasia, Iran, and Turkmeniya (USSR) to Aral Sea and L. Balkhash. Resident and migrant, wintering mainly in central Asia and from N Arabia to NW India, also Africa.

Passage migrant through Egypt, wintering abundantly near Khartoum, fairly commonly in arid parts of N Sudan south to 13°N, also in N and NE Ethiopia (flock of *c.* 30 in rocky coastal hills: Smith 1957), with 1 record in Djibouti, Nov (Welch and Welch 1986). Vagrant, Namibia (1, Swakopmund, Sept 1930), probably an escaped cagebird (Brooke 1988).

Regarded as a major agricultural pest of sorghum near Khartoum, Sudan (Beshir 1978, Schmutterer 1969). 200,000–250,000 Bimaculated Larks were estimated to winter on sorghum fields near Khartoum (Nov 1975), of which *c.* 10% were thought to have been killed by aircraft application of fenthion pesticide (Beshir 1978).

Description. *M. b. bimaculata* (Ménétries): NE Turkey, S Transcaucasia, Iranian mts east to Elburz and Kerman; winters N Sudan. ADULT ♂: forehead, crown, mantle, sides of neck and cheeks grey-brown, with evenly distributed small dark brown streaks; nape mealy, paler than crown; back, rump and uppertail-coverts noticeably darker brown than head – brown or grey-brown or faintly rufescent brown, mottled with dark brown. Scapular feathers with very dark brown outer webs, blackish midlines, and contrasting pale buff inner webs: the effect in some birds is of broad contrasting dark brown and buffy lines on mantle and scapulars. Tail slightly shorter than in Calandra Lark *M. calandra*, and concealed by uppertail-coverts to within 5–10 mm of tip; tail-feathers blackish brown including central tail-feather, which is broadly fringed pale rufous especially towards base; T6 with outer web fringed buffy; T2–T6 with white tips 2 mm deep on outer webs and 5 mm deep on inner webs. Lores whitish; long, broad, whitish or pale buff superciliary stripe from base of bill to well behind eye; ear-coverts dark brown dorsally, merging through rufescent brown to mid-brown ventrally, shading to whitish below eye. Chin and throat white; sides of upper breast with large black mark; breast buffy or very pale grey with scattered small rufous streaks and larger rufescent brown dappling at sides. Belly off-white or creamy; flanks buff; undertail-coverts white. Remiges brown to dark brown, all except P9 and P8 and tertials with 1–2 mm wide buff tips and with narrow buffy fringe to outer webs; upperwing-coverts with blackish brown centres, merging through pale rufous to very pale buff fringes and tips. Axillaries, underwing-coverts and underwing grey-brown. Upper mandible dark horn or blackish brown; lower mandible pale horn, yellower at gape; eye light to dark brown; legs and feet bright pink-brown, flesh-brown or yellowish brown. Sexes alike. SIZE: wing, ♂ (n = 16) 119–128 (125), ♀ (n = 9) 110–118 (113). WEIGHT: (various races, Asia, May) ♂ (n = 13) 48–59 (54·1), ♀ (n = 7) 48–62 (54·1).

IMMATURE: upperparts somewhat paler than adult, each feather with white tip, dark brown subterminal mark, and pale rufous fringe, giving white-spotted, scaly effect. Underparts like adult but with pale brownish smudge in place of adult's black breast mark.

M. b. rufescens (Brehm): S Turkey, Syria, Lebanon and Iraq; winters NE Ethiopia. Upperparts more olive, less greyish, and ear-coverts and sides of breast behind the black patch substantially more rufous-washed than in nominate race. SIZE: wing, ♂ (n = 19) 119–129 (123·5), ♀ (n = 10) 111–117 (114); tail, ♂ (n = 17) 49–57 (53·2), ♀ (n = 11) 47–53 (49·6); bill to skull, ♂ (n = 19) 20·4–23·2 (21·3), ♀ (n = 18) 19·2–21·4 (20·1), bill to feathers av. 4·2 shorter; tarsus, ♂ (n = 17) 26–27·5 (27·0), ♀ (n = 11) 25·5–27·5 (27·0).

Field Characters. A medium or large lark wintering in NE Africa, readily distinguished from all other larks except Calandra Lark *M. calandra* and Greater Short-toed Lark *Calandrella brachydactyla* by conspicuous black patch on side of upper breast. Very similar to Calandra Lark but lacks white trailing edge in wing, and with white tip to tail (Calandra Lark has white sides to tail). Much larger than Greater Short-toed Lark, with larger bill, long white superciliary stripe and white-tipped tail.

Voice. Tape-recorded (73, 90, LEO). Song very like that of Calandra Lark q.v., but less rich, louder, more varied, less sustained and with fewer harsh grating sounds (Cramp 1988). Recalls song of Greater Short-toed Lark. Mimics other birds. Flight call a cheery trill, 'prrp' or 'tchup-turrup', like Eurasian Skylark *Alauda arvensis*, giving rise to onomatopoeic name *Jaghiagha* in Sudan.

General Habits. On breeding grounds a high-altitude bird inhabiting cultivated plateaux, dry heath, stony wastes, semi-desert, bare steppe and gravelly areas up to 2150 m. In Africa frequents harvested and evidently standing cereal crops (Sudan), dry short grass plains, and bare stony hills (Eritrea), also beaches and barren semi-desert.

Gregarious on wintering grounds. Forms tight-packed flocks at watering places (Meinertzhagen 1951). Foraging flocks of up to 500 birds near Khartoum are loose and not well integrated (Beshir 1978). Mixes with other larks including Lesser Short-toed Lark *C. rufescens*. Roosts at night communally in open desert with scattered clumps of grass, *Cymbopogon nervatus*, each bird in flat pit *c.* 40 mm deep which it digs with its feet; av. density of roost holes was 'about 230/100 sq. meters' (but av. distance between holes given as 20–50 m: Beshir 1978). Birds also roost in self-made flat pits in sorghum fields in hottest time of day, in shade of plants. Birds defaecate in them (Beshir 1978). Sand- and dust-bathes. Song-flight exactly like that of *M. calandra*, q.v.

Forages by walking and pecking, like other larks. In Sudan eats guinea-corn *Sorghum* grain by nipping it from low seed-heads when seed still soft and by perching on heads of tall plant (Beshir 1978); also digs up freshly-sown grain with bill (Schmutterer 1969).

Visits Sudan Oct–Mar; Ethiopia and Djibouti mainly Nov (1 early Aug, Eritrea).

Food. In Palearctic, wide variety of insects, and seeds of *Amaranthus*, *Malcolmia*, legumes, *Hypericum*, *Galium*, *Heliotropium* and grasses (Cramp 1988). In Africa only *Sorghum* seeds recorded (see above).

References
Beshir, E. S. A. (1978).
Cramp, S. (1988).

Genus *Ammomanes* Cabanis

Small, pale, plain-backed larks of open desert, with cryptically-coloured plumage matching substrate. Bill short, conical and strong, deeper than broad at nostrils, slightly curved (**A**); nostrils concealed by tuft of plumes. Outer primary small (*A. cincturus*, *A. grayi*) or large (*A. deserti*, *A. phoenicurus*). Tail somewhat short to medium-length, rufous with variable black pattern on outer half, or (*A. grayi*) black with white base and tip. Legs and feet somewhat weak, not robust; hind claw curved, equal in length to or shorter than hind toe.

A

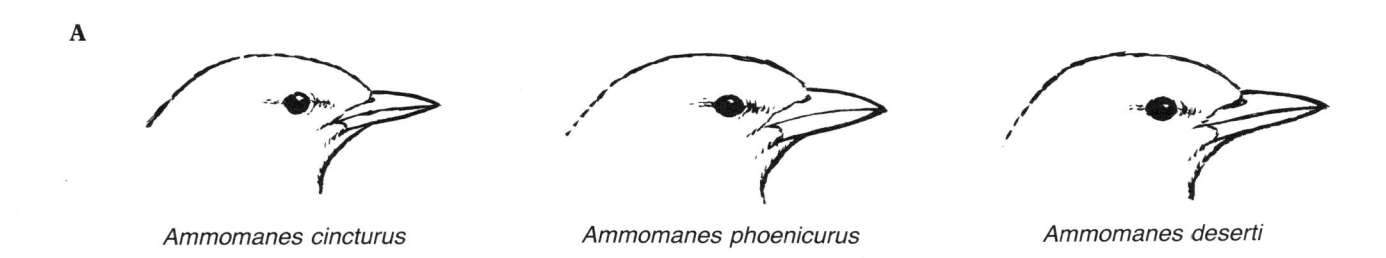

Ammomanes cincturus   *Ammomanes phoenicurus*   *Ammomanes deserti*

Feed on seeds and invertebrates; apparently dependent on metabolic water and invertebrates for primary sources of water; seldom drink. Tend to crouch and hide if threatened.

Africa and Asia; 4 species, 3 in N Hemisphere, 1 in SW Africa. 3 species in Africa, of which 1 endemic.

Ammomanes cincturus (Gould). Bar-tailed Lark. Ammomane élégante.

Melanocorypha cinctura Gould, 1841. *In* Darwin, Zool. Voy. 'Beagle', pt. 3, p. 87; São Tiago, Cape Verde Is.

Forms a superspecies with *A. phoenicurus* (India).

Plate 2 (Opp. p. 33)

Range and Status. Cape Verde Is, N Africa, Arabia, Iraq, Iran, Afghanistan, Pakistan. Resident; dispersive.

Resident or local migrant in Sahara, commoner in W than in E, occurring in all African countries south to 18°N in west and 15°N on Nile. Morocco, common in south. Algeria, occurs south of line Chegga–Bou Saada–Messaad–Laghouat–Aïn Sefra, and frequent in winter in Hoggar and Tassili. Libya, widespread in Tripoli desert, probably commoner than *A. deserti*, and in Fezzan extends from Jebel Soda to Ghat (perhaps absent from sand seas of Oubari and Mourzouk); common in N Cyrenaican desert, extending south at least as far as sand sea at 30°N, and occurs in Libyan desert west of Jaghbub; Egypt, common near Salum (Siwa), otherwise uncommon and local – sparse records between Dakhla and Kharga and elsewhere. Mauritania, local in N and widespread in S, common in Nouakchott area but erratic and sometimes absent for 2 months. Mali, frequent in desert pasture in W-centre, in Irrigui and Awana. Niger, frequent in Ténéré; other records from Agadès, Arlit, Abangarit and Tin Telloust (Giraudoux *et al.* 1988). Chad, common along Oum Chalouba–Fada road between 16° and 17°N but in Wadi Rimé–Wadi Achim Reserve only in lava-strewn Mortcha area, between Oum Chalouba and Wadi Sofaya. Sudan, widespread and locally common from Darfur, Wad Medani and the Eritrean border northward.

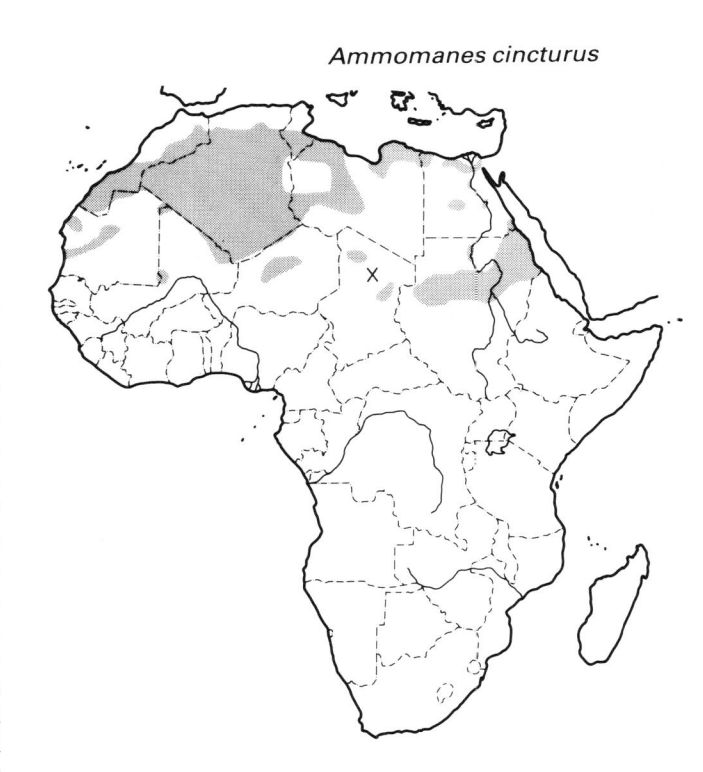

Description. *A. c. arenicolor* (Sundevall) (only race in Africa): Sahara, Middle East and Arabia. Includes '*pallens*' (Agadès, Aïr and elsewhere), distinguishable only in series by being slightly less pinkish above. ADULT ♂: upperparts uniform, soft yellowish or pinkish brown, very slightly greyer on crown and mantle, yellower on nape and pinker on rump. Forehead and crown speckled with minute soft dark spots and streaks; lores and featherlets around eye white; ear-coverts soft pinkish brown, sometimes slightly dappled; sides of neck same as nape. Uppertail-coverts and tail pale rufous-brown; all rectrices including central pair with distal one-third to one-quarter brownish black (surrounded with narrow fringe of buff). Chin and throat white with greyish feather bases showing through; breast very pale buff with greyish or pinkish wash especially towards sides, and a few soft, barely-discernible darkish streaks in centre; belly and undertail-coverts white or extremely pale pinkish buff; flanks pinkish buff. Remiges pale rufous-brown, outer primaries with dark blackish ends surrounded by narrow buff fringe, tertials pale rufous with broad buff fringes (the rufous and buff merging); upperwing-coverts with dark brown centres grading to broad buffy fringe. Axillaries and underwing-coverts creamy. Upper mandible pink or yellowish grey, lower mandible pinker, with grey tip; eye yellowish brown or dark brown; legs and feet yellowish grey, whitish, or greyish horn with pink tinge. Sexes alike. SIZE: wing, ♂ (n = 16) 91–100 (95·4), ♀ (n = 19) 86–93 (89·6); tail, ♂ (n = 9) 48–56 (53·5), ♀ (n = 11) 45–51 (48·3); bill to skull, ♂ (n = 9) 13·4–14·8 (14·1), ♀ (n = 11) 11·6–13·8 (12·7), bill to feathers av. 3·0 shorter; tarsus, ♂ (n = 9) 21–23 (22·0), ♀ (n = 9) 20–22·5 (22·0). WEIGHT: (Algeria) ♂ (n = 10) 21–23·5; (other African weights) ♂ (n = 2) 14, 17, ♀ (n = 2) 16, 18.

IMMATURE: like adult but slightly paler and buffier; blackish ends to primaries and tail less extensive and less sharply defined (may be absent in primaries).

NESTLING: unknown.

Field Characters. This species and its congeners are the least variegated of African larks and their very plainness serves to identify them as *Ammomanes* spp. (Other plain-backed larks, i.e. hoopoe-larks *Alaemon*, horned larks *Eremophila*, Thick-billed Lark *Ramphocoris clot-bey* and ♀ Black-crowned Sparrow-Lark *Eremopterix nigriceps*, are all, except the last, easy to distinguish by other characters.) A small, neat, pale purplish or pinkish rufous desert lark with white underparts, unmarked above or below except by blackish end to tail and blackish tips to outer primaries, both features showing well in flight and distinguishing it from congeners and ♀ Black-crowned Sparrow-Lark. Often recalls bunting *Emberiza* (Cramp 1988). Pale races extremely similar to Desert Lark *A. deserti* at rest and when walking. Seen from behind, tail of Bar-tailed Lark shows dark tip at rest as well as in flight, while tail of Desert Lark appears mainly rufous because only central feathers are showing. When tail of Desert Lark partly spread, brown centres of other feathers become visible, and whole tail appears mixture of dark brown and rufous, not rufous with dark tip.

Voice. Tape-recorded (60, 63, 73, 78, JOHN). ♂ song a short phrase of clipped trills and single notes, 'turr-ee tre-le tree-tree-you', with 2nd half lower-pitched than 1st. Song-flight is steeply undulating, a series of steep ascents and plunges; 'turr-ee tre-le' given during plunge and 'tree-tree-you' at peak of ascent. 'Tree-le' may give way to sustained trill or to 'zoo-ee' or 'zoo-it' sounds and high-pitched 'seeeeeee' notes (Cramp 1988). Flight call a thin, high, descending whistle 'peeyu'. Various other calls in literature seem to be variants of above (Cramp 1988).

General Habits. The most desert-adapted of all Saharan larks, inhabiting areas without water or any relief from sun: wide, flat or sloping stone or sand deserts with only the most sparse vegetation, hammada, sandy wadis in mountain valleys and desolate lava fields.

Solitary or in pairs when breeding, otherwise in flocks usually of < 8, occasionally up to 50 (W Sahara). Associates with other larks, in N Africa with both horned larks *Eremophila* spp., both crested larks *Galerida* spp., and the Lesser Short-toed Lark *Calandrella rufescens*. Forages in flock by running forward 5–6 steps then pausing; picks from and digs into surface, sometimes opening wings, making little jump, and dashing after passing insect (Cramp 1988). Rather wild and shy, but adults with young out of nest were approachable (Tunisia). Runs fast, looking long-legged; flights fast and lengthy. Flying flock twists and turns in unison. Sings in flight, also from bush, low eminence, or ground. Makes loafing hollow in sand by vigorous kicking, then prostrates itself in it with head outstretched, wings spread, tail flattened (Cramp 1988). In worst heat up to 5 birds shelter together in shade of rock slab (Saudi Arabia) crouching (**A**).

Resident, but can disperse far after breeding. Recorded in Tassili and Hoggar Mts (S Algeria) only Dec–Mar. In drought years moves to less arid regions after breeding (Heim de Balsac and Mayaud 1962).

Food. 6 birds (S Morocco) contained seeds (including *Aizoon*) and insects; 2 (Algeria) contained seeds; 4 (Cape Verde Is) held seeds, other plant material, grit, grasshoppers, a beetle and other insects.

Breeding Habits. Solitary territorial breeder. Song-flight deeply undulating, bird like a yo-yo, flying 25–40 m high with head raised and chest puffed out, or in high circular sweep. Density: song-flights *c.* 100 m apart (Libya); on Cape Verde Is. *c.* 10 nests in 2 km².
NEST: shallow depression lined with vegetation, often with rim of stones on exposed side (Cramp 1988), sited in lee of tussock or small stone.
EGGS: 2–4, av. of 27 clutches (Tunisia, Algeria) 2·8. Subelliptical; glossy; white, lightly spotted with black, grey and purple. SIZE: (n = 60) 19·3–23·7 × 13·2–16·9 (21·5 × 15·3). WEIGHT: *c.* 2·6.
LAYING DATES: N Africa, Feb–May, mainly Feb–Apr; Mali, Jan–Apr.

Reference
Cramp, S. (1988).

Ammomanes deserti (Lichtenstein). Desert Lark. Ammomane isabelline.

Plates 2, 4
(Opp. pp. 33, 49)

Alauda deserti Lichtenstein, 1823. Verz. Doubl. Zool. Mus. Berlin (1823); Upper Egypt (= Aswan).

Range and Status. N Africa and Middle East to Afghanistan, NW India.

Resident, much of Africa north of 15°N, from Mauritania and Morocco east through Mali, Algeria, N Niger, Libya, N Chad, Egypt, Sudan, NE Ethiopia, Djibouti and N Somalia. Common but local.

Description. *A. d. geyri* Hartert (including '*janetti*' and '*bensoni*'): Mauritania to Nigeria (Kano) and SE Algeria (Hoggar Mts). ADULT ♂: crown, nape, hindneck, mantle, back, uppertail-coverts and wing-coverts sandy grey-brown, rump pink-brown. Tail-feathers darker brown with rufous outer margins and narrow pale tips, central pair paler than rest of tail, outer ones darker. Chin and throat whitish, remainder of underparts sandy buff. Flight feathers dark brown with rufous outer margins. Bill pale brown with darker tip; eye hazel; legs and feet yellowish brown. Sexes alike. SIZE: wing, ♂ (n = 23) 92–107 (99·0), ♀ (n = 16) 89–100 (95·0); tail, ♂ (n = 23) 59–70 (64·0), ♀ (n = 16) 58–68 (63·0); bill, ♂ (n = 22) 14·4–17·3 (15·9), ♀ (n = 16) 13–16·8 (15·1); tarsus, ♂ (n = 23) 20–24 (21·8), ♀ (n = 16) 20–23·4 (21·6). WEIGHT: ♂ (n = 4) 24·1–26·6 (25·1), 1 ♀ 21·1, 2 unsexed 20, 24·9.

IMMATURE: feathers of underparts, wing-coverts and flight feathers with narrow, pale tips; a few spots on chest.
NESTLING: down white, greyish white or buff-white, long and dense. Inside of mouth orange-yellow with 1 dark spot either side of back of anterior tongue, and 1 dark spot on tips of mandibles.

A. d. payni Hartert (including '*monodi*'): Morocco south of Great Atlas Mts and SW Algeria. Upperparts dark pinkish grey, otherwise as *geyri*.

A. d. algeriensis Sharpe: Algeria, Tunisia and W Libya, south to Tibesti, Chad. Upperparts grey-brown, underparts buff.

A. d. whitakeri Hartert: SW Libya. Darker than *geyri*, upperparts grey-brown, undertail-coverts browner.

A. d. mya Hartert: W and central Sahara. Upperparts pale sandy cinnamon, tail-coverts rufous-brown. Tail brown, central rectrices more rufous; underwing-coverts cinnamon. Larger: (2 ♂♂, 1 ♀) wing, ♂ 107, 108, ♀ 109; tail, ♂ 64, 69, ♀ 69; bill, ♂ 18·1, 18·3, ♀ 16·7; tarsus, ♂ 22·5, 24, ♀ 21.

A. d. isabellina (Temminck): Egypt west of Nile Valley and Dakhla Oasis. Paler and more sandy brown than *mya*. Size similar to *geyri*.

A. d. deserti (Lichtenstein) (including '*borosi*'): S. Libya, Egypt east of Nile to Red Sea and south to Sudan. Upperparts dark grey-brown.

A. d. kollmanspergeri Niethammer: NE Chad and Sudan (Darfur). Upperparts darker and more reddish brown than *geyri*, underparts more rufous.

A. d. erythrochrous Reichenow: Sudan from Dongola south to Kordofan and west to Chad (Ndjamena). Upperparts paler and more pinkish cinnamon than *geyri*.

A. d. samharensis Shelley: Red Sea coast of Sudan and Ethiopia (Eritrea) south to Massawa. Upperparts darker and more grey-brown, underparts sandy buff. Smaller: wing, ♂ (n = 11) 91–102 (98·0), ♀ (n = 5) 88–93 (91·0).

A. d. assabensis Salvadori: Ethiopia and NW Somalia. Upperparts dark smoky grey, underparts dark buff, marks on chest. Smaller: wing, ♂ (n = 3) 86–87 (86·6), ♀ (n = 2) 85, 86.

A. d. akeleyi Elliot: N Somalia. Upperparts pale sandy grey, rump and tail-coverts dark pink-buff. Tail brown, outer tail-feathers tipped sandy. Throat whitish, underparts pinkish buff. Smaller: wing, ♂ (n = 11) 85–93 (90·0), ♀ (n = 3) 80–85 (83·0).

Field Characters. A small to medium-sized, unstreaked and very plain brown lark, upperparts varying racially but generally pale; flight feathers darker, with rufous margins. Care must be taken to distinguish pale races from Bar-tailed Lark *A. cincturus*, especially at rest; habitats usually mutually exclusive. In flight latter shows contrasting wedge of black on pale rufous tail, deepest on central feathers, dark tips to primaries contrasting with pale rufous wing, whereas in Desert Lark all except central tail-feathers are dark brown with rufous margins. For further details, see Bar-tailed Lark.

Voice. Tape-recorded (60, 63, 73, 78, 90). Not very vocal. A variety of simple soft twittering and liquid notes (Simmons 1952): (1) Flight song a rolling, trilled 'trrreee' or 'trreeoo', a fast repeated 'willip-chillip'; a fast repeated rising and falling 'chee-lit, chu-lu-lit'; a moderately paced 'chucle-chucle, cheelee'. (2) Contact call when foraging, 'chu' in a soft quiet undertone; sometimes a slurred 'chee-lu'. (3) Flight calls: 'chu' or 'wu', rapidly repeated; 'chee-wit'; 'chwee-u, chwee-u, chwee-u'; 'chu-lit'; 'chlit, chee-wu, chee-wu'; 'chu-lu-weet, chu-lit'. (4) Excitement calls: 'chee-u, chee-u, chee-u, chee-wit'; 'chul-wit-chutle' repeated 4–5 times. (5) A fast liquid twitter when chasing. For further variations, see Cramp (1988).

General Habits. Occurs on stony flats in arid savannas, in arid open scrub, on bare clays, rocky or stony hillsides, lava fields and basalt-derived gravels and stony patches, and in sandy or stony wadis on desert margins. Occurs in the central Sahara, but restricted to hills and oases and well-vegetated wadis, avoiding dunes and sandy plains and absent from predominantly sandy desert. Also occurs in savanna woodlands on stony slopes. In general, favours hills and higher parts of desert, density decreasing with altitude. In central Sahara 2·17 birds/km^2 recorded at 2250–2700 m, 1·8 birds/km^2 at 1800–2250 m and 0·83 birds/km^2 at 1000–1200 m. On desert margins occurs in low-lying areas and at bases of hills and cliffs. Common on Red Sea shore. In desert found up to 6 km from water. Seldom drinks in habitats on desert margins, but has been recorded drinking in central Sahara, between 07·00 h (1 h after sunrise) and 11·00 h.

Solitary, in pairs or loose aggregations, seldom in flocks; in breeding season also in family parties. Tame and easily approached. Loose foraging flocks move slowly over substrate, maintaining cohesion with contact calling. Birds maintain individual distance of *c.* 2 m. Individual suddenly running or flying may release flight pursuit reaction in nearest neighbour, resulting in low, zig-zag aerial chase over varying distances, accompanied by fast liquid twitterings. Chases terminated by leading bird alighting. Sometimes whole group may fly after leading bird. Chases occur in flight when individual distances are transgressed (Simmons 1952). Loose flocks break up Feb–Mar.

Forages among stones, woodland detritus and ungulate droppings. Breaks up dung to find seeds, hammering dung and large seeds with bill. Breaks open plant buds to eat parasites. Insects are taken from ground and among stones; also takes insects from low bushes either by reaching up or by gleaning. Feeds on flying insects up to 6 m above ground (Cramp 1988).

Food. Invertebrates, including grasshoppers (Orthoptera), beetles (Coleoptera), buprestid beetles, bugs (Hemiptera), ants (Formicidae); spiders; seeds of *Astragalus*, *Artemisia* and other shrubs; green shoots of grasses. Young fed grasshoppers and seeds (Cramp 1988). Stomach contents often include sand and grit.

Breeding Habits. In brief, undulating display flight, ♂ ascends steeply to 5–10 m, descends vertically; may fly horizontally in between; also undulating horizontal flight between 2 high points, with deep, floppy wing-beats, with equal periods of closed and open wings; may dip 2 m in undulations. May sing throughout display or only after attaining maximum height (Cramp 1988).

NEST: thick-walled cup of grass stems, fibres and other plant material, lined with plant down and wool, placed on ground against small shrub, tuft or rock. Side of nest away from backing built up with stones 10–15 mm diam.; if in open, nest is surrounded by small stones; built by both ♂ and ♀.

EGGS: 3–7, mean (87 clutches) 3·8. Clutch size varies with habitat, 4 on desert margins, 3 in centre of desert. Ovate; pinky white, usually entirely covered with lilac, grey, brown and sienna speckles and spots, sometimes lightly marked only on broad end. SIZE: (n = 150, Tunisia) 21·5–25 × 16–17; (n = 3, Hoggar Mts, *mya*) 21–22 × 15·3–16·8 (21·6 × 15·9); (n = 10, Somalia, *akeleyi*) 20–21 × 15–16 (20·2 × 15·7).

LAYING DATES: Morocco, Apr; Algeria and Tunisia, late Mar–May; Libya, Feb–Mar (♂♂ with enlarged testes); Mauritania, Jan–Mar; Sudan, Feb–Apr; Niger, June; Ethiopia, Apr–May; Somalia, Apr–June.

INCUBATION: by ♀ only; period unknown.

DEVELOPMENT AND CARE OF YOUNG: young fed by both parents; probably leave nest before they are able to fly, still dependent on parents.

References
Cramp, S. (1988).
Simmons, K. E. L. (1952).

Ammomanes grayi (Wahlberg). Gray's Lark. Ammomane de Gray.

Alauda grayi Wahlberg, 1855. Oefv. K. Sv. Vet.-Akad. Förh. 12, p. 213; between Kuiseb and Swakop rivers, Damaraland.

Range and Status. Endemic resident in Namib desert from S Angola (Pico do Azevedo) south to Namibia (Aus). Restricted to desert and margins. Locally common.

Description. *A. g. grayi* (Wahlberg): W Namibia from Cape Cross south to S edge of Namib Desert north of Aus. ADULT ♂: entire upperparts, including sides of face, wing-coverts, scapulars and central pair of tail-feathers, plain pinkish grey-buff, becoming more rufous on tertials and rump. Tail-feathers brownish black with white outer webs and tips. Whitish streak over eye. Underparts white washed with buff on breast and belly; a few dark feathers on sides of breast. Flight feathers plain brown; underside of wing whitish buff. Bill grey with dark tip; eye olive-brown; legs and feet greyish flesh. Sexes alike. SIZE: (20 ♂♂, 18 ♀♀) wing, ♂ 78–84 (81·0), ♀ 73–82 (77·0); tail, ♂ 51–61 (55·0), ♀ 46–57 (51·0); bill, ♂ 13·9–16·3 (15·1), ♀ 13·2–15·5 (14·6); tarsus, ♂ 21–24 (22·1), ♀ 19–23·9 (21·7). WEIGHT: ♂ (n = 37) 19·3–26·5 (22·7), ♀ (n = 35) 17·5–23 (20·2).

IMMATURE: upperparts obscurely mottled.

NESTLING: flesh pink, with grey-white down on crown, back, scapulars and flanks. Inside of mouth deep yellow with 2 black lateral spots almost at back of anterior tongue, and 2 black spots on tips of upper and lower mandibles.

A. g. hoeschi Niethammer: N Namibia from about Cape Cross to S Angola. Darker, more brownish grey on mantle, and redder, less fawn on secondaries. Larger: wing, 2 ♂♂ 83, 85, ♀ (n = 10) 77–81 (79·0). WEIGHT: 2 ♀♀ 20, 21·7.

Field Characters. A small, unstreaked lark with pale upperparts and white underparts. Unlike any lark within its range but at a distance, especially when perched on stone, can be mistaken for Tractrac Chat *Cercomela tractrac*, which is larger and greyer with longer, thinner bill and white rump conspicuous in flight. Usually in small groups (chats found singly or in pairs).

Voice. Tape-recorded (88, F, GIB). Flight song a combination of high-pitched, very sharp tinkling notes with a loud up-slurring whistled note, accompanied by loud, reedy whirring of wings. Alarm call, high-pitched 'tseet'. Other calls: continuous mellow 'tew', and contact call of 3 piping notes. Adults alarmed or wary at the nest keep vocal contact with soft 'tew'.

General Habits. Present on bare whitish, pink and grey gravel plains with sparse vegetation and scattered rocks. In Namib Desert most abundant between 600 and 1000 m (Willoughby 1971). Absent from sand desert and coastal dunes. North of Swakopmund common on gravel flats just inland from the beach. Avoids well-vegetated wadis, but present in numbers on adjacent gravel plains after good rains when there is sparse but even grass cover.

Throughout the year occurs in loose flocks or groups of up to 30, including juveniles. Mean group size (n = 121) 3·3. Individuals not strongly associated within groups, which may partially break up and join transient groups. Tends to crouch when alarmed; when flushed seldom flies far.

Foraging flocks spread out over a hectare or more. Takes food items from surface of ground; also uncovers buried items by flicking sand sideways with bill, leaving a distinctive raked pattern. Forages around zebra and antelope droppings and entrances of rodent burrows. May forage actively during hot part of day; between foraging bouts perches on stones or twigs above gravel surface and exposes thinly feathered underwings and sides while facing into prevailing wind, to lose body heat. Shelters from heat in rodent burrows. Metabolic water and invertebrates are apparently primary water source; seldom recorded drinking.

No regular movements; nomadic and opportunistic to some extent, but remains within desert.

Food. Invertebrates and vegetable matter, relative proportions varying with rainfall. Mean throughout year: seeds 56% (grass seeds *Stipagrostis* spp., forbs, including *Cleome* spp. and *Monsonia umbellata*), vegetable matter 1% (basal nodes of grasses, stomach contents of herbivorous invertebrates) and invertebrates 43% (termites *Hodotermes*, ants, beetles, stick insects, locusts, flies, moths, spiders, solifugids). Nestlings fed exclusively on invertebrates (small locust nymphs, spiders, beetle larvae and flies).

Breeding Habits. Monogamous. Courtship and pairing take place within groups; paired birds leave group to nest; rejoin group as soon as young can fly. Display of ♂ more vocal than visual, consists of undulating flight c. 7 m above ground accompanied by singing and wing-whirring; usually performed for c. 30 min after sunset and for 2 h before sunrise, occasionally during the day.

NEST: thick-walled cup of soft, fine grass inflorescences, usually *Stipagrostis* spp.; int. diam. 60, ext. diam. 110, depth 35–50, placed in depression dug in gravel c. 130 diam., 75 deep, lined around inner perimeter with a layer of round pebbles 5–10 mm diam. Top of cup flush with or slightly above surrounding gravel. Nest often sited next to 1–2 large stones, solitary grass tuft or shrub, to provide shade.

EGGS: 2–3, mean (19 clutches) 2·4. Ovate; dull white finely speckled with pinkish brown, grey-brown and slate, often concentrated at obtuse end. SIZE: (n = 21) 19·8–22·4 × 14·6–16 (21·2 × 15·3).

LAYING DATES: Namibia, Mar–July, exceptionally Sept, Jan. Apparently breeds opportunistically after rains; no clear-cut gonadal cycle in ♂♂.

INCUBATION: period (n = 2) 12, 13 days.

DEVELOPMENT AND CARE OF YOUNG: both adults feed nestlings. When bringing food they run towards nest, stop abruptly, stand still and erect, then run towards nest again. Small nestlings fed twice in 30 min. Nestlings leave nest before they are able to fly, at c. 10 days; still dependent on parents; when able to fly but still under parental care, join small foraging flocks.

BREEDING SUCCESS/SURVIVAL: from 22 eggs laid in 10 nests, 11 young fledged. No data on subsequent survival. Adults preyed on by Lanner Falcon *Falco biarmicus* and Red-necked Falcon *F. chicquera*.

References
Niethammer, G. (1969).
Willoughby, E. J. (1971).

Genus *Calandrella* Kaup

Small to medium-sized larks of open, well-grazed grasslands. Wing more pointed than *Mirafra*, outer primary much reduced. Tail dark and square, outer rectrices white. Bill strong and short, nostrils concealed by bristles or plumelets. Nest cup-shaped, without dome, on foundation of coarse material, often with ramp of stones or mud-flakes.

7 species worldwide, 4 in Africa, 1 endemic. 2 superspecies in Africa, *C. brachydactyla*/*C. cinerea* and *C. rufescens*/*C. somalica*.

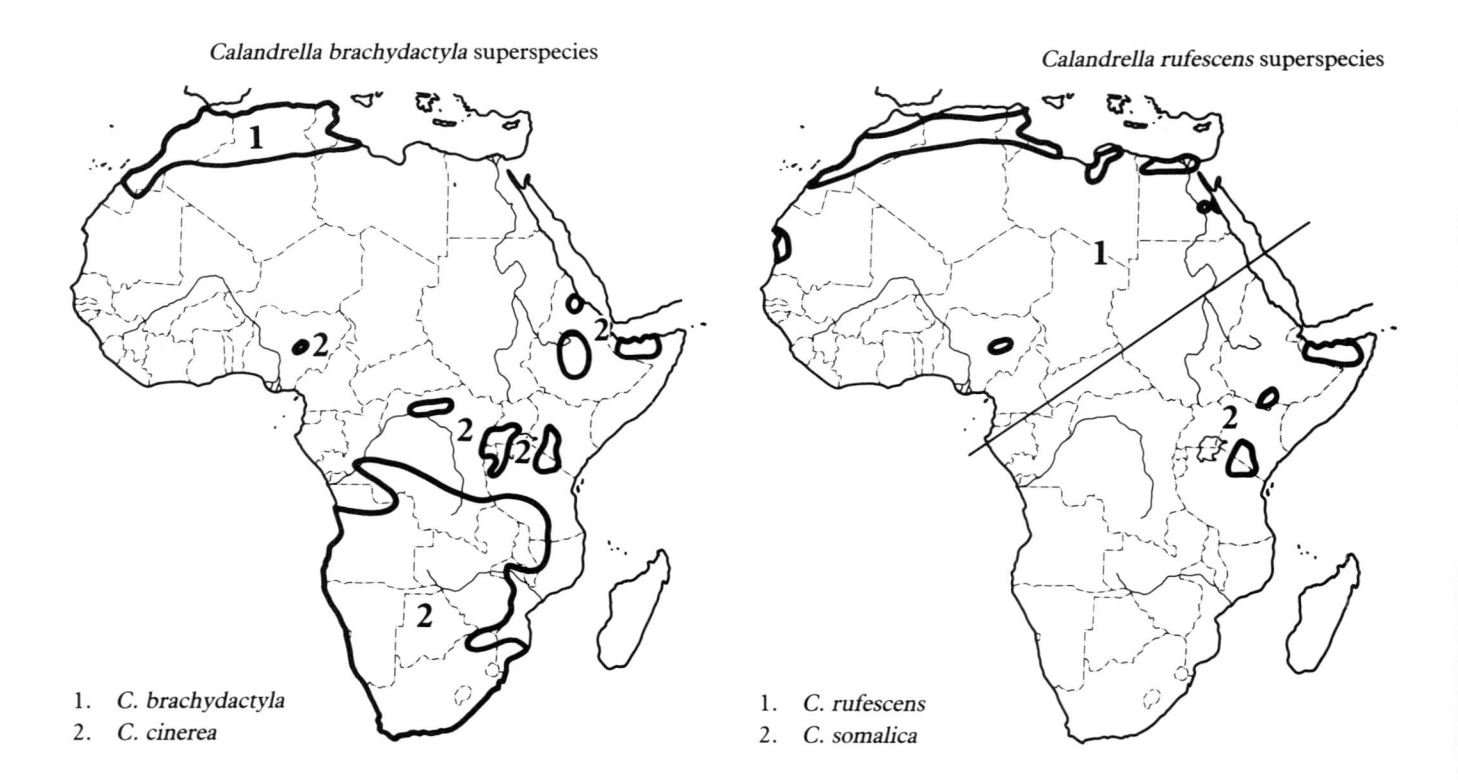

Calandrella brachydactyla superspecies

1. *C. brachydactyla*
2. *C. cinerea*

Calandrella rufescens superspecies

1. *C. rufescens*
2. *C. somalica*

Calandrella brachydactyla (Leisler). Greater Short-toed Lark. Alouette calandrelle.

Alauda brachydactila [sic] Leisler, 1814. Ann. Wetterauischen Gesell., 3, pl. 19, p. 357; Montpellier, France.

Forms a superspecies with *C. cinerea* and Asian *C. acutirostris*.

Range and Status. Breeds Africa, Palearctic to N Mongolia and NE China; winters Africa, Middle East, central Asia.

Resident and Palearctic migrant. Breeds Morocco, N Algeria, Tunisia (many localities in coastal and southern areas), NW Libya (rather scarce); has bred NE Egypt (no recent records). Winters sparsely central Algeria, extreme S Tunisia; widespread in Sahel, from Mauritania, N Senegal, Mali (locally abundant), Niger (widespread), Nigeria (common in extreme NE), Chad, Sudan (abundant on short-grass plains), NE Ethiopia (common) and Somalia (several records). Single records from Cameroon (Efulen), Zaïre (Iyonda), and Djibouti (Ras Siyan); 2 records Kenya (Athi R., Mombasa).

Description. *C. b. brachydactyla* (Leisler): breeds Europe from Mediterranean north to central France, Yugoslavia and S Rumania; winters N Africa and south of Sahara from Mauritania to Somalia. ADULT ♂: forehead, crown, mantle and back streaked, feathers with broad greyish black centres shading through rufous to pink-buff, buff or cinnamon-buff edges. Rump and uppertail-coverts tawny buff or cinnamon-buff, feathers darker along shafts, tail-coverts with blackish streaks along shafts. Tail-feathers blackish brown, outer pair mainly white, bases of both webs and outer portion of inner webs grey-brown, next pair with broad off-white outer edge, central pair with broad rufous-cinnamon margins. Sides of neck and foreneck off-white or buffy, speckled and streaked darker brown, ear-coverts brown, with thin pale streaks. Lores and broad supercilium creamy white, short buff streak below eye. Chin and throat cream or white. Breast pale buffy brown, feather centres sometimes darker rufous-buff in centre of breast and brown on sides; at sides of upper breast a prominent black or dark olive-brown crescent mark; occasionally a few narrow black spots across upper breast forming narrow mottled gorget; rest of underparts creamy white. Flight feathers dark brown with narrow buff outer edges and tips, outermost primary with broader buff edge; tertials dark grey-brown, broadly margined pink-cinnamon or buff; scapulars and upperwing-coverts dark olive-brown with broad cinnamon-buff outer edges. Underwing-coverts and axillaries greyish or whitish buff, occasionally rufous-buff. Upper mandible horn-brown, darkest at base and tip, lower mandible and cutting edge yellowish horn; eye brown; legs and feet brownish flesh. SIZE: wing, ♂ (n = 37) 92–102 (96·2), ♀ (n = 17) 86–94 (90·1); tail, ♂ (n = 11) 54–59 (56·9), ♀ (n = 9) 50–57 (53·9); bill, ♂ (n = 5) 13·2–15·4 (14·3), ♀ (n = 5) 13·1–14·6 (13·7); tarsus, ♂ (n = 7) 20·4–21·9 (21·1), ♀ (n = 7) 19·3–21·2 (20·2). WEIGHT: unsexed (Morocco, on spring migration) (n = 161) 16–24 (20·7).

IMMATURE: like adult, but all feathers of upperparts, including upperwing-coverts, dark brown with buff edges and whitish tips. Underparts white, breast buff with rounded dark grey or brown spots. Outer primary longer than adult, with broadly rounded tip, marginated pale buff.

NESTLING: skin bluish black above, orange-red below, with long pale buff down on upperparts and flanks. Inside of mouth deep yellow with orange palate, 2 large black spots on each side of tongue, 1 near tip of tongue, 1 on tip of upper mandible. Gape-flanges yellowish white. Bill horn-brown.

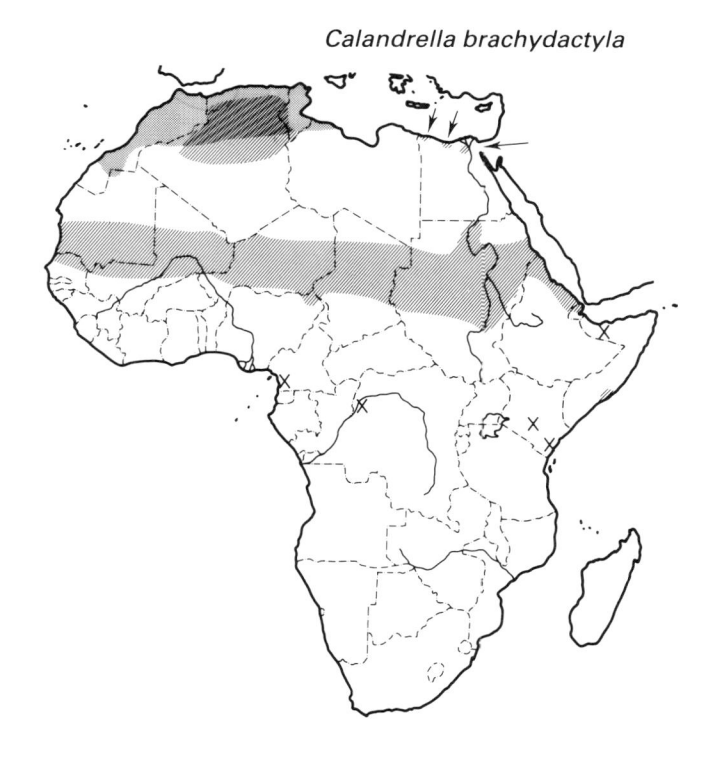

Calandrella brachydactyla

C. b. rubiginosa Fromholz: breeds NW Africa, Malta; winters N Africa and probably also south of Sahara. Overall more rufous, especially on crown.

C. b. longipennis (Eversmann): plains north of Caucasus and perhaps Ukraine to N Mongolia and NE China. Palearctic migrant; listed for N Africa, Mali, Chad, Somalia and Kenya but status uncertain (see below). Crown, hindneck and rump more sandy grey, streaks of upperparts narrower. Bill small: ♂ (n = 11) 12·7–13·7 (13·3), ♀ (n = 3) 12·4–13·9 (13·2).

TAXONOMIC NOTE: identification of races of this species in winter is very difficult; exact distribution of various races remains to be worked out, and some may occur which are not listed here.

Field Characters. A small, pale lark with short, pale bill, no crest, and unstreaked white underparts; tail dark with white outer feathers. Characteristic small blackish patch at sides of neck, not always easy to see, distinguishes it from Lesser Short-toed Lark *C. rufescens*, which also has streaked breast. Distinct white supercilium outlines dark top of head producing capped effect; cap rufous in N African birds (brown in Lesser Short-toed). Tertials almost reach wing-tips (longer than in Lesser). Desert Larks *Ammomanes* spp. have paler, unstreaked upperparts, different tail patterns. Voice distinctive, q.v.

Voice. Tape-recorded (62, 73, 78, CHA – 76, 89, 90). Flight song a series of discrete phrases *c.* 2 s long with pauses of 2–3 s between phrases; phrase frequently starts with a few staccato notes which accelerate into a tinkling trill, but many phrases lack these introductory notes and consist of 2–3 trills at different pitches and speeds; notes not fluty or melodious but bubbling and pleasant. Imitations of other species frequently incorporated. Ground song similar but quieter; descending muscial trill also given from ground. Calls include loud 'pee-yi', given singly or as first note of song, and a 'chup' or 'chirrup' in flight. In Morocco wintering birds gave a variety of calls, including 'dzi', 'ti', 't-t-t-t' and 'tur'. For other calls and song variants, see Cramp (1988).

General Habits. Occurs in dry open plains and uplands, terraces, slopes and undulating hills, on sandy or clay soils or on gravels, and in fallow fields with stubble or in cereal croplands where there are bare areas and dirt tracks; also in croplands with mosaic of small patches of bare soil. Winters in semi-arid and arid plains, on cultivated fields, croplands immediately after the harvest, well-grazed and sparsely covered grasslands, airstrips and muddy margins of flood-plains.

Highly gregarious when not breeding, occurring in flocks of a few to several hundred on migration, occasionally aggregations of thousands (10,000 in Mali: Lamarche 1981).

Forages on ground and takes food from ground and low plants; digs with bill for buried items, and occasionally takes flying prey. Drinks water regularly.

Migrates south across Mediterreanean coast mid-Aug–Oct, peaking late Aug–mid-Sept when flocks may number several thousands. Spring passage north late Jan–Apr, early in S Morocco (probably N African breeding birds), continuing to early Mar and Apr when birds arrive at breeding areas.

Food. Invertebrates: ants (*Tetramorium*, *Messor*), grasshoppers, termites, bugs, adult, pupal and larval Lepidoptera, flies Bibionidae, Hymenoptera, adult and larval beetles (Carabidae, Histeridae, Cerambycidae, Chrysomelidae, Bruchidae, Curculionidae), spiders, woodlice, snails; also seeds (grasses *Panicum*, *Setaria*, *Eleusine*, *Eragrostis*; forbs Polygonaceae, Chenopodiaceae, Amaranthaceae, Papaveraceae, Cruciferae, Leguminosae), plant leaves and stems. Nestlings fed only invertebrates.

Breeding Habits. Territorial, probably monogamous (Cramp 1988). Territories usually clustered in neighbourhood groups of 10–20 pairs, 10–20 km between groups. Territory 40–50 m diam., density *c.* 2–3 pairs/ha, up to 4–5 pairs/ha (Hungary). ♂ begins display soon after arrival at nesting area; bird climbs steeply with rapidly beating wings to 8–15 m, frequently in spiral, giving short introductory notes; thereafter gives main trilling song while rising to 30–50 m, then extends wings, stays motionless in air on final note of song-phrase, closes wings and descends, or may open wings to effect gliding descent; before reaching ground, beats wings a few times to climb again for a few m, then drops down and repeats major song-flight sequence (Cramp 1988).

NEST: cup of grass leaves, stems and rootlets, ext. diam. 87, int. diam. 55–80, depth 30–60; lined with softer vegetation, feathers, plant down and wool, in scrape in ground; frequently surrounded by small clods of earth or stones; built mainly by ♀ in 7–10 days after she chooses site and excavates scrape or prepares nest hollow.

EGGS: 3–5, mean (69 clutches) 3·5 (Tunisia and Algeria). Sub-elliptical; whitish, creamy-white, greenish or greyish, either heavily but evenly mottled with pale brown and lavender grey, or spotted and blotched darker brown, with some pale purplish grey, marks larger and denser and forming cap at obtuse end. SIZE: (n = 130) 17–22·8 × 13·4–15·6 (19·8 × 14·8). WEIGHT: (n = 30) 2·21 g.

LAYING DATES: Morocco, Apr–May; Algeria, Apr–May; Tunisia, Apr–June.

INCUBATION: by ♀ only; begins with last egg; period 13 days. Brood hatches in 24 h.

DEVELOPMENT AND CARE OF YOUNG: young fed by both adults, from dawn to dusk, average 10–13 times per h, rate may drop to 6 times per h around midday. Young leave nest at 9–10 days; fledging period 12–13 days. Young of 1st brood become independent soon after leaving nest, those of 2nd brood remain with parents for some weeks.

BREEDING SUCCESS/SURVIVAL: no data for Africa. European nests predated by snakes and destroyed by sheep.

Reference
Cramp, S. (1988).

Plates 4, 6
(Opp. pp. 49, 97)

Calandrella cinerea (Gmelin). **Red-capped Lark. Alouette cendrille.**

Alauda cinerea Gmelin, 1789. Syst. Nat., 1, pt. 2, p. 798; Cape Town.

Forms a superspecies with *C. brachydactyla* and Asian *C. acutirostris*.

Range and Status. Africa, Arabia.

Resident, partly nomadic. An isolated population in N Nigeria (Jos Plateau, uncommon and very local); W and SE Ethiopia, central N Somalia (fairly common), central and SW Kenya (locally common), SW Uganda (east to Entebbe), NW and S Tanzania, Zaïre (scattered localities in NE, N-central, SW and S), Rwanda (scarce and local), Angola (central plateau from N Bihé south to

Humpata, coastal plain at Benguela and flood-plains of larger rivers from Kwanza south to Cunene), N and E Zambia (locally common on dry plains), Malawi (common on bare or recently burnt ground), S Mozambique on flood-plains of Limpopo R., Zimbabwe (widespread, locally common nomad), widespread but local in Botswana and Namibia. In South Africa, common resident and local nomad in S Transvaal, W and S coastal Natal, widespread in Orange Free State, NE, E and W Cape Province, local in Karoo and S coast. Vagrant Abd-el-Kuri.

Description. *C. c. cinerea* Gmelin (including '*witputzi*'): S Namibia, South Africa (Cape Province east to Grahamstown, W Orange Free State). ADULT ♂: crown rufous, with erectile crest usually laid flat; nape and hindneck brown, sides of neck and foreneck brown streaked whitish, ear-coverts brown, lores buffy white, supercilium broad and white, chin and throat white. Mantle brown, back brown mottled with darker brown, rump reddish brown, uppertail-coverts rufous. Tail-feathers dark brown, outer pair with white outer webs, adjacent pair with narrow whitish outer edge, central pair with buffy margins. Breast brownish with rufous patches at sides, with variable amount of brown or rufous extending onto flanks; remainder of underparts, including undertail-coverts, white. Flight feathers dark brown, outer primary with buffy white outer web, rest with narrow paler brown outer edges; scapulars and tertials brown with paler brown margins, upperwing-coverts brown with darker centres. Underwing-coverts and axillaries light brown. Bill black; eye brown; legs and feet dark brown to black. Sexes alike. SIZE: (8 ♂♂, 12 ♀♀) wing, ♂ 93–99 (95·0); ♀ 85–97 (89·5); tail, ♂ 56–62 (58·0), ♀ 47–58 (53·1); bill, ♂ 13·8–15·8 (14·6), ♀ 13–14·9 (14·2); tarsus, ♂ 20–23 (21·3), ♀ 19–23 (21·0). WEIGHT: ♂ (n = 19) 21·5–26·3 (24·1), ♀ (n = 14) 20–26 (23·1).
IMMATURE: darker than adult; upperparts dark brown, crown, back and flight feathers with subapical blackish bar and whitish tip, breast and lower breast brown, spotted irregularly dark brown, belly buff.
NESTLING: skin blackish with grey-white down on crown, wings, back and thighs.
C. c. anderssoni (Tristram): South Africa (E Cape Province, Natal, Transvaal), NE Namibia, N Botswana, SW Zambia (north to Monze). Upperparts browner, streaking and mottling more blackish and contrasting. Crown and sides of breast darker rufous-brown. Intergrades with *cinerea* in Natal and E Cape.
C. c. spleniata (Strickland) (including '*ongumensis*'): W Namibia on margins of Namib Desert. Upperparts pale yellowish sandy with faint streaking, much paler than *anderssoni*.
C. c. millardi Paterson: S Botswana (Tsane, Tsabong, Kakia, Mumpswe). Upperparts whitish grey, lacking any sandy tinge.
C. c. saturatior Reichenow: Zimbabwe, Zambia, Malawi, Angola, Shaba, Zaïre to lower Congo R., east to Tanzania, Uganda and W Kenya. Upperparts brighter and more rufous than *anderssoni*.
C. c. williamsi Clancey: Kenya (Naivasha, Kinangop, Athi and Kapiti Plains). Upperparts greyer and colder than *anderssoni*, darker and more heavily streaked than *cinerea*, rufous on crown and sides of breast dull.
C. c. erlangeri (Neumann) (including '*fuertesi*'): central Ethiopia. Differs from other races by large black patches at sides of breast; crown deep rufous, forehead dusky blackish, black streaking on hindcrown; upperparts dull sandy grey with heavy black streaking, chest and flanks pinkish apricot.
C. c. blanfordi (Shelley): N Ethiopia (Eritrea). Considerably paler above than *erlangeri*, more sandy, less blackish, streaks narrower and brown, not black, nape and hindneck pale sandy with reduced streaking, cap paler red; below, dark breast patch present but much smaller, less pronounced, rufous areas paler.
C. c. daroodensis White: central and N Somalia. Paler above than *blanfordi*, greyish sandy, crown paler and less sharply defined, some streaking still present on nape and hindneck. Below mainly white but still with slight pinkish wash on breast and flanks. Blackish breast patches like *blanfordi*, small but distinct. Underwing-coverts whiter than *blanfordi*. Smaller than *blanfordi*.
TAXONOMIC NOTE: Ethiopian and Somali races are distinguished from those further south in having black, not red, patches on sides of breast; patches are especially large and conspicuous in *erlangeri*, reduced but still present in the other 2. Peters (1960) treated Eritrean and Somali birds as a distinct species, *blanfordi*, while keeping *erlangeri* as a race of *cinerea*, although it is at least as different from other *cinerea* races as *blanfordi*. White (1960a) split Eritrean and Somali birds into 2, describing Somali birds as *daroodensis*. Hall and Moreau (1970) acknowledged the distinctness of these populations, suggesting that both *erlangeri* and the *blanfordi* group (*blanfordi*, *daroodensis* and birds from Arabia) were incipient species. White (1961) treated Ethiopia/Somali populations as races of *cinerea*; we prefer to do the same until more is known of their life histories, while at the same time noting their distinctness.

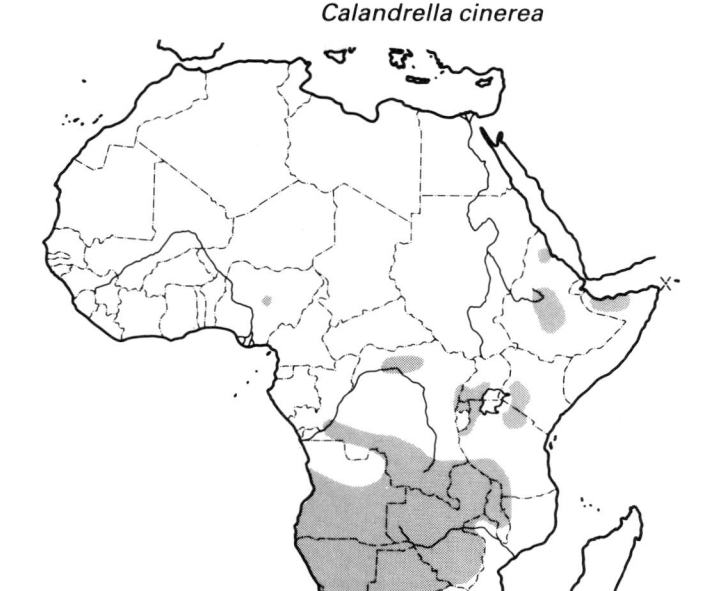

Field Characters. A medium-sized lark, readily distinguished by red breast-patches and red crown outlined by conspicuous white stripe. Underparts unstreaked; in flight, dark tail contrasts with brown body and rufous rump. Flight said to be pipit-like (Maclean 1985).

Voice. Tape-recorded (86, 88, C, F, LUT, PAY). Flight song, a medley of high-pitched notes and trills, some melodious, some sibilant or harsh, including imitations of other species: 'tree-tree-chew-chew-chew-chew-tip-tip-tip-tschreet-tschreet-lululu . . .'. Flight calls, 'drelit', 'chirrup' or hard 'pt'; other calls include sharp 'tsree' and harsh 'tshweep'.

General Habits. Present in short, dry grasslands, particularly where grazed short or recently burnt, ploughed lands, croplands after the harvest, fallow fields, cattle paddocks, airfields and bare areas in dwarf shrublands.

Outside breeding season occurs in flocks of 5–20, sometimes hundreds. Forages on ground, walking quickly about, occasionally running; pecks food from soil surface or gleans foliage of grasses or shrubs. Breaks up antelope droppings to obtain seeds, and probes in cattle dung for fly and beetle larvae. Drinks water regularly. Raises crest when anxious or hot.

Has definite but irregular movements into both breeding and non-breeding areas; nomadic rather than a regular migrant. Some birds resident.

Food. Invertebrates (ants, mantids, termites, adult and larval Lepidoptera, bugs Pentatomidae, scale insects *Margarodes*, grasshoppers, wasps, weevils) and seeds (grasses *Brachiaria*, *Setaria*, *Urochloa*, *Eleusine*, *Panicum*; sedges *Fimbristylis*; forb *Gisekia*). Nestlings fed invertebrates.

Breeding Habits. Territorial; territory size $c.$ 0·5 ha in SW Cape Province. ♂ in display ascends almost vertically into air to height of $c.$ 50 m, then flies into wind flapping slowly, rising and dipping alternately, singing (see *Voice*). Remains in air 7–10 min, singing at short intervals, gradually descending to 30–40 m, when it closes its wings and drops; opens wings just above ground and flies up again to repeat display, or else drops down to perch on ground.

NEST: cup of dry grass and roots, int. diam. $c.$ 50, depth $c.$ 25, lined with finer material, placed in excavation in ground against grass tuft, shrub, stone, or large clod of earth, or out in open; hoof-print or other existing hollow may also be used. Rim of nest just above or level with ground surface; ramp of small stones or pieces of earth often present on open side of nest. Built mainly by ♀, in $c.$ 5 days.

EGGS: 2–4, mean (204 clutches) 2·15. Sub-elliptical; white, creamy white or pale cream, spotted and speckled with grey and some brown, sometimes markings indistinct and cloudy, sometimes concentrated into dark zone or cap on obtuse end. SIZE: (n = 102) 17–23·3 × 13·2–16·8 (21·2 × 15·1).

LAYING DATES: Ethiopia, May–June, Oct; Zaïre, May, July–Aug; E Africa: Region A, Mar–May, Oct, Dec; Region B, Apr; Region D, Feb–July, Nov–Dec; Malaŵi, July, Sept–Oct; Zambia, May–Sept; Zimbabwe, Mar–Oct, Dec; Namibia, Mar, July, Sept–Oct; South Africa: SW Cape, E Cape and Karoo, Aug–Dec; Orange Free State, Aug–Jan; Natal, July–Nov; Transvaal, July–Feb, Apr.

INCUBATION: period 13–15 days.

DEVELOPMENT AND CARE OF YOUNG: young fed by both adults, but mainly by ♀; rate varies in SW Cape from 2 to 23 feeds/h, several feeds being given at rate of 1 per min, followed by gap of up to 35 min. Faecal sacs removed by adults after feeding. Young brooded up to 7 days, shaded on hot, still days. Nestling period (n = 25) 9–18 days (12·0).

BREEDING SUCCESS/SURVIVAL: of 135 eggs in 66 nests, 108 eggs hatched and 82 young left nests in SW Cape. Small clutches more successful than large, mean production 1·52 young/nest from clutch of 2, 2·0 young/nest from clutch of 3; nests early in breeding season marginally more successful than later nests (Winterbottom and Wilson 1959).

References
Borrett, R. P. and Wilson, K. J. (1971).
Winterbottom, J. M. and Wilson, A. H. (1959).

Plate 2
(Opp. p. 33)

Calandrella rufescens (Vieillot). **Lesser Short-toed Lark. Alouette pispolette.**

Alauda rufescens Vieillot, 1820. Tabl. Enc. Methorn 1, p. 322; Tenerife, Canary Is.

Forms a superspecies with *C. somalica* and Asian *C. raytal*.

Range and Status. Africa, Canary Is, Iberian peninsula, Middle East and central Asia to N China (Manchuria).

Resident, Palearctic visitor and intra-African migrant. Breeding resident Morocco, Algeria, Tunisia (mainly coastal, Enfidaville, Hergla, Monastir, Kairouan to Sfax; central areas north to Tunis), N and NW Libya (southern limits not clear), Egypt (Nile delta). Winters in breeding range with some southward expansion in S Morocco, Tunisia (frequent in area west and south of breeding area), and Egypt (Nile valley, Red Sea Coast); also in Mauritania (rare) and N Nigeria (Jos Plateau, rare). Locally common in breeding areas, sparse and seldom recorded in southern parts of wintering range.

Description. *C. r. minor* (Cabanis): N Africa (except Nile Delta), Middle East from N Sinai north to S Turkey, east to W and central Iraq. ADULT ♂: upperparts pale rufous-cinnamon when fresh, greyish sandy buff or pink-buff when worn. Crown narrowly streaked black, hindneck and sides of neck more faintly streaked dark grey; mantle, back, scapulars and uppertail-coverts mottled blackish, feather centres with blackish brown triangular marks; rump uniform pink-cinnamon to greyish sandy buff, sometimes faintly streaked dark grey. Tail-feathers blackish brown or black, outer pair white or buffy except for basal half of inner webs, next pair with white outer webs, central pair grey-brown with broad rufous-buff margins and blackish shafts, rest of feathers with narrow rufous-buff margins. Lores, narrow supercilium, area in front of eye and narrow eye-ring buffy white. Ear-coverts brown streaked buffy; dark line from gape below eye to ear-coverts; line of dark spots along sides of chin to lower ear-coverts. Chin and throat white to pale pinkish buff, often with faint grey streaks. Breast and flanks cinnamon-buff or pale buff, breast with band of distinct blackish streaks, contrasting with white of throat, streaks edged rufous-cinnamon on lower breast, olive-brown on flanks; belly and undertail-coverts cream or white. Flight feathers greyish black, outer primary edged white, rest pale buff, margins indistinct on inner webs and on tips of secondaries, tertials olive-brown with broad cinnamon or buff margins. Upperwing-coverts dark olive-brown marginated pale cinnamon or buff, especially on tips, which form 2 pale bars across wing. Underwing-coverts and axillaries off-white to pale buff. Bill horn-grey, upper mandible darker, base of lower mandible yellowish; eye brown; legs and feet flesh-brown or yellowish brown. Sexes alike. SIZE: wing, ♂ (n = 18) 87–93 (91·1), ♀ (n = 13) 82–89 (84·9); tail, ♂ (n = 16) 50–56 (53·2), ♀ (n = 12) 46–53 (49·2); bill, ♂ (n = 14) 11·2–13·2 (12·3), ♀ (n = 12) 10·1–13 (12·0); tarsus, ♂ (n = 15) 19·5–21·3 (20·4), ♀ (n = 11) 18·4–20·7 (19·7). WEIGHT: unsexed (n = 5) 19–21 (19·8), (n = 7) 20–25.

IMMATURE: feathers of upperparts, including wing-coverts and tail-feathers (except outer pair) with pink-buff, cream or white margins, longer feathers submarginally bordered black; breast with small and indistinct grey spots. Outer primary longer and more rounded at tip than adult, tipped buffy. Bill horn, tinged flesh.

NESTLING: pale buff or pale yellow-brown down on upperparts. Inside of mouth flesh, 1 black spot either side of tongue and 1 at tip, 1 black spot on both mandibles, gape yellow.

C. r. nicolli Hartert: Egypt (Nile delta). Upperparts darker than *minor*, greyish cinnamon-brown, with broad and heavy black streaks. Chest pale buff, heavily marked.

C. r. heinei (Homeyer): NE Rumania, Ukraine, N Caucasus and Turkmeniya east to NW Altai Mts and Kirgiziya, N Afghanistan. Winters Iraq, Iran, Egypt. Similar to *minor*, but upperparts pale olive-grey, slightly tinged cinnamon when fresh, rump sandy grey, streaking slightly heavier and more distinct, although narrower than *nicolli*.

Field Characters. A small lark with short, pale bill, streaked upperparts, breast and sometimes flanks, dark tail with white outer feathers. Very similar to Greater Short-toed Lark *C. brachydactyla*; colour of upperparts varies locally in both species so no help in identification, except that Lesser never has rufous on crown. Lesser distinguished by streaked breast, narrower supercilium (less capped effect), shorter tertials (not reaching wing-tip), but best distinction is voice, q.v.

Voice. Tape-recorded (62, 73 – 90, ROC). Song a continuous jumble of rattles, trills, churrs, warbles, whistles, squeaks and buzzy and grating notes, much of it imitations of other species, including songs of Greater Short-toed Lark and Eurasian Skylark *Alauda arvensis*. Pauses are irregular and brief, material between pauses varying from a single call note to lengthy trilling, quite different from regular, repetitive format of Greater Short-toed Lark. Performance has bustling, insistent, effervescent quality. Renderings of contact-alarm calls include 'prrrt', 'tchchrrr', 'chirrick' and 'chit-it-it-it'. For more details of songs and calls, see Cramp (1988).

General Habits. Occurs in Mediterranean and semi-desert areas on sparsely vegetated dry saline clays and gravels, and in grasslands on steppes and stony plains. Winters in grasslands (Nigeria).

Gregarious when not breeding, often in large flocks, sometimes numbering thousands. Forages on ground, taking food items from soil surface and low plants, and occasionally flying prey. Drinks dew from grass; can go without drinking water for months, but will drink available water, even if brackish or saline.

Migrates south Sept–Nov; large movements recorded Morocco (Tiznit) Nov; reports may be confused by local movements of post-breeding flocks. Recorded Mauritania (Banc d'Arguin) moving south in flocks of up to 15 in Sept and Nov, single birds Dec; several birds near Jos, Nigeria, Dec (Elgood 1982). Moves north Mar in Morocco and NW Algeria, Apr in Mauritania (Nouadhibou).

Food. Invertebrates: damselflies Odonata, grasshoppers, termites, bugs (Scutelleridae, Lygaeidae, Cicadidae), larval Lepidoptera (Pyralidae, Noctuidae, Geometridae), flies, including Tabanidae, ants (Formicidae and other

Hymenoptera), adult and larval beetles (Cicindelidae, Carabidae, Histeridae, Geotrupidae, Buprestidae, Tenebrionidae, Chrysomelidae, Curculionidae), spiders (Thomisidae, Lycosidae, Gnaphosidae); also seeds (Chenopodiaceae, Amaranthaceae, Cruciferae, Rosaceae, Malvaceae, Zygophyllaceae, Lythraceae, Boraginaceae, grasses Gramineae, cereal grains). Food intake up to 6·8 g dry weight of caterpillars/day.

Breeding Habits. Monogamous, territorial. Territory size varies widely according to quality of territory. In display, ♂ sings mostly in flight but also from low perches. In song-flight bird ascends steeply and rather jerkily with abrupt beating of wings and spreading of tail, usually giving series of rhythmless calls, followed by continuous and varied song with many imitations of other birds' calls; at 20–30 m, bird heads into light wind and flutters in undulations while gaining height, then glides down with shallow, shivering wing-beats, singing; repeats this several times while moving forward, then descends; alternate ascending and gliding similar to display of Greater Short-toed Lark. Known also to display on ground (**A**) (I. Willis, pers. comm.).

NEST: cup of dry vegetation in scrape in ground in shelter of tuft of grass; built by ♀ only.

EGGS: 2–5, mean (84 clutches) 3·6. Sub-elliptical; variably coloured whitish, yellowish or buff, more or less spotted and blotched with dark brown. SIZE: (n = 80, *C. r. heinei*) 17–21·3 × 13–15·7 (19·1 × 14·7).
LAYING DATES: Algeria and Tunisia, Apr–June.
INCUBATION: by ♀ only, or with ♂ assisting.
DEVELOPMENT AND CARE OF YOUNG: young fed by both parents; young leave nest at 9 days.

Reference
Cramp, S. (1988).

Plate 4
(Opp. p. 49)

Calandrella somalica (Sharpe). Somali Short-toed Lark. Alouette roussâtre.

Alauda somalica Sharpe, 1895. Proc. Zool. Soc. Lond., p. 472; Haud, Somaliland.

Forms a superspecies with *C. rufescens* and Asian *C. raytal*.

Range and Status. Endemic resident. 3 disjunct populations: (1) N Somalia west of 49°E and north of 8°N, and adjacent E Ethiopia (Haud); (2) S Ethiopia (Mega, Yavello, Negelli); (3) Kenya/Tanzania from Naivasha and Nairobi to Crater Highlands, Arusha and Tsavo Nat. Parks, and sight record from Samburu.

Description. *C. s. perconfusa* (see White 1960b): central and W plateau of N Somalia. ADULT ♂: feathers of top of head and upperparts, including scapulars and upperwing-coverts, dark brown broadly fringed pale sandy to reddish buff. Tail-feathers blackish brown with narrow buff edges and tips, outer pair with outer web and adjacent portion of inner web buff, next pair with buff confined to outer web. Pre-orbital region, broad eye-ring and supercilium whitish to buff, cheeks and ear-coverts brown, darker towards rear, with narrow pale streaks, bordered behind by pale buff half-collar; line of dark streaks down side of neck. Breast pale brown to buff with dark streaks which sometimes coalesce at sides of upper breast to form blackish patches; rest of underparts whitish with variable amount of buff on flanks. Primaries and secondaries dark brown with narrow pale sandy or buff edges and tips, outer primary with outer web entirely buff; tertials long, richer dark brown with much broader pale fringes; axillaries and underwing-coverts pale brown to reddish buff. Bill, upper mandible horn or pale horn, lower mandible whitish flesh, overall

Calandrella somalica

'somewhat reddish' (Archer and Godman 1961); eye dark brown; legs and feet flesh-white or greyish flesh. Sexes alike. SIZE: (11 ♂♂, 11 ♀♀) wing, ♂ 83–90 (85·7); ♀ 78–91 (85·2); tail, ♂ 43–50 (47·7), ♀ 41–51 (45·9); bill, ♂ 13–15 (13·8), ♀ 12·5–14·5 (13·5); tarsus, ♂ 19·5–21 (20·5), ♀ 18·5–21·5 (20·3).

IMMATURE and NESTLING: unknown.

C. s. somalica (Sharpe) (including '*vulpecula*'): area of red sands from Burao, N Somalia, to Haud, E Ethiopia. Edges to feathers of upperparts, wings and tail bright vinous red, entire underparts washed bright pink.

C. s. megaensis Benson: S Ethiopia. More heavily streaked above than *perconfusa*, feathers edged reddish, red-brown wash on breast and flanks.

C. s. athensis (Sharpe): Kenya/Tanzania. Upperparts with heavy dark streaks, feathers edged whitish or pale sandy, underparts white with pale brown wash on breast and flanks.

TAXONOMIC NOTE: we follow Hall and Moreau (1970) in treating Afrotropical populations as a full species, distinct from *C. rufescens*.

Field Characters. A smallish lark that travels in flocks when not breeding. Upperparts reddish to brown or sandy with moderate dark streaking (Ethiopia/Somalia) or pale brown overlaid with heavy dark streaks (Kenya/Tanzania), some buff on outer tail-feathers, no red in wings. Similar-sized White-tailed Bush-Lark *Mirafra albicauda* and Singing Bush-Lark *M. cantillans* have similar streaked upperparts and breast but red patches in wings and much white in tail. Larger Gillett's Lark *M. gilletti* in Somalia has no red in wing or white in tail but is more uniform red-brown above, less streaked, with breast streaks red-brown, not black.

Voice. Not tape-recorded. Call, low-pitched 'piri-pip'; flight song in breeding season.

General Habits. Inhabits open grassy plains, acacia short-grass savanna, tall grass in glades; in E Africa, 1200–1650 m. In non-breeding season forms large flocks which fly fast and low over the ground; fairly tame then, but single birds in breeding season rather shy. Said to be sedentary in Somalia (Archer and Godman 1961) but evidently a wanderer in Kenya/Tanzania; one caught in nets at night at Ngulia (Kenya), Dec.

Food. Unknown.

Breeding Habits. Territorial. Flight song in breeding season.

NEST: deep depression in ground forming cup, lined with grass, placed against grass tuft out in open plain.

EGGS: 4–5 (*perconfusa*), 2–3 (*athensis*). Ivory-white, sometimes pale olive-brown, whole surface closely speckled with dark brown, olive-brown or ochreous over patches of ash-grey, somewhat more heavily at large end. SIZE: av. 19·5 × 13·7 (*perconfusa*), 18 × 14·5 (*athensis*).

LAYING DATES: Somalia, June; E Africa: Region D, Mar–June, during long rains (gonadal data Dec).

Genus *Spizocorys* Sundevall

Small, gregarious larks of open, well-grazed grasslands, arid dwarf shrublands and gravel plains. Plumage streaked on crown and back; wings relatively long, outer primary less reduced than *Calandrella*; bill short, strong. Nest a simple cup, without dome; often with rim extended out over substrate.

5 species, all endemic to Africa.

Spizocorys conirostris (Sundevall). Pink-billed Lark. Alouette à bec rose.

Alauda conirostris Sundevall, 1850. Oefv. K. Sv. Vet.-Akad. Förh. 7, p. 99; Vetchkop, Orange Free State.

Plate 6 (Opp. p. 97)

Range and Status. Endemic resident. W Zambia (common but local); N and SE Namibia, south to Kalkrand, south and east to Kalahari Gemsbok Nat. Park (locally common nomad); NE, E and S Botswana (locally common); South Africa from S Transvaal (locally common nomad) to NW Natal (uncommon and local), central and N Orange Free State (very local, uncommon to rare), and N, NE and E Cape (uncommon and local).

Description. *S. c. conirostris* (Sundevall): South Africa (interior Natal, S Transvaal, E Orange Free State). ADULT ♂: crown, nape, hindneck, mantle and back rufous-brown, feathers darker in centre, giving a streaked effect. Rump brown, less streaked than back, feather centres darker, uppertail-coverts buffy brown, less streaked than rump. Tail-feathers brown, outer webs of outer pair buff, adjacent pair with narrow buff outer edges. Sides of neck brown, foreneck brown with darker markings; lores and supercilium buffy white, dark line from bill through eye onto ear-coverts, white crescent below eye, under which is thin black line; thin black malar streak, ear-coverts brown becoming paler towards eye. Chin and throat white, breast to undertail-coverts rufous-buff, breast with teardrop-shaped dark brown spots. Flight feathers brown, outer primary with whitish buff outer edge, secondaries with narrow buff outer edges, all with inner webs edged buff. Scapulars, tertials and upperwing-coverts dark brown edged buff; underwing-coverts, axillaries and edges of inner webs of underside of flight feathers rufous-buff. Bill pink; eye pinkish to yellowish brown; legs and feet pink. Sexes alike. SIZE: (18 ♂♂, 17 ♀♀) wing, ♂ 72–80 (76·1), ♀ 70–77 (72·5); tail, ♂ (n = 15) 38–45 (41·9), ♀ 35–42 (38·6); bill, ♂ (n = 17) 11–13·6 (12·2), ♀ 11–12·8 (12·1); tarsus, ♂ 16–19 (17·9), ♀ 16–20 (18·2). WEIGHT: ♂ (n = 11) 12·5–15·5 (14·0), ♀ (n = 13) 12–16·5 (14·6).

IMMATURE: darker than adult, speckled above and below, crown feathers tipped buffy, back feathers tipped whitish. Flight feathers and tail edged buffy. Wing-coverts dark brown, edged light brown, tipped white. Feathers of chest tipped white. Bill blackish.

NESTLING: blackish above, pink below, with down on crown, back, rump, scapulars and wings. Gape pale slate, inside of mouth yellow, with 2 spots at tips of upper and lower mandibles and 3 spots on anterior tongue, 1–2 spots at tip, 2 laterally about halfway along.

S. c. transiens (Clancey): South Africa (E and N Cape Province, east to E Transvaal). Upperparts less heavily streaked.

S. c. barlowi Roberts: South Africa (NW Cape Province), S Botswana (Kakia, Tsabong), S Namibia (north to Kalkrand). Similar to *transiens* but paler, underparts much paler, belly whitish.

S. c. damarensis Roberts: NW Namibia (Swakop R. to Ovamboland). Paler than *barlowi*, especially below; less streaked.

S. c. crypta (Irwin): NE Botswana (Makgadikgadi Pan, L. Dow). Upperparts with greyish feather edges.

S. c. makawai (Traylor): Zambia (Liuwa and Mutala Plains). Blackish brown above, feathers edged pinkish buff; underparts pinkish chestnut.

S. c. harti (Benson): Zambia (Matabele Plain). Paler above and below than *makawai*, feathers above edged pale grey.

Spizocorys conirostris

Field Characters. A small lark, widespread in the dry interior, with dark, streaked upperparts, pinkish underparts, and short, conical pink bill. Stark's Lark *Eremalauda starki* is much paler, with white underparts and a crest; similar to Sclater's Lark *S. sclateri* and colour of underparts variable so no help in identification; best distinguished by less heavily-marked face, without dark teardrop below eye characteristic of Sclater's, and shorter, all-pink bill without dark tip. Similar to Botha's Lark *S. fringillaris*, q.v. for differences.

Voice. Tape-recorded (75, F, GIB, LEM). Flight call a musical trill or chirp of 2–3 syllables 'see-see-see' or 'tri-tri-tri', repeated often (Maclean 1985). Alarm call similar, but accented on the first syllable. Birds on the ground call a faint, insect-like 'trrr-krik-krik'.

General Habits. Occurs in open, short grassland, in croplands immediately after the harvest and in fallow fields, also on burnt ground. In Kalahari found on dunes with fairly dense grass cover.

When not breeding occurs in groups of 5–20 or more. Shy and not easy to see. Forages by walking or running in short bursts on open ground. Drinks water regularly.

Food. Seeds and invertebrates.

Breeding Habits. Monogamous. Flight song in display similar to the call.

NEST: cup of grass, ext. diam. 80–90, diam. of cup 50–60, 35–40 deep; lined with rootlets and hair, sometimes placed on slight foundation of sticks, placed in scrape in ground against tuft of grass or forb, or sometimes out in the open; rim of cup expanded outwards to form apron on exposed side. Cup of nest constructed first, then apron, then lining added. Pieces of grass up to 100 mm long used in construction; nests contain 700–800 pieces of material. In Kalahari most nests faced east, southeast or northeast, in E Transvaal most faced east or southeast.

EGGS: 2–3, mean (68 clutches) 2·1. Ovate; white, usually densely speckled with grey-brown or rusty brown over slate, markings sometimes concentrated at obtuse end. SIZE: (n = 32) 16·1–20·6 × 12·7–14·4 (18·2 × 13·6).

LAYING DATES: South Africa: Transvaal, Nov, Jan, Apr–May; Natal, Feb; Orange Free State, Sept, Nov–May; N Cape, Feb, Apr–Sept.

INCUBATION: begins with first egg; period 12 ± 1 days.

DEVELOPMENT AND CARE OF YOUNG: young fed by both adults; leave nest when *c.* 10 days old, before able to fly.

BREEDING SUCCESS/SURVIVAL: of 75 eggs laid in 37 nests, 36 young hatched, 11 left nest (Maclean 1970). Nests in open sites less successful than those in concealed sites (Maclean 1970).

Reference
Maclean, G. L. (1970).

Spizocorys fringillaris (Sundevall). Botha's Lark. Alouette de Botha.

Alauda fringillaris Sundevall, 1850. Oefv. K. Sv. Vet.-Akad. Förh. 7, p. 99; Leeuspruit, Vredefort, Orange Free State.

Plate 6
(Opp. p. 97)

Range and Status. Resident, endemic to South Africa, scattered localities in SE Transvaal (area bordered by Hendrina-Wakkerstroom-Volksrust-Standerton) and Orange Free State (west of Wesselbron, Allemanskraal Dam, Warden, Verkykerskop). Uncommon and local, but possibly not as rare as previously supposed; minimum population probably 1000, possibly as many as 20,000 (Allan *et al.* 1983). No evidence of decrease in population, but is no longer found at type-locality or vicinity.

Description. ADULT ♂: crown, nape, hindneck, mantle and back brown, streaked blackish. Sides of neck buffy brown, foreneck paler brown; narrow loral line buff, narrow dark line from bill to eye, pale patch above and behind eye, narrow dark malar streak; chin and throat greyish white. Rump greyish brown, feathers with dark centres, uppertail-coverts brown, unstreaked. Tail-feathers blackish brown, outer pair with broad white outer edges, next pair similar, but with dark tips. Breast and flanks heavily streaked blackish brown, breast, upper belly and flanks rufous-buff, belly and undertail-coverts pale buff to white. Flight feathers dark brown, primaries with narrow buff outer edge, secondaries similar but with paler tips. Scapulars dark brown edged paler brown; tertials dark brown; wing-coverts brown, shafts darker. Underwing-coverts and axillaries rufous-buff. Bill pink; eye brown; legs and feet pink. Sexes alike. SIZE: (5 ♂♂, 4 ♀♀) wing, ♂ 77–83 (79·6), ♀ 76–83 (79·0); tail, ♂ 38–42 (40·0), ♀ 40–42 (41·0); bill, ♂ 12–14·7 (13·5), ♀ 12–14 (12·8); tarsus, ♂ 19–22 (19·8), ♀ 19–20 (19·7). WEIGHT: ♂ (n = 6) 15·7–19·2 (17·9), ♀ (n = 3) 20·2–21 (20·5).

IMMATURE: crown and back feathers tipped whitish, flight feathers and tail marginated buffy; underparts lack blackish streaking, breast spotted with rufous-brown. Bill horn; legs and feet pink.

NESTLING: skin grey with long buff down on crown, back, rump and scapulars. Inside of mouth deep yellow, with 3 black spots on tongue, 1 apical and 2 lateral. Bill horn with creamy white egg tooth and gape; legs and feet pink.

Field Characters. An uncommon and local lark with a very restricted range. Small and dark, heavily streaked above, below streaked on flanks as well as breast. Very similar to Pink-billed Lark *S. conirostris* with which it is found; differs in more slender, dark-tipped bill, white belly contrasting with buff breast and flanks (underparts uniform pinkish in Pink-billed), streaked flanks and more white in tail.

Voice. Tape-recorded (88, F, ALDG, GIB, LEM). A melodius two-syllabled 'tcheree' repeated several times, both when perched and in flight (Allan *et al.* 1983), and a sharp 'chuk' repeated either rapidly or sporadically, only given in flight, probably a contact call.

General Habits. Patchily distributed in well-grazed grasslands on plateaux and uplands, where grass short and well-drained. Avoids the longer grass in valley bottoms, poorly drained areas, planted pastures, croplands and rocky areas.

Usually solitary, in pairs or small groups of 3–6. Forages by walking about briskly, looking down, pausing at intervals to look around; sometimes darts after prey or pursues prey into air. Does not dig for food. Dependent on surface water and drinks regularly.

Food. Invertebrates (beetles, moths) and seeds.

Breeding Habits. Monogamous. Has no flight display.
NEST: cup of dry grass blades, ext. diam. 78–100, diam. of cup 51–63, depth 31–45 (n = 18); lined with fine grass and occasionally sheep's wool and hair; placed in excavation in ground between grass tufts.

EGGS: 2–3, mean (8 clutches) 2·25. Ovate; cream, heavily marked with fine speckles of dark brown and grey-brown, often concentrated on obtuse end. SIZE: (n = 12) 17·7–19·7 × 12·6–14·5 (18·6 × 13·6).

LAYING DATES: SE Transvaal, Nov–Dec; Orange Free State, Dec.

DEVELOPMENT AND CARE OF YOUNG: young fed by both adults.

BREEDING SUCCESS/SURVIVAL: of 17 eggs laid in 8 nests, 9 young were fledged; no data on subsequent survival.

Reference
Allan, D. G. *et al.* (1983).

Spizocorys sclateri (Shelley). Sclater's Lark. Alouette de Sclater.

Calandrella sclateri Shelley, 1902. Bds Afr., 3, p. 136; Hauntop R., Maltahohe, Great Namaqualand.

Range and Status. Resident, endemic to S Namibia, (Maltahohe, Kalkrand) and Cape Province, South Africa (Bushmanland south to Kliprand, and Great Karoo east to Hopetown, Philipstown, Murraysburg, south to Rietbron).

Description. ADULT ♂: crown, nape, hindneck, mantle, back and rump brownish buff, streaked darker brown, feathers with dark centres and buff-brown edges; uppertail-coverts light brown, indistinctly streaked darker. Tail brown, outer web of outer 2, sometimes 3 feathers whitish. Sides of neck brownish, foreneck greyish, speckled dark brown, lores whitish, indistinct buff supercilium, dark line from base of bill through eye and around ear-coverts; white crescent below eye, under which is dark teardrop-shaped mark; pale patch between this mark and dark brown ear-coverts. Chin and throat white with a few brown speckles, rest of underparts rufous-buff, somewhat paler on upper breast and undertail-coverts; breast streaked brown. Flight feathers brown, outer primary with whitish buff outer edge, secondaries with pale tips; scapulars brown with rufous-buff inner edge, tertials dark brown, outer webs paler, wing-coverts brown with buff fringes. Underwing-coverts and axillaries rufous-buff. Bill brownish pink to brownish horn; eye brown; legs and feet light brown. Sexes alike. SIZE: (10 ♂♂, 12 ♀♀) wing, ♂ 76–87 (82·9), ♀ 76–84 (80·3); tail, ♂ 40–45 (42·5), ♀ 39–45 (42·1); bill, ♂ 13–15·5 (14·2), ♀ 13·2–15·8 (14·4); tarsus, ♂ 15–18 (17·0), ♀ 14–18 (16·6). WEIGHT: 1 ♂ 21, 1 unsexed 17.

IMMATURE: feathers of crown, mantle and back tipped whitish; mantle appears lightly barred. Wing-coverts and tertials tipped whitish, tertials also marginated buff. Primaries and secondaries marginated buff, tipped white. Facial markings less distinct. Breast and belly feathers tipped whitish. Tail-feathers tipped buff.

NESTLING: skin bright pink, darker on feather tracts; long pale buff down on crown, back, wings and rump; inside of mouth orange-yellow, tongue spots present.

TAXONOMIC NOTE: *Spizocorys sclateri* is regarded here as not subspecifically divisible; we do not recognize races *capensis* (Ogilvie-Grant) and *theresae* (Meinertzhagen) (see Dean and Colston 1988).

Field Characters. An uncommon and local lark of the Karoo and interior dry country. Told from Stark's Lark *Eremalauda starki* by darker upperparts, pinkish underparts, lack of crest and darker bill; does not overlap with Botha's Lark *S. fringillaris*. Similar to Pink-billed Lark *S. conirostris* but with proportionately larger head and longer, wedge-shaped bill, giving big-headed appearance; bill brownish, not pink. Bold face pattern with dark 'teardrop' below eye distinctive.

Voice. Tape-recorded (F). Calls include 'peew', 'trew', 'tew', 'trit, 'tchweet', 'trit-trit' and 'trit-tew-tew-trit' given in flight or on perch. Chirps and twitters while foraging.

General Habits. Found in arid stony plains with scattered grasses and poorly drained sparse dwarf shrublands on clays. Commonest in grassy N Karoo and Bushmanland.

When not breeding, occurs in groups of up to 25. Nomadic and opportunistic, moving into areas where rain has fallen, then moving on after breeding. Drinks water regularly (Steyn and Myburgh 1989).

Food. Invertebrates: lepidopteran larvae, weevils, small beetles, ants (*Messor capensis*, *Monomorium* spp.); seeds of grasses (*Brachiaria glomerata*, *Schmidtia kalahariensis*, *Stipagrostis*), shrubs (Leguminosae) and forbs (*Polygonum*). All stomachs examined (n = 7) contained pieces of gravel up to 4 mm diam.

Breeding Habits. Monogamous. Birds arrive at breeding area in small, loose groups, pairs form within group.

NEST: cup of leaves, stems and feathery awns of grasses *Stipagrostis obtusa* or *S. ciliata*; cup diam. 50–60, depth 25–30; in scrape *c*. 80 diam., 30–40 deep, excavated by bird using its feet and scattering soil by wing-fluttering; in hard stony soil next to stone or tuft, or in open gravel patch. Several scrapes made before one is selected for nest. Scrape lined with small pebbles before material placed in it; cup often ringed by 'apron' of grass or small pebbles.

EGGS: 1 only (20 clutches). White or off-white tinged brownish, finely spotted with greyish brown, or speckled with brown and a few brown blotches over underlying slate speckles, markings usually concentrated on obtuse end. SIZE: (n = 3) 20–22 × 13·9–15 (20·0 × 14·4).

LAYING DATES: N Cape, May, Aug, Sept, Nov. Breeds opportunistically, when conditions suitable.

INCUBATION: in short spells (*c.* 15 min) by both sexes; period not less than 13 days.

DEVELOPMENT AND CARE OF YOUNG: newly hatched young closely brooded; fed by both parents.

BREEDING SUCCESS/SURVIVAL: nests sometimes swamped by sudden rainfall (Steyn and Myburgh 1989).

References
Dean, W. R. J. and Colston, P. R. (1988).
Hockey, P. A. R. and Sinclair, J. C. (1981).
Steyn, P. and Myburgh, N. (1989).

Spizocorys obbiensis Witherby. Obbia Lark. Alouette d'Obbia.

Spizocorys obbiensis Witherby, 1905. Ibis, p. 514; Obbia, Somaliland.

Plate 4
(Opp. p. 49)

Range and Status. Resident, endemic to narrow strip of coastal Somalia from Obbia to Hal Hambo, 30 km south of Mogadishu, 570 km long and max. 16 km wide, total area 1200–1500 km^2; generally abundant; not endangered (Ash 1981).

Description. ADULT ♂: crown light grey-brown to buff-brown, feather centres dark brown, nape and hindneck similar but feathers edged whiter; sides of neck grey-brown, feathers with dark centres giving mottled appearance, foreneck whitish; cheeks whitish with dark moustachial streak and another dark streak below eye; dark line from bill through eye; white mark above and behind eye. Chin and throat white. Mantle buffy brown, feathers with dark centres, back similar but feathers with dark central spot shading to lighter brown and buff edges. Rump and uppertail-coverts pale buff-brown, feathers with dark centres. Tail dark brown, outer feathers with narrow white outer edge, adjacent pair similar but white edge narrower, central pair with greyish buff margins. Rest of underparts white or off-white, breast to upper belly and flanks streaked dark brown, sides of breast and flanks with brown tinge. Flight feathers brown, outer primary with narrow whitish outer edge, scapulars and upperwing-coverts brown, edged pale buffy brown, tertials brown, marginated pale buffy brown to off-white. Underwing-coverts and axillaries pale buffy brown with pinkish tinge. Bill pinkish brown, lower mandible paler and more yellowish; eye brown; legs and feet pink to pale brown. Sexes alike. SIZE: (5 ♂♂, 1 ♀) wing, ♂ 69·5–70 (69·9), ♀ 64; tail, ♂ 37–39 (37·9), ♀ 35·7; bill, ♂ 13·5–15·5 (14·2), ♀ 13·2; tarsus, ♂ 19·1–19·6 (19·4), ♀ 17·2. WEIGHT: ♂ (n = 5) 13·7–15·6 (14·8), 1 ♀ 12·3, 1 unsexed 14·7.

IMMATURE and NESTLING: not known.

Field Characters. A small, short-tailed greyish lark, heavily streaked on breast and flanks. Bill fairly stout, legs pink. When foraging has distinctive 'hunched' posture. Unlike any other lark in its very limited range.

Voice. Not tape-recorded. Flight call, 'tip-tip'.

General Habits. Occurs on sand dunes and plains with patches of low, heavily grazed halophytic scrub and forbs, usually close to the sea. Restless and very active; forages briefly in one place, then flies off 100 m or more to another. Runs about actively and clambers over low vegetation while foraging. In pairs or small parties, occasionally in flocks as large as 30.

Food. Unknown.

Breeding Habits.

NEST: a cup, thin- to thick-walled, of dry rootlets, dry vegetation, woolly seeds, dead leaves, sometimes pieces of string, in scrape or depression in sand, ext. diam. 80–90, int. cup diam. 50–60, depth of cup 30–50. In open site next to forb, grass or sedge clump, or beneath sparse cover of creeping *Indigofera* or *Zygophyllum*, on dunes close to sea.

EGGS: 2–3, mean (6 clutches) 2·1. Blunt pyriform; greyish or creamy white, variably speckled with brown and grey, with occasional larger speckles or blotches and hairlines of brown, markings sometimes concentrated on broad end; some eggs erythristic. SIZE: (n = 6) 17·5–19·2 × 13–14 (18·2 × 13·5).

LAYING DATES: Somalia, May–July, possibly also Nov–Dec (gonadal data).

Reference
Ash, J. S. (1981).

Plate 4
(Opp. p. 49)

Spizocorys personata Sharpe. Masked Lark. Alouette masquée.

Spizocorys personata Sharpe, 1895. Proc. Zool. Soc. Lond., p. 471; Milmil, Ogaden.

Range and Status. Endemic resident. S and E Ethiopia (Yavello, Ogaden), N-central Kenya. Locally common.

Description. *S. p. personata* Sharpe: E Ethiopia. ADULT ♂: crown, nape and hindneck grey-brown with a few inconspicuous streaks and spots of darker brown giving slightly scaly appearance; sides of lower neck whitish, upper neck grey-brown, foreneck whitish. Black 'mask' from around eye to base of bill and down across cheeks to form broad malar stripe, which at base extends out into throat, forming a sort of inverted 'T'. Ear-coverts grey-brown, separated from mask by small pale patch. Chin and throat white. Mantle grey-brown, back greyish brown with indistinct dark streaks, rump and uppertail-coverts greyish brown. Tail-feathers dark brown, outer pair rich buff with narrow dark brown inner edges, next pair with narrow pale rufous outer edges, central pair rather paler brown. Breast greyish brown, shading to more rufous-brown on belly, undertail-coverts buff. Flight feathers brown, outer primary with buff outer web. Scapulars and upperwing-coverts dark brown with pale fringes, tertials dark brown edged buff. Axillaries and underwing-coverts grey-brown; carpal region pale brownish pink. Bill pale yellowish horn, lower mandible paler; eye dark brown to hazel; legs and feet flesh white. Sexes alike. SIZE: (2 ♂♂, 2 ♀♀) wing, ♂ 84, 87, ♀ 80, 83; tail, ♂ 48, 49, ♀ 43, 44; bill, ♂ 14·6, 16·3, ♀ (n = 1) 16; tarsus, ♂ 20, 21, ♀ 20, 21. WEIGHT: 1 unsexed 19·7.
   IMMATURE and NESTLING: unknown.
   S. p. yavelloensis Benson: S Ethiopia, N Kenya (Didi Galgalla Desert). Upperparts dark and greyish, breast dark and greyish, contrasting with rufous belly. SIZE: (5 ♂♂) wing, 80–87 (83·8); tail, 48–51 (49·6); bill, 13·5–15·5 (14·5); tarsus, 19–20 (19·7).
   S. p. intensa (Rothschild): N-central Kenya (Isiolo, Chanler's Falls). Upperparts rich brown, breast darker and greyer, belly deeper reddish. SIZE: (3 ♂♂) wing, 80–85 (83·3); tail, 48–51 (49·9); bill, 14–16 (14·7); tarsus, 19–19·5 (19·3).
   S. p. mcchesneyi (Williams): N Kenya (Marsabit Plateau). Upperparts darker than *personata*.

Field Characters. A rather dark, dull grey-brown desert lark with a large yellowish pink bill and diagnostic black 'mask' on face and throat. Plumage colour varies somewhat with race, but cinnamon belly always contrasts with grey breast. Outer tail-feathers pale buff.

Voice. Unknown.

Spizocorys personata

General Habits. In Kenya, occurs in dwarf shrub grasslands with sparse cover and scattered bushes on black cotton and black lava soils, and in black lava desert; in SE Ethiopia noted on bare ground covered with small lava boulders, the only vegetation being scanty short grass and a few scattered thorn bushes (Benson 1946).

Food. Invertebrates (grasshopper egg cases), grass seeds, bulbs and corms.

Breeding Habits. Unknown.
   LAYING DATES: E Africa: Region D, May; Ethiopia (breeding condition July, including ♀ with egg yolking in ovary).

Reference
Benson, C. W. (1946).

Genus *Eremalauda* Sclater

Small larks of arid and semi-arid shrublands and grasslands. Plumage buffy to pale brown, streaked above, whitish below, lightly or more heavily streaked on breast. Face boldly patterned in *dunni*, less so in *starki*; both have white eye-ring. Bill short and stout. Nest an open cup; lacks either apron or ramp.
   2 species, Africa and Asia, 1 endemic to Africa.

Eremalauda dunni (Shelley). Dunn's Lark. Alouette de Dunn.

Calendula dunni Shelley, 1904. Bull. Br. Orn. Club 14, p. 82; Ogageh Wells, Kordofan.

Range and Status. Africa, Israel, Lebanon, Jordan, central and S Arabia.

Resident and nomadic. Fragmented distribution along S edge of Sahara from Mauritania (widespread but irregular, common to uncommon), Mali (between 14° and 18°N; Tombouctou, In Alahy, Taberreshat; numbers fluctuate locally), Niger (Takouhout, Tanout, Marandet, Lagane), Chad, W and central Sudan (uncommon).

Description. *E. d. dunni* (Shelley): only race in Africa. ADULT ♂: upperparts sandy to pink-isabelline streaked red-brown, more heavily on crown than mantle; collar around hind-neck paler, with little streaking. Folded wing, tertials, scapulars, greater coverts and bases of flight feathers darker, with rufous tone. Area between bill and eye whitish, leading to broad pale eye-ring and short pale streak behind eye; ear-coverts pale sandy. Tail red and black; 2 central pairs of feathers pale red-brown, rest black, outer webs and tips of outer pair reddish buff, rest variably tipped buff according to wear. Underparts white to cream, sides of breast pale cinnamon, unstreaked. Primaries and secondaries rufous-brown, outer webs cinnamon, tips of outer primaries dusky. Underwings pale cinnamon. Bill dull yellowish white tinged brown on tip; eye light brown; legs and feet pale flesh, pink-white or sandy white. Sexes alike. SIZE: wing, ♂ (n = 13) 82–87 (83·7), ♀ (n = 4) 76–81 (78·6); tail, ♂ (n = 10) 50–55, ♀ (n = 5) 48–51; bill, ♂ (n = 12) 13·8–15·8 (14·8), ♀ (n = 4) 13·8–14·6 (14·2); tarsus, (10 ♂♂, 5 ♀♀) 19–20.

IMMATURE: similar to adult but darker, feathers on crown, mantle and scapulars tipped white, those of hindneck and back to uppertail-coverts narrowly edged off-white; flight feathers plain rufous-cinnamon, without black on tips of primaries.

NESTLING: unknown.

Field Characters. A small lark, overall very similar to larger Desert Lark *Ammomanes deserti* and same-sized Bar-tailed Lark *A. cincturus*, with both of which it occurs. Differs from both by streaked crown and upperparts, larger head and stouter bill giving big-headed appearance, broad, rounded wings and broad tail. Tail pattern also different, black with rufous centre rather than rufous with dark tip (Bar-tailed) or rufous with dark centre (Desert). For differences from Kordofan Lark *Mirafra cordofanica*, see that species.

Voice. Tape-recorded (78). Flight song variable, described as 'a scratchy warbling, interspersed with melodious but melancholy whistling . . .', or 'series of short, sweet, rambling phrases . . .', 'chee chee cheeree cheer-eer cheeree cheee' or 'pee-pee-pee-pooo' (Cramp 1988). Song from ground, 'a series of quick rambling notes, each phrase lasting 2–4 s, and a shorter phrase ending with a long "cheee"'. Song from ground taped by Chappuis, 'chit-tit-woodly-widdly-wit'. Contact calls include single or repeated 'ziup' or 'chiup chiup' in flight, and shrill chirp 'chrruit-chrruit'; also 'chleep' or 'cheelip', 'cheep', 'chleek' and 'cheet-cheet'. Alarm calls include loud ringing 'chee-oop', 'tu-wep', 'chup-chup-chee-oo', 'chip-chip-twee' and 'chipchipchipchipchip-chip-chip'. Fledged young have repeated, high-pitched, soft plaintive call, 'pee-pee' or 'peeuw-peeuw-peeuw'. For further details, see Cramp (1988).

General Habits. Inhabits flat arid lowlands at edge of desert and dry grassy plains (e.g. Sudan); in Mauritania, steppe with grass *Aristida plumosa*. Avoids savanna and mountainous areas.

When not breeding occurs in small groups, sometimes up to 20, often with other species of larks, especially sparrow-larks *Eremopterix*. Tame and allows close approach. Forages by picking food from substrate and digging for seeds, leaving distinctive holes in ground. Also forages at bases of grass tufts and shrubs, and jumps up to take items from foliage of plants. Pecks at animal droppings to extract seeds. Active when foraging, alternately running about and pausing to peck at food. Shelters from heat during day by keeping to shade under overhanging rocks or under bushes where roosting hollows are made; has been recorded roosting during day in lizard burrows.

Mainly sedentary, but may become nomadic during droughts. Local fluctuations in numbers occur in S Sahara.

Food. Invertebrates, grass seeds *Panicum turgidum* and other small seeds. Nestlings fed on green caterpillars.

Breeding Habits. Monogamous, territorial. In breeding season, territories are *c.* 5 ha (Mauritania). Courtship display flight performed from start of breeding season:

90 ALAUDIDAE

♂ rises into wind to a height of 30 m, sometimes 50 m or more; while singing, remains more or less stationary in air, but swings from side to side with slow wing-beats, presenting a floppy appearance. Ends flight by descending gradually on outstretched wings, or plunging to ground to chase another bird. Display flight may last 4–5 min.

NEST: scrape lined with fresh vegetation, on ground next to tuft of grass (de Naurois 1974).

EGGS: 2–3 (n = 2). Ovate; white with blackish or lavender spots and blotches.

LAYING DATES: Mauritania, Oct–Jan; Mali (young, Apr–May); Sudan, Jan–Feb.

DEVELOPMENT AND CARE OF YOUNG: adults perform rodent-run and injury-feigning distraction displays to lure intruders away from nests with young.

References
Cramp, S. (1988).
de Naurois, R. (1974).

Plate 6
(Opp. p. 97)

Eremalauda starki (Shelley). Stark's Lark. Alouette de Stark.

Calandrella starki Shelley, 1902. Bds Afr., 3, p. 135; Wilson Fountain, Great Namaqualand (= Wilsonfontein, Damaraland).

Range and Status. Endemic resident and nomad. Coastal SW Angola from Benguela south (local and uncommon); widespread in Namibia, particularly in S and SE, where locally abundant; S and SW Botswana; W Cape Province, South Africa to about Prieska in E, south to Brandvlei (locally common).

Description. ADULT ♂: crown to back streaked blackish and buff, crown feathers relatively long and erectile; sides of neck and foreneck buff, eye-ring, lores and supercilium whitish, cheeks buff grading to brown on ear-coverts; chin and throat whitish. Rump slightly pinkish buff, feathers with dark centres, uppertail-coverts plain pinkish buff. Tail-feathers brown, outer pair white with dark outer half of inner webs, adjacent pair with white outer webs, central pair paler brown with buff margins and tips. Breast brownish buff with a few dark brown speckles, shading to buff lower breast and white on belly, flanks washed brownish buff, undertail-coverts buff. Flight feathers brown with narrow buff outer edge, outer primary edged white. Scapulars and tertials brown edged buff. Underwing-coverts and axillaries off-white to rich buff. Bill whitish horn with dark tip; eye brown; legs and feet pinkish white. Sexes alike. SIZE: (8 ♂♂, 7 ♀♀) wing, ♂ 80–85 (83·0), ♀ (n = 6) 77–83 (79·6); tail, ♂ 43–48 (45·6), ♀ 40–45 (41·7); bill, ♂ 13·2–14·5 (13·5), ♀ 12·6–13·8 (13·1); tarsus, ♂ 19–22 (20·3), ♀ 17–20 (19·1). WEIGHT: ♂ (n = 40) 16–21·2 (18·5), ♀ (n = 30) 15·5–22·5 (18·6).

IMMATURE: like adult but feathers of upperparts with whitish tips, giving spotted effect.

NESTLING: mauvish pink, darker on bill tip and feather tracts, eyes purplish black; head and back with long greyish buff down. Gape pale yellowish white, inside of mouth yellow with 3 black spots on tongue, 1 either side and 1 on tip, and 1 black spot at tip of upper and lower mandibles.

TAXONOMIC NOTE: *Eremalauda starki* is regarded here as not subspecifically divisible; we do not recognize the race *gregaria* (Clancey).

Field Characters. A smallish, pale lark with streaked crown and upperparts but only lightly streaked breast, underparts mainly white. Dark-tipped pale bill looks rather heavy for size of bird; crest diagnostic when raised, but even when flat gives head peaked appearance. Sclater's Lark *Spizocorys sclateri* and Pink-billed Lark *S. conirostris* lack crest and have darker upperparts, tawny underparts; Pink-billed lacks dark tip to bill, Sclater's has bold face pattern.

Voice. Tape-recorded (F, GIB). Song a long series of soft churring chirps and whistles (Maclean 1985): 'chirr-chirr-chirr-chree-chree-chirr-chirr-chirr-chree-chree' or 'prrr-prrr-prrr-preee-preee-prrr-prrr-preee-preee' given on ground or in air; also flight call 'chree-chree', 'trree-trree', 'churr-churr', 'churr-cheer-churr' or 'cheer-churr' repeated irregularly.

General Habits. Occurs on semi-arid and arid sparsely grassed stony plains, sparsely vegetated flat, stony limestone and in open savanna woodlands on sands. In Namib Desert occurs on thinly grassed gravel flats to limit of vegetation.

When not breeding, occurs in groups of 4–5, sometimes up to several hundred. Forages by creeping along with body close to ground, pecking at ground or occasionally uncovering buried items by sideways jerks of bill. When it finds food, raises head and squats down to nibble and swallow it. Drinks water regularly in dry season, less often after rains; individuals probably do not drink every day.

Avoids heat by crouching in shade of overhanging rocks or vegetation. Sometimes perches on rocks or low shrubs while facing into prevailing wind, holding folded wings away from body to expose thinly feathered sides.

Nomadic, moving into areas where rain has fallen. Abundant some years in Namib Desert after rain, when hundreds nest on thinly grassed gravel plains.

Food. In Namib Desert, arthropods 20% (termites *Hodotermes*, ants, beetles, bugs, flies, spiders, solifugids), seeds 75% (grasses *Stipagrostis* spp., forbs *Cleome* spp., *Monsonia umbellata*), green vegetation 5% (grass culm bases, flowers of *Tribulus* sp.) (Willoughby 1971).

Breeding Habits. Territorial. Breeding density in Namib Desert estimated at 5–12 birds/ha (Willoughby 1971). In display, ♂ sings while rising steadily straight up from ground to height of 20–200 m; there it hovers against the wind, or circles slowly if no wind, and sings continuously for several min; then closes wings and dives vertically to ground, where it perches or chases conspecific. Sometimes sings while flying slowly with legs dangling and wings fluttering for short distances 1–2 m above ground. ♂ displaying on ground faces ♀ and sings continuously, with body horizontal, crest raised, head bobbing.

NEST: cup of white, silky inflorescences of grasses *Aristida* and *Stipagrostis*; cup diam. 45–64, depth 25–38, built on foundation of pieces of earth and a few stones in excavation in gravel, stony soil or in sandy patch on limestone flats, against rock or shrub, or sometimes out in open (in winter); sometimes has sparse rim of sand and spider webs; rim usually just above or level with substrate; faces east, southeast or south in Kalahari, south, south-southeast or southeast in Namib.

EGGS: 2–3, mean (20 clutches) 2·3. Sub-elliptical; white or pinky white, spotted with pale pinky brown and some grey markings sometimes concentrated at obtuse end. SIZE: (n = 13) 19–20·4 × 13·4–14·8 (19·7 × 14·3).

LAYING DATES: Namibia, Mar, May, Aug; Namib Desert, Apr–May; South Africa, Kalahari, Aug–Nov; season varies according to rainfall.

INCUBATION: period 12 ± 1 days.

DEVELOPMENT AND CARE OF YOUNG: both adults brood and feed young. Feeding call, quiet 'chop chop', releases food-begging in young, which are fed largely on green grass seeds. Young leave nest at age of *c.* 10 days, before they can fly. Adults react to disturbance at nest with alarm display, flying up a few m calling 'tree', then landing on ground, repeating display as long as necessary. Another alarm reaction is to flip wings in bouncing flight, then suddenly dive with wings partly folded and held high, sometimes snapping them. Gives injury-feigning display in response to distress calls from young: bird flies towards intruder, flops down and flutters away along ground with soft 'zzz' call.

BREEDING SUCCESS/SURVIVAL: of 12 eggs laid in 6 nests, 4 young hatched but none survived (Maclean 1970). Adults preyed on by Red-necked Falcon *Falco chicquera*.

References
Maclean, G. L. (1970).
Willoughby, E. J. (1971).

Genus *Chersophilus* Sharpe

A single species of mid-sized NW African lark of uncertain affinities, but with distinctly *Mirafra*-like plumage (especially of juvenile), long thin decurved bill, moderately robust legs and elongated hind claw. Upperparts rufous or dark brown, all feathers broadly pale-fringed. Underparts white; chin and breast heavily streaked rufous or dark brown. Rectrices somewhat pointed, blackish, the outer one white. P10 minute (unlike *Mirafra*). Nostrils concealed.

Chersophilus duponti (Vieillot). Dupont's Lark. Sirli du Dupont.

Alauda duponti Vieillot, 1820. Faune Franç., p. 173, pl. 76; Provence.

Plate 2
(Opp. p. 33)

Range and Status. N Africa and Spain. Resident. Has bred Portugal and S France, accidental Italy, Malta.

Resident N Africa but sparse and uncommon almost throughout (locally common in Spain, however). Morocco, uncommon in NE between Midelt and Aïn Béni Mathar and between Wadi Moulouya and Algerian border, but locally common on *Stipa* grass and *Artemisia* steppe north of Plaine de Tafrata, west to Itzer–Midelt road; 10, Aïn Béni Mathar, Algeria, Jan, nowhere common; most records from northwest towards Moroccan border; bred commonly in Tebessa, near Tunisian border, in 1935–37. Tunisia, rare and local breeder in grass and bush plains from Feriana southeast to Libyan border (Gafsa, Sidi Mansour, Gabès, Mahares, Kebili, Douz, Matmata, Tatahouine). Libya, thinly distributed in sand and stone desert in NW, east to 15°E, south to 31°N; rare and local in Cyrenaica, in semi-desert from Jebel Akhdar to Al Adem (Tobruk). Egypt, rare near coast from Libyan border east to Arabs Gulf; no recent records. Density (Spain): 8·9 birds per 10 ha in *Gypsophilla hispanica*, 20·5 birds per 10 ha in *Linum/Genista* (Suárez et al. 1982).

Description. *C. d. duponti* (Vieillot): Morocco, N Algeria, NW Tunisia (also Spain). ADULT ♂: upperparts dark brown, strongly variegated with pale buff. Forehead dark rufous with indistinct line of dark-rufous-fringed feathers above lores; crown blackish, browner or more rufous at sides, and with irregular whitish or very pale buff median stripe, usually more prominent towards forecrown than on hindcrown; lores pale; narrow whitish superciliary stripe; nape and sides of nape greyer than crown. Mantle, scapulars, back and rump dark brown, the feathers broadly fringed buff, and overall brown colour varying from blackish brown to warm brown; rump greyish, mottled dark brown; uppertail-coverts chiefly dark rufous. Tail-feathers pointed, central 2 pairs dark brown; outer tail-feather entirely white; T5 with black inner web and white outer web; T4 and T3 black. Chin white; throat white, strongly streaked dark brown, ear-coverts dark brown mottled with white, with 2 thin broken lines of black spots below eye; breast white, strongly streaked with rufescent dark brown. Belly, flanks and undertail-coverts white. Wings brown to dark brown; all feathers (especially tertials, primary coverts and secondary coverts) broadly fringed whitish or pale buff, and primary and secondary coverts with dark subterminal mark between pale buff fringe and brown centre of feather; lesser and median coverts rufescent with whitish inner edge. Axillaries and underwing-coverts mainly greyish brown. Bill dark brown with lower mandible and cutting edges of upper mandible pink or horn-brown; eye brown; legs flesh-brown. Sexes alike. SIZE: wing, ♂ (n = 15) 99–106 (103), ♀ (n = 10) 88–95 (91·7); tail, ♂ (n = 8) 58–66 (62·5), ♀ (n = 6) 52–58 (54·6); bill to skull, ♂ (n = 14) 22·5–25·1 (23·9), ♀ (n = 9) 19·3–23·2 (21·6), bill to feathers av. 3·2 shorter; tarsus, ♂ (n = 8) 23–25 (24·0), ♀ (n = 6) 22–23·5 (23·0); hind claw (n = 52) 8–12·5 (10·2) (Cramp 1988). WEIGHT: (Spain, unsexed, n = 33) 32–47 (39·4).

IMMATURE: like adult but pale feather-fringes broader (up to 2 mm), much less black at centre of feather; pale median crown-stripe indistinct or absent; 2 rows of minute spots on ear-coverts indistinct.

NESTLING: unknown.

C. d. margaritae (Koenig): southern slopes of Algerian Atlas Mts from Biskra east to SE Tunisia, Libya and Egypt. Like nominate race but rufous or rufous-brown where nominate race is dark brown on head, back, wings, breast and tail. Immature like adult but with fluffy plumage, paler above and below, the feathers more conspicuously white-fringed, each with small blackish mark between white fringe and rufous-brown centre, especially conspicuous in tertials; underparts less streaked and less strongly marked than in adult; forehead, crown and nape more strongly marked, with speckled appearance. Bill longer, tail shorter than nominate race: bill to skull, ♂ (n = 29) 24·5–29·2 (26·6); tail, ♂ (n = 8) 55–58 (56·9).

Chersophilus duponti

Field Characters. A small or medium-sized secretive N African lark, without crest and with long, thin, markedly decurved bill, plumage rufous (*margaritae*) or rather indistinctive and like Eurasian Skylark *Alauda arvensis* (nominate *duponti*). Pale median crown-stripe; pale buffy crescent below speckled dark brown cheeks. Wings rather short – tips barely show below tertials. Song unlark-like: resembles twittering of Brown Linnet *Carduelis cannabina*.

Voice. Tape-recorded (63, 73, CHAR, HAZ, ROC). Song a linnet-like twittering, with light, fluty tremolos or fast liquid bubbling, many distinctive nasal or twangy 'whee-ur-wheeee' or 'hoo-hee-ur-hoo-eeee' phrases on rising pitch, and slow 'wzeep' or 'pizzeep' notes (like Reed Bunting *Emberiza schoeniclus*), ventriloquial 'dwur-dee' notes (like Eurasian Bullfinch *Pyrrhula pyrrhula*) and occasional rasping or buzzing trills (Cramp 1988). Call, from ground and in flight, an easily-imitable whistle 'hoo-hee', 2nd syllable rising in pitch. Other notes, 'dweeje' (like European Greenfinch *Carduelis chloris*); alarm a quiet 'tsii' (Cramp 1988).

General Habits. Inhabits flat or shelving (never steep or dissected) open country from sea level to >1000 m (Morocco), with clumped growth 30–50 cm tall of feather-grass *Stipa tenacissima* and *Artemisia* and expanses of bare soil between the vegetation. In non-breeding season, inhabits cereal fields, particularly barley and oats.

Solitary, in pairs, or small flocks. Mixes with Eurasian Skylarks and Calandra Larks *Melanocorypha calandra* in winter foraging flocks. Secretive; an inveterate runner; conceals itself behind vegetation and can be almost impossible to watch on foot, but tolerates vehicles. Shy and retiring. When flushed flies for *c.* 50 m, drops and runs into thickest available cover; very hard to flush a second time. Runs very fast; can look like a rodent. Stands high in slim, erect posture to observe person from behind tuft of vegetation (**A**). Never recorded to drink. Wholly terrestrial, but climbs up small plants (Cramp 1988). Feeds by digging with bill into ground and around bases of *Artemisia* tufts; also by splitting open balls of horse dung (Smith 1965). Flight fast, with more flapping than flight of Eurasian Skylark, for example. Sings throughout moonlit Feb nights, before dawn and incessantly for 2 h after dawn, but less during daytime (Smith 1965). Sings from small eminence on ground, and in flight.

Not migratory, but sometimes disperses far, occurring in Algerian desert well south of breeding range.

Food. Insects, seeds. 2 birds (Spain) contained a caterpillar, 12 beetles of 5 families; 2 others contained ants, grasshoppers, 6 beetles, and 90 *Asphodelus* seeds (Cramp 1988).

Breeding Habits. Territorial, solitary nester. Song-flight: rises to 100–150 m (higher than Eurasian Skylark) and sings for up to 30 (possibly 60) min; then makes sudden, vertical, very fast descent to ground. Also has short, low and straight song-flight during which it gives bursts of wing-claps: bird rises silently from ground at 45° angle to *c.* 10 m, then, while still rising, gives *c.* 10 wing-claps followed by a few more interspersed with short version of song, after which it descends silently (Hazevoet 1989). Wing-claps clearly audible for up to 50 m. Song-flight begins as early as 05·00 h, increases in frequency until *c.* 11·00 h, then tapers off. Song resumes at sunset but without wing-clapping.

NEST: deep scrape lined with rootlets, small twigs, vegetable fibres and animal hair, set into tussock, under a bush, or in lee of large stone. Ext. diam. 94–96, int. diam. 69–71, ext. depth 41, depth of cup 32 (Spain). Nest building takes 3–4 days (SE Spain).

EGGS: 3–4; in Algeria and Tunisia 57 clutches of 3, and 18 of 4 (mean 2·76). Subelliptical; glossy; white or pink, densely spotted with reddish brown on purple-grey undermarkings. SIZE: (n = 14, *C. d. duponti*) 22·4–24·6 × 16·6–18 (23·6 × 17·2); (n = 10, *margaritae*) 22·6–24·2 × 16·6–18 (23·4 × 17·2). WEIGHT: *c.* 3·6.

LAYING DATES: Morocco, Feb; Algeria, N Tunisia, Mar–June; Libya, Mar.

INCUBATION: period 12–13 days (n = 4) (Cañadas *et al.* 1988).

DEVELOPMENT AND CARE OF YOUNG: fledging period *c.* 10–11 days (Spain).

BREEDING SUCCESS/SURVIVAL: of 17 eggs in 4 nests, SE Spain, 15 hatched but only 3 nestlings fledged, others being eaten by snake *Malpolon monspessulanus* (Cañadas *et al.* 1988).

References
Cañadas, S. *et al.* (1988).
Cramp, S. (1988).
Suárez, F. *et al.* (1982).

Genus *Pseudalaemon* Phillips

Bill stout and strong, nostrils concealed by bristles; legs strong, hind claw straight and shorter than hind toe. Tail short and square, wing with outer primary short and slender. Nest cup-shaped, without dome.

Endemic to Africa; monospecific.

ALAUDIDAE

Plate 4
(Opp. p. 49)

Pseudalaemon fremantlii (Phillips). Short-tailed Lark. Cochevis à queue courte.

Calendula fremantlii Phillips, 1897. Bull. Br. Orn. Club 6, p. 46; Gedais.

Range and Status. Endemic resident. S Ethiopia (Mega), Somalia (fairly common and widespread south to 5°N, and south along coast); N and S Kenya (Marsabit and plateau south of central highlands, from Athi Plains and Nairobi Nat. Park to Konza and Simba), Tanzania (Arusha west to Crater Highlands and Serengeti). Locally common, sometimes abundant (Somalia).

Description. *P. f. fremantlii* (Phillips): Somalia, SE Ethiopia. ADULT ♂: crown buff, streaked dark brown; nape and hindneck whitish streaked dark brown, forming pale collar; sides of neck and foreneck pale buff with a few brown markings. Lores and broad supercilium buff; dark brown streak from bill through eye onto upper ear-coverts; white streak below eye, separated from white cheeks by narrow black line; cheeks white crossed by diagonal broad dark streak; narrower dark streak beside chin and upper throat; ear-coverts brown; chin and throat white. Feathers of upperparts dark brown edged and tipped buffy or grey-brown on mantle, edged pale greyish on back and rump, uppertail-coverts with dark centres and pale greyish edges, less heavily streaked than back. Tail-feathers dark brown, outer pair with white outer webs, next pair with narrow white outer edges, central pair grey-brown, marginated buffy. Upper breast whitish with a few teardrop-shaped darker brown spots, often coalescing to form dark patch at sides; lower breast pale rufous with a few indistinct streaks. Flight feathers greyish brown with narrow pale outer edges; scapulars and upperwing-coverts brown with pale narrow outer edges, broad buff inner edges, tertials brown narrowly marginated whitish buff. Underwing-coverts and axillaries light brown to buff. Bill dark grey, lower mandible paler; eye brown; legs and feet flesh white. Sexes alike. SIZE: (10 ♂♂, 11 ♀♀) wing, ♂ 82–87 (84·9); ♀ 79–88 (83·2); tail, ♂ 37–47 (40·8); ♀ 37–45 (39·5); bill, ♂ 18–20 (18·8), ♀ 17·4–20·6 (18·5); tarsus, ♂ (n = 9) 18–21 (19·7), ♀ (n = 9) 17·5–20·5 (19·1). WEIGHT: ♂ (n = 5) 23–26 (25·0), ♀ (n = 4) 20·5–25 (21·9), unsexed (n = 2) 18·5, 20·4.

IMMATURE: feathers of upperparts, secondaries and primaries tipped white; faint brown spots on breast.

NESTLING: unknown.

P. f. megaensis Benson: S Ethiopia (Mega), N Kenya. Darker than *fremantlii*, more heavily streaked blackish, feathers of upperparts edged more rufous, breast and sides more reddish brown.

P. f. delamerei Sharpe: S Kenya, Tanzania. Similar to *megaensis*, but feathers of upperparts edged greyer.

Field Characters. A medium-small lark, with a long bill which gives it a big-headed appearance; short, square tail noticeable in flight. Complex, bold face pattern and dark breast-patches characteristic. Sympatric Crested Lark *Galerida cristata* and Thekla Lark *G. malabarica* larger, with conspicuous crests, plain faces, shorter bills and longer tails.

Voice. Not tape-recorded. Said to have 'a curious, sharp unmistakable note on rising' (Mackworth-Praed and Grant 1955).

General Habits. Occurs on short-grass plains, coastal grasslands and open shrublands, particularly where there are gravels and bare patches, and in light, open woodlands with rocky outcrops. Often near human habitation and even enters buildings.

Solitary or in pairs in breeding season; at other times may form flocks of 10–30, sometimes 45–60, and over large areas scattered birds may total hundreds (Somalia). Not shy; often perches on rocks. Shelters in shade of bushes during heat of day.

Partly sedentary; locally nomadic.

Food. Seeds, grass culm bases and corms.

Breeding Habits.

NEST: in tuft of grass.

EGGS: 3–4 (n = 2). Ivory-white, spotted with sienna or umber and underlying shades of ink-grey. SIZE: (n = 5) 19–21·5 × 15·5–16 (20·6 × 15·9).

LAYING DATES: Ethiopia (young bird c. 2 months old, Sept); Somalia, May (fledgling, Jan); E Africa: Region D, May–June.

Genus *Galerida* Boie

Not satisfactorily distinguishable from *Alauda* or *Lullula* (see family diagnosis). Small to medium larks with short to long crest arising in middle of crown, rather short to rather long bill, large and strong in *G. magnirostris*, fairly stout in *G. modesta*, more slender in *G. cristata/malabarica*, nostrils concealed. 10th primary shorter than, equal to, or longer than primary coverts; tertials short to long; legs robust, hind claw long and not strongly decurved.

5 species; 4 in Africa, 2 endemic (*G. magnirostris*, *G. modesta*), and 2 sibling species in N Africa, Europe and Asia (*G. cristata*, *G. malabarica*).

Galerida modesta Heuglin. Sun Lark. Cochevis modeste.

Plate 4
(Opp. p. 49)

Galerida modesta Heuglin, 1864. J. Orn., p. 274; Bongo, Bahr-el-Ghazal, Sudan.

Range and Status. Endemic resident. Senegambia, Guinea, Sierra Leone, Ivory Coast, Mali (fairly common in S), Burkina Faso (north to Fadd N'Gourma), N Ghana (Lawra, Wa, Gambaga), Nigeria (locally not uncommon, Zaria, Kano, south to Igbetti, Ilorin to Wukari in dry season), Cameroon (Adamawa Plateau, Benué Plain, and E), Central African Republic, W and S Sudan (uncommon in W, fairly common in S), extreme NW Uganda (Nile Valley), Zaïre (Ubangi R., S Ubangi-Shari, Uelle basin). Locally common.

Description. *G. m. modesta* Heuglin (including '*giffardi*'): Burkina Faso, N Ghana, N Nigeria, east to Sudan (Darfur) and NW Uganda (confined to narrow belt). ADULT ♂: upperparts sandy rufous, crown, nape and back heavily streaked blackish brown, hindneck less streaked, thus forming paler collar; rump and uppertail-coverts with cinnamon tinge and fewer streaks. Lores and supercilium whitish; dark line from bill through eye over and around brown ear-coverts; cheeks buffy crossed by dark diagonal streak; chin and throat whitish, some dark spots at side of throat. Tail blackish brown, outer webs of feathers pale rufous. Breast pale rufous heavily streaked blackish, rest of underparts including undertail-coverts buff. Flight feathers dark brown, narrowly edged rufous, tertials dark brown with narrow rufous margins, scapulars and upperwing-coverts blackish brown with broader pale rufous margins, tipped whitish. Underwing-coverts and axillaries rufous-buff. Bill blackish, lower mandible whitish; eye dark brown; legs and feet light brown. Sexes alike. SIZE: (5 ♂♂, 5 ♀♀) wing, ♂ 72–84 (80·2), ♀ 76–82 (78·2); tail, ♂ 42–50 (46·8), ♀ 41–44 (42·8); bill, ♂ 11·5–13·5 (12·4), ♀ 11·5–14 (12·3); tarsus, ♂ 15–18 (16·0), ♀ 17–19 (18·0). WEIGHT: 2 ♂♂ 21·5, 22, 1 ♀ 18.
IMMATURE: unknown.
NESTLING: skin blackish, with long, white, hair-like down on crown, back, wings and thighs; inside of mouth orange-yellow; 3 black spots on tongue, 1 at tip and 1 either side, 1 black spot at tip of upper and lower mandibles (Serle 1943).
G. m. bucolica Hartlaub: N Zaïre (Ubangi R.) east to W Uganda. Darker than *modesta*.
G. m. strumpelli Reichenow: Cameroon (Foumban, Tibati, Ngaoundéré). Darker than *bucolica*, more heavily streaked with black and with broader rufous feather edges. Larger: wing 86–90.
G. m. nigrita Grote: Senegambia, Sierra Leone, Mali. Darker above than *strumpelli*, feathers of upperparts with narrower rufous edges. Smaller: wing 78.

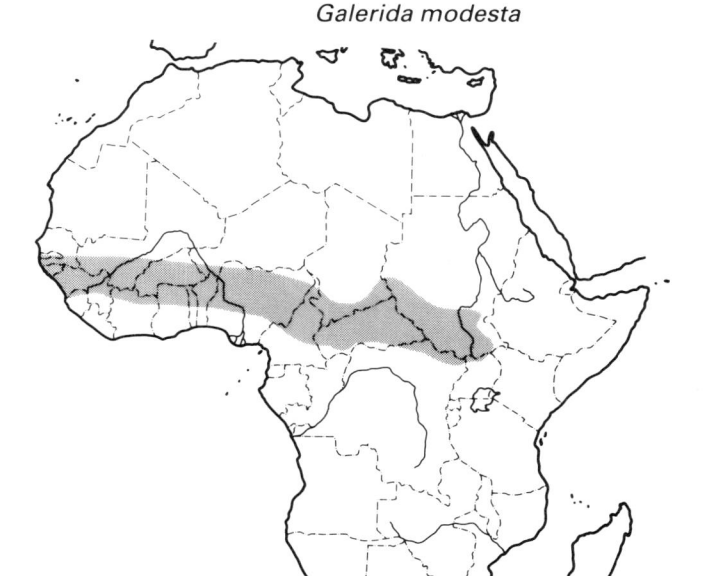

Field Characters. A small, dark lark, upperparts and breast heavily streaked black; wings show some rufous in flight. On ground most likely to be confused with Flappet Lark *Mirafra rufocinnamomea*, which has very variable plumage but usually much more rufous below, breast lightly spotted, not heavily streaked, and pale tips to feathers of upperparts give scaled rather than streaked appearance.

Voice. Tape-recorded (78, BRU, CHA). Song of 2 types: (1) given on ground and in flight, a rapid tuneless jingle of notes lasting 1–2 s, repeated every few s; (2) given in air, probably a display song (Chappuis 1985), a continuous stream of notes, some melodious, some grating and squeaky, incorporating imitations of other species, including Common Bulbul *Pycnonotus barbatus* and perfect rendition of Red-rumped Swallow *Hirundo daurica* (Chappuis tape, 1985).

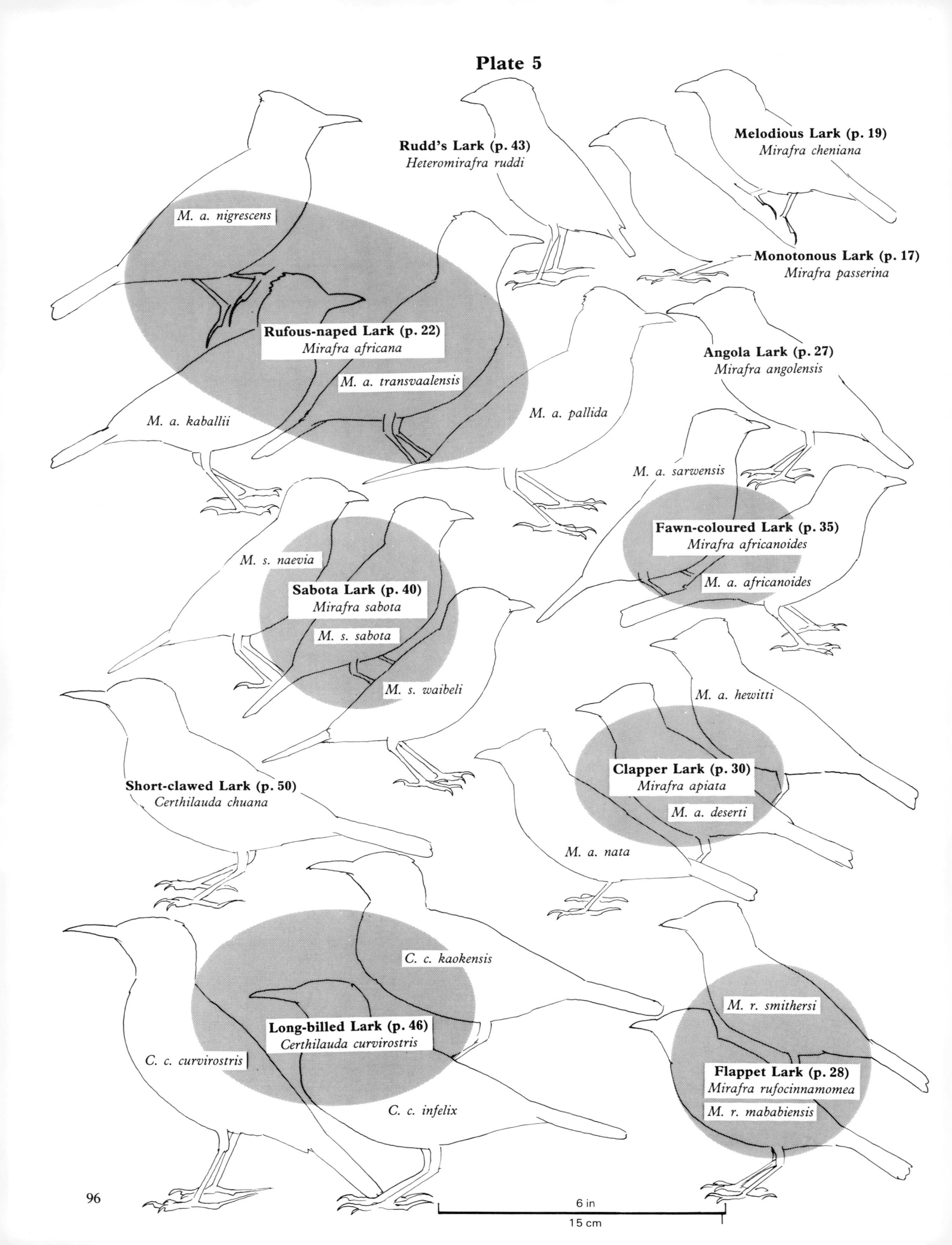

Plate 5

Rudd's Lark (p. 43)
Heteromirafra ruddi

Melodious Lark (p. 19)
Mirafra cheniana

M. a. nigrescens

Monotonous Lark (p. 17)
Mirafra passerina

Rufous-naped Lark (p. 22)
Mirafra africana
M. a. transvaalensis

M. a. kaballii   *M. a. pallida*

Angola Lark (p. 27)
Mirafra angolensis

M. a. sarwensis

Fawn-coloured Lark (p. 35)
Mirafra africanoides
M. a. africanoides

M. s. naevia

Sabota Lark (p. 40)
Mirafra sabota
M. s. sabota

M. s. waibeli

M. a. hewitti

Clapper Lark (p. 30)
Mirafra apiata
M. a. deserti

Short-clawed Lark (p. 50)
Certhilauda chuana

M. a. nata

C. c. kaokensis

Long-billed Lark (p. 46)
Certhilauda curvirostris

C. c. curvirostris

C. c. infelix

M. r. smithersi

Flappet Lark (p. 28)
Mirafra rufocinnamomea
M. r. mababiensis

96

6 in
15 cm

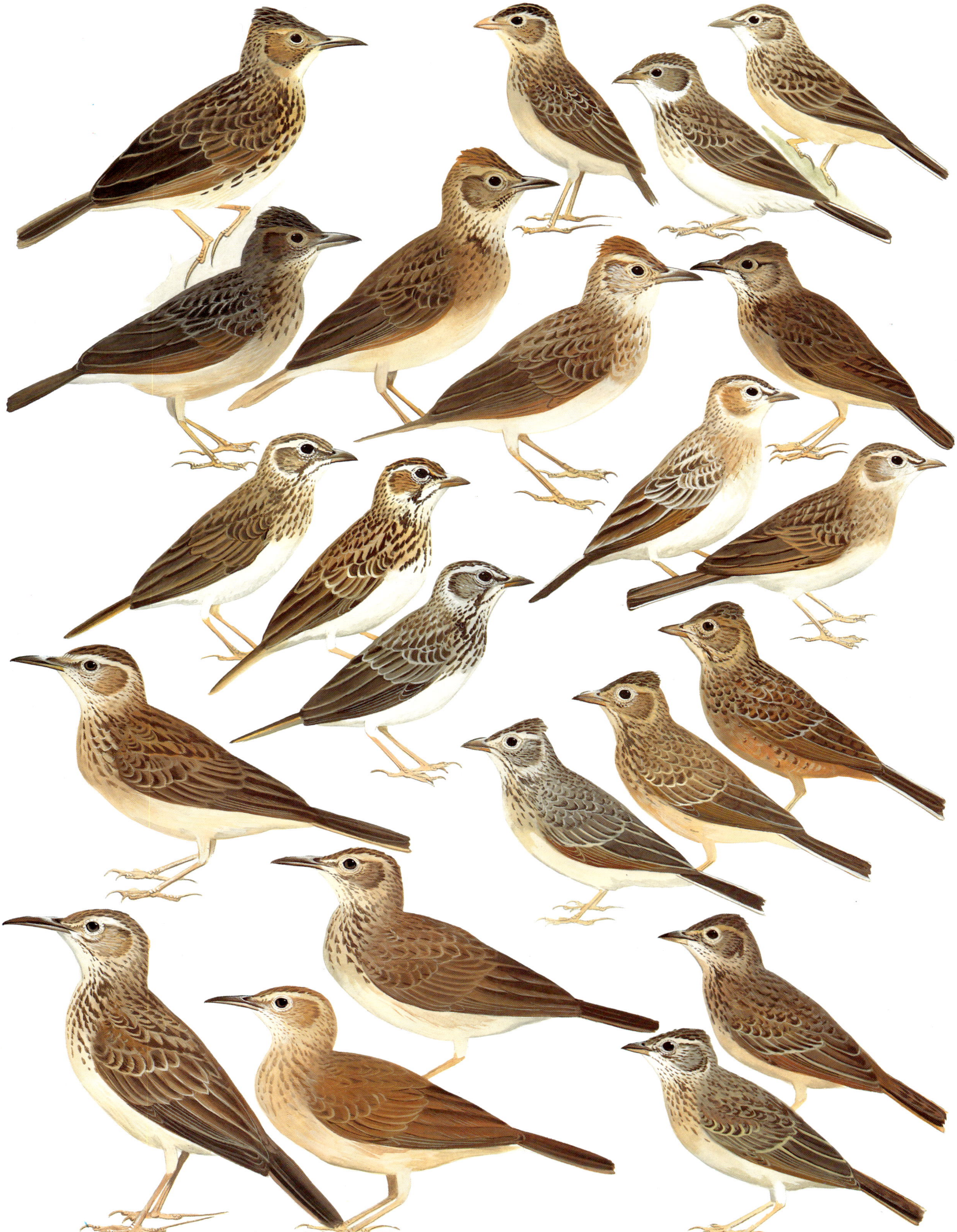
MWW1987

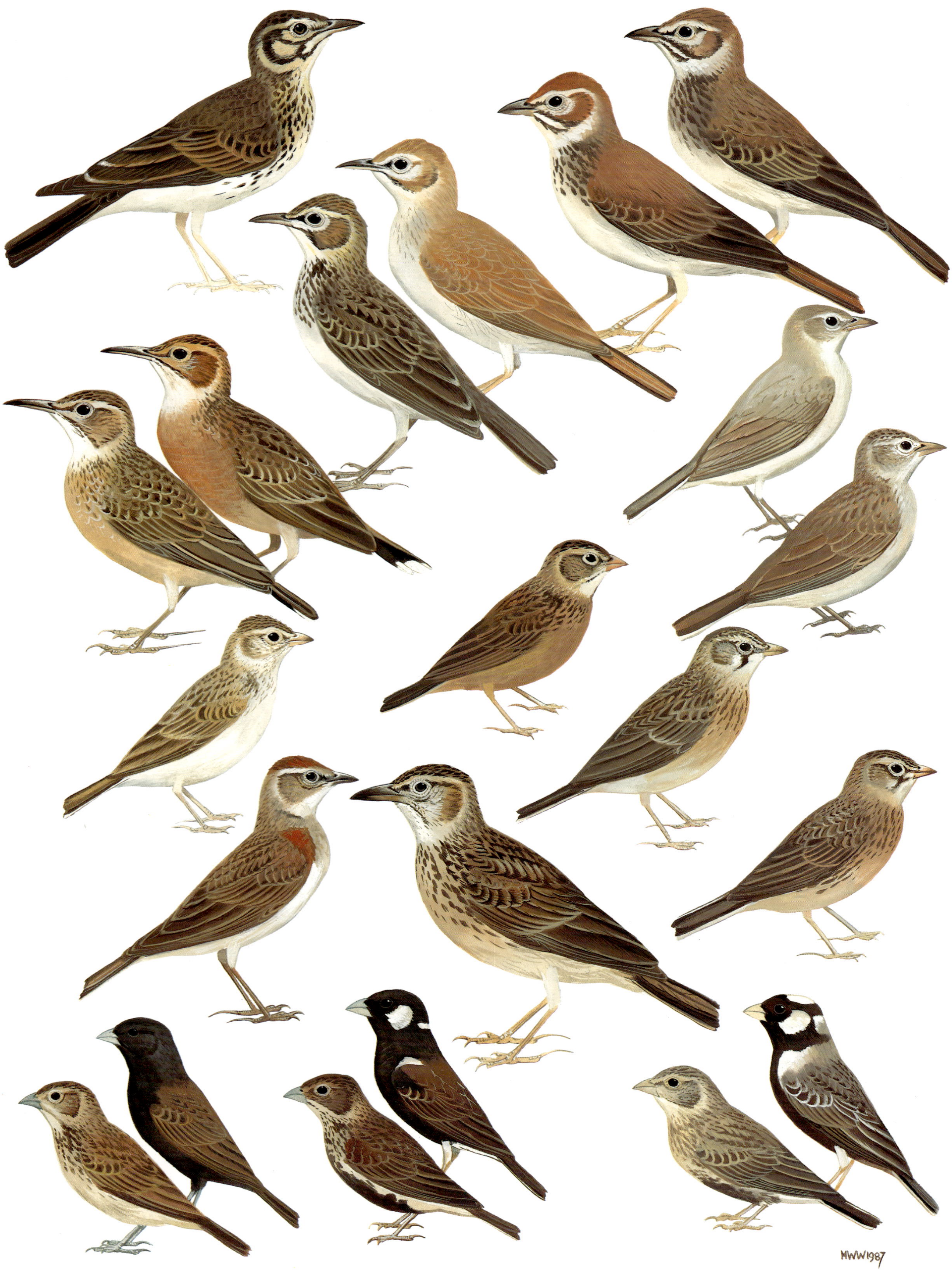
MWW1987

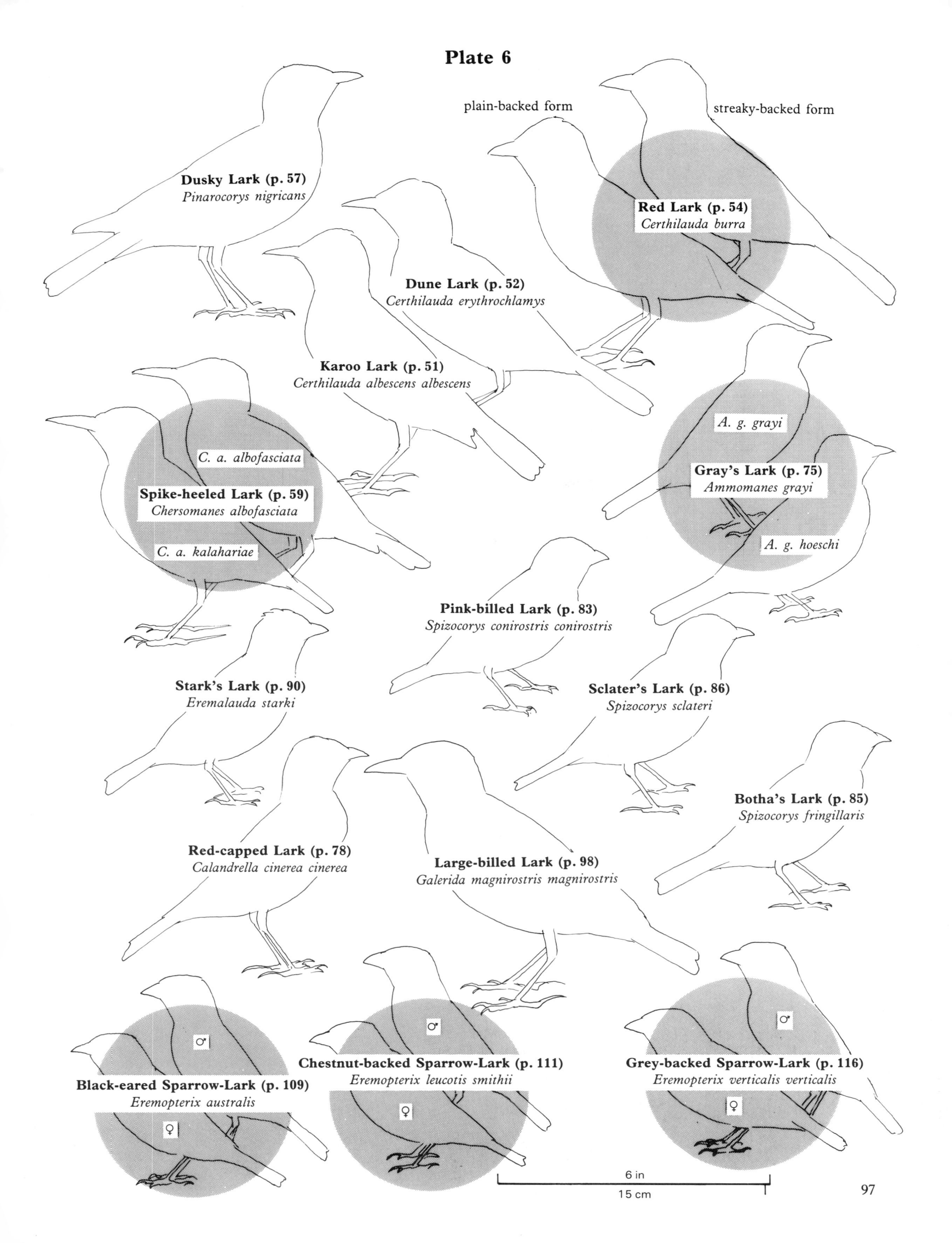

Plate 6

Dusky Lark (p. 57)
Pinarocorys nigricans

Dune Lark (p. 52)
Certhilauda erythrochlamys

plain-backed form   streaky-backed form

Red Lark (p. 54)
Certhilauda burra

Karoo Lark (p. 51)
Certhilauda albescens albescens

C. a. albofasciata

Spike-heeled Lark (p. 59)
Chersomanes albofasciata

C. a. kalahariae

A. g. grayi

Gray's Lark (p. 75)
Ammomanes grayi

A. g. hoeschi

Pink-billed Lark (p. 83)
Spizocorys conirostris conirostris

Stark's Lark (p. 90)
Eremalauda starki

Sclater's Lark (p. 86)
Spizocorys sclateri

Botha's Lark (p. 85)
Spizocorys fringillaris

Red-capped Lark (p. 78)
Calandrella cinerea cinerea

Large-billed Lark (p. 98)
Galerida magnirostris magnirostris

Black-eared Sparrow-Lark (p. 109)
Eremopterix australis

Chestnut-backed Sparrow-Lark (p. 111)
Eremopterix leucotis smithii

Grey-backed Sparrow-Lark (p. 116)
Eremopterix verticalis verticalis

6 in / 15 cm

General Habits. Preferred habitat seems to be a mixture of grassland and rocky areas, e.g. grassy inselbergs, but occupies wide selection of habitat types: hillsides with thin grass cover, open savannas, grassy patches in woodland, always where there is some grass; readily takes to man-made habitats (cultivation, fallow fields, pastures) and quite at home with man, occurring on bare ground in and around villages, on football pitches, airfields and in suburbs.

Solitary or in pairs in breeding season, in groups of 3–6 at other times.

Some evidence for local movements, possibly regular migration, e.g. Sudan, Nigeria. In Nigeria moves south in dry season, and present in some areas only at this time.

Food. Invertebrates (insects) and seeds (grasses, cereal grains, other hard seeds).

Breeding Habits. In display, ♂ hovers in air while singing (see *Voice*).

NEST: flimsy cup of grass and rootlets, int. diam. *c.* 50, sparsely lined with grass and roots, placed in scrape in ground, at base of tuft of grass, shrub or stone; built by ♀, accompanied by ♂ (Serle 1943).

EGGS: 1 only (2 clutches). Creamy white, densely covered with reddish brown spots. SIZE: (n = 1) 19·9 × 14·7.

LAYING DATES: Mali, May–June, Nov–Jan; Ghana, Dec–Jan (song-flight June); Nigeria, Feb–Mar, Oct–Nov; Sudan, July.

References
Serle, W. (1943).
Shuel, R. (1938).

Plate 6
(Opp. p. 97)

Galerida magnirostris (Stephens). **Large-billed Lark. Cochevis à gros bec.**

Alauda magnirostris Stephens, 1826. Shaw's Gen. Zool. 14, p. 26; South Africa (= near Cape Town).

Range and Status. Endemic resident. S Namibia (in extreme S, along W Orange R.); South Africa: Cape Province (widespread and locally common in Namaqualand, Bushmanland south of 29°S, south to SW and Agulhas Plain, east through Little and Great Karoo to E and NE), Orange Free State (mainly in W, scattered records in N and E), and extreme SW Natal; W, S and SE Lesotho. Locally common.

Description. *G. m. magnirostris* (Stephens): SW Cape north along coast to Piquetberg. ADULT ♂: crown, nape, hindneck, mantle, back and rump light brown streaked blackish, less heavily on hindneck and rump (feathers with blackish brown centres, brown edges); uppertail-coverts grey-brown with dark centres. Tail-feathers dark brown, outer pair with narrow buffy white outer edges, central pair medium brown with pale margins. Crown feathers long and erectile. Sides of neck and foreneck brown, streaked whitish. Lores and supercilium whitish; dark line from base of bill through eye onto upper ear-coverts; ear-coverts brown, cheeks whitish, indistinct moustachial streak, chin and throat whitish. Some individuals have cheeks and sides of chin and throat lightly speckled. Underparts whitish, sometimes tinged yellowish on belly and undertail-coverts; breast heavily marked with broad dark streaks, lower breast and flanks more sparingly marked with narrow streaks. Flight feathers dark brown, often with pale outer edges and tips; scapulars dark brown, edges paler, tertials with dark brown centres shading to broad greyish brown margins; upperwing-coverts dark brown edged paler. Underwing-coverts and axillaries light brown, feathers along bend of wing barred dark and pale brown. Bill dark brown, basal half of lower mandible and base of upper mandible yellow; eye brown; legs and feet pinkish brown. Sexes alike. SIZE: (7 ♂♂, 8 ♀♀) wing, ♂ 99–107 (104), ♀ 93–103 (98·0); tail, ♂ 58–61 (59·8), ♀ 51–59 (55·3); bill, ♂ 20–22 (20·7), ♀ 19–21·6 (20·3); tarsus, ♂ 23–26 (24·5), ♀ 21–26 (24·7). WEIGHT: ♂ (n = 3) 41–48 (45·0), ♀ (n = 2) 35, 43, unsexed (n = 2) 35, 43.

IMMATURE: feathers of upperparts tipped buff. Primaries with narrow buff edge, broader buff tip, secondaries with broad buff margins. Breast spotted rather than streaked; feathers with spots also have buff tips.

NESTLING: skin brownish black above, pinkish brown to reddish on sides and below. Down long, buff to yellowish buff, on crown, back, scapulars, rump and thighs. Gape pale butter yellow.

G. m. harei (Roberts): Namaqualand and Karoo east to E Cape and W Orange Free State. Underparts, especially belly, with yellowish tinge. Wing larger: ♂ (n = 46) 102–111 (106), ♀ (n = 27) 95– 104 (99·8); bill smaller: ♂ (n = 52) 18·9–23·5 (21·4), ♀ (n = 24) 18·5–22·3 (20·6).

G. m. montivaga (Vincent): Lesotho and SW Natal. Upperparts darker, feathers edged paler buff. Bill smaller: ♂ (n= 10) 18·7–20·2 (19·4), ♀ (n = 4) 17·1–19·6 (18·4).

Field Characters. A large, stocky, heavy-looking lark, boldly streaked blackish above and below (**A**). Tail dark, square, and relatively short. Crest often held flat but diagnostic when raised, e.g. in alarm or during song. Told from same-sized Long-billed Lark *Certhilauda curvirostris* by thick, yellow-based bill.

Voice. Tape-recorded (75, 88, F, GIB). Song given from perch or in display flight, a quick, lilting phrase of 3–8 notes repeated over and over: 'too-toodle-oo', 'tweedle-doo', 'tree-triddly-pee', 'whit-titwiddliddle-widdly'; thin and high-pitched but not unmusical, with a ringing quality. Also has lengthier song in which it gives perfect imitations of other species interspersed between song-phrases (at least 13 species recorded: Maclean 1985).

General Habits. Present in montane semi-arid sparse grassland, semi-arid succulent and non-succulent dwarf shrublands, coastal macchia, croplands before and after the harvest and fallow fields. Commonest in succulent Karoo.

Solitary or in pairs, exceptionally in family group of 3–4 birds. May hold territories year-round. Forages by walking about, searching bases of plants and pecking food from bare ground. Digs for bulbs and seeds, and perches on shrubs to feed on fruits. Breaks open antelope, goat and sheep droppings to extract seeds. In dwarf shrubland takes seeds directly from low plants by pecking downwards into the plant. Drinks water regularly if available. Often on prominent perch (on top of rock, tree, shrub, fence post), especially when singing.

Food. Invertebrates: termites (*Hodotermes mossambicus*, *Microhodotermes viator*), roaches (Blattidae), adult and larval beetles, larval Lepidoptera, ants (*Monomorium*) and other insects; seeds (grasses *Sorghum*, *Stipagrostis*, cereals (oats, wheat, rye), sedges, forbs *Atriplex semibaccata*, *Tetragonia echinata*, *Hermannia*, *Arctotis*, *Aizoon*, *Malephora*, *Aridaria*, *Polygonum*, legumes), fruits (*Lycium*) and stems, basal nodes of grasses and dicotyledon leaves, and bulbs. All stomachs contained pieces of gravel up to 5 mm diam. Nestlings fed beetle larvae.

Breeding Habits. Monogamous; territorial. In display flight, ♂ rises to height of 15–50 m, then circles with fluttering flight, singing. Every few seconds, closes wings momentarily and dips down, then continues circling flight. At end of flight display, closes wings and descends in long curving dive, opens wings just above shrub layer, and lands on perch. Display lasts *c.* 2–5 min; repeated at 5–10 min intervals.

NEST: open cup of twigs, coarse grass and rootlets, int. diam. 60–70, depth *c.* 30, lined with finer material, including wool, feathers and plant down, built on foundation of small sticks, placed in excavation at base of small shrub or tuft of grass or out in open; usually faces south in Karoo.

EGGS: 2–4, mean (81 clutches) 2·5. Ovate or sub-elliptical; white or pale cream, densely spotted, speckled or blotched with brown, grey-brown and grey. SIZE: (n = 62) 20·9–25·7 × 15·4–17·4 (23·2 × 16·6).

LAYING DATES: South Africa: SW Cape, July–Dec; E Cape, Aug–Dec, Feb–Apr; Karoo, Aug–Nov, Feb; N Cape, Oct; Orange Free State, Oct–Dec, Apr.

DEVELOPMENT AND CARE OF YOUNG: both adults feed young; feeding rate (2 young) 24 feeds in 313 min. Adults bringing food fly towards nest, land on ground some distance away and walk to nest, giving 'characteristic call' when near nest (David 1971). Nest sanitation performed by both adults after feeding young; 5 faecal sacs removed in 313 min. Young up to 5 days old are shaded by adults. ♀ displaced from nest gave distraction display by fluttering close to ground as though wounded.

Reference
David, J. H. M. (1971).

Plates 2, 4
(Opp. pp. 33, 49)

Galerida cristata (Linnaeus). Crested Lark. Cochevis huppé.

Alauda cristata Linnaeus, 1758. Syst. Nat. (10th ed.) 1, p. 166; 'in Europae viis' = Vienna.

Galerida cristata

Range and Status. Africa, Iberia to S Sweden, east to Arabia, E India and Korea. Resident and partial migrant.

One of the commonest larks throughout much of its African range: Morocco, N Algeria, Tunisia, N Libya, N Egypt and entire Nile valley, Mauritania (local in N), Senegambia and Guinea, across central Mali, S Niger (north to Arlit, 19°N), N Togo, N Nigeria, across central Chad and Sudan, NE and coastal Ethiopia, Djibouti, and coastal N Somalia. Also N Ghana (Mole Game Res.), N Benin, N Cameroon, N Central African Republic (Manovo-Gounda-Saint Floris Nat. Park), N Kenya and SE Ethiopia (locally common).

Description. *G. c. kleinschmidti* Erlanger: NW Morocco south to Rabat and Azrou, east to Rif. ADULT ♂: forehead and crown brown, streaked blackish, elongated feathers on hindcrown being mainly black; remaining upperparts brown to dark brown variegated with streaks and blotches, most feathers having dark brown centres and paler brown margins. Tertials and central rectrices greyish brown or olivaceous brown. Outermost rectrix slightly paler than remainder of tail and outer web of outer rectrix pale rufous; remainder of tail blackish. Lores pale buff; weakly developed superciliary stripe pale buff; ear-coverts brown-and-buff streaked; sides of neck pale buff with blackish streaks. Chin and throat white or off-white in centre, speckled with dark brown at sides (especially of chin), the speckles forming indistinct moustachial mark. Breast off-white or very pale creamy, more or less heavily streaked with blackish brown; belly and undertail-coverts creamy. Remiges and upperwing-coverts brown or dark brown, narrowly fringed with pale buff; some coverts broadly fringed pale buff and with dark brown centres. Underwing-coverts and axillaries pale cinnamon. Bill dark brown or horn brown, base of lower mandible paler; eye dark brown; legs and feet pale brown or flesh-brown. Sexes alike. SIZE: wing, ♂ (n = 7) 102–108 (106); bill to nares, ♂ (n = 7) 13–15 (14.0); see also *arenicola*, below. WEIGHT: (*G. c. pallida*, Spain) ♂ (n = 20) 36–47 (41.4).

IMMATURE: differs from adult in having all feathers of upperparts white-tipped or broadly white-fringed with a dusky subterminal line between white tip and rufous-brown centre, giving bird white-spotted appearance like young thrush. Dark streaks on breast are sparse and obscure.

NESTLING: down long and dense on upperparts and flanks, sparse on underparts, whitish or yellowish; mouth dull orange-yellow with black spot inside tip of upper and lower mandible and large V-shaped spot on tip of tongue; gape-flanges white or pale yellow; bare skin purple-blue, legs and feet pink (Cramp 1988).

TAXONOMIC NOTE: numerous races have been described (Crested Larks have 'proved a source of trinomial effusion almost amounting to a jest in the ornithological world': R. Meinertzhagen), differing in ground colour, streaking, size and bill shape. However, differences are slight, and in general apparent only in series. The races recognized below, and descriptions and measurements, are based largely on Cramp (1988) and White (1961).

G. c. riggenbachi Hartert: W Morocco from Casablanca to Sous valley. Rich cinnamon-brown upperparts, cinnamon-buff underparts, heavily streaked crown, mantle and breast; intergrades with *kleinschmidti* between Rabat and Casablanca. Wing, ♂ (n = 5) 104–114 (109); bill to skull, ♂ (n = 5) 21–22 (22.0).

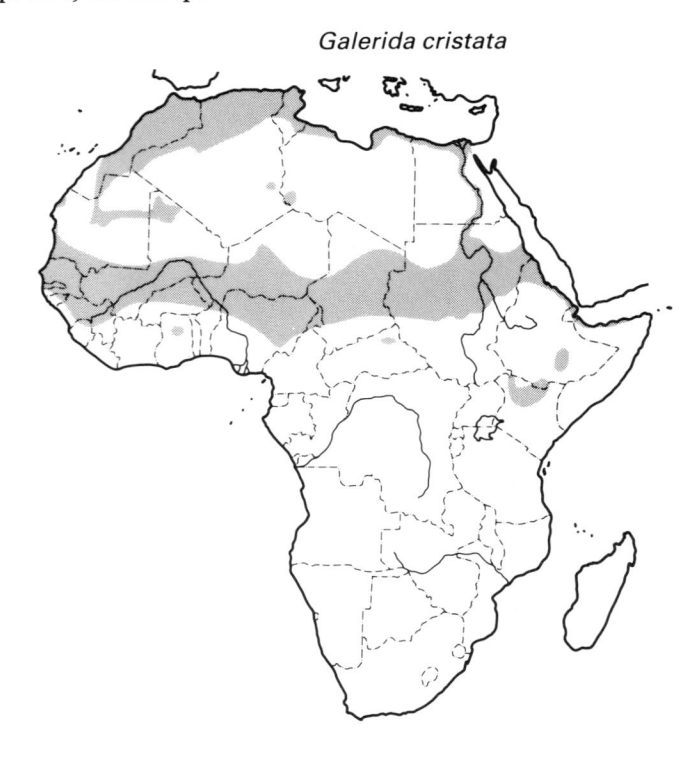

G. c. carthaginis Kleinschmidt and Hilgert: NE Morocco (coastal plains from Tellian Atlas west to Oujda) to Tunisia (Sous). Buffier than *riggenbachi*, streaks narrower, underparts paler; wing longer: unsexed (n = 12) 105–113 (110); bill to nares, unsexed (n = 12) 13.2–16 (15.7), bill to skull, ♂ (n = 5) 21–22 (21.0).

G. c. randonii Loche: E Morocco (from Moulouya valley, west to Marrakesh?) to Algeria (Hauts Plateaux). Like *carthaginis* but more heavily streaked above and much larger: wing, unsexed (n = 6) 113–117 (116); bill to nares, unsexed (n = 6) 16–19.5 (17.3), bill to skull, ♂ (n = 5) 23–25 (24). WEIGHT: ♂ (n = ?) 46–54.

G. c. macrorhyncha Tristram: Algeria (Atlas Saharien and N Sahara from Laghouat and Ghardaïa west to Figuig and Béchar) and N Mauritania (Atar). Like *randonii* but streaks pale olive- or rufous-brown, upperparts appearing uniform. Wing, unsexed (n = 10) 105–114 (109); bill to nares, unsexed (n = 10) 15–18 (16.3), bill to skull, ♂ (n = 5) 23–26 (24.5). WEIGHT: ♂ (n = 16) 37–42 and (n = 8) 44–54 (47.6).

G. c. arenicola Tristram: Algeria (Biskra and Ouargla eastwards), Tunisia (Gafsa and Sfax southwards) and Libya (Tripolitania). Like *macrorhyncha* but smaller, bill shorter. SIZE: (11 ♂♂, 8 ♀♀) wing, ♂ 107–112 (110), ♀ 101–106 (104); tail, ♂ 60–64 (62.0), ♀ 56–61 (58.0); bill to skull, ♂ 21.3–24 (22.5), ♀ 20.4–22.7 (21.4), bill to feathers av. 3.2 shorter, bill to nares 14–17.5 (16.0); tarsus, ♂ 25–27 (26.0), ♀ 25–27 (26.0).

G. c. festae Hartert: Libya (Cyrenaica: Benghazi to Tobruk). Upperparts cinnamon-rufous with sharp, narrow streaks, underparts cinnamon-buff with heavy streaks on breast. Wing, ♂ (n = 15) 105–109 (107), ♀ (n = 11) 98–102; bill to nares, unsexed (n = 15) 13.5–15.5 (14.7).

G. c. brachyura Tristram: NE Libya (Cyrenaica, south of *festae*) to NE Egypt (east to Wadi Natrun, Red Sea coast, also Sinai) (and Israel to Iraq). Like *altirostris* but upperparts greyer, less brown. Wing (n = 10) 104–110 (106); bill to nares (n = 10) 14–15 (14.5).

G. c. helenae Lavauden: SE Algeria (Illizi, north of Tassili); probably SW Libya. Upperparts uniform red-brown, underparts rufous-buff with large brown spots on breast.

G. c. jordansi Niethammer: Niger (northwest of Agadès). Like *helenae* but breast with small, faint spots. WEIGHT (*helenae* with *jordansi*): ♂ (n = 3) 37–42 (39·7), ♀ 38.

G. c. nigricans C. L. Brehm: Egypt (Nile delta north of barrage, east to Dumyat). Upperparts dark olive-grey, heavily streaked, underparts pale cinnamon with broad sharp black streaks. Wing, ♂ (n = 12) 99–106 (103); bill to nares, ♂ (n = 12) 12·5–17 (13·9).

G. c. maculata C. L. Brehm: Egypt (Nile valley from Cairo to Aswan; El Faiyum; borders of Nile delta). Like *nigricans* but streaks narrower above and below. Wing, ♂ (n = 24) 102–110 (105); bill to nares, ♂ (n = 24) 13–15·5 (14·0).

G. c. halfae Nicoll: Egypt (Nile valley south of Aswan) to N Sudan (Wadi Halfa). Paler and greyer than *maculata*. Intergrades with *maculata* and *altirostris*. Wing (n = 10) 100–107 (103); bill to nares (n = 10) 12–14 (13·3).

G. c. altirostris Brehm: Sudan (Nile valley between Dongola and Berber, east to Red Sea) and Ethiopia (Eritrea). Paler sandy-brown than *halfae*.

G. c. somaliensis Reichenow: N Somalia, SE Ethiopia, N Kenya. Duller and greyer than *altirostris*; very like *brachyura*. WEIGHT: ♀ (n = 21) 26–39 (32·9).

G. c. balsaci De Keyser and Villiers: coastal Mauritania. Upperparts dull greyish brown, underparts pale buff; heavily streaked.

G. c. senegallensis Muller: Senegambia and Guinea to Sierra Leone and Niger. Like *balsaci* but darker below.

G. c. alexanderi Neumann: Nigeria to W Sudan. Upperparts cinnamon-brown, not heavily streaked; underparts like *senegallensis*.

G. c. isabellina Bonaparte: Sudan (Kordofan to Nile valley). Like *alexanderi* but upperparts more cinnamon, less streaked.

Field Characters. Geographically variable, grey-brown to cinnamon above, cinnamon, buff or whitish below, with heavy or thin black streaks on upperparts, breast and flanks, or almost unstreaked. A medium-sized, rather bulky lark, with long, spiky, much-used forecrown crest. Bill strong. Perches freely on ground, rocks and low bushes, with half-upright posture. Vocal; main call a slurred, plaintive whistle 'klee-tree-weeoo'. Flight flapping and floating, with rather broad wings and shortish tail with cinnamon sides. Cinnamon underwing-coverts often conspicuous in flight. For distinctions from extremely similar Thekla Lark *G. malabarica*, see that species. Erect, spiky crest distinguishes it from all sympatric larks except Thekla. Note, however, that crest *can* lie quite flat. Some Palearctic larks in N Africa are crested (Eurasian Skylark *Alauda arvensis*, Woodlark *Lullula arborea*) and so are some larger Afrotropical *Mirafra* spp. (Rufous-naped Lark *M. africana* and allospecies), but their crests are all of uniformly short feathers.

Voice. Tape-recorded (78, C, McVIC, MOR). ♂ song loud, with long, soft, fluty whistles, double notes, tremolos, twitters and sometimes mimicry of other birds; aptly described as 'utz-utz-utz-zwee-du-du' or 'wee-too-wee-too-chee-too-twee-too-chee-twee-chee'. Series of linked phrases lasts 4–12 s; *c.* 12 series given in song-flight with intervals of 3–5 s. Song from perch, with pauses, up to *c.* 22 s long. Courtship song a swallow-like, quiet rambling twitter. Main call a pure, liquid, musical, lilting 'peeleevee', 'dji-dji-djii' or 'twee-tee-too'. For twittering, flight, rattle, distress, copulation and feeding calls, see Cramp (1988).

General Habits. Over most of range characteristically inhabits farmland, with plenty of open sandy or loamy soil soft enough to be probed for food. Commonly near human habitations and artefacts: urban wasteland, parks, railways, playing-fields. Natural habitat is flat, dry, shelving or gently undulating lowland plains, steppe, and bushy or tree savannas with vegetation covering only about half of soil surface. In Ghana bovals (shallow soil with short grass and low shrubs); in Somalia salt-pans, lava-plains, roadsides, and open sea-shores, nesting in damp irrigated meadows. Avoids hills, much-dissected land, woods and mobile sands; forages on firm substrates, and on very hard laterites if there are dust pockets permanent enough to contain such food as ant-lion larvae.

Nearly always in pairs, all year round, sometimes solitary, seldom more than *c.* 4 together and even then associated only loosely. Larger flocks may occur on migration or at drinking places. Tame. In general does not associate with other larks, although recorded with Greater Short-toed and Lesser Short-toed Larks *Calandrella brachydactyla* and *C. rufescens* (Tunisia, Libya) and with Thekla Larks (Spain); also with Ruddy Turnstones *Arenaria interpres* (coastal Somalia).

Forages by pottering along ground, walking, turning aside, pecking and probing at soil surface and low vegetation, occasionally running, often stopping to spend minute or more digging vigorously with bill into soil. Digs with blows of bill to left and right, pushing, pecking and stabbing, to make tunnel-shaped hole 20 mm deep and 20 mm wide (Cramp 1988). Persistent at a dig, returning to it soon after being disturbed. Digs where dung is lying, and into hoe-lines or side of sowing furrow, exposing and eating new-sown seeds. In Nigeria 1 moved systematically from 1 ant-lion larva sand funnel to another, making quick stabs into centre of each and evidently eating larvae in about half of 80 funnels in *c.* 10 m² of sandy laterite (Fry 1966). Extracts seeds from seed-heads on ground and from low plants; turns over dead leaves on ground; pursues insects and beats larger ones against ground, their wings often falling off (Cramp 1988). Hammers snails against stone 'anvils' (Morocco), in manner of Song Thrush *Turdus philomelos* (Riley 1989).

♂ sings in flight or from low perch, telegraph wire or from ground. Sometimes sings quietly whilst feeding.

Resident. Small numbers of *Galerida* sp. crossing Straits of Gibraltar into Africa in autumn are probably mainly *G. cristata* (Cramp 1988).

Food. Seeds, shoots, leaves; beetles and other invertebrates. For European foods, see Cramp (1988). In Africa: ants, grain and snails (Morocco); seeds, grain and grass (Tunisia); grain, grass- and *Chenopodium* seeds, ants, larvae and spiders (Egypt); seeds and insects

(Niger); ant-lion larvae (Nigeria); shoots of *Suaeda* (Somalia); seeds in 8 stomachs, insects including beetles and hawk-moth larvae in 9 other stomachs (Kenya).

Breeding Habits. Solitary, territorial, monogamous breeder – but Lynes (1912) often saw 3 birds together in breeding season and found record 8 eggs in a nest (Cramp 1988). ♂: ♀ ratio 4·3 : 1 (USSR); ♂♂ associate closely in winter (Tunisia: Niethammer 1954). Av. size of territory 0·58 ha (Denmark); nests may be only 50 m apart (Egypt). In song-flight ♂ takes off from elevated point, ascends at angle into wind, and starts singing at height of c. 30–70 m, sometimes continuing to 100–200 m; uses slow, fluttering wing-beats, with much hovering; flies in wide circles; av. song duration (Europe) 3·4 min (Cramp 1988). Chases conspecifics from territory; 2 birds chase around area of c. 30 m diam., land and face each other c. 1 m apart; both sing vigorously, 1 crouching, 1 erect, then crouching bird suddenly chases other away (Nigeria: Mundy and Cook 1972). Rebuffs territorial intrusions by Hoopoe Lark *Alaemon alaudipes* and sparrow larks *Eremopterix* spp. (Sudan).

Pair-bonding display: ♂ pursues ♀ in zigzag flight, calling. On ground, ♂ erects crest vertically, fans raised tail, half crouches, and ruffles plumage (**A**), minces towards ♀ and dances in same posture in front of her. ♀ usually crouches silently; if she moves away, ♂ chases her (Cramp 1988). Courtship ends abruptly. Copulation may follow pursuit-flight; ♀ drops to ground, lies flat or crouches, with wings trembling violently, tail cocked and cloaca exposed (**B**); ♂ climbs rather than flies onto her. Copulation repeated, with wing-shivering between acts.

NEST: shallow depression in ground with untidy lining of straws and other vegetation. Sometimes domed. Ext. diam. av. 135, int. diam. 63, depth of cup 79, walls av. 36 thick. Sited in open, in shelter of low shrub or tussock, or under bank. Built by ♀, taking 1–4 days (Cramp 1988).

EGGS: 3–5, 115 clutches av. 3·9 eggs (N Africa); 3–6 (Somalia). Subelliptical; glossy; whitish or grey, finely spotted and speckled buff and grey, sometimes making zone at broad end. SIZE: (n = 16, *G. c. arenicola*) 20–23·4 × 16–17·8 (21·9 × 17·0). WEIGHT: c. 3·3.

LAYING DATES: N Africa, Apr–June; Senegambia, Sept–June; Mali, Apr–May; Togo (juv., Feb); Nigeria, Nov–Mar, May; Ethiopia, Dec–Mar; Somalia, Apr–May; E Africa: Region D, Mar.

INCUBATION: by ♀ only, but ♂ may stand over eggs shading them from sun when ♀ away. Begins with last or 3rd egg. Incubation spells of c. 1 h; in breaks of 10–20 min ♀ preens, ♂ sings nearby, both may move 400 m away. ♀ lands 1–15 m away from nest, looks around, and walks or runs to it, making good use of any cover (Cramp 1988). Hatching takes 2 days. Period: 11–13 days.

DEVELOPMENT AND CARE OF YOUNG: young brooded by ♀ for 5–7 days. Until young 2 days old, ♀ leaves nest every 20–30 min to collect food; at 3–4 days, ♀ broods in bouts of 15–20 min (Denmark). At first ♀ eats faeces; after 3–5 days, ♀ carries faecal sacs (and regurgitated pellets) away. Eyes open at 4–6 days. Leave nest at c. 8–11 days, sometimes returning to it until they can fly. First flight at 15–16 days; sustained flight at 20 days. Young fed by both parents before and after fledging.

Reference
Cramp, S. (1988).

A

B

Galerida malabarica (Scopoli). Thekla Lark; Short-crested Lark. Cochevis de Thékla.

Alauda malabarica Scopoli, 1786. Del Flor. et Faun. Insubr., fasc. 2, p. 94; China = Malabar Coast.

Plates 2, 4
(Opp. pp. 33, 49)

Range and Status. Africa, Spain, Balearic Is., India.

Resident, frequent to common; locally commoner than Crested Lark *G. cristata*. Morocco, frequent and widespread except in south; recent records from Ouezzane, Sefrou, Itzer, Zeïda, Midelt, Settat, Sidi Bou Othmane, Ijoukak, Bouarfa, Tinejdad, Boumalne du Dadès and in Sous (Massa, Taroudant, Aoulouz). Algeria, coast north of Eastern Erg, Ouargla, south of Mzab, El Goléa, north of Western Erg, Béchar; up to 2300 m on Jebel Mahmel, Aurès. Tunisia, common throughout, but absent Kroumirie, and breeds along coasts only between Sfax and Gabès; scattered in central plateau; numerous in mountains from Kasserine to Gafsa and south of El Hamma-Kebili to Remada, Djeneien. Libya, locally common in Tripoli; all of Jebel Nafusa east to Homs; also Jebel Waddan (16°E) and Sirte; Cyrenaica, from Ajadabia to Jebel Akhdar – common in hills. NW Egypt: was common near Salum in 1920 but has not been recorded since 1928 (Cramp 1988). Ethiopia, frequent in W Highlands, SE Highlands and S Ethiopia. Somalia, abundant and widespread south to 5°N, extending down E coast to 1°N. Kenya, locally common at 400–1300 m from Huri Hills and Turbi to Marsabit and Loiyengalani, also extending all along E shore of L. Turkana (A. D. Forbes-Watson, pers. comm.) and numerous in Dida Galgalla Desert.

Description. *G. m. erlangeri* Hartert: N Morocco (Tangier south to Oulmès and Azrou, east to Algerian border). ADULT ♂: plumage greyish brown to dark brown above, heavily streaked on forehead, crown, mantle, back and upperwing-coverts with blackish brown. Outer tail-feather pale rufous; T5 with black inner web and pale rufous outer web and tip; T1 dark brown, fringed greyish; remaining tail-feathers blackish. Lores pale buff; featherlets around eye pale buff; ear-coverts brown, streaked with dark brown; sides of face below eye blackish, joining poorly-developed blackish moustachial streak. Centre of chin and throat off-white or very pale buff; sides of chin and throat mottled blackish; breast off-white or very pale buff, lightly or heavily marked with brownish black streaks or blotches; remainder of underparts very pale buffy white. Remiges and upperwing-coverts brown or dark brown, narrowly fringed with pale buff; some coverts broadly fringed pale buff and with dark brown centres. Bill dark brown with pale pinkish base to lower mandible; eye hazel to dark brown; legs and feet flesh-brown or yellow-flesh. Sexes alike. SIZE: wing, ♂ (n = 10) 101–107 (104), ♀ (n = 10) 94–100 (97·0); (*G. m. ruficolor*) wing, ♂ (n = 25) 102–110 (106), ♀ (n = 11) 95–102 (99·2); tail, ♂ (n = 10) 58–64 (60·7), ♀ (n = 8) 48–58 (53·6); bill to skull, ♂ (n = 15) 16·6–19·2 (18·1), ♀ (n = 8) 16·3–18·5 (17·4); tarsus, ♂ (n = 10) 24–27 (26·0), ♀ (n = 8) 24–26 (25·0). WEIGHT: (*G. m. theklae*, Spain) unsexed (n = 23) 34–41 (36·8).

IMMATURE: ground colour of upperparts buffer than adult, all feathers with narrow pink-cinnamon fringe on sides of broader white tip, with dusky subterminal line between white tip and rufous-brown centre; crest pale-tipped. Dark streaks on breast are sparse and obscure.

NESTLING: down long and dense on upperparts and flanks, sparse on underparts, whitish or yellowish; mouth dull orange-yellow with small black spot inside tip of both mandibles, single spot on each side of base of tongue, and 3 tiny spots and thin black line near tip of tongue (Cramp 1988).

G. m. ruficolor Whitaker: Morocco (south of *erlangeri* to High Atlas), N Algeria and N Tunisia (south to 36°N). Feathers of upperparts with broad black centres and pale greyish rufous fringes. SIZE: (see above). WEIGHT: ♂ (n = 5) 32–39 (36·2).

G. m. aguirrei Cabrera: Morocco (south of Anti-Atlas Mts) to NW Mauritania. Like *ruficolor* but dark feather-centres of upperparts and breast narrower; upperparts more rufous.

G. m. superflua Hartert: E Morocco (west to Moulaya R.) and N Algeria (Hauts Plateaux, Atlas Saharien) to E Tunisia. Sandier above than *erlangeri* and much less heavily streaked (feather-centres of upperparts narrow, dark olive-brown with greyish fringes); rump and central tail-feathers pale rufous; whiter below than *ruficolor*, much whiter than *erlangeri*, with smaller, more discrete breast streaks. Wing, ♂ (n = 11) 105–109 (107), ♀ (n = 13), 99–103 (101). WEIGHT: ♂ (n = 21) 33–43 (39·6).

G. m. carolinae Erlanger: Algeria (N Sahara), Morocco (Figuig), central Tunisia, N Libya, NW Egypt. Like *superflua* but spots on breast narrower. Wing, ♂ (n = 11) 99–105 (102), ♀ (n = 8) 94–98 (95·9). WEIGHT: ♂ (n = 3) 38–39·5 (38·9).

G. m. deichleri Erlanger: Algeria (NE Sahara), S Tunisia (Douz). The most distinctive race: very pallid above, with crown and nape whitish or pale rufous or somewhat streaked pale brown, and remaining upperparts very pale rufous without dark marks; below white, with pinkish wash and with small, scattered dark streaks on breast.

G. m. praetermissa (Blanford): highland and coastal Ethiopia. Like *erlangeri* but underparts buffer and breast speckles smaller.

G. m. huei Erard and de Naurois: Ethiopia (Bale Mts, Arussi). Like *praetermissa* but streaking of upperparts blacker and more intense; underparts white, washed pale buff. Wing, ♂ (n = 4) 101–107 (104), ♀ (n = 3) 95·5–97 (96·5); bill to skull, ♂ (n =

4) 16–18 (17·0). WEIGHT: ♂ (n = 4) 34–37 (35·0), ♀ (n = 3) 32–34 (33·3).

G. m. huriensis Benson: N Kenya, SE Ethiopia. Like *praetermissa* and *erlangeri* above but nape and upper mantle paler, and crown striping more obvious since blackish feathers have very pale edges; chin and throat pure white (i.e. whiter than *erlangeri* or *praetermissa*); belly pinkish white.

G. m. ellioti Hartert: Somalia. Above like *huriensis* but less saturated, sandier and slightly more rufescent; underparts like *huriensis*.

G. m. harrarensis Erard and Jarry: Ethiopia (Jigjigga, Harrar). Like *ellioti* but darker, less sandy; dark streaks wider above and on breast. Bill short and fine. Wing, ♂ (n = 2) 100, 101, ♀ 93; bill to skull 14·5.

G. m. mallablensis Colston: Somalia (Mallable, 30 km northeast of Mogadishu). Like *ellioti* but upperparts grey-tinged; feathers edged pale whitish, greyish or buffy (rather than warm sandy-brown); lacks cinnamon wash on nape, uppertail-coverts and ear-coverts (present in *ellioti*); breast more heavily streaked dark brown; back, rump and flanks greyer. Wing, 1 ♂ 96; tail 51; bill to skull 16·5; tarsus 22. WEIGHT: 1 ♂ 33·8 (Colston 1982).

Field Characters. Same as Crested Lark *G. cristata* q.v., from which adult and juvenile (but not nestling) almost indistinguishable except by smaller size, finer bill, and sometimes by habitat. Geographically variable, greyish brown, rufous, sandy or pale rufescent cream, heavily or lightly streaked above and on breast, or virtually plain. Distinctions from Crested Lark: Thekla Lark slightly smaller, with short, more fanned, less spiky crest; bill usually shorter and looks broad-based and straighter, less decurved than in most Crested Larks; breast spots better-defined (like Dupont's Lark *Chersophilus duponti*) but stop short of ear-coverts, leaving whitish half-collar from throat to side of neck; underwing-coverts greyish (not cinnamon-buff); uppertail-coverts pale rufous-grey, contrasting with grey-brown rump (no contrast in Crested Lark) (Hollom *et al.* 1988).

Voice. Tape-recorded (63, 73, JOHN, ROC). Like voice of Crested Lark q.v., but repertoire more restricted. Song attractive, musical, with many thin but loud and drawn-out whistles, rattling like *Sylvia* warbler, thrush-like phrases (*Turdus*), finch-like twittering, buzzing and mewing: 'sweee-swee-eee-sweeeoo', 'chew-wheeeeeeoo', 'tweet-a-tweet-a-twee', with 5 bursts in 43 s including pauses of 3–5 s; gentle, unstressed, with pulsating sounds like Eurasian Skylark *Alauda arvensis*, but less shrill and vibrant (Cramp 1988). A versatile mimic of other birds, including Woodchat Shrike *Lanius senator*, Sardinian Warbler *Sylvia melanocephala*, Great Tit *Parus major* and Chaffinch *Fringilla coelebs*. Contact-alarm call a soft, melodious, fluty whistle of 2–4 notes, rising then falling in pitch. Flight call, uttered commonly, a buzzy, twittering 'twititititititi-tiree-tweeee' or 'twirr-twirr-twirr-tweeeerrrr-tweeeeee', last note high-pitched (D. A. Zimmerman, pers. comm.).

General Habits. Generally uses rockier, more hilly ground than Crested Lark, where such ground abuts onto bare, dry, open soil, walled cereal fields, or wadi beds with oleander *Nerium*; heath, fallow arable land, abandoned farmland, *Stipa* grass steppe, coastal dunes, plantations of *Euphorbia cactoides*; also around villages, where ♂♂ sing from rooftops. In Ethiopia at 1550–1770 m in stony acacia steppe and on plateaux above 2300 m and as high as 3200 m; in Kenya in dwarf scrub grassland on rocky ground or lava at 400–1300 m; and in Somalia in open grassy and lightly-bushed areas, dissected ground and dunes.

In pairs. Not shy. Flock of *c.* 20 (once, Morocco, Nov), and more prone to flock than Crested Lark (Morocco, Eritrea). Sometimes associates with Crested Lark and with flocks of Lesser Short-toed Larks *Calandrella rufescens* and Temminck's Horned Larks *Eremophila bilopha*. Forages on ground, where gait and feeding actions much like Crested Lark but less hunched-looking, and nimbler. Perches on low plants. Searches under stones by pushing bill under and flipping stone over with quick sideways jerk (Cramp 1988). Does not dig for food (cf. Crested Lark). Leaps up to 1 m to snatch insect in flight or from vegetation (Abs 1963). Roosts on ground amongst rocks, and in captivity scratches shallow pit in loose sand to roost in it at night. Comes to waterholes to drink, but can do without water.

Food. Insects and seeds. For Spanish diet, see Cramp (1988). 9 birds (Morocco) contained seeds of *Echium* grass, and Portulacaceae, and ants, weevils, fly pupae and eggs; 18 birds (Tunisia) contained grass seeds, grain, and small snails and beetles; 14 birds (Libya) contained grain and many grasshoppers (Cramp 1988). A millipede brought to nestlings (Morocco). In Bale Mts, Ethiopia, takes numerous insects (beetles, fly pupae are main prey) and some seeds (Dorst and Roux 1972).

Breeding Habits. Solitary, territorial, monogamous breeder. 4·7 pairs per km^2 (degenerate woodland, Morocco: Thévenot 1982), 10–18 pairs per km^2 in N Algeria. Av. 32 birds in nine 22-km motor transects and av. 50 birds in eight 24-km transects (Blondel 1962). ♂ sings from ground eminence or low vegetation or in flight. Takes off, and may start singing at 10–12 m; ascends very high, flying in wide circles with slow, weak, fluttering action with wings and tail fully spread. Sings up to 7 min, av. 3 min. Ceases singing at considerable height, then plunges vertically to perch on ground.

NEST: construction and site very like that of Crested Lark q.v. (Abs 1963).

EGGS: 3–7, av. of 21 clutches (N Algeria) 4·6 eggs, av. of 65 clutches (E Algeria, Tunisia) 3·9 eggs. 2 broods. Subelliptical; glossy; whitish or grey, finely spotted and speckled buff and grey, sometimes making zone at broad end. SIZE: (n = 35, *G. m. theklae*, Spain) 21·6–25 × 15·4–17·7 (22·8 × 16·5). WEIGHT: *c.* 3·2.

LAYING DATES: W Morocco, Feb–May; Algeria, Tunisia, Apr–June; Ethiopia, Feb–Aug, Oct; Somalia (n = 21), May–June; E Africa: Region D, May.

DEVELOPMENT AND CARE OF YOUNG: adult feeding 2 young in nest (Morocco) approached it in stages, hovered briefly and dropped onto or next to nest (Pasteur 1958). Young leave nest at *c*. 9 days, *c*. 6 days before they can fly (Abs 1963).

BREEDING SUCCESS/SURVIVAL: 12 found dead of exposure on Iknioun Piste, Morocco, on 1 day in Dec (Thévenot *et al.* 1982).

References
Abs, M. (1963).
Archer, G. F. and Godman, E. M. (1961).
Cramp, S. (1988).
Erard, C. and Jarry, G. (1973).
Erard, C. and Naurois, R. de (1973).
Pasteur, G. (1958).

Genus *Lullula* Kaup

Not satisfactorily distinguishable from *Alauda* or *Galerida* (see family diagnosis). A single species of lark, with medium-length crest arising in middle of crown, rather short, slender and tapering bill, small 10th primary, short tertials, hind claw longer than in all *Galerida* spp., and short tail. Plumage relatively rich and variegated. Arboreal and terrestrial. W Palearctic including N Africa.

Lullula arborea (Linnaeus). Woodlark. Alouette lulu.

Alauda arborea Linnaeus, 1758. Syst. Nat. (10th ed.) 1, p. 166; Europe.

Plate 2
(Opp. p. 33)

Range and Status. NW Africa, Europe north to 62°N, Asia Minor. East to 56°E in USSR and *c*. 57°E in Elburz Mts, Iran. Partial migrant, birds from NE half of range vacating it to winter in SW half. Major decline in most western areas of range during 20th century. Accidental Kuwait, Faeroes, Iceland.

Rather sparse and local resident in N Morocco, N Algeria and NW Tunisia. In Morocco widespread from sea level up to 2160 m in Rif; occurs in pre-Rif and Middle Atlas Mts, and High Atlas up to Oukaïmeden (2200 m), Yagour (2400 m) and B'Lu (2600 m). In Tunisia a rare resident breeder in Kroumirie and along Algerian border south to Bou Chebka, and perhaps in central plateau in north. Winter records from Tunis and Zarza. Scarce and irregular winter visitor to Tripoli coast, NW Libya, Nov–Mar and to Nile delta and NW Egypt coast. May have bred near Tarhuna and Garian, Libya, 1923.

Description. *L. a. pallida* Hilgert (only race in Africa): NW Africa and S Europe east to Iran. ADULT ♂: forehead and crown heavily streaked black and buff; mantle, back, scapulars and upperwing-coverts greyish brown or rufescent brown, heavily streaked with black; rump and uppertail-coverts uniform grey-brown. Central tail-feather brown; outer tail-feather pale greyish or buffy; T5 black with large white spot at end and with outer web narrowly edged white; T4 and T3 black with white spot at end; T2 all black. Lores pale; broad pale buffy eye-stripe; ear-coverts blackish brown, contrasting with pale greyish sides of neck. Chin whitish or very pale buff in centre with moustachial line of black speckles at side; throat very pale buffy white with small but discrete black spots; breast very pale buff, heavily marked with discrete black streaks; belly and undertail-coverts very pale buffy white. Primaries dark brown, narrowly edged white, the white showing conspicuously in closed wing at point of emargination of P8, 7 and 6; greater primary coverts black with broad whitish or buffy tip, and lesser and median primary coverts white. Secondaries dark greyish brown with buff outer edges and narrow pale tips, tertials dark brown with buffy white fringes, broader on outer webs, and tipped with browish buff triangular wedge; axillaries, underwing-coverts and broad inner margins to underside of primaries and secondaries white, rest of underside of primaries and secondaries with greyish sheen, giving pale appearance to whole of underwing. Upper mandible dark brown, lower mandible horn-brown with pinkish base; eye dark brown; legs and feet yellowish brown. Sexes alike. SIZE: (10 ♂♂, 5 ♀♀, Iberia) wing, ♂ 94–98 (95·6), ♀ 88–94 (91·1); tail, ♂ 48–53 (50·4), ♀ 45–48 (46·3); bill to skull, ♂ 13·6–14·9 (14·4), ♀ 14·2–15·1 (14·7), bill to feathers av. 2·9 shorter; tarsus, ♂ 21–23 (22·0), ♀ 20–22·5 (22·0). WEIGHT: (Spain) unsexed (n = 19) 23–29 (26·1).

IMMATURE: like adult but crown blacker; eye-stripe shorter; marks on breast dusky (not black) and rounder – looks spotted rather than streaked; more extensive white or pale buff in T5 and T6.

NESTLING: white or buffy grey down, long and dense on head and upperparts, shorter and scantier on underparts.

Field Characters. A NW African lark more likely to be confused with more common Eurasian Skylark *Alauda arvensis* than with any other. Medium-sized with short tail, broad rounded wings, fine bill, long, distinct, white superciliary stripe meeting on nape, rufescent cheeks, richly marked plumage with small black streaks on rufescent brown or buffy ground above and on pale buff breast, buffy white chin and belly, crest (often not apparent in field), characteristic black-and-white mark in carpal area of folded wing, and distinctive song. Smaller than Eurasian Skylark; legs pink-brown (yellow-brown in Eurasian Skylark); in flight tail looks blackish with whitish tips but without white edge.

Voice. Tape-recorded (CHA – 62, 73, B). ♂ song sweet, clear, liquid and mellow, delivered as 3–5 s phrases separated by 1–3 s pauses. Each phrase is largely a repetition of 1 note, or 2 notes alternating, typically with fluty 'lu-lu-lu' element (origin of name *Lullula*), starting quietly, followed by crescendo, diminuendo and accelerando in falling pitch. Each phrase seems different from the last. For analyses, see Cramp (1988). Some mimicry of other birds. Main call, 'd'lui', 't'luiii' or 'tew-leet'. ♀ song a twittering variation upon 'd'lui' call theme.

General Habits. Inhabits wooded, sunny hills and lowlands on well-drained soils, with scattered large trees, bushes and plenty of open grassy places; edges of cork oak *Quercus suber* and holm oak *Q. ilex* woods on stony hillsides, macchia and open woods up to 3000 m in Atlas Mts, lowland woods on dunes, felled woodlands, open pine-heath, and long-fallow farmland.

In pairs and family parties; occasionally flocks (> 20) in winter. Feeds on ground, bustling about quietly and inconspicuously. Perches freely in bushes and trees, walking along large limbs. Dust-bathes. Roosts on ground, in grass or stubble, near bushes. Flight undulating and hesitant, with jerky, wavering progress. Sings freely in flight, on ground or in tree. Sings in Morocco from 1st week Apr.

Not a conspicuous migrant since not very gregarious. Ringing recoveries in Europe show strong southwesterly component from summer in Baltic and N Europe to winter in SW France. No recoveries from Africa, where populations appear to be resident; but some northern birds winter on Egyptian and Libyan coastal plains, and movements across Straits of Gibraltar into Morocco occur mid-Oct to mid-Nov (38 birds in 1976, 133 in 1977: Tellería 1981).

Food. Mainly insects and spiders in summer and seeds in winter. In Europe arthropods include dragonflies, grasshoppers, thrips, larval ant-lions, adult, larval and pupal Lepidoptera and Diptera, Hymenoptera, beetles and grubs, spiders, centipedes and snails; and plant material includes seeds of pine, knotweed, hemp, pinks, poppies, legumes, composites and grass; also some leaves and buds (Cramp 1988).

Breeding Habits. Solitary, territorial, monogamous breeder. Not studied in Africa. Territory size 1–3 ha (Poland), 4–5 ha (Britain). In Germany 25 singing ♂♂ in woods of 115 km² and 26–30 in pinewood of 165 km². Song-flight is fluttering, looping, spiral ascent to 50–80 m, cruising aloft in wide circles, and descent sometimes with final plummet.

NEST: neat, substantial, deep cup made in deep depression in ground, of stems, leaves, pine needles, grass and moss, lined with finer grasses and some hair. Cup 66 wide, 59 deep; walls up to 100 thick at top and 22 at base. Nest and depression formed by both sexes in 6–7 days; in shelter of scrub, bracken or heather, often near base of sapling.

EGGS: 1–6, usually 3–4 (Britain). Subelliptical; slightly glossy; white with fine brown speckles and blotches, often forming band at broad end. SIZE (n = 100, *L. a. arborea*, Britain) (eggs of *pallida* same size) 19·7–23·5 × 15·1–17·4 (21·6 × 16·3). WEIGHT: *c.* 3·4.

LAYING DATES: N Africa, end Mar to end May.

INCUBATION: by ♀ only, beginning with last egg. For incubation behaviour (Poland), see Mackowicz (1970). Period: 12–15 days.

DEVELOPMENT AND CARE OF YOUNG: young hatch at 2·8 g; grow rapidly to days 9–11, when *c.* 21 g, then more slowly. Fledging period 10–13 days, but flight weak for some days thereafter. ♀ broods for much of time to day 4, ♂ feeding young; then brood fed by ♂ and ♀ equally. Young leave nest by walking. For further details, see Cramp (1988).

References
Cramp, S. (1988).
Mackowicz, R. (1970).

Genus *Alauda* Linnaeus

Not satisfactorily distinguishable from *Galerida* or *Lullula* (see family diagnosis). Small to medium larks with short to medium-length crest arising in middle of crown, short or medium tapering bill, minute 10th primary, tertials longer than those of most but not all *Galerida* spp., hind claw medium-long, not strongly decurved.

4 species, Palearctic: 1 endemic to Cape Verde Is (*A. razae*), 1 in S Asia (*A. gulgula*), 1 in Japan (*A. japonica*) and 1 throughout Palearctic including N Africa (*A. arvensis*). The last 3 form a species group.

Alauda arvensis Linnaeus. Eurasian Skylark. Alouette des champs.

Alauda arvensis Linnaeus, 1758. Syst. Nat. (10th ed.) 1, p. 165; Sweden.

Plate 2
(Opp. p. 33)

Range and Status. NW Africa, Europe north to 71°N, Asia between *c.* 40°N and Arctic Circle, east to E Siberia. Migratory, wintering in Europe west of 13°E, Mediterranean basin including N African coasts and Maghreb, N and SE Arabia, N India, and China between *c.* 21° and 45°N.

Breeding resident NE Morocco, N Algeria and NE Tunisia, and non-breeding winter visitor from N Morocco to N Egypt. Rather scarce breeder in cultivated lowlands of Morocco, Algeria and Tunisia, more numerous in coastal regions and on high plateaux. Winter visitor in large numbers in N Morocco and N Algeria, widespread and locally numerous, but uncommon in Tunisia, mainly in coastal lowlands south to Bahiret el Bibane. Particularly abundant in Le Rharb area, Morocco, and on Middle Atlas moorlands. Common in coastal and steppe areas of Tripoli (Libya) east to Sirte, and in Cyrenaican coastal zone; scarcer inland. Frequent in N Egypt.

Description. *A. a. harterti* Whitaker: NW Africa. ADULT ♂: forehead and crown brown with heavy black stripes, in some birds irregular, in others forming *c.* 5 rather neat black rows; hindneck and sides of neck greyer and paler brown with blackish stripes; rest of upperparts brown or greyish brown, heavily mottled blackish brown with rufescent brown margins and paler whitish tips. Uppertail-coverts long. Tail dark brown; T6 all white, T5 with outer half white and inner half blackish (end of inner half sometimes white). Lores and narrow superciliary stripe whitish buff; ear-coverts pale brown, streaked dark brown. Chin and throat buffy white, throat with fine, short, dark brown streaks; breast buffier, washed rufous, strongly streaked black-brown; flanks the same but much less streaked; belly and undertail-coverts whitish. Wing feathers dark brown with tips, inner webs and outer webs of outer primaries fringed buffy white, outer webs of secondaries and inner primaries fringed rufescent-buff and innermost secondaries and tertials fringed pale buff distally; all upperwing-coverts dark brown with blackish centres and well-defined pale buff tips. Axillaries and underwing-coverts grey, washed buff. Upper mandible dark horn-brown, lower mandible pale horn; eye dark brown; legs and feet yellowish brown. Sexes alike. SIZE: (Morocco) wing, ♂ (n = 7) 111–114 (112), ♀ (n = 2) 101–103.

IMMATURE: upperparts like adult but each feather white-tipped and buff-edged; the tertials, crown and some mantle feathers with narrow black subterminal line between buff fringe and brown interior. Breast spotting diffuse. Feathers of underparts fluffy.

NESTLING: skin blackish dorsally, whitish ventrally with wispy, straw-coloured down on forehead, nape, scapulars, rump and flanks. Down sparse, up to 12 mm long on rump.

TAXONOMIC NOTE: extent to which various races reach N Africa in winter is unclear and race of great majority of non-breeding visitors unknown. Nominate *arvensis* (Azores and Britain to Urals, south to Alps) has been taken in Tunisia; *cantarella* (NE Spain, S Europe, Mediterranean islands, east to Caucasus) said to reach N Africa from Balkans and USSR (Cramp 1988); *intermedia* (E Asia) said to reach Algeria and Tunisia (Etchécopar and Hüe 1967) but is most unlikely to occur there (Cramp 1988); numerous birds crossing Straits of Gibraltar and wintering in Morocco may be *arvensis*, or *cantarella* and *sierrae* (S Iberia) (Cramp 1988). Strength of migration to W and SW Europe from USSR east to at least 50°E makes it likely that these and other races all enter Africa.

A. a. arvensis Linnaeus: (see above). Darker than *harterti*, with heavier streaks on upperparts. SIZE: (Netherlands, Sept–Feb) wing, ♂ (n = 226) 109–123 (115·5), ♀ (n = 193) 99–112 (105); tail, ♂ (n = 37) 62–70 (66·2), ♀ (n = 19) 59–64 (61·5); bill to skull, ♂ (n = 38) 14·1–17·2 (15·6), ♀ (n = 20) 13·6–16·1 (14·8), bill to feathers av. 3·2 shorter; tarsus, ♂ (n = 37) 23·5–27 (25·0), ♀ (n = 21), 22·5–25·5 (24·0). WEIGHT: (Nov) ♂ (n = 90) 34–55 (44·1), ♀ (n = 68) 28–47 (37·6).

A. a. cantarella Bonaparte: (see above). Like *harterti* but more olive. Smaller than *arvensis*: wing, ♂ (n = 18) 109–117 (112), ♀ (n = 8) 99–106 (103).

A. a. sierrae Weigold: (see above). Like *arvensis* but ground colour pinker or sandier; same size as *cantarella*.

Field Characters. Medium-sized lark, streaky brown above, buffy white below, with closely-streaked breast rather sharply demarcated from white belly; with short but often-used crest, whitish trailing edge to wings, white tail-sides, characteristic song and flight calls, sustained song-flight, and habit of hovering briefly before landing. Most likely to be confused, in northern Maghreb, with Crested Lark *Galerida cristata*, Lesser Short-toed Lark *Calandrella rufescens*, Woodlark *Lullula arborea* and Dupont's Lark *Chersophilus duponti*. Crested Lark has longer crest and warm buff sides of tail; Lesser Short-toed Lark is small, not crested, with short pale bill and short hind claw; Woodlark has short, blackish, white-tipped tail and black-and-white carpal mark in closed wing; and Dupont's Lark has long decurved bill. When flushed, gives 'chirrup' call and shows characteristic white outer tail-feathers and white trailing edge to wing.

Voice. Tape-recorded (CHA – 62, 73, B). Song, delivered with spirit, is loud, modulated, shrill warble, delivered without pause for 3–5 min while bird hovers 50–100 m up in air. Song is lilting, with undulating pitch pattern (Cramp 1988), trills, tremolos, and frequent repetition of notes and phrases. Sometimes mimics other birds. Flight call a liquid rolling or rippling 'chirrup', 'tschirit' or 'tschrit-ju'. For several other calls, see Cramp (1988).

General Habits. Inhabits cultivated land, pasture, open grass sward of playing-fields and airfields, edges of marshes, dunes, saltings, limestone downs, and grassy plateaux and moors in Atlas Mts.

Solitary or in pairs and family parties on breeding grounds, gregarious in winter. Walks with legs well flexed; crouches when uneasy. Feeds by walking over ground, taking items from soil or low plants; occasionally digs with bill in loose soil, but not deep or persistently. Does not drink in central Europe but may do so in Africa. Usually calls on taking flight. Flight over short distance fluttering or wavering; frequently hovers 1–2 m above ground. Prolonged flight strong and a little undulating, with floppy wing action (Witherby *et al.* 1938). Perches on low bush or wall, but avoids trees. Often dust-bathes. Roosts in short or tussocky grass.

Autumn migration is west or southwestward across Europe, from N and central USSR (east to *c.* 50°E) to W and SW Europe, with numerous recoveries in W France and Iberia. Birds from SW USSR and SE Europe move southwest to Italy and SE France; neither group has yet produced recoveries in N Africa. Winter visitors present in Libya and Egypt Oct–Apr. Inland in Libyan desert south to Serir, 27° 20′N: 30 in Mar, Apr, 20 in early May; and in Morocco in absolute desert at Tiznit, Nov; and 30 in cultivated Dadès valley, Feb (Smith 1965).

Food. Not studied in Africa. In Europe feeds in stubble mainly on grain, in cereal fields on grain, in ploughed land on weed seeds, in winter wheat fields on wheat leaves, in leas and vegetable fields on weed seeds, and in other young crops on insects, grass flowers, leaves and seeds. For list of large variety of invertebrates and plant materials, see Cramp (1988).

Breeding Habits. Few African data. Solitary, monogamous, territorial breeder. Territorial boundaries often contiguous in Europe; territory usually 0·5 ha in size (limits, 0·25 ha, *c.* 20 ha). In song-flight ♂ takes off into wind, and at 10–20 m starts steep spiralling ascent accompanied by singing, with tail spread and wings fluttering; at height of 50–100 m, bird hovers (10–20 beats per s), alternating with slow horizontal circling over territory; after several min it descends in slow spirals, still singing, wings spread but not flapping, either gliding down to land or suddenly stopping singing at 10–20 m and plummeting to ground (Cramp 1988).

NEST: shallow depression in ground with foundation of grass leaves and stems, lined with finer grasses and sometimes a little hair. Sited in open or among tussocky grass or young crops. Built probably by ♀ only.

EGGS: 3–7, usually 3–5 in Europe, probably fewer in Africa (Etchécopar and Hüe 1967). Subelliptical; glossy; grey-white, often tinged greenish, thickly spotted with olive or brown. SIZE: (n = 11, NW Africa) av. 23·2 × 17·2; (n = 300, nominate *arvensis*) 20·5–26·5 × 15·3–18·5 (23·4 × 16·8). WEIGHT: 3·35.

LAYING DATES: NW Africa, Apr–May.

INCUBATION: by ♀ only. Period: 11 days.

DEVELOPMENT AND CARE OF YOUNG: clutch hatches within 8 h; feeding by ♀ begins a few min after 1st young hatches, by ♂ 2 h later. ♀ broods up to day 5. Eyes open on day 4, then young fed by ♂ and ♀ equally. Faecal sacs at first swallowed by parent, later carried away in bill and dropped in flight. Young spend av. 8·5 days in nest (Britain) and leave spontaneously within 4 h. Parental feeding declines after day 12. Fledging period 15–20 days. Young starts feeding itself as soon as it can fly well.

BREEDING SUCCESS/SURVIVAL: of 319 eggs (Britain), 67% hatched and 46% fledged; av. annual adult mortality (Britain) 33·5% (Delius 1965). Oldest ringed bird 8 years 5 months.

References
Cramp, S. (1988).
Delius, J. D. (1963, 1965).

Genus *Eremopterix* Kaup

Small, gregarious larks of open grassland, savannas, shrublands and sparsely vegetated desert plains. Plumage patterned, sexes not alike; ♂♂ with much black on head and underparts, ♀♀ mainly brown and streaky, sparrow-like. Bill short and strong, nostrils concealed by plumelets; feet and legs small. Nest cup-shaped, without dome, with unusual felted spider web and sand rim in some species.

7 species world wide, 6 in Africa, 5 endemic. 1 superspecies in Africa, *E. verticalis/E. signata*; *E. leucotis* and *E. nigriceps* form a species group.

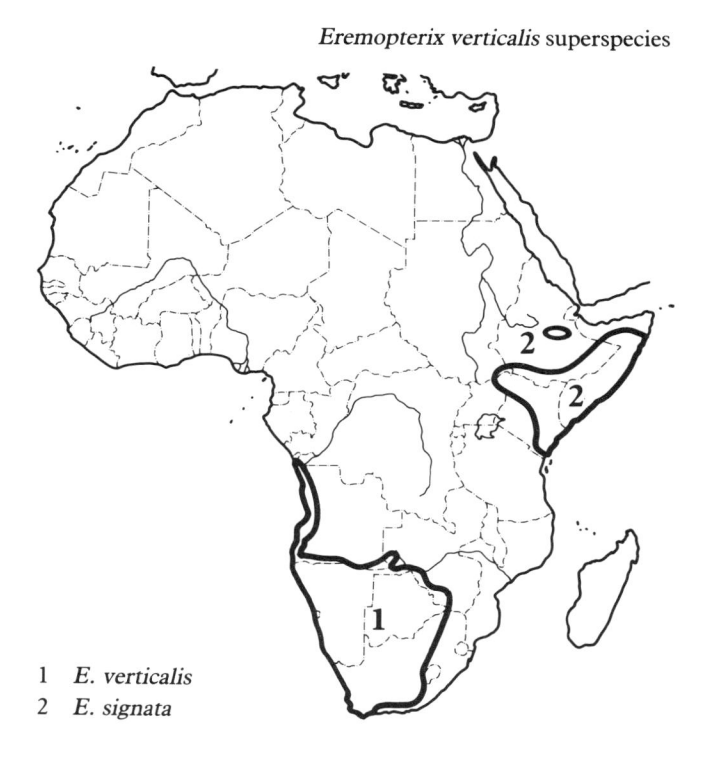

Eremopterix verticalis superspecies

1 *E. verticalis*
2 *E. signata*

Eremopterix australis (Smith). Black-eared Sparrow-Lark. Alouette-moineau à oreillons noirs.

Megalotis australis A. Smith, 1836. Rep. Exped. Expl. Cent. Afr., p. 49; country along the Orange R. (= Colesberg).

Plate 6
(Opp. p. 97)

Range and Status. Endemic resident, nomadic. S Namibia south of 24°S; S Botswana in S and SW Kalahari; South Africa: Cape Province, from N Namaqualand south and east across Great Karoo to Prince Albert, Cradock and Tarkastad, exceptionally to Oudtshoorn; widespread in N Karoo (locally common); vagrant W Transvaal; scattered localities in W and S Orange Free State (exceptionally to Excelsior in E, generally rare). Locally common, abundant at times.

Description. ADULT ♂: head, neck and underparts entirely black except for crown, which is dark sooty brown, shading to brown on nape and hindneck; sides of neck and foreneck sooty brown. Mantle dark brown, back reddish brown or dark chestnut, rump brown, uppertail-coverts dark brown. Tail blackish brown, central pair of feathers brown with paler fringes. Flight feathers dark brown or blackish; scapulars reddish brown or chestnut with dark centres, tertials brown, with narrow buffy inner and broader outer edge (edged rufous in some birds), wing-coverts brown edged pale brown to rufous. Underwing entirely blackish. Bill bluish white; eye orange-red to reddish brown; legs and feet brownish to greyish white. ADULT ♀: crown, nape and hindneck dull rufous-brown to chestnut, streaked darker brown, sides of neck and foreneck brown, some streaking on foreneck, cheeks and ear-coverts brown, indistinct pale supercilium and eye-ring, chin and throat whitish speckled brown. Mantle brown, back russet, rump and uppertail-coverts darker than back, all streaked darker brown. Tail dark brown, outer feathers with outer webs buff, adjacent pair with narrower buffy edges, central pair brown with broad paler brown margins. Underparts whitish or greyish white, streaked from chin to undertail-coverts with dark brown or blackish; sometimes a blackish patch on lower breast. Flight feathers dark brown, primaries narrowly edged buff; tertials dark brown, broadly edged rufous-brown, scapulars and wing-coverts rufous-brown with darker shafts. Underwing dark brown. SIZE: (10 ♂♂, 6 ♀♀) wing, ♂ 76–80 (78·0), ♀ 74–78 (75·3); tail, ♂ 41–45 (43·1), ♀ 39–44 (41·3); bill, ♂ 10–12 (10·8), ♀ 10–12 (11·1); tarsus, ♂ 15–18 (16·4), ♀ 16–19 (17·5). WEIGHT: ♂ (n = 6) 12–15·5 (14·0), 1 ♀ 15, 1 unsexed 15.

IMMATURE: similar to adult ♀, but above barred with buff and blackish; below mottled dusky and off-white on breast and flanks.

NESTLING: skin bright pinkish flesh, deep purplish black on feather tracts, eyes blackish; sparse light grey down on head and back. Gape pale yellowish white, inside of mouth yellow with 3 black spots on tongue, 1 on either side and 1 at tip; 1 black spot at tip of upper and lower mandibles.

Field Characters. A dark sparrow-lark. ♂ has head and underparts entirely black, distinguishing it from Chestnut-backed Sparrow-Lark *E. leucotis*, which has white face, nuchal collar and belly. Rounded black wings conspicuous in flight. ♀ has upperparts dull rufous with dark streaks, underparts mottled black and white, without the black patch on belly of sympatric sparrow-larks. Further distinguished from ♀ Chestnut-backed by brown rump concolorous with back, not contrastingly pale.

Voice. Tape-recorded (F, GIB). Short, twittering song given on ground, alarm call sharp 'tsee' or 'dzee', or 'preep'; distress call 'chee-chee-chee' (Maclean 1985). Song in display flight consists of buzzy notes.

General Habits. Present in sparse grassland on sands, dwarf shrublands on sands, clays and stony soils, and in dwarf shrublands on poorly drained clays.

Gregarious even when breeding, usually in groups of 5–10; when not breeding, flocks may number 50–100 birds. Forages in open areas among shrubs and stones; creeps about picking food items from ground surface or bases of plants. Does not drink water, and is totally dependent on metabolic water.

Nomadic over most of range, following rainfall.

Food. Invertebrates: harvester termites (*Hodotermes*), beetles, ants (*Tetramorium*), leaf-hoppers (Homoptera); seeds of grasses (*Schmidtia*, *Stipagrostis*) and forbs (*Gisekia*, *Chenopodium*, *Hypertelis*, *Hermannia*) and fruits of *Lycium*. Gravel up to 3 mm diam. usually present in stomach. Young fed exclusively invertebrates (harvester termites, other insects).

Breeding Habits. Monogamous, territorial. In display, ♂ flies over territory in large circles a few m above ground, giving a buzzy call; wing-beats exaggerated, 'butterfly-like', effect accentuated by broad black wings.

NEST: built by ♀ only, in 4–5 days; cup of fine grass and rootlets, int. diam. 52, depth of cup 33 (n = 50), on foundation of sticks in excavation in ground. Rim of nest usually a little above surrounding substrate, covered with felted mixture of sand and spider-webs (taken from nests of *Seothyra* sp. spiders) brought by ♂, applied by ♀. Sited at base of shrub or grass tuft, occasionally out in open. Most nests in Kalahari faced east (Maclean 1970).

Eremopterix australis

EGGS: 2–4, mean (60 clutches) 2·1. Sub-elliptical; white, finely speckled with pinky brown, concentrated to form zone at obtuse end. SIZE: (n = 103) 16·7–20·6 × 12·3–14·9 (18·3 × 13·4).

LAYING DATES: South Africa: N Cape, Jan–Feb, May–July (varies according to rainfall); Karoo, Feb, July, Sept–Nov; Orange Free State, Oct.

INCUBATION: by both sexes; period 12 ± 1 days.

DEVELOPMENT AND CARE OF YOUNG: young fed, brooded and shaded by both adults; ♂ does major share of brooding and shading. Begging released in young by feeding call, a single short 'preep'. Nest sanitation by both adults. Young leave nest at 7–10 days, led away by ♀ using feeding call. Both adults perform elaborate distraction displays if intruder near nest; ♂ gives 'butterfly' flight over territory, also hovers and flutters over nest; display accompanied by buzzing 'dzee' alarm call, and 'preep'. ♂ may perch on shrub and display with spread wings before resuming butterfly flight. Both adults perform injury-feigning display; swoop towards intruder with loud wing-beats (in ♂♂ this is high-pitched 'whoop'), flop down on ground and flutter away with a soft 'tik tik tik tik' call (♂), or raspy 'che chee chee chee' (♀). Also wing-snaps in alarm.

BREEDING SUCCESS/SURVIVAL: of 78 eggs in 37 nests, 25 hatched and 8 young left nests (Kalahari: Maclean 1970).

Reference
Maclean, G. L. (1970).

Eremopterix leucotis (Stanley). Chestnut-backed Sparrow-Lark. Alouette-moineau à oreillons blancs.   Plates 4, 6
(Opp. pp. 49, 97)

Loxia leucotis Stanley, 1814. *In* Salt's Travels in Abyssinia, App., p. 60; coast of Eritrea.

Range and Status. Endemic; resident and wanderer or intra-African migrant. Widespread in sahelian and soudanian savannas from Senegambia, S Mauritania and Mali from *c.* 10 to 17°N (though common only to 14°N) across to E Ethiopia and NW Somalia, extending farther north in Nile valley (Sudan) to *c.* 20°N; then rather more local and sparse south to NE Uganda (Mt Kamalinga), drier areas of Kenya, extreme S Somalia, and N Tanzania, south to Tabora, Mbeya and coast at Dar-es-Salaam. 350 km further south, reappears from central Malaŵi, SE Zambia and Mozambique to N Namibia, S Angola, Botswana, Zimbabwe, and NE South Africa. Extends through Transvaal and Orange Free State to N Natal and Colesberg area of Cape Province. Northern populations occur from sea level up to at least 1800 m, but southern ones appear to be largely absent above *c.* 1200 m. Common, seasonally abundant, in flocks on drier plains.

Description. *E. l. melanocephala* (Lichtenstein): Senegambia to Nile R. ADULT ♂: crown, face, around eye, sides of neck, chin, throat, breast, belly, undertail-coverts and underwing very dark blackish chocolate-brown; cheeks and ear-coverts white; nape has white patch extending onto sides of neck but not continuous with cheeks; mantle, back and scapulars chestnut with variable number of feathers edged or tipped white or grey; rump and uppertail-coverts pale pinkish buff; tail brown, with outermost feather paler (outer web almost white), and central pair with rufous edges; flanks greyish white; primaries brown with outer edges tinged rufous and inner edges paler; secondaries brown with broader rufous edges and pale tips; tertials dark brown with broad rufous edges and tips; wing-coverts mainly white, with a few scattered chestnut feathers. Bill pale horn to white; eye brown; legs and feet horn to greyish flesh. ADULT ♀: like ♂ except as follows: crown brownish chestnut, feathers with dark centres which may be fairly prominent; nape paler; mantle and back brownish chestnut, some feathers with dark centres but usually paler and duller than ♂; white patch on face mottled with brownish grey; chin, throat and breast very pale buff with dark brown or chestnut centres to feathers giving mottled appearance; centre of belly dark brown paling to buff on flanks; wings like ♂ but chestnut much duller. SIZE: (10 ♂♂, 10 ♀♀) wing, ♂ 76–81 (78·0), ♀ 73–80 (76·1); tail, ♂ 48–55 (49·4), ♀ 43–47 (45·9); bill, ♂ 10–11·5 (11·3), ♀ 10–11·5 (10·8); tarsus, ♂ 15–16·5 (15·7), ♀ 15–16·5 (15·7). WEIGHT: (Nigeria, n = 5) 12·3–13·8 (13·2), 1 immature 11·9.

IMMATURE: like adult ♀, but head, mantle and all wing-coverts dark brown with broad white tips to feathers; breast like adult ♀; chin, throat and belly mainly greyish white with some brown centres to feathers.

NESTLING: skin black, with feather tracts tan; eye black; gape white; inside of mouth orange with papillae on palate; tongue orange with 3 black spots.

E. l. leucotis (Stanley): E and S Sudan to Ethiopia (including Eritrea). ♂ with shoulder black or dark brown (not chestnut and white); mantle and wings darker than in ♂ *melanocephala*; ♀ very much darker generally than ♀ *melanocephala*.

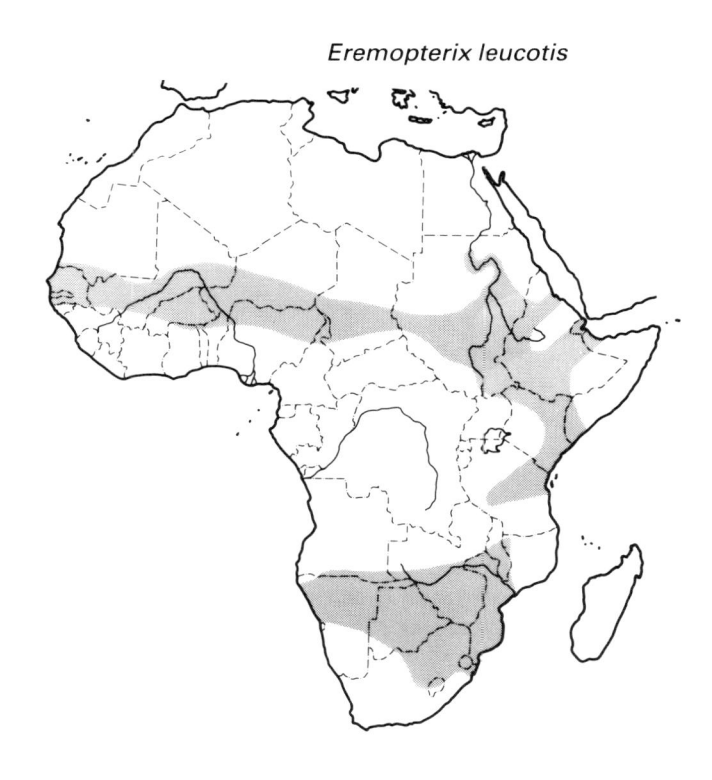

E. l. madaraszi (Reichenow): NE Uganda, Kenya and N Tanzania, and a separate population in N Malaŵi and Mozambique. ♂ richer red above and with larger bill (5 ♂♂ 11–14 (12·3)), and ♀ duller, less reddish on upperparts and whole throat and foreneck, as well as middle of belly, dull black.

E. l. smithii (Bonaparte): S Malaŵi and S Zambia south through Zimbabwe (except west), and E Botswana to South Africa. ♂ paler and more tawny than *leucotis* and *madaraszi*, and black on shoulder more extensive; ♀ paler overall, with black restricted to middle of lower breast and belly. Larger: wing, 10 ♂♂ 78–87 (81·9).

E. l. hoeschi White: N Namibia, S Angola, N and NE Botswana east to W Zimbabwe. Back of ♀ colder and greyer, and both ♀ and immature paler below than other races.

Field Characters. A dark sparrow-lark. ♂ has head and neck entirely black except for white face-patch and nuchal collar; back rich chestnut, belly white, pale rump shows in flight. Only other sparrow-lark with chestnut back is Black-eared *E. australis*, but ♂ easily distinguished by lack of any white in plumage and brown rump. ♀ similar to ♀ Black-eared Sparrow-Lark but has black belly, pale rump (rump of ♀ Black-eared same colour as back). Overlaps with several other species of sparrow-lark but these are all paler, with grey or light brown upperparts, ♂♂ with heads patterned white with brown or black, ♀♀ without mottled chestnut upperparts of ♀ Chestnut-backed.

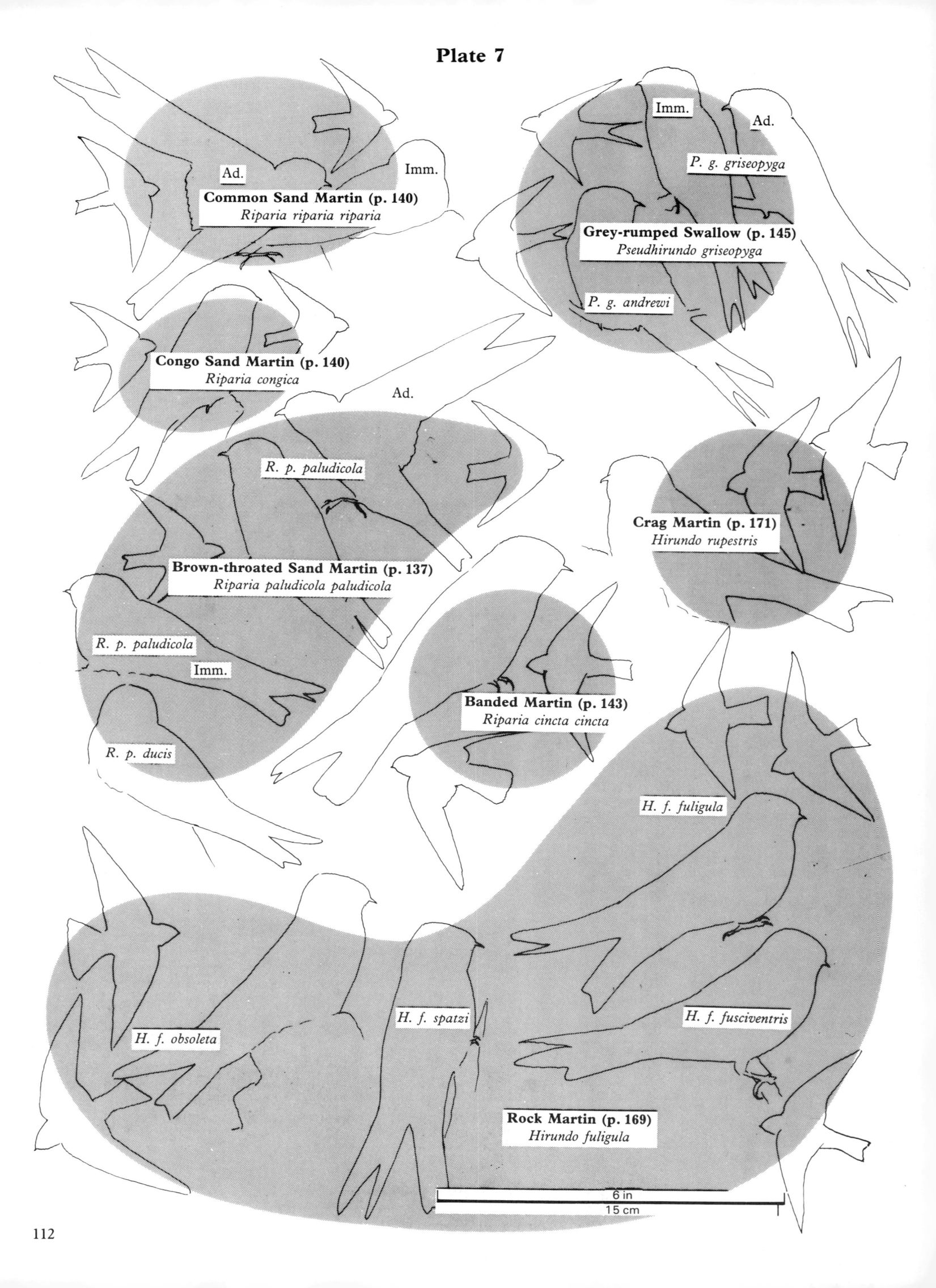

Plate 7

Common Sand Martin (p. 140)
Riparia riparia riparia
Ad. / Imm.

Congo Sand Martin (p. 140)
Riparia congica

Brown-throated Sand Martin (p. 137)
Riparia paludicola paludicola
R. p. paludicola
Ad.
R. p. paludicola
Imm.
R. p. ducis

Banded Martin (p. 143)
Riparia cincta cincta

Grey-rumped Swallow (p. 145)
Pseudhirundo griseopyga
Imm. / Ad.
P. g. griseopyga
P. g. andrewi

Crag Martin (p. 171)
Hirundo rupestris

Rock Martin (p. 169)
Hirundo fuligula
H. f. obsoleta
H. f. spatzi
H. f. fuligula
H. f. fusciventris

6 in / 15 cm

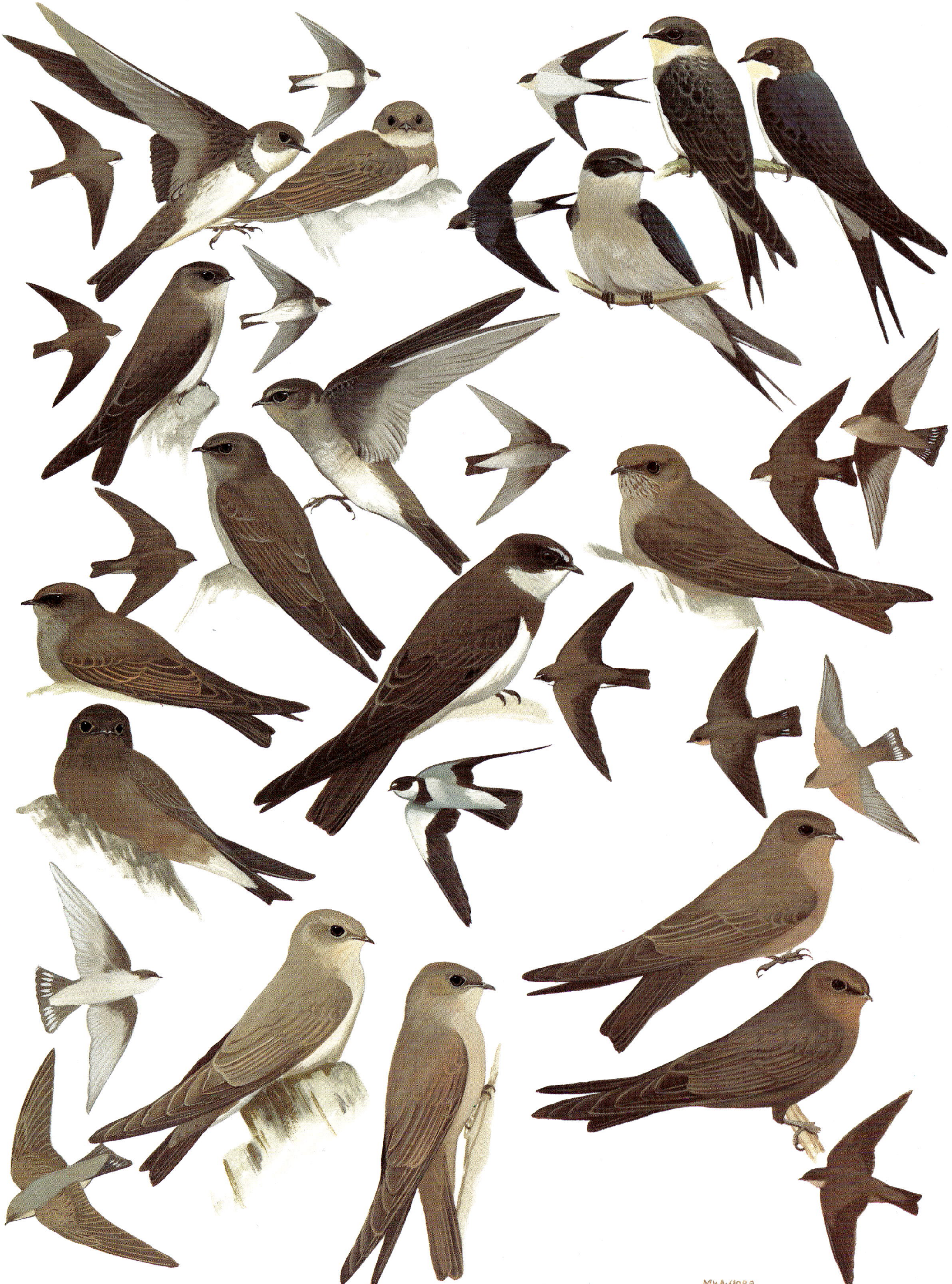
MWW1988

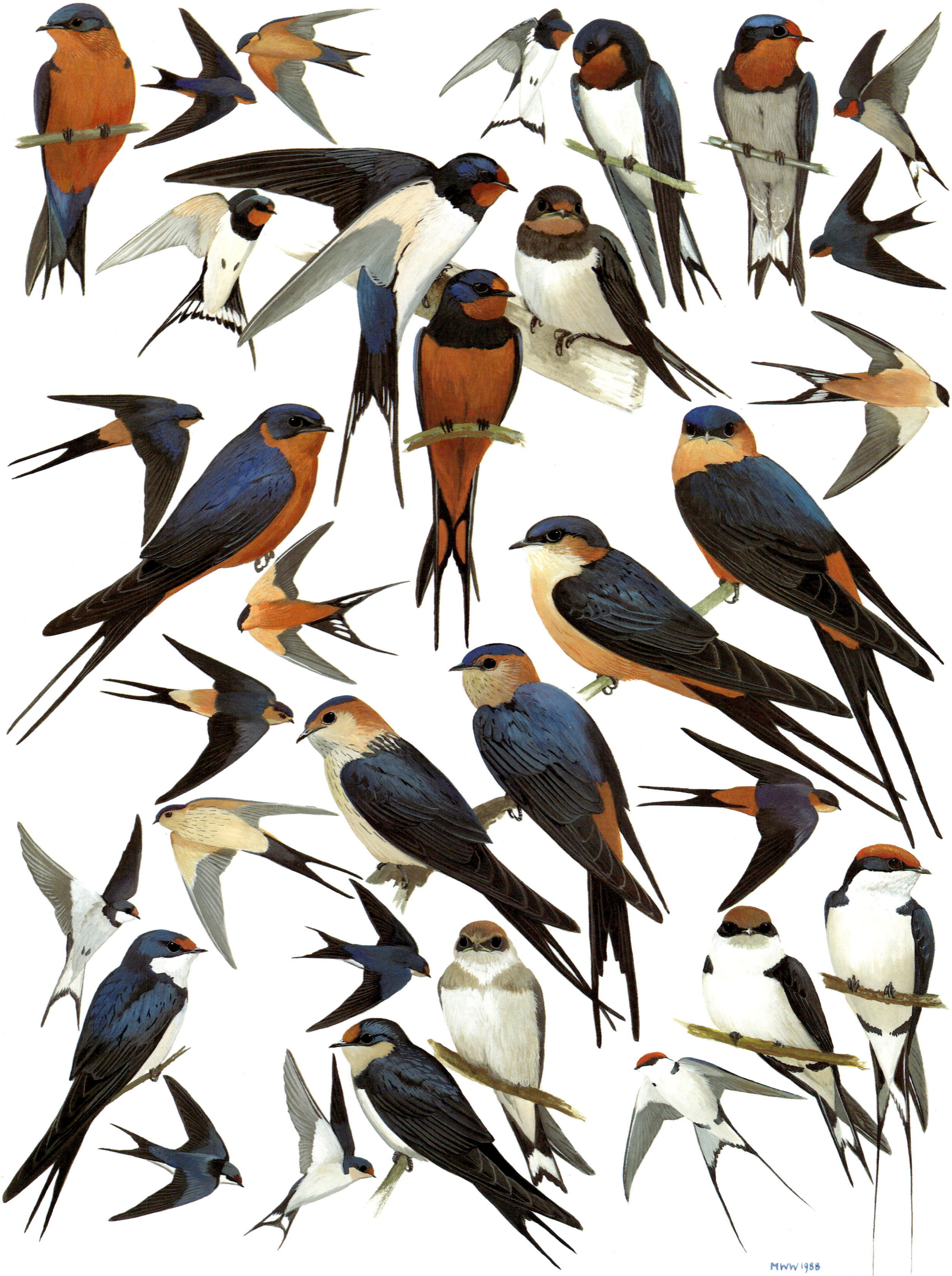
MWW 1988

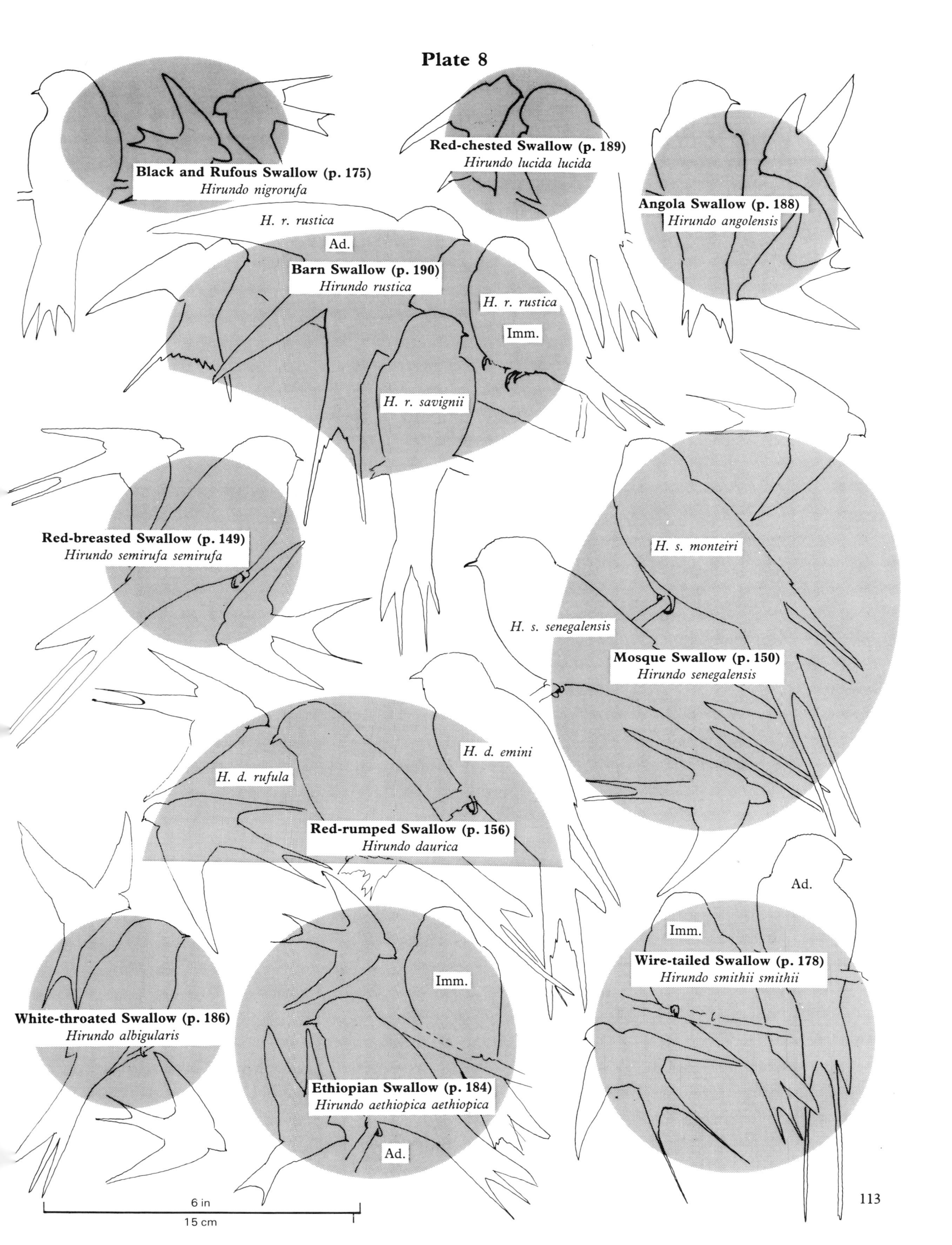

Plate 8

Black and Rufous Swallow (p. 175)
Hirundo nigrorufa

Red-chested Swallow (p. 189)
Hirundo lucida lucida

Angola Swallow (p. 188)
Hirundo angolensis

H. r. rustica
Ad.
Barn Swallow (p. 190)
Hirundo rustica
H. r. rustica
Imm.
H. r. savignii

Red-breasted Swallow (p. 149)
Hirundo semirufa semirufa

H. s. monteiri
H. s. senegalensis
Mosque Swallow (p. 150)
Hirundo senegalensis

H. d. emini
H. d. rufula
Red-rumped Swallow (p. 156)
Hirundo daurica

Ad.
Imm.
Wire-tailed Swallow (p. 178)
Hirundo smithii smithii

White-throated Swallow (p. 186)
Hirundo albigularis

Imm.
Ethiopian Swallow (p. 184)
Hirundo aethiopica aethiopica
Ad.

6 in
15 cm

113

Voice. Tape-recorded (78, McVIC, PAY). Song complex, and formed of 2 series of notes: 2–4 rather hoarse, base notes 'kree', and 2 softer, sweeter notes 'hu-hu', the first rather higher in pitch. Whole song lasts *c.* 10 s, and carries only a short distance. Call a short, rattling, not very distinctive 'chip-chip'.

General Habits. A widespread bird on open stony or sandy ground, grassy savannas, airports, fields and farmland; abundant on bare soils and especially on recently burnt ground. Prefers low to medium rainfall areas and in some areas has a special affinity with black cotton and black lava soils. Feeds on ground, sometimes at animal droppings.

Occurs in pairs when breeding (pairs sometimes nest very close together), and in flocks at other times. Flocks generally of 5–50 birds, occasionally a few hundreds; sometimes mixes with other sparrow-larks, especially Chestnut-headed Sparrow-Lark *E. signata*.

On ground walks with shuffling gait when foraging, or runs. Occasionally perches on bushes or trees and, when hot, on overhead wires. When alarmed, flocks may circle around at some height; but single birds usually fly low and drop to ground very suddenly. Drinks regularly (Skead 1975), but often occurs well away from water. Sings from ground and in flight.

Some populations sedentary; others migratory or nomadic. All across northern edge of range moves south from more northern parts in dry season after breeding in rains (Elgood *et al.* 1973). In E Africa irregularly nomadic; appears in Tsavo East Nat. Park (Kenya) only soon after rain (Lack 1985). An influx during dry season into Southern Province, Zambia, probably from the southwest; population in middle Zambezi and Luangwa valleys resident, but species is dry season visitor onto adjacent plateaus, apparently with little mixing of the 2 populations (Irwin 1982). Almost everywhere numbers fluctuate considerably from year to year.

Food. Seeds, especially small grass seeds, and some insects (mainly when breeding). Some larger insects, especially long, thin grasshoppers, fed to young (Morel and Morel 1984). Feeds in cultivated cereal crops and visits stubble fields.

Breeding Habits. Evidently monogamous; nests solitarily or in small, concentrated groups, with nests as little as 2–3 m apart. Sings during display flight, circling at up to 10 m above ground. Performs bouncing flight over nest when disturbed. In courtship on ground, ♂, puffed up and with wings outspread, prances after ♀. Chases conspecifics and other *Eremopterix* species; faces larger birds by puffing up, but crouches on nest to be 'invisible' in hottest weather.

2 broods may possibly be produced in a season: young bird incubated egg in nest when ♀ left; ♀ returned 8 min later with long insect; young ate it with difficulty; when ♂ returned, ♀ chased off young (Morel and Morel 1984). Also, young reared early may start to breed late in same season: ♂ of pair in late Mar was not in complete plumage (Morel and Morel 1984).

NEST: shallow scrape in soil lined with dry grass and rootlets, diam. 50, depth 30. Not domed (unlike many larks' nests), but usually placed in shade of clod, stone or grass tuft (35 of 41 nests, Senegal: Morel and Morel 1984). 14 of 15 nests in E Transvaal were on south, southeast or east of protective object (Chittenden and Batchelor 1977); those in Senegal faced east or northeast to avoid the sun after mid-morning. Both sexes build. Further material and lining often added after first egg laid.

EGGS: 1–2, occasionally 3; mean clutch size in Zimbabwe (n = 32), Transvaal (n = 15) and Malaŵi (n = 23) all 1·9, but in W Africa usually only 1 (Serle *et al.* 1977), and always 1 (n = 60) in Senegambia (Morel and Morel 1984). Colour varies even within a clutch; most eggs off-white with brownish and bluish grey spots mainly at larger end. SIZE: (n = 16, southern Africa) 17·6–19·9 × 12·9–14·2 (18·9 × 13·5); (n = 3, Senegambia) 18·8 × 13·5.

LAYING DATES: Mauritania, Mar; Senegambia, Oct–Mar; N Nigeria, Oct–Mar (especially Jan); Ethiopia, Jan, Mar, May; Sudan, Nov–Mar; E Africa: Region D, Mar, May; Zimbabwe, all months, but mainly Feb–Sept; Malaŵi, Apr–Sept; Mozambique, May; Transvaal, Feb–July. In regions liable to flooding breeds in dry season, but in low rainfall areas breeds in rains (Benson 1963).

INCUBATION: by both sexes by day, but only ♀ at night and until 1 h after dawn, reliefs every 0·5–3 h during day (Chittenden and Batchelor 1977), or never more than 1 h (Morel and Morel 1984). At nest-relief ♀ solicits ♂ bringing food, which is regurgitated 10–15 s before ♂ gives it to her. Number of visits never more than 15 in a day; off-duty parent perches on nearby tree or termitarium with wings open in hottest part of day, but farther away at other times. Sits tightly on eggs in cooler periods, but simply shades them at hottest time. Period: 11 days.

DEVELOPMENT AND CARE OF YOUNG: both parents feed young, mainly with insects; growth is rapid. Parent approaches nest carefully, rushes the last 0·5 m or so, then departs quickly. Nest kept very clean, parents always removing faecal sac before leaving. Amount of brooding depends on age of young and outside temperature. Food given to young often appears too large for gape; young hold beak and neck upwards and parent forces food in; maximum of 20 visits per h.

References
Chittenden, H. N. and Batchelor, G. R. (1977).
Irwin, M. P. S. and Lorber, P. (1983).
Morel, G. J. and Morel, M.-Y. (1984).

Eremopterix nigriceps (Gould). Black-crowned Sparrow-Lark. Alouette-moineau à front blanc.

Pyrrhulauda nigriceps Gould, 1841. *In* Darwin, Zool. Voyage 'Beagle', pt. 3, p. 87; São Tiago, Cape Verde Is.

Plate 2
(Opp. p. 33)

Range and Status. Africa, Cape Verde Is., Arabia, India.

Resident S Morocco, Egypt (common breeder on SE coastal plain, vagrant north to Wadi el Natrun), Mauritania (throughout except Majabat, locally common), Senegal (but not Gambia), Mali (throughout Sahel, especially common in E), Niger (throughout except N), central Chad, central Sudan, NE Ethiopia, Djibouti (locally abundant), Somalia (very common and widespread, south to 1°N), Socotra. Has recently appeared in N Nigeria (fairly common in Nguru-Gashua-Gogorem area), possibly a result of desertification (Ash 1990).

Description. *E. n. albifrons* (Sundevall): Mauritania, Mali, Chad, W Sudan (Darfur and Nile valley). ADULT ♂: forehead, nape, hindneck, cheeks, ear-coverts, sides of neck and foreneck white; crown, broad band through eye from crown to base of bill, chin, throat and broad bar from lower throat across sides of neck black. Mantle and back greyish brown, feathers edged paler, rump pale grey-brown shading into buffy uppertail-coverts. Tail blackish brown, outer feathers broadly edged buffy white, central pair edged whitish grey-brown. Entire underparts black except for white patch on sides of breast; thighs white. Flight feathers dark greyish brown, narrowly edged buffy on outer edge, tertials, scapulars and upperwing-coverts grey-brown with broad pale grey-buff margins. Underwing-coverts and axillaries black. Bill pale horn-brown or bluish white; eye brown; legs and feet pale flesh, greenish white or whitish horn. ADULT ♀: upperparts pale rufous-cinnamon, crown feathers with dark shafts, hindneck and uppertail-coverts often paler and buffy. Sides of neck and patch around eye pink-buff, cheeks and ear-coverts deeper cinnamon-buff. Underparts pale cream-buff, occasionally a few black or grey specks on chin, band across breast pale cinnamon-buff, sometimes with a few narrow brown streaks. Tail like ♂, but central pair of feathers rufous-cinnamon. Flight feathers, tertials and greater upperwing-coverts grey-brown, edged rufous-cinnamon, lesser and median coverts greyish brown, edged pale grey-buff. Underwing-coverts black. SIZE: *E. n. albifrons*: wing, ♂ (n = 15) 79–82 (80·3), ♀ (n = 9) 74–80 (76·6); remainder of measurements *E. n. nigriceps* (Cape Verde Is.): tail, ♂ (n = 14) 41–48 (43·6), ♀ (n = 15) 40–45 (42·2); bill, ♂ (n = 13) 12·4–13·6 (12·9), ♀ (n = 11) 11·9–14·3 (13·4); tarsus, ♂ (n = 13) 15·6–17·4 (16·5), ♀ (n = 12) 15·7–17·3 (16·6). WEIGHT: *E. n. albifrons*: ♂ (n = 3) 14–16, 1 ♀ 12.

IMMATURE: similar to adult, but feathers of crown, mantle, scapulars and upperwing-coverts have contrasting pale buff or warm buff edges, each feather with subterminal narrow dull grey spot or arc, tertials cinnamon with broad off-white fringes, fine and indistinct dusky spots on rump and uppertail-coverts. Breast heavily marked with poorly defined black spots. Flight feathers narrowly and evenly edged pink-cinnamon.

NESTLING: pale sandy down on upperparts.

E. n. melanauchen (Cabanis): E Sudan (east of Nile valley), Ethiopia, Somalia, Red Sea coast, Socotra. ♂ has crown more extensively black, upper mantle black, white band across hindneck; white patch on forehead smaller than in *albifrons*, rest of upperparts grey. ♀ not as rufous above, more greyish buff, band across breast grey.

E. n. forbeswatsoni Ripley and Bond: Socotra. Upperparts more rufous, less brown, than *melanauchen*, particularly on edgings of inner secondaries and middle rectrices; white patch on forehead extends further back, from base of upper mandible to crown.

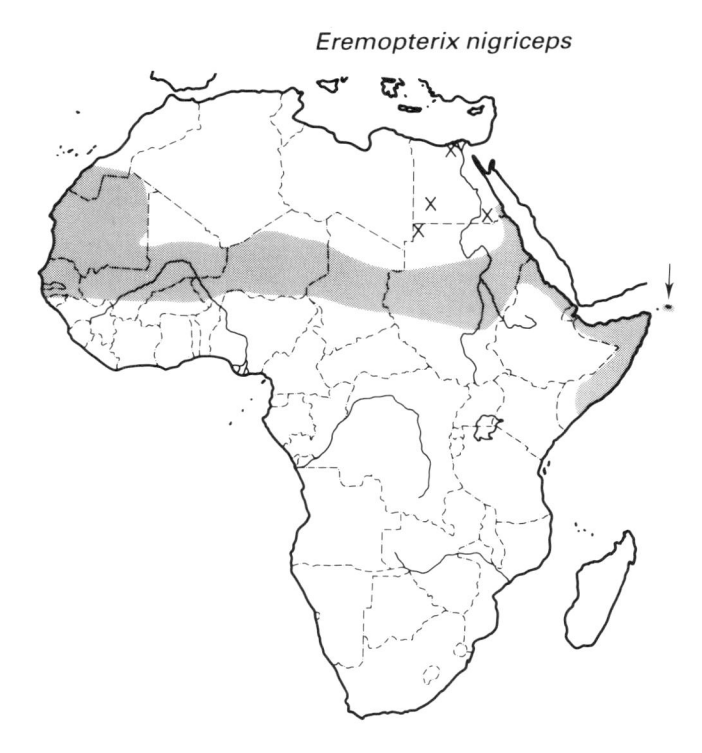

Field Characters. A small, dumpy lark appearing square-headed in the field. ♂ easily told from all other larks except sparrow-larks *Eremopterix* by black and white head pattern and black underparts. Only sympatric sparrow-lark is Chestnut-backed *E. leucotis*, which is even smaller; ♂ has head black except for white cheeks (no white on forehead), chestnut back, white belly. ♀ not unlike similarly-coloured desert larks *Ammomanes* spp. and Dunn's Lark *Eremalauda dunni* but smaller and chunkier, with short, mainly black tail and black underwings. ♀ Chestnut-backed is smaller and darker, upperparts mottled chestnut and blackish, breast streaked, black patch on belly.

Voice. Tape-recorded (78, HAZ, HOL). Song, high, thin, pure whistles given in discrete phrases a few s apart; singer repeats same phrase over and over. Number of notes and form of phrase vary individually and geographically; song-phrases from Chappuis (1985) include 'chi-WEE-chu', 'chi-wee-wee' and a more complex 'chit-CHIVY-chiwee-CHER-wit'. Call, a harsh, ringing, down-slurred 'CHEEEoo'. For other songs and calls, see Cramp (1988).

General Habits. Occurs mainly in Sahel savannas, especially in vegetation along streams and wadis, coastal grasslands with succulents and open sparsely-vegetated desert margins.

Solitary or in pairs, more often gregarious, occurring in flocks of 20–30, sometimes up to 60. Forages on ground by running a few steps and pausing before running on. Picks seeds from low bushes directly or by hovering; chases aerial prey. Drinks water regularly, including brackish or saline water. Avoids heat by roosting in shade of bushes and rocks, excavating small hollow in which it lies with spread wings, often panting. Wings frequently held away from body while foraging, and legs often dangle in flight on hot days.

Food. Invertebrates: grasshoppers, beetles, bugs (Reduviidae), larval Lepidoptera, spiders (Araneae); seeds (grasses *Panicum* spp., e.g. *P. turgidum*, cereal grains and forbs including *Zygophyllum simplex*). Young fed invertebrates.

Breeding Habits. Monogamous, territorial. Density in area of *c.* 75 ha of sparse grassland and shrubland in Saudi Arabia, 0·3 pair/ha in prime habitat, 0·1 pair/ha in sub-optimal habitat (Morgan and Palfery 1986). ♂♂ establish territories where they display to unmated ♀♀. In ground display, ♂ runs towards ♀ with head down, crown feathers raised, wings held away from body; on reaching her, gives excited call, turns away and leaps back to starting point. Display may be repeated up to 6 times; frequently follows courtship chase. In song-flight, ♂ rises from ground, singing, and with rapidly beating wings climbs steeply to 6–10 m; at that point he sings for <1 min, while circling in stiff-winged fluttering flight, often with legs dangling. At end of flight, ♂ descends in series of swooping glides with wings held stiffly spread at slight dihedral. Also has aerial courtship chase in which ♂ chases ♀ in rapid, switchback flight low over ground for 10–15 s.

NEST: site selected by ♀ attended by ♂; built by ♀ alone in *c.* 1 day. Cup of grass on base of twigs, lined with grass, hair, feathers, wool or any soft material available, placed in excavation in ground at base of shrub, grass tuft or stone; may have rim of small pebbles. ♀ excavates nest hollow by pressing down on substrate with her breast and turning with spread wings. Next, small pebbles are placed around rim, then twig foundation in hollow, finally finer material and lining.

EGGS: 2–3, mean (18 clutches) 2·6. Subelliptical; greyish or dirty white evenly speckled and blotched light brown over underlying purplish grey spots. SIZE: (n = 11) 17·5–18·8 × 13–14·7 (18·3 × 13·9).

LAYING DATES: Morocco, Apr; Egypt, Feb; Mauritania, Mar–Apr north of 18°N, Aug–Oct south of 18°N; Senegal, Oct–Mar; Mali, July–Oct; Ethiopia, Jan, Mar, Sept; Sudan, Nov–Mar; Somalia, Apr–May.

INCUBATION: by both sexes, mainly by ♀; probably begins with 2nd egg; period (n = 5) 11–12 days.

DEVELOPMENT AND CARE OF YOUNG: young fed by both adults, by ♂ more than ♀; shaded or brooded more by ♀; 261 feeding visits in 1 day, 29·5% of visits 05·00–08·00 h, 37% 16·00–18·00 h, only 3% 11·00–13·00 h. Food delivered to young in bouts lasting 9–18 min. Both adults remove faecal sacs from nest after feeding. Young leave nest at 8 days; brood split between parents, usually 1 young each.

BREEDING SUCCESS/SURVIVAL: mean brood size 1·8 young/nest (n = 10), with estimated overall productivity 0·18–0·36/nest attempt in Saudi Arabia (Morgan and Palfery 1987).

References
Cramp, S. (1988).
Morgan, J. H. and Palfery, J. (1986).

Plate 6
(Opp. p. 97)

Eremopterix verticalis (Smith). **Grey-backed Sparrow-Lark. Alouette-moineau à dos gris.**

Megalotis verticalis A. Smith, 1836. Rep. Exped. Expl. Cent. Afr., p. 48; country on both sides of the Orange R. (= Colesberg).

Forms a superspecies with *E. signata*.

Range and Status. Endemic resident and nomad. Angola, on coastal plain from Cabinda to Benguela, and arid SW; Zambia west of Zambezi; Zimbabwe mainly in W and NW, sporadically on central plateau; widespread in Botswana, commonest in S; widespread, but with considerable local movements, in Namibia; in South Africa present erratically but often abundantly W and NW Transvaal and W Orange Free State, widespread and common in Cape Province, mainly in Karoo, but regularly present in SW, exceptionally to Cape Peninsula. Common to abundant.

Description. *E. v. verticalis* (Smith): South Africa (Cape Province, Orange Free State, W Transvaal), Botswana, Zimbabwe (Bulawayo, Hwange), Zambia. ADULT ♂: top of head black except for white patch from centre of crown to nape; black collar from throat across side of neck around hindneck, separated from black of head by white nuchal collar (variable in extent). Black of crown continues round eye and base of bill to join black chin and throat; large white patch on cheeks and ear-coverts. Mantle dark grey, back lighter grey, feathers with dark grey centres, broad paler grey margins, rump and uppertail-coverts rather paler grey than back. Tail-feathers dark greyish brown to blackish, outer pair with white outer webs and tips, light brown inner webs, central pair medium grey-brown with paler fringes. Underparts entirely black except for pale greyish or greyish white flanks, pale grey or white patch on sides of breast, thighs white. Flight feathers greyish brown, secondaries narrowly edged white; scapulars, tertials and wing-coverts brown, broadly edged buffy or off-white. Underwing sooty brown to black. Bill pearly grey to bluish grey; eye brown; legs and feet pale grey to pinkish grey. ADULT ♀: crown, nape and hindneck streaked light and dark brown, sides of neck and foreneck lighter brown, buffy eye-ring, cheeks, chin and throat whitish, speckled dark brown, ear-coverts brown with narrow pale streaks. Mantle and back brown streaked darker brown, rump brown, uppertail-coverts buffy white with darker centres. Tail-feathers like ♂ except central pair more broadly marginated white, rest narrowly edged white. Breast and flanks buff-brown, lightly speckled brown, belly black. Flight feathers brown, scapulars with dark brown centres, lighter brown edges, tertials and upperwing-coverts brown, broadly edged pale brown. Underwing dark brown to black. SIZE: (14 ♂♂, 12 ♀♀) wing, ♂ 78–83 (80·4), ♀ 75–82 (77·7); tail, ♂ 40–45 (42·5), ♀ 39–44 (41·5); bill, ♂ 11·3–13·4 (12·4), ♀ 11·4–13 (12·3); tarsus, ♂ 16–18 (17·0), ♀ 16–19 (17·5). WEIGHT: ♂ (n = 27) 14–19·8 (17·0), ♀ (n = 30) 12·5–21·2 (17·0).

IMMATURE: mantle and back pale reddish brown, feathers with buffy white edges, coverts with similar pale edges. Ear-coverts mottled chestnut brown, cheeks buffy chestnut. Throat and breast buff, with indistinct darker chestnut centres to feathers, belly dark sooty brown, flanks buff.

NESTLING: bright pink, purplish on feather tracts, eyes blackish; head and back with pale grey down. Gape pale yellowish white, inside of mouth yellow with 3 black spots on tongue, 1 either side and 1 at tip, and 1 black spot at tip of upper and lower mandibles.

E. v. damarensis Roberts: South Africa (NW Cape), Namibia, NW Botswana, coastal Angola and Cabinda, Zambia (Liuwa Plain, Livingstone, Lochinvar Ranch, Loma, Matabele Plain). Overlaps with *E. v. khama* on Liuwa Plain, Livingstone, Lochinvar Ranch and Loma. ♂ paler above, upperparts more sandy, feathers marginated sandy buff; ♀ buffier above, feathers with reddish brown margins; below more buffy.

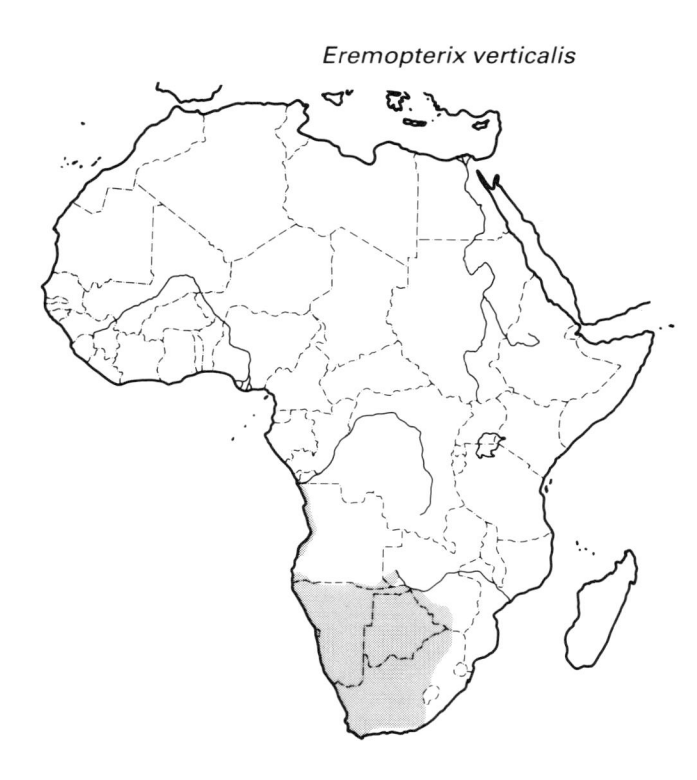

E. v. khama Irwin: NE Botswana (Makgadikgadi Pan), Zimbabwe, Zambia (Luachi R., Liuwa Plain, Mutala Plain). ♂ greyish white above, feathers with indistinct whitish margins, secondaries and coverts broadly margined white; ♀ pale greyish brown above, whitish below. Immature, above pale brown, not reddish, feathers edged white, ear-coverts mottled brown, cheeks whitish, throat, breast and flanks whitish.

E. v. harti Benson and Irwin: Zambia (Liuwa Plain, Kalabo, Senanga, Siloana Plains). May occur in Caprivi Strip, Namibia (Winterbottom 1971). Overlaps with both *E. v. damarensis* and *E. v. khama* on Liuwa Plain. ♂ and ♀ similar to *khama* but darker, less greyish above, wing-coverts not so conspicuously marginated white.

Field Characters. The only grey-backed sparrow-lark in southern Africa; the other 2 species (Black-eared *E. australis* and Chestnut-backed *E. leucotis*) both have chestnut backs. ♂ with strongly patterned black and white head very like species from north of the equator but unlike local birds, which have either no white on head (Black-eared) or white cheeks and nuchal collar only (Chestnut-backed). ♀ less distinctive but has pale grey-brown upperparts (no chestnut), black belly (lacking in ♀ Black-eared, present in ♀ Chestnut-backed).

Voice. Tape-recorded (F, GIB). Song in flight, a sharp 2–3 note simple, unmelodious 'twip-twip-chik' repeated over and over; chirping call, 'chruk-chruk'; alarm, 'prink'.

General Habits. Occurs in semi-arid or arid open short grass or sparsely grassed plains on Kalahari sands and gravels, grasslands with scattered bush, dry pans, burnt grassland, croplands after the harvest, fallow fields and dwarf shrublands on sands and clays. Present in Namib Desert beyond limits of vegetation, and on isolated red sand dunes with sparse vegetation along Kuiseb R. In Kalahari, most numerous on limestone flats along rivers.

Gregarious, even when breeding; usually seen in groups which vary in size from 5 or 6 to many hundreds. Forages in loose flocks, creeping along and pecking at food on surface of ground, occasionally uncovers buried items with sideways jerks of bill. Breaks open inflated fruits of *Atriplex lindleyi* to extract seed. Drinks water frequently and regularly, but can maintain body weight without free water (Willoughby 1968) and in Namib Desert probably depends on metabolic water. Roosts in shade of small bush during heat of day.

Nomadic and opportunistic; moves into semi-arid and arid areas where there has been rain; also has extensive movements in dry season (Benson and Irwin 1965).

Food. In Namib, invertebrates 8%, including harvester termites, small beetles, ants; seeds 91%, including grass *Stipagrostis* spp., forbs *Cleome* spp., *Monsonia umbellata*; green vegetation 1% (grass culm bases, sprouting grass seeds). In Karoo, commonly feeds on seeds of alien *Atriplex lindleyi*. Young fed exclusively invertebrates (grasshoppers, harvester termites, larval Lepidoptera).

Breeding Habits. Monogamous. Breeding density in Namib Desert 5–12 birds/ha; in grassy, gravel plains of

Eremopterix verticalis

N Cape Province 20–25 birds/ha in optimal habitat. In song-flight, ♂ circles, often with legs dangling, over territory at 15–30 m in series of undulations giving simple, unmelodious song; ends display with vertical dive to ground (Willoughby 1971).

NEST: built by ♀ only, in 4–5 days; cup of dry grass and rootlets, int. diam. 54, depth of cup 30 (n = 139), lined with finer material, usually awns from grasses (*Stipagrostis*), fluffy seeds (*Eriocephalus* spp.) or sometimes wool; placed in excavation in ground, or on foundation of small stones or mud pellets where ground too hard to dig, at base of stone, tuft of grass or shrub. In Kalahari summer, nests on east, southeast or south side of shelter; winter nests often out in open away from shelter.

EGGS: 2–3, mean (204 clutches) 2·1. Sub-elliptical; dull white, speckled with yellowish brown, brown and grey, often more densely marked at obtuse end. SIZE: (n = 314) 16·9–21 × 12·3–15·7 (19·3 × 13·9).

LAYING DATES: Zambia, Feb; Namibia, Mar–June, Aug; South Africa: Transvaal, Aug, Dec; N Cape, July–Oct, Jan–June; Karoo, July–Feb, May; E Cape, Sept–Oct, Mar–Apr. Breeding season in arid areas varies according to rainfall.

INCUBATION: by both sexes; period 12 ± 1 days.

DEVELOPMENT AND CARE OF YOUNG: young fed by both adults, who release begging in young by calling 'tsee ree'. Adults remove faecal sacs from nest after feeds. Young leave nest at 7–10 days, led by ♀.

BREEDING SUCCESS/SURVIVAL: in Kalahari, of 243 eggs laid in 114 nests, 77 hatched and 42 young left nests (Maclean 1970). No data on subsequent survival. Adults preyed on by Red-necked Falcon *Falco chicquera*.

References
Maclean, G. L. (1970).
Willoughby, E. J. (1968).
Willoughby, E. J. (1971).

Plate 4
(Opp. p. 49)

Eremopterix signata (Oustalet). **Chestnut-headed Sparrow-Lark. Alouette-moineau d'Oustalet.**

Pyrrhulauda signata Oustalet, 1886. Bibl. Ecole Haut. Etud., 31, art. 10, p. 9; N Italian Somaliland.

Forms a superspecies with *E. verticalis*.

Range and Status. Endemic; resident and wanderer. Extreme SE Sudan, SE Ethiopia, Somalia south of 10°N, and across Kenya north of 3°N in west, about 100 km south of L. Turkana and down to coast at *c.* 2°S. Since 1960s has extended south along coast and inland to Tsavo East and West Nat. Parks (SE Kenya), apparently replacing Fischer's Sparrow-Lark *E. leucopareia* (Lack *et al.* 1980). 1 record at Diani Beach, 25 km south of Mombasa. SW limit to range largely defined by higher rainfall areas of highlands. Common to abundant.

Description. *E. s. signata* (Oustalet): range of species except SE Sudan and west of L. Turkana. ADULT ♂: head, chin and throat chestnut, except centre of crown, cheeks and ear-coverts, which are white; hindneck white; dark brown smudge on upper mantle; mantle and back sandy brown; tail brown, with central pair of rectrices paler and with broad, faintly reddish brown edge to proximal half; outermost pair has outer web and distal third of inner web white; breast, belly and flanks dirty white, with broad dark chestnut line down centre becoming broader and blacker towards tail; undertail-coverts dark chestnut-brown; remiges like mantle, but inner primaries and all secondaries and tertials have paler tips and edges; P8, P7, P6 and P5 partially emarginated; scapulars and all wing-coverts brown with paler edges; underwing very dark brown. Bill light grey to greyish white; eye brown; legs and feet pinky flesh. ADULT ♀: like ♂ but lacks most of black and chestnut; top of head like mantle but more streaked than ♂, with dark centres to feathers; indistinct orange-chestnut stripe over eye; face pale with indistinct darker spots; chin to undertail-coverts variable dirty white, with a hidden dark line down centre, especially on belly; wings and tail like ♂; underwing dark chestnut. SIZE: (19 ♂♂, 9 ♀♀) wing, ♂ 75–82 (79·1), ♀ 75–81 (77·9); tail, ♂ 43–56 (49·4), ♀ 42–53 (46·9); bill, ♂ 11·5–14 (12·9), ♀ 10–12·5 (11·7); tarsus, ♂ 16–18 (17·1), ♀ 16–18 (16·9). WEIGHT: (Kenya, May) 2 ♂♂ 15, 16, 1 ♀ 16.

IMMATURE: like adult ♀ but underparts paler.
NESTLING: undescribed.

E. s. harrisoni (Ogilvie-Grant): SE Sudan, NW Kenya west of L. Turkana. Upperparts more greyish brown and less sandy.

Field Characters. A pale sparrow-lark characteristic of semi-arid country, the only one in most of its range, but meets Fischer's *E. leucopareia* in the south, and at times occurs in the same places as Chestnut-backed *E. leucotis* and Black-crowned *E. nigriceps*. ♂ Chestnut-headed is a much brighter and neater looking bird than ♂ Fischer's, with well-defined patches of white and chestnut on the head, whiter underparts without brownish suffusion; black line down underparts much broader. ♀ has paler and usually greyer upperparts than ♀ Fischer's, black line on underparts usually partly or wholly concealed, and paler face with pale orange-buff supercilium (no supercilium in ♀ Fischer's). Chestnut-backed is a much darker bird altogether, ♂ Black-crowned has white patch on forehead (not on centre of crown) and both sexes are more ash brown than Chestnut-headed.

Voice. Tape-recorded (McVIC). Song a simple twitter, usually from ground but also in flight. Call a sharp 'chip-up' (Tomlinson 1950); also various 'tssp' calls.

General Habits. Prefers short grass plains and stony semi-desert scrub, although also in desert areas. Rare above 2000 m. Usually on ground, sometimes perching on rocks or low bushes.

Solitary when breeding, but outside breeding season occurs in flocks of tens or occasionally hundreds, sometimes with a few Chestnut-backed or Black-crowned Sparrow-Larks.

Feeds while walking about on ground, taking seeds mainly from the ground, rather than from vegetation. Drinks regularly, and in N Kenya desert noted coming to waterhole in morning with other larks and sandgrouse (S. Keith, pers. comm.). May move to waterholes in flocks, but also in twos and threes.

Sedentary in some areas, but many wander extensively and irregularly, e.g. many appeared in Tsavo East Nat. Park Apr–May and Sept–Oct 1975, Feb and May–Sept 1976 (Lack 1985).

Food. Largely granivorous, mainly small grass seeds. 3 times seen to take insects to feed young in nest (Lack 1985).

Breeding Habits. Advertises territory with fluttering song flight *c.* 6 m above ground.

NEST: cup of fibrous grasses, in hollow in ground next to tuft of grass, rock or fallen log.

EGGS: 3–5, usually 4; white or pale buff, minutely speckled over whole surface with various shades of brown. SIZE: (n = 3, Somalia) 18·6 × 12·5.

LAYING DATES: Somalia, about mid-June (Archer and Godman 1961); Sudan, May, June; E Africa: Kenya (adults carrying food, June, July).

Reference
Lack, P. C. (1985).

Eremopterix leucopareia (Fischer and Reichenow). Fischer's Sparrow-Lark. Alouette-moineau de Fischer.

Plate 4
(Opp. p. 49)

Coraphaites leucopareia Fischer and Reichenow, 1884. J. Orn., p. 53; Little Arusha, NE Tanganyika Territory.

Forms a superspecies with Indian *E. grisea*.

Range and Status. Endemic; resident and wanderer. NE Uganda (Mt Moroto and E Karamoja) east to Meru Nat. Park (Kenya) and south through Kenya and Tanzania except for West Lake Province, the southwest corner, and east of line from Mt Kilimanjaro to Dar-es-Salaam to Songea; around northern end of L. Tanganyika; and south to N Zambia (Lundazi) and Malaŵi (south to Mzimba and Kasungu districts). Occupies gap in range of Chestnut-backed Sparrow-Lark *E. leucotis* q.v., with the only overlap in central and E Tanzania. First seen at Lundazi, Zambia in 1976 and not recorded there previously (Dowsett 1979). Recorded until at least early 1960s in Tsavo East Nat. Park, Kenya, but by mid-1970s had apparently been replaced by Chestnut-headed Sparrow-Lark *E. signata* (Lack *et al.* 1980). Common.

Description. ADULT ♂: crown, nape and neck variably chestnut, tawny, or sometimes grey with darker centres to feathers; face and around eye black, and a blackish ring around hindneck; cheeks and ear-coverts white to buff; chin and throat very dark brown; sides of neck dark brown, the lower part fringed chestnut; mantle and back greyish brown with dark centres to feathers; rump paler; uppertail-coverts pale brown with white edges; tail dark brown, except that outermost feather has distal half of inner web and distal two-thirds of outer web rufous-brown, and the next inner pair has distal half of outer web rufous-brown; breast buff, belly white, both with broad, very dark brown band down centre; undertail-coverts very dark brown; basal half of outer edge of outer web of primaries rufous; outer edges and tips of secondaries and tertials rufous; remiges otherwise dark brown; scapulars dark brown with rufous-brown edges and tips to feathers; wing-coverts rufous-brown with dark brown centres to feathers; underwing very dark brown. Bill greyish black to greyish white with darker tip; eye brown; legs and feet pale brown. ADULT ♀: crown and head grey-brown with dark centres to feathers; stripe over eye, sides of neck and half-hidden ring around back of neck orange-brown; cheeks and lores buffy brown; chin white; throat and breast buff to tawny, grading into white on flanks and belly; belly with variably wide dark brown band down centre; mantle, back and wings like ♂ but paler, and wings have less rufous on outer edges; outer web of outer tail-feather nearly white; otherwise like ♂. SIZE: (25 ♂♂, 19 ♀♀) wing, ♂ 74–79 (76·1), ♀ 71–78 (74·1); tail, ♂ 39–47 (43·1), ♀ 37–49 (40·9); bill, ♂ 11–13 (11·9), ♀ 11–12·5 (11·6); tarsus, ♂ 16–18·5 (17·3), ♀ 16–18·5 (17·1).

IMMATURE: ♂ like adult ♀ but some adult ♂ head markings may be visible. ♀ like adult ♀ but with whitish tips to feathers of upperparts, all wing-coverts and wings.

No subspecies recognized, but many birds in southern populations are darker than northern ones.

Field Characters. The only sparrow-lark in much of its range but does just overlap with Chestnut-backed *E. leucotis* and Chestnut-headed *E. signata*. Chestnut-backed is much darker, ♂ with mainly black head (no brown), chestnut back; ♀ has upperparts chestnut with dark mottling, black patch on belly but no black line down centre. ♂ Fischer's has same colour combination (chestnut, black and white) as Chestnut-headed but is much dingier-looking, with no white on crown or hindneck, white face-patch and breast suffused with brown; also chin and throat blackish, not chestnut, and black on underparts confined to line down centre. ♀ similar to ♀ Chestnut-headed but darker and browner above, dark line on underparts broader and more conspicuous (not concealed). ♂ Fischer's readily distinguished from other bird families (though somewhat suggestive of ♂ House Sparrow *Passer domesticus*), but ♀ away from ♂♂ can be confusing, often passed over among the hordes of look-alike streaky brown birds that crowd the plains of E Africa. Black line on underparts (if you know to look for it!) diagnostic but can be hard to see; orange suffusion on breast and collar a useful clue, also short tail and chunky shape.

Voice. Tape-recorded (McVIC). Song a brief warble, uttered from ground. Calls of flock a low 'twee-eez' and a soft twittering.

General Habits. Inhabits dry, short grass plains or bare patches in grassy country, in low and medium rainfall areas from sea level to 1800 m. In Malaŵi also occurs in gardens, fields, small areas of cultivation in woodland and grassy dambos, once on an airstrip (Dowsett 1983; Karcher and Medland 1989). Occurs very often on roads, in groups of 5–8. Forages by walking about between grass tussocks, pecking at ground. Regularly seeks shade of isolated tree or bush in hottest part of day. When nesting, sometimes raises feathers to increase heat loss (Reynolds 1977).

Eremopterix leucopareia

Mainly sedentary, but some irregular movements, not clearly related to rainfall patterns, but presumably in response to food availability.

Food. Unknown.

Breeding Habits
NEST: shallow scrape in ground, lined with dry grass.
EGGS: 2–3; creamy white or greyish white, with variable amount of chocolate-brown or sepia spots and blotches, and with mauve-grey undermarkings. SIZE: 19 × 12.
LAYING DATES: E Africa: Kenya, Apr–July, Dec; Tanzania, Mar; Region B, May–June; Region C, Apr–Aug; Region D, Feb–July, Dec. (Regions B and D in rains, Region C in dry season). Malaŵi, probably Apr–May. Breeds in drier seasons in areas of higher rainfall where liable to flooding, but in rains in lower rainfall areas (Benson 1963).

Genus *Eremophila* Boie

Characterized by strongly marked black and white or black and yellow facial patterns and elongated feathers at edge of crown forming wispy 'horns'; remaining upperparts unstriped. Bill small and rather weak; nostrils concealed below fringe of stiff black feathers. Legs neither long nor short; hind claw somewhat elongated and decurved.

2 species, comprising a superspecies, the 'horned' larks, one (*E. bilopha*) confined to N Africa and N Arabia, the other (*E. alpestris*) Holarctic. Being the only New World representative of its Old World family, *E. alpestris* occupies a wide range of lowland lark habitats in N America, but in the Old World its habitat is Arcto-alpine; *E. bilopha* inhabits lowland desert steppe.

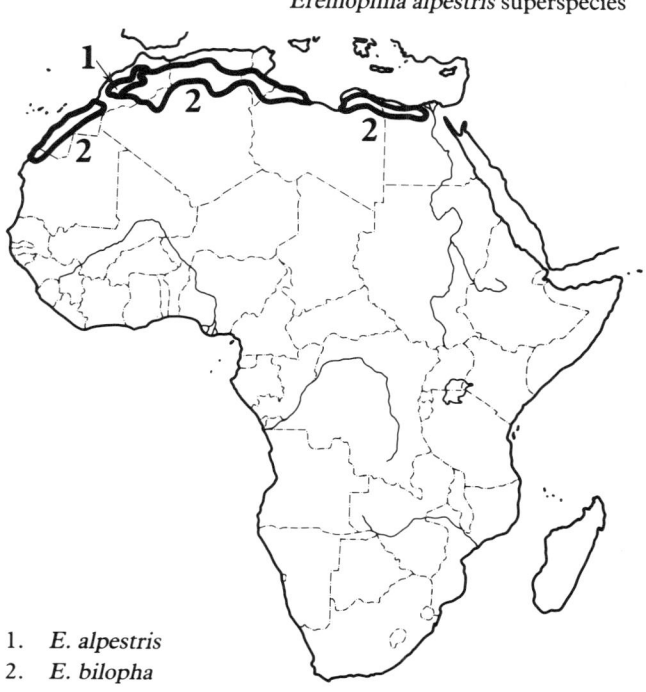

Eremophila alpestris superspecies

1. *E. alpestris*
2. *E. bilopha*

Eremophila alpestris (Linnaeus). Horned Lark; Shore Lark. Alouette hausse-col.

Plate 2 (Opp. p. 33)

Alauda alpestris Linnaeus, 1758. Syst. Nat. (10th ed.) 1, p. 166; N America = coast of South Carolina.

Forms a superspecies with *E. bilopha*.

Range and Status. N America, Colombian Andes, NW Africa, Arctic Eurasia, and Balkans to Himalayas, W Manchuria, N Mongolia and S Siberia (Stanovoy Mts). Northern populations migratory, southern ones mainly resident but with vertical migrations.

Resident and altitudinal migrant in Morocco; occurs in Middle Atlas between Azrou and Itzer, in High Atlas at Oukaïmeden, Yagour, Zaouïa Ahansal, Tabant, Tizi n' Tichka and Imilchil. Probably widespread, breeding between 2000 and 3500 m; somewhat lower in winter, recent records south of High Atlas at Azib d'Iriri, El Kelaa des Mgouna, Tizi n' Tazazert and Boumalne du Dadès.

Description. *E. a. atlas* Whitaker (only race in Africa): endemic to Morocco. ADULT ♂: nasal feathers and lores black; forehead pale yellow; superciliary stripe yellow, joining pale yellow area behind black cheek; forecrown black, forming band up to 8 mm deep; hindcrown pinkish brown. Black of forecrown extends back at sides above yellow superciliary stripe as narrow line of curving black feathers, sometimes standing out from side of crown to form 'horns' up to 16 mm long behind eye (from eye to tip of horn). Nape and upper mantle pinkish brown; lower mantle, back, scapulars and upperwing-coverts plain grey-brown; rump brown with indistinct dark brown stripes; uppertail-coverts brown or pinkish brown, sometimes pale tipped. Central tail-feather dark brown with broad pale brown fringe; rest of tail black but outer tail-feather with outer web mainly white and T5 with narrow outer fringe near tip. Lores and ear-coverts black. Chin and throat pale yellow, confluent with pale yellow area behind black ear-coverts; breast black; centre of belly white or whitish; lower breast, sides of breast and flanks washed grey-brown; undertail-coverts white. Remiges brown or dark brown; inner primaries and secondaries narrowly fringed and tipped grey-brown; outer primary (P9) with whitish outer web; upperwing-coverts grey-brown. Underwing silvery grey, undertail black. ADULT ♂ (non-breeding): same, except that black lores and ear-coverts are peppered with pale yellow. Bill brown-black, lower mandible with grey base; eye brown; legs and feet black. ADULT ♀: like ♂, differing as follows: black of forecrown, lores, cheeks, throat and breast somewhat restricted (in fresh plumage with broad pale feather-tips); hindneck and upper mantle yellower (pale, not deep, vinous pink); lower mantle, scapulars and back appear less uniform (Cramp 1988). Bill slightly paler than ♂. SIZE: wing, ♂ (n = 14) 109–118 (114), ♀ (n = 11) 102–108 (106); *E. a. flava* (slightly smaller N European race, 40 ♂♂, 30 ♀♀), wing, ♂ 108–116 (112), ♀ 100–107 (103); tail, ♂ 64–74 (68·4), ♀ 56–67 (60·7); bill to skull, ♂ 13·6–15·5 (14·7), ♀ 12·6–14·8 (13·8), bill to feathers av. 3·6 shorter; tarsus, ♂ 21–24 (22·5), ♀ 21–23 (22·0). WEIGHT: (*E. a. atlas* not known; *E. a. flava*, E Germany, winter) ♂ (n = 16) 36–45 (39·9), ♀ (n = 21) 30–44 (36·9).

IMMATURE: rufescent brown with yellowish tinge around face. Forehead, crown, mantle, back, scapulars and upperwing-coverts profusely spotted white, each feather with white tip, blackish penultimate mark and brownish or rufescent-brown centre. Breast and flanks with indistinct dull bars and crescents. Wings yellowish brown or rufescent brown; all remiges and especially tertials with 1–2 mm white or buffy tips and fringes, the tertials with dark penultimate band. Outer rectrices buffy where adult is white.

NESTLING: (*E. a. flava*) long dense tufts of white or straw-yellow down, on upperparts only; bare skin brown, mouth and gape flanges yellow, with 3 black spots at base of tongue and 1 at tip.

Field Characters. A medium-sized NW African lark, adult rather uniformly pink-brown above and white below, strikingly patterned on face and breast with black and yellow, thin black 'horns' generally quite evident in summer, less so in winter. Only similar lark is Temminck's Horned Lark *E. bilopha*, which is white where Horned Lark is yellow; moreover the 2 spp. are entirely segregated in Morocco by altitude and habitat, at least in summer, *E. alpestris* in mountains above 2000 m and *E. bilopha* in desert steppe below 1000 m. Juvenile lacks black face and breast-bands, but has yellowish face and is heavily spotted with white above and scalloped with blackish below.

Voice. Tape-recorded (CHA – 62, 73). Song and calls lack volume and vehemence of most larks and are sibilant. Call a shrill 'tsip' or 'tseep' like small pipit, also 'tsissup' and 'tsweerrp' like White and Yellow Wagtails *Motacilla alba* and *M. flava*. Song a weak warble, in various races (not *atlas*) described as 'tsip-tsip-tsee-didi', 'chit-chi-chiddle-chee-la' and 'tsee-tsee-(tsee)-tee-lee-lee' (Cramp 1988).

General Habits. In NW Africa inhabits bare high plateaux and upland ranges, and barren slopes with unmelted patches of snow, also areas of alfa grass *Stipa tenacissima* lower down (Smith 1965). Forages on ground amongst stones and short grass, pecking at ground and vegetation. Solitary, in pairs or family parties in breeding season; at other times gregarious, forming flocks of up to 15–20 birds (e.g. Morocco).

Food. Mainly insects in summer and seeds in winter (for lists of foods, see Cramp 1988). Only Moroccan record: stonefly carried to young (Lynes 1920).

Breeding Habits. Very little known in Africa; for details of breeding habits outside Africa, see Cramp (1988). Song-flight: ♂ ascends silently in undulations to 80–250 m, where it sings, gliding with wings and tail widely spread or beating wings slowly and deeply, heading into wind and remaining almost stationary; at conclusion, closes wings and drops head-first, vertically, with whizzing sound (Cramp 1988). Solitary, territorial.

NEST: natural or excavated depression in ground, lined with small twigs, rootlets, grass stems and leaves, with vegetable down and finer grasses inside, and surrounded with small stones. Diam. av. 64, depth 42. Built by ♀ only, digging depression with bill and feet, in 1–4 days.

EGGS: 2–4, usually 4. Subelliptical; glossy; greenish white with heavy yellowish brown spotting and brown hair-streaks. SIZE: 20·5–26 × 14·7–18 (22·7 × 16·2). WEIGHT: *c.* 3·15.

LAYING DATES: Morocco, May.

INCUBATION: by ♀ only. Period: 10–11 days.

DEVELOPMENT AND CARE OF YOUNG: young leave nest at 9–12 days, and fly at 16–18 days.

Reference
Cramp, S. (1988).

Eremophila alpestris

Eremophila bilopha (Temminck). Temminck's Horned Lark. Alouette bilophe.

Alauda bilopha Temminck, 1823. Pl. col., livr. 41, pl. 244, fig. 1; deserts of Aqaba, Arabia.

Plate 2
(Opp. p. 33)

Forms a superspecies with *E. alpestris*.

Range and Status. NW and N Africa, Sinai, Jordan, Syria, W Iraq, N and NE Saudi Arabia. Resident and short-distance migrant. Accidental Lebanon, Malta.

Resident from N Mauritania to N Egypt, with some internal migration in Libya. Morocco: locally common in S; abundant in E on Hauts Plateaux, and more local to southeast of Atlas Mts in Sahara but common on plains south and east of Boumalne (flock of 100 between Bouarfa and Figuig: Smith 1965). Vehicle transects gave 46 birds in 3 km of pure stand of succulent shrub *Aizoon*, 6 in 2 km of *Salsola*, > 100 birds in 70 km of *Salsola* and grass with *Aizoon* clumps, and 0·6 per km of adjacent barren plains (Valverde 1957). Algeria: frequent and widespread, north to Barika, Aïn Oussera and Mecheria and south to near Ouargla, Ghardaïa, Béchar, Guir and Touggourt; 50 birds in 16-km transect, Laghouat (Haas 1969). Tunisia: scarce from Bou Hedma and the Chotts southward, frequent in Chott Rharsa. Libya: widespread in Tripoli in northern steppe and semi-desert, east to 16°E; may breed in Jebel Waddan, Jebel Soda and Hammada el Homra; in Cyrenaica extends from Ajadabia east to Egypt, in south probably no further than 100 mm isohyet; N Egypt: locally frequent from Libyan border to 30½°E (Siwa, Bir Qattara, Qaret el Tarfaya, Minqar Abu Dweiss, El Moghra, Nahlet el Balah, Bir Bouweib and near Bahariya).

Description. ADULT ♂: like *E. alpestris* but a cleaner cut bird with face black and white, not black and yellow, with longer 'horns', pink-brown rather than grey-brown upperparts, and rufous tertiaries and central tail feathers. Nasal feathers, lores and ear-coverts black, the last extending downwards towards black breast patch. Forehead white, joining white superciliary area and silky white patch behind black ear-coverts, also joining white chin and throat. Forecrown black in band *c*. 7 mm deep, extending backwards at side of crown to form curving black horn often standing out from side of head; horn up to 22 mm long (from hind edge of eye). Crown, nape, sides of neck, mantle, back and rump pink-brown; uppertail-coverts pale rufous. Central tail-feather pale rufescent brown with paler edges, outer tail-feather black with outer web white, T5 with very narrow white outer edge, T4 with even narrower white edge towards tip, T3 and T2 all black. Chin and throat white; breast black, forming large crescentic patch with tips of crescent near bend of wing; sides of breast behind black patch pinkish brown; flanks pinkish brown; belly and undertail-coverts white. Remiges dark brown with 2–3 mm broad whitish or buffy tips to both webs; outer primary (P9) with narrow outer web white, and remaining primaries with proximal half of outer webs pink-brown up to point of emargination; tertials rufous; upperwing-coverts pale rufous, fringed and tipped white. Axillaries, underwing-coverts and underwings silvery grey; underside of tail black, with white outer web to T6. Bill bluish grey or greyish dark brown, base of lower mandible paler; eye brown or dark brown; legs and feet pinkish grey or dark purplish brown. ADULT ♀: less difference between the sexes than in *E. alpestris*; horns av. shorter; black bands on forehead, under eye and on breast slightly narrower than in ♂ and less sharply demarcated. SIZE: wing, ♂ (n = 50) 96–106 (100), ♀ (n = 24) 89–96 (92·5); tail, ♂ (n = 18) 62–69 (65·6), ♀ (n = 9) 57–64 (60·0); bill to skull, ♂ (n = 17) 14·2–16·1 (15·1), ♀ (n = 9) 12·7–14·9 (13·9), bill to feathers av. 2·7 shorter; tarsus, ♂ (n = 16) 20–22 (21·0), ♀ (n = 10) 18·5–21·5 (20·0). WEIGHT: (Algeria, 2 ♂♂) 38, 39.

IMMATURE: lacks black face and breast marks of adult. Upperparts sandy-cinnamon, all feathers with whitish tips; forehead, eye-ring and narrow supercilium cream-white; ear-coverts sandy-cinnamon; chin and throat creamy, breast sandy-cinnamon, belly creamy.

NESTLING: at *c*. 5 days, covered with minute feathers breaking through fluffy whitish down on head and upperparts (none on underparts); thick, pale gape-flanges (from photographs in Shannon 1974).

Field Characters. A medium-sized N African desert lark, the adult rather uniformly cinnamon-rufous above and white below, strikingly patterned on face and breast with black and white, the long wispy 'horns' generally quite evident (**A**, p. 124). In flight, undersides of wings prominently white, and tail looks black with white edges. Only similar lark is Horned Lark *E. alpestris*, which has yellow (not white) forehead, supercilium, cheeks and throat, and is allopatrically restricted to High Atlas Mts in Morocco. Juvenile lacks black face and breast-band, and is heavily spotted and scalloped with white above; tail like adult's but black duller and edges buff (not white).

A

Voice. Tape-recorded (60, 62, 73, 78). Very like voice of Horned Lark q.v., but less vigorous. Disconnected bursts of soft, melodious twittering, or quiet, fine warbling, or monotonous repetition of 'dee-dee-eeee', 'chep-seee-eee' or 'chep-ep-seeee'. Thin, drawn-out whistles alternating with richer chirrups and dry rattles; on ground gives quiet 'see(y)oo', metallic 'chee-u' or 'chee-yoo' and 'sweet-teee-ooo', with soft nasal 'tzew' when flushed (Cramp 1988). Flight call a loud 'tsip' or 'sweeeep' like wagtail *Motacilla* sp.

General Habits. Inhabits flat wormwood *Artemisia* steppe, plains with thin wispy grass or the chenopod *Anabasis*, plateaux with *Stipa* grass, gravel hammada and semi-desert, silty wadi beds, and slopes carrying low cover of succulent *Aizoon*. Prefers areas with compacted soil and avoids soft sand.

Solitary or in pairs in breeding season, at other times in small flocks, av. of 4 birds, Morocco, but up to 100 there, and flocks up to 20 frequently reported in Algeria and Libya. Mixes with other larks including Bar-tailed Lark *Ammomanes cincturus* (Libya). Tame, at least in still conditions, less so if windy, and flocks very shy. Feeds on ground, walking, running and pecking at earth and vegetation in manner of other larks; feeding flock keeps together in tight bunch, birds turning over stones many times their own weight in search for food (Smith 1965). Birds shot in Iraq in early morning had crops full of water and may have been drinking dew (Cramp 1988). Sings on ground, and in song-flight (see below).

Mainly resident, but many winter records in Chott Rharsa, Tunisia, may be migrants from Algeria, and in Libya occurs south in Fezzan to Jebel Soda in winter, and recorded on Gulf of Sirte in winter where not known to breed.

Food. Seeds, insects and larvae. Seeds include *Aizoon* (Morocco); fruits of *Ephedra transitoria* (Jordan: Clarke 1980); also fleshy tips of desert xerophytes.

Breeding Habits. Evidently solitary, strongly territorial, breeder. Displaying ♂ sings from air; song-flight like that of Horned Lark, but ascent rather feeble and never to any great height, up to 25 m in Jordan but only 2–4 m in S Morocco, bird facing wind and dropping back almost vertically to or slightly ahead of starting point. On ground ♂ adopts courtship posture with head up, horns raised, feathers ruffled, tail spread and wings drooped, and dances with rapid steps in front of ♀ (Cramp 1988).

NEST: shallow depression in ground lined with twigs, grasses and rootlets, with deep inner cup lined with soft grass-heads; once, a mud lining with rag and wool embedded in mud (Jennings 1980); almost invariably with rampart of small stones or flakes of dry mud at one side (Shannon 1974). Ext. diam. (Algeria) 110, int. diam. 70, ext. depth 40, depth of cup 25. Sited under tussock, by tuft of grass, or in open.

EGGS: 2–4, av. of 17 clutches (N Africa) 2·88. Subelliptical; glossy; greenish white with heavy yellowish brown spotting and brown hair-streaks. SIZE: (n = 40) 19·7–24·7 × 14·8–16·4 (22·0 × 15·3). WEIGHT: *c*. 2·65.

LAYING DATES: Morocco, Feb–Apr; Algeria, Apr–May.

INCUBATION: during dust storm, incubating bird became agitated, restlessly moving about on eggs and repeatedly leaving nest (Shannon 1974).

DEVELOPMENT AND CARE OF YOUNG: young fed by both parents, in nest and for some time afterwards. Parent shades young from hot sun. One brood left nest at *c*. 16 days (probably atypically late: Jennings 1980).

References
Cramp, S. (1988).
Shannon, G. R. (1974).
Valverde, J. A. (1957).

Family HIRUNDINIDAE: swallows and martins

A highly specialized group of passerines adapted for aerial feeding on invertebrates. Small; body slender and streamlined, neck short, wings long, narrow and pointed, tail sometimes deeply forked, with long streamers. Plumage glossy dark blue or green above, whitish, buff or rufous below, sometimes streaked; or rather plain brown throughout; often with conspicuous pale rump, and white 'windows' in tail. Bill short, broad, dorso-ventrally compressed; gape wide; a few species with weak rictal bristles. Nostrils either open, rounded and dorsally directed or oval and operculate. 10 primaries, with P10 extremely reduced; 2 genera (*Psalidoprocne* in Africa, and *Stelgidopteryx*) have outer primaries serrated. Tail with 12 rectrices. Tarsi short, sharply ridged at rear; feet weak, front toes more or less united at base, claws strong; tarsi unfeathered (except in *Delichon*). Sexes alike.

Many species highly migratory. Efficient fliers with excellent aerial manoeuvrability. Most species gregarious when not breeding, some also when breeding. Nests are simple cup of feathers and grass in chamber at end of burrow (usually self-made) in bank or in flat ground, or half-cup or retort-shaped construction of mud pellets attached to rock. Eggs white, spotted or unspotted. Chick scantily covered with down at hatching; replaced by denser 2nd down after a week.

Worldwide except for polar regions and most oceanic islands. 15 genera, 74 species (Turner and Rose 1989). In Africa, 7 genera (2 endemic, another restricted to Africa and Madagascar) and 38 species (29 endemic). 6 species breed in Eurasia, N Africa and subsaharan Africa; 2 breed in Eurasia and N Africa and winter in subsaharan Africa; and 1 Malagasy species winters in SE Africa.

Structurally and biologically one of the most distinctive of songbird families. Affinities uncertain, but DNA evidence suggests closest alliance is with bulbuls, Old World warblers and 8 other families ('Superfamily Sylvioidea': Sibley and Ahlquist 1990).

Subfamily PSEUDOCHELIDONINAE: river martins

Glossy black swallows with robust, deep, broad bill, central rectrices with short or long filamentous rachis, and bright-coloured soft parts (bill orange-and-yellow or green-and-black, eye and orbital ring red or white, legs and feet pink). Bronchial rings incomplete (more or less complete in Hirundininae).

2 species, *P. eurystomina* in Africa and *P. sirintarae* in Asia (breeding grounds unknown; some winter in Thailand: Thonglongya 1968), sometimes distinguished generically (Brooke 1972), although they are surprisingly alike in morphology and plumage (Zusi 1978). African species nests in burrow in ground.

Genus *Pseudochelidon* Hartlaub

Pseudochelidon eurystomina **Hartlaub. African River Martin. Hirondelle de rivière.**

Pseudochelidon eurystomina Hartlaub, 1861. J. Orn., p. 12; Gabon.

Plate 10
(Opp. p. 161)

Range and Status. Endemic, equatorial migrant. Breeds Jan–May on middle Congo R. and lower Oubangui R. from Lukolela (1° 7'S) to Betou (3° 3'N) (Zaïre–Congo border) and Ikengo (0° 8'S), and on upper Congo R. (Zaïre) from Bumba (22° 28'E) to Basoko (23° 35'E), perhaps also near Umangi (21° 25'E). Occurs May–Nov near Gabon–Congo coast, from L. Onangué (1°S) and lower Nyanga R. (3°S) (Gabon) to mouth of Kwilu-Niari R. (4° 30'S, Congo). Strong local belief that they also breed from lower Nyanga R. to Kwilu-Niari R. has recently been vindicated: *c.* 800 birds nested in 5 colonies at Gamba (2° 45'S) on Gabon coast in Oct 1988 (P. Christy and P. D. Alexander-Marrack, pers. comm.). Only localities where migration between coast and middle Congo R. has been detected, are M'Passa and Makokou, NE Gabon (Brosset and Erard 1986), where in Feb immature birds were commoner than adults (P. Christy, pers. comm.). Whether the same birds breed twice a year, once in Gabon and once in Zaïre, or whether each population 'winters' in the breeding region of the other, is unknown.

Pseudochelidon eurystomina

Description. ADULT ♂: black. Whole head and underside black, matt or with slight bluish gloss, the matt black of nape divided very abruptly from oily green-glossed mantle, back, wings and tail. Rump and uppertail-coverts matt black without green gloss. Underside of remiges and rectrices shiny black. Central pair of rectrices slightly pointed, the rachis projecting 1 mm when feather worn. Bill bright yellow, merging to orange at cutting edges and base; eye scarlet, eyelids dull pink, black pupil small; legs and feet brown-pink, claws pale buff. Sexes alike. SIZE: (10 ♂♂, 5 ♀♀) wing, ♂ 119–128, ♀ 118–127; tail, ♂ 44–48, ♀ 44–47; bill to feathers, ♂♀ (n = 15) 11–12·5, bill to skull, 1♂ 14·5, 1♀ 15, 1 unsexed 15·3, width at gape, ♂♀ (n = 3) 9·5–11·2 (10·3); tarsus, ♂♀ (n = 15) 14–16.

IMMATURE: blackish brown above, greyish brown below; feathers pale-fringed, particularly undertail-coverts.

NESTLING: scant whitish down on crown, humeral tracts and lower back.

Field Characters. A peculiar black swallow with large, wide, round head, robust yellow bill, red eye, pink legs and rather short, rounded tail. Very gregarious; usually on great rivers. Actions in flight and on ground are swallow-like, but a few African River Martins in flock of *Hirundo* swallows are readily recognized by voice and 'peculiar shape' (Brosset and Erard 1977) (**A**).

Voice. Tape-recorded (ERA). Flight call 'chip-ip-ip' or 'cheer, cheer, cheer . . .', with rasping quality, like tern *Sterna*. Call also described as jingling. Resting flock makes chattering sound, like weavers *Ploceus* (but without any long wheezing notes). In aerial pursuits said to emit 'little jingling song' (Brosset and Erard 1977).

A

General Habits. Inhabits lakes, lake-like expanses of great rivers, river mouths and swampy coastal areas, nesting in flat sandbars by rivers, or (Gamba, Gabon) in grassy sand in plain 2–3 km from coast.

Gregarious. Flight energetic, with brief periods of gliding, more like Purple Martin *Progne subis* than typical *Hirundo* swallow. Dense flock of up to 1000 birds moves near nesting ground with co-ordinated precision of waders or European Starlings *Sturnus vulgaris*; but when foraging, birds scatter in all directions (Chapin 1953). Migrants at M'Passa (Gabon) forage around buildings in morning, above large river and adjacent forest for much of day, and over grassy airfield in late afternoon. In evening, flock circles high in air, with much calling, and 2–3 birds often involved in aerial pursuits. Hovers into wind. Flocks rest on bare ground, telephone wires, corrugated iron roofs, slender branches of bushes, and bare branches at top of tall trees. Quite mobile on ground, resembling a small pratincole more than a swallow (Brosset and Erard 1977).

Present on middle Congo R. breeding grounds (Zaïre) from mid-Jan to May, and on Gabon coastal breeding grounds from late May to at least Nov. Migrants were not detected at Makokou (NE Gabon) until 1977, when daily passage of tens or hundreds of birds from mid-Jan to mid-Mar (max. 1000 in early Feb). Subsequently passage noted there late Dec to late Mar, and early June to early Sept (Erard 1981). In 1985, thousands passed eastwards (some northwards) in 2nd half of Feb (Brosset and Erard 1986). Movements evidently determined by river water levels. Spring migrants at Makokou include numerous juveniles, but in autumn all are adult (P. Christy, pers. comm.).

Food. Airborne insects: mainly ants, also small butterflies (many Lycaenidae), Hemiptera, Homoptera, beetles and flies.

Breeding Habits. Colonial nester. Mating system unknown. 1 colony of *c.* 400 nests (Congo R.); 5 colonies *c.* 200 m apart in a line (Gabon) of *c.* 20, 100, 100, 150 and 400 birds.

In display on ground, shortly before breeding season, 2 birds hold body horizontally with wing-tips drooping, wrists held away from breast, neck stretched vertically up, head held horizontally. They run parallel with one another, then face each other and perform 'a kind of pivoted dance' with lateral displacements and jingling calls. In another display, bird in 'vertically-stretched' posture approaches another, then digs ground with bill (Brosset and Erard 1977). In a variant of the first display, interpreted as pair-bonding courtship, a bird ran across ground towards others, with wings quivering and bill open, as if begging. Once, when one bird stopped displaying to another, the latter started to display similarly to the first one (P. D. Alexander-Marrack, pers. comm.).

NEST: cavity with a few dead leaves, twigs and copal-tree seed-pods on floor, dug at end of oblique tunnel *c.* 1 m long, in flat sand bar (Zaïre). Birds at Gamba colony (Gabon) brought grass into nest burrows soon after digging them, sometimes collected on ground a few m away. Burrows, *c.* 0·5 m apart, are excavated by martins themselves, but one colony was on site formerly occupied by Rosy Bee-eaters *Merops malimbicus*, whose burrows may have been re-used by the martins. A few bee-eaters were still entering nests in late Oct, but aggressive incident between martin and bee-eater observed only once. Gamba colonies sited on old beach ridges thinly covered with short grass; ridges 1·5 m higher than surrounding grassy savanna plain, 0·5 × 2 km in extent, surrounded by swamp forest, 2–3 km from coast (P. D. Alexander-Marrack, pers. comm.).

EGGS: 3; white. SIZE: (n = 8) 21·9–26 × 16·4–18·2 (24·1 × 17·4).

LAYING DATES: Congo, Zaïre, Feb–Mar; Gabon, about Sept–Nov (nests excavated Sept–Oct, mainly late Oct; adults entering with food late Nov).

BREEDING SUCCESS/SURVIVAL: in the 1930s colonies were raided at night, before and during the breeding season, by fishermen who took adult birds and nestlings back to their villages to eat (Chapin 1953). Evident high incidence of this practice may well decimate population. Breeding colonies also vulnerable to flooding; middle Congo R. sand bars are exposed in Feb–Mar and July–Aug; fresh eggs found at end of Mar give little time for rearing before waters rise in Apr (Chapin 1953).

References
Bannerman, D. A. (1939).
Brosset, A. and Erard, C. (1977).
Chapin, J. P. (1953).

Subfamily HIRUNDININAE: typical martins and swallows

Brown or glossy blue or green hirundines with bill short but broad, rather weak (robust in American *Progne*); black soft-parts; tail square or shallowly forked ('martins') or deeply forked with long, attenuated outer rectrix ('swallows'); bronchial rings more or less complete (unique among Passeriformes).

HIRUNDINIDAE

Genus *Psalidoprocne* Cabanis

Medium-sized swallows with outer edge of outer primary rough or saw-edged (in ♂♂ only), due to the barbs being curved backwards i.e. towards front of wing. Glossy black, blue or green (unglossed brown in 2 species), underwing-coverts often white or grey. Tail deeply forked (square in 1 species), outer tail-feathers not as filamentous as in *Hirundo*. Nest, a tunnel and egg-chamber, in dry perpendicular bank (4 species), or on ledge on cliffs (1 species). Flight slow and fluttering; perches much less than do *Hirundo* spp.

Endemic; 5 species (12 species recognized in Peters 1960, who treated 8 races of *pristoptera* as specifically distinct; but see White 1961, and Hall and Moreau 1970). 2 superspecies, *P. obscura/P. pristoptera* and *P. albiceps/P. fuliginosa*.

Psalidoprocne obscura superspecies

1. *P. obscura*
2. *P. pristoptera*

Psalidoprocne albiceps superspecies

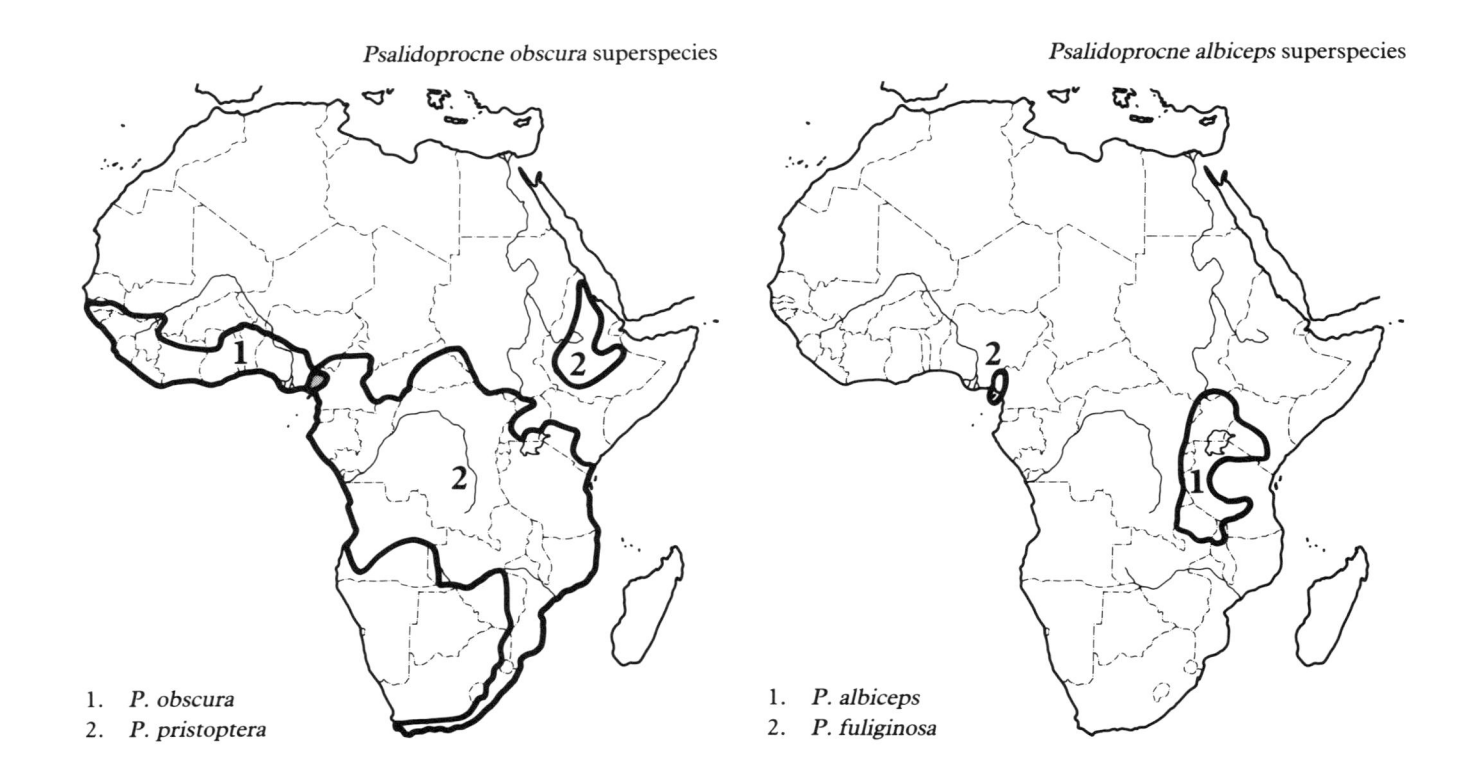

1. *P. albiceps*
2. *P. fuliginosa*

Plate 9
(Opp. p. 160)

Psalidoprocne nitens (Cassin). **Square-tailed Saw-wing. Hirondelle à queue courte.**

Atticora nitens Cassin, 1857. Proc. Acad. Nat. Sci. Philadelphia, sig. 2, p. 38; Muni R., Gabon.

Range and Status. Endemic resident in lowland forests of western Africa, in Guinea (Kakoulima, 50 km northeast of Conakry), Sierra Leone, Liberia (Mt Nimba, 900–1300 m; also Paul Town, Zorzor), Ivory Coast (coastal to central regions, Abidjan, Mt Toukini, Comoé Nat. Park), Ghana (forest belt from Tarkwa to Tafo; also Mole Nat. Park in N), SE Nigeria (Calabar, Mambilla and Owerri), S Cameroon (south of *c*. 5°N in forest region), and N and central Zaïre (east to Semliki valley, south to Ganda Sundi) south through Gabon to Cabinda. Common Liberia, Gabon and NE Zaïre; uncommon to frequent elsewhere.

Description. *P. n. nitens* (Cassin): Guinea to central Zaïre. ADULT ♂: upperparts including head dark blackish brown glossed with dull green. Tail not forked, blackish brown with purple-blue wash. Lores black, ear-coverts blackish brown. Chin and throat dark grey-brown; remaining underparts dark brown glossed with green. Primaries, secondaries and upper-wing-coverts blackish brown, outer webs with indistinct glossy green edges; axillaries and underwing-coverts blackish brown faintly glossed with dull green; rest of underwing dark brown, unglossed. Bill black; eye brown; legs black. Sexes alike. SIZE: (9 ♂♂, 16 ♀♀) wing, ♂ 92–98 (95·2), ♀ 86–95 (89·4); tail, ♂ 38–51 (47·9), ♀ 40–49 (45·7); bill to feathers, ♂ 4–6 (4·4), ♀ 5–6 (5·1); tarsus ♂ 6–10 (8·6), ♀ 6–10 (8·5). WEIGHT: (Liberia) ♀ (n = 13) av. 10·0, unsexed (n = 7) av. 9·6.

IMMATURE: like adult but plumage with only indistinct dull green gloss.

NESTLING: dark chocolate-brown above, paler brown below.

P. n. centralis Neumann: NE Zaïre from Tshuapa to Semliki valley. Like nominate form but throat blackish brown, slightly glossed with green.

Field Characters. A small, all-dark swallow with square tail. Very similar to Forest Swallow *H. fuliginosa*, which has slightly forked tail, rufous wash on throat. Other saw-wings *Psalidoprocne* spp. have long, forked tails; their young are brown, with shorter but still forked tails. Young White-throated Blue Swallow *H. nigrita* is dark brown with square tail but has white throat-patch, white spots in tail.

Psalidoprocne nitens

Voice. Tape-recorded (C, CHA, ERA). Calls, a rattle and a soft 'sip'.

General Habits. Inhabits primary and secondary lowland forest, including edges and clearings.

Occurs in pairs or flocks of up to 30, often with Fanti Saw-wing *P. obscura* and Pied-winged Swallow *H. leucosoma*, flying above and within forests, sometimes close to ground in clearings and villages. Less common in modified habitats, e.g. around villages, where often replaced by Black Saw-wing *P. pristoptera*. Flight slow and fluttering, often interspersed with periods of perching on small trees.

Mainly sedentary, but moves to Comoé Nat. Park, central Ivory Coast, in July.

Food. Airborne insects, including wood-boring beetles (bostrichids), termites and ants.

Breeding Habits. Solitary nester, territorial. During courtship, presumed ♀ makes many loops and zigzags in flight, sometimes holding wings rigid and below horizontal; presumed ♂ follows her, holding wings below horizontal and vibrating them rapidly to produce noise audible at short distance. Both then glide, after which ♂ turns and approaches ♀; turning his head up high, he seems to touch ♀'s bill (courtship feeding?), then one or both produce a rattling call (Brosset and Erard 1986).

NEST: chamber padded with lichens and moss, at end of tunnel in perpendicular bank. Tunnel 0·3–2 m long, 6 cm in diam., rises somewhat towards egg-chamber.

EGGS: 2, once 4 (A. Brosset, pers. comm.). Long ovals; white. SIZE: (n = 2) 16–19 × 11–14.

LAYING DATES: Sierra Leone (breeding condition July); Cameroon, Jan, July; Gabon, July–Mar (dry season to beginning of rains); Angola (Cabinda), Aug; NE Zaïre, Feb, May, July–Oct.

Reference
Brosset, A. and Erard, C. (1986).

Psalidoprocne obscura (Hartlaub). Fanti Saw-wing. Hirondelle fanti.

Hirundo obscura Hartlaub, 1855. J. Orn. 3, p. 355; Dabocrom, Gold Coast.

Forms a superspecies with *P. pristoptera*.

Range and Status. Endemic resident and intra-African migrant. From S Senegambia and extreme SW Mali through Guinea-Bissau, Guinea, Sierra Leone, Liberia, Ivory Coast (south of c. 9°N), Ghana, Togo (southern forest region), Benin (coast; also Pendjari and Arli Nat. Park along Burkina Faso border), and S and central Nigeria (along Niger R. north to 11°N) to W Cameroon (Victoria, Kumba). Uncommon, sometimes common, especially Sierra Leone (commonest swallow, Freetown), Ivory Coast and Nigeria. Vagrant Central African Republic (Manovo-Gounda-Saint Floris Nat. Park).

Description. ADULT ♂: entire head and body plumage glossy dark bottle green, with slight bluish sheen on rump and uppertail-coverts; lores black. Tail very long, strongly forked, black glossed with dark green, T5 and T6 with gloss on outer webs only. Primaries and secondaries blackish, outer webs with dull green gloss, inner webs with brown border; greater coverts similar but without brown border; lesser and median coverts dark glossy green; axillaries and underwing-coverts dark brown, unglossed. Bill black; eye brown; feet black. ADULT ♀: like ♂ but tail shorter. SIZE: wing, 1 ♂ 97, ♀ (n = 2) 88, 94, unsexed (n = 4) 85–97 (91·0); tail, 1 ♂ 100, ♀ (n = 2) 88, 91, unsexed (n = 3) 67–90 (78·2); depth of fork, 1 ♂ 57, ♀ (n = 2) 41, 63, unsexed (n = 2) 40, 66; bill to

Plate 9
(Opp. p. 160)

feathers, 1 ♂ 4·5, ♀ (n = 2) 4, 5, unsexed (n = 4) 4–5 (4·5); tarsus, 1 ♂ 9, 1 ♀ 9, unsexed (n = 4) 6–8 (6·3). WEIGHT: (Ghana, Nigeria) 2 unsexed 10, 8·8.

IMMATURE: similar to adult but brown, with variable amount of glossy on upperparts.

NESTLING: unknown.

Field Characters. An all-dark saw-wing with very long, forked tail, glossy green body, dark underwings. Replaces Black Saw-wing *P. pristoptera* in W Africa though just meets race *petiti* in Cameroon; *petiti* very different, with shorter tail, unglossed brown body, white underwing-coverts. Might just meet Mountain Saw-wing *P. fuliginosa* in Cameroon or E Nigeria, but latter also has unglossed brown body, and shorter tail with shallow fork.

Voice. Tape-recorded (CHA). Call, a soft 'seep'.

General Habits. Occurs along forest and woodland fringing grassland, savanna and rivers.

Occurs in pairs or small parties, sometimes flocks of up to 50. Flight buoyant; also hesitant and flickering (Field 1974). Hawks for insects in clearings in woodland and savanna, and along borders of gallery forest. Sometimes uses tree as vantage point, flying out to hunt from it.

Sedentary in S, partly migratory in N of range: Gambia mainly July–Nov; Sierra Leone, 'immense' flocks 12 Dec, scarce later, Freetown (Bannerman 1939); Ivory Coast south of 7° 30'N, common Oct–May, rarer in rains; in N present Apr–May to Sept; in S Ghana (inland from Cape Coast) largely a dry season visitor end Sept–end Mar with a few in May–Sept, in N mainly wet season June–Sept; and Nigeria, resident (Lagos, Warri) in S, and breeding visitor in N (south *c.* 11°N) May–Oct.

Food. Airborne insects.

Psalidoprocne obscura

Breeding Habits.

NEST: rounded egg-chamber lined with grass and moss at end of horizontal tunnel, *c.* 50 cm long; in river bank, pit or road cutting; excavated by both sexes.

EGGS: 2 (3 clutches). Elongated, oval; not glossy; white. SIZE: (n = ?) 17·2–19·5 × 12·3–13·5 (18·8 × 12·9).

LAYING DATES: Senegambia, Sept; Sierra Leone, Sept–Oct (at end of rains) (breeding condition July); Nigeria, May–Oct (in rains).

Plate 9
(Opp. p. 160)

Psalidoprocne pristoptera (Rüppell). Black Saw-wing. Hirondelle hérissée.

Hirundo (Chelidon) pristoptera Rüppell, 1836. Neue Wirbelt., Vögel, p. 105, pl. 39, fig. 2; Simen Province, Ethiopia.

Forms a superspecies with *P. obscura*.

Range and Status. Endemic resident and intra-African migrant. E Nigeria (Obudu and Mambilla plateaux), Cameroon (north to Adamawa Plateau), N Central African Republic (Bamingui-Bangoran Nat. Park) east to SW Sudan (Bahr el Ghazal, Western Equatoria), Ethiopia (W and SE Highlands and central and S Rift above 1200 m), Uganda (Mt Morongole in N, Kabale, Kigezi and Mbarara in SW, Mt Elgon in E), and Kenya (mainly highlands south of *c.* 2°N and west of 38·5°E; also SE coast at Ribe, Sokoke Forest and Shimba Hills). Gabon, Congo, Zaïre, Rwanda, Burundi and Tanzania south to Angola (south to W Huila and N Moxico), Namibia (E Caprivi), Zambia (from Mankoya eastwards, mainly north of 14°S), E Zimbabwe (from Marandellas and Zimbabwe ruins east and south to Sabi valley), Mozambique, Swaziland, E Transvaal (including Escarpment region, Lowveld along Crocodile R. and SE Highlands), Natal and Cape Province (mainly coastal,

Psalidoprocne pristoptera

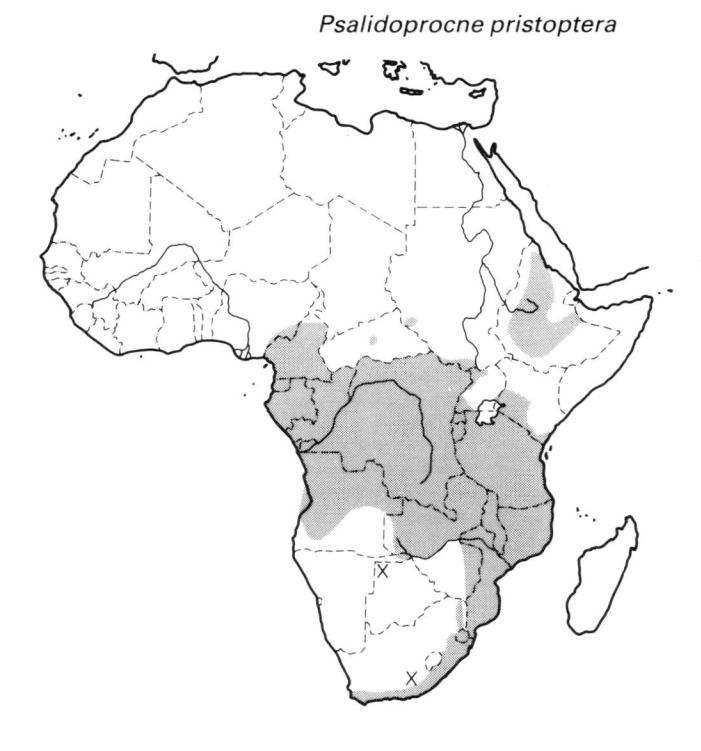

west to Cape Town; inland record at Swinside Dam, *c.* 32°S, 26°E). Vagrant Botswana (Shakawe). Frequent to abundant in highlands; rare to uncommon in lower Congo basin, W Sudan, plains of E Africa, Zambia south of 14°S, and central Zimbabwe (population of *holomelaena* a few hundred individuals at most: Irwin 1981).

Description. *P. p. pristoptera* (Rüppell): W Highlands of N Ethiopia. ADULT ♂: head, entire body, and lesser and median upperwing-coverts glossy purplish blue; lores black. Tail blackish brown with dull green gloss, strongly forked, outer tail-feather broad and blunt. Primaries, secondaries and greater coverts blackish brown, outer webs with faint blue wash; axillaries and underwing-coverts white, under primary coverts with dusky tips, rest of underwing dark brown, under primary shafts straw-coloured. Bill black; eye brown; legs purplish. ADULT ♀: like ♂ but tail less forked. SIZE: (10 ♂♂, 10 ♀♀) wing, ♂ 99–108 (102), ♀ 92–107 (98·7); tail, ♂ 54–77 (63·6), ♀ 51–71 (58·5), depth of fork, unsexed (n = ?) 20–25; bill to feathers, ♂ 4–5 (4·6), ♀ (4–6) (5·0); tarsus, ♂ 6–9 (6·9), ♀ 6–8 (7·5). WEIGHT: (Ethiopia) 1 ♂ 12, 1 ♀ 12.

IMMATURE: similar to adult but dark brown, without gloss.
NESTLING: naked except for tufts of grey down on lower flanks, shoulders and back of head.

P. p. blandfordi (Blundell and Lovat): central part of W Highlands of Ethiopia from Blue Nile and Bilo to Addis Ababa. Greenish wash over purplish blue on head, mantle and upperwing-coverts; below darker blue than nominate race with only a slight green wash. Tail fork 20–25.

P. p. oleaginea (Neumann): SW Ethiopia and SE Sudan. Glossy rich oily green; outer tail-feathers less broad than nominate race. Tail fork 24–34.

P. p. antinorii (Salvadori): central and S Ethiopia from Addis Ababa and Harar south to Yavello, Alghe, Burgi and N end L. Turkana. Gloss purplish bronze; outer tail-feathers not quite as broad as in nominate race. Tail fork 25–37.

P. p. mangbettorum (Chapin): Sudan west to NE Zaïre. Gloss rich oily green; outer tail-feathers more attenuated than in nominate race, tail more forked; white of axillaries and underwing-coverts extends to sides of breast. Outer tail-feather 79–94; tail fork 33–54.

P. p. chalybea (Reichenow): N and central Cameroon to NE Zaïre; also W Sudan (9°–10°N, 24°E). Gloss rich oily green; outer tail-feathers more attenuated than in nominate race; tail more forked and underwing-coverts and axillaries grey, eye blackish and legs brown. Outer tail-feather 85–94; tail fork 38–54.

P. p. petiti Sharpe and Bouvier: Angola (Cabinda), Gabon, S Cameroon and E Nigeria. Gloss bronzy brownish black; underwing-coverts white with grey tinge. Outer tail-feather 70–82; tail fork 25–35. WEIGHT: (Cameroon) ♂ (n = 3) 10–11 (10·6), ♀ (n = 3) 10–12 (11·0).

P. p. reichenowi (Neumann): N and central Angola, S Zaïre and W Zambia. Gloss greenish, less rich than in *oleaginea*; underwing-coverts greyish white, outer tail-feather not quite as broad as in nominate race (72–78); tail fork 21–33.

P. p. orientalis (Reichenow): S Tanzania (north to Iringa) and E Zambia to E Zimbabwe (border highlands to Sabi valley) and Mozambique (south to Beira). Gloss dull greenish, less rich than in *oleaginea*; underwing-coverts white (grey in immature birds); tail longer, more deeply forked and outer tail-feather narrower (76–90); tail fork 37–48.

P. p. holomelaena (Sundevall): South Africa, coastal region from Cape Town to Natal and S Mozambique west to NE Transvaal, SE Zimbabwe (Sabi Valley west to Zimbabwe ruins and Hawau), and S Malaŵi (Nchalo). Gloss dull greenish, less rich than in *oleaginea*, underwing-coverts sooty grey or blackish. Outer tail-feather 82–90; tail fork 38–49. WEIGHT: (Zimbabwe) unsexed (n = 14) 10–12 (11·2); (South Africa) ♂ (n = 9) 10–13 (11·6), ♀ (n = 7) 11–13 (11·4).

P. p. massaica (Neumann): N and central Tanzania (south to Mwanihana and Magombera Forests) and Kenya. Like *holomelaena* but wing longer (110–119); outer tail-feather 85–99; tail fork 39–57. WEIGHT: (Uganda) ♂ (n = 11) 10–14 (12·1), ♀ (n = 3) 12–15 (13·5); (Kenya) ♂ (n = 9) 10–13 (11·6), ♀ (n = 5) 7–11 (9·4).

P. p. ruwenzori (Chapin): E Zaïre (highlands from Rwenzori to Mt Kabobo). Like *holomelaena* but tail shorter (73–85); tail fork 28–35.

Field Characters. A medium-sized swallow with all-dark body and forked tail; in much of its range the only swallow of this description. Widespread, with much geographic variation. Most races have some degree of body gloss, various shades of dark green, dark blue or blackish, underwing-coverts white, grey or dark brown. Replaced in W Africa by Fanti Saw-wing *P. obscura*, q.v. for differences. Nigerian race *petiti* (white underwing-coverts, unglossed brown body) meets Mountain Saw-wing *P. fuliginosa*; latter distinguished by heavier build, shorter tail with shallower fork, brown underwings. Brown immature similar to immature brown ♀ White-headed Saw-wing *P. albiceps*, but latter has heavier body with broader head, and shorter, less forked tail.

Voice. Tape-recorded (CHA, ERA, GIB, McVIC, STJ). Main call, a high-pitched sibilant 'see-see' or 'see-see-seeu'; apparent racial difference illustrated in Maclean (1985); call of *holomelaena* 'chirr-chirr-cheeeu', that of *orientalis* 'tseeu-tseeu-tsee-ip'. Other calls include low 'chirp' (alarm) and mellow 'hui' in courtship flight.

General Habits. Occurs in variety of habitats at all elevations but mainly in forest and woodland, especially in clearings and at edges; also moorland, grassy highland, tall grass savanna, thornbush, river valleys and around villages; usually not far from water.

Occurs in pairs or small groups; sometimes in flocks of 6–15. Flight slow, buoyant, undulating, fluttering and gliding; not swift. Often forages fairly close to ground, e.g. in clearings and below canopy in miombo woodland. Feeds usually at dawn and dusk but active any time during cloudy days. Roosts on base of twigs of tall trees, sometimes on tall grass.

Movements complicated, not well understood. Sedentary, altitudinal migrant and in some countries a migrant, breeding in rains and absent in dry season. Mainly sedentary Gabon, Sudan, Ethiopia and E Africa (partially migratory Ethiopia and Tanzania ?); partial migrant in Malaŵi (hundreds seen in non-breeding season), Zambia (most Oct–Apr, few July–Aug), Mozambique and South Africa. Altitudinal migrant in Zimbabwe (*orientalis*). Migrant in Nigeria (present only in rainy season), Angola, S Zaïre and Zambia (*reichenowi*) (absent during dry season), and South Africa and Zimbabwe (*holomelaena*) (present mainly Sept–Apr, probably in Mozambique in non-breeding season). Of 134 birds ringed, Malaŵi (Nchalo), 6 recaptured on same site, one 4 years later (Hanmer 1989).

Food. Airborne insects, including bostrichid beetles, ants and flies.

Breeding Habits. Monogamous; solitary nester; once (Gabon) 10 pairs were excavating nests on face of sandy cliff. During courtship, presumed ♂ pursues ♀ with wings moving stiffly and kept mostly below horizontal plane of body.

NEST: chamber, 75 × 130 × 70, lined with *Usnea* lichens and dry grass, at end of horizontal tunnel, *c.* 45 cm long and 2·5 cm in diam., in sandy cliff, earth bank or pit, occasionally in stream-bank (Short and Horne 1985). Excavated by both sexes who remove bits of earth with bill and push them out of tunnel with feet; takes up to 3 weeks. Sometimes hole abandoned by bee-eaters used as nest. Before entering hole, birds wheel by it several times, then dive straight into it.

EGGS: 1–3, usually 2, av. (n = 27 clutches, Cameroon, Ethiopia, E Africa, southern Africa) 2·03. Long, oval; white. SIZE: (n = 16) 17·9–20·7 × 11·9–14·3 (18·7 × 12·8). Double brooded.

LAYING DATES: Cameroon, Apr, June, Oct; Gabon, Apr–July; Congo (breeding condition Nov); Zaïre, Feb–Sept; Ethiopia, Feb–July; E Africa: Kenya, Feb; Tanzania, Jan–Mar, June–Dec; Region A, Feb, June, Dec; Region B, Feb; Region C, Dec; Region D, Jan–Mar, May–Dec; Angola, Mar, May; Zambia, Jan–Mar; Malaŵi, Feb–Mar, Aug, Dec; Mozambique, Oct–Mar, May; Zimbabwe, *holomelaena* Mar–Apr, *orientalis* July–Apr (July 1, Aug 2, Sept 3, Oct 3, Nov 4, Dec 4, Jan 2, Feb 3, Mar 1, Apr 3 clutches); South Africa, Aug, Oct–Apr.

INCUBATION: by ♀ (also ♂ ?); usually in spells of 10–50 min. Period: 19 days.

DEVELOPMENT AND CARE OF YOUNG: eyes begin to open at 5 days; primary quills 3–10 mm and tail-feathers 0–5 mm long at 10 days; body covered with contour feathers at 17 days. Young fed by both parents, with intervals between feeds of 4–20 min at first, up to 2 h when nearer to fledging. Brooded much of time when small, thereafter at variable intervals; leave nest at 24–27 days.

References
Moreau, R. E. (1940).
van Someren, V. G. L. (1956).
White, C. M. N. (1961).

Plate 9
(Opp. p. 160)

Psalidoprocne albiceps Sclater. **White-headed Saw-wing.** Hirondelle à tête blanche.

Psalidoprocne albiceps Sclater, 1864. Proc. Zool. Soc. Lond., p. 108, pl. 14; Uzinza, Tabora district, Tanganyika Territory.

Forms a superspecies with *P. fuliginosa*.

Range and Status. Endemic resident and partial intra-African migrant. Extreme SE Sudan (upper Nile), E Zaïre (from Sudan border, Lendu Plateau, Semliki and Ruzizi-Kivu District south to Zambian border), Uganda, SW and W Kenya (mainly L. Victoria basin; also north to 2°N and east to 38°E), Rwanda, Burundi, Tanzania (mainly L. Victoria region, and south in W to Kigoma and Mahari Mt, to Ufipa Plateau and Iringa in S, and Babati, Kolo, Manyara and Ngurus in N and E), Zambia (Northern and Luapula provinces south to *c.* 12°S) and N Malaŵi south to Mzimba and Viphya Mts; also N Angola (Cacolo, Malange District). Frequent to common in most of range; locally abundant E Zaïre. Vagrant Ethiopia (L. Abaya, lower Omo R./north end L. Turkana), central and S Zaïre (Kasai, W Shaba), and W and S Zambia (Sesheke, Ndola, Lusaka, Chipata).

Psalidoprocne albiceps

Field Characters. A dark brown saw-wing, appearing matt (unglossed) in the field. Body broader and tail shorter, less deeply forked than Black Saw-wing *P. pristoptera*. White-headed adults unmistakable, especially ♂; ♀ sometimes does not have much white on crown but throat always white. All-brown immature ♀ very similar to immature Black Saw-wing, distinguished by shape (broader body, shorter tail with shallow fork).

Voice. Tape-recorded (McVIC).

General Habits. Occurs in miombo woodland, riparian, evergreen and moist montane forest, especially edges and clearings, grassland, savanna (E Zaïre), and occasionally over papyrus (Zambia).

Occurs in pairs or groups of up to 4, rarely 40 (Zaïre) when feeding. Flight slow and hesitant; feeds low over bush or grassland. Perches on bushes, grass stems and tree tops.

Mainly sedentary, but partly migratory: present NE Zambia early Oct to early May, Malaŵi Oct–Apr, and probably partial migrant in E Africa.

Food. Airborne insects, including winged ants, small beetles and flies.

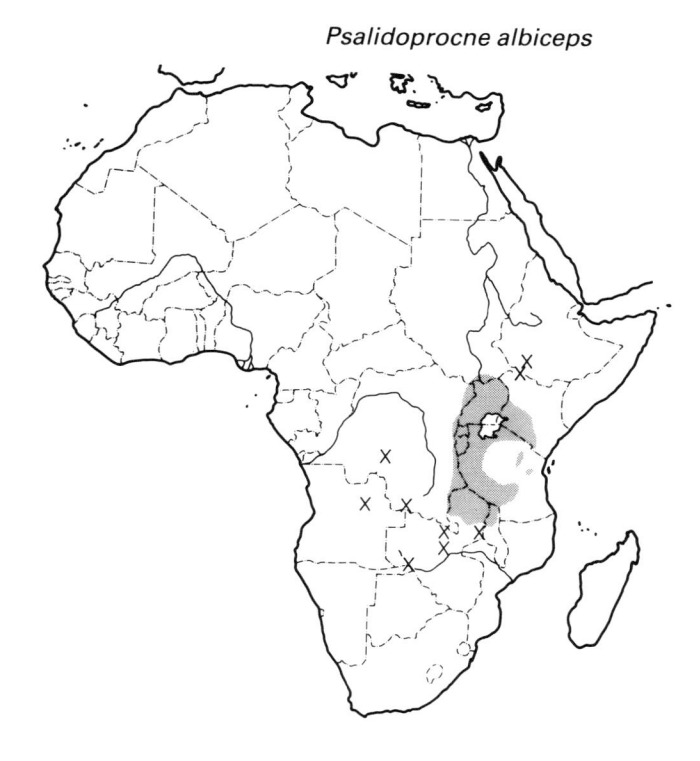

Description. *P. a. albiceps* Sclater: range of species except N Angola. ADULT ♂: forehead to nape, chin and throat white; black stripe from base of bill through eye and ear-coverts. Body, upperwings and tail blackish brown with hint of oily green gloss, tail forked; axillaries and underwing-coverts paler, greyish brown. Bill black; eye brown; legs dark grey-brown to black. ADULT ♀: like ♂ but top of head with variable amount of dark mottling, throat also with some mottling around the edges, less pure white; tail shorter. SIZE: (10 ♂♂, 9 ♀♀) wing, ♂ 95–106 (101), ♀ 91–106 (97·5); tail, ♂ 49–71 (61·1), ♀ 50–72 (60·0), depth of fork, ♂ 19–29 (24·7), ♀ 15–37 (19·4); bill to feathers, ♂ 4–6 (5·0), ♀ 4–5 (4·3); tarsus, ♂ 6–8 (7·4), ♀ 6–8 (7·2). WEIGHT: (Uganda) ♂ (n = 17) 6–14 (11·7), ♀ (n = 6) 7–14 (10·6); (Kenya) 1 ♀ 11, 2 unsexed 11, 13.

IMMATURE: somewhat lighter and browner, ♂ with pale grey chin and upper throat, and with age some white feathers starting to appear on crown; ♀ entirely brown.

NESTLING: unknown.

P. a. suffusa Ripley: N Angola (Cacolo, Malange District). White area on head reduced to small cap; throat greyish; axillaries, underwing-coverts and ear-coverts greyish brown.

Breeding Habits. Solitary nester. ♂ commonly chases ♀ during courtship.

NEST: chamber lined with grass, lichens and a few feathers, at end of tunnel, with slight upward incline, 25–60 cm long, in roadside bank or similar dry site (not in river banks).

EGGS: 2–4. Rather pointed; white. SIZE: (n = 2) 19–19·2 × 13 (19·1 × 13·0). Sometimes double brooded (Zaïre).

LAYING DATES: Sudan, Aug–Sept; E Africa: Kenya, Nov–Mar; Uganda, Apr–June, Aug; Region A, Apr, Aug; Region B, Apr–June, Dec; Region C, Nov–Jan; Rwanda, Mar–Apr, Nov (mainly rains); Zaïre, May–Oct; Zambia, Dec–Jan; Malaŵi, Feb (enlarged testes Oct).

Psalidoprocne fuliginosa Shelley. Mountain Saw-wing. Hirondelle brune.

Plate 9
(Opp. p. 160)

Psalidoprocne fuliginosa Shelley, 1887. Proc. Zool. Soc. Lond., p. 123; Mt Cameroon.

Forms a superspecies with *P. albiceps*.

Range and Status. Endemic resident, Cameroon (Mt Cameroon) and Bioko; several sight records E Nigeria (Obudu Plateau) (Ash *et al.* 1989). Common.

Description. ADULT ♂: lores blackish; rest of head and entire body plumage dull dark chocolate-brown, without gloss, rather blacker on rump, uppertail-coverts, belly, flanks and undertail-coverts. Flight feathers blackish, becoming paler on margins of

inner webs; coverts chocolate-brown; axillaries and underwing-coverts greyer and paler, under primary coverts tipped black. Tail forked. Bill black; eye brown; legs brown. Sexes alike. SIZE: (4 ♂♂, 4 ♀♀) wing, ♂ 101–106 (105), ♀ 96–101 (98·8); tail, ♂ 60–64 (62·6), ♀ 48–58 (53·1), depth of fork, ♂ 20–26 (21·6), ♀ 19–25 (21·2); bill to feathers, ♂ 5–6 (5·6), ♀ 5–6 (5·8); tarsus, ♂ 7–9 (8·1), ♀ 7–9 (8·1). WEIGHT: ♂ (n = 9) 11·5–14 (12·6), ♀ (n = 6) 11–14 (12·5), unsexed (n = 10) 10·5–12·5 (11·7).
IMMATURE: like adult but paler brown.
NESTLING: unknown.

Field Characters. A dark brown, rather heavy-bodied swallow. Similar to race *petiti* of Black Saw-wing *P. pristoptera*; both have brown bodies without gloss, but *petiti* is slimmer, with more deeply forked tail, whitish underwing. Fanti Saw-wing *P. obscura* is black with green gloss and larger, with longer and much more deeply forked tail.

Voice. Tape-recorded (C, CHA, DYE). Song, a distinctive, soft, melodious 'dju-dju-diob-djuob-djuob', said to be quite different from any other swallow (Stuart 1986); calls, 'see-su' and 'tchuk-tchuk'.

General Habits. Inhabits forest interior and clearings from sea level to mountain peaks, also montane grassland, farmland and human settlements. On Bioko occurs from *c.* 300 m upwards; on Mt Cameroon, S and SE slopes from sea level to 3000 m and N slopes from 600 to 2900 m (Stuart 1986).
 Occurs in pairs or small parties, occasionally in groups of over 50; associates with other species of swallows and swifts. Flight unhurried; sometimes appears to pause in mid-air before swooping downward (Stuart 1986); sometimes flies close to ground. Perches on exposed branches on edges and clearings of forests, sometimes on tops of tall grass.
 Mainly sedentary, but records away from Mt Cameroon and Obudu Plateau suggest some seasonal movements.

Food. Airborne insects.

Breeding Habits. Usually solitary nester; once, 10–15 pairs found nesting beside waterfall; also 4–5 pairs inspecting nest-site in ravine (Stuart 1986).

Psalidoprocne fuliginosa

NEST: pad, *c.* 9–10 cm in diam., of tightly woven strands of lichens and moss, placed on ledge in hole on lava cliff, in deep ravine, cave beside waterfall, once inside shed.
 EGGS: 2. Slightly glossy; white. SIZE: (n = 2) 18·8–18·9 × 12·2–13·2 (18·9 × 12·7).
 LAYING DATES: Mt Cameroon, Oct, Dec–Mar (1 clutch recorded each month).

References
Eisentraut, M. (1963).
Serle, W. (1981).
Stuart, S. N. (1986).

Genus *Phedina* Bonaparte

2 small, brown-and-white martins, striped below, with nearly square tails, forming a superspecies, with one species in SW Congo Basin and the other breeding in the Malagasy Region, partly wintering (also breeding?) adjacently in Africa. Morphologically, the genus is barely distinguishable from *Hirundo* (Hall and Moreau 1970) or *Riparia*. We would merge *Phedina* with *Riparia* on the basis of tunnel-nesting of *P. brazzae*, were it not that *P. borbonica* is said to build saucer nests of twigs on rock ledges and houses (Delacour 1932, Rand 1936).

Phedina brazzae Oustalet. **Brazza's Martin. Hirondelle de Brazza.**

Phedina brazzae Oustalet, 1886. Naturaliste 8, p. 300; below Kwamouth.

Forms a superspecies with *P. borbonica*.

Range and Status. Endemic resident (or migrant), Congo, SW Zaïre and NE Angola. Ranges from about Djambala (Congo), Bolobo and Kwamouth (mouth of Kasai R., Zaïre) southeastwards to Kananga (= Luluabourg), Lubilashi R. and Lubishi R. (Zaïre). In Angola known only from Camissombo and Dundo (Lunda).

Description. ADULT ♂: forehead and crown very dark brown; nape, sides of neck, cheeks, mantle, back, rump, uppertail-coverts, scapulars and tertials dark brown. Wings and tail very dark brown; tail square or rounded. Lores very dark brown, or pale brown with buffish wash. Chin and throat silky white, profusely marked with small dark brown stripes; breast and flanks silky white, profusely marked with larger long dark brown stripes; breast sometimes suffused greyish brown; belly and undertail-coverts silky white with sparse, long, thin, dark brown stripes. Underwing-coverts dark brown; undersides of remiges and rectrices silvery dark brown. Bill black with whitish area at gape; eye brown; legs and feet black. SIZE: (4 ♂♂, 4 ♀♀) wing, ♂ 99–103 (101), ♀ 96–100 (98·0); tail, ♂ 45–47 (46·0), ♀ 44·5–46·5 (45·9); bill to skull, ♂ 8·4–9·4 (8·9), ♀ 8·1–8·9 (8·4).

IMMATURE: like adult but upperwing-coverts, tertials, and feathers of mantle, back and rump with whitish edges and rufous tips (greater upperwing-coverts especially rufous-tipped); underparts have less profuse and less distinct stripes than in adult; chin and throat greyish, with brown stripes; belly nearly plain white. Bill black with conspicuous white gape-flanges; mouth fleshy white.

NESTLING: not described.

Field Characters. A dark brown martin with silky white underparts heavily streaked with brown, from chin to undertail-coverts. Underside of wings dark brown. Tail square or slightly rounded – never forked. Riverine, in SW Congo Basin; gregarious. Very like Mascarene Martin *P. borbonica* (which is less heavily striped below and has slightly forked tail), but the 2 are not sympatric. The 3 small sand martins *Riparia congica*, *R. paludicola* and *R. riparia* occur in range of Brazza's Martin but have unstriped, plain brown and plain white underparts. Tail-moulting Lesser Striped Swallow *Hirundo abyssinica* has same streaked breast as Brazza's Martin, but has rufous cap and rump and white wing linings.

Voice. Unknown.

General Habits. Inhabits forested rivers in lush savanna woodlands in SW part of Congo Basin, nesting in sand banks. Breeds in small, loose colonies; general habits probably like those of sand martins *Riparia* spp., but almost nothing on record. Foraging flocks mingle with Lesser Striped Swallows (Chapin 1953). All records east of 20°E appear to be July–Nov (when breeding), so may be intratropical migrant.

Food. Unknown.

Breeding Habits. Colonial nester. One colony was of 'half a dozen tunnels, rather widely scattered' (Chapin 1953). Other tunnels 'sometimes closely grouped'.

NEST: tunnel *c.* 50 cm long, in sandy bank of river or drainage ditch, with terminal egg chamber containing fine grass, cotton and a few feathers.

EGGS: single clutch known, of 3 eggs, perhaps incomplete; white. SIZE: (n = 3) 18·4–18·7 × 12–12·8 (18·5 × 12·5).

LAYING DATES: Zaïre (Gandajika, 6° 44′S, 23° 7′E), July, 'nesting continues until October'; juvenile Sept.

Reference
Chapin, J. P. (1953).

HIRUNDINIDAE

Plate 10
(Opp. p. 161)

Phedina borbonica **(Hartlaub). Mascarene Martin. Hirondelle des Mascareignes.**

Phedina madagascariensis Hartlaub, 1860. J. Orn., p. 83; Madagascar.

Forms a superspecies with *P. brazzae*.

Range and Status. Breeding visitor, Mauritius and Reunion (*P. b. borbonica*) and Madagascar (*P. b. madagascariensis*); latter winters eastern Africa; migrant Aldabra Is; vagrant Amirantes Is (Feare 1977).

Status in Africa poorly known. Plentiful ('hundreds') L. Chilwa (Malaŵi) late June to late July 1944; 8 Chididi (Malaŵi), Apr 1959; 'large numbers' Inhaminga (18° 24'S, 35° 03'E, Manica e Sofala, Mozambique) June–July 1968; scattered records of 1–3 birds Pemba I. (Tanzania) Nov–Mar (once 2 birds Aug–Sept) in 1920s; 16 at L. Jipe (Kenya) June 1978 (Zimmerman 1978). Sight records from Chiromo (Malaŵi) July 1985, Tamboharta Pan, Lundi R. (Zimbabwe), and Klipplaatdrift (South Africa) Feb 1986.

Present Madagascar Aug–Apr, breeding Oct–Nov; that the species is present on Pemba at same season suggests that it may also breed there or on mainland (Moreau 1966).

Description. *P. b. madagascariensis* (Hartlaub) (only subspecies in Africa). ADULT ♂: forehead, crown, nape and ear-coverts brownish grey; mantle to rump and scapulars greyish brown, faintly streaked or dappled with darker brown. Uppertail-coverts, tail and wings dark brown; rectrices extremely narrowly edged white; tail slightly forked. Most birds from Madagascan breeding grounds have tertials plain dark brown, but migrants in fresh plumage in Africa have tertials with 1–2 mm wide white tips. Chin and throat white, profusely streaked with small brown stripes; breast greyish white with small or large, distinct or diffuse, brown streaks; flanks streaked; belly and undertail-coverts white. Underwing-coverts dark grey, most marginal coverts with broad whitish or pale buff tips. Bill black; eye brown; legs and feet black. Sexes alike. SIZE: (Madagascar) wing, ♂ (n = 5) 114–118 (116), ♀ (n = 5) 109–117 (113); tail, ♂ (n = 5) 50·5–52·5 (51·5), ♀ (n = 5) 48–54 (51·6); bill to skull, ♂ (n = 5) 10·3–13·9 (12·2), bill to feathers, ♂♀ (n = 9) 6·2–8·3 (7·7); tarsus, ♂♀ (n = 9) 11–14 (12·4). (Mozambique) wing, ♂ (n = 5) 101–114 (108), ♀ (n = 6) 102–116 (108); tail, ♂ (n = 5) 50·4–58 (55·2), ♀ (n = 6) 49·5–60 (53·2) (R. A. Earlé, pers. comm.). WEIGHT: (Mozambique) ♂ (n = 4) 21·3–22·9 (22·1), ♀ (n = 5) 17·9–22·7 (20·3).

IMMATURE: like adult but secondaries broadly tipped white.

Field Characters. A rather nondescript dark brown and white martin which might be confused with sand martins *Riparia* spp., and possibly tail-moulting Lesser Striped Swallow *H. abyssinica*. Upperparts dusky brown, wings blackish above and below; tail blackish, slightly forked; underparts greyish, boldly mottled or streaked blackish; belly and undertail-coverts white. Voice distinctive. Distinctive languid flight. Very like Brown-throated Sand Martin *R. paludicola* but larger, stockier, with streaked brown (not plain brown) breast. Common Sand Martin *R. riparia* is same size as Mascarene Martin, and dark-winged, but has white underparts with brown breast-band. Tail-moulting Lesser Striped Swallow has same streaked breast as Mascarene Martin, but has rufous cap and rump and white wing linings.

Voice. Tape-recorded (GI, HOR). Call note 'phree-zz' (Clancey *et al*. 1969).

Phedina borbonica

General Habits. Flock in Mozambique inhabited recently-logged woodland and immediately adjacent cultivated clearings. Birds foraged low over treetops in poor weather, higher on warm, calm days, mixing with Lesser Striped Swallows, Black Saw-wings *Psalidoprocne pristoptera*, African Palm Swifts *Cypsiurus parvus* and Bat-like Spinetails *Neafrapus boehmi*. Flight languid, rather slow and fluttering, with much gliding on set wings. Perches on dead upper twigs of trees. Not shy. Mozambique birds were moulting remiges June–July (suggesting that they were winter residents); several were quite fat.

Food. Airborne insects.

Reference
Clancey, P. A. *et al*. (1969).

Genus *Riparia* Forster

Sand martins. Large to small swallows with uniform brown upperparts, brown and white underparts; plumage rather soft; no metallic colours. Feet unfeathered; tail only slightly forked, without streamers. Eggs white. Nest in self-dug burrows in sandy ground.

4 species, 2 endemic to Africa, 1 Palearctic (wintering throughout Africa), and 1 widespread in Africa and S Asia. 2, *R. paludicola* and *R. congica*, are very closely allied and may be conspecific (*R. congica* poorly known); they form a superspecies with the larger *R. riparia*.

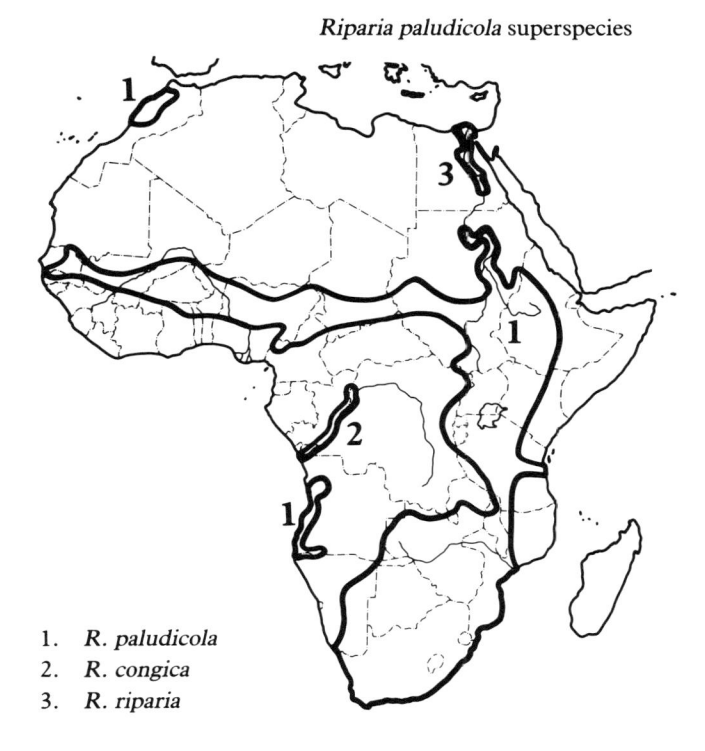

Riparia paludicola superspecies

1. *R. paludicola*
2. *R. congica*
3. *R. riparia*

Riparia paludicola (Vieillot). Brown-throated Sand Martin; African Sand Martin. Hirondelle paludicole.

Plate 7 (Opp. p. 112)

Hirundo paludicola Vieillot, 1817. Nouv. Dict. d'Hist. Nat., 14, p. 511; South Africa.

Forms a superspecies with *R. riparia* and *R. congica*.

Range and Status. Africa, Madagascar, N India, Burma, N Vietnam, Taiwan, N Philippines. Resident and partial migrant. Vagrant Israel, Saudi Arabia and Oman.

Morocco, resident, common and widespread in lowlands west of Atlas Mts, between 30° and 35°N; range recently extended to Oued Bou-Regreg, Rabat. Remainder of African range is subsaharan. Senegambia, occurs on middle Senegal R., 5 records Gambia R., elsewhere rare, mainly in winter; no breeding records. Mali, sparse, north to Tamesna, south to Ségou, but common in W in Baoulé Nat. Park. Guinea, recorded only in NE, probably breeding (Walsh 1987). Ghana, uncommon in N, Oct–May (overlooked June–Sept?); occurs in White Volta R. valley. Niger, rare, records all months except July–Sept and Jan, in or near Niger R. valley. Nigeria, sparse resident, locally common, in valleys of major rivers north to Sokoto and L. Chad; south of Benué R. recorded only on Mambilla Plateau and at Ibadan. Cameroon, sparse; records at 6 localities. Chad, 'strictly sedentary' and common in Shari R. valley southeast of L. Chad, also in Zakouma. Central African Republic, recorded in Manovo-Gounda-Saint Floris Nat. Park and near Bangui. Sudan, fairly common along Nile up to Khartoum and Blue Nile, also in Darfur and Bahr el Arab, and sparse in far S in Dec–Feb (Nikolaus 1987). Ethiopia, frequent to common, locally abundant, throughout except SE and Rift Valley. E Africa, widespread, locally abundant, mainly between 1200 and 3000 m but also common on lower Rufiji R. Zaïre, known only from Kivu highlands, where frequent mainly between 1500 and 2300 m, but occurs up to 3700 m on Mt Karisimbi; probably occurs also in SE on upper Luapula R. Zambia, locally common resident, confined to vicinity of larger rivers and lakes. Malaŵi, uncommon, confined to perennial rivers; up to 2000, Shire R., Jan; not on L. Malaŵi. Zimbabwe, mainly in middle Zambezi, Sabi and Limpopo valleys, scarce on L. Kariba; elsewhere uncommon. Angola, S Cuanza Norte and Condo on Cuanza R.; records at Massangano and Quissama Nat. Park; also on lower Cunene R.; common at Foz. Namibia, vagrant west to Zais. Botswana, common in N near water, elsewhere sparse. South Africa, widespread resident, dispersing after breeding, in Transvaal commonest in highveld; common throughout Natal, and Cape Province except NW.

Riparia paludicola

Description. *R. p. mauritanica* (Meade-Waldo): Morocco. ADULT ♂: upperparts from forehead to back, uppertail-coverts, tail, wings, cheeks, sides of neck and lores almost uniform brown; rump slightly paler; rectrices and remiges slightly darker. Outer 3 rectrices extremely narrowly edged whitish. Tail shallowly forked. Chin, throat and breast greyish brown, merging gradually into white belly and undertail-coverts; flanks washed brown. Undertail-coverts long. Outer edge of outer primary, inner edges of inner secondaries, and tertials, extremely narrowly edged pale brown. Underwing-coverts and axillaries grey-brown. Tips of folded wings extend 5 mm beyond tail. Bill black; eye dark brown; legs and feet black. Sexes alike. SIZE: (7 ♂♂, 7 ♀♀) wing, ♂ 97–104 (101), ♀ 98–103 (100); tail, ♂ 38–45 (41·5), ♀ 40–44 (42·1); bill to skull, ♂ 8·7–10·3 (9·4), ♀ 9–10·5 (9·6), bill to distal corner of nostril, ♂ 4·7–4·9 (4·8), ♀ 4·4–5·1 (4·7); tarsus, ♂ 10·4–11·8 (10·9), ♀ 9·9–11·5 (10·7).
IMMATURE: like adult but slightly paler and greyer; feathers narrowly fringed buff, particularly on forehead, back, rump, uppertail-coverts and upperwing-coverts.
NESTLING: unknown.
R. p. paludicola (Vieillot): Angola and South Africa north to Zambia, Malaŵi and S Tanzania (Njombe Highlands). Like *mauritanica* but larger: wing, unsexed (n = 62) 97–114 (104); tail, longest rectrix, unsexed (n = 15) 49–59, shortest rectrix (n = 15) 41–50. WEIGHT: ♂ (n = 3) 11·5–13·5 (12·4), ♀ (n = 3) 11·5–12·6 (11·9), unsexed (n = 111) 11–16 (13·5) (R. A. Earlé, pers. comm.).
R. p. paludibula (Rüppell) (= *minor* (Cabanis): Brooke 1975): Senegambia to Sudan and NE Ethiopia. Darker than *paludicola*. Small: wing, unsexed (n = 10) 95–103 (97·0).
R. p. schoensis (Reichenow): highlands of Ethiopia. Darker than *paludibula*. Large: wing (♂, ♀) 109–117.
R. p. newtoni (Bannerman): Cameroon (Bamenda Highlands) and adjacent Nigeria (Mambilla Plateau). Darker than *schoensis* and much darker than *paludicola*; throat and breast greyer; belly more extensively white.
R. p. ducis (Reichenow): central Tanzania and Zaïre (Kivu) north to Uganda and Kenya. Dark (like *newtoni*) with crown blackish brown; but little or no white on belly. WEIGHT: unsexed (n = 19, Nairobi, Kenya) 8–10·5 (9·2) (D. J. Pearson, pers. comm.).

Field Characters. A small, gregarious, mid-brown or rather dark brown martin with white belly and undertail-coverts (whitest in Cameroon, less extensively white, or pale brown, in E Africa). Quiet, not very vocal; occurs mainly near permanent water, but also far away from water in cool highlands and hot lowlands. Very like Common Sand Martin *R. riparia*, and same size (both species vary racially, through same range of weights and wing-lengths), but distinguished by brown throat, concolorous with head (Common Sand Martin has white throat contrasting with brown cheeks, and brown breast-band well demarcated from white in front and behind). Even more similar to Congo Sand Martin *R. congica* q.v., but allopatric. Readily distinguished from Banded Martin *R. cincta*, which is larger, with white throat, white eyebrow and well-demarcated breast-band. Somewhat similar to small, dark tropical races of Rock Martin *Hirundo fuligula*, which told by white spots in tail, rocky habitat, rufescent tones in plumage, stockier shape (broader head and body), and less fluttering flight. 20% smaller than Crag Martin *H. rupestris*, which has white spots in tail, blackish underwing-coverts contrasting with silvery tail-feathers (underwing of Brown-throated Sand Martin is more uniform dark brown), and is thick-set. In Cameroon, white belly readily distinguishes it from Forest Swallow *H. fuliginosa*. Otherwise, likely to be confused only with *Phedina* martins, which are, however, heavily striped below.

Voice. Tape-recorded (88, GIB, GREG, McVIC). Song a soft twittering. ♂ and ♀ digging nest-burrow work silently but utter low 'chee, wer-chi-cho, wer-chi-cho' as they change shifts. Contact call a thin 'sree-sree' or 'svee-svee', like Common Sand Martin but clearer in tone; also 'skrrr' from feeding flocks. ♂ arriving at nest with young calls 'chi-choo' (van Someren 1956).

General Habits. Inhabits permanent rivers, lakes and environs, in breeding season where there are sandbanks and cliffs; more widespread at other times, occurring over grassland, bushland, coastal sand-dunes (Angola), cold mountain moors (Zaïre), ponds, dams, sewage farms, reedbeds and estuaries.

Occasionally in ones and twos but generally in flocks, typically of c. 20, occasionally up to 1000. Flight fluttering, rather stiff-winged, lacking momentum, without long swoops and gliding of most swallows. Feeds in flight, commonly over water and waterside vegetation, mixing with Common Sand Martins, Banded Martins, Barn Swallows *H. rustica* and Black Saw-wings *Psalidoprocne pristoptera*; in still weather generally forages within 3 m of water surface. Continues feeding until late dusk. Often perches on reeds and waterside grasses and bushes. Roosts gregariously at night on slanting reeds and papyrus stems, huddling in small groups, often with greater numbers of Common Sand Martins and Barn Swallows. Near Harare (Zimbabwe) flocks tend to roost at different sites from Common Sand Martins; the 2 species also tend to feed at separate sewage works (Tree 1986a).

Disperses quite widely after breeding, seldom amounting to migration. Occurs near Harare in summer months only (Irwin 1981); in Transvaal occurs year-round in highveld but mainly Feb–Aug in bushveld and lowveld (Tarboton *et al*. 1987); visitor to Njombe area (Tanzania), possibly non-breeding; all S Sudan records are Dec–Feb. 3 Zimbabwe birds recovered 7, 11 and 55 km from place where ringed.

Food. Airborne insects, probably mainly very small, including mosquitoes, syrphid and muscid flies, midges Chironomidae, ants *Pheidole*, small beetles, and a grasshopper, retrieved from water surface when dropped (South Africa: Taylor 1942); also dragonflies up to 25 mm long.

Breeding Habits. Breeds solitarily or (usually) colonially. Colony generally of c. 6–12 nest burrows; one of 137 nests (South Africa), and up to 500 burrows reported. At Massa (Morocco) groups of 3–6 pairs dispersed along 2–3 km of river (Cramp 1988).

In display, 2 birds sat close together, circled around each other, then ♂ crouched with head down and stretched forward and nibbled ♀'s breast and cloacal feathers. ♀ crouched; ♂ mounted her, climbing up from side, and copulated with open wings and spread tail (Broekhuysen and Stanford 1954). Copulation also sometimes attempted in flight.

NEST: saucer of grass on pad of small feathers, in chamber at end of burrow c. 35 mm wide and 30–80 cm long, more or less straight and slightly ascending. Chamber c. 87 × 112 × 75 mm; another 127–140 mm high. Chamber floor c. 12 mm lower than highest point of tunnel. Burrows dug by ♂ and ♀, in sandy or friable earth (not stony soil or clay); sited in low, perpendicular sandbank by river or road, in gorge, quarry or abandoned mine dump. Bird sometimes uses old nest-hole of Pied Starling *Spreo bicolor*, or half-dug, abandoned nest-hole of bee-eater *Merops*. Pair starts digging at several places for up to 15 cm, before abandoning all but one place. Small bits of earth loosened with bill and pushed back towards entrance with feet. Pair takes 3 weeks (once 2 weeks) to finish burrow, digging alternately in short spells, then preening and sunning together for long spells.

EGGS: 2–4, av. (n = 14 clutches) 2·9. Slightly glossy; white. SIZE: (n = 50, southern Africa) 15·3–18·7 × 10·5–13·8 (16·9 × 12·2).

LAYING DATES: Morocco, Oct, Dec–Mar (plenty of other evidence suggests breeding Sept–May; breeding season evidently variable, depending on duration of summer rains: Thévenot *et al*. 1981); Guinea (new burrows, Nov); Nigeria, Oct–Feb; Chad, Jan–Feb; E Africa: Region A, June, Region C, June–July, Region D, Mar–July, Sept (mainly Mar–June); Zambia, June–Sept (mainly Aug); Malaŵi, June–July; Zimbabwe, June–Sept (mainly June and Aug); Botswana, July; South Africa: Natal, Mar–Nov, Transvaal, Feb–Oct, SW Cape, Aug–Feb (in dry season, at low water).

INCUBATION: mainly by ♀; sometimes ♂ and ♀ in nest together by day, or both absent for 10 min; ♂ and ♀ roost in nest at night. Period: c. 12 days.

DEVELOPMENT AND CARE OF YOUNG: young fed by ♂ and ♀. Nestling period c. 25 days (Kenya). Young fly well at first appearance.

BREEDING SUCCESS/SURVIVAL: greatest longevity of ringed bird c. 5 years (Zimbabwe).

References
Cramp, S. (1988).
van Someren, V. G. L. (1956).

Plate 7
(Opp. p. 112)

Riparia congica (Reichenow). Congo Sand Martin. Hirondelle de rivage du Congo.

Cotile congica Reichenow, 1887. J. Orn., p. 300; Manyanga.

Forms a superspecies with *R. riparia* and *R. paludicola*.

Range and Status. Endemic, resident, middle and lower Congo River, Zaïre, from Mbandaka (equator) to below Boma (near mouth); also lower Oubangui R. upstream to about Impfondo (1° 40′N). Abundant, at least locally. 1 record Likati R. near Muma (3° 30′N, 23° 20′E).

Riparia congica

Description. ADULT ♂: upperparts from forehead to rump, also cheeks and sides of neck, uniform greyish brown; lores, rectrices and remiges slightly darker brown. Chin and throat white; breast greyish brown, forming band, distinct in most birds, less so in others (immatures?); belly and undertail-coverts white. Underwing-coverts brown. Undersides of remiges and rectrices glossy dark grey-brown. Tail slightly forked. Bill black; eye dark brown; feet dusky brown. Sexes alike. SIZE (6 ♂♂, 3 ♀♀) wing, ♂ 90–98 (94·0), ♀ 89·5–95·5 (92·3); tail, ♂ 42–44 (43·2), ♀ 41–47 (44·0), depth of fork, ♂ 3–7 (5·0), ♀ 4·3–5·5 (4·8); bill to feathers, ♂♀ 5–5·5 (5·25), bill to skull, ♂ 9–9·5 (9·3), ♀ 8·5–9 (8·8); tarsus, ♂ 9–11 (10·3), ♀ 10·5–11; unsexed (n = 2) wing, 93, 94.

IMMATURE: like adult but tertials and feathers of mantle and back very narrowly tipped whitish; brown breast-band less distinct.

NESTLING: unknown.

Field Characters. A small brown riverine martin, with white underparts and brown breast-band. Confined to Oubangui/Congo Rivers, where abundant. Very like Common Sand Martin *R. riparia* but much smaller and relatively short-winged. Very like allopatric Brown-throated Sand Martin *R. paludicola*, particularly race *mauritanica*, differing from it only in having paler throat and darker breast-band and in being slightly smaller.

Voice. Unknown.

General Habits. Occurs along Congo R. at high and low water, but also feeds over forest several km away from river. Perches on trees and on sand near nest-holes, particularly when nesting on sandy islets exposed at low water. Not migratory.

Food. Unknown, but doubtless small airborne insects.

Breeding Habits. Nests colonially, making rows of tunnels in any low declivity on sandy islets exposed by falling waters, e.g. near Lukolela (Zaïre) in Feb and Mar. Nests often close to edge of water, and sometimes near nest-holes of Grey-rumped Swallow *Pseudhirundo griseopyga* (Chapin 1953). Nothing further known.

Reference
Chapin, J. P. (1953).

Plate 7
(Opp. p. 112)

Riparia riparia (Linnaeus). Common Sand Martin; Bank Swallow. Hirondelle de rivage.

Hirundo riparia Linnaeus, 1758. Syst. Nat. (10th ed.), p. 192; Sweden.

Forms a superspecies with *R. paludicola* and *R. congica*.

Range and Status. Breeds Palearctic, south to Mediterranean, lower Nile valley, Mesopotamia, Gujarat and Fukien. Winters in Africa, S and SE Asia.

Breeds Egypt, mainly in Nile valley south to Aswan, and widespread in and near Nile delta, and has bred NW Africa, in Morocco in 1969 (Ruthke 1971); may have bred Tunisia in 19th century (Cramp 1988). Passage migrant throughout N Africa, between Europe and subsaharan Africa in spring and autumn, widespread and sparse, but locally abundant. Abundant N Morocco, Apr (not every year); frequent Algeria and Tunisia, mainly near coast but often crosses desert; common Tripoli (Libya) Mar–May, sometimes abundant (1000) at coastal wadis July–Aug; in Cyrenaica and Fezzan (Libya) uncommon to frequent, occasionally flocks of up to 700;

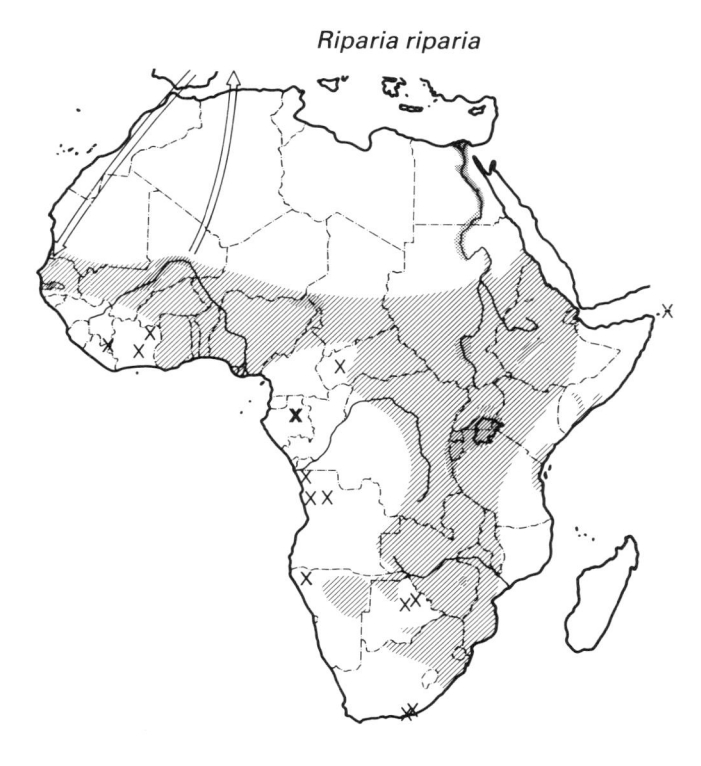

Riparia riparia

rare in Libyan desert (but once 100, Apr–May, at Serir). Occasionally summers NW Africa, and rarely winters Morocco.

In subsaharan Africa uncommon to common and locally abundant, from Senegambia to Ethiopia (mainly in interior on great lakes and rivers, or in highlands) and southward (mainly between 25° and 35°E) to NE South Africa. Evidently winters sparsely throughout that vast region (see map), but literature often fails to distinguish wintering from passage-migrant flocks. Greatest wintering concentration is in equatorial E-central Africa: Burundi and L. Victoria basin (common to locally very abundant). Absent from, or only casual in: Guinea, Sierra Leone, Liberia and Ivory Coast; Cameroon, Gabon, Congo and Angola; Somalia, E Kenya, E Tanzania and NE Mozambique; W Namibia and SW South Africa.

Mauritania, sparse, sometimes abundant, Sept–Dec, Feb–Apr. Senegambia, abundant L. Guier, once thousands, Mar, on middle Gambia R. Mali, 20,000–30,000 L. Faguibine and L. Horo, Nov–Dec, 50,000 Mar–Apr. Nigeria, 2000 Nguru Dec–Jan, 100,000–175,000 moving north *per hour* in late Mar to early Apr afternoons, Malamfatori (L. Chad), and 1 million estimated per day (Ash *et al.* 1967). Sudan, 'very common, locally abundant ... Nile Sudd is a major winter quarter' (Nikolaus 1987). Ethiopia, frequent throughout, locally abundant in Rift Valley. Socotra, 3 in May, E Africa, locally abundant winter visitor and passage migrant Sept to early May, particularly at L. Kyogo, L. Victoria and L. Tanganyika; less common at Rift Valley lakes and L. Turkana. Zaïre, commoner in E half. Burundi, very abundant throughout, from late Sept to early May, particularly on Ruzizi R. plain and in eastern savannas. Zambia, uncommon to frequent, sporadic, sometimes low hundreds, once 2000 (Ndola, Mar), and thousands, Victoria Falls, Jan. Malaŵi, regular but sparse, once 3000 (Chiromo, Jan) and 2000 (Nchalo, Nov). Zimbabwe, commoner than in 1950s, flocks of 100 on Mashonaland Plateau Oct–Mar, once 1800 (Harare, Feb), but estimated to be 12,000 (Mar 1975) and 20,000 (Dec 1981) in huge roost of Barn Swallows *Hirundo rustica* near Harare (Tree 1986a). South Africa, widespread but sparse in Transvaal bushveld and highveld, particularly May, and frequent on Natal littoral, Oct–Apr. First records Cape Province 1977: Grahamstown and Cape Recife.

Description. *R. r. riparia* (Linnaeus): Europe and Asia, wintering in Africa. ADULT ♂: upperparts from forehead to back, uppertail-coverts, tail, wings, cheeks, sides of neck and breast almost uniform brown; forehead and rump slightly paler than remainder of upperparts, and rectrices and remiges slightly darker. Outer 3 rectrices extremely narrowly edged whitish. Tail shallowly forked. Lores dark brown. Chin white or cream-white, throat white; breast brown, forming pectoral band *c.* 8 mm deep in midline and up to 20 mm deep at sides of breast; flanks washed brown; lower breast, belly and undertail-coverts white. Undertail-coverts very long. Outer edge of outer primary, inner edges of inner secondaries, and tertials, extremely narrowly edged pale brown. Underwing-coverts and axillaries grey-brown. Tips of folded wings extend 10 mm beyond tail. Sexes alike. SIZE: wing, ♂ (n = 25) 103–111 (107), ♀ (n = 16) 103–110 (106·5); tail, ♂ (n = 24) 48–54 (51·0), ♀ (n = 17) 48–54 (51·0), depth of fork, ♂ (n = 24) 7–13 (9·1), ♀ (n = 17) 7–13 (9·7); bill to skull, ♂ (n = 10) 9·2–10·5 (9·9), ♀ (n = 15) 8·9–11·2 (9·9), bill to feathers av. 4·1 shorter; tarsus, ♂ (n = 13) 9·6–10·8 (10·3), ♀ (n = 18) 9·8–11·5 (10·6). WEIGHT: Zimbabwe (Harare roost) means (n = 300) fall from 14 g in Oct to 12 g in Dec–Jan, then rise to 13 g in Mar (Tree 1986a); (Zambia, Feb), unsexed (n = 43) 9–13 (11·4); (Uganda, Nov–Feb), unsexed (n = 107) 10·5–14·4 (12·8), (Mar, n = 50) av. 13·6, (Apr, n = 385) av. 14·4, (May, n = 4) av. 17·9 (Pearson 1971); (Nigeria, L. Chad, Mar–Apr), unsexed (n = 32) 10·8–20·4 (13·2) (Fry *et al.* 1970; heaviest birds accumulate up to 28% fat/body weight just before departure); (Morocco, spring), unsexed (n = 255) 8·9–17·3 (11·3) (Ash 1969); (Germany, July–Aug), ♂ (n = 19) 11·4–16·5 (12·7), ♀ (n = 32) 11·2–15·6 (13·6).

IMMATURE: like adult, but feathers of upperparts (particularly forehead, back, rump and uppertail-coverts) fringed pinkish or cream-white at tips, and lores, cheeks, sides of neck, chin, throat and breast washed rufous-buff.

NESTLING: skin pink, mouth yellow, gape-flanges pale yellow; down pale grey, rather short but plentiful.

R. r. shelleyi (Sharpe): Egypt (Nile delta and valley south to about 23°N), wintering south to Sudan and NE Ethiopia (N Red Sea coast). Paler above than nominate race. Smaller: wing, ♂ (n = 7) 92–96 (94·6), ♀ (n = 6) 88–98 (93·3).

R. r. diluta (Sharpe and Wyatt): Asia, south of range of nominate race; winters in India. 1 record, Egypt. Upperparts paler than in nominate race; breast-band paler, narrower and less clear-cut.

Field Characters. A small, gregarious, mid-brown or rather dark brown martin with white underparts crossed by distinct, brown breast-band. Underwing uniform dark brown. Very like Brown-throated Sand Martin *R. paludicola*, and same size, but distinguished by white chin and throat, and brown breast-band well demarcated from white in front and behind (Brown-throated Sand Martin has brown throat and breast, not clearly demarcated from white belly). Even more similar to Congo Sand Martin *R. congica* q.v. Banded Martin *R. cincta* is larger, with deeper breast-band, distinct white eyebrow, and white underwing-coverts. Crag Martin *H. rupestris* is larger, lacks a breast-band, has white spots in tail, and blackish underwing-coverts contrasting with silvery tail-feathers, and is thickset. Otherwise likely to be confused only with *Phedina* martins, which are, however, heavily striped below.

Voice. Tape-recorded (B, C, 62, 73). Song a harsh twittering, with notes like contact call but slightly harder: 'ch-cher ch-cher cher chi-chi-chi-chi-chi-ch-chi-chi-chi-i-i-i-i'. Contact call a harsh or hoarse, grating 'tschrd' or 'tschr', used at perch and in flight. For excitement, warning, alarm, distress and juvenile calls, see Cramp (1988).

General Habits. Inhabits low airspace over most types of terrain except forest and built-up areas; occurs above or near grassland, swamps, reedbeds, rivers and lakes, and in breeding season near sandy or loamy nesting banks. In Zimbabwe occurs over open water, vleis, moist grassland bordering dams, irrigated pastures, sugar-cane fields, and sewage disposal works (open ponds, sludge beds, filter tanks). Gregarious at all seasons; often mixing with other swallows when feeding. All food taken on wing; av. foraging height 15 m; occasionally follows ploughing tractor. Forages up to 6 km from nest, but mainly within 0·25 km. Roosts in nest (sometimes ♀ alone), and in non-breeding season gregariously in reed, reedmace and papyrus beds, commonly mixing with Barn Swallows, also with wagtails *Motacilla* and Brown-throated Sand Martins. Often greatly outnumbered at roosts by Barn Swallows; 'small numbers' roost with c. 1 million Barn Swallows near Eldoret, Kenya (Best 1977), and near Harare, Zimbabwe, Common Sand Martins form 4–13% of large Barn Swallow roosts (Tree 1986a). Once flock roosted on bare ground, with bee-eaters *Merops* (Jordan). Goes to roost later at dusk than do most other hirundines. Arrival of new flock excites birds already settled at roost, leading to aerial evolutions (L. Chad, also Zaïre: Verheyen 1952). Bathes by dipping onto surface of water in fluttering flight; also bathes in dew. Sometimes comes to ground to perch, sunbathe and dust-bathe. 300 once sunbathed together on ground, lying on sides and spreading one or both wings. Perches freely on variety of low vantage-points: fences, tops of bushes, bare side-branches of trees, low roofs. In breeding season commonly plays with feathers, dropping and catching them again in flight.

Post-breeding moult suspended during autumn migration (Dowsett 1971). In Kenya and Uganda remex moult is from Oct/Nov to mid-Mar/mid-Apr (Ginn and Melville 1983), and in Zambia and Zimbabwe from mid-Oct/early Dec to mid-Feb/early Apr (duration 120–135 days: Tree 1986a). Duration of moult in W Africa c. 141 days. For plumage sequence of moult, see Bub et al. (1981).

Main immigration to N tropics is in Sept–Oct and emigration in Apr, with a few lingering to early May. Season in southern Africa shorter: mainly Oct–Mar; extreme dates, Transvaal, 10 Sept and 20 Apr, and Zambia, 13 Aug and 31 May. Numerous ringing recoveries between NW Europe and Morocco and Senegambia show routes of Swedish, German and British birds (Persson 1973, Bub and Klings 1968, Mead and Harrison 1979). British birds leave in Aug, cross to Biscay coast (France), follow Ebro valley (Spain), cross W Mediterranean to Morocco and reach N Senegambia by Oct–early Nov. Evidently they move eastwards in winter to Niger Inundation Zone (Mali), then in spring cross Sahara to N Algeria, overfly Mediterranean further east than in autumn, and return to Britain via W-central Europe. Families appear to keep together and to follow traditional routes; however, adults seem to travel faster than young birds on average. They fly by day, and at night roost gregariously, usually in reedbeds (Cramp 1988). Birds from Sweden and Germany also winter mainly in Senegambia. Vast flocks at L. Chad, late Mar–early Apr 1967 and 1968 (1 million birds on some days) were bound probably for E Europe and USSR. Only E African ringing recovery is from Uganda (Entebbe) to east of Kuybyshev (USSR, 52° 26′N, 53° 11′E). In west, recoveries are from Nigeria (L. Chad) to Tunisia, Malta and Czechoslovakia (1 each), from Morocco to Britain (4), from Britain to Morocco (33), Algeria (11), Tunisia (2), Libya (1) and Senegambia (27), from Sweden to Algeria and Mali (1 each), from Denmark to Algeria, Nigeria (L. Chad) and W Central African Republic (1 each), from Netherlands to Algeria (1), and from W Germany to Algeria, Libya and Burkina Faso (1 each).

Food. Small airborne insects (and a few spiders), mainly soft-bodied. In Europe main groups fed to nestlings are mayflies, dragonflies, stoneflies, grasshoppers, bugs, caddis-flies, dipteran flies of 27 families, small hymenopterans, and beetles. In one study (Scotland, n = 2623 items), 69% were Diptera; in another (USSR, n = 579 items), 48% were leafhoppers and 37% Diptera; and in a third (USSR, n = 2488 items), 50% were Diptera and 10% leafhoppers. At L. Chad (Nigeria) one bird had eaten only small demoiselles, another only the minute midge *Tanytarsus spadiceonotatus* (dry weight of 1 midge, 0·00022 g), and another, small beetles and bugs. 32 birds there contained 40% by dry weight of *Tanytarsus* and other chironomid and cecidomyid midges, and 60% larger insects: 14 dragonflies, 13 bugs, 15 hymenopterans, and 20 beetles (most < 5 mm long) (Fry et al. 1970).

Breeding Habits. Monogamous, colonial, double-brooded. Av. colony size, N Europe, *c.* 80 nest-burrows but *c.* 40 pairs. Habits not studied in Africa; for European data on nest-site selection, pair-bonding behaviour, mate-guarding and copulation, see Cramp (1988).

NEST: saucer of feathers, grass and leaves in cavity at end of burrow usually 46–90 cm long, dug by ♂ and ♀ (taking *c.* 14 days: Europe) in perpendicular river bank, quarry or sea sand-cliff. In Egypt burrow sometimes sited in flat ground. 2nd clutch usually laid in same nest, 7·5 days after 1st brood flies.

EGGS: 2–6, av. (Britain) 4·8. Subelliptical; smooth; white. SIZE: (n = 25, *R. r. shelleyi*, Egypt) 15·2–19 × 11·7–13 (16·9 × 12·2). WEIGHT: *c.* 1·37.

LAYING DATES: Egypt, Mar–Apr.

INCUBATION: by ♂ and ♀, in equal shares by day, mainly ♀ at night; nest-relief very rapid. Period: 14–15 days.

DEVELOPMENT AND CARE OF YOUNG: brooded constantly on first day, by ♂ and ♀ alternately, but hardly at all by day after day 7. Parents swallow nestlings' faeces for first 4 days, then carry them away. At 9 days young are mobile and noisy, and defaecate from burrow entrance. Young fed by ♂ and ♀; up to 200 food-visits a day; brood given *c.* 7000 insects (dry weight *c.* 7 g) daily, in food-balls of 100–200 mg wet weight (USSR). Young roost in nest for *c.* 1 week after flying; by day they loaf together on wires or twigs, fed there (and in flight) by parents.

BREEDING SUCCESS/SURVIVAL: low mean weight at northern edge of Sahara, spring, suggests that many birds may perish in desert crossing; some found dead there in cold spell, Apr (Morocco: Ash 1969). Av. annual 1st-year mortality, Britain, 77%, and adult mortality 65% (Mead 1979); av. annual adult mortality, Sweden, ♂♂ 54·4%, ♀♀ 60·9% (Cramp 1988). Oldest bird, 9 years.

Reference
Cramp, S. (1988).

Riparia cincta (Boddaert). Banded Martin. Hirondelle à collier.   Plate 7
(Opp. p. 112)

Hirundo cincta Boddaert, 1783. Tabl. Pl. enlum., p. 45; Cape of Good Hope.

Range and Status. Endemic resident and intra-African migrant. Widespread resident from Ethiopia to Malaŵi and Zambia (except that it vacates Kenya in Dec), widespread breeding visitor to Zimbabwe and South Africa Sept–Apr. Breeds in Angola and Zaïre, but sparse and poorly known there, and picture complicated by non-breeding migrants from south and east. Fairly common resident around Gamba, Gabon (D. E. Sargeant, pers. comm.). To north and west of Zaïre, mainly a sparse non-breeding visitor June–Oct, but some presumptive evidence that a few may breed in e.g. N Nigeria and spend dry season (Nov–Mar) near coast. Further details under races and movements.

Generally sparse; only 54 records in Zambia in one year and 53 records in another 2 years. Uncommon and local in Zimbabwe and South Africa. In E Africa, locally common in L. Victoria basin and W Rift; elsewhere uncommon and local. Frequent in Ethiopia. Uncommon to rare throughout W Africa, singly or in flocks of up to 10 (Ghana) and 70 (Nigeria), but abundant on islands in L. Chad (Nigeria), late June.

Description. *R. c. cincta* (Boddaert): South Africa, lowland Lesotho, SW Zimbabwe, wintering north at least to Angola and S Zaïre. ADULT ♂: entire upperparts from forehead to rump and tail, with wings, cheeks and sides of neck, almost uniform dark greyish brown. Lores black; narrow line of

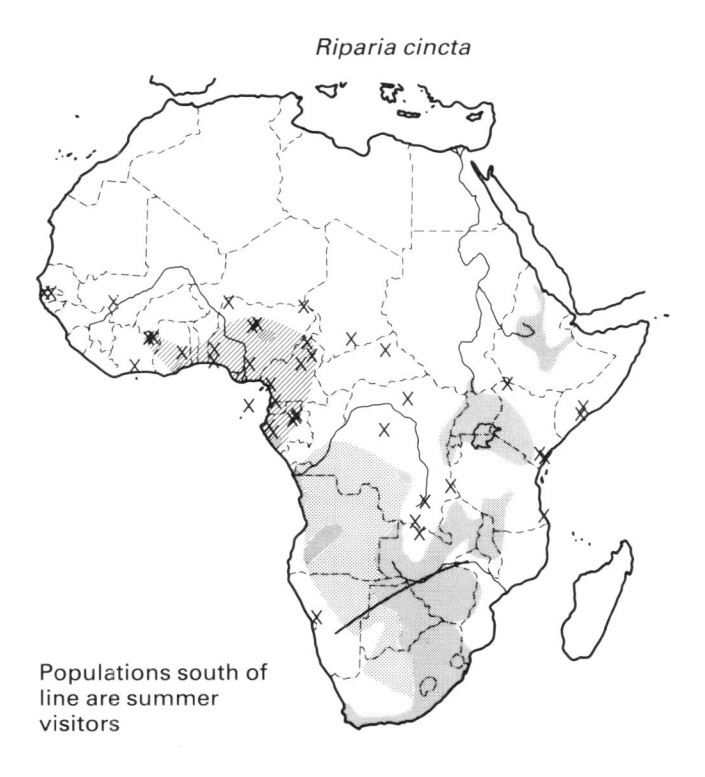

Populations south of line are summer visitors

feathers immediately above nostril running at sides of forehead above lores to eye forms short, narrow, whitish superciliary stripe. Rump rather paler than mantle and back. Rectrices and remiges darker than mantle and back. Tail not forked. Tips of folded wings do not extend beyond tail. Chin and throat silky white, occasionally with very slight buff wash; breast brown (same shade as mantle) forming pectoral band of uniform width from midline to sides of breast, c. 15 mm deep; belly and undertail-coverts white; upper belly sometimes with a few brown feathers in midline. Underwing-coverts white, some marginal coverts brownish; undersides of wing and tail silvery dark brown. Bill black; eye dark brown; legs dark greyish brown. Sexes alike. SIZE: (26 ♂♂, 19 ♀♀) wing, ♂ 116–140 (129), ♀ 117–140 (127·5); tail, ♂ 50·4–69 (59·9), ♀ 53–66·2 (60·1); bill to feathers, ♂ 7·6–9·9 (8·7), ♀ 7·4–10 (8·7), bill to skull, ♂ 11–14 (11·6); tarsus, ♂ 11–13 (12·1), ♀ 10·3–14 (12·1). WEIGHT: (southern Africa) ♂ (n = 4) 20·5–24·5 (22·3), ♀ (n = 5) 22–28·7 (24·6) (R. A. Earlé, pers. comm.).

IMMATURE: like adult but upperwing-coverts and tertials broadly edged rufous-buff (2 mm wide on tertials).

NESTLING: unknown.

R. c. xerica Clancey and Irwin: N Botswana, N Namibia, wintering grounds unknown, perhaps W Africa; has occurred S and W Angola. Uppertail-coverts slightly darker than in cincta; underparts paler and breast-band narrower (depth c. 12 mm).

R. c. suahelica van Someren: Zimbabwe, W Mozambique, Zambia, Malaŵi, SE and E Zaïre, S and NW Tanzania, W Kenya, Uganda, Sudan (Juba and Lado only: probably this race); breeding visitor to Zimbabwe; elsewhere mainly resident; wintering grounds unknown – perhaps W Africa. Upperparts darker than in cincta and xerica; breast-band c. 18 mm deep; short white superciliary stripe better defined. Smaller: wing, ♂ (n = 10) 126–138 (130·5), ♀ (n = 5) 123–129 (127). WEIGHT: ♀ (n = 5) 19–30·5 (25·7).

R. c. parvula Amadon: range poorly known – probably central and N Angola, extreme NW Zambia, and SW Zaïre east to 24°E. Winters north to Cameroon. Same colour as xerica, but breast-band broad (c. 18 mm).

R. c. erlangeri Reichenow: Ethiopia; resident. A few records of visitors to Kenya. Same colour as cincta, but breast-band shallower; irregular line of brown feathers in midline extends from breast-band to upper belly. Larger: wing, unsexed, 135–146.

Field Characters. A medium-sized hirundine (about size of Barn Swallow *Hirundo rustica*) with square tail lacking streamers, easily told by dark brown upperparts and white underparts with well-demarcated brown breast-band and short white superciliary stripe. Last feature distinguishes it from all other African martins and swallows; otherwise, similar to Common Sand Martin *R. riparia* but larger and more strongly marked. Characteristic delicate 'soft' flight. Quiet. Not markedly gregarious.

Voice. Tape-recorded (88, C, GIB, McVIC). Song a short, subdued, squeaky warble, 'chirip cherip chee chirup', sometimes ending in trill 'chiruiiii' (van Someren 1956); also described as loud and chattering. Flight calls: 'chip', 'chrip' and 'kip'.

General Habits. Inhabits airspace low over grassland, marshes, vleis, small bodies of water, lightly-bushed pasture, and even over rocky shore and surf (van der Merwe 1986).

Solitary, in pairs or small flocks (tens); after nesting, several families flock together; rarely, hundreds gather together on migration. Foraging flight slow, silent, graceful and seemingly effortless, with slow, deliberate, tern-like wing-beats. Skims grasstops, and forages above herds of cattle, zebras and antelopes; attracted to grass fires. Mixes freely with other foraging swallows and swifts. Skims still water surface with scarcely a ripple. 1–2 birds rest on fences, mounds, flat ground, and bare twigs at top of bush. Suns itself by lying on ground with wings and tail spread.

Breeding visitor to South Africa Sept–Apr (extreme dates 10 Sept, 8 May) and to Zimbabwe Sept to Apr, May or occasionally June, with pre-breeding passage Sept–Oct (flocks of 20–100) and main departures from late Mar to mid-Apr. Rarely, winters in Zimbabwe. Flock of c. 1000 on L. McIlwaine, early May, probably consisted of South African migrants. 1 ringing recovery, Zimbabwe, 4 years and 5 km away. Occurs in Zambia mainly Oct–Apr (85% of all records), with a few records May, June and Sept (13%) but hardly any in July–Aug (2%) (Aspinwall 1983). Recorded Malaŵi all months except June, Aug. Resident, with ill-defined local movements, Angola, Zaïre, E Africa, Ethiopia. Rare, eastern seaboard, mainly Aug (SE Somalia, Kenya, Tanzania). Breeding visitor to 3000 m in Mau Narok (Kenya), mainly Feb–July (Sessions 1966); wandering flocks occur elsewhere in Kenya until Nov, but there are no records from late Nov (30 at L. Turkana) to late Jan (Lewis and Pomeroy 1989, Lewis 1989). 2 records S Sudan. Irregular passage migrant, interior of Gabon, late Aug to early Sept and late Oct to Feb.

W African records, from Gambia to Principe I., S Chad and N Central African Republic, are May–Oct, except for Gambia Jan and Apr, NE Ivory Coast Mar–July, Ghana (near Winneba) Nov–May, SW Nigeria Feb–Mar and central Nigeria (mainly Jos Plateau) early Mar to mid-Nov.

Food. Insects. Those fed to nestlings include small beetles (Scarabaeidae, Silphidae), moths, flies, mantis nymphs, lace-wings and moth caterpillars; caterpillars taken 'probably from grass flower-heads' (van Someren 1956).

Breeding Habits. Nests solitarily.

NEST: dish-shaped structure of old coarse grass, lined with fine grass and feathers, in offset chamber c. 15 cm diam., at end of straight, slightly inclining burrow 60–90 cm long. Burrow excavated by bird near top of perpendicular bank of donga, gulley, river channel, embankment, dam or pit, near to water; also (once) in roof of burrow of antbear *Orycteropus afer* 800 m from water (Vincent 1944), in colony of Pied Starlings *Spreo bicolor* (Stark and Sclater 1900), and in plastic drainage pipe (van der Merwe 1986).

EGGS: 2–5, in southern Africa 2–4, av. (n = 15 clutches) 3·2. White. SIZE: (n = 26, southern Africa) 19–23·9 × 14·1–16 (21·1 × 15·0).

LAYING DATES: Nigeria (c. 20 birds seen entering holes in river bank, Kaduna, month?; pair carrying grass and entering bank hole, Shendam, June); Ethiopia, May–Aug; E Africa: Region A, June, Region B, Feb–June, Aug–Sept (breeds in rains); Gabon (young food-begging, Jan); Zaïre, July; Angola (said to breed Nov–Dec); Malaŵi (nest-building Jan); Zambia, Aug, Oct, Dec–Jan; Zimbabwe, Sept, Nov–Feb (mainly Jan–Feb); Botswana, Sept–Oct; South Africa: Natal, Nov–Feb, S Cape, Aug–Oct.

INCUBATION: incubating bird fed by mate.

DEVELOPMENT AND CARE OF YOUNG: young fed by both parents, with one visit every 10 min in forenoon and from 15·00 to 18·00 h (Kenya). Parent carries large quantity of insects in mouth, and gives parts of mouthful to each nestling in turn. Chicks' droppings evidently eaten or removed from nest by parents. Nestling period: between 21 and 24 days (n = 1). Fledglings accompany parents on wing for days or weeks.

BREEDING SUCCESS/SURVIVAL: 1 caught by thigh-feathers in sticky weed *Becium angustifolium* (Critchley 1975). Greatest longevity 4 years (bird ringed as adult, Harare).

Reference
van Someren, V. G. L. (1956).

Genus *Pseudhirundo* Roberts

Very small swallow (< 10 g), differing from *Hirundo* spp. in nesting in rodent burrows and having immaculate white eggs, and from *Riparia* spp. in its glossy blue back and deeply-forked tail; moreover, *Riparia* martins dig their own nest burrows. Probably more closely allied with Australian White-backed Swallow *Cheramoeca leucosterna* and with Neotropical burrow-nesting swallows *Atticora* spp. than with other African hirundines, and perhaps congeneric with one or both of them.

Endemic, 1 species.

Pseudhirundo griseopyga (Sundevall). Grey-rumped Swallow. Hirondelle à croupion gris.

Hirundo griseopyga Sundevall, 1850. Oefv. K. Sv. Vet.-Akad. Förh., 7, p. 107; Natal.

Plate 7
(Opp. p. 112)

Range and Status. Endemic, resident and migrant. In general rather sparse, particularly in W Africa, but frequent and locally common to abundant in other regions.

Transvaal, frequent in eastern lowveld, scarce and local elsewhere but once 200 in Kruger Nat. Park; Natal, formerly along littoral plain, now only in NE. 3 in S Atlantic at 15° 16'S, 2° 46'E, late Nov 1964 (Acland 1966, Donnelly 1966, Took 1967). Zimbabwe, breeding visitor Apr–Oct, but partly resident in low-rainfall years; the commonest breeding swallow on Mashonaland Plateau; scarce in dry regions, probably does not breed on Kalahari Sands; occurs up to 2100 m at Inyanga; assembles in flocks of 2000–3000 at end of breeding season. Botswana, sparse, but commoner in N; all months, but mainly Apr–Oct. Angola, central Huila to Huambo, N Bihé east to Luacano, N Moxico; Cunene R.; Zaïre basin (Malanje, Cangandala). Zambia, frequent to common throughout, above 950 m (once, 1950 m). Malaŵi, frequent, from below 1600 m up to 2300 m (Dedza Mts, Nyika Plateau). Equatorial Guinea and Gabon, scarce, coastal areas only; breeding colonies on lower Ogoué R. E Congo and Zaïre, common resident on middle Congo R., and frequent and widespread in NE, E and SE Zaïre. E Africa, local and uncommon, 900–

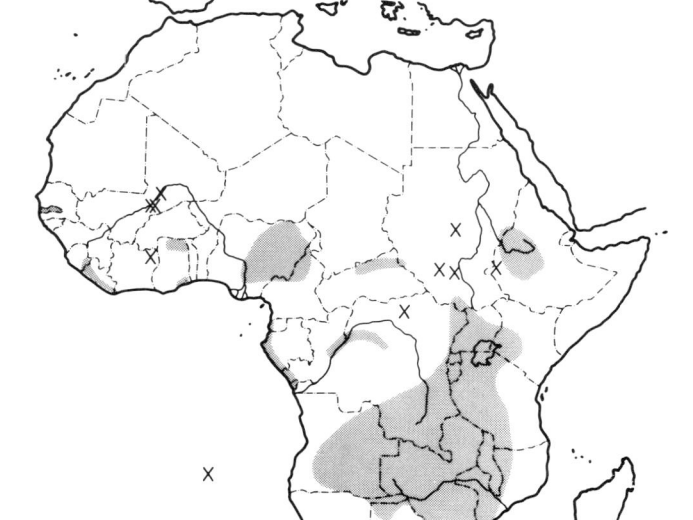

2200 m, with marked seasonal fluctuations; widespread in W Kenya, more local in L. Victoria basin, but in flocks of *c.* 100 in Kibondo (NW Tanzania), and sometimes numerous near north of lake (Uganda) June–Sept; also frequent in interior of S Tanzania. Ethiopia, frequent to locally abundant, central and western W Highlands and Rift Valley. Sudan, rare; breeding colonies on upper Blue Nile R. (Fazogli). N Central African Republic, frequent in Bamingui-Bangoran and Manova-Gounda-Saint Floris Nat. Parks. NW Cameroon, rare. Nigeria, scarce and highly local, sporadic, perhaps resident; breeds in Niger R. valley; flock recorded Mambilla (Green 1990). Ghana, locally frequent in N, 1 record in S in forest zone, and once a strong passage on Accra plains. Ivory Coast, 1, Comoé. Mali, 3 records, Oct–Jan. E Sierra Leone; coastal parts of Liberia, and common on Mt Nimba. Senegambia, Gambia R. only, uncommon, flocks of up to 300 in some years.

Description. *P. g. griseopyga* Sundevall: range of species, except *melbina* (below). ADULT ♂: forehead, crown, nape, hindneck and sides of neck dark brown, feathers fringed slightly paler brown; sometimes slight, pale superciliary stripe; lores blackish brown, ear-coverts very dark brown. Brown crown and neck sharply demarcated from glossy blue-black mantle, back and wings; rump and uppertail-coverts pale grey-brown; tail dark brown with bluish gloss, especially on elongated outer rectrices; inner rectrices fringed pale buff when new. Entire underparts white, chin and belly sometimes creamy white; chin, throat and breast with distinct salmon-pink wash in fresh plumage, but colour quickly disappears. Underwing-coverts white; marginal coverts brownish; undersides of remiges and rectrices dark blackish grey. Bill black; eye black-brown; legs and feet pale brown to dark brown. Sexes alike. SIZE: wing, ♂ (n = 39) 92–103 (97·5), ♀ (n = 38) 87–103 (95·8); tail, outer feather, ♂ (n = 35) 51–87·5 (74·4), ♀ (n = 35) 56·4–85 (71·2), inner feather, ♂ (n = 37) 33–46·5 (39·3), ♀ (n = 36) 33·5–47 (39·9); bill to feathers, ♂ (n = 27) 4·2–6·1 (5·1), ♀ (n = 28) 4·2–5·8 (5·0); tarsus, ♂ (n = 39) 9·4–13 (11·1), ♀ (n = 37) 9·2–12·2 (11·0). WEIGHT: (southern Africa) ♂ (n = 10) 9–12·5 (10·6), ♀ (n = 11) 7·7–11·5 (9·8) (R. A. Earlé, pers. comm.); (Kenya) ♀ (n = 5) 6–7 (6·2). W African birds (*P. g. 'gertrudis'*) clinally smaller: wing, unsexed (n = 5) 90–96 (93·3); weight (Liberia) 1 ♂ 10, 1 ♀ 9·6, 1 unsexed 10·3.

IMMATURE: forehead and crown feathers rather more obviously pale-fringed than in adult, forming frosting on forehead and slightly more distinct superciliary stripe; mantle and back feathers fringed pale buff or whitish, giving scalloped effect. Little or no blue gloss. Tertials with *c.* 1 mm whitish fringes. Chin and throat at first buffy, and after a few weeks creamy white with salmon pink wash becoming greyer towards breast and at sides of neck and sometimes forming a distinct greyish patch (Fry 1973). Belly and flanks grey; undertail-coverts white, washed pinkish buff.

NESTLING: down pale grey; bill purplish grey, gape-flanges cream, inside mouth yellow, eye grey, tarsus grey-brown.

P. g. melbina J. and E. Verreaux: Gambia R., E Sierra Leone, Liberia; coastal region from Equatorial Guinea to Cabinda. Rump dark greyish brown or dark brown.

P. g. andrewi (Williams): breeding grounds unknown; several at L. Naivasha, Kenya, Apr 1965. Described as a distinct species (Williams 1966), but see Dowsett (1972). 1 ♀ collected: browner than *melbina*, underparts pale brown, wing 100 mm, weight 10 g.

Field Characters. Small (lightest-weight African hirundine), slimly built, somewhat gregarious swallow; quiet, with rather weak, fluttering flight. Upperparts dark brown, back slightly glossy blue-brown, rump grey or greyish brown; tail deeply forked, without any white; underparts uniform white or greyish white. In habits resembles sand martin *Riparia* rather than typical swallow *Hirundo*. Told from all other hirundines by combination of deeply-forked tail and grey rump; but bird moulting tail streamers could be confused with cliff swallow, particularly white-throated Preuss's Cliff Swallow *Hirundo preussi*, which is best distinguished by its white-spotted tail.

Voice. Tape-recorded (86, GIB, MOY). Weak hissing twitter, or nasal 'wha', 'pa' or 'jeew', audible only within 20 m. Voice little used.

General Habits. Inhabits dry grasslands, sheltered plains, edges of vleis with short grass and bare ground, particularly when burnt; also woodland with plenty of short grass and bare earth spaces; often near water; breeds on alluvium, fallow farmland, polo fields (Zimbabwe), airfields (Malaŵi), golf courses, and flat sand bars in rivers (Nigeria).

In pairs or small loose flocks; after breeding and before migrating forms flocks of up to 200, occasionally 3000. Forages low over short-grass sward with patches of bare soil; fond of burnt grassland. Mixes freely with other martins and swallows (including Banded Martin *Riparia cincta*, Brown-throated Sand Martin *R. paludicola*, Red-throated Cliff Swallow *Hirundo rufigula*, Angola Swallow *H. angolensis*, Red-rumped Swallow *H. daurica* and Lesser Striped Swallow *H. abyssinica*). Follows grass-cutting tractor, to feed on disturbed insects. About 30 hunted all day from central tree in marsh-side scrub, Nigeria, resting on tree between forays. On Mt Nimba, Liberia, hunts low over 700 m ridges, sweeping up from one slope, crossing ridge and diving 30–50 m down next slope, repeating manoeuvre again and again (Colston and Curry-Lindahl 1986). Flight 'soft' and fluttering, quiet, without long glides – like flight of Common Sand Martin *R. riparia*. Flies directly into narrow nest-hole in flat ground, evidently without pausing or perching at entrance. Often sits on ground, and perches on woody twig, waterside willows, or grass stem. Roosts at night in reeds; once several roosted on bare ground (Schmidl 1982). Moults almost entirely out of breeding season (Earlé 1987c).

Evidently migratory over much of range, although no clear picture emerges. Transvaal, recorded all months, but mainly Apr–Sept (when it breeds). Zimbabwe, breeding visitor arriving Apr/May and departing Sept/Oct (some Nov); many stay all year in dry years. Zambia, mainly dry-season breeding visitor – N Kafue basin Mar–Sept and near Lusaka early Apr to late Sept; in Dec–Feb only one-third as many records/month as Apr–Nov; some evidence suggests that migrants may winter in lowland Mozambique (Aspinwall 1980a, Tree 1976). Great numbers appear in Akagera Nat. Park, Rwanda, Mar–Sept, particularly from late May to mid-

Aug. In E Africa, marked seasonal fluctuations in most localities, partly because of preference for burnt grassland (Britton 1980). Resident on middle Congo R. Not regarded as migrant in Ethiopia. In Bamingui-Bangoran Nat. Park, Central African Republic, June–Sept only. Strong northwestward migration near Accra, Ghana, late Jan. Perhaps partially migrant in Liberia, but pattern difficult to discern. Dry-season visitor to Gambia, extreme dates 25 Nov and 21 Feb.

Food. Airborne insects.

Breeding Habits. Nests are solitary, or c. 3–10 grouped in loose colony. Possible display: a few birds once packed very close together and flew over colony uttering weak, concerted cries (Serle 1957).

NEST: loose, bulky pad of soft or coarse, dry grass blades, stems and heads, in rounded chamber 'the size of a clenched fist' (Serle 1957), at end of horizontal or slightly declining, straight or curved burrow c. 60 cm long; burrow dug '2 to 3 feet into the ground at a fairly deep angle' (Jourdain and Shuel 1935) in flat ground or low, shelving bank. Substrate firm sand, or hard dry soil with sparse growth of stunted grass, often burnt-over. Burrow generally an old rodent hole; occasionally uses old kingfisher and bee-eater burrows, and funnel in terrestrial termite-mound. 'Nest in holes made . . . by the birds themselves' (Chapin 1953, referring to sources including Jourdain; and Shuel 1935, who did not actually make that claim).

EGGS: 1–4, mainly 2–3 near equator and 4 in Malaŵi. Very delicate; short ovate or pointed ovate; slightly glossy; white. Shape and size vary even within clutch. SIZE: (n = 17, southern Africa) 14·3–17·6 × 11–12·9 (16·0 × 11·9); (n = 5, Nigeria) 15·5–17·9 × 11·8–12·1 (17·05 × 11·8).

LAYING DATES: Liberia (juv. June); Ghana (juv. Jan); Nigeria, Niger R. valley, Nov–May; Sudan, Jan; Ethiopia, Mar–Apr, Dec; E Africa: Region B, July, Region C, May–July, Sept, Region D, Mar, Sept–Oct; Gabon (nestlings Aug); Zaïre (digging Jan–Apr, July–Aug; nestling Aug); Angola, Aug–Sept; Zambia, July–Sept (mainly Aug); Botswana, July; Malaŵi, July–Oct; Zimbabwe, July–Nov, with 65% of 59 clutches in Aug; South Africa, May–Dec (80% in June–July: Earlé 1987c). Everywhere nests in dry (winter) season.

BREEDING SUCCESS/SURVIVAL: nests often flooded during rains.

References
Earlé, R. A. (1987c).
Serle, W. (1957).

Genus *Hirundo* Linnaeus

Small to rather large swallows; back usually glossy blue and tail deeply forked, but several species plain brown and square-tailed. Rump same colour as back, or white or rufous. Outer tail-feather sometimes greatly elongated and attenuated, and outer feathers often with oblong white spots on inner webs. Nostrils oval, with lateral membrane. Build cup- or retort-shaped nest using mud pellets, on ledge, rock face or under overhang.

About 34 species, 24 in Africa (19 endemic). In *Hirundo sensu stricto* we recognize 1 superspecies, *H. angolensis/H. lucida/H. rustica*, on the basis of similar plumage and nests and association with human habitation. 5 species, *H. semirufa*, *H. senegalensis*, *H. daurica*, *H. abyssinica* and *H. cucullata*, have sometimes been included in the genus *Cecropis* because of their red rump, deeply forked tail and nest built with mud pellets and with an entrance tunnel (Chapin 1953, Brooke 1972). We prefer to take a conservative approach and have included *Cecropis* with *Hirundo*, following Hall and Moreau (1970), Cramp (1988) and J. Farrand, Jr (pers. comm.). *H. semirufa/H. senegalensis* and *H. abyssinica/H. cucullata* have been considered to be superspecies (Hall and Moreau 1970, Short et al. 1990), but we treat them as species groups because of the large overlaps in range.

11 species, the cliff swallows, have often been separated generically, as *Petrochelidon*. They comprise 2 distinct species, the African *H. (P.) fuliginosa* (which may not be a cliff swallow at all), the Australasian *H. (P.) nigricans*, and 9 species which are closely allied and have been treated as a single superspecies (Hall and Moreau 1970). The 9 species comprise 3 American, 1 Asiatic, 1 Australasian and 4 African forms. The last have been treated as a superspecies but, consequent upon the recent discovery of *H. perdita*, we prefer to recognize 2 superspecies, *H. spilodera/H. perdita* and the smaller *H. preussi/H. rufigula* (see Fry and Smith 1985). We are more impressed by similarities than by differences between cliff swallows and other martins (cf. Brooke 1972) and thus follow Phillips (1973) and Turner and Rose (1989) in merging *Petrochelidon* with *Hirundo*.

3 species, the crag martins, form a superspecies, with 1 member in Africa and SW Asia, 1 in N Africa, S Europe and Asia, and 1 in India. Crag martins are brown swallows associated with rocky hills and buildings, and are often separated as *Ptyonoprogne* because they lack glossy blue plumage and tail streamers. We include them in *Hirundo*, since they have (slightly) forked tails with white spots, cup-shaped mud nests, spotted eggs, and voice and incubation behaviour like *H. rustica* (Mayr and Bond 1943). Taxonomic treatment is very variable. Tropical crag martins are small and dark (African *fusciventris* group of races, Indian *H. concolor* group), and higher-latitude forms are large (dark southern African *fuligula* group, pale Sahara *obsoleta* group, and the not-so-pale Eurasian *rupestris*). Much of the variation is

clinal, but restriction to rocky hills ensures some clinal discontinuities. Size difference between *fusciventris* and *fuligula* groups is abrupt, but their shades intergrade. *Fusciventris* and *obsoleta* groups and *rupestris* intergrade in size, but we separate *H. rupestris* specifically from the *obsoleta* group because breeding ranges evidently overlap in N Algeria, NW and W Arabia, and more extensively in SW Asia (Cramp 1988, p. 253 and maps pp. 249, 255). There are some grounds for recognizing the *fuligula*, *fusciventris* and *obsoleta* groups as 3 separate species. However, we follow Cramp (1988) in treating them as a single, highly polytypic species, *H. fuligula*, while realizing that the species then embraces much greater morphological variation than occurs between it and *H. rupestris*. *H. fuligula* and *H. rupestris* form a superspecies.

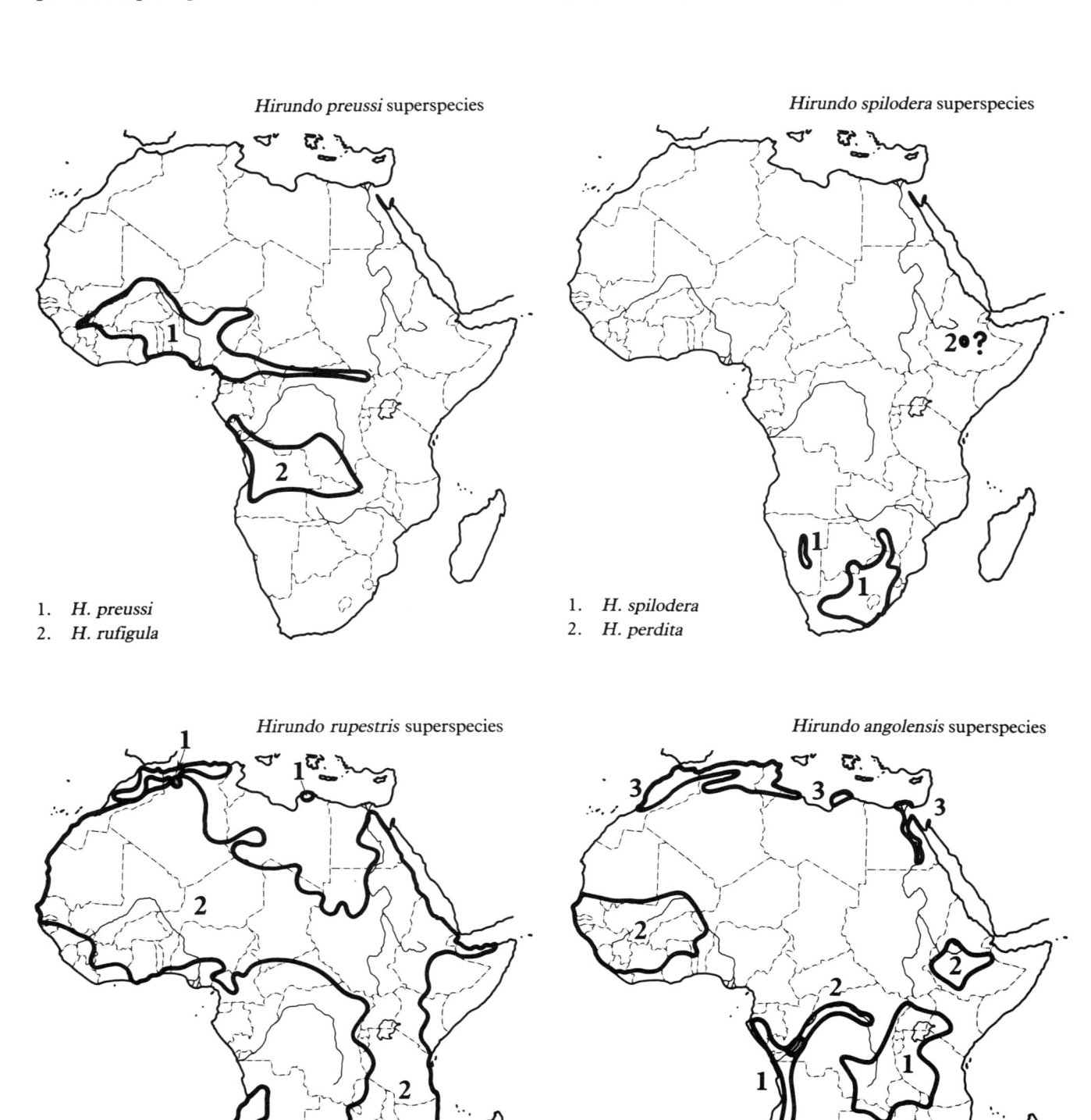

Hirundo preussi superspecies

1. *H. preussi*
2. *H. rufigula*

Hirundo spilodera superspecies

1. *H. spilodera*
2. *H. perdita*

Hirundo rupestris superspecies

1. *H. rupestris*
2. *H. fuligula*

Hirundo angolensis superspecies

1. *H. angolensis*
2. *H. lucida*
3. *H. rustica*

Hirundo semirufa Sundevall. Red-breasted Swallow. Hirondelle à ventre roux.

Hirundo semirufa Sundevall, 1850. Oefv. K. Sv. Vet.-Akad. Förh., 7, p. 107; Magliesburg, Transvaal.

Plate 8
(Opp. p. 113)

Range and Status. Endemic resident and intra-African migrant; from Senegambia (not N), Mali (south of *c*. 13·5°N), SW Niger (Niamey east to Maradi and Tessaoua), Guinea, Sierra Leone, Liberia, Ivory Coast (mainly forest zone but occasionally north to northern border), SE Burkina Faso (also Ouagadougou area), Ghana (mainly forest zone but north to Mole and Bolgatanga), Togo, Benin and Nigeria (except extreme NE), to Cameroon, SW Chad, SW Central African Republic (Lobaye Préfecture, Bangui area) and Sudan (upper Nile north to *c*. 6°N; also breeds Kulme in W); also Gabon, Congo, Zaïre (except upper Uelle in NE), Uganda, Kenya west of the Rift, east to Narok, north to *c*. 2°N, with wanderers to Hell's Gate, Mt Suswa and Ewaso Ngiro marshes; NW and N Tanzania (Nyaruonga, Ngare Dovash, Tarime, Shirati), and Rwanda, south to Angola (except S and SW), Zambia (except SW), Malaŵi (west of Rift east to Lilongwe, Vwazu Marsh and Blantyre), NE and E Botswana, Zimbabwe, central and S Mozambique, South Africa (NE Cape, Orange Free State, Transvaal, interior Natal), W Swaziland, and N Lesotho; also Namibia (Etosha area, E Kavango). Vagrant SE Kenya (Tsavo East); South Africa (Cape Agulhas). Uncommon (especially in western W Africa) to locally common. Has increased its range and population southwards and eastwards in recent decades, at least in southern part of range, probably because more nesting habitats available due to road construction combined with clearing and fragmentation of woodland (Earlé and Brooke 1989).

wing, ♂ (n = 10) 110–121 (115). WEIGHT: (Ghana) 2 unsexed 13, 15; (Cameroon) 1 ♀ 21; (Uganda) 2 ♂♂ 25, 26, 2 ♀♀ 20, 23, 1 unsexed 22.

Description. *H. s. semirufa* Sundevall: SE Angola, S Zaïre, Zambia and W Malaŵi to South Africa, Botswana and Namibia. ADULT ♂: upperparts from forehead to lower back glossy blue-black, rump orange-chestnut, uppertail-coverts blue-black. Tail-feathers blue-black, less glossy, outer pair long and attenuated, white spot on inner web, decreasing in size inwardly, always present on T4–T6, sometimes on T3, not on 2 central pairs. Lores, area around eyes and ear-coverts black, latter with a little gloss. Chin, throat, malar region, small wedge on sides of neck, and underparts orange-chestnut, rather paler on throat and deeper on flanks. Wings black washed with dark blue, more on coverts than flight feathers; axillaries and underwing-coverts pale rufous. Bill black; eye brown; legs black. ADULT ♀: like ♂ but outer tail-feathers shorter. SIZE: (10 ♂♂, 10 ♀♀) wing, ♂ 124–138 (130), ♀ 125–138 (130); tail, ♂ 106–140 (121), ♀ 96–142 (115), depth of fork, ♂ 61–93 (77·3), ♀ 42–82 (61·3); bill to feathers, ♂ 7–9 (7·6), ♀ 6–8 (7·3); tarsus, ♂ 11–16 (14·0), ♀ 12–15 (13·2). WEIGHT: ♂ (n = 22) 26–36 (30·9), ♀ (n = 28) 25–40 (29·5).
IMMATURE: brown above with some blue on feather-tips, chin and throat much paler, outer tail-feathers shorter, tips to wing-coverts and secondaries buff.
NESTLING: skin pink; naked but for a few long greyish white downy plumes on head and back; gape creamy yellow.
H. s. gordoni (Jardine) (including '*neumanni*'): Senegambia to Sudan, Kenya, N Angola and N Zaïre. Paler below, smaller:

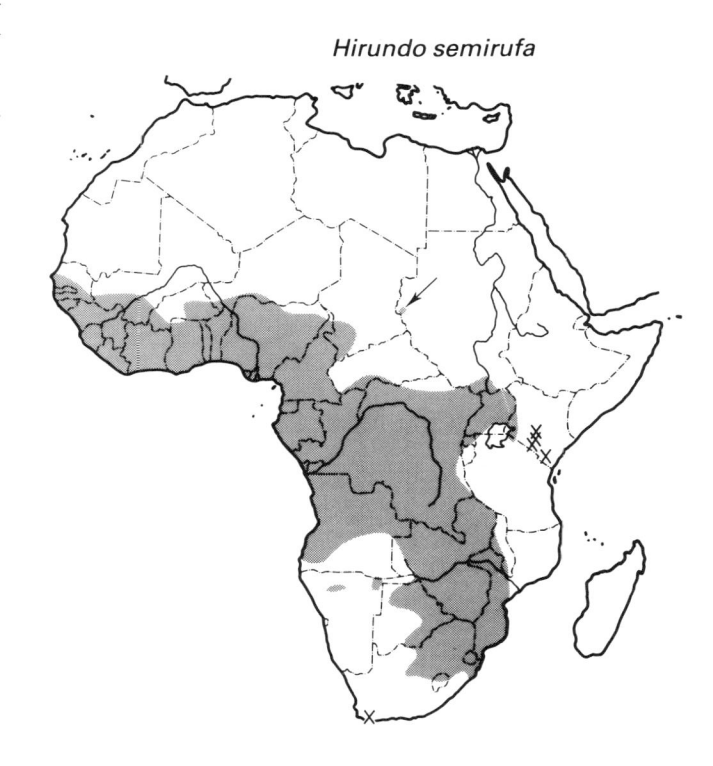

Field Characters. A large blue and red swallow with red rump, tail with white spots and long streamers; dark cap extends to below eye and onto ear-coverts. Similar Mosque Swallow *H. senegalensis* is even larger, though tail-streamers shorter; dark cap extends only to level of eye, chin to upper breast whitish, underwing-coverts white. Red-rumped Swallow *H. daurica* has rufous hind collar separating cap from back, dark undertail-coverts, no white spots in tail.

Voice. Tape-recorded (86, 88, C, F). Distinction between song and calls unclear. Song said to be 'soft and gurgling' (Maclean 1985) but not illustrated there. Vocalizations are of 3 main types: (1) phrases consisting of chuckling, gurgling notes introduced by a few short chips and ending with a long note that is either higher or lower than the central chuckle: (a) higher: 'chip-cherp-purrrrter-eee'; lower: 'chip-chip-cheedle-urrr', 'chip-chili-widdlyiddly-oh'; (2) long down-slurred whistle, 'peeeeurrr'; (3) individual short high-pitched notes, 'chip', 'chissick', 'weet-weet'; last said to be alarm (Maclean 1985).

General Habits. Inhabits grassland, semi-arid scrub, and open savanna, often near water, sometimes clearings in woodland or forest but avoids forested areas.

Occurs in pairs, or groups of up to 8, or in large numbers when hawking for insects over grass fires. Sometimes pair remains together during non-breeding season (Gabon). Associates with other swallows. Flight slow, graceful; glides over grassland, sometimes briefly hovering. Perches on telegraph wires, leafless twigs and occasionally roads.

Largely resident near equator but migratory at higher latitudes. Breeds in N savannas in rainy season, and winters near equator in dry season, mixing with resident populations there. Resident S Ghana, S Nigeria and Gabon. Present Gambia Apr–July, Ghana, Nigeria (north of c. 8°N) May–Sept; Cameroon May–Sept; and Sudan Apr–Oct; Tanzania Nov–May; present Zambia, Zimbabwe, Botswana and South Africa July–Mar and largely absent in coldest months Apr–July. Marked passage noted in latter half of Feb, Chirawmoo, Zimbabwe (Tree 1987) and southern Africa where 1 recovered more than 50 km from site of ringing (Earlé 1987g).

Food. Airborne insects, including black ants.

Breeding Habits. Monogamous, solitary nester. Once 2 ♀♀ and 1 ♂ at 1 nest with 8 eggs (each ♀ laid 4 eggs: Earlé 1987d, 1989); after chicks hatched, 2nd clutch of 3 eggs (laid by only 1 ♀: Earlé 1989).

NEST: gourd- or retort-shaped structure, made of mud pellets, with long tubular entrance and inner chamber lined with grass, feathers, and sheep wool; plastered to overhanging bank or occasionally side or roof of culvert, house, electricity tower, hollow tree trunk, antbear burrow, warthog hole in termite mound; most built less than 1 m above ground level. Tunnel entrance c. 5 cm in diam., tunnel, horizontal (n = 18, Bloemfontein) 10–37 (25·5) cm long. Size and shape of nest varies but widest outside measurements (n = 18) 16–22 (20·0) cm; nest walls 0·5–2 cm thick, usually thicker walled than those of Lesser Striped Swallow *H. abyssinica* and Greater Striped Swallow *H. cucullata* (R. A. Earlé, pers. comm.). 2 nests took 13 and 16 days to construct: egg chamber and entrance tunnel each took half the time.

EGGS: 1–6, usually 3, av. (n = 199 clutches, Bloemfontein, South Africa) 3·4. Slightly glossy; white. Laid at 24-h intervals. Double-brooded; second clutch laid 16–30 (20·9) days after successful fledging of first brood. SIZE: (n = 124, *H. s. semirufa*) 19·6–25·2 × 14–16·6 (22·1 × 15·5), (n = 7, *H. s. gordoni*) 18·8–20·3 × 13·5–15·3 (19·3 × 14·0). WEIGHT: 1st clutch (n = 18) mean 3·1, 2nd clutch (n = 12) mean 2·8, eggs lose 15–23% (18·5%) of weight during incubation.

LAYING DATES: Senegambia, May, July; Sierra Leone, June; Ivory Coast (breeding condition, June); Mali, July–Aug; Burkina Faso, July–Aug; Ghana, Apr, June, Aug; Togo, Sept; Nigeria, Apr–Sept; Cameroon, Mar–July (peak June); Gabon, Jan, Apr, June, Sept–Oct; Zaïre, Apr–July, Oct–Dec; Sudan, July–Aug; E Africa: Uganda, May–June; Kenya, Apr–June, Aug; Region B, May–June; Zambia, Oct–Feb; Malaŵi, Oct–Jan, Mar; Mozambique, Aug, Nov; Zimbabwe, Aug–Apr (Aug 2, Sept 22, Oct 34, Nov 58, Dec 57, Jan 34, Feb 23, Mar 3, Apr 2 clutches); Botswana, Sept, Feb; South Africa: Oct–Mar (mainly Oct–Dec), Transvaal, Oct–Mar (Oct 2, Nov 12, Dec 15, Jan 11, Feb 4, Mar 9 clutches).

INCUBATION: starts after clutch complete; by ♀ only, with ♀ spending as little as 14% of daylight hours incubating (Earlé 1989). Period: 16 days.

DEVELOPMENT AND CARE OF YOUNG: black dots of future feathers under skin at day 4; eyes start to open at day 5; sheath of tail-feathers visible at day 6; eyes fully open at day 10; tail-feathers break through sheath at day 11; reaches adult weight at day 18. Young remain in nest until days 23–25 (23·9). On leaving nest, tail c. 40 mm and wing 20 mm shorter than adult. Young sleep in nest for up to 15 nights after first departure.

BREEDING SUCCESS/SURVIVAL: of 175 eggs (48 nests, South Africa, 3 successive seasons), 80% hatched, of 140 young, 74% fledged (total success rate 59%); failure due to desertion, eggs falling from nests, and chick starvation; 82% of 1st clutches and 44% of 2nd clutches were successful. Zimbabwe, 4 ringed birds lived 2·6–5·6 years (Irwin 1981); South Africa, 1 unsexed 5·75 years (Earlé 1987g).

References
Aspinwall, D. R. (1980d).
Earlé, R. A. (1989).
Earlé, R. A. and Brooke, R. K. (1989).
Lewis, A. D. (1982).

Plate 8
(Opp. p. 113)

Hirundo senegalensis Linnaeus. **Mosque Swallow. Hirondelle des mosquées.**

Hirundo senegalensis Linnaeus, 1766. Syst. Nat. (12th ed.), 1, p. 345; Senegal.

Range and Status. Endemic resident and partial intra-African migrant. Subsaharan Africa from S Mauritania (middle and upper Senegal R. valley), Senegambia, Mali (south of 14°N), Burkina Faso, Ivory Coast (north of 9°N), SW Niger (Parc du W, Ayorou, Filingue, Dosso), Ghana (N savannas; coastal region from Sekondi east to Accra Plains and Akwapim Hills; also forest region at Akim Oda), Benin (mainly north; also south to coast),

Hirundo senegalensis

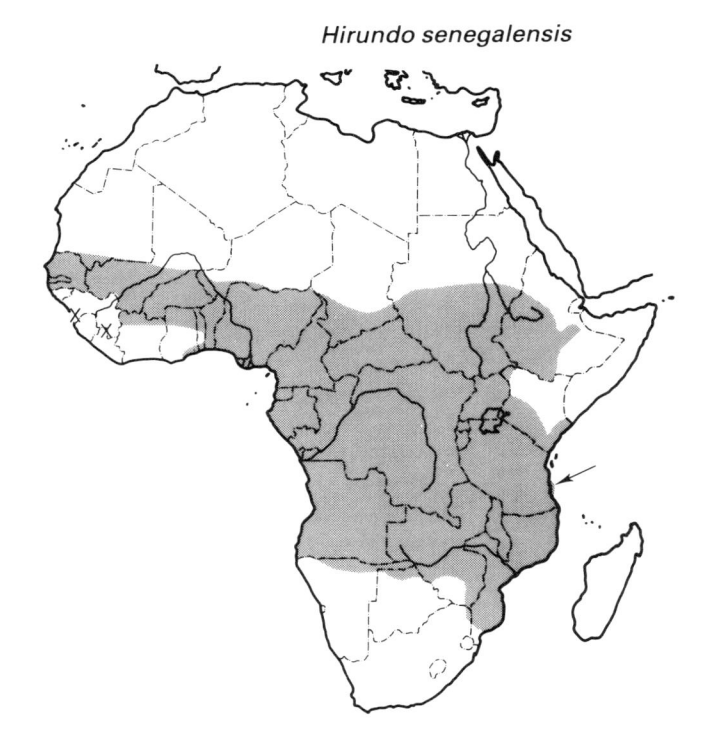

Nigeria, Cameroon to S Chad (mainly south of 10°N), Sudan (south of 14°N), W and S Ethiopia (east to central W Highlands and SE Highlands), W and central Kenya south of 2°N and mainly west of 38°E; also Moyale in N and coastal region north to Somalia border) and Somalia (1 record near border at 1° 30′S, 41° 30′E). Then south through equatorial and central Africa to S Angola (south to Cunene R.), N Namibia (south to c. 18°S), N Botswana, Zimbabwe (N and E from N Kalahari Sands east and south to Chipinga uplands, Sabi valley and Limpopo R. to Chikwarakwara), Mozambique (except extreme S) and NE Transvaal (mainly Kruger Nat. Park). Vagrant Guinea (Conakry), Liberia (Voinjama), Mafia I. Rare to locally common (rare: Benin, Central African Republic, N Kenya (Moyale), N Botswana; uncommon: Mali, coastal Angola, Malawî, South Africa; frequent: Ivory Coast, Nigeria, Gabon, rest of Angola, Ethiopia, Zambia, Mozambique; common: Ghana, Chad, Sudan, Uganda, Kenya, Tanzania, Congo and Zimbabwe).

Description. *H. s. senegalensis* Linnaeus: Mauritania and Senegambia to N Ghana, N Nigeria and N Cameroon, S Chad and W Sudan. ADULT ♂: forehead to lower back, and uppertail-coverts, glossy blue with purple reflections, rump dark rufous. Tail-feathers glossy blue-black without white on inner webs; outermost pair very long. Lores and stripe under eye white; supra-loral line and short streak behind eye black; cheeks and ear-coverts white to pale rufous, becoming darker rufous on sides of neck, rufous extending onto sides of head behind eye and continuing as hind collar below nape; collar narrowest on hind neck, sometimes broken by blue feather-tips, but feather bases always rufous. Chin to lower throat variable, whitish to pale rufous, upper breast same or deeper rufous, rest of underparts chestnut-rufous, darker posteriorly. Wings black, flight feathers with slight bluish gloss, upperwing-coverts with more pronounced gloss, axillaries and underwing-coverts pale fawn to white, rest of underwing dusky black. Bill black; eye brown; legs black. Sexes alike. SIZE: (10 ♂♂, 10 ♀♀) wing, ♂ 138–156 (144), ♀ 138–151 (143); tail, ♂ 81–109 (93·9), ♀ 75–110 (95·2), depth of fork, ♂ 57–83 (66·6), ♀ 32–73 (59·4); bill to feathers, ♂ 9–10 (9·5), ♀ 8–11 (9·4); tarsus, ♂ 15–19 (16·3), ♀ 14–18 (16·0).

IMMATURE: upperparts darker than adult, dark brown with dark blue gloss, rump pale rufous, sides of neck and nuchal collar much paler rufous, almost whitish; underparts much paler, brown patch at sides of breast extends to form incomplete breast-band; uppertail-coverts, inner secondaries and tertials with narrow pale rufous tips.

NESTLING: unknown.

H. s. saturatior Bannerman: from S Ghana to S Cameroon, N Zaïre to S Sudan, Ethiopia, Kenya, Uganda, Rwanda and Burundi. Deeper rufous below than nominate race. WEIGHT: (Kenya) ♂ (n = 7) 41–50 (43·4), ♀ (n = 5) 41–54 (44·8); (Uganda) ♂ (n = 3) 45–50 (47·7).

H. s. monteiri (Hartlaub): Angola, S Zaïre and Tanzania southwards. Differs from nominate race in having white spots on inner webs of all but central tail-feathers; darker and often longer-tailed. WEIGHT: (Tanzania) ♀ (n = 3) 39·5–43·7 (42·1); (South Africa) ♂ (n = 7) 41–51 (47·4), ♀ (n = 9) 38–47 (43·4).

Field Characters. The largest and heaviest swallow in Africa; at a distance might even be mistaken for a small falcon. Larger than similar Red-breasted Swallow *H. semirufa* but tail-streamers shorter; further differs in having pale face and throat and partial rufous hind collar, whiter underwing-coverts. Western and central races further distinguished by lack of white spots in tail, but southern African race *monteiri* does have white tail spots. Much smaller Red-rumped Swallow *H. daurica* has paler and more uniform underparts (little contrast between throat and rest of underparts), dark undertail-coverts, no white spots in tail. Mosque Swallow's 'tin trumpet' call diagnostic, and flap-and-sail flight characteristic.

Voice. Tape-recorded (86, 88, C, BRU, GIB). Song a guttural, fast 'chuckle-chuckle-chuckle'; calls include nasal, drawn-out 'naaaah', piping 'pyuuuu', high-pitched 'mew', 'weh' and nasal 'ya'.

General Habits. Occurs in a variety of habitats including grassland, savanna, acacia bush with baobabs, light woodland, villages, cultivated areas, forest edges, clearings and river valleys; avoids closed forest and arid regions; in Zambia associated with culverts (rather than bridges) (D. Aspinwall, pers. comm.). Prefers more wooded areas than Red-breasted Swallow, especially in south of range.

Solitary or in pairs or groups of c. 6, but up to 100 congregate to hawk prey over grass fires. Flight slow, with much sailing, then burst of shallow flapping. Forages 20–30 m above tree tops, sometimes with other species of swallows or swifts. Roosts on tops of tall trees; also on tall grasses and bushes.

Sedentary or (sometimes nearby) partly migratory. North of equator resident coastal Ghana, central Nigeria, Gabon; present Benin Mar–June; S Nigeria (Lagos) Jan–May; Cameroon montane districts usually Sept–Oct. More migratory at high latitudes north of equator: present Senegambia largely May–Oct, N Nigeria Feb–

July, Mali Mar–Oct. South of equator, appears to be sedentary, although breeding season visitor in parts of Zimbabwe (Mashonaland, present Aug–Apr, sedentary elsewhere).

Food. Airborne insects, including ants, termites and flies.

Breeding Habits. Monogamous, solitary nester.
NEST: gourd- or retort-shaped with long entrance, made of mud pellets, nest chamber lined with grass and feathers, in hole in tree (often baobab) or house, occasionally in bank of river or old termite mound. Built by both sexes, usually in early morning or late afternoon, taking up to 1 month.
EGGS: 2–4, av. (n = 15 clutches, Uganda) 2·7. Pure white. SIZE: (n = 11) 21–22·1 × 14–15 (21·7 × 14·7). Double-brooded.

LAYING DATES: Mauritania, July–Sept; Senegambia, July–Oct, Dec; Burkina Faso, July; Ghana, Jan, Apr–July, Oct (but mainly wet season); Nigeria, May, Aug; Cameroon, Feb–Sept; Gabon, Apr; Sudan, W, June–Sept, Nov, S, Dec–Feb; Ethiopia, Apr, July, Nov; E Africa: Kenya, Apr–May, Dec; Uganda, Feb–July, Oct–Dec; Tanzania, Jan–Feb, Apr, Oct, Dec; Region A, Mar–June; Region B, Mar–June, Sept, Nov; Region C, Apr–June, Dec; Region D, Mar–June, Oct–Nov; Region E, Apr–May, Dec (most breeding in long rains); Rwanda, Jan, Mar–Apr (most breeding in long rains); Zaïre, Apr–July; Angola, Nov–Jan; Zambia, Aug–Apr; Malaŵi, Jan–Feb, May, July, Sept–Dec; Zimbabwe, Aug–Apr; South Africa, Nov–Jan.

References
Aspinwall, D. R. (1980b).
Pitman, C. R. S. (1931).

Plate 10
(Opp. p. 161)

Hirundo abyssinica Guérin-Méneville. Lesser Striped Swallow. Hirondelle striée.

Hirundo abyssinica Guérin-Méneville, 1843. Rev. Zool., p. 322; Ethiopia.

Range and Status. Endemic resident and intra-African migrant. Subsaharan Africa from central Senegambia, Guinea and Sierra Leone to S Mali (north to Ansongo), Liberia, Ivory Coast, Ghana, N Togo, Burkina Faso, extreme SW Niger (Gaya, Yeri), Nigeria (south of *c.* 13°N), Cameroon, and SW Chad to Central African Republic (Lobaye Préfecture and Bangui Aua in SW, Bamingui-Bangoran and Manovo-Gounda-Saint Floris Nat. Parks in NE), Sudan (Darfur, Bahr El Ghazal; also extreme S), Ethiopia (east to W and SE Highlands and Ogaden above 900 m) and Somalia (south of *c.* 5°N and west of 46°E). Then south through equatorial, central and E Africa to N Namibia, N and NE Botswana, E South Africa (E Cape east of Jansenville, Steytlerville, Tarkastal; extreme E Orange Free State; Transvaal; and Natal), and Swaziland. Frequent to locally abundant, but uncommon Senegambia, Sudan and Congo basin forest belt except along major rivers and in Orange Free State. Vagrant Oman (1987). Increasing in numbers due to use of man-made structures for nest-sites.

Description. *H. a. abyssinica* Guérin-Méneville: Ethiopia and E Sudan. ADULT ♂: forehead, crown, nape, hindneck, side of neck and ear-coverts light reddish chestnut, mantle and back glossy blue-black; rump and uppertail-coverts paler rufous chestnut (but longer uppertail-coverts dull blue-black, edged pale chestnut); rufous band from sides of rump onto vent. Tail-feathers brownish black washed with dark metallic blue; T6 and T5 with a prominent white spot, T4 and T3 with either reduced white or no white, T6 very long. Lores grey to greyish white, black spot in front of eye. Cheeks, chin, throat and underparts white, undertail-coverts with buffy tinge; boldly streaked black, most heavily on throat and upper breast; streaks on undertail-coverts narrow and brown. Wings brownish black, primaries and secondaries with dull dark blue wash, tertials and upperwing-coverts with glossy blue wash; axillaries and underwing-coverts buff, under primary coverts with dark tips.

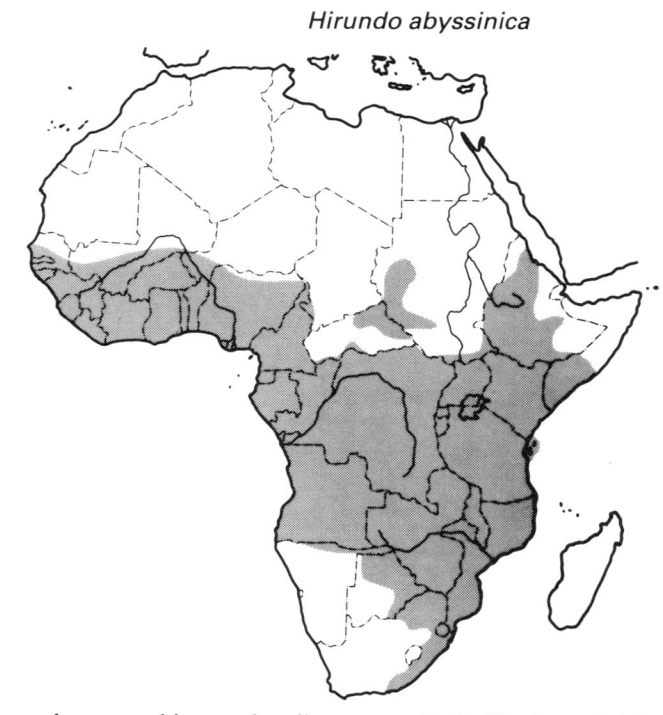

Bill black; eye brown; feet black. ADULT ♀: like ♂ but outer tail-feather shorter. SIZE: (10 ♂♂, 10 ♀♀) wing, ♂ 103–110 (108), ♀ 100–111 (104); tail, ♂ 74–94 (86·6), ♀ 52–81 (68·1), depth of fork, ♂ 50–100 (64·6), ♀ 32–62 (45·4); bill to feathers, ♂ 5–7 (6·1), ♀ 5–7 (5·8); tarsus, ♂ 10–12 (10·5), ♀ 9–12 (10·3).

IMMATURE: similar to adult but top of head speckled with brownish black, rump paler, breast sometimes with more buff, and innermost secondaries and wing-coverts with tawny tips.

NESTLING: naked but for some long, greyish white down; bill greyish, gape white.

H. a. unitatis (Sclater and Mackworth-Praed): S Sudan and Zaïre west to Gabon, central Angola, E Zambia, Zimbabwe (not NW), N and E Botswana, Transvaal, E Cape and Mozambique. More heavily streaked below and larger than nominate race. Wing 107–117, tail streamers 80–120. WEIGHT: (Angola) ♂ (n = 6) 13–15 (14·5); (Uganda) ♂ (n = 8) 16–27 (18·3), ♀ (n = 2) 18, 18; (Kenya) ♂ (n = 2) 15, 16, ♀ (n = 2) 12, 13; (Zimbabwe) 1 ♂ 20; (Botswana and South Africa) unsexed (n = 4) 17–19 (17·5).

H. a. ampliformis Clancey: S Angola to W Zambia, and NW Zimbabwe to N Namibia. Similar to *unitatis* but shaft streaking on underparts coarser and blacker. Larger: wing 112–123, tail streamers up to 162.

H. a. maxima (Bannerman): SE Nigeria and S Cameroon to SW Central African Republic. Darker and more heavily streaked below than *unitatis*. Streaking of underparts similar to *ampliformis* but crown less tawny, more chestnut. Larger than nominate race: wing, ♂ (n = 10) 105–114 (110). WEIGHT: (Cameroon) 1 ♂ 23, 1 ♀ 22.

H. a. puella Temminck and Schlegel: Senegambia to NE Nigeria and N Cameroon. More finely streaked and smaller than nominate race: wing 96–106. WEIGHT: (Liberia) ♂ (n = 5) av. 15·2.

H. a. bannermani Grant and Mackworth-Praed: NE Central African Republic, SW and W Sudan (Darfur). Striping below much finer and narrower than in other races, and crown pale rufous.

Field Characters. A small blue and white swallow with rufous head and rump, the only one outside of southern Africa with striped underparts except the 2 *Phedina* martins, which are brown with short, square tails. In southern Africa confusion possible with Greater Striped Swallow *H. cucullata*, which also has rufous head and rump and striped underparts, but is considerably larger, with much finer streaks on underparts, giving it a paler appearance, and entire underparts buffy (not just undertail-coverts), cap darker, not extending onto ear-coverts, flight slower and more measured. Frequently-given nasal song of Lesser Striped Swallow diagnostic.

Voice. Tape-recorded (5, 72, 81, 86, 88, C, F). Very vocal. Song consists of variable number of introductory notes, squeaky, liquid or buzzy, 'tsli', 'tsi-tsi', 'chewp', 'brrrrt', leading into main body of song, 3–6 nasal tinny notes, typically descending the scale, sometimes turning up at the end: 'nyee-nyee-nay-naah-noh-naw'; 'nyee-nay-naw-nay'; notes loud and accentuated, with slight pause between each.

General Habits. Inhabits open grassy areas; also occurs in savanna, woodland, cultivated areas, forest edge and clearings, around granitic hills, cliffs, and human habitation, and over water; occasionally mangroves and river valleys in forests but avoids dense forest; in Zambia associated largely with bridges (not culverts) (D. Aspinwall, pers. comm.).

Solitary or in pairs; often in flocks of 5–20, sometimes 100, often in company with swifts and swallows. Flight agile and darting, often 10–20 m above tree tops, with much flapping and gliding. Feeds around cattle and other large mammals, catching insects disturbed by them; sometimes hovers over and takes caterpillars from standing crops; perches to feed on fruits of pigeon wood tree (*Trema orientalis*) and *Acacia cyclops*, balancing with flapping wings (Every 1988; McLean 1988). Also perches on wires and leafless twigs. Vocal at perch and in flight. Sometimes (Somalia) roosts in sugar-cane fields in large numbers (200–300: P. Becker, pers. comm.).

Mainly sedentary within c. 10° of equator; mainly migratory at higher latitudes, where it breeds in the rains, returning to winter at lower latitudes. Sedentary S Ghana, Gabon, Zaïre, Angola and Zambia; probably also Chad, Central African Republic, Ethiopia and E Africa. Apparently breeding visitor in rains Senegambia, Sierra Leone, N Ghana (mid-Mar to late July), Togo (no records Sept–Jan), Nigeria (Mar–July), N Cameroon (June–Oct), Zimbabwe (July–Mar), Botswana (July–Mar), Mozambique (July–Mar) and South Africa (July–Mar). Situation complicated in Zimbabwe because some present all year with birds leaving after breeding to unknown destinations (probably north) and being replaced by other birds (from South Africa ?); and in Zambia and Botswana because *ampliformis* in W breeds in winter in dry season while *unitatis* in E breeds in summer in wet season.

Food. Airborne insects, including alate termites, ants, bees, flies and beetles; also caterpillars, fruits and seed pods.

Breeding Habits. Monogamous, usually solitary nester, but colonial Zambia (usually 3–4 nests together, once a colony of 50 nests: Bowen 1979b). Single bird that had lost its mate became helper at adjacent nest of Greater Striped Swallow (Oates 1991).

NEST: gourd- or retort-shaped, made of mud pellets slightly smaller than those of Red-breasted Swallow *H. semirufa* (R. A. Earlé, pers. comm.); spherical egg-chamber with tubular entrance, bowl lined with grass and feathers; plastered to overhang in cave, under roof, rock, culvert, verandah, bridge or large branch, or in crevice or hollow of baobab. Tunnel horizontal, 6–25 cm long, egg-chamber 15 cm in diam., 10 cm long. Built by both sexes, taking up to 7 weeks. Sometimes old nest of other swallow used as base for nest, with pair modifying it and adding tunnel to it. Once a pair (Angola) evicted Wire-tailed Swallow *H. smithii* from nest and used it as base for nest.

EGGS: 2–4, usually 3, av. (n = 97 clutches, Cameroon, E Africa, southern Africa) 2·86. Slightly glossy; immaculate white, occasionally with pale reddish spots and speckles. SIZE: (n = 72) 17·3–21·8 × 12·5–15·4 (19·7 × 13·9). Double-brooded.

154 HIRUNDINIDAE

LAYING DATES: Senegambia, Mar, July; Sierra Leone, Mar–May; Burkina Faso, Aug; Ghana, S, Mar–July, N, May–Aug; Togo, Mar–June; Nigeria, Mar–July (mainly May–July); Cameroon, Dec–May; Gabon, June–Mar, probably all year; Sudan, July–Aug (young Dec); Ethiopia, Mar–Aug; E Africa: Kenya, Feb–Aug, Oct–Dec; Uganda, Feb–Aug, Oct–Dec; Tanzania, Jan–Aug, Oct-Dec (mostly Oct–Dec); Zanzibar and Pemba, Jan–Dec; Region A, May; Region B, Feb–June; Region C, Dec–May; Region D, Jan–Sept, Nov–Dec; Region E, Apr, July–Aug, Dec (confined to rains, especially long rains); Rwanda, Mar–June (long rains); Angola, Jan–Apr, June–Nov; Zaïre, July–Oct; Namibia, Sept, Nov–Mar; Malaŵi, July–Feb; Mozambique, May–Aug, Oct–Dec; Zambia, May–Aug (*ampliformis*), Sept–Apr (*unitatis*); Zimbabwe, nearly all months, mainly Sept–Jan (Jan 15, Feb 9, Mar 5, May 1, June 1, July 4, Aug 10, Sept 49, Oct 63, Nov 90, Dec 34 clutches); Botswana, Sept–Jan (*unitatis*); South Africa, Aug–Apr (with 70% Oct–Dec: Earlé 1988a).

INCUBATION: by ♀ only; ♂ often perches nearby. Period: 14–21 days (probably 18–19).
DEVELOPMENT AND CARE OF YOUNG: young fed by both parents, 2·6–9 times per h (Moreau 1947). Young leave nest at 17–28 days.
BREEDING SUCCESS/SURVIVAL: nests usurped by Mocking Cliff-Chat *Myrmecocichla cinnamomeiventris* and White-rumped Swift *Apus caffer*. 1 ringed bird recovered after 6 years (Earlé 1987g).

References
Aspinwall, D. R. (1981a).
Clancey, P. A. (1969).
Earlé, R. A. (1988a).
Moreau, R. E. (1947).

Plate 10
(Opp. p. 161)

Hirundo cucullata Boddaert. Greater Striped Swallow. Hirondelle à tête rousse.

Hirundo cucullata Boddaert, 1783. Table Pl. enlum., p. 45; Cape of Good Hope.

Range and Status. Endemic intra-African migrant, breeding in Angola (highlands of Huila and Humpato), central Namibia (c. 19–26°S, 16–17°E from S Ovamboland to N Great Namaqualand), E Botswana (Francistown, Gaberones), Zimbabwe (above 1200 m from Matopos to Inyanga, Melsetter and Chimanimani Mts), W Swaziland, Lesotho and South Africa (SW Cape including Cape Town region north to 32°S; E Cape from c. 23°E; Orange Free State, central and S Transvaal and Natal). Winters central Africa north to N Angola (coastal Mossamedes, Cuanza Sul, Lunda districts), S Zaïre (Kasai, Shaba), Zambia, SW Tanzania (north and east to Morogoro) and Malaŵi (north end L. Malaŵi, also Nsanje District in passage). Frequent to common.

Description. ADULT ♂: forehead, crown, nape and hindneck chestnut with some blue-black streaks, becoming paler on sides of neck and hindneck, producing incipient pale collar; mantle and back glossy blue-black; rump and anterior uppertail-coverts tawny, lower coverts blue-black tipped tawny. Tail brownish black washed with dull metallic blue; a white patch on inner web of T3–T6, largest on T6 and becoming progressively smaller inwards, T1–T2 without white, T6 greatly elongated. Lores dusky; cheeks, ear-coverts, chin, throat and underparts light buff, narrowly streaked dark brown; a small wedge of dark blue extends from mantle across sides of lower neck; undertail-coverts white tinged buff. Wings brownish black, upperwings including greater coverts washed dull metallic blue, lesser and medium coverts glossy blue, underside of flight feathers browner and paler; axillaries and underwing-coverts buffy white, under primary coverts with narrow dark streaks. Bill black; eye brown; legs black. ADULT ♀: like ♂ but outer tail-feather shorter. SIZE: wing, ♂ (n = 10) 100–132 (123), ♀ (n = 10) 117–130 (122); tail, ♂ (n = 10) 41–115 (88·0), ♀ (n = 10) 74–94 (87·1), depth of fork, ♂ (n = 3) 56–59 (56·9), ♀ (n = 5) 43–53 (49·9); bill to feathers, ♂ (n = 10) 7–8 (7·2), ♀ (n = 10) 6–8 (6·9); tarsus, ♂ (n = 10) 10–14

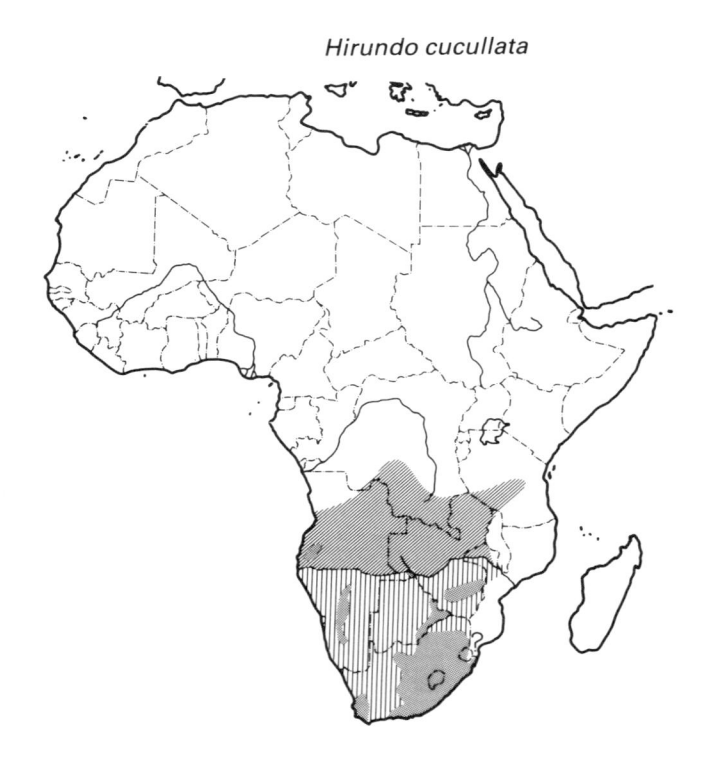

(11·9), ♀ (n = 10) 10–15 (13·2). WEIGHT: (southern Africa) ♂ (n = 23) 23–31 (27·0), ♀ (n = 24) 21–31 (27·2), unsexed (n = 34) 19–35 (27·0).

IMMATURE: similar to adult but duller, chestnut on top of head mixed with black, rump paler, innermost secondaries and wing-coverts tipped tawny.

NESTLING: skin pinkish red; greyish white down on head and back, bill-flange white.

Field Characters. A large swallow, blue and buff rather than blue and white, with chestnut cap and rump. Underparts streaked, but streaks fine, only visible at close range, bird appearing plain at a distance. Only other swallow in its range with streaked underparts is Lesser Striped Swallow *H. abyssinica*, which is considerably smaller, with more agile, darting flight, and has white underparts with much heavier dark streaking, visible at a distance. The Lesser Striped Swallow also has paler and more uniform rufous cap which extends onto ear-coverts, rufous band across vent and characteristic nasal song.

Voice. Tape-recorded (81, 88, LUT). Song a brief (2–3 s) phrase of 1–2 low, clipped notes, a double note and a gurgling trill: 'chip-cherker-kurrrrrrrr'; in between phrases, a sharp 'chep'. Flight call, 'chissick'.

General Habits. Occurs in treeless country including montane grassland; sometimes over water; avoids woodland and forest.

Solitary or in pairs, flocks of up to 30 in non-breeding season; groups sometimes formed on breeding grounds before migrating (R. A. Earlé, pers. comm.). Flight slow and leisurely, with gliding between wing-beats. Forages over grassland and water, often with other swallows and swifts. When feeding on *Acacia cyclops* seed pods, hovers before and/or clings to pods, then plucks out ripe seeds (Every 1988). Often perches on trees and wires, or on ground in early morning.

Intra-African migrant, present on breeding grounds Aug–Apr; occasionally arrives July. In Mar–Apr moves north to spend non-breeding season in central Africa, probably mainly west of 25°E and north of Cunene-Zambezi rivers (D. Aspinwall, pers. comm.). 1 ringed South Africa (Helderberg Nature Reserve, Somerset West, Cape), recovered Kasanza, Zaïre, 3156 km away (Earlé 1987g).

Food. Airborne insects; occasionally mulberries and seeds of *Acacia cyclops*; also shell grit (Roos and Roos 1989).

Breeding Habits. Monogamous and solitary. One pair were helped at the nest by a Lesser Striped Swallow from an adjacent nest which had lost its mate; the Lesser Striped helped build the nest and feed the incubating ♀

and chicks (Oates 1991). In courtship, ♂ throws head back and puffs out throat feathers when singing, both on perch and in flight (Maclean 1985).

NEST: bowl with horizontal tunnel 7–23 cm long, made of mud pellets (**A**), bowl lined with grass and feathers; plastered to roof or eaves of building, rock overhang, culvert, bridge, underside of fallen tree, or old mine. Entrance to tunnel 38–51 wide, 35–50 high. Built by both sexes, usually in morning and afternoon; takes 2–3 weeks. Same site may be used year after year, by one or both birds (♂ used same nest for 3 seasons, South Africa: Schmidt 1962). Largest recorded nest 600 in height, 300 in width with centre hollow, bowl and tunnel at top of entire structure (Pringle 1989). Once used old Blue Swallow *H. atrocaerulea* nest as base for own nest (Zimbabwe: Snell 1963).

EGGS: 2–4, usually 3, av. (n = 31 clutches, southern Africa) 2·97. Laid in early morning. Often 2–3 broods per year. Slightly glossy; white, rarely with brownish spots. SIZE: (n = 104) 19·1–24·5 × 13·5–16·5 (21·7 × 15·1).

LAYING DATES: Angola, Oct; Namibia, Nov–Mar; Botswana, Nov–Apr; Zimbabwe, Aug–Mar (Aug 3, Sept 6, Oct 11, Nov 7, Dec 15, Jan 4, Feb 2, Mar 5 clutches); South Africa: Aug–Apr, Transvaal, Oct–Apr (Oct 7, Nov 26, Dec 14, Jan 14, Feb 23, Mar 16, Apr 2 clutches).

INCUBATION: by ♀ only, although both birds sleep in nest at night. Period: (n = 17) 16–20 (17·6) days.

DEVELOPMENT AND CARE OF YOUNG: young hatch on successive days; wing-feather sheaths visible at 4 days, wing-feathers break through skin at 6 days, eyes half open at 8 days, some wing- and tail-feathers break through sheath and unfold at 11 days, fully feathered, with pinkish grey legs, at 20 days. Young fed by both parents; leave nest at 23–30 days but return to nest for another 9–22 days.

BREEDING SUCCESS/SURVIVAL: SE Cape, successive broods of 2 and 3 all fledged. Nests usurped by Southern Anteater-Chat *Myrmecocichla formicivora*, Mocking Cliff-Chat *M. cinnamomeiventris*, Red-breasted Swallow *H. semirufa* and White-rumped Swift *Apus caffer* (once, 4 of 10 swallow nests robbed by swifts including some that had half-grown young: Schmidt 1962). Other losses due to bad weather, scarcity of food, infestation of hippoboscid flies (*Ornithomyia*), and nests breaking off corrugated roofs (Schmidt 1962). 5 birds (South Africa) lived for 5, 5·5, 6, 6·4 and 6·4 years (Earlé 1987g).

Reference
Schmidt, R. K. (1962).

HIRUNDINIDAE

Plate 8
(Opp. p. 113)

Hirundo daurica Linnaeus. Red-rumped Swallow. Hirondelle rousseline.

Hirundo daurica Linnaeus, 1771. Mantissa, p. 528; Siberia.

Range and Status. Africa, SW Arabia, S Europe and Middle East to Sri Lanka, China, Japan and Ussuriland. Resident south of 20°N, mainly migratory north of 20°N. Accidental in Europe north to Britain, Norway, Sweden and Finland.

Breeding summer visitor to NW Africa, spring passage migrant throughout N Africa; resident and partial intratropical migrant in, and in winter Palearctic visitor to, northern tropical and E Africa.

Frequent and widespread in Morocco north of 30° 30'N. In Algeria rare as breeding visitor in extreme W coastal region, and has nested at Tigzirt and L. Télamine; passage migrant in small numbers. In Tunisia scarce passage migrant in spring only. Rarely, winters in Morocco and Tunisia. Spring passage migrant in Libya, quite common in some years on Tripoli coast, rare in Cyrenaica and interior, but flocks of up to 100 sometimes in Ghat, Ajjial, Oubari and Brak. In Egypt, locally numerous spring migrant in Nile valley and Western Desert, in flocks of hundreds, but scarce in autumn; scarce spring migrant in Eastern Desert and on Red Sea coast.

Widespread, uncommon, frequent or common from Senegambia to Ethiopia, south to Malaŵi. Mauritania, scarce winter visitor near Nouakchott, Oct–Nov, Mar–Apr and June (probably Palearctic *rufula*). Senegambia, locally common resident in Gambia and Casamance, scarce on coast, frequent in towns and villages. Guinea, records at Kankan, Badela and Beyla district. Mali, mainly passage migrant, in spring; rare in autumn (probably *rufula*). Sierra Leone, resident Birwa Plateau. Burkina Faso, resident Arli Nat. Park. Niger, records in SW at Yéri, Niamey and Dosso. Ivory Coast, frequent resident between $7\frac{1}{2}°$ and 10°N. Ghana, frequent and widespread wet-season visitor in N (but resident at Tumu reservoir) and mainly resident in S. Nigeria, common resident in soudanian and N guinean savannas, penetrating S guinean savannas in Niger R. valley; has bred at Obudu Cattle Ranch (Ash 1990); partial migrant, much less common at height of rains than at other times; records in S Nigeria mainly Mar–May. *H. d. rufula* not known in Ghana or Nigeria. Cameroon, resident in N, south to Adamawa Plateau (*domicella*) and at Manenguba, Bamenda, Oku, Kumbo, Ndop and Djang (*kumboensis*). Chad, common Zakouma and Bahr Salamat in Jan and probably resident (*domicella*); a few records in SW; regular autumn visitor to Ouadi Rimé–Ouadi Achim (*rufula*). Central African Republic, recorded in Manovo-Gounda-Saint Floris Nat. Park only. Sudan, common resident Darfur, Bahr el Ghazal and Equatoria (*domicella*, *emini*), uncommon to north and northeast of Sennar (*rufula*), chiefly as erratic autumn migrant in Nile valley in flocks of up to 500 (Hogg et al. 1984). Ethiopia, winter visitor to W Highlands and NE, common in towns (*rufula*); frequent or common resident in W and SE Highlands, NE, S and SE Ethiopia and Rift Valley (*melanocrissa*) and in W Ethiopia (*domicella*). Somalia, 5 records of *rufula* in N mountains, May, and 1 in S (race?). Uganda, resident from Mt Lonyili and Mt Morongole in NE, south in rocky hills to SW and SE borders. Resident central Kenya north to Mt Nyiro and Mt Marsabit, but very sparse below 1600 m in W Kenya and L. Victoria basin. Widespread resident in highlands of N and E Tanzania, also in SW Zaïre, resident between 1200 and 2150 m in hills in extreme E and SE. Zambia, Nyika Plateau, Sunzu, Kambole, perhaps Diwa Hill and Chipata, once in middle Zambezi valley. Malaŵi, not uncommon above 1200 m, on Nyika Plateau, Viphya Mts, Mafinga Mts and Misuku Hills. Zimbabwe, 2 sight records near Harare, Feb and Mar.

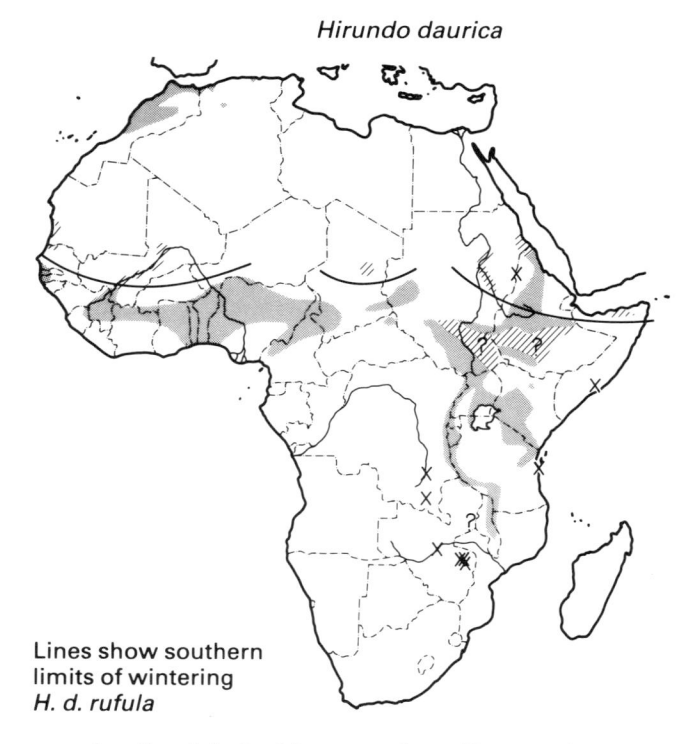

Description. *H. d. rufula* (Temminck): N Africa, S Europe, through Middle East to USSR (Tien Shan Mts) and Kashmir. Winters mainly in subsaharan Africa. ADULT ♂: crown, mantle, scapulars and back glossy dark blue. Forehead, line above lores, narrow supercilium, nape, sides of neck and hindneck rufous, forming well-demarcated rufous half-collar between crown and mantle. Lores dusky, cheeks and ear-coverts buffy. Upper rump rufous, lower rump very pale rufous or buff; uppertail-coverts glossy dark blue. Tail bluish black, slightly glossy, deeply forked, with tapering outer rectrices, curving towards each other so that tips meet or even cross. Chin to vent uniformly buffy cream, with minute dark shaft-streaks, particularly on chin and breast; flanks washed slightly more rufous. At sides, cream of chin and throat merge into buff cheeks and rufous sides of neck. Undertail-coverts black with bluish gloss. Median and greater wing-coverts dark blue, tips of primaries glossy bluish; remainder of upperwing sooty blackish. Underwing-coverts buffy cream; undersides of remiges and rectrices glossy blackish. Bill black, mouth pale yellow-grey;

eye dark brown; legs and feet black or red-brown with flesh-pink tinge. Sexes alike. SIZE: wing, ♂ (n = 31) 120–128 (124·5), ♀ (n = 22) 118–127 (121); tail, ♂ (n = 29) 94–107 (102), ♀ (n = 18) 89–99 (95·6), fork depth, ♂ (n = 30) 53–64 (59·3), ♀ (n = 18) 46–56 (52·0); bill to skull, ♂ (n = 13) 10·4–12·1 (11·2), ♀ (n = 13) 9·3–12·5 (10·9), bill to feathers av. 4·4 less; tarsus, ♂ (n = 13) 13·5–14·7 (14·1), ♀ (n = 10) 13·2–14·3 (13·8). WEIGHT: (S Spain, summer) ♂ (n = 27) 19–29 (22·1), ♀ (n = 25) 20–28 (22·4); (SE Morocco, early Apr) unsexed (n = 7) 15·2–19·9 (17·1).

IMMATURE: like adult but less blue gloss on crown; back feathers blackish without any gloss; rump much paler rufous than in adult; lesser, median and greater wing-coverts narrowly tipped pale buff; tertials and inner primaries broadly tipped pale buff; and underparts without any black shaft-streaks.

NESTLING: skin pink, bill dark grey, gape-flanges yellow-white, legs greyish; down long and soft, restricted to tufts on forehead, neck and back.

H. d. melanocrissa (Rüppell): highlands of Ethiopia. Forehead glossy dark blue, like crown; rufous stripe over eye narrower than in *rufula*; rump variable, either dark rufous or buffy, but uniform and not 2-toned as in *rufula*; underparts slightly buffier than in *rufula*; minute dark shaft-streaks on breast but not on belly.

H. d. domicella (Hartlaub and Finsch): N tropics from Senegambia and Guinea to Sudan (east to Baro R.), but not Sierra Leone highlands, Cameroon highlands, or SE Sudan. Like *melanocrissa* but underparts from chin to vent white, or white with faintest pink-buff wash, with (at most) vestigial darkish shaft-streaks.

H. d. emini (Reichenow): SE Sudan, Uganda, E Zaïre, W Kenya, Tanzania, N Zambia, Malawi. Like *melanocrissa* but underparts pale rufous, darkening towards belly and flanks, and without dark shaft-streaks except for vestiges on chin. Rump medium or dark rufous, never pale rufous. WEIGHT: ♂ (n = 2) 23, 30, ♀ (n = 3) 24, 28, 29; unsexed (n = 5) 22·5–25·6 (24·3) (D. J. Pearson, pers. comm.).

H. d. kumboensis (Bannerman): Birwa Plateau, Sierra Leone, and Bamenda Hignlands, Cameroon. Like *melanocrissa* and *emini* but underparts intermediate in tone. Small: wing 110–122.

Field Characters. Medium-sized swallow with dark blue cap, wings, back and tail, and pale or dark rufous rump and underparts. Undertail-coverts black, combining with black under surface of tail to give bird seen from below a dipped-in-ink appearance which distinguishes it at once from all other blue-and-rufous African hirundines. Upperparts very like those of Mosque Swallow *H. senegalensis*, with rufous rump, deeply-forked blue-black tail, and rufous hind collar dividing blue cap from blue back, but Mosque Swallow much larger, with more measured, flap-and-sail flight. Also resembles Red-breasted Swallow *H. semirufa*, but told from it by black (not rufous) undertail-coverts, rufous-buff (not blue) ear-coverts, and buffy (not white) underwing-coverts. Fine striations on breast of some races not usually apparent in field. Good character is caliper-like tail shape. Flight lacks elegance of e.g. Barn Swallow *H. rustica*.

Voice. Tape-recorded (C, CHA, GIB, GREG, LEM, MOR). Not as vocal as many other swallows. ♂ song a sweet, subdued twitter, 1·3–2·3 s long, starting or ending with dry rattle, and with clicks and nasal sounds. Contact call a complaining, sparrow-like chirp, 'djuit' or 'tchreet', given singly or twice in quick succession, often repeated. Also gives long-drawn, nasal 'djuui', 'gultsch' or 'quitsch' like European Greenfinch *Carduelis chloris* (Cramp 1988), and mewing note like kitten or Common Buzzard *Buteo buteo*. Alarm a short, sharp 'ee', 'kit' or 'kier'.

General Habits. A bird of rocky hills and dissected, dry country, in Palearctic much less dependent on human habitations and anthropogenic land than is Barn Swallow *H. rustica*, but south of Sahara common in towns. In N Africa inhabits plains and mountains, sea-cliffs, farmland, pasture and rocky hillsides; sometimes near buildings and ruins of massive construction. In W Africa occurs widely over dissected, lateritic, grassy and wooded savannas, in Ghana particularly near reservoirs in N savannas, and in Nigeria seldom far from 2–3 m high earthen cliffs by erosion gulleys and streams (its main nest sites there); also at salt pans, granitic inselbergs, and over cattle pasture and marshes. In Sudan occurs in open woodland and bushy grassland near low-lying rivers. Common in townships in Ethiopia. In E Africa frequents rocky hillsides at 1200–3000 m, but occurs also in hot lowlands of L. Victoria basin, and down to 500 m near coast. More montane a bird in southern than in N Africa.

Solitary, in pairs, or flocks of up to 20; forages mainly on its own, also with other hirundines, e.g. Barn Swallow, Red-breasted Swallow and Banded Martin *Riparia cincta*; on migration also mixes with Common Sand Martin *R. riparia* and Common House Martin *Delichon urbica*. Flight less elegant than that of Barn Swallow, with flat-winged glides, soaring ascents, slow turns and sweeps, and much less rapid wing-beats; patrols along rock-face or other regular route rather like Crag Martin *H. rupestris* or Rock Martin *H. fuligula* (Cramp 1988). Lands on ground, to collect mud for nests and evidently (rarely) to feed. Commonly attends grass fires. Perches on telegraph wires and small leafless trees. Roosts in reeds, with Barn Swallows and Common Sand Martins (Mali), also in tall trees along rivers, and in breeding season in nest. Flock of 900 hirundines including 300 Red-rumped Swallows once found on ground in morning after heavy rain (Sierra Leone).

Arrives Morocco late Mar to early May and departs Oct; pronounced migration across Straits of Gibraltar, Apr and Sept; spring passage Tunisia late Mar to late May (Thiollay 1977, Riols 1978), and in Egyptian Western Desert late Mar to mid-Apr. Ghana, occurs in N savannas at Bolgatanga Apr–Sept and at Mole July–Aug. Nigeria, occurs at Abeokuta in SW in Apr–May, and in N less common in May–July than in Aug–Apr. Similarly at Jebel Marra, Sudan, birds move north in Apr–May and south in Oct (*domicella*: Lynes 1925). Bird ringed in Nairobi, Kenya, recovered 2 months later 240 km to WNW.

Food. Airborne insects. Not studied in Africa, but in France 25-day nestlings were fed 255 items, 94% winged ants and 6% other Hymenoptera, Diptera, carabid and curculionid beetles and a cockroach (Prodon 1982). Recorded eating flies at camel dung (Arabia), taking termites as they emerged from ground (Nepal), and settling on *Artemisia* to take insects from them (USSR).

Breeding Habits. Monogamous; breeds solitarily or in colonies of 3–4 pairs (and 20–50 pairs in Near East). ♂ defends territory around nest from other ♂♂, independent young, and other swallows including Barn Swallows (Zaïre). Birds form pairs shortly before nesting; in courtship, 2 birds fly alongside each other with synchronized, shallow wing-beats; ♂ also flies slowly, singing, from time to time vibrating wings, then dives steeply with wings closed (Cramp 1988). Double-brooded. Pair breeds in long and short rains each year (Kenya: Dyson 1976). One nest used for 5 years (Cyprus).

NEST: bowl attached usually beneath horizontal surface, with entrance tunnel attached similarly, made of mud pellets reinforced with dry grass stems, bowl lined with plant down, wool and feathers. Bowl (n = 132, Spain: de Rebollo 1980) externally 7·6–22·4 (18·1) long, 4·7–22·1 (17·5) wide, 6·2–15 (10·4) high; tunnel (n = 128) 4·9–14·2 (10·0) long, 4·5–9·9 (7·1) wide and 4·3–8·2 (5·7) high. Nest weight (n = 19) 712–1410 (975) (Cramp 1988). Built by ♂ and ♀ in 8–15 days. Sited in culverts and concrete drainage pipes, under bridges, stone and masonry eaves, and overhangs in sheer rockfaces; in roof of cave, also brick buildings and a verandah (Malaŵi).

EGGS: 4–5, av. (n = 65, Spain) 4·5. Subelliptical; slightly glossy; white with fine red-brown speckling. SIZE: (n = 237, Spain, Morocco) 18·2–22·5 × 13·1–15·5 (20·3 × 14·2). WEIGHT: 2·01.

LAYING DATES: Morocco, late Apr–July; Senegambia (Gambia, Casamance), Feb–May, Oct–Nov; Ghana (nest-building near coast, Mar–Apr); Nigeria, Oct–Jan (Zaria, Kari, Plateau Province), July–Oct (Kainji); Ethiopia, Jan, Apr–Aug, perhaps Dec and Mar; Zaïre (nest-building Luofu, Mar, Lendu Plateau, Aug); E Africa: Region A, Mar–July, Region B, Mar–Apr, June, Region C, Sept, Dec–Mar, Region D, Feb–Sept; Malaŵi, July, Dec–Jan. Breeds in the rainy season.

INCUBATION: by both sexes, ♀ more than ♂, beginning with last egg or 1–2 days after clutch completed. Period: (n = 21, Spain) 13–16 (14·5) days.

DEVELOPMENT AND CARE OF YOUNG: av. weight 1·8 on 1st day and 20·9 on days 20–21 (Cramp 1988). Fed by both parents; av. 9·4 nest-visits per h on day 5, 22·8 visits on day 10, and 24·5 on day 15 (Bulgaria). When 24 days old, young given live insects *c.* 14–20 mm long. Fledging period 26–27 days, occasionally 22–25 days (Prodon 1982). After leaving nest, young accompany parents for 8–9 days and are fed by them for at least 5 days.

BREEDING SUCCESS/SURVIVAL: of 558 eggs (Spain) 80% hatched and 74·4% survived to fledging of young. 2nd clutches survive rather better than 1st ones, and occasional 3rd clutches better than 2nd ones (de Rebollo 1980). Oldest ringed bird 8 years (Nairobi; ringed as adult, recaptured alive after 7 years 10 months: Campbell 1977).

References
Cramp, S. (1988).
Prodon, R. (1982).
Rebollo, F. de L. (1980).

Plate 9
(Opp. p. 160)

Hirundo fuliginosa (Chapin). **Forest Swallow; Dusky Cliff Swallow. Hirondelle de forêt.**

Lecythoplastes fuliginosa Chapin, 1925. Ibis, p. 149; Lolodorf, Cameroon.

Range and Status. Endemic resident or intratropical migrant, West African forests. E Nigeria (Oban Hills Forest Reserve: Ash 1990); W Cameroon, known from 9 localities north to Yoko and west to Mt Cameroon; Equatorial Guinea (2 localities) and N, central and E Gabon. In Gabon an irregular migrant, abundant on passage at Makokou in Jan 1963; 4 at M'Passa, Mar 1977; juvenile netted July 1963 (Brosset and Erard 1986). Not uncommon but poorly known. Sight records from Gambia (Jensen and Kirkeby 1980) almost certainly refer to Rock Martins *Hirundo fuligula*.

Description. ADULT ♂: forehead and crown black glossed with blue. Remaining upperparts, wings and tail very dark blackish brown, unglossed or with the slightest gloss on wing-coverts. Chin dark rufous; throat dark rufescent brown; remaining underparts, including undersides of wings and tail, very dark brown. Breast washed with rufescent brown. Breast and belly unglossed, but remiges and rectrices somewhat glossy below. Upper mandible black; lower mandible paler, with yellowish white base; eye dark brown; legs and feet brownish, grey brown, pale brown or dark brown. Sexes alike. SIZE: (12 ♂♂, 7 ♀♀) wing, ♂ 84–92 (87·3), ♀ 86–91 (89·4); tail, ♂ 40–49·5 (44·9), ♀ 40·4–45·5 (44·0); bill to skull, ♂ 7·8–9·3 (8·2), ♀ 8·9–9·8 (9·1), bill to feathers, ♂ 5·4–6·5 (6·1), ♀ 5·4–7·1 (6·1); tarsus, ♂ 9·2–11·2 (10·1), ♀ 9·9–11·2 (10·6). (7 more unsexed birds fall within these ranges.)

IMMATURE: labels of some skins state that in life gape is yellow and base of lower mandible dirty white. These may be immature birds, in which case immature plumage is otherwise indistinguishable from that of adults.

Field Characters. A small, not usually very gregarious, uniform blackish brown, rain forest swallow with rusty brown throat. Tail square or slightly forked. Rather silent. The only all-dark African swallow except for saw-wings *Psalidoprocne* and African River Martin *Pseudo-*

Hirundo fuliginosa

General Habits. Inhabits clearings in rain forest. Mainly in pairs, but up to 5 together on breeding grounds (Serle 1954), and once abundant at Makokou (Gabon), Jan, when thought to be migrating (Brosset and Erard 1986). A pair in Kumba (Cameroon) roosted at night by clinging upright or aslant on smooth cement outside wall of house under verandah (Serle 1954). Moults between June and Oct, the first 2 primaries falling simultaneously (R. A. Earlé, in press). Roosts in nest at night, at least in breeding season (Chapin 1948).

Food. Unknown.

Breeding Habits. Solitary breeder. Courtship song (see Voice).
NEST: 1 was hemispherical chamber made of mud pellets, with 2 very short spouted openings set diametrically opposite each other and close to rock surface; chamber floor with pad of white vegetable down; fixed to horizontal roof of cave at base of 15 m cliff flanking forest ravine (Serle 1954). Other nests found 'under (between) rocks in virgin forest' (Cameroon: R. A. Earlé, in press), and fixed to ceilings in houses (Chapin 1948). 2 separate nests near Nkoebissay (Cameroon) were in immediate proximity of breeding pairs of Grey-necked Rockfowl *Picathartes oreas* (R. D. Etchécopar, pers. comm.). 1 nest, built by ♂ and ♀, took 2 weeks to construct; birds began lining nest with vegetable down whilst still adding mud pellets; twice, at this stage, 4 birds slept in nest (Chapin 1948), which suggests cooperative breeding.
EGGS: 2–3. Ovate; smooth, glossless; immaculate white (or white very sparingly and lightly spotted with pale orange-brown). SIZE: (n = 13) 17–20 × 12·5–13·2 (18·9 × 12·9).
LAYING DATES: Cameroon, Apr–June (young Nov, Jan); Gabon (immature July).

chelidon eurystomina. Readily distinguished from last by habitat, bill and eye colour, and from sympatric Black Saw-wing *P. pristoptera* by tail shape. Forest Swallow is very similar to sympatric Square-tailed Saw-wing *P. nitens* and to Mountain Saw-wing *P. fuliginosa* (sympatric on Mt Cameroon only), both of which are square-tailed. Saw-wings have weaker flight, high above clearings; Forest Swallow probably has fast, swooping flight of its genus, low and high in clearings. Square-tailed Saw-wing has grey (not rufous-brown) throat. Mountain Saw-wing is larger, with dark brown throat and grey underwings. Forest Swallow also resembles darker races of Rock Martin which are dark brown rather than blackish, and have pale russet throat and white marks in tail.

Voice. Not tape-recorded. Evidently rather silent. A courting pair, clinging side by side for several s, uttered a soft, sweet song (Serle 1954). Flight call, not often given, a soft 'wheet'.

References
Chapin, J. P. (1925, 1948).
Earlé, R. A. (1989).

Hirundo preussi (Reichenow). Preuss's Cliff Swallow. Hirondelle de Preuss.

Plate 10
(Opp. p. 161)

Lecythoplastes preussi Reichenow, 1898. Orn. Monatsber, p. 115; Sanaga River near Edea, Cameroon.

Forms a superspecies with *H. rufigula*.

Range and Status. Endemic resident and migrant in northern tropics. Locally common from entire Niger R. valley in Mali to L. Chad (Nigeria) and Cameroon; also in NE Guinea (Bafing, Kankan), N Togo, Central African Republic (Lobaye Préfecture, Bangui) and NE Zaïre (at 3° 41'N, 29° 8'E near Dungu R.). In Sierra Leone 2 colonies between Falaba and Gberia Timbako (Tye 1985). In Mali common but local from 17°N to Bamako; Ivory Coast, now known from 8 localities, south to Katiola (Demey and Fishpool 1991); Ghana,

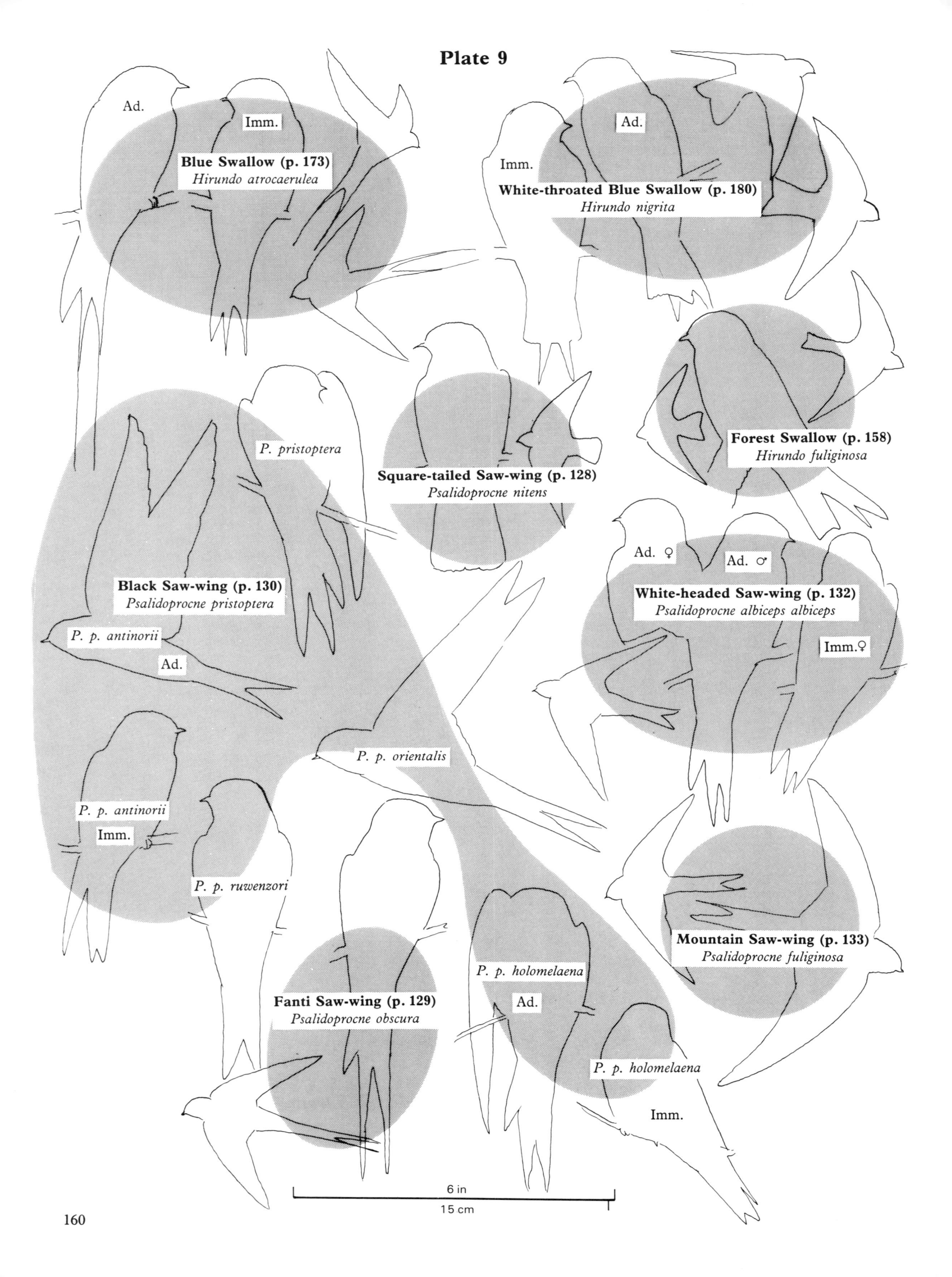

Plate 9

Blue Swallow (p. 173)
Hirundo atrocaerulea
Ad.
Imm.

White-throated Blue Swallow (p. 180)
Hirundo nigrita
Ad.
Imm.

Square-tailed Saw-wing (p. 128)
Psalidoprocne nitens

Forest Swallow (p. 158)
Hirundo fuliginosa

Black Saw-wing (p. 130)
Psalidoprocne pristoptera
P. pristoptera
P. p. antinorii
Ad.
P. p. antinorii
Imm.
P. p. ruwenzori
P. p. orientalis

White-headed Saw-wing (p. 132)
Psalidoprocne albiceps albiceps
Ad. ♀
Ad. ♂
Imm. ♀

Fanti Saw-wing (p. 129)
Psalidoprocne obscura

P. p. holomelaena
Ad.
P. p. holomelaena

Mountain Saw-wing (p. 133)
Psalidoprocne fuliginosa
Imm.

6 in / 15 cm

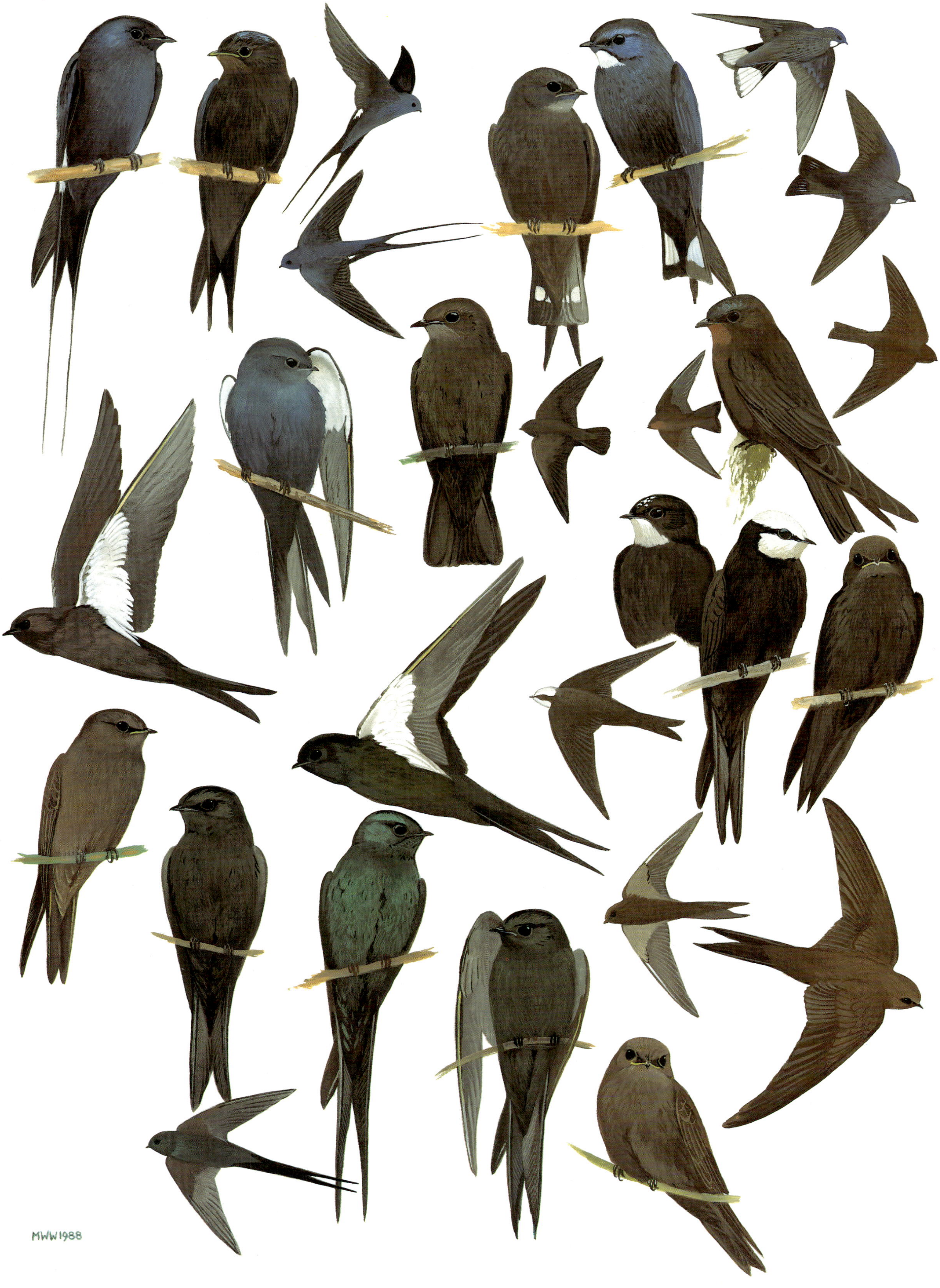
MWW 1988

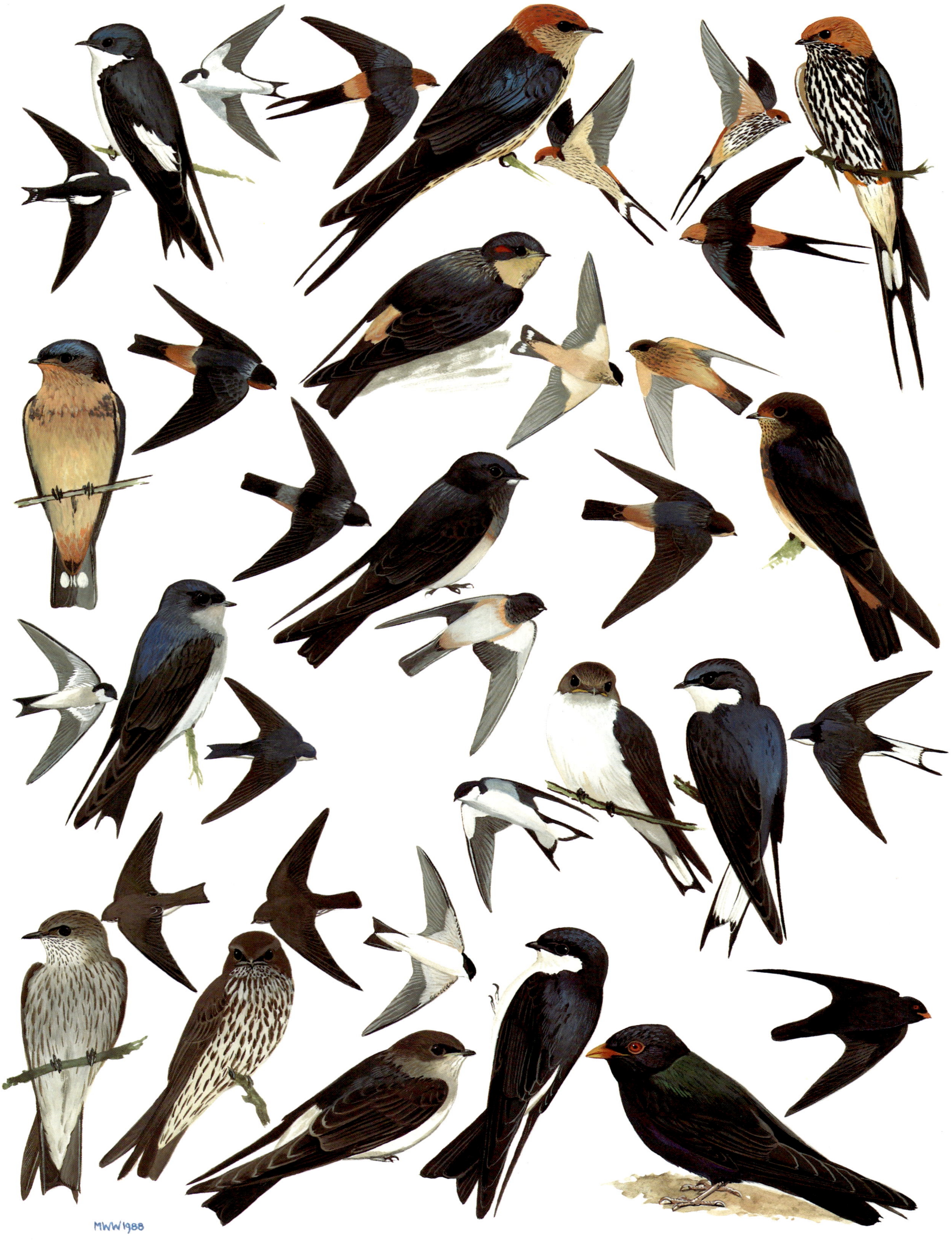
MWW 1988

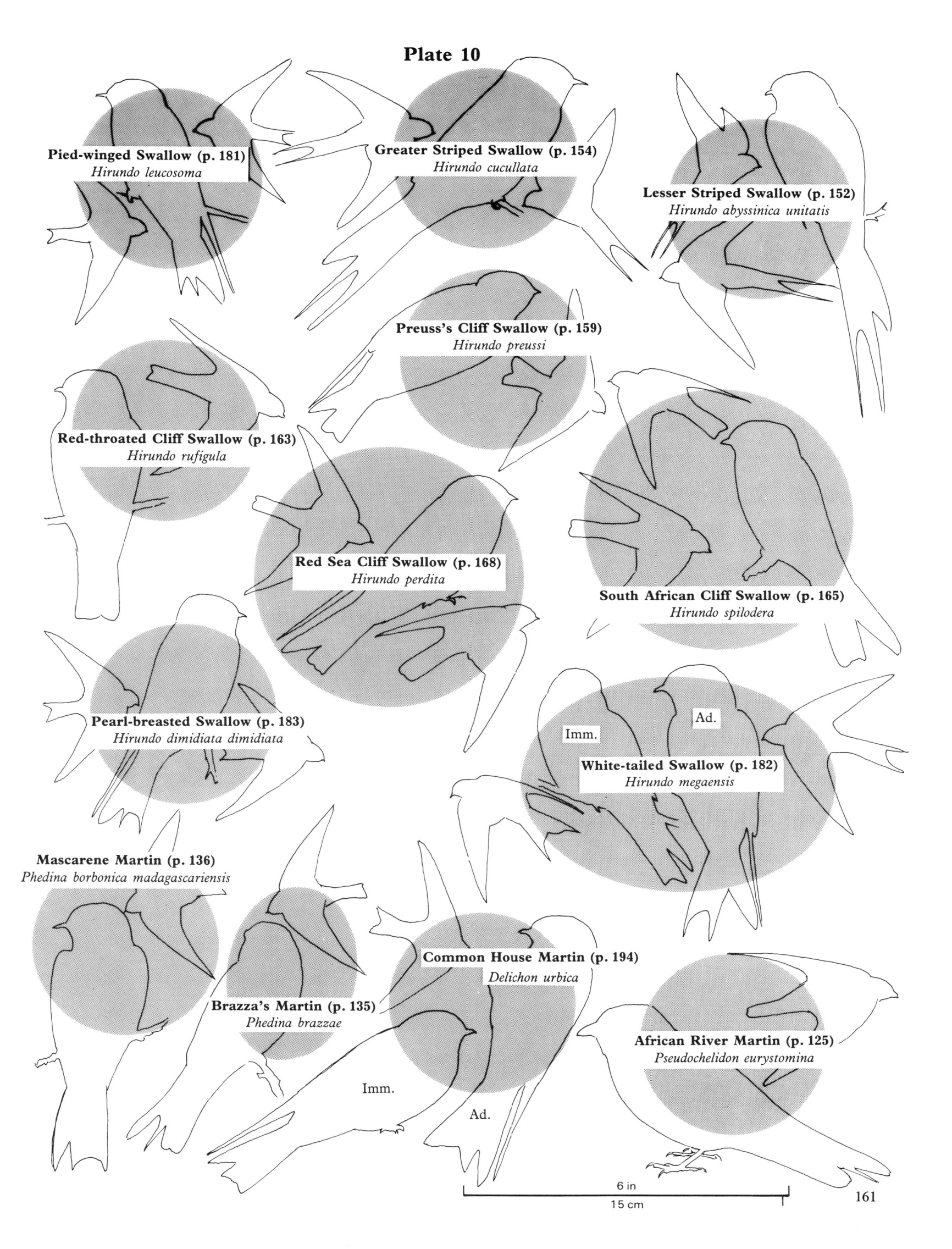

Plate 10

Pied-winged Swallow (p. 181)
Hirundo leucosoma

Greater Striped Swallow (p. 154)
Hirundo cucullata

Lesser Striped Swallow (p. 152)
Hirundo abyssinica unitatis

Preuss's Cliff Swallow (p. 159)
Hirundo preussi

Red-throated Cliff Swallow (p. 163)
Hirundo rufigula

Red Sea Cliff Swallow (p. 168)
Hirundo perdita

South African Cliff Swallow (p. 165)
Hirundo spilodera

Pearl-breasted Swallow (p. 183)
Hirundo dimidiata dimidiata

Imm. Ad.

White-tailed Swallow (p. 182)
Hirundo megaensis

Mascarene Martin (p. 136)
Phedina borbonica madagascariensis

Brazza's Martin (p. 135)
Phedina brazzae

Common House Martin (p. 194)
Delichon urbica

African River Martin (p. 125)
Pseudochelidon eurystomina

Imm.
Ad.

6 in
15 cm

Hirundo preussi

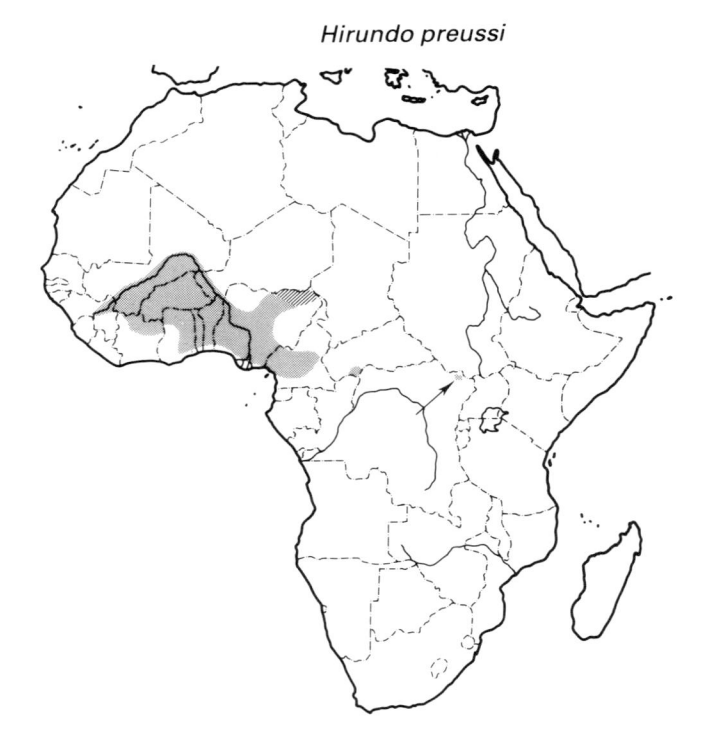

locally frequent along streams with overhanging rock-cliffs north of 8°N, and recorded near coast on Accra Plains and in forest zone; Niger, occurs all year in 'W' Nat. Park, and has bred Niamey airport; Togo, locally common north of 9°N; Nigeria, locally common, and now known from Serti and Obudu Plateau (Ash *et al.* 1989), but breeding colonies sparser in N than in S; Cameroon, colonies at Batchenga, Akonolinga, Garoua-Boulai and Wum. Highly local in SW Central African Republic and NE Zaïre, but probably breeds in both countries.

Description. ADULT ♂: forehead, crown and nape glossy blue-black; mantle, scapulars and back glossy blue-black with irregular grey-white streaks, particularly on mantle where they may form ruff around hindneck. Rump pale buff. Tail and wings blackish brown; tail forked, outer rectrix dark brown with minute pale mark on inner web, next 3 rectrices with large oval white mark on inner web, middle 2 rectrices unmarked black or blackish brown. Lores and areas below eyes buff. Upper cheeks and sides of head blue-black, concolorous with crown and enclosing small acute-triangular dark rufous patch behind eye. Chin to vent uniform buff or pinkish buff, grey feather bases showing through on chin, throat and to some extent on breast, giving these areas mottled greyish appearance. Shorter undertail-coverts very pale buff, longer ones dull blackish with pale buff fringes. Bill black; eye very dark brown; legs and feet brown. SIZE: (9 ♂♂, 8 ♀♀) wing, ♂ 92–99 (95·2), ♀ 93–99 (96·3); tail, ♂ 49–54 (51·5), ♀ 50–54 (52·0), central tail feathers av. 9·5 shorter (♂) and 11 shorter (♀); bill to skull, ♂ 7·5–9·5 (8·9), ♀ 8·7–9·8 (9·1); tarsus, ♂♀ 11. WEIGHT (Cameroon) 1 ♂ 12, 2 ♀♀ 14, 14.
IMMATURE: dark brown where adult is navy blue; crown feathers tipped fulvous; fulvous collar; chin, throat and breast warm pinkish brown; belly silky white with pink wash; undertail-coverts grey-brown.
NESTLING: not known.

Field Characters. A small martin with blue-black upperparts and pale buffy or whitish rump and underparts. Mantle streaked with white (generally visible only in close view, but can be surprisingly conspicuous). Tail forked, black, with white spots visible when bird banks in flight. Wedge of dark rufous behind eye, usually hard to discern. Highly gregarious, sometimes flocking with similar-looking Common House Martin *Delichon urbica*. Voice quite like that of Common House Martin, from which distinguished by buffier or greyer rump and underparts, less cleanly pied appearance, streaked mantle, whitespotted tail and rufous patch behind eye. Not unlike juvenile and tail-moulting adult Grey-rumped Swallow *Pseudhirundo griseopyga*, but much more sociable. Nests sometimes indistinguishable from those of Lesser Striped Swallow *Hirundo abyssinica* (Serle 1940).

Voice. Tape-recorded (CHA). Flight call 'prrp-prrrp.' When alarmed near nesting site, flock falls silent, then a few birds emit loud 'pseep' (Ashford 1968).

General Habits. Inhabits savannas with streams, woods and cultivation, spending about half of year near nesting site, generally a bridge. Highly gregarious, flocking (often densely) at all seasons. Occurrence sporadic, with flocks of up to 1000 suddenly appearing near rivers, inselbergs or towns, then not seen again; even when nesting, birds disappear *en masse* for hours at a time. Forages low down but also commonly very high, bird barely visible even when viewed with 10 × binoculars (probably accounting for 'disappearance'). Feeding flocks mix with Common House Martins, also with Barn Swallows *H. rustica*, Lesser Striped Swallows, Common Sand Martins *Riparia riparia*, Fanti Saw-wings *Psalidoprocne obscura* and Mottled Swifts *Tachymarptis aequatorialis* (Tye 1985). Known to alight on bare ground, and collects mud for nests from ground. Sometimes roosts with other swallows; also roosts in own half-built nest (Serle 1965).

After breeding disperses widely; in Nigeria the only Aug–Sept records are all from extreme NE, where species is frequent as rains migrant in June–Sept.

Food. Unknown.

Breeding Habits. Densely colonial nester; breeding colonies in Mali of 1000 (Ansongo) and 1500–2500 nests (Bamako); in Sierra Leone 50, Ghana 100, Niger 100, Togo 20, Nigeria 50, 100, 150 and up to 400 nests, and in Cameroon 200 (Wum) and 500 nests (Lom R.).

Breeding protracted; at 1 Nigerian colony of 65 old nests (early Jan), repair and new building started in late Jan, 93 nests and egg-laying in Mar, 150 nests in June, and some nestlings still being fed in late July (Ashford 1968).
NEST: flask-shaped chamber of mud consisting of roughly hemispherical egg-chamber and long, downward-

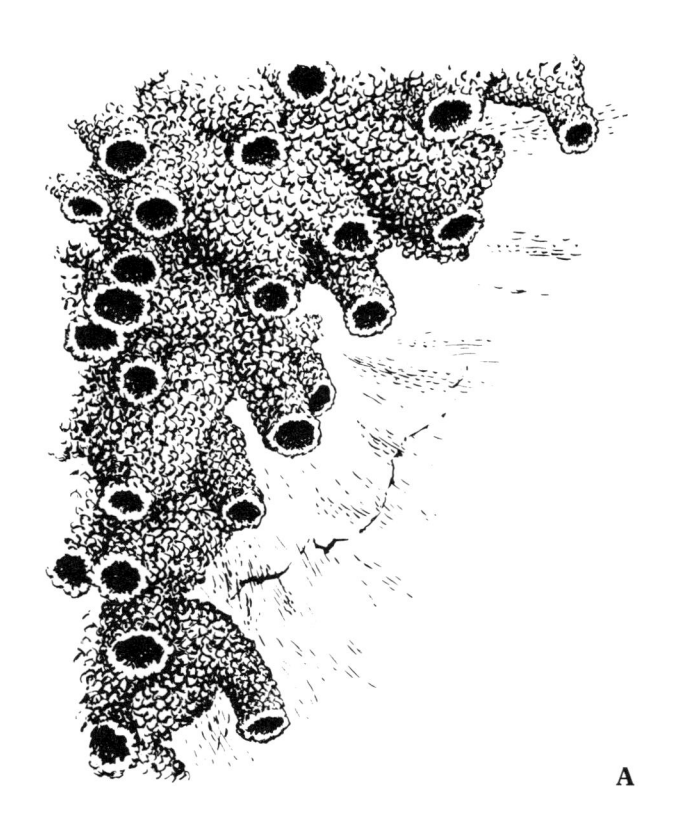

pointing entrance tunnel (**A**). Chamber scantily lined with dry grasses, feathers, hirsute seeds, and sand loosened from walls. External diam. of chamber 12–20 cm; funnel leaves chamber horizontally then curves downward (sometimes sideways); funnel 2–15 cm long, diam. of external opening 30–40 mm (Serle 1940) or 45–50 mm (Ashford 1940). Nests packed together on under surface of overhanging rock, often above water (Ghana: Grimes 1987); those under arcade surrounding house were plastered in angle between wall and corrugated iron roof (Chapin 1953). In Nigeria breeds mainly under concrete bridges over water, but also high on cliffs flanking Niger R., in concrete culverts under railway, under thatched roof of native hut, and on ceiling of 3 m high tunnel at Oyo dam. Nest usually built under horizontal surface, but sometimes on vertical cliff or mud wall of house; on wall below eaves nests placed in 2 tiers (Serle 1965).

Most nests are contiguous and sometimes so densely packed that they can be counted by the funnels but not by the chambers. 150 nests in 1 colony were in 7 clusters spanning 5·5 m of concrete tunnel, the largest cluster of 70 nests. Nest evidently built often 2 months before laying begins (Nigeria). At 1 colony mud was gathered at lakeshore 100 m away, but only from small patches in 1 m^2 area of fine, wet silt; removal of mud forms patch in ground 5–7 cm in diam. and *c.* 25 mm deep (Broadbent 1969). At another colony mud was gathered 20 m away (Tye 1985). Chamber built first, then funnel. Nest built by both sexes, often working simultaneously. Bird arriving at nest expresses blob of mud from each side of gape, and works it onto nest with bill. From time to time all nest-building birds leave colony in a flock, returning later. A few pairs of Lesser Striped Swallows sometimes nest in same colony.

EGGS: 2–3, av. (n = 26 clutches, Nigeria, Feb) 2·08, av. (n = 10 clutches, Nigeria, May) 2·9; all of 9 clutches (Cameroon) were of 3 eggs. Ovate, long ovate or broad ovate; smooth, glossy; ground white with pale violet shell marks. 2 types: speckled type finely, closely and evenly speckled with pink-brown or red-brown; spotted type has sparser, bolder red-brown markings forming ring or cap (Serle 1940). In clutch of 3, one egg often much more lightly marked than other 2. SIZE: (n = 23, Nigeria) 15·3–18·7 × 12·2–13·4 (17·3 × 12·7); (n = 54, Nigeria) 17–21·5 × 11·9–13·4 (18·8 × 12·9); (n = 27, Cameroon) 16·9–20·2 × 11·9–13·1 (18·6 × 12·6).

LAYING DATES: Sierra Leone, May; Mali, thought to breed all year at Ansongo and Nov–July at Bamako (Lamarche 1981); Ghana, May–June, Dec–Jan, probably also Feb–Apr; Niger, June; Togo, Mar–Apr, June; Nigeria, Mar–July; Cameroon, Apr–May.

INCUBATION: incubating bird sits close. When breeding colony alarmed, all birds leave, wheel around in compact flock, then suddenly return together, twittering excitedly.

BREEDING SUCCESS/SURVIVAL: parts of huge colony at Bamako (Mali) destroyed every year by rising waters of Niger R. (Lamarche 1981). Colony at Kete Kratchi (Ghana) regularly swept away by floods, but after breeding completed (Grimes 1987).

References
Ashford, R. W. (1968).
Serle, W. (1940, 1965).
Tye, A. (1985).

Hirundo rufigula Bocage. **Red-throated Cliff Swallow.** Hirondelle à gorge fauve.

Hirundo rufigula Bocage, 1878. J. Sci. Nat. Lisboã 6, p. 256; Caconda.

Forms a superspecies with *H. preussi*.

NOMENCLATURAL NOTE: see note (2) under *Hirundo fuligula*.

Range and Status. Endemic, resident and intratropical migrant: central and N Angola and adjacent parts of Congo, Gabon, Zaïre and Zambia. Locally on central Angolan plateau from N Huila and Huambo to W Malanje

Plate 10
(Opp. p. 161)

and S Lunda; from Caconda and Pungo Andongo north to Congo R. below Matadi (Cabinda) and Tchibanga and Mouila (Gabon); northeast to Lubilashi and Sankuru rivers (Zaïre) and southeast to Kitwe (Zambia). Locally common in Angola. In Zambia probably spreading; in early 1970s known only from Chikata Rapids in Kabompo and at Kitwe (300–400 nests in 1972), but by late 1970s established in 6 localities: Kitwe, Chikata Rapids, Chingola, Ndola, Solwezi and Mwombezhi R. (Bowen 1979a, Taylor 1979); large numbers seen on Mundwiji Plain, Mwinilunga, May. Evidently colonized S Gabon in 1970s, with construction of bridges over Nyanga and Ngounié rivers (Christy 1984).

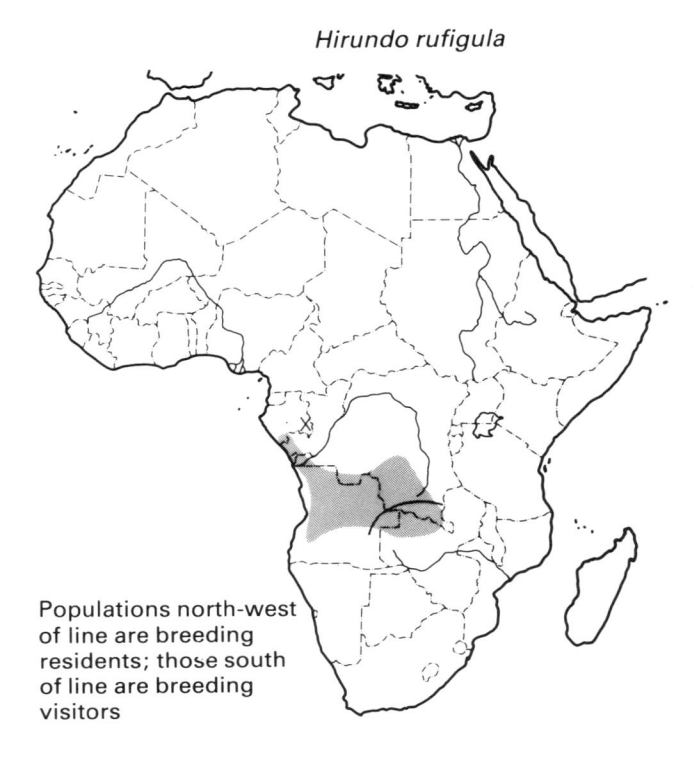

Hirundo rufigula

Populations north-west of line are breeding residents; those south of line are breeding visitors

Description. ADULT ♂: forehead, crown and nape glossy blue-black with narrow line of buff feathers at base of nostrils connecting with lores. Mantle, scapulars and back glossy blue or blue-black, with irregular white streaks, particularly on mantle where they may form ruff around hindneck. Rump rufous, sharply demarcated from back and tail. Tail and wings blackish brown, tail shallowly forked; outer feather dark brown with small pale mark on inner web; next 3 feathers with oval white mark on inner web; middle 2 rectrices unmarked black or blackish brown. Chin, throat and upper breast rufous; lower breast, belly and flanks pinkish buff or very pale orange; shorter undertail-coverts rufous, longer ones blackish with pale buff fringes. Bill black; eye red-brown; legs and feet dusky. Sexes alike. SIZE: wing, ♂ (n = 3) 98–99 (98·7), ♀ (n = 6) 94–98 (96·3); unsexed (n = 43) 90–99 (95·8); tail, ♂ (n = 3) 48–52 (50·3), ♀ (n = 6) 47–52 (50·0), unsexed (n = 40) 50–59 (54·2), central tail feathers av. 7·0 (♂, n = 3), 8·5 (♀ n = 6) and 8·4 (unsexed, n = 26) shorter; bill to skull, unsexed (n = 41) 7–10 (8·3); tarsus, unsexed (n = 41) 12–14 (12·7).
IMMATURE: dark brown where adult is navy blue; belly silky white with pink wash; rufous parts paler than adult.
NESTLING: not described.

Field Characters. A small martin with blue-black upperparts, rufous rump, chin, throat, upper breast and undertail-coverts and pale buff belly. Mantle streaked with white (visible only in close view). Tail blackish with concealed white spots, shallowly forked. Overlaps, and easily confused with, South African Cliff Swallow *H. spilodera*, which is larger, with rufous forehead, and whitish throat heavily speckled with black.

Voice. Tape-recorded (GIB, STJ). Call, 'prrrt' and 'tchirrup', rather like Common Sand Martin *Riparia riparia* (Tree 1964); also described as grating, hoarse 'tre-tre-tre' with accent on 1st syllable (Christy 1984).

General Habits. Feeds gregariously over farmland, sewage ponds and open dry country, sometimes mixing with other hirundines including Lesser Striped Swallows *Hirundo abyssinica* and Grey-rumped Swallows *Pseudhirundo griseopyga*, also with Little Swifts *Apus affinis*. Forages up to 30 km from breeding site. Habits very like those of Preuss's Cliff Swallow *H. preussi* (Chapin 1953).

Migrants present in Zambia early Apr to early Dec, mainly May–Nov.

Food. Unknown.

Breeding Habits. Densely colonial nester; one colony of *c.* 200 birds (Dean 1974); another of 'several hundred' birds (Chapin 1953); Zambian colonies of *c.* 20, 25, 50, 80, 100, 140 and 300–400 birds (Taylor 1979), and Gabon ones of *c.* 100 and 400 birds. About 60 birds copulated gregariously on ground, ♂♂ flying above perched ♀♀ and landing on their backs (Christy 1984).
NEST: evidently much like that of Preuss's Cliff Swallow and South African Cliff Swallow q.v.; a half-sphere of mud with wide, short entrance spout, lined with fine grass, without feathers. Nearly always built under man-made structures, usually concrete road bridges. Building in a given colony lasts at least 5 weeks (Zambia: Tree 1964). Old nests often taken over by Little Swift (Dean 1974) and White-rumped Swift *Apus caffer* (Hall 1960, Cannell 1968). White-rumped Swifts also expropriate partially-built nests, lining them untidily with feathers; 8–9 swallow nests expropriated by swifts in one colony of *c.* 70 nests (Tree 1964).
EGGS: 1–2, usually 2. Immaculate pinkish white, or white speckled with reddish, rufous or violet, sometimes forming cap at broad end. SIZE: (n = 5) 19–20 × 12·5–13·5 (19·5 × 12·7).
LAYING DATES: Angola, July–Oct (nest-building May); Zaïre, Apr (Boma), June (Kouilou); Zambia, Sept–Oct.

References
Christy, P. (1984).
Tree, A. J. (1964).

Hirundo spilodera Sundevall. South African Cliff Swallow. Hirondelle sud-africaine.

Hirundo spilodera Sundevall, 1850. Oefv. K. Sv. Vet.-Akad, Förh. 7, p. 108; Valsch River, Orange Free State.

Forms a superspecies with *H. perdita*.

Range and Status. Endemic; migratory. Visits South Africa Aug–Apr, breeding commonly mainly between 25° and 31°S and 24° and 31°E, and sparsely in central Cape Province west to 20°E. Bred for a few years near Bulawayo (Zimbabwe) and near Windhoek (Namibia) in early 1960s. Distribution depends on availability of nesting sites, which are nowadays almost all artificial; range has extended through E Cape with establishment of towns and construction of large buildings and bridges. Major extension of range to Namibia, Zimbabwe and west in Cape Province in and after the very wet summer of 1961–62 (Rowan 1963). Largest colonies, 975 and 1306 birds (Willows, Driekloof, Earlé 1987b). Winters in W Zaïre. Vagrant Malawî (Medland 1985), E Gabon (Leconi): D. E. Sargeant, pers. comm.

Description. ADULT ♂: forehead and forecrown dark brown; small dark rufous feathers above nostrils; hindcrown and nape dark greyish brown with slight bluish gloss. Lores rufous; cheeks and sides of head dull black; neck and sides of neck blue-black. Mantle, back and scapulars blue-black with irregular greyish white streaks, especially on mantle; rump rufous; uppertail-coverts blue-black, fringed rufous. Tail and wings blackish brown with slight gloss on exposed tips of primaries and tertials. Chin and upper throat pale rufous, variably but usually heavily covered with minute black spots which become more contiguous towards cheeks and sides of breast; breast rufous with coarse, rather diffuse, blue-black spots; lower breast pinkish or buffish white with irregular rufous marks; belly and flanks very pale buff or pinkish white, with rufous streaks on flanks; undertail-coverts rufous or (longest ones) blue-black with broad rufous fringes. Underwing-coverts pale rufous; undersides of remiges and rectrices silvery blackish. Sexes alike; ♂ and ♀ about same size, but largest individuals are ♀♀ (Earlé 1985d). SIZE: wing, ♂ (n = 226) 105–118 (111), ♀ (n = 136) 105–119 (112); tail, ♂ (n = 172) 45–56 (50·5), ♀ (n = 99) 45·4–57·2 (51·5); bill to feathers, ♂ (n = 47) 7·4–9·2 (8·1), ♀ (n = 29) 7–9·1 (8·1); tarsus, ♂ (n = 127) 11·7–16·8 (13·9), ♀ (n = 77) 12·3–17·8 (13·8). WEIGHT: ♂ (n = 500) 16–24·2 (20·4), ♀ (n = 500) 16·5–26 (20·7). Birds (unsexed, n = 50) av. lightest at dawn (20·3) and heaviest at dusk (22·5) (Earlé 1986c).

IMMATURE: upperparts browner than adult, lacking blue gloss. Rump pale pink at fledging, becoming very pale buff or creamy whitish within a few weeks. Underparts not conspicuously mottled or streaked – rather, chin and throat pale pinkish buff becoming dark grey-brown towards cheeks and on upper breast; lower breast pale buff; belly white. Late fledglings have decided pinkish wash to breast and belly; bill yellowish, turning black 6–8 weeks after fledging (Earlé 1985d).

NESTLING: hatchling blind, pink, with yellow gape and orange inside of mouth. At 5 days has short, very sparse white down on head and back, and very swollen gape-flanges. Feathered nestling like immature but retains down until it flies (Earlé 1985f), and dark feathers buff-fringed.

Field Characters. A small, gregarious swallow with square, unmarked tail, blue-black upperparts, sooty crown, rufous or buffy rump, and buffy white underparts and wing linings. Forehead and lores pale rufous; mantle streaked whitish; throat and breast variably spotted with black, often heavily (can look dark-throated in field); undertail-coverts rufous. More likely to be mistaken for Rock Martin *Hirundo fuligula* than blue *Hirundo* swallow; but combination of blue-black upperparts, unspotted square tail, and speckled pinkish breast is diagnostic. On wintering grounds (Zaïre) rufous forehead and lores and lack of white spots in tail distinguish it from sympatric Red-throated Cliff Swallow *H. rufigula*.

Voice. Tape-recorded (88, F, GIB, GIL, EAR). Lacks any long-duration song. Adult has 8 distinct calls (Earlé 1986d): (1) chatter-call of series of warbling or twittering notes, 'chor-chor-chor-chor', duration 0·4–0·5 s, given by several birds together at all times of day; (2) threat, the same but notes quicker and harsher, duration up to 1·2 s; (3) nest-relief, 2 simultaneous warbles, at 2 and 4 kHz, repeated 5–10 times, each lasting 0·4–0·5 s; given by both sexes, usually on wing, also at nest; (4) low-intensity alarm, a double 'chick' of 0·05 s duration, composed of one element at 2–3 kHz and another at 3–4 kHz (thought to be produced by opposite sides of syrinx); (5) high-intensity alarm, usually given immediately after low-intensity alarm call, a note of 0·025 s duration at 4–5 kHz; (6) distress, 2 simultaneous harsh notes; (7) contact, 1–2 whistles, duration 0·12 s, with up to 7 whistles in < 1 s, followed by pause of a few s. Nestling has food-soliciting call; and contact call like (8) but raspy.

Experiments show strongly developed homing ability (Earlé 1987a).

General Habits. Breeding habitat is airspace around large bridges and buildings, small culverts and quarries in flat grassveld and semi-arid karoo areas. Wintering habitat unknown.

Highly gregarious; in flocks of 10–1000. Forages on wing at mean 2·2 m above ground (n = 202 measurements), 44% within 1 m of ground, 10% >6 m high. Flushes prey from vegetation by hovering 2–3 cm above bush, then flies back and forth above bush seizing fleeing moths, before hovering over another bush. 300 birds once seen feeding on ground, probably on harvester termites *Hodotermes mossambicus*. Follows ploughs, often settling on earth to pick up insects. Feeds in tightly-knit flock over grazing sheep and cattle and over foraging flock of Helmeted Guineafowl *Numida meleagris* (Skead 1979). Attracted to bush fires to feed on fleeing insects (Bilby 1957). Flight straight and fast with continuous beats, interspersed with short glides when foraging. Foraging flight slower, with twists and turns, than e.g. nest-building flight. On ground, walks. Main comfort movements: 1-wing stretch, wing-and-leg stretch, 2-wing stretch, jaw-stretch, preening on loafing perch, scratching (direct, under wing), wing-and-tail shake, and 'bumble-bee' shake (in flight, after long spell of incubation, bird puffs out trunk feathers and shakes itself with characteristic whirring sound) (Earlé 1985e). Bathes by flying low over still water and falling in head-first, then preening at perch; rain-bathes at perch by continuously shuffling plumage and preening. Flight speed 9·2 m per s (Earlé 1985f).

Roosts in nest all South African summer. Before breeding, colony is deserted for 4 h every morning; after foraging, flock re-forms in air and enters nests all at once. Social activity stimulated by rain, inducing excited chasing. Mud-gathering for nest-building is social, up to 40 birds gathering mud from one small part of large muddy area available. Occasionally feeds socially on ground, 10–50 birds taking termites. Before nest-building commences, bird encountering feather or leaf in air grabs it, and other birds play by trying to take it from its bill; several birds also try to pick up feather or leaf from surface of water (Earlé 1985e).

Small proportion of population starts moulting in South Africa Mar–Apr; species completes moult in Zaïre by late Aug (Earlé 1987b).

Present South Africa Aug–Apr; arrives in Transvaal a few days earlier than in Orange Free State (Broekhuysen 1974). Winters probably mainly in W Zaïre and SE Congo, where known from 8 localities, Apr–Sept; *c.* 30 wintered near Bloemfontein, July 1988, and a few seen in Kalahari Gemsbok Park July 1988 (Herholdt 1989). 4 recoveries in Zaïre of Transvaal-ringed birds, from Feshi, Kansanza, Kobo and Tshikapa, 2311–2528 km from place where ringed, one after only 15 days (Earlé 1987b). Locality and date records from intervening Botswana (Penry 1986) and NW Zambia suggest direct migration between breeding and non-breeding grounds.

Food. Large variety of airborne, grounded and some flightless insects, mainly Coleoptera (particularly Scarabaeidae and Curculionidae), Diptera (particularly Drosophilidae and Muscidae), Hymenoptera and Hemiptera. Muscid flies and *Onthophagus* and barine beetles each occurred in >50% of 142 stomachs (South Africa: Earlé 1985a). Flightless insects include carabid beetles, lepidopterous and beetle larvae. Also spiders, and stones (gastroliths) up to 5 mm diam. Thysanoptera, found in suction-trap samples, are too small (<1 mm) to be eaten by Cliff Swallows (Earlé 1985a).

Breeding Habits. Monogamous, colonial breeder. Defends territory around nest entrance, extent determined by reach of bill (Earlé 1985e).

Courtship and pair-bonding period very short; acquisition of nest-site by ♂ is prerequisite. ♂ clings or squats repeatedly at same spot, defending it by pecking and threatening neighbours; squatting ♂ is harassed by other birds hovering close to it and alighting on its back. 'Squatting' ♂ builds mud ledge (see below), chasing away other birds alighting on it but reacting by wing-quivering to chosen ♀. Paired ♂ and ♀ apparently soon recognize each other's voices. Unmated ♂♂ start to build nest, trying to attract ♀ by occupying it. Threat posture consists of (1) raising crown feathers, (2) calling, (3) wing-quivering, then (4) pecking or chasing intruder. Nest-owning ♂, returning to site and finding intruder, flies directly at it, strikes it on nape or head with its bill and clings to its back with its claws. Fighting birds may fall to the ground, where attacker may try to copulate with its victim. Fight lasts up to 5 min and attracts other birds which all attempt to mate. Pairs show very little aggression to birds in adjacent nests (Earlé 1985c).

NEST: hemispherical chamber of mud, containing pad of wool, grass, plant down and sometimes feathers (Earlé 1985c). Sited in clusters on vertical surface or under overhang on concrete bridges, rock-cliffs, buildings, large culverts, water towers and quarries, often over water. Chamber has short spout, with entrance facing obliquely down. Nest made of *c.* 1300–1800 mud pellets, gathered 6–300 m away. When gathered 20 m away, pellets added to nest at rate of 1 per min, bird spending 4–12 s (av. 11·4, n = 64) at mud-gathering site. Collects nest-lining material by hovering above flower-heads of Compositae and pecking at seed plumes; collects feathers opportunistically on wing; takes sheep's wool from barbed-wire fences. Dimensions: total length (n = 6) 140–210 (184); outside depth (n = 8) 77–126 (90·5); inside length (n = 7) 109–154 (131); inside depth (n = 7) 65–110 (81·9); inside width (n = 7) 111–155 (133); entrance width (n = 11) 25–37 (29·7); entrance height (n = 11) 34–57 (43·0); mud mass (dry, n = 9) 731–979 (840) g. Nest built by both sexes, stimulated by rain which forms puddles. 2 nests took 5 and 7 days to build.

Building is in 6 stages: (1) mud pellets applied to vertical surface at position of bird's foot-hold, to form long horizontal ledge, (2) nest-base formed on ledge, (3) nest-body formed on base, (4) upcurving side-walls added, (5) roof and entrance tunnel added, and (6) soft pad added in depression on floor of mud bowl (illustrated in Earlé 1985c). Undamaged nests used repeatedly, year by year. Slightly damaged nests immediately repaired during current nesting, and may be repaired and re-used next year; nests with large breaks have new entrance added. Nest is built in deep crevice merely by closing mouth of crevice with mud (containing entrance spout).

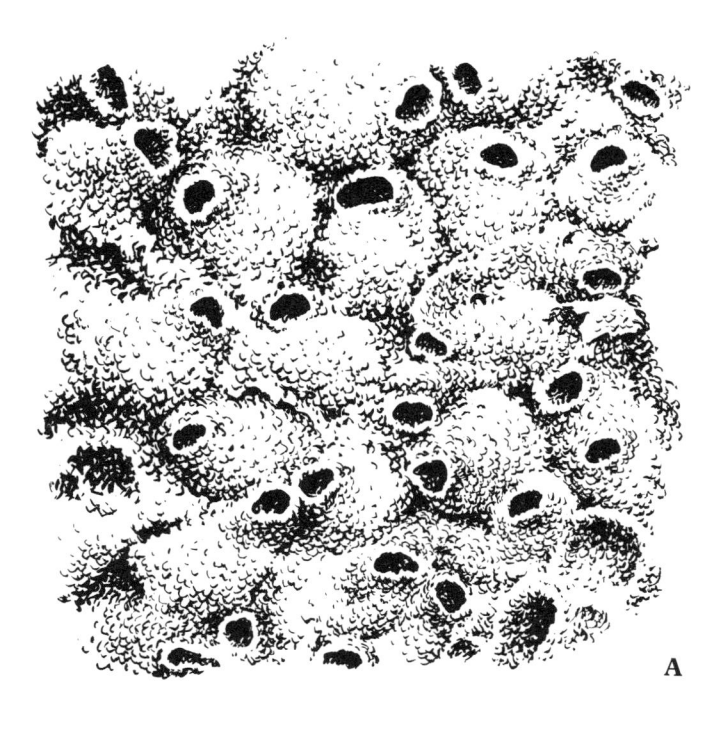

Contiguous nests with entrances close together have entrances facing in different directions (**A**), so that adult birds perching at each entrance are not in mutual pecking range. 1 nest had 2 entrances. As with bee-eaters, c. 16% of nests (n = 177 clutches) are parasitized by other ♀♀ in the colony (Earlé 1985f).

EGGS: 1–4, usually 3, av. of 455 clutches at one colony 2·3 eggs, and of 230 at another, 2·7 eggs. Overall av. (n = 1226 clutches) 2·4 (Earlé 1985f). Laid at 24 h intervals, at night or in early morning. Subspherical; white, spotted and blotched with red-brown, purple and grey, mainly at broad end. Small proportion of pairs have second broods. SIZE: (n = 199) 18·3–22·9 × 13·1–15·2 (20·3 × 14·1). WEIGHT: (n = 26) 1·5–2·6 (2·2).

LAYING DATES: Zimbabwe, South Africa, Sept–Mar.

INCUBATION: starts with 1st egg (parent roosts in nest at night in any event). Period: (n = 16) 14–16 (14·6) days. Incubation by both sexes, for av. bout of 9·2 min. Sitting bird leaves only when mate arrives and clings to nest. Feathers brought to nest during (and after) incubation period.

DEVELOPMENT AND CARE OF YOUNG: eyes start to open at day 4, fully open by day 9. Primary feather sheaths break open on days 9–10 and rectrices on days 11–12. Weight increases nearly linearly from 1·5 to 3 g at day 1 to as much as 31 g about days 19–22, then declines to 23 g at day 24 when nestling flies. Period: 24·55–25·42 days (Burgerjon 1964), or av. 24·1 days (Earlé 1985f); chick can fly at 23 days when wing is 93 long. Chicks fed by both parents equally; food always carried in bill (not regurgitated by parent); 2–9 feeds per young per h, highest in afternoon. Fledglings do not return to nest for at least 4 days after leaving. Buff feather-fringes on back soon lost (but tertials remain buff-fringed for months). Yellow gape regresses at 4–6 weeks after leaving nest.

BREEDING SUCCESS/SURVIVAL: at 2 large colonies near Bloemfontein, main predators of eggs and chicks falling from nests were yellow mongooses *Cynictis penicillata* (one of which patrolled colony up to 5 times per day) and ants *Techmomyrmex albipes* and *Acantholepis capensis* (Earlé 1985b). Other predators of chicks were Black-headed Heron *Ardea melanocephala* and of fledglings and adults Steppe Buzzard *Buteo buteo*, Spotted Eagle-Owl *Bubo africanus* and Fiscal Shrike *Lanius collaris*. One year the 2 ant species attacked chicks in nest and colony was abandoned. Another colony abandoned after people took chicks for food and wantonly destroyed site. Several nests robbed by monitor lizard (Grobler and Jacobs 1985). Adult birds frequently get jammed in nest entrance and die; others die entangled in sheep's wool nest lining (de Swardt 1988). House Sparrows *Passer domesticus* and Cape Sparrows *P. melanurus* remove swallow eggs, then expropriate nests (Earlé 1985b). Nests also taken over by White-rumped Swifts *Apus caffer* and Little Swifts *A. affinis* for breeding and by Pied Barbets *Tricholaema leucomelaina* for roosting. 3 ectoparasites are specific to South African Cliff Swallows, the flea *Xenopsylla trispinis*, tick *Ornithodoros peringueyi*, and louse-fly *Ornithomya inocellata*; the last may affect chick survival (Earlé 1985b). Many nestlings killed by mite *Macronyssus bursa* (Burgerjon 1964). One bird survived 12 min floating in river, and flew away 2 min later (Penry 1987).

Of 310 clutches, 8% of C/1, 74% of C/2 and 69% of C/3 hatched; of 216 broods (419 nestlings), 80% of broods of 1, 69% of broods of 2, and 81% of broods of 3 fledged; overall success rate 48·8% (303 fledglings from 621 eggs) (Burgerjon 1964). From 2245 eggs in 951 clutches, 1268 young fledged, a 56·5% overall success rate (Earlé 1985f).

References
Burgerjon, J. J. (1964).
Earlé, R. A. (1985a,b,c,e,f, 1986a,b, 1987b).

Plate 10
(Opp. p. 161)

Hirundo perdita Fry and Smith. Red Sea Cliff Swallow. Hirondelle de la Mer Rouge.

Hirundo perdita Fry and Smith, 1985. Ibis 127, pp. 1–6; Sanganeb Light House, Red Sea.

Forms a superspecies with *H. spilodera*.

Range and Status. Known only from a single specimen, found dead at Sanganeb light-house (19° 43′N, 37° 26′E), 20 km northeast of Port Sudan (Sudan), on 9 May 1984. It had evidently been caught up in a major hirundine migration (500 Barn Swallows *H. rustica* and 100 Common Sand Martins *Riparia riparia* per h) and may have originated in Red Sea Hills of Sudan or Ethiopia (Fry and Smith 1985). Pale-rumped swallows, perhaps this species, were seen in June 1985 at 3300 m altitude in Bale Mts (7° 50′N, 39° 22′E), Ethiopia (G. Nikolaus, pers. comm.), and in Nov–Dec 1988 in Awash Nat. Park (6 birds) and around cliffs by L. Langano (20–24 birds), Ethiopia (Madge and Redman, in press).

Description. (From photograph, Fry and Smith 1985; only wings and tail preserved). ADULT: forehead, lores, crown, nape, sides of neck and mantle dark brown or blackish brown with dark steel-blue gloss; gloss appears strongest on mantle (but that might be owing to light incidence in the photograph). Base of lower neck and mantle feathers white or whitish, with some suggestion that some white marks may show in life. Feathers of forecrown and hindneck and, particularly, sides of neck, apparently with small whitish or pale grey patch affecting a few barbs near mid-length of each feather. Scapulars, tertials and rectrices dark brown with faint bluish gloss (no gloss on inner webs or rectrices); scapulars with barely discernible fringed effect. Rump feathers dark grey with white bases and white fringes; most feathers with white fringe (1–2 mm wide near rachis); uppertail-coverts dark glossy brown, one feather looking fresh and with dark blue gloss; lateral coverts with white fringes. Discrete white spot on chin, same width as base of mandible, i.e. *c.* 9 mm, and 3–4 mm deep in midline; a few minute buffy feathers at sides of white spot; throat and upper breast bluish black, most feathers evidently narrowly fringed pale grey or bluish grey; lower feathers of upper breast broadly fringed buff and buff-grey; lower breast and belly silky white, strongly demarcated from upper breast by narrow warm buff line *c.* 1 mm wide; vent pale grey, strongly washed pale orange. Proximal undertail-coverts buffy pale grey, next ones and also lateral ones glossy black or bluish black with very broad (*c.* 6 mm) tip pale grey or buffy grey or pale orange-buff; single distal undertail-covert, projecting *c.* 8 mm beyond others (loose?), glossy black or very dark brown with distinct narrow greyish tip. Wings dark brown or blackish brown, slightly glossy, with suggestion (in some lights) of faintest greenish gloss. Tips of inner greater wing-coverts and of outer lesser wing-coverts a slightly paler brown, giving poorly emphasized dappled effect near wrist. A single inner greater or tertial covert in right wing is strongly glossed dark blue. Underwing-coverts silky white, the marginal underwing-coverts with dark grey bases showing through (giving grey-and-white striped effect), greater under primary coverts mainly plain dove grey. Some middle and most inner underwing-coverts washed very pale orange. Axillaries very pale orange; rather long (tip 12 mm short of tip of tertials). Underside of remiges and rectrices dark greyish brown with very strong pale greyish or silvery sheen, particularly at inner margins of primaries. Remex and rectrix rachis undersides white, more apparent in wing than in tail. Bill, legs, feet and claws black. Wing, 114, 115 mm; 9th primary longest, P10 78 shorter, P8 1, P7 7, P6 15, P5 22, P4 28–30, P3 36 P2 41–42 and P1 47–48 shorter. No emargination or notching. Tail square, not forked. Freed middle rectrix 57 mm long, outer one 59 mm long.

Field Characters. Probably very like those of South African Cliff Swallow *H. spilodera* q.v., from which Red Sea Cliff Swallow differs in having blue-black cap (not dark brown), grey rump (not pale rufous), blackish lores and area around eye (not buffy orange), white chin spot, and black throat and upper breast (not pale rufous, speckled with black). A small, square-tailed swallow, probably gregarious and inhabiting cliffs, with blue-black upperparts, head and throat, grey rump, white wing linings, chin-spot and belly, and rufous-buff undertail-coverts.

Birds seen in Ethiopia in 1988 resembled South African Cliff Swallow more than Red Sea Cliff Swallow: rump pinkish buff or rufous, centre of throat pale, and ill-defined dusky partial breast-band (Madge and Redman, in press). They foraged over grassland with other hirundines and around cliff faces, swirling about with fluttering, frequent turns and gliding, sometimes alighting briefly.

Breeding Habits. Condition of plumage suggests that pre-breeding moult was complete and that bird was about to breed (May). Nothing further known.

References
Fry, C. H. and Smith, D. A. (1985).
Madge, S. C. and Redman, N. J. (in press).

Hirundo fuligula Lichtenstein. Rock Martin; Pale Crag Martin. Hirondelle isabelline.

Hirundo fuligula Lichtenstein, 1842. Verz. Saugth. Vög. Kaffernl., p. 18; Baviaan River, Cape Province.

Plate 7
(Opp. p. 112)

Forms a superspecies with *H. rupestris* and *H. concolor* (India).

NOMENCLATURAL NOTES: (1) Pale races (*obsoleta, pallida, presaharica, spatzi, arabica, buchanani*) are called Pale Crag Martins (Hirondelle du désert), and dark ones (the remainder, sometimes including *pusilla*) Rock (or African Rock) Martins (Hirondelle isabelline).

(2) Readers should beware of confusion between *Hirundo rufigula* Bocage, 1878 (Red-throated Cliff Swallow) and *Cotyle rufigula* Fischer and Reichenow, 1884 (African Rock Martin). The latter name has been widespread in the literature, usually as *Ptyonoprogne* (*Hirundo*) *fuligula rufigula* (e.g. Cramp 1988), but is preoccupied by *rufigula* Bocage and should be replaced by the synonym *fusciventris* Vincent, 1933.

(3) In French, all 3 species have been known as Hirondelle de rochers (e.g. Christy 1984).

Range and Status. Africa, Arabia, Iran to Pakistan.

Common and widespread resident over greater part of Africa, in vicinity of rocky hills, cliffs and gorges, and of buildings and towns, from coast up to 3000 m. Some northern and southern African populations partially migrant. Winter visitor to N Somalia (probably from Arabia), N Senegambia and SW Mauritania. Absent from Tunisia, Guinea-Bissau and Guinea (?), Rio Muni, Congo, Zaïre (except montane E border), Botswana (except far E), and from all islands except Socotra; 1 record Gabon (Leconi): D. E. Sargeant, pers. comm.

Hirundo fuligula

Line shows approximate southern limit of wintering Palearctic populations

Description. *H. f. fusciventris* (Vincent) (= *H. f. rufigula* (Fischer and Reichenow)): S Chad, Central African Republic, S Sudan, SW Ethiopia, Uganda, W Kenya, south through E Africa to 20°S in Zimbabwe and 22°S in Mozambique. ADULT ♂: forehead, crown, nape, sides of neck and ear-coverts dark brown; mantle and scapulars to uppertail-coverts dark brown with indistinct greyish wash. Tail dark brown, square or slightly forked; middle and outer rectrices plain, 4 intervening ones with large white spot on inner web. Lores blackish. Chin and throat pinkish fawn; breast rufescent dark brown; belly, flanks and undertail-coverts dark brown, the last sometimes pale-tipped. Wings dark brown; axillaries and underwing-coverts dark brown, some with rufous inner edges. Bill black, mouth pink-grey; eye dark brown; legs and feet dusky pinkish brown. Sexes alike. SIZE: (Sudan, Kenya, Tanzania, Zambia, N Zimbabwe): wing, ♂ (n = 45) 109–125 (114), ♀ (n = 33) 106–124 (114); tail, ♂ (n = 43) 43–50 (46·3), ♀ (n = 35) 40–50 (46·4), bill, ♂ (n = 13) 7–8 (7·5), ♀ (n = 8) 7–8 (7·5); tarsus, c. 11. WEIGHT: (Kenya) unsexed (n = 4) 10–15 (13·8). Size rather uniform from Sudan to N Zimbabwe (Irwin 1977), but S Zimbabwe birds (Matopos) larger: wing (n = 4) 120–125 (123).

IMMATURE: like adult but tertials, greater, median and lesser wing-coverts, and feathers from mantle to rump, fringed buff at tip.

NESTLING: down light grey, rather long; on inner and outer supraorbital, occipital, spinal, humeral and crural tracts.

H. f. bansoensis (Bannerman): Cameroon and Nigeria west to Sierra Leone and probably lower Gambia R. Like *fusciventris* but even darker brown, particularly on upperparts, which are glossy.

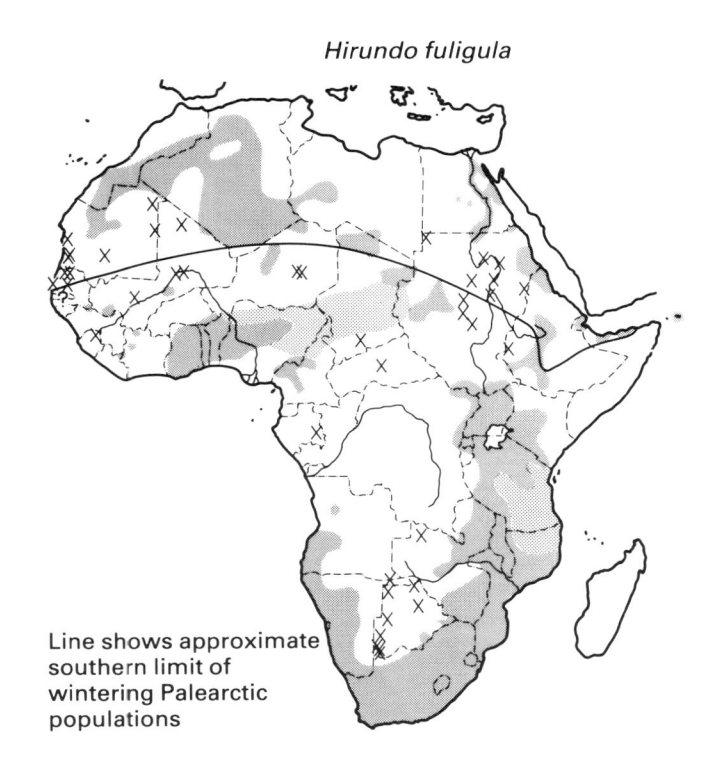

H. f. pusilla (Zedlitz): central Ethiopia and Eritrea west through central Sudan, central Chad and S Niger to S Mali. Like *fusciventris* but browns paler; breast pink-grey (rather than rufescent brown). Upperparts not (or barely) glossy.

H. f. buchanani (Hartert): Niger (Aïr massif). Like *fusciventris* but browns paler and greyer; slightly paler than *pusilla* (same shade as Crag Martin *H. rupestris*). Wing, unsexed (n = 6) 111–115 (113).

H. f. arabica (Reichenow): N Sudan west to Ennedi (Chad), south to N Darfur, north to Gebel Elba, east to Red Sea; coastal Ethiopia (Eritrea), N Somalia, Socotra; also SW Arabia. Paler and greyer than above races; throat pale pinkish, breast pale grey-brown washed with pink.

H. f. spatzi (Geyr von Schweppenburg): S Algeria (Ahagger), Tassili (N'Ajjer), SW Libya (Fezzan). Like *arabica* but paler; underparts with only slight suggestion of pink wash. Wing, unsexed (n = 8) 117–121 (119). WEIGHT: (n = 9) 14–16·5.

H. f. presaharica (Vaurie): Algeria from Atlas Saharien south to Timimoun and El Goléa, west to S Morocco and N Mauritania. Like *spatzi* but paler. Wing, unsexed (n = 9) 117–123 (121). WEIGHT: 16·2, 24·5.

H. f. obsoleta (Cabanis): Egypt (Nile valley except delta, east to Red Sea Hills); also Sinai, N Arabia, Iran. Like *presaharica* but paler. Upperparts paler and greyer than in *presaharica*, underparts off-white (except for dusky undertail-coverts).

H. f. pretoriae (Roberts): S Mozambique (Lebombo range), SW Zimbabwe (intermediate with *fusciventris*) and E South Africa (Transkei, East Griqualand, Natal, Zululand, Swaziland, E Orange Free State, E Transvaal). Paler than *fusciventris*; brown shade above near that of *pusilla* but greyer; chin, throat and breast pinkish cinnamon; flanks, belly and undertail-coverts dark greyish brown. Bill stout. Large: wing, ♂ (n = 15) 127–139 (132), ♀ (n = 12) 127–137 (133); tail, ♂ (n = 15) 50–56 (54·0), ♀ (n = 12) 51–57 (54·0). Birds from SE Zimbabwe (Selungwe Hills) dark above, bright cinnamon below; size between *fusciventris* and *pretoriae* (Irwin 1977).

H. f. fuligula (Lichtenstein): S Namibia, South Africa (except range of *pretoriae* in east), SE Botswana. Like *pretoriae* but fractionally smaller and darker: wing, ♂ (n = 12) 128–135 (132), ♀ (n = 8) 126–135 (131). WEIGHT of southern African birds (race?): ♂ (n = 5) 16·3–24·3 (20·2), ♀ (n = 4) 16–18·3 (17·1) (R. A. Earlé, pers. comm.); unsexed (n = 7) 16–30 (22·4), unsexed (n = 20) 18·5–23·6 (21·1).

H. f. anderssoni (Sharpe and Wyatt): SW Angola, Namibia except Great Namaqualand. Paler above and below than *fuligula*, and (particularly in north of range) greyer and slightly smaller: wing, ♂ (n = 22) 123–138 (131), ♀ (n = 20) 123–139 (130).

Field Characters. A robust grey-brown or dark reddish brown swallow, closely associated with bare crags and buildings. Thick-set, with broad head, trunk, rump, tail and wings; quiet, with graceful but slow and stable flight. Geographically variable in size and colour, with at least 11 races in Africa, in 3 main groups. All have round white spots in tail, usually quite conspicuous as bird banks in sailing cliff-side flight. Saharan populations either very like Crag Martin *H. rupestris* (but somewhat smaller, paler and greyer, with paler underwing-coverts), or much paler, with pale mantle, back and rump contrasting sharply with dark brown wings and tail, and underwing-coverts and undertail-coverts much paler than in Crag Martin. Chin and throat always unstreaked (finely streaked in Crag Martin). Tropical populations small (barely larger than Common Sand Martin *Riparia riparia*), uniformly very dark brown above, rufescent dusky brown below, with clearer pinkish or rufous throat. In poor light bird appears blackish, lighter throat only discernible at close quarters. Southern African birds (SW Angola and S Zimbabwe to Cape) large (size of Banded Martin *R. cincta*; larger than South African Cliff Swallow *H. spilodera*), with uniform dark greyish brown upperparts and rufous or pinkish brown throat and breast. For distinction from Crag Martin, see that species.

Voice. Tape-recorded (88, GIB, HOL, HRS, MOY). Song a soft, liquid, melodious twittering. Flight calls, 'wit', 'wik-wik-wik', rapid high-pitched 'chi' sounds sometimes ending in wagtail-like 'chis-wick' (Cramp 1988). Song and calls little used and very quiet. Bird made clapping noise (with bill?) as it flew close past head of person intruding at nest (Schmidt 1964).

General Habits. Highly aerial, inhabiting warm, sheltered situations from sea level up to at least 3000 m, around inselbergs (kopjes) and craggy hillsides and mountains, in Saharan massifs (common in Atlas, Ahagger, Aïr, Tibesti, Ennedi, Red Sea Hills), desert ravines, and Egyptian monuments (Abu Simbel), in savanna and rainforest zones of W Africa, and in great range of habitats in E and southern Africa. Occurs in villages and towns in forest zone (Ghana, Nigeria), nesting on buildings and rocky outcrops. Large expanses of sheer rock faces are necessary habitat component, particularly for nesting. Nests under small overhangs on perpendicular rock faces, dam walls, tall bridges, occasionally on verandahs or under house eaves, and on high city buildings (e.g. in Harare). Inhabits coasts (probably breeds in Dar-es-Salaam) and sometimes forages over rocks at low tide (South Africa: Skead 1966). Forages mainly within 10 m of cliff or ground; at Amani, Tanzania, all insects were caught by pair within 40 m of their nest and within 10 m of ground (Moreau 1939a). Attracted to grass fires, perhaps from afar, mixing with swifts and other swallows and keeping to edge of smoke. Seldom drinks; does so by swooping low over still water, like other swallows. Becomes inactive in even slight rain.

In pairs, breeding often loosely colonial; occasionally in flock of up to 50. Quiet. Flight leisurely, without swoops and speed-changes of most congeners; spends much time gliding slowly back and forth within few m of cliff face or building, but cruises easily at 80 km per h (Maclean 1985). Active at dawn and dusk; on wing in half-light. Out of breeding season roosts on buildings; hundreds sleep in rows on Cape Town Parliament buildings in winter.

Moult protracted, in all months but mainly in 3 months prior to principal breeding season (E and southern Africa); moult sometimes arrested (Earlé 1988b).

Tropical populations resident. Northern populations partially migratory (e.g. in Oman). *H. f. presaharica* occurs out of breeding range, in Ahagger (S Algeria) in winter; in NW Africa some breeding grounds are vacated Oct–Feb. Scarce winter visitor to N Senegambia and coastal and SW Mauritania. *H. f. arabica* occurs in NE and SE Ethiopia in winter (Jan–Mar), and *H. f. obsoleta* is non-breeding visitor to N Somalia (probably from Arabia or Iran, rather than from Egypt). *H. f. fuligula* has also occurred outside its breeding range, in Zimbabwe (Bulawayo), July (Irwin 1977). Some altitudinal movement in southern Africa; at Lydenburg (Transvaal) occurs in winter only (Tarboton *et al.* 1987).

Food. Airborne insects, including Hemiptera.

Breeding Habits. Nests isolated or in loose colony of up to 40 pairs. Density of *c.* 5 pairs per km² (Cape Town: Siegfried 1968). Nest used repeatedly; broods reared up to 4 times in quick succession, perhaps by same pair. Same site used for years, once for 14 years (Ussher 1944). Pair breeds in long and short rains each year (Kenya: Dyson 1976). Nests are built on rock-faces, and also commonly on rural and urban buildings throughout range. Nested on buildings in South Africa (East London) at turn of century (Batten 1943); 60 years ago species had already 'largely forsaken the rock-dwelling habit for European buildings' in Tanzania (Sclater and Moreau 1933).

NEST: usually a half-cup stuck to vertical surface, also sometimes a full cup on ledge. Built of mud pellets (once 960 pellets) using 1–2 types of mud, with small twigs and fibres stuck to inside, lined with soft vegetable down, also fine grass and feathers. Half-cup nests used repeatedly; whole-cup nests sometimes built on top of old ones. Built by ♂ and ♀ equally, taking up to 40 days (but most of nest built in *c.* 20 days: Brooke and Vernon

1961). Unoccupied nests sometimes taken over by Little Swifts *Apus affinis* (Carr 1984).

EGGS: 2–3, laid at 24 h intervals. Elongated, sub-elliptical; smooth, slightly glossy; white, spotted and speckled with black, purple-grey, yellow-brown or chestnut, often forming wreath at broad end. One nest contained 5 eggs, with 2 birds incubating at once (Uganda: T. Butynski and J. Kalina, pers. comm.), another contained 6 eggs (Schmidt 1964). SIZE: (n = 15) 17·1–19·5 × 12–14 (18·5 × 13·0). WEIGHT: *c.* 1·65.

LAYING DATES: Morocco, Algeria, (Jan)–Feb–Apr; Ivory Coast, Feb; Ghana, May–July, Oct–Dec; Niger, July–Aug, Sept, Dec; Togo, July, Oct; Nigeria, May–June, Oct–Dec; Chad, July–Aug; Ethiopia, Jan–Mar, May, July–Aug, Oct–Nov; Djibouti, Nov; E Africa: Region A, Dec–Feb, Apr, June, Region C, all months except May, Region D, all months (of 86 records, 22% in May, 17% in Apr, 15% in Nov, 9% in Mar; breeds in rains or just after); Malaŵi, Jan–Feb, Apr, Sept; Zambia, Feb–Apr, July–Nov; Zimbabwe, Aug–Nov, Jan–Apr (mainly Sept); Namibia, Aug–June (peak in Feb); Botswana, Sept; South Africa: Transvaal, July–Jan (62% in Oct), elsewhere mainly Aug–Apr, but in Kalahari almost any month after rain.

INCUBATION: by ♂ and ♀, in shifts of 2–12 min, eggs covered for 50–90% of day (Tanzania: Moreau 1939a). Period: 16·5–18 days (n = 5 nests). Nest-relief rapid, birds changing over smoothly, with single movement.

DEVELOPMENT AND CARE OF YOUNG: in 1 brood young born 8 h and 13·5 h apart (Schmidt 1964). Young brooded and fed by both parents equally. Young brooded assiduously for 3–4 days after hatching and for about 20% of time in following week. Av. feeding rates (feeding visits per nestling per 200 min – 280 h of observation) 44 (brood of 1), 37 (brood of 2), 36 (brood of 3). When 3 nestlings were transferred into another nest 5 km away containing 1 young of same age, foster parents' feeding rate immediately rose from 44 to 74 per young per 200 min (Moreau 1939a). Nestling period 25–30 days – hard to evaluate, since young flutter out of nest to nearby perches and back again over period of several days (Brooke and Vernon 1961). Whole brood may also leave nest together, decisively (Hull 1944), roosting in nest for several nights thereafter (once for 23 nights). Young accompanied and fed by parents for up to 19 days after leaving nest, but all driven away within 9 months.

BREEDING SUCCESS/SURVIVAL: several reports in literature, of most or all fledglings dying at nests studied, indicate low success rate. Oldest ringed bird 6 years 9 months.

References
Irwin, M. P. S. (1977).
Moreau, R. E. (1939a).

Hirundo rupestris **Scopoli. Crag Martin. Hirondelle de rochers.**

Plate 7
(Opp. p. 112)

Hirundo rupestris Scopoli, 1769. Ann. 1, Hist.-Nat., p. 167; Tyrol.

Forms a superspecies with *H. fuligula* and *H. concolor* (India).

Range and Status. NW Africa, Cyrenaica, Iberia, S France and Switzerland to Turkey, Iran, central Asia east to 100°E and Himalayas east to 120°E; also W Arabia.

In NW Africa common and widespread resident; in N Morocco south to 29–30°N, and in N Algeria mainly north of 35°N. Scarce Tunisia, north of 35°N; has bred and probably still does. Resident Cyrenaica, Libya. Winter visitor to NW Africa from Europe, large numbers crossing Straits of Gibraltar in Oct and Mar; scarce winter visitor to Senegambia, Egypt and N Sudan (Nile valley and Red Sea coast) and in Ethiopia south to Rift Valley.

Description. ADULT ♂: forehead, crown, nape, sides of neck, ear-coverts and lores brown or greyish brown; mantle and scapulars to uppertail-coverts slightly paler greyish brown. Tail brown, slightly forked; middle and outer rectrices plain brown, 4 intervening ones with large white spot on inner web. Chin white with numerous minute pale brown spots and sometimes with buff wash; throat white; breast very pale buffy grey gradually merging through buffy or pinkish buff lower breast to pale donkey-brown flanks, belly and undertail-coverts.

Wings brown to dark brown; lesser and greater wing-coverts and tertials fringed very pale buff. Axillaries and underwing-coverts dark greyish brown with buffy tips; underside of remiges and tail dark silvery grey. Bill black, base of lower mandible greyish or reddish; eye dark brown; legs and feet pale flesh-brown. Sexes alike. SIZE: wing, ♂ (n = 21) 127–134 (131), ♀ (n = 12) 126–136 (131); tail, ♂ (n = 14) 52–57 (54·5), ♀ (n = 8) 52–56 (54·3); bill to skull, ♂ (n = 15) 10·7–12·2 (11·4), ♀ (n = 8) 10·5–12·2 (11·5); tarsus, ♂ (n = 15) 11–12·5 (11·7), ♀ (n = 8) 10·5–12 (11·4). WEIGHT: (Gibraltar, unsexed) (n = 1300) 17·7–33 (av. 20·4 Mar, 25·6 Nov dawn, 27·1 Nov dusk) (Elkins and Etheridge 1974, 1977).

IMMATURE: upperparts slightly more variegated than adult, all feathers being edged pale buff, conspicuously so on rump and tertials; greater wing-coverts edged warm buff. Underparts like adult but breast and belly and undertail-coverts darker, the feathers buff-fringed.

NESTLING: down short, dense, dark brown-grey; mouth yellow, gape pale yellow; legs and feet pink.

Field Characters. An almost uniformly brown swallow (shade of Common Sand Martin *Riparia riparia*) closely associated with cliffs and crags; underwing-coverts blackish; throat finely mottled, tail with white spots. Thick-set, with broad head, body, tail and wings; flight graceful but slow and stable. Very like some races of Rock Martin *H. fuligula*, but usually considerably darker than its parapatric Saharan populations, and further distinguished from *H. f. obsoleta* by having rump concolorous with mantle (not paler than mantle), mottled throat, dark brown (not pale brown) axillaries, and dark brown (not grey-brown) belly and undertail-coverts.

Voice. Tape-recorded (CHA – ROC, SEL). Song a quiet, persistent throaty twittering, in flight, in breeding season but also in winter. Contact call, 'prrrit' and variations: 'pit-pit', 'ti', 'pritit', 'drit' and 'zi'. Considerable variety of alarm calls: 'dschirr', 'chiupi', 'sfooee', 'zwidi' and pure loud whistling 'siu'. Also excitement and annoyance calls.

General Habits. Highly aerial, inhabiting warm, dry, sheltered situations from sea level up to 2150 m (Europe) and 4000 m (Turkey), in craggy hills and mountains, sea-cliffs, and around buildings, foraging mainly near precipitous cliffs and in gorges but also over adjacent pine woods and villages and (Morocco) swamps and lakes. Mainly uses airspace below level of dominant neighbouring land surfaces (Cramp 1988), but also forages in thermals high above mountain peaks.

Flight graceful but less energetic than in most congeners, with much slow gliding back and forth within a few m of cliff face, where bird forages almost exclusively when cliffs are sunlit. When cliffs not sunny, e.g. in early morning, forages over open country elsewhere. One fed on ground, on abundant gnats (Aden). Not usually associated with marshes, but in winter hundreds feed over L. Rincon marshes, Morocco (Smith 1965). Thousands roosting in caves in Gibraltar probably feed within 16 km, or even 5 km, of cliffs; elsewhere in Europe forages well within 1·5 km of roost or nest (Cramp 1988).

Generally in pairs or small, loose flocks. Winter foraging flocks of up to 400. By day often loafs on exposed, sunny rock-face or ledge. Roosts at night on cliffs or in caves. In limestone sea-caves in Gibraltar 1500–2000 sleep on ledges and 'chimneys' among stalactites (Elkins and Etheridge 1974). Very aggressive to its own species at roosts.

Food. Airborne insects: flies, stoneflies, caddis-flies, beetles, bugs, butterflies, wasps (Spain: Cramp 1988).

Breeding Habits. Nests are solitary or in small colonies. In colony nests av. 30·5 m apart (Switzerland) or 14–16 m apart (Spain). Pair defends territory around nest and close to cliff-face. Sometimes colony includes Common House Martins *Delichon urbica* and (Israel) Rock Martins. Courtship-feeding evidently rare. During nest-building, flying bird deliberately drops feather or grass, retrieving it in air then presenting it to mate, who uses it to line nest. For further details, see Cramp (1988).

NEST: half-cup of mud, lined with fine grasses and feathers. Av. size (n = 4): length 14·4 cm, width 10 cm, height 8·2 cm, thickness 21 mm. Built by both sexes, taking 9–20 days (Spain). Generally built on precipitous rock-face, under overhang or in fissure or small cavity (Mallorca); in Alps often on churches, castles, hotels and silos.

EGGS: 1–5, av. (Switzerland) 3·2. Long, subelliptical; slightly glossy; white, sparsely spotted with red and grey, mainly at broad end. SIZE: (n = 105) 17·2–23·2 × 12·7–15·4 (20·2 × 14·0). WEIGHT: 2·2.

LAYING DATES: Morocco, few conclusive data, but evidently Mar–June (Thévenot et al. 1982).

INCUBATION: by ♀ (89% of time) and ♂ (5%); changeover accompanied by calling but involves no display. ♀ continues to line nest during incubation. Period: (n = 9, Spain) 13–17 days.

DEVELOPMENT AND CARE OF YOUNG: ♀ broods young 2–3 times as much as ♂ (Italy: Farina 1978). Fledging period (n = 8, Spain), 24–27 days. Both sexes feed nestlings, in and out of nest. Young fed by parents for 14–21 days after flying.

Reference
Cramp, S. (1988)

Hirundo atrocaerulea Sundevall. Blue Swallow. Hirondelle bleue.

Hirundo atrocaerulea Sundevall, 1850. Oefv. K. Vet.-Akad. Förh., 7, p. 107; lower Caffraria: Umvoti, Natal.

Plate 9
(Opp. p. 160)

Range and Status. Endemic resident and intra-African migrant, some perhaps sedentary. Breeds SW Tanzania (Kitulo Plateau, also Mbeya, Mufindi, Iringa), Malaŵi (N and central montane grasslands; also Mt Mulanje and highlands in SE near boundary of Mwanza and Ntcheu districts); NE Zambia (Nyika Plateau); Zimbabwe (Inyanga Highlands south to Chimanimani Mts near Mozambique border); W Mozambique (mts near E Zimbabwe and SE Malaŵi borders); W Swaziland; and South Africa (upper Natal, E Transvaal); also SE Zaïre (Marungu: Ketendwe and Sambwe). Non-breeding season visitor north to NE Zaïre (Lendu Plateau west of L. Albert), Uganda (north shore of L. Victoria, from Entebbe area eastwards through Kyagwe; also in Tororo area and Queen Elizabeth Nat. Park) and W Kenya (Mumias, Mungatsi). In passage Zimbabwe (Chipinga). Vagrant Zimbabwe (Harare), Mozambique (mouth of Zambezi R.) and NW and NE Tanzania (Kifanya, Korogwe). Frequent to common, Malaŵi ('large' populations on 1000 km² of grassland, Nyika Plateau) and Tanzania; uncommon, Uganda and Kenya. Uncommon to rare, Zambia, Zimbabwe, Swaziland and South Africa, declining in recent years and close to extinction in Swaziland (*c.* 12 pairs 1987: Allan 1988) and South Africa (Transvaal, 29 pairs, 1987; Natal, 22 pairs, 1987: Allan *et al.* 1987, Allan 1988) and vulnerable in Zimbabwe due to afforestation and spread of exotic trees and other plants which are destroying its specialized nest-sites.

Description. ADULT ♂: entire head, body and upperwing-coverts metallic blue-black with variable white patch on flanks and occasionally a few white streaks on rump. Tail-feathers black washed metallic greenish blue, underside of shafts whitish, outermost pair greatly elongated. Primaries and secondaries brownish black with slight greenish blue tinge; shafts of primaries whitish. Bill black; eye brown; legs and feet black. ADULT ♀: like ♂, but outer feathers (T6) much shorter. SIZE: wing, ♂ (n = 10) 110–117 (114), ♀ (n = 10) 105–110 (107); tail, ♂ (n = 10) 123–140 (135), ♀ (n = 10) 57–76 (69·3), depth of fork, ♂ (n = 8) 70–113 (83·0), ♀ (n = 4) 21–79 (38·0); bill to feathers, ♂ (n = 10) 5–7 (6·2), ♀ (n = 10) 5–7 (5·9); tarsus, ♂ (n = 10) 6–8 (6·6), ♀ (n = 10) 5–8 (6·6).
IMMATURE: similar to adult but sooty black with some glossy blue above: throat brownish; outer tail-feather shorter; bill and gape yellow.
NESTLING: 'almost naked' (Snell 1970), otherwise unknown.

Field Characters. A slim-looking swallow with almost wholly blue body and extremely long tail-streamers. Small amount of white on flanks and primary shafts very hard to see in the field. Can only be confused with Black Saw-wing *Psalidoprocne pristoptera* subspp., which have all-dark bodies and forked tails; blue-bodied races of latter are confined to Ethiopia, and they and some black-bodied races also have white or pale grey underwings. Black-bodied races with dark underwings distinguished by shorter and broader outer tail-feathers (long streamers of Blue Swallow so spindly that bird can appear tail-less in flight).

Voice. Tape-recorded (GIB). Advertising call or song a high-pitched, somewhat nasal 'peep-peep-peep-peep-peep' or 'tee-tee-tee-tee' given by ♂ during courtship and while foraging and caring for young. Commonest call, a soft 'choop-choop'; other calls include 'peep-peep' in alarm or when threatening; a quiet, unmusical 'chip-chip' in flight; and a quick 'chip-pu-pu', 'rip-rip' and 'reap-reap'.

General Habits. Inhabits open grassland, often at high altitudes with fairly heavy rainfall (e.g. Transvaal, 1220–1760 m and where av. annual rainfall exceeds 1400 mm: Allan *et al.* 1987). Also found along edges of woods and swamps; in non-breeding range in Uganda and Kenya, open grassy areas, often with bushes and trees (M. Carswell, pers. comm.).

Solitary, in pairs, or small family parties; not very gregarious, but sometimes in groups of up to 40 (Dyer 1988). Flight swift, graceful, without gliding; swoops over meadows and up and down gullies and valleys in search of prey, usually close to ground, although may fly to and fro over valley for considerable time at 50–100 m. Often settles on isolated bush near ground or on grass stem.

Intra-African migrant; some may be resident in Zimbabwe. On breeding grounds mainly Oct–Apr (earliest late Aug: Malaŵi, Zimbabwe), latest departure 26 May (Zambia). Moves northwards at end Apr; spends off-season in NE Zaïre, S Uganda (Mar–Sept) and W Kenya (Apr–Sept). Moults in off-season. Tanzania population probably also moves north for non-breeding season.

Food. Airborne insects.

Breeding Habits. Monogamous, solitary nester, territorial; density (South Africa) varies from 1 pair/800 ha to 1 pair/30 ha with nests 2500 m to 250 m apart; largest concentration in one area South Africa (1987) 10 pairs (Kaapsehoop, Transvaal). Occasionally gregarious: 2 nest-sites, escarpment of E Transvaal, 1–3 ♂♂ and up to 6 ♀♀ flying together at same time around and into shaft opening (Dunning 1989). Some birds often return year after year to same place to nest (Snell 1979). ♂♂ arrive at breeding areas c. 4 days before ♀♀ (Zimbabwe: Snell 1970). When in breeding area, ♂ and ♀ hawk for prey and fly in and out of depressions and pot holes that sometimes housed previous nests. Sometimes ♂ singing on post fluffs out white flank feathers (territorial or courtship display (?): R. A. Earlé, pers. comm.). In courtship (Dyer 1988), both ♂ and ♀ perform display flight which consists of 3 consecutive phases: (1) *Approach flutter flight* (**A**) when ♂ flies up in a direct line behind ♀; as he approaches her, he changes his wing-beats from a 'leisurely swallow pattern' to 'a series of shallow and very rapid wing-beats' (Dyer 1988). (2) *Dorsal glide* (**B**) when ♂ next flies directly over ♀'s back, then slightly depresses and widely spreads his tail, and at the same time 'extends his wings upward and outwards in a "Y"' (Dyer 1988); he may glide in this position for 1–2 s over her back. (3) *Head drop* (**C**) when ♂ with tail and wings spread next 'extends his neck vertically downwards so the tip of his bill almost touches her head' (Dyer 1988) or actually touches her bill (D. Allan, pers. comm.); when in this position he gives the 'peep-peep-peep-peep-peep' or 'tee-tee-tee-tee' call. Thereafter, he flies away with normal wing-beats (Dyer 1988) or with wings extended upwards, parachutes slowly almost to the ground before darting off (D. Allan, pers. comm.).

NEST: open cup of mud mixed with grass and applied evenly, not in pellets; lined with grass, hair and usually with large white feathers; not supported from below; attached to side or roof of pot hole, old mine, or burrow in ground, including that of aardvark *Orycteropus afer*, usually 0·3–5 m below surface of ground; also under overhang of bank or rock, in culvert, and under roof of house or garage (Zimbabwe). Often has a clear approach, free of obstructing vegetation. Built by both sexes with onset of rains; one in Zimbabwe took 28 days (Snell 1970). At 2 nest-sites (E Transvaal), apparently only ♀ constructs nest with up to 3 ♂♂ carrying nest material into same shaft; not known if they are involved in co-operative building of 1 nest or in building several nests (Dunning 1989). Sometimes, previous year's nest repaired; same nest may be used in successive years (1 used for 18 years in Zimbabwe, but not known if by same birds: Snell 1981). Occasionally, uses old nest of White-throated Swallow *H. albigularis* and Greater Striped Swallow *H. cucullata*.

EGGS: 2–3, rarely 4, av. (n = 17 clutches, Zimbabwe) 2·82, av. (n = 15 clutches, South Africa) 2·9. White, marked with fine brown and purplish brown spots and streaks. SIZE: (n = 9) 17·7–19·5 × 12·5–13·8 (18·3 × 13·2); (n = 6) 17·3–19·6 × 12·7–13·4 (18·8 × 13·0). 2–3 broods per year.

LAYING DATES: Tanzania, Nov–Dec, Feb (breeding in rains); Zaïre, Feb; Malaŵi, Oct–Jan; Zambia, Nov–Jan; Mozambique, Dec; Zimbabwe, Sept–Feb (Sept 1, Oct 6, Nov 8, Dec 16, Jan 13, Feb 6 clutches); South Africa: Sept–Feb (peak Dec–Jan), Natal, Nov–Mar, Transvaal, Nov–Mar (Nov 13, Dec 6, Jan 10, Feb 4, Mar 1 clutches).

INCUBATION: by ♀ only. Period: 15–16 days.
DEVELOPMENT AND CARE OF YOUNG: gape of young yellow at first, white by the time it leaves nest. Young brooded largely by ♀ but fed by both parents; leave nest at 20–26 days, when they can fly; return to nest for a few nights after leaving it.

BREEDING SUCCESS/SURVIVAL: many nests and broods destroyed by heavy rain, flooding, adverse weather and fallen vegetation over opening to nest-site. Some birds also driven from nest by Greater Striped Swallow. Of 37 eggs in 14 nests in Zimbabwe, 78% hatched and 93% chicks fledged, giving a total success of 73% (Snell 1969, 1970, 1979).

References
Allan, D. G. (1988).
Allan, D. G. *et al.* (1987).
Dunning, J. B. (1989).
Dyer, M. (1988).
Earlé, R. A. (1987e).
Snell, M. L. (1969, 1970, 1979).

Hirundo nigrorufa Bocage. Black and Rufous Swallow. Hirondelle rousse-et-noire.

Hirundo nigrorufa Bocage, 1877. J. Sci. Math. Phys. Nat., Acad. Sci. Lisboa 6, no. 22, p. 158; Caconda, Angola.

Plate 8
(Opp. p. 113)

Range and Status. Endemic resident and partial local migrant, Angola (central plateau from N Huila and Cuanza Sul east to S Lunda and N Moxico; also N at Kwango) east to S Zaïre (S Shaba) and N Zambia (Mwinilunga, Luapula and W Northern districts east to Mporokoso and Kasama but not Ndola). Uncommon Angola, uncommon to locally common Zaïre and Zambia.

Description. ADULT ♂: upperparts including top of head down to ear-coverts and upperwing-coverts glossy violet-blue; tail-feathers blackish with reduced gloss, and variable narrow white edge to inner web, outer pair almost matt and slightly elongated. Preorbital region blackish, chin, throat and underparts rich tawny red; small patch at side of upper breast and undertail-coverts glossy violet-blue. Flight feathers blackish with slight purplish gloss; primary shafts pale, especially on underside. Bill black; eye dark brown; legs blackish horn to black. ADULT ♀: like ♂ but slightly paler and tail shorter. SIZE: wing, ♂ (n = 10) 100–112 (108), ♀ (n = 5) 102–108 (105), unsexed (n = 2) 105, 110; tail, ♂ (n = 10) 41–60 (52·9), ♀ (n = 5) 38–50 (47·1), unsexed (n = 2) 47, 56, depth of fork, ♂ (n = 5) 14–24 (18·4), ♀ (n = 3) 4·5–12 (5·8), unsexed (n = 2) 10·5, 20·5; bill to feathers, ♂ (n = 10) 5–7 (6·1), ♀ (n = 5) 6–7 (6·4), unsexed (n = 2) 6, 6·5; tarsus, ♂ (n = 10) 7–9 (7·8), ♀ (n = 4) 7–8 (7·4). WEIGHT: (Angola) 1 ♂ 14·5, 1 ♀ 14.

IMMATURE: very similar to adult but slightly paler above and below; outer tail-feather not elongated.

NESTLING: unknown.

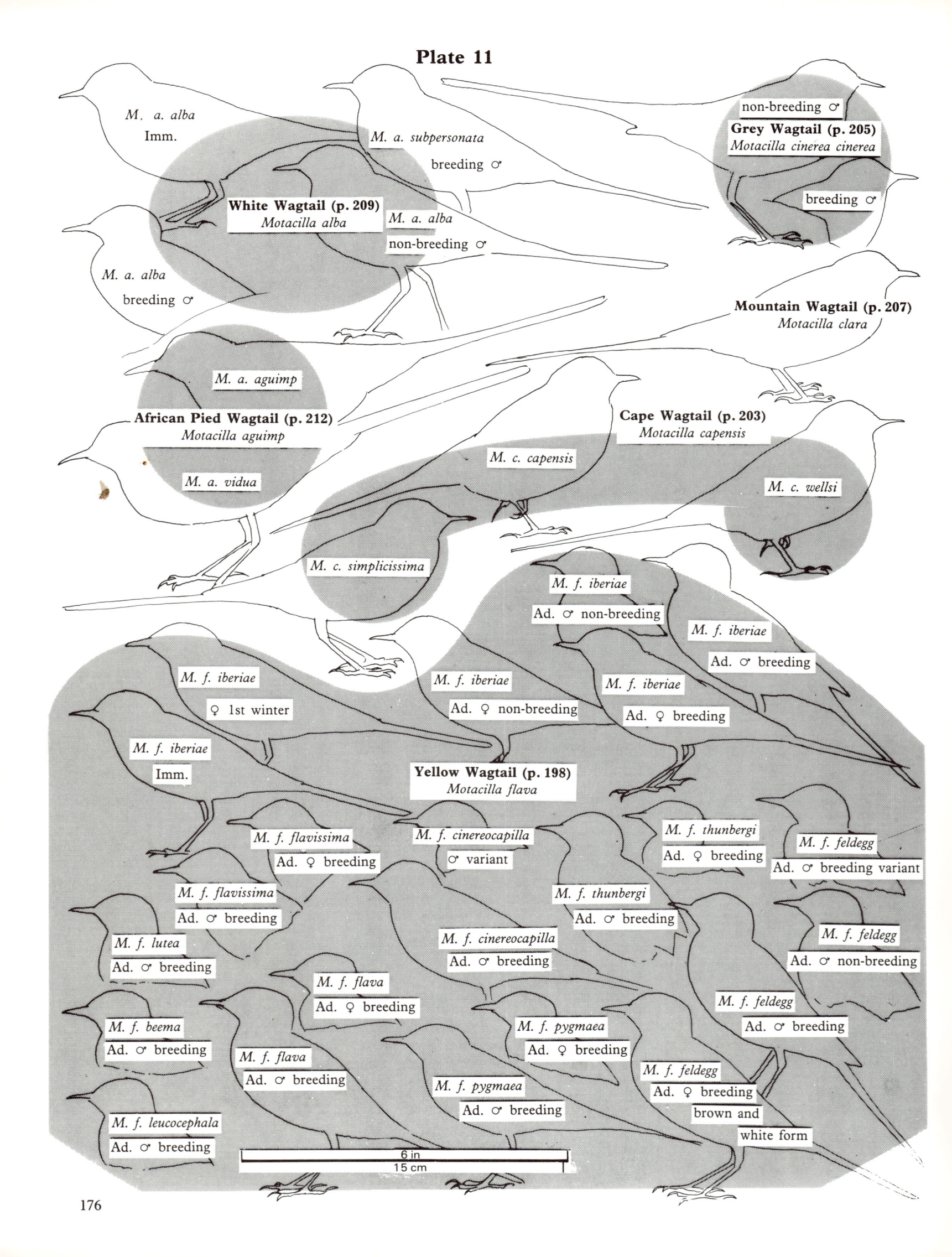

Plate 11

M. a. alba Imm.

M. a. subpersonata breeding ♂

White Wagtail (p. 209)
Motacilla alba

M. a. alba non-breeding ♂

M. a. alba breeding ♂

non-breeding ♂
Grey Wagtail (p. 205)
Motacilla cinerea cinerea

breeding ♂

Mountain Wagtail (p. 207)
Motacilla clara

M. a. aguimp

African Pied Wagtail (p. 212)
Motacilla aguimp

M. a. vidua

M. c. capensis

Cape Wagtail (p. 203)
Motacilla capensis

M. c. wellsi

M. c. simplicissima

M. f. iberiae Ad. ♂ non-breeding

M. f. iberiae Ad. ♂ breeding

M. f. iberiae ♀ 1st winter

M. f. iberiae Ad. ♀ non-breeding

M. f. iberiae Ad. ♀ breeding

M. f. iberiae Imm.

Yellow Wagtail (p. 198)
Motacilla flava

M. f. flavissima Ad. ♀ breeding

M. f. cinereocapilla ♂ variant

M. f. thunbergi Ad. ♀ breeding

M. f. feldegg Ad. ♂ breeding variant

M. f. flavissima Ad. ♂ breeding

M. f. thunbergi Ad. ♂ breeding

M. f. lutea Ad. ♂ breeding

M. f. cinereocapilla Ad. ♂ breeding

M. f. feldegg Ad. ♂ non-breeding

M. f. flava Ad. ♀ breeding

M. f. beema Ad. ♂ breeding

M. f. pygmaea Ad. ♀ breeding

M. f. feldegg Ad. ♂ breeding

M. f. flava Ad. ♂ breeding

M. f. pygmaea Ad. ♂ breeding

M. f. feldegg Ad. ♀ breeding brown and white form

M. f. leucocephala Ad. ♂ breeding

6 in
15 cm

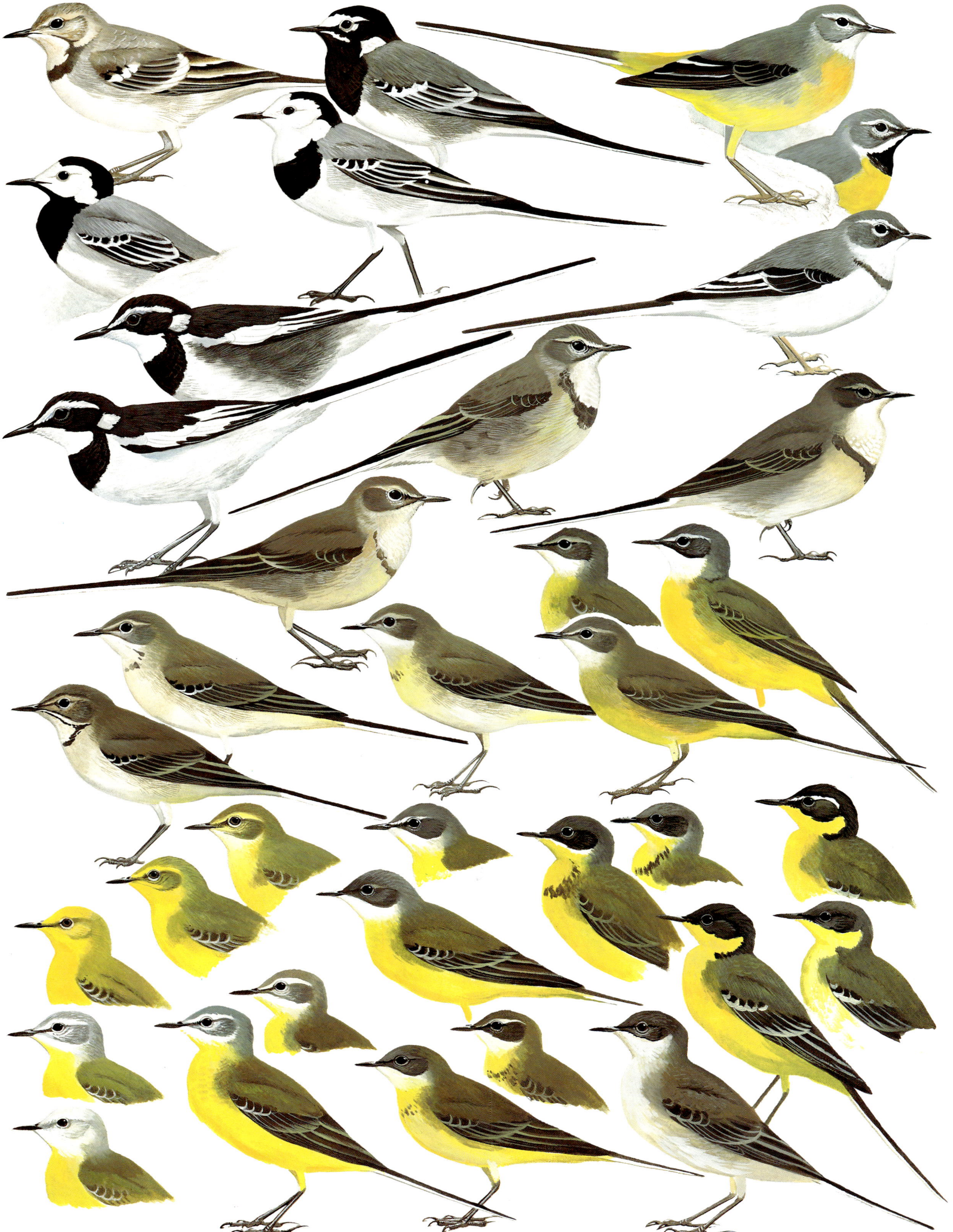

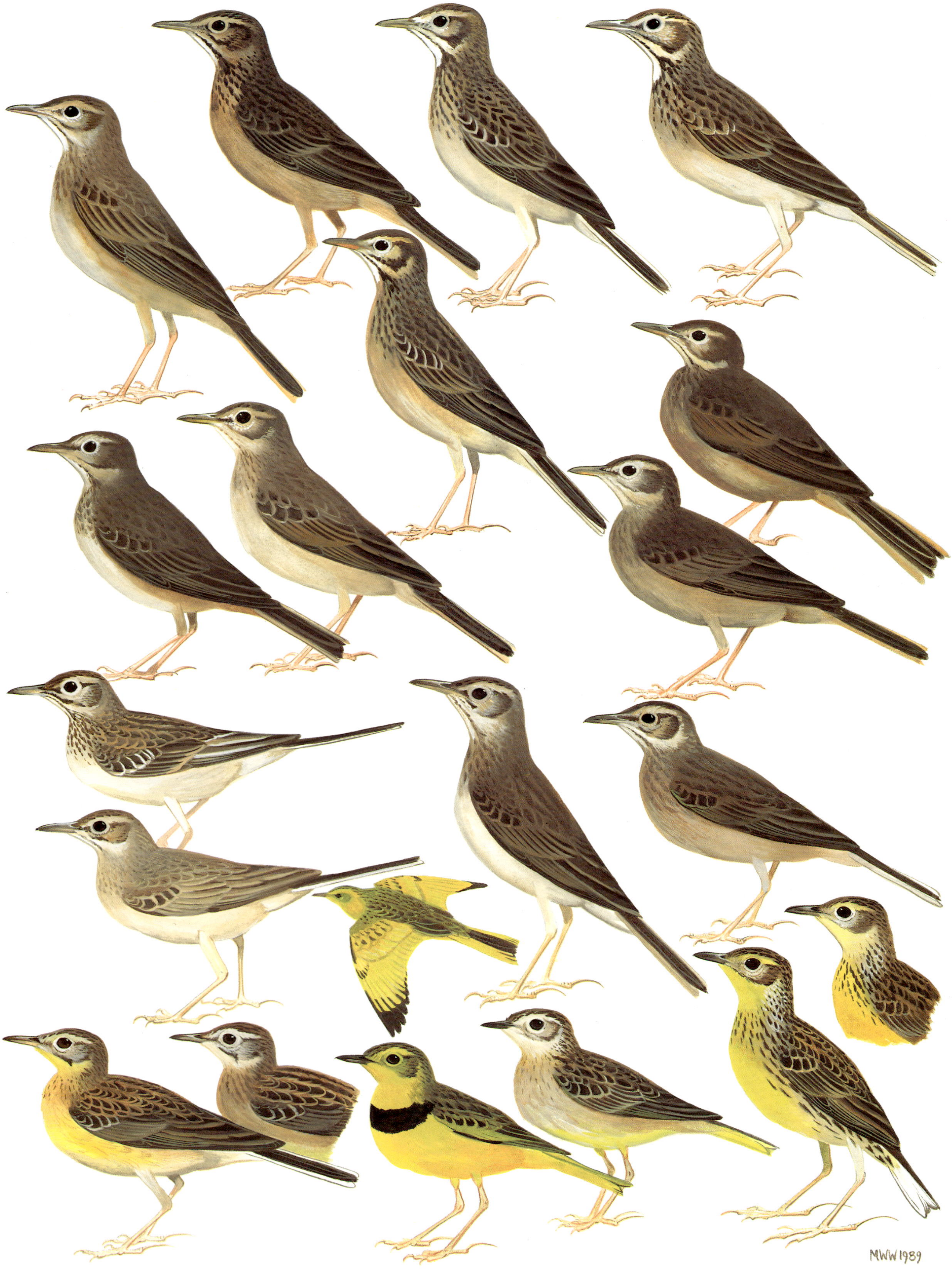
MWW 1989

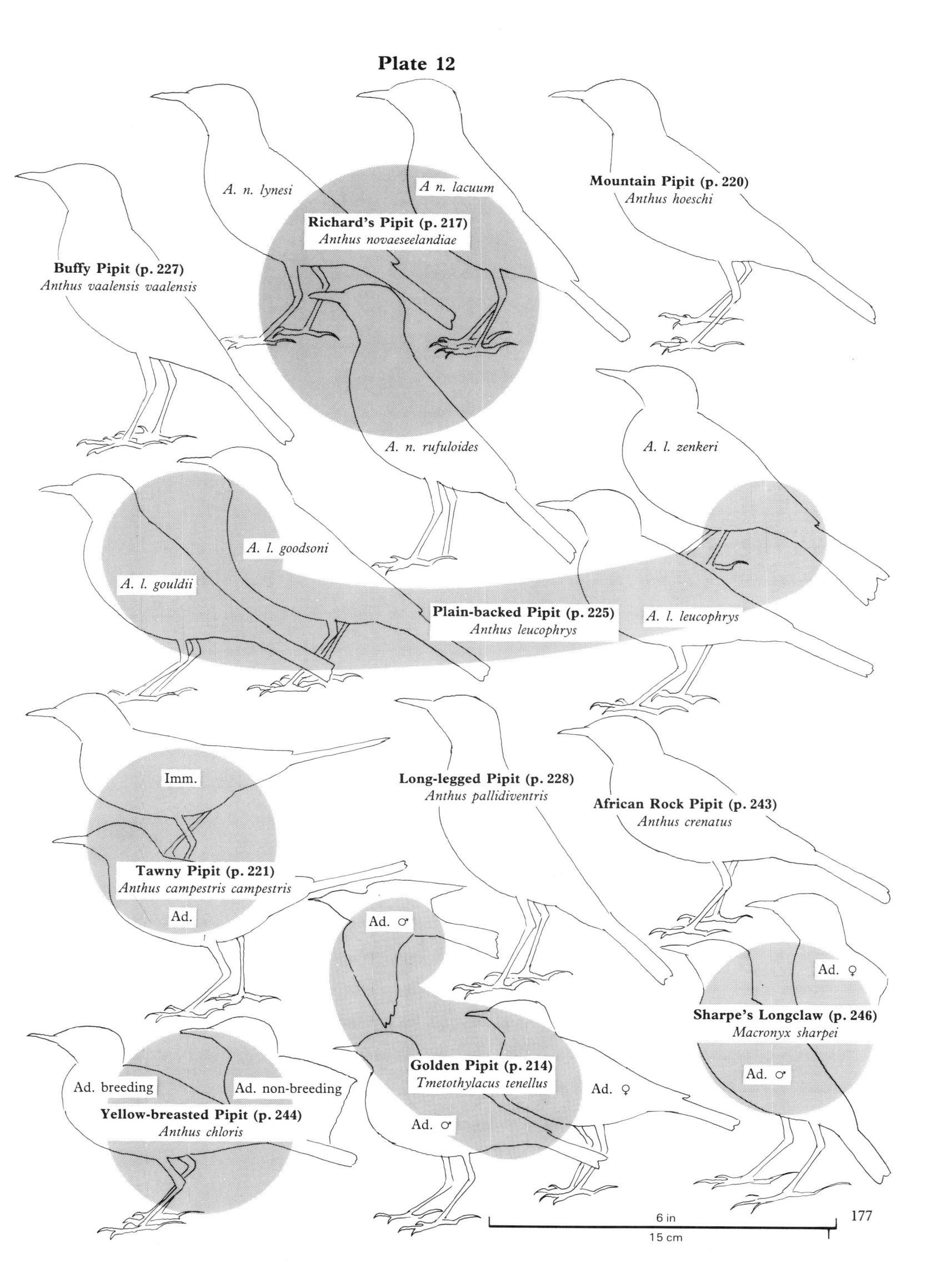

Plate 12

Buffy Pipit (p. 227)
Anthus vaalensis vaalensis

A. n. lynesi

A n. lacuum

Richard's Pipit (p. 217)
Anthus novaeseelandiae

Mountain Pipit (p. 220)
Anthus hoeschi

A. n. rufuloides

A. l. zenkeri

A. l. goodsoni

A. l. gouldii

Plain-backed Pipit (p. 225)
Anthus leucophrys

A. l. leucophrys

Imm.

Long-legged Pipit (p. 228)
Anthus pallidiventris

African Rock Pipit (p. 243)
Anthus crenatus

Tawny Pipit (p. 221)
Anthus campestris campestris
Ad.

Ad. ♂

Ad. ♀

Sharpe's Longclaw (p. 246)
Macronyx sharpei

Ad. breeding   Ad. non-breeding

Yellow-breasted Pipit (p. 244)
Anthus chloris

Golden Pipit (p. 214)
Tmetothylacus tenellus

Ad. ♀

Ad. ♂

Ad. ♂

6 in
15 cm

177

Field Characters. A small swallow with upperparts blue (including rump), underwing-coverts and entire underparts except undertail-coverts rufous, tail with shallow fork. Other fork-tailed species with rufous underparts larger, with deeply forked tail, rufous rump, pale underwing. Red-throated Cliff Swallow *H. rufigula* has short, square tail with white spots, rufous rump, pale underwing.

Voice. Tape-recorded (ASP). Song buzzy and shrill, given in flight. Call, short, strident 'eeek', decreasing slightly in pitch and given *c.* once per s.

General Habits. Inhabits open grasslands, particularly those seasonally flooded and bordering streams; also grassy margins of swamps, clearings in open woodland and occasionally man-made ponds.

Solitary, in pairs, or parties of up to 5; sometimes in groups of up to 20 in non-breeding season. Flight low and swooping, usually 1–2 m above grass. Perches frequently, usually near ground on grass stems, low bushes and small termite mounds.

Mainly sedentary, with some seasonal movements in NW Zambia where absent from breeding grounds during rains (Mwinilunga, late Nov to late Feb/early Mar); not known where birds go (Bowen 1983b).

Food. Airborne insects, including flies and beetles.

Breeding Habits. Monogamous, solitary nester, territorial; nests *c.* 100 m apart along river (Zambia). At beginning of breeding season, ♂ chases ♀ while giving buzzy call, and both hold wings rigidly up, like Blue Swallow *H. atrocaerulea*.

NEST: open cup of mud and rootlets, lined with grass and a few feathers; attached under overhang of bank of stream or river; once in side of pit (Zambia: Benson 1956); well hidden. Usually less than 0·5 m above water surface – 9 nests, Zambia, 0·2–0·7 (0·4) m. Dimensions (n = 1) width 80, int. diam. 60, depth 45 (Zambia). Built in *c.* 4 weeks, probably by both sexes.

EGGS: 2–3, usually 3, av. (n = 13 clutches, Zambia) 2·69. Usually 1 brood per season, rarely 2 broods because of shortness of breeding season July/Aug–Oct (Bowen 1983b). Blunt, long ovals; somewhat glossy; white to creamy, marked with peppery speckling or spotting of warm brown, dark brown or chocolate-brown. SIZE: (n = 18, Zambia) 16·5–19·1 × 12·3–13·4 (17·6 × 12·9).

LAYING DATES: Angola (immature *c.* 2 months old Dec); Zaïre, July; Zambia, Aug–Oct (Aug 2, Sept 9, Oct 1 clutches) (but building as early as 1 July).

References
Bowen, P. St. J. (1983b).
Bowen, P. St. J. and Colebrook-Robjent, J. F. R. (1984).

Plate 8
(Opp. p. 113)

Hirundo smithii Leach. **Wire-tailed Swallow. Hirondelle à longs brins.**

Hirundo smithii Anonymous = Leach, 1818. In Tuckey's Narr. Exped. Expl. Zaire, app., p. 407; Chisalla Island, Congo.

Range and Status. Subsaharan Africa, Afghanistan, Turkestan, Indian subcontinent, Burma, Thailand, Laos and Vietnam.

Resident and partial intra-African migrant. S Mauritania (upper Senegal and Karakoro river valleys near Gourage, Ould Yenji, Saboussire and Kidira, *c.* 14·5–15°N, 11·5–12·5°W); S Senegambia, Guinea-Bissau, N and central Guinea, NW Sierra Leone, Mali (Niger R. drainage north to *c.* 17°N; also rivers of Boucle de Baoulé Nat. Park), Burkina Faso, Ivory Coast (northern guinea belt south to *c.* 8°N; also southern Lakes Yamoussoukro and Ayamé south to *c.* 5·5°N); Ghana (N savanna region south and east to Nyanyanu on coast and Accra Plains), Togo, Benin, SW Niger (Niger R. drainage; also north and east to Maradi and Tessaoua), Nigeria (N and central savanna region north to Kano and Kari and south to Ibadan and Serti); N and central Cameroon (mainly Adamawa Plateau and savanna region south to *c.* 5°N); S Chad, Central African Republic (except Sahel in extreme NE), Sudan (Nile valley north to *c.* 20°N; also north to *c.* 10° in W), Ethiopia (east to arid lowlands in N, E and S). Also NE Gabon, Zaïre (lower Congo R., Shaba, and N and E below 1500 m including shores of Lakes Albert, Kivu and Tanganyika); Uganda, Kenya (except arid N

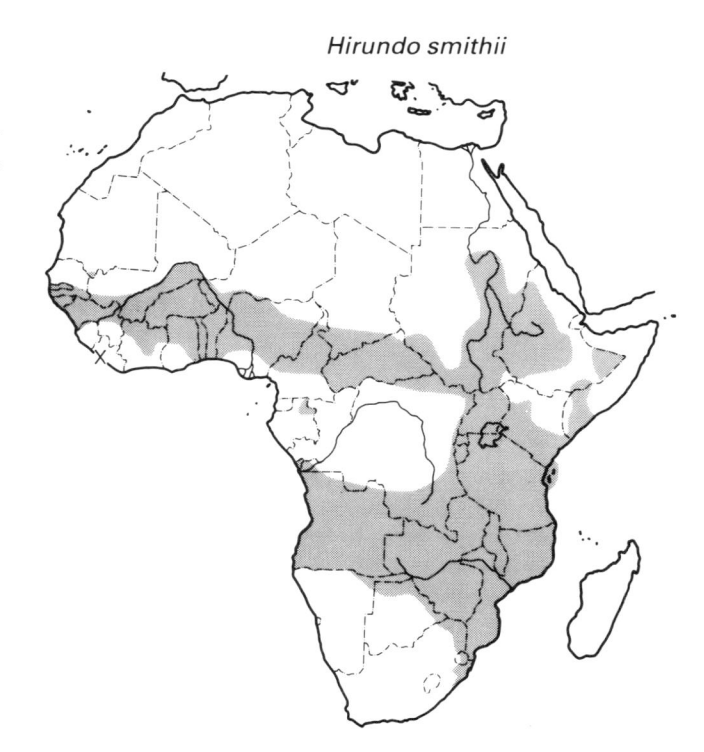

Hirundo smithii

and E), Somalia (Juba valley and coastal area south of *c.* 2°N), Tanzania (including Zanzibar and Pemba), Rwanda, Burundi, Angola, Namibia (NE, and estuary of Cunene R.), Botswana (Okavango; also N and E north of *c.* 20°S and east of *c.* 27°E), Zambia, Malaŵi, Mozambique, Zimbabwe, South Africa (E Transvaal, mainly Lowveld region west to Palala R.; Zululand and Natal coast south to *c.* 30°S) and Swaziland (absent in W). Vagrant Liberia (lower Junk R.). Avoids forests of W Africa and Congo basin, and arid regions. Uncommon or local in western W Africa (except common Baoulé, Mali), Gabon, central Sudan and S Somalia; frequent to locally abundant elsewhere. Has spread considerably during last 100 years with building of bridges and other man-made structures next to water.

Description. *H. s. smithii* Leach: only race in Africa. ADULT ♂: forehead and crown bright reddish chestnut; rest of upperparts and upperwing-coverts glossy violet-blue. Tail-feathers black edged purple, inner webs of all but central pair with white patch; outer pair greatly elongated, slender and wire-like. Lores and feathers around eye black, ear-coverts black with violet-blue wash. Chin, throat and underparts white; throat and breast sometimes with pale salmon-pink flush; small blue-black patch extending from side of neck to side of upper breast; blue-black band from lower flanks across undertail-coverts, not quite meeting in centre. Primaries and secondaries brownish black washed dull bluish purple; axillaries and underwing-coverts white. Bill black; eye brown; legs black. ADULT ♀: like ♂ but outer tail-feathers much shorter. SIZE: (10 ♂♂, 10 ♀♀) wing, ♂ 104–114 (109), ♀ 99–111 (106); tail, ♂ 34–110 (77·6), ♀ 40–75 (49·5), depth of fork, ♂ 40–85 (62·2), ♀ 10–26 (19·2); bill to feathers, ♂ 6–8 (7·0), ♀ 6–8 (6·9); tarsus, ♂ 6–9 (7·4), ♀ 6–8 (6·9). WEIGHT: (Ethiopia) unsexed (n = 195) 10–15 (12·4); (Kenya, Tanzania and Zimbabwe) ♂ (n = 3) 11–13 (12·0), ♀ (n = 4) 9–13 (11·1); (South Africa) ♂ (n = 10) 12–15 (12·8), ♀ (n = 12) 10–14 (12·6).

IMMATURE: similar to adult but duller, with little gloss; top of head ashy brown; underparts (especially breast) washed with buff, outer tail-feathers only slightly elongated.

NESTLING: naked except for some grey down.

Field Characters. A small, slim, fast-flying swallow appearing pure white from below, including underwing and much of underside of tail. Tail-streamers very thin and long (somewhat shorter in ♀), almost invisible at any distance; entire top of head reddish. Other swallows with white underparts are larger and lack needle-thin streamers and chestnut cap; White-throated Swallow *H. albigularis* has dark breast-band, chestnut patch on forehead only (not conspicuous in field); Pearl-breasted Swallow *H. dimidiata* has incomplete breast-band, short tail with shallow fork, dark blue cap; Ethiopian Swallow *H. aethiopica* has chestnut on forehead only, incomplete blue breast-band, tail without streamers.

Voice. Tape-recorded (F, CHA, GIB, MOY, McVIC). Song a twittering 'chirrikweet, chirrik-weet' (Maclean 1985); call, 'twit', 'pwee' and 'che'. Usually silent.

General Habits. Occurs in open country, towns and villages, often near rivers and lakes in savanna areas, from sea level to 2750 m (Ethiopia); avoids forest, dry woodland and desert.

In pairs, or small groups especially when feeding; some pairs remain together for at least a year, roosting in nest area even when not breeding (lower Zambezi R.: Hanmer 1976). Sometimes flocks with Lesser Striped Swallow *H. abyssinica* and Grey-rumped Swallow *Pseudhirundo griseopyga*. Flight swift and graceful, rapid wingbeats alternating with swoops and glides. Hawks for prey in open areas around houses, gardens, fields and over water. Tends to fly 10–20 m above tree line with Lesser Striped Swallows and African Palm Swifts *Cypsiurus parvus*. Perches on dead trees, telegraph wires and railings, and road surfaces of bridges.

In W Africa sedentary in Mauritania (present Jan, Mar, Apr, Sept), Gambia, Central African Republic, and Nigeria except in E (Serti) where dry season visitor Nov–Jan; migratory in Benin, Burkina Faso (present Pendjari and Arli Nat. Parks only July–Jan), Ghana and Gabon (the only individuals noted in NE, Oct–Jan, were not breeding). In E Africa, sedentary in Zanzibar but locally migratory in Kenya (1 ringed Nairobi, recovered 75 km to north). In central and southern Africa, sedentary on lower Zambezi R. (Mozambique), but elsewhere partly migratory, present mainly Mar/Apr to Oct. Zambia, present all year, common Mar and July–Nov, scarce Dec–Jan and May–June, many departing at onset of rains and returning when rains ease off and water levels recede; Zimbabwe, scarce Apr–July; Natal, mainly present early Mar to mid-Oct; and Mozambique, common mid-Mar to early Oct, few records Nov–Feb, with most appearing to leave as soon as young able to accompany adults.

Food. Airborne insects, including ants, beetles, termites, bugs, butterflies, moths and flies.

Breeding Habits. Solitary nester (once 4 nests together: Angola). Once a new pair appeared to claim ownership of territory abandoned 24 h previously by another pair (Zimbabwe: Nyandoro 1987).

NEST: small open cup of mud, lined with grasses, stems, rootlets and a few feathers; generally rather flat, and placed close to ceiling of verandah or house, inside water tower, under bridge, under overhang or in cave (Zanzibar, Ethiopia). Not supported from below, but sometimes old swallow nests used as base. Built 30 cm to 15 m above water or ground, by both sexes in *c.* 7 days. Birds use same nest for successive broods, repairing it before starting next brood.

EGGS: 2–4, usually 2–3, av. (n = 9 clutches, South Africa) 2·9, av. (n = 84 clutches, Malaŵi) 2·71, av. (n = 10 clutches, Kenya) 2·7. Blunt and ovate; slightly glossy; white with reddish or dark brown spots. SIZE: (n = 32) 17–19·5 × 12·5–14·2 (17·0 × 19·5). Double brooded (Zanzibar) or 2–3 broods (Gambia), and up to 4 broods in 12 months on lower Zambezi R.

LAYING DATES: Senegal, Jan–May, July–Dec; Gambia, Jan–Dec (peaks Jan–Mar and July–Aug); Guinea-Bissau, Sept; Guinea, Feb–Mar; Ivory Coast, Feb; Burkina Faso, Jan; Ghana, July (nest building Oct); Togo, Apr–May, Oct; Nigeria, Oct–May; Sudan, Feb–Apr, Sept, Nov; Ethiopia, Jan, Mar–May, Aug–Nov; E Africa: Tanzania, Jan–May, Sept–Oct; Region A, May,

Sept–Oct; Region B, Jan–Feb, Apr–July, Sept; Region C, Jan–July, Sept–Oct, Dec; Region D, Feb–Dec; Region E, Apr–May (breeds mainly in rains: peak Apr–May, also Oct); Zaïre, Feb–Apr, Aug–Oct; Angola, June–July, Sept; Zambia, Jan–Oct (Jan 1, Feb 3, Mar 13, Apr 5, May 1, June 1, July 3, Aug 17, Sept 6, Oct 11 clutches); Malawi, all months, with 2 peaks Feb–Apr and Aug–Oct (Jan 1, Feb 22, Mar 15, Apr 14, May 5, June 2, July 6, Aug 27, Sept 25, Oct 11, Nov 1, Dec 1 clutches); Zimbabwe, all months, with 2 peaks Feb–Apr and Aug–Oct (Jan 2, Feb 19, Mar 26, Apr 21, May 10, June 6, July 7, Aug 42, Sept 39, Oct 36, Nov 9, Dec 5 clutches); Mozambique, June–Sept; Botswana, Aug (6 clutches); South Africa: Apr–May, July, Sept–Oct, Transvaal, Mar–Oct (Mar 1, Apr 2, May 1, July 2, Aug 2, Oct 1 clutches).

INCUBATION: by ♀ only with ♂ often perched close by; begins when clutch complete. Period: 13–19 days.

DEVELOPMENT AND CARE OF YOUNG: young leave nest at 15–24 days; return at night to sleep, for up to 3 weeks. ♀ usually broods small young for most of first 3 days, thereafter ♂ helps to variable extent. ♀ feeds young rather more than does ♂. Young receive up to 16 feeds per h; intervals between feeds more prolonged as young grow. Both parents remove faecal sacs for first *c.* 7 days.

BREEDING SUCCESS/SURVIVAL: Common Kestrel *Falco tinnunculus* takes young (Shaw 1979). Tanzania (Serengeti), 2 nests (5 eggs) produced 2 fledglings; Zimbabwe, 1 nest, 3 fledglings. 2 ♀♀ lived for over 8 years and 9·5 years (Hanmer 1989).

References
Aspinwall, D. R. (1980c).
Moreau, R. E. (1939b).
Scott, A. J. (1986).
Shaw, J. R. (1979).

Plate 9
(Opp. p. 160)

Hirundo nigrita Gray. White-throated Blue Swallow. Hirondelle noire.

Hirundo nigrita G. R. Gray, 1845. Gen. Birds, 1, pl. 20; Niger R.

Range and Status. Endemic resident, from Sierra Leone (except dry north), Liberia, Ivory Coast (forest region from *c.* 8°N south), Ghana (forest regions especially in SW; also Kete Kratchi in guinean belt), S Benin (forest region including Mono R. near Togo border), Nigeria (forest region north to 7°N at Olokemeji), Cameroon (forest regions from *c.* 6°N south), SW Central African Republic (Lobaye and Haute Sangha préfectures), Gabon, Congo, Zaïre (mainly Congo basin east and north to Upper Uelle District and south to Kasai; absent Shaba) and N Angola (Cabinda; Lunda Province: Luachimo R., Cassai R. and L. Carumbo; also Cuanza Norte Province, *c.* 9°30'S, 14°30'E on Luacala R. between Dondo and N'Dala Tando). Frequent to locally common.

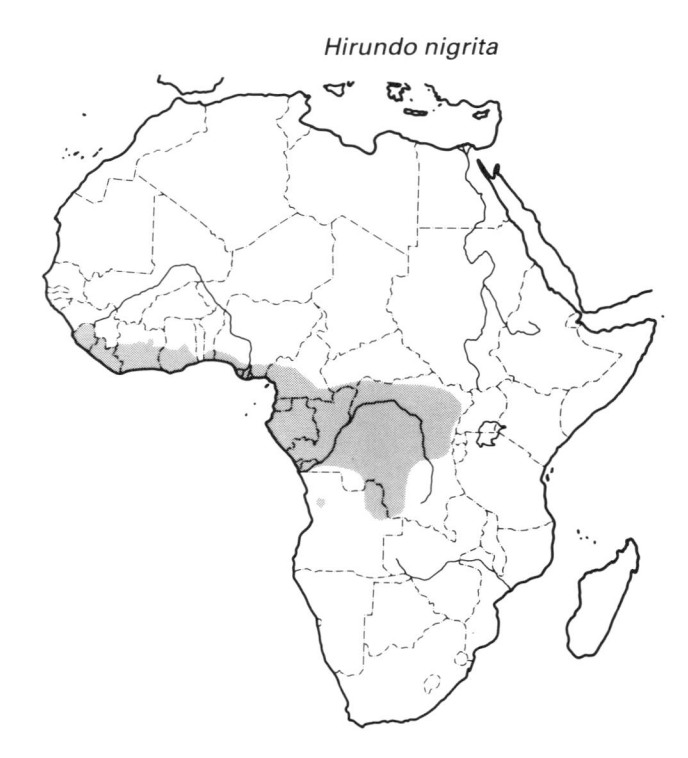

Description. ADULT ♂: upperparts glossy blue-black washed with purple. Tail short, fork shallow, blackish with variable blue gloss and with white patches on inner webs of all except central feathers. Lores black; chin, sides of throat and cheeks glossy purplish blue, middle of throat white; underparts glossy blue-black. Primaries and secondaries black with dull purple wash; tertials and upperwing-coverts with black centres and glossy purple borders; axillaries and underwing-coverts blackish with glossy purple tips. Bill black; eye dark brown; legs black. Sexes alike. SIZE: (10 ♂♂, 10 ♀♀) wing, ♂ 104–109 (106), ♀ 103–111 (106); tail, ♂ 28–43 (36·7), ♀ 34–41 (37·9); bill to feathers, ♂ 7–9 (7·9), ♀ 8–9 (8·3); tarsus, ♂ 7–8 (7·2), ♀ 7–9 (7·7). WEIGHT: (Liberia) 1 ♂ 16·6, 1 ♀ 18·5; (Gabon) 1 ♂ 16·6; (Angola) ♂ (n = 3) 17–22 (18·9), ♀ (n = 3) 15·5–22 (18·9).

IMMATURE: like adult but duller and browner; underparts with only an indication of blue gloss.

NESTLING: scantily covered with blackish down on crown, nape, humeral tracts, and lower back; inside of mouth yellow; no tongue spots; gape-flanges cream; bill flesh, tipped grey.

Field Characters. A small swallow of rivers in forest, entirely blue-black except for white throat-patch and white spots in tail. Tail short with very shallow fork, almost square. No other all-blue swallow in its range.

Voice. Not tape-recorded. Call, a low twittering.

General Habits. Occurs along streams, rivers and lakes with forested banks in rain forest; also mangroves.

Usually in pairs, sometimes in groups of up to 8, once a flock of over 40 (Cameroon: Serle 1954). Flight rapid with almost continuous wing-beats and little gliding; flies from perch to catch insects above river surface or vegetation overhanging it. Regularly feeds on swarming ants and termites. Perches on rocks and branches of fallen trees in mid-stream, on snag or open ground on bank, and on house up to 100 m from river.

Sedentary throughout range, but possibly migratory Sierra Leone (present only Dec–July: Harding and Harding 1982).

Food. Airborne insects, including tabanid flies, Odonata, termites and ants.

Breeding Habits. Solitary nester, territorial; 1 nest per km of river, NE Gabon. In courtship (Gabon: C. Erard, pers. comm.), presumed ♂ alights c. 10 m from presumed ♀, then crouches with head upright and body horizontal, appearing more slender than usual. ♂ also spasmodically raises and lowers wings; ♀ appears to rest; then ♂ and ♀ alternately look at and face away from each other (♂ looks at ♀ while ♀ looks away, then vice versa).

NEST: open cup of mud pellets, lined with grass and feathers, usually built against or under rock or fallen tree overhanging river, 10 cm to 2 m above water; also on cliffs over water, in upside-down dugout canoe, under bridge and under roof of house close to river. Dimensions (n = 1), ext. diam. 110, int. diam. 75; ext. depth 55, int. depth 30. Takes 3 weeks to construct; sometimes pair repairs and uses old nest (Brosset and Erard 1986).

EGGS: 2–3, rarely 4, av. (n = 18 clutches) 2·5. Takes up to a week to complete clutch. Ovate; smooth, almost glossless; rosy white, strongly spotted with reddish brown to grey-brown, especially at larger end. SIZE: 17·3–18·8 × 12·4–13·4 (18·0 × 12·9). Sometimes 2 broods per year (Gabon).

LAYING DATES: Sierra Leone, Mar; Liberia (enlarged testes and fresh breeding plumage, Aug); Ghana, Jan, June; Nigeria, Jan–Feb, Aug (most Jan–Feb); Cameroon, Jan–Feb, June (when water level low); Gabon, Jan–Mar 16 clutches, June–Aug 6 clutches, nesting in dry season; Zaïre, Jan–Feb, July, Dec (probably nests all months); Angola, Mar, Aug.

INCUBATION: by ♀ only, ♂ remaining nearby to defend territory. Period: at least 15 days.

DEVELOPMENT AND CARE OF YOUNG: young covered with filamentous blackish down at 4–5 days; leave nest at 17–18 days; cared for by both parents.

BREEDING SUCCESS/SURVIVAL: Gabon, 1 nest, 3 fledglings (Sept), 1 nest, 2 fledglings (Feb).

Reference
Brosset, A. and Erard, C. (1986).

Hirundo leucosoma Swainson. Pied-winged Swallow. Hirondelle à ailes tachetés.

Hirundo leucosoma Swainson, 1837. Birds of West Afr., 2, p. 74; W Africa.

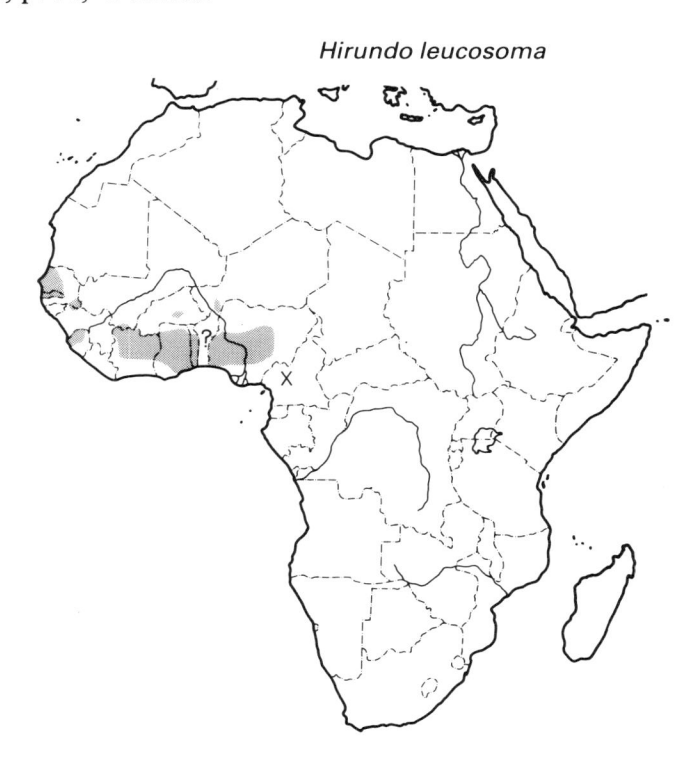

Plate 10
(Opp. p. 161)

Range and Status. Endemic resident in savannas of W Africa. Some seasonal movements in parts of range. W and SE Senegambia, N Guinea (Fouta-Djalon), NW Sierra Leone, S Mali (north to c. 11°N at Bougouni), Burkina Faso (Ouagadougou, Koubri), Ivory Coast south to c. 7·5°N (Béoumi, Bouaké); Ghana (mainly northern savanna but south to coast from Cape Coast to Accra Plains); Togo (from Binaparba in N to Nuatja in S); Nigeria (from Badagri in W on coast north to c. 11°N and east to c. 10°E); also SW Niger (Parc du W, Tapoa). Vagrant, Cameroon (1, Mbanti). Uncommon to frequent.

Description. ADULT ♂: upperparts from forehead to uppertail-coverts and small patch at sides of breast bright glossy steel-blue. Tail-feathers black with blue and green gloss; central pair with no white, T2 with a small white spot on inner web, T3–T6 with large white spot on inner web, increasing in size outwardly; outer pair elongated but not long, tapering to a point. Lores and under eye black; chin, throat and underparts white. Primaries and outer secondaries blackish with duller blue gloss, inner secondaries mainly white with only tips and shafts glossy blue; tertials with white outer and blue inner webs; inner greater upperwing-coverts white; rest of upperwing-coverts glossy blue; axillaries and underwing-coverts white. Bill black;

eye brown; legs black. ADULT ♀: like ♂ but tail less forked. SIZE: wing, ♂ (n = 10) 95–100 (98·5), ♀ (n = 10) 93–100 (96·1); tail, ♂ (n = 10) 36–49 (45·7), ♀ (n = 10) 39–47 (42·0), depth of fork, ♂ (n = 9) 13–18 (15·2), ♀ (n = 7) 7–10 (8·5); bill to feathers, ♂ (n = 10) 6–7 (6·3), ♀ (n = 10) 5–6 (6·0); tarsus, ♂ (n = 10) 6–8 (6·7), ♀ (n = 10) 6–7 (6·8).

IMMATURE: duller than adult, with brown head and much less blue gloss on upperparts.

NESTLING: unknown.

Field Characters. A small blue and white swallow with large white patch on wings, conspicuous both in flight and at rest; tail fairly short, with white spots, shallow fork. The only swallow in Africa with white wing-patches.

Voice. Tape-recorded (CHA).

General Habits. Inhabits mature, undisturbed tree savanna; also occurs around villages and towns, and occasionally along rivers in forests and forest clearings.

Usually in pairs, sometimes solitary. Typically flies low, between trees, with buoyant and jerky flight. Perches on telegraph wires, defoliated trees and bushes.

Mainly sedentary, but present Senegal only Dec (Dakar), and late Apr (in SE); Niger, present Nov to early Apr; Ghana, mainly wet season (early Mar–July Accra Plains, Nov–May Tumu, wet season Mole Nat. Park). Most northerly records in Nigeria are in rains.

Food. Airborne insects.

Breeding Habits. Solitary nester and territorial, although sometimes nests alongside Red-breasted Swallow *H. semirufa* (Bannerman 1939).

NEST: open cup lined with fibres and built on support of roof in house, well, or water tower. Built by both sexes; takes 15–30 days.

EGGS: 4 (2 clutches); white. SIZE: (n = 2) 17–17·5 × 12–13 (Holman 1947); (n = ?) 19–20·5 × 12·5–13·5 (19·7 × 13·1).

LAYING DATES: Gambia, Apr–June; Sierra Leone, Apr–June; Ghana, May–June (wet season); Nigeria, Apr–June.

Plate 10
(Opp. p. 161)

Hirundo megaensis Benson. **White-tailed Swallow.** Hirondelle à queue blanche.

Hirundo megaensis Benson, 1942. Bull. Br. Orn. Club 63, p. 10; 10 miles west of Mega, 4000 feet, Ethiopia.

Range and Status. Endemic resident, Ethiopia, with range of only 10,000 km² around Yavello and Mega, Sidamo Province (50 km north and 15 km northeast of Yavello and 50 km east of Mega). Frequent to common on 60-km stretch of road between Yavello and Mega: 15–20 birds seen per day, 1971, 25 on 25 Jan 1975; on 35-km stretch of same road, 14 on 21 Feb 1989 (Ash and Gullick 1989). Classified as 'Rare and Threatened' in the Red Data Book; clearing of bush for cultivation and increase in grazing noted in 1989 (Ash and Gullick 1989).

Description. ADULT ♂: top of head and upperparts glossy steel-blue; hindneck and mantle with some white mottling. Central pairs of tail-feathers (T1–T3) pure white with thin black shaft streak; T4 white with pale grey edge of outer web extending 10–12 mm from tip, short (2–3 mm) grey edge on tip of inner web; T5 slightly elongated (c. 5 mm), outer web blackish blue, inner web pure white with grey tip (1–1·5 mm wide); T6 elongated (c. 20–25 mm), outer web blackish blue, inner web white. Lores, narrow band over base of bill, cheeks and ear-coverts glossy blue; underparts pure white. Wings glossy blue-black, inner web of innermost greater covert white; axillaries and underwing-coverts white with some brown mottling along bend of wing. Bill black; eye brown; legs black. ADULT ♀: less strongly blue than ♂; lores, narrow band over base of bill and ear-coverts blackish, white on tail-feathers limited to distal half of inner web, tips dark; outermost pair shorter and with less white on inner web than ♂. SIZE: wing, ♂ (n = 6) 99–105 (101), ♀ (n = 4) 97–103 (99·1); tail, ♂ (n = 6) 53–58 (54·3), ♀ (n = 4) 40–51 (44·5), depth of fork, ♂ (n = 4) 20–25 (23·0), 1 ♀ 16; bill to feathers, ♂ (n = 6) 6–7 (6·0), ♀ (n = 4) 5–6 (5·7); tarsus, ♂ (n = 6) 8–11 (9·7), ♀ (n = 4) all 10. WEIGHT: 1 ♂ 11.

IMMATURE: similar to adult ♀ but browner, especially on head, ear-coverts and rump; some feathers of crown tipped glossy blue. Tail-feathers, flight feathers, some tertials, and upper tail-coverts with distinct narrow pale fringes. Throat and rest of underparts pure white.

NESTLING: unknown.

Field Characters. A small blue and white swallow with much white in tail. Range very restricted; only other blue and white swallow in area is Ethiopian Swallow *H. aethiopica*, which is larger and longer-tailed, white in tail confined to row of white spots, forehead chestnut, underparts off-white, not pure white, buffy throat and breast, incomplete blue breast-band. Brown immatures similar but Ethiopian has brown breast-band, buffy throat and breast.

Voice. Unknown.

General Habits. Occurs in open, arid acacia and short grass savanna, mainly at 1600–1725 m (Ash and Gullick 1989). Feeds on insects associated with flowering trees (Collar and Stuart 1985).

Reported to be sedentary (noticed throughout period of June 1941 to Mar 1942: Benson 1946) although possibly with some local movements since absent Mega/Yavello area 14–16 Nov 1988 (R. T. Wilson, pers. comm.) but 'lots' present 21–25 Feb 1989 (J. S. Ash, pers. comm.).

Food. Small beetles 3–5 mm long.

Breeding Habits. Suspected of nesting in holes in tall 'chimney-stack' termitaria Jan–Feb (Benson 1942; Collar and Stuart 1985), also possibly in culverts along Mega–Yavello road (Ash and Gullick 1989).

References
Ash, J. S. and Gullick, T. M. (1989).
Benson, C. W. (1942, 1946).
Collar, N. J. and Stuart, S. N. (1985).

Hirundo dimidiata Sundevall. Pearl-breasted Swallow. Hirondelle à gorge perlée.

Plate 10
(Opp. p. 161)

Hirundo dimidiata Sundevall, 1850. Oefv. K. Sv. Vet.-Akad. Förh., 7, p. 107; Upper Caffraria: Leroma, Transvaal.

Range and Status. Endemic resident and intra-African migrant. Angola (Cuanza Norte to Huila and Cunene, east to Bié, Lunda and Moxico districts); SE Zaïre (Shaba), SW Tanzania (1 record, Mbeya), N and E Namibia, Zambia, Malaŵi (west of Rift Valley south to Ntchew and Nsanje); Botswana, Zimbabwe, W Mozambique, W Swaziland, W Lesotho; South Africa (Transvaal, Orange Free State, NW Natal and S and E Cape west and south to Cape Town; also N Cape). Uncommon to frequent; but locally common Angola, South Africa, Lesotho, Zambia and Malaŵi. In Cape has increased considerably this century (Tree 1986a); but in Zimbabwe some decline in recent years (Irwin 1981).

Description. *H. d. dimidiata* Sundevall: Cape to NE Namibia, Botswana, Orange Free State, Lesotho, S Transvaal, Swaziland and NW Natal. ADULT ♂: entire upperparts including forehead glossy steel-blue or purplish blue. Tail blackish with reduced gloss, outer tail-feathers almost matt, and longer, somewhat tapered and pointed. Lores and ear-coverts dull blue-black. Chin, throat and underparts dull white, breast washed grey; small blackish patch at sides of breast; a few dark shaft-streaks on lower breast and belly; undertail-coverts with dark tips and sometimes additional dark bar separated from tip by white bar. Wings brownish black washed with dark metallic blue; lesser and median underwing-coverts grey-brown, axillaries and greater underwing-coverts white. Bill black; eye brown; legs black. ADULT ♀: like ♂ but with gloss greener and outer tail-feathers shorter. SIZE: wing, ♂ (n = 10) 95–105 (99·2), ♀ (n = 10) 96–101 (99·5); tail, ♂ (n = 10) 42–64 (53·8), ♀ (n = 10) 43–59 (51·2), depth of fork, ♂ (n = 10) 11–20 (14·7), ♀ (n = 4) 9–13 (11·1); bill to feathers, ♂ (n = 10) 6–7 (6·2), ♀ (n = 10) 6–7 (6·0); tarsus, ♂ (n = 10) 7–8 (7·1), ♀ (n = 10) 6–8 (7·1). WEIGHT: (Namibia) 1 ♂ 13, 1 unsexed 13·8; (South Africa) ♂ (n = 2) 11, 12, unsexed (n = 3) 10–12 (11·0); (N Cape) unsexed (n = 4) 11–15 (12·5).

IMMATURE: similar to adult, but dusky black above and duller, lacking much of metallic gloss; innermost secondaries with white tips; tail shorter.

NESTLING: skin pink, naked except for some grey down on head and back; gape yellowish white.

H. d. marwitzi Reichenow: Angola, S Zaïre, Zambia, Malaŵi,

SW Tanzania, Zimbabwe, N Transvaal and Mozambique. Differences from nominate form slight in Zimbabwe (Irwin 1981, Tree 1986a). Elsewhere darker and greyer below. Smaller: wing 94–107 vs 100–108. WEIGHT: (Zambia) unsexed (n = 3) 10–12 (11·2); (Zimbabwe) unsexed (n = 9) 11–12 (11·5).

Field Characters. A small swallow, with blue upperparts, including cap and rump, tail moderately forked, with no white spots. Similar Common House Martin *Delichon urbica* and Grey-rumped Swallow *Pseudhirundo griseopyga* have contrasting pale (grey or white) rumps, greyish underwing (mainly white in Pearl-breasted). White-throated Swallow *H. albigularis* has chestnut forehead, dark breast-band, white spots in tail.

Voice. Tape-recorded (88). Song a series of harsh, sharp notes, 'chip-chip-chip-cher-cher-chip-chip-cher ...' broken into 1–2 notes per s; calls, a guttural 'kss-kss' during courtship, 'twit' in flight, also 'chik', 'chut' and 'chuchut'.

General Habits. Inhabits grassland and bush, often in vicinity of water; also edges and clearings of miombo woodlands, human settlements and cultivated areas, and occasionally swampy ground and edges of ponds.

In pairs or small groups, occasionally in flocks of up to 40 or even 100. Flight quick and agile; sometimes forages below canopy of miombo trees. Flock of 100 hawking low over kraal at head height, feeding on swarms of insects; some birds rested on cow-pats between feeds (Cape Province: Spearpoint 1990). Often perches on dead twigs on tree tops, or on stones or open ground.

Populations north of *c*. 26°S (*marwitzi*) probably mainly sedentary, with some nomadic movements, while those south of there (*dimidiata*) are migratory, present on breeding grounds Aug to Mar–Apr (although some overwinter in flocks of up to 20 in Cape and Transvaal), and spend non-breeding season in N Botswana, Zimbabwe, Zambia and Zaïre May–Oct (dry season). Observations in Zambia: Jan (12), Feb (9), Mar (3), Apr (7), May (57), June (27), July (50), Aug (18), Sept (20), Oct (34), Nov (7), Dec (8) (D. Aspinwall, pers. comm.).

Food. Airborne insects; possibly grass seeds (Every 1988); pieces of white-wash (= dried remains of a mix of lime-rich paint) once found in stomach (Dean 1989).

Breeding Habits. Monogamous; solitary breeder but reported to nest in 'small groups' in Zaïre. Courtship: ♂ flutters in front of perched ♀; both open bills wide and make guttural 'kss-kss' sounds, then ♂ attempts to mount ♀.

NEST: cup of mud pellets, lined with grass, hair and feathers; stuck to wall of building, culvert, well, mine shaft, rock face or antbear burrow. Ext. diam. 125–135, ext. depth 45; int. diam. 95, int. depth 30. Built by both sexes (but apparently mainly by ♂) in about 3–4 weeks. Sometimes used for 2nd brood, also in following year.

EGGS: 2–4, usually 2–3, once 6 (Claasen 1991); av. (n = 9 clutches, South Africa) 2·4. Laid every 24 h, usually in early morning. Glossy white. SIZE: (n = 38) 14·8–19·5 × 11·3–14 (17·3 × 12·6). 2 broods per year, South Africa.

LAYING DATES: Zaïre, Aug–Oct; Zambia, Aug–Sept; Malaŵi, Aug–Sept; Mozambique, Sept (feeding young Oct); Namibia, Oct–Jan; Zimbabwe, Aug–Feb (Aug 4, Sept 13, Oct 6, Nov 2, Dec 2, Jan 1, Feb 2 clutches); South Africa, Aug–Mar (peak Sept, with 77% Sept–Nov).

INCUBATION: starts when clutch complete; by ♀ only, with ♂ perching nearby. Period: 16–17 days.

DEVELOPMENT AND CARE OF YOUNG: wing quills appear at 7 days, sheaths break at 11 days; eyes open at 10 days. Young leave nest at 20–23 days when they can fly; return to nest for next 4 nights. Brooded by ♀ for periods of 1–5 min for at least 7 days; fed by both ♀ and ♂ from time of hatching until 20 days after they leave nest; 1 brood fed 8 times in 10 min (Spearpoint 1990).

References
Aspinwall, D. R. (1979b).
Benson, C. W. (1949).
Schmidt, R. K. (1959).
Tree, A. J. (1986a).

Plate 8
(Opp. p. 113)

Hirundo aethiopica Blanford. Ethiopian Swallow. Hirondelle d'Ethiopie.

Hirundo aethiopica Blanford, 1869. Ann. Mag. Nat. Hist. (4), 4, p. 329; Bakakit, Tigré, northern Ethiopia.

Range and Status. Endemic resident and partial intra-African migrant. Senegambia (Basse-Casamance, Niokolo-Koba, and Siné Saloum Nat. Parks), Guinea (Kindia area), Ivory Coast (now numerous in E, expanding: Demey and Fishpool 1991), Ghana (Cape Coast to Accra Plains, Tamele and Sugu), Togo (1 record: 30 at Fazao), Benin (N and central savannas), E Mali (eastern Niger R. drainage including Gao, Menaka, Tamesua), Burkina Faso (extreme N), S Niger (Parc du W and Niamey east to Maradi-Tanour), Nigeria (throughout, except in SE south of Ogoja), N Cameroon (N montane districts and Adamawa Plateau north to L. Chad), central and S Chad (north to *c*. 15°N), NE Central African Republic (Manovo-Gounda-Saint Floris Nat. Park; uncommon), central and SE Sudan (north to *c*. 16°N), Ethiopia, Djibouti, Somalia (except NE); south to N and central Uganda (Murchison Falls, Kidepo and Rwenzori Nat. Parks; Entebbe), N, W-central and coastal Kenya and extreme NE Tanzania (Tanga); also vagrant Masasi, Moshi and Zanzibar. Frequent to locally abundant; range extending southwards in E Africa, and westwards, to W Africa in recent decades.

Description. *H. a. aethiopica* Blanford: range of species except E Ethiopia, Somalia and NE Kenya. ADULT ♂: forehead and forecrown deep chestnut, rest of upperparts glossy steel-blue. Tail blue-black, all but central tail-feathers with white spot on inner web; outer tail-feathers elongated. Lores sooty black, ear-coverts blackish glossed with blue. Chin, throat and upper breast buffy; sides of breast dark glossy steel-blue, forming incomplete breast-band; rest of underparts white with dusky wash. Primaries matt black with only a hint of blue; secondaries and tertials broadly edged blue; underside of primary shafts white; axillaries and underwing-coverts white. Bill black; eye dark brown; legs black. Sexes alike. SIZE: wing, ♂ (n = 10) 97–110 (105), ♀ (n = 10) 99–112 (105); tail, ♂ (n = 10) 39–64 (54·5), ♀ (n = 10) 39–62 (48·9), depth of fork, ♂ (n = 7) 16–30 (23·6), ♀ (n = 3) 17–23 (21·0); bill to feathers, ♂ (n = 10) 6–8 (6·9), ♀ (n = 10) 6–7 (6·7); tarsus, ♂ (n = 10) 7–9 (8·3), ♀ (n = 10) 7–10 (7·8). WEIGHT: (Chad) unsexed (n = 87) 11–15 (13·0); (Ethiopia) unsexed (n = 39) 11–17 (14·0); (Kenya) ♂ (n = 4) 10–16 (14·0), ♀ (n = 5) 14–17 (15·3).

IMMATURE: similar to adult but crown and forehead brown, upperparts ashy brown with some steel-blue gloss; breast-patch brown.

NESTLING: unknown.

H. a. amadoni White: E Ethiopia, Somalia and NE Kenya. Differs from nominate race in having throat and upper breast white without buff.

Field Characters. A medium-sized swallow, rather smaller than Barn Swallow *H. rustica* and Red-chested Swallow *H. lucida*, but sharing chestnut forehead and ring of white spots in tail. Throat not red but white or pale buff, blue breast-band incomplete, tail-streamers fairly short; from below, e.g. perched on wire, appears all white with narrow black ring starting across sides of breast, not at all like immature Barn Swallow which has similar short streamers but red-ochre throat and broad, complete, brown breast-band. Underwing-coverts whitish or pale buff, vs. deeper buff in Barn or grey-brown in Red-chested. Wire-tailed Swallow *H. smithii* is smaller, slimmer, with complete chestnut cap, long thin tail-streamers.

Voice. Tape-recorded (CHA, McVIC). Song is quite loud, sustained and melodious squeaking and twittering. Call, 'chip', 'cheut' or 'cheep-cheep'.

General Habits. Occurs in grasslands, savannas, bush and open woodlands; also in villages and towns, over inland waters and marshes, and along coastal rocks and cliffs. Avoids forest, although sometimes in clearings. From sea level to 2750 m (Ethiopia).

In pairs or family parties; flocks of up to 500 in non-breeding season. Flight a little less graceful, wing-beats slightly faster with less gliding, than longer-winged Barn Swallow; often flies for extended periods very low, only 30 cm or so above ground (L. Grant, pers. comm.). Sometimes (Somalia) roosts in sugar-cane fields in large numbers (2000–3000: P. Becker, pers. comm.). After breeding (Kenya) remains in vicinity of nesting area. Forages over surrounding bush, especially open areas; also takes food from surfaces such as walls and ceilings. Returns at dusk to roost on or near nest; members of pair usually vary roosting positions (L. Grant, pers. comm.), although 1 pair roosted in same position throughout the following non-breeding season while the young roosted elsewhere (Grant and Lewis 1984).

Mainly sedentary (e.g. Laikipia Plateau, Kenya; most of Nigeria) but locally migratory. Present in rains (breeding), then sparse or absent in dry season in e.g. coastal Ghana (present mainly Mar–Oct), N Ghana (Mar–July), Benin, SE Nigeria (mainly May–Sept), Chad (absent Abéché Dec to early Mar), Central African Republic (present Aug–Dec), Sudan (mainly May–July), and Ethiopia (Eritrea) (present all year but commonest Apr–May). Where these populations go in the non-breeding season is not known.

Food. Airborne insects, including flies, termites, winged ants, bees and cicadas.

Breeding Habits. Monogamous; loosely colonial. During courtship, ♂ and ♀ perch side by side, then from time to time presumed ♂ gives loud song and presumed ♀ gives softer, twittering subsong (♂ also gives this song in flight). Every few min both leave perch, wheel around, and return to perch; then ♂ (and occasionally ♀) makes series of brief flights, sometimes hovering over nest-site. On return to perch, ♂ squawks loudly and hovers over ♀ before alighting. In Kenya (Laikipia Plateau), ♂ may start to sing during short rains (Nov), several months before nesting begins, but no nesting takes place until long rains (Mar–May), when ♂ sings regularly; later, during 2nd and 3rd broods (July–Dec, usually July) he rarely sings.

NEST: cup-like structure made of sticky wet mud pellets brought one at a time, reinforced with dry grass and rootlets, lined with grass, hair, feathers, coconut or palm leaf fibres, twine and strands of plastic; diam. 100 × 80, depth 130 (n = 1, Kenya, rather larger than average); placed under verandah or eaves of house, under (occasionally on top of) beam, 8–10 m down well (SE Ethiopia), in cave, under exposed dead coral overhang or bridge (Ghana, Kenya). Mud pellets (0·15–3·35 mm diam.) usually stuck to surface, with normally no support from below; nest flattened where attached to beam. Built by both sexes; material brought to nest-site usually in morning; completed in about 2 weeks. New nest may be built on old or broken nest of swallow or swift, often by addition of new mud to birds' own nest of previous season (L. Grant, pers. comm.). Nest always relined before being used for 2nd brood. Nest sometimes built but not used. Both birds roost at or near nest while it is still under construction.

EGGS: 2–4, rarely 1, av. (n = 14 clutches, Kenya coast) 2·7, av. (n = 7 clutches, Chad and inland Kenya) 3·14, av. (n = 4 clutches, Somalia) 3·75; laid every 24 h. Glossy; white, faintly tinged with pink, with speckles and blotches of rufous and chestnut varying in intensity and density, mainly in wide belt around broad end. SIZE: (n = ?) 17·5–20·1 × 12·2–14 (18·7 × 13·2), (n = 5) 17–18·5 × 12–13 (av. ?). 2 broods per season, occasionally 3.

LAYING DATES: Senegambia, Mar, July–Aug, Dec–Jan; Guinea, Dec–Jan; Ghana, Apr–June; Benin, Aug; Niger, Aug; Nigeria, Mar–Nov (mostly May–July); Chad, July–Nov (in wet and early cool seasons); Sudan, Mar–July; Ethiopia, Apr–Oct (birds at nest, Jan, Danakil); Somalia, Apr–May, Sept; E Africa: Uganda, Mar; Kenya, Apr–July, Dec–Jan (mainly Apr–July, sometimes other months when it rains); Region D, Apr–May; Region E, Mar–June, Nov (mainly in long rains).

INCUBATION: probably mainly by ♀; begins with last egg. Period: c. 14 days.

DEVELOPMENT AND CARE OF YOUNG: fledging period 25 days; young often cling to edge of nest for some days before leaving. After fledging, young roost on or beside nest for 7 days or more. Both parents feed small young, almost continuously from dawn to dusk; 10 feeds in 25 min (Grant and Lewis 1984). As young grow, parents are absent for periods of up to 30 min. Parents feed newly hatched young with regurgitated material, and older young with small insects held in their bills. Sometimes both parents arrive at nest at same time; they swallow some faecal sacs, carry others away in bills. Young learn to defaecate over edge of nest when less than 1 week old, but parents may still continue to carry away sacs for some time afterwards (L. Grant, pers. comm.). Young of earlier broods show interest in later brood, but not known whether they feed them.

BREEDING SUCCESS/SURVIVAL: adults sometimes desert nest and young due to persistent chasing by Little Swift *Apus affinis* and White-rumped Swift *A. caffer* (Kenya: Laikipia Plateau). Sometimes, White-rumped Swifts knock young out of nest, then are attacked by parents in the air and knocked to the ground (L. Grant, pers. comm.). Parents also chase Lesser Striped Swallows *H. abyssinica*, which sometimes take over their nests. Entire broods of young, from 1 week old to fully fledged, die from disease and/or parasites; 4 nestlings died when caught in hair loop, and 1 probably when trampled by older young (L. Grant, pers. comm.).

Reference
Grant, L. and Lewis, A. (1984).

Plate 8
(Opp. p. 113)

Hirundo albigularis Strickland. White-throated Swallow. Hirondelle à gorge blanche.

Hirundo albigularis Strickland, 1849. In Jardine's Contrib. Orn., p. 17, pl. 15; South Africa (= Cape Peninsula).

Range and Status. Endemic resident and intra-African migrant. Breeds South Africa (throughout except NW Cape), Lesotho, Swaziland, Botswana (west to c. 22°E), Zimbabwe (mainly on central plateau and in E), Mozambique (throughout except N), and Malaŵi (in S near Mt Dzonze, Liwonde and Nchalo; possibly also Mt Mulanje). Common throughout except uncommon Botswana, uncommon to frequent Malaŵi. Also Angola – 1 breeding record, central plateau (12°S, 16°E) but no details (Hall and Moreau 1970) – and Zambia – Mansa (11° 11'S, 28° 53'E) and in Luanshya District (13° 8'S, 28° 25'E). Non-breeding visitor Namibia (Cunene R. estuary, Skeleton Coast Park; Windhoek area; Welverdiend Farm (25° 50'S, 19° 57'E), E Kavango and E Caprivi Strip, uncommon), Angola (central and E regions, locally frequent), Zaïre (Shaba), Zambia (throughout except NE, locally uncommon to abundant: once a flock of c. 1000 at Blue Lagoon), and central and S Malaŵi (from Bunda/Lilongwe and Mt Dzonze south to Nsanje District). Vagrant, Cameroon (1, Kumbo, 6° 12'N, 10° 40'E); Tanzania (1, L. Jipe, 3° 35'S, 37° 45'E, late July).

Description. ADULT ♂: forehead chestnut; rest of upperparts glossy dark blue. Tail-feathers dull blue-black, all except central pair with large white patch on inner web, outermost pair elongated. Lores, around eye and ear-coverts dull blue-black. Chin and throat pure white bordered below by glossy blue-black breast-band, narrow in centre; rest of underparts off-white to pale brownish grey. Wings brownish black, primaries and secondaries washed dull blue, coverts broadly edged glossy blue, axillaries and greater underwing-coverts white. Bill black; eye brown; legs black. ADULT ♀: like ♂ but tail less deeply forked. SIZE: (10 ♂♂, 10 ♀♀) wing, ♂ 123–

133 (129), ♀ 124–137 (129); tail, ♂ 51–77 (67·6), ♀ 57–70 (63·2), depth of fork, ♂ 31–40 (36·7), ♀ 12–28 (20·1); bill to feathers, ♂ 8–10 (8·5), ♀ 8–10 (8·9); tarsus, ♂ 8–10 (8·7), ♀ 9–11 (9·9). WEIGHT: (Angola) 1 ♂ 20, 1 ♀ 19; (Namibia) 1 ♀ 22; (Mozambique) ♂ (n = 10) 20–29 (22·8); (South Africa) unsexed (n = 4) 19–21 (20·2).

IMMATURE: like adult, but chestnut area on forehead smaller and paler (forehead sometimes all blue-black); breast-band brown to dull black; upperparts duller and flight feathers with paler edges.

NESTLING: unknown.

Field Characters. A medium-sized barn swallow type with chestnut forehead, blue breast-band and white spots on fairly deeply forked tail; quickly distinguished from Barn Swallow *H. rustica*, Angola Swallow *H. angolensis* and Red-chested Swallow *H. lucida* by white throat. Other swallows with white underparts lack chestnut forehead and white tail spots, and either lack breast-band (Pearl-breasted Swallow *H. dimidiata*) or have brown breast-band and upperparts (Common Sand Martin *Riparia riparia* and Banded Martin *R. cincta*).

Voice. Tape-recorded (88, GIB). Song, a disjointed series of harsh, sharp, fast twittering notes, 'chit-chit ...', broken into groups, 2–3 per s; call, similar notes given individually, also some chittering, squeaking and nasal notes.

General Habits. Inhabits open country and grassland, often near water; also flood plains, open woodlands, human habitation; open water near nest-sites on bridges and dams.

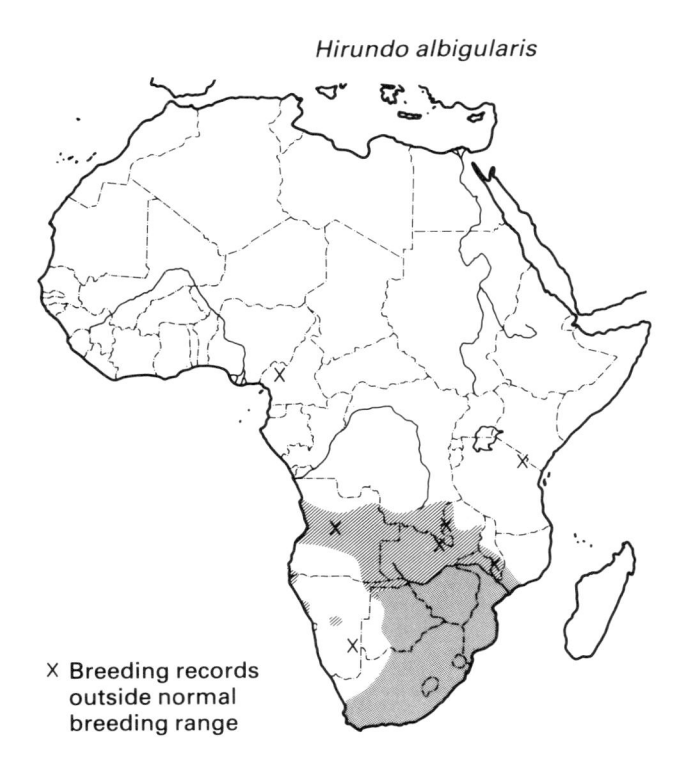

Solitary or in pairs or family groups; on migration in flocks of 20–1000. Flight quick and agile; often hawks for prey over water or open grassland. Often in company with Black Saw-wing *Psalidoprocne pristoptera*, Blue Swallow *H. atrocaerulea*, Rock Martin *H. fuligula* and Barn Swallow. Once followed within a few m of a man and 2 dogs, apparently catching insects flushed by them (Vernon and Dean 1988). Perches on posts or stumps in water.

Mainly migratory, breeding south of Zambezi R. in southern summer, and wintering north of Zambezi. Arrives at breeding areas Aug and departs late Mar/Apr. On wintering grounds, e.g. Zambia, arrivals are from early Mar, main arrivals late May to early June, and departures late Sept/Oct (latest 1 Dec). No distant ringing returns. Rarely overwinters on breeding grounds; a few remain all year Zimbabwe. Some return annually to same place to breed (Zimbabwe, 1 ringed pair present for 3 years, another for 2 years: Irwin 1981). 3 recovered more than 21 km from ringing site in South Africa (Earlé 1987g).

Food. Airborne insects.

Breeding Habits. Solitary nester. ♂♂ usually arrive first, ♀♀ a couple of weeks later.

NEST: cup of mud pellets, lined with fine grass, rootlets, hair and feathers; plastered onto wall, eaves or ceiling of verandah, under bridge, in culvert or under overhang of rock by stream. Often uses same nest-site in subsequent breeding season, and a pair often rears broods from several nests at a site in rotation (R. A. Earlé, pers. comm.).

EGGS: 2–5, usually 3, av. (n = 79 clutches, South Africa) 2·8. Pinkish white or buff with spots and speckles of reddish brown and slate. SIZE: (n = 176) 18·1–23 × 12·9–15·8 (20·0 × 14·2). Double-brooded in Zimbabwe, multiple-brooded in South Africa.

LAYING DATES: Zambia, Oct (1 record, but immatures all months Luanshya); Malaŵi, Oct (in breeding plumage, possibly carrying food, Dec); Mozambique, Oct; Botswana, Aug–Sept; Zimbabwe, Aug–Mar (Aug 5, Sept 27, Oct 23, Nov 21, Dec 18, Jan 14, Feb 8, Mar 6 clutches); South Africa: Natal/Zululand, Sept–Mar (Sept 8, Oct 7, Nov 3, Dec 19, Jan 5, Feb 4, Mar 3 clutches), Cape, Aug–Mar (peak Oct), Transvaal, Aug–Mar (Aug 1, Sept 53, Oct 58, Nov 60, Dec 40, Jan 39, Feb 51, Mar 20 clutches).

INCUBATION: long spells on nest by day (R. A. Earlé, pers. comm.). Period: 15–16 days.

DEVELOPMENT AND CARE OF YOUNG: young fed by both parents; leave nest at 20–21 days; return to nest for at least another 12 days before final departure.

BREEDING SUCCESS/SURVIVAL: 3 birds South Africa lived 8–10 years (Earlé 1987g).

Hirundo albigularis

X Breeding records outside normal breeding range

Hirundo angolensis Bocage. Angola Swallow. Hirondelle d'Angola.

Hirundo angolensis Bocage, 1868. J. Sci. Math. Phys. Nat., Acad. Sci. Lisbôa 2, no. 5, p. 47; Huila, Angola.

Forms a superspecies with *H. rustica* and *H. lucida* (and in S Asia and Australasia with *H. tahitica* and *H. neoxena*).

Range and Status. Endemic resident and partial local migrant. W Gabon (Mouila, Tchibanga), SW Congo, Angola (Cabinda and in W from Ulge and W Malange south to Huila and Cunene), and inland along Congo R. to *c.* 16°E at Kwamouth. Also NE Namibia (Caprivi Strip), N Zambia (Mwinilunga and Ndola areas), Zaïre (from E Upper Uelle District to Kasai and N Shaba), Uganda north to Bunyoro, Teso and Mt Elgon, W Kenya (mainly Nyanza but east to Nyahururu, Timau, Naro Moru, Mogotio, Mara R. and Nairobi), Rwanda, Burundi, Tanzania (L. Victoria basin to Iringa and Songea) and N Malaŵi. Seasonally abundant Uganda, Kenya and W Tanzania; frequent, Angola; rare to locally common, Zambia; uncommon, Malaŵi. Vagrant, E Tanzania (Uluguru Mts: Morningside and Bunduki).

Description. ADULT ♂: forehead dark chestnut, rest of upperparts glossy blue-black; lores, stripe under eye, ear-coverts and sides of neck black. Tail-feathers blue-black, all except central pair with large white patch on inner web, outer pair slightly elongated. Chin, throat, foreneck and upper breast reddish chestnut; glossy blue-black band across breast, broadest at sides, narrow or even partly absent in centre; rest of underparts dull ash-brown, undertail-coverts tipped white. Wings black, primaries and secondaries with dull blue gloss, tertials and upperwing-coverts with glossy blue edges, axillaries and underwing-coverts ashy brown. Bill black; eye brown; legs brownish black. ADULT ♀: like ♂ but slightly smaller, and outer tail-feathers shorter. SIZE: (10 ♂♂, 10 ♀♀) wing, ♂ 113–127 (118), ♀ 114–122 (117); tail, ♂ 48–59 (52·6), ♀ 41–58 (50·4), depth of fork, ♂ 12–21 (17·3), ♀ 10–16 (12·4); bill to feathers, ♂ 7–10 (8·1), ♀ 7–9 (8·2); tarsus, ♂ 7–9 (8·3), ♀ 8–10 (8·4). WEIGHT: (Uganda) ♂ (n = 7) 12–18 (15·6), ♀ (n = 6) 16–20 (18·0); (Malaŵi) unsexed (n = 4) av. 17·8.

IMMATURE: like adult but upperparts duller with less gloss; forehead and throat paler chestnut; outer tail-feathers shorter.
NESTLING: unknown.

Field Characters. A barn swallow type with red forehead and throat, blue breast-band and white spots in tail; tail-streamers short, like Red-chested Swallow *H. lucida*. Best distinguished from other group members by drab appearance (grey-brown underparts and underwing), and from Barn Swallow *H. rustica* also by narrow breast-band and red of throat extending to breast.

Voice. Tape-recorded (C). Song a weak twittering.

General Habits. Poorly documented but evidently like those of Barn Swallow. Inhabits open country, grasslands, plains, cultivated fields and air space above open water; also bush country, edges of swamps and forests, and villages. In pairs or groups of 3–5; often in large flocks (up to 500) in non-breeding season. Flight rather sluggish. Rests on trees or telegraph wires.

Partially migratory, e.g. in E Africa, 500 on tiny Magogo Is in L. Victoria at midday on 29 July (Britton 1980); Malaŵi, present May–Oct; in Zambia largely absent during rains (Dec–Mar), although a few present all year (Aspinwall 1982).

Food. Insects, including winged ants, small black wasps (Impenetrable Forest, Uganda: R. M. Glen, pers. comm.), mayflies and small beetles.

Breeding Habits. Solitary nester, occasionally colonial (Gabon, 1 colony of *c.* 150 nests on government building and 6 pairs in thatched house: Rand *et al.* 1959). In courtship, ♂ (?) arches and quivers wings while making weak twitters.

NEST: open cup built with mud and grass, lined with grass and feathers, not very bulky, plastered onto overhang of bank, cliff or large rock, roof of cave, bridge support, sometimes house wall, verandah or under eaves of building.

EGGS: 2–3, av. (n = 9 clutches, Uganda) 2·89 (T. Butynski, pers. comm.). Laid on consecutive days over 3-day period. White, spotted with rusty brown. SIZE: (n = ?) 18·3–19·5 × 13·1–13·5 (18·8 × 13·4). Triple-brooded (Uganda: T. Butynski, pers. comm.).

LAYING DATES: Gabon, May, Sept; Angola, Aug–Sept; E Africa: Uganda, Jan–July, Oct–Dec (pair in Bwindi Forest Preserve hatched young Oct, Dec and Aug: T. Butynski, pers. comm.); Kenya, May; Tanzania, Nov–Feb; Region A, Mar–June; Region B, Jan–June, Oct–Nov; Region C, Jan, Aug, Oct–Dec;

Region D, Apr–May (throughout E Africa breeding largely confined to rains); E Zaïre, Mar–May, Aug, Oct, Dec; Malaŵi, Oct–Dec; Zambia, May–Dec.

INCUBATION: begins after last egg laid. Period: c. 17–18 days (7 nests, Uganda).

DEVELOPMENT AND CARE OF YOUNG: young remain in nest 22–27 (23) days (11 nests, Uganda); return to nest at night for at least 3 weeks after fledging (T. Butynski, pers. comm.).

BREEDING SUCCESS/SURVIVAL: of c. 55 eggs (20 nests), 40 hatched and c. 90% fledged (from 17 nests, mean 2·12 fledglings) (T. Butynski, pers. comm.).

Hirundo lucida Hartlaub. Red-chested Swallow. Hirondelle à gorge rousse.

Plate 8
(Opp. p. 113)

Hirundo lucida Hartlaub, 1858. J. Orn. 6, p. 42; Casamance R., Senegal.

Forms a superspecies with *H. rustica* and *H. angolensis* (and in S Asia and Australasia with *H. tahitica* and *H. neoxena*).

Range and Status. Endemic resident with some movements in non-breeding season, in 3 widely separated populations: (1) in W Africa from S Mauritania (Senegal R. valley; also Maghmouda and Tilemsi), Senegambia, Guinea, Sierra Leone, Mali (north to c. 17·5°N in Sahel), Burkina Faso, NW Liberia, N and central Ivory Coast (south to Lamto) to SW Niger (Niamey, Saga), Ghana (N savannas including Bolgatanga, Mole, Tamali and Tumu), Togo south to 8°N and Benin (Pobé, lower Oneme R. valley); vagrant W Nigeria (1, Jebba, 9° 8′N, 4° 49′E); (2) in Congo and Zaïre (Congo R. valley from Brazzaville to Kisangani); and (3) Ethiopia (W and SE Highlands). Common to abundant; may be increasing as bridge-building makes further nest-sites available.

Description. *H. l. lucida* Hartlaub (including '*clara*'): W Africa. ADULT ♂: forehead dark chestnut, rest of upperparts glossy blue-black, lores, stripe under eye and ear-coverts black, latter with some blue gloss. Tail-feathers blue-black, all but central pair with extensive white patch on inner web; outer pair slightly elongated (**A**, p. 190). Chin, throat and upper breast reddish chestnut, bordered below by glossy blue-black band, broadest at sides, narrowest in centre, usually complete, sometimes broken; rest of underparts white. Wings brownish black, primaries and secondaries washed blue, tertials and upperwing-coverts broadly edged blue, underwing-coverts dusky, under primary coverts tipped dark blue, axillaries whitish. Bill black; eye brownish black; legs brown. ADULT ♀: like ♂ but outer tail-feathers shorter. SIZE: (10 ♂♂, 10 ♀♀) wing, ♂ 103–118 (110), ♀ 101–121 (109); tail, ♂ 48–69 (56·4), ♀ 46–60 (54·1), depth of fork, ♂ 18–31 (21·9), ♀ 9–25 (13·8); bill to feathers, ♂ 7–8 (7·3), ♀ 7–9 (7·4); tarsus, ♂ 6–8 (7·3), ♀ 7–9 (7·7).

IMMATURE: similar to adult but upperparts and breast less glossy, forehead and throat paler chestnut, and outer tail-feathers shorter than adult ♂.

NESTLING: unknown.

H. l. subalaris Reichenow: Congo and Zaïre. Forehead with chestnut darker and more extensive than in *lucida* and throat darker rufous.

H. l. rothschildi Neumann: Ethiopia. Differs from *lucida* in being more violet, less blue above; axillaries white (not dusky). WEIGHT: 1 unsexed 15·1.

Field Characters. A medium-sized barn swallow type with red forehead and throat, blue breast-band and white spots in tail. Very like Barn Swallow *H. rustica* but smaller, with short tail-streamers; red band on forehead narrower, red of throat extends to upper breast, breast-band narrower (always broad and complete in Barn Swallow), underwing-coverts dark, not buff, white tail spots larger. Immatures hard to distinguish since tails same length and plumage often scruffy-looking, but immature Barn Swallow should always have broader brown breast-band, smaller tail spots, buffy underwing.

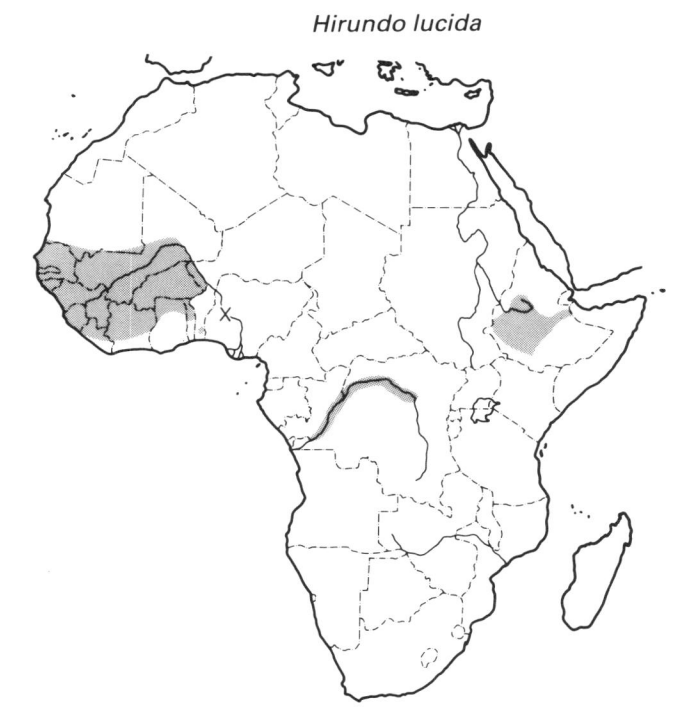

Voice. Tape-recorded (MOR). A weak twittering warble, similar to Barn Swallow but clearer and louder.

General Habits. In W Africa occurs in villages, around bridges, over marshes, and in bush and open country; in Congo and Zaïre also in forest clearings and along river banks; in Ethiopia in grassland and over large freshwater lakes, rivers and highland streams and marshes at 1800–2750 m.

Occurs singly or in pairs, or in flocks of 6–10, occasionally 40 or more. Flight swift and darting. Perches on telegraph wires, bare branches, and rocks and posts in rivers. Roosts communally in large numbers at night.

In W Africa migratory status uncertain; populations both sedentary and migratory: resident Gambia and N Ghana (Tumu, 10° 55′N); but wet season visitor central Ghana (Tamale 9° 20′N), Togo (several localities between July and Aug, once 11° Dec); and S Benin (common Pobé, Mar–June, Oct). Apparently sedentary in Ethiopia, Congo and Zaïre.

Food. Insects, including flying termites.

Breeding Habits.
NEST: open cup of grass and mud, lined with grass and feathers; placed under eaves, bridge or rock; also in verandah, hut and termite mound.
EGGS: 3–4, white with reddish spots. SIZE: (n = 4) 17·5–18 × 12·5–15 (17·6 × 13·4), (n = ?) 16·8–20·9 × 12·5–14·2 (18·9 × 13·2). Double-brooded (Walsh *et al.* 1990).
LAYING DATES: Mauritania, July–Aug; Senegal, Mar–July, Dec; Gambia, Mar–June; Mali, Aug–Sept; Niger, Sept–Oct; Liberia, Feb–Mar; Ghana, Mar, July (visiting nest-sites); Togo, Mar–June, Dec (apparently raised a brood). Zaïre, Feb–May, Aug–Sept (in rains); Ethiopia, Apr–June, possibly Feb.

Hirundo rustica

Hirundo lucida

A

Plate 8
(Opp. p. 113)

Hirundo rustica **Linnaeus. Barn Swallow. Hirondelle de cheminée.**

Hirundo rustica Linnaeus, 1758. Syst. Nat. (10th ed.), p. 191; Sweden.

Forms a superspecies with *H. lucida*, *H. angolensis* and (in Asia and Australasia) *H. tahitica* and *H. neoxena*.

Range and Status. Holarctic: breeds N Africa, whole of Europe, and nearly all of Asia and N America between 20° and 67°N (70°N in Norway). Winters in subsaharan Africa, India, SE Asia to Papua New Guinea, and South America. Summering and wintering grounds exclusive, except for S Spain, S Iran, and Hainan; Nile valley race is resident. Accidental Iceland, Azores and Spitzbergen.

Breeding summer visitor to N Africa (east to Cyrenaica); has bred in Hoggar Mts, S Algeria. Resident and partial migrant in Egypt; passage migrant between Eurasia and subsaharan Africa, throughout N half of Africa; and winter visitor to entire subsaharan Africa south of a line from Sierra Leone to Djibouti (but common at S end of L. Chad, Jan). Scarce west of 14°W (implying that species does not follow coast but crosses W Sahara).

Common to abundant, but less (or only locally) so in arid regions and at high elevations. 4–8 million pairs breeding in W Europe produce 14–28 million fledglings (Cramp 1988), so in autumn 22–44 million Barn Swallows enter Africa to winter there, with perhaps twice that number additionally from E Europe and Asia. Concentrations of hundreds or thousands occur in most African countries, with greatest concentrations at reed-bed roosts, e.g. tens or hundreds of thousands at Somerset West (Cape), Mashu (Natal) and Rosherville and Vrischgewaard (Transvaal), c. 160 000 at Kitwe and at Choma and 2–3 million near Ndola (all Zambia: Penry 1979, Taylor 1982), c. 1 million at Eldoret, Kenya (Best 1977) and also near Tafo, Ghana, in rain forest zone. Great flocks roost also in mealie fields in southern Africa, e.g. 100 000 (Zimbabwe).

Hirundo rustica

Description. *H. r. rustica* Linnaeus: N Africa east to Libya, and Eurasia east to W Altai Mts; winters throughout subsaharan Africa (also in India, SE Asia). ADULT ♂ (breeding): forehead chestnut; crown to uppertail-coverts, sides of neck and upperwing-coverts glossy blue-black with violaceous tinge. Tail blackish brown with strong blue gloss, inner web of all feathers except central pair with large round or oval white patch (a long oval or bar in outermost rectrix: **A**, p. 190); T6 long and attenuated. Lores blackish, chin and throat chestnut; upper breast steely blue-black forming band c. 15 mm wide, sharply demarcated from lower breast; lower breast, belly and undertail-coverts white. Remiges blackish brown with exposed surfaces of feathers glossed blue (occasionally violaceous or greenish blue). Axillaries and underwing-coverts pale buff; underside of remiges and rectrices dark grey-brown, rectrices showing the white marks. Bill black; eye dark brown; legs and feet black, tinged brown. ADULT ♂ (non-breeding): in fresh plumage (autumn) white underparts washed with pink. ADULT ♀: like ♂ in breeding and non-breeding plumages, but tail averages shorter. SIZE: (Europe, summer) wing, ♂ (n = 64) 120–129 (124), ♀ (n = 26) 118–125 (122); tail, outer feather, ♂ (n = 73) 89–119 (103), ♀ (n = 30) 70–99 (86·9); depth of fork, ♂ (n = 75) 47–78 (61·9), ♀ (n = 31) 27–57 (44·2); bill to skull, ♂ (n = 25) 11·2–12·8 (12·2), ♀ (n = 11) 11·7–12·8 (12·3), bill to distal corner of nostril, ♂ (n = 24) 5·3–6·3 (5·7), ♀ (n = 10) 5·4–6·2 (5·8); tarsus, ♂ (n = 20) 10·3–11·5 (10·8), ♀ (n = 12) 10·2–11·4 (10·8). WEIGHT: (Uganda, Zambia) Oct (n = 23) 15·2–19·1 (17·0), Nov (n = 81) 14·5–20·8 (17·7), Dec (n = 100) 13·9–21 (17·6), Jan–Feb (n = 51) 16–20 (17·9), Mar (n = 39) 16·5–22·2 (19·1), Apr (n = 3) 17·2–17·9 (17·6); (South Africa, n = 1147) avs: Oct 21·5, Nov 20·8, Dec 20·0, Jan–Feb 19·2, Mar 20·2, Apr 19·7; (Morocco) Mar–Apr (n = 2357) 11–28·2 (16·0); (W Germany) summer, ♂ (n = 125) 16·1–21·4 (18·8), ♀ (n = 137) 16–23·7 (19·4). All African weights above are of unsexed adults. Adults starved to death in poor weather: (South Africa) Nov (n = 51) 11·3–15·4 (13·0) (Skead and Skead 1970), Apr (n = 28) 11·7–15·8 (13·6) (Broekhuysen 1953), (Morocco) Apr (n = 31) av. 12·8 (Ash 1969).

IMMATURE: like adult but duller, and lacks long outer rectrices. Forehead, chin and throat pale rufous; upperparts, wings and tail dark brown, glossed blue-black; some wing-coverts narrowly pale tipped, giving scaly appearance; lores, ear-coverts, sides of neck and breast-band dark blackish brown.

NESTLING: upperparts with sparse tufts of long, pale grey down; underparts bare; mouth yellow, gape-flanges pale yellow. Second coat of short, woolly down grows between sprouting feathers, in second week.

H. r. savignii Stephens: Nile valley and delta, Egypt; resident. Lower breast, belly and undertail-coverts deep rufous. Immature differs from immature *H. r. rustica* in having chin and throat dark rufous and belly and undertail-coverts dark rufous-buff.

H. r. transitiva (Hartert): Israel, Lebanon, S Syria; winters in (or migrates through) Egyptian W Desert; records claimed in Tunisia and Sudan. Underparts pale rufous.

H. r. gutturalis Scopoli: E Asia, wintering mainly in SE Asia; identified wintering in small numbers in Botswana, Zimbabwe, Transvaal, Natal and S Mozambique. Like nominate race but blue-black upper breast band narrow in centre, so that chestnut throat often connects to buffy lower breast. Smaller; wing, ♂ (n = 15) 112–122 (116), ♀ (n = 13) 110–117 (113).

Field Characters. The most widespread, and often the commonest, swallow in Africa. Uniform blue-black upperparts, rufous throat, blue-black breast-band, white underparts and wing linings, long tail-streamers and white windows in tail distinguish it from all other swallows except the following. Egyptian residents dark rufous (not white) below. Red-chested Swallow *H. lucida* very similar, but is smaller, with shorter tail-streamers, more white in tail (**A**, p. 190), narrower rufous forehead band, narrower breast-band, rufous of throat continuing onto breast. For distinction of immatures, see *H. lucida*. Ethiopian Swallow *H. aethiopica* is smaller still, with shorter tail-streamers, wider rufous forehead band, white or buffy (not rufous) throat, and broken breast-band. Angola Swallow *H. angolensis* q.v. differs mainly in having ash brown (not white) underparts and wing linings.

Voice. Tape-recorded (33, 86, 88, B – 62, 73). Song (♂ only) is simple but pleasing, a warbling twitter 3–15 s long including several dry rattles. Contact call, used freely all year, is a quite loud 'witt-witt', sometimes run together, e.g. 'tswit-tswit-tswit-tswit'. For follow-contact, enticement, courtship, copulation, warning, alarm and juvenile calls, see Cramp (1988).

General Habits. In N Africa breeding birds inhabit farmland, fields, pasture, open leafy suburbs, rural villages, and other man-formed open countryside with sheds and outhouses for nesting, and nearby rivers, lakes or ditches for drinking and foraging. On migration occurs over all habitats, including sea-straits, desert, woods and forest, concentrating at oases, along rivers, coasts and other leading topographical lines. In subsaharan Africa, occupies wide variety of man-made and natural habitats: arid, open and bushy grassland, woodland, cultivated land, pasture, open water, forest edge and, particularly, swampy country. At all altitudes from sea level up to 2200 m (Zimbabwe) and 3000 m (E Africa).

Solitary or locally gregarious (e.g. at drinking pool) when breeding; at other times more or less strongly gregarious, foraging in loose flocks of tens or hundreds, commonly with many other species of hirundines, and roosting at night in flocks of hundreds to millions (see above). Forages in flight, mainly low over ground or water and near herds of large mammals; recorded foraging around party of Ruffs *Philomachus pugnax* feeding on ground, taking flies flushed by them from low vegetation. Also takes insects direct from vegetation, e.g. *Locris* bugs from lawn grass by swooping, and insects from shrubs by hovering (Brooke 1956). In South Africa flocks of up to 50 swallows eat seeds of *Acacia cyclops* trees (see Food), by flying slowly upwind to tree 1–5 m tall, hovering momentarily at pod, plucking entire seed, and continuing in flight; some swallows simply perch in tree to feed (Hofmeyr 1989). *Acacia cyclops* is an introduced Australian species, so this habit must have developed recently (J. M. Lock, pers. comm.). Not attracted to grass fires (or attracted much less than other hirundines). Flight elegant, with easy wing-beats, short but graceful glides, dashing chases after insects, lifts and abrupt changes of pace and direction. Can hover momentarily. Perches readily on grasses, twigs, wires and buildings. Frequently calls; one of the more vocal swallows. Drinks in glide at surface of unruffled water, with lower mandible leaving long wake.

Birds in huge flock gathering to roost perform mass movements high in air above roost site for *c.* 10 min before plunging vertically in small groups into reeds or mealie field, when light has almost faded (Webber 1975); one huge flock perched on ground before going to roost, with groups rising to fly with reckless speed in undulating follow-my-leader fashion (Penry 1979). Another cloud of birds flying high condensed and swooped down to maize field after dark, with much twittering; at sunrise they ascended 'like a great smoke spiral' until out of sight above low clouds (Hornby 1973); may leave roosts flying upward in waves, over period of a few min. Also roosts in elephant grass *Pennisetum*, occasionally in bushes, trees and other crops and once on ground (tropical Africa:

B

Cramp 1988). 250 birds slept in rows or huddling in clusters in stone-built barn in cold weather, South Africa (**B**), many dying (Broekhuysen 1961). Birds come to roost site from 30 to 40 km away (Namibia: Becker 1974); roosting flights can look like migration (Roberts 1989). Juveniles predominate in large roosts; at 2 Kenya roosts 68 and 80% of birds were juveniles (Reynolds 1971, Best 1977), and in Zaïre proportion of juveniles at a roost increased through winter to 100% in Mar (de Bont 1962).

H. r. savignii starts moulting Apr–June, finishes Sept–Nov. NW African *rustica* starts moulting on breeding grounds; moult half completed in Ivory Coast in Sept and terminated Nov–Dec. Some European *rustica* start moulting there Aug–Oct, but most delay commencement until arrival in wintering grounds, where (southern Africa) moult begins with P1 from mid-Sept to mid-Nov and finishes with P10 from late Jan to late Mar, i.e. duration *c.* $4\frac{1}{2}$ months (Stresemann and Stresemann 1968, Dowsett 1966, Piper 1974, Loske and Lederer 1988, and Cramp 1988 reviewing 8 other sources).

Egyptian population resident; NW African one winters probably in Ivory Coast; European and W Asian birds winter in subsaharan Africa. Travels in small or vast flocks by day and (at least some over Mediterranean) by night. Crosses Mediterranean and Sahara on broad front, but migrating flocks tend to follow coasts and similar leading lines. Present in NW Africa mainly mid-Mar to mid-Sept, S Ghana Sept to early May (earliest 29 July, latest 3 June), E Africa and South Africa Aug–May, mainly Sept–Apr, rarely oversummering (June, July) in both areas. Main passage through tropical Africa is mid-Sept to early Nov, with exodus from southern Africa in Feb (heavy passage in Zambia from mid-Feb), and spring migration through N Africa in mid-Mar to late Apr. Adult ♂♂ arrive on breeding grounds first, then adult ♀♀, then 1-year-olds. Some birds already paired on arrival, and pair-bonding sometimes starts on or before migration: occasionally sings in winter (Togo, Jan: Douaud 1957) and commonly sings and displays on spring passage through N Africa.

Some 2 million Barn Swallows spending part of their

lives in Africa have been ringed (e.g. 980,000 in Britain, 144,000 in southern Africa, 14,000 in E Africa, up to 1987). 260,000 ringed in E Asia have not resulted in any African recoveries (Kang 1971). Numerous recoveries indicate that: British and Irish birds winter mainly in eastern South Africa (but extended their range westward in South Africa after 'winter' of 1962-63: Rowan 1968, Mead 1979); Scandinavian birds winter mainly in Congo (Cramp 1988); N European birds winter mainly in W Congo basin; central European birds winter somewhat further south in Congo basin; many German birds winter in S Ghana; birds ringed in Transvaal breed mainly west of Urals; birds ringed in Cape Province and Natal breed mainly east of Urals.

Ringed abroad, recovered in Africa: (a) from Britain and Ireland to Tunisia (11), Algeria (45), Western Sahara (1), Mauritania (3), Senegambia (1), Liberia (2), Ghana (6), Nigeria (20), Cameroon (6), Principe I. (1), Bioko (1), Uganda (2), Congo (4), Zaïre (17), Angola (1), Namibia (8), Botswana (2), Zimbabwe (4), Mozambique (2) and South Africa (348) (Mead and Clark 1987, 1988); (b) from NW Europe (France, Luxembourg, Belgium, Holland, Germany) to NW Africa (31), Mauritania (3), Senegambia (2), Sierra Leone, Ivory Coast, Ghana, Nigeria and Cameroon (27), N Zaïre (3), Gabon (1), near lower Congo R. (12), S Zaïre (3), Zambia (1) and South Africa (11) (Zink 1970); (c) from central Europe (Germany, Czechoslovakia, Switzerland) to NW Africa (15), Libya (1), Ghana (3), Nigeria (4), W and S Congo basin (18) (Zink 1970); (d) others from Germany to Egypt (1) and Tanzania (1), Poland to Uganda (2), Kenya (2) and South Africa (1), Estonia to Malawi (1) and South Africa (2), USSR to South Africa (3), Belgium to Ghana, Ivory Coast, Congo, Zaïre and South Africa (5), Holland to Ivory Coast, Zambia and Namibia (4), Denmark and Finland to Namibia (3), Czechoslovakia, Finland, Hungary, Greece and USSR to Kenya (7), Spain to Ghana, and Switzerland to Nigeria, Zaïre and South Africa (1 each) (see maps in Davis 1965, Mead 1970, Zink 1969, Loske 1986).

Ringed and recovered in Africa: Uganda to Kenya (1), Zimbabwe to Uganda (1), Transvaal to Zimbabwe, Kenya, Uganda and Nigeria (1 each).

Ringed in Africa, recovered abroad: (a) from W Cape Province to Britain (7) and USSR between 34° and 62°E (17) (mainly between 53° and 57°N); (b) from Transvaal to Britain and Ireland (14), Europe between 5° and 25°E (20) (north to 66°N), USSR between 28° and 40°E (33) and USSR between 40° and 93°E (10); (c) from Natal to USSR between 23° and 45°E (12) (north to 65°N) and USSR between 62° and 82°E (13); (d) Zimbabwe and Zambia to France (1), Czechoslovakia (1), Finland (1), Leningrad (1), Ukraine (6), and 86°E near Leninsk-Kuznetskiy (1); (e) others from Libya to Czechoslovakia (1), Namibia and Botswana to Britain (3), Czechoslovakia (1) and Poland (2), Zaïre to Czechoslovakia (1), Ukraine (1) and N Kazakhstan (1), Uganda to Turkey (1), and Kenya to Lebanon (1), Iraq (1), Ukraine (1), Kabardino-Balkarian ASSR (1) and Dagestanian ASSR (1) (see map in Rowan 1968, and see also de Bont 1960, Ingram 1974, Oatley 1983, Loske 1986, Backhurst 1981, 1988).

2 rapid recoveries, from Cape Town to 42° 15'N, 42° 53'E near Kutaisi, USSR, 8500 km in 34 days, and from Johannesburg to 54° 39'N, 86° 10'E near Leninsk-Kuznetskiy, 12,000 km in 34 days, suggest av. transit of 300-400 km per day.

Food. Small airborne insects, mainly ants, flies and beetles (25 birds, Uganda); also spiders, amphipods, and in South Africa arillate seeds of rooikrans *Acacia cyclops* (Every 1988, Hofmeyr 1989). Plucks boll worms and army worms from vegetation in flight (*idem*), cercopid bugs *Locris transversa* from lawn (Zimbabwe: Talbot 1974), and amphipods from beach (South Africa: Broekhuysen 1952). 4 birds, Orange Free State, each contained 17-120 ants (van Ee 1988). 2 road kills, Kenya, contained 29 beetles of c. 11 spp. 4·5-10·0 mm long, 22 winged ants, 4 wasps 2-6 mm long, 9 flies, a bug, and lepidopterous wings (Lack and Quicke 1978). Birds feeding around flock of Ruffs were evidently taking minute midges *Polypedilum lobiferum* disturbed by the waders from grasses (Taylor 1964). Casualties at a Zimbabwe roost had stomachs packed solid with small flying ants (Hornby 1973). In Europe, takes great variety of insects; numerically, 78% of adult diet in Poland (355 items) were Hymenoptera and 14% Coleoptera; 62% of nestling diet in Czechoslovakia (4606 items) were Diptera and 28% Hymenoptera (Cramp 1988).

Arillate seeds of alien *Acacia cyclops* form significant component of diet in Cape (Port Elizabeth, Betty's Bay); in Jan. 80% of swallows present in area eat them, each bird taking av. 3·4-7·3 seeds daily. Black seed and attached red aril are swallowed whole, the aril digested and the seed voided or regurgitated. 70 arils weigh 1 g and protein content is 13% (20% in swallows' insect prey) (Hofmeyr 1989).

Breeding Habits. Few African data, but exhaustively studied in Europe (Cramp 1988). Monogamous, occasionally bigamous. Solitary nester, or sometimes a loose colony of c. 5 pairs; rare instances of up to 280 pairs in 1 farm and 95 pairs in one stable (Germany). Mean number of pairs nesting together (n = 141) 2·9. Density of 1·5-11 pairs per m² over much of N Europe. Pair defends territory of 4-25 m² around nest. Pair remains faithful to nest-site for years. Only c. 1% of 1-year-olds return to nest-site, but most breed within 10 km (Germany) or 30 km (Britain) of birthplace (Cramp 1988). In Morocco 2, perhaps 3, broods raised in same nest (Mayaud 1986). ♂ selects site, shows it to ♀ by landing on site, singing and giving enticement call; when ♀ accepts site she does most of the nest-building while ♂ perches nearby, singing. Mated ♂ may refurbish or build nest before mate arrives in the area; otherwise, pair formation is completed when ♀ begins roosting at nest site. ♂ guards ♀ for many days, when neighbouring ♂♂ chase and harass her (particularly in large colony). Courtship feeding does not occur. ♂ solicits ♀ by vigorous singing, hovers above her, mounts and copulates with his wings flailing (Cramp 1988).

NEST: shallow cup or half-cup, made of mud pellets usually mixed with some plant material, lined with feathers; *c.* 20 cm wide and 10 cm deep. Built mainly by ♀, with ♂ helping to bring material, taking *c.* 8 days (plus 2 days to add lining). Sited 2–5 m above ground, on vertical stone, masonry or wooden surface, with narrow or considerable ledge below nest, in airy outhouse or farm building.

EGGS: 2–7, mainly 4–5. Laid daily, within 2 h of sunrise. Long ellipses; smooth, glossy; white, lightly marked with red-brown and some purple or grey spots. SIZE: (n = 250, Europe) 16·7–23 × 12·3–14·8 (19·7 × 13·6).

LAYING DATES: Morocco, Feb–June (mainly Apr–May); Algeria, Tunisia, Mar–June or July.

INCUBATION: mainly or entirely by ♀, beginning with last egg. Period: (n = 93) 11–19 (15·25) days.

DEVELOPMENT AND CARE OF YOUNG: eyes open at 4–9 days. Young first fed by ♀ within 1 h of hatching, then brooded assiduously by her for 3 days. ♂ feeds young during that time, and takes greater share in brooding for next few days. Young start to beg at 3–4 days. Each food bolus given to 1 young, and not shared. Parents remove faecal sacs until young 12–14 days old. Period: (n = 110) 18–23 (19·5) days. Brood fledges usually over 2 days, then fed by parents for 3–12 (av. 5·3) days.

BREEDING SUCCESS/SURVIVAL: in European studies, breeding success rates of 72, 76, 87 and 88% measured (total of 3088 eggs); also 86% (3941 1st-clutch eggs), 92% (2651 2nd-clutch eggs) and 85% (54 3rd-clutch eggs) (Cramp 1988). Main egg losses are 8·5% to predation and 7·2% to infertility.

In Europe, av. annual 1st-year mortality 70, 74 and *c.* 80%, 2nd-year mortality 47 and 69%, and av. annual adult mortality 43, 60–65 and *c.* 70% (Cramp 1988). Oldest ringed bird 15 years 11 months.

Many migrants die during Saharan crossing, and when encountering cold weather; also when blown out to sea. Common prey of many falcons.

References
Cramp, S. (1988).
Dowsett, R. J. (1966).
Mead, C. J. (1970).
Mendelsohn, J. M. (1973).
Møller, A. P. (1984b).
Rowan, M. K. (1968).

Genus *Delichon* Horsfield and Moore

House martins. 2 closely-allied species, one west and one east Palearctic, forming a superspecies, and a third species in the Himalayas. *Delichon* is anagram of *Chelidon*, Greek for swallow. Small, black-and-white, colonial martins, tails forked but without streamers, distinguished from other hirundines by thickly white-feathered tarsi and toes. Nest a half-cup, built of mud, attached commonly to house walls. Winter in tropics. 1 species breeds south to NW Africa and winters widely in Africa. Rarely, hybridizes with Barn Swallow *Hirundo rustica*; one such hybrid netted in Zambia (Dowsett 1978).

Plate 10
(Opp. p. 161)

Delichon urbica **(Linnaeus). Common House Martin. Hirondelle de fenêtre.**

Hirundo urbica Linnaeus, 1758. Syst. Nat. (10th ed.) 1, p. 192; Europe.

Forms a superspecies with *D. dasypus* (E Palearctic).

Range and Status. W Palearctic from NW Africa and Britain to central Asia and Kashmir, north to 70°N. Migrant to subsaharan Africa.

Breeding summer visitor to Maghreb (Morocco to Tunisia), where common; may breed on Jebel Nafusa, Libya, and breeding suspected at Sebha, Fezzan, 1959. Non-breeding 'winter' visitor to entire subsaharan Africa except Somalia and arid parts of Kenya, NE Tanzania and W South Africa. Sight records Socotra, May (Ripley and Bond 1966). Widespread at least over higher pected – flies very high and easily overlooked. 90 million estimated to enter Africa each autumn; high feeding flocks of 20–300 often seen in winter in e.g. Nigeria, Cameroon, but whereabouts of the millions remain mysterious. 2000–3000 in valley on Mt Elgon (Kenya) and later estimate of 50 000 there (Rolfe and Pearson 1973) is largest recorded African assemblage. Winters south to Cape where locally common Oct–Apr. Highest counts in southern Africa: Zaïre *c.* 1000 Feb; Zambia 1000 Oct (Mwenze), 1000 Nov (Lusaka), 2000 Dec

(Rufunsa), 1000 Jan (Nyika); Zimbabwe 450 and 860 Nov (Nyabasanga); South Africa 800 Dec and Feb (Cape Province), 2450 and 3000–4000 Apr (Transvaal). Isolated nests or small breeding colonies in Namibia (Otjiwarongo 1928, Windhoek 1985) and South Africa: Cape Town 1892, 1893, Somerset West 1969, Transvaal 1928, E Cape Province (Keiskammahoek for several years up to 1946, Kokstad 1967).

Description. *D. u. urbica* (Linnaeus): only race known from Africa (although the smaller southern races *fenestrarum* (central Europe) and *meridionalis* (S Europe) must also winter there). ADULT ♂ (breeding): forehead, crown, nape, mantle, back, scapulars and inner wing-coverts steely blue-black, glossy. Rump snowy white, sometimes with very narrow brown shaft-streaks; white patch *c.* 30 mm deep. Uppertail-coverts glossy blue-black. Tail blackish brown; underside brown, slightly glossy. Lores black, ear-coverts blue-black, sides of neck glossy blue-black, all sharply defined from snowy white chin and throat. Breast, belly and undertail-coverts white, flanks greyish white; undertail-coverts with dark shaft-streaks. Feathers of legs and toes white. Wings blackish brown, inner remiges very narrowly pale-fringed; underside silvery pale brown; axillaries and underwing-coverts greyish pale brown. Bill black; eye dark brown; legs and toes feathered white (skin pink). ADULT ♂ (non-breeding): like breeding ♂ but rump, cheeks, chin, throat and flanks buffy brown or brownish grey. Sexes alike. SIZE: (Morocco, Algeria, unsexed, n = 32) wing, 99–108 (104); av. wing length increases northward by 1 mm every 2·2° latitude; (Netherlands, Apr–Aug) wing, ♂ (n = 33) 106–115 (111), ♀ (n = 30) 105–116 (110); tail, ♂ (n = 26) 58–65 (61·1), ♀ (n = 25) 56–64 (60·4), depth of fork, ♂ av. 20·2, ♀ av. 18·6; bill to skull, ♂ (n = 10) 9·3–10·8 (9·9), ♀ (n = 18) 8·7–10·7 (9·8), bill to feathers av. 3·4 shorter; tarsus, ♂ (n = 13) 11–12 (11·5), ♀ (n = 18) 10·5–12 (11·4). WEIGHT: (Morocco, Apr, unsexed, n = 49) 12–17 (14·4); (Morocco, other months, unsexed, n = 252) 10–20 (14·5); (South Africa, n = 3) 18–19 (18·4); (South Africa, exhausted birds Nov, unsexed, n = 107) 10–15 (12·1); (South Africa, exhausted birds Apr, unsexed, n = 45) 12–16 (13·3); (NW Europe, May–June, unsexed, n = 30) 16–21 (18·4).

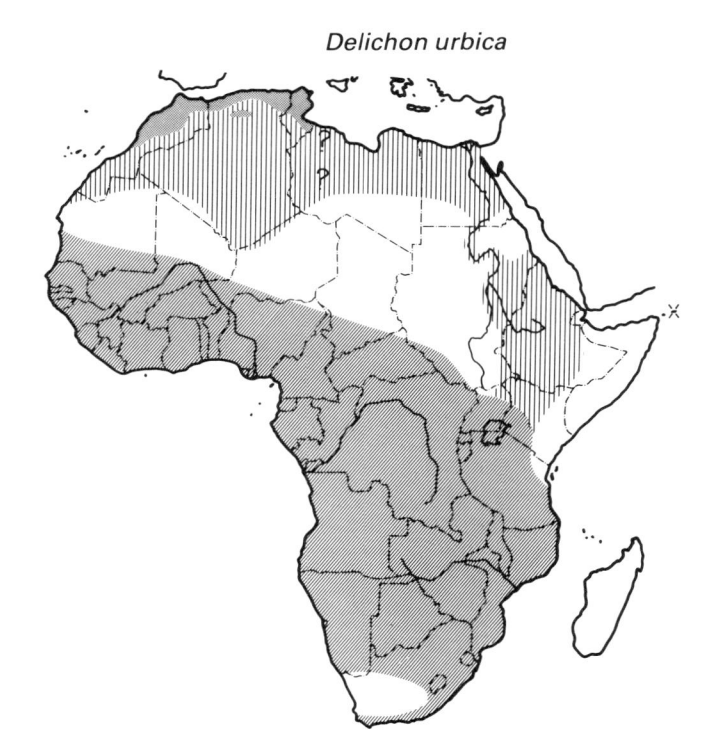

Delichon urbica

IMMATURE: dark greyish brown where adult is blue-black; some glossy blue feathers in mantle; rump off-white rather than snowy white, the patch only 20 mm deep; uppertail-coverts with white-fringed tips; underparts off-white, chin, throat and breast diffused with brownish grey; inner secondaries with white tips up to 2 mm deep.
NESTLING: in 1st week only a few tufts of long pale grey down on upperparts; skin pink, mouth yellow, gape-flanges pale yellow; in 2nd week short, dense, whitish down develops all over (except belly), the feathers growing through it.

Field Characters. A small, blue-black-and-white, high-flying martin; flight not graceful or swooping but flap-and-glide, on rather short, triangular wings. Blue-black above, with white rump, and white below. Rump and throat sometimes brownish in autumn and winter. Always a clear line of demarcation between black and white, just under eye. Tail with shallow fork, lacking streamers and white spots. Characteristic hard grating call often given in flight. Often associates with cliff swallows (e.g. Preuss's Cliff Swallow *H. preussi*, Red-throated Cliff Swallow *H. rufigula*), which are similar in size, shape and plumage, but lack capped appearance (with sharp demarcation between black cheeks and white throat) or have rufous in plumage. For distinctions from Pearl-breasted Swallow *H. dimidiata*, see that species.

Voice. Tape-recorded (GIB – B, C, 62, 73). Vocal. Song a sweet, very soft twittering. Call a hard 'prt' or 'pr-prt' like 2 pebbles being rubbed together, ♀'s call higher-pitched and more melodious than ♂'s. For other calls, see Cramp (1988).

General Habits. Inhabits airspace above breeding colonies, which throughout range are now almost entirely on urban and rural buildings and bridges; also forages far from nest, over sea-cliffs, forests and precipitous mountains up to at least 2 km from nest. In Nigeria associates with similar-looking Preuss's Cliff Swallow *Hirundo preussi*, both species foraging at least to height limit of vision using 10 × binoculars, and in Zambia with Red-throated Cliff Swallow *H. rufigula*. Throughout tropical Africa, frequents cliffs and valleys at 1600–3000 m, but also occurs over open country at lower elevations. In Zimbabwe strongly associated with hillsides, cliff-faces and similar rocky places, feeding so high in good weather as to be out of sight. Tolerates wet, windy and cool conditions but vulnerable to extreme heat and cold.

Feeds almost entirely on insects taken in aerial pursuits: flies up fast to catch prey from below, then glides down again. Occasionally hunts around bush fires or follows tractor ploughing herbaceous vegetation; takes caterpillars suspended from trees on silk.

In breeding season roosts in nest, occasionally in trees; after breeding, roosts communally in old nests – exceptionally up to 14 birds in a nest. Considerable circumstantial evidence from Germany and USSR that some birds roost on wing at great altitude. In Africa reported roosting in reeds (Zaïre, Verheyen 1952, Curry-Lindahl 1963; Tanzania, Sassi and Zimmer 1941; Zambia); also in trees and possibly in cliffs; but paucity of African roosting records suggests that Common House Martins sleep on the wing there also (Harper and Harper 1974).

120 adult and immature Common House Martins fluttered around nests containing young in active colony of Red-throated Cliff Swallows, Zambia, Oct. They mixed freely with the swallows, without aggression, and clung to outside of some (unbroken) nests; they entered the many broken nests with exposed grass-lined interiors, resting quietly in them, rearranging grass lining, pecking at broken mud edges, and gently bill-pecking each other. Similarly, Common House Martins entered broken nests of Lesser Striped Swallows *H. abyssinica*, Zambia, Oct (Fellowes 1971). Roosting in the nests was not proven in either instance.

Whole Palearctic population winters in subsaharan Africa; autumn migration in Europe and N Africa mainly Sept–Oct, with many late-breeders moving in Nov; return migration begins in Africa Mar, first European arrivals late Mar to early Apr, main arrivals Apr–May. Migrations within Africa poorly documented. Ringing recoveries: from Sweden, Germany and France (2) to Morocco, from Britain, Sweden, Germany and Netherlands to Algeria, from France (3), Germany and Poland to Tunisia, from Poland to Libya, from Netherlands to Nigeria, from Algeria to S Mali, from Switzerland to Cameroon, from Germany to Angola, from USSR (Rybachii, 55° 11′N) to South Africa (Ceres, Cape Town), from Germany to Botswana and Zambia (2), from Finland to Zimbabwe, from Norway and Sweden to Transvaal, from Kenya to USSR (Caucasus), and from Morocco to Tunisia. Occurs in southern Africa Oct–May, mainly Nov–Apr; occasionally 'oversummers' Transvaal May–Oct; 1 oversummering record E Africa (Samburu, Kenya, late June).

Food. Flying insects. Almost nothing known about diet in Africa (ants in Zaïre), but many studies in Europe detail wide range of small airborne insects down to 1·1 mm long (see Cramp 1988). Also takes a few airborne spiders.

Breeding Habits. Monogamous; breeds gregariously in dense colonies and sub-colonies, usually under house eaves. In colony, mud walls of adjacent nests often contiguous. Max. size 500 nests, but most colonies on houses are <5 nests; av. of 4·85 nests in 126 town colonies (Germany). Cliff colonies larger: av. of 21 nests in 38 cliff colonies (Britain). Selection of nest-site and mate intimately linked; a ♀ may choose mate by accepting his nest-site (Lind 1960, Cramp 1988). ♂ inspects several sites, and flies around chosen one, returning to it repeatedly; pair-bonding occurs when ♀ repeatedly visits site also. Courtship-feeding of ♀ by ♂ occurs. ♂ takes initiative in soliciting copulation; he 'shows' nest to ♀, lands, sings, and runs toward ♀ in crouched posture. Copulation takes 20–30 s; after copulation, birds preen or allopreen. Pair spends some time in nest before laying; laying often starts when nest only half-built; remainder of mud wall and soft lining added during laying period (Cramp 1988).

NEST: half-cup of mud built against vertical wall and overhanging roof, with small oval entrance at top; lined with down and feathers collected on the wing. Placed under eaves, bridges, culverts and natural overhangs in sea- and inland cliffs. Mud brought in single wet, almost liquid, pellets; in 10 nests 690–1495 (av. 1074) pellets. Pellets gathered up to 150 m away, from side of muddy puddle, bird taking 6–8 s to collect each one. Nest built by both sexes; pellet fixed in 30–60 s with rapid, shivering movements. Av. size of 5 nests: 178 wide, 143 high, 106 from front to back; entrance 24 high, 66 wide (Lind 1960, Cramp 1988).

EGGS: NW Africa 4–6; N Europe 1–9, mainly 4–5, av. 4·4. Eggs laid daily. Subelliptical; slightly glossy; white. SIZE: (n = 80, W Mediterranean) 16–19·6 × 12·2–13·5 (17·9 × 12·7); (n = 14, NW Africa) av. 19·5 × 13·5. WEIGHT: 1·68.

LAYING DATES: Morocco, Algeria, Tunisia, Mar–May; South Africa, Sept, Oct, Dec, Jan, May.

INCUBATION: by both sexes, with frequent changeovers: ♀ bout duration av. 13·1 min, ♂ 8·7 min. Begins with last egg. Period: 14–16 days.

DEVELOPMENT AND CARE OF YOUNG: weight of young at hatching 1·8 g, peaks at 24·9 g on day 16, reduces to 18·3 g at fledging on c. day 27. Young fed within 2 h of hatching. Brooded by both sexes, ♀ more than ♂; broods of 1–3 young brooded for up to 11 days, larger broods for only 5 days. Faecal sacs removed by parents for 4 days; from day 5 young start defaecating out of nest entrance and parents no longer enter nest fully to feed them. Eyes open at 8–9 days. Young poke head out of entrance from day 9. Parents usually alternate feeding visits, bringing bolus of food in throat, or mass of larger, leggy insects held in bill.

Parents start luring young to leave nest from day 15, more so in days 19–24. Luring behaviour important – young will not fledge unless lured. Parents fly slowly past repeatedly with rapidly-beating wings, hovering in front of nest, not bringing food. Up to 20 other birds join in luring. ♀ escorts 1st young to fledge; ♂ attends remaining young in nest. On the wing young do not follow parents closely; they soon become independent. Fledging period: 22–32 days (depends on weather).

BREEDING SUCCESS/SURVIVAL: of 919 eggs (Germany) 79% hatched, of which 91% fledged (72% success). Av. annual mortalities 36–67% (Europe). >100, mainly juveniles, died in unseasonable cold weather, South Africa, Nov 1968 (Skead and Skead 1970) and >50 there in Apr 1953 (Broekhuysen 1953). Abundant desiccated corpses of trans-Saharan migrants found 50–100 km north of Ténéré, N Niger, Nov (Heu 1961). Oldest ringed bird 14 years 6 months.

References
Bryant, D. M. (1975).
Cramp, S. (1988).
Harper, J. and Harper, L. (1974).
Lind, E. A. (1960).
Lockhart, P. S. (1970).

Family MOTACILLIDAE: wagtails, pipits and longclaws

Slim, insectivorous, ground-living, 9-primaried oscines. Bill slender with nasal operculum and rictal bristles. Tarsus acutiplantar, sides covered with more or less unbroken sheath; acrotarsium scutellate. Plumage mainly streaky brown (pipits), cryptic browns above and yellow below (longclaws), or black and white, often yellow below (wagtails). Wing and tail often moderately long. Neck short. Sexes alike, except among wagtails. Forage on ground by walking. Nest on ground or (wagtails) in stonework. Sing in flight or from eminence. Many species migratory and rather gregarious. On DNA–DNA hybridization evidence, treated by Sibley and Ahlquist (1990) as Subfamily Motacillinae in Family Passeridae, other subfamilies being Passerinae (sparrows), Prunellinae (accentors), Ploceinae (weavers) and Estrildinae (waxbills).

54 species, in 3 main genera and 2 monotypic genera (*Dendronanthus*, India; *Tmetothylacus*, Africa). Old World grasslands and wetlands, except that 9 spp. of *Anthus* occur in New World.

Genus *Motacilla* Linnaeus

Wagtails: plumage grey, black and white, black and yellow, or grey and yellow; sexes alike or (usually) different. Bill fine; legs long and slender, hind claw generally long and decurved; tail moderate to long, outer feathers white. Old World, ground-feeding insectivorous birds associated with grass meadows, watersides and human habitation. Constantly wag long tail up and down when foraging. Nests on ground or in cavity in stonework. Most species migratory and rather gregarious.

11 species; in Africa, 3 endemic and 3 Palearctic ones breeding in NW Africa and wintering in part in subsaharan Africa. 1 species endemic to Madagascar and 1 to Japan. A 12th wagtail, the oriental *Dendronanthus indicus*, is closely related to *Motacilla*.

Contains 3 groups, characterized as follows. (1) Yellow wagtails, *M. flava*/*M. citreola*/*M. capensis*/*M. flaviventris*, respectively in Eurasia, Asia, southern Africa and Madagascar; water meadow and waterside birds with outermost 2 tail-feathers mainly white. *M. citreola*, which is almost indistinguishable from *M. flava* in voice and habitat but is longer-legged, would form a superspecies with it were it not for extensive overlap with *M. flava* in Asia. *M. capensis* and *M. flaviventris* comprise a superspecies; leg and tail lengths are nearer to *M. alba* than to *M. flava* (B. Wood, pers. comm.) but the superspecies is more closely allied to *M. flava*. Juveniles of *M. citreola* and *M. flava* are extremely similar in plumage to adult *M. capensis*. (2) Stream wagtails, *M. clara*/*M. cinerea*, a superspecies: fast-flowing rivers, often montane; tails long, outermost 3 tail-feathers white. (3) Pied wagtails, *M. alba*/*M. lugens*/*M. aguimp*/*M. madaraspatensis*/*M. grandis*, forming a superspecies; respectively Palearctic, African, Indian and Japanese; habitats waterside and suburban; plumage lacks any yellow; outermost 2 tail-feathers white.

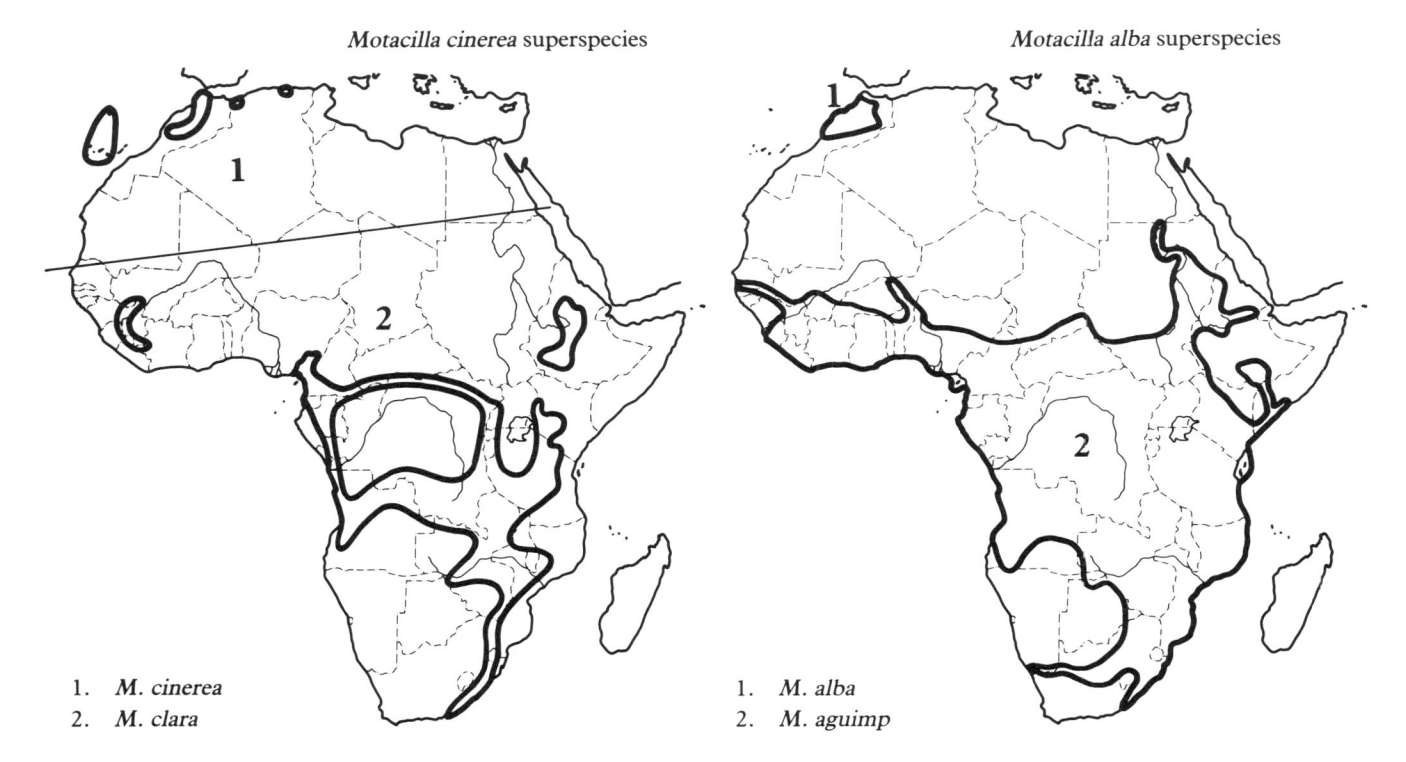

Motacilla cinerea superspecies

Motacilla alba superspecies

1. *M. cinerea*
2. *M. clara*

1. *M. alba*
2. *M. aguimp*

Plate 11
(Opp. p. 176)

Motacilla flava Linnaeus. Yellow Wagtail. Bergeronnette printanière.

Motacilla flava Linnaeus, 1758. Syst. Nat. (10th ed.) 1, p. 185; southern Sweden.

TAXONOMIC NOTE: partly on the basis of claimed sympatry, the Yellow Wagtail complex has been treated multi-specifically by many authors (e.g. Grant and Mackworth-Praed 1952, Knox 1989); in particular, *M. flavissima*, *M. lutea*, *M. thunbergi* and *M. feldegg* have often been regarded as separate from *M. flava*. But many populations hybridize and intergrade, and here we follow currently accepted practice and treat them all as conspecific (whilst acknowledging that some, especially *feldegg*, are biologically distinctive).

11 W and central Palearctic races, and 3 hybrid populations sometimes named as races, occur in Africa. The blue- or grey-headed *flava* can be regarded as the principal, central one, and it forms an axis with the similar grey-headed *cinereocapilla* and *iberiae* to the southwest and *beema* to the east (the endemic Egyptian *pygmaea* is like *cinereocapilla*). The axis separates 2 very similar yellow-headed races, *flavissima* (west of *flava*) and *lutea* (east of *flava*), and also *lutea* (north of *beema*) from the white-headed *leucocephala* (south of *beema*). It also separates the black-headed *feldegg* and *melanogrisea* to the south from the slaty-headed *thunbergi* to the north; these 3 races are somewhat alike (Sharrock and Dale 1964). In Africa, usually the only birds which can be identified confidently as to race are spring ♂♂. Some commonly-used English names are given below.

Range and Status. NW Africa, Egypt (Nile valley), Europe, Asia north of Himalayas and China; Sakhalin; W Alaska. Winters in Transcaucasia, S and SE Asia, Indonesia, W New Guinea, and throughout subsaharan Africa.

Many races occur in Africa, where in spring they are distinctive and some behave as incipient species with distinct habitats and migrations. In continental overview, status can be considered in 4 parts. (1) In NW Africa, locally common breeding visitor (wintering in W Africa; some may winter in NW Africa), sparse winter visitor from Europe, and frequent to abundant passage migrant between Europe and W Africa. Breeds Naïr, Toufat and Cheddid Is and at Cap Timiris, Mauritania (Altenburg *et al.* 1982). In Morocco breeding race (*iberiae*) widespread from sea level up to 1000 m, with local concentrations of several dozen pairs (e.g. at Merja Zerga and Moulouya). Wintering birds mainly *iberiae* and *cinereocapilla*, with some *flava* and *flavissima*. All of these races, and *thunbergi*, occur on passage, with foraging and roosting flocks often of hundreds. In Algeria, *iberiae* breeds on plateaux from Boughzoul to Macta and Fetzara, Arzew, Wadi Rhiou, Dahras, Reghaïa and perhaps Laghouat; *flava* occurs commonly on passage and frequently winters; *iberiae*, *cinereocapilla* and *thunbergi* occur regularly on passage; 4 records of *flavissima*; some spring records of migrant *feldegg* in Hoggar, Tassili, Aïn Oussera and Aïn Amenas. In Tunisia, a scarce breeding visitor (*iberiae*), rare but regular winter visitor and common passage migrant. (2) In Egypt, *M. f. pygmaea* is endemic resident, breeding in Nile valley from 24°N to Nile delta, also in Faiyoum and Wadi Natrun. Abundant at L. Manzala, May, in loose colonies of 15–20 pairs per ha (Meininger *et al.* 1986). *M. f. cinereocapilla*, *flava*, *flavissima*, *thunbergi*, '*dombrowskii*' and *feldegg* are passage migrants, the last abundant in spring in Western Desert and along Nile, with a few in Gebel Elba (SE Egypt). (3) Frequent to abundant passage migrant throughout rest of N and Saharan Africa; e.g. in Libya *flava* abundant (*feldegg*, *thunbergi* and *cinereocapilla* much less so) in coastal wadis near Tripoli in spring and common in autumn, common in Cyrenaica in spring but scarce in autumn, one of the commonest spring migrants through Fezzan, at all oases, with 3000 once at Sebha, and in Libyan Desert common in spring at Jalo, Jaghbub, Kufra and Serir, with small flocks in autumn at Jalo, Serir and 'Awainat. (4) Frequent to abundant passage migrant and winter visitor to subsaharan Africa, wintering north to about 17°N, but absent from S Namibia, S Botswana, Orange Free State (except N), Lesotho, and Cape Province, except coastal areas from Cape Town to Natal. Flocks of tens and hundreds commonplace; reed-bed roosts often of thousands. Roost of 50,000 at L. Chad, Nigeria, Mar, and perhaps many times that number nearby (Fry *et al.* 1972); roosts of 20,000 at Vom and 57,000 at Kano (Nigeria: Ashford 1970), 30,000 at Mopti and 40,000 at L. Horo (Mali).

Thumbnail winter ranges: throughout subsaharan Africa, *flava*, *thunbergi*; W Africa, *flavissima* (east to Mali, Liberia), *cinereocapilla* (Mali to Nigeria), *iberiae*

(Senegambia to Chad, Zaïre); Egypt, *pygmaea*; N tropics, *feldegg*, '*dombrowskii*' (Nigeria to Sudan), *melanogrisea* (vagrant, L. Chad); NE Africa, '*superciliaris*', '*perconfusus*' (Egypt to Ethiopia); E Africa, *beema*, *lutea* (Sudan to Cape: *beema* west to Chad), *leucocephala* (vagrant south to Malaŵi). Further information below.

In Nigeria (country where racial wintering ranges best known), *flava* occurs throughout and is commonest race, *thunbergi* winters in S and sparsely in centre (Jos Plateau), *cinereocapilla* winters north of 12°N, abundantly, *feldegg* sparsely in N and commonly in NE, and probably *iberiae* in W (Wood 1975). In South Africa, only *flava* winters in any numbers; it appears to have been much commoner in Transvaal 100 years ago than today (Tarboton *et al.* 1987).

Description. *M. f. flava* Linnaeus (Blue-headed Wagtail): central Europe, east to Urals. Winters throughout species' subsaharan range. ADULT ♂ (breeding): forehead, crown, nape and hindneck blue-grey, often with few olive-green feathers on crown; long, narrow, white superciliary stripe from nostril to above hind end of ear-coverts; lores, under eye, and ear-coverts like crown but darker, and finely marked with white; mantle, back and rump yellowish olive-green; tail brownish black, narrowly fringed olive, T6 white, T5 white but with proximal quarter of inner web and much of shaft blackish; wings brownish black, secondaries narrowly and coverts and tertials broadly fringed whitish, making 2 white bars in closed wing; entire underparts bright yellow. Bill black (greyish or yellowish at base of lower mandible); eye dark brown; legs and feet black. ADULT ♂ (non-breeding): head (except for chin, throat and superciliary stripe), mantle, back and rump greyish olive-brown; chin, throat and breast white or greyish white, shading to yellow on belly and undertail-coverts. Whitish edges to wing-coverts and remiges much narrower than in summer. ADULT ♀ (breeding): much like non-breeding adult ♂, but upperparts slightly brighter, and breast, flanks and belly rather more washed with yellow. Often an irregular pectoral band of dark brown spots. ADULT ♀ (non-breeding): like non-breeding ♂, but supercilium yellowish and underparts paler, usually with irregular 'necklace' of brown spots, like breeding ♀. SIZE: (Netherlands) wing, ♂ (n = 127) 77–86 (81·8), ♀ (n = 67) 74–82 (78·5); tail, ♂ (n = 51) 65–76 (69·4), ♀ (n = 33) 62–73 (67·1); bill to skull, ♂ (n = 34) 15·1–16·5 (15·9), ♀ (n = 21) 14·7–16·2 (15·6); tarsus, ♂ (n = 36) 23·3–25 (24·0), ♀ (n = 26) 22·5–24·8 (23·5). WEIGHT: (Morocco, spring) ♂ (n = 129) 12·3–20·4 (14·9), ♀ (n = 61) 11·2–17·9 (13·9); (mainly *flava*, Nigeria) ♂ (n = 465), samples rise from av. 16·1 in early Nov to 18·4 in late Mar, ♀ (n = 574), samples rise from av. 15·5 in early Nov to 16·3 in late Mar (dusk weights; dawn weights *c.* 1·0 g lower in autumn and *c.* 2·0 g lower in spring: Wood 1978), range, ♂♂ 14·5–19·7, ♀♀ 13·7–19·2 (dusk); (mainly *flava*, Uganda) ♂ (n = 224) from 14·1–21·5 (17·0) in Oct to 17·5–20·4 (19·3) in Apr, ♀ (n = 230) 13·8–19·1 (15·8) in Oct to 15·6–21·6 (18·5) in Apr. Increase in weight in first half of Apr, throughout Nigeria, very rapid (Fry *et al.* 1970).

IMMATURE: like non-breeding adult, but upperparts from forehead to rump dark brown; supercilium, chin and centre of throat and breast buffy; sides of throat from base of lower mandible to breast, and band across breast, dark brown; belly and undertail-coverts yellowish buff.

NESTLING: skin yellow-red, gape-flanges pale yellow, tip of bill dark; skin becomes pink and flanges deep yellow; legs grey; down long, plentiful, yellowish or buffy-white; on upperparts only.

M. f. iberiae Hartert (Spanish Yellow Wagtail): NW Africa, Spain, Portugal, SW France; winters from Senegambia to Cameroon, less commonly from Chad to Zaïre. Breeding ♂: like ♂ *M. f. flava* but chin and throat white, not yellow. Small: wing, ♂ (n = 13) 78–84 (80·6), ♀ (n = 6) 73–77 (75·1). WEIGHT: (Morocco, spring) ♂ (n = 29) 12·4–16·5 (14·3).

M. f. cinereocapilla Savi (Ashy-headed Wagtail): Italy, Sicily, Sardinia, NW Yugoslavia; winters from Mali to Nigeria, east to L. Chad; records from Uganda, L. Turkana and Kenya highlands could refer to *thunbergi*. Breeding ♂: like ♂ *iberiae*, but white superciliary stripe vestigial or absent. Intergrades with *iberiae* in Corsica and S France, and perhaps Tunisia and Algeria. Large: wing, ♂ (n = 49) 79–87 (83·0), ♀ (n = 10) 77–82 (79·6). WEIGHT: (Morocco, spring) ♂ (n = 37) 12·6–18·2 (14·9).

M. f. pygmaea (A. E. Brehm) (Egyptian Yellow Wagtail): Egypt. Resident; vagrant to Khor Arbat, Sudan, 1908 (Goodman and Atta 1987). Breeding ♂: like ♂ *cinereocapilla* (dark grey crown and ear-coverts; no white superciliary stripe; white chin and throat) but smaller: wing, ♂ (n = 10) 74–79 (76·2), ♀ (n = 5) 70–75 (72·4). Non-breeding ♂: like breeding ♂ but forepart colours brighter, and centre of chin and throat yellow.

M. f. beema (Sykes) (Sykes's Wagtail): USSR – lower Volga R. east to near L. Baikal, and from 60°N in SW Siberia to N Kazakhstan and Altai; winters E Africa from Sudan to Tanzania and sparsely in Malaŵi, Zimbabwe and South Africa; migrants in spring occur west to Nigeria (L. Chad); occurs on E coast of Somalia in Mar–Apr. Breeding ♂: like ♂ *M. f. flava* but forehead, crown, hindneck, lores and ear-coverts paler ash-grey; white superciliary stripe broader.

M. f. flavissima Blyth (British Yellow Wagtail): Britain and adjacent continental coast; winters from Senegambia to Liberia (abundant on Mt Nimba: Colston and Curry-Lindahl 1986), Mali (large roosts) and probably Ivory Coast; claimed in Nigeria (L. Chad), Kenya, Zaïre, Botswana and Zimbabwe (probably *lutea*: Pearson and Backhurst 1973, Clancey 1980). Breeding ♂: hindcrown, nape, hindneck and border of ear-coverts yellow-olive; lores blackish; rest of head bright yellow.

M. f. lutea (S. G. Gmelin) (Eastern Yellow-headed Wagtail): USSR – lower Volga basin north to Kazan' and Perm', and east through Kazakhstan to L. Chany and L. Zaysan; winters in dry E African plains from Sudan to Malaŵi, Zimbabwe and South Africa. Commonest race wintering in Somalia and Zimbabwe. Recorded on Socotra I., and in Zaïre at L. Suzi (S Marungu). Breeding ♂: entire head including lores bright yellow, concolorous with underparts; some indistinguishable from ♂ *flavissima*. WEIGHT: (Kenya) ♂♂ increase from av. 16·4 in Oct to 17·0 in Jan, 17·5 in Mar, 18·6 (max. 26·4) in Apr; ♀♀ increase similarly 15·3–15·7–16·0–16·9 (max. 22·6).

M. f. leucocephala Przevalski (White-headed Wagtail): NW Mongolia, N Sinkiang and adjacent USSR. Vagrant in Africa south to Sudan (Suakin), Uganda (Kampala), Kenya (several, Nairobi), Tanzania (Arusha Chini), Zambia (Kanyamwe in Feira), and Malaŵi. Breeding ♂: centre of throat yellow and hindneck sometimes greyish, otherwise head entirely white, including lores.

M. f. thunbergi (Billberg) (Grey-headed Wagtail): Norway east to N Siberia; winters almost throughout subsaharan Africa; in Senegambia on passage only; uncommon in Sudan. Breeding ♂: forehead, lores, forecrown and ear-coverts blackish or dark slaty grey, shading to dark blue-grey hindcrown, nape and hindneck; centre of chin and throat yellow, with broad white stripe below lores and ear-coverts. Same size as *cinereocapilla*. Intergrades with *beema* and nominate *flava*.

M. f. feldegg Michahelles (Black-headed Wagtail): Yugoslavia, Bulgaria, Greece, Turkey, E Mediterranean, Iraq, S Ukraine, Crimea, W Caspian, Iran, Afghanistan; winters in swampy country from Nigeria to N Zaïre, Uganda and Sudan, and south to Ituri R. and L. Kivu (Zaïre), and Burundi (6 records); in 1977–78 many wintered in millet and sorghum fields at Arada, E Chad (Newby 1980). 1 winter record Somalia where, however, large passage in north in Mar–Apr. Vagrant South Africa (SW Cape, ?Transvaal). Frequent on passage in Sudan; abundant in E Chad; common at L. Chad; rare in Mali; occasional in W Niger. 1 record Morocco (Smith 1968), rare in Algeria, but common on passage on N Africa coast from Libya eastwards. Breeding ♂: head black, sharply defined from bright yellow chin and throat. Same size as *cinereocapilla*.

M. f. melanogrisea (Homeyer): Volga delta and E Caspian east to Tarbagatay and Ili R. basin; a few spring migrants at L. Chad, Nigeria (Fry *et al.* 1972). Breeding ♂: like ♂ *feldegg* but upperparts brighter green, and a white stripe between yellow throat and black cap (like *thunbergi*).

M. f. 'superciliaris' (Brehm): S Iran (?) to Turkestan and lower Volga; winters from Egypt to Sudan and Ethiopia; on cultivated land. Breeding ♂: like ♂ *feldegg* but with narrow white or yellow superciliary stripe. Perhaps a hybrid between *feldegg* and *flava* (Cramp 1988).

M. f. 'dombrowskii' (Tschusi): Romania, Yugoslavia; winters from Nigeria to Ethiopia and Sudan; vagrant to Zaïre (Buta). Breeding ♂: like ♂ *M. f. flava* but forehead darker grey, and ear-coverts blackish; chin sometimes white. 'Dombrowskii' birds are hybrids between *feldegg* and nominate *flava*, with wide geographical zone of hybridization.

M. f. 'perconfusus' Grant and Praed: a hybrid or variant recorded from Holland, Denmark and Germany; also from Sudan and Ethiopia. Breeding ♂: like ♂ *M. f. flava*, but superciliary stripe yellow.

Field Characters. A rather short-tailed wagtail, common and widespread in dry or wet short-grass country, often near cattle; generally with some yellow in plumage, which distinguishes it from all other African wagtails except Grey Wagtail *M. cinerea*. Vocal, with distinctive monosyllabic call (but see Citrine Wagtail *M. citreola*). Grey Wagtail readily told by black throat (♂ only), long tail, swift-stream habitat, and disyllabic call. Some ♀ and immature (autumn) Yellow Wagtails, entirely without yellow, can be extremely similar to Cape Wagtail *M. capensis* and even to immature White Wagtail *M. alba*, and are best told by voice.

Highly variable. Most breeding-plumage (spring) ♂♂ readily identifiable to race (see descriptions above, and Plate 11). Spring ♀♀ of most races usually have same distribution of yellow in underparts as ♂ of their race but more white on chin or throat, and obscure version of ♂'s head marks (dark or pale cheeks, presence or absence of superciliary stripe, etc.). *M. f. feldegg* and *melanogrisea* ♀♀ have dusky, black-mottled heads and bright wing marks; ♀♀ of *iberiae* and *cinereocapilla* have dark cheeks and whitish chins; ♀ *thunbergi* has dark crown, cheeks and upperparts, and poorly-marked supercilium; and ♀ *flavissima* is dark brown above with pale underparts. Other ♀♀, and certainly juveniles, are seldom safely identifiable as to race even when (as often happens) several races occur together. Lone ♀♀ best left unidentified.

Voice. Tape-recorded (89, GIB, MOR – 62, 73, C). Call, freely given at rest and in flight all year, is quite loud, sibilant monosyllable 'psie'. Sometimes faintly disyllabic: 'psüip' or 'psiib'; occasionally disyllabic 'tsi-weep'. Call of southern races lower-pitched and harsher than in northern races. Song, on breeding grounds only, a prolonged but not loud twittering, with variable repetition of 'psie' call note: a sweet, high-pitched 'tsee-wee-sirr tsee-wee-sirr' (*flavissima*). Also an unmusical, hard, rattling trill, 'trrizzz' (*pygmaea*) (Cramp 1988).

General Habits. On breeding grounds inhabits water-meadows, drier parts of saltmarshes, riversides, lakesides, and similar moist, unobstructed areas. On wintering grounds occurs in variety of habitats: marshes, rice fields, cereal fields, cassava and banana plots, floating vegetation and lakeside debris, pans, dams, forest clearings, sports- and airfields, parks, coastal mudflats, open bushland, and grazed grassland up to 3000 m. Associated with grazing cattle, zebras and antelopes. *M. f. feldegg* keeps to swamps, lakesides and other wet places; *lutea* predominates in dry plains. Around Nairobi, Kenya, *lutea* occurs mainly on sewage farms, *flava* in cattle pasture, *thunbergi* and *beema* equally in both habitats, and *feldegg* around game reserve waterholes (Reynolds 1974). On Mt Nimba (Liberia), deforested by iron ore mining, *flavissima* forages on paved roads being recolonized by vegetation, which zig-zag across entire mountain (Colston and Curry-Lindahl 1986). For roost habitats, see below.

Breeding birds strongly territorial. On wintering grounds, some birds territorial, others markedly gregarious; hence sometimes occurs singly but elsewhere in small or huge flocks. In Nigeria, single bird (n = 6 territories) defends 60–130 (av. 82·5) m of stream; territory 0·15–0·36 (0·25) ha (Wood 1976). In Gabon flock of 3–5 birds occupies same 500–800 m² territory all winter, defending it aggressively (Brosset and Erard 1986). A bird will temporarily defend 1 water trough. Territorial flocks of 4–6 also common in Zaïre. Seldom forms flocks in E Africa except at roost and when migrating. Races mix freely; species also commonly mixes with other wagtails, and pipits. Migrants can forage densely: > 1000 per ha or along 500 m of shoreline (L. Chad, spring). Territory-holders join communal roost, arriving earlier and departing later than non-territorial birds (Wood 1976).

Roosts, often in traditional sites, are in tall grass (near Nairobi, *Pennisetum purpureum*), sugarcane fields, reedmace *Typha*, reeds *Phragmites*, clumps of shrubs and trees in shallow water, and trees, especially fig *Ficus dekdekena*. Birds congregate in their hundreds, or tens of thousands. Catchment area for roosts of > 20,000 at Vom, Nigeria, 270 km² in Nov and 825 km² in Feb (Wood 1976). 3000 birds at Ibadan, Nigeria, roosted in area of only 400 m² of 2 m tall grass, taking at least 28 min to arrive in flocks of 2–60, mainly *c.* 20 birds (Broadbent 1969). When nearly dark, birds fly to roost *en masse*, circling then diving down steeply. In morning, they depart in waves. Roost site is often changed, probably because predators quickly learn to exploit it. In evenings before site is abandoned, flocks fly high towards roost area, calling, and first assemble on open ground nearby, preening, feeding and calling; when assembly nearly complete, calling is audible from afar as twittering murmur (B. Wood, pers. comm.).

Feeds by walking briskly, picking items from ground or water surface, or making quick run at prey, or making short flight up from ground to seize insect in mid-air. Occasionally takes insect from low vegetation by hovering, or flies low over water to snatch insect from surface. In Nigeria averages 6·5 paces between successive prey capture, late Nov, but 22·6 paces in Mar when food scarcer; Nov–Dec feeding rate highest at dawn and dusk, and low at midday, but in Jan–Mar feeds at constant rate throughout day (Wood 1976). During heat of day, spring birds at L. Chad rest in shade under vegetation for short periods; otherwise they stop feeding only to wash or preen.

Spring and autumn migrations are on broad front, over savannas, deserts and seas around N half of Africa. Numerous spring and autumn records at Saharan oases. In autumn most birds evidently overfly N African coast. *M. f. flava* is common along Moroccan Atlantic coast Sept–Oct, and occurs up to 100 km offshore. Migrates in parties of *c*. 20, up to *c*. 80, also in ones and twos. Arrives in northern tropics in late Sept (E Africa) or during Oct (W Africa), at various Nigerian stations with great regularity (e.g. Ibadan, 2–4 Oct in 4 years). Marked fidelity to winter quarters, both to general area and roost (Nigeria: Kano) and to specific territory (Ghana). Arrives in southern Africa late Oct, becoming widespread mid-Nov (Zimbabwe). Occasionally winters in Transvaal (June, July). Leaves Zimbabwe mid-Mar to mid-Apr, N tropics late Apr to early May.

During 2–3 weeks before spring departure from L. Chad and other Nigerian localities, weight increases by one-third as fat is laid down; this suggests that Sahara is crossed non-stop, in up to 3 nights and 2 days (Fry *et al.* 1970, Wood 1982). Emigration from L. Chad mainly in late afternoon. Evacuation of northern tropics is progressive, from Mar in S to May in N. In Zaïre and Nigeria *feldegg* departs first, then *flava*, then *thunbergi*. Emigration very regular; for 8 years last departures at Ibadan (Nigeria) were between 28 April and 9 May. Spring migration of nominate *flava* and *flavissima* further east in NW Africa than autumn migration, but in NE Africa east European races show opposite tendency. Passage in N Africa is more obvious in spring than in autumn; in Mar–May, peaking in early May. *M. f. thunbergi* appears to have loop migration, entering Africa further east than it departs (Gatter 1987a, b).

Over 250 recoveries in Africa of birds ringed in W Palearctic and vice versa. All recoveries up to 1975 mapped analytically by Zink (1975), including *c*. 80% of the following: 36 recoveries in Morocco, 9 in Algeria and 3 in Tunisia, of birds ringed mainly in S England, Belgium, S Sweden and N Poland; 41 recoveries in Western Sahara and near coast of Mauritania, Senegambia and Guinea-Bissau, mainly from S England, Belgium, Holland and W Germany; 26 recoveries in W Africa between 10°W and 10°E (Mali to Nigeria), mainly from around S Baltic; also 1 from Denmark and 2 from Italy to Nigeria. Several recoveries from Finland to Greece show a N European migratory divide, but up to 1989 only 1 recovery from Finland to Egypt (and 1 from Egypt to Czechoslovakia). 1 from Finland to W Central African Republic.

Recoveries from Senegambia to Morocco (1), Spain (3), France (2), Belgium (1) and England (1); from S Ghana to Libya (1), Malta (1), Italy (3) and Finland (2). 38 recoveries from Kano, N Nigeria, to S Ghana (1), SE Nigeria (1), N Algeria (2), Tunisia (2), N Libya (1), Malta and Sicily (6), Italy (16), Greece (1), Hungary (1), Czechoslovakia, Germany and Poland (4), Finland (2) and USSR (31°E, 51°N) (1). 33 recoveries from Vom, central Nigeria, to S Algeria (1), N Algeria (1), Morocco (1), Tunisia (1), Libya (2), Malta and Sicily (6), Italy (10), Crete (1), Turkey (1), Greece (2), Romania (1), Poland (1), Estonia (1) and Finland (4). From L. Chad, NE Nigeria, to Italy (1) and Crete (1). 36 recoveries from Kariobangi, Nairobi, Kenya, to Tanzania (1), Uganda (2), W Saudi Arabia (5), Bahrein and Qatar (2), Iran (3), Azerbaijan (4), Astrakhan (1), Dagestan ASSR (about 47°E, 43°N) (1), Bashkirian ASSR (about 56°E, 55°N) (1), USSR between 40° and 48°E and 50° and 57°N (15), and Siberia (about 81°E, 59°N on Ob R.) (1) (Backhurst 1977, 1988). From Harar, Ethiopia, to USSR (about 41°E, 51°N) (1).

Food. Insects, a few other invertebrates, and a little plant matter. For large variety of insects recorded in diet in Europe, see Cramp (1988). In Nigeria 45% of diet (by number) Hemiptera, 23% beetles, 15% Hymenoptera, 14% grasshoppers, 2% flies and 1% other insects (89 birds: Wood 1976); also termites, spiders, grasshopper nymphs, beetle larvae, and a few berries of salt-bush *Salvadora persica* (L. Chad: Fry *et al*. 1972). In spring, wagtails at L. Chad subsist mainly on the minute midge *Tanytarsus spadiceonotatus*; at L. Victoria (Uganda) main food is also small Diptera. Other African records: phasmids, cockroaches, shield bugs, hawk-moth caterpillars, longicorn beetles, crustaceans, and small seeds; and in Gabon observed taking ants, termites, small orthopterans, beetles and butterflies.

Breeding Habits. Solitary, monogamous, territorial breeder. Territorial ♂♂ highly aggressive, threatening and chasing intruding conspecifics and other songbirds. ♂ sings from perch or in display-flight; at perch, hunches body, draws in neck, puffs out breast, tilts head back, and droops wings and tail, then ascends steeply for a few m and continues singing with tail fanned, legs dangling and toes clenched, then descends with rapidly fluttering wings (Cramp 1988). Rival ♂♂ threaten each other, one bird in 'singing' posture, strutting about and swaying, other sitting back on tail with yellow breast thrust out and head tilted far back.

NEST: grass cup lined with hair, placed in shallow scrape in soil. Built by ♀, with ♂ nearby, taking 4–20 days (England).

EGGS: 4–6, laid in early mornings. Subelliptical; smooth, glossy; grey-white to buff, densely spotted and hair-streaked in shades of brown. SIZE: (n = 200, *M. f. flava*) 17–20 × 12·8–15·3 (18·5 × 14·0). WEIGHT: 1·8.

LAYING DATES: Morocco, Algeria, Tunisia, Egypt, Apr–May.

INCUBATION: by ♀ for 67% of time and ♂ for 17% (eggs unattended for 17%). Begins with last egg. Hatching synchronous. Period: 11–13 (12·4) days (Britain).

DEVELOPMENT AND CARE OF YOUNG: young cared for by ♀ and ♂. Nestling period: 10–14 (11·1) days. Fledglings stay with parents for several weeks, and may migrate with them.

BREEDING SUCCESS/SURVIVAL: 67 and 51% overall (99 and 808 eggs, Britain). Losses due to desertion, predation, farming, weather and trampling by cattle (Cramp 1988). Oldest ringed bird 8·5 years.

References
Cramp, S. (1988).
Dittberner, H. and Dittberner, W. (1984).
Fry, C. H., Ferguson-Lees, I. J. and Dowsett, R. J. (1972).
Smith, S. (1950).
Wood, B. (1975, 1976, 1978, 1979, 1982).

Not illustrated

Motacilla citreola Pallas. Citrine Wagtail. Bergeronnette citrine.

Motacilla citreola Pallas, 1776. Reise versch. Prov. Russ. Reichs 3, p. 696; E Siberia.

Range and Status. Central Palearctic, where common, wintering in India and SE Asia. Signs of westward spread in USSR (Cramp 1988). Breeds west to Byelorussia; vagrant to 22 other European countries, annually in Britain. Common at Eilat (Israeli Red Sea), Mar, making 'their first landfall' (Hammond 1985), so probably commoner in NE Africa than records suggest. Common winter visitor to, and probably passage migrant through, coastal S Oman.

Vagrant. 1 Djibouti (Oeud Petite Douda, late Feb 1990: Welch and Welch 1990). 1 Zaranik, Egypt (Sinai), 100 km east of 'Africa' border, early Apr 1983 (C. H. Fry).

Description. *M. c. citreola*: northern Palearctic (3 subspecies, but subspecies in Africa not known). ADULT ♂ (breeding): forehead, crown and nape feathers bright yellow, tipped dusky and tinged with olive; rest of head bright yellow. Hindneck black; mantle and scapulars dark grey; back and rump grey; uppertail-coverts black; tail black with white edges (T5, T6). Sides of neck and entire underparts bright yellow. Wing black, with white or greyish white fringes to primaries, secondaries and tertials, and broad white tips to median and greater coverts and primary coverts. Axillaries and underwing-coverts pale brownish grey. Bill black; eye black; legs and feet brownish black. ADULT ♂ (non-breeding): as breeding, but hindneck dark grey, washed with olive; forehead, forecrown and ear-coverts yellow, mottled olive-green; broad yellow supercilium, and mottled ear-coverts surrounded by bright yellow; undertail-coverts white. ADULT ♀ (breeding): like non-breeding ♂, but white tips to median and greater coverts narrower. ADULT ♀ (non-breeding): centre of forehead, crown and nape same dark olive-grey as hindneck and mantle; otherwise like breeding ♀, but yellows paler and undertail-coverts white. SIZE: (17 ♂♂, 7 ♀♀) wing, ♂ 85–90 (87·9), ♀ 80–85 (82·3); tail, ♂ 74–81 (76·6), ♀ 70–76 (72·6); bill to skull, ♂ 16·7–18·8 (17·5), ♀ 16·1–17·4 (16·8); tarsus, ♂ 25·3–27·7 (26·3), ♀ 23·8–25·6 (24·6). WEIGHT: USSR (May–June) ♂ (n = 13) 20–24 (22·2), ♀ (n = 8) 18–25 (21·1).

IMMATURE: forehead buff, uppertail-coverts black, rest of upperparts dull olive-brown; supercilium very pale buff; lores, ear-coverts and cheeks dull olive-brown, streaked with white; underparts buffy, tinged with olive on sides of breast. Bill grey-pink; eye dark brown; legs greyish flesh.

NESTLING: unknown.

Field Characters. Very like Yellow Wagtail *M. flava* in plumage, voice and habitat, and the 2 species often flock together. Longer-legged than Yellow Wagtail, sometimes noticeably so. Adults distinguished by having mantle to rump dark grey in Citrine Wagtail, bright olive in Yellow Wagtail; hindneck is black in ♂ Citrine, dark grey in ♀ Citrine, bright olive in Yellow. Juveniles and intermediate-plumaged winter birds of both species very alike, but with practice Citrine Wagtail can be told by its yellow-and-green-mottled ear-coverts with darker margin and yellower interior, surrounded by clear yellow, its wider whitish edges to tertials, and longer legs.

Voice. Tape-recorded (HOL, SVEN). Contact call 'sreep', very like Yellow Wagtail but said to be harsher.

General Habits. Inhabits marshes, edges of lakes, and wet tussocky grassland. In S Arabia mainly coastal, on edges of brackish lagoons, sewage ponds and in irrigated farmland. Forages on moist open ground at edge of water, and on beds of water hyacinth *Eichhornia*, in company of Yellow Wagtails and White Wagtails *M. alba*. Picks prey from ground while walking, and from water surface while wading in shallows; snatches insects from low vegetation and in brief upward flutter. Rests and preens on ground, or in low bush. Loosely gregarious; not particularly timid.

Migratory; Asiatic breeding and wintering grounds practically exclusive; winters to S India and central Malaysian peninsula. Regular spring migrant at Eilat, Israel, from unknown wintering grounds (see above).

Food. Wide variety of small insects, larvae and spiders (Asia).

Reference
Cramp, S. (1988).

Motacilla capensis Linnaeus. Cape Wagtail. Bergeronnette du Cap.

Motacilla capensis Linnaeus, 1766. Syst. Nat. (12th ed.) 1, p. 333; Cape of Good Hope.

Forms a superspecies with *M. flaviventris* (Madagascar).

Plate 11
(Opp. p. 176)

Range and Status. Endemic resident, with seasonal movements in Transvaal and Mozambique. 3 populations, with ecological (and plumage) differences. (1) Local and mainly uncommon in Kenya between 1500 and 3000 m; commoner in SW Uganda from Entebbe and Masaka to Ankole and Kigezi; scarce in E Zaïre, Rwanda, Burundi and NW Tanzania. (2) Population in Angola, SE Zaïre, Zambia, NE Namibia (Caprivi Strip) and N Botswana is common in W-central Angola, at mouth of Cunene R., in Balovale (Zambia) west of Zambezi, around Mweru and Bangweulu swamps, and locally in Chobe and Okavango valleys (Namibia, Botswana); otherwise sparse but widespread in Angola and Zambia. (3) In Zimbabwe breeds sparsely from 1200 to 2200 m; sparse in N Transvaal; abundant in winter in E Zululand and Mozambique plain; sparse in NW Namibia; common in S Namibia and throughout South Africa south of 23°S, where far more a bird of human habitations, farms, homesteads and villages than to north of Limpopo. Sometimes sedentary for years (Skead 1954).

Where commonest, in SW Cape, occurred in half of all vleis counted in 1950s, with monthly median to 14 birds, and maximum of 49 in one vlei. 36 at Athlone sewage farm May–Oct 1962 and up to 70 in Nov–Dec (Winterbottom 1964). 25–50 in Durban Bay, Natal, Apr–July. Hundreds congregate along shores of highveld pans, Transvaal, in winter. Breeding density of one pair per 3–9 ha, E Cape. May have declined in Zimbabwe since 1950s; reportedly declining in many other areas of southern Africa.

Description. *M. c. capensis* Linnaeus (includes *bradfieldi* Roberts: Namibia): South Africa to S and W Namibia, Zimbabwe (Mashonaland Plateau) and Mozambique. ADULT ♂: forehead, crown, nape, mantle, back, scapulars, sides of neck, and ear-coverts olivaceous dark grey (crown and neck less olivaceous than other parts). Tail brownish black with 2 outermost feathers (T5, T6) white; T5 with brown border to inner web proximally; T3 sometimes with very narrow white end and edge to outer web; T1 usually slightly pale-edged. Lores dark brown or blackish; narrow buffy superciliary stripe from base of nostrils to behind eye. Chin and throat creamy; throat bordered by dusky pectoral band *c.* 5 mm deep in centre and 2 mm deep at sides of breast; sides of breast and flanks olivaceous grey; centre of lower breast and belly yellowish cream or creamy white; undertail-coverts white. Wings blackish brown; all remiges and upperwing-coverts narrowly edged grey or buffy; tertials conspicuously pale-edged; axillaries grey, most other underwing-coverts whitish. Bill black; eye dark brown; legs and feet blackish. Partial albino known (Broekhuysen 1969). Sexes alike. SIZE: (14 ♂♂, 11 ♀♀) wing, ♂ 77–87 (82·5), ♀ 77–87 (82·1); tail, ♂♀ 77–91; bill to feathers, ♂♀ 13–15; tarsus, ♂ 21–24·5. WEIGHT: unsexed (n = 184) 17–25 (20·9).

IMMATURE: like adult but warm brown above rather than olivaceous grey. Secondary coverts broadly tipped pale buff; tertials and other wing-coverts tipped and edged buffy. Tail shorter than that of adult. Underparts, especially belly, usually washed very pale yellowish.

NESTLING: skin red-orange, mouth orange. Down light grey, 10 mm long, on head, wings, back, flanks, thighs and abdomen.

M. c. simplicissima Neumann: NE Namibia (Caprivi Strip), Botswana, Angola, SE Zaïre, Zambia, W Zimbabwe. Upperparts more olivaceous, especially crown; outer rectrix edged with black on inner web; dusky breast-band reduced to irregular spot in centre of breast; underparts washed slightly yellower than in nominate race.

M. c. wellsi Ogilvie-Grant: E Zaïre to Kenya. Upperparts much darker than in nominate race; outer rectrix never has any black on inner web; pectoral band darker, slightly larger and better defined; underparts with salmon pink wash. WEIGHT: ♂ (n = 8) 19–25 (22·5), ♀ (n = 6) 21–23·5 (22·8).

Field Characters. A somewhat nondescript, southern African wagtail, brownish grey above, off-white to buff below, with brown wash on flanks, blackish pectoral band in 2 races (*capensis*, *wellsi*), reduced to breast spot in *simplicissima*. Tail blackish, with conspicuous white sides. Range overlaps those of 4 and 5 other wagtails in S and E Africa, respectively. They are distinguished from Cape Wagtail as follows: Grey Wagtail *M. cinerea* always has markedly yellow belly and long tail, and inhabits swift-flowing streams; African Pied Wagtail *M. aguimp* is strikingly black-and-white in all plumages, with white in wing. White Wagtail *M. alba* (barely penetrates south of equator) is generally boldly pied but juveniles can be very like Cape Wagtail; and Yellow Wagtail *M. flava* usually has yellow underparts and lacks pectoral band, but some can look like Cape Wagtail. White and Yellow Wagtails can, however, confidently be distinguished by voice, q.v. For distinctions from Mountain Wagtail *M. clara*, see that species.

Voice. Tape-recorded (72, 81, 88, F, LEM, PAY). (1) A canary-like song, irregular repetitions of call note interspersed with a slightly less piping and lower-pitched 'tweep tweep' and with short, pleasant, finch-like twitter or warble 'ti-ti-ti-ti-ti-ti-ti-ti' lasting 0·9 s. One song of 30 s duration contained *c.* 20 'tsee' and 5 'tweep' notes and 3 twitters. (2) Flight call, a short, sharp, loud 'tsiep', 'tsee' or 'tseet', sometimes a double 'tsee tsee'. (3) A longer 'tseeep' or 'tsayp', commonly when foraging. (4) A long-drawn, lazy whistle, 'where-you-are?'.

General Habits. South of Limpopo R. occupies wide range of habitats, and is characteristically the wagtail of vicinity of dwellings, villages and large expanses of lawn in towns. Inhabits tree-lined rivers, sea coasts, offshore islands, sand dunes, saltmarshes, vleis, pastures, farmyards and gardens. In Barberspan, W Transvaal, *c.* 80% of birds live near buildings (Farkas 1962). North of Limpopo, much less urban (replaced in towns by African Pied Wagtail), and inhabits mainly slow-flowing streams with stretches of rock and intermittent pools with muddy edges, drainage lines through vleis, muddy edges of dams and sewage farms, also lawns and homesteads. In Zimbabwe breeds from 1200 to 2200 m. In E Africa inhabits shallow bays of lakes, edges of swamps, grassland, forest clearings and cultivation at 900–3000 m.

In pairs, family parties, and in winter in loose flocks of up to 50 birds. Forages by walking purposefully, pecking ground and low vegetation, making running dash or small jump into air after insect. If undisturbed, and food supply good, forages over grass sward for up to *c.* 1 h without flying, but territorial, with frequent minor conflicts and short aerial chases with congeners when present. Constantly wags tail vertically whilst foraging; on alighting, wags tail through wide arc a few times. Often stops to loaf and preen; tail wagging then less frequent. Often feeds amongst cattle. Continues to feed in light rain. Readily walks in shallow water covering feet, pecking at insects in and just above it. Generally quite tame in gardens and suburbs. Calls on taking flight (when tail spread and white outer feathers conspicuous). Sings from low eminence, tree or building.

Usually roosts gregariously in palms, pine, plane or eucalyptus trees, or in reeds; sometimes solitarily in hedge or (once) old swallow's nest. Communal roosts in reedbeds involve up to 170 birds (Knysna, South Africa, Sept), which fly in at twilight over period of 20–30 min (but mostly in 10 min), arriving in ones, twos or small parties, flying lower in windy than in calm weather; they circle over reedbeds, calling, then dive down, zigzagging or directly, entering the reeds (Winterbottom 1964). Once several mobbed African Marsh Harrier *Circus ranivorus* at roost site. Roosts are often in busy city centres.

Evidently sedentary in much of South Africa (ringing studies); but partial migrant in east, vacating parts of Orange Free State in winter and visiting Durban and Ndumu, also Mozambique, only in winter. In Zambia moves away from Mwinilunga, and into Luangwa valley, in rains (Berry 1981). Mainly sedentary in E Africa, but with some local wandering.

Food. Mainly insects. Insects (including Diptera, moths and beetles) in all 17 birds in a Uganda sample. In South Africa food includes *Musca domestica*, butterfly *Catopsilia florella*, dragonfly *Pantala flavescens*, dragonfly nymphs, ants, termites, mosquitoes and larvae, chironomid midges, aphids, cerambycid larvae, sandhoppers, fish up to 2 cm long, tadpoles, fiddler crabs *Uca annulipes* (Begg 1981), and possibly ticks and small chameleons. Garden birds eat raw meat, fat, suet, cheese, hard-boiled egg, mealie meal, cake and bread. Snails recorded once (Uganda). Caterpillars feature importantly in items fed to nestlings (Winterbottom 1964).

Breeding Habits. Monogamous, territorial, solitary breeder. One pair held territory of *c.* 40 ha; elsewhere 3 pairs divided up 40 ha area. ♂ may persistently peck at his reflection in house windows in territory. In pair-bonding display, all year, ♂ presents bill-full of twigs to ♀, who takes them one by one and drops them. ♂ places scraps of nest material in front of incubating ♀, who ignores them. Rarely, ♂ runs to meet ♀ with his wings outspread and quivering; and ♂ and ♀ flutter breast-to-breast from elevated perch to ground. No courtship display in association with copulation (Skead 1954). Evidently pairs for life.

NEST: deep cup of fine grass stems and leaves and rootlets, neatly lined with fine rootlets, cattle-hair, wool and sometimes a few feathers, built on or into roughly constructed, bulky foundation of grasses, weeds, roots, reed strips, and pieces of rag. Also recorded: pine needles, grass runners, twigs, string, fluff and seed-pods. Cup 65 mm internal diam., 50 mm deep. Sited on ground or up to 6 m high in vegetation but usually only 30–40 cm high. Of 214 nests in South Africa, 40% in grass (half of them overhanging water), 33% in or under bushes, usually by water, 12% in clumps of reeds, 3% in trees, 3% on sand dunes and 3% in crevices in walls and buildings. Nest usually well hidden in vegetation. Other sites: under rock, holes and ledges in rock, crevice in bank of stream or donga, hedges, creepers, eaves, and trucks. Nest material placed concurrently at several different sites (at least 3) before one is selected (Skead 1954). Built by ♂ and ♀, taking 2–10 days. New nest generally built for each clutch, but same nest sometimes used for up to 4 clutches.

EGGS: 1–4, exceptionally 5–7, mean clutch size (South Africa, 4 localities, n = 534 clutches) 2·76–2·96. Individual ♀♀ can lay av. clutch of 3·3 over several years (Skead 1954, Earlé 1986). Eggs laid at 24 h intervals, in early morning. Dull yellowish, clouded or indistinctly speckled or mottled at broad end with brown; occasionally immaculate; once pure white (Herholdt 1986). SIZE: (n = 240, southern Africa) 18·6–25·5 × 14–16·8 (21·0 × 15·4). Several broods per season: in South Africa often only 1 clutch per year, but up to 6 clutches per season known.

LAYING DATES: E Africa: Region B, Mar, June–Sept, Nov–Dec, Region D, Jan, Apr–May, July, Sept; Angola (enlarged gonads Aug); Zambia, July–Aug; Zimbabwe, Aug–Dec, Feb (mainly Sept–Oct); South Africa: Transvaal, all months but 60% Sept–Nov, Natal, July–Jan.

INCUBATION: by ♂ and ♀; at night mainly by ♀. Period: (W Cape, n = 15 nests) 13·7 ± 1·6 days, (E Cape, n = 13) 14·3 ± 0·9 days, (Transvaal, n = 10) 13·4 ± 1·0 days. When alarmed, incubating bird flies off nest low and silently, keeping tail closed.

DEVELOPMENT AND CARE OF YOUNG: eyes half open by day 4 and fully open by day 8. Quills well grown by day 6, and feathers start to break out on day 8. Alert and half-feathered, with primary feathers starting, on day 9. First fear reaction on day 11. Nest too small (for 3 chicks) on day 12. By day 15 young are smooth-feathered, with little down remaining. Young brooded by ♀ and ♂ assiduously for 2 days; one parent covers them or roosts beside nest for 6–10 nights. On very hot days, parent spreads wings over nest to protect chicks. Young fed by ♂ and ♀ equally, once with 18 nest visits in 60 min (SW Cape). ♂ stops feeding them when they fly, after which ♀ feeds them alone. Parents remove faecal sacs; later, young deposit uncapsulated faeces on nest-edge, parents remove them. Faecal capsules dropped by parent in flight or on alighting. Fledging period: 15·2–15·9 days (averages at 13, 10 and 10 nests, S Africa); exceptionally up to 19–20 days.

Parents with flightless chick out of nest, pursued by boomslang *Dispholidus typus*, feigned injury by adroitly fluttering just out of reach; later it advanced with spread wings at observer (Oatley 1988).

BREEDING SUCCESS/SURVIVAL: one marked ♀ laid 24 eggs in 8 nests in 11 months (Skead 1954); another, in Bloemfontein city centre, laid 50 eggs in 15 clutches in 48 months, with 29 young fledging (58% success) (Earlé 1981). Parasitized by Red-chested Cuckoo *Cuculus solitarius* and Diederik Cuckoo *Chrysococcyx caprius*. Nestlings parasitized by lice and fly *Passeromyia heterochaeta*. Drastic decline of Cape Wagtails at East London, 1939–42, attributed to tarsal joint disease like poultry scaly-leg (Winterbottom 1964). Oldest ringed bird, 5·5 years, but a white-headed one lived for 11 years (Skead 1954).

References
Skead, C. J. (1954).
Winterbottom, J. M. (1964).

Motacilla cinerea Tunstall. Grey Wagtail. Bergeronnette des ruisseaux.

Motacilla cinerea Tunstall, 1771. Orn. Brit., p. 2; Yorkshire, England (Clancey 1946).

Forms a superspecies with *M. clara*.

Plate 11
(Opp. p. 176)

Range and Status. NW Africa, Azores, Madeira, Canaries; Europe east to line from Trondheim to Danube delta; Turkey, Caucasus, Iran; Asia from Ural Mts to Kamchatka and Japan. Winters in W Europe, Mediterranean, Africa, coastal Arabia, S and SE Asia.

Resident and Palearctic visitor. Common and widespread in N Morocco in summer, from sea level up to 3100 m; common in High and Middle Atlas Mts in Nov–Dec. Rare breeder in Algeria (Djurdjura, Tlemcen, Kherrata, Lakhdaria), but frequent and widespread winter visitor in N, with desert records at Ouargla, Béni Abbès, Bordj Omar, Touggourt, Laghouat and Mitidja. Uncommon winter visitor to Tunisia. In Libya small numbers winter along coast in NW and NE and in jebels up to 100 km inland; several desert records of passage migrants, with 10 and 30 at Serir in spring 1969 and 1970. Uncommon to frequent winter visitor to N Egypt coast, Nile delta and valley, and Western Desert.

Mauritania, regular on passage near Nouakchott; 1 in N (Zouerate). Senegambia, 3 winter records in NW and 4 at mouth of Gambia R. Mali, quite common and widespread winter visitor in SW; records on Niger R. at Ansongo and L. Korientzé. Niger, 4 at Mari, 1985.

Commoner winter visitor in E Africa. In Sudan, local and uncommon passage migrant in N and winter visitor in S. Frequent to common in W and SE Highlands of Ethiopia, also in Rift Valley and NE; mainly between 1500 and 3000 m, but some down to 1000 and up to 4000 m; below 500 m near Dancalia, Eritrea. Uncommon passage migrant in Somalia. Recorded in N Central African Republic (Manovo-Gounda-Saint Floris Nat. Park), and in NE Zaïre (Avakubi, Faradje, Mahagi Port, Bambesa). In Uganda widespread in W from West Nile and Acholi south to Ankole and Kigezi, and in E frequent on Mt Elgon. From there occurs frequently through Kenya highlands to N Tanzania (Arusha Nat. Park, North Pare Mts, Mt Kilimanjaro, East Usambaras) with some on passage in N and SE Kenya. A few winter on Zomba Mt, Malaŵi (first record, 1964). Vagrant Zambia (Chinkuli Dam, Lusaka, May 1979), Zimbabwe (Nyanga, 3 in Dec 1983; E highlands, date?; Hwange Nat. Park, Mar 1984); Namibia (Swakopmund, 3 records, 1971–1984; Windhoek, Jan–Feb 1984); Transvaal (Woodbush, 3 records 1983–1989; Natal (Mkuzi, Sept 1974, Mtunzini, June 1977, Karkloof, Mar 1979, between Dundee and Nqutu, Mar 1987); Cape Province (Cape of Good Hope Nature Reserve, Apr 1989, Port Elizabeth, July 1989) (Every 1990).

Density of 6–8 birds on 6–7 km of river, Ethiopia; once 8 in 800 m of small wadi stream (Tyler and Ormerod 1986, 1987).

Motacilla cinerea

Field Characters. A slim, graceful, long-tailed wagtail of tumbling mountain streams and brisk lowland brooks; breeding ♂ mid-grey above with olive yellow rump and white superciliary stripe, and bright yellow below with black chin and throat, separated from dark grey cheeks by broad white stripe. ♀, winter ♂ and juveniles are buffier and lack black throat. Loud, hard, disyllabic call-note, very like that of Mountain Wagtail *M. clara*. Shape and habits very like Mountain Wagtail, which is more blue-grey above with blackish cheeks and narrow black necklace on white (never yellow) underparts. The 2 species often occur together. Tail wagged deeply up and down; underside of closed tail appears pure white.

Voice. Tape-recorded (62, 73, 78, 89, B). Song a musical 'tzi-tzi-tzi-tzi, tzee-ree-ree-ree' with trilling variants (see Cramp 1988). Call a clipped, rather high-pitched, metallic 'chissik' or 'tzitzi' or a harder 'tchichipp'.

General Habits. Habitat is primarily fast-running, clean montane streams, secondary lowland rivers and lake shores, with shingle, gravel, or boulders, plenty of trees and dense herbage, and (in summer) cliff ledges and recesses for nesting. Uses ornamental waters (Egypt), but mainly in coastal wadis. In Ethiopia inhabits rocky streams on steep, wooded escarpments, rivers flowing through deforested grassland, and river beds with trickle of water and a few deep pools; also sewage farms, cattle farms with manure and slurry, and forest tracks (Tyler and Ormerod 1987); in Eritrea mainly on dams and streams, but often far from water in *Combretum* and dry acacia woodland (Smith 1957).

Solitary or in pairs, or family parties in NW Africa. Forages along stream or up to 100 m away in grassy places and on paths (Ethiopia); walks and runs, pecking

Description. *M. c. cinerea* Tunstall (only race in Africa): NW Africa, Europe, east to Iran. Winters in W Europe, NW, W and E Africa. ADULT ♂ (breeding): head and upperparts mid- to dark grey, except for narrow but distinct white superciliary stripe, narrow white eye-ring, black lores, white moustachial stripe, and olive-yellow rump and uppertail-coverts. Tail black, outer 3 rectrices white. Wing-coverts olive-grey; remiges black, inner ones (and particularly tertials) with white edges. Chin and throat black, sharply demarcated from breast. Breast and remaining underparts bright yellow. Bill black, mouth grey-pink; eye dark brown; legs and feet pink-brown or dark brownish flesh, soles yellower. ADULT ♂ (non-breeding): the same, but chin and throat white or buffy white, and superciliary stripe indistinct, buffy. ADULT ♀ (breeding): like breeding ♂, but chin and throat buffy white, mottled with black (mainly on chin), sometimes as black as in ♂. ADULT ♀ (non-breeding): like non-breeding ♂, but underparts paler yellow and breast buffy. SIZE: wing, ♂ (n = 13) 82–89 (85·0), ♀ (n = 10) 80–86 (83·8); tail, ♂ (n = 14) 94–104 (97·9), ♀ (n = 10) 92–99 (95·8); bill to skull, ♂ (n = 29) 15·1–16·9 (16·1), ♀ (n = 24) 15·1–16·7 (15·9); tarsus, ♂ (n = 35) 19·4–21·8 (20·8), ♀ (n = 31) 19·7–21·6 (20·6). WEIGHT: (central Europe) ♂ (n = 87) 15–22 (18·0), ♀ (n = 26) 14–20 (17·2).

IMMATURE: like non-breeding ♀ but pale markings buffier and dark markings more olive.

NESTLING: skin, including bill and legs, yellow-flesh; gape-flanges pale yellow; mouth bright yellow; down, on upperparts only, long, dense, pale grey-yellow or golden buff.

at water and rock surfaces and amongst vegetation; rarely, turns over leaves, jumps into air for an insect, makes aerial pursuit of one, or catches a resting butterfly. Flycatching flight from ground is steep rise up to 6 m (Schifferli 1972). Peck rate 14·8 per min (Tyler and Ormerod 1987).

Autumn passage across Straits of Gibraltar into Morocco mainly mid-Sept to mid-Oct (at least 200 birds: Telléria 1981), and near Suez, Egypt, 39 birds in the same period (Cramp 1988). One of the earliest Palearctic migrants to Ethiopia, arriving N Eritrea mid-Aug and Addis Ababa mid-Sept. Visits E Africa Sept–Mar. May–June records in southern Africa (see above) suggest some regular 'oversummering'. An altitudinal migrant in Morocco (Cramp 1988).

Recoveries of birds ringed in Europe and found in Africa: Germany to Morocco (8), Algeria (3), Tunisia (1), Mauritania (1), Mali (1), Senegambia (6) and Ghana (2) (Schloss 1982; although the last 2 are likely to refer to Yellow Wagtails *M. flava*); USSR (43° 45′N 40° 0′E) to S Sudan (Imatong Mts).

Food. Mainly aquatic insects. Of 177 items (Ethiopia), adult chironomid and empidid flies were 34% (numerically) and their larvae, including simuliids, 11%, adult caenid mayflies 6%, and ephemerellid and baetid mayfly nymphs 29%, adult caddis-flies 2%, stoneflies 1%, beetles 11% and larval beetles 5% (Tyler and Ormerod 1987). One bird caught a *Pholotis* butterfly (Ethiopia). Diet in Europe similar, and includes caterpillars, damselflies, dragonflies and their larvae, and a few small molluscs, crustaceans, amphibia and fish. Estimated to take 19,800 items per day (Britain, summer).

Breeding Habits. Not studied in Africa. Monogamous, territorial, solitary breeder. For courtship behaviour, see Cramp (1988).

NEST: cup of grass, roots, moss and small twigs, lined with hair. Sited on cliff ledge or in crevice in wall, in earthen bank or among tree roots. Both sexes gather material and build nest, but probably only ♀ adds lining.

EGGS: 3–7, laid daily. Subelliptical; smooth, glossy; white, buff or grey, mottled with brown or grey and often with dark streaks. SIZE: (n = 200, Europe) 17–21·7 × 13–15·5 (19·0 × 14·3). 2 broods (rarely 3).

LAYING DATES: NW Africa, Mar–May.

INCUBATION: by ♀ and ♂. Begins with last egg. Period: 11–14 days.

DEVELOPMENT AND CARE OF YOUNG: young brooded assiduously by day by ♀ and ♂, and at night by ♀ only. Both parents feed young and remove faecal sacs, dropping latter in running water. Fledging period: 11–17 (13·5) days. Young may return to nest for few days after leaving; start to feed themselves 4–5 days after first departure from nest; parents may feed them for a further 1–2 weeks.

BREEDING SUCCESS/SURVIVAL: in 2 European studies (338 and 590 eggs), successes from eggs laid to fledglings reared were 42 and 57% (Cramp 1988).

References
Cramp, S. (1988).
Tyler, S. J. and Ormerod, S. J. (1986, 1987).

Motacilla clara Sharpe. Mountain Wagtail; Long-tailed Wagtail. Bergeronnette à longue queue.

Plate 11
(Opp. p. 176)

Motacilla clara Sharpe, 1908. Ibis, p. 341, *nom. nov.* for *Motacilla longicauda* Rüppell, 1840; Simen, Abyssinia (preoccupied).

Forms a superspecies with *M. cinerea*.

Range and Status. Endemic resident; nomadic on seasonal streams. More or less continuously distributed on high ground in eastern Africa, from Ethiopia to E Cape, occurring west of 25°E in S Zaïre, N Zambia and Angola (west to Atlantic coast). Separate populations in E Nigeria/Cameroon/Rio Muni/Gabon and in Liberia/Sierra Leone/Ivory Coast.

Sparse to frequent in Sierra Leone (upper Rokel R., Bintumane, Sefadu/Gandahun); present in lowland W Liberia, and frequent on Mt Nimba; W Ivory Coast (Comoé, Bandama, Meno and Cavally rivers); SW Mali (upper Bafing valley) and probably Mandingo Mts; not yet reported from Guinea, but must surely occur there. Ghana, 1 record Begoro. In Nigeria sparse to frequent on Obudu Plateau; recorded in Oban Hills (300 m), Serti (Mambilla Plateau) and Calabar (near sea level). Frequent and widespread in hills and lowlands of W Cameroon. In Gabon common in savannas and gallery forests near Booué, recorded at Loa-Loa waterfalls (Ivondo basin) Nov–Dec; once Makokou. Central African Republic, recorded Lobaye Préfecture. Resident and frequent in Ethiopia at 1500–3000 m, common in S Tigré Province but rare or absent in N Tigré and Eritrea; frequent in Tana basin, Tacazze R. and W escarpment of Begemder and Simien Provinces (Tyler and Ormerod 1986).

In E Africa scarce in lowlands; frequent in highlands from Mt Elgon, through Kenya (east to Mara R., Narok, Nairobi and Naro Moru) to Tanzania (from Mt Kilimanjaro to Arusha Nat. Park, Njombe, Mt Rungwe and Ufipa Plateau). Scarce in Burundi, between 1400 and 1600 m near Kumoso and Makamba. In S Malaŵi, from 70 to 1600 m in Nsanje and Chikwawa districts; also adjacently in Mozambique including Mt Gorongoza.

Local and uncommon in Zambia but widespread on Eastern Province plateau; absent from Western and Barotse Provinces. Occurs up to 1800 m in Zimbabwe, in E highlands and from Mashonaland Plateau to Victoria Falls; local in Sabi valley; absent from Matopos and Limpopo valley. Local in Angola (Cabinda, E Benguela and Huambo Highlands, Mossamedes Escarpment, Cuanza Sul, Luhanda, Dundo, Quela, Gabela). In South Africa restricted to Escarpment region and E edge of highveld in Transvaal, where only locally at all common; and in Swaziland, Zululand, Transkei, Natal and E Cape.

Densities of 6–7 pairs on 1·6 km of river, 9–12 pairs on 2·5 km, and 13–14 pairs on 2·4 km, Ethiopia (Tyler and Ormerod 1987), and in southern Africa *c*. 1 pair per 0·5 km of pristine river and 1 pair per 1·0 km of unclean river (Maclean 1985); density of 11·1–25 (16·9) pairs per 10 km on Palmiet R., Natal (Piper and Schultz 1988).

Motacilla clara

Description. *M. c. torrentium* Ticehurst: E Uganda, Kenya and Rwanda to Angola and South Africa. ADULT ♂: forehead, crown, nape, sides of neck, mantle, back, scapulars, rump and uppertail-coverts uniform grey, somewhat darker on top of head. Pre-orbital area and stripe under eye blackish, becoming dark grey on ear-coverts; narrow white superciliary stripe from nostril above lores to slightly behind eye. Central tail-feathers (T1) black; T2 black with irregular white marks at tip and sometimes with narrow white margin to outer web; T3 white with black outer web and broad black edge to inner web; T4 white, sometimes with narrow black line proximally on outer web; T5 and T6 entirely white. Wing blackish, most remiges and coverts extremely narrowly edged white, but tertials and tertial coverts broadly edged white. Axillaries and underwing-coverts white, but outer under primary coverts black. Chin and throat white; narrow blackish breast-band, *c*. 8 mm deep in centre and 2 mm deep at sides of breast; remaining underparts white, greyer on flanks; undertail-coverts white. Bill and mouth black, base of lower mandible sometimes grey; eye dark brown; legs and feet pinkish or greyish brown. Sexes alike. SIZE: wing, unsexed (n = 37) 72–83 (79·0); tail, unsexed (n = 37) 90–104; bill, unsexed (n = 11) 14–15·5; tarsus, unsexed (n = 11) 20–22 (21·0). WEIGHT: (Kenya) ♂ (n = 3) 17–21 (19·3), ♀ (n = 1) 15.

IMMATURE: like adult but blackish breast-band vestigial or absent; underparts sometimes suffused buffy.

NESTLING: not described.

M. c. clara Sharpe: Ethiopia. White borders of tertials and tertial coverts much less broad than in *torrentium*. Wings longer, tail proportionately shorter. Wing, unsexed (n = 12) 83–91; tail, unsexed (n = 12) 94–103.

M. c. chapini Amadon: W Africa to W Uganda. A poor race, not certainly distinct from *torrentium*, but in series upperparts said to be darker. WEIGHT: (Uganda) ♂ (n = 5) 15–21·5 (17·1), ♀ (n = 5) 15–24·5 (18·7).

Field Characters. A slim, graceful, long-tailed wagtail of swift, rocky streams; clear blue-grey above, white below, with black cheeks, black 'necklace', and black tail with much white in sides. Loud, hard, disyllabic call-note. Shape and habits very like Grey Wagtail *M. cinerea* which, however, always has yellow belly. Cape Wagtail *M. capensis* has overall browner appearance, with dull olive-grey or olive-brown upperparts, underparts off-white to buff and often rather blotchy-looking, with brown flanks; breast-band either almost lacking (race *simplicissima*) or broader than in Mountain Wagtail and less clear-cut. Brown or grey-brown cheeks not contrastingly dark. Tail of Cape Wagtail shorter and less white (2, not 3, white outer feathers), and in flight shows less white in wings than Mountain Wagtail. Cape Wagtail also has different habitat and call.

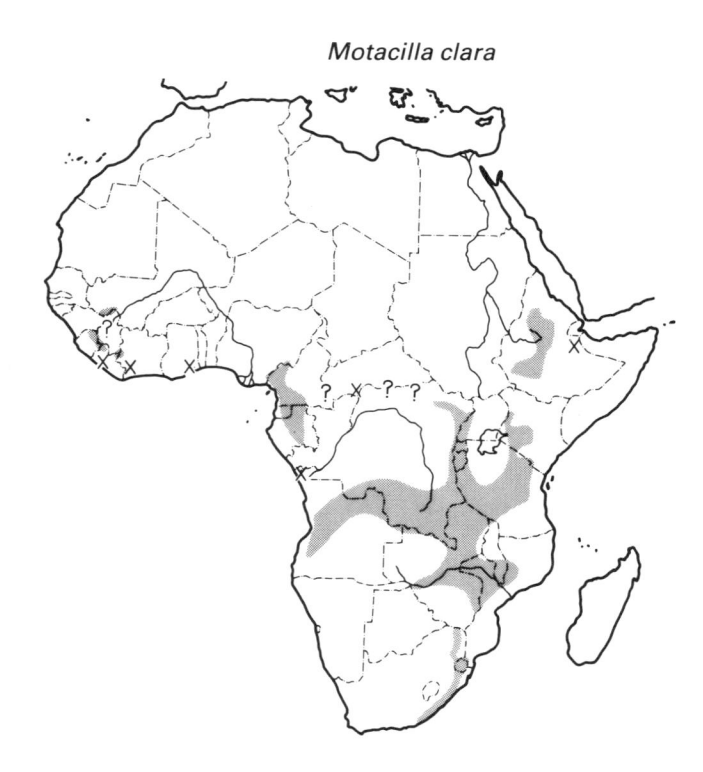

Voice. Tape-recorded (72, 78, F, GIB, GREG, McVIC, WALK). Song a sibilant, quiet and sustained medley of 4 phrases. One passage of 75 s consisted of rather regularly repeated phrases in the sequence A A B A B A A B A A B A, C C B C C C C D D D D. Phrase A is decelerating, falling trill of *c*. 21 notes, 'trrrrrip'; B contains 3 short notes, 'ts tswe tsu', at successively lower pitch but 'tswe' rising in pitch; C has 3 longer notes, 'tse tse teee', 'tse' at 3·5 kHz and 'teee' at 4·4 kHz; D has 4 notes, 'tu tseeeu tu tseeeu'. Phrase A lasts 1·1 s, B 0·8 s, C 1·0 s and D 1·8 s; pause of 1–5 s between phrases. Call, usually given on taking flight, is loud, hard 'chissik'; also has high-pitched contact call, 'teep'. Alarm: 'phweep' or 'chweep'.

General Habits. Occurs along fast-flowing, turbulent, bouldery streams in lowland and montane forest; often near (and behind) waterfalls; also on swift streams in *Brachystegia* and acacia woodland (Angola). Streams may be thickly overhung with woody vegetation or almost deforested (Ethiopia: Tyler and Ormerod 1987). Found occasionally on river beds where water reduced to series of deep pools with narrow trickle over algae-covered rocks. Usually in more open situations in montane than in lowland country.

In pairs, feeding along edge of dashing or quiet water, along fallen tree trunks, on stony reaches, shingle, rocks, and grassy banks. Pecks at prey on water and rock surfaces and from amongst vegetation. Rarely (1% of feeding methods) jumps up or flies to catch winged insect (Tyler and Ormerod 1987). Catches butterflies resting on ground (Chapin 1953). Larger prey beaten against rock before being carried to young. Forages within 1 m of Grey Wagtails (Ethiopia) without evident conflict, but may be interspecifically territorial with them elsewhere. Threatens conspecifics with puffed body plumage and raised tail (Maclean 1985). A lively bird, constantly on the move, working streamside assiduously, tail wagging deeply, then flying a few 100 m to new foraging area. Often feeds foot-deep in water, sometimes belly-deep. Flies low over water, fanning tail and calling as it takes off. Flight markedly undulating.

Sedentary on permanent streams. No evidence of any seasonal movements to lower altitudes in Ethiopia (Tyler and Ormerod 1986), but regarded as somewhat nomadic in E and southern Africa. All Ivory Coast records in dry season (Oct–Mar) and all Mali records Dec–May.

Food. Mainly aquatic insects. Of 304 items (Ethiopia), adult chironomid and empidid flies were 53% (numerically), and their larvae (including simuliids) 4%, adult caenid mayflies 6·5% and 3 families of mayfly nymphs 20%, adult hydropsychid caddis-flies 3%, perlid stoneflies 3%, and Lepidoptera and Coleoptera (including a few larvae) 11% (Tyler and Ormerod 1987). Also recorded: tadpoles and large dragonfly nymphs (Ethiopia), butterflies (Zaïre) and mosquitoes (South Africa). Southern African diet mostly dragonfly larvae (Maclean 1985).

Breeding Habits. Monogamous, territorial, solitary breeder. Territory strictly linear, being a stretch of river and narrow zone either side of it. Territory length, 200–700 m (Ethiopia) and 400–900 (590) m on Palmiet R. (Natal). Territory is all-purpose, defended by pair all year; pair keeps together. In territorial display, pair stands on rock, calling loudly, fluffing plumage, lowering and vibrating wings, raising tail vertically and fanning it; often followed by hectic aerial chases with much calling (Piper and Schultz 1989). Courtship display: bird flies erratically, chasing mate which flies with tail fanned; sometimes they fly vertically upwards – a 'spiral dance'. At other times bird walks towards mate with drooping wings, body plumage fluffed, and tail raised and fanned (Piper 1989).

NEST: neat, deep cup of fine rootlets and hair, built into coarse and bulky foundation of grasses, leaves and moss. Birds submerge foundation materials in water first, making them soggy and pliable. Nest is kept damp throughout construction, by adding wet material. Inner bowl of fine rootlets added after nest dries. Ext. diam. 120–140, height 70–150. Sited in niche in stream bank, often concealed behind overhanging vegetation (85% of 48 nests), under bridges (10%) and in trees (5%); up to 5 m above water surface and 9 m from river side. Pair uses same nest-site year after year, and uses same nest if it can be refurbished. Av. of 2·55 breeding attempts per year (Natal: Piper and Schultz 1988).

EGGS: 1–4, usually 2–3, laid on successive days (Ethiopia, Natal). Ovate; dull greyish, spotted with brown, mainly at broad end. SIZE: (n = 27, southern Africa) 17·4–21·9 × 14·2–16 (19·9 × 15·2). WEIGHT: 2·45.

LAYING DATES: Ethiopia, Jan–Apr, Sept–Nov; Tanzania, all year; Malaŵi, Mar, July–Nov; Zimbabwe, Aug–Nov (mainly Sept–Nov); South Africa: Transvaal, Aug–Dec, May (mainly Oct–Nov); Natal and E Cape Sept–Dec.

INCUBATION: by ♂ and ♀. Period: 13–14 days.

DEVELOPMENT AND CARE OF YOUNG: young fed by ♂ and ♀, av. 4·9 visits to nest per h (Tanzania). For growth, see Piper (1989). Fledging period c. 14 days (Natal) or 15–16 days (Ethiopia). After leaving nest, young dependent on parents for 14–30 days (first brood) or up to 60 days (later broods).

BREEDING SUCCESS/SURVIVAL: on Palmiet R., Natal, av. of 5·5 eggs per pair per year produce 1·55 fledglings (28% success), and adult survival rate is 95%. It implies that 70% of fledglings disperse away from Palmiet R. (a 'prime' area) each year (Piper and Schultz 1988). Oldest ringed bird 10 years (Piper 1987).

References
Moreau, R. E. (1949).
Piper, S. E. (1989).
Piper, S. E. and Schultz, D. M. (1988, 1989).
Tyler, S. J. and Ormerod, S. J. (1987).

Motacilla alba Linnaeus. White Wagtail. Bergeronnette grise.

Plate 11
(Opp. p. 176)

Motacilla alba Linnaeus, 1758. Syst. Nat. (10th ed.) 1, p. 185; Sweden.

Forms a superspecies with *M. aguimp*, *M. madaraspatensis* (India), *M. grandis* (Japan) and *M. lugens* (E Asia).

Range and Status. NW Africa, whole of Europe and Asia south to Iran, Himalayas and Vietnam. Winters in SW Europe, Africa and S Asia.

Resident and Palearctic migrant. As breeding bird, confined to Morocco, where race *subpersonata* is endemic; widespread but rather sparse in lowlands (wadis Guir, Ziz, Rheriss, Mgoun, Sous, Nfis, Oum er Rbia, Sebou, Loukkos), also in High Atlas (Ziz gorge, Rich, Oukaïmeden, Ourika valley). Evidently a vertical migrant. Marked recent range expansion in SW Morocco. May occasionally breed in Algeria.

Frequent to abundant Palearctic winter visitor to N

Africa, south to Sierra Leone, central Chad and Kenya. Autumn immigrants must number several million, since 1 million were estimated to have passed a boat 200 km north of Egypt in 4·5 h in spring (Larmuth 1973). Abundant in Morocco and Algeria, wintering from coast up to 2200 m, with several roosts of thousands each. Scarce to common in Tunisia. In Libya abundant in Tripoli, common in Fezzan and Cyrenaica, scarce to common in winter at oases in desert. Common to abundant in Egypt. Common in NW and SW Mauritania and NW Senegal; the commonest and most widespread of all Palearctic birds in Mauritania, ubiquitous in Nouakchott. Common on Gambian coast, scarce on Gambia R. Uncommon in Sierra Leone. Frequent to common, and widespread, in Mali south of 17°N. Uncommon to frequent in S Niger and very local in N. Uncommon to frequent in N Nigeria, south to Zaria and Jos Plateau; very rare at L. Chad. Frequent in Chad where mapped, very common in Ouadi Rimé–Ouadi Achim Faunal Reserve. Common to abundant in N Sudan, uncommon in S. Common throughout Ethiopia, much less so in Somalia, where found mainly in NW and SE. Winters commonly on Socotra I., on Hadibu plain and Goahal valley; also on adjacent Abd-el-Kuri I. In Uganda small numbers winter south to Murchison Falls Nat. Park, Soroti and Mbale, and rarely to Kampala, Rwenzori Nat. Park and Kitwe. In Kenya common only on L. Turkana; further south rare to frequent, with up to 30 on Kariobangi sewage works. Vagrant to Tanzania (L. Manyara, Dar-es-Salaam), Malawî (Mangochi, 'winter' 1932–33), Cameroon (c. 4, Waza, Ndjamena), Togo (1 at Oti Toutionga, Jan 1987: Walsh *et al.* 1990). Central African Republic (Manovo-Gounda-Saint Floris Nat. Park; Sangha Préfecture), Gabon (M'Passa, Dec 1972), and Zaïre (8, before 1953, at Avakubi, Niangara and Faradje; perhaps frequent in extreme NE); sight records from Victoria Falls, Zambia.

Description. *M. a. subpersonata* Meade-Waldo: Morocco. ADULT ♂ (breeding): forehead and short superciliary stripe white, sharply demarcated from adjacent black areas. Crown, nape and hindneck black. Mantle, back, rump, uppertail-coverts and scapulars mid-grey. Tail-feathers black with white sides; T1 black with narrow white outer fringe, T2–T4 black (T4 sometimes with a little white on outer fringe), T5 and T6 white, with long dark wedge basally on inner webs. Lores, cheeks and ear-coverts black, eye-ring white, with small crescent of white featherlets below eye; black ear-coverts adjoin black hindneck and black throat. Chin, throat and breast black, moustachial streak white, and white oblong on side of neck below ear-coverts. Black breast sharply demarcated from rest of underparts, which are white with grey wash on sides of breast and flanks. Wings greyish black, median and greater coverts broadly fringed white and forming 2 wing-bars, and tertials broadly fringed white. Bill black, mouth black; eye black-brown; legs and feet black. ADULT ♂ (non-breeding): centre of crown and nape washed olive-grey; chin white; throat mottled black and white. ADULT ♀ (breeding): like ♂ but head usually less contrastedly pied: a few grey feathers on crown and white ones on chin and ear-coverts; white moustache not clear-cut. ADULT ♀ (non-breeding): like breeding ♀, but crown and hindneck greyer, chin white, throat mottled black and white. SIZE: (*M. a. alba*, Europe) wing, ♂ (n = 44) 87–96 (91·6), ♀ (n = 18) 85–92 (88·2); tail, ♂ (n = 80) 80–90 (85·0), ♀ (n = 42) 77–87 (82·0); bill to skull, ♂ (n = 40) 15·2–17·2 (16·2), ♀ (n = 21) 14·9–16·9 (16·0), bill to feathers, ♂ (n = 18) 8·5–9·8 (9·0), ♀ (n = 14) 8·5–10·1 (9·2); tarsus, ♂ (n = 38) 22·7–25 (23·8), ♀ (n = 23) 22·2–24·4 (23·2). WEIGHT: (*M. a. alba*, Belgium) ♂ (n = 25) 20–24·6, ♀ (n = 16) 17·6–21·9, laying ♀ (n = 4) 24·2–27·9.

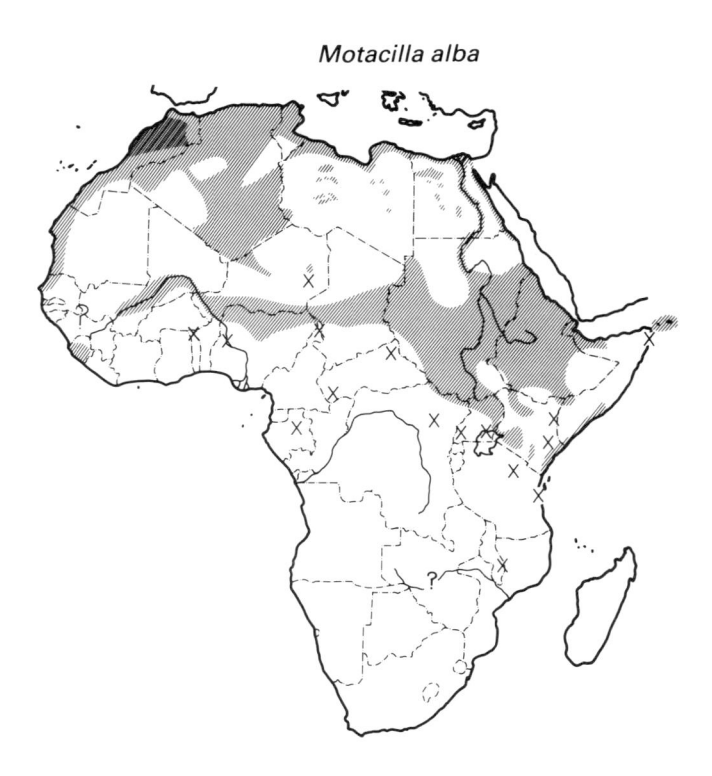

IMMATURE: head brownish grey; dusky moustachial stripe; chin and throat greyish white; narrow dark grey-brown gorget; rest of underparts grey-white, sometimes with buff tinge.

NESTLING (*M. a. alba*): bill, legs and skin flesh-pink; mouth orange-yellow, tongue reddish, gape-flanges pale yellow; down, on upperparts only, long and grey.

M. a. alba Linnaeus: Iceland, continental Europe east to Urals, Asia Minor, Levant; winters from Netherlands, Bulgaria and Turkey south to Senegambia, Kenya and Arabia. Like *subpersonata* but lores, cheeks and ear-coverts always white, joining white oblong on side of neck. Mantle a clearer grey or bluish grey. Median and greater coverts whiter than in *subpersonata*, forming more conspicuous wing-bars; tertials more conspicuously white-fringed. Measurements given above.

M. a. yarrellii Gould: Britain and locally on adjacent continental coast; winters south to NW Africa. Less frequent than *alba* in Morocco (Thévenot *et al.* 1980), wintering near Atlantic coast. 2 records Tunisia. Like *alba* but mantle, back, rump, uppertail-coverts and scapulars black in ♂, greyish black in ♀.

M. a. dukhunensis Sykes: SW USSR to central Siberia, Altai Mts and Yenisei, south to N Iran; winters in S Asia. Frequent on Socotra I., where no other race recorded; vagrant Ethiopia. Geographically variable; like *alba* but white tips of median and greater coverts broader, mantle and scapulars to uppertail-coverts usually paler, bluer grey; closed wing usually looks paler.

Field Characters. The common, familiar winter wagtail of N Africa. Juveniles of this (and most) wagtail species can be difficult to identify – they are best told by accompanying adults. In Palearctic Africa adult White Wagtail readily distinguished from Grey Wagtail *M. cinerea* by lack of yellow, shorter tail with 2 (not 3) white outer tail-feathers, and habitat. Wintering birds broadly sympatric with African Pied Wagtail *M. aguimp* in Mali, Niger, Sudan, Ethiopia, Somalia, Uganda and Kenya, with Mountain Wagtail *M. clara* in Ethiopia, and with Cape Wagtail *M. capensis* in Kenya. Adult always distinguishable from adult African Pied Wagtail by its white (not black) forehead and face, lack of white wing-patch; juveniles can be almost identical, although African Pied Wagtail usually has whiter wings. Distinguished from Mountain Wagtail by white face, broad black breast-band, shorter tail with white on 2 (not 3) outer feathers, and habitat. Adult distinguished from Cape Wagtail by strongly pied plumage and voice, but juveniles can be practically indistinguishable.

Voice. Tape-recorded (HAZ – 62, 73, 78, B, C). Call, given commonly in flight and whilst foraging, a loud, shrill 'tschizzik' or more musical 'tziwirrp'. Alarm: 'chik'. Song a lively, warbling twitter, consisting mainly of slurred repetitions of the 'tschizzik' call.

General Habits. Breeding birds in Morocco inhabit rocky and sandy reaches of rivers in fertile valleys, and banks of clear-running streams flowing into sandy and rocky desert south of Atlas Mts. Other races wintering in Africa depend far more on human habitation and agriculture, and frequent lake and river edges, dams, sewage works, marshes, farmland, gardens, villages, cities, parks, fields, and well-sites and nomadic settlements in the Sahara.

Solitary, in pairs, small flocks, and large concentrations on migration and at roosts. Forages by walking and picking items from ground; walks in toe-deep water, picking at surface; often runs for 1 m to make stab at insect; sometimes makes short flight upward to catch prey in mid-air; occasionally hovers to snatch food from water surface or vegetation. Forages along 'tide-line' by puddles and larger waters, taking items washed up. On seashores, searches tide-line for insects and crustaceans in seaweed. Feeds on crumbs around houses; forages near, and occasionally upon, cattle. Searches carcasses for maggots. Follows the plough, and feeds on weevils in open-air grain stores (Egypt). Usually tame and confiding.

Territorial in winter, usually but not always in pairs. Once 49 birds studied, each in its own feeding territory (Egypt: Simmons 1965). Typically, pair maintains 1 main and 1–2 subsidiary feeding territories, defending them from other wagtails. Territories of different pairs are well spaced, and not necessarily confluent. ♂ tends to dominate ♀, and to chase her away aggressively if food is short. ♂ sometimes sings in territory (Egypt).

Large numbers roost in exotic neem trees *Azadirachta indica* (E Chad). Much larger roosts form further north: 2000 birds in Egypt and several thousand in Morocco. In Egypt roosts are usually in sugar cane, but one was in a row of small trees on bank of Nile in Gezira. Birds fly to roost individually or in well-separated small parties; first they form a preliminary gathering nearby, calling, feeding and chasing each other (Simmons 1965).

Abundant on autumn passage along whole of African Mediterranean coastal regions, where winter visitor Oct–Apr. Present SW Mauritania mid-Sept to early Apr, N Nigeria early Nov to late Mar, and Kenya Nov to early Mar. At most localities arrives 2–3 weeks later than Yellow Wagtail *M. flava*, and in spring departs 1–2 weeks earlier. Begins to assume breeding dress in Feb. Abundant on spring passage through Atlas Mts, in Maghreb coastal lowlands and over straits of Gibraltar in Mar to mid-Apr; also through Nile delta and across Egyptian Mediterranean coast Mar to early Apr. During a night of mass migration across E Mediterranean in late Mar, thousands died on a ship (Larmuth 1973).

Ringing recoveries: Senegambia to Morocco (1) and Spain (1), Morocco to Tunisia (1), Austria (1), France (1), Germany (4), Switzerland (1) and Netherlands (1). Iceland to Morocco (1); circumstantial evidence that some birds wintering in Mauritania and Senegambia may be from Iceland (Morel and Roux 1973). Britain to Morocco (7), Algeria (1), Mauritania (1). Further recoveries in Morocco from Netherlands (15), Belgium (3), W Germany (9), E Germany (5), France (1) and Switzerland (1). Recoveries in Algeria from Sweden (1), W Germany (4), Switzerland (2), Poland (5). Poland to Tunisia (4); Sweden to Algeria (1) and Tunisia (1). Norway and Sweden to Egypt (11); Finland to Egypt (19). To Sudan from Sweden (2), Latvia (2), Lithuania (1), Finland (1) and USSR (1) (57° 45′N, 39° 40′E).

Food. Flies and other insects, supplemented with crumbs, worms, snails and crustaceans. For detailed diet, see Cramp (1988). Not studied in Africa, where only flies, snails, maggots and weevils recorded.

Breeding Habits. Barely studied in Africa, very well known in Europe. Monogamous, solitary, territorial. For territorial and courtship behaviour, see Cramp (1988).

NEST: rough cup of twigs, grass stems, leaves, rootlets and moss, neatly lined with hair and feathers. Sited in hole or crevice, in Europe commonly in man-made structure (stone wall, bridge).

EGGS: 3–8, mainly 5–6. Laid daily. Smooth, glossy; white or grey-white with fine, even, grey-brown speckling. SIZE: (n = 250, *alba*, Europe) 18–22 × 14·2–16·3 (20·2 × 15·1).

LAYING DATES: Morocco, Apr–June.

INCUBATION: by ♀ and ♂; mainly ♀ by day, solely ♀ at night. Period: 11–16 (12·6) days.

DEVELOPMENT AND CARE OF YOUNG: young fed by ♂ and ♀. Fledging period: 11–16 (13·7) days. Fed by parents for 4–7 days after leaving nest.

BREEDING SUCCESS/SURVIVAL: Finland, 79% hatching and 77% fledging survival (n = 1147 eggs); Britain, 64% hatching and 83% fledging survival (n = 5747 eggs) (Cramp 1988).

References
Cramp, S. (1988).
Simmons, K. E. L. (1965).

212 MOTACILLIDAE

Plate 11
(Opp. p. 176)

Motacilla aguimp Dumont. African Pied Wagtail. Bergeronnette pie.

Motacilla aguimp Dumont, 1821. Dict. Sci. Nat. éd. Levrault, 21, p. 266; lower Orange River.

Forms a superspecies with *M. alba* (Eurasia), *M. madaraspatensis* (India), *M. grandis* (Japan) and *M. lugens* (E Asia).

Range and Status. Endemic resident. Wanders locally; migratory in South Africa. Vagrant Israel. Common to frequent in humid regions and near water and human habitations, from Sierra Leone and S Mali through Chad (mainly south of 10°N) to Ethiopia, Kenya (except NE) and permanent rivers in S Somalia. From these countries southward throughout tropics to Angola, N and E Botswana, Zimbabwe, Mozambique and South Africa.

Range in peripheral countries: Senegambia, several records Fatoto, Gambia R. and Kédougou, SE Senegal; Niger, at 7 localities on or near Niger R. Vagrant, São Tomé. Chad, records at L. Chad, frequent in Bahr Salamat, common in S. In Sudan, locally common along rivers and low-lying streams and swamps; Egypt, probably breeds along shores of L. Nasser; 1 record Aswan. Ethiopia, range poorly understood; scarce in highlands, commoner at low elevations in S. In E Africa a widespread resident and wanderer from sea level to 3000 m; very few records in Kenya north of 2°N in west and equator in east. Somalia, confined to Juba R. and lower Shebelle R. Occasional non-breeding visitor to Zanzibar and Pemba. Range in S Angola may be greater than shown. In Botswana, scarce in Okavango and Tuli area; commoner in Caprivi Strip (Namibia); reaches mouth of Cunene R. (Angola/Namibia). In South Africa a scarce to common resident; breeding migrant to E Cape; mainly non-breeding visitor to Transvaal.

Motacilla aguimp

Birds south of line are breeding visitors

Description. *M. a. vidua* Sundevall: range of species except range of *aguimp*. ADULT ♂ (breeding): crown, centre of forehead, nape, lores, area under eye, ear-coverts, mantle, scapulars and back black. Broad, well-defined white superciliary stripe from nostrils to well behind eye. Rump greyish black; uppertail-coverts black, lateral ones with white outer webs. Tail-feathers black; outer 2 pairs white (T5 with inner web edged black); central pair sometimes with 1 mm wide white outer edge. Chin and throat white; broad, black pectoral band about 12 mm deep in midline and 5 mm deep at sides of breast; sides of neck white, forming square patch surrounded by black; rest of underparts white. Primaries black, narrowly tipped white, with large area of white on base of both webs, largest on P5 and P4; secondaries black with broader white tip and white edge to entire outer web; tertials black with broad white edges to outer web; primary coverts and alula black; secondary and median coverts white; lesser coverts black; underwing-coverts white. Bill black; eye dark brown; legs and feet black. ADULT ♂ (non-breeding): upperparts slaty grey rather than black. ADULT ♀ (breeding): dark parts of plumage slightly less intense black than in breeding ♂. ADULT ♀ (non-breeding): upperparts dark olive-grey. SIZE: (NE Africa) wing, ♂ (n = 10) 93–102 (96·4), ♀ (n = 11) 88–96 (91·3); tail, ♂ (n = 10) 86–94 (91·1), ♀ (n = 11) 81–93 (87·0); bill to skull, ♂ (n = 10) 17·6–19·1 (18·3), ♀ (n = 11) 16·9–18·5 (17·9), bill to nostrils, ♂ (n = 9) 10–11·4 (10·8), ♀ (n = 10) 10·1–11·5 (10·8); tarsus, ♂ (n = 9) 23·9–26·4 (25·2), ♀ (n = 11) 23·2–25·6 (24·2). Liberian birds smaller: wing, ♂ (n = 3) 91–93 (91·8), 1 ♀ 85·5. WEIGHT: (Kenya, Uganda) ♂ (n = 13) 22–33 (27·3), ♀ (n = 6) 23–30 (26·1); (Zambia) unsexed (n = 55) 23–30·5 (27·0).

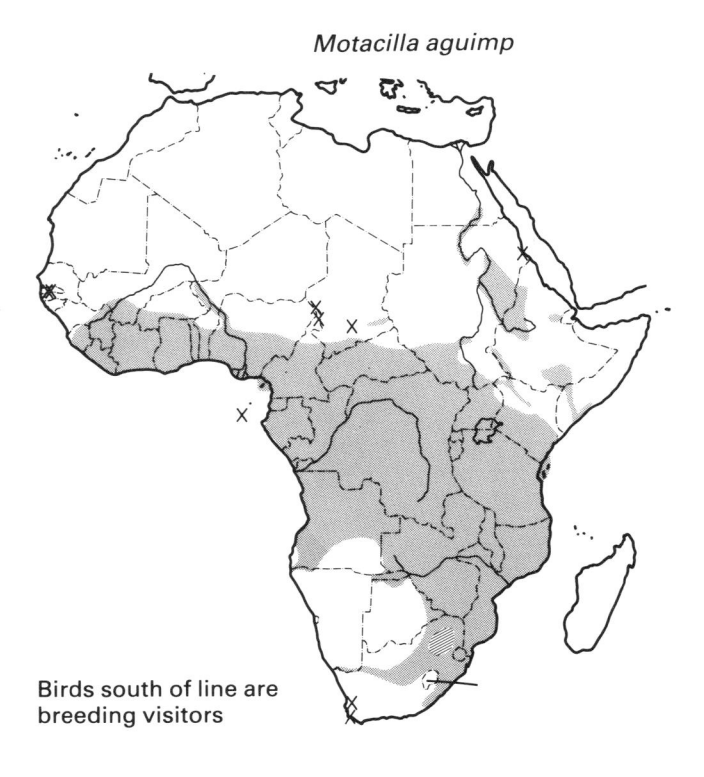

IMMATURE: somewhat resembles adult Cape Wagtail *M. capensis*, but has far more white in wings. Remiges and rectrices black, like adult *M. aguimp*, but all other parts black in adult are dark olivaceous grey-brown. Except in remiges and rectrices, all parts white in adult are heavily suffused with buff. Greater and median primary coverts white, tipped buffy.
NESTLING: naked, bill and skin yellow, with dusky down on crown, back and axillary region.
M. a. aguimp Dumont: Orange R. and lower Vaal R. in Cape/Namibia, Orange Free State, Lesotho, SW Transvaal; resident. Black breast-band larger than in *vidua* with its hind border less clear-cut (owing to a mixture of black and white feathers, and to feathers of upper belly having grey bases which show through). Sides of breast (concealed below wing) and flanks with large irregular patches of black. Outer tail-feather has black border to inner web, and T5 has greater extent of black than in *vidua*.

Field Characters. A tame and familiar bird with striking, boldly pied plumage. Only likely to be confused with White Wagtail *M. alba*, from which distinguished by black of crown extending to bill, long white superciliary stripe above black face-patch, and black mantle. Folded wing shows large solid white patch on black, whereas wing of *alba* is dark with many narrow white stripes (feather edgings). In flight, African Pied Wagtail shows conspicuous white patch on wing. Readily distinguished from Mountain Wagtail *M. clara* by black upperparts, white wing-patch, and broad black pectoral band. Immature quite like adult Cape Wagtail, but has far more white in wings, and different voice.

Voice. Tape-recorded (5, 22, 78, 81, 88, C, F). Song a sustained medley of mellow piping and high-pitched notes, quite melodious, and recalling Island Canary *Serinus canaria* in rhythm and phrasing: 'tsip weet-weet, twip-twip-twip, weep-weep, tip-tip-tip, weet-woo-woo'. Imitates other birds, including Common Bulbul *Pycnonotus barbatus* and Chattering Cisticola *Cisticola anonyma* (Brosset and Erard 1986). Call, usually given on taking flight, loud 'chizzit', not as emphatic or clearly disyllabic as in White Wagtail; also described as short whistle, 'tuwhee'.

General Habits. In humid tropics a common resident near human habitation: villages, parks, playing fields, lawns and large gardens in cities, on roofs and paved roads, farms, forest clearings, camp sites, dams, sewage ponds and along sandy margins of rivers and streams; also on sand banks, rocks, rapids, dry river beds, and coastal lagoons. In drier country far more restricted to riversides. Frequently enters houses. In Liberia and Cameroon prefers rushing torrents to sluggish water, but in Nigeria occurs by torrents, and also rivers which have dwindled to series of small pools. From sea level up to 1600 (Zimbabwe) and 3000 m (E Africa).

Occurs singly, in pairs, family parties and, out of breeding season, in loose aggregations where feeding is good (up to 100 birds at Blantyre sewage farm, Malaŵi: Nhlane in press). At such times, flocks of 5–10 birds walk and forage together. Walks about briskly whilst foraging, looking leggy, tail almost constantly wagging up and down (deeply and vigorously when bird alights or after running); often makes short, quick run to seize insect from ground 1 m away; also occasionally stands on tiptoe to reach for insect on vegetation, and jumps up or flies after passing insect. Occasionally hovers over water to pick floating insect from surface. Once, 2 settled on deep water to wash (Bannerman 1936). Peck rates on lawns av. 4·5 per min, on sewage percolation filters 21 and sewage drying beds 22 per min (Malaŵi: Nhlane in press). Usually tame. Bathes frequently. Flight deeply undulating. Usually calls during short flight, and often also on ground; commonly uses rooftop as singing post.

Hundreds roost in trees lining streets in central Nairobi, Kenya. 200–400 roosted at night in small yachts anchored off Entebbe, Uganda, many sleeping under floorboards (Pitman 1966). Breeding wagtails roosted on small boats at Watamu, Kenya, where they were nesting (Donnelly 1977). 50 roost in Blantyre, Malaŵi, in factory roof or trees in middle of city.

Generally sedentary, with some local wandering, but in South Africa breeding visitor to E Cape Sept–Mar; in Transvaal, common resident on lowveld rivers but absent from highveld except at major waters, where non-breeding migrant.

Food. Insects. Tabanid and other flies, dragonflies, orthopterans, moths, butterflies, small beetles, termites and ants (Gabon: Brosset and Erard 1986). 19 birds (Uganda, Kenya) all contained insects, including flies, beetles and larvae. Also takes grasshopper nymphs and worms. At Blantyre sewage works (Malaŵi), feeds almost entirely on 2 species of fly larvae (Nhlane in press). Attracted to mealworms *Tenebrio molitor*; also eats crumbs and meat scraps in gardens. Twice seen to take small fish from shallow water (Zimbabwe: Edwards 1988). 1 bird seen dipping large insect, usually dismembered cricket, into water on 7 occasions; each insect was dipped up to 17 times, then bird flew off with it in bill, perhaps to nestlings (Liggitt 1985).

Breeding Habits. Monogamous, territorial, solitary breeder. Pair defends nesting territory, typically *c.* 100 m diam., by a river, and an adjacent or nearby feeding territory of 1–2 ha, elongated, up to several hundred m long. Feeding territory defended by ♂ and ♀. Aggressive to conspecifics, including own young as soon as they become independent; commonly attacks its reflection in windows and car wing-mirrors. In what was evidently a territorial display, directed at Common Sandpiper *Actitis hypoleucos*, crouched with bill raised and closed tail held almost vertically (Tree 1963). Courtship not described. Strong circumstantial evidence that birds pair for life and remain faithful to a particular spot for years – at least 5 years (van Someren 1956, Donnelly 1976–1978, Niven 1981).

NEST: rough, bulky cup, loosely constructed of soft dry grass, plant stems, rootlets, feathers, dead leaves, plant down and occasional bits of string and rags; neatly lined with fine dry grass, hair, small feathers, fine rootlets and (Kenya coast) strands of coconut fibre and soft seaweed. Cup 58 mm int. diam., 38 mm deep. Built by both sexes; once material brought with 13 visits in 44 min. Sited on ground in clump of reeds, long grass, weeds, niche in river bank, pile of flood debris, or (often) cleft in rocks in middle of river; on rock shelf in cave, among creepers, in hollow end of decayed branch, in cavity of dead tree in flood water, in roofing thatch (often), on top of pillar, on a constantly-used chair in house (nest built in 13 h: Niven 1981), behind house rainwater pipe, shelf in out-house and, commonly, in glove-box or similar niche in small moored boats.

EGGS: 2–7, usually 3–4 (E Africa, most of southern Africa; but av. of 2·2 in Malaŵi (n = 30 clutches). 1 bird in Kenya laid 6 clutches of 4 and one of 3 eggs in 1976 (av. 3·9) and 8 of 3 and one of 4 eggs in 1977 (av. 3·1). Laid at 24 h intervals. White, clouded with pale grey, evenly freckled or densely flecked with warm sepia brown. SIZE: (n = 45, southern Africa) 19·4–24·9 × 14·6–16·6 (21·8 × 15·9). 2–3 broods a year, but exceptionally breeding can be almost non-stop. What were probably the same 2 birds throughout, laid 16 clutches in 24 months (Watamu, Kenya: Donnelly 1978).

LAYING DATES: Mali, Feb–May; Niger, Apr; Ghana (fledglings May, June); Nigeria, Oct–Apr; Gabon, Aug, Dec–Feb; Ethiopia, ?May; E Africa: Region A, Jan, Mar–May, Sept, Oct, Region B, Feb–July, Sept, Oct, Dec, Region C, Feb, Mar, May–July, Oct–Dec, Region D, every month (mainly Apr, July, Sept), Region E, Dec–Oct (breeds mainly in rains in A and C, peaks in long rains in B); Malaŵi, Feb–May, Aug–Nov (mainly Mar and Sept–Oct); Zambia, Feb–Dec (mainly Sept); Zimbabwe, Feb–Mar, Apr–Dec (mainly Aug–Oct);

South Africa: Transvaal, all months except May (June 2, July 14, Aug 33, Sept 57, Oct 80, Nov 48, Dec 38, Jan 12, Feb 7, Mar 14, Apr 2 clutches).

INCUBATION: by ♀ (mainly) and ♂. Period: 13–14 days. In several instances, parents adapted immediately to removal of nest and eggs or young by man to new site up to 20 m away (Donnelly 1977, Niven 1981).

DEVELOPMENT AND CARE OF YOUNG: young fed by both parents. One brood of two 3-day-old chicks fed equally by ♂ and ♀ 8.2 times per hour, with beakful of insects or with single grasshopper, caterpillar or butterfly. Young brooded by ♀ alone at first; later in nestling period ♀ broods only at night. Weights and measurements of growing young plotted by Nhlane (1990). Fledging period: 15–16 days. Some young can fly quite well when they fledge; parents accompany them and 'teach' them to fly; fly well at 17–20 days. Soon thereafter (once at age 27–28 days: Donnelly 1977) young are vigorously chased away by parents which can start laying new clutch 3 days later.

BREEDING SUCCESS/SURVIVAL: the Watamu (Kenya) 'pair' (see above) reared 39 fledglings to independence from 55 eggs in 16 clutches (Donnelly 1978). Heavily parasitized by Red-chested Cuckoo *Cuculus solitarius*, Diederik Cuckoo *Chrysococcyx caprius* and Black and White Cuckoo *Oxylophus jacobinus*: E African native legend calls African Pied Wagtail 'the father of the hawk' (= cuckoo). Suffers from diseases of foot: 10% of 102 birds handled at Entebbe (Uganda) had swollen, scaly legs; toes, or whole foot, sometimes missing (Pitman 1966).

References
Brosset, A. and Erard, C. (1986).
Nhlane, M. E. D. (1990, in press).
van Someren, V. G. L. (1956).
Vincent, A. W. (1946).

Genus *Tmetothylacus* Cabanis

Bill slender and pointed. Legs and toes long, with long hind claw; lower part of tibiotarsus unfeathered. Tail medium-length; wing bluntly pointed. Sexes distinct. ♂ mainly bright yellow and black, ♀ streaked brown with yellow-tinged underparts, tail-feathers and flight feathers. Feed on ground for insects; perch on small trees and bushes.

Monotypic; endemic.

Plate 12
(Opp. p. 177)

Tmetothylacus tenellus (**Cabanis**). **Golden Pipit. Pipit doré.**

Macronix tenellus Cabanis, 1878. J. Orn. xxvi, p. 205; Taita, SE Kenya.

Range and Status. Africa; vagrant Oman.

Endemic resident, local migrant and probably intra-African migrant. W and S Somalia (west of 47°E), SE Ethiopia (N Ogaden west to Sassabanch and S Ogaden west to Doua R.), Kenya north and east of the highlands (west to Chalbi, Baringo, Nairobi and Olorgesaillie) and NE Tanzania lowlands (west to Arusha and south to Ruaha, Rufiji R. and Dar-es-Salaam); often common to locally abundant. Also SE Sudan (east of Kapoeta) to NE Uganda (Kidepo valley); uncommon. Vagrant W Kenya (Macalder), S Tanzania (Mikindani), Zimbabwe (1, Wankie Game Reserve, Mar 1972) and South Africa (Transvaal: 1, Irene, Jan 1906, 1, Rust der Winter Dam, Jan–Feb 1986 and 1 between Shingwedzi and Letaba, Kruger Nat. Park, Nov 1988).

Description. ADULT ♂: top and sides of head, mantle and scapulars streaked blackish brown and yellowish green, back and rump yellowish green with fewer streaks. Central tail-feathers dark brown, T2–T4 yellow with dark brown tips and distal outer edges, T5 and T6 wholly yellow. Supercilium broad, yellow; lores and ear-coverts greenish, streaked dusky.

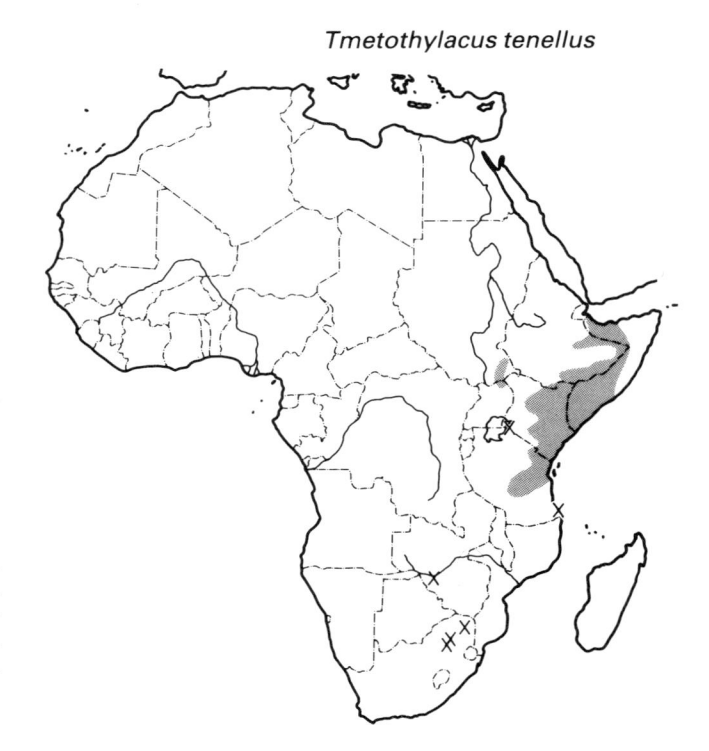

Broad black band across breast; rest of underparts bright canary yellow. Primaries yellow, outer five feathers blackish distally, inner four with black absent or reduced to small spot at tip; secondaries wholly yellow; tertials dark brown with buff fringes and yellow at base; primary coverts and alula yellowish tipped black; outer greater and outer median coverts mainly yellow, with base of outer webs dark brown; lesser coverts dark brown, fringed yellow; inner greater and inner median coverts dark brown fringed buff; bend of wing, underwing-coverts and axillaries yellow. Bill brown; eyes brown; legs and feet pale brown, lower part of tibiotarsus unfeathered. ADULT ♀: buffish brown above with dark brown streaking, less distinct on rump. Tail dark brown, outer feathers yellowish buff, T5 with some yellowish buff on inner web. Supercilium buff. Underparts buff, yellower on centre of belly. Wing dark brown, primaries with narrow yellow edging, coverts fringed buff; underwing-coverts and axillaries yellow; underside of primaries with broad yellow edges to inner webs. SIZE: (10 ♂♂, 6 ♀♀) wing, ♂ 80–87 (84·1), ♀ 78–82 (80·5); tail, ♂ 61–64 (62·8), ♀ 57–62 (59·8); bill, ♂ 15–17 (16·2), ♀ 15–17 (16·2); tarsus, ♂ 25–29 (27·1), ♀ 27–29 (27·6). WEIGHT: ♂ (n = 14) 18–21·5 (20·1), ♀ (n = 10) 18–22 (20·0).

Hind claw rather long and curved, 10–13. Wing-tip bluntly pointed; P7–P9 longest and about equal, P6 1 mm shorter, P5 4–5 mm shorter; P6–P8 emarginated on outer web.

IMMATURE: like adult ♀, but breast spotted; bases of secondaries yellow in immature ♂.

NESTLING: unknown.

Field Characters. A medium-sized pipit with long legs and toes. ♂ unmistakable, with brilliant golden-yellow underparts, bold black breast-band, broad golden-yellow band across wing and yellow rump and sides to tail (**A**). ♀ and immature at rest quite like *Anthus* pipits, with streaked brown upperparts, but underparts rich buff with variable amount of pale yellow usually visible on belly (lacking in some individuals); in flight, yellow wing-edging, underwing and sides to tail apparent.

Voice. Tape-recorded (GIB, McVIC). ♂ gives a series of soft, thin sibilant notes 'dji-dji-dji-dji ....', at rate of about 5 per s, interspersed every 1–1·5 s by lower notes 'dur-dur'; from bush or tree, or during fluttering display flight; otherwise silent.

General Habits. Occupies low arid or semi-arid grasslands with bushes or small trees; from sea level to *c*. 1000 m, occasionally to 1500 m. Sometimes solitary, but often in family groups or loose parties. Frequently perches on top of bush or small tree, wagging tail like wagtail *Motacilla*. Forages with deliberate walking gait, taking prey from ground or grass stems.

Numbers fluctuate seasonally in most areas, with marked influxes during rains. Movements may be partly local, but birds often absent from SE Kenya during June–Oct dry season, and Tanzania records mainly Dec–Mar; also, vagrants have occurred far to south of normal range in Jan and Mar, and to the northeast in Oman in June, so longer movements may occur within African range. Frequently attracted to lights at night at Ngulia, SE Kenya, Nov–Jan (Backhurst and Pearson 1977).

Food. Insects.

Breeding Habits. Monogamous, territorial. In songflight display, ♂ rises from small tree or bush top on fluttering wings (**B**), then planes to ground with wings held in V over back and tail raised and spread.

NEST: cup of thick stems and grasses, lined with rootlets, hidden low down in bush.

EGGS: 2–4, white or greenish white, heavily spotted dark brown. SIZE: *c*. 20 × 14·5.

LAYING DATES: Ethiopia, May; Somalia, May; E Africa: Kenya, July (coast), Nov–Dec, Apr (southeast); Tanzania, Nov; Region D, Dec–Jan, during rains.

Reference
Lewis, A. and Pomeroy, D. (1989).

Genus *Anthus* Bechstein

Bill slender and pointed. Legs rather long. Tail medium to longish or (*A. brachyurus*) rather short. Wing blunt or bluntly pointed; tertials long, often equal to longest primaries. Wing formula, particularly length of P6 and P7, in some cases diagnostic at the species level. Plumage brown above, usually streaked. Underparts buff or whitish, streaked on breast and flanks; 2 species, *A. chloris* of southern Africa and *A. lutescens* of South America, are yellow below. Most of outer tail-feather and distal part of adjacent feather usually white or pale buff; colour and pattern important in identification. Sexes alike. Mainly ground-feeding insectivorous birds of open country; some inhabit open woodlands or clearings in forest, a few feed in trees.

40 species worldwide: 19 in Africa, of which 11 endemic. 13 breed in Afrotropics, 1 in N Africa (which is also a Palearctic migrant); 5 are entirely Palearctic migrants. *A. novaeseelandiae*/*A. hoeschi* and *A. spinoletta*/*A. petrosus* form superspecies, and *A. vaalensis*/*A. leucophrys*/*A. pallidiventris* and the migrants *A. pratensis*/*A. trivialis*/*A. cervinus* form species groups. We follow Knox (1988) in regarding *A. spinoletta* and *A. petrosus* as distinct species because of pronounced differences in plumages and in habitat preference. We also follow Hall (1961) in treating African members of the *A. novaeseelandiae* complex as conspecific with Asian and Australasian birds; differences of colouration, structure and voice between Asian *richardi* and some African races appear to be less marked than between African forms themselves. In view of its segregated breeding in areas of sympatry with *A. novaeseelandiae rufuloides*, and its different calls and display (Mendelsohn 1984, Dowsett-Lemaire 1989), we follow Clancey (1984a) and Maclean (1985) in treating the high-altitude Lesotho breeding form as a separate species *A. hoeschi*. However, we prefer to treat the little known form *latistriatus* as a race of *novaeseelandiae*. A pipit collected near L. Chad and identified as *A. godlewskii* (White 1957; see also Hall 1961, Moreau 1972, Cramp 1988) would appear to have been a local race of *A. novaeseelandiae*. In the *A. similis* complex, habitat differences and small structural differences exist between adjacent forms in southern Africa (Hall and Moreau 1970, Clancey 1985a), but pending further information on ecology and behaviour we treat this complex as a single species. While recognizing two plain-backed pipit species in southern Africa, *A. leucophrys* and *A. vaalensis*, we accept that intergradation occurs between darker and sandier forms in E Africa, and thus, following Hall (1961), we recognize only *A. leucophrys* in this region. Cooper (1985) has advanced arguments for associating *A. chloris* and the longclaw *Macronyx sharpei* in a genus *Hemimacronyx*, and this has been followed by Clancey (1990). However, in its bill, leg and foot structure, tail pattern, flight and behaviour, *chloris* resembles typical pipits rather than *sharpei* and we retain it in *Anthus*. (See also Taxonomic Note under *A. similis*, *A. novaeseelandiae* and *A. hoeschi*.)

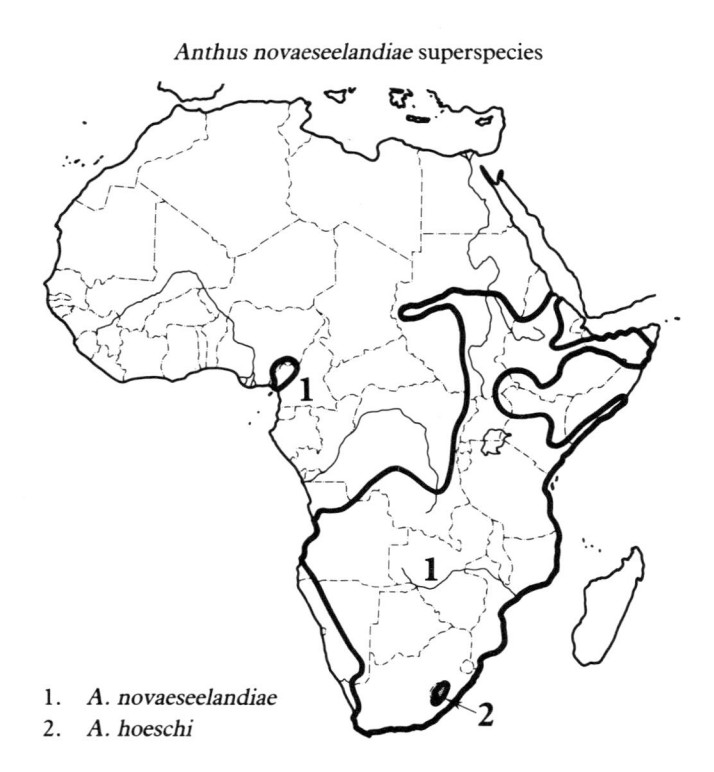

Anthus novaeseelandiae superspecies

1. *A. novaeseelandiae*
2. *A. hoeschi*

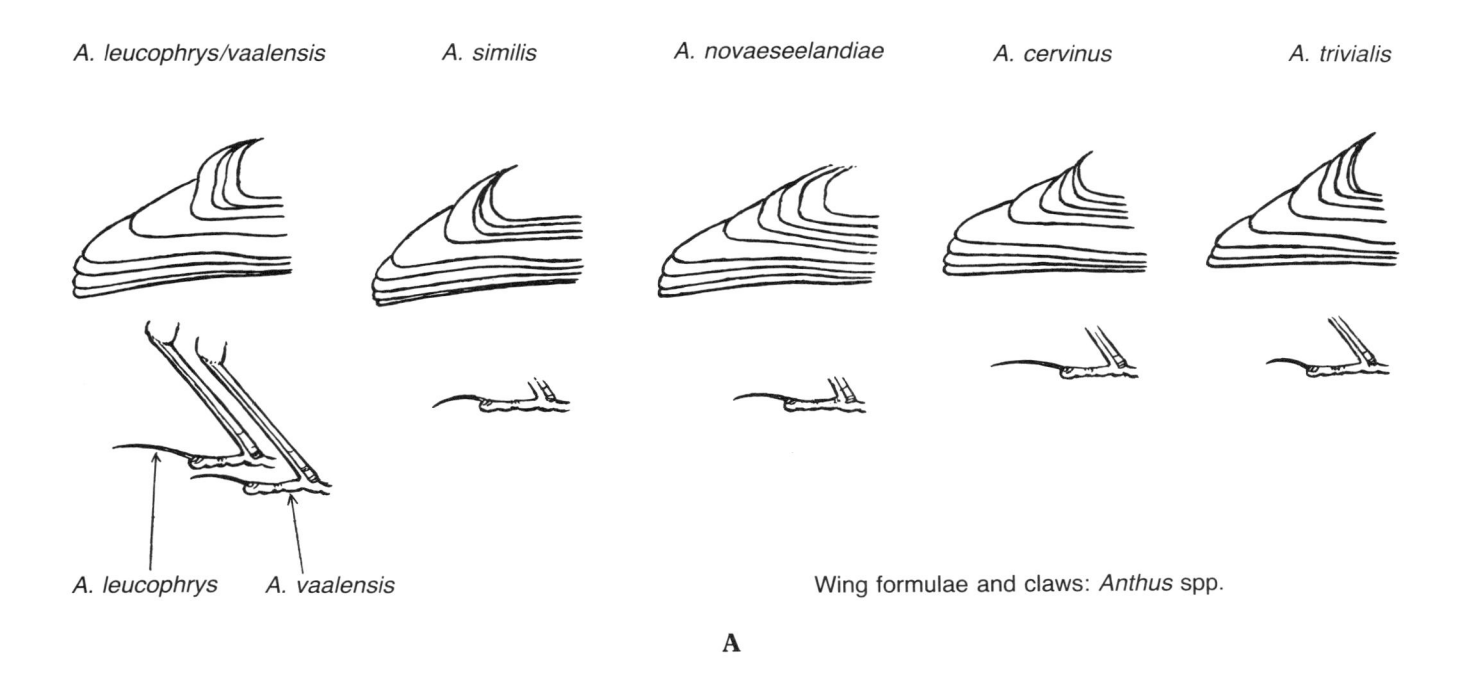

| A. leucophrys/vaalensis | A. similis | A. novaeseelandiae | A. cervinus | A. trivialis |

A. leucophrys A. vaalensis

Wing formulae and claws: *Anthus* spp.

A

Anthus novaeseelandiae (Gmelin). Richard's Pipit. Pipit de Richard.

Alauda novae Seelandiae Gmelin, 1789. Syst. Nat. I, pt. 2, p. 799; Queen Charlotte's Sound, South Is., New Zealand.

Forms a superspecies with *A. hoeschi*.

Plate 12
(Opp. p. 177)

Range and Status. Breeds Africa, SW Arabia, central, E and S Asia to Australia and New Zealand. Palearctic populations winter from India and S China to Malaysia; rare Europe.

Resident and local migrant; rare Palearctic visitor N Africa. In W Africa distribution highly restricted: locally common Cameroon highlands and adjacent SE Nigeria (*cameroonensis* and *lynesi*); *lynesi* also recorded N Nigeria (once, Kano), Chad (once, N'Djamena) and W Sudan (Darfur, uncommon). In E and central Africa, widespread and common to abundant in open grassy country at all altitudes, in central and SE Sudan, Ethiopian highlands, coastal Ethiopia to Djibouti and N Somalia (north of 7°N), Uganda (except NE), W and central Kenya, coastal S Somalia and Kenya, Tanzania, highlands of Rwanda, Burundi and E and S Zaïre; southern Africa, throughout except for N and SW Angola, W Namibia and NW Cape Province, South Africa. In N Africa, Palearctic winter visitor, rare to vagrant Morocco, N Algeria and Libya; rare to uncommon in Egypt (mainly Nile delta and Faiyoum) (*richardi*). Rare but widespread coastal and SW Mauritania and Mali (Sahel) (race uncertain, perhaps *richardi*). Birds from NE Nigeria and W Chad, originally assigned to *richardi*, apparently a local unnamed population; and racial identity of birds from Liberia (Zwedru), Ghana (once, north of Srogboe) and Central African Republic (Manovo-Gounda-Saint Floris Nat. Park) uncertain.

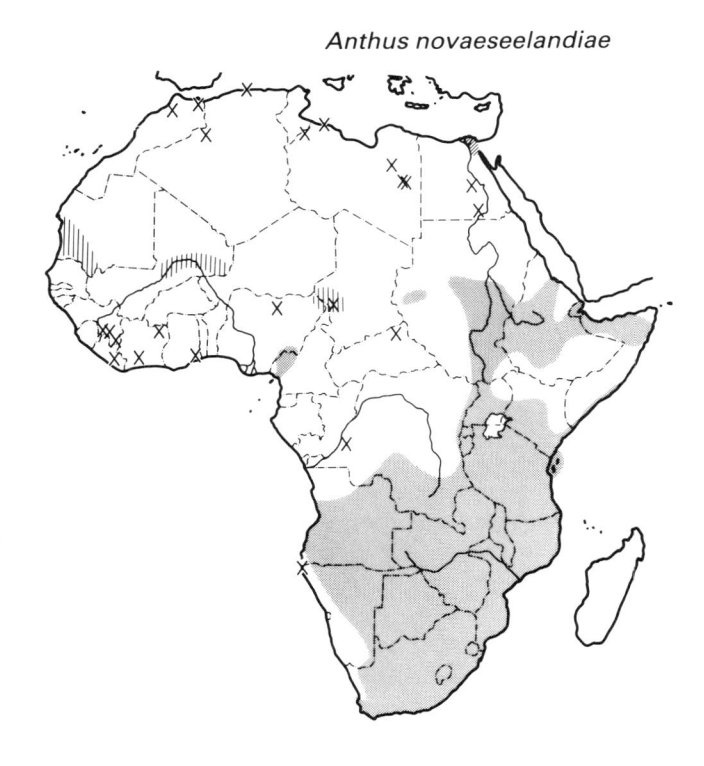

Description. *A. n. rufuloides* Roberts: South Africa (except N Cape Province), Swaziland and Lesotho lowlands; in non-breeding season to Namibia, N Botswana, Zimbabwe and S Mozambique. ADULT ♂: upperparts warm buffish brown, top of head and mantle with diffuse dark brown streaking, hind-neck paler, back and rump almost uniform. Tail dark brown, central feathers edged pale brown, outer pair white on outer web and most of inner web, T5 with outer web white distally and long narrow white wedge against shaft on inner web. Supercilium broad, buffish; lores buff; ear-coverts buff streaked dark brown; stripe through eye and narrow moustachial stripe dark brown; malar streak blackish. Underparts pale buff, deeper buff on breast and flanks, whiter on chin, upper breast with short dark brown streaks. Primaries and primary coverts blackish brown, narrowly edged buff; secondaries, tertials and greater coverts blackish brown, broadly edged and tipped rich buff; median coverts blackish with more contrasting pale buff edges; lesser coverts buffish brown; underwing-coverts and axillaries buff. Bill blackish brown, base of lower mandible yellowish; eye dark brown; legs and feet pinkish flesh. Sexes alike. SIZE: (10 ♂♂, 10 ♀♀) wing, ♂ 85–95 (90·0), ♀ 82–90 (86·0); tail, ♂ 61–64 (62·5), ♀ 58–60 (59·3); bill, ♂ 16–18 (17·2), ♀ 16–18 (16·7); tarsus, ♂ 26–28 (26·8), ♀ 25–26 (25·7). WEIGHT: unsexed (n = 10) 21–27 (23·5).

Hind claw longish and rather straight, 11–15; wing-tip bluntly pointed; P7–P9 longest and about equal, P6 1–2 mm shorter, P5 7–10 mm shorter; P6–P8 emarginated on outer web.

IMMATURE: more strongly streaked above, feather centres darker; tertials, greater coverts and median coverts sharply edged whitish.

NESTLING: unknown.

A. n. bocagei Nicholson: South Africa (N Cape Province), Namibia (except N saline pans), Botswana (except N saline pans) and S and W Angola; in non-breeding season to Zimbabwe and S Mozambique. Paler above than *rufuloides*, more sandy. Smaller: wing, ♂ (n = 10) 82–88 (86·0).

A. n. grotei Niethammer: saline pans of N Botswana and N Namibia; in non-breeding season to Zimbabwe and S Zambia. Paler and greyer than *bocagei*.

A. n. lichenya Vincent (including '*katangae*'): Malaŵi (except S), adjacent highland NW Mozambique, Zimbabwe plateau, N and E Zambia, NE Angola, S Zaïre, W Tanzania, Rwanda, Burundi, W Uganda. Darker and more heavily streaked above than *rufuloides*; richer buff below with sharper streaking extending lower on breast and sometimes to flanks; white on T5 often vestigial or absent. Wing, ♂ (n = 10) 88–92 (90·7).

A. n. lacuum Meinertzhagen: W, central and S Kenya, SE Uganda and N and central Tanzania. More strongly streaked above than *rufuloides*, wing-feather edgings and upperparts tinged more cinnamon; chest streaks short and sparse; white in outer tail-feathers usually tinged buff. Wing, ♂ (n = 10) 84–90 (87·3). WEIGHT: unsexed (n = 27, Kenya) 20–25 (22·9).

A. n. spurium Clancey: coastal lowlands of Mozambique to S Malaŵi and SE Tanzania; also NE Namibia and N Botswana (Okavango drainage and delta) and adjacent Zambezi valley. Similar to *lacuum* but colder and greyer above; breast streaking blacker. Smaller: wing, ♂ (n = 7, Mozambique) 84–88 (86·0).

A. n. annae Meinertzhagen: N Ethiopia (Eritrea), Djibouti, N Somalia and lowlands of SW Somalia, E Kenya and NE Tanzania. As *lacuum* but paler, with greyer tinge to upperparts. Slightly smaller: wing, ♂ (n = 6) 82–88 (85·7).

A. n. cinnamomeus Rüppell: W and SE Highlands of Ethiopia. Similar to *lacuum* but warmer above and richer cinnamon on wing-feather edgings and upperparts; below more strongly buff with breast spotting blacker; white in outer tail-feathers tinged buff. Larger: wing, ♂ (n = 10) 90–99 (93·4). WEIGHT: unsexed (n = 38) 18–31 (27·0).

A. n. stabilis Clancey: central and SE Sudan. Less warmly tinged than *lacuum*, more buffish, streaking on crown and mantle more pronounced; streaks below heavier and blacker, extending to lower breast; white in outer tail-feathers tinged buff. Large: wing, ♂ (n = 4) 91–95.

A. n. lynesi Bannerman and Bates: Cameroon (east of Bamenda and Banso Highlands), SE Nigeria (Mambilla and Obudu plateaux); Chad (N'Djamena) and W Sudan (Darfur). Darker above than *lacuum* with heavier, blacker streaking and contrasting cinnamon-buff wing-feather edging; deep cinnamon-buff below with streaking confined to upper breast. Pale areas of outer tail-feathers buffish white. Large: wing, ♂ (n = 6) 95–100 (97·5); hind claw rather short, 9–10.

A. n. camaroonensis Shelley: Cameroon (Mt Cameroon and Mt Manenguba). Like *lynesi* but less reddish; more buff below, wing-feather edging pale buff. Wing, ♂ (n = 4) 92–95; hind claw 9–11.

A. n. itombwensis Prigogine: E Zaïre (Itombwe Highlands and Mt Kabobo). Darker above than *lacuum* or *lichenya*, crown and mantle with heavy blackish streaking; buff below with heavy streaking extending to mid-breast and flanks. White in T5 usually reduced as in *lichenya*. Large: wing, ♂ (n = 28) 90–98 (93·4); hind claw 10–16.

A. n. latistriatus Jackson: W Kenya (Kavirondo Nov 1894), SW Uganda (Ankole Nov 1925, L. Edward Jan 1927), E Zaïre (L. Edward May 1927, Itombwe Sept 1950), NW Tanzania (Bukoba Dec 1909); breeding area unknown (Prigogine 1981, Clancey 1984c). Very dark above, blacker than *itombwensis*; rich cinnamon-buff below with contrasting whitish throat and strong breast marking, with streaks extending onto flanks; fresh wing-coverts and tertials with rich buff edging; T6 greyish buff, T5 with greyish buff spot at tip. Resembles *lynesi* but more extensively streaked below. Size similar: wing, ♂ (n = 2) 94, 94, (1 ♀) 88; but hind claw slightly shorter, 7–10, and rather straight; wing formula as *rufuloides* and other African races.

A. n. richardi Vieillot: breeds W Siberia; winters S Asia and in small numbers S Europe; rare to vagrant N Africa. Like *lacuum* but less warmly tinged; more buffish above with pale feather edges more prominent, especially on head; whiter below; wing-feather edging more sandy buff; outer tail-feathers pure white. Larger with longer legs and hind claw than Afrotropical races: wing, ♂ (n = 23) 97–102 (99·9); tarsus, unsexed (n = 9) 29–33 (30·7); hind claw 15–20.

It is not clearly established that Palearctic birds cross the Sahara. Birds collected L. Chad Mar–June and Oct–Nov are close to *richardi* in colouration (though slightly paler and sandier) but are rather smaller with shorter tarsi and hind claw, and white outer tail-feathers tinged buff. Wing, ♂ (n = 2) 98, 94, ♀ (n = 4) 88–94 (92·3); tarsus, unsexed (n = 6) 25–29 (27·3); hind claw 10–13. They are close to *stabilis* and presumably represent an undescribed local population. A June ♀ had enlarged ova, and a May ♂ had growing testes. The racial identity of birds from Mauritania, Mali, Liberia, Ghana and Central African Republic also remains uncertain.

TAXONOMIC NOTE: African pipits of the *novaeseelandiae* complex have been treated as a separate species *A. cinnamomeus* by Prigogine (1981) and Clancey (1984a, 1986a, 1990). However, *A. n. richardi* closely resembles many African forms in plumage, and is identical in wing structure and similar in voice and display. We therefore follow Hall (1961) and Hall and Moreau (1970) in treating African birds as conspecific with extralimital races. Racial variation in Africa is complex, and distributions obscured by non-breeding movements. Some previously recognized forms (*lichenya*, *lacuum*, *spurium* and *annae*) were synonymized by White (1957) and Hall (1961) with *A. n. cinnamomeus*, but their plumage characters appear distinct and consistent, and we follow Clancey (1986a) in retaining them as subspecies. The Lesotho form first described as *A. richardi editus* shows segregated breeding where it occurs with *A. n. rufuloides*, and has different calls and displays (Mendelsohn 1984, Dowsett-Lemaire 1989), and is treated here as a separate species *A. hoeschi*. When more is known of their breeding distribution, movements and ecology, other dark montane populations may prove to warrant specific status. At present, however, we prefer to treat these, including the distinctive *lynesi* and *camaroonensis* of W Africa, and the enigmatic *latistriatus* of central Africa, as forms of *novaeseelandiae*.

Field Characters. The standard pipit in eastern and southern Africa, to which all others should be compared. Medium large, rather long-legged and longish tailed (**A**). Buffish or warm brown above with clear dark streaking, and buff or whitish below with short streaks across chest. Face pattern bold, with pronounced pale supercilium and dark eye-stripe, moustachial stripe and malar streak. Outer tail-feathers usually white (buffish in W African highland birds), legs pinkish, lower mandible yellowish. Some montane populations are much darker and more heavily streaked on breast. Similar to Long-billed Pipit *A. similis*, but more distinctly streaked, and typically slightly smaller, shorter billed and shorter tailed; best distinguished by voice, and in hand by longish hind claw (most populations) and shorter fifth primary (**A**, p. 217). Differs from immature Tawny Pipit *A. campestris* in stouter bill, more rounded crown, broader build, longer legs and longer hind claw. Plain-backed Pipit *A. leucophrys* and Buffy Pipit *A. vaalensis* are larger, and plainer above, with buff outer tail-feathers and plainer face patterns.

Voice. Tape-recorded (78, 88, C, GIB, LEM, PEAR, STJ). Typical song, a 3–4 note phrase, 'pli-pli-pli-pli' (range 2–11: Dowsett-Lemaire 1989), repeated every few seconds from low perch or during upward loop of cruising display flight; also gives continuous string of notes, 'pli-pli-pli-pli-pli ...' (range 12–26 notes per song: Dowsett-Lemaire 1989) during steep dive at end of display flight. Intervals between cruising songs 1·5–3 s. Call, a sharp 'chip', usually given when flushed, and a loud, sharp 'psiu' in flight.

General Habits. Occupies a variety of habitats including short grassland, grassy lake edges, open savanna and cultivation, in flat country or on gentle slopes; in Kenya most abundant in short grass trampled and denuded by stock or game and littered with droppings, but also occupies richer pastures of higher plateaux (van Someren 1956). Ranges from sea level to above 3400 m in E Africa, 2400 m in Zimbabwe and 2700 m in South Africa. In W Africa mainly montane, occurring up to 3700 m in Cameroon.

Quite gregarious; commonly forms scattered groups, and sometimes flocks of up to 100 during non-breeding season. Attracted to recently burnt ground, along with other small passerines such as larks and wheatears. Makes short runs, pausing to stand upright; dips tail occasionally during pauses. Also walks with strutting gait. Perches on posts and bushes, especially when disturbed. Forages from ground or short vegetation, or from crevices in rocks; breaks up dung to secure larvae and digs for food at roots of grass clumps; sometimes leaps or makes short pursuit flights to catch aerial prey. Flight powerful with marked undulations and strong take-off.

Southern populations distinctly migratory, birds from South Africa, Namibia and Botswana moving north and east in non-breeding season. Other populations mainly sedentary; extent of local movements unclear. Palearctic birds regularly occur well to west of normal breeding range, and some apparently winter S Europe and occur N Africa, especially Egypt; extent of penetration of Africa not clear, but passage Mauritania Sept–Oct and Mar–Apr and occurrence in Mali Sahel Sept–Apr suggest these are Palearctic birds.

Food. Insects; also other arthropods and occasionally seeds. In Zimbabwe, largely short-horn grasshoppers and a variety of beetles Oct–Apr, with caterpillars an important addition Jan–Apr; also termites when available. Of 67 stomachs (Zimbabwe), 76% contained Coleoptera, 69% Orthoptera, 34% Hymenoptera, 18% Hemiptera, 19% Lepidoptera, 24% Isoptera, 3% Diptera, 4% Odonata, 6% Myriapoda, 3% Dictyoptera, 6% Arachnida, 7% seeds, 6% feathers and 5% grit (Borrett and Wilson 1971). Larvae of flies and dung beetles form large proportion of food in Kenya (van Someren 1956).

Breeding Habits. Monogamous; territorial, nests 200–400 m apart in densely occupied areas, South Africa; up to 25 nests in 16 ha, Kenya. In display ♂ rises obliquely to 30–75 m, cruises with undulating flight and repeated song-phrase (see *Voice*) for 100–150 m, then dives steeply to land on ground or bush. ♂ perches and calls from rock or bush near nest. ♀ returning to nest drops to ground a few m away, then approaches with quick walk.

220 MOTACILLIDAE

NEST: a neat, shallow cup, int. diam. c. 70, depth c. 35, of dry grass lined with fine rootlets, fibre and hair, set into soil at base of small rock, grass tuft or low shrub, or on low sloping bank, usually partly concealed; often close to open space or path.

EGGS: 2–4, av. (n = 66 clutches, South Africa) 2·7, av. (n = 36 clutches, Kenya) 2·6, av. (n = 17 clutches, E Tanzania) 2·9; pale cream, closely freckled and spotted with grey and shades of brown, sometimes with heavy circle near broad end. SIZE: (n = 182, South Africa) 19–23 × 14–18 (20·7 × 15·3); (n = 11, Kenya) 20–21 × 15–16 (20·6 × 15·1).

LAYING DATES: Ethiopia, Apr; Kenya, Mar–July, Nov–Dec; S Tanzania, Mar–May, Sept–Oct; Zambia, Aug–Oct; Zimbabwe and South Africa, Sept–Dec; Cameroon highlands, Feb–Apr. Usually at end of dry season, but during rains in W and central Kenya and on E African coast.

INCUBATION: by both sexes, mainly by ♀. Period: 12–13 days.

DEVELOPMENT AND CARE OF YOUNG: young remain in nest up to 17 days if undisturbed; brooded at first by ♀; fed by both sexes; squat on ground among stones and vegetation after leaving nest.

BREEDING SUCCESS/SURVIVAL: in Kenya many clutches taken by rats and other ground predators; many nests trampled by stock and game (van Someren 1956).

References
Borrett, R. P. and Wilson, K. J. (1971).
Clancey, P. A. (1986a).
Cramp, S. (1988).
van Someren, V. G. L. (1956).

Plate 12
(Opp. p. 177)

Anthus hoeschi Stresemann. **Mountain Pipit. Pipit du Drakensberg.**

Anthus hoeschi Stresemann, 1938. Orn. Monatsber. xlvi, p. 151; Erongo Mts, N Namibia.

Forms a superspecies with *A. novaeseelandiae*.

Range and Status. Intra-African migrant; endemic to southern Africa. A breeding visitor to South Africa/Lesotho, where locally common on high plateaux of Drakensberg massif: in Lesotho, in the Thaba Pitsoa Mts of the Liqaleneng, Malutsenyene, and in the Drakensberg in the Sangebethu valley, the Lekalebalele valley and the Sehlabathebe Nat. Park; in NE Cape Province at Naude's Nek. Occurs in non-breeding season NW Zambia (Balovale, Mwinilunga and Kabompo), E Angola (N Luanda) and S Zaïre (L. Kabwe and Palenge, Shaba Province); apparently uncommon. Also, probably on passage, N Cape Province (Kimberley), Namibia (Erongo, near Karibib and Okahandja) and Botswana (Francistown).

Description. ADULT ♂: upperparts brown, strongly streaked blackish brown, less distinctly so on rump. Tail blackish brown, central feathers edged pale brown, outer pair buffish white except at base of inner web, T5 buffish white at tip. Supercilium broad, buffish; streak through eye and narrow moustachial stripe dark brown. Underparts pale buff, whiter on chin, upper breast with short bold blackish brown streaks; narrow malar streak blackish. Primaries and primary coverts blackish brown, narrowly fringed buff; secondaries, tertials, greater coverts and median coverts blackish brown with broad buff edges and tips; lesser coverts buff; underwing-coverts and axillaries buff. Bill dark brown, base of lower mandible flesh; eyes dark brown; legs and feet orange-brown. Sexes alike. SIZE: wing, ♂ (n = 9) 94–98 (95·1), ♀ (n = 6) 89–92 (89·7); tail, ♂ (n = 9) 63–72 (67·2), ♀ (n = 6) 61–67 (64·1); bill, ♂ (n = 9) 18–19 (18·6), ♀ (n = 6) 16–19 (17·5); tarsus, ♂ (n = 3) 28–30 (29·0), ♀ (n = 2) 28, 28. WEIGHT: unsexed (n = 13) 23·5–31 (27·0).

Hind claw longish, 9–14. Wing-tip bluntly pointed; P7–P9 longest and about equal, P6 1–2 mm shorter, P5 8–13 mm shorter; P6–P8 emarginated on outer web.

IMMATURE AND NESTLING: unknown.

TAXONOMIC NOTE: formerly considered a race of Richard's Pipit *A. novaeseelandiae*. The breeding population of the high altitude Lesotho Plateau, described as *A. richardi editus* by Vincent (1951), was considered by Clancey (1978) to be the same as the population occurring in Zambia in the non-breeding season and named *A. r. lwenarum* by White (1946). Clancey later (1984a) decided that 2 birds from Namibia, described as *hoeschi* by Stresemann (1938) were also non-breeding members of this population. The population retains *hoeschi* as the oldest name, *editus* and *lwenarum* becoming synonyms. *A. hoeschi* is considered specifically distinct because of its extensive breeding overlap with *A. n. rufuloides*, with clear vocal and behavioural differences between the two (Mendelsohn 1984, Dowsett-Lemaire 1989).

Field Characters. Very similar to local races of Richard's Pipit *A. novaeseelandiae* and Long-billed Pipit *A. similis*. Chunkier than Richard's Pipit, mantle more heavily streaked, breast usually more heavily streaked, outer tail-feathers buff, not white, base of bill pinkish flesh rather than yellowish; song similar but deeper and more varied. Long-billed Pipit has similar bill and tail colour, but less distinctly streaked above, underpart streaking narrower but more extensive, less confined to chest; in hand shows longer sixth primary.

Voice. Tape-recorded (GIB, LEM). Song given during cruising display flight or return to ground contains strings of notes of *c.* 1·5–2·5 s duration, 'tchrit-chritchritchri', 'tsi-plu-tsi-plu-tsi-plu' and a lower pitched 'chiri-chiri-chiri-chiri', separated by intervals of 1–1·5 s (Dowsett-Lemaire 1989); when perched, a soft disyllabic 'tuchit-tuchit'; flight calls, 'psiu' and 'chiri'; alarm at nest, a repeated 'twit . . . twit . . . twit'.

General Habits. Breeds on rocky hillsides, heathland and short montane grassland at 2000–2700 m, usually separated altitudinally from Richard's Pipit, but the two may overlap on flat high grassland. Flies down frequently to below 2000 m to feed on flatter grassland and ploughed fields among Richard's Pipits. May perch on rock or fence.

Moves off the Drakensbergs in Mar, returning in Oct. Identical looking birds occur and moult in dambos and damp meadows of the Zaïre–Zambia watershed May–Oct, and these probably represent the same population. Presumed migrants have occurred N Cape Province (Apr), Namibia (Mar) and E Botswana (Oct).

Food. Unknown.

Breeding Habits. Monogamous and territorial; nests 500 m to less than 100 m apart. Performs display flight similar to Richard's Pipit, rising obliquely to 15–50 m and cruising a short distance with slight undulations before diving steeply to ground. Sings during display flight (see *Voice*); also from large rock or fence.

NEST: cup of grass on ground, tucked well under a tuft.

EGGS: 3–4; whitish, spotted brown and grey, spots forming dense ring round broad end. SIZE: (n = 3) 21–22 × 15 (Maclean 1985).

LAYING DATES: South Africa and Lesotho, late Nov to early Jan.

References
Brooke, R. K. (1984).
Clancey, P. A. (1978, 1984a).
Dowsett-Lemaire, F. (1989).
Mendelsohn, J. (1984).

Anthus campestris (Linnaeus). Tawny Pipit. Pipit rousseline.

Plate 12
(Opp. p. 177)

Alauda campestris Linnaeus, 1758. Syst. Nat. (10th ed.), p. 166; Sweden.

Range and Status. Breeds Africa, continental Europe and Asia from Black and Caspian Sea areas east to Mongolia and south to Afghanistan and N Tibet; winters Africa, SW Asia and India.

Palearctic winter visitor; also breeding summer visitor NW Africa. Breeds Morocco (Rif, Middle Atlas, High Atlas and near Mediterranean coast); N Algeria (south to Hauts Plateaux and the Aurès Massif); and NW Tunisia; uncommon to frequent. Possible breeding NW Libya (singing birds May), SW Mauritania (Nouakchott, carryiny food July and Sept) and NW Somalia (Bacaadaweyn, War Idaad, enlarged gonads May: Ash and Miskell 1990). Winters south of Sahara, across Sahel and soudanian savanna belts, from SW Mauritania and Senegambia through central and S Mali, S Niger, N Nigeria and central Chad to central Sudan, mainly north of 12°N but south to N Ivory Coast, N Ghana and N Togo; also Ethiopia to NW Somalia (and once Abd-el-Kuri), and south to N and E Kenya and S Somalia; frequent to common Senegambia to N Nigeria and Sudan to Ethiopia, rare to uncommon elsewhere. Rare to vagrant further south, in N Liberia (Voinjama), Ivory Coast (Ferkessedougou), Ghana coast (Accra), E Nigeria (Yola), S Sudan (Juba), Uganda (Entebbe) and SE Kenya (Tsavo). In Sahara, winters S Algeria (Hoggar and Tanezrouft, frequent); also recorded mid-winter NE Chad (Yebjyeba) and Egypt (Nile delta, Faiyoum, Aswan). On passage N Africa, frequent to common

Anthus campestris

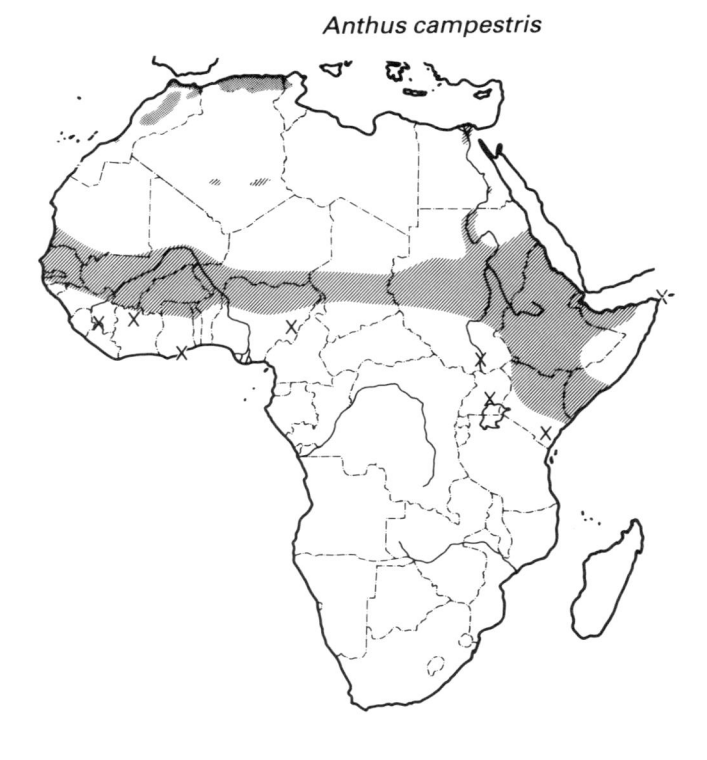

autumn Morocco (especially W coast) and Egypt; uncommon to frequent Algeria, Tunisia and Libya; generally more common spring, with records in Saharan oases.

Description. *A. c. campestris* (Linnaeus): breeds throughout range except central Asia; winters Africa and Arabia. ADULT ♂: upperparts pale sandy brown, often tinged greyish, all except rump with indistinct blackish brown streaking. Tail dark brown, central feathers broadly edged pale buff, outer pair white, T5 with outer web and much of distal inner web also white. Supercilium prominent, creamy; lores and ear-coverts brown, flecked creamy. Underparts creamy white, breast more buff, usually with a few short dark brown streaks, flanks buff; malar streak narrow but distinct, dark brown. Primaries, secondaries and primary coverts dark brown, narrowly edged whitish; tertials and greater coverts brown, broadly edged and tipped sandy buff; median coverts with blacker centres and broad contrasting buffish white tips; lesser coverts sandy brown; underwing-coverts and axillaries creamy buff. Bill dark brown, base of lower mandible flesh; eyes blackish brown; legs and feet bright yellowish flesh. Sexes alike. SIZE: (10 ♂♂, 10 ♀♀) wing, ♂ 90–95 (92·8), ♀ 86–90 (88·5); tail, ♂ 69–73 (71·3), ♀ 66–70 (68·0); bill, ♂ 17–19 (18·7), ♀ 16–19 (17·9); tarsus, ♂ 24–27 (24·7), ♀ 24–26 (24·6). WEIGHT: unsexed (n = 1, Ethiopia) 23·8, (n = 1, Algeria) 23, (n = 7, Malta, spring and autumn) 26–32 (27·8).

Complete moult in Palearctic July–Sept; in Africa, partial moult Feb to early Apr involves body, tertials and often central tail-feathers. Hind claw short to medium, curved and quite strong, 9–12. Wing-tip bluntly pointed; P7–P9 longest, P6 2–3 mm shorter, P5 10–14 mm shorter, P6–P8 emarginated on outer web.

IMMATURE: scalloped or strongly streaked above and with sharp dark streaking across upper breast. First winter differs from adult in being more distinctly streaked above and on breast.

NESTLING: covered with pale buff down above, long and dense; mouth orange-red, gape-flanges yellow.

A. c. griseus Nicoll: E Iran and SW Kazakhstan; migrates through Middle East and Egypt. Paler, greyer above and whiter below than nominate.

Field Characters. A fairly large, pale, slim pipit with pale legs and rather long tail with white outer feathers. Adult almost uniform sandy brown above and usually unstreaked below, with prominent supercilium and bold wing-covert bars. Immature distinctly streaked and closely resembles Richard's Pipit *A. novaeseelandiae*, which has similar tail pattern and wing formula. Best distinguished by wagtail-like build and carriage, with finer bill and shorter legs, and by voice; also in hand by short hind claw. Some races of Long-billed Pipit *A. similis* are rather plain, with equally long tail, but are darker and greyer, with heavier build, longer legs, buff-tinged outer tail-feathers and much blunter wing.

Voice. Tape-recorded (78–62, 73, B). Song a repetition of a loud 'chiree' or 'chireeo' (with pitch descending at end of note), notes separated by pauses of 2–3 s, usually during song-flight; calls, a loud 'tseuc', often on take-off or landing, and 'tzeep', recalling Yellow Wagtail *Motacilla flava*, given in flight; also, in alarm near nest, a shrill 'sree' and a high-pitched 'tji-tji-tji'.

General Habits. Winters in short dry grassland, open or with scattered bushes and small trees; in Senegal in acacia-dominated sahelian wooded grassland; in Ethiopia up to 2000 m. Breeds NW Africa on dry mountain slopes and plateaux up to 2600 m, occurring in Atlas above tree line; also on stony plains, coastal sand dunes and vineyards.

Usually solitary outside breeding season but often in small groups on migration, occasionally flocks of up to 30. In Darfur, small flocks in excited chases around bush tops shortly before spring migration (Lynes 1924). Sometimes quite tame. Feeds on ground or among low herbage. Occasionally takes flying insects, including termites, by leaping up or making short pursuit flights. Carriage most horizontal of all large pipits (Cramp 1988); walks quickly or runs with frequent pecks and alert pauses, recalling a wagtail. Wags tail strongly and frequently. Perches freely on rocks, walls and occasionally trees, typically higher than Richard's Pipit. Flight like wagtail, with less powerful take-off and shallower undulations than Richard's Pipit.

Nominate *campestris* winters largely in Sahel zone. Main reception areas appear to be Senegambia/Mali in W and Sudan/Ethiopia in E. Eastern birds probably also include *griseus*. Autumn passage N Africa occurs late Aug–Nov; apparently mainly through Morocco, with strong movement along Atlantic coast, and through Egypt. Birds arrive Senegambia from early Oct, Mali Sept–Oct, Nigeria from mid-Oct, Sudan and Ethiopia Oct–Nov. Return movement begins early, N Morocco late Jan–Feb and Egypt from mid-Feb; a Jan influx

Senegambia perhaps involves birds from further east. However, many birds remain in winter quarters until Mar–Apr (once a bird with enlarged gonads in mid-May N Somalia). Spring passage N Africa apparently on a broader front, mainly mid-Mar to Apr with last birds early May. 2 birds recovered Morocco Sept (ringed W Germany and Sweden).

Food. Mainly insects. Winter diet includes locusts, grasshoppers, termites, Hymenoptera, beetles, caterpillars and grass seeds.

Breeding Habits. Monogamous; solitary and territorial. Territory size usually 10–30 ha in Europe (Cramp 1988). ♂ sings throughout breeding season, in Europe Apr–July; usually in flight, also from tree or hummock. In song-flight rises silently and almost vertically on fluttering wing-bets to 20–30 m, then performs a series of undulations before plummeting steeply with wings and tail raised; sings during horizontal phase but most strongly during descent (Cramp 1988). Hectic aerial chases of ♀ by ♂ during pre-laying period.

NEST: cup of grasses, roots and dead weeds, lined with some hair; placed in sheltered depression, often under grass tuft; built mainly by ♀; ext. diam. 105, cup diam. 70, depth 50.

EGGS: in Europe 4–5, occasionally 6. Whitish, spotted or heavily blotched with brown and purplish grey. SIZE: (n = 137, Europe) 19–24 × 15–17 (21·9 × 15·7). Often double-brooded.

LAYING DATES: Morocco, Algeria and Tunisia, late Apr to May; *c.* 2 weeks later on high ground.

INCUBATION: by ♀, beginning with last egg; ♂ may help. Period: 12–14 days.

DEVELOPMENT AND CARE OF YOUNG: young remain in nest 13–14 days; brooded at first by ♀; fed by both parents; fed for *c.* 2 weeks after leaving nest.

BREEDING SUCCESS/SURVIVAL: in Europe (Sweden and Netherlands) *c.* 80–89% of eggs hatch with 60–74% of young fledging; up to 40% of young lost soon after fledging (Cramp 1988).

Reference
Cramp, S. (1988).

Anthus similis Jerdon. Long-billed Pipit. Pipit à long bec.

Anthus similis Jerdon, 1840. Madras J. Lit. Sci. II, p. 35; Jalna, Nilgiri Hills, India.

Plate 13
(Opp. p. 240)

Range and Status. Breeds Africa, Socotra, S Arabia, mountains of Near East and Asia; Asian birds winter in central India and Burma.

Resident. Range disjunct in W Africa: central Niger (Aïr, Monts Bagzans); highlands of Sierra Leone, Guinea, N Liberia, W Ivory Coast (Mt Nimba), SE Ghana (Togo range), central Nigeria (Jos), SE Nigeria (Obudu Plateau and Gashaka-Gumti Game Reserve) and W Cameroon, locally frequent to common; central and E Mali (L. Aougoundou, Anderamboukane) and SW Mali (Bamako, Bougouni), status uncertain; Chad (1, Abéché). In eastern Africa, in broken hilly or arid rocky country: in W and central Sudan, NE Sudan to N Ethiopia and Somalia north of *c.* 8° 30′, Socotra, central and S Ethiopia to NE Uganda, W, central and S Kenya and N Tanzania, locally frequent to common; coastal Kenya and S Somalia, status uncertain; vagrant Djibouti (1, Arta). In central and southern Africa, more widespread: highlands of SW Uganda, Rwanda, Burundi and E Zaïre; *Brachystegia* woodland from S Congo and S Zaïre to central Angola, NE Namibia (Caprivi), Zambia, Malaŵi and SW Tanzania; montane grassland in SW Tanzania and N Malaŵi; woodlands from Zimbabwe and adjacent Mozambique to NE Botswana; above tree line on Mt Moco, central Angola; frequent to common; rocky and mountainous areas from South Africa, Swaziland and Lesotho to S Botswana, Namibia and SW Angola; locally common.

Description. *A. s. nicholsoni* Sharpe: South Africa (NE and N Cape Province, W Orange Free State and S Transvaal) and SE Botswana. ADULT ♂: upperparts brown, indistinctly streaked, most strongly on head. Tail dark brown, central feathers tinged buff, outer pair with buffish white outer web and large distal wedge on inner web, T5 with buffish white confined to outer edge and tip of inner web. Supercilium prominent, buff; lores and ear-coverts brown, streaked buff. Underparts warm buff, deeper on breast and flanks, whiter on chin and throat. Dark brown streaking on entire breast, extending to flanks. Fine malar streak dark brown. Primaries and primary coverts dark brown, narrowly edged warm buff; secondaries, tertials and median and greater coverts dark brown with broad warm buff fringes and tips; lesser coverts buffish brown; underwing-coverts and axillaries warm buff. Bill dark horn, base of lower mandible pale pinkish; eyes dark brown; legs and feet flesh brown. Sexes alike. SIZE: (10 ♂♂, 10 ♀♀) wing, ♂ 95–102 (99·0), ♀ 89–97 (93·6); tail, ♂ 69–73 (71·3), ♀ 67–71 (69·2); bill, ♂ 18–21 (19·4), ♀ 18–20 (19·0); tarsus, ♂ 26–27 (26·3), ♀ 25–27 (26·0). WEIGHT: unsexed (n = 5) 22–25 (23·0).

Hind claw strong, rather short and curved, 7–12. Wing bluntly pointed; P6–P9 longest and about equal, P5 usually <5 mm shorter; P5–P8 emarginated on outer web.

IMMATURE: darker with scalloped plumage above, and well-defined spotting rather than streaking on breast.

NESTLING: unknown.

A. s. petricolus Clancey: Lesotho and South Africa south of *nicholsoni* (mountains of W and S Cape, W Natal, E Orange Free State and E Transvaal). Slightly darker and colder than *nicholsoni*, less buffy.

A. s. leucocraspedon Reichenow: South Africa (NW Cape), SW Botswana and S and W Namibia to SW Angola. Paler, more sandy above than *nicholsoni*; underparts paler buff, with short fine breast streaks; whiter on chin and throat.

A. s. palliditinctus Clancey: desert edges from extreme NW Namibia to SW Angola. Greyer above than *leucocraspedon*, even lighter below.

A. s. frondicolus Clancey: Zimbabwe Plateau and adjacent Mozambique to NE Botswana. A woodland form. Colder above than *nicholsoni*, streaking more prominent. Bill shorter (unsexed, n = 7) 17–18, tail shorter (unsexed, n = 7) 63–70, with extensive pale wedges on inner web of T5.

A. s. schoutedeni Chapin (including '*chorsophilus*'): S Zaïre to S Congo, the Angolan Plateau, W Zambia and NE Namibia. A miombo woodland form. Streaking above more prominent than in *nicholsoni*; below, breast streaking sharp, throat and belly rather white. Pale areas of outer tail-feathers whiter and more extensive than in *nicholsoni*, with large white wedge on inner web of T5. Smaller than *nicholsoni*: wing, ♂ (n = 7) 90–98 (92·1); bill relatively short, 15–18; tail relatively short, 60–70; hind claw 7–10.

A. s. nyassae Neumann (including '*winterbottomi*'): E Zambia, SW Tanzania, Malaŵi and adjacent NW Mozambique. Mainly a miombo form. Like *schoutedeni* but richer warmer brown above, and more vinous buff below. Birds from above 1800 m in Malaŵi and SW Tanzania are streaked blacker above and deeply suffused below, and usually have reduced white in T5. Measurements similar to *schoutedeni*. Wing formula as in *nicholsoni* and other races of *A. similis*.

A. s. moco Traylor: Mt Moco, central Angola, above the tree line. Less distinctly streaked than *schoutedeni*, from which separated altitudinally; tail longer, without extensive white in T5.

A. s. dewittei Chapin (including '*hallae*'): highlands of SE and E Zaïre, Rwanda, Burundi and SW Uganda. Similar to *moco* but darker and colder above and more lightly streaked below. Smaller and generally darker than *hararensis*.

A. s. hararensis Neumann: Ethiopian Highlands. Darker, colder and greyer above than *nicholsoni*, streaking rather more distinct; wing-feather edging and underparts cinnamon-buff. Tail pattern similar to *nicholsoni*. Large: wing, ♂ (n = 10) 94–103 (97·9); tail, 69–78; bill, 19–21; hind claw 7–10. WEIGHT: unsexed (n = 68) 21·5–29 (25·3).

A. s. chyuluensis van Someren: N, W and S Kenya to N Tanzania. Like *hararensis* but throat and belly paler.

A. s. nivescens Reichenow: NE Sudan (Red Sea Hills), N Ethiopia (Eritrea), Djibouti and N Somalia. Paler above than *hararensis* with narrow dark streaking; paler below. Smaller than *hararensis*: wing, ♂ (n = 10) 90–98 (93·4); bill rather long, 20–22.

A. s. sokotrae Hartert: Socotra. As *nivescens*, but streaking above blacker, more distinct. Wing smaller, (6 ♂♂) 83–88, but bill relatively long, (6 ♂♂) 21–22.

A. s. jebelmarrae Lynes: W Sudan (Darfur) and central Sudan. Upperparts paler and warmer, streaking indistinct; wing-feather edging cinnamon-buff; underparts with strong cinnamon wash, breast streaking much reduced. Wing, ♂ (n = 10) 89–97 (94·4).

A. s. asbenaicus Rothschild: central Niger (mountains of Aïr); recorded central and E Mali. Much paler, more sandy buff than other races, spots on breast practically absent.

A. s. bannermani Bates (including '*josensis*'): Sierra Leone, Guinea, Liberia, Ivory Coast, Ghana, Nigeria and Cameroon; recorded SW Mali. Darker, with blackish streaking above, very bold streaking on breast and flanks; underparts and wing-feather edging cinnamon-buff. Smaller: wing, ♂ (n = 10) 85–92 (88·8); hind claw 7–10. Wing formula as in *nicholsoni*.

TAXONOMIC NOTE: the racial situation of *A. similis* in Africa is complex. *A. s. moco* from Mt Moco, W Angola, differs markedly from *schoutedeni* present in woodland around its base (Traylor 1962, Hall and Moreau 1970), and appears to have affinities with other central African montane populations, and with birds of South Africa/Namibia. Clancey (1985a) separated the southern African woodland forms *schoutedeni*, *nyassae* and *frondicolus* as a full species *A. nyassae*. He also separated *bannermani* and dark birds from the highlands of Malaŵi, NE Zambia and SW Tanzania (originally assigned to *nyassae*) and associated these with *latistriatus* (Clancey 1985a, 1986d, 1990). This last form is known only from a few non-breeding specimens and is treated here in *A. novaeseelandiae*. The situation is still unclear, however, and pending further biological information on W African birds and on adjacent forms in southern Africa, we treat the *similis* complex as a single species.

Field Characters. A large robust pipit similar to Richard's Pipit *A. novaeseelandiae* and best distinguished from it by voice. Bill longer than Richard's Pipit and most races larger with longer tail. Plumage varies geographically, but Long-billed Pipit shows more diffuse streaking above than Richard's Pipit, and less distinct streaks on breast. Facial pattern like Richard's Pipit but usually less bold; outer tail-feathers buffy (most races), legs pale or pinkish brown, base of lower mandible pinkish. In hand, all races separable from Richard's Pipit by short curved hind claw and longer and emarginated 5th primary (**A**, p. 217). Birds of less patterned races could be confused with Plain-backed Pipit *A. leucophrys*, but this lacks dark malar streak and has buffy brown outer tail-feathers and shorter-looking bill.

Voice. Tape-recorded (75, 78, 88, GIB, LEM, STJ). Song simple and unmusical, a random sequence of varied disjointed notes, delivered at rate of 1 per 1–2 s, 'chreu, shreep, chew-ee', on ground or in fluttering display flight; tone more ringing and notes more varied than in similar songs of Plain-backed Pipit and Buffy Pipit *A. vaalensis*; call-notes, sharp 'wheet' and loud, ringing 'che-vlee'.

General Habits. Preferred habitat varies geographically: in W Africa mainly montane grasslands, open areas in forest and rocky slopes above 1900 m; in E and NE Africa lightly grassed stony hillsides or arid rocky gullies with scattered bushes or small trees, from sea level up to 2600 m; across central southern Africa, light miombo woodland with an understorey of short grass, especially on edges among scattered trees and bushes (also montane grasslands at 1800–2000 m in NE Zambia, N Malaŵi and SW Tanzania); in South Africa, lightly grassed or shrubby hillsides or mountainous terrain with rocks and boulders.

Usually solitary or in pairs, but sometimes forms small parties. Forages on ground, walking with half upright carriage; stands erect when alert. Takes food from grass stems; also occasionally from tree branches. Flight strong and undulating. Usually flies some distance when disturbed, often landing on bush or small tree where it perches upright with bill pointing slightly upwards. Also perches on rocks and mounds. Miombo birds fly up to trees, and habitually walk about on branches like Tree Pipit *A. trivialis*.

Food. Mainly insects; occasionally spiders, scorpions and snails. In Zimbabwe, mainly grasshoppers, crickets, beetles, termites, wasps and spiders (Borrett and Wilson 1969).

Breeding Habits. Monogamous, territorial. Male sings from rock, tree or bush; also in flight. Ascends from perch on fluttering wings while singing, often to considerable height, and may circle, with undulations, before gliding down to perch on ground with wings and tail widely spread.

NEST: deep cup of dry grass lined with rootlets and finer grass; on ground in shelter of rock, clod or tuft of vegetation, often on slope; well hidden.

EGGS: 2–3, occasionally 4, av. (n = 31 clutches, South Africa) 2·3, av. (n = 19 clutches, E Africa) 2·5. Stone or whitish, densely spotted brown, grey and violet. SIZE: (n = 62, South Africa) 20–24 × 15–18 (21·9 × 15·7); Sierra Leone, *c*. 19 × 15.

LAYING DATES: Sierra Leone, Mar–Apr; Cameroon, Mar–Apr; Niger (eggs in ovaries, June); Sudan (Darfur) June–Oct; Sudan coast, May; Kenya, Apr–May; N Tanzania, Oct–Apr; Malaŵi and Zambia, Sept–Nov; Zimbabwe, Jan–Feb, July–Nov (mostly Sept–Nov); South Africa, Oct–Dec. Mostly in the early rains, but miombo birds in the dry season.

INCUBATION: by ♀ only. Period: 13–14 days in Israel.

DEVELOPMENT AND CARE OF YOUNG: fledging period *c*. 2 weeks.

References
Clancey, P. A. (1985a).
Cramp, S. (1988).

Anthus leucophrys Vieillot. Plain-backed Pipit. Pipit à dos roux.

Anthus leucophrys Vieillot, 1818. Nouv. Dict. d'Hist. Nat. xxvii, p. 502; Cape of Good Hope.

Plate 12
(Opp. p. 177)

Range and Status. Endemic resident; some populations nomadic or locally migratory. Widespread north of equator, uncommon to locally common: small numbers in extreme S Mauritania (Senegal R. valley and delta); Senegambia and Guinea Bissau to Sierra Leone, Liberia and Ivory Coast, S Mali (north in delta to *c*. 15°N), Ghana, Togo, Nigeria (north to Zaria and L. Chad, absent SE); N Cameroon (south to *c*. 5°N), S Chad, Central African Republic (except SW) and S Sudan (east to Nile) to N Zaïre borders, Uganda, W, SW and central Kenya and W, S and central Ethiopia to NW Somalia. South of equator, frequent to common from Rwanda, Burundi, NW Tanzania, Gabon (rare, 1 pair, Leconi), Lower Congo R. and S Zaïre (Kasai and Shaba) to central, S and E Angola, Zambia (except Luangwa and Middle Zambezi valleys), N Malaŵi, SW and S Tanzania (Songwe R. to Rukwa and Iringa District), NE Namibia (Ovamboland-Kavango-Caprivi), NW Botswana (especially Okavango) and extreme NW Zimbabwe (Kasungula); and South Africa (Transvaal highveld to Natal and S and SW Cape), Lesotho lowlands and W Swaziland to SW Mozambique borders (Lebombo Mts).

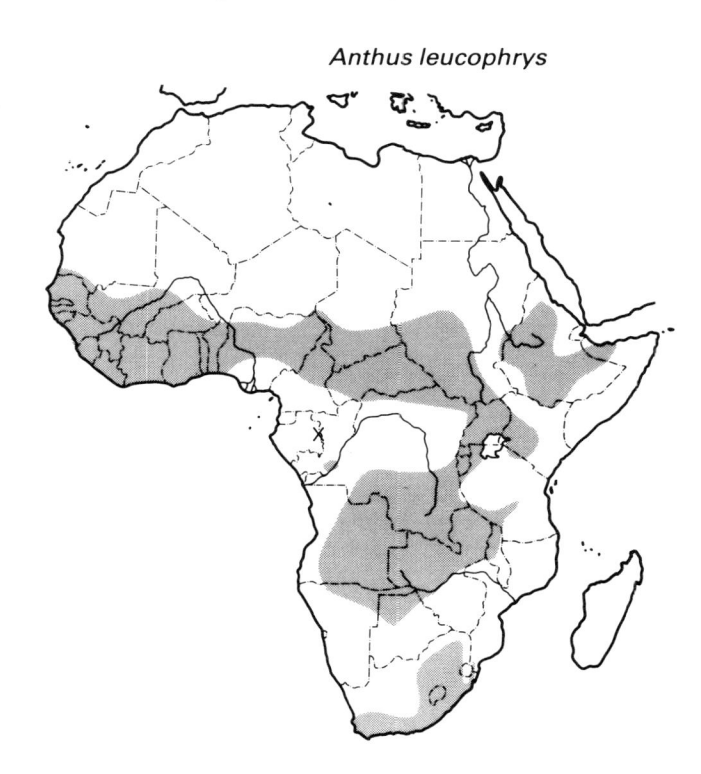

Description. *A. l. leucophrys* Vieillot (including '*enunciator*'): South Africa, Lesotho, Swaziland, Mozambique. ADULT ♂: upperparts uniform, rather olive brown, top of head faintly streaked. Tail blackish brown, outer feathers pale brown, T5 with pale brown outer web and pale tip to inner web extending as wedge along shaft. Supercilium buffish; lores brown, ear-coverts mottled brown and buffish, darker eye-stripe and moustachial stripe; malar streak faint. Underparts pale buff, browner on breast, paler on chin and throat; indistinct brown streaks across upper breast. Remiges and larger wing-coverts blackish brown, secondaries broadly edged and tertials, greater coverts and median coverts broadly edged and tipped sandy brown; lesser coverts brown; underwing-coverts and axillaries warm buff. Bill blackish brown, base of lower mandible yellowish flesh; eye dark brown; legs and feet pale or flesh brown. Sexes alike. SIZE: (10 ♂♂, 10 ♀♀) wing, ♂ 92–102 (97·2), ♀ 91–103 (95·1); tail, ♂ 65–72 (67·2), ♀ 62–67 (64·0); bill, ♂ 17–20 (18·7), ♀ 17–19 (17·9); tarsus, ♂ 30–31 (30·6), ♀ 28–31 (29·5). WEIGHT: unsexed (n = 6) 21–24 (23·2).

Hind claw long, straight and strong, 11–17. Wing-tip bluntly pointed; P6–P9 longest and about equal, P5 5–10 mm shorter; P6–P8 emarginated, P5 slightly so.

IMMATURE: browner, upperparts, wing-coverts and tertials with narrow buffish white fringes, breast with well-marked short dark brown streaks.

NESTLING: unknown.

A. l. tephridorsus Clancey: S Angola, SW Zambia, NE Namibia and NW Botswana. Paler, more greyish above.

A. l. bohndorffi Neumann (including '*prunus*'): S Zaïre to N and central Angola, N and E Zambia, N Malaŵi and Tanzania. Upperparts browner, less olive than nominate, underparts paler.

A. l. zenkeri Neumann: S Mali to Ghana, Nigeria, Cameroon, Central African Republic, S Sudan, N Zaïre, Uganda, NW Tanzania, Rwanda/Burundi and W Kenya. Upperparts browner above as in *bohndorffi*, but wing-feather edges more cinnamon; deeper buff than *bohndorffi* below, with cinnamon tinge, pale pattern on T5 reduced to spot near tip. Wing, ♂ (n = 10, Kenya and Uganda) 91–97 (94·9). WEIGHT: unsexed (n = 5, Kenya) 24–28 (26·2). Hindclaw shorter, 8–12.

A. l. gouldii Fraser: Sierra Leone, Liberia and Ivory Coast. Darker above than *zenkeri* with less cinnamon on wings; paler, more creamy below with more distinct breast streaks. Smaller: wing, ♂ (n = 5) 87–95 (91·2).

A. l. ansorgei White: S Mauritania, Senegambia and Guinea Bissau. Like *gouldii* but paler and greyer above. Wing, ♂ (n = 6) 92–96 (93·8).

A. l. omoensis Neumann: N, central and W Ethiopia and adjacent SE Sudan. Like *zenkeri* but darker, richer cinnamon-buff below and more heavily streaked on breast. Larger: wing, ♂ (n = 10) 97–109 (102). WEIGHT: unsexed (n = 8) 23–32 (26·9).

A. l. saphiroi Neumann: SE Ethiopia (N Ogaden-Harar) to NW Somalia. Slightly paler above than *zenkeri*, with sandier wing-feather edging; size similar.

A. l. goodsoni Meinertzhagen: central and SW Kenya to extreme N Tanzania. Paler and sandier above than *zenkeri* and *saphiroi*, paler, more sandy below. Larger: wing, ♂ (n = 9) 95–102 (98·9). Hind claw 8–10.

TAXONOMIC NOTE: this and the paler, sandier *A. vaalensis* exist as 2 good species in southern Africa. In E Africa, however, birds very similar to *A. vaalensis* (*goodsoni* and *saphiroi*) meet and intergrade with typical *A. leucophrys*; all northern populations are therefore treated here as races of *leucophrys* (*vide* Hall 1961, Hall and Moreau 1970).

Field Characters. A large, robust pipit with unstreaked upperparts and indistinct, diffuse streaks on breast (more defined in races *gouldii* and *ansorgei*). Upperparts normally dark brown (paler in *saphiroi* of Ethiopia and *goodsoni* of Kenya), underparts warm buff, washed cinnamon in some races, and wings edged rich buff. Tail broad-looking, outer feathers buff. Pale supercilium prominent; shows dark eye-stripe and moustachial stripe but lacks distinct malar streak. Richard's Pipit *A. novaeseelandiae* and Long-billed Pipit *A. similis* have streaks on upperparts (though indistinct on some races of Long-billed Pipit), streaks or spots on breast and conspicuous dark malar streak. In southern Africa, overlaps with very similar plain Buffy Pipit *A. vaalensis*, whose plumage is variable but paler above and paler, less richly coloured below. Plain-backed Pipit further distinguished from Buffy Pipit by yellowish (not pink) base to lower mandible, and in hand by longer, straighter hind claw (**A**, p. 217).

Voice. Tape-recorded (78, 88, C, GRI, LEM, MOR). Song simple and unmusical, a rather monotonous series of single or alternating notes, 'chree, cheup, chree, cheup, ...', delivered at rate of about 1 per s; from ground or low perch; calls, thin 'tsissik' in flight, and 'trt-tit' at take-off.

General Habits. Frequents flat or hilly short grass country, flood plains and cultivation; especially well-grazed areas with patches of bare ground and scattered bushes, trees and termite mounds. Often breeds on recently burnt ground. Generally prefers moister habitat in southern Africa than Buffy Pipit. Mainly at low to medium altitude, but occurs to 2200 m in E Africa.

Usually solitary or in pairs, occasionally small groups. Walks with rather upright stance; takes prey from ground. Pauses and wags tail deliberately. Perches occasionally on mound or bush. When flushed usually flies low for short distance, landing again on ground, flight heavy-looking and undulating.

Some populations nomadic or locally migratory: occurs N Nigeria mainly Jan–May; reaches SW Mozambique as non-breeding visitor July–Sept; dry season visitor Zambia to Middle Zambezi and Kafue.

Food. Insects and their larvae; seeds. Caterpillars and mantids included in diet of nestlings, South Africa (McCleland 1987).

Breeding Habits. Monogamous and territorial; several pairs sometimes nest close together. Displaying ♂ flies high in air, flutters in circle, then drops headlong to ground.

NEST: deep cup of grass, lined with rootlets and sometimes hair, feathers or fur; ext. diam. *c.* 120, diam. of cup *c.* 70, depth of cup *c.* 50; set in soil at base of overhanging grass tuft.

EGGS: 2–3, sometimes 4, av. (n = 19 clutches, South Africa) 2·7, av. (n = 6 clutches, Kenya) 2·3. Whitish or greyish, closely speckled brown and dark and light grey.

SIZE: (n = 54, South Africa) 19–25 × 14–17 (21·5 × 15·6), (n = 6, Kenya) 19–20 × 14–16 (19·9 × 15·5).

LAYING DATES: Senegambia, Apr–June; Sierra Leone, Jan–May; Burkina Faso (juv. July); Ghana, Feb–June; Nigeria, throughout year in S (e.g. Ilaro Jan, May, Sept, Dec), mainly Mar–May in N; Kenya, Mar–June, Aug; SW Tanzania, Dec; Zambia, Sept, Nov; South Africa: Transvaal, Sept–Dec, Natal and Cape, Oct–Dec. Breeds mainly during rains.

DEVELOPMENT AND CARE OF YOUNG: young fed by both parents.

Anthus vaalensis Shelley. Buffy Pipit. Pipit du Vaal.

Plate 12
(Opp. p. 177)

Anthus vaalensis Shelley, 1900. Bds Afr. 2, p. 311; Newcastle, Natal.

Range and Status. Endemic resident and intra-African migrant. Confined to central and southern Africa, mainly plateau areas: Angola (central plateau, reaching coast at Luanda); S Zaïre (Shaba, also Middle Kasai valley); Zambia (mainly in W, absent from Luangwa valley); SW Tanzania (Ufipa and Tukuyu to Songea); Malaŵi (except extreme S); Zimbabwe and adjacent Mozambique; Botswana; NE and central Namibia; South Africa (Transvaal south to inland Natal and S and W Cape; 1, Dududu near Natal coast); lowland Lesotho; and W Swaziland. Widespread and locally common; becoming increasingly so in overgrazed stock-farming areas. Dry season breeding visitor to Zambia and Malaŵi.

Description. *A. v. vaalensis* Shelley (including '*daviesi*'): South Africa to S Botswana. ADULT ♂: upperparts uniformly sandy brown, top of head faintly streaked darker. Tail dark brown, outer feathers pale buff on outer web and distal part of inner web, T5 with buff confined to tip and outer edge. Supercilium broad, buffish; lores dusky; ear-coverts sandy brown with darker moustachial streak; malar streak fine, dark brown. Underparts pale sandy buff, deeper on breast, paler on chin and throat; indistinct greyish brown streaks on upper breast. Primaries, secondaries and primary coverts brown; tertials, greater coverts and median coverts brown, broadly edged and tipped buff; lesser coverts sandy brown; underwing-coverts and axillaries pale buff. Bill dark brown, base of lower mandible pinkish; eye dark brown; legs and feet pale yellowish brown. Sexes alike. SIZE: (8 ♂♂, 8 ♀♀) wing, ♂ 102–110 (106), ♀ 97–107 (101); tail, ♂ 76–80 (78·1), ♀ 72–76 (74·1); bill, ♂ 18–21 (19·4), ♀ 18–20 (19·1); tarsus, ♂ 28–32 (30·1), ♀ 28–30 (29·4). WEIGHT: unsexed (n = 18) 23–31 (27·0).

Hind claw short and weak, 8–12 (**A**, p. 217). Wing-tip bluntly pointed; P6–P9 longest and about equal, P5 5–10 mm shorter; P6–P8 emarginated on outer web. P5 slightly so.

IMMATURE: browner above, with pale buff feather edging, breast distinctly spotted.

NESTLING: unknown.

A. v. exasperatus Winterbottom: salt pans and peripheral arid country in NE Botswana; in non-breeding season to W Zimbabwe. Colder and greyer above than nominate. Smaller: wing, ♂ (n = 4) 98–101.

A. v. chobiensis (Roberts): Zimbabwe and adjacent Mozambique to N Botswana, NE Namibia (Caprivi), S Zaïre, Zambia, Malaŵi and SW Tanzania. Less sandy, more vinaceous above than nominate, wing-coverts and underparts tinged more cinnamon-buff. Smaller: wing, ♂ (n = 6) 96–102.

A. v. namibicus Clancey: NE and central Namibia (Okavango R. from Rundu to Andara; Tsumkwe; Karibib). Darker, more blackish above; more vinaceous buff, less tawny buff below than nominate. Smaller: wing, ♂ (n = 7) 95–103.

A. v. neumanni Meinertzhagen: Angolan plateau. Slightly darker and browner above than nominate; deeper buff below. Smaller: wing, ♂ (n = 4) 94–101.

Field Characters. A large, pale, plain-backed pipit of southern Africa, with indistinct breast streaks, buff outer tail-feathers and a prominent supercilium. Very similar to Plain-backed Pipit *A. leucophrys*, but paler above and below, and slightly larger and heavier; for further differences see Plain-backed Pipit.

Voice. Tape-recorded (GIB, STJ). Song a series of rather dry, monotonous notes, 'chrep, chiri, chree, chreu', delivered at rate of about 1 per s, slightly lower pitched than similar song of Long-billed Pipit *A. similis*; call-note sharp, di-, tri- or monosyllabic, 'chipip', 'chipipip', 'chik', recalling that of White Wagtail *Motacilla alba*.

General Habits. Frequents bare or sparsely grassed ground, typically with rocks or termite mounds for perching, especially in areas overgrazed by domestic stock; generally in drier country than Plain-backed Pipit, though both may be found together.

Solitary or in pairs; sometimes in small groups, and attracted in numbers to burnt ground. Takes food from ground. Walks or runs; pauses very upright; and wags tail deliberately. Restless and rather shy. Perches on rocks, termite mounds or furrows when disturbed. Flies briefly to trees and wires.

Dry season breeding visitor to Zambia and Malaŵi Apr–Oct. Local and long-distance movements in Zimbabwe; in some years leaves Mashonaland Plateau entirely during rains. In South Africa, subject to nomadic movements, especially in winter (Maclean 1985).

Food. Insects; also other arthropods and occasionally seeds. In Zimbabwe, largely short-horn grasshoppers and a variety of beetles Oct–Apr, with caterpillars an important addition Jan–Apr; also large numbers of termites whenever available. Of 67 stomachs (Zimbabwe), 76% contained Coleoptera, 69% Orthoptera, 34% Hymenoptera, 18% Hemiptera, 19% Lepidoptera, 24% Isoptera, 3% Diptera, 4% Odonata, 6% Myriapoda, 3% Dictyoptera, 6% Arachnida, 7% seeds, 6% feathers and 5% grit (Borrett and Wilson 1971).

Breeding Habits. Monogamous, territorial.
NEST: bulky, loosely built cup of coarse grass stems, lined with fine grass and rootlets; well hidden in depression under grass tuft or against stone; diam. of cup $c.$ 60, depth $c.$ 40.
EGGS: 2–3. Dull white, closely speckled with shades of grey and light brown. SIZE: (n = 21, South Africa) 20–24 × 15–17 (22·0 × 15·3).
LAYING DATES: Zambia, June–Oct; Zimbabwe, July–Feb (mainly Sept–Dec); South Africa, Aug–Dec.

Reference
Borrett, R. P. and Wilson, K. J. (1971).

Plate 12
(Opp. p. 177)

Anthus pallidiventris Sharpe. Long-legged Pipit. Pipit à longes pattes.

Anthus pallidiventris Sharpe, 1885. Cat. Bds Br. Mus. 10, p. 560; Gaboon.

Range and Status. Endemic resident; confined to equatorial Africa. Locally common Equatorial Guinea and W and S Gabon to S Congo and the Lower and Middle valley of the Congo R. (upstream to Mbandaka, Zaïre); also NW Angola (Cabinda, Lower Cuanza valley from Luanda to Barraca and N Malange); once SW Cameroon (Avelé).

Description. *A. p. pallidiventris* Sharpe: Equatorial Guinea to NW Angola including Lower Congo R. valley inland to Brazzaville; once Cameroon. ADULT ♂: upperparts rather plain greyish brown, slightly streaked on head. Tail blackish brown, T6 buffish brown, T5 buffish brown with paler outer edge and tip. Supercilium prominent, whitish; lores and ear-coverts brown, flecked pale buff, with darker moustachial stripe; underparts white, breast washed greyish buff and indistinctly streaked. Primaries and secondaries dark brown; tertials, greater and median coverts dark brown, edged and tipped buffish; lesser coverts brown; underwing-coverts and axillaries brownish white. Bill brown, lower mandible with much yellow at base; eye blackish; legs and feet light brown. Sexes alike. SIZE: (9 ♂♂, 7 ♀♀) wing, ♂ 93–99 (96·0), ♀ 89–95 (91·6); tail, ♂ 64–70 (65·6), ♀ 59–66 (63·7); bill, ♂ 18–21 (19·7), ♀ 18–19 (18·4); tarsus, ♂ 31–35 (32·2), ♀ 29–32 (31·0).
Hind claw long and strong, 13–16. Wing-tip bluntly pointed; P6–P9 longest, P5 5–10 mm shorter; P6–P8 emarginated on outer web, P5 slightly so.
A. p. esobe Chapin: middle basin of Congo R. from Kunungu upstream to Mbandaka and Lukolela. Darker and greyer brown than nominate and more heavily streaked on breast.
IMMATURE AND NESTLING: unknown.

Field Characters. A large, plain-backed pipit with long legs, long-looking bill and prominent supercilium. Similar to local races of Plain-backed Pipit *A. leucophrys* and Buffy Pipit *A. vaalensis* with which it overlaps in southern part of range. Upperparts somewhat greyer, underparts whiter, less buffy, breast streaks more distinct; outer tail-feathers light brown, wing-edging without rich tone of Plain-backed Pipit; base of lower mandible yellow, like Plain-backed Pipit. Race *esobe* more distinctive, with darker upperparts and more pronounced breast streaks.

Voice. Tape-recorded (78, ERA). Song, 'tuit-tidii', repeated every 2–3 s; contact call between members of pair in tall vegetation, 'psip'; flight call, a liquid 'pouititit'; also gives a single husky 'ptic-ptic' (Brosset and Erard 1986).

General Habits. Inhabits small areas of natural grassland, including larger grassy clearings in forest; also man-made habitats, including pastures, airports, bare ground near houses, and cultivation. In pairs or small parties; stands upright, constantly wagging tail. Forages on ground; runs after insects, making small jumps to catch those flying low. Only occasionally perches in trees.

Food. Arthropods, including small Orthoptera and other small flying insects, and spiders.

Breeding Habits. Unknown.
LAYING DATES: Gabon, May (breeding condition Mar); Zaïre, probably May–July, at end of rains.

Reference
Brosset, A. and Erard, C. (1986).

Anthus melindae Shelley. Malindi Pipit. Pipit de Malindi.

Plate 13
(Opp. p. 240)

Anthus melindae Shelley, 1900. Bds Afr. 2, p. 305; Malindi, Kenya.

Range and Status. Endemic resident. Coastal Kenya from Mombasa northwards, including lower Tana valley, and coastal Somalia north to 3°N and inland along Jubba valley to Bu'aale and Webi Shabeele to Mahaddeillen. Locally common to abundant; rare in Kenya south of Sabaki R.

Description. *A. m. melindae* Shelley: Kenya coast and S Somalia north to Mogadishu and Webi Shabeele valley. ADULT ♂: upperparts earth brown, top of head streaked dark brown, mantle with indistinct dark brown mottling. Tail blackish brown, T6 pale brown with whitish tip and outer edge, T5 also with whitish tip. Supercilium whitish; lores, moustachial area and stripe through and behind eye brown; broad line below eye to ear-coverts mainly buff; prominent malar streak dark brown. Underparts whitish, suffused greyish buff on breast and flanks; brown streaks across breast and a few on flanks. Remiges and upperwing-coverts dark brown, tertials and greater coverts edged rich buff, median coverts and alula edged whitish; underwing-coverts and axillaries greyish brown. Bill dark horn, base of lower mandible bright yellowish flesh; eye dark brown; legs and feet bright yellowish flesh. Sexes alike. SIZE: (9 ♂♂, 3 ♀♀) wing, ♂ 81–89 (84·3), ♀ 81–83 (82·3); tail, ♂ 57–63 (61·2), ♀ 55–60 (57·7); bill, ♂ 17–19 (17·9), ♀ 16–18 (17·0); tarsus, ♂ 25–28 (26·8), ♀ 24–26 (25·3). WEIGHT: unsexed (n = 8) 19–27 (22·5).
Hind claw medium, fairly straight, 10–11. Wing-tip blunt; P6–P9 longest and about equal, P5 *c.* 3 mm shorter; P5–P8 emarginated on outer web.
IMMATURE: somewhat darker and browner above than adult, with narrow whitish edges to feathers of upperparts and wing-coverts.
NESTLING: covered with grey down.
A. m. mallablensis Colston: Mallable, 30 km northeast of Mogadishu, Somalia. Paler ash brown above than nominate, edges of tertials and greater coverts paler; malar stripe fainter, streaking below paler, finer and less extensive, belly whiter. Smaller: wing, ♀ (n = 3) 76–79 (77·7). WEIGHT: ♀ (n = 3) 19–20·5 (19·6).

Field Characters. A rather robust, long-legged pipit similar in size and structure to Richard's Pipit *A. novaeseelandiae*. Overall tone greyer and darker than local race of Richard's Pipit, which is sandy buff; upperparts, including top of head, much plainer looking, with diffuse mottling rather than dark streaks; underparts white with greyish tone, sometimes lightly washed buff on breast and flanks, not warm sandy buff; streaks on breast extend to flanks and upper belly. Tail has less white, legs brighter, more orange, lower mandible with prominent orange-yellow base; in hand, wing blunter with P5 emarginated. Local race of Long-billed Pipit *A. similis* larger, longer-billed and longer-tailed, with fainter streaking below, warmly tinged wing-coverts and underparts and buffish outer tail-feathers.

Voice. Tape-recorded (McVIC). Song a repeated jangling 'creer' or 'kurree', uttered about once per s from perch, usually 1–1·5 m above ground; calls, 'tsweep', uttered in flight, much like Yellow Wagtail *Motacilla flava*; also 'tweet-tweet' and 'tirrip-tirrip-tirrip' (Britton and Britton 1978).

General Habits. In Kenya, confined to low-lying short coastal or riparian grassland subject to seasonal flooding. In Somalia, more widespread on short grassland, edges of drying flood pans and cultivation; *A. m. mallablensis* occurs on coastal sand-dunes. Usually singly or in pairs, sometimes in small groups. Walks or runs with rather upright carriage, pausing to stand erect; rarely wags tail. Perches on sticks and low bushes. Takes food from ground or base of grass stems; catches small locusts in the air. Escape flight usually low and for short distance.

Food. Insects (including small locusts) and other arthropods.

Breeding Habits. Monogamous and territorial, with nesting density in N Kenya coastal grasslands 2–4 pairs per ha (Britton and Britton 1978).

NEST: deep substantial cup of grasses attached to growing grass, usually in tussock; cup diam. 70–80, depth *c.* 50.

EGGS: 2–3; off-white with chocolate brown markings and grey undermarkings, concentrated at broad end. SIZE: (n = 5) 18–21 × 14–16 (19·8 × 15·3).

LAYING DATES: Kenya, Apr–June, during rains.

Reference
Britton, P. L. and Britton, H. A. (1978).

Plate 13
(Opp. p. 240)

Anthus brachyurus Sundevall. Short-tailed Pipit. Pipit à queue courte.

Anthus brachyurus Sundevall, 1850. Oefv. K. Sv. Vet.-Akad. Förh. 2, p. 100; Upper Umlaas R., Natal.

Range and Status. Endemic resident, local migrant and possibly intra-African migrant. Distribution disjunct, obscured by sporadic occurrence and local movements. Occurs SW Uganda (Ankole, Toro), Rwanda, N Burundi, adjacent NW Tanzania and E Zaïre; S Gabon (west and north to Mouila), S Congo and S Zaïre to NE Angola (Kasai R. drainage) and Zambia (west to 30°E); S Tanzania (Iringa highlands); S Mozambique (locally, near Beira); and South Africa (S Natal, E, W-central and SW Transvaal); generally uncommon to frequent. Recorded also Zimbabwe (Beit Bridge), Orange Free State (1, *c.* 29°S, 26°E: Brooke 1984), Lesotho (1, *c.* 29°S, 29°E: Brooke 1984).

Description. *A. b. leggei* Ogilvie-Grant (including '*eludens*'): throughout range except South Africa. ADULT ♂: upperparts blackish, feathers edged dull olive-brown, rump plain dark olive-brown. Tail-feathers dark brown, outermost pair greyish white, T5 pale at tip. Supercilium indistinct; lores and ear-coverts dark olive. Underparts pale yellowish buff, with bold blackish streaking on breast and some streaking extending also to throat, flanks and belly. Wing blackish brown, tertials and wing-coverts edged pale olive-buff; underwing-coverts and axillaries pale yellowish buff. Bill dark brown above, flesh below; eye brown; legs and feet flesh. Sexes alike. SIZE: (10 ♂♂, 10 ♀♀) wing, ♂ 63–68 (66·2), ♀ 62–66 (64·2); tail, ♂ 34–39 (37·4), ♀ 34–39 (36·4); bill, ♂ 13–15 (13·9), ♀ 12–15 (13·5); tarsus, ♂ 17–19 (17·6), ♀ 17–18 (17·3).

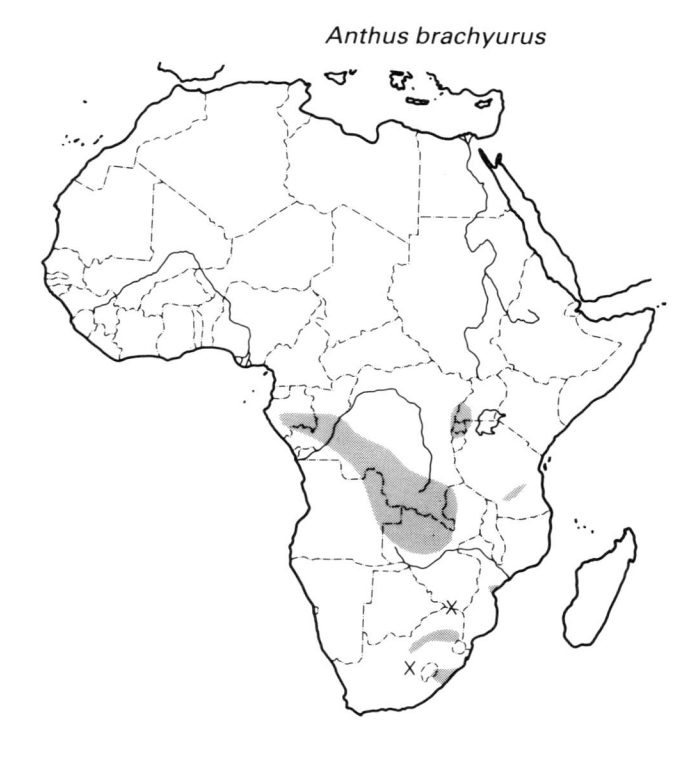

Hind claw short and curved. Wing-tip blunt; P6–P9 longest, P5 1–3 mm shorter; P6–P8 emarginated on outer web.

IMMATURE: upperparts, including rump, dark brown with narrow warm buff feather fringes; streaking below broader than in adult; fringes to tertials and coverts warm buffish brown and more sharply defined.

NESTLING: unknown.

A. b. brachyurus Sundevall: South Africa. Upperpart feathers more strongly edged with olive-buff than in *leggei*, streaking below slightly less heavy.

Field Characters. A small dark pipit with a very short tail, shorter and thinner than in other small pipits; when flushed 'flies off speedily, resembling a large *Cisticola*' (Sinclair 1984). Similar to Bush Pipit *A. caffer*, but upperparts darker and more olive, mottled-looking rather than streaked, underparts more extensively streaked. Lack of supercilium gives plain-faced look. Habitat and voice differences useful.

Voice. Tape-recorded (GIB). Song during cruising display flight, a series of nasal notes, 'bzip, bzeent, bzeeu', delivered at rate of about 1 per 1·5 s; also a rolling 'tip-pee-reep-pip-rip' (Maclean 1985); other calls include soft, nasal 'tseep'.

General Habits. Most terrestrial of the small pipits. Frequents short open grassland, usually moist but less than 15 cm high. In Zambia prefers sparsely vegetated ground, on sandy soils or after recent burning (D. R. Aspinwall, pers. comm.). Typically occurs between 800 and 1500 m, although recorded above 1800 m in S Tanzania and near sea level in Mozambique. Winters on Natal coast on short grass with sandy substrate.

Usually solitary, but occasionally several together in loose concentration. Shy, walks long distances on ground and difficult to flush. Flight jerky and dipping, in wide circles a few m to just above ground; does not land on perch, but drops into grass 20–30 m from take-off and runs along ground in crouching attitude. May perch on stone or low termite mound.

Highly mobile, but movements poorly understood. Appearances unpredictable locally, but conform to broad seasonal pattern in many areas. May undertake considerable intra-African movements. Recorded S Zaïre Aug–Nov (1 June, 1 Dec), Zambia Sept–Mar (1 Apr, 1 Aug), Malaŵi Nov, S Tanzania Nov–Feb, Angola Dec–Jan; by contrast, in Gabon June–July, Rwanda/Burundi May–Aug, E Zaïre May–June, Uganda May–June (but also Feb) (D. R. Aspinwall, pers. comm., T. Harris, pers. comm.). Thus, most records from tropics during or near rains, those from 4°–16°S mainly Sept–Mar, those from central Africa mainly May–Aug. Birds in NE Rwanda in dry season only (June–Aug) possibly from S Uganda (L. Edward, Akagera: Vande weghe 1981). Recorded Zimbabwe Nov; South Africa all months but mainly Sept–Jan. Probably an altitudinal migrant Natal, breeding on lower slopes of Drakensberg escarpment in summer and regular Zululand coast in winter (J. C. Sinclair, pers. comm.). Birds near Beira probably migrants from E Zimbabwe highlands.

Food. Insects and their larvae; grass seeds (Maclean 1985).

Breeding Habits. In display, ♂ cruises a few m above ground in wide circles (diam. 100 m or more), giving nasal calls (see *Voice*) accompanied by buzzing wing-snaps.

NEST: small deep cup of coarse grass, lined with fine material; ext. diam. 90–110, diam. of cup 60–70, depth of cup *c.* 35; placed on ground between grass tufts.

EGGS: 2–3, av. (n = 3 clutches, South Africa) 2·7. White, creamy or pinkish white, finely speckled light brownish and purplish grey, often more densely at broad end. SIZE: (n = 7, South Africa) 16–18 × 13–14 (17·3 × 13·3) (Maclean 1985); (Zaïre) 15–16 × 13.

LAYING DATES: S Zaïre, Nov; Zambia (breeding condition Dec); S Tanzania, Nov–Dec; Angola, Dec; South Africa, Sept–Jan.

Reference
Brooke, R. K. (1984).

Anthus caffer **Sundevall. Bush Pipit; Bushveld Pipit. Pipit cafre.**

Plate 13
(Opp. p. 240)

Anthus caffer Sundevall, 1851. Oefv. K. Sv. Vet-Akad. Förh. 7, p. 100; Mohapoani, Transvaal.

Range and Status. Endemic resident, partly nomadic. Distribution disjunct: S Ethiopia (highlands east of the Rift between Gatela and Asheba); S Kenya and N Tanzania (Lolgorien, Loita, Ngong and Simba south to Serengeti, Arusha, Kidugallo and Dar-es-Salaam); W Angola (plateau in Huambo, N Bié and N and W Huila); range more continuous from N Malaŵi (Edingeni), N Zambia (Mpika) and extreme SE Zaïre to central Zambia (Copper Belt west to Kasempa and south to Barotse and Lusaka); and from central Zimbabwe plateau (Harare and Rusape south and southwest to Masvingo, Bulawayo and N Matopos) to E Botswana (Nata to Gaborone), South Africa (Transvaal and NE Natal), E Swaziland and S Mozambique (south of the

Limpopo). Local; frequent Angola and Zimbabwe southward; uncommon Kenya, Tanzania and Zambia; rare Malawi; status uncertain Ethiopia.

Anthus caffer

Description. *A. c. caffer* Sundevall: SE Botswana, SW Zimbabwe, Transvaal, W Swaziland and adjacent South Africa (N Natal). ADULT ♂: upperparts warm brown, head to mantle and scapulars boldly streaked dark brown, rump and uppertail-coverts almost plain. Tail dark brown, T6 white except at base of outer web. Supercilium buffish and indistinct; lores and ear-coverts dark brown, flecked buff. Underparts suffused pale buff, breast and flanks streaked dark brown. Wing dark brown, tertials and upperwing-coverts broadly edged and tipped warm buff; underwing-coverts and axillaries buffish white. Bill dark brown above, pale horn below; eye brown; legs and feet light brown. Sexes alike. SIZE: wing, ♂ (n = 13) 74–78 (75·4), ♀ (n = 8) 72–75 (73·3); tail, ♂ (n = 8) 47–53 (51·0), ♀ (n = 10) 44–53 (48·6); bill, ♂ (n = 3) 15, ♀ (n = 5) 14–16 (15·0); tarsus, ♂ (n = 3) 17–19 (18·0), ♀ (n = 5) 17–19 (18·0). WEIGHT: 1 unsexed 16.

Hind claw short and curved, 6–8. Wing-tip blunt; P6–P9 longest, P5 2–4 mm shorter; P6–P8 emarginated on outer web, P5 slightly so.

IMMATURE: paler than adult, more spotted above.

NESTLING: unknown.

A. c. traylori Clancey: S Mozambique and adjacent South Africa (E Transvaal and extreme NE Natal). Less warm above than nominate, light streaking on hindneck paler, rump more olivaceous and distinctly streaked dark brown; whiter below. Smaller: wing, ♂ (n = 14) 70–75 (72·0).

A. c. mzimbaensis Benson: NE Botswana to Zimbabwe Plateau, occurring also N Zambia and W Malawi. Like *traylori* but larger: wing, ♂ (n = 7) 75–80 (77·0).

A. c. blayneyi van Someren: Kenya and Tanzania. Differs from nominate in having upperparts more sandy buff, rump streaked and belly whiter. Smaller: wing, ♂ (n = 2) 67, 71.

A. c. australoabyssinicus Benson: Ethiopia. Paler than *blayneyi*, streaking above finer; streaking below extending further onto throat; outer tail-feathers tinged buff.

TAXONOMIC NOTE: some Zambian birds and the population of Angola probably represent an undescribed race, being nearer *blayneyi* than *caffer*, but smaller than *mzimbaensis* and less white below (Clancey 1989).

Field Characters. A small warm brown pipit with well-streaked upperparts and breast, white outer tail-feathers and rather plain face. Most similar to Short-tailed Pipit *A. brachyurus* but tail longer and broader. Upperparts paler, warmer and browner than Short-tailed Pipit, with light and dark streaking more contrasting, streaking on underparts more restricted to breast. Behaviour, habitat and voice different.

Voice. Tape-recorded (88, GIB, McVIC, STJ). Song a repetition of phrase of 2–3 nasal notes, first note lower, 'werrp-cheer, werrp-cheer, werrp-cheer-chirr . . .', given from tree. Call, a distinctive 'see-ip' or 'bzhzhzht', often given on alighting on branch after flushing (D. R. Aspinwall, pers. comm.).

General Habits. Frequents open woodland with patchy ground cover; short grassland with sandy patches and scattered acacia or broad-leaved trees; edges of *Brachystegia* woodland. In E Africa, ranges from near sea level to 2200 m; in Zimbabwe mainly above 1200 m.

Solitary, in pairs or in small scattered flocks in non-breeding season. Usually associates with mixed bird parties. Not particularly shy, but easily overlooked and difficult to see when perched. Forages on ground, walking and picking among grass and leaf litter. Flies to bush or low tree branch when disturbed, with quick erratic flight. May walk along tree branch.

In Transvaal, probably nomadic, and numbers fluctuate. Birds from Zimbabwe may move north to Zambia and Malawi. Appearances in Kenya are sporadic, mainly Mar–Sept.

Food. Insects.

Breeding Habits.
NEST: small thick-walled cup of grass, lined with rootlets, set into ground under tuft of grass; well concealed.
EGGS: 2–3, av. (n = 28 clutches, South Africa) 2·5, av. (n = 3 clutches, Kenya) 2·7. White, with small clear reddish brown and slate-grey spots. SIZE: (n = 65, South Africa) 17–21 × 13–15 (18·5 × 14·0); (n = 3, Kenya) 18–19 × 13–14 (18·8 × 13·2).
LAYING DATES: Kenya, Mar–Apr; Zimbabwe, Nov, Jan; South Africa: Transvaal, Nov–Feb (with 77% Nov–Dec), Natal, Oct–Mar.

References
Clancey, P. A. (1989).
Maclean, G. L. (1985).

Anthus sokokensis **van Someren. Sokoke Pipit. Pipit de Sokoke.**

Anthus sokokensis van Someren, 1921. Bull. Br. Orn. Club 41, p. 124; Sokoke, Kenya.

Range and Status. Endemic to coastal East Africa; resident. Confined to Sokoke Forest and Gedi, Kenya, and Kiono Forest (near Sadani) and Pugu Hills, Tanzania. Population estimates for Sokoke Forest Reserve 3000–5000 pairs (Britton and Zimmerman 1979) and 2000+ pairs (Kelsey and Langton 1984); uncommon Kiono Forest; rare Pugu Hills. Population at Moa, N coastal Tanzania, discovered in 1931, probably extinct. Status threatened by disturbance of forest habitat.

Description. ADULT ♂: entire upperparts broadly streaked warm buff and blackish brown, tinged rufous on rump and uppertail-coverts. Tail-feathers blackish brown and pointed, central pair broadly edged rufous-buff; outer pair white except base of inner web, T5 tipped white. Supercilium narrow and buffish white; lores buffish; ear-coverts streaked pale buff and dark brown. Underparts white, tinged yellowish buff; upper breast broadly streaked and lower breast, flanks and belly more finely streaked blackish. Primaries and secondaries blackish brown, narrowly edged buff; tertials and inner greater coverts dark brown, broad pale rufous or creamy buff fringes forming panel on closed wing; median and outer greater coverts broadly tipped buffish white; lesser coverts dark brown, broadly fringed buff; underwing-coverts and axillaries white. Bill blackish horn above, pinkish below with brown tip; eye dark brown; legs and feet fleshy pink. Sexes alike. SIZE: (3 ♂♂, 3 ♀♀) wing, ♂ 68–71 (69·3), ♀ 68–71 (69·6); tail, ♂ 46–52 (48·3), ♀ 49–50 (49·7); bill, ♂ 15–17 (16·0), ♀ 15–16 (15·7); tarsus, ♂ 19–20 (19·3), ♀ 18–20 (19·0). WEIGHT: unsexed (n = 10) 12–17 (15·0).

Hind claw short and straight, 7–8. Wing-tip blunt; P6–P9 longest, P5 *c.* 1 mm shorter; P6–P8 emarginated on outer web, P5 slightly so.

IMMATURE AND NESTLING: unknown.

Field Characters. The only small pipit in its very restricted range. Richly coloured, with contrasting reddish buff and black streaks on upperparts, and white underparts with broad blackish spots on breast and streaks on upper belly and flanks. Shows rufous panel on closed wing, two whitish wing-bars and white outer tail-feathers.

Voice. Tape-recorded (CART). A loud, high-pitched 'sweer', given from perch or on ground.

General Habits. Inhabits ground stratum of coastal evergreen forest. In Sokoke, prefers dense, uncleared *Afzelia*-dominated forest, occurring at densities of up to 1 pair/2 ha (Britton and Zimmerman 1979); elsewhere at lower densities in open *Brachystegia* woodland, *Cynometra-Manilkara* forest and low land rain forest. Lives mainly on the forest floor, where shy and difficult to observe. Usually seen singly, and often located by call. Feeds among sparse grass. When flushed, may fly only a few m before dropping to ground, but often flies to high perch, where easily lost to sight.

Food. Insects, including white ants and beetles.

Breeding Habits. Unknown.

References
Britton, P. L. and Zimmerman, D. A. (1979).
Collar, N. J. and Stuart, S. N. (1985).
Kelsey, M. G. and Langton, T. E. S. (1984).

Plate 13
(Opp. p. 240)

Anthus trivialis **(Linnaeus). Tree Pipit. Pipit des arbres.**

Alauda trivialis Linnaeus, 1758. Syst. Nat. (10th ed.), p. 166; Sweden.

Range and Status. Breeds Eurasia east to *c.* 145°E, north to N Scandinavia and *c.* 63–65°N in USSR, and south to N Mediterranean, N Asia Minor, Caucasus, N Iran, central Asia and Himalayas; winters Africa and India.

Plate 13
(Opp. p. 240)

Palearctic winter visitor. Winters south of Sahara in a belt from S Mauritania (mainly south of 17°N, also Nouakchott), Senegambia, Guinea, Sierra Leone and Liberia east through central and S Mali, Ivory Coast, Ghana, central and S Nigeria, Cameroon and Central African Republic, to N Gabon and N Zaïre; also from S Sudan (north to c. 8°N) and W and central Ethiopia (north to c. 16°N) southwards through Uganda, Kenya (except low E areas), E and SE Zaïre, Rwanda, Burundi and Tanzania (except SE) to Zambia, Malaŵi, Zimbabwe, W Mozambique, E Angola, NE Namibia, N and E Botswana and South Africa (Transvaal); frequent to locally abundant, but uncommon Angola, Botswana and Transvaal. Vagrant Natal (Durban), Namibia (Möwe Bay, Naukluft). On passage, frequent to common throughout Sahel and NW Somalia (in Chad and Sudan mainly autumn); in N Africa, frequent to common autumn (uncommon Libya), common spring. Rare winter records Morocco, Algeria, Egypt (Nile delta and valley, Faiyum), Niger (Zinder) and E Sudan probably include passage stragglers. Singing N Algeria (Djudjura) June (possible breeding?).

Anthus trivialis

× Vagrant
+ Mid-winter record

Description. *A. t. trivialis* (Linnaeus): Europe, N Asia Minor, N Iran and Siberia east to c. 140°E; only subspecies wintering in Africa. ADULT ♂: upperparts olive-brown, top of head to mantle and scapulars streaked dark brown, back lightly streaked, rump plain. Tail dark brown, outermost feathers white on outer web and most of inner web, T5 with small white spot or wedge at tip of inner web. Supercilium and lores buff; ear-coverts brown and buff. Underparts whitish, washed pale buff, especially on breast and flanks, breast and sides of throat with heavy blackish brown streaks, malar stripe blackish brown. Primaries and secondaries dark brown, narrowly edged pale buff; tertials, greater and median coverts brown, edged whitish; lesser coverts brown edged buff; underwing-coverts and axillaries dusky grey fringed pale buff. Bill dark brown, base of lower mandible pale flesh; eye blackish brown; legs and feet pale brownish flesh. Sexes alike. SIZE: (10 ♂♂, 10 ♀♀, Europe) wing, ♂ 83–90 (85·8), ♀ 82–89 (85·0); tail, ♂ 57–60 (58·6), ♀ 53–57 (55·0); bill, ♂ 15–17 (15·8), ♀ 15–17 (15·7); tarsus, ♂ 21–23 (21·9), ♀ 21–23 (22·3). WEIGHT: unsexed (n = 51, Nigeria Oct–Feb) 18–24 (21·4), (n = 108, Ethiopia Oct–Feb) 18·8–24·8 (21·6), (n = 35, Kenya Nov–Feb) 18·2–25·3 (21·7), (n = 39, Kenya Mar–Apr) 19·7–33 (23·6), (n = 13, Ethiopia Mar–Apr) 19·2–33·1 (22·4), (n = 46, Nigeria Mar–May) 21–39 (28·9), (n = 51, Morocco Mar–May) 15–23 (18·2).

Complete moult July–Aug in Palearctic; in Africa, body plumage, tertials and central and outer tail-feathers moulted Jan–Mar. Hind claw short and curved, 8–10. Wing-tip bluntly pointed; P8–P9 longest, P7 0–2 mm shorter, P6 3–6 mm shorter; P7–P8 emarginated on outer web.

IMMATURE: first winter bird like adult.

Field Characters. A sleek, medium-sized pipit with distinctly streaked olive-tinged brown upperparts, pale buffy underparts with strong blackish streaks across breast, white outer tail-feathers, narrow blackish malar streak and prominent buff wing-bars. Similar in size to Meadow Pipit *A. pratensis* and Red-throated Pipit *A. cervinus*, but slightly more attenuated; more warm olive and buff with thicker bill, pinker legs, streaking below more confined to breast, and more noticeable pale eye-ring. Easily separated from Meadow Pipit, and with experience from Red-throated Pipit, by voice. Smaller than Richard's Pipit *A. novaeseelandiae*, more olive, with shorter legs and bill and different call. Distinguished from Meadow, Red-throated and Richard's pipits in hand by short hind claw; also by shorter 6th primary (**A**, p. 217).

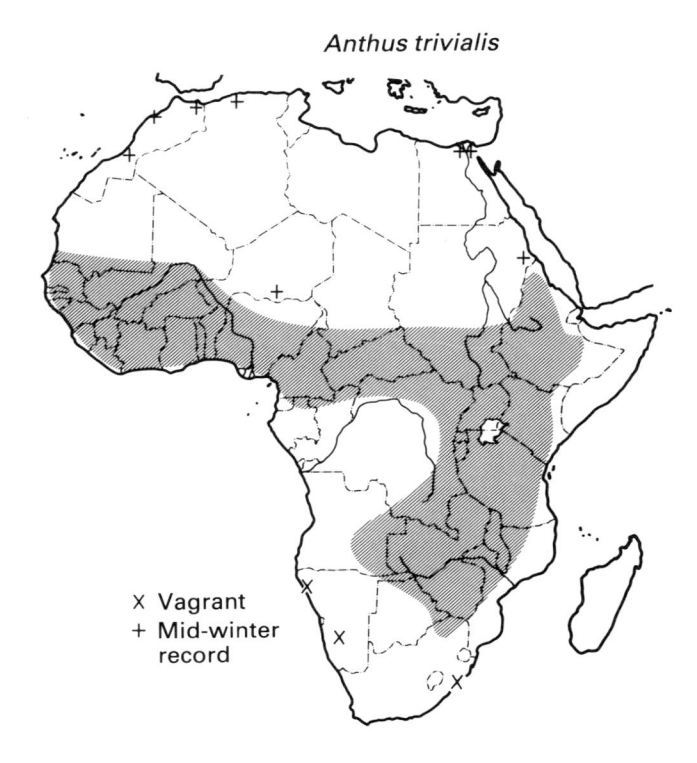

Voice. Tape-recorded (GIB – 78, B, C). Song a loud canary-like chatter, lasting c. 5–8 s, comprising several contrasting sections 'chew-chew-chew-chew . . . chi-chi-chi-chi-chwer . . . chwee-chwee-chwee . . .' and ending with three drawn-out notes, '. . . chee-er-chee-er-chee-er'. Calls, single 'teez' or 'tseep' in flight, usually given when flushed, shorter and more nasal than similar call of Red-throated Pipit; high-pitched alarm, 'sip-sip-sip'.

General Habits. Frequents forest edge, open woodland and wooded savanna. In W Africa found mainly in guinea savanna, forest-grassland mosaic, forest clearings and derived habitat. In E and southern Africa most common in montane habitat, including forest edge, cultivation, plantations and grassy hillsides with scattered bushes; also ranges to lower, drier, open woodland. In central and E Africa mainly at 1500–3000 m; in southern Africa at 450–2400 m.

Not markedly gregarious. Usually in ones and twos or small groups, but may gather in larger parties to roost in tussocky grass. Rather unobtrusive. Forages on ground in low-bellied posture, but carriage rather less horizontal than Meadow or Red-throated Pipits. Gait a deliberate walk, with occasional short runs; frequently wags tail during pauses. Settles habitually on low tree branches, to which it generally flies when disturbed. Flight strong, bounding. Song given from perch or in song-flight, in which it rises at angle of *c*. 50° to *c*. 30 m, beginning song near top of ascent, then descends in parachute glide with wings horizontal, tail raised and legs dangling.

Birds from Europe, including Finland and NW USSR, move southwest to Iberia and W Mediterranean basin. N African passage data, however, indicate broad-front movement to and from Africa. First birds Aug, but main passage protracted, Morocco mid-Sept to late Oct, Egypt early Sept to early Nov. Autumn movement relatively light, with only 1 European recovery (see below), suggesting many birds overfly Mediterranean area as well as Sahara. South of Sahara, passage evident coastal Mauritania (Oct), sometimes several thousands; N Nigeria (Sept–Oct), Chad (from mid-Sept, mainly early- to mid-Oct), N Gabon (Nov), Rwanda and Kenya (Oct to early Dec). Arrives winter quarters W Africa Oct, and Kenya south to Zambia mainly Nov. Vacates winter quarters throughout continent mid-Mar to mid-Apr. First spring birds appear N Africa late Feb, but main movement Mar to mid-May (peak mid-Apr), much heavier than in autumn except in Mauritania, where only 'dozens' Mar–May (Lamarche 1988). Birds recovered Morocco (4, spring, ringed France, Belgium, Netherlands and West Germany; 1, autumn, ringed W Germany); Algeria (1 early Jan, ringed Denmark; 3, spring, ringed Britain, Belgium and Finland); Tunisia (3, spring, ringed France, Belgium and Finland); Mauritania (1, Oct, ringed Britain); and Ivory Coast (1, Dec, ringed Belgium). Birds ringed Tunisia (Apr) recovered Italy (1) and USSR (1, *c*. 57°N, 29°E). Birds wintering E and southern Africa, and on passage in Egypt, are presumably from SW Asia or USSR north and northeast of Caspian Sea.

Food. Chiefly insects; also spiders; seeds. Winter diet W Africa includes grasshoppers, beetles and millet grains; in Zimbabwe, mainly beetles, also moths, bugs (Hemiptera) and termites.

Reference
Cramp, S. (1988).

Anthus pratensis (Linnaeus). Meadow Pipit. Pipit des prés.

Alauda pratensis Linnaeus, 1758. Syst. Nat. (10th ed.), p. 166; Sweden.

Plate 13
(Opp. p. 240)

Range and Status. Breeds W and N Europe to W Siberia; also Iceland and SE Greenland; winters W Europe, Mediterranean, Africa and SW Asia.

Palearctic winter visitor. Winters north of Sahara, in Morocco (south to High Atlas and Oued Sous), N Algeria (south to borders of Sahara), N and central Tunisia, coastal Libya (Tripolitania and locally Cyrenaica) and N Egypt (Nile delta, Faiyoum, Wadi el Natrun, Suez Corridor); common to very abundant Morocco and Algeria, frequent Tunisia, common Libya and Egypt; also coastal Mauritania (south to Nouakchott) and inland in Tagant and Karakoro; occurs in desert oases in Algeria (south to the Hoggar, where common), Libya (south to Sebha and Serir) and Egypt (south to Dakhla and along Nile to Asyut).

Description. *A. p. pratensis* (Linnaeus): throughout range except Ireland and W Scotland; only subspecies in Africa. ADULT ♂: upperparts greenish olive-brown, top of head to mantle, scapulars and back broadly streaked blackish brown, rump and uppertail-coverts almost uniform. Tail dark brown, central feathers fringed greenish olive, outer pair with whitish outer web and large white distal wedge on inner web, T5 with smaller white distal wedge on inner web and white tip to outer web. Supercilium pale greyish or yellow; ear-coverts brown or yellowish brown. Underparts whitish, washed grey or yellow-buff, sides of throat, breast and flanks with broad blackish brown streaking; narrow malar streak blackish. Primaries and secondaries blackish brown, narrowly edged greenish or olive; tertials and upperwing-coverts with pale olive or whitish edges and tips; underwing-coverts and axillaries buffish white. Bill dark brown, base of lower mandible pale brownish flesh; eye blackish brown; legs and feet pale yellowish brown or flesh-brown. Sexes alike. SIZE: (10 ♂♂, 10 ♀♀, Europe) wing, ♂ 76–83 (78·9); ♀ 75–81 (77·9); tail, ♂ 54–57 (55·7), ♀ 52–56 (54·1); bill, ♂ 14–17 (15·6), ♀ 14–16 (15·4); tarsus, ♂ 21–23 (21·8), ♀ 21–23 (22·1). WEIGHT: unsexed (n = 13, Malta autumn and winter) 14·5–22 (17·5).

Complete moult July–Sept in breeding area; in Africa, body plumage, tertials and some tail-feathers moulted Jan–Mar. Hind claw fairly long and weak, 9–15. Wing-tip bluntly pointed; P7–P8 longest and equal, P9 0–1 mm shorter, P6 0–2 mm shorter; P6–P8 emarginated on outer web.

IMMATURE: first winter like adult.

Field Characters. A medium-sized pipit with streaked brown upperparts, pale buff or whitish underparts with streaked breast, and white outer tail-feathers. Upperparts more olive- or greenish-tinged than Tree Pipit *A. trivialis*, Red-throated Pipit *A. cervinus* or Water Pipit *A. spinoletta*. Otherwise plumage most like Tree Pipit; underparts duller, less buffy, with streaks narrower and extending more to flanks, legs usually browner, bill finer, face pattern less distinct; in hand, shows longer, straighter hind claw and longer 6th primary. Best told from Tree and Red-throated Pipits by call. Much less heavily streaked below than non-breeding Red-throated Pipit, rump unstreaked. For distinctions from Water Pipit and European Rock Pipit *A. petrosus*, see those species.

Voice. Tape-recorded (62, 73, 76, 89, B). Call a diagnostic thin, squeaky 'tsip' when flushed, usually repeated rapidly 3 or more times.

General Habits. Frequents open country; in N Africa, seashores, salt- and freshwater marshes, cultivation, olive groves, forest glades and edges, tussocky pasture and moorland up to above 3000 m. Usually in small parties or loose-knit flocks; sometimes hundreds together. Forages on ground with distinctive creeping gait, moving forward slowly and steadily on flexed legs, picking insects from vegetation; only occasionally walks or runs in more upright posture. Occasionally pounces on low-flying or disturbed insects. Wags tail but less frequently than Tree Pipit.

Flight fluttering with hesitant, jerky action, usually high and erratic when bird is flushed. Perches freely on posts, wires and bushes and even in trees, but less habitually than Tree Pipit.

Resident or partial migrant in W Europe, but N and E populations are all medium-distance migrants (Cramp 1988). Autumn movements within Europe are southwest, birds from NW Europe migrating towards Iberia, and many from central Europe and east of Baltic into Italy. N African wintering extends along Mediterranean coastal belt, but penetration into Africa greatest in W, especially Morocco and Mauritania. Birds arrive NW Africa Oct–Dec, Egypt from late Sept, and remain until Mar to early Apr (latest Egypt 19 Apr). Marked passages on Mediterranean coasts Morocco and Algeria Oct to early Dec and late Feb to Mar. Birds recovered Morocco (over 50, ringed Iceland, Britain, France, Belgium, Netherlands, W Germany and Denmark); Algeria (several, ringed Britain, Belgium, Netherlands, Switzerland, Finland and Lithuania); and Tunisia (1, ringed Lithuania). Birds wintering Libya and Egypt presumably all from USSR.

Food. Insects and their larvae (mainly flies, beetles and ants), spiders, small earthworms and snails; occasionally seeds.

Reference
Cramp, S. (1988).

Plate 13
(Opp. p. 240)

Anthus cervinus (Pallas). **Red-throated Pipit. Pipit à gorge rousse.**

Motacilla cervinus Pallas, 1811. Zoographica Rosso-Asiat., p. 511; Kolyma, Siberia.

Range and Status. Breeds N Scandinavia east through arctic USSR to Bering Strait; winters Africa, S and SE Asia.

Palearctic winter visitor. In N Africa, winters Egypt (throughout but mainly Nile delta and valley and Mediterranean and Red Sea coasts, common to abundant); NW Libya and SE Tunisia (Kebia, Medenine, Gabès; frequent to common near coast); in small numbers W Morocco; rarely Algeria. South of Sahara, winters mainly in a belt from S Mauritania (north to Nouadhibou) and Senegambia east through central and S Mali, Ivory Coast (south to Abidjan and Adiopodoumé), N Ghana, S Niger, N Nigeria (south mainly to Zaria and Jos Plateau; also coast in W), S Chad, and central, S and coastal Sudan to Ethiopia, and south to Kenya (largely W and S but also drier N and E in vicinity of permanent water); frequent to common in W, locally common to very abundant from Niger eastwards. Small numbers also winter south to Sierra Leone, coastal Ghana, SW Nigeria, NE Zaïre, Rwanda, Uganda and NE Tanzania (south to Dar-es-Salaam); also S Somalia (south of 6°N). Vagrant Zambia (3 records Ndola) and South Africa (1, Natal: Umvoti R. mouth). On passage, frequent Morocco, Algeria and Tunisia (including desert oases, mainly spring); common Libya (especially coast but also desert oases); widespread, often very abundant, Egypt, Sudan and Chad. Spring movement evident also in Senegambia and coastal Mauritania, heavy in N Somalia and coastal Ethiopia.

Description. ADULT ♂ (breeding): entire upperparts including rump and uppertail-coverts pale brown, broadly streaked blackish brown. Tail blackish brown, central feathers fringed buff, outer pair dusky white on outer web and distally on most of inner web, T5 tipped white on inner web. Lores and ear-coverts buffish brown; supercilium and chin to throat and upper breast deep buffish pink, upper breast often with a few narrow blackish streaks, sometimes more heavily streaked; rest of underparts buffish white, boldly streaked blackish on lower breast and flanks. Primaries and secondaries blackish brown, narrowly edged buff; tertials, greater and median coverts dark brown tipped and edged buffish white, lesser coverts dark brown edged paler brown; underwing-coverts and axillaries greyish white. Bill dark brown, base of lower mandible yellowish flesh; eye dark brown; legs and feet yellowish or brownish flesh. ADULT ♂ (non-breeding): like breeding ♂, but supercilium and chin to breast buffish, sometimes with pink tinge; whole breast heavily streaked and sides of throat bordered by blackish malar streak. ADULT ♀ (breeding): like breeding ♂, but pink on breast and head usually less intense and extensive, upper breast usually buff and more heavily streaked. ADULT ♀ (non-breeding): like non-breeding ♂, but lacking pink tinge on supercilium and throat. SIZE: (10 ♂♂, 10 ♀♀, N Europe and Siberia) wing, ♂ 82–88 (85·2), ♀ 79–86 (82·4); tail, ♂ 54–59 (55·9), ♀ 50–55 (53·3); bill, ♂ 14–16 (15·2), ♀ 14–16 (15·0); tarsus, ♂ 21–23 (21·7), ♀ 20–22 (21·0). WEIGHT: unsexed (n = 34, Ethiopia Oct–Feb) 17·2–21·8 (19·3), (n = 54, Kenya Oct–Feb) 16·7–21·5 (19·3), (n = 101, Ethiopia Mar–Apr) 16·4–29·3 (20·9), (n = 22, Kenya Mar–Apr) 16·9–27·6 (21·0).

Complete moult Aug–Sept in Palearctic; in Africa some pink feathers grown on head Dec–Jan, but main moult of body and head feathers, coverts, tertials and central and outer tail-feathers Feb–Mar. Hind claw long, rather weak and straight, 9–13. Wing-tip bluntly pointed; P7–P9 longest and about equal, P6 1–4 mm shorter; P7–P8, emarginated, P6 usually slightly so.

IMMATURE: first winter like non-breeding adult ♀.

Field Characters. A medium-sized pipit, size of Tree Pipit *A. trivialis* and Meadow Pipit *A. pratensis*, with pronounced streaks on brown upperparts, heavy blackish streaks on white underparts and white outer tail-feathers. Distinctive breeding plumage acquired in Africa, starting in Jan with pink on throat, and by Mar–Apr face to breast bright, almost uniform orange-pink. In non-breeding plumage distinguished from Tree and Meadow Pipits by streaked rump, bolder streaking on underparts, broader white edges to wing-coverts and by call. Bill thinner than Tree Pipit, tail rather shorter, legs browner and general appearance more brown and white (rather than olive and buff); hind claw longer and straighter (**A**, p. 217).

Voice. Tape-recorded (McVIC – 78, 89). Call a high-pitched 'tseeeaz', more drawn out and more piercing than call of Tree Pipit; typically given in flight and usually when flushed, but occasionally from ground.

General Habits. Frequents wet, grassy edges of lakes, dams and pools, cattle-trampled mud, short grassland, ditches, irrigation, cultivation, seashores and high altitude moorland. In E Africa, ranges commonly to 2500 m, occasionally to above 3000 m.

Quite gregarious. Usually in small loose-knit groups, but sometimes in concentrations of many hundreds. Often associates in wet habitat with Yellow Wagtails *Motacilla flava*, and may roost communally with them. Forages on ground with rather horizontal posture and creeping walk like Meadow Pipit, probing among vegetation; often makes short runs; wags tail when pausing. Tends to move behind cover when disturbed and flushes late. Often flies high and for some distance, flight less bounding than Tree Pipit, less jerky than Meadow Pipit. May perch on bushes, fences and occasionally trees.

The westernmost populations, from N Scandinavia, migrate mainly east of Baltic, but some birds through S Sweden. Large numbers occur on passage central and E Mediterranean, but few further west. A number winter NE Africa north of Sahara, but majority pass on to sahelian and soudanian savanna belts south of the desert, and to E Africa. Main influx occurs Egypt late Sept to mid-Nov; passage Libya mid-Sept to Nov; Sudan coast from mid-Sept. Concentrations in sahel tend to move south as floods and pools disappear Oct–Jan. Reaches S Sudan late Oct, Kenya from late Oct but mainly Nov, Ivory Coast from Nov, but N Ghana and N Nigeria mainly Jan. Main departure from tropical wintering areas occurs late Mar to mid-Apr, with last birds south of Sahara early May. Spring movements begin Egypt late Feb and peak late Mar to mid-Apr, with latest birds 17 May and 7 June; passage Libya Apr to mid-May, Tunisia late Mar to mid-May. Many Scandinavian birds presumably winter in W or central Sahel. Birds recovered Apr NE Algeria (1, ringed Finland) and Tunisia (1, ringed Sweden).

Food. Insects and their larvae (especially ants, beetles, flies (Diptera)), small worms, freshwater molluscs and grass seeds.

Anthus cervinus

Reference
Cramp, S. (1988).

238 MOTACILLIDAE

Plate 13
(Opp. p. 240)

Anthus spinoletta (Linnaeus). Water Pipit. Pipit spioncelle.

Alauda spinoletta Linnaeus, 1758. Syst. Nat. (10th ed.), p. 166; Italy.

Forms a superspecies with *A. petrosus* and N American *A. rubescens*.

Range and Status. Breeds in mountains of central and SW Europe, the Caucasus and central Asia; winters W Europe, Mediterranean basin including N Africa, and S and E Asia.

Palearctic winter visitor. Winters north of Sahara: in Morocco, common near E coast, frequent Atlantic coast south to Oualidia, uncommon Oued Sous and higher terrain in Middle and High Atlas; in Algeria, locally common near coast (mainly Oranie), frequent inland south to Saharan Atlas with records on borders of Sahara from Ghardaia, Ouargla, Touggourt and El Oued; frequent near coast in Tunisia and W Libya (Tripolitania); in Egypt, locally common in N (Nile delta, Wadi el Natrun, Faiyoum, Suez Corridor), frequent Bahig oasis, uncommon along Nile south to Aswan and on Red Sea coast.

Description. *A. s. spinoletta* (Linnaeus): breeds in mountains of central and SW Europe; winters Europe and NW Africa. ADULT ♂ (breeding): entire upperparts greyish brown. Tail blackish brown, outer feathers with whitish outer web and distal part of inner web, T5 with small white tip. Supercilium broad and whitish; lores and ear-coverts greyish brown. Underparts pale pinkish buff, usually unstreaked on breast. Primaries, secondaries and primary coverts blackish brown, narrowly edged warm brown; tertials and upperwing-coverts broadly edged warm brown; underwing-coverts and axillaries white. Bill blackish brown; eye blackish brown; legs dark or blackish brown. ADULT ♂ (non-breeding): upperparts warm brown with indistinct darker streaking (although rump plain); underparts whitish with dark brown streaks across breast, extending to flanks; dark brown malar streak. ADULT ♀: like ♂, but breeding plumage less pinkish below and frequently with a few breast streaks. SIZE: (10 ♂♂, 10 ♀♀) wing, ♂ 89–93 (91·1), ♀ 81–91 (85·3); tail, ♂ 60–68 (63·3), ♀ 58–62 (59·5); bill, ♂ 17–18 (17·4), ♀ 16–18 (17·1); tarsus, ♂ 23–25 (24·2), ♀ 23–24 (23·4). WEIGHT: unsexed (n = 11, Malta winter) 19·5–23 (21·8).

Complete moult July–Sept in breeding area; breeding plumage acquired by partial moult Jan–Mar. Hind claw longish, weak, rather curved, 10–14. Wing-tip bluntly pointed; P6–P9 longest and about equal, P5 9–13 mm shorter; P6–P8 emarginated on outer web.

IMMATURE: first winter like non-breeding adult except upperparts browner, with mottling more obvious; streaking on underparts slightly darker.

A. s. coutellii (Audouin): breeds mountains of E Turkey to N Iran; winters S Asia and Egypt. Breeding plumage paler, sandier grey above than nominate, more strongly vinous and whitish below. Smaller: wing, ♂ (n = 10) 85–90 (88·0).

Field Characters. A largish pipit with dark legs, white outer tail-feathers and broad whitish supercilium. Breeding plumage (seen in Africa Mar–Apr) distinctive, with plain brownish grey upperparts and whitish underparts with pink flush to breast. In non-breeding plumage, has mottled brown upperparts and whitish underparts with streaked breast. Larger and less olive than Meadow Pipit *A. pratensis*; also distinguished from this and Red-throated Pipit *A. cervinus* by more pronounced supercilium, larger bill, longer and darker legs, fuller tail and more upright stance. Much less boldly streaked than Red-throated Pipit. Call distinctive (see below). Tawny Pipit *A. campestris* is also plain above, but paler, more sandy, with shorter, pale legs and more horizontal stance. For distinctions from similar European Rock Pipit *A. petrosus*, see that species.

Voice. Tape-recorded (62, 73, 76, 89, B). Call a sharp 'phisst', similar to that of Meadow Pipit, but lower, less feeble and squeaky, and usually given singly.

General Habits. Winters on coastal saltings, by lagoons, along wadis and in wet cultivation and marshy habitat inland up to 2000 m. Usually solitary, but forms loose-knit groups for feeding, especially in late winter and spring. Forms flocks for roosting; up to several hundred together in reeds and tall grass, Egypt. Feeds on ground, occasionally leaping to catch flying insects. Frequents open mud more than Meadow Pipit and feeds on floating vegetation. Hops, walks and runs, gait less creeping than in Meadow Pipit. Wags tail frequently and strongly when alarmed. Rather shy. Typically flushes at c. 40–50 m range in rather zigzag fashion; longer flights quite high. Flight smooth; lacks hesitant flutter of Meadow Pipit, but not bounding like larger pipits (Cramp 1988).

Nominate birds disperse in autumn from mountains of Europe. Some move to NW Europe, but many migrate southwest to W Mediterranean basin; a proportion of these reach NW Africa. Early arrivals Morocco and

Algeria late Aug, but main influx from mid-Oct. Departure Mar to mid-Apr with late birds to early May. *A. s. coutellii* wintering in Egypt presumably come from mountains of Turkey; main arrival from early Oct (earliest 20 Sept); departure Mar to mid-Apr (latest 14 May).

Food. Insects (especially beetles, flies (Diptera), bugs (Hemiptera) and larvae of Lepidoptera), amphipods, spiders, freshwater snails, annelids; also algae and other plant material.

Reference
Cramp, S. (1988).

Anthus petrosus (Montagu). European Rock Pipit. Pipit maritime.

Plate 13
(Opp. p. 240)

Alauda petrosa Montagu, 1798. Trans. Linn. Soc. Lond. 4, p. 41; Coast of Wales.

Forms a superspecies with *A. spinoletta* and N American *A. rubescens*.

Range and Status. Breeds coasts of Britain and Ireland, Faroes, NW France, Fenno-Scandia and NW USSR; winters W European coasts south to Iberia; rarely to Italy, Malta and Morocco.

Palearctic winter visitor. Rare NW Moroccan coast (Tangier–Oualidia).

Description. *A. p. littoralis* (C. L. Brehm): breeds Fenno-Scandia (including Danish islands) and NW USSR; probably this race Morocco. ADULT ♂ (breeding): upperparts dark olive-grey, head lightly streaked, mantle with indistinct darker mottling, rump plain, more olive. Tail dark brown, central feathers with broad olive-brown fringes, outer pair with pale greyish outer web and tip to inner web, T5 with small grey spot at tip. Supercilium creamy buff; lores and ear-coverts grey-brown. Chin creamy; rest of underparts pinkish buff, breast and flanks with faint brown streaks; faint brown malar streak. Primaries, secondaries and primary coverts blackish brown, narrowly edged olive-grey; tertials and greater coverts blackish brown, broadly edged olive-grey; median coverts blackish with olive-grey tips; lesser coverts dark grey with broad olive-grey tips. Underwing-coverts and axillaries dusky grey, fringed creamy. Bill blackish brown; eye blackish brown; legs horn or dark brown. ADULT ♂ (non-breeding): mottling above more pronounced; underparts creamy without pink tinge, chest, flanks and sometimes sides of belly more heavily streaked dusky olive-brown. Sexes alike. SIZE: (10 ♂♂, 10 ♀♀) wing, ♂ 87–92 (89·6), ♀ 80–88 (84·0); tail, ♂ 58–65 (60·7), ♀ 54–59 (56·5); bill, ♂ 17–19 (18·2), ♀ 17–18 (17·7); tarsus, ♂ 22–24 (23·3), ♀ 22–23 (22·7). WEIGHT: unsexed (n = 59, Netherlands Oct) 18–27 (22·4).

Complete moult in breeding area July–Sept; partial moult Jan–Mar. Hind claw longish, curved, 9–12. Primaries 6–9 longest, P5 9–11 mm shorter; P6–P8 emarginated on outer web.

IMMATURE: first winter like non-breeding adult except upperparts browner, with mottling more obvious; streaking on underparts slightly darker.

Field Characters. A fairly large pipit with mottled dark brown upperparts, streaked buff underparts and dark brown legs. Distinguished from Water Pipit *A. spinoletta* and Meadow Pipit *A. pratensis* by grey, not white, outer tail-feathers. More olive-toned than non-breeding Water Pipit, less white below, with supercilium less pronounced. Larger than Meadow Pipit with longer bill, darker legs and more upright stance; call slightly different.

Voice. Tape-recorded (B, CHA, SEL). Call, like that of Water Pipit, a sharp 'phisst', usually given singly.

General Habits. Coastal, rarely penetrating more than a short distance inland. Winters on rocky shores, estuaries and salt marsh. Usually solitary, but may form small groups on migration. Feeds on ground, on mud or among rocks and seaweed. Walks with fairly upright stance; occasionally runs. Often rather tame. Escape flight usually low and short, often ending in curving sweep (Cramp 1988); easy, without hesitant action of Meadow Pipit.

Scandinavian *littoralis* leave breeding areas and move to coasts of W Europe late Oct to Nov, remaining until Mar. Populations of Britain and Ireland (nominate *petrosus*) basically resident.

Food. Insects (especially flies (Diptera) and beetles), crustaceans, small worms, small fish; also seeds.

Reference
Cramp, S. (1988).

Plate 13

Tree Pipit (p. 233)
Anthus trivialis trivialis

Ad. breeding ♂

Meadow Pipit (p. 235)
Anthus pratensis pratensis

Red-throated Pipit (p. 236)
Anthus cervinus

Ad. ♀ Non-breeding

Short-tailed Pipit (p. 230)
Anthus brachyurus brachyurus

Ad. breeding ♀

Sokoke Pipit (p. 233)
Anthus sokokensis

Bush Pipit (p. 231)
Anthus caffer caffer

A. s. leucocraspedon   *A. s. schoutedeni*

Long-billed Pipit (p. 223)
Anthus similis

A. s. nicholsoni   *A. s. hararensis*

Striped Pipit (p. 242)
Anthus lineiventris

Water Pipit (p. 238)
Anthus spinoletta spinoletta

Ad. non-breeding

Malindi Pipit (p. 229)
Anthus melindae melindae

European Rock Pipit (p. 239)
Anthus petrosus petrosus

Ad. non-breeding

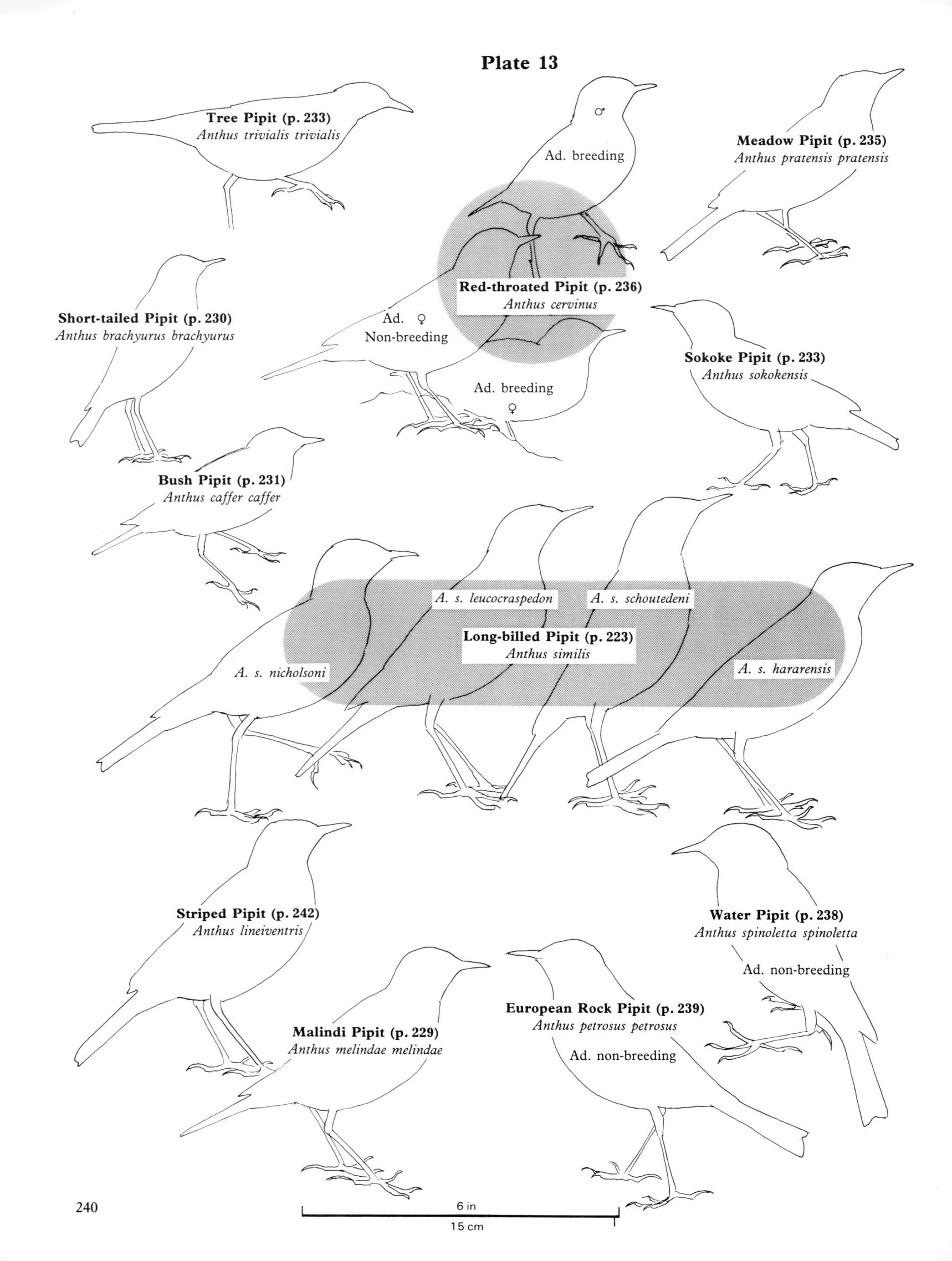

6 in
15 cm

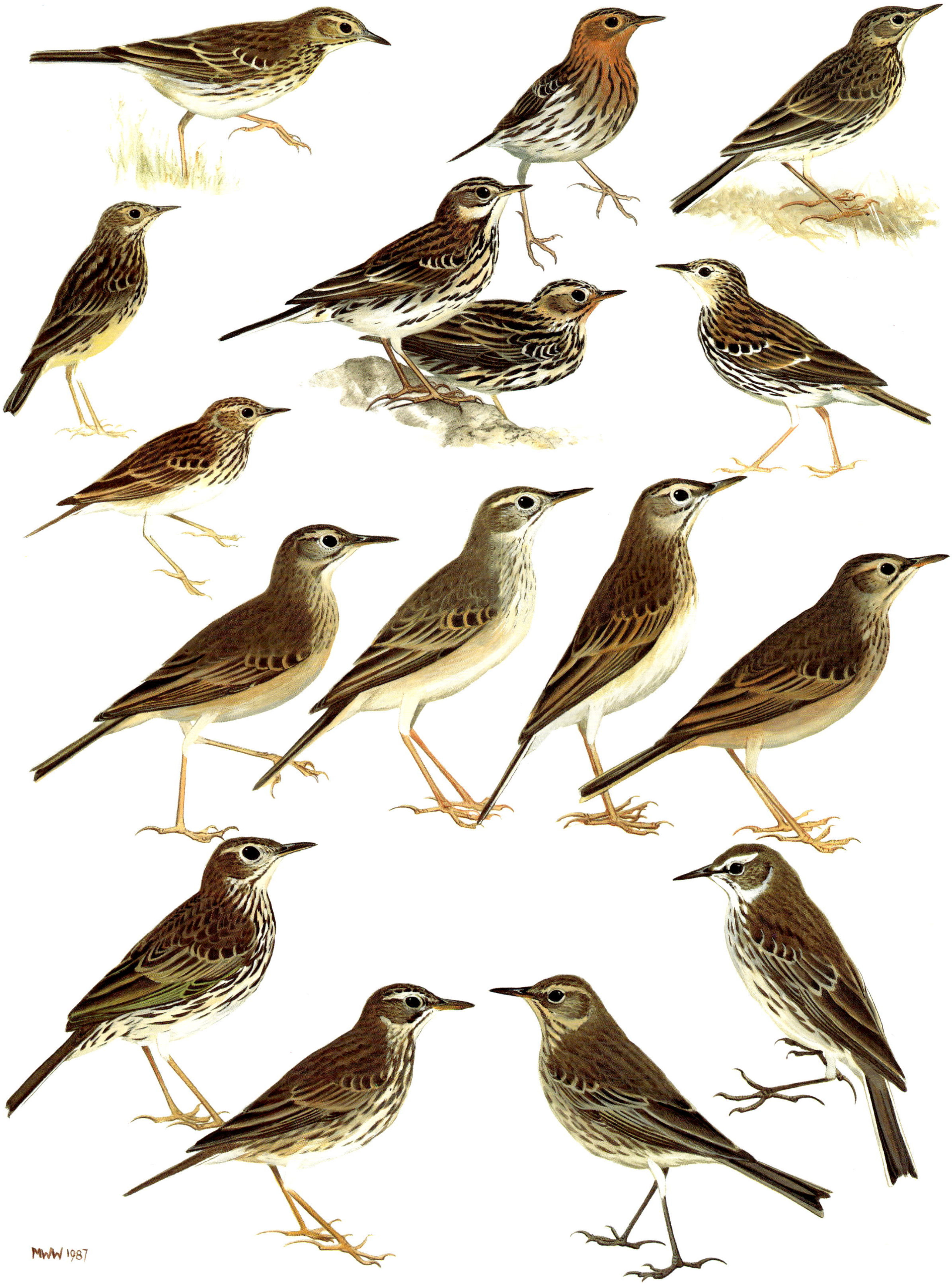
MWW 1987

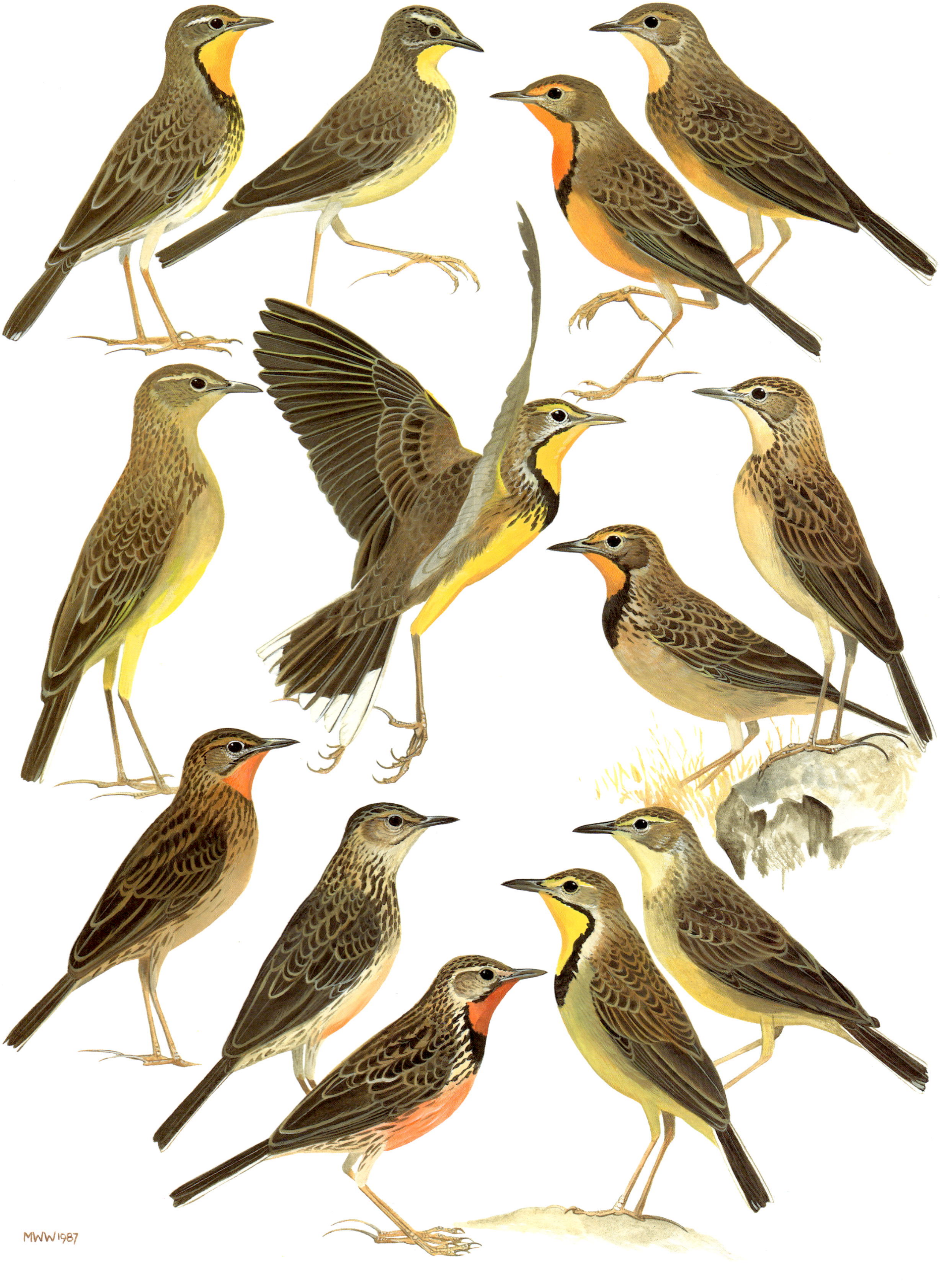
MWW 1987

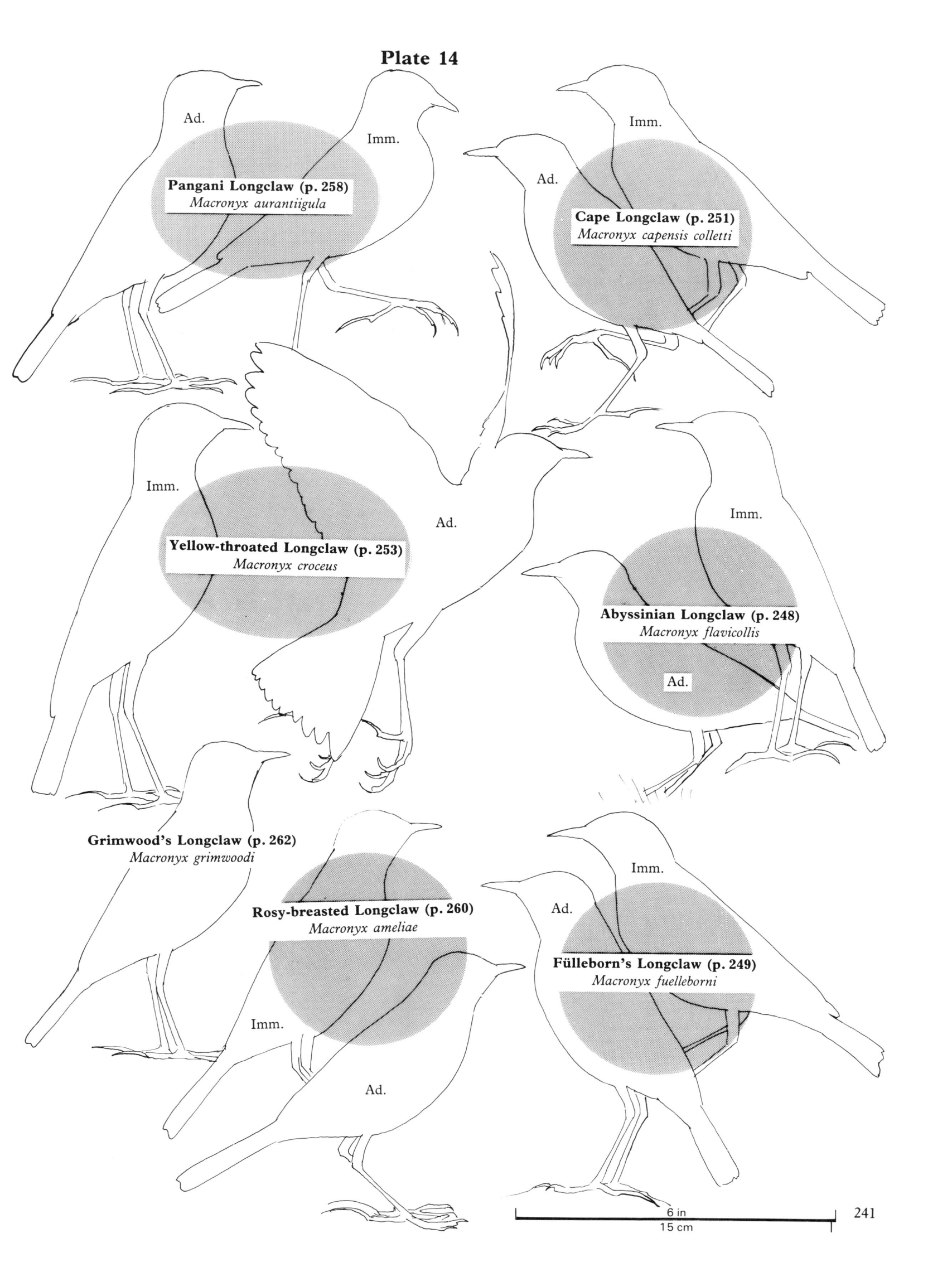

Plate 14

Pangani Longclaw (p. 258)
Macronyx aurantiigula

Cape Longclaw (p. 251)
Macronyx capensis colletti

Yellow-throated Longclaw (p. 253)
Macronyx croceus

Abyssinian Longclaw (p. 248)
Macronyx flavicollis

Grimwood's Longclaw (p. 262)
Macronyx grimwoodi

Rosy-breasted Longclaw (p. 260)
Macronyx ameliae

Fülleborn's Longclaw (p. 249)
Macronyx fuelleborni

6 in / 15 cm

241

Plate 13
(Opp. p. 240)

Anthus lineiventris Sundevall. Striped Pipit. Pipit strié de Sundevall.

Anthus lineiventris Sundevall, 1850. Oefv. K. Sv. Vet.-Akad. Förh. 7, p. 100; Limpopo R.

Range and Status. Endemic resident. Distribution somewhat fragmented: highlands of SE Kenya (Chyulus and Taitas) and E Tanzania (Kilimanjaro, Mbulu escarpment and Pares south to Songea and Tunduru); W Tanzania (Kibondo); SE Zaïre (Upemba Nat. Park); Zambia (eastern N and Eastern provinces, Muchinga and Middle Zambezi escarpments west to Victoria Falls; also E Copperbelt Province, Kafue, N Mwinilunga and Balovale), Malaŵi and highlands of N and W Mozambique; Zimbabwe (E highlands and central plateau north to Zambezi) through E Botswana to South Africa (Transvaal and E Natal), E Swaziland and SW Mozambique (Lebombo Mts); also W Angola (N Malange south to Mt Moco and highlands of W Huila). Local; uncommon to frequent.

Description. ADULT ♂: upperparts pale olive-brown, streaked dark brown from head to mantle and scapulars, less distinctly on back and rump. Tail dark brown, outer feathers white on outer web and all except base of inner web, T5 with white distal third to inner web, T4 with white tip to inner web. Supercilium long, narrow and buffish; lores and ear-coverts dark brown, flecked buff; malar stripe narrow. Underparts pale buff with broad dark brown streaks across breast, finer streaks at sides of throat and long fine streaks on flanks, sides of belly and undertail-coverts. Primaries, secondaries, primary coverts and alula dark brown edged yellowish green; tertials and rest of upperwing-coverts dark brown, edged and tipped olive-brown.

Underwing-coverts and axillaries greyish buff with darker streaking, tinged bright yellowish in carpal region. Bill dark brown above, horn below; eye brown; legs and feet light flesh brown. Sexes alike. SIZE: (10 ♂♂, 10 ♀♀) wing, ♂ 85–90 (87·5), ♀ 82–91 (83·9); tail, ♂ 63–69 (66·5), ♀ 59–67 (63·4); bill, ♂ 19–21 (20·1), ♀ 18–20 (19·0); tarsus, ♂ 28–30 (29·3), ♀ 26–29 (27·5). WEIGHT: unsexed (n = 12, Zimbabwe and South Africa) 33–38 (35·0).

Hind claw short and curved. 9–10. Wing blunt, P6–P9 longest, P5 1–3 mm shorter; P5–P8 emarginated on outer web.
IMMATURE: paler than adult, more spotted above.
NESTLING: unknown.

Field Characters. A large olive-brown pipit with long bill and strong legs, long streaks over most of underparts and white outer tail-feathers. Yellowish green edging to wing- and tail-feathers diagnostic but not easy to see in the field. Often the only pipit in its rocky woodland habitat.

Voice. Tape-recorded (75, 81, 88, CHA, STJ, WALK). Largely silent, but song during breeding season loud and melodious; composed of whistled phrases lasting 2–5 s, separated by *c.* 1 s pauses; varied, but phrases usually repeated 2–3 times; 'wip-chew-chew-whitty-chew-chee-chee-wip-chew-chew', and a more extended 'ch-ch-wee-wee-wee-cheew-cheew-cheew . . .'.

General Habits. Confined to rocky terrain with trees. In E Africa, occurs on hillsides at 1000–2000 m; in Zambia, Malaŵi and Zimbabwe, among rocks under *Brachystegia* at 500–2200 m; in South Africa, on rocky ground in thin woodland or thornbush, steep slopes of gorges, hills and mountains; in coastal hinterland of Natal as low as 150 m; mainly replaced by African Rock Pipit *A. crenatus* above 1000 m.

Solitary or in pairs, sometimes in small groups in non-breeding season. Mainly silent and retiring but not especially shy. Forages from ground, among stones and sparse vegetation. Runs rapidly with horizontal carriage and legs well flexed, crossing low, flat rocks. Flies to nearest low tree branches when disturbed. Walks about on branches like Tree Pipit *A. trivialis*. Often perches motionless parallel to branch with bill pointing upward.

Food. Insects, including grasshoppers.

Breeding Habits. Monogamous and territorial. When defending territory, ♂ often sings from low tree branch.
NEST: substantially-built cup of dry grass, twigs and dead leaves, lined with soft dry grass and rootlets; on ground against rock or tuft of grass, partially covered by

overhanging grass, entrance at side; ext. diam. *c.* 120, int. diam. *c.* 90, depth of cup *c.* 50.

EGGS: 2–3, av. (n = 9 clutches, South Africa) 2·8. Creamy white, closely speckled reddish brown, with underlying spots of dark and light grey. SIZE: (n = 29, South Africa) 20–25 × 15–18 (22·5 × 16·4).

LAYING DATES: Tanzania, Nov; Malaŵi, Oct; Zambia (breeding condition Nov); Zimbabwe, Sept–Jan; South Africa: Natal, Sept–Nov, Transvaal, Dec–Jan.

Reference
Maclean, G. L. (1985).

Anthus crenatus Finsch and Hartlaub. African Rock Pipit. Pipit des rochers.

Plate 12
(Opp. p. 177)

Anthus crenatus Finsch and Hartlaub, 1870. Vög. Ost-Afr., p. 275; near Cape Town.

Range and Status. Endemic resident on South African mountains. Distribution patchy, from SW, S and E Cape Province to E and S Orange Free State, W Natal and SE Transvaal; also Lesotho and W Swaziland. Locally frequent to common.

Description. ADULT ♂: upperparts olive-brown, top of head streaked slightly darker, sides of neck greyer. Tail-feathers dark brown, narrowly edged yellowish green, T6 with pale outer edge and whitish tip to inner web. Supercilium prominent, buffish, extending to well behind eye, lores and ear-coverts olive-brown, finely streaked buff. Chin to throat whitish, bordered by narrow dark brown malar stripe; rest of underparts rich buff, with short indistinct brown streaks across breast; undertail-coverts dark brown, broadly tipped buff. Wing brown, primaries, secondaries and alula edged yellowish green, tertials and greater coverts edged olive-green; underwing-coverts and axillaries buff, broadly tipped yellowish green. Bill dark brown, basal half of lower mandible yellowish; eye hazel; legs and feet pale brown. Sexes alike. SIZE: (7 ♂♂, 5 ♀♀) wing, ♂ 87–92 (89·3), ♀ 84–89 (86·8); tail, ♂ 59–65 (62·9), ♀ 60–65 (62·1); bill, ♂ 20–22 (20·9), ♀ 18–22 (20·4); tarsus, ♂ 29–31 (29·7), ♀ 28–30 (29·0).

Hind claw short and curved, 10–12. Wing-tip blunt; P5–P9 about equal; P5–P8 emarginated.

IMMATURE AND NESTLING: unknown.

Field Characters. A large stocky pipit with heavy legs and bill. Plain, with unstreaked upperparts and fine streaks on breast visible only in good light and at close range. Supercilium bold, outer tail-feathers buffish. Richard's Pipit *A. novaeseelandiae*, Mountain Pipit *A. hoeschi* and Long-billed Pipit *A. similis* all have streaked upperparts and breast, whiter underparts. Best told from plain-backed Buffy Pipit *A. vaalensis* and Plain-backed Pipit *A. leucophrys* by rocky habitat and distinctive voice. Yellowish wing-edging and touch of yellow at bend of wing visible only at very close range.

Voice. Tape-recorded (88, GIB, LEM). Song a sweet phrase of two syllables, the second somewhat trilled, 'whee-tsrreeee, whee-tsreeee', repeated several times, falling in pitch on second note; also a slower more quavering 'whee-pr-rreeu' (Maclean 1985).

General Habits. Confined to hills, mountains and lower slopes of escarpments, mainly above 1000 m; prefers open areas with rocky outcrops, boulders and grass clumps or small shrubs (Clancey 1964). Usually occurs singly. Does not consort with other pipits. Shy and unapproachable, easily overlooked when not calling. Forages on ground, creeping in grass or running among stones. Flies far when flushed.

Food. Insects; also spiders and seeds.

244 MOTACILLIDAE

Breeding Habits. Monogamous. Maintains large territory, apparently throughout year. ♂ conspicuous in breeding season, singing from regular rock perch in very upright posture with bill pointing skywards; also sings during aerial display.

NEST: cup of grass and roots, lined with finer material, on ground by grass clump or sheltering stone (Maclean 1985).

EGGS: 3–4, white or greyish white with dense brown and grey speckling concentrated at broad end (Maclean 1985). SIZE: (n = 15) 21–22 × 15–16 (21·5 × 15·3).

LAYING DATES: South Africa, Nov–Jan, Transvaal, Dec.

Reference
Maclean, G. L. (1985).

Plate 12
(Opp. p. 177)

Anthus chloris Lichtenstein. Yellow-breasted Pipit. Pipit à gorge jaune.

Anthus chloris Lichtenstein, 1842. Verz. Samml. Kaffernl., p. 13; Kaffirland (= Likwa Vaal and Modder rivers).

Range and Status. Endemic to southern Africa. Resident and partial migrant. Breeds at high altitudes E South Africa and Lesotho on the Stornberg and Drakensberg; from NE Cape Province through E Lesotho, the high interior of Natal and E Orange Free State to SE Transvaal; locally frequent to common Natal and Transvaal, rare to uncommon NE Cape and Lesotho. Some migrate to lower elevations in non-breeding season to reach Natal coast, E Cape Province, Swaziland and Vaal basin; 1 old record S Cape at Swellendam. Vulnerable; habitat threatened by burning and grazing and by afforestation. Fewer recent records of migrants at middle and lower altitudes suggests numbers decreasing.

Description. ADULT ♂ (breeding): top of head and mantle blackish brown with broad yellowish olive edges to feathers producing a broadly streaked effect; nape, rump and uppertail-coverts greyer. Tail-feathers blackish brown, edged buff, T6 white except basal fringes of outer web, T5 tipped white on inner web. Supercilium narrow, yellowish; lores and ear-coverts streaked rich brown, whiter area below eye, bordered below by thin blackish malar streak. Chin to upper belly bright yellow, tinged brown on flanks, usually with a few fine brown streaks on upper breast; lower belly and undertail-coverts greyish white. Primaries, secondaries and primary coverts blackish brown, narrowly edged yellow; tertials and greater coverts broadly edged rich buff; median coverts edged whiter; lesser coverts and carpal region edged yellow; greater underwing-coverts blackish tipped white; lesser underwing-coverts and axillaries yellow. Bill dark horn, base of lower mandible yellowish; eye dark brown; legs and feet yellowish flesh. Sexes alike. ADULT ♂ (non-breeding): differs from breeding ♂ in having feathers of upperparts edged warm buffish brown; supercilium warm buff; lores and ear-coverts more uniform warm buffish brown, malar streak indistinct. Below, chin to throat pale buffish brown; breast warm buffish, usually with a few fine brown streaks; yellow confined to patch on centre of belly. SIZE: (9 ♂♂, 6 ♀♀) wing, ♂ 81–90 (85·0), ♀ 81–85 (82·8); tail, ♂ 61–69 (65·3), ♀ 59–67 (64·2); bill, ♂ 16–17 (16·7), ♀ 16–18 (16·8); tarsus, ♂ 24–26 (24·9), ♀ 24–25 (24·7). WEIGHT: 1 unsexed 25.

Hind claw long, weak, 12–17. Wing-tip blunt; P6–P9 longest and about equal, P5 c. 2 mm shorter; P5–P8 emarginated on outer web.

IMMATURE: blackish brown above, head streaked and mantle and back scalloped creamy buff. Below buffy, breast more dusky and with short fine brown streaks, centre of belly with slight yellow wash. Tertials and wing-coverts with broad and sharply-demarcated creamy buff edges.

NESTLING: unknown.

Field Characters. A fairly large pipit. In breeding plumage the only *Anthus* with yellow underparts, but these are only conspicuous in front views when bird is displaying; easily overlooked as Richard's Pipit *A. novaeseelandiae* when flushed from cover. Shows greyish white face (dark in most other pipits) and fine malar streak. Outer tail-feathers white. Superficially not unlike immature Cape Longclaw *Macronyx capensis*, with similar dark-streaked upperparts, but latter is larger and bulkier with short, stout bill, orange-buff rather than yellow underparts and longclaw flight. Non-breeding bird is warm buffish below with yellow patch on belly, and has warmer brown upperparts and rather plain face. Yellow wing-linings visible when wings raised.

Voice. Tape-recorded (58, 75, F, GIB). Song, a repeated double 'see-chick ... see-chick ...' separated by short pauses, first note sibilant and soft, the 'chick' abrupt and metallic (Vernon 1983); calls, a single whistled 'tseeu' and a rapid continuous 'chick, chick, chick ...' in alarm, quieter than display call.

General Habits. Breeds at 1400–2400 m, in lush flat grassland, also on grassy slopes in mountainous country; frequents areas with tussocks; avoids recently grazed or burnt areas. In non-breeding season some descend to lower-altitude grassland or bush savanna, while others spend winter on snowline (J. C. Sinclair, pers. comm.).

Usually in pairs, but forms small flocks in non-breeding season. Skulking and furtive; creeps through grass and runs across more open spaces. Tends to freeze when approached, or to run, keeping behind cover of tall grass. Flushes only when approached closely and usually flies far.

Most move off highland breeding grounds Apr, returning Sept.

Food. Insects, including mantids and small beetles.

Breeding Habits. Monogamous, territorial. Several pairs may nest close together in suitable sites. Territorial ♂ bolder, perching on grass tufts or other low eminences. Sings from stone or termite mound, or in display flight, in which it climbs to 20–25 m with floppy wing-beats, cruises for about 100 m, then dives vertically into grass (Vernon 1983).

NEST: neat cup of stalks, grass blades and rootlets, lined with fine rootlets and often hair; set into ground under tussock or in grass clump, usually well concealed.

EGGS: usually 3, av. (n = 8 clutches) 2·9. White to greenish white, with brown and lilac-grey spots, often concentrated at broad end. SIZE: (n = 17) 19–22 × 15–17 (20·7 × 15·8).

LAYING DATES: South Africa: Natal, Nov–Dec, Transvaal, Nov–Jan; during summer rains.

References
Brooke, R. K. (1984).
Clancey, P. A. (1985).
Maclean, G. L. (1985).

Genus *Macronyx* Swainson

Most species larger and more robust than pipits, with more upright stance; some more lark-like in appearance. Dorsal plumage cryptic but underparts brightly coloured; 6 species also have blackish necklace across breast; tail variably edged and tipped white. Colour changes caused by rapid abrasion and sun bleaching of plumage, especially upperparts, complicate study of regional plumage variation. Legs and toes long, hind claws very long, enabling birds to walk easily over tussocky grass. Rictal bristles generally pronounced.

Widespread in grasslands over subsaharan region; sympatric species exhibit wider ecological tolerance in areas where no congeners are present. Flight normally jerky, glides interspersed with short periods of flaps. Voices distinctive; whistled melodious calls and simple songs, often given in flight. Bear a strong resemblance, in appearance, habitat, nest and habits, to meadowlarks *Sturnella* (Icteridae) of the New World (Friedmann 1946): a striking example of convergent evolution. Breed mostly during or just after rains; grass cover for nest concealment probably an important factor in timing of breeding.

Endemic; 8 species. We broadly adopt the species groups proposed by Cooper (1985), i.e. (1) *M. flavicollis*, *M. fuelleborni* and *M. capensis* (the last 2 constituting a superspecies), (2) *M. croceus* and *M. aurantiigula* and (3) *M. ameliae* and *M. grimwoodi*. We do not accept Cooper's reinstatement of the genus *Hemimacronyx* to include *M. sharpei* and *Anthus chloris*, and retain *sharpei* in *Macronyx* because it has so many typical longclaw characteristics.

Macronyx fuelleborni superspecies

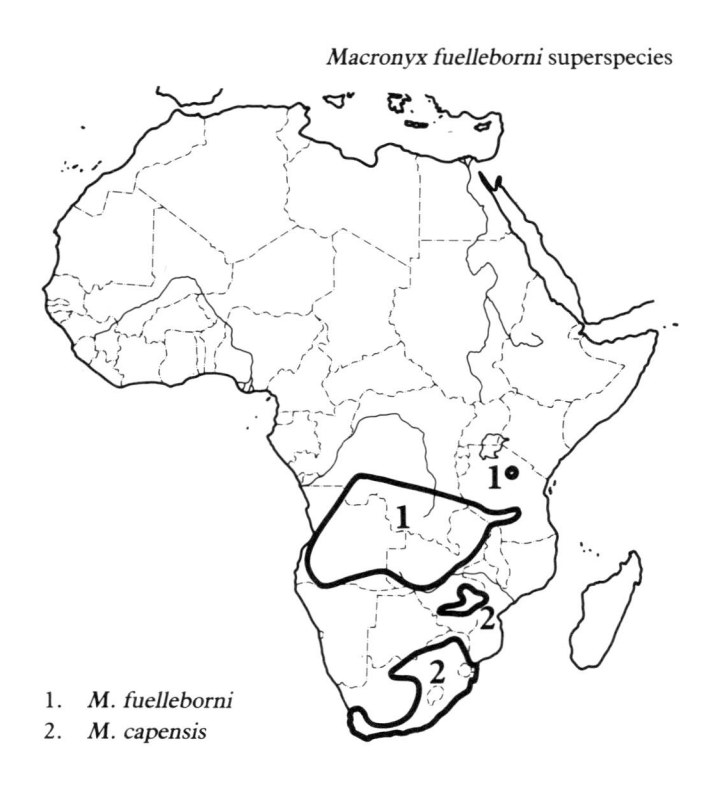

1. *M. fuelleborni*
2. *M. capensis*

Plate 12
(Opp. p. 177)

Macronyx sharpei Jackson. Sharpe's Longclaw. Sentinelle de Sharpe.

Macronyx sharpei Jackson, 1904. Bull. Br. Orn. Club 14, p. 74; Mau Plateau, Kenya.

Range and Status. Resident, endemic to highland grasslands in W and central Kenya, from slopes of Mt Elgon east to highlands on both sides of the Rift, including Mau and Il Kinangop plateaux, Aberdare Mts and Mt Kenya. Widespread and locally common. Not yet threatened but habitat being reduced. At S end of Kinangop Plateau occurs in grassland planted with *Eucalyptus* saplings; such areas will become unsuitable as trees grow. Other grassland habitat in central Kenya may be similarly threatened by afforestation or agriculture.

Description. ADULT ♂: feathers of upperparts brownish black, fringed or edged cinnamon to yellow-ochre giving well-patterned appearance. Forehead to nape with almost equally broad dark and pale streaks; hindneck with smaller dark feather centres and more prominent pale edges; sides of neck pale cinnamon with narrow dark olive-brown streaks. Lores pale cream, tinged deep lemon-yellow; face pale cream; ear-coverts washed cinnamon-brown, rear border darker; supercilium lemon-yellow with small dull olive-brown streaks; narrow moustachial stripe dull dark olive-brown. Feathers of upperparts, including uppertail-coverts and scapulars, fringed pale cinnamon to yellow-ochre. Uppertail-coverts very long and acuminate. Tail-feathers dark olive-brown; T1 fringed buff; T2–T4 narrowly edged buff on outer web, T4 with small white tip; distal third of T5 white with dark olive-brown shaft streak, proximal half of outer web narrowly edged pale yellow; T6 white except for dark olive-brown shaft streak and proximal third of inner web (**A**). Chin to centre of belly deep lemon-yellow; line of small dark spots below moustachial stripe, and breast-band of blackish streaks; flanks cinnamon with brownish black streaks; thighs, sides of belly and undertail-coverts pale yellow to pale horn with olive-brown streaks. Primaries olive-brown, outer webs narrowly edged pale yellow, tips pale when fresh. Secondaries dark olive-brown, inners narrowly fringed buff, outers with pale yellow edges to outer webs; tertials brownish black, fringed buff to cinnamon. Upperwing-coverts dark olive-brown with pale fringes; marginal coverts, lesser coverts and alula fringed deep lemon-yellow, median and greater coverts fringed pale cream-white to yellowish buff. Axillaries deep lemon-yellow; underwing-coverts white with olive-brown centres; underside of remiges pale olive-brown. Upper mandible pale brown to blackish; lower mandible usually paler, often pale horn, greyish or pale brown, with dark tip; cutting edges pale; gape pale horn; eye brown to very dark brown; legs and feet horn, pale to deep flesh, brownish flesh or yellow-ochre. Hind claw less curved than in other longclaw species. ADULT ♀: similar to ♂ but yellow of underparts duller and variably tinged pale buff on breast; supercilium buff, tinged yellow; white in tail duller. SIZE: (27 ♂♂, 9 ♀♀) wing, ♂ 84–92 (88·0), ♀ 83–86 (84·4); tail, ♂ 58–71 (64·5), ♀ 57–66 (61·6); bill, ♂ 15–19 (17·6), ♀ 16–18 (16·9); tarsus, ♂ 27–32 (30·0), ♀ 26–31 (29·1); hind claw, ♂ 13–19 (14·8), ♀ 13–16 (14·7).

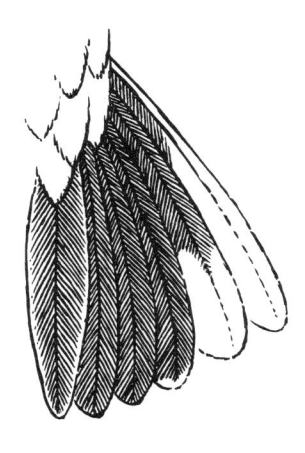

IMMATURE: upperparts paler than adult: feather centres dull olive-brown to brownish black, edged or fringed pale horn to buff; supercilium buff. Underparts dull whitish cream with pale yellow wash only on lower breast and centre of belly; breast heavily spotted dull olive-brown. All upperwing-coverts fringed whitish cream.

JUVENILE: like immature but feathers of upperparts sometimes fringed paler, whitish buff.

NESTLING: unknown.

Field Characters. The smallest longclaw, shape more slender and pipit-like, bill smaller and finer. Tail proportionately as long as Rosy-breasted Longclaw *M. ameliae* but long wings make it appear relatively short. Adults distinguished from Yellow-throated Longclaw *M. croceus* by lack of solid black necklace or heavily-marked dark breast-band, all ages by more boldly patterned upperparts. In flight, separable from Yellow-throated Longclaw by white in tail being mainly at sides (almost completely white T6, white distal third of T5 and tip of T4), from Rosy-breasted Longclaw by lack of noticeably pale area on lesser upperwing-coverts but pale fringes to greaters. Does not overlap with Pangani Longclaw *M. aurantiigula*. Told from sympatric pipits by yellow underparts, voice, and distinctive longclaw flight and display.

Voice. Song tape-recorded (C, F, McVIC). Song of ♂, given in flight or when perched, a thin, rather plaintive series of whistles rising in pitch: 'yo yo teu-tee' or 'yoo si-si-si-si', or a more complex series of variably-pitched notes, e.g. 'tee s-si-si ya yee yo-yo'; also gives repeated thin 'sip' notes. Calls: (1) a high-pitched 'tsit' or 'tsip' (in flight when breeding), (2) a louder, sharper 'tswit', (3) a long, plaintive metallic 'weee', (4) a richer rising 'yooee' or descending 'eeoo', (5) a thin 'eeeu' (P. B. Taylor, pers. obs.). Some notes resemble those of Ayres's Cisticola *Cisticola ayresii*, with which it occurs in central Kenya.

General Habits. Occurs on open, treeless high-altitude grassland and rolling downland, with short, often tussocky, grass up to *c*. 30 cm high, often grazed by domestic stock. Habitat may become moist during rains, with temporary pools or waterlogged patches, and may dry out completely during non-breeding season. Altitudinal range 1850–3400 m.

Occurs singly, in pairs or in family groups. Normally found on ground; shy, reluctant to fly, hiding in grass at approach of observer or escaping by running; uses grass tussocks as observation posts. When flushed often flies low and straight for some distance, then makes long, shallow descent and alights, but sometimes rises steeply, calls and then drops sharply after a very short flight. Easier to approach in cold and windy weather, and singing males perched on fence posts may be approached closely. Flight action irregular, flaps interspersed with short pauses, but not markedly undulating. Forages in grass or on small open patches of ground.

Normally sedentary, but makes very limited seasonal local movements away from areas which become very dry in non-breeding season (P. B. Taylor, pers. obs.).

Food. Insects, especially beetles and grasshoppers.

Breeding Habits. Apparently monogamous; solitary nester. ♂ sings all year but more frequently during breeding season, when singing occurs erratically throughout the day. ♂ normally sings in display flight, rising vertically to a height of 10–45 m and then flying in level circles, alternating equal short periods of flapping flight and gliding; singing begins during each period of flaps and ends during next glide; bird then drops rapidly, with dangling legs, straight down to ground or fence post. Also sings from perch on fence post, sometimes spreading and shivering tail so that white is visible; song may be repeated monotonously for several min. ♂ also makes parabolic flight, rising to $c.$ 20 m and uttering 'sip' calls. Breeding pair circle $c.$ 20 m up with erratic flight action and 'tsit' or 'tsip' calls, then descend to nesting area; also make short rapidly-fluttering flight together, rising only $c.$ 1 m from ground and descending quickly (P. B. Taylor, pers. obs.).

NEST: deep, solidly-constructed cup of dry grass, lined with rootlets and usually well-hidden at base of grass tussock or larger herbaceous plant; also in grass under tiny bush, on side of low anthill or on ploughed land under lump of earth, sometimes in hollow scooped out by bird. Diam. of cup $c.$ 70, depth $c.$ 60.

EGGS: 2–3, av. (n = 18 clutches) 2·2. Subelliptical; pale greenish white to buff, obscurely mottled or finely speckled pale to dark brown, or yellowish brown undermarked with grey; markings often more numerous at blunt end. SIZE: (n = 18) 22–25·5 × 16–17·3 (23·9 × 16·5).

LAYING DATES: Mar–June, Sept–Oct, Dec; breeds during or just after rains.

DEVELOPMENT AND CARE OF YOUNG: on close approach of intruder, adult with newly hatched young flew heavily from nest for short distance and then ran as if injured before flying again (P. H. B. Sessions, E. Afr. Nat. Hist. Soc. nest record card).

Plate 14
(Opp. p. 241)

Macronyx flavicollis Rüppell. Abyssinian Longclaw. Sentinelle d'Abyssinie.

Macronyx flavicollis Rüppell, 1840. Neue Wirbelth. Vögel, p. 102; Simen, Ethiopia.

Range and Status. Endemic to highlands of Ethiopia except extreme N and W, from Simien south to Kaffa and Sidamo. Widespread, locally scarce to abundant; commonest between 1800 and 2750 m. Numbers apparently reduced recently through increase in cultivated land and grazing (Ash and Gullick 1989).

Description. ADULT ♂: feathers of upperparts brownish black, edged or fringed buff to cinnamon, giving well-patterned appearance. Pale feather edges narrow on forehead and crown, broader on nape and hindneck; sides of neck buff to cream. Lores and cheeks whitish to cream; ear-coverts buff to cream; broad eye-stripe olive-brown; supercilium saffron yellow to above eye, buff to cream behind eye. Mantle feathers and scapulars broadly fringed cinnamon or buff; feathers of back, rump and uppertail-coverts with paler centres than rest of upperparts, dull olive-brown, and fringed cinnamon. Tail-feathers dark olive-brown with paler shafts; T1 fringed buff; T2–T5 narrowly edged buff on outer web; T3 with small white spot at tip of inner web; distal quarter of inner web and tip of outer web of T4 white; distal third of inner web and tip of outer web of T5 white; basal half of T6 dark olive-brown, distal half white, outer web often white almost to base (**A**). Moustachial stripe dark olive-brown, continuing as an unbroken necklace across upper breast. Chin, throat, malar region and foreneck deep saffron yellow, enclosed by dark necklace. Sides of breast and lower breast buff to cinnamon; breadth of dark necklace variable, often occupying entire centre of breast; variable blackish brown streaks at sides of breast. Rest of underparts buff to cinnamon, paler on flanks and thighs; centre of belly variably washed yellow. Primaries and secondaries olive-brown, outer webs of primaries narrowly edged pale yellow or yellowish buff, tips whitish when fresh. Secondaries narrowly fringed buff; tertials brownish black with broad cinnamon fringes. Upperwing-coverts brownish black with pale fringes: marginal coverts and alula fringed saffron yellow, lesser coverts fringed white to buff, median and greater coverts fringed buff

A

to cinnamon. Axillaries white; underwing-coverts whitish with olive-brown centres; underside of remiges pale olive-brown, tinged cinnamon. Upper mandible black (occasionally dark brown or dark grey), lower mandible grey or blue-grey; eye dark brown; legs and feet pale brown to mid-brown, tinged flesh or yellow, soles yellow. ADULT ♀: like ♂, but may have saffron yellow of chin to neck paler and duller; less or no yellow on belly. SIZE: (32 ♂♂, 19 ♀♀) wing, ♂ 87–96 (91·0), ♀ 84–91 (87·4); tail, ♂ 51–66 (60·3), ♀ 50–65 (58·8); bill, ♂ 16–20 (17·5), ♀ 16–18 (17·1); tarsus, ♂ 30–34 (32·5), ♀ 27–34 (31·5); hind claw, ♂ 11–20 (14·8), ♀ 12–15 (13·8).

IMMATURE: like adult, but feathers of upperparts with darker tawny brown fringes; chin to neck buff to cinnamon, usually paler than rest of underparts; moustachial stripe and necklace browner, necklace composed of spots.

JUVENILE: like immature, but feathers of upperparts fringed even darker tawny brown, fringes especially broad on tertials. Lower breast, flanks and belly washed deep cinnamon; centre of belly washed yellow.

NESTLING: unknown.

Field Characters. A small longclaw with yellow throat, solid dark necklace, strongly patterned upperparts and more white in outer portion of tail than any except Sharpe's Longclaw *M. sharpei*. No other longclaw occurs within its range. Separated from pipits by longclaw shape, colour of underparts, voice and behaviour.

Voice. Not tape-recorded. Song clear and trilling, call note piping (Mackworth-Praed and Grant 1960).

General Habits. Inhabits open grassland from 1200 to 4100 m. At higher elevations (3800–4100 m) occurs in tussocky grass of moorlands; at 1800 m recorded from short-grassed edge of swamp (Benson 1946). Occurs singly, in pairs or in family groups. Apparently tame and confiding; often stands on vantage point such as lump of earth, rock, small bush or fence, from which it calls.

Food. Unknown.

Breeding Habits. Little known.
NEST: cup of dry grass, lined with plant fibres, rootlets and sometimes horsehair; either on ground or up to 10 cm above it; in grass tuft, once in young growing corn.
EGGS: 2–4. Subelliptical to ovate; glossy; pale greenish white, flecked dull mid-brown. SIZE: (n = 11) 23–24 × 16·5–17·5 (23·4 × 17·0).
LAYING DATES: several records during rains (June–Aug), one in dry season (Jan) in SW (Doko) (von Erlanger 1907).

Macronyx fuelleborni Reichenow. Fülleborn's Longclaw. Sentinelle de Fülleborn.

Macronyx fuelleborni Reichenow, 1900. Orn. Monatsber., p. 39; Unika Highlands, S Tanganyika.

Forms a superspecies with *M. capensis*.

Plate 14
(Opp. p. 241)

Range and Status. Endemic resident, with occasional local movements. Locally common central and S Zaïre (especially S Kwilu), south throughout Kasai and Shaba and east to Marungu Highlands. Common over much of Angola, from Cuanza Sul and W Malanje, east across central plateau and south to Bié and Huila; in SW extends into Namibia in extreme N Ovamboland, where occasional. Locally common Zambia but absent from dry country in extreme SW and not recorded east of Luangwa valley. Extends north into S Tanzania as far as Uzungwe and in SW to Ufipa Plateau; also recorded in N at Mbulu. Not recorded Malaŵi but almost certainly occurs in extreme north (Chitipa District), adjacent to Tanzania/Zambia borders.

Description. *M. f. ascensi* Salvadori: range of species except for nominate race. ADULT ♂: forehead to nape very dark olive-brown to brownish black, feathers narrowly fringed greyish horn to dull cinnamon, giving a slightly scaly effect; hindneck and sides of neck greyish horn with slightly darker feather centres, forming greyish collar. Supercilium deep lemon-yellow; lores white to buff; ear-coverts cinnamon-brown. Feathers of mantle, scapulars and uppertail-coverts very dark olive-brown to brownish black, fringed dull cinnamon to greyish horn; back and rump greyish horn, sometimes tinged dull cinnamon, with slightly darker feather centres. Tail-feathers dark olive-brown; T1 narrowly fringed pale buff; T2–T6 narrowly edged yellow to yellow-buff on outer web and with white area at tip of inner web and shaft becoming progressively more extensive from T2 (often absent) to T6 (distal third); on T4–T6 outer web is also white at tip, but less

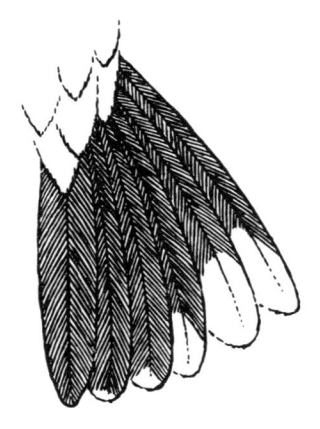

A

extensively than inner web (**A**). Moustachial stripe black or brownish black, extending as an unbroken necklace across upper breast; white or buff-white line running from lores above moustachial stripe and broadening to small whitish patch at side of lower neck, sometimes continuing as small buff-white area at side of breast below necklace. Chin, throat, malar region and foreneck deep lemon-yellow, enclosed by dark necklace. Sides of breast yellow-buff to dull lemon-yellow; breast to belly and thighs dull lemon-yellow; breast sometimes with very faint narrow dark olive-brown streaks; flanks yellow-buff; undertail-coverts buff to yellow-buff. Primaries and secondaries dark olive-brown; outer webs of primaries narrowly edged pale yellow; primaries tipped pale buff when fresh. Secondaries and tertials fringed buff, palest at tips; tertials darker than secondaries, more brownish black. Upperwing-coverts dark olive-brown with pale fringes: alula, marginal and outer lesser coverts fringed bright yellow, inner lesser coverts fringed yellowish white to buff; median and greater coverts fringed buff to greyish horn. Axillaries white or yellowish white, with olive-brown centres; underwing-coverts pale olive-brown with whitish fringes; underside of remiges pale olive-brown. Upper mandible dark horn to black, lower mandible grey, blue-grey or pale horn, tip blackish; eye dark brown; legs and feet horn to brown, with yellowish tinge. ADULT ♀: like ♂, necklace sometimes browner and less well developed. SIZE: wing, ♂ (n = 15) 100–112 (105·5), ♀ (n = 24) 95–110 (99·8); tail, ♂ (n = 7) 66–74 (71·1), ♀ (n = 16) 59–72 (67·4); bill, ♂ (n = 9) 20–22 (21·7), ♀ (n = 17) 19–24 (21·4); tarsus, ♂ (n = 8) 39–44 (41·4), ♀ (n = 16) 38–42 (40·1); hind claw, ♂ (n = 15) 19–25 (22·1), ♀ (n = 23) 18–25 (20·2). WEIGHT: ♂ (n = 9) 53–64 (57·9), ♀ (n = 7) 46–60 (52·6).

IMMATURE: upperparts like adult, sometimes darker overall with duller olive-brown tone; supercilium duller. Necklace browner, narrower and more diffuse at edges; chin to foreneck buff with deep lemon-yellow flecks; rest of underparts less bright yellow than in adult, washed deep buff to dull yellow.

JUVENILE: upperparts often have less prominent pale feather fringes, giving overall darker effect. Chin to foreneck buff to cinnamon; necklace of dull olive-brown spots (and sometimes streaks), often ill-defined; rest of underparts deep buff to cinnamon, washed yellow on centre of lower breast and belly; yellow on wings duller.

NESTLING: natal down black; gape yellow; inside of mouth orange (A. J. Scott, pers. comm.).

M. f. fuelleborni Reichenow: extreme S Tanzania. Rump brighter, more cinnamon-tinged; whitish to pale buff wash on sides of breast below necklace, sometimes extending to centre of breast; occasionally a narrow whitish line immediately above necklace; yellow of underparts less extensive, flanks, thighs and sides of lower breast washed dull buff. Smaller: wing 93–103.

Field Characters. A large longclaw, very similar to slightly smaller Yellow-throated Longclaw *M. croceus*. Adult and immature Fülleborn's Longclaw distinguished by lack of streaks on breast, centre of necklace narrower, duller yellow underparts, less pronounced yellow supercilium, and different voice. Streaking on upperparts usually less contrasting. Juveniles very similar but Yellow-throated Longclaw more heavily streaked on breast and flanks. Ecologically separated from Yellow-throated Longclaw which prefers richer vegetation types. In flight, less white on outer tail-feathers and more on tips of inner feathers than Rosy-breasted Longclaw *M. ameliae*, giving effect of white band across end of tail rather than white outertail. Cape Longclaw *M. capensis* allopatric but ranges come close in E Zambia; adults have orange throat and supercilium, dull orange belly contrasting with brownish cinnamon flanks, immature similar but without orange throat.

Voice. Tape-recorded (CHA, MOY, STJ). Song a distinctive 2-syllabled trill 'chree-er'; also gives a repeated and monotonous whistled 'jee-o-wee' both when perched and when flushed. Usual calls are a metallic 'weee' and a chirping 'chwee'.

General Habits. Occurs in grass on plains and downs, dambos, marshy meadows and hollows, including areas with scattered bushes or small trees; usually on moist to seasonally shallowly inundated ground, but also in dry grass near water. Not confined to large tracts of grassland, occurring e.g. along drainage lines in woodland. Grass usually short but occasionally moderately long, and cover may be sparse, tussocky or fairly dense. In Zambia also found in wet grass at sewage treatment works. Generally occupies drier ground than Rosy-breasted and Grimwood's Longclaws *M. grimwoodi* where all occur

Macronyx fuelleborni

together (e.g. NW Zambia); in only area of sympatry with Yellow-throated Longclaw (Pungu Andongo, NW Angola) occupies grassland in *Brachystegia* woodland, Yellow-throated Longclaw occupying richer vegetation of coastal plain. Altitudinal range 730–2450 m.

Occurs in pairs and family parties. Fairly shy, hiding in grass when approached, but flushes quite easily; does not fly far. Sometimes perches in bushes or on small trees, but much less frequently than Yellow-throated Longclaw; also perches on large flower heads of thatch grass *Hyparrhenia*. Flight stiff-winged with irregular action. Normally forages on ground but catches termite alates in flight.

Moult probably consists of complete post-breeding and partial pre-breeding moult (Traylor 1965), rather than 2 complete moults described by Verheyen (1953).

Mainly sedentary throughout the year, but makes very local seasonal movements in response to habitat changes. At Ndola, Zambia, commonest in breeding areas Oct–Apr, wandering away locally in May–Sept, when birds move to damper grassland near permanent wetland, but reappearing in breeding areas occasionally throughout dry season (P. B. Taylor, pers. obs.).

Food. Insects, especially grasshoppers and beetles; spiders.

Breeding Habits. Monogamous, solitary nester.

NEST: shallow, wide, neat cup of dry grass, lined with rootlets or fine grass; ext. diam. *c*. 90, depth of cup *c*. 30; usually on ground, sometimes up to 15 cm above ground; alongside or within grass tussock (once in small bush) in 30–60 cm high grass; usually well hidden by overhanging grass but occasionally open to view from above; often on moist or inundated ground.

EGGS: 2–3, av. (n = 40 clutches) 2·7. Subelliptical; off-white, cream or greenish grey speckled red-brown, grey-brown or violet. SIZE: (n = 20) 22–27 × 17–19 (24·4 × 18·1).

LAYING DATES: Zaïre, Sept–Feb, May; Angola, Sept–Feb; Zambia, Oct–Feb, once June; Tanzania, Nov–May. Breeds mainly during rains, when ground cover good.

DEVELOPMENT AND CARE OF YOUNG: feathered nestlings fed about every 10 min by ♀ accompanied by ♂ (Vincent 1946); young also fed by both parents. Adult flushed from nest gives distraction display, fluttering helplessly through grass before flying to nearby bush (A. J. Scott, pers. comm.). Adults with young wary, flying to tree or bush when disturbed and uttering sharp alarm notes; they also hover and circle low before landing at nest and sometimes perch on nearby low eminence before approaching it.

Macronyx capensis (Linnaeus). Cape Longclaw. Sentinelle du Cap.

Plate 14
(Opp. p. 241)

Alauda capensis Linnaeus, 1766. Syst. Nat. (12th ed.) 1, p. 288; Cape of Good Hope, South Africa.

Forms a superspecies with *M. fuelleborni*.

Range and Status. Endemic resident. E Zimbabwe, from Inyanga Highlands south to the Vumba, and in adjacent Mozambique; Mashonaland Plateau along central watershed from Chinhoyi, Harare and Mangula south and west to Great Zimbabwe and the N fringe of Matobo Hills. South Africa from Transvaal highveld to Orange Free State, Natal and E and S Cape Province, west to winter rainfall district of S and SW Cape; also in Swaziland, Lesotho, extreme S Mozambique (S Maputo District and the plateau of S and central Lebombos) and extreme SE Botswana. Frequent to locally common.

Description. *M. c. colletti* Schouteden (including '*latimerae*' and '*stabilior*'): range of species except for nominate race. ADULT ♂: forehead to nape dark olive-brown, feathers fringed tawny-olive to cinnamon; hindneck and sides of neck greyish horn to dull cinnamon with slightly darker feather centres on hindneck, forming greyish collar. Supercilium bright orange; lores and line below eye to base of ear-coverts white or washed pale yellow; ear-coverts greyish horn to dull cinnamon. Mantle and scapulars dark olive-brown, feathers fringed tawny-olive to cinnamon; back, rump and uppertail-coverts greyish horn to dull cinnamon with slightly darker feather centres. Tail-feathers dark olive-brown; T1 narrowly fringed buff; T2–T6 tipped white on both webs, white becoming progressively more extensive from T2 (may be less than 5 mm) to T6 (distal third of feather white); outer webs of T2–T6 narrowly edged pale yellow to pale orange; outer web of T6 white except at base (**A**). Moustachial stripe brownish black, extending as unbroken necklace across upper breast, broadening in centre. Chin, throat, malar region and foreneck bright deep orange, enclosed by dark necklace. Sides of breast, flanks, thighs and undertail-coverts cinnamon to tawny-olive, sometimes pale grey on thighs and undertail-coverts; centre of breast and belly orange-yellow. Primaries and secondaries dark olive-brown; outer webs of P1–P9 narrowly edged orange-yellow; outer web of P10 edged cream; primaries tipped buff when fresh. Secondaries broadly edged dull orange-yellow, tipped buff; tertials dark olive-brown with buff to dull cinnamon fringes. Upperwing-coverts dark olive-brown, fringed paler; alula and marginal coverts fringed bright orange, lesser coverts fringed dull orange-yellow, median and greater coverts fringed dull cinnamon. Axillaries white, edged bright orange; underwing-coverts and underside of remiges pale olive-brown. Bill brown to blackish, base of lower mandible grey or blue-grey; eye hazel to dark brown; legs and feet light brown with flesh, reddish or yellow tinge. ADULT ♀: like ♂, but supercilium dull orange only to above eye; often duller than ♂ on underparts, with browner and less well-defined necklace and more greyish horn to dull cinnamon on breast and sides of belly. SIZE: (35 ♂♂, 20 ♀♀) wing, ♂ 96–106 (101), ♀ 89–102 (95·6); tail, ♂ 61–73 (67·1), ♀ 59–69 (64·2); bill, ♂ 20–23 (21·4), ♀ 19–23 (20·9); tarsus, ♂ 34–41 (36·7), ♀ 33–40 (35·6); hind claw, ♂ 15–25 (18·5), ♀ 15–23 (18·7). WEIGHT: ♂ (n = 7) 49·2–55 (52·5), ♀ (n = 11) 45·2–52·2 (47·7), unsexed (n = 13) 40–54·5 (45·0).

IMMATURE ♂: feathers of upperparts with darker centres; upperwing-coverts and remiges fringed pale cinnamon to cream, giving more patterned appearance; hindneck with darker feather centres and paler fringes; chin to foreneck cream to pale orange, with some bright orange feathers; incomplete necklace of dark brown spots extending over entire breast; orange-yellow of underparts paler, duller and confined to centre of lower breast and belly; rest of underparts heavily washed dull cinnamon to dark grey. IMMATURE ♀: like ♂ but paler on underparts.

JUVENILE: like immature but even less orange-yellow on underparts; spots of necklace less numerous, forming narrow band across upper breast.

NESTLING: unknown.

M. c. capensis (Linnaeus): S and SW Cape Province. Underparts darker and duller: sides of breast, flanks, thighs, undertail-coverts and often entire breast below necklace dull cinnamon to cinnamon-brown; breast usually obscurely streaked olive-brown to buffy brown; orange-yellow of underparts restricted to belly and centre of lower breast.

Field Characters. A large, chunky, short-tailed longclaw with orange throat and supercilium, orange-yellow breast and belly contrasting with brownish cinnamon flanks; immatures lack orange throat. Has characteristic upright stance on ground (**B**); mewing alarm call distinctive. Slightly larger Yellow-throated Longclaw *M. croceus* has yellow throat, supercilium and underparts, white on face and neck outside necklace (usually lacking in Cape Longclaw); immatures have pale buffy to yellowish underparts, yellow (not buff) edges to wing-feathers.

Voice. Tape-recorded (58, 75, 81, 88, F, WALK). Advertising call a series of clear, piping, far-carrying whistles 'dweet' or 'deweet', sometimes given in duet with 'meew' call, e.g. 'deweet, meew, deweet, meew', usually given in flight; contact call a high-pitched far-carrying whistle 'tsweet', given on ground or in flight; loud, cat-like 'meew' or 'meyew' alarm call, given in flight or when perched; a hard 'chack', possibly also an alarm call; loud sparrow-like chirps 'chi-rup' or 'cheeerr-up'; loud staccato 'choi', 'chio', 'chi-cho' and 'choik', often run together 'choichoichoik' and given with alarm and advertising calls. One hand-reared nestling mimicked the calls of Crowned Eagle *Stephanoaetus coronatus*, Cape Wagtail *Motacilla capensis* and Rufous Sparrow *Passer motitensis* (Farkas 1962).

General Habits. Inhabits short, often dense, grassland in cool climates, e.g. montane grassland on highveld plateau in N part of range and at lower elevations in cooler south, descending to sea level in S Natal and S Cape. Where sympatric with Yellow-throated Longclaw, usually occurs at higher elevations and also occupies moister areas, e.g. marshes, moist to seasonally flooded ground and margins of vleis with rushes and thin scrub; also in dry grassland beside rivers, lakes and lagoons. Avoids areas with dense growth of bushes or scattered trees. Occurs on agricultural land (pasture, fallow or cultivated), but in SW Cape avoids ploughland and requires grass or equivalent. In Cape Province also occurs in short, thin fynbos (upland to coastal) and even in intertidal zone of rocky seashores (Skead 1966). In Natal, occurs on *Themeda* and *Aristida* sourveld; in seasonally burnt grassland near Pietermaritzburg, at 715 m, prefers short grass (height less than 20 cm, mean height 6·5–11·9 cm); not present in tall (90 cm) rank grass; in short grass, tolerates tall grass component only if sparse and with mean height of 52 cm or less; grasses are *Aristida, Themeda, Heteropogon, Eragrostis, Hyparrhenia,* and *Microchloa* (Bowland 1984). Altitudinal range sea level (Cape) to 2300 m (Zimbabwe).

Occurs in pairs or small loose groups, in winter sometimes with pipits, especially on recently burnt ground. Normally on ground, but sometimes stands on low eminence such as grass tussock, mound, rock, fence or (occasionally) low bush. Crouches low when alarmed; flight direct, short bursts of stiff wing-beats alternating with glides on stiff wings, tail usually spread. Forages mostly on ground, walking with long strides; scratches open termite tunnels with feet; feeds among ashes on recently burnt ground, and in intertidal zone of Cape beaches among rocks and freshly washed-up seaweed; in E Orange Free State often follows the plough to pick up insect larvae; associates with domestic stock, feeding on insects disturbed by them; in winter occasionally forages around cowsheds. Chases insects on ground or hawks them in air; beats larger prey on ground before swallowing it; catches termite alates in flight from perch in bush or small tree.

Normally sedentary, but post-breeding altitudinal movements by some South African populations of *colletti* (Clancey 1980).

Food. Adult and larval insects, including beetles, termite alates, reduviid bugs, hippoboscid flies, and especially grasshoppers; also grass seeds.

Breeding Habits. Monogamous, territorial, solitary nester. In display flight rises *c*. 10 m above ground, fluttering and singing, then drops to grass again; also sings from perch on low mound or fence. Droops wings and fans tail in threat posture; pairs chase and fight other pairs, both in air and on ground, giving repeated alarm and advertising calls.

NEST: deep cup of grass, lined with rootlets or fine grass; int. diam. 83–89, depth 57–80; in dry grassland, often near water; on ground beside, between or within grass tussocks, beside broad-leaved plant, or in thick grass or weeds, sometimes in shallow depression in soil; usually well hidden.

EGGS: 2–5, av. (n = 34 clutches) 3·0. Subelliptical to ovate; white, creamy white or buff, with brown to dark brown markings, usually small freckles, over entire surface; most also with sparse to dense heavy blotches, sometimes concentrated at thick end. SIZE: (n = 75) 22–27·1 × 16·7–19·4 (23·8 × 17·7).

LAYING DATES: Zimbabwe, Sept–Mar (most Nov–Feb); Mozambique, from Sept; Transvaal, Aug–Apr with 69% of records Nov–Jan; Natal, Sept–Mar (most Nov–Dec); W Cape, Aug–Dec; breeds mainly during rains.

References
Bowland, A. E. (1984).
Irwin, M. P. S. (1981).

Macronyx croceus (Vieillot). Yellow-throated Longclaw. Sentinelle à gorge jaune.

Alauda crocea Vieillot, 1816. Nouv. Dict. d'Hist. Nat. 1, p. 365; Senegal.

Plate 14
(Opp. p. 241)

Range and Status. Endemic resident; the commonest and most widespread longclaw, found over much of subsaharan Africa. Scarce in Senegambia; not uncommon Guinea-Bissau and Guinea; rare and localized Mali, north to *c*. 15°N at Plateau Dogon. Widespread and locally frequent or common from Sierra Leone to central Cameroon and Chad, extending north to extreme SW Niger (W Nat. Park); much restricted north of 8°N in Ivory Coast and 10° 30′N in Ghana and Nigeria; local in N Cameroon but occurs near L. Chad and Mokolo; in Chad extends east to Bol (L. Chad) and along shores of Bahr az Zoum, Chari and L'Aouk; in Central African Republic, common in Bamingui-Bangoran Nat. Park. Widespread and not uncommon from S half of Gabon to NW Angola (Cabinda, S Cuanza Norte and adjoining Malanje (Pungo Andongo)). From NE Congo/NW Zaïre border extends across savannas of N Zaïre and adjoining S Central African Republic to S Sudan and E Africa, where widespread and locally common. Occurs throughout Uganda, much of W and central Kenya, and NW Tanzania south to Mpanda. 1 old record, SE coast Somalia (Ash and Miskell 1983). Reappears in coastal lowlands from Kenya (Witu) to Mozambique, and inland across S-central Tanzania to Songea, N Iringa and Kilosa, Malaŵi (throughout), Zambia (only east of Luangwa valley) and central and SE Mozambique

Macronyx croceus

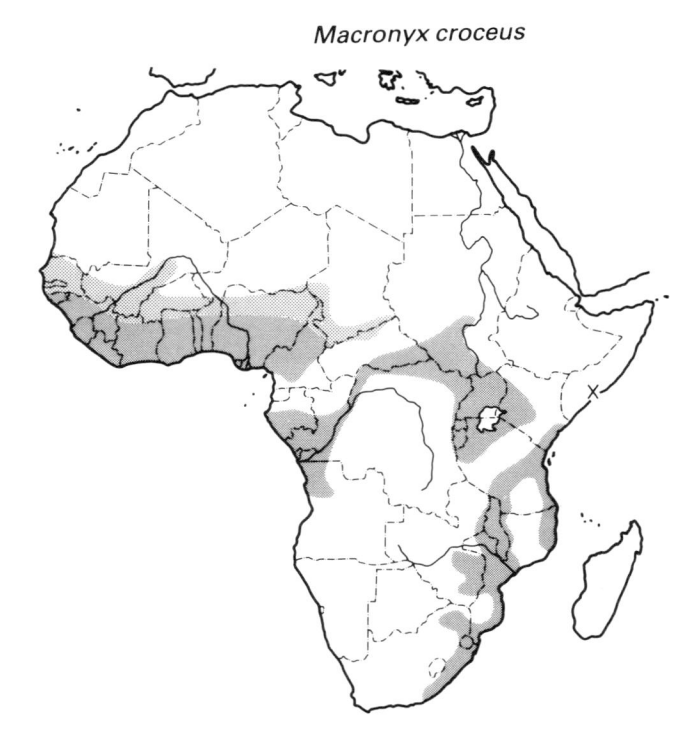

(common). Widespread in E, N and central Zimbabwe, northwest across Mashonaland Plateau and south to Mt Buhwa and Sabi–Lundi confluence. In South Africa relatively common but confined to lower altitudes, from E Transvaal through Natal to E Cape Province.

Description. (Includes '*hygricus*', '*tertius*' and '*vulturinus*'.) ADULT ♂: forehead to nape dark olive-brown, feathers narrowly fringed paler olive-brown giving a moderately streaked effect; hindneck and sides of neck with broader pale fawn-brown to smoke grey feather fringes and less pronounced dark centres. Supercilium deep lemon-yellow; lores white or tinged yellow; ear-coverts cinnamon-brown. Scapulars and mantle to uppertail-coverts very dark olive-brown, paler on rump and uppertail-coverts, feathers fringed dull to pale cinnamon on mantle, dull fawn to smoke grey elsewhere. Tail-feathers dark olive-brown with pale shafts; T1 broadly fringed yellow-buff to buff; T2 narrowly fringed whitish to buff; T3 and T4 with distal quarter and third respectively of inner web and tip of outer web white, outer web narrowly fringed buff to pale yellow; distal half of T5 white or tinged dusky on outer web, which is narrowly fringed deep lemon-yellow; distal half of T6 white, outer web whitish with deep lemon-yellow edge (**A**). Moustachial stripe dull dark brown to black, extending as an unbroken necklace across upper breast, broadening markedly in centre; area below eye and ear-coverts to side of neck white or cream, bordering black necklace and continuing as cream-tinged area at sides of breast. Chin, throat, malar region and foreneck deep lemon-yellow, enclosed by dark necklace. Sides of upper breast washed cream, of lower breast washed lemon-yellow with buff tinge; feather centres dark olive-brown; centre of breast to thighs and belly deep lemon-yellow, variably streaked dull dark brown below necklace; flanks and undertail-coverts yellow-buff, with darker feather centres to coverts and dark streaks on flanks. Primaries and secondaries dark olive-brown; primaries tipped whitish when fresh, outer webs edged deep lemon-yellow. Outer secondaries tipped white and outer webs edged yellow-buff; inner secondaries and tertials broadly fringed buff to dull cinnamon. Upperwing-coverts dark olive-brown with broad pale fringes; alula and marginal coverts fringed deep lemon-yellow, lesser coverts fringed dull yellow-buff, median coverts fringed dull yellow-buff to dull cinnamon, greater coverts fringed dull cinnamon. Axillaries pale yellow; underwing-coverts white with olive-brown centres; underside of remiges pale olive-brown. Upper mandible brown to blackish, lower mandible grey or pale blue, sometimes tinged horn or tipped dark, rarely slate-green; eye brown to dark brown, rarely black, grey-brown or steel-blue; legs and feet pale brown to dull yellow, sometimes tinged flesh, straw or even greenish yellow. ADULT ♀: like ♂ but underparts less bright, yellow often duller, washed buff to olive-yellow from breast downwards; necklace dull dark olive-brown and less well-defined; often less white on face and neck. SIZE: (111 ♂♂, 124 ♀♀) wing, ♂ 93–107 (99·6), ♀ 87–103 (96·4); tail, ♂ 66–86 (76·2), ♀ 64–83 (73·4); bill, ♂ 17–23 (20·3), ♀ 18–23 (20·1); tarsus, ♂ 33–40 (36·9), ♀ 32–41 (36·7); hind claw, ♂ 16–27 (21·1), ♀ 16–32 (20·8). WEIGHT: ♂ (n = 11) 43·8–54·6 (49·2), ♀ (n = 16) 43–51·6 (48·0), unsexed (n = 28) 38–55 (45·8).

IMMATURE ♂: duller; upperparts fringed darker, more dull cinnamon; white of head and neck replaced by dull buff; yellow of head and underparts duller, washed dull buff or olive-yellow; streaking on breast, flanks and thighs dark olive-brown and more extensive; necklace dark olive-brown and ill-defined, broken into spots and streaks; supraloral area and supercilium dull yellow. Central tail-feathers broadly fringed dull cinnamon; white on all tail-feathers duller; yellow of wings and tail duller. IMMATURE ♀: like immature ♂ but duller; supercilium largely dull buff with hardly any yellow; necklace of narrow streaks and spots, paler brown. WEIGHT: unsexed (n = 4) 37·5–46·7 (41·9).

JUVENILE: upperparts like immature; yellow of underparts replaced by dull buff, palest on chin and throat where dark cream, darkest across centre of breast, on flanks and undertail-coverts; trace of dull yellow on centre of breast and belly. Moustachial line of small spots, paler and very indistinct; necklace of small olive-brown spots; lores and supercilium buff; yellow on wings dull and pale.

NESTLING: skin blackish, gape yellow; at 5 days dull brownish pink with dull grey down, mouth orange, gape yellow.

Field Characters. The standard longclaw over much of Africa, with which others should be compared. Large and chunky with bright yellow throat and underparts separated by well-defined black necklace, streaked breast, white band outside necklace on face and neck,

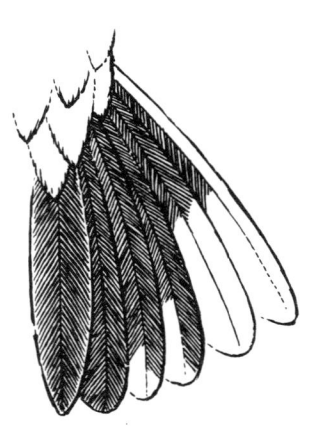

A

prominent yellow supercilium and moderately patterned upperparts. Sympatric with Pangani, Cape and Rosy-breasted Longclaws (*M. aurantiigula*, *M. capensis*, *M. ameliae*), and marginally with Fülleborn's Longclaw *M. fuelleborni* and Sharpe's Longclaw *M. sharpei*, which replace it in central Africa and Kenya highlands respectively. In flight, tail pattern similar to Fülleborn's, Pangani and Cape Longclaws (white across tip), except in Fülleborn's Longclaw white is less extensive on outermost feather; for further differences see these species.

Voice. Tape-recorded (7, 58, 75, 78, C, F). Much regional and individual variation. Typical song, given throughout range, is a rapid, rolled 'tir-tri-tri-tri' ('what do I see'), the 'tri' sometimes repeated up to 5 times; regional variants include: (1) a slurred 'trrrry tree-tree-tree' (Zimbabwe), (2) a rapid 'tritritri-tri-tri-tri' (Natal) or 'trooee-tri-tri-tri' (Kenya), the 'tri' repeated up to 12 times. Other multiple calls, which may also be songs or advertising calls, are (1) a descending whistled 'te-ter-to' followed by repeated 'ter-to' notes (Natal), (2) a whistled 'tyorry teer-teer-teer-teer', the 'teer' repeated up to 11 times (Kenya), (3) a repeated loud 'tyor ti-ti' (South Africa). Alarm call a loud piping 'twip pipipipipi', the 'pi' repeated up to 9 times, sometimes preceded by a whistled 'teeoo' or 'twi-tee'. A hard 'ker kwee-kwee-kwee' or rapid 'keer-keer-keer' (Kenya) may also indicate alarm. Single whistled call-notes, given by both sexes and possibly with both contact and alarm functions, are (1) 'tyoo-eoo' or 'tyo-EEEoo', (2) 'te-OEE', (3) 'tsirreoo', (4) 'twee-ee' and (5) a slurred 'tsrrrry' (Kenya); calls (1) and (4) are also given by ♂ during display flight. Songs and multiple calls are given from perch or in display flight; simple calls are often given in flight and both sexes may call together.

General Habits. Occurs in short to long, dense or rank grass, in open to thickly-bushed grassland, thornveld and lightly wooded areas; also meadows, old cultivation, clearings in heavily-wooded areas and (W Africa) the seashore. Not confined to the vicinity of water but often found in moist to wet areas, e.g. vleis, along drainage lines, at swamp and marsh edges and near dams, lakes and rivers. In areas of sympatry, Cape Longclaw occupies moist habitats, Yellow-throated Longclaw occurring in drier areas, among bushes and trees and in denser and longer grass. Altitudinal range sea level to 2350 m (Mau Ravine, Kenya); away from tropics largely confined to warmer coastal areas.

Occurs in pairs or family parties. In non-breeding season generally silent and unobtrusive but confiding and easy to approach. Often perches on vantage point such as rock, telegraph wire, top of bush or tree, up to 12 m above ground. Flight normally slow and jerky, with glides and flaps and spread tail. Runs rapidly; perches with difficulty on slender twigs, fluttering wings to keep balance. When disturbed on ground first crouches; as intruder comes closer, stands upright and motionless with bill pointing upwards, then usually flies to top of bush or tree but sometimes makes short flight and drops into grass. Follows predator (domestic cat, serval, mongoose) through grass, flitting above it.

Normally diurnal, but during breeding season may call up to 30 min before daybreak (Kenya). Roosts on ground in grass. When grass is wet, suns itself on termite mound or other eminence. Forages on ground, often in pairs, among grass or in open areas, walking with long strides; associates with cattle, feeding on insects disturbed by them.

Mainly sedentary, but with some local seasonal movements: at Juba, Sudan, occurs Aug–Jan, leaving when habitat dries out (G. Nikolaus, pers. comm.); at Blantyre, Malaŵi, moves into airport grass and adjacent bush during rains; in South Africa, regularly moves away from some breeding areas June–Jan, especially in Natal and E Cape (G. L. Maclean, pers. comm.).

Food. Mainly adult and larval insects, especially grasshoppers, beetles and Lepidoptera; also mantids, ants, millipedes, molluscs, and some plant material. Nestlings fed mainly on larvae of Lepidoptera, grasshopper nymphs and mantids.

Breeding Habits. Monogamous, territorial, solitary nester. Becomes wary and more difficult to approach during breeding season; pair occupies territory, giving songs and display flights, up to a month before breeding commences (Serle 1949). Display flight (often circular) fluttering, with spread tail, accompanied by song; also sings from perch on fence, bush or low tree. ♂ chases ♀ in flight; also hovers up to 3 m above ♀ giving single whistled calls.

NEST: cup-shaped, usually deep and well-constructed but sometimes shallow and/or loosely made, rarely very little material; of dead grass stems and blades, lined with fine rootlets and fibres; mud occasionally incorporated into foundation; ext. diam. 100–145, depth to 120; int. diam. 85–100, depth of cup 30–60. Placed on ground among grass stems or against base of tussock, bush or sapling, up to 15 cm above ground in tussock (usually in moist places or areas liable to flooding) with approach ramp, once 60 cm above ground in acacia bush (van Someren 1956); sometimes overhung by grass blades, or under matted grass or broad-leaved herb and reached by a side entrance; occasionally built in depression such as hoof-mark. Built by ♀, with ♂ in attendance, often uttering whistled calls from nearby perch on bush or tree.

EGGS: 2–4, av. (n = 119 clutches) 3·0. Subelliptical to ovate; grey-white, grey-buff, pink-buff or white to pale blue, usually densely and finely spotted and streaked brown (pale to chocolate or red-brown), markings either evenly distributed or concentrated at blunt end; sometimes with pale grey undermarkings. SIZE: (n = 178) 20–26 × 16–19 (23·8 × 17·3). Sometimes double-brooded.

LAYING DATES: Sudan, Oct; Senegambia, June–Aug, Dec; Mali, Apr–Oct; Sierra Leone, May–June; Liberia, June; Ghana, June–Aug; Togo, July–Aug; Nigeria, Apr–Oct; Kenya, all months but mostly during rains; Uganda, Mar–Nov (most in rains); Tanzania, Nov–June; Zaïre, Aug–Nov; Zambia, Dec, June; Malaŵi,

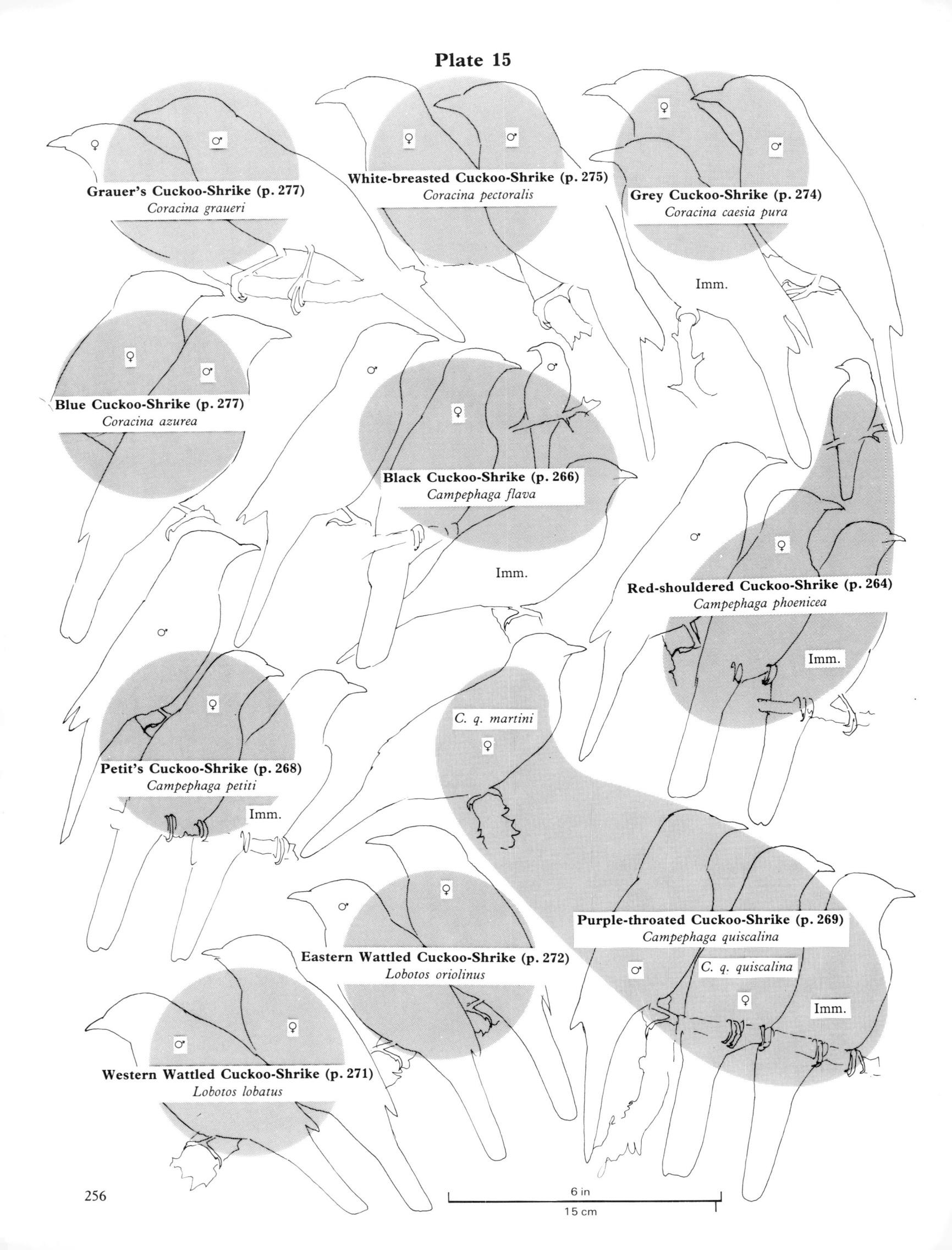

Plate 15

Grauer's Cuckoo-Shrike (p. 277)
Coracina graueri

White-breasted Cuckoo-Shrike (p. 275)
Coracina pectoralis

Grey Cuckoo-Shrike (p. 274)
Coracina caesia pura

Blue Cuckoo-Shrike (p. 277)
Coracina azurea

Black Cuckoo-Shrike (p. 266)
Campephaga flava

Red-shouldered Cuckoo-Shrike (p. 264)
Campephaga phoenicea

Petit's Cuckoo-Shrike (p. 268)
Campephaga petiti

Purple-throated Cuckoo-Shrike (p. 269)
Campephaga quiscalina

C. q. martini
C. q. quiscalina

Eastern Wattled Cuckoo-Shrike (p. 272)
Lobotos oriolinus

Western Wattled Cuckoo-Shrike (p. 271)
Lobotos lobatus

6 in
15 cm

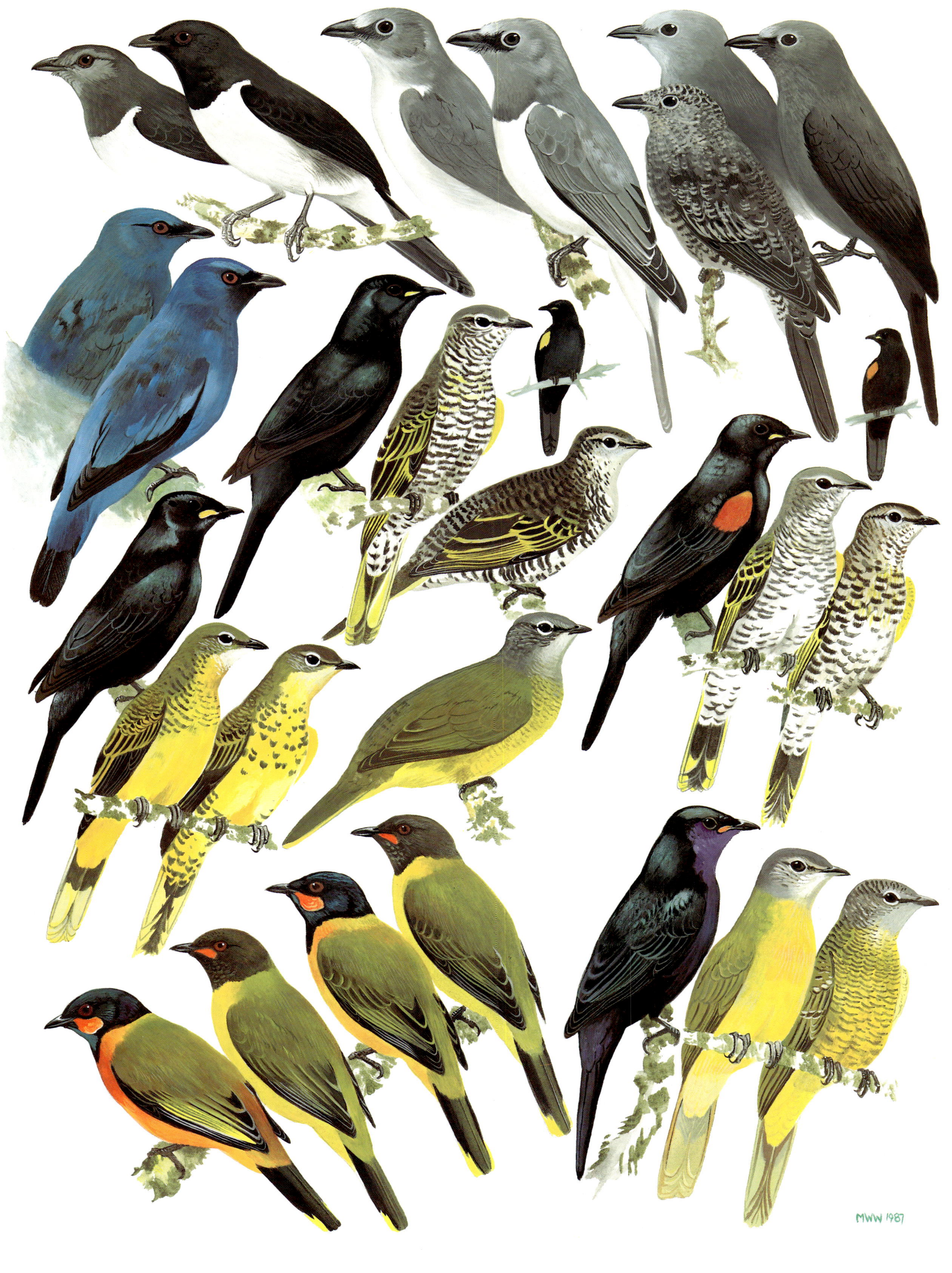
MWW 1987

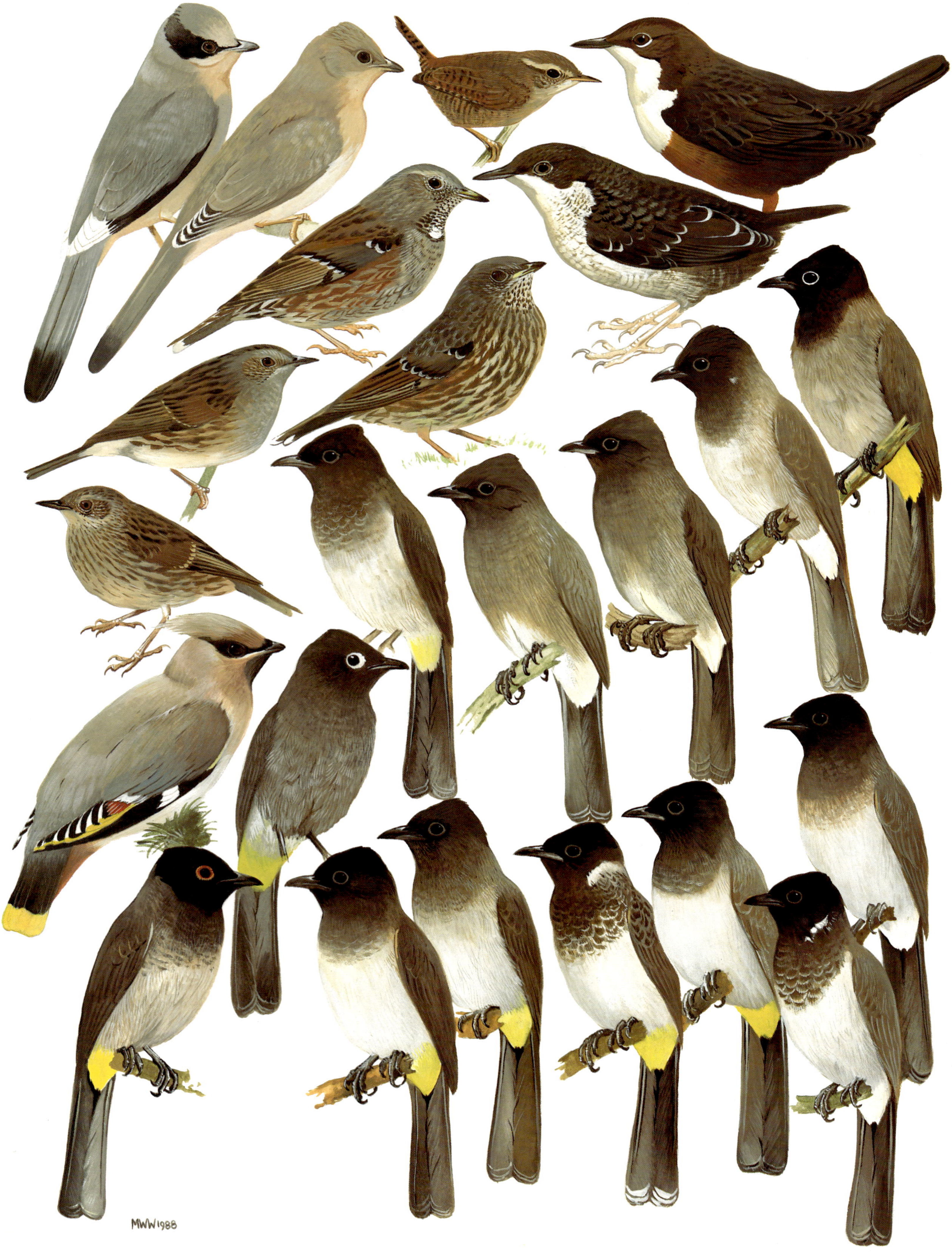
MWW 1988

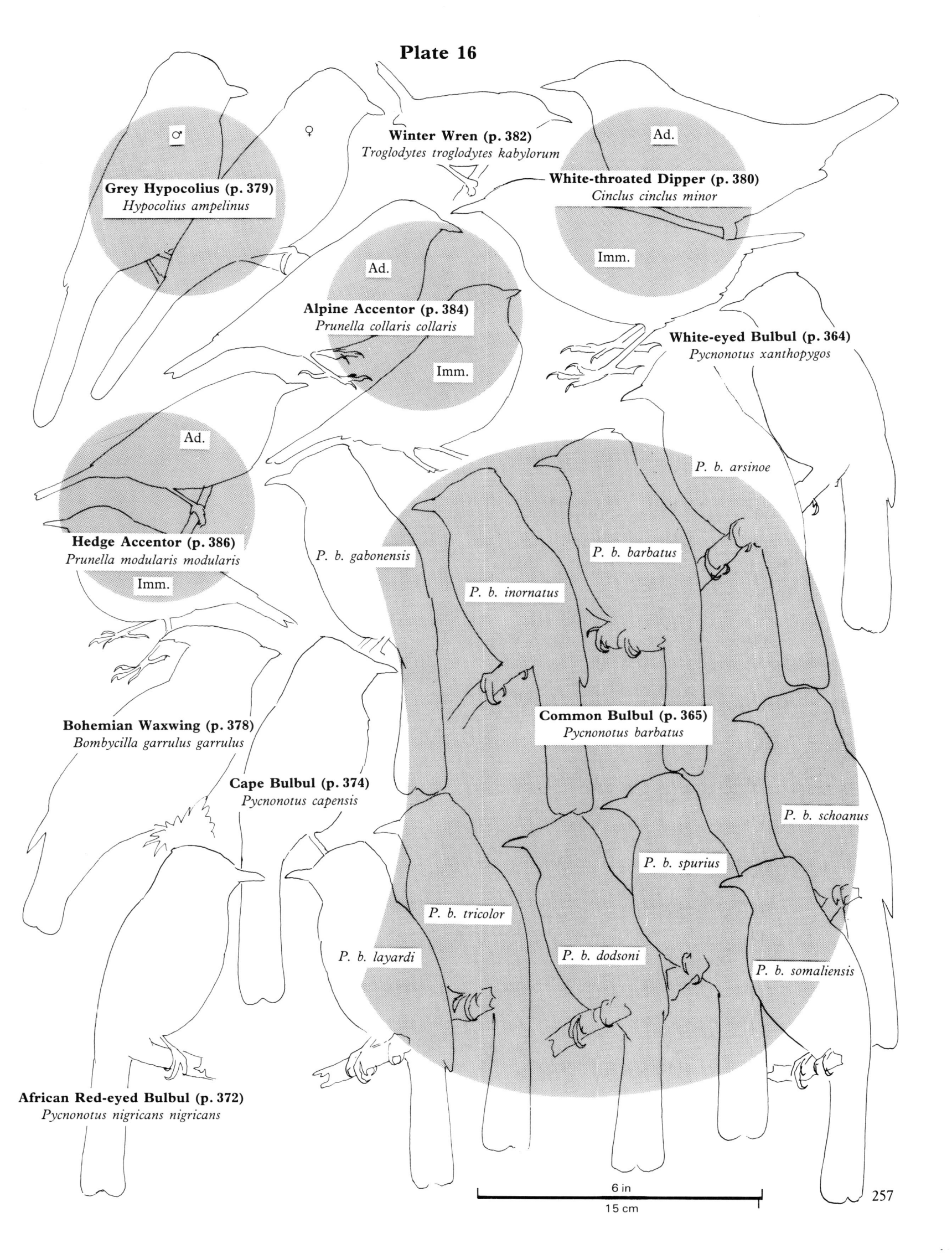

Plate 16

Grey Hypocolius (p. 379)
Hypocolius ampelinus
♂ ♀

Winter Wren (p. 382)
Troglodytes troglodytes kabylorum

White-throated Dipper (p. 380)
Cinclus cinclus minor
Ad. Imm.

Alpine Accentor (p. 384)
Prunella collaris collaris
Ad. Imm.

White-eyed Bulbul (p. 364)
Pycnonotus xanthopygos

Hedge Accentor (p. 386)
Prunella modularis modularis
Ad. Imm.

Bohemian Waxwing (p. 378)
Bombycilla garrulus garrulus

Cape Bulbul (p. 374)
Pycnonotus capensis

Common Bulbul (p. 365)
Pycnonotus barbatus

P. b. arsinoe
P. b. barbatus
P. b. gabonensis
P. b. inornatus
P. b. schoanus
P. b. spurius
P. b. tricolor
P. b. layardi
P. b. dodsoni
P. b. somaliensis

African Red-eyed Bulbul (p. 372)
Pycnonotus nigricans nigricans

6 in
15 cm

257

Nov–Apr; Mozambique, Nov–May; Zimbabwe, Sept–Mar (before and during rains); South Africa: Transvaal, Aug–Apr (most Nov–Jan), Natal, Oct–Feb. Breeds mainly during or just after rains.

INCUBATION: solely or mostly by ♀. Period 13–14 days. Incubating bird approaches nest by circuitous routes; in E Africa bird flushes only on close approach, but in Natal birds flush early, sometimes when intruder over 100 m from nest (G. L. Maclean, pers. comm.). One did not desert when grass was cut around nest area, and when flushed by dog gave successful distraction display, flying weakly 0·5 m above ground in front of dog for c. 40 m (L. H. Brown, E. Afr. Nat. Hist. Soc. nest record cards).

DEVELOPMENT AND CARE OF YOUNG: young fed by both parents, in nest and after departure; nestling period 16–17 days. Parents approach nest on foot; often whistle when bringing food to young, which utter thin begging calls 'schreee-schreee'; adults once pulled apart large locust between them before feeding it to young (van Someren 1956). When leaving nest, young are feathered but hardly able to fly; they remain with parents for some time after independence; 1 bird c. 6–7 weeks old begged for food for 10 days and was sometimes fed by 1 parent, at a time when both parents were feeding second brood (Marchant 1942). Intruders are drawn away from young by distraction display, adult fluttering helplessly through grass before flying off (A. J. Scott, pers. comm.).

References
Irwin, M. P. S. (1981).
van Someren, V. G. L. (1956).

Plate 14
(Opp. p. 241)

Macronyx aurantiigula Reichenow. Pangani Longclaw. Sentinelle dorée.

Macronyx aurantiigula Reichenow, 1891. J. Orn., p. 222; Pangani R., Tanganyika.

Range and Status. Endemic resident. SE Somalia south of 3°N (uncommon) and coastal Kenya south to Mombasa; common in central and S Kenya, from Meru, Athi Plains and Hola south through Tsavo region to NE Tanzania (Ngulu, Katesh).

Description. ADULT ♂: upperparts well patterned, dull dark olive-brown to brownish black, feathers of forehead to hindneck edged buff to pale cinnamon, paler, even whitish buff, on nape and hindneck, feathers of mantle to uppertail-coverts broadly fringed buff to pale cinnamon. Sides of neck buff with dull dark olive-brown feather centres; lores whitish to pale orange-yellow, supraloral line dull dark olive-brown; supercilium orange-yellow to above eye, extending as an ill-defined whitish streak behind eye; ear-coverts cinnamon-brown. Tail-feathers dark olive-brown with paler shafts; T1 broadly fringed buff to whitish buff; T2–T5 narrowly fringed pale buff (or pale yellow on T5) and with white tip to inner web becoming progressively more extensive, occupying distal quarter of T5; outer web of T2–T5 tipped whitish; distal third of inner web of T6 white, outer web buffish white almost to base (**A**). Moustachial stripe blackish, extending as an unbroken narrow necklace across upper breast; a whitish line runs from lores above moustachial stripe to side of upper breast. Chin, throat, malar region and foreneck orange-yellow or deep yellow, enclosed by dark necklace. Sides of breast buff; centre of breast to centre of belly deep yellow; flanks, sides of belly and undertail-coverts white to whitish buff; thigh feathers often fringed deep yellow; centre of breast with narrow area of brownish black streaks below necklace, sides of breast, flanks and thighs with larger dull olive-brown streaks; undertail-coverts with narrow dull olive-brown streaks. Primaries and secondaries olive-brown, outer webs of primaries narrowly edged pale yellow; primaries tipped whitish when fresh. Secondaries and tertials fringed buff, latter darker, with broader fringes. Upperwing-coverts dull dark olive-brown with pale fringes; marginal coverts, lesser coverts and alula fringed deep yellow, median and greater coverts fringed buff. Axillaries white with olive-brown centres; underwing-coverts pale olive-brown with paler fringes; underside of remiges pale olive-brown. Bill dark grey-brown to black, lower mandible often paler, grey to pale horn with darker tip; upper eyelid orange-yellow, lower paler (like lores); eye mid- to dark brown; legs and feet pale brown, tinged flesh. ADULT ♀: like ♂ but less yellow on supercilium, chin to foreneck sometimes paler orange-yellow or yellow, necklace sometimes narrower. SIZE: (8 ♂♂, 9 ♀♀) wing, ♂ 89–94 (92·5), ♀ 87–95 (88·7); tail, ♂ 68–73 (70·6), ♀ 66–73 (69·4); bill, ♂ 19–21 (19·8), ♀ 19–21 (20·1); tarsus, ♂ 33–37 (34·5), ♀ 31–

37 (33·1); hind claw, ♂ 15–20 (16·8), ♀ 15–18 (16·4). WEIGHT: 1 ♂ 36·6, 2 ♀♀ 42·5, 42, unsexed (n = 8) 36·7–43 (39·1).

IMMATURE: like adult but blackish necklace incomplete, of spots; chin to foreneck buff with some orange or orange-yellow spots; rest of underparts buff with yellow wash on breast; streaking of underparts less heavy than adult; yellow on wings often paler, yellowish white. Legs and feet flesh-grey.

JUVENILE: like immature but fringes of feathers of upperparts more sandy; chin to foreneck entirely buff.

NESTLING: unknown.

Field Characters. A medium-sized longclaw, generally resembling somewhat larger Yellow-throated Longclaw *M. croceus*. Adults differ from Yellow-throated Longclaw in more orange-yellow throat contrasting with yellow breast (especially in ♂, but contrast often difficult to see), narrower dark necklace, heavier and more extensive streaking on breast and flanks, whitish sides to breast and flanks, paler streaking on upperparts; immatures have paler underparts with heavier streaking, orange spots on throat, paler streaking on upperparts.

Voice. Tape-recorded (GREG, McVIC). Song, usually given when perched, descending 'teeoo', rapidly repeated. Calls given when perched or in flight: (1) repeated high-pitched, sibilant, plaintive 'seeee', (2) repeated rising whistled 'weee' or 'ooeee', (3) series of rapid 'chrry' notes, musical but rather chirruping. Anxiety call when breeding, plaintive 'k-lee', uttered in flight.

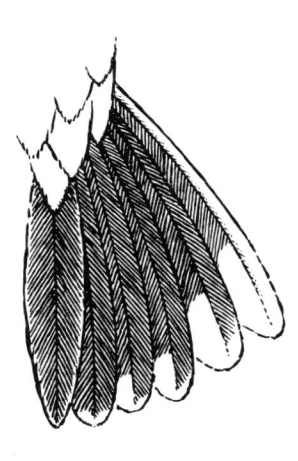

A

General Habits. Occurs in grassland with acacia and other bushes, typically in semi-arid plateau country below 1000 m, but extends to higher, more verdant areas in NE Tanzania; also in more open, littoral grasslands. Not associated with water; in thornbush country prefers areas with taller grass and thick cover. Altitudinal range sea level to 1800 m. Considered allopatric with Yellow-throated Longclaw, which normally prefers more open, richer grassland in less arid areas, but both occur in *Dobera/Albizzia* parkland along lower Tana R., Kenya (Britton 1980) and both occur and nest in acacia-bushed grassland near Mombasa, Kenya (P. B. Taylor, pers. obs.).

Occurs in pairs or family parties; wary and difficult to approach. Regularly perches on vantage point (bush or small tree), also at tops of grass stems; takes refuge on ground when disturbed. Flight rather weak, stiff-winged with irregular action. Forages on ground, including short-grassed tracks and other open areas; also catches insects in flight, *c.* 1 m above ground, and jumps up to pick insects from grass stems.

Mainly sedentary, but in SE Kenya absent from breeding habitat at Mombasa in driest period (Jan–Apr) (P. B. Taylor, pers. obs.) and more often seen in Tsavo E Nat. Park in wet season (Lack 1985).

Food. Insects, especially beetles, grasshoppers, termites, mantids and small moths and their larvae; also spiders and small snails. Heads are removed from termites fed to nestlings (van Someren 1956).

Breeding Habits. Monogamous, solitary nester.

NEST: cup-shaped, of dry grass blades and stems, lined with fine rootlets and fibres; on ground in dense grass or up to 10 cm above ground within grass tussock and with approach ramp; sometimes in deep scrape; surrounding grass less than 1 m tall; sometimes well concealed by overhanging grass blades. Built by ♀ with ♂ in attendance (P. B. Taylor, pers. obs.).

EGGS: 2–4, av. (n = 10 clutches) 2·5. Subelliptical to ovate; white with numerous small spots of grey-brown, deep buff, or rich reddish brown, sometimes with a few larger blotches; markings often more numerous at thick end. SIZE: (n = 10) 23–25·5 × 15–17·3 (24·0 × 16·2).

LAYING DATES: Kenya and Tanzania, all months except Aug, Oct and Dec; also breeding condition Tanzania, Oct; nest-building Kenya, Dec.

INCUBATION: probably by ♀ only. Period: 13 days.

DEVELOPMENT AND CARE OF YOUNG: young fed by both parents. When disturbed near nest, adults fly around giving repeated anxiety calls; ♂ leads intruder away from nest with short flights and plaintive calling, dropping to ground between flights (van Someren 1956).

Reference
van Someren, V. G. L. (1956).

260 MOTACILLIDAE

Plate 14
(Opp. p. 241)

Macronyx ameliae de Tarragon. **Rosy-breasted Longclaw; Pink-throated Longclaw.**
Sentinelle à gorge rose.

Macronyx ameliae de Tarragon, 1845. Rev. Zool., p. 452; Port Natal (= Durban).

Range and Status. Endemic resident, with some local movements. Distribution disjunct. W Kenya (locally common) to central Kenya highlands and N Tanzania (local, south to L. Manyara); single record near Tabora, central Tanzania; also SW Tanzania, near Malawi border. Widespread but local throughout most of Zambia and adjacent SE Zaïre (Shaba); in Angola local on E plateau (N and E Moxico, Lunda, west to N Bié) and also in extreme S Huila; local in extreme N Malawi (Songwe R. mouth), but recent records near Lilongwe and in SW by Mozambique border. Zimbabwe, recent colonist at L. Kariba (near Charara), and very local on central watershed of Mashonaland Plateau; also in extreme NW, and across N Botswana to NE Namibia. Coastal S Mozambique (Same R.) to Zululand, formerly also coastal Natal south to Durban but now almost extinct there (1 recent record, S Natal); vulnerable in coastal areas because of habitat destruction or degradation of grasslands.

Macronyx ameliae

☐ Formerly resident

Description. (Includes 'altanus' and 'wintoni'.) ADULT ♂: feathers of upperparts blackish, broadly edged cinnamon to warm buff from forehead to back and paler buff to whitish on rump and uppertail-coverts; scapulars similar but with brighter, often paler, edges. Sides of neck pale cinnamon with dark olive-brown streaks. Supraloral area pink, supercilium whitish to buff; lores and cheeks buffish white; ear-coverts cinnamon-brown. Tail-feathers dark olive-brown, shafts pale, T1 fringed and T2–T4 narrowly edged buff to white; T4 with small whitish tip to inner web; T5 white on distal third of inner web and variably from distal third to almost all of outer web; T6 white except for dark olive-brown proximal quarter of inner web (**A**). Ill-defined blackish moustachial stripe extends to join sides of blackish necklace across upper breast; chin, throat, foreneck and malar region burnt orange, pinkish scarlet, geranium pink or peach-red with an orange tinge. Upper breast cream to buff with broad blackish streaks and heavy blotches forming broken necklace around red of neck, broadest and most complete in centre. Sides of breast buff to whitish, strongly streaked blackish; lower breast to belly like throat but slightly paler, or almost dull pale orange, extent variable.

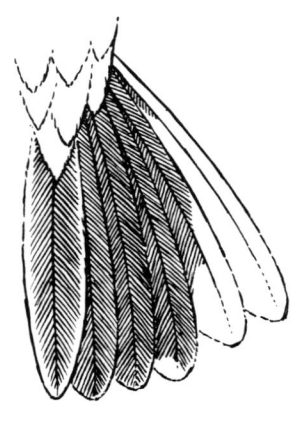

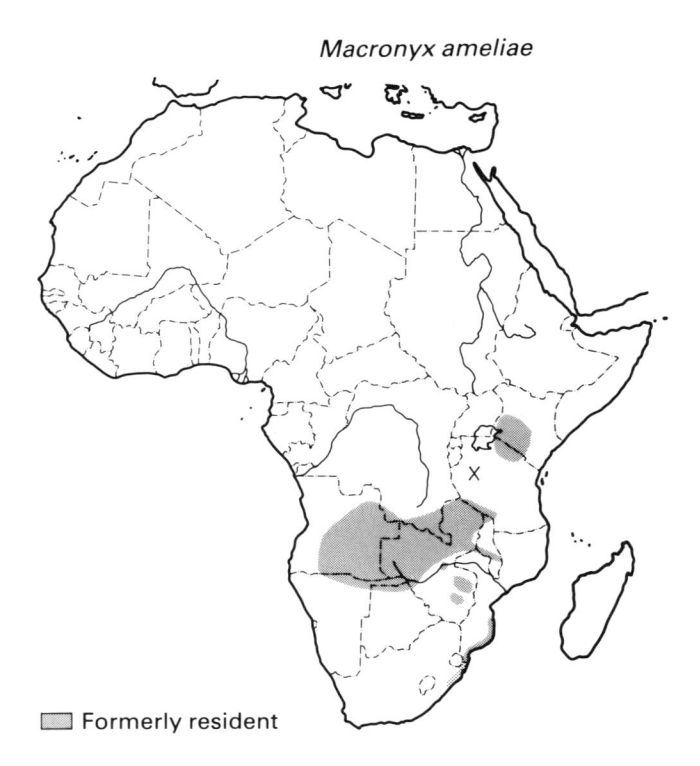

Flanks, thighs and undertail-coverts dull buff, strongly streaked brownish black. Primaries and secondaries bright olive-brown, primaries narrowly edged whitish to pinkish on outer web, secondaries more broadly edged whitish buff; both with pale tips, which soon abrade on primaries. Tertials brownish black, broadly fringed buff to pale cinnamon, paler at tips. Upperwing-coverts dark olive-brown with pale fringes; alula and marginal coverts fringed pink, red or orange (like lower breast), lesser coverts fringed white, median and greater coverts edged buff and tipped whitish. Axillaries and underwing-coverts white, axillaries tinged pink, all feathers basally olive-brown. Underside of remiges pale olive-brown. Upper mandible brownish horn to dark brown (darkest on culmen), sometimes flesh or grey; lower mandible paler, often flesh or grey to whitish, especially at base; inside of mouth dark red-brown, small yellow gape-flanges; eye dark brown to blackish; legs and feet light yellow-brown or brownish flesh to mid-brown, sometimes with orange tinge; soles pale yellow. ADULT ♀: upperparts like ♂ but supraloral area buff. Chin to foreneck buff with variable, often patchy, wash of red or pink or with just a few reddish spots; upper breast buff with brownish black streaks; lower breast to belly duller and less extensively reddish or orange, and washed buff. SIZE: wing, ♂ (n = 36) 85–95 (90·5), ♀ (n = 22) 81–93 (87·3); tail, ♂ (n = 34) 67–86 (74·9), ♀ (n = 15) 61–76 (69·7); bill, ♂ (n = 32) 16–21 (18·8), ♀ (n = 18) 17–20 (18·7); tarsus, ♂ (n = 33) 28–34 (30·5), ♀ (n = 18) 27–32 (29·7); hind claw, ♂ (n = 33) 13–20 (16·0), ♀ (n = 18) 12–19 (15·3). WEIGHT: ♂ (n = 9) 30·3–39·8 (34·4), ♀ (n = 4) 30·8–34 (32·6).

IMMATURE ♂: after post-juvenile moult, similar to adult ♂ on upperparts but supraloral area buff; red or pink of underparts paler and less extensive, often patchy and mixed with buff, especially on throat and foreneck. Necklace incomplete, of blackish streaks and large blotches; complete necklace may not develop until second year; some ♂♂ with fully adult red

A

underparts have only streaked necklace. White on tail-feathers duller. IMMATURE ♀: like adult ♀ but less extensively red or pink on underparts; white on tail-feathers duller. Birds may breed before assuming full adult plumage.

JUVENILE: like adult ♀ but sometimes with feathers of upperparts variably fringed and edged white to buffish white, especially on head, neck, mantle and scapulars; outer web of T6 partly to wholly dusky; chin to foreneck greyish white or tinged buff, with variable small dark spots on neck and throat; rest of underparts buff to whitish buff, sometimes with salmon or pink wash; breast with heavy blackish streaks. White on tail-feathers washed olive-brown. Upperwing-coverts, and sometimes tertials and inner secondaries, fringed either whitish or dull buff.

NESTLING: skin flesh-brown, gape yellow, mouth pale salmon-yellow.

Field Characters. A medium-sized longclaw with rather slender, pipit-like shape; tail relatively longer than other species, flight less stiff-winged. Easily distinguished from all longclaws except Grimwood's Longclaw *M. grimwoodi* by pink throat and underparts; ♀ has streaks on breast instead of black necklace of ♂ and throat may be mainly buff, but always some pink on belly; immature like ♀ but with even less pink. In flight shows pale patch on upperwing formed by whitish fringes to lesser coverts, and more white in outertail, less on tips, than other species, i.e. lacks white terminal band. For distinctions from Grimwood's Longclaw, see that species.

Voice. Tape-recorded (F, 88, CHA, GIB). Song, given in flight or when perched: (1) a series of squeaky whistled notes, 'tee-yoo', (2) one or more 'tee-yoo' notes followed by several hard notes, 'trip' or 'tyip'; both (1) and (2) usually end with drawn-out wheeze; (3) series of 4–6 'tee' notes followed by drawn-out 'yoooo'. Usual call, single, sharp but rather plaintive 'chuit', sometimes longer 'chiteet', often repeated; also metallic, descending 'teeang'.

General Habits. Occurs in short or tussocky grassland (occasionally in longer grass), permanently or seasonally moist (sometimes flooded or dry); near swamps, marshes, rivers, pans and open water, on floodplains or highland downs, by vleis and dambos; on coast also in flooded cultivation, fallow ricefields and wet pasture. At L. Kariba, Zimbabwe, it is a recent colonist on shore in beds of grass *Panicum repens* (Irwin 1981). In Zululand, prefers edges of vleis or lakes with fine grass (*Sporobolus subtilis, S. virginicus* and *Paspalum vaginatum*), especially where grazed short (G. L. Maclean, pers. comm.). Grassland usually treeless but at Ndola, Zambia, with scattered small acacia trees. Although normally ecologically segregated from sympatric congeners by its preference for moister habitat, sometimes occurs alongside them on drier ground. Altitudinal range sea level to 2200 m.

Occurs in pairs or family parties. Found mainly on ground or low perch (grass tussock, termite mound, roadside laterite dump, fence post, low bush, or tall grass stem at level of flower heads where it is concealed from view). Seldom uses tree or bush as singing or observation post, except in Zululand and at Ndola, Zambia, where regularly perches on acacia bushes up to 2 m high. Usually shy and difficult to approach, except for perched singing males. Crouches when disturbed on ground; reluctant to fly, either hiding in grass or running; when flushed rises steeply, often not flying far, and may run after landing. Flight more pipit-like, less stiff-winged, than other longclaws. Seldom calls in flight, unlike other species, unless giving territorial song. Forages on ground, in grass or on earth tracks; also catches insects, especially termite alates, in flight; digs out insect larvae from edges of decaying grass tufts.

♂♂ have extended pre-breeding body moult, continuing throughout breeding period (Clancey 1967).

Mainly sedentary throughout the year but moves locally in response to seasonal habitat changes such as drying out or burning of grassland. Generally sedentary in Zimbabwe but has local movements and numbers may fluctuate considerably, increasing after a series of good rainy seasons. At Ndola, Zambia, 1974–80, recorded in all months but regularly present in moist breeding areas at swamp edges only Oct–Apr; departed when grass completely dry, and was then confined to permanently moist areas, sometimes with tall grass, up to 5 km from breeding grounds; this off-season habitat was not occupied during the rains, when the grass was tall and under water (P. B. Taylor, pers. obs.).

Food. Insects, up to the size of grasshoppers and small locusts, also small frogs.

Breeding Habits. Monogamous, solitary nester. Territorial; ♂♂ fight in air (Kenya). ♂ sings either from top of bush (Zambia and Kenya, song type 3) or in display flight: he rises high, sometimes almost out of sight, and circles or hovers while repeating plaintive song, legs often dangling; glides to earth, legs still dangling, and often hovers over grass before landing. At Ndola, Zambia, singing heard only end-Oct to end-Apr, most Dec–Mar; known breeding Feb–Mar (P. B. Taylor, pers. obs.).

NEST: deep, compact cup of thin dry grass, neatly lined with rootlets or fine dead grass, usually well concealed by overhanging tuft, sometimes with short approach run through dense grass; ext. diam. *c.* 100, int. diam. *c.* 75, depth of cup *c.* 60; on or slightly above ground, against or between grass tussocks, occasionally inside crown of old tuft. Built by ♀, with ♂ in attendance (P. B. Taylor, pers. obs.).

EGGS: 2–4, av. (n = 30 clutches) 2·8. Subelliptical to ovate, slightly pointed at one end; very pale green or blue to white, heavily spotted and blotched brown and violet, grey, pale lilac or mauve; markings sometimes concentrated at thick end. SIZE: (n = 32) 20–23·8 × 15–17·3 (21·8 × 16·2).

LAYING DATES: Kenya, Apr–June, Dec; Tanzania, Dec–Jan, Mar, May (gonads active Nov); Zambia, Sept, Nov–Dec, Feb–Mar, June; Zaïre, Dec–Jan; Malaŵi, Dec–Jan, Mar; Mozambique, Mar; Zimbabwe, Nov–Apr; Botswana (breeding condition Oct); South Africa, Oct–Dec; breeds mainly during or after rains.

MOTACILLIDAE

INCUBATION: by ♀ only. Period: 13–14 days. Incubating bird leaves nest readily until incubation advanced, when it may sit very close.

DEVELOPMENT AND CARE OF YOUNG: young fed by both parents, or by ♀ with food taken from ♂ (van Someren 1956); nestling period *c*. 16 days. On leaving nest, young are feathered but hardly able to fly. ♂ gives distraction display, usually *c*. 100 m from nest, when intruder approaches: perches on bush or termite mound, uttering loud 'chuit' calls, or hovers and flutters jerkily at height of *c*. 10 m with dangling legs; leads intruder away by making short low flights, pausing to repeat calls from perch.

References
Irwin, M. P. S. (1981).
van Someren, V. G. L. (1956).

Plate 14
(Opp. p. 241)

Macronyx grimwoodi Benson. Grimwood's Longclaw. Sentinelle de Grimwood.

Macronyx grimwoodi Benson, 1955. Bull. Br. Orn. Club 75, p. 102; Chitunta Plain, Mwinilunga District, N Rhodesia.

Range and Status. Endemic resident from SW Zaïre (Gungu, near Kwilu R.) to central and E Angola (N Bié, N Moxico and Lunda provinces) and extreme NW Zambia (Mwinilunga District, south to Kampanda). Local, frequent to common.

Description. ADULT ♂: feathers of upperparts cinnamon with prominent blackish centres, giving boldly patterned appearance: forehead and crown almost black, with narrow cinnamon streaks; hindneck and nape with broader cinnamon edges and less prominent brownish black centres; sides of neck cinnamon with blackish spots or streaks. Lores and short moustachial streak black; supercilium, face and malar region cinnamon; ear-coverts cinnamon-brown. Mantle and scapulars with broad blackish centres and cinnamon to tawny edges; back and rump with broader, rich tawny fringes; uppertail-coverts fringed buff to cinnamon. Central 4 tail-feathers (T1–T4) dull dark olive-brown, shafts pale; T1 and T2 with narrow buffy brown fringes; T3 with 7–10 mm white spot at tip of inner web; distal third of inner web of T4 white; proximal half of T5 dull dark olive-brown, distal half of inner web white and of outer web pale brown-grey; inner web of T6 white almost to base, outer web pale brown-grey (**A**). Chin, throat and foreneck peach-red to burnt orange; breast to undertail-coverts dull cinnamon, tinged tawny buff and sometimes pinkish buff on centre of belly; entire breast, from immediately below red of neck, streaked or spotted blackish, markings heaviest at sides; flanks with long, narrow, brownish black streaks; thighs and undertail-coverts with broader, brownish black streaks. Primaries and secondaries olive-brown, with cinnamon edges to outer webs and whitish tips in fresh plumage; tertials blackish with broad buff to cinnamon fringes. Upperwing-coverts brownish black with broad pale fringes, on leading edge of outer wing and alula pinkish buff, on lesser coverts white, buff to cinnamon on other coverts. Axillaries white, underwing-coverts pale olive-brown, fringed white; underside of remiges pale olive-brown. Bill dark horn, lower mandible often paler, even whitish, or with grey to whitish base; eye dark brown; legs and feet horn or brown, often with reddish or pinkish tinge. ADULT ♀: like ♂, but chin and throat paler, flesh to flesh-ochre, and feathers of leading edge of wing and alula less markedly fringed pinkish buff. SIZE: (7 ♂♂, 3 ♀♀) wing, ♂ 90–100 (94·6), ♀ 89–97 (94·0); tail, ♂ 66–76 (70·6), ♀ 66–73 (70·0); bill, ♂ 21–24 (21·9), ♀ 17·5–22 (20·2); tarsus, ♂ 35–40 (37·0), ♀ 31–40 (35·7); hind claw, ♂ 17–21 (19·0), ♀ 18–21 (19·7).

IMMATURE AND NESTLING: unknown.

Field Characters. Larger than Rosy-breasted Longclaw *M. ameliae*, with longer, heavier bill. Upperparts with broader blackish streaks and brighter cinnamon feather edges, pink of underparts confined to throat, breast streaked (no necklace), flank streaks narrower and less extensive. Similar pale wing-patch (white fringes to lesser coverts) shows in flight but is more contrasting, since rest of wing darker.

Voice. Unknown.

General Habits. Occurs, often alongside Rosy-breasted and Fülleborn's Longclaws *M. fuelleborni*, in moist short grass on plains, near streams and rivers, along drainage lines (dambos) and in grassy hollows. Where all 3 species occur together (e.g. in NW Zambia), Grimwood's Longclaw may occupy the wettest areas and Fülleborn's Longclaw the driest, but there is often no apparent ecological difference between Grimwood's and Rosy-breasted (Benson *et al.* 1971). Altitudinal range 800–1500 m. Behaviour resembles that of other grassland longclaws but has not been studied. Will perch on tall grass stems in open areas.

Food. Small beetles, weevils, grasshoppers, pentatomid bugs.

Breeding Habits. Almost unknown. Song-flight resembles that of Fülleborn's Longclaw (Lippens and Wille 1974). Nest and eggs undescribed, but one nest with 2 eggs found on ground at Gungu (SW Zaïre) on 18 Feb 1958 during rains (Lippens and Wille 1976).

Family CAMPEPHAGIDAE: cuckoo-shrikes

Small to medium-sized birds. Bill broad at base, often rather stout, notched and slightly hooked. Nostrils round, obscured or nearly so by frontal feathering and well-developed rictal bristles. Tarsi short, scutellated in front, smooth behind; feet quite strong. Wings rather long and pointed, with outermost primary (P10) about half length of P9, P7–P8 longest. Tail with 12 rectrices, usually moderately long, and rounded. Feathers of lower back and rump thick and partially erectile, with stiffened pointed shafts; easily shed.

Sexually dimorphic, in some genera markedly so (especially *Campephaga*). Nestling plumage usually white, replaced by plumage similar to adult ♀ at first moult. Inhabit tree canopy and secondary growth; feed mainly on insects, including hairy caterpillars. The minivets of Asia (*Pericrocotus*) are brightly coloured, occurring in lively parties; other genera less active, usually solitary, most with grey, black and white plumage. One long-range migrant in E Asia (*Pericrocotus brevirostris*); other species move locally in Africa and Australia.

10 genera, 72 species. Old World, mainly tropics; 3 genera in Africa (2 endemic), with 10 species, all endemic. DNA–DNA hybridization studies indicate close relationship with orioles (Oriolidae) (Sibley and Ahlquist 1990).

Genus *Campephaga* Vieillot

Slim, small- to medium-sized birds. Bill quite small, but broad at base; tail rather long, and rounded, with outer feathers considerably shorter (up to 30 mm) than T5. Thick, partially erectile feathering often gives rounded appearance to back. Sexes very different: ♂♂ mostly glossy black with prominent gape wattles, ♀♀ olive-yellow and white, most with strong blackish barring. Mainly solitary, unobtrusive and silent; glean and snatch insects in forest and woodland canopy and foliage of moist thicket.

4 species, endemic to Africa; 2 resident in forest, 2 resident and locally migrant in forest edge, woodland and thicket.

Because *C. phoenicea*, *C. flava* and *C. petiti* are broadly allopatric, we consider that they form a superspecies. However, they present a complex taxonomic situation. *C. phoenicea* and *C. flava* are very similar in plumage, occupy similar habitats and are locally migratory. Their breeding ranges overlap only in SW Ethiopia and (rarely) W Kenya. No intermediate ♂♂ are known, but some ♀♀ from central and E Kenya (east of the range of *C. phoenicea* ♂♂) have yellow in tail-feathers intermediate in extent between *C. phoenicea* and typical *C. flava* (Hall and Moreau 1970).

CAMPEPHAGIDAE

C. petiti is a resident forest bird, and while ♂♂ closely resemble all-black *C. flava*, ♀♀ are quite different from those of *C. phoenicea* and *C. flava*, having underparts bright yellow with little barring. Where *C. petiti* and *C. phoenicea* overlap geographically in Cameroon and NE Zaïre, they are ecologically separated. Where *C. petiti* and *C. flava* overlap in W Kenya, E Zaïre and NW Angola, they are largely separated, with *C. petiti* in true forest, *C. flava* in woodland and forest edge. However, ♀♀ or young ♂♂ from L. Kivu and Ruzizi valley in the Rwanda Rift, and from NW Angola, have underparts that are barred but are also strongly yellow, suggesting that they are hybrids between *C. petiti* and *C. flava* (Hall and Moreau 1970, Prigogine 1972, Vande weghe 1988). Because *C. petiti* and *C. flava* in the Rwanda Rift and NW Angola and *C. phoenicea* and *C. flava* in central and E Kenya show some intermediate characters, further studies are needed to determine to what extent *C. phoenicea*, *C. flava* and *C. petiti* are genetically isolated from each other.

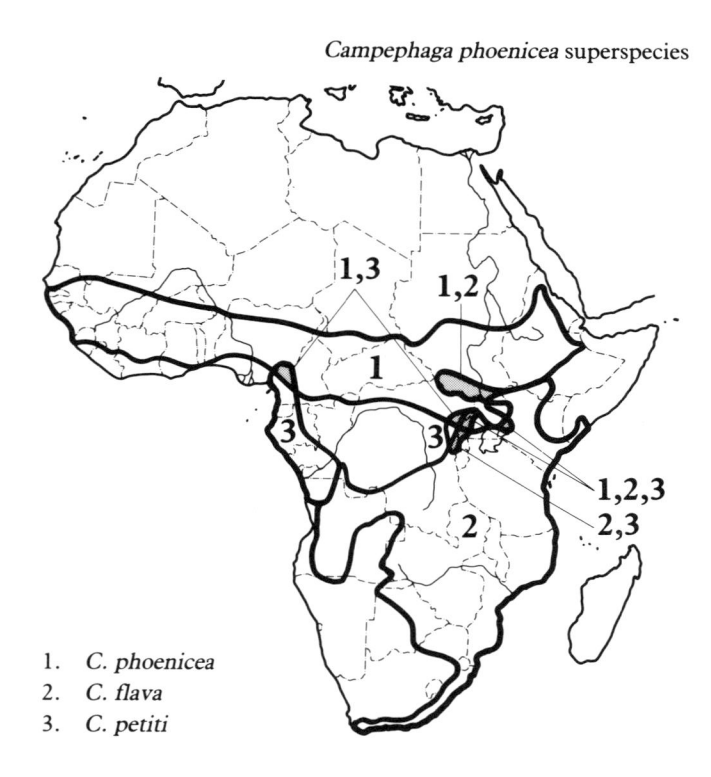

Campephaga phoenicea superspecies

1. *C. phoenicea*
2. *C. flava*
3. *C. petiti*

Plate 15 *Campephaga phoenicea* (Latham). Red-shouldered Cuckoo-Shrike. Echenilleur à épaulettes rouges.
(Opp. p. 256)
Ampelis phoenicea Latham, 1790. Ind. Orn. 1, p. 367; Gambia.

Forms a superspecies with *C. flava* and *C. petiti*.

Range and Status. Endemic resident and intra-African migrant, widely distributed in savanna north of equator to 12°–14°N (16°N in Ethiopia); from extreme S Mauritania (S Guidimaka: Karakoro), E and S Senegambia, S Mali (Kami, Bamako), Guinea, N and central Sierra Leone, N and central Ivory Coast, Ghana (throughout except SW), Togo, SW Niger (west of Niger R., Park du W), N and central Nigeria (also along coast in W), Cameroon (4–5°N northwards), N Congo (Impfondo), S Chad (south of *c*. 10°N), and Central African Republic to Sudan (north to Darfur, Kordofan and the Upper Blue Nile), Ethiopia (W and central areas north to 16°N and SE Highlands), N Zaïre (south to Bomokandi R. and Irumu), Uganda (NW and central, south to Ankole) and extreme W Kenya (east to Kakamega). Uncommon to locally common. Vagrant W Zaïre (near Lukolela) and NE Rwanda (J. P. Vande weghe, pers. comm.). Wet season visitor to northernmost parts of range.

Description. ADULT ♂: entire body black with greenish blue gloss. Bristly feathers of forehead and pre-orbital area black without gloss. Tail-feathers black, edged greenish blue. Wings largely black; secondaries and tertials edged greenish blue; lesser and median coverts bright red with yellow bases, sometimes entirely orange or orange-yellow; rest of upperwing, underwing-coverts and axillaries black with blue gloss; inner webs of primaries and secondaries with slight yellow tinge. Bill black; gape yellow; eye black or dark brown to grey; legs and feet black. ADULT ♀: upperparts brown to grey-brown, rump and uppertail-coverts sometimes greyer, sometimes richer brown; black barring on rump and uppertail-coverts, variably extending onto back and scapulars. Tail-feathers blackish

A

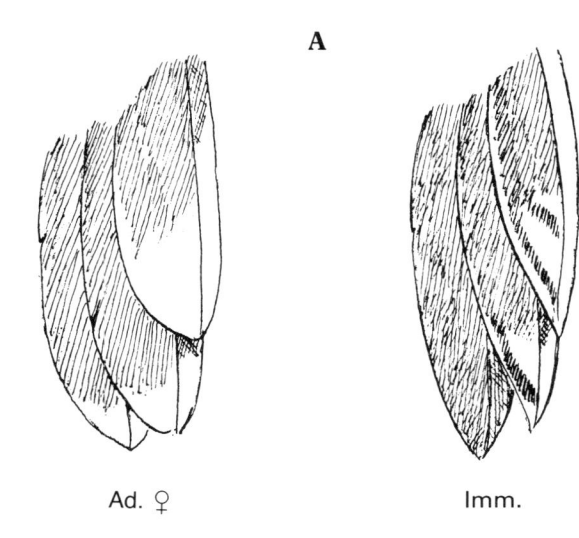

Ad. ♀   Imm.

brown, central pair tinged green; outer pair with outer web entirely yellow, outer portion of inner web yellow (up to c. 30 mm) (**A**, left), T5 with outer web either all yellow or distal half yellow, less yellow on tip of inner web (c. 20 mm), T4 with distal portion of outer web yellow, inner web with still less yellow at tip (c. 10 mm). Supercilium pale grey or whitish, lores and streak through eye dark brown, ear-coverts greyish brown streaked white. Underparts white with black bars and crescentic marks, strong yellow tinge to sides of breast, often extending to centre of breast and flanks, tibial feathers yellow barred black. Wings blackish brown; primaries, alula and primary coverts narrowly edged yellow; distal part of secondaries, tertials and rest of upperwing-coverts edged and tipped yellow; underwing-coverts and axillaries bright yellow; inner webs of secondaries and primary bases with broad yellow edge, giving generally yellow appearance to underwing. SIZE: (10 ♂♂, 10 ♀♀) wing, ♂ 94–102 (97·8), ♀ 95–102 (97·4); tail, ♂ 82–90 (85·2), ♀ 84–93 (88·3); bill, ♂ 16–20 (17·9), ♀ 18–21 (19·4); tarsus, ♂ 18–19 (18·4), ♀ 18–19 (18·9). WEIGHT: (Liberia) 1 immature ♂ 26; (Ghana) 1 immature ♂ 27·5, ♀ (n = 4) 25·5–30 (28·0), unsexed adult (n = 7) 28·1–30 (29·5); (Ethiopia) ♂ (n = 44) 23·1–32·5 (27·0), ♀ (n = 44) 23–32·1 (27·2); (Kenya/Uganda) ♂ (n = 3) 29–31 (29·7), 1 ♀ 26.

IMMATURE: feathers of top and sides of head and upperparts with white terminal and black subterminal bands, underparts more heavily barred than in ♀, bars broader, those on belly short and spade-shaped, producing blotchy, almost spotted appearance. Tail-feathers pointed, outer webs as in ♀, tips of inner webs not plain yellow like ♀ but patterned, with broad diagonal stripe of dull brownish yellow, crossed by narrower diagonal brown stripe separating paler edge from yellower interior (**A**, right).

NESTLING: unknown.

Field Characters. Glossy blue-black ♂ with red or orange shoulder-patch readily distinguishable from black ♂♂ of other species; however, some individuals have orange-yellow shoulder-patch which almost matches yellow patch of some Black Cuckoo-Shrikes *C. flava*; shoulder-patch of latter usually but not invariably smaller, confined to lesser coverts, but a few individuals probably not safely distinguishable. ♀ with black-barred white underparts extremely similar to ♀ Black Cuckoo-Shrike but greyer above, without olive-green tinge, and underside of tail of perched birds shows much less extensive yellow (outer half yellow, basal half dark).

Immatures also very similar but immature Black Cuckoo-Shrike has more yellow in tail, without striped tips.

Voice. Tape-recorded (BRU, CHA, ERA). Song (duet of ♂ and ♀) a bustling medley of high-pitched squeaks, churrs and whistles continuing without pause for 5–10 s; sounds rather like a high-class sunbird. Calls of pair in flight, a burbled 't-chew' 'ti-chulu' or 'ti-chew-cher', often repeated; contact calls between ♂ and ♀, a sharp repeated 'tchit' or 'tchit-tchiwi-tchiwi'.

General Habits. Frequents thickets in savanna, wooded grassland, moist secondary growth, forest patches including edges and clearings; generally in thicker habitat than Black Cuckoo-Shrike in Kenya and Uganda. Ranges from sea level to about 1600 m.

Usually singly or in pairs. Unobtrusive and mainly silent, keeping to bushes and small trees and flying low from one to the next with undulating flight. Forages mainly within foliage, gleaning from leaves and twigs; occasionally drops to ground to take prey and makes aerial flycatching sallies.

Migratory in W Africa east of c. 8°E, and in Sudan, moving north to breed during the rains Mar/Apr–Sept/Oct, and returning to lower latitudes to edge of forest belt in Nov–Feb dry season. In Nigeria present in S guinea zone all months, moving north to Zaria May–Nov and south to coast in west Nov–Apr. In Sudan present Darfur (c. 12°N) in rains. In Uganda resident with increase in population in Jan–Apr probably due to non-breeding migrants from N tropics. Elsewhere apparently wanders randomly (Britton 1973).

Food. Caterpillars and other insects (Orthoptera, Hemiptera).

Campephaga phoenicea

Breeding Habits. Monogamous.

NEST: small shallow cup of moss and lichens, bound with spider webs; well concealed in fork of leafless tree c. 10 m high (Uganda: Jackson and Sclater 1938).

EGGS: 2 (1 clutch Uganda, 1 Kenya). Yellowish green, spotted and mottled with purplish. SIZE: 21 × 16.

LAYING DATES: Gambia (carrying nesting material, Aug); Nigeria, May–Sept; W Sudan (Darfur), July; S Sudan, June; Uganda, Mar–Apr; Kenya (building, June).

Reference
Britton, P. L. (1973).

Plate 15
(Opp. p. 256)

Campephaga flava Vieillot. Black Cuckoo-Shrike. Echenilleur noir.

Campephaga flava Vieillot, 1817. Nouv. Dict. d'Hist. Nat. 10, p. 49; South Africa.

Forms a superspecies with *C. phoenicea* and *C. petiti*.

Range and Status. Endemic resident and intra-African migrant, widely distributed in E, central and southern Africa; from SW Ethiopia (north to southern W highlands and southern Rift Valley), Sudan (extreme S only), Kenya (throughout except NE), and S Somalia (Middle and Lower Jubba Valley and coast) south through Uganda (south of c. 1°N), Tanzania (including Zanzibar and Mafia), Rwanda, Burundi, E and S Zaïre (north to Luluaborg, Kasongo, L. Kivu and L. Albert), Zambia, Malaŵi, Mozambique and Zimbabwe to Angola (S Huila to SE Cuanza Norte, Malange and Lunda), Namibia (extreme N and Caprivi, also in W near Karibib), N and E Botswana, South Africa (N and E Transvaal, E Natal to S and SW Cape) and Swaziland. Uncommon to frequent, though seasonally common Kenya, Zambia, S Mozambique and Transvaal (Nylsvlei) where density c. 1 pair/30 ha (Tarboton *et al.* 1987). Mainly a non-breeding visitor SW Zaïre, N Tanzania, Rwanda, Uganda, Kenya and S Sudan.

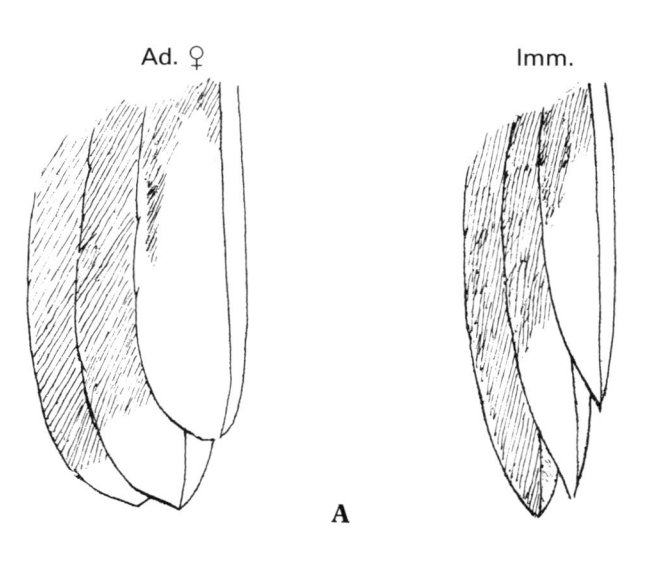

Description. ADULT ♂: entire body black with greenish blue gloss. Bristly feathers of forehead and pre-orbital area black without gloss. Tail-feathers black, edged greenish blue. Wings largely black, secondaries, tertials and upperwing-coverts edged greenish blue; lesser coverts often yellow, forming patch near bend of wing; underwing-coverts blue-black or blue-black and yellow; axillaries blue-black; inner webs of primaries and secondaries washed yellow, giving yellowish grey appearance to underwing. Bill black; gape yellow; eye dark brown; legs and feet black. ADULT ♀: olive-green to olive-brown above shading to grey or brown on rump and uppertail-coverts, which are barred black, barring variably extending onto back and scapulars. Tail blackish brown, central feathers tinged green, outer pair entirely yellow except for basal one-third of inner web, T5 with all except base of outer web and distal half of inner web yellow, T4 with distal half of outer web and 10–15 mm tip to inner web yellow (**A**, left). Supercilium pale; lores and streak behind eye dark brown; ear-coverts streaked brown and white. Underparts white with blackish bars and crescentic marks; yellow tinge on sides of breast, often extending to centre of breast and flanks, tibial feathering yellow with black bars. Wings blackish brown; primaries, alula and primary coverts narrowly edged yellow; distal part of secondaries, tertials and rest of upperwing-coverts broadly edged and tipped yellow; underwing-coverts and axillaries bright yellow; inner webs of secondaries and primary bases with broad yellow edge, giving generally yellow appearance to underwing. SIZE: (10 ♂♂, 10 ♀♀) wing, ♂ 100–111 (104), ♀ 99–108 (102); tail, ♂ 86–99 (93·2), ♀ 85–96 (92·2); bill, ♂ 16–20 (17·7), ♀ 18–21 (19·5); tarsus, ♂ 19–21 (20·1), ♀ 20–21 (20·5). WEIGHT: (Zambia, Botswana, Zimbabwe) ♂ (n = 5) 30–33 (31·6); (Zambia, Botswana, South Africa) ♀ (n = 3) 34–40 (36·8); (Kenya, Uganda) ♂ (n = 3) 29–34 (32·3), ♀ (n = 7) 27–34 (30·9).

IMMATURE: like ♀ but top and sides of head and upperparts entirely barred, feathers with black subterminal and white terminal bands; underparts more heavily barred, bars broader, those of belly short and spade-shaped producing blotchy or spotted appearance. Tail-feathers pointed, with yellow areas as in ♀ (**A**, right).

NESTLING: skin deep purplish black on top, reddish below; bill black, mouth yellow; eyes closed; body covered with white down (42 mm long) (Skead 1966) (**B**).

Field Characters. All-black ♂ similar to Petit's and Purple-throated Cuckoo-Shrikes *C. petiti* and *C. quiscalina*; Petit's has same blue-green gloss, best separated by more extensive and conspicuous yellow gape; Purple-throated has strong rich purple gloss on face, neck and underparts. For separation of yellow-shouldered ♂♂ from Red-shouldered Cuckoo-Shrike *C. phoenicea*, see that species. Drongos *Dicrurus* are also black with bluish sheen, but have square or forked tails (tails rounded in cuckoo-shrikes), loud voices, aggressive behaviour and use conspicuous perches. ♀ extremely similar to ♀ Red-shouldered but upperparts olive-green rather than grey or brown, underside of tail of perched bird more extensively yellow. Immatures also similar, but immature Red-shouldered has striped tip to underside of tail.

Voice. Tape-recorded (86, 88, F, GIB, MOY, PAR). A high-pitched, penetrating trill with a rather ringing quality, not dry; lasting *c.* 1 s, often immediately repeated to form a double trill. Also a very different, harsh, down-slurred 'chweeu', a hissing note, a short 'chip' and various squeaking notes.

General Habits. Frequents a wide range of habitats including miombo, mopane and acacia woodland, riparian forest, edge of evergreen forest, semi-arid bushland, scrub, savanna, gardens and exotic plantations. In E Africa ranges from coast to montane forests up to 1800 m (exceptionally to 3000 m), but breeds above 900 m. In S-central Africa breeds mainly in woodland at 1000–2200 m, but occurs along rivers and in forest edge as low as 50 m in non-breeding season. Breeds near sea level in Cape Province.

Occurs singly, in pairs or small groups usually dominated by ♀♀/immatures; once 12 ♂♂ together (Benson and Benson 1977); common in mixed bird parties. Mainly unobtrusive and rather silent, foraging in foliage of canopy and edges, searching leaves, lichens and bark and catching prey in fluttering hops or short flights; occasionally feeds on ground. Usually moves on quickly from one tree to the next. ♂ may become active when excited, flying back and forth with trilling or harsh churring calls, and flicking wings over back (Skead 1966). Raises rump feathers when alarmed, and directs rounded back towards source of danger. Flight rather weak-looking, with slow regular wing-beats.

A wanderer; local appearances tend to be irregular, although pair may occupy site for some months (Skead 1966). Many populations partly or wholly migratory. Vacates plateau country of southern Africa after breeding, and moves to lower altitudes; many birds move north to the equator. Occurs Kenya, N Tanzania, Uganda and Rwanda, mainly as non-breeding visitor Apr–Sept; Somalia June–Oct; S Sudan as non-breeding visitor during rains. In Malaŵi, leaves plateau for lower elevations Apr/May–Sept; in Zambia present mainly Oct–Apr with some overwintering east of 25°E; in Zimbabwe, present central plateau Sept–Apr; in Botswana present Oct–May; in Transvaal, present mainly Oct–May with some overwintering; in Natal, present mid-Sept–June; in Cape Province all months. Attracted to lighted area at night SE Kenya (Ngulia) Nov–Jan.

Food. Insects (mainly caterpillars but also moths, katydids, flying termites); also spiders, fruit and tree seeds.

B

Breeding Habits. Monogamous. Pair often moves about in breeding season, ♂ following close behind ♀. Excited ♂ may perform fluttering moth-like flight near ♀ (Skead 1966).

NEST: small shallow cup of moss and lichens (**B**), lined with finer fibres, hair and leaves, and bound and secured to branch with spider webs; outer diam. 70–100, diam. of cup 40–45, depth 13–20, in flat fork or saddle, usually 5–18 m high in tree, well camouflaged, often far out on branch; in Zambia and Zimbabwe, usually in *Brachystegia* trees. ♀ builds nest; takes up to 10 days, working 2 h at a time (van Someren 1956); ♂ accompanies ♀ during building.

EGGS: 1–3, usually 2, av. (n = 37 clutches, southern Africa) 1·9. Pale green, yellowish green or greyish green, speckled with dark brown and with underlying violet-grey marks. SIZE: (n = 42, southern Africa) 22–26 × 16–18 (23·6 × 17·3).

LAYING DATES: Ethiopia, Apr (possibly breeding, Feb); Kenya, Mar–May; N Tanzania, Jan–Feb, Apr–May, Dec; Zanzibar, Oct; SE Zaïre, Oct–Jan; Zambia, Oct–Jan; Malaŵi, Oct–Jan (enlarged gonads, Sept); Zimbabwe, Sept–Feb (Sept 5, Oct 13, Nov 44, Dec 30, Jan 6, Feb 3 clutches); Mozambique, Sept–Jan; South Africa: Transvaal, Oct–Jan (Oct 4, Nov 18, Dec 17, Jan 6 clutches); E Cape, Oct, Jan.

INCUBATION: by ♀ only; period c. 20 days; ♂ feeds ♀ on nest and defends area near it.

DEVELOPMENT AND CARE OF YOUNG: on day 5 young have feathers emerging on back, flanks and breast (but not head, nape, crurals or tail), eye one-quarter open; on day 6, feathers on abdominal tracts and head, primaries and tail-feathers small (Skead 1966). Brooded by ♀ only, fed by both parents from hatching on. Leave nest 20–23 days after hatching.

BREEDING SUCCESS/SURVIVAL: 1, Zimbabwe, recovered at place of ringing c. 3 years later (Irwin 1981); 1 ♀ 5–5·5 years old (Hanmer 1985).

References
Britton, P. L. (1973).
Skead, C. J. (1966).
van Someren, V. G. L. (1956).
Vincent, A. W. (1949).

Plate 15
(Opp. p. 256)

Campephaga petiti Oustalet. Petit's Cuckoo-Shrike. Echenilleur de Petit.

Campephaga petiti Oustalet, 1884. Ann. Sci. Nat. Zool. Ser. (6), 17, art. 8, p. 1; Landana.

Forms a superspecies with *C. flava* and *C. phoenicea*.

Range and Status. Endemic resident, mainly equatorial evergreen forest. Several sight records SE Nigeria (Gotel Mts and Obudu Plateau: Ash *et al*. 1989); SW Cameroon (forest/savanna mosaic of montane district, between c. 6·5–5·5°N, 10–11°E); Gabon, W Congo, W Zaïre (Lower Congo), NW Angola (Cabinda, S Cuanza Norte and Cuanza Sul); also NE Zaïre (E edge of the rain forest and L. Kivu), SW Uganda (Impenetrable, Maramagambo, Kalinzu, Kashoya-Kitomi, Kibale and Itwara forests) and W Kenya (Kakamega and Nandi forests). Uncommon to locally frequent, although locally common Kakamega.

Description. ADULT ♂: head and body black with greenish blue gloss. Bristly feathers of forehead and pre-orbital area, centre of belly and thighs without gloss. Tail-feathers black, edged greenish blue. Wings largely black; secondaries, tertials and upperwing-coverts edged glossy greenish blue; underwing-coverts and axillaries black; inner webs of primaries and secondaries grey. Bill black; large gape wattles yellow; eye dark brown; legs and feet black. ADULT ♀: top of head to upper mantle olive-green; scapulars, lower mantle and back barred black and yellow with strong olive wash which fades out on lower back; rump and uppertail-coverts with narrower, wavy, less distinct barring. 2 central pairs of tail-feathers dark olive-green, narrowly tipped yellow; rest blackish, narrowly edged olive. T3 with c. 7–8 mm yellow tip, T4 with distal half of outer web and c. 15 mm tip of inner web yellow, T5 with whole of outer web and c. 25 mm of inner web yellow, outer pair (T6) all yellow except basal half of inner web. Nasal feathering, supraloral line and supercilium buff; lores and short streak behind eye blackish brown; broken eye-ring white; ear-coverts olive streaked white or pale yellow. Chin and upper throat white or pale yellow, rest of underparts bright rich yellow with variable amount of barring, some with almost none or with just a trace on sides of breast and neck, most typical birds with barring across breast and upper flanks, a few with entire underparts barred except centre of belly. Wings black, distal part of primaries, alula and primary coverts narrowly edged yellow; distal part of secondaries, tertials and rest of upperwing-coverts broadly edged and tipped yellow; underwing-coverts and axillaries yellow; inner webs of secondaries and primary bases with broad yellow edge, giving generally yellow appearance to underwing. SIZE: (8 ♂♂, 10 ♀♀) wing, ♂ 92–101 (97·1), ♀ 94–102 (97·3); tail, ♂ 82–92 (89·2), ♀ 83–93 (88·9); bill, ♂ 18–19 (18·4), ♀ 18–20 (18·7); tarsus, ♂ 18–21 (19·2), ♀ 19–21 (19·6). WEIGHT: (Cameroon, Uganda, Kenya) ♂ (n = 16) 26–37 (32·0); (Cameroon, Uganda) ♀ (n = 3) 28–36 (30·7).

IMMATURE: top and sides of head dark green with indistinct black barring, mantle and upper back olive-green with dark bars and very narrow white terminal fringe; lower back yellow barred black, rump and uppertail-coverts yellow with more indistinct black barring. Sides of neck and breast with dark bars, rest of underparts yellow with large, spade-shaped black marks, heaviest on breast, usually with an area of unmarked clear yellow on both sides of lower breast and belly. Tail-feathers pointed, otherwise much like ♀ except yellow somewhat less extensive.

NESTLING: unknown.

Field Characters. ♂ glossy blue-black, very similar to ♂ Black Cuckoo-Shrike *C. flava*, but yellow gape wattles much more conspicuous; never has yellow shoulder-patch. ♂ Purple-throated Cuckoo-Shrike *C. quiscalina* has purple face, throat and breast, smaller gape wattles. ♀ told from Black and Red-shouldered *C. phoenicea* Cuckoo-Shrikes by greener upperparts with black and yellow barring, bright yellow underparts; ♀ Purple-throated has similar (but unbarred) yellow underparts but grey head and face, unbarred green upperparts, plain green, not black and yellow, wings. Immature additionally told from immature Purple-throated by black spots on underparts.

Voice. Tape-recorded (McVIC).

General Habits. Inhabits moist evergreen primary and secondary forest including secondary growth and patches and strips in forest/savanna mosaic, mainly at 1000–2000 m but near sea level along Lower Congo and Angola. In Kibale Forest, Uganda, commonest in logged areas with dense undergrowth and sparse canopy (J. Kalina, pers. comm.). Usually occurs singly or in pairs. Silent and unobtrusive; forages in higher and middle canopy of trees, but also sometimes in undergrowth. Perches motionless and rather upright, sometimes on exposed twig.

Food. Caterpillars; also flying insects and grasshoppers.

Breeding Habits.
NEST: cup of lichens and mosses, placed on branch fork *c.* 13 m above ground near top of *Markhamia platycalyx* tree (Uganda: Skorupa 1982).
LAYING DATES: Zaïre, Feb–May; Uganda, June (1 record); Kenya (enlarged gonads and courting June–July).

Campephaga quiscalina Finsch. Purple-throated Cuckoo-Shrike. Echenilleur pourpré.

Campephaga quiscalina Finsch, 1869. Ibis p. 189; Fanti.

Plate 15
(Opp. p. 256)

Range and Status. Endemic resident with fragmented distribution in evergreen forest in equatorial and S central Africa; from Sierra Leone, S Guinea, N Liberia (Mt Nimba, Ganta), S Mali ('south of Selengue' 11° 37'N, 8° 14'W), Ivory Coast (5–8°N, east to 5°W), Ghana (Akropong, Amedzofe, Ejura, Subri River Forest Reserve, between *c.* 6–7·5°N, 1°W–1°E), W Togo (Ahoué-houé), Nigeria (Nindam Forest Reserve), W and SW Cameroon, Gabon (W and NE including Makokou region) and SW Central African Republic (Lobaye Préfecture) to NW and W Angola (Cabinda, Cuanza Norte Province, Chouzo and Chingoroi); also S Sudan (Imatong Mts and Aloma Plateau) and NE Zaïre to central Uganda (Malabigambo Forest, Mubende, Mabira, Mt Elgon), Kenya (W and central highlands east to Mt Kenya and Nairobi and south to Nguruman), Tanzania (Bukoba and Minziro in NW, Loliondo and Oldeani in N, Ulugurus, Mwanihana Forest and Mahenge in E), Zambia east of 24°E, north of 14°S and the Muchinga Escarpment and west of 32°E; and SE Zaïre (S Shaba including Kambove). Uncommon to frequent, except locally common Kenya, Zambia and Angola.

Description. *C. q. quiscalina* Finsch: Guinea and Sierra Leone to Lower Congo and Angola. ADULT ♂: entire body black, ear-coverts, cheeks, chin, throat and breast glossed purple, grading into blue-purple on belly, hindneck blue-purple, otherwise glossed steel-blue. Bristly feathers of forehead and pre-orbital area black without gloss. Tail-feathers black, edged steel-blue. Wings black; secondaries, tertials and upperwing-coverts edged glossy blue; underwing-coverts and axillaries black glossed purple; underside of primaries and secondaries brownish black. Bill black; gape yellow to deep orange; eye dark brown or red-brown; legs and feet dark brown or blackish.

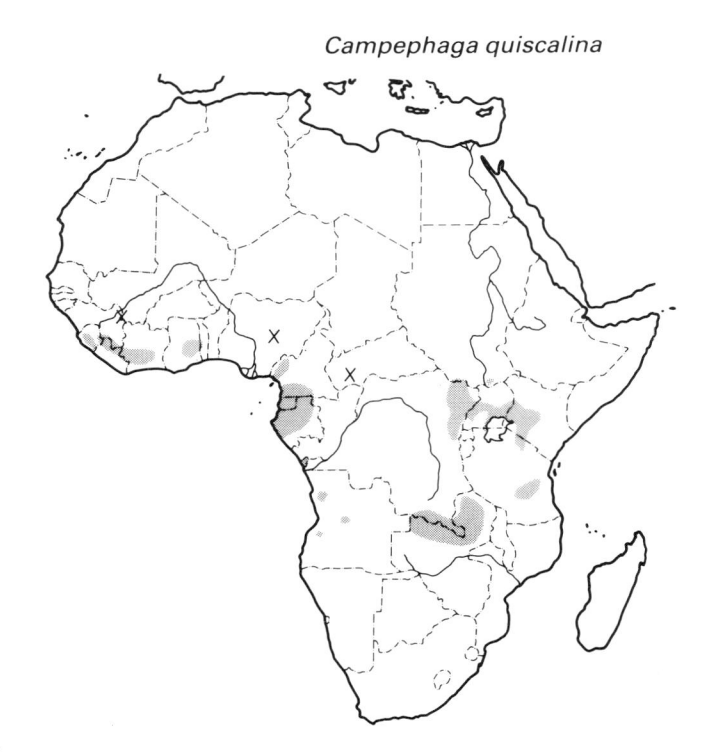

ADULT ♀: top of head to hindneck and upper mantle grey; supraloral line and indistinct supercilium grey-white, lores and streak behind eye black, broken eye-ring white. Cheeks and ear-coverts with variable amount of white streaking, streaks broad to narrow. Upperparts including scapulars yellowish green, somewhat brighter on rump and uppertail-coverts. Tail-feathers brown with strong green wash, T3 and T4 with narrow yellow tips, T5 with broader (c. 10 mm) tip and edges of distal one-third of inner and outer webs yellow, outer pair (T6) with entire outer web, broad (c. 20 mm) diagonal tip and narrow fringe to all of inner web yellow. Chin and throat white, rest of underparts bright yellow or orange-yellow. Wings brown, feathers edged green, outer webs of smaller coverts and tertials wholly green; underwing-coverts and axillaries yellow; yellow edge to inner webs of primaries and secondaries giving generally yellow appearance to underwing. Soft parts as adult ♂ but lacks noticeable gape. SIZE: (10 ♂♂, 8 ♀♀) wing, ♂ 95–105 (100), ♀ 96–99 (97·3); tail, ♂ 72–77 (73·7), ♀ 71–79 (76·8); bill, ♂ 18–20 (19·2), ♀ 19–21 (19·8); tarsus, ♂ 19–21 (19·7), ♀ 17–20 (19·1). WEIGHT: (Liberia) ♂ (n = 9) mean 39·7, ♀ (n = 4) 32–43 (38·3); (Kenya) 1 ♀ 30; (Zambia) 2 ♀♀ 34, 35.

IMMATURE: like adult ♀ but top of head browner, feathers of head, upperparts, tertials and upperwing-coverts with variable amount of black and white barring (black subterminal and white terminal bands), tail-feathers pointed, patterned yellow areas of T4–T6 with brown stripe just inside edges, most pronounced at tips. Chin and throat pale grey, yellow underparts with dark barring, bars broad and blotchy in younger birds, increasingly narrow with age.

NESTLING: unknown.

C. q. martini (Jackson): Zambia, SE and E Zaïre to Kenya, N Tanzania and S Sudan. Differs from nominate race in having dusky grey barring on upper breast of ♀. WEIGHT: (Uganda) ♂ (n = 8) 26–39 (32·9), ♀ (n = 5) 29–40 (34·8).

C. q. munzneri Reichenow: E Tanzania. Like *martini* but throat of ♂ glossed steel-blue.

Field Characters. ♂ distinguished from all-dark ♂♂ of Black and Petit's Cuckoo-Shrikes *C. flava* and *C. petiti* by strong purple gloss on face, throat and breast, continuing as wash on belly and hindneck; also slightly heavier looking. ♀ with yellow underparts similar to ♀ Petit's from below but quite different above, with grey head, unbarred green upperparts, plain green wings. Immature additionally differs from immature Petit's by barred rather than spotted underparts.

Voice. Tape-recorded (ERA, GREG, HOR, McVIC). Song a modulated whistle; contact call (♂ only), 'whit-whit-whit'; young near nest give high-pitched, penetrating 'tseeoo' (Brosset 1972).

General Habits. Inhabits moist evergreen forest, both primary and secondary, especially edges and clearings; mainly in lowlands, but ranging to 1500 m in Zambia and confined in Kenya to intermediate and montane levels, 1500–2500 m. In Gabon, occurs in scattered large trees, shrubs and tall bushes, and forest regenerating after clearance (Brosset and Erard 1986); in Zambia also found in riparian and dry forest (D.R. Aspinwall, pers. comm.). Moves around slowly, foraging in creepers and foliage, capturing prey under leaves and on trunks; tends to keep higher in the canopy than Black Cuckoo-Shrike. Mainly silent and unobtrusive, although ♂ (defending territory?) emits 'long whistles' at regular intervals (Brosset and Erard 1986). Often perches motionless on branch in upright stance. Sedentary at least in Gabon and Zambia.

Food. Insects, mainly caterpillars (sometimes 95% of diet at the nest: Brosset and Erard 1986), Orthoptera, Hemiptera, Hymenoptera and moths; also spiders.

Breeding Habits. Monogamous; probably territorial (Brosset and Erard 1986).

NEST: small, shallow cup of lichen threads, bound with spider webs, and lined with finer fibres and moss; outer diam. 80, diam. of cup 50–55, depth 23; bound to horizontal branch of tree 6–30 m above ground (in Gabon *Pentaclethra* and *Harungana* trees); built by ♀ though ♂ accompanies ♀ during building.

EGGS: 2. Greenish blue with light rusty brown, brown-red or dark brown speckling and mottling, more concentrated at broad end; 1 clutch with underlying slate purple and ashy markings. SIZE: (n = 1, Zambia) 24 × 18; Gabon c. 23 × 17.

LAYING DATES: Sierra Leone (breeds at end of rains); Liberia (enlarged gonads Feb); Ghana ('about to breed' Feb); Gabon, Aug, Nov, Dec; Angola, Mar, Sept; Zaïre, May, July, Aug (in rains); Sudan, Apr–May; Kenya, June, Oct, Dec (carrying food July, Nov); Zambia, Nov, Jan (breeding condition and nest building Oct).

INCUBATION: by ♀ only; period 20 days. ♀ may be gone up to one-third of daylight hours; ♂ guards nest in her absence; also calls to her with special contact call (see *Voice*).

DEVELOPMENT AND CARE OF YOUNG: young first brooded by ♀ and fed mainly by ♂, later fed equally by both parents; droppings removed by ♀; leave nest 22 days after hatching.

References
Brosset, A. (1972).
Brosset, A. and Erard, C. (1986).
Madge, S. G. (1972).

Genus *Lobotos* Bates

Bill quite small; tail rounded, rather less long than in *Campephaga*. Predominantly green above and yellow or orange-yellow below with blackish head and large yellow wattles below eye; ♀♀ similar to ♂♂ but duller. Inhabit treetop canopy in primary rain forest.

2 species, very similar in appearance, endemic to Africa. The wattled cuckoo-shrikes have often been included in *Campephaga*, but in view of their colourful plumage, with ♀♀ similar to ♂♂, and their facial wattles, we prefer to follow Chapin (1953) and Brosset and Erard (1986) and place them in a separate genus. The two allopatric forms are sometimes regarded as races, but ♂♂ have distinctly different plumages, and we treat them as members of a superspecies.

Lobotos lobatus superspecies

1. *L. lobatus*
2. *L. oriolinus*

Lobotos lobatus (Temminck). Western Wattled Cuckoo-Shrike. Echenilleur à barbillons.

Plate 15
(Opp. p. 256)

Ceblepyris lobatus Temminck, 1824. Pl. Col. livr. 47, p. 279; Gold Coast.

Forms a superspecies with *L. oriolinus*.

Range and Status. Endemic resident W Africa in rainforest zone, mainly west of Dahomey Gap. Known from few localities: E Sierra Leone (Gola Forest); Liberia (near Mano and Loffa rivers in W, Mt Nimba, and near Zwedru, Juarzon and Kaobli in E: W. Gatter, pers. comm.); W Ivory Coast (Tai Nat. Park, reasonably abundant: Gartshore and Carson 1989), and Ghana (old records 'interior of Fanti' *c*. 1875, Ashanti probably near Kumasi 1939, Mampong 1937: Grimes 1987). Sight record E Nigeria (Ikpan 5° 5′N, 8° 40′E) (Ash *et al.* 1989). Uncommon to frequent locally Sierra Leone and Liberia. Threatened by exploitation and clearance of primary forest.

Description. ADULT ♂: entire head, nape and throat black, glossed dark green; bristly feathers of lores and frontal band black, without gloss. Ill-defined yellow collar at base of nape, shading into olive-yellow mantle, back and scapulars; rump and short uppertail-coverts orange-chestnut; long uppertail-coverts olive-yellow. Central pair of tail-feathers dark green, next pair black, narrowly tipped yellow; rest black with distal portion yellow, increasing progressively to cover over two-thirds on outermost feather (**A**, left, p. 272). Breast and flanks bright orange-yellow; belly and undertail-coverts yellow. Remiges, alula and primary coverts black, primaries narrowly edged white, secondaries edged yellow, tertials with most of outer web washed greenish yellow; rest of upperwing-coverts olive-green; underwing-coverts, axillaries and bend of wing bright yellow. Bill black; large orange to velvety orange gape

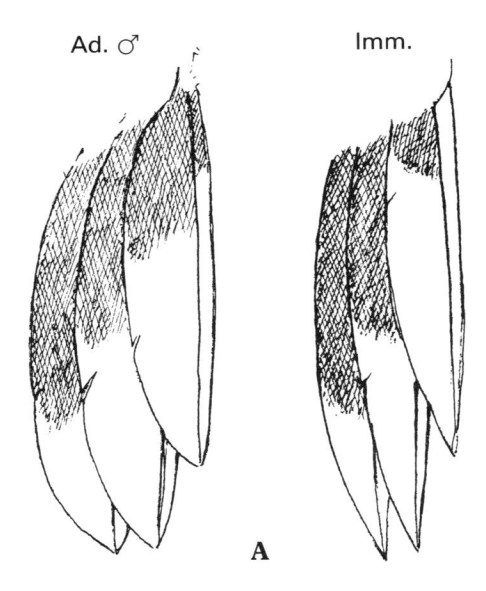

272   CAMPEPHAGIDAE

Ad. ♂   Imm.

A

wattles; eye dark brown or red-brown; legs black. ADULT ♀: similar to ♂, but head dull black without gloss, forehead and crown dusted green; mantle and uppertail-coverts wholly olive-green; breast to undertail-coverts yellow; gape wattles smaller.
SIZE: (10 ♂♂, 3 ♀♀) wing, ♂ 93–100 (97·6), ♀ 93–95 (93·7); tail, ♂ 74–79 (76·0), ♀ 73–80 (76·7); bill, ♂ 16–19 (17·4), ♀ 16–18 (17·3); tarsus, ♂ 19–21 (20·2), ♀ 19–20 (19·3).
WEIGHT: (Liberia) unsexed (n = 21) 29–37 (33·1).

IMMATURE: like adult ♀ but remiges and primary and greater coverts tipped white, and upper- and underparts with slight dark barring; tail feathers similar but narrower and pointed (**A**, right).

NESTLING: unknown.

Field Characters. See under Eastern Wattled Cuckoo-Shrike *L. oriolinus*.

Voice. Not tape-recorded. Flight-call, 'zit', like Song Thrush *Turdus philomelos* (W. Gatter, pers. comm.).

General Habits. Confined to primary lowland rain forest; in Liberia close to rivers (W. Gatter, pers. comm.). Generally silent and inconspicuous, perched among dense high foliage, typically 20–45 m above ground (W. Gatter, pers. comm.); usually in ones or twos, often with mixed bird party. Usually seen flying from tree to tree or above canopy.

Lobotos lobatus

Food. Caterpillars, grasshoppers, mantids and small seeds.

Breeding Habits.
LAYING DATES: Liberia (breeding condition Feb, Aug–Nov, fully grown immature birds May–July).

Plate 15
(Opp. p. 256)

Lobotos oriolinus Bates. **Eastern Wattled Cuckoo-Shrike. Echenilleur loriot.**

Lobotos oriolinus Bates, 1909. Bull. Br. Orn. Club 25, p. 14; Assobam.

Forms a superspecies with *L. lobatus*.

Range and Status. Endemic resident in rain forest. S Cameroon (Sangmelima, Dja R., Assobam, Molundu), SW Central African Republic (Nola, Lobaye Préfecture), Gabon (Makokou, M'Passa, Mimongo, Camma R.) and E Zaïre (west of L. Albert and L. Kivu from Medje and Lendu Plateau south through Angumu and Lundjulu to Bukavu and Kamituga). Range possibly quite extensive but recorded from few localities. Uncommon to rare.

Description. ADULT ♂: entire head, nape, throat and upper breast black, glossed dark green; bristly feathers of lores and frontal band black without gloss; upper mantle bright yellow, merging into yellowish green lower mantle, back and scapulars; rump and uppertail-coverts yellower green. Lower breast to undertail-coverts bright yellow with saffron tinge. Central pair of tail-feathers dark green, rest black with yellow tips, yellow distal portion increasing progressively from *c*. 10 mm on T2 to < 30 mm on outer pair. Remiges, alula and primary and greater coverts black; primaries narrowly edged white, secondaries narrowly edged yellow, tertials and greater coverts with most of outer web washed yellowish green; rest of upperwing-coverts yellowish green, like back; underwing-coverts, axillaries and bend of wing bright yellow, inner webs of primaries and secondaries edged whitish. Bill black; large orange gape wattles; eye dark brown; legs and feet black. ADULT ♀: similar to ♂ but head duller, black with green tinge especially on forehead and above eye; underparts yellow without saffron tinge; gape wattles smaller. SIZE: (18 ♂♂, 8 ♀♀) wing, ♂ 92–105 (100), ♀ 94–102 (98·1); tail, ♂ 68–91·5 (85·3), ♀ 73–85·5 (79·6); bill, ♂ 15·5–18 (17·2), ♀ 15·5–18 (16·7); tarsus, ♂ 17·5–20 (18·5), ♀ 18·5–19 (18·6) (M. Louette and P. Colston, pers. comm.).

IMMATURE ♂ (♀ not described): frontal band dull yellow; forehead and forecrown dull green, feathers with narrow black fringes; crown and nape dull, slightly greenish black, a few feathers with whitish fringes; golden collar around hindneck, rest of upperparts like adult but lightly barred, feathers with pale terminal and black subterminal bands; barring similar but more pronounced on scapulars and lesser coverts; tail-feathers like adult but pointed, T3 with single small blackish bar 1 mm from tip. Supercilium and ear-coverts, joining broad band from sides of neck round lower throat, bluish green, lores blackish, indistinct pale broken eye-ring; chin buff merging into greenish grey throat, rest of underparts yellow like ♀ but with narrow dusky bars on breast. Wings like adult but primary coverts with narrow white tips, greater coverts with white terminal and black subterminal bands.

NESTLING: unknown.

Field Characters. The 2 wattled cuckoo-shrikes, which are allopatric, are most distinctive. With their blackish heads, green backs, yellow underparts, black and yellow tails and black and green wings, they seem to mimic the 2 black-headed orioles in their ranges, Western Black-headed *Oriolus brachyrhynchus* and Black-winged *O. nigripennis*. Best distinguished by short, small black bill (heavy red bill in orioles), and pronounced orange gape wattles. Other marks are blue gloss to black head, saffron or orange-chestnut underparts of ♂, and somewhat smaller size.

Voice. Unknown.

General Habits. Inhabits primary forest high in canopy (Gabon, 25 m high: Brosset and Erard 1986). Most W African records from near rivers between sea level and 700 m; in E Zaïre at 850–1300 m. Solitary or in pairs; perches quietly, and hunts like a flycatcher, sallying to take insects from under leaves. Follows parties of insectivores.

Lobotos oriolinus

Food. Small caterpillars, grasshoppers and other insects.

Breeding Habits.
LAYING DATES: Zaïre, Dec–Jan, Apr–June, Aug (A. Prigogine, pers. comm.).

Genus *Coracina* Vieillot

Medium-sized birds. Bill rather stout; tail medium, usually almost square, outer feathers only slightly shorter. Plumage predominantly grey or grey and white, although one African species (*C. azurea*) is brilliant blue. ♀♀ differ only slightly from ♂♂, being paler, less boldly patterned. Inhabit forest and woodland canopy; insectivorous. Mainly solitary and unobtrusive, perching quietly in upright pose on shady branch.

40 species, widely distributed in warmer regions of Old World; 4 in Africa, all endemic.

C. caesia and *C. pectoralis* overlap geographically, but are usually separated, the former in higher-altitude forest and the latter in lower-altitude woodland. However, since their adult plumages are different and immature plumages very different, we treat them as members of a species group rather than a superspecies.

Coracina caesia (Lichtenstein). Grey Cuckoo-Shrike. Echenilleur gris.

Ceblepyris caesia Lichtenstein, 1823. Verz. Doubl. Zool. Mus. Berlin, p. 51; Galgenbosch, E Cape Province.

Range and Status. Endemic resident, mainly in montane forests. Main range discontinuous through eastern Africa; also another population in W Africa. In W Africa, SE Nigeria (Obudu Plateau), W Cameroon (Mt Cameroon to Mt Oku and Banso Mts, sea level to 2100 m, mainly above 570 m) and Bioko. In eastern Africa, from Ethiopia (central and S highlands), SE Sudan (Imatong, Dongatona and Didinga Mts), Kenya (W and central highlands, 1600–3000 m, north to Mt Nyiru, Mt Marsabit and the Matthews Range; also Mrima Hill near sea level in SE), Uganda (in E, Mts Lonyili, Morongole, Moroto, Kadam and Elgon; in SW, Rwenzoris and Kigezi), E Zaïre (highlands at 1430–3000 m, from Mahagi to mountains west of Baraka, and Mt Kabobo), W Rwanda, W Burundi and Tanzania (Mahari Mt in W; and Ngorongoro, Mbulu, Kilimanjaro, Usambaras, Ulugurus and Mwanihana Forest south to Njombe and Rungwe Mt) to SE Malaŵi (Thyolo Mt; wanderers to Chiradzulu/Lisau, Malaŵi Hill), W Mozambique (Chimanimani, Chipinga and Chiperone Mts), Zimbabwe (E highlands from Inyangani Mt to Mt Selinda; also Emberengwa Mt in SW), South Africa (E Transvaal and Natal to E and S Cape), Swaziland and S Mozambique (Lebombo Range). Locally frequent to common in north of range, mainly uncommon from Malaŵi and Zimbabwe southwards. Density in Transvaal: 6 birds per 4·5 ha. Vagrant Central African Republic (Manovo-Gounda-Saint Floris Nat. Park).

Description. *C. c. caesia* (Lichtenstein): Zimbabwe and Mozambique to South Africa. ADULT ♂: body and most of head uniform medium grey; forehead, forecrown and area round eye paler, bristly feathers of nostrils and pre-orbital region black, eye-ring white. Central tail-feathers blackish grey fringed medium grey, rest blackish, outer pair browner, becoming progressively paler towards tip. Remiges, alula and primary coverts black, primaries, secondaries and primary coverts with grey outer edges, tertials with grey outer webs; rest of upper wing-coverts grey; underwing-coverts and axillaries pale grey; underside of secondaries and primary bases with broad greyish white edge to inner web. Bill black; eye dark brown or black; legs and feet black. ADULT ♀: like ♂ but paler below, especially on chin and throat, pre-orbital region grey. SIZE: (10 ♂♂, 8 ♀♀) wing, ♂ 127–134 (131), ♀ 121–130 (127); tail, ♂ 112–116 (114), ♀ 107–115 (111); bill, ♂ 21–25 (23·1), ♀ 20–23 (21·5); tarsus, ♂ 23–25 (24·0), ♀ 23–24 (23·9).

IMMATURE: dark brown above with white fringing and barring; white below, barred dark greyish brown. Tail black, outer 2 pairs of feathers with white tips, rest with white distal outer edging, upperwing-coverts fringed and barred white; underwing pale grey.

NESTLING: skin black, with some grey down.

C. c. pura (Sharpe): W Africa and Ethiopia south to Malaŵi and adjacent NW Mozambique. Smaller than nominate race: wing (10 ♂♂) 116–124 (122), (10 ♀♀) 113–122 (116). WEIGHT: (Uganda) ♂ (n = 12) 42–63 (48·1), ♀ (n = 12) 40–50 (46·5).

Field Characters. Fairly large for a cuckoo-shrike, especially southern birds (nominate race). More or less uniform grey except for blackish wings and tail, pale forehead and eye-ring, blackish area in front of eye. The 2 other grey cuckoo-shrikes, White-breasted *C. pectoralis* and Grauer's *C. graueri*, have white underparts, and White-breasted does not occur in same habitat. Greyish barred immature also unlike any other cuckoo-shrike.

Voice. Tape-recorded (20, 88, F, McVIC, STJ). A thin, high-pitched, down-slurred whistle 'seeea', lasting *c.* 0·5 s.

General Habits. Frequents evergreen forest, forest patches and well-wooded watercourses with tall trees. Typically at higher altitudes, above 1000 m in W Africa, 1500–3000 m in central and E Africa, though as low as 900 m in E Tanzania and exceptionally near sea level in coastal Cameroon and Kenya. In S at lower elevations, mainly below 1200 m in Zimbabwe, and to sea level in coastal bush in South Africa. Occurs singly, in pairs or family groups of up to 7. Forages in tree tops, catching insects in air or hopping and picking up prey on trunks and branches; moves from branch to branch with quick flutter of wings. Unobtrusive; perches motionless for long periods in shade below foliage.

Some movements to lower elevations after breeding: in Cameroon, may migrate to lower elevations after breeding (Stuart 1986); in Transvaal, undergoes altitudinal migration from higher forests to base of escarpments and lowvelds in winter (Tarboton *et al.* 1987); Natal, winter visitor only, Feb–July, Oct (Bisley

Valley, Pietermaritzburg: Vernon 1972). Some movements also in Zimbabwe as indicated by occurrence of 1 individual at Emberengwa Mt, 300 km west of normal range (Irwin 1981), and Kenya where 1 occurred on coast, Mar, 100 km northeast of Usambaras.

Food. Insects, including caterpillars, crickets, grasshoppers and some beetles.

Breeding Habits. Monogamous.

NEST: shallow bowl of lichens bound with spider webs; in fork or on branch of tree.

EGGS: 1–2. Pale bluish or greenish, spotted and streaked with olive. SIZE: (n =3, South Africa) 26–30 × 19–20 (27·6 × 19·3).

LAYING DATES: Cameroon (breeding condition Nov); Sudan, Aug, Dec; Ethiopia (possible breeding Feb–Mar); E Africa: Region A, Mar, Aug, Oct; Region B, Jan; Region D, Dec, Feb (avoids breeding in wettest months).

INCUBATION: by both sexes; period unknown.

DEVELOPMENT AND CARE OF YOUNG: young remain with parents until next breeding season (Maclean 1985).

Coracina pectoralis (Jardine and Selby). White-breasted Cuckoo-Shrike. Echenilleur à gorge blanche.

Graucalus pectoralis Jardine and Selby, 1828. Illustr. Orn. 2, pl. 57; Sierra Leone.

Plate 15
(Opp. p. 256)

Range and Status. Endemic resident, widely distributed north and south of the forest belt; from S Senegambia, Guinea (throughout but mainly in woodland in NE, north of Fouta-Djalon massif), N Sierra Leone, S Mali, N Ivory Coast, Ghana (in N, and dry E border south to Ho), Nigeria (mainly guinea savanna zone, north to Zaria, south to Ibadan and Enugu) and N Cameroon (Adamawa Plateau and Benué Plains, largely 6–10°N), east to Central African Republic (Lobaye Préfecture, Manovo-Gounda-Saint Floris Nat. Park), N Zaïre (Upper Uelle District to L. Albert), W and S Sudan (north to Darfur, Wadweil, Rumbek and Boma Plateau), W and S Ethiopia (W border areas north to 14°N and through southern Rift Valley to *c.* 39°E), N and E Uganda (south to Kabalega Nat. Park, Serere and Bukedi) and extreme W Kenya (Elgon, Mangiki, Kakamega); also from W and S Tanzania (north to Ngara, Tabora, Chunya, Songea and Soga), NE Rwanda (Akagera), E and S Burundi and SE Zaïre (Shaba) to Zambia and Malawi (throughout), Mozambique, interior Angola, NE Namibia (Kavango and Caprivi), N Botswana (south to Okavango and Makgadikgadi), Zimbabwe and NE South Africa (lowveld N and NE Transvaal). Vagrant Mauritania (Gouraye, 15° 30′N, 12° 00′W), SE Kenya (Tsavo), N Tanzania (Arusha Nat. Park) and NW Namibia (Skeleton Coast Park: Hoanib and Ombonde rivers). Frequent to common except uncommon Namibia to South Africa and rare Kenya and Rwanda. Occurs at lower altitudes than Grey Cuckoo-Shrike *C. caesia* in areas of geographical overlap.

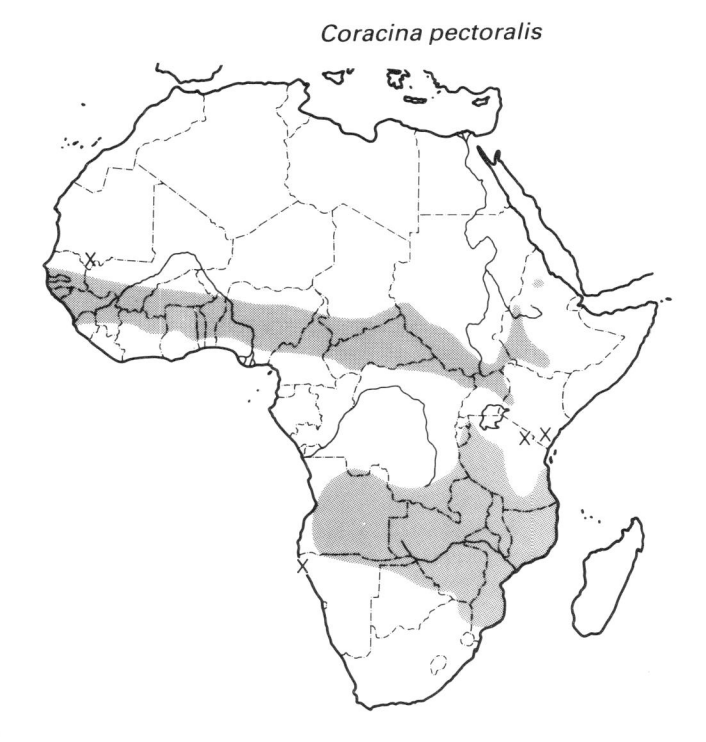

Description. ADULT ♂: upperparts and head to upper breast grey, paler on forehead, darkening through forecrown to grey of crown, paler above and behind eye; pre-orbital region black, extending to part way under eye; eye-ring white; in some birds a small, pale grey patch separating eye from black pre-orbital region; cheeks, ear-coverts and chin to upper breast somewhat darker grey than upperparts. Central pair of tail-feathers very dark grey, rest blackish brown, outer pair gradually paling to narrow white tips with variable amount of narrower pale edging on both webs, absent in some birds. T5 blackish up to white tips without gradation, tips even narrower than in T6, a few birds with a little pale edging on outer portion of both webs. Lower breast to undertail-coverts white. Remiges, alula and primary coverts black, primaries and their coverts narrowly edged grey, secondaries more broadly so, tertials grey on whole outer web; rest of upperwing-coverts grey; underwing-coverts and axillaries white; underside of secondaries and primary bases white on inner web. Bill black; eye dark sepia; legs and feet black. ADULT ♀: differs from ♂ in having chin and upper throat white, lower throat, sides of neck and upper breast pale grey; black pre-orbital region more restricted, some birds with partial ring of pale feathers outside eye-ring; and lower breast and flanks sometimes with faint grey barring. SIZE: (10 ♂♂, 9 ♀♀) wing, ♂ 139–146 (142), ♀ 134–143 (138); tail, ♂ 107–122 (115), ♀ 105–112 (110); bill, ♂ 22–25 (23·2), ♀ 19–21 (20·3); tarsus, ♂ 22–24 (23·2), ♀ 21–24 (22·3). WEIGHT: 1 unsexed 58; (Cameroon, Angola) 2 ♂♂ 58, 49; (Cameroon) 1 ♀ 58.

IMMATURE: (full size but still with some downy feathers) feathers on top of head to hindneck white with dark brown arrow-shaped marks, those of upperparts dark brown with white fringes and bases, with overlay of some grey adult feathers, underparts white with dark brown spots (spade-shaped on long undertail-coverts). Primaries and secondaries edged white distally, alula, upperwing-coverts and tertials edged and tipped white, wing-coverts with black subterminal band. With age, grey replaces barred and spotted upperparts and spots on underparts disappear; white wing-edgings retained longer than other juvenile characters.

NESTLING: skin steel-grey with coarse white down on upperparts (**A**).

A

Field Characters. A large cuckoo-shrike with a relatively heavy bill, ♂ with upperparts, throat and upper breast grey contrasting with white underparts, ♀ similar but throat to breast much paler. The only grey cuckoo-shrike outside forest.

Voice. Tape-recorded (86, CHA). A rather loud, long, high-pitched down-slurred whistle with a buzzy, burry quality, not pure, lasting 1–1·5 s. Also thin, weak, high-pitched calls, 'tsip-tsip', 'tsitsitsi', 'sweet-sweet'. ♂ said to give a soft, whistled 'duid-duid' (Maclean 1985).

General Habits. North of equator inhabits mature savanna woodland and bushy savanna with open groves of tall leafy trees, especially *Acacia*; in southern Africa, well-developed woodland, including *Brachystegia*, *Baikiaea* and *Colophospermum mopane*. Generally below 1500 m.

Usually solitary or in pairs, keeping mainly to foliage of upper branches. Gleans trunks, branches and leaves; hawks for insects in air; occasionally takes prey from ground. Moves with long hops from branch to branch; flight slow, with flap-and-glide.

Mainly resident, although some seasonal movements; Sudan, common in dry season with some moving north in rains, recorded Jan–May, July, Sept; Central African Republic, present Feb-Mar, June, Aug–Sept, apparently absent early in dry season, Oct–Jan; southern Africa, some movement to drier savanna at end of rains after breeding; some also to lower elevations in Malaŵi and present mainly in winter in Transvaal.

Food. Insects, especially caterpillars; also Orthoptera, Hymenoptera and ants.

Breeding Habits. Monogamous and territorial.

NEST: shallow cup of weed stems covered with lichens (**A**), lined with soft lichens and firmly secured to branch with spider webs, outer diam. *c.* 110, diam. of cup 80, depth 40; placed 6–15 m above ground on branch of tall tree, usually across wide flat fork; built in *c.* 6 days (Whittingham 1964), usually by ♀, sometimes by ♂; they travel up to 300 m or more to collect material.

EGGS: 1–2 (n = 8 clutches, Malaŵi, all 2). Pale bluish green or dull green, heavily speckled with sepia-brown and underlying ash-grey, sometimes concentrated in zone at broad end (Vincent 1949). SIZE: (n = 13, South Africa) 23–30 × 19–20 (27·3 × 19·5).

LAYING DATES: Gambia, June; Ghana, Jan; Malaŵi, Oct–Feb, May; Zambia, Aug–Dec (Aug 1, Sept 5, Oct 2, Nov 1, Dec 1 clutches); SE Zaïre, Sept; Angola, Sept; Zimbabwe, Aug–Dec (Aug 3, Sept 63, Oct 53, Nov 21, Dec 7 clutches); South Africa: Transvaal, Nov.

INCUBATION: by both sexes; period 23 days.

DEVELOPMENT AND CARE OF YOUNG: young fed by both parents; leave nest 24 days after hatching but dependent upon parents for another 2–3 months; remain with parents until next breeding season.

References
Vincent, A. W. (1949).
Whittingham, A. P. (1964).

Coracina graueri Newman. Grauer's Cuckoo-Shrike. Echenilleur de Grauer.

Coracina graueri Newman, 1908. Bull. Br. Orn. Club 23, p. 11; 90 km west of Lake Edward.

Range and Status. Endemic resident in E Zaïre montane forests. Local and uncommon in highlands along W side of Albertine Rift, from Djugu and Mongbwalu west of L. Albert south to Lutunguru west of L. Edward; also Mt Kahuzi south to Elila R. west of Uvira.

Description. ADULT ♂: upperparts and entire head to upper breast dark slaty grey; lores, around eye and chin to breast blacker, sides of neck, upper breast and upperparts with slight greenish blue gloss. Lower flanks and thighs medium grey, rest of underparts white. Tail-feathers black with bluish sheen, 2 outer pairs browner with very narrow white tips. Wings mainly black, secondaries, tertials and upperwing-coverts dark slate with glossy blue-green edges; underwing-coverts and axillaries white, underside of flight feathers light grey. Bill black; eye dark brown; legs and feet blackish grey. ADULT ♀: similar to ♂ but slightly paler slate above; sides of head and neck and chin to upper breast medium grey, lores darker grey, undertail-coverts pale buff. SIZE: (15 ♂♂, 16 ♀♀) wing, ♂ 110–116 (112), ♀ 105–113 (110); tail, ♂ 86–104 (103), ♀ 84–107 (103); bill, ♂ 19–19·5 (19·2), ♀ 19–20·5 (19·8); tarsus, ♂ 21–23 (22·4), ♀ 20·8–23·5 (22·3) (P. Colston and M. Louette, pers. comm.).

IMMATURE: resembles ♀: forehead light grey, lores with some black spots, throat grey with some darker blackish feathers, undertail-coverts beige (A. Prigogine, pers. comm.).
NESTLING: unknown.

Field Characters. A rare forest cuckoo-shrike confined to E Zaïre. Dark head and throat contrast strikingly with white underparts. Sympatric Grey Cuckoo-Shrike *C. caesia* is entirely grey; White-breasted Cuckoo-Shrike *C. pectoralis* also has white underparts and ♂ has grey throat, but it is allopatric, found only in woodland.

Voice. Unknown.

General Habits. Inhabits montane and transitional forest at 1400–1900 m, but as low as 1140 m on W edge of Itombwe highlands. Mainly allopatric or parapatric with Grey Cuckoo-Shrike, typically at lower altitudes (Prigogine 1980). Forages quietly among lower boughs and in high tree tops; in pairs or small groups, also in undergrowth as member of mixed bird party. Little known.

Food. Caterpillars.

Breeding Habits.
LAYING DATES: Zaïre, Jan, May, June (A. Prigogine, pers. comm.).

Reference
Prigogine, A. (1980).

Plate 15
(Opp. p. 256)

Coracina azurea (Cassin). Blue Cuckoo-Shrike. Echenilleur bleu.

Graucalus azureus Cassin, 1852. Proc. Acad. Nat. Sci. Philadelphia 5, p. 348; Sierra Leone.

Range and Status. Endemic resident, with fragmented range in W Africa and Congo basin; from Sierra Leone, Liberia (throughout), SW Ivory Coast (Mt Nimba and Tai Nat. Park to San Pedro, Gagnoa and Lamto), S Ghana (c. 5–7°N, 0° 30′–3°W including Atewa Forest Reserve, Bia Nat. Park, Fumso, Kumasi, Kwahu Tafo, Prahsu and Tarkwa) and S Nigeria (mainly west of Lower Niger R.; also Calabar) to SW Cameroon (between Mamfu and Kumba, also Ndian), N and W Gabon (Momokou to Belinga area), SW Central African Republic (Lobaye Préfecture), N and central Zaïre (scattered localities south to Luebo/Kasai and east to Ituri) and NW Angola (Cabinda). Locally frequent to common: 5–7 pairs/km^2 Gabon (M'Passa: Brosset and Erard 1986).

Description. ADULT ♂: top of head, ear-coverts, sides of neck, entire upperparts including scapulars, and breast to undertail-coverts brilliant glossy blue; forehead, pre-orbital region and

Plate 15
(Opp. p. 256)

narrow band of feathers above eye black; chin bluish black, shading into dark ultramarine throat. Tail-feathers black, central pair edged bright blue. Primaries and primary coverts black, narrowly edged blue; secondaries black with broader blue edges; tertials and rest of upperwing-coverts, underwing-coverts and axillaries blue; underside of flight feathers grey. Bill black; eye dark brown or red-brown; legs and feet black. ADULT ♀: similar to ♂ but greener, less brilliant blue, and throat like rest of underparts. SIZE: (9 ♂♂, 6 ♀♀) wing, ♂ 110–112 (111), ♀ 100–108 (105); tail, ♂ 82–92 (89·0), ♀ 82–88 (85·2); bill, ♂ 19–22 (20·8), ♀ 19–20 (18·3); tarsus, ♂ 20–21 (20·4), ♀ 20–21 (20·4). WEIGHT: (Liberia) ♂ (n = 10) mean 46, ♀ (n = 3) 43–51 (46·6).

IMMATURE: like adult ♀, but belly feathers edged white, secondaries broadly edged white and mottled blue-grey; tertials with distal black and white barring; primary coverts edged white; tail tipped white.

Coracina azurea

Field Characters. When seen well, brilliant blue ♂ and only slightly duller ♀ truly unmistakable, not only unlike any other cuckoo-shrike but unlike any other forest bird in Africa. Often appears blackish, however, when observed from below in shady canopy.

Voice. Tape-recorded (C, CHA, KEI). Song in Gabon a clear, loud, down-slurred. 'PEEEOO' prefaced by a short 'chup' –– 'chup-PEEEOO'; calls well spaced, with intervals often punctuated by conversational notes, 'chup-puji' or 'wee-chup', or a 5–6 note chatter 'chachacha ...'. Song in Ivory Coast has rather more nasal quality, with up-slurred as well as down-slurred notes, 'POOEET-POOIT-POOEET-PEEOO' or POOI-PEEOO-POOI-PEEOO', sometimes just repeated 'PEEEOO'. Intervals between phrases punctuated by sharp chatter, faster than that in Gabon, and nasal 'wer-tit'.

General Habits. Inhabits lowland primary and secondary forests; also open woodlands and occasionally clearings; below 1190 m in E Zaïre where separated altitudinally from Grauer's Cuckoo-Shrike *C. graueri* (Prigogine 1980). Mainly frequents upper canopy and tops of tall trees, singly, in pairs or sometimes small family groups of up to 6; only rarely descends to undergrowth and edges. Hunts like a flycatcher, snatching prey in air and from foliage. Runs along branches and vines and gleans undersides of leaves. Unobtrusive and easily overlooked unless calls are known; perches for long periods in vertical posture in shade below foliage. Joins mixed-species bird parties, especially in dry season.

Food. Caterpillars, grasshoppers, termites and beetles; occasionally small snails and *Croton* fruit.

Breeding Habits. Monogamous and apparently territorial (powerful calls uttered by ♂ probably involved in territorial competition).

NEST: loosely constructed bowl of lichens and spider webs, placed on horizontal branch *c.* 6 m above ground (Brosset and Erard 1986).

LAYING DATES: Liberia (display Jan, feeding fledglings Mar); Ghana (signs of breeding Feb, ♀ feeding fledglings May); Nigeria (signs of breeding Feb); Gabon (♀ with egg in oviduct Jan, ♂ in full sexual activity Nov, main song period Oct–Feb).

References
Brosset, A. and Erard, C. (1986).
Prigogine, A. (1980).

Family PYCNONOTIDAE: bulbuls

Small to medium-sized birds with short neck, rather slender body, and short, broad, rounded wings; 10 primaries, P10 often short, half the length of P9; tail with 12 feathers, medium to longish; square, rounded or (a few species) graduated. Bill short to medium, decurved toward tip, slender (except Asian *Spizixos*), notched, with nasal operculum. Rictal bristles usually well developed. Legs and feet rather short and weak. Body feathers long, soft and fluffy, especially on lower back; hindneck partly bare of feathers (only evident when neck is stretched); always a few bristles on nape, short and hard to see in some species, long and evident in others, e.g. *Criniger*. Some species have crown feathers elongated, forming crest, or erectile, giving head peaked, shaggy appearance when raised. Plumage mainly dull: olive, dark green, brown and grey, some species with yellow underparts, brightly coloured vent or distinctive marks on face; bright green in *Chloropsis* (nectar-adapted Asiatic bulbuls: Olson 1989). Sexes alike; ♂♂ usually larger. Immature like adult; unspotted.

Most in tropics, a few in temperate zone; inhabit mainly forest, also open country or scrub with trees, some adapted to man-made habitats. Arboreal, a few partly terrestrial. Eat fruit and insects, also a little nectar and pollen. Territorial (except *Andropadus latirostris*, a lekking species). Some species form groups which defend large home range. Helpers at nest in at least 2 species. Voices distinctive, often best means of distinguishing similar species, typically with burry, gravelly quality, many chattery, some with whistles; very few are musical (beautiful songs noted by early writers doubtless came from other birds). Sedentary or locally nomadic; cool-climate species partly migratory.

The family characteristically has a thin sheet of bone or some trace of ossified connective tissue covering the posterior margin of the nostril (Olson 1989). This is lacking in *Nicator*, which Olson returns to its original position in the bush-shrikes Malaconotidae, and in *Neolestes*, which he considers *incertae sedis* near *Prionops* (Prionopidae). We follow this arrangement, and the latter 2 genera will be treated in a later volume.

The position of Pycnonotidae within Passeriformes is uncertain. Placed next to cuckoo-shrikes Campephagidae by earlier authors, but not by Sibley (1970) who thought that the shared character of abundant rump feathers was convergent. Egg-white proteins of bulbuls suggest relationship to drongos Dicruridae and possibly starlings Sturnidae and orioles Oriolidae (Sibley 1970). A completely different arrangement is offered by Sibley and Ahlquist (1990), who place bulbuls in Superfamily Sylvioidea of the Parvorder Passerida, in between kinglets Regulidae and African warblers Cisticolidae, far removed from drongos, cuckoo-shrikes and orioles (in Parvorder Corvida). We place bulbuls next to cuckoo-shrikes not out of conviction but because we follow the conventional classification of Campbell and Lack (1985) until clear evidence indicates otherwise.

Africa, Asia, NW Australasia, Malagasy Region. 123 species; 52 in Africa, of which all but one are endemic, in 9 endemic genera and 2 (*Pycnonotus*, *Criniger*) shared with Asia.

Genus *Andropadus* Swainson

Bill short to medium, black or light brown, the upper mandible sometimes with small serrations behind a terminal notch; nostrils are partly feathered slits. Head without crest. Wings and tail moderately long. Legs and feet medium-sized, blackish, brownish yellow or flesh. Plumage mainly uniform olive, green or brown, underparts paler, belly sometimes with some yellow, some species with grey head, 1 with yellow moustachial streak.

Confined to forest. Voices unimpressive, with grating or burry quality; songs generally subdued, some including whistles. Some species seldom sing, but *A. virens*, *A. latirostris* and *A. importunus* sing loudly and persistently throughout the day.

United with *Pycnonotus* by Delacour (1943) and Rand (1958) but separated by White (1962) and Hall and Moreau (1970). Pterylography of the 2 genera is almost indistinguishable (Heyman and Morlion 1980), though they differ not only in plumage but also in ecology and behaviour. We consider *Andropadus* a group of primitive African forest bulbuls. *Pycnonotus* is a primarily Asian genus; the few African representatives are members of a widespread Asian superspecies which has recently invaded Africa via the Middle East.

Endemic; 11 species. No superspecies. *Contra* Chapin (1953), White (1962) and Hall and Moreau (1970), *A. montanus* is probably not close to *A. masukuensis*; we agree with Stuart and Jensen (*in* Stuart 1986) and Dowsett-Lemaire and Dowsett (1989) that because of vocal and other differences, *A. montanus* is an independent species.

Andropadus masukuensis Shelley. Shelley's Greenbul. Bulbul des Masuku.

Andropadus masukuensis Shelley, 1897. Ibis, p. 534; Masuku Range, 7000 ft.

Range and Status. Endemic resident. Middle-elevation highlands: E Zaïre, from extreme NE at edge of Ituri Forest (Arebi, Bondo Mabe, Mt Wago) south along W edge of Rift to Mt Kabobo; extends east of Rift to E side of Rutshuru Plain and Impenetrable and Kalinzu forests in Uganda, Nyungwe Forest in Rwanda, and E side of L. Tanganyika at Kigoma and Kungwe-Mahari Mts, but apparently absent Rwenzoris and volcano regions of Kivu and Rwanda. W Kenya (S Mt Elgon to Nandi and Kakamega forests), E Tanzania from S Pare and Usambara Mts to Njombe, Songea, Mt Rungwe and Isoko, and extreme N Malaŵi (Masuku Hills). Fairly common to locally abundant, but density difficult to assess since song spasmodic and not far-carrying (Dowsett-Lemaire 1989)

Description. *A. m. masukuensis* Shelley: S Tanzania (Rungwe, Ukinga), Malaŵi. ADULT ♂: top and sides of head olive, indistinct pale eye-ring, rest of upperparts, including scapulars and wing-coverts, brighter, more greenish olive, especially on rump and uppertail-coverts. Tail-feathers olive, outer edges greener; shafts blackish brown above, whitish below. Chin and throat grey-olive, rest of underparts dull olive, paler than upperparts; centre of breast and belly palest and with slight yellowish cast. Primaries black-brown, outer webs bright greenish olive; outer webs of secondaries greenish olive, inner webs brown with greenish wash and pale buff inner edging; on underwing, inner webs of secondaries broadly edged yellow-buff, base of inner webs of primaries edged olive-buff, under-wing-coverts and axillaries yellow-buff. Upper mandible dark brown to black, lower mandible grey, blue-grey or slate; eye brown, dark brown, red-brown or chestnut; legs and feet slaty blue, blue-grey, lead-grey, greyish olive, grey-green or green, sometimes with yellowish cast, soles olive-yellow. Sexes alike. SIZE: (13 ♂♂, 4 ♀♀) wing, ♂ 77–86 (82·1), ♀ 75–84 (80·0); tail, ♂ 73–83 (77·9), ♀ 74–76 (75·0); bill, ♂ 15–17 (15·9), ♀ 15·5–16 (15·9); tarsus, ♂ 20–24 (21·8), ♀ 21–23 (22·1). WEIGHT: ♂ (n = 3) 28·8–29·6 (29·3), 2 ♀♀ 27·5, 30·4, 1 juvenile with incomplete tail, 23·8. Diurnal weight change (n = 1) 14·1%.

IMMATURE: somewhat duller than adult; breast deeper olive, crown and nape more olive.

NESTLING: unknown.

A. m. roehli Reichenow: E Tanzania except range of nominate race. Top of head somewhat darker and duller, face and sides of head greyer, pale eye-ring more pronounced; throat grey, rest of underparts duller, greyer olive except undertail-coverts which are as bright as in nominate race. SIZE: (7 ♂♂, 3 ♀♀) wing, ♂ 76–86 (79·3), ♀ 74–78 (76·0); tail, ♂ 71–79 (74·4), ♀ 67–77 (71·7); bill, ♂ 14–17 (15·7), ♀ 15–16 (15·7); tarsus, ♂ 19–22 (20·6), ♀ 19–22 (20·3). WEIGHT: ♂ (n = 6) 23–30 (25·9), ♀ (n = 3) 23–31·8 (26·6).

A. m. kakamegae (Sharpe): Zaïre, Uganda, Kenya, W Tanzania (Kigoma). Top and sides of head dark grey, pale eye-ring not pronounced. Chin and throat pale grey, lower throat darker; rest of underparts brighter, more yellowish olive than in nominate race, with gingery cast, especially on flanks and undertail-coverts. SIZE: (5 ♂♂, 6 ♀♀) wing, ♂ 78–81 (79·0), ♀ 70–79 (75·0); tail, ♂ 72–76 (73·2), ♀ 66–73 (69·0); bill, ♂ 13·5–16·5 (14·7), ♀ 12–16 (14·3); tarsus, ♂ 19–21 (20·2), ♀ 19–22 (20·3). WEIGHT: ♂ (n = 15) 22–28 (24·2 ± 1·8), ♀ (n = 9) 20–25 (23·5).

A. m. kungwensis Moreau: Kungwe-Mahari Mts, Tanzania. Top of head dark grey, feathers edged blackish giving slightly mottled effect. Lores and facial area grey streaked white, eye-ring white, throat paler grey than face; upperparts as in nominate race, breast pale grey with greenish wash, becoming greener on belly and flanks. (Originally described as a race of *A. tephrolaemus*, later reclassified by Hall and Moreau (1964) as a race of *masukuensis*.)

TAXONOMIC NOTE: 2 groups of subspecies, a grey-headed western group (*kakamegae/kungwensis*) and a green-headed nominate group (*masukuensis/roehli*), each an incipient species (Hall and Moreau 1970). Relationships at the species level are unclear. It was merged with *A. montanus* by White (1962), but treated independently by Stuart and Jensen (in Stuart 1986). Hall and Moreau (1970) thought that *A. masukuensis* and *A. montanus* form a superspecies. We follow Stuart (1986) and Dowsett-Lemaire (1989) in keeping *masukuensis* independent; it is probably more closely related to W African *A. tephrolaemus*.

Field Characters. A very common but generally silent bulbul, which usually draws attention to itself by its foraging behaviour, q.v. Upperparts bright green, underparts dull olive, head and neck grey (*kakamegae*), olive-grey (*roehli*) or olive (*masukuensis*); pale eye-ring not broad but visible in the field. Grey-headed races very similar to race *kikuyuensis* of Mountain Greenbul *A. tephrolaemus* but underparts duller, less yellow; occurs at lower elevations. Other races of Mountain Greenbul have grey head and underparts (olive in Shelley's Greenbul from same area). Stripe-cheeked Greenbul *A. milanjensis* is larger, with brighter, greenish yellow underparts and pale-streaked dark face-patch.

Voice. Tape-recorded (GREG, LEM, McVIC, STJ). A quiet bird, voice little known. One taped in Malaŵi (*masukuensis*) gave a rather soft, lilting, pleasant song of 8 notes, 'chip, wa-da-tee, chee-tu, ti-wew', first note introductory, clipped, next phrase ('wa-da-tee') ascending scale, next 2 notes falling; same phrase repeated with little variation. One heard in N Tanzania (*roehli*) gave a simple 'ke-kew-ke-kew-ke', fairly loud, not really reminiscent of Malaŵi birds (F. Dowsett-Lemaire, pers. comm.). Tonal quality of song reminiscent of Ansorge's Greenbul *A. ansorgei*. Call, subdued 'wit-wit-wit'. Said to give nasal 'kwew-kwa-kwew' while moving in dense thickets (Sclater and Moreau 1932).

General Habits. Inhabits montane and intermediate-level rain forest, also gallery forest and riverside scrub; altitudinal range 1000–2300 m Zaïre, 900–2300 m E Africa, 1600–2000 m Malaŵi.

Typically occurs singly or in pairs; larger groups assemble at fruiting trees, and numbers may also join bird parties. Forages at all levels though less frequently in canopy than ground and mid-stratum. Often in leafy vine tangles and shrubby ground thickets (*Dracaena*, *Acanthopale*); spends much time at outer edge of foliage, examining dense clusters of fruiting vines, eating many berries but also searching for insects. Systematically gleans large tree-trunks and smaller branches, clinging to them like woodpecker, looking for small invertebrates in crevices, epiphytes and patches of moss. Often uses foraging method of treecreeper *Certhia*, starting at base of large tree, working its way up in short hops for 6–10 m, then flying down to base of next tree to resume climbing (Zimmerman 1972). Once attended ant swarm for 1 h as member of mixed-species flock (Willis 1983). Irregular visitor to bird bath in shady corner of garden, S Tanzania (Beakbane 1983).

Duration of primary moult, W Kenya (n = 1) 143 days (Mann 1985).

Records from 500 m in Usambaras (Stuart and Jensen 1981) and from gallery forest in Malaŵi at 1200 m may represent off-season downward movement.

Food. Insects, small fruits (e.g. *Maesa*, *Trema*, *Urera*), seeds.

Breeding Habits.
NEST: shallow cup, light but strong, built of a variety of fibrous material, including vine tendrils (e.g. Ampelidaceae, *Vitis* sp.), small roots of saprophytic plants, veins from skeletonized leaves, flower stalks of *Acalypha* sp., a black fibre resembling horsehair (*Marasmius*?), sometimes a little moss, unlined or lined with the same black fibre; slung in fork of small sapling in closed or open forest, 1·5–5 m above ground; 1 nest, ext. diam 8·4 cm, int. diam. 5·2 cm; another, ext. diam. 8·9 × 7·6 cm.

EGGS: 1–2. Pinkish buff or pinkish white, densely spotted and blotched with brown, chocolate and black over grey undermarkings, especially at thick end. SIZE: (n = 3) 15·5–16 × 24–25 (24·7 × 15·8).

LAYING DATES: Zaïre, Feb, Sept–Nov, possibly May; E Africa: Uganda (breeding condition Mar, Oct–Nov); Kenya (breeding condition June); Region D, Jan–Feb, July–Aug, Oct–Dec; Malaŵi, Oct–Nov.

INCUBATION: incubating bird (sex?) sits tight, allowing approach of human observer to within 12 m without leaving nest.

References
Sclater, W. L. and Moreau, R. E. (1932).
Zimmerman, D. A. (1972).

Andropadus montanus **Reichenow. Cameroon Montane Greenbul. Bulbul concolore.**

Plate 17
(Opp. p. 304)

Andropadus montanus Reichenow, 1892. J. Orn. 40, p. 188; Buea, 950 m, Mt Cameroon.

Range and Status. Endemic resident, Nigeria (Obudu Plateau, uncommon, and Mt Gangirwal) and Cameroon (Mt Oku and Ndu through Bamenda Highlands to Mt Manenguba, Mt Kupé, Rumpi Hills and Mt Cameroon). Not on Bioko, and an old record from Togo now thought to be an error (Stuart 1986). Scarce to very common.

Though widely distributed in Cameroon, commonest in the most threatened forests (in N) and rarer in better preserved forests in S.

Description. (Includes '*concolor*'.) ADULT ♂: entire plumage very uniform; yellowish olive-green, somewhat paler below; rump and uppertail-coverts a little brighter than rest of upperparts, underparts more yellowish, especially on belly and undertail-coverts, with almost brownish cast, chin and throat slightly paler. Shafts of tail-feathers dark brown above, yellowish white below. Primaries dark brown, outer webs olive-green, secondaries similar but inner webs progressively greener inwardly; on underside, inner webs of both edged straw yellow, underwing-coverts and axillaries sulphur yellow. Bill black; eye red or red-brown; legs grey or grey-black. Sexes alike. SIZE: (9 ♂♂, 5 ♀♀) wing, ♂ 77–84 (81·2), ♀ 73–82 (77·4); tail, ♂ 74–80 (76·8), ♀ 73–80 (75·4); bill, ♂ 17–18·5 (17·7), ♀ 16·5–18 (17·3); tarsus, ♂ 21·5–25 (23·2), ♀ 21–24 (22·9). WEIGHT: ♂ (n = 5) 30–34 (32·2), 1 ♀ 30, (4 ♂♂, 1 ♀) 29–35 (32·7), unsexed (n = 10) 29·3–36·7 (33·1).

IMMATURE: 'probable imm. ♂' (Eisentraut 1968) had grey-brown eye, yellowish grey feet.

NESTLING: unknown.

Field Characters. The only montane bulbul in its restricted range with uniform plumage. Mountain Greenbul *A. tephrolaemus* has grey head, brighter yellow underparts, brighter, greener back; Cameroon Olive Greenbul *Phyllastrephus poensis* has brown upperparts, rusty tail, brown and white underparts, wing-flicking habit.

Voice. Tape-recorded (CHA, LEM). Song, a short, low-pitched, husky babble with a few higher-pitched squeaky notes, lasting 2–3 s; rises slightly in pitch, ends with lower, rapid, chuckled 'ta-da-da-da'. Calls, a scolding 'ker-ker-ker . . . ', a low 'purrrr' and 'a loud, descending chatter, totally unlike that of Shelley's Greenbul *A. masukuensis*' (Stuart 1986).

General Habits. Inhabits montane forest, second growth, forest clearings and scrubby undergrowth. Occurs singly or in pairs. Similar to Mountain Greenbul in actions, general behaviour and habitat but more unobtrusive; keeps fairly low in vegetation, occasionally foraging up to 10 m.

Moves to lower levels in non-breeding season, when found down to 500–550 m on SE slopes of Mt Cameroon, corresponding with lower level of mist belt (Serle 1964)

Food. Insects, fruit, fruit pulp, seeds.

Breeding Habits. Unknown.
LAYING DATES: Cameroon (breeding condition Jan–Mar, Nov; bird with brood patch, Mar; juveniles Mar–Apr).

Andropadus montanus

Reference
Stuart, S. N. (1986).

Plate 17
(Opp. p. 304)

Andropadus tephrolaemus (Gray). Mountain Greenbul. Bulbul à gorge grise.

Trichophorus tephrolaemus Gray, 1862. Ann. Mag. Nat. Hist. 10 (3), p. 444; Mt Cameroon, 7000 ft.

Range and Status. Endemic resident. E Nigeria (Obudu Plateau, Gotel Mts, Mambilla Plateau), Cameroon (widespread in highlands from Banso Mts and Mt Oku to Rumpi Hills, Mt Kupé and Mt Cameroon); Bioko; highland and montane areas of E Zaïre from Rwenzoris to Mt Kabobo; Rwanda, Burundi; W and SW Uganda (Rwenzoris, Kigezi, Ankole), Mt Elgon, Kenya highlands from Cheranganis to Mau Narok and from Aberdares to Kikuyu, Mt Kenya and Nyambenis; Nguruman and Taita Hills. E Tanzania (range under races); Malaŵi (N highlands south to Viphya Mts, and in S, Zomba and Mt Mulanje; mainly above 1800 m except in S where down to 1500 m); extreme NE Zambia (Nyika Plateau, Mukutu and Mafinga Mts); W central Mozambique (Mulanje and Namuli Mts). Common to abundant; density in Malaŵi up to 2–3 pairs/ha.

Description. *A. t. kikuyuensis* (Sharpe): Zaïre to central Kenya. ADULT ♂: top and sides of head, hindneck and upper mantle slate grey, cheeks, ear-coverts and under eye with pale streaks, broken eye-ring white. Mantle to rump and uppertail-coverts, scapulars, upperwing-coverts and outer webs of flight feathers uniform bright olive-green, somewhat paler and brighter on rump and uppertail-coverts. Tail-feathers olive with brighter green edges, inner webs of all but central pair narrowly edged pale greenish; shafts brown on upperside, whitish on underside. Chin, throat and upper breast light grey, rest of underparts bright olive-yellow, somewhat darker on breast and flanks, belly almost pure yellow. Primaries and secondaries dark brown, underside of inner webs edged pale yellow; axillaries and underwing-coverts yellow. Bill black; eye brown, hazel, red-brown or dark red; legs and feet greenish grey, blue-grey, greenish or greenish brown. Sexes alike, but ♂♂ slightly larger. SIZE: (15 ♂♂, 13 ♀♀) wing, ♂ 84–94 (88·5), ♀ 83–88 (85·8); tail, ♂ 78–90 (83·7), ♀ 73–86 (80·8); bill, ♂ 16·5–19 (17·6), ♀ 15·5–18 (16·8); tarsus, ♂ 21–26

(23·3), ♀ 21–25 (23·0). WEIGHT: ♂ (n = 14) 29–41 (33·1), ♀ (n = 9) 29–32·5 (30·3), unsexed (n = 37) 28·5–35·5 (33·1 ± 2·0).

IMMATURE: duller; head and neck feathers tipped olive, underparts darker, dull olive with little yellow, even on belly. Eye brown or grey-brown without reddish tinge of adult.

NESTLING: unknown.

A. t. tephrolaemus (Gray): Mt Cameroon and Bioko. Pale streaks on face faint or lacking; eye-ring narrower; grey of head not extending to breast or mantle. SIZE: wing, unsexed (n = 16) 80–89 (84·1). WEIGHT: unsexed (n = 16) 29·2–40·2 (36·3).

A. t. bamendae Bannerman: Nigeria, Cameroon except range of *tephrolaemus*. Like nominate but larger; underparts duller, much less yellow, breast with dark olive wash. Variation clinal: birds from N (Mt Oku) clearly *bamendae*, but closer to Mt Cameroon (Rumpi Hills, Mt Kupé) intermediate between this and nominate. SIZE: wing, all unsexed (n = 33, Mt Kupé) 78–92 (84·6), (n = 12, Manenguba) 85–96 (88·7), (n = 7, Rumpi Hills) 78–89 (84·0), (n = 22, Mt Oku) 83–97 (89·0). WEIGHT: all unsexed (n = 33, Mt Kupé) 31·8–46·7 (38·3), (n = 12, Mt Manenguba) 32·4–44·5 (39·1), (n = 7, Rumpi Hills) 34–44·7 (38·1), (n = 22, Mt Oku) 31–44·1 (37·9).

A. t. nigriceps (Shelley): S Kenya (Nguruman Hills) and N Tanzania (Mt Kilimanjaro, Crater Highlands, Mbulu, Mt Hanang). Forehead and crown blackish, becoming dark grey on hindneck; broken eye-ring greyish white, pale streaks on grey face. Upperparts, wings and tail duller and darker green than *kikuyuensis*; throat and underparts mainly grey, paler in centre of belly where washed yellowish, flanks and thighs with yellow-olive wash, undertail-coverts yellow-olive.

A. t. usambarae (Grote): SE Kenya (Taita Hills) and NE Tanzania (S Pare and W Usambara Mts). Underparts as *nigriceps* but top of head same grey as *kikuyuensis*, with narrow black border from above lores over and behind eye, separated from ear-coverts by grey post-ocular streak. WEIGHT: ♂ (n = 2) 31, 32, ♀ (n = 2) 28, 29.

A. t. neumanni (Hartert): Uluguru Mts. Top of head black as *nigriceps* but no grey eye-ring or eye streak.

A. t. fusciceps (Shelley): S part of range north to S Tanzania (Umalila, Isoko, Mt Rungwe, Poroto Mts). Crown clear grey but dull, with slight brownish wash; blackish area above lores and in front of eye, continuing as line over eye; face unstreaked or with faint streaking, broad broken white eye-ring. Green of upperparts duller than in other races. Underparts dull grey with brownish cast, paler on belly, lower flanks grey-olive, undertail-coverts dull yellowish olive with pale tips giving mottled effect; underwing yellow-buff, not yellow. WEIGHT: Malawi, ♂ (n = 40) 32·2–40 (38·4), ♀ (n = 90) 30–41 (36·8). Individuals change weight within 3 days by 2 g, or 4·7% of sample mean (n = 10, Malawi: Dowsett 1983).

A. t. chlorigula (Reichenow): E Tanzania from Kiboriani, Ngurus and Ukagurus to Iringa, Njombe and Songea. Top and sides of head grey with brownish cast, face unstreaked or faintly streaked, broad broken white eye-ring, blackish area before and behind eye. Chin pale grey; broad olive-yellow patch on throat becoming olive-green on sides of breast; rest of underparts grey, paler in centre of belly; lower flanks and undertail-coverts yellow-olive, undertail-coverts with pale tips giving mottled effect. WEIGHT: 1 ♀ 33, unsexed (n = 2) 39, 41.

('*A. t. kungwensis*' Moreau is a race of *A. masukuensis*, q.v. (Hall and Moreau 1964)).

TAXONOMIC NOTE: inclusion of all of the above populations in this species *tephrolaemus* is tentative. They are usually divided into 3 groups: (1) *tephrolaemus/bamendae/kikuyuensis* (grey head and breast, yellow belly); (2) *nigriceps/usambarae/neumanni/fusciceps* (variable amount of black on grey head, underparts grey with olive flanks); (3) *chlorigula* (like *nigriceps* but with yellow throat).

Hall and Moreau (1970) regard these 3 groups as forming incipient species. Dowsett-Lemaire and Dowsett (1989) and Stuart (1986) consider birds from Cameroon and Nigeria specifically distinct. Voice and plumage suggest *kikuyuensis* (Zaïre–central Kenya) may represent an incipient species (Dowsett-Lemaire and Dowsett 1990). Until the relationships between the various groups and between *tephrolaemus sensu lato* and *masukuensis* have been worked out, we prefer to leave the populations treated here as races of *tephrolaemus*.

Field Characters. An abundant bulbul of montane forests. Wide range of calls and variety of racial plumages: Cameroon races distinguished from Cameroon Montane Greenbul *A. montanus* by grey head; *kikuyuensis* very similar to race of Shelley's Greenbul *A. masukuensis kakamegae*, which also has grey head and green upperparts, but underparts of Mountain Greenbul much brighter and yellower, pale eye-ring more conspicuous; usually found at higher elevations. All other races told from other races of Shelley's Greenbul by grey head and underparts (dull olive in Shelley's). Northern races of Stripe-cheeked Greenbul *A. milanjensis*, with bright yellow underparts, overlap only grey-bellied races of Mountain Greenbul; grey-headed nominate race of Stripe-cheeked Greenbul closer to *nigriceps* but has green, not grey, underparts and striped cheeks.

Voice. Tape-recorded (32, 53, 86, C, LEM, McVIC). A wide variety of songs and calls, differing regionally. *A. t. chlorigula* has a 5-note song, 'chip-chop-cho-chilly', or 'it's not so chilly', first note short and clipped, last note loudest; intervals between phrases c. 10 s. Song of *fusciceps* has same form but last 2 notes more rounded, 'chip-chop-wicho-weeyo', and a shortened phrase 'chop-wi-choda'. Call like that of Stripe-cheeked Greenbul, a chattering 'twuk' repeated up to 12 times, also 'pichu' or 'hor-pi-chu'. N Tanzania races said to have 3 types of song: (1) a conversational 'kwew-ki-kwew-ki-kwew', very like Shelley's Greenbul; (2) a more deliberate 'kwo, kwer-kwer, kwee, kwo', slightly falsetto, especially on note 4, reminiscent of *chlorigula*; and (3) rich, passionate warbling (Moreau and Sclater 1938) – the last is almost certainly an error. Song of *kikuyuensis* from Rwanda (Nyungwe) is of 2 types: (1) a bustling continuum of low, nasal, husky notes all on one pitch – notes run into one another, unlike distinct notes and well-defined phrases of Tanzanian birds; (2) closer to *nigriceps/chlorigula* – a more defined phrase of 6–8 distinct notes, repeated with minor variations: 'jur-jitjur-dajur-jur-jerry', last note upslurred and not unlike last note of Tanzanian song but not as loud. Cameroon birds have large variety of calls and chatters, all very similar to Shelley's Greenbul of E Africa, with very little resemblance to those of E African races of Mountain Greenbul (Stuart 1986). Song of *bamendae* recorded in Nigeria is continuous series of 7-note phrases with hardly a pause between them, same form repeated: second note lower than first, notes 3–5 rising in pitch, last 2 dropping off, last on same pitch as first, e.g. 'chep, chop, chup, chap, chip, chap, chep'; notes dry and somewhat burry, with something of quality of *nigriceps* group but weaker, less definite, and with different form; call, a nasal 'dzut-dzuwi', second note upslurred. Song of nominate race from Mt Cameroon very like *bamendae* (F. Dowsett-Lemaire, pers. comm.)

General Habits. Inhabits montane forest of all types, including primary forest, second growth, gallery forest, edges and clearings; not confined to unbroken stretches of forest, occupying small isolated patches on plateaux (including cedar on Mt Mulanje), even bracken-briar; on high mountains extends into bamboo where often numerous, and in recently deforested areas (e.g. Nyungwe (Rugege) Forest, Rwanda) typical of roadside brush and bushes. Has wide altitudinal range, reaching tops of many mountains, occurring up to 3300 m in bamboo, down to 1350 m in E Africa and 1000 m in Cameroon; 2 records of juveniles at sea level in Cameroon (Serle 1964) considered unusual.

Occurs singly, in pairs, or in groups of up to 6, often family parties with juveniles; large numbers gather at fruiting trees, especially figs. Often in mixed-bird flocks with e.g. Shelley's Greenbul and Abyssinian Hill Babbler *Alcippe abyssinica*. Forages at all levels, from undergrowth to canopy; catches insects in air and by gleaning foliage and bark, hopping and probing along moss- or lichen-laden branches, clinging to trunks like woodpecker. Attends swarms of army ants (*Anomma, Dorylus*), especially those climbing small trees, to catch insects and spiders: waits in low vegetation, drops to forest floor and darts back up; snatches prey from low trunks, saplings, foliage and in air. Feeds higher, in mid-stratum and canopy, when insect abundance greatest (Malaŵi: Dowsett and Dowsett-Lemaire 1984). After morning feed, birds rest in thick clumps of woody undergrowth, ♂♂ singing. Inquisitive and noisy; calls continuously while feeding. Visits garden bird baths (Beakbane 1983).

Only adults moult remiges and rectrices; juveniles of 3 months moult only body feathers. On Nyika Plateau, Malaŵi, mean duration of moult in 3 years was 110–131 days. Moults during rains over a 6-month period, beginning when young become independent. Weights remain high in all but earliest stages of moult, suggesting long, 'easy' moult period (data from Dowsett and Dowsett-Lemaire 1984).

Mainly sedentary. Banded birds (n = 59) in Malaŵi (Nyika) moved as much as 2500 m (mainly 250–500 m). 2 birds dispersed 2·5 km from one forest patch to another (Dowsett 1985). Some altitudinal movement during cool season in Tanzania (Uluguru Mts) and Malaŵi, down to 700–1000 m, once 500 m.

Food. Insects, including bark beetles, and spiders; fruits: at least 63 species, diam. 3–15 mm, recorded Malaŵi (Dowsett-Lemaire 1988); in Rwanda, genera include *Agelaea, Coccinia, Harungana, Ilex, Maesa, Polyscias, Rutidea, Schefflera, Trema* and *Urera* (Dowsett-Lemaire 1990).

Breeding Habits. Monogamous, territorial. Mean territory diam. (n = 39) 80 m (Dowsett 1985). Density in forest–grassland mosaic 2–3 pairs/ha, in pure forest *c*. 1 pair/0·5 ha; breeds in forest patches of only 0·12–0·16 ha, visiting adjacent bush (second growth) or edge of next forest patch; density only 1 pair/2 ha where Stripe-cheeked Greenbul equally common (Dowsett-Lemaire 1989). In Malaŵi 31% of breeding adults (35/113) held same territory more than 1 season.

NEST: a shallow, open cup, light but strong and well-knit, neatly built of wide range of materials including twigs, tendrils, plant stems, thick, coarse grass blades, dried leaves, spider webs and especially moss, which frequently covers exterior; sometimes unlined, more often lined with fine grasses and hair-like fibres; ext. diam. 90–120, int. diam. 60–75, ext. depth 35–95, int. depth 30–45; 1 nest oval, not round: ext. diam. 130 × 55, int. diam. 70 × 30. Placed 1·2–4 m above ground in undergrowth or low shrub, where slung between horizontal twigs; in tree fern, and once in fork formed by crossed bamboo stems.

EGGS: 1–3, usually 2. Truncated ovate; smooth, with slight gloss; white or cream, sometimes with pinkish or purplish tone; heavily freckled, especially at broad end, and sometimes almost obscuring ground colour, with blackish brown, olive-brown, tawny or purplish chocolate over grey-lilac undermarkings. SIZE: (n = 9) 23·2–27·7 × 16·2–18·3 (25·2 × 17·3).

LAYING DATES: Cameroon, May–June, Aug, Nov–Mar, mainly in dry season but also in rains, and possibly all year; Zaïre, Mar–Aug, Oct–Nov, possibly all year; E Africa: Region A, Nov; Region B, Jan, Mar; Region C, Sept–Dec; Region D, Sept–Mar; avoids breeding in wettest months which are cold and misty (Brown and Britton 1980). Malaŵi, Aug–Nov (Aug 8, Sept 22, Oct 25, Nov 5 clutches). Breeds at end of dry season and beginning of rains (peak period for both fruit and insects: Dowsett and Dowsett-Lemaire 1984). Zambia, Sept–Oct (breeding condition Nov).

INCUBATION: by ♀ only; ♂ brings food to nest.

BREEDING SUCCESS/SURVIVAL: in exceptionally dry season, Malaŵi, none of 16 nesting attempts was successful, and only 2 juveniles were ringed where 20 were ringed in previous season (Dowsett-Lemaire 1985). Average annual mortality rates (Malaŵi) 51·4 and 48·5% in ♂♂, 76·5 and 59·1% in ♀♀ (high values may be in part due to birds moving away from the area). Lives at least 7·9 years; av. further life expectancy 0·8–1·6 years (Dowsett 1985). Several nest records, Zaïre, list single chick, though original clutch size not known. A one-footed adult managed to survive (Prigogine 1971).

References
Dowsett, R. J. (1985).
Dowsett, R. J. and Dowsett-Lemaire, F. (1984).
Dowsett-Lemaire, F. (1985).
Stuart, S. N. (1986).

Andropadus milanjensis (Shelley). Stripe-cheeked Greenbul. Bulbul montagnard.

Xenocichla milanjensis Shelley, 1894. Ibis, p. 9, pl. 1, fig. 1; Milanji Hills, Nyasaland.

Plate 17
(Opp. p. 304)

Range and Status. Endemic resident from S Kenya (Ol Doinyo Orok, Chyulu and Taita Hills) through highlands of E Tanzania (west to Monduli Mt, Mbulu, Uzungwa Mts, Mt Rungwe and Isoko) and Malaŵi (Misuku Hills and Nyika Plateau to Shire Highlands and Mt Mulanje, but many gaps in centre of range), to Mozambique (Unangu, Namuli and Chiperone Mts, highlands along Zimbabwe border and pockets of suitable habitat on Manica Plateau, especially common on Mt Gorongoza), and Zimbabwe (Inyanga Hills to Mt Selinda, very common only above 1400 m; less common at 950 m in Honde valley, irregular below 600 m at Haroni–Lusitu confluence). Common to locally abundant; density N Malaŵi at least 5 pairs/10 ha (Dowsett-Lemaire 1989).

Description. *A. m. striifacies* (Reichenow and Neumann) (including '*chyulu*'): Kenya and Tanzania south to Iringa District. ADULT ♂: top of head and upperparts, including wing-coverts and edges of flight feathers, more or less uniform bright olive-green, rump and wing edgings somewhat brighter but head uniform with mantle. Outer webs of tail-feathers similar, inner webs brown with very narrow pale margins; shafts brown above, whitish below. Patch from lores and base of bill broadly around eye to cheeks and ear-coverts blackish, streaked white on cheeks and ear-coverts. Chin pale grey with olive wash, rest of underparts from throat to undertail-coverts olive-yellow, washed darker on breast and flanks. Inner webs of primaries and secondaries dark brown, edged pale yellow on underside, broadest on secondaries; axillaries and underwing-coverts olive-yellow. Bill black, inside of mouth yellow; eye grey, brown or yellow; legs and feet grey, greyish brown, brown or sepia, soles yellow or white. Sexes alike, but ♂♂ slightly larger. SIZE: wing, ♂ (n = 14) 92–102 (96·6), ♀ (n = 9) 89–97 (92·7); tail, ♂ (n = 14) 83–96 (88·4), ♀ (n = 9) 77–88 (84·8); bill, ♂ (n = 13) 18–21 (19·6), ♀ (n = 9) 18–21 (19·4); tarsus, ♂ (n = 13) 22–25 (23·0), ♀ (n = 9) 21–24 (22·4). WEIGHT: Kenya, ♂ (n = 11) 39–50 (45·1 ± 3·1), ♀ (n = 5) 37–43 (39·8); Tanzania, ♂ (n = 19) 39·1–47·8 (42·7 ± 2·3), ♀ (n = 3, non-breeding) 39·4–44·8 (41·4), ♀ (n = 5, breeding) 40–47 (42·3).

IMMATURE: cheeks and ear-coverts greenish, white stripes dull and inconspicuous; underparts duller, greyish wash on breast and belly; bill sepia, gape wattle yellow; eye dark brown or sepia; legs light dusky grey, soles brighter yellow than in adult.

NESTLING: unknown.

A. m. olivaceiceps (Shelley): S Tanzania (Rungwe), Malaŵi (except Mulanje Mt), Mozambique (Unangu). Similar to *striifacies*, with dark top of head, but overall darker and duller; upperparts somewhat greener, less yellow; below greener, yellow areas paler, less rich, wash on breast and flanks olive-green rather than yellow-brown. Eye dark reddish or pale yellowish. WEIGHT: ♂ (n = 15) 37–46·5 (41·5), ♀ (n = 3) 44·2–48·3 (45·8).

TAXONOMIC NOTE: eye colour said to differ consistently without regard to sex or season between the 2 populations in Malaŵi, northern birds (Nyika Plateau, Misuku Hills) having brown eyes, southern ones (Ntchisi to Thyolo) having yellow eyes (Dowsett 1974). However, birds in Thyolo reported to change plumage and eye colour in Sept (beginning of breeding season): crown, moustache (?) and upperwing-coverts become darker, eye changes from orange to pale yellow (Johnston-Stewart 1982). A pair displayed in this plumage. Normal plumage resumed in Nov.

A. m. milanjensis (Shelley) (including '*disjunctus*'): Malaŵi (Mulanje Mt), Mozambique (except Unangu), Zimbabwe. Forehead to nape, sides of head, chin and upper throat grey, merging to yellow-olive on lower throat; narrow eye-ring white, broadening to short crescent on upper eyelid; upperparts somewhat duller and darker than in *olivaceiceps*, underparts duller and greener, washed yellowish on belly only. Eye plae grey or brown. Immature has crown greener than adult, eye grey-brown. SIZE: wing, ♂ (n = 29) 84–100 (93·7), ♀ (n = 12) 85–94 (88·6); bill, ♂ (n = 29) 19·6–23·5 (21·7), ♀ (n = 10) 20·7–23·2 (21·5). WEIGHT: ♂ (n = 45) 33·2–46·6 (42·5), ♀ (n = 26) 34·6–47·4 (40·5).

Field Characters. A common bulbul of eastern highland forests, the only one in its range with dull golden underparts. Closely related Mountain Greenbul *A. tephrolaemus* is a little smaller but same shape, with similar vocal quality, but sympatric races have grey underparts (*chlorigula* with yellow on throat only). Blackish face-patch, with pale streaks visible at close range, is also distinctive. Eye colour too variable (brown, grey or yellow) to be useful field character. In Malaŵi, duller race *olivaceiceps* overlaps with similar Shelley's Greenbul *A. masukuensis*, but is larger, with heavier bill; Shelley's Greenbul has plain olive head and throat without dark face-patch. Voice has similar quality to Mountain Greenbul but song has different rhythm.

Voice. Tape-recorded (32, 88, F, GIB, STJ, WALK). Song, 8–12 dry, clipped, gravelly notes all on one pitch, with a distinct rhythm which varies regionally. In Zimbabwe (nominate race), 4 measured notes followed by 3 run together, ending with a short trill: 'chop-chap-chip-chip, chipa-chip, prrrr'; in N Malaŵi (*olivaceiceps*) song slower, with lethargic quality, 4–5 measured notes, sometimes fewer: 'cho, cha, chip, cho, chip', 'cherp, cho, cher, chip', 'cherp, chorp, chip' or just 'were-chop'; some of these 'phrases' may qualify as calls, distinction between the 2 blurred; in E Africa (*striifacies*), 3 measured notes, then 3 run together, ending with a double note instead of a trill: 'chip-chop-chop, peechopee, chochop'. In both cases, 'partial' songs frequent, with beginning or ending notes left off, as is typical of many bulbuls, but tone and general form remain the same throughout range. Song may have additional conversational note or 2 tacked on at end: 'wert', 'wer-tidi', 'gwik'. Contact and general call, a loud, rapidly repeated 'cha', and said to have a 'company call' with different birds answering in turn (van Someren 1939). In courtship, throaty 'twer-twi'.

General Habits. Inhabits primary rain forest, mainly montane and submontane, both interior and edge, and riverine forest; ventures out into nearby scrub and thickets and even to bushes in *Brachystegia*. Locally enters plantations of conifers and hardwoods.

Sometimes solitary, often in small, noisy parties; gathers in large numbers at fruiting trees. Forages at all levels, most often in middle strata and canopy. Hops and creeps along branches in search of insects, also hawks for them, especially when sun comes out after rain, flying out from forest in aerial pursuit; occasionally attends ant swarms. Can appear shy, skulking in dense vegetation, but fairly bold at times, emerging from forest to feed on berries in nearby bushes. Readily came to trap baited with guava (Swynnerton 1907). When singing said to sidle along branch with little hops, but also sings while stationary on perch (S. Keith, pers. obs.). Noisy but also furtive around bird baths, S Tanzania (Beakbane 1983).

Mainly sedentary but in cold season descends to lower levels in Tanzania (down to 150 m in Usambaras, 250 m in Ulugurus), Malaŵi (to 600–900 m on Mt Mulanje) and Zimbabwe (to 350 m, Haroni-Lusitu).

Food. Insects and fruit; 25 fruit species listed for Malaŵi (Dowsett-Lemaire 1988). Of 9 stomachs, Zimbabwe, 2 contained only small fruits, 4 mainly fruit and some insects, 4 insects only (Swynnerton 1907), whereas in Tanzania diet mainly berries (Sclater and Moreau 1932). Insects include beetles and their larvae, earwigs and caterpillars; fruits include *Loranthus*, *Lantana*, *Ficus*, *Rapanea*, *Sapium manianum*, *Clidemia hirta* and jasmine. Worms and seeds also taken.

Breeding Habits. Monogamous, territorial. Birds said to give throaty 'twer-twi' when courting (Sclater and Moreau 1932).

NEST: a neatly rounded cup, often so thin and fragile that it can be seen through from below; built of roots, twigs and grass, covered on outside with moss, cocoons and lichen fibres, lined with fine grass and tree fern roots and fibres; ext. diam. 90–100, int. diam. 64, ext. depth 64, int. depth 20; placed 4–7 m above ground, typically in upright fork at top of small sapling, also in horizontal fork towards end of branch or in bush or creeper; may have spray of 'old man's beard' (*Usnea* sp.?) draped over fork for concealment (Swynnerton 1907).

EGGS: usually 2, sometimes 1. Rather elongate and bluntly ovate; white or off-white, sometimes buffy, densely spotted, blotched, marbled, streaked, lined and scrawled with shades of brown and chocolate on undermarkings of grey, blue-grey or lilac-grey; markings concentrated in cap or broad band around large end. SIZE: (n = 6, southern Africa) 24·3–26 × 17·5–18·1 (25·1 × 17·8).

LAYING DATES: E Africa: Region C, July; Region D, Dec–Feb, May (building nest Nov); Malaŵi, Sept–Oct; Mozambique, Nov–Dec (breeding condition Aug); Zimbabwe, Oct–Mar, mainly Nov–Dec (Oct 4, Nov 13, Dec 20, Jan 1, Mar 2 clutches).

INCUBATION: by ♀ only. Incubating bird may sit tight and not leave until nest tree is shaken, or may leave nest on approach of observer and quickly return.

BREEDING SUCCESS/SURVIVAL: nest destroyed by baboons (Swynnerton 1907).

References
Dowsett, R. J. (1974).
Sclater, W. L. and Moreau, R. E. (1932).
van Someren, V. G. L. (1939).
Swynnerton, C. F. M. (1907).

Andropadus virens Cassin. Little Greenbul. Bulbul verdâtre.

Andropadus virens Cassin, 1858. Proc. Acad. Nat. Sci. Phil., 1857, p. 34; Cape Lopez and Muni R., W Africa.

Plate 18
(Opp. p. 305)

Range and Status. Endemic resident, mainly sedentary, locally nomadic. The most widespread forest bulbul in Africa, abundant everywhere except at periphery of range. Range continuous in W and central Africa from Gambia (mouth of Gambia R.) and SW Senegal (Basse Casamance) to S Central African Republic and extreme S Sudan, north to extreme SW Mali (Mandingo Mts), 10°N in Nigeria, Adamawa Plateau (Cameroon) and extreme SW Chad, south to N Angola (Cuanza Norte, Malanje, Lunda, N Moxico) and N Zambia (south to N half Western Province, Mkushi R. and Muchinga Mts). In Zaïre almost throughout, but apparently absent from highlands on W side of L. Tanganyika from Mt Kabobo to Marungu Plateau, and from volcano region; also only single records Rwanda (extreme N end Akagera Nat. Park) and Burundi (Kigwena Forest). Reappears in forests of W and S Uganda, extending north to Budongo, Sezibwa R. and Mt Elgon, east into W Kenya (to Kakamega, Kisumu, Lolgorien and Tarang'anya), and south into NW Tanzania at Bukoba. After another gap, occurs on Kenya coast from Kilifi and Rabai through Shimba Hills into Tanzania, south on coast to Dar-es-Salaam, inland to Usambaras and Arusha and through SE highlands from Ngurus and Ulugurus through Uzungwas to Tukuyu and Songea, also in SW at Kitungulu near Zambian border; continuing into Malaŵi (south in highlands to Nkhotakota District, and after gap in highlands east of Rift) and highlands of central N Mozambique. Bioko, Zanzibar, Mafia I.

Density in Gabon variable, up to 500 individuals/km² (Brosset and Erard 1986); in Nigeria, 1 bird/2·5 ha (Johnson 1989); in Malaŵi, 2–6 pairs/10 ha locally (Dowsett-Lemaire 1989).

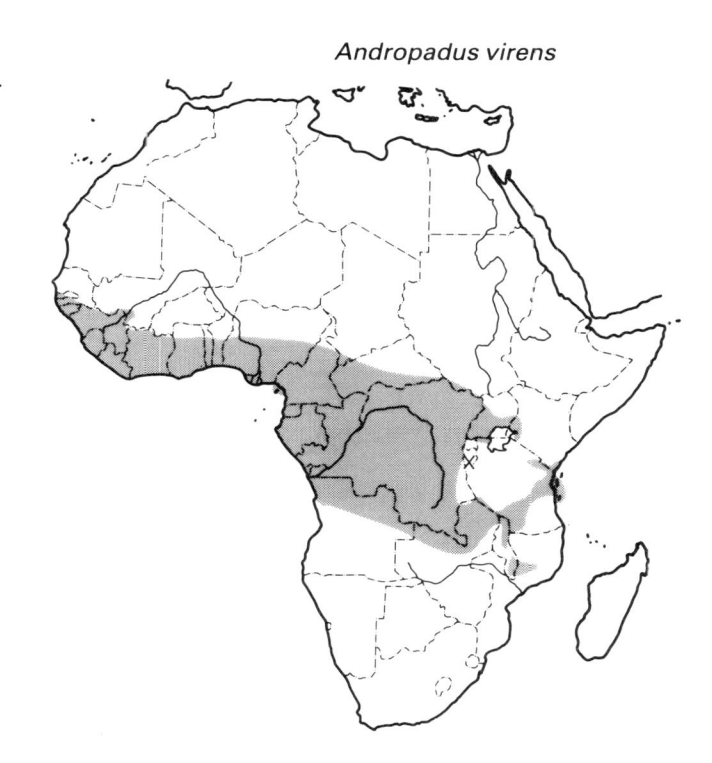

Description. *A. v. virens* Cassin (including '*holochlorus*'): Cameroon and Bioko to Sudan, W Kenya and Angola. ADULT ♂: top and sides of head and upperparts, including scapulars and upperwing-coverts, dark greenish olive, becoming browner on rump; uppertail-coverts and tail brown with a rusty wash, shafts of tail-feathers dark maroon above, pale yellow below. Ear-coverts variable, in some birds with somewhat paler shafts and centres producing slightly striped effect (only visible in the hand). Chin and throat dull pale olive with yellowish cast, becoming darker olive on breast and flanks and pale yellow in centre of belly, undertail-coverts pale brown with pale yellow tips. Primaries and secondaries dark brown edged olive, inner secondaries with slight coppery tinge on inner webs; on underside, inner primaries and secondaries edged pale yellowish buff on inner webs; underwing-coverts pale yellow becoming brighter yellow on axillaries and under primary coverts. Bill blackish or dusky, lower mandible browner, gape bright yellow (variable in extent, almost lacking in some birds); distal half of upper mandible serrated, with progressively deeper notches towards tip; eye grey-brown; legs and feet sometimes yellowish orange, usually yellowish brown, soles yellow. Sexes alike. SIZE: (12 ♂♂, 12 ♀♀) wing, ♂ 70–82 (75·7), ♀ 65–80 (72·3); tail, ♂ 62–72 (66·6), ♀ 58–70 (64·8); bill, ♂ 13–16 (14·3), ♀ 13–15·5 (14·2); tarsus, ♂ 18–21 (19·3), ♀ 18–21 (19·6). WEIGHT: Zaïre, ♂ (n = 14) 23–32 (28·5 ± 2·6), ♀ (n = 13) 22–30 (26·2 ± 1·9); Kenya and Uganda, ♂ (n = 9) 21–26 (24·1 ± 1·4), ♀ (n = 6) 20–21 (20·3 ± 0·5). Diurnal weight change 10·4–23·2% (W Kenya: Mann 1985).

IMMATURE: upperparts like adult but top of head slightly darker, wings browner with rusty wash on upperwing-coverts and secondaries; below, darker and more uniform olive, yellow in middle of belly paler and less extensive, undertail-coverts light olive, lower mandible, tip of upper mandible and gape yellow. An earlier (juvenile?) plumage described as 'underparts from chin to belly grey, undertail-coverts tinged isabelline and brown, upperparts olive-green washed with russet, mainly on back and upperwing-coverts; inner webs of inner secondaries edged isabelline; tail feathers more narrow and pointed, tips of primaries more rounded' (Verheyen 1953).

NESTLING: unknown.

TAXONOMIC NOTE: birds from Uganda and W Kenya ('*holochlorus*') average slightly larger and yellower below, but because there is much individual variation and intergradation, we do not consider them a separate race.

A. v. erythropterus Hartlaub: Gambia to Nigeria. Darker and greyer above than nominate race, with almost no green; pale area on centre of belly a narrow line of dull whitish yellow; flight feathers, including wing edgings, less olive, more brown, with slight rusty tinge. WEIGHT: Liberia, ♂ (n = 9) 20·7–30·7 (26·0 ± 1·8), ♀ in breeding condition (n = 5) 19·6–23·4 (21·8).

A. v. zombensis Shelley (including '*marwitzi*', '*shimba*'): extreme SE Zaïre (SE Shabsa), N Zambia and SE Kenya to Malaŵi and Mozambique, also Mafia I. Paler and more greyish green above than nominate race, olive of underparts paler and with greyish tinge, yellow on belly paler. Northern populations (*marwitzi*, *shimba*) tend to be somewhat browner above. WEIGHT: Kenya (coast), ♂ (n = 12) 20–25 (23·3 ± 1·6), ♀ (n = 5) 18–22 (20·2); N Tanzania, ♂ (n = 26) 22–30 (26·1

± 1·8), ♀ (n = 3) 22·8–31·9 (28·3); N Zambia, ♂ (n = 7) 22·1–26·5 (24·9).

A. v. zanzibaricus Pakenham: Zanzibar. Duller, less green and somewhat darker above than *zombensis* on opposite mainland, more or less intermediate between *zombensis* and nominate; greyer and less yellow below than either.

Andropadus hallae was described by Prigogine (1972), based on a single specimen, an adult ♂ collected 16 Sept 1970 at Nyamupe, E Zaïre (3° 21'S, 38° 8'E). It resembles the Little Greenbul *A. virens* but is overall darker, 'in particular, crown and back dark olive-green, uppertail-coverts a little lighter, rectrices olive-green washed with dark brown, chin and throat dark green without any yellowish wash, remainder of underside dark green slightly yellowish in middle of abdomen, undertail-coverts greenish olive with lighter tips, remiges olive-green washed with dark brown, interior of remiges dark grey becoming paler towards the margins, axillaries pale green, upper and lower mandibles almost black, tarsus and feet black, claws blackish'. It is the same size as Little Greenbul (wing 75, tail 68, culmen from skull 15, tarsus 18), but the tail appears to be less graduated. It was collected in primary forest at 990 m, in a place which the collector insisted was not the normal habitat of Little Greenbul. It is geographically sympatric with Little Greenbul, though possibly with ecological separation, and Prigogine (loc. cit.) concluded they were a pair of sibling species.

Collecting has continued in the area where the bird was taken and no similar specimens have been found. A. Prigogine (pers. comm.) now believes the bird to be a melanistic specimen of *A. virens*, even though he has examined over 1000 specimens of *virens* and none show any evidence of melanism.

Field Characters. An abundant and ubiquitous bulbul, whose presence is easily established by its bustling song, given all day, even at the hottest hours, and all year. Much less easy to see, however, as it prefers to remain hidden in thick brush and undergrowth; even at eye level it remains infuriatingly just out of sight, singing from an invisible perch, or briefly glimpsed as a shape in the shadows. When called up by playing back its voice it darts across a track or other opening and dives into cover again without affording a proper look. Adults mainly dull olive with some yellow in centre of belly, with short and broad dark bill, no eye-ring, dull grey-brown eyes, yellowish brown legs; a size larger than similar-looking Little Grey Greenbul *A. gracilis* and Ansorge's Greenbul *A. ansorgei*, but hard to tell from same-sized Plain Greenbul *A. curvirostris*, q.v. Young similar to very young Yellow-whiskered Greenbul *A. latirostris*, which lacks yellow whiskers but has dusky malar stripe separating cheeks from paler throat, centre of throat to breast darker and browner, undertail-coverts pale tawny, not olive, feet bright orange-yellow rather than brown or yellow-brown.

Voice. Tape-recorded (32, 53, 86, C). Cheerful, bustling song is one of the 'basic' sounds of lowland rain forest; it consists of 2 basic parts, with optional additions in between: several low, gravelly introductory notes on same pitch, followed by a pause; then a long, unmusical jumble of bubbling notes, some grating but mainly rather squeaky, continuing on one pitch until just before the end when they rise into a shrill finale. In many cases, this is the complete song, but in others the initial pause is followed by a sort of musical interlude, with up to 10 tuneful whistles, interspersed with a few gravelly notes and often repeated before proceeding to the bubbling conclusion. Whistled themes vary from bird to bird but an individual may repeat its own theme (see cuts 2 and 3 in Keith and Gunn 1971). Scolding call, a repeated, dry 'cuk-cuk-cuk . . . , sometimes rising in pitch.

General Habits. Inhabits all types of forest, but prefers second-growth and edge situations. In primary forest occupies undergrowth and dense vine tangles, sometimes ascending to middle levels, occasionally to canopy; more typical of clearings, open spaces around fallen trees, thick brush at forest edge (e.g. in Gabon, *Aframomum* thickets under parasol trees *Musanga*: Brosset and Erard 1986). Extends out beyond main forest bloc along gallery forest and into forest-savanna mosaic, invading isolated forest patches and readily flying from one to another. Partial to waterside vegetation, occurring around ponds and along streams, also in swamp forest (Nigeria, Lagos) and mangrove forest (Zanzibar). Sometimes enters man-made habitats (rubber and cocoa plantations) and even gardens. Where outnumbered by Yellow-whiskered Greenbul, keeps more to edge habitats while latter occupies forest interior; where more numerous, invades interior as well (Gabon: Brosset and Erard 1986). Mainly lowland forest, ascending locally to lower edge of montane forest: to 2200 m in Cameroon, 1830 m in Zaïre, 1800 m in E Africa, 2000 m in Zambia, 2050 m in Malaŵi.

Occurs singly and in pairs; generally not sociable, though gathers in some numbers at fruiting trees with other frugivores. Found in only 4% of undergrowth bird parties, Gabon (Brosset and Erard 1986), although reportedly common in bird parties E Zaïre (Prigogine 1971) and noted in party with 3 bulbul species and many other passerines, Rwanda (Vande weghe 1984). Shy and skulking; flight quick and jerky between patches of cover. Forages mainly within leafy cover, but comes to ground to catch insects disturbed by ant columns, and exploits swarming termites. Emerges in evening to bathe in shaded streams and pools, later preening itself before going to roost.

Tends to wander locally; rarely territorial except when nesting. Of 114 birds ringed over period of 12 years, Gabon, only 1 recaptured (twice, 8 and 23 months later) (Brosset and Erard 1986).

Undergoes complete moult Nov–Feb (Zaïre, Verheyen 1953). Primary moult lasting 314 days recorded W Kenya (Mann 1985).

Food. Insects: Orthoptera, Neuroptera, Lepidoptera, Diptera, Coleoptera (especially Curculionidae); Phasmidae, shield bugs, caterpillars, termites, ants. Spiders. Berries and small fruits, e.g. *Allophylus, Musanga, Trema, Solanum, Ficus, Macaranga, Maesopsis berchemioides*; also exotics *Lantana* and *Clidemia hirta*. Seeds. Often stated to be largely or entirely a fruit eater, but this may be true only seasonally. Of 27 stomachs, Cameroon, 18 contained fruit, 9 insects (Serle 1954).

Breeding Habits. Territorial while breeding. Seasonal gatherings of 6–10 singing birds in small area may be connected with breeding. In display at singing post beats wings, lowers tail slightly, lowers head and puffs out feathers of throat, belly, flanks and especially upper back, giving it a humped appearance. In probable pre-copulatory behaviour, 1 bird (♂?) perched on branch 4 m above ground and sang while mate flew silently from branch to branch nearby, rapidly beating wings and showing other signs of excitement (Germain et al. 1973).

NEST: shallow cup of dead leaves, grasses, twigs and pieces of bark (**A**), sometimes bound with cobwebs and bark strips, lined with plant fibres and fine roots and grasses, also fibres of fungus *Marasmius*; other materials used as locally available, e.g. nest near field built with strips of maize leaves; ext. diam. 10–11·4 cm, int. diam. 6·4 cm, ext. depth 7 cm, int. depth 2–3·3 cm; placed low down, from just above ground to 1·5 m, in small shrub or tangled mass of bracken or under large leaves (1 suspended between 3 twigs), generally concealed but sometimes in fairly exposed position.

EGGS: 2, occasionally 3. Somewhat glossy; ground colour varying from dull white or pale greyish to beige, pink, pale mauve or pale violet, profusely (or sometimes indistinctly) speckled with various shades of red-brown, sometimes salmon or violet, with purplish slate undermarkings, usually in cap around large end. SIZE: (n = 5) 20–23 × 14·5–16 (21·9 × 15·2).

LAYING DATES: Senegambia, June–July; Liberia (breeding condition June, Aug); Ghana, Feb, Apr–May; Nigeria (breeding condition May–June, Sept); Cameroon, Jan–Mar (breeding condition May–June, Sept), probably breeds all year (Germain et al. 1973); Gabon, Oct–Mar, especially Jan–Feb (short dry season), occasionally July–Aug; breeding delayed when short dry season delayed; Congo, Nov–Dec; Zaïre, all months near the equator, elsewhere during the rains (e.g. Itombwe, all months except dry season June–Aug); Angola (breeding condition Aug); Central African Republic (breeding condition May–June); Sudan, Aug–Sept; E Africa: Uganda (breeding condition June–July, Nov); Zanzibar (breeding condition Nov–Feb); Region B, Jan–May, Sept–Oct; Region D, Feb; Malawi, Oct; Zambia, Aug–Oct (breeding condition Nov), mainly in hot, pre-rains season.

INCUBATION: by ♀ only. Period: at least 13 days. Incubating bird can be approached to within 3 m.

DEVELOPMENT AND CARE OF YOUNG: fledging period: 11–12 days.

BREEDING SUCCESS/SURVIVAL: 60% of nests predated, Gabon. Bird caught in trap set for squirrel *Funisciurus*, probably a not uncommon occurrence since they live in same habitat (Malbrant and McClatchy 1952). Bird in Kakamega, Kenya, lived 7 years 9 months (Mann 1985).

References
Brosset, A. and Erard, C. (1986).
Zimmerman, D. A. (1972).

Andropadus gracilis Cabanis. Little Grey Greenbul. Bulbul gracile.

Andropadus gracilis Cabanis, 1880. Orn. Centralb. 5, p. 174; Angola.

Plate 18
(Opp. p. 305)

Range and Status. Endemic resident. W African forest belt from Sierra Leone and SE Guinea through Liberia and Ivory Coast (north to Sipilou, Bouaké, Dabakala) to Ghana (not uncommon, widespread) and Togo; and from Nigeria (not uncommon, extending north in forest belt to Ibadan, Ife, Enugu, and isolated population at Nindam Forest Reserve, also Obudu Plateau), Cameroon (mainly in S, north in highlands to Mt Kupé, Rumpi Hills and Korup Nat. Park), Gabon, Congo, SW Central African Republic, N and central Zaïre (south in E to Mt Kabobo), Angola (Cabinda and extreme N Lunda (Dundo)) and Uganda (common in forests of S and W, north to Budongo, east to Entebbe, Kifu and Mabira), and a disjunct population in W Kenya (Kakamega). Density in Gabon varies from 10 to 12 pairs/km^2 in old second growth and riverine forest to 5–6 pairs/km^2 in areas of human habitation where habitat more broken up (Brosset and Erard 1986); density in Nigeria 1 bird/3 ha (Johnson 1989).

Description. *A. g. ugandae* van Someren: E Zaïre to W Kenya. ADULT ♂: top and sides of head greyish olive, lores darker, narrow eye-ring white. Mantle, back, scapulars and upper-wing-coverts olive-green, becoming yellowish green and then gingery brown on rump, uppertail-coverts rich brown with rufous wash; tail-feathers brown with slight rufous wash, very narrowly edged olive and with pale tips when fresh; shafts brown to red-brown above, pale brown below. Chin, throat and upper breast light olive-grey, rest of underparts rather variable: brownish or greenish olive with variable amount of yellow from lower breast to lower belly, palest in centre of belly; lower flanks with gingery wash, undertail-coverts pale brown. Primaries and secondaries dark brown edged greenish olive, on underside with pale yellow edges to basal half of inner webs; underwing-coverts and axillaries lemon yellow, greater under primary coverts light brownish grey. Bill dark brown to black; eye brown, olive-brown or brownish grey; legs and feet olive-green, brownish horn, blue-grey. Sexes alike. SIZE: (13 ♂♂, 8 ♀♀) wing, ♂ 66–77 (73.6), ♀ 68–78 (72.1); tail, ♂ 58–65 (61.1), ♀ 57–65 (60.5); bill, ♂ 14–16 (14.9), ♀ 13.5–15.5 (14.5); tarsus, ♂ 17–20 (18.2), ♀ 16–18 (17.3). WEIGHT: ♂ (n = 36) 18–23 (20.9), ♀ (n = 15) 19–28 (22.4).

IMMATURE: (*A. g. gracilis*) ♂ with partly ossified skull brighter green above and brighter yellow below than adult, grey throat washed with green, eye-ring yellow, not white. Bill black, gape and base of bill yellow; eye grey-brown; legs and feet olive.

NESTLING: unknown.

A. g. gracilis Cabanis: SE Nigeria to central Zaïre and Angola. Considerably duller and greyer below, with very little green or yellow.

A. g. extremus Hartert: Sierra Leone to SW Nigeria. Very similar to *ugandae*; yellow of underparts somewhat brighter. WEIGHT: Liberia, ♂ (n = 2) 16.9, 17.1, ♀ in breeding condition (n = 2) 15, 17.1; Nigeria, unsexed (n = 11) 20–27 (24.0).

Field Characters. An extremely small greenbul, smaller than all others in its range except Ansorge's Greenbul *A. ansorgei*, with which it sometimes shares the same niche. Plumage much more yellow and green than Ansorge's Greenbul, which appears brown and grey in the field, with brown upperparts, grey-brown underparts, gingery wash on flanks; both have same narrow white eye-ring. Song quite similar to Ansorge's Greenbul in general tone but phrases longer and more complex (see *Voice*). Not particularly secretive; often sings from exposed perch (Button 1964a). Plumage extremely similar to Plain Greenbul *A. curvirostris*, which has similar yellow on underparts and gingery flanks but has more pronounced brown breast, sharply contrasting with grey throat rather than merging into it. Plain Greenbul is also larger and heavier, with larger bill which appears decurved, reddish rather than brownish eyes, and only a trace of an eye-ring, not visible in the field; voice different, lower and slower. Often in same habitat as Little Greenbul *A. virens*, whereas close relatives Ansorge's and Plain Greenbuls more often in true forest, though all 3 can occur together.

Voice. Tape-recorded (53, ERA, McVIC). Song, variable phrases of 4–6 whistled notes, pure and fairly high-pitched but not loud, and easily overlooked: 'clip, wor, ta-chi-ta-chi', 'we were, to-wee', 'ta-cher, ta-wer-chee'. About same pitch as song of Ansorge's Greenbul and similar in tone, but more notes per phrase and less slurring; the 2 can easily be confused. Unlike song of Plain Greenbul (*contra* Lippens and Wille 1976), which is slower and lower-pitched. Call, 'twit' or 'tchick'.

Andropadus gracilis

General Habits. Inhabits primary and secondary forest, including isolated patches in savanna; riverine forest, seasonal swamp forest, parkland and various man-made situations: abandoned cultivation, overgrown plantations, especially where vegetation is in final stages of regeneration; small patches of trees around villages. Prefers second growth, clearings and forest edge to main body of forest; more typical of small clumps of trees and tall, dense patches of brush and shrubs. Lowland and intermediate forest, up to 1200 m in Cameroon, 1550 m in Zaïre, 1700 m in E Africa.

Found singly, in pairs or in small family groups, also mixed-species flocks of insectivores, although less tied to bird parties than other bulbuls; in Cameroon, noted in party with Rufous-vented Paradise Flycatcher *Terpsiphone rufocinerea* and Cassin's Malimbe *Malimbus cassini*. Forages at all heights, including canopy, but mainly at mid-level, 4–18 m, both on open limbs and twigs, among leaves (especially gleaning underside), and in tangles of lianas and clusters of dead leaves; sometimes catches insects in air. Seeks out fruiting trees. Active; fairly tame and easily observed.

Primary moult duration, Kenya (n = 1) 207 days (Mann 1985).

Sedentary, at least in Gabon.

Food. Insects: Orthoptera, small beetles, moths, caterpillars and other insect larvae, small mantids, black ants; spiders; small snails (once: Johnson 1989). Berries and small fruits of e.g. *Rauwolfia*, *Trema*, *Croton*, *Musanga*; seeds.

Breeding Habits. Monogamous, territorial.

NEST: only 2 described (Erard 1981). Open cup with foundation of dead leaves closely pressed together and strengthened with a binding of thin, dry, twisted vines, with a smooth, compact lining of fine stems and vegetable fibres; more like a small and more carefully-built nest of Common Bulbul *Pycnonotus barbatus* than that of an *Andropadus*. Green leaves included in outer binding of nest aid concealment. Ext. diam. 99, int. diam. 65, ext. depth 70, depth of cup 40. Site, 2·9 m above ground in foliage at end of horizontal branch of avocado tree. A second nest in early stages of construction was a loose cup of dead leaves bound with cobwebs, 8 m above ground at junction of many twigs inside bunch of leaves on horizontal branch of tree at edge of clearing. Built by ♀ alone; ♂ accompanied her in silence, perched 12 m away while she was building nest.

EGGS: 2. Creamy white with rosy tinge, large end thinly covered with tiny brown and mahogany spots over a layer of faint pale grey and pale lilac markings.

LAYING DATES: Liberia (breeding condition July); Ghana (breeding condition June, Jan); Nigeria (breeding condition Mar, July); Cameroon, May (breeding condition Dec); Gabon, Aug (nest construction Jan, ♀ soliciting copulation Mar, breeding condition Nov–Dec); Zaïre, Sept–June; Uganda (breeding condition Jan–Feb).

INCUBATION: by ♀ only.

References
Brosset, A. and Erard, C. (1986).
Erard, C. (1981).
Zimmerman, D. A. (1972).

Andropadus ansorgei Hartert. Ansorge's Greenbul. Bulbul d'Ansorge.

Plate 18
(Opp. p. 305)

Andropadus ansorgei Hartert, 1907. Bull. Br. Orn. Club 2, p. 10; Degama, S Nigeria.

Range and Status. Endemic resident. Sierra Leone to Liberia (fairly common Mt Nimba) and Ghana (forest zone; 1 record Tafo); status in Ivory Coast uncertain: not listed by Thiollay (1985) but recent sight records from Yapo Forest (Demey and Fishpool 1991). Nigeria: east of Lower Niger R., probably also forest zone of SW; Cameroon, Equatorial Guinea (Rio Muni), Gabon; central and NE Zaïre, many localities but with large apparent gaps; W Kenya (now only certainly in Kakamega Forest, formerly Mt Elgon and Nyarondo). No records from Angola, but occurs on Zaïre border at Kasongo Lunda. Uncommon to locally fairly common, but status and distribution uncertain in many areas due to confusion in the field with more widespread Little Grey Greenbul *A. gracilis*. Some range gaps may be real, not apparent; lack of records from well-collected Ugandan forests suggests real gap.

Description. *A. a. kavirondensis* (van Someren): Kenya. ADULT ♂: top of head to hindneck brownish olive grey, becoming somewhat paler on sides of head and neck; narrow whitish eye-ring. Mantle and back olive-brown, becoming gingery rufescent on rump, uppertail-coverts rufous brown; tail brown with slight rufescent wash, shafts same colour above, pale brown below. Chin and throat pale brownish grey, rest of underparts dull greyish brown with gingery wash on flanks, centre of belly paler with slight yellowish tinge only visible in the hand, undertail-coverts pale yellowish brown. Primaries and secondaries olive-brown edged paler, undersides with pale edges to inner webs; axillaries and underwing-coverts pale buffy white, under primary coverts olive-grey. Bill black to brownish horn with black base and culmen; eye dark brown, brown or chestnut brown, occasionally red-brown; legs and feet greyish olive or dark olive. Sexes alike, but bill of ♀ dull blackish horn, legs and feet of ♀ greenish grey. SIZE: (11 ♂♂, 9 ♀♀) wing, ♂ 72–78 (74·1), ♀ 68–73 (70·0); tail, ♂ 59–65 (62·3), ♀ 58–62 (59·6); bill, ♂ 13–16 (14·7), ♀ 13·5–15 (14·6); tarsus, ♂ 15–18·5 (16·9), ♀ 14·5–18 (16·3). WEIGHT: Liberia, ♂ (n = 4) 19–20·9 (20·0), 1 ♀ 20·1; Kenya, ♂ (n = 5) 17–19 (17·9), ♀ (n = 5) 17–23 (18·2).

IMMATURE: full-grown ♀ with unossified skull like adult but head and breast slightly greyer; cutting edge and tip of bill pale brown, gape yellow; legs and feet light olive.

NESTLING: unknown.

A. a. ansorgei Hartert: Sierra Leone to Zaïre. Breast and flanks browner, centre of belly less contrastingly pale and without yellow wash, undertail-coverts rufous-brown.

Field Characters. A tiny, short-tailed greenbul; looks grey below except for gingery brown wash on flanks. Same size as Little Grey Greenbul and with similar pale eye-ring, but latter has pronounced yellow on belly, different voice. Appears extremely small in the field, much smaller than similar-looking Plain Greenbul *A. curvirostris*; at a distance might be mistaken for a Palearctic warbler.

Voice. Tape-recorded (53, C, ERA, HOR, STJ). Song, soft, whistled phrases of 3 notes, of which 1 is always slurred, sometimes 2: 'twee-tui-twit', 'wee-tui-wee', 'teeu-tee-wuee', 'tiu-hit-wee' (or 'you hit me'). Notes well spaced, unhurried, pure and quite sweet, but soft and easily lost among other voices. Sings all year in Gabon. Call, a short, sibilant, rattling trill, 'reminiscent of flight call of Brown-headed Cowbird *Molothrus ater*' (Zimmerman 1972).

General Habits. Inhabits lowland and intermediate forest, exceptionally to lower edge of montane forest (Zaïre, Mt Nyombe, 1450 m). Primary forest: chiefly in canopy in Gabon, with a special liking for tangles of lianas; middle levels in Liberia, lower and middle levels in Kenya, where forages along smaller limbs and branches; thickets and undergrowth within forest in Ghana. Also in secondary forest, strips of forest along streams in farmland, late stages of regrowth in formerly cultivated areas, and on large islands in rivers. Usually singly or in pairs, sometimes family parties; does join mixed-species foraging flocks but more independent of them than most bulbuls.

Sedentary, at least in Gabon.

Food. Insects, including Coleoptera, Orthoptera, caterpillars, moths; also small fruits, e.g. *Allophylus*, *Macaranga*, *Musanga*, and seeds.

Breeding Habits. Nest and eggs not described.
LAYING DATES: Nigeria, May, July, Oct–Dec; Gabon (singing intensifies at beginning of rains which follow the long dry season, and during short dry season); Zaïre (juveniles Nov, breeding condition Mar, July, Nov, near breeding condition Feb, Aug); in Ituri District, 'probably breeds through whole rainy season' (Chapin 1953)); Kenya (breeding condition June).

References
Brosset, A. and Erard, C. (1986).
Zimmerman, D. A. (1972).

Plate 18
(Opp. p. 305)

Andropadus curvirostris Cassin. Plain Greenbul; Cameroon Sombre Greenbul. Bulbul curvirostre.

Andropadus curvirostris Cassin, 1860. Proc. Acad. Nat. Sci. Phil., 1859, p. 46; Camma R., W Africa.

Range and Status. Endemic resident. Forest, from Sierra Leone, SE Guinea and Liberia (fairly common Mt Nimba) to Ivory Coast (common, north to Sipilou, Maraoué), Ghana (widespread, fairly common) and Togo; and from S Nigeria (not uncommon in lowland forest west of Lower Niger, scarce east of river, to Owerri), Cameroon (common to abundant), Bioko and Gabon (common) to SW Central African Republic, Zaïre (common and widespread in central and N, absent from SE), Angola (Cabinda and forested areas in Cuanza Norte and N Lunda), extreme S Sudan (common), Uganda (common to abundant in W and central forests, from Bwamba, Budongo and Impenetrable to Mt Elgon) and W Kenya (Elgon, Kakamega and Nandi forests). Density in Gabon probably 10–12 pairs/km^2 (Brosset and Erard 1986).

Description. *A. c. curvirostris* Cassin: forest zone of central and S Ghana to Kenya, Angola, Zaïre. ADULT ♂: top and sides of head dark grey-olive, somewhat paler on face and ear-coverts, lores darker; indistinct grey eye-ring. Upperparts, including scapulars, tertials and upperwing-coverts, brown-olive, rump somewhat brighter, with rusty wash, some feathers with yellowish tips; uppertail-coverts reddish brown, tail darker red-brown, feathers sometimes narrowly edged olive (variable) and with narrow pale tips when fresh, shafts red-brown above, pale brown below. Chin and throat light olive-grey, clearly demarcated from olive-brown breast; belly olive-yellow, becoming pale yellow in centre of lower belly; flanks olive-brown with gingery wash, undertail-coverts pale brownish with pale yellow tips. Primaries and secondaries dark brown edged olive-green; on underside, inner webs with narrow pale yellow margins, broadest on secondaries, axillaries and underwing-coverts lemon yellow, greater under primary coverts grey-brown. Bill black, distal half of cutting edges olive; eye chestnut brown to red-brown; legs pale brownish or olive, olive-green, dark green or dark greenish-grey, feet somewhat greyer. Sexes alike, but bill of ♀ dark grey-brown with paler cutting edges and tip. SIZE: wing, ♂ (n = 40) 74–82 (77·3); ♀ (n = 15) 71–75 (73·3); tail, ♂ (n = 11) 69–77 (73·0); ♀ (n = 12) 68–75 (71·5); bill, ♂ (n = 11) 14·5–16 (15·4), ♀ (n = 12) 15–16 (15·3); tarsus, ♂ 18–20 (18·9), ♀ 17–21 (19·0). WEIGHT: Uganda, ♂♀ (n = 19) 21·5–36; Kenya, ♂ (n = 13) 22–29 (25·0 ± 2·1), ♀ (n = 10) 22–28 (25·0 ± 2·1). Diurnal weight change of 10·4% (Kenya: Mann 1985).

IMMATURE: belly yellower than adult, flanks dull olive without gingery wash; gape yellow.

NESTLING: unknown.

A. c. leoninus Bates: Sierra Leone to central Ghana (Goaso and Mampong in Ashanti), in thickets north of forest/savanna boundary. All upperpart colours a shade darker than in nominate race; tail blacker brown, less reddish; duller and slightly greyer below, with less yellow. WEIGHT: Liberia, ♂ (n = 7) 22–26·6 (24·1 ± 1·7), ♀ in breeding condition (n = 9) 20·3–22·9 (21·5 ± 0·7).

Field Characters. A nondescript olive greenbul with narrow black bill, red-brown eye, grey eye-ring, greenish legs; fairly small, but noticeably larger than tiny Little Grey Greenbul *A. gracilis* and Ansorge's Greenbul *A. ansorgei*, from which it also differs in lack of eye-ring, darker, more slender bill with noticeable curve, and feeding niche (usually forages within leaf clusters rather than on open branches). Underparts very similar to Little Grey Greenbul, with yellow on belly and gingery wash on flanks. Useful character (in good light only) is neatly defined grey throat sharply demarcated from brown breast and contrasting with dark head. Sings from low, concealed perch (Button 1964a). Very like same-sized Little Greenbul *A. virens* and often in same bushy situation; throat of Little Greenbul barely defined, merges into breast, and Little Greenbul has broad, browner bill, yellowish brown legs, and bustling song given constantly.

Voice. Tape-recorded (53, C, ERA, STJ, Z1M). Song, a series of varied phrases of 3–4 slurred whistles, low and mellow, almost pure, with no bulbul burr: 'wee-du-duee', 'wee-du, wee-dooer', 'tyu-wee-dyu-drrreee' (fourth, upslurred note introduced by a chatter). Pitch lower than songs of Little Grey and Ansorge's Greenbuls, delivery slower, with a lazy, casual quality, as if it were almost too much of an effort to sing. Not silent, as often claimed, but subdued song tends to be submerged among louder voices and singer frequently not in view. Calls, a harsh chatter, a dry, rather muted 'prrrt' or 'trrrrrrrr', and a double 'twit-twit'.

General Habits. Inhabits primary and secondary forest, also gallery forest, final stages of regenerating forest in old cultivation; clearings and edges; reaches islands in rivers. Mainly in lowlands, locally in intermediate forest, to 1800 m in Zaïre, 1600 m in Sudan, 2300 m in E Africa. Chiefly in understorey, 2–15 m, ascending into lower trees and sometimes into canopy, especially in dry season.

Occurs singly, in pairs or small territorial family groups; regularly joins mixed-species flocks, especially at forest edge on border between primary forest and second growth, and joins other bulbuls at fruiting trees. Forages mainly in leafy tops of higher shrubs and in vine tangles. Gleans insects from foliage, sometimes while hovering. Common but often overlooked because concealed in foliage. Perches for long periods on bare branches in low levels of mature bush (Button 1964a).

Primary moult duration of up to 140 days (W Kenya: Mann 1985).

Sedentary, at least in Gabon.

Food. Caterpillars, mainly 4–7 cm long; Orthoptera, mantid larvae, moths. Fruits, especially *Rauwolfia*, also *Musanga*, *Heisteria*, *Macaranga*, *Allophylus*, *Agelaea*, *Trema*, *Urera*. Small seeds. Of 42 stomachs, Cameroon, fruit in 29, insects in 11, snails in 2 (Serle 1965).

Breeding Habits. Territorial; advertises territory with song. ♀ solicits copulation in horizontal posture with beating wings. Nest and eggs not described.

LAYING DATES: Sudan, Jan, Aug, Oct; Ghana, Feb; Nigeria (♂ with slightly enlarged testes Aug); Cameroon (breeding condition Aug–Oct, Dec–Jan); Zaïre (breeding condition Sept–May); Gabon (increased singing late Sept–Feb, ♀ soliciting copulation Dec–Jan, during short dry season); Angola (immature Feb, breeding condition Feb–Mar); East Africa: Region B (breeding condition Jan–Feb, Apr, July).

BREEDING SUCCESS/SURVIVAL: 2 birds in Kakamega, Kenya, lived 9 years 11 months and 10 years 8 months, and a third 19 years 6 months, apparently a record for any tropical species (Zimmerman 1986a).

References
Brosset, A. and Erard, C. (1986).
Zimmerman, D. A. (1972).

Plate 17
(Opp. p. 304)

Andropadus gracilirostris Strickland. Slender-billed Greenbul. Bulbul à bec grêle.

Andropadus gracilirostris Strickland, 1844. Proc. Zool. Soc. Lond., p. 101; Fernando Po.

Range and Status. Endemic resident. SW Senegal (Basse Casamance), Guinea and Sierra Leone to extreme SW Mali (uncommon, Mandingo Mts), Liberia (common), Ivory Coast (from coast north to Man, Sifie, Dabakala), Ghana (common) and Togo; and from Nigeria (locally common, north to Ibadan, Ife, Nsukka), Cameroon (mainly in S, north to Mt Kupé), Bioko, Gabon (widespread and fairly common), Congo, SW and SE Central African Republic, extreme S Sudan and Zaïre (almost throughout except extreme NW and SE) to Angola (Cabinda, Cuanza Norte, N Malanje, N Lunda), Rwanda, Burundi, Uganda (fairly common in S and W, to Budongo, Mabira, Mt Elgon), Kenya (W and central highlands north to Kapenguria and Ndoto Mts, east to Meru, south to NW Mara Game Reserve and Nairobi) and Tanzania (NW and Kungwe-Mahari Mts). Density in Gabon 7–9 pairs/km^2 (Brosset and Erard 1986).

Description. *A. g. percivali* (Neumann): highlands of central Kenya, intergrading with nominate race in W Kenya. ADULT ♂: top of head to hindneck dull olive, rest of upperparts, including scapulars and upperwing-coverts, somewhat brighter and greener, especially on rump and uppertail-coverts. Vestigial grey line from lores to over and just behind eye, not present in all individuals; barely perceptible narrow grey eye-ring; ear-coverts dull olive-brown, feathers with paler centres and darker tips, giving slightly mottled effect. Tail brown washed with olive, somewhat paler on underside, shafts maroon above, pale brown below. Chin, throat and cheeks creamy white, rest of underparts pale grey becoming creamy white in centre of belly and pale buff on thighs and undertail-coverts. Primaries and secondaries dark brown edged greenish olive; on underside, inner webs broadly edged yellow-ochre, broadest on secondaries; axillaries and under primary coverts orange-yellow, underwing-coverts yellow-ochre. Bill black or blackish; eye brown, red-brown, red or deep red; legs and feet brown, dark brown, blackish or slaty black. Sexes alike. SIZE: (10 ♂♂, 10 ♀♀) wing, ♂ 82–90 (86·3), ♀ 81–87 (85·1); tail, ♂ 72–82 (78·2), ♀ 72–80 (75·9); bill, ♂ 17–20 (18·3), ♀ 17–20 (18·4); tarsus, ♂ 19–22·5 (21·2), ♀ 19–22 (20·8). WEIGHT: ♂ (n = 43) 24–37 (30·7), ♀ (n = 32) 26–40 (31·8).
   IMMATURE: ('chagwensis') very young (tail 14): upperparts, wings and tail rich dark brown with no trace of olive or green; below, darker and duller grey with brownish wash, lower flanks and undertail-coverts cinnamon. Later, browner above than adult, wash on breast and flanks brown, not olive, undertail-coverts and narrow margins on inner webs of underside of tail-feathers cinnamon.
   NESTLING: unknown.
   A. g. gracilirostris Strickland (including 'congensis', 'chagwensis'): Guinea to Sudan, W Kenya and Angola. Upperparts browner, underparts darker and greyer, some birds with olive wash on breast and flanks, others with a few creamy yellow streaks; 'congensis' said to be darker but this is barely perceptible; 'chagwensis' said to be like 'congensis' below but lighter olive, tail lighter and browner, but this is only true for some individuals.

Field Characters. A drab bulbul living in the canopy. In typical view, all one sees is plain grey underparts of bird feeding among foliage at top of tall tree. Olive upperparts often hard to see at this height, but not necessary for identification. Underpart colour distinctive; other forest bulbuls have underparts brown, olive, yellow or white, none are uniform light grey. Whistled song distinctive, although at heights of 30 m it can be hard to tell which species is involved.

Voice. Tape-recorded (53, C, BRU, KEI, MOR, ZIM). Normal song, a series of single pure, high-pitched, down-slurred whistles lasting *c.* 1·5 s, given at intervals of 2–4 s; quality reminiscent of European Starling *Sturnus vulgaris*. Sometimes whistles are shorter, with a more burry quality, or both types can be given in succession. Also has a 3-part song, 'chip-cheeo-wu', accent on second (highest) note. Contact call between foraging birds, soft 'frue', 'fu-ue' or 'fu-wuwu' (Dowsett-Lemaire 1990). Alarm scold, an agitated 'chewi'.

General Habits. Inhabits primary and especially secondary forest, gallery forest, abandoned cultivation with forest in various stages of regeneration, tall, leafy trees left standing in agricultural land, patches of dense woodland in savanna; fond of forest edge, clearings and roadside vegetation. Lowland to montane forest, up to 2300 m in Zaïre, 2500 m in Rwanda, 1800 m in Sudan, 2400 m in E Africa.
   Can occur at all levels in forest but usually only in canopy, where it works through leaf clusters searching for fruit and gleaning insects from leaves, 'foraging like sunbird' (Gee and Heigham 1977). Occurs singly, wandering individuals sometimes constantly on the

move, or in pairs or family parties. Numbers may gather at fruiting trees and vines: 20 birds in liana, Rwanda (Dowsett-Lemaire 1990). Often joins other fruit-eaters like Common Bulbul *Pycnonotus barbatus* and Yellow-whiskered Greenbul *A. latirostris*. Also joins mixed-species foraging flocks; in S Cameroon noted in party with Zenker's Honeyguide *Melignomon zenkeri*, Velvet-mantled Drongo *Dicrurus (adsimilis) modestus*, Little Green Sunbird *Anthreptes seimundi* and Yellow-capped Weaver *Ploceus dorsomaculatus* (Germain *et al.* 1973). Sometimes descends to lower levels, e.g. Ghana, where common in small bushes and undergrowth (Lowe 1937), and Kenya (Kakamega), where it feeds at 4 m in canopy of plantation of small trees just outside forest. Wanders within large territory (home range?) of 8–12 ha (Gabon: Brosset and Erard 1986). Although common, its quiet manner and rather undistinguished whistle often make it inconspicuous.

Sedentary, at least in Gabon.

Food. Insects: Orthoptera, including grasshoppers, Hemiptera, moths, winged ants, caterpillars, mantid larvae. Spiders. Fruit species recorded in Gabon include *Musanga, Morinda, Macaranga, Croton, Heisteria, Ficus, Xylopia, Allophylus, Rauwolfia*; in Rwanda (Nyungwe), *Agelaea, Bridelia, Harungana, Ilex, Musanga, Ocotea, Podocarpus, Syzygium, Polyscias, Rutidea, Sapium, Schefflera* (2 species), *Trema, Urera*. Seeds.

Breeding Habits. Territorial. ♀ solicits copulation in horizontal posture, rump feathers erect, tail lowered and spread, wings lowered and quivering.

Nest and eggs not described. ♀ seen on several occasions carrying nest material (dead leaves and vegetable fibres) from forest edge to canopy of tall tree in forest 200 m away, accompanied by singing ♂ (Brosset and Erard 1986).

LAYING DATES: Ghana (breeding condition Feb); Nigeria (breeding condition June, feeding young July); Cameroon, Sept, Nov; Zaïre (evidence of breeding nearly all months, probably nests all year); Rwanda, Nov–Dec; Central African Republic (breeding condition June); Sudan, Aug; Uganda (breeding condition May, Nov); Kenya (breeding condition June, fledgling being fed by adult Nov); E Africa (building nest Feb).

BREEDING SUCCESS/SURVIVAL: bird killed by Red-thighed Sparrowhawk *Accipiter erythropus* (Brosset and Erard 1986).

References
Brosset, A. and Erard, C. (1986).
Zimmerman, D. A. (1972).

Andropadus latirostris Strickland. Yellow-whiskered Greenbul. Bulbul à moustaches jaunes.   Plate 18

Andropadus latirostris Strickland, 1844. Proc. Zool. Soc. Lond., p. 100; Fernando Po.   (Opp. p. 305)

Range and Status. Endemic resident, both sedentary and nomadic. Extreme SW Senegal (Basse Casamance), Sierra Leone, SE Guinea and Liberia to Ivory Coast (north to Korhogo), Ghana (north at least to Mampong) and Togo. Nigeria (widespread in S) to Cameroon (north to Mt Oku), Bioko, Gabon and S Congo. Rather local and uncommon in Angola (Cabinda, Cuanza Norte, Gabela, N Lunda (Dundo)); Zaïre (abundant in N and E, absent SE and many parts of S). Extreme SE Central African Republic, S Sudan (common in mountains), Rwanda, N Burundi and adjacent areas of NW Tanzania; W and S Uganda north to Budongo, Mabira, Elgon; Kenya highlands, in W from Kapenguria to Nguruman Hills, in E from central region north on montane islands to Mt Nyiru, and on Tanzania border at Ol Doinyo Orok; W Tanzania from Kigoma to Kungwe-Mahare Mts, and on Ufipa Plateau. Common to abundant nearly everywhere, often the most abundant bulbul or even the most abundant bird species (e.g. Gabon, where it comprised 28% of 8624 birds captured: Brosset and Erard 1986). Density very variable but up to 500 birds/km² Gabon; in Nigeria, 1 bird/0·7 ha (Johnson 1989).

Description. *A. l. latirostris* Strickland (including '*australis*', '*pallidus*', '*eugenius*', '*saturatus*'): range of species except race *congener*. ADULT ♂: top and sides of head, including lores, ear-coverts and cheeks, dark sooty olive, becoming dark greenish olive on rest of upperparts, rump with rufous tinge, uppertail-coverts dark russet. Tail dark reddish brown, feathers narrowly edged greenish brown, shafts very dark red-brown above, off-white below. Bright yellow moustache stripe on sides of throat, rest of underparts dull olive-green, becoming lighter and browner on flanks and pale yellow on centre of lower breast and belly, undertail-coverts pale brown tipped pale brownish yellow. Primaries and secondaries blackish brown, outer webs lighter brown edged greenish olive, upperwing-coverts dark olive-brown; underside of flight feathers with greyish cast, inner webs of secondaries and base of inner webs of primaries edged pale buffy grey, axillaries and underwing-coverts dull yellowish olive. Bill black or dark brown, cutting edges, tips and gape pale yellow; eye dark brown, brown or grey-brown; legs and feet yellow-brown to brownish flesh or dark brownish orange. Sexes alike. SIZE: (12 ♂♂, 12 ♀♀) wing, ♂ 83–91 (87·1), ♀ 77–83 (80·3); tail, ♂ 77–83 (79·8), ♀ 69–80 (74·5); bill, ♂ 15·5–17 (16·2), ♀ 14–16 (15·2); tarsus, ♂ 20–22 (21·2), ♀ 19–21 (20·7). WEIGHT: W Kenya (Kakamega), ♂ (n = 15) 19–32 (27·9 ± 2·5), ♀ (n = 15) 22–29 (25·1 ± 1·7), unsexed (n = 82) 22·7–33·5 (27·6 ± 2·3); Kenya and Uganda, ♂ (n = 14) 25–35 (28·9 ± 3·1), ♀ (n = 11) 23–32 (26·0 ± 2·1). Diurnal weight change 10·1–21·6% (W Kenya: Mann 1985).

IMMATURE: upperparts like adult but with rufous wash, underparts mainly dingy brown without olive tone, some yellowish white in centre of belly, undertail-coverts like adult but washed pale tawny. Yellow malar stripe lacking in youngest birds, then starts as a trace and grows in gradually; at time of first moult, when brown of underparts being replaced by olive, stripe evident but not full adult width; dark area between cheeks and incipient pale malar stripe appears as dark malar stripe. Basal half of lower mandible yellow, inside of mouth bright orange-yellow; legs and feet orange-yellow.

NESTLING: naked, pinkish brown above, pink below; gape yellow, mouth bright orange; soles bright yellow.

TAXONOMIC NOTE: '*eugenius*' (E Zaïre to S Sudan and W Kenya, where it intergrades with '*saturatus*') said to differ from nominate race by slightly greener upperparts, yellower underparts, brighter yellow malar stripes which meet below bill to form yellow chin; '*saturatus*' (W and E Kenya, SW Tanzania) said to have greener upperparts and underparts than '*eugenius*', yellow malar stripes slightly more confluent at anterior end, and to be larger (Rand 1958). We find these differences so insignificant as to be untenable.

A. l. congener Reichenow: Senegal to SW Nigeria. Upperparts much darker, duller and greyer; tail dark grey-brown, appearing almost black; underparts paler and duller, more greyish olive, centre of belly whitish yellow. Slightly smaller: wing, ♂ (n = 48) 75–87 (81·7 ± 3·1), ♀ (n = 40) 74–85 (77·1 ± 2·7). WEIGHT: Liberia, ♂ (n = 10) 24–31·5 (27·3 ± 2·2), ♀ (n = 4) 19·3–23·2 (21·5), ♀ in breeding condition (n = 6) 21·5–25 (23·1 ± 1·2), immature ♂ (n = 4) 21·9–27 (25·3); Nigeria, unsexed (n = 47) 24–33 (28·0).

TAXONOMIC NOTE: '*australis*' (Ufipa Plateau, Tanzania), with rather paler upperparts, may be a sibling species (Ripley and Heinrich 1966). Voice is very distinctive. Usual vocalization is 'zik', repeated several times in rapid succession, 'zik zik zik zik zik zik', very similar to call of Scarlet-chested Sunbird *Nectarinia senegalensis*; another call is a monosyllabic 'tjeurr . . . tjeurr . . . ', uttered once or a few times at intervals. Other characters (foot colour, ecology, behaviour) cited by Ripley and Heinrich are less convincing: foot colour is variable (see Zimmerman 1972).

Field Characters. The 'basic' bulbul of most forests, abundant and widespread, often the dominant species and rivalled only by Little Greenbul *A. virens*. Tuneless chattery song heard all day. Shy when in undergrowth but often feeds at higher levels where somewhat easier to see, although tends to sing from concealed perch. Adults with yellow whiskers, which are often puffed out, unmistakable, but dingy immature without whiskers can be confused with similar-sized adult Little Greenbul (for differences, see that species).

Voice. Tape-recorded (10, 13, 32, 53, C, BROS). Song, a string of dry, tuneless notes jumbled together, 'chop, chip, chirrup, chup, prip, prip, chup, chirrup, chop . . .', lasting 4–8 s, usually 6–7 s; speed and pitch remain constant but volume increases; notes at first almost inaudible, then gradually become louder; volume greatest at end of song, which often closes with a loud 'KICK-KICK'. Intervals between songs variable, from 5 to 30 s, but at times of intense singing one song may lead into another with hardly any interruption. Call, repeated 'chip, chip, chip . . .'. Alarm rattle, a series of rapid 'dit' notes; buzzy 'cherrrt' given in intraspecific aggressive encounters (Willis 1983).

General Habits. Inhabits all types of primary and secondary forest, both interior and edge, including degraded forest and regenerating young second growth; primary forest habitat in Liberia characterized by *Gilbertiodendron preussi*, *Heritiera utilis*, *Chrysophyllum albidum*, *Lophira alata* and *Uapaca* spp. (Mattes and Gatter 1989). Outside forest spreads into deforested areas overgrown with thick brush, plantations (e.g. coffee forest, Angola), abandoned cultivation, gallery forest, isolated thickets in forest–grassland mosaic, ravines overgrown with tree ferns, bamboo thickets, even large gardens. Lowland and highland forest, up to 2200 m in Cameroon, 2300 m in Zaïre, 2700 m in Sudan, 3000 m in E Africa.

Solitary and generally unsociable; said to join bird parties, commonly in second growth (Prigogine 1978), but this is not the norm; found in only 13% of undergrowth parties, Gabon (Brosset 1969). Gathers with congeners while bathing and with other species at fruiting trees but usually travels singly, not in family parties or even in pairs, though on occasion large numbers of wandering individuals concentrate at forest edge – 104 caught in 3 days in mist net at same spot, Gabon (Brosset and Erard 1986). Ecological niche broad; found at all levels, from ground to canopy. On ground, hunts for invertebrates among fallen logs and branches. Catches insects disturbed by ant columns, sitting in vegetation just above ants and darting to ground to capture prey (Lambert 1984); when displaced by bristlebills *Bleda* spp. from ground, takes prey from undergrowth above ants or from nearby trunks and lianas, some in air (Willis 1983). Digs in loose soil; often hovers while plucking fruit or gleaning insects from foliage. Generally rather shy, especially at lower levels; even when feeding in fruiting tree, where most species highly visible, prefers to fly up from undergrowth, gorge on berries and return to cover, rather than remain in view for long periods. Crouches on approach of hawk (e.g. *Accipiter tachiro*) or person, flitting wings and giving alarm rattle. In intraspecific disputes birds raise yellow malar stripes, give buzzy notes and chase each other (Willis 1983). Bathes in water repeatedly; sometimes sunbathes, with spread wings and tail. Song continues all day, an ever-present background sound of many forests; often only bird singing in heat of afternoon.

Duration of primary moult Kakamega, Kenya (n = 22) 90–218 (145·9) days; for 2 individuals, 306 and 335 (Mann 1985). In Liberia moult begins soon after nestlings fledge; moulting season mainly Jan–May (Mattes and Gatter 1989).

Sedentary and nomadic. Mainly nomadic in high-density areas where lek system used (see below). In Gabon, 90% of individuals wander, small minority sedentary; of 791 individuals ringed in 12 years, only 82 (10·4%) were recaptured (recapture rate in other bulbuls 70%). Rate higher in W Kenya (Kakamega), where 18·4% of 375 ringed birds recaptured, 6·31% a year or more later (Mann 1985), and much higher in Nigeria (Ogba Forest), where population considered sedentary (Johnson 1989). (♂♂ more sedentary than ♀♀ – over twice as many ♂♂ recaptured). Some individuals remain sedentary while in immature plumage, but no young ringed on nest were recaptured (Brosset and Erard 1986). Sedentary birds have small ranges; most recaptured in exact spot where ringed. Range not defended; tolerates congeners, numbers occurring in same area without fighting or aggressive behaviour; ringed (sedentary) and unringed (nomadic) birds often caught in same net (Brosset 1981a). Likewise, no individual aggression shown when birds gather in loose flocks at food concentrations. Reasons for sedentary behaviour in minority of individuals in largely nomadic population not clear; not related to age, sexual activity or physical condition; possibly genetic (Brosset 1981a).

Food. Omnivorous. Eats many fruits and berries, e.g. *Allophylus, Bridelia, Capsicum, Cissus, Dacryodes, Ficus, Galiniera, Harungana, Heisteria, Ilex, Maesa, Morinda, Musanga, Ocotea, Polyalthia, Polyscias, Prunus, Rapanea, Rutidea, Sapium, Schefflera, Solanum, Trema, Tricalysia, Urera, Vitis*; also arils of *Agelaea, Coelocaryon, Pycnanthus*. Takes a wide variety of invertebrates: molluscs (Cyclostomidae, Helicidae), woodlice, spiders; insects: Orthoptera, including large grasshoppers; Diptera, Coleoptera (Geotrupidae, Staphylinidae); caterpillars, moths, mantid larvae, winged termites, ants. A few small vertebrates (frogs, geckos).

Breeding Habits. Social system flexible, varying regionally.

(A) In high-density areas, e.g. Gabon, polygamous and non-territorial; uses leks; some evidence of cooperative breeding. In months before breeding season, ♂♂ assemble to sing in chorus at traditional lek sites; size of lek 100–200 m long, 40–50 m wide, often along road or river; spaces between leks 600–1200 m (Gabon: Brosset 1981d). Within lek ♂ selects song post in patch of thick foliage; post may be at any height, from undergrowth to canopy. Space between singers 10–20 m. Leks frequented during singing period only (Sept–Apr in Gabon), birds silent for rest of year (except for young of year which give subsong outside breeding season at time of first moult: Brosset 1981b). Lek behaviour seasonal, coinciding with breeding season. As ♂ comes into breeding condition he defends an area c. 20 m diam. around song post, from which he chases off intruders by flying around with flailing wings; to any bird that fails to fly off (presumed ♀) he gives display very like that of Common Bulbul *Pycnonotus barbatus*, lowering and spreading tail and rapidly beating wings which are drooped and slightly curved, but with body more stretched out and yellow moustaches prominently displayed. ♂♂ and ♀♀ gather on lek at 08·00–12·00 h and 16·00–17·00 h; ♀ chooses mate and copulation takes place; unclear whether ♀ chooses ♂ which has best site or a particular individual regardless of site (Brosset 1981d). Rapid turnover of both ♂♂ and ♀♀ at lek.

♀ takes over all duties of raising young. ♀♀ show complete lack of territoriality even in vicinity of nest; ♀♀ incubate within sight of one another, 2–3 nests in loose group only a few m apart, while large areas of adjacent similar habitat untenanted; in a given group clutches often at same stage of incubation, showing synchrony of breeding (Brosset 1981b). On approach of danger, several ♂♂ join in defence of nest.

Nests only in short dry season in Gabon (unlike other bulbuls in same forests which may nest all year); season linked to breeding system and food supply (Brosset 1981b). Best months for food (both insects and fruit) are Oct–Nov, when ♀♀ become fat before laying, and Mar–Apr, when young become independent of parental feeding (immediately on leaving nest) and need to find food easily. In intervening dry months (Dec–Feb), food supply somewhat reduced but nests and incubating birds keep dry, and since only ♀ feeds young she can leave nest without young getting wet.

Within populations using lek system, some ♂♂ sing in isolation away from lek, possibly ejected from lek by dominant ♂♂ (birds caught at centre of lek all large and heavy, perhaps dominant). They are irregularly distributed, locations varying daily, so probably nomadic, i.e. not on territory (Brosset 1981d); not known if they take any part in rearing young.

(B) Where less abundant, monogamous and territorial. Pairs remain on same territory in successive years. 12 pairs on 16 ha, Kenya, i.e. av. territory size 1.3 ha (van Someren 1956).

NEST: rather loosely built cup, typically round, sometimes oval or extended to shape of hammock; outer layer of twigs, rootlets, strips of bark and plant stems, inner layer of dry leaves often bound together with spider webs and fibres of fungus *Marasmius*, lined with fine rootlets, plant fibres, dry grass or *Marasmius*; extra layer of dead leaves sometimes added to outside; ext. diam. 90–120, int. diam 50–80 (65 × 45 in oval cup), ext. depth 40–70, int. depth 30–40. 'Very similar in construction to nest of

Phyllastrephus; less untidy and thus looks smaller; has same general appearance of an accidental collection of debris caught up in a fork' (van Someren 1956). Placed 0·3–1·9 m above ground in tangled forest undergrowth (shrub, small bush or sapling), often in fork, attached to arms of fork with vine tendrils. Higher sites used in dry year when forest undergrowth wilted, 2–7 m above ground under canopy of small sapling or bushy tree, both inside and outside forest; 1 also in leafy *Begonia* in outdoor orchid house (van Someren 1956). May take up to 14 days to build original nest, but for second brood nest can be built in only 5 days.

EGGS: 1–4, usually 2; Gabon mean (? clutches) 2·43. Laid at 1-day intervals, occasionally 2. Destroyed clutch can be replaced in 7 days; new clutch laid either in old nest or in new one a few m away. Eggs broad ovate; smooth, glossy; white, greyish white or shades of pale pink, with markings of brown, purple-brown, maroon, or wine red, sometimes dark pink or carmine, and underlying spots of pale to dark grey, grey-violet, lilac-grey or sometimes pale beige; usually concentrated around large end, sometimes evenly but sparingly spotted all over; 'like *Pycnonotus* but more glossy – different from other forest bulbuls' (van Someren 1956).
SIZE: (n = 19) 19–25 × 14–16·5 (22·2 × 15·8).

LAYING DATES: Liberia, Oct–Mar (mainly Nov–Dec), early in dry season when the most food is available (Mattes and Gatter 1989); Nigeria, May; Cameroon, June, May, Aug (breeding condition Mar); Gabon, Dec–Apr (short dry season); Central African Republic (breeding condition June); Zaïre (all months, especially in rainy season); Rwanda, Jan, Mar, Sept–Nov; Angola (breeding condition Oct, immature Feb); Sudan, Jan–Feb, May–July; E Africa: Uganda, Sept–Oct (breeding condition Apr, July–Aug, Nov); Kenya, Oct, Dec; Tanzania (immature Dec); Region A, June, Dec; Region B, Feb, Apr–Oct; Region C, Jan, June–July; Region D, Mar–June, Aug, Dec. Seasons ill-defined in Regions A–C, most Region D records in long rains.

INCUBATION: by ♀ only. Period: 12–14 days. Incubating bird surprised at nest either gives distraction display or freezes, even allowing itself to be picked up with no attempt at escape. In display, drops to ground from nest, lies there fluttering, then trails off with drooping wings and spread tail; sometimes runs off noisily through leaves like rodent as soon as it hits ground. One bird followed by observer fluttered on ground for 15 m, flew into bushes and continued fluttering for another 10 m, then disappeared; when observer returned to nest, bird was already back on it; another bird gave distraction display for 20 m while observer remained at nest instead of following her; when bird saw display was not working she flew back to nest and resumed incubating, even though observer was still standing by it! (van Someren 1956).

DEVELOPMENT AND CARE OF YOUNG: young tended by ♀ alone in lek systems, elsewhere fed by ♂ as well. Brooded closely by ♀ for first week, thereafter at intervals until almost feathered; fed only insects for first 4 days, then mixture of fruit and insects, and during last week in nest only fruit; fed on av. every 40 min. ♂ may feed chicks directly or hand food to ♀ to give them. When ♀ brings food, gives low chirrup, at which chicks stretch necks and open mouths; when ♂ brings food while ♀ brooding, gives brief call from tree above nest, ♀ answers and hops onto side of nest to let him give food. Fledging period: 12–15 days; young independent almost on leaving nest, do not follow parents around.

BREEDING SUCCESS/SURVIVAL: reproductive success varies widely, along with predation level; 45–85% of nests predated annually, av. 75%. Complete cessation of breeding for 2 successive years, Gabon, while environment unchanged and other birds bred normally, probably adopted as anti-predator strategy (Brosset 1981b): in year breeding resumed, predation of nests down to 46%, population jumped 50%, 2·43 young fledged per nest; in following 4 years, predation of nests increased to 78%, young fledged per nest decreased to 1·95. Cessation of breeding believed to break 'search image' in older predators, inhibit its formation in younger ones, thus predation drops sharply (Brosset 1981b). Identified predators include monkeys, mongooses, shrikes, puff-back-shrikes *Dryoscopus* and carnivorous ants; nests parasitized by Emerald Cuckoo *Chrysococcyx cupreus*. Maximum life expectancy, Gabon, 7–8 years (Brosset 1981c); the only birds recaptured 2 years after being ringed were large, apparently ♂♂, so large size apparently increases chances of survival (Brosset and Erard 1986). 8 birds in Kakamega, Kenya, lived from 2 years 7 months to 9 years 7 months (Mann 1985).

References
Brosset, A. (1981a, b, d).
Brosset, A. and Erard, C. (1986).
Mattes, A. and Gatter, W. (1989).
van Someren, V. G. L. (1956).

Plate 18
(Opp. p. 305)

Andropadus importunus (Vieillot). **Sombre Greenbul; Zanzibar Sombre Greenbul. Bulbul importun.**

Turdus importunus Vieillot, 1818. Nouv. Dict. d'Hist. Nat., nouv. ed., 20, p. 266; Anteniquoi (= Outeniqua) Forest (Cape Province).

Range and Status. Endemic resident. Abundant on coast of E and S Africa, extending inland to varying degrees. Extreme S Ethiopia (Dolo), Somalia south of c. 4°N, chiefly along Juba and Webbe Shebelle rivers. Kenya/Tanzania, inland on Tana R. to Garissa, further south to Ngulia, Emali, Kibwezi, Makueni, Kitui, slopes of Kilimanjaro, Kilosa, and central Kenya highlands from Nairobi and Thika to Meru and Ndoto Mts; up to 1220 m in Chyulu Hills, 2000 m in highlands. Zanzibar and Mafia Is but absent Pemba. Mozambique, almost throughout except NW; in N, widespread on littoral and extending inland along river valleys, but not in open woodland or on central plateau; common on offshore islands, including Margaronque; status on Bazaruto unclear: said to be absent by McCulloch (1967) but present by Clancey (1971). Malaŵi, upper Shire valley and S littoral of L. Malaŵi to Makanjila, Bua R. mouth and Bana, with population 16 km N and S of Nkhata Bay, confined to rim of lake; Malaŵi Hill escarpment to *c.* 600 m; plains at this level between Mulanje and Zomba; isolated record from Chisunga, 1200 m. Zambia, only in SE, extending inland along river valleys – Zambezi (to N shore L. Kariba), Kafue (to Kafue), and lowest reaches of Luangwa, Lukushasha and Lunsemfwa. Possible sight record Namwala, 150 km west of Kafue (Bruce-Miller and Bruce-Miller 1975), although not repeated, might indicate an isolated population. Zimbabwe, occurs Middle Zambesi valley from Kanyemba to Mana Pools and Chirundu; in E, in Honde valley to 900 m, around Umtali and in Lusitu-Haroni area, southwest over Chipinga Uplands (common) to Mt Selinda and Chikore; in Sabi valley upstream to Mutema, and on Lundi R. to Chipinda Pools; along Limpopo R. to Chikwarakwara, and may be on Nuanetsi R. and elsewhere in SE lowveld. Transvaal, Zoutpansberg, and E highlands and lowlands (uncommon resident Kruger Nat. Park), thence south through Swaziland, Natal and coastal regions of Cape Province to Cape Town, where it occurs in winter rainfall area (but absent Hottentots-Holland District).

Description. *A. i. insularis* (Hartlaub) (including '*subalaris*' and '*somaliensis*'): N part of range south to Tanzania except SE. ADULT ♂: top and sides of head and upperparts, including wing-coverts, brownish olive; tail-feathers darker, outer edges greener, sometimes with very narrow whitish yellow margins to inner webs and/or tips. Throat, breast and flanks lighter brownish olive, somewhat paler on chin; belly pale yellow, undertail-coverts and thighs richer, brownish yellow. Flight feathers dark olive-brown edged greener; axillaries, underwing-coverts and inner webs of flight feathers bright ochreous yellow. Bill black to slaty; eye white to pale yellow; legs and feet dark olive-brown, grey, slaty or black. Sexes alike, but ♀♀ slightly smaller. SIZE: (14 ♂♂, 9 ♀♀) wing, ♂ 82–91 (85·9), ♀ 77–92 (82·8); tail, ♂ 70–85 (76·4), ♀ 69–81 (73·6); bill, ♂ 15·5–19 (17·6), ♀ 15–16·5 (16·1); tarsus, ♂ 19·5–23 (21·1), ♀ 20–22 (20·6). WEIGHT: Kenya (coast), ♂ (n = 7) 25·1–30 (27·3 ± 1·7), ♀ (n = 9) 22–27·4 (25·5 ± 1·6), unsexed (n = 10) 25–30·9 (29·2 ± 1·2); Tanzania (Amani), ♂ (n = 3) 26·5–29·5 (27·5). ♂♂ increase weight with latitude (Kenya to South Africa: Britton 1972). Lightest just before breeding, heaviest during and after moult (Malaŵi: Hanmer 1978).
IMMATURE: like adult but slightly more olivaceous above; eye-ring yellow; eye brown. See also *Development and Care of Young*.
NESTLING: orange gape-flanges, brown eye, yellow eye-ring.
A. i. importunus (Vieillot) (including '*errolius*' and '*noomei*'): Cape Province, Natal, S Zululand, W Swaziland, highlands of N and E Transvaal. Much darker than other races. Upperparts dark olive-green, underparts uniform olive-grey except for slight paling on belly (a few feathers with pale yellow tips).

Andropadus importunus

Underwing as in *insularis* but usually pale yellow without ochreous tinge. WEIGHT: 1 ♂ 34·6, unsexed (n = 14) 30·3–39·2 (35·7 ± 2·5); southern Africa (may include some *oleaginus* and *hypoxanthus*), ♂ (n = 24) 26–40 (31·4), ♀ (n = 16) 24–30 (26·9), unsexed (n = 113) 26–44·5 (32·6). IMMATURE: yellow ring round eye, eye grey, bill horn, gape yellow.
There is a poorly marked cline of increasing yellow in underparts from SW to NE, and upperparts become slightly greener.
A. i. oleaginus Peters (= '*mentor*' Clancey, see Rand 1958): N Zululand, E Swaziland, Mozambique south of Save R., lowland E Transvaal, Zimbabwe along Limpopo R. Upperparts slightly paler than in *importunus*; throat, upper breast and flanks greyish yellowish olive, lower breast, belly and undertail-coverts bright, rich yellow. SIZE: (22 ♂♂, 14 ♀♀) wing, ♂ 81–91 (86·8), ♀ 80–88 (83·0); bill, ♂ 17–19·5 (18·4), ♀ 16·1–18·8 (17·5).
A. i. hypoxanthus Sharpe (including '*loquax*'): Mozambique north of Save R., Zimbabwe except along Limpopo R., Zambia, Malaŵi, SE Tanzania including Mafia I. Like *oleaginus* but upperparts greener, yellow of underparts even brighter. Birds from Mafia I., usually included in this race, have rather paler yellow underparts and browner upperparts, intermediate between *hypoxanthus* and *insularis*. WEIGHT: Zimbabwe, ♂ (n = 51) 29·5–40·5 (33·9 ± 2·7), ♀ (n = 40) 24·3–33·3 (28·5 ± 2·4); Mozambique, unsexed (n = 63) 23–40 (29·2 ± 3·4); Malaŵi, unsexed (n = 275) 23–36 (29·5 ± 2·4).
[*A. i. fricki* Mearns of central Kenya highlands (like *insularis* but with yellow eye-ring) now almost certainly referable to immature *insularis* (D. J. Pearson, pers. comm.).]

Field Characters. An abundant but seldom-seen bird that constantly advertises its presence with its distinctive song. Song persistent to the point of annoyance, hence the name *importunus*, 'importunate'. Hard to see; skulks in thick vegetation; pale eye distinctive. Uniform olive-green South African birds can be confused only with Cape Bulbul *Pycnonotus capensis*, which is however bold and familiar, with different voice, greyer plumage, white eye-ring (as opposed to eye) and yellow vent. In north, could be confused with larger Yellow-bellied Greenbul *Chlorocichla flaviventris*, which is less skulking, often moving about in small groups, has different voice, brighter yellow underparts, bright yellow (not brownish) throat, dark reddish eye and broken white eye-ring.

Voice. Tape-recorded (10, 14, 32, 58, 81, 86, 88). Kenya: full song, a series of loud, clear, strong notes, 'ti-ti-wer cheeo-cheeo cheeo-wer chi-wee chi-wer chi-wee', fast and run together, with these phrases hard to make out; final up-slurred 'che-wee' may be replaced by down-slurred 'wee-chuk'; second song, perhaps also used as call, a loud 'preeto' or 'willie', followed by a jumble of notes in an undertone, inaudible except at close range. Southern Africa: song, loud 'willie' followed by jumble of notes that are equally loud, not subdued, followed by some separate notes and often ending with either a wheezing note or a down-slurred whistle. Also, a loud 'plee-plee-plee-plee . . .' (alarm call?).

General Habits. Typical habitat is dense coastal scrub and thickets, including dune scrub. Spreads inland in dense riverine bush and forest; in some areas confined to this habitat, elsewhere occupies other types of bush, scrub and woodland, e.g. thickets in thornveld. In southern Africa ascends mountain slopes along wooded valleys and narrow ravines, reaching evergreen forest in mist belt. In Mozambique often in reeds at river's edge (Vincent 1947), and once recorded in mopane (Malaŵi). Prefers natural vegetation to man-made habitats, but in Zanzibar found in overgrown cultivated plots, in coastal E Africa in various types of cultivation, in W Cape in woodlands of mixed exotic trees, and in Malaŵi sometimes nests in village mango trees.

Usually occurs singly or in pairs, although large flocks sometimes form in non-breeding season (Hanmer 1984). Generally shy, keeping to dense foliage (but can be 'squeaked up' by person). Dives quickly into cover when alarmed. Emerges at intervals to sing on top of bush or on exposed branch, and even perches on telephone wires. A most persistent vocalist, singing all day, even at midday, and all year. Flight rapid and direct, with fast wing-beats. Forages in trees and bushes, clambering about on branches and in creepers; in forest generally keeps to upper stratum, in scrub feeds in undergrowth or on ground among leaf litter.

Primary moult starts about middle of breeding season and takes 80–120 days; immatures frequently interrupt moult.

Sedentary.

Food. Insects, small snails, fruit, berries. In Tsavo Nat. Park, Kenya, ate only fruit (25% *Commiphora*), taken from bushes, not trees (Lack 1985).

Breeding Habits. Monogamous, territorial. ♂♂ spar, giving double whistle. In pre-copulatory display, pair once perched facing each other on horizontal branch of tree *c.* 2 m above ground; ♂ rapidly raised and lowered head in elliptical arc, bill slightly open; after 8–10 such movements he slowly turned through a full circle, constantly head-dipping, with crown and nape feathers raised, wings opened slightly and quivered rapidly. ♀ bobbed head, out of phase with ♂: when his head at top of arc, hers at bottom. Throughout display both birds called with soft keening note. Display lasted *c.* 5 min, then they copulated (Longrigg 1978).

NEST: a cup, usually flimsy, thin and shallow, even semi-transparent, sometimes bulky and solid; made of twigs, weed stems, rootlets, tendrils, fibres and coarse, dry grass, sometimes with a few dead leaves, once with string; lined with finer grasses and fibres, also pine needles; diam. of cup *c.* 7 cm. Situated 1–4 m (usually 1·5–2·5 m) above ground in fork, either in main stem of slender tree or bush, or at end of branch where supported by leaves and twigs; sometimes in dense cover, with larger trees overhead, sometimes at edge of clearing or thicket, where it may be poorly concealed.

EGGS: 1–3; southern Africa mean (96 clutches) 1·9; 9 clutches in Tanzania/Malaŵi all of 2. Oval; somewhat glossy; dull whitish, cream or deep cream, rather evenly marked with spots, streaks and twirls of umber, red-brown, purple, olive-brown or greenish brown, with undermarkings of grey, grey-green or ashy. SIZE: (n = 59, southern Africa) 20·7–26·8 × 15·3–18·2 (23·1 × 16·7).

LAYING DATES: E Africa: Region D, Jan, Mar, May, Aug; Region E, Apr–Oct, Dec; Zanzibar, mainly Sept–Jan, also May–June; Malaŵi, Jan–Apr, June, Oct–Dec (breeding condition Sept); Zambia, Sept, Nov; Zimbabwe, Oct–Feb (Oct 4, Dec 2, Jan 1, Feb 1 clutches); Mozambique, Sept–Mar; South Africa: Transvaal, Oct–Jan; Natal Oct–Apr, especially Oct–Dec (Oct 12, Nov 17, Dec 16, Jan 9, Feb 4, Mar 3, Apr 2 clutches); Cape Province, Oct–Jan.

INCUBATION: by ♀ only. Period: 15–17 days. Sits tight even when branch pulled down by man; one slipped off at last moment and fluttered to nearby perch, where soon joined by mate responding to shrill whistles of alarm (Vincent 1947). Readily deserts nest if disturbed.

DEVELOPMENT AND CARE OF YOUNG: for 3 months young have orange gape-flanges, brown eye, yellow eye-ring; at 3–6 months yellow gape-flanges, brownish eye, yellow eye-ring; at 6–9 months yellowish gape-flanges, yellowish eye, yellow eye-ring; at 9–12 months very faint yellowish gape-flanges, yellow eye, vestigial yellow eye-ring; at 12 months indistinguishable from adult. Pneumatization of skull 25% at 3 months, *c.* 50% at 6, *c.* 75% at 9, complete at 12 months (Hanmer 1978). Fledging period (southern Africa): 14–16 days.

BREEDING SUCCESS/SURVIVAL: nests parasitized by Jacobin Cuckoo *Oxylophus jacobinus* (n = 21). Mean survival rate 0·47, mean life expectancy 1·39 years; no difference in rates between adults and immatures (Malaŵi: Hanmer 1984). 9 birds lived >7·5–12 years (Hanmer 1989).

References
Hanmer, D. B. (1978).
Vincent, A. W. (1947).

Genus *Calyptocichla* Oberholser

Bill straight, slender, laterally compressed, flesh-pink; operculum completely feathered; wings long; tail rather short; underparts golden yellow. Distinctive whistled song and calls. Forest, mainly in canopy.
Endemic; monotypic.

Calyptocichla serina (J. and E. Verreaux). Golden Greenbul. Bulbul doré.

Criniger serinus J. and E. Verreaux, 1855. J. Orn. 3, p. 105; Gabon.

Plate 21
(Opp. p. 352)

Range and Status. Endemic resident, both sedentary and nomadic. Sierra Leone, SE Guinea, Liberia (common Mt Nimba), Ivory Coast (local, north to Man, Abengourou), S Ghana (rare); S Nigeria (across forest zone, rare) and Cameroon (widespread in S, north to Rumpi Hills) to Gabon (common and widespread), S Congo, Angola (Cabinda), SW Central African Republic (Bangui area) and extreme W Zaïre (Tshela). Uncommon and local N and NE Zaïre, but common in highlands of Itombwe (Kivu). Bioko. Widespread but nowhere abundant; locally common to rare.

Description. ADULT ♂: top of head, upperparts and upperwing-coverts yellowish olive-green, somewhat brighter on rump; tail similar but a little duller, shafts brown above, yellow below. Cheeks and ear-coverts olive with pale streaks; chin and upper throat off-white becoming yellow on lower throat; breast and flanks yellowish olive-green, feathers of breast and upper flanks with yellow shafts, giving streaked effect; lower breast, belly and undertail-coverts bright golden yellow. Flight feathers dark olive-brown, outer edges of primaries and outer webs of secondaries yellowish olive-green, inner webs of secondaries progressively greener inwardly; inner webs of underside of flight feathers edged pale yellow, axillaries and underwing-coverts yellow tinged ochre. Bill light brown to pale pinkish brown, tip darker, base of lower mandible whitish; eye dark grey to brown or pale grey-brown; legs and feet variable – grey, olive-grey or dark bluish grey to dark blue, greenish blue or dull dark red. Sexes alike. SIZE: (10 ♂♂, 10 ♀♀) wing, ♂ 86–94 (89·7), ♀ 82–93 (87·8); tail, ♂ 70–78 (73·2), ♀ 69–78 (72·7); bill, ♂ 19–21 (20·4), ♀ 19–21 (20·2); tarsus, ♂ 19–21 (20·4), ♀ 19–21·5 (19·6). WEIGHT: ♂ (n = 6) 36·4–42 (39·8), ♀ (n = 7) 35–42·7 (37·7).
IMMATURE: like adult but olive of underparts duller, throat pale yellow, lower mandible pinkish flesh.
NESTLING: unknown.

Field Characters. A medium-sized greenbul of lowland forest with white throat, bright golden underparts, long, thin, pale bill noticeable in the field, and un-bulbul-like 2-note whistle. Best distinguished from other lowland forest greenbuls by canopy habitat – others, like bristle-bills *Bleda* and Icterine and Xavier's Greenbuls *Phyllastrephus icterinus* and *P. xavieri* live in undergrowth and lower mid-levels. Grey-headed Greenbul *P. poliocephalus* is often in canopy but lives at higher elevations; Joyful Greenbul *Chlorocichla laetissima* feeds up to middle storeys but is also in intermediate forest, and has yellow throat and bubbly song.

Voice. Tape-recorded (53, ERA, KEI, LEM). Call (song?), a high-pitched, clear double note, 'tyup-teeyu' or 'tyup-tyuyu', note 1 brief, note 2 longer and downslurred; repeated at intervals of 1–6 s; reminiscent of Waller's Chestnut-winged Starling *Onychognathus walleri* which has similar clear tone. In between calls, a few notes may be given in an undertone, a 2–3 note chatter or a soft 'tuyu'. Contact calls of feeding birds include thin, high-pitched descending whistle, 'tseeyu', and 1–5 lower-pitched, short, clipped notes: 'tyip', 'tyip-tyu', 'tyip-tyu-tyip', 'chop-tee-chop', 'chop-ti-chip-ti-chop'.

General Habits. Inhabits primary forest, especially around clearings, but more often second growth and relict forest patches; also plantations, e.g. cacao, with shade trees, regenerating forest on abandoned farmland;

in Gabon fond of young second growth with *Sarcophrynium*, secondary forest with Burseraceae and Myrtiscaceae (Brosset and Erard 1986). Lowland forest, to 1300 m (Cameroon).

Keeps mainly to upper levels and canopy in forest, down to 20 m; occasionally descends to fruiting bushes, e.g. *Rauwolfia*. Solitary, in pairs or family parties of 3–4. Eats fruit, and in Mayombe Forest, Congo, often associates with Slender-billed Greenbul *Andropadus gracilirostris* in fruiting trees (F. Dowsett-Lemaire, pers. comm.). Gleans leaves for insects, and sometimes joins mixed-species flocks; catches termites in flight.

Nomadic, constantly on the move, covering large areas; sedentary only while nesting.

Food. Fruit, including *Ficus, Heisteria, Macaranga, Morinda, Musanga, Rauwolfia, Xylopia*; small seeds; insects, including Orthoptera, termites, large caterpillars.

Breeding Habits. Territorial. Boundary squabbles occur during breeding season; sings only while breeding.

NEST: only 1 described (but others seen: Brosset and Erard 1986); built of dry leaves and plant fibres; situated in fork at end of branch, 30–40 m up in crown of tall tree at edge of forest or near village, sometimes concealed by leaves. Built by ♀, accompanied by singing ♂.

EGGS: unknown.

LAYING DATES: Nigeria (breeding condition, Aug); Cameroon Mar, Aug; Gabon, Dec–Feb (breeding condition Sept); Zaïre, Jan, Nov, probably May.

INCUBATION: by ♀ only.

DEVELOPMENT AND CARE OF YOUNG: young fed by both parents; given caterpillars.

Reference
Brosset, A. and Erard, C. (1986).

Genus *Baeopogon* Heine

Bill short and broad, broader than deep at base, nostrils narrow, slit-like, half covered with feathers; rictal bristles weak; upperparts olive-green, underparts grey or buff; tail short, T3–T6 whitish, giving birds a superficial resemblance to honeyguides Indicatoridae. Loud, whistled calls. Forest, mainly in canopy.

Endemic; 2 species.

Plate 21
(Opp. p. 352)

Baeopogon indicator (**Verreaux**). **Honeyguide Greenbul. Bulbul à queue blanche.**

Criniger indicator J. and E. Verreaux, 1855. J. Orn. 3, p. 105; Gabon.

Range and Status. Endemic resident. NW Sierra Leone (Kilimi), SE Guinea and Liberia (north to Mt Nimba) to Ivory Coast (north to Sipilou, Korhogo and Comoé Nat. Park), Ghana and Togo. Nigeria (widespread, north to Kagoro), central and S Cameroon (north to Mt Cameroon), Gabon, E and S Congo, Zaïre (widespread from lower Congo R. to E highlands, extending south to Upemba Nat. Park and S Lualaba Province on Angolan border at Dilolo), crossing into adjacent NW Zambia (Salujinga to Sakeji and Lisombo Stream). Angola (Cabinda, and rare and local in forests of Cuanza Norte and Malanje); SW and SE Central African Republic, S Sudan, forests of Uganda from Impenetrable, Bwamba and Budongo to Mabira, extending to W Kenya (Kakamega and Nandi; old records from Elgon and upper Yala R.). Fairly common and widespread; density in Gabon 6–8 pairs/km^2; in Nigeria (Ogba Forest), 1 bird/10 ha (Johnson 1989).

Description. *B. i. indicator* (Verreaux) (including '*chlorosaturatus*'): range of species except *leucurus*. ADULT ♂: crown and upperparts, including scapulars and upperwing-coverts, yellowish olive-green, somewhat darker on crown and washed greyish on nape and hindneck. Central 2 pairs of tail-feathers blackish, T3 with inner web and broad (*c.* 18 mm) tip to outer web blackish, rest of outer web white, T4–T6 white edged yellowish with dark tip progressively shorter outwardly, from *c.* 13 mm on T4 to *c.* 7 mm on T6. Lores, around eye, cheeks and ear-coverts grey with olive wash; chin and throat grey with hint of olive; breast and flanks mixed grey and olive; lower belly, centre of upper belly, thighs and undertail-coverts creamy buff. Primaries and secondaries blackish brown edged golden green, edges becoming broader inwardly from extremely narrow on outer primaries to whole of outer web on inner secondaries, outermost primary short, without edging. Axillaries and underwing-coverts yellowish olive, feathers on underside of bend of wing creamy yellow. Sexes alike except for eye colour. Bill black to dark grey; eye, ♂ white to creamy or greyish white, ♀ brown to grey; legs and feet grey to slate. SIZE: (12 ♂♂, 7 ♀♀) wing, ♂ 96–103 (100), ♀ 91–97 (93·9); tail, ♂ 68–75 (71·1), ♀ 65–70 (68·1); bill, ♂ 18–19·5 (18·8),

♀ 17–19·5 (17·9); width of bill at base very variable, 3·5–7·5 mm (Friedmann and Williams 1971); tarsus, ♂ 18–22 (20·4), ♀ 19–21 (19·7). WEIGHT: ♂ (n = 13) 44–53 (47·8), ♀ (n = 6) 40–48 (43·9).

IMMATURE: like adult but somewhat duller above, crown browner, belly grey-white; outer 3 pairs of tail-feathers entirely white, T3 with white on outer web more extensive (dark tip only c. 8 mm). Eye of ♂ dull greyish buff.

NESTLING: unknown.

B. i. leucurus (Cassin) (including '*togoensis*'): Sierra Leone to Togo. Underparts greyer, less olive, chin and throat paler grey with no olive, belly and undertail-coverts paler and brighter, usually creamy white, sometimes pale buff.

Field Characters. A medium-sized bulbul that spends most of its time in canopy foliage; easy to overlook unless calling. Greenish above, greyish below, which with white tail make it look like large honeyguide; distinguished by stocky bulbul shape and large head (honeyguides are slim and small-headed), lack of moustachial stripe or other patterning on head or underparts, direct flight (honeyguides have undulating flight, often flicking and spreading tail), and behaviour – bulbuls glean foliage while honeyguides perch quietly for long periods. White eye of ♂ Honeyguide Greenbul is distinctive – all honeyguides have dark eyes (for distinctions from Sjöstedt's Honeyguide Greenbul *B. clamans*, see that species).

Voice. Tape-recorded (32, 53, 86, BRU, ERA, HOR). Song, a variable, lilting phrase of 8–10 high-pitched, pure whistles, the last usually drawn-out and down-slurred; 3 lower notes of a rather buzzy quality, 'tyer-yer-yer', are sometimes tacked on after a slight pause. Interval between song-phrases, 4–6 s. In Liberia and Ivory Coast, a more often heard variant consists of 4 short introductory notes and 2 upslurred whistles, 'ti-ti-tu-wi-TOOY-TOIY', repeated at same intervals with very little change in form; 2–3 additional whistles may be added after a slight pause. The common call is a long 'wolf whistle', rising and then falling; though independent of the song, it may also be interspersed between song-phrases. Less common is a nasal, mewing call.

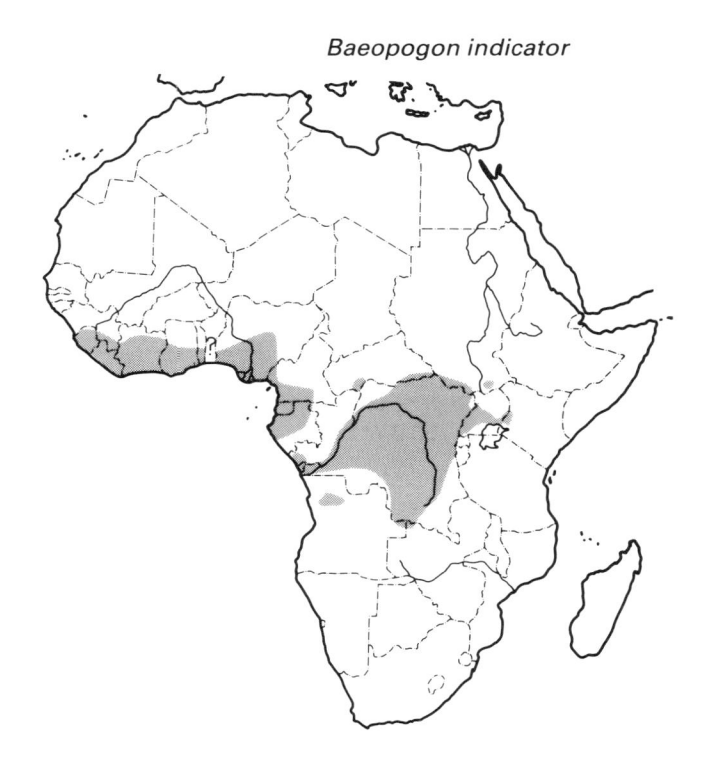

Claim that song is 'directly followed by female's loud semi-musical call' (Chapin 1953) has not been confirmed by recent workers; all calls are probably given by ♂♂. Likewise, the 'monotonous keeto-keeto-keeto-keeto' of Mackworth-Praed and Grant (1952) must refer to some other species.

General Habits. Inhabits primary forest, especially around clearings, but prefers more open habitats: forest edge, second growth with *Musanga* and *Aframomum*, gallery forest, strips of forest along streams in cultivated areas, coffee forest and other tree-covered plantations. Lowland and intermediate forest; up to 450 m in Cameroon, 1700 m in Zaïre, 2000 m in E Africa.

Sometimes solitary but usually in pairs or groups of up to 7. Prefers leafy canopy and emergents, where it forages in thick creeper tangles, often hovering. Also occurs at lower levels, especially in search of fruit; in Liberia (Mt Nimba) eats berries in low bushes in grassy areas 50 m from forest (J. R. Karr, pers. comm.). Sometimes joins mixed-species foraging flocks despite diet being mainly fruit. Occasionally follows ant swarms, catching flushed insects in air or on trunks and vines. Active, moving from tree to tree, and vocal, often singing throughout the day. Flashes white tail-feathers, probably an intraspecific recognition signal (Brosset and Erard 1986). Sedentary.

Food. Berries and small fruits (*Ficus, Musanga, Heisteria, Rauwolfia, Croton, Xylopia*; *Trema* especially favoured); also insects, including caterpillars, large black ants, beetles, termite alates; and spiders.

Breeding Habits. Territorial. In Gabon defends area of 15–18 ha with agonistic displays and voice, especially the 'wolf whistle' call (Brosset and Erard 1986); in Kenya (Kakamega) territory at least 4 ha (Zimmerman 1972).

NEST: only 1 described, a loosely-built cup of dry twigs and dead leaves, lined with a bed of thin yellow plant stems whose colour contrasts sharply with the dull outer portion; ext. diam. 75, int. diam. 55, depth 50, depth of cup 40; securely fastened to 2 small twigs c. 8 m above ground on branch of orange tree close to road and c. 20 m from building. Another seen but not described, in slender branches in canopy of forest tree.

EGGS: not recorded. Since 2 small young found in the above nest, clutch may be 2.

LAYING DATES: Sierra Leone (breeding condition Feb); Ghana, Feb (building nest Aug); Nigeria (breeding condition Sept–Mar, i.e. dry season); Zaïre, June (breeding condition in rains, probably breeds most of year); Gabon, Nov; E Africa: Region B (building nest Mar); Zambia (breeding condition Aug, Oct–Nov).

INCUBATION: by ♀ only.

Reference
Brosset, A. and Erard, C. (1986).

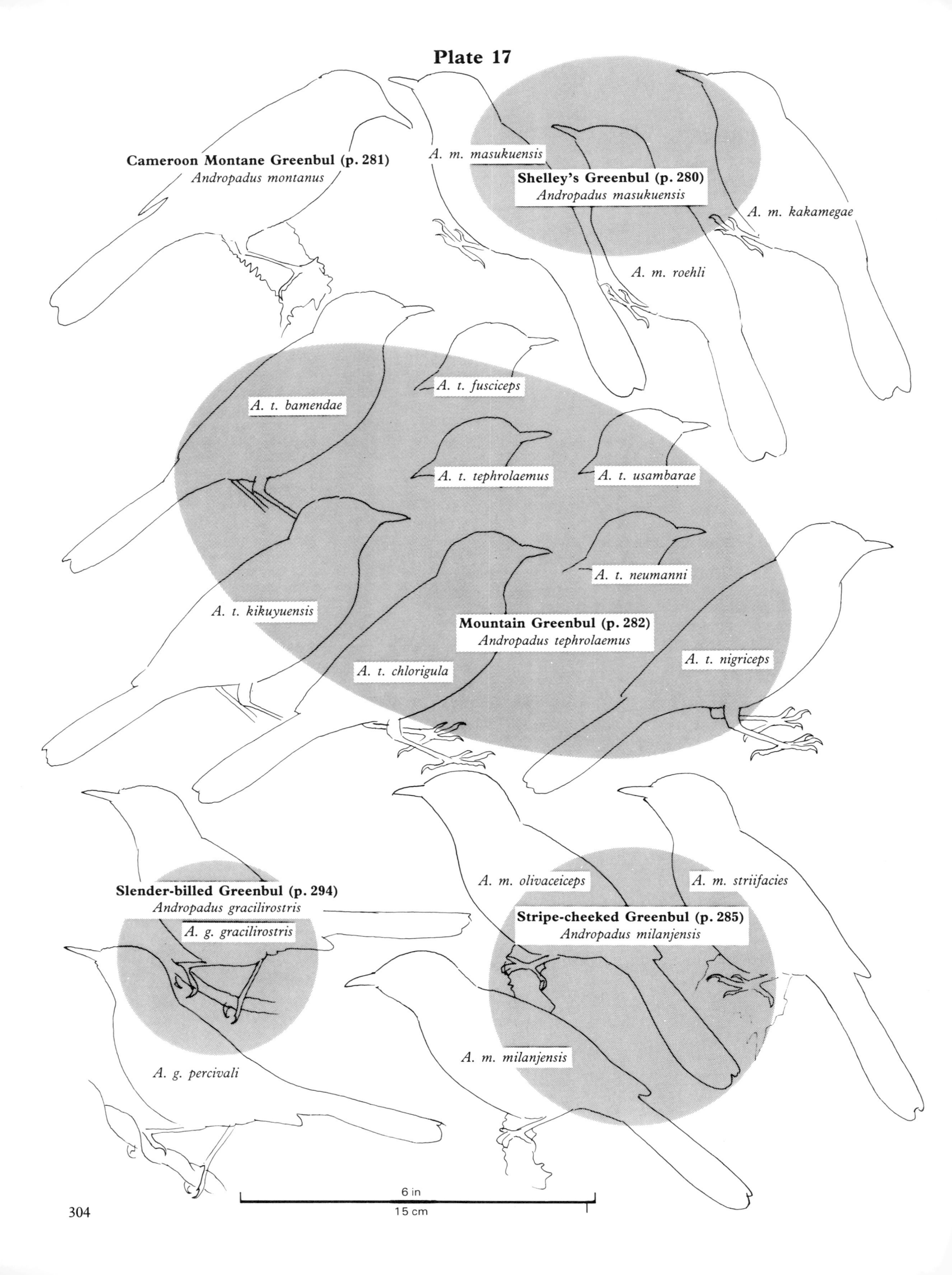

Plate 17

Cameroon Montane Greenbul (p. 281)
Andropadus montanus

Shelley's Greenbul (p. 280)
Andropadus masukuensis

A. m. masukuensis
A. m. kakamegae
A. m. roehli

A. t. bamendae
A. t. fusciceps
A. t. tephrolaemus
A. t. usambarae
A. t. neumanni
A. t. kikuyuensis

Mountain Greenbul (p. 282)
Andropadus tephrolaemus

A. t. chlorigula
A. t. nigriceps

Slender-billed Greenbul (p. 294)
Andropadus gracilirostris

A. g. gracilirostris
A. g. percivali

A. m. olivaceiceps
A. m. striifacies

Stripe-cheeked Greenbul (p. 285)
Andropadus milanjensis

A. m. milanjensis

6 in
15 cm

304

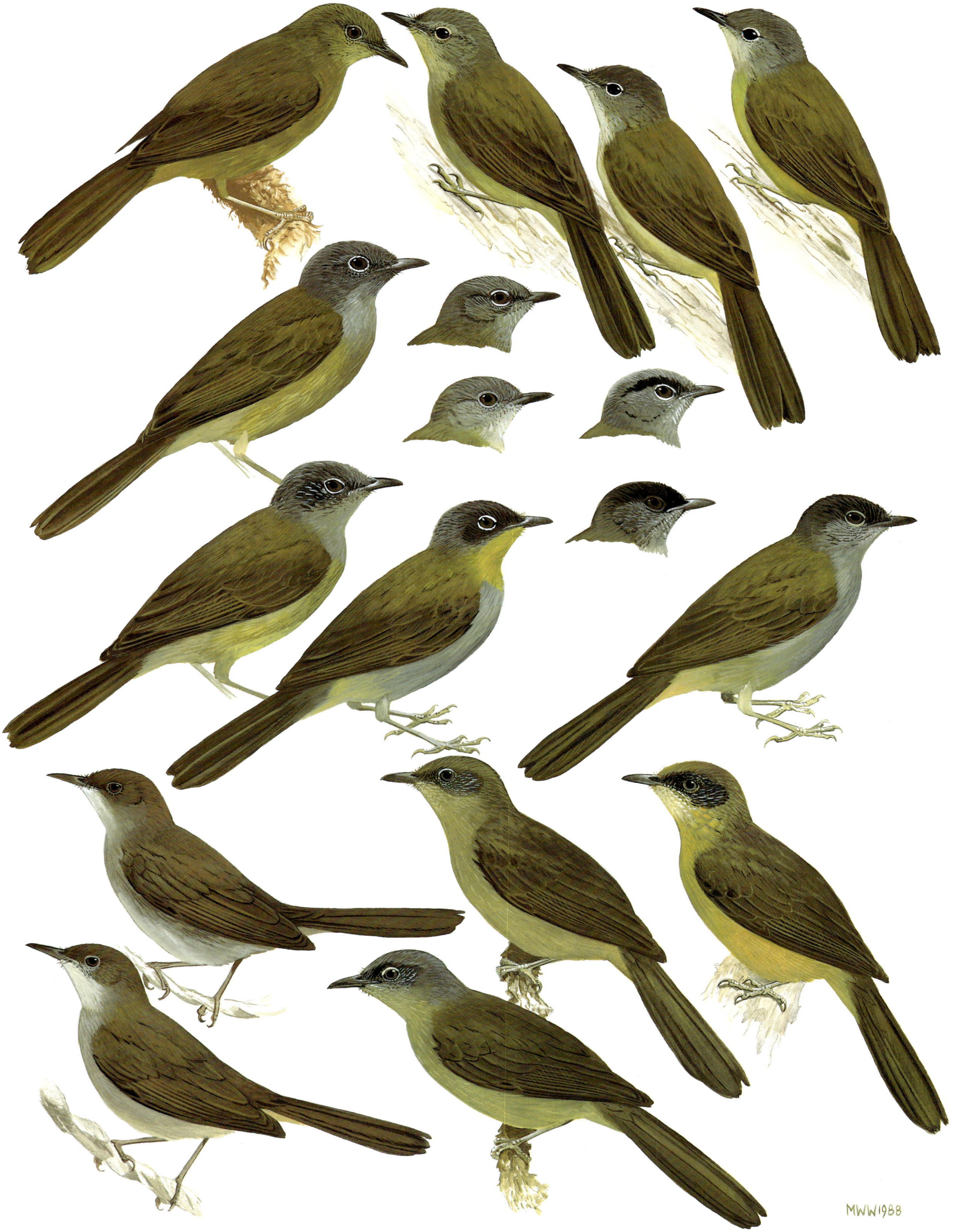
MWW1988

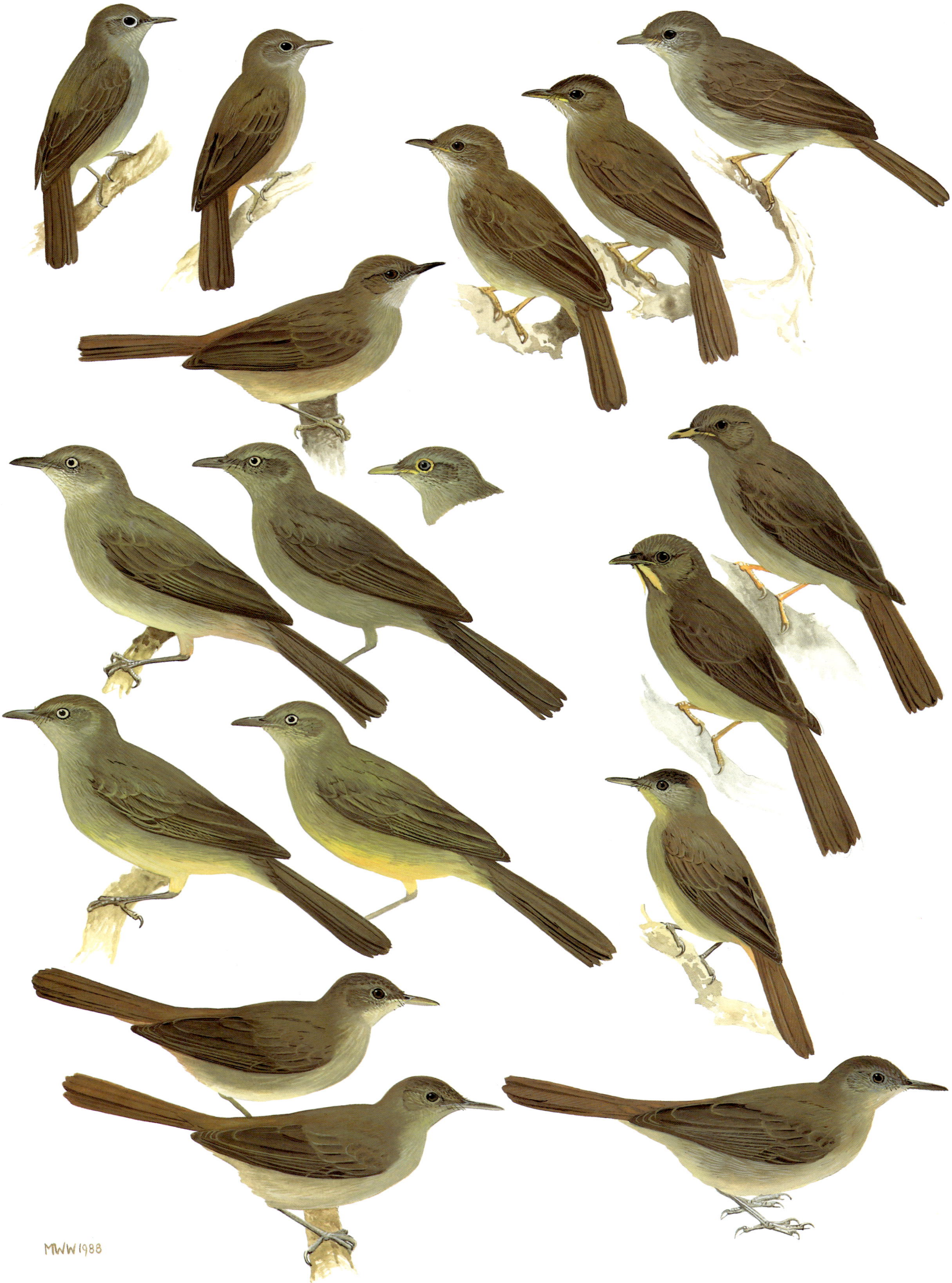
MWW 1988

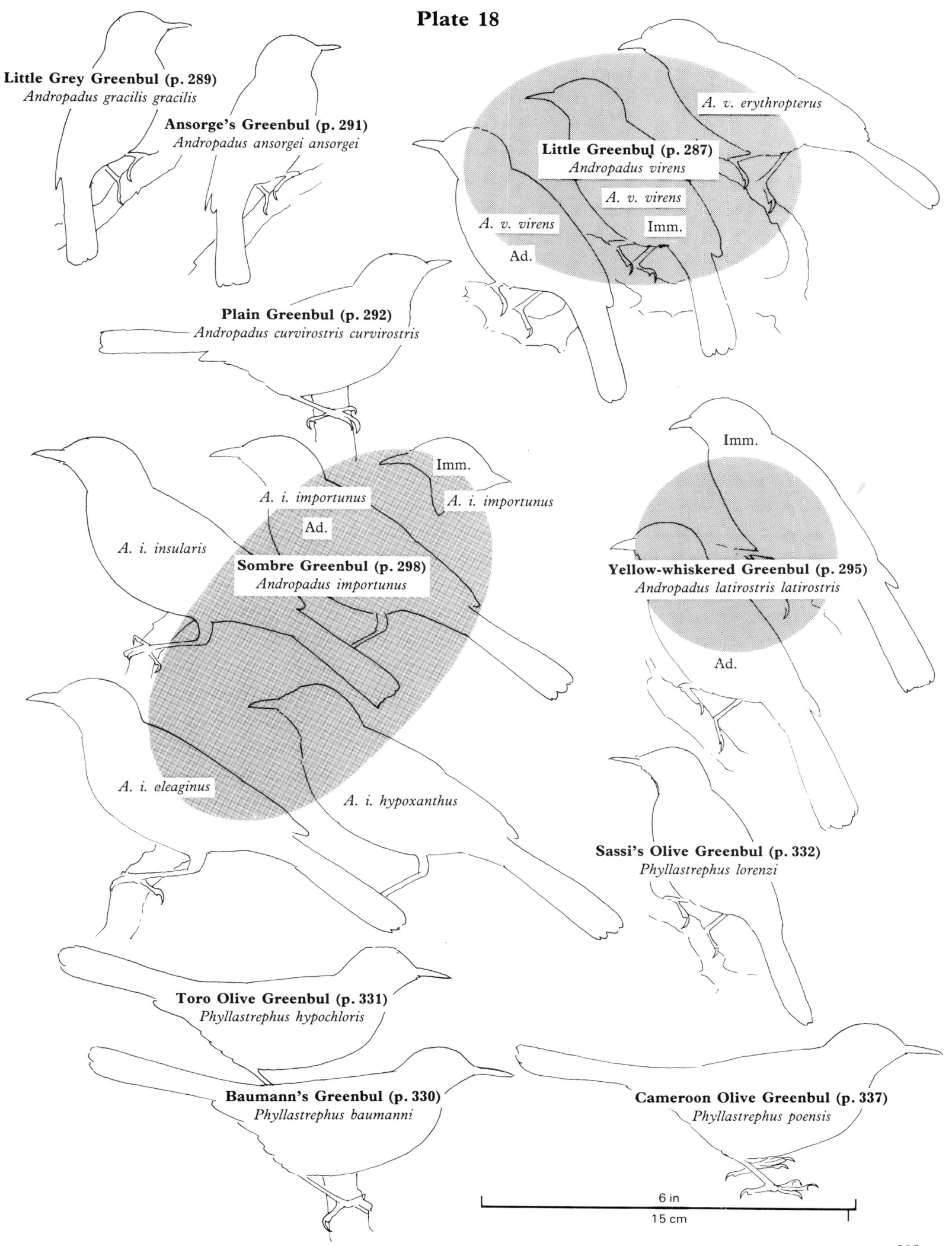

Plate 18

Little Grey Greenbul (p. 289)
Andropadus gracilis gracilis

Ansorge's Greenbul (p. 291)
Andropadus ansorgei ansorgei

Little Greenbul (p. 287)
Andropadus virens
A. v. erythropterus
A. v. virens Imm.
A. v. virens Ad.

Plain Greenbul (p. 292)
Andropadus curvirostris curvirostris

Sombre Greenbul (p. 298)
Andropadus importunus
A. i. importunus Ad.
A. i. importunus Imm.
A. i. insularis
A. i. eleaginus
A. i. hypoxanthus

Yellow-whiskered Greenbul (p. 295)
Andropadus latirostris latirostris
Imm.
Ad.

Sassi's Olive Greenbul (p. 332)
Phyllastrephus lorenzi

Toro Olive Greenbul (p. 331)
Phyllastrephus hypochloris

Baumann's Greenbul (p. 330)
Phyllastrephus baumanni

Cameroon Olive Greenbul (p. 337)
Phyllastrephus poensis

6 in
15 cm

305

Baeopogon clamans (Sjöstedt). Sjöstedt's Honeyguide Greenbul. Bulbul bruyant.

Plate 21
(Opp. p. 352)

Xenocichla clamans Sjöstedt, 1893. Orn. Monatsber. 1, p. 28; Ekundu, Cameroon.

Range and Status. Endemic resident. 4 disjunct populations: (1) Extreme SE Nigeria (Ikpan Block, Oban West), Cameroon from Korup Nat. Park, Kumba and Mt Cameroon along coast to Campo on border of Equatorial Guinea, and coastal Gabon to Fernan Vaz district, inland in Cameroon to Bitye and Sangmelima, in Gabon to Makokou; (2) S Congo (Mayombe: Dowsett-Lemaire and Dowsett 1989); (3) NE Zaïre in Uele and Ituri districts, from Libokwa and Poko to Ngayu and Beni; (4) E Zaïre in small area of Maniema and Kivu, 2° 16′–3° 30′S, 26° 7′–28° 17′E. Locally fairly common; density in Gabon about that of *B. indicator* (6–8 pairs/km^2) (Brosset and Erard 1986).

Description. ADULT ♂: forehead grey with olive wash, crown to hindneck olive-green, feathers with variable amount of grey on outer half and narrow dark fringes giving scaly effect. Lores, around eye, cheeks and ear-coverts grey. Rest of upperparts, including scapulars, tertials and upperwing-coverts golden green. 2 central pairs of tail-feathers blackish brown, T3 same but outer web white to within *c.* 10 mm of tip; 3 outer pairs white, outer webs broadly edged sulphur-yellow. Chin and centre of throat grey bordered with band of buff around lower throat, extending onto sides of neck as incipient collar and in some individuals extending up sides of throat, separating it from grey cheeks. Breast and upper belly greyish buff washed yellowish olive-green, flanks grey-olive, lower belly and undertail-coverts rich buff. Primaries and secondaries blackish brown edged golden green, narrowly on outer primaries, progressively broader on secondaries; outermost primary short, without edging. Axillaries and underwing-coverts ochre with variable amount of greyish; feathers along underside of bend of wing bright ochre; inner webs of underside of primaries and secondaries edged greyish buff. Bill black to dark grey or dusky, cutting edges and tip paler; eye dark reddish brown to dull red; legs and feet grey to blue-grey. Sexes alike. SIZE: (9 ♂♂, 7 ♀♀), wing, ♂ 93–107 (101), ♀ 91–99 (94·4); tail, ♂ 67–76 (72·4), ♀ 66·5–73 (69·3); bill, ♂ 18–20 (18·6), ♀ 16·5–18 (17·4); tarsus, ♂ 19–21·5 (19·8), ♀ 18·5–22 (19·7).

IMMATURE: like adult but duller, buffy areas paler, less rich, throat with light buffy wash.

NESTLING: unknown.

Field Characters. Very like Honeyguide Greenbul *B. indicator*; underparts paler and buffier, throat paler than breast rather than concolourous with it. Other characters only partially helpful; dark red eye distinguishes it from ♂ Honeyguide Greenbul (with pale eye) but not from ♀ (dark eye); lack of dark tips to white tail-feathers distinguishes it from adult Honeyguide Greenbul but not from immature. Best character is harsh call, frequently given, very different from pleasant whistles of Honeyguide Greenbul.

Voice. Tape-recorded (53, ERA). Call, a single, grating, down-slurred note, 'chyaa', 'pyow', 'peeyu', loud, carrying up to 300–400 m, and repeated for long periods; 1 bird called for 15 min without pause (Chappuis 1975). Call note often runs into song, an irregular jumble of scratchy, nasal notes and a few whistles, rather like latter part of song of Little Greenbul *Andropadus virens*.

General Habits. Inhabits primary forest and old second growth, with a preference for water, occurring along river banks and small streams inside forest; mainly lowlands, but ascends to 1500 m in E Zaïre. Ecologically separated from Honeyguide Greenbul (Brosset and Erard 1986), preferring interior of forest to open areas and foraging at intermediate levels and lower canopy. Sometimes descends to undergrowth.

Occurs singly, in pairs or in small groups; sometimes joins mixed-species flocks. Gleans leaves and searches for fruit. Ecology partly tied to small black wasp *Polybioides melaina* which builds large paper nests in trees along watercourses (see *Food*). Calls near to nests and tears them apart, despite the wasps' being 'very courageous and venomous' (Chapin 1953). Calls continuously, spreading tail to display white feathers. Territorial all year; defends territory of 8–10 ha with song and aggressive displays, responding strongly to playback of taped song (Brosset and Erard 1986).

Food. Fruits and berries, e.g. *Musanga, Heisteria*; insects and their larvae: moths, caterpillars, and especially contents of nests of *Polybioides melaina* and other Hymenoptera. Young fed exclusively on wax, larvae and pupae of wasps.

Breeding Habits. Territorial (see *General Habits*).

NEST: only 1 described (Erard 1977); loosely-built cup of dead leaves bound together with interwoven rootlets and vine stems, suspended between small branches like a hammock, attached with spider webs. Leaves and twigs hang down from underside, attached by threads of fungus *Marasmius*, making nest look like bunch of dead leaves caught among branches. A typical bulbul nest, like large version of nest of Yellow-whiskered Greenbul *Andropadus latirostris*. Situated 15 m above ground in leaves at end of branch of 25 m tree, in small clearing on bank of river.

EGGS: unknown.

LAYING DATES: Gabon, Feb (carrying nest material Jan), but singing season suggests that breeding encompasses 2 rainy seasons and intervening short dry season (Brosset and Erard 1986); Zaïre (breeding condition Feb, Mar, July).

DEVELOPMENT AND CARE OF YOUNG: young fully feathered at c. 15–16 days (Erard 1977). Fed by ♀ only; when 2 young in nest, fed on average every 27 min (= every 54 min per young); when only 1 young left, fed every 33 min (Erard 1977). When ♀ brings food, ♂ keeps watch nearby. ♀ arrives silently, or gives a few short notes; ♂ answers, then sings, moving about near nest, gradually approaching it, singing continuously, sometimes answered by ♀. He stops a few m from nest and ♀ feeds young, but if alarmed both stay away and sing and give alarm calls together. If raptor flies close to nest, birds immediately stop calling and hide among leaves; if ♀ is incubating she flattens herself on nest and freezes.

References
Brosset, A. and Erard, C. (1986).
Erard, C. (1977).

Genus *Ixonotus* Verreaux

Upperparts and wings brown-olive, with white spots on lower back and wings; in tail, T4–T6 pure white; underparts creamy white; bill slender and weak, nostrils slit-like, partly feathered; tail fairly long. Inhabit forest canopy; travel in noisy groups giving un-bulbul-like metallic chatter.

Endemic; monotypic.

Ixonotus guttatus Verreaux. Spotted Greenbul. Bulbul tâcheté.

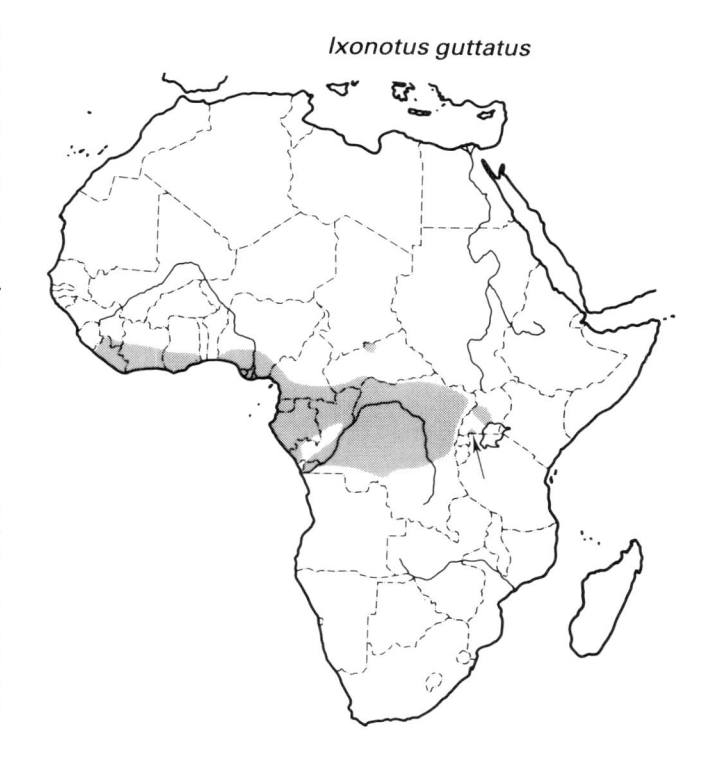

Plate 21
(Opp. p. 352)

Ixonotus guttatus J. and E. Verreaux, 1851. Rev. Mag. Zool. Paris, p. 306; Gabon.

Range and Status. Endemic resident. Liberia (north to Mt Nimba), SE Guinea, Ivory Coast (scarce, north to Man, Agnibilekrou), Ghana (uncommon and local, north to Sunyani). S Nigeria (local but not uncommon), Cameroon (north to Meket Mbeng, Baseng, Korup Nat. Park), Gabon (common), Congo, Zaïre (widespread in N, south to Kwango, Kasai, Maniema and Kivu), Central African Republic (Bamingui-Bangoran Nat. Park, and in SW), W-central Uganda (Budongo, Bugoma, Mpigi and Kifu forests) and forests either side of Kagera R. in S Uganda (Malabigambo) and NW Tanzania (Minziro). Locally common to scarce; density in Nigeria (Ogba Forest) 1 bird/2 ha (Johnson 1989).

Description. (includes '*bugoma*'). ADULT ♂: crown to hindneck greyish olive, feathers fringed blackish producing scaly effect. Mantle and upper back olive; feathers of lower back, rump, uppertail-coverts, scapulars, upperwing-coverts and inner secondaries dark brown-olive (blackish on rump) tipped with large white or yellowish spots. 2 central pairs of tail-feathers blackish brown, 3 outer pairs pure white, T3 with outer web white, inner web blackish brown, but variable at tip, where sometimes white extends onto inner web, sometimes dark onto outer web. Lores and moustachial area whitish separated by thin black line; ear-coverts and area around eye grey-brown with variable amount of pale streaking. Chin and throat white, breast to undertail-coverts white variably washed with cream or sulphur-yellow, most pronounced on lower belly and flanks. Primaries and secondaries blackish brown edged brownish olive; axillaries and underwing-coverts cream to pale yellow, inner webs of underside of primaries and secondaries edged whitish. Upper mandible dark grey, lower mandible grey, pinkish grey or grey-brown; eye, ♂ dark brown to deep red-brown, ♀ greyish white; legs and feet grey to dark grey or bluish grey. Sexes alike. SIZE: (10 ♂♂, 6 ♀♀) wing, ♂ 86–96 (90·9), ♀ 84–91 (87·2); tail, ♂ 70–80 (74·7), ♀ 67–74 (70·7); bill, ♂ 16–18 (17·0), ♀ 16–17 (16·5); tarsus, ♂ 17·5–19·5 (18·1), ♀ 18–19·5 (18·3). WEIGHT: Liberia, ♂ (n = 3) 35·2–36·2 (35·8), ♀ (n = 2) 31·4, 35·4; Uganda, ♂ (n = 15) 31–40 (35·2), ♀ (n = 3) 34·1–38 (36·0).

IMMATURE: full-grown immature identical to adult; one very small bird (wing 52, tail 21), still with white down on underparts, has top of head and back brown, not olive, otherwise mainly like adult; white marks on wings and rump already in place but tinged buff, outer tail-feathers pure white.

NESTLING: unknown.

Field Characters. Social, travelling in groups from tree to tree and keeping up a constant chatter. White spots on wings and rump distinctive but hard to see since birds are usually in canopy. From below appears almost pure white: underparts, underwing and underside of broad tail all white. Can be mistaken for Western Nicator *Nicator chloris*, which also has white spots on wing, but is a large, solitary, rather sluggish bird of undergrowth and mid-levels with completely different calls; it also differs in duller underparts, yellow undertail-coverts, smaller wing spots and heavy, shrike-like bill.

Voice. Tape-recorded (53, C, ERA, MOY). Contact call a continuous chirruping, the notes thin and high-pitched, with a metallic quality like the chink of silver coins; typically a double note, 'tsi-tsrit', 'tsi-trrr', 'tsi-tsup', 'tsrrr-tit'; sometimes 3 or 4 notes together, or just a single 'tsit'. Song unknown. Quality and tone of call completely unlike that of any other bulbul; sounds more like a large sunbird.

General Habits. Inhabits primary forest, keeping mainly to canopy, less often in middle and lower levels; and second growth, frequenting crowns of tall trees and tops of nearby bushes and shrubs. Also occurs in clearings, forest edge, gallery forest, forest–savanna mosaic, and man-made habitats: trees shading plantations, open cultivation, even near houses. Lowland and lower intermediate forest, to 1250 m in Uganda, 1300 m in Zaïre.

The most gregarious bulbul, in groups all year; noisy bands of 5–50 birds work through trees, constantly on the move, keeping contact with sharp clicking calls. Group does not spend long in any one tree; even when they find one with a large fruit crop they gorge themselves, then fly off and rest for a while before flying on to next tree. One group of 6 moved through dense foliage 6 m above forest floor, keeping very close together; birds 'moved around quite rapidly, following each other in short flights across gaps in foliage or between trees; as birds at the back caught up with those in front, up to 5 birds would sit very close together on a branch, pressed side to side, stretching up or peering about, flicking wings, usually one at a time, and flicking tail. Then group split up as individuals moved off, only to reform further away. One bird while foraging constantly raised first one wing then the other, flicking it quite mechanically and keeping it well bent at the bend (**A**). Wing-flicking apparently a regular activity at all times, even when birds close together' (Tanzania: M. W. Woodcock, pers. comm.). Rump feathers also sometimes raised and fluffed out. Group is monospecific, but may mingle with other species at fruiting trees, e.g. Yellow-spotted Barbet *Buccanodon duchaillui* (Friedmann and Williams 1969). Once in mixed-species flock (Germain *et al.* 1973). Actively gleans leaves and branches, sometimes hanging upside down. Favourite feeding site is underside of branch, from which it usually collects food items by hovering (Johnson 1989).

Sedentary for part of year, nomadic at other times. In Gabon, largest groups form in long dry season when fruiting trees are scattered; group uses area of several dozen ha (Brosset and Erard 1986).

(N.B.: Young's notes (1946) on *I. guttatus* in fact refer to Western Nicator.)

Food. Fruit, mainly of *Ficus*, also *Macaranga*, *Rauwolfia*, *Musanga* and *Heisteria*; insects, including ants gleaned from bark, caterpillars and swarming termites. Nestlings fed both fruit and caterpillars.

Breeding Habits. A cooperative breeder; young fed by group of 4–6 birds, both in nest (Chapin 1953) and after leaving it (Brosset and Erard 1986).

NEST: open, shallow, rounded cup, rather frail and flimsy, crudely built of dry leaves and leaf stems, rootlets and bits of bark, bound with spider webs; with thin lining of small epiphytic orchid and other rootlets; regurgitated seed pellets often attached to exterior, to reinforce and camouflage it (Herroelen 1955). Round nests, ext. diam. 70–73, int. diam. 60–65; oval nest, ext. diam. 75 × 85, int. diam. 60 × 70, ext. depth 25–60, int. depth 20–26; placed 2·5–4·5 m above ground in fork of small tree or bush, wild or cultivated, often near road or building; once inside knot hole on upper surface of horizontal branch 10 m up in kola-nut tree in farmland choked with second growth (Serle 1965).

EGGS: 2. Ovate; smooth, glossy; pale yellowish, cream or pale brown, thickly and boldly blotched and spotted with blackish brown, chocolate, olive-brown and yellow-brown over greyish undermarkings; markings usually concentrated around large end but may obscure most of yellow undercoat. SIZE: (n = 2) both 23·5 × 16.

LAYING DATES: Liberia (breeding condition Aug); Nigeria (building nests June–Sept); Cameroon, Mar–May, June, Aug (young Oct); Zaïre, Dec–Feb, Apr, June–July, Oct (probably almost all year); Gabon (flying young Feb, Oct).

INCUBATION: by ♀ only.

DEVELOPMENT AND CARE OF YOUNG: young fed by both parents and by helpers, before and after fledging. As party of adults leaves nest after feeding young, one sometimes remains behind to brood them (Chapin 1953).

Reference
Serle, W. (1965).

Genus *Chlorocichla* Sharpe

Bill long and stout, culmen and gonys gently curved, nostrils oblong, operculum partly feathered. Rictal bristles well developed, nuchal hair short. Tail rounded or slightly graduated, same length as wing. Legs rather long, feet strong. Throat white or yellow, forming usually distinct patch.

Inhabit thickets and scrub, mainly outside forest; also forest edge and clearings. Often travel in noisy family groups; voices chattery, with nasal, burry quality (except *C. laetissima*).

Endemic; 6 species. No superspecies. We consider *C. flaviventris* and *C. falkensteini* too different to form a superspecies (cf. Hall and Moreau 1970). *C. laetissima* is vocally aberrant. *C. prigoginei* closely resembles *C. laetissima*, but they are sympatric and may form a pair of sibling species (De Roo 1967). *C. simplex* and *C. flavicollis* have similar voices. *C. flavicollis* was placed by Sclater (1930) in a monotypic genus *Atimastillas*; it forms a link with *Pyrrhurus* and *Thescelocichla*.

Chlorocichla laetissima (Sharpe). Joyful Greenbul. Bulbul joyeux.

Andropadus laetissimus Sharpe, 1899. Bull. Br. Orn. Club 10, p. 27; Nandi, Equatorial Africa (= Kenya).

Plate 21
(Opp. p. 352)

Range and Status. Endemic resident. Sudan (Imatong Mts: fairly common but local); E Zaïre (Bondo Mabe and Lendu Plateau to highlands west of L. Edward; Mt Kabobo; Marungu Highlands); extreme N Zambia (Kasangu, 8° 25′S, 29° 22′E); W Uganda (Toro, Ankole and Impenetrable Forest, Kigezi); sight record S Uganda (Mabira); W Kenya (Mt Elgon and Cherangani Hills to Sotik, Mundiri R., Olooloo Escarpment and Lolgorien). Common locally but distribution patchy; absent from many areas where it might be expected.

Description. *C. l. laetissima* (Sharpe): range of species except for *schoutedeni*. ADULT ♂: top and sides of head, cheeks, ear-coverts, mantle to uppertail-coverts, scapulars and upperwing-coverts, bright yellowish olive-green; feathers of top of head with narrow dark centres, giving scaly effect; feathers from nostrils to lores and narrow band across base of upper mandible brownish yellow, and semi-erect, as are those of forehead; vestigial short supercilium yellowish, indistinct broken eye-ring yellow. Tail-feathers olive-green edged yellowish, shafts dark maroon above, straw below. Chin and throat bright yellow, rest of underparts greenish yellow to golden green, darker on sides of breast and flanks, brighter on centre of belly, undertail-coverts tipped yellow. Primaries and secondaries dark brown, outer webs yellowish olive-green, inner webs with narrow pale edges; axillaries and underwing-coverts greenish yellow, margins of inner webs of underside of primaries and secondaries yellow-buff. Bill black, slaty black or dark brown, with brown cutting edges and tip; eye brown to chestnut, brown-red or bright russet; legs and feet black, blue-grey, greenish grey or greyish brown, soles pale yellow. Sexes alike. SIZE: (12 ♂♂, 12 ♀♀) wing, ♂ 101–109 (106), ♀ 97–109 (103); tail, ♂ 95–104 (99·8), ♀ 94–102 (97·4); bill, ♂ 21–23 (21·8), ♀ 20–23 (21·5); tarsus, ♂ 22–25 (23·8), ♀ 18–25 (23·1). WEIGHT: Kenya, ♂ (n = 5) 47–54 (50·6), ♀ (n = 8) 43–55 (50·0).

IMMATURE: back washed brownish, underparts greener.
NESTLING: unknown at hatching; for later stages, see *Breeding Habits*.

C. l. schoutedeni Prigogine: E Zaïre (Mt Kabobo, Marungu Highlands), Zambia. Darker above, especially on nape and hindneck; face darker and more olive, less green; underparts overall darker (heavier olive wash on breast and flanks).

Field Characters. One of the most brightly coloured bulbuls, the only one entirely green and yellow (yellow-green above, golden green below, bright yellow throat). Forages in groups with happy, bubbling chatter. Smaller Golden Greenbul *Calyptocichla serina*, a solitary treetop bird with a descending whistle, has similar plumage but pink (not black) bill, grey face, white throat. Other bulbuls with yellow underparts are not uniform, having e.g. grey heads, white throats, darker or browner upperparts, and most have yellow of underparts paler and browner, not golden (for distinctions from Prigogine's Greenbul *C. prigoginei*, see that species).

Voice. Tape-recorded (32, 86, C, GREG, HOR, LEM, Z1M). 2-part, bubbling song lasts 4–5 s; first part (*c.* 3 s) starts with low notes, gets louder, ends with sharp 'chik-chik'; pause of 0·5 s; second part (*c.* 1·5 s) fades away with low notes, almost an afterthought. Pleasant and energetic but notes not clear or melodious, still have burry bulbul tone. (Description of loud, beautiful song in Jackson and Sclater (1938) clearly refers to some other species.) Call, a sharp 'chik' or 'chak'.

General Habits. Inhabits primary and secondary forest, especially more open parts; forest edge, gallery forest, isolated forest patches; at altitudes from 1050 to 2300 m.

Chlorocichla laetissima

Sociable, usually in monospecific parties of 4–8; sometimes joins mixed-species flocks. Forages mainly at low and intermediate levels, less often in dense undergrowth or canopy.

Food. Berries and seeds.

Breeding Habits. Unknown in the wild. Nest built in captivity was open cup of Spanish moss, rootlets and fine grasses, ext. diam. 228, depth of cup 64 (Everitt 1964). Eggs not seen. At 7 days chick almost fully feathered, gape bright orange with pale yellow flanges; 6 days later plumage green above, washed brown, yellow below, washed green; at 6 months identical to parents. Fed by both parents with 'live food' (including a moth) and fruit (Everitt 1964a). Fully independent at 31 days, but had begun to feed itself before then.

LAYING DATES: Sudan, Mar; Kenya, July.

Reference
Everitt, C. (1964a).

Plate 21
(Opp. p. 352)

Chlorocichla prigoginei De Roo. Prigogine's Greenbul. Bulbul de Prigogine.

Chlorocichla prigoginei De Roo, 1967. Rev. Zool. Bot. Afr. 75 (3/4), pp. 392–395; Maboya, Kivu, Zaïre.

Range and Status. Endemic resident. Known only from 2 areas in E Zaïre: on Lendu Plateau at 1675 m between Djugu (1° 55'N, 30° 30'E) and Nioka (2° 9'N, 30° 40'E); and between Beni (0° 28'N, 29° 28'E) and Butembo (0° 8'N, 29° 17'E) (localities Butembo, Maboya, Kiwira, Mutaka, Kabasha). Rare, and seriously threatened by forest destruction (Collar and Stuart 1985).

Description. ADULT ♂: top of head and upperparts, including scapulars and upperwing-coverts, yellowish olive-green; ear-coverts similar, with hint of pale streaking; lores and pre-orbital region pale grey, merging with broad grey-white eye-ring; feathers from nostril to narrow band over base of upper mandible brownish grey, and semi-erect, as are those of forehead. Chin white, throat bright yellow. Tail-feathers olive edged yellowish olive-green, shafts deep maroon above, straw

Chlorocichla prigoginei

Sexes alike. SIZE: wing, ♂ (n = 5) 91–98 (94·0), ♀ (n = 3) 86–90 (88·3); tail, ♂ (n = 4) 89–95 (92·0), ♀ (n = 2) 87, 88; bill, 1 ♂ 18, 1 ♀ 17; tarsus, ♂ (n = 5) 21–22·5 (21·6), ♀ (n = 3) 21–22·5 (21·5).

IMMATURE: 2 birds with partly ossified skulls have plumage like adult but pale leg scales.

NESTLING: unknown.

Field Characters. Very similar to sympatric Joyful Greenbul *C. laetissima* but smaller, with white chin, grey area between bill and eyes, pale grey eye-ring.

Voice. Unknown.

General Habits. Inhabits isolated forest patches, gallery forest and thickets at 1350–1800 m.

Wing moult apparently slow and often interrupted, especially in secondaries (De Roo 1967).

Food. Stomach of 1 bird contained seeds, fruit pulp and a small green caterpillar.

below. Underparts greenish yellow, sides of breast and flanks darker and greener, undertail-coverts yellowish brown with yellow tips. Primaries and secondaries dark brown edged yellowish olive-green, more broadly inwardly, inner webs with narrow pale margins; axillaries and underwing-coverts bright yellow; inner webs of underside of primaries and secondaries edged pale yellow. Bill blackish; eye deep red; legs and feet dull bluish grey, dark horn or blackish, claws grey-brown.

Reference
De Roo, A. (1967).

Chlorocichla flaviventris (Smith). Yellow-bellied Greenbul. Bulbul à poitrine jaune.

Trichophorus flaviventris Smith, 1834. S. Afr. Quart. J., 2nd Ser., 2, p. 143; near Port Natal (i.e. Durban).

Plate 21
(Opp. p. 352)

Range and Status. Endemic resident in coastal and low-lying areas of E and central Africa from S Somalia (locally very common), Kenya (south from Boni Forest, inland to E edge of highlands from Meru to Ngong: common), Tanzania (inland to Mt Kilimanjaro, L. Manyara, Mt Hanang, Shinyanga, Tabora, Iringa and L. Malaŵi, and in SW at Sumbawanga: common) and SE Zaïre through Zambia (almost throughout, locally scarce or absent), Malaŵi (almost throughout) and Zimbabwe (widespread but absent from NW and parts of Mashonaland Plateau; vagrant Harare: Woodall 1971) to S half Angola (and north to Luanda on coast), N Namibia (all along N border from Swartboois Drift to Zambezi), NW Botswana (Okavango, Chobe R.), N and E Transvaal (lowland river valleys), Mozambique (almost throughout, generally common) and Natal below 300 m.

Description. *C. f. occidentalis* Sharpe (including '*zambesiae*' and '*ortiva*'): W Tanzania to Zaïre, Angola, Botswana, Transvaal and Mozambique. ADULT ♂: upperparts including wing-coverts and tail light olive-green; top of head somewhat browner and darker, rump somewhat paler and brighter. Tail-feathers edged brighter green, with narrow pale tips when fresh; shafts brown on upperside, pale yellow on underside. Feathers of head often raised to form shaggy crest. Lores and facial area dull olive-brown. Chin and throat pale yellow, chin sometimes almost white; rest of underparts bright rich yellow with brownish wash on breast. Remiges olive-brown, outwardly edged green above, inwardly edged yellow below; bend of wing, axillaries and underwing-coverts yellow. Bill purplish slate, horn or horn-black, cutting edges paler; broken eye-ring white, upper 'half' broader and more conspicuous than lower; eye dark brown, red-brown or red; legs and feet grey, slate-grey or blackish. Sexes alike, but ♂♂ larger. SIZE: wing, ♂ (n = 27) 95–110 (103), ♀ (n = 21) 89–102 (96·6); tail, ♂ (n =

17) 88–99 (93·5), ♀ (n = 11) 87–92 (89·5); bill, ♂ (n = 27) 19·5–24 (21·2), ♀ (n = 21) 17·5–21·5 (20·1); tarsus, ♂ (n = 17) 22·5–26 (24·6), ♀ (n = 11) 22–25·5 (24·2). WEIGHT: Zaïre, ♂ (n = 8) 43–50 (46·5 ± 2·1); ♀ (n = 3) 38–41 (39·7); Zambia, unsexed (n = 18) 32–46 (37·3 ± 2·8); Malaŵi, unsexed (n = 10) 36–46 (39·8 ± 3·2); Zimbabwe, ♂ (n = 21) 34·5–51·2 (42·0 ± 3·8), ♀ (n = 12) 31·9–39·6 (36·9 ± 2·5).

IMMATURE: like adult but duller and paler, head same colour as back, more olive-brown than in adult; belly pale yellow, tail-feathers more pointed, tips of outer primaries more rounded; plumage more downy. Eye grey, becoming red at 6 months.

NESTLING: naked, dark greyish brown on back, yellowish on belly; gape yellow, mouth cadmium, bill brown; feet grey-brown.

C. f. centralis Reichenow: Somalia, Kenya, E Tanzania, N Mozambique. Darker and browner above than *occidentalis*, especially top of head, which contrasts more with back; rump brown, not noticeably paler than back; wing edgings more olive, less bright. WEIGHT: Kenya, ♂ (n = 14) 43–54·2 (49 ± 3·1), ♀ (n = 21) 36–50·4 (42·4 ± 3·2); Tanzania, ♂ (n = 2) 53, 55·5.

C. f. flaviventris Smith: Natal and S Mozambique. Browner above than *occidentalis*, but rather richer brown than *centralis* and not as dark, also head same colour as back, not contrasting. Uppertail-coverts and tail with reddish tinge, shafts red-brown above, whitish below; wing edgings light brown. Yellow of underparts paler, less rich, wash on breast greener and underwing-coverts paler, sulphur yellow.

Brightness and richness of underparts varies within as well as between all races.

Field Characters. A common, noisy bulbul of eastern lowlands, told from most others in same habitat by bright yellow underparts. Most resembles Sombre Greenbul *Andropadus importunus* but is larger, throat yellow instead of brown, bill longer, eye red-brown, broken white eye-ring, very different voice. Fischer's Greenbul *Phyllastrephus fischeri* has darker, browner upperparts, whitish underparts; brown-headed race *alfredi* of Yellow-streaked Greenbul *P. flavostriatus* has yellow streaks on whitish underparts but darker, browner upperparts, different voice.

Voice. Tape-recorded (14, 32, 58, 75, 86, 88, C, F). Song (Kenya), rather lengthy phrases consisting of 6 slow, low, quiet introductory notes followed by 4 long ones with buzzy, nasal quality: 'ruk ... tuk, tuk; raa, tuktuk; raaa-caaa-caaa-caaa'. In Zambia and southern Africa, song notes have same quality but different rhythm, 4–6 notes on same pitch, slowly and methodically delivered: 'jit-jar-jui-jer', 'jit-jar-jeer-jeer-jar', 'jit-jeer-jar-jui-jeeo', 'jit-jeer-jar-jeeo-jui-jeer'; 3–4 s intervals between phrases, sometimes punctuated by a few chuckles. Contact call 'quar-tooa, quar-toar', changing to 'kerr-quar' when bird finds food and summons mate. Scolding call, a loud, repeated 'pow', 'paa' or 'neh'; also gives variety of chuckling and chattering calls. Soft 'chaacha' in display. Call of young, 'cherup, cherup'.

General Habits. Inhabits forest, preferring thick undergrowth around clearings and at edge to interior; shrubby second growth, gallery forest, riverine bush, coastal scrub and dune forest; *Cryptosepalum* forest in NW Zambia; occasionally *Brachystegia* woodland, and anthill thickets in mopane; bushy tangles and thickets, including dense bush in wooded savanna; in Zimbabwe also drier thickets on hillsides and even semi-arid savanna woodland. In Sokoke Forest, Kenya, common in all 4 forest types, including *Brachystegia* (Britton and Zimmerman 1979); recently established as garden bird in Bulawayo, Zimbabwe (Feather 1986). Mainly in lowlands, reaching 1700 m in Malaŵi, 2100 m in E Africa; in Malaŵi above 1000 m restricted to riparian forest. Typically in undergrowth, but in forest also middle strata and sometimes canopy.

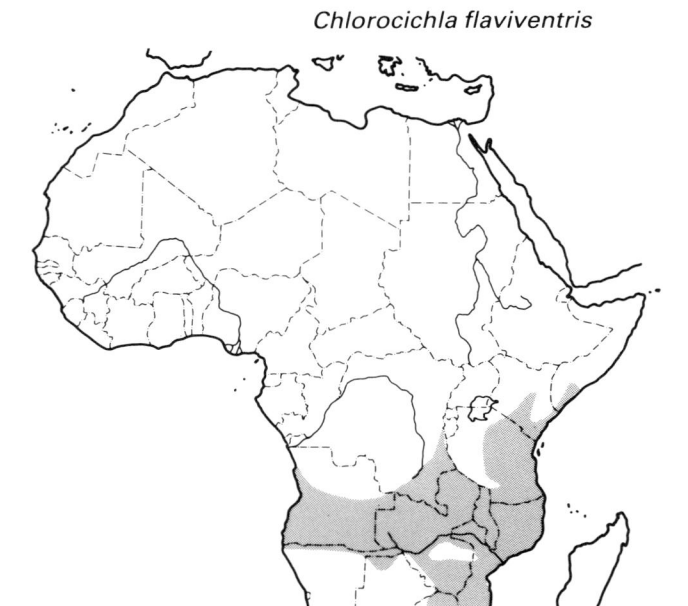

Forages in pairs or groups of up to 6. Often joins bird parties: in Mozambique with Yellow-streaked Greenbul, Kretschmer's Longbill *Macrosphenus kretschmeri*, flycatchers *Batis fratrum*, *B. molitor* and *Trochocercus cyanomelas*, and puffback *Dryoscopus cubla*; travelled up to 30 m behind party, on same route and at same speed (Vincent 1935). Hawks insects in flight, and flocks with other bulbuls at fruiting trees. Frequently lands on antelope and grooms its head, ears and even eyes (for ectoparasites?), usually with antelope's acquiescence (Vernon 1972, Steyn 1975, Chalton 1976). Once briefly joined other birds at swarm of army ants *Dorylus* (Willis 1983). Noisy, often the most vociferous member of a party, yet skulking and hard to see, slipping away into thicker vegetation on approach of observer. In South Africa apparently emerges from forest only when sunning itself on winter mornings (Maclean 1985).

Moults during breeding season (Zaïre, Malaŵi, Mozambique). 2 complete moults a year, Zaïre (Dec–Mar and May–July).

Food. Insects; berries and other fruit, both soft and with hard seeds (seeds sometimes ingested); flowers, seeds. Fruits of *Lantana*, *Carissa*, *Cissus*, *Allophylus*, *Clausena* and *Teclea* fed to young.

Chlorocichla flaviventris

Breeding Habits. Monogamous; territorial. 4 pairs in 16 ha, central Kenya. In pre-copulatory display in vegetation 3 m above ground, one bird chased the other, both stopped, ♀ crouched and fanned wings. ♂ then faced ♀, both erected crown feathers, opened bills and rotated heads elliptically and out of phase. After 2 min ♂ tried to mount ♀ but she snapped at him. Birds exchanged positions on branch and continued display for c. 2 min, then resumed chase. Throughout display both gave soft 'chaacha' call (Frost 1979).

NEST: a small, shallow, loosely built cup of twigs, pieces of vine, tendrils, plant stems and grass, lined with creeper tendrils, fine rootlets, grass blades and dead leaves, sometimes so thin that eggs can be seen from below; ext. diam. 10 cm, int. diam. 7·5 cm, ext. depth 5 cm, depth of cup 2·5 cm; placed 1–4 m above ground on outer twigs of horizontal branch or in fork in centre of small shrub, usually fairly well protected and concealed by foliage; also noted in open without concealment, sitting bird clearly visible at several paces (Vincent 1935). Built by both sexes; 1 took 7 days to complete (Kenya).

EGGS: usually 2; southern Africa mean (26 clutches) 2·1. Elongate; glossy; white, cream, pale ochreous or pale olive, heavily spotted, freckled and blotched with olive, brown, umber and sepia over grey or yellowish grey, markings often concentrated at big end, occasionally covering most of egg; general appearance not unlike eggs of some nightjars (Vincent 1935). SIZE: (n = 33, southern Africa) 21·6–26·9 × 15·9–18 (24·8 × 17·1).

LAYING DATES: Zaïre, all year; Angola (breeding condition Nov); E Africa: Region C, Oct–Jan; Region D, Mar–June, Dec; Region E, May, July–Aug (breeds during rains); Malaŵi, Oct–Mar; Zambia, Sept–Nov; Zimbabwe, Sept–Mar, mainly Oct–Dec (Sept 1, Oct 31, Nov 44, Dec 23, Jan 12, Feb 1, Mar 2 clutches); Mozambique Oct–Jan (breeding condition Mar); South Africa: Natal, June, Sept–Jan, mainly Oct–Dec.

INCUBATION: begins with laying of first egg; by ♀ only, fed by ♂ on nest. Period: 14 days. Incubating bird sat close, periods off nest brief, longest 10 min; hopped off when nest inspected by man but remained in nest tree, protesting loudly, flirting tail, opening and closing wings, raising and lowering crest; returned to nest when man left (van Someren 1956). Another ♀ was more timid, slipped off nest at person's approach, sat tight only when eggs about to hatch, then on leaving nest feigned injury – tumbled off nest, lay fluttering on ground, and crept away with one wing drooping; returned to nest as soon as man left (van Someren 1956).

DEVELOPMENT AND CARE OF YOUNG: feathers grow rapidly: at day 14, 2 young overflowed nest and ♀ could barely cover them with wings when brooding them at night. Chicks brooded closely by ♀ for 2 days after hatching; ♀ fed by ♂ while brooding. On day 3, ♂ fed chicks, then brooded them while ♀ went for food; ♀ returned and fed them, then both left nest. For first 3 days chicks fed with larvae only, then both larvae and berries (♀ brought only larvae, ♂ only berries); fruits with hard seeds had seeds removed before being fed to young. On day 16 young able to stand up, stretch and preen; induced to leave nest by parents with berries who called to them from branch 0·5 m from nest. Young at first remained in nest and answered parents with call, a low, double 'cherup, cherup'; then hopped and flapped out of nest and onto branch. On subsequent feeds, young tempted further away from nest (van Someren 1956). Juvenile accompanied by both adults was fed by ♀ with red berries (Vincent 1935).

References
Maclean, G. R. (1985).
van Someren, V. G. L. (1956).
Vincent, J. (1935).

Chlorocichla falkensteini (Reichenow). Yellow-necked Greenbul; Falkenstein's Greenbul. Bulbul de Falkenstein.

Plate 20 (Opp. p. 321)

Criniger Falkensteini Reichenow, 1874. J. Orn. 22, p. 458; Chinchoxo, Loango Coast.

Range and Status. Endemic resident. 4 apparently disjunct populations: (1) S Cameroon (north to Kumba, Yaoundé, Bitye, east to Nguilili) and N Gabon (south to about equator); (2) near Bamingui, Central African Republic; (3) S Congo, Angola (Cabinda), extreme W Zaïre (lower Congo R.); (4) W Angola from Cuanza Norte (Uige) south along escarpment to Lubango (Sa da Bandeira). Locally common to rare; in Gabon density up to 12–15 birds/10 ha in some areas, rare elsewhere.

Description. ADULT ♂: top and sides of head, including cheeks and ear-coverts, and entire upperparts, including scapulars and upperwing-coverts, yellowish olive-green, slightly brighter on rump; lores and pre-orbital region grey. Tail-feathers olive-green with brighter green edging, shafts dark maroon above, straw below. Chin and throat sulphur yellow, extending as small wedge from lower throat onto sides of neck; rest of underparts mainly grey, lower belly and centre of upper belly creamy, sometimes with a few yellow feathers in centre, thighs washed olive-yellow, undertail-coverts pale grey to pale brown, often washed olive-yellow. Primaries and secondaries dark brown, edged yellowish olive-green, more broadly inwardly, outer webs of inner secondaries entirely yellowish olive-green, inner webs duller olive-green; inner webs of basal half of primaries and all of secondaries with pale yellow edging above and below, brighter and more extensive below, axillaries and underwing-coverts sulphur yellow. Bill black to dark greenish grey; eye deep brick red, blood red, crimson or orange-red; legs and feet blue, slate-blue or blue-grey. Sexes alike. SIZE: (8 ♂♂, 7 ♀♀) wing, ♂ 83–93 (89·3), ♀ 83–90 (86·0); tail, ♂ 77–84 (79·9), ♀ 73–85 (78·0); bill, ♂ 18–21 (19·1), ♀ 17–19 (17·9); tarsus, ♂ 20–23 (21·0), ♀ 19–21 (20·5). WEIGHT: Cameroon, 1 ♂ 29, ♀ (n = 3) 29–31 (30·0).

IMMATURE: plumage like adult, eyes light brown, gape yellow.

NESTLING: unknown.

Field Characters. A skulking bulbul of thick brush outside forest, with bright green upperparts, yellow throat, grey underparts, and gravelly voice. Yellow-bellied Greenbul *C. flaviventris* has greenish upperparts (although browner than Yellow-necked) but underparts entirely yellow; race *flavigula* of larger Yellow-throated Leaf-love *C. flavicollis* has yellow throat but brown upperparts and underparts; Simple Greenbul *C. simplex* has white throat, brown upperparts.

Voice. Tape-recorded (53, C, ERA). Song, phrases of 6–10 notes, 1–2 loud and up- or down-slurred, others brief and clipped: 'kip-kop-ko-KWEE-ko-witawit-kup', 'kip-kop-ko-KWEER-ko-KWAY'. Burry, nasal quality very similar to that of Yellow-bellied Greenbul and Simple Greenbul. Chattering calls also similar.

General Habits. Inhabits dense vegetation, including copses and small isolated forest patches, and early stages of regenerating forest in cutover areas, but not in forest proper except in clearings and around villages. Readily occupies man-made habitats, especially those overgrown with *Musanga* and thick brush (e.g. Solanaceae): neglected orchards, banana plantations, coffee forest, fallow fields, abandoned villages, even gardens (Zaïre, Kinshasa).

Occurs in pairs or family parties; occasionally joins mixed flocks of insectivores. Forages among leaves of bushes and trees of medium height, often hidden and seldom seen despite frequent calls. Territorial all year.

Sedentary in Gabon.

Food. Berries and fruits: *Trema, Solanum, Rauwolfia, Ficus*, small peppers; 'large berries, possibly coffee' (Angola: Hall 1960); also caterpillars and termite alates.

Chlorocichla falkensteini

Breeding Habits. Territorial.

NEST: only 1 described (Bates 1909). Shallow cup of shreds of bark and leaf stems, lined with fine grasses, in fork of bush; int. diam. 55 × 65. Built by ♀? (bird carrying nest material was accompanied by presumed ♂, singing: Brosset and Erard 1986).

EGGS: 2. Elongated pointed ovals; slightly glossy; pale greenish to pale stone, with irregular lines and scrolls of umber brown over olive and grey, giving marbled appearance, concentrated around large end. SIZE: 22–25·5 × 16–17.

LAYING DATES: Cameroon, May, July; Gabon, Nov (flying young June, carrying nest material Mar, song intensifies Sept–Feb).

INCUBATION: by ♀ only.

DEVELOPMENT AND CARE OF YOUNG: young just out of nest fed caterpillars by both parents (Brosset and Erard 1986).

Reference
Brosset, A and Erard, C. (1986).

Plate 20
(Opp. p. 321)

Chlorocichla simplex (Hartlaub). **Simple Greenbul. Bulbul modeste.**

Tricophorus simplex Hartlaub, 1855. J. Orn. 3, p. 356; Rio Boutry, Gold Coast.

Range and Status. Endemic resident. Guinea-Bissau (Catio), coastal and S Guinea, Sierra Leone (fairly common), Liberia (locally abundant, from coast to Mt Nimba), Ivory Coast (north to c. 10°N, fairly common), Ghana (common, north to Mole Game Reserve), S Togo, S Benin, Nigeria (common, north to Ilorin, Enugu, Serti, Gashaka-Gumti Game Reserve), Cameroon (common, north to Rumpi Hills, Kumba and Tombel), Gabon, Congo, Zaïre (widespread except SE, east to base of E highlands), SW Central African Republic, extreme S Sudan (Bangangai Forest), Uganda (Bwamba) and N Angola (south to Cuanza Norte, N Lunda (Dundo)).

Common to locally abundant; density in Gabon 4–6 pairs/10 ha (Brosset and Erard 1986); in Cameroon, 5 birds/30 ha (Decoux and Fotso 1988).

Description. ADULT ♂: entire top of head to hindneck, cheeks and ear-coverts dark brown, loral area slightly blacker, cheeks and ear-coverts with indistinct narrow pale streaks, broken eye-ring white, broader above than below eye, each part shorter than in other bulbuls. Upperparts, including scapulars and upperwing-coverts, olive-brown, slightly paler on rump; tail-feathers brown with olive margins, outer 2 pairs with narrow pale tips when fresh, shafts maroon above, straw below. Chin and throat white, white just extending onto sides of neck. Breast and flanks pale olive-brown, lower flanks washed cinnamon; belly creamy white, sometimes with pale yellowish wash; thighs and undertail-coverts pale cinnamon. Flight feathers dark brown, outer webs edged olive-brown, more broadly inwardly, basal half of inner webs edged paler, shafts maroon above, straw below; axillaries, underwing-coverts and inner margins of underside of flight feathers cinnamon. Bill black to greenish black, cutting edges grey; tongue pale yellow; eye dark brown; legs and feet dark grey to bluish or slate. Sexes alike, but ♀ eye deep red-brown. SIZE: (12 ♂♂, 12 ♀♀) wing, ♂ 97–109 (103), ♀ 91–101 (95·5); tail, ♂ 87–100 (92·7), ♀ 84–92 (87·8); bill, ♂ 19·5–23 (22·3), ♀ 19–21·5 (20·3); tarsus, ♂ 22·5–26·5 (25·0), ♀ 20–26 (24·1). WEIGHT: ♂ (n = 8) 40·7–54·5 (47·8 ± 4·6), ♀ in breeding condition (n = 4) 40·4–49·2 (45·1).

IMMATURE: unknown.
NESTLING: inside of mouth flesh-red, gape-flanges yellow.

Field Characters. A dull brownish bulbul of brushy areas outside true forest; parties skulk in undergrowth – usually located by chatter. White throat and broken eye-ring (especially upper half) conspicuous in the field. Distinguished from white-throated populations of Yellow-throated Leaf-Love *C. flavicollis* by eye-ring, cinnamon undertail-coverts and underwing, more secretive habits and different voice, and from brown forest bulbuls by non-forest habitat.

Voice. Tape-recorded (53, CANE, CART, ERA, GRI, KEI). Song, a variable phrase of 8–12 notes, alternately sharp and burry, lasting 3–4 s; starts low and gets louder, third note from the end typically highest and loudest: 'wido, wida-wit-do, wida-KWIT-to-kwair'; final note 'kwair' has a peculiar nasal quality very like that of Yellow-bellied Greenbul *C. flaviventris*. Basic call a chattering 'kurru-kurru-kurru . . .', lower-pitched than that of Yellow-throated Leaf-love, less sharp, and with a burry quality, probably used as contact call by members of pair or group; other calls include a nasal 'tjcaaa', low 'chukudu' and sharp 'chichikak'.

General Habits. Inhabits brushy areas outside true forest, including dense shrub in savanna, orchard bush, coastal and swampy thickets, gallery forest and forest edge; occupies regenerating forest on old cultivation and undergrowth in open secondary forest, and readily adapts to cultivated areas, fallow fields, recently cleared land, roadside brush and other man-made habitats, spreading along forest roads to colonize villages. Lowlands, up to *c*. 1200 m.

Occurs in pairs or small family groups; sometimes joins mixed-species flocks. Shy and furtive, skulking in foliage and located mainly by voice, but sometimes emerges to perch in the open. Territorial at all seasons, reacting strongly to playback of recorded voice.

Sedentary in Gabon.

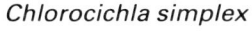

Food. Fruit of *Solanum*, *Trema*, *Rauwolfia*, palms, red peppers; seeds; insects (caterpillars, termite alates); spiders.

Breeding Habits. Territorial. Defends area of 1–2 ha.
NEST: small, shallow cup, thin but strong and compact, of twigs, rootlets, plant stems, dry grasses, dry vines and other fibrous material, bound with cobwebs, unlined (**A**); ext. diam. 80–110, int. diam. 50–70, depth

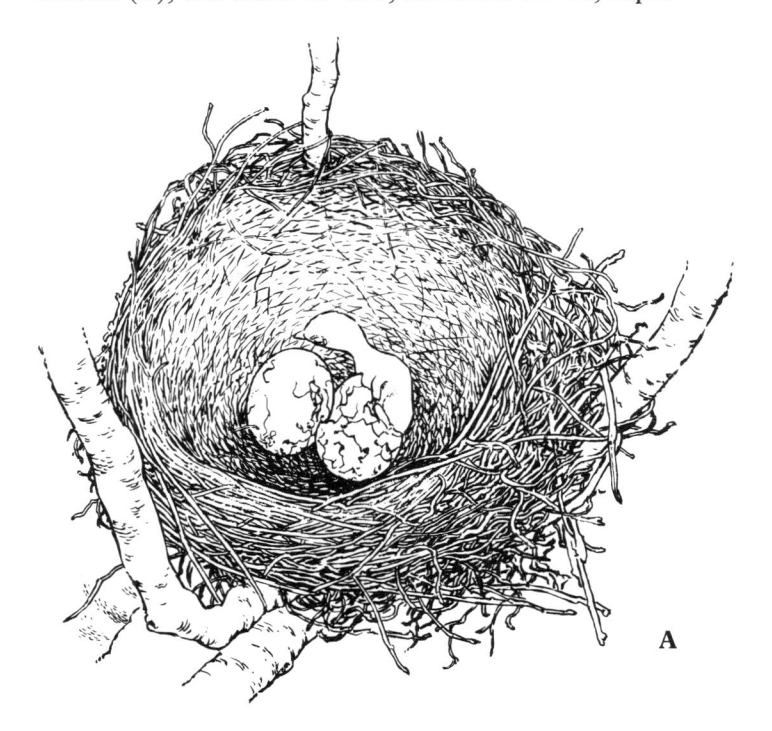

80–100, depth of cup 40; 'a thinner version of *Pycnonotus barbatus* nest' (Brosset and Erard 1986). Placed 0·9–2 m above ground in fork of dense clump of low bushes, saplings or shrubs, wild or cultivated; of 12 nests, Gabon, 8 were in *Solanum*, 1 in *Trema*, 1 in *Harungana*, 1 in manioc, 1 in ornamental hedge around huts (Brosset and Erard 1986); elsewhere in cassava and *Triumfetta*.

EGGS: 2 (32 clutches), sometimes 1 (2 clutches). Regular to long ovals; glossy; creamy white to beige, pale olive or pale reddish grey, with spots and scrolls of dark red-brown and sepia over pale grey, violet-grey or pale mauve: 'like eggs of *Emberiza cia*' (Brosset and Erard 1986); sometimes unspotted (Prigogine 1961). SIZE: (n = 21) 22–26 × 16–19.

LAYING DATES: Sierra-Leone, Mar; Liberia (breeding condition Feb, July–Sept); Ghana, Feb, Apr (building nest July, Dec); Nigeria, Jan, June–Sept; Cameroon, Jan, Mar, June–Sept, Dec; Gabon, Sept–Feb, once July in unusual year (Sept 1, Oct 2, Dec 1, Jan 6, Feb 2 clutches); Zaïre, Aug, Oct–Mar.

INCUBATION: by ♀ only. Period: 14 days. ♀ slips quietly off nest on approach of person.

BREEDING SUCCESS/SURVIVAL: 3 of 24 eggs infertile, 7 of 11 nests robbed (Gabon: Brosset and Erard 1986).

Reference
Brosset, A. and Erard, C. (1986).

Plate 20
(Opp. p. 321)

Chlorocichla flavicollis (Swainson). Yellow-throated Leaf-love. Bulbul à gorge claire.

Haematornis flavicollis Swainson, 1837. Birds of West Afr. 1, p. 259; W Africa.

Range and Status. Endemic resident. S Senegal, Gambia, Guinea, N half Sierra Leone, N Ivory Coast (common south to Toumodi), N Ghana, Togo, Benin; broadly across Nigeria from Zaria, Birnin Kudu and Kari south to Abeokuta, Ibadan, Enugu and Serti, avoiding drier areas in N and coastal and forest areas in S. Central and S Cameroon (north to Adamaua Plateau), Gabon, Congo, SW Central African Republic, Zaïre (throughout except heavily forested central region), Angola south to Huambo and Moxico; widespread in N and W Zambia, south to c. 14°S in central areas and to Senanga in W, but not in Luangwa valley. Rwanda, Burundi, extreme S Sudan, W-central Ethiopia from Didessa R. valley, Angar, Gutin, Nekempte and Argio to Dembidollo, Bure and Gilo R. at Mungum; central and S Uganda north to Bunyoro and Mbale; humid areas of W Kenya west of Rift from L. Victoria inland to 2000 m, east to Soy near Eldoret; N and NW Tanzania around L. Victoria and south to Kibonde. Absent from Malaŵi (Benson 1947, Benson and Benson 1977; cf. Chapin 1953, Hall and Moreau 1970). Common to locally abundant.

Description. *C. f. flavigula* (Cabanis) (including '*pallidigula*'): Angola, SE Zaïre (north to S Kasai, S Maniema) and E Zaïre (Kivu) to Kenya, Tanzania, Zambia. ADULT ♂: forehead to nape and sides of head including cheeks and ear-coverts greyish olive, feathers of top of head with darker centres giving slightly scaly effect, especially on forehead and forecrown. Hindneck to rump, scapulars and upperwing-coverts greenish olive, back feathers shading to grey at tips; tail-feathers somewhat darker and browner, edged greenish olive, shafts dark brown to dark maroon above, straw-coloured below. Chin and throat sulphur yellow, variable, brighter in E of range; rest of underparts variable, brownish or greenish olive, greyer in W of range; each breast feather with white shaft bordered by variable stripe of pale grey, more pronounced in E of range, feather tips grey, paler in E. Variable-width stripe up centre of belly pale brown to creamy or pale sulphur; undertail-coverts medium to light brown, paler toward tips. Primaries and secondaries dark olive brown edged greenish olive, more broadly inwardly; axillaries and underwing-coverts light yellowish brown, underside of bend of wing yellow-ochre. Bill black or greyish black; eye grey-buff; legs and feet grey or dark grey, faintly tinged blue or olive. Sexes alike, but eye of ♀ grey-white, and some ♀♀ have paler, creamy white throat. '*Pallidigula*' (Zaïre to Kenya) has deeper yellow throat and browner underparts in E. SIZE: (12 ♂♂, 10 ♀♀) wing, ♂ 100–113 (108), ♀ 98–113 (103); tail, ♂ 94–109 (101), ♀ 91–105 (95·9); bill, ♂ 20–23 (21·5), ♀ 20–23 (21·3); tarsus, ♂ 24–27 (25·6), ♀ 24–27·5 (25·4). WEIGHT: Zaire, ♂ (n = 5) 49–60 (53·4), ♀ (n = 2) 38, 47, immature ♂

(n = 2) 45, 53, 1 immature ♀ 48; Uganda and Kenya, ♂ (n = 5) 46–58 (50·9), 1 ♀ 37, unsexed (n = 17) 42–53 (47·3 ± 3·0).

IMMATURE: very like adult but darker and browner, throat whiter, tail-feathers narrower and more rounded at tip.

NESTLING: inside of mouth orange, gape-flanges yellow-white.

C. f. soror Neumann (including '*simplicicolor*'): N-central Cameroon (Tibati) to Congo, Zaïre (south to N Kasai, forested Maniema), Central African Republic, Sudan, Ethiopia. Throat white with only faint tinge of sulphur; underparts paler and greyer than other races. Crown and upperparts greyer in N (Cameroon), browner in S (Congo). WEIGHT: unsexed (n = 7, Ethiopia) 37·7–45·8 (41·7).

C. f. flavicollis (Swainson): Senegal to N Cameroon (Adamaua Plateau). Throat bright yellow; underparts olive-brown with scattered yellow streaks on centre of breast and belly, darker, browner and more uniform than *soror*, usually not paler on belly (though some are); lower belly and undertail-coverts brown, as dark as rest of underparts, not pale.

TAXONOMIC NOTE: in N Cameroon, yellow-throated *flavicollis* on Adamaua Plateau separated by only 50 km from white-throated *soror* at Tibati; no intergrades are known and they may meet as separate species (Rand 1958).

Field Characters. A large bulbul of thick vegetation away from forest, moving about in noisy, chattering parties. Puffs out yellow or white throat, which contrasts with darker head and underparts. One of the few non-forest bulbuls, likely to be confused only with white-throated Simple Greenbul *C. simplex* where it overlaps range of *soror* (for differences, see Simple Greenbul).

Voice. Tape-recorded (53, 86, C, HOR, KEI, MOR, MOY). Usual calls are a loud, nasal, twangy 'chow' or 'kyow', fairly high-pitched, and a more abbreviated 'kyip' or 'chik'. Given by group members while foraging, doubtless for contact; several birds often call together. 'Chow' combines with shorter notes to form a loose 'song', e.g. 'chow, chow, ka-chika-chow, chow', 'chitto, ka-chika chow, chow, chit'. These notes all have a harsh, grating quality; the 'thrush-like warbling' of Mackworth-Praed and Grant (1952) is probably an error.

General Habits. Occurs mainly in brushy and wooded habitats away from forest: clumps of bushes and low trees along streams and rivers; thickets, vine tangles and other low, dense vegetation; orchard bush, gallery forest, open woodland. Penetrates second growth, but not in true forest except around clearings; avoids large forest blocks. Adapts to man-made habitats of all kinds, including plantations, abandoned cultivation, *Lantana* scrub, orchards, parks, wooded gardens, even eucalyptus. Prefers mango trees in Sudan, where known as Mango Bulbul. In Rwanda, regularly in papyrus (Vande weghe 1981). Lowland and intermediate elevations, up to 1800 m in Cameroon and 2000 m in Zaïre.

Occurs singly, in pairs or (more often) in family groups. Forages low down in bushes and on ground, but rises to middle levels where trees taller. Rather shy and usually well hidden, but emerges from vegetation from time to time to look around. Easily located by continuous chattering. Scolds intruders; flicks wings and puffs out throat. Party members often move in succession, one landing on perch just vacated by another. Sometimes joins mixed-species foraging flocks.

Undergoes 2 complete moults Dec–July, covering breeding period (Zaïre, Upemba Nat. Park).

Sedentary in Gabon.

Food. Fruit of *Solanum*, *Trema*, *Rauwolfia*, guava, mango, red pepper; also some insects (caterpillars, beetles).

Breeding Habits. Territorial. Probably a cooperative breeder (Elgood 1982). Groups of 3 (not necessarily a pair and a young bird) jointly defend territories (Brosset and Erard 1986). ♂ and ♀ chase each other in display.

NEST: cup of small twigs, rootlets, leaf stems and dry leaves, bound by cobwebs; one nest in garden was lined with casuarina needles. Situated 2–10 m above ground in tree fork formed by small branches or junction of palm leaves, or slung between 2 prongs of fork on horizontal branch, usually in thick-foliaged tree such as lime, mango or fig. Readily nests in garden or compound, even near buildings. Nest built by 1 bird, accompanied by its mate (♂?); material collected from under bush while mate watched; both birds noisy, following each other from tree to tree; later presumed ♂ remained in tree and drove away all birds approaching nest; construction complete in *c.* 10 days; first egg laid 2 days later (Uganda: Jackson and Sclater 1938).

EGGS: 2 (4 clutches). Pointed ovals: slightly glossy; pale creamy white or dirty white to pale buff or pale pink with spots and blotches of dark to light brown and red-brown over dark grey, lilac-grey and pale purple, concentrated at large end. SIZE: (n = 4) 24–25 × 16·5–17 (24·5 × 16·7).

LAYING DATES: Gambia, July, Nov–Dec; Mali, Nov (unconfirmed reports in rainy season, June–Sept); Nigeria, May–Aug; Gabon (young out of nest, Oct, singing increases Aug–Feb); Congo, Dec; Zaïre, Dec–July (breeding condition Sept–Oct); Rwanda, Oct–Dec; Angola (breeding condition Jan, Aug); Sudan, Sept; E Africa: Uganda, Apr–June; Region B, Dec–June, Aug–Sept (with peaks both in long rains and dry season); Zambia, Sept–Oct.

DEVELOPMENT AND CARE OF YOUNG: after leaving nest young accompanied by both parents. Parent protects threatened nestling by silently dropping to ground and fluttering off, flopping about with wings extended. Later it scolds from nearby bush. Second parent present while other gives distraction display.

BREEDING SUCCESS/SURVIVAL: parasitized by Red-chested Cuckoo *Cuculus solitarius*.

Reference
Jackson, F. J. and Sclater, W. L. (1938).

PYCNONOTIDAE

Genus *Thescelocichla* Oberholser

Bill similar to *Chlorocichla*; legs and feet stout; tail long and graduated, T4–T6 with broad white tips; face grey, streaked and spotted white; breast grey with white streaks, rest of underparts yellowish white. Inhabit swampy forest and second growth, typically with palms. Forage in noisy groups which often chatter in chorus; voice loud and raucous.

Endemic; monotypic.

Plate 21
(Opp. p. 352)

Thescelocichla leucopleura (Cassin). Swamp Palm Bulbul; White-tailed Greenbul. Bulbul des raphias.

Phyllostrophus leucopleurus Cassin, 1856. Proc. Acad. Nat. Sci. Phil., 1855, 7, p. 238; Moonda R., Gabon.

Range and Status. Endemic resident, sedentary and nomadic. Single records W Gambia (Abuko Nature Reserve) and SW Senegal (Basse Casamance); Guinea Bissau, SE Guinea, Sierra Leone, Liberia (widespread), Ivory Coast (abundant, north to 9° 20′N), Ghana (widespread, locally common), Togo, S Benin, Nigeria (locally common, north to Jos Plateau), Cameroon (common in S, north to Rumpi Hills), Gabon (locally abundant), Congo (widespread); Angola (Cabinda, common); SW Central African Republic; in Zaïre, most of lowland forest areas, extending into savanna in gallery forest, in E from near Sudan border south to base of Itombwe Highlands; W Uganda (Bwamba). Generally common and widespread.

Thescelocichla leucopleura

Description. ADULT ♂: top and sides of head brownish olive to greyish olive, feathers of top of head fringed black giving mottled effect. Upperparts, including upperwing-coverts, olive brown, warmer brown on rump and uppertail-coverts. Tail-feathers brown, edges washed olive, outer 4 pairs with broad white tips increasing in depth outwardly, from 15 on T3 to 32 on T6; shafts red-brown above, cream below. Feathers of lores, eye-ring, cheeks, ear-coverts and around base of bill with white shafts and centres, giving streaked and spotted effect to whole facial area; some birds have short white streak extending back from over eye. Chin and throat off-white; breast feathers olive with white shafts; rest of underparts, axillaries and underwing-coverts creamy white to pale creamy yellow. Primaries and secondaries brown, undersides with pale sheen. Bill black, lower mandible paler, tinged greyish or bluish; eye brown, red-brown or reddish (in Bwamba, Uganda, eye whitish to cream with pale pinkish tinge: Friedmann and Williams 1971); legs and feet dark grey-brown, grey or bluish grey. Sexes alike. SIZE: (12 ♂♂, 12 ♀♀) wing, ♂ 110–117 (114), ♀ 103–111 (106); tail, ♂ 97–107 (102), ♀ 91–102 (96·9); bill, ♂ 22–25 (23·4), ♀ 21–25 (23·0); tarsus, ♂ 23–28 (25·6), ♀ 23–26 (24·8). WEIGHT: Liberia, ♂ (n = 2) 60·6, 67·4, 1 ♀ 58, 1 ♀ in breeding condition 59; Cameroon, 1 ♀ 60; Uganda, ♂ (n = 13) 55–70 (62·5).

IMMATURE: browner above, throat and breast pale yellow, tail-feathers more pointed.

NESTLING: unknown.

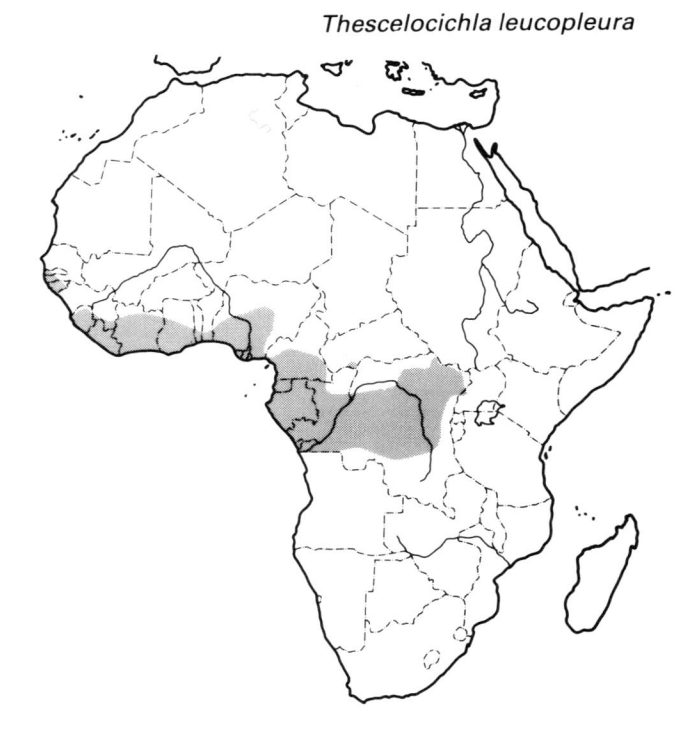

Field Characters. A large bulbul with creamy belly and much white in tail; travels in noisy groups which keep up continual raucous chatter. Only other bulbuls with white in tail are much smaller Spotted Greenbul *Ixonotus guttatus*, which has outer tail-feathers all white, short bill, entire underparts white, white spots on wing and rump, and honeyguide greenbuls *Baeopogon* which are smaller and plumper, with dark underparts; all have different calls and habitat.

Voice. Tape-recorded (40, 53, B, ERA, GRI, KEI). Foraging flock gives unstructured babbling; low, grating notes begin almost in an undertone, then increase in volume and become interspersed with higher, shriller ones, e.g. 'kokoko . . . eeko-RIT-toto-TREE-tu-TRIT-toro'; shorter phrases frequent; most notes are brief and clipped, lower ones have buzzy, nasal timbre. When one bird starts to call, it often sets off others, resulting in group chorus. These bursts end abruptly, are often followed by 10–15 s of silence before next bird starts up; the silent interlude may be punctuated by a few desultory chuckles and sharp notes. Song is simply a more formal, structured arrangement of these same notes, repeated by single bird from crown of palm or mid-level perch in forest. It is sometimes in 2 parts, with a slight pause in between: 'ti-tu-to-TRIT-to . . . ti-to-REE-to-RAA'; songs tend to work up to a climax, last notes loudest. Scolding call is a rapid, loud, insistent nasal chatter, all on one pitch: 'cha-cha-cha-cha-cha . . .'. Loud rattle 'chack-ack-ack' accompanies intraspecific chases (Willis 1983).

General Habits. Prefers swampy areas with palm trees, especially *Raphia*, also oil palm *Elaeis*, both inside and outside forest. Also found in drier parts of primary and secondary forest, especially around clearings, and in degraded edge situations, e.g. sparse second growth mixed with elephant grass; sometimes in *Pandanus*. Extends outside forest bloc into gallery forest and forest–grassland mosaic; adapts well to man-made habitats – cacao plantations, stands of bananas and other fruit trees, gardens, mango trees around old villages. Lowland forest, up to 1050 m in Cameroon and 1110 m in Zaïre.

Travels in noisy bands of 3–8, sometimes 10–12; minimum social unit 3 adults. Members maintain contact with babbling chatter; group defends territory by loud chorusing, accompanied by displays of spread wings and tail. Territory size varies from a few ha in marshy area inside forest to 12–15 ha in cutover areas and forest edge (Gabon: Brosset and Erard 1986). In Gabon apparently never joins mixed-species flocks, but in Zaïre joins flocks when they move through its habitat (Prigogine 1971), and in Uganda sometimes joins up with Leaf-love *Pyrrhurus scandens*, which also travels in noisy groups, the resulting din being considerable (van Someren and van Someren 1949). Forages at all levels from ground to canopy, moving freely from one to another; in canopy fond of vine tangles in openings created by tree falls; clings to trunks and gleans bark; probes into ferns and other plants growing on branches; takes prey from trunks, foliage and vines while hovering; seen fluttering about on path feeding on column of black ants (Malbrant and Mclatchy 1952). In Gabon follows swarms of ants *Dorylus wilverthi* in vegetation, including canopy (Willis 1983). Restless, constantly on the move. Mobs snakes, owls and hawks, e.g. *Accipiter tachiro*, often accompanied by squirrel *Funisciurus anerythrus* which shares same habitat (Brosset and Erard 1986). Not shy but often hard to see because hidden in thick vine and leaf tangles; loud calls give away its presence. Known as 'talky-talky bird' in Ghana, where children 'refuse to eat the flesh of these birds, fearing that if they do they will never stop talking' (Lowe 1937).

Partly nomadic. Appearance irregular in Bwamba forest, Uganda, where sometimes common, sometimes absent, and a similar situation noted in Nigeria (Bannerman 1936).

Food. Insects, including caterpillars, ants and emerging termites; fruit: *Ficus*, *Heisteria*, *Morinda*, *Musanga*, *Trichilia*; peeled palm nuts, royal palm berries, bananas; seeds.

Breeding Habits. Evidently a cooperative breeder: 3–4 birds came and gave alarm calls at nest on approach of observer; young leaving nest attended by 4 adults (Brosset and Erard 1986).

NEST: only 1 described: made of long strips of half-rotten leaves, 2 m above ground in banana plant in abandoned plantation (Brosset and Erard 1986); another (no details) in swamp palm (Bates 1930).

EGGS: 2 (1 clutch, and 2 nests with 2 young each). Ochre, spotted with brown-olive and grey-violet.

LAYING DATES: Liberia (breeding condition June); Ghana, Feb; Cameroon, Feb, May, July (breeding condition Oct), in rainy season; Gabon, Jan, Mar–Apr, Aug–Sept; Zaïre (breeding condition Nov).

INCUBATION: by ♀ only.

DEVELOPMENT AND CARE OF YOUNG: young fed by both parents, also by 4 other birds on leaving nest. In another case 2 individuals were carrying food accompanied by a third which was singing (Brosset and Erard 1986). Young fed with soft insects.

Reference
Brosset, A. and Erard, C. (1986).

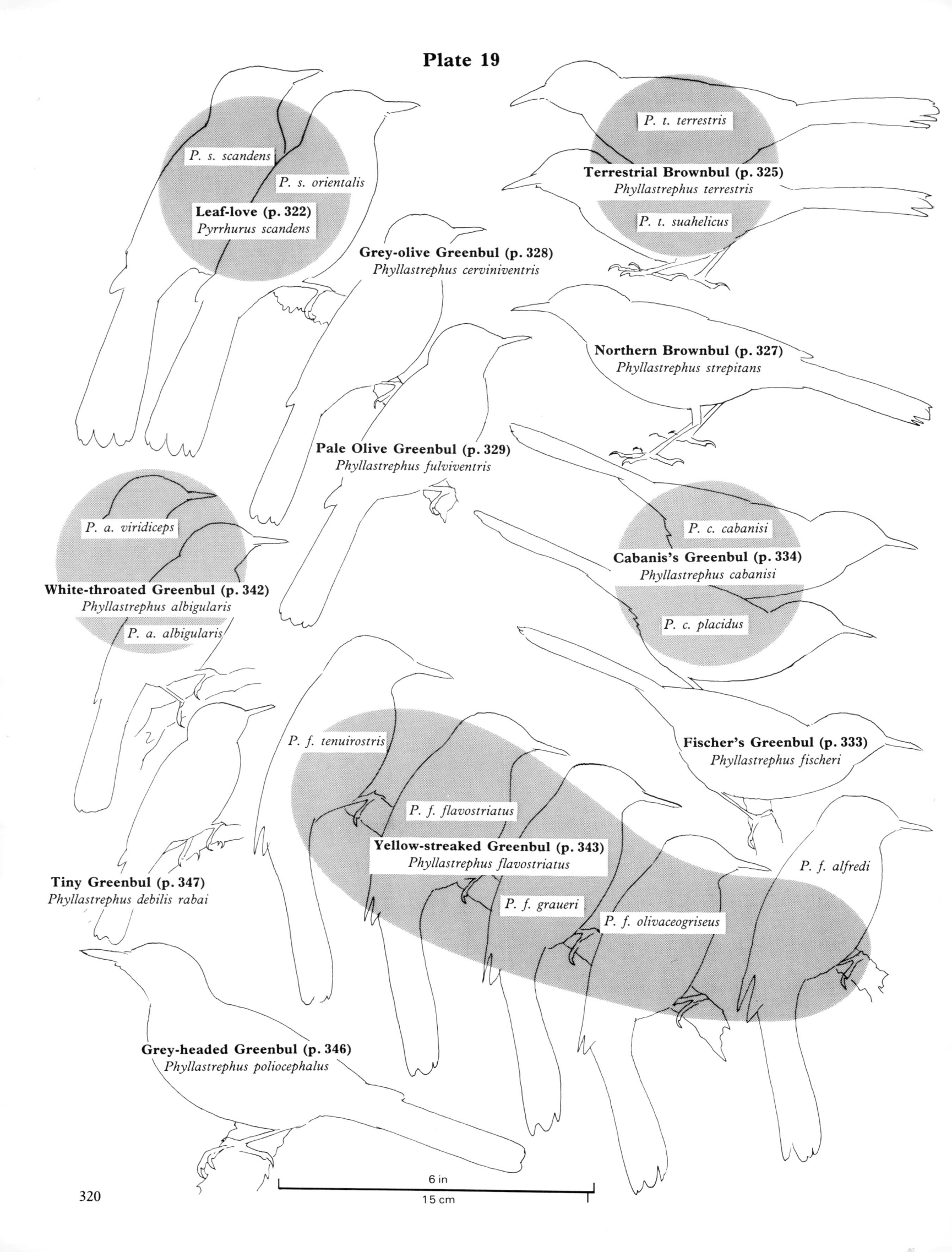

Plate 19

P. s. scandens
P. s. orientalis
Leaf-love (p. 322)
Pyrrhurus scandens

P. t. terrestris
Terrestrial Brownbul (p. 325)
Phyllastrephus terrestris
P. t. suahelicus

Grey-olive Greenbul (p. 328)
Phyllastrephus cerviniventris

Northern Brownbul (p. 327)
Phyllastrephus strepitans

Pale Olive Greenbul (p. 329)
Phyllastrephus fulviventris

P. a. viridiceps
White-throated Greenbul (p. 342)
Phyllastrephus albigularis
P. a. albigularis

P. c. cabanisi
Cabanis's Greenbul (p. 334)
Phyllastrephus cabanisi
P. c. placidus

P. f. tenuirostris
P. f. flavostriatus
Yellow-streaked Greenbul (p. 343)
Phyllastrephus flavostriatus
P. f. graueri
P. f. olivaceogriseus
P. f. alfredi

Fischer's Greenbul (p. 333)
Phyllastrephus fischeri

Tiny Greenbul (p. 347)
Phyllastrephus debilis rabai

Grey-headed Greenbul (p. 346)
Phyllastrephus poliocephalus

6 in
15 cm

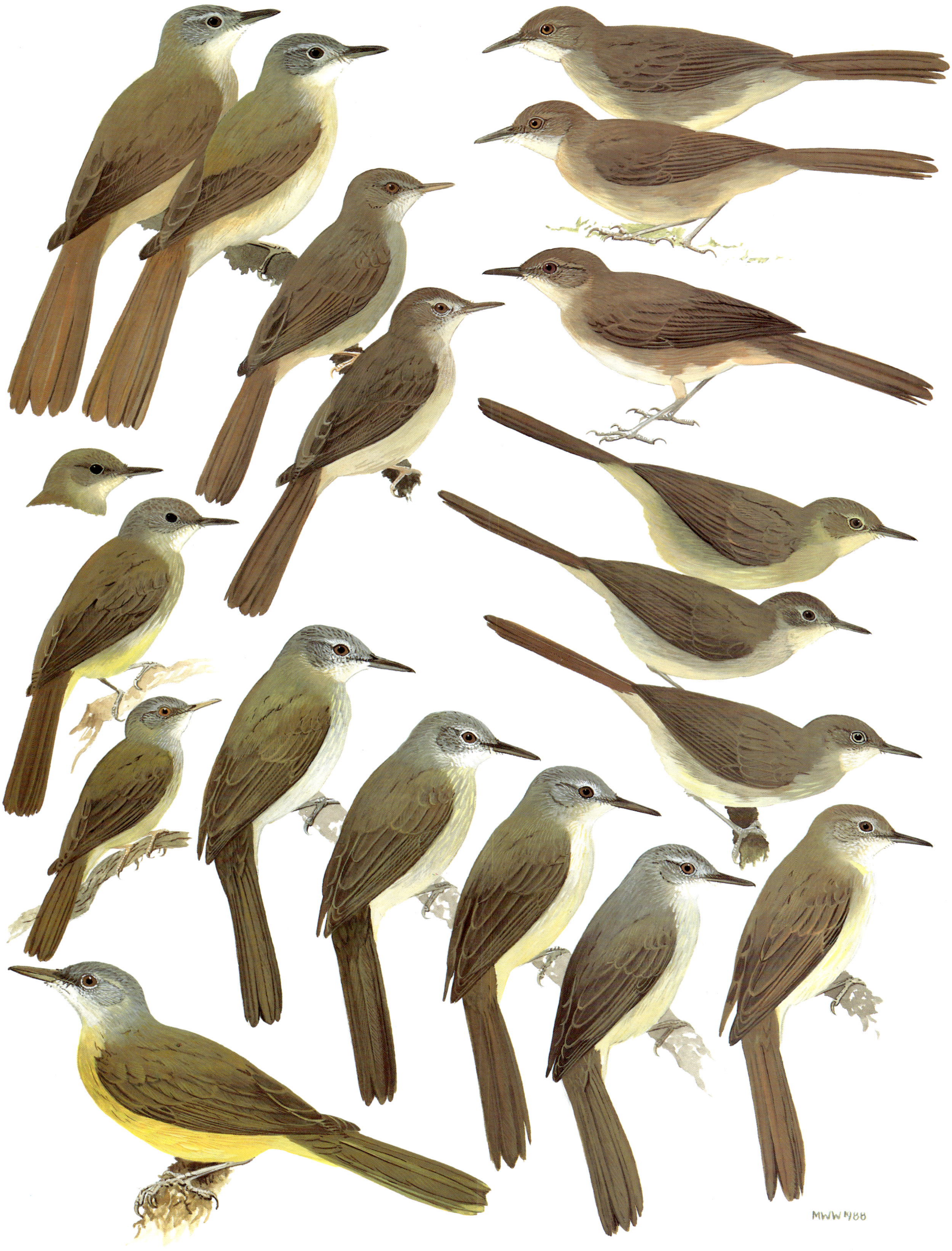
MWW 1988

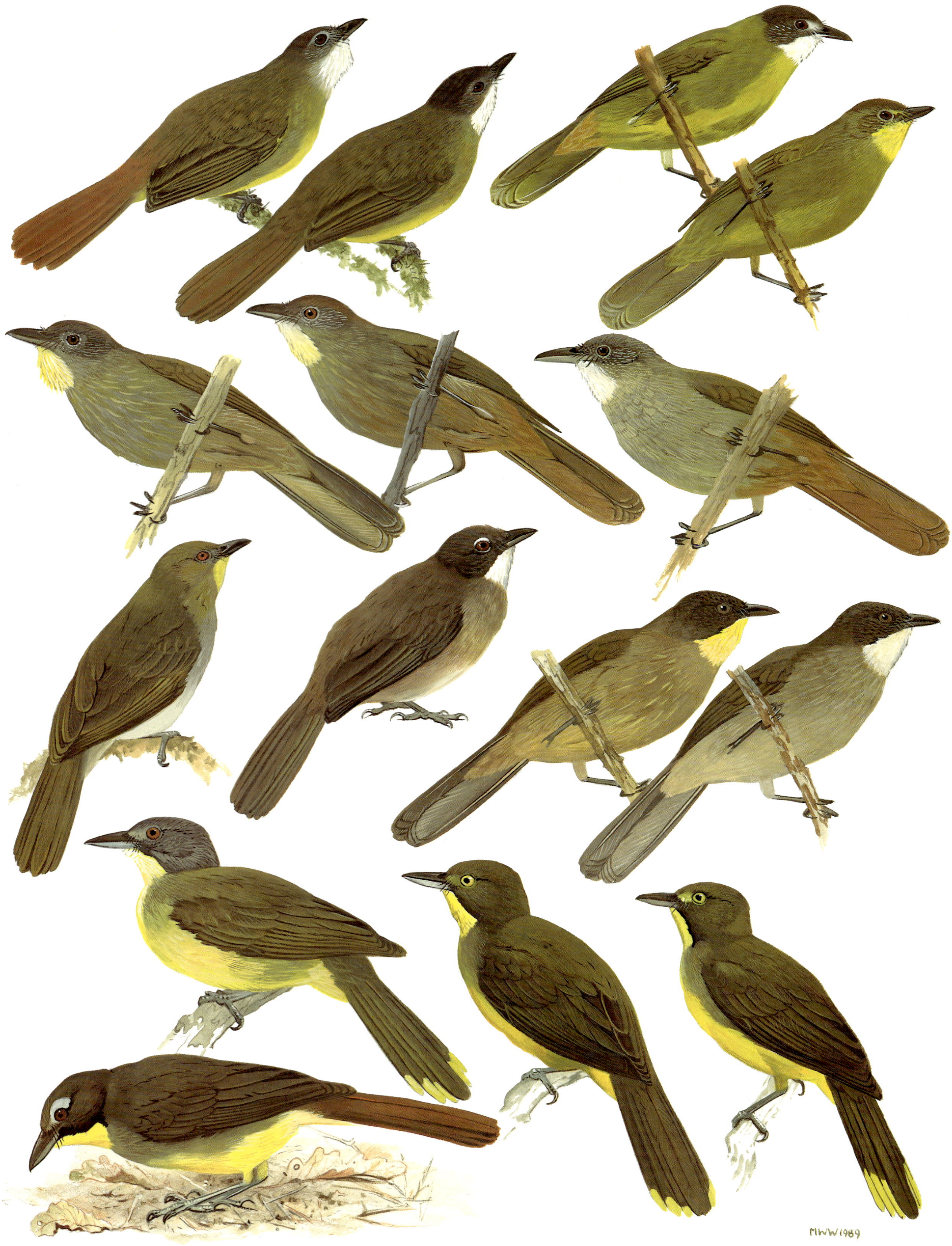

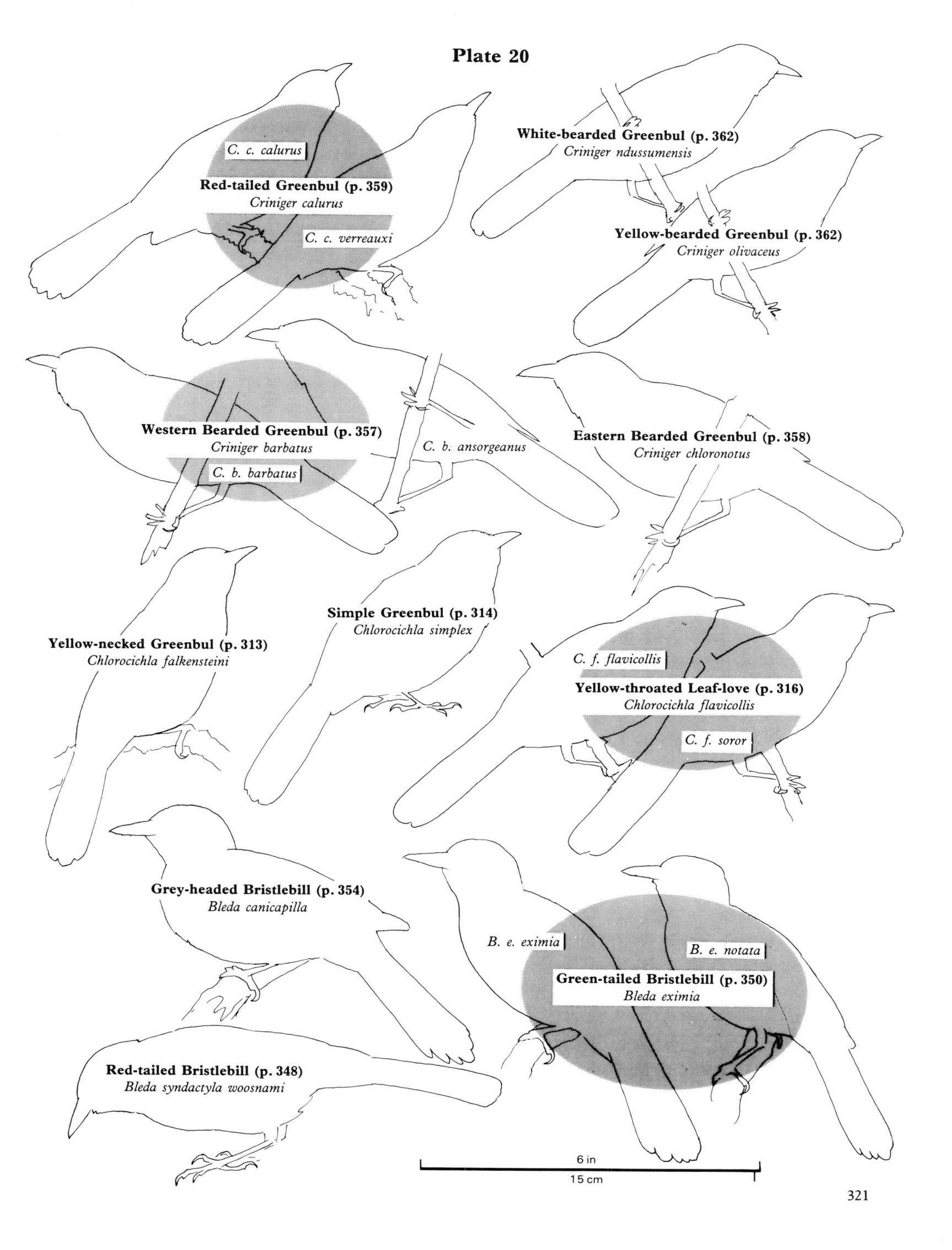

Plate 20

Red-tailed Greenbul (p. 359)
Criniger calurus
C. c. calurus
C. c. verreauxi

White-bearded Greenbul (p. 362)
Criniger ndussumensis

Yellow-bearded Greenbul (p. 362)
Criniger olivaceus

Western Bearded Greenbul (p. 357)
Criniger barbatus
C. b. ansorgeanus
C. b. barbatus

Eastern Bearded Greenbul (p. 358)
Criniger chloronotus

Yellow-necked Greenbul (p. 313)
Chlorocichla falkensteini

Simple Greenbul (p. 314)
Chlorocichla simplex

Yellow-throated Leaf-love (p. 316)
Chlorocichla flavicollis
C. f. flavicollis
C. f. soror

Grey-headed Bristlebill (p. 354)
Bleda canicapilla

Green-tailed Bristlebill (p. 350)
Bleda eximia
B. e. eximia
B. e. notata

Red-tailed Bristlebill (p. 348)
Bleda syndactyla woosnami

6 in
15 cm

321

PYCNONOTIDAE

Genus *Pyrrhurus* Cassin

Bill medium length, not deep but broad at base, not laterally compressed; rictal bristles strong. Frontal feathers stiff, scale-like; tail long, broad, rounded, bright red. Inhabit lower levels of forest. Foraging behaviour and voice very similar to *Thescelocichla* with which they often associate; travel in noisy groups, which often call in chorus; voice loud and raucous.

Usually placed in *Phyllastrephus*, but bill shape is much closer to *Thescelocichla* and *Chlorocichla flavicollis* which, voice and behaviour suggest, are its 2 closest allies, and its plumage is distinct. We follow Sclater (1930) in retaining *scandens* in *Pyrrhurus*.

Endemic to Africa; monotypic.

Plate 19
(Opp. p. 320)

Pyrrhurus scandens (Swainson). Leaf-love. Bulbul à queue rousse.

Phyllastrephus scandens Swainson, 1837. Birds of West Afr. 1, p. 270, pl. 30; W Africa.

Range and Status. Endemic resident. Extreme S Sudan (uncommon), W Gambia (common), SW Senegal (Santiaba Forest), Guinea-Bissau, N and E Guinea, NE Sierra Leone, N Liberia, SW Mali (uncommon) and presumably intervening parts of Guinea; Ivory Coast (throughout except extreme N); Ghana (not uncommon, north at least to Mole Game Reserve); Togo (north to Mo and Alejdo Forest where common); Nigeria (locally common, north to Kainji, Kaduna and Kagoro, also Obudu Plateau); Cameroon (widespread, north to Adamawa Plateau), Gabon (rare and local), S Congo (Mah), Central African Republic north to Bamingui-Bangoran Nat. Park. In Zaïre occurs away from main forest block, in W along Congo and Oubangui rivers, in S in scattered localities in Kwango, Kasai and Katanga; reappears in NE where fairly widespread from Beni north. Uganda (common Bwamba, elsewhere only sight records Mabira, Mukono and Kifu); extreme NE Rwanda (along Akagera R.), and W Tanzania (Ujiji, Basondo, Mahari Mt, Kibwesa). Locally common and range extensive but distribution fragmented, with many apparent gaps.

Description. *P. s. orientalis* Hartlaub (including '*acedis*', '*upembae*'): range of species except nominate race. ADULT ♂: forehead and crown grey, somewhat paler on forehead, shading to olive-grey on nape and hindneck, feathers with dark fringes giving slightly scaly effect; nape with a few black bristles. Mantle, back, scapulars and lesser coverts greyish olive-green, progressively paler and more tawny on rump and bright tawny-rufous on uppertail-coverts; feathers of lower back and rump long and fluffy. Tail pale rufous, outer feathers paler toward tips, shafts pale rufous above, straw below. Lores and preorbital area pale buff, sometimes joining in narrow pale band across base of upper mandible; eye-ring white; cheeks and ear-coverts light brownish grey with fine pale streaks; prominent black rictal bristles. Chin and throat white, breast and flanks mixed buff and pale grey, lower flanks washed pale cinnamon, lower breast and belly creamy white to whitish yellow, thighs and undertail-coverts pale cinnamon. Primaries and secondaries dark brown, edges of primaries and outer webs of secondaries greenish rufous, inner webs of inner secondaries paler, more rufous-brown; median and greater coverts and tertials brown; axillaries and most underwing-coverts pale yellowish cinnamon, under primary coverts darker and browner, inner webs of underside of primaries and secondaries margined pale cinnamon. Bill, upper mandible dark horn, olive-green or bluish grey, lower mandible whitish, pale grey or yellowish grey, cutting edges and tips pale grey or whitish; eye reddish chestnut, greyish brown, bluish grey or pale lemon-yellow, rim of eyelids light greenish; legs and feet blue-grey, pale pinkish grey or pale olive-green. Sexes alike. SIZE: (5 ♂♂, 9 ♀♀) wing, ♂ 100–109 (105), ♀ 91–108 (98·8); tail, ♂ 94–101 (98·0), ♀ 87–102 (92·2); bill, ♂ 22–24·5 (23·0), ♀ 20–22·5 (21·6); tarsus, ♂ 24–28 (26·2), ♀ 23·5–25·5 (24·7). WEIGHT: ♂ (n = 10) 43–53 (47·2), ♀ (n = 11) 33–51 (40·7).

IMMATURE: forehead to nape, back, rump and upperwing-coverts grey-olive washed russet, especially on back and rump; chin and underparts silky white, throat washed grey, flanks and undertail-coverts pale red-brown; tail-feathers narrower and more pointed, tips of primaries more rounded.

NESTLING: unknown.

P. s. scandens Swainson: Senegal to N Cameroon. Grey top of head with olive wash, upperparts browner olive, less grey, underparts darker and richer, more heavily washed cinnamon and pale yellow. Larger: wing, ♂ 100–118, ♀ 99–110; tail, ♂ 96–109, ♀ 92–108. WEIGHT: ♂ (n = 2) 60·6, 67·4.

TAXONOMIC NOTE: in '*acedis*' and '*upembae*', tail colour and amount of grey wash on breast are variable, and difference in colour of upperparts is very slight; we do not consider them valid.

Field Characters. A large, noisy, rather pale bulbul with grey head, rusty tail and buffy yellow underparts. Tail is brighter and paler than other rufous-tailed species and more conspicuous, since frequently fanned. Best identified by loud, raucous voice and group chorusing by foraging birds. Cabanis's Greenbul *Phyllastrephus cabanisi* is darker, with brown head, no buff on dull yellow underparts, very different voice. Voice quite similar to Swamp Palm Bulbul *Thescelocichla leucopleura* with which it often occurs, but latter very different-looking, with yellow-white underparts, white in tail.

Voice. Tape-recorded (32, 53, CHA, ERA, HOR, KEI). In Uganda (Bwamba), contact call between pairs or group members is low, hoarse but powerful, 1–3 notes, 'kawp', 'kawpa' or 'kawpakawp'; this regularly speeds up and becomes group chorus, during which high, sharp notes are added, 'chap' or 'chapachap'. Birds from Liberia (Mt Nimba) have rather different chorus; 1 or more birds keep up a constant sharp, high-pitched 'chip-chip-chip-chip-chip . . .', while others join in at intervals with hoarse 'karp', 'kurp' or 'koko' (S. Keith, tape and notes).

General Habits. Inhabits forest and thickets, with a distinct preference for water. In forest interior likes swampy places with luxuriant vegetation and palms (*Raphia*, *Phoenix reclinata*) or *Pandanus*; in clearings and at edge (especially along banks of rivers and streams) frequents dense liana tangles. Outside forest occupies forest–savanna mosaic, especially gallery forest, also isolated forest patches, thickets in woodland, orchard bush; forest strips in cultivation, thickets bordering open ground. Lowlands, to 1200 m.

Travels in pairs or small groups, which chorus to defend communal territory, like Swamp Palm Bulbul (Brosset and Erard 1986). Groups regularly join mixed-species flocks. Forages at all levels from ground to canopy; gleans foliage, opening up curled leaves; clings to tree trunks and branches, probing surfaces; clambers among vines and creepers; often hangs upside down under twig or leaf clump. Makes short sallies into air after flying termites. On ground searches exposed roots and matted clumps of dead leaves. Bathes in streams, dropping down from perch above water, splashing itself, flying back to perch to shake and preen, then dropping down for another dip.

Moult of primaries and tail-feathers begins with innermost and proceeds outwardly. 2 complete moults a year, May–June and Oct–Nov (Zaïre, Upemba Nat. Park: Verheyen 1953).

Sedentary, at least in NE Gabon.

Food. Insects and their larvae, including beetles, grasshoppers, caterpillars, flying termites, small snails; berries and seeds.

Breeding Habits. Territorial.

NEST: cup-shaped, of dry leaves, grass and strips of palm leaf, unlined; int. diam. (n = 1) 60 × 70, nest appearing small for size of bird; slung between twigs of bush or tree overhanging stream, attached by cobwebs and black hair-like fibres which form net around outside.

EGGS: 2–3. Slightly pointed ovals; somewhat glossy; dull creamy white, buff, greyish or pale stone, with small brown dots and scrolls, purple-brown mottling around large end, and suffused greyish undermarkings.

LAYING DATES: Sudan, Oct; Liberia, Jan; Ghana, Dec–Feb; Togo (carrying nest material, Mar); Nigeria (breeding condition Oct–Nov); Cameroon, Dec–Jan; Zaïre, Oct–Nov, probably May–June (in dry season in Upemba Nat. Park); Central African Republic (breeding condition June); Uganda (breeding condition Apr, July).

INCUBATION: by ♀ only.

Genus *Phyllastrephus* Swainson

Bill long, slender, laterally compressed; culmen almost straight, curved only near tip; gonys slightly convex; nostrils narrow, slit-like; rictal bristles short to long. Plumage mainly brown or olive-green, some species with yellow underparts. Tail rounded or graduated, often reddish. Sexual size dimorphism greater than in other bulbul genera.

Inhabit forest, mainly undergrowth and lower levels. Voices with grating, burry quality, generally not loud, songs not well developed, call often a chatter; no group chorusing, but duetting in 2 species.

PYCNONOTIDAE

Endemic; 17 species. Olson (1989) removed Madagascar '*Phyllastrephus*' spp. from the Pycnonotidae altogether. 4 superspecies: *P. cerviniventris/P. fulviventris*, *P. baumanni/P. hypochloris*, *P. fischeri/P. cabanisi* and *P. flavostriatus/P. poliocephalus*. *P. lorenzi* is partly sympatric with *P. hypochloris* and so must remain an independent species (cf. Hall and Moreau 1970). Based on plumage and voice, *P. poensis* may be closest to Cabanis's Greenbul *P. cabanisi*, of which it may be the ecological counterpart (Stuart 1986).

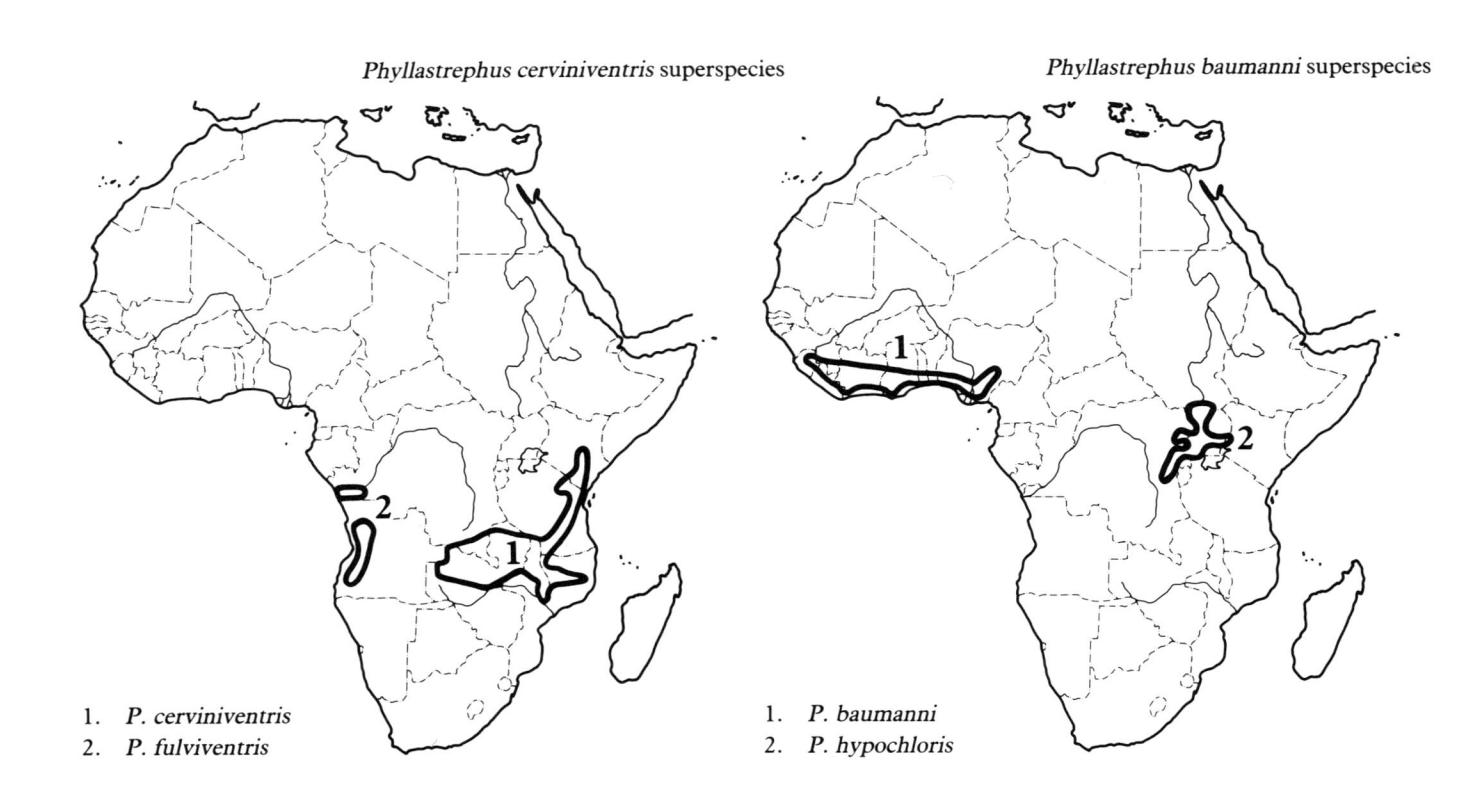

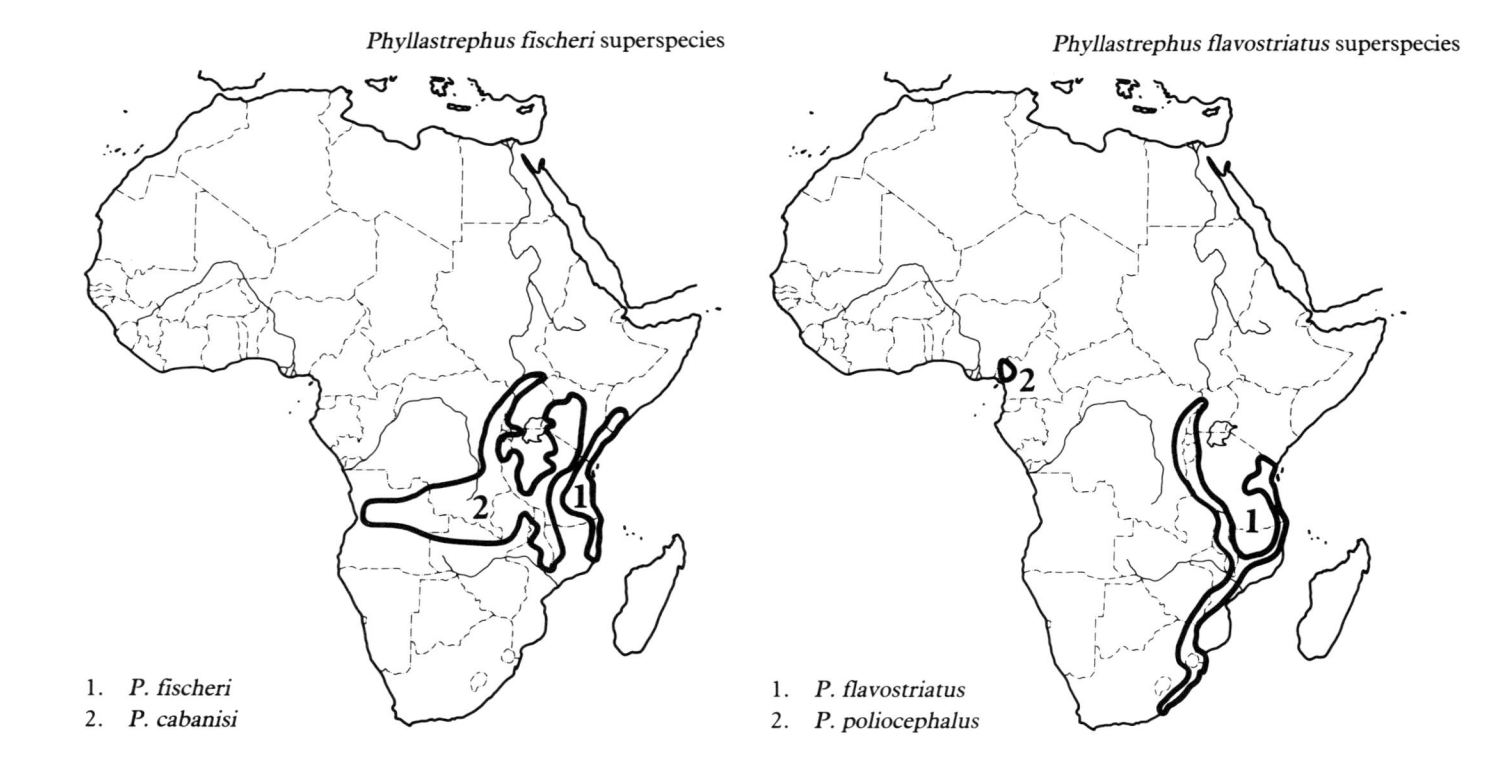

Phyllastrephus cerviniventris superspecies

1. *P. cerviniventris*
2. *P. fulviventris*

Phyllastrephus baumanni superspecies

1. *P. baumanni*
2. *P. hypochloris*

Phyllastrephus fischeri superspecies

1. *P. fischeri*
2. *P. cabanisi*

Phyllastrephus flavostriatus superspecies

1. *P. flavostriatus*
2. *P. poliocephalus*

Phyllastrephus terrestris Swainson. Terrestrial Brownbul. Bulbul jaboteur.

Phyllastrephus terrestris Swainson, 1837. Birds of West Afr. 1, p. 270; Anteniquoi (*ex* Levaillant) (= Outeniqua, Cape Province).

Plate 19
(Opp. p. 320)

Range and Status. Endemic resident. S Somalia (lower Juba R., rare), coastal Kenya (and isolated population '*bensoni*' in Meru and Chuka forests, Mt Kenya); coastal Tanzania, and locally inland, also in SW at Rukwa and Kasanga; widespread in Malaŵi, common but local in Zambia. Extreme SE Zaïre between L. Tanganyika and L. Mweru, and along Luapula R. (Kiniama) on Zambian border; isolated population in Angola in central and S Huila where frequent, mainly along Cunene R. In Namibia confined to riverine bush in the Caprivi Strip and Okavango R.; in Botswana in Okavango and along Chobe and Botletle rivers, and in SE along Limpopo R. Zimbabwe, widespread and common in major river valley systems, ascending to central plateau in suitable habitat; sparse in E highlands, uncommon above 1200 m; recent record Harare Botanic Gardens. Almost throughout Mozambique, although status in NE not known; generally common, especially in lowlands, also on offshore islands (Inhaca, Bazaruto). In Transvaal, common in Escarpment region and Soutpansberg-Blouberg, in nearly all forest patches; local in lowveld, restricted to larger river systems (Subie, Letaba, Olifants, Levuvhu), also Limpopo R. and its larger tributaries. Zululand and Natal along coast to W Cape Province, as far as Swellendam and Montagu. Generally frequent to common, sometimes local.

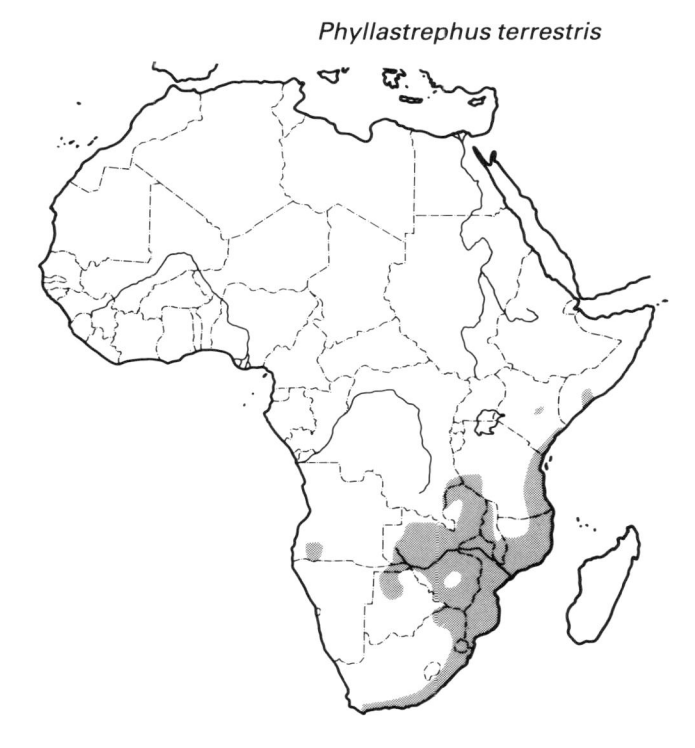

Description. *P. t. intermedius* Gunning and Roberts (including '*robertsi*' and '*katangae*'): E Zululand and S Mozambique to Angola, Zaïre and Zambia, intergrading with *suahelicus* in Malaŵi and N Mozambique. ADULT ♂: upperparts, including scapulars, wing-coverts and outer margin of flight feathers, brownish olive, somewhat darker on top of head; uppertail-coverts, rump and tail brown tinged rusty, shafts of tail-feathers brown above, sometimes whitish at base, white below, tips of tail-feathers variably tipped pale. Lores blackish, cheeks and ear-coverts brownish olive with a few thin, pale streaks. Throat white, upper breast and flanks pale brownish olive (whiter in centre of breast), becoming pale buffy olive on lower flanks and undertail-coverts; lower breast and belly white. Flight feathers dark olive-brown, inner margins on underside pale buffy white; underwing-coverts and axillaries pale buffy white. Bill blackish horn, lower mandible often paler, sometimes whitish at base; narrow white eye-ring; eye red to red-brown or golden brown; legs and feet slate-grey, blue-grey, brownish horn or pale silvery grey with fleshy tint, soles yellowish. Sexes alike. SIZE: wing, ♂ (n = 73) 76–95 (86·9), ♀ (n = 48) 73–93 (78·5); tail, ♂ (n = 8) 83–97 (90·8), ♀ (n = 5) 77–90 (82·4); bill, ♂ (n = 68) 18·5–25·7 (22·8), ♀ (n = 47) 19–25·5 (21·1); tarsus, ♂ (n = 8) 21–24 (22·9), ♀ (n = 5) 21·5–24 (22·9). WEIGHT: ♂ (n = 51) 29·5–40·5 (33·9 ± 2·7), ♀ (n = 40) 24·3–33·3 (28·5 ± 2·4).

IMMATURE (*suahelicus*): paler and brighter above than adult; top and sides of head yellowish brown, paler on forehead; sides of back rufous-brown, rump and uppertail-coverts tawny-orange, tail-feathers paler and richer with tawny edges; throat and underparts mainly pale yellowish brown, becoming buff on mid-belly and pale tawny on undertail-coverts. Wing-coverts and flight-feathers edged tawny. Bill pinkish brown, paler on lower mandible; eye dark grey; legs and feet pale brownish flesh.

NESTLING: yellowish white gape-flanges, pale greyish brown eye.

P. t. terrestris Swainson: Cape Province to W Zululand, Swaziland, E and N Transvaal. Upperparts considerably darker, including top and sides of head; lower back to rump warmer, gingery brown; below, sides of breast and flanks darker and more contrasting, and white of breast and belly with creamy yellow wash. Larger: wing, ♂ (n = 36) 73–100 (89·1), ♀ (n = 27) 73–88 (82·3); bill, ♂ (n = 35) 19·1–29·1 (25·0), ♀ (n = 27) 19·5–24·8 (22·5). WEIGHT: unsexed (n = 16) 33–47 (39·6 ± 4·8).

P. t. suahelicus Reichenow (including '*bensoni*'): Kenya/Tanzania, grading clinally with *intermedius* in N Mozambique. Breast and flanks somewhat browner, less grey than *intermedius*, white of underparts tends to have slight buffy wash; upperparts with slight reddish wash, rump and tail somewhat redder and darker. WEIGHT: ♂ (n = 19) 31–40 (36·1 ± 2·5), ♀ (n = 17) 25–34 (29·1 ± 2·7).

Field Characters. A brown and white bulbul of eastern coastal and riverine undergrowth, separated from most others in same habitat by lack of green or yellow in plumage. Slightly larger and with longer, heavier bill than Northern Brownbul *P. strepitans*; upperparts slightly darker and duller, and lacks buffy wash on underparts. Best separated by voice, more skulking habits, and usually by thicker, more forested habitat, though both birds can occur together. Can be distinguished from Fischer's Greenbul *P. fischeri* only with extreme care; they occur together in dark undergrowth. Both have dark cap contrasting with conspicuous white throat; Terrestrial Brownbul is browner, less olive above and lacks yellowish wash on underparts, but eye colour is best character (dark in Terrestrial Brownbul, pale in Fischer's Greenbul).

Voice. Tape-recorded (7, 14, 58, 75, 86, 88, CHA, KEI, LEM). Song (E Africa), a series of phrases of 4–10 low, burry, husky notes all on one pitch, e.g. 'chrrt-chrrt-chrrt-chrrt'; 'churr-jit-jit-chaaa-chaaa-chaaa-jiho-jit-jiho'; 'churr-jit-jit-jiho'; 'chrrt-chrrt-chrrt-chrrt-chrrt-jihojit-jihojit-jiho'; 'jiho-jit-chrrt-chrrt-chrrt-chrrt'. The 'ho' part of the double note 'jiho' has a buzzy, nasal quality. Intervals between phrases 4–10 s, sometimes punctuated by a quiet, conversational 'chut' or 'chudut'. The grating 'chrrt' and 'chaaa' notes are also used as calls, but one bird at Gedi, Kenya, incorporated them into formal song (above) given regularly and continuously from perch (S. Keith, tape and notes). Song in southern Africa said to be warbled 'wicherwer-wicherwer' (Maclean 1985). Groups of foraging birds give continuous chattering, a jumble of grating notes: 'rakakak', 'churr', 'jijijijijit', 'ruka', 'chaaa', 'rukukukut', 'grrrrr', 'churrpa-churrpa'. A foraging pair may from time to time break into a rough duet; the grating notes speed up and rise in pitch and turn into a short, harsh, grating trill interspersed with some pure, shrill notes. These bursts of song are variable in form and usually short (3–4 s) in duration, and are given at irregular intervals. Similarly, 3 birds performed a 'trio', intermittently speeding up grating contact calls and adding different notes to produce a loose 'song' (Malaŵi: C. Chappuis, tape and notes).

General Habits. Inhabits lower storey of primary and secondary evergreen forest, typically lowland but locally also montane, in Natal reaching mistbelt evergreen forest at 1500 m and in Kenya occurring in forest at the SE base of Mt Kenya; also deciduous thickets. Equally partial to riparian forest, riverine bush, dense thickets and tangles in ravines and along watercourses, and also found away from water in thick thorny scrub, dense woods and bushy areas, including weedy tangles bordering cultivation. Altitudinal range Malaŵi 50–1750 m, exceptionally to 2050 m; at higher elevations favours drier habitats (Dowsett-Lemaire 1989). In Mozambique ascends to 1220 m on Mt Gorongoza.

Moves about in small groups of 3–6; joins mixed-species flocks, and several times noted in company of Arrow-marked Babbler *Turdoides jardinei*. Usually forages on or near ground, scratching among leaf litter with bill and feet, rustling leaves, probing into decaying vegetable matter; also feeds in trees, even to just below the canopy, wherever there are accumulations of decaying leaves and vegetable debris trapped in tangles of creepers and lianas (Irwin 1980). Seen attending swarm of army ants Dorylini, sitting in vegetation just above ants and darting down to capture prey items (Lambert 1984). Fairly noisy, but skulking and hard to see; moves away with a chatter of alarm if approached. Members of party cross path or other open area reluctantly, one by one.

Essentially sedentary; of 62 birds ringed, Zimbabwe, 38 (61·3%) recaptured in same place, several 7–8 times (Manson 1985). Birds at exceptionally high elevation (2050 m), Malaŵi, may have been wanderers (Dowsett-Lemaire 1989).

Moult starts near beginning of breeding season; primary moult lasts 3–4 months. Immatures start moulting at 4–5 months; interrupted moult found in some.

Food. Arthropods, including insects (termites, beetles, grasshoppers), especially ants; also snails, small lizards, fruits and seeds.

Breeding Habits. Monogamous; territorial.

NEST: a shallow cup or bowl, flimsy and frail, sometimes so thin that the eggs can be seen from below, untidily built of twigs, plant stems and fibres, sometimes strengthened on the outside with leaves, roots, grass and moss; lined with rootlets, fine fibres and a few leaves; int. diam. 5 cm, int. depth 2·5 cm; situated 1–2 m above ground on branch or slung between twigs, usually near edge of bush or thicket.

EGGS: 1–4, usually 2, mean (47 clutches) 2·1. Glossy; white, sometimes pale brown, streaked, blotched and spotted with dark olive-brown and sepia over grey undermarkings, often concentrated at thick end. SIZE: (n = 41, southern Africa) 20·9–26·9 × 15·5–18 (24 × 16·9). Possibly double-brooded (Hanmer 1978).

LAYING DATES: E Africa: Region D, Jan; Region E, Aug; Zaïre, 'mainly during rainy season' (Lippens and Wille 1976); Angola (breeding condition Nov–Dec); Zambia, Mar–Apr (breeding condition Sept–Oct); Malaŵi, Nov–June; Mozambique, Oct–Apr; Zimbabwe, Sept–Feb, Apr, mainly Nov–Dec (Jan 2, Feb 1, Apr 1, Sept 1, Oct 2, Nov 17, Dec 14 clutches); Botswana, Dec; South Africa: Transvaal, Nov–Dec; Natal, Oct–Jan; Cape Province, Nov–Jan.

INCUBATION: incubating bird (sex?) slipped off nest and moved away into undergrowth at slightest alarm, but later, on close approach of human observer, became bold, hopping about and calling within 1–2 m (Swynnerton 1907).

DEVELOPMENT AND CARE OF YOUNG: young up to 5 months have yellowish white gape-flanges, pale greyish brown eye; 5–9 months, yellowish gape, orange–brown eye; 9–24 months, horn gape, reddish eye, same as adult.

BREEDING SUCCESS/SURVIVAL: mean survival rate 0·52, mean life expectancy 1·59 years; no difference in rates between adults and immatures (Malaŵi: Hanmer 1984). 3 birds lived > 8·5–9 years, 1 > 12 years (Hanmer 1989).

References
Maclean, G. L. (1985).
Vincent, J. (1935).

Phyllastrephus strepitans (Reichenow). Northern Brownbul. Bulbul brun.

Criniger strepitans Reichenow, 1879. Orn. Centralb. 1, p. 139; Malindi, E Africa (Kenya Colony).

Range and Status. Endemic resident. SE Sudan (Boma Hills and borders of Ethiopia, Kenya and Uganda, north to Lado, and isolated population in Jebel Marra region); SW, S and SE Ethiopia and Rift Valley; Somalia (locally common south of 5°N); N Uganda south to Wadelai, Kidepo Valley Nat. Park, Mt Moroto; Kenya (common and widespread, almost throughout except driest parts of NW and E, and humid SW) and Tanzania south to Dar-es-Salaam and slopes of Ulugurus, inland to Moshi-Arusha area and country west of L. Natron. May occur in NE Zaïre but evidence conflicting. Common to locally abundant.

Description. ADULT ♂: top of head and upperparts, including scapulars and upperwing-coverts, reddish brown, becoming brighter on rump and rufous on uppertail-coverts. Tail dark reddish brown, somewhat graduated: T6 12–16 shorter than T1; shafts red-brown above, straw below. Indistinct loral line, pre-orbital area and cheeks pale grey-brown, eye-ring white, ear-coverts pale brown with very fine pale shaft streaks. Chin and throat white to very pale buff; breast and flanks grey-brown with variable buffy wash, becoming richer and buffier on lower flanks; lower breast and belly white, thighs and undertail-coverts buff. Primaries and secondaries dark brown, edged pale brown on primaries, becoming reddish brown on secondaries; axillaries and underwing-coverts rich buff, inner webs of underside of primaries and secondaries pale buff. Bill black, sometimes tinged olive, lower mandible paler, especially on underside; eye brown, red-brown or pale stone, colour not corresponding to size or sex, though red-brown in heaviest birds of both sexes (eye colour may be related to dominance status, or merely dimorphism not related to other characteristics: Wood 1989); legs and feet deep bluish black to grey. Sexes alike. SIZE: (19 ♂♂, 19 ♀♀) wing, ♂ 75–85 (80·4), ♀ 72–82 (75·6); tail, ♂ 78–90 (84·9), ♀ 74–88 (81·3); bill, ♂ 18–21·5 (19·9), ♀ 16–20·5 (18·4); tarsus, ♂ 20·5–23 (22·2), ♀ 20·5–24 (21·7). WEIGHT: ♂ (n = 9) 19–29 (24·5 ± 3·2), ♀ (n = 13) 17–24 (20·8 ± 2·0), unsexed (n = 45) 21·4–32·9 (27·1 ± 3·5).

IMMATURE: like adult but top of head and back slightly warmer in tone, eye duller.

NESTLING: a very small bird (wing 50; tail-feathers just sprouting from sheaths; bill 9; tarsus 20·5) has feathered upperparts, red-brown like adult but paler on top of head, underparts covered with buffy down, paler in centre of breast and belly.

Field Characters. A scrub-country bulbul, often the only one present in driest parts of range, except very different Common Bulbul *Pycnonotus barbatus*. Most likely to be confused with Terrestrial Brownbul *P. terrestris*; upperparts more rufous, face brown rather than olive, underparts with buffy wash, undertail-coverts cinnamon, bill shorter and darker. Terrestrial Brownbul also has brighter, whiter throat-patch, contrasting more with breast. Grey-olive Greenbul *P. cerviniventris* has much greyer upperparts, darker underparts, pale bill and pale pinkish legs; Sombre Greenbul *Andropadus importunus* and Yellow-bellied Greenbul *Chlorocichla flaviventris* are green above, yellow below.

Voice. Tape-recorded (CHA, GREG, HOR, KEI, McVIC, STJ). Song, given by pair in duet, a harsh, shrill chattering, often preceded by some slower, grating notes; bouts of chattering last up to 10 s and are often repeated with only a slight break, forming a lengthy series; they have a noisy, insistent quality, like a babbler *Turdoides*, and may speed up and get louder in the middle, recalling wood-hoopoe *Phoeniculus*. Lower, more conversational contact notes given by foraging group.

General Habits. Inhabits scrub, undergrowth and thickets, both deciduous and evergreen, in thornbush, dry, open woodland, wooded grassland and forest; coastal scrub, groundwater forest, evergreen scrub in juniper woods (Ethiopia), *Salvadora* thickets, thick riverine brush; *Afzelia*, *Brachystegia* and *Cynometra* forest in Sokoke, Kenya (Britton and Zimmerman 1979).

Travels in pairs and small noisy parties; rather shy, although less so than Terrestrial Brownbul; when crossing open space between bushes, birds fly one at a time, in follow-my-leader fashion. Flicks wings constantly, while foraging and when perched. Keeps very low down; mean foraging height of 19 birds, Tsavo East Nat. Park, Kenya, 1·2 m; insects taken from leaves (70%), stems (9%), ground (6%) or air (7%) (Lack 1985). 2–3 birds observed apparently anting on patch of sun-lit sand where many ants present, on path in bush, 'crouched, wings spread, apparently in semi-entranced attitude' (Brown and Newman 1974).

Food. Fruit and insects. In Tsavo East Nat. Park, Kenya, ate only insects (n = 45 items: Lack 1985).

Breeding Habits. Almost unknown. 1 nest found 'in an acacia' but not described (E African nest record card); eggs not known.

LAYING DATES: Sudan, Jan–Feb; Ethiopia, May, possibly Dec; Uganda (breeding condition May); Kenya, June; Tanzania (breeding condition Mar).

Plate 19
(Opp. p. 320)

Phyllastrephus cerviniventris (Shelley). Grey-olive Greenbul. Bulbul vert-olive.

Phyllostrophus [sic] *cerviniventris* Shelley, 1894. Ibis, p. 10, pl. 2; Zomba and Tschiromo, Nyasaland.

Forms a superspecies with *P. fulviventris*.

Range and Status. Endemic resident. Scattered localities in S Kenya (Meru, Tharaka, Thika, Lolterish, Mzima Springs, Kitovu Forest, Bura); occurrence near coast at Rabai and Shimoni needs confirmation. N and E Tanzania (L. Manyara, around bases of Mt Meru and Mt Kilimanjaro, W and E Usambaras and Ulugurus, to Luwegu R. and Songea); Zambia (widespread but local, absent only from N between L. Tanganyika and L. Mweru and areas close to Tanzania border, and from Barotse and most of Southern provinces); extreme SE Zaïre; Malaŵi (widespread, almost throughout) and N Mozambique. Locally fairly common but generally rather scarce.

Description. ADULT ♂: top of head and upperparts, including scapulars and upperwing-coverts, grey-olive tinged brownish, slightly colder and darker on head, becoming paler on lower rump and rich rufous on uppertail-coverts. Tail-feathers rufous-brown with narrow pale tips when fresh, shafts dark red-brown above, straw below. Faint pale loral line (sometimes lacking); pre-orbital area, cheeks and ear-coverts greyish olive-brown with narrow pale streaks. Chin and throat white or very pale buff; sides of neck, breast and flanks greyish olive-brown, with tawny wash on lower flanks and thighs, centre of breast with a few buffy streaks, becoming creamy buff on belly and tawny on undertail-coverts. Primaries and secondaries dark brown edged olive; axillaries and underwing-coverts pale tawny, inner margins of underside of primaries and secondaries pale buff. Upper mandible dark horn, horn-grey or horn-brown, cutting edges whitish, lower mandible pale horn or whitish; eye golden brown to ochre, pale orange, yellow or creamy white; legs and feet pale flesh, brownish white or pinkish grey. Sexes alike. SIZE: (8 ♂♂, 5 ♀♀) wing, ♂ 79–93 (85·0), ♀ 76–79 (77·6); tail, ♂ 76–87 (82·3), ♀ 73–77 (75·0); bill, ♂ 19–22 (20·8), ♀ 19–20 (19·3); tarsus, ♂ 21–24 (22·5), ♀ 21–22 (21·4). WEIGHT: ♂ (n = 2) 28·8, 32·9, 1 ♀ 23·5, unsexed (n = 4) 23·5–28·5 (26·0).

IMMATURE: browner, less olive above, with tawny edges to secondary coverts. Eyes grey in juvenile and immature, then greyish green, then yellow in adult (Vincent 1935).

NESTLING: unknown.

TAXONOMIC NOTE: *P. c. schoutedeni* Prigogine, described from 4 specimens from R. Kisanga, Zaïre (27° 35′E, 11° 43′S), is apparently darker and greyer below, especially throat and breast, belly and undertail-coverts darker rufous, tail-feathers dark brown, less rufous than nominate race. A bird from Kakanda near the Lualaba R. is 'intermediate' between these Kisanga birds and the nominate (Prigogine 1969); but that seems unlikely since '*schoutedeni*' is geographically intermediate. Until the situation becomes clearer, we prefer not to recognize *schoutedeni*.

Phyllastrephus cerviniventris

Field Characters. A smallish bulbul, with overall similarity to somewhat larger sympatric congeners Northern Brownbul *P. strepitans*, Terrestrial Brownbul *P. terrestris* and Cabanis's Greenbul *P. cabanisi*; has same rusty rump and tail but upperparts paler and greyer, underparts buffer, undertail-coverts tawny; pale underside of bill and whitish or pinkish legs and feet noticeable in the field (Turner and Zimmerman 1979). The 2 brownbuls have brown upperparts, whitish underparts, Cabanis's Greenbul has dark olive upperparts, creamy yellow throat and belly.

Voice. Tape-recorded (86, CART, CHA, GREG, LEM, McVIC). Calls include a loud, insistent, nasal scold 'kaaa-kaaa-kaaa-kaaa . . .', a curious quick, short, descending trill, almost musical, unlike any other bulbul sound, and a variety of chips, chucks and gravelly notes; all of these may be blended together into a loose, conversational 'song'.

General Habits. Inhabits thickets and matted growth of all kinds, with a distinct preference for streamside vegetation; gallery forest, gullies in open woodland, groundwater forest, bamboo thickets, lake littoral (Malaŵi); sometimes enters true forest, e.g. Malaŵi, where occupies *Metarungia* thickets on forest floor (Dowsett-Lemaire 1989). Mainly lowlands, 400–1900 m.

In pairs and small parties; constantly on the move, flicking wings and tail in typical *Phyllastrephus* fashion; rather shy and secretive. Forages in low tangles of vines and shrubby vegetation, up to 4 m, gleaning bark and leaves, and on ground, where it moves with long hops; occasionally hawks for insects. In Zambia very fond of feeding on logs, especially ones that have fallen across streams and ravines (C. Carter, pers. comm.). Expanded niche on Chipata Mt, Malaŵi, may be related to absence of Cabanis's Greenbul (Dowsett-Lemaire 1989).

Food. Insects, including grasshoppers and beetles; some fruit.

Breeding Habits.
NEST: irregularly-shaped, untidy cup, fairly deep, of fine rootlets, moss and dead leaves, lined with more rootlets, ext. diam. 100, ext. depth 55 (n = 1); placed in fork or suspended between branches of fork in streamside bush; one was 1 m above surface of stream.
EGGS: 2 (3 clutches). Grey with tinge of cream, well covered with dull blue-grey and brown marks. SIZE: (n = 2) both 23·5 × 16.
LAYING DATES: Kenya, Dec; Zambia, Apr (breeding condition Oct); Malaŵi, Apr–May, Oct–Nov; Mozambique (flying juvenile and breeding condition Apr).
DEVELOPMENT AND CARE OF YOUNG: fledged young out of nest fed by both parents (Lewis and Pomeroy 1989).

Phyllastrephus fulviventris (Cabanis). Pale Olive Greenbul. Bulbul à ventre roux.

Phyllostrephus fulviventris Cabanis, 1876. J. Orn. 24, p. 92; Chinchoncho, Loango Coast.

Forms a superspecies with *P. cerviniventris*.

Plate 19
(Opp. p. 320)

Range and Status. Endemic resident. Extreme W Zaïre (Manyanga and Ngombe Lutete on lower Congo R.) and Angola: Cabinda (Loango coast), and from Cuanza Norte (Ndala Tando, Massangano, Dondo, Pungo Andongo) and Luanda (Kisama Nat. Park) south along escarpment to NW Huila (Mt Bongo) and N Mossamedes (Capangombe). Locally common (e.g. Ndala Tando, where commonest forest bulbul (Dean *et al.* 1988)), but little known and status unclear.

Description. ADULT ♂: forehead to hindneck dark olive-brown, becoming slightly paler and more olive on mantle, back, scapulars and upperwing-coverts, rump and uppertail-coverts rufous. Tail-feathers dark reddish brown edged olive, with narrow pale tips when fresh, shafts red-brown above, straw below. Broken eye-ring white, merging with pale loral line and very short pale streak just above and behind eye; line of short white feathers below eye-ring, separated from pale lores by dark line; cheeks and ear-coverts light olive-brown with whitish shaft streaks. Chin and throat creamy white, rest of underparts pale dull yellow with variable tawny wash, breast washed lightly and flanks more heavily with pale olive-brown, undertail-coverts pale yellowish tawny. Primaries and secondaries dark brown edged olive, axillaries and underwing-coverts creamy yellow, inner webs of underside of primaries and secondaries margined pale greyish buff. Upper mandible dark brown or dark grey, lower mandible paler, grey to slate-grey, pinkish grey or whitish horn, cutting edges whitish; inside of mouth cadmium yellow; eye dark brown to red-brown; legs and feet flesh to bluish grey or greenish grey, toes yellowish grey. Sexes alike. SIZE: (12 ♂♂, 9 ♀♀) wing, ♂ 84–95 (89·9), ♀ 77–84 (81·3); tail, ♂ 83–95 (90·4), ♀ 76–83 (79·6); bill, ♂ 21–25 (22·5), ♀ 18·5–23 (20·8); tarsus, ♂ 22–25 (23·3), ♀ 21–24 (22·4). WEIGHT: 1 ♂ 37, 1 ♀ 30.
IMMATURE AND NESTLING: unknown.

Field Characters. An undergrowth bulbul with olive-brown upperparts, pale tawny yellow underparts and rusty tail. Range approaches but does not overlap that of Terrestrial Brownbul *P. terrestris* (browner upperparts,

330 PYCNONOTIDAE

whitish underparts) (Traylor 1962); Cabanis's Greenbul *P. cabanisi sucosus* has greener upperparts, darker, olive and yellow underparts; White-throated Greenbul *P. albigularis* has green upperparts, yellow underparts with contrasting white throat; Yellow-bellied Greenbul *Chlorocichla flaviventris* has similar upperparts but green tail, bright yellow underparts.

Voice. Not tape-recorded. Alarm call said to be a loud, sharp 'tsik-tschirr-tschirr' (Heinrich 1958).

General Habits. Inhabits forest patches in savanna, gallery forest with tangled lianas, thickets in dry *Croton* woodland, thick bush dotted with baobabs, coffee forest; habitat said to be like that of Terrestrial Brownbul (Heinrich 1958). Forages at all levels, including canopy, but most often low in bushes, also on ground among dry leaves, like a thrush (Heinrich 1958).

Food. Only insects recorded.

Breeding Habits. Unknown. In Angola, breeding condition Aug, juvenile July.

Plate 18
(Opp. p. 305)

Phyllastrephus baumanni (Reichenow). Baumann's Greenbul. Bulbul de Baumann.

Phyllostrephus baumanni Reichenow, 1895. Orn. Monatsber. 3, p. 96; Misahöhe, Togo.

Forms a superspecies with *P. hypochloris*.

Range and Status. Endemic resident. N and E Sierra Leone, N and E Liberia (uncommon), Ivory Coast (rare, scattered localities from coast (San Pedro) north to Beoumi and Mt Nimba area), Ghana (known from 3 localities, Cape Coast, Bosum east of Mpraeso, Ejura) but probably more widespread and abundant than records suggest (Grimes 1987); S Togo (Ahoué-houé, Ounabé, Misahöhe); Nigeria (coast (Lagos) north to Ilaro, Ibadan, Ife and Enugu, and at Pandam at S edge of Jos Plateau; not uncommon, possibly less common in SE). An undergrowth species previously reported as uncommon to rare, but recent use of mist nets is proving it more common than originally supposed. Density in Nigeria (Ogba Forest) 1 bird/5 ha (Johnson 1989).

Description. ADULT ♂: top of head to hindneck dark olive-brown; upperparts, including scapulars and upperwing-coverts, olive-brown with gingery wash, becoming rufous on rump and uppertail-coverts. Tail-feathers rufous with variable narrow ginger edging, shafts dark maroon above, straw below. Feathers of forehead with pale fringes giving slightly scaly effect; lores, cheeks, ear-coverts and eye-ring grey-brown, cheeks and ear-coverts with narrow pale streaks. Chin and throat pale grey with slight greenish tinge, feathers with pale shafts; rest of underparts greyish olive, somewhat browner on flanks, breast feathers with pale grey shaft streaks, undertail-coverts dull light brown with pale tips. Primaries and secondaries dark brown with narrow olive edging, axillaries and underwing-coverts pale yellowish olive, inner webs of underside of primaries and secondaries edged pale buff. Upper mandible black to dark brown, lower mandible bluish white to pale grey or pinkish horn, cutting edges cream; eye brown to hazel, chestnut, tawny or yellowish brown; legs and feet grey to blue-grey or dusky grey. Sexes alike, but ♂♂ larger. SIZE: (10 ♂♂, 4 ♀♀) wing, ♂ 71–82 (77·9), ♀ 69–72 (70·5); tail, ♂ 75–83 (80·2), ♀ 66–75 (70·0); bill, ♂ 20–22 (20·6), ♀ 17·5–21 (19·0); tarsus, ♂ 21·5–24 (23·1), ♀ 21–23 (21·8). WEIGHT: ♂ (n = 5) 27·3–32·9 (30·3), ♀ (n = 4) 23·5–26·5 (24·9), unsexed (n = 9) 26–31 (29·0).

IMMATURE AND NESTLING: unknown.

Phyllastrephus baumanni

Field Characters. A medium-sized greenbul, brown above and grey below, with long, rusty tail and no contrasting throat-patch. Skulks in undergrowth, where most likely to be confused with Little Greenbul *Andropadus virens*, which has darker and greyer upperparts, darker and less rusty tail, olive-brown underparts and reddish wash on wings; 2-tone bill much longer than Little Greenbul's, which appears uniform dark.

Voice. Not tape-recorded. Said to be noisy, giving a continuous clucking call similar to that of Little Greenbul, also a loud swearing note, and 1 or 2 fluty notes similar in quality to typical calls of Common Bulbul *Pycnonotus barbatus* (Marchant 1953), which may be the same as the 'thrush-like "seer-seer"' of Button (1964b).

General Habits. Inhabits thick undergrowth in primary and secondary forest and forest edge; also gallery forest, small woods in savanna, parklands with tall trees, thick brush along streams in orchard-bush country. Forages among foliage in dense creeper tangles. Vertical foraging niche 3–20 m (Ilaro, Nigeria: Button 1964b). In pairs or groups of 3–5; joins mixed-species flocks. Rather shy; perches and moves quietly.

Food. Insects, including Orthoptera, Hemiptera, Lepidoptera and their larvae, Coleoptera and ants; spiders; also berries, small seeds and a few leaves.

Breeding Habits. Nest and eggs unrecorded. Birds in breeding condition Liberia, Jan, Apr, Oct–Nov; Nigeria, Dec.

Phyllastrephus hypochloris (Jackson). Toro Olive Greenbul. Bulbul du Toro.

Plate 18 (Opp. p. 305)

Stelgidillas hypochloris Jackson, 1906. Bull. Br. Orn. Club 19, p. 20; Kilirau, Toro, Uganda.

Forms a superspecies with *P. baumanni*.

Range and Status. Endemic resident. Extreme S Sudan (Imatong Mts to Aloma Plateau near Yei); E Zaïre in region of Beni (Beni-Mawambi, Beni, Semliki valley, Maboya (common)), in central Kivu (Irangi, 850 m, west of Mt Kahuzi) and S Kivu (Itombwe, around Kamituga). Widespread in forests of W and S Uganda, from Impenetrable to near shores of L. Victoria (Buddu, Malibigambo (common)), and north to Bwamba, Bugoma and Budongo, east to Mabira, Busoga and Elgon. W Kenya at Kakamega (old records from Elgon and Lerundo). Common in a few forests but generally scarce.

Description. ADULT ♂: top and sides of head, including lores, cheeks and ear-coverts, brownish olive-green, feathers of forehead to nape with narrow dark fringes giving slightly scaly effect, cheeks and ear-coverts with narrow pale streaks. Upperparts, including scapulars, olive-green becoming gingery on rump; uppertail-coverts dull rufous. Tail-feathers dark reddish brown narrowly edged greenish, with narrow pale tips when fresh, shafts maroon above, straw below. Chin and throat pale grey with yellowish wash; rest of underparts pale olive with indistinct greenish yellow streaks, flanks a little darker and browner, centre of belly pale cream to pale yellow, undertail-coverts pale brown with rufous tinge, tips pale yellow. Primaries and secondaries dark brown, broadly edged olive-green; axillaries and underwing-coverts pale yellowish olive to lemon yellow, inner webs of underside of primaries and secondaries edged greyish buff. Bill of ♂ black, gonys pale flesh becoming yellow at tip, gape pale yellow to dull yellowish flesh; bill of ♀ brownish black, tomia olive to yellowish horn, gonys pale olive-flesh, gape olive to yellowish horn; eye, ♂ brownish orange, ♀ rust-brown; legs and feet greenish grey or bluish grey to pale grey-brown, soles dull yellow, claws brown. Sexes alike. SIZE: (12 ♂♂, 9 ♀♀) wing, ♂ 72–82 (76·7), ♀ 67–73 (69·4); tail, ♂ 67–81 (74·6), ♀ 64–71 (67·2); bill, ♂ 17–20 (18·4), ♀ 16–19 (17·7); tarsus, ♂ 20–24 (21·4), ♀ 19–21 (20·2). WEIGHT: Uganda, ♂ (n = 14) 19–30 (25·6), ♀ (n = 11) 17–24 (20·9); Kenya, ♂ (n = 10) 19·8–24 (22·1 ± 1·3), ♀ (n = 7) 18–24 (20·4 ± 1·7).

IMMATURE: top of head paler and browner, underparts slightly paler, throat yellower, undertail-coverts pale rufous, fresh tail-feathers lack pale tips. ♀ has bill dark brown with yellow tomia and tip, feet pale bluish, edges of scutes, claws and soles dull pale yellow.

NESTLING: unknown.

TAXONOMIC NOTE: sometimes considered a subspecies of *P. baumanni*. However, compared to the often very slight differences between full species in Pycnonotidae, the differences between these 2 are quite considerable (Zimmerman 1972). Little is known of their habits, and we agree with Zimmerman that it would be premature to consider them conspecific.

Field Characters. A small, dingy, nondescript bulbul with a rather long bill, very hard to identify in the field except by voice (q.v.). Occurs in same undergrowth habitat as Cabanis's Greenbul *P. cabanisi sucosus*, which is paler below, with yellowish throat and belly. Shorter, stouter bills distinguish extremely similar Cameroon Sombre Greenbul *Andropadus curvirostris* (more rufous brown above, darker below) and Little Greenbul *A. virens* (darker and duller above, darker throat and breast, yellow belly); in hand, paler, greyer, yellow-streaked underparts of Toro Olive make separation easier. At a distance looks like small Slender-billed Greenbul *A. gracilirostris* but occurs in undergrowth, not canopy (Turner and Zimmerman 1979).

Voice. Tape-recorded (GREG, McVIC). Song, a loud, sharp, tuneless phrase 'titiwah', short and clipped, third note lower; sometimes 4 notes 'titutawah'; rapidly repeated 2–4 times. Call, a harsh, shrill chatter, which often precedes song: 'chrrrrrrrrtitiwah'. Quality of notes reminiscent of Little Greenbul *Andropadus virens*.

General Habits. Inhabits primary forest in lowland and transition zones, to 1440 m in Zaïre, 1600 m in Sudan, 1800 m in Uganda. Prefers dense forest undergrowth, and only once not in this habitat at Kakamega (Zimmerman 1972); elsewhere (e.g. Bwamba: van Someren and van Someren 1949) ascends to middle strata, and fond of dense growth along forest streams, thick, tangled vegetation in clearings, at forest edge and in adjoining cultivation; also seasonal dry swamp forest.

Occurs in pairs or small (family?) parties; when one bird was caught in mist net, others came in response to its cries (Friedmann and Williams 1969). Sometimes joins mixed-species flocks, with e.g. Cabanis's Greenbul and Dark-backed Weaver *Ploceus bicolor* (Chapin 1953). Forages in undergrowth, on lower branches of trees, and in tangles of ferns and creepers on large tree trunks.

Food. Insects, including beetles, and their larvae; fruit.

Breeding Habits. Nest and eggs not described. Pair seen building nest in creeper-covered tree (no details: Friedmann and Williams 1971). Breeding indications from gonadal data: Sudan, May; Zaïre (Itombwe) Dec–Mar; elsewhere data for most months (Lippens and Wille 1976).

Reference
Zimmerman, D. A. (1972).

Plate 18
(Opp. p. 305)

Phyllastrephus lorenzi Sassi. Sassi's Olive Greenbul. Bulbul de Lorenz.

Phyllastrephus lorenzi Sassi, 1914. Anz. K. Akad. Wiss. Wien, Math.-Naturw. Kl. 51, p. 309; Moera, near Beni, E. Congo.

Range and Status. Endemic resident. Rare, confined mainly to narrow strip in E Zaïre. Single record from Uele District (Bambera); Kibali-Ituri District from Bondo Mabe and Irumu to Beni, including Ituri and Semliki Forests, just crossing into Uganda at Bwamba (1 record); narrow zone around Kamituga, S Kivu (Itombwe), 2° 49'–3° 54'S, 28° 9'–28° 22'E. Recent record from Hombro (2° 13'S, 28° 42'E), west of Mt Kahuzi, suggests Beni and Kamituga populations may meet (Prigogine 1980).

Description. ADULT ♂: top of head mottled black and olive, blacker on crown, more olive on forehead; patch on hindcrown rich brown; lores, cheeks, ear-coverts and behind eye dull olive, with faint pale streaking on ear-coverts, continuing as broad, darker olive collar around nape and hindneck. Upperparts to scapulars and rump dark olive-brown tinged rusty, becoming more rufous on upperwing-coverts and uppertail-coverts; tail dark red-brown, shafts whitish on underside. Rictal bristles black. Chin and throat olive-yellow, rest of underparts yellowish olive-green, becoming olive-yellow on centre of belly, undertail-coverts cinnamon. Primaries and secondaries blackish brown, outer margins of primaries light greenish brown becoming progressively darker and more rusty inwardly and on secondaries; inner margins of secondaries pale buff, broader on underside; inner webs of primaries also edged

pale buff on underside; axillaries, underwing-coverts and bend of wing olive-yellow. Bill blackish brown, lower mandible, cutting edges and tip paler; eye brown; legs and feet greyish to pale greyish brown. Sexes alike. SIZE: wing, ♂ (n = 6) 71–78 (75·2), ♀ (n = 8) 68–72 (70·4); tail, ♂ (n = 6) 66–74 (69·0), ♀ (n = 8) 58–64 (61·0); bill, ♂♀ (n = 16) 16·5–20 (17·9). WEIGHT: 1 ♀ 19.
   IMMATURE: like adult but blackish area on crown poorly developed.
   NESTLING: unknown.

Field Characters. A forest bulbul with typical colouring (brownish above, yellow-green below, dark reddish tail) except for curious and distinctive black cap; moves slowly, allowing good views of the cap.

Voice. Unknown.

General Habits. Inhabits primary forest in transition and lower montane zones, in rather narrow altitudinal band, 1060–1820 m. Occurs in small groups at lower and middle levels; joins mixed-species flocks.

Food. Unknown.

Breeding Habits. Unknown. Birds in breeding condition Jan, Mar–May, Aug.

Reference
Prigogine, A. (1980).

Phyllastrephus fischeri (Reichenow). Fischer's Greenbul. Bulbul de Fischer.

Criniger fischeri Reichenow, 1879. Orn. Centralb. 1, p. 139; Muniuni (= Muniumi), E Africa (near mouth of Tana R.).

Forms a superspecies with *P. cabanisi*.

Range and Status. Endemic resident. Extreme S Somalia (lower Juba R.); coastal Kenya south from lower Tana R. (Makeri) and Witu, inland to Rabai and Shimba Hills, coastal Tanzania inland to foothills of E Usambaras, E Ulugurus (Kimboza and Kibungo forests), and Uzungwa Mts (Mwanihana Forest); N Mozambique (Netia). Common to abundant.

Description. (Includes 'grotei', 'munzneri', 'sokokensis' and 'P. alfredi itoculo' Vincent). ADULT ♂: top of head olive-brown to brownish olive, mantle and back same but slightly greener, becoming richer, gingery brown on rump and dull rufous on uppertail-coverts. Tail-feathers dull reddish brown with narrow buffy fringes to tips and inner webs, outer webs narrowly edged olive-brown, shafts dark red-brown above, straw below. Cheeks and ear-coverts light olive-brown with fine pale streaks, eye-ring white. Chin and throat white, breast and flanks greyish to brownish olive, browner on lower flanks, lower breast with a few creamy yellow streaks, belly creamy white to pale yellowish, undertail-coverts pale brown to pale cinnamon. Primaries and secondaries dark brown, outer webs of primaries pale greenish brown, outer webs of secondaries brown, upperwing-coverts brown, axillaries, underwing-coverts and inner margins of underside of primaries and secondaries buff. Bill black or brownish black, cutting edges pale horn to whitish; eye white to creamy or whitish yellow; legs and feet grey with bluish, lavender or fleshy tinge. Sexes alike. SIZE: wing, ♂ (n = 27) 77–96 (87·1), ♀ (n = 23) 73–85 (79·7); tail, ♂ (n = 16) 77–93 (87·2), ♀ (n = 15) 74–85 (80·5); bill, ♂ (n = 12) 21–24 (22·7), ♀ (n = 11) 19–22 (20·3); tarsus, ♂ (n = 6) 21–25 (22·5), ♀ (n = 5) 21–22 (21·8). Southern birds average larger. WEIGHT: Kenya (coast north of Mombasa) ♂ (n = 25) 29–34·5 (31·2 ± 1·5), ♀ (n = 26) 22–31·1 (26·4 ± 2·2), (coast south of Mombasa) ♂ (n = 10) 34–39 (35·7 ± 1·8), ♀ (n = 10) 24–33 (28·3 ± 3·1); Tanzania (Usambara Mts, 300–600 m) 1 ♂ 38, 1 ♀ 30, 1 immature ♀ 25, immature ♂ (n = 3) 33·5–34 (33·8). Data suggest N–S cline of increasing size and weight (Britton 1972).

   IMMATURE: like adult but upperparts, breast and flanks somewhat darker, belly yellower, tail paler with more extensive pale fringes, eye grey.
   NESTLING: unknown.
   TAXONOMIC NOTE: the specific recognition of *fischeri* was suggested by Ripley and Heinrich (1966) and accepted by Benson *et al.* (1970), Britton (1972) and Dowsett (1972). It differs from *P. cabanisi* in plumage, eye colour and longer bill with straighter culmen.

Field Characters. A bulbul of coastal lowland forest and scrub, common but shy and secretive, usually on ground or in low undergrowth. Upperparts more olive, less brown than the 2 brownbuls, Northern *P. strepitans* and Terrestrial *P. terrestris*, and underparts with yellow wash, but such plumage differences not much use in shadowy undergrowth. Best distinguished by pale eye (dark in brownbuls).

Voice. Tape-recorded (CHA, KEI). Call, a series of harsh, scolding churrs, interspersed with an explosive 'pidi-tit'. Churrs frequently accelerate into a distinctive 4–5 note form, in which first note is longest, loudest and highest, the following notes on a descending scale: 'WREEE-ga-ga-ga-ga'; possibly a contact call since often answered by neighbour.

General Habits. Inhabits lowland evergreen forest and dense coastal thickets; in Sokoke Forest, Kenya, occupies all 4 forest types: lowland rain forest, *Afzelia*, *Brachystegia* and *Cynometra-Manilkara*. Lowlands, mainly below 300 m, occasionally to 600 m (Usambara and Uzungwa Mts, Tanzania).

Occurs in pairs or small groups; joins mixed-species flocks. Keeps to ground stratum, foraging in low bushes and tangled vegetation, and on ground itself, turning over leaves and debris while searching for arthropods. Shy and secretive, but presence often detected by loud contact call (see *Voice*); when one bird calls, it is often answered by others.

Food. Arthropods.

Breeding Habits. Unknown. Birds in breeding condition, E Africa (Region E), Apr, Dec.

Plate 19
(Opp. p. 320)

Phyllastrephus cabanisi (Sharpe). **Cabanis's Greenbul. Bulbul de Cabanis.**

Criniger cabanisi Sharpe, 1881. Cat. Bds Br. Mus. 6, p. 83; Angola.

Forms a superspecies with *P. fischeri*.

Range and Status. Endemic resident. Central plateau of Angola from W Malanje, Cuanza Sul and Huambo to Lunda and N Moxico; Zaïre in SE lowlands (Kasai, Katanga) and E highlands from Lendu Plateau to Mt Kabobo and Marungu; Rwanda, W Burundi; mountains of extreme SE Sudan; Uganda in forests of W from Kibale to Impenetrable, and formerly Mubende Hills, reappearing on Mt Elgon and in extreme NE on Mt Lonyili (Kidepo Nat. Park). Widespread in highlands of W and central Kenya, including montane islands north to Marsabit and Mt Kulal, and in SE to Taita Hills and Mt Kasigau. Throughout highlands of E Tanzania from N border to Mbeya and down E side of L. Malaŵi to Litembo; also in NW at Bukoba and in W from near Burundi border (Kahambwe R., Kasulu) south to Ufipa Plateau. N half of Zambia; highlands of N and S Malaŵi, and scattered localities in between; highlands of N Mozambique along Malaŵi border, south to Namuli Mts. Common to abundant. The 3 point plots in Hall and Moreau (1970) around the N end of L. Victoria may be in error due to confusion in the literature, since the Uganda race is montane. Localities in this region listed by Jackson and Sclater (1938) for *P. fischeri succosus* [sic] and by van Someren (1922) for *P. cabanisi hypochloris* may all refer to *P. hypochloris*.

Description. *P. c. placidus* (Shelley): Kenya east of the Rift, N, E and S Tanzania (west in S to Rungwe Mts), extreme NE Zambia, Malaŵi, Mozambique. ADULT ♂: top of head and upperparts to rump, including scapulars, brownish olive-green, top of head darker and browner; in some individuals head feathers have dark fringes producing slightly scaled effect. Uppertail-coverts rufous, sometimes with slight greenish wash, tail-feathers darker rufous with trace of greenish on outer edges, shafts red-brown above, straw below. Lores greyish brown, slightly paler than forehead, eye-ring white, cheeks and ear-coverts brown with fine pale streaks. Chin and throat variable, whitish to creamy, neither yellowish as in other races nor pure white as in Fischer's Greenbul *P. fischeri*. Rest of underparts duller creamy white, brightest on belly, with variable greyish olive wash on breast and flanks, undertail-coverts pale dull brown, sometimes washed cinnamon, thighs darker brown. Primaries and secondaries dark brown, outer webs olive-brown with rufous tinge, tertials uniform olive-brown, upperwing-coverts brown, axillaries and underwing-coverts creamy white, underside of inner webs of primaries and secondaries broadly edged pale buff. Bill horn-brown, horn-grey or yellowish grey, lower mandible much paler except for dark patch along sides of distal half, gape whitish; eye light to dark brown, grey-brown, grey, dull yellow-grey or dull brownish ochre; legs and feet slaty grey or slaty blue, soles yellow, olive-yellow or dull orange, claws pale red-brown. Sexes alike. SIZE: wing, ♂ (n = 29) 78–92 (85·0), ♀ (n = 35) 74–85 (78·6); tail, ♂ (n = 31) 78–88 (82·1), ♀ (n = 30) 71–84 (76·5); bill, ♂ (n = 30) 16·5–21·5 (19·6), ♀ (n = 34) 17–20 (18·6); tarsus, ♂ (n = 10) 22–23·5 (22·8), ♀ (n = 10) 21·5–23·5 (22·3). WEIGHT: ♂ (n = 58) 21–32 (27·5), ♀ (n = 66) 19–31 (23·5). Diurnal weight change of 11·7% (Kenya: Mann 1985).

IMMATURE: juvenile still with a few downy feathers (wing 63; tail 22; bill 13; tarsus 20) has top of head, upperparts and upperwing-coverts dark red-brown, uppertail-coverts and tail like adult, underparts blotchy olive-brown and white, paler and creamier on belly, lower flanks and undertail-coverts cinnamon, wings and tail like adult; bill black; eye warm sepia. Full-grown birds have medium red-brown upperparts, olive feathers starting to appear on lower back and rump, underparts still more or less like juvenile, blotchy brown and creamy, but lower flanks brown, less cinnamon, thighs and undertail-coverts cinnamon, tail and wings like adult except tail-feathers pointed, narrower, gape white.

NESTLING: pinkish orange above, paler below, gape yellow, inside of mouth deep yellow without markings.

P. c. sucosus (Reichenow): Kenya west of Rift, Uganda, Sudan, E Zaïre south to Mt Kabobo, Rwanda, NW Tanzania (Bukoba). Top of head and upperparts greener; chin, throat and underparts creamy yellow, sides of neck, breast and flanks washed greenish olive rather than greyish olive, pale eye-ring tinged olive, gape and inside of mouth yellow, eye sometimes with olive tinge. WEIGHT: W Kenya, ♂ (n = 25) 22–31 (26·5 ± 2·0), ♀ (n = 26) 19–26 (23·0 ± 2·2). Immature like immature *placidus* but more olive above, belly yellowish.

Birds from Oldeani and Mbulu, N Tanzania, are intermediate between *sucosus* and *placidus*. Birds from Mt Kabobo are *sucosus* (Dowsett and Prigogine 1974), not *cabanisi* (Prigogine 1960). Birds from Marungu Plateau are intermediate between the 2.

P. c. cabanisi (Sharpe): Angola, SE Zaïre, Zambia except extreme NE (Mafinga Mts), W and SW Tanzania to Ufipa Plateau. Similar to *sucosus* but yellow of underparts somewhat less bright, though considerably brighter than *placidus*. Large: wing, ♂ (n = 49) 84–98 (92·5), ♀ (n = 28) 78–88 (84·7). WEIGHT: ♂ (n = 34) 33–42 (37·1), ♀ (n = 18) 27–39 (31·9).

P. c. nandensis Cunningham-van Someren and Schifter: N Nandi Forest, W Kenya. Back greener than *sucosus*, uppertail-coverts and tail browner, less chestnut, throat and belly brighter yellow, breast and flanks darker, outer webs of remiges buffy citrine, inner webs black without any brown.

P. c. ngurumanensis Cunningham-van Someren and Schifter: Nguruman Hills, SW Kenya. Like *nandensis* but underparts less yellow, breast-band less pronounced, wings browner.

TAXONOMIC NOTE: Dowsett (1972) proposed that *placidus* and *cabanisi* be treated as separate species, because of differences in size, colour pattern and habitat; but Dowsett and Dowsett-Lemaire (1980) showed that vocally the 2 are in fact closely related. Further, Malaŵi *placidus* reacts strongly to playback of Zambia *cabanisi* song (F. Dowsett-Lemaire, pers. comm.). For these reasons we treat *placidus* as a race of *cabanisi*.

Field Characters. A common and widespread bird whose grating churrs are a familiar sound in forest undergrowth. Plumage rather undistinguished, with greenish upperparts, rufous tail, throat and underparts creamy (*placidus*) to pale yellow (*sucosus, cabanisi*). Best distinguished by pale throat and belly contrasting with dark breast. Duet distinctive, but bird does not sing much; with practice, churr can be distinguished from those of other bulbuls.

Voice. Tape-recorded (KEI, LEM, STJ). (a) Race *placidus*: contact call a churr, harsh and grating but low and usually not strident; often drowned out by other louder forest sounds. Song often takes form of a duet, one bird doing the churr, the other a whistled series of continuous double notes, 'chirry, chorry, chirry, chorry, chorry, chirry . . .', preceded by and sometimes followed by a subdued churr. Single bird may also sing without accompaniment, giving a similar whistle, 'chweeu, chweeu . . .'. Call of ♂ during pre-copulatory display, low 'pruit-pruit- pri-pri-pri, pri-pri-cho', without normal harsh tone (van Someren 1956); anxiety/alarm, a high-pitched 'squii, squii' or 'ziew-ziew-ziew-ziew'; also gives bubbling 'chut-ut' or 'chut-ut-ut'. (b) Race *cabanisi*: generally very similar to *placidus*. At start of song, both birds churr, then one breaks into the 'chirry, chorry . . .' while the other gives a loud, grating 'cheee-cheee-cheee . . .'. Another type of call, quite different from the churr, consists of a few short notes followed by a few whistled ones: 'tititi-twer-tee-choo', 'tititi-twee', 'titititi-chooee'; sometimes many short notes followed by a single whistle, run together and accelerated, with a scolding quality reminiscent of scold-call of Little Greenbul *Andropadus virens*.

General Habits. Inhabits primary and secondary forest of all types (lowland, transition, montane), forested ravines in highlands, lowland gallery forest, thick, tangled streamside brush with lianas; in SE Zaïre occupies woodlands rather than equatorial forest (Chapin 1953); in Zambia, also in *Cryptosepalum* forest and fringing *Syzygium*. Race *cabanisi* usually at lower elevations than others, e.g. in Zambia, where *cabanisi* in lowlands, *placidus* confined to montane forest of NE, and Zaïre, where *sucosus* in E highlands, 1300–2100 m, *cabanisi* in SE lowlands. In Sudan, *sucosus* found at 1400–2000 m, in Rwanda (Nyungwe) up to 2400 m; in E Africa, *placidus* ranges up to 2700 m, occasionally descending to 600 m (Sigi Forest, Amani, Tanzania: Stuart and Jensen 1981); in Malaŵi occurs at 1050–2000 m, occasionally down to 700 m.

Travels in pairs, family parties or small groups of adults (in adult groups paired birds remain close together). During breeding season, non-breeding birds join together to roam through gallery forest (Zaïre). Groups may attach themselves to mixed-species flocks, with e.g. babblers, cuckoo-shrikes, flycatchers, weavers and other bulbuls, and join other birds at a plentiful food supply such as an exposed termite trail, but more often feed on their own. Forages mainly at lower levels, in undergrowth and on ground, sometimes higher, including canopy of small trees. Where presumed competitor Yellow-streaked Greenbul *P. flavostriatus* absent, forages up to 6–7 m, sometimes up trunks and lianas to 16 m (Malaŵi: Dowsett-Lemaire 1989). Creeps through thick vegetation and vine tangles, gleans bases and lower sections of tree trunks, clinging to surface and fluttering, hops about on ground, probing among decaying leaves and moss. Small groups sometimes follow army ants *Dorylus* as ants pass through underbrush where bulbuls can perch low; bulbuls leave when ants enter open understorey (Willis 1983). Once followed swarm for 5 h 15 min. During minor intraspecific disputes around ants, raises head and throat feathers and gives growling rattle 'raaaagh'. Flicks tail and wings (one wing at a time), especially while calling but also while foraging and at rest; alarm calls accompanied by slight upward jerks of tail and one or both wings (Willis 1983). Scratches head over wing. Shy but also curious, responding to 'squeaking' as well as song-playback. Joins other birds to scold and mob owl or snake, even chameleon. Much chattering given for 1 h before sunset as birds settled into roost (van Someren 1939).

Partial moult noted after breeding, Kenya. 2 complete moults a year, SE Zaïre, in Mar–May and Oct–Dec; birds able to breed even during moult of primaries and tail-feathers, including ♀ with egg in oviduct. ♂ with large testes was moulting in all tracts, Zambia (Payne 1969). Duration of primary moult, W Kenya (n = 2) 104 and 190 days (Mann 1985).

Generally very sedentary; same party found in same small area of forest every day in Kakamega, Kenya (Zimmerman 1972); same territory occupied for 7 years, Malaŵi (Dowsett-Lemaire 1989). Some local wandering, Malaŵi, also downward movement noted during cold season.

Food. Chiefly arthropods: weevils and other beetles, moth larvae, mantids, nymphal grasshoppers, ants, spiders; also small molluscs and a few berries.

Breeding Habits. Monogamous; mated pair remains together for many seasons. In pre-copulatory display, ♂ hops about in undergrowth intermittently flicking wings and fanning tail, raises head feathers, puffs out throat, fluffs up rump feathers; moves in circle, giving special call (see *Voice*).

NEST: shallow, elliptical cup of bark fibres, tendrils, dry sedges, dead leaves and moss, bound together with cobwebs, often with additional outer layer of dead leaves, lined with fine fibres and rootlets; bits of bark and leaves hang down from outside for camouflage, giving impression of a mass of leaf debris caught in fork; ext. diam. 89 × 64, int. diam. 57 × 44, ext. depth 51, int. depth 19 (n = 1). Placed 1·5–5 m above ground, slung in lateral fork or placed in vertical fork in woody undergrowth or low bush, sometimes between 2 loops of creeper; many nests well shaded, others exposed. Built by both sexes in 7–10 days.

EGGS: 2, sometimes 3. Long ovals; pinkish white or pink, with broad wreath around large end of bold spots, blotches and scrolls of dark brown or black, rest of surface sparsely marked with small spots and thin scrolls of dark brown, undermarkings grey or grey-mauve. SIZE: (n = 2) 21·3–22·3 × 14·8–15·2; (n = ?) av. 24 × 16. Double-brooded.

LAYING DATES: Angola (breeding condition June, Aug–Sept); Zaïre, Dec–Jan, Mar–May, July (probably all year); Rwanda, Sept–Jan; Sudan, Dec–Mar; E Africa: Uganda (breeding condition Oct–Dec); Region C, Jan; Region D, Dec–July; mainly in rains; Zambia, Mar, Sept (breeding condition Aug, Oct–Nov); season extensive, starting perhaps late Aug and continuing throughout rains (Benson and Irwin 1965); Malaŵi, Oct–Jan (breeding condition and juveniles Sept).

INCUBATION: by both sexes, mainly by ♀. Period: 11–12 days, longer in cold weather. Incubating bird, especially ♀, quite tame; when approached carefully, sits tight, can be lifted to give view of eggs, even stroked; if surprised, slips off nest, drops to ground, flutters, lies on side or back with wing half open and head lying limp, i.e. playing dead; if intruder still present, turns upright and staggers a few m with trailing wing and spread tail, then topples over and lies panting; if this fails, goes a few more m with broken wing, flutters and collapses again, playing dead (van Someren 1956).

DEVELOPMENT AND CARE OF YOUNG: young fed with invertebrates only: moth larvae, carabid larvae and pupae, nymphal long-horned grasshoppers, glow-worms, spiders. Most food gathered within 20 m of nest. Brooding bird may be tame: one took larva from hand of man, then used man's hand as perch from which to feed young! (van Someren 1956). Fledging period: 16–18 days. Young remain with parents long after fledging, even while parents are building new nest, moving away only when the new clutch is laid.

BREEDING SUCCESS/SURVIVAL: exposed nests predated by mongoose, genets, goshawks, shrikes, snakes; during 5-year period, total population on census tract, central Kenya, did not increase (van Someren 1956). 3 ringed birds recaptured after intervals of $7\frac{1}{2}$–$8\frac{3}{4}$ years (Kenya: Mann 1985).

References
Dowsett, R. J. (1972).
van Someren, V. G. L. (1956).

Phyllastrephus poensis (Alexander). Cameroon Olive Greenbul. Bulbul olivâtre.

Phyllostrophus [sic] *poensis* Alexander, 1903. Bull. Br. Orn. Club 13, p 35; Bakaki, Fernando Po.

Plate 18
(Opp. p. 305)

Range and Status. Endemic resident. SE Nigeria (Obudu Plateau, Danko Forest Reserve, Mambilla Plateau, common), central Cameroon highlands (Mt Oku, Bamenda, L. Bambulue, Foto near Dschang, Mt Manenguba, Mt Nlonako, Mt Kupé, Rumpi Hills, Mt Cameroon) and Bioko. Not uncommon on Bioko when discovered at the turn of the century, but apparently no subsequent records; not found despite thorough search in 1962–63 (Eisentraut 1973). Widely distributed in Cameroon and locally not uncommon; not under immediate threat (Stuart 1986).

Description. ADULT ♂: top of head to hindneck dark olive-brown; feathers of forehead greyer, with dark fringes giving slightly scaly effect. Mantle, back, upper rump, scapulars and upperwing-coverts greenish olive-brown, becoming rufous on lower rump and uppertail-coverts. Tail-feathers dark rufous, margins paler, slightly greenish, shafts dark maroon above, straw below. Lores, pre-orbital region and feathers around eye grey, ear-coverts grey-brown; line of short white streaks from base of bill under eye and onto anterior ear-coverts. Chin and throat greyish white, breast and flanks light olive-brown, richer on lower flanks; belly whitish with creamy yellow streaks, some extending onto lower breast; undertail-coverts pale rufous-brown. Primaries and secondaries dark brown, outer webs greenish olive-brown tinged rusty; axillaries and underwing-coverts yellowish olive, inner webs of underside of primaries and secondaries margined pale yellowish buff. Upper mandible black or brownish black, lower mandible blackish, underside whitish horn, cutting edges and tips whitish; eye brown, grey-brown or hazel; legs and feet grey, blue-grey or bluish slate, soles yellowish. Sexes alike. SIZE: (13 ♂♂, 8 ♀♀) wing, ♂ 85–91 (88·1), ♀ 72–80 (75·5); tail, ♂ 77·5–88 (84·5), ♀ 72–77 (74·5); bill, ♂ 21–24 (22·6), ♀ 19–21 (19·9); tarsus, ♂ 21·5–26·5 (24·7), ♀ 21–23·5 (22·1). WEIGHT: ♂ (n = 5) 28–32 (30·0), ♀ (n = 3) 24, 24, 25, unsexed (n = 23) 23·8–37·1 (28·9).

IMMATURE: a small ♀ (wing 69; tail 47; bill 15; tarsus 22) has upperparts, wings and tail dark brownish chestnut, darker and duller on top of head and tail, rump tawny; chin and throat dirty white, dull brown band across breast, rest of underparts mainly dirty white, some rusty feathers on lower belly and undertail-coverts. Culmen dark brown, rest of bill mixed yellow and grey; eye grey-brown; feet grey, soles yellow. Another, slightly smaller, bird (wing 64; tail 29; bill 13; tarsus 23) is similar, but rusty feathers on flanks, not on belly, undertail-coverts all rusty.

NESTLING: (eyes just open, skin transparent, no feathers showing) bill greyish, edges and tip yellow, gape yellow, inside of mouth orange-yellow, tongue dusky, feet fleshy grey.

Field Characters. A medium-sized, undistinguished-looking bulbul, brown above, pale below, with white throat and rather long dark reddish tail; in looks, voice and behaviour more reminiscent of Cabanis's Greenbul *P. cabanisi* of eastern Africa than of any local species. Few bulbuls share its montane habitat and none resemble it; Cameroon Montane Greenbul *Andropadus montanus* is entirely olive-green; Mountain Greenbul *A. tephrolaemus* is green with grey head, yellowish belly; Grey-headed Greenbul *P. poliocephalus* has bright yellow underparts.

Voice. Tape-recorded (LEM). Song, a series of dry, unmusical notes, 'chewp, chop, chop, chip, chip, chewp, chip, cher . . .'. Contact call, a low, grating 'churrr', 'charrr' or 'chorrr', reminiscent of churring calls of Cabanis's Greenbul.

General Habits. Inhabits montane forest and forest-filled ravines on plateaux at 1000–2200 m. Prefers higher elevations than Grey-headed Greenbul but absent from highest elevation forest on Mt Cameroon and Bamenda Highlands (Stuart 1986).

Occurs in pairs and small (family?) parties. Forages in thick foliage, especially in tangles of creepers, at all levels from undergrowth to canopy, and around clearings. Phlegmatic and unobtrusive, moving slowly and easily overlooked. Gleans leaves, tree trunks and moss-covered branches, frequently flicking wings and tail. Joins mixed-species flocks at army ant swarms.

Food. Insects and small fruits.

Breeding Habits.
NEST: only 2 described (Serle 1954). One was an open, regular cup of loosely assembled moss and a few leaf skeletons, with an inner layer of fine tendrils and lining of fibres; ext. diam. 85, int. diam. 55, ext. depth 90, int. depth 40; placed 1·5 m above ground in low bush in forest, resting on 2–3 moss-covered twigs, well camou-

flaged (moss of exterior matched moss of twigs); the other was built of leaves, lined with fibres, slung between horizontal twigs of shrub with cobwebs and rootlets.

EGGS: 2 (1 clutch). Long ovals; smooth, glossy; pale pink, largely obscured by grey mottling and intricate pattern of darker grey bunting-like scrawls, overlain with blotches and scrawls of dark brown, mainly in wreath around large end. SIZE: (n = 2) 25·1 × 14·9, 25 × 15·1.

LAYING DATES: Nigeria, Dec (immature June); Cameroon, Nov–Apr, mainly Nov–Jan, in dry season; Bioko, Nov.

References
Serle, W. (1954).
Stuart, S. N. (1986).

Plate 21
(Opp. p. 352)

Phyllastrephus icterinus (Bonaparte). Icterine Greenbul. Bulbul ictérin.

Trichophorus icterinus 'Temm.', Bonaparte, 1850. Consp. Gen. Av. 1, p. 262; Guinea.

Range and Status. Endemic resident. Sierra Leone, Liberia (north to Mt Nimba), Ivory Coast (abundant, north to Sipilou, Comoé Nat. Park), S half Ghana, S Nigeria (forest zone only), S Cameroon (very common, north to Mamfe, Ngoumé). Equatorial Guinea, Gabon (abundant), Congo, N Angola (Cabinda), SW Central African Republic, Zaïre (common to abundant throughout main central forest bloc, also in lower Congo (Mayombe)); Uganda forests in W (Bwamba, Bugoma, Budongo) and S (Malabigambo). Density in Nigeria (Ogba Forest) 1 bird/5 ha (Johnson 1989).

Description. ADULT ♂: top of head and upperparts, including scapulars and lesser coverts, olive-green (median and greater coverts same but browner), paler olive band on lower rump; uppertail-coverts with slight greenish rufous tinge; rump feathers long and fluffy. Tail-feathers dark rufous narrowly edged greenish, shafts pale red-brown above, yellowish straw below. Indistinct pale line from lores over and behind eye; inconspicuous eye-ring dull yellow; cheeks and ear-coverts olive-green with a few fine pale shaft streaks. Chin and throat sulphur yellow; sides of breast and flanks greenish or yellowish olive, paler than upperparts, centre of breast yellow with olive wash, lower breast, belly and undertail-coverts sulphur yellow, latter sometimes with olive wash. Primaries and secondaries dark brown, primaries edged dull pale olive, secondaries with outer webs brownish olive-green, axillaries and underwing-coverts sulphur yellow, inner webs of underside of primaries and secondaries margined pale yellowish buff. Bill, upper mandible black to greenish or brownish black, lower mandible grey, greenish grey or greenish yellow, cutting edges and tip yellow; gape pale yellow; eye brown, brown-ochre, ochre-grey or light brownish grey; legs and feet bluish grey or bluish green, soles pale green or greenish yellow. Sexes alike. (Birds from Cameroon to Uganda, '*tricolor*', average greener above and duller yellow below; a specimen from Sierra Leone is also greener above, and we concur with Louette (1981) that no races are distinguishable.) SIZE: (12 ♂♂, 12 ♀♀) wing, ♂ 70–79 (74·7), ♀ 64–70 (67·6); tail, ♂ 64–71 (67·5), ♀ 56–63 (60·3); bill, ♂ 17–20·5 (18·2), ♀ 16–19 (17·3); tarsus, ♂ 17·5–19·5 (18·4), ♀ 17·5–19·5 (18·4). WEIGHT: ♂ (n = 28) 16·5–24·5 (20·3), ♀ (n = 19) 15–22 (17·6).

IMMATURE: like adult but greener above, slight brownish suffusion on breast and throat; tail-feathers more pointed.

NESTLING: unknown.

Field Characters. A small greenbul with green upperparts, yellow underparts and rusty tail, often one of the commonest bulbuls in its area and a component of most mixed-species flocks. Distinguished from all yellow-bellied species except Xavier's Greenbul *P. xavieri* by small size and voice. Almost identical in plumage to Xavier's Greenbul and often found in same flock, although tends to forage at somewhat lower levels. Smaller overall, but ♂♂ are larger than ♀♀ in both species, and ♀ Xavier's overlaps in size with ♂ Icterine. Large ♂ Xavier's and small ♀ Icterine might be distinguishable by direct size comparison. Uppertail-coverts reddish (greenish in Xavier's), and tail brighter red.

Voice. Tape-recorded (53, C, CART, ERA). Typical call an insistent, repeated 'gur-guk' or 'gur-gur-gur'; also some hard trills. Contact call 'chip-paaa', 'chip-chipaaa', 'gur-gur-gaaa', or 'naa-jaa-jaa-jaa-jaa-jaa-jaa'; the 'paaa' or 'gaaa' sound has a characteristic nasal quality.

General Habits. Inhabits primary and old secondary forest, mainly in undergrowth of interior, but follows other species to edges and clearings; also in dry seasonal swamp forest, forest–grassland mosaic, and cocoa plantations (Bioko). Lowland and lower transition zones, to 1100 m in Cameroon, 1250 m in Uganda.

Travels in family parties of 3–5, probably a pair with young of last brood, sometimes in groups of up to 12. Family groups stay together for considerable periods; 1 bird ringed as juvenile recaptured with same 2 adults 10 months later (Brosset and Erard 1986). Cohesion of group maintained by nasal contact call, frequently given. Alarm call of bird caught in net gathers whole group. Group defends territory, mainly by song, but when 2 groups come in contact they also display with spread tail and wings, and sometimes actual fighting takes place. Sometimes forages in monospecific flocks but much more typically in mixed-species flocks, of which it is one of the most constant members; noted in 183 of 271 such flocks (Gabon: Brosset and Erard 1986). In mixed flock may act as follower, eating insects flushed by others, or as leader, discovering concentrations of caterpillars or butterflies and attracting other species to it. Forages in undergrowth and shrubs, 2–8 m above ground; gleans leaves, twigs, small branches and creeper tangles; captures fleeing prey in fluttering flight. Follows diurnal mammals such as squirrels *Funisciurus*, rats *Hybomys* and small antelopes, catching insects they displace while moving through undergrowth. Sometimes attends army ants *Dorylus*, remaining in foliage 1–10 m above swarm, usually briefly, once for an hour (Willis 1983). Readily mobs potential predator (owl, snake, pangolin, human); alarm calls cause other bird species in flock to flee or freeze. Bathes frequently during heat of day, also in late afternoon, in hollows in trees. One bathed in same spot as Xavier's Greenbul and Green-tailed Bristlebill *Bleda eximia*; individuals bathed in order of decreasing size (Brosset and Erard 1986).

Sedentary in Gabon; of 64 birds ringed, 36 recaptured at least once in same spot; some ringed adults and immatures recaptured many times in up to $7\frac{1}{2}$ years (Brosset and Erard 1986). Some birds observed up to 800 m from point of ringing.

Food. Chiefly insects: Orthoptera, Hemiptera, Neuroptera, flies, beetles, moths, caterpillars (mainly under 30 mm, not over 60 mm: Gabon, Brosset and Erard 1986), mantid larvae, termite alates, ants; spiders; once a snail (Johnson 1989); also fruit and small seeds. Nestlings fed larvae, including caterpillars, some Orthoptera and spiders.

Breeding Habits. Territorial, monogamous. Pair bond long-lasting; both members of pair recaptured in same net several times, once after 5 years 1 month (Gabon: Brosset and Erard 1986).

NEST: small cup of dry leaves with a few twigs and rootlets, bound by strands of fungus *Marasmius*, sometimes also lined with it, with dead leaves hanging down from outside to break up shape (**A**); ext. diam. from 50 × 60 to 80 × 90, thickness *c*. 10, ext. depth 40–70, int. depth 30–60; slung like hammock with cobwebs and *Marasmius*, in horizontal fork toward end of branch, sometimes in open, sometimes under large leaf, 0.65–11 m above ground in bush, shrub or liana.

EGGS: 2 (n = 41), occasionally 1 (n = 2). Glossy, reddish or brownish pink, finely speckled brown or carmine, with characteristic more or less solid band of dark brown, 4–5 mm wide, around middle of egg towards large end. SIZE: 17–20·9 × 12–14·3. (Note: description and measurement of eggs by Bates (1927) may not be of this species: Prigogine 1984.)

A

LAYING DATES: Liberia, (breeding condition Aug, Oct); Ghana (breeding condition Sept); Nigeria, (breeding condition Nov, Jan; building nest Jan); Cameroon, Jan, June, Sept (breeding condition Nov–Dec); Bioko, Nov; Gabon, Oct–Apr (fledged young just out of nest June, Aug); probably all year, with peak at end of rainy season and in short dry season; Zaïre, all year; Uganda (breeding condition Jan–Feb).

INCUBATION: by ♀ only. Period: 14 days. When surprised on nest, incubating bird falls to ground, does rodent-run distraction display. Some pairs unafraid of man during incubation and rearing of young, others remain wild.

DEVELOPMENT AND CARE OF YOUNG: single young in nest fed alternately by parents 18 times in 3 h (Brosset and Erard 1986); in other nests, 70 feeds in 18 h (Brosset 1971b). Orthoptera broken up by being beaten on branch before fed to young. Young grow fast, leave nest at 12 days when still unable to fly, remain hidden among leaves where fed by parents; at 20–25 days join parents in mixed-species flocks.

BREEDING SUCCESS/SURVIVAL: of 38 nests, Gabon, 27 destroyed by predators. Of 11 clutches which produced flying young, one egg was infertile in 4 cases, 1 young disappeared in 3 cases; in 8 of 11 nests, only 1 young raised (Brosset and Erard 1986). Lives at least $7\frac{1}{2}$ years.

Reference
Brosset, A. and Erard, C. (1986).

Plate 21
(Opp. p. 352)

Phyllastrephus xavieri (Oustalet). Xavier's Greenbul. Bulbul ictérin tâcheté.

Xenocichla Xavieri Oustalet, 1892. Naturaliste, ser. 2, 6, p. 218; Bangui, on Ubangi R.

Range and Status. Endemic resident. Cameroon (fairly widespread in S, north to c. 5°N in region north and west of Mt Cameroon); Gabon (uncommon), E and S Congo (Gamboma, Mayombe), SW Central African Republic, main forest bloc of E central Zaïre (very common in Itombwe). Fairly widespread in W and S Uganda from Maramagambo and Kalinzu forests north to Bwamba and Budongo, east to Butambala and Malabigambo and adjacent Minziro Forest in NW Tanzania. Locally common but numbers difficult to assess due to confusion with Icterine Greenbul *P. icterinus*.

Description. *P. x. xavieri* (Oustalet) (including '*P. icterinus sethsmithi*'): range of species except *serlei*. ADULT ♂: top of head and upperparts, including scapulars and lesser coverts, olive-green (median and greater coverts same but browner), paler olive band on lower rump; uppertail-coverts olive or greenish with variable light reddish wash; rump feathers long and fluffy. Tail-feathers rufous washed greenish, shafts dull red-brown above, pale straw below. Indistinct pale line from lores over and behind eye; inconspicuous eye-ring dull yellow; cheeks and ear-coverts olive-green with a few fine pale shaft streaks. Chin and throat sulphur yellow; sides of breast and flanks greenish or yellowish olive, paler than upperparts, centre of breast yellow with olive wash, lower breast, belly and undertail-coverts sulphur yellow, latter sometimes with olive wash. Primaries and secondaries dark brown, primaries edged dull pale olive, secondaries with outer webs brownish olive-green, axillaries and underwing-coverts sulphur yellow, inner webs of underside of primaries and secondaries margined pale yellowish buff. Upper mandible black to dark brown, lower mandible light grey, greenish grey or yellowish, cutting edges yellow, tip light grey-green, gape yellow; eye brown to grey, pale yellowish grey or sandy; legs and feet bluish or greenish grey, brown or flesh, soles yellowish. Sexes alike, but ♂♂ larger. SIZE: (12 ♂♂, 12 ♀♀) wing, ♂ 82–91 (86·7), ♀ 71–76 (73·8); tail, ♂ 75–85 (79·8), ♀ 65–71 (68·3); bill, ♂ 20–24 (22·0), ♀ 17–21 (18·4); tarsus, ♂ 20–23 (21·8), ♀ 17–19·5 (18·6). WEIGHT: ♂ (n = 28) 23–31 (26·3), ♀ (n = 27) 16–22 (19·5).

IMMATURE: somewhat darker above, paler yellow below, more heavily washed olive on breast, lower mandible mainly yellow, tips of both mandibles more extensively yellow, gape broadly yellow.

NESTLING: unknown.

P. x. serlei Chapin: north and west of Mt Cameroon. Marginal; somewhat paler, duller, less bright yellow below, with more pronounced olive wash across breast.

Phyllastrephus xavieri

Field Characters. A larger version of Icterine Greenbul, q.v., although ♀ Xavier's and ♂ Icterine overlap in size. In the field, throat of Xavier's Greenbul appears paler than breast, while in Icterine Greenbul throat and breast same colour (Brosset and Erard 1986), although this is not apparent in skins. Colour of uppertail-coverts of Xavier's Greenbul appears overall greenish rather than reddish, and tail less red; these differences are also visible in the field (Brosset and Erard 1986). Tends to forage at higher levels, but the 2 species may flock together.

Voice. Not yet described, but the voices of *P. xavieri* and *P. icterinus* are being studied by F. Dowsett-Lemaire (pers. comm., Oct 1990). Vocalizations appearing under this species on Chappuis's disc #5 (1975) are now known to have been made by the Golden Greenbul *Calyptocichla serina* (C. Chappuis, pers. comm.).

General Habits. Inhabits primary and secondary forest, including seasonal dry swamp forest, both interior and edge, also large trees left standing in cultivated clearings. Lowland and transition zones, to 900 m in Cameroon, 1470 m in Zaïre, 1500 m in Uganda.

Travels in pairs and family groups of 4–6; regularly joins mixed-species flocks of insectivores. Forages at all levels, from undergrowth to canopy, but with a distinct preference for upper levels, above 15 m. Works actively through foliage, gleaning underside of leaves and capturing insects in twisting flight.

Sedentary, at least in Gabon, where bird ringed as juvenile recaptured in same spot 5 years 8 months later (Brosset and Erard 1986).

Food. Mainly insects: beetles, moths, caterpillars, Orthoptera and their eggs, mantid larvae; spiders; also a few small berries and seeds.

Breeding Habits. Territorial.
NEST: cup of dry leaves lined with fine rootlets; ext. diam. 80, int. diam. 55–60, ext. depth 55–60, int. depth 35–40; slung in fork of thin branch or placed between 2 small branches, 1–2·5 m above ground in bush (e.g. *Alchornea*) in dense undergrowth.

EGGS: 2. Bright pink to rose-beige, with spots and streaks of dark brown or blackish over ashy undermarkings, concentrated in conspicuous band around base of large end which is less strongly marked. SIZE: (n = 5) 20·3–22·8 × 15·6–16·1.

LAYING DATES: Cameroon (breeding condition Aug); Gabon, Feb (singing increases during rains which follow long dry season and during short dry season); Zaïre, Nov–June in Itombwe, probably all year at equator; Uganda (breeding condition Feb, Apr, June–July, juvenile May).

BREEDING SUCCESS/SURVIVAL: lives a minimum of 5 years 8 months.

Reference
Brosset, A. and Erard, C. (1986).

Phyllastrephus leucolepis Gatter. Liberian Greenbul. Bulbul ictérin tâcheté.

Plate 21
(Opp. p. 352)

Phyllastrephus leucolepis Gatter, 1985. J. Orn. 126, pp. 155–156; Zwedru, Liberia.

Range and Status. Known only from 2 isolated patches of rain forest 20 km northwest of Zwedru, near Cavalla R., 6° 12′N, 8° 11′W, Grand Gedeh County, Liberia; several birds observed there, 1981–84.

Description. Sex of single existing specimen unknown. Forehead, top of head, nape, hindneck, back and scapulars olive-green, rump cinnamon washed olive-green. Tail cinnamon. Lores, ring of small feathers around eye, cheeks and ear-coverts olive with some fine yellow streaking. Chin, throat and underparts sulphur yellow, breast and flanks washed olive. Wing feathers dark brown, outer webs olive-brown. On the primaries, secondaries, greater coverts, alula, and medium and greater primary coverts is a subterminal band of white, 7–12 mm broad on primaries and secondaries, 3–4 mm broad on coverts. Upper mandible black, lower mandible, cutting edges and tip horn, base of lower mandible blackish; eye colour unknown; legs and feet grey-horn. SIZE: wing, fresh 74–75, dry 71–72; tail unknown; bill 17, with distinct curve near tip; tarsus 23.

IMMATURE AND NESTLING: unknown.

Field Characters. Combination of yellow underparts and white wing spots renders it unmistakable. Spotted Greenbul *Ixonotus guttatus* has pure white underparts and travels in noisy groups giving distinctive call; no call yet heard from Liberian Greenbul, which moves about in pairs or singly. Western Nicator *Nicator chloris* also has white wing spots but is large and sluggish, with heavy bill and white underparts except for yellow throat and undertail-coverts.

Voice. Unknown.

General Habits. Inhabits rain forest, known so far only in transition zone between evergreen and semi-deciduous forest. Habitat includes the following trees and shrubs (from Gatter 1985): (a) in undergrowth up to *c.* 20 m high (principal habitat): *Diospyros ?manni, D. gabonensis, Scotellia coriacea, Coula edulis, Necepsia afzelia, Cola buntingii*; (b) in mid-levels (20–35 m high): *Gilbertiodendron preussi, Heritiera utilis, Coula edulis, Uapaca* sp.; (c) in upper levels (35–50 m high): *Gilbertiodendron preussi, Heritiera utilis, Chrysophyllum albidum, Lophira alata*. In 12 ha of habitat there were 108 tree species with chest-high diam. of > 10 cm.

Joins mixed-species flocks of other bulbuls, shrikes *Malaconotus*, sunbirds *Nectarinia* and malimbes *Malimbus*. Forages mainly 4–8 m above ground in branches, hopping about and flicking both wings at the same time; white bands conspicuous when wings flicked, suggesting this may act as visual signal in dark forest. Ecological niche appears to lie between those of Icterine Greenbul *P. icterinus* and sunbirds (leaves and twigs) and of Yellow-bearded and Western Bearded Greenbuls *Criniger olivaceus* and *C. barbatus* (thick branches, trunks, lianas). Frequents branches close to trunk; Icterine Greenbul prefers smaller outer branches.

Food. Unknown.

Breeding Habits. Unknown.

Reference
Gatter, W. (1985).

Plate 19
(Opp. p. 320)

Phyllastrephus albigularis (Sharpe). **White-throated Greenbul. Bulbul à gorge blanche.**

Xenocichla albigularis Sharpe, 1881. Cat. Bds Br. Mus. 6, p. 103, pl. 7; Fantee.

Range and Status. Endemic resident. Extreme SW Senegal (Basse Casamance), Sierra Leone, SE Guinea, Liberia, Ivory Coast (coast north to Mt Nimba, Maroué Nat. Park); Ghana (coast north at least to Bia Nat. Park and Mampong); S Togo, S Benin, Nigeria (widespread in S, also on Jos Plateau); Cameroon (rather rare), Gabon (widespread, common to uncommon), S Congo (Mayombe: Dowsett-Lemaire and Dowsett 1989), Angola (rare: Cuanza Norte (Canzele), Cuanza Sul (Gabela)); Zaïre (rare in W and S, known from Duma (NW) and Kunungu (lower Congo R.); and from Kasai and Sankuru districts east to Kivu (scarce), becoming common only in NE); SW and SE Central African Republic (common on Ouossi R.), extreme S Sudan (common). Widespread and common to abundant in W and S Uganda north of equator, from Budongo, Bugoma and Kibale across to Mt Elgon. Distribution curiously patchy; common in some areas, apparently rare or absent in others where it might be expected. Density in Gabon 10–12 pairs/km^2; in Nigeria (Ogba Forest) 1 bird/10 ha (Johnson 1989).

Description. *P. a. albigularis* (Sharpe): range of species except for *viridiceps*. ADULT ♂: forehead to hindneck dark brownish or greyish olive, feathers variably tinged darker, giving slightly scaly effect only visible at close range. Upperparts, including scapulars and upperwing-coverts, greenish olive, becoming paler on lower rump, uppertail-coverts rufous. Tail-feathers dark reddish brown edged greenish, shafts red-brown above, straw below. Lores and area around eye grey, indistinct narrow eye-ring grey-white, cheeks and ear-coverts grey-brown with fine pale streaks, prominent black rictal bristles. Chin and throat white; breast and flanks greenish olive, centre of breast paler with some yellow streaks; lower breast and upper belly pale yellow mixed with some pale greyish or olive feathers, lower belly pure sulphur yellow, undertail-coverts yellow tinged olive. Primaries and secondaries dark brown edged olive-green, axillaries and underwing-coverts pale yellow, sometimes tinged olive, inner margin of undersides of primaries and secondaries pale yellowish buff. Bill black, dark grey or dark brown, cutting edges and tip pale, lower mandible paler, especially on underside; eye dark brown to pale brown, pale grey or whitish; legs and feet grey, lead grey or bluish flesh. Sexes alike, but ♂♂ larger. SIZE: (12 ♂♂, 12 ♀♀) wing, ♂ 80–89 (83·5), ♀ 66–76 (72·8); tail, ♂ 72–82 (77·0), ♀ 61–68 (65·1); bill, ♂ 20–22 (21·2), ♀ 18–20 (19·0); tarsus, ♂ 19·5–22 (20·4), ♀ 18·5–20·5 (19·4). WEIGHT: ♂ (n = 23) 24–31 (27·5), ♀ (n = 24) 17–28 (21·9).

IMMATURE: full-grown bird is like adult, but gape, cutting edges and tips of mandibles yellow, eye light brownish grey, rim of eyelids greenish grey or pale greenish, feet light bluish grey, toes and claws yellowish. A younger ♀ (wing 63; tail 63; bill 18; tarsus 19·5) has breast darker and browner, lower mandible entirely yellow.

NESTLING: unknown.

P. a. viridiceps Rand: Angola. Top of head washed green, upperparts greener, throat washed pale yellowish buff.

Field Characters. A small, slender greenbul with proportionately rather long bill; greenish, with pale yellow underparts and white throat contrasting with darker breast. Red-tailed Greenbul *Criniger calurus* also has conspicuous white throat, greenish upperparts and reddish tail, but is larger and heavier, underparts mixed dark green and brighter, more orange-yellow, voice very different; Cabanis's Greenbul *P. cabanisi sucosus* has similar upperparts but yellowish throat, duller, olive-yellow underparts. Xavier's and Icterine Greenbuls *P. xavieri* and *P. icterinus* have entire underparts yellow, including throat.

Voice. Tape-recorded (53, ERA). Song, a series of short (2–3 s), bustling phrases of rapidly-repeated notes, variable in form and speed but usually rising in pitch and intensity in the middle and then trailing off; notes not musical but clear, without burry bulbul quality. Calls, harsh 'jerr-it' and a variety of hard, grating churrs and trills.

(*Note*: the 2 cuts from Gabon appearing under this species on Chappuis's disc #5 (1975) are now known to have come from the Icterine Greenbul; only the cut from Ivory Coast comes from White-throated Greenbul: C. Chappuis, pers. comm.)

General Habits. Inhabits primary and secondary forest, mainly lowland, but up to 1850 m in Uganda. Occurs in interior of forest but prefers edge situations where there are matted layers of lianas and other tangled vegetation, around clearings and places where fallen trees have produced openings in forest, and at edge of forest bordering rivers. Within forest mainly in tangled undergrowth, but also in middle and upper levels, as far as lower canopy, especially in long dry season (Gabon: Brosset and Erard 1986).

Occurs in pairs and small family parties; regularly joins mixed-species flocks. Quick and agile as it works through vine tangles, gleaning leaves and stems; also gleans tree branches. Sexual size dimorphism so great that ♂ and ♀ may behave as 2 'ecological species' (Dyer *et al.* 1986). Territorial all year; reacts violently to playback of song, returning again and again to source of sound even when human present (Chappuis 1975). Flicks wings excitedly while scolding.

Sedentary in Gabon. Ringed birds recovered at same spot, 1 after 11 months, another after 6 years (Brosset and Erard 1986).

Food. Mainly insects and their larvae: Orthoptera, Neuroptera, Hemiptera, Coleoptera, Lepidoptera and their larvae, ants; also spiders, small gastropods and some fruit.

Breeding Habits. Territorial.

NEST: nest (not described) situated in low bush in forest, in fork of branch or slung between 2 branches.

EGGS: 2. Ovate or long elliptical ovals; high gloss; buff, pinkish beige or red, with spots and scrawls of brown, violet or purplish over fine grey undermarkings, concentrated around large end. SIZE: (n = 4) 21·2–24·7 × 15·2–15·9 (22·8 × 15·5).

LAYING DATES: Nigeria (breeding condition Aug–Nov); Cameroon (breeding condition Aug); Gabon (juvenile July, breeding condition Dec, singing intensifies at beginning of rainy seasons and during short dry season); Zaïre (breeding condition Aug; 'mainly toward end of rainy season': Lippens and Wille 1976); Angola (breeding condition Feb); Central African Republic (breeding condition May); Uganda, Mar–May.

INCUBATION: by ♀ only.

BREEDING SUCCESS/SURVIVAL: lives to a minimum of 6 years.

References
Brosset, A. and Erard, C. (1986).
Harrison, C. J. O. and Parker, S. A. (1965).

Phyllastrephus flavostriatus (Sharpe). Yellow-streaked Greenbul. Bulbul à stries jaunes.

Andropadus flavostriatus Sharpe, 1876. Ibis, p. 53; Macamac, Transvaal.

Forms a superspecies with *P. poliocephalus*.

Range and Status. Endemic resident. SE Kenya (2 old specimens, Fort Hall and Taita Hills; current status unknown); Tanzania: coast (Pugu Hills, Lindi, Mikindani), E highlands (S Pares, Usambaras, Nguru, Ukagurus, Ulugurus, Uzungwa), SW (Ufipa Plateau) and W (Kungwe-Mahari Mt.); SW Uganda (Impenetrable Forest and Kigezi District); W Rwanda, N Burundi and highlands of E Zaïre from west of L. Albert to Mt Kabobo; south to extreme NE Zambia (Nyika Plateau, Mukutu and Mafinga Mts), N and SE Malaŵi (not west of Rift south of Ntchisi), Mozambique (N highlands adjacent to S Malaŵi, near coast at Netia, central highlands at Inhamitanga and Gorongoza and along Zimbabwe border, and in extreme S in Lebombo Mts), Zimbabwe (E highlands from Inyanga to Mt Selinda), Transvaal (Escarpment region from Soutpansberg to Kangwane), E Swaziland (Lebombo Mts), Zululand (Ngoye Forest, isolated records L. St Lucia) and E Cape Province (coastal Transkei). Common to locally abundant; density in N Malaŵi 3–4 pairs/10 ha (Dowsett-Lemaire 1989).

Phyllastrephus flavostriatus

Description. *P. f. olivaceogriseus* Reichenow (including '*itombwensis*', '*ruwenzorii*'): Rwenzori Mts, SW Uganda, W Rwanda and N Burundi (Nyungwe and Bugoie Forests), and west of Rift Valley in highlands northwest of L. Tanganyika (Itombwe) and Mt Kabobo. Absent from Virunga Volcanos (Prigogine 1973). ADULT ♂: top and sides of head and hindneck grey, lores and narrow eye-ring whitish, fine pale streaks on ear-coverts. Mantle, back, rump and scapulars grey with variable amount of olive-green wash; uppertail-coverts olive-green; tail-feathers brown with olive-green wash, most pronounced on edges, all but inner pair with pale yellow margins to inner webs, shafts brown on upperside, white on underside. Chin and throat white, rest of underparts, including thighs and undertail-coverts, white, with grey wash on breast and flanks, lightly streaked yellow from lower breast to undertail-coverts (sometimes heavily streaked). Wings, including coverts, bright greenish, inner webs of flight feathers brown, edged whitish yellow on underside, broadest on secondaries; axillaries and underwing-coverts lemon yellow. Bill dark brown or black, base of lower mandible greyish, cutting edges yellow; inside of mouth yellow; eye dull grey, blue-grey or light grey, sometimes with tinge of olive or lavendar; legs and feet light grey to lead grey. Sexes alike, but ♂♂ larger. SIZE: wing, ♂ (n = 27) 94–106 (99·1), ♀ (n = 30) 78–90 (86·1); tail, ♂ (n = 27) 81–100 (92·8), ♀ (n = 30) 70–88 (81·5); bill, ♂ (n = 27) 20·5–24 (21·8), ♀ (n = 30) 17·5–21 (19·2); tarsus, ♂ (n = 6) 24–27 (25·3), ♀ (n = 10) 22–26 (23·2). WEIGHT: ♂ (n = 13) 32–42 (36·8), ♀ (n = 16) 23–34 (28·9).

IMMATURE: breast dusky olive-green, centre of belly pale yellow.

NESTLING: unknown.

TAXONOMIC NOTE: '*itombwensis*' (Prigogine 1973) has more olive or olive-brown wash on back, producing greater contrast with grey crown; chin and throat off-white (not white), grey on underparts darker, yellow streaking somewhat reduced; '*ruwenzorii*' (Prigogine 1973) has upperparts like *itombwensis* but slightly yellower underparts, especially on undertail-coverts. Since individual variation is considerable, however, we prefer not to assign them names.

P. f. graueri Neumann: Zaïre in highlands west of lakes Albert, Edward and Kivu. Head like *olivaceogriseus* but back greener, rump and uppertail-coverts gingery brown, tail-feathers richer brown although with same green edgings. Wing-coverts and edges of flight feathers greenish rufous, not green. Less grey wash on underparts, lower flanks olive; yellow streaking on breast and upper belly heavier, coalescing on lower belly to become solid yellow; underwing brighter lemon yellow.

P. f. kungwensis Moreau: Kungwe-Mahari Mts, W Tanzania. Very like *graueri*; grey of crown a little darker, nape tinged olive; belly and undertail-coverts paler yellow, latter with brownish wash. (Differences sufficiently small for it to be merged with *graueri* were their ranges contiguous, but they are separated by *olivaceogriseus* and so subspecific status best retained.)

P. f. tenuirostris Fischer and Reichenow: Kenya and E Tanzania to lowlands of Mozambique north of Zambezi. Above like *kungwensis* but wings duller – brownish olive not greenish rufous; yellow of underparts considerably paler, flanks olive-green, undertail-coverts brownish yellow. WEIGHT: Tanzania, ♂ (n = 12) 30·9–39·6 (34·7 ± 2·5), ♀ (n = 8) 24·2–30·1 (26·8 ± 1·8).

P. f. vincenti Grant and Mackworth-Praed: highlands of SE Malawi east of Shire R. valley and adjacent N Mozambique. Generally darker and greyer below than *tenuirostris*, throat duller white, yellow streaks narrower, less bright, undertail-coverts more olive than yellow.

P. f. flavostriatus (Sharpe) (including '*distans*', '*dendrophilus*', '*dryobates*'): South Africa, Zimbabwe and lowlands of Mozambique south of Zambezi. Upperparts duller and browner than above races; brownish wash on grey of head; wings and tail mainly brown with only hint of green edging; wash on underparts grey-brown, yellow streaks pale and narrow, about as in *olivaceogriseus*. SIZE: wing, ♂ (n = 23) 90–100 (94·3), ♀ (n = 40) 75–94 (82·3); bill, ♂ (n = 22) 22·7–27·1 (24·8), ♀ (n = 39) 19·2–24·1 (21·9). WEIGHT: Zimbabwe, ♂ (n = 15) 30·8–39·8 (33·9 ± 2·5), ♀ (n = 11) 26·2–30·1 (29·0 ± 1·2).

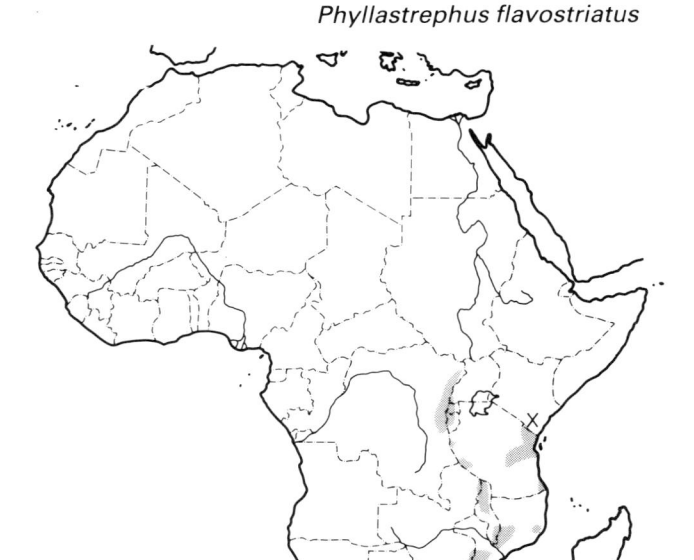

P. f. uzungwensis Jensen and Stuart: Mwanihana Forest, Uzungwa Mts, E Tanzania. Identical to *tenuirostris* except that yellow streaks merge on belly to form bright yellow patch, conspicuous in the field.

P. f. alfredi Shelley: SW Tanzania, Zambia, N Malaŵi. Differs from all other races in having top of head and hindneck brown rather than grey, upperparts brown with an olive wash, uppertail-coverts dark rufous, tail-feathers reddish brown with greenish brown outer margins, wing-coverts and outer margins of secondaries and inner primaries rufous-brown. Below, paler and whiter than other races; closest to *kungwensis* but patch on sides of breast brown, not grey, lower belly whitish, undertail-coverts pale brownish. WEIGHT: ♂ (n = 11) 28·1–35·8 (31·8), ♀ (n = 14) 22–32·4 (25·2).

TAXONOMIC NOTE: *P. f. alfredi* is distinctive; it has been treated as a full species, and is regarded by Hall and Moreau (1970) as an incipient species. However, its eggs are very similar to those of *P. f. vincenti* and there is no difference in general habits or voice (Benson 1951, confirmed by tapes of F. Dowsett-Lemaire and S. Keith); we regard it as a race of *P. flavostriatus*.

Field Characters. A common, noisy bulbul, not at all shy, frequently seen in mixed bird parties. Underparts look pale in the field, yellow streaking not apparent except in very yellowest races in west of range, and in race from Uzungwa Mts, Tanzania. Best distinguished by foraging methods and wing-flicking habit. Grey crown and greenish upperparts give superficial similarity to Tiny Greenbul *P. debilis*, but much larger and longer-billed, with darker eye, different voice; bolder and more conspicuous.

Voice. Tape-recorded (14, 32, 86, 88, GRI, McVIC, WALK). Song, a series of measured, loud, down-slurred notes, clear but with a slight overtone of burry bulbul quality, descending the scale and sometimes speeding up at the end. Number of notes very variable, typically 6–9, range 4–11. In typical E African song, 4–5 down-slurred notes are followed by 3–4 somewhat faster ones all on one pitch – 'chip, cheer, cheer, choy, chewy, tu-tu-tu'. Songs of southern birds tend to have longer series of down-slurred notes and lack faster ones at the end. Song almost always introduced by a variety of conversational and call notes; e.g. a rattling trill, 'tirrrrrit', 'tirrr-tit' or 'tititititit'; loud 'chap', 'chip' or 'chop'; low, confiding notes, 'cher-dut' or 'cher-da-da'. Sometimes only a few introductory notes but often a considerable preamble, leading straight into the song proper without a break. Intervals between songs vary, and, combined with irregular introduction, give casual quality to the performance. Has wide variety of chattering calls; typically a loud, repeated 'chip' and the rattles described above.

General Habits. Inhabits rain forest, chiefly montane, including small, isolated patches; also forest edge, riverine forest and thick bush; sometimes bamboo. Wide altitudinal range, from near sea level in S to 3000 m in central Africa.

Occurs singly or in pairs, but more often in groups, typically 5–6 (once 15 noted in bird party: Long 1961). Group may forage on its own but frequently joins mixed-species flocks, which include other bulbuls, cuckoo-shrikes *Coracina*, *Campephaga*, babblers *Trichastoma*, warblers *Macrosphenus*, flycatchers *Batis*, *Trochocercus*, shrikes *Dryoscopus*, drongos *Dicrurus*, white-eyes *Zosterops* and weavers *Ploceus*. A small group regularly followed a pair of Tullberg's Woodpeckers *Campethera taeniolaema* (Rwanda: Kunkel and Kunkel 1969). Forages at all levels, most often in middle and upper strata but sometimes in undergrowth and on rotten logs. Gleans leaves and bark, especially bark covered in moss and lichen, clinging to trunks and working its way round them like woodpecker; runs along branches, probing into epiphytes; often hangs upside down; climbs among creepers, searching for food among bundles of dead leaves and other accumulated debris. Sometimes works its way up from low bushes to canopy, progressing in short hops, then flies down to next patch of undergrowth to repeat the process. Repeatedly flicks wings, one at a time, flashing the yellow underwing-coverts.

Estimated moult duration 90 days, Tanzania (Moreau *et al.* 1947), where moult of population as a whole takes 5 months, some birds moulting 2 months later than others; some overlap between end of breeding season and beginning of moult in population. Duration of 145 days recorded Malaŵi, where high weight suggests lengthy, energetically 'easy', moulting period (Dowsett and Dowsett-Lemaire 1984).

Sedentary. 14 ringed birds moved only 20–380 m (234 ± 88) (Dowsett 1985); but occasional bird visits isolated forest patches up to 10 km from nearest regular habitat, one bird must have travelled at least 32 km across woodland (Dowsett-Lemaire 1989). Some evidence for altitudinal movement, Malaŵi.

Food. Mainly insects, including eggs and larvae; grasshoppers, beetles, bugs, caterpillars, moth eggs; spiders, small snails. A few fruits and berries also taken, but captive bird that ate insect larvae and grasshoppers refused bananas and other fruits (Swynnerton 1907).

Breeding Habits. Monogamous; territorial. Average territory size Malaŵi (Nyika), 2–3 ha (Dowsett-Lemaire 1983). Site fidelity shown in Malaŵi, where half of breeding adults (11/22) held same territory more than 1 season, as follows: 2 seasons, 7 birds; 3 seasons, 2 birds; 4 seasons, 1 bird; 10 seasons, 1 bird (Dowsett 1985). Surplus of non-breeding birds noted in population (Dowsett-Lemaire 1985).

NEST: an open cup or bowl, rather fragile, built of twigs, tendrils, rootlets, creepers, dry grass, dead leaves, thin bark; outside often covered with moss, sometimes a few feathers, and floor reinforced with additional plant fibres; the whole bound together with spider webs; unlined, or lined with fine grass and fibres. Ext. diam. 100–120, int. diam. 50–64; sometimes elliptical: int. diam. 45 × 64; ext. depth 50–64, int. depth 25–38. Placed among clumps of dry leaves, bracken fronds, moss, creepers and other dense vegetation, or suspended in horizontal fork of twig; at Mt Selinda, Zimbabwe, typically slung from broad leaves of *Dracaena* at junction with stem (Swynnerton 1907); usually low, 1–2 m above ground.

EGGS: usually 2; Zimbabwe (12 clutches) all 2; southern Africa (6 clutches) 2–3 (2·3). Elongate-ovate; smooth, slightly glossy; purple-pink, brown-pink, pale brownish flesh with rich pink suffusion, mauve, grey, occasionally pure white; with spots, streaks, scrawls and pencillings of brown, red-brown, light chocolate, rich chestnut or purple, and undermarkings of slate-grey, blue-grey or purplish slate; typically in band around broad end, sometimes covering whole end; markings like 'line of Arabic written with purple ink, hastily blotted, then written over again with brown' (Sclater and Moreau 1932). SIZE: (n = 11, southern Africa) 21–23·9 × 15·4–17 (22·7 × 16·5). Double-brooded.

LAYING DATES: Zaïre, Feb–Oct (juvenile Jan); E Africa: Uganda, July, Nov (nest-building Jan); Region C, Aug; Region D, Oct–Jan; Zambia, Jan, Sept, Nov; Malaŵi, Sept–Jan; Mozambique, Sept, Nov (breeding condition July, building nest Jan); Zimbabwe, Oct–Jan, usually Nov–Dec (Oct 9, Nov 21, Dec 30, Jan 5 clutches), also Feb; South Africa: Transvaal, Oct–Mar; Natal, Oct–Jan.

INCUBATION: by ♀ only (n = 2, Zaïre: Prigogine 1953). Sitting birds tame, even allowing observers to touch their bills. One remained on nest, occasionally raising 1 wing or giving faint sound of protest, while photographer broke off surrounding twigs, moved camera to within 0·5 m; another left nest on approach of observer but quickly returned, remained on nest facing observer 1 m away; third bird sitting on empty nest refused to move until touched, then remained within a few m, scolding (Swynnerton 1907).

BREEDING SUCCESS/SURVIVAL: average annual mortality rates in Malaŵi (Nyika) calculated as 34·8% in ♂♂, 60% in ♀♀; minimum age of oldest individual 9·9 years, average further life expectancy 1·2–2·4 years (Dowsett 1985). Egg taken by snake *Dasypeltis medici* (Macdonald and Dean 1978).

References
Dowsett-Lemaire, F. (1989).
Prigogine, A. (1973).
Sclater, W. L. and Moreau, R. E. (1932).
Swynnerton, C. F. M. (1907).
Vincent, J. (1935).

Plate 19
(Opp. p. 320)

Phyllastrephus poliocephalus (**Reichenow**). Grey-headed Greenbul. Bulbul à ventre jaune.

Xenocichla poliocephala Reichenow, 1892. J. Orn. 40, pp. 189, 220; Buea, Mt Cameroon, 1200 m.

Forms a superspecies with *P. flavostriatus*.

Range and Status. Endemic resident. Submontane forest in E Nigeria (Obudu Plateau) and W Cameroon: Mt Cameroon, 500–1500 m (very common); Rumpi Hills, 1250–1500 m (fairly common); Mt Kupé, common 950–1600 m, much rarer 1600–2000 m; Mt Nlonako, 1050–1800 m (common); S slope Bamenda Highlands at Foto.

Lives within fairly narrow altitudinal range in which some forest clearance has taken place, but not seriously threatened at present (Collar and Stuart 1985).

Description. ADULT ♂: top of head and hindneck grey, sides of head light grey, indistinct pale loral line, narrow white eye-ring, pronounced black rictal bristles. Mantle grey-green, shading to green on back and scapulars and brighter green on rump and uppertail-coverts. Tail-feathers olive-brown edged green, tips and inner margins of all but central pair pale yellow, shafts brown above, whitish below. Chin and throat white, rest of underparts, including flanks, thighs and undertail-coverts bright yellow. Upperwing-coverts olive-brown, primaries and secondaries brown, outer webs yellow-olive, inner margins of undersides pale yellow, broadest on secondaries; axillaries and underwing-coverts bright yellow. Upper mandible grey to blackish, lower mandible mixed grey and yellowish, tip and cutting edges pale horn; eye dark brown, grey-brown or grey; legs and feet grey. Sexes alike, but ♂♂ larger. SIZE: (10 ♂♂, 8 ♀♀) wing, ♂ 96·5–108 (103), ♀ 86–95·5 (89·1); tail, ♂ 89–102 (97·1), ♀ 83–89 (86·3); bill, ♂ 22·5–26 (24·6), ♀ 20–23 (21·5); tarsus, ♂ 25–29 (27·1), ♀ 24–27·5 (25·8). WEIGHT: ♂ (n = 2) 64·0, 54·5, 1 ♀ 37·8, 1 unsexed (probably ♂) 55·5.

IMMATURE AND NESTLING: unknown.

TAXONOMIC NOTE: shows vocal and behavioural similarities with Yellow-streaked Greenbul *P. flavostriatus*, though lacking its characteristic song, and has been considered conspecific (e.g. by White 1962); however, it is much larger, with plumage well beyond the range of variation shown in Yellow-streaked Greenbul, and geographically distant, and we prefer to follow Hall and Moreau (1970) and Dowsett-Lemaire and Dowsett (1989) in considering the 2 as allospecies of a superspecies.

Phyllastrephus poliocephalus

Field Characters. Largest bulbul in Cameroon montane forest; should be easily told from other local species by white throat, bright yellow underparts, but can be confused with Mountain Greenbul *Andropadus tephrolaemus*, which also has grey crown and green upperparts, but differs in its grey throat and much duller, olive-yellow underparts. Voice and foraging methods different.

Voice. Tape-recorded (CHA, LEM). Call, a loud, full but rather harsh 'churp', 'chewp' or 'chup', sometimes double or triple, 'cher-cherp' or 'chuchuchup'; tone and quality very similar to calls of Yellow-streaked Greenbul. Calls interspersed with a quick, sharp trill, 'tidididit'. Never heard to give song of Yellow-streaked Greenbul despite numerous visits (C. Chappuis, pers. comm.).

General Habits. Inhabits tall, mature rain forest at intermediate elevations.

Occasionally found singly, but almost always a conspicuous member of noisy mixed-species flocks where, like closely-related Yellow-streaked Greenbul, often predominant. Noted in company of Elliott's Woodpecker *Dendropicos elliotii*, Tullberg's Woodpecker *Campethera taeniolaema*, Mountain Greenbul, Gilbert's Babbler *Kupeornis gilberti*, Pink-footed Puffback *Dryoscopus angolensis*, Waller's Chestnut-winged Starling *Onychognathus walleri* and Dark-backed Weaver *Ploceus bicolor*. Moves actively in foliage and on tree trunks and lianas, flicking wings frequently; in canopy and middle levels, sometimes on low trunks and in undergrowth.

Moult may overlap with breeding season: ♂♂ moulting primaries at same time as ♀ found with brood patch (Stuart 1986).

Food. Insects.

Breeding Habits. Breeding indications: Cameroon, egg in oviduct Feb, testes enlarged Nov, ♀ with brood patch Feb, immatures with skull ossification just starting, Jan, Feb.

Reference
Stuart, S. N. (1986).

Phyllastrephus debilis (Sclater). Tiny Greenbul; Slender Bulbul; Smaller Yellow-streaked Greenbul. Bulbul minute.

Plate 19
(Opp. p. 320)

Xenocichla debilis Sclater, 1899. Ibis, p. 284; north of Inhambane, Portuguese E Africa.

Range and Status. Endemic resident; coastal Kenya and Tanzania south from Tana R. mouth to Rufiji R., inland to Rabai and Shimba Hills, Usambaras, Ulugurus and Ngurus; SE Tanzania at Ndungidi and Liwale; Mozambique south to Limpopo, commonest in N; Zimbabwe in Haroni-Lusitu area and upstream on tributaries to 800 m. Common.

Description. *P. d. rabai* Hartert and van Someren (including '*shimbanus*'): Kenya and lowland Tanzania south to Rufiji R. ADULT ♂: top of head grey, paling slightly on sides, sometimes with small amount of greenish wash on hindneck; upperparts bright olive-green; tail-feathers olive-brown edged yellow-green, shafts brown on upperside, whitish on underside; chin and throat white, underparts pale grey-white, grey wash on breast and flanks, variably streaked pale yellow; lower belly yellowish white, undertail-coverts light greenish yellow. Primaries and secondaries dark brown edged bright green, inner web of underside of secondaries broadly, and primaries narrowly, edged yellow-white, underwing-coverts and axillaries pale yellow. Bill grey, pale blue-grey, greyish horn or dusky brown, lower mandible and cutting edges paler, lacking hook at end of upper mandible characteristic of genus; mouth pale yellow; eye colour variable – white, cream, yellow-white, pale golden, red-brown or dark grey-brown, but usually pale; legs and feet black, brown, lead grey or pale blue-grey, soles sandy. Sexes alike. SIZE: wing, ♂ (n = 12) 61–70 (64·5), ♀ (n = 6) 59–68 (63·2); tail, ♂ (n = 11) 57–64 (59·9), ♀ (n = 5) 57–63 (59·6); bill, ♂ (n = 12) 14·5–16·5 (15·3), ♀ (n = 5) 15–16 (15·5); tarsus, ♂ (n = 11) 17–20 (18·4), ♀ (n = 5) 17–19 (17·9). WEIGHT: ♂ (n = 32) 13·5–17 (15·4 ± 0·9), ♀ (n = 25) 11·6–15 (13·3 ± 1·0).

IMMATURE: top of head greener, almost concolorous with back, lores and forehead light green, not grey.
NESTLING: unknown.

P. d. albigula (Grote): Usambara and Nguru Mts, Tanzania, above 600 m. Forehead grey but crown and hindneck with strong olive-green wash; upperparts somewhat darker; throat with greyish wash, rest of underparts darker and greyer with very little yellow streaking, flanks greyish green, lower belly greyish white, undertail-coverts light yellow-olive. Upper mandible blackish, tip, cutting edges and lower mandible horn, yellow-horn or pale horn; mouth orange; eye yellow; legs and feet grey to dark grey, soles yellow, yellow-brown or brown. WEIGHT: ♂ (n = 12) 13·7–16·6 (15·1 ± 1·4), also 1 ♂ 19, ♀ (n = 13) 12·9–16·6 (14·5 ± 1·1).

P. d. debilis (Sclater): SE Tanzania, Zimbabwe, Mozambique. Top of head closest to *rabai* but with light olive wash; upperparts intermediate in tone between other 2 races; underparts similar to *rabai*, less grey on breast and flanks but yellow less bright, more washed out; considerably paler and yellower than *albigula*. Lower belly and undertail-coverts light yellow-olive. SIZE: wing, ♂ (n = 19) 64–70 (66·9), ♀ (n = 11) 61–65 (63·0). WEIGHT: Zimbabwe, ♂ (n = 12) 13·3–16·5 (14·6 ± 0·9), ♀ (n = 7) 12·5–14·2 (13·1 ± 0·6); Mozambique, ♂ (n = 15) 13·4–17 (15·3), ♀ (n = 7) 12·9–14·9 (13·9).

Field Characters. Smallest African bulbul; more likely to be taken for a leaf-gleaning warbler. Half the size of Yellow-streaked Greenbul *Phyllastrephus flavostriatus*, with tiny, paler bill, pale eye, paler grey or green head, brighter green back.

Voice. Tape-recorded (GIB, GREG, KEI, McVIC, STJ, WALK). Song, a variable series of 2 types of notes, a short, clipped 'tut' or 'dit' and a low, nasal 'daa'; basic motif is in 2 parts separated by slight pause, notes of second part given slightly faster: 'tut-daa-dit . . . daadidaa'; the first part may begin with up to 4 short notes: 'tutututut-daa-dit . . . ', while the 3 notes of the last part may go up the scale, the last almost falsetto: 'daadaray'. Pauses between songs may be interrupted by short rattle, 'rititit'. The song as described by Sclater and Moreau (1932), a 'brief, loud, sweet warble, chilly and pure as a Robin's', clearly refers to some other species. Call, a series of sharp, grating notes with a definite pattern: 3–4 short ones, the last highest and sharpest, 'kukukick' or 'kukukukick', a brief pause, then 1–2 longer series that include a grating rattle, e.g. 'kukukukick . . . . krrrr-kuku-krrrr, kukukukukick'; sometimes immediately preceded by a dry 'reee-cheer', the 'cheer' down-slurred (from second bird?); call may be repeated regularly from perch as if it were a song (cf. Terrestrial Brownbul *P. terrestris*).

General Habits. Inhabits thick coastal scrub, lowland evergreen forest and fringing second growth, forest edge, riverine forest, forest–woodland mosaic, sometimes woodland, including miombo; up to 1500 m in E Africa. In Sokoke, Kenya, common in 3 forest types – lowland rain forest, *Cynometra-Manilkara* and *Afzelia*.

Sometimes solitary, usually in pairs or small parties, sometimes with other species. Forages both in canopy and lower levels; creeps about in tangled undergrowth and lianas, gleaning insects from leaves in agile manner like warbler *Phylloscopus*. In Sokoke, regularly travels in parties of *c*. 10 birds which constantly call to each other as they forage; sings from canopy (15 m) in very early morning only (S. Keith, pers. obs.).

Food. Insects.

Breeding Habits. Monogamous; territorial.

NEST: small neat cup of fibres, grass heads, ferns, dry leaves and lichens, bound together with spider webs; 2–5 m above ground in fork of sapling or on outer branches of bush.

EGGS: 2. Light blue, blunt end covered with spots and blotches of dark and light brown over grey. SIZE: (n = 4, southern Africa) 18·4–18·9 × 13·4–13·7 (18·6 × 13·5).

LAYING DATES: E Africa: Region D, Dec; Region E, May; Zimbabwe, Oct–Jan.

Genus *Bleda* Bonaparte

Bill deep, laterally compressed; culmen straight, strongly hooked at tip, gonys convex; nostrils oval, rictal bristles well developed. Legs long, legs and feet stout; middle and outer toes fused as far as the first joint. Upperparts green, underparts bright yellow; tail red in 1 species, outer feathers tipped yellow in 2 species. ♂♂ larger than ♀♀. Inhabit forest undergrowth; attend army ant swarms. Songs whistled, some musical, often quite powerful.

Endemic; 3 species. New evidence suggests *Bleda e. eximia* is a monotypic species; other races of *B. eximia* may belong with *B. canicapilla* (C. Chappuis, pers. comm.).

Plate 20
(Opp. p. 321)

Bleda syndactyla (Swainson). **Red-tailed Bristlebill. Bulbul moustac.**

Dasycephala syndactyla Swainson, 1837. Birds of West Africa 1, p. 261; Sierra Leone.

Range and Status. Endemic resident. Sierra Leone, Liberia, Ivory Coast (frequent north to 7°N), S Ghana (rare), S Nigeria (uncommon); common S Cameroon, north to Korup Nat. Park and Ngoumé; Equatorial Guinea, Gabon (common), Congo; Angola (Cabinda, Cuanza Norte, Cuanza Sul, Lunda), extreme NW Zambia; SW and SE Central African Republic, Zaïre (almost throughout except SE, south to Lualaba (Kasaji) and in E to Mt Kabobo); extreme S Sudan (Lotti Forest); widespread W and S Uganda; W Kenya from Elgon and Kitale to N Nandi and Iruru forests; NW Tanzania at Minziro Forest and Bukoba. Common to locally abundant; density in Gabon 15–18 pairs/km^2; apparent rarity in some areas may be due to extremely secretive behaviour.

Description. *B. s. woosnami* Ogilvie-Grant: E Zaïre (west to Kasai, Lualaba, Upper Congo), Sudan, Kenya, Uganda, Zambia. ADULT ♂: top of head and upperparts including scapulars greenish olive; lower rump greenish rufous, some individuals with lower layer of olive-yellow feathers forming band partly obscured by rufous feathers; rump feathers loose and fluffy; uppertail-coverts bright rufous. Tail bright rufous, a shade darker than uppertail-coverts, shafts red-brown above, straw below with rufous tinge. 4 long black bristles on either side of gape. Pre-orbital region pale yellow-olive, ring of bare skin around eye light blue, ear-coverts greenish olive becoming darker on cheeks, which merge into very dark stripe from base of lower mandible down sides of throat. Chin, throat and underparts bright rich yellow, breast, flanks and thighs washed brownish olive, undertail-coverts rather paler yellow. Primaries and secondaries dark brown edged olive-brown, tertials more or less uniform olive-brown, upperwing-coverts olive-brown, axillaries and underwing-coverts pale dull yellow, brighter at bend of wing, inner edges of undersides of primaries and secondaries pale greyish buff. Upper mandible black to dark brown or dark grey, lower mandible pale grey or bluish, cutting edges paler; eye red-brown to brown or grey-brown; legs and feet pale grey or bluish grey to red-brown or greyish pink. Sexes alike. SIZE: (12 ♂♂, 12 ♀♀) wing, ♂ 101–112 (107), ♀ 95–103 (98·8); tail, ♂ 86–99 (92·5), ♀ 79–93 (84·4); bill, ♂ 24–28 (25·8), ♀ 21–24 (22·5); tarsus, ♂ 23–27 (24·8), ♀ 24–26 (25·0). WEIGHT: ♂ (n = 47) 36–56 (45·3), ♀ (n = 44) 35–48 (47·0). Diurnal weight changes (n = 5) 10·2–23·2% (W Kenya: Mann 1985).
  IMMATURE: entirely russet. Later plumage like adult, bill black, tomia and outer half of both mandibles yellow, gape yellow, bare skin round eye yellow or pale greenish, eye dark brown, feet dull yellow or ochreous yellow.
  NESTLING: unknown.
  B. s. syndactyla (Swainson) (including '*multicolor*'): Sierra Leone to W Zaïre (east to Bangui, Lukolela, Tshuapa). Yellow of underparts paler, less rich. Heavier. WEIGHT: Angola, ♂ (n = 4) 49–52 (50·5), ♀ (n = 8) 40–50 (45·6); Cameroon unsexed (n = 8) 45·7–57 (50·3); Liberia, ♂ (n = 5) 45·5–52·6 (50·2), ♀ (n = 4) 41·8–49·5 (44·8).
  B. s. nandensis Cunningham-van Someren and Schifter: N Nandi Forest, W Kenya. Darker and greener above than *woosnami*, brighter below. Slightly larger than birds in nearby Kakamega: wing (mean) ♂ 109, ♀ 101·5.

Field Characters. A large bulbul with bright yellow underparts, rather long rufous tail, often flicked, and ring of pale blue skin around eye. Common but extremely shy and suspicious: one of the hardest African bulbuls to observe. Its plaintive descending whistles are a familiar background sound in many forests, yet it seldom comes into the open, even in response to song playback. The other 2 bristlebills are same size, with same yellow underparts, and often occur in same forest, but have yellow-tipped green tails and lack eye-ring; all 3 species have different voices. Smaller Xavier's Greenbul *Phyllastrephus xavieri* has green upperparts and yellow underparts and slight rusty wash on tail, but travels with mid-level bird parties and is easy to see, not shy. Bearded bulbuls *Criniger* spp. are nearly as large but have dark olive underparts with yellow confined to mid-line, and 2 species have white throats; voices different.

Voice. Tape-recorded (53, 86, C, ERA, HOR, KEI). Song, a series of plaintive, quavering whistles with 'nostalgique' quality (Chappuis 1975); typically descending (and increasing in volume) but in variant, Gabon, whistles went up the scale, the series ending in 2 short notes (Chappuis 1975). Whistles are preceded by 1 or 2 or sometimes a series of chattering notes, and occasionally break into further chatter at the end. Intervals between songs sometimes punctuated by a few conversational notes, in typical bulbul fashion. Variety of other calls includes loud, nasal 'kur' repeated up to 10 times, scolding chatter, and sharp 'chip' or 'chup'. Alarm, loud 'yamp' and rapidly repeated 'chert'. Begging call of young, a peeping 'euh'.

General Habits. Principal habitat primary and old secondary forest; also gallery forest, dense moist thickets and forest–grassland mosaic (Liberia); occasionally old abandoned farms overgrown with dense second growth. Mainly in lowlands, but also transition zone in E, reaching 1550 m in E Zaïre, 2150 m in E Africa; on Mt Cameroon, 200–950 m (Stuart 1986). Almost always in undergrowth, in thickest tangles of lianas and other choking vegetation, e.g. around windfalls in forest. In Gabon sometimes ascends to higher strata, especially during long dry season (Brosset and Erard 1986).
  Occurs in pairs or small family parties. A regular member of mixed-species flocks, including one centred around foraging squirrels (Brosset 1969). Forages in low vegetation and on ground, where it turns over leaves,

often together with ant-thrushes *Neocossyphus* spp. and alethes (Dowsett and Dowsett-Lemaire 1989). Large and aggressive, with a powerful beak; occupies niche of shrike, feeding on small vertebrates as well as large insects. Sometimes takes fruit from vines at forest edge, together with other frugivores (Liberia: Willis 1983). Regularly follows ants (e.g. *Dorylus wilverthi*), mainly on or close to ground, sometimes higher; clings to vertical saplings crosswise, with 2 outer toes on lower foot together, exactly as in similarly syndactyl Neotropical antbirds (Willis 1983). Captures prey on ground and in air, or makes short flights to glean it from trunks, vines and leaves. Dominates other species at ant swarms except Red-tailed Ant-Thrush *Neocossyphus rufus*. Intraspecifically aggressive; in disputes gives loud chips and chatters and some song. Chases frequent; attacks with bill raised and white throat puffed out. Fights between individuals may be to establish dominance hierarchy rather than defend territory (Brosset and Erard 1986). Territorial all year but territories overlap. In areas where ants are common, birds may have home ranges rather than defended territories. When ant swarm and attendant flock enters home range, owners become dominant, keeping the best vantage points for themselves. 2 pairs or families present at different ends of large colony, foraging near each other but avoiding contact (Gabon: Willis 1983). Banded birds did not follow ants moving 200 m away (see also Green-tailed Bristlebill *B. eximia*). Members of pair stay together all year, often captured in same net; however, considerable turnover of partners noted in Gabon, where 18 individuals had new partners within 1 year of being ringed (Brosset and Erard 1986). Site fidelity marked: of 20 ringed over 2-year period, Gabon, 16 recaptured in same place during next 6 months, one 6 years later (Brosset and Erard 1986).

Post-juvenile moult begins before departure from nest. Primary moult duration, W Kenya (n = 4) 90–159 (133·5) days (Mann 1985).

Sedentary, at least in Gabon.

Food. Small vertebrates, including amphibians; arthropods, often large, including centipedes, millipedes, spiders, roaches, ants, termite alates, mantids, caterpillars, moths (large Attacidae), Coleoptera, Orthoptera (grasshoppers up to 35 mm long); some fruit (including bramble) and seeds.

Breeding Habits. Monogamous, territorial.

NEST: either a shallow cup, or built in form of broadened mouth of trumpet; outer layer of pliant twigs and tendrils bound together with cobwebs, inner layer of large dead leaves, lined with smaller twigs, rootlets and tendrils; easily mistaken for clump of dead leaves; fungus *Marasmius* grows among leaves and helps bind them; 1 nest, diam. 100 × 90, depth *c.* 50. Placed in upper foliage layer of low bush in forest undergrowth, 1–3·5 m above ground, sometimes on top of bunch of dead leaves.

EGGS: 2 (23 clutches), occasionally 1 (2 clutches). Rather glossy; olive-yellow, pale olive or pale buff, irregularly but heavily spotted with pale to dark brown and dark olive, often covering most of egg and giving overall dark appearance. SIZE (n = 3) 24·5–27·3 × 17·7–19·8 (26·1 × 18·5).

LAYING DATES: Sudan, Jan, Mar, June, Aug, Oct; Liberia (breeding condition Mar–Apr); Cameroon, Oct–Dec; Gabon, Oct–Mar, especially Jan–Feb; Central African Republic (breeding condition May–June); Zaïre, Sept–Oct (breeding condition July); Angola (breeding condition Sept–Oct, Feb, juveniles Jan–Feb); E Africa: Uganda, Apr (fledgling Dec, breeding condition Jan–Feb, June); Region B, Apr–May, Aug, mainly in long rains.

INCUBATION: by ♀ only.

BREEDING SUCCESS/SURVIVAL: 11 of 14 nests destroyed and 3 fledglings killed (Brosset and Erard 1986). 2 birds lived 5 years 10 months and 8 years 8 months (W Kenya: Mann 1985).

References
Brosset, A. and Erard, C. (1986).
Willis, E. O. (1983).

Plate 20
(Opp. p. 321)

Bleda eximia (Hartlaub). Green-tailed Bristlebill. Bulbul à queue verte.

Trichophorus eximius Hartlaub, 1855. J. Orn. 3, p. 356; Dabocrom, Gold Coast.

Range and Status. Endemic resident. Sierra Leone, SE Guinea, Liberia, Ivory Coast (north to Mt Nimba), S Ghana (rare, north to Bia Nat. Park and Sukuma Forest Reserve). Extreme SE Nigeria (Oban area, Boshi-Okwangwo Forest Reserve), Cameroon (very common and widespread in S, north to Korup Nat. Park and Ngoumé), Equatorial Guinea (including Bioko), Gabon (very common and widespread), Congo, N Angola (Cabinda), SW and SE Central African Republic, extreme S Sudan (Bengengai Forest), Zaïre (lower Congo and most of central forest bloc), forests of W and S Uganda from Bwamba and Budongo to Malabigambo and east to Sezibwa R. and Mabira, and extreme NW Tanzania (Minziro Forest). Common to locally abundant in most areas; density in Gabon, 30 family units/km^2; with addition of unattached birds = 12 birds/10 ha (Brosset and Erard 1986).

Description. *B. e. ugandae* van Someren: Zaïre (except lower Congo R.), Sudan, Uganda. ADULT ♂: top of head and upperparts, including scapulars and upperwing-coverts, olive-green; tail-feathers olive-brown edged olive-green, outer 4 pairs with broad yellow tips, depth increasing from 12–15 mm on T3 to 22–32 mm on T6; shafts maroon above, straw below. Lores and pre-orbital spot dull greenish yellow; cheeks and ear-coverts dark olive-green, slightly darker in moustachial area which contrasts sharply with yellow throat. Chin, throat and underparts bright, rich yellow, sides of breast, flanks and thighs washed olive-green, light wash sometimes continuing across centre of breast. Primaries and secondaries dark brown with olive-green outer webs; axillaries and underwing-coverts bright yellow, brightest at bend of wing; inner margins of undersides of primaries and secondaries paler yellow. Upper mandible black, dusky or dark grey, lower mandible light blue or blue-grey; eye dark brown or grey-brown to chestnut, yellow or whitish; legs and feet pale grey, blue-grey or brown-grey to bluish white or whitish flesh. Sexes alike. SIZE: (12 ♂♂, 12 ♀♀) wing, ♂ 99–106 (103), ♀ 91–100 (94·8); tail, ♂ 87–102 (91·8), ♀ 80–89 (84·1); bill, ♂ 22–24 (22·9), ♀ 19–22·5 (21·0); tarsus, ♂ 23·5–26 (25·0), ♀ 22·5–25·5 (24·1). WEIGHT: ♂ (n = 9) 37–48 (42·4), ♀ (n = 9) 33–39 (36·0).

IMMATURE: largely dark russet-brown, downy breast feathers whitish with tips washed brown; bill dusky brown, tomia and gape yellow, rim of eyelids greenish yellow, eye dark brown, legs brown-yellow, feet very pale yellow. Brown feathers gradually replaced by plumage which is very similar to adult but duller, especially on head; bill (including lower mandible) brownish black, rim of eyelids and gape yellow, eye dull grey, legs light grey, feet pale buff.

NESTLING: unknown.

B. e. notata (Cassin): Nigeria and Bioko to lower Congo. Lores and pre-orbital spot bright yellow, yellow on tips of tail-feathers less extensive, *c.* 5–10 mm less on each feather. WEIGHT: Cameroon, ♂ (n = 2) 33, 35, 1 ♀ 34, unsexed (n = 6) 33·9–41·8 (39·3).

B. e. eximia (Hartlaub): Sierra Leone to Ghana. Lores and pre-orbital spot yellowish green, indistinct; yellow tips to tail-feathers narrower, *c.* 10 mm on outer pair, absent on T3 (or just thin yellow fringe). Heaviest race. WEIGHT: ♂ (n = 4) 50·1–51·4 (50·7), ♀ (n = 2) 43·5, 45·8, ♀ in breeding condition (n = 6) 40·4–48·3 (45·2 ± 2·7).

TAXONOMIC NOTE: *ugandae* and *notata* may be races of *Bleda canicapilla* (C. Chappuis, pers. comm.).

Field Characters. A large bulbul with green upperparts, yellow underparts and yellow-tipped green tail. Red-tailed Bristlebill *B. syndactyla* has darker upperparts, red tail, blue bare skin around eye, different voice; Grey-headed Bristlebill *B. canicapilla* also has yellow-tipped green tail, but distinguished by grey head, pale grey (not yellow) area in front of eye and different voice.

Voice. Tape-recorded (CHA). Common call, a sharp, loud 'chip-chi-chup-chup', or just 'chip-chi'; this leads into 2 song types in Gabon (race *notata*, Chappuis recording): (a) 8–10 down-slurred whistles, high-pitched at first but descending the scale, each a little longer and slower than the last; early notes sharp, later ones softer, burry, but tuneless, without the mournful quality of Red-tailed Bristlebill, but nevertheless with some generic resemblance; (b) quite different, a brief refrain of 5–6 pleasant, fluty notes without set form. Groups give loose chorus at nightfall, like Common Bulbul *Pycnonotus barbatus*, a jumble of call notes and snatches of both song types. Also has a hard rattle, attached to regular call, 'chip-chi-trrrrrr ...'. Contact calls, 'tsik', 'pit', 'tchek' and 'pieu'. In Bwamba, Uganda (race *ugandae*), song/call quite different, a series of loud, un-bulbul-like clear, whistled notes all on same pitch, repeated rapidly, with longer introductory note, 'wheee ... tu-tu-tu-tu-tu ...', number of notes varying (S. Keith, pers. obs.).

General Habits. Inhabits primary forest and old second growth, in some areas also forest–grassland mosaic. Mainly in undergrowth < 3 m. Lowlands only, to 650 m E Zaïre, 1250 m Uganda.

Social unit is a pair and 1 young, which stays with its parents for at least 8 months. Pair bond does not last more than 1 year; ringing shows that partners often change (Brosset and Erard 1986). Regularly joins mixed-species flocks: observed in 84 of 271 flocks, Gabon. Always part of mixed flocks following army ant swarms, where restless and agile; waits on ground or sits on branch above swarm, darting to ground and back again. Keeps ahead or to sides of swarm, avoiding the numerous thrushes (*Neocossyphus*, *Alethe*) that stay very close to ants. Catches prey mainly on ground (sometimes tossing leaves), less often in short aerial sallies, occasionally gleans it from vines and trunks (Willis 1983). Dominated by Red-tailed Bristlebill *B. syndactyla* and Red-tailed Ant-Thrush *Neocossyphus rufus*. Also seen to stay briefly among ants, apparently anting (Lippens and Wille 1976). Shy and secretive but also noisy.

Bleda eximia

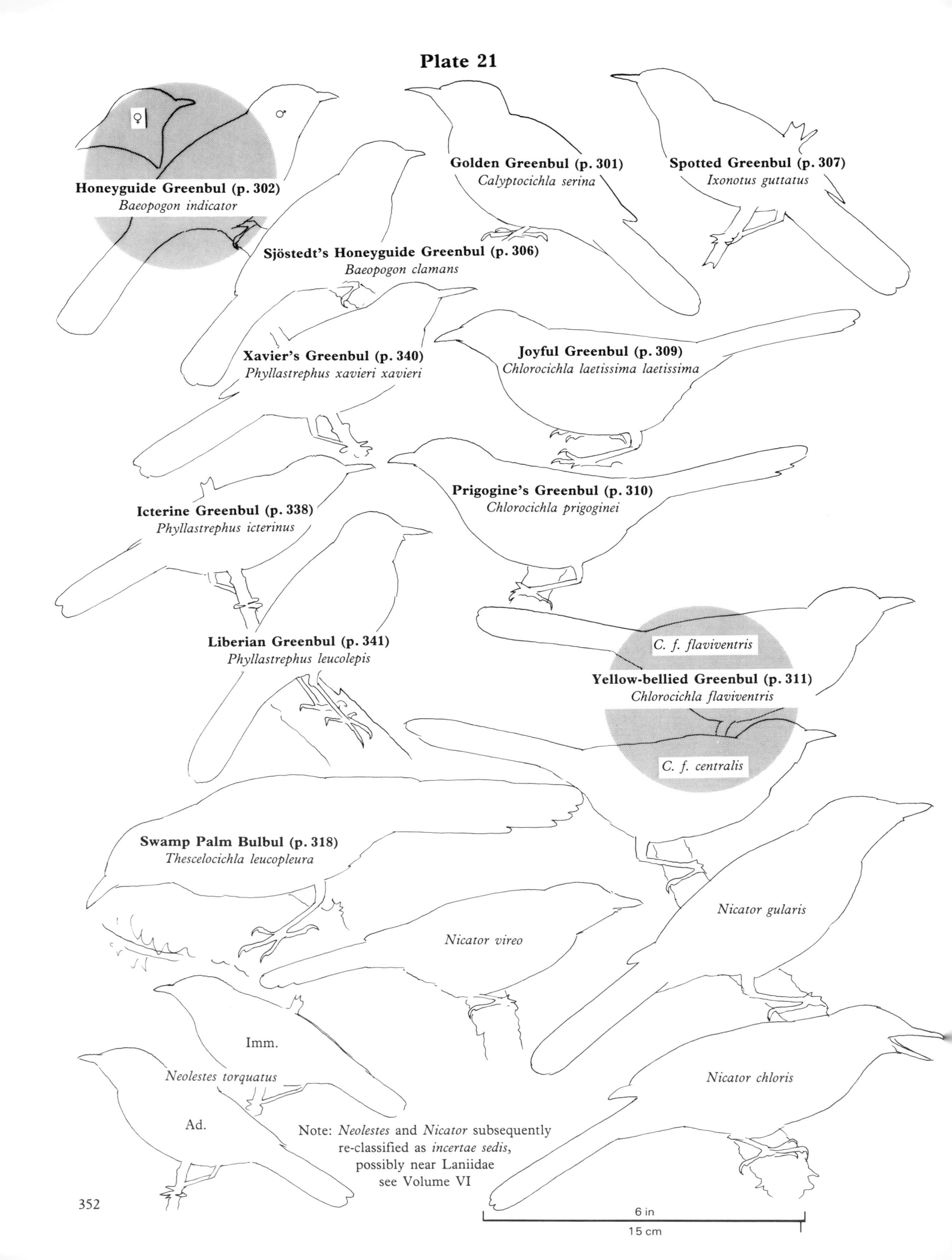

Plate 21

Honeyguide Greenbul (p. 302)
Baeopogon indicator

Golden Greenbul (p. 301)
Calyptocichla serina

Spotted Greenbul (p. 307)
Ixonotus guttatus

Sjöstedt's Honeyguide Greenbul (p. 306)
Baeopogon clamans

Xavier's Greenbul (p. 340)
Phyllastrephus xavieri xavieri

Joyful Greenbul (p. 309)
Chlorocichla laetissima laetissima

Icterine Greenbul (p. 338)
Phyllastrephus icterinus

Prigogine's Greenbul (p. 310)
Chlorocichla prigoginei

Liberian Greenbul (p. 341)
Phyllastrephus leucolepis

C. f. flaviventris

Yellow-bellied Greenbul (p. 311)
Chlorocichla flaviventris

C. f. centralis

Swamp Palm Bulbul (p. 318)
Thescelocichla leucopleura

Nicator vireo

Nicator gularis

Imm.

Neolestes torquatus

Nicator chloris

Ad.

Note: *Neolestes* and *Nicator* subsequently re-classified as *incertae sedis*, possibly near Laniidae see Volume VI

352

6 in
15 cm

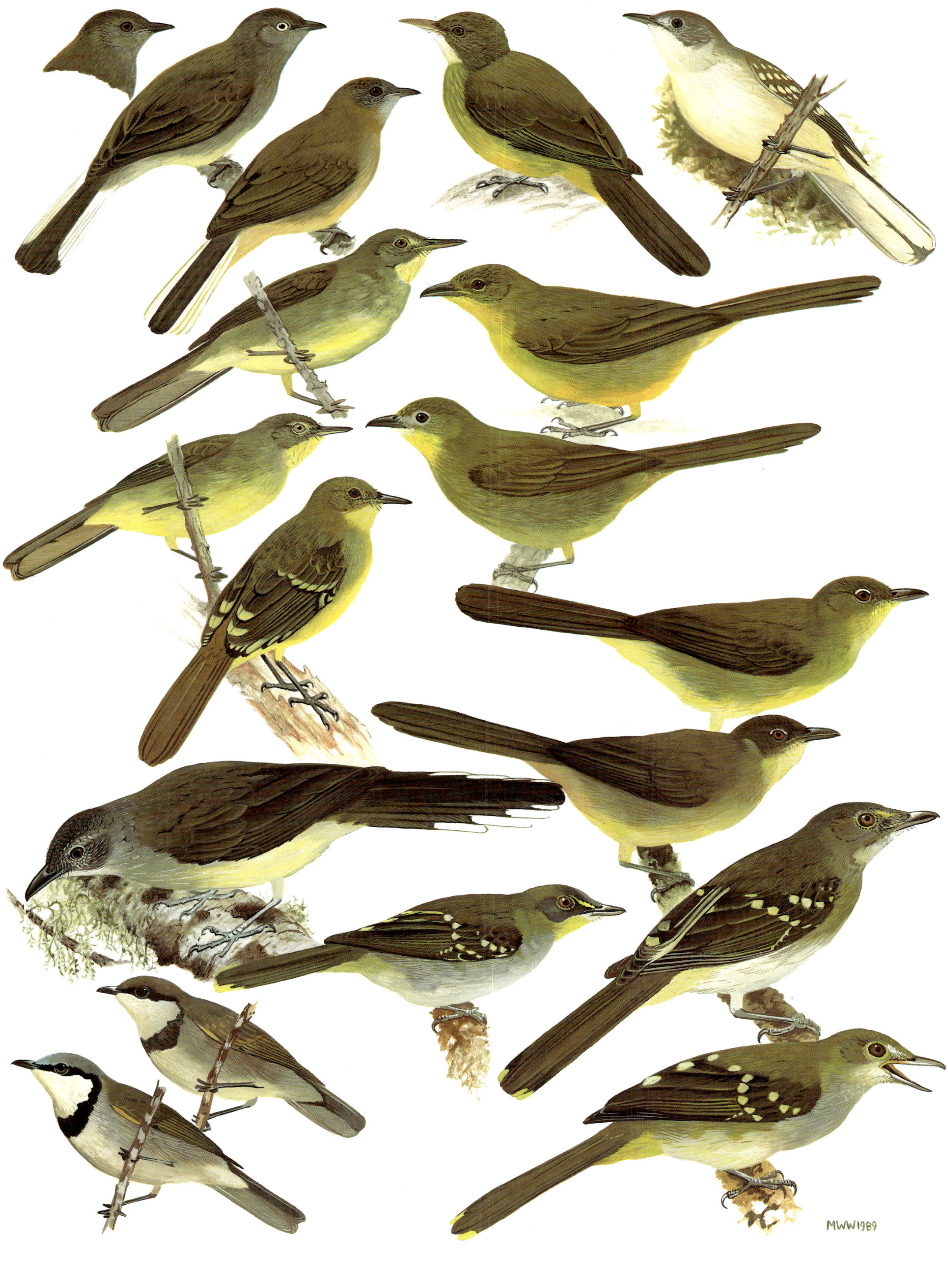
MWW1989

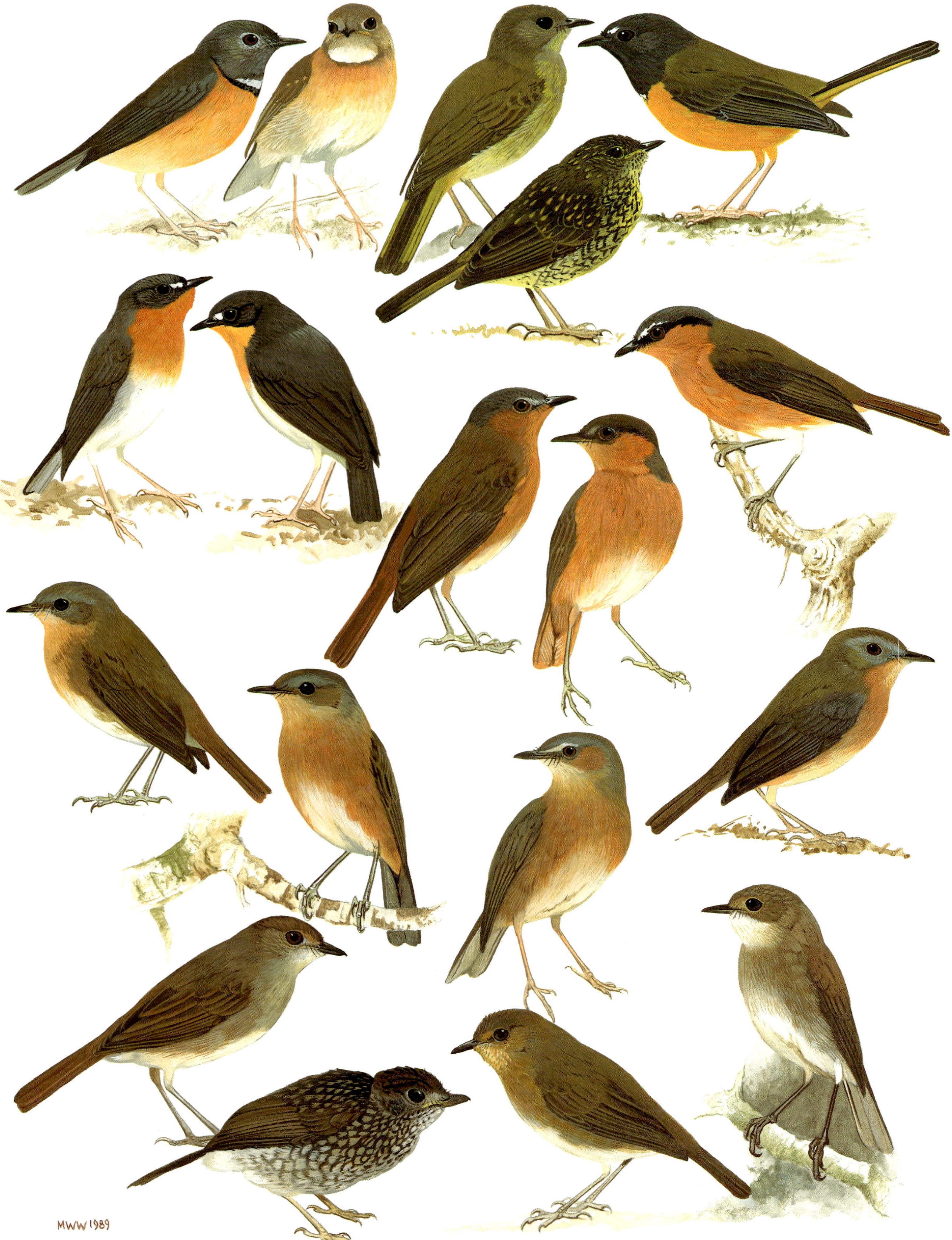
MWW 1989

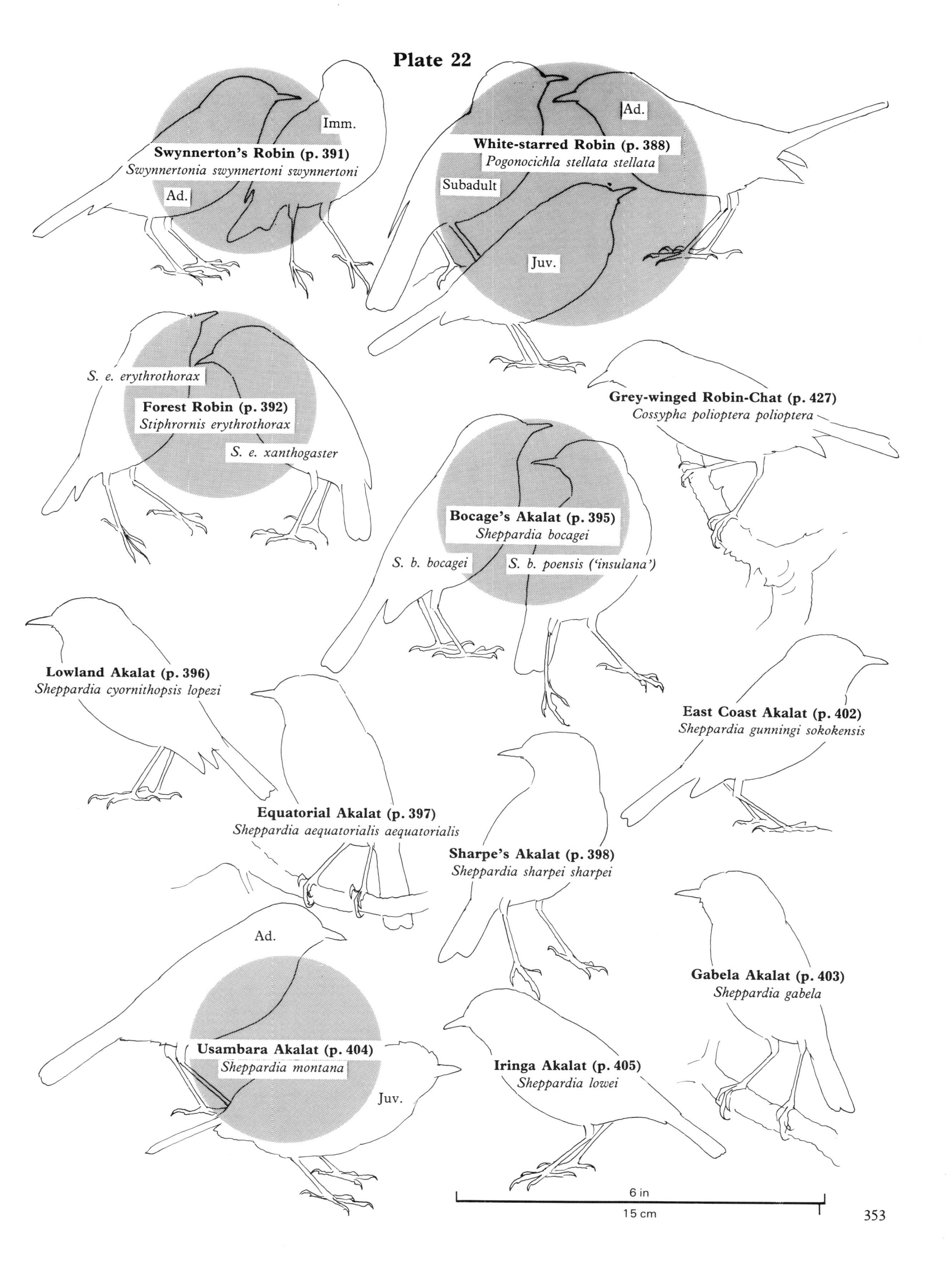

Plate 22

Swynnerton's Robin (p. 391)
Swynnertonia swynnertoni swynnertoni
Imm.
Ad.

White-starred Robin (p. 388)
Pogonocichla stellata stellata
Ad.
Subadult
Juv.

Forest Robin (p. 392)
Stiphrornis erythrothorax
S. e. erythrothorax
S. e. xanthogaster

Bocage's Akalat (p. 395)
Sheppardia bocagei
S. b. bocagei
S. b. poensis ('insulana')

Grey-winged Robin-Chat (p. 427)
Cossypha polioptera polioptera

Lowland Akalat (p. 396)
Sheppardia cyornithopsis lopezi

Equatorial Akalat (p. 397)
Sheppardia aequatorialis aequatorialis

Sharpe's Akalat (p. 398)
Sheppardia sharpei sharpei

East Coast Akalat (p. 402)
Sheppardia gunningi sokokensis

Gabela Akalat (p. 403)
Sheppardia gabela

Usambara Akalat (p. 404)
Sheppardia montana
Ad.
Juv.

Iringa Akalat (p. 405)
Sheppardia lowei

6 in
15 cm

354 PYCNONOTIDAE

At alarm calls of flock members may freeze, crouch or flee silently to low vine tangles or treefalls, but often gives loud alarm at approach of *Accipiter tachiro*. Different pairs attended different ends of large colony (Gabon: Willis 1983), but birds banded at site of colony were replaced by unbanded birds when colony moved 200 m, suggesting birds do not follow ants beyond their territory. Does not defend territory but has overlapping home ranges. Noisy disputes erupt around ant swarms, with loud singing and spreading of tail-feathers to show yellow tips, also chasing and violet fights, birds gripping each other and refusing to let go until almost touched by human observer.

Sedentary in Gabon, where site fidelity very marked. 76% of ringed birds recaptured in same spot; even young do not disperse far, remaining in the territory where they were born (Brosset and Erard 1986). However, birds in canopy at Cape Coast, Ghana, may have been migrants (Grimes 1987).

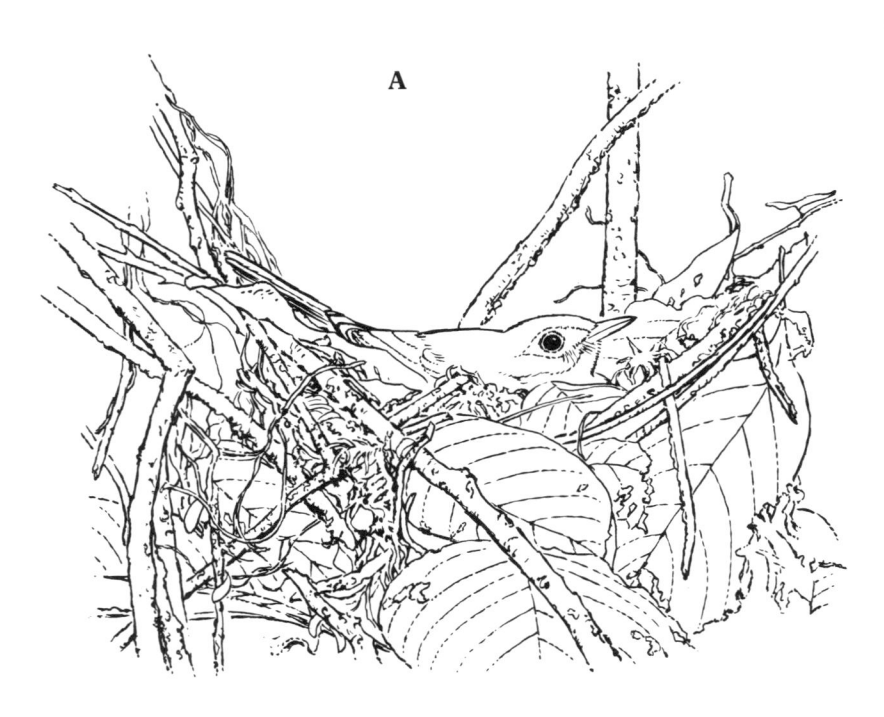

Breeding Habits. Monogamous.

NEST: rather rudimentary, with base of plant debris on which are placed several large dead leaves curled round to form funnel, bound together with strands of fungus *Marasmius* (**A**), lined with leaf skeletons, thin plant fibres and more *Marasmius*; attached to supporting branch by mycelium of fungus; usually concealed in clump of dry leaves, but clump exposed, sometimes visible at 30 m; placed 0·5–3 m above ground in small, isolated bush in open part of forest, away from main path taken by tree mammals, to decrease chances of predation (Brosset 1971b). Built by both sexes.

EGGS: 2 (66 clutches), sometimes 1 (2 clutches); laid at 1-day intervals. Green, ochre or light brown, heavily spotted with dark brown, red-brown, violet-brown or dark olive, sometimes concentrated around broad end, sometimes diffuse. SIZE: (n = 2) 22–24 × 16–17.

LAYING DATES: Liberia (breeding condition June–July); Cameroon, May, July–Aug, probably all year; Gabon, all year except long dry season (June–Sept), especially Jan–Feb (short dry season). Zaïre, all year; Uganda (breeding condition Feb).

INCUBATION: by ♀ only. Period: 14 days. Bouts of incubation 2 h, absences *c.* 40 min. Shy, readily flushed from nest; takes up to 1 h to return if observer still nearby, both birds calling from concealed posts.

DEVELOPMENT AND CARE OF YOUNG: young fed by both parents, 14 visits in 20 h of observation (n = 1, Gabon). Young in nest 13–14 days; both parents take turns brooding.

BREEDING SUCCESS/SURVIVAL: usually only 1 young raised; 15 nests which produced flying young averaged 1·3 young per nest; 82% of nests destroyed (Gabon: Brosset and Erard 1986). Longevity considerable: 5 birds recaptured 8 years after being ringed, 1 after 10 years.

Food. Large insects and their larvae: beetles, Orthoptera (including cicadas, katydids), butterflies, large caterpillars; also ants, small frogs and some fruit. Young fed Orthoptera, Neuroptera, caterpillars, spiders.

References
Brosset, A. (1971b, 1981c).
Brosset, A. and Erard, C. (1986).
Willis, E. O. (1983).

Plate 20
(Opp. p. 321)

Bleda canicapilla (Hartlaub). **Grey-headed Bristlebill. Bulbul fourmilier.**

Trichophorus canicapillus Hartlaub, 1854. J. Orn. 2, p. 25; Sierra Leone.

Range and Status. Endemic resident. W Gambia, SW Senegal, Guinea-Bissau, Guinea, SW Mali (rare), Sierra Leone, Liberia, Ivory Coast (very common, north to 9° 30′N); Ghana (widespread and common), Togo, Benin; Nigeria (common in S, and also at Nindam Forest Reserve; possible record in gallery forest at Aliya (11°

10′N, 10° 55′E). Common except at western limits of range. Density in Nigeria (Ogba Forest) 1 bird/10 ha (Johnson 1989).

Description. *B. c. canicapilla* (Hartlaub): Guinea-Bissau to Nigeria. ADULT ♂: top of head to hindneck, cheeks, ear-coverts and sides of neck grey, lores and pre-orbital region paler grey; 3–4 stiff black bristles at base of bill. Upperparts, including scapulars, tertials and upperwing-coverts olive-green. Tail-feathers olive-green, edges somewhat brighter, outer 3 pairs with broad (12–15 mm) yellow tips, outer feather only slightly broader than T3, not broadening outwardly as in Green-tailed Bristlebill *B. eximia*; shafts maroon above, straw below. Chin white, throat pale yellow, rest of underparts bright yellow, sides of breast, narrow line across centre of breast and lower flanks washed olive-green. Primaries and secondaries dark brown edged olive-green; axillaries, underwing-coverts and inner margins of undersides of primaries and secondaries yellow. Upper mandible blackish, dark horn or dark olive-green, lower mandible greenish blue, blue-grey or bluish horn; small bare patch behind eye blue-grey; eye brown to red; legs and feet blue to greenish blue, slate or pearly grey. Sexes alike. SIZE: (8 ♂♂, 8 ♀♀) wing, ♂ 98–109 (102), ♀ 93–102 (95·5); tail, ♂ 84–101 (92·0), ♀ 82–98 (86·9); bill, ♂ 21–24 (22·4), ♀ 19·5–22 (20·5); tarsus, ♂ 24–28 (25·8), ♀ 22–26 (25·1). WEIGHT: ♂ (n = 6) 39·4–46·5 (42·7 ± 3·6), ♀ (n = 4) 36–40·5 (37·9), ♀ in breeding condition (n = 3) 38–41·3 (39·3).
IMMATURE AND NESTLING: unknown.

B. c. moreli Erard: Senegal and Gambia. Paler than nominate race; head pearl grey, back brownish or greyish olive, less green, without yellow tone; yellow of underparts, including underwing, paler, more sulphur-yellow; yellow tips to tail feathers very pale and washed-out, almost white on outer pair; bill shorter and thicker.

Bleda canicapilla

Field Characters. A large bulbul with green upperparts, yellow underparts and yellow-tipped green tail. Less of a deep forest bird than is very similar Green-tailed Bristlebill, more often in edges and secondary growth; latter also has yellow-tipped green tail, but distinguished by green head, yellow spot in front of eye, different call note.

Voice. Tape-recorded (53, ERA, KEI, MOR, MOY). Song extremely variable; typically begins with a few soft, almost inaudible warm-up notes, then introduced by variable number of regular call notes, loud 'TSIK' and especially 'CHEEwo', followed by lilting refrain of sweet whistles; every song different, both in form and in number of notes (usually 8–14, sometimes as few as 4 or as many as 20), but with general tendency to work down the scale, often ending with a low 'sign-off' note 'chor'. A few notes are down-slurred, like those of Green-tailed Bristlebill, with tell-tale family burry quality, but most are pure and fluty, some quite powerful, often like human whistle. A very fine singer for a bulbul, perhaps the best in Africa; some individuals extremely imaginative, changing tone and direction of notes in an almost whimsical way reminiscent of Musician Wren *Cyphorhinus arada* of the Neotropics. Not all songs are brilliant, however; one variant is a series of bubbling, bustling notes reminiscent of Little Greenbul *Andropadus virens*, with few whistles, while others consist mainly of 'CHEEwo' calls with a few whistles tacked on at the end. 'CHEEwo' call (sometimes 'chuWEE-CHEEwo') very characteristic, and best means of distinguishing voice from that of Green-tailed Bristlebill, some of whose songs are not unlike certain Grey-headed Bristlebill variations. Said to give an excited 'chityu-chityu-chityu-wirra-wirra' at dusk (Button 1964b), perhaps an evening chorus as in other members of the genus. Alarm on approach of person, a rattling 'di-i-i-i-i-i-i-i . . .'; on approach of *Accipiter tachiro*, multiple chips 'dit-dit-dit . . .'; begging call of young just out of nest, continuous 'chish' at rate of 2 notes per s (Willis 1983).

General Habits. Inhabits undergrowth of primary forest and second growth but typically at edges and around clearings rather than in deep forest; also occupies forest–grassland mosaic, riverine forest, thickets in dense woodland, swamp forest, gardens, palm trees bordering tall rank swamp vegetation.

Travels in small groups that forage in low vegetation close to ground; catches insects in air or gleans them from leaves; often joins mixed-species flocks. Attends ant columns, standing on ground or perching in low vegetation (sometimes up to 5 m), from which it drops to ground (mainly) or takes prey from trunks or foliage or catches it in air (Willis 1983). Usually only 1 pair or family attends a swarm or joins a mixed-species flock; young may quiver wings and give begging call. Rather shy and secretive; on approach of person flashes yellow corners of tail, gives alarm rattle, retreats and whistles from a distance (Willis 1983). When members of mixed flock give alarm, retreats to far side of flock, hides in liana tangles or crouches, flicking wings and tail; also wipes bill, sleeks body and peers under outstretched wing; leaves ants if no cover nearby. Around ants interacts aggressively with Brown-chested and Fire-crested alethes *Alethe poliocephala* and *A. diademata*, Red-tailed Bristlebill *B. syndactyla*, White-tailed Ant-Thrush *Neocossyphus poensis* and Little Greenbul *Andropadus virens*. Territorial; in intraspecific disputes birds give bursts of whistled song, flick wings, flash tail spots and chase one another (Willis 1983). Button (1964b) lists as typical behaviour, 'frenzied chases through barer levels of lower bush', possibly squabbles over dominance as in other *Bleda* species.

Apparently subject to local movements: 1 found flying about inside building in Ghana (Legon) Oct.

Food. Insects, including Orthoptera, Isoptera, caterpillars, flies, termites, black ants; also millipedes, spiders, ticks, small frogs, some fruit and a few seeds.

Breeding Habits. Very little known.
NEST: only 1 found: open, flimsy cup of dry, dead leaves, on base of dry twigs arranged in criss-cross fashion, lined with black fibres; ext. diam. 110, int. diam. 65, ext. depth 90, int. depth 25; rather insecurely placed among trailing vines and branches hanging from low bush in forest, 1·6 m above ground.

EGGS: only 1 clutch (of 2) described. Ovate; smooth, somewhat glossy; white tinged olive-brown, covered with bold, irregularly-shaped blotches and spots of dark brown and ash-grey and purple-grey undermarkings, forming broad wreath around large end. SIZE: (n = 2) 23·4–24 × 16·4–16·9.

LAYING DATES: Senegal, Jan; Sierra Leone (breeding condition May, Aug–Sept, Dec); Liberia (breeding condition July–Aug); Ivory Coast (breeding condition Mar, June); Nigeria, June.

Reference
Willis, E. O. (1983).

Genus *Criniger* Temminck

Bill stout, deep, thicker at base than near tip; tip strongly hooked; culmen curved, gonys almost straight; nostrils oval. Rictal bristles well developed. Legs and feet short and stout. Wings and tail rounded. Plumage soft, feathers long; some species crested; long hair-like feathers on nape, resting on mantle; conspicuous white or yellow throat-patch, feathers long and frequently puffed out. Inhabit forest, mainly lower levels. Songs in Africa rather weak and tuneless, calls loud, whistled.

11 species, Africa and Asia. 5 species in Africa, all endemic; 2 form a superspecies (*C. barbatus*/*C. chloronotus*).

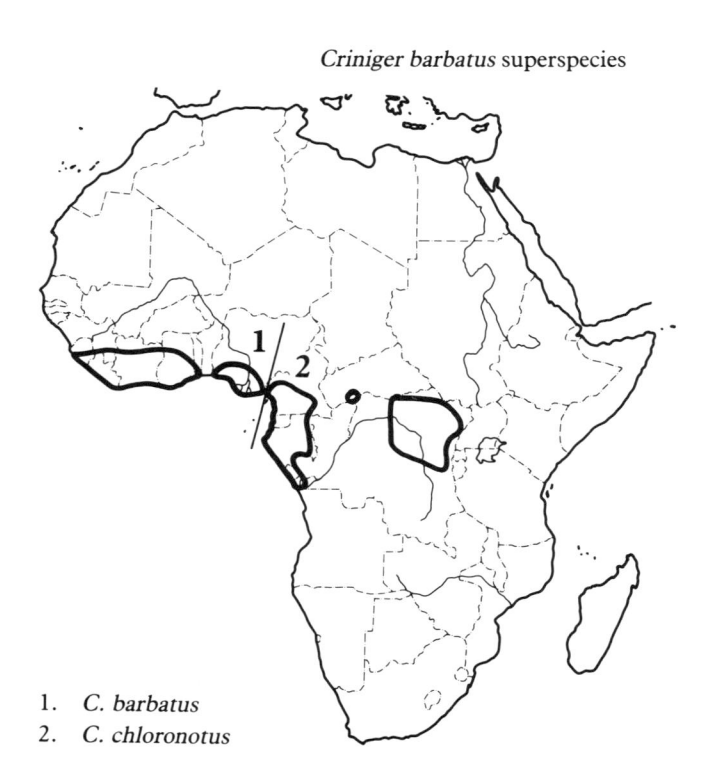

Criniger barbatus superspecies

1. *C. barbatus*
2. *C. chloronotus*

Criniger barbatus (Temminck). **Western Bearded Greenbul. Bulbul crinon occidental.**

Trichophorus barbatus Temminck, 1821. Pl. col. 3, livr. 15, pl. 88; Sierra Leone.

Forms a superspecies with *C. chloronotus*.

Plate 20
(Opp. p. 321)

Range and Status. Endemic resident. Sierra Leone, SE Guinea, Liberia (common), Ivory Coast (common north to 9° 30′N); S Ghana (not uncommon), Togo, S Nigeria (not uncommon but most numerous east of lower Niger R.).

Criniger barbatus

Description. *C. b. barbatus* (Temminck): Sierra Leone to Togo. ADULT ♂: forehead brownish grey, feathers with indistinct pale shaft streaks; crown and nape brownish grey, feathers edged olive-green, those of back of head elongated, sometimes raised to form short bushy crest; a few long black bristles on nape, stretching onto mantle. Mantle to uppertail-coverts yellowish olive-green, mantle feathers with grey centres, uppertail-coverts with slight rufous wash in centre. Tail-feathers dull reddish brown, central pair margined olive-green on both webs, rest margined olive-green on outer webs, buff on inner webs; shafts blackish above, yellowish cinnamon below. Pre-orbital area and around eye grey, narrow eye-ring white; cheeks and ear-coverts brownish olive with prominent white shaft streaks; rictal bristles black. Chin and throat bright yellow; breast feathers grey with olive-green edges and pale shaft streaks producing somewhat streaked effect; belly, flanks and undertail-coverts olive-yellow with buffy tinge. Primaries and secondaries dark brown with olive-green outer webs, inner secondaries darker olive-green on both webs, upperwing-coverts olive-green; axillaries olive-yellow, underwing-coverts and margins of inner webs of undersides of primaries and secondaries yellowish buff. Upper mandible black or dark horn, lower mandible grey to bluish, cutting edges yellow; eye dark brown to chestnut-brown; legs and feet black, grey or blue-grey. Sexes alike. SIZE: wing, ♂ (n = 11) 103·5 ± 3·3, ♀ (n = 6) 93–99 (96·3); tail, ♂ (n = 11) 90·4 ± 2·1, ♀ (n = 6) 84–94 (88·8); bill, ♂ (n = 11) 23·3 ± 1·1, ♀ (n = 6) 19·5–21 (20·3); tarsus, ♂ (n = 3) 22–24 (23·0), ♀ (n = 6) 20–23·5 (22·4). WEIGHT: ♂ (n = 9) 41·2–48·1 (44·4 ± 2·7), ♀ (n = 5) 37·8–43 (40·8), ♀ in breeding condition (n = 3) 40·3–47·3 (43·7), immature ♂ 47·2, immature ♀ (n = 4) 33·5–47·2 (40·6).

IMMATURE: an almost fully grown bird (wing 74; tail 85; bill 19·5; tarsus 24·5) has top of head brown with a few grey feathers growing through, back olive-brown with a hint of greenish, uppertail-coverts dull cinnamon; tail paler than adult, especially on underside; face with very little streaking, throat paler yellow, band across breast plain brown without streaks, flanks pale brown, belly pale yellow. Primaries with light olive-green outer margins, secondaries with pale buff inner margins. Full-size bird with unossified skull identical to adult in plumage; upper mandible black, lower mandible and cutting edges yellow, gape yellow; eye brown; legs and feet bluish grey.

NESTLING: unknown.

C. b. ansorgeanus Hartert: Nigeria. Top of head browner grey, upperparts somewhat paler, especially on rump; uppertail-coverts and tail bright rufous, tail-feathers narrowly margined olive-yellow, shafts brown above, yellow-brown below. Chin white, throat very pale yellow; underparts browner, undertail-coverts yellowish cinnamon; axillaries and underwing-coverts pale dull cinnamon. Bill, upper mandible dark horn-grey or black, lower mandible pale bluish or greyish, cutting edges pale blue; eye blood-red, mahogany-red or red-brown; legs and feet pale blue-grey.

Field Characters. A large, noisy bulbul, with yellow throat which is frequently puffed out and contrasts with dull, dark underparts. Similar-sized Red-tailed Greenbul *C. calurus* distinguished by white throat, yellow underparts; Yellow-bearded Greenbul *C. olivaceus* is smaller, with brighter yellow throat, olive-yellow underparts, different feeding habits, and is quiet and unobtrusive.

Voice. Tape-recorded (53, C). Commonest song has 2 low introductory notes with slightly throaty quality, followed by 2 louder, higher, clear ones, 'toro-TWEE-yer'; a longer version of this song begins with 2 high rather than low notes, 'chi-chi-tay-EE-TWEE-yer'. A second song is quite different, a series of up to 10 loud, clear, down-slurred whistles on same pitch, melodious and un-bulbul-like. Call, a loud 'WEEP' or sharp 'tyip'. Also said to give a 'loud, throaty, concerted babbling' (Serle 1957).

General Habits. Inhabits lowland primary and secondary forest, forest–grassland mosaic, and gallery forest. Forages at all levels from thick undergrowth to lower canopy, but most often fairly low down. Gleans leaves and plucks berries in fluttering flight; follows ant swarms. Sometimes solitary but mainly gregarious; a prominent and noisy member of mixed-species flocks, often associating with Red-tailed Greenbul. Frequent puffing out of throat not connected to calling or display, possibly a contact signal.

Food. Insects, including moths and small beetles; fruits, including liana berries and *Musanga*; and seeds. 1 stomach, Liberia, contained 20% insects and 80% seeds (M. F. Carter, pers. comm.).

Breeding Habits. Unknown.
LAYING DATES: Sierra Leone (breeding condition Dec–Apr); Liberia (large juvenile Oct, breeding condition Mar–Apr); Ghana (juveniles July, breeding condition Mar).

Plate 20
(Opp. p. 321)

Criniger chloronotus (Cassin). Eastern Bearded Greenbul. Bulbul crinon oriental.

Trichophorus chloronotus Cassin, 1860. Proc. Acad. N. Sci. Phil. 1859, p. 43; Camma R., W. Africa (= Gabon).

Forms a superspecies with *C. barbatus*.

Range and Status. Endemic resident. Widespread in Cameroon north to Korup Nat. Park and Mt Manenguba, east in lowlands to Mieri and Bitye; Gabon (common), S Congo, extreme W Zaïre (Mayombe), N Angola (Cabinda), Central African Republic (Lobaye Préfecture), N and E Zaïre, south in E to Itombwe; extreme W Uganda (Bwamba). Generally fairly common; density in Gabon 7–8 groups/km^2.

Description. ADULT ♂: forehead to hindneck slate-grey, feathers with blackish centres and narrow black fringes producing slightly mottled effect; crown feathers long, over lying mantle, sometimes raised to give head a shaggy look; a few black bristles on hindneck, stretching onto mantle. Mantle to rump, scapulars and upperwing-coverts yellowish olive-green; uppertail-coverts and tail bright rufous, all tail-feathers except central pair with very narrow greenish margins to outer webs, shafts maroon above, reddish buff below. Pre-orbital area grey with tiny pale spots and streaks; cheeks and ear-coverts dark grey with white shaft streaks; narrow pale eye-ring. Chin and throat white; breast light grey with brownish wash, feathers of upper breast with narrow dark tips giving slightly scaly effect; flanks greenish yellow, belly pale dull yellow, undertail-coverts yellowish cinnamon. Primaries and secondaries dark brown, outer webs yellowish olive-green, inner secondaries with inner webs dull green; axillaries and underwing-coverts dull olive-yellow, inner margins of undersides of primaries and secondaries greyish buff. Bill, upper mandible dark horn to dark olive-green with blue-grey cutting edges, lower mandible pale, blue, blue-grey, bluish green or greenish grey; eye dark brown, red-brown or red; legs and feet blue-grey, purplish blue or greyish violet. Sexes alike. SIZE: (12 ♂♂, 4 ♀♀) wing, ♂ 98–111 (105), ♀ 99–102 (101); tail, ♂ 84–96 (90·4), ♀ 86–93 (89·8); bill, ♂ 21·5–25 (22·9), ♀ 21·5–22·5 (22·0); tarsus, ♂ 22–25 (23·3), ♀ 22–24 (23·0). WEIGHT: unsexed (n = 2, Cameroon) 38·8, 45·3.
IMMATURE: like adult except lower mandible greenish, cutting edges yellowish, gape yellow.
NESTLING: unknown.
TAXONOMIC NOTE: originally described as a full species, this form has been considered a race of *C. barbatus* by recent authors. However, we agree with Chappuis (1975) that vocal differences warrant its treatment as a separate species.

Field Characters. A large bulbul with grey head and breast, bright red tail, white throat frequently puffed out; yellow on underparts pale, confined to belly; characteristic quavering 2-note song. Sympatric Red-tailed Greenbul *C. calurus* also puffs out white throat but is smaller, with dull red tail, olive breast, bright yellow underparts, very different voice.

Voice. Tape-recorded (53, ERA). Song very different from Western Bearded Greenbul *C. barbatus*: 2 quavering notes, second one pitched about a fifth higher; notes last *c*. 1 s and follow each other without a break; repeated at 5–10 s intervals. Not loud; no variation; quality rather mournful. Alarm, a weak chatter.

General Habits. Inhabits primary lowland rain forest; in E Zaïre also intermediate and lower montane forest. Sea level to 1800 m.
Occurs in groups of 3–5, in dry season up to 7–8; sometimes solitary or in pairs. Groups seem to be held together by song of leader. Home range of group apparently large: of 23 birds ringed in 10 years in Gabon, only 5 were recaptured, after intervals varying from 1 to 49 months (Brosset and Erard 1986). Group territory

advertised by song, defended by communal displays in which 'beard' puffed far out.

Occupies lower and middle levels, mainly below 15 m, but ascends to lower canopy in dry season. Forages especially in matted vines and tangles of small branches, gleaning insects from leaves and probing bark. Sometimes joins mixed-species flocks. Follows ant swarms. Suspicious and secretive but noisy; presence indicated by characteristic song (see *Voice*).

Sedentary, at least in Gabon.

Food. Mainly insects and their eggs and larvae, including Orthoptera and Coleoptera; also some fruit. Young fed Orthoptera and large larvae.

Breeding Habits. Territorial. Group habits suggest co-operative breeding.

NEST: shallow cup, solid and well constructed (**A**); outer wall of interlocking pliant twigs and moss, inner wall of thin plant stems, lined with strands of fungus *Marasmius*; invariably decorated with stems of the epiphytic fern *Microgramma owariensis*, c. 30 cm long, bound together around rim, a feature unique to this nest (Brosset 1971b); leaves on stems remain green throughout incubation and raising of young; 1 nest, ext. diam. 92, diam. of cup 70. Typically situated on bunch of green leaves forming crown of low shrub in open understorey, e.g. *Alchornea floribunda*, 0·5–2 m above ground; nest is set on base of bits of wood and decaying leaves on which fungus soon grows, fixing base to leaves on which it rests; it becomes so firmly attached that it can be turned upside down and shaken vigorously without being dislodged from leaves (Brosset and Erard 1986). Less often situated in exposed position on upper side of bare branches of bush. (For comparison with nest of sympatric Red-tailed Greenbul, see Brosset 1971a.)

EGGS: normally 2 (25 clutches), sometimes 1 (2 clutches). Glossy; very variable; ground colour grey-beige, rose, wine, red-brown or chocolate; unmarked or with spots and scrolls of grey-violet, dark brown or black, sometimes dark crown on large end; some have 2 parallel bands of dark olive-brown around centre and a few large, tear-shaped marks on large end.

LAYING DATES: Cameroon (breeding condition Mar); Gabon, Aug, Nov–Mar (Aug 3, Nov 1, Dec–Mar 42 clutches), i.e. mainly during short dry season; Zaïre, Dec–May, probably all year (Chapin 1953).

INCUBATION: by ♀ only. Period: at least 14 days.

DEVELOPMENT AND CARE OF YOUNG: single young in nest fed 12 times in 4 h; of 2 young in another nest, one had av. 2 feeds per h and one appeared to be ill (rickets?: A. Devez *in* Brosset 1971).

BREEDING SUCCESS/SURVIVAL: in 9 nests with 2 young, both young reached flying stage in only 3 nests; second young usually smaller, stops being fed and dies at age of 5–6 days. 26 of 35 nests (74%) destroyed by predators; 2 raided by mandrills *Papio mandrillus* (Brosset and Erard 1986).

References
Brosset, A. (1971b).
Brosset, A. and Erard, C. (1986).

Criniger calurus (Cassin). Red-tailed Greenbul. Bulbul à barbe blanche.

Plate 20
(Opp. p. 321)

Trichophorus calurus Cassin, 1857. Proc. Acad. N. Sci. Phil., 1856, 8, p. 158; Moonda R., W Africa (= Gabon).

Range and Status. Endemic resident, from extreme SW Senegal, Guinea-Bissau, Guinea, extreme SW Mali (Bafing-Makana), Sierra Leone, and Liberia to Ivory Coast (local, north to Nimba, Maraoué, Abengourou) and Ghana (not uncommon, north to Bia Nat. Park, Ejura and Amedzofe); and from S Nigeria (common), Cameroon (very common, north to Korup Nat. Park and Dimako), Gabon (very common), Congo, Angola (Cabinda, and in extreme NE at Dundo) and Zaïre (throughout forested areas, very common) to SW and SE Central African Republic, extreme SW Sudan (Bengengai) and Uganda (widespread, from Budongo, Bwamba and Impenetrable Forests east to Mabira and Elgon). Bioko. Single record from Bukoba, Tanzania, reported by Harvey (1979) refers to Yellow-throated Leaf-love *Chlorocichla flavicollis* (Baker 1990). Density in Gabon, 12 groups/1 km^2; in Nigeria (Ogba Forest), 1 bird/5 ha (Johnson 1989).

Description. *C. c. emini* Chapin: W-central Zaïre (Lukolela) and NE Angola to Uganda and Tanzania. ADULT ♂: top of head to hindneck olive-brown, slightly greyer on hindneck; a few long black bristles on hindneck (**A**), lying flat on mantle and upper back. Mantle to rump olive-green, paler on lower rump; uppertail-coverts dull reddish green. Tail-feathers dull brownish red with olive-green outer margins and all except central pair with yellowish buff inner margins, shafts reddish above, straw below. Lores, around eye, cheeks and ear-coverts dark grey, cheeks and ear-coverts with thin white streaks. Sides of breast, narrow band across centre of breast, and flanks olive-green, rest of underparts bright, rich yellow, undertail-coverts of some birds lightly washed with pale cinnamon. Primaries and secondaries dark olive-brown, outer webs olive-green, brighter and paler on primaries, inner webs of secondaries progressively less dark inwardly, upperwing-coverts olive-green, edges of outer webs brighter; axillaries and underwing-coverts yellowish olive, small feathers along bend of wing yellow, inner margins of undersides of primaries and secondaries yellowish buff. Bill grey or pale grey to blue-grey, culmen black or dark brown; eye red-brown, brown or dark crimson, sometimes grey, rim of eyelids dark grey; legs and feet dark grey to blue-grey, soles tinged yellowish. Sexes alike. SIZE: (12 ♂♂, 9 ♀♀) wing, ♂ 86–95 (90·1), ♀ 82–88 (84·6); tail, ♂ 75–89 (82·8), ♀ 71–81 (76·8); bill, ♂ 19–21 (20·0), ♀ 18·5–20·5 (19·1); tarsus, ♂ 20–22·5 (20·9), ♀ 20–22·5 (21·3). WEIGHT: Uganda and Central African Republic, ♂ (n = 6) 22–34 (28·4), ♀ (n = 4) 25–30 (28·3); Nigeria, unsexed (n = 4) 32–35 (34·0).

IMMATURE: like adult but upperwing-coverts with dull cinnamon fringes, tail-feathers more pointed. Bill blackish; eye dark brown; feet blue-grey. Very young bird (wing 53; tail 15; bill 12), with some downy feathers remaining on underparts, especially lower belly and undertail-coverts, has upperparts dull olive-green, top of head with brownish wash, tail-feathers cinnamon, tips paler; chin and throat buffy white, sides of breast, narrow band across breast, and flanks dull olive, breast and belly sulphur yellow washed ochre on lower belly, undertail-coverts ochre; wings as adult.

NESTLING: unknown.

C. c. calurus (Cassin): S Nigeria (Benin) to extreme W Zaïre (lower Congo); Bioko. Like *emini* but uppertail-coverts and tail brighter and redder.

C. c. verreauxi Sharpe: Senegal to SW Nigeria. Top of head darker than other races, upperparts darker green, uppertail-coverts about like back, with only hint of reddish wash, tail less red than other races, mainly brownish olive-green, dark red confined to narrow stripe down centre of each feather. WEIGHT: Liberia, ♂ (n = 9) 41·2–48·1 (44·4 ± 2·7), ♀ (n = 4) 37·9–42·8 (40·3), ♀ in breeding condition (n = 3) 40·3–47·3 (43·7), 1 immature 47·2.

Apparently intergrades with *calurus* but line of intergradation has not been determined (Elgood 1982).

Field Characters. A common bulbul of lower-level forest bird parties, with reddish tail and bright yellow underparts contrasting with olive flanks. The white beard, frequently puffed out (**A**), is conspicuous, especially in dark undergrowth. The only *Criniger* in W Africa with a white throat; further, Yellow-bearded Greenbul *C. olivaceus* has olive-green underparts, Western Bearded Greenbul *C. barbatus* grey underparts. Eastern Bearded Greenbul *C. chloronotus* (Cameroon–Zaïre) has white throat but same grey underparts as Western Bearded Greenbul except for some yellow on belly (for distinctions from White-bearded Greenbul *C. ndussumensis*, see that species).

Voice. Tape-recorded (32, 53, C, ERA, GRI, MOR). Song rather weak and tuneless, typically a 5-note phrase, 'chit-chiro-chiro' or 'cher-chiro-cheelu', often preceded by a few short, dry chips. A 3-note variant, 'chit-der-chyit', the notes more punctuated (less run together) also given, and sometimes a form combining both types, 'chit-der-chiro-cheelu'. Bird responding to playback of its song gave an agitated 'der-der-der . . .', 8–10 notes ending with a double 'chup-chyip'; the series was repeated several times (Chappuis 1975). More common and noticeable than songs is the clear, loud whistled call note, 'peeyu' or 'pueeyu'. Also has a variety of low-pitched calls, including a hoarse 'cr-r-mmm' or 'urr-churr'. Alarm call almost identical to that of monkey *Cercopithecus cephus* which lives in same habitat (Brosset and Erard 1986). Chorusing 'like *Criniger barbatus* but with more whistled notes than rasping cries' described by Marchant (1953).

Criniger calurus

General Habits. Inhabits mainly primary and old secondary forest, but extends out along broader stretches of gallery forest and into forest–grassland mosaic. Within forest occurs in tangled vegetation along roads and tracks as well as in interior. Occupies mainly undergrowth and mid-levels, but in Gabon also ascends to canopy during long dry season (Brosset and Erard 1986). Lowland and transition forest, sometimes lower montane, up to 950 m in Cameroon, 1750 m in Zaïre, 1500 m in Uganda.

Occurs in groups of 3–12. Groups are territorial, maintaining same territory for considerable periods; of 50 birds ringed, Gabon, 22 recaptured in same place, 1 after 25 months, another after 4 years (Brosset and Erard 1986). Territory defended by group, all members giving monosyllabic call, one individual, clearly the leader, giving elaborate song. One of the most consistent members of mixed-bird flocks, and often the leader; present in 127 out of 271 parties, Gabon (Brosset and Erard 1986). Forages from low to upper middle levels (2–25 m), mainly by gleaning, creeping around tree trunks and along branches, extracting insect larvae from crevices, searching leaf surfaces, including underside, prying into creeper tangles and bunches of dead leaves caught in forks. In Sierra Leone typically takes food in fluttering flight, plucking berries and seizing insects from leaves (Field 1979). In mixed flocks keeps close to any *Campethera* woodpecker (*C. caroli*, *C. cailliautii*, *C. nivosa*), even though these feed mainly on ants. Sometimes attends ant swarms. Flicks wings regularly, sometimes also fans tail. At night observed sleeping while hanging head down from end of branch (Brosset and Erard 1986).

Sedentary, at least in Gabon.

Food. Mainly insects and their larvae, including beetles, grasshoppers, leaf hoppers, bugs, Lepidoptera (including caterpillars), Diptera, ants and ant pupae; also spiders and some fruits, berries and small seeds, including peppers and oil palm fruits; once, a lizard (Johnson 1989). Young fed caterpillars and adult Lepidoptera.

Breeding Habits. Territorial.

NEST: cup of rootlets, plant stems and moss on base of small twigs and hard dry leaves, with outer layer of thin tendrils (**B**); one was lined with fine blackish fibres (*Marasmius?*) making interior of nest look almost black, matching dark eggs (see below); 1 nest, ext. diam. 110, int. diam. 65. Situated in bush, 1·8–2·8 m above ground, on support formed by junction of several leaf petioles and attached to them by spider webs; one was placed only 0·6 m above forest floor on bunch of fallen sticks near small stream.

EGGS: 2 (4 clutches). Pyriform; glossy; usually brightly coloured (rose, olive-green, brown-violet), with dark cap on large end, rest of surface spotted olive, chocolate or purple; one clutch in dark nest (see above) had pink ground colour almost completely obscured by dense chocolate-brown marks, presenting uniform dark surface. SIZE: (n = 4) 22–23·5 × 15·1–16 (22·8 × 15·6).

LAYING DATES: Liberia (immature Nov); Cameroon, Feb, Apr–Aug, Dec (breeding condition Sept); Bioko, Feb–Mar, Nov (breeding condition Aug), probably no well-defined breeding season (Eisentraut 1973); Gabon, Dec–Feb (short dry season), also breeding condition Aug; Zaïre, all year; Central African Republic (breeding condition June); Uganda, June (breeding condition May, Nov, juveniles Apr, July).

INCUBATION: by ♀ only.

DEVELOPMENT AND CARE OF YOUNG: young fed by both parents, at rate of 20 feeds in 8 h (1 nest); fledge at 14 days. 1 young observed both feeding itself and also being fed frequently by members of its group (Gabon: Brosset and Erard 1986).

BREEDING SUCCESS/SURVIVAL: of 4 nests, Gabon, 3 produced 2 young each, 1 was destroyed by predator.

B

Reference
Brosset, A. and Erard, C. (1986).

Plate 20
(Opp. p. 321)

Criniger ndussumensis Reichenow. White-bearded Greenbul. Bulbul de Reichenow.

Criniger verreauxi ndussumensis Reichenow, 1904. Vög. Afr. 3, p. 383; Ndussuma country: Kinyawanga, near Beni (= Zaïre).

Range and Status. Endemic resident. SE Nigeria, SW Cameroon, Gabon, N Angola (Cabinda), SW Central African Republic (Lobaye Préfecture); N Zaïre east to Semliki and Itombwe. Much rarer than Red-tailed Greenbul *C. calurus* in Itombwe (Prigogine 1971), and apparently less common throughout its range.

Criniger ndussumensis

Description. Differs from Red-tailed Greenbul only as follows: bill more slender and slightly shorter; more pronounced greyish white spot in front of eye; rictal bristles and nuchal hairs less well developed; flanks darker (more heavily washed olive), undertail-coverts cinnamon-buff rather than yellow. SIZE: wing, ♂ (n = 8) 88–96 (91·2), ♀ (n = 8) 81–88 (84·5); tail, ♂ (n = 5) 78–86 (81·3), ♀ (n = 8) 72–78 (74·3). Measurements completely overlap those of sympatric *C. calurus emini* (see Prigogine 1971).

TAXONOMIC NOTE. Status uncertain. It hybridizes with *C. calurus emini* in NE Zaïre and was treated as a subspecies of *C. calurus* by Chapin (1953). But no such hybridization is known in Cameroon or Gabon, where the birds behave as 2 species. White (1956, 1962), Rand (1958, and *in* Peters 1960), Hall and Moreau (1970) and Prigogine (1971) treated them as distinct species. Rand, whom we follow, called this form *C. ndussumensis*; White and Prigogine believed it to be a race of *C. olivaceus*; Hall and Moreau regarded it as a full species forming a superspecies with *C. olivaceus*.

It is extremely similar to *C. calurus*; Brosset and Erard (1986) found no differences in voice, behaviour or ecology in Gabon; their data could refer to either form but are all listed under *C. calurus*. Chapin (1953) found no differences in habits or voice in E Zaïre. Stuart (1986) could not separate them either in the field or in the hand in Cameroon, and his data may refer to either species. However, Chappuis (1975) and Dowsett-Lemaire and Dowsett (1989) found vocal differences (see *Voice*).

Field Characters. Probably not safely distinguishable in the field from Red-tailed Greenbul (but see *Description* and *Voice*).

Voice. Tape-recorded (53, C, LEM). Song of 3 notes, the last lower and slightly down-slurred, 'chuk-ker-chyer', similar to 3-note song of Red-tailed Greenbul but tonal quality quite different, burry and low-pitched rather than whistled (singing birds not examined in the hand; this song *might* have been given by Red-tailed Greenbul: Chappuis 1975). Call, a hard trill, different from Red-tailed Greenbul (Dowsett-Lemaire and Dowsett 1989).

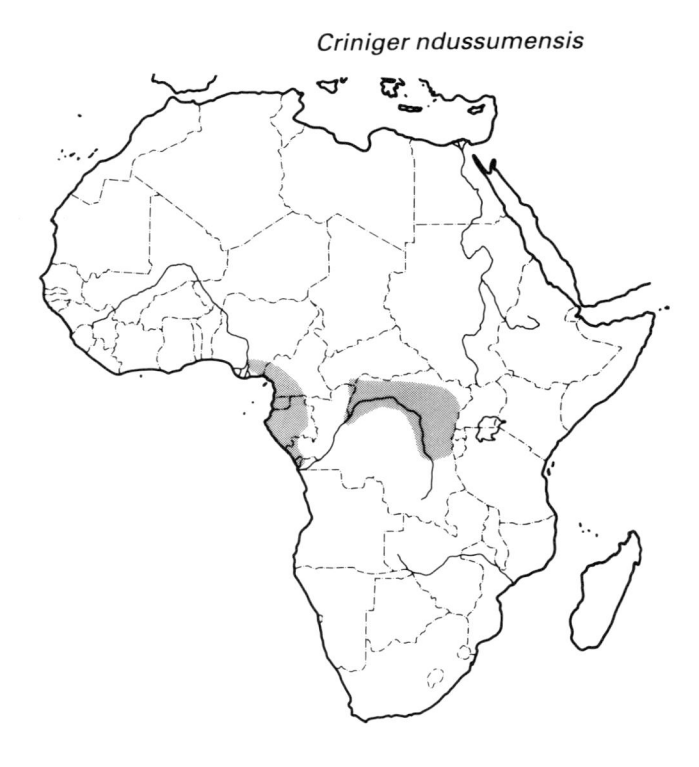

General Habits, Food, Breeding Habits. Laying Dates in E Zaïre (Itombwe), Feb–Mar. No further data can be assigned with certainty to this form due to problems of identification (see above).

Reference
Rand, A. L. (1958).

Plate 20
(Opp. p. 321)

Criniger olivaceus (Swainson). Yellow-bearded Greenbul. Bulbul à barbe jaune.

Tricophorus olivaceus Swainson, 1837. Birds of West Afr. 1, p. 264; W Africa.

Range and Status. Endemic resident with disjunct distribution west of Dahomey gap. Senegambia (Casamance); extreme SE Guinea (N'Zérékoré); extreme SW Mali, south of Mandingo Mts: Sagabari, Bafing-Makana, Kangaba; Sierra Leone (moderately common in Gola Forest; also in Nimini Hills, Kono District); W, N and SE Liberia (Lofa-Mano Forest, Mt Nimba, Sapo Nat. Park, Greenville, Grand Gedeh County); Ivory Coast (Tai Nat. Park, Yapo Forest); Ghana, 2 old records from Fanti but intensive searches in 1970s failed to find it (M. A. Macdonald, *in* Collar and Stuart 1985). Locally not uncommon but generally rare; threatened by habitat destruction; classed as 'Vulnerable' in IUCN/ICBP Red Data Book.

Description. ADULT ♂: upperparts, including scapulars and upperwing-coverts, rather dark olive-green, slightly brighter on rump; top of head either darker and browner or same colour as back. Some long, black, hair-like bristles on hindneck stretching over mantle. Tail-feathers brown with slight reddish tinge, margins olive-green. Thin loral line pale olive, narrow eye-ring white; from base of bill below eye to ear-coverts dark olive-green streaked yellowish white; chin and throat pale bright yellow, underparts dark olive-green becoming yellowish in centre and forming pale midline from lower breast to undertail-coverts. Flight feathers dark brown, outer margins olive-green, inner margins olive-buff; underwing-coverts and axillaries dull yellowish olive, bend of wing greenish yellow. Bill bluish grey to blackish grey, cutting edges horn; eye deep brown to red-brown; legs and feet blue-grey. Sexes alike. SIZE: (7 ♂♂, 6 ♀♀) wing, ♂ 87–91 (88·6), ♀ 78–88 (82·2); tail, ♂ 76·5–84 (79·6), ♀ 71–80 (75·0); bill, ♂ 17·5–19 (18·4), ♀ 17–18·5 (17·7); tarsus, ♂ 21·5–26 (22·9), ♀ 20·5–23 (21·9). WEIGHT: ♂ (n = 8) 25·7–31·7 (28·7 ± 2·2), ♀ (n = 2) 23·8, 29·8, ♀ in breeding condition (n = 2) 26·4, 27·3.

IMMATURE: said to have top of head brownish olive, mantle browner, below, duller yellow (Mackworth-Praed and Grant 1973), but a freshly collected immature ♀, fully grown but with unossified skull, is indistinguishable from adults.

NESTLING: unknown.

Field Characters. A medium-sized greenbul, entirely olive-green except for yellow throat and slight rusty wash on tail. Smaller than sympatric congeners Western Bearded Greenbul *C. barbatus* and Red-tailed Greenbul *C. calurus*. Western Bearded has yellow throat but greyish head and grey-brown underparts; Red-tailed has green and yellow underparts but white throat; both are bold and noisy, conspicuous in bird parties, while Yellow-bearded is quiet and unobtrusive, with different foraging techniques (see below).

Voice. Not tape-recorded. Generally silent; only known call an undistinguished little 'chup' (Field 1979).

General Habits. Inhabits lowland primary forest; not in second growth. Sometimes solitary but usually joins mixed-species flocks. Forages at all levels from ground stratum to canopy; gleans bark, clinging to trunks, peering into crevices, searching undersides of branches; feeds more like *Phyllastrephus* than *Criniger* (Field 1979).

Food. Mainly insects, also small fruits.

Criniger olivaceus

Breeding Habits. Unknown.
LAYING DATES: Guinea (breeding condition May); Sierra Leone (young being fed by parents Apr); Liberia (breeding condition Nov).

References
Collar, N. J. and Stuart, S. N. (1985).
Field, G. D. (1979).

Genus *Pycnonotus* Boie

Bill comparatively short, culmen curved and not sharply ridged, gonys almost straight, nostrils oval or elongated; rictal bristles weak. Wings short, about same length as tail; tail square, rounded or slightly graduated; legs short to medium, usually black. Crown feathers erectile and sometimes elongated. Plumage more variable than other genera. Subgenus *Pycnonotus*, to which all African species belong, characterized by red, orange, yellow or white vent generally contrasting with rest of underparts.

A few species confined to forest, but most inhabit open country with scattered trees and scrub, readily adapting to man-made habitats, especially gardens. Vociferous, but song short and undistinguished.

36 species; 4 in Africa, 3 endemic. Asiatic *P. jocosus* successfully introduced in several parts of the world. The 3 endemic species (*P. barbatus/nigricans/capensis*) are almost identical behaviourally (Irwin 1958) and differ mainly in eye-ring colour. They are broadly parapatric, with and without ecological segregation where they overlap. The few instances of hybridization have occurred only where man has altered the natural habitat, and we follow Liversidge (1983) in treating them as separate species in a superspecies.

We place *Pycnonotus* last in the family because we agree with Short *et al.* (1990) that it is a recently expanding, speciating, derived genus that should follow other groups.

Pycnonotus barbatus superspecies

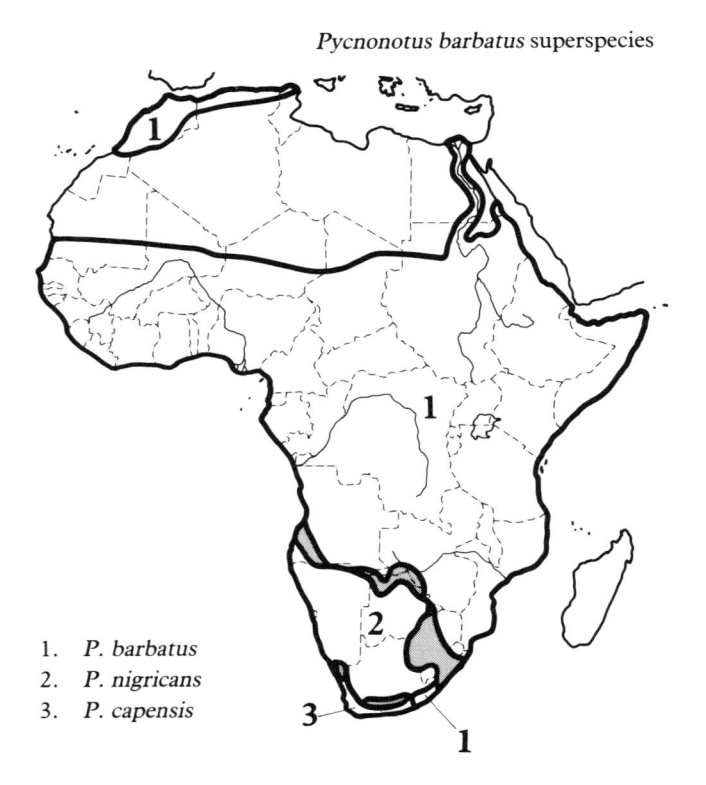

1. *P. barbatus*
2. *P. nigricans*
3. *P. capensis*

Plate 16
(Opp. p. 257)

Pycnonotus xanthopygos (Ehrenberg). White-eyed Bulbul. Bulbul des jardins.

Ixus xanthopygos Ehrenberg, 1833. *In* Hemprich and Ehrenberg, Symb. Phys., Aves, fol. bb.; Arabia = Muwailah on coast near north end of Red Sea.

Forms a superspecies with *P. barbatus*, *P. nigricans* and *P. capensis*, also Asian *P. leucotis*, *P. leucogenys* and possibly *P. cafer* and *P. aurigaster*.

Range and Status. Resident Turkey, Lebanon, Israel, W Syria, Jordan, Sinai and Arabian peninsula.

Only 1 certain record; ♂ collected Egypt (near Cairo) was apparently paired with a Common Bulbul *Pycnonotus barbatus* and had enlarged testes (Meinertzhagen 1930). Reported occurrence at Suez (Wimpfheimer *et al.* 1983) was an error for Common Bulbul (S. M. Baha el Din *in* Goodman and Meininger 1989).

Description. ADULT ♂: entire head to nape, including face, chin and upper throat, brownish black; hindneck, sides of neck and mantle to uppertail-coverts, scapulars and upperwing-coverts uniform greyish brown; tail dark brown narrowly tipped pale brown; lower throat and upper breast dark brown; lower breast, upper belly and flanks pale grey-brown, becoming whitish on lower belly; undertail-coverts bright lemon-yellow. Wings medium brown, axillaries, underwing-coverts and inner margins of underside of flight feathers pale grey-brown, bend of wing creamy. Bill grey-black; bare skin around eye white; eye brown to dark brown; legs and feet black or slate-blue, soles dark grey or flesh-grey. SIZE: wing, ♂ (n = 18) 96–103 (99·4), ♀ (n = 19) 90–95 (92·8); tail, ♂ (n = 15) 84–95 (89·1), ♀ (n = 15) 80–88 (84·6); bill, ♂ (n = 15) 19·7–22 (21·0), ♀ (n = 15) 19·3–21·2 (20·3); tarsus, ♂ (n = 14) 22·7–24·9 (23·6), ♀ (n = 15) 21·8–23·7 (22·8). WEIGHT: unsexed (n = 56, Israel) 35–46 (44·0).

IMMATURE: like adult but feathers softer and looser, especially on rump, vent and undertail-coverts; tail-feathers narrow, rather pointed; P10 longer, with more broadly rounded tip.

Field Characters. A robust bulbul, slightly larger and stouter-billed than parapatric race *arsinoe* of Common Bulbul. Distinguished by blacker head (contrasting more with back and breast) and grey-brown underparts, whitish only in centre of belly; Common Bulbul has more extensive brown breast changing abruptly to white belly. Also has white eye-ring and bright yellow vent (vent white in Common Bulbul), and tail often looks conspicuously pale-tipped (C. H. Fry, pers. comm.).

Pycnonotus xanthopygos

Voice. Tape-recorded (HOL). Song, disjointed series of short phrases, some notes mellow and fluty, others with husky bulbul quality, rendered e.g. 'twur-tu-TWEE-teeroo' (Cramp 1988) or 'buli-buli-buli-buli' (Hollom *et al.* 1988). Alarm call 'tscheck' or 'trratsch'; other calls include 'wit-wit-wit . . .' and 'teewit'. For further details, see Cramp (1988).

General Habits. Habitat similar to Common Bulbul: almost anywhere with trees, bushes or scrub, including oases, river valleys, wadis and coastal thickets; also palm and citrus groves, orchards, banana plantations and gardens.

Generally occurs in pairs; pairs remain together all year, often for several years. Becomes gregarious where food abundant; in high-density areas, hundreds gather in post-breeding flocks. Some birds hold territories all year, but vacate them temporarily to join feeding flocks (Cramp 1988). Active and noisy; squabbles break out among flock members; sings all year, although less regularly in early winter. Normally shy, but becomes tame around humans. Catches insects in flight; also gleans them from leaves, sometimes hovering.

Food. Mainly fruit, including peaches, plums, strawberries, tomatoes and prickly pear *Opuntia*; also seeds, leaves, nectar, insect larvae, flies, moths, flying ants, bees, wasps, mole-crickets, locusts, worms and snails.

Reference
Cramp, S. (1988).

Pycnonotus barbatus (Desfontaines). Common Bulbul; Dark-capped Bulbul; Black-eyed Bulbul. Bulbul commun.

Plate 16
(Opp. p. 257)

Turdus barbatus Desfontaine, 1789. Hist. Acad. R. Sci. Paris (1787), p. 500, pl. 13; 'Côtes de Barbarie' (= near Algiers).

Forms a superspecies with *P. nigricans* and *P. capensis*, also Asian *P. xanthopygos*, *P. leucotis*, *P. leucogenys* and possibly *P. cafer* and *P. aurigaster*.

Range and Status. Endemic resident. Perhaps the most widespread and abundant bird in Africa, found from extreme N Morocco (Tangiers) and N Egypt to Cape Province, South Africa. Morocco, widely distributed and locally common: in coastal regions from Tangiers south to Goulimine and Bou Izakarn, east to Plaine des Trifa and to 800 m in Mts des Beni Snassene. Ascends Moyen Atlas to 900 m, Haut Atlas to 2300 m, descending toward desert in valleys of Ziz R. to Tafilalet and Dra R. to Zagora, and to near Fezzou. Southern limits, oases in Djebel Bani, e.g. Akka and Tarhjijt. Algeria, very common on coast and inland to watershed of coastal Atlas Mts, up to 1000 m; further south, records from Batna, Biskra, Baniane and El Goléa. Tunisia, scarce, mainly in N in Kroumerie and valley of Medjerda R.; records from Sousse, Kairouan, Gabès and Nefta. Egypt, throughout Nile valley and delta, west on coast to Bahig, and in Fayoum and oasis of Wadi Natrun; in SE occurs in Gebel Elba region (abundant on coast). Since 1944 has colonized Suez Canal area.

South of Sahara found everywhere except driest parts of Ethiopia and Somalia, and in SW. Mauritania, fairly common in S Sahel and especially along Senegal R.; several records from Adrar region. Mali to Sudan, ranges north to 16°–18° N, in Niger also in Aïr, in Chad to Ennedi in E, also in Tibesti; in Sudan follows Nile north to Egypt, and present all along Red Sea coast. Common to abundant Ethiopia up to 3000 m except Danakil Desert and driest areas of E; in Somalia and N Kenya penetrates desert along rivers and wherever there is a little vegetation, absent only from the most treeless areas. Common to abundant in every other country in its range; absent SW Angola, Namibia except extreme NE (chiefly along Okavango R. and in Caprivi Strip) and S Botswana (occurs south through Okavango region to L. Ngami, and along N and E fringe, mainly Nata R., Shashi R. and around Macloutsie). In South Africa occurs south to Transvaal (absent only in SW), E Orange Free State, Natal, Transkei and E Cape Province to Sundays R. Present Zanzibar and Mafia Is; absent Pemba and Gulf of Guinea Is.

Pycnonotus barbatus

Description. *P. b. layardi* Gurney (including '*micrus*', '*tenebrior*', '*pallidus*' and '*naumanni*'): SE Kenya (Taveta), coastal Tanzania (inland to Arusha, Songea), and E and S Zambia to NE Botswana and South Africa. ADULT ♂: top of head, face and chin black, grading to dark brown on throat and ear-coverts but sharply demarcated from mantle; feathers on back of head slightly elongated to form short crest; upperparts and wing-coverts drab greyish brown, tail darker brown, feathers with narrow pale tips which disappear with wear. Breast brown, feathers with narrow pale fringes; belly and flanks dull white, latter washed with pale grey-brown; undertail-coverts and vent bright yellow. Flight feathers dull sooty brown, undersides with pale basal half to inner webs; axillaries and underwing-coverts dull white to pale brown, bend of wing cream-coloured. Bill black, inside of mouth yellow; narrow eye-ring black, inner surface of eyelid yellow, eye dark brown; legs and feet black. Sexes alike, but ♂♂ larger and heavier. Leucistic individuals not uncommon. SIZE: wing, ♂ (n = 24) 91–101 (97·0), ♀ (n = 25) 89–99 (92·8); tail, ♂ (n = 10) 83–88 (86·0), ♀ (n = 10) 82–87 (84·1); bill, ♂ (n = 24) 16–21·8 (19·5), ♀ (n = 25) 16–21·2 (18·8); tarsus, ♂ (n = 10) 20–23 (21·7), ♀ (n = 10) 20–24 (21·7). WEIGHT: ♂ (n = 59) 34–49·5 (40·4), ♀ (n = 46) 29–42 (34·0), unsexed (n = 609) 23–52·9 (38·9). Large samples showed little or no seasonal weight change (Hanmer 1978, Crowe *et al.* 1981). Species as a whole small in hot regions, larger in cool ones (Crowe *et al.* 1981; cf. Britton 1972, Hanmer 1978).

IMMATURE: duller and paler than adult, upperparts with rusty wash, especially on rump; feathers softer and looser, especially on rump, vent and undertail-coverts; tail-feathers narrow, rather pointed; P10 longer, with more broadly rounded tip. Weighs less than adult (Hanmer 1978).

NESTLING: naked, flesh-brown or chestnut above, slightly paler below; gape yellow, mouth yellow-pink.

P. b. barbatus Desfontaine: Morocco to Tunisia. Head and upperparts brown with little contrast, underparts more uniform than other races, less contrast between breast and belly, brown throat grading into grey-brown breast and off-white belly; vent white. Much closer to *inornatus* than to *arsinoe* of Egypt. SIZE: wing, ♂ (n = 17) 100–107 (103), ♀ (n = 18) 93–100 (96·8).

P. b. inornatus (Fraser) (including '*goodi*'): Senegal to Ghana, N Niger (Aïr), N Nigeria, N Cameroon, W Chad, intergrading with '*nigeriae*' (= *gabonensis*) in Nigeria. Very like *barbatus*, but upperparts greyer, belly whiter, undertail-coverts white or cream, breast feathers plain with only hint of pale fringing. WEIGHT: Ghana, unsexed (n = 35, July–Sept) 31–41 (35·6), one bird increased from 31·6 in Aug (wet season) to 38 in Nov (dry season); Nigeria, unsexed (n = 26, Nov–Mar) 36–46 (40·2), (n = 13, Apr–June) 34–42 (38·2); L. Chad, unsexed (n = 161, season?) 28–47 (36·0 ± 3·1).

P. b. arsinoe (Lichtenstein): Egypt and Sudan south to Darfur, Kordofan, 10°N in Nile valley and L. No, and E Chad. Head and throat blackish, brown on breast ends abruptly, contrasting with white belly, not grading into it as in *inornatus*, breast feathers with pronounced pale fringes, undertail-coverts white. Variable white mark on ear-coverts in most birds. Meets yellow-vented *tricolor* without intergradation. SIZE: wing, ♂ (n = 21) 91–97 (93·6), ♀ (n = 12) 86–91 (88·7). WEIGHT: Chad, unsexed (n = ?) 20–35 (29·0), (Ennedi) 1 ♂ 26, 1 ♀ 28.

P. b. schoanus Neumann: highlands of Ethiopia from Eritrea south to NW Sidamo-Borama and east to Dire Dawa and Harar; extreme SE Sudan (Boma Hills). Fairly similar to *arsinoe*, with white undertail-coverts, but upperparts darker, head blacker, pale spot on ear-coverts indistinct, breast darker. SIZE: wing, ♂ (n = 17) 86–97 (91·3), ♀ (n = 12) 83·5–97 (89·4). WEIGHT: 1 ♂ 33, 1 ♀ 34.

P. b. somaliensis Reichenow: Djibouti, NW Somalia, SE Ethiopia (NE Bale, SE Arussi). Like *arsinoe*, with white undertail-coverts, but upperparts slightly paler, white spot on ear-coverts more conspicuous, breast feathers with more pronounced pale fringes, producing spotted effect where breast meets belly.

P. b. spurius Reichenow: S Ethiopia from S Bale (Gedel Mts) to N Sidamo-Borama (Alghe, Sagan R.). Like *schoanus* but undertail-coverts deep yellow. SIZE: wing, ♂ (n = 7) 84–87 (85·1), ♀ (n = 7) 80–86 (83·1).

P. b. dodsoni Sharpe (including '*peasei*', '*chyulu*' and '*teitensis*'): Somalia (except NW) and adjacent lowlands of Ethiopia west to Mega and Yavello, lowland N and E Kenya south to Chyulu Hills, Tsavo and Kipini; hybridizes in E and S Kenya with highland populations of *tricolor*. Like *somaliensis*, but undertail-coverts yellow, breast spotting more pronounced, extending to upper belly, pale tips to tail-feathers broader; feathers of upperparts with pale fringes producing somewhat scaly effect. Small: wing, ♂ (n = 14) 78–84 (81·3), ♀ (n = 9) 70–81 (78·1). WEIGHT: unsexed (n = 29, Kenya) 21–33 (26·5 ± 2·4).

P. b. gabonensis Sharpe (including '*nigeriae*'): central Nigeria and central Cameroon to Gabon and S Congo. Head, upperparts and breast darker than *inornatus*, with abrupt transition from breast to belly. Undertail-coverts with variable amount of yellow, bridging gap between *inornatus* and *tricolor*, paler in north, yellower in south. WEIGHT: Cameroon (Yaoundé) 2 ♂♂ 42, 46, 1 ♀ 34, unsexed (n = 3) 36·8, 46·8, 43·8.

P. b. tricolor (Hartlaub) (including '*minor*', '*fayi*', '*vaughan-jonesi*' and '*ngamii*'): E Cameroon, Congo, Zaïre, S Sudan and W and central Kenya south to Angola (except SW), N Namibia, NW Botswana (Okavango) and Zambia north and west of *layardi*. Like *layardi*, but crown dark brown, not contrasting with upperparts which are darker and browner than *layardi*; face and chin dark brown rather than black. SIZE: wing, ♂ (n = 12) 98–105 (101), ♀ (n = 5) 92–96 (94·0). WEIGHT: Zaïre (Upemba Nat. Park), ♂ (n = 8) 37–41 (39·6 ± 1·2), ♀ (n = 5) 34–41 (36·8 ± 2·4); Uganda, ♂♀ (n = 50) 31–42; Kenya, ♂♀ (n = 428) 27–46; Zambia, unsexed (n = 46, Balovale) 29·5–41 (34·6 ± 2·5), (n = 8, Mwinilunga) 36–42·5 (39·7 ± 1·9); Botswana, unsexed (n = 14) 32–40·9 (37·4).

TAXONOMIC NOTE: we follow White (1962) in recognizing only the most well-marked races of this highly polytypic species, and prefer not to retain names of intermediate populations. Nominate *barbatus* of NW Africa is far more like *inornatus* of W Africa than *arsinoe* of Egypt, suggesting recent colonization of NW Africa, making it unlikely that nominate *barbatus* is a good species (cf. Hall and Moreau 1970).

Field Characters. Probably the most familiar bird in Africa. Size of thrush, with dark, slightly crested head, grey-brown upperparts and breast, white belly and white or yellow vent. Bold, noisy and conspicuous; often lives close to man. In SW Africa, dark eye-ring distinguishes it from closely-related African Red-eyed Bulbul *P. nigricans* (red eye-ring) and Cape Bulbul *P. capensis* (white eye-ring). For further distinctions, see those species.

Voice. Tape-recorded (5, 53, 58, 69, 86, C, F). Song, phrases of 3–6 notes, somewhat abrupt and disjointed, e.g. 'quick, chop, toquick', 'quick, quake, koyo', 'quick, chu, kway-ko', 'kooee, ti-ti-kwitta', 'chip, chop, chip, kweeko'. Very varied over the range of the species, though individual birds tend to have limited repertoire; pleasant, mellow but not rich or sustained. Many popular renderings, the best being 'quick, doctor, quick!'. Intervals between phrases vary from 2 to 10 s, typically *c.* 5 s; intervals may be silent or contain chattering notes. Sometimes raises wings when singing. Common call is loud, somewhat shrill and querulous 'quick, quick', 'klip, klip' or 'tit-tit-tit'; may indicate mild alarm or annoyance, and a more strident version used when scolding predator, together with nasal 'scaaa' and other harsh notes. Loud chattering given at dawn and dusk, often in chorus. Low, husky 'jer-dut' or 'jut' given when at rest. A low 'cheedle-cheedle-cheedlelit' given as greeting at nest (van Someren 1956). Calls of young include a piping whistle (bird lonely), 'tweet-tweet' (food-begging), angry scolding twitter (hawk in nearby tree) and a purring comfort noise (Preston 1975).

General Habits. Occupies almost any wooded or bushy habitat; absent only from large tracts of unbroken forest or woodland, open grassland with no bushes, and treeless desert. Partial to water; especially common in riverine habitats, penetrating desert, forest and woodland along streams. Lives at edge of forest, including cutover areas overgrown with parasol trees *Musanga*, thickets of Solanaceae and other plants of first stage of forest regeneration. Enters old second-growth and more open parts of virgin forest, even temporarily entering heavy forest when attracted by a special food source; found in canopy of Kakamega Forest, Kenya, when attracted by fruiting trees (Zimmerman 1972), and in NW Zambia often in canopy of evergreen forest patches. Lives in swamp forest. Quick to follow man's invasion of forest, along roads and railway lines and around camps, even flying over it to occupy isolated camps and clearings within a few years of their creation. In Nigeria invaded cattle ranch and new hotel on Obudu Plateau; at Mt Nimba (Liberia), as forest was cleared, 1963–82, successively invaded all types of habitat at all elevations, especially common at forest edge and in clearings; population increase enormous, and now the most common bird in the Nimba region (Colston and Curry-Lindahl 1986). Sparse or absent in virgin stands of *Brachystegia*, *Baikiaea* and mopane woodland, especially the drier parts, but tolerates more open areas where there are bushes and shrubs, and in thicker parts found in dambos and open ground along streams; in Zambia partial to thickets on woodland anthills. Ascends to 3000 m or more, in bamboo thickets, juniper woods (Ethiopia), bracken-briar thickets (Malaŵi) and dense *Leucosidea* scrub (South Africa). Found in all types of wooded savanna and parkland, thickets, coastal and other scrub; in N Africa, macchia; in Serengeti, vegetation on kopjes and isolated trees on plains. Penetrates semi-arid areas in vegetation along streams and wadis; colonizes oases; in Sahara found almost anywhere there is water and a few trees; characteristic of palm groves and wells; in SE Egypt lives in *Dracaena ombet* forest (Goodman and Meininger 1989). Occurs up to 2500 m on Jebel Marra (Sudan). Lives in stands of papyrus mixed with *Miscanthidium* and other marsh plants, and even in pure papyrus several km out in lake from shore (Vande weghe 1981). Enters plantations of pine, *Eucalyptus* and other introduced trees; orchards, citrus groves; cultivated and fallow land; parks, gardens; particularly fond of scrubby vegetation around native cultivation and in and around villages.

Usually in pairs; in groups of 3–4 after breeding season (probably family parties), sometimes up to 30 (Ghana), once >70 (Zimbabwe). Congregates at fruiting trees, mixing with green pigeons, mousebirds, Olive Thrush *Turdus olivaceus*, white-eyes and other frugivores: 28 counted in a single fig tree. Not territorial outside breeding season; wanders about feeding opportunistically. Forages in trees and bushes; also takes food from ground. Once broke up 7·5 cm lizard on ground – first bit off head, then ate the viscera and flew off with the head (Sharland 1975). Plucks berries, sometimes hanging head-down from small branch like weaver *Ploceus*, and swallows them whole. Pecks into larger fruits; makes hole in orange on tree to eat flesh inside. Probes flowers for nectar. Gleans insects from leaves, sometimes hovering. Takes termites from galleries on branches accidentally broken and exposed. Once entered *Brachystegia* woodland, not its usual habitat, to exploit outbreak of caterpillars (Vernon 1975). Often sallies for insects on wing; agile, twisting and turning in pursuit of prey. Returns to same perch (top of bush or small tree) like

Plate 23

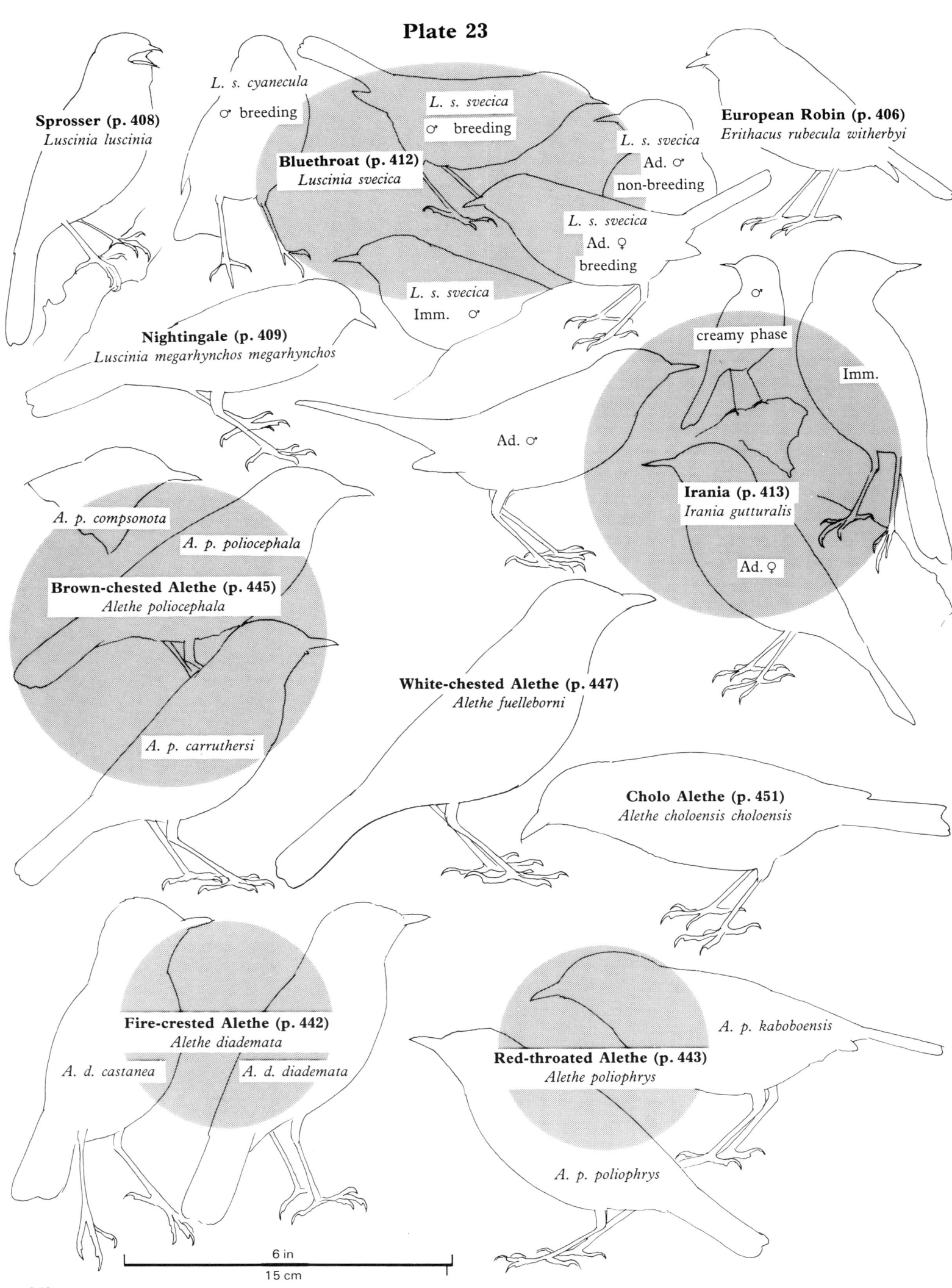

Sprosser (p. 408)
Luscinia luscinia

L. s. cyanecula
♂ breeding

L. s. svecica
♂ breeding

Bluethroat (p. 412)
Luscinia svecica

L. s. svecica
Ad. ♂
non-breeding

European Robin (p. 406)
Erithacus rubecula witherbyi

L. s. svecica
Ad. ♀
breeding

L. s. svecica
Imm. ♂

Nightingale (p. 409)
Luscinia megarhynchos megarhynchos

♂
creamy phase

Imm.

Ad. ♂

Irania (p. 413)
Irania gutturalis

Ad. ♀

A. p. compsonota

A. p. poliocephala

Brown-chested Alethe (p. 445)
Alethe poliocephala

A. p. carruthersi

White-chested Alethe (p. 447)
Alethe fuelleborni

Cholo Alethe (p. 451)
Alethe choloensis choloensis

Fire-crested Alethe (p. 442)
Alethe diademata

A. d. castanea   *A. d. diademata*

Red-throated Alethe (p. 443)
Alethe poliophrys

A. p. kaboboensis

A. p. poliophrys

6 in / 15 cm

368

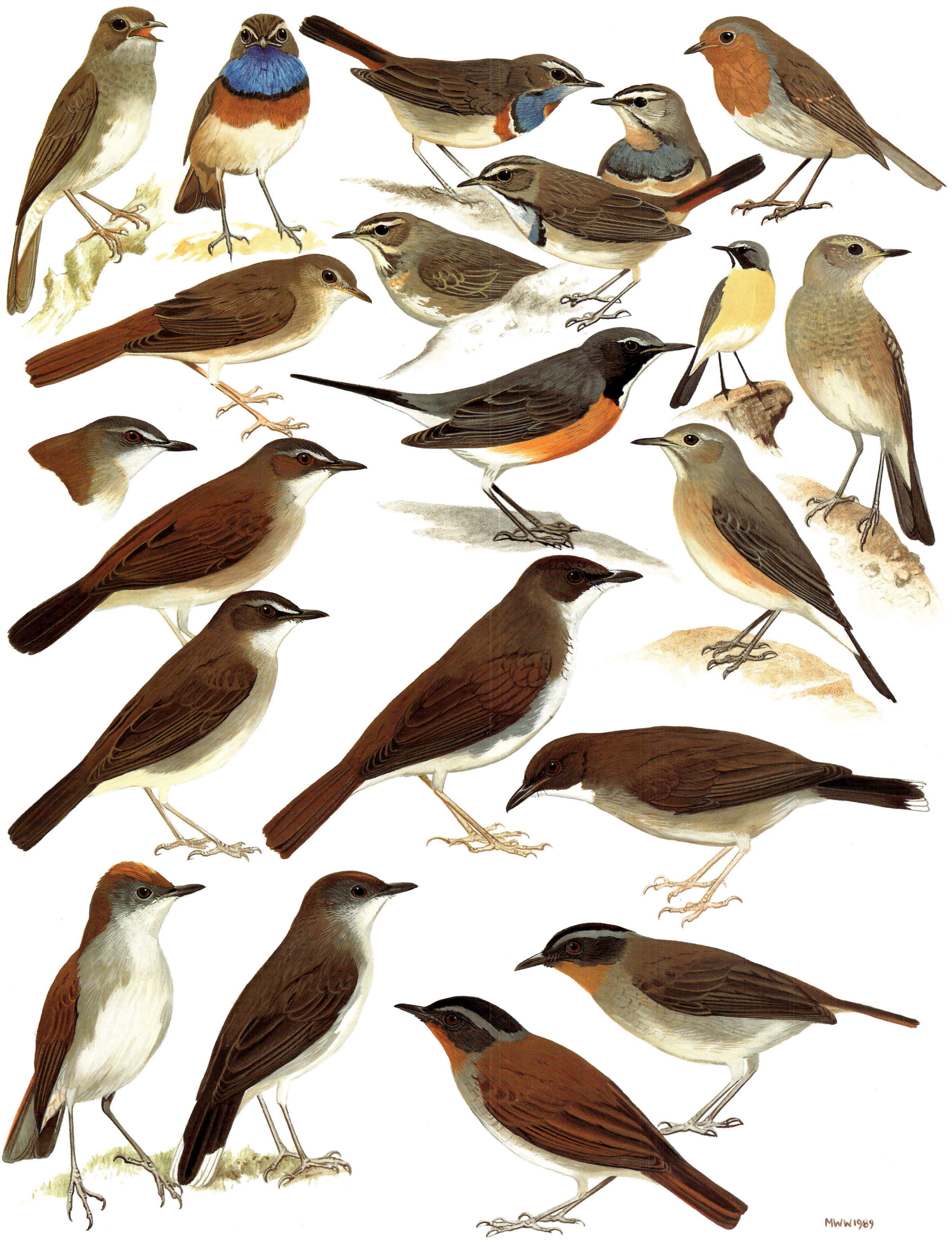
MWW 1989

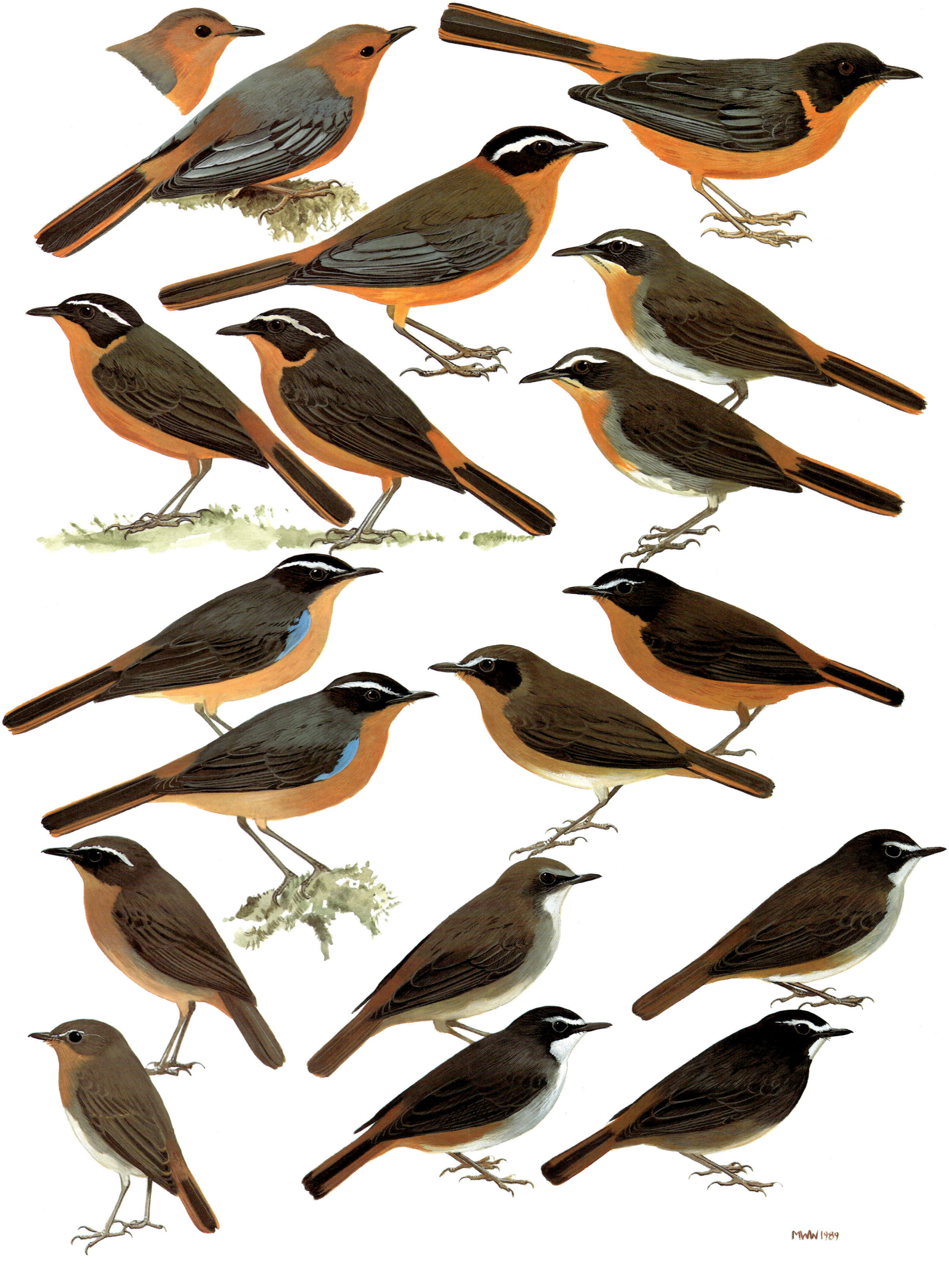
MWW 1989

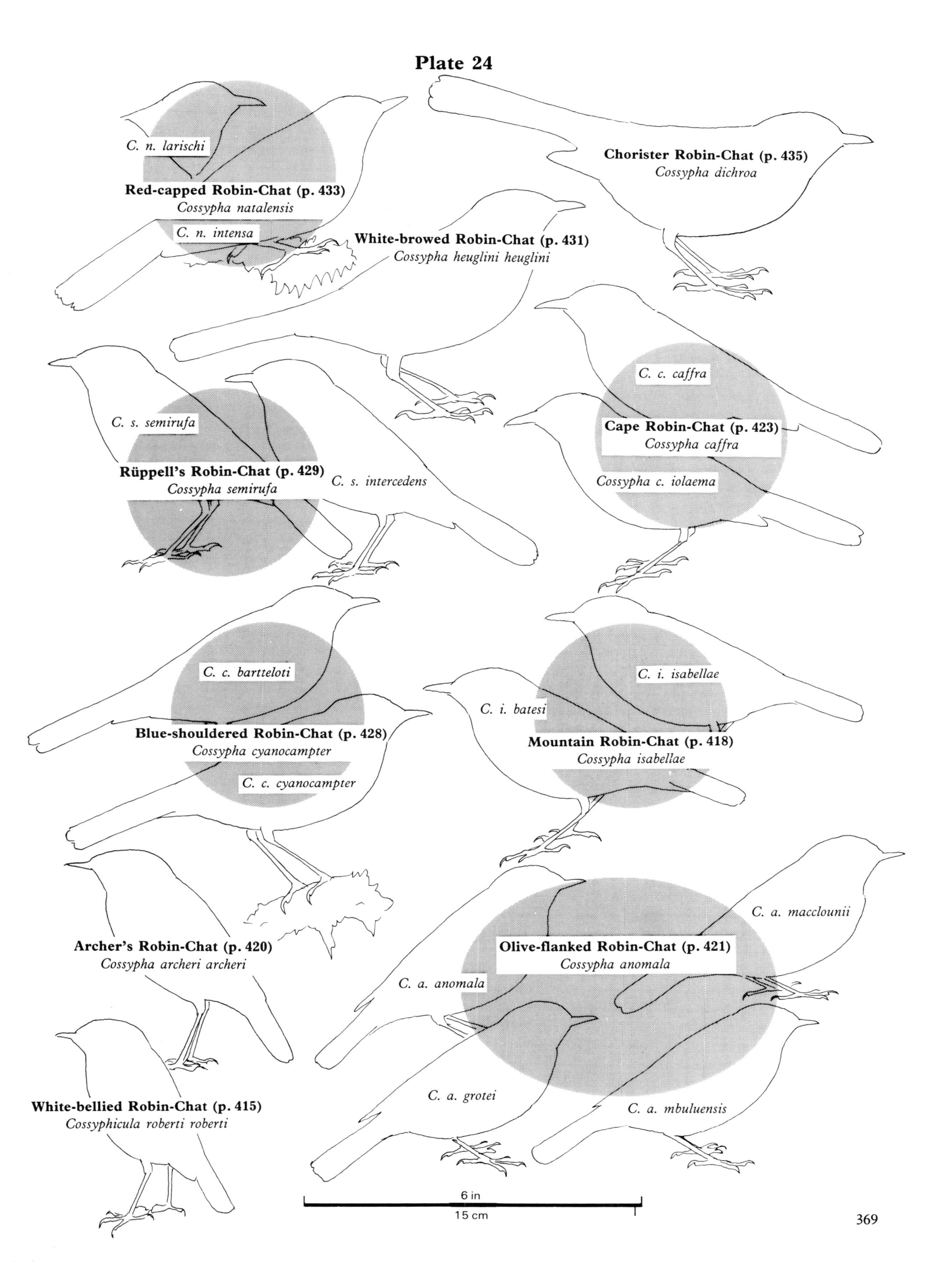

Plate 24

C. n. larischi

Red-capped Robin-Chat (p. 433)
Cossypha natalensis

C. n. intensa

White-browed Robin-Chat (p. 431)
Cossypha heuglini heuglini

Chorister Robin-Chat (p. 435)
Cossypha dichroa

C. s. semirufa

Rüppell's Robin-Chat (p. 429)
Cossypha semirufa

C. s. intercedens

C. c. caffra

Cape Robin-Chat (p. 423)
Cossypha caffra

Cossypha c. iolaema

C. c. bartteloti

Blue-shouldered Robin-Chat (p. 428)
Cossypha cyanocampter

C. c. cyanocampter

C. i. isabellae

C. i. batesi

Mountain Robin-Chat (p. 418)
Cossypha isabellae

Archer's Robin-Chat (p. 420)
Cossypha archeri archeri

C. a. macclounii

Olive-flanked Robin-Chat (p. 421)
Cossypha anomala

C. a. anomala

C. a. grotei

C. a. mbuluensis

White-bellied Robin-Chat (p. 415)
Cossyphicula roberti roberti

6 in
15 cm

369

flycatcher, or uses different perches. Catches moths, mayflies and grasshoppers, but mainly termites, taken as they fly from nests. Attends swarms of ants *Dorylus* at edges of forest and woodland. Takes fallen fruit, sometimes becoming intoxicated on fermenting fruit, when it can easily be caught by hand. Eats beeswax put out at honeyguide feeding station, Kenya; 1 even 'defended' site against Lesser Honeyguides *Indicator minor*; when it was collected, its stomach contained only wax (Horne and Short 1990). Accepts food from man at feeders, picnic tables and campgrounds.

One of the earliest birds to sing in the morning, starting before dawn and continuing for *c.* 2 h. Often congregates to sing before sunset. Communal singing noted following feeding at Kakamega, Kenya, birds gathering in top of tall, thinly-foliaged tree (Zimmerman 1972). Fond of bathing, using rivers, pools, gutters, bird baths, and rainwater in foliage. May congregate morning and evening around water source. In drier areas drinks regularly; in wetter habitats gets water from vegetation. Seen to climb down reed *Phragmites* to drink water; catches drips from leaking tap (Brooke 1981). Flies rather like a thrush *Turdus*; moves through foliage with great ease. Quarrelsome while feeding, drinking or bathing but not overly aggressive. Quick to spot snake, owl or predatory mammal, several gathering to scold noisily, attracting other birds. Scolded stuffed Spotted Eagle-Owl *Bubo africanus* in garden, even scolding space where owl had been after it was removed (Vernon 1975). Recorded anting, using typical passerine postures; ants not seen in bill, but there were large camponotine ants on ground (Brown and Newman 1974). Bold but wary of man. At sunset calls and preens on top of bush or exposed branch. Roosts in thick foliage, e.g. of fig or orange tree (Morocco); 1 pair roosted together in same patch of foliage every night for 2 months (Gabon: Brosset 1971a). Communal roosting reported in non-breeding season. One of the last birds to retire at night, with much twittering and fuss.

Moult overlaps with breeding season. Duration of primary moult variable, 4–5 months in Malaŵi and Mozambique, 5–9 months in Kenya. 8% of birds in Kenya had interrupted moult in breeding season (Britton 1972); none did so in Malaŵi/Mozambique (Hanmer 1978). In Zaïre (Upemba Nat. Park), 2 moults a year, Nov–Dec and Apr–June (Verheyen 1953).

Sedentary, or wanders locally, particularly in winter (Tangiers: Pineau and Giraud-Audine 1979; Algeria; southern Africa: Lawson 1962). In Senegambia leaves the most arid parts of Ferlo region in dry season. Strictly non-migratory in Nigeria, but in Chad both resident and migrant at Ouadi Rimé–Ouadi Achim Faunal Reserve: influx of migrants from south during rains, when residents restricted to wadi habitat.

Food. Wild and cultivated fruits; also flowers, nectar, insects and some seeds. In North Africa, fruits of *Olea europaea, Arbutus unedo, Balanites aegyptiaca, Pistacia lentiscus, Jasminum fruticans, Ziziphus*, oleander *Nerium*, oleaster *Elaeagnus*, asphodel (Liliaceae); figs, pomegranates, apricots, dates, citrus. Thick and fleshy parts of flowers, e.g. asphodel, oleander. Insects include termites Isoptera. In Afrotropics, eats any fruit available; over 50 species recorded by single observer (Vernon 1975), including figs *Ficus* (especially preferred), wild asparagus, avocado, guava, mango, papaya, banana, apple, peach *Prunus persica*, pear, plum, wild blackberry *Rubus, Boscia senegalensis, Commiphora*, spinous cucumber *Cucumis*, wild gourds, cultivated *Duranta* berries, *Erythrococca, Lantana, Morinda*, oil palm *Elaeis guineensis*, parasol tree *Musanga*, Persian lilac *Melia azedarach, Rapania, Rauwolfia*, red pepper *Capsicum, Solanum* (fruits up to 13 mm diam.), *Tetrorchidium didymostemon, Trema*, Virginia creeper *Parthenocissus quinquefolia*. Less favoured, eaten only when no others available, were *Antedesma venosa, Cordia ovalis, Ilex mitis, Dovyalis zeyheri*, and *Pittosporum viridiflorum* (Vernon 1975). In Malaŵi, fruits 3–15 mm in diam. of 21 species recorded by Dowsett-Lemaire (1988); principal genera are *Afrocrania, Bridelia, Croton, Cussonia, Ficus* (3 species), *Harungana, Ilex, Macaranga* (2 species), *Maesa, Myrica, Polyscias, Prunus, Rauwolfia, Sapium, Syzygium* (2 species), *Trema, Viscum* and *Rhus*. Eats flowers and their parts: *Acacia* spp., *Aloe aculeata*, bottlebrush *Calothamnus quadrifidus, Celtis integrifolia, Eucalyptus, Grevillea, Pinus, Prosopis juliflora, Protea*; young leaves (*Acacia, Prosopis*); nectar from *Aloe, Greyia sutherlandi*; seeds of bramble; crystallized *Acacia* gum. In Ghana and elsewhere commonly eats fruits of neem *Azadirachta indica* (J. M. Locke, pers. comm.).

Insects include termites (especially alates) and flying ants, also aphids, beetles, dragonflies, fly grubs, small black wasps, Lepidoptera (adults and larvae), including butterfly *Belenois mesentina*, hawkmoths, *Acraea, Laphygma exempta*; mantises, mayflies, grasshoppers, stick insects; insect eggs. Small lizards and geckos also taken. In Kenya (Tsavo Nat. Park), diet (n = 139 items) was insects (14%), fruit (72%) and flowers (14%); *Commiphora* only 25% of fruit eaten (Lack 1985). Omnivorous in captivity, consuming almost any human food (see list in Preston 1975). Even eats beeswax (Horne and Short 1990).

Breeding Habits. Monogamous; said to mate for life (van Someren 1956). Pair bond maintained by allopreening and by greeting ceremony with duet (see *Voice*), accompanied by slow wing-flapping and spreading of tail, after which birds sit side by side; also, while feeding young, both hunt for food and return to nest together. Territorial only during breeding season. In Morocco, av. 1·6 pairs/km^2 in undisturbed woodland, 14·8 pairs/km^2 in degenerate woodland, *c.* 21–24 pairs/km^2 in semi-arid scrub in lowlands and in high valleys (Thévenot 1982). 2 nests once found 15 m apart, Sierra Leone. Density 10 pairs/100 ha in mixed woodland, Zimbabwe (Vernon 1985). Territoriality variable; territory defended weakly or violently. In Zaïre (Upemba Nat. Park), 2 fought on ground with beaks and claws, singing between attacks. ♂♂ flight fiercely over ♀ (Priest 1935). In Morocco, many pairs on territory apparently do not breed (Brosset

1961). In Gabon, populations unstable, territories irregularly occupied; pairs may remain together for only 1–2 nestings, continually forming and breaking up where vegetation changes rapidly throughout the year, e.g. in clearings around villages, fields, overgrown abandoned agricultural land, but are more likely to stay together where vegetation more stable, e.g. streamside brush, woodland edge and gardens (Brosset 1971a, Brosset and Erard 1986).

NEST: broad, oval, shallow cup, usually rather thin and fragile but carefully constructed; built of twigs, plant stems (e.g. banana and *Pennisetum*: Prigogine 1961), grass stems, moss, bits of lichen, rootlets and leaves, often bound together with spider webs; lined with fine grasses and roots, creeper tendrils, plant fibres, wool or animal hair, sometimes twine and thread; often attached to support with spider webs. Ext. diam. 75–120, int. diam. 60–70, ext. depth 50–60, int. depth 37·5–50 – small for size of bird. Usually 1–5 m above ground, sometimes up to 13 m, once 0·3 m in bracken (van Someren 1956); in almost any kind of bush, shrub or tree, wild or cultivated, placed in fork or on branch or twigs to which attached with spider webs; occasionally suspended between twigs in fork. Sites include: top of bunch of bananas (common); crown of palm or surface of palm leaf (one on leaf of doum palm had no base, just a ring of grass and hair to keep the eggs in place); head of papyrus *Cyperus papyrus* in swamps. In Zanzibar often in undergrowth in coconut plantations. Often nests close to man, by path in garden, in creeper on wall, in tree over building; once on wire roll inside shed (Marshall 1969), once under rag in leafless garden bush (Clarke 1985). Usually screened by vegetation; sometimes in open. Constructed in 8–10 days (n = 1 observation: Julliard 1986); by ♀ only. During construction bird seen clinging to palm trunk and pulling off fibres (Vincent 1947).

EGGS: 2–5; Morocco 2–4 (12 clutches/2, 48 c/3, 2 c/4); Egypt 3–5, usually 3; Cameroon 2 (18 clutches); Gabon 1–3 (6 c/1, 39 c/2, 5 c/3); Zaïre, 2–3; E Africa, 2 eggs 'common', 3 eggs 'frequent', 4 eggs 'exceptional' (van Someren 1956); Malaŵi/Zambia/Zimbabwe 1–4 (21 c/1, 136 c/2, 179 c/3, 2 c/4); Natal/Zululand 1–3 (14 c/1, 28 c/2, 70 c/3); South Africa (n = 178) 2–3 (2·6); 1 clutch of 6 probably laid by 2 ♀♀. Laid at 1-day intervals, occasionally 2-day; clutch of 4 laid in 2 days, Ghana, must have been laid by 2 birds (Bannerman 1936). Subelliptical, more pointed at one end, some bluntly rounded; smooth, glossy; variable colour: ground colour brown, russet, vinaceous, pink, pinkish grey, violet-rose, or white, lightly to heavily covered with spots and freckles, and sometimes a few streaks and hair-lines, of brown, red-brown, maroon, red, violet-red, violet or purple, occasionally black, over grey, ashy, pale lilac or mauve shell marks; markings vary from large and conspicuous to small and faint, sometimes dense throughout, often concentrated at blunt end, occasionally almost absent. SIZE: (n = 31, N Africa) 23–28 × 16·2–21·3; (n = 34, Cameroon) 21·7–26·1 × 15·1–17·6 (23 × 16·4); (n = 27, Zaïre) 21·4–24 × 15·7–17·2 (23 × 16·4); (n = 221, southern Africa) 19·8–25·6 × 15·2–18·1 (22·4 × 16·3). Double brooded.

LAYING DATES: Morocco, Mar–Sept, mainly May–July (date of earliest eggs varies by year and region); Algeria and Tunisia, May–Aug; Egypt, Mar–Aug; Mauritania, Feb–Apr in N, July, Dec in S; perhaps all year (Lamarche 1988); Senegambia, all months, mainly June–Aug; Mali, Feb–Apr, July–Nov, perhaps all year; Niger, Dec–Jan; Liberia (Mt Nimba), probably all year; Ghana, probably most of year, mainly July–Dec; Burkina Faso, Dec–Jan (juveniles and nest-building Aug); Togo, Jan–Apr; Nigeria, all months, peaking in rains Apr–July; Sudan, Jan, Mar–Oct; Ethiopia: *schoanus* Jan–July, possibly Aug, Dec; *somaliensis* Jan; *spurius* Mar–May, Nov; *dodsoni* Jan, possibly Feb, Apr–June, Oct–Dec, Somalia, Feb, June, Aug; Cameroon, all months except Sept, but with altitudinal variation: in lowlands all months except Aug–Sept, peaks in latter half of dry season and middle of rains; montane areas Dec–Aug (gonadal data Oct–Nov), breeding mainly during dry season (Eisentraut 1972, Serle 1981); Gabon, all year, peaking in short dry season and at beginning of rains; Zaïre, all year; Rwanda, Sept–May; E Africa: Region A, Mar–July, Oct–Nov; Region B, Mar–Jan; Region C, Jan–Feb, Apr–Dec; Region D, all months; Region E, Apr–Sept (breeds during both long and short rains, also just before and just after rains); Angola, nearly all year, peak Sept–Nov; Zambia, July–Mar, peaking Sept–Nov (July 1, Aug 5, Sept 28, Oct 38, Nov 17, Dec 10, Jan 4, Feb 6, Mar 1 clutches); Malaŵi, Sept–Jan, peak Oct–Nov (Sept 5, Oct 31, Nov 22, Dec 15, Jan 10 clutches); Zimbabwe, July–Apr, mainly Sept–Dec (July 2, Aug 17, Sept 140, Oct 404, Nov 208, Dec 103, Jan 47, Feb 17, Mar 11, Apr 1 clutches); Mozambique, Sept–Feb; South Africa: Transvaal, Sept–Apr; Natal/Zululand, Sept–Apr, peak Oct–Dec (Sept 4, Oct 24, Nov 36, Dec 28, Jan 14, Feb 5, Mar 4, Apr 1 clutches).

INCUBATION: begins just after laying of first egg; usually by ♀ only, sometimes also by ♂ which always remains nearby. Period: 13 days (Morocco), 12–13 days (E Africa), 14 days (Gabon), 12–14 days (southern Africa). Incubating bird crouches low, just bill and tail visible above rim; gives distraction display when danger present, pretending to be wounded. ♂ carries food to incubating ♀.

DEVELOPMENT AND CARE OF YOUNG: young naked for first week, then grow feathers quickly. From hatching to 3 months gape yellow, 3–5 months slightly yellowish, after 5 months blacks as in adult. For first 3 months skull 25% pneumatized, 3–6 months 25–50%, 6–9 months 50–75%, 9–11 months 75–99%, 12–23 months 100%, bone pink and hard; from 24 months bone white and hard (Hanmer 1978).

Young brooded constantly for first few days, by both parents; at 4–5 days brooded 58% of time, at 10–11 days 24%. ♀ often fed on nest by ♂; she eats berries but gives insects to young. Young fed only insects for first week, especially thoraxes of Orthoptera, also caterpillars, butterflies, larval and imago noctuids, mantids, winged termites, flying ants *Pheidole*, fly grubs. After first week diet augmented by fruits and berries of e.g. *Lantana*, *Trema*, *Ficus*, *Cissus*, *Clausena*. Fed by both parents, who often visit nest together, one or both with food. Frequency of feeding varies: in N Africa the same throughout the day; during first part of fledging period 6·5 visits per h, during latter part 8·2; av. 7·2 (18 h of observation: Julliard 1986); in E Africa more frequent, 8–12 visits per 30 min, but with frequent breaks when ♀ broods (van Someren 1956); in Gabon only 3 visits per h (20 h of observation), but prey larger (Brosset 1971a). At midday parents sit side by side on edge of nest, shading young.

♀ seen to preen, also sunbathe, lying on her side, spreading one wing and tail, raising crest fully and fluffing rump feathers to let sun penetrate (van Someren 1956). Faeces carefully removed from nest (sometimes apparently swallowed) up to time of fledging, nest and ground below it remaining completely clean. Fledging period: 10–12 days (South Africa); 15–17 days (Zimbabwe); about 13–16 days elsewhere. Young can fly only short distance at first, remain bunched together in foliage for a few days before following adults, loudly begging for food. Parents care for them for c. 1 month. At 15–20 days young start to feed themselves, but are fed by parents up to day 27 and beg until day 40 (Julliard 1986). Feeding stops when second nesting under way; 2 young stopped begging when incubation of second clutch began, 20 days after they left their nest, but even though they were able to feed themselves, they were not truly independent and were tolerated near second nest. Young fledged early Sept remained with parents until following Feb. Helpers sometimes attend nest and assist with feeding of young.

Other species not tolerated near nest (Julliard 1986). Usually calls loudly on approach of human intruder, though nearby intruder in full view once tolerated even while young being tended (van Someren 1958).

BREEDING SUCCESS/SURVIVAL: great loss of young squeezed out of small nests (Tanzania: Sclater and Moreau 1933). Morocco (n = 5 nests) 10 of 13 eggs hatched, 5 young fledged; (n = 8 nests), 9 young fledged = 1·12/nest; Gabon (n = 64 nests) 0·82 young fledged per nest. Losses by predation 42·2%, Gabon, all to army ants. Eggs on palm leaf washed away by rain shower. Parasitized by cuckoos: Jacobin *Oxylophus jacobinus* (n = 135), Emerald *Chrysococcyx cupreus* (n = 12), Klaas's *C. klaas* (n = 3), Diederic *C. caprius* (n = 2) and Red-chested *Cuculus solitarius* (n = 3). Killed by shrikes, mongooses, often by cats, once by monkey *Cercopithecus aethiops* (Leck 1977). Excitable and noisy parents advertise location of both nest and fledged young. Mean survival rate in the wild 0·47 (Malawi: Nchalo), mean life-expectancy 1·39 years; no significant difference between survival rates of adults and immatures (Hanmer 1984). 9 birds lived for >7·5–11·5 years (Hanmer 1989), 1 bird for 12 years, South Africa (T. B. Oatley, pers. comm.) and 1 bird for 18 years in Chad (R. J. Dowsett, *fide* T. B. Oatley, pers. comm.). In urban areas, Zimbabwe, mean life expectancy 2·02 years, max. 9·92 years (Irwin 1981). Captive bird lived for 26 years 5 months (Dryden 1981).

References
Brosset, A. (1971a).
Brosset, A. and Erard, C. (1986).
Cramp, S. (1988).
Julliard, J.-P. (1986).
van Someren, V. G. L. (1956).
Vincent, A. W. (1947).

Plate 16
(Opp. p. 257)

Pycnonotus nigricans (Vieillot). African Red-eyed Bulbul; Black-fronted Bulbul. Bulbul brunoir.

Turdus nigricans Vieillot, 1818. Nouv. Dict. d'Hist. Nat. 20, p. 253; banks of the Orange R. in Namaqualand, restricted to Goodhouse, NW Cape Province, by MacDonald, Contr. Orn. West. S. Afr., 1957, p. 116.

Forms a superspecies with *P. barbatus* and *P. capensis*, also Asian *P. xanthopygos*, *P. leucotis*, *P. leucogenys*, and possibly *P. cafer* and *P. aurigaster*.

Range and Status. Endemic resident and partial local migrant. Breeds from SW Angola (north to Benguela, Gambos, Chingoroi), Namibia (except extreme SW desert), and Botswana (except Okavango) to Cape Province (Orange R. mouth, N and E Karoo to E Griqualand), Orange Free State, W Lesotho (lowlands, to 2000 m) and SW Transvaal. Non-breeding visitor to extreme S Zambia (Livingstone area and Sesheke District), NW Zimbabwe (Umgusa Forest Reserve to Victoria Falls, mainly Hwange Nat. Park), central Transvaal, Natal and Zululand. Status in Zimbabwe uncertain: dry season visitor to Hwange Nat. Park, a few believed resident around Victoria Falls, and a possible breeding record from Bulawayo. Vagrant Cape Peninsula (Zwaanswyk: Martin 1990). Locally common to abundant.

Description. *P. n. nigricans* (Vieillot) (including '*grisescentior*'): Angola and Zambia to NW and N Cape Province and NW Transvaal, intergrading with *superior* along middle Orange R. and in Griqualand West. ADULT ♂: top and sides of head black, grading into blackish brown on chin and throat. Rest of upperparts, including wings, grey-brown, feathers variably fringed paler; tail-feathers blackish brown with very narrow pale tips which disappear with wear. Sides of neck and breast brown, feathers fringed whitish, giving scaly effect. Lower breast white with a few brown streaks; belly white, flanks pale brown, vent and undertail-coverts bright yellow. Axillaries and underwing-coverts whitish; underside of flight feathers dark grey-brown, basal half of primaries with pale inner margins. Bill black; bare skin round eye orange to orange-red; eye brick red to orange; legs and feet black. Sexes alike, ♀ slightly smaller. 1 individual with both ovary and testis reported (Storey and Harrison 1969), and leucistic individuals occur. SIZE: wing, ♂ (n = 31) 91–103 (94·8), ♀ (n = 28) 84–98 (91·2); tail, ♂ (n = 10) 74–82 (77·0), ♀ (n = 10) 74–78 (75·6); bill, ♂ (n = 31) 16–22 (19·1), ♀ (n = 28) 16–20·1 (18·6); tarsus, ♂ (n = 10) 19–24 (21·7), ♀ (n = 10) 20–23 (21·4). WEIGHT: unsexed (n = 316) 21·6–37·4 (30·8).

IMMATURE: duller and somewhat paler above than adult; eye-ring pink, eye brown.

NESTLING: naked, gape bright pink.

P. n. superior Clancey: E Cape Province, Orange Free State, Lesotho, S Transvaal. Slightly darker above; blackish brown of throat extends further toward breast, brown area on breast larger. Larger, especially tail: ♂ 85·5–95.

Field Characters. Very similar to Common Bulbul *P. barbatus*, from which best distinguished by red eye wattle, even though it is sometimes not easy to see except at close range (Holliday 1965). Slightly darker head and throat produce a little more contrast with white belly. Immature told from immature Common Bulbul by pink eye-ring. Cape Bulbul *P. capensis* has white eye-ring and more or less uniform grey underparts.

Voice. Tape-recorded (88, F, GIB, HEL, LUT, PAY). Extremely similar to that of Common Bulbul, q.v. Notes tend to be a little richer and deeper and delivered more slowly. Bird on the Gillard tape seems to be singing 'The Red-eyed Bulbul'! Both sexes sing, ♂ having slightly wider repertoire (Everitt 1964b).

General Habits. Occupies rather drier country than Common Bulbul, replacing it in arid parts of Botswana. Dry woodland, acacia savanna, brushy hillsides, semi-arid scrub, riverine bush; also orchards and gardens. Lives in shrubberies and gardens around dry-country farms. Often in open and rather treeless districts, where found in bush along dry watercourses, but always dependent on water and within reach of rivers, wells and other surface water. As long as there is water, it can exist in semi-arid country, e.g. in Botswana, where found even at remote springs (Irwin 1956).

Usually in pairs or small, loose groups; in Botswana, forms flocks in winter (Beesley and Irving 1976). Tame and familiar, the dry-country equivalent of the Common Bulbul; at home with man, living in gardens and visiting feeders and bird baths. Usually forages in trees, sometimes hanging upside down, like Common Bulbul; not often on ground. Noted repeatedly hawking for 'small insects' in air over small thorn tree and arum-fringed stream (Powett 1963). Noisy, calling and singing frequently, usually from top of tree or bush. Flicks wings and flirts tail while singing and when alarmed (Maclean 1985). Anting recorded.

Some dry season movement north from Botswana to Zambesi R. and NW Zimbabwe. Only Zambia specimens are from dry season, June–Oct; sight records at other times of year open to question (Aspinwall 1975), so not yet known if a few are breeding residents. Classed as 'nomadic' (regular, temporary visitor) in Kalahari-Gemsbok Nat. Park, N Cape Province (Mills 1976). In winter some birds from interior highlands of South Africa move to lower levels in E Transvaal, Natal and Zululand.

Food. Fruits, including mulberry, grapes, wild figs *Ficus*, *Lantana* and muchinga *Popowia abovata*; and insects, including beetles, aphids and ants. Also nectar. Study of captive birds showed each bird ate an av. 6·73 g dry weight of food per day, which was 21·55% of bird's own weight (Liversidge 1970).

Breeding Habits. Monogamous; territorial.

NEST: a circular, shallow cup of dry grass, fine twigs, rootlets and fibres, lined with finer grass stems; ext. diam. 90, int. diam. 60, depth of cup 50 (n = 1, Namibia: Hoesch and Niethammer 1940); 1–4 m up in fork in bush or tree, often well hidden. Captive pair included Spanish moss *Usnea* among construction materials, and placed nest in one of topmost branches of crabapple tree. Built by both sexes.

EGGS: typically 3, rarely 4; (9 clutches) 2–3 (2·3). Laid at 1-day intervals. Oval to elliptical; pinkish white, pale pink or white tinged reddish, marked, often copiously, with spots, speckles and blotches of dark red, red-brown, brown, chestnut and purplish grey, with underlining of grey or blue-grey; often thickly spotted at blunt end, rest of egg largely unmarked. SIZE: (n = 42) 19·7–25·4 × 14·7–17·5 (22·5 × 16·3). Double-brooded.

LAYING DATES: southern Africa, Sept–Mar (timing varies with rainfall); Namibia, Sept–Feb, peak Dec–Jan; Botswana, Nov–Feb; SW Transvaal Dec–Jan.

INCUBATION: begins with laying of first egg; by both sexes (in captivity) or by ♀ only (in wild, n = 2 nests). Period: 12–13 days (in captivity), 11–12 days (in wild).

DEVELOPMENT AND CARE OF YOUNG: at 5 days, eyes open, quills showing in wings; at 10 days nearly fully feathered, dark greenish brown above, head darker; leave nest at 12–13 days, when head, neck and upperparts as adult, underparts clear yellowish greyish white, unmarked; no coloured eye-ring. Fully able to fly 5–7 days after leaving nest. At 3 months eye-ring yellowish white, gradually changing to pink a month later; at 8 months, exactly as adult.

Young fed by both parents; they feed themselves 2 weeks after leaving nest but still accept food from parents. Food of young not recorded, but a juvenile Jacobin Cuckoo *Oxylophus jacobinus* was seen being fed by both foster parents on porridge from a workman's pot! (V. Simpson *in* Beesley and Irving 1976).

BREEDING SUCCESS/SURVIVAL: in captivity, out of 3 nests, 3 eggs hatched out of 6 laid. Frequently parasitized by Jacobin Cuckoo (12 records). Bulbul pair gave vigorous battle with persistent cuckoo pair; cuckoos eventually driven off, ♀ cuckoo seen to drop egg into flower bed (Clarke 1974). Unidentified trematode worm found in several captive birds that died in the same cage (Liversidge 1970).

References
Everitt, C. (1964b).
Maclean, G. R. (1985).

Plate 16
(Opp. p. 257)

Pycnonotus capensis (Linnaeus). Cape Bulbul. Bulbul du Cap.

Turdus capensis Linnaeus, 1766. Syst. Nat. (12th ed.), 1, p. 295; Cape of Good Hope.

Forms a superspecies with *P. nigricans* and *P. barbatus*, also Asian *P. xanthopygos*, *P. leucotis*, *P. leucogenys* and possibly *P. cafer* and *P. aurigaster*.

Range and Status. Resident, endemic to Cape Province, South Africa. SW Cape north to Orange R. and east to Sundays R. near Port Elizabeth; mainly within 100 km of the coast, but in man-made habitats penetrates into edge of Karoo, e.g. to Laingsburg, Prince Albert, Graaff-Reinet and Pearston. Formerly extended east to King William's Town and East London. Common to locally abundant. Breeding density 13–19 pairs/20 ha.

Description. ADULT ♂: head, upperparts, tail, wings and breast sooty brown; crown darker, feathers slightly raised to give a square, peaked effect; chin, lores and face blackish brown. Upper belly and flanks paler brown, lower belly whitish, undertail-coverts and vent lemon-yellow. Underwing as upperwing except for variable creamy white area on primary coverts at edge of wing, sometimes extending inwards as brown and white mottling. Bill black; bare skin round eye white; eye red-brown; legs and feet black. Sexes alike. Albinistic birds recorded. SIZE: wing, ♂ (n = 9) 90–102 (95·6), ♀ (n = 10) 83–97 (90·8); tail, ♂ (n = 9) 72–87 (80·1), ♀ (n = 11) 74–88 (80·2); bill, ♂ (n = 9) 18–21·7 (20·2), ♀ (n = 10) 19–21 (20·0); tarsus, ♂ (n = 9) 21–24 (22·2), ♀ (n = 11) 19–24 (21·5). WEIGHT: unsexed (n = 176) 28·2–47·3 (38·6).

IMMATURE: like adult; white eye-ring takes 2–3 months to develop.

NESTLING: naked skin transparent at birth, soon thickens and becomes opaque; after 24 h purplish brown. Gape pink, tip of tongue black.

Field Characters. The only *Pycnonotus* bulbul in its range except at extreme edges. Distinguished by broad white eye-ring from African Red-eyed Bulbul *P. nigricans* (red eye-ring) and Common Bulbul *P. barbatus* (brown eye-ring); also by brown of underparts extending onto flanks and belly, and head same colour as rest of plumage, not contrastingly darker. Immature has brown eye-ring for first few months, when distinguished by plumage colour (same as adult).

Voice. Tape-recorded (81, 88, F, LEM, WALK). Song, a series of short phrases, 'chip-chee-woodely', 'chupwheet-churry-up', 'ship-shape-worta-weetawoo', 'what-yer-rightaway', 'whetcha-willya'; quality very like that of Common Bulbul. Common call, loud 'piet-my-jol'. Excited chatter during wing-flicking display, repeated 'chop-doodle-it'. Contact calls, 'chirrup', 'key-link' or more anxious 'chit-tee' or 'chit-tee-tee'; alarm, a low 'churr' or 'churrt'.

General Habits. Inhabits scrub, coastal heath, dune forest, riverine bush, farmlands, plantations and gardens; particularly fond of exotic wattle *Acacia cyclops*; mainly below 500 m. Prefers moist conditions, i.e. areas of winter rainfall or where rain is evenly spread throughout the year; generally does not enter drier regions with sparse vegetation on desert soils (Liversidge 1970), although near Prince Albert occurs in *Acacia karroo* woodland along drainage lines, together with African Red-eyed Bulbul (W. R. J. Dean, pers. comm.).

Occurs singly or in pairs all year, but in winter also often travels in small, noisy, fast-moving foraging flocks, typically of 5–9 birds; up to 40 gather at fruiting trees. Eats a wide range of berries but normally only one type at a time; concentrates on certain bushes while neglecting similar ones nearby. Hawks insects, including termites, on the wing; also takes them from flowers. Comes to bird feeders. In winter sometimes roosts near food source. Allopreens at any time of year: bird sidles up to its mate until their bodies meet, then each in turn preens the back of the other's head. Sings especially loudly and for longest periods (10–23 min) during nest-building and early incubation; uses 1 or 2 well-defined prominent song perches within its territory.

Moult takes place Dec–May, to a more limited extent in other months; it may vary by several months from one year to the next. Timing of moult for each bird varies individually. There is no correlation between moult and the breeding season (Liversidge 1970); some birds moult while breeding.

Mainly sedentary, with local wandering in non-breeding season. Some evidence that birds return to same foraging areas in successive winters.

Food. Fruit, seeds, nectar and insects. In one study *c.* 90% of items were vegetable matter, mainly fruit, and 10% insects (Liversidge 1970). Of 25 plant species eaten, the most important were: *Rhus crenatus* (28·6% of number of identified items), *Acacia cyclops* (23·4%), *Chrysanthemoides monilifera* (16·5%), *Mundia spinosa* (7·6%) and *Rapanea gilliana* (7·0%). Adults increase intake of insects somewhat in Nov–Dec. 42% of items fed to young are insects. Dry weight of food eaten daily is 22–30% of body weight (captive birds: Liversidge 1970).

Breeding Habits. Monogamous; usually but not always pairs for life. Pair remains together all year. Territorial only during breeding season, and even then territory not strongly defended nor actively advertised. Defence in order of priority is for mate, nest-site and song perch (Liversidge 1970). Territory size 1·1–2·2 ha; same territory used in successive years.

Threat displays are of 4 types: (1) crest flattened and head lowered so crown and back form straight line, wings slightly raised so carpal joints prominent (**A**); at higher intensity, tail spread and depressed. (2) Crest slightly raised, wings pressed close to body, tail raised, undertail-coverts fluffed out (**B**); at higher intensity, tail also spread. (3) Like (1) but beak opened to show red gape. (4) Like (1) but wings spread farther, back and rump feathers raised to give 'larger than life' look (**C**); usually

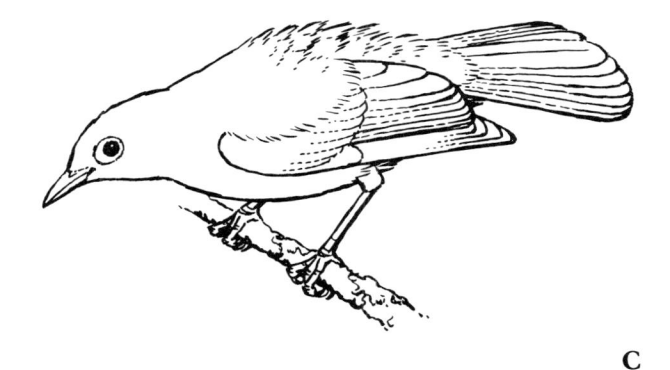

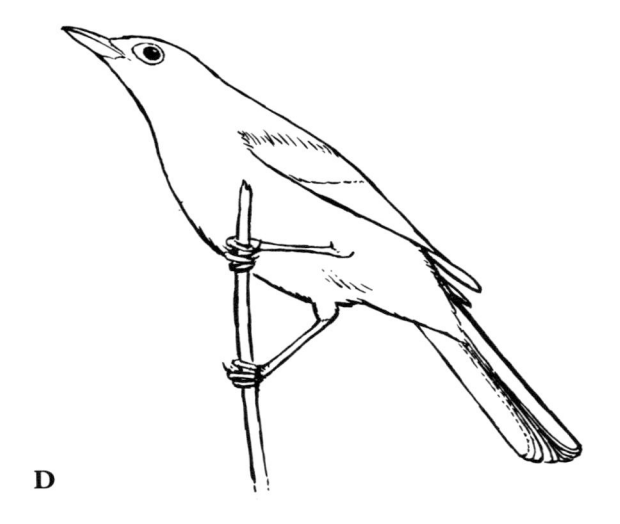

376 PYCNONOTIDAE

directed at other bird species. In submissive/appeasement display (**D**), crest flattened, neck drawn in, head held high, bill pointing upwards, wings held to sides as at rest.

In early stages of courtship, ♂ circles ♀ in exaggerated fluttering flight with tail well spread. In display on perch, ♂ flattens crest, bends head forward and down, tilts back forward, fluffing up lower back and rump feathers, spreads primaries and tail so that outer tail-feathers come close to primaries (**E**). Sometimes he sways from side to side or turns around. He is usually on higher perch than ♀, so yellow undertail-coverts conspicuous. In pre-copulatory display, ♂ raises crest slightly, fans tail and raises it just above the horizontal; ♀ lowers head and flattens crest, keeping body horizontal, and rapidly flicks wings a short distance out from sides. Incubating ♀ displays to returning ♂ by standing upright, lowering tail and flicking wings rapidly up and down, raising them to vertical position over head (**F**) and snapping them back to sides, giving chattering call (see *Voice*).

NEST: cup of twigs (some retaining leaves), rootlets and grass stems on foundation of coarse twigs (**G**); lined with sheep's wool, cobwebs, paper, string and cotton wool, with fine roots and vegetable fibres on top. Placed in fork of tree or shrub, or on outer horizontal branch 5–20 mm in diam., just within outer leaf layer (once in creeper hanging over dead bush), 0·3–4·0 m above ground (37·5% >1·5 m); 63 species of nesting bush identified, most popular being *Acacia cyclops*; also in pergola in garden. Site selected and nest built by ♀ (in later stages sometimes helped by ♂); ♂ accompanies her and sings while she gathers nest material. Birds call loudly after depositing nest material. Nest built in 2–7 days (once 10 days), but can be built rapidly – 1 was 75% completed in 10 h (Liversidge 1970); most building activity takes place between 07·00 and 12·00 h. Builds new nest for second brood, sometimes incorporating material from first nest (van der Merwe 1987).

EGGS: 2–3, rarely 4 or 5; mean (74 clutches) 2·6. White with pinkish tinge, spotted with dark red and purplish grey. SIZE: (n = 44) 21·7–26 × 15·9–18 (23·7 × 16·9). Laid at 1-day intervals, usually very early in morning. Often double brooded; second clutch can be larger or smaller than first one.

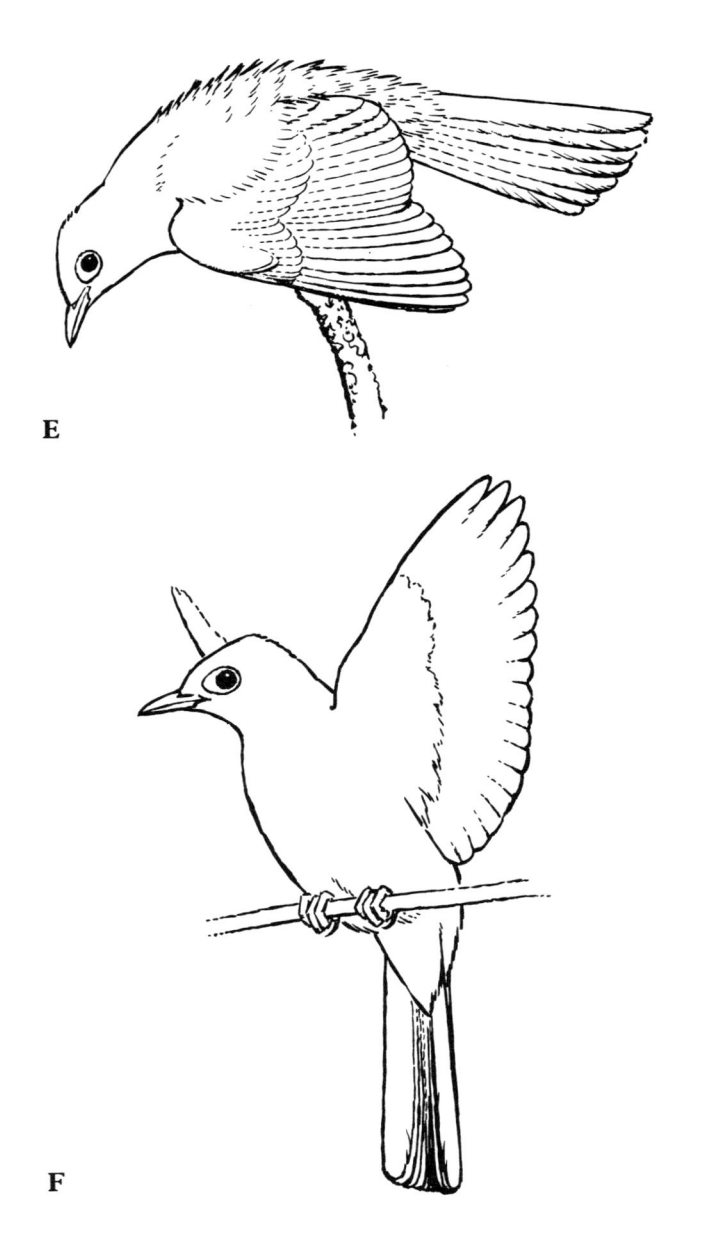

LAYING DATES: Aug–Mar, mainly Sept–Nov.

INCUBATION: by ♀ only. Usually begins when last egg laid (Liversidge 1970), but in 2 clutches of 3, began with first egg (Roberts 1987, van der Merwe 1987). Period: 11·5–14 days. ♀ leaves at intervals to feed herself; ♂ sometimes brings her food.

DEVELOPMENT AND CARE OF YOUNG: eyes closed at hatching, start to open on day 3, fully open day 6 or 7. At day 3 or 4, feather tracts begin to show through skin as dark lines, quill tips break surface 1 day later; wing quills appear first, grow rapidly, break open on day 6; tail-feathers also appear on day 6. By day 8–9, body almost covered with quills, little skin visible; next day tips of quills break into feathers; by day 11 nestling fully feathered. Gape orange by day 4 in one bird, whitish by day 6 in another; later, always yellow. Black tongue tip soon disappears, completely by day 5. Weights: day 1, 2·7–3·9; day 2, 3·3–5·9; day 3, 4·5–9·5; day 4, 6·7–12·5; day 5, 13·2–16·6; day 8, 17·6–23·6; day 10, 22·3–26·7; day 13, 35·8–45·1. Young fledge after 11–15 days, usually 12–14; tail is then half grown, power of flight not strong. 2 weeks later tail two-thirds grown, flight strong; 3–4 weeks out of nest fledging full size but crest not well developed, eye-ring still brown. Fully mature and able to breed within 1 year.

Young fed by both parents, in nest and after leaving it; given mainly insects up to day 6 and mainly fruit thereafter. Brooded mainly by ♀ when small, occasionally by ♂. Faeces swallowed by parents up to day 8, after which sacs removed and dropped 5–20 m from nest. Fledglings fed for at least 2 weeks after leaving nest; after 19 days one was being fed at rate of 9 times in 45 min (Liversidge 1970); still occasionally fed after 1 month, but not after any second brood hatches, although tolerated near nest. Fully independent 40 days after leaving nest, when they start to call and sing. Young of the year gather in loose flocks.

BREEDING SUCCESS/SURVIVAL: from 233 eggs laid, 63 (27%) birds fledged. Survival rate of fledglings 18% after 1 month, 10% after 1 year. Annual adult mortality rate 25%; average life-expectancy of adult 3·1 years. Heavily parasitized by Jacobin Cuckoo *Oxylophus jacobinus*: in one area over a 4–year period, 12–73% (av. 39%) of nests were parasitized (Liversidge 1970). However, since majority of parasitized nests were pillaged by predators anyway, cuckoos were responsible for only a 7% loss of the bulbul population. Predators caused 60% of the loss of nests with eggs or young; of these, 70% were mammals (monkey *Cercopithecus aethiops*, genet *Genetta felina*, domestic cat *Felis domesticus*, mongooses and especially rodents, which ate many eggs); 16% were birds (African Marsh Harrier *Circus ranivorus*, White-browed Coucal *Centropus superciliosus*, Southern Boubou *Laniarius ferrugineus*); others included puff adder *Bitis arietans* and egg-eating snake *Dasypeltis scabra*. A few nests were destroyed by storms.

Reference
Liversidge, R. (1970).

Family BOMBYCILLIDAE: waxwings, Grey Hypocolius and silky flycatchers

Medium-sized, plump-bodied birds with dense, soft, silky plumage in blending tones, crested, typically with crest defined by blackish line running from eye to eye behind it (Delacour and Amadon 1949), with small but quite stout bill, wide at base, nostrils covered by short, dense, velvety feathers, and long or short, patterned tail. Leg short or very short. Berry-eaters.

Family sometimes thought to include West Indian Palmchat *Dulus dominicus* (Sibley and Ahlquist 1990). Apart from *Dulus*, 5 genera, perhaps mutually more closely related than different appearances suggest: waxwings *Bombycilla*, Grey Hypocolius *Hypocolius ampelinus*, and American silky flycatchers (*Phainopepla*, *Phainoptila*, *Ptilogonys*).

Genus *Bombycilla* Vieillot

Gregarious, migratory, grey-and-rufous bombycillids with short, yellow- or red-tipped square tail; 2 species with waxy red tips to most secondary feathers. Long-winged; outermost primary minute.

3 species, Nearctic and Palearctic, 1 a vagrant in Africa.

378 BOMBYCILLIDAE

Plate 16
(Opp. p. 257)

Bombycilla garrulus (Linnaeus). Bohemian Waxwing. Jaseur de Bohême.

Lanius garrulus Linnaeus, 1758. Syst. Nat. (10th ed.) 1, p. 95; Europe.

Range and Status. Holarctic: breeds Sweden east to Kolyma R. and Kamchatka, and from N Alaska to Hudson Bay. Winters regularly south to central and SE Europe, 40°N in central Asia, S Kiangshu Province (China), Oregon and Montana; but strongly irruptive, every few years extending much further south. Accidental Malta, Cyprus, Israel.

Accidental Africa: single occurrence – flocks near Blida (Algeria), late autumn 1841.

Description. *B. g. garrulus* (Linnaeus) (only race vagrant to Africa): Europe to Siberia. ADULT ♂: plumage peculiarly soft. Forehead bordering bill velvety black; forehead rufous, merging to pink-brown on crested crown; nape, mantle, back and wing-coverts drab brown, rump and uppertail-coverts pale grey; tail black with bright yellow tip. Lores and thin streak behind eye black; chin black; ear-coverts pink-brown (whitish in narrow line next to black eye-streak), throat and sides of neck grey-brown, breast pink-brown, belly buff or pale brown, undertail-coverts rufous. Remiges and primary coverts black, coverts tipped white, primaries tipped white and distally edged yellow (forming a V on each feather and a neat yellow line in closed wing), and inner secondaries tipped white and with bright red, waxy appendage at tip of each shaft. Underside of wing pale grey; underside of tail black, tipped yellow. Bill black, base blue-grey, small patch at base of lower mandible white; eye dark brown; legs and feet black. ADULT ♀: very like ♂, differing in number of minor features, mainly: waxy red tips to secondaries average smaller and fewer (mode 7 in ♂♂, 6 in ♀♀); yellow tail-tip narrower; fewer primaries with pale V-marks. SIZE: wing, ♂ (n = 11) 114–125 (118), ♀ (n = 9) 114–122 (118); tail, ♂ (n = 11) 60–67 (62·4), ♀ (n = 8) 57–64 (60·6); bill to skull, ♂ (n = 28) 16·8–19·5 (17·8), ♀ (n = 24) 17–18·9 (17·8); tarsus, ♂ (n = 25) 20–22 (21·0), ♀ (n = 25) 19–21 (20·0). WEIGHT: Hungary (Dec–Apr) ♂ (n = 27) 45–65 (56·0), ♀ (n = 13) 50–70 (60·0).

IMMATURE: differs from adult mainly in having darker, grey plumage (particularly head), shorter crest, darker breast, streaked belly, and in averaging 2–3 fewer red tips to secondaries.

Field Characters. A plump, crested, vinaceous-brown bird, size of European Starling *Sturnus vulgaris*. Chin black, rump grey, undertail-coverts bright rufous, tail conspicuously yellow-tipped black. Closed wing usually looks strongly pied, with some bright yellow and red. Rump feathers often puffed out, and crest raised high. Dumpy; gregarious, berry-eating; perches unmoving on berry-bush, except for occasionally tugging at and swallowing fruit. Arboreal, tame. Characteristic voice.

Voice. Tape-recorded (76, 89). Rather silent. Usual note, most often given in flight, a thin, high trill, 'sirrrrrr'.

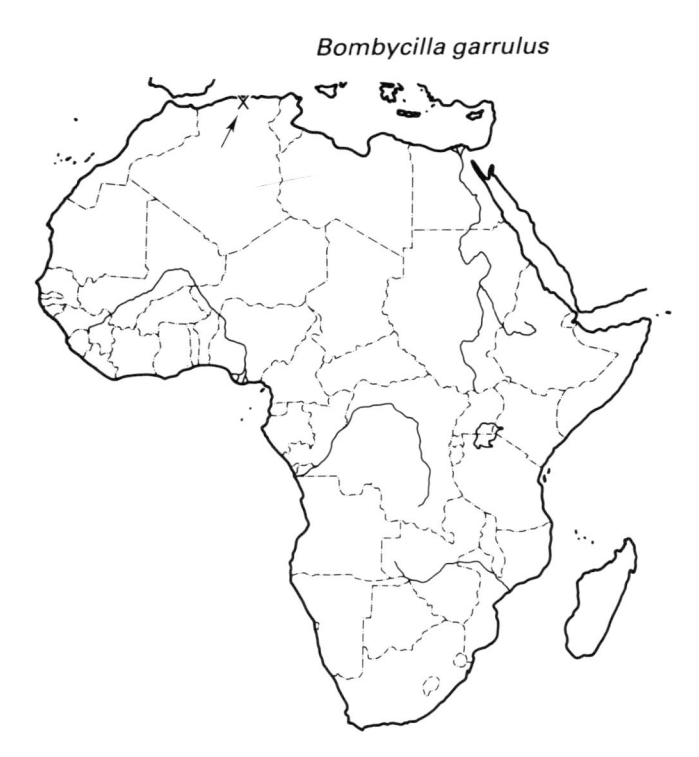

General Habits. In winter quarters inhabits open places with scattered ornamental and berry-bearing trees, tall hedges, thickets of sea-buckthorn *Hippophae rhamnoides*, *Cotoneaster*, lines of trees by busy roads, parks and gardens.

Gregarious in winter, seldom occurring singly; usually in flocks of 10–100, all flying together and alighting on one or a few neighbouring bushes to feed. Sluggish, sitting immobile for long periods. Flight direct, not fast nor undulating.

Baltic and Russian birds migrate southwest in autumn. In the great irruption of 1965–66, numerous birds ringed in Oulu, Finland, recovered mainly to southwest, particularly in S France.

Food. In winter berries of *Berberis*, *Cotoneaster*, *Pyracantha*, juniper, privet, yew, holly, elder, buckthorn, hawthorn and hips of dog-rose and guelder-rose.

Reference
Cramp, S. (1988).

Genus *Hypocolius* Bonaparte

A gregarious, migratory, grey, desert bombycillid. Waxwing-like but lacks bright colours; relatively short-winged; long-tailed. Single species, *H. ampelinus*, Iraq to Afghanistan, accidental in NE Africa.

Hypocolius ampelinus Bonaparte. Grey Hypocolius. Hypocolius gris.

Hypocolius ampelinus Bonaparte, 1850. Consp. Av. 1, p. 336; Abyssinian coast.

Plate 16
(Opp. p. 257)

Range and Status. Mesopotamia north to 37°N in Iraq, and Iranian coastal lowlands east to about Hamun-e Jaz Murian and W Sistan. Locally in S Afghanistan and in Murgab valley, S Turkmeniya. Short-distance migrant, wintering sparsely in Arabian peninsula south to Hejaz Mts (Saudi Arabia), where locally common, and Bahrain. Vagrant NE Africa, United Arab Emirates, Oman, Pakistan, NW India.

Accidental Ethiopia (Massawa, 1850), NE Sudan (Gebel Elba, 3 records: Goodman 1984) and Egypt (1, Wadi Shallal: El Negumi 1949).

Description. ADULT ♂ (breeding): forehead bordering bill velvety black; rest of forehead greyish buff; crown buffy grey. Ear-coverts black overlain with silky dark grey; lores black. Broad black line connects eyes across hindcrown; nape, sides of neck, mantle, back, scapulars and upperwing-coverts soft French grey, paling on rump and uppertail-coverts. Tail grey with black end *c.* 20 mm deep. Chin and throat pale pinkish buff, merging to French grey on sides of neck and breast, which merges to pinkish buff on belly, flanks and undertail-coverts. Primaries black with clear-cut white ends, the white patch longest on outermost primary (*c.* 25 mm long); secondaries with grey outer webs and dusky inner webs; all upperwing-coverts fringed grey. Axillaries pale grey; remaining underwing-coverts pale grey or creamy; underside of remiges like upperside. Bill black or dark horn with paler base; eye dark brown; legs and feet reddish flesh. ADULT ♂ (non-breeding): plumage the same, legs and feet yellowish flesh.

A

ADULT ♀: like ♂ but entire plumage buffier. Lacks black line from base of bill through lores, ear-coverts and around hindcrown. Tip of tail brown rather than black, less distinctly demarcated than in ♂, *c.* 15 mm deep, and outer rectrix without any dark tip at all. Underparts buffier than ♂, breast lacking grey wash of ♂'s breast. Primaries with outer webs grey, white tips of both webs 3 mm deep, and subterminal blackish mark near end of outer web and more broadly on inner web (which is dusky). Secondaries and upperwing-coverts plain, concolorous with back. Bill horn or dark horn with paler base. SIZE: wing, ♂ (n = 15) 97–106 (102), ♀ (n = 17) 97–105 (101); tail, ♂ (n = 14) 109–122 (110·5), ♀ (n = 17) 96–109 (104); bill to skull, ♂ (n = 14) 19·1–20·6 (19·9), ♀ (n = 15) 18·3–21·0 (19·7); tarsus, ♂ (n = 14) 23–25 (24·0), ♀ (n = 14) 23–25 (24·0). WEIGHT: (SW Afghanistan) 1 ♂ 48, 3 ♀♀ 49, 50, 55.

IMMATURE: like adult ♀ but entire head buffier, and remiges (before first moult) plain brown without any variegation.

Field Characters. A plump-bodied, long-tailed, grey or fawn bird of date-palms in semi-desert. About size of Great Grey Shrike *Lanius excubitor*, and ♂ of somewhat similar appearance with black mask, which is, however, triangular (**A**) rather than oblong, and with strongly patterned wing. Readily told from shrike by white-tipped primaries, which in flight contrast strongly with black distal and grey proximal half of wing. Gregarious; generally quiet; rather sluggish; easily overlooked. Behaviour much like Bohemian Waxwing *Bombycilla garrulus*.

Voice. Tape-recorded (BUN, HOL). Quiet in winter, but flocks occasionally make pleasant or plaintive mewing or whistling sounds and trilling or rolling notes recalling European Bee-eater *Merops apiaster*. Main calls: (1) mewing 'meee', loud and repeated, fluty 'peeeooo' or 'wheeoo' like ♂ Eurasian Wigeon *Anas penelope*; (2) when disturbed, 'meew' and 'kirrr' notes intermingled at 60–80 notes per min; (3) loud, trilling, bee-eater-like 'kirrrek' (Bunni and Siman 1978).

General Habits. Breeding and wintering habitat is palm-groves, oases, broad-leaved subdesert scrub and watered gardens; also (Iraq) impenetrable acacia scrub, tamarisk, willow and poplar. Likes date-palms and poplar stands. Arboreal, feeding gregariously in trees and bushes, keeping to outsides when feeding but diving into middle when alarmed; often sits on top, with slightly drooping wing (**A**, p. 379). Flight strong, direct, whirring, without undulations; flocks fly high, gliding down to perch. Tame, sitting tight in bush; if disturbed, often flies far. Hops and clambers about in vegetation; feeding movements deliberate, stretching and balancing to reach berries (Cramp 1988). Occasionally drops to ground for insect.

Food. Mainly fruit: dates, figs, and berries of *Lycium barbarum* and *Morus alba* (mulberry) (Iraq). Also leaves of *Lycium*, shoots of tamarisk and (Saudi Arabia) berries of *Solanum incanum*, *Withania somnifera*, *Lycium arabicum* and *Ochradenus baccata*; in India berries of *Zizyphus* and shrub *Salvadora persica*. Takes a few beetles and other insects, especially to feed to young.

Reference
Cramp, S. (1988).

Family CINCLIDAE: dippers

Wings short, outermost primary short but well developed; tail short, square or slightly rounded, with 12 rectrices; plumage blackish, sometimes with grey, white and rufous, feathers copious, with under-covering of down. Tarsus long, strong, booted; claws stout. Sexes alike. Feed on aquatic insects, by walking and swimming underwater. Nest roofed, sited by stream or behind waterfall. Eggs white, unspotted.

1 genus, with 5 species in Andes, Rockies and mountainous regions of Palearctic. 1 species in Africa.

Genus *Cinclus* Borkhausen

Plate 16
(Opp. p. 257)

Cinclus cinclus (**Linnaeus**). **White-throated Dipper. Cincle plongeur.**

Sturnus cinclus Linnaeus, 1758. Syst. Nat. (10th ed.) 1, p. 168; Europe.

Range and Status. Africa, Spain, Britain and Norway to central China and E Siberia. Extinct in Cyprus. Vagrant Malta, Iraq.

Resident Morocco, rare Algeria, vagrant Tunisia, possibly Libya. In Morocco widespread but uncommon in the Rif and Middle Atlas, commoner in High Atlas from 900 to 2600 m (wadis Zar, Ourika, Nfis; Assif Melloul; Oukaïmeden, L. Ifni, Setti Fatma) and very common further east (Zaouïa Ahansal, Tabant, Imilchil); south to Todra. Near Asni (31° 17′N, 7° 58′W), 24 birds on total of 11·5 km of 3 rivers (Tyler and Ormerod 1991). Records at 5 Algerian localities (mainly Hodna and Aurès Mts in NE); 2 records Tunisia (Aïn Draham, 1941, 1943); 1 claim (1916) from E Libya.

Description. *C. c. minor* Tristram: endemic to Africa. ADULT ♂: forehead, crown, nape, lores, ear-coverts and sides of neck dark brown, sharply demarcated from white throat, less sharply from slaty mantle. Lower eyelid white. Mantle to tail and wings dark slate, shiny, with feathers margined matt black-brown. Chin, throat, breast pure white; upper belly pale rufous, sharply demarcated from breast and merging into black belly, flanks and undertail-coverts. Centre of belly sometimes suffused with rufous: amount and intensity of rufous on underparts somewhat variable. Underside of wings and tail dark brown. Bill brown-black; eye rich deep brown; legs and feet brown-black. Sexes alike. SIZE: (9♂♂, 3♀♀) wing, ♂ av. 103 (SD 1·9) (live birds; skins are 97–102, av. 99·0), ♀ av. 94·3; bill to feathers, ♂ av. 13·7 (SD 0·7), ♀ av. 12·9; tarsus, ♂ av. 36·1 (SD 0·8), ♀ av. 34·9 (Tyler and Ormerod 1991). WEIGHT: ♂ (n = 9) av. 76·0, ♀ (n = 3) av. 68·3.

IMMATURE: entire upperparts, including forehead to nape, dark slate grey; lores, ear-coverts and sides of neck dark brown-grey, mottled; chin white; throat, breast and upper belly creamy white, with brownish feather-tips giving scaly appearance; upper belly suffused with buff, darkening to slate grey lower belly and undertail-coverts. Remiges and upper-wing-coverts pale-tipped.

NESTLING: (*C. c. aquaticus*): skin pink-orange; underparts bare, upperparts with long, dense, dark grey down; mouth pink or bright orange.

Field Characters. A robust, tubby, blackish songbird with strikingly white breast, confined to fast-running clearwater streams. Short-tailed; otherwise rather like a small thrush *Turdus* sp. Walks and swims freely in streams, feeding at edge or under water. Flight low, fast and direct. White lower eyelid often conspicuous as eye is blinked. Frequent 'zit' call attracts attention; easily heard over noise of running water.

Voice. Tape-recorded (62, 73, B). Usual call, in flight and at perch, a loud 'zit' or 'tze', 2–3 times in succession, occasionally up to 100 per min. Also, loud rattling 'r-r-r-r' or 'zur-r-r-r'. Song (Europe) a sweet warble given by both sexes.

General Habits. Little information in NW Africa. In Europe inhabits swift-flowing, cold, clearwater streams, 1–20 m wide, in mountains and foothills, dispersing in winter, occasionally to lowland waters. Requires bouldery, stony or gravelly shallows for feeding, and does not feed near or otherwise require vegetation.

Solitary, or a pair not very close together, or in dispersed family party. Territorial, patrols stream by walking; occasionally runs for an insect; walks unhesitatingly into water until submerged; swims, or rather bobs, at surface like miniature duck. Occasionally dives from surface, or jumps into deeper water from rock, submerging immediately. Feeds whilst walking on stream bed, usually totally submerged, progressing always upcurrent, moving pebbles with bill and swallowing insects found underneath. Submerges for up to 10 s (30 s in captivity), av. 3·2 s; dives up to 12 times per min, surfacing for as little as < 1 s before diving again. Most prey eaten under water; larger items (e.g. fish) brought to surface. Sometimes feeds on ground beside stream, turning over stones and tossing debris; also catches airborne insect in short flight. Cocks tail. Flies readily when disturbed or changing feeding station; rather shy. Flight low over water (often cuts corners, over land) and direct, whirring, fast, without undulation. Roosts at night solitarily; rarely, a few together; often under bridges.

In NW Africa sedentary, or with some altitudinal movements (Chaworth-Musters 1939); Moroccan records outside habitual areas also suggest some local movement (Thévenot *et al.* 1982).

Food. Aquatic invertebrates, mainly insects. Of 1211 items (Asni, Morocco), 2 were not aquatic (1 fly, 1 caterpillar) and 3% were not insects (hydras, worms, molluscs, Crustacea). Of the remainder, caddis-fly larvae (Trichoptera: Hydropsychidae) were 52% numerically (88·5% by weight) and mayfly larvae (Ephemeroptera: Bactidae) 22% numerically (1·7% by weight) (Tyler and Ormerod 1991). In Europe mainly aquatic insects, particularly caddis-fly larvae (Trichoptera) and mayfly nymphs (Ephemeroptera); also spiders, amphipods, isopods, millipedes, centipedes, molluscs, fish and plant material.

Breeding Habits. Little known in NW Africa; following data mainly European (from Cramp 1988). Monogamous, nests solitarily, territorial. Breeding territory *c.* 4000 m², sometimes separate from winter territory. ♀♀ usually breed at age 1, ♂♂ may not. Adjacent nests hundreds of m apart (rarely < 100 m; often > 1 km).

Pair-formation begins with advertising display. Advertising bird (either sex) stands upright and sings for several s with head thrown back and turned slowly from side to side, bill pointing up, breast ruffled, wings half-spread sideways or upwards, and tail fanned. ♂ runs (or swims) to ♀, giving advertising display, until their bills touch, runs dipping and blinking around ♀, often wing-quivering. If ♀ departs, ♂ chases her, singing in flight; they may then both fly high. One or both may land directly in water, and swim ashore. Courtship feeding of ♀ by ♂ frequent. ♀ solicits by crouching, with wings half open and quivered, often gaping, singing or calling. If ♂ stimulated to mount, he does so without further ado, hopping or flying straight onto her back; during copulation, which lasts 2–7 s, both birds beat wings.

NEST: ball built mainly of moss, also grass leaves, with wide entrance pointing to side or down, and inner cup of fine roots, stems and hair; greatest ext. diam. *c.* 220–260. Built in hole in streamside wall or culvert, on rocky ledges, under banks, and behind waterfalls, usually within 1 m horizontally and 2 m vertically of water. Built by ♂ and ♀; only ♀ completes cup; moss

Cinclus cinclus

gathered up to 30 m from site and (if dry) wetted before use. Construction takes 18–28 days.

EGGS: 3–5. Subelliptical; glossy; white. SIZE: (n = 7, Morocco) av. 26 × 18; (n = 200, *C. c. aquaticus*, Europe) 23·4–28·4 × 16·5–20·1 (25·6 × 18·8). WEIGHT: *c.* 4·55.

LAYING DATES: Morocco, Mar.

INCUBATION: by ♀ only, beginning with last (or penultimate) egg. Period: (n = 26) 12–18 (16) days.

DEVELOPMENT AND CARE OF YOUNG: young brooded by ♀ until 12–13 days old (not by ♂); fed by both parents, up to 20 items per meal at 5 days, but later fewer, larger items per meal; fed every 2–16 min (Europe). Small young given single items, older ones food-balls. After chick fed, it retreats to back of nest, allowing another to take its place. Young lean out of entrance at 11–12 days, begging loudly; sometimes elicit feeding by Grey Wagtails *Motacilla cinerea* with nearby nest (Yoerg and O'Halloran 1991). Parents remove faecal sacs from nest up to day 9; thereafter chicks defaecate out of nest and parents collect sacs and drop them in flight into stream. Fledging period: (n = 15) 20–24 (22) days (Britain). Brood takes 1–4 days to leave nest; young immediately start dipping in water. Parents, mainly ♀, continue feeding young for 1–2·5 weeks after young leave nest.

BREEDING SUCCESS/SURVIVAL: 61% of 1986 eggs hatched and 51% of young survived to 14 days (Britain). Av. annual mortality 64% (Britain). Oldest bird 7 years 10 months.

References
Cramp, S. (1988).
Tyler, S. J. and Ormerod, S. J. (1991).

Family TROGLODYTIDAE: wrens

Very small, small and medium-sized non-migratory, insectivorous songbirds mainly with barred, brown plumage, thick and soft, short tails (often cocked), short but well-developed outermost primary, nostrils partly covered with membrane, strong legs and feet, scutellate tarsi and large claws.

60–65 species, in 12–14 genera, including *Donacobius atricapillus* (until recently thought to be a mockingbird). Neotropical and Nearctic, with 1 species, *Troglodytes troglodytes*, also widespread in Palearctic, including N Africa.

Genus *Troglodytes* Vieillot

Small to very small; the smallest wrens. Bill short, slightly curved, without notch, nostrils oval; tail with 12 feathers, very short, usually cocked; wings short and rounded; claws strong, curved, hind toe large. Plumage brown above, brown, grey or white below, with barring on wings and tail (also on underparts in 2 species).

9 species in New World, 1 also in Old World.

Plate 16
(Opp. p. 257)

Troglodytes troglodytes (**Linnaeus**). **Winter Wren; Wren. Troglodyte mignon.**

Motacilla troglodytes Linnaeus, 1758. Syst. Nat. (10th ed.) 1, p. 188; Europe.

Range and Status. Holarctic: N America, Newfoundland, Iceland, Africa, Europe north to 70°N, Near East, S Asia south to Taiwan and north to Sakhalin; Japan, Kurile and Aleutian Is. Resident and partial migrant.

Common resident in N Africa in 2 populations, *kabylorum* in Maghreb and *juniperi* in Libya. Former is widespread in Morocco from Rif, Tangier and Casablanca inland to Middle, High and Anti-Atlas ranges up to 2600 m, east to about Talsint, and on coastal plain and foothills from about Ujda east through Algeria to N Tunisia. In Tunisia widespread but sparse, south to Jugurtha, Le Kesra and Enfida; commonest in N and NE Kroumirie (NW Tunisia). Race *juniperi* confined to Jebel Akhdar, Libya, from Tocra Pass east to Derna; single winter records Benghazi and Tobruk.

Description. *T. t. kabylorum* Hartert: NW Africa, S Spain. ADULT ♂: forehead, crown and nape dark brown, greyish and suffused with russet towards nape; mantle, back, rump, wings and tail russet, with dark brown bars indistinct on back and distinct on tail (*c.* 8 dark bars on tail and uppertail-coverts). Lores speckled brown, sometimes with suggestion of superciliary stripe; whitish or buffy superciliary stripe conspicuous behind eye, with dark brown line below, on upper ear-coverts;

lower ear-coverts, cheeks and sides of throat finely streaked brown-on-white; chin, throat and breast pale buff or rufous-brown; centre of belly pale buff; sides of breast and belly, and particularly flanks, buff-brown with irregular, narrow dark brown bars. Undertail-coverts barred pale buff and blackish brown. Wings more conspicuously dark banded, tertials and upperwing-coverts tipped buff, and outer primaries closely barred black and buffy white. Underside of wings and tail pale grey-brown. Upper mandible blackish brown, lower mandible horn or yellowish, darker towards tip; eye dark brown, narrow whitish eye-ring; legs and feet light brown. Sexes alike. SIZE: (6 ♂♂, 6 ♀♀) wing, ♂ 47–49 (48·3), ♀ 43–47 (45·4); tail, ♂ 29–33 (31·0), ♀ 29–31 (30·5); bill to skull, ♂ 13·4–15·4 (14·1), ♀ 12·9–13·8 (13·3); tarsus, ♂ 17·3–18·4 (17·7), ♀ 16·5–17·2 (16·9). WEIGHT: unknown; unsexed (n = 13, *T. t. troglodytes*, Portugal) 7·1–10·2 (8·8).

IMMATURE: like adult but ground colour paler, duller; barring less distinct.

NESTLING: skin pink; underparts covered with short, sparse, dark grey down; gape-flanges pale yellow, mouth bright yellow, legs and feet pink.

T. t. juniperi Hartert: Cyrenaica (Libya). Like *kabylorum* but slightly smaller, with longer bill: wing, ♂ (n = 15) 46–48; bill to skull, ♂ (n = 9) 15·5–16·5 (16·0).

Field Characters. A very small, but robust and active, thin-billed songbird, readily distinguished by blackish-barred rufescent wings, tail and flanks, pale eye-stripe relieving otherwise rather drab brown plumage, and by permanently-cocked very short tail. Vocal – sings all day and nearly all year. Skulks in low vegetation; much more often heard than seen; presence usually first apparent from alarm scold.

Voice. Tape-recorded (CHA – 62, 73, B). Song a loud, vehement, shrill rattling warble of 4–6 s duration, well structured but regionally variable. Song continuous, but usually consists of 5–10 parts, some sweet and melodious, others fast ticking trills. Often ♂ sings once only, but at other times song repeated after few s *ad lib*. Song sometimes abruptly curtailed in middle. For courtship-song and whisper-song, see Cramp (1988), where numerous calls are described. Main call a single or double loud 'tek tek', usually an alarm scold, becoming a churr as alarm increases.

General Habits. Inhabits sunny and dense pine woods, keeping to understorey and ground cover, open rocky hillsides with bracken or macchia, hedges, mossy or dry-stone walls and outbuildings, particularly when overgrown with scrambling vegetation, sea-cliffs, crops, fallen timber, gardens, orchards, stream valleys, and mountains up to treeline. Commonly enters crevices in rocks, walls and masonry.

Solitary, not shy, but spends much time within dense ground vegetation or concealed from view within drystone wall, where creeps and makes small leaps or flies for as little as 1 m with whirring wings, constantly searching for food on surfaces and in small crevices in wood and stone. Gleans leaves, leaf-litter and bark; feeds mainly within 2 m of ground. Hops on ground with legs bent, tail cocked and flicked. Flies swiftly and directly, but for only short distances. Roosts at night solitarily, or 2–3 together, ♂♂ and ♀♀ separately before and after breeding season or together in winter. In severe weather, 30–60 cram together in e.g. a nest-box, and 96 once recorded roosting in loft (England).

Resident, probably mainly sedentary, but montane birds may descend in winter. Partially migratory in Europe, with 1 ringed nestling moving from Gotland (Sweden) to Granada (Spain), but no evidence yet that migrants penetrate Africa.

Food. Not studied in Africa. In Europe chiefly arthropods, particularly beetles and spiders, also caterpillars, aphids, fly and caddis-fly larvae. Most prey < 10 mm long. Occasionally takes food from water (crustaceans, tadpoles, small fish) and eats small soft fruits.

Breeding Habits. Little known in N Africa. In Europe mainly polygynous, ♂♂ strongly territorial with up to 150 territories per km². In Morocco 23 territories per km² in degenerate woodland and up to 74 in humid forest (Thévenot 1982). Starting early in season, ♂ builds numerous nests in his territory: av. (n = 25 by same ♂ in 3 years, England) 6·3 nests; av. (n = 88, W Germany) 2·5 nests, ratio of nests used for breeding to unused nests varying from 1·6:1 to 1:6 (Dallmann 1977).

NEST: a neat, unlined ball with side entrance, made of moss, dead leaves and grasses and a little woodier vegetation; situated in crevice in rock, building, old wall or tree-hole, in other birds' nests, thatch, holes in banks, and in dense, usually thorny vegetation (Europe). ♂ builds several such nests; ♀ adds lining of hair and feathers to the one she chooses. Building takes 1–5 days. SIZE: (n = 116, Germany) av. 145 long × 130 broad × 113 high; int. width 62, int. height 56.

EGGS: 4–5 (Morocco). Subelliptical; glossy; white, sometimes with dark speckles at broad end. SIZE: (n = 150, *T. t. troglodytes*, Europe) 14·7–18·4 × 11·6–13·8 (16·4 × 12·6). WEIGHT: av. 1·32.

LAYING DATES: Morocco, Mar–June; probably double-brooded.

INCUBATION: in Europe by ♀ only, for 10–15 min periods alternating with 7–9 min periods off nest. Period: (n = 43) 12–20 (16) days. Clutch takes 1–3 days to hatch.

DEVELOPMENT AND CARE OF YOUNG: eyes open at 5–7 days. Brooded by ♀, assiduously at first, waning over 1st week. Young fed mainly by ♀; ♂ either does not feed them in nest, or gives token help, or feeds them equally with ♀ (Europe: varies regionally with mating system). After being fed, nestlings void faecal sac; parents start removing sacs on about day 4. Young fledge in morning, then huddle on branch for 1–2 days; fledglings escorted and fed by ♂ and ♀ equally. Fledging period: (n = 42) 14–19 (17·3) days (England: Armstrong 1955).

BREEDING SUCCESS/SURVIVAL: no African data. Success rate 65% (Britain): 71% of eggs in 825 clutches hatched, 92% fledged. Av. annual mortality 63% (Britain). Lives up to 6 years 7 months.

References
Armstrong, E. A. (1955).
Cramp, S. (1988).

Family PRUNELLIDAE: accentors

Small, inconspicuous, brownish ground-foraging songbirds with thin but hard bill, wide at base, nostrils free but covered with membrane, short but quite strong legs, rounded wings, thick and rather hard plumage. Sexes nearly alike. Eggs blue. Most species montane, resident. Eat insects, some seeds. Possess crop and muscular gizzard, and related not with other insectivores (except wagtails Motacillidae), but evidently with sparrows and weavers. On DNA-DNA hybridization evidence, treated by Sibley and Ahlquist (1990) as Subfamily Prunellinae in Family Passeridae, other subfamilies being Motacillinae (pipits), Passerinae (sparrows), Ploceinae (weavers) and Estrildinae (waxbills).

Single genus, with 14 species, Palearctic, one breeding in NW Africa, another a scarce winter visitor to N Africa.

Genus *Prunella* Vieillot

Plate 16
(Opp. p. 257)

Prunella collaris (Scopoli). Alpine Accentor. Accenteur alpin.

Sturnus collaris Scopoli, 1769. Ann. 1, Hist.-Nat., p. 131; Carinthia.

Range and Status. Mountains of NW Africa, central and S Europe, Asia Minor, central and S Asia, Japan and Taiwan, from 2000 m up to snowline (3000 m, Europe; 8000 m, Asia). Altitudinal and latitudinal migrant, in W Palearctic wintering from Spain to Sicily and Greece and sparsely in NW Africa.

Resident in, and probably winter visitor to, Atlas Mts, NW Africa. Breeds Morocco in W High Atlas southeast of Marrakesh, where common at 2900–3600 m on Jebel Angour (Toubkal range) and occurs down to 2400 m in Jan (Taddert), and E High Atlas on Midelt (3400 m); also probably in Middle Atlas (Jebel Bou Naceur, Jebel Bou Iblane). Breeds in Algeria on Jebel Babor (2000 m); sight records from 4 other Algerian localities within 70 km of coast (Ledant *et al*. 1981). In Tunisia a rare visitor Oct–Mar, regular on Jebel Kornine and Jebel Ressas (Tunis), otherwise only 6 records.

Description. *P. c. collaris* (Scopoli): NW Africa, Europe east to Carpathians and N Yugoslavia; winters to south. ADULT ♂: forehead and crown dull grey with olive-brown tinge; nape, sides of neck, mantle, scapulars and back grey with olive-brown wash and indistinct but broad dusky streaks; outer webs of scapulars rufescent; uppertail-coverts greyish olive-brown, each with black shaft-streak, white tip, and black subterminal bar; tail black-brown, feathers with narrow grey fringes, grey-brown spot at end of outer webs and large pink, buff or white spot at end of inner webs; underside of tail dark brown, white-tipped. Lores and ear-coverts grey, speckled whitish, and surrounded by clear bluish grey on supercilium, cheeks and breast. Chin and throat white with black speckles or broken black bars, forming gorget; sides of breast grey, tinged rufous; belly pale buffy grey in centre and rufous at sides; flanks with large blackish or rufescent patches; undertail-coverts white, heavily spotted black. Remiges black-brown, primaries narrowly and secondaries broadly edged rufous, tertials broadly rufous-edged and pale-tipped; lesser coverts grey, median

coverts black with small white tips, greater coverts black with warm brown base and triangular white spot at tip, lesser primary coverts black spotted with white, greater primary coverts black with short white streak at tip of outer web. Underside of wing buffy grey, some coverts white-tipped. Bill brown-black at tip and yellow at base; eye reddish brown; legs and feet reddish brown. Sexes similar but ♀ slightly duller.
SIZE: wing, ♂ (n = 18) 101–110 (105), ♀ (n = 10) 95–100 (97·4); tail, ♂ (n = 17) 60–70 (65·1), ♀ (n = 7) 55–64 (58·7); bill to skull, ♂ (n = 13) 16·8–18 (17·4), ♀ (n = 4) 17·3–17·6 (17·4); tarsus, ♂ (n = 15) 24–27 (25·0), ♀ (n = 4) 24–25 (25·0). WEIGHT: Spain, (May) ♂ (n = 8) 38·5–43, ♀ (n = 2) 36·5, 40.

IMMATURE: like adult but browner – head and breast brown, speckled blackish, where adult blue-grey; gorget less well defined, chin and throat unspotted or spotted (rather than barred) with black, breast brown with black-brown spots; lower breast, sides of breast and flanks extensively marked with rufous.

NESTLING: upperparts and thighs with long but scanty dark grey down; other parts bare. Mouth red with 2 black spots at base of tongue; gape-flanges white.

Field Characters. Very high altitude; an unobtrusive but well-marked, robust, thin-billed bird often first seen when flushed from boulders near old snow, when white-tipped tail is good feature. General shape and stance pipit-like or like Eurasian Skylark *Alauda arvensis*. Head and breast mainly blue-grey, bill yellowish, back streaky brown, forewing black with 2 rows of white spots, whitish gorget, sides rufous and flanks dark-spotted, and undertail pied (often conspicuously so). Retiring; when flushed flies short distance and dives for cover; but feeds in open, moving with shuffling gait, flicking wings and tail.

Voice. Tape-recorded (62, 73). Song like that of Hedge Accentor *P. modularis* q.v., but slower, lower-pitched, richer and more musical, somewhat recalling Eurasian Skylark. Duration *c.* 8·5 s. Main call even more Skylark-like, a quiet, rippling 'tchirririp'; also husky 'churrp'.

General Habits. Inhabits sunniest available sites between treeline and snowline: stony or boulder-strewn flats and slopes with patches of alpine grass sward, with no or few shrubs. Often feeds around mountain chalets, ski-lifts and picnic places; seldom below 2000 m even in winter.

In pairs in summer, solitary, slightly or strongly gregarious in winter. Sometimes 10 or more associate loosely at 1 winter locality, and may feed in dense flocks like Eurasian Skylark. 40 together at 3050 m on Jebel Angour, Morocco, Dec (Juana and Santos 1981). Gait a rapid short-stepped walk with little runs and hops, sometimes creeping like Hedge Accentor, but usually with more erect carriage. Constantly flicks wings and tail. Flight undulating and fluent; often calls in flight. Sings in flight, from top of boulder or, occasionally, low shrub. Feeds on ground, particularly at melting edge of snow, pecking at wet ground, moss, lichens and grass; clings to rock faces and pecks into cracks in rock; sometimes makes short aerial sally and briefly hovers.

Food. In Europe mainly small insects, also spiders, snails, worms and variety of seeds; some berries and small leaves.

Breeding Habits. Not studied in Africa. Territorial and essentially monogamous, but pair 'helped' by 1–2 non-breeders, and locally may be freely promiscuous (Cramp 1988). Density: 14 birds (5+ pairs) in 2 km, and 18 birds (6+ pairs) in *c.* 3 km (Switzerland). Song-flight a brief flutter upward, hover, and downward glide (Cramp 1988).

Pair-bonding involves much aerial chasing, calling and undulating display-flights. Courting ♂ gives ripple-calls and sings, and hops around ♀ with wings drooped, plumage ruffled and body trembling. ♀ adopts soliciting posture, crouching with flattened body, wings half-spread and quivering, tail closed, steeply raised and moved rapidly from side to side, exposing bright red cloaca. ♂ stands motionless behind soliciting ♀ for several s, jerking head, then suddenly leaps on her. ♀ leaps away at almost same instant. Film analysis shows copulation lasts 0·15 s: successful impregnation probably helped by ♂ having 12 × 8 mm sperm duct either side of cloaca; sperm may be expressed onto ♀'s cloaca during fleeting copulation (Cramp 1988).

NEST: a cup loosely made of grass leaves and stems, neatly lined with moss, hair and feathers; in rock cleft or on ground between boulders. Nest 170 × 120 × 80 high, cup 80 wide, 53 deep.

EGGS: 3–4, laid daily. Subelliptical; glossy; pale blue. SIZE: (n = 70, Europe) 20·5–26·7 × 15–17·1 (23·0 × 16·6). WEIGHT: *c.* 3·4.

LAYING DATES: Algeria, pair feeding young out of nest in June.

INCUBATION: by ♂ and ♀. Period: 14–15 days.

DEVELOPMENT AND CARE OF YOUNG: 1 brood hatched over 16 h. Max. weight of young 38–40 g by day 12.

Plate 16
(Opp. p. 257)

Prunella modularis (Linnaeus). Hedge Accentor; Dunnock. Accenteur moucher.

Motacilla modularis Linnaeus, 1758. Syst. Nat. (10th ed.) 1, p. 184; Europe.

Forms a superspecies with *P. immaculata* (Himalayas) and *P. rubida* (Japan).

Range and Status. W Palearctic: Spain and Britain to W slopes of Ural Mts and Transcaucasia; breeds north to 70°N, south to 40°N. Resident, partial migrant, and migrant, wintering south to Gibraltar, Mediterranean islands and Iraq. Accidental Iceland, Faeroes and Africa.

Rare visitor to Morocco and Algeria – 1 or 2 seen most years, e.g. at Tétouan and Tiztoutine (1979) and Irherm and Ifrane (1981) (Morocco); may have bred recently in Algeria (E.D.H. Johnson *in* Cramp 1988). In Tunisia a rare but possibly regular visitor mid-Oct to mid-Mar, in northern forests and on passage at Cap Bon; in Libya 5 records, Jan–Mar 1965, west of Tripoli town; Egypt, very small numbers Oct–Mar on N coast from Matruh to Bahig, also 2 records from Giza and 1 from Suez.

Description. *P. m. modularis* (Linnaeus): N and central Europe, wintering in S Europe and commonly in central and W Mediterranean islands; rare, Morocco to Egypt. ADULT ♂ (non-breeding): forehead, crown and nape olive-grey; mantle and scapulars rufous-brown, broadly streaked black; back, rump and uppertail-coverts olive-brown (sometimes slightly streaked dusky); tail blackish olive-brown. Lores speckled dark brown; ear-coverts dark brown streaked with olive- or rufous-brown and whitish buff. Sides of head around lores and ear-coverts plain bluish grey (supercilium, sides of neck, chin and throat). On breast bluish grey merges into greyish white belly and undertail-coverts. Flanks and sides of breast olive-brown, with dull dark olive-brown streaks on flanks. Wing blackish brown, coverts broadly fringed rufous-brown; greater coverts rufous-brown with blackish along shafts and whitish buff fringe at tip of outer web, lesser coverts olive-brown, primary coverts and primaries black-brown with narrow rufous fringes, secondaries and tertials black-brown with broad rufous fringes. Underside of wings and tail dull grey. Bill black, base of upper mandible and (more so) lower mandible pale pinkish; eye light brown or bright reddish brown; legs and feet light reddish brown. ADULT ♀ (non-breeding): like ♂, but sides of face grey rather than blue-grey, and upperparts less heavily streaked. SIZE: (Netherlands) wing, ♂ (n = 78) 69–74 (71·1), ♀ (n = 48) 65–72 (68·6); tail, ♂ (n = 30) 55–62 (58·0), ♀ (n = 25) 53–59 (55·7); bill to skull, ♂ (n = 27) 13·9–15·5 (14·6), ♀ (n = 23) 13·6–15·6 (14·4); tarsus, ♂ (n = 29) 20–22 (21·0), ♀ (n = 26) 19–21 (20·0). WEIGHT: Netherlands (Oct–Feb) ♂ (n = 7) 18·3–23 (20·7), ♀ (n = 5) 16–20 (18·4).

IMMATURE: like adult ♀ but buffier, without clear greys on head, and whole underparts from chin to flanks and belly dark-streaked.

P. m. obscura (Hablizl): E Turkey, Caucasus, Iran, wintering to south; 3 records Egypt (Goodman and Watson 1983). Above, duller rufous-brown than *modularis*; below, ground colour paler grey, with more extensive olive-brown wash. Smaller: wing, ♂ (n = 34) 67–71 (69·0), ♀ (n = 34) 65–70 (67·5). WEIGHT: (Turkey) ♂ 20, ♀ 22, (Egypt) ♂ 17·8.

Field Characters. An unobtrusive streaky-brown bird with sparrow-like plumage, but bluish grey face and rufescent mantle and wings. Thin-billed and slim-bodied. Keeps low down in thick vegetation and feeds on ground, where it has characteristic creeping gait, with short, shuffling hops on flexed legs, the body horizontal. Constantly twitches wings.

Voice. Tape-recorded (62, 73, 89, B). May sing occasionally in Africa (sings most of year in Europe). Song a short, thin, sibilant warble at one pitch, unmusical but not unpleasant, a jangling 'sissi-weeso' and variants lasting 2–4 s. Not far-carrying. Song repeated in bouts of c. 8. Main call a shrill, piping 'seep'. Several other calls (see Cramp 1988).

General Habits. Uses wide variety of habitats on both breeding and wintering grounds: gardens, orchards, scrubland, coppiced woods, pedunculate oak and other broad-leaved forest with plenty of ground vegetation, hedgerows, parks, farms, shrubby dunes and sea-cliffs, pine woods, plantations, and shrubby wetland.

Solitary. Sings from top of low bush, but otherwise unobtrusive, feeding mainly on ground concealed under vegetation, or in open but within 5 m of hedge or shrubs. Creeping gait on ground (see above); constantly flicks wings. Flies low, seldom far; without undulation.

Agonistic and sexual encounters occur all year.

Great number of ringing returns show that NW European population (Scandinavia, Germany) winters in S Europe and Spain south to Gibraltar. Single African recovery, from Brunswick (W Germany) to near Ouezzane, N Morocco.

Food. Small arthropods and soft invertebrates in summer but almost entirely seeds in winter (*Spergula, Polygonum, Atriplex, Lamium, Rumex, Ranunculus, Stellaria, Geranium, Plantago*).

Reference
Cramp, S. (1988).

Family TURDIDAE: thrushes

Small to medium-sized songbirds; bill slender to fairly stout, rictal and nasal bristles present; tongue non-tubular, with extensive blade-like, reedy tip; syrinx with turdine 'thumb'; 10 primaries, first often very short; tail medium or short, long in a few species, rounded or square, occasionally graduated, with 12 feathers (exceptionally 10 or 14); tarsus long and strong, usually booted, although some chats have scutellate tarsi when young, feet acutiplantar, strong and well-developed. Plumage mainly grey, brown or rufous, a few blue or green; young spotted or scaled in most species. Single annual moult in autumn; in spring, partial body moult by a few species, otherwise moult is by abrasion.

Mainly terrestrial, some arboreal; typically inhabit woodland and forest, some species in grassland, moorland and rocky habitats, including cliffs; many adapted to cultivation, gardens and human habitation. Eat fruit, insects, worms and snails. Female builds cup-shaped nest in bush or crevice, incubates and broods young; eggs usually pale and speckled. Many are extremely fine singers. Often migratory.

Almost worldwide; about 300 species. 125 species in Africa, of which 89 are endemic, 23 are non-endemic breeders (some also Palearctic migrants), and 13 are Palearctic migrants only. It is extremely difficult to define this family, as its boundaries are constantly changing. Most recent authors now treat it as a subfamily (Turdinae) within the Muscicapidae; Sibley and Monroe (1990) retain the 'true thrushes' (*Turdus, Zoothera, Monticola, Neocossyphus*) and *Alethe* in Turdinae, also *Chaetops* which may be a babbler or a warbler (Olson 1984, 1989); other African thrush genera they place in the tribe Saxicolini ('chats') within the subfamily Muscicapinae. Olson (1989) proposes a subfamily Myadestinae within Muscicapidae, which would include *Neocossyphus, Modulatrix* and *Pinarornis*, but Jensen (1990) believes *Modulatrix* (including *Arcanator*) is a babbler (Timaliidae) and *Neocossyphus* is best treated as a separate family. Olson (1984) shows that the sugarbirds (Promeropidae) have the turdine thumb and considers them to be thrushes that have evolved the highly specialized morphology connected with nectarivory. We prefer to retain Turdidae as a family until its relationships with sister groups are worked out, while recognizing that this is probably in part an artificial grouping. As a general rule, we use the genera and species of Peters (1964), with modifications suggested by recent research; the sequence of genera and species within the family is our own. Our classification of the small forest thrushes (genera *Pogonocichla, Swynnertonia, Stiphrornis, Sheppardia, Cossypha, Xenocopsychus* and *Alethe*) is based on that of Jensen (1990), with some modifications. Treatment of *Turdus* and *Zoothera* follows Irwin (1984), except that we agree with Short *et al.* (1990) that *Psophocichla* is a valid genus.

Genus *Pogonocichla* Cabanis

Small; bill broad and flycatcher-like, rictal bristles pronounced. Plumage unlike any other African thrush: head and neck slate blue, upperparts olive-green, tail black and yellow, underparts bright yellow. Some races have an unspotted olive-green plumage stage intermediate between spotted young and adults. Inhabit montane forest undergrowth; catch some food on the wing, also glean foliage and forage on the ground. Partly migratory.

Endemic; monotypic. An isolated genus, considered primitive by Jensen (1990).

Plate 22
(Opp. p. 353)

388 TURDIDAE

Pogonocichla stellata (Vieillot). **White-starred Robin. Akalat étoilé.**

Pogonocichla stellata (Vieillot), 1818. Nouv. Dict. d'Hist. Nat. 21, p. 468; Plettenberg Bay, southern Cape Province, South Africa.

Range and Status. Endemic resident in montane forest of eastern Africa from extreme SE Sudan (common in Imatong Mts, uncommon elsewhere), E Zaïre (Rwenzori Mts to Mt Kabobo, including volcano region), W Rwanda, W Burundi, W Uganda (Rwenzoris and Kigezi), and Kenya (from Mt Elgon, Cheranganis, Laikipia Plateau, Mt Kulal, Mt Nyiru and Mt Uaraguess through central highlands to Ol Doinyo Orok, Chyulu Hills, Kasigau, Taita Hills and Mrima Hill) to Tanzania (Mahari Mt and Ufipa Plateau in W, and throughout E from crater highlands, Mt Kilimanjaro, Pare Mts and Usambaras to Njombe, Songea and Mt Rungwe); Malaŵi (throughout); Mozambique (N-central mts east to Ribaue, also Mt Gorongoza and along Zimbabwe border, and in S in Lebombo Mts and in coastal bush (Macia, Maputo)); Zimbabwe (E highlands from Inyanga to Chipinga); Swaziland and South Africa: Transvaal (Escarpment Region), Natal, Cape Province west to Swellendam. Common to abundant. Density in Transvaal 5 pairs/4·5 ha; a 25-ha forest patch on Nyika Plateau, Malaŵi, held 36–40 pairs (Dowsett-Lemaire 1983).

Description. *P. s. stellata* (Vieillot) (including '*margaritata*'): South Africa (Zululand south of Umfolosi R., Natal, Transkei, and S Cape west to Knysna and George). ADULT ♂: entire head and neck dark bluish slate apart from variable, usually distinct silky white supraloral 'star' and usually concealed white spot in centre of throat. Mantle, back, rump and uppertail-coverts green; central pair of tail-feathers black, distal half of outer web of next pair black, remainder bright yellow with 10 mm-deep black terminal band. Underparts bright golden yellow; primaries bluish grey, outer webs pale grey, secondaries, tertials and upperwing-coverts olive-green; underwing-coverts pale yellow. Bill black; eye dark brown; legs and feet pinkish. Sexes alike but ♀ smaller. SIZE: (5 ♂♂, 1 ♀) wing, ♂ 84–89 (86·4), ♀ 76; tail, ♂ 66–79 (71·6), ♀ 57; bill, ♂ 16–19 (16·8), ♀ 16; tarsus, ♂ 26–28 (26·8), ♀ 26. WEIGHT: ♂ (n = 143) 19–25 (21·4), ♀ (n = 90) 18–23 (20·2).

Uniquely for a robin, 2 distinct immature plumages: a spotted juvenile plumage moulted after 3 months into a plain olive subadult plumage which, in southern populations at least, is retained for 12 months until next annual moult.

SUBADULT: upperparts from forehead to rump dusky olive; central pair of tail-feathers dark greyish brown, remainder dull dusky yellow; face dusky olive with no white spots above lores or on throat; sometimes a pale orbital ring; underparts from chin to undertail-coverts dull pale yellow, variably marbled with grey on belly; wings olive-grey, outer webs of primaries not differently coloured; flight feathers and some yellowish-tipped upperwing-coverts are retained from juvenile plumage; wings and tail av. 2 mm shorter than in adults.

JUVENILE: crown to rump dark blackish brown, feather-shafts and -tips pale yellow; chin and throat pale straw-yellow, feathers edged black, giving scaly appearance, especially on chest; belly yellowish white, not mottled.

NESTLING: skin pinkish orange; long grey down on dorsal feather tracts; inside of mouth pale ochre-yellow; legs and feet yellowish pink.

Pogonocichla stellata

TAXONOMIC NOTE: both subadult and adult plumages vary geographically. 2 racial groups, one with simple piping call (*ruwenzorii*, *intensa*, *guttifer*, *orientalis* and *stellata*), the other with complex piping call (*macarthuri*, *helleri* and *transvaalensis*; call of *elgonensis* unknown).

P. s. intensa Sharpe: Sudan, highlands of N and central Kenya, N Tanzania (Loliondo, crater highlands, Mt Meru). Back olive-green with bronze wash; rump and uppertail-coverts yellow; primaries and secondaries edged slate-grey, tertials edged green; black band in tail 14–17 mm deep. Juvenile has 4 central tail-feathers blackish, yellow of the others bright. Subadult has upperparts olive-green, underparts yellow mottled with olive.

P. s. ruwenzorii (Ogilvie-Grant) (including '*friedmanni*'): Rwenzori Mts and Kigezi in SW Uganda to W Ruanda and Zaïre (Kivu and Mt Kabobo). Deeper yellow below than *intensa*; secondaries and tertials edged olive-green; black band in tail up to 20 mm deep; white supraloral and throat spots smaller than *intensa* and *guttifer*. Juvenile and subadult birds similar to *intensa*; some apparently moult directly from juvenile to adult plumage.

P. s. elgonensis (Ogilvie-Grant): Mt Elgon, Uganda. Like *intensa* but tail all black. Juvenile and subadult have all tail-feathers blackish with bases variably yellow.

P. s. guttifer (Reichenow and Neumann): Mt Kilimanjaro, Tanzania. Back darker olive than above races; primaries and secondaries all edged blue-grey; black tail-band 14–17 mm deep. Juvenile has 4 central tail-feathers blackish, yellow of the others bright. Subadult plumage undescribed.

P. s. pallidiflava Cunningham-van Someren and Schifter: Imatong Mts, S Sudan. Underparts very pale yellow; tail brighter yellow than *guttifer* (Primuline Yellow *vs* Spanish Yellow).

P. s. macarthuri (van Someren): Chyulu Hills, Kenya. Paler yellow below than other northern races; back greener, less olive, more like *orientalis*; primaries and secondaries edged blue-grey, without green wash; rump and uppertail-coverts golden yellow; black tips of tail-feathers 10 mm deep. Juvenile has 2 central tail-feathers blackish, others rather duller yellow than in adult. Subadult olive below with narrow yellow streaking.

P. s. helleri Mearns: Kenya (Taita Hills, Kasigau), Tanzania (Pare Mts, Losogonoi). Back green; outer edges of primaries and secondaries blue-grey with green wash; rump and uppertail-coverts yellow; central pair of tail-feathers black, black tips to others 7–10 mm deep. Juvenile has less contrasting upper- and underparts than northern races; subadult dull greyish olive above, dull pale yellow below, mottled grey.

P. s. orientalis (Fischer and Reichenow): Tanzania (Kungwe Mt to Usambaras) to Malaŵi and central Mozambique (Namuli and Chiperone Mts). Mantle, back and rump green, uppertail-coverts yellowish green; primaries olive-grey, secondaries, tertials and upperwing-coverts olive-green; black tail-band 10 mm deep. Juvenile and subadult like *helleri*.

P. s. transvaalensis (Roberts) (including 'chirindensis', 'hygrica' and 'lebombo'): E highlands of Zimbabwe; Mozambique (Mt Gorongoza); N Transvaal (Blouberg, Zoutpansberg and Woodbush to Barberton); high-altitude forests of Zululand south to Umfolozi R. Back, rump and uppertail-coverts green; outer edges of primaries and secondaries silvery blue-grey; black tip of tail narrow, up to 10 mm deep. Juvenile tail with 2 central feathers blackish, others dingy yellow; subadult dull olive-green above and dull pale yellow below with only faint greyish mottling; primaries edged olive.

Field Characters. A distinct robin with dark bluish head, green back and bright yellow underparts; some races with yellow rump, all except *elgonensis* with yellow in tail, producing flash of yellow in retreating birds. White 'stars' (supraloral spots) often concealed. The only robin with yellow underparts (others red or orange). Dark head and yellow underparts give superficial resemblance to Forest Weaver *Ploceus bicolor*. Dark green subadult plumage unique among robins; spotted juveniles told by yellow in tail. Very common in highland forests but shy; plaintive song, quiet but persistent, often the first clue to its presence. Subadults seldom call (only a soft 'chut' when alarmed) and are always inconspicuous. For differences from Swynnerton's Robin *Swynnertonia swynnertoni*, see that species.

Voice. Tape-recorded (58, 75, 86, 88, C, F, FEE, KEI, MAN, OAT). Song quiet; in nominate race a soft warbling, beginning and ending with slurred 'wheeyoo'; in *P. s. ruwenzorii* much faster, similar to song of Lesser Double-collared Sunbird *Nectarinia chalybea* (but see also Willis 1985). *P. s. orientalis* in Usambaras whistles 'fur-fee-fur-fee, for-her-for-her' over and over again at 4 notes per s. Both sexes sing. Adults of nominate race utter a quiet warbling subsong at end of moult period in Apr–May, inaudible beyond 6 m. Rattle call, used at all ages, soft 'chut', given singly or in rapid succession to form ratchet-like sound used just before and after roosting, and by adults when nestlings in danger. Loud piping, audible at 100 m or more, is predator-warning, used in breeding season by adult (seldom by subadult, never by juvenile); it varies geographically; simple, 2-syllabled 'too-twee', or complex 'ter-wheh dada weeyoo' (*P. s. transvaalensis*) or 'wheh cheeyoo wheh-ter-wheh te cheeyoo' (*P. s. helleri*). 'Too-twee' uttered 4–5 times in 2–2·5 s. Location call of fledglings, high-pitched 'seep'.

General Habits. Breeds in primary and secondary montane forest, forest edge, thickets in cutover areas and overgrown fields, bamboo, bracken-briar, and moorlands in stands of tree heath and giant lobelia; at 1600–3300 m in Zaïre, above 2000 m in Sudan, mainly 1600–3300 m in E Africa, progressively lower further south, at sea level in Cape Province. In non-breeding season, occurs in coastal bush and riparian growth, exceptionally under mango trees and in understorey of pine plantations.

Flight silent; hops when on ground. Bathes regularly in streams overhung with thickets, mainly in last hour of daylight. Forages at all levels in forest, most frequently in undergrowth. Tosses leaf litter, gleans foliage, twigs and bark; makes aerial sallies to capture flying prey; ascends into canopy for small fruits. Associates with driver ant columns. At army ant columns, seldom perches on ground but waits like small flycatcher 1–5 m above ants on ground (occasionally up to 25 m above them); the most generalized forager of all African ant followers (Willis 1985). Appears at any disturbance in forest, especially clearing of undergrowth. Intraspecifically aggressive, but supplanted by other ant-following robin species. Adults of southern races retain olive subadult plumage for 2 years and do not breed until 2 years old.

An altitudinal migrant, descending in dry season from montane forests to lowland riverine and moist evergreen forests, down to 300 m in E Africa and to sea level in Mozambique and South Africa. Usually does not move far but *P. s. transvaalensis* visits forests near Limpopo R. mouth in Mozambique, several hundred km from nearest breeding site. Altitudinal movement involves some adults of both sexes and many subadults, some starting their journeys while still in juvenile plumage. In southern Africa, birds descend in full moult in Mar–Apr and ascend in Sept, moving by day through evergreen cover along wooded drainage lines. On Nyika Plateau, Malaŵi, breeding ♂♂ strictly sedentary but all ♀♀ depart in non-breeding season (Dowsett 1985) and move 20–100 km.

Food. Small invertebrates; small berries and drupes eaten when available. In 214 faecal samples in Natal there were beetles in 83%, moths in 58%, ants in 43%, spiders in 34%, flies in 24%, caterpillars in 23%, amphipods in 20%, bugs and wasps each in 18%, crickets and other orthopterans in 11% and centipedes in 4%. Fruit of *Canthium*, *Cassipourea*, *Ficus*, *Hedychium*, *Ilex*, *Kiggelaria* and *Rhus* were in 34% of samples. In lowland habitats also feeds on termites. Remains of small frogs *Arthroleptis xenodactyloidea* and molluscs recorded in Zimbabwe.

Breeding Habits. Monogamous; territorial. Both sexes sing to advertise territory. ♂ occupies territory throughout year in Natal but sings mainly in Oct–Dec. All territories have at least one dense thicket of undergrowth. Territory size 0·5–0·75 ha (Natal), with av. of 0·6–0·7 ha on Nyika Plateau (Malaŵi), where some pairs breed in isolated forest patches of 1 ha. Size of territory is related to audible range of song – *c.* 45 m. Threatening adult lowers and sleeks head and fluffs back. In courtship display, ♂ sways slowly and rhythmically with out-stretched neck, clinging to side of branch on which ♀ sits passively; he may then give display flight with rapidly-fluttering wings and fully-fanned tail, accompanied by loud 'weee-weee-weee-weee' notes which merge into normal song.

NEST: domed structure with side entrance; extent of dome varies from barely overlapping inner rim of cup to a porch-like overhang; constructed of dead leaves, rootlets, tendrils and moss, lined with soft, skeletonized leaves and fine plant inflorescences (such as *Galopina*). SIZE: (n = 85, Natal): ext. diam. 110–230 (150), int. diam. 50–85 (65), depth of cup 40, entrance hole (circular: diam. 50; elliptical: 55 × 45). Well concealed; live stems of fern and *Asparagus* frequently incorporated in walls of dome. Dry weight of 1 nest from Natal, 47 g. Placed usually on sloping ground, often against base of small tree or mossy rock; 23% of Natal nests were above ground level in hollows in banks, on fallen tree trunks, or in herbaceous growth on top or side of boulder. Built by ♀ in 7 days.

EGGS: 2–3, usually 2 but 3 south of Zambezi R.; mean clutch size (n = 78, Natal) 2·9, (n = 26, Malaŵi) 2·3; laid on consecutive days. Oval to long oval; white (rarely pale green), freckled and blotched with pinkish brown and lilac, usually on larger two-thirds of egg, often coalescing to form cap or ring at large end. In clutch of 3, last egg often paler and smaller. SIZE: (n = 138, Natal) 20–24 × 15–17 (22·0 × 16·0). WEIGHT: (n = 18) 2·7–3·2 (3·0).

LAYING DATES: Zaïre, Aug–May; Rwanda, Aug; E Africa: Uganda (fledglings Mar–Apr, Sept, carrying nest material Feb); Region A, Jan–Feb, June, Aug–Oct; Region B, Dec–Jan; Region C, Dec–Feb, May, Oct; Region D, all months. Breeds in dry season at high altitudes, in both wet and dry seasons at lower altitudes, preferring wet. Zambia, Sept–June; Malaŵi (Nyika Plateau), Sept–Jan (Sept 5, Oct 64, Nov 120, Dec 63, Jan 6 clutches: Dowsett and Dowsett-Lemaire 1984); Mozambique, Nov; Zimbabwe, Oct–Jan (Oct 3, Nov 13, Dec 16, Jan 3 clutches); Transvaal and Natal, Oct–Dec.

INCUBATION: begins at or soon after completion of clutch. By ♀ only. Period: 16–18 (16·5) days.

DEVELOPMENT AND CARE OF YOUNG: eyes begin to open on days 4–5, fully open by day 7; feathering on flanks by day 7, primaries bursting from quills by day 9; fully feathered by day 12. Nestling weights peak on days 13–14 (8 chicks from 3 broods, Natal, were up to 14% heavier than fledglings). Nestling period: (n = 9, Natal) 14–16 (14·3) days; in wet weather, brood may delay leaving nest for a day or more. Nestlings fed by ♂ and ♀; feeding visits av. 10 per h; faecal sacs removed and discarded 30+ m from nest. Each parent cares for at least 1 fledgling; fledglings hide in dense cover and move about warily, giving location call. Adult pipes loudly if danger threatens. Parental care lasts 40–42 days after fledging, possibly less when brood fledges late in season.

BREEDING SUCCESS/SURVIVAL: in Natal 35 of 60 nests produced young (91 chicks from 179 eggs, i.e. 51%). Success rate of 11 pairs whose nests were inspected by careful researcher (18%) was significantly lower than rate of 32 pairs with unchecked nests (78%). Of 72 breeding pairs in Malaŵi, 46 successful ones produced 83 young (av. 1·8 young per pair: Dowsett-Lemaire 1985). Survival rate of breeding birds over 9-year period (Natal): ♂♂ 0·77–0·90, ♀♀ 0·68–0·84. Av. annual mortality (Malaŵi): ♂♂ 16·7–20·6%, ♀♀ 25·8–40·4% (Dowsett 1985). 1 ♂ lived at least 14 years. Parasitized by Red-chested Cuckoo *Cuculus solitarius*.

References
Dowsett, R. J. (1985).
Dowsett-Lemaire, F. (1985).
Moreau, R. E. (1951).
Oatley, T. B. (1982a, b and c).
Willis, E. O. (1985).

Genus *Swynnertonia* Roberts

Very small; superficially similar to *Pogonocichla* but bill long and narrow, not broad and flycatcher-like, rictal bristles absent, tail and wings much shorter and more rounded; lacks erectile white tufts in front of the eyes, tail uniform grey, underparts orange, not yellow; sexually dimorphic (sexes alike in *Pogonocichla*). Feeding behaviour different (Jensen 1990). Inhabits montane and mid-altitude forests; sedentary.

Endemic; monotypic. Sometimes placed in *Pogonocichla* (White 1962, Hall and Moreau 1970), but we follow Irwin and Clancey (1974) in maintaining it as a separate genus.

Swynnertonia swynnertoni (Shelley). Swynnerton's Robin. Akalat de Swynnerton.

Erythacus swynnertoni Shelley, 1906. Bull. Br. Orn. Club xvi, p. 125; Chirinda Forest, Mt Selinda, S Rhodesia.

Plate 22
(Opp. p. 353)

Range and Status. Endemic resident. 4 isolated populations: (1) Usambara Mts (Mtai forest block), N Tanzania (pers. comm. to A. Tye from G. Anderson and T. Evans). (2) E scarp of Uzungwa Mts, S Tanzania (Mwanihana Forest and Chita, and very recently 80 miles south of there (M. W. Woodcock, pers. comm.)); (3) E highlands of Zimbabwe from Stapleford to Chirinda; (4) Mt Gorongoza, Mozambique. Common at Chirinda where locally 4–6 pairs/ha, and on Mt Gorongoza; scarce in Mwanihana Forest, less so at Chita; listed as rare in ICBP/IUCN Red Data Book because of its extremely limited range.

Description. *S. s. swynnertoni* (Shelley) (including '*umbratica*'): Zimbabwe and Mozambique. ADULT ♂: entire head and hindneck dark grey; mantle to uppertail-coverts olive-brown; tail uniform slate-grey; chin and throat dark grey, white crescent across lower foreneck, edged with black on lower side; breast yellowish orange, flanks tinged olive; belly, vent and undertail-coverts white. Wings slate-grey, secondaries and tertials edged on outer webs with olive-brown; upperwing-coverts olive-brown; underwing-coverts dull yellow. Bill black; eye dark brown; legs and feet pale greyish pink. ADULT ♀: similar to ♂ but duller; crown, face and neck tinged olive; chin and upper throat buffy grey. SIZE: wing, ♂ (n = 106) 66–73 (70·4), ♀ (n = 41) 65–70 (67·3); tail, ♂ (n = 98) 43·5–52 (49·8), ♀ (n = 37) 41–47 (44·2); bill, ♂♀ (n = 164) 12–16 (14·0); tarsus, ♂♀ (n = 164) 23–28 (26·1). WEIGHT: (Zimbabwe, Jan–Dec) ♂ (n = 105) 13·8–20·4 (15·8), ♀ (n = 44) 14·4–19·5 (16·3).

IMMATURE: brown above, feathers of crown, face, nape, mantle and upperwing-coverts tipped buffy yellow; underside like adult but markedly paler; chin light buff, crescent on foreneck drab greyish white and brown, breast pale yellow, feathers with brown tips, belly mottled grey and white; underwing-coverts dark brown. Spangled juvenile feathers of upperparts replaced with olive-brown feathers in adult, so immatures of 4–15 months resemble adult ♀ above but are paler than adults on breast and belly. Bill horn-coloured, grey tip to lower mandible, sometimes persisting for several months; eye tinged grey up to about 4 months; legs and feet medium flesh-pink, paler than adult.

NESTLING: skin pink with dark grey down, gape bright yellow-ochre.

S. s. rodgersi Jensen and Stuart: Mwanihana Forest and Chita (Uzungwa Mts), Tanzania, at 1000–1700 m altitude. Underparts yellower than nominate race, crown more olive, back greyer, less olive.

Field Characters. A small robin with very restricted range. Not unlike White-starred Robin *Pogonocichla stellata* but smaller, with orange breast and flanks, white belly, no yellow on rump or tail. Black-bordered white breast-patch large and visible (small and hard to see in White-starred Robin). Tail usually carried at a 45° angle, making white undertail-coverts conspicuous from behind.

Voice. Tape-recorded (32, 88, CHA, GIB, MAN, OAT, STJ, WALK). Song varies geographically; at Chirinda (Zimbabwe) it is remarkably loud for so small a bird; of 3 or 4 notes; 4-note version sounds very like song of White-browed Scrub-Robin *Cercotrichas leucophrys*: 'tee-werche-woo' or 'tee-terwer-choo'; 3-note version 'cha-chee-roo' often used when 2 birds call alternately. At Vumba (Zimbabwe) song is only moderately loud, 'woot-it-chweeeee'. At Seldomseen (Zimbabwe) song is 'zitt-zitt-slurr', the 'slurr' 2 tones lower and rising; sometimes 'slurr-zitt-zitt'. Soft quiet musical warbling lasting 3–4 s is given by displaying birds and when several birds present at ant column.

Commonest call is descending, squeaky trill or purr, often ending with 2–3 harsh, squeaky notes, 'trrrrrrrrrrr-wee-twaw-twee'; used as alarm call, uttered incessantly by ♂ and ♀ when nest threatened, or when person enters territory, and very often when several birds gather at driver-ant column; it is quiet, not audible beyond *c*. 25 m. At Seldomseen (Zimbabwe) alarm is high-pitched monotonous purr. Plaintive, high-pitched, descending 'seeeep' given by ♂ and ♀ when anxious; much like calls of Cape Robin-Chat *Cossypha caffra* and Brown Scrub-Robin *Cercotrichas signata*. Short, low chirp used by parents when approaching nest with newly-hatched chicks or chicks about to fledge (Manson 1990).

General Habits. Inhabits undergrowth of evergreen forest; 1000–1700 m in Tanzania, 850–1750 m in Mozambique, 900–1800 m in Zimbabwe. At Chirinda and Mt Gorongoza, favours stands of *Dracaena fragrans*, especially during breeding season.

Usually in pairs, but can be solitary in non-breeding season (Manson 1985). Likes areas of open forest floor with plenty of dead leaves. Seldom perches over 2 m above ground. Several gather at driver ant columns. Allows close approach by observers in dry season, but wary when breeding. Flight can be silent or audible. Forages mainly on forest floor where ground cover absent, by hopping, vigorously tossing leaf litter with bill; at ant columns perches on low vantage point and pounces on invertebrates flushed from litter by ants.

Food. Mainly insects; in 25 stomachs and 4 faeces (Chirinda, Zimbabwe), there were beetles in 93%, ants in 55%, wasps in 38%, caterpillars and moths in 21%, plant bugs in 21%, spiders in 21%, flies in 17%, and orthopterans, termites, centipedes, woodlice, small forest frogs *Artholeptis xenodactyloides* and small fruits in 10%. In 20 stomachs, beetles and ants made up 72% of identified food items (Dick 1981).

Breeding Habits. Monogamous, territorial; solitary nester. In territorial defence puffs out chest, making black and white foreneck conspicuous.

NEST: moderate to deep cup of dead leaves, leaf midribs and moss, lined with rootlets and tendrils, often with shiny, hair-like vegetable strands. Small and compact, or bulky and asymmetrical: ext. diam. 80–140, int. diam. 50–60, depth of cup 30–40. Sites (n = 52): *Dracaena* leaf axils 34%, hollow stumps or rot-holes 25%, forked stems of shrubs 19%, crevices in tree trunks 12%, among epiphytes on sloping trunk 10%. Height above ground (n = 57) 0·3–1 (0·9) m.

EGGS: 2 (once 3). Laid early in the day at *c.* 30-h intervals. Glossy; blue-green, blotched and spotted with shades of brown and mauve, or pale buff, speckled with rufous; markings often form cap or ring at large end. SIZE: (n = 19) 20–22 × 14–16 (20·9 × 14·9).

LAYING DATES: Zimbabwe, Oct–Jan (79% Nov–Dec).

INCUBATION: by ♀ only. Apparently starts with first egg: one nestling larger and more active than other (in 4 nests: Manson 1990). 44 incubation stints lasted 5–23 min (av. 14 min); 41 absences were of 5–23 min (av. 10 min). Period: 15–16 days.

DEVELOPMENT AND CARE OF YOUNG: ♀ broods chicks for 7–9 (av. 8) days; brooding bird regularly draws back into crouching position to inspect chicks. ♂ and ♀ feed them, ♂ passing food to brooding ♀ who gives it to chicks. Feeding rate increases throughout nestling period; at 1 nest, 57 feeds in first 3 days and 135 in last 3 days of period. Period: 14 days. Chicks may jump from nest before fledging and hide in leaf litter if encouraged by parent (Manson 1990).

BREEDING SUCCESS/SURVIVAL: occasionally parasitized by Red-chested Cuckoo *Cuculus solitarius*.

References
Collar, N. J. and Stuart, S. N. (1985).
Jensen, F. P. and Stuart, S. N. (1982).
Manson, A. (1990).

Genus *Stiphrornis* Hartlaub

Very small. Plumage somewhat resembles *Sheppardia* (orange throat and breast), but tail shorter, bill very fine, rictal bristles vestigial; feeding behaviour also different (Jensen 1990). Inhabit undergrowth of lowland forests.
Endemic; monotypic.

Plates 22, 26
(Opp. pp. 353, 401)

Stiphrornis erythrothorax **Hartlaub. Forest Robin. Robin. Rougegorge de forêt.**

Stiphrornis erythrothorax Hartlaub, 1855. J. Orn., p. 355; Dabocrom, Gold Coast (probably near 5° 20′N, 1° 30′W).

Range and Status. Endemic resident. SE Guinea (borders of Sierra Leone and Liberia), Sierra Leone, Liberia (common), Ivory Coast (widespread, north to Sipilou and Maraoué), S Ghana (widespread and not uncommon), S Togo, Nigeria (north to *c.* 10°N), Cameroon (north to *c.* 6°N), Bioko, SW and SE Central African Republic, extreme SW Sudan (Bengengai Forest), Gabon, S Congo (Mayombe), N half of Zaïre (south in E to Itombwe); W Uganda from Budongo to Malabigambo Forest, east to Mabira; NW Tanzania (Minziro Forest). Vagrant W Kenya (Kipkabus, 2500 m). Frequent to locally abundant; density in Gabon 12–15 pairs/km².

Description. *S. e. erythrothorax* Hartlaub: Sierra Leone to S Nigeria. ADULT ♂: forehead black with white supraloral spots; crown slate grey with olive wash; nape to uppertail-coverts and upperwing-coverts olive-green, wings and tail dark brown with olive wash; lores and ear-coverts black; chin, throat, foreneck and upper breast bright orange, silky, plush-like feathers enhancing colour; flanks dusky grey; rest of underparts and underwing-coverts white. Bill black; eye dark brown; legs and feet greyish-mauve. ADULT ♀: like ♂ but orange throat less vivid. SIZE: (20 ♂♂, 9 ♀♀) wing, ♂ 62–67 (63·8), ♀ 60–65 (62·2); tail, ♂ 31–40 (33·6), ♀ (n = 8) 32–35 (33·3); bill, ♂ 12·5–15 (14·1), ♀ 13·5–14·5 (14·0); tarsus, ♂ 21–23 (21·5), ♀ 21–23 (21·7). WEIGHT: (Liberia) ♂ (n = 17) 14·9 ±

1·4, ♀ (n = 7) 14·5 ± 1·4; unsexed (n = 15, Cameroon) 15·6–20 (17·0).

IMMATURE: white supraloral spots variable, becoming conspicuous before loss of all juvenile speckled feathers; sides of face grey, chin and throat whitish; rufous terminal spots on upperwing-coverts and mantle; upper breast mottled dull orange and dusky.

S. e. gabonensis Sharpe: Gabon and Bioko. Upperparts sooty grey, ♀ with faint olive wash.

S. e. xanthogaster Sharpe (including '*mabirae*'): interior of Cameroon, along Congo watershed to Ituri, Uganda and W Kenya. Like *gabonensis* above; chin, throat and breast much paler and less clearly demarcated from belly; belly and flanks less white, with buffy orange wash. In E, more olive-washed above and orange on throat more distinct. WEIGHT: unsexed (Uganda) 15–17.

Field Characters. A diminutive, short-tailed robin of lowland forest with bright yellow-orange throat and breast, white belly and black face with white supraloral spot which is permanently visible, unlike that of White-starred Robin *Pogonocichla stellata*. Dull olive and grey upperparts blend so well with background that bird is almost invisible from behind, even when moving. Normally carries tail at a 45° angle. Lowland and Equatorial Akalats *Sheppardia cyornithopsis* and *S. aequatorialis* are larger, with orange-red underparts, rufous tail, blue-grey face with no white spot, different voices; Forest Robin's plaintive, repetitive song characteristic.

Voice. Tape-recorded (32, 57, C, CART, ERA, LEM, ZIM). Song high-pitched, fast-paced, repetitive: 'heeyie heir eeyie heir ee weeyie' (Sierra Leone), 'tee air hrai eees' (Gabon), 'terwee chee chee chee chee cha-why' (Zaïre, Kivu) or 'ter-ter tweee ter ter churrri' (Uganda, Bwamba). The Kivu song has 2–6 'chee' notes. In W Africa sings a plaintive 4-note phrase 'sw-i-i-wi'. Call a harsh, grating 'kweg'; also a quiet but sharp, ratchet-like 'grrrek' (like alarm/roosting call of other robins). Pair sings in duet, ♂ giving trills and whistles and ♀ rasping and grating notes (Brosset and Erard 1986).

General Habits. Inhabits undergrowth of primary lowland rain forest, and in W Africa riverine forest and dense thickets in savanna. Ranges into transitional montane forest to 1400 m (E Zaïre); shuns dense secondary growth of ecotones and felled areas. Often occurs in thickets of Marantaceae, creeping and flitting about among the stems.

Solitary, in pairs or groups of 3–4 (family unit?). Wary and difficult to approach, often retreating 20–30 m away in dense cover, perching behind large leaf and stretching neck to peer at observer over it. Forages in dense thickets, on open leaf-covered ground and around mossy fallen tree trunks. Regularly attends army ant columns and makes short, flycatcher-like sallies to capture flying insects. Pair in Gabon accompanied army ant column for 100 min, hopping on low lianas and ground 10–30 m ahead of ants, tossing leaves and pecking at tiny insects (Willis 1986). Flight normally silent, but when several birds gather at driver ant column, there is much supplanting and chasing with whirring wings and bill-clicking, also calling and wing-flicking – wings are spread and raised vertically, often in time with final 'cha-why' of song. Also raises closed tail. Hops, but bird once seen to walk down sloping branch. Often perches on vertical stems. Roosts 1–2 m above ground, often on slender branch protruding from thicket. Sings at any time of day but especially at dawn and dusk, usually in undergrowth, from perch 1–2 m above ground.

Food. Small insects, including beetles, and their larvae, ants, termites, caterpillars and parasitic wasps.

Breeding Habits. Monogamous; solitary nester, territorial. Territory size in primary forest (estimated from number of singing birds) 6–7 ha, large for so small a bird (Gabon: Brosset and Erard 1986).

NEST: open cup of green moss on foundation of dead leaves, lined with fine fibrous strips or hair-like fungal mycelium strands or blackish rootlets; ext. diam. *c.* 90, int. diam. 65, depth of cup 38–40. Usually on ground at base of tree, often in hollow between bole and buttress root, also in hollow in upturned roots of fallen tree. One nest was 1 m above ground on top of decaying tree trunk and another on top of termite mound in what looked like refurbished nest of Fire-crested Alethe *Alethe diademata*.

EGGS: 2. Ovate to bluntly ovate; glossy; green to greenish blue with spots and blotches of reddish browns of varying hues and intensities, sometimes also with mauve and lilac, coalescing to form ring or cap at large end, or so dense as to mask ground colour. SIZE: (n = 4) 20·7–21·7 × 14·3–14·7 (21·2 × 14·5).

LAYING DATES: Liberia (breeding condition Apr, June, juvs Apr, Aug); Ghana, Feb; Nigeria, Mar; Cameroon, Feb, May, Sept (juvs June, Oct, Nov); (Bioko (flying young Aug); Central African Republic (breeding condition May–June); Gabon, Oct–Mar (juvs June, Aug–Sept); Zaïre, Aug–Jan (breeding condition Feb, Apr–July), probably all year; Uganda, Feb–Mar.

INCUBATION: by ♀ only, with nest absences of 20–50 min. Period: (n = 1, Gabon) 16 days.

DEVELOPMENT AND CARE OF YOUNG: young fed by both sexes.

BREEDING SUCCESS/SURVIVAL: most nests robbed by predators or (n = 3) parasitized by cuckoos (probably Dusky Long-tailed Cuckoo *Cercococcyx mechowi*). 1 ♂ lived 5 years, 1 ♀ ringed as juvenile lived 8 years.

Reference
Brosset, A. and Erard, C. (1986).

Genus *Sheppardia* Haagner

Small, with broad bill, well-developed rictal bristles, long, thin tarsi, fairly short tail; diagnostic character is combination of unicoloured tail and pale (white or cinnamon) tuft of erectile feathers over the eye. These feathers are normally overlain, except just in front of the eye where they form a small supraloral spot, but they are raised to form a bright pale stripe when birds are excited.

Inhabit forest and moist woodland, mainly in undergrowth; feed on ground, especially around army ants, also flycatch from perch.

Endemic, mostly in tropics; 8 species, in 2 superspecies, 1 containing *S. montana* and *S. lowei* and the other containing the rest.

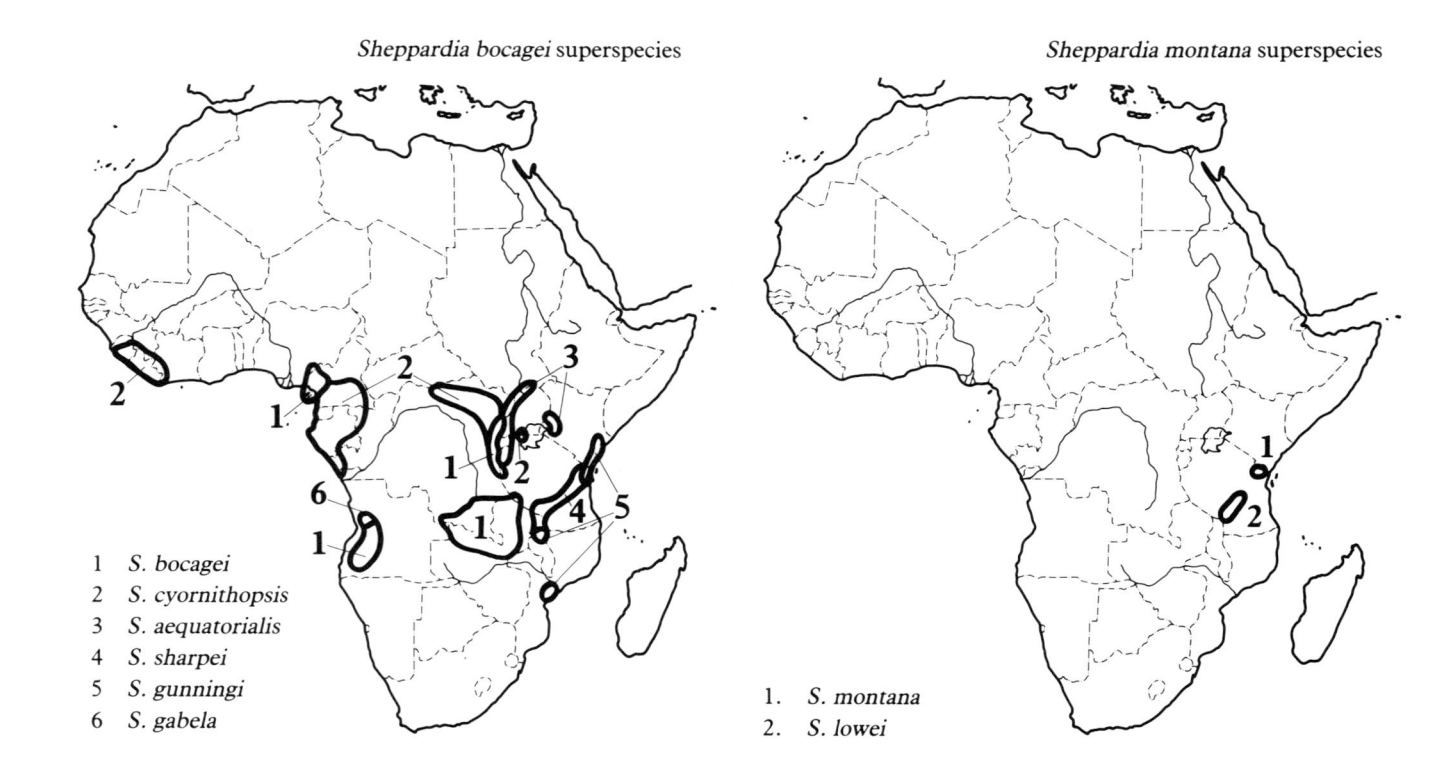

Shepppardia bocagei superspecies

1 *S. bocagei*
2 *S. cyornithopsis*
3 *S. aequatorialis*
4 *S. sharpei*
5 *S. gunningi*
6 *S. gabela*

Sheppardia montana superspecies

1. *S. montana*
2. *S. lowei*

Sheppardia bocagei (Finsch and Hartlaub). Bocage's Akalat. Akalat à joues rousses.

Plates 22, 26
(Opp. pp. 353, 401)

Cossypha bocagei Finsch and Hartlaub, 1870. Vög. Ost-Afr., p. 284; Mossamedes Province, Angola.

Forms a superspecies with *S. cyornithopsis*, *S. aequatorialis*, *S. sharpei*, *S. gunningi* and *S. gabela*.

Range and Status. Endemic resident. Distribution disjunct. Highlands of SE Nigeria (Obudu Plateau), Cameroon (Mt Nlonako, Mt Kupé, Rumpi Hills, Mt Cameroon) and Bioko; Zaïre in E (mts west of L. Edward, Itombwe, Mt Kabobo) and SE (Katanga: Haut Lomami, Lualaba, Haut Luapula, Upemba Nat. Park); Tanzania in W (Mahari Mt) and extreme SW (Chito Hill, Kitungulu); E Angola (highlands of Cuanza Sul and Huambo, escarpment from E Benguela to Huila and Leba); Zambia (Northern and Luapula provinces south to Muchinga Escarpment, Copperbelt, Northwestern Province south to Mombwezhi R., sight record from south of Kasempa). Frequent to locally abundant; density in Zambia, 3–4 birds in forest patch of <1 ha.

Description. *S. b. bocagei* Finsch and Hartlaub: Angola. ADULT ♂: forehead, crown, nape and lores dark grey; sides of neck and entire underparts rusty orange, except centre of belly white; mantle, scapulars and rump tawny-olive; uppertail-coverts tawny; tail tawny-brown; primaries and secondaries dull brown, edged olive on outer webs; outer webs of primary-coverts olive, inner webs dull brown; other wing-coverts and tertials with outer webs tawny-olive, inner webs dull brown; underwing-coverts and axillaries rusty orange. Bill black; eye dark brown; legs pinkish grey. Sexes alike. SIZE: (3 ♂♂, 2 ♀♀) wing, ♂ 80–82 (80·7), ♀ 73, 76; tail, ♂ 69–72 (70·0), ♀ 60, 63; bill, ♂ 16–17 (16·7), ♀ 15, 16; tarsus, ♂ 27–28 (27·7), ♀ 25, 25.

IMMATURE: of this race unknown; see *S. b. granti*.
NESTLING: unknown.

S. b. chapini (Benson): Zambia, SE Zaïre. Crown with slight olive wash; upperparts less tawny; underparts slightly paler. SIZE: ♂ (n = 7) wing 80–82 (81·1), tail 61–69 (65·1), bill 16–17 (16·3), tarsus 25–27 (25·9).

S. b. ilyai (Prigogine): W Tanzania, east of Mt Kungwe. Like *chapini* but crown more olive-grey; flanks washed olive. SIZE: (1 ♂, 2 ♀♀) wing, ♂ 83, ♀ 75, 76; tail, ♂ 61, ♀ 54, 57; bill, ♂ 16, ♀ 15, 16; tarsus, ♂ 25, ♀ 24, 24.

S. b. kaboboensis (Prigogine): Mt Kabobo (E Zaïre). Like nominate but has black line under eye; upperparts less tawny; central rectrices duller brown. Smaller? SIZE: (1 ♀) wing 70; tail 53; tarsus 26.

S. b. kungwensis (Moreau): Mt Kungwe (W Tanzania). Like *kaboboensis* but crown olive-grey, more olive than *chapini*; lores and ear-coverts slate-grey; black line under eye narrower; whole tail duller than nominate. Smaller than nominate. SIZE: wing, ♂ (n = 3) 71–78 (74·7), ♀ (n = 3) 69–76 (73·3); tail, ♂ (n = 3) 52–58 (55·3), ♀ (n = 3) 50–58 (54·7); bill, ♂ (n = 1) 17, ♀ (n = 3) 17 (17·0); tarsus, ♂ (n = 3) 25–27 (26·3), ♀ (n = 3) 26–27 (26·7).

S. b. schoutedeni (Prigogine): mountains west of L. Edward to Kivu. Crown slaty black without olive tinge.

S. b. granti Serle: W Cameroon. Very like *kungwensis*, including line below eye (but this paler in some individuals); tail darker red-brown; legs paler. Smaller than nominate, size similar to *kungwensis*. SIZE: wing, ♂ (n = 6) 69–74 (72·0), ♀ (n = 5) 66–72 (67·8); tail, ♂ (n = 5) 51–56 (53·6), ♀ (n = 5) 45–54 (47·8); bill, ♂ (n = 5) 16–17 (16·6), ♀ (n = 5) 16–17 (16·4); tarsus, ♂ (n = 5) 24–25 (24·6), ♀ (n = 5) 22–24 (23·6). WEIGHT: (Cameroon) 1 ♂ 17·5, unsexed (n = 16) 16·4–20·9 (18·3).

IMMATURE: as adult except upperparts black spotted with rufous; underparts black spotted with rufous-buff; centre of belly dirty white. Spots remain longest on breast, neck, nape and head.

S. b. poensis (Alexander) (='*insulana*' Grote): Bioko. As *granti* but crown darker, black with olive wash; lores, ear-coverts and line below eye darker, slate; underparts richer rufous. SIZE: (2 ♂♂, 2 ♀♀) wing, ♂ 72, 75, ♀ 65, 66; tail, ♂ 51, 54, ♀ 44, 45; bill, ♂ 16, 17, ♀ 15, 16; tarsus, ♂ 25, 27, ♀ 23, 23.

TAXONOMIC NOTE: as noted by Hall and Moreau (1970), there are 2 groups of populations in this species. Southern birds (*bocagei* group), from Angola, Zambia and SE Zaïre, live in lowland evergreen forest and are morphologically different from northern birds (*insulana* group), which are confined to montane forest. Prigogine (1987) believed these differences (wing and tail length, wing–tail ratio, colour of lower eyelid, degree of contrast between head and back colour) warranted the classification of northern birds as a full species, *Cossypha insulana*. We prefer to wait until the biology of the birds is better known before taking such a step, and believe like Moreau and Benson (1956), White (1962) and Hall and Moreau (1970) that the morphological differences between the 2 groups of populations are at the subspecies level.

Field Characters. A fairly large akalat, very similar in appearance to slightly larger Grey-winged Robin-Chat *Cossypha polioptera* (bright orange underparts, uniform rufous tail), but latter has prominent white supercilium, black stripe through eye and loud robin-chat song. Although marginally sympatric in NW Zambia and E Zaïre (Mt Kabobo), the 2 species do not occur in the same forests. For differences from Mountain Robin-Chat *Cossypha isabellae*, see that species.

Voice. Tape-recorded (57, 86, CART). Song a simple refrain of 7–10 high-pitched sweet whistles, last few notes highest: 'deee da, doe-doo-doo-doo, dee da dee', or 'my pretty peggy-lee, pretty bo-peep'. Other calls include quiet twittering and a sibilant ratchet note reminiscent of Dusky Flycatcher *Muscicapa adusta*. Alarm, 1–3 high-pitched 'seeep' notes.

General Habits. Inhabits transitional and montane evergreen forest, sometimes bamboo, 600–1700 m in Cameroon, 1250–2450 m in E Zaïre, 1000–2100 m in Angola, 1300–2400 m in W Tanzania. In N Zambia and S Zaïre, frequents gallery forest along drainage lines in *Brachystegia* woodland at 1300–1500 m, and abundant in 'mushitu' (swamp forest with dense sapling growth).

Often in pairs, communicating with soft, sibilant calls. Feeds extensively on ground, dropping from low perch to forage in leaf litter. Attends ant columns. In half light, especially after sunset, leaves forest to feed in low *Afromomum* growth in surrounding woodland, giving twittering contact calls. Mainly sedentary but young bird at 30 m at foot of Mt Cameroon was presumably post-breeding wanderer.

Food. Insects, especially wasps (Hymenoptera, including Braconidae); also beetles (Tenebrionidae, Staphylinidae), ants (*Camponotus*, *Pheidole*), flies (Diopsidae, Tipulidae), bugs (Psyllidae, Reduviidae), termites, cockroaches, moths, lacewings (Mantispidae) and spiders. 1 bird ate at least 20 small wasps.

Breeding Habits.
NEST: only 1 described: cup of moss lined with black vegetable fibres, sited 0·75 m above ground in hollow stump.
EGGS: 2. Cream, thickly clouded with reddish brown and ash-grey. SIZE: 21–22 × 16.
LAYING DATES: Cameroon, Dec–Mar; S Zaïre, Nov–Feb; E Zaïre, Jan–May; Zambia, Nov (breeding condition Sept, Nov; spotted juv. Jan); Angola (breeding condition Aug–Sept).

Reference
White, C.M.N. (1945).

Plates 22, 26
(Opp. pp. 353, 401)

Sheppardia cyornithopsis (Sharpe). Lowland Akalat. Merle rougegorge.

Callene cyornithopsis Sharpe, 1901. Bull. Br. Orn. Club 12, p. 4; Efulen, Cameroon.

Forms a superspecies with *S. bocagei*, *S. aequatorialis*, *S. sharpei*, *S. gunningi* and *S. gabela*.

Range and Status. Endemic resident. S Guinea (borders of Sierra Leone and Liberia), N Liberia, E Ivory Coast (Mt Nimba to Tai Nat. Park and Cavally R.); Cameroon east of Sanaga R., and 1 record from between Foumban and Banyo; Gabon; S Congo (Mayombe); N Zaïre east to Ituri District, south in E lowlands to Itombwe; Uganda (Bwamba, lakeshore forest at Sango Bay, Malabigambo Forest) and nearby Minziro Forest in NW Tanzania. Common to locally abundant.

Description. *S. c. houghtoni* Bannerman: Sierra Leone, Liberia. ADULT ♂: entire upperparts, lores, ear-coverts, scapulars and sides of breast olive, slightly darker on crown (crown feathers edged darker, giving scalloped appearance); uppertail-coverts fringed tawny; tail-feathers dull brown with broad tawny fringes; belly white; remainder of underparts rusty orange, paler on chin and throat; primaries and secondaries dull brown, outer webs fringed olive, inner webs fringed buff; tertials and wing-coverts olive-brown; underwing-coverts and axillaries pale rusty orange. Bill black; eye brown; legs grey. Sexes alike. SIZE: (9 ♂♂, 3 ♀♀) wing, ♂ 72–79 (74·9), ♀ 65–69 (67·3); tail, ♂ 55–61 (58·2), ♀ 48–53 (51·0); bill, ♂ 15–17 (15·9), ♀ 14–16 (15·0); tarsus, ♂ 24–26 (24·9), ♀ 22–23 (22·3). WEIGHT: (Nimba, Liberia, May–Sept) ♂ (n = 9) 18–21·2 (19·6), ♀ (n = 2) 17·2, 17·3.
IMMATURE: as adult except upperparts, face, chin, throat, breast and wing-coverts dark brown with rusty spots, the spots

remaining longest on head; belly buff, scalloped dark brown.

S. c. cyornithopsis (Sharpe): S Cameroon, Gabon. Upperparts slightly browner; flanks washed with olive; undertail-coverts paler, even white, but no consistent difference in shade of other orange areas of underparts.

S. c. lopezi (Alexander): E Zaïre, W Uganda. As nominate but flanks olive-brown. WEIGHT: (Uganda) ♂ (n = ?) 14–17, ♀ (n = ?) 12–17.

Field Characters. A small robin with orange underparts (except white belly) and rusty rump and tail. Western birds (nominate race) almost indistinguishable from Equatorial Akalat *S. aequatorialis*; however, eastern race *lopezi*, which just overlaps Equatorial Akalat in E Zaïre, is somewhat smaller, with orange confined to throat and breast, flanks brown, and sometimes a brown band separating orange breast from white belly. Best distinguished from Equatorial Akalat by voice and habitat (lowland *vs* highland forest).

Voice. Tape-recorded (CHA). Song varied, a mixture of chuckles, churrs and whistles, usually starting low and ending on 2 high notes, e.g. 'chew-chew-chuckle-twaa-twaa-gurrrr-tweee-tweee', lasting 2–3 s; somewhat reminiscent of Nightingale *Luscinia megarhynchos* and quite unlike simple 1-note song of Equatorial Akalat. Identification of this song is questionable (Chappuis 1975); it is certainly very different from songs of other akalats. Call a short whistled 'piee'; alarm a low 'trrr'.

General Habits. Inhabits undergrowth of primary forest, secondary forest (especially where *Musanga* and *Afromomum* are dominant), forest–grassland mosaic (Liberia) and seasonal swamp forest. Lowlands, sometimes lower montane forest, to 900 m in Liberia, 1500 m in Zaïre, 1200 m in Uganda.

Occurs singly or in pairs; sometimes joins mixed-species flocks. Foraging bird in Liberia perched 0·7–2 m above ground, making flycatching sallies in air or dropping to ground; sometimes sat motionless on perch for several min; wary but curious: approached within 2 m of person, inspecting him while giving subdued low 'trrrr' note indicating caution, and lifting tail (S. Keith, pers. obs.). 1 bird accompanied column of safari ants *Dorylus wilverthi* for 60 min (Willis 1986). At dusk, emerges into forest clearings; occasionally perches on roots or stumps next to road (Zaïre: Chapin 1953).

Food. Insects: beetles, ants, termites, caterpillars, small orthopterans; also spiders and small millipedes.

Breeding Habits.
NEST: only one described (Prigogine 1984). Shaped like a purse, of moss, twigs and black fibres, lined with black fibres; diam. 90–100, ext. depth 130, int. depth 80, depth below side entrance 55; placed 0·5 m above ground in dense undergrowth, attached to 3 twigs.

EGGS: 2 (only 1 clutch known). 1 egg brownish rose lightly spotted with red around large end, the other olive-beige with brown spotting, darker around large end. SIZE: 21·2–21·6 × 15·3.

LAYING DATES: Liberia (♂♂ in breeding condition June–Sept, recently-fledged juv. Oct); Gabon (juv. attended by parents mid-Dec); Zaïre, Aug–May; Uganda (breeding condition and juvs Jan–Feb).

INCUBATION: by ♀ only.

References
Brosset, A. and Erard, C. (1986).
Prigogine, A. (1984).

Sheppardia aequatorialis (Jackson). Equatorial Akalat. Merle rougegorge équatorial.

Plates 22, 26
(Opp. pp. 353, 401)

Callene aequatorialis Jackson, 1906. Bull. Br. Orn. Club 16, p. 46; Kericho, Lumbwa, W Kenya.

Forms a superspecies with *S. bocagei*, *S. cyornithopsis*, *S. sharpei*, *S. gunningi* and *S. gabela*.

Range and Status. Endemic resident, S Sudan (Imatong Mts); highlands of E Zaïre (Lendu Plateau, Kivu between L. Edward and L. Kivu, Itombwe); W Rwanda, W Burundi; SW Uganda (Impenetrable, Kalinzu and Kibale forests); W Kenya (Elgon, Nandi and Kakamega to Sotik, Molo and Mau Forest). Common to locally abundant.

Description. *S. a. aequatorialis* Jackson (including '*pallidigularis*'): range of species except *acholiensis*. ADULT ♂: top of head olive-brown; mantle, back, scapulars, lesser and median coverts brighter olive-brown, lower back with cinnamon wash, rump and uppertail-coverts tawny. Tail-feathers dark red-brown narrowly edged tawny. Pre-orbital region dark grey, supercilium, around eye and small patch behind eye dull bluish grey, ear-coverts olive-brown. Belly white, rest of underparts brownish orange, paler on chin, throat and undertail-coverts.

Primaries, secondaries and greater coverts dark brown edged cinnamon, axillaries and underwing-coverts pale orange, inner edges of underside of primaries and secondaries whitish or pale buff. Bill blackish brown; eye dark brown; legs and feet dull light olive, pale greenish, greyish horn or olive-grey, soles yellowish, claws olive-grey or pale greenish. Sexes alike. SIZE: (8 ♂♂, 6 ♀♀) wing, ♂ 67–73 (69·9), ♀ 63–68 (66·0); tail, ♂ 45–53 (50·0), ♀ 42–49 (44·5); bill, ♂ 13–15 (14·0), ♀ 13·5–14 (13·8); tarsus, ♂ 20–22·5 (21·6), ♀ 20·5–22 (21·3). WEIGHT: (Kenya/Uganda) ♂ (n = 7) 13–17 (14·9), ♀ (n = 7) 12–16 (14·5). Diurnal weight change 10·9–20·1% (13·9%) (Kenya: Mann 1985).

IMMATURE: very dark; head blackish, streaked pale orange; upperparts and upperwing-coverts dark brown with large pale orange spots; rump and uppertail-coverts tawny, unspotted; tail like adult. Pre-orbital region black; throat feathers pale buff edged with black; breast and sides of belly heavily mottled black and pale buffy orange, flanks washed with orange, lower belly and vent off-white, feathers dark at base and narrowly fringed with black; undertail-coverts pale orange, unspotted.

S. a. *acholiensis* Macdonald: S Sudan (Imatong Mts). Distinguished from other races by olive-brown (not russet-brown) head and mantle.

TAXONOMIC NOTE: formerly considered a subspecies of *S. cyornithopsis* (e.g. Chapin 1953), but the 2 are sympatric in E Zaïre (Prigogine 1971).

Sheppardia aequatorialis

Field Characters. A small robin with orange underparts (except white belly) and rusty rump and tail. Plumage unpatterned, dull blue-grey face without white spots or stripes, tail uniform, without black. In flight, rusty tail looks bright, almost orange, against background of forest undergrowth. Similar-sized and sympatric White-bellied Robin-Chat *Cossyphicula roberti* also has orange throat and breast and white belly but no orange on flanks, and further differs in black-and-red tail, white loral stripe, and voice. Flycatcher-like; makes short sallies and often returns to same perch where it periodically flicks its wings. Common but unobtrusive, easily overlooked by observer unfamiliar with its monotonous single-note song. For distinctions from very similar Lowland Akalat *S. cyornithopsis*, see that species.

Voice. Tape-recorded (57, GREG, HOR, STJ). Song insignificant for a robin, a single low, mournful, quavering note, 'prrru', 'prruyu' or 'prrrer', repeated at intervals of 1–2 s. Aggressive calls in excitement or in answer to tape, 5–6 note 'wer-di-di-dit ... cho', with pause before last note which is lower in pitch; also a dry 'tchac' (Dowsett-Lemaire 1990). Alarm, especially when young nearby, a dry, guttural rattle like call of Common Stonechat *Saxicola torquata*. Soft 'whet' or 'whit' while foraging or in flight.

General Habits. Inhabits mainly montane forest; 1800 m in Sudan, 1150–1870 m in Zaïre, 1100–1200 m in Burundi, up to 2430 m in Rwanda (commoner below 2000 m), 1600–2500 m in E Africa. In mature forest avoids areas of dense understorey, preferring mid-strata and open lower canopy. In dry season occurs in dense thickets of regenerating forest in lava fields north of L. Kivu, Zaïre (altitudinal migrants?).

Solitary, although several may gather at ant swarms. Forages mainly on or near ground; perches on saplings 1–1·5 m above ground and drops onto prey; catches insects in air and gleans them from logs and trunks; sometimes flycatches in canopy. Attends ant swarms regularly. Usually keeps to shade during day, venturing into open glades and onto paths only at dusk and dawn; however, observed by path in bright sunlight in mid-morning, making short sallies for insects in low vegetation (Kakamega, Kenya: S. Keith, pers. obs.).

Food. Mainly insects: moths, wasps, flies, beetles, ants (but not doryline army ants), orthopterans and hemipteran bugs; also spiders and small millipedes.

Breeding Habits. Monogamous; solitary nester.

NEST: only 3 described (Chapin 1978, Kalina and Butynski 1989; T. Butynski, pers. comm.); small cup of decayed leaves and green moss lined with fine fibres, sunk into ground on leaf-strewn slope or 0·5 m above ground in buttress of large tree; diam. of cup 60, depth of cup 57.

EGGS: 2. Pale green, speckled with reddish brown, thinly at small end, heavily at large end where markings form cap. SIZE: (n = 2) 21·8–21·9 × 15·3.

LAYING DATES: Sudan, Jan, Apr, Nov; Zaïre, Oct–May; Uganda (breeding condition Apr); Kenya (juvs June, breeding condition June–July).

BREEDING SUCCESS/SURVIVAL: ringed bird in Kakamega, Kenya, lived at least 7 years 9 months (Mann 1985).

References
Chapin, R. T. (1978).
Kalina, J. and Butynski, T. M. (1989).
Oatley, T. B. (1961).

Plates 22, 26
(Opp. pp. 353, 401)

Sheppardia sharpei (Shelley). **Sharpe's Akalat. Akalat de Sharpe.**

Callene sharpei Shelley, 1903. Bull. Br. Orn. Club 13, p. 60; Masisi Hill, N Nyasaland.

Forms a superspecies with *S. bocagei*, *S. cyornithopsis*, *S. aequatorialis*, *S. gunningi* and *S. gabela*.

Range and Status. Endemic resident. Restricted to E and SW Tanzania (Usambaras, Ngurus, Ulugurus, Dabaga Highlands, Poroto Mts, Mt Rungwe and Umalila) and N Malaŵi (eastern escarpment of Nyika Plateau, Musisi). Frequent to locally common; population of 25–30 pairs in 75-ha Manyenjere Forest, Malaŵi.

Description. *S. s. sharpei* (Shelley): Malaŵi, S Tanzania. ADULT ♂: entire upperparts brownish olive, browner on uppertail-coverts; supercilium creamy white; tail dull brown; chin, throat and sides of neck rusty orange; breast and flanks paler rusty orange, heavily washed olive; belly white; undertail-coverts pale orange-buff; remiges and wing-coverts dull brown, fringed olive on outer webs, buff on inner webs; underwing-coverts and axillaries pale rusty orange. Bill black; eye sepia; legs pale purplish grey. Sexes alike. SIZE: wing, ♂ (n = 2) 66, 69, ♀ (n = 3) 62–66 (63·7); tail, ♂ (n = 2) 52, 57, ♀ (n = 3) 45–50 (48·0); bill, ♂ (n = 1) 14, ♀ (n = 2) 14, 15; tarsus, ♂ (n = 2) 22, 23, ♀ (n = 3) 21–22 (21·7).

IMMATURE: as adult except upperparts and wing-coverts black, streaked rusty orange; no supercilium; underparts buff, more orange on breast, scalloped black. Streaks remain longest on ear-coverts.

S. s. usambarae MacDonald: Usambara and Nguru Mts, Tanzania. Orange of underparts paler, orange-buff; supercilium less prominent, greyish white; white of belly extends further onto breast. Immature with spots on upperparts paler, broader than on nominate immature. SIZE: (5 ♂♂, 2 ♀♀) wing, ♂ 68–73 (70·6), ♀ 65, 66; tail, ♂ 53–59 (55·2), ♀ 49, 51; bill, ♂ 14–16 (15·0), ♀ 14, 15; tarsus, ♂ 21–22 (21·8), ♀ 20, 21. WEIGHT: (Tanzania) ♂ (n = 3) 14·5–15 (14·8), ♀ (n = 3) 12–14·2 (12·7).

Field Characters. A rather small akalat, confined to highlands of Tanzania and Malaŵi. Orange of underparts rather dull and washed-out compared to, for example, Equatorial Akalat *S. aequatorialis*. Tail same colour as back, not contrastingly rufous; white loral line continues to above eye. Usambara and Iringa Akalats *S. montana* and *S. lowei* are larger, grey or brown without orange.

Voice. Tape-recorded (FAR, LEM, STJ). Song a simple thin whistle, 'chee-chiddely chiddely' or variant. Warning call a series of high-pitched, metallic 'pink' notes ending with a guttural 'cheeyurr-churrr'. Birds chasing one another gave long series of mixed 'eek' and 'jazz jazz jazz' notes, also subsong 'hee hi see weeyou' or 'see heh chiwee chu' (Willis 1985).

General Habits. Inhabits dense undergrowth of montane forests and bamboo, especially those regularly in cloud; altitudinal range 600–2600 m in E Africa, 1500–2100 m in Malaŵi.

An elusive, little-known species which keeps to thick cover, but (like other akalats) comes into open to feed on paths and roads, especially after sunset. Forages mainly on or near ground; one pair foraged cooperatively, one bird turning over leaf litter, the other pouncing on exposed prey (Sclater and Moreau 1933). Pair followed army ant column (Zambia) sallying from low perches to the ground, also to foliage. Stance rather upright; raises closed tail; snaps bill (Willis 1985); similar foraging observed in Malaŵi (Dowsett-Lemaire 1983).

Some cold season altitudinal movements in Tanzania: descends escarpment of E Usambaras to Amani-Sigi Forest at 600 m.

Food. Insects.

Breeding Habits. In courtship display one bird sat on twig while other hovered above it 'uttering thin robin-like squeaks' (Moreau and Moreau 1937).

NEST: deep, open cup of dead leaves, lined with leaf midribs and skeletal leaf fragments; int. diam. c. 45–50, depth c. 45–50. Commonly sited in coppice growth on lopped sapling or stump, or among root sprouts of trees (e.g. *Parinari*, *Ocotea*), usually 1–1·3 m above ground. Built by ♂ and ♀; one bird (probably ♂) collects material, mate takes it, with wings fluttering, and incorporates it into nest.

EGGS: 2. Pale pink-buff ground colour with dark red-brown markings concentrated at large end, often forming band; markings may fade perceptibly during incubation. SIZE: (n = 4, Tanzania) 19–21 × 14–16 (20·3 × 15·0).

LAYING DATES: E Africa: Region C, Oct–Nov; Region D, Dec (♂ and ♀ in breeding condition N Tanzania Feb, S Tanzania Sept–Nov); Malaŵi, Oct–Dec (Oct 5, Nov 1, Dec 1 clutches: Dowsett and Dowsett-Lemaire 1984).

BREEDING SUCCESS/SURVIVAL: parasitized by Barred Long-tailed Cuckoo *Cercococcyx montanus*.

References
Moreau, R. E. and Moreau, W. M. (1937).
Sclater, W. L. and Moreau, R. E. (1933).

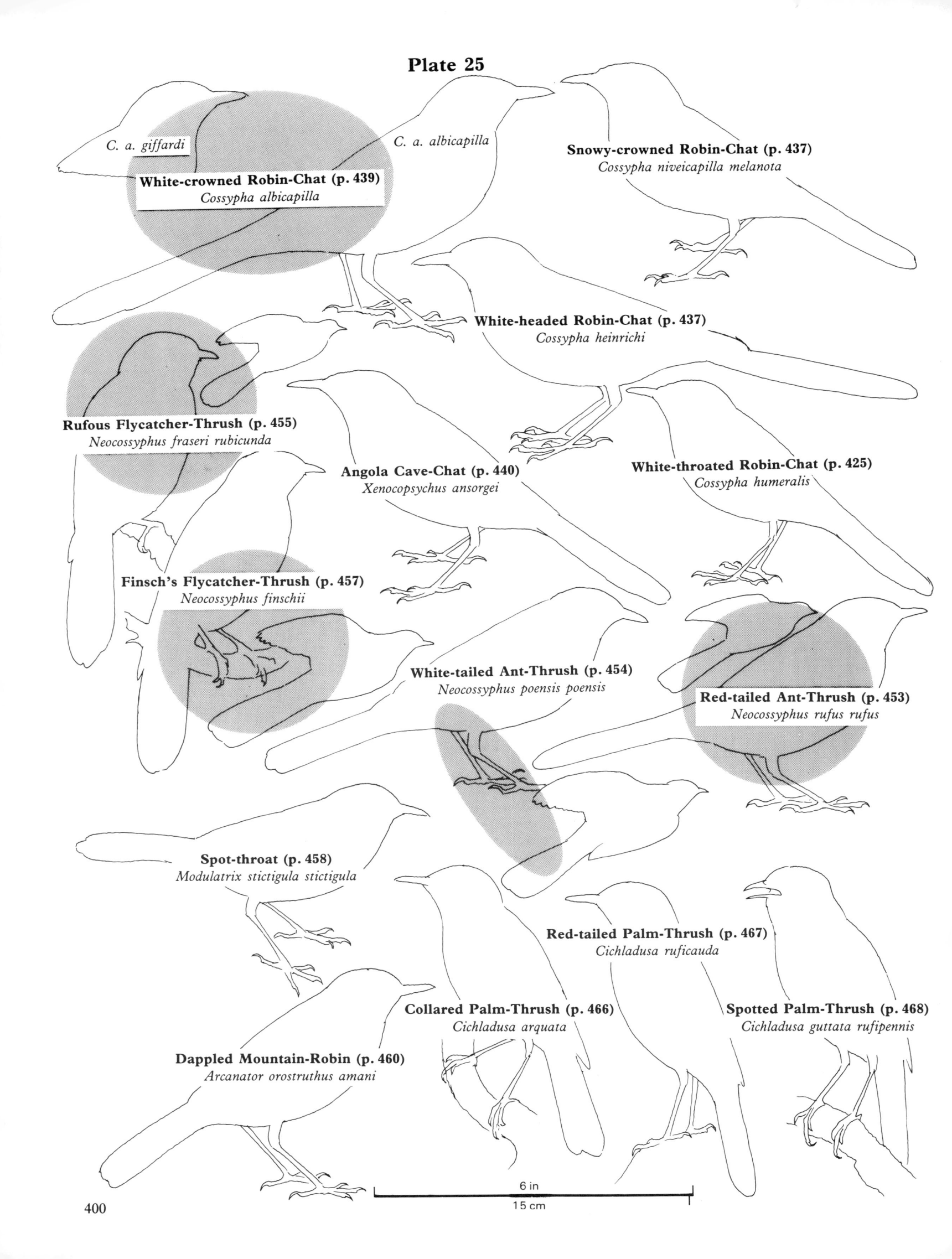

Plate 25

C. a. giffardi

C. a. albicapilla

White-crowned Robin-Chat (p. 439)
Cossypha albicapilla

Snowy-crowned Robin-Chat (p. 437)
Cossypha niveicapilla melanota

White-headed Robin-Chat (p. 437)
Cossypha heinrichi

Rufous Flycatcher-Thrush (p. 455)
Neocossyphus fraseri rubicunda

Angola Cave-Chat (p. 440)
Xenocopsychus ansorgei

White-throated Robin-Chat (p. 425)
Cossypha humeralis

Finsch's Flycatcher-Thrush (p. 457)
Neocossyphus finschii

White-tailed Ant-Thrush (p. 454)
Neocossyphus poensis poensis

Red-tailed Ant-Thrush (p. 453)
Neocossyphus rufus rufus

Spot-throat (p. 458)
Modulatrix stictigula stictigula

Red-tailed Palm-Thrush (p. 467)
Cichladusa ruficauda

Collared Palm-Thrush (p. 466)
Cichladusa arquata

Spotted Palm-Thrush (p. 468)
Cichladusa guttata rufipennis

Dappled Mountain-Robin (p. 460)
Arcanator orostruthus amani

6 in
15 cm

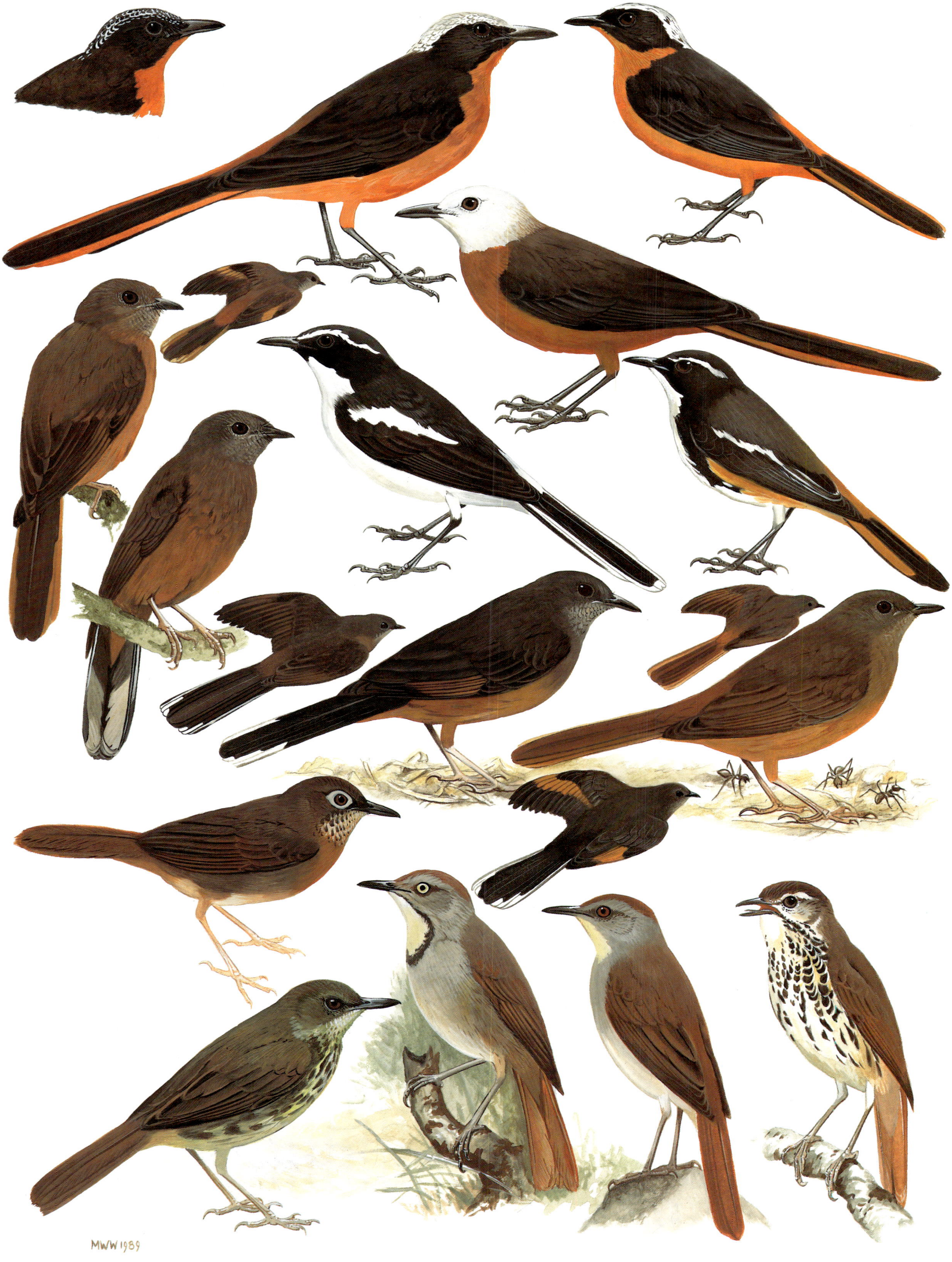
MWW 1989

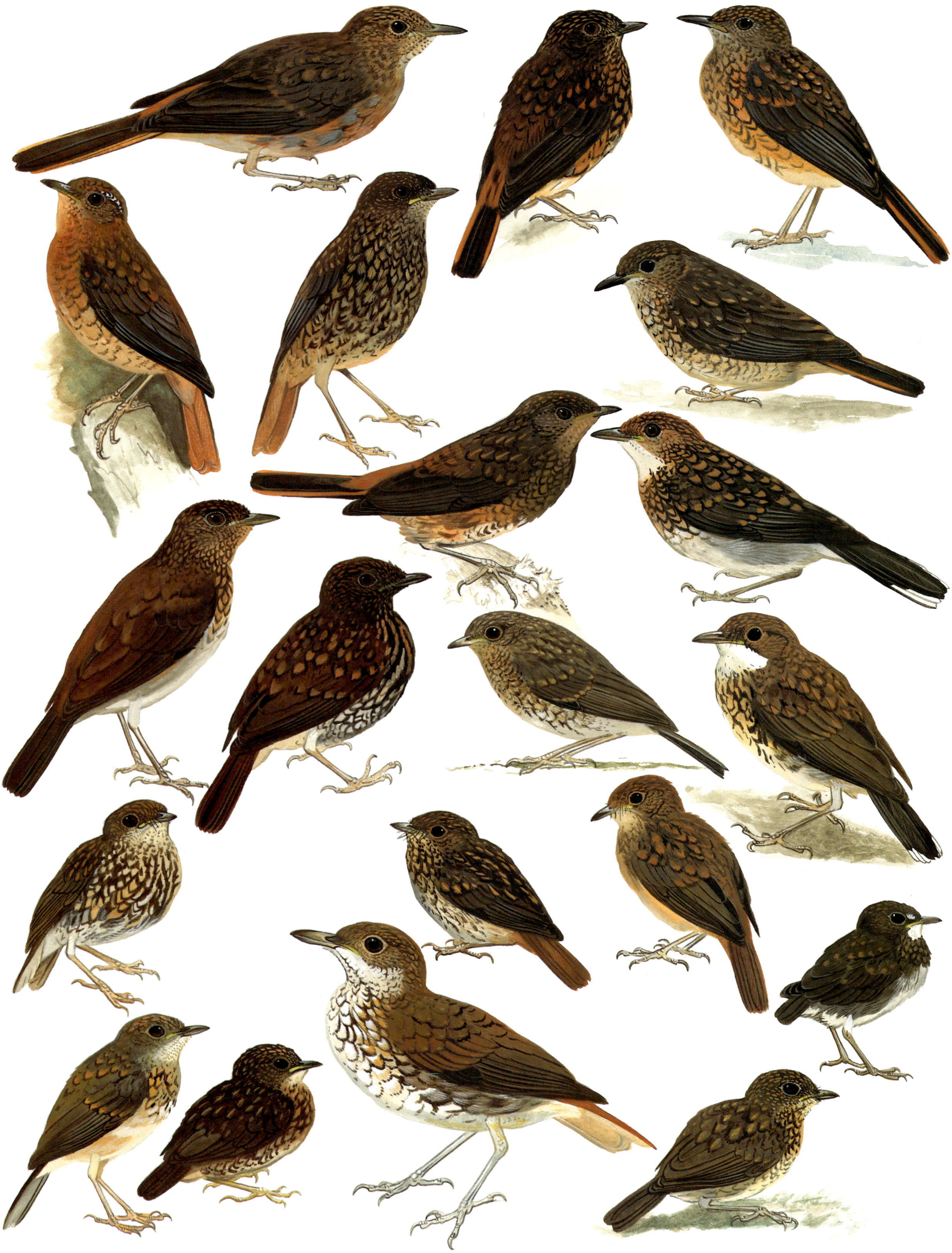
MWW1989

Plate 26

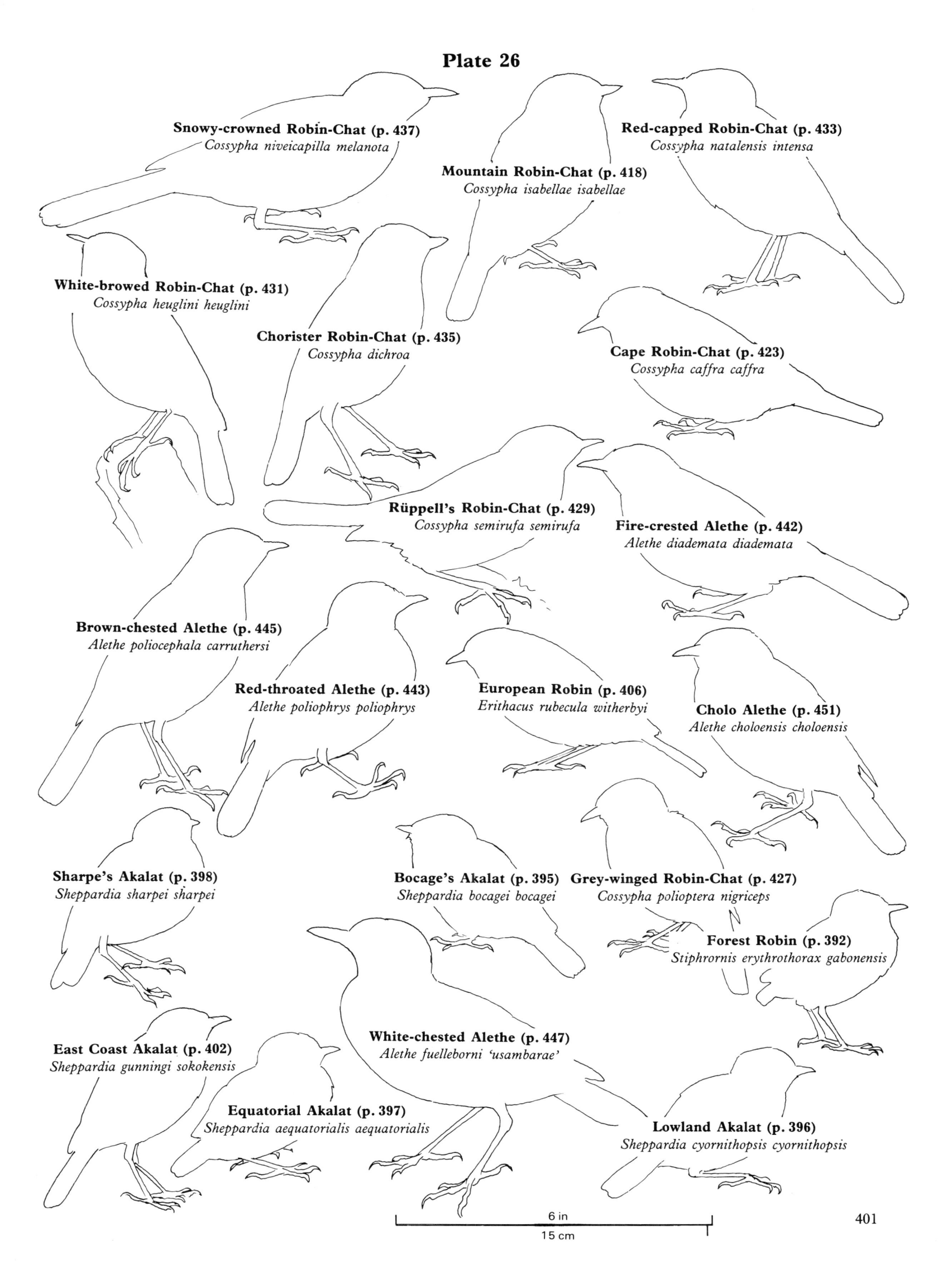

Snowy-crowned Robin-Chat (p. 437)
Cossypha niveicapilla melanota

Red-capped Robin-Chat (p. 433)
Cossypha natalensis intensa

Mountain Robin-Chat (p. 418)
Cossypha isabellae isabellae

White-browed Robin-Chat (p. 431)
Cossypha heuglini heuglini

Chorister Robin-Chat (p. 435)
Cossypha dichroa

Cape Robin-Chat (p. 423)
Cossypha caffra caffra

Rüppell's Robin-Chat (p. 429)
Cossypha semirufa semirufa

Fire-crested Alethe (p. 442)
Alethe diademata diademata

Brown-chested Alethe (p. 445)
Alethe poliocephala carruthersi

Red-throated Alethe (p. 443)
Alethe poliophrys poliophrys

European Robin (p. 406)
Erithacus rubecula witherbyi

Cholo Alethe (p. 451)
Alethe choloensis choloensis

Sharpe's Akalat (p. 398)
Sheppardia sharpei sharpei

Bocage's Akalat (p. 395)
Sheppardia bocagei bocagei

Grey-winged Robin-Chat (p. 427)
Cossypha polioptera nigriceps

Forest Robin (p. 392)
Stiphrornis erythrothorax gabonensis

East Coast Akalat (p. 402)
Sheppardia gunningi sokokensis

White-chested Alethe (p. 447)
Alethe fuelleborni 'usambarae'

Equatorial Akalat (p. 397)
Sheppardia aequatorialis aequatorialis

Lowland Akalat (p. 396)
Sheppardia cyornithopsis cyornithopsis

6 in
15 cm

Sheppardia gunningi Haagner. East Coast Akalat. Akalat de Gunning.

Plates 22, 26
(Opp. pp. 353, 401)

Sheppardia gunningi A.K. Haagner, 1909. Ann. Tvl. Mus. 1, 3, p. 180; Mzimbiti, near Beira, Mozambique.

Forms a superspecies with *S. bocagei*, *S. cyornithopsis*, *S. aequatorialis*, *S. sharpei* and *S. gabela*.

Range and Status. Endemic resident in 3 separate areas: (1) Kenya (lower Tana R. south from Makeri, Sokoke Forest, Rabai, Shimba Hills and Shimoni) and NE Tanzania (Pugu Hills); (2) Malaŵi west of L. Malaŵi in Nkhata Bay District, from Choma Mt south to Kuwilwe and west to near Chikangawa; (3) Cheringoma Plateau of coastal Mozambique between Beira and Zambezi R. including Chineziwa and Inhamitanda Forest. A 1932 sight record from Netia, N Mozambique coast. Common; density of *c.* 6 pairs in 7·5 ha forest on Choma Mt, Malaŵi. Total Malaŵi population probably over 3000 pairs.

Description. *S. g. gunningi* Haagner: Mozambique. ADULT ♂: entire upperparts from crown to uppertail-coverts bronze-brown. Tail greyish brown. Lores and supercilium bluish grey, ear-coverts olive-brown; chin and throat light orange-yellow; flanks olive-yellow; belly white, undertail-coverts buff. Primaries, secondaries and tertials brown, outer webs of primaries greyish; upperwing-coverts bluish grey, underwing-coverts pale orange. Bill blackish brown, base of lower mandible yellowish; eye dark brown; legs and feet greyish pink, soles primrose yellow. Sexes alike. SIZE: wing, ♂ (n = 13) 72–76 (74·1), ♀ (n = 4) 65–68 (67·2); tail, ♂ (n = 12) 52–58 (54·3), ♀ (n = 4) 46–51 (48·6); bill, ♂ (n = 4) 15–17 (16·0), ♀ (n = 3) 15–16 (15·2); tarsus, ♂ (n = 2) 21, 22, ♀ (n = 2) 20, 20. WEIGHT: ♂ (n = 8) 17–19 (18·0); ♀ (n = 3) 16–17 (16·7).

IMMATURE: mottled; olive-brown upperparts and wing-coverts with pear-shaped buff spots centred on feather shafts; underparts buffy, feathers of throat and breast edged with reddish brown.

NESTLING: unknown.

S. g. sokokensis (van Someren): Kenya and Tanzania. More richly coloured than nominate race. Outer webs of primaries bluish grey; lores and supercilium bluer; underside from throat to breast rich orange, lower breast and belly white. Small. SIZE: wing, ♂ (n = 11) 67–72 (70·5), ♀ (n = 7) 63–72 (65·6). WEIGHT: ♂ (n = 2) 13, 16, ♀ (n = 4) 12–15 (13·5).

S. g. bensoni Kinnear: Malaŵi. Like *S. g. sokokensis* but white on underside restricted to belly. Large. SIZE: wing, ♂ (n = 8) 72–78 (74·8), ♀ (n = 3) 68–70 (69·0).

Field Characters. A small akalat of eastern lowland forests, with bright yellow-orange underparts, blue-grey supercilium and wings contrasting with olive cheeks and back, tail plain brown without rufous, and thin akalat song. Flicks wings and cocks tail like other robins. The only akalat in its range, and far smaller than sympatric robin-chats.

Voice. Tape-recorded (CART, FAR, GIB, HOR, LEM, McVIC, OAT). Song fast, simple, high-pitched: several 4- or 5-note phrases, lasting 1–2·5 s (av. 2 s), e.g. 'too-yoo-twee-yoo, tootoo-yoo-twee-yoo'. Number of notes varies but pattern and duration are characteristic. One bird, responding to playback of its song, uttered descending 3-syllable call like that of Red-chested Cuckoo *Cuculus solitarius*, but an octave higher. Alarm a single, piping 'seeep' in series, frequently interspersed with guttural clicks; not loud or far-carrying. Click sometimes rapidly repeated to form ratchet call very like that of sympatric Red-capped Robin-Chat *Cossypha natalensis*.

General Habits. Inhabits lowland forest; below 300 m in Kenya, Tanzania and Mozambique, and at 475–1750 m in Malaŵi. Favours moist valley bottoms and dense understorey; less common in drier, more open growth, e.g. *Cynomitra-Manilkara* and *Afzelia* forests (Kenya: Britton and Zimmerman 1979). Fond of mossy logs (Irwin 1963), and often in dense thickets formed around fallen forest tree.

Solitary; skulks; ♂ discloses presence by repetitive song during breeding season; sings all day in first half of rainy season, but only sporadically at dawn and dusk in dry season. Ventures into open glades, footpaths and tracks at dusk. Forages on ground, in undergrowth and mid-strata of forest, and exploits driver ant columns, usually perching near van of column and sallying down to snatch fleeing invertebrates. Feeds in association with Red-capped Robin-Chat *Cossypha natalensis* and Eastern Bearded Scrub-Robin *Cercotrichas quadrivirgata*.

May vacate high altitudes on Choma Mt (Malaŵi) in non-breeding season.

Food. Insects: beetles, ants, grasshoppers, moths, caterpillars, termites, crickets, plant bugs; also very small frogs (Mozambique, Malaŵi: Irwin 1963, Oatley 1970a).

Breeding Habits. Unknown.
LAYING DATES: probably Nov–Dec in most areas.

Kenya (juv. Apr); Tanzania (breeding condition Oct–Nov); Malaŵi, Dec (juv. Feb); Mozambique (♂♂ with fully enlarged testes and singing, Nov).

References
Benson, C. W. (1946).
Collar, N. J. and Stuart, S. N. (1985).

Sheppardia gabela (Rand). Gabela Akalat. Akalat de Gabela.

Muscicapa gabela Rand, 1957. Fieldiana Zool. 39, pp. 41–43; 15 km south of Gabela (= Amboim), Angola.

Forms a superspecies with *S. bocagei*, *S. cyornithopsis*, *S. aequatorialis*, *S. sharpei* and *S. gunningi*.

Plate 22
(Opp. p. 353)

Range and Status. Endemic resident. Known only from patches of secondary forest within 40 km of Gabela, Angola, at Londa (= Conda), Conde and Assango. Probably locally frequent, although whole population may be restricted to an area of <1000 km² (Hall and Moreau 1962).

Description. ADULT ♂: forehead and crown dark brownish olive, feathers edged darker on crown to give faint scaling; nape, hindneck and face brownish olive; lores and orbital ring buffy; chin and throat whitish, bordered by 15 mm wide brownish olive, mottled breast-band; mantle, back, rump and uppertail-coverts brownish olive, becoming rusty on rump and uppertail-coverts; tail brownish black with brownish olive edging to feathers. Lower breast, belly and undertail-coverts white; flanks brownish olive; thighs olive; wing-feathers brownish black, primaries with outer edges brownish olive, inner webs edged grey; upperwing-coverts brownish olive, underwing-coverts buffy white. Bill black; eye dark; legs and feet dark horn. Sexes alike. SIZE: (1 ♂, 1 ♀) wing, ♂ 66, ♀ 60·5; tail, ♂ 52, ♀ 46; bill, ♂ 14; tarsus, ♂ 22, ♀ 22.

Field Characters. A small akalat confined to a limited area around Gabela, Angola. Nondescript, brown above, white below with brown breast-band; tail without rufous. Only other small forest thrushes in its range are Bocage's Akalat *S. bocagei* (orange underparts), Brown-chested Alethe *Alethe poliocephala* (larger, with grey head, white eyebrow, rufous back) and Forest Scrub-Robin *Cercotrichas leucosticta* (patterned face, white in wings and tail).

Voice. Unknown.

General Habits. Inhabits primary forest, where undergrowth densest, and tangled growth of secondary forest. Often at edge of forest near clearings and coffee plantations, early and late in day. Gleans insect prey from leaves and branches of undergrowth. Remains motionless for long periods. Probably exclusively insectivorous, although actual food unknown.

Breeding Habits. Nothing known. Birds in breeding condition Sept (Pinto 1962).

Reference
Collar, N. J. and Stuart, S. N. (1985).

Sheppardia montana (Reichenow). Usambara Akalat. Cossyphe des Usambaras.

Alethe montana Reichenow, 1906. Orn. Monatsber. 15, p. 30; Usambara, Tanganyika.

Forms a superspecies with *S. lowei*.

Range and Status. Endemic to W Usambara Mts, Tanzania, where resident in montane forest from 1600 to 2300 m at Shume, Shagayu and Mazumbai. Common to locally abundant; especially numerous above 2000 m. Population density in Shume and Shagayu forests (90 km^2) averages 2–3 birds per ha, indicating total population *c.* 28 000 (van der Willigen and Lovett 1981). Threatened by habitat clearance in its tiny range.

Description. ADULT ♂: crown to nape dark brownish olive, becoming slightly paler on mantle to rump; uppertail-coverts and tail dark reddish brown, tail-feathers brighter on outer web. Ear-coverts and sides of neck pale olive-grey. Concealed streak of orange-rufous feathers with brownish olive tips from base of upper mandible to above eye; pre-orbital area dull white. Chin and throat off-white; breast and sides of belly pale olive-grey, becoming darker on flanks; centre of belly and vent off-white; undertail-coverts very pale buff. Flight feathers dark brown, outer webs ranging from dark brownish olive to rufescent brown, inner edges pale greyish buff, palest on primaries. Upperwing-coverts and scapulars dark brownish olive. Axillaries pale olive-grey; underwing-coverts pale buff. Bill hooked, black; noticeable black rictal bristles; eye dark brown to dull brown; legs and feet lead grey, brown or purplish with silver wash. Sexes alike. SIZE: wing, ♂ (n = 5) 76–80 (77·0), ♀ (n = 4) 70–80 (75·0); tail, ♂ (n = 5) 54–60 (57·0), ♀ (n = 4) 53–59 (56·0); bill from skull, ♂ (n = 5) 15·5–17 (16·5), ♀ (n = 3) 16–17 (16·5); tarsus, ♂ (n = 5) 29–32 (30·0), ♀ (n = 4) 28–32 (30·0). WEIGHT: ♂ (n = 4) 23–26 (24·2), ♀ (n = 6) 19–22·5 (21·3), unsexed (n = 95) 15·5–25·5 (21·8).

IMMATURE: upperparts very dark blackish brown, with centres of feathers pale yellowish buff, giving mottled appearance (much more so on mantle to uppertail-coverts and upperwing-coverts than on forehead to nape). Chin, throat, breast and flanks dark brown with off-white centres to feathers; breast strongly mottled. Flight feathers, tail, belly and vent as in adult. No concealed stripe above eye. Upper mandible blackish with horn tip; lower mandible bright yellowish horn.

NESTLING: unknown.

Field Characters. A stocky, dull brown robin, pale below, with very long legs and rufous spot in front of eye (visible when bird is excited, e.g. at ant swarms). Of other sympatric robins, Sharpe's Akalat *S. sharpei* is smaller with orange underparts and whitish grey stripe over eye, and White-starred Robin *Pogonocichla stellata* has blue-grey head, brilliant yellow underparts, and black and yellow tail. Juveniles are larger than Sharpe's Akalat, generally darker, and with longer legs; juvenile White-starred Robin has black and yellow tail. Occurs with Sharpe's Akalat only around 1600 m.

Voice. Not tape-recorded. Song (or subsong?) a series of long, thin notes, all on one pitch, 'ee seabee ee hee-hee ee seabee hee lichee seabee'. Alarm a nasal scold 'jahh jah jah jah jah'. At ant swarms, birds frequently interact with a faint rattle 'chr-r-r-ree', rapid snapping of beak, faint 'chup' or 'cheu', and song-like 'eek' or 'eeleek' notes.

General Habits. A bird of montane forest undergrowth, thickets and degraded forest with some remaining canopy, from 1600 to 2300 m. Replaces Sharpe's Akalat at higher altitude and in drier forest.

Active and tame. Spends much time perching just above ground; also perches up to 4 m in bushes. Known to 'ant': bird holds ant in bill, rubs it on tips of wing-feathers, then eats it. Scratches head over wing; stretches one leg, wing and tail on same side; sometimes stands on one leg. Forages mainly on forest floor; also on trunks, lianas, limbs, and by sallying in mid-air. Catches most prey after short sally or rush; searches in leaf litter, tossing leaves. Habitually associates with driver ant swarms, up to 12 birds gathering together; sometimes has to flutter to remove ants. They dispute occasionally, chasing, flicking tails up to 35° above horizontal (showing off whitish undertail-coverts), and flitting wings. Also interacts with White-starred Robin and White-chested Alethe *Alethe fuelleborni*. Sedentary.

Food. Tiny insects, particularly those flushed by driver ants.

Breeding Habits. Unknown. In Shume, Tanzania, brood patches and presence of juveniles indicate breeding season from Oct to Mar, with peak in Nov–Dec.

References
Collar, N. J. and Stuart, S. N. (1985).
Sclater, W. L. and Moreau, R. E. (1933).
Willis, E. O. (1985).

Sheppardia lowei (Grant and Mackworth-Praed). Iringa Akalat. Cossyphe d'Iringa.

Plate 22
(Opp. p. 353)

Alethe lowei Grant and Mackworth-Praed, 1941. Bull. Br. Orn. Club 61, p. 61; Njombe, Iringa Region, Tanganyika.

Forms a superspecies with *S. montana*.

Range and Status. Endemic to Tanzania. Resident in montane forest patches above 1450 m from Dabaga, Mufindi and Chita in Uzungwa Mts, south to the Njombe area and Mdando Forest in Livingstone Mts. Uncommon to locally common, but threatened by habitat destruction.

Description. ADULT ♂: crown to nape dark brownish olive, becoming slightly paler on mantle to uppertail-coverts, and on face and ear-coverts. Concealed white streak from base of bill to above eye, tipped pale buff. Concealed, small white spot below eye. Tail brown. Chin and throat warm pale buff, becoming more olive on breast, flanks and undertail-coverts, and merging into off-white belly. Flight feathers dark brown with brownish outer webs and paler inner edges (palest on primaries). Upperwing-coverts and scapulars brownish olive. Axillaries and underwing-coverts pale olivaceous buff. Bill hooked, black; black rictal bristles; eye brown; legs and feet brownish grey. Sexes alike. SIZE: wing, 1 ♂ 75, ♀ (n = 2) 69, 70, unsexed (n = 7) 67–74 (71·5); tail, 1 ♂ 58, 1 ♀ 53; bill from skull, 1 ♂ 16, 1 ♀ 17; tarsus, 1 ♂ 31. WEIGHT: ♂ (n = 6) 16·5–21·2 (19·3), 1 ♀ 17·5, unsexed (n = 15) 16·2–20 (18·1).

IMMATURE: upperparts very dark blackish brown, with centres of feathers pale buff, giving mottled appearance. Feathers of upperparts have blackish margins, with centres off-white on chin, throat and belly, and pale buff on face, breast and flanks, giving mottled appearance. Flight feathers, tail and undertail-coverts as in adult. No concealed spots above or below eye. Bill black.

NESTLING: unknown.

Field Characters. An elusive, dull-coloured robin, very similar to Sharpe's Akalat *S. sharpei*, with which it is sympatric at some localities, but larger, stockier, with longer legs and duller, buff underparts; stripe over eye shorter, less distinct, and pale buff instead of whitish grey. Juveniles of the 2 species very similar, but Iringa Akalat is larger with longer legs. Juvenile also like juvenile White-starred Robin *Pogonocichla stellata*, but tail brown instead of black and yellow. Juvenile Olive-flanked Robin-Chat *Cossypha anomala* is much larger with buff or rufous uppertail-coverts.

Voice. Not tape-recorded. Song (or subsong?) a long series of loud whistles, 'whee' or 'wree', 2 per s, or tinkling trills 'her hee her-her-her-her here her here her here her'. Alarm a nasal rattle 'ra-a-a-a-a-a-a-a-ah', or 'tak' notes. When birds fight at ant swarms, submissive one gives tinkling trills 'che-e-e-e-e che-e-ree-e-e-e-her', with song-like variations, and aggressor gives 'tak' calls and snaps its bill.

General Habits. Inhabits montane forest and thickets, usually quite dry, from 1450 m to at least 2450 m. Generally increases in abundance with altitude, replacing the very similar and partially sympatric Sharpe's Akalat in higher and drier areas.

Forages mainly by hopping on forest floor; less often searches among debris, and gleans trunks, vines and branches, and makes aerial sallies; catches most prey after short sallies or hop-runs; searches in leaf litter, tossing leaves. Habitually attends driver ant swarms, where birds often fight; during intraspecific chases, chased bird crouches and flutters wings, while aggressor flits wings and raises tail. When alarmed by humans, raises tail and flits wings, but soon becomes tame. Believed to be sedentary.

Food. Tiny insects, often flushed by driver ants.

Breeding Habits. Unknown. Enlarged gonads and presence of juveniles indicate breeding season between Sept and Feb.

References
Collar, N. J. and Stuart, S. N. (1985).
Willis, E. O. (1985).

TURDIDAE

Genus *Erithacus* Cuvier

Closely allied to *Luscinia*, but plumage softer and more copious, and tail square. Shorter-winged, less migratory and less robust than *Luscinia*; song less powerful and less varied. Sexes alike; young spotted.

3 species, 1 in W Palearctic, including N Africa, 2 in Japan.

Plates 23, 26
(Opp. pp. 368, 401)

***Erithacus rubecula* (Linnaeus). European Robin. Rougegorge familier.**

Motacilla rubecula Linnaeus, 1758. Syst. Nat. (10th ed.) 1, p. 188; Europe.

Range and Status. W Palearctic, from NW Africa, Portugal, Britain and Norway to Urals and W and S shores of Caspian Sea; also east of Urals to 84°E. Resident, partial migrant, and migrant.

Resident in Morocco, Algeria and Tunisia; winter visitor to N Africa. In Maghreb resident populations restricted to montane forest where widespread and locally common. Winter visitors often abundant near Maghreb coast; densities of 62 per km^2 in Algerian hillside macchia and 250 per km^2 in mixed woodland (Rooke 1947), and 130 per km^2 in mixed countryside near Casablanca (Morocco: Cramp 1988). In Libya common Oct–Apr north of 32°N in Tripolitania, frequent but local in Cyrenaican coastal zone, and along Egyptian coast and throughout Nile delta (up to 50 per km^2 in coastal W Sinai). Migrants are scarce south of N African coastal zone, but reach 28°N on Moroccan coast, 29–30°N in Algeria (where reports from Ahoggar at 24°N, Jan, need confirming: Cramp 1988); in Fezzan and Libyan desert recorded at El Hammam, Marada and Jaghbub (29–30°N) and Serir (27½°N); in Egypt south on Nile to 30°N. Also, 5 records coastal Mauritania, Nov–Dec and Mar.

Erithacus rubecula

Description. *E. r. rubecula* (Linnaeus): range of species except Britain, some Canary islands, Algeria, Tunisia, Crimea, Transcaucasia and east of Urals; winters south to N Africa. ADULT ♂: entire upperparts from forehead to tail and wings plain brown, warmest on rump. Lores, sides of forehead, ear-coverts, cheeks, chin, throat and breast orange. Narrow line between brown and orange above eye, and broad line behind orange from side of neck to side of breast, bluish grey. Belly white in centre, fawn at sides; flanks pale or quite dark buff; undertail-coverts pale buff or creamy. Greater wing-coverts often with rufous spot at tip, forming pale bar in closed wing; but spots soon wear away. Underside of wing buff. Bill brown-black with base of lower mandible paler; eye black-brown; legs and feet brown-black to flesh-brown. Sexes alike. SIZE: (Netherlands, Sept–Mar) wing, ♂ (n = 66) 70–77 (72·8), ♀ (n = 23) 68–74 (71·4); tail, ♂ (n = 99) 53–60 (56·4), ♀ (n = 38) 51–59 (55·1); bill to skull, ♂ (n = 92) 13·7–15·6 (14·4), ♀ (n = 35) 13·5–15·1 (14·2); tarsus, ♂ (n = 96) 24–27 (25·0), ♀ (n = 37) 24–27 (25·0). WEIGHT: W Morocco (Mar) unsexed (n = 9) av. 16·1; Netherlands (Nov–Feb) ♂ (n = 12) 16·8–24·5 (19·0), ♀ (n = 7) 15·8–20 (18·3).

IMMATURE: brown above and buff below, heavily spotted with pale buff from crown to rump, and with pale spots and dark crescents from breast to flanks. Buff wing-bar formed by tips to greater coverts. No orange or blue-grey in plumage. Soft parts black.

NESTLING: underparts bare; upperparts with long, scant grey down (down black by 7 days); mouth yellow, gape-flanges pale yellow.

E. r. witherbyi Hartert: Tunisia, Algeria (grading with *rubecula* in NE Morocco and S Spain). Like *rubecula* but smaller: wing, ♂ (n = 10) 66–73 (69·5), ♀ 67·5.

Field Characters. Adult readily distinguished by 'red' face and breast. Bluish grey border to orange breast can be hard to see if bird in shade. Heavily buff-spotted brown juvenile like juvenile of several other chats and thrushes; distinguished by humid woodland habitat, ticking alarm call, very slender black legs and upright stance.

Voice. Tape-recorded (CHA – 62, 73, 76, 89, B). Highly vocal. Song, uttered all year, a short, bittersweet medley of shrill and liquid notes with sad, haunting quality. Song, particularly common in autumn, is succession of 1–3 s phrases given at rate of *c.* 10 per min; each phrase contains av. 4·8 motifs; song highly variable. Call a short, sharp 'tic', usually repeated (alarm, territorial defence); also thin 'tswee' (contact-alarm). For other calls, see Cramp (1988).

General Habits. Breeding habitat in NW Africa mainly montane coniferous forest with humid leaf-litter and dead wood. Winter migrants occupy wide variety of habitats: woodland, farmland, dense macchia, large gardens, orchards and plantations.

Solitary (winter) or in pairs; keeps to open undergrowth, feeding largely on ground in woods or in open but near cover. Stance upright; often flicks tail up when hopping on ground; bows and bobs foreparts if anxious. Pugnacious and strongly territorial. Sings all year, from low perch; both sexes sing. In winter (S Spain) hunts chiefly from low perch, dropping and falling forward onto prey with flick of tail; foraging by hopping along ground provides only 4–6% of diet (Herrera 1977).

Migrants arrive in N Africa late Sept (sometimes late Aug), with main passage in Oct. Spring passage begins Feb, with main passage from mid-Mar to early Apr; N Africa vacated by late Apr, earlier in west than east (Cramp 1988). Most records at Saharan oases late Dec to Mar (once Nov). Ringing recoveries: singles to Algeria from USSR (Rybachiy), Switzerland, W Germany and Czechoslovakia. No evidence for migration or vertical displacement of resident NW Africa populations. As well as N European birds, Mediterranean ones may move into N Africa.

Food. Small beetles and other hard and soft invertebrates, and in winter small fruits and seeds. In S Spain: in holm oak *Quercus ilex* woods, Oct–Feb, plant material (bits of acorns) varied monthly from 16·5% by volume (17 birds) to 78% (16 birds), and of >1900 invertebrates 76% by number were ants, 12% beetles up to 8 mm long, and the rest larvae up to 20 mm long, and earwigs (Herrera 1977); in farmland, Nov–Jan, plant food (berries) varied monthly from 26 to 42% by volume and of >300 invertebrates 73% by number were ants and 15% beetles. At a highland locality in 4 winters, fruits of *Pistacia lentiscus* were 31% of plant diet, *Phillyrea latifolia* 19% and *Viburnum tinus* 6%; at a lowland site *P. lentiscus* was 89% (Herrera 1984).

Breeding Habits. Solitary, monogamous, territorial breeder. Strongly territorial all year, ♂ and ♀ defending separate territories; size in Britain (av. of averages, n = 221) 0·43 ha, and in N Africa about the same (with up to 300 birds per 100 ha). *c.* 10% of British ♂♂ bigamous. Breeding territories larger: Britain (av. of averages, n = 163) 0·84 ha, Canary Is (n = 7) 0·9–1·5 ha, Morocco <3·9 ha in forest (25·8 pairs per 100 ha: Thévenot 1982).

In pair-bonding ♀ has to overcome ♂'s aggression. ♀ approaches ♂, contact-calling; at first ♂ drives her away, later he treats any contact-calling bird as if a ♀. As ♀ approaches, ♂ pecks ground and retreats, resulting in hot-pursuit aerial chases around territory, both birds singing. ♀ spends more time with ♂, until paired. In sedentary birds, pairing is up to 4 months prior to nest-building. Courtship-feeding rare before nest-building, but ♂ gives ♀ 30–50 feeds per day over hatching period. 65% of >1400 copulations follow courtship-feeds (Cramp 1988).

NEST: cup of fine grasses, leaves and moss, lined with finer fibres and hair, built on bulky base of dead leaves; diam. 130, height 45. Sited in bank, among tree roots, in crevice, hollow stump or nest-box, usually well within 5 m of ground. Built by ♀ in *c.* 4 days.

EGGS: 4–6, av. NW Africa (n = 15) 4·2. Subelliptical; dull white or bluish, with faint pink freckles and blotches. SIZE: (n = 28, Morocco, Algeria) 17·5–20·5 × 14·5–16·5; mean (n = 200, Europe) 19·6 × 15·0. WEIGHT: *c.* 2·4.

LAYING DATES: NW Africa, Apr–May.

INCUBATION: by ♀ only, starting with last (sometimes penultimate) egg. Period: (n = 105, Europe) 12–21 (13·7) days.

DEVELOPMENT AND CARE OF YOUNG: in Britain hatches at 1·85 g; at day 1 average weights 2·2–3·2; day 7, 13·1–13·9; day 10, 17·8; day 13, 18·2–18·5 (Cramp 1988). Brooded by ♀ for 7 days. Fed by ♂ and ♀, or ♂ alone if ♀ begins 2nd brood. Young fed within 2 h of hatching by ♂ more than ♀, ♂ often transferring food to ♀ first. Eyes open at 4–6 days. Faecal sacs removed by ♂ or ♀, sometimes eaten but after day 5 carried away. Fledging period: (n = 125) 10–18 (13·4) days (Cramp 1988). All young fledge within 18 h; for first 4–5 days incapable of sustained flight and spend most time on or near ground; brood division between parents not uncommon.

BREEDING SUCCESS/SURVIVAL: in Britain 71% of >1400 eggs hatched, of which 77% fledged (55% success); annual adult mortality 62% (Cramp 1988).

References
Cramp, S. (1988).
Herrera, C.M. (1977, 1984).
Kersten, M. *et al.* (1983).
Rooke, K. B. (1947).

Genus *Luscinia* Forster

An assemblage of small, skulking Old-World thrushes. Bill slender, rictal bristles inconspicuous, legs rather long; tail medium length, slightly rounded; wings long and bluntly pointed. Sexes alike in some species, olive or warm brown above and greyish white below; in others, ♂♂ have brightly patterned heads and throats and/or dark blue upperparts, while ♀♀ are much duller, olive and whitish. Tail plain or with rufous or white patches. Inhabit leafy scrub, woodland thicket and marsh edges, and feed largely on ground. Songs rich and powerful.

11 species worldwide, mostly Asian; 3 in Africa as Palearctic migrants, 1 also breeding NW Africa.

408 TURDIDAE

Plate 23
(Opp. p. 368)

Luscinia luscinia (Linnaeus). Sprosser; Thrush Nightingale. Rossignol progné.

Motacilla luscinia Linnaeus, 1758. Syst. Nat. (10th ed.) 1, p. 184; Sweden.

Luscinia luscinia

Range and Status. Breeds NE Europe and USSR between 45° and 65°N to *c.* 90°E; winters E and southern Africa.

Palearctic winter visitor. Winters mainly south of 7°S from SW Tanzania (Rukwa) to Zambia (except NW), extreme SE Zaïre, Malaŵi, NW Mozambique (Zambezi valley) and N Zimbabwe; locally common to abundant. Also S Zimbabwe, NE Namibia (Caprivi Strip), N Botswana (N of 21°S) and Transvaal; uncommon to rare. North of 7°N in central Ethiopia (Awash Nat. Park), Kenya east of the highlands (Machakos, Kitui, Kibwezi and Voi-Taita) and N Tanzania (near L. Manyara); locally common. On southward passage occurs in Egypt, N and central Sudan (from Nile eastwards), Ethiopia (mainly in and west of Rift), N Kenya (Marsabit), central and inland SE Kenya (from E edge of highlands to *c.* 50 km from coast), and inland NE, central and S Tanzania; locally common to very abundant. Also on southward passage SW Somalia (middle and lower Jubba valley), frequent; SE Sudan, W Kenya, and coastal Tanzania (Dar-es-Salaam), uncommon to rare. On northward (return) passage, SE Kenya (including coast) and central Ethiopia, frequent to common; S Somalia (3 records, Mar). Vagrant Libya (Wadi Kaam, Tripoli, Apr), Nigeria (Kano, Sept–Oct), Sudan (Sanganeb, spring), Namibia (Windhoek); winter Natal records require confirmation (Cyrus and Robson 1980).

Description. ADULT ♂: upperparts dark earth brown, typically tinged olivaceous, although some tinged rufous especially on head and uppertail-coverts, while others paler and greyer, especially on mantle and scapulars. Tail dark brown, tinged rufous. Supercilium indistinct; lores brownish white; ear-coverts dark brown with paler streaking. Chin and centre of throat greyish white, bordered by broad brown moustachial streak; sides of neck, breast and flanks greyish brown with variable amount of dark mottling and spotting; belly dull white; undertail-coverts pale brownish with variable sparse barring. Primaries, secondaries, primary coverts and alula brownish black with brown outer edgings; rest of wing-coverts and tertials dark earth brown. Underwing-coverts and axillaries creamy buff with brown bases. Bill dark brown, base of lower mandible paler; eyes dark brown; legs and feet pale or pinkish brown. Sexes alike. SIZE: (10 ♂♂, 10 ♀♀, Europe/USSR) wing, ♂ 85–92 (88·4), ♀ 84–90 (87·5); tail, ♂ 65–71 (68·6), ♀ 67–69 (68·1); bill, ♂ 17–18 (17·4), ♀ 16–17 (16·7); tarsus, ♂ 27–29 (27·8), ♀ 25–28 (26·9). WEIGHT: unsexed (n = 33, Egypt, Aug–Oct) 20·6–29·9 (24·4); (n = 84, E Sudan, Aug–Sept) 14–23 (19·3); (n = 602, Kenya, Nov–Jan) 16·6–30·7 (22·2); (n = 172, Zimbabwe and Malaŵi, Dec–Mar) 19·5–33 (24·2); (n = 96, Kenya, Mar–Apr) 18·8–36·8 (24·7); (n = 4, Ethiopia, Mar–Apr) 34–36 (34·8).

A single complete moult occurs on breeding grounds July–Aug. Wing-tip bluntly pointed; P8 longest, P7 2–5 mm shorter, P6 7–10; P9 2–5 mm shorter, about equal to P7; P10 minute and narrow, much shorter than primary coverts; only one primary outer web (P8) emarginated.

IMMATURE: first winter like adult, but wing-coverts, tertials and flight feathers tinged more rufous, and tertials and greater coverts with pale buff tips.

Field Characters. Very like Nightingale *L. megarhynchos* but upperparts darker and duller, less warm brown, with greyish or olivaceous tone; tail darker, brownish rufous, contrasting less with back. Also has noticeable dark malar stripe bordering white chin and throat (lacking in Nightingale), dark mottling on breast and flanks, and in hand, very small first primary and more pointed wing-tip (**A**). Voice similar to Nightingale but can be distinguished with practice.

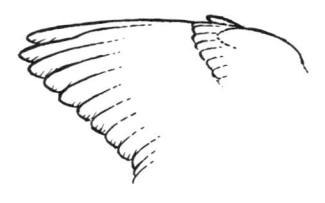

Luscinia luscinia

Luscinia megarhynchos

A

Voice. Tape-recorded (15, 69, 86, 88, C, F). Song similar in structure to that of Nightingale, but phrases longer (*c.* 5 s) and separated by pauses of 3–5 s; speed of delivery slower, with units more separated; phrases typically in three sections, beginning with whistling 'wheet-wheet-wheet ...', followed by deep fluty 'chook-u-chook-u ...' or 'chiddy-ock-chiddy-ock-chiddy-ock ...' and ending with deep rattle; slow introductory crescendo whistles typical of Nightingale are lacking. In Africa, full song often heard in winter quarters Feb–Mar; at other times, and on southward passage, phrases less organized and distinct, units more disjointed, interspersed with croaks, strung together over periods of up to 20 s or more, the whole effect much more guttural, less liquid, than song of Nightingale. Contact-alarm calls, low croaking 'krrrk', guttural 'tuc' or 'tuc-tuc' and whistled 'wheeet', sharper and higher-pitched than corresponding call of Nightingale.

General Habits. Frequents moist leafy scrub and thicket, woodland undergrowth and overgrown stream beds. Sometimes shares wintering habitat with Nightingale but more confined to rank, moist cover. Utilizes drier and more varied habitat on passage: scrub, hedges, bushland thicket, and mangroves on Sudan coast. Winters typically at 500–1500 m, but ranges up to 2200 m on passage through E Africa.

Occurs singly or in small loose groups when migrating. Skulking, but not especially shy; sometimes closely approachable. Perches low down and feeds largely on ground, hopping and disturbing leaves to search. Also feeds in herb and shrub layer and recorded taking flying insects. Cocks tail and flicks it frequently when excited; also fans it to one side or the other. Territorial in winter quarters, also for periods of days or weeks on autumn stopover in NE Africa and on passage through Kenya. In Zimbabwe, 3 territorial birds *c.* 100 m apart from one another, a fourth bird *c.* 40 m away (Smith 1951). Territorial defence includes chasing and other aggressive interactions. Especially active for short periods at dawn and dusk, when whistled 'wheeet' call characteristic. Sings regularly in winter quarters (in Zimbabwe from early Jan) and intermittently on southward passage, usually from sunrise to midday, often in considerable heat. Song-duels reported in Zambia.

Western populations breeding W Baltic and central Europe migrate southeast in autumn to cross E Mediterranean. Enters Africa east of 25°E, mainly across northern Red Sea. Southward passage initially rapid, through Egypt and NE Sudan into W and central Ethiopia, but onward movement through central and SE Kenya and NE and central Tanzania is 2 months later. Main movement Egypt and coastal Sudan mid-Aug to early Oct (first birds early Aug); central Ethiopia mid-Sept to early Nov; Kenya and Tanzania early Nov to mid-Dec (first birds mid-Oct). Most birds enter southern Africa on front *c.* 250 km wide immediately east of Mt Kilimanjaro, changing migratory direction from S to SW. Arrives main wintering areas late Nov to Dec; remains until mid- to late Mar. Return route through Africa similar to southward route, though birds occur Kenya coast, and avoid Sudan, presumably using more eastern route through Ethiopia. Main spring passage Kenya late Mar to mid-Apr; Ethiopia early to mid-Apr (latest 27 Apr); less protracted and less heavy than in autumn. Last birds Egypt early May. Birds recovered Egypt (7, Aug–Nov, ringed Denmark, S Sweden and S Finland; 1, Mar, ringed Denmark); and South Africa (Natal, 1, Nov, ringed Denmark); also Yemen (1, May, ringed Sweden). Birds ringed Kenya, Nov–Dec, recovered Lebanon (1, Apr; 1, Sept); Syria (1, Apr); Israel (1, Apr); Finland (1, June); and USSR (Ukraine) (1, June; 1, July).

Food. Mainly arthropods, including ants, beetles, small grasshoppers, adult, pupal and larval Lepidoptera, adult and larval Diptera, millipedes; also small molluscs; sometimes fruit and occasionally seeds.

References
Ash, J. S. (1973).
Cramp, S. (1988).
Nikolaus, G. and Pearson, D. J. (1982).
Pearson, D. J. (1984).

Luscinia megarhynchos C. L. Brehm. Nightingale. Rossignol philomèle.

Plate 23
(Opp. p. 368)

Luscinia megarhynchos C. L. Brehm, 1831. Handb. Nat. Vög. Deutsch., p. 356; Germany.

Range and Status. Breeds NW Africa, W, central and SE Europe, Asia Minor, the Caucasus, N Iran, N Afghanistan and S USSR east to *c.* 85°E; winters tropical Africa south to about the equator.

Palearctic winter visitor and intra-African migrant. Breeding visitor N and W Morocco south to the High Atlas, N Algeria and N Tunisia (Kroumirie to Cap Bon); common to abundant. Winters in Senegambia (uncommon) and Guinea (uncommon) to Sierra Leone, Liberia and Ivory Coast, thence east through Ghana, Nigeria (uncommon in N), N Cameroon, S Chad, Central African Republic, N Zaïre and S Sudan to N and E Uganda, extreme W Kenya and central and S Ethiopia; also central and SE Kenya to S Somalia (Jubba and Webi Shabeele valleys) and extreme NE Tanzania (south to Dar-es-Salaam). In W Africa, mainly in guinea savanna and savanna–forest mosaic south of 10°N where common to locally abundant; in E Africa, locally common to abundant. On passage throughout N Africa (uncommon in autumn, common in spring), and the Sahel (frequent to common autumn, less common spring); uncommon Saharan oases, Sudan coast and N Ethiopia; common to abundant on southward passage, S Somalia.

Luscinia megarhynchos

Description. *L. m. megarhynchos* C. L. Brehm: breeds NW Africa, Europe and Asia Minor; winters W and central Africa east to Sudan, Uganda and W Ethiopia. ADULT ♂: upperparts dark olive-brown or warm rufous-brown, uppertail-coverts brighter chestnut-brown. Tail dark chestnut-brown. Supercilium indistinct; lores brownish white; ear-coverts brown with paler streaking. Sides of neck, sides of throat, breast and flanks pale greyish brown; rest of underparts dull white, undertail-coverts washed creamy buff. Primaries, secondaries, primary coverts and alula blackish brown with rufous-brown outer edgings; tertials and rest of wing-coverts rufous-brown; underwing-coverts and axillaries creamy buff with brown bases. Bill dark brown; eye dark brown; legs and feet pale pinkish to pale grey-brown. Sexes alike. SIZE: (10 ♂♂, 10 ♀♀, Europe/NW Africa) wing, ♂ 79–89 (84·7); ♀ 80–88 (83·1); tail, ♂ 65–72 (68·9), ♀ 65–70 (68·3); bill, ♂ 17–19 (17·8), ♀ 16–18 (17·1); tarsus, ♂ 27–28 (27·6), ♀ 26–28 (27·3). WEIGHT: unsexed (n = 24, Egypt, Aug–Sept) 20·0–26·9 (22·6); (n = 19, N Nigeria, autumn) 19·1–22·4 (19·4); (n = 8, S Nigeria, Nov–Feb) 17–24 (20·2); (n = 8, N Nigeria, spring) 23·2–33 (29·1); (n = 276, Morocco, Mar–May) 12·2–24·6 (18·3).

A single complete moult occurs on breeding grounds July–Aug. Wing-tip bluntly pointed; P8 longest (sometimes equal to P7), P7 0–2 mm shorter, P6 4–6; P9 5–9 shorter, between P5 and P7; P10 very small, between 2 shorter and 4 longer than primary coverts; outer webs of P7 and P8 emarginated.

IMMATURE: juvenile mottled blackish and warm brown above; whitish below with dark brown feather-tips. Wing and tail as adult but with buff spots at tips of tertials and greater coverts. First winter as adult but with buffish tips remaining on tertials and greater coverts.

NESTLING: down grey-black, on upperparts only; inside of mouth orange, no tongue spots; gape-flanges white shading to pale yellow at angle of gape.

L. m. africana (Fischer and Reichenow): breeds Caucasus and E Turkey to Iran; winters E Sudan, Ethiopia, Somalia, Kenya and Tanzania. Like nominate race but slightly larger, with rufous tinge mainly confined to head and uppertail-coverts, upperparts and edges of wing-feathers more earth brown. Wing, ♂ (n = 10) 85–91 (87·4). WEIGHT: unsexed (n = 207, Kenya, Nov–Dec) 16–25 (19·8).

L. m. hafizi Severtzov: breeds central Asia from Aral Sea to border of W Sinkiang; winters E Sudan, Ethiopia, Somalia, Kenya, Tanzania and once Uganda. Larger and paler than *africana*, greyish brown above with contrasting chestnut tail and uppertail-coverts; whiter below than other races; supercilium more distinct. Intergrades with *africana*. Wing, ♂ (n = 10) 90–95 (92·2). WEIGHT: unsexed (n = 116, Kenya, Nov–Dec) 15·8–26·1 (20·8); (n = 6, *africana/hafizi*, Kenya, Mar–Apr) 21·5–39·2 (27·3).

Field Characters. A small, rather featureless thrush, with warm brown upperparts, pale greyish brown underparts and rufous rump and tail. Skulking and hard to see; rufous tail is often the only feature seen well as the bird flies off and dives into cover. Loud song and calls best indicators of its presence. Race *hafizi* somewhat larger, paler and greyer, with longer tail. For distinctions from Sprosser *L. luscinia*, see that species.

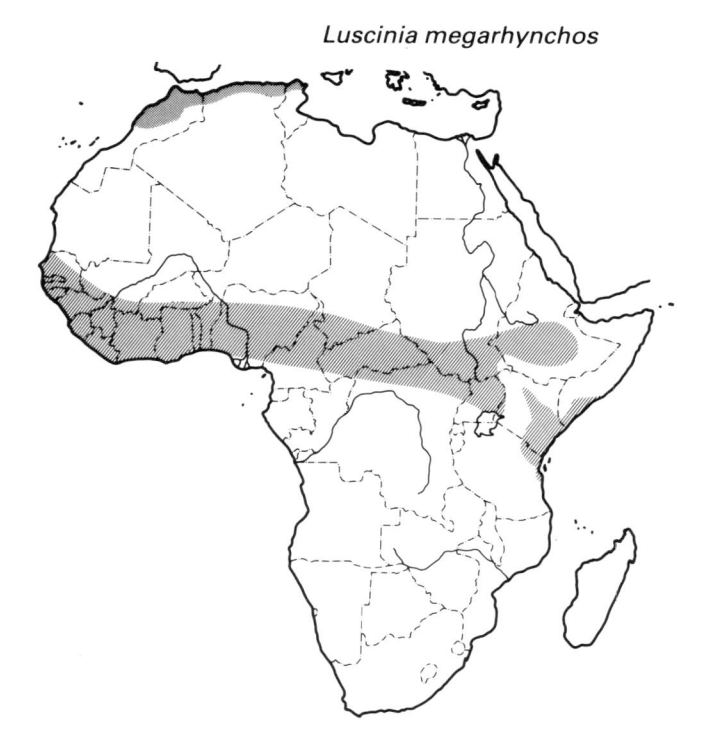

Voice. Tape-recorded (5, 53, C, MOR, PEA – 89). Song powerful, rich and varied, consisting of succession of full-sounding phrases 2–4 s long, separated by pauses of about same length. Each phrase involves repetition of a basic unit, usually with a change of unit type in mid-phrase and a terminal flourish. Some units clear, rich and piping, others bubbling or rattling. Individuals produce a succession of different phases: 'cheeoo-cheeoo-cheeoo-chi', 'pichu-pichu-pichu-pichurrrrrrrr-chi', 'wheet-chook-chook-chook-chook-chook-chi'. Phrases often introduced by slow whistling crescendo notes, 'wheet-wheee-wheeeee-wheeeeee ...'. Daytime song usually shorter and less varied than nocturnal one, with whistling introductory notes less frequent (Cramp 1988). Winter song similarly structured and varied, though less powerful. 3 song types recognized by Cramp (1988): (1) territorial-song, loud and continuous, given by ♂ mainly at night for advertisement, also by day in interactions with rivals; (2) courtship-song, quieter, given by ♂, especially in vicinity of nest; (3) contact-song, short and fragmentary, given by both ♂ and ♀ anywhere in territory. Contact/alarm calls, a grating 'krrr', hard 'tec-tec', whistled 'wheet' or 'wheet-krrr', harsh 'kaarr'; in courtship a bleating 'hä-hä-hä-hä'. For further details, see Cramp (1988); for distinctions from Sprosser *L. luscinia*, see that species.

General Habits. On NW African breeding grounds inhabits low thickets, coastal maquis, woodland and bushy lower slopes up to 1300 m. In non-breeding season in tropical Africa occurs in dense forest edge and secondary growth, riverine and woodland thickets, thorny scrub, edges of cultivation and garden hedges, from sea level to 1600 m in Ethiopia and Kenya. In E Africa, in drier scrub and woodland undergrowth than wintering Sprosser, less confined to valleys and small watercourses.

Solitary, but local concentrations occur on passage and in winter quarters. Secretive but not especially shy; sometimes approachable and confiding. Keeps mainly to ground and low perches; flight flitting and low. Frequently flicks wings and cocks tail when excited. Moves on ground by short or long hops on rather long legs, with erect carriage, head and tail held up and wings drooped. Forages mainly within or around thick cover, taking prey from leaf litter, bare ground or leaves and twigs. Also drops onto insects from perch or catches them in flight. Commonly sedentary and territorial on wintering grounds; much interaction between individuals and chasing, especially at dawn and dusk. Often sings from thick cover during day, Nov–Mar, usually in morning. Winter territories apparently small (e.g. Togo, 3–4 birds singing c. 10 m apart in same thicket: Douaud 1957).

Two main migratory pathways, *megarhynchos* from Europe over a broad front across N Africa to Senegambia east to Uganda and W Ethiopia; and *africana* and *hafizi* across Arabia mainly to Ethiopia, Kenya and NE Tanzania. *Megarhynchos* departs Europe late July to early Sept; movement through Europe broadly southwest, birds occurring throughout Mediterranean region but commonest in W (Cramp 1988). Scarcity of autumn records N Africa (e.g. few birds seen, mainly mid-Aug to early Oct; only 2 recoveries, see below) suggests most fly directly from Europe to N tropics, where passage occurs Senegambia mainly Oct–Nov (earliest 20 Aug), N Nigeria Sept to early Nov (peak late Oct), Chad mainly Oct (earliest 6 Sept), W and central Sudan Sept–Oct. Main arrival wintering grounds W Africa and S Sudan late Oct to early Dec. *Megarhynchos* departs for N mainly mid-Mar to early Apr; recorded Senegambia to early May; main passage N Nigeria late Mar to early Apr, N African coast (with a few inland Algeria and Libya) late Mar to mid-May, peak mid-Apr, movement on broad front. Birds recovered Morocco (2, Sept, ringed France); several, Mar–Apr, ringed Netherlands, France, Belgium, W Germany); Libya (1, spring, ringed Italy); Nigeria (1, Mar, ringed Tunisia); and Cameroon (1, Apr, ringed Hungary). Birds ringed Morocco (May) recovered France (1) and Netherlands (1); ringed Tunisia (Apr) recovered Italy (3) and Hungary.

Hafizi and *africana* common on autumn passage S Oman (Sept to early Oct), central Ethiopia (late Sept to early Nov), S Somalia and SE Kenya (Nov). Winters Kenya mainly Nov – early Dec to late Mar – early Apr with *hafizi* most numerous on coast, *africana* inland. Latest record Ethiopia 20 Apr. Bird ringed Ethiopia, Sept, recovered USSR (c. 40°N, 46°E).

Food. Mainly insects, including ants and their pupae, adult and larval beetles, small Lepidoptera, caterpillars, flies (Diptera), small Orthoptera, termites; also spiders and earthworms; occasionally berries.

Breeding Habits. Few African data; well studied in Europe (Cramp 1988). Monogamous; a few reports of bigamy (1 ♂, 2 ♀♀: Davis 1975). Solitary and territorial; territory 0·13–1·9 ha (Europe, USSR); serves for pair formation, nesting and most feeding. ♂ arrives on breeding grounds a few days before ♀; establishes and defends territory by singing territorial-song (mid-Apr to July, Mediterranean) from low undergrowth and low branches of trees; also by chasing and threatening with bowing posture. ♂ courts ♀ with courtship-song, usually at night for c. 2 weeks until pair formed; in display, ♂ sings quietly on ground or on branch near ♀, fanning and raising tail, sometimes drooping wings and exchanging harsh alarm calls with ♀; he then pursues her in whirring flight with bleating calls, lands beside her and sings again but more intensely, dancing with tail fanned, wings spread and quivering, and head lowered; copulation may follow. ♂ also gives courtship-song while ♀ collects nesting material. Courtship-song of ♂ replaced by increased contact calls as pair-bond strengthens.

NEST: a bulky cup of dead leaves and grass lined with finer grasses, feathers and hair; occasionally domed; on or close to ground in herbage or low creepers. Nest-site selected and nest built by ♀.

EGGS: in NW Africa usually 4–5. Bluish with light reddish brown speckling. SIZE: NW Africa, 18–24 × 15–17. Often double-brooded.

LAYING DATES: Morocco, Algeria and Tunisia, late Apr to May.

INCUBATION: by ♀ alone; begins with first egg. Period: 13–14 days. ♀ sits tightly on nest from day 9.

DEVELOPMENT AND CARE OF YOUNG: young brooded by ♀ for first few days, when food brought by ♂; thereafter brooded less regularly and fed by both parents; both parents perform nest sanitation. Young leave nest 10–12 days after hatching, dispersing into surrounding cover until able to fly 3–5 days later; self-feeding 3–9 days after fledging, but partly fed by parents for another 6–20 days, often beyond territory; parents may divide brood between them. When pair starts second clutch, ♂ may depart with first brood while ♀ begins incubating. After young hatch, parents give alarm calls on approach of intruders, most intensely when young just out of nest; ♀ may perform distraction display.

References
Ash, J. S. (1973).
Cramp, S. (1988).
Pearson, D. J. (1984).

TURDIDAE

Plate 23
(Opp. p. 368)

Luscinia svecica (Linnaeus). Bluethroat. Gorgebleue à miroir.

Motacilla svecica Linnaeus, 1758. Syst. Nat. (10th ed.), p. 187; Sweden.

Range and Status. Breeds N Scandinavia, central, NE and SW Europe, Caucasus, N Iran and N and central Asia; also W Alaska; winters Mediterranean basin to N tropical Africa and SW, S and SE Asia.

Palearctic winter visitor. In N Africa, winters locally and in small numbers near coasts of W and N Morocco, N Algeria and N Tunisia; in some Saharan oases in Morocco and Algeria; and in Egypt (N coast, Nile delta and valley, south at least to Aswan, W desert oases and Red Sea). Winters more widely south of Sahara, from S Mauritania to Sudan, south to c. 12–13°N (including S Mali, N Ghana, SE Niger and N Nigeria), and in Ethiopia south to c. $7\frac{1}{2}$°N; locally frequent to common. Vagrant NW Ivory Coast (Bako), S Ghana (Kumasi), S Nigeria (Ibadan) and S Ethiopia (lower Omo R.). Frequent to common in N Africa on both passages.

Description. *L. s. svecica* (Linnaeus): breeds Scandinavia east through N USSR; winters N Africa and S Asia. ADULT ♂ (breeding): top of head dark brown with blackish streaks forming dark lines on each side of crown; rest of upperparts dark brown with pale buff fringes to feathers, especially on rump; uppertail-coverts with bright chestnut bases. Tail blackish brown, all but central feathers with basal half bright chestnut. Prominent supercilium buffish; lores and ear-coverts dark brown, mottled and streaked buff; sides of neck tinged greyish. Chin, throat and upper breast metallic blue, bordered below by narrow black and broader chestnut bands; spot on centre of lower throat chestnut; rest of underparts buffish white, flanks suffused greyish. Remiges and upperwing-coverts dark brown with pale edging; axillaries and underwing-coverts dark brown fringed buff. Bill dark brown, paler at base of lower mandible; eye dark brown; legs and feet yellowish brown. ADULT ♂ (non-breeding): like breeding ♂, but chin and upper throat buff; broad blackish line along side of lower throat, and above this a blue malar streak; chestnut throat feathers with whitish bases; colours of throat and breast generally obscured by buff fringes. ADULT ♀: like non-breeding ♂, but with little or no blue on upper breast or malar region, and few if any chestnut feathers on throat; chin and centre of throat mainly buff, bordered by broad blackish lines at sides and narrow blackish breast-band; lacks chestnut on lower breast. SIZE: (10 ♂♂, 10 ♀♀, N Europe/Siberia) wing, ♂ 74–79 (76·8), ♀ 70–75 (72·7); tail, ♂ 53–58 (55·5), ♀ 51–55 (52·9); bill, ♂ 16–17 (16·1), ♀ 15–17 (15·8); tarsus, ♂ 27–28 (27·7), ♀ 26–28 (27·1). WEIGHT: unsexed (n = 20 on migration, Germany) 17–22 (19·8); (n = 19, *L. s. svecica* or *cyanecula*, Ethiopia) 14·9–21·4 (17·2).

Complete moult on breeding grounds, but ♂♂ acquire breeding plumage by moulting chin to upper breast Feb–Mar. Wing-tip bluntly pointed; P7 and P8 longest, P6 0·5–2 mm shorter, P5 2–4; P9 4–8 shorter, between P3 and P5; P10 very small; outer webs of P6–P8 emarginated.

IMMATURE: first winter bird has primary and outer greater coverts and tertials tipped buff; otherwise like non-breeding adult, except that ♂ has less blue in malar streak, less chestnut on throat and paler chestnut breast-band.

L. s. cyanecula (Meisner): breeds Europe south of 60°N; winters N Africa and Iberia. ♂ differs from nominate race in having white spot on lower throat, occasionally no spot at all; ♀ and immature not certainly distinguishable. Wing, ♂ (n = 10) 74–78 (76·2). WEIGHT: unsexed (n = 14 Morocco, Mar–May) 12·3–16·9 (14·2); (n = 2, Nigeria, Dec, probably *cyanecula*) 15·5, 15·5.

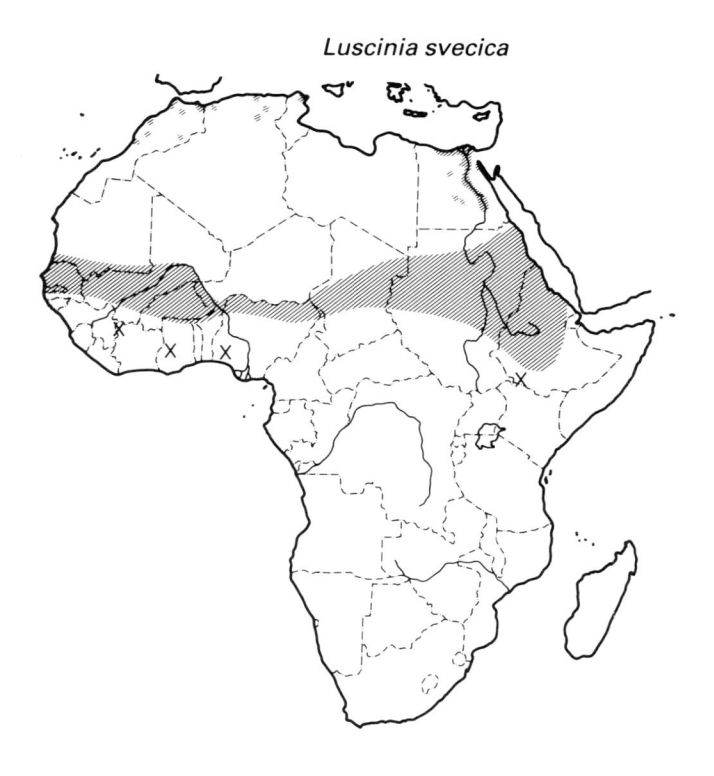

Luscinia svecica

L. s. magna (Sarudny and Loudon): breeds Caucasus to Iran, wintering Middle East, Sudan and Ethiopia. Paler than nominate, greyer above; gorget paler blue, throat spot absent or very small and white. The largest race: wing, ♂ (n = 10) 79–83 (80·7).

L. s. volgae (Kleinschmidt): breeds central Russia, wintering Middle East and Egypt. ♂♂ individually variable, intermediate between nominate race and *cyanecula*.

Field Characters. A small, shy thrush, dark brown above and whitish below, generally seen as it flits from one patch of cover to another, when rufous sides to base of blackish tail (all plumages) are diagnostic. Breeding ♂ with blue throat and breast, red and black breast-bands, unmistakable; other plumages very variable but all birds have conspicuous white supercilium, white or buffy throat bordered by blackish moustachial stripe and broken black breast-band.

Voice. Tape-recorded (62, 73, B). Song loud, sweet and exceptionally varied; consists of phrases c. 10–20 s long, separated by pauses. Phrase begins typically with slow and hesitant sweet whistled notes 'djip-djip-djip …', accelerating into a variety of motifs, with much repetition (Cramp 1988). Lacks full rich quality of song of Nightingale *L. megarhynchos* and at times suggests an *Acrocephalus* warbler; sweet musical notes and liquid trills alternate with harsh discordant sounds; twanging 'trr-trr-trr-trr' is characteristic. Mimics calls of other birds, crickets, tree frogs and other sounds. Apparently no pronounced difference between nominate *svecica* and *cyanecula*. Contact-alarm calls, a hard staccato 'tacc-tacc', croaking 'turrc', plaintive 'hweet'. For further details, see Cramp (1988).

General Habits. Usually frequents waterside habitats: reeds, clumps of grass, tamarisks, edges of rice fields, sugar cane, tidal flats with saltmarsh vegetation, thickets along watercourses, and oases; in W Africa, typically in *Mimosa pigra* thickets; in Ethiopia, in rank streamside herbage at over 1900 m. Usually solitary, but small concentrations may occur on passage and in winter quarters. Unobtrusive; skulks close to ground, moving furtively in thick cover. When flushed flies low and for short distance, diving quickly back into cover. Emerges to feed in open if undisturbed, hopping, pausing and briefly running. Carriage erect, with tail frequently raised, giving bird a long-legged appearance. Feeds on ground and in low vegetation, turning over leaves and soil; also catches flying insects. Gives snatches of song in winter quarters.

European ringing recoveries indicate that *cyanecula* from central Europe and W USSR move southwest. Some nominate *svecica* from Scandinavia move south, but many migrate southeast to Asia. Both races occur throughout African range, but their distribution and relative abundance is unclear. In general, *cyanecula* is commoner in the west, *svecica* in the east. In N Africa, *cyanecula* predominates on passage, and almost all wintering birds are this race. Nearly all birds wintering in Senegal and most in Mali and Chad are *cyanecula*; birds reaching N Nigeria are mainly *svecica*. The two races occur in about equal numbers in Sudan and Ethiopia. Most *cyanecula* appear to winter in tropical Africa, but only a small proportion of nominate *svecica* do so. In Africa, *magna* from SW Asia occurs only in small numbers in NE Sudan and Ethiopia; *volgae* is known only from Egypt.

Arrival and passage N Africa and Red Sea coast is diffuse, mainly mid-Sept to Nov. South of Sahara, birds reach Senegal from mid-Sept, Mali Oct–Nov, Nigeria early Dec, Chad Oct (*svecica* in Nov); departure mainly Mar. Return passage begins early, in Feb Morocco and Egypt, but main N African movement Mar, continuing to late Apr; recorded Ethiopia to early May and exceptionally Senegambia late May. Birds recovered Morocco (5, Oct–Feb, ringed Belgium and Netherlands), N Algeria (1, Mar, ringed Germany), Algerian Sahara (1, Apr, ringed Norway) and Tunisia (1, Feb, ringed Belgium).

Food. Mainly insects, including caterpillars, beetles, ants, grasshoppers; also small snails, earthworms, small frogs; occasionally seeds and fruit.

Reference
Cramp, S. (1988).

Genus *Irania* de Filippi

Bill slender, nostrils in front of frontal feathering; rictal bristles inconspicuous; legs strong and fairly long; tail medium length; wings long, bluntly pointed. Sexes distinct. ♂ patterned black, grey, white and orange-rufous; ♀ mainly greyish brown. Tail black in both sexes. Frequent thickets. Feed partly on ground.

Monotypic; Palearctic migrant in Africa.

Irania gutturalis (Guérin). Irania; Persian Robin. Iranie à gorge blanche.

Cossypha gutturalis Guérin, 1843. Rev. Zool., p. 162; N Abyssinia.

Plate 23
(Opp. p. 368)

Range and Status. Breeds SW Asia, from S and E Turkey, the Caucasus and E Mediterranean to Iran, N Afghanistan and Soviet Turkestan; winters E Africa.

Palearctic winter visitor. Winters inland Kenya north, east and south of the highlands, from Isiolo and Meru south to Kitui, Tsavo and the Taitas, west to Nairobi, Namanga and (rarely) Mara; also N Tanzania west to Serengeti and south to Dodoma, and once S Tanzania (Mbarali, $8\frac{1}{2}°$S); locally frequent to common. On southward passage, coastal, central and probably E Ethiopia, and Kenya in plateau country north and east of the highlands, locally common to abundant; NW Somalia, uncommon; Sudan coast (Khor Aba'at), rare. On return passage, SE and E Kenya and central Ethiopia, frequent to common; NW Somalia and Djibouti, rare.

Description. ADULT ♂: entire upperparts grey. Tail black. Supercilium white; lores and sides of chin and throat black; ear-coverts dark grey; centre of chin and throat white; black sides of throat occasionally connected by black bar across upper breast. Breast and flanks usually deep orange-rufous, sometimes pale orange or pale cinnamon-buff; belly and undertail-coverts whitish. Remiges and primary coverts blackish, broadly edged grey; rest of upperwing-coverts grey; underwing-coverts and axillaries deep orange-rufous to pale cinnamon-buff. Bill black; eye hazel; legs and feet black. ADULT ♀: upperparts brownish grey. Indistinct buffish supercilium; lores and ear-coverts brownish; sides of neck, chin and throat brownish grey; centre of chin and throat greyish white; breast warm buff or rufous-buff barred grey and whitish. Flanks, underwing-coverts and axillaries orange-buff. Tail, wings and soft parts as ♂. SIZE: (10 ♂♂, 10 ♀♀) wing, ♂ 94–102 (96·6), ♀ 90–98 (93·5); tail, ♂ 72–81 (78·6), ♀ 70–79 (74·1); bill, ♂ 18–20 (19·1), ♀ 18–21 (19·5); tarsus, ♂ 25–27 (25·6), ♀ 24–27 (25·5). WEIGHT: unsexed (n = 500, Kenya, Oct–Feb) 17·6–26·7 (21·6); (n = 10, Kenya, Mar–Apr) 22·8–29·7 (24·8).

Complete moult on breeding grounds; body and head moult in Africa Nov–Dec. Wing-tip bluntly pointed; P8 longest (sometimes = P7); P7 0–1 mm shorter, P6 1·5–4, P5 6–11; P9 4–10 shorter, between P5 and P7; P10 very small; P6–P8 emarginated on outer web.

IMMATURE: first winter as adult ♀, but wing-feathers dark brown, edged pale brown, greater coverts with whitish tips. First summer plumage like adult, acquired by full head and body moult Nov–Dec, but juvenile wing- and tail-feathers retained.

Field Characters. A small thrush, slightly larger than Nightingale *Luscinia megarhynchos*, and with longer bill. ♂ has striking combination of grey upperparts, white supercilium and throat, black face and orange breast, giving it a strong superficial resemblance to robin-chats *Cossypha*, but distinguished by all-black tail and orange underwing. ♀ duller and browner with rather uniform head, but has diagnostic orange-buff flanks and underwing and black tail. Lives in dry, scrubby habitat where no robin-chats occur.

Voice. Tape-recorded (McVIC, PEA – 73, ROC). Song quite loud and melodious; consists of vigorous warbling phrases, sometimes sustained for over 10 s, separated by short pauses. Phrases comprise a mixture of fluty whistles and throaty grating notes, but lack much variety; recall song of a *Sylvia* warbler. Full song given in winter quarters; also a brief quiet warbling sub-song by wintering and passage birds. Calls, a hard 'tec' and dry grating 'turrr' or 'eet-trr', similar to calls of *Luscinia* spp.; also a noisy 'tzi-lit', recalling note of White Wagtail *Motacilla alba*.

General Habits. Frequents scrub and thickets in semi-arid areas, especially dry acacia and *Commiphora* woodland, margins of cultivation, cover along gullies and watercourses, sometimes gardens; typically in drier habitats than *Luscinia* spp. Winters mainly at 300–1500 m.

Generally solitary, but often in loose wintering concentrations, and small groups may occur on migration. Shy and elusive, spending much time perched low down in dense bushes or small trees. Usually dives into thicket if flushed, but sometimes flies a considerable distance. Feeds among low twigs; also on ground, turning over leaves to search, sometimes in the open. Gait and stance on ground similar to Nightingale, moving with long hops, carriage erect, wings drooping and tail usually cocked high. Carriage typically more horizontal than Nightingale when perched. When alarmed, extends legs and raises head so that white throat prominent; tail raised slowly to *c*. 45°, spread and slowly lowered. Wintering birds apparently territorial, some occupying same thicket for many weeks. ♂ often sings Jan–Mar, typically from bush or small tree branch 1–2 m above ground. Song-duets recorded in Kenya between ♂♂ *c*. 30 m apart, in which singing bird had head extended, back flattened, wings drooped and tail spread.

Lack of records in Egypt and Sudan suggests that birds from Asia Minor initially move southeast. All populations apparently cross Arabia to enter Africa on a narrow front through Eritrea and central/E Ethiopia in late Aug to mid-Oct. Onward movement is delayed, with main passage E and SE Kenya Nov to early Jan (earliest 2 Nov). Southward movement continues through early part of winter, birds moving on from Tsavo (SE Kenya) as bush dries out. Main movement into Tanzania is east of Mt Kilimanjaro, but considerable numbers winter in L. Manyara area, presumably migrating west of the mountain. Wintering sites occupied to end March or beginning of Apr. Return route through Africa and Arabia apparently similar to the autumn route, with passage through SE Kenya and central Ethiopia in late Mar to mid-Apr.

Irania gutturalis

Food. Mainly insects, including adult and larval beetles, ants, small grasshoppers; also centipedes, spiders; occasionally berries.

References
Cramp, S. (1988).
Pearson, D. J. (1984).

Genus *Cossyphicula* Grote

Very small; bill broad at base, rictal bristles copious and stiff, feet small. Plumage like *Sheppardia* except for red and black tail; white stripe in front of eye does not include the erectile feathers characteristic of *Sheppardia*. Catches most food on the wing, unlike *Cossypha*, and song very similar to *Stiphrornis*. Since it has characteristics of both *Cossypha* and *Sheppardia*, we follow Chapin (1953) and Dowsett-Lemaire (1990) in retaining it in a separate genus.

Endemic; monotypic.

Cossyphicula roberti (Alexander). White-bellied Robin-Chat. Cossyphe à ventre blanc.

Plate 24
(Opp. p. 369)

Callene roberti Alexander, 1903. Bull. Br. Orn. Club 13, p. 37; Fernando Po.

Range and Status. Endemic resident. 2 populations: (1) E Nigeria (Obudu Plateau), central Cameroon (from Foto near Dschang to Mt Nlonako, Muambong, Mt Kupé, Rumpi Hills and Mt Cameroon), and Bioko; (2) E Zaïre from mountains west of L. Edward to Itombwe, SW Rwanda (Nyungwe) and SW Uganda (Impenetrable Forest). Frequent.

Description. *C. r. roberti* (Alexander): Bioko and W Cameroon. ADULT ♂: entire upperparts, scapulars, tertials and wing-coverts olive-brown; lores black with narrow white line above, from bill to eye; ear-coverts rusty brown. Central pair (sometimes two pairs) of tail-feathers black; next pair with black inner web, orange-rufous outer web; others orange-rufous; outermost pair with dark grey smudge of variable extent on outer web near tip (**A**). Chin, throat and breast orange-rufous, flanks washed olive; belly and undertail-coverts white; primaries and secondaries dull brown, fringed olive on outer web, buff on inner web; underwing-coverts and axillaries orange-buff. Bill black; eye brown; feet grey. Sexes alike. SIZE: (15 ♂♂, 4 ♀♀, Bioko, Cameroon) wing, ♂ 66–70 (68·5), ♀ 66–68 (66·5); tail, ♂ 43–51 (46·0), ♀ 41–46 (44·3); bill, ♂ 14–16 (15·2), ♀ 14–15 (14·5); tarsus, ♂ 18–21 (19·7), ♀ 18–21 (19·1).
IMMATURE: like adult except upperparts and wing-coverts black, streaked orange-rufous; underparts buff, washed orange on breast, and scalloped with black.
C. r. rufescentior Hartert: L. Edward to Kivu (E Zaïre). Flanks, sides of belly and undertail-coverts orange-rufous.

Field Characters. Smallest robin-chat; size of akalat *Sheppardia*, with plaintive song atypical of robin-chats, but with tell-tale red-and-black robin-chat tail. Otherwise very similar to sympatric Equatorial Akalat *Sheppardia aequatorialis*, with same orange throat and breast and white belly, but lacks orange on flanks and has white supraloral line, different voice. Bocage's Akalat *S. bocagei* larger, with entirely orange underparts, grey crown, orange of neck extending onto ear-coverts, and uniform red tail. Spotted juveniles have red-and-black tail (tail uniform red in young akalats).

Voice. Tape-recorded (LEM). Song is high-pitched and plaintive; in Rwanda (Nyungwe), a series of whistled notes given in a rapid, continuous flow, lasting 1 min or more without a break; often preceded by a lengthy (up to several min) introduction of subdued phrases; song increases in intensity as singer works itself up (Dowsett-Lemaire 1990). In Nigeria (Obudu Plateau), song short (1·5 s) and melancholy, of 6-note motifs. Rwanda song so similar to that of Forest Robin *Stiphrornis erythrothorax*, that latter reacted strongly to tape (see sonagrams in Dowsett-Lemaire 1990).

General Habits. Inhabits primary and secondary forest; 650–2000 m in Cameroon, 1150–2200 m in Zaïre, 1600 m in Uganda and up to 2000 m in Rwanda (Nyungwe). In Nyungwe prefers damp hollows and flat areas with dense undergrowth, especially of *Mimulopsis arborescens* (Acanthaceae), often along streams. Usually occurs singly; gleans insects from leaves and catches them in air, darting out from perch; feeds at all levels from ground to over 13 m, typically at 2–4 m just above shrub layer.

Food. Mainly insects; also small fruits (*Galmiera*).

Breeding Habits. Nest and eggs unknown. Birds in breeding condition Cameroon, Mar, Zaïre, Jan–May, Uganda, Mar–Apr, June.

References
Dowsett-Lemaire, F. (1990).
Stuart, S. N. (1986).

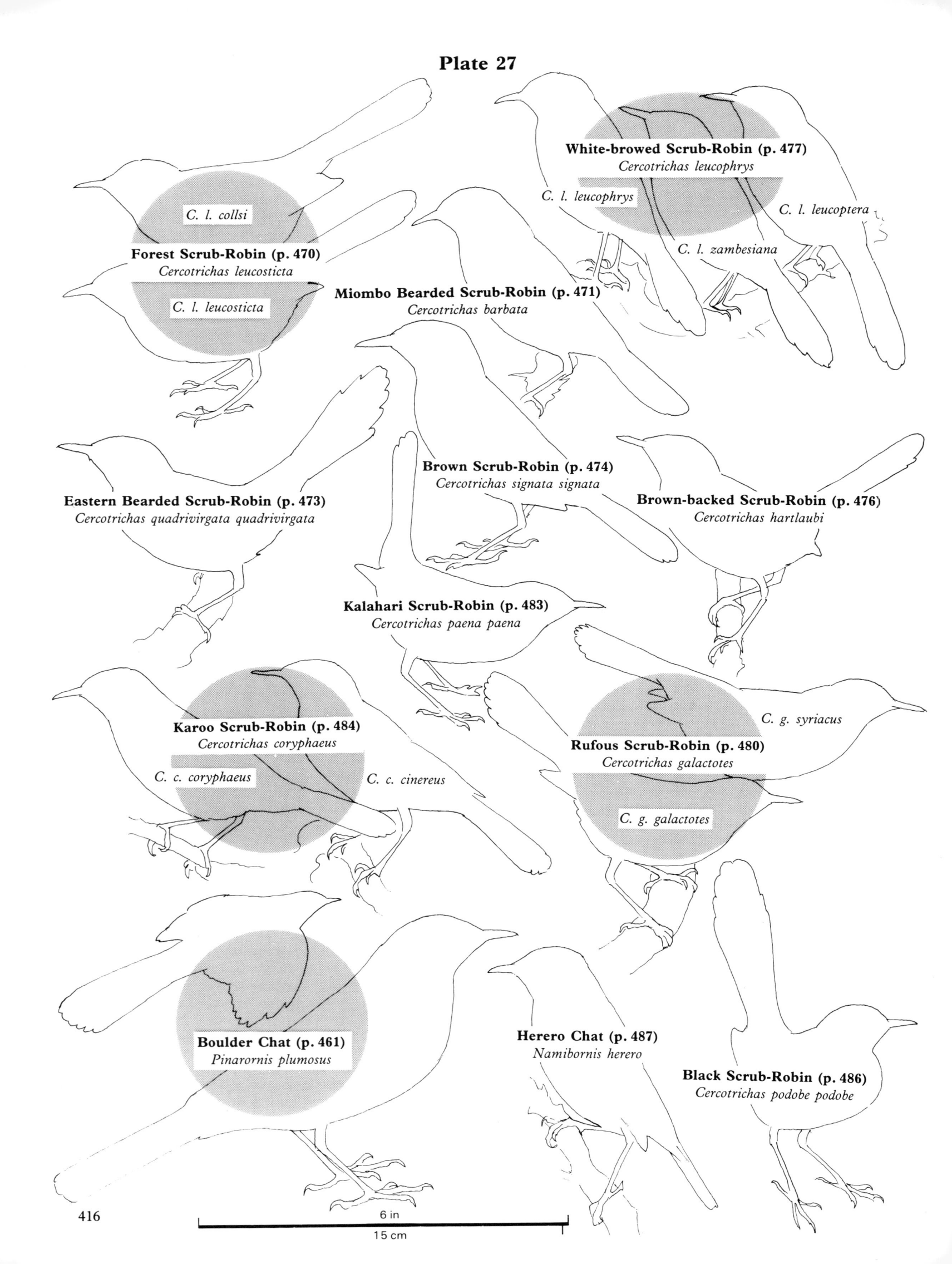

Plate 27

White-browed Scrub-Robin (p. 477)
Cercotrichas leucophrys

C. l. leucophrys

C. l. leucoptera

C. l. zambesiana

C. l. collsi

Forest Scrub-Robin (p. 470)
Cercotrichas leucosticta

C. l. leucosticta

Miombo Bearded Scrub-Robin (p. 471)
Cercotrichas barbata

Eastern Bearded Scrub-Robin (p. 473)
Cercotrichas quadrivirgata quadrivirgata

Brown Scrub-Robin (p. 474)
Cercotrichas signata signata

Brown-backed Scrub-Robin (p. 476)
Cercotrichas hartlaubi

Kalahari Scrub-Robin (p. 483)
Cercotrichas paena paena

Karoo Scrub-Robin (p. 484)
Cercotrichas coryphaeus

C. c. coryphaeus

C. c. cinereus

Rufous Scrub-Robin (p. 480)
Cercotrichas galactotes

C. g. syriacus

C. g. galactotes

Boulder Chat (p. 461)
Pinarornis plumosus

Herero Chat (p. 487)
Namibornis herero

Black Scrub-Robin (p. 486)
Cercotrichas podobe podobe

6 in / 15 cm

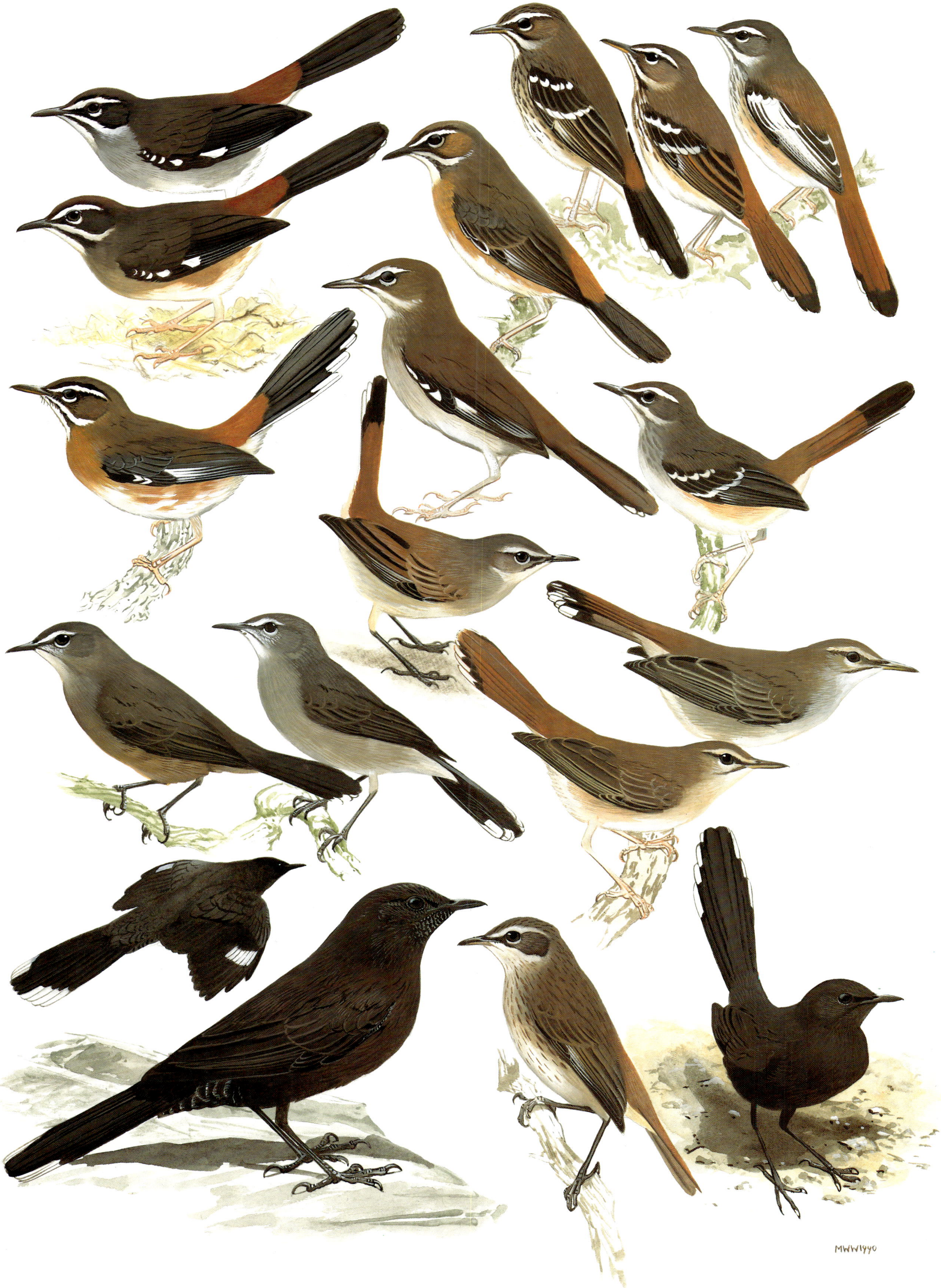
MWW1990

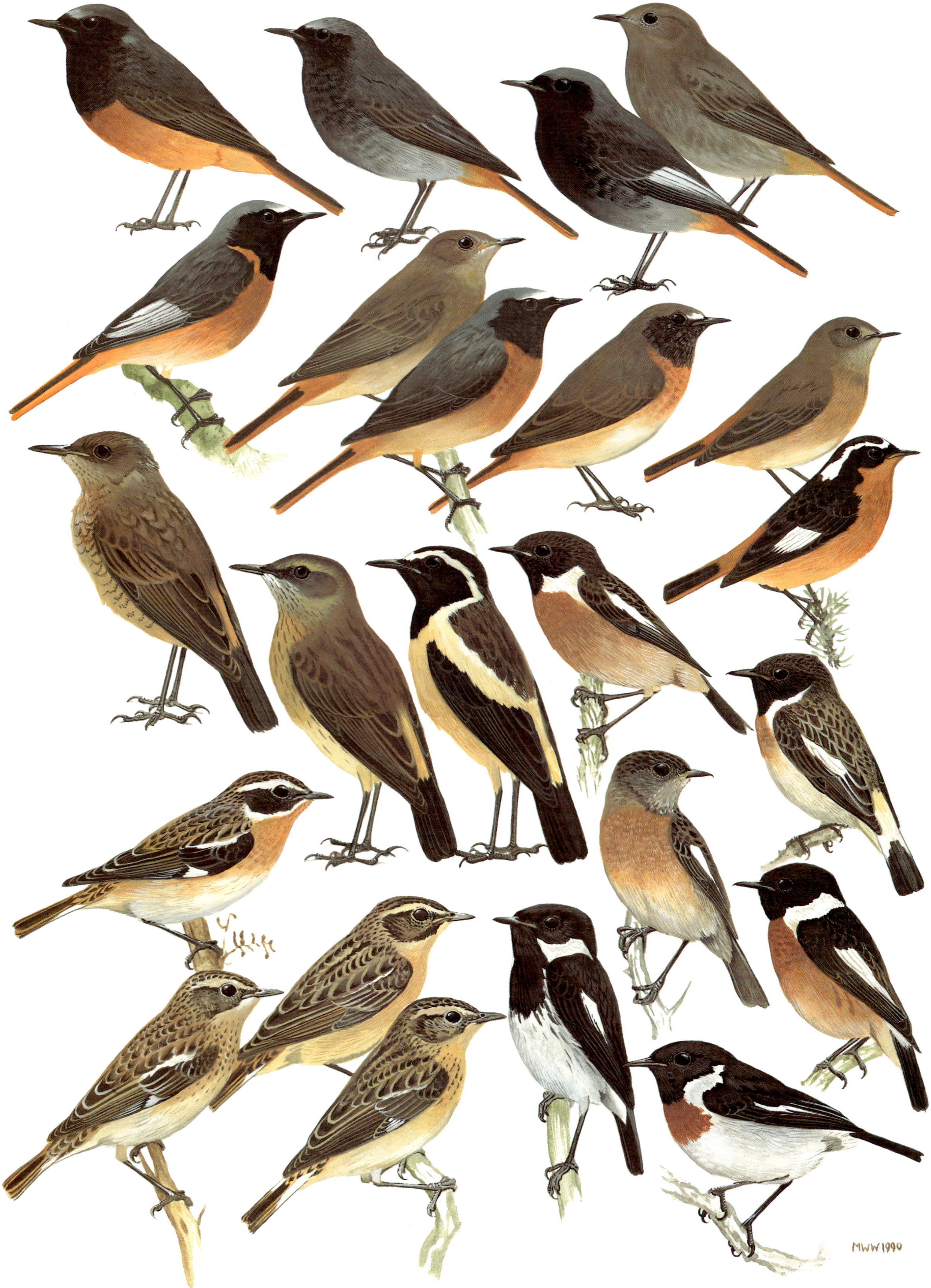
MWW1990

Plate 28

Black Redstart (p. 489)
Phoenicurus ochruros

P. o. phoenicuroides Ad. ♂ breeding

P. o. gibraltariensis Ad. ♂ non-breeding

P. o. gibraltariensis Ad. ♂ breeding

P. o. gibraltariensis Ad. ♀ breeding

Common Redstart (p. 491)
Phoenicurus phoenicurus

P. p. samamisicus Ad. ♂ breeding

P. p. phoenicurus Ad. ♀ breeding

P. p. phoenicurus Ad. ♂ breeding

P. p. phoenicurus Ad. ♂ non-breeding

Juv.

Moussier's Redstart (p. 493)
Phoenicurus moussieri

Ad. ♀

Ad. ♂

Buff-streaked Chat (p. 499)
Saxicola bifasciata

Ad. ♂

Ad. ♀

S. t. rubicola Ad. ♂

S. t. variegata Ad. ♂

S. t. torquata Ad. ♀

Common Stonechat (p. 494)
Saxicola torquata

S. t. torquata Ad. ♂

Whinchat (p. 497)
Saxicola rubetra

Ad. ♂ breeding

Ad. ♀ breeding

Ad. ♂ non-breeding

♂ 1st winter

S. t. albofasciata Ad. ♂

S. t. axillaris Ad. ♂

6 in
15 cm

417

TURDIDAE

Genus *Cossypha* Vigors

Small to medium-sized thrushes with long tarsi, some larger species with long, graduated tail; most with much red in plumage, all with some red, most with white eyebrow, tail red with dark central feathers (except *C. polioptera*, *C. archeri* and some races of *C. anomala*).

Inhabit forest, woodland, thickets, agricultural areas and gardens; feed mainly on ground; most have loud, distinctive songs with brilliant mimicry and much variety and inventiveness; some of the world's finest songsters.

Endemic; 14 species. Among African genera closest to *Cossyphicula* and *Sheppardia*, but relationships with these genera and within *Cossypha* not well understood; only 1 clear superspecies (*C. cyanocampter/ C. heuglini/ C. semirufa*). *C. heinrichi*, *C. niveicapilla* and *C. albicapilla* form a species group; *C. dichroa* is also included in this group by Jensen (1990), but it is probably more closely related to *C. natalensis* (cf. Hall and Moreau 1970). *C. polioptera* lacks black in the tail and has concealed tufts of erectile feathers above the eye, and for these reasons was placed in *Sheppardia* by Jensen (1990); however, in plumage it resembles several members of *Cossypha*, and it has a typical *Cossypha* song, loud and full of mimicry. A second species group comprising *C. archeri*, *C. anomala*, *C. caffra* and *C. humeralis* is listed by Jensen (1990), but S. Stuart (pers. comm.) believes *C. humeralis* is clearly independent and *C. archeri* and *C. anomala* unrelated. Chapin (1953) believed *C. humeralis*, with its white wing-bar and black and white plumage, to be closest to *Xenocopsychus ansorgei* (which he placed in *Cossypha*), but T. B. Oatley (pers. comm.) believes that despite plumage differences it is closest to *C. caffra*, based on build and proportions, slender tarsus, colour of belly and undertail-coverts, similar warning call and shared egg colour. *C. archeri* and *C. isabellae* are similar in size, proportions, general colour and voice, and form an east–west species pair; *C. anomala* is well named, with no clear relatives; these last 3 may not belong in *Cossypha* at all.

Cossypha cyanocampter superspecies

1 *C. cyanocampter*
2 *C. semirufa*
3 *C. heuglini*

Plates 24, 26
(Opp. pp. 369, 401)

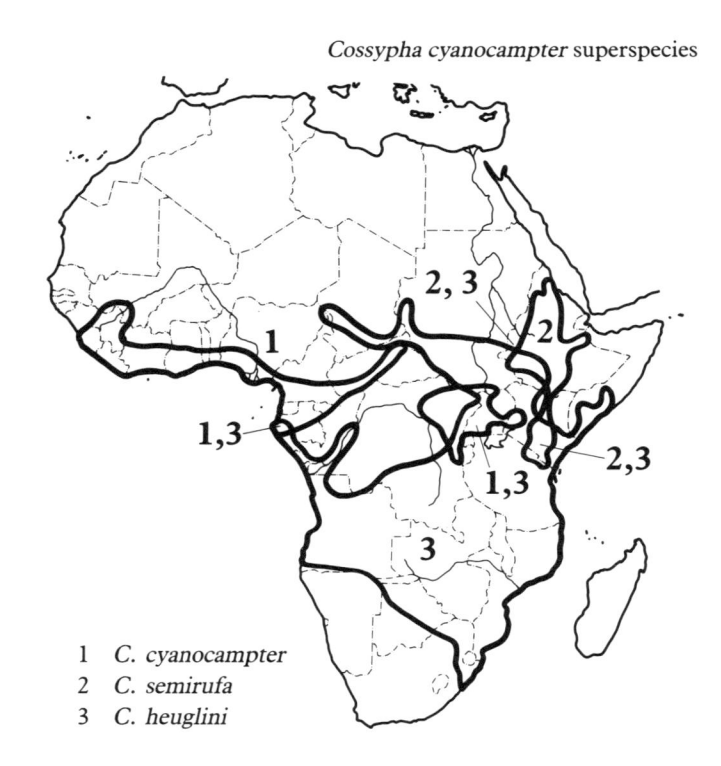

Cossypha isabellae Gray. Mountain Robin-Chat. Cossyphe d'Isabelle.

Cossypha isabellae Gray, 1862. Ann. Mag. Nat. Hist. (3) 10, p. 443; Mt Cameroon.

Range and Status. Endemic resident in montane forest in W Cameroon from Bamenda Highlands southwest to Mt Cameroon, west and north to E Nigeria (Obudu Plateau, Gotel Mts on Mambilla Plateau, and Chappal Hendu (Green 1990)). From 800 m (Mt Cameroon) to 2700 m, though in most locations does not occur below 1100 m. Common to abundant, most numerous at highest elevations.

Description. *C. i. batesi* (Bannerman): range of species except Mt Cameroon. ADULT ♂: forehead, crown, and ear-coverts olive-brown, merging into slightly paler mantle and back, and reddish brown rump. Uppertail-coverts orange-rufous; central pair of tail-feathers dark brown, the rest orange-rufous, tipped with dark brown, outermost pair also with broad dark brown outer margin along distal half. Supercilium and supra-loral line (continuing over base of bill) white; pre-orbital area and cheeks dark greyish black. Chin, throat, breast, flanks and undertail-coverts orange-rufous, becoming paler on upper belly and off-white on lower belly. Flight feathers dark brown with olive-brown outer margins. Upperwing-coverts and scapulars olive-brown. Underwing-coverts pale orange-buff. Bill hooked, black; black rictal bristles; eye brown; legs and feet grey, sometimes brownish or olive-grey. Sexes alike. SIZE: wing, ♂ (n = 10) 73–81 (77·0), ♀ (n = 9) 71–76 (73·0); tail, ♂ (n = 8) 54–60 (58·0), ♀ (n = 7) 51–54 (53·0); bill to skull, ♂ (n = 10) 17–19 (18·1), ♀ (n = 8) 17–18 (17·6); tarsus, ♂ (n = 9) 27–32 (29·9), ♀ (n = 9) 27–30 (28·6). WEIGHT: unsexed (n = 41) 19·5–28·2 (24·3).

IMMATURE: forehead to rump, upperwing-coverts and face very dark brown with pale rufous centres to feathers, giving head a streaked and back a mottled appearance. Chin, throat, breast and flanks very dark brown with very pale buff centres to feathers, giving mottled appearance, merging into dirty off-white centre of belly. Upper mandible blackish with horn tip; lower mandible horn.

NESTLING: unknown.

C. i. isabellae Gray: Mt Cameroon. Forehead, crown, face and ear-coverts blackish olive-brown, merging into dark olive-brown mantle, back and upperwing-coverts. Rump and uppertail-coverts bright rufous. Central pair of tail-feathers blackish (not dark brown), the rest orange-rufous, tipped blackish, outermost pair also with blackish outer margin along distal half. Flight feathers very dark blackish brown, secondaries and tertials with dark olive-brown outer margins, and primaries with pale brown outer margins. Compared with nominate form, slightly richer orange underparts, and buff rather than off-white on belly.

Field Characters. A smallish *Cossypha* of montane forest undergrowth, with black face, white lores and supercilium, red and black tail, and underparts either uniform orange (nominate *isabellae*) or with whitish belly (*batesi*).

Sympatric with two similar-looking species, White-bellied Robin-Chat *Cossyphicula roberti* and Bocage's Akalat *Sheppardia bocagei*. Race *isabellae* distinguished from both by orange belly, but White-bellied Robin-Chat is very similar to race *batesi*, including white eyeline, but is very much smaller; its tail is red and black like *isabellae*; tail of *batesi* is less contrasting (centre browner). Bocage's Akalat, which generally replaces Mountain Robin-Chat at lower altitudes, is intermediate in size between the other 2; it lacks the white eyeline, and orange of neck extends onto ear-coverts; tail darker red and more uniform (less contrast between centre and sides). Immature Mountain Robin-Chat distinguished from immatures of the other 2 by larger size and buff (not whitish grey) belly; also, tails of immatures are similar to those of their adults.

Voice. Tape-recorded (57, LEM). Generally very quiet. Song a rather tuneless 2-note trill 'tree-treeeeee', repeated up to 5 times in rapid succession, giving 10-note phrase. Call a quiet, snoring 'drrr'.

General Habits. Inhabits understorey and ground stratum of montane forest, usually replacing marginally sympatric Bocage's Akalat at higher altitudes. Perches on or near ground, on tree stumps, fallen trees, shrubs, or less commonly on branches up to 9 m from ground. Occasionally leaves forest cover to feed in cultivated areas, often among coco-yams. Solitary, retiring, inconspicuous, but tame. Often feeds on insects around driver ant swarms, where 4–5 birds can be seen together.

Probably sedentary.

Food. Insects, including beetles; also small seeds.

Breeding Habits. Courtship and territorial behaviour unknown.

NEST: only 2 known. 1 cup-shaped and bulky with rudimentary dome of leaf skeletons and moss; bulk of nest composed of moss and a few leaf skeletons, with an intermediate layer of moss and roots, and lining of fine grass stems; situated 1·6 m above ground, wedged into cleft in bole of tall forest tree (Serle 1950). The other was in hollow, dead stump, a cup of green moss and loose lichen, with a moss dome (F. Dowsett-Lemaire, pers. comm.).

EGGS: 2. Pale greenish blue, one with faint rusty speckling. SIZE: 22·3–22·6 × 15·4–15·5.

LAYING DATES: Cameroon, Sept. Evidence from gonads, incubation patches, primary moult and presence of juveniles suggests an ill-defined breeding season, with records in most months.

References
Eisentraut, M. (1973).
Serle, W. (1950).
Stuart, S. N. (1986).

420 TURDIDAE

Plate 24
(Opp. p. 369)

Cossypha archeri Sharpe. Archer's Robin-Chat. Cossyphe d'Archer.

Cossypha archeri Sharpe, 1902. Bull. Br. Orn. Club 13, p. 9; Ruwenzori Mts, Uganda.

Range and Status. Endemic resident in montane forest around Albertine Rift. Ranges from northwest of L. Edward south to Mt Kabobo (Zaïre) and from Mt Ruwenzori (Uganda and Zaïre) south through W Rwanda to W Burundi (Ijenda). Common to abundant.

Description. *C. a. archeri* Sharpe (including 'albimentalis'): range of species except Mt Kabobo. ADULT ♂: forehead, crown, nape and ear-coverts dark brown, merging into rich amber-brown on mantle, back and rump; dark reddish brown uppertail-coverts. Supercilium white. Greyish black face, chin and around eye. A small, variably sized white spot in centre of chin; throat orange-rufous; breast dark orange-rufous, becoming paler on flanks and sides of belly, and pale buff to off-white in centre of belly and vent; undertail-coverts orange-rufous. Flight feathers dark brown, outer margins of primaries pale brown, of secondaries and tertials reddish brown. Upperwing-coverts and scapulars rich amber-brown. Underwing-coverts and axillaries orange-brown. Bill hooked, black; eye dark brown; legs and feet brown. Sexes alike. SIZE: wing, ♂ (n = 5) 74–82 (78·0), ♀ (n = 3) 70–74 (73·0); tail, ♂ (n = 5) 59–62 (61·0), ♀ (n = 3) 53–56 (55·0); bill from skull, ♂ (n = 5) 17–18 (17·6), ♀ (n = 2) 17; tarsus, ♂ (n = 5) 31–33 (32·1), ♀ (n = 3) 28–31 (29·5). WEIGHT: ♂ (n = 4) 20–25, ♀ (n = 3) 21–26.

IMMATURE: upperparts similar to adult. Indistinct supercilium of orange-buff feathers with dark outer margins. Feathers of underparts have same basic colour as in adult (i.e. darkest rufous on breast) but have dark olive-brown tips giving mottled appearance, especially on throat and breast. Flight feathers and tail as in adult. Bill dark brown, lower mandible usually paler.
NESTLING: unknown.

C. a. kimbutui (Prigogine): Mt Kabobo, E Zaïre. Forehead and face very dark greyish black, merging into blackish olive on crown, nape and ear-coverts; mantle, back, rump, scapulars and upperwing-coverts olive-brown; tail dark brown. Flight feathers dark brown, outer margins of primaries greyish, of secondaries and tertials olive-brown.

Field Characters. A rather small, plump robin-chat with brown upperparts, uniform reddish tail, grey face, white eyebrow and orange underparts. A high-altitude species, the only robin in most of its range except the very different Cape Robin-Chat *Cossypha caffra* (mainly grey, with orange throat and breast, red and black tail). At lower elevations must be distinguished with care from several similar-looking species. Slightly larger Bocage's Akalat *Sheppardia bocagei* has grey top of head, rufous of throat extending to eye and ear-coverts, and lacks white eye-stripe; slightly smaller Equatorial Akalat *S. aequatorialis* has olive upperparts, bluish grey face with no white eye-stripe, pale orange chin and throat extending almost to eye; smaller White-bellied Robin-Chat *Cossyphicula roberti* has similar blackish pre-orbital area and white supercilium but orange of chin and throat extends to eye and onto ear-coverts, and tail is red with black centre, not uniform red. Immature Archer's Robin-Chat more rufous below than immatures of any of these 3; further distinguished from White-bellied Robin-Chat by plain brown tail (latter already has black and red tail of adult), and from Bocage's and Equatorial Akalats in lacking spots on underparts. Monotonous plaintive song distinctive.

Voice. Tape-recorded (32, C, CART, LEM). Song a series of high-pitched melancholy, unmusical notes 'ee chee chee cheeyie yee', frequently repeated; notes trilled, not pure. Call a frog-like double or triple grunt 'tonk-tonk' or 'cop-cop-cop'. Sings throughout the day for several months of the year.

General Habits. Inhabits montane forest undergrowth, bamboo and giant heath, from 1600 m to upper limit of *Senecio* (groundsel) around 4300 m. Common to abundant along stream edges, usually in dense vegetation close to ground, occasionally up to 10 m high in forest. In very moist forests occurs throughout understorey; in drier forests usually restricted to vicinity of streams.

Occurs singly or in pairs. Feeds on forest floor and near the ground on moss-covered fallen logs. Joins mixed-species flocks; sometimes follows driver ant swarms, but less habitually than other robins; showed no interest in ants in Nyungwe Forest, Rwanda (Dowsett-Lemaire 1990). When alarmed, it flicks tail to angle of 45° above horizontal.

Believed to be sedentary.

Food. Insects, including beetles, ants, caterpillars and bugs.

Breeding Habits. Territorial; territory size in optimal habitat less than 1 ha (Nyungwe Forest, Rwanda: Dowsett-Lemaire 1990).

NEST: only 2 known (Masterson, 1981; J.-P. Vande weghe, pers. comm.). One, a neat cup of rootlets and tendrils, in depression 1·2 m up on side of large mass of moss clinging to trunk of giant *Senecio*; no strong exterior, moss provided its only support. Second situated 1 m from ground in dead branches along stream at 1750 m.

EGGS: 2. Pale blue, freckled with a few small pale brown spots, mostly around centre and towards the small end, not forming any distinct ring. SIZE: (n = 2) 23·5 × 16, 24·4 × 15·7.

LAYING DATES: Rwanda, Jan, Apr, Oct, Dec (Nyungwe Forest); Uganda, Jan; Zaïre (enlarged gonads, juveniles present, Oct–Nov (Mt Kabobo), and Oct, Mar and May (Itombwe)). These data suggest ill-defined breeding season.

References
Chapin, J. P. (1953).
Prigogine, A. (1960, 1971).

Cossypha anomala (Shelley). Olive-flanked Robin-Chat. Cossyphe à flancs olives.

Callene anomala Shelley, 1893. Ibis (6)5, p. 14; Mt Mulanje, Nyasaland.

Plate 24
(Opp. p. 369)

Range and Status. Endemic resident in E African montane forest. An isolated population in N Tanzania (Mbulu); main range is highlands of E and S Tanzania, from Ukagurus and Ulugurus southwards, higher mountains of Malaŵi south to Mulanje, N Mozambique (Mt Namuli) and NE Zambia (Nyika Plateau). Occurs from 1000 to at least 2600 m. Uncommon to common; density in Malaŵi 10 pairs/10 ha (Dowsett-Lemaire 1989).

Description. *C. a. grotei* (Reichenow): highlands of E Tanzania from Ukagurus and Ulugurus southwest to Njombe and Songea. ADULT ♂: forehead white or pale grey, merging into white supercilium, which becomes pale grey behind eye. Crown, nape and ear-coverts dark slaty grey, becoming slightly paler on mantle, and tinged with rufous-brown on back and scapulars. Rump dull rufous-brown, merging into bright orange-rufous uppertail-coverts. T1 and T2 blackish brown with reddish tinge; rest of tail-feathers orange-rufous, tipped with dark brown, T6 also has dark brown outer margin along distal half. Face and moustachial stripes black. Chin and throat white. Centre of breast pale grey, sides dark grey, merging into olive-rufous flanks and white belly and vent; undertail-coverts pale buff or pale rufous. Flight feathers dark grey-brown, secondaries and tertials with olive outer margins, primaries with pale grey-brown outer margins. Upperwing-coverts dark slaty grey. Underwing-coverts and axillaries pale grey to pale orange-buff. Bill black; eye dark brown; legs and feet brown or grey-brown. Sexes alike. SIZE: wing, ♂ (n = 5) 76–80 (78·0), ♀ (n = 4) 73–83 (77·0); tail, ♂ (n = 5) 63–67 (65·0), ♀ (n = 4) 60–68 (62·0); bill to skull, ♂ (n = 5) 18–19 (18·6), ♀ (n = 3) 17–19 (18·2); tarsus, ♂ (n = 4) 31–34 (32·2), ♀ (n = 3) 32–33 (32·4). WEIGHT: ♂ (n = 7) 26–28 (26·9), ♀ (n = 2) 22·5, 25.

IMMATURE: feathers of rump, upperwing-coverts and underparts brown with buff centres, often tipped dark brown, giving speckled appearance; centre of belly and vent pale buff with very few speckles. Uppertail-coverts buff or rufous. Flight feathers and tail dark brown. Bill paler than adult, especially lower mandible.

NESTLING: unknown.

C. a. anomala (Shelley): Mt Mulanje, S Malaŵi, and Mt Namuli, N Mozambique (includes *C. a. gurue* from Namuli, said to have darker upperparts). Grey forehead, supercilium and face, merging into brown on crown, sides of neck and nape. No dark moustachial stripe. Chin white, merging into very pale grey throat. Mantle, back, scapulars and upperwing-coverts brown, becoming more rufous on rump and reddish brown on uppertail-coverts. Tail dull reddish brown without darker central tail-feathers. Breast grey, becoming much paler on belly. Olive-rufous flanks duller than in *grotei*. Flight feathers as in *grotei* but browner, less grey.

C. a. macclounii (Shelley): Tukuyu District, S Tanzania south to Viphya Plateau, Malaŵi. Like *grotei* but white forehead and supercilium less pronounced, and slightly more olive wash on slate-grey crown, back and upperwing-coverts. Rump and uppertail-coverts much duller rufous. Tail dull reddish brown, like nominate form. Underparts and flight feathers as in *grotei*. WEIGHT: ♂ (n = 6) 22–26·5 (24·3), ♀ (n = 2) 24, 26·5.

C. a. mbuluensis (Grant and Mackworth-Praed): Mbulu Highlands, N-central Tanzania. Similar to *grotei*, but white supercilium very pronounced, continuing behind eye. Upperparts, face, moustachial stripe and wings even darker; head almost black. Tail pattern as in *grotei* but black and orange strongly contrasted (rather than blackish brown and orange-rufous). Breast and flanks dark grey, flanks with olive-rufous wash. Belly grey; vent and undertail-coverts rufous. Axillaries grey. Legs and feet dark brown.

Field Characters. A typical *Cossypha* in most characteristics (although *anomala* and *macclounii* do not have the typical *Cossypha* tail pattern). Races *grotei*, *macclounii* and *mbuluensis* readily distinguished from all related robins by white supercilium and throat, black moustachial stripe, pale grey underparts and olive-rufous flanks. Nominate *anomala* has much less contrast in head pattern, but has the pale grey underparts and olive-rufous flanks. All other *Cossypha* spp. have at least some orange or orange-rufous on underparts. Sympatric White-chested Alethe *A. fuelleborni* and Cholo Alethe *A. choloensis* have rich brown upperparts and are much whiter below, and lack olive-rufous flanks and distinctive head pattern; White-chested Alethe is much larger, Cholo Alethe is somewhat larger, with white tips to outer tail-feathers. Immature Olive-flanked Robin-Chat best separated from other immature robins by size, and from other robin-chats by brown tail; it lacks the white belly of sympatric immature *Alethe* spp.

Voice. Tape-recorded (86, LEM). Song a series of 4–7 thin warbling or whistling notes, 'er see hurry per yoh' or 'tuo-tjrrio-tu', with much regional variation. First 3 and last 3 notes are usually on a descending scale, with antepenultimate note highest. Song of *mbuluensis* quite different, a short whistle followed by a long one, with emphasis on second note, 'fe-fuuuur'. Alarm a short, loud, harsh 'har', 'chop' or 'wump', repeated irregularly.

General Habits. Inhabits montane forest, forest patches, edges and nearby gardens; often low in dense vines or shrubs in forest openings, along streams, among stands of tree ferns; favours places where it can feed on soft ground.

Occurs singly or in pairs. Flicks wings, raising and slightly lowering tail at the same time. Usually keeps to ground, or to twigs, vines or logs very close to ground, but perches at 3–5 m when singing, and ascends to 7 m when driver ants are climbing vegetation. Feeds on ground or on logs, trunks, lianas and twigs, gleaning surfaces or taking insect after short sally or hopping run. Often associated with ant swarms, but also occurs in forest patches without ants. Aggressive towards smaller birds at ant swarms (e.g. White-starred Robin *Pogonocichla stellata* and Iringa Akalat *Sheppardia lowei*), but displaced by larger ones (e.g. White-chested Alethe).

Some evidence of movement to lower altitudes outside breeding season (at least in S Malaŵi where occurs at only 900 m on Mt Mulanje in Aug), but probably mainly sedentary. Strictly sedentary on Nyika Plateau, with no homing ability after translocation (Dowsett and Dowsett-Lemaire 1986). Retrapped birds moved 0–620 (150 ± 139) m from ringing site in same forest patch (Dowsett 1985).

Food. Insects.

Breeding Habits. Territorial. On Nyika Plateau (Malaŵi), territories occupied by pairs or unmated birds. Territory size 0·25–2·5 ha, smaller where there is dense vegetation along streams, larger in dry forests on ridges with sparse ground cover (Dowsett-Lemaire 1983). Of 26 birds, 9 held same territory for 2 years, 9 for 3 years, 2 for 4 years, 1 for each successive number of years, up to 10 (Dowsett 1985).

NEST: 1 was deep cup of dead leaves and moss lined with black vegetable fibres and layer of red-brown vegetable matter, placed in hollow 1·3 m above ground in dead tree; ramp of green moss in front of entrance; another was placed on horizontally growing stems of canes in a thicket; a third was on the ground, 1 m up a stream bank, with a 70 mm diam. open cup; a fourth was built into dried grass draping the top of a stump 0·7 m above ground.

EGGS: 2. Pyriform; putty-coloured, some with mottling and darker shading at large end. SIZE: 25–27 × 15–17.

LAYING DATES: Tanzania, July, Oct–Dec; Malaŵi (Nyika Plateau), Nov–Jan (Nov 22, Dec 20, Jan 6 clutches).

BREEDING SUCCESS/SURVIVAL: of 37 pairs on Nyika Plateau, 16 (43%) successfully raised young; productivity was 0·7 young per pair, 1·7 per successful pair (Dowsett-Lemaire 1985). Average annual mortality rate 13·7% in ♂♂, 43·3% in ♀♀; lives at least 10 years (Dowsett 1985).

References
Dowsett-Lemaire, F. (1983).
Dowsett, R. J. and Dowsett-Lemaire, F. (1984).
Willis, E. O. (1985).

Cossypha caffra (Linnaeus). Cape Robin-Chat. Cossyphe du Cap.

Motacilla caffra Linnaeus, 1771. Mantiss. Plantar., p. 527; Cape of Good Hope.

Plates 24, 26
(Opp. pp. 369, 401)

Range and Status. Endemic resident. Extreme S Sudan (common Imatong Mts, rare elsewhere); highlands of E Zaïre from S end of Albert Nat. Park (Kibati, Tshumagassa, Mt Gahinga) to Itombwe, and on Marungu Plateau; volcano region of extreme E Kivu, SW Uganda and NW Rwanda; SE Rwanda (Nyungwe); Uganda in SW (Kigezi, Ankole), NE (Mt Lonyili, Mt Morongole) and E (Mt Elgon); throughout W and central highlands of Kenya and on outlying mountains (Kulal, Nyiru, Marsabit, Matthews Range, Chyulu and Taita hills); Tanzania, in W at Ngara, Mahari Mt and Ufipa Plateau, and throughout N and E highlands south to Songea and Malaŵi border; NW Zambia (Mafinga Mts and Nyika Plateau); Malaŵi, common above 1500 m, south in N to Viphya Mts, and in S from Dedza and Mangoche to Mwanza and Mulanje; Mozambique (N-central mountains east to Mt Namuli, mountains along Zimbabwe border, Mt Gorongoza and probably elsewhere on Manica Platform); Zimbabwe (E highlands from Inyanga south to Chimanimani Mts and in non-breeding season to Mt Selinda, inland on watershed of Mashonaland Plateau to Marandellas); W Swaziland; lowlands of Lesotho; South Africa in Transvaal (almost throughout except E Lowfeld and drier NW), Orange Free State, Natal (throughout but in lowlands only in winter) and Cape Province (throughout except dry W-central region); extreme SW Namibia (north to Great Fish R.). Some winter in Lebombo Mts (Transvaal–Mozambique border, E Swaziland) and coastal lowlands of Natal. Common.

Description. *C. c. iolaema* Reichenow: Malaŵi and N Mozambique to Sudan, including E Zaïre (Marungu). ADULT ♂: upperparts from forehead to rump, including scapulars, olive-grey, more strongly olive-brown on rump, less so on crown; uppertail-coverts tawny-rufous; central tail-feathers grey-brown, others rufous, variably tipped grey-brown; supercilium white, extending behind eye; lores and ear-coverts black; moustachial streak white, malar black; sides of neck and breast grey; chin, throat and centre of breast orange-rufous; front and sides of belly grey, centre white; thighs olive-rufous; undertail-coverts orange-buff; remiges and wing-coverts dull brown, edged olive-brown; alula and outer primary coverts tipped white, producing a white spot distal from bend of wing; underwing-coverts and axillaries orange-buff. Bill black; eye dark brown; feet black. Sexes alike. SIZE: wing, ♂ (n = 11) 81–90 (85·2), ♀ (n = 11) 73–84 (79·2); tail, ♂ (n = 11) 72–83 (77·6), ♀ (n = 11) 69–79 (73·6); bill, ♂ (n = 10) 17–19 (18·3), ♀ (n = 10) 16–19 (17·8); tarsus, ♂ (n = 10) 27–31 (29·2), ♀ (n = 10) 28–31 (29·1). WEIGHT: (Malaŵi) ♂ (n = 35) 26·2–34 (28·9), ♀ (n = 23) 25·4–32 (28·0).

IMMATURE: like adult but upperparts and wing-coverts black, streaked and spotted orange-buff; underparts orange-buff, scalloped black. These marks remain longest on nape, face, rump and lower breast.

NESTLING: skin pale pinkish orange, iron-grey down on head, flanks and rump.

C. c. kivuensis Schouteden: SW Uganda to Kivu (E Zaïre) and northwest of L. Tanganyika.

C. c. caffra: SW Cape Province to Zimbabwe, where intergrades with *iolaema* in E. As *iolaema* except upperparts paler, more olive-brown; orange and grey of underparts paler. Slightly larger. SIZE: (5 ♂♂, 5 ♀♀) wing, ♂ 85–92 (88·8), ♀ 76–82 (80·2); tail, ♂ (79–86 (82·8), ♀ 75–78 (76·4); bill, ♂ 18–20 (19·4), ♀ 18–19 (18·6); tarsus, ♂ 29–31 (30·0), ♀ 29–31 (29·8).

C. c. namaquensis Sclater: Namibia to Orange Free State, north of last race. Like nominate but white supercilium better developed behind eye and broader in many individuals. Still larger. SIZE: (5 ♂♂, 1 ♀) wing, ♂ 90–95 (92·6), ♀ 85; tail, ♂ 83–90 (87·2), ♀ 79; bill, ♂ 19–20 (19·6), ♀ 17; tarsus, ♂ 30–31 (30·2), ♀ 31.

Field Characters. A familiar garden bird of eastern and southern Africa. The only robin-chat with a grey belly (orange confined to throat, breast and undertail-coverts). Song less powerful and more repetitive than other robin-chats, with fewer imitations. Often sings in full view from prominent perch. On alighting flicks wings and partly fans tail; when moving about on ground or in thicket, frequently jerks tail up and lets it fall slowly.

Voice. Tape-recorded (5, 13, 15, 20, 24, 36, 46, 86, C, F). Song melodious; deliberate, high-pitched, whistled phrases of 5–7 notes, commencing with down-slurred single note (inaudible at a distance). Mimics other birds but less than most other robin-chats; skilfully assimilates borrowed notes without changing nature of its song. Some individuals never imitate; others incorporate calls of up to 20 different birds; sometimes also inserts its own alarm call. Song of northern races faster than that in S.

General Habits. In E and central Africa inhabits highland tea and coffee plantations, woodland, gardens, giant heath, edges of bamboo, forested ravines and bracken-briar fringes of large forests; seldom penetrates forest interior. South of Limpopo R. occurs in most areas with dense cover 2 m high and with scattered trees or other prominent song-posts. In dry W of southern Africa, largely confined to trees along watercourses and vicinity of human settlements. Common in gardens from Transvaal to Cape Province. Occurs above 1600 m in Sudan, 1800–3475 m in Zaïre, above 1800 m in Zambia. In E Africa seldom resident below 1500 m and ranges up to 3400 m, exceptionally down to 500 m; in South Africa occurs from sea level (only as winter visitor along Natal coast) to 2800 m.

Solitary or in pairs. Has hopping gait. Short flights silent and agile, but long flights often audible and with a shallow undulation. Bathes regularly in well-sheltered streams or garden bird baths; drinks frequently. Calls at dusk when preparing to roost and when predator seen. Flits wings, and jerks tail upwards 2–3 times. Roosts in leafy site 2–8 m above ground level; uses same perch nightly for weeks at a time.

Forages mainly on ground, mates frequently feeding within sight of each other, under shrubs and garden hedges; often perches on low branch, rock or other eminence, flying down to capture prey; prefers to disturb prey in leaf litter by hopping through it rather than by whisking leaves with bill. Follows army ant columns. Gleans foliage, twigs and tree trunks; makes short sallies; freely ascends to canopy where caterpillars abundant, especially if feeding fledgling Red-chested Cuckoo *Cuculus solitarius*. Visits bird tables, taking bone meal, grated cheese and other tidbits, raids pets' food bowls, and even enters houses to snatch butter.

A marked altitudinal migrant in southern Africa north to Malaŵi, some movements also in E Africa. Non-breeding birds move down from mountains, scarps and plateaux in dry season (e.g. Natal: Cyrus 1989); extent of movements not known.

Food. Insects; also earthworms, crustaceans, molluscs; occasionally frogs, small lizards, small drupes and berries. In 17 stomachs and 87 faecal samples (mainly from Natal), ants occurred in 88% of samples, beetles (Coccinellidae, Curculionidae, Melolonthidae, Scarabaeidae, Staphylinidae, Tenebrionidae) in 67%, fruit in 65%, moths and caterpillars in 46%, termites in 35%, parasitic wasps in 17%, bugs (Naucoridae, Tingidae) in 15%, spiders in 12%, crickets and grasshoppers in 10%, flies (Asilidae, Tabanidae) in 9%, centipedes and pseudoscorpions in 4%. Termites often predominate in diet of birds in acacia woodlands. In SW Cape Province, eats seeds (with attached oil-rich funicles) of exotic *Acacia cyclops*, also fruits of *Asparagus*, *Celtis*, *Cestrum*, *Ficus*, *Halleria*, *Hedychium*, *Ilex*, *Kiggelaria*, *Maytenus*, *Morus*, *Olea*, *Physalis*, *Psidium*, *Rhus*, *Rubus* and *Solanum*.

Breeding Habits. Monogamous, territorial. Territory size 0·05–0·75 ha. In South Africa pairs for life; mated pair remains in territory all year; lost mate is quickly replaced from substantial non-breeding population, without disruption to neighbouring pairs. Territorial skirmishes between single birds and mated pairs occur throughout year except during moult. ♂ advertises territory from prominent song-post (or several in large territory). In territorial display, ♂ faces rival, raises fully-fanned tail and sings loudly; then he chases intruder, often accompanied by ♀. Such pursuits may go on intermittently all day if trespasser persistent. In precopulatory display, ♂ sings from low horizontal perch c. 1 m above ground and struts sideways along it, drooping wings and fanning depressed tail while gradually increasing song intensity; when ♀ appears, ♂ stands upright, singing with lowered, fanned tail, then bows in different directions; in bowing movement head is held horizontal and lowered smoothly until below level of back; this position is maintained for 1–2 s. ♀ approaches, flicking tail and flexing drooped wings; ♂ leaves perch and flies with fanned, lowered tail towards her, singing loudly, and copulation ensues, after which ♂ flies up above canopy or vegetation-level uttering loud, ecstatic whistles and varied, incomplete song-phrases. ♀ remains at perch for a while, lowering wings spasmodically before resuming normal activities.

NEST: substantial, often bulky structure of twigs, bark fragments, dead leaves, dried bents and grass stems, enclosing cup of finer material; cup lined with fine rootlets, other soft vegetable material, and often animal hair; deep, sometimes 5 mm deeper than wide; diam. 60–70 (66), depth av. 52, ext. diam. of nest 150. 30% of 966 South African nests were on ground; higher nests placed in leafy vegetation, branch and flood debris, recesses in stream banks, open buildings, or discarded artefacts overgrown with vegetation; av. height above ground 1·1 m (highest 3·66 m). Many nests made visible by untidy ramps of long grass stems; ground nests much harder to find. Built by ♀ in 6–15 days; ♂ sometimes escorts ♀ but does not carry material or aid in construction.

EGGS: 2–3, usually laid on consecutive days, although 40–60 h lapses between 2nd and 3rd egg are known. In SW Cape, 83% of 515 clutches had 2 eggs. Ratio of C/2 to C/3 in Transvaal (106 clutches) 1 : 1·08; in Natal (210 clutches) 1 : 1·06. South African mean (975 clutches) 2·3. 2-egg clutches predominate in Malaŵi and 3-egg clutches in Kenya. Av. clutch in Malaŵi (23 clutches) 2·1 eggs. Ovate to elliptical ovate; white, cream, buff, pale green or blue, variably freckled, spotted and blotched with salmon-pink, russet or chocolate, often forming cap at large end; markings sometimes uniformly dense, making

Alarm call, guttural 'wur-de-dur' or 'gar-ga-garg'; faster north of Zambezi R. and sometimes of 4–5 syllables. Anxiety note, given when nest or dependent young threatened, is soft, plaintive, descending whistle 'peeeeooo', like call of Grey Cuckoo-Shrike *Coracina caesia*. Location call of fledgling, quiet 'seeep' at intervals, becoming strident squeaking when bird fed by parent. On occasion adult utters alarm call and song in flight.

whole egg brown or pink. Blue eggs (rare) may be immaculate. SIZE: (n = 559, South Africa) 19–26 × 14–18 (23·0 × 16·5). Most pairs probably attempt to rear 2 broods and some 3; 3rd broods sometimes successful after completion of post-breeding moult (South Africa).

LAYING DATES: Sudan, Jan, Mar; Zaïre (breeding condition May, July); E Africa: Region A, Jan–June, Aug, Oct–Nov; Region C, Jan, Mar, May; Region D, all months (peaking in long rains, with smaller peak in short rains); Zambia, Oct–Jan; Malawi, Oct–Jan (Oct 4, Nov 9, Dec 9, Jan 4 clutches); Zimbabwe, Sept–Dec (Sept 4, Oct 10, Nov 11, Dec 7 clutches); South Africa: Transvaal, Aug–Feb (Aug 14, Sept 30, Oct 66, Nov 52, Dec 36, Jan 6, Feb 1 clutches); Natal, Aug–Dec; Orange Free State, Sept–Jan, Apr–May; E Cape, July–Jan, May; Karoo, Aug–Dec; SW Cape, June–Nov.

INCUBATION: begins with laying of last egg, or up to 2 days later. In South Africa only ♀ incubates; in E Africa ♂ said to incubate for short spells in late afternoon and to bring food to incubating ♀ when eggs close to hatching (van Someren 1956). Period: (n = 38) 13–19 (16·1) days. Periods of >16 days probably result from late start of incubation.

DEVELOPMENT AND CARE OF YOUNG: chick gains 1·5–2 g daily up to day 10, less thereafter; fully feathered by day 12, when it invariably shows fear.

Chicks brooded regularly by ♀ for first 4–5 days, especially in early morning or bad weather; fed by both parents; nest visits increase from av. 6 per h per chick for first 4–5 days to 8 per h per chick for rest of period; overall av. 10 visits per h for nests with 1 chick and 16 per h for broods of 2 or 3; food may be given to more than 1 chick during a visit. Parent carries away or swallows faecal sacs; fledglings may perch on rim of nest for long periods and defaecate there. Nestling period: (n = 32) 15–18 (16·5) days. Periods of 15 days probably short due to disturbance; wet weather delays fledging.

After leaving nest young can scarcely fly; they hide in bushes where their dark, buff-spotted plumage conceals them well. ♂ and ♀ continue to feed them for 5–7 weeks (much less if another clutch is started within 10–20 days, in which case only ♂ feeds them). Post-juvenile moult begins 11–18 weeks after hatching.

BREEDING SUCCESS/SURVIVAL: 20·8% of eggs laid over 2 years in acacia savanna in Natal produced fledged young; 10 fledglings raised from 19 nests; predation was greatest factor in reducing breeding success (Earlé 1981). Commonly parasitized by Red-Chested Cuckoo *Cuculus solitarius* in South Africa, especially in Natal and Transvaal, where 16–22% of nests affected. Several birds lived for over 7 years in South Africa (max. 7 years 8 months).

Reference
Rowan, M. K. (1969).

Cossypha humeralis (Smith). **White-throated Robin-Chat. Cossyphe à gorge blanche.**

Dessonornis humeralis A. Smith, 1836. Rep. Exped. Expl. Cent. Afr., p. 46; Marikwa (= Marico) R., W Transvaal.

Plate 25
(Opp. p. 400)

Range and Status. Endemic resident. From Zambezi-Limpopo watershed (Zimbabwe: northern limits on Zambezi tributaries above 900 m) through Limpopo basin to E and S Botswana (east of a line from Plumtree to Kanye), S Mozambique south from Buzi and Save rivers, including Bazaruto I., Transvaal (except Highveld), Swaziland, N Natal (south to Tugela R.) and extreme NE Cape Province. Common.

Description. (Includes '*crepuscula*'). ADULT ♂: forehead, crown, nape, hindneck, mantle and back slate-grey; tail-feathers tawny-orange, central pair blackish brown, remainder broadly tipped black. Lores, ear-coverts and sides of neck black; white supercilium extending to rear of ear-coverts; chin to belly immaculate white; rump and uppertail-coverts buffy orange; flanks and undertail-coverts buffy orange. Flight feathers black, outer webs of primaries and secondaries edged grey, outer webs of tertials entirely white; lesser coverts and inner secondary coverts white, rest of upperwing-coverts black. Marginal underwing-coverts black, rest of underwing-coverts and axillaries white. Bill black; eye dark brown; legs and feet black, soles grey. Sexes alike, but ♀ slightly duller. SIZE: wing, ♂ (n = 5) 79–83 (81·2), ♀ (n = 7) 76–82 (78·4); tail, ♂ (n = 5) 74–79 (75·8), ♀ (n = 7) 70–78 (73·0); bill, ♂ (n = 5) 19–20 (19·6), ♀ (n = 6) 17–19 (18·5); tarsus, ♂ (n = 5) 26–28 (26·8), ♀ (n = 7) 26–28 (27·1). WEIGHT: (Zimbabwe) ♂ (n = 14)

20·2–24·7 (22·4 ± 1·5), ♀ (n = 8) 19–23·1 (20·1 ± 1·5) (Jackson 1989); (Zimbabwe/Transvaal) ♂ (n = 14) 22–26 (24·2), ♀ (n = 13) 19–25 (21·0), ♂♀ (n = 54) 17–28 (22·9).

IMMATURE: mottled; upperparts blackish brown, feathers with orange-buff tips; underparts buff, feathers tipped black. No wing-bar in birds under 8 weeks old; tail as in adult.

NESTLING: hatchling pink-skinned with long, grey down which persists until after fledging; gape yellow-ochre, legs and feet pinkish yellow, bill pinkish horn.

Field Characters. A woodland robin with unique colour combination (grey upperparts, rufous rump, red-and-black tail, white supercilium, black wings with white flash, underparts immaculate white except orange lower flanks). Superficially similar Angola Cave-Chat *Xenocopsychus ansorgei* lacks rufous and is allopatric. When dashing for cover, gives impression of a small black-and-white bird with a red tail (Sinclair 1984). On brief glimpse in bush, black-and-white plumage might suggest Southern Boubou *Laniarius ferrugineus*, which is larger, with heavy bill, no white supercilium, black tail.

Voice. Tape-recorded (15, 24, 81, 88, C, F, KEI, OAT). Song, by ♂ and ♀, a sustained, high-pitched jumble of whistles, warbles and trills, mixed with much mimicry; phrases last up to 10 s. Mimics many resident bird species, also e.g. European Bee-eater *Merops apiaster*, Common Greenshank *Tringa nebularia* and Willow Warbler *Phylloscopus trochilus*. At age 9 months a hand-reared bird mimicked 10 locally common birds. Basic call, guttural 'burg', used at roost and as high-intensity alarm when nest threatened (like 'wur' in 'wur-de-dur' note of Cape Robin-Chat *C. caffra*). Contact call, uttered by ♂ and ♀ throughout year, is plaintive 'seep-chweeyu', the high-pitched 'seep' being used also as alarm (very like call of Familiar Chat *Cercomela familiaris*). May use alarm calls of other birds (e.g. Kurrichane Thrush *Turdus libonyanus*) when danger threatens young.

General Habits. A bird of thickets in woodland and tree savanna; dense growth in dry gullies, along watercourses and on termitaria; gardens of rural and suburban homesteads with adequate cover.

Usually in pairs. Sings all year, but only ♂ in June–July (mainly in early morning). In heat of day, ♂ sings quietly from thickets; at other times usually from elevated perch on leafy tree top. Singing increases in frequency and intensity after rain. Depresses and fans tail while singing.

Forages mainly on ground; hawks flying insects, especially just before dusk when ventures out into open tracks and paths. Picks up freshly fallen berries from ground or plucks them *in situ* from perch or in hovering flight. Bathes regularly in puddles and by flying through garden sprays. Roosts singly or in pairs 1–2·5 m above ground; does not necessarily use same roost every night; often chooses protruding branch in sheltered place (e.g. side of hedge in lee of wind). ♂ and ♀ roost close to each other but not on same perch.

Sedentary.

Food. Arthropods and some fruits. In 38 samples (12 stomachs and 26 faeces) there were beetles in 63%, ants in 55%, termites in 42%, caterpillars and moths in 37%, spiders in 18%, orthopterans in 11%, plant- and assassin-bugs in 11%, flies in 5%, arachnids (other than spiders) in 5% and centipedes in 3%. Fruit was found in 13% of samples and included *Antidesma venosum*, *Capparis tomentosa*, *Euclea divinorum*, *E. schimperi* and *Grewia microthyrsa*.

Breeding Habits. Monogamous; territorial; solitary nester. In presence of rival tail raised over back and fanned, and bird facing towards or away from rival. Aggressive pursuits of conspecifics and other species accompanied by loud bill clicking.

NEST: moderate to deep cup of dead leaves, coarsely lined with leaf fragments, midribs, stalks and rootlets, set in main body of dried grass, bents and twigs; cup rimmed with twigs of up to 10 mm diam., often making ramp at one side or platform all around; ext. diam. 150 (av. of 50 nests), diam. of cup 60, depth of cup 40. Most nests built on ground and very well concealed, nest rim level with surface; 8 of 129 nests were sited up to 2 m above ground in holes in banks, trees and hollow stumps; often near edge of thicket, frequently (in Zululand) under sloping *Sanseviera* leaf; garden nests often in tins and pots.

EGGS: 3; 10% of records are C/2 but these clutches probably incomplete or reduced by predation. Eggs laid on consecutive days, in morning or afternoon. Oval to elliptic oval; dull white, greyish white or creamy white, variably clouded (except at small end) by small pinkish brown spots and blotches, concentrated into cap or ring at large end and combined with spots or blotches of lilac and mauve; 3rd egg in clutch usually paler and more finely marked. SIZE: (n = 137) 18–23 × 13–16 (20·9 × 15·1).

LAYING DATES: Zimbabwe, Sept–Dec (90% of clutches Oct–Nov); Mozambique, Sept–Oct; Transvaal, Sept–Dec (Sept 3, Oct 23, Nov 13, Dec 5 clutches); Natal, Sept–Nov (22% of clutches Sept, while most Nov clutches are replacements).

INCUBATION: begins with 3rd egg or shortly thereafter, by ♀ only. Period: 14–15 days. ♂ roosts and spends much time during day on low perch close to incubating ♀.

DEVELOPMENT AND CARE OF YOUNG: all down lost by days 15–17; juvenile plumage begins to appear at about 30 days with underwing-coverts and flanks; white humeral feathers appear at *c.* 40 days; central tail-feathers usually 2–3 mm shorter than rest until *c.* 56 days; white wing-bar fully developed at 60 days; moult to adult plumage (excluding flight feathers which are retained for a further 12 months) complete at *c.* 80 days. Hand-reared birds made first singing attempts at 48 days and first imitations at 54 days.

♂ starts bringing food as soon as first egg hatches; eggshells and faecal sacs are removed and dropped 5–10 m from nest. Nestling period: (n = 1) 13–14 days.

Young fed by both parents until 6–7 weeks old; still solicit food with wing-shivering at 42 days.

BREEDING SUCCESS/SURVIVAL: of 26 nests in NE Natal, 23% robbed of eggs, 19% robbed of chicks; 58% (15 nests) fledged a total of 37 young. Parasitized by Red-chested Cuckoo *Cuculus solitarius* (Rowan 1983).

Cossypha polioptera Reichenow. Grey-winged Robin-Chat. Cossyphe à sourcils blancs.

Plates 22, 26
(Opp. pp. 353, 401)

Cossypha polioptera Reichenow, 1892. J. Orn., p. 59; Bukoba, Lake Victoria.

Range and Status. Endemic resident. Distribution disjunct. Mts of E Sierra Leone; Mt Nimba (Liberia and Ivory Coast) and nearby Mt Tonkui (near Man, Ivory Coast: abundant); N-central Nigeria, very local, from Birnin Gwarri to Loko and Pandam; E Nigeria (Mambilla Plateau and Gashaka–Gumti Game Reserve) and adjacent Cameroon (Bamenda Highlands and Adamawa Plateau); NW Angola (Duque da Braganza, Ndala Tando); E Angola (Luacano), extreme S Zaïre (Katanga: Kolwezi) and extreme NW Zambia (Salujinga to source of the Zambezi, Lisombo Stream and Sakeji Stream); extreme S Sudan (fairly common but local); E Zaïre (Lendu Plateau, Mt Kabobo); riverine forest in S Burundi, 1200–1460 m, and adjacent Tanzania (Kibondo); NW Tanzania at Bukoba; W and S Uganda from Impenetrable and Lugalambo forests to Mubende, Mabira and Mt Elgon; W Kenya from Mt Elgon, Cherangani Mts, Kakamega, Kaimosi and Nandi to Sotik, Mau Forest, Rapogi and Tarang'anya. Locally common but generally rather scarce.

Description. *C. p. polioptera* Reichenow (including '*grimwoodi*': differences described by White are not evident in BMNH specimens, including the type): S Sudan, Uganda, W Kenya, E Zaïre, NW Tanzania, N Angola, NW Zambia. ADULT ♂: forehead, crown and nape dark slate-grey; supercilium from bill to above or slightly behind eye white; lores, ear-coverts and narrow stripe below supercilium and above eye black; chin, throat, sides of neck and entire underparts rusty, paler (near-white) on belly; mantle, scapulars and rump olive-brown; uppertail-coverts tinged rufous; tail reddish brown; primaries and secondaries dull brown edged (broadly on secondaries) olive-brown; tertials olive-brown; wing-coverts brownish slate; underwing-coverts and axillaries pale rufous, feathers basally grey. Bill black; eye dark brown; feet grey-brown. ADULT ♀: as ♂ but crown often tinged olive. SIZE: wing, ♂ (n = 3) 78–82 (80·3), ♀ (n = 5) 74–79 (75·6); tail, ♂ (n = 3) 60–70 (65·0), ♀ (n = 5) 55–60 (58·6); bill, ♂ (n = 2) 16, 17, ♀ (n = 5) 16 (16·0); tarsus, ♂ (n = 2) 25, 26, ♀ (n = 5) 23–25 (24·0). WEIGHT: (Kenya) ♂ (n = 4) 17–20 (18·5), ♀ (n = 3) 15–17 (16·0).

IMMATURE: as adult ♀ but no white supercilium; crown and face spotted rufous (spotting later restricted to hindcrown and ear-coverts); wing-coverts edged rufous.

NESTLING: unknown.

C. p. nigriceps (Reichenow): Sierra Leone to N Cameroon. Crown black; supercilium extends further behind eye, meeting rufous of sides of neck; hindneck rufous, forming a collar. SIZE: wing, ♂ (n = 5) 79–83 (81·2), ♀ (n =' 3) 74–76 (75·0); tail, ♂ (n = 5) 64–70 (67·4), ♀ (n = 3) 60–62 (61·0); bill, ♂ (n = 5) 16–19 (17·2), ♀ (n = 3) 15–18 (16·7); tarsus, ♂ (n = 4) 25–28 (25·7), ♀ (n = 3) 24–26 (24·7). WEIGHT: (Nimba, Liberia, Oct–Jan) ♂ (n = 5) 19–23·2 (21·7), ♀ (n = 3) 18–20·3 (19·1).

C. p. tessmanni (Reichenow): E Cameroon. Said to be darker than *nigriceps*, especially on the cheeks (White 1962).

Field Characters. A rather small *Cossypha*; underparts reddish orange, tail uniform rufous, white supercilium, broad black stripe through eye. When it forages on ground, supercilium breaks outline and makes bird hard to detect among leaf litter. Grey wing shared by other robin-chats and akalats, not a field character. Regularly flicks wings and tail when perched. Differs from most robin-chats in having no black in tail. Very similar Bocage's Akalat *Sheppardia bocagei* is slightly smaller, with almost plain grey face (small white supraloral spot, indistinct grey eye-line). Tends to forage at higher levels than Bocage's Akalat.

Voice. Tape-recorded (57, 86, GREG, McVIC, OAT). Song similar to that of Red-capped Robin-Chat *Cossypha natalensis* but higher-pitched, rich whistles and a medley of imitations of calls of pigeons, cuckoos, bee-eaters, orioles, bulbuls, thrushes, flycatchers, shrikes and others. Soft 'gut' uttered by bird when suspicious; alarm a sibilant 'kwick-kweeek-kwick kwick kwick'. Contact call 'till-ull-tweee', like a call of White-browed Robin-Chat *C. heuglini*, but much quieter; also 'tee-ta-too too tweee', first 3 syllables very quiet.

General Habits. Inhabits edges and clearings of middle elevation forest, gallery forest in plateau woodlands or open grasslands, lakeside thickets, forested ravines and 'mushitu' (NW Zambia). Altitudinal range from 400 m (W Africa) to 1800 m in Zaïre and Sudan, 1100–2130 m in E Africa.

Hops on ground; flies silently and rapidly, weaving adroitly through thickets. Forages on ground, also in middle strata and lower canopy.

Generally sedentary, but possible local migrant July–Oct, Sudan.

Food. Beetles (Carabidae, Tenebrionidae), ants (Camponotidae), moths and caterpillars up to 20 mm long, bugs (Pentatomidae, Lygaeidae), ichneumon wasps, tipulid flies, orthopterans, scorpions and spiders.

Breeding Habits. Almost unknown. Nest undescribed; eggs said to be olive-green.

LAYING DATES: Liberia (breeding condition Jan); Rwanda, Oct; Uganda, Apr, June, Oct (juvs Aug–Sept); Zambia Sept–Oct.

BREEDING SUCCESS/SURVIVAL: bird in Kakamega, Kenya, lived at least 3 years 5 months (Mann 1985).

Reference
Oatley, T. B. (1970a).

Plate 24
(Opp. p. 369)

Cossypha cyanocampter (Bonaparte). **Blue-shouldered Robin-Chat. Cossyphe à ailes bleues.**

Bessonornis cyanocampter Bonaparte, 1850. Consp. Av. i, p. 301; Dabakom, Gold Coast.

Forms a superspecies with *C. semirufa* and *C. heuglini*.

Range and Status. Endemic resident. Extreme SW Mali (Bafing Makana, Sagabari, Kangaba), E Guinea, Sierra Leone, Liberia, Ivory Coast (north to Sipilou, Bouaké), Ghana (north to Bia Nat. Park, Mampong), S Togo; S Nigeria, S Cameroon (north to Mt Cameroon, Kumba), Central African Republic in SW (Lobaye Préfecture) and NE (Manovo-Gounda-Saint Floris Nat. Park); NE Zaïre west to *c.* 23°E, south to Itombwe; extreme S Sudan (Imatong Mts); forests of W and S Uganda from Budongo, Bwamba and Impenetrable to Mabira and Malabigambo, and Minziro Forest in NW Tanzania; W Kenya from Elgon to Kakamega, Kaimosi and Nandi. Frequent to locally common; density of *c.* 7–8 pairs/km² of secondary forest (Gabon).

Description. *C. c. cyanocampter* (Bonaparte): Sierra Leone to Gabon. ADULT ♂: centre of forehead, crown, nape, lores, line under eye, ear-coverts and sides of neck black; supercilium from sides of forehead to sides of neck white; mantle, back and scapulars olive-green; rump olive-rufous; uppertail-coverts rufous; central tail-feathers black, others rufous edged black on outer webs (on outermost pair almost whole of outer web black); entire underparts rufous, paler on belly; remiges, greater coverts and primary coverts dull grey-brown, edged steel-blue on outer webs; lesser and median coverts blue; underwing-coverts and axillaries rufous. Bill black; eye dark brown; feet brownish grey. Sexes alike. SIZE: wing, ♂ (n = 11) 81–91 (87·5), ♀ (n = 11) 79–84 (81·8); tail, ♂ (n = 11) 73–87 (80·2), ♀ (n = 11) 69–79 (72·4); bill, ♂ (n = 10) 19–21 (19·6), ♀ (n = 10) 18–19 (18·4); tarsus, ♂ (n = 8) 26–30 (28·1), ♀ (n = 10) 25–28 (26·6). WEIGHT: (Nimba, Liberia, July–Oct) ♂ (n = 3) 28–32·2 (30·6), 1 ♀ 29.

IMMATURE: like adult except face and crown spotted rufous; remiges edged dull pale brown; wing-coverts and tertials tipped rufous.

NESTLING: down rusty orange.

C. c. bartteloti (Shelley) (including '*pallidiventris*'): NE Zaïre to S Sudan and W Kenya. Mantle, back and scapulars darker, blackish olive. Smaller. SIZE: wing, ♂ (n = 6) 80–86 (83·3); tail, ♂ (n = 6) 66–76 (71·2); bill, ♂ (n = 5) 17–19 (17·8); tarsus, ♂ (n = 6) 29 (29·0). WEIGHT: (Kakamega, Kenya, June) 1 ♂ 31; unsexed (n = 27) av. 28·7; diurnal weight change 11·5–13·8% (Mann 1985).

Field Characters. A medium-sized robin-chat with white supercilium, orange underparts and prominent pale blue shoulder-patch contrasting with blackish wing. A great skulker but a superb songster and mimic; its presence is usually revealed when calls of canopy species emanate from the undergrowth! More of a forest bird than similar-sized robin-chats. Range overlaps with White-browed Robin-Chat *C. heuglini* and Red-capped Robin-Chat *C. natalensis*, but these are more in forest edge and undergrowth outside forest. Red-capped (all-red head) is also a fine singer and mimic, but similar-looking White-browed (larger, uniform greyish wing without shoulder-patch) has totally different song with set form and no mimicry.

Voice. Tape-recorded (32, 57, C, ERA, GREG, STJ, ZIM). A superb mimic; song includes many perfect imitations of other birds' calls, especially those of cuckoos, African Green Pigeon *Treron calva* and birds of prey, e.g. Crowned Eagle *Stephanoaetus coronatus*; mimicked call may be repeated 2 or more times, but sooner or later the robin-chat's own deep, deliberate notes are interpolated (as in song of Red-capped Robin-Chat). Song battle with human recorded in which bird answered recordist's whistle with superior whistles of its own (Keith and Gunn 1971). Capable of singing at 2 frequencies simultaneously, producing a base accompaniment to treble variations. A 'curious dry, croaking rattle' (Chapin 1953) which often initiated song in Ituri (Zaïre), thought to be imitation of frog, was probably the alarm note.

General Habits. Inhabits dense undergrowth at edges of and around clearings in primary forest, thickets of secondary regenerating forest, gallery forest, wooded ravines, seasonal swamp forest. Mainly in lowlands, from near W African seaboard to 900 m in Liberia, 2000 m in Sudan, over 1700 m in Zaïre, 2000 m in E Africa. Absent from large tracts of apparently suitable habitat in Zaïre.

An exceptionally adroit flier, able to pursue twisting, weaving rivals through dense, tangled vegetation with seemingly reckless speed. Forages on or near ground in dense cover from which (unlike other robin-chats) it seems never to emerge, even at dusk. Frequently sings in complete darkness after nightfall (but not throughout night).

Sedentary; but partial migration suggested by observation of bird in Gabon perfectly imitating Nightingale *Luscinia megarhynchos* which never occurs there (Brosset and Erard 1986).

Food. Arthropods: beetles, ants, termites, cicadas, grasshoppers, moths, caterpillars, millipedes and spiders.

Breeding Habits. Monogamous, solitary nester. Territory size 3–6 ha in secondary forest in Gabon (Brosset and Erard 1986).
NEST: variable; either a bulky, loose construction of stout twigs, moist dead leaves, decomposing grass blades and moss, thickly lined with moss (Cameroon), or once a neat cup fashioned wholly of fine grass fibres interwoven with live green moss filaments growing on top of decaying log on which nest was sited (Uganda). SIZE: (1 nest, Cameroon) ext. diam. 120, cup diam. 65, depth 45. Height above ground (3 nests) 0·5–1 m; nests respectively on mossy log, suspended in dense *Afromomum* growth, and in fork of pollarded tree.
EGGS: few clutches found; all were of 2 eggs. Elliptical ovate; slightly glossy; olive-green ground colour overlaid with brown, with discrete mauve or violet markings in brown overlay, forming cap at large end. SIZE: (n = 6) 22–25 × 15–17 (23·4 × 16·1).
LAYING DATES: Liberia (breeding condition Aug–Sept); Nigeria (breeding condition May); Cameroon, Mar, July, Oct; Gabon, Jan; Sudan, Oct; Zaïre, Feb–Sept, Nov–Dec; Uganda, Apr (breeding condition Jan–Feb, July).
BREEDING SUCCESS/SURVIVAL: oldest ringed wild bird lived 8 years 5 months (Kenya: Mann 1985), but captive ♂ still alive and breeding (1–2 broods annually) after 24 years in aviary (Curio 1989).

References
Brosset, A. and Erard, C. (1986).
Curio, E. (1989).
Mann, C. (1985).

Cossypha semirufa (Rüppell). Rüppell's Robin-Chat. Cossyphe de Rüppell.

Plates 24, 26
(Opp. pp. 369, 401)

Petrocinchla semirufa Rüppell, 1840. Neue Wirbelth., Vögel, p. 81; Abyssinia.

Forms a superspecies with *C. cyanocampter* and *C. heuglini*.

Range and Status. Endemic resident. Ethiopia (common throughout highlands and Rift valley); SE Sudan (Boma); Kenya at Moyale on Ethiopia border, on Mt Marsabit, and in highlands in or east of Rift valley south of *c*. 0° 30′N, including Chyulus and Taita Hills; N Tanzania (Crater Highlands, Longido, Essimingor, Arusha Nat. Park, Pare Mts). Common to locally abundant.

Description. *C. s. semirufa* (Rüppell): Ethiopia exept SE, to SE Sudan and N Kenya. ADULT ♂: centre of forehead, crown, nape, lores, line under eye, ear-coverts and sides of neck black; supercilium from sides of forehead to sides of neck white; mantle, back and scapulars olive-brown; rump and uppertail-coverts rufous; central pair of tail-feathers black, others rufous edged black on outer webs; entire underparts dark rufous; remiges and wing-coverts dull brown, edged olive on outer webs; underwing-coverts and axillaries rich rufous. Bill black;

eye dark brown; feet dark brown. Sexes alike. SIZE: (10 ♂♂, 10 ♀♀) wing, ♂ 79–82 (81·0), ♀ 73–82 (76·2); tail, ♂ 67–75 (70·9), ♀ 55–69 (63·2); bill, ♂ 16–19 (17·6), ♀ 17–18 (17·5); tarsus, ♂ 29–31 (30·2), ♀ 27–30 (28·7).

IMMATURE: as adult except no supercilium; crown black, spotted olive; mantle, back and scapulars olive-brown, scalloped black; underparts rufous-buff, scalloped black (heavier on breast). Later, as supercilium appears, crown black, streaked rufous; wing-coverts spotted rufous (these rufous markings not evident in newly-fledged juveniles). Juvenile spotting lasts longest on head and wing-coverts.

NESTLING: unknown.

C. s. donaldsoni (Sharpe): SE Ethiopia (Harar) and adjacent Somalia, intergrading with last on Arussi Plateau. Rufous of underparts extends up sides of neck to form a near-complete collar; mantle, back and scapulars darker, blackish olive. Bill more robust. ♂ larger. SIZE: (2 ♂♂, 2 ♀♀) wing, ♂ 88, 91, ♀ 78, 81; tail, ♂ 83, 86, ♀ 75, 76; bill, ♂ 19, 19, ♀ 17, 18; tarsus, ♂ 29, 33, ♀ 29, 29.

C. s. intercedens (Cabanis): central and SE Kenya, N Tanzania. Larger than *donaldsoni*. SIZE: (6 ♂♂, 6 ♀♀) wing, ♂ 88–96 (92·3), ♀ 84–90 (87·7); tail, ♂ 77–87 (83·0), ♀ 71–81 (77·5); bill, ♂ 19–20 (19·7), ♀ 18–20 (19·0); tarsus, ♂ 30–31 (30·8), ♀ 26–31 (28·7).

Field Characters. A highland forest robin-chat with a powerful and beautiful song. Plumage very similar to White-browed Robin-Chat *C. heuglini*, which is larger, with blue-grey wings and olive (not black) central tail-feathers, and has different song and habitat (thickets rather than forest). Only other robin-chat in its range is Cape Robin-Chat *C. caffra*, which has grey belly, different song and lives outside forest.

Voice. Tape-recorded (10, 27, 32, 56, B, C, LOW, ROC, STJ, ZIM). Song mellow yet powerful, composed largely of mimicry of other birds' calls mixed with robin-chat's own notes. Learns to mimic lengthy phrases from tunes whistled by people; also mimics small mammals (dog, squirrel). Whistled song-call consists of series of low notes, 'hoo-hoo-hoo ...', concluding with loud double note 'heeyo', ending with an octave drop. Alarm a guttural rattle.

General Habits. A bird of evergreen forest and forest margins, tangled thickets flanking streams and dry gullies, and gardens with dense shrubberies; in Ethiopia also juniper and *Podocarpus* forests and edges of woodland. Unlike White-browed Robin-Chat, inhabits forest interior; especially common in forests of Chyulu Hills (Kenya). Altitudinal range 1000–3200 m in Ethiopia, 1400–2300 m in E Africa.

Rather pugnacious; drives other species from bird-baths. Forages extensively on ground. Accompanies driver ant columns to catch invertebrates that they flush; adept at robbing ants of captured insect larvae. Frequently associates with small forest antelopes such as suni *Nesotragus moschatus*, catching insects disturbed as antelope moves about. Often emerges from cover to feed in paths and roadways after sunset, and after rain at other times of day.

Food. Mainly insects, including beetles, moths, caterpillars, mantids and grasshoppers.

Breeding Habits. Monogamous. ♂♂ advertise territories with song, vying with each other for extended periods throughout the day.

NEST: open cup of rootlets, fibres and moss on foundation of dead, decaying leaves and twigs; cup lined with vegetable fibres and some moss; built by ♂ and ♀ 1–2·5 m above ground in variety of sites: recess in earthen bank of bushy gully, hollow in stump or tree, collection of debris, thick vegetation (leafy creeper or shrub); in gardens may build in hanging basket of fern or other plants.

EGGS: 2–3, laid at 1–2 day intervals (usually daily). Glossy; olive to olive-brown. SIZE: 21–24 × 15–17 mm.

LAYING DATES: Ethiopia, Mar–Aug; E Africa: Region D, Mar–June, Dec. Most breeding commences in the long rains.

INCUBATION: mainly by ♀, with short spells by ♂. Period: 12–13 days.

DEVELOPMENT AND CARE OF YOUNG: observations at 1 nest showed ♀ stayed in vicinity after eggs hatched, occasionally brooding nestlings; ♂ foraged further afield and brought food to ♀ which then fed young; for first few days faecal sacs usually swallowed by parent, thereafter taken away and dropped; when nestlings 1 week old, once fed 31 times between 10·30 and 15·00 h, 26 times by ♀. Nestling period: 15–16 days.

BREEDING SUCCESS/SURVIVAL: in Kenya highlands often parasitized by Red-chested Cuckoo *Cuculus solitarius*.

Reference
van Someren, V. G. L. (1956).

Cossypha heuglini Hartlaub. White-browed Robin-Chat. Cossyphe d'Heuglin.

Cossypha heuglini Hartlaub, 1866. J. Orn., p. 36; Wau, Bahr-el-Ghazal, SW Sudan.

Plates 24, 26
(Opp. pp. 369, 401)

Forms a superspecies with *C. cyanocampter* and *C. semirufa*.

Range and Status. Endemic resident. Nigeria, sight record in extreme NE on Cameroon border; extreme N Cameroon and SW Chad from valleys of Shari and Logone rivers near L. Chad to Fort Archambault; SW and NE Central African Republic (Lobaye Préfecture, Manovo-Gounda-Saint Floris Nat. Park); Sudan (Darfur, and widespread south of *c*. 10°N); SW and S Ethiopia; Somalia south of 3°N, very uncommon; widespread in W, central and S Kenya, north on outlying mountains to Loima Hills, Mt Nyiru and Mt Marsabit, and in E along Tana R. and on coast; E and S Zaïre and along lower Congo R.; S Congo and W Gabon; and through most of central Africa south to S Angola, NE Namibia (Caprivi), NW Botswana (Okavango), Zimbabwe (except dry NW), N and E Transvaal and N Natal. Very common.

Description. *C. h. heuglini* (Hartlaub): Central African Republic, S Sudan and S Ethiopia to E Angola, Zimbabwe and E Transvaal. ADULT ♂: centre of forehead, crown, nape, lores, line under eye, ear-coverts and sides of neck black; supercilium from sides of forehead to sides of neck white; mantle, back and scapulars olive-green to olive-grey; rump and uppertail-coverts rufous; central pair of tail feathers olive-grey, next pair with inner web olive-grey, outer rufous, or wholly rufous, other feathers rufous except outermost pair which has outer web olive-grey; entire underparts rufous; remiges dull grey-brown; wing-coverts brownish slate; underwing-coverts and axillaries rufous. Bill black; eye dark brown; feet dark brown. Sexes alike. SIZE: wing, ♂ (n = 10) 99–106 (101·9), ♀ (n = 10) 90–102 (94·8); tail, ♂ (n = 10) 83–95 (91·3), ♀ (n = 10) 76–91 (83·9); bill, ♂ (n = 9) 19–22 (20·7), ♀ (n = 10) 18–21 (19·3); tarsus, ♂ (n = 10) 29–33 (31·3), ♀ (n = 10) 28–32 (30·0). WEIGHT: (Zimbabwe) ♂ (n = 14) 30·5–44·1 (37·5 ± 3·7), ♀ (n = 10) 29·1–35·9 (32·3 ± 1·9).

IMMATURE: as adult except no supercilium; crown black, spotted rufous; rest of upperparts rufous-olive, scalloped black; underparts rufous, scalloped black.

NESTLING: skin pinkish orange, down grey; bill, legs and feet yellow; weight at hatching 2·4 g.

C. h. intermedia (Cabanis): SE Sudan, coastal E Africa to Zululand. Smaller. SIZE: (6 ♂♂, 4 ♀♀) wing, ♂ 89–94 (91·2), ♀ 81–91 (86·5); tail, ♂ 73–85 (80·3), ♀ 76–83 (78·2); bill, ♂ 19–21 (20·2), ♀ 19–20 (19·5); tarsus, ♂ 27–30 (29·3), ♀ 28–29 (28·2).

C. h. subrufescens (Bocage): Gabon to W Angola. Like nominate but central pair of tail-feathers black.

Field Characters. A common bird of scrubby undergrowth, whose loud, crescendo song, one of the most distinctive in Africa, is its best field character. Large, with black head, long white supercilium, grey wings and red-orange underparts. Plumage similar to somewhat smaller Rüppell's Robin-Chat *C. semirufa*, which has darker upperparts, blackish wings, black (not olive) central tail-feathers; their ranges overlap but they are ecologically separated, Rüppell's living mainly in forest. Main difference is song; Rüppell's has typical unstructured robin-chat song with much mimicry. For differences from Blue-shouldered Robin-Chat *C. cyanocampter*, see that species.

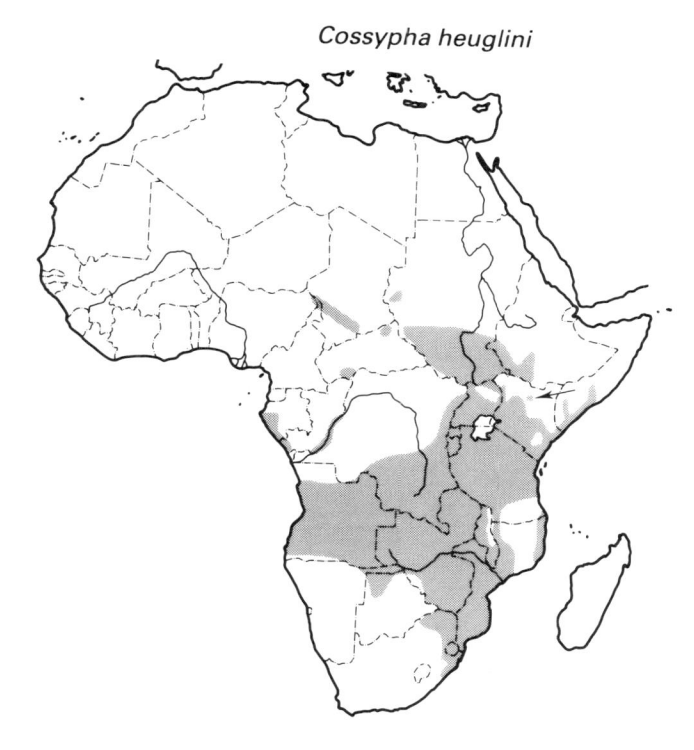

Voice. Tape-recorded (25, 33, 42, 51, 56, 57, 58, B, C, F, KEI). Crescendo song is highly distinctive – a simple motif of a few syllables, 'woot woot chero-cheee' repeated 5–6 times, beginning quietly, increasing in volume and rising in pitch, very loud at peak volume; after 2–5 repetitions of song, motif changes, either between or during songs. In early morning (occasionally at other times), ♂ and ♀ sing antiphonally, ♀ adding high-pitched 'tsreeeu' after each of last 3 motifs of ♂'s song. In absence of ♀, ♂ can add ♀'s notes in the right places without otherwise altering his song. ♂ and ♀ sometimes sing at same time; ♂'s song is then not crescendo; such duets are frequently the first song of the day (at roost), and may be sung at intervals all day in response to duets by neighbouring pair. Song does not incorporate mimicry of other birds (but hand-reared bird can become accomplished mimic). Mimicry incorporated into sub-song and rattle call, which includes alarm or mobbing calls of other birds, e.g. 'tru-wick' call of Common Bulbul *Pycnonotus barbatus*, especially when robin-chat young are threatened. Begging call of fledglings, also used as location call, is clear-cut 'zett', very like call of Black-cheeked Waxbill *Estrilda nonnula*.

Alarm, staccato rattle of 'tsreck' notes, also used when going to roost at dusk. Contact call varies geographically; a loud, fast-paced 'pit-porleee' or 'cheeritter-porleee' in Kivu; 'da-da-teee' 3 times then 'da-teee' 3 times in Malaŵi; 'chickle-ter-tweeep' in NW Zambia; 'dont-you-do-eeet' in Zambesi valley; 'putt-poo-leeee', 'pep-pep poo-leee' in NE Transvaal; 'tit tut-tut terweee' in NE Natal. Common component is high-pitched, upward-slurred final syllable (at a distance the only note audible).

General Habits. A bird of riverine forest and evergreen bush clumps. Avoids interior of large forests although sometimes at edges; requires understorey cover in form of evergreen shrubs; favours termitaria thickets in woodlands, and suburban gardens with dense shrubberies (especially of *Bougainvillea*). In discontinuous fringing forests, ranges 50–100 m into adjacent *Phragmites* reedbeds. Wary and elusive in rural habitats but not so around human habitation. Occurs from sea level to 2200 m in E Africa, 1800 m in Ethiopia, 1750 m in Zaïre, 1500 m in Zimbabwe, but not above 1000 m south of Limpopo R.

Sings from high perches, also from undergrowth. At loudest point of song, bird pumps notes out with beak open wide, tail jerking in unison with each note, neck and breast inflated. Sings at any time of day in breeding season. Flight silent and adroit. Spends much time on ground; on open ground (river sandbank, lawn), progresses 1–2 m by rapid, long hops, then pauses. Forages mainly on ground, whisking leaf litter beneath thickets; catches flying termites on the wing, sallying out from cover of bush clumps. Leaves cover after sunset to feed on open ground. Aggressive; hops sideways (presenting side view) to drive off thrushes or *Laniarius* shrikes; if they do not take flight leaps onto them, pecking viciously. Bathes frequently, singly or in pairs, in shallow edges of rivers or bird baths; also sunbathes, leaning away from sun to expose sides, raising body feathers.

Food. Mainly insects, especially beetles and ants. In 19 stomachs and 9 faeces (Zaïre, Zambia, Malaŵi, Zululand) there were: ants (Ponerinae, Dorylinae, Camponotinae, Myrmicinae) in 86%, beetles (Anthicidae, Buprestidae, Carabidae, Coccinellidae, Cocujidae, Curculionidae, Lagriidae, Lampyridae, Tenebrionidae) in 75%, caterpillars and moths in 32%, termites and bugs (Anthocoridae, Reduviidae) each in 21%, grasshoppers and hunting wasps (Ichneumonidae) each in 14%, millipedes in 11%, woodlice and flies each in 7%, spiders and a small frog each in 4%. 25% of samples contained unidentified fruits.

Breeding Habits. Monogamous; territorial. Pair bond strong; pair remains on territory all year. Territory size (southern Africa) 0·3–2·0 ha; territories linear in low rainfall areas, where dense evergreen cover confined to drainage lines. In territorial defence, pair approaches trespasser, perches high, circles intruder, singing loudly. In courtship display, ♂ sings continuously with tail fully fanned and depressed and feathers of underparts raised (a greeting posture); he follows ♀ until she responds with fanned tail, upstretched neck and quiet twittering song. After copulating, birds move to high perch and give loud, antiphonal song.

NEST: moderately deep cup in substantial foundation of dead leaves and coarse twigs, lined with compound-leaf raches, leaf midribs, rootlets or even very fine twigs; sometimes only a pad of rootlets in small rot-hole. Constructed from locally available material, e.g. elephant grass (*Pennisetum*) where common; ext. diam. av. 150, cup diam. av. 66; cup depth av. 41. Situated commonly in niche or hollow in tree, on coppiced stump, in branches of bush or small tree or suspended in roots under overhang of river bank; sometimes (3·8%) on ground. Av. height of 184 sites in Zimbabwe and Zululand 1·6 m; highest 8 m, in bamboo clump. Built by ♀ in less than a week; old nests sometimes re-used, ♀ adding new lining to cup.

EGGS: 2–3, mainly 2; C/3 only 5% of records; clutches of 1 probably incomplete. Laid at daily intervals; 2-day interval between 1st and 2nd eggs reported (van Someren 1956). Ovate to elliptical ovate; fairly glossy; olive, sometimes cream or beige, usually almost obscured by brown clouding on which reddish brown markings sometimes superimposed; immaculate blue eggs once recorded. SIZE: (n = 98, southern Africa) 21–26 × 15–18 (23·0 × 16·6). Sometimes double-brooded.

LAYING DATES: Gabon (♂ with enlarged testes, Apr); Zaïre, Oct–Jan; Rwanda, Jan–Apr, Aug–Nov; Sudan, Aug–Sept; Ethiopia, June (possibly Dec); E Africa: Region B, Jan–Dec; Region C, Oct–Feb, May, Aug; Region D, Jan, Apr–July, Nov; Malaŵi, Sept–Apr; Zambia, Sept–Feb; Angola (breeding condition Oct–Nov); Zimbabwe, Aug–Jan (Aug 2, Sept 34, Oct 56, Nov 77, Dec 48, Jan 9 clutches); S Mozambique, Transvaal and Natal, Sept–Jan. Everywhere breeding peak is in wet season, although it breeds all year in equatorial areas; in E Africa most clutches commenced in Apr–May (43%) and Nov–Jan (29%).

INCUBATION: begins with last egg or up to 2 days later, by F only. Period: 12–17 (14·5) days.

DEVELOPMENT AND CARE OF YOUNG: eyes start to open on day 4; legs and feet start turning pink and feather shafts appear on all tracts by day 5; eye round by end of day 7, primary quills feathering on day 8; bill horn-coloured and mantle feathered by day 10; weight 25–28 g on day 11. At 3 weeks, tail 4 cm long and flight well developed; in week 5 tail reaches 60 and orange underwing-coverts appear; in week 6 adult feathers appear in breast and mantle; by week 9 underparts are almost immaculate orange and white feathers appear in supercilium; adult plumage complete by week 12.

In cool weather, ♀ broods frequently during first 4–5 days. ♂ and ♀ feed nestlings and remove faecal sacs, sometimes swallowing them. Nestling period: 13–17 (15·0) days. Initially fledglings unable to flutter more than a few m, and they hide low down in thickets; at this period adults incorporate alarm calls of other bird species in their rattle calls and readily mob and pester potential predators, including tree snakes; ♂ in Zululand flew at head and pecked back of neck of 1·2 m boomslang *Dispholidus typus*, almost dislodging it from tree. Young bird utters rattle calls from 11 days after hatching, and fledgling ♂♂ may attempt to sing as early as 19 days of age. Juveniles prefer to sit on leafy ground rather than perch on low branches, but fly up at approach of parent to beg with shivering wings; they also start to peck at objects on ground and soon actively pursue invertebrates such as small crickets. Period of parental care not known, but probably 3–4 weeks after fledging.

BREEDING SUCCESS/SURVIVAL: much nest predation, especially by boomslangs and cats in NE Zululand where only 4 of 15 nests produced fledglings. Multiple unsuccessful attempts are commonplace, even though 2 broods can be reared in a season. Occasionally parasitized by Red-chested Cuckoo *Cuculus solitarius*, E Africa to Zimbabwe. Territorial adults survive well: 2 ♂♂ lived at least 11 and 12 years, 1 ♀ at least 8 years (Hanmer 1989).

Reference
Farkas, T. (1973).

Cossypha natalensis A. Smith. Red-capped Robin-Chat. Cossyphe à calotte rousse.

Cossypha natalensis A. Smith, 1840. Illustr. Zool. South Africa, Aves, pl. 60; Port Natal (= Durban).

Plates 24, 26
(Opp. pp. 369, 401)

Range and Status. Endemic resident. Distribution patchy north of equator; widespread further south but status unclear in many areas due to confusion between residents and migrants. Nigeria (Nindam Forest Reserve, where fairly common; Gashaka-Gumti Game Reserve, uncommon on Chappal Waddi above 1400 m (Green 1990)); Cameroon (Yaoundé); SE Central African Republic (Ouossi R.); forests of S Sudan from Bengengai to Boma Hills; Ethiopia south of *c.* 7°N; Somalia, common along Juba R. valley and on coast south of 3°N; coastal Gabon south of Libreville, S Congo (Mayombe), coastal Cabinda (Angola); Congo/Zaïre along middle Congo R. from Bolobo to Kinshasa; widespread S Zaïre south of *c.* 5°S (north in E to Mt Kabobo), and south through W Angola (Cuanza Norte, Malanje and Cuanza Sul) to Huila (Leba), and in E to N Lunda and Moxico; Rwanda, Burundi; Uganda north to Murchison Falls Nat. Park and east to Jinja; Kenya in scattered sites around periphery of W and central highlands and along Tanzanian border and on coast; almost throughout Tanzania and widespread in Zambia except in Western and Eastern Provinces; extreme E Namibia (E Caprivi: Branfield 1990); locally common Malaŵi below 1500 m but local west of Rift escarpment; coastal N Mozambique, more widespread south of Zambezi; Zimbabwe in E and along Zambezi valley; E Transvaal (lowveld), Swaziland, lowlands of Natal and E Cape Province west to Great Fish R. Also Zanzibar, Mafia, Bazaruto and Inhaca Is. Scarce to locally very common or numerous, e.g. commonest ground bird at Gedi (Kenya coast) in May–Nov. Density of 1–2 pairs/ha on L. Malaŵi.

Description. *C. n. intensa* Mearns (including '*garguensis*' (= '*tennenti*') and '*seclusa*'): former fall within the range of variation of *intensa*, whereas latter are variably coloured intergrades between *intensa* and *larischi*): S Somalia and S Sudan through E Angola and E Africa to Zimbabwe, E Transvaal and Mozambique. ADULT ♂: upperparts from forehead to mantle olive-rufous, variably mottled dark grey (feather bases grey); back dark blue-grey, variably mottled olive-rufous; scapulars dark blue-grey; rump olive-rufous; uppertail-coverts rufous; central pair of tail-feathers black, others rufous, outermost pair with black fringe on outer webs; face, sides of neck and entire underparts dark rufous; remiges, greater coverts and primary coverts dark brown, fringed blue-grey on outer web; median and lesser coverts blue-grey; underwing-coverts and axillaries rufous. Bill black; eye brown; feet pinkish grey to brown. Sexes alike. SIZE: wing, ♂ (n = 10) 87–98 (94·0), ♀ (n = 10) 85–95 (90·1); tail, ♂ (n = 10) 69–81 (77·5), ♀ (n = 10) 71–79 (74·5); bill, ♂ (n = 10) 18–20 (18·8), ♀ (n = 10) 17–22 (18·4); tarsus, ♂ (n = 9) 24–29 (26·7), ♀ (n = 10) 24–27 (25·9).

WEIGHT: (Sokoke, Kenya, Apr) ♂ (n = 2) 29, 32; (Zimbabwe) ♂ (n = 25) 28·3–39·8 (31·5 ± 2·4), ♀ (n = 14) 24·4–33 (28·7 ± 2·3).

IMMATURE: like adult except entire upperparts and wing-coverts black, spotted rufous; entire underparts rufous, scalloped black. Spots persist longest on wing-coverts and head.

NESTLING: skin pink, with large grey down feathers; gape orange, flanges whitish yellow; legs and feet pinkish, soles yellow.

C. n. natalensis A. Smith: E Cape Province to Natal. Rufous of upperparts darker, more olive-brown.

C. n. larischi Meise: N Angola to Nigeria. Darker above than nominate, especially cap. Tail shorter: ♂♀ (n = 10) 67–76 (68·9).

Field Characters. The only robin-chat with an all-orange head (some races have brown cap). Combination of beady black eye set in orange face, orange underparts and blue-grey wings reminiscent of New World Prothonotary Warbler *Protonotaria citrea*. Characteristic movement of perched bird is upward jerk of tail together with wing flick; tail sinks slowly back down and may be briefly fanned. Bird hard to see, and in partial view can be confused with Chorister Robin-Chat *C. dichroa* and

White-browed Robin-Chat *C. heuglini* in same habitat. White-browed has same bluish wings and orange underparts, but black face with white eyebrow, and loud crescendo song with no imitations. For differences from Chorister Robin-Chat, see that species.

Voice. Tape-recorded (17, 22, 24, 57, 58, C, F, OAT). 3 main vocalizations: song, whistled social call and guttural ratchet call. Song, mainly in breeding season, is rich medley of imitations of other bird calls with some original notes and very human whistling ('*vox humana*': Farkas 1969); not as loud or powerful as that of Chorister Robin-Chat, and best distinguished by whistles being slurred, as if bird the worse for drink. An excellent mimic; often repeats mimicked call many times as if to rehearse it. In 20 min of continuous song, bird imitated 17 other bird species. Also imitates mammals, e.g. whining, yapping dog. Use of certain birds' calls (e.g. African Fish-Eagle *Haliaeetus vocifer*) by territorial ♂ often stimulates neighbours to employ same call: all ♂♂ within earshot can 'fish eagle' at once. In Malaŵi (Thyolo), imitated Green-headed Oriole *Oriolus chlorocephalus* so well that the oriole answered back and a duet went on for some min (Johnston-Stewart 1984). Out of breeding season, song is quieter and less varied; bird seldom sings in coolest, driest months (South Africa). Social call is plaintive 2-syllabled 'tiuh-tah' or 'see-saw', given all year, particularly by territorial birds before breeding. Ratchet call, given when birds going to roost, also when young threatened, is like scold of Terrestrial Brownbul *Phyllastrephus terrestris* but more drawn out. Nestlings utter 'sreee' begging call from day 12; when they fledge it is shortened and acts as location call.

General Habits. Occurs in variety of forest habitats: dune, riverine, gallery and ravine forests; also sand forest, and seasonally dense deciduous thickets in bushveld and woodland; in Angola, dense *Brachystegia* woodland. Shuns extensive forests in high rainfall areas and usually cool montane forest, but in Ethiopia found in olive/*Podocarpus*/juniper forests to 2000 m. Seldom above 1200 m south of Limpopo R., but up to 2000 m in Sudan, 1670 m in Zaïre and 2200 m in E Africa (although mainly below 1500 m). Breeds in rural and suburban gardens with plenty of trees and shrubs; common in greater Durban and all towns along Natal coast.

Flight silent, fast and adroit; catches flying prey with ease. Has quick, hopping gait; fledglings walk and hop. Bathes regularly in pools, water-filled hollows in stumps and trees, and commonly in bird baths; bathes at any time of day or season, especially when moulting. Seldom drinks. Raises and quivers wings while singing (Lippens and Wille 1976). Forages mainly on ground, vigorously whisking away leaf litter; breaks open termite galleries on fallen wood or tree boles; catches flying termites. Attends driver ant columns and opportunistically attends mammals and birds, especially mole-rats *Cryptomys*, whose mound-building activities flush invertebrates, also elephant-shrews *Rhynchocyon cirnei* and Spotted Ground-Thrush *Zoothera guttata*. Ascends freely to mid-strata and canopy for berries and drupes; gleans leaves. In half-light comes onto paths, tracks, glades and gardens, foraging later than other robins, akalats or robin-chats.

Partial migrant. In Natal, moves into coastal evergreen forests from interior in winter; individuals are faithful to same wintering ground; some birds permanently resident in coastal forests. In several regions rare or absent in certain seasons, e.g. S Ethiopia and S Sudan, absent June–Oct; coastal Kenya, absent Dec–Apr when whereabouts remain a mystery; Rwanda, Burundi and Tanzania W of 33°, absent Dec–May; in W Angola, Katanga (Zaïre) and Zambia north of 12°S mainly a dry season visitor (Britton 1971). First-year ♂♂ prone to wander; passage occurs on W shore of L. Kivu (E Zaïre) in secondary *Bridelia* scrub; in May birds are abundant one day, absent the next; extremes of size and plumage in different individuals show that they are from different populations. May be breeding migrant to some places north of Congo basin; but resident in Nindam, Nigeria (Stuart and Gartshore 1986).

Food. A wide spectrum of invertebrate prey; also small fruits. In 33 stomachs and 14 faecal samples (E Zaïre, Zambia, Malaŵi, Mozambique, Natal) there were: beetles (8 families, including Scarabaeidae, Curculionidae and Elateridae) in 79%, ants in 77%, moths (adults and larvae, including Sphingidae) and orthopterans each in 34%, termites in 26%, spiders (Attidae, Salticidae) in 21%, bugs (Fulgoridae, Pentatomidae) in 17%, wasps and centipedes each in 13%, other insects in 6%, isopods, flies and scorpions each in 2%. 1 bird ate a small crab (Zaïre). Fruits including *Erythroxylon*, *Euclea* and *Halleria* were in 13% of above samples.

Breeding Habits. Monogamous; territorial; solitary nester. Breeding territories in E Transvaal 0·21–0·84 ha; ♂ and ♀ both aggressive in territorial defence against intruders of their own sex. Courtship display of ♂ involves erect stance with flattened plumage, fanned tail and partially-lifted wings which give bird hump-backed appearance (Farkas 1969); wings are vibrated rapidly while bird dances alternately left and right, gradually approaching or drawing away from mate with vocal accompaniment of fast-paced, rather quiet song made up entirely of imitations. This display may be followed by a fluttering display flight with audible wing-beats. Mating (of captive birds: Roddis 1964): ♂ raises wings and erects tail above quivering body, ♀ droops wings and also quivers body; after coition ♂ rubs back of head under ♀'s throat.

NEST: open cup of dead leaves and fine twigs, supplemented with other material (grass, moss, lichen and even dried hippopotamus dung); usually lined with leaf midribs, sometimes also with dead leaves, tendrils and (rarely) animal hair; ext. diam. 100–250 (135), cup diam. 55–75, cup depth 30–50. Built by ♂ and ♀ in up to 5 days; site varies with habitat: 78% of forest nests are in hollows, rot-holes or crevices in stumps and trees; in deciduous thickets, 95% are in hollows in earthen or shale sides of drainage gullies. 5% built on flat ground, where skilfully concealed among leaf litter and field layer. Mean height above ground of 93 nests (Natal) 1·2 m.

EGGS: 2–4, normally 3 (all 1-egg and many 2-egg clutches are incomplete); av. (135 clutches, southern Africa) 2·9. Laid on consecutive days, usually in early morning. Ovate to elliptical ovate; highly glossy; turquoise-blue, olive-green or dark brown (all eggs basically blue but may have secondary and tertiary overlays of olive-green and brown). Majority between olive and brown; of 81 clutches, 5 were blue, 20 olive, 39 olive-brown and 17 brown. SIZE: (n = 216, Natal) 20–25 × 15–18 (22·4 × 16·5).

LAYING DATES: Nigeria, June; Rwanda, Jan, Mar; Central African Republic (breeding condition June); Sudan, Aug; E Africa: Region B, Apr–May, Region E, July; Zambia and Malaŵi, Nov–Dec; southern Africa (south of Zambezi R.), Sept–Jan; 50% of clutches commenced in Nov; Dec–Jan clutches are mainly replacements. Throughout range, breeding usually coincides with rainy season.

INCUBATION: begins with completion of clutch; by ♀ only. ♀ often fed by ♂ (Farkas 1969). Period: 14–15 (14·4) days.

DEVELOPMENT AND CARE OF YOUNG: eyes open by day 7; dorsal tract feathered from day 8, feathering widespread after day 9. Moult of juvenile plumage begins in week 7, orange feathers appearing at sides of breast; by week 10 only head remains mottled and by week 12 bird is in adult plumage; flight feathers retained for a further 12 months.

Young fed by ♂ and ♀. ♀ may brood intermittently for first 7–8 days. Nestling period: (n = 2) 12 days. Fledglings at first stay on or near ground under dense cover, beg with wing-shivering and neck stretched stiffly towards parent; ♂ and ♀ attend them for 28–42 days; young start to feed themselves in week 4 after hatching; adept at catching food on wing by week 12, although they forage mainly on ground.

BREEDING SUCCESS/SURVIVAL: in 62 clutches, 43% of eggs produced fledged young. Oldest birds, ringed as adults, 8 years (Natal, Malaŵi).

References
Britton, P. L. (1971).
Britton, P. L. and Rathbun, G. B. (1978).
Farkas, T. (1969).
Oatley, T. B. (1959).
Stuart, S. N. and Gartshore, M. E. (1986).

Cossypha dichroa (Gmelin). **Chorister Robin-Chat. Cossyphe choriste.**

Muscicapa dichroa Gmelin, 1789. Syst. Nat. 1(2), p. 949; South Africa.

Plates 24, 26
(Opp. pp. 369, 401)

Range and Status. Resident and altitudinal migrant, endemic to South Africa. From Zoutpansberg Range in N Transvaal south along NE escarpment through W Swaziland to Natal, extreme E Orange Free State, E and S Cape Province to the isolated Ruitersbos Forest, Mossel Bay District. Common.

Description. *C. d. dichroa* (Gmelin) (including '*haagneri*'): range of species south of c. 24°S. ADULT ♂: forehead, crown, nape, hindneck, mantle and back dark slate-grey, shading to buffy orange on rump and uppertail-coverts; lores, ear-coverts and moustachial area black; chin, throat, sides of neck and rest of underparts including underwing-coverts orange. Tail with 2 central feathers and outer webs of 2 outermost feathers blackish slate, otherwise orange. Wings dark slate-grey with lighter slate-blue edgings to primaries and secondaries and shoulder area. Bill black; eye dark brown; legs and feet brownish pink, soles grey. Sexes alike. SIZE: wing, ♂ (n = 24) 99–110 (104); ♀ (n = 25) 97–103 (99·7); tail, ♂ (n = 5) 80–93 (87·7), ♀ (n = 4) 80–84 (82·3); bill, ♂ (n = 5) 19–22 (21·0), ♀ (n = 6) 19–21 (20·0); tarsus, ♂ (n = 5) 29–31 (30·4), ♀ (n = 6) 30–31 (30·7). WEIGHT: (Natal, July–June) ♂ (n = 28) 42–56 (47·6), ♀ (n = 34) 37–53 (44·1).

IMMATURE: mottled; dusky brown feathers of upperparts have buff spots, buff feathers of underparts edged with sooty black. Wings and tail like adult.

NESTLING: pink-skinned with long, sooty grey down feathers. Inside of mouth bright orange.

C. d. mimica Clancey: Zoutpansberg and Woodbush, NE Transvaal. Like nominate race but smaller (wing av. 5 mm and tail av. 8·5 mm shorter) with no overlap in size.

Field Characters. A large robin-chat confined to evergreen forest in South Africa. Darker than closely-related Red-capped Robin-Chat *C. natalensis* and larger, with blackish back and wings and mainly black head. Song very similar to Red-capped, which is often in same habitat, but contact calls different; Red-capped has characteristic 'see-saw'. Altitudinal migrants may occur within range of White-browed Robin-Chat *C. heuglini*, which has white supercilium and different song.

Voice. Tape-recorded (14, 20, 24, 50, 56, 58, 75, 81, 88, C, F). Song (Sept–May) composed of loud whistled phrases liberally interspersed with imitations of calls of other bird species; single robin mimics up to 26 different species: Emerald Cuckoo *Chrysococcyx cupreus* and Black-headed Oriole *Oriolus larvatus* are favourite subjects. Contact call, uttered by ♂ and ♀ all year (mainly Oct–May), a whistled 'toy-toy' or, in north, a 3-syllabled 'peep-borrow'. 1 breeding pair regularly used antiphonal call of Southern Boubou *Laniarius ferrugineus* as contact call. Basic call a guttural, rachet-like churr, used all year when birds are going to roost and at dawn; also serves as high-intensity alarm call when adults or young are threatened by predator. Imitated alarm calls sometimes incorporated into basic ratchet churr when danger threatens dependent young. Fledglings utter a penetrating 'chip' location call and a 2-syllable 'chis-sick' when hungry.

General Habits. A solitary bird of moist evergreen forest. Breeds at sea level from Cape Province to S Natal, but in north effectively montane, breeding from *c.* 1400 to 1800 m. North of *c.* 30°S occurs below 600 m only in winter.

In drier months, Apr–Sept, forages in orchards and gardens adjacent to stands of forest; in Oct–May forages mainly in mid-strata and lower canopy of forest. In winter attracted by activities of doryline ants and mole-rats *Cryptomys* spp., which flush earthworms and insects from soil and leaf litter. On ground hops; flies silently. Bathes regularly in forest streams and occasionally in water-filled rot-holes in trees.

Migrates altitudinally in Apr and Sept. Leapfrog movements take place, some birds crossing others in intervening habitat to spend winters at lower levels. Some winter within breeding range, others in fragmented patches of forest on Natal coast, and in Gwaliweni Forest (Lebombo Mts, NE Zululand) where they do not occur in Oct–Mar. Absent from coastal plain of Zululand. Most migrants are 1st year birds, but adults also move. Adult ♂ returns to same winter territory, defended from time of arrival in Apr.

Food. Mainly insects. In 11 stomachs and 33 faecal samples from Natal and Transvaal there were: beetles in 73%, ants in 61%, moths and caterpillars in 34%, centipedes and millipedes in 16%, orthopterans in 9%, plant bugs and assassin bugs in 9%, flies in 9%, woodlice in 9%, termites in 7%, spiders in 5%, other arachnids in 5%, wasps in 2% and other insects in 2%. Also eats earthworms. 1 stomach contained 59 caterpillars and another held 79 driver ants. Fruit remains (in 57% of above samples, higher than in any other robin studied: Oatley 1970a) were of *Asparagus*, *Burchellia*, *Celtis*, *Kiggelaria*, *Maytenus*, *Scolopia*, *Scutia*, *Solanum*, *Vepris*, *Xymalos*, and exotic *Rhus*, *Psidium*, *Hedychium*, *Lantana* and *Rubus*.

Breeding Habits. Monogamous; territorial. Usually pairs for life.

NEST: shallow cup or pad of rootlets or leaf ribs, with similar but finer material sometimes used as lining; green moss sometimes worked into rim; cup diam. av. 70, cup depth av. 40, ext. diam. av. 90; shape conforms to shape of rot-hole or crevice in tree, which is usual site, 1·5–12·5 m above ground (av. of 32 nests, 4·9 m). Built by ♀, mainly in early morning and late afternoon.

EGGS: 3, laid on successive mornings (2- and 1-egg clutches due to egg loss after start of incubation). Broad ovals; highly glossed; either immaculate green (dark or light olive, apple- or blue-green) or brown (dark chocolate, maroon-brown, reddish brown or copper); all brown eggs are 'green' eggs overlaid with a suffusion of brown. Intermediates occur; green, brown/green and brown egg clutches occur in approximate ratio 4:1:5. SIZE: (n = 56) 23–29 × 18–20 (24·5 × 18·8). WEIGHT: (n = 3) 4·6–4·7.

LAYING DATES: South Africa, Oct–Dec. 70% of clutches commenced in Nov; Dec clutches are mainly replacements.

INCUBATION: commences at moment of, or on same day as, laying of 3rd egg; by ♀ only. Period: not determined but within range of 15–19 days.

DEVELOPMENT AND CARE OF YOUNG: hatching weight *c.* 4 g; wing quills begin to erupt on day 4; eyes on day 6; dorsal tract feathering by day 8, general feathering by day 9; weight 30–35 g at day 10. Moult out of mottled plumage begins *c.* 35 days after hatching, beginning at sides of breast and ending with head; complete adult plumage attained by about day 80. Flight feathers are retained for a further 12 months. Moult period for all birds, Jan–Mar.

Young fed by ♂ and ♀. Nestling period: (n = 1) 14 days. Basic churr call can be uttered by nestlings from day 11. Fledglings sit concealed in dense cover while parent forages in vicinity. ♂ and ♀ attend young for *c.* 6 weeks after fledging; towards end of that period young bird moves about more and starts to forage for itself. Family, although split up during day, roosts together at dusk.

BREEDING SUCCESS/SURVIVAL: 69% of eggs in 24 clutches hatched; 31% of eggs in 12 clutches for which final outcome known produced fledged young. Annual survival rate of adults (Natal) estimated to be 70% for ♀♀ and 87·5% for ♂♂; oldest known bird lived for at least 26 years.

Reference
Oatley, T. B. (1969, 1970a).

Cossypha heinrichi Rand. **White-headed Robin-Chat. Cossyphe à tête blanche d'Angola.**

Cossypha heinrichi Rand, 1955. Fieldiana Zool. 34, pp. 327–329; Duque de Braganza, Angola.

Range and Status. Endemic resident. Known from only 2 small areas: *c.* 30 km northeast of Duque de Braganza in N Angola (12 specimens, 1954, 1957), and 500 km to the north in Bombo-Lumene Forest Reserve 4° 30'S, 16° 8'E and at nearby Nkiene in W Zaïre (4 specimens, 1975, 1980). Birds have been seen at 4 other sites within 15 km radius of Bombo-Lumene and a bird 'almost certainly of this species' was glimpsed 550 km further north near Mbandaka (equator, 18° 24'E) (Harrison 1977).

Description. ADULT ♂: entire head and neck white; back greyish olive; rump and uppertail-coverts deep rufous-orange; central tail-feathers black, next 2 orange-chestnut with black inner webs and orange tips, next 3 orange-chestnut, T6 with outer webs black distally, chestnut basally. Underparts from breast to undertail-coverts and thighs rich orange-rufous. Primaries and secondaries black, narrowly edged with greyish olive; upperwing-coverts greyish olive; underwing greyish black, coverts tipped with chestnut. Bill black; eye purplish red (Angola) or reddish brown to sandy brown (Zaïre); legs and feet slate. SIZE: wing, ♂ (n = 4) 120–121 (120), ♀ (n = 3) 110–117 (114); tail, ♂ (n = 2) 141, 147, 1 ♀ 128; bill, ♂ (n = 2) 20, 23, 1 ♀ 21; tarsus, ♂ (n = 2) 37, 38, 1 ♀ 37. WEIGHT: ♂ (n = 2) 61, 69, ♀ (n = 2) 56, 56, subadult (n = 2) 45, 53.

IMMATURE: feathers of head, neck and breast buffy ochraceous with sooty black margins giving scaled appearance; back feathers buffy rufous with broad black margins; belly plain yellowish buff; tail as in adult but feathers more pointed.

Field Characters. The only forest undergrowth bird in Africa with all-white head and neck. Large, with red-orange underparts and long, graduated tail. Wary and skulking, keeping to dense cover.

Voice. Unknown.

General Habits. Frequents undergrowth of gallery forests in Angola and evergreen forest clumps of 1–325 ha in tree savanna in Zaïre.

Singly or in pairs, but up to 4 at driver ant columns. Forages on ground or in undergrowth to heights of 3–4 m; attracted to driver ants and will forsake forest cover to follow ant columns out into open savanna.

Food. Principally ants, including doryline driver ants; 1 stomach was crammed with big-headed driver ant soldiers; also beetles.

Breeding Habits. Very little known; nest and eggs undescribed. Both parents feed young.

LAYING DATES: Angola (juvs Nov, and post-juv. moult Apr, suggest egg-laying Oct); W Zaïre (attended juv. with yellow gape late Nov, and post-juv. moult Nov, suggest egg-laying Feb and Sept).

References
Collar, N. J. and Stuart, S. N. (1985).
Harrison, I. D. (1977).
Ripley, S. D. and Heinrich, G. H. (1966).

Plate 25
(Opp. p. 400)

Cossypha niveicapilla (Lafresnaye). **Snowy-crowned Robin-Chat. Petit Cossyphe à tête blanche.**

Turdus niveicapillus Lafresnaye, 1838. Essai nouv. manière grouper Passereaux, p. 16; Sénégal.

Range and Status. Endemic resident. Widespread in W Africa south of 14°N, and further north to extreme S Mauritania (upper Senegal R. valley, Karakoro) and Niger R. delta in Mali. In Chad occurs at Ndjamena, but northern limit elsewhere probably 10°N; in Central African Republic occurs north to Manovu-Gounda-Saint Floris Nat. Park. Sparse in main forest blocs of W and central Africa, where only at edges and in clearings; occurs sparingly in Gabon, S Congo, N Angola (Cabinda, also N Lunda (Dondo)) and W Zaïre (Lukula, Kinshasa). Locally fairly common in Sudan north to *c.* 13°N; common to frequent in W Ethiopia in lowlands and SW highlands. Ranges widely in L. Victoria basin, east in W Kenya highlands to Nandi and Kipteget R., south in

Plates 25, 26
(Opp. pp. 400, 401)

Tanzania to Kasulu and Gombe Stream Game Reserve, north in Uganda to Murchison Falls Nat. Park and Gulu; Rwanda, Burundi; widespread NE Zaïre west to c. 24°E, south to Itombwe. Common.

Description. *C. n. niveicapilla* (Lafresnaye): W Africa to Nigeria and Sudan. ADULT ♂: crown and nape white, variably mottled black; forehead, lores, supercilium, line under eye, ear-coverts and sides of neck black; mantle, back and scapulars slate-grey, washed olive; rump olive-rufous; uppertail-coverts rufous; central pair of tail-feathers black, others rufous, outermost pair with black outer fringe; entire underparts rufous; remiges very dark brown, fringed blue-grey; wing-coverts very dark brown, broadly fringed blue-grey; underwing-coverts and axillaries rufous. Bill black; eye dark brown; feet dark brown to black. Sexes alike. SIZE: (10 ♂♂, 10 ♀♀) wing, ♂ 96–106 (103), ♀ 88–96 (92·9); tail, ♂ 88–102 (96·8), ♀ 82–91 (87·2); bill, ♂ 18–21 (20·1), ♀ 18–20 (19·1); tarsus, ♂ 28–31 (29·3), ♀ 26–30 (28·0).

IMMATURE: like adult except entire upperparts including wing-coverts very dark brown, spotted rufous; entire underparts rufous, scalloped very dark brown. Spots last longest on head.

C. n. melanota (Cabanis): 'pure' *melanota* restricted to L. Victoria basin; intergrades in size occur from Rwenzori Mts northwest; intergrades in colour occur in broad band from S Ghana and S Nigeria to N Cameroon and eastwards; S Cameroon birds are *melanota* in colour, nominate in size. Like nominate except mantle, back and scapulars black, with no or only slight olive wash; remiges and wing-coverts black, sometimes with slight blue-grey fringing. Larger. SIZE: (Victoria basin: 6 ♂♂, 4 ♀♀) wing, ♂ 101–108 (105), ♀ 98–104 (100); tail, ♂ 92–109 (101), ♀ 87–99 (92·7); bill, ♂ 21–22 (21·3), ♀ 20–21 (20·2); tarsus, ♂ 29–31 (30·2), ♀ 28–32 (30·0).

Field Characters. A large, long-tailed robin-chat, distinguished from all except White-crowned *C. albicapilla* by having white crown instead of white supercilium. Distinguished from larger White-crowned by smaller white crown-patch with no scaling, not reaching base of bill and not extending as far down sides of head; also by rufous (not black) chin, rufous hind-collar and, at close range, blue-grey edging to primaries and brown (not red) eye. A much better singer than White-crowned; song more powerful, with much variation and imitation.

Voice. Tape-recorded (10, 40, 56, 57, C, ERA, GREG, HOR, KEI). Song, loudest at sunset, a rich medley of imitations of other birds' vocalizations and human whistles, distinctive for its fast-paced delivery and manner in which portions of borrowed phrases are strung together to form a varied output of tantalizingly familiar but not always identifiable versions of local bird calls. Can sing continuously for up to 15 min (North and McChesney 1964). Quickly learns new calls, at first uttering straight imitations, then trying them in different keys, and finally working them into a medley. Contact call a whistled 'wheeeoo wheeeoo wheeeoo'. Alarm note a guttural, ratchet-like churr, similar to that of several other robin-chats, heard especially at dusk.

General Habits. In W Africa and N Zaïre typically a bird of thickets that develop around single large woodland trees, termite mounds, large boulders or inselbergs; also frequents gallery and fringing forests; in E Africa commonly inhabits forest edges (and sometimes interior), thickets, rank growth along streams and gardens with dense shrubberies. From near sea level up to 1400 m through most of range; to 2000 m in Cameroon, between 900 and 1400 m in Uganda and Tanzania and up to 2000 m in W Kenya.

Wary and skulking, usually only seen briefly in flight between thickets. Singly or in pairs (seldom possible to tell whether single bird is one of a pair or not). Several birds may gather in one large thicket in otherwise open woodland; much chasing may then ensue, with birds flying through and over tree tops. Forages mainly on ground and at low levels in dense vegetation.

Food. Invertebrates and small fruits; beetles, ants, termites, caterpillars, mantids, small millipedes and tiny snails; berry seeds were found in 4 of 16 stomachs in Zaïre.

Breeding Habits. Monogamous.

NEST: open, rather shallow cup of dead leaves, soft twigs, leaf stems, dried grass and tendrils or moss, lined with rootlets; ext. diam. c. 90, depth of cup 30. Sites include forks or rot-holes in trees, tops of coppiced stumps, bases of palm fronds, and dense foliage; height above ground (n = 3 nests) 1–1·6 m.

EGGS: 2–3. Ovate; glossy; olive-green variably overlaid with suffusion of chocolate or reddish brown. SIZE: (n = 11) 21–25 × 16–18 (22·0 × 16·7).

LAYING DATES: Senegambia, June–Sept; Mali, Mar; Niger, July; Burkina Faso, July–Aug; Nigeria, May–Aug; Sudan, Mar, July; N Zaïre, Feb–Nov; E Zaïre, Mar; Rwanda, Mar, Sept–Dec; E Africa: Uganda, May–June; Region B, Jan, Mar–July, Oct. Most E African records are during long rains.

Cossypha niveicapilla

Cossypha albicapilla (Vieillot). White-crowned Robin-Chat. Grand Cossyphe à tête blanche.

Turdus albicapilla Vieillot, 1818. Nouv. Dict. d'Hist. Nat. 20, p. 254; Sénégal.

Plate 25
(Opp. p. 400)

Range and Status. Endemic. A locally frequent resident or partial migrant from Gambia to Bamingui R. (Chad/Central African Republic) and in SE Sudan and SW Ethiopia.

Uncommon throughout Gambia but common near Kédougou; occurs in Niokola-Koba Nat. Park, Senegal, and in NE Guinea. Old sight records from N Sierra Leone. Common in Mandingo Mts, Mali, and widespread north to 13° 30′N, at least in July–Oct. Frequent in dense guinean woodland in N Ivory Coast, with scattered records south to 8°N; once at Toumodi, 6° 34′N. Status in Ghana poorly known – probably a not uncommon resident south to Mole, 9° 22′N. Recorded in N Togo and N Benin, also in 'W' Nat. Park in Niger in Oct–Mar. In Nigeria frequent and widespread in guinean woodland north to 11° 15′N and south to middle Niger and Benue R. valleys (8°N). In Cameroon restricted to Adamawa Plateau. An old record from Bamingui R. across Chad/Central African Republic border, but no recent records from either country. In Sudan occurs only in Boma Hills, and in Ethiopia a poorly known bird of the SW highlands.

Description. *C. a. albicapilla* (Vieillot): Senegal to Guinea. ADULT ♂: forehead and crown feathers white with very narrow blackish tips which soon wear off, and concealed black bases that make white look greyish rather than snowy; nape and hindneck the same but feathers slightly elongated. Lores, chin, ear-coverts, cheeks, sides of neck, mantle, upper back, scapulars and wings glossy brownish black; lower back and rump dark rufous-orange; uppertail-coverts bright rufous; tail long, T1 blackish, T2–T6 bright rufous, graduated, with outer web of T6 blackish. Apart from chin, entire underparts rich orange-rufous, including axillaries and underwing-coverts. Underside of tail dull rufous. Bill black; eye red; legs and feet black. Sexes alike. SIZE: (13 ♂♂, 6 ♀♀) wing, ♂ 124–131, ♀ 111–118; tail, ♂ 126–136, ♀ 112–119; bill to feathers, ♂ 17–20, ♀ 15–19; bill to skull, ♂ (n = 6) 23–26 (24·5), 1 ♀ 21; tarsus, ♂ 34–38, ♀ 32–36. WEIGHT: unknown.

IMMATURE AND NESTLING: unknown. Subadult like adult except face, upperparts and wings paler, browner; wing-coverts spotted rufous.

C. a. giffardi Hartert: Ghana (or further west) east to Cameroon. Subspecific identity of populations between Guinea and Ghana and on Bamingui R. unknown. Like *albicapilla* but feathers from forehead to nape with less white, so that crown is effectively black with crescentic white feather-tips; mantle, wings and T1 black (not brownish black); underparts paler than in *albicapilla*. WEIGHT: (n = 3, Ghana) ♂♀ 57–60 (58·0).

C. a. omoensis Sharpe: SE Sudan and SW Ethiopia. Like *giffardi* but underparts darker.

Field Characters. A large, long-tailed robin-chat (largest in the genus) of thickets in guinean woodlands, readily distinguished from all sympatric congeners except Snowy-crowned Robin-Chat *C. niveicapilla* by its white or black-and-white-speckled crown. Snowy-crowned Robin-Chat has white restricted to broad band from centre of forehead to nape; sides of forehead and crown and band across base of bill black. White-crowned has much broader area of white, from base of bill to further back on head and extending down sides of crown and forehead, always with some black scaling, narrow west of Mali, broader further east. White-crowned lacks narrow rufous hindneck collar of Snowy-crowned (difficult to see in the field). Although White-crowned keeps to thick cover, a cautious approach usually ensures fleeting or quite good views of it skulking on the ground, when its much greater size alone serves to distinguish it: length *c.* 27 cm (Snowy-crowned, *c.* 20 cm). For further distinctions, see Snowy-crowned Robin-Chat.

Voice. Tape-recorded (57, MEES). Song a sustained varied warbling, with trills and low scratchy notes, somewhat recalling song of Reed Warbler *Acrocephalus scirpaceus* or even Nightingale *Luscinia megarhynchos*. Delivery varies from measured and hesitant to somewhat hurried or even lively. Does not imitate. Song pleasant, but not as musical or melodious as songs of Snowy-crowned Robin-Chat or Red-capped Robin-Chat *C. natalensis*, for instance. Alarm (?) a high thin 'sweowee'.

General Habits. Inhabits gallery forests in dry guinean savanna woodland ('kurmis'), thickets, dense plantations, riverine bush, and similar areas with spaced timber trees and a dense undergrowth casting deep shade. Also occurs in large gardens with thick shrubberies and unkempt hedges, e.g. in Bathurst. From sea level up to 950 m.

Usually seen singly but probably lives in pairs. Keeps mainly to the ground, foraging by tossing leaf litter aside with swipes of the bill, and searching stems at head-height and the ground surface. When disturbed, at first retreats by hopping or running or flying further into undergrowth, but not difficult to watch and if approached with caution bird soon returns and resumes foraging, noisily turning dead leaves aside, only a few m from observer. Occasionally climbs up through vegetation with strong hops, exposing itself momentarily on top of thicket to peer around, then drops out of sight again.

Mainly resident, but may be a partial migrant in Mali (and elsewhere), where said to be commonest in N in rains (Apr–Oct) and in S in dry season (Nov–Mar).

Food. Unknown. Birds watched foraging on ground below thickets in Nigeria appeared to be taking small insects. In Cameroon sometimes caught in traps baited with termites and ants.

Breeding Habits. Barely known; appears to be monogamous and territorial.

NEST: only one described was a scant cup of rootlets and decaying leaves with peripherally-arranged tendrils, in hollowed-out top of small tree stump 1·2 m high, the hollow 110 in diam. at top; ♀ was incubating. Site was thick low vegetation on bank of small stream in grass-woodland country (Serle 1940).

EGGS: 2 (1 clutch only). Oblate ovals; smooth, slightly glossy; pale grey-green ground with ashy violet blotched shell-marks, nearly obscured by reddish brown spots, blotches and clouds. SIZE: (n = 2) 27·3 × 17·3, 27 × 17·9.

LAYING DATES: Gambia, June, Aug–Sept, Dec; Niger, Dec; Nigeria, July; Burkina Faso, Dec.

Genus *Xenocopsychus* Hartert

Fairly large; tail long and graduated; plumage entirely black and white. Inhabit cliffs and caves and nearby forest patches. Closest to *Cossypha* and included in *Cossypha* by White (1962); considered by Chapin (1953) to be close to *Cossypha humeralis*. However, all members of *Cossypha* have at least some red in the plumage, and most have a distinctive tail pattern (red with black centre). Differences in proportions, colour pattern and preferred habitat justify upholding it as a separate genus (Jensen 1990).

Endemic to Angola; monotypic.

Plate 25
(Opp. p. 400)

Xenocopsychus ansorgei Hartert. Angola Cave-Chat. Cossyphe des grottes.

Xenocopsychus ansorgei Hartert, 1907. Bull. Br. Orn. Club 19, p. 81; Lobango, Huila, Angola.

Range and Status. Endemic resident, confined to rocky hills in 3 areas in W Angola: near Ndala Tando, Cuanza Norte; Mt Soque, Huambo; and highlands of Lobango-Humpata-Leba region of W Huila. Very local; frequent to common.

Description. ADULT ♂: forehead to rump, including scapulars, black, washed grey on rump; uppertail-coverts white; central pair of tail-feathers black; next pair with inner web and tip of outer web white, rest of outer web black; remaining pairs white, outermost pair with black outer fringe near tip; supercilium from bill to sides of neck white; lores, chin, sides of face, ear-coverts and sides of neck black; entire underparts white; remiges and most wing-coverts black; tertiary coverts and some lesser and median coverts white, producing broad white shoulder-streak; axillaries and inner underwing-coverts white; outer underwing-coverts black. Bill black; eye dark brown; legs and feet black. SIZE: (3 ♂♂, 5 ♀♀) wing, ♂ 96–104 (99·3), ♀ 90–98 (93·0); tail, ♂ 98–115 (107), ♀ 94–104 (99·8); bill, ♂ 21–24 (22·7), ♀ 21–23 (22·0); tarsus, ♂ 29–31 (30·3), ♀ 28–30·5 (29·3).

Field Characters. Black-and-white plumage and long tail more suggestive of Asian magpie-robin *Copsychus*

than any African thrush. Specialized rocky habitat and restricted distribution (W Angola) further preclude confusion with any other species.

Voice. Not tape-recorded. A musical song, 'dülülü dülü dülülü', similar to song of European Woodlark *Lullula arborea* (Braun 1956). Alarm call a harsh, 2-syllabled 'birr-djarr'. Also a quiet 'ui ... ti ... ti ...' (Pinto 1962).

General Habits. Frequents rocky hills and gorges, especially where there are jumbles of weathered sandstone boulders with adjacent forest patches; prefers (but not restricted to) areas with recesses and open cave mouths where it takes shelter.

In pairs. Forages on quartzite rock outcrops, probing lichens. Moves and behaves like a *Cossypha* (Hall 1960).

Food. Insects, including larvae and adult beetles (Tenebrionidae, Staphylinidae), weevils *Blosyrus* and ants.

Breeding Habits. Monogamous; almost certainly territorial.

NEST: open cup of twigs and bents lined with finer plant material, placed on rock ledge under overhang.

EGGS: 2. White, clouded with brown and reddish speckling. SIZE: (n = 2) 23·1 × 18, 21·3 × 16·2.

LAYING DATES: Sept–Nov.

Reference
Braun, R. H. (1956).

Genus *Alethe* Cassin

A very homogeneous group of medium-sized thrushes with relatively short, rounded tails and long legs, and uniform plumage (red-brown upperparts and wings, whitish underparts). 2 species have white spots in tail (*A. choloensis*, nominate race of *A. diademata*), 1 has red throat (*A. poliophrys*). Whistled songs and calls have generic similarity.

Inhabit ground stratum of evergreen forests. Habitually attend army ant swarms. Mainly sedentary; some altitudinal migration at higher latitudes.

Endemic to Africa; 5 species, 2 comprising a superspecies (*A. fuelleborni*/*A. choloensis*). *A. diademata* has longer tail and shorter legs than the others, and has orange crown-patch and short crest. The other 4 were put in a single superspecies by Hall and Moreau (1970); however, *A. poliocephala* and *A. poliophrys* are partly sympatric, e.g. in Bururi Forest, Burundi, where they coexist at all levels from 1700 to 2300 m with no indication of interbreeding (Vande weghe 1988; cf. Prigogine 1980, 1984). Songs of *A. choloensis* and *A. fuelleborni* are very similar, while that of *A. poliocephala* has a different form.

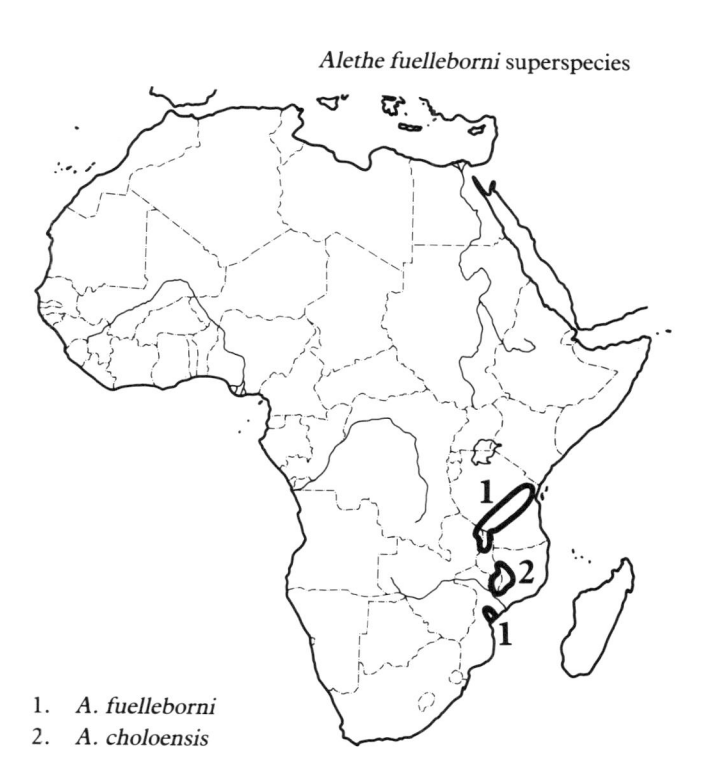

Alethe fuelleborni superspecies

1. *A. fuelleborni*
2. *A. choloensis*

Plates 23, 26
(Opp. pp. 368, 401)

Alethe diademata (Bonaparte). Fire-crested Alethe. Alèthe à huppe rousse.

Bessonornis (Turdus) diadematus Bonaparte, 1851. Consp. Av. 1 (1850), p. 302; Guinea.

Range and Status. Endemic resident. Senegal (Basse Casamance), Guinea-Bissau, Guinea, Sierra Leone, Liberia, Ivory Coast (north to Korhogo), Ghana, Togo; S Nigeria, Cameroon (north to Korup Nat. Park and Mt Kupé), Bioko, extreme S Central African Republic, extreme SW Sudan (Bengengai Forest), Gabon, Congo, Angola (Cabinda, Lunda), Zaïre (throughout lowland forest zone), forests of W and S Uganda from Budongo and Jinja to Impenetrable and Malabigambo, NW Tanzania (Minziro Forest). Common to abundant; density in Gabon 13–16 pairs/km^2.

Description. *A. d. woosnami* Ogilvie-Grant: Zaïre (except lower Congo R.) to Sudan and Uganda. ADULT ♂: forehead grey-brown, becoming rich brown along sides of crown to hindneck; centre of crown pale orange, feathers tipped brown so orange patch partly concealed when feathers not raised; crown feathers form short erectile crest. Mantle to uppertail-coverts and scapulars chestnut. Tail uniform blackish brown. Preorbital area, supercilium, cheeks, ear-coverts and sides of neck grey; chin to undertail-coverts white; sides of breast and flanks and thighs grey. Primaries and secondaries blackish brown, secondaries with dull chestnut outer margins that increase in extent inwardly, inner secondaries with outer web and part of inner web chestnut; upperwing-coverts with dark brown inner webs, chestnut outer webs; axillaries and underwing-coverts grey with white margins. Bill and inside of mouth black; eye brown, hazel or red; legs and feet pale blue, pale grey, grey-brown or grey-black. Sexes alike. SIZE: (12 ♂♂, 12 ♀♀) wing, ♂ 90–97 (93·2), ♀ 84–88 (86·6); tail, ♂ 66–80 (71·0), ♀ 59–69 (65·2); bill, ♂ 17–20 (18·4), ♀ 17–19·5 (18·1); tarsus, ♂ 23–26·5 (25·4), ♀ 22–25 (23·9). WEIGHT: ♂ (n = 38) 28–38 (32·6), ♀ (n = 36) 22–39 (29·8).

IMMATURE: crown black with pale orange streaks; upperparts including lesser and median coverts black with large pale orange spots. Pre-orbital area and eye-ring orange, cheeks and ear-coverts mottled black and orange. Chin and throat pale orange with fine black streaks; breast streaked black and orange, belly to undertail-coverts orange, feathers of upper belly with narrow black fringes. Wings and tail like adult but innermost secondaries with small pale spots at tip, greater coverts tipped with large orange spots narrowly fringed black. Later, streaks remain on crown and some orange spots on face but spots disappear from upperparts and most of wings; upperparts chestnut like adult; throat and breast mixed white and orange, rest of underparts white except for black fringes to feather tips. Bill brownish black; gape yellow; inside of mouth orange; eye greyish brown; legs and feet dull greenish buff.

NESTLING: skin dark covered with long sooty down; gape white.

A. d. castanea Cassin: Nigeria to Bioko and W Zaïre (lower Congo R.). Forehead chestnut, not grey-brown; upperparts brighter chestnut; outer webs of tail-feathers edged dull chestnut; undertail-coverts fringed buff; primaries as well as secondaries edged chestnut; upperwing-coverts mainly chestnut (marginally darker on inner webs), underwing-coverts white without grey centres. WEIGHT: unsexed (n = 39, Cameroon) 28·4–37·6 (32·3). Immature has paler crown and back than *woosnami*, black feathering less extensive.

A. d. diademata (Bonaparte): Senegal to Togo. Upperparts and wing-edgings cinnamon-brown, much duller than other races; tail black with large white spots on outer 3 feathers, increasing in size outwardly. Bill black or greenish black; eye chestnut-red, amber or brownish orange; legs and feet blue, slate, grey or purplish grey. WEIGHT: (Liberia) ♂ (n = 13) 32·1 ± 2·4, ♀ (n = 7) 31·8 ± 0·9.

TAXONOMIC NOTE: nominate *diademata* formerly treated as a separate species, but habits and voice almost identical to those of *castanea*.

Field Characters. A common alethe whose monotonous call is a familiar sound in lowland forest. Shorter-legged and with more horizontal pose than sympatric Brown-chested Alethe *A. poliocephala*, which lacks crown-stripe and has pale supercilium, duller and browner underparts, no white in tail, and is largely silent. Spotted young very similar to Brown-chested Alethe but larger.

Voice. Tape-recorded (53, ERA, KEI, MOR). Song (nominate race), 3 sweet, liquid whistles, each one higher than the last, representing 'do-re-sol' on the sol-fa scale: 'hah-her-huee'; second note may be slightly down-slurred, last note may be slurred up or down; singer sometimes only reaches 'fa'. Thin 'peep' occasionally added after 'so'. Also has conversational subsong, a mixture of whistles and rattles, e.g. 'wee-were-chiwowo-huee-her-dzit-what-your-titic' (S. Keith, pers. obs.) or 'herdy-hear-were-turder-tatata-hear-her-do' (Willis 1986). Call, a monotonous series of single low whistles, like first note of song, repeated on same pitch at regular intervals of 2·5–3 s. Alarm, harsh, clicking chatter; when attacking conspecific gives grunt 'chahh' and 4–10 chip notes, while supplanted bird gives whimpering squeak 'eeh-eeh-eeh-eeh' (Willis 1986). Bird near person gave low, nasal 'taaaaa' like bush-shrike *Laniarius* (Liberia: S. Keith, pers. obs.). Song of *castanea* has 2 notes only, 'hoe-fear', second higher than first and down-slurred, like first and third notes of song of nominate race, and identical in tone, pitch and quality. In reaction to playback it sings, or gives faint short whistle 'hoo' at rate of 1 per s.

General Habits. Inhabits primary and old secondary forest, forest regenerating on old cultivation with *Musanga* and *Afromomum*, gallery forest, forest-grassland mosaic and cocoa plantations (Bioko). Mainly in lowlands, reaching 1500 m in Cameroon and E Africa, 1460 m in Zaïre.

Habitually attends swarming army ants *Dorylus*; seen at 69 out of 90 swarms, Gabon (Brosset and Erard 1986). When swarms abundant, only a few birds follow each one; when swarms few, up to 25 or 50 congregate at one (Willis 1986). Social unit is pair with young. Sings at dawn, then leaves territory to join ant swarm along with congeners; may also wander in early morning and in evening, visiting inactive ant colonies; follows ants wherever they go, not just in its own territory. Aggressive towards conspecifics, chasing them with grunts and chipping notes, ousting them from best perches around ants; bird being chased gives little squeaky notes, flicks wings downward. In display raises crest, fluffs belly feathers, gapes to display colour of open mouth, puffs out white throat and gives subsong; display sometimes ends in combat, birds gripping one another with feet, raising wings above their backs to show white underwing. Dominated by bristlebills *Bleda* and ant-thrushes *Neocossyphus*; spotted young often dominate older birds. Forages on ground and in low vegetation; stands on ground ahead of ants or perches above them on slender horizontal branches, not vertical stems which are often climbed by ants. Captures ground prey in hopping rush or short flight, or uncovers it by tossing leaves; gleans prey from logs, trunks, branches, lianas, leaves and tangles of debris; occasionally catches it in air; large items chewed and pounded on ground. Snatches prey rapidly from among ants, but often attacked by them and has to remove them from plumage. When ants not present, e.g. during dry season, joins mixed-species flocks. When nervous it flicks wings and tail, spreads tail to show white corners; on approach of hawk or when birds overhead give alarm notes, flattens crest and looks up.

Sedentary, at least in Gabon.

Food. Insects, especially grasshoppers and crickets, also flies, moths, caterpillars, beetles, roaches, termites, black ants, army ants; spiders; small snails and small frogs.

Breeding Habits. Territorial; territory advertised by song; size (Gabon) 5–6 ha.

NEST: shallow cup without rim, either compact or loose and rudimentary, of strips of bark, rootlets, dead twigs, moss, dead leaves and a little earth, lined with small black rootlets, strands of fungus *Marasmius* and a few leaves; ext. diam. 80–110, int. diam. 40–50, ext. depth 30–40, int. depth 10–20. Placed up to 3·3 m above ground in large cavity in living tree trunk, dead stump, or rotting log on ground, in termite mound, or on ground under end of decaying log.

EGGS: 2 (usually) or 3. Long ovals; almost no gloss; green, beige, pinkish ochre or pinkish white, heavily spotted brown, chestnut, rich maroon and light red over violet-grey and dull lilac, especially at large end, sometimes covering most of shell. SIZE: (n = 4) 24·6–28·2 × 16·7–18·5 (26·8 × 17·8). Sometimes double-brooded.

LAYING DATES: Liberia (breeding condition June–Sept, recently-fledged juveniles Oct–Nov); Cameroon, May, Aug, Oct (juveniles Nov); Gabon, Dec–Mar, especially Dec–Jan, mainly in short dry season (juveniles seen Jan–Aug, suggesting season extended); Zaïre, Apr–Dec in N, Sept–May in E highlands (Itombwe); Angola (breeding condition Feb); E Africa: Region B, Mar.

INCUBATION: by ♀ alone.

DEVELOPMENT AND CARE OF YOUNG: nestling period 12–13 days.

BREEDING SUCCESS/SURVIVAL: 1 nest destroyed by mandrills *Papio mandrillus*. Many live > 4 years, 1 lived > 8 years (Brosset and Erard 1986).

References
Brosset, A. and Erard, C. (1986).
Willis, E. O. (1986).

Alethe poliophrys Sharpe. Red-throated Alethe. Alèthe à gorge rousse.

Plates 23, 26
(Opp. pp. 368, 401)

Alethe poliophrys Sharpe, 1902. Bull. Br. Orn. Club 13, p. 10; Ruwenzori.

Range and Status. Endemic resident. Mountains of E Zaïre (Rwenzoris to Mt Kabobo), W Uganda (Rwenzoris, Impenetrable Forest), W Rwanda and W Burundi. Common.

Description. *A. p. poliophrys* Sharpe: range of species except Mt Kabobo. ADULT ♂: centre of forehead, crown and nape black, encircled by broad grey line from sides of forehead over lores and eye and around hindneck; pre-orbital region, cheeks and ear-coverts grey-black; eye-ring black. Mantle to rump, scapulars, and lesser and median coverts bright reddish chestnut; tail-feathers dark brown, outer webs edged chestnut. Chin dark grey, throat and foreneck orange-rufous, rest of underparts pale brown to off-white, almost pure white on centre of belly, breast and upper flanks with variable grey wash, lower flanks with variable brown wash; variable faint narrow brown barring on pale underparts, sometimes complete from breast to undertail-coverts, sometimes almost absent. Primaries and secondaries dark brown, inner primaries and all secondaries edged reddish chestnut; greater coverts with reddish chestnut outer webs, brown inner webs; axillaries and underwing-coverts off-white. Bill black or brownish black; eye red-brown to dark brown; legs and feet whitish to pale pink, greyish pink or brownish grey, claws grey-pink. SIZE: (10 ♂♂, 8 ♀♀) wing, ♂ 95–104 (98·9), ♀ 88–102 (93·5); tail, ♂ 64–74 (68·6), ♀ 59–69 (62·5); bill, ♂ 19·5–21 (20·3), ♀ 19–21 (19·7); tarsus, ♂ 28·5–31·5 (29·9), ♀ 28–30·5 (29·2). WEIGHT: ♂ (n = 18) 30–40 (35·3), ♀ (n = 16) 30–45 (35·1).

IMMATURE: head, mantle, upper back, throat and breast black with broad orange streak in centre of each feather (paler orange on breast); lower back to rump reddish chestnut; belly and undertail-coverts white, feathers of upper belly fringed black with pale orange subterminal band; wings like adult except tips of coverts have orange spot and blackish fringe.

NESTLING: unknown.

A. p. kaboboensis Prigogine: Mt Kabobo. Upperparts to rump, upperwing-coverts and wing-edgings olive-brown, much less rufous; tail-feathers with little rufous except at edges; orange-rufous of throat paler. Larger: (8 ♂♂, 4 ♀♀) wing, ♂ 100–104 (102), ♀ 96–98 (97·2); tail, ♂ 70–73·5 (71·7), ♀ 67–69 (68·0).

Field Characters. Restricted to mountain forests bordering central African Rift valley. Does not overlap Fire-crested Alethe *A. diademata* of lowland forests, but in some areas sympatric with Brown-chested Alethe *A. poliocephala carruthersi*. Larger than Brown-chested, with broad grey (not narrow white) supercilium, brighter red upperparts and wings, reddish tail (blackish in Brown-chested), and red throat. Spotted young similar to Brown-chested Alethe; best distinguished by redder upperparts and tail, throat with orange and black streaks (not pale), and larger size.

Voice. Tape-recorded (CART, LEM). Song poorly structured, even more so than in other members of the genus (Dowsett-Lemaire 1990); consists of a single whistled note, pure or slightly burry, downslurred ('peeeyoo' or 'peeeyurr') or on one pitch with slight rise in the middle ('wooiyoo'); same note repeated at intervals of 2–3 s, or several different notes mixed together in more continuous song, including upslurred 'fueee'. 1 bird imitated 2-note call of Black-tailed Oriole *Oriolus percivali* (Dowsett-Lemaire 1990). Whistled note also serves as contact call in mixed-species flock (J. Kalina, pers. comm.). Scold/alarm, soft 'raa-chaa-chaa-chaa-chaa', unlike rattles of congeners; also gives faint chirp 'wer'.

Alethe poliophrys

General Habits. Inhabits undergrowth of dense montane forest, high-altitude gallery forest and wooded ravines, and lower edge of bamboo zone, 1300–3000 m.

Occurs singly or in pairs or small groups. Follows columns of army ants, e.g. *Dorylus nigricans*; joined by Brown-chested Alethe in zone of overlap (Burundi: Vande weghe 1988). Hops on ground ahead of ants or waits in low vegetation. Displaces smaller thrushes (*Pogonocichla*, *Sheppardia*). Captures prey mainly on ground, in short flights or rushes, sometimes by tossing leaves; less often flies up to glean it from trunks or lianas. Sometimes wanders through semi-open understorey away from ants. Also joins mixed-species flocks, especially in dry season; flocks in Uganda (Impenetrable Forest) include Narina's Trogon *Heterotrogon narina*, Emerald Cuckoo *Chrysococcyx cupreus*, Lemon-rumped Tinkerbird *Pogoniulus bilineatus*, bulbuls (*Andropadus*, *Pycnonotus*), thrushes (*Pogonocichla*, *Neocossyphus*), flycatchers (*Terpsiphone*, *Melaenornis*, *Muscicapa*), warblers (*Prinia*, *Bathmocercus*), babblers *Alcippe*, white-eyes, sunbirds, waxbills *Estrilda*, orioles and starlings *Onychognathus* (J. Kalina, pers. comm.). Generally very shy; when person approaches, hops away rapidly on ground and disappears behind dense vegetation; also very curious, sometimes moving closer to inspect intruder if he keeps quiet. When attracted by imitation of its whistle, raises 1 or both wings to show white linings, flicks tail, and scolds or chirps (Willis 1985).

Food. Insects, including beetles, flies and army ants (60–80 in 1 stomach: Chapin 1953); also spiders, earthworms and snails.

Breeding Habits. Has extensive home range in Rwanda (Nyungwe), rather than proper territory, because army ants move fast and swarms there are widely spaced (1–2 km apart) (Dowsett-Lemaire 1990); singing mostly associated with sexual activity and breeding. Only 1 nest and 1 clutch of eggs found (Kalina and Baranga 1991).

NEST: cup of green moss, lined with dry brown moss stems; int. diam. 70 × 60, int. depth 45; 1·05 m above ground, in cleft in moss-covered tree buttress.

EGGS: 2. Dark green with faint brown speckles around large end. SIZE: (n = 2) 25 × 18.

LAYING DATES: Zaïre, Sept–July, probably all year; Rwanda, Sept–Oct; Uganda, Mar.

References
Dowsett-Lemaire, F. (1990).
Kalina, J. and Baranga, J. (1991).
Willis, E. O. (1985).

Alethe poliocephala (Bonaparte). Brown-chested Alethe. Alèthe à poitrine brune.

Plates 23, 26
(Opp. pp. 368, 401)

Trichophorus poliocephalus Bonaparte, 1850. Consp. Av. 1, p. 262; Dabocrom, Gold Coast (see Mees 1988).

Range and Status. Endemic resident. Sierra Leone, Liberia, Ivory Coast (north to Comoé, Sipilou), S Ghana; S Nigeria (Lagos to Obudu Plateau and north in E to Gashaka-Gumti Game Reserve), Cameroon (throughout forest zone, north to Korup Nat. Park and Mt Oku), Gabon, S Congo (Mayombe), Angola (Cuanza Norte (Quicolungo), Cuanza Sul (Gabela)). Reappears in SE Central African Republic and Zaïre (NE, E south to Mt Kabobo, and SE (Upemba Nat. Park)), east to extreme SE Sudan (Imatong Mts), W and S Uganda north to Budongo and Mabira, and separate population in NE on Mt Lonyili; Kenya west of the Rift from Mt Elgon to forests of Kakamega, Nandi and Mau, and east of the Rift from Meru to Nairobi; E Rwanda, SW Burundi; and Tanzania in NW (Minziro Forest) and W (Mahari Mt, Ufipa Plateau). Common to abundant; density in Gabon 6–7 pairs/km².

Description. *A. p. carruthersi* Ogilvie-Grant: E Zaïre, Uganda, W Kenya. ADULT ♂: forehead to nape dark grey-brown; white line from base of bill over eye, becoming grey behind eye. Pre-orbital area blackish, becoming browner on cheeks and ear-coverts. Mantle and sides of neck cinnamon-brown, grading into reddish chestnut, dull on back, scapulars and upperwing-coverts, brighter on rump and uppertail-coverts. Tail uniform blackish brown. Chin black, throat white, breast and flanks pale dull brown, greyer on flanks, belly and undertail-coverts creamy white. Primaries and secondaries blackish brown edged cinnamon-brown; axillaries and underwing-coverts creamy white. Bill black; eye red-brown or chestnut; legs and feet pinkish flesh, sometimes with grey or brown tinge. Sexes alike. SIZE: (12 ♂♂, 9 ♀♀) wing, ♂ 82–88 (84·4), ♀ 80–86 (82·9); tail, ♂ 48–57 (52·0), ♀ 48–54 (51·6); bill, ♂ 15·5–19 (17·4), ♀ 16–18·5 (17·3); tarsus, ♂ 24–27 (26·0), ♀ 25–26·5 (25·7). WEIGHT: ♂ (n = 6) 23–32 (27·3), ♀ (n = 7) 22–30 (25·8). A ringed ♀ weighed exactly the same (30) when recaptured as it did 9 years earlier (Gichuki and Schifter 1990).

IMMATURE: a young bird (wing 49, tail 10, bill 13, tarsus 25; downy feathers on belly) has top of head black streaked pale orange, mantle, back, lesser and median coverts black with large pale orange spots, rump and tail orange with black fringes to feathers, cheeks and ear-coverts dull pale orange-brown with dark mottling, chin and throat buff, breast and flanks pale dull orange with dark blotches on breast, belly and undertail-coverts creamy with orange suffusion. Primaries and secondaries like adult, greater coverts blackish with broad subterminal orange band and narrow black tip, underwing-coverts and axillaries not yet grown except along bend of wing (bases of flight feathers bare, uncovered).

With age, orange spots disappear from upperparts, tail and most of wing, but orange streaks on head remain; underparts whiter with no orange suffusion, breast blotched black and orange; in final stage head loses streaks, last few spots leave wings and breast blotches disappear. Fully grown but still immature birds resemble adults. Bill dark brown, basal half of lower mandible yellowish, gape yellow; eye dark brown or greyish brown; legs and feet pinkish white.

NESTLING: unknown.

A. p. compsonota (Cassin): Bioko, Nigeria and Cameroon to NW Angola (Quicolungo). Top of head and face greyer and darker; upperparts brighter reddish chestnut. Larger: wing, ♂♀ 85–102. WEIGHT: unsexed (n = 64, Cameroon) 32·5–46·2 (37·3).

A. p. poliocephala (Sharpe): Sierra Leone to Ghana. Like *compsonota* but ear-coverts brownish chestnut. WEIGHT: (Liberia) ♂ (n = 9) 32·7 ± 2·8, ♀ (n = 7) 31·9 ± 1·6.

A. p. akeleyi Dearborn: Kenya east of Rift. Head and face greyer than *carruthersi*, upperparts somewhat duller, less rich. Larger than *carruthersi*: wing, ♂♀ 94–100. WEIGHT: ♂ (n = 4) 30–32 (31·0), ♀ (n = 5) 30–38 (33·8).

A. p. hallae Traylor: near Gabela, Cuanza Sul, Angola. Crown reddish olive, not grey or brown; ear-coverts light chestnut-brown, like *poliocephala*; back like *akeleyi*.

A. p. nandensis Cunningham-van Someren and Schifter: N Nandi Forest, Kenya. Head sooty black; back darker and less reddish than *carruthersi* or *akeleyi*; tail black; breast-band faint, deep olive-buff; flanks greyish. Immature darker than *akeleyi*, spots rufous, not light buff. WEIGHT: ♂ (n = 25) 25·5–32·5 (27·8), ♀ (n = 15) 26–31 (28·5).

A. p. giloensis Cunningham-van Someren and Schifter: Sudan. Top and sides of head blackish, ear-coverts browner; back to rump cinnamon-brown; distinct light drab breast-band; belly washed grey-brown, flanks darker greyish brown. WEIGHT: 1 ♂ 27.

A. p. kungwensis Moreau: W Tanzania (Mt Kungwe). Upperparts darker and redder than *carruthersi* or *akeleyi*, about like nominate; top and sides of head grey, not blackish, with olive wash.

A. p. ufipae Moreau: SW Tanzania (Ufipa Plateau) and SE Zaïre (Upemba Nat. Park). Like *carruthersi* but upperparts slightly redder, top and sides of head olive-brown, not grey. Mt Kabobo birds intermediate between *carruthersi* and *ufipae* (Prigogine 1984).

A. p. vandeweghei Prigogine: Rwanda, Burundi. Differs from *carruthersi* in paler back, less pronounced dark breast-band. Larger: wing, ♂ 98·5–101 (99·5), ♀ 93–96 (94·3). Back duller, less reddish than *kungwensis* and *ufipae*; browner, darker and more olive than *akeleyi*.

Field Characters. A plump, upright, long-legged bird of the forest floor and low undergrowth. Rather nondescript, with dark grey-brown head, white eyebrow, chestnut back and pale underparts; brown wash on breast often indistinct, not a good field mark. Often occurs with Fire-crested Alethe *A. diademata* around ant swarms; latter is larger, with orange stripe on crown (not always easy to see), purer white underparts with grey, not brown, flanks and sides of breast, and no white supercilium; western race *diademata* also has white spots on tail and brown back. Spotted young very similar, best told by size; young of western race of *A. diademata* have white spots in tail.

Voice. Tape-recorded (53, C, GREG, LEM, McVIC). Generally silent. Song a series of 5–8 pure, rather mournful whistles, descending the scale and becoming less intense towards the end. Also has buzzy subsong, 'reiz-eiz-serrt-serr-riz-sez-seees' or 'razz-raah-zaeid-zerr' (Willis 1986). Contact-call between pair, 'tu-iit'; aggressive calls at ant swarms, a growl 'raagh' and 'seiz-seiz-seiz-seiz'; alarm, a rising whistle, 'seeeleeeeeh'. Once imitated chatter of Grey-headed Bristlebill *Bleda canicapilla* when latter approached its nest (Willis 1986). Bird ringed and released gave a 'rather quiet, scratchy, finch-like song of *c*. 20 notes' (Mann 1974).

General Habits. Lives mainly in primary forest, sometimes old secondary growth and forest–grassland mosaic; in undergrowth and on forest floor, especially near streams. Mainly in lowlands but also in transition and lower montane forest, to 900 m in Liberia, 1600 m in Zaïre, 2200 m in Cameroon, 2500 m in Sudan, 2800 m in E Africa.

Generally occurs singly, in pairs or small family parties, but up to 12 may gather at ant swarm. Forages on ground and fallen logs and in low, dense vegetation; drops to ground from perch to catch insect, then returns to perch; tosses leaves. Regularly joins bird parties following ants. Waits on ground or on low perches 5–15 m ahead of swarm, catching arthropods fleeing from the ants. Takes most prey on ground, either directly or after a few quick hops or short flights; also gleans prey from logs, trunks, lianas and foliage. Around ants it interacts aggressively with Fire-crested Alethe, ant-thrushes *Neocossyphus* spp. and bristlebills *Bleda* spp.; frequently chases conspecifics, giving growl or snapping bill; flashes one or both wings after successful chase; in between chases gives faint subsong. Young birds especially aggressive, attacking adults and often supplanting them (Willis 1986). Visits inactive ant colonies and waits for ants to swarm. Less attracted to ants than Fire-crested Alethe; joins mixed-species flocks not following ants. Pairs have broadly overlapping home ranges rather than defined territories; areas with best food resources occupied by dominant birds. At dusk bathes in forest streams; one preened and flashed wings while giving medley of song, scolds and chatters (Willis 1986).

Sedentary, at least in Gabon and W Kenya; bird ringed in Kenya was recaptured 9 years later only 100 m from original site (Gichuki and Schifter 1990).

Food. Invertebrates, especially insects and their larvae (beetles, termites, ants, mole-crickets, mantids), millipedes, spiders. Also small molluscs and tiny frogs.

Breeding Habits.
NEST: cup of moss, often deep, sometimes strengthened with framework of rootlets and strands of fungus *Marasmius*, which may also form lining; ext. diam. 65–70, int. diam. 60–65, ext. depth 60, int. depth. 40; placed 1–7 m above ground in cavity, e.g. in rot-hole or crack in tree trunk, old woodpecker hole, on top of dead stump, on termite mound against side of tree; once in crack between 2 tree buttresses, resting on a knob, and once plastered to side of fork.

EGGS: 1–3, laid at 1-day intervals. Long ovals; green, greenish brown or chocolate, heavily spotted with chestnut and violet-grey, especially around large end. SIZE: (n = 4) 20–28 × 15–19·5 (23·0 × 16·9).

LAYING DATES: Sierra Leone, July; Liberia (breeding condition May–June, juveniles June–Jan); Ghana (breeding condition Sept); Nigeria (breeding condition June); Cameroon (juveniles Apr, Nov, birds with brood patch Dec–Jan); Gabon, Dec–Mar (juveniles June–Aug), breeding possibly continuous Oct–June; Central African Republic (breeding condition June); Zaïre, Sept–Apr (juveniles July); Angola (breeding condition Aug); Sudan, Mar; E Africa: Kenya, Mar, Nov; Region B, Apr–May; Region D, Oct.

INCUBATION: by ♀ only; begins with 2nd egg. Period: 17 days. ♀ leaves nest every 40–50 min to feed; very shy, only returning to nest after careful watching and circling nest in surrounding vegetation.

BREEDING SUCCESS/SURVIVAL: of 4 nests with eggs, Gabon, all destroyed by predators. Bird in W Kenya lived at least 9 years (Gichuki and Schifter 1990).

References
Brosset, A. and Erard, C. (1986).
Willis, E. O. (1986).

Alethe fuelleborni Reichenow. White-chested Alethe. Alèthe à poitrine blanche.

Alethe fuelleborni Reichenow, 1900. Orn. Monatsber. 8, p. 99; Peroto-Ngosi, Tandalla, Tanganyika.

Forms a superspecies with *A. choloensis*.

Plates 23, 26
(Opp. pp. 368, 401)

Range and Status. Endemic resident. Widespread in mts of E Tanzania from S Pares and Usambaras through Ngurus, Ukagurus, Ulugurus and Uzungwas to Mahenge, Livingtone Mts and Mt Rungwe; extreme NE Zambia (Mafinga Mts, Nyika Plateau); N Malaŵi south to S Viphya Plateau; central Mozambique (Mt Gorongoza, coast near Sofala (Beira)). Common; density in continuous forest, Malaŵi, 2 pairs/10 ha (Dowsett-Lemaire 1989).

Description (includes '*usambarae*', '*xuthura*'). ADULT ♂: forehead and supraloral line grey; pre-orbital area, cheeks and ear-coverts blackish; crown to hindneck grey-brown, mantle olivaceous to cinnamon-brown, becoming increasingly reddish on back, scapulars and median and lesser upperwing-coverts, rump and uppertail-coverts cinnamon-rufous. Tail-feathers chestnut, outer webs edged reddish chestnut, shafts maroon above, straw below. Chin and throat white, bordered by dark grey sides of neck which continue as patch across sides of lower throat forming incipient half-collar; rest of underparts white, variable patch of cinnamon-brown at sides of breast, flanks mixed cinnamon-brown and grey; breast feathers sometimes tipped darker, giving scaly effect (scales weak or absent in north of range). Primaries and secondaries blackish brown, primaries edged cinnamon-brown; secondaries edged cinnamon-rufous, inner webs of inner secondaries cinnamon-brown; greater coverts with outer web cinnamon-rufous, inner web blackish brown; axillaries and underwing-coverts white. Bill black; eye red-brown to dark brown; legs and feet pale, grey or greyish pink. Sexes alike. SIZE: (decreases from N to S: Jensen *et al.* 1985) wing (Malaŵi), ♂ (n = 17) 106–112 (107·9 ± 1·7), ♀ (n = 11) 103–107 (104·6 ± 1.12), N Tanzania mean, ♂♀ (n = 22) 110; tail, ♂ (n = 4) 78–82 (79·8), ♀ (n = 5) 70–76 (73·4); bill, ♂ (n = 4) 23–26 (24·5), ♀ (n = 5) 20·5–24 (22·3); tarsus, ♂ (n = 4) 32–35 (33·3), ♀ (n = 5) 32–34 (32·8). WEIGHT: (Malaŵi) ♂ (n = 17) 41·6–56·5 (49·0), ♀ (n = 10) 44–58 (49·7). Heavy at beginning of breeding season, light while feeding young, heavy when moulting (Dowsett and Dowsett-Lemaire 1984).

IMMATURE: top of head dark brown, feathers with pale orange shaft-streaks; rest of upperparts blackish brown with pale orange spots; with age, spots disappear, upperparts like adult but some feathers, especially on back, retain blackish tips. Throat dark grey-brown with some white, changing to off-white with narrow dark barring. Breast black with pale orange spots; belly white, feathers with variable amount of narrow dark fringing, giving scaly effect, grey of flanks washed pale orange, undertail-coverts pale orange.

NESTLING: unknown.

TAXONOMIC NOTE: much individual variation in back colour within populations, and size variation is clinal; scaling on breast feathers generally lacking in northern birds ('*usambarae*'), but weak scaling present in some individuals. We treat the species as monotypic.

Field Characters. A large alethe of E African highland forests, the only one in its range, with reddish upperparts and tail, white underparts and no eye-stripe. From behind might be mistaken for Spot-throat *Modulatrix stictigula*, but Spot-throat has red-brown underparts, spotted throat.

Plate 29

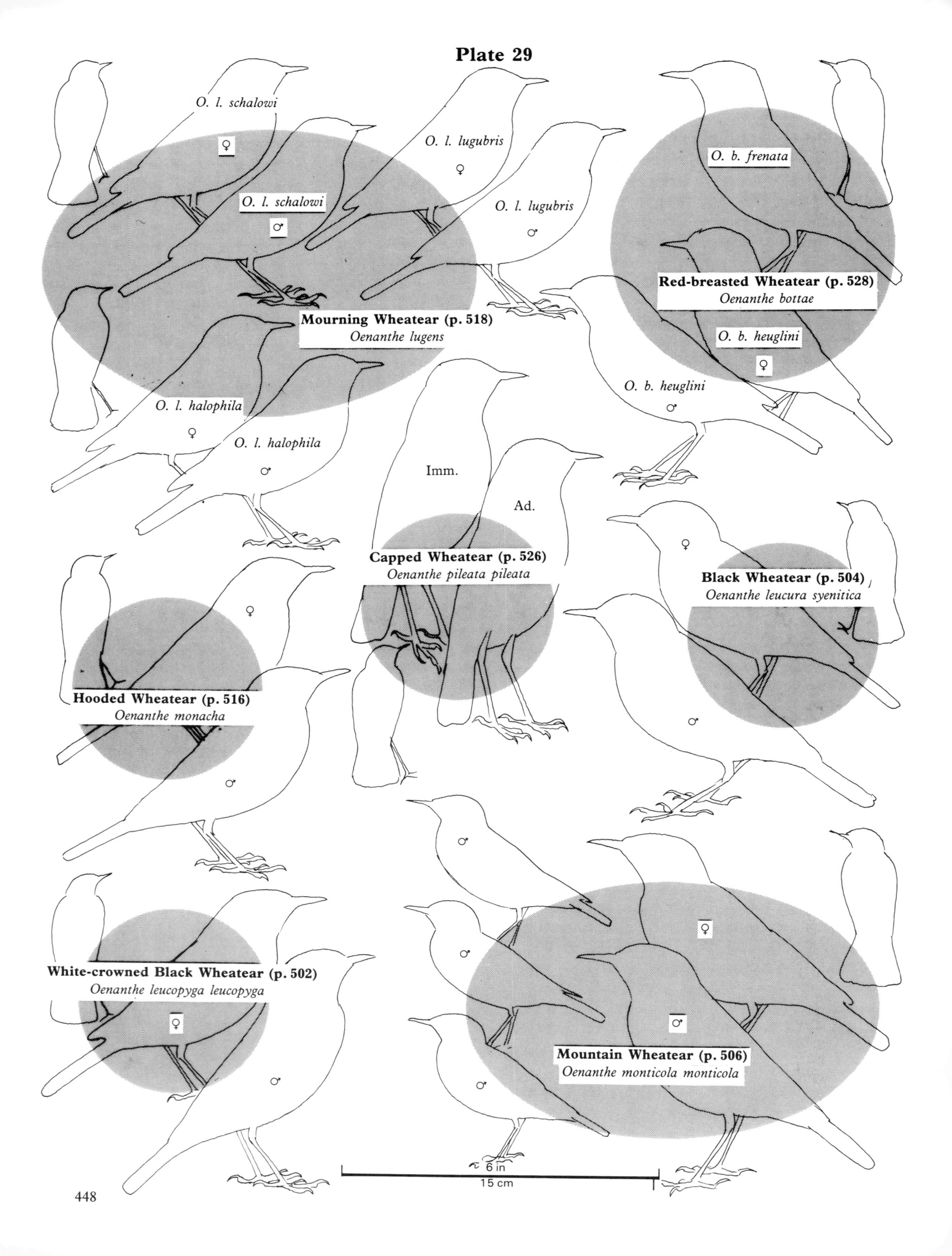

O. l. schalowi ♀
O. l. schalowi ♂
O. l. lugubris ♀
O. l. lugubris ♂
O. l. halophila ♀
O. l. halophila ♂

Mourning Wheatear (p. 518)
Oenanthe lugens

O. b. frenata
O. b. heuglini ♀
O. b. heuglini ♂

Red-breasted Wheatear (p. 528)
Oenanthe bottae

Imm. Ad.

Capped Wheatear (p. 526)
Oenanthe pileata pileata

♀
♂

Black Wheatear (p. 504)
Oenanthe leucura syenitica

♀
♂

Hooded Wheatear (p. 516)
Oenanthe monacha

♀
♂

White-crowned Black Wheatear (p. 502)
Oenanthe leucopyga leucopyga

♀
♂

Mountain Wheatear (p. 506)
Oenanthe monticola monticola

6 in
15 cm

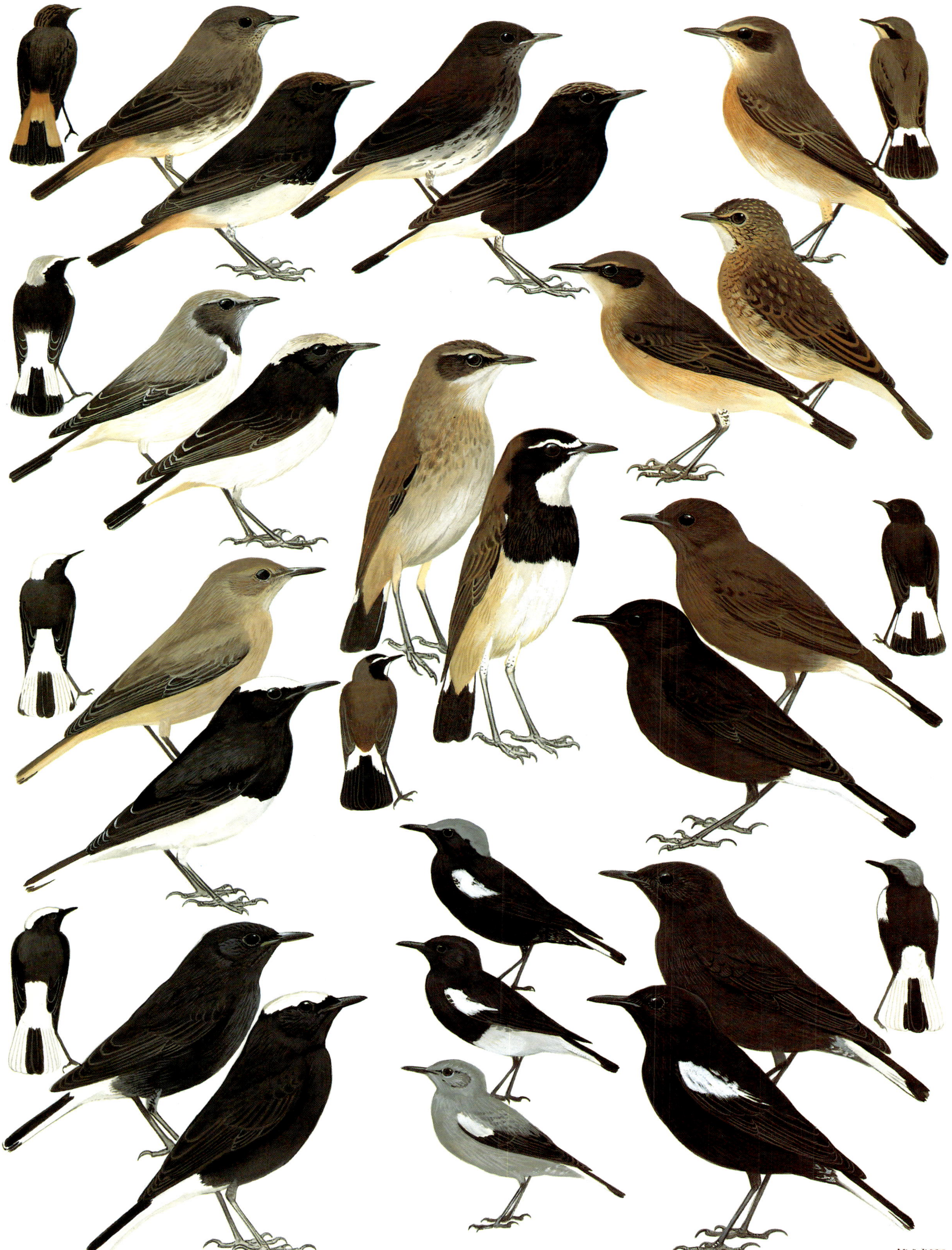
MWW1990

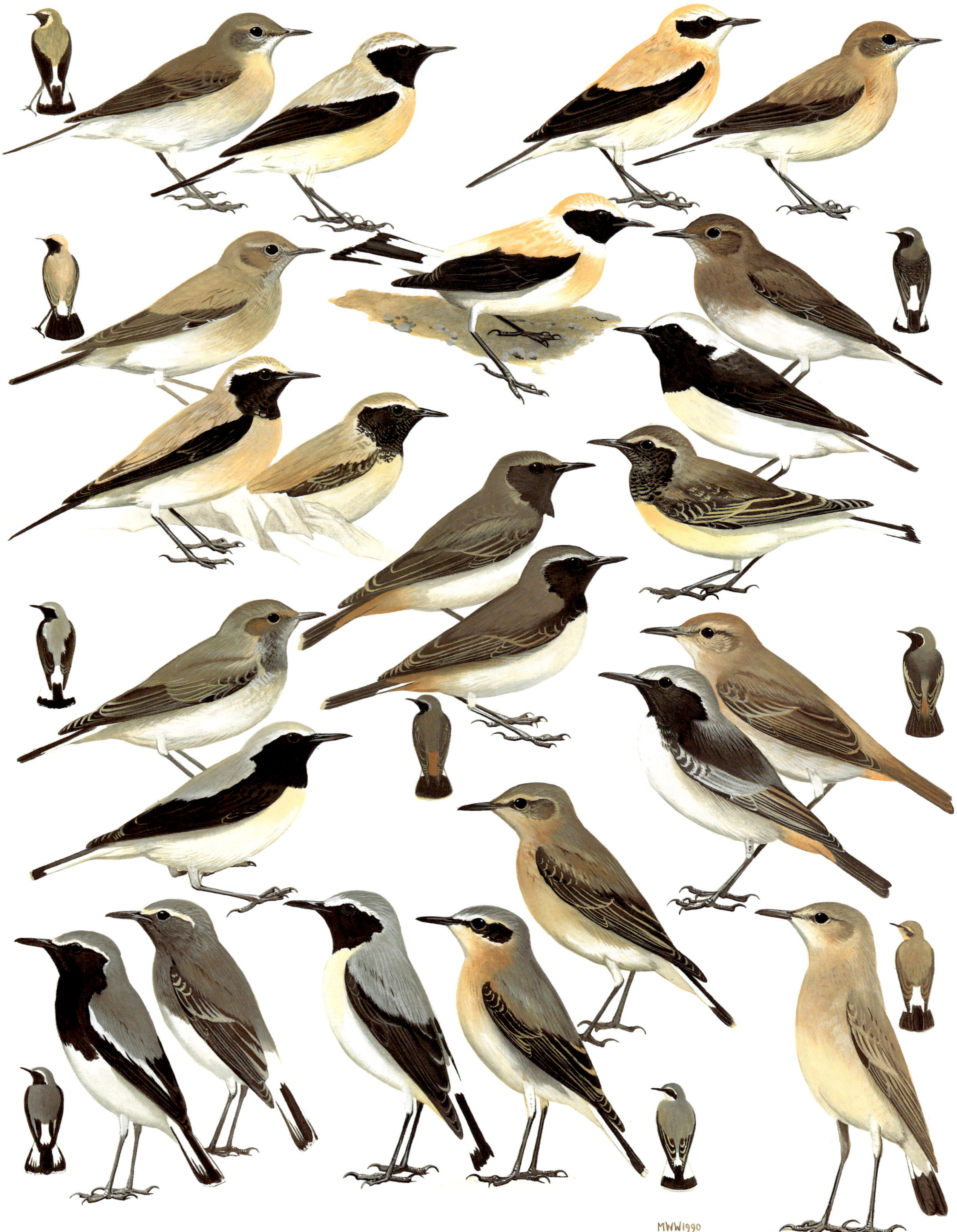
MWW1990

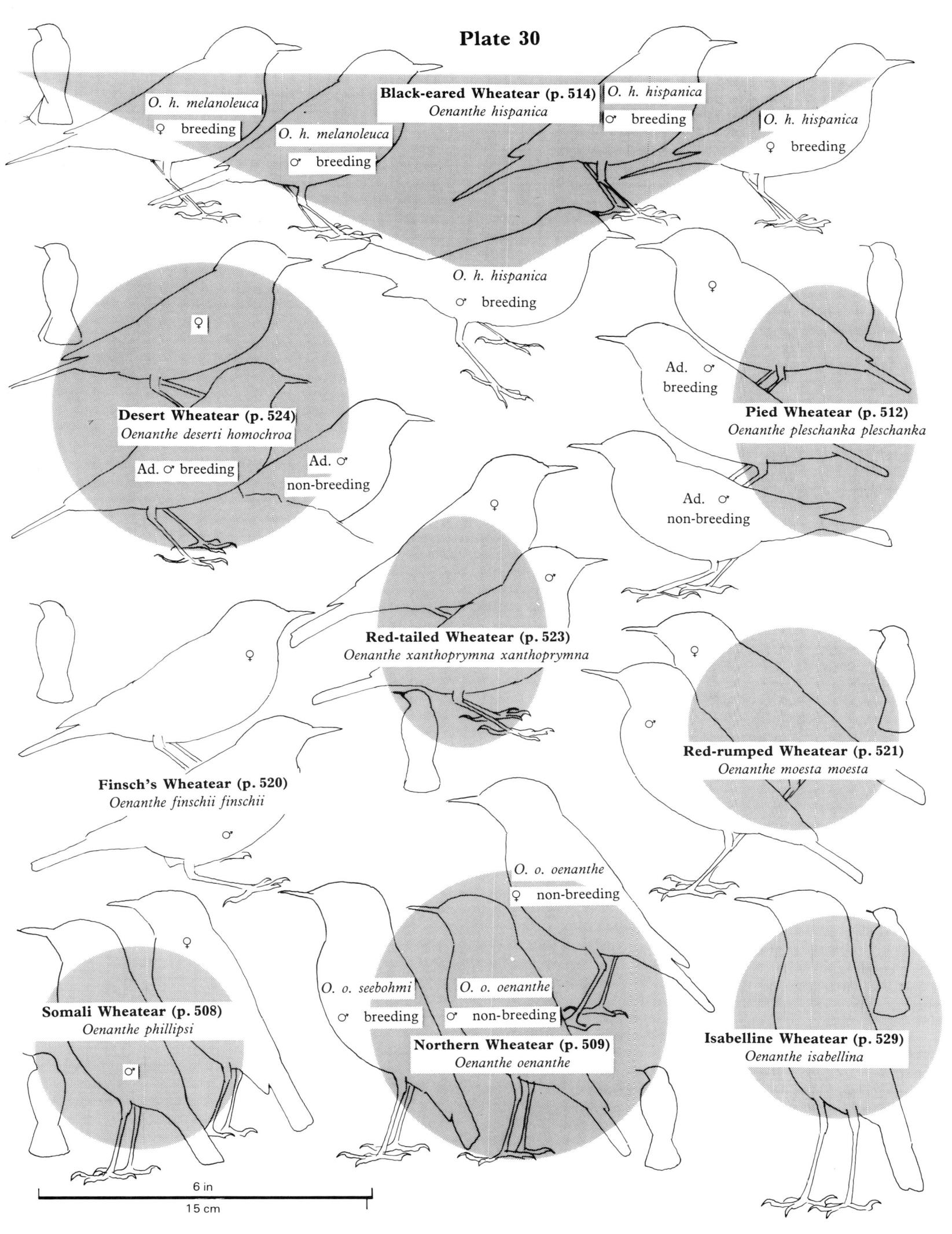

Plate 30

Black-eared Wheatear (p. 514)
Oenanthe hispanica

- *O. h. melanoleuca* ♀ breeding
- *O. h. melanoleuca* ♂ breeding
- *O. h. hispanica* ♂ breeding
- *O. h. hispanica* ♀ breeding
- *O. h. hispanica* ♂ breeding

Desert Wheatear (p. 524)
Oenanthe deserti homochroa
- ♀
- Ad. ♂ breeding
- Ad. ♂ non-breeding

Pied Wheatear (p. 512)
Oenanthe pleschanka pleschanka
- ♀
- Ad. ♂ breeding
- Ad. ♂ non-breeding

Red-tailed Wheatear (p. 523)
Oenanthe xanthoprymna xanthoprymna
- ♀
- ♂

Finsch's Wheatear (p. 520)
Oenanthe finschii finschii
- ♀
- ♂

Red-rumped Wheatear (p. 521)
Oenanthe moesta moesta
- ♀
- ♂

Somali Wheatear (p. 508)
Oenanthe phillipsi
- ♀
- ♂

Northern Wheatear (p. 509)
Oenanthe oenanthe
- *O. o. oenanthe* ♀ non-breeding
- *O. o. seebohmi* ♂ breeding
- *O. o. oenanthe* ♂ non-breeding

Isabelline Wheatear (p. 529)
Oenanthe isabellina

6 in / 15 cm

Voice. Tape-recorded (86, 88, GIB, LEM). Contact call a loud 'wooeeooo', rising in pitch in the middle, carrying 200–300 m; song often starts with contact call, then a series of varied whistles, mainly pure, some slightly trilled, given at intervals of c. 1 s, e.g. 'wooeeoo, peeoo, weeowip, torweeyurr, pweeurr, wurrweeyu, wiptowee, prrreee' (Malaŵi); song of Tanzanian birds has rather more burry quality, common component an upslurred 'wurrrreee'. Alarm, a loud rattle 'shreeeeeeeh'; in aggression, a whistled 'seeeeh', and fragments of subsong, e.g. 'wi-wecher-seet-wipwipwip-eup-shree' or 'seee-herrr-reeeeeeih-wheroe' (Willis 1985). Food-begging call of young, a sharp, very high-pitched 'srrreeeeee', like cricket or grasshopper.

General Habits. Inhabits tall forest with open understorey, with saplings but no grass or other ground plants; mainly montane forest, 1600–2400 m in Malaŵi, 1800–2100 m in Zambia, up to 2600 m in Tanzania, but sometimes breeds as low as 500 m, and in cold season occurs at lower levels, down to 250 m in Tanzania and in coastal forest in Mozambique.

Regularly attends army ant swarms. Stands on ground just ahead of ants, waiting for invertebrates flushed by them; if overtaken by ants, stamps feet, fluffs and shakes plumage and removes them with bill. Hops and runs to capture prey on ground, or makes short aerial sally; gleans it from trunks, low branches, sometimes leaves, seldom above 2 m. When ants are inactive (especially in dry season) it forages on its own or occasionally joins mixed-species flocks. Fairly aggressive toward conspecifics around ants; chases are frequent, accompanied by rattles, whistles and bill-snapping, sometimes some subsong (Willis 1985). Local ♂ dominates other ♂♂, although sometimes supplanted by his own young; he also dominates smaller thrushes (*Cossypha*, *Pogonocichla*, *Sheppardia*) and the Mountain Greenbul *Andropadus tephrolaemus*. Shy and elusive, very hard to see except around ant swarms; stays close to ground in low vegetation; when alarmed flicks wings and flees; flight rapid and low. Sings only during breeding season. Song apparently has sexual function, linked to breeding activity, and has no strong territorial use (Dowsett-Lemaire 1987); there is no countersinging between neighbours, and song playback causes little reaction except close to singing bird or within 10–20 m of nest.

Mean duration of individual moult (n = 17) 176 days (Malaŵi: Dowsett and Dowsett-Lemaire 1984). Adults moult wing- and tail-feathers after breeding; juveniles moult only body feathers. Moult is timed to end before cold weather, when food in short supply, although it may be delayed by birds which have dependent young at a late date (Dowsett and Dowsett-Lemaire 1984). In the species as a whole, moult begins before breeding is complete, but that has not been shown for any individual.

Mainly sedentary, but some altitudinal migration in non-breeding season in Tanzania, Malaŵi, and probably Mozambique, apparently mainly by ♀♀. 39 retrapped birds moved 0–480 (246 ± 127) m from ringing site (Malaŵi: Dowsett 1985); birds in small patches moved shorter distances. Within a single breeding season, only one-third of 28 adults moved further than the mean territory diam. (240 m) of the species, maximum move 370 m.

Food. Mainly insects, including beetles, moths, ants and their larvae, and army ants *Dorylus*; also spiders, millipedes, worms, snails and small amphibians; a few small berries.

Breeding Habits. Monogamous; territorial. Territory size in small forest patches 0·5–4 ha. Of 22 territorial adults, 4 held same territory for at least 2 years, 9 for at least 3 years, 2 for 4 years, 4 for 5 years, and 1 each for 6, 7, 8 years (Dowsett 1985). Successful breeding depends on constant source of army ant activity; all breeding territories contain an active ant colony (Dowsett-Lemaire 1989). Rarely breeds in successive years since ants not dependable and may move away. During breeding season, ♂♂ visit unoccupied forests to search for ant colonies. Birds can breed in forest patch devoid of ants if nearby patch has ants; parents cross over to get food, and when young are fledged the whole family moves across; 6–8 birds can feed at large ant swarm despite intraspecific chasing by territory holder. Non-breeding adults are present in breeding areas during breeding season.

NEST: thick cup of green moss, lined with fine rootlets; 1 nest diam. 80, depth 50; placed 1–8 m above ground in fork of tree, in liana tangle, or on top of stump; 1 nest on stump exposed, covered only by a few creepers, another in fork was hidden in base of fern clump.

EGGS: 2 (only 1 clutch described: Jensen *et al.* 1985). Pale green with brown to dark green spots. SIZE: (n = 2) 25·5 × 18·2, 26·9 × 17·7. Lays again if first attempt fails (Dowsett-Lemaire 1985).

LAYING DATES: Tanzania, Oct–Mar; Zambia, Nov; Malaŵi, Oct–Jan (Oct 2, Nov 19, Dec 12, Jan 6 clutches) (end of dry season, beginning of rainy season).

INCUBATION: by ♀ only; ♂ may sing nearby.

DEVELOPMENT AND CARE OF YOUNG: young fed by both parents, in nest and for at least 6 weeks after leaving it. 1 ringed young still present in territory 2 months after independence.

BREEDING SUCCESS/SURVIVAL: average annual mortality rate 13·2% in ♂♂, 13·3% in ♀♀; lives at least 7·9 years (Dowsett 1985).

References
Dowsett-Lemaire, F. (1987).
Jensen, F. P. *et al.* (1985).
Willis, E. O. (1985).

Alethe choloensis Sclater. Cholo Alethe. Alèthe du Mont Cholo.

Alethe choloensis Sclater, 1927. Bull. Br. Orn. Club 47, p. 86; Mt Cholo, Nyasaland.

Forms a superspecies with *A. fuelleborni*.

Plates 23, 26
(Opp. pp. 368, 401)

Range and Status. Endemic resident. S Malaŵi east of Shire R. from Namizimu Hills and Mangoche Mt through Chikala Hill, Malosa and Zomba Mts, Mauze Hill, and Shire Highlands (Chiradzulu, Ndirande, Bangwe, Soche, Malabvi) to Mulanje Mt and Cholo Mt; central Mozambique (Namuli and Chiperone Mts). Density in Malaŵi (Cholo) 1 pair/5 ha; locally common but total population only 1500 pairs; classed as endangered in I.C.B.P. Red Data Book (Collar and Stuart 1985); its habitat is being reduced every year by deforestation, especially on Mt Mulanje.

Description. *A. c. choloensis* Sclater: Malaŵi, Mozambique (Mt Chiperone). ADULT ♂: forehead to hindneck and mantle cinnamon-brown, becoming increasingly rufous on back, rump and scapulars; uppertail-coverts cinnamon-brown. Tail-feathers dark brown edged cinnamon-brown, T2 with tiny white spot at tip of shaft, T3–T6 tipped with larger white spots increasing outwardly from 6 mm on T3 to 16 mm on T6. Lores dark brown, cheeks and ear-coverts brown; chin and throat white; small brown patch at sides of breast, becoming grey towards centre of breast; flanks mixed brown and grey, centre of lower breast and belly white, with buff wash on breast and flanks and all except centre of belly, undertail-coverts rich buff. Primaries and secondaries dark brown, primaries narrowly and secondaries more broadly edged rufous; upperwing-coverts with rufous outer webs, dark brown inner webs; axillaries and underwing-coverts white, some dark marks on under primary coverts. Bill black; eye dark brown; legs and feet pale flesh. Sexes alike. SIZE: (9 ♂♂, 3 ♀♀) wing, ♂ 90–101 (97·0), ♀ 93–99 (96·7); tail, ♂ 64–72 (67·8), ♀ 63–69 (66·7); bill, ♂ 18·5–21 (19·4), ♀ 18·5–21 (20·2); tarsus, ♂ 28·5–31 (30·2), ♀ 30–31 (30·5).

IMMATURE: upperparts from top of head to rump and wing-coverts dark brown, mottled orange (feather shafts orange broadening to spot at tip); throat feathers dirty white with narrow dark tips giving scaly effect); breast blackish with pale orange spots, rest of underparts dirty white mottled blackish, less mottling on lower belly and undertail-coverts, flanks washed pale orange; wings and tail like adult. Bill black, keel of lower mandible yellow, gape yellow; legs and feet pale brown.

NESTLING: unknown.

A. c. namuli Vincent: Mozambique (Mt Namuli). Breast and belly paler and whiter, with very little buff wash, undertail-coverts white.

Field Characters. Mountains of S Malaŵi and Mozambique only. Upperparts and wings reddish brown, tail dark with white spots at tips of outer feathers, underparts white with grey and buff wash on breast and flanks (nominate race) or mainly white (*namuli*). Similar to closely-related White-chested Alethe *A. fuelleborni* but allopatric; no other alethe in its range. Other small thrushes in its habitat are smaller Olive-flanked Robin-Chat *Cossypha anomala* (reddish tail without white spots, grey face, pale supercilium, grey breast, russet flanks, brown-orange undertail-coverts, different song), and smaller Eastern Bearded Scrub-Robin *Cercotrichas quadrivirgata* (black and white striped face, orange breast and flanks, white flashes in wing, much white in tail, sweet warbling song).

Voice. Tape-recorded (LEM). Contact call a loud, down-slurred 'peeeyoo'. Song, phrases of 3–4 notes (usually beginning with contact call) lasting 2–2·5 s, less often 2 notes (1·5 s): 'tyerr .... wor-tyer-chee' (second note lowest); 'wor ... tee-tyer-tyurr' (first note lowest, second highest, third and fourth descending); 'tyerrr-tuwee-tyurrr'; 'werrr-cheeya'. Rather softer, higher-pitched and more variable than White-chested Alethe; some notes pure, many with rolling, burry quality. Rattle call like that of White-chested Alethe; alarm, thin 'seeee'.

General Habits. Inhabits ground stratum in mid-altitude evergreen forest, often quite close to edge; in patches of weeds or dense tangles of bushes and branches. Breeds mainly at 1000–1700 m (but at 720 m on Mt Mulanje: Johnston-Stewart 1989); in summer months ascends to 1900 m.

Solitary, or in groups of up to 5 around ants. Regularly attends ant swarms; waits on ground or perches on log or in low bush; darts down to ground from perch, snatches prey, hurriedy returns to perch, often with much wing-flapping and bill-snapping (Johnston-Stewart 1982). Also tosses leaves on ground, or gleans prey from trunks and foliage in short flight. Immatures learning to feed around ants were inefficient, catching little, while adults called nearby (Johnston-Stewart 1989). Chases other species, e.g. White-starred Robin *Pogonocichla stellata*, giving rattle call. Sometimes forages far from ants with other species, e.g. Cabanis's Greenbul *Phyllastrephus cabanisi*. Shy and elusive; when alarmed, flicks wings and tail and flashes white tail spots, dives into cover.

Mainly sedentary, but some altitudinal migration in Malaŵi, up to 1900 m and down to 680 m.

Food. Insects, including army ants, beetles and 'grubs'.

Breeding Habits. Almost unknown; only 1 nest found (Benson and Benson 1947).

NEST: cup of green moss lined with fine dry tendrils; ext. diam. 140, int. diam. 95, depth 50; 4 m above ground in fork of tree.

EGGS: 3. Glossy; green, heavily mottled with dull and light chestnut over lilac and pale grey. SIZE: (n = 2) 25·9 × 19·2, 25·5 × 19·1.

LAYING DATES: Malaŵi, Sept, Nov, Jan; Mozambique, Sept.

Reference
Dowsett-Lemaire, F. (1987).

Genus *Neocossyphus* Fischer and Reichenow

Medium-sized birds, legs long or short, tail relatively long, bill broad or narrow, wings and tail rounded. Plumage rather dense, feathers with downy bases, somewhat fluffy on rump; predominantly rufous, with variable rufous wingbar across primaries and secondaries. Juvenile plumage unspotted.

Genus comprises 2 ground-dwelling 'thrushes', with long legs and narrow bills, one in Lower Guinea, one in both Upper and Lower Guinea, and 2 arboreal 'flycatchers' (formerly genus '*Stizorhina*'), with short legs and broad bills, one in Upper and one in Lower Guinea. In each area the 2 sympatric species have almost identical plumage, although structure, voice and ecology differ; one behaves like a flycatcher and the other like a thrush. While members of each pair resemble one another, plumage differences between the 2 pairs are considerable: the Lower Guinea pair (*rufus/fraseri*) is almost uniformly rufous, while the Upper Guinea pair (*poensis/finschi*) has white in the tail and darker upperparts.

How this remarkable evolutionary phenomenon arose is not clear. Perhaps the ancestral stock first split geographically, into Upper and Lower Guinea, and after evolving plumage differences split again in each area, and the separate populations evolved structural differences. Later the populations merged, by which time isolating mechanisms had evolved allowing the 2 populations to live together as good species. Fry (1964) thought it more likely that 'a *Neocossyphus* and a *Stizorhina* say in Central Africa (= Lower Guinea) each speciated on migrating to Upper Guinea', but that would be asking rather a lot of coincidence; it assumes that each bird retained its structural characters while 'happening' to evolve identical plumage.

The genus *Neocossyphus* has had a chequered history and its systematic position is still unresolved. While generally considered a thrush (e.g. by Peters 1964), its syringeal morphology is not turdine (Ames 1975). Juveniles lack the spotted plumage characteristic of most young thrushes, but several other turdines have unspotted young (e.g. *Cercotrichas*, *Myrmecocichla*); in the hand *Neocossyphus* spp. do not 'feel' like true thrushes (Jensen 1990), and on this and other evidence Jensen concluded the genus does not belong in the subfamily Turdinae and is apparently not closely related to other families, and is perhaps best listed as a separate family. Olson (1989) proposed that the genera *Myadestes* (including *Phaeornis*), *Stizorhina*, *Neocossyphus*, *Modulatrix* and *Pinarornis* be combined in a subfamily Myadestinae in the larger family Muscicapidae, with the Myadestinae 'the primitive sister-group of the much larger clade defined by the derived turdine condition of the syrinx'. We retain *Neocossyphus* in Turdidae, pending further studies.

Endemic to tropical Africa; 4 species. Sympatry and obscure evolutionary history prevent any superspecies groupings; the 4 are best classed together loosely in a species group.

Neocossyphus rufus (Fischer and Reichenow). Red-tailed Ant-Thrush. Grive fourmilière à queue rousse.

Pseudocossyphus rufus Fischer and Reichenow, 1844. J. Orn., p. 58; Pangani.

Plate 25
(Opp. p. 400)

Range and Status. Endemic resident. SE Cameroon (from Mt Cameroon eastwards), Gabon and Congo (common) east to NE Zaïre (Uele, Kisangani, Kivu) and W Uganda; and in coastal Kenya and Tanzania from Tana R. to Mikindani; also Zanzibar. Single record Somalia (Ola Uager, 1° 18'N, 41° 40'E: Roche 1987). Rare to common; av. density 3–4 pairs/km^2 in NE Gabon, where commoner at higher altitudes (800–1000 m).

Description. *N. r. gabunensis* Neumann: range of species except coastal Kenya and Tanzania. ADULT ♂: forehead, crown, nape, lores, ear-coverts and hindneck olive-brown washed grey. Mantle and scapulars olive-brown tinged tawny. Rump and uppertail-coverts deep rufous-chestnut. Tail orange-rufous to pale cinnamon-rufous, with 2 central pairs of rectrices darker, almost deep chestnut. Chin, throat and upper breast pale greyish olive-brown with a slight greenish tinge; rest of underparts bright tawny-cinnamon, darker on flanks and undertail-coverts. Marginal and lesser upperwing-coverts olive-brown tinged tawny; median and greater upperwing-coverts dark olive-brown with almost rufous outer webs. Primaries blackish brown on inner webs and shafts, becoming bright rufous on outer webs; outer secondaries like primaries but outer webs duller, more olive, less deep rufous; inner secondaries entirely rufous olive-brown. Underwing-coverts, axillaries, fringes to inner webs of underside of flight-feathers and whole underside of tail-feathers orange-rufous to pale tawny-cinnamon. Bill brownish black; eye brown; legs and feet horn or flesh colour. Sexes alike. SIZE: (7 ♂♂, 10 ♀♀) wing, ♂ 112–121 (118), ♀ 109–117 (114); tail, ♂ 92·5–100 (95·0), ♀ 88·5–98 (93·5); bill, ♂ 19·5–21 (20·2), ♀ 19–21 (20·1); tarsus, ♂ 27·5–31 (29·4), ♀ 28–30 (28·9). WEIGHT: ♂ (n = 5) 64·5–72 (68·7), ♀ (n = 7) 55–71·5 (63·9).

IMMATURE: like adult but duller, underparts more olive and mantle browner.

NESTLING: unknown.

N. r. rufus (Fischer and Reichenow): coastal Kenya and Tanzania. Paler on throat and upper breast, without (or with ill-defined) greyish tinge on breast. WEIGHT: ♂ (n = 10) 56–80 (64·1), ♀ (n = 6) 52–70 (59·0).

Field Characters. A shy, rufous bird, with size, proportions and stance of typical thrush. Usually seen around swarms of army ants. Lack of white in tail distinguishes it from White-tailed Ant-Thrush *N. poensis* and Finsch's Flycatcher-Thrush *N. finschii*. Separated from Rufous Flycatcher-Thrush *N. fraseri* by rufous tail, lack of wing-bar and underwing pattern, larger size, longer and more slender bill, stouter legs and especially by thrush-like rather than flycatcher-like gait.

Voice. Tape-recorded (57, GREG, HOR, KEI, McVIC, STJ). Advertising call a sibilant, descending whistle, 'pseeyew', sometimes followed by a lower one at constant pitch 'pseeyew-pyeeew'. Song consists of these 2 whistles and a long, descending, rustling trill which begins on a high pitch and drops gradually at first and then abruptly while accelerating. Series of short crackling trills 'prrr ... prrr' around army ants are commonest call, also used as flight call on take-off. Probable contact call given when moving in forest, a high-pitched whistle 'pseee'. A sharp, rasping 'treet' and a thin, high-pitched 'tseem' are alarm or anxiety calls near nest. Birds in hand give vigorous, rhythmic bursts of bill clattering.

General Habits. In western part of range inhabits primary lowland forest, rarely old second growth. In east also occurs in other forest types, including riverine (Tana R.) and rich coastal (Sokoke). Widespread in forested areas, up to 1400 m in Congo Basin, to 900 m in eastern part of range. Frequents mainly forest floor and lowest understorey; also moves through open middle levels. In fragmented habitat may appear in open; seen perching briefly on wire (Kenya, Diani Beach: S. Keith, pers. comm.)

Social unit is the pair; each pair has an extensive home range which overlaps those of other pairs; inside part of this area, pair dominates conspecifics, outside it they dominate the pair. Regularly looks for and follows swarms of army ants, at least in Gabon. Positions itself in front of swarm, close to ants, driving off White-tailed Ant-Thrushes to either side and Rufous Flycatcher-Thrushes overhead; dominates all smaller species. Birds maintain an individual distance of at least 2·5 m. Tail-up postures with body fluffing, like Eurasian Blackbird *Turdus merula*, are used in agonistic display; tail-up postures with body sleeking given when about to flee. Tail spread and lifted somewhat during pauses in disputes with conspecifics. Follows and scolds Pel's Fishing-Owl *Scotopelia peli*. Forages like shrike or *Melaenornis* flycatcher: drops to ground from slanting perch in front of or among ants to snatch up prey. On ground gait thrush-like (hops).

Sedentary in NE Gabon.

Food. Mainly insects (grasshoppers, beetles, ants, caterpillars); also spiders, small millipedes and other arthropods flushed by swarming army ants; occasionally tiny snails.

Breeding Habits. Monogamous, solitary nester.

NEST: only 3 found. 2 situated in large knot-holes, 4 and 10 m high in dead trees, 1 in jagged hollow in top of broken-off hardwood stump *c.* 1·5 m high. In the last, the nest was a slender pad of tendrils and fine plant fibres, with a few dead leaves and some rodent fur; the cavity nests were a mixture of dry rootlets and flower stems on a layer of dry leaves above a mass of fine blackish rootlets.

EGGS: 2 (2 clutches). Elongated oval; slightly glossy; whitish, more or less completely obscured by red-brown or rufous and brown spots and blotches, which may coalesce into cap at broad end. An oviduct egg was pale greenish white, freckled with reddish and with red-brown blotches. SIZE: (1 clutch, Zaïre) 27·1 × 18·3 and 27·5 × 18·4; (oviduct egg) 26·5 × 19·5.

LAYING DATES: Cameroon, Nov; Gabon, Dec, Jan; Zaïre, Apr, May (♀ with ovary enlarged, Sept); E Africa: Zanzibar, Sept; Kenya, Apr; Uganda (♀ in breeding condition Oct).

INCUBATION: by ♀ only.

DEVELOPMENT AND CARE OF YOUNG: as soon as young leave nest they follow adults in vicinity of army ants, but approach the swarms closely only when able to fly well.

References
Brosset, A. and Erard, C. (1986).
Brown, L. H. (1970).
Willis, E. O. (1986).

Plate 25
(Opp. p. 400)

Neocossyphus poensis (Srickland). **White-tailed Ant-Thrush.** Grive fourmilière à queue blanche.

Cossypha poensis Strickland, 1844. Proc. Zool. Soc., p. 100; Clarence, Fernando Po.

Range and Status. Endemic resident. Occurs from Sierra Leone to S Nigeria, on Bioko, and from S Cameroon and Gabon south to NW Angola (Canzele), and east to S Central African Republic, N Zaïre (Kinshasa and Equateur to Ituri and Kivu), Rwanda, W Uganda and W Kenya (Kakamega and Nandi forests). Frequent to common; av. density 6–7 pairs/km² in NE Gabon.

Description. *N. p. poensis* (Strickland): Sierra Leone to Cameroon, Gabon and S Congo (Loango Coast). ADULT ♂: forehead, crown, nape and hindneck slaty brown; lores, ear-coverts, mantle, rump, uppertail-coverts, scapulars, wing-coverts and inner secondaries similar but tinged brownish. Tail blackish, with large distal white spots decreasing in size from T6 to T4. Chin and throat smoke grey with a cinnamon tinge, more olive on upper breast. Rest of underparts, axillaries and underwing-coverts bright cinnamon-rufous to ferruginous tawny. Primaries and secondaries sooty brown; outer webs of inner primaries and outer secondaries with dark chestnut bases producing patch on closed wing; inner webs from P8 to S6 have bright orange-rufous or cinnamon-rufous bases which produce a conspicuous wing-bar in fluttering flight. Bill blackish; eye dark brown; legs and feet pale buffy flesh colour. Sexes alike. SIZE (6 ♂♂, 5 ♀♀): wing, ♂ 102–112 (106), ♀ 97–109 (102); tail, ♂ 84·5–89 (86·0), ♀ 76–85·5 (80·7); bill, ♂ 18·5–20 (19·0), ♀ 18–20·5 (19·1); tarsus, ♂ 26·5–27·5 (27·0), ♀ 25·5–27·5 (26·7). WEIGHT: ♂ (n = 41) 45–60 (53·1), ♀ (n = 21) 43–60 (50·5).

IMMATURE AND NESTLING: unknown.

N. p. praepectoralis Jackson: N Angola, Central African Republic, W Zaïre (Lukolela, Lusambo), to Uganda. Upperparts browner, especially on mantle; white on outer pair of tail-feathers restricted to inner web and extreme tip of outer web.

N. p. kakamegoes Cunningham-van Someren and Schifter: Kakamega Forest, W Kenya. 'Head, with scaly appearance, and back Chaetura Drab. Tail black, white spot on tail 31–34. Throat buffy brown, sharply defined from upper breast which is near brownish olive. Lower breast tawny olive, belly richer rufous but not as intense as in *praepectoralis* (Uganda birds) or birds from North Nandi. Outer web of primaries Rood's Brown' (Cunningham-van Someren and Schifter 1981).

N. p. nigridorsalis Cunningham-van Someren and Schifter: North Nandi Forest, W Kenya. 'Crown (scaly) and back to upper tail-coverts Blackish Brown. Tail Jet Black, white spot on outer rectrix 30–34. Throat slightly paler than breast, Tawny Olive, slightly scaly in appearance. Belly Dresden Brown with flanks Antique Brown, not as rich rufous as *praepectoralis*. Outer web of primaries Walnut Brown compared with Russet of *praepectoralis*. Undertail-coverts each with a dark shaft and narrow dark fringe of Antique Brown' (Cunningham-van Someren and Schifter 1981).

Field Characters. A dark thrush, almost always seen around driver ants, fluttering near ground, displaying large white tail-tips and rapidly emitting squeaky, ticking and sparkling calls. Easily separated from Rufous Flycatcher-Thrush *N. fraseri* and Red-tailed Ant-Thrush *N. rufus* by very dark upperparts (rufous restricted to underparts) and white on tail. Distinguished from Finsch's Flycatcher-Thrush *N. finschii* by much darker, less brown upperparts and tail-feathers, slaty, not russet rump, narrow bill, and turdine rather than muscicapine stance.

Voice. Tape-recorded (57, ERA, GREG, HOR, McVIC, MOY). Advertising call is a single note 'hewiiit' (prolonged ascending whistle, with a short inflection at beginning and end) which can be repeated several times. Bird easily attracted by imitation of this call. Around ants occasionally gives *Turdus*-like song, long but subdued, like (but richer than) courtship song of Rufous Flycatcher-Thrush. During intraspecific chases gives loud 'heeer her-hih'. Normal call around army ants, on take-off, or in flight through forest understorey, a ticking 'prrt-prrt', a series of short squeaky trills sounding like splitting husks. Birds in hand clatter mandibles rhythmically.

General Habits. Widespread in primary lowland forest; also in old second growth, forest edge, large clearings with fallen trees, degraded forest on islands in rivers. Sea level to 1500 m (E Zaïre, Uganda) or 1700–1900 m (Kenya, Rwanda), once at 2520 m, Rwanda (Uwinka: Dowsett-Lemaire 1990). Because of its dependence on army ants, it inhabits mainly the ground and lowest understorey.

Social organization as in Red-tailed Ant-Thrush q.v. Has very large home range (ringed bird was netted at points 800 m apart). In front of army ant swarm it dominates Fire-crested and Brown-chested Alethes *Alethe diademata* and *A. poliocephala*, but is driven off by Red-tailed Ant-Thrush and Red-tailed Bristlebill *Bleda syndactyla*. Flashes white tail spots during interactions with other species and in slow fluttering flight. Slowly raises and lowers tail when perched, alerted or during agonistic encounters. Fairly aggressive interspecifically; chases involve tail spreading and short flights towards the opponent on the ground. Foraging behaviour like that of Red-tailed Ant-Thrush. Also seen joining mixed-species flocks and feeding on swarming winged termites and ants.

Sedentary (at least in NE Gabon).

Food. Mainly small arthropods, especially insects, larvae and pupae (ants, including army ants; termites, beetles, grasshoppers); also millipedes, spiders and other invertebrates driven out by raiding army ants; exceptionally very small frogs and fruits.

Breeding Habits. Monogamous, probably solitary nester. Nest and eggs undescribed.

LAYING DATES: Nigeria (fledglings June); Cameroon (young Sept, Nov); Zaïre, June, Nov (breeding condition, young Apr–May, Oct); Angola (oviduct egg Feb).

References
Brosset, A. and Erard, C. (1986).
Willis, E. O. (1986).

Neocossyphus fraseri (Strickland). Rufous Flycatcher-Thrush; Rufous Flycatcher. Grive fourmilière rousse.

Plate 25
(Opp. p. 400)

Muscicapa fraseri Strickland, 1844. Proc. Zool. Soc., p. 101; Fernando Po.

Range and Status. Endemic resident. Bioko, S Nigeria (Calabar), S Cameroon, S Central African Republic, N Zaïre and SW Sudan (Zande District and Aloma Plateau), S and W Uganda, NW Tanzania (Minziro, Bukoba), south to NW Zambia (N Mwinilunga) and N Angola (south to Cuanza Sul). Common to locally abundant. In NE Gabon density averages 20·8 pairs/km² in primary forest, probably higher in old second growth.

Description. *N. f. rubicunda* (Hartlaub): Nigeria and Central African Republic to Angola and Zambia, east in Zaïre to Lualaba R. ADULT ♂: forehead, crown, nape, lores, ear-coverts and hindneck greyish olive-brown. Mantle and scapulars russet cinnamon-brown; rump and uppertail-coverts bright ferruginous chestnut; tail-feathers dark olive-brown with outer pair, distal half of next and tip of third pair rusty buff or pale tawny. Chin and throat pale clay-colour, washed smoke grey, rest of underparts rich tawny. Upperwing-coverts olive-brown, outer webs russet cinnamon-brown. Flight-feathers dark olive-brown, outer webs edged russet cinnamon-brown; secondaries have rusty bases (from basal quarter on 1st to basal third on 6th), P1–P9 have rusty buff spots on inner webs which, with bases of secondaries, create conspicuous diagonal bar on underside of wings. Underwing-coverts pale tawny or rusty buff. Bill blackish, base of lower mandible and gape paler; eye olive chestnut-brown; legs and feet horn or light brown. Sexes alike. SIZE: (16 ♂♂, 13 ♀♀) wing, ♂ 90·5–105 (98·0), ♀ 90·5–105 (97·2); tail, ♂ 74–85 (79·9), ♀ 73·5–86 (80·4); bill, ♂ 14·5–16·5 (15·6), ♀ 14·5–16·5 (15·3); tarsus, ♂ 19–22 (20·2), ♀ 18·5–22 (20·0). WEIGHT: ♂ (n = 35) 30–44 (35·8), ♀ (n = 21) 27–39 (35·5).

IMMATURE: like adult but much duller and darker with margins to upperwing-coverts more rusty, head, breast and upperparts more smoky.

NESTLING: skin black with sparse but long blackish grey down.

N. f. fraseri (Strickland): Bioko. Slightly darker, especially on rump and undertail-coverts (barely distinguishable from *rubicunda*).

N. f. vulpina (Reichenow): S Sudan, Uganda and NE Zaïre, east of *rubicunda*. Tail more extensively rufous-tawny than other subspecies.

Field Characters. A rufous bird the size of a small thrush but resembling a large flycatcher, with upright posture, short legs, rather plump appearance and a peculiar habit of rapidly fanning the outer tail-feathers. Solitary; usually seen calling from top of stump, liana or thick branch. Distinguished from White-tailed Ant-Thrush *N. poensis* and Finsch's Flycatcher-Thrush *N. finschii* by lack of white in tail. Easily separated from larger Red-tailed Ant-Thrush *N. rufus* by short, wide bill, short legs, flycatcher shape, tail-flicking habit, and in flight by striking wing-bar and especially underwing pattern (pale orange-rufous underwing-coverts and band across remiges contrasting with dark brown underwing).

Voice. Tape-recorded (32, 57, C, MOY, STJ). Varied repertoire. Commonest calls are territorial and consist of series of 4–5 ascending whistles: (1) with high-pitched, pure tone 'wee-wee-wee-wee' for advertising when patrolling territory; (2) shorter, more rapidly emitted, with some vibration, 'weet-weet-weet-weet' during counter-singing at a distance; (3) 5 lower-pitched notes with more scolding, even hoarser quality, with interval between last 2 notes 'swit-swit-swit-swit-swit' to drive neighbour off; (4) still louder, more scolding and hoarser notes in series of 3, sometimes 2, when fighting. Contact call between partners during take-off, short 'seet'. Courtship song is a short subdued, *Turdus*-like phrase. Partners recognize one another by a slow series of 3–4 whistled, slightly ascending notes 'weeeee-ee-eee-eeeee' which are highly individual and constant from year to year. Feeble, ascending, vibrated notes 'sweet' indicate anxiety; when louder and repeated at regular intervals, show alarm; when hoarser and more scolding 'tswit-tswit-tswit', with variable rhythm (last note usually louder), used as mobbing calls like those of Shining Drongo *Dicrurus atripennis*. Birds in hand usually clatter mandibles vigorously. Young in nest peep and chirp; outside nest utter a long trill, almost a wail, recalling notes of Grey-throated or Grey Tit-Flycatchers *Myoparus griseigularis* or *M. plumbeus*.

General Habits. Widespread in lowland primary forest and old second growth, sea level to 1500 m. Essentially an understorey bird, although also in open mid-strata; found in dense places with numerous lianas but seems to prefer windfall clearings with decaying, broken-off trunks 5–20 m high; less frequent in 'cathedral forest'. Present along streams in secondary forest, in old cultivation (i.e. cacao, coffee and banana plantations) with important tree cover, and in disused agricultural land overgrown with tall stands of old parasol trees *Musanga* and herbaceous gingers *Aframomum*. Avoids recently logged clearings, cultivated land and young plantations, but may come close to human dwellings when tree cover and dense vegetation are present. In NE Gabon occupies vertical space from ground to canopy, though with a seasonable preference for lower strata during rains, high understorey and lower canopy during dry season.

Social unit consists of monogamous territorial pair. Partners often move separately in territory, which is defended by ♂, though ♀ frequently participates in advertising, especially when mixed-species flocks approach. Defenders rapidly flick wing-tip and spasmodically fan outer tail-feathers. During territorial disputes, intimidation posture includes fluffing breast while flattening feathers of back and underparts, erecting crown feathers, drooping wings, and spreading wrists to show wing-bar; in threat posture, stance more upright, plumage sleeked, wrists spread wider and tail raised. In true fight posture, wings held half-open and raised to display underwing pattern, tail-feathers spasmodically fanned and flicked.

Forages like flycatcher, using 4 principal methods (NE Gabon): sallies from perch (large bough, arched liana, or top of rotting stump), snaps up prey, especially on bark (31·4%) and on or under leaves (36·8%), less frequently in air between tops of adjacent shrubby trees (18·2%) or on ground (1·6%) where it hops. Rarely gleans in foliage. Frequently – especially during dry season – joins mixed-species flocks of foraging insectivorous birds and mammals (35·2% of observations in NE Gabon), particularly around swarming army ants, though it is not a habitual ant-follower.

Sedentary (at least in NE Gabon).

Food. Mainly small insects (av. 10·3 mm, n = 46, NE Gabon): especially ants, beetles, termites and Diptera, also moths, caterpillars, small Orthoptera and cicadas; sometimes small fruits; occasionally small snails and millipedes.

Breeding Habits. Monogamous; solitary nester. Territorial; in NE Gabon av. territory size 4·0 ha (n = 36). In courtship behaviour ♂ approaches ♀ as if in territorial dispute, then hops on perch in lateral asymmetry posture (with wing raised most on the side turned towards ♀ and body tilted away from her), crouches with quivering wings, ruffles lower mantle and rump and raises tail. Interrupted courtship flight follows with 'turdine' song; ♂ may then preen ♀'s wings and flanks. Crouching ♀ solicits ♂ for copulation while quivering wings and raising bill. At nest ♂ displays in various postures including slow wing flapping and showing wing-marks.

NEST: usually a rudimentary, very shallow cup of thin rootlets, including some moss and pieces of rotting leaves; sometimes more bulky; outer diam. *c.* 10 cm; placed in open cavities (e.g. knot-holes in trunks or slanting limbs, opened up old woodpeckers' or barbets' holes near the top of rotted, broken-off trunks, openings in root systems of epiphytes), once in hollow at top of old cut-off banana stump, and in old nest of African Thrush *Turdus pelios*. Built by ♀.

EGGS: undescribed. Clutch said to be 1–2; and oviduct egg pale blue blotched reddish brown and with dark reddish-purple cap markings (Mackworth-Praed and Grant 1963), but source of this information unclear.

LAYING DATES: Cameroon, Aug, Oct, Dec; NE Gabon, Sept–Mar (especially at beginning of rains); Zaïre, Sept (♂♀ with enlarged gonads in almost every month Jan–Oct, but especially during rains); E Africa: Region B, Apr, Sept; Zambia, Oct; Angola, Mar.

INCUBATION: by ♀ only; incubation bouts last 30–60 min in morning and evening, 45–90 min in afternoon, alternating with 30–45 min of feeding. ♂ guards nest, greets ♀ with courtship song and brief flights. Period: unknown.

DEVELOPMENT AND CARE OF YOUNG: young leave nest when 14 days old. ♀ broods chicks intermittently for first 7 days. Both adults feed young, more frequently as they get older. Adults, especially ♂, give alarm when squirrels *Aethosciurus poensis* approach nest-site, freeze and stay away from nest when African Goshawk *Accipiter tachiro* is in vicinity. They vigorously mob potential predators, human observer included, when young are out of nest. Period of dependence on parents unknown, but at least 1 month; young stay in parents' territory until next breeding cycle.

BREEDING SUCCESS/SURVIVAL: in NE Gabon, 2 birds ringed as adults recaptured 8 years later.

References
Brosset, A. and Erard, C. (1986).
Eisentraut, M. (1973).
Erard, C. (1987, 1990).
Willis, E. O. (1986).

Neocossyphus finschii (Sharpe). **Finsch's Flycatcher-Thrush. Grive fourmilière de Finsch.**

Plate 25
(Opp. p. 400)

Cassina finschi Sharpe, 1870. Ibis, p. 53; Fantee.

Range and Status. Endemic resident. Sierra Leone, Liberia, S Ivory Coast (south of a line Sipilou–Gagnoa–Abengourou), S Ghana, S Benin, S Nigeria (south of Ibadan, Ife and Mamu Forest, but reported from Kagoro). Rare to common.

Description. ADULT ♂: forehead, crown and hindneck slightly greyish olive-brown, becoming paler on lores and ear-coverts. Mantle and scapulars olive-brown, rump and uppertail-coverts greyish chestnut-brown; tail slaty brown with large distal white spots on outer 3 feathers, decreasing in size inwardly. Chin and throat pale clay colour, upper breast olive-tawny, rest of underparts rich rufous-tawny. Upperwing-coverts and inner secondaries greyish olive-brown. Flight-feathers slaty brown, outer edge of inner primaries russet cinnamon-brown; wing-bar formed by rusty buff basal third of S1–S6 and spots on base of inner webs of P1–P7 (P8). Axillaries and underwing-coverts pale tawny or rusty buff. Bill blackish, paler below; eye brown; legs and feet flesh colour. Sexes alike. SIZE: (9 ♂♂, 5 ♀♀): wing, ♂ 101·8 ± 3·2, ♀ 97 ± 5·3; tail, ♂ 74 ± 2·1, ♀ 70·8 ± 1·3; bill, ♂ 16·7 ± 0·5, ♀ 17·2 ± 0·4. WEIGHT: ♂ 37·7 ± 2·4, ♀ 36·3 ± 1·8.

Field Characters. A solitary, brown and rufous, flycatcher-like bird of forest mid-strata. Distinguished from Red-tailed Ant-Thrush *N. rufus* and Rufous Flycatcher-Thrush *N. fraseri* by white spots on outer tail-feathers, and from White-tailed Ant-Thrush *N. poensis* by flycatcher shape and foraging behaviour, browner, less slaty upperparts and central tail-feathers, russet rump, shorter tarsi and, when seen from below, broader bill.

Voice. Tape-recorded (57, GRI). Repertoire reminiscent of Rufous Flycatcher-Thrush q.v., but slower and

458 TURDIDAE

on a lower pitch. Territorial calls are 4-note 'tswe-tswe-tswe-tswe', rapidly uttered, and full, throaty and deliberate 'tsw-tswee ... tsweeee'. A long drawn-out and plaintive double or treble whistle 'wee ... weeee-eee' also often given. Pair scolds African Goshawk *Accipiter tachiro* with a loud buzzy 'word-word-word'. Rufous Flycatcher-Thrush completely ignores playback of songs and calls of this species.

General Habits. Widespread in primary or secondary lowland evergreen or semi-deciduous forest, sea level to 1500 m. Prefers thickest parts of forest. Lives mainly in understorey and lower canopy. Not very active; sits motionless on branch or liana. Social unit is territorial pair. General habits like those of Rufous Flycatcher-Thrush, q.v. Joins mixed-species flocks, sometimes around raiding army ants, though less dependent on them than Red-tailed and White-tailed Ant-Thrushes. Forages like true flycatcher, sallying or sweeping down from perch to catch insects on or under leaves or on bark, often hovering, or in open space. Frequently changes perches but does not range over large area. When alarmed, often slips away in quick, direct flight.

Food. Small insects: ants, termites, beetles, grasshoppers, Diptera, Hymenoptera, Hemiptera and Lepidoptera.

Breeding Habits. Monogamous, territorial, probably solitary nester. Nothing known of reproductive behaviour nor breeding season, except that a bird in S Nigeria collected nest material at base of epiphytes in Mar; nest could not be found. Birds with enlarged gonads Liberia, Aug–Sept.

Neocossyphus finschii

References
Marchant, S. (1942).
Walker, G. R. (1939).
Willis, E. O. (1986).

Genus *Modulatrix* Ripley

An endemic genus of uncertain affinities, sometimes grouped with babblers (Timaliidae). Size of small thrush, with narrow bill, long tarsus and toes, and short, rounded wings (7th primary longest). Tail slightly rounded, with 12 somewhat pointed feathers. Turdine characteristics include tail shape and general plumage appearance. Timaliid characteristics include lack of 'turdine thumb' in the syrinx (Ames 1975), lack of spotted juvenile plumage, and general behaviour. An additional species, *orostruthus*, is frequently included in *Modulatrix*, but we follow Irwin and Clancey (1986) in assigning this to a monotypic genus, *Arcanator*. In view of the uncertainties still surrounding the systematic position of *Modulatrix*, we adopt the conservative position of retaining the genus in the Turdidae, pending further investigations.

1 species, restricted to montane forest in eastern Africa.

Plate 25
(Opp. p. 400)

Modulatrix stictigula (Reichenow). **Spot-throat. Grive à gorge tachetée.**

Turdinus stictigula Reichenow, 1906. Orn. Monatsber. 14, p. 10; Mbaramo, Usambara, Tanganyika.

Range and Status. Endemic resident, montane forests of NE, E and S Tanzania and N Malaŵi, from Usambaras southwest to N Malaŵi (Misuku Hills). Ranges from 900 to 2700 m, principally above 1200 m. Uncommon to abundant (especially numerous at higher altitudes); density in Malaŵi c. 1 singing bird per ha (Dowsett-Lemaire 1989).

Modulatrix stictigula

Description. *M. s. pressa* (Bangs and Loveridge): E and S Tanzania (from Ukaguru Mts southwards), N Malaŵi. ADULT ♂: forehead dark reddish brown; ear-coverts brownish olive with tawny centres to feathers; eye-ring off-white; crown, nape, hindneck, sides of neck, mantle, back and rump brownish olive; uppertail-coverts dark reddish brown; tail dark redbrown. Chin and throat off-white, merging into tawny breast, with dusky brown feather-tips giving speckled appearance; sides of breast and flanks brownish olive; belly off-white in centre, tawny at sides; vent and undertail-coverts tawny. Scapulars and wings brownish olive, except for dark reddish brown outer webs of primaries; axillaries brownish olive; underwing slightly paler than upperwing. Upper mandible blackish; lower mandible greyish brown; eye dark brown; legs and feet dark brown with pale soles. Sexes alike. SIZE: wing, ♂ (n = 8) 78–83 (80·5), ♀ (n = 5) 76–79 (78·0), unsexed (n = 22) 75–85 (80·0); tail, ♂ (n = 5) 69–75 (72·0), ♀ (n = 4) 67–69 (68·0); bill to skull, ♂ (n = 5) 19–21 (20·2), ♀ (n = 4) 18–20 (19·5); tarsus, ♂ (n = 4) 31–33 (31·9), ♀ (n = 4) 30–32 (31·4). WEIGHT: ♂ (n = 3) 30·5–37·5 (33·5), 1 ♀ 32·5, unsexed (n = 27) 26–36 (30·5).

IMMATURE: like adult but throat less markedly speckled, underparts slightly paler.

NESTLING: hatchling salmon-pink; gape yellow with red furrow in centre of palate.

M. s. stictigula (Reichenow): Usambaras and Ngurus, NE Tanzania. Spots on chin and throat smaller, paler and less distinct. Slightly larger: wing, ♂ (n = 5) 80–83 (82·0), ♀ (n = 4) 78–82 (80·0). WEIGHT: ♂ (n = 5) 30–38 (34·6), 1 ♀ 33·5, unsexed (n = 49) 27–34 (30·5).

Field Characters. A shy bird of forest undergrowth, mainly dull brown, with spotted throat and tawny underparts. Easily confused in poor light with Dappled Mountain-Robin *Arcanator orostruthus*, Cabanis's Greenbul *Phyllastrephus cabanisi*, or Pale-breasted Illadopsis *Trichastoma rufipennis*. Dappled Mountain-Robin has pale yellow underparts with broad dark streaks, Cabanis's Greenbul is olive with yellowish white underparts, Pale-breasted Illadopsis has pale greyish underparts. Orange Ground-Thrush *Zoothera gurneyi* is larger and bulkier, with unspotted bright orange throat and breast contrasting with pure white belly, and conspicuous white spots in wings. Spot-throat is a noisy bird usually seen on the forest floor, and reluctant to fly; voice distinctive.

Voice. Tape-recorded (KEI, LEM, STJ). Song is loud series of rich, clear, high-pitched fluty notes and thrush-like warbles, 'eit, seeyee, heet, eeree, reeze, peeyer', but with enormous regional and individual variation. Call typically a 2-note whistle, second note pitched higher than first; it varies regionally, tending to be rich and fluty in north of range (e.g. 'hooooree' in W Usambaras) and to be a thinner whistle in south (e.g. 'seeeeyee' in Iringa Highlands). In E Usambaras, 2-note call is preceded by an additional note at intermediate pitch. In Misuku Hills 2 types of whistle, either rising or falling in pitch. Alarm is usually 3 harsh notes 'jeir jeir jeir' or 'chew churri'. Noisy throughout year, but especially when breeding.

General Habits. Occurs in montane forest, including dense, high-altitude thickets, degraded forest, thick undergrowth near forest patches, and gardens.

Solitary, in pairs or in groups of 3–4 (probably family parties). Keeps on or very close to the forest floor. Hops on ground with tail somewhat elevated; looks back and forth constantly rather than keeping still, and pecks or runs and pecks at tiny prey. Turns over fallen leaves and debris in search of food. Frequently rests on tops of hummocks, logs, lianas or bases of tree-ferns, close to forest floor. After landing on perch flicks tail up and flits wings. Sometimes visits ant swarms (but does not forage regularly near ants). Shy and nervous. Flight weak; prefers to run to cover if alarmed, or more usually a mixture of flying and running along forest floor. Flees immediately if other bird species give alarm call or if it sees a person.

One record in dry bushland at 600 m in Dec may indicate some altitudinal movements; otherwise appears to be sedentary.

Food. Insects; once berries.

Breeding Habits. Courtship behaviour unknown.

NEST: (only 1 known: Sclater and Moreau 1932). Neat cup of twigs, mostly 120–150 long, lined with felt of leaf skeletons, 1·6 m above ground in fork of forest sapling.

EGGS: 2. Heavily blotched and scrawled with crimson-lake on a pinkish ground. SIZE: 23·5 × 18.

LAYING DATES: E Usambara Mts, Nov. Gonads and brood patches indicate breeding season in Tanzania Oct–Mar, with peak Nov–Dec. In N Malaŵi, frequency of singing suggests breeding peak in Feb–Mar.

INCUBATION: sitting bird very shy and wary.

References
Ripley, S. D. and Heinrich, G. H. (1969).
Sclater, W. L. and Moreau, R. E. (1932).

TURDIDAE

Genus *Arcanator* Irwin and Clancey

An endemic genus of uncertain affinities, possibly belonging with babblers (Timaliidae). Size of small thrush, with long narrow bill, fairly long tarsus and toes, and short, rounded wings (5th primary longest). Tail slightly rounded, with 12 somewhat pointed feathers. Turdine characteristics include tail shape and general plumage appearance. Timaliid characteristics include lack of a 'turdine thumb' in the syrinx (Ames 1975, F. P. Jensen, pers. comm.), and lack of spotted juvenile plumage. Frequently included in *Modulatrix*, but we follow Irwin and Clancey (1986) in assigning it to a monotypic genus, *Arcanator*. In view of the uncertainties still surrounding the systematic position of both *Arcanator* and *Modulatrix*, we adopt the conservative position of retaining them in the Turdidae, pending further investigations.

1 species, restricted to montane forest in eastern Africa; very rare.

Plate 25
(Opp. p. 400)

Arcanator orostruthus (Vincent). Dappled Mountain-Robin. Grive tachetée.

Phyllastrephus orostruthus Vincent, 1933. Bull. Br. Orn. Club 53, p. 133; Mt Namuli, N Mozambique.

Range and Status. Endemic resident, in 3 isolated populations in lower montane forest: NE Tanzania (E Usambara Mts); E Tanzania (E scarp of Uzungwa Mts from Mwanihana southwest to Chita); and N Mozambique (Mt Namuli). Ranges from 900 to 1700 m. Rare, density very low, but more common at Chita. Population estimates in E Usambaras range from a few hundred to a few thousand. Threatened by habitat destruction.

Description. *A. o. amani* (Sclater and Moreau): E Usambara Mts, Tanzania. ADULT ♂: forehead to neck brownish olive-green, becoming somewhat brighter and greener from mantle to rump and on scapulars, upperwing-coverts and tertials; rump with rufescent tinge, uppertail-coverts cinnamon-brown. Tail-feathers dark olive-brown with reddish tinge on outer webs, shafts red-brown above, straw below. Lores dark olive, cheeks and ear-coverts brownish olive. Chin and throat yellowish white, feathers tipped olive giving smudgy spotted appearance; rest of underparts pale yellow, with broad olive stripes on breast, upper belly and flanks (feathers olive with yellow edges); sides of breast and lower flanks more solid olive (yellow on feathers reduced); lower belly and undertail-coverts unstreaked, with variable ochre wash, quite pronounced in some birds. Primaries and secondaries dark brown, outer webs richer brown, paler and brighter on primaries; axillaries and underwing-coverts pale greyish, with a little dark mottling on underside of primary coverts, rest of underwing with pale sheen. Bill brown to black, base of lower mandible whitish; eye brown or red-brown; legs and feet pinkish grey. Sexes alike. SIZE: wing, ♂ (n = 4) 84–90 (88·0), 1 ♀ 84, unsexed (n = 7) 85–92 (88·0); tail, ♂ (n = 4) 66–74 (71·0), 1 ♀ 63·5; bill to skull, ♂ (n = 6) 19–22 (20·3), 1 ♀ 21, unsexed (n = 6) 20–21 (20·5); tarsus, ♂ (n = 4) 24–26 (25·0), 1 ♀ 25, unsexed (n = 7) 26–28 (27·2). WEIGHT: ♂ (n = 3) all 30, 1 ♀ 32, unsexed (n = 6) 30–35 (31·8).

IMMATURE: similar to adult (without speckled plumage of typical immature turdines), but dull olive-green on breast with very little yellow, giving uniform olive appearance.

NESTLING: unknown.

A. o. orostruthus (Vincent): Mt Namuli, N Mozambique. Forehead, crown, flight feathers and especially tail much more reddish brown. Breast less dappled. Eye sepia. SIZE: wing, 1 ♂ 84; tail, 1 ♂ 78; bill to skull, 1 ♂ 21·7; tarsus, 1 ♂ 29.

A. o. sanjei Jensen and Stuart: Uzungwa Mts, Tanzania. Tail dark brownish olive with only very slight reddish tinge. Breast more strongly dappled than in other races; flanks and sides of belly dark olive-green with no dappling. Bill much broader at base: 1 ♂ 7·8, compared with 1 ♂ 6·9 for nominate form and ♂ (n = 4) 5·9–7·0 (6·6) for *amani*. Slightly larger. SIZE: wing, ♂ (n = 4) 89–92 (90·5), ♀ (n = 2) 87, 89, unsexed (n = 8) 87–94 (91·0); tail, ♂ (n = 4) 73–76 (75·0), ♀ (n = 2) 69, 72; bill to skull, 1 ♂ 23·3; tarsus, 1 ♂ 29. WEIGHT: ♂ (n = 4) 30·5–34 (32·5), ♀ (n = 2) 69, 72, unsexed (n = 8) 28·5–38 (34·5).

Field Characters. An elusive, seldom-seen species of montane forest undergrowth, whose very rounded wings are obvious in flight (M. W. Woodcock, pers. comm.). Easily identified if distinctive dappling and streaking on pale yellow underparts is seen, otherwise easily confused in dim light with 3 sympatric species of similar size and proportions and similarly-coloured upperparts and wings,

Spot-throat *Modulatrix stictigula*, Pale-breasted Illadopsis *Trichastoma rufipennis*, and Cabanis's Greenbul *Phyllastrephus cabanisi*. Spot-throat has tawny underparts, with spots confined to throat; Pale-breasted Illadopsis has unspotted pale grey underparts; Cabanis's Greenbul has unspotted yellowish white underparts. Immature Orange Ground-Thrush *Zoothera gurneyi* has heavily spotted breast but underparts mainly pale orange, upperparts red-brown, white spots on wings.

Voice. Tape-recorded (WOOD). Song, phrases of 5–8 piping notes, delivered rather rapidly. Notes are pure and song is pleasant but not rich or melodic. Singer takes phrase and repeats it many times with little variation. 3 phrases recorded by M. W. Woodcock in Tanzania: (1) 5 notes, 3rd loudest, given twice with almost no pause: 'tu-tu-WEEtaloo, tu-tu-WEEtaloo'; (2) 8 notes, last loudest: 'tu-tupeeter-tu, ter-tu-WEE; (3) 8–9 notes, last loudest: 'pu-piddly-per-tuwer-PEE' or PEEWEE.

General Habits. Inhabits undisturbed, very moist evergreen rain-forest at 900–1700 m (generally at lower altitudes than Spot-throat, although the 2 are sympatric in several areas). Occurs in dense undergrowth, especially along streams (M. W. Woodcock, pers. comm.). Circumstantial evidence suggests that it forages on the ground. Nothing else known.

Food. Insects.

Breeding Habits. Unknown. A juvenile in Nov in E Usambara Mts, Tanzania, and bird with well-developed incubation patch in Uzungwa Mts, Tanzania, in Nov are the only indications of breeding season.

Reference
Collar, N. J. and Stuart, S. N. (1985).

Genus *Pinarornis* Sharpe

Distinctive single species of rock-dwelling chat endemic to SE Africa. Plumage sooty, with white marks in wing and tail; throat speckled. Long-bodied, long-tailed and quite long-legged. Syrinx lacks turdine thumb. Humerus with well-developed second tricipital fossa. Thought to be related to *Stizorhina*, *Neocossyphus*, *Modulatrix* and the New World genus *Myadestes* (Olson 1989).

Pinarornis plumosus Sharpe. Boulder Chat. Merle des rochers.

Plate 27
(Opp. p. 416)

Pinarornis plumosus Sharpe, 1876. *In* Layard and Sharpe's Birds of South Africa, p. 230. Victoria Falls (in error for Matopos: Irwin 1957).

Range and Status. Endemic resident. Restricted to bouldery granitic hills and outcrops, from W Malaŵi and E Zambia, through Zimbabwe, to E Botswana. Frequent to common and widespread in Eastern Province, Zambia, and Matopos Hills, Zimbabwe, but elsewhere highly localized. In Zambia occurs from Chipengali, Kalikali and Tamanda to Chipata, Chadiza and west to Nyanje and Nyakolwe Hills, with records between Luangwa R. and Lukasashi at 14° 9′S, 30° 24′E, and at Serenje; but absent from Muchinga Escarpment. In Malaŵi known only from Mchinja, Mt Majete and rocky knolls in Dzalanyama Forest. In Zimbabwe occurs on *c.* 21 kopjes apart from Matopos Hills (map in Irwin 1981), on hills southeast of the watershed, in Limpopo R. and Sabi R. valleys, in NE at Mtoko, Mt Darwin and Mt Mwenji, and on granitic fringes of SE lowveld. Locally common in Botswana from Francistown to Tuli.

Density of 12 breeding pairs in 10 km², Matopos Hills, Zimbabwe (Grobler and Steyn 1980).

Description. ADULT ♂: head and body very dark brownish grey, gradually becoming darker towards rear; rump, upper-tail-coverts, belly and undertail-coverts brownish black; but chin and throat feathers narrowly tipped white, giving brindled greyish effect; and undertail-coverts sometimes narrowly

Pinarornis plumosus

and inconspicuously white-tipped. Tail black or very dark brownish black; central pair of feathers (T1) wholly black or sometimes very narrowly white-tipped; T2 and T3 wholly black, or sometimes broadly white-tipped on both vanes; white tip of T3 up to 10 mm deep; T4, T5 and T6 black with increasing amounts of white at end, up to 25 mm deep in T6; white tips of T6 sometimes invaded with black on edge of outer web and towards tip; underside of tail glossy black with white patches showing at tips. Wings blackish brown, with bold white mark half way along inner web of all primaries except outermost (P10) and of outer 4 secondaries; white extends onto outer webs of P1, P2 and P3, and to a lesser extent onto outer webs of P4 and the outer 3 secondaries. Axillaries sooty; underwing-coverts sooty, broadly white-tipped; underside of primaries shiny brown with the white areas showing. Bill black; eye dark brown; legs and feet black. Sexes alike. SIZE: (4 ♂♂, 5 ♀♀) wing, ♂ 116–120 (118), ♀ 112–125 (116); tail, ♂ 109–125 (119), ♀ 107–124 (118); bill to skull, ♂ 22·5–26 (24·0), ♀ 21–26 (23·5); tarsus, ♂ 29–32 (31·3), ♀ 27–30 (28·8).

IMMATURE: like adult but forepart of body slightly paler; trunk plumage soft and fluffy, especially on rump.

NESTLING: crown, wings, back and thighs with blackish down; gape-flanges orange (W. R. J. Dean and I. A. W. Macdonald, pers. comm.).

Field Characters. A dull black, thrush-like bird confined to broken granite hill country, from Malaŵi to Botswana, with distinctive 'squeaky wheel' call. Long-tailed, round-winged, rather long-legged; total length 23–27 cm. Corner of tail white, conspicuous in flight. Bold white bar in wing, not visible in resting bird but very conspicuous in flight. Throat and undertail-coverts often barred with white.

Voice. Tape recorded (86, 88, CHA, WALK). Song a medley of high-pitched, thin but quite sweet notes, reminiscent of European Robin *Erithacus rubecula*. Main vocalization (song?, contact call?) a slightly variable phrase of 3–4 pure, high-pitched notes: 'tweeweep teeeep teeweeeeeeeet or 'tsuui tsu tsu p'loo'. Each note lasts *c*. 0·5 s and is followed by 0·5 s pause; notes given regularly, making phrase sound like 3–4 turns of squeaky wheel. Each note is at one pitch, or rises or falls in pitch; last note generally lower pitched (2–3 kHz) than first one (4–7 kHz). Entire phrase lasts 2·5–3·0 s and is repeated almost without break, and with only minor variation, for up to 17 min.

General Habits. Inhabits lower slopes of granitic hills and outcrops with large boulders and plenty of scattered trees. Outcrops may be as small as <0·5 ha, or mountainous kopjes; and surrounded by either humid forest or semi-arid savanna. Also steep, rocky, dry stream beds.

Occurs mainly on or among boulders under wooded canopy (Irwin 1981), and avoids large, precipitous rock faces (of the sort frequented by Mocking Cliff-Chat *Myrmecocichla cinnamomeiventris*, from which Boulder Chat is thereby segregated, although the 2 sometimes nest within 300 m of each other: Medland 1985).

In pairs or family parties. Active when foraging, moving over large boulder in series of strong hops, sometimes assisted with a few quick wing-flaps, then changing station a short distance with a strong flapping flight uphill or a swooping glide downhill. Also runs along branches and rocks; and walks. Raises long tail up high immediately on landing, and quickly lowers it again. Lost to view, amongst boulders, for much of the time. When not foraging, sits quietly on rock, in full view, evidently soaking up sun; squats down on tarsi, stands and fluffs up rump, then squats again (Maclean 1985). Vocal; frequently gives 'squeaky wheel' call, particularly when nest is approached or fledglings are about; gives warbling song with bill pointing upward; ♂ (and ♀?) occasionally sings from prominent perch.

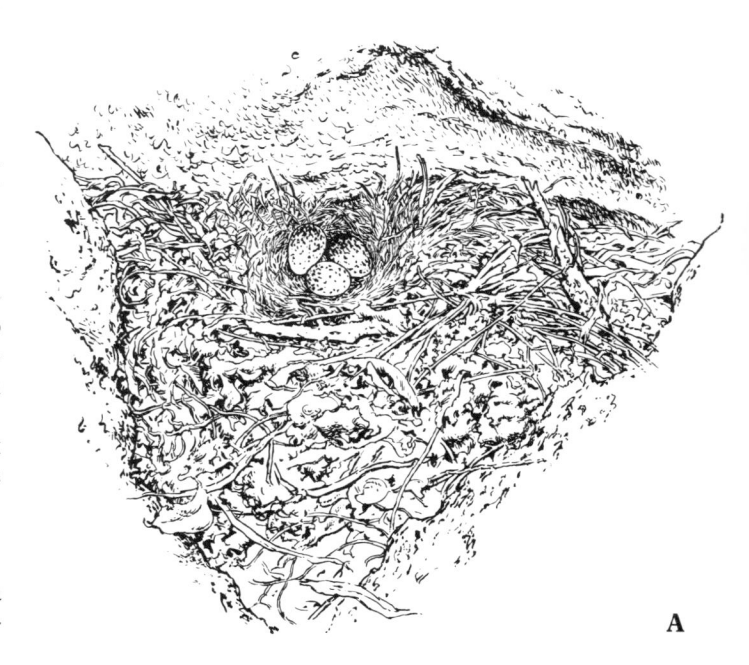

A

Gleans food from leaves and bark; hawks termites in flights. Evidently completely sedentary.

Food. Insects and small lizards.

Breeding Habits. Monogamous, solitary nester. One pair of birds made 3 breeding attempts in 1 season (W. R. J. Dean and I. A. W. Macdonald, pers. comm.).

NEST: neat unlined cup formed of loosely-knitted leaf petioles, on dead leaves and vegetable debris which have already accumulated in site, supported around outside by collection of bits of bark, small earth clods and twigs; cup diam. 88–110, cup depth 35–54. Sometimes with considerable base of earth clods (**A**). Sited on ground, half hidden under and between boulders, under log, or in horizontal cleft in rock, near base of kopje.

EGGS: 2–3, av. (22 clutches, Zimbabwe) 2·8. Long ovals; glossy; greenish white, with numerous red-brown or warm brown speckles, concentrated at broad end. SIZE: (n = 40, Zimbabwe) 25·0–32·7 × 17·6–19·9 (27·0 × 18·9). WEIGHT: 5·5 when fresh; (n = 4) 3·9–4·8 (4·6) near hatching.

LAYING DATES: Malaŵi, Nov; Zambia, Dec; Zimbabwe, Sept–Dec (Sept 11, Oct 46, Nov 44, Dec 6 clutches).

DEVELOPMENT AND CARE OF YOUNG: at 1–2 days young of 1 brood weighed 5·6, 6·3 and 7·1. Eyes begin to open and quills to sprout at 5 days. Fed by both parents, with variety of soft-bodied insects. Parents swallow faecal sacs and evidently remove egg-shells to distance of *c*. 15 m. Young leave nest at 16–20 days, before they can fly, and hide in rock crevices if approached.

BREEDING SUCCESS/SURVIVAL: 5 out of 33 clutches parasitized by Red-chested Cuckoo *Cuculus solitarius*, which has a '*Pinarornis*' gens of spotted eggs (W. R. J. Dean and I. A. W. Macdonald, pers. comm.); 8 out of a further 14 clutches similarly parasitized (Matopos, Zimbabwe: Grobler and Steyn 1980). Of 22 eggs, 14 hatched, but only 4 young fledged.

References
Grobler, J. H. and Steyn, P. (1980).
Medland, R. D. (1985).

Genus *Cichladusa* Peters

Close to *Cossypha* and *Alethe*, with similar plain tail, but differs from both and from all African forest robins in building mud nest. Bill stronger than *Cossypha*, similar to *Alethe*, nostrils rounded, rictal bristles weak. Frontal feathers short, stiff and sub-erectile. Tail long, about same length as wing. 2 species closely associated with palms.

Endemic to Africa. 3 species, of which 2 (*C. arquata*/*C. ruficauda*) form a superspecies.

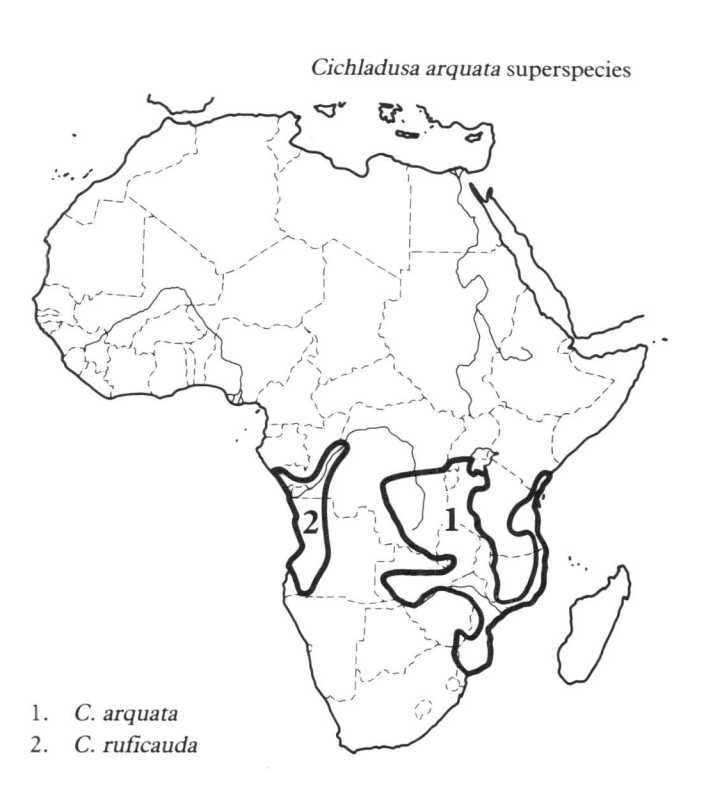

Cichladusa arquata superspecies

1. *C. arquata*
2. *C. ruficauda*

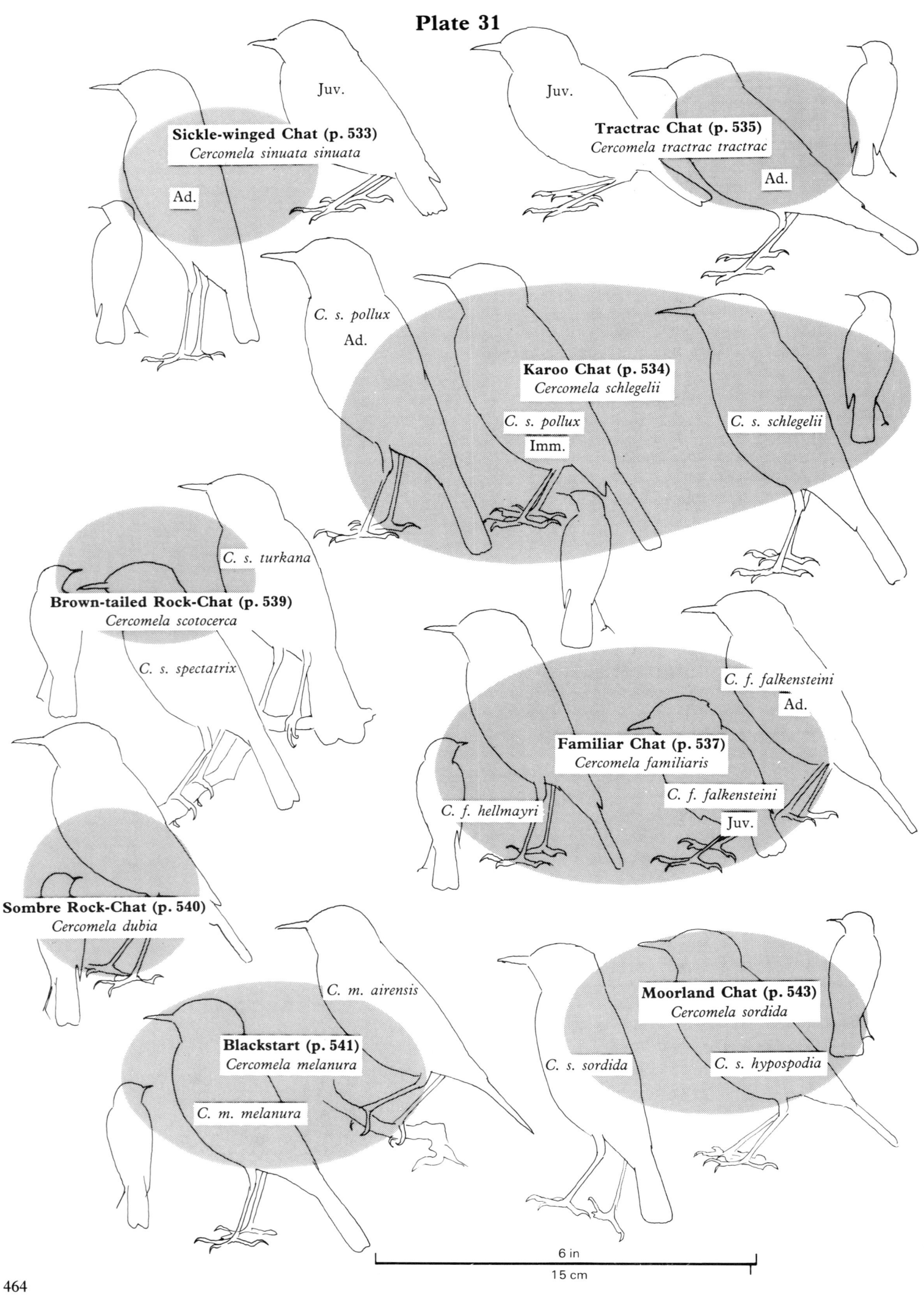

Plate 31

Sickle-winged Chat (p. 533)
Cercomela sinuata sinuata
Ad.
Juv.

Tractrac Chat (p. 535)
Cercomela tractrac tractrac
Juv.
Ad.

C. s. pollux
Ad.

Karoo Chat (p. 534)
Cercomela schlegelii

C. s. pollux
Imm.

C. s. schlegelii

C. s. turkana

Brown-tailed Rock-Chat (p. 539)
Cercomela scotocerca

C. s. spectatrix

C. f. falkensteini
Ad.

Familiar Chat (p. 537)
Cercomela familiaris

C. f. hellmayri

C. f. falkensteini
Juv.

Sombre Rock-Chat (p. 540)
Cercomela dubia

C. m. airensis

Blackstart (p. 541)
Cercomela melanura

C. m. melanura

Moorland Chat (p. 543)
Cercomela sordida

C. s. sordida

C. s. hypospodia

6 in
15 cm

464

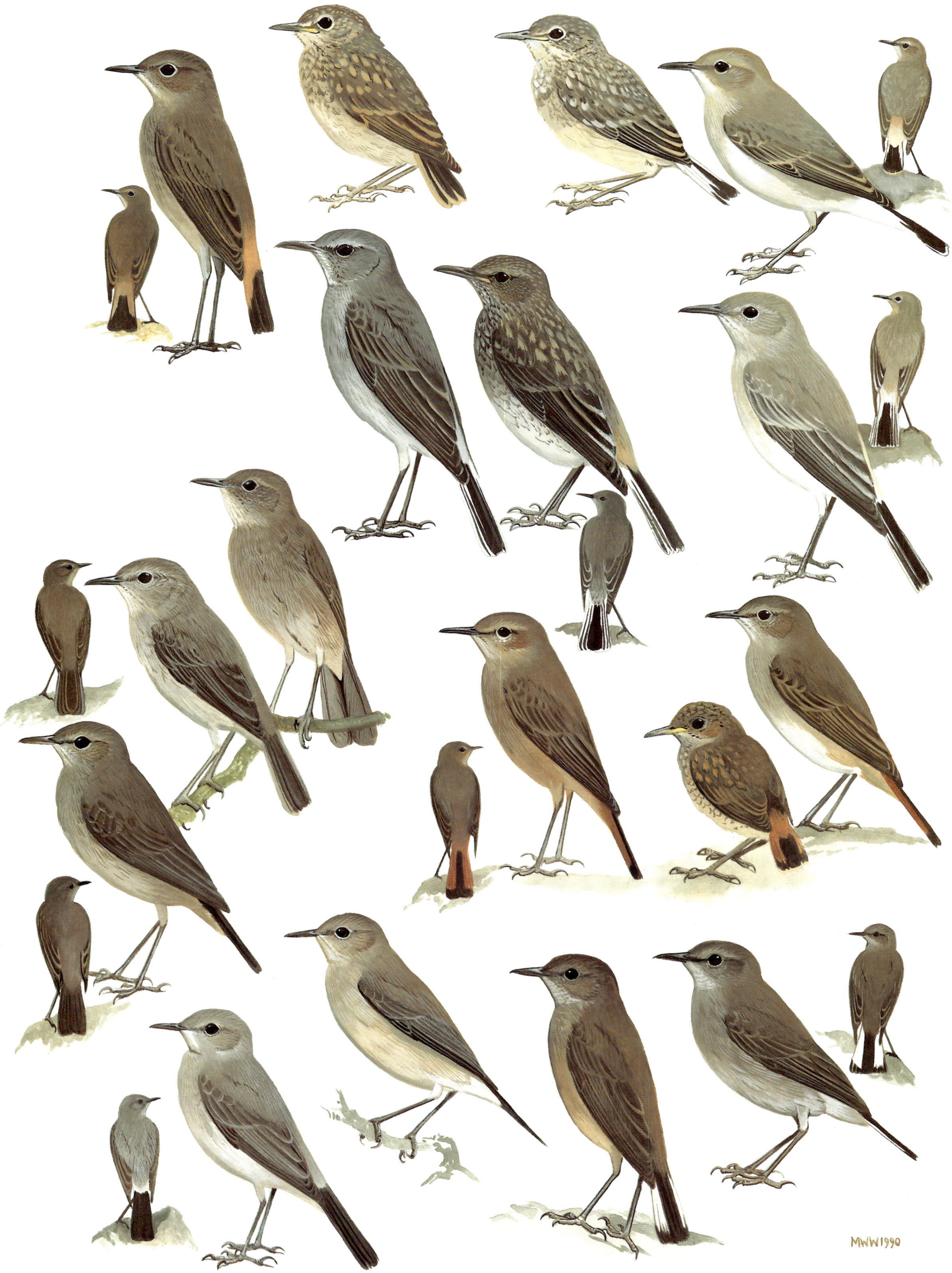
MWW1990

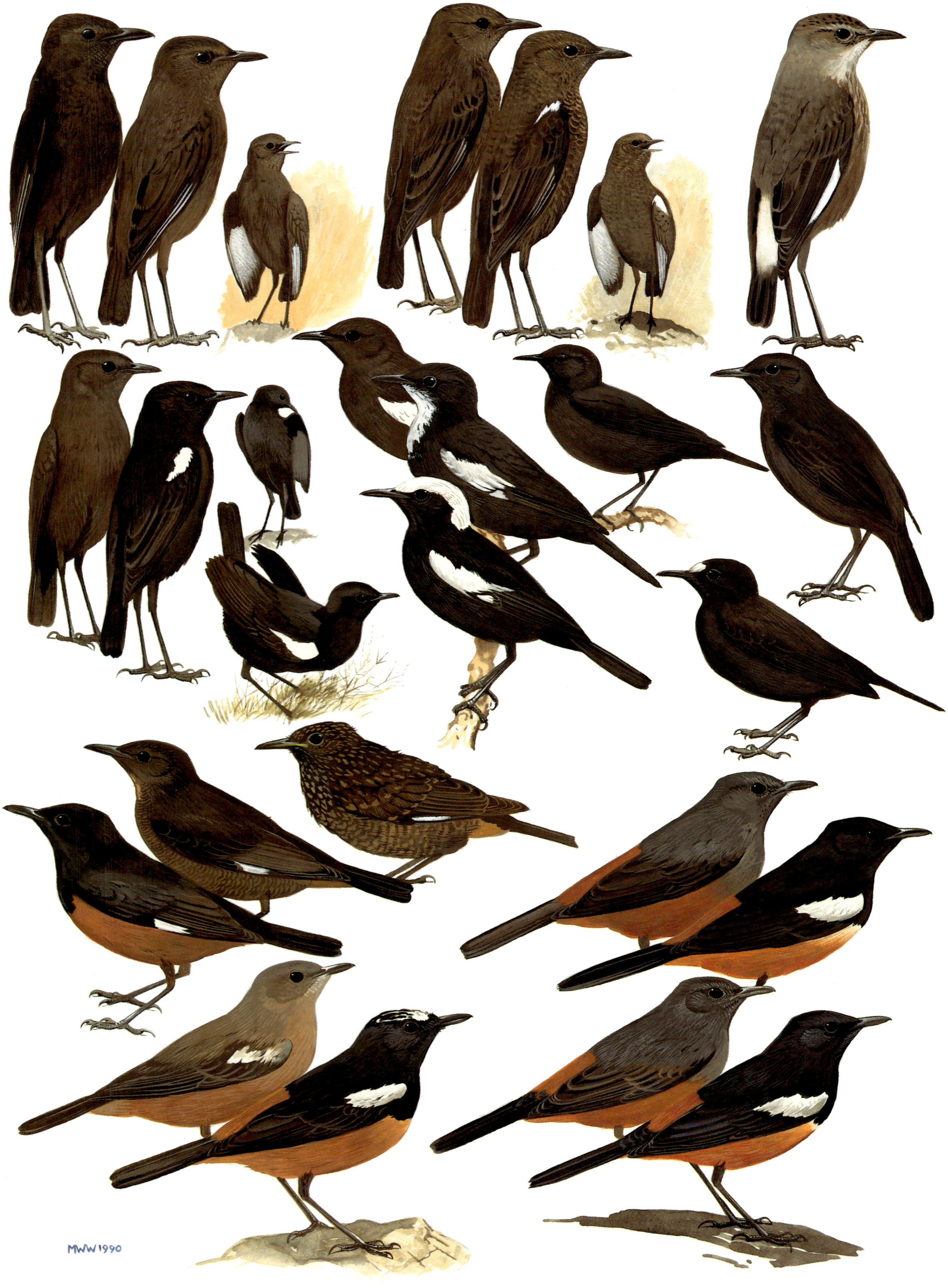
MWW 1990

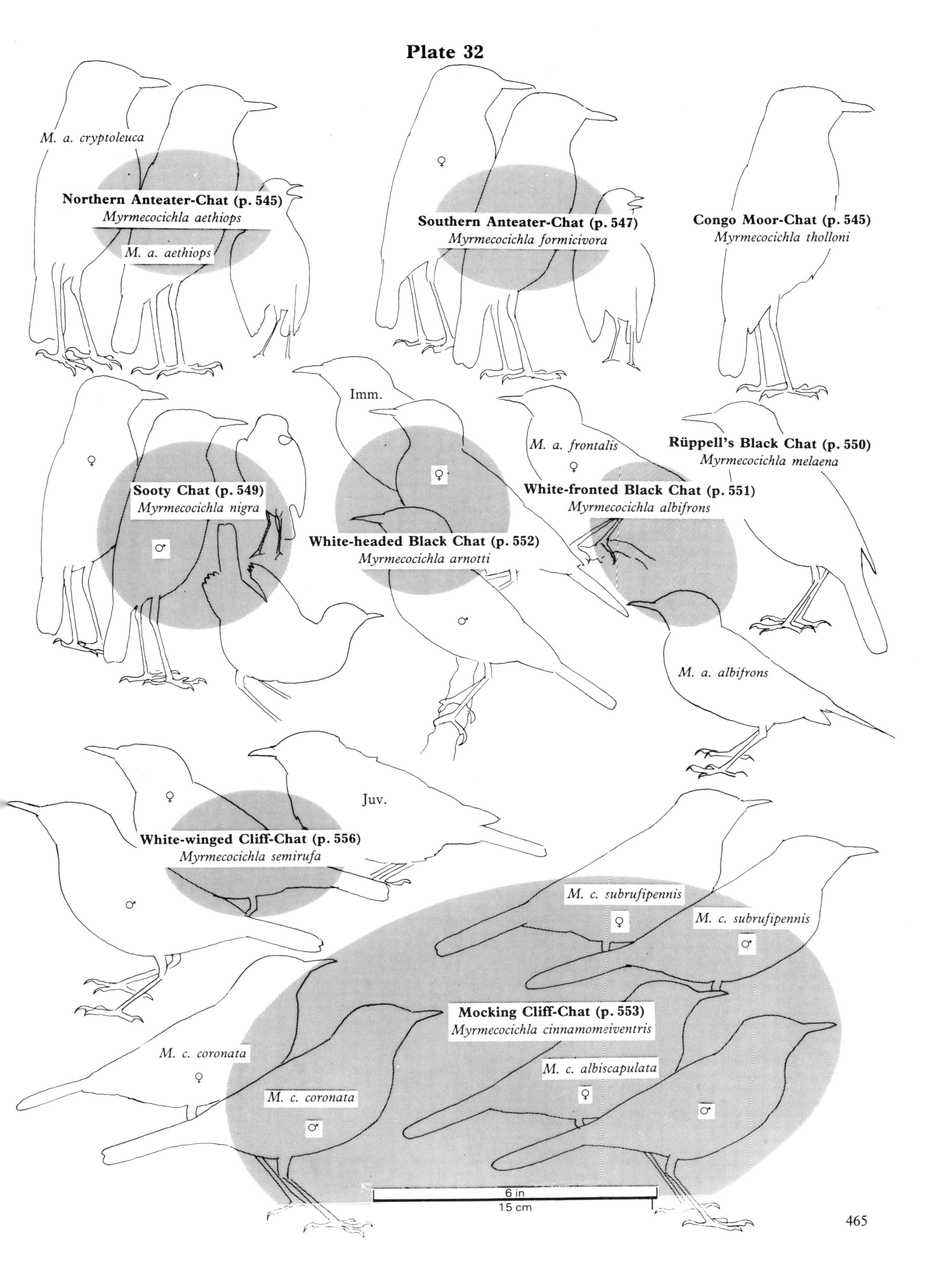

Plate 32

M. a. cryptoleuca

Northern Anteater-Chat (p. 545)
Myrmecocichla aethiops

M. a. aethiops

♀

Southern Anteater-Chat (p. 547)
Myrmecocichla formicivora

Congo Moor-Chat (p. 545)
Myrmecocichla tholloni

Imm.

♀

Sooty Chat (p. 549)
Myrmecocichla nigra

♂

♀

M. a. frontalis
♀

White-headed Black Chat (p. 552)
Myrmecocichla arnotti

♂

Rüppell's Black Chat (p. 550)
Myrmecocichla melaena

White-fronted Black Chat (p. 551)
Myrmecocichla albifrons

M. a. albifrons

♀

White-winged Cliff-Chat (p. 556)
Myrmecocichla semirufa

Juv.

♂

M. c. subrufipennis
♀

M. c. subrufipennis
♂

M. c. coronata
♀

M. c. coronata
♂

Mocking Cliff-Chat (p. 553)
Myrmecocichla cinnamomeiventris

M. c. albiscapulata
♀

♂

6 in
15 cm

465

466 TURDIDAE

Plate 25
(Opp. p. 400)

***Cichladusa arquata* Peters. Collared Palm-Thrush; Morning Warbler. Cichladuse de Peters.**

Cichladusa arquata Peters, 1863. Monatsber. Kon. Akad. Wiss. Berlin, p. 134; Sena, Zambesi R., Mozambique.

Forms a superspecies with *C. ruficauda*.

Range and Status. Endemic resident. Extreme S Uganda (upper Kagera R., Ankole) and coastal Kenya (south from Garsen, inland to Mariakani and Shimba Hills) through W Tanzania (locally to Mwanza and Kibondo and Rukwa), Rwanda, Burundi, E and SE Zaïre (Ruzizi valley south to Luapula R. and east to Sankuru District and Kasai), and Zambia (widespread, mainly in major river systems (Zambezi, Kafue, Luangwa, Luapula) and swamps (Mweru Marsh)), to Zimbabwe (in Zambezi valley but only upstream from Victoria Falls, and in Sabi valley, on Lundi R. and at Haroni–Lusitu confluence); and through E Tanzania (coastal and inland to Selous Game Reserve, Ruaha Nat. Park and Kilosa) and Malaŵi (mainly Shire valley and L. Malaŵi littoral, locally elsewhere) to Mozambique (littoral south at least to Save R., inland in Zambezi and Save valleys and to Gorongoza Nat. Park, and along Limpopo). Sight record from N Transvaal (Letaba camp, Kruger Nat. Park). Locally common but generally sparse, confined to special habitat.

Description. ADULT ♂: crown, nape and hindneck gingery brown; supercilium, sides of face and neck grey, joining to form collar round upper mantle. Chin, throat and centre of breast buff with uneven black border. Mantle greyish brown, back brown becoming orange-red on lower back, rump and uppertail-coverts; tail rufous. Sides of breast grey, flanks grey with buffy wash, lower flanks tawny; belly buff, undertail-coverts rufous-buff. Wings mainly rufous; primaries brown with rufous bases and outer webs, outer secondaries brown on outer half of inner webs; underwing-coverts tawny. Bill black; eye pale yellow to grey or greyish white; legs and feet slate-grey to brown. Sexes alike. SIZE: wing, ♂ (n = 36) 85–92 (89·0), ♀ (n = 37) 81–90 (86·0); tail, ♂ (n = 36) 81–91 (85·0), ♀ (n = 37) 79–91 (82·0); bill, ♂ (n = 32) 17·1–20·4 (19·5), ♀ (n = 33) 17–20·4 (18·8); tarsus, ♂ (n = 33) 22–27 (25·0), ♀ (n = 33) 22–27 (24·4). WEIGHT: ♂ (n = 5) 33·6–38·3 (36·2), ♀ (n = 8) 30–35·5 (33·1), unsexed (n = 41) 28·5–38 (34·2).

IMMATURE: throat, chest and underparts spotted or streaked brownish. Crown and nape brown, with dark streaks. Eye brown.
NESTLING: unknown.

Field Characters. About size of robin-chat *Cossypha*, with rufous wings, rump and tail, grey from face down sides of neck to breast. Buff throat with black border unique among African birds; allopatric Red-tailed Palm-Thrush *C. ruficauda* has buff throat but no black border. Flicks and fans tail or waves it slowly up and down; flicks wings on ground. Voice distinctive.

Voice. Tape-recorded (35, 39, 86, C, F). Song, a melodious, twice-repeated 'de dee doodle-oo deedee'; also 'da de da dee da, da de da dee doo', incorporating harsh guttural notes and more complex variations. Alarm notes, an abrasive, chattery call and harsh 'churr churr'.

Cichladusa arquata

Mimics calls of other species. Wings often clapped in mid-flight producing loud 'prrrup prrrup'.

General Habits. Peculiarly locally and discontinuously distributed. Inhabits riparian ivory palm *Hyphaene ventricosa* thickets along rivers, and bush clumps in ivory and borassus palm *Borassus aethiopicum* savannas. Also occurs in gardens some distance from its usual habitat, and is common in some E African coastal towns.

Occurs in pairs or family parties. Forages on ground, mainly in thicket understorey but also in open. Prey usually picked from ground, but sometimes drops from low perch onto insects on ground. Hops on ground with tail raised; when disturbed moves up into bushes or palms. Sometimes flies between trees or thickets in prolonged glide with wings outstretched.

Food. Invertebrates and small vertebrates (frogs). 1 stomach contained 93 insects of 11 spp.; *Nezara viridula* (38%), *Forficula* spp. (25%), Blattidae sp. (17%), *Heteronychus* spp. (6%), remainder Lygaeidae, Notonectidae, Corixidae, Melolonthidae, Scarabaeidae, Elateridae, Acrididae. Young fed centipedes, long-horned grasshoppers, larval Elateridae, caterpillars, moths and small frogs.

Breeding Habits. Monogamous, territorial.

NEST: like a truncated cone, diam. at base 150, diam. at top 85–105, height *c*. 50, depth of cup 25; of grass bound with mud, mud often predominating in the walls, lined with fine grass stems and strips of palm leaf. Placed on palm leaf near base of frond, against or away from trunk; in cavity where frond has broken away; in the leaf mass of dragon tree *Dracaena*; or on rafters and beams under eaves of pump stations or boat houses, once on capital of pillar in verandah (Fuggles-Couchman 1986), once on functioning air-conditioner (Hanmer 1989b). Built by both sexes; birds carry material along same route back and forth, calling at staging posts on the way.

EGGS: 2–3, mean (14 clutches) 2·3. Elliptical ovals; pale greenish white, plain or faintly speckled or sparingly spotted with reddish and chestnut in zone around large end, overlying pale lilac-purple and pale slate marks. SIZE: (n = 6) 24–27 × 15–18 (25·7 × 16·5).

LAYING DATES: Zimbabwe, Oct–Dec, Feb–Mar; Zambia, Oct–Nov, Feb–Mar; Malawi, Nov–Mar; Zaïre, Dec–May; Tanzania, May.

INCUBATION: by both sexes, at irregular intervals; period (n = 1) 13 days.

DEVELOPMENT AND CARE OF YOUNG: nestling period (n = 1) 20 days. Young almost fledged at 13 days when heavily striped on throat, chest and underparts. At 16 days striping begins to disappear from throat, and collar first appears; at 20 days collar fairly well marked. Young utter a continuous cricket-like wheeze, 'zzeezzee'; stretch wings and flap at 13 days. Fed by both adults, one immediately following the other; they use same route to and from nest, stopping at staging posts and calling; final approach to nest direct. Fed once every 10 min in morning, less frequently at midday. Faecal sacs removed from nest and dropped well away from it.

BREEDING SUCCESS/SURVIVAL: 1 record; full clutch of 3 young reared.

References
Benson, C. W. and Benson, F. M. (1947).
Donnelly, B. G. (1967).
Wilson, K. J. (1964).

Cichladusa ruficauda (Hartlaub). **Red-tailed Palm-Thrush; Red-tailed Morning Warbler.** Cichladuse à queue rousse.

Plate 25
(Opp. p. 400)

Bradyornis ruficauda Hartlaub, 1857. Syst. Orn. Westafr., p. 66; Gabon.

Forms a superspecies with *C. arquata*.

Range and Status. Endemic resident. S Gabon coast south through Cabinda (mainly coastal, in palm savannas) and coastal and inland Angola (common throughout escarpment zone and in lowland dry forest, locally common in scattered localities in W Guinea Forest zone (Mavoio, Maquela-do-Zomba, Uige, south to about Malange), W central plateau, coastal lowlands to Benguela and lower Kunene R. valley (absent from arid SW); in Congo and Zaïre along Congo R. valley north and east to Mbandaka. Locally common.

Description. ADULT ♂: crown, nape and hindneck rufous-brown, lores dark brown, supercilium and ear-coverts greyish. Chin and throat yellowish buff shading to grey on breast and sides of neck; mantle greyish, back, upperwing-coverts, scapulars and tertials rufous-brown, rump and uppertail-coverts orange-rufous. Tail-feathers rufous, central pair browner. Upper belly and flanks grey, shading to pale brown-buff on lower flanks and undertail-coverts; lower belly buff. Primaries and secondaries dark brown, outer webs fringed rufous (except on outer primaries). Underwing-coverts pale brown-buff. Bill black; eye red to reddish brown; legs and feet blue-grey. Sexes alike. SIZE: wing, ♂ (n = 8) 85–91 (88·0), ♀ (n = 7) 84–89 (86·0); tail, ♂ (n = 8) 72–85 (78·0), ♀ (n = 7) 71–85 (75·0); bill, ♂ (n = 7) 17·7–20 (19·0), ♀ (n = 7) 18·6–20 (19·0); tarsus, ♂ (n = 8) 23–25 (24·3), ♀ (n = 7) 23–25 (24·0). WEIGHT: 1 ♂ 30, 1 ♀ 30, 1 unsexed 28.

IMMATURE: crown streaked and brownish; back feathers with dusky tips; throat and breast grey, feathers of breast with dusky tips giving mottled appearance. Eye brown.
NESTLING: unknown.

Field Characters. A small thrush not much bigger than a Sprosser *Luscinia luscinia*, from which distinguished by rufous upperparts, longer and brighter rufous tail, buff throat, grey and buff belly and song. Like closely-related Collared Palm-Thrush *C. arquata*, with similar wing- and tail-flicking movements, but lacks black border to throat and ranges do not overlap.

Voice. Tape-recorded (C, F). Loud, clear song, uttered most frequently at dawn and dusk; similar to Collared Palm-Thrush, interspersed with churring notes, and a rich, throaty babbling. Mimics calls of other species.

General Habits. Inhabits ivory palm *Hyphaene ventricosa* thickets (coastal Cabinda), oil palm *Elaeis guineensis* groves, plantations, and thickets in forest and gallery forest, bush clumps in low-altitude, dry *Adansonia digitata – Acacia welwitschii* woodland and banana plantations. Frequents houses and gardens.

Occurs in pairs or small parties. Forages on ground and in low bushes; when disturbed, moves upwards to tops of palms or other trees. More often heard than seen. Roosts in palms.

Food. Invertebrates, including beetles (adult and larval stages) and spiders; occasionally oil palm fruits.

Breeding Habits. Monogamous, territorial.
NEST: thick-walled mud cup, shaped like truncated cone, with base *c.* 100 diam., height *c.* 50, lined with grass and finer material; either placed at base of ivory or oil palm frond, built into niche in main stem of baobab or other large tree, or placed on rock ledge. Frequently nests on buildings, e.g. pump or boat houses in gallery forest, on window or other ledges on brick buildings, commonly on rafters. When building, adults use series of staging perches to and from source of nest material (mud) to nest, and call at each stop.
EGGS: 1–4, mean (4 clutches) 2·2. Elliptical ovate; pale greenish white, freckled with pinky or dull rufous speckles, markings heavier or blotched at obtuse end. SIZE: (n = 2) 23·7–24·1 × 16·2–16·5.
LAYING DATES: Angola (including Cabinda), Sept–Oct, Dec–Jan, Apr; Zaïre, Oct–Apr.
DEVELOPMENT AND CARE OF YOUNG: both adults feed nestlings, using staging posts where they stop and sing while carrying food to nest; same route used repeatedly.

References
Dean, W. R. J. (1976).
Serle, W. (1955).

Plate 25
(Opp. p. 400)

Cichladusa guttata (Heuglin). Spotted Palm-Thrush; Spotted Morning Warbler.
Cichladuse à poitrine tachetée.

Crateropus guttatus Heuglin, 1862. J. Orn. 10, p. 300; middle course of the Bahr-el-Abiad (White Nile).

Range and Status. Endemic resident. S Sudan (Bahr el Ghazal and Equatoria, common); SW and S Ethiopia (common); S Somalia (coast from about Mogadishu south, inland along Juba and Webi Shebelle river valleys; common); throughout Kenya except N deserts and in SW around L. Victoria; E and central Tanzania south to Wembere, Ruaha Nat. Park, Morogoro and Dar es Salaam; Uganda south to Butiaba, Buruli and L. Opeta; extreme NE Zaïre (W shores of L. Albert).

Description. *C. g. guttata* (Heuglin): Sudan, Zaïre, Uganda and NW Kenya west of L. Turkana. ADULT ♂: crown, nape and hindneck brown, faint streaking on crown; sides of neck and foreneck whitish. Ear-coverts brown, streak over eye and cheeks whitish, malar streaks blackish. Mantle, back and rump brown; uppertail-coverts orange-rufous, tail rufous-brown. Underparts creamy white, with buffy wash on flanks and undertail-coverts, sometimes on throat; teardrop shaped black spots on sides of breast extending down flanks and across belly. Wings brown, flight feathers edged rufous. Underwing-coverts buff. Bill black; eye brown; legs and feet black. Sexes alike. SIZE: wing, ♂ (n = 15) 81–90 (85·0), ♀ (n = 6) 80–87 (84·0); tail, ♂ (n = 15) 72–86 (78·0), ♀ (n = 6) 73–81 (77·0); bill, ♂ (n = 14) 17·4–20 (18·6), ♂ (n = 6) 17·3–19·7 (18·3); tarsus, ♂ (n = 15) 22·5–26·5 (24·5), ♀ (n = 6) 24–26 (25·1). WEIGHT: ♂ (n = 17) 17–30 (24·4), ♀ (n = 17) 16–30 (21·9).

IMMATURE: breast off-white; spotting browner, less sharp; malar streaks broader and browner.

NESTLING: unknown.

C. g. intercalans Clancey: SW Ethiopia in E Gamo-Gofa, Sidamo and W Bale, south, east of L. Turkana, to highlands of Kenya; interior of Tanzania from L. Natron and L. Manyara, Arusha and the Pare Mts, south to about 7° 30'S in the Great Ruaha drainage, and marginally in E Zaïre. Redder and darker above than *guttata*, with heavier dark shaft-streaking. Flight feathers redder, as in *rufipennis*. Below buffy, less white, spotting on breast and sides heavier and blacker. Size as *guttata*.

C. g. rufipennis Sharpe: littoral of S Somalia, Kenya and Tanzania, as far south as Dar-es-Salaam. Similar to *intercalans*, but below less deep buff and with smaller spotting on breast and sides. Smaller: (7 ♂♂, 5 ♀♀) wing, ♂ 76–78 (77·0), ♀ 70–77·5 (75·0).

Field Characters. White underparts and spotted breast easily distinguish it from Collared Palm-Thrush *C. arquata*. More likely to be mistaken for White-browed Scrub-Robin *Erythropygia leucophrys*, which is considerably smaller, with streaked, not spotted breast, 2 white wing-bars and broad black subterminal band to rufous tail, and less powerful song.

Cichladusa guttata

Voice. Tape-recorded (5, 27, 33, 36, B, C). Song loud and clear as in other palm-thrushes; 'usually introduced by a chortle' (North 1958). Main component typically 3 sweet, slurred notes, 'tee-it, too-yu, roo-y-o', but a wide variety of warbling and other notes, calls and churrs are included; also mimicry of other birds. Song battle with wintering Nightingale *Luscinia megarhynchos* recorded by North (1958). Alarm call a churr. Most vocal at dawn and dusk but also sings during the day and on moonlit nights.

General Habits. Inhabits thick scrub and thickets along streams and rivers, bush clumps in dry savannas, and evergreen scrub in *Juniperus* woodland in S Ethiopia. Not restricted to palms. Mainly found at lower altitudes, seldom above 1500 m, though higher in S Ethiopia.

Frequents gardens in towns and villages. Occurs in pairs or family parties; shy, keeping to cover. Forages on ground, where it flicks wings and waves tail slowly.

Food. Invertebrates, including small snails; *Cordia* fruits. Young fed long-horned grasshopper nymphs, beetle larvae, Lampyrinae larvae, lepidopteran larvae.

Breeding Habits. Monogamous, territorial.

NEST: made entirely of mud, bound with a few pieces of grass or leaves, lined with fine rootlets and other material; *c.* 65 diam. at base and *c.* 65 high, cup 30–40 deep; placed on thick branch of tree, usually 2–3 m above ground; built by both adults, mostly by ♀, in *c.* 9 days.

EGGS: 2–3, mean (18 clutches) 2·1. Ovate; unmarked bright turquoise or blue-green. SIZE: (n = 21) 19·2–23·2 × 14·9–15·8 (21·9 × 15·3).

LAYING DATES: Sudan, Sept; Zaïre, Sept; E Africa: Region A, Mar–Apr, June, Aug; Region B, May; Region D, Feb–June, Nov–Dec; Region E, Apr.

INCUBATION: almost entirely by ♀; period *c.* 12 days.

DEVELOPMENT AND CARE OF YOUNG: young fed by both adults.

References
Benson, C. W. (1946).
North, M. E. W. (1958).
Pitman, C. R. S. (1930).
van Someren, V. G. L. (1956).

470 TURDIDAE

Genus *Cercotrichas* Boie

Small, bushland chats with rather long, graduated, white- or pied-tipped tails which are commonly cocked high over back. Rectrices broad. Plumage mainly brown and rufous, often variegated; one species mainly black (*C. podobe*). Outermost primary longer than coverts. Legs long; tarsus scutellate in front. Sexes alike. Juvenile not spotted, but like adult. Nest open, eggs spotted.

10 species, all native to Africa, 8 endemic (and resident), 1 also in Europe and Asia (a migrant), 1 also in Arabia. 4 compose a superspecies (*C. barbata*/*C. quadrivirgata*/*C. leucosticta*/*C. signata*), and 5 rufous species are mutually closely allied, essentially allopatric, but with some overlapping.

Cercotrichas leucosticta superspecies

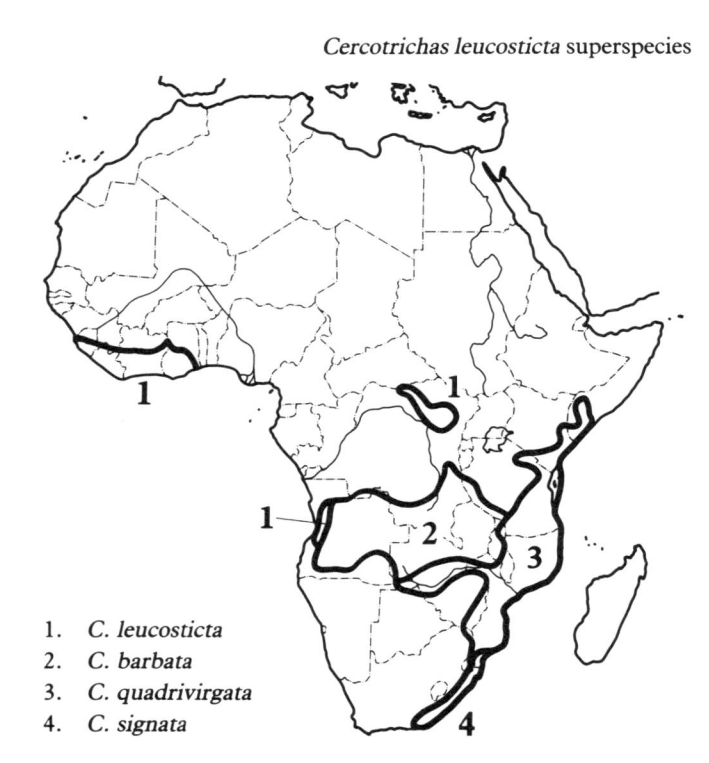

1. *C. leucosticta*
2. *C. barbata*
3. *C. quadrivirgata*
4. *C. signata*

Plate 27
(Opp. p. 416)

Cercotrichas leucosticta **(Sharpe). Forest Scrub-Robin. Robin-agrobate de Ghana.**

Cossypha leucosticta Sharpe, 1883. Cat. Birds Brit. Mus. vii, p. 44; Accra, Gold Coast.

Forms a superspecies with *C. barbata*, *C. quadrivirgata* and *C. signata*.

Range and Status. Endemic resident. 3 disjunct populations: (1) upper guinea forests from Sierra Leone (Kambui Hills) and Liberia to Ivory Coast (north to Mt Nimba, Tai Nat. Park, Comoé Nat. Park) and Ghana (north to *c.* 7°N). (2) Angola from Cuanza Norte, lower Cuanza R. and Gabela south along escarpment to Huila (Chingoroi, 13° 37′S). (3) SE Central African Republic (Ouossi R.) and NE Zaïre from lower Uele R. to Ituri District and across Semliki R. to Bwamba Forest, Uganda, and south to Itombwe. Uncommon to frequent.

Description. *C. l. leucosticta* (Sharpe): Ghana. ADULT ♂: forehead, crown, nape, back and scapulars olive-brown, more or less scalloped blackish brown on crown; crown edged by black line above white supercilium (from bill to above ear-coverts); lores and spot under eye dark grey; ear-coverts paler grey; moustachial stripe white (to sides of neck); malar streak dark grey (meeting grey of breast); chin and throat white; rump and uppertail-coverts dark rufous; rectrices dull brown edged olive-brown, with white spot at tip of outer three pairs; sides of breast olive-grey, centre greyish buff; rest of underparts white, strongly washed with rufous-buff on belly and lower flanks, more olive on upper part of flanks; primaries and secondaries dark brown, secondaries edged grey, P4–P7 narrowly edged white on emarginated part of outer web, P4–P8 with basal one-third of outer web white (producing a white wing-patch); tertials olive-brown, tinged rufous at edges; primary coverts dark brown, other wing-coverts olive-brown, feathers of alula with white spot at tip (together with white marks on primaries, this produces 3–5 white spots on closed wing); underwing-coverts and axillaries silvery white. Colours of soft parts of this race unknown (see below). ♀ and immature of this race unknown. SIZE (1 unsexed, probably ♂) wing, 84; tail, 70; bill, 18; tarsus, 26.

C. l. colstoni Tye: Liberia (and presumably Sierra Leone). ADULT ♂: upperparts slightly darker, especially on rump which is dark rufous-brown; breast-band broader, mottled pale and dark grey (feathers with pale shaft-streaks and darker edges) and continuous across breast; rest of underparts dirty white washed with dull grey-brown, tinged rufous-buff on flanks only; tail tipped white on outer two or three pairs. Bill black; eye brown; legs pale pink. Perhaps smaller. Sexes alike. SIZE: (9 ♂♂, 7 ♀♀) wing, ♂ 76–83 (78·7), ♀ 71–75 (72·9); tail, ♂ 66–71 (68·0), ♀ 59–64 (61·6); bill, ♂ 18–20 (18·6), ♀ 17–19 (18·1); tarsus, ♂ 25–28 (26·4), ♀ 24–27 (25·3). WEIGHT: (Nimba, Liberia) ♂ (n = 9) 23·2–31·2 (26·6), ♀ (n = 7) 21·5–28 (24·3).
IMMATURE: like adult except upperparts from crown to back scalloped black; lesser wing-coverts spotted rufous; underparts rufous-buff, scalloped black.

C. l. collsi (Alexander): NE Zaïre and adjacent Central African Republic. Like nominate except rump and uppertail-coverts rufous-olive (much less strongly rufous); breast-band pure grey (not mottled); flanks olive-grey; rest of underparts white. WEIGHT: (Central African Republic) ♂ (n = 3) 26–28 (27·0), 1 ♀ 27.

C. l. reichenowi (Hartert): NW Angola. Like *collsi* but upperparts paler, purer olive; breast-band paler, French grey; flanks olive-buff; more white on tail (outer 4 pairs tipped white and tips on outer 3 pairs larger).

Field Characters. The only scrub-robin found in the interior of large forests, with sombre plumage as befits its habitat. Dark brown above, with chestnut rump, white-tipped black tail without red, distinct blackish malar stripe, grey breast-band, grey wings with white flashes. Extremely wary; jerky up-and-down tail movement probably the best means of spotting it in dim light. White-browed and Brown-backed Scrub-Robins *C. leucophrys* and *C. hartlaubi* inhabit forest clearings but not interior; both are smaller and have different songs; also, White-browed has paler upperparts, rich rufous rump and tail-edges, brown-streaked breast; Brown-backed looks more like Forest Scrub-Robin, with grey-brown upperparts and smudgy grey streaks on breast, but has much rufous on rump and tail, indistinct malar stripe, different wing pattern.

Voice. Tape-recorded (53, C, VIEL). Song, sweet, high-pitched whistles, particularly melodious, variable and inventive, with individual phrases lasting 3–5 s; loudness varies; does not mimic other birds. Considered to be finest songster of Zaïre forests (Chapin 1953). Alarm a high-pitched, rapid 'chit-chit-chit'; monosyllabic 'chuck' uttered at intervals from ground by foraging bird, perhaps a contact call.

General Habits. Inhabits woody undergrowth of lowland and escarpment forests; avoids dense herbaceous growth. Exceptionally wary, slipping away at first sign of disturbance. Forages on ground; makes short run, pauses, then runs on, like other scrub-robins; often feeds in association with driver ant columns.

Food. Invertebrates, mainly insects, including beetles, black ants, termites, caterpillars, grasshoppers; also small millipedes and amphibians and very small snails.

Cercotrichas leucosticta

Breeding Habits. Monogamous.
NEST: only 1 described (Chapin 1953); shallow cup of rootlets and dead leaves situated *c.* 1·6 m above ground in rot-hole in bole of large forest tree.
EGGS: not described.
LAYING DATES: Liberia, Aug–Sept; Zaïre, Mar, Sept–Oct (♂♀ with enlarged gonads, Feb, June); Central African Republic (breeding condition June).

Reference
Chapin, J. P. (1953).

Cercotrichas barbata (Hartlaub and Finsch). Miombo Bearded Scrub-Robin. Robin-agrobate barbu du Miombo.

Plate 27
(Opp. p. 416)

Cossypha barbata Hartlaub and Finsch, 1870. Vög. Ost-Afr., p. 864; Caconda, Huila, Angola.

Forms a superspecies with *C. leucosticta*, *C. quadrivirgata* and *C. signata*.

Range and Status. Endemic resident in *Brachystegia* woodland; central plateau of Angola from W Malanje south to central Huila and Bihe, north to S Zaïre in Kwango (Gungu) and widely in SE Zaïre west to Dilolo and Kolwezi and north to Manyema; W Tanzania from Kigoma to Tukuyu; widespread in Zambia except in Zambezi and Luangwa valleys; N and central Malaŵi below 1500 m. Frequent to common.

Description. Monotypic (supposed differences of '*thamnodytes*' not supported by long series at BMNH). ADULT ♂: crown, nape, back and scapulars grey-brown, tinged earth-brown when fresh; crown underlined by dark brown line above white supercilium (from bill to behind eye); lores dark brown; sometimes narrow ring under eye white; ear-coverts brown; moustachial stripe white (extending to rear of ear-coverts), malar dark brown; chin and throat white; rump and uppertail-coverts rufous. Tail: central pair of feathers dark brown; next 2 pairs dark brown with or without triangular white tip (extending up shaft); outer 3 pairs with progressively larger white area (outermost pair three-quarters white). Breast and flanks rich buff, darker on breast; belly and undertail-coverts white; remiges dull brown, fringed grey-brown (almost white on middle primaries), outer web of P3–P7 basally white, producing wing-patch; wing-coverts dull brown, tipped white on some lesser coverts; alula dark brown, broadly fringed white; underwing-coverts and axillaries white. Bill black, often horn at base of lower mandible; eye brown; feet pale brown. Sexes alike. SIZE: (10 ♂♂, 10 ♀♀) wing, ♂ 81–88 (84·3), ♀ 77–84 (79·9); tail, ♂ 65–75 (70·1), ♀ 62–70 (65·0); bill, ♂ 17–19 (17·9), ♀ 17–19 (17·6); tarsus, ♂ 23–27 (25·4), ♀ 25–27 (25·9).

IMMATURE: top and sides of head, sides of neck and upperparts brown with orange tinge, each feather fringed black producing scalloped effect. Rump and uppertail-coverts like adult but some scalloping on rump; wing-coverts with broad orange-buff subterminal band and narrow black tip. Underparts mainly buff, incipient orange-brown band across breast with a few dark streaks, lower flanks tinged orange-brown. Bill horn.

NESTLING: unknown.

Field Characters. The scrub-robin of miombo woodland. This and closely-related Eastern Bearded Scrub-Robin *C. quadrivirgata*, which it partly overlaps in some areas of Zambia, Malaŵi and Tanzania, are the only scrub-robins with unstreaked cinnamon underparts. Cinnamon much more extensive in Miombo Bearded Scrub-Robin, covering all except centre of belly; ear-coverts and white face stripes also washed cinnamon; upperparts paler and greyer than Eastern Bearded, but tail with more white; voices similar. Some races of White-browed Scrub-Robin *C. leucophrys* have a touch of cinnamon on sides of breast and flanks, but are smaller, with streaked breast, brown back, much rufous in tail, different song.

Voice. Tape-recorded (56, 86, CA, LEM). Song is loud, varied and melodious, made up of rising and falling cadences and repeated short phrases; differs from song of Eastern Bearded Scrub-Robin mainly in its continuity, unbroken by frequent pauses, although in Malaŵi songs of the 2 species are very similar. Alarm call 1–2 sharp notes followed by drawn-out buzzing churr: 'chek-chek-kwezzzzzzz', indistinguishable from that of Eastern Bearded.

General Habits. Inhabits ground stratum and thickets in *Brachystegia* (miombo) woodland and understorey of dry, evergreen *Cryptosepalum* forest and *Marquesia* thickets. Occurs between 1300 m and 1700 m throughout range except in east where descends to 500 m at edge of L. Malaŵi.

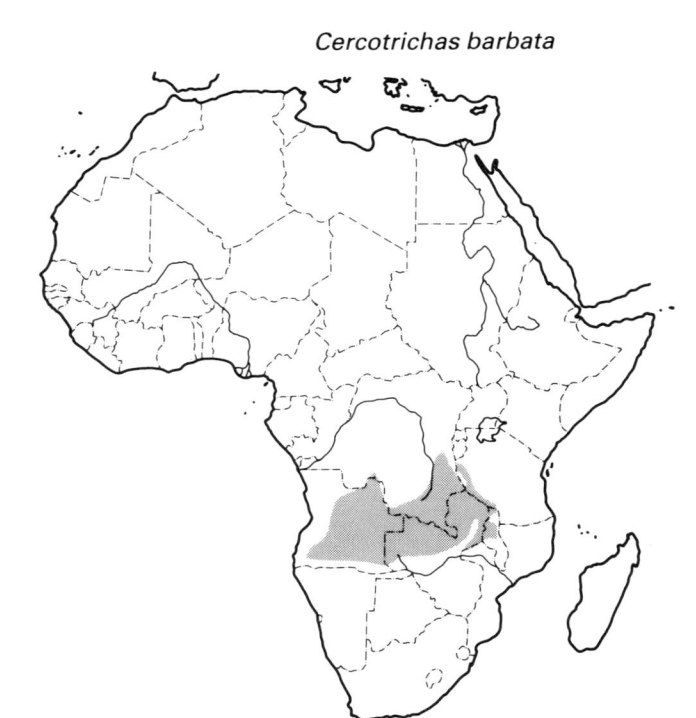

Usually in pairs. Flight silent. Sings throughout day at onset of breeding, at other times sporadically in early morning, sometimes after sunset. Forages on ground under thickets, low scrub and in short grass, especially near large termite mounds.

Food. Insects, including termites (*Microcerotermes* and *Odontotermes*), beetles (Carabidae, Curculionidae), ants, small grasshoppers and plant bugs; also spiders (Oxyopidae, Salticidae).

Breeding Habits. Monogamous; territorial; similar to Eastern Bearded Scrub-Robin in all known aspects. ♂ sings loudly and frequently throughout day at start of breeding season.

NEST: open cup of dried grass, rootlets and leaf stalks, sited in hollow stump or hole in tree, from ground level to at least 1 m high.

EGGS: 2–3, laid on successive days. Ovate; glossy; pale green, spotted and blotched with dark brown, russet and lilac, especially at large end. SIZE: (n = 6, Zambia, Malaŵi) 19–21 × 14–16 (20·3 × 15·5).

LAYING DATES: Zaïre, Aug–Nov; Angola, Sept–Nov; Zambia, Aug–Nov; Malaŵi, Oct–Dec. Mainly in early part of rainy season.

INCUBATION: by ♀ only. Period unknown.

DEVELOPMENT AND CARE OF YOUNG: young fed by both parents.

Reference
Benson, C. W. and Irwin, M. P. S. (1966).

Cercotrichas quadrivirgata (Reichenow). Eastern Bearded Scrub-Robin. Robin-agrobate barbu oriental.

Plate 27
(Opp. p. 416)

Thamnobia quadrivirgata Reichenow, 1879. Orn. Centralb., p. 114; Kipini, Kenya.

Forms a superspecies with *C. leucosticta*, *C. barbata* and *C. signata*.

Range and Status. Endemic resident. Somalia south of 3°N along Juba R. valley and on coast; coastal Kenya, inland up Tana R. to Garissa and in Ngaia Forest (Meru), and in south to Taita and Kitovu Forest; Tanzania on coast and inland to North Pare Mts, L. Manyara, Itigi, Morogoro and Songea; Malaŵi below 1000 m, widespread in S north to Dedza and probably Salima, and L. Malaŵi littoral further north, especially between Chinteche and Nkhata Bay; along Zambezi R. basin in Mozambique, Zambia, Zimbabwe and E Namibia (Caprivi); in Zambia also in plateau country of Southern Province, Kafue basin and Luano valley, in Zimbabwe in SE from Umtali and Sabi R. valley to Chibi, Mt Buhwa and Limpopo; widespread in Mozambique on coastal lowlands and inland on Zambezi, Save, Limpopo and other rivers; E Swaziland, Zululand, Transvaal along Limpopo in N and other rivers in E; Botswana south to Maun, NE Namibia and SE Angola. Zanzibar and Mafia Is. Frequent to common.

Cercotrichas quadrivirgata

Description. *C. q. quadrivirgata* (Reichenow): mainland Africa. ADULT ♂: forehead and line above supercilium very dark brown; supercilium white; lores very dark brown; ear-coverts brown; moustachial streak white, malar dark brown; chin and throat white; crown, nape, back and scapulars dull olive-brown, washed rufous when fresh; rump dark rufous; central 2 pairs of tail-feathers dark brown, others with white mark near tip which is progressively larger on outers (outermost pair half white); breast and flanks rufous-buff; belly and undertail-coverts white; primaries dark brown narrowly edged white, and inner primaries basally white, producing wing-patch; secondaries dark brown narrowly edged olive-brown; tertials olive-brown, edged rufous; alula and outer wing-coverts dark brown, alula and lesser coverts tipped white; inner wing-coverts grey-brown; underwing-coverts and axillaries white. Bill black; eye dark brown; feet pale brownish pink. Sexes alike. SIZE: (no evidence for cline suggested by White (1962): outlying large birds present in all regions of range) wing, ♂ (n = 18) 77–89 (81·7), ♀ (n = 13) 76–81 (77·9); tail, ♂ (n = 17) 68–86 (75·5), ♀ (n = 13) 65–76 (69·7); bill, ♂ (n = 18) 17–20 (18·5), ♀ (n = 13) 18–20 (18·4); tarsus, ♂ (n = 18) 24–28 (26·1), ♀ (n = 13) 23–28 (25·5). WEIGHT: (Zimbabwe) ♂ (n = 20) 23·4–30·7 (26·6 ± 2·4), ♀ (n = 12) 21·2–31·2 (25·6 ± 2·9).

IMMATURE: as adult but upperparts, face and wing-coverts scalloped rufous and very dark brown; underparts scalloped very dark brown.

NESTLING: naked at hatching (no down); skin dark pinkish mauve to blackish, gape dull yellow-ochre; legs and feet light pink.

C. q. greenwayi Moreau: Mafia and Zanzibar Is. Upperparts greyer brown; rump less strongly rufous, more olive-rufous; breast and flanks much paler, washed buff rather than rufous. Larger. SIZE: wing, ♂ (n = 5) 83–90 (85·8), 1 ♀ 84; tail, ♂ (n = 5) 74–84 (78·4), 1 ♀ 79; bill, ♂ (n = 4) 19–21 (19·5); tarsus, ♂ (n = 5) 25–28 (26·4), 1 ♀ 27.

Field Characters. A scrub-robin of coastal forest and riverine scrub, with strongly marked black-and-white face, cinnamon breast and flanks, white flashes in wing, much white in tail. Rather shy, spending much time feeding quietly on ground where it blends perfectly with leaf litter, almost invisible until it moves. Extremely fine, sweet song distinguishes it from all except Miombo Bearded Scrub-Robin *C. barbata*. Larger Brown Scrub-Robin *C. signata* (partly sympatric south of Limpopo R.) has less strongly marked face, grey breast and flanks, less white in tail and different song. Churring alarm like that of Brown Scrub-Robin but preceded by 2 sharp 'chek' notes. Sympatric but ecologically separated White-browed Scrub-Robin *C. leucophrys* has streaked breast, much red in tail, different wing pattern and voice.

Voice. Tape-recorded (5, 15, 33, 58, 72, 86, 88, B, C, F, KEI). Song is individually varied series of sweet whistled phrases, series lasting 12–62 s, intervals between series 2–19 s. Duration of series and intervals characteristic for each individual ♂, which can be identified by them even though the songs themselves may be alike in other respects. Each series is preceded by 3 slow whistles, 'why wooo weee'; many phrases repeated, not consecutively, and sequence varies; song may include some mimicry of other birds. Much regional and individual variation, the more versatile ♂♂ being among the finest

474 TURDIDAE

songsters of African forests. Alarm, 1–2 sharp notes followed by drawn out, buzzing churr, 'chek-chek kwezzzzzz'; commonly given as warning and when going to roost. Anxiety call, uttered when young are threatened, is high-pitched descending whistle, not very loud.

General Habits. Inhabits xeric sand forest, woodland thickets and riverine growth from S Somalia to Natal (L. St Lucia); also lowland evergreen forest, especially north of range of Brown Scrub-Robin; in mesic areas usually avoids damp valleys and frequents ridges where forest drier. Ranges generally from sea level to 1300 m; in E Africa usually below 1000 m, exceptionally to 1800 m.

In pairs all year. Flight silent. On ground hops with head held low in line with body; hops 14–20 cm long. In dry habitats, sand-bathes every evening, choosing open areas such as vehicle tracks where sand has been warmed by sun; sometimes digs shallow hole by whisking away sand with bill, then lies in hole, slightly on one side, flapping free wing and stretching head and neck out on sand. Once observed bathing in water-filled rot-hole (Baker 1983). Forages mainly on ground, whisking leaf litter and digging into soil with bill; breaks open termite galleries on twig trash and makes vertical sallies from ground to heights of 1 m or more to catch flying insects.

Food. Mainly beetles, ants and termites. In 16 stomachs and 5 faecal samples (Malaŵi, Natal) there were: beetles (Buprestidae, Curculionidae, Carabidae, Elyteridae, Melolonthidae, Scarabaeidae, Staphylinidae, Tenebrionidae) in 76%, ants in 71%, termites in 48%, crickets, grasshoppers and mantids in 29%, spiders in 24%, moths and caterpillars (Sphingidae) in 19%, wasps in 14%, plant bugs and flies each in 10%, millipedes in 5%. Not known to eat fruit.

Breeding Habits. Monogamous; territorial. At start of breeding season ♂ sings at intervals throughout day, frequently but very quietly in undergrowth, and at intervals ascends to canopy of tallest trees where it sings loudly, with slightly drooped wings, continuously raising and lowering fully-fanned tail.

NEST: deep, open cup composed of dead leaves, small twigs, dry grass and, in sand forest habitats, *Usnea* lichen, lined with rootlets, tendrils, leaf midribs and often bush pig *Potamochoerus porcus* hair. Some nests in confined site are little more than pad of rootlets; 1 semicircular cup was backed by wooden wall of hollow in which it was built. SIZE: (n = 20, Natal) ext. diam. 90, cup diam. 57, cup depth 45. Usual site is dry rot-hole or hollow stump top, nest rim a little below rim of hollow so that sitting ♀ can see out without raising herself from nest; other sites include crevice formed by dividing trunks; 1 nest placed between loose bark and bole of dead tree. Hole usually < 100 in diam.; mean height above ground 1·47 m, but one at 18 m in sub-canopy of 32 m high tree (Sokoke, Kenya: Short and Horne 1985).

EGGS: 2–3, usually 3 in Natal. Laid on successive days. Ovate to elliptical ovate; glossy; pale green, white, bluish white, blue or aquamarine, boldly blotched and freckled with pale to very dark brown and greyish mauve, usually forming cap or ring at large end. SIZE: (n = 45, Natal) 18–22 × 14–16 (20·4 × 14·9).

LAYING DATES: E Africa: Region D, Feb; Region E, Dec–Jan; Malaŵi, Oct–Jan; Zambia, Sept–Nov; Zimbabwe, Sept–Dec; South Africa: Natal, Sept–Nov, Transvaal, Oct–Dec.

INCUBATION: begins at dusk after laying of last egg; by ♀ only. Period: (n = 3) 11–14 days.

DEVELOPMENT AND CARE OF YOUNG: eyes open on day 6; feathers of dorsal tract predominantly rufous, tipped with sooty black; wing-coverts tawny. Fed by ♂ and ♀; at one nest young fed every 20–30 min; parents often carried faecal material from nest and dropped it as they flew off (Short and Horne 1985). ♀ broods chicks intermittently for first few days after they hatch. Nestling period: (n = 2) 15–17 days.

BREEDING SUCCESS/SURVIVAL: only 17% of eggs laid in 15 nests (Natal) produced fledged young; 5 of 22 nests (22·7%) were parasitized by Red-chested Cuckoo *Cuculus solitarius*.

Reference
Oatley, T. B. (1970b).

Plate 27
(Opp. p. 416)

Cercotrichas signata (Sundevall). Brown Scrub-Robin. Robin-agrobate barbu brun.

Cossypha signata Sundevall, 1850. Oefv. K. Sv. Vet.-Akad. Förh. vii, p. 101; Umslango (= Umhlanga), Natal.

Forms a superspecies with *C. leucosticta*, *C. barbata* and *C. quadrivirgata*.

Range and Status. Endemic resident. Southern Africa, from Limpopo R. mouth (Mozambique) along littoral to coastal forests of Alexandria and Springmount, E Cape Province (South Africa), with small outlying populations in Uitenhage and Humansdorp districts; also inland, in some NE Transvaal escarpment forests, from Woodbush (Tzaneen District) to Ngome, Nkandhla and Ngoye (Zululand) and Amatole Forest (E Cape). Inland distribution disjunct, with inexplicable absences from some forests. Common only in the south, from coastal Transkei to Alexandria Forest.

Description. *C. s. signata* (Sundevall) (including '*reclusa*' and '*oatleyi*'): Woodbush Forest, Transvaal, south to Ngoye, and coastal forests from Tugela R. southwards. Has hybridized with Eastern Bearded Scrub-Robin *Cercotrichas quadrivirgata* (L. St Lucia). ADULT ♂: centre of forehead and crown, nape, ear-coverts, hindneck, side of neck, mantle and back dark olive-brown; a cline in mantle tone from dull greyish brown in E Cape to darker reddish brown in high forests of Zululand and Transvaal; rump and uppertail-coverts paler olive-brown; central 2 pairs of tail-feathers dark olive-brown, others blackish, broadly tipped white. Lores black; sides of forehead and crown blackish brown; supercilium and semi-circle or spot below eye white; chin and throat white; malar stripes and sides of foreneck grey; upper breast and flanks grey; lower breast white marbled with grey; belly and undertail-coverts white. Primaries black with white edging to outer webs, basal 10 mm of each feather white, forming a white bar; alula black, tipped with white; secondaries, tertials and scapulars greyish brown; upperwing-coverts dark grey, flecked with white on lesser coverts; underwing-coverts white with blackish bases. Bill black; eye dark brown; legs and feet flesh-coloured to greyish pink, soles pale yellow. Sexes alike. SIZE: (20 ♂♂, 10 ♀♀) wing, ♂ 83–91 (86·1), ♀ 77–85 (80·5); tail, ♂ 73–84 (77·9), ♀ 71–75 (73·2); bill, ♂ 19–25 (22·5), ♀ 20–24 (22·3); tarsus, ♂ 25–28 (26·8), ♀ 25–28 (26·5). WEIGHT: (n = 5, Natal) 36–42 (38·2).

IMMATURE: forehead and crown dark brown, feathers tipped dull buff; nape and mantle buff, feathers tipped black; rump rufous-brown, feathers edged with black. Underparts buffy white, breast feathers edged dusky, giving scaled appearance. Wing-coverts dark brown with apical buff spots and thin edging of black; tertials thinly edged with rufous buff. Gape yellow, legs and feet pinkish cream, soles cream-coloured.

NESTLING: unknown.

C. s. tongensis (Roberts): Mozambique and Zululand coast from Limpopo R. mouth to St Lucia estuary. Distinctive; pale, short-billed. Upperparts medium brown; underparts whiter than in nominate race with grey areas replaced by buffy grey or buff, especially on flanks; malar stripe more distinct than in *C. s. signata*. Bill, ♂♀ 18–22.

Field Characters. A dark scrub-robin of coastal forests of South Africa and Mozambique, with brown upperparts, grey breast and flanks, white supercilium and crescent under eye, white wing-flashes. No rufous on rump or tail. White spots on dark brown tail are often all that is glimpsed of the retreating bird. When disturbed perches 2–3 m high and remains quiet, allowing a person to walk underneath. Alarm call distinctive; song somewhat like higher-pitched version of Orange Ground-Thrush *Zoothera gurneyi*.

Voice. Tape-recorded (14, 56, 58, 75, 88, F). Song is melodious and varied, high-pitched, sweet, pure whistles arranged in short phrases which end with high-pitched chirps and buzzes; another fine, inventive singer. Commonest call, 6–10 sibilant 'zeeet-seeet-seeet-seeet' notes, sometimes with slight downward cadence, uttered stridently when bird alarmed; also given quietly as contact call. When fledglings are in danger this call may be preceded by a high-pitched, soft, descending whistle, very like that used similarly by Cape Robin-Chat *Cossypha caffra*. Song of *tongensis* similar but somewhat less varied; alarm call less strident, a single, drawn-out 'schizzzzzzz', used in breeding season when one bird is chasing another. Call may also be incorporated in song-phrases as first and last syllables.

General Habits. A bird of evergreen forest, especially dune forest; seldom moves far from dense tangles of undergrowth or stands of *Isoglossa* to which it retreats if alarmed.

Solitary or in pairs. Flight silent. Nominate race bathes (less frequently than most other robins) in quiet pools; *C. s. tongensis*, which seldom has access to surface water, foliage-bathes in early morning on dew-laden leaves, and dust-bathes in sandy forest tracks after sunset. Gait a fast run with head, body and tail held horizontally; on stopping, bird adopts upright stance and tail is jerked up over back. Song is usually uttered from crown of tree 6–8 m high, but occasionally from lower perches or ground. Forages entirely on ground; whisks leaf litter away to expose soil and concealed invertebrates; both races associate with mole-rats *Cryptomys* spp., whose tunnelling and mound-building activities disturb invertebrates; one individual spent nearly all of 2 consecutive days on a few m² of ground where a mole-rat was active. Foraging bird occasionally 'marks time' quite rapidly, paddling like gull foraging on sandy beach; this habit unknown in other scrub-robins. After foraging, bird flies up to perch *c*. 1 m high, bill-wipes vigorously and starts preening. In between foraging it may hop onto protruding root and sit for long periods with head sunk into shoulders and bill pointing slightly upwards. In breeding season sings at any time of day; at other times primarily an early-morning songster.

Food. Arthropods; some seeds. 21 stomachs and 6 faecal samples indicated that ants (63%), beetles (59%) and millipedes (48%) are most frequent prey; orthopterans and moths each occurred in 30% of samples, bugs and flies in 7%, wasps, spiders and other arachnids in 4% (Oatley 1970). Seeds or other fruit remains occurred in 7% of samples and were probably picked up from ground. No obvious difference in diets of long-billed and short-billed races.

Breeding Habits. Monogamous. Appears to remain paired throughout year but solitary birds are encountered in non-breeding season.

NEST: loosely-formed cup of dead leaves, rootlets, fibres and sometimes moss, lined with fine rootlets and animal hair. 1 nest: cup diam. 60, cup depth 57. Placed 1–2·5 m above ground in hole 10–45 cm deep in tree trunk.

EGGS: 2–3, laid on successive mornings. Ground colour white to pale bluish green, blotched and spotted with various shades of brown, mauve and grey; markings sometimes coalesce to form cap or ring at large end; small end frequently immaculate. SIZE: (n = 13) 21–24 × 15–17 (22·5 × 16·2).

LAYING DATES: South Africa: E Cape and Natal, Oct–Dec; Transvaal (nest-building Oct).

INCUBATION: (n = 2) 14–15 days. Remarkably silent at nest; ♂ sings anywhere but near nest, and never warns incubating ♀ of approaching danger.

DEVELOPMENT AND CARE OF YOUNG: nestling period: (n = 1) 16 days. Young fed and attended by both parents, both in and out of nest. 'Schizzz' alarm note used frequently after young have left nest and for as long as family party remains together.

Reference
Oatley, T. B. (1970a).

Plate 27
(Opp. p. 416)

Cercotrichas hartlaubi (Reichenow). **Brown-backed Scrub-Robin. Robin-agrobate à dos brun.**

Erythropygia hartlaubi Reichenow, 1891. J. Orn., p. 63; Mutjora in upper Semliki Valley, southwest of Ruwenzori Mts.

Range and Status. Endemic resident. Lowlands of S Cameroon, and in highland valleys north to Bamenda; SW Central African Republic; a small population in NW Angola (Cuanza Norte: Salazar, Quiculungo); widespread NE Zaïre, west to c. 25°E, south in lowlands to Itombwe (Kamituga, Mulembe); Rwanda, Burundi; W and S Uganda north to Masindi and Elgon; NW Tanzania south to Bukoba, Ngara. Rather scarce in highlands of Kenya, from Kaimosi and Nandi south to Loita Plains and east to Mt Kenya, Meru and Nairobi; apparently declining east of the Rift (Lewis and Pomeroy 1989). Scarce to common.

Description. (Includes 'kenia'). ADULT ♂: crown and nape dark chocolate-brown, crown outlined somewhat darker; supercilium from above bill to above ear-coverts white; lores dark brown; ear-coverts and sides of neck grey-brown, chin and throat white; mantle and scapulars grey-brown, becoming olive-brown on back and upper rump; lower rump and uppertail-coverts dark rufous; tail-feathers basally rufous (about half on outers, two-thirds on inners), then dark brown, and tipped white (more broadly on outers); breast mottled buffish grey and white; rest of underparts white, washed buff to a variable extent (usually more so on flanks and undertail-coverts); remiges and wing-coverts dark chocolate-brown, greater and some lesser coverts broadly tipped white, producing double wing-bar; remiges narrowly fringed buff when fresh; underwing-coverts and axillaries mottled grey and white. Bill black; eye dark brown; feet pinkish or bluish grey. Sexes alike. SIZE: (all except Cameroon birds) wing, ♂ (n = 9) 68–70 (69·1), ♀ (n = 4) 62–66 (63·7); tail, ♂ (n = 8) 64–69 (66·7), ♀ (n = 4) 57–62 (60·5); bill, ♂ (n = 8) 17–18 (17·7), ♀ (n = 4) 17–18 (17·5); tarsus, ♂ (n = 9) 25–28 (26·1), ♀ (n = 4) 24–26 (25·0). Cameroon birds are smaller: SIZE: (8 ♂♂) wing, 65–68

(66·0); tail, 61–65 (63·1); bill, 17–18 (17·7); tarsus, 24–26 (25·0). WEIGHT: ♂ (n = 2) 20, 21, ♀ (n = 3) 17–20 (18·3).

IMMATURE: top and sides of head, sides of neck, upperparts and wing-coverts dark brown with pale orange-brown streaks; rump and tail like adult; chin and throat off-white, underparts buff, breast with dark scalloping.

NESTLING: skin dark pinkish brown; no down feathers.

Field Characters. A small scrub-robin of central Africa. Uppertail-coverts and inner half of tail bright rufous, contrasting with dingy brown upperparts and dark outer-tail; all except central tail-feathers tipped white. Moves tail incessantly, even while preening; sings from prominent position at top of bush, tree or elephant-grass stem, raising and lowering fanned tail; also sings in flight. Broadly sympatric with White-browed Scrub-Robin *C. leucophrys*, but generally ecologically separated, preferring lusher habitats. Sympatric races of White-browed Scrub-Robin have brighter, red-brown backs, rufous of tail extending further towards tip, breast with sharp brown instead of smudgy grey streaks, wings with rufous edgings. Songs similar in form. For differences from Forest Scrub-Robin *C. leucosticta*, see that species.

Voice. Tape-recorded (53, C, GREG, HOR, KEI, McVIC). Song, loud for size of bird, is similar to White-browed Scrub-Robin but somewhat richer and mellower, a series of clear, tuneful, whistled phrases, e.g. 'cher weee ter, cher weee ter too' and 'chur whip chee-cheee chittereeeyoo cha-cha' or 'keyup cheechee weee-yoo chuprep weeeyoo chuprep'; ♂ sings continuously for 20–30 min at a time; ♀ sometimes duets with ♂, using simpler phrases like 'see-lit-see, cheee-why, lit-see, see-see'. Takes a phrase and repeats it, like White-browed Scrub-Robin. Anxiety call, used when feeding chicks of fledglings, 'piri' or 'pri-prit'.

General Habits. Inhabits tall grass savanna; especially favours elephant grass, and so occurs often in and about villages; also inhabits patchwork mosaics of old fields, banana plantations, millet fields and regenerating secondary growth. Shuns interior of closed forest, but occurs in large clearings in forest zone. Altitudinal range from near sea level in Cameroon to 2200 m in E Africa.

Usually in pairs; forages on ground or low down in shrubs and among stems of tall grass.

Sedentary; no known movements.

Food. Insects: beetles and their larvae, moths, flies, grasshoppers and lantern flies (Homoptera, Fulgoridae); also small millipedes.

Breeding Habits. Monogamous, solitary nester. Breeding territories small.

NEST: open, deep cup on ground or near it in creepers, woody shrub or grass; constructed largely of dry grass and rootlets. Ground nests often have rim flush with surrounding leaf litter and cup sunken; rim frequently thickened at one side.

EGGS: 2–3, rarely 4. Creamy, sometimes pinkish, freckled and spotted with shades of brown, most densely near middle, sometimes forming a ring. SIZE: av. about 21 × 14 (van Someren 1956).

LAYING DATES: Angola (breeding condition Nov); Rwanda, Apr, Oct; E Africa: Region B, Feb, Aug; Region D, Apr–May, Nov; E Zaïre, Mar–May, possibly July.

INCUBATION: by ♀ only; ♂ often perches near sitting ♀ and sings quietly. Period: 12 days.

DEVELOPMENT AND CARE OF YOUNG: quills appear after day 4. Young fed by ♂ and ♀, which approach nest via high point to scan surrounding area for predators; parent visits nest every 3–5 min (less often during 1st half of afternoon) when chicks 1 week old; swallows or carries away faecal sacs. Adults' tail posturings particularly extreme when feeding nestlings. Nestling period: 14–15 days.

BREEDING SUCCESS/SURVIVAL: 40% success rate (E Africa: van Someren 1956).

Reference
van Someren, V. G. L. (1956).

Cercotrichas leucophrys (Vieillot). White-browed Scrub-Robin. Robin-agrobate à dos roux.

Plate 27
(Opp. p. 416)

Sylvia leucophrys Vieillot, 1817. Nouv. Dict. d'Hist. Nat. 11, p. 191; Gamtoos R., E Cape Province.

Range and Status. Endemic resident. Ethiopia, in Rift valley and around base of E highlands into Somalia, where widespread except NE, south into Kenya (throughout except extremely arid areas in N and E) and S Sudan (rare). Tanzania, Uganda, Rwanda, Burundi, Zaïre around periphery of main forest bloc and in cleared areas within it; S Gabon, S Congo, and throughout most of central and southern Africa except arid SW; south to central Namibia, Botswana except SW, Transvaal, extreme NW Orange Free State (needs confirmation: Earlé and Grobler 1987); NE and SE Cape Province (along coast to George). Frequent to abundant. Density in Transvaal 45 pairs/100 km² in acacia, 8 pairs/km² in broad-leaved woodland.

Description. *C. l. leucoptera* (Rüppell): Ethiopia, N Somalia, SE Sudan, N Kenya south to Nairobi. ADULT ♂: forehead, crown and nape grey-brown; supercilium from above bill to above ear-coverts white; lores dark brown; ear-coverts buffy brown; chin and throat white; mantle, back and scapulars earth-brown (variable in shade); rump and uppertail-coverts rufous-brown; tail-feathers rufous-brown with subterminal dark brown band, tipped white (very narrowly on inners, up to a quarter of outermost), outer web of outermost pair dark brown to base; breast and sides of neck mottled grey and white (or white streaked with grey); flanks buff; belly and undertail-coverts white; remiges except inner secondaries dark brown, fringed buff when fresh; wing-coverts and inner secondaries dark brown, broadly fringed white; underwing-coverts and axillaries mottled grey and white. Bill black, base of lower mandible yellow-horn; eye brown; feet grey or brown. Sexes alike. SIZE: (10 ♂♂, 10 ♀♀) wing, ♂ 70–74 (71·7), ♀ 62–70 (66·9); tail, ♂ 68–75 (71·5), ♀ 63–71 (67·3); bill, ♂ 17–19 (17·6), ♀ 16–18 (17·2); tarsus, ♂ 23–26 (24·5), ♀ 22–26 (23·8). WEIGHT: (Turkana, Kenya, May) 1 ♂ 17·5, 1 ♀ 16.

IMMATURE: like adult except upperparts rufous-buff, scalloped dark brown; no supercilium; face, chin, throat and breast scalloped brownish grey.

NESTLING: see under *C. l. zambesiana*.

C. l. eluta Bowen: S Somalia, NE Kenya. Averages paler above.

C. l. vulpina (Reichenow): E Kenya, E Tanzania. Dark brown and white areas of tail less extensive (narrower bands); mantle and scapulars more rufous; crown and nape less grey, more olive-tinged.

C. l. brunneiceps (Reichenow): Kenya south and west of Nairobi to L. Natron and NE Tanzania west of Kilimanjaro to Loliondo; intergrades with *vulpina* in Simba-Makindu area. As *leucoptera* but crown and nape darker, blackish olive, wearing to blackish grey; mantle and scapulars slightly more olive-tinged; slightly less white on wing-coverts (usually separates into two wing-bars rather than a single patch); breast buffy white; sides of neck and breast more heavily streaked very dark brown. Larger. SIZE: wing, ♂ (n = 8) 73–78 (75·2), ♀ (n = 2) 70, 73; tail, ♂ (n = 8) 69–77 (73·5), ♀ (n = 2) 70, 74; bill, ♂ (n = 8) 18–22 (19·2), ♀ (n = 2) 18, 19; tarsus, ♂ (n = 8) 26–28 (27·0), 1 ♀ 25.

C. l. sclateri (Grote): central Tanzania from Shinyanga, Mbulu and Lolkisale south to Kilosa and Iringa. Like *brunneiceps* but upperparts brighter brown, less olive; underparts less heavily streaked brown (narrower, paler streaks). Smaller. SIZE: (5 ♂♂, 3 ♀♀) wing, ♂ 67–72 (69·4), ♀ 66–68 (67·0); tail, ♂ 60–68 (64·0), ♀ 61–65 (63·3); bill, ♂ 17–19 (18·0), ♀ 17–19 (18·0); tarsus, ♂ 23–25 (24·2), ♀ 24–25 (24·7).

C. l. zambesiana (Sharpe): S Sudan, Uganda, W Kenya, W and S Tanzania, N and E Zaïre, Malawi, N Mozambique, E and S Zambia, E Zimbabwe. Colour of upperparts intermediate between those of *sclateri* and *brunneiceps*; lacks white edges to inner secondaries; in southern populations, brown streaks on underparts average broader but paler than in *sclateri*, while northern populations (including '*ruficauda*') are similar to *sclateri* in this respect. Smaller than *sclateri*. SIZE: wing, ♂ (n = 6) 63–70 (67·5), ♀ (n = 5) 61–66 (64·2); tail, ♂ (n = 6) 56–66 (62·5), ♀ (n = 5) 55–62 (58·4); bill, ♂ (n = 4) 16–17 (16·5), ♀ (n = 5) 16–17 (16·2); tarsus, ♂ (n = 6) 22–25 (23·3), ♀ (n = 5) 21–24 (22·6). WEIGHT: (Zimbabwe) ♂ (n = 14) 12·9–20·3 (17·0 ± 2·1), ♀ (n = 5) 15–17·9 (16·0 ± 1·1).

NESTLING: naked (no down); skin black, gape yellow.

C. l. munda (Cabanis): R. Congo to central Angola and W Katanga (Zaïre), intergrading in SE Zaïre and NW Zambia with *zambesiana*. Like *zambesiana* but upperparts average slightly duller; tail less rufous (feathers often dark brown to base) though in SE Zaïre intermediates occur with rufous-tinged dark brown tail, or with tail pattern of preceding races but with dark brown area more extensive. Not sexually dimorphic in size. SIZE: (5 ♂♂, 5 ♀♀) wing, ♂ 65–68 (66·0), ♀ 63–69 (65·4); tail, ♂ 59–61 (59·8), ♀ 56–64 (60·0); bill, ♂ 15–17 (16·0), ♀ 15–17 (16·0); tarsus, ♂ 22–25 (23·6), ♀ 22–24 (23·2).

C. l. ovamboensis (Neumann): S Angola, N Namibia, SW Zambia, N Botswana, W Zimbabwe. Like *munda* but upperparts greyer, less rufous; less contrast between crown and back; tail lacks rufous apart from fringing near feather-bases; breast less streaked, in this character integrating with *munda* in S Angola and N Botswana and with *leucophrys* (such intergrades including '*makalaka*') in NE Botswana and W Zimbabwe; primaries and secondaries with paler (often white) fringes. Larger than *munda* but, like it, not sexually dimorphic. SIZE: wing, ♂ (n = 5) 67–74 (70·8), ♀ (n = 3) 68–73 (70·7); tail, ♂ (n = 5) 64–72 (68·6), ♀ (n = 3) 63–72 (67·0); bill, ♂ (n = 5) 17–18 (17·6), ♀ (n = 2) 17, 18; tarsus, ♂ (n = 5) 23–26 (24·6), ♀ (n = 3) 25 (25·0).

C. l. leucophrys (Vieillot) (including '*pectoralis*'): S Zimbabwe and Transvaal, intergrading with *zambesiana* in S Mozambique. Like *ovamboensis* but underparts more heavily streaked with very dark brown (as in *zambesiana*). Not, or only slightly, sexually dimorphic in size. SIZE: wing, ♂ (n = 10) 65–73 (69·4), ♀ (n = 6) 64–72 (69·2); tail, ♂ (n = 10) 62–73 (66·4), ♀ (n = 6) 63–68 (65·2); bill, ♂ (n = 9) 17–19 (18·0), ♀ (n = 6) 17–18 (17·3); tarsus, ♂ (n = 10) 23–25 (24·3), ♀ (n = 6) 23–26 (24·5).

Field Characters. The common scrub-robin of much of east and central Africa, with which all others should be compared; prefers drier habitat than most. Draws attention to itself by exaggerated tail movements (**A**), rattling alarm call and frequently-uttered song, often given from exposed perch. Races differ widely, but all have basic pattern of reddish brown upperparts, rufous rump and uppertail-coverts, black-and-white striped face, 2 white wing-bars, streaked breast. Tail varies from all-black to red with black subterminal bar; white tips to all but central feathers. Underparts mainly white, but some races have cinnamon wash on breast and flanks; breast streaking varies from sharp and copious to smudgy and faint. Northern *leucoptera* group ('White-winged Scrub-Robin') has red back, wing-bars coalescing into almost solid white patch. When comparing it with other species, one must be careful to select the correct races. Overlaps in NE Africa with Rufous Scrub-Robin *C. galactotes*, which is plain-looking with no breast streaks, only buffy edgings to wing feathers (no white patch or bars), rufous tail with black subterminal and white terminal bands. For differences from other scrub-robins, see those accounts.

Voice. Tape-recorded (7, 11, 12, 21, 38, 53, 58, 69, 75, C, F, KEI). Pleasant, rather high-pitched song of pure whistles; takes a phrase and repeats it several times, then switches to another phrase, e.g. 'wee-choo-choo-tiddly, widdly-widdly-widdly-widow, skoo-be-doo'; in southern Africa, 'peep-bo-go' or (commonly) 'willie-dee bedee bedee bedeoo'; 3–10 syllables to a phrase, typically 4–5, variations almost endless. Alarm a staccato, scolding rattle, used when bird preparing for roost; when danger threatens young, alarm drawn out into loud ratcheting, persistently uttered by both ♂ and ♀. In Kenya, 'chee-ip' when young threatened. Location call of fledglings, 'chissik'.

General Habits. Inhabits almost any kind of bushy or wooded country, including overgrown cultivation. Characteristic of thornveld in E half of continent. Occurs in arid acacia steppe in NE, but in SW largely replaced in this habitat by Kalahari Scrub-Robin *C. paena*; where sympatric, White-browed usually inhabits dense cover along dry watercourses. Along coast in dune scrub and edges of dune forest. Sparse in *Brachystegia* woodland – usually in disturbed areas with discontinuous ground over. In large clearings in lowland forests of the Congo R. basin in Zaïre. Ranges from sea level to 1400 m on wooded plateaux, where common; occurs up to 2200 m in E Africa.

Flies silently with short rapid undulations. On leaf litter usually hops; on open ground (sandy tracks and footpaths) usually runs, footprints 75–100 mm apart, forming a straight line. After run of *c.* 1 m bird stops, stands erect, raises tail and lowers wings in spasmodic jerks. If disturbed, flies off low, settles quickly and scuttles off very like a mouse, dodging between stems and tussocks with great speed. Bathes in rain puddles and bird baths, also in wet foliage, usually in early morning. Also dust-bathes.

Song normally uttered from top of a bush or low branch of tree; perch often exposed but always close to dense cover. In breeding season sings continuously in moonlight for an hour before dawn; under bright moon sings briefly at any time of night. Forages mainly on ground; searches leaf litter under thickets with sideways whisks of bill, and darts after mobile invertebrates on open ground. Regularly breaks open earthen termite galleries built around leaf and twig trash; also searches rhino middens and droppings of other large herbivores for dipteran larvae and scarab beetles. Sometimes catches termite alates on the wing. Eats small drupes of *Vitex*, and probes aloes for nectar in winter (when birds can become orange-faced with pollen).

Food. Insects. Termites important, but bird usually ignores large harvester termites *Hodotermes* which are common in thornveld. In 23 stomachs and 26 faecal samples from southern Africa there were: termites in 69%, ants in 67%, beetles (Scarabaeidae, Chrysomelidae, Carabidae, Curculionidae) in 59%, moths and caterpillars in 31%, plant bugs in 27%, crickets and grasshoppers in 18%, spiders (Salticidae) in 12%, millipedes and parasitic wasps each in 4%, flies and ant lions each in 2%. Fruit in 10% of samples. Also nectar.

Breeding Habits. Monogamous; territorial. At onset of breeding season sings with increasing vigour; territorial adult ♂♂ often chase young ♂♂; fighting and grappling commonplace, with both birds falling to the ground. During territorial and courtship display, loud song is accompanied by almost erect, fully-fanned tail with white-tipped feathers showing prominently.

NEST: open, deep cup of dry grass blades and stems lined with ginger-coloured grass rootlets; frequently with substantial ramp. SIZE: (n = 50, South Africa) ext. diam. 110, cup diam. 55, cup depth 42. Sited in dry grass or tussock at base of small tree or edge of shrub;

A

often in hanging, dead leaf skirts of large aloe; 4–50 cm above ground, av. (n = 53) 20 cm; exceptionally as high as 1·5 m. Built in 4–5 days, once in only 8 h, by both ♂ and ♀.

EGGS: 2–4, av. 2·7; 68% of 125 clutches in South Africa with 3 eggs. Laid on consecutive days, usually in early morning. Ovate to elliptical ovate, rarely pyriform. In southern Africa pure white, often glossy, spotted and blotched with shades of brown and with greyish mauve, markings commonly forming ring and cap at large end; in E Africa creamy to buff, with spots and blotches of browns and grey fairly evenly distributed (van Someren 1956). SIZE: (n = 100, South Africa) 18–22 × 13–16 (20·1 × 14·3). WEIGHT: (n = 5, Natal) 1·9–2·1 (2·0).

LAYING DATES: Ethiopia, Mar–May; Zaïre, all year; Rwanda, Oct–Jan, Mar–Apr; Angola, Oct–Nov; E Africa: Region A, Apr–May, July; Region B, Apr–May, Sept; Region C, Oct–Dec, Feb, May; Region D, Dec–Feb, Apr–June; Region E, Aug, Oct, Dec, Mar; 50% of E African eggs laid in Apr–May; Malaŵi, Oct–Jan; Zambia, Oct–Dec; Mozambique, Dec–Feb; Zimbabwe and South Africa, Sept–Jan (69% of clutches Oct–Nov). Breeds in rains throughout range.

INCUBATION: begins with last egg; by ♀ only. Period: 12 days.

DEVELOPMENT AND CARE OF YOUNG: some feathering at 7–8 days, when weight 7–8 g; white wing-bars prominent by day 10; weight at fledging c. 15 g. Nestling period: (n = 4) 11–12 (11·5) days.

Fed by both ♂ and ♀; ♀ broods intermittently for first few days and shelters young from direct sunlight. Young can fly well at fledging; at first they hide in dense cover and give location call, but after a few days each chick tends to follow one or other foraging parent. ♂ sings regularly at this stage; 1 fledgling with full-length tail sang long and well.

BREEDING SUCCESS/SURVIVAL: 93 eggs in 34 clutches (Natal) produced 32 fledglings (34% success). Slender mongoose *Herpestes sanguineus* is frequent nest predator. Oldest known bird (Natal) collided with car 11 years and 9 months after being ringed as adult. In southern Africa parasitized at very low level by Red-chested Cuckoo *Cuculus solitarius* and Diederik Cuckoo *Chrysococcyx caprius*.

Reference
van Someren, V. G. L. (1956).

Plate 27
(Opp. p. 416)

Cercotrichas galactotes (Temminck). Rufous Scrub-Robin; Rufous Bush Robin. Agrobate roux.

Sylvia galactotes Temminck, 1820. Man. d'Orn., ed. 2, 1, p. 182; Algeciras.

Range and Status. Africa, SW and SE Europe, SW Asia and SE Arabia. Resident and migrant.

4 distinct populations in Africa. (1) Nominate *galactotes* is frequent, locally common, breeding summer visitor to N Africa (Morocco to Egypt), Spain, and from Sinai to S Syria; breeding grounds vacated in winter, when whole population in N tropical Africa (probably mainly in W Africa). (2) Races *syriacus* and *familiaris* (Near and Middle East) winter in Ethiopia and Kenya (vagrant to Malaŵi (1), Tanzania (2), Uganda (1) and Zaïre (1)). (3) Endemic *minor* is a frequent to common resident from S Mauritania and N Senegambia through sahelian and sudanian savannas to Red Sea coast of Sudan, N Ethiopia and N Somalia. (4) Endemic race *hamertoni* is resident in N and central Somalia.

Status of nominate *galactotes* wintering in W Africa unclear owing to confusion with *minor*. *Syriacus* and *familiaris* evidently broadly overlap on E African wintering grounds. Scarce in Tunisia, but throughout rest of range species is common or locally common as breeding bird (e.g. Morocco, Libya, Egypt, Mali, Nigeria, Chad, Sudan, Ethiopia, Somalia); it can be very common on passage, e.g. 40 in small area in Morocco, Apr, and 24 in a wadi, Djibouti, Nov (Welch and Welch 1986).

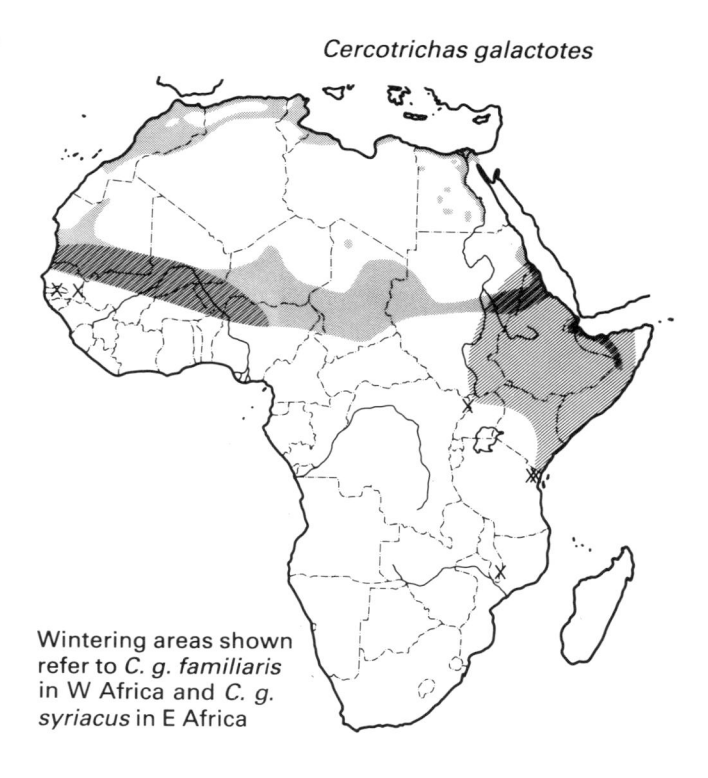

Cercotrichas galactotes

Wintering areas shown refer to *C. g. familiaris* in W Africa and *C. g. syriacus* in E Africa

Description. *C. g. galactotes* (Temminck): N Africa, Spain, Near East; winters W Africa. ADULT ♂: forehead to scapulars and back rufous-brown; rump, uppertail-coverts and tail brighter – fox red. Tail graduated; central pair of feathers darker towards tip, other 5 pairs boldly marked with terminal white and subterminal black patches. White mark on T2 1 mm deep (sometimes absent), on T3 2–4, T4 5–8, T5 10–14 and T6 15–20 deep; black patch 7–12 deep on T2 but only 5 deep on T6. Sides of forehead and crown creamy, forming long, distinct superciliary stripe. Lores and upper ear-coverts dusky, forming narrow dark line through eye. Lower ear-coverts pinkish brown, sides of neck rufescent brown. Chin and throat creamy white, breast pale greyish rufescent brown; belly, flanks and undertail-coverts pinkish cream or whitish. Wings rufescent brown; remiges and tertials dark brown, with rufous fringes to outer webs, narrow on primaries and secondaries, broad on tertials; tips of secondaries and inner primaries narrowly fringed buffy white; lesser coverts rufous-brown, median coverts rufous with browner centres, greater coverts dark brown with rufous-brown outer-edge fringes and whitish tips. Bill black-brown, base of lower mandible horn-brown or whitish; eye brown; legs and feet flesh-brown or pale pinkish grey-horn. Sexes alike. SIZE: wing, ♂ (n = 17) 84–92 (87·7), ♀ (n = 16) 81–89 (86·1); tail, ♂ (n = 18) 64–73 (68·9), ♀ (n = 16) 65–70 (67·5); bill to skull, ♂ (n = 16) 19–21·6 (20·2), ♀ (n = 14) 18·1–20·1 (19·2); tarsus, ♂ (n = 17) 26–28 (27·0), ♀ (n = 16) 25–28 (27·0). WEIGHT: unsexed (n = 9, Morocco, Apr) 21–25 (22·7).

IMMATURE: very like adult; body feathers shorter; browns slightly greyer; rufous parts slightly sandier; underparts paler; black and white marks in rectrices both smaller than in adult (see Cramp 1988).

NESTLING: naked; skin black; mouth yellow-orange, gape-flanges white.

C. g. minor (Cabanis): Senegambia to N Somalia. Like nominate *galactotes* but upperparts warmer pink-brown; subterminal black band in tail narrower; wing-tip rounder. Smaller: wing, ♂ (n = 7) 79–85 (82·1), ♀ (n = 4) 78–81 (79·2); tail, ♂ (n = 6) 61–71 (66·8), ♀ (n = 4) 64–66 (65·1). WEIGHT: unsexed (n = 6, Nigeria, Nov–Mar) 17–19·5 (18·4).

C. g. hamertoni (Grant): E Somalia. Upperparts darker than above races. Small: wing (unsexed, n = 2) 70, 71.

C. g. syriacus (Hemprich and Ehrenberg): SE Europe, Turkey and Syria south to Lebanon, wintering E Africa. Forehead to back and wings drab grey-brown; wing-tip more pointed than in nominate race. Wing longer: wing, ♂ (n = 17) 85–92 (88·3), ♀ (n = 6) 84–87 (85·6).

C. g. familiaris (Ménétries): Transcaucasia and Iraq to Kazakhstan and Afghanistan, wintering E Africa. Like *syriacus* but slightly paler and greyer above, whiter below. WEIGHT: unsexed (n = 26, *familiaris* or *syriacus*, Kenya, Oct–Jan) 17·5–24 (20·3).

Field Characters. A slender but robust, strong-billed, strong-legged largely terrestrial chat, rufous or grey-brown and rufous above, pale below, with well-marked pale supercilium and thin black line through eye, and long, foxy tail, frequently cocked over back and fanned, graduated, with conspicuous white tip and subterminal black band. On breeding grounds unlikely to be mistaken, except in N Somalia where overlaps with White-browed Scrub-Robin *C. leucophrys*, which is distinguished by either small black streaks forming moustache and on breast, or grey head, foxy back, and broad white fringes to upperwing-coverts and tertials. Wintering birds penetrate much further into E African range of *C. leucophrys* and overlap geographically with Brown-backed and Eastern Bearded Scrub-Robins *C. hartlaubi* and *C. quadrivirgata*. Former has dark brown back and wings, 2 white wing-bars, and tail with rufous base and blackish end, lacking white spots; latter has moustache, brown breast, and white-tipped black tail. However, unlikely to occur on same ground as last 2 since they have different habitats: wet bushland (Brown-backed) and forest (Eastern Bearded).

Voice. Tape-recorded (53 – 62, 73). Song is of indefinite duration, consisting of regularly-repeated but variable fluty note or twitter, every 2 s, thrush-like (*Turdus*) in quality or tremulous, like European Robin *Erithacus rubecula*, either with pause between notes or with fainter, sibilant or murmuring notes interposed to give unbroken sound. Rate, 24 notes/phrases in 45 s. Effect is of sustained, quite far-carrying, sweet, liquid warbling, but repetitive and rather unmusical. Main call a hard 'tek tek' or 'chack chack'; contact call a sibilant 'tseeeet' or 'zip'; alarm, 'zi-zi-zi . . .'; warning, fluty 'piu' (Cramp 1988).

General Habits. Inhabits dry bushy country, varying from thick woodlands of southern soudanian zone with heavy grass layer and 5 m trees forming almost continuous canopy, to open sand dunes with well-spaced clumps of shrubs and only occasional trees. Farmland, wooded pasture, palm groves, large gardens (but not suburban), orchards, plantations, groves of orange, *Pistacia*, *Rhus* and *Retama*; tangles of *Opuntia*, tamarisk, *Salvadora* and *Hyphaene*; open *Acacia tortilis*, *Prosopis*, *Zizyphus* and *Combretum* woods with fallen timber and undercover; well-vegetated wadis, oases, hedged cattle enclosures, and scrub around farmland and human habitations. In Tunisia a marked partiality for dry, sandy places and dusty roadsides with *Opuntia* and aloes (Whitaker 1905).

Solitary or (usually) in pairs or family parties; loosely gregarious on migration. Feeds mainly on ground, moving with strong hops in shade of vegetation or within a few m of cover. Skulking but not shy; rather, bold and inquisitive. Perches freely in trees, but seldom above c. 3 m high; spends much time perching on very low bough or sticks and dead branches lying on ground. Feeds by probing soil, flicking dead leaves, and taking insects from low flowers. On bush or ground constantly moves tail slowly up and down, fanning it against ground, and cocks it in series of jerks higher and higher, until almost lying along back; half-cocked tail widely fanned and wagged jerkily through small arc; wing-tips droop when tail fanned or cocked, and wing-tips frequently flicked forwards, the wrists turning accordingly.

Sings for long periods from large shaded bough or from bush-top; also telegraph wire, generally 2–5 m high. Only ♂ sings. Dust-bathes frequently in captivity and probably in wild (necessary to maintain condition of plumage: Beven 1970).

C. g. galactotes arrives S Algeria, S Tunisia and Egypt early Apr, and rest of N African range late Apr/early

May, and departs to south late Sept/early Oct (mid-Sept in Egypt). Crosses Fezzan and Libyan desert in spring and presumably crosses Sahara on both passages, particularly in west (although few records, race generally not determined). Winters on Red Sea coast, Sudan, with a few west to Nile (Nikolaus 1987). In W Africa common migrant (presumably race *galactotes*) Sept–Oct and Apr and winter visitor Mauritania and Mali. Wintering *galactotes* identified in Senegambia, Mali, Niger and N Nigeria (1 record, Kano: Wilkinson 1979), but none recorded from Chad or W Sudan.

C. g. syriacus and *familiaris* winter equally commonly in Somalia Sept–May; in Ethiopia *syriacus* common, *familiaris* uncommon; in N and E Kenya Nov–Apr, with passage in Nov and Mar–Apr; spring passage through Somalia mid-Mar to mid-May. Most or all Kenyan birds are evidently *familiaris* (Britton 1980). 1 record Uganda, Oct (Kidepo Valley Nat. Park), 2 in extreme NE Tanzania, and 1 (attributed to *syriacus*) in NE Zaïre, Dec (Ishwa plain, Mahagi).

Food. Mainly insects. 3 Moroccan birds ate 5 large caterpillars, 2 beetles, other insects, and 6 *Nitraria* fruits (Valverde 1957). Otherwise diet known much better out of Africa. Insects include dragonflies, grasshoppers, earwigs, mantises, cicadas, ant-lion larvae, caterpillars, moths, flies, grubs, ants, wasps, bees and their larvae, beetles and larvae; also spiders, millipedes, centipedes, earthworms and occasionally fruits (Cramp 1988).

Breeding Habits. Monogamous, territorial, solitary breeder; territories contiguous. Av. territory size (n = 5, SE Spain) 2–8 ha; density 2·8 pairs per km^2 (Lopez Iborra 1983); singing ♂♂ 200 m apart, W Sahara, but only av. 2 birds per 24 km of wadi in N Sahara (Blondel 1962, Valverde 1957). In song-flight ♂ flies up 2–3 m from bush then glides to another bush or flies with slow, shallow beats. On ground threatens rival with wings open and tail fanned and cocked (**A**); rival may adopt forward-threat posture (**B**) (Cramp 1988).

NEST: loosely-built, untidy, rather flat cup of grass, rootlets, fine twigs, and bits of bark, lined with silky vegetable fibres, wool, hair, feathers and commonly (Tunisia, Palestine) snake skins; ext. diam. 130 × 200, height 65–100; int. diam. 70–90, cup depth 40–65. Built by both sexes, 0·5–2·5 (av. 2) m above ground in palm thicket, *Opuntia* or thorn-bush (*Acacia, Zizyphus*).

EGGS: 3–5 N Africa, mainly 2–3 in Afrotropics; laid daily; double-brooded. Subelliptical; glossy; white or very pale bluish or greenish grey heavily speckled with brown, purple-brown or purple-grey, with some streaks. SIZE: (n = 29, *C. g. galactotes*, Tunisia) 19·5–26 × 14·3–18·0 (22·5 × 16·3); (n = 2, *C. g. minor*, Niger) 21·0–21·3 × 15–16. WEIGHT: *c*. 3·2.

LAYING DATES: Morocco, Algeria, Tunisia, mainly mid-May to mid-June; Libya, May; Mauritania, June; Niger (Aïr) June, Aug (nestlings June, July); Togo, May; Nigeria, Apr; Ethiopia, May.

INCUBATION: by ♀ only, beginning with last egg. Period: 13 days.

DEVELOPMENT AND CARE OF YOUNG: eyes open at 6 days; fed by ♂ and ♀. May leave nest as early as day 10; fledglings at first perch separately, but when they can fly better they perch together. By day 25 young ceaselessly follow parent (brood sometimes divided between ♂ and ♀), begging and shivering wings; young start feeding themselves *c*. 3 weeks after fledging (Cramp, 1988). Fledging period: 12–13 days.

Reference
Cramp, S. (1988).

Cercotrichas paena (Smith). Kalahari Scrub-Robin. Agrobate du Kalahari.

Erythropygia paena A. Smith, 1836. Rep. Exped. Expl. Cent. Afr., p. 45; between Latakoo and the Tropic (= north of Kuruman), N Cape Province.

Plate 27
(Opp. p. 416)

Range and Status. Endemic resident. Angola in dry coastal plain of Benguela and Mossamedes and along Namibian border in S Huila and probably Cuando-Cubango; Namibia except extreme arid regions of coast and S; Botswana; SW Zimbabwe from Gwaai and Umgusa Forest Reserve south to Botswana border and along Limpopo; common in W and central Transvaal, becoming more local and patchily distributed further east, as far as Burgersfort, vagrant in Lowveld; W and central Orange Free State and N Cape Province. Common.

Description. *C. p. paena* (Smith): Botswana, N Cape Province. ADULT ♂: upperparts brownish grey, washed with earth-brown on mantle, back, scapulars and rump; crown underlined dark grey, above white supercilium (from bill to above ear-coverts); lores and upper ear-coverts dark brown; sides of neck and lower ear-coverts buffish grey; chin and throat white; uppertail-coverts rufous; tail-feathers basally rufous, with subterminal very dark brown band (about one-third of feather and extending up shaft); all feathers tipped white, broadest (one-quarter of feather) on outermost, progressively less on inners, with central pair having only tiny white tip; breast, flanks and undertail-coverts greyish buff; belly paler buff (often white); remiges and wing-coverts dark brown, fringed buff (broadly on inner secondaries, tertials and inner wing-coverts); underwing-coverts white; axillaries buff. Bill black; eye dark brown; feet dark grey-brown. Sexes alike. SIZE: wing, ♂ (n = 9) 70–73 (71·7); ♀ (n = 10) 69–73 (70·7); tail, ♂ (n = 8) 63–69 (66·5), ♀ (n = 10) 64–70 (66·0); bill, ♂ (n = 9) 19–21 (20·0), ♀ (n = 10) 17–20 (18·7); tarsus, ♂ (n = 9) 25–27 (25·8), ♀ (n = 10) 24–27 (24·9). WEIGHT: (n = 21) 17–23 (19·7).

IMMATURE: like adult except upperparts browner and scalloped with black; underparts scalloped black; fringes on remiges richer, deeper buff.

NESTLING: entirely naked; skin black, gape yellow.

C. p. oriens (Clancey): Orange Free State, Transvaal and S Zimbabwe. Upperparts darker, browner, underparts darker, brownish buff rather than greyish buff. SIZE: (5 ♂♂, 5 ♀♀) wing, ♂ 71–74 (72·2), ♀ 70–73 (71·6); tail, ♂ 65–71 (67·0), ♀ 66–71 (67·6); bill, ♂ 18–20 (19·2), ♀ 18–19 (18·6); tarsus, ♂ 24–26 (24·8), ♀ 24–25 (24·6).

C. p. damarensis (Hartert): Namibia. Like *paena* but upperparts paler, greyer, especially on head.

C. p. benguellensis (Hartert): SW Angola. As *damarensis* but upperparts paler still (head grey, rest of upperparts brownish grey, rump and tail paler rufous); underparts paler, less buff. Slightly smaller (wing and tail). SIZE: (6 ♂♂, 3 ♀♀) wing, ♂ 68–69 (68·5), ♀ 66–71 (68·3); tail, ♂ 60–64 (62·8), ♀ 60–64 (61·3); bill, ♂ 19–20 (19·5), ♀ 19–21 (20·0); tarsus, ♂ 23–26 (24·7), ♀ 25 (25·0).

TAXONOMIC NOTE: these four races form a cline, with palest birds in NW, darkest in E.

Field Characters. A pale scrub-robin of Kalahari sand vegetation with distinctively patterned tail, rufous with broad black subterminal band, graduated white terminal band, broadest on outer feathers, narrowest on innermost. No wing-bars or breast streaks. When alighting, jerks tail up and flicks wings away above body several times, a movement so characteristic that bird can be identified even in silhouette. Keeps tail cocked even when foraging on ground, especially if wary of observer. Partly overlaps with White-browed Scrub-Robin *C. leucophrys* towards edges of range. White-browed has darker, redder upperparts, broad white wing-bars; sympatric races have black tail (rufous confined to rump and uppertail-coverts), cinnamon wash on breast and flanks, breast streaks present but reduced, sometimes faint. When perched, Kalahari Scrub-Robin raises and lowers tail smoothly (White-browed does so jerkily). For differences from Karoo Scrub-Robin *C. coryphaeus*, see that species.

Voice. Tape-recorded (35, 38, 88, F, GIB, LUT, OAT). Song more musical and varied than those of White-browed and Karoo Scrub-Robins; a pleasing medley of whistles and warbling chirps, some syllables repeated several times, then augmented and repeated again: 'weeyoo, weeyoo, cheeeep, weerip, willerip, willerip, cheeeoo, cheecheeoo'. Has a sparrow-like anxiety note 'seeeup', which also serves as a contact call. Alarm, a rasping 'zeee', reminiscent of scolding note of some prinias.

General Habits. Inhabits a range of vegetation communities on Kalahari sand, including *Combretum*, *Terminalia*, *Colophospermum* and *Acacia*, particularly *A. mellifera*, fringing ephemerally-flooded depressions; at edge of range frequents clearing or open areas in *Baikiaea* woodland and bushveld, and favours heavily grazed and browsed areas with low secondary scrub; also old cultivation.

Solitary or in pairs; spends much time on ground where it runs or occasionally hops with tail held slightly above horizontal; when wary, closes tail and holds it vertically and droops wings until tips almost touch ground. Frequently suns itself on topmost branches of bush or tree in early morning. Sings from highest available perch (*contra* Maclean 1985), such as telegraph wire or top of camelthorn tree *Acacia erioloba*, especially after rain. Forages mainly on bare ground; after completing search, flies up to top of nearby bush or tree; in old cultivation forages right out in the open. Quick to retreat to cover if disturbed.

Evidently some seasonal movements in Zimbabwe, where some birds vacate ungrazed areas with tall grass growth.

Food. Insects, especially termites; regularly eats harvester termite workers (Hodotermitidae), which other robins seldom feed on. In 8 stomachs (Botswana, Transvaal) there were: termites (Hodotermitidae, Termitidae) in 100%, beetles (Coccinellidae, Curculionidae and Tenebrionidae) in 62%, ants (Ponerinae, Myrmicinae) in 62%, bugs (Pentatomidae), moths and caterpillars each in 50%, grasshoppers and mantids each in 37%, spiders in 12%. 4 stomachs contained berry seeds.

Breeding Habits. Monogamous, solitary nester.

NEST: open cup, neatly and compactly lined with rootlets and tendrils and sometimes animal hair; outer construction, often untidy, is of dried grass, leaves and sometimes fine twigs; ext diam. 100, cup diam. *c.* 55. Some nests (on ground) are no more than a roughly-shaped cup of dead leaves and twigs. Sited low down in thorny bush, often hidden by scant grass, although many nests not well concealed; sometimes on ground. 1 was built in discarded jam tin in garden. Height of 73 sites (4 on ground) av. 0.33 m, max. 1.52 m.

EGGS: 2–4, av. (n = 50 clutches) 2·3; laid at daily intervals. Ovate to elliptical ovate; glossy; white, rarely pale green, marked with spots and freckles of yellowish brown and russet-brown and some smears of greyish purple and lilac, often forming ring or cap at large end. SIZE: (n = 60) 18–23 × 14–16 (20·2 × 14·6).

LAYING DATES: Angola (breeding condition June–July); Namibia, Oct–Mar; Zimbabwe, Sept–Dec; South Africa: Transvaal, July–Jan; Orange Free State, Aug–Dec; N Cape, Feb, Apr, July, Sept, Nov–Dec. At E end of range in Orange Free State and Transvaal, laying peaks at onset of summer rains in Oct–Dec; in arid areas of N Cape, Botswana, Namibia and Angola, where rain is unpredictable, breeds after rainfall in any season.

INCUBATION: begins as soon as clutch completed. Period: (n = 1) 13 days.

DEVELOPMENT AND CARE OF YOUNG: young fed by ♂ and ♀. Nestling period (n = 1) 14 days. ♂ and ♀ feed fledglings; duration of post-fledging parental care unknown.

BREEDING SUCCESS/SURVIVAL: 1 record of parasitism by Diederik Cuckoo *Chrysococcyx caprius* (Transvaal).

Plate 27 (Opp. p. 416)

Cercotrichas coryphaeus (Lesson). **Karoo Scrub-Robin. Agrobate du Karoo.**

Sylvia coryphaeus Lesson, 1831. Traité Orn., p. 419; E Cape Province.

Range and Status. Endemic to southern Africa in SW Namibia (Great Namaqualand), Cape Province north to Orange R. and West Griqualand, east along S coast to Great Fish R. (but absent between Stilbaai and Gamtoos R.), and in NE to *c.* 28° E; S, W and central Orange Free State; lowlands of Lesotho. Common.

Description. *C. c. coryphaeus* (Lesson): Namibia, NW Cape Province east and south to Port Elizabeth, W Orange Free State. ADULT ♂: entire upperparts, sides of neck and ear-coverts dark greyish olive, slightly greyer, less olive on crown and nape; tail-feathers black, broadly tipped white on outer 4 pairs; narrow supercilium white (from bill to just behind eye); lores dark brown; narrow ring under eye white; malar area mottled brown and white (often forming a dark brown malar streak); chin and central throat-streak white; sides of throat grey; breast greyish olive-buff; rest of underparts and under-wing dull brownish buff, sometimes olive-tinged; remiges and wing-coverts brown, fringed slightly paler when fresh. Bill black; eye dark brown; feet black. Sexes alike. SIZE: wing, ♂ (n = 9) 73–78 (75·2), ♀ (n = 10) 69–79 (73·8); tail, ♂ (n = 9) 68–79 (74·8), ♀ (n = 10) 70–76 (72·1); bill, ♂ (n = 8) 16–19 (17·4), ♀ (n = 10) 16–19 (17·1); tarsus, ♂ (n = 9) 26–28 (26·9), ♀ (n = 10) 25–30 (26·8). WEIGHT: (n = 7) 19–22 (20·2).

IMMATURE: like adult except upperparts and underparts barred with dark brown.

NESTLING: skin pink, dark grey down or head, flanks and rump (other scrub-robins lack down feathers); inside of mouth yellowish orange, flanges cream-coloured, bill dark horn.

C. c. cinereus (Macdonald): SW Cape Province from Orange R. south to Bredasdorp. Upperparts and underparts slightly greyer, paler. SIZE: wing, ♂ (n = 6) 73–77 (74·3), 1 ♀ 69; tail, ♂ (n = 6) 70–76 (72·2), 1 ♀ 63; bill, ♂ (n = 6) 16–18 (17·0); 1 ♀ 16; tarsus, ♂ (n = 5) 26–28 (27·2), 1 ♀ 28.

Field Characters. A bird of karoo and other scrubby habitats in SW South Africa and S Namibia. More or less uniform dull grey-brown; viewed from in front, presents oval white throat-patch flanked by thin dark grey malar stripes; from behind, black tail with small white spots on outer 4 feathers; also has white supercilium and crescent below eye. Does not cock tail, unlike partly sympatric Kalahari Scrub-Robin *C. paena*; when excited, jerks tail from side to side and flicks wings rapidly. Kalahari is pale, with much rufous on upperparts and tail, and has different voice.

Voice. Tape-recorded (75, 88, GIB, LEM). Song a noisy, not very melodious, series of chirps and chattering, churring and grating notes interspersed with some whistles, reminiscent of reed warbler *Acrocephalus* or Tit-Babbler *Parisoma subcaeruleum*. Breeding ♂ includes imitations of other birds' calls in its song and breeding ♀ sometimes accompanies ♂ song with scolding. Call a sharp 'cheeup', frequently followed by 'switip-switip-tweety'.

General Habits. Widespread in arid shrub habitats in SW Africa, particularly in places where damp soil or shelter from wind allows shrubs at least 1 m high. Ubiquitous in waterless strandveld of coastal Namaqualand; totally independent of surface water. Frequents thorny growth along gullies, using tallest trees as songposts. Avoids mesic areas of winter rainfall region but common in dry fynbos; in E Cape extends into thornveld where open spaces between bush clumps have a light cover of low-growing scrub; occupies homestead gardens and farmyards.

Occurs in pairs. Perches on top of shrub or fence-post, then flies short distance and plunges into base of another shrub, or lands on ground and runs under and around shrubs (for up to 25 m if pursued). Spends much time near ground, where it often spots mongoose, cat or snake and loudly and persistently scolds it, hopping up and down excitedly on top of nearest vantage point. Runs and hops; hops while foraging undisturbed on small patches of bare ground, e.g. floor of dry gully.

Flies up to low vegetation to glean invertebrates from twigs and leaves or to pluck small fruits. Visits beaches and forages actively on piled seaweed wrack in intertidal zone.

Food. Insects, including ants, beetles (Chrysomelidae, Curculionidae, Elateridae, Scarabaeidae, Tenebrionidae), flies (Muscidae, Sepsidae), small parasitic wasps, small grasshoppers; also seeds (ingested with ants?) and fruits of *Lycium*.

Breeding Habits. Monogamous; territorial. Rarely, a helper at nest (SW Cape). Breeding territory small, defended by ♂ and ♀; forages widely outside defended area (Farkas 1988). Territorial encounters evoke excited staccatos of scolding, much wing- and tail-flicking, dancing up and down on shrubs and scurrying about on ground. At occupied nest often perches within arm's length of observer.

NEST: deep symmetrical cup, often on platform of dead twigs (which help to conceal it in twig trash); twigs too heavy for bird to carry are dragged to nest site; lined with dried rootlets, leaf fragments, hair (especially sheep's wool) or feathers; overall diam. 100–160, diam. of cup 60–75, cup depth 50. Nearly always sited on ground (n = 357), usually under shrub or dense herbaceous cover, sometimes in hollow under side of rock or in old tin can or other discarded artefact; rarely (n = 17) 30–100 cm above ground in centre of thorny bush, recess in bank or hole in tree. Built by ♀ in 5–20 days; ♂ feeds ♀ on occasion during building period. Replacement nests are common.

EGGS: 2–4, av. (n = 358) 2·8; 3-egg clutches are usual and comprise 76% of all clutches in winter rainfall area of SW Cape and in E Cape, 67% in Orange Free State and 62% in Karoo. Eggs laid on consecutive days. Ovate to elliptical ovate; greenish blue, variably marked with spots of brown, reddish brown and purplish grey, sometimes so thickly as almost to obscure ground colour; markings concentrated at large end, sometimes at equator. SIZE: (n = 106) 17–22 × 14–16 (20·0 × 14·8). Often rears 2 broods per season.

LAYING DATES: Namibia, Jan; South Africa: Orange Free State, Sept–Dec; Cape (Karoo), July–Jan, Mar, May; E Cape, Aug–Dec; SW Cape, July–Dec. Breeds in autumn in Karoo after good rains.

INCUBATION: by ♀ only; ♂ feeds her in and out of nest. Period: (n = 7) 13–15 (14·5) days.

DEVELOPMENT AND CARE OF YOUNG: ♀ broods chicks for first few days; both parents feed nestlings, with caterpillars and other insects, found at least 100 m from nest. Chick gives 'cheeup' call from *c*. day 10. Nestling period: (n = 4) 10–15 (12) days. Adults feed fledged young for at least 3–4 weeks.

BREEDING SUCCESS/SURVIVAL: nest failure common; pair in Orange Free State made 5 breeding attempts in 1 season. Diederik Cuckoo *Chrysococcyx caprius* parasitizes nests in E Cape (3 records).

Reference
Farkas, T. (1988).

486 TURDIDAE

Plate 27
(Opp. p. 416)

Cercotrichas podobe (Müller). **Black Scrub-Robin; Black Bush Robin. Merle podobe.**

Turdus podobe Müller, 1776. Syst. Nat., Suppl., p. 145; Senegal.

Cercotrichas podobe

Range and Status. Africa, SW Arabia (north to 160 km north of Mecca) and central Arabia (Riyadh). Accidental S Israel (1981: Eames 1986).

Resident or possibly partial migrant in subdesert steppe from Mauritania and N Senegambia to Red Sea coast of Sudan and Ethiopia (Eritrea); also Djibouti, adjacent E Ethiopia and N Somalia. Most northerly occurrences, Afar, 21°N, Mauritania, where resident, and Tamanrasset, Algeria, Feb. Frequent, Aïr (Niger). Frequent to very common; 22 birds in one small wadi (Yemen); sea level (Somalia coast) up to at least 1500 m.

Description. *C. p. podobe* (Müller): endemic to Africa. ADULT ♂: entirely greyish black, except for wings, rump, undertail-coverts and tail. Rump black; long, graduated tail black, all feathers except central ones with sharply-demarcated white end – white patch progressively larger from T2 to T6 where 15–19 deep; undertail-coverts black with broad white chevron tips (**A**). Remiges black, all except S1–2 and P1–2 with basal two-thirds of inner webs rufous; lesser and median coverts black or greyish black, tertials and all other upperwing-coverts blackish brown; alula and marginal coverts with small white tips which soon wear off. Underside of wing black. Worn plumage with brownish tinge throughout. Bill black; eye hazel or dark brown; legs and feet black or dark horn-brown. Sexes alike. SIZE: wing, ♂ (n = 15) 90–102 (93·1), ♀ (n = 15) 85–92 (88·7); tail, ♂ (n = 12) 104–119 (110), ♀ (n = 10) 97–105 (101), outermost rectrix 26–32 shorter than innermost; bill to skull, ♂ (n = 13) 19·0–21·3 (20·0), ♀ (n = 11) 17·8–20·1 (18·9), bill to feathers av. 4·5 shorter; tarsus, ♂ (n = 13) 28·1–30·4 (29·3), ♀ (n = 11) 26·7–29·4 (28·1). WEIGHT: (Niger, Chad) ♂ (n = 3) 24–27 (25·7), ♀ (n = 4) 24–26 (25·0).

IMMATURE: like adult but plumage fluffy, loose and much browner; white tips to rectrices smaller, only 10–12 deep on T6 and often lacking altogether on T2 and T3; undertail-coverts all black, or with very narrow white tips; rufous on inner webs of remiges more extensive than in adult, and tertials and greater coverts tinged rufous. Soft parts browner than in adult.

NESTLING: naked; skin black, mouth deep orange, gape-flanges whitish.

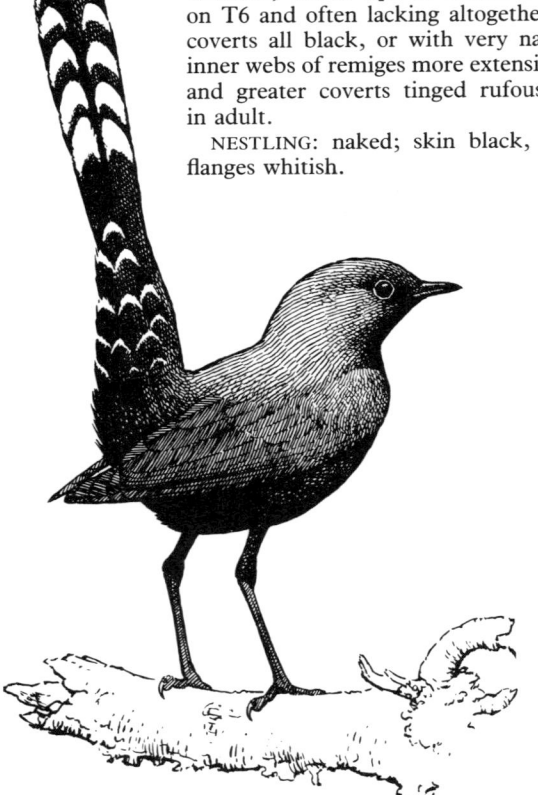

Field Characters. A conspicuous, readily-identified songster of sahelian and soudanian savannas, slightly built, sooty black with long, graduated, white-tipped tail which is steeply cocked and flirted, and bold white chevrons in undertail-coverts. In flight wings sometimes show rufous (inner webs of remiges). In, on and under bushes; looks leggy; song far-carrying, sings most of year.

Voice. Tape-recorded (53, MOR). Song, quite loud and far-carrying, of 2 types. (1) Unhurried delivery of variable phrase, beginning with sweet note, often with trill in middle, and usually ending with harsh or scratchy note, 'sheeea-tititi-sjra' or 'see-see-seeta-treea', 1st note and final notes each falling; phrase lasts 1–2 s and is followed by 1–2 s pause; song of 16 phrases delivered in 50 s. (2) Sustained babbling medley of sweet, fluty, rolled, squeaky and scratchy notes, without pauses, lasting at least 47 s, with characters of sylviid subsongs, and somewhat reminiscent of subsong of Olivaceous Warbler *Hippolais pallida*. Song (1) very like that of Rufous Scrub-Robin *C. galactotes*.

General Habits. Inhabits areas with scattered acacia bushes or spaced clumps of doum- and date-palms, tamarisks or *Salvadora* bushes. Occurs in hot, arid subdesert with widely-scattered low, leafless thorn shrubs but not trees, and also in well-timbered wadis, with bordering thickets. Particularly characteristic of wadis in Chad (Newby 1980). Often in hedges of living and cut thorn surrounding goat and cattle enclosures. Seldom perches in any tall trees, but keeps mainly on or within 2 m of ground, singing from top of shrub or other conspicuous perch and spending much time on ground under or near bushes.

A

In pairs all year round; territorial and rather aggressive to its own species. ♂ and ♀ often well apart, but in vocal contact. Active; forages mainly on ground, probing seemingly bare sand or soil under bushes with some leaf and woody litter. Also feeds in bushes. Often moves from ground up into bush, and back again. On ground cocks tail very high and proceeds with long hops; in shrub tail carried vertically, even forwards from perpendicular, or held down, in axis of body. Fans tail jerkily, and droops wing-tips when tail cocked. Not particularly shy, can be quite tame.

No good evidence of migration in W Africa, and thought to be strictly sedentary in N Chad (Newby 1980); but evidently a non-breeding visitor to Somalia, where all records are Nov–Mar (Ash and Miskell 1983). In Saudi Arabia and N Yemen breeds up to 1500 m, and several records up to 2400 m may refer to migrants (Brooks *et al.* 1987).

Food. Unknown.

Breeding Habits

NEST: a flat, rather ragged cup of dry grass leaves, palm fibres, scraps of cloth, twigs and rootlets, lined with hair and fine grass; sited *c.* 1–2 m high in young date palm (in between young vertical fronds), in *Zizyphus* bush, in crevice in tree trunk and within thick palm undergrowth; once in roof of shed and once in a grain silo.

EGGS: 2–4. Ground pale greenish or grey-white, covered with olive, blue-grey or red-brown speckles, often forming zone around large end. SIZE: (n = 3) 23–24·5 × 16·5.

LAYING DATES: Senegambia, Mar–July; Mali, Feb–Sept; Nigeria/Chad (L. Chad), Apr–June; Sudan (Setit), May; Ethiopia, May, July–Aug(?).

References
Archer, G. and Godman, E. M. (1961).
Bannerman, D. A. (1936).
Cramp, S. (1988).
Heim de Balsac, H. and Mayaud, N. (1951).

Genus *Namibornis* Bradfield

Single species of chat- or flycatcher-like bird endemic to arid SW Africa. Brown above with reddish tail, white below with streaky breast; plumage and proportions recall *Cercomela* chats and *Bradornis* flycatchers. Resembles chats more than flycatchers in habits, adult call notes, and fledglings, but flycatchers more than chats in nest construction and site, eggs and begging call. Song a complex warble, rather robin-like. Unpublished studies of anatomy and feather structure suggest closer affinity with chats than flycatchers (see Jensen and Jensen 1971).

Namibornis herero (de Schauensee). Herero Chat. Traquet du Herero.

Bradornis herero de Schauensee, 1931. Proc. Acad. N. Sci. Phil. 83, p. 449; Karibib, Damaraland, SW Africa.

Plate 27
(Opp. p. 416)

Range and Status. Endemic, confined to escarpment zone hills and mountains from extreme SW Angola (Iona Peak, S Mossamedes Province, 16° 54′S, 12° 34′E) to central Namibia (Naukluft Mts). Localities include Gamsberg, Spitzkop, Uis and mountains of the Kaokoveld. Rare to locally uncommon; apparently not threatened despite restricted distribution (Collar and Stuart 1985).

Description. ADULT ♂: top of head to mantle grey with brownish wash, shading to pinkish brown on back and tawny on rump and uppertail-coverts; forehead to hindneck with distinct dark streaks, mantle and back more faintly streaked. Tail-feathers blackish brown edged tawny, proportion of tawny on each feather increasing outwardly until outer pair entirely pale rufous except for dark shaft streak and basal half of inner web. Supraloral line and supercilium white; dark brown mask from base of bill and lores through and below eye onto ear-coverts; very faint pale eye-ring. Chin and upper throat pure white, unstreaked; lower throat to belly silky white with dark streaks, with brownish wash on breast and stronger, pinkish brown wash on flanks; undertail-coverts white, unstreaked. Flight feathers dark brown, inner primaries narrowly edged buff, secondaries edged tawny and tipped pale buff, tertials

488 TURDIDAE

with broader buff outer edges and tips, scapulars tawny brown with dark centres, underwing-coverts dark brown broadly edged and tipped tawny, axillaries and upperwing-coverts white to pale buff, under primary coverts with dark mottling. Bill black; eye brown to dark grey-brown; legs and feet black. Sexes alike. SIZE: wing, ♂ (n = 7) 90–95 (91·9), ♀ (n = 8) 87–92 (89·9); tail, ♂ (n = 4) 68–70 (68·6), ♀ (n = 3) 63–68 (65·7); bill, ♂ (n = 4) 18–18·5 (18·3), ♀ (n = 3) 19–20 (19·5); tarsus, ♂ (n = 4) 22–22·5 (22·1), ♀ (n = 3) 23–24 (23·3). WEIGHT: ♀ (n = 2) 26·5, 28.

IMMATURE: above, mottled dark brown, rusty and buff, tail redder and shorter than adult, face mask well-defined; underparts whitish, chin to upper belly mottled blackish brown (Maclean 1985).

NESTLING: unfeathered young not described; fairly well-feathered nestling had yellow gape, orange inside of mouth, black bill, dusky face wash already apparent.

Field Characters. An elusive bird of the escarpment country and adjacent semi-desert hills. Most noticeable feature is dark face mask contrasting with white supercilium and throat. Streaked underparts also distinctive, although not as easy to see except at fairly close range. From behind could be mistaken for Familiar Chat *Cercomela familiaris*, which also has brown upperparts, rufous rump and black-centred rufous tail, but Familiar Chat lacks face mask and eye-stripe, has uniform grey-brown, unstreaked underparts, and inverted 'T' tail pattern.

Voice. Tape-recorded (GIB). Song 'a set of beautiful, mellow, warbled and trilled, jumbled short phrases, distinctly robin-like and very similar to some song phrases of the Damara Rockjumper *Achaetops pycnopygius*' (Jensen and Jensen 1971). 'Breeding contact call' of 3 notes, given during breeding season, is 'a rather subdued, though quite penetrating, trilled "ji-ju-juu"' (Jensen and Jensen 1971). Alarm, a long, subdued but harsh 'churrrr'. Begging call of nestling a loud, slightly trilled 'tsrrp'.

General Habits. Inhabits rocky hills, mountains and inselbergs with some vegetation cover, on slopes, in ravines and dry watercourses, and on level ground around their bases. Not on cliffs or bare slopes; some vegetation always required, typically mixed *Acacia–Commiphora* associations with low trees (3–4 m high) thinly scattered over rock-strewn slopes. Ground cover absent except for short periods during rains; av. rainfall 75–250 mm/year. In Angola occurs in *Colophospermum–Commiphora* savanna on rocky mountain slopes (Dean *et al*. 1988).

Mainly in pairs, but in dry season in groups of up to 5 (last brood?). Hunts from low perch in bush or tree, dropping down to pick food from ground. Flies low, and flicks wings on alighting, like Familiar Chat. Rather shy. ♀ with enlarged ovary was moulting wing- and tail-feathers (Hoesch and Niethammer 1940).

Mainly sedentary, but some local movement west into Namib Desert after good rains.

Food. Mainly insects and their larvae, including ants; also berries (e.g. *Commiphora saxicola*) and picnic scraps at campsites.

Breeding Habits. Territorial. In probable courtship display, bird perched in tree and made 'wallowing' movements as if shaping nest; second bird hopped onto same spot as first flew off, repeated performance. First bird then adopted food-begging posture, fluttering wings but remaining silent, second bird fed it an insect (Jensen and Jensen 1971).

NEST: bulky but otherwise flycatcher-like cup of dry, fine grass, vegetable fibres and a few rootlets and thin plant stems, rather crudely bound together (and connected to supporting twigs) with cobwebs and fine silky seed fibres (**A**); lined with similar fine materials and some animal hair; 1 nest ext. diam. 80, int. diam. 55, ext. depth 60, int. depth 35; placed 1·3–4 m above ground in outer twigs of low branch, in multiple fork or (once) in clump of *Loranthus*.

EGGS: 1–4, usually 2. Pale greenish white with rather sparse covering of discrete small dark red-brown speckles, especially at broad end. SIZE: 21–23·2 × 15·1–16·6 (22·5 × 16·1). Double-brooded in good seasons; when conditions are right, probably triple-brooded.

LAYING DATES: Namibia, Feb–May.

DEVELOPMENT AND CARE OF YOUNG: young fed by both parents. Single nestling shaded by parents standing or crouching over it for periods of 15 min between feedings during hottest 6 h of day.

References
Hoesch, W. and Niethammer, G. (1940).
Jensen, R. A. C. and Jensen, M. K. (1971).

A

Genus *Phoenicurus* Forster

Small, slim bodied Old World thrushes. Bill slender and pointed, nostrils in front of frontal feathering; rictal bristles distinct. Legs longish and slender. Tail medium length, slightly rounded; wings fairly long, tip rounded or bluntly pointed. Plumage soft. Sexes distinct. ♂♂ strikingly patterned black, grey, white and chestnut, ♀♀ brownish; rump and tail mainly bright chestnut in both sexes. Perch in trees and bushes and on rocks; feed mainly by sallying to ground; nest in holes and crevices.

11 species, mostly Asian; 3 in Africa, of which 1 endemic, 2 Palearctic migrants which also breed in NW Africa.

Phoenicurus ochruros (Gmelin). Black Redstart. Rougequeue noir.

Motacilla Ochruros Gmelin, 1774. Reise d. Russeland VII, p. 101; N Iran.

Plate 28 (Opp. p. 417)

Range and Status. Breeds Europe north to c. 55°N, Africa, and from Near East and Caucasus through N Iran and Afghanistan to mountains of central and S Asia; winters S Europe, Africa and S Asia.

Resident and Palearctic winter visitor. Uncommon local breeder Morocco (Rif, Middle Atlas, High Atlas) and N Algeria (Djudjura Mts). Winters Morocco, N Algeria and N Tunisia south to N borders of Sahara; coastal Libya (mainly in W); Egypt (Mediterranean coast, Nile delta and valley south to Luxor, and Suez corridor); NE Sudan (Red Sea Hills), N Ethiopia (Eritrea), Djibouti and N Somalia (south to c. 9°N); mainly frequent to common. Also coastal Mauritania (Nouadhibou-Nouakchott), Saharan oases of Algeria (Ahaggar), Libya (Hon, Serir) and Egypt (Siwa, Bahariya, Dhakla); central Sudan (near Khartoum) and central Ethiopia (Awash valley), rare to uncommon. Vagrant S Mali (Bamako) and Chad (north of Kanem).

Description. *P. o. gibraltariensis* (Gmelin): breeds Europe and NW Africa; winters W and S Europe, N Africa and Middle East. ADULT ♂: top of head, mantle and scapulars dark grey; face and chin to chest black; rump and uppertail-coverts bright chestnut. Central tail-feathers blackish brown, rest bright chestnut. Flanks grey; belly greyish white; undertail-coverts orange-buff. Remiges and primary and secondary coverts blackish brown, outer webs of secondaries and especially tertials broadly edged white, forming conspicuous wing panel; median and lesser coverts blackish; underwing-coverts and axillaries blackish tipped grey. Bill black; eye blackish brown; legs and feet black. ADULT ♀: upperparts dark greyish brown, sides of head, chin to breast and flanks greyish brown, belly brownish white; wing-feather edgings browner than ♂ and wing panel absent. Rump, uppertail-coverts and tail like ♂. SIZE: (10 ♂♂, 10 ♀♀, Europe) wing, ♂ 84–89 (86·3), ♀ 80–85 (82·5); tail, ♂ 63–70 (65·4), ♀ 59–66 (62·0); bill, ♂ 14–16 (15·0), ♀ 14–16 (15·0); tarsus, ♂ 23–24 (23·7), ♀ 22–24 (23·3). WEIGHT: unsexed (n = 50, Malta, winter) 13–20 (16·5).

Breeding plumage acquired by abrasion; complete moult on breeding grounds Aug–Sept. Adult ♂ in fresh plumage has black face and breast partly obscured by grey feather-tips. Wing-tip rounded; P6 and P7 longest; P5 2–5 mm shorter; P8 1–2 shorter, P9 9–13, usually between P3 and P4; P10 very small; P5–P8 with emarginated outer webs.

IMMATURE: juvenile dull brown above with faint darker mottling, rump and uppertail-coverts chestnut. Chin to breast and flanks brown, with dark feather-tips; belly white. Tail- and wing-feathers as adult ♀. First winter and first summer ♂ and ♀ like adult ♀, but upperparts rather browner, some first summer ♂♂ with throat and breast blacker.

NESTLING: down dark grey; mouth rich yellow, gape-flanges ivory.

P. o. phoenicuroides (Horsfield and Moore): breeds mountains of central Asia; winters S and SW Asia, N Somalia, Djibouti, N Ethiopia, NE Sudan and Egypt (once Giza). ♂ has lower breast to undertail-coverts, axillaries and underwing-coverts chestnut; often a whitish band on forecrown; no white wing panel. ♀ paler brown, less greyish than *gibraltariensis*, especially below. Rather smaller: wing, ♂ (n = 10) 81–86 (83·3), ♀ (n = 10) 76–83 (78·8), WEIGHT: unsexed (n = 11, NE Sudan, Nov) 12–15 (13·3).

Field Characters. ♂ of N African race distinctive, with dark grey upperparts, black face and breast, grey flanks and belly, white patch on wing and chestnut rump and tail. ♂ of race *phoenicuroides* (Sudan–Somalia) with chestnut underparts and pale band on forecrown, resembles Common Redstart *P. phoenicurus*, but white on forehead less extensive, black of throat extends to breast, belly chestnut, not whitish. Lacks white patch on wing whereas *P. p. samamisicus* (visiting Sudan–Ethiopia) does have white wing-patch; ♀ and immature (both races) darker and greyer than Common Redstart, especially on underparts. In the hand, wing rounder than Common Redstart, with P9 shorter (**A**). Voice and habitat different.

A

Phoenicurus ochruros

Phoenicurus phoenicurus

Voice. Tape-recorded (CHA – 62, 73, 76, 89). Song consists of short high-pitched phrases (*c.* 1–2 s) separated by pauses of a few seconds; two types, one a jinging 'jirr-jirr-ji-ji-ji-ji', the other a harsh scratchy 'trchch-titutili', sometimes connected to form a longer (3–4 s) song-phrase; contact-alarm calls, 'tsip', 'ticc-ticc' or 'tsip-ticc-ticc'.

General Habits. *P. o. gibraltariensis* breeds on rocky hills and mountain slopes up to the snowline; in Morocco at 2000–3500 m, in Algeria at *c.* 1700 m; winters from sea level up to 2500 m on rough stony hillsides with scattered trees and bushes, edges of cultivation, rocky deserts, towns, villages and ruins. *P. o. phoenicuroides* winters mainly on rocky slopes with scattered trees and in woodland glades above 1300 m.

Solitary, sometimes in twos and threes; rather unobtrusive. Markedly more terrestrial than Common Redstart. Perches upright on rocks and buildings, sometimes on low bushes and tree boughs; quivers tail. Feeds mainly on ground, moving with a brisk hop; also makes quick runs, pausing in alert posture. May dig 2–4 cm into ground to obtain larvae. Also drops onto prey from perch, catches insects in flight and hovers to take items from walls and foliage. Calls less frequently than Common Redstart. Sings occasionally in autumn and winter. Apparent courtship and incipient nest building recorded in Egyptian winter quarters (Moreau and Moreau 1928).

P. o. gibraltariensis winters mainly in Mediterranean basin. In Europe, birds breeding west of *c.* 13°E migrate southwest in autumn, those east of 13° migrate southeast; S European birds apparently sedentary. Morocco mountain populations descend in January to 2000 m or below. In N Africa winters mainly in Morocco, W Algeria and Egypt; main arrival Oct, departure Mar to early Apr. Birds recovered Morocco (13 ringed France, 10 Belgium, 2 Netherlands, 16 W Germany and 2 E Germany); Algeria (4 ringed France, 7 Belgium, 1 Netherlands, 20 W Germany, 4 E Germany and 1 Austria); and Egypt (1 ringed W Germany); most birds wintering Egypt presumably from central and E Europe. *P. o. phoenicuroides* winters SW Asia and NE Africa in S Red Sea and Gulf of Aden hinterland. Arrival in Sudan late Oct to Nov; present Somalia Oct–Mar.

Food. Mainly insects: beetles, ants, bugs (Hemiptera), grasshoppers, adult and larval Lepidoptera, flies (Diptera); also spiders, millipedes, earthworms, small crustaceans, and occasionally fruit.

Breeding Habits. Usually monogamous; ♂ sometimes with 2 ♀♀ (Cramp 1988). Solitary and territorial; territory defended by ♂ alone. ♂ sings from high rock or exposed perch, sometimes in flight, from nest building to hatching of second brood, though not while feeding first brood in nest; mainly in morning and evening; may have more than one regular song post in territory (Cramp 1988). In courtship, ♂ chases ♀, singing constantly; ♀ settles and ♂ flies in front of her, singing and twisting from side to side.

NEST: loosely constructed cup of dry grasses, moss and fibre, lined with hair and feathers; ext. diam. 120–150, int. diam. 60, depth of cup 40–50; in fissure or hole in rocks; in Europe usually on ledge or in hole in building. Built by ♀; site usually selected by ♀, sometimes assisted by ♂.

EGGS: 4–6. Glossy; white, rarely tinged blue. SIZE: (Europe) 17–22 × 13–16 (19·4 × 14·4); calculated weight 2·2. 2 broods usual in Europe.

LAYING DATES: Morocco and Algeria, late Apr–June.
INCUBATION: by ♀ alone, beginning with last egg. Period: 12–13 days.

DEVELOPMENT AND CARE OF YOUNG: young brooded by ♀ only; mainly fed in nest by ♀ though ♂ may pass food to ♀ to feed to young. Leave nest 12–17 days after hatching, and seek safety among rocks or vegetation; sometimes not fully able to fly; may return to nest to roost for first few days. ♂ may take brood away while ♀ incubates next clutch. Young may be independent by 11 days after fledging, or family may stay together for 3–4 weeks (Cramp 1988).

Reference
Cramp, S. (1988).

Phoenicurus phoenicurus (Linnaeus). Common Redstart. Rougequeue à front blanc.

Motacilla phoenicurus Linnaeus, 1758. Syst. Nat. (10th ed.), p. 187; Sweden.

Plate 28
(Opp. p. 417)

Range and Status. Breeds Europe and Africa, and through Siberia east to L. Baikal; also Asia Minor and Caucasus to Iran; winters Africa, SW Europe and S Arabia.

Resident and Palearctic winter visitor. Uncommon and local breeder in N Morocco (Rif and Middle Atlas); NE Algeria (Djebel Babor, Dj. Chelia, Chrea and Forest of Méridja); formerly bred N Tunisia (Cap Bon) but no recent evidence. Winters north of Sahara, regular Jan–Feb N Morocco and N Algeria, uncommon; Tunisia, Libya and Egypt, rare. Winters south of Sahara in a belt from SW Mauritania (delta region), Senegambia, N Guinea, S Mali (south of *c.* 15°N), N Ivory Coast, Burkina Faso and N Ghana (open wooded savannas south to about Tamale) east through N and central Nigeria, S Niger, N Cameroon (south to Touroua), S Chad, NE Central African Republic and S Sudan (south of *c.* 12°N) to W, N and central Ethiopia; also south to NE Zaïre, E Rwanda, Uganda, extreme NW Tanzania and W Kenya (west of 37°E); generally common to abundant locally in soudan and guinea savanna belts and in N and E Uganda; a few overwinter in Sahel; uncommon Rwanda, SW Uganda and W Kenya. Vagrant Liberia (north of forest area), S Ivory Coast (near Abidjan), S Ghana (Shai Hills), S Nigeria (Ibadan, Bututu), S Somalia (Jubba valley), NE Kenya (Moyale) and S Kenya (Kajiado, Ngulia, Voi); also South Africa (Transvaal: 1 Potchefstroom, 1 Vaal Reefs). On passage, both seasons in Sahel, in N Africa, on Red Sea coast, and in saharan oases, common to abundant; Djibouti and NW Somalia, uncommon. Population declines noted during recent cycle of sahelian droughts.

Description. *P. p. phoenicurus* (Linnaeus): breeds Europe, Siberia and NW Africa; winters throughout African range. ADULT ♂: broad band from forehead to above eye white; crown to back and scapulars French grey; rump and uppertail-coverts bright chestnut. Central tail-feathers blackish brown with bases and most of outer webs chestnut, rest chestnut. Lores, chin, ear-coverts, throat and sides of neck black; breast and flanks bright orange-chestnut; belly white; undertail-coverts rufous-buff. Remiges and wing-coverts brown, greater, median and lesser coverts with grey edging. Underwing-coverts and axillaries chestnut. Bill black; eye dark brown; legs and feet black. ADULT ♀: forehead to mantle and scapulars brown; sides of head buff-brown; chin and whole throat pale buff, mottled brown; breast, flanks and underwing-coverts orange-buff. Rump, uppertail-coverts and tail like ♂. SIZE: (10 ♂♂, 10 ♀♀, Europe/USSR) wing, ♂ 78–81 (79·1), ♀ 73–79 (76·3); tail, ♂ 56–60 (57·7), ♀ 54–59 (57·2); bill, ♂ 14–16 (15·1), ♀ 14–16 (15·1); tarsus, ♂ 22–24 (22·8), ♀ 21–23 (21·9). WEIGHT: unsexed (n = 60, Egypt, Aug–Oct) 11·5–20·7 (16·9); (n = 38, N Nigeria, autumn) 11·7–16 (13·8); (n = 93, Ethiopia, Oct–Feb) 11–19·1 (13·7); (n = 50, Ethiopia, Mar–May) 10·6–23·3 (15·3); (n = 14, central Nigeria, spring) 12–20 (15·7); (n = 28, N Nigeria, spring) 12·6–21·9 (14·7); (n = 110, Morocco, Mar–May) 10·6–17·1 (13·1).

Breeding plumage acquired by abrasion. Complete annual moult on breeding grounds, July–Aug. Freshly moulted birds have tertials and wing-coverts edged warm brown. Adult ♂♂ Oct–Dec have white forehead and supercilium largely obscured by grey feather-tips; black throat with whitish tips; mantle feathers and scapulars tipped brown, tertials and greater coverts with broad grey edgings. Wing-tip bluntly pointed; P7 and P8 longest; P6 1–3 mm shorter, P5 5–10; P9 5–10 shorter, between P4 and P6; P10 very small; P6–P8 emarginated on outer webs.

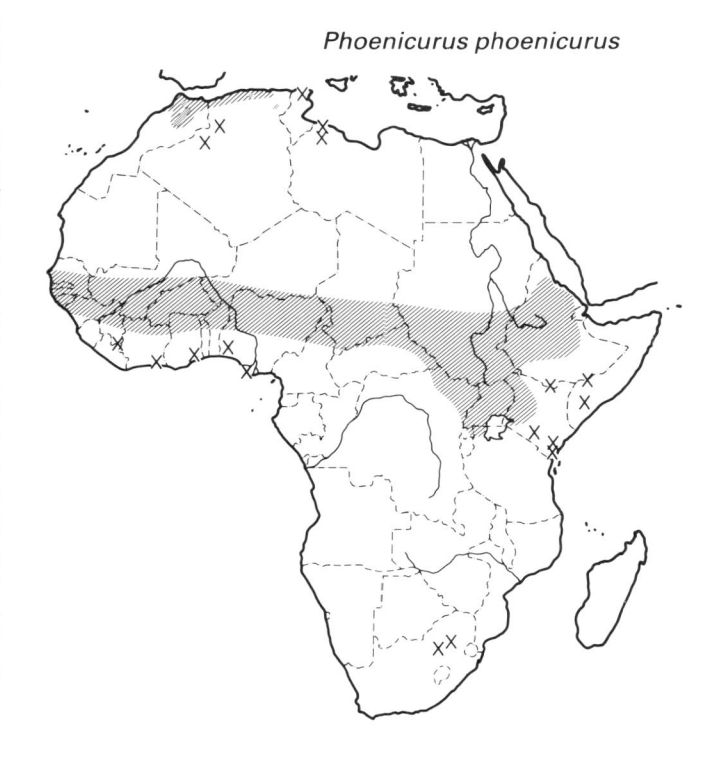

IMMATURE: juvenile spotted blackish and buff above; buffish white below with dark brown feather-tips. Wing- and tail-feathers like adult ♀, but median and lesser coverts with buff tips. First winter ♂ like adult ♂ but with broader brown feather-tips which largely obscure head pattern, and tertials and greater coverts edged browner like adult ♀. First summer ♂ resembles breeding adult ♂ but retains some brown feather-tips above and white feather-tips on throat. First winter and summer ♀ like adult.

NESTLING: down dark grey, fairly long; mouth orange, gape-flanges yellow.

P. p. samamisicus (Hablizl): breeds Asia Minor and Caucasus to Iran; winters Arabia and Ethiopia to S Sudan; on passage Egypt and N Sudan; vagrant Algeria. Adult ♂ has outer webs of primaries narrowly and of secondaries broadly edged white to produce conspicuous wing panel; ♀ and immature paler, greyer above than nominate.

Field Characters. ♂, with grey upperparts, black face and throat, conspicuous white forehead and bright chestnut breast and tail, can only be confused with race *phoenicuroides* of Black Redstart *P. ochruros*, q.v. ♀ browner and paler than ♀ Black Redstart, especially on underparts. In hand, wing more pointed than Black Redstart, P9 longer (see **A**, p. 490). For differences from Moussier's Redstart *P. moussieri*, see that species.

Voice. Tape-recorded (53, C, CHA, MOR – B). Song sweet and melancholy; consists of 2-part phrases of c. 2 s duration separated by pauses of several seconds: a pure toned 'ji-gju-gju-gju . . .' followed by a variable squeaky or mechanical warble including hard strangled sounds, in which other species may be mimicked, and ending in a flourish. Contact-alarm notes, plaintive 'hweet' and liquid 'twick' or 'hwee-tick-tick'.

General Habits. Breeds in old oak and cedar forest at medium to high altitudes (1400–2200 m in Middle Atlas). Usually winters mainly in semi-arid habitats, thickets in thorn steppe, dry open woodland, riverine acacia and gardens, up to 2000 m. In N Ethiopia, any woodland above 600 m, nominate race generally at higher altitudes than *samamisicus*. Often in more open scrubby areas on passage.

Usually solitary; sometimes in parties or local concentrations on migration. Active and restless, calling frequently and constantly shivering tail; rather shy. Forages from bushes or lower branches of trees, flying out to catch prey on ground, usually returning to eat it; makes short flights to catch insects in air, and picks items from trunks, branches and leaves, sometimes hovering near foliage (Cramp 1988). Typical stance on open perch half upright, more horizontal in foliage. Moves on ground with more upright stance and quick, light hop. Flight buoyant and agile with slight undulation and sweep up to perch. On wintering grounds may be aggressive to other Palearctic migrants (Cramp 1988). Sings occasionally in winter in Africa.

Birds from W, central and N (including Baltic) Europe move southwest towards Iberia, and enter Africa through Morocco and Algeria. A few remain throughout winter in W Mediterranean basin, but the great majority cross Sahara. Passage occurs along whole N African coast mid-Sept to Oct, Libyan and Egyptian migrants presumably coming from USSR. Saharan crossing apparently on a broad front, with birds common in oases. South of Sahara, main passage occurs Senegal from mid-Sept, Chad late Sept to Oct and N Sudan late Sept to Oct (though *samamisicus* passes mainly in Sept). Arrives main wintering areas Oct, reaching Uganda by Nov; departs again late Mar to early Apr. Passage rather more conspicuous in spring than autumn. Loose concentrations of hundreds occur Mali Mar–Apr, and prominent movement of both nominate race and *samamisicus* N Ethiopia (Eritrea) Apr. Strong passage N Africa, mainly mid-Mar to mid-May, with ♂♂ passing earlier than ♀♀. Some birds return to Europe via a more easterly route, through Tunisia and Libya. Birds recovered Morocco (many autumn and spring; also 7 in Jan–Feb and 4 in June–July; ringed W, central and N Europe); Algeria (several autumn and spring; also 6 in Jan–Feb; ringed W, central and N Europe); Tunisia (several, spring, ringed France, Belgium, Netherlands, W Germany, E Germany, Sweden and Finland); Libya (6, spring, ringed Britain, France, Netherlands, W Germany, Sweden and Poland); Mauritania (1, Oct, ringed France; 1, Nov, ringed Finland); Senegal (1, Jan, ringed Britain); Chad (1, Nov, ringed W Germany); Zaïre (1, Mar, ringed Finland). Birds ringed Tunisia (May) recovered France, Italy, Hungary, Poland and USSR (c. 57°N, 41°E); ringed Niger (May) recovered Switzerland (1); ringed Nigeria (Oct) recovered Algeria (1); and ringed Uganda (Mar) recovered Iraq (1, Apr). Nominate birds wintering NE Africa are probably all from USSR.

Food. Insects, including grasshoppers, beetles, flies, ants and bees; millipedes, molluscs and earthworms; also occasionally fruit.

Breeding Habits. Mainly monogamous, but occasionally 1 ♂ with 2 ♀♀ (Cramp 1988); solitary and territorial. Territory size in Europe 0·1–1 ha; defended by ♂ who shows aggression to intruding ♂♂. ♀ arrives several days after ♂. ♂ sings from high, exposed perches, from a few days after arrival to hatching of last brood; also has circling song-flight a few m above ground. Displaying ♂ adopts greeting posture in front of ♀, crouching slightly with wings spread and tail fanned; perches near ♀ and beats wings deeply (Cramp 1988).

NEST: cup of grasses, roots and moss, lined with hair and feathers; usually in hole in tree, stump or wall. Built by ♀, although ♂ usually selects site.

EGGS: 5–7; NW Africa mean (7 clutches) 5·9. Pale blue, usually unmarked, occasionally with fine dark red-brown speckles. SIZE: (n = 250) 16·6–21·5 × 12·3–15·2 (18·7 × 13·8), calculated weight 1·9. Frequently double-brooded in Europe.

LAYING DATES: Morocco, Algeria and Tunisia, May–June.

INCUBATION: mainly by ♀. Period: 12–14 days, beginning with last egg.

DEVELOPMENT AND CARE OF YOUNG: young brooded by ♀ only; fed in nest by both parents though ♀ takes larger share; leave nest 12–15 days after hatching; remain in care of parents 10–14 days after fledging.

BREEDING SUCCESS/SURVIVAL: no information from N Africa. Of 479 eggs in 76 nests (Finland), 81·2% hatched, 92·3% of these fledged (Cramp 1988).

Reference
Cramp, S. (1988).

Phoenicurus moussieri (Olph-Galliard). **Moussier's Redstart. Rubiette de Moussier.**

Erithacus Moussieri Olph-Galliard, 1852. Ann. Soc. Agric. Lyon 4, p. 101; Algeria.

Plate 28
(Opp. p. 417)

Range and Status. Endemic to NW Africa. Resident, dispersive and perhaps migratory over relatively short distances. Breeds commonly at middle to upper altitudes and on high summits: in Morocco, in the Rif, Middle Atlas and High Atlas (reaching the coast near Tamri), and in small numbers in plains areas of the Atlantic seaboard, around the mouth of the Tensift, Essaouira, Agadir and Massa; in N Algeria, in the Atlas Tellien, the Aurès Mts and the Saharan Atlas around Bou Saada, Djelfa, Aflou, El Bayadh and Ksours; in N Tunisia, in higher mountain regions north to Djebel Ichkeul but avoiding the coast; in NW Libya suspected at Jebel Nefoussa. In non-breeding season many move to lower altitudes. In Morocco and Algeria, reaches Mediterranean coast and fringes of Sahara south to Goulimine, Ksar-es-Souk, Béni Abbès and Touggourt; in Tunisia, coastal areas from Cap Bon to Gabès and deserts south to Nefta. In Libya, fairly common in NW from Sabratha east to Wadi Kaam and in Jebel Nefoussa; recorded once Nov Benghazi. Vagrant Italy, Malta, Britain.

Description. ADULT ♂: forehead, lores and ear-coverts black; broad white band across forecrown and above eye extending back to join white patches on sides of upper mantle; rest of crown to hindneck, mantle and scapulars black; back black, feathers tipped buffish; rump and uppertail-coverts bright orange-chestnut. Central tail-feathers dark brown; rest orange-chestnut with outer webs edged dark brown distally. Entire underparts orange-chestnut, washed with buff on belly. Remiges and upperwing-coverts black, white basal areas to outer webs of inner primaries and secondaries forming large prominent wing-patch; underwing-coverts and axillaries bright chestnut. Bill black; eye dark brown; legs and feet black. ADULT ♀: forehead to back and scapulars brown; sides of head buffish brown; entire underparts orange-buff, richest on breast and washed with brown on sides of breast and throat. Wings blackish brown without white patch; upperwing-coverts tipped and secondaries broadly edged buff; underwing-coverts and axillaries pale orange-buff. Rump, uppertail-coverts and tail like ♂.
SIZE: (10 ♂♂, 10 ♀♀) wing, ♂ 65–71 (67·4), ♀ 63–69 (64·8); tail, ♂ 48–53 (50·2), ♀ 46–53 (49·7); bill, ♂ 14–16 (14·9), ♀ 14–15 (14·9); tarsus, ♂ 24–26 (24·7), ♀ 23–26 (24·3). WEIGHT: Algeria (Nov) ♂ (n = 2) 14·5, 15, ♀ (n = 2) 15, 15.
Breeding plumage acquired by abrasion. Complete moult after breeding late June–Aug. Adult ♂ in autumn has black upperparts obscured by grey-brown feather-tips, and white head-band mottled grey. Wing-tip rounded; P6 and P7 longest; P5 0·5–2 mm shorter, P4 3–5; P8 0·5–2 shorter; P9 6–9, usually shorter than P3; P10 small, c. 35% P9; P5–P8 emarginated on outer webs.
IMMATURE: juvenile dark brown above, mottled warm buff, buff below with narrow dark scaling, especially on breast. Rump, uppertail-coverts and tail like adult. Wing like adult but browner and with broad warm buff fringes to tertials and greater coverts; ♂ with white wing-patch. First winter ♂ like fresh-plumaged adult but supercilium less broad, and tertials and greater coverts with buff fringes.
NESTLING: unknown.

Field Characters. A small, rather plump redstart with short wings and tail. ♂ in breeding plumage distinctive, with black head, upperparts and wings, sharply contrasting broad white stripe around crown, large white wing-patch, and bright orange-rufous underparts and tail. In non-breeding plumage, feathers have pale fringes, giving greyish mottled effect, but white marks still present. ♀ like ♀ Common Redstart *P. phoenicurus* but more compact, with rufous tinge above and below; immature like Common but more rufous, spots and scales less distinct, ♂ with some white in wing. Longer, slimmer shape of Common also useful.

Voice. Tape-recorded (60, 73). Song a thin reedy warbling composed of 2–6 s phrases. Each phrase 8–10 repetitions of a buzzy motif, rendered 'derdzu-derdzu-derdzu . . .' (P. A. D. Hollom *in* Cramp 1988); contact-alarm note, a high-pitched 'eep, eep', often followed by a rasping 'tr-rr-rr'.

General Habits. In east of breeding range inhabits dry, grassy, stony or rocky slopes with scrub and low bushes, and old or degraded forests with scattered pines, oaks and cedars, mainly at 1500–2300 m. In Atlas region occurs at forest base and on stony summits and denuded plateaux; in High Atlas, up to 3000 m among bushes and xerophytic plants. In winter, moves to lower, flatter country, occurring in scrub and bushes along wadis and in oases, and in *Zizyphus* scrub on plains.
Usually solitary in non-breeding season. Confiding and at times inquisitive. ♀ more shy than ♂. Perches on low bushes or tree branches, flying to catch prey from ground; sometimes digs with bill into ground for food. Occasionally pursues flying insects. Shivers tail con-

stantly like other redstarts, but flicks wings like a *Saxicola* chat (Cramp 1988). Flight low and rapid with fluttering wing-beats.

Movement out of higher breeding areas occurs Oct to avoid snow, ♂♂ departing before ♀♀; birds return late Feb to early Apr. Extent of movement within general breeding range not well known: in NE Morocco large numbers move into valley of Moulouya in winter; in Algeria movements in winter into semi-desert from around Monts des Ksours and Atlas Saharien; and in Tunisia from higher mountains in N to coastal regions and deserts to the S. One nestling ringed Morocco (near Tetouan) recovered following Dec c. 12 km to northwest.

Food. Insects, mainly ants, beetles, grasshoppers and larvae; also some plant material.

Breeding Habits. Monogamous; solitary and territorial. Song delivered by ♂ from tree, bush, rock or ground. ♂ gives soft thin song when courting and chasing ♀.

NEST: cup of dry coarse grass, lined with feathers and hair, usually on ground protected by bush or tuft of vegetation; sometimes in hole in wall or trunk of old oak, or under thatched roof.
EGGS: 3–6 (usually 4–5); mean (143 clutches Algeria and Tunisia) 4·2. Glossy; pale blue or white, unmarked. SIZE: (n = 83) 17–20 × 13–16 (18·2 × 14·0). 1 brood, possibly 2.
LAYING DATES: Algeria and Tunisia, Apr to mid-June; Moroccan Sahara from mid-Mar.
DEVELOPMENT AND CARE OF YOUNG: young fed by both sexes, remain with parents after leaving nest.
BREEDING SUCCESS/SURVIVAL: sometimes parasitized by European Cuckoo *Cuculus canorus* (Cramp 1988).

References
Cramp, S. (1988).
Heim de Balzac, H. and Mayaud, N. (1962).

Genus *Saxicola* Bechstein

Small to medium-sized, compact chats. Bill short and strong, rather broad at base; rictal and nasal bristles well developed. Legs longish, rather slender. Tail shortish and square; wing-tips rounded or bluntly pointed. Sexes distinct: ♂♂ mainly black and white, most with white patch on inner wing-coverts, some with reddish breast; ♀♀ streaked brown above, brown or buff below. Frequent open country; perch upright on tops of bushes and other plants, or on rocks; feed by aerial or ground sallying.

12 species, mostly Asian. 3 in Africa: 1 endemic, 1 resident and Palearctic migrant, 1 Palearctic migrant only. The Buff-streaked Chat *S. bifasciata* is better placed in *Saxicola* than in *Oenanthe*; it is quite unlike any wheatear, but resembles the 2 geographically closest *Saxicola* spp. (*S. torquata*, *S. rubetra*) in many details of plumage, structure and ecology (Tye 1989).

Saxicola torquata (Linnaeus). Common Stonechat. Traquet pâtre.

Plate 28
(Opp. p. 417)

Motacilla torquata Linnaeus, 1776. Syst. Nat. (12th ed.), p. 238; Cape of Good Hope = Cape Flats.

Range and Status. Breeds Africa, Madagascar, Europe, and most of Asia south to the Himalayas, S China and Japan. African and W Palearctic populations resident or not highly migratory; central and E Palearctic populations highly migratory, wintering NE Africa, India, SW and SE Asia.

Resident and local migrant; also Palearctic winter visitor to N Africa. Breeds N Africa in Morocco (south on coast to c. 30°), N Algeria and N Tunisia (central high plateaux south to Lekef and east to Cap Bon), uncommon to locally common. In W Africa, NW Senegal, central Mali and SW Niger, uncommon; highlands of E Sierra Leone, SE Guinea, N Liberia and W Ivory Coast (Mt Nimba), locally common; highlands of SE Nigeria (Obudu, Mambilla), Cameroon (W and Adamawa Plateau) and Bioko, common; at lower altitudes from S Gabon to Lower Congo, common. In eastern Africa mainly above 1500 m in W Sudan (Darfur), common; W and central Ethiopia; extreme S Sudan to NE Uganda; from E Zaïre, W and S Uganda, Rwanda and Burundi to NW Tanzania, from E Uganda to central and S Kenya (east to c. 38°E) and N Tanzania, common to abundant. In central and southern Africa mainly above 1000 m in Angola, S Zaïre, Zambia, S Tanzania, Malaŵi, NW and W Mozambique, N Botswana, Zimbabwe, NE Namibia; all levels in South Africa except dry interior of Cape Province; common to abundant. Non-breeding visitor to lowland S Malaŵi and S Mozambique. Populations localized in N and W Africa, more widespread elsewhere, numbers increasing with spread of cultivation.

Palearctic winter visitor: N Morocco, N Algeria (south to c. 29°N) and N Tunisia, common to abundant; Libya (coastal Tripoli and Cyrenaica provinces), locally common; Egypt (N coast and Nile delta and lower valley, locally common; western oases and Red Sea coast, uncommon to frequent), and from Sudan (N and central east of Nile) to Djibouti and Ethiopia (south to c. 8°N), uncommon to frequent. Irregular in coastal Mauritania, 1 record Aleg (Lamarche 1988). Vagrant N Mali (near Arawan, El Kseib Ounan), N Niger (Arrigui), Libyan desert (Jaghbub, Serir), Chad (Djebel, Abougoudam), SW Sudan (near Wau) and NW Somalia (Berbera).

Description. *S. t. torquata* (Linnaeus): South Africa (SW and S Cape to Natal, E Orange Free State and E Transvaal highlands) and W Swaziland; non-breeding visitor S Mozambique. ADULT ♂: entire head (including throat), back and scapulars black, feathers variably tipped brown; rump feathers blackish tipped white; uppertail-coverts white. Tail black. Sides of neck white; breast and flanks deep chestnut-red; belly and undertail-coverts white; in some, narrow white patch on sides of breast joins belly to white collar. Remiges and upperwing-coverts black, with prominent patch formed mainly by white inner greater and inner median coverts, tertials tipped buff and edged whitish when fresh; axillaries and underwing-coverts blackish, tipped white. Bill black; eye dark brown; legs and feet black. ADULT ♀: streaked dark brown and buff above, including scapulars and sides of neck; uppertail-coverts white. Tail dark brown. Lores and ear-coverts dark brown; breast and flanks suffused warm rich buff; belly and undertail-coverts pale buff. Wing blackish brown, coverts and tertials edged and tipped buff; white patch smaller than in ♂. SIZE: (10 ♂♂, 10 ♀♀) wing, ♂ 68–75 (73·4), ♀ 68–72 (70·2); tail, ♂ 49–54 (52·3), ♀ 48–53 (50·7); bill, ♂ 14–17 (15·8), ♀ 15–16 (15·8); tarsus, ♂ 23–25 (23·7), ♀ 23–25 (23·5). WEIGHT: unsexed (n = 7) 13–17·1 (15·2).

Wing-tip rounded; P5–P8 longest, P4 3–5 mm shorter; P9 6–8 shorter, between P2 and P4; P10 small; P4–P8 emarginated on outer webs.

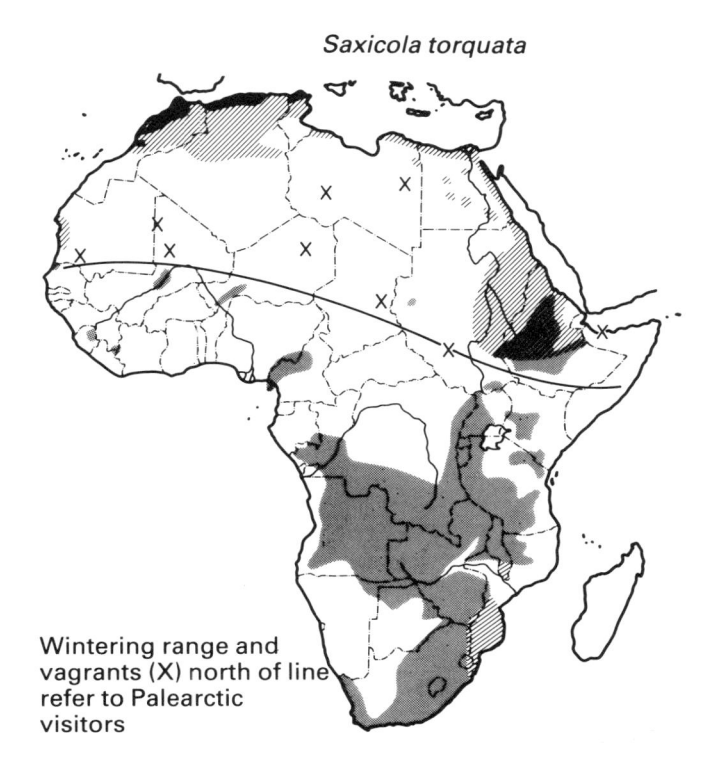

IMMATURE: above, including top and sides of head, dark brown, spotted pale buff; rump pale rufous-buff; throat greyish; rest of underparts warm buff, mottled dusky, especially on breast. Tail dark brown. Wing-feathers dark brown with buff fringes; inner median and greater coverts white.

NESTLING: down brownish grey, fairly long; mouth yellow, gape-flanges pale yellow.

S. t. clanceyi Latimer: South Africa (coastal W Cape). ♂ whiter on flanks and sides of breast than nominate, chestnut more restricted to centre of breast; ♀ paler on abdomen.

S. t. stonei Bowen: interior South Africa (N and W Transvaal, N Cape, W Orange Free State), E and N Botswana, NE Namibia, the Zimbabwe plateau, S and E Angola, S Zaïre, Zambia and SW Tanzania; occurs Mozambique coast as non-breeding visitor. ♂ paler chestnut below than nominate, abdomen tinged tawny.

S. t. oreobates Clancey: highlands of Lesotho, wintering at lower altitudes and reaching E Zimbabwe. ♂ with chestnut of underside darker, more vinous, than in nominate, abdomen tinged buff; ♀ darker, less buffish above, and darker below on throat, breast and flanks. Large: wing, ♂ (n = 7) 71–76.

S. t. promiscua Hartert: highlands of E Zimbabwe, N and W Mozambique, Malaŵi and S and SE Tanzania. ♂ similar to nominate but chestnut on breast more restricted; ♀ paler on abdomen. Slightly smaller: wing, ♂ (n = 10) 64–71 (68·4).

S. t. axillaris (Shelley): highlands of Kenya, N and W Tanzania including L. Victoria basin, W and S Uganda, Rwanda, Burundi and E Zaïre. ♂ with deep chestnut on breast much more restricted than nominate, with white at sides; ♀ darker above. Smaller: wing, ♂ (n = 10) 63–71 (67·7). WEIGHT: unsexed (n = 13) 13–20 (15·5).

S. t. albofasciata (Rüppell): highlands of Ethiopia, S Sudan and NE Uganda. ♂ like *axillaris* but lacks chestnut below, black of throat extending over upper breast and upper flanks; ♀ like *axillaris* but even darker. Wing, ♂ (n = 10) 68–74 (70·6). Probably an incipient species.

S. t. salax (Verreaux): NW Angola, lower Congo, Gabon, Cameroon, SE Nigeria and Bioko. ♂ with chestnut paler than nominate and more restricted on breast, with white at sides; ♀ darker than nominate above, but whiter on abdomen. Smaller: wing, ♂ (n = 3) 65–66.

S. t. moptana Bates: inner Niger delta, Mali and Senegal delta. Like *salax* but slightly smaller, chestnut on breast still paler and more restricted; underwing-coverts and axillaries mainly white.

S. t. nebularum Bates: highlands of Sierra Leone, Guinea, Liberia and W Ivory Coast. Like *salax* but ♂ with more extensive chestnut below. Smaller than nominate: wing, ♂ (n = 5) 66–70.

S. t. jebelmarrae Lynes: W Sudan (Darfur). Similar to *stonei* but ♂ with more extensive pale chestnut below. Wing, ♂ (n = 10) 67–71 (68·8).

S. t. rubicola (Linnaeus): breeds locally N and W Morocco, N Algeria and N Tunisia; also Europe to Caucasus; winters Morocco–Egypt to N edge of Sahara; also to Middle East. ♂ differs from nominate in having more buff tipping and streaking above; rump and uppertail-coverts streaked rusty brown with little white; below more buffy on belly; wings browner with more buff edging to coverts and tertials. ♀ streaked brown on rump and uppertail-coverts. Wing, ♂ (n = 10) 65–69 (66·8). WEIGHT: unsexed (n = 50, Malta, winter) 13–19 (15·3).

S. t. armenica Stegmann: breeds S Caspian area; winters Egypt, E Sudan and N Ethiopia, also Iraq and Arabia. Differs from Afrotropical races and from *rubicola* in having white at base of tail-feathers. ♂ more buffy above and paler below than *rubicola*; ♀ also paler; both sexes have extensive buffy white rump and uppertail-coverts. Wing longer with tip more pointed; P7 and P8 longest, and P9 longer than P4 as opposed to shorter in Afrotropical races; wing, ♂ (n = 10) 75–78 (76·6).

Saxicola torquata

Wintering range and vagrants (X) north of line refer to Palearctic visitors

S. t. variegata (Gmelin): breeds N Caspian area; winters Egypt, E Sudan, N Ethiopia and rarely NW Somalia. ♂ with one-third to two-thirds base of tail white. Otherwise similar to *armenica* but more buffy, sometimes rufous, on abdomen. Slightly smaller: wing, ♂ (n = 10) 67–74 (71·1).

Field Characters. A small chat with large, round head, plump form, and short tail, usually seen sitting upright on prominent perch, frequently flicking wings and tail. Much geographic variation, but ♂ always has blackish head and tail, dark back, white patch on side of neck and (except in Ethiopian race) chestnut breast. ♀ has mottled brown head and upperparts, duller chestnut underparts. Both sexes have white patch on inner wing-coverts, rump white (orange-brown with white mottling in European/ N African race). Wintering ♂♂ of W Palearctic races browner above than resident birds, while Asian visitors are paler and slimmer, with white at base of tail. Can only be confused with Whinchat *S. rubetra* (for differences, see that species).

Voice. Tape-recorded (53, 86, 88, C, F). Song a variable canary-like warble, consisting of short (*c*. 1–1·5 s) phrases separated by intervals of at least 2 s; sometimes in short scrappy bouts, but often in long sustained sequences. Shriller and more monotonous than song of Whinchat, with more repetition of phrases and smaller pitch changes. Calls, a plaintive 'wiet' and a repeated 'trrk-trrk' or 'wiet-trrk-trrk' contact-alarm.

General Habits. Inhabits open country, especially with scattered scrub and low bushes; alpine moorland, grassy hillsides, montane forest edge, cultivation, rank herbage, marshy areas and swamp edges. Mainly a montane species in tropical Africa, but race *moptana* inhabits lowland floodplain, and birds range to low levels in Gabon and Lower Congo valley. Typically at 1800–3000 m in S Sudan and Ethiopia and at 1500–3000 m in Kenya and Uganda. Further south, occurs mainly at 1000–2000 m; at Cape to sea level, with race *clanceyi* resident in coastal sand dunes. NW African residents inhabit macchia, thorn hedges and cultivation from sea level to above 2000 m.

Usually in pairs or family groups, foraging within territory. Palearctic wintering birds typically solitary or in pairs. Perches conspicuously with bold upright stance, on low bush, tall herbage, fence or telegraph wire, often along road sides. Inquisitive but wary. Restless, and frequently flicks wings and tail. Darts down to ground to take prey, returning to same perch or one nearby; sometimes hawks aerial insects. Flight close to ground, jerky, with white rump and wing-patch prominent; sometimes hovers; can turn quickly, going into cover or landing on perch. Gait, a bouncing hop (Cramp 1988). Associates with Black-lored Cisticola *Cisticola nigriloris* in high grasslands of SW Tanzania (Harpum 1978).

In southern Africa some seasonal movement to lower altitudes. Birds from E South Africa reach S Mozambique in non-breeding season; Lesotho highland breeders disperse to winter as far as E Zimbabwe; some Malawi birds descend from breeding area at 1000–2000 m to as low as 450 m (as far south as Chididi). Palearctic birds occur N Africa mainly Oct–Mar (a few Sept and Apr), wintering in coastal scrub, on bushy hillsides and in wadis and desert oases. European *rubicola* winters Morocco to Egypt with notable concentration near Algerian coast. Birds ringed France (8), Belgium (15), Holland (5), W Germany (1) and Luxemburg (1) recovered Morocco; ringed France (3), Belgium (6), Holland (5), W Germany (5), Switzerland (2) and Britain (2) recovered Algeria; ringed Italy (2) recovered Tunisia; ringed Morocco (1) recovered Spain. SW Asian *variegata* and *armenica* reach Egypt, E Sudan and N Ethiopia, with *variegata* apparently commoner.

Food. Small or medium-sized insects and their larvae; also woodlice, spiders, snails, small earthworms, small fish and lizards; seeds and fruit. In Kenya, food brought to nestlings included small moths and their larvae, nymphs of short-horn grasshoppers, glow-worms, nymphs of acridids, small weevils, melolonthid beetles, termites, a lacewing, a small butterfly, a small millipede and a spider (van Someren 1956).

Breeding Habits. Monogamous, although very occasionally 1 ♂ with 2 ♀♀, and once 1 ♀ with 4 ♂♂ in succession (Cramp 1988). Territorial; often remain paired in territories all year round. Change of mate occurred in only 7 of 34 pairs over 18 months in Kenya (Dittami and Gwinner 1985). ♂ sings alone or near mate, from elevated perch, with head raised and white patches exposed, and in song-flight, when he rises almost vertically from perch with slow jerky action, feet dangling, occasionally hovering briefly. When defending territory, ♂ chases invading ♂♂, repeatedly flicks wings on landing, exposing white patches, and flicks and fans tail. When courting ♀, ♂ chases her, hovers above her or crouches low before her with primaries stiffly lowered, bend of wing buried under white collar, white rump and wing-patches conspicuous (Cramp 1988).

NEST: deep cup of rootlets, grass and plant stems, lined with finer rootlets and sometimes with hair, wool and feathers; ext. diam. *c*. 100, cup diam. 55–70, depth 25–40; on ground or on low bank, well hidden at base of grass tuft or under small bush; built by ♀.

EGGS: 4–6 (NW Africa), 2–5 (subsaharan Africa), laid at daily intervals; Kenya mean (30 clutches) 3·0; southern Africa mean (50 clutches) 3·2. Bluish green, variably spotted or freckled reddish brown, markings concentrated at larger end. SIZE: (n = 87, southern Africa) 17–20 × 13–15 (18·9 × 14·2); (n = 22, Kenya) 18–21 × 14–16 (19·8 × 14·9); (n = 24, S Zaïre) 17–21 × 14–15 (18·5 × 14·1). Single-brooded in subsaharan Africa, but 2–3 broods usual in Europe.

LAYING DATES: Morocco, Algeria and Tunisia, Mar–June; Senegambia, Mar; Sierra Leone, Apr; Nigeria, Jan; Cameroon, Mar; Sudan (Darfur), May; S Sudan, Apr–May; Ethiopia, Apr–May; Uganda, Feb–Mar; Kenya, Jan–June (mainly Mar–May); N Tanzania,

Sept–Feb; S Zaïre, Aug–Sept; S Tanzania, Malaŵi, Zambia, Aug–Nov; Zimbabwe, July–Dec (mainly Aug–Nov); South Africa: Transvaal, Aug–Dec (mainly Aug–Oct); Natal, Aug–Dec; Lesotho, Sept–Dec.

INCUBATION: by ♀ only. Period: 14–15 days; begins when clutch complete.

DEVELOPMENT AND CARE OF YOUNG: young brooded by ♀; fed by both parents; remain in nest 13–16 days. Fledglings remain with parents for 3–4 months in Kenya, first breed at 1 year old. In Europe, young fed by both parents for 3–4 days after fledging; ♀ then starts to build nest for next brood, while ♂ continues to feed fledglings for 5–10 more days (Cramp 1988).

BREEDING SUCCESS: of 97 eggs in 39 clutches (Kenya), 70% hatched and 58% produced fledged young; only 5/39 nests failed to produce fledged young (Dittami and Gwinner 1985). In Europe (Jersey), of 332 eggs in 61 clutches, 81% hatched and 64% produced fledged young (Cramp 1988).

References
Cramp, S. (1988).
Dittami, J. P. and Gwinner, E. (1985).
van Someren, V. G. L. (1956).
von Hecke, P. (1965).

Saxicola rubetra (Linnaeus). Whinchat. Traquet tarier.

Motacilla rubetra Linnaeus, 1758. Syst. Nat. (10th ed.), p. 186; Sweden.

Plate 28
(Opp. p. 417)

Range and Status. Breeds W, central and N Europe and east through the USSR between 50° and 65°N to W Siberia, also Caucasus to N Iran; winters tropical Africa.

Palearctic winter visitor. Winters south of Sahara, apparently two main areas: from S Mauritania to Guinea, Sierra Leone, Liberia, S Mali, Ivory Coast, Ghana and Nigeria (except extreme N) to Cameroon and N Gabon; and NE Zaïre and S Sudan (south of 6°) to SW Ethiopia, Uganda and W and central Kenya, south through Zaïre (E of Congo R.), Rwanda/Burundi and W Tanzania to Malaŵi (south to 13°S) and N Zambia (south to Muchinga escarpment and Kafue at 16°S, west to Copper Belt). Common to abundant, especially in guinea savanna and savanna/forest mosaic, less common in soudan belt; frequent to uncommon east of L. Victoria and south of about 3°S. Vagrant SE Kenya (Ngulia, Mombasa), Namibia (Swakopmund, Jan 1925), Botswana (Okavango delta: Qhaaxwha, Sept 1984, Xaxaba, Mar 1987) and South Africa (Transvaal: Diepgazet, March 1987). Individuals occasionally recorded in mid-winter, N Africa: S Morocco (Massa, Irherm), W Algeria (Béni Abbès), N Algeria (La Macta, El Kala) and Tunisia (Tozeur, Sousse, Sfax, Gafsa). On passage throughout N Africa (uncommon to locally common autumn, common to abundant spring) and in Sahel (common to abundant Mauritania and Senegal, frequent elsewhere); also Central African Republic (Feb–Mar), Ethiopia (except Ogaden, mainly spring), Djibouti and NW Somalia.

+ Northern mid-winter records

Description. ADULT ♂ (breeding): top of head to mantle and scapulars blackish brown, feathers edged rufous-brown; rump and uppertail-coverts rufous-brown streaked blackish. Tail blackish brown, basal quarter of central feathers and basal half of other feathers white, outer web of outermost pair mainly white. Supercilium broad and white; lores and ear-coverts black, marked with olive-brown. Chin and sides of throat white, forming broad stripe extending back to white patch on side of neck; centre of throat, breast and flanks orange-buff; belly and undertail-coverts pale buff. Primaries and secondaries blackish brown, edged buff, broadly so on secondaries; bases of inner primaries and outer secondaries white, forming small visible wing-spot. Primary coverts blackish with white bases; greater, median and lesser coverts black with narrow buff edges; innermost greater, median and lesser coverts mainly pure white, forming a larger wing-spot. Underwing-coverts and axillaries buffish white with dusky bases. Bill black; eye blackish brown; legs and feet black. ADULT ♂ (non-breeding): like breeding male, but supercilium buffish, dusky moustache less well-defined, sides of throat less white, underparts less bright, more buff, with light streaking on breast; white spot on inner wing-coverts smaller. ADULT ♀ (breeding): like breeding ♂ but supercilium, chin and throat buffish, moustachial streak browner, underparts paler; pale spot on inner wing-coverts less distinct. ADULT ♀ (non-breeding): like non-breeding ♂ but pale spots on coverts indistinct. SIZE: (10 ♂♂, 10♀♀, Europe) wing, ♂ 73–77 (75·6), ♀ 72–76 (73·5); tail, ♂ 45–48 (46·4), ♀ 44–50 (46·4); bill, ♂ 13–15 (14·3), ♀ 13–15 (14·4); tarsus, ♂ 22–23 (22·2), ♀ 21–24 (22·4). WEIGHT: unsexed (n = 12, Nigeria, Oct–Feb) 13·5–15 (14·4); (n = 68, Nigeria, Mar–Apr) 13·5–26·5 (19·7); (n = 29, Uganda, Oct–Mar) 12·9–18·3 (15·5).

Breeding plumage acquired by partial moult in Africa Jan–Feb; complete moult occurs in Palearctic July–Aug. Wing-tip bluntly pointed: P8 longest (sometimes = P7); P7 0–1 mm shorter, P6 2–3, P5 6–9; P9 2–4 shorter, between P5 and P6; P 10 minute; P6–P8 emarginated on outer web.

IMMATURE: first winter like adult but little white at base of primaries and secondaries; white on primary coverts confined to bases; first winter ♂ with some white at bases of inner wing-coverts like non-breeding ♀; almost lacking in first winter ♀. First summer like adult but retaining reduced white at bases of primaries and secondaries and on primary coverts.

Field Characters. Same size as Common Stonechat *S. torquata* but slimmer, head less rounded, stance less upright; mottled upperparts, buff underparts (tinged orange in ♂), white patch on blackish wing like Common Stonechat, but with second, smaller white patch at base of primaries. ♂ has blackish face-patch outlined by broad white supercilium and white malar stripe; in ♀, face-patch brown and stripes buffy but same pattern evident, whereas ♀ Common Stonechat has plain brown head without stripes. In flight, both sexes show white sides to base of blackish tail, but rump same colour as back.

Voice. Tape-recorded (CHA, McVIC – 53, 73, B, C). Song a fluty warble, consisting of short (1–1·5 s) phrases given in lengthy bouts; song-phrases richly varied and mimetic, including whistles and rattling notes; frequency range broader than song of Stonechat, with more low-pitched sounds; contact-alarm call, a repeated 'tuc-tuc' or 'hüü-tuc-tuc'.

General Habits. Inhabits moist open areas with perches and access to ground; open or lightly bushed grassland, grassy marsh and forest edges, cultivation, open areas recently cleared of crops or bush. In Uganda, up to 2300 m, especially in tall grassland; in Sierra Leone, occasionally among small mangroves.

Solitary in winter, but may form groups of up to 30 on migration. Perches low on tall plant, grass stem or other prominent vantage point; less upright than Stonechat. Hunts from perch; takes prey from ground or vegetation, sometimes in flight. Stance on ground half upright; gait a rapid hop, with occasional run. Flight usually low, level and fast, from one perch to next. Flicks wings and tail, bobs when alarmed. Sometimes establishes territory on wintering grounds for a few weeks. Sings only occasionally in Africa.

Birds leave European breeding grounds late Aug to Sept. Autumn movements from NW and central Europe are southwest, towards Iberia. Passage in N Africa light though protracted, late Aug to Nov. Passage also light in N tropics (Senegambia late Sept to mid-Nov, N Nigeria Sept, S Sudan Oct–Dec), suggesting long unbroken flights from S Europe and S USSR to low tropical latitudes. Saharan crossing apparently on broad front, though with fewer birds in central sector. First birds reach main wintering areas south to Uganda mid- to late Sept, but main arrival Oct–Nov; arrival Zambia and Malaŵi late Oct to Dec. Departure Zambia mainly Feb; Zaïre, Uganda and Rwanda mid-Mar to early Apr; W Africa mainly early Apr. Return passage Senegambia mid-Mar to early May, Ethiopia mid-Apr, but numbers N tropics again small. Substantial spring passage N Africa, mainly late Mar to mid-May. Birds recovered Morocco (5, Sept–Nov, ringed France, Belgium, Netherlands and Finland; 12, Mar–May, ringed Britain, Netherlands, Belgium and Finland); Algeria (8, Mar–May, ringed Britain, Finland and Austria); Tunisia (1, Nov, ringed Finland; 1 Apr, ringed Britain); Ivory Coast (1, Apr, ringed Finland); Ghana (1, Mar, ringed Finland); and Togo (1, Dec, ringed Tunisia Apr). Birds ringed Tunisia (Apr–May) recovered Italy (4), Poland (1), Norway (1) and USSR (1, *c.* 58°N, 47°E); 1 ringed Nigeria (Apr) recovered Libya (Apr). Birds wintering E central and southern Africa and those on passage Egypt and Ethiopia presumably from USSR.

Food. Insects, including ants, beetles, Hemiptera, grasshoppers, adult and larval Lepidoptera; woodlice, millipedes, centipedes, spiders, snails and earthworms; also seeds.

References
Cramp, S. (1988).
Zink, G. (1980).

Saxicola bifasciata Temminck. Buff-streaked Chat. Traquet bifascié.

Saxicola bifasciata Temminck, 1829. *In* Temminck and Laugier, Nouv. Rec. de Pl. Col. d'Ois. (4), 79, pl. 472, fig. 2; 'Caffrerie' = E. Cape Province.

Plate 28
(Opp. p. 417)

Range and Status. Endemic resident in South Africa, Lesotho and Swaziland. Occurs in E Cape Province north and east of a line from Grahamstown to Graaf-Reinet, northeast to the Drakensberg Mts of Transkei, Natal, Lesotho and E Orange Free State, north through W Swaziland and E Transvaal to the Soutpansberg; isolated populations in mountains north of Potgietersrus and in central and W Waterberg; no longer in Magaliesberg. Usually above 1000 m but down to sea level on Natal–Transkei border. Generally frequent, locally common, especially in Natal Drakensberg.

Description. ADULT ♂: forehead and broad supercilium rich buff; crown and nape dark brownish black, lightly streaked with light brown; face, chin, throat, sides of neck and upper breast black, mottled buff in a few (1st year birds?); mantle and back mottled dark- and cinnamon-brown with white feather-edgings when fresh; rump, uppertail-coverts and scapulars variable, white to orange-buff, or white washed with orange-buff; tail black, fading to dark brown; lower breast and belly rich cinnamon-buff; primaries and secondaries dark brown, fading to pale brown; tertials black, edged and tipped buff, fading and wearing to brown, untipped; wing-coverts black, sometimes edged cinnamon-buff when fresh; underwing-coverts black or dark grey-brown, underside of remiges silver-grey; axillaries white. Bill black; eye dark brown or black; legs and feet black, soles brown. ADULT ♀: forehead, crown, nape and ear-coverts brown, streaked sepia; lores sepia; indistinct supercilium grey-brown; mantle and back warm brown, mottled with dark brown; rump and uppertail-coverts cinnamon-buff; tail black; entire underparts from chin to belly rich cinnamon-buff, breast feathers with sepia shaft-streaks; primaries and secondaries dark brown; tertials and wing-coverts dark brown or black, edged and tipped buff, fading to brown and losing tips; underwing and soft parts as ♂. SIZE: wing, ♂ (n = 42) 88–98 (93·2), ♀ (n = 10) 83–92 (87·5); tail, ♂ (n = 45) 57–66 (61·3), ♀ (n = 10) 56–60 (57·7); bill, ♂ (n = 43) 18–22 (20·1), ♀ (n = 10) 18–20 (19·4); tarsus, ♂ (n = 45) 30–34 (31·9), ♀ (n = 10) 30–33 (31·2). WEIGHT: 1 ♂ (Aug) 37·5, ♀ (n = 3) 32·7–34·8 (33·7).

IMMATURE: like adult ♀, but crown, mantle, back and face speckled with buff; chin, throat and breast scalloped buff and dark brown.

NESTLING: unknown.

Field Characters. ♂ like Common Stonechat *S. torquata* but larger (**A**); distinguished from it by broad white eyebrow and from other chats by black throat and tail, buff V from rump through shoulders on brown back, and orange or white rump and underparts. ♀ distinguished from *Cercomela* chats by general orange-brown appearance with brown back, orange-buff rump and black tail. Immature recognized by buff speckling on back and dark scalloping on breast.

Voice. Tape-recorded (20, 72, 88, GIB, ROC, TYE). Song consists of short phrases 2–6 s long, separated by pauses of 2–10 s. Phrases often start with 1–4 'chacks', followed by a series of whistling, trilled or harsh notes, rather similar to those of other chats. Pauses between song-phrases are often interrupted by 'chack' calls. Sometimes mimics other birds. Both sexes sing, in both breeding and non-breeding seasons. Call-notes (both sexes), 'chack' and a short squeaky whistle 'weet', uttered singly or in combination; used in territorial defence and in alarm; 'chack' rather commoner. Will countersing with whistling human (K. Gamble, pers. comm.).

General Habits. Frequents boulder-strewn hillsides and rock outcrops in grassland, often with aloes and low, scattered bushes and trees. Seldom occurs away from rocky ground but enters cattle kraals, fallow fields and roadside areas, especially when not breeding. Rarely below 900 m; in Natal Drakensberg found especially in boulder fields on montane grassland just below the Cave Sandstone escarpment of the 'Little Berg' at *c.* 1500–1700 m, and only rarely above the Cave Sandstone.

Usually solitary or in pairs. Apparently territorial when breeding and partially so when not. Members of pair often remain together when not breeding, sometimes with fledged young, forming small family parties. Defends non-breeding territory by displays, song and chasing; displaces *Cisticola* spp., Long-billed Pipit *Anthus similis*, Familiar Chat *Cercomela familiaris* and Common Stonechat *Saxicola torquata*; is displaced by Fiscal Shrike *Lanius collaris* (Tye 1988). Sings briefly during disputes; in non-breeding season frequently sings quietly during foraging, while watching for prey.

Perches primarily on boulders, also on low trees and bushes. Flies low between perches. May flick wings and tail on alighting or when disturbed. Normally wary but grows tame near human habitation. When not breeding spends most of time (*c.* 70%) foraging, *c.* 20% resting, often in shade, and remainder on alarm, preening, interactions with other birds and singing (Tye 1988). Reacts to overflying raptors by looking up, bobbing head and flicking wings.

Uses 4 main foraging techniques, but mostly aerial sallying (70% of attempts) and ground sallying (15%) from boulders up to *c.* 5 m tall, in areas with grass up to *c.* 60 cm high. Sallies to ground mostly 0·5–2 m from perch; flies up to 20 m in pursuit of aerial prey and to heights of 10 m (Tye 1988). Also uses dash-and-jab on large flat rocks and in grass shorter than 30 cm, where moves with a bounding gait, jumping rather than hopping; runs on rocks and where grass very short (Tye 1988). Occasionally makes short upward sallies to pick prey from vegetation (e.g. tree trunks) without landing on the trunk. Often beats large prey on rock before swallowing.

Generally sedentary, although wanders somewhat outside breeding season, and birds from highest altitudes may move lower in winter.

Food. Primarily insects. In one study in montane grassland, Apr, ate mainly grasshoppers, which were extremely abundant, also beetles, spiders, ants (especially winged ♂♂); but when alate termites emerged after showers, switched to eating them almost exclusively (Tye 1988a). Half-grown nestlings given grasshoppers, caterpillars and other insects (W. R. Tarboton, pers. comm.). Adults encouraged to feeding tables will take fat and maize-meal mixture (K. Gamble, pers. comm.).

Breeding Habits. Monogamous, territorial, hole-nester. In breeding season, ♂ sings loudly from prominent rocks.

NEST: a large and untidy cup, of soft, dry grasses, lined with animal hair, rootlets and finer grass, on a bulky foundation of grass and roots; int. cup diam. *c.* 70–80, depth 40–50; on ground under downslope side of rock or boulder or in crevice in rocks or wall; often concealed by surrounding grass tufts. Built mainly by ♀.

EGGS: 2–4, usually 3; mean (18 clutches) 2·8. Sub-elliptical; smooth, slightly glossy; creamy white or buff, occasionally tinged pale bluish, heavily freckled with lilac and red-brown, normally with a cap or ring of deeper colour or speckles at big end. SIZE: (n = 20) 21–25·2 × 15·1–17·3 (23·1 × 16·5); (n = 9) 22–24·4 × 15·5–17 (23·1 × 16·4).

LAYING DATES: mainly Sept–Dec (peaking Oct in Natal), occasionally to Feb; probably often double-brooded.

INCUBATION: apparently by ♀ alone (W. R. Tarboton, pers. comm.).

DEVELOPMENT AND CARE OF YOUNG: young fed by both sexes.

References
Tye, A. (1988, 1989).
Vincent, A. W. (1947).

Genus *Oenanthe* Vieillot

A rather poorly defined genus, merging with *Cercomela* through the southern African species of the latter, and also close to *Myrmecocichla* and *Saxicola*. Distinguished from all these except some *Cercomela* species by white or orange tail-coverts and basal part of tail, with black or brown distal part, although one race of *O. monticola* has a wholly-black tail. The plumage pattern of *O. monticola* suggests a close link between *Oenanthe* and *Myrmecocichla*.

Species and subspecies differentiated by variations in black-and-white or sandy plumage. Plumage characters labile; many species polymorphic in crown, throat or belly colour; degree of polymorphism or of sexual or age dimorphism may vary between races of the same species. Characters involved in polymorphism or dimorphism in some species are responsible for subspecific or even specific differentiation in others (see Mayr and Stresemann 1950). Sandy plumage cryptic in arid habitat; black-and-white can be cryptic on broken ground; also said to signify low palatability (Cott 1985).

Live predominantly in arid or semi-arid zones; terrestrial insectivores, hunting ground-living or aerial prey. Three major foraging techniques, used to varying degrees by different species and in different habitats: (1) dash-and-jab – running a short distance, pausing to scan, then either pecking at prey, running to peck at prey or running on to next pausing position; (2) aerial sallying – flying up from perch to capture airborne prey; (3) ground sallying – flying down from perch to capture terrestrial prey. Some species also occasionally sally-glean, i.e. fly up from ground to pick prey from vegetation. More robust species with longer tarsi and stronger bill tend to use dash-and-jab more. Often move by an asymmetric hop (Alexander 1985), intermediate between running and hopping. Often capture large prey relative to body size, food niche overlapping that of *Lanius* shrikes.

Nest normally a cup, placed in cavity, built mainly or entirely by ♀; eggs incubated by ♀; young fed by both parents. Eggs white to pale green or blue, spotted with red-brown or unspotted. Young normally spotted, but unspotted in *leucopyga*, *leucura* and *monticola*.

Alarm notes a harsh 'chack', like knocking two stones together, and/or a whistle.

Many species coexist to an unusual degree for congeners; some niche separation based on habitat, leading to latitudinal zonation in N Africa and among Palearctic species wintering in Sahel (Lack 1971, Smith 1971) but food preferences generally broad, with much overlap between species. Interference competition, through interspecific territoriality, is the major mechanism bringing about ecological isolation in many circumstances.

Afrotropical and Palearctic, 1 species extending to Nearctic in Alaska, Greenland and NE Canada. 18 species, 16 in Africa of which 12 breed there, including 3 endemics. African populations of 3 other species are at least incipient endemic species. Intrageneric relationships are debatable (see Tye 1989b). The genus is at present centred on the Middle East and N Africa and may have originated there.

O. leucura, *O. leucopyga* and extralimital *O. alboniger* form a superspecies of large, predominantly black-and-white plumaged birds. *O. monticola* links them with a superspecies comprising *O. phillipsi* and *O. oenanthe*; ♂ black-phase *monticola* very similar to ♂ *O. leucopyga*, and ♀ to ♀ *O. leucura*; grey-phase *monticola* similar to *O. phillipsi*. *O. phillipsi* and *O. monticola* share some unusual plumage characteristics, including pale wing-coverts; these 2 and *O. oenanthe* are the only predominantly grey species. *O. monacha* is an aerially-feeding desert derivative of the *alboniger* superspecies.

A third superspecies comprises *O. pleschanka* and *O. hispanica*, two closely-allied, near-parapatric species; there is limited hybrid introgression around zones of parapatry (Haffer 1977) and sympatry (Loskot 1986).

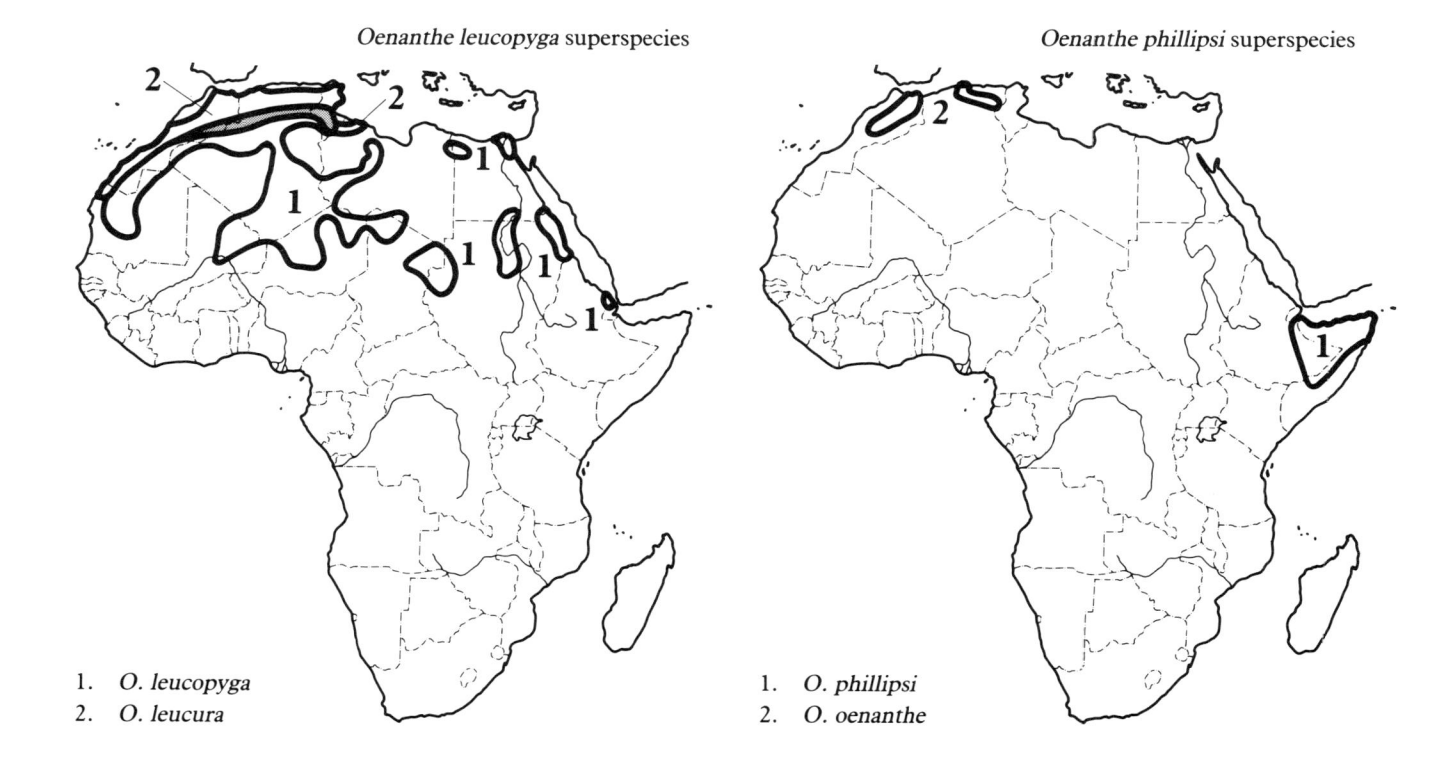

Oenanthe leucopyga superspecies

Oenanthe phillipsi superspecies

1. *O. leucopyga*
2. *O. leucura*

1. *O. phillipsi*
2. *O. oenanthe*

502 TURDIDAE

O. lugens, *O. finschii* and extralimital *O. picata* form a species-group, with *lugens* and *picata* sometimes formerly regarded as conspecific. *O. moesta* is usually placed in a superspecies with *O. xanthoprymna* but may be closer to the *lugens* species-group which, like *moesta* and *xanthoprymna*, also shows a tendency towards a red tail, especially in the African *lugubris* group. *O. moesta* also shares further characteristics with *O. finschii* and *O. lugens*. *O. deserti* and *O. xanthoprymna* are difficult to place: they seem close to the *lugens* species-group, but they may be more closely related to the following superspecies.

O. isabellina, *O. bottae* and *O. pileata* form a final superspecies. They are the most sandy brown species, with *isabellina* and *bottae* almost lacking black on the body plumage. Despite *pileata*'s unique black necklace, it shares face pattern and colour of upperparts with *bottae* and general colouration with *bottae* and *isabellina*.

Oenanthe pileata superspecies

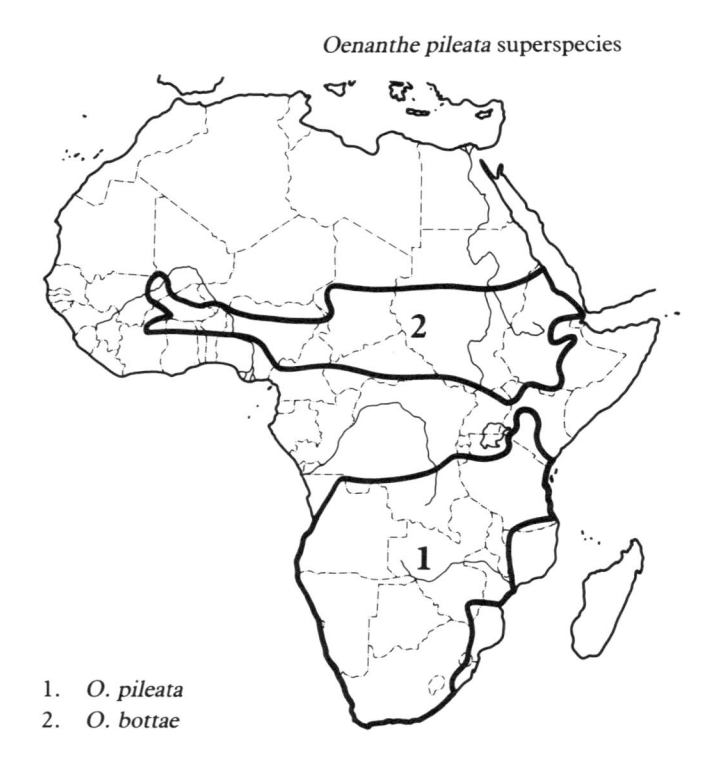

1. *O. pileata*
2. *O. bottae*

Plate 29
(Opp. p. 448)

Oenanthe leucopyga (C. L. Brehm). White-crowned Black Wheatear. Traquet à tête blanche.

Vitiflora leucopyga C. L. Brehm, 1855. Der vollstandige Vögelfang, p. 225; Korosko, Egypt.

Forms a superspecies with *O. leucura* and extralimital *O. alboniger*.

Range and Status. N Africa, Sahara, Arabia, Middle East to Iran.

Resident and local migrant. Breeds in rocky areas in N Africa from Mauritania to Egypt, Sudan and possibly Ethiopia (Eritrea), and in central Saharan massifs. Not near coast in W, and up to 2750 m in Hoggar. Sometimes rather local, but generally abundant in suitable habitat. In Morocco, 0·6 pairs/km^2, and up to 1 bird/km of transect (Blondel 1962); at Tamanrasset, Algeria, *c.* 60 birds in 2 km^2 oasis (Gaston 1970). Frequent winter visitor to Somalia north of 11°N. Breeding range extended north in Algeria and Tunisia in 1920s, and in 1960s and 1970s, the latter possibly correlated with recent desertification (Heim de Balsac 1975, Ledant *et al.* 1981). Vagrant N Nigeria (Malamfatori in Aug, Maiduguri).

Description. *O. l. aegra* Hartert: NW Africa, W Egyptian oases, central Saharan massifs, Darfur. ADULT ♂: forehead and supercilium black; crown white, extending to variable degree onto nape; lower rump, tail-coverts and vent white; rest of upperparts, face and underparts entirely black; rectrices white, with distal half of central pair very dark brown; other rectrices often with brown smudges on one or both webs near tip; all rectrices tipped white when fresh; remiges and wing-coverts black or very dark brown; underwing-coverts and axillaries black; underside of remiges greyish brown. Bill, legs and feet black; eye very dark brown. Sexes alike. SIZE: wing, ♂ (n = 35) 97–108 (104), ♀ (n = 21) 92–103 (98·1); tail, ♂ (n = 34) 65–72 (69·1), ♀ (n = 20) 62–70 (65·2); bill, ♂ (n = 30) 19–22 (20·6), ♀ (n = 21) 19–23 (20·1); tarsus, ♂ (n = 35) 24–28 (26·1), ♀ (n = 21) 24–26 (25·1). WEIGHT: Niger (Aïr, June–July) ♂ (n = 5) 22·6–26·7 (25·4), ♀ (n = 2) 21·2, 22, newly-fledged juveniles (n = 6) 16·6–29 (23·5); Algeria (Nov–Jan) ♂ (n = 5) 25–39 (29·6), ♀ (n = 3) 24–26 (25·0).

FIRST WINTER: like adult but crown black, becoming white at first post-nuptial moult; intermediates occur, often young birds with 1 or 2 white feathers or adults with 1 or 2 black; remiges and wing-coverts paler brown, with narrow pale tips when fresh; brown smudges on outer rectrices often larger.

JUVENILE: unspotted; like first-winter but body plumage dark brown, not black.

NESTLING: unknown.

O. l. leucopyga: Nile valley of Egypt, E Sudan, Eritrea, Djibouti. Plumage variable, ranging in appearance from that of typical *aegra* to that of *ernesti*. Larger. SIZE: wing, ♂ (n = 35) 101–112 (106), ♀ (n = 29) 95–108 (101); tail, ♂ (n = 35) 64–76 (69·5), ♀ (n = 28) 62–71 (65·3); bill, ♂ (n = 32) 20–24 (22·1), ♀ (n = 27) 20–23 (21·7); tarsus, ♂ (n = 34) 23–28 (25·5), ♀ (n = 27) 23–27 (24·9).

O. l. ernesti: breeds Sinai and Middle East; may enter Africa in winter but individuals not distinguishable from some *leucopyga*. Black of plumage with strong blue gloss. Larger, but with shorter tarsi. SIZE: wing, ♂ (n = 16) 105–115 (108), ♀ (n = 10) 101–106 (103); tail, ♂ (n = 17) 68–76 (72·2), ♀ (n = 11) 65–70 (66·7); bill, ♂ (n = 17) 21–24 (22·7), ♀ (n = 11) 20–22 (21·3); tarsus, ♂ (n = 15) 23–27 (25·6), ♀ (n = 11) 24–27 (24·8).

Field Characters. More black on underparts than other wheatears except Black Wheatear *O. leucura*. White-capped adult resembles Hooded Wheatear *O. monacha* but has black belly. Black-capped immature like ♂ Black Wheatear but smaller, less robust and without black band across end of tail.

Voice. Tape-recorded (53, 60, 63). Full song broken into short phrases 1·5–2 s long, by pauses of 3–9 s; distinctive, rich and fluty, of whistled, slurred or slightly trilled notes; unlike other wheatears, with no or few harsh or chacking sounds. May include mimicry. Quieter, more continuous song or subsong not broken into phrases; similar in quality but with more chuckling or chacking notes, as in songs of other wheatears. Will countersing with and imitate a whistling human (Fischman 1977). ♀♀ and young occasionally sing. Many calls, including a harsh 'chack', loud and quiet whistles, and a grating call like running fingernail over comb teeth, similar to buzzy begging call of young. Most song at dawn and dusk, though bird occasionally sings at night; sometimes sings in winter.

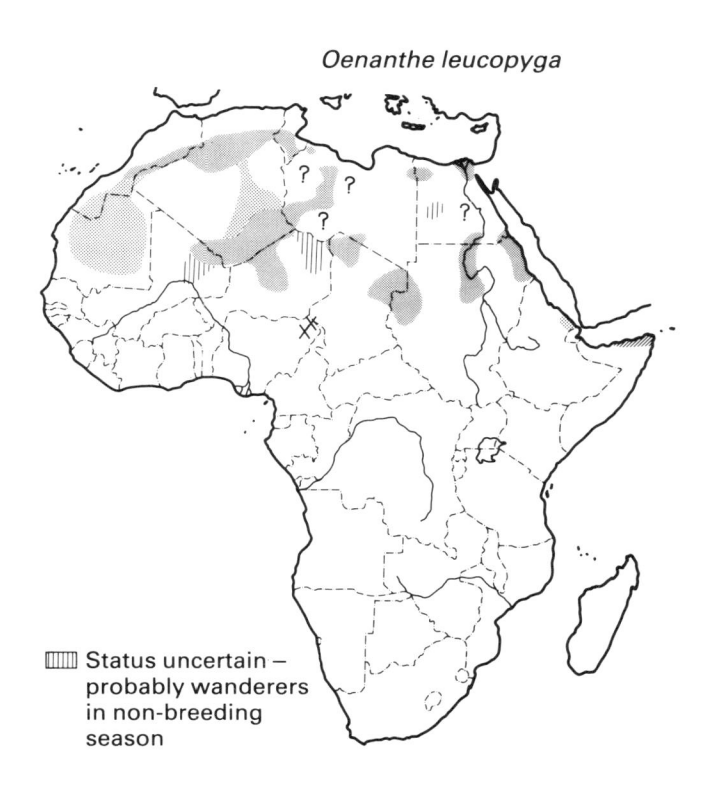

General Habits. A bird of true desert, but locally in areas receiving up to 150 mm rain annually; also found among human habitation. In desert occupies pebbly and rocky areas, ravines, wadis, cliffs, lava-fields; all with minimal vegetation. Occasionally recorded in sandy desert. In oases inhabits palm groves, cultivation and buildings. At all altitudes in Saharan massifs. Independent of water but will drink if water available.

Territorial; when not breeding occurs singly or in pairs; pair may stay in breeding territory all winter; defends territory at migration stopover points for up to 2 weeks. Sometimes interspecifically territorial with other wheatears, including Pied *O. pleschanka* and Mourning *O. lugens*, and with Blue Rock-Thrush *Monticola solitarius*. Individual autumn feeding territories average 2·4 ha (n = 6); sometimes apparently feeds outside territory. Defence initially by display: perches on rock, lowers and spreads tail and chacks or sings or whistles loudly; or crouches with lowered or raised spread tail, shuffling about. May fight briefly with territorial neighbours, or chase in flight. Aggressive encounters may end with sleeked posture, bill pointing skyward. Song usually from high perch on tree or rock, often accompanied by jerky motion, causing front part of body to rock back and forth; also sings in slow flight with much gliding and tail spread.

Tame, bold and curious, visiting houses, stopped vehicles, people on foot. Protected by and sacred to some Bedouin. Bobs head and wags tail in response to potential predator. Attacks snakes, cats and people near nest, sometimes mobbing in groups. Breeding birds call and hover over predator, leading it away from nest. In response to falcon, flies to shadow, where plumage cryptic.

Perches on rocks, posts, telegraph poles, bushes. Forages by ground- and aerial-sallying or by dash-and-jab; occasionally catches insects and plucks berries in bushes. If perches unavailable may hover over feeding area. Also digs for larvae, throwing earth sideways. Flicks wings when it runs. Flight floating, somewhat swallow-like. Active all day unless temperature high, when it spends hottest hours perched in shade of rocks or in lower branches of acacias, occasionally sallying to ground. Feeds until well after sunset. In Arabia and Israel roosts under stones, on cliff ledges or in high, narrow rock crevices, members of a pair roosting separately. Shelters from rain or dust storms under rocks.

Mainly sedentary, though NW African birds may move south in winter. Some evidence for shift from poorest habitats in dry season (Destre 1984).

Oenanthe leucopyga

▥ Status uncertain – probably wanderers in non-breeding season

Food. Mainly insects: beetles (including Buprestidae), tenebrionid larvae, ant lions, small Hymenoptera, grasshoppers (including adult locusts *Schistocerca gregaria*), caterpillars up to 6·4 cm and, especially, ants and flies; occasionally berries. Beats large caterpillars and beetles before swallowing. In more desolate country, takes larger prey, particularly locusts and lizards (including Lacertidae and Gekkonidae, eating tails of latter). Eats scraps and vegetable matter, e.g. meat thrown from table (sometimes fed by Bedouin), pomegranate seeds, grass seeds, dates and flour. In Arabia and Israel also recorded eating crickets, butterflies, aphids, worms, ticks and other parasites picked from camels, insects found in camel dung (Fischman 1977). Young fed especially larvae (Palfery 1988).

Breeding Habits. Monogamous cavity nester, helpers at some nests. Normally territorial, expelling passage migrant wheatear spp. and Common Redstart *Phoenicurus phoenicurus*, but breeding birds in Egypt permitted passage migrant Northern Wheatear *O. oenanthe* and Desert Wheatear *O. deserti* to intrude, without showing aggression towards them (Meinertzhagen 1954). Other species, e.g. larks, chased from vicinity of nest or feeding bird. Defence of breeding territory mainly by ♂. Territory size *c.* 0·4 ha in Israel; av. 6·6 ha (n = 7, range 4·8–9·1) in Arabia. Sings much in early breeding season. In courtship display, ♂ leans forward so body nearly horizontal, shuffles to within a few cm of ♀, which remains still. ♂ chacks or sings quietly for a few s, then pair separate or mate; copulation very brief. In dancing display ♂ performs rapid flight loops in front of perched ♀; ♂ may also fly up and down like a yo-yo (**A**: I. Willis, pers. comm.). ♂ may chase ♀ in twisting, rapid flight; sometimes a third bird joins chases.

NEST: of dried grass mixed with slender twigs, or acacia bark, lined with hair, wool or plant down; rather loose, and sometimes very large. Placed in rock crevice, hole in bank or wall, building (where nest may be exposed, e.g. on mantlepiece); one found in base of abandoned nest of Golden Eagle *Aquila chrysaetos*, 15 m above ground (Valverde 1957). Often has 'pathway' or heap (up to several kg) or small (up to 10 g) pebbles built on outer edge or partly supporting nest. Pebbles collected over a period of weeks; function unknown but may help keep nest contents cool (George 1978). Nest and pebble heap built mainly by ♀; site may be re-used in successive years.

A

EGGS: 3–5, rarely 2 or 6; NW Africa mean (31 clutches) 3·9; smaller clutches in drier years. Subelliptical; almost without gloss; creamy white or very pale greenish or bluish, speckled with red-brown, mainly at large end, often forming a crown; occasionally with deep blotches of pinkish lilac. SIZE: (n = 9, N Africa) mean 23·8 × 16·8; (n = 3, Aïr) 20·2–20·5 × 14·8–15·3 (20·4 × 14·9); (n = 7, Morocco) 21·6–23 × 15·7–16·8 (22·3 × 16·1). Sometimes double-brooded.

LAYING DATES: Mauritania, Jan–Feb; N Africa, Sudan late Jan–May, possibly with peak Mar–Apr in Morocco, Algeria; central Sahara, Mar–July.

INCUBATION: by ♀; period 2 weeks.

DEVELOPMENT AND CARE OF YOUNG: young fed by both sexes. ♀ parent may lay second clutch while first brood still in nest, when ♂ alone may feed latter. Juveniles quickly develop immature plumage; able to fly well within 3 days of emergence; remain near hiding places initially; after few days follow parents. Fledglings fed by adults for *c.* 3 weeks; large broods may be divided between parents. In Arabia, one first brood stayed in parents' territory after becoming independent and at least 1 bird carried food to, and faecal sacs from, second-brood nestlings (Palfery 1988). Independent juveniles may set up individual or group territories in or outside parental territory. At least some breed in second calendar year, when crown still black; these ♀♀ may lay smaller eggs.

References
Arnault, C. (1926).
Cramp, S. (1988).
Heim de Balsac, H. (1926).
Palfery, J. (1988).
Pasteur, G. (1956).
Tye, A. (1987).
Valverde, J. A. (1957).

Plate 29
(Opp. p. 448)

Oenanthe leucura (Gmelin). **Black Wheatear. Traquet rieur.**

Turdus leucurus Gmelin, 1789. Syst. Nat. 1, p. 820; Gibraltar.

Forms a superspecies with *O. leucopyga* and extralimital *O. alboniger*.

Range and Status. SW Europe and NW Africa.

Resident, with some local movements. Breeds in rocky and mountainous areas from Western Sahara to W Libya. Common to abundant up to 2000 m in Morocco, in Algeria on Hauts Plateaux and Aurès south to foot of Saharan Atlas and to coast in Algeria and SW Morocco.

One breeding record Djebel Moussa (Tangier Peninsula, Morocco). Common and mainly coastal in Western Sahara; rare NW Mauritania; common in central Tunisia and Jebel Nafusa in Libya, frequent in massifs further north in Tunisia. In Morocco, density 2·6 birds/km² (rocky habitat), 2 per 5 km (wadi) (Blondel 1962). Vagrant E Libya (Salum) and Egypt (Quseir, sighting). Mainly lives north and west of 100 mm isohyet.

Description. *O. l. syenitica* (Heuglin) (including '*riggenbachi*'): only subspecies in Africa. ADULT ♂: entire plumage dull brownish black except tail-coverts and tail; wing-feathers somewhat browner than body plumage; tail-coverts white. Rectrices basally white, with distal half of central pair and distal one-sixth of others brownish black; all tipped white when fresh. Bill, legs and feet black; eye dark brown. ADULT ♀: like ♂ but dark areas of plumage browner; underparts mottled brown and dark sepia (feathers sepia, tipped paler brown). SIZE: wing, ♂ (n = 23) 96–105 (99·2), ♀ (n = 18) 90–99 (93·9); tail, ♂ (n = 23) 63–72 (67·3), ♀ (n = 18) 61–70 (65·1); bill, ♂ (n = 23) 20–23 (21·8), ♀ (n = 17) 19–23 (21·4); tarsus, ♂ (n = 23) 25–27 (26·0), ♀ (n = 18) 24–27 (25·7). WEIGHT: Algeria (Oct–Nov) 1 ♂ 44, 1 unsexed (first winter) 37·5.
JUVENILE: upperparts like adult ♀; underparts as upperparts, i.e. more uniform dark brown than adult ♀, paler brown than adult ♂. Sometimes has some faint, pale-brown spotting but often unspotted.
FIRST WINTER: like juvenile but underparts darker, closer to colour of adult ♂.
NESTLING: down on upperparts grey, long and dense; underparts bare. Mouth yellow; no tongue spots; gape flanges pale yellow.

Field Characters. Distinguished by entirely black plumage (except for white around base of tail) from all wheatears of the region except immature White-crowned Black *O. leucopyga*, from which told by broad black band across end of tail. Large-bodied appearance, resembling a rock-thrush *Monticola*, distinguishes it from extra-limital *opistholeuca* morph of Eastern Pied Wheatear *O. picata*.

Voice. Tape-recorded (62, 73 – B). Song in phrases mostly 2–4 s long, interrupted by pauses of 2–8 s (Spain); phrases occasionally > 12 s. A rich warble, often begun and ended with harsher churring or chacking notes. ♀ sometimes sings, song like ♂ but harsher. Alarm call a loud 'pee-pee-pee', often repeated; also 'chack', quiet 'chut', and shrill 'zweer', all uttered when bird disturbed. Begging call of young, a similar 'sweer'. Perches on and sings from rocks, bushes, trees and roofs; also sings in fluttering flight with tail spread.

General Habits. Found in rocky and mountainous places, including sea-cliffs, gorges and scree, in semi-desert and areas completely lacking vegetation. Also in flatter country with rock outcrops and poor grass cover. Occasionally among ruins or houses; also orchards, where recorded hunting insects on piles of rotting vegetation (Valverde 1957). Generally at higher altitudes than White-crowned Black Wheatear; ranges of these 2 species rarely overlap. Up to 3000 m in Atlas, down to base where White-crowned Black Wheatear absent; reaches coast in places where ground rocky.
Usually solitary or in pairs; sometimes in small groups (family parties?). Shy; when agitated spreads wings and tail. Not as active as some wheatears, behaving more like a *Turdus* thrush. Forages mainly by dash-and-jab and ground sallying, but also captures some prey by aerial sallying or sally-gleaning from vegetation. Searches around rocks and under bushes; pecks into ground with strong bill. Forages until quite dark in evening, when crepuscular beetles available. Beats large prey items on ground and may spread wings when handling them. Regurgitates pellets.
May descend from highest altitudes in Atlas in winter.

Food. Mainly insects, especially beetles and ants, also grasshoppers, larvae; seeds and an olive stone once found in stomach. Occasionally captures lizards. In Spain also recorded eating scale insects, bees, wasps, mantises, adult and larval Lepidoptera, flies, spiders, millipedes, scorpions and berries.

Breeding Habits. Apparently monogamous, loosely territorial, cavity-nester. In Spain pairs had overlapping home ranges of *c.* 14 ha; in Western Sahara 3 pairs in 1 km (orchards), 6 pairs in 7 km (bare slopes). Aggressive in breeding season, in Spain chasing rock-thrushes *Monticola* spp. and other birds. Song and aerial chases form part of courtship; ♂ has tail-shivering display. Sings much during breeding season, beginning about Feb. ♂ leads ♀ to potential nest-sites.

Oenanthe leucura

NEST: bulky, of coarse grass and rootlets, neatly lined with much hair, wool or feathers. Placed in rock crevice or hole under rock or tussock, in wadi bank or wall. Erects a partial barrier of pebbles at hole entrance, or partly under nest, or leading like path to nest; can contain several kg of stones, perhaps accumulated over several seasons. In Spain, individual stones weighed up to 28 g, mean 7 g. Stones brought and nest built by both sexes.

EGGS: 3–5, NW Africa mean (101 clutches) 4·0. Subelliptical; smooth, almost glossless; white to pale green or pale blue, sparsely spotted with red-brown or deep violet, mainly at big end. SIZE: (Tunisia) average 24 × 17; (Algerian Sahara) 21·7–25·3 × 16·7–18·2.

LAYING DATES: throughout NW Africa mainly Mar–May but from Feb in S, and occasionally June. Double-brooded.

INCUBATION: by ♀; period c. 16 days.

DEVELOPMENT AND CARE OF YOUNG: young fed by both parents; by ♂ more at first, while ♀ brooding chicks (France). Leave nest at c. 15 days, when barely able to fly. Fledglings hide under rocks; fed by parents for c. 2 weeks but begin self-feeding after 1 week.

BREEDING SUCCESS/SURVIVAL: in Spain at least 60% of nests failed, probably mainly through losses to snakes and lizards.

References
Cramp, S. (1988).
Richardson, F. (1965).
Valverde, J. A. (1957).
Whitaker, J. I. S. (1905).

Plate 29
(Opp. p. 448)

Oenanthe monticola Vieillot. Mountain Wheatear; Mountain Chat. Traquet montagnard.

Oenanthe monticola Vieillot, 1818. Nouv. Dict. d'Hist. Nat. 21, p. 234; Namaqualand.

Forms a link between the superspecies *O. alboniger/O. leucopyga/O. leucura* and the superspecies *O. phillipsi/O. oenanthe*.

Range and Status. Endemic resident. Swaziland, South Africa, Lesotho, Namibia and Angola. Frequent to common W Swaziland, Transvaal highveld, W Natal, Orange Free State, lowland Lesotho, NE Cape Province, east through Karoo to SW Cape, north through Namibia to coastal SW Angola. Isolated population (*O. m. nigricauda*) in highlands of Huambo and S Cuanza Sul provinces of Angola. In mountains in E half of range but descends to coast along Atlantic seaboard. Absent from Kalahari sandveld and E Namibia.

Description. *O. m. monticola* (including '*griseiceps*'): Swaziland and Natal to Great Namaqualand. ADULT ♂: polymorphic in general body colour, and in colour of crown and belly, with any combination of crown and belly colour possible in 'black phase' ♂♂. *Black phase*: forehead, crown, nape and hindneck black, dark ashy grey or pale French grey; some black-crowned birds have a faint pale line from bill to above eye; sides of neck, face, chin, throat, breast, mantle, back, rump and scapulars black; uppertail-coverts white. Central pair of rectrices black; next pair black with small white patch at base; others white with distal part black (or sometimes wholly white), the black least extensive on pairs 4 and 5 from centre (three-quarters white). Belly and undertail-coverts white, black or mottled black-and-white. Remiges black or very dark brown; lesser and median coverts white; other wing-coverts black or very dark brown; underwing-coverts and axillaries black; underside of remiges smoke-grey. Bill, legs and feet black; eye dark brown. *Grey phase*: entire body plumage and underwing French or ash-grey, paler on underparts, becoming paler through belly to white on undertail-coverts; belly somewhat variable, whiter in some individuals. Uppertail-coverts, lesser and median wing-coverts white; remiges and other wing-coverts very dark brown. Body plumage wears to show dark shaft-streaks, producing streaky grey and black appearance. ADULT ♀: entire plumage dark blackish brown or earth brown, wearing to rather mottled earth brown and sepia; tail as ♂; uppertail-coverts white; undertail-coverts often partly white. SIZE: wing, ♂ (n = 91) 107–122 (113), ♀ (n = 32) 103–115 (109); tail, ♂ (n = 91) 68–82 (74·7), ♀ (n = 31) 65–79 (73·0); bill, ♂ (n = 88) 20–25 (22·2), ♀ (n = 30) 20–24 (21·8); tarsus, ♂ (n = 89) 29–34 (31·4), ♀ (n = 30) 28–33 (30·2). WEIGHT: unsexed (n = 3, South Africa) 30·8, 32·6, 35.

JUVENILE: unspotted; like adult ♀, but often with more black in tail, sometimes black almost to base on inner webs.

FIRST-WINTER ♂: like juvenile, or as follows (seems to moult from juvenile plumage at any time from early winter to the following breeding season). *Black phase*: like adult but always has black crown; upperparts brownish black; wings very dark brown; lesser and median coverts initially as rest of wing. *Grey phase*: like adult but body darker grey; lesser and median coverts grey, as rest of body; uppertail-coverts often grey; fringes on secondaries and tertials and tips on tail-feathers white in fresh plumage.

NESTLING: patches of long grey down on crown, back, flanks and wings.

O. m. atmorii (Tristram): Namibia, from Damaraland north. Like *O. m. monticola* but black-phase ♂♂ with black crown often have pronounced white line from bill to above eye; black-phase ♂♂ with grey crown have crown often paler, greyish white; grey-phase ♂♂ much paler, with body plumage greyish white. Smaller. SIZE: wing, ♂ (n = 15) 105–111 (108), ♀ (n = 11) 99–107 (103); tail, ♂ (n = 15) 65–76 (70·2), ♀ (n = 11) 62–73 (67·5); bill, ♂ (n = 14) 20–23 (21·6), ♀ (n = 11) 20–24 (21·5); tarsus, ♂ (n = 14) 28–32 (29·9), ♀ (n = 11) 28–30 (28·8).

O. m. albipileata (Bocage): coastal Benguela Province, Angola. Like last race but ♀ paler, dull brown on upperparts, sandy brown on underparts with much white on belly. Smaller. SIZE: (7 ♂♂, 6 ♀♀) wing, ♂ 99–107 (102), ♀ 100–106 (101); tail, ♂ 66–69 (67·7), ♀ 65–70 (67·0); bill, ♂ 19–21 (20·0), ♀ 20 (20·0); tarsus, ♂ 29–30 (29·6), ♀ 28–29 (28·8). WEIGHT: ♂ (n = 4, Apr) 28–35 (30·7); 1 ♀ (Aug) 30.

Polymorphism of above 3 races varies geographically: black-crowned and black-bellied forms of black-phase ♂ commoner further south and east.

O. m. nigricauda Traylor: highlands of Huambo and S Cuanza Sul, Angola. ADULT ♂: similar to black-phase *monticola*. Forehead, crown and nape dark ashy grey or black; lower rump and uppertail-coverts white; lower belly (or entire belly) and undertail-coverts greyish white mottled with black; lesser and median wing-coverts black or (in immature?) white; rest of plumage black, moderately glossy; tail black with dark grey fringes on basal three-quarters of outer webs. ADULT ♀: paler than *monticola* and with lower belly white. SIZE: wing, ♂ (n = 4) 100–101 (100·8), 1 ♀ 99; tail, ♂ (n = 2) 70, 72; bill, ♂ (n = 4) 20–21 (20·3); tarsus, ♂ (n = 2) 28, 30.

TAXONOMIC NOTE: '*griseiceps*' as originally described referred to ♂ ashy grey-headed, black phase; used by later authors to refer to a supposed larger eastern race, but no size difference exists.

Field Characters. ♂ distinguished from other chats of the region by combination of black or grey plumage with black-and-white tail and white shoulder-patches. White-headed Black Chat *Myrmecocichla arnotti* is similar to black-phase ♂ but has wholly black rump and tail. Darker wings and tail and white shoulder-patches distinguish grey phase from similar *Cercomela* spp. ♀ and young distinguished by white in tail from Southern Anteater-Chat *Myrmecocichla formicivora*; much darker than ♀ Northern Wheatear *O. oenanthe* and young Capped Wheatear *O. pileata*.

Voice. Tape-recorded (88, F, GIB, HEL, ROC). Song loud, stronger and more musical than most wheatears; a jumble of whistles and trills, in phrases mostly 1–2 s, up to 7 s long, interrupted by pauses of 1–8 s. Also includes mimicry of wide variety of birds and a bull (Plowes 1948). ♀ song includes more harsh notes, especially a twanging 'cherr'. Song used in disputes and chases is less obviously broken into phrases, and includes more harsh notes. Alarm note 'chit-chit'.

General Habits. Generally in arid, hilly country remote from human habitation, though will adapt to human presence. Within these limits, habitat rather varied, but usually grassy and sometimes with bushes, and always with rocks or mounds; includes open grassland on boulder-strewn slopes; rock outcrops and kopjes in grassveld; quarries, old mine workings; also riverine bush, farmyards and gardens. In Angola on rocky slopes of high peaks with short grass and scattered stunted trees, and in arid coastal zone; in Namibia in semi-desert and hills; in Natal mostly above 1200 m. In Lesotho inhabits sandstone hill slopes up to 2500 m, staying in river valleys below this altitude in the highlands. In Namibia reported to avoid sandstone, preferring granite and limestone (Hoesch and Niethammer 1940).

Usually in pairs or solitary, sometimes small parties. Territorial, less strongly in non-breeding season. Wary but conspicuous, perching on boulders, termite mounds, bushes, trees, or any other eminence. Often flicks wings and tail. Flies low between perches, showing white patches; from high perch glides short distance, drops suddenly, then flies low. Moves with long hops over rocks; forages by dash-and-jab, sallying to ground and into air, often returning repeatedly to same perch (Andersson 1872). May grow tame near human habitation and will visit bird tables, where not aggressive. Sings mostly early and late in day, often 2–3 h before sunrise, or all night if moonlit. Usually sings while perched, but occasionally in song-flight. Occasionally mobs Fiscal Shrike *Lanius collaris*.

Food. Normally insectivorous, but will take table scraps, including bread crumbs, suet, porridge, grated coconut. Diet includes grasshoppers, spiders, ants. One bird caught moths at a light at night, and visited a *Cactoblastis* rearing shed to catch escaped moths, having learnt the times when workers came to open the cages (Taylor 1946). Cutworms and other larvae, grasshoppers, crickets, spiders, small beetles and one butterfly given to nestlings (Taylor 1946, Plowes 1948).

Breeding Habits. Monogamous, territorial, cavity-nester. In early breeding season ♂ displays with twisting flight down hill-slope, with brief song.

NEST: bulky and roughly-built of dry grass, bark, flower heads, moss and other plant material, lined with fibres and hair. Often on a foundation of spiders' web, earth, twigs or small stones. Base of one nest made almost entirely of *c.* 250 trapdoor-spiders' nest entrances, with sand still adhering to them (Plowes 1948). Normal nest-site under rock on hillside or in cleft in rocks, also in hole in side of erosion gully, in stone wall, on gutter, under verandah roof and in nest box. One nest in old nest-tunnel of Pied Starling *Spreo bicolor* (Vincent 1947).

Sometimes caked mud or twigs used to disguise front of nest. Built by ♀ in 4–14 days, bringing material every 5–10 min. One nest required an estimated 400 beakloads (Plowes 1948). Lining material worked in by scratching with feet and bill while bird turns round. Same nest used for consecutive broods and repaired between them; may use nest of previous year. Occasionally uses old nest of Familiar Chat *Cercomela familiaris*. Eggs laid *c*. 3 days after nest completed.

EGGS: 2–4, southern Africa mean (50 clutches) 2·8; Natal and Zululand (8 clutches) 2·9; 1 laid per day. Subelliptical; slightly glossy; pale blue to greenish blue with lilac and pinkish rufous specks, usually densely concentrated at big end. SIZE: (n = 204, southern Africa) 20·9–26·4 × 15·8–19 (23·5 × 16·8).

LAYING DATES: throughout range June–Mar (peak Sept–Nov, possibly later (Jan–Feb) in Namibia); opportunistically after rain in arid areas. Usually 2–3, occasionally perhaps 4 broods per season; *c*. 12 days between fledging and re-laying.

INCUBATION: by ♀; period 13–14 days. ♂ briefly visits ♀ on nest *c*. 6 times per day, when ♀ may leave nest but ♂ does not then incubate (Plowes 1948).

DEVELOPMENT AND CARE OF YOUNG: newly-hatched young covered with patches of long grey fluff; first primaries appear at *c*. 7 days; throat bright orange; eyes open, primaries begin breaking sheaths, rump feathers begin to grow, and young begin to churr when fed, at *c*. 10 days. Nestling period *c*. 16–17 days; young fed by ♂ and ♀, all day until dark; if ♀ at nest, ♂ may give food to her to pass to young. 5-day-old young given *c*. 90 feeds/day (30 for each chick), at 2·5 each/h (Plowes 1948). ♀ often broods nestlings after bringing food; ceases brooding at night when young *c*. 10 days old. Adults swallow faecal sacs for initial *c*. 10 days, later carry them off. Food brought very infrequently one day each side of fledging date. Newly-fledged young can fly a little but stay under cover, hide among rocks. Adults with dependent fledglings call loudly and fly close to potential predators near nest.

BREEDING SUCCESS/SURVIVAL: very hot summer days can kill broods in nests under roofs.

References
Davies, C. G. (1910).
Plowes, D. C. H. (1948).
Taylor, J. S. (1946).
Vincent, A. W. (1947).

Plate 30
(Opp. p. 449)

Oenanthe phillipsi (Shelley). **Somali Wheatear. Traquet somali.**

Saxicola phillipsi Shelley, 1885. Ibis, ser. 5, 3, p. 404, pl. 12; mountains near Berbera.

Forms a superspecies with *O. oenanthe*.

Range and Status. Endemic; locally common, widespread resident in Somalia north of 4°N and SE Ethiopia (Ogaden). Altitudinal range 600–2000 m.

Description. ADULT ♂: forehead and supercilium white; crown, nape, back and scapulars pure French grey; sides of neck, ear-coverts, lores, chin, throat and breast black; lower rump and uppertail-coverts white. Central pair of rectrices black with white edges near base; other rectrices white with distal three-quarters of shaft black, broadening on outer web to a black wedge which extends across inner web near tip; all rectrices tipped white when fresh. Belly and undertail-coverts white. Remiges very dark brown or black; secondaries and tertials narrowly tipped white and, when fresh, outer web of tertials and inner secondaries fringed white; greater primary coverts black; other wing-coverts pale grey; underwing-coverts and axillaries black; underside of remiges smoke-grey. Bill black; eye dark; legs and feet black. Sexes alike. SIZE: wing, ♂ (n = 12) 80–86 (83·9), ♀ (n = 6) 79–83 (80·7); tail, ♂ (n = 13) 44–52 (48·9), ♀ (n = 6) 44–49 (46·0); bill, ♂ (n = 13) 17–18 (17·5), ♀ (n = 6) 17–18 (17·3); tarsus, ♂ (n = 13) 23–26 (25·2), ♀ (n = 5) 22–25 (23·8).

IMMATURE: like adult except crown, nape, back and scapulars grey tinged with brown; sides of neck, ear-coverts, lores, chin, throat and breast black, mottled or scalloped to variable extent with grey, sometimes entirely grey; black areas

Oenanthe phillipsi

of tail browner and broader than in adult; primaries dark brown, initially tipped buff; secondaries and tertials with broader buff fringes than adult; greater primary coverts brown, with narrow cream edgings progressively worn away; lower mandible often basally horn.

NESTLING: early stages undescribed; later, upperparts above rump and underparts down to belly mottled brown and dark grey; rump and uppertail-coverts white; belly and undertail-coverts white smudged with dark brown; tail as immature; wing-coverts buff; bill dark brown, basally horn.

TAXONOMIC NOTE: sometimes regarded as a subspecies of *O. oenanthe*, but see Hall and Moreau (1970), Tye (1986).

Field Characters. Tail pattern and white rump identify it as a wheatear. Distinguished from other wheatears in its range by combination of black throat and grey back. Immature identified by grey mottling on black areas of underparts. Told from NW African race of Northern Wheatear *O. oenanthe seebohmi* by grey wing-coverts.

Voice. Not tape-recorded. Song unknown. Calls, a low, drawn-out whistle, uttered repeatedly, and a double buzzing sound, accompanied by wing-flicking (Miskell in press); also said to make a 'sharp metallic "clicking" sound common to most of its kind' (i.e. as in other wheatears) (Archer and Godman 1961).

General Habits. Inhabits mainly open, stony ground, grasslands, burnt grass, semi-desert, light bush. May occur far from water. Solitary or in pairs. Perches on stones, tops of small bushes, telegraph wires but apparently not trees. Restless, flicks tail up and down, sometimes quivers wings. Never flies far. Confiding, permitting close approach. Hunts by sallying to ground.

Food. Insects, including grasshoppers, mantids, beetles, ants, larvae.

Breeding Habits. Apparently monogamous ground-nester; probably territorial.

NEST: of grass seed, fine grass and hair, on ground under bush, rock or fallen branch.

EGGS: 3–4, mean (3 clutches) 3·3. Subelliptical; smooth, slightly glossy; turquoise-blue, entire surface with red-brown freckles or small spots which sometimes coalesce to form crown at big end. SIZE: (n = 7) 18·5–21 × 15–15·5 (19·7 × 15·4).

LAYING DATES: Apr–June.

DEVELOPMENT AND CARE OF YOUNG: both parents bring food. Adult plumage gained progressively during first year by combination of wear and moult: final traces lost at first full moult, at age of about 12 months.

References
Archer, G. and Godman, E. M. (1961).
Tye, A. (1986).

Oenanthe oenanthe **(Linnaeus). Northern Wheatear. Traquet motteux.**

Plate 30
(Opp. p. 449)

Motacilla oenanthe Linnaeus, 1758. Syst. Nat. (10th ed.) 1, p. 186; Europe (= Sweden).

Forms a superspecies with *O. phillipsi*.

Range and Status. Breeds NW Africa, Eurasia, Greenland and NE Canada, Alaska and adjacent NW Canada; winters Africa, mainly south of Sahara.

Intra-African and Palearctic migrant. Breeds (*O. o. seebohmi*) Morocco, Algeria in Middle and High Atlas and Aurès; this race winters S Morocco (Ouarzazate), Western Sahara, Mauritania (in SW, but mainly east of 14° 30′W, where perhaps at least 50,000 individuals winter: Browne 1982); uncommon to rare Senegal, Mali; vagrant Cameroon (De Greling 1972), SE Libya. Birds from outside Africa common to abundant on passage N Africa, uncommon to rare in winter SW Morocco, Algeria south of Atlas, extreme S Tunisia, coastal SE Egypt north to Quseir. Common winter visitor W Mauritania, perhaps more common in SW (total population south of 18° 2′N *c.* 150,000 birds in 77,000 km²; birds seen/h 0·59, per km² 2·0: Browne 1982); common to abundant winter visitor Sahel zone in N Senegal (up to 10 per 25 ha, once 17 per 25 ha: Morel 1968), Mali (mainly 14–18°N, further south in dry or degraded habitat), Niger, common to abundant to 16°N, frequent Aïr; frequent SE, uncommon SW Burkina Faso; in Chad abundant on passage, common winter visitor Ennedi and

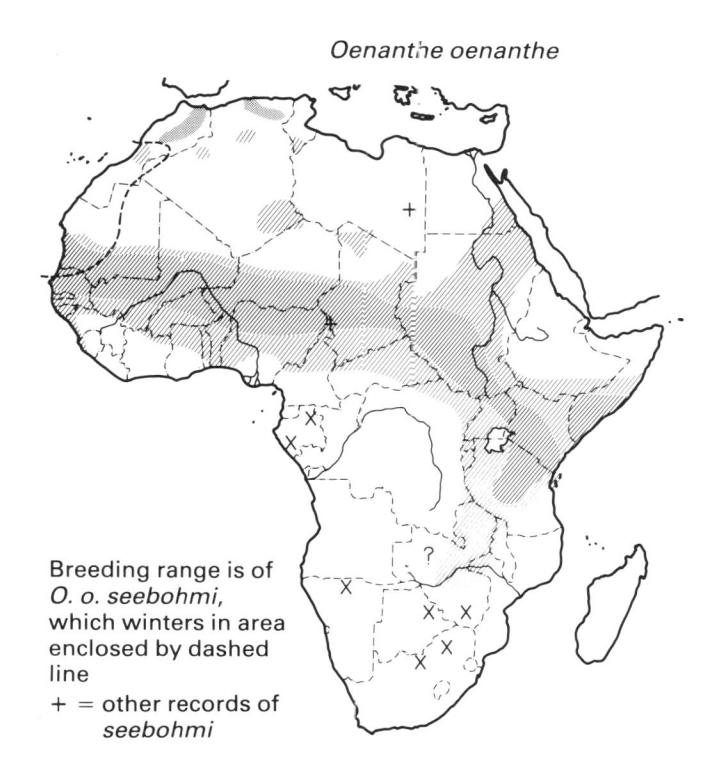

Breeding range is of *O. o. seebohmi*, which winters in area enclosed by dashed line
+ = other records of *seebohmi*

Sahel, frequent saharan and soudanian zones (mainly north of 11°N and more common west of 20°E), one record June, Tibesti; in Sudan abundant on passage, common to very abundant winter visitor west of White Nile, abundant Bahr-el-Ghazal, common Darfur, elsewhere frequent to common (prefers S sahelian and soudanian zones, mainly south of 13°N); frequent to common on passage Ethiopia, wintering frequently in S; rare NW Somalia, locally common to abundant winter visitor in S. Frequent passage migrant, uncommon winter visitor Gambia and Casamance (often near coast, but recorded throughout); rare coastal and E Guinea and coastal Sierra Leone; vagrant NE Ivory Coast, N Ghana; frequent NW Benin; common to abundant N Nigeria (uncommon south of 10°N, straggling to coast especially in W); frequent to common W Cameroon (N Bamenda Highlands) and N Cameroon; 2 records Gabon: Nyanga, Nov (D. E. Sargeant, pers. comm.), Makokou, Mar; frequent in winter N central Central African Republic and NE Zaïre (Uele District); frequent on passage, rare winter visitor Rwanda; frequent to common N and E Uganda, W and S Kenya (to coast in SE), Tanzanian highlands and uplands; uncommon S and W Uganda, N and E Kenya; 2 single records oversummering birds, Tsavo; vagrant Zanzibar; uncommon to frequent Zambia, Malaŵi in winter; one record July, Zambia, and occasional records all months, Malaŵi; rare N Zimbabwe; vagrant Zimbabwe south of 18°S, Namibia (Etosha Nat. Park), Botswana and Transvaal. Perhaps 124 million individuals winter in Africa (Moreau 1972).

Description. *O. o. oenanthe* (including '*integer*', '*rostrata*', '*oenanthoides*'): breeds N Europe, N Asia to Siberia and Alaska, winters Sahel and E Africa (but some breeding N Europe have characters of next race, and birds with characters of both races breed side by side in Siberia and Mongolia). ADULT ♂: forehead and narrow supercilium creamy white or buff; upperparts from crown to upper rump, including scapulars, grey more or less tinged with olive-brown, strongly so on upper rump; lores and ear-coverts black; lower rump and uppertail-coverts white. Rectrices basally white, with distal two-thirds of central pair and distal third of others black, all tipped buff when fresh. Underparts buff, richer on throat and breast, paler on chin and belly. Remiges and wing-coverts black, tipped and (except primaries) fringed buff when fresh; underwing-coverts, axillaries and underside of remiges dark smoke-grey. Bill, legs and feet black; eye dark brown. ADULT ♀: forehead and narrow supercilium buff; upperparts olive-brown, sometimes tinged grey and occasionally purer grey like ♂; lores dark brown; ear-coverts bronze; rump and tail as ♂; underparts dull yellow-buff, deeper on throat and breast; wing as ♂ but black replaced by dark brown; underwing-coverts and axillaries grey. SIZE: wing, ♂ (n = 53) 94–104 (98·8), ♀ (n = 31) 91–98 (94·3); tail, ♂ (n = 52) 51–63 (56·5), ♀ (n = 31) 47–60 (53·1); bill, ♂ (n = 51) 16–20 (18·1), ♀ (n = 30) 17–21 (18·2); tarsus, ♂ (n = 51) 25–29 (26·9), ♀ (n = 31) 25–28 (26·3). There is a cline: birds small in W, large in E. WEIGHT: Senegal (Jan) 1 ♀ 24, 1 unsexed 24; Algerian Sahara (spring migration) unsexed (n = 3) 21·4–29·3 (24·7); (Ennedi, Sept) unsexed (n = 2) 20·5, 24·5; NE Nigeria (Mar) unsexed (n = 10) 20–33 (25·7); Aldabra (Mar) 1 ♀ 18·5; Kenya, Zambia, Zimbabwe (Nov–Jan) ♂ (n = 3) 22·2–27 (24·7), unsexed (n = 5) 20·8–24·9 (22·8).

FIRST WINTER: ♂ and ♀ like adult ♀ but remiges of both sexes dark brown; belly richer buff when fresh; wing shorter than adult of respective sex.

FIRST SUMMER: ♂ like adult ♂ but wing dark brown; often browner on upperparts.

JUVENILE: upperparts olive-brown, scalloped with sepia and spotted with buff; underparts buff, scalloped with dark brown; wing brown with broad deep buff-brown fringes.

NESTLING: patches of grey down on crown, spine, and shoulders; pterylae on back, nape, head, flanks and wings visible as dark patches.

O. o. libanotica (Hemprich and Ehrenberg) (including '*nivea*', '*virago*', '*argentea*'): breeds S Europe from Spain to Middle East, S and central Asia to W Turkestan (but some breeding in this area have characters of last race), and Siberia (where mixes with last race); winters Sahel and E Africa. ADULT ♂: like ♂ *oenanthe* but forehead and supercilium pure white and often broader; upperparts purer grey, less tinged olive-brown; upper rump strongly tinged brown (not olive); underparts paler, almost white, tinged buff on throat and upper breast; black tail-band averages narrower, usually half of central pair and one-quarter of others. ADULT ♀: like ♀ *oenanthe* but upperparts average purer brown or grey-brown, less tinged olive; underparts average paler buff. Ranges of colour variation in the 2 subspp. overlap. SIZE: wing, ♂ (n = 26) 93–102 (97·7), ♀ (n = 13) 90–99 (94·0); tail, ♂ (n = 26) 52–60 (56·8), ♀ (n = 13) 50–58 (53·8); bill, ♂ (n = 26) 17–20 (18·6), ♀ (n = 13) 15–19 (17·5); tarsus, ♂ (n = 25) 25–29 (27·1), ♀ (n = 13) 25–27 (25·9). Size cline, as in last race.

O. o. leucorrhoa (Gmelin) (including '*schioeleri*'): breeds Iceland, Greenland, NE Canada; winters W Africa. ADULT ♂: like ♂ *oenanthe* but throat and breast average richer buff (broad overlap between the 2 races). ADULT ♀: no consistent plumage differences. Both sexes larger (overlap with western populations of *oenanthe* and *libanotica* is minimal). SIZE: (12 ♂♂, 6 ♀♀) wing, ♂ 96–107 (103), ♀ 97–102 (99·0); tail, ♂ 55–61 (58·6), ♀ 53–57 (55·2); bill, ♂ 17–19 (18·3), ♀ 18–19 (18·3); tarsus, ♂ 27–29 (27·8), ♀ 27–29 (27·7). Icelandic '*schioeleri*' fall within the lower part of the size range.

O. o. seebohmi (Dixon): breeds Atlas Mountains, winters NW and W Africa. ADULT ♂: like ♂ *libanotica* but sides of neck, chin and throat black (often meeting black of wings), very rarely buff, sometimes only partly black; underwing-coverts and axillaries near-black. ADULT ♀: like ♀ *libanotica* but upperparts vary from brown to grey (like ♂); chin, throat, sides of neck and lores variable, often dusky grey but can be typical ♀ buff, or mottled black and buff, or almost as black as ♂. SIZE: wing, ♂ (n = 5) 94–99 (97·0), ♀ (n = 5) 89–95 (91·8); tail, ♂ (n = 5) 58–64 (60·0), ♀ (n = 5) 50–57 (54·0); bill, ♂ (n = 5) 17–20 (18·6), ♀ (n = 4) 17–20 (18·5); tarsus, ♂ (n = 5) 26–28 (26·8), ♀ (n = 5) 25–28 (26·2).

FIRST WINTER: ♂ as adult ♀ but throat often with some dark mottling.

JUVENILE: as *oenanthe*.

TAXONOMIC NOTE: *O. o. seebohmi* is an incipient species.

Field Characters. Adult ♂ of pale-throated races distinguished from all other wheatears by grey upperparts and black eye-stripe. Adult ♂ *seebohmi* separated from other black-throated wheatears except allopatric Somali Wheatear *O. phillipsi* by grey back. ♀ and first-winter birds of both sexes resemble other white-rumped 'brown' wheatears, most readily recognized by tail pattern: black band broader than Hooded *O. monacha*, Black-eared *O. hispanica* and Pied *O. pleschanka* (though approached by some ♀♀ of eastern race *melanoleuca* of Black-eared, but latter more slender, flick wings more and have different call note); narrower than Isabelline *O. isabellina*, ♀ Desert *O. deserti*, Red-breasted *O. bottae* and immature Capped *O. pileata*. Approached most closely by Isabelline, immature Capped and Red-breasted but body plumage paler than last. Distinguished from Isabelline by darker wing, contrasting more with body plumage, and by head-bob (less emphatic) and tail-wag (see below); from immature Capped by more uniform, unspotted upperparts. Tail pattern almost identical with ♀ Finsch's Wheatear *O. finschii*, but some ♀ Finsch's have dark throat and most have paler greyish upperparts.

Voice. Tape-recorded (60, CHA – 53, 62, 73, B). Song comprises brief phrases *c.* 1 s long, separated by pauses of *c.* 4–8 s; or longer, more continuous song. Phrases include chacks, creaks, whistles, trills and mimicry; often begin with a few, quiet chacks. ♂ sometimes sings in winter quarters. Alarm calls 'chack' and a squeaky whistle 'weet', given singly or in combination. A buzzing call, like drawing fingernail over comb and resembling begging call of juveniles, occasionally used when territory threatened by dominant species, e.g. Isabelline Wheatear.

General Habits. Inhabits mainly short-grass acacia steppe and degraded savanna, also open cultivated land, wadis, barren rocky hills, burnt ground. Prefers areas with rocks, anthills or bushes for perches. On migration often in subdesert. Altitudinal range sea level to 3000 m; corresponds with habitat availability, e.g. not below 1000 m in Malaŵi, and on montane grasslands in W Cameroon. In W Africa occupies a belt in Sahel south of, but overlapping, range of Isabelline Wheatear; in E Africa often at higher altitudes or in hillier, less arid and more wooded country than Isabelline. Where range overlaps that of Capped Wheatear, prefers less arid areas until latter departs southward. *O. o. seebohmi* breeds in rocky terrain with low, sparse scrub, usually above 1500 m; at lower altitudes adapts to degraded bushy country.

Usually solitary and, when not breeding, both sexes individually territorial, even at migration stopovers; occasionally in pairs. Territory size *c.* 2–4 ha (Senegal, Zambia); radius *c.* 90 m (Kenya). Territories defended interspecifically against other wheatears, including dominant Isabelline, Capped, Mourning *O. lugens* and subordinate Pied, Black-eared; occasionally overlaps territory with Isabelline, Pied. Subordinate to shrikes *Lanius* spp., Mountain Rock-Thrush *Monticola saxatilis* and Spotted Flycatcher *Muscicapa striata*; displaced or evicted by them, but often has overlapping territories. Dominance relationships may be reversed by circumstances: established Common Stonechat *Saxicola torquata* evicts passage Northern Wheatears but latter dominant when both resident. Defence includes song, calls, display (tail-spreading, wing-lowering revealing white rump), aerial chases, fights on ground and in air (beating with wings, grappling with feet). Attempts defence against dominant birds (including shrikes) by harassing them. Subordinate individuals (♀♀, first-winter birds) or all birds in presence of dominant species (especially Isabelline Wheatear) avoid conspicuous behaviour and prominent perches.

Perches on stones, anthills, fences, bushes, low branches of trees; flies low between perches. Wags tail after landing or running, bobs head when disturbed. Tail-wag quick down, slow up-and-down; head bob rapid, <0·1 s (faster than Isabelline). Forages by dash-and-jab, ground and aerial sallying; occasionally sally-gleans. Relative use of 3 main foraging techniques depends on habitat structure, particularly length of vegetation (dash-and-jab more when short) and availability of elevated perches. In Senegal 80% of perches used for sallying were ≤1 m high (Tye 1984). When disturbed often flies off to elevated perch, stands erect; evades pursuit by flying low and turning sharply behind objects; reluctant to be driven from territory. Reacts to overflying raptors by looking up; to falcons by crouching on ground or moving to lower perch; and to bird-hunting species (e.g. Red-necked Falcon *Falco chicquera*) by flying to cover in bush or hole. Active before sunrise and until near-dark; rests in shade around midday in hot areas (>30°C in Senegal: Tye 1982). Can be caught at night, when dazzled by lights, at edges of dirt roads where possibly roosting (Sudan, Zambia); also roosts on ledges in sides of ditches, and possibly in rodent holes.

O. o. seebohmi leaves Atlas early Sept to late Oct; on passage Zouérat (N Mauritania), Tindouf (S Morocco) early Apr; first appears in breeding areas in Atlas late Mar, but some still in winter quarters, N Senegal, Apr. Main route of *O. o. leucorrhoa* into Africa is from Spain to Morocco/Algeria and vice versa; 23 ringed in Britain recovered Morocco (20), Algeria (3). *O. o. oenanthe* and *libanotica* pass into Africa on broad front across Mediterranean. *O. o. oenanthe* ringed in Britain, Netherlands, Belgium, Germany, Switzerland, Denmark, Norway, Russia, recovered Morocco; ringed Britain, Belgium, Sweden, Latvia, recovered Algeria; ringed Belgium, recovered Tunisia. Passage through Egypt over open desert as well as Nile Valley; in autumn appears to migrate E–W through Sudan; in E Africa main route is Rift Valley and east of L. Victoria (hardly occurs west of lake). The only Palearctic wheatear to winter commonly south of equator. Main passage, arrival and departure dates: Western Sahara, end Jan to Apr; Morocco, end Aug to mid-Nov (common mid-Sept to mid-Oct), end Feb to late May (peak end Mar to early May); Algeria, *O. o. oenanthe* Aug–Oct, *leucorrhoa* mainly Nov–Dec (though some *leucorrhoa* reach Sahel by Oct), *oenanthe* Jan to end Apr, peak Mar–Apr (east of 0°, mainly Mar onward, i.e. 1 month later), *leucorrhoa* until May; Tunisia, Aug–Oct, late Feb–May (peak mid-Mar to Apr); Libya, Egypt, mid-Aug to mid-Nov (peak Sept), late Feb to late May (peak late Mar to mid-Apr). Mauritania, arrives Aug (Sept in SW), departs Jan–Apr (latest June); Senegal, arrives Sept (*leucorrhoa* perhaps not until late Oct), departs late Jan to early May; Gambia, arrives Sept (mainly Nov), departs Feb–Mar; Mali, arrives Oct (*leucorrhoa* Oct–Nov), departs Apr to early May; SW Niger, Sept–May; Chad, arrives late Aug to Oct (common from end Sept), departs Jan–May (peak Mar); Nigeria, arrives end Sept (common Nov), departs by late Mar (possible peak early Mar, L. Chad); Zaïre, arrives Oct, departs by end Mar (possible peak mid-Mar); Sudan, arrives late Sept (peak late Sept to Oct), departs early May (peak Mar to early May); Ethiopia, arrives late Aug (common from mid-Sept, peak Sept–Oct); Somalia, present late Aug and Feb to mid-May. Rwanda, Uganda, Kenya, arrives mid-Sept (peak Oct), departs Mar to mid-Apr; Tanzania, arrives Sept–Nov (peak Oct in N), departs Feb–Mar; Zambia, arrives Oct, departs Mar–May; Malaŵi, Zimbabwe, arrives late Sept to Oct, departs Mar. Autumn passage appears heavier in Algeria, Kenya, Uganda, with no obvious spring peak in Kenya, Uganda; both passages about equal in Egypt. Spring migration commences very early: passage lasts *c.* 55 days in Libya, *c.* 5 months SW Algeria, but individuals cross Sahara quickly. *O. o. oenanthe* and *leucorrhoa* both reach N Africa with *c.* 6 weeks to spare before breeding. ♂♂ migrate in spring ahead of ♀♀.

Passage birds in Tunisia Mar and Aug very fat, but in Sept not fat (Heim de Balsac and Blanchet 1951–1957); in Nigeria birds on passage or about to depart had body weights 20–33 g, fat content 2–33% (Ward 1963). 3 shot on migration in Algerian desert had body weights 21·4–29·3 g, fat index (% fat/lean dry weight) 53·5–105·3, water index (% water/l.d.w) 146–178, while 2 found freshly dead had body weights 12·8, 13·9, fat index 3·0, 6·6, water index 139, 164; i.e. died when fat, not water, exhausted (Haas and Beck 1979). In N Nigeria fat accumulated when insect abundance declining and high temperatures force rest in shade for 6 of 12 daylight h; but energy expenditure low at that time (Ward 1963).

Some evidence for local movements related to rainfall: in Serengeti first appears in areas which have had recent rain, and movement Sept–Oct depends on distribution of local storms; confined in early Oct to areas which had at least 25 mm rain, i.e. mostly woodland, spreading later onto plains as Capped Wheatear begins to depart. Leaves many areas of central Kenya as grass grows after Nov rains, numbers increasing again Dec–Jan as grass dries out and burnt patches appear.

Food. Mainly insects; in Africa including termites, ants, bees, beetles; also millipedes, small snails; rarely, seeds and berries. In 3 stomachs, Zimbabwe, termites, beetles (especially weevils) and ants predominated; preference for ants and termites may account partly for attraction to burnt areas: these insects not greatly affected by fire but more easily seen on bare burnt ground.

Breeding Habits. Following details apply to *O. o. seebohmi*; for *O. o. oenanthe*, see Cramp (1988). Monogamous, territorial, ground or crevice-nester.

NEST: placed under rock, in wall or pile of stones, in deep cleft, under dense bush or cushion of decaying vegetation.

EGGS: 4–6, NW Africa mean (7 clutches) 5·3. Very pale blue or almost white; more or less indistinguishable from eggs of *O. o. oenanthe*. SIZE: (n = 23, Morocco/Algeria) mean 23 × 16·3; (n = 10, NW Africa) 20·5–22·6 × 15·6–16·2 (21·6 × 15·9).

LAYING DATES: Morocco, Algeria, Apr–June (not until May in High Atlas).

References
Borrett, R. P. and Jackson, H. D. (1970).
Cramp, S. (1988).
Mackworth-Praed, C. W. and Grant, C. H. B. (1951).
Mayaud, N. (1951).
Tye, A. (1984).

Plate 30
(Opp. p. 449)

Oenanthe pleschanka (Lepechin). **Pied Wheatear. Traquet pie.**

Motacilla pleschanka Lepechin, 1770. Novi. Comm. Acad. Sci. Petropolitanae 14, p. 503, pl. 24; Saratov on Volga R.

Forms a superspecies with *O. hispanica*.

Range and Status. Breeds Cyprus and Eurasia from Black Sea to Mongolia and NW India; winters Africa, Arabia and India.

Palearctic migrant. Common to abundant winter visitor to Sudan, Ethiopia up to 2500 m, Somalia, E, N, central and S Kenya; frequent W Kenya, Uganda, Tanzania south to *c.* 4°S; uncommon E-central Chad, Zanzibar; rare Libya (records NW Mar, SE Oct, Mar); vagrant South Africa (Mtunzini). Frequent on passage in Egypt (including Cyprus race), some probably wintering in S. Cyprus race apparently winters Sudan, Ethiopia.

Description. *O. p. pleschanka*: breeds E Europe, Asia. ADULT ♂: forehead, crown, nape and hindneck dull brown (autumn), wearing to white (spring), with partly-worn plumage mottled; indistinct supercilium cream; lores, ear-coverts and sides of neck black, just meeting wings; chin, throat and upper breast black or rarely ('*vittata*' phase) creamy white merging into buff on breast; mantle, back and scapulars mottled brown and black, wearing through a scalloped stage to black; rump and uppertail-coverts white, occasionally tinged buff. Rectrices basally white, distal half to two-thirds of central pair and band across others near tip black, band broader on outer pairs and extending up outer web; amount of black on tail somewhat variable; all rectrices tipped white when fresh. Breast, belly and undertail-coverts buff, richer on breast, which may be partly obscured by a dull brown wash. Remiges and wing-coverts very dark brown, when fresh tipped buff; secondaries and tertials also fringed buff when fresh; underwing-coverts and axillaries black; underside of remiges silver-brown. Bill black; eye dark brown; legs and feet black, soles grey. ADULT ♀: like ♂ but upperparts mottled dull brown with pale feather-edgings; lores, ear-coverts, chin, throat and breast greyish fawn, becoming mottled with dark grey on chin and throat through wear; remiges and wing-coverts browner; underwing-coverts and axillaries brownish grey. SIZE: wing, ♂ (n = 49) 89–102 (94·7), ♀ (n = 26) 85–98 (92·2); tail, ♂ (n = 47) 55–67 (61·4), ♀ (n = 26) 53–63 (59·3); bill, ♂ (n = 49) 15–19 (17·1), ♀ (n = 26) 15–19 (17·0); tarsus, ♂ (n = 49) 21–25 (22·5), ♀ (n = 24) 20–23 (22·0). WEIGHT: Kenya (Nov) 1 ♂ 14·3; E Africa (winter) ♂ (n = 2) 17·5, 20, ♀ (n = 2) 20, 20.

FIRST WINTER: like adult but ♂ has remiges brown (like adult ♀) and chin and throat more heavily scalloped with buff.
O. p. cypriaca (Homeyer): breeds Cyprus, winter range unclear, apparently throughout winter range of nominate race. ADULT ♂: like black-throated ♂ of nominate race but much smaller; both subspecies have a similar range of variation in extent of black on tail and in colour of underparts and crown but cypriaca has reduced black on tail more rarely than nominate does. ADULT ♀: like ♂ but buff scalloping on chin and throat in fresh dress is broader (blacker on throat than nominate ♀); crown does not wear to white but remains dull brown with greyish buff spots near nape; back does wear, as in ♂ (darker than nominate ♀), but remains browner than ♂. SIZE: wing, ♂ (n = 17) 82–89 (85·2), ♀ (n = 7) 81–85 (82·3); tail, ♂ (n = 17) 53–60 (57·2), ♀ (n = 7) 53–61 (56·6); bill, ♂ (n = 16) 15–18 (16·5), ♀ (n = 7) 16–18 (16·4); tarsus, ♂ (n = 16) 21–23 (21·6), ♀ (n = 7) 21–22 (21·6). WEIGHT: Chad (Mar) 1 ♂ 23.

TAXONOMIC NOTE: the 'Cyprus Wheatear' O. p. cypriaca has been regarded as a separate species but the differences (in song, sexual dimorphism, and size) are only at the level used to separate subspecies in other wheatear species.

Field Characters. ♂ in fresh plumage distinguished from other wheatears except Black-eared O. hispanica by dark brown wash on back; from latter by greater extent of black on throat (in black-throated phase) and by black background to brown on back, not white/sandy. In worn plumage closely resembles Mourning Wheatear O. lugens, but undertail-coverts white, not rufous-tinged. Hooded Wheatear O. monacha has more black on breast and nape. ♀ of Cyprus race similar to ♂ (both races), but nominate ♀ in fresh plumage indistinguishable in field from ♀ of eastern race of Black-eared Wheatear (melanoleuca, q.v. for further distinguishing features), but in worn plumage some can be identified by mottling on breast; ♀ O. h. hispanica is usually paler, sandier on upperparts. ♀ darker, smaller and more slender, with less black on tail, than Isabelline O. isabellina and ♀ Northern O. oenanthe Wheatears; closely resembles ♀ Finsch's Wheatear O. finschii, but more slender.

Voice. Tape-recorded (McVIC – 53, 62, ROC). Song of nominate race very brief (c. 0·25 s) and monotonous; c. 3 notes, the first 2 simple chirps, the last trilled: 'tri tri trree'. Slight variations occur, e.g. 4 notes, middle notes trilled. Sometimes includes mimicry. Song repeated many times, with 3–5 s gaps between phrases. Apparently sometimes emits 2 notes simultaneously (Jackson and Sclater 1938). Song of Cyprus race cicada-like, a rapid series of buzzes, c. 3–6 per s, a series lasting c. 2–10 s: 'bizz-bizz-bizz . . .' (Sluys and Van den Berg 1982). Call a harsh 'zack'. Alarm note like drawing fingernail rapidly over comb; similar to begging call of nestling wheatears.

General Habits. Prefers less arid country than many wheatears, with more trees and bushes, though has no absolute requirement for them. Frequents plains with bushy areas and scattered trees, montane moorland, rocky hills, woodland fringes, cultivation. Sometimes attracted to burnt areas. Mainly 1200–2000 m in Somalia and Kenya, and commonest wheatear below 2500 m on high plateaux of Ethiopia; avoids lowland semi-desert in Eritrea. Solitary in Africa, though ♂ and ♀ may occasionally defend joint territory. Territorial, even at migration stopover points; also defends against other wheatears including White-crowned Black O. leucopyga, Mourning, Isabelline and Northern. Territories occasionally overlap those of Northern or Isabelline. Territory radius in Kenya c. 50 m (Leisler et al. 1983). Consistently subordinate to Isabelline and Northern (both in Kenya and outside Africa) and to 'Schalow's' O. lugens schalowi and Capped O. pileata (in Kenya). Aggressive towards Fiscal Shrike Lanius collcris, Whinchat Saxicola rubetra, Barn Swallow Hirundo rustica, Yellow Wagtail Motacilla flava and Mountain Rock-Thrush Monticola saxatilis. Sometimes sings in Africa (Feb, Mar). Hovers c. 60 cm above ground, probably over mammalian or reptilian predator (cf. other wheatears), uttering alarm note. Perches on trees and bushes more than many wheatears, mainly at height of 1–3 m. Forages mainly by ground sallying, also uses aerial sallying and dash-and-jab. At Tsavo, Kenya, 64% of 100 prey taken on ground, 30% in flight, 6% from bushes (Lack et al. 1980). Hops or runs; hops more in longer vegetation. Recorded roosting at night on ledges in sides of ditch (Morrison 1945). Present in winter quarters late Sept to May, mainly Oct–Mar. Autumn passage peak late Sept to Oct in Somalia. One ringed in Ethiopia retrapped same site following year.

Food. Insects; in Africa including beetles, ants, other Hymenoptera, Dipteran larvae, Hemiptera, ant lions, Lepidoptera (adults and caterpillars), grasshoppers; also mites, spiders and some seeds.

References
Christensen, S. (1974).
Cramp, S. (1988).
Sluys, R. and van den Berg, M. (1982)

514 TURDIDAE

Plate 30
(Opp. p. 449)

Oenanthe hispanica (Linnaeus). Black-eared Wheatear; Spanish Wheatear. Traquet oreillard.

Motacilla hispanica Linnaeus, 1758. Syst. Nat. (10th ed.) 1, p. 186; 'Hispania' (= Gibraltar).

Forms a superspecies with *O. pleschanka* (see Haffer 1977 and Loskot 1986).

Range and Status. Breeds around W Mediterranean and east through S Europe to Caucasus and Iran; winters in Sahel and Horn of Africa.

Resident, Palearctic and intra-African migrant. Breeds N Africa from Morocco south to Atlas, Algeria south to Saharan Atlas, Tunisia (common) south to *c.* 34°N, NW Libya (frequent), mainly near coast. Winters in Sahel: frequent N Senegal; rare Gambia; frequent to common Mauritania north to 18° 30'N; in Mali, 12° 30'–18°N; frequent extreme SW Niger, rare further north; Chad to *c.* 15°N; rare NE Burkina Faso and Niger; uncommon in Nigeria north of 12°N; common to abundant Sudan, 9–14°N in W, 10–15°N in E, vagrant *c.* 31°E, 5°N; possibly some winter as far north as S Egypt. Uncommon to frequent Ethiopian highlands; abundant N Somalia; vagrant Kenya (Athi R.). In Mauritania south of 18°N, 190,000 birds estimated to winter in 76,000 km^2, with 0·75 birds seen per hour, 2·5 per km^2 (Browne 1982); up to 3 birds per 25 ha, Senegal (Morel 1968). Common to abundant passage migrant N Africa from Western Sahara to Egypt.

Oenanthe hispanica

Description. *O. h. hispanica*: breeds W Mediterranean to Yugoslavia, winters Sahel east to Mali. ADULT ♂: upperparts rich russet-buff, head wearing to cream, progressively from forehead; lores and ear-coverts black; chin and throat polymorphic, black or cream (percentage of black throats increases towards E), occasionally intermediate; rump and uppertail-coverts white, occasionally tinged buff on rump. Rectrices basally white, distal three-quarters of central pair and distal one-quarter to one-third of others black, with more black on outer webs and outer feathers; all tipped white when fresh. Breast rich buff; belly and undertail-coverts buff. Remiges and wing-coverts very dark brown or black; remiges narrowly tipped buff when fresh, wing-coverts tipped and fringed buff when fresh; tertials more broadly tipped and fringed buff; scapulars black, fringed buff when fresh; underwing-coverts and axillaries black; underside of remiges smoke-grey. Bill, eye, legs and feet black. ADULT ♀: like ♂ but upperparts sandy brown; lores and ear-coverts bronze-brown; chin and throat cream or dusky (dark feather-bases showing to variable extent); black areas of wing and tail replaced by dark brown. SIZE: wing, ♂ (n = 90) 85–97 (90·6), ♀ (n = 34) 84–91 (87·4); tail, ♂ (n = 88) 55–66 (61·3), ♀ (n = 34) 52–63 (58·4); bill, ♂ (n = 88) 15–20 (17·1), ♀ (n = 32) 16–19 (16·9); tarsus, ♂ (n = 85) 20–25 (22·6), ♀ (n = 32) 21–24 (22·3). WEIGHT: Algeria (Apr) 1♂ 14·5; Kenya (Mar) 1 ♂ 18; Morocco (Sept) 1 unsexed 20 (Apr) unsexed (n = 2) 14·6, 21·5.

IMMATURE: like adult ♀ but upperparts and underparts grey-brown scalloped with brown, less so on belly; buff edgings on scapulars and wings broader; bill and feet horn, darkening with age.

NESTLING: (*c.* 2 days old) tufts of down on nape, spine, flanks and above eyes.

O. h. melanoleuca (Güldenstadt): breeds S Italy to Caucasus; winters Sahel west to Mali, and in Horn of Africa. ADULT ♂: like ♂ *hispanica* but upperparts russet-brown, heavily washed or mottled with grey (autumn), brown fading and grey wearing to buffy white, with grey mottling remaining to a variable extent (spring); chin and throat buff, or black, which often extends onto upper breast (percentage of black throats increases further east); black feather-bases may show on upper rump; black areas of tail often more extensive; breast powdery cinnamon-buff, fading to rich buff. ADULT ♀: like ♀ *hispanica* but upperparts darker, greyish brown; breast powdery buff or pale cinnamon-brown; underwing black or greyish.

IMMATURE: like adult ♀ but with broader buff tips and fringes on remiges; black-throated ♂♂ have chin and throat heavily scalloped with buff.

Field Characters. ♂ distinguished from similar ♂ Pied Wheatear *O. pleschanka* by pale, not black, back; pale-throated ♂ from all other wheatears by combination of black stripe through eye and bright, sandy, white or mottled brown upperparts. Black-throated ♂ also told from other black-throated wheatears except Desert *O. deserti* and Finsch's *O. finschii* by colour of upperparts; from Desert by much white in sides of tail, and black scapulars (sandy buff in Desert); from Finsch's by less extensive black on sides of neck, not (or barely) meeting black on wing; also less black in tail. ♀ similar to ♀ of several other wheatears, but slender build and tail pattern of inverted T distinguish it from many; narrower terminal tail-band (especially in *O. h. hispanica*), with more extensive black on outer pair of rectrices than Northern *O. oenanthe*. ♀ *O. h. melanoleuca* not certainly distinguishable in the field in fresh plumage (difficult or impossible in the hand) from ♀ Pied since ranges of plumage variation in the 2 species overlap, but in worn plumage ♀ Pied tends to show mottling on breast; ♀ *O. h. hispanica* has paler, sandier upperparts than ♀ Pied.

Voice. Tape-recorded (53, 62, 73). One song type high-pitched, of short phrases 1–3 s long, separated by 2–7 s pauses; each phrase begins haltingly, with 1–3 notes leading into a rapid scratchy jumble of whistles; mimicry sometimes included. Another type longer, not broken into phrases, consisting of similar notes mixed with call-notes; this type given in winter quarters and on breeding grounds. Calls, a sharp 'tack' or rapid series of tacks, and a plaintive whistle, often given in combination; also a buzzing 'beeezh' call like juvenile begging call, given when threatened.

General Habits. Breeds in dry, stony country, with shrubs, olive and cactus groves, vineyards, rocky hills, broken ground; up to 1200 m (higher outside Africa). Avoids dense forest. Although it does not require presence of trees or bushes, generally winters in less arid country, usually with more trees, than Northern and Isabelline *O. isabellina* Wheatears. Winter habitat mainly acacia savanna (often denser than that used by other wheatears); also wadis, cultivated ground, gardens, oases, rocky hills, burnt ground; up to 2000 m in Ethiopia. The two phases said to prefer different habitats in Chad (Salvan 1968).

On passage and in early spring, sometimes occurs in small parties, sometimes with Northern Wheatears. Otherwise normally solitary, apparently territorial. Interspecific territoriality or aggression recorded in Africa with Pied (Kenya), Northern and Isabelline Wheatears (Senegal). Occasionally sings in winter.

Perches habitually higher than Northern and Isabelline Wheatears, on bushes, trees, stones, fences, telephone wires. Flicks tail and wings. Flies low; sometimes hovers higher, especially in breeding season. Forages by ground and aerial sallying from perches up to 3 m high and by dash-and-jab. In Chad forages mainly from dawn to 08·00 h and from 15·00 to 17·00 h, resting 11·00–15·00 h (Salvan 1964). In Sudan, most active in evening, until 1 h after sunset.

Migrates across Mediterranean on broad front, then over open desert, as well as Nile Valley. Autumn passage in N Africa and Sudan mainly late Aug to Sept, stragglers to Nov. Reaches Mauritania Sept, Senegal mid-Sept to mid-Oct, Mali late Aug (but mainly Oct–Nov), Niger Oct, Chad late Sept to Nov, Somalia mid-Oct. Spring migration Morocco, end Feb to May; Algeria–Libya, late Mar to early May; Sudan and Egypt, Mar–Apr. Spring passage more marked than autumn in Morocco, Libya, Egypt. Departs Senegal Apr, Mauritania Apr–May, Mali, Niger Mar–Apr (straggling to May), Chad Jan–Mar, Somalia early Mar. Individuals may return to same winter home range in successive years (Sharland 1967). ♂♂ migrate in spring ahead of ♀♀.

Food. Mainly insects; in Africa includes ants, beetles, Hemiptera, mayflies (caught at swarms), larvae, grasshoppers; also small snails, worms, woodlice, mites, berries and seeds. Hemiptera eaten include strong-smelling species.

Breeding Habits. Apparently monogamous, territorial, ground- or hole-nester. Territorial defence mainly by ♂, which generally perches higher than ♀, probably to facilitate defence (Santos and Suárez 1985). Sings much on arrival in breeding areas, declining thereafter. ♂ has circling, fluttering song-flight with tail spread; sings also from perch. Defence also includes aerial chases, fights, displays with tail-spreading, head-raising, wing-lowering and flicking. Territory defended against congeners in Asia. Courtship by intruding ♂ recorded (Pineau and Giraud-Audine 1979); displaying ♂ flew rapidly in horizontal figure-of-8, amplitude *c.* 10 m, within 1 m of ground, in front of ♀, then landed near ♀, wings vibrating; display ended when ♀'s mate appeared. This display typical of courtship in Asia. Courtship also includes aerial chases.

NEST: rather flat cup, loosely constructed of dry grass and fibres, lined with hair, wool or down; in Asia ext. diam. *c.* 120, int. diam. 60, depth 20, weight 6 g; placed under rock or tussock, in crevice in wadi side, scree or ruin, or in burrow; usually in wide, shallow crevice; one found in old jam-jar, partly sunken in ground. In Eurasia, ♀ builds in *c.* 1 week, but ♂ may accompany her and mandibulate material.

EGGS: 4–5, occasionally 3 or 6; Morocco, Algeria and Tunisia mean (55 clutches) 4·6. Subelliptical; smooth, slightly glossy; rich greenish blue, spotted with brick red, mainly at big end, sometimes forming a crown, and sometimes with a few deep lilac patches. SIZE: (n = 27, NW Africa) 18·5–21·5 × 14·5–15·5 (20·0 × 15·0); (n = 8, Algeria) 20–21·6 × 15–15·6 (20·9 × 15·4).

LAYING DATES: N Africa, Apr–June. Often double-brooded; may commence building second nest while first brood still dependent.

INCUBATION: in Eurasia by ♀; period in N Africa *c.* 13 days.

DEVELOPMENT AND CARE OF YOUNG: in Asia all eggs hatch within 24 h; young fed by both parents; eyes open about day 5; wait at nest entrance from about day 10; emerge days 11–12; can fly by day 12; self-feed by days 14–16, including aerial sallying from day 18; independent days 20–22.

References
Brosset, A. (1961).
Cramp, S. (1988).

Plate 29
(Opp. p. 448)

Oenanthe monacha (Temminck). Hooded Wheatear. Traquet à capuchon.

Saxicola monacha Temminck, 1825. *In* Temminck and Laugier, Nouv. Rec. de Pl. Col. d'Ois. 60, pl. 359, fig. 1; Nubia (= Luxor).

Range and Status. Egypt and Arabia to Afghanistan and NW Pakistan.

Resident and partial intra-African migrant. Egypt east of Nile south to Aswan, and uncommon west of Nile; rare (no breeding records) N Sudan in Nile valley and on Red Sea coast south to Suakin; vagrant Djibouti and Egypt/Libya border near coast.

Description. ADULT ♂: forehead, crown and nape white; lores, ear-coverts, sides of neck, chin, throat, breast, mantle, back and scapulars black (breast and mantle finely scalloped white in fresh plumage); rump, tail-coverts and belly white. Rectrices white, with distal half of central pair black or very dark brown and extending up shaft (but sometimes only smudged brown on outer web); other rectrices smudged to a variable extent with black or sepia near tip, especially on outer web; all tail-feathers tipped white when fresh. Remiges and wing-coverts black or very dark brown, tipped white when fresh, especially secondaries; underwing-coverts and axillaries black or very dark brown; underside of remiges greyish sepia. Bill, legs and feet black; eye brown. ADULT ♀: entire upperparts and sides of neck greyish fawn; entire underparts sandy buff; rump and uppertail-coverts pinkish or cinnamon-buff; rectrices basally pinkish or cinnamon-buff, distal half to two-thirds of central pair brown and distal portion of others brown to a variable extent – up to one-third of outer web on outermost – with wedge across inner web; remiges and wing-coverts greyish brown, with paler fringe on basal half to three-quarters of inner web of remiges; wing-feathers tipped and fringed buff when fresh; underwing-coverts and axillaries cream and grey; underside of remiges smoky grey-brown. Bill, legs and feet black; eye brown. SIZE: wing, ♂ (n = 16) 101–109 (105), ♀ (n = 8) 98–105 (101); tail, ♂ (n = 15) 67–75 (70·4), ♀ (n = 8) 67–76 (70·5); bill, ♂ (n = 16) 22–24 (22·6), ♀ (n = 8) 21–23 (22·1); tarsus, ♂ (n = 16) 22–24 (22·9), ♀ (n = 8) 21–23 (22·1). WEIGHT: Iran (winter) 1 ♂ 22·5, (June) 1♀ 18.

IMMATURE: like adult ♀ but upperparts paler sandy fawn; rump paler buff; fringes to remiges and wing-coverts richer buff; light areas of tail white, dark areas brown and cinnamon; bill horn.

FIRST WINTER: ♂ like adult ♂ but black of mantle and underparts scalloped with white; tail brown, like ♀; remiges and wing-coverts like adult ♀, contrasting markedly with colour of upperparts; underwing-coverts mottled cream and brown.

NESTLING: unknown.

Field Characters. Combination of very little black on outer tail-feathers and white belly distinguishes ♂ from all other wheatears; White-crowned Black *O. leucopyga* has similar tail but black belly. ♂ most closely resembles Mourning Wheatear *O. lugens*, but has more black on breast, less on tail, longer bill and shorter legs. Buoyant, butterfly-like flight distinctive. ♀ distinguished from most wheatears by very pale buff plumage and pinkish tail with little black; Red-tailed *O. xanthoprymna* and Red-rumped *O. moesta* have deeper, rufous tail with broad black subterminal band, while Mourning Wheatear has cinnamon-buff only on undertail-coverts.

Voice. Not tape-recorded. Song a sweet, subdued warble (Meinertzhagen 1930) lasting *c.* 2 s, apparently only by ♂, from post or in flight. A rattling call heard in Middle East (Bundy and Sharrock 1986). Calls, harsh 'zack' and alarm call 'wit wit'.

General Habits. Inhabits desolate, unvegetated desert wadis and ravines, preferring those with steep sides. Also recorded in grassland with acacia trees (Sudan) and a mountain monastery (Egypt: D. Merrie, pers. comm.). Outside Africa also frequents buildings near cultivation. Usually solitary or in pairs. Both sexes defend large individual territories outside breeding season; usually interspecifically territorial with Mourning Wheatear, but sometimes ousted from winter territory by later-arriving Mourning Wheatears (Hartley 1949). Shy, diving for cover among rocks when alarmed. Usually silent and elusive. Hunts mainly by aerial sallying, up to 100 m high; also sallies to ground. In Iran used aerial sallying more in middle part of day, perhaps because terrestrial prey was hiding from heat. In Middle East recorded visiting water-troughs to eat large ticks from camels and other livestock, also perching on Nubian Ibex (A: I.

A

Willis, pers. comm.). Seen to capture and eat large dragonflies (Hartley 1949).

Mainly sedentary but may wander away from breeding areas in winter.

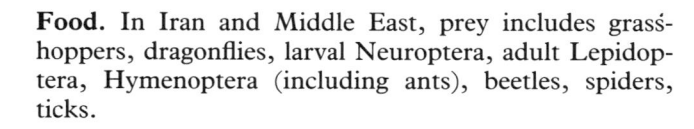

Oenanthe monacha

B

Breeding Habits. Monogamous and territorial. Breeding territory perhaps *c*. 1 km² in Iran. Aggressive, evicting conspecifics, other wheatears, Sand Partridge *Ammoperdix heyi*, Scrub Warbler *Scotocerca inquieta* and others. When breeding only ♂ sings, while perched or in circular song-flight, which is fluttering, usually circular, with tail spread (**B**). Nest and eggs not certainly described; one in Iran in hole in wadi bank.

EGGS: those attributed to this species are subelliptical; slightly glossy; pale blue, unspotted or with sparse, tiny red-brown specks concentrated at big end. SIZE: 21·6 × 15·6.

LAYING DATES: Egypt, Mar–Apr.

DEVELOPMENT AND CARE OF YOUNG: nestlings in Iran fed throughout day at 3–7 feeds/h, by both parents; in one day ♂ made 13 feeding visits and removed 1 faecal sac, ♀ 49 visits and 5 faecal sacs. Fledglings accompanied by parents.

Food. In Iran and Middle East, prey includes grasshoppers, dragonflies, larval Neuroptera, adult Lepidoptera, Hymenoptera (including ants), beetles, spiders, ticks.

References
Cornwallis, L. (1975).
Cramp, S. (1988).
Goodwin, D. (1957).

[*Oenanthe picata* (Blyth). Eastern Pied Wheatear. Traquet pie d'Orient.

Not illustrated

Saxicola picata Blyth, 1847. J. Asiat. Soc. Bengal 16, p. 131; between Sind and Ferozepure.

Range and Status. Breeds Iran east to Turkestan, Afghanistan and Kashmir. Winters S Iran to India, recently found wintering regularly (*O. p. picata*) in small numbers in Oman and United Arab Emirates (Fry and Erikson 1989). There are no firm records for this species in Africa: a 19th-century specimen of *O. p. opistholeuca* in the British Museum (Natural History), whose label mentions the word 'Nubia' but not as a definite locality, probably came from Asia, and a recent sight record on the Red Sea coast of Egypt (Farrow 1990) gives no detail of subspecies (important for identification of this polytypic and polymorphic species) or description, so the possibility of confusion with other species cannot be eliminated.]

Plate 29
(Opp. p. 448)

518 TURDIDAE

Oenanthe lugens (Lichtenstein). Mourning Wheatear. Traquet deuil.

Saxicola lugens Lichtenstein, 1823. Verz. Doubl. Zool. Mus. Berlin, p. 33; Nubia.

Range and Status. N and E Africa, Arabia and S Iran.
Resident and partial intra-African migrant and Palearctic migrant. Breeds NW Africa from Algeria to W Egypt, in narrow band along N edge of Sahara south of High Atlas and Saharan Atlas; status in S Morocco uncertain, reported present May but no definite breeding records. Generally locally common to abundant. Common and evenly distributed S Tunisia; in Libya mainly south of 100 mm isohyet and north of 30°N, though breeds sparsely in arid areas nearer coast; frequent E Egypt. Frequent to common resident in 3 isolated populations above 1200 m in mountains of Ethiopia, Somalia north of 10°N and S Kenya–N Tanzania. Vagrant Niger (Aïr) Nov, Mauritania (Cap el Zass, Zouérate, Kediet ej Jill).

Oenanthe lugens

h = *halophila*
l = *lugens*
ls = *lugubris*
p = *persica*
v = *vauriei*
s = *schalowi*

Description. *O. l. halophila* (Tristram): Morocco to NW Egypt. ADULT ♂: forehead, crown, nape and hindneck white, more or less obscured with pale grey-brown, especially on crown which is also sometimes lightly streaked dark grey; sides of neck, lores, ear-coverts, chin and throat black (joining black of wings); mantle, back and scapulars black; rump and uppertail-coverts white. Rectrices basally white, with distal half to two-thirds of central pair and one-fifth to one-quarter of others black, the black extending up the shaft; all tipped white when fresh. Underparts white, tinged pale orange-buff on undertail-coverts; remiges and wing-coverts very dark brown, tipped white when fresh; underwing-coverts and axillaries black; underside of remiges smoke grey, with white fringe on basal half of inner web, not reaching shaft. Bill, legs and feet black; eye brown. ADULT ♀: upperparts pale sandy brown, often with brown shaft-streaks on crown; lores, ear-coverts, sides of neck, chin and throat vary from black, lightly scalloped with buff, to lores and ear-coverts sandy brown, sides of neck mottled buff and grey, chin and throat buff. Underparts pale buff with faint orange tinge to undertail-coverts; rump, uppertail-coverts, tail and wings as ♂, but black replaced by brown; underwing-coverts and axillaries dark greyish brown. SIZE: wing, ♂ (n = 21) 89–96 (92·3), ♀ (n = 19) 83–91 (87·3); tail, ♂ (n = 21) 56–66 (60·7), ♀ (n = 19) 55–64 (58·4); bill, ♂ (n = 20) 18–21 (18·9), ♀ (n = 18) 18–19 (18·7); tarsus, ♂ (n = 18) 24–26 (25·2), ♀ (n = 18) 22–26 (23·9). WEIGHT: (Algeria, Nov–Dec) ♂ (n = 3) 22–25 (24·0), ♀ (n = 3) 19–22 (21·0).
IMMATURE: like adult ♀ but lores, ear-coverts, sides of neck, chin and throat as palest ♀♀; tips on wings and tail broader, and buff; underparts richer buff.
NESTLING: unknown for *O. l. halophila*, but see *O. l. schalowi*.
O. l. lugens: E Egypt, Middle East. Sexes alike: like ♂ *halophila* but crown and nape usually less tinged buff, purer grey or white; white on inner web of remiges broader, usually reaching shaft. Marginally larger than respective sexes of *halophila*. SIZE: wing, ♂ (n = 16) 90–99 (93·9), ♀ (n = 9) 83–93 (88·7); tail, ♂ (n = 16) 56–66 (60·8), ♀ (n = 9) 54–61 (58·2); bill, ♂ (n = 15) 18–21 (19·1), ♀ (n = 9) 18–20 (18·8); tarsus, ♂ (n = 16) 23–29 (25·6), ♀ (n = 9) 23–26 (24·6).
O. l. persica (Seebohm): breeds S Iran, winters Arabia, Sudan, possibly Egypt, but individual birds not certainly distinguishable from *lugens*. ADULT ♂: like *lugens* but crown and nape sometimes buffier or browner (like *halophila* or still browner); white on inner web of remiges intermediate between *halophila* and *lugens*, not usually reaching shaft, and differently shaped; undertail-coverts sometimes richer orange-buff.
ADULT ♀: nearly identical with ♂ but black of back and throat often replaced by very dark brown. Slightly larger. SIZE: wing, ♂ (n = 6) 90–99 (96·0), ♀ (n = 7) 90–95 (92·0); tail, ♂ (n = 5) 62–66 (64·0), ♀ (n = 7) 59–62 (60·1); bill, ♂ (n = 6) 18–20 (19·2), ♀ (n = 6) 19–20 (19·2); tarsus, ♂ (n = 5) 25–26 (25·2), ♀ (n = 7) 24–26 (24·7).
O. l. lugubris (Rüppell): Ethiopia. ADULT ♂: like ♂ *halophila* but forehead, crown and nape dirty brown, heavily streaked black; black of underparts extends to breast and flanks, to variable degree; belly dimorphic, either white, with mottling at junction with black of breast, or black. Mantle, back, rump and scapulars black; uppertail-coverts pale orange-buff. White of tail replaced by orange-buff; black on outer tail-feathers broader, one-third of length, and not extending up shaft; remiges and wing-coverts black, without white fringe.
ADULT ♀: entire upperparts, lores, ear-coverts and sides of neck dark sepia; chin, throat and breast greyish buff, heavily streaked and mottled with very dark brown, streaks extending onto belly and flanks; belly and undertail-coverts buff; uppertail-coverts pale orange-buff. Tail as ♂ or with black replaced by very dark brown; remiges and wing-coverts very dark brown, narrowly fringed brown when fresh; indistinct grey fringe on inner web of primaries; underwing-coverts and axillaries very dark brown. Smaller. SIZE: wing, ♂ (n = 12) 83–89 (85·2), ♀ (n = 4) 78–81 (79·0); tail, ♂ (n = 12) 53–61 (57·8), ♀ (n = 4) 53–55 (53·8); bill, ♂ (n = 11) 17–19 (18·1), ♀ (n = 4) 17–18 (17·5); tarsus, ♂ (n = 11) 21–25 (23·1), ♀ (n = 3) 21–23 (22·0).
O. l. vauriei Meinertzhagen: Somalia. ADULT ♂: as ♂ *lugubris* (white-bellied form) but forehead, crown, nape and hindneck pale greyish buff, streaked black; black of underparts extends only onto upper breast; narrow white fringe on inner webs of remiges. ADULT ♀: like ♀ *lugubris* but upperparts dirty brown mottled with sepia; underparts streaked as *lugubris* but

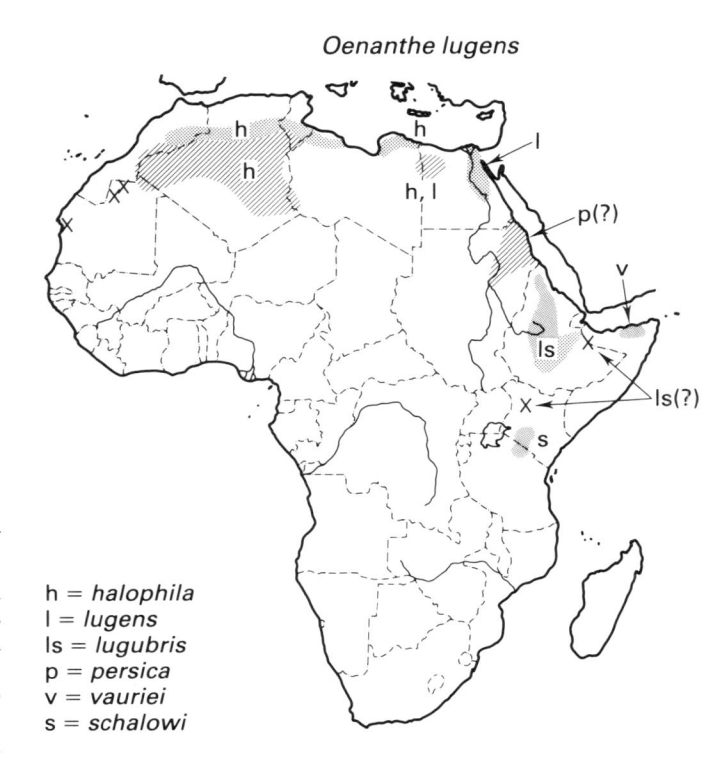

buff, less grey. SIZE: wing, ♂ (n = 5) 81–86 (83·6), ♀ (n = 5) 81–84 (81·8); tail, ♂ (n = 5) 52–58 (54·5), ♀ (n = 5) 53–56 (54·6); bill, ♂ (n = 5) 18–20 (19·2), ♀ (n = 4) 19–20 (19·5); tarsus, ♂ (n = 5) 23–24 (23·4), ♀ (n = 5) 20–23 (22·2).

O. l. schalowi (Fischer and Reichenow): S Kenya–N Tanzania. ADULT ♂: as white-bellied ♂ *lugubris* but back very dark brown, less pure black; crown less streaked, more uniform dirty brown in many birds. ADULT ♀: as ♀ *lugubris* but upperparts mottled sepia and dark brown. Slightly larger than *lugubris*. SIZE: wing, ♂ (n = 12) 82–90 (86·9), ♀ (n = 5) 82–86 (83·6); tail, ♂ (n = 12) 56–63 (58·9), ♀ (n = 5) 55–59 (57·0); bill, ♂ (n = 11) 17–20 (18·6), ♀ (n = 5) 18–19 (18·4); tarsus, ♂ (n = 12) 22–25 (24·0), ♀ (n = 5) 22–24 (23·0).

NESTLING: dark brownish pink with greyish tufts of down on head, base of wings and pelvic area, and shorter tufts on back.

TAXONOMIC NOTE: highly polytypic, with isolated populations representing at least incipient species, including: *schalowi*, *lugubris/vauriei*, extralimital *lugentoides/boscaweni* (Arabia) and *lugens/persica/halophila*. Although *schalowi* at one extreme appears strikingly different from *halophila* at the other, extremes are linked by a chain of subspecies along which characters change one by one, so best regarded as members of a single species.

Field Characters. The many races may be considered in 2 groups for purposes of field identification: '*lugens*' group (*lugens*, *halophila*, *persica*) and '*lugubris*' group (*lugubris*, *vauriei*, *schalowi*). ♂ *halophila* and both sexes of *lugens* and *persica* resemble several other N African wheatears. Distinguished from Hooded *O. monacha* by less black on breast and nape, more on tail; from Finsch's *O. finschii* by black back; from Pied *O. pleschanka* by rufous-tinged undertail-coverts. ♀ *halophila* told from other ♀ wheatears by throat (often dark) and undertail-coverts (tinged rufous); last feature differentiates it from dark-throated ♀ Finsch's. The *lugubris* group distinguished from other E African wheatears by cinnamon-buff rump and base of tail; larger than Red-tailed *O. xanthoprymna* and Red-rumped *O. moesta*, with more black or dark brown on body. Tail-wag and head-bob characteristic (see below).

Voice. Tape-recorded (60, C, CHA, GREG, JOHN, McVIC). Song variable: in breeding season normally loud and broken into 5 s phrases by 3–4 s pauses; in winter similar but may also be continuous, shrill or quieter. One or several phrases sometimes repeated many times. Often resembles songs of other wheatears; includes whistles, chacks, grating, squeaking and twanging notes. Both sexes sing in winter and spring. Contact and territorial call a soft but sharp 'chut, chut', louder when alarmed. Begging call of young a buzzing or rasping 'brrzh', as in other wheatears.

General Habits. Inhabits wild, rocky, desert country, including wadis, cliffs, kopjes, boulder-strewn slopes and scree, rough stony ground and shingle ridges, buildings and other masonry constructions; poorly vegetated or with scrub. In Ethiopia also on moors and ploughed land; in N Ethiopia (Eritrea) and N Africa sometimes in villages. Found in mountains throughout range; to over 4000 m in Ethiopia.

Occurs singly or in pairs. Territorial; in winter sometimes defends against *Ammomanes* larks and other wheatears, including Hooded, Red-tailed and White-crowned Black *O. leucopyga*, but sometimes ignores intruders, including Pied, Northern *O. oenanthe* and Capped *O. pileata*. Succeeds in evicting Hooded from latter's recently-established territories. Strength of interspecific defence not clearly related to breeding cycle. On arrival in Egyptian wintering areas, much song and fighting occurs, with over 65% of song incidents and 85% of territorial encounters occurring in first 3 of the 7–8 months spent in winter quarters. Winter territories held by single birds in Egypt *c.* 1·5 ha, larger in more barren areas (Hartley 1949). In winter territorial displays, perches upright on rock with tail horizontal and wings depressed, flicking wings and bobbing head, or spreading wings to show white patch while turning back on opponent; may call or sing. More active defence includes aerial chases and brief, fluttering fights. Sings mostly from rock, sometimes in fluttering flight with tail fanned; in autumn sings most in morning and evening; in winter most in mid-afternoon (Hartley 1949). Perches on and sings from rocks, telegraph wires, buildings, walls and banks. In autumn in hottest part of day selects shaded perches. Territories consistently contain broken ground, providing shade. Forages by dash-and-jab, aerial sallying, mainly within 1 m of the ground, and ground sallying. Quiet, sometimes very tame, rarely flies far. Has very emphatic head-bob, like that of Isabelline Wheatear *O. isabellina* (q.v.); tail-wag like that of Northern Wheatear: slow movement down then up, quick down. Takes cover from hawks under boulders. Although often lives far from water, will drink on occasion. Active shortly before sunrise and until *c.* 20 min after sunset. May sleep in hottest part of day.

Some movement of N African populations in winter, visiting Western Sahara, S Algeria (Hoggar), S and NE Libya, W Egypt. Birds in winter in Sudan may be N African or Asian breeders.

Food. Mainly ants; also beetles, grasshoppers and other insects. Outside Africa also recorded eating ant lions, termites, Hemiptera, other Hymenoptera, flies, spiders and plant material. In Iran, 76% of prey 1–5 mm long, 19% 5–10 mm, 4% 10–15 mm, 1% 15–20 mm (Cornwallis 1975). Young in Israel given termites, locusts, beetles, flies, larvae, wood lice, scorpions, berries (Zachai 1984).

Breeding Habits. Monogamous, territorial, cavity-nester; helpers at some nests. Breeding territories defended against other wheatears, including migrant Black-eared *O. hispanica* and Desert *O. deserti* and breeding Hooded. During courtship display, May, ♂ stood upright and hopped around crouching, wing-shivering ♀ before attempting to mount, then rapid aerial chase before landing and copulating (Bundy and Morgan 1969). ♀♀ may move between territories of ♂♂ (Haas 1986).

NEST: loosely built, flat cup of dry grass, stems and

roots, lined with rootlets, hair or wool. In Israel ext. diam. *c.* 110, height *c.* 50, depth of cup *c.* 25. Placed deep in narrow fissure in rocks, cliff or wall, or under rock, or in tunnel in bank. One nest in river bank consisted of 15 cm tunnel terminating in spherical cavity lined with dry grass and hair (*schalowi*). Built by ♀ (N Africa), in *c.* 2 weeks (Israel). Often (N and E Africa) builds rampart of small stones under and around nest, number varying with size of cavity; may make a wide entrance narrow by this means.

EGGS: N Africa 3–6, mean (26 clutches) 4·2; E Africa 1–3. Subelliptical; unglossed or slightly glossy; pale grey-blue to dull greenish blue, spotted with red-brown and violet, mainly at big end, often forming a crown; in *schalowi* spots finer, fainter, covering entire egg, or sometimes absent. SIZE: (Algerian Sahara) 21·7–23·8 × 15·8–16·3; (n = 21, Algeria) 19–20·8 × 14·5–16·4 (20·2 × 15·6); (n = 42, N Africa) 19·5–21·5 × 15–16·5; (Kenya) 20–20·5 × 15–15·5. Normally double-brooded (Egypt, E Africa).

LAYING DATES: throughout NW Africa, Mar–Apr; Egypt, Feb–June; Ethiopia, Mar–Aug. E Africa: Kenya, Apr–July; Region D, Oct–July with possible peak Mar.

INCUBATION: period 13–14 days (E Africa).

DEVELOPMENT AND CARE OF YOUNG: young in nest 15–16 days (E Africa). Fed by both sexes, and occasionally by more than 2 adults; first-brood ♂♂ may feed second-brood young, while young ♀♀ leave parental territory soon after fledging (Haas 1986). In Israel, nestlings 5–6 days old given 4·4 feeds/h by ♀, 4·1 by ♂; fledglings 15–18 days old given 9·3/h by ♀, 2·1 by ♂ (Zachai 1984).

References
Cramp, S. (1988).
Hartley, P. H. T. (1949).
Jackson, F. J. and Sclater, W. L. (1938).
van Someren, V. G. L. (1956).
Vaurie, C. (1950).

Plate 30
(Opp. p. 449)

Oenanthe finschii (Heuglin). Finsch's Wheatear. Traquet de Finsch.

Saxicola finschii Heuglin, 1869. Orn. Nordost Afr. 1, p. 350; Syria.

Range and Status. Breeds Asia Minor, Middle East to Afghanistan; winters Egypt to India.

Palearctic migrant; uncommon Nile delta to Suez Canal, rare Siwa/Qattara depression.

Oenanthe finschii

Description. *O. f. finschii*: only subspecies in Africa. ADULT ♂: forehead, crown and nape pale grey, more or less mottled with darker grey, sometimes washed with buff; lores, ear-coverts, very narrow supercilium, chin, throat, sides of neck and upper breast black (meeting black of wings); mantle and back very pale grey or creamy white, more or less washed with buff; rump and uppertail-coverts white. Rectrices basally white, distal half of central pair and distal one-fifth of others black, the black extending up shaft and outer edges; all tipped white when fresh. Lower breast, belly and undertail-coverts white; scapulars black. Remiges very dark brown or black, narrowly tipped (tertials also fringed) white when fresh; wing-coverts, underwing-coverts and axillaries black; underside of remiges smoke-grey. Bill, legs and feet black; eye dark brown. A rare variant exists with crown and/or underparts washed with red-ochre. ADULT ♀: like adult ♂ but upperparts greyish brown; lores and ear-coverts brown; chin and throat mottled grey and buff, usually darker lower on throat; black of tail and wing replaced by dark brown; underparts buffish white; underwing buffish grey. SIZE: wing, ♂ (n = 17) 84–94 (88·8), ♀ (n = 12) 84–87 (85·2); tail, ♂ (n = 16) 56–64 (59·1), ♀ (n = 12) 54–58 (55·9); bill, ♂ (n = 17) 17–19 (18·2), ♀ (n = 12) 16–20 (17·9); tarsus, ♂ (n = 17) 24–26 (24·9), ♀ (n = 11) 23–25 (24·0).

IMMATURE: as adult, but remiges of ♂ like adult ♀.

Field Characters. ♂ distinguished from all other wheatears except ♂ of eastern race of Black-eared, *O. hispanica melanoleuca*, by extent of white on upperparts: continuous from crown to tail. Told from eastern Black-eared by black extending onto upper breast, narrower white area on back and larger size. ♀ sometimes resembles ♂ in having dark throat, but has upperparts duller brown; this phase readily distinguishable from other wheatears except Desert *O. deserti*; latter has broad black band on tail. Pale-throated ♀ similar to other ♀ wheatears, especially Pied *O. pleschanka*, Black-eared and Northern *O. oenanthe*, and often difficult to distinguish in the field; tail-band broader than most Pied and Black-eared; upperparts often paler than Northern.

Voice. Tape-recorded (CHA, PAN). Song given by ♂, less frequently by ♀; varied, including whistles and harsh sounds, some mimicry; in winter short, subdued. Call when disturbed, 'chack'.

General Habits. Inhabits desert edge and neighbouring cultivation. Mostly solitary and individually territorial, on migration and in winter quarters. Little known in Africa. Winter territories 3–6 ha in Iran, where birds sometimes returned to same territory in successive winters; defended against all wheatears except Hume's *O. alboniger*. In Iran 97% of prey caught on ground, by dash-and-jab and ground sallying (Cornwallis 1975); elsewhere also uses aerial sallying. Rarely bobs head, but flicks wings and wags tail up, then down.

Food. In Asia mainly insects, especially ants and beetles, also seeds and other plant material.

References
Cornwallis, L. (1975).
Cramp, S. (1988).
Ticehurst, C. B. (1927).

Oenanthe moesta (**Lichtenstein**). Red-rumped Wheatear; Tristram's Wheatear. Traquet à tête grise.

Saxicola moesta Lichtenstein, 1823. Verz. Doubl. Zool. Mus. Berlin, p. 33; Egypt.

Plate 30
(Opp. p. 449)

Range and Status. N Africa and Middle East.
Resident and partial local migrant. Breeds coastal Western Sahara and SW Morocco, in Hauts-Plateaux, and in narrow zone south of Atlas from NE Morocco to central Tunisia, also in narrow (c. 40 km) zone near or along coast from Libya to W Egypt, with gap in drier Gulf of Sirte area. S limit in Libya well north of 100 mm isohyet; in Morocco N limit is at about 200 mm isohyet. Vagrant Nile valley (Egypt), Jan. Patchily distributed but locally common. 3 road transect counts Morocco, Sept, gave 1 bird/10 km, 1 per 4 km, 1 per 6·5 km (Knight *et al.* 1973); 9 transects Apr–June gave 2 birds/ 22 km (Blondel 1962).

Description. *O. m. moesta* (including '*theresae*'): only subspecies in Africa. ADULT ♂ (Algeria): forehead and broad supercilium creamy white; crown, nape and hindneck pinkish grey, tinged rufous when fresh; lores, ear-coverts, sides of neck, chin, throat, scapulars and back dull black, tinged grey when fresh; rump and uppertail-coverts cinnamon-buff. Rectrices basally rufous, distal three-quarters of central pair and one-third to half of others dark brown. Breast and belly greyish white, tinged pinkish when fresh; undertail-coverts pale cinnamon-buff. Remiges brown, tipped and fringed buff when fresh; wing-coverts black, broadly tipped buff when fresh; underwing-coverts and axillaries black; underside of remiges silver-grey. Bill, legs and feet black; eye brown. There is an east–west colour cline: SW Moroccan '*theresae*' has back more intense black, crown darker grey, rump, tail and undertail-coverts deeper rufous; Tunisian birds have back more grey-brown, crown paler pinkish grey, rump and undertail-coverts buff, tail paler cinnamon; Libyan and W Egyptian birds have rump and tail still paler. A few Libyan birds have strong rusty wash on underparts, crown and supercilium. ADULT ♀: forehead, crown, nape, sides of neck, lores and ear-coverts rich rusty brown; back and scapulars grey-brown, tinged rufous;

rump and uppertail-coverts rufous; tail and remiges as ♂; underparts pale pinkish buff, richer on undertail-coverts; wing-coverts brown, broadly tipped rufous-buff when fresh; underwing-coverts and axillaries grey. No general colour cline, but SW Moroccan *'theresae'* are darker, more grey-brown above and duller, browner, less russet on crown. SIZE: clinal; Morocco to Tunisia: wing, ♂ (n = 22) 90–99 (93·5), ♀ (n = 14) 84–91 (88·5); tail, ♂ (n = 22) 60–70 (65·0), ♀ (n = 14) 58–66 (62·1); bill, ♂ (n = 21) 19–22 (20·5), ♀ (n = 14) 19–21 (19·9); tarsus, ♂ (n = 22) 26–29 (27·8), ♀ (n = 14) 25–28 (26·8); Libya to W Egypt: wing, ♂ (n = 9) 86–93 (89·4), ♀ (n = 7) 84–89 (86·4); tail, ♂ (n = 9) 58–67 (62·4), ♀ (n = 7) 58–63 (60·1); bill, ♂ (n = 9) 20–21 (20·3), ♀ (n = 7) 19–21 (19·6); tarsus, ♂ (n = 8) 25–28 (27·0), ♀ (n = 7) 25–26 (25·4).

IMMATURE: like adult ♀, but upperparts more uniform grey-brown, tinged with rufous (no contrast between crown and back). Adult plumage attained at post-juvenile moult.

NESTLING: unknown.

Field Characters. Both sexes distinguished from other chats of the region except Red-tailed Wheatear *O. xanthoprymna* by rufous rump and broad black tail-band. Red-tailed Wheatear is smaller, slighter; black-throated race *xanthoprymna* separated from ♂ Red-rumped by brown upperparts, lacking pale crown. ♂ most closely resembles Mourning Wheatear *O. lugens*, but has prominent pale fringes on wings. ♀ further distinguished from other ♀ wheatears and from pale-throated race of Red-tailed (*chrysopygia*) by bright rufous crown. Song distinctive.

Voice. Tape-recorded (53, 60, JOHN). Typical song a highly characteristic series of *c*. 12–13 whirring trills of successively higher pitch, often preceded by 2–4 'tlik' notes, the whole song lasting *c*. 6–8 s and repeated after pauses of *c*. 4–6 s: 'tlik, tlik, trooooooee trooooooee trooooooee …'. Courtship song like brief version of this, consisting of 3–4 whistles followed by a trilling, rising warble; lasts 1–2 s and repeated after pauses of *c*. 5–9 s. Calls, 'chack', 'tlik', or a rapid series of them forming a rattle, and whistles. Grating juvenile begging call typical of wheatears.

General Habits. Inhabits sandy, stony or clay semi-desert with sparse, low bushes, also in chotts and sebkha (salt flats). Prefers flat ground; avoids cultivation, rocky areas and broken ground. Habitat similar to that of Desert Wheatear *O. deserti*, but where both species occur, Red-rumped said to take 'poorer' areas (Heim de Balsac and Mayaud 1962). Habitat always contains elevated perches and rodent holes.

Solitary or in pairs; some pairs maintain joint territory all year. Flights short, buoyant, undulating, butterfly-like. Perches on bushes, walls, telegraph wires; not shy. When alarmed, adult may dive into burrow and escape from another exit. Forages by dash-and-jab, less often by sallying to ground.

Mainly resident but most in E Morocco shift south in winter.

Food. Mainly insects, including beetles, caterpillars (up to 5 cm), grasshoppers and ants; occasionally green plant material. Beetles often beaten and elytra rejected. Most prey brought by one pair to their 3 young were pale form of scorpion *Scorpio maurus* (Jarry 1969).

Breeding Habits. Monogamous, territorial, hole-nester. ♂ seen driving off larks and a Desert Wheatear from near nest-hole (Smith 1971); ♀ may also defend territory. Disputes involve tail-fanning and aerial chases. Spring (typical) song (uttered from Feb on) often given in flight, with tail spread; also from ground or perch. ♀ sings, but not as frequently as ♂. Courtship includes a wing-fanning display, from Dec onward; ♂ and ♀ leap-frog with bounds of 5–10 m, giving antiphonal duet of courtship song.

NEST: large, of coarse grass and rootlets, lined with hair, down, feathers (of other birds), often snake skins, bits of paper, rag, string. Built by ♀, at end of burrow of jird *Meriones*, sand-rat *Psammomys* or other rodent; up to 2 m from entrance; in bank or flat ground. Burrow sometimes curved and branching and may have more than one entrance. Reputed to make second exit as route to escape from monitor lizards. Entrance often under bush. Occasionally builds in pile of stones or mud wall.

EGGS: 4–5, occasionally 2 or 3 (incomplete clutches?); NW Africa mean (6 clutches) 4·0. Subelliptical; smooth, slightly glossy; very pale greenish blue or grey-blue, thinly spotted with red-brown, mainly at big end; some unspotted, some nearly white (in same clutch as more 'normal' ones). SIZE: (n = 19, Algeria) 22–25·5 × 16–18 (24·2 × 16·8); (Tunisia) 23–24 × 15·5–16.

LAYING DATES: throughout NW Africa, Feb to early June. Those laying early may experience snow during incubation. First brood often immediately followed by second while first-brood fledglings still dependent; eggs often laid in same nest without nest being repaired. Possibly up to 3 broods raised per year: this may be the most fecund passerine of N African subdesert zone, in terms of length of breeding season and number of eggs laid (Heim de Balsac and Mayaud 1962).

INCUBATION: probably by ♀ alone; often sits tight and can be caught on nest; sometimes calls from nest-hole if approached.

DEVELOPMENT AND CARE OF YOUNG: young fed by both sexes. Fledglings stay near nest-hole, beg by calling. Well-grown juveniles continue to take refuge in rodent holes, even when independent.

Reference
Cramp, S. (1988).

Oenanthe xanthoprymna (Hemprich and Ehrenberg). Red-tailed Wheatear. Traquet à queue rousse.

Saxicola xanthoprymna Hemprich and Ehrenberg, 1833. Symb. Phys. Aves, fol. d.d.; Nubia.

Range and Status. Breeds Armenia, N Iraq, Iran, Afghanistan; winters NE Africa, Arabia, India.

Palearctic migrant; common in coastal hills of SE Egypt from Quseir south; frequent coastal Sudan, lowland W Ethiopia, N Ethiopia (Eritrea), Somalia north of 11°N; vagrant Nile valley (Egypt and Sudan), NW Sudan, SE Libya.

Description. *O. x. xanthoprymna*: breeds Armenia, N Iraq, NW Iran; African vagrants are this race, otherwise winter ranges of 2 subspecies in Africa are similar. ADULT ♂: crown, nape, mantle and back greyish brown (browner when fresh); forehead and narrow supercilium white; lores, ear-coverts, sides of neck, chin and throat black, sharply cut off from back, and just meeting black on wing; rump and uppertail-coverts pale orange-rufous. Rectrices white, with distal half to two-thirds of central pair, and distal one-third of others, very dark brown; all tipped rufous-buff when fresh. Breast and belly silvery white, merging into rufous undertail-coverts; remiges and greater wing-coverts brown, tipped buff, and tertials fringed buff, when fresh; lesser and inner median coverts very dark brown or black; underwing-coverts and axillaries black. Bill black; eye dark brown; legs and feet black, soles brown. ADULT ♀: as ♂ except black less extensive on sides of neck, not meeting black of wing, and merging with grey of back; lesser and inner median coverts brown or dark brown, more like rest of wing. SIZE: wing, ♂ (n = 9) 91–97 (95·0), ♀ (n = 5) 86–94 (90·4); tail, ♂ (n = 8) 55–64 (60·7), ♀ (n = 5) 52–62 (56·6); bill, ♂ (n = 9) 18–21 (18·9), ♀ (n = 5) 18–20 (18·6); tarsus, ♂ (n = 9) 23–26 (24·4), ♀ (n = 5) 23–25 (23·6).

TAXONOMIC NOTE: *O. x. 'cummingi'* is a probable hybrid, found in zone of overlap between nominate race and *O. x. chrysopygia*, in W Iran, and is intermediate in plumage, having black throat and underwing of *xanthoprymna* and red tail of *chrysopygia*. Measurements included in the above.

O. x. chrysopygia (De Filippi): breeds Iran to Afghanistan; winters India, Arabia, NE Africa. ADULT ♂: upperparts greyish brown; indistinct supercilium pale grey or buff; lores and ear-coverts greyish brown, ear-coverts tinged bronze; lower rump, tail-coverts and tail like *xanthoprymna* but white parts and feather tips replaced with deep orange-rufous; underparts pale buffish grey, richer buff on breast and flanks and merging into pale orange-buff undertail-coverts; wing like ♀ *xanthoprymna*; underwing-coverts and axillaries silvery white, basally dark grey. Larger. Sexes alike. SIZE: wing, ♂ (n = 10) 91–99 (96·2), ♀ (n = 18) 88–100 (93·2); tail, ♂ (n = 10) 60–66 (62·2), ♀ (n = 18) 56–62 (58·9); bill, ♂ (n = 10) 18–22 (19·4), ♀ (n = 17) 18–21 (19·5); tarsus, ♂ (n = 9) 24–26 (25·6), ♀ (n = 18) 24–27 (25·3).

Field Characters. *O. x. chrysopygia* distinguished from all other wheatears except ♀ Red-rumped *O. moesta* by combination of red in tail and pale throat. Told from ♀ Red-rumped by small size, slender build and lack of rufous on crown. *O. x. xanthoprymna* and *'cummingi'* distinguished from all except ♂ Red-rumped by red rump. Both subspecies differ from ♂ Red-rumped in having uniform grey-brown, not grey and black, upperparts, and by small, slender build, contrasting with robust appearance of Red-rumped. Distinguished from redstarts *Phoenicurus* by behaviour, black tail-band and colour of upperparts. Tail-wag distinctive: very slowly up, almost cocked, then rapidly down.

Voice. Tape-recorded (CHA, HOL). Song reported to be strongly mimetic (Dementiev and Gladkov 1954) although this not substantiated by existing recordings. Said to be silent in winter quarters (Smith 1971), but sings at spring migration stopovers in Kuwait (A. Tye, pers. obs.). Song in Kuwait like that of other non-breeding wheatears; continuous, not broken into separate phrases, mostly of whistles and trills with very few harsh notes, and included distinctive phrase 'pip-pip-pip, pip-pip-pip, preeeew', with each series of 3 'pips' rising, and 'preeeew' initially rising briefly then sliding down scale. This phrase also occurs in song of Isabelline Wheatear *O. isabellina* recorded by Chappuis (1975).

General Habits. Frequents Red Sea coastal plain and rocky hills. Mostly in arid habitats: cliffs, gorges, acacia grassland, annual grassland, rocky desert. Preference for rocky, hilly ground may separate it from other, plains-dwelling wheatears (Isabelline and Desert *O. deserti*). Solitary and territorial in winter: defence against Isabelline Wheatear recorded; one chased by Mourning Wheatear *O. lugens* (Short and Horne 1981). Often perches on bushes, rocks, trees, telephone wires. On flatter ground forages mainly by dash-and-jab; also uses ground sallying and occasionally sally-gleaning and aerial sallying, which perhaps predominate on cliffs and broken ground. Wags tail on landing; often flicks wings. Habits little known, especially in Africa. Present in African winter quarters mid-Oct to Mar.

Food. Insects; in Asia eats mainly ants, beetles, termites, caterpillars, also wide variety of other insects; occasionally small lizards, seeds, fruits.

References
Bates, G. L. (1935).
Cornwallis, L. (1975).
Cramp, S. (1988).

Plate 30
(Opp. p. 449)

Oenanthe deserti (Temminck). Desert Wheatear. Traquet du désert.

Saxicola deserti Temminck, 1825. *In* Temminck and Laugier, Nouv. Rec. de Pl. Col. d'Ois. 60, pl. 359, fig. 2; Nubia.

Range and Status. Breeds N Africa, Middle East, Central Asia to Mongolia; winters Africa and Middle East to India.

Resident, intra-African and Palearctic migrant. Breeds from N Western Sahara and S Morocco in band south of Atlas to Tunisia and along coast to Egypt; has bred Mauritania; N African populations partially migratory, moving south to winter both within breeding range and in Sahara and N Sahel from Senegal to Sudan and Somalia; mostly north of 17°N in W, mainly north of 16° Chad, 15–17°N in Sudan, south to 8°N and around Mogadishu in Somalia. Winter visitors from outside Africa occur in eastern half of winter range. Generally common to abundant in breeding range and in winter in N Sahel, Somalia and Socotra; frequent to rare winter in S Sahel, Aïr, Ethiopian highlands and N Africa. In Mauritania, 220,000 estimated to winter in 67,000 km^2 (between 17° 26'N and 19°N); 1·0 birds seen per h, 3·3 per km^2 (Browne 1982). Questionable sight records from Gambia and Kenya.

Description. *O. d. homochroa* (Tristram): breeds Western Sahara to W Egypt, winters N and S fringes of Sahara. ADULT ♂: forehead and narrow supercilium cream; crown and nape greyish sandy; lores, ear-coverts, chin, throat and sides of neck black, throat scalloped with white when fresh; back and scapulars deep sandy buff, feathers tipped greyish when fresh; lower rump and uppertail-coverts white, cream, or buff. Tail distally black, basal third white, tipped buff when fresh; breast, belly and undertail-coverts buff, richer on breast. Primaries dark brown or black, tipped buff when fresh, with margin of basal half of inner web white; secondaries like primaries but tip and entire margin of inner web white, and fringed white on outer web when fresh; tertials dark brown or black, fringed sandy buff; wing-coverts black, fringed white when fresh; lesser coverts broadly fringed white or buff; axillaries and underwing-coverts black, tipped white; underside of remiges silver-grey, inner webs fringed white or cream. Bill, legs and feet black; eye dark brown. ADULT ♀: like ♂ but black areas of face usually replaced by dusky buff (feathers basally grey, distally sandy buff, grey showing to variable extent, sometimes appearing black scalloped with buff); black of tail and wings browner; axillaries and underwing-coverts dark grey, broadly tipped buff. SIZE: wing, ♂ (n = 48) 82–93 (89·6), ♀ (n = 19) 83–90 (86·8); tail, ♂ (n = 48) 51–67 (61·3), ♀ (n = 19) 53–61 (57·8); bill, ♂ (n = 46) 16–19 (17·6), ♀ (n = 19) 16–18 (17·6); tarsus, ♂ (n = 46) 22–27 (24·6), ♀ (n = 18) 23–26 (24·5). WEIGHT: Algeria (Jan) 1 ♂ 26.

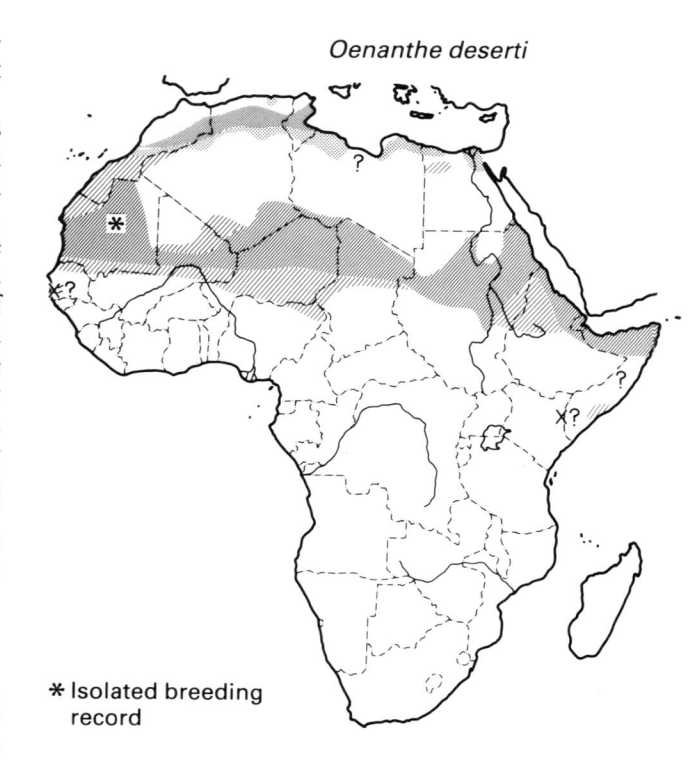

Oenanthe deserti

∗ Isolated breeding record

IMMATURE: like adult ♀, but entire upperparts greyish sandy brown, scalloped with dark brown; entire underparts greyish buff, scalloped on breast with brown; edges and tips of wing and tail-feathers sandy brown. First-winter ♂ has browner remiges than adult ♂, and throat more broadly scalloped with buff.
NESTLING: down pale grey.
O. d. deserti (including '*atrogularis*': supposed plumage differences not supported by long series of specimens): breeds E Egypt to central Asia, winters NE Africa to India, vagrant Niger, Chad. Back and scapulars tinged greyer, uniform with crown and nape. Difference more evident in worn plumage when grey tips of *homochroa* lost. Intermediates and birds with characters of both races occur in Egypt, Sudan, Eritrea. Slightly larger, with body size and amount of white on inner webs of remiges increasing in a cline from W to E across Asia. SIZE: wing, ♂ (n = 25) 90–96 (92·3), ♀ (n = 16) 86–91 (88·6); tail, ♂ (n = 25) 55–67 (60·5), ♀ (n = 15) 53–62 (57·8); bill, ♂ (n = 23) 17–19 (17·7), ♀ (n = 14) 16–19 (17·7); tarsus, ♂ (n = 23) 23–26 (24·8), ♀ (n = 14) 23–26 (24·2).

O. d. oreophila (Oberholser): breeds Tibet, Ladakh, and Kunlun Shan range, winters S Arabia, Socotra and Abd-al-Kuri, possibly S Somalia. Like deserti but grey of crown wears less than back, so crown often greyer in worn dress; white on inner web of remiges usually extends to shaft. Intergrades with last race in Afghanistan. Larger. SIZE: wing, ♂ (n = 14) 97–104 (100), ♀ (n = 12) 92–100 (95·8); tail, ♂ (n = 14) 59–68 (63·8), ♀ (n = 11) 59–64 (62·1); bill, ♂ (n = 14) 17–21 (18·6), ♀ (n = 11) 16–19 (17·8); tarsus, ♂ (n = 13) 25–28 (26·3), ♀ (n = 11) 24–27 (25·9).

Field Characters. ♂ distinguished from other black-throated wheatears by broad black band on tail below white rump, and sandy back. ♀ distinguished from other pale-throated wheatears except Isabelline O. isabellina by the same characters, and from Isabelline by dark wing and much smaller size; some races show white on primaries in flight. Tail-wag distinctive (see below); even more emphatic than Isabelline (q.v.).

Voice. Tape-recorded (53, 60). Song unlike other wheatears, a plaintive, descending whistle of 2–4 notes, lasting 0·3–1·5 s; often rendered 'see-dool-a-dol', trilled towards the end and often followed by a grating churr of c. 0·5 s like running a fingernail over a wooden comb. Repeated at fairly regular 4 s intervals, with slight variation in form of whistle. Also has a more typical wheatear song of longer phrases, incorporating this main song-phrase, but also including plaintive whistles, chacks and churrs. Often sings in winter, loudly or quietly; ♀ occasionally sings, with 'more uniform, rolling phrases' than ♂ (Chappuis 1975). Calls, a whistle 'sweee' and a hard 'tuk', sometimes (when alarmed) combined, or a series of 'tuks' forming a rattle.

General Habits. Breeds in stony and bushy, or sandy subdesert, sebkha (salt flats), dry wadis, tamarisk woodland; prefers flat ground; avoids coast (except in Western Sahara) and broken rocky ground. In sandy terrain, requires hard substrate (e.g. hillocks) for nesting in burrows, or bushes or stone slabs for nesting under. Absent from driving dunes and stony desert bare of vegetation. Up to 1500 m on Hauts Plateaux of Atlas. S limit in Cyrenaica may correspond with 100 mm isohyet. Winters in similar habitat, but also sandier areas. Overlaps broadly in N Africa with Red-rumped Wheatear O. moesta.

In winter usually solitary, defending individual territories; also defends against Red-rumped Wheatear. Winter territories 0·3–0·8 ha in Iran. Sings much in winter. Wags tail loosely and with great amplitude, 4–8 times rapidly after a run or after landing on perch; flicks wings. Perches on stones, bushes, grass-stems, trees, telephone wires. Often confiding, but flies far when alarmed; escapes by flying over obstacle and dodging to one side when out of sight. Frequently hovers. Forages by dash-and-jab and sallying to ground from perches 25–150 cm high; sometimes uses aerial sallying; occasionally hovers into wind and dives on prey. Sits at dune crests, searching for food in crumbling sides. One ♂ became tame, feeding on locust hoppers which escaped from cages (Smith 1971).

Spreads out after breeding, moving inland in Western Sahara. Leaves high altitudes in winter in N, but remains up to 2000 m in Somalia, 1300 m in N Ethiopia (Eritrea). Some NW African breeders move south c. 1000 km Aug–Sept and return Mar–Apr, though NE African populations probably less migratory. Arrives Mauritania, Mali and central Sahara Sept, departs Mar–Apr. Palearctic migrants arrive Sudan, Eritrea, Somalia Aug–Nov (peak Somalia Oct), depart Feb–Mar.

Food. Mainly insects; in Africa including beetles, ants, grasshoppers, caterpillars, Dipteran larvae; also some seeds. One bird ate early instar locusts but, although it tried, could not deal with live 4th and 5th instars (Smith 1971). In Asia also other insects, wood lice, millipedes, spiders, worms and small lizards.

Breeding Habits. Monogamous, territorial, hole- or ground-nester. Breeding territories large, c. 4 ha. Interspecifically territorial with Red-rumped Wheatear. Subordinate to Red-rumped and (in Asia) to Isabelline, Northern O. oenanthe, Finsch's O. finschii, and Mourning O. lugens Wheatears, but dominant to Pied O. pleschanka. Courtship and pair-formation recorded on spring migration; includes aerial chases; also ♂ fans and wags tail. Sings much on arrival in breeding areas. Song-flight vertically ascending or circling; also sings from perch, sometimes from ground.

NEST: rather bulky, of dry grass and root fibres, lined with wool, hair or feathers; in Asia ext. diam. c. 130, int. diam. 60–100, cup depth 20, height 50. Placed up to 0·5 m down old rodent burrow or other hole in ground, mound, masonry of culvert or bridge, road, railway or wadi bank, or on ground under stone slab, outcrop, bush or old box. Built mainly by ♀, whom ♂ may accompany or aid.

EGGS: 3–5, occasionally 6; Morocco mean (9 clutches) 4·3; Algeria/Tunisia mean (48 clutches) 4·1. Subelliptical; smooth, slightly glossy; greenish blue, spotted with red-brown and violet, especially near big end; some unspotted, some nearly white. SIZE: (n = 62, Tunisia) 18–22 × 15–16·8; (n = 43, NW Africa) 18·6–21·5 × 14–16 (19·9 × 14·9). Often double-brooded.

LAYING DATES: Mar–May throughout N African range.

INCUBATION: by ♀; period 13–14 days (Asia).

DEVELOPMENT AND CARE OF YOUNG: young fed by both parents. Fledglings may be fed for 3 weeks after leaving nest.

References
Ash, J. S. (1981).
Cramp, S. (1988).
Valverde, J. A. (1957).

Plate 29
(Opp. p. 448)

Oenanthe pileata (Gmelin). Capped Wheatear. Traquet du Cap.

Motacilla pileata Gmelin, 1789. Syst. Nat. 1, p. 965; Cape of Good Hope.

Forms a superspecies with *O. bottae* and *O. isabellina*.

Range and Status. Endemic to E and southern Africa; resident and intra-African migrant. Common to abundant, and apparently resident, Kenya highlands and Serengeti region of Tanzania. Common dry-season breeding visitor to lowland Kenya north to Mombasa, and Tanzania (except L. Victoria basin and extreme W). In Serengeti, 1–180 per km^2, weighted mean 76 (Folse 1982). Vagrant Zanzibar, S Somalia and W Uganda (Ruwenzori and W Nile). Frequent to locally common dry-season breeding visitor to Malaŵi, Zambia, S Zaïre, Zimbabwe, Tete District of Mozambique, Transvaal highveld. In Zimbabwe mainly in highlands, uncommon in river valleys and SE lowveld. Irregular Botswana. Frequent to common partial migrant Orange Free State, Natal, Cape Province, S Namibia (less common in winter); further north in Namibia mostly a locally common breeding visitor in the rains, absent June–Oct. Status uncertain in Angola, where generally sparse but locally abundant SW and NE.

Description. *O. p. pileata*: South Africa excluding NE Cape Province, Namibia north to *c*. 26°N. ADULT ♂: broad forehead band and narrow supercilium white; forecrown black, merging on hindcrown into rich brown of nape, back and scapulars, chin and throat white; lores, ear-coverts, sides of neck and broad breast-band black; rump cinnamon-brown; uppertail-coverts white. Rectrices basally white, with distal three-quarters of central pair and distal half of others very dark brown or black; all tipped white when fresh. Belly white; flanks and undertail-coverts buff. Remiges and wing-coverts very dark brown or black, tipped and fringed cinnamon-buff when fresh, broadly so on tertials; underwing-coverts and axillaries grey, fringed white; underside of remiges silver-grey. Bill, legs and feet black; eye dark brown. Sexes alike. SIZE: wing, ♂ (n = 34) 91–106 (98·5), ♀ (n = 13) 91–104 (95·8); tail, ♂ (n = 33) 57–70 (62·5), ♀ (n = 12) 55–66 (60·6); bill, ♂ (n = 33) 18–23 (20·2), ♀ (n = 13) 18–21 (19·3); tarsus, ♂ (n = 34) 29–36 (32·4), ♀ (n = 12) 29–33 (31·8). WEIGHT: South Africa, 1 unsexed 32·5.

IMMATURE: like adult but upperparts, lores and ear-coverts rich brown, spotted with pale brown; tips and fringes on wing-feathers broader; chin and throat off-white, spotted with brown; breast and belly yellow-buff, mottled and scalloped with brown, more so on breast, forming incipient band. Adult plumage attained at post-juvenile moult.

NESTLING: unknown.

O. p. livingstonii (Tristram) (including '*albinotata*'): central and E Africa from NE Cape Province, Botswana, Zimbabwe and N Mozambique north. Like *pileata* but forehead band averages narrower, like supercilium in width; upperparts darker, blackish brown; rump darker, richer brown but less cinnamon-tinged. Averages smaller. SIZE: wing, ♂ (n = 39) 87–101 (93·5), ♀ (n = 17) 83–94 (90·2); tail, ♂ (n = 37) 54–64 (58·3), ♀ (n = 17) 51–60 (55·5); bill, ♂ (n = 38) 18–21 (19·3), ♀ (n = 16) 18–21 (19·1); tarsus, ♂ (n = 39) 28–33 (30·7), ♀ (n = 17) 27–33 (30·2). WEIGHT: Zambia (June) 1 ♂ 32·5, immatures (n = 3, Dec–Jan) 24–30·7 (27·7); Serengeti, 1 unsexed 29.

O. p. neseri MacDonald: Namibia north of *c*. 26°N, Angola. Intermediate between *pileata* and *livingstonii* in colour of upperparts; amount of white on forehead more similar to *pileata*; size closer to *livingstonii*. SIZE: wing, ♂ (n = 9) 92–97 (94·6), ♀ (n = 2) 90, 90; tail, ♂ (n = 9) 56–62 (60·0), ♀ (n = 2) 54, 55; bill, ♂ (n = 8) 18–20 (19·4), ♀ (n = 2) 18, 19; tarsus, ♂ (n = 9) 29–33 (31·2), ♀ (n = 2) 30, 30.

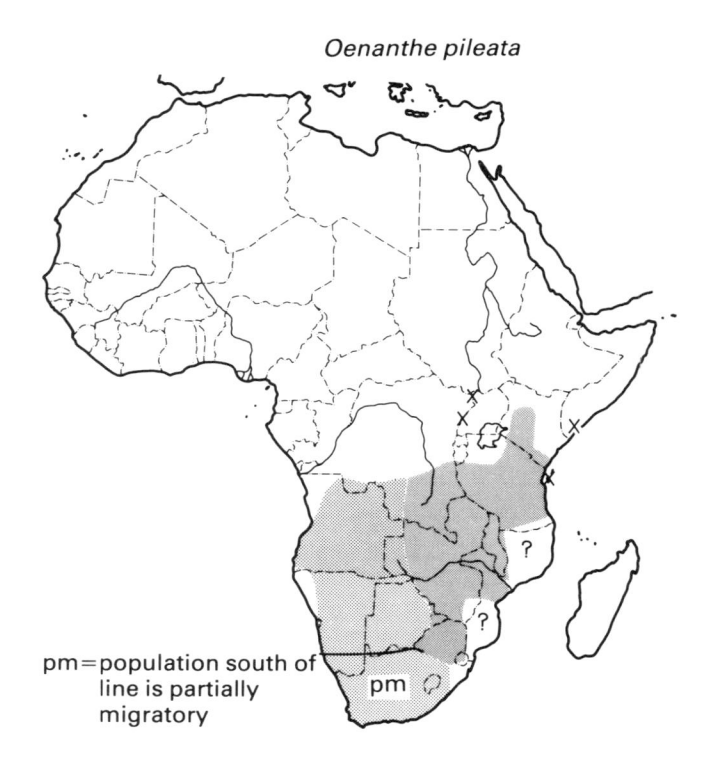

Field Characters. Adult distinguished from all other chats by unique black 'necklace' (**A**). Immature similar to other predominantly brown wheatears; tail pattern most similar to Isabelline *O. isabellina* and Red-breasted *O. bottae*; tail and/or darker upperparts separate it from Black-eared *O. hispanica*, Isabelline and ♀ Desert *O. deserti*; larger, with paler upperparts and breast than Red-breasted *O. b. heuglini*. General appearance closest to ♀ and first-winter Northern *O. oenanthe*, but black tail-band broader (cf. Isabelline); also initially has buff-spotted upperparts and wing-coverts and broad chestnut edges to tertials and inner secondaries, which are less prominent in Northern; when older shows traces of the dark breast-band.

Voice. Tape-recorded (53, 60). Song of brief phrases, mostly 1–2 s, occasionally up to 7 s long, interrupted by pauses typically 5–30 s long. Phrases varied, a jumble of whistles and trills with comparatively few harsh churring and clicking notes; often imitative, including notes of African Wattled Lapwing *Vanellus senegallus*, Crowned Lapwing *V. coronatus*, Temminck's Courser *Cursorius temminckii*, Little Bee-eater *Merops pusillus*, Yellow-throated Longclaw *Macronyx croceus*, Groundscraper Thrush *Turdus litsitsirupa*, Fork-tailed Drongo *Dicrurus* *adsimilis*, dog bark, goat bleat and other animals (Benson 1940). Calls, a thin whistle 'weet', often incorporated in song-phrases; a longer, louder rising 'sooeet', and clicks; all used as alarm calls.

General Habits. Found mainly on dry grassy plains, especially where ground almost bare or overgrazed, or grass very short or recently burnt; often with scattered bushes or ant- or termite-hills. In Angola also in *Brachystegia* savanna. Occasionally in cultivation, including stubble, ploughed land; cattle pens. Arrives E Africa when grass deteriorating and being burnt; particularly attracted to burnt ground in early dry season when alternative bare areas unavailable. Departs E Africa when grass becomes too long at start of rains. In Malaŵi mainly 500–1500 m, occasionally higher; in Zimbabwe breeds at 1700 m, occasionally found up to 2200 m; in E Africa breeds mainly above 1400 m.

Usually solitary and territorial, sometimes in pairs or family groups. Sings until dusk, and often at night, especially when disturbed, e.g. by grass fires. Hops on ground; usually perches low, on stones, ant hills, low bushes, posts, wire fences, buildings, low trees. Flies low between perches. Perches upright; bobs head emphatically on landing or when alarmed, wags tail and flicks wings. Forages by dash-and-jab; may also spot prey while flying low, and hover over it a few s before landing.

Movements of migratory populations complex and not fully understood. E African populations are among the few birds to migrate towards equator to breed in cool, dry season (mostly May–Nov), then south or southwest for rainy season to moult in drier areas (cf. Red-breasted Wheatear). Apparently resident Kenya highlands and N central Tanzania but dry-season breeding visitor to remainder of E Africa from SE Kenya and SE Zaïre to Transvaal highveld and Swaziland. In SE Kenya and Tanzania adults present mainly Apr–Sept, immatures until Jan. In rest of SE African range (Malaŵi, Zambia, SE Zaïre, Zimbabwe, Mozambique, Transvaal) adults present mainly late Apr to early Nov; except in SE Zaïre (whence they depart Nov) immatures remain until Feb when they undergo a delayed post-juvenile moult into adult plumage, and probably remain all year (through rains) (Irwin 1971). However, it is a summer breeding visitor (Nov–July) to parts of Kalahari, and some May–Nov visitors to Zimbabwe may be non-breeders, having bred further south or southwest (e.g. Kalahari). Present all year at junction of southern (Nov–July) breeding area and northern (May–Nov) breeding area in N Botswana, where perhaps breeds June–Oct. Breeding visitor Nov–July in SW Botswana, irregular in SE (frequent all year if dry but absent during rains in wet years – usually between Nov and July); one breeding record (Mar) in this area (Beesley and Irving 1976). Conceivably, migratory populations could breed twice a year, in dry season in central and E Africa and in rains in arid Kalahari (Benson 1982). Mainly resident *O. p. pileata* exhibits some post-breeding movement, and is uncommon in SE Cape Province in winter.

Food. Insects, especially ants, also flies, locusts, beetles, bugs, other Hymenoptera, termites, arachnids, lucerne butterflies, caterpillars. In Serengeti, apparently purely insectivorous, specializing on ants (Folse 1982).

Breeding Habits. Monogamous, territorial, hole-nester. Territories defended against wintering Palearctic migrant wheatears including Northern, Pied *O. pleschanka* and 'Schalow's' *O. lugens schalowi*. In display flight ♂ flies or hovers *c.* 3 m above ground, tail spread, wings fluttering rapidly, singing. Courtship display includes subdued song by one bird (♂?), including mimicry, with tail-fanning showing white rump; flexes wings and runs crouching round other bird *c.* 50 cm away, interspersed with short flips into air to *c.* 2·5 m high with tail fanned, head held high and wings fluttering as in song-flight. Sometimes other bird also flies briefly.

NEST: of straw, dry grass, leaves, rootlets, lined with finer material including hair, feathers; placed 0·3–1 m down old rodent burrow in open, flat ground or low bank; sometimes in an enlarged chamber which may be lined with nest material. Occasionally nests in termite mound; one found under old metal railway sleeper, Zimbabwe (James 1947); one burrow had 2 exits, both used.

EGGS: 3–4, sometimes 5, 6; also 2 or 1 (incomplete?); Zimbabwe, Zambia, Malaŵi mean (23 clutches) 3·0. Subelliptical; slightly glossy; pale translucent greenish white or bluish white, occasionally white, unmarked or faintly speckled with yellowish pink. SIZE: (n = 44, southern Africa) 20·9–27 × 15·5–20·4 (23·4 × 17·4).

LAYING DATES: Cape Province, Natal, Sept–Jan; Transvaal, Aug–Sept; SW Botswana, Nov–Jan; SE Botswana, Mar; N Botswana, June–Oct(?); Zimbabwe, June–Jan (peak Aug); Zambia, July–Oct (peak Sept); Malaŵi, Aug–Oct; Namibia, Feb, July; Angola, July–Oct. E Africa: Kenya, Uganda, Apr–June, Dec–Jan; Region A, Apr; Region C, Apr–May, Aug–Sept; Region D, Apr–July, Sept, Dec–Jan; records suggest breeding end of rains and in dry season (Apr–Sept) Region C,

528 TURDIDAE

and mainly in rains (Apr–July, Oct–Dec) Region D, possibly because in Region C high rainfall floods nests, and Region D too dry to provide sufficient food except in rains (Brown and Britton 1980).

DEVELOPMENT AND CARE OF YOUNG: fledglings accompanied by ♂ or ♀.

References
Irwin, M. P. S. (1981).
Vincent, A. W. (1947).
White, C. M. N. (1961).

Plate 29
(Opp. p. 448)

Oenanthe bottae (Bonaparte). Red-breasted Wheatear. Traquet à poitrine rousse.

Campicola bottae Bonaparte, 1854. Compt. rend. hebdom. des séances de l'Acad. des Sci. 38, p. 7; Yemen.

Forms a superspecies with *O. isabellina* and *O. pileata*.

Range and Status. Africa and SW Arabia.

Resident and intra-African migrant. *O. b. frenata* uncommon resident Eritrea and S Ethiopia, locally common to abundant resident W, central and SE highlands of Ethiopia; vagrant Djibouti (or *O. b. bottae* from Arabia). *O. b. heuglini* uncommon resident W Eritrea and W highlands of Ethiopia; intra-African migrant between N Guinea and Sahel zones; possibly partially resident in Mali, Nigeria, Chad, Central African Republic, Sudan; vagrant Ivory Coast, rare Burkina Faso, uncommon Mali and S Mauritania, rare to frequent Niger, N Ghana, N Togo, N Benin, seasonally and locally frequent to common central and N Nigeria, N Central African Republic, Cameroon highlands, widespread S Chad, locally and seasonally common to abundant Darfur and Nile valley of Sudan, seasonally frequent N Uganda (Karamoja) and probably extreme NW Kenya, vagrant S Kenya.

Description. *O. b. heuglini* (Finsch and Hartlaub) (including 'campicolina'). ADULT ♂: forehead, crown, nape, back and rump very dark brown, mottled with rufous-buff when fresh (May–July); narrow supercilium white; lores black, ear-coverts black or very dark brown, forming stripe through and under eye; cheeks, chin and throat white to rufous-buff; uppertail-coverts white. Rectrices white with distal three-quarters of central pair and distal half of others black, tipped white when fresh. Breast, flanks and undertail-coverts rich rufous-buff when fresh, paler on belly, and changing to orange-brown; scapulars, remiges and wing-coverts very dark brown, edged and tipped rufous-buff when fresh, more broadly on tertials and wing-coverts; underwing-coverts and axillaries pinkish buff, fading to silver-grey. Bill black; eye very dark brown; legs and feet black, soles brown. Sexes alike. SIZE: wing, ♂ (n = 60) 83–94 (87·6), ♀ (n = 27) 80–88 (84·1); tail, ♂ (n = 59) 49–63 (54·5), ♀ (n = 28) 50–60 (52·6); bill, ♂ (n = 56) 16–20 (17·3), ♀ (n = 24) 15–18 (16·9); tarsus, ♂ (n = 59) 24–30 (27·2), ♀ (n = 28) 23–29 (26·4). WEIGHT: Nigeria (Apr–May) 1 unsexed adult 19·6, juvenile (n = 3) 19·6–20·6 (19·8); Chad, unsexed (n = 6) 17–23 (20·0).

IMMATURE: like adult but mantle, back and ear-coverts very dark brown spotted with dark rufous-buff; chin, throat, breast and flanks scalloped with dark brown; fringes on wing-feathers broader.

NESTLING: unknown.

O. b. frenata (Heuglin): paler, greyer above; supercilium narrower, cream; chin and throat white; breast richer rufous, more clearly distinct from throat; underwing-coverts and axillaries cream. Much larger: wing, ♂ (n = 15) 95–105 (101), ♀ (n = 6) 96–102 (99·3); tail, ♂ (n = 14) 59–65 (62·1), ♀ (n = 6) 56–65 (60·3); bill, ♂ (n = 14) 18–22 (20·1), ♀ (n = 6) 20–21 (20·5); tarsus, ♂ (n = 14) 30–35 (33·0), ♀ (n = 6) 33–35 (33·5).

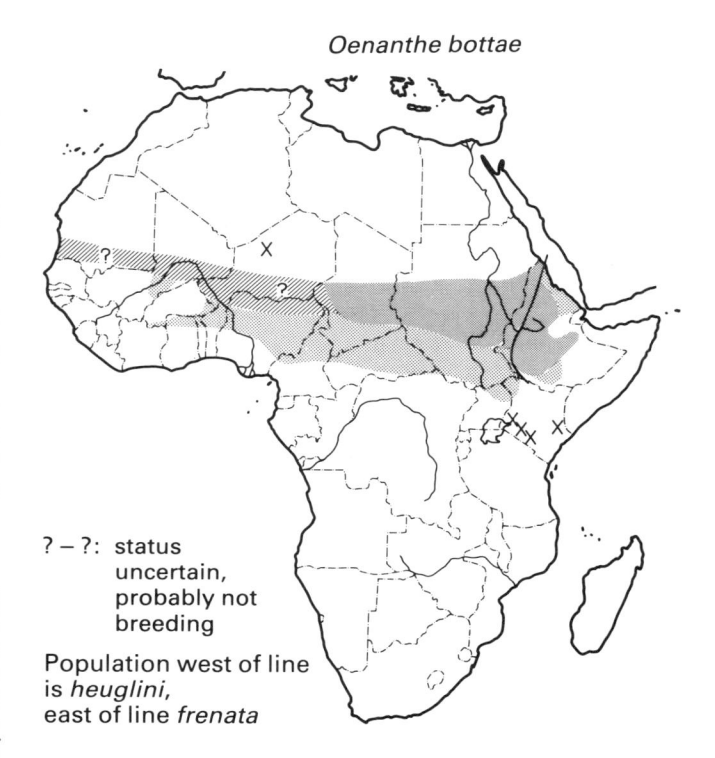

TAXONOMIC NOTE: in Africa this species comprises two well-marked parapatric incipient species, with almost no overlap in size and habitat.

Field Characters. Distinguished from other wheatears and chats by combination of lack of black on underparts with dark brown upperparts and white rump. Darker than Isabelline Wheatear *O. isabellina* and pale-throated morph of Black-eared *O. hispanica*; further distinguished from latter by broad black band on tail. May be confused with ♀ and first-winter Northern *O. oenanthe*, especially

Oenanthe bottae

? – ?: status uncertain, probably not breeding

Population west of line is *heuglini*, east of line *frenata* after this species has been dust-bathing, but tail pattern distinctive. Closest in appearance to ♀ Desert *O. deserti* but *frenata* is much larger, and *heuglini* darker brown above, more rufous below.

Voice. Tape-recorded (53, LEM). Song, extended phrases of 30–60 s, interrupted by brief pauses up to 2 s long; phrases separated by longer pauses of *c.* 10 s, which may be interrupted by whistles or chacks. Phrases very varied, containing chacks, whistles, jumbled trills, chuckling, churring and chirping notes, and mimicry. Incorporates notes of White-bellied Bustard *Eupodotis senegalensis* and Greenshank *Tringa nebularia* (tape-recorded), African Wattled Lapwing *Vanellus senegallus*, Green Sandpiper *Tringa ochropus*, Wood Sandpiper *T. glareola*, falcon sp., Black Kite *Milvus migrans*. Mimicry often near beginning of song. Usual call a hard 'chack'.

General Habits. *O. b. heuglini* frequents inselbergs, flat ground or boulder-strewn rocky hillsides with dry, short (≤5 cm) grass; sometimes on fallow cultivated ground; in Central African Republic inhabits *Syzygium–Adina* riparian woodland. Prefers, and breeds on, bare, overgrazed and, especially, burnt areas. *O. b. frenata* occurs in montane habitats 1800–4100 m, including tussock-grass and giant heath moorlands and highland grass savanna.

Usually in pairs; in parties of up to 10 when not breeding. May sing at night, on moonlit nights or when disturbed. 'Chack' call and mimicked alarms of other species used as alarm calls, with drooping wings and flicking tail. Perches on rocks, trees, bushes, termite mounds, grass stems. Generally shy, not easily approached, but when alarmed does not fly far.

O. b. heuglini apparently resident Ethiopia, but at least partial migrant elsewhere. Breeds late dry season in N guinean and soudanian zones, shifting north at start of rains (about Apr) to soudanian and sahelian zones, where moults, returning about Nov. Migration related to grass height. Such movement away from equator after breeding unusual (cf. Capped Wheatear *O. pileata*); but reduces overlap with Palearctic migrant wheatears. Latitudinal migration obscured by local movements to burnt ground.

Food. Insects, including beetles and ants.

Breeding Habits. Apparently monogamous, territorial, hole- or ground-nester. Song used in territorial advertisement, usually in hovering or slow, fluttering, often rising song-flight. During song-flight, tail spread and depressed, exposing tail pattern. Sings most during territory establishment; generally silent when feeding young (Bates 1927). In courtship, ♂ hops around ♀, raising and spreading tail, fluttering wings and uttering snatches of song.

NEST: of coarse, dry grass, lined with fine grass, with or without inner lining of fur; usually in old burrow, or hole in termites' nest, or rock crevice; placed up to 1 m into hole.

EGGS: 2–4, usually 3; one laid per day. Slightly glossy; very pale blue to greenish blue, sometimes with small red-brown flecks, especially at big end. SIZE: (n = 2, Nigeria) 21 × 15, 21·5 × 15. Usually double-brooded, using same nest.

LAYING DATES: late dry season: Nigeria, Cameroon, Nov–Mar; Sudan, Mar; Ethiopia, Feb–June.

INCUBATION: apparently by ♀; period 14 days (1 second nest). When a first-brood juvenile approached second nest, ♀ called and ♂ chased it away; ♂ (and ♀ after eggs hatched) may also chase away conspecifics and other birds within 5 m of nest, including Speckled Pigeon *Columba guinea*, Whinchat *Saxicola rubetra*, larks (MacGregor 1950).

DEVELOPMENT AND CARE OF YOUNG: nestlings emerge from nest-hole after 15 days; almost immediately begin to peck at ground and make short vertical flights. Bill yellow and tail short at first. 4 days after emerging, begin to move away from nest-hole; ♀ may lead chick back. Return to nest-hole to roost. Both parents feed nestlings; as they grow, adults may rest for shorter period than usual at midday. Dependent fledglings dive into hole, or crouch, in response to adult alarm calls. Adults gave alarm call at Pallid Harrier *Circus macrourus* only after chick emerged (MacGregor 1950).

References
Ebbutt, D. P. (1967).
Elgood, J. H. *et al.* (1973).
Lynes, H. (1924–25).
MacGregor, D. E. (1950).

Oenanthe isabellina (Temminck). Isabelline Wheatear. Traquet isabelle. Plate 30
(Opp. p. 449)

Saxicola isabellina Temminck, 1829. *In* Temminck and Laugier, Nouv. Rec. de Pl. Col. d'Ois. (4) 79, pl. 472, fig. 1; Nubia.

Forms a superspecies with *O. bottae* and *O. pileata*.

Range and Status. Breeds Eurasia from Bulgaria, NE Greece, Turkey and Syria to E USSR and Manchuria; range expanding to NW; winters India, Arabia and Africa.

Palearctic migrant, wintering mostly south of Sahara in Sahel zone; common to abundant all winter from S Mauritania and N Senegal to S Egypt, Sudan, Ethiopia, Djibouti, Somalia and Socotra; less common Niger and W Chad. West of Sudan, mainly 16–18°N, where often the most common *Oenanthe*. In east, common, locally abundant, south to N Uganda and N Tanzania; straggles to Rwanda, Zaïre, N Zambia, with one doubtful record Angola ('Kakindo', untraced, possibly Cacondo). A few winter N Africa and Sahara but mainly passage migrant there. Frequent to common on passage Egypt to Tunisia and S Algeria, straggling to Morocco. Greater frequency of records in NW Africa in recent years may correlate with expanding breeding range. Some evidence for winter range changes related to rainfall: may have shifted south in W Africa in last 25 years (cf. Morel and Roux 1966, Browne 1982, Gee 1984, Morel 1972 and pers. comm.), while in Kenya certain areas support this species in dry years, but Northern Wheatears *O. oenanthe* in wet years (D. J. Pearson, pers. comm.).

Oenanthe isabellina

Description. ADULT ♂: upperparts sandy brown or isabelline (some crown feathers often with dark brown centres, giving mottled appearance); narrow frontal band and indistinct supercilium (both sometimes absent) cream; lores sandy brown, cream, dusky or black; ear-coverts sandy brown, like crown or slightly darker; chin buffy white, occasionally like underparts; lower rump and uppertail-coverts white (sometimes strongly tinged orange-brown on rump). Rectrices white with distal one-third to half very dark brown or black, except central pair which have distal half to two-thirds very dark brown or black; all tipped buff when fresh. Entire underparts pinkish buff, middle of belly paler, sometimes flanks and upper breast darker. Remiges and wing-coverts pale sandy brown to dark brown with buff or rusty buff tips and (except primaries) outer margins, broader on tertials; all fringes reduced by wear; alula like primaries or darker; underwing-coverts silver-grey or white; axillaries grey with broad pale fringes. Bill, legs and feet black; eye brown or dark hazel. ADULT ♀: as ♂ except frontal band usually absent; lores less often dark. SIZE: wing, ♂ (n = 76) 94–107 (99·7), ♀ (n = 46) 92–107 (96·8); tail, ♂ (n = 77) 50–64 (56·9), ♀ (n = 47) 52–63 (55·2); bill, ♂ (n = 73) 16–21 (18·4), ♀ (n = 44) 16–21 (18·1); tarsus, ♂ (n = 77) 27–35 (30·2), ♀ (n = 47) 26–35 (30·0). WEIGHT: Kenya (Nov) unsexed (n = 3) 20·8–26·7 (23·8); Senegal (Jan) ♂ (n = 2) 27·1, 27·2.
IMMATURE: reaches Africa after post-juvenile moult, when similar to adult, except retains duller brown juvenile remiges.

Field Characters. Larger than most wheatears. Overall appearance pale, plumage uniform, lacking contrast. Distinguished from most wheatears by combination of white rump and lack of black on body; much paler than Red-breasted *O. bottae*. Difficult to tell from ♀ of Northern (stance, proportions and size not distinctive in W Africa where most Northern are large Greenland race *O. o. leucorrhoa*), Desert *O. deserti*, Black-eared *O. hispanica* and Pied *O. pleschanka*; best distinguished by tail pattern and (from last 3) by size; also less contrast between wing feathers and rest of plumage. Dark lores reputedly distinctive but absent from many Isabelline and present in most Northern; i.e. no consistent difference. Underwing light grey or white (dark grey in Northern, ♀ Desert and Pied, black in Black-eared). Head-bob and tail-wag distinctive and may be diagnostic (see below).

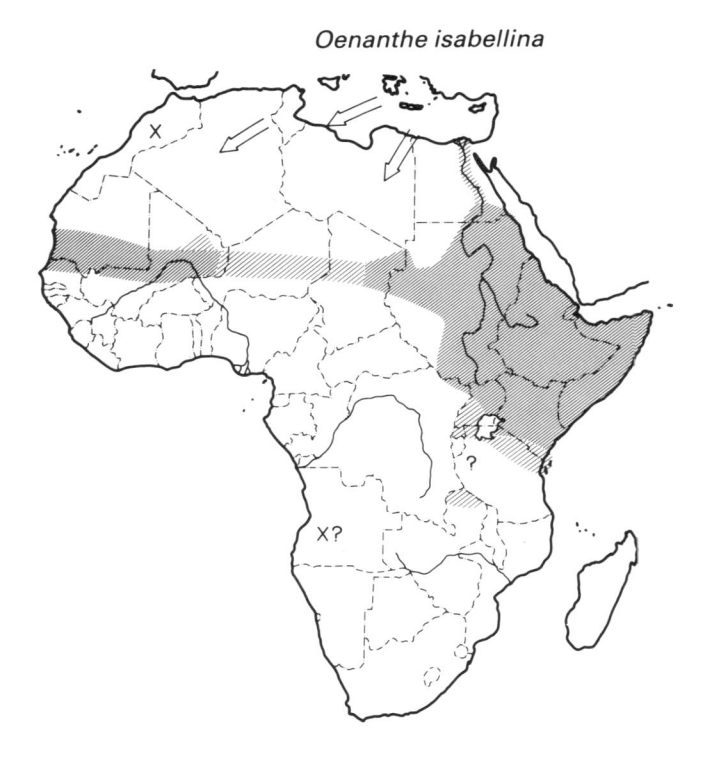

Voice. Tape-recorded (McVIC, TYE – 53, 62, 73, ROC). Song a jumble of whistles and harsh twanging or chacking notes, uttered in phrases *c*. 3–7 s long with pauses of *c*. 2–6 s between phrases. Phrases resemble those of other wheatears but perhaps with greater proportion of whistles and nasal twanging notes. Each phrase consists of a few note-types repeated, or a wide variety of notes, and may include call-notes and mimicry. Phrases may be repeated or slightly modified. Common call, a harsh 'chack'; also a loud whistle 'wheew', uttered singly or in combination.

General Habits. Found in steppe and sub-desert habitats on sandy or stony ground, usually with some short grass cover. Prefers flat plains or gentle hillsides to broken ground. Usually in lowlands but up to 2500 m in Ethiopia. Territories usually, but need not, include a few bushes or trees. Sometimes attracted to freshly-burnt ground. Enters villages, oases, sewage farms and old cultivation. May occur in bushier country when on migration. Usually in more arid regions than Northern Wheatear, but their ranges overlap broadly in W Africa, where they defend mutually exclusive territories.
Solitary and normally territorial, advertising with song, and defending foraging area against conspecifics,

other wheatears and other species which are probably competitors for food (e.g. Tawny Pipit *Anthus campestris*); however, territories overlap with those of *Lanius* shrikes (Tye 1984), and sometimes with other wheatears (Leisler *et al.* 1983). Territory radius c. 80 m in Kenya (Leisler *et al.* 1983), area 2–3 ha in Senegal (A. Tye, pers. obs.). Generally dominant over other, smaller wheatear species, especially Northern and Pied, but subordinate to and displaced by *Lanius* shrikes, often in circumstances suggesting interference competition; sometimes harasses shrikes (Tye 1984). One bird attacked its image in a mirror, another defended a cage of locust hoppers from competitors (Smith 1971). Aggressive behaviour includes calls, song, chasing in flight and fighting. Displays used in Africa include song-flight (circular flight from and back to elevated perch, often with head held higher than horizontal body); hovering over potential predators or competitors; 2 birds hovering close to one another (related to fighting); walking in parallel with neighbour along boundary; tail-spreading on ground or in flight (e.g. accompanying parallel walk). Often sings 'unsolicited' and briefly during displays and inter- and intra-specific territorial disputes. Some individuals (♂♂?) spend up to 10% of their time singing in winter quarters, both at subdued volume and in full song; others (♀♀?) do not sing (Tye 1988b). Sing from elevated perch, during song-flight or, occasionally, from ground. Calls used as alarm or challenge notes in response to invasion of territory by humans, conspecifics, other wheatears and certain other birds (e.g. shrikes *Lanius* spp.).

Perches on rocks, bushes, trees, mounds and occasionally buildings. Often bobs head and wags tail when alarmed, when halting after a run, or when territory threatened. Head-bob is a rapid down–up movement, often accompanied by raising tail and flicking both wings. The bob takes 0·2–0·3 s and is more emphatic than in some wheatear species, often revealing white rump over the head when seen from front. On ground, tail-wag usually consists of 3 or 4 jerky, evenly-spaced down–up movements, rather different from those of Northern Wheatear (q.v.); pattern less consistent when on bush or other perch (Tye and Tye 1983). Flies low and direct, rising late to land on elevated perch. Rests in shade of bush or tree-crown in hottest part of day, when most singing and preening take place. Forages mostly in cooler early and late hours, often until almost dark, and uses dash-and-jab, ground and aerial sallying. Although often largely terrestrial (has short wing, long tarsus), the relative use of each foraging technique depends on habitat structure (availability of perches, vegetation height). In Tsavo, Kenya, 86% of 268 prey items captured by dash-and-jab, 8% by ground sallying, 4% by aerial sallying, 1% in bushes (Lack 1985); but in Senegal acacia savanna, ground and aerial sallying more common than dash-and-jab (A. Tye, pers. obs.). 80% of sallies to ground in Senegal were from perches ≤ 1 m high (Tye 1984); sallies further from higher perches. Often appears to roost in holes in ground, e.g. mammal burrows; sometimes roosts by roadside where it can be caught when dazzled by lights (P. B. Taylor, pers. comm.). In Senegal, reacted to overflying birds of prey by ceasing activity and watching predator, and to falcons by crouching on ground or moving to lower perch in bush. Habitual bird-eating species, e.g. Red-necked Falcon *Falco chicquera*, stimulated birds in the open to fly to cover (Tye 1984).

Migrates mainly by night, entering Africa heading southwest on broad front from Red Sea to Tunisia. May fly west along Mediterranean before crossing Sahara. Main autumn passages in Tunisia late Aug to late Nov, Egypt Aug–Oct. In Sudan, passage mainly east of Nile. Main arrivals in winter quarters (some earlier) are: Mauritania and Senegal late Oct, Mali Oct, Sudan, Eritrea and Somalia Sept, inland Ethiopia and Kenya Oct, Tanzania Dec. Main departures Feb–Mar, although movement begins Jan in Senegal (but see below); stragglers remain until Apr or May. Spring passage in Algeria, Tunisia and Libya Feb–May, peak Mar; Egypt Mar and Apr. Spring passage appears more marked in Algeria and Tunisia, but autumn passage appears heavier in Sudan; this probably reflects the attractiveness of these areas as refuelling stops at the respective seasons. Some evidence for local movements in winter quarters: in Senegal, birds moving in Jan could have come from further north as the dry season progresses; in Kenya, vanishes from many areas after Oct–Nov rains, often reappearing as conditions dry out (D. J. Pearson, pers. comm.).

Food. Primarily insects, prey size ranging from c. 2 to 40 mm, caught on ground or in flight. In Africa beetles, grasshoppers and ants (including *Messor barbarus*) especially important; also Diptera and caterpillars. Mites and small seeds also recorded.

References
Aspinwall, D. R. (1977).
Cramp, S. (1988).
Tye, A. (1984, 1988b).

532 TURDIDAE

Genus *Cercomela* Bonaparte

A somewhat diverse assemblage of small, plain-coloured, Old World, ground-feeding chats forming a link between the even more terrestrial wheatears (*Oenanthe*) and the bushchats (*Saxicola*). Upperparts blackish brown or grey with rumps concolorous or rufous or white; tails black or (1 species) rufous or (4 species) with white edges; ear-coverts always slightly darker than surrounding plumage, and somewhat rufescent and glossy; underparts plain, grey, buff or brown. Bill fine. Legs slender and rather long. Tail medium-length or (*C. sordida*) short. 3 SW African species have the outer (9th) primary attenuated, markedly so in 2 of them (**A**).

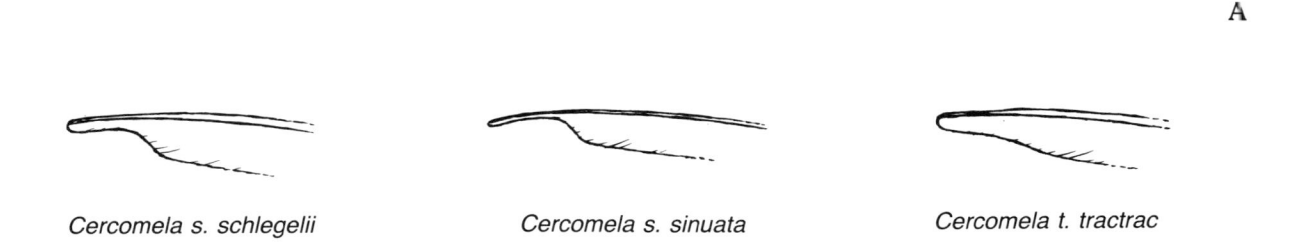

A

Cercomela s. schlegelii   *Cercomela s. sinuata*   *Cercomela t. tractrac*

9 species: 1 Indian (*C. fusca*), 1 Arabian and NE African (*C. melanura*), and 7 African endemics. 3 SW African species, the sickle-winged chats, appear to form a natural group, although they embrace the largest rock-chat, *C. schlegelii*, a grey wheatear-like species with white rump and white outer tail, and the smallest species, *C. sinuata*, a brown bird with rufous rump. Further north there is a brown-tailed superspecies, *C. familiaris*/*C. scotocerca*, and 3 independent black-tailed spp., *C. melanura*, *C. dubia* and *C. fusca*, also a distinctive, short-tailed, long-legged, montane bird, *C. sordida*, which might warrant generic separation (*Pinarochroa*). Some Oriental chats customarily placed in the genus *Saxicola* (*S. caprata*, *S. ferrea*, *S. jerdoni*, *S. melanoleuca*) may be more closely allied with *Cercomela* than with the stonechats (*S. torquata*, *S. rubetra*, *S. bifasciata*, *S. insignis*, *S. macrorhyncha*, and *S. dacotiae*). Although ♂♂ of these Oriental chats, which are pied birds, look quite different from *Cercomela*, the ♀♀ with black tails or rufous rumps are very similar to African *Cercomela* spp.

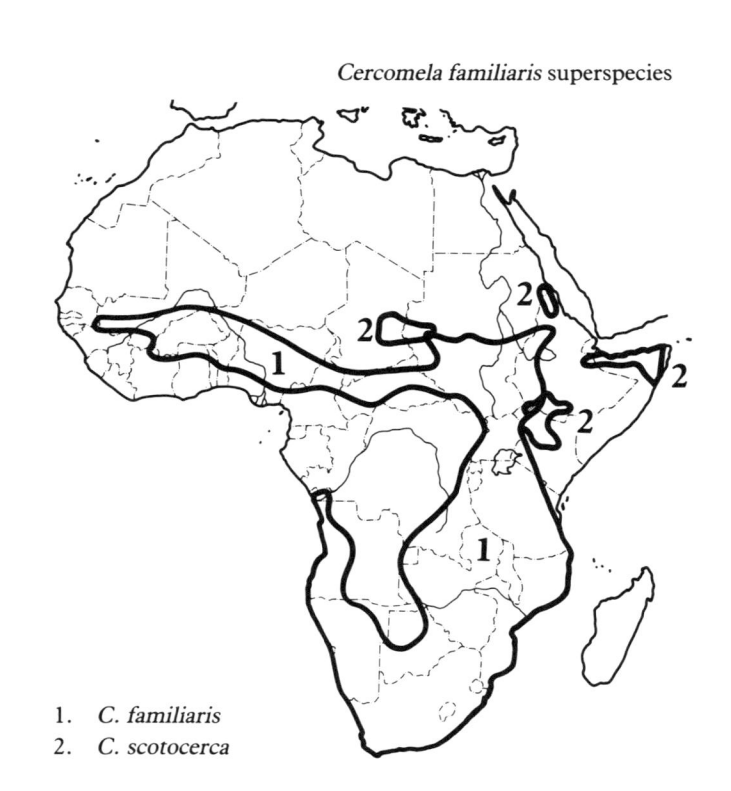

Cercomela familiaris superspecies

1. *C. familiaris*
2. *C. scotocerca*

Cercomela sinuata (Sundevall). Sickle-winged Chat. Traquet de roche brun.

Lusc.(inia) sinuata Sundevall, 1858. Oefv. K. Sv. Vet.-Akad. Förh. 2(3), p. 44; Cape Town.

Plate 31
(Opp. p. 464)

Range and Status. Endemic, South Africa, Lesotho and S Namibia; resident and vertical migrant. Common. In agricultural habitats, about twice as common on ploughed land as in pasture, and twice as common in cereal fields as on ploughed land (Winterbottom 1968a).

Description. *C. s. sinuata* (Sundevall): W Cape, east through Karoo to Great Fish River. ADULT ♂: forehead and crown brown; nape, mantle, back and scapulars brown, barely perceptibly paler than crown; brown back grades into pale rufous rump and dark rufescent brown uppertail-coverts. Tail dark brown with concealed pale rufous base, visible at sides when tail spread: T5 with proximal half rufous, T6 with proximal three-quarters of outer web pale rufous (whitish at edge). Narrow ring of whitish featherlets around eye; lores brown; indistinct paler brown superciliary stripe from lores to well behind eye; ear-coverts darkish rufescent brown, slightly darker than crown. Chin, throat and breast pale grey-brown; belly pale grey-brown or buffy brown, paler in centre and buffier towards flanks; undertail-coverts very pale buff or off-white. Wings dark brown; inner secondaries and tertials broadly fringed rufescent buff and secondary coverts broadly fringed buff, forming buffy stripe in closed wing. Tip of P9 greatly attenuated, in some specimens forming sickle 1 mm wide and 10 mm long (see **A**, p. 532). Underwing-coverts buffy, speckled dark brown. Bill black; eye dark brown; legs and feet black. Sexes alike. SIZE: (8 ♂♂, 7 ♀♀) wing, ♂ 78·5–85 (80·6), ♀ 72–78·5 (76·1); tail, ♂♀ 48–57; bill, ♂♀ 13–16; tarsus, ♂♀ 26–30.

IMMATURE: like adult but heavily spotted with buff above. Upperwing-coverts and tertials with broad rufescent buff margins, and underparts (especially breast) heavily mottled with dark greyish. P9 sickle-tipped.

NESTLING: unknown.

C. s. ensifera Clancey: Namaqualand to N and NE Cape, W Orange Free State and SW Transvaal. Paler and warmer brown than *sinuata*. Tail, unsexed (n = 18) 46–54 (50·3).

C. s. hypernephela Clancey: Lesotho (Natal in winter). Like *ensifera*, but slightly darker and cooler brown. Tail longer: ♂ (n = 2) 57, 58.

Field Characters. A small, brown chat which can be confused with Familiar Chat *C. familiaris* and immatures of Capped Wheatear *Oenanthe pileata* and Buff-streaked Chat *Saxicola bifasciata* (the latter being parapatric). Upperparts plain smoky greyish brown, underparts greyish buff, paling to creamy undertail-coverts; ear-coverts rufescent, slightly darker than head. Wings dark brown, feathers with rufous fringes. Rump and base of tail pale rusty brown, with solid blackish triangle at end of tail, pointing forwards (**A**, opposite). Familiar Chat (usually in different habitat) has black in tail shaped like an inverted 'T' (like many wheatears), with oblong block at side of tail, and rump, rufous; its upperparts and underparts are the same tone, but Sickle-winged Chat has underparts markedly paler than upperparts, and overall appears much paler in the field; also has longer legs and more upright stance. Only other chat with similarly-shaped blackish triangle is Karoo Chat *C. schlegelii*, a grey bird with white edges to tail.

Voice. Tape-recorded (GIB). Rather silent. Usual call a quiet 'chak-chak', quite high-pitched.

General Habits. Inhabits karoo, drier fynbos, shrubby semi-desert, montane grasslands, cultivated land and well-grazed pastures, and (Lesotho) alpine slopes.

Solitary or in pairs, perching on ground, low shrub or fence. Forages from low perch, dropping onto ground to catch prey. Flicks wings on landing, but less deliberately than does Familiar Chat. Flights mainly low and short. Can use very upright perching posture, looking long-legged (**B**, p. 534).

Race *hypernephela*, breeding up to 2300 m in Lesotho, is partial migrant to Natal in winter (Clancey 1964).

Food. Insects.

A

534 TURDIDAE

Breeding Habits. Evidently a solitary, territorial, monogamous breeder.

NEST: a neat cup with foundation of coarse grass, moss and twigs, lined with soft dry grass and plant down. 1 nest was built entirely of reddish down from *Protea* flowers (Vincent 1947). Sited on open ground under tuft of grass or tiny shrub or in lee of stone, or in crevice in wall or rock.

EGGS: 2–3, av. of 8 clutches 2·5. Pale blue-green, indistinctly speckled with pale rust, mainly at broad end. SIZE: (n = 47) 18·5–22·5 × 13·6–15·6 (20·1 × 14·7).

LAYING DATES: South Africa, Aug–Mar (mainly Oct–Jan).

Reference
Maclean, G. L. (1985).

Plate 31
(Opp. p. 464)

Cercomela schlegelii (Wahlberg). Karoo Chat. Traquet du Karoo.

Erithacus schlegelii Wahlberg, 1855. Oefv. K. Sv. Vet.-Akad. Förh. 12, p. 213; Damaraland.

Range and Status. Endemic resident, SW Angola (north to Mossamedes; common on gravel plains of Iona Nat. Park), W and S Namibia (sparse to frequent) and South Africa (common, Cape Province south to Little Karoo and east to SW Orange Free State and the Drakensberg and Matatiele). Sight record Barberspan, Transvaal, 1985; records near Vredefort, N Orange Free State, 1989. 190 birds in 850 ha in karroid broken veld, Klaarstroom, S Great Karoo; the commonest bird species in succulent karoo (Winterbottom 1966, 1968a).

Description. *C. s. pollux* (Hartlaub): W Cape north of Olifants R., to E Cape, Griqualand West and W Orange Free State. ADULT ♂: forehead, crown, nape, sides of neck, mantle, back, rump and scapulars uniform dark grey (lower rump whitish in some specimens); uppertail-coverts slightly paler grey, some feathers whitish. Tail-feathers blackish brown; T6 with outer web white, T5 with outer web white except at tip, T4 with outer web white except for 11 mm at tip, and T3 with outer web white except for 22 mm towards tip; T2 and T1 all blackish. Lores mid- or dark grey; featherlets around eye grey; ear-coverts brownish grey. Chin, throat and breast pale grey, grading to whitish on belly and white undertail-coverts; flanks grey. Primaries blackish brown, P9 sickle tipped (see generic account: **A**, p. 532); inner secondaries and tertials narrowly fringed greyish white; upperwing-coverts blackish brown, broadly fringed brownish grey. Axillaries and underwing-coverts white. Bill black; eye dark brown; legs and feet black. Sexes alike. SIZE: (4 ♂♂, 4 ♀♀) wing, ♂ 101–111 (107), ♀ 97–101 (99·4); tail, ♂ 75–76 (75·5), ♀ 66–71 (65·2); bill to feathers, ♂ 14·5–16 (15·25), ♀ 14–17 (15·25); tarsus, ♂ 32–35 (33·0), ♀ 31–34 (32·4). WEIGHT: immatures (n = 3) 15–17·5 (16·2).

Cercomela schlegelii

IMMATURE: upperparts browner (less grey) than adult, and all feathers, especially on mantle and back, with large buff spots, so that upperparts look heavily spotted; fringes of tertials broader and paler, throat and breast browner, heavily mottled with dark grey or blackish, belly lightly mottled with dark brown.

NESTLING: unknown.

C. s. schlegelii (Wahlberg): Namibia (W Damaraland, east to Erongo Mts). Above much paler grey – almost pinkish grey; rump and uppertail-coverts white. Below white or creamy white with throat and breast washed greyish. Much smaller. SIZE: wing, ♂ (n = 6) 94·5–97 (95·6), ♀ (n = 5) 83·5–90 (86·7); tail, ♂ (n = 6) 65–68, ♀ (n = 5) 59–62; bill, ♂♀ (n = 11) 13–16·5; tarsus, ♂♀ (n = 11) 26–30.

C. s. benguellensis (Sclater): NW Namibia, edge of Kaokoveld, SW Angola. Slightly darker than nominate race. Smaller: wing, 85–90.

C. s. namaquensis (Sclater): NW Cape (Richtersveld, N Little Namaqualand and Bushmanland) and S Namibia (Rehoboth, Damaraland). Like *pollux* but slightly paler and smaller. Extent of white in rump and uppertail-coverts variable. Size between *pollux* and *schlegelii*: wing, ♂ (n = 9) av. 102, ♀ (n = 5) av. 95; tail, ♂ (n = 9) av. 70, ♀ (n = 5) av. 67; bill, ♂ (n = 9) av. 19, ♀ (n = 5) av. 18.

Field Characters. A small to medium dark grey chat, with paler grey breast paling to white on undertail-coverts, and with blackish brown wings and tail. Black in tail forms a solid triangle, with acute angle pointing forwards (meeting grey or white rump); sides of tail white, forming conspicuous panels (**A**). In Angola and Namibia small (size of Familiar Chat *C. familiaris*) and white-rumped; in South Africa larger (size of Mountain Wheatear *Oenanthe monticola*) and rump grey, concolorous with mantle. White-rumped Namibian birds resemble Tractrac Chat *C. tractrac*, which is whiter and has only end of tail black; they also resemble grey-phase ♂ Mountain Wheatear, which has white shoulder or wrist patch. ♀ Mountain Wheatear (without white in wing) is much darker than Karoo Chat, and black in tail is oblong not triangular. In South Africa distinguished from congeners by greater size, longer tail, more robust build, and dark grey rump meeting white-sided black tail.

Voice. Tape-recorded (GIB). Calls, a rattling 'tirr-tit-tat' or 'tirr-tit-tat-tut'; also 'chak-chak' or 'trrat-trrat'.

General Habits. Inhabits karoo, scrubby and bushy plateaux and stony hillsides. In Little Namaqualand avoids coastal plains and lower Orange R. valley. Further north its preference for high ground with scrubby vegetation is more marked (Tractrac Chat mainly in lower, stonier habitats: Macdonald 1957).

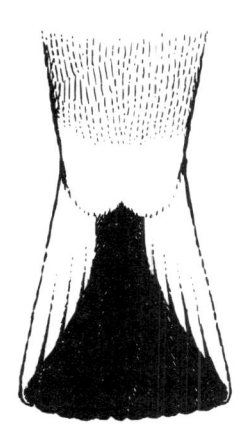

Solitary or in pairs. Perches on top of bush, rock, fence or telegraph wire, often fluttering to get its balance. Forages on ground; keeps look-out from low perch, dropping to ground for insect. Flicks wings (not as frequently as Familiar Chat). Flight undulating; flights mainly low and short.

Food. Insects and seeds.

Breeding Habits. Evidently a solitary, territorial, monogamous breeder.

NEST: cup; foundation of twigs, lined with dry grass and vegetable down. Sited on ground under tuft of grass, bush or rock.

EGGS: 2–4, av. of 10 clutches 2·6. Pale greenish blue, heavily freckled with rusty red. SIZE: (n = 40) 19·1–24 × 14·3–16·5 (20·7 × 15·4).

LAYING DATES: South Africa, Aug–Mar; season may vary with rainfall.

References
Macdonald, J. D. (1957).
Maclean, G. R. (1985).

Cercomela tractrac (Wilkes). Tractrac Chat. Traquet de roche pâle.

Plate 31
(Opp. p. 464)

Motacilla tractrac Wilkes, 1817. Encycl. Londinensis 16, p. 89; Auteniquois country (Orange R.?).

Range and Status. Endemic resident, SW Angola (desert coast of Mossamedes and compact desert of Iona Nat. Park), W and S Namibia and South Africa (Cape Province south to Great Karoo and east to SW Orange Free State, about Aliwal). Frequent; density of 13 birds in 850 ha in karroid broken veld, Klaarstroom, S Great Karoo (Winterbottom 1986b). Common and widespread in Skeleton Coast Park, Namibia (Ryan *et al.* 1984).

Description. *C. t. tractrac* Wilkes: South Africa (Karoo to Aliwal). ADULT ♂: forehead, lores, crown, nape, sides of neck, mantle, back and scapulars uniform greyish brown or brownish grey, becoming rufescent on rump; uppertail-coverts white. Tail brownish black with creamy white or buffy white base concealed below uppertail-coverts, and with increasing amounts of white in outer webs, such that T5 has proximal half white, distal half black, with white in outer web to within 17 mm of tip, and T6 has completely white outer web. Ear-coverts slightly browner than crown. Chin and throat greyish white in centre, merging to pale grey on sides; breast pale grey; belly whitish or greyish white in centre, merging to buffy brown on flanks; undertail-coverts white. Wings dark brown with outer web of outer primaries narrowly fringed buff towards base, and inner secondaries and tertials narrowly or quite broadly fringed buff all around (forming narrow whitish trailing edge to inner part of wing). Shape of P9 varies: tip rounded, attenuated or emarginated (Meinertzhagen 1950). Upperwing-coverts grey-brown, broadly fringed buff. Underwing-coverts and axillaries white. Bill black; eye dark brown; legs and feet black. Sexes alike. SIZE: (4 ♂♂, 4 ♀♀) wing, ♂ 83–86 (84.2), ♀ 82–89 (85.2); tail, ♂ 48–53.5 (50.0), ♀ 48–54 (50.0); bill to feathers, ♂ 13–14 (13.5), ♀ 13.5–15 (13.7); tarsus, ♂ 26–29.5 (27.0), ♀ 29–31 (29.5).

IMMATURE: dark brown, heavily mottled buff. Forehead and crown dark brown, heavily spotted buff; nape, mantle and back buffy brown with rufescent wash. Tail-feathers like adult but with buffy tips. Throat and breast buffy, mottled rufescent dark brown. Primary coverts, remaining upperwing-coverts, inner secondaries and tertials broadly fringed pinkish buff.

NESTLING: unknown.

C. t. albicans (Wahlberg): Namibia (W Damaraland, N Great Namaqualand). Much whiter than nominate race. Upperparts grey; feathers whitish-tipped, so that upperparts look mealy; rump as well as uppertail-coverts white or creamy. Entire underparts creamy white. Upperwing-coverts, remiges and tertials broadly fringed creamy white. Larger. SIZE: (3 ♂♂, 3 ♀♀) wing, ♂ 92–94 (92.7), ♀ 88–96 (91.3); tail, ♂ 47–51 (49.4), ♀ 47–49 (48.1); bill to feathers, ♂ av. 17, ♀ av. 15.5; tarsus, ♂ av. 29.7, ♀ av. 29.

C. t. barlowi (Roberts): Namibia (central and S Great Namaqualand, Aus, Bushmanland). Upperparts and underparts intermediate in tone between nominate *tractrac* and *albicans*. Rump white; barely any trace of rufescent or pinkish brown wash in lower back adjacent to rump. Throat and belly off-white, breast very pale grey. Wing, unsexed (n = 7) 79–88 (85).

C. t. hoeschi (Niethammer): NW Namibia, SW Angola. Darker than *barlowi*. Wing 90–97.

C. t. nebulosa Clancey: coastal sand dunes, SW Namibia and W South Africa. Warmer and buffier than *barlowi*; colder and greyer than *albicans*, and smaller: wing, ♂ (n = 9) 81.5–89 (85.5), ♀ (n = 7) 79.5–83.5 (80.3).

Field Characters. A small, subdesert chat with white underparts and wheatear-like tail pattern – tail white with solid black end forming obtuse triangle with apex in centre of tail (**A**). In Angola and Namibia has pale grey upperparts, white-fringed wing-feathers, white rump, and is same size as Karoo Chat *C. schlegelii* there. In South Africa smaller than local Karoo Chat, and head, back and wings dark brown; distinguished from Karoo Chat by having white rump and tail-base, whiter underparts and shorter tail.

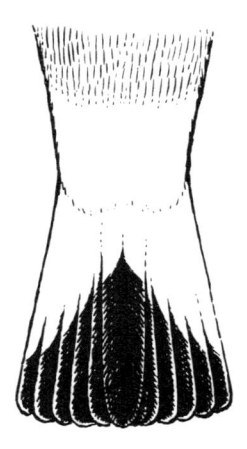

Voice. Tape-recorded (GIB). Rather silent. Alarm a sharp 'trac-trac'.

General Habits. Inhabits open shrubby plains in semi-desert, karoo and stubbly grassveld, straggly bushes on coastal sand dunes, and dry watercourses (Maclean 1985). Occurs in fur seal colony *Arctocephalus pusillus* at Wolf Bay, Namibia (Shaughnessy and Shaughnessy 1987). Prefers flat to hilly or dissected country.

Solitary or in pairs; up to 10 separated birds in wide area can be in view at once. Perches on low bush or rock and flies down to ground when insect seen. Frequently flicks wings and jerks tail (less deliberately than Familiar Chat *C. familiaris*). Runs fast. Tame and inquisitive; when alarmed hovers briefly to improve view (Maclean 1985).

Food. Insects, including tenebrionid beetles dead on roads.

Breeding Habits. Loosely territorial (Maclean 1985).

NEST: neat cup; foundation of twigs, lined with soft grass and hair. Sited on ground under shrub or boulder; once inside gnarled desert cabbage *Welwitschia mirabilis*.

EGGS: 2–3. Greenish blue, immaculate, or finely speckled with red-brown. SIZE: (n = 14) 19.5–24.2 × 15–16.9 (22.1 × 16.1).

LAYING DATES: South Africa, Aug–Apr (mainly Sept–Oct, but season varies with rainfall).

Reference
Maclean, G. L. (1985).

Cercomela familiaris (Stephens). **Familiar Chat; Red-tailed Chat. Traquet de roche à queue rousse.**

Saxicola familiaris Stephens, 1826. Gen. Zool. 13, p. 241; southern Africa.

Forms a superspecies with *C. scotocerca*.

Plate 31
(Opp. p. 464)

Range and Status. Endemic resident, in W Africa rather local, uncommon to common, from SE Senegambia to mid Central African Republic, and in E and southern Africa frequent to common from Sudan and N Ethiopia to South Africa, north to Angola and Cabinda, as follows: Senegambia, common in Kédougou region (Morel 1985); Mali, fairly common and widespread in S, mainly in rocky hills; Ivory Coast, on northern inselbergs (Niangbo, Korhogo, Boundiali, Odienné); Ghana, local, uncommon (Gambaga, Bongo Hills, Mole); Nigeria, locally common on bare rocky ground and in erosion gullies, from Sokoto to Igbetti, Ogoja, Jos and Obudu Plateaux; Cameroon (Cameroon montane district, Adamawa, Sakdjé); Central African Republic, frequent to common, Lobaye Préfecture and Bamingui-Bangoran Nat. Park); Sudan, frequent in Kordofan and Boya Hills (Equatoria) and Boma Hills, and local and uncommon elsewhere south of 13°N (Nikolaus 1987); Ethiopia, uncommon, locally frequent, in NW and SW; NE Uganda, N Kenya, common and widespread in L. Turkana basin and Karamoja; some scattered records in E Uganda and W Kenya may refer to wanderers, but common all year at Olooloo escarpment, NW Mara Game Reserve, SW Kenya (Pearson 1990); widespread in SW Uganda in Ankole and Kigezi regions. From NW and NE Tanzania (Arusha, Mara, Tabora, Kigoma, West Lake regions) extends south, becoming common in S Tanzania. Widespread in Malaŵi, between 90 and 1550 m, scarce up to 1850 m. In Zambia rather local and uncommon in E, scarce in W, absent from Barotse Province, and in North-Western Province known only at Kasempa, Zambezi Rapids, Balovale and Chavuma Hill. Throughout Zimbabwe: sparse in parts of Mashonaland, commoner in dry country, sparse or absent in NW Matabeleland; more widespread south than north of Limpopo R. (Campbell 1988). Remainder of southern Africa, widespread and common resident, particularly in dry areas, but uncommon in moist E littoral. Angola, coastal plain from Cunene R. to Luanda, central plateau from W and N Huila to S Cuanza Sul, Malanje and N Bihé, also Cabinda.

Description. *C. f. familiaris* (Stephens): South Africa (S Cape east to Great Kei R.). ADULT ♂: forehead, crown, nape, sides of neck, mantle, back and scapulars uniform dark greyish brown merging to bright rufous on rump and uppertail-coverts. Central pair of tail-feathers uniform dark brown, rest rufous with dark brown tips *c.* 9 mm deep, T6 with outer web entirely dark brown. Ear-coverts rufescent brown, contrasting with adjacent plumage. Chin, throat, breast and flanks pale greyish buff; belly and undertail-coverts pale greyish buff, washed rufous or pinkish brown. Wings dark brown; all upperwing-coverts and remiges narrowly edged and tipped buffy. Underwing-coverts grey. Bill black; eye dark brown; legs and feet black. Sexes alike. SIZE: wing, ♂ (n = 22) 85–92 (88·0), ♀ (n = 24) 79–85 (82·0); tail, ♂♀ (n = 46) 55–68; bill, ♂♀ (n = 54) 13–19; tarsus, ♂♀ (n = 42) 22·5–26. WEIGHT: unsexed (n = 18) 17–28·8 (20·8).

IMMATURE: like adult but upperparts buff-spotted and underparts with dusky feather-tips, giving scaly appearance, particularly on breast.

NESTLING: blind; head and upperparts covered with dense grey down, not long, gape-flanges yellow (from photograph in Steyn 1966). Later, bill dark brown, gape-flanges whiter, tongue and middle of palate orange-yellow, rest of mouth yellow (Serle 1940).

TAXONOMIC NOTE: the following 4 races are identifiable in series, but differences between them and nominate *familiaris* are of shade only and are very slight. *C. f. falkensteini* and *omoensis* are more distinct.

C. f. galtoni (Strickland): South Africa (NW and N Cape) to Namibia (Damaraland) and W Botswana. Upperparts like nominate race but from forehead to back slightly paler. Underparts paler, pinker and less ochreous than in nominate race.

C. f. angolensis Lynes: N Namibia (Etosha Pan, Ovamboland, Kaokoveld, N Namib) and W Angola. Upperparts very slightly paler than in *galtoni* and rufous parts slightly less intense.

C. f. hellmayri (Reichenow): South Africa (NE Cape, Orange Free State, Transvaal), SE Botswana, Zimbabwe (plateau), Swaziland, S Mozambique. Undertail-coverts and belly very slightly paler than in nominate race. Albino (see Webb 1976).

C. f. actuosa Clancey: South Africa (Drakensbergs, Transkei, W Natal, Lesotho). Browns darker and warmer than in nominate race and rufous parts less vinaceous.

C. f. falkensteini (Cabanis): Zambezi valley to NW Ethiopia and W Africa, except range of *omoensis*. Differs from all previous races in having much greyer underside, whiter in centre of belly, and lacking warm buffy tinge. Throat and breast grey. WEIGHT: unsexed (n = 19) 13–16 (14·6) (J. S. Ash, pers. comm.).

Cercomela familiaris

C. f. omoensis Neumann: SW Ethiopia, SE Sudan, NW Kenya, NE Uganda. Upperparts slightly darker than in *falkensteini*; underparts greyer and darker.

Field Characters. A slim, tame, confiding chat with orange-rufous rump and sides of tail; centre and tip of tail blackish brown, forming dark inverted 'T' on orange background (**A**). Rest of plumage nondescript brown-grey, belly paler. In pairs, in broken or rocky country with shrubs; perches on shrubs and ground, flicking wings on landing. Most similar to Brown-tailed Rock-Chat *C. scotocerca* (overlap in Kenya/Uganda), which has similar dingy brown plumage but lacks rufous rump and tail. Race *turkana* of Brown-tailed Rock-Chat (in zone of overlap) does have rufescent wash on rump, narrow rufous-buff edges to tail-feathers, but these are hard to see in the field, whereas rufous of Familiar Chat stands out clearly. In N tropics, ♀♀ and immatures of Common Redstart *Phoenicurus phoenicurus* and Black Redstart *P. ochruros* have brown plumage, rufous rump and tail, but tail brighter, and all except central pair rufous to tip; tail constantly shivered, and wings not flicked. For distinction from Sickle-winged Chat *C. sinuata* in southern Africa, see that species.

Voice. Tape-recorded (53, 86, 88, GIB, HEL, MOY). Song quiet, a nondescript random series of soft whistles interspersed with subdued chattering (like alarm note): 'tsweep-tsweep-tsweep-cher-cher-tsweep-tsweep-cher-cher-cher-tsweep-tsweep' (Steyn 1966), also described as 'peep-churr-churr, peep-chak-chur, peep-peep . . .' (Maclean 1985). Alarm a harsh ratchet-like 'cher-cher' or 'chuck-chuck', or staccato series of rapid notes with high 'whee' whistles: 'whee-chuck-chuck-chuck, whee-chuck'.

General Habits. Inhabits shrubland or light woodland, particularly *Brachystegia*, generally with scattered rocks or dissected by small erosion gullies; also rocky outcrops, inselbergs (kopjes), wooded valleys and hillsides, escarpments, cultivation, and tree-lined stream beds in dry country. Readily adapts to human settlements; at Essexvale (Zimbabwe) common around a school village but uncommon in surrounding countryside (Steyn 1966).

In pairs or up to 5 birds 'not necessarily family parties' (Vincent 1947); tame. Forages by keeping watch at perch on bush-top, low outer branch of tree, fence, rock, termitarium, wall or building, and flying down to ground to catch insect. Flicks wings on landing 3–4 times, rather more deliberately than do most other chats; hops on ground, flicking wings at each stop; eats insect on ground or flies with it back to elevated perch, flicking wings on return to perch and slowly raising tail once or twice. 2 observations of 1–2 birds flying up and clinging to a klipspringer *Oreotragus oreotragus*, searching its ears and body as it lay down, and following it apparently to feed on disturbed insects, suggest that the association may be regular (Steyn and Hosking 1988).

Mainly sedentary (e.g. Essexvale, Zimbabwe), but in Mali thought to occur north to 15° 31′N in the rainy season only (June–Oct) (Lamarche 1981).

Food. Mainly insects; also bread, and mulberries taken from ground and swallowed whole (Taylor 1936, Steyn 1966). Mulberries, an earthworm, and harvester termites *Hodotermes mossambicus* brought to nestlings (Zimbabwe). Said to eat pine seeds and to damage nursery pine seedlings (Vernon 1967).

Breeding Habits. Solitary, monogamous breeder, probably territorial.

NEST: neat, firm, thick cup of hair, wool, feathers and soft plant material, with deep, roughly-made base, on foundation of stones, clods of earth and bits of bark. Stones are 'used to reduce the crevice in which they are nesting to the desired size' before nest-building (Earlé 1981). 1 nest (in nest-box) had foundation of 147 stones, with *c.* 50 more added the following season; stones 11–20 × 38–48 mm, most weighing 5–6 (max. 7·5) g (Steyn 1966). Another (in house wall) had foundation of 361 stones, 15 three-inch nails, and *c.* 250 items including pieces of bark, wood, bamboo, tin, lime plaster, mud, dung, rag, bone, paint skin, glass and rubber: total weight 1163 g (James 1929). Non-artificial sites also reinforced with pebbles (Plowes 1943). Sited in hole, in ground, among tree-roots, in wall of erosion gulley, rock-face, building, 20 cm inside old burrow of White-fronted Bee-eater *Merops bullockoides*, old nest chamber of Sociable Weaver *Philetairus socius*, nest-box and tin; once in a constantly-used customs road-boom (Steyn 1967). In dry stream-beds in open country, nest sited in area with very little cover (Vincent 1947). Nest-site inspected for 3 days before building starts. Nest built (by both birds?) in 2–2½ and 3–4 days, with 1st egg laid 1–4 days later (Steyn 1966). Some nests have dry plant stems and twiglets piled against outer side (Vincent 1947).

EGGS: 2–4, usually 3; av. of 95 clutches, southern Africa, 3·1. Laid on successive days in early morning (before 07:00 h). Bright greenish blue, sparsely speckled with red-brown, concentrated around broad end. Sometimes a 2nd clutch laid in same nest *c.* 2 weeks after 1st brood fledges. SIZE: (n = 186, southern Africa) 18·0–23·6 × 14·1–16·6 (20·4 × 15·1).

LAYING DATES: Mali, Mar; Ivory Coast (nest-building Feb); Ghana, Dec; Nigeria, Mar–May; E Africa: Region C, Mar; Angola (breeding condition Sept–Oct); Zambia, Sept–Dec, mainly Oct; Malaŵi, Sept–Nov; Zimbabwe, July–Jan (July 4, Aug 31, Sept 67, Oct 55, Nov 26, Dec 12, Jan 1 clutches); South Africa, Sept–Feb (mainly Oct–Nov).

INCUBATION: not known whether both sexes incubate. 1st and 2nd eggs are brooded overnight before 3rd egg laid (Steyn 1966). Incubating bird leaves nest to chase away starlings and to mob shrikes. Period: (Zimbabwe) 13–13·5 days. Eggs hatch in same sequence as they were laid.

DEVELOPMENT AND CARE OF YOUNG: weight: (n = 3) 2·5–3 at day 1, 3·5–5 at day 2, 9·5–13 at day 5, 16·5–20·5 at day 8, 17·5–19 at day 11 and 19–21 at day 14. Feathers first break from quills on body at 6 days, when down sparse and eyes half open; at 11 days well-feathered, with little down remaining, and abdomen no longer swollen. Bill growth 95% complete by day 7 and tarsus growth 95% complete by day 9; wing grows at constant rate from days 3 to 14; tail grows slowly to day 5 and accelerates thereafter (Steyn 1966). Young fed by both parents; fledging period 15–18 days ± 1 day. Both parents performed injury-feigning distraction display when a nestling, disturbed by observer, flew prematurely to ground; they ran along ground with wings half-opened and dragging. Fledglings tended by both parents, remaining in vicinity of nest or moving few hundred m away.

BREEDING SUCCESS/SURVIVAL: bird ringed as nestling, Zimbabwe, trapped where ringed after 6 years 2 months. Dead bird evidently impaled on *Opuntia rosea* (Dean and Dean 1987).

Reference
Steyn, P. (1966).

Cercomela scotocerca (Heuglin). Brown-tailed Rock-Chat. Traquet de roche à queue brune.

Plate 31
(Opp. p. 464)

Saxicola scotocerca Heuglin, 1869. Orn. Nordost. Afr. 1, p. 363; near Keren, Bogosland.

Forms a superspecies with *C. familiaris*.

Range and Status. Endemic resident in 4 isolated populations: (1) W Sudan (Darfur), common on Jebel Marra up to 2460 m, and Chad east of 19°E, between 11° and 15°N, locally abundant – 10–20 per ha at Kilingen (Salvan 1967); (2) E Sudan and N Ethiopia (Red Sea Hills), common between 1100 and 1700 m; (3) E-central Ethiopia (Awash valley) to N Somalia, fairly common north of 9°N and very local south to 7°N; and (4) SW Ethiopia, N Kenya, and NE Uganda, widespread but uncommon, from Karamoja and Turkana to L. Baringo, N Uaso Nyiro R., Garba Tula and Marsabit (old record from Wajir).

Description. *C. s. scotocerca* (Heuglin): E Sudan, N Ethiopia. ADULT ♂: forehead, crown, nape, sides of neck, mantle, back and scapulars plain greyish brown; rump rufescent brown; uppertail-coverts dark brown with rufescent brown or buffy fringes. Tail uniformly dark brown. Ear-coverts rufescent brown in contrast with grey-brown sides of neck and with crown and throat; circle of whitish featherlets around eye, especially whitish above it. Chin and throat pale greyish or buffy white; breast very pale brown, belly pale greyish or buffy white; undertail-coverts off-white; flanks pale brown. Wings dark brown; remiges, tertials, primary coverts and other coverts narrowly and rather inconspicuously fringed buffy. Axillaries and underwing-coverts pale brown, under primary coverts somewhat darker, inner edges of undersides of flight feathers pale buffy brown giving underwing pale appearance. Bill black; eye dark brown; legs and feet black. Sexes alike.

SIZE: (5 ♂♂, 2 ♀♀) wing, ♂ 68–71 (70·2), ♀ 67, 67; tail, ♂ 51·5–55 (52·7), ♀ 50, 51; bill, ♂ 12–15 (13·4), ♀ 10·5, 12·5; tarsus, ♂ 20·5–22 (21·5), ♀ 18·5, 19.

IMMATURE AND NESTLING: unknown.

C. s. furensis Lynes: Sudan (Darfur). Barely distinguishable from nominate *scotocerca*; slightly warmer shade of brown below.

C. s. spectatrix S. Clarke: Ethiopia (Awash valley), Somalia south to 9°N. Upperparts much paler and greyer than nominate race; underparts paler and greyer. Large: unsexed (n = 4) wing 80–85; tail 60–62; bill 11·5–13; tarsus 25–26. WEIGHT: 1 unsexed 20·6 (J. S. Ash, pers. comm.).

C. s. validior Berlioz and Roche: Somalia between 7° and 9°N. Like *spectatrix* but undertail-coverts yellowish. Larger: unsexed (n = 3) wing 83–90; tail 62–67; bill 13–14; tarsus 25–28. *C. s. spectatrix* and *validior* treated as specifically distinct from *scotocerca* and *turkana* by Berlioz and Roche (1970).

C. s. turkana van Someren: Kenya, Uganda, S Ethiopia. Very slightly darker than nominate race; rump rather more rufescent (at least in some specimens); outer edges of rectrices rufescent brown; underparts darker than in nominate race and more uniformly grey-brown or ochreous brown. Larger: wing, ♂ (n = 3) 77–80 (78·0), ♀ (n = 3) 71–78 (75·2); tail, ♂ (n = 3) 56–60 (57·7), ♀ (n = 3) 54–58 (55·7). WEIGHT: 1 ♀ 16 (S. Keith, pers. comm.).

Field Characters. An undistinguished-looking rock- and bush-dwelling chat, grey or grey-brown with dark brown wings and tail, stocky upright stance, noticeably large eye and typical chat behaviour (tame, perching low, dropping onto ground for insects). Should be compared carefully with Familiar Chat *C. familiaris* in zone of overlap in N Kenya/Uganda; for differences see that species. Altitudinally and ecologically distant from Moorland Chat *C. sordida*, confined to high moorlands. In Sudan likely to be confused only with Blackstart *C. melanura* (black tail, in sharp contrast with grey-brown back). In E Ethiopia and N Somalia overlaps with Blackstart and Sombre Rock-Chat *C. dubia*. Latter is darker and browner than pale grey sympatric race of Brown-tailed Rock-Chat (and is almost identical in plumage to Red Sea Hills race of Brown-tailed Rock-Chat, but is longer-legged than it, much longer-winged, relatively short-tailed, and is not known to be large-eyed).

Voice. Tape-recorded (GREG, McVIC). A sweet strong 'chuke-chuke' and a short sweet trill (Mackworth-Praed and Grant 1960).

General Habits. Inhabits rocky scarps, sandstone or boulder-strewn wadis, with scattered low shrubs (Somalia); arid rocky country with plenty of acacia bushes (Sudan, Ethiopia), keeping to bushes more than to rocks; also bushed grassland (Kenya/Uganda).

Solitary. Fearless, but if approached too close, hops off boulder perch and disappears amongst rocks and undergrowth (Archer and Godman 1961). Flicks wings and tail. Sings from top of vegetation.

Food. Ants and termites in 2 stomachs, grains in 1.

Breeding Habits. Solitary breeder.

NEST: like that of Blackstart; foundation of grass, lined with hair, sited in hole or deep crevice in rocks with a little slope of pebbles outside (Mackworth-Praed and Grant 1960).

EGGS: not known.

LAYING DATES: Ethiopia (Eritrea), Apr.

Reference
Archer, G. and Godman, E. M. (1961).

Plate 31
(Opp. p. 464)

Cercomela dubia (**Blundell and Lovat**). **Sombre Rock-Chat. Traquet de roche sombre.**

Myrmecocichla dubia Blundell and Lovat, 1899. Bull. Br. Orn. Club 10, p. 22; Fontaly, Abyssinia.

Range and Status. Endemic, probably resident, E-central Ethiopia (upper Awash valley eastwards, about 10°N) and Somalia (1 old record from northern mountains). Rare and little-known; perhaps locally frequent.

Description. ADULT ♂: upperparts brown or dark greyish, with very slight rufous wash on rump; uppertail-coverts dark brown. Tail uniformly dark brown with outer web of T6 fringed palish in one specimen. Lores pale grey; sides of neck grey brown; ear-coverts rufescent brown. Chin and throat pale grey merging into pale greyish brown breast; centre of belly whitish or pale grey merging into browner thighs and greyer flanks;

undertail-coverts greyish or brownish white (3 specimens) or brownish black (type, paratype). Outer webs of remiges and wing-coverts narrowly fringed lighter brown or greyish brown. Underwing-coverts grey-brown. Bill black; eye dark brown; legs and feet black. Sexes alike. SIZE: (2 ♂♂, 1 ♀) wing, ♂ 79, 83, ♀ 76; tail, ♂ 55, 55, ♀ 54; bill, ♂ 15, 15, ♀ 14.5; tarsus, ♂ 23.5, 24, ♀ 23.

IMMATURE AND NESTLING: unknown.

Field Characters. Nondescript. Very like Blackstart *C. melanura* but tail dark brown rather than black, undertail-coverts greyish or brownish white (not cream-white), and bill rather stouter. Also like Brown-tailed Rock-Chat *C. scotocerca*, local race of which is, however, pale grey (with dark wings and tail). Posture said to be upright; habits like those of Blackstart.

Voice. Unknown.

General Habits. Type specimen was collected at dusk, from a flock of 12–15 birds, which were making short flights, alternating between bushes and rocks, in a directional manner which suggested migration (Ogilvie-Grant 1900).

Food. Unknown.

Breeding Habits. Unknown.

Reference
Ogilvie-Grant, W. R. (1900).

Cercomela melanura (Temminck). Blackstart; Black-tailed Rock-Chat. Traquet de roche à queue noire.

Plate 31
(Opp. p. 464)

Saxicola melanura Temminck, 1824. Pl. Col. 43, pl. 257, fig. 2; Arabia (Sinai).

Range and Status. Africa and Middle East (Sinai, Israel and W Jordan to Aden, W and central Saudi Arabia, S Oman). Accidental Syria, Kuwait.

Resident, frequent to common. Mali, Hombori foothills and in E in Azzawakh Mts, hills of Agalam-Galam, and south of Adrar des Iforras. Niger, Tillia, Arlit, Aïr, Massif de Djada, Filingué, Tibiri, Dallol Bosso. Chad, widespread in north, common Tibesti and south to Largeau, 3 records south to Mortcha and Elela. Sudan, locally common from Darfur to Red Sea. Egypt, 6 records from Gebel Elba region where probably breeding resident, possibly only summer resident (Goodman and Atta 1987), and sight records as follows: Giza, 9 Oct (Meinertzhagen 1930); Aswan, 16 Oct (Vuilleumier 1979); Sakkara, 22 Feb (Keith 1980); pair, Abu Simbel, Mar 22 (Welch and Welch 1983). Ethiopia, common in NE and Danakil Desert, also on Dahlac, Dancalia and Nocra Is. Djibouti, quite common in Forêt du Day and Mabla Mts, 14 seen at Garab on 1 day and 13 at Goula; also Tadjoura, As Eyla and Loyada (Welch and Welch 1986). Somalia, common north of 7°N. Vagrant (?) Gambia, 4, Fatoto, Feb 1974.

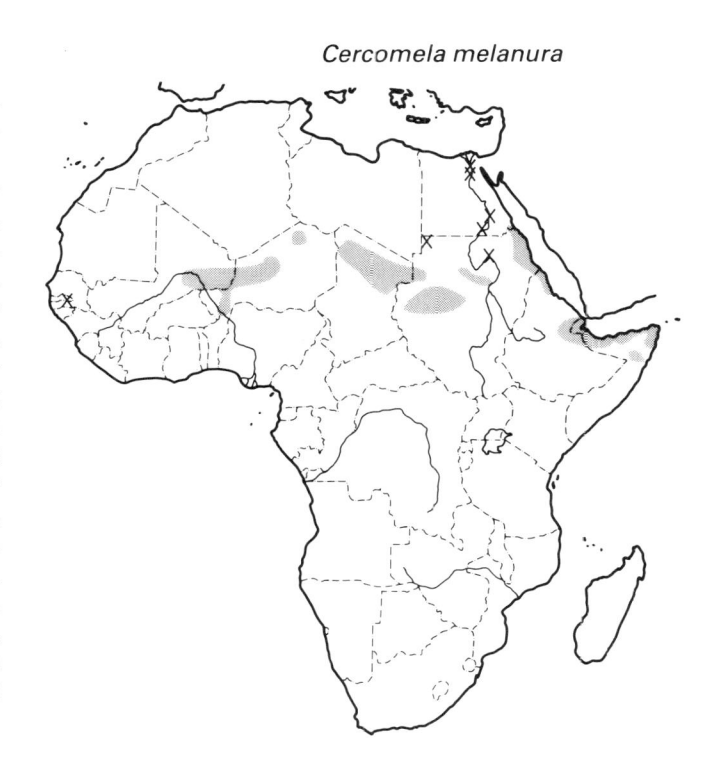

Description. *C. m. lypura* (Hemprich and Ehrenberg): SE Egypt, NE Ethiopia, and Sudan from Nile to Red Sea coast. ADULT ♂: forehead to rump uniform pale sandy grey-brown; uppertail-coverts black, shorter ones at sides tipped pale grey; tail black. Lores dusky; ear-coverts pale greyish brown; cheeks pale brownish grey. Chin, throat and breast pale brownish grey; belly, flanks and undertail-coverts cream-white, tinged pink in fresh plumage. Remiges and tertials dark brown or brown-grey with outer webs narrowly edged pale grey; lesser wing-coverts pure grey, other upperwing-coverts greyish brown. Axillaries and underwing-coverts pale grey-brown. Bill black or brown-black; eye dark brown; legs and feet black. Sexes alike. SIZE: (*C. m. airensis*, Sudan and Niger, 9 ♂♂, 8 ♀♀) wing, ♂ 78–82, ♀ 74–77; tail, ♂ 57–58, ♀ 53–55; bill to skull, ♂ 15·7–16·8, ♀ 16·0–16·8, bill to feathers, ♂♀ 12–13; tarsus, ♂ 23–24, ♀ 22–23.

IMMATURE: like adult but browner, less grey; uppertail-coverts brown-tipped; breast creamy; remiges with broader, less clear-cut, pale fringes.

NESTLING: unknown.

C. m. aussae Thesiger and Meynell: Ethiopia (Danakil Desert), Djibouti, Somalia. Like *lypura* but darker and greyer.

C. m. airensis Hartert: Niger (Aïr) to Sudan (Kordofan). Browner than foregoing races, not grey, but with sandy or cinnamon tinge. Birds in Tibesti and Ennedi darker than those in Aïr. WEIGHT: ♂ (n = 4) 14–15, ♀ (n = 2) 15, 15.

C. m. ultima Bates: Mali, W Niger. Brown, not grey; like *airensis* but darker and without cinnamon tinge. However, doubtfully distinct from *airensis* since colour variable in this species (Lamarche *in* Mayaud 1988).

C. m. melanura (Temminck): Egypt (Sinai), Israel, Jordan, Saudi Arabia. Paler and greyer than other races. Not certainly recorded in Africa, but sight records from N Egypt may refer to this race since it breeds in Sinai.

Field Characters. A plain pale grey-brown terrestrial desert chat with 2 distinctive features: tail is all-black, and bird constantly flicks tail open, often drooping then flicking wings half-open at same time.

Voice. Tape-recorded (53, HOL, JOHN). Song given as 'chi koo chri-ki chiu-teoo' (*aussae*, Ethiopia, lasting 0·9–1·0 s), and a 'weedley' song with scratchy and liquid notes (*ultima*, Niger, lasting 1·3–1·5 s) (Cramp 1988). In Arabia song described as 'chree chrew chitchoo chirri chiwi' or 'chee-yu-chwe' or 'chup t'chee chup ter tutcher'. Contact call a loud, liquid 'chura lit' or 'tyoo-trit'; alarm a high-pitch whistle 'feefee' or 'whee'.

General Habits. Inhabits rocky hills with or without cover but usually with sparse acacia, dissected rocky subdesert, thorny bushes in rocky ravines, screes, steep hillsides with rockfalls and scattered shrubs (bases of inselbergs), sandstone scarps, and sandy dry river beds. Readily takes to environs of human habitation.

In pairs in breeding season, in pairs or solitary in winter. Tame and fearless. Spends most of time in restless flitting between rocks, ground and shrubs or small trees; sits in, rather than on, bush. Perches and forages freely around stone-built houses, huts, stock-enclosures, low walls, and manure yards. Frequently flirts tail and wings, half-opening them in a measured way (**A**). Feeds by scanning from low perch, and dropping onto insect on ground; also searches vegetation, hovers briefly, and pursues insect on wing. Forms loose foraging flock with warblers *Sylvia melanocephala*, *S. curruca* and *S. leucomelaena* (Egypt: Goodman and Meininger 1989). Sings from tree-top, house, or rocky eminence, for most of year but infrequently in autumn (particularly Oct). In winter sings mainly in early morning, early forenoon, and evening (Hartley 1952). In aggressive display faces away from opponent, depresses and spreads tail to show black plumage.

Sight records in non-breeding season in Egypt (see above) indicate some local movement.

Food. Insects. In Arabia caterpillars and winged insects (Bundy 1986) and berries, including *Lycium shawii*.

Breeding Habits. Evidently solitary and territorial, probably monogamous; c. 4 pairs per km of desert wadi (Oman: Bundy 1986). Aerial chasing occurs early in breeding season, both birds continually calling.

NEST: shallow cup or pad of grass, hair, leaves and bits of rag, lined with finer vegetation and hair. Sited up to 0·5 m into crevice in rock-face or hole in high earthen bank of dry watercourse, cavity in stone wall, expansion-gaps in road bridges, or under house eaves. Rock-hole nests may have perimeter or platform of pebbles (Mackworth-Praed and Grant 1960; not noted by other authors, however). Nest built by ♂ and ♀.

EGGS: 3–4. Subelliptical; smooth, glossy; very pale blue, finely speckled or heavily freckled with chestnut-brown, speckles sometimes concentrated at broad end. SIZE: (n = 26) 18·0–21·5 × 14·0–16·5 (19·6 × 14·7). WEIGHT: c. 2·3.

LAYING DATES: Mali, Apr–June; S Sahara, Mar–June; Niger, May, (nestlings June, July); Sudan (fledglings May); Ethiopia, Feb–Apr; Somalia, May–June (probably also Mar–Apr: Archer and Godman 1961).

DEVELOPMENT AND CARE OF YOUNG: fledglings attended by both parents.

References
Bundy, G. (1986).
Cramp, S. (1988).

Cercomela sordida (Rüppell). Moorland Chat; Hill Chat. Traquet afroalpin.

Saxicola sordida Rüppell, 1837. Neue Wirbelt., Vögel, p. 75, pl. 26, fig. 2; Simen, Abyssinia.

Plate 31
(Opp. p. 464)

Range and Status. Endemic resident, mountains of Ethiopia, E Uganda, Kenya and N Tanzania. In Ethiopia common to locally abundant in highlands on both sides of Rift Valley, but rare below 2100 m; in E Africa common in alpine moorland from upper edge of forest to 4400 m, on Mt Elgon, the Cheranganis, Aberdares and Mt Kenya, scarce as low at 2300 m in forest glades; the commonest bird on Mt Kilimanjaro above 3400 m, and seen up to nearly 5200 m there (Moreau 1935); also occurs at Engamat, Olosirwa and Ololmoti, and said to occur on Kinangop Plateau (Kenya), Mt Meru (Tanzania) and Rwenzoris (W Uganda) (Britton 1980).

Description. *C. s. sordida* Rüppell: Ethiopia. Forehead, crown, nape, ear-coverts, sides of neck, mantle, back, rump and scapulars uniform dark grey-brown; uppertail-coverts dark brown. Central 2 pairs of tail-feathers (T1, T2) brownish black; T3–T5 white with dark brown tip 5 mm deep, and T6 white with dark brown tip and distal half of outer web dark brown. Indistinct greyish superciliary stripe from nostril across top of eye and ear-coverts; lores brown; ear-coverts slightly less greyish than surrounding plumage. Entire underparts buffy or pale greyish brown. Wings dark brown; remiges very narrowly fringed buffy but inner secondaries and tertials broadly fringed buffy brown; upperwing-coverts dark brown, broadly fringed buffy brown. Underwing-coverts buffy white. Arussi Plateau birds, '*schoana*', said to be browner above and more fulvous below; Ghera region birds darker than '*schoana*' and rustier below than Kenyan and Tanzanian races (Desfayes 1975). Bill black; eye dark brown; legs and feet black. Sexes alike. SIZE: wing, ♂ (n = 3) 72–78 (74·7), ♀ (n = 3) 71–73 (72·0), unsexed (n = 11) 67–78. WEIGHT: 2 ♀♀ 18, 19; (unsexed, n = 10) 19–23 (21·5) (J. S. Ash, pers. comm.).

IMMATURE: like adult but upperparts indistinctly barred, and breast speckled with dark brown.

NESTLING: unknown.

C. s. ernesti (Sharpe): Kenya, Uganda. Upperparts slightly darker brown than nominate race. WEIGHT: ♀ (n = 12) 15–24 (18·75).

C. s. hypospodia (Shelley): Tanzania (Mt Kilimanjaro). Upperparts darker than in *ernesti* and crown blackish brown; underparts greyer; chin, throat and breast mid-grey, upper belly pale buffy grey, lower belly, flanks and undertail-coverts olivaceous brown. SIZE: (3 ♂♂, 3 ♀♀) wing, ♂ 73–74 (73·3), ♀ 71–74 (72·3); tail, ♂ 45–47 (46·3), ♀ 47–52 (49·0); bill, ♂ 13–15 (14·0), ♀ 14–15 (14·7); tarsus, ♂ 29–34 (31·0), ♀ 29–30 (29·7).

C. s. olimotiensis (Elliott): Tanzania (Crater Highlands). Upperparts like those of *ernesti*, underparts like those of *hypospodia*.

Field Characters. A small, dark brown chat with wheatear-like tail, white with dark brown tip and central feathers; one of the very few bird species resident above 3500 m, and one of the commonest at that altitude. Short-tailed, but behaviourally a typical chat, flicking tail and wings, perching low, and making small pounce onto insect on ground.

Voice. Not tape-recorded. Song not known. 'Silent creatures' (Moreau 1935). Call a pleasant chirping or metallic 'werp-werp', and a more sibilant alarm (Mackworth-Praed and Grant 1960).

General Habits. A very common resident in high alpine grasslands and rocky moors, mainly at 3400–4400 m, among tree heaths above bamboo zone; marshy bamboo-forest glades; bare slopes with no vegetation 'but a few scraps of half-dead grass' (Moreau 1935); open sward of low-growing matted grass; tussock-grass; *Artemisia*-

covered subalpine scrub. Cultivated fields in Ghera region at 2150 m in Ethiopia, including cultivated clearings, without rocks, in thick rain forest (Desfayes 1975).

In pairs and small parties. Sits on sprigs of heath and everlasting (*Helichrysum*) flower heads and stems and sticks projecting from alpine meadows, and flies or jumps down to take insects. Alert and active; flits about amongst rocks and boulders. Often perches with legs splayed (**A**, p. 543); in freezing conditions can look dumpy, like European Robin *Erithacus rubecula* (**B**, p. 543). Flicks tail and wings, fanning tail, and bows down. Extremely tame, approaching a person to within a few cm.

Food. Mainly small beetles and their larvae, also caterpillars and small shells.

Breeding Habits. Solitary breeder.

NEST: large, untidy grass cup lined with moss, lichen, plant down, fur (from hyrax?) and a few feathers; 1 nest ext. depth *c.* 56, ext. diam. 105 × 118, cup depth *c.* 18, cup diam. 68. One was sited in cleft at top of bole of giant groundsel; another in side of grassy tussock.

EGGS: 3. Blue, lightly spotted and streaked with black and violet. SIZE: (n = 1) *c.* 23 × 16·5.

LAYING DATES: Ethiopia, Feb, Mar, May, July; E Africa: Region D, Jan, June, July, Sept (dry months following each rainy season).

References
Jackson, F. J. and Sclater, W. L. (1938).
Pollard, J. R. T. (1946).

Genus *Myrmecocichla* Cabanis

Small to medium-sized, black or grey chats with patches of white in plumage, with marked sexual dichromatism. In shape, smaller species rather like wheatears *Oenanthe* and larger ones like robin-chats *Cossypha*. Thin-billed, bill shape typically turdine; wings round; tail rather short or moderately long; legs long, legs and toes robust. Non-migratory. Defined biologically by diet, mainly of ants and termites (in most species, so far as known). Mostly ground-dwelling and earth-hole nesting; but the 2 largest species (*M. cinnamomeiventris* and *M. semirufa*, with rufous bellies, sometimes separated as '*Thamnolaea*'), inhabit rocky hillsides and nest in old swallows' nests, and 2 small species (*M. arnotti* and *M. albifrons*, which we treat with some reservation as a superspecies, formerly genus '*Pentholaea*') are somewhat arboreal and nest in tree holes. The remaining 5 species comprise: *M. melaena*, with characteristics intermediate between '*Thamnolaea*' and '*Pentholaea*'; *M. nigra* and the grey *M. tholloni*; and a superspecies (*M. aethiops* and *M. formicivora*), to which *M. tholloni* may belong.

Endemic to subsaharan Africa; 9 species.

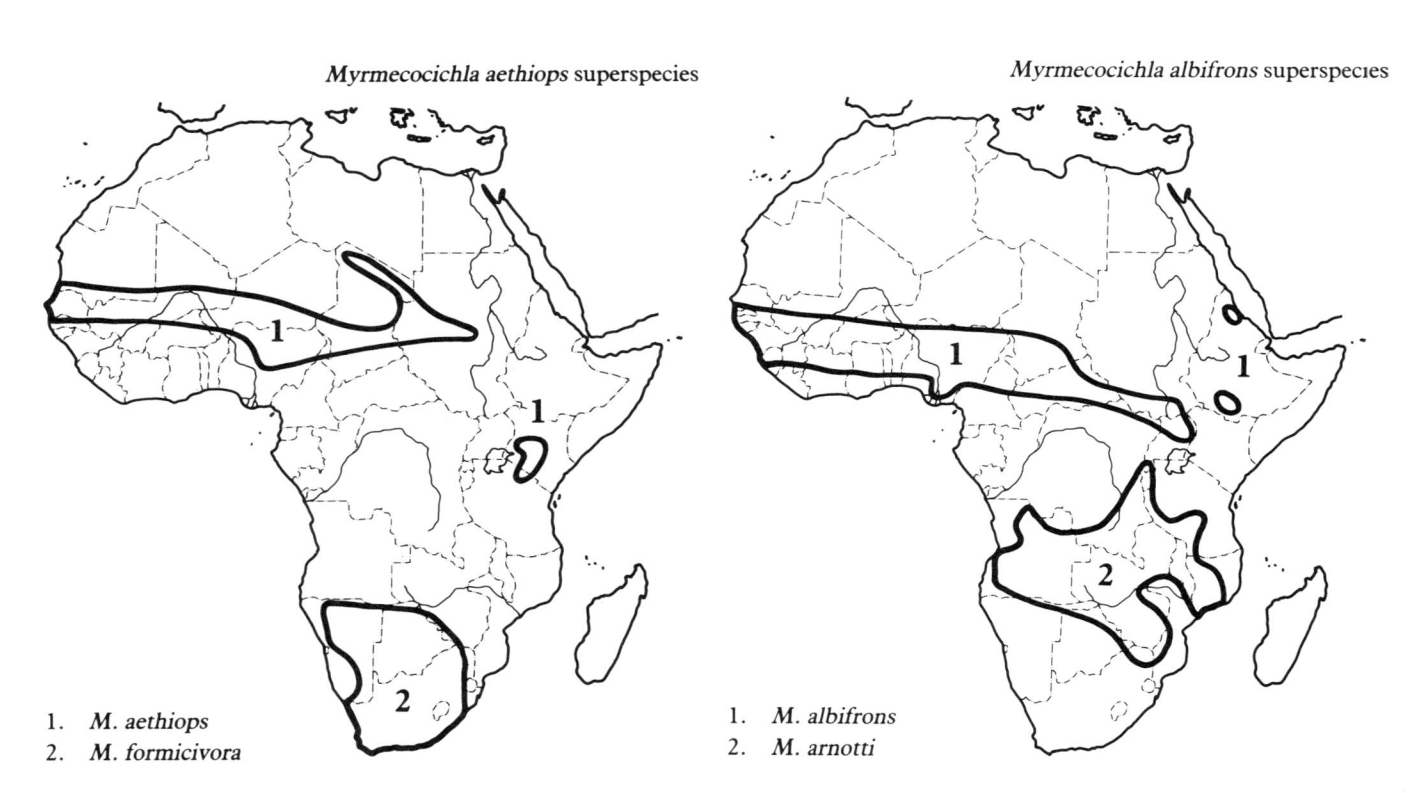

Myrmecocichla aethiops superspecies

1. *M. aethiops*
2. *M. formicivora*

Myrmecocichla albifrons superspecies

1. *M. albifrons*
2. *M. arnotti*

Myrmecocichla tholloni (Oustalet). Congo Moor-Chat. Traquet-fourmilier du Congo.

Plate 32
(Opp. p. 465)

Saxicola tholloni Oustalet, 1886. Naturaliste 8, p. 300; Leketi, Alima R., Congo.

Range and Status. Endemic resident. In open grass country in Central African Republic (Manovo-Gounda-Saint Floris Nat. Park), Congo. Zaïre east to *c.* 22°E, and Angola (local in Cuanzo Sul, Huambo, Lunda and Bié south to Lusasinga R. Locally common).

Description. ADULT ♂: forehead and crown grey with dark brown centres to feathers; nape paler grey with dark brown centres giving appearance of paler ring around neck; face grey, mottled with darker areas; chin, throat, cheeks and ear-coverts dirty greyish white; mantle and back dark brown; rump and uppertail-coverts white; tail dark brown; breast dirty greyish white, but brown centres of feathers give streaked appearance; belly and flanks greyish brown; undertail-coverts white; primaries blackish brown with basal third of P4–P9 all white, but only basal third of inner web of P1–P3 white; secondaries, scapulars, tertials and all wing-coverts blackish brown; underwing white. Bill black, mouth dark flesh; eye dark brown; legs and feet black. Sexes alike. SIZE: (5 ♂♂, 5 ♀♀) wing, ♂ 103–110 (105), ♀ 98–105 (100); tail, ♂ 62–68 (65·0), ♀ 56–59 (58·0); bill, ♂ (n = 4) 21–23 (22·0), ♀ 19–22 (20·6); tarsus, ♂ 32–36 (34·4), ♀ 32–33·5 (32·7).

IMMATURE: more uniform, darker brown on head and mantle than adult, underparts greyer. No white spot in wings, but has pale tips to secondaries, tertials and wing-coverts.

NESTLING: unknown.

Field Characters. A medium-sized, greyish chat with contrasting white patch in wings, white chin and throat, and conspicuously white rump and uppertail-coverts. Grey parts can appear blackish in poor light. The only chat in the genus with a white rump. Partially sympatric Sooty Chat *M. nigra* has white on shoulders, not in wings or on rump. Flight characteristic (see below).

Voice. Not tape-recorded. Song unknown. Alarm call a shrill but weak 'peep', very like that of Northern Anteater-Chat *M. aethiops*.

General Habits. Occurs in open grasslands on sandy or marshy ground, usually with a few bushes or small trees, and especially at roadsides. Lives mainly on ground, but often uses small bush or tree, or even telegraph wire, as perch to look for food. Occurs in pairs or small (family?) parties. Very wary, with characteristic shivering or 'flittering' flight (Lynes 1938). Feeds by flying out from elevated perch to take insects from the ground; may return again and again to the same few perches on bushes or low trees (Hall 1960). May use holes in ground as refuge (Hall 1960).

Food. Mainly insects, especially beetles and grasshoppers; fragments of a small lizard noted in one stomach (Hall 1960).

Breeding Habits. Has characteristic display-flight, flying near to the ground in a circle around nest-site with vibrating wings. Hops up and down on one place when disturbed near nest (Lippens and Wille 1976).

NEST: very similar to nest of Sooty Chat q.v., in sandpit.

EGGS: 2 young seen in brood; eggs undescribed.

LAYING DATES: Zaïre, (young July in dry season); Angola, (breeding condition, Aug); evidently breeds at end of dry season (Lynes 1938).

Myrmecocichla aethiops Cabanis. Northern Anteater-Chat. Traquet-fourmilier brun du nord.

Plate 32
(Opp. p. 465)

Myrmecocichla aethiops Cabanis, 1850. Mus. Hein. 1, p. 8; Senegal.

Forms a superspecies with *M. formicivora*.

Range and Status. Endemic resident; 2 isolated populations, Senegambia to Sudan, and Kenya/N Tanzania. In west, virtually restricted to sahelian and soudanian savannas between about 12° and 18°N, from Mauritania and Senegambia (perhaps only vagrant, Gambia), east through S Chad to Sudan (common up to 1700 m in Darfur, rather less common in Kordofan north to *c.* 14°N); and from Toumbouctou (Mali) and Niger delta in N, south to Jos Plateau in Nigeria. Records from Ennedi and Tibesti (Chad). Widespread but only locally common; large areas where not recorded. In Kenya and NW Tanzania, from Mt Elgon and Matthews Range south to Ngong, Loita and Mara, and Crater Highlands of N Tanzania, also locally in Serengeti Nat. Park (Stronach 1990). Common at 1500–3000 m (mainly above 1700 m).

Myrmecocichla aethiops

Description. *M. a. aethiops* Cabanis: Senegambia to Chad. ADULT ♂: crown and nape dark brown, sometimes with paler edges to feathers, giving streaky appearance; rest of upperparts, underwing, belly, undertail-coverts, and tail uniform dark brown; chin and throat uniform dark brown or, usually, dark brown with pale edges to feathers; breast dark brown, upper breast feathers sometimes pale-edged; wings entirely dark brown except that basal two-thirds of inner webs of primaries are white (above and below). Bill black; eye dark brown; legs and feet black. Sexes alike, though ♀ may be slightly lighter brown. SIZE: (19 ♂♂, 9 ♀♀) wing, ♂ 105–117 (112), ♀ 108–115 (111); tail, ♂ 73–81 (77·3), ♀ 71–84 (77·8); bill, ♂ 21·5–25 (22·8), ♀ 20–23·5 (22·6); tarsus, ♂ 32–36·5 (34·6), ♀ 32·5–35·5 (34·1).
IMMATURE: like adult but warmer brown, and recognizably darker up to about 6 months; throat and breast feathers have buff edges. Gape yellow (disappears after 3 months).
NESTLING (*M. a. cryptoleuca*): hatches naked, down greyish brown, gape swollen and pale yellow to whitish. Eyes open about 7th day.
M. a. sudanensis Lynes: Sudan. Paler, less blackish on brown. Smaller: wing, ♂ (n = 10) 103–109 (105), ♀ (n = 14) 97–107 (102).
M. a. cryptoleuca Sharpe: Kenya, Tanzania. Darker, sootier brown than *aethiops*. Slightly larger: wing, ♂ (n = 20) 111–122 (117), ♀ (n = 11) 109–115 (114); tail, ♂ (n = 14) 61–70 (66·6), ♀ (n = 8) 54–66 (63·0); bill, ♂ (n = 19) 23–27 (25·1), ♀ (n = 10) 22–26 (24·6); tarsus, ♂ (n = 15) 34–39 (36·4), ♀ (n = 7) 33–37 (35·3). WEIGHT: (Jan) ♂ (n = 20) 51–66 (58·5), ♀ (n = 11) 47–58 (54·7).

Field Characters. A dark brown, terrestrial chat, looking black at a distance, with largely white primaries showing conspicuously in flight. Tame and conspicuous, with weak flight; stance upright; appears rather short-tailed and long-legged. Sooty Chat *M. nigra* is distinctly smaller, has no white in wings, and ♂ has white shoulder-patch. White-fronted Black Chat *M. albifrons* is smaller still and browner, has a large white patch in wing rather than silvery sheen, and ♂ has white forehead.

Voice. Tape-recorded (30, 53, C, GREG, McVIC, MOR). A prolonged, varied mixture of clear whistles and more guttural notes, recalling an *Acrocephalus* warbler. Quality less fluty than its allospecies. Subsong often uttered by young ♂♂. Call (sometimes included in song) is loud, sharp, slightly rasped 'tsui' or 'tseeo'. Begging call of fledgling, soft 'brrt', also sometimes given by adults. Alarm call, high-pitched 'pit'.

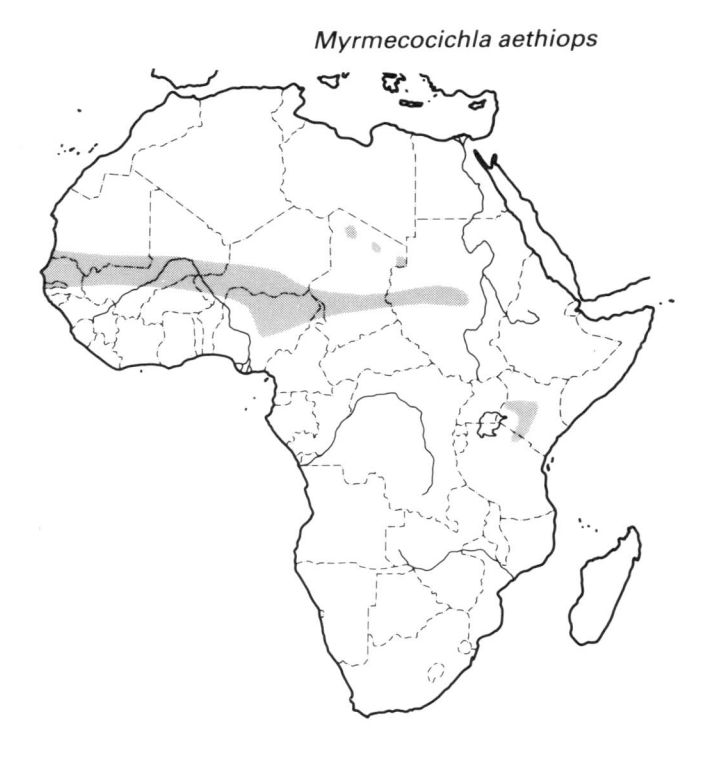

General Habits. Inhabits short grass country, usually with bushes, and almost always with termitaria. Requires unlined wells or other earth holes for nest. Often seen on road verges, in quarries, in or near villages and in cultivated areas. Also in open acacia woodland and montane grassland in highlands of Kenya and N Tanzania.
Usually in pairs or small family parties of up to 15 birds. Perches and feeds on ground; perches on and sings from termitaria, bushes, small trees and telegraph wires. Small parties seen feeding on harvested millet fields near Niger R. in Mali (Duhart and Deschamps 1963). Frequently enters termitaria and aardvark dens, where it roosts all year and nests. Digging of potential nest/roost holes occurs all year. Song (and sometimes calls) accompanied by drooping of wings – exposing white of primaries – and spreading of tail. 6–8 birds sometimes engage in display with head and neck stretched up, accompanied by shrill squawking (Elliott and Fuggles-Couchman 1948); aggressive during group encounters at territory borders and towards intruders. Submissive birds crouch on ground with head and tail bent down, head, body and tail-feathers spread. Often associates with Chestnut-bellied Starling *Spreo pulcher* in humid sahelian zones (Mauritania: Lamarche 1988).
Moult usually starts towards end of breeding season and takes about 6 months. Post-juvenile moult includes head and body plumage and several primaries and secondaries (Haas 1986).
In Kenya sedentary, but in Mauritania may move north in rains (Gee 1984).

Food. Insects (especially moths and termites, also beetles, ants, grasshoppers, caterpillars), spiders, small vertebrates and occasionally fruits of *Withania somnifera*.

Breeding Habits. Evidently monogamous, with pair-bond remaining for several years, but in family groups any mating system appears possible and inbreeding not uncommon. Regular cooperative breeder in Kenya. In courtship display, ♂ hops and wing-flicks between 2 spots a few times in front of ♀, which trembles with drooped wings (Bates 1927). Highly territorial. Groups advertise by singing from elevated perches. Singing bird often droops wings and raises and fans tail.

NEST: flat grass cup lined with fine rootlets, at end of straight, horizontal tunnel up to 1·5 m long, dug by both sexes in termite mound, aardvark den or side of earth well or bank.

EGGS: 2–5; Nigeria mean (n = 6 clutches) 3·3; Kenya mean (n = 74 clutches) 3·6. Slightly glossed; pure white. SIZE: (n = 20, Nigeria) 21–26 × 16·5–18 (23·4 × 17·6); (n = 7, E Africa) 25·5–29·3 × 18·4–19·3 (27·0 × 18·8); (n = 6 eggs, 3 clutches, Kenya) 23–27 × 18–20 (25·5 × 19·1). WEIGHT: (n = 18, Nigeria) 4·0.

LAYING DATES: Mauritania, July, Sept; Senegambia, July; Mali, July–Sept; Nigeria, June–July; Sudan, June–Aug; E Africa: Region A, Mar, Region C, Jan, Apr, May, Sept; Region D, all months, mainly during long rains Mar–June. Almost all in early to mid-rainy seasons.

INCUBATION: begins with 3rd to 4th egg, only by ♀. Period: 14–16 days. Only rarely found to be incubating in Nigeria (Mundy and Cook 1974).

DEVELOPMENT AND CARE OF YOUNG: nestling period 21–23 (16–20) days; both parents and helpers feed young; each young fed av. 6 times per h; each adult brings food 3–5 times per h; young are brooded in Kenya but apparently not in Nigeria, even at night when parents roost in nest-hole (Mundy and Cook 1974). Young of previous brood, especially ♂♂, stay with parents and help to feed subsequent broods until able to join a strange group. Not more than 2 helpers seen at nest in Nigeria (Mundy and Cook 1974), but up to 7 helpers seen at nest in Kenya.

BREEDING SUCCESS AND SURVIVAL: 56 broods failed due to parasitism by Greater Honeyguide *Indicator indicator*. Annual adult mortality about 30%.

References
Haas, V. (1986).
Mundy, P. J. and Cook, A. W. (1974).

Myrmecocichla formicivora (Vieillot). Southern Anteater-Chat. Traquet-fourmilier brun du sud.

Plate 32
(Opp. p. 465)

Oenanthe formicivora Vieillot, 1818. Nouv. Dict. d'Hist. Nat. 21, p. 421; Sunday's R., Cape Province, South Africa.

Forms a superspecies with *M. aethiops*.

Range and Status. Endemic resident, southern Africa. Namibia (except Namib Desert and Caprivi Strip), Botswana (except extreme N), SW Zimbabwe (2 records), and most of South Africa except for NE Transvaal, coastal Natal, and coastal SE and SW Cape Province. Occurs as far south as Somerset West and reaches coast in both central S and W Cape Province. Rarer in eastern part of range, including Lesotho.

Description. ADULT ♂: entire upperparts dark greyish brown, most feathers (especially on head) with pale edges; tail dark brown; chin, throat, breast and upper belly dark sandy brown with darker centres to feathers; lower belly and undertail like upperparts; wings brown, but basal third of inner web of primaries white; wing-coverts uniform dark brown except for a small white patch on lesser coverts. Bill black; eye dusky; legs and feet black. ADULT ♀: like ♂ but without white patch on wing-coverts. SIZE: (21 ♂♂, 13 ♀♀) wing, ♂ 97–104 (101), ♀ 92–100 (96·2); tail, ♂ 61–69 (64·2), ♀ 56–64 (60·9); bill, ♂ 19·5–24 (21·9), ♀ 19·5–22 (20·9); tarsus, ♂ 31·5–36 (33·7), ♀ 31–34·5 (32·9). Size cline from S of range (large) to NW (small). NW Namibia birds, *M. f. 'minor'*, are smallest: (2 ♂♂, 6 ♀♀) wing, 87–98 (92·4); tail, 57–65 (59·9); bill, 19–21·5 (20·1); tarsus, 28–32 (30·0). WEIGHT: southern Africa (n = 63) 35–51·1 (42·2); Namibia (Oct) 1 ♂ 42, 1 ♀ 38.

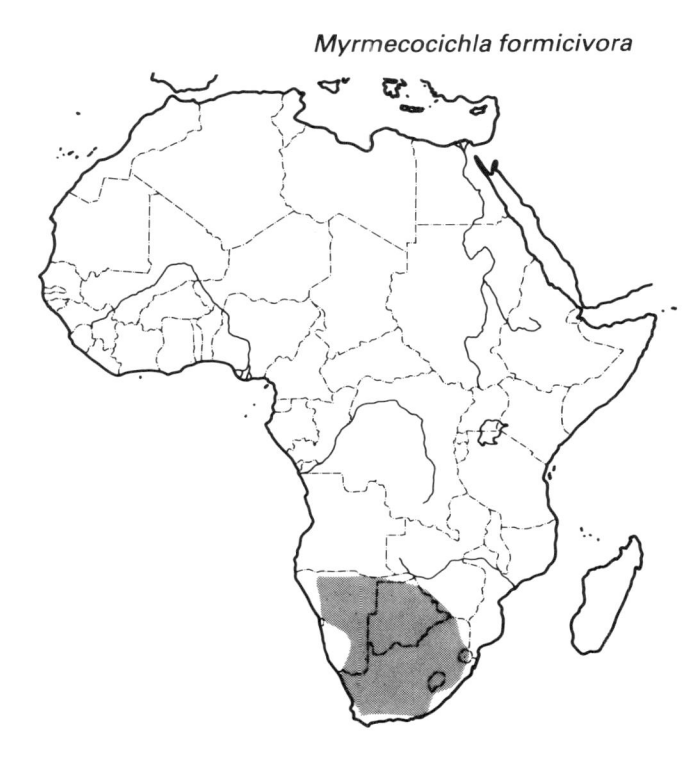

IMMATURE: generally dark brown, washed with rusty. Juvenile blackish brown with paler spots. White on wing of ♂, though less extensive than in adult, is present from very young stage.

NESTLING: newly hatched young naked; skin dark reddish orange, gape cream.

Field Characters. A dark brown, terrestrial chat, looking black at a distance, with largely white primaries showing conspicuously in flight. ♂ has small white shoulder-patch, often concealed. Stance upright. Rather short-tailed and long-legged. Has characteristic 'twinkling' flight. Perches on ground, termite hills, low shrubs.

Voice. Tape-recorded (72, 81, 88, LUT, ROC). Song a varied mixture of whistles and guttural notes, typically 'tee-a-u, oo-oo, tee-a-u', with the 'oo-oo' at a lower pitch and the second 'tee' rather drawn out. Quality of notes very similar to Eurasian Blackbird *Turdus merula*, but often repeats a phrase 2–3 times, and song may be fairly sustained. Occasionally mimics other birds. Call a clear plaintive whistle, 'weeei'.

General Habits. Inhabits open grassy country, especially rolling hills; also semi-desert scrub. Commonest in drier country with a few bushes and many termitaria. Often occurs by roadsides. Sea level to over 1300 m.

In pairs or family groups of up to *c.* 6 birds. Groups split up into pairs in late winter or spring. Groups slightly larger in Dec (summer) than July (winter): 2·85 *vs* 1·81. Perches freely on bushes, roadside posts, telegraph wires and termitaria. Feeds mostly by dropping onto ground or into vegetation from elevated perch (occasionally after hovering), but also whilst hopping or running along ground. Has been recorded foraging from loose soil excavated by aardvarks or porcupines, in whose burrows it often nests (Skead 1974). Often hovers before tumbling to low perch or to ground. Mobs a predator by hovering above it, calling, or perches close by with hanging wings, raising and lowering tail continuously (Herholdt 1987). At sunset preens on ground then suddenly flies off in one direction, lands briefly and then flies off in another direction, until too dark to follow (Herholdt 1987). Roosts in the nest-hole (♀ always and ♂ sometimes when with eggs or small chicks) but also evidently elsewhere.

Entirely sedentary.

Food. Insects, especially ants and termites. In 33 birds near Bloemfontein, South Africa, Hymenoptera (almost all Formicidae) dominated numerically in summer, and they and termites (entirely *Hodotermes*) in winter. Termites and orthopterans (Acrididae) were 77% of dry weight in winter. Solifugae, taken only in summer, were then largest proportion of diet by weight. Hemiptera, Dermaptera, various beetles (especially Tenebrionidae), caterpillars (including lucerne caterpillars) and millipedes also recorded, and fruit taken Feb–Apr (Earlé and Louw 1988). *Trinervitermes* termites taken when mounds are broken.

Breeding Habits. Most details from a study in Orange Free State (Earlé and Herholdt 1988). Monogamous. Helpers not recorded at any of 22 first broods, but first-brood fledglings help at subsequent nests. Not aggressive to conspecifics (Herholdt 1987). Apparently no territorial behaviour, but pair have defined home range. Ranges stable, but turnover of birds high, with main dispersal occurring in late winter and spring (Aug–Oct). Adjacent pairs forage together when feeding young. Both sexes sing, but ♀ only when excited. Song used almost entirely for courtship. In display ♂ flies with stiff wings, tilting sideways to show off white shoulder-patches; may also fly rapidly up to 40–50 m and hover for several s. All displays accompanied by vigorous singing. Usually 2 broods (19 of 22 ♀♀, including all 10 successful with first brood) with ♀ laying new clutch av. 29 days after previous chicks fledge (n = 5), but 7 days (n = 6) after loss of eggs and 19 days (n = 3) after loss of chicks. Second clutch never in same burrow (though once in same swallow nest), but third clutch 4 times in same nest as first.

NEST: cup of dry grass and fine rootlets at end of tunnel; length of tunnel (n = 17) 0·35–1·5 (0·92) m, width at opening 5·5–12 (8·0) cm, height at opening 6–11 (8·0) cm. Tunnel dug with bill by both adults, usually in 8–10 days depending on soil type and recent rainfall (Herholdt 1987), in sandy bank by stream, roadside, ditch, quarry or in entrance or roof of aardvark, hyaena or porcupine burrow. Often uses burrow dug in previous years, but never burrow of another bird species. If no suitable sites available, uses hole in concrete bridge or nest of swallow, e.g. Greater Striped Swallow *Hirundo cucullata* in culvert under tarred road. Once evicted chicks of swallow and took over nest (Earlé 1987).

EGGS: 1–5, usually 3–4; southern Africa mean (n = 14 clutches) 3·2; Orange Free State mean (n = 51 clutches) 3·7. Well pointed; glossy; white, occasionally with a few brownish spots. SIZE: (n = 18, Orange Free State) 22·1–25·5 × 16·6–19·4 (23·6 × 18·0); (n = 37, southern Africa) 21·7–28·5 × 16·5–18·8 (24·0 × 17·9). WEIGHT: (n = 10, Orange Free State) 3·8–4·2 (4·1).

LAYING DATES: South Africa: Aug–Jan (mainly Oct–Dec), Mar, Apr; Orange Free State, Sept–Feb (mainly Sept–Nov); SW Cape Province, Aug–Nov; Natal, Sept–Nov; Barberspan, Sept–Mar.

INCUBATION: by ♀ only, starting with 4th or last egg. Period: 14–15 days.

DEVELOPMENT AND CARE OF YOUNG: young hatch naked; eyes start opening on day 4, fully open at day 9; pterylae clearly visible on day 4 or 5; tarsus reaches adult length by about day 10; fledging weight of 40–45 g reached by day 11; tail and wing at fledging *c.* 20 mm shorter than in adult.

Fledgling sex ratio: *c.* 0·6 ♂ to 1 ♀ (n = 44) (difference not significant). Period: 15–18 days. Young of first brood fed by parents, but ♂ and ♀ fledglings of first brood contribute *c.* 10% of food items to 2nd and 3rd broods. Young fed mainly before noon (9·4 feeds per h before, 4·3 feeds per h after) and helpers only fed before noon. After fledging, young continue to be fed for at least 7 days. Fledged juveniles use nest-hole for roost, shelter, and bolt-hole when alarmed.

BREEDING SUCCESS: Orange Free State: of 170 eggs, 52% hatched, of which 81% fledged (overall 42% success). 5 failures caused by brood parasitism by Greater Honeyguide *Indicator indicator* (Earlé and Herholdt 1987), predation by striped mouse *Rhabdomys pumilio* (2), desertion of chicks due to mite infestation, and rain-induced burrow collapse.

References
Earlé, R. A. and Herholdt, J. J. (1988).
Earlé, R. A. and Louw, S. v. d. M. (1988).
Herholdt, J. J. (1987).

Myrmecocichla nigra (Vieillot). Sooty Chat. Traquet-fourmilier noir.

Plate 32
(opp. p. 465)

Oenanthe nigra Vieillot, 1818. Nouv. Dict. d'Hist. Nat. 21, p. 431; Malimbe, Portuguese Congo.

Range and Status. Endemic resident. Rather irregularly distributed in savannas of Nigeria (less than 10 localities, from Sokoto and Zaria to Mambilla Plateau in SE, where frequent and widespread), montane grasslands of Cameroon, across N edge of forest belt of Zaïre, through Central African Republic to extreme S Sudan, Uganda (not NE) and SW Kenya to N Serengeti Nat. Park, Tanzania; occurs on S side of L. Victoria and ranges through W Tanzania into S Zambia east to Lavushi Manda and Mpika; S half of Zaïre to Kasai, S Congo, S Gabon, and much of Angola. Range meets that of Northern Anteater-Chat *M. aethiops* but area of overlap very small. Common locally, especially at higher altitudes. Seems to be absent from some areas within geographical range which appear suitable.

Description. ADULT ♂: entirely glossy black except for lesser and median wing-coverts and base of greater coverts, which are white, and primaries, which are dark brown. Bill black; eye dark brown; legs and feet black. ADULT ♀: whole plumage nearly uniform dark brown. SIZE: (22 ♂♂, 18 ♀♀) wing, ♂ 93–107 (99·0), ♀ 88–101 (93·3); tail, ♂ 61–70 (65·5), ♀ 56–67 (61·8); bill, ♂ 17·5–21 (19·2), ♀ 17·5–20 (18·5); tarsus, ♂ 27·5–32·5 (30·4), ♀ 28–32 (29·7). WEIGHT: Uganda (Feb) unsexed (n = 2) 41, 42·5; Cameroon (Feb) ♂ (n = 2) 37, 37, (May) ♂ (n = 3) 44–46 (44·7), ♀ (n = 3) 37–38 (37·7), unsexed (n = 1) 40; Congo (Sept) unsexed (n = 1) 39.
IMMATURE: ♂ like adult ♂ but less glossy; ♀ like adult ♀.
NESTLING: unknown.

Field Characters. ♂ is glossy black with white shoulder-patch; ♀ is very dark brown throughout. In flight readily distinguished from Northern Anteater-Chat and Congo Moor-Chat *M. tholloni* by lack of white 'window' in primaries. Very similar to immature White-headed Black Chat *M. arnotti*, which has black head, but Sooty Chat has smaller white patch in wing-coverts (greater coverts only, not alula), and shorter tail (extending less far beyond wings); it is also found in more open country, with scattered bushes and trees.

Voice. Tape-recorded (53, 86, C, CART, MOR, MOY, WALK). A quiet, thrush-like, rather wild-sounding song, 'teei-teeo teei teei' (pause) 'teeeo', 2nd note stressed and last a long drawn-out whistle; but several variations. Song simpler than that of White-fronted Black Chat *M. albifrons*. Both sexes sing, and subsong-like song heard once from ♀ (V. Haas, pers. comm.). Call a variety of whistles. Occasionally mimics other birds.

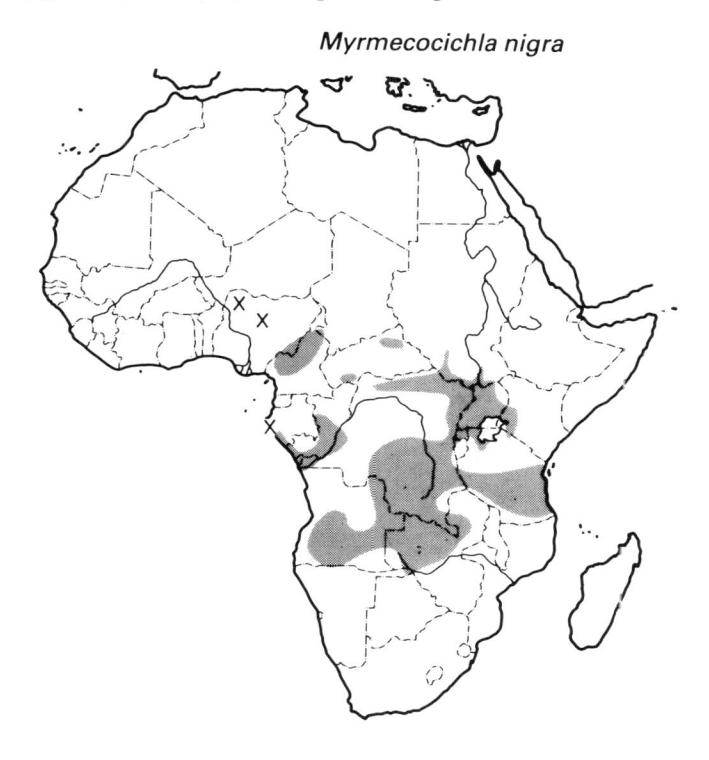

General Habits. Inhabits, almost exclusively, country with short grass and termitaria, with or without scattered bushes. Common along roads in some parts, in farmed areas, and on recently burnt land. At least in E Africa and Cameroon occurs mainly between 1200 and 1700 m.

Solitary, territorial and sedentary. Occurs in pairs all year, although one group of 2 adult ♂♂ and 1 ♀ seen. In another territory, of a pair observed over 12 h in Oct in Kenya, a second ♀ seen repeatedly. One aggressive encounter between 2 ♀♀ seen near territory boundary. Submissive ♀ behaved like submissive Northern Anteater-Chat, q.v. (V. Haas, pers. comm.) Flies mainly only short distances (< 100 m): only one 300 m flight (while singing) seen in 12 h (V. Haas, pers. comm.).

Sits conspicuously on top of bush or termitarium, slowly moving tail up and down, especially just after alighting. When hopping about on ground, carries tail cocked up. Sunbathes. Roosts in nest hole.

Food. Insects, including caterpillars, termites, ants and Orthoptera; occasionally seeds.

Breeding Habits. Solitary and territorial. Territory centred around nest hole. Sings on the wing, especially in fluttering flight from bush to bush, or from perch on bush or termite mound, when it sways body back and forth. While singing in air, flies around with undulating flight and rapidly beating wings. Occasionally responds to song from neighbours (V. Haas, pers. comm.). Sometimes starts to sing before dawn.

NEST: neat, grass-lined cup built in hollow of termite mound, sandbank, aardvark den or occasionally stump, or at end of burrow up to 1 m long dug by birds themselves into sandbank. Burrow usually slopes slightly upward.

EGGS: 2–5. Oval; white, usually with faint bluish tinge and sometimes with fine black spots. SIZE: (n = 2, Uganda) 25·5–26·3 × 16·1–16·3 (25·9 × 16·2).

LAYING DATES: E Africa: Region A, Feb–Apr, Region B, Aug–June; Uganda, May–June, Oct; Zambia, Mar (gonads active Oct); Angola, Aug–Nov; NE Zaïre (Uele), Mar–June; SE Zaïre, Sept–Feb, July. Season often ill-defined but peak breeding usually in rains.

Plate 32
(Opp. p. 465)

Myrmecocichla melaena (Rüppell). Rüppell's Black Chat. Traquet noir d'Abyssinie.

Saxicola melaena Rüppell, 1837. Neue Wirbelt., Vögel, p. 77; Agami Province, N Abyssinia.

Range and Status. Endemic resident, highlands of Ethiopia. Occurs only above *c*. 1800 m, on plateau of S Eritrea, south through Wollo and Tigré to Shoa, Gojjam and Gonder Provinces. Not in SE highlands or N part of W highlands. Uncommon but locally frequent.

Description. ADULT ♂: entire plumage black or very dark brown except for primaries, which have inner webs with basal two-thirds white and distal third dark brown, outer webs nearly black. Bill black; eye brown; legs and feet black. Sexes alike. SIZE: (4 ♂♂, 5 ♀♀, 5 unsexed) wing, ♂ 87–93 (89·3), ♀ 85–92 (89·6), unsexed 90–93 (91·4); tail, ♂ 61–65 (62·0), ♀ 59–68 (63·4), unsexed 64–67 (65·4); bill, ♂ 17–19 (18·0), ♀ (n = 4) 18–19 (18·6), unsexed 17·5–18·5 (17·9); tarsus, ♂ 27–29 (28·3), ♀ 27·5–30 (28·5), unsexed 28·5–39 (28·9).

IMMATURE AND NESTLING: unknown.

Field Characters. A large, black chat with white wings conspicuous in flight, inhabiting rocky gorges at high altitudes. Mocking Cliff-Chat *M. cinnamomeiventris* and White-winged Cliff-Chat *M. semirufa* have chestnut on underparts. Smaller White-fronted Black Chat *M. albifrons* occurs mainly at lower altitudes, is more arboreal, and ♂ has white forehead.

Voice. Unknown.

General Habits. Lives among cliffs, ravines, gorges and wet rocks above 1800 m, preferring the tops of ravines, especially near waterfalls. Occurs mainly on bare rocks; usually in pairs, sometimes small (family?) parties. When standing, flirts tail high over back.

Food. Unknown.

Breeding Habits. Nests in cracks in cliff-faces; nest-building June (Smith 1957).

Myrmecocichla albifrons (Rüppell). **White-fronted Black Chat. Traquet noir à front blanc.**

Saxicola albifrons Rüppell, 1837. Neue Wirbelt., Vögel, p. 78; Temben Province, N Abyssinia.

Forms a superspecies with *M. arnotti*.

Plate 32
(Opp. p. 465)

Range and Status. Endemic resident. Occurs in savannas from extreme S Mauritania, Senegambia (rather scarce), and Sierra Leone to Central African Republic, extreme NE Zaïre, and S Sudan west of Nile R. as far north as Juba, Yei and Moru Districts; extends south to Murchison Falls Nat. Park, L. Kyoga and Teso (Uganda); a probably separate population in lowlands of SW Ethiopia; and another in N Ethiopia (Eritrea), up to *c.* 2000 m. Common locally.

Description. *M. a. frontalis* (Swainson): Senegambia to Cameroon. ADULT ♂: entirely black, with slight gloss, but forehead and usually front of crown white; primaries and secondaries dark brown with undersides of remiges almost white. Bill black; eye dark grey-brown; legs and feet black. ADULT ♀: like ♂, but lacks white on forehead. SIZE: (10 ♂♂, 7 ♀♀) wing, ♂ 77–85 (80·7), ♀ 71–80 (76·0); tail, ♂ 53–68 (64·9), ♀ 52–60 (56·3); bill, ♂ 14·5–17·5 (16·2), ♀ 14·5–17 (15·9); tarsus, ♂ 22–25 (23·5), ♀ 21·5–24 (23·1). WEIGHT: Ghana (July–Sept) 1 ♂, 20·5, 1 ♀, 18·5, (Nov–May) unsexed (n = 3) 19·5–24·9 (21·8); Cameroon (Feb) 1 ♂ 20.

IMMATURE: like adult except that all body feathers and wing-coverts are dark blackish brown with pale subterminal spots.

NESTLING: covered in black down; mouth-lining bright orange.

M. a. limbata (Reichenow): E Cameroon. Amount of white on wing-coverts intermediate between last and next races; ♂ with a few white feathers on wing-coverts; ♀ dark brown all over.

M. a. clericalis (Hartlaub): S Sudan and Uganda. ♂ with large white patch on wing-coverts.

M. a. pachyrhyncha (Neumann): SW Ethiopia. ♂ like *frontalis* but white of crown extends further back. ♀ greyer on head than ♀ *frontalis*, with darker centres to feathers, giving faintly streaked appearance; chin and throat dirty grey.

M. a. albifrons (Rüppell): N Ethiopia south to Yavello. Like *frontalis* but inner webs of wing-feathers whitish. ♀ has chin and throat paler than rest of body, with dark midribs to feathers.

Field Characters. ♂ all black with white forehead and white in wings, some races with white shoulder-patch; ♀ lacks white forehead. ♂ distinguished from Northern Anteater-Chat *M. aethiops* by less white in wings; and from Sooty Chat *M. nigra*, Rüppell's Black Chat *M. melaena* and Northern Anteater-Chat by conspicuous white forehead. ♀, with no white on head, very similar to ♀♀ of all 3 but they are all larger; Sooty Chat and Northern Anteater-Chat are browner, and the latter and Rüppell's Black Chat have white in wings. ♀ White-fronted Black Chat also resembles Black Tit *Parus leucomelas* in both appearance and behaviour, but latter has white patch on wing-coverts and outer edge of primaries.

Voice. Tape-recorded (53, MOR). Much less noisy than congeners. Song a short, sharp, distinctive series of complex notes; sometimes mimics other birds. Call a series of chattering notes and a pleasant 'tweet' note.

Myrmecocichla albifrons

General Habits. A conspicuous, fairly common bird of open savanna woodland, clearings and bushy country, especially in rocky areas and those with erosion gullies. Fond of recently burnt bare stony ground with stumps or bushes which it uses as hunting posts. Also in bare, stony semi-arid areas, and in cultivated areas where some stumps remain. Spends much of time on top of small bushes and less time than congeners on ground. Occurs up to 2000 m in Ethiopia; in pairs or small family parties throughout year. Feeds from vantage points, pouncing to ground like a shrike. Often flicks tail conspicuously when feeding and singing.

Mainly sedentary, but possibly intra-African migrant in Mauritania (Lamarche 1988).

Food. Insects.

Breeding Habits. Monogamous, territorial. ♂ has butterfly-like display flight.

NEST: grass cup lined with small roots or cobwebs, placed in crevice, under boulder or in hollow stump up to 1 m above ground.

EGGS: 2–3. Pale greenish or bluish, liberally spotted with rufous and brown. SIZE: *c.* 25·5 × 19.

LAYING DATES: Mali, said to be Nov–Feb; Ghana, Jan; Nigeria, Dec–Apr; NE Zaïre, Jan–Feb. In most places breeds in dry season, mainly after fires have burned off the grass.

552 TURDIDAE

Plate 32
(Opp. p. 465)

Myrmecocichla arnotti (Tristram). **White-headed Black Chat; Arnott's Chat. Traquet d'Arnott.**

Saxicola arnotti Tristram, 1869. Ibis, p. 206; Victoria Falls, Southern Rhodesia.

Forms a superspecies with *M. albifrons*.

Range and Status. Endemic resident. All across N and central Angola, S Zaïre, across Zambia, throughout Zimbabwe except east of *c.* 33°E, Rwanda, Burundi (2 records), W and S Tanzania north to Tabora and east to Morogoro, Malaŵi, N Mozambique, NE Namibia (Caprivi Strip), NE third of Botswana and NE South Africa (NE Transvaal). Locally common but inexplicably absent from some areas, and has suffered from habitat destruction in others, e.g. Mashonaland Plateau, Zimbabwe.

Description. *M. a. arnotti* (Tristram): range of species except Angola. ADULT ♂: crown to upper nape white; remainder of upperparts, entire underparts, remiges, tail and underwing glossy black; scapulars glossy black with outer edge of outer webs white; all wing-coverts white, greater and primary coverts with black tips; alula black. Bill black; eye dark brown; legs and feet black. ADULT ♀: black duller than ♂, sometimes with brownish tinge; crown black; chin, throat, breast and sides of neck white with small black tips to feathers, especially on chin and throat; extent of white down breast variable; otherwise like ♂. SIZE: (25 ♂♂, 20 ♀♀) wing, ♂ 96–108 (102·5), ♀ 93–105 (97·4); tail, ♂ 70–82 (74·7), ♀ 65–77 (69·2); bill, ♂ 17–22 (19·5), ♀ 17–20·5 (18·9); tarsus, ♂ 27·5–31 (28·9), ♀ 25·5–30·5 (27·9). WEIGHT: 1 ♀ 35·2.
 IMMATURE: like adult but usually lacks white on head, although ♂♂ sometimes have whitish bases to crown feathers, and ♀♀ sometimes have a few whitish feathers on chin. Mouth orange-yellow, gape yellow.
 NESTLING: unknown.
 M. a. harterti (Neunzig): Angola. ♂ has white on head restricted to forehead and (narrow) superciliary stripe; white on wings like *arnotti* but primary coverts mainly black, with white bases. ♀ like ♀ *arnotti*, but has small white triangular patch on throat (tips of feathers dark, giving dirty white appearance).

Field Characters. A thick-set, black chat with large white patch in wing, conspicuous in flight and at rest, and pure white cap (♂) or white throat (♀). Fairly conspicuous; hops about near ground, typically in *Brachystegia* woodland. ♂ resembles black form of Mountain Wheatear *Oenanthe monticola* but lacks white rump and undertail-coverts. The only sympatric congener, Sooty Chat *M. nigra*, has no white on head in either sex, although immature is very similar (q.v.). Largely allopatric Congo Moor-Chat *M. tholloni* is mainly grey, not black, and Southern Anteater-Chat *M. formicivora* is browner, with no white on head.

Voice. Tape-recorded (35, 86, CHA, GIB, WALK). A rather quiet mixture of complex musical notes, interspersed with a few thin, high-pitched whistles 'feee', rising then falling in pitch. Does not have the clear strident whistles of some congeners, and has fewer guttural notes than most of them. On its own the 'feee' is a contact and alarm call. ♂ and ♀ may sing together and both are versatile and accurate mimics of other species (7 listed by Barbour 1972). Another alarm call is a quiet, musical 'fik'.

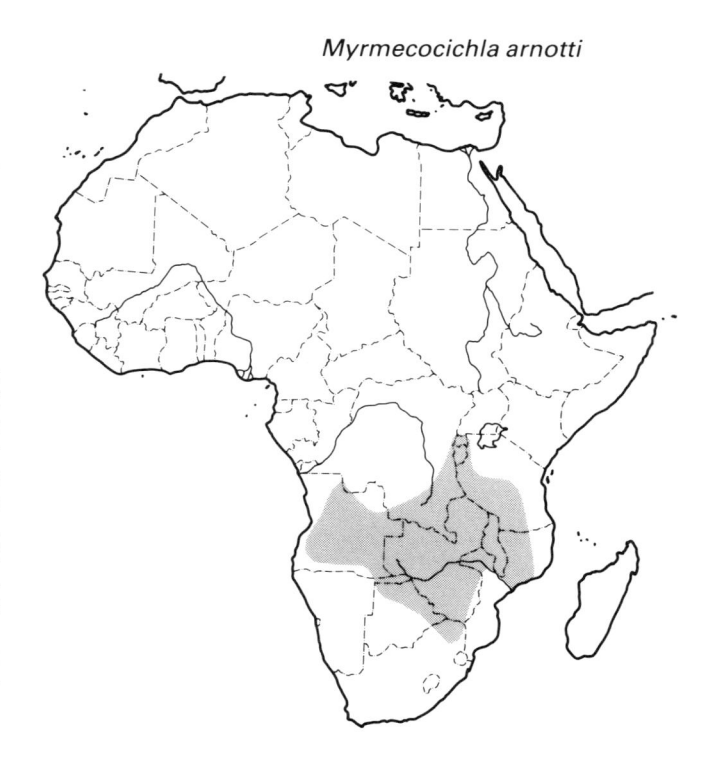

General Habits. Inhabits well-developed *Baikiaea* and, especially, *Brachystegia* woodland, mainly below 1500 m. Prefers areas where understorey is fairly open, with little grass. Occasionally occurs in cultivated areas but absent from extensive areas of secondary growth. Inexplicably absent from some areas. Thought to have been more common and widespread in Zimbabwe in the past, before the cutting down of extensive areas of *Brachystegia* woodland (Irwin 1981).
 Occurs in pairs or small groups. Shy but fairly noisy and conspicuous. Behaviour rather *Turdus*-like, hopping about low down amongst tree trunks. Searches for food on trunks and on ground and may hang sideways; swallows gravel. Pounces to ground for food from perch up to 3 m high. Characteristically raises and lowers tail when alighting. In flight wing-beats audible. Song often accompanied by wing-flicking (Barbour 1972).
 Sedentary.

Food. Insects and spiders, especially ants; also beetles. Larvae, green caterpillars, moths and a mole cricket (all large) fed to large nestlings (Barbour 1972).

Breeding Habits. Monogamous. Both sexes sing.
 NEST: a grass cup usually lined with soft grass, roots, and/or feathers, in hole in tree, usually 2–4 m above

ground; int. diam. 80, depth 40. One nest in hole in stump filled to rim with woody droppings of beetles and dry petioles. Occasionally under thatched roof (Lippens and Wille 1976).

EGGS: 2–4. Oval; pale bluish or greenish, variably spotted with pale brown and with mauve undermarkings, often forming cap at larger end. SIZE: (n = 12, southern Africa) 20·9–24 × 15·2–17·5 (22·6 × 16·6); (n = 3, Zambia) 22·0 × 15·8; (n = 5, Zambia and Zimbabwe) 22·3–23·4 × 15·2–17·5 (22·7 × 16·1).

LAYING DATES: E Africa: Region C, Jan, July, Sept, Oct; Zambia, Aug–Nov, Jan; Zimbabwe, Aug–Dec (mainly Oct–Nov); Malaŵi, Sept–Nov; Angola, Aug; S Zaïre, Aug–Dec. Mainly end of dry season and early rainy season.

DEVELOPMENT AND CARE OF YOUNG: chicks fed by both adults.

Reference
Barbour, D. (1972).

Myrmecocichla cinnamomeiventris (Lafresnaye). Mocking Cliff-Chat; Mocking Chat. Traquet de roche à ventre roux.

Turdus cinnamomeiventris Lafresnaye, 1836. Rev. Zool., pl. 55; Cape Province.

Plate 32
(Opp. p. 465)

Range and Status. Endemic resident. Local; restricted to cliffs, kopjes, inselbergs, gorges and similar boulder-strewn areas. Locally common from E South Africa through eastern Africa to W Ethiopia. West of Nile, range far more fragmented: absent from numerous rocky localities which look suitable.

South Africa, locally common; vertical migrant. Zimbabwe, common and widespread but absent from parts of NW Matebeleland; up to 2440 m in Chimanimani Mts (Mt Binga). E Botswana; W Mozambique. Malaŵi, common, from L. Malaŵi up to 2150 m; absent from Mulanje. Zambia, widespread in E, west to Victoria Falls, about Mkushi R., several localities northwest of Ndola, Kafue Nat. Park (Kaindabaila only), and a sight record in E Mwinilunga. Tanzania, widespread in E, west to rocky shores of L. Malaŵi, Iringa, Tukuyu, central Tabora, Mwanza, Kibondo, Ngara and Bukoba. Kenya and Uganda, rather more local, 600–2200 m, absent from some suitable-looking areas. NE Zaïre, Gaima Hills (Nzoro), Aba, Piagga (Faradje). Rwanda, Nyarubuye, Ntaruka, Mutumba, Mushushu, and Mabari Massif; Burundi, abundant between Nyakuzu and Mpinga. Ethiopia, frequent and widespread in W and in hills between Dire Dawa and Jejega, rather uncommon between 1200 and 2500 m in Eritrea. Sudan, uncommon and local in SE from upper Blue Nile to Boma Hills and Fazughli; locally in Kordofan and Darfur (Jebel Marra). Chad, rare, known only from Koulbouss (14° 22′N, 22° 27′E), Kilingen (Abéché), Goz Beida, and Kyabe in S. Central African Republic, in Manovo-Gounda-Saint Floris and Bamingui-Bangoran Nat. Parks. Cameroon, Adamawa Plateau, Benué plain, Meri (Mandara). Nigeria, common and widespread on granitic hills north of Niger and Benué rivers, less common north of Sokoto and Kano; south of the rivers only at Igbetti and Eruwa in W, and in E on Chappal Hendu and Chappal Tale in Gashaka-Gumti Game Reserve, and on Obudu Plateau. Togo, Péwa (9° 16′N, 1° 14′E). Ghana, Shai Hills on Accra Plains (a population established since 1950s: Grimes 1987), and in N Nakpanduri and Konkori Scarp (Mole). Benin and Burkina Faso, Arli-Pendjari Nat. Park (5 localities within 90 km).

Ivory Coast, Touba and Boundiali only. Mali, widespread in Mopti and Mandingo Mts, north to Hombori (15° 16′N, 1° 40W). Mauritania, Assaba Escarpment (Djouk, 17° 9′N, 12° 12′W); Senegambia, common near Kédougou (Ibel, Dinndéfélou).

Description. *M. c. cinnamomeiventris* (Lafresnaye): South Africa (east to *c.* 32°E), lowland Lesotho, W Swaziland. ADULT ♂: glossy black except as follows: rump orange, merging to dark rufous on uppertail-coverts; narrow transverse white line between breast and belly; upper belly pale orange, grading through orange lower belly to dark rufous undertail; median and lesser coverts white, forming large patch in wing; some lesser coverts white with black tip. Bill black; eye dark brown;

legs and feet black. ADULT ♀: entire head, breast, mantle, back and wings very dark grey; rump, uppertail-coverts, belly and undertail-coverts very dark rufous with upper belly greyish. Demarcation between colours of breast and belly less precise than in ♂. Tail and remiges black, like ♂. SIZE: wing, ♂ (n = 15) 109–122·5 (115), ♀ (n = 13) 103–115 (109); tail, unsexed (n = 29) 90–100; bill, unsexed (n = 30) 18–22; tarsus, unsexed (n = 30) 27·5–31·5. WEIGHT: unsexed (n = 5) 40·7–51 (47·7).

IMMATURE: ♂ like adult ♂, and ♀ like adult ♀, but rufous feathering duller and much fluffier, particularly on breast. Black parts drab, not glossy, white patch in wing narrower; primary coverts and greater wing-coverts edged grey. Gape whitish.

NESTLING: skin bright pink, with sparse tufts of long black down.

M. c. autochthones Clancey: S Africa (coastal Transkei, Natal, E Transvaal lowveld), E Swaziland, S Mozambique. Like *cinnamomeiventris* but no white line between breast and belly. Smaller: wing, ♂ (n = 7) 104–111 (108).

M. c. subrufipennis (Reichenow): Zambia and Malaŵi to Uganda, SE Sudan and SW Ethiopia. ♂ like nominate ♂ but upper belly paling through pale orange to small area of white abutting black breast. Median upperwing-coverts completely white, never black-tipped. ♀: head, mantle, back, upperwing-coverts and breast paler grey than in nominate ♀, forming contrast between grey areas and black remiges. Belly (especially upper belly) paler than in nominate ♀, more orange than dark rufous. Similarly, rump and uppertail-coverts paler than in nominate ♀, pale orange rather than dark rufous.

M. c. odica Clancey: E Zimbabwe. ♂ like ♂ *subrufipennis*; ♀ like *subrufipennis* but belly and flanks tawny chestnut. Same size as *cinnamomeiventris*.

M. c. albiscapulata (Rüppell): Ethiopia. Adult ♂ like ♂ *subrufipennis* but uppertail-coverts glossy blue-black, not rufous. ♀ like ♂ *albiscapulata* (hence totally unlike ♀♀ of other races), but differs from ♂ in lacking white wing patch.

M. c. kordofanensis (Wettstein): Sudan (Kordofan, Nuba Hills). Adult ♂ like ♂ *subrufipennis*, but forehead and crown snowy white. ♀ like ♀ *subrufipennis* but mantle, back and scapulars browner; crown slightly greyer; chin, throat and breast brown with olivaceous or rufescent wash (rather than dark grey).

M. c. coronata (Reichenow): Togo and SE Burkina Faso to W Sudan (Darfur). ♂ like ♂ *kordofanensis* but white forehead and crown feathers very narrowly fringed with blackish, so crown looks white speckled with black; forehead and crown sometimes black with a few irregular white feathers (e.g. Bamenda, Cameroon; Bauchi, Nigeria). Some ♂♂ lack white in upper belly (Cameroon). ♀ like ♀ *kordofanensis* but underparts yellower (throat yellowish grey and belly bright orange rather than dark rufous).

M. c. cavernicola Bates: Mali (Mopti). ♂ like ♂ *kordofanensis* but forehead and crown black not white; also upper belly orange, without infusion of white (hence *cavernicola* is almost indistinguishable from ♂ of nominate race, except that its median coverts are wholly white without any black tips). ♀ like ♀ *coronata*.

M. c. bambarae Bates: Mali (Mandingo Mts). ♂ like ♂ *cavernicola* (i.e. cap black not white), but rump darker and with smaller extent of orange, uppertail-coverts blue-black (not rufous) and white wing-patch much smaller (restricted to lesser coverts; median coverts black, not white). ♀ like ♀ *cavernicola* but rufous rump patch smaller (uppertail-coverts black, not rufous).

TAXONOMIC NOTE: black-crowned ♂♂ like ♂ *cavernicola* occur alongside white-crowned ♂ *coronata* at several localities, namely Arli-Pendjari Nat. Park (Burkina Faso and Benin, Green 1980; within 10 km of each other), Péwa (Togo: Cheke 1982; at same locality), Obudu Plateau (Nigeria: Fry 1966; in same area but with 350 m vertical separation), Bauchi (Nigeria) and Bamenda (Cameroon) (skins in Brit. Mus. Nat. Hist., at same localities as *coronata*). Whether these are morphs of *M. c. coronata* or are a separate species has not been determined.

Field Characters. ♂ unmistakable, a rather long-tailed, slender thrush-like large chat of cliffs and ravines. Glossy black with dark orange belly and undertail-coverts and large white patch in wing (median and lesser coverts); rump orange (black in Ethiopia); crown black (white from Togo east to Nile); upper belly pales to whitish band bordering black breast (Zambia to SW Ethiopia). ♀ patterned like ♂, but dark grey where ♂ is black (and in Ethiopia lacks white wing-patch) and could be mistaken for other species, but is nearly always closely accompanied by ♂. Loud, rich song.

Voice. Tape-recorded (22, 42, 51, 53, C, F). Voice powerful and far-carrying. One of Africa's finest songsters, with rich, sustained song of 2 types, each given by both sexes: (1) a clear, loud, melodious, fluty warbling, slow and strong, without imitations; (2) a less loud, rambling, rapid song or subsong consisting mainly of imitations of other birds. Over 30 imitated species listed by Benson (1940) and Farkas (1961), including several francolins and other non-passerines, thrushes, robins and other chats; Mountain Nightjar *Caprimulgus poliocephalus* 'perfectly imitated' (Ethiopia, Cheesman 1935). Songs often given in duet. Call a variable, sibilant 'ist-sisissi'; alarm a harsh 'kraat'. For most of year 1 bird singing or calling is generally answered immediately or after a few s by its mate; singing may then take form of duet. Begging call of nestling a high-pitched 'zerrr' or 'sreeee'. Fledgling alarm: 'tschree-tschreee'.

General Habits. Inhabits granitic hills in savanna and rising above forest, escarpments, cliffs, gorges, boulder-strewn slopes with precipices at least 6 m deep, wooded boulder-strewn lake shores. Requires leafy, woody vegetation, particularly with fig trees *Ficus*, but avoids dense, shady forest, although in Ethiopia (Eritrea) sometimes lives amongst rocks with no vegetation (Smith 1957). Sometimes adapts to brick and stone houses (e.g. on Zomba Mt, Nkhata Bay and Livingstonia, Malaŵi; often associated with houses in Ethiopia). Also uses road culverts and bridges (Malaŵi, Ethiopia). In Transvaal highveld favours wild, rugged ravines with rock-falls, roomy and wide, with rocky stream in middle, rock-faces and slabs full of seams, holes and clefts, and large *Ficus ingens* trees (Farkas 1961).

In pairs (or, transiently, family parties). Never gregarious. Shy but inquisitive and becomes tame around human habitations. Flight strong, flapping and with sweeping glides; on landing, raises tail slowly and exaggeratedly. Walks nimbly on flat slabs and ledges of rock, stopping from time to time and slowly wagging tail. Uses passages and tunnels amongst boulders, disappearing from view then reappearing some m away. Flies often across cliff face, dropping with half-spiral downward dive, or flying straight up perpendicular cliffs and overhangs, for several m. ♀ commonly follows ♂, flying a few m or seconds behind him. ♂ has a few favourite perches in territory: jagged peaks, isolated boulders, dry branches at tree top. Forages in trees and on ground. In fig tree makes long jumps from branch to branch or short flights within canopy, to feed on fruits. Uses thick branch as look-out for insects on ground; catches insect by dropping onto it from tree perch, or by searching amongst rocks, under trees, and in clearings between tussocks near water, hopping and running for a few strides with folded tail held straight up. When bird sitting on branch, tail moves up and down constantly, and when excited wagging is powerful, through c. 180°. Tail half-spread on upswing and closed tight on downswing. Both sexes and all ages swing tail. Juveniles have playful confrontations with dassies *Procavia capensis*, running at one, stopping in front of it, jumping into air and snapping bill, then jumping over it and worrying it from behind. Afrikaans name, Dassievoel, suggests that such interactions may be regular features of birds' life.

Resident in tropics, but a vertical migrant at higher latitudes; vacates Transvaal highveld mid-May to mid-Sept.

Food. Figs (several *Ficus* spp. including *F. ingens*), insects, including flying ants, and spiders.

Breeding Habits. Solitary, territorial, monogamous. Paired ♂ and ♀ remained together for 2 seasons. Territory 'fairly extensive' (Farkas 1961). Territory defended by (a) aggressive, precisely synchronized song duet by ♂ and ♀ of pair, and (b) by threat posture. Head and back plumage is flattened, but breast and belly feathers are strongly erected; tail bobs slowly, spread wide and, with body held erect, bird approaches opponent by running to left and to right, zigzagging, whilst singing.

In courtship, ♂ has song-flight with frequent changes in direction, zigzagging between trees, cliff-faces and bushes, making white shoulders conspicuous, and singing song-type (1); flight lasts 10–20 s. Song-flight most frequent at time of nest-building and egg-laying (Farkas 1966). Additionally, courting ♂ approaches ♀ in low shrubs in series of crouching runs, with wings part-opened and drooping to show white shoulders, tail partly fanned and half raised, and singing song-type (2). If ♀ responds, ♂ cocks tail vertically, almost imperceptibly vibrates wings, and utters 'zerrr' nestling-begging call; copulation follows (Farkas 1961, 1966).

NEST: untidy but compact substrate of feathers, down, fine dry grass, leaves, twigs, dead leaves, small roots and mammal hair (dog, antelope, dassie), lined with cobwebs, soft plant fibres and mammal hair which becomes felt-like. Generally built in bowl of mud nest of Lesser Striped Swallow *Hirundo abyssinica*, occasionally of Greater Striped Swallow *H. cucullata* (Vincent 1947) and of swifts and probably African Rock Martin *H. fuligula* (Smith *et al.* 1966), also in crevice in rock and sometimes crevice in masonry, brick or stone-walled outhouse. Nest built by ♂ and ♀ in c. 7 days (up to 3 weeks if weather unfavourable, South Africa). Pair appropriates swallow's nest by driving owners away, and ♂ removes all swallow's nest lining and contents (eggs and young). Pair uses same swallow's nest for several successive broods, until the nest tunnel disintegrates. In swallow's nest substrate is built at first on floor of cavity opposite entrance, in strip from deepest part of bowl up wall to roof. Further strips are laid down adjacently, until entire bowl lined; then felt-like padding of hair is added.

EGGS: 2–4, in southern Africa av. of 36 clutches, 2·8 eggs. In Zaïre always 2 eggs. Eggs laid at 24 h intervals, usually between 08·00 and 09·00 h. White, cream, pale green, blue, or bluish green, spotted with pale red-brown and mauve. SIZE: (n = 28, southern Africa) 24–27·6 × 17–19·1 (25·6 × 18·2).

LAYING DATES: South Africa, Aug–Dec (mainly Sept–Nov); Zimbabwe, Aug–Dec (Aug 1, Sept 17, Oct 43, Nov 37, Dec 7 clutches); Malaŵi, Oct–Dec; Zambia, Nov; Zaïre (bird in breeding condition Feb); E Africa: Region A, Apr, June, Region B, Mar, Region C, Mar, May, Nov, Dec, Region D, Mar, Apr, Dec (all E African breeding in rainy season); Ethiopia, Feb, Apr–June; Nigeria, Dec–May; Ghana, (nestlings May, fledglings July).

INCUBATION: by ♀ only, sitting all night (South Africa) and for long periods by day in cool weather and during 2–3 days before hatching. Period: 14–15 days.

DEVELOPMENT AND CARE OF YOUNG: chicks hatch in morning. Subcutaneous feathers visible by days 6–7, when rectrices and remiges break out. Contour feathers grow from days 8 to 9, on underparts first. Eyes open on days 10–11, when bill and legs darken. Sexual dimorphism in plumage apparent by day 13. Young leave nest on days 19–21 (once on days 15–16), with wings and tail well developed but head feathering scant. Underwing-coverts and axillaries develop from weeks 5 to 7 or 8. Young normally fed by both parents, although at one nest week-old young were fed by ♀ only, every 10 min; after giving food, ♀ waited for faecal sac to be produced, then flew off with it. On fledging at about 20 days, young can fly well for 8–10 m, gaining height. Fledgling moves on ground with triple strides, or (longer distances) with clumsy hops; 4–5 days after fledging it can run rapidly. Family keeps close together for some time.

References
Farkas, T. (1961, 1966).
Smith, V. W. *et al.* (1966).

556 TURDIDAE

Plate 32
(Opp. p. 465)

Myrmecocichla semirufa (Rüppell). **White-winged Cliff-Chat. Traquet de roche d'Abyssinie.**

Saxicola semirufa Rüppell, 1837. Neue Wirbelt., Vögel, p. 74; L. Tana.

Range and Status. Endemic resident, Ethiopia. Locally frequent to common, 1550–1850 m in Yavello Hills, usually above 2100 m elsewhere, but in Sidamo and Borana mainly between 1500 and 1800 m in hilly downland; uncommon, Eritrea, between 1250 and 2500 m, from Guna Guna north to Adi Quala.

Description. ADULT ♂: entirely glossy black, except as follows: belly, flanks and undertail-coverts orange; large white patch in wing formed by white bases to all primaries except outermost (short P10). Bill, eye, legs and feet black. ADULT ♀: forehead, lores, crown, ear-coverts, hindneck, mantle and upperwing-coverts black. Remaining upperparts, tail and wings brownish black, sometimes with rump feathers faintly fringed brown and with same white wing-patch as in adult ♂. Chin and throat blackish, with irregular pale buff stripe down centre; breast and belly blackish with feathers more or less broadly fringed buffy or orange; undertail-coverts orange, centre of belly sometimes orange or dusky orange. SIZE: wing, ♂ (n = 6) 111–117 (115), ♀ (n = 6) 102–111 (106); tail, ♂ (n = 6) 77–88·5 (80·2), ♀ (n = 6) 68·5–77·5 (74·0); bill, ♂ (n = 3) 18–22 (20·0); tarsus, ♂ (n = 3) 28–34 (31·0), ♀ (n = 3) 30–31 (30·5).

IMMATURE: black, profusely spotted above with small, buffy orange spots. Tips of rump feathers and uppertail-coverts broadly orange; upperwing-coverts similarly blackish with buff spots. Same white patch in wing as adult. Chin, throat and breast black with buff spots becoming more profuse and larger posteriorly, so that belly is buffy mottled with black and undertail-coverts buffy or dull orange. Older birds like adult but blackish underparts heavily and finely banded buff anteriorly and more orange posteriorly.

NESTLING: unknown.

Field Characters. A striking bird of highland Ethiopia, ♂ like none other except Mocking Cliff-Chat *M. cinnamomeiventris*. ♂ glossy black with orange belly and undertail-coverts and white patch at base of primaries (showing conspicuously when wing folded as well as in flight). Also characterized by cliff habitat and loud, clear whistling calls. ♀ less striking, brownish black, with median buff stripe on throat, fine orange-buff bars on underparts, orange undertail-coverts, and same white patch in wing as ♂. Mocking Cliff-Chat occurs in same habitat and altitudes, but ♂ has white shoulders and no white in primaries and ♀ no white in wing at all.

Voice. Not tape-recorded. Said to have a modulated, fluty song.

Myrmecocichla semirufa

General Habits. Inhabits borders of forest, slopes of small valleys, rocks and scrub-covered mountain gorges, and ravines choked with large boulders, also degraded park-like *Podocarpus* forest, rocky outcrops and stony ground on fringes of farms. Occurs in pairs or small parties; shy and retiring.

Food. Immature birds feed on newly-emerged termites (Addis Ababa: Urban 1978).

Breeding Habits.
NEST: compact construction of grass stems and moss, lined with hair and feathers; sited in rock crevice or hole in stone wall.
EGGS: 3. Glossy; white or greenish white, covered with fine pale rufous speckling. SIZE: *c.* 25 × 19.
LAYING DATES: Ethiopia, June–Aug (young Aug).

BIBLIOGRAPHY

The bibliography is in three parts: (1) general and regional references, (2) references for each family and (3) acoustic references. Titles in (1) are usually not repeated in (2). These two lists cover all of the works cited in the text and all other significant works consulted. If a reference cited in the text does not appear in the appropriate family list (2), it will be found in (1).

1. General and Regional References

Allport, G., Ausden, M., Hayman, P., Robertson, P. and Wood, P. (1989). The conservation of the birds of Gola Forest, Sierra Leone. *ICBP Study Report* No. 38. Int. Council for Bird Preserv., Cambridge.

Ames, P. L. (1971). The morphology of the syrinx in passerine birds. *Peabody Mus. Nat. Hist. Bull.* 37.

Archer, G. and Godman, E. M. (1961). 'The Birds of British Somaliland and the Gulf of Aden', Vols 3 and 4. Gurney and Jackson, London.

Ash, J. S. (1990). Additions to the avifauna of Nigeria, with notes on distributional changes and breeding. *Malimbus* 11, 104–116.

Ash, J. S. and Miskell, J. E. (1983). Birds of Somalia, their habitat, status and distribution. *Scopus*, Special Suppl. 1, 1–97.

Ash, J. S., Dowsett, R. J. and Dowsett-Lemaire, F. (1989). New ornithological distribution records from eastern Nigeria. *Tauraco Res. Rep.* No. 1, 13–27.

Bannerman, D. A. (1936, 1939). 'The Birds of Tropical West Africa', Vols 4–5. Crown Agents, London.

Basilio, A. (1963). 'Aves de la isla de Fernando Po'. Editorial Coculsa, Madrid.

Bates, G. L. (1930). 'Handbook of the Birds of West Africa'. John Bale, Sons and Danielsson Ltd, London.

Becker, P. (1974). Beobachtungen an paläarktischen Zugvögeln in ihrem Winterquartier in Südwestafrika. *S.W.A. Wissenschaftliche Gesellschaft*, Windhoek, 1–86.

Beesley, J. S. S. and Irving, N. S. (1976). The status of the birds of Gaborone and its surroundings. *Botswana Notes Rec.* 8, 231–261.

Benson, C. W. and Benson, F. M. (1977). 'The Birds of Malaŵi'. Montfort Press, Limbe, Malaŵi.

Benson, C. W., Brooke, R. K., Dowsett, R. J. and Irwin, M. P. S. (1971). 'The Birds of Zambia'. Collins, London.

de Bie, S. and Morgan, N. (1989). Les oiseaux de la Reserve de la Biosphère 'Boucle du Baoulé', Mali. *Malimbus* 11, 41–60.

Bonde, K. (1981). An annotated checklist to the birds of Lesotho. Unpubl. MS, 91 pp., Maseru.

Bouet, G. (1955–1961). 'Oiseaux de l'Afrique Tropicale'. Off. Rech. Scient. Tech. Outre-Mer, Paris.

Bowen, P. St J. (1983). Palaearctic migrants in Mwinilunga District, North-Western Province. *Bull. Zamb. Orn. Soc.* 13–15, 43–84.

Boyer, H. J. and Bridgeford, P. A. (1988). Birds of the Naukluft Mountains: an annotated checklist. *Madoqua* 15, 294–315.

Braine, S. (1990). Records of birds of the Cunene River estuary. *Lanioturdus* 25, 4–21.

Britton, P. L. (Ed.) (1980). 'Birds of East Africa'. East Africa Natural History Society, Nairobi.

Britton, P. L. and Zimmerman, D. A. (1979). The avifauna of Sokoke Forest, Kenya. *J. E. Afr. Nat. Hist. Soc.* 169, 1–16.

Brooke, R. K. (1984). South African Red Data Book – Birds. *S. Afr. Natn. Sci. Progr. Rep.* 10, 1–213. CSIR, Pretoria.

Brosset, A. (1961). Ecologie des oiseaux du Maroc Oriental. *Trav. Inst. Sci. Chérif.*, sér. Zoologie, no. 22.

Brosset, A. (1969). La vie sociale des oiseaux dans une forêt équatoriale du Gabon. *Biol. Gabon.* 5, 29–69.

Brosset, A. and Erard, C. (1986). 'Les Oiseaux des Régions Forestières du Nord-Est du Gabon', Vol. 1. Société Nationale de Protection de la Nature, Paris.

Brown, C. J. (1990). Birds of the West Caprivi Strip, Namibia. *Lanioturdus* 25, 22–37.

Brown, L. H. and Britton, P. L. (1980). 'The Breeding Seasons of East African Birds'. East Africa Natural History Society, Nairobi.

Browne, P. W. P. (1982). Palaearctic birds wintering in southwest Mauritania: species, distributions and population estimates. *Malimbus* 4, 69–92.

Brunel, J. (1958). Observations sur les oiseaux du Bas-Dahomey. *Oiseau et R.F.O.* 28, 1–38.

Bundy, G. (1976). 'The Birds of Libya'. British Ornithologists' Union Check-list No. 1. BOU, London.

Campbell, B. and Lack, E. (1985). 'A Dictionary of Birds'. T. and A. D. Poyser, Calton.

Carroll, R. W. (1988). Birds of the Central African Republic. *Malimbus* 10, 177–200.

Carswell, M. (1986). Birds of the Kampala area. *Scopus*, Special Suppl. 2, 1–89.

Cave, F. O. and Macdonald, J. D. (1955). 'Birds of the Sudan'. Oliver and Boyd, Edinburgh.

Chapin, J. P. (1953). The birds of the Belgian Congo, Vol. 3. *Bull. Am. Mus. Nat. Hist.* 75A.

Chappuis, C. (1975, 1978, 1985). Illustration sonore de problèmes bioacoustiques posés par les oiseaux de la zone éthiopienne. *Alauda* 43 (4), 427–474; 46 (4), 327–355; 53(2), 115–136.

Chappuis, C. (1986). Revised list of sound-recorded Afrotropical birds. *Malimbus* 8, 25–39.

Cheke, R. A., Walsh, J. F. and Sowah, S. A. (1986). Records of birds seen in the Republic of Togo during 1984–1986. *Malimbus* 8, 51–72.

Clancey, P. A. (1964). 'The Birds of Natal and Zululand'. Oliver and Boyd, Edinburgh.

Clancey, P. A. (1971). A handlist of the birds of southern Moçambique. Lourenço Marques: *Inst. Invest. Científica Moçambique*, Ser. A, 11, 1–167.

Clancey, P. A. (Ed.) (1980). 'Checklist of Southern African Birds'. Southern African Ornithological Society, Johannesburg.

Clancey, P. A. (1985). 'The Rare Birds of Southern Africa'. Winchester Press, Saxonwold.

Clancey, P. A. (1986). Endemicity in the southern African avifauna. *Durban Mus. Novit.* 13(20), 245–284.

Clancey, P. A. (Ed.) (1987). 'Checklist of Southern African Birds: First Updating Report'. Southern African Ornithological Society, Johannesburg.

Collar, N. J. and Stuart, S. N. (1985). 'Threatened Birds of Africa and Its Islands'. ICBP/IUCN Red Data Book, 3rd ed., pt. 1. Int. Council for Bird Preserv., Cambridge.

Collar, N. J. and Stuart, S. N. (1988). 'Key Forests for Threatened Birds in Africa'. *ICBP Monograph* No. 3. Int. Council for Bird Preserv., Cambridge.

Colston, P. R. and Curry-Lindahl, K. (1986). 'The Birds of Mount Nimba, Liberia'. British Museum (Natural History), London.

Cramp, S. (Ed.) (1988). 'Handbook of the Birds of Europe, the Middle East and North Africa. The Birds of the Western Palearctic', Vol. 5. Oxford University Press, Oxford.

Curry-Lindahl, K. (1981). 'Bird Migration in Africa', Vols 1, 2. Academic Press, London.

Cyrus, D. and Robson, N. (1980). 'Bird Atlas of Natal'. University of Natal Press, Pietermaritzburg, South Africa.

Dean, W. R. J. (1971). Breeding data for the birds of Natal and Zululand. *Durban Mus. Novit.* **9**, 59–61.

Dean, W. R. J. (1974). Breeding and distributional notes on some Angolan birds. *Durban Mus. Novit.* **10**, 109–125.

Dean, W. R. J. and Huntley, M. A. An updated list of the birds of Angola. Unpubl. ms.

Dean, W. R. J., Huntley, M. A., Huntley, B. J. and Vernon, C. J. (1988). Notes on some birds of Angola. *Durban Mus. Novit.* **14**, 43–92.

Dejonghe, J.-F. and Czajkowski, M. A. (1983). The lifespan of birds ringed in France, French departments abroad, and countries with French influence. *Alauda* **51**, 27–47.

Demey, R. and Fishpool, L. D. C. (1991). Additions and annotations to the avifauna of Côte d'Ivoire. *Malimbus* **12**, 61–86.

Desfayes, M. (1975). Birds from Ethiopia. *Rev. Zool. Afr.* **89**, 505–535.

Donnelly, B. (1985). The birds of the Matobo (formerly Matopos) National Park, Zimbabwe. *Honeyguide* **31**, 11–23.

Dowsett, R. J. (1985). Site fidelity and survival rates of some montane forest birds in Malaŵi, south-central Africa. *Biotropica* **17**, 145–154.

Dowsett, R. J. (Ed.) (1989). Enquête faunistique dans la forêt du Mayombe et check-liste des oiseaux et des mammifères du Congo. *Tauraco Res. Rep.* No. 2.

Dowsett, R. J. and Dowsett-Lemaire, F. (1984). Breeding and moult cycles of some montane forest birds in south-central Africa. *Terre Vie* **39**, 89–111.

Dowsett, R. J. and Prigogine, A. (1974). The avifauna of the Marungu Highlands. *Hydrobiol. Survey of L. Bangweulu – Luapula R. Basin*, Vol. XIX.

Dowsett-Lemaire, F. (1983). Ecological and territorial requirements of montane forest birds on the Nyika Plateau, south-central Africa. *Gerfaut* **73**, 345–378.

Dowsett-Lemaire, F. (1985). Breeding productivity and the non-breeding element in some montane forest birds in Malaŵi, south-central Africa. *Biotropica* **17**, 137–144.

Dowsett-Lemaire, F. (1989). Ecological and biogeographical aspects of forest bird communities in Malaŵi. *Scopus* **13**, 1–80.

Dowsett-Lemaire, F. (1990). Eco-ethology, distribution and status of Nyungwe Forest birds (Rwanda). In 'Survey of the Fauna and Flora of Nyungwe Forest, Rwanda' (Dowsett, R. J. Ed.). *Tauraco Res. Rep.* No. 3, pp. 31–85.

Dowsett-Lemaire, F. and Dowsett, R. J. (1990). Zoogeography and taxonomic relationships of the forest birds of the Albertine Rift Afromontane region. In 'Survey of the Fauna and Flora of Nyungwe Forest Birds, Rwanda' (Dowsett, R. J. Ed.). *Tauraco Res. Rep.* No. 3, pp. 87–109.

Dupuy, A. (1969). Catalogue ornithologique du Sahara Algérien. *Oiseau et R.F.O.* **39**, 225–241.

Dyer, M., Gartshore, M. E. and Sharland, R. E. (1986). The birds of Nindam Forest Reserve, Kagoro, Nigeria. *Malimbus* **8**, 2–20.

Earlé, R. A. and Grobler, N. J. (1987). 'First Atlas of Bird Distribution in the Orange Free State'. National Museum, Bloemfontein.

Eisentraut, M. (1963). Die Wirbeltiere des Kamerungebirges. P. Parey, Hamburg & Berlin, 353 pp.

Eisentraut, M. (1973). Die Wirbeltierfauna von Fernando Poo und Westkamerun. *Bonn. zool. Monogr.* **3**, 1–400.

Elgood, J. H. (1982). 'The Birds of Nigeria'. British Ornithologists' Union Check-list No. 4. BOU, London.

Elgood, J. H., Sharland, R. E. and Ward, P. (1966). Palaearctic migrants in Nigeria. *Ibis* **108**, 84–116.

Elgood, J. H., Fry, C. H. and Dowsett, R. J. (1973). African migrants in Nigeria. *Ibis* **115**, 375–411.

Etchécopar, R. D. and Hüe, F. (1967). 'The Birds of North Africa'. Oliver and Boyd, Edinburgh.

Feather, P. J. (1986). The Bulawayo garden bird survey 1973–1982. *Honeyguide* **32**, 13–33.

Field, G. D. (1974). 'Birds of the Freetown Peninsula'. Fourah Bay College Bookshop, Freetown.

Frade, F. (1951). 'Catálogo das Aves de Moçambique'. Ministerio das Colonias, Lisbon.

Frade, F. and Bacelar, A. (1955). Catálogo das Aves da Guiné Portuguesa. *Anil. Junta. Invest. Ultram.* **10**, 4.

Friedmann, H. (1930). Birds collected by the Childs Frick expedition to Ethiopia and Kenya Colony. *Bull. U.S. Nat. Mus.* **153**, 1–516.

Friedmann, H. (1978). Results of the Lathrop Central African Republic Expedition 1976, ornithology. *Los Angeles County Mus. Contrib. Sci.* No. 287, 1–22.

Friedmann, H. and Williams, J. G. (1971). The birds of the lowlands of Bwamba, Toro Province, Uganda. *Los Angeles County Mus. Contrib. Sci.* No. 211, 1–70.

Fry, C. H. (1985–1986). Coded bibliography of African ornithology, 1985–1986. *Malimbus* **7** (Suppl.) 1–41; **8**, 1–38.

Fuggles-Couchman, N. R. (1984). The distribution of, and other notes on, some birds of Tanzania. *Scopus* **8**, 1–17, 73–78, 81–92.

Gartshore, M. E. (1989). An avifaunal survey of Tai National Park, Ivory Coast. *ICBP Study Report* No. 39. Int. Council for Bird Preserv., Cambridge.

Gatter, W. (1988). The birds of Liberia (West Africa). *Verh. orn. Gesell. Bayern* **24**, 689–723.

Gaugris, Y., Prigogine, A. and Vande weghe, J.-P. (1981). Additions et corrections à l'avifaune du Burundi. *Gerfaut* **71**, 3–39.

Germain, M., Dragesco, J., Roux, F. and Garcin, H. (1973). Contribution à l'ornithologie du Sud-Cameroun II. Passeriformes. *Oiseau et R.F.O.* **43**, 212–259.

Ginn, P. J. (1976). Birds of Makgadigadi: a preliminary report. *Wagtail* **15**, 21–96.

Ginn, P. J. (1979). 'Birds of Botswana'. Chris van Rensburg Publ., Johannesburg.

Giraudoux, P., Degauquier, R., Jones, P. J., Weigel, J. and Isenmann, P. (1988). Avifaune du Niger: état des connaissances en 1986. *Malimbus* **10**, 1–140.

Goodman, S. M. and Meininger, P. L. (Eds) (1989). 'The Birds of Egypt'. Oxford University Press, Oxford.

Gore, M. E. J. (1981). 'The Birds of The Gambia'. British Ornithologists' Union Check-list No. 3. BOU, London.

Green, A. A. (1983). The Birds of Bamingui-Bangoran National Park, Central African Republic. *Malimbus* **5**, 17–30.

Green, A. A. (1990). The avifauna of the southern sector of the Gashaka-Gumti Game Reserve, Nigeria. *Malimbus* **12**, 31–51.

Green, A. A. and Sayer, J. A. (1979). The birds of Pendjari and Arli National Parks (Benin and Upper Volta). *Malimbus* **1**, 14–28.

Grimes, L. G. (1987). 'The Birds of Ghana'. British Ornithologists' Union Check-list No. 9. BOU, London.

Günther, R. and Feiler, A. (1986). Zur Phänologie, Ökologie und Morphologie angulanischer Vögel (Aves), II. Passeriformes. *Faun. Abhandl. Staatliches Mus. Tierkunde Dresden* **14**, 1–29.

Hall, B. P. (1960). The ecology and taxonomy of some Angola birds. *Bull. Brit. Mus. (Nat. Hist.), Zool.* **6**, 367–463.

Hall, B. P. and Moreau, R. E. (1970). 'An Atlas of Speciation in African Passerine Birds'. British Museum (Natural History), London.

Harvey, W. G. and Howell, K. M. (1987). Birds of the Dar es Salaam area, Tanzania. *Gerfaut* **77**, 205–258.

Heim de Balsac, H. and Mayaud, N. (1962). 'Les Oiseaux du Nord-Ouest de l'Afrique'. Paul Lechevalier, Paris.

Heinrich, G. (1958). Zur Verbreitung und Lebensweise der Vögel von Angola, II and III. *J. Orn.* **99**, 322–362, 399–421.

Hines, C. J. H. (1985–1987). The birds of eastern Kavango, SWA/Namibia. *J. SWA Soc.* **40–41**, 115–147.

Hoesch, W. and Niethammer, G. (1940). 'Die Vogelwelt Deutsch Südwestafrikas'. Deutschen Ornithologischen Gesellschaft, Sonderheft, Berlin.

Hogg, P., Dare, P. J. and Rintoul, J. V. (1984). Palaearctic migrants in the central Sudan. *Ibis* **126**, 307–331.

Howells, W. W. (1985). The birds of the Dande Communal Lands, Middle Zambezi Valley, Zimbabwe. *Honeyguide* **31**, 26–48.

Irwin, M. P. S. (1981). 'The Birds of Zimbabwe'. Quest Publ., Harare.

Irwin, M. P. S., Niven, P. N. F. and Winterbottom, J. M. (1969). Some birds of the lower Chobe River area, Botswana. *Arnoldia (Rhod.)* **4**, 1–40.

Jackson, F. J. and Sclater, W. L. (1938). 'The Birds of Kenya Colony and the Uganda Protectorate'. Gurney and Jackson, London.

Jensen, J. V. and Kirkeby, J. (1980). 'The Birds of The Gambia'. Aros Nature Guides, Århus.

Johnston-Stewart, N. G. B. (1982). Evergreen forest birds in upper Thyolo. *Nyala* **8**, 69–84.

Keith, S. and Gunn, W. W. H. (1971). 'Birds of the African Rain Forests'. Sounds of Nature #9, Fed. Ontario Naturalists and Amer. Mus. Nat. Hist., New York.

Keith, S., Twomey, A., Friedmann, H. and Williams, J. (1969). The avifauna of the Impenetrable Forest, Uganda. *Am. Mus. Novit.* 2389.

Kemp, A. C. (1976). The distribution and status of the birds of the Kruger National Park. *Koedoe Monog.* **2**, 1–130.

Lack, P. (1985). The ecology of the land-birds of Tsavo East National Park, Kenya. *Scopus* **9**, 2–23, 57–96.

Lamarche, B. (1981). Liste commentée des oiseaux du Mali. Pt 2: Passereaux. *Malimbus* **3**, 73–102.

Lamarche, B. (1988). Liste commentée des oiseaux de Mauritanie. *Etudes Sahar. et Ouest-Afr.* **1**(4), 1–162.

Ledant, J. P., Jacob, J. P., Jacobs, P., Malher, F., Ochando, B. and Roche, J. (1981). Mis à jour de l'avifaune Algérienne. *Gerfaut* **71**, 295–398.

Lewis, A. and Pomeroy, D. (1989). 'A Bird Atlas of Kenya'. A. A. Balkema, Rotterdam.

Lippens, L. and Wille, H. (1976). 'Les Oiseaux du Zaïre'. Lannoo, Tielt.

Liversidge, R. (1985). Alien bird species introduced into South Africa. *Proc. Symp. Birds and Man*, pp. 31–44.

Louette, M. (1981). 'The Birds of Cameroon: an Annotated Check-List'. Paleis der Academiën, Brussels.

Lynes, H. (1924, 1925). On the birds of north and central Darfur. *Ibis* **11**(6), 648–719; **12**(1), 71–131.

Macdonald, J. D. (1957). 'Contribution to the Ornithology of Western South Africa'. Trustees of the British Museum, London.

Mackworth-Praed, C. W. and Grant, C. H. B. (1957, 1960). 'Birds of Eastern and North Eastern Africa'. Longmans, London.

Mackworth-Praed, C. W. and Grant, C. H. B. (1962, 1963). 'Birds of the Southern Third of Africa'. Longmans, London.

Mackworth-Praed, C. W. and Grant, C. H. B. (1970, 1973). 'Birds of West Central and Western Africa'. Longmans, London.

Maclean, G. L. (1985). 'Roberts' Birds of Southern Africa', 5th ed. Trustees of the John Voelker Bird Book Fund, Cape Town.

Mann, C. F. (1985). An avifaunal study in Kakamega Forest, Kenya, with particular reference to species diversity, weight and moult. *Ostrich* **56**, 236–262.

Mayaud, N. (1986, 1988). Les oiseaux du nord-ouest de l'Afrique. Notes complémentaires. *Alauda* **54**, 213–229; **56**, 113–125.

Mayr, E. and Greenway, J. C. (1956). Sequence of passerine families (Aves). *Breviora, Mus. Comp. Zool.* **58**, 1–11.

Moreau, R. E. (1950). The breeding seasons of African birds. 1. Land birds. *Ibis* **92**, 223–267.

Moreau, R. E. (1966). 'The Bird Faunas of Africa and Its Islands'. Academic Press, London and New York.

Moreau, R. E. (1972). 'The Palaearctic-African Bird Migration Systems'. Academic Press, London and New York.

Morel, G. J. and Morel, M. Y. (1988). Liste des oiseaux de Guinée. *Malimbus* **10**, 143–176.

Morel, G. J. and Morel, M.-Y. (1990). 'Les Oiseaux de Sénégambie (avec Cartes de Distribution)'. Off. Rech. Scient. Tech. Outre-Mer., Paris.

Morel, G. J. and Roux, F. (1966). Les migrateurs paléarctiques au Sénégal. *Terre Vie* **113**, 19–72, 143–176.

Morel, G. J. and Roux, F. (1973). Les migrateurs paléarctiques au Sénégal: notes complémentaires. *Terre Vie* **27**, 523–550.

Newby, J. E. (1980). The birds of the Ouadi Rime – Ouadi Achim Faunal Reserve: a contribution to the study of the Chadian avifauna. *Malimbus* **2**, 29–50.

Newman, K. (1983). 'Newman's Birds of Southern Africa'. Macmillan South Africa, Johannesburg.

Nikolaus, G. (1987). Distribution atlas of Sudan's birds with notes on habitat and status. *Bonn. zool. Monog.* 25.

Olson, S. L. (1989). Preliminary systematic notes on some Old World passerines. *Riv. ital. Orn. Milano* **59**, (3–4), 183–195.

Pakenham, R. H. W. (1979). 'The Birds of Zanzibar and Pemba'. British Ornithologists' Union Check-list No. 2. BOU, London.

Peters, J. L. (1960, 1964, 1979). 'Check-List of Birds of the World', Vols 8, 9 and 10. Museum of Comparative Zoology, Cambridge, Massachusetts.

Pineau, J. and Giraud-Audine, M. (1979). Les oiseaux de la péninsule tingitane. *Trav. Inst. Sci. Chérif.*, sér. Zoologie, **38**, 1–147.

Priest, C. D. (1935) 'The Birds of Southern Rhodesia', Vol. 3. William Clowes, London and Beccles.

Prigogine, A. (1953). Contribution à l'étude de la faune ornithologique de la région à l'ouest du Lac Edouard. *Annl. Mus. r. Congo belge*, Sér. 8, 24.

Prigogine, A. (1960). La faune ornithologique du Massif du Mont Kabobo. *Annl. Mus. r. Congo belge*, Sér. 8, 85.

Prigogine, A. (1971, 1978, 1984). Les oiseaux de l'Itombwe et de son hinterland. *Mus. Roy. Afr. Cent. Ann.* Sér. 8, 185, 223, 243.

Prigogine, A. (1980). Etude de quelques contacts secondaires au Zaïre oriental. *Gerfaut* **70**, 305–384.

Prigogine, A. (1984). Secondary contacts in central Africa. *Proc.V Pan-Afr. Orn. Congr.*, pp. 81–96.

Raikow, R. J. (1982). Monophyly of the Passeriformes: test of a phylogenetic hypothesis. *Auk* **99**, 431–445.

Rand, A. L. (1951). Birds from Liberia. *Fieldiana Zool.* **32**(9), 558–653.

Rand, A. L., Friedmann, H. and Traylor, M. A. (1959). Birds

from Gabon and Moyen Congo. *Fieldiana Zool.* **41**, 221–411.
Ripley, S. D. and Bond, G. M. (1966). The birds of Socotra and Abd-el-Kuri. *Smithsonian Misc. Coll.* **151**, 1–37.
Ripley, S. D. and Heinrich, G. H. (1966). Comments on the avifauna of Tanzania, I. *Postilla* **96**, 1–45.
da Rosa Pinto, A. A. (1970). Um catálogo das aves do Distrito da Huila (Angola). *Mem. Trab. Inst. Invest. Ci. Angola* **6**, 5–160.
da Rosa Pinto, A. A. (1972). Contribuciao para o estudo da avifauna do Distrito de Cabinda (Angola). *Mem. Trab. Inst. Invest. Ci. Angola* **10**, 9–90.
da Rosa Pinto, A. A. and Lamm, D. W. (1953–1960). Contribution to the study of the ornithology of Sul do Save (Mozambique), pts. 1–4. *Mem. Mus. Alvaro de Castro* **2**, 65–85; **3**, 125–159; **4**, 107–167.
Salvan, J. (1968–1969). Contribution à l'étude des oiseaux du Tchad. *Oiseau et R.F.O.* **38**, 249–273; **39**, 38–69.
Schmidl, D. (1982). 'The Birds of the Serengeti National Park, Tanzania'. British Ornithologists' Union Check-list No. 5. BOU, London.
Schönwetter, M. (1979). 'Handbuch der Oologie', Vol. II. Akademie Verlag, Berlin.
Schouteden, H. (1954–1955). De vogels van belgisch Congo en van Ruanda-Urundi, III. *Ann. Mus. Congo belge*, C. Zoologie Sér. IV, **4**, fasc. 1–2.
Sclater, W. L. (1930). 'Systema Avium Aethiopicarum', pt. 2. BOU, London.
Sclater, W. L. and Moreau, R. E. (1932–1933). Taxonomic and field notes on some birds of north-eastern Tanganyika Territory. *Ibis* **11**(1), 656–683; (2), 1–33, 187–218, 399–439.
Serle, W. (1940). Field observations on some northern Nigerian birds. *Ibis* **14**(4), 1–47.
Serle, W. (1943). Further field observations on northern Nigerian birds. *Ibis* **85**, 413–437.
Serle, W. (1948–1949). Notes on the birds of Sierra Leone. *Ostrich* **19**, 187–199; **20**, 70–85.
Serle, W. (1950). A contribution to the ornithology of the British Cameroons. *Ibis* **92**, 343–376, 602–638.
Serle, W. (1954). A second contribution to the ornithology of the British Cameroons. *Ibis* **96**, 47–80.
Serle, W. (1957). A contribution to the ornithology of the Eastern Region of Nigeria. *Ibis* **99**, 371–418.
Serle, W. (1965). A third contribution to the ornithology of the British Cameroons. *Ibis* **107**, 60–94.
Serle, W. (1981). The breeding seasons of birds in the lowland rain forest and in the montane forest of West Cameroon. *Ibis* **123**, 62–74.
Serle, W., Morel, G. J. and Hartwig, W. (1977). 'A Field Guide to the Birds of West Africa'. Collins, London.
Short, L. L., Horne, J. F. M. and Muringo-Gichuki, C. (1990). Annotated check-list of the birds of East Africa. *Proc. West. Found. Vert. Zool.* **4**(3).
Sibley, C. G. and Ahlquist, J. E. (1990). 'Phylogeny and Classification of Birds. A Study in Molecular Evolution'. Yale University Press, New Haven.
Sibley, C. G. and Monroe, B. L. Jr. (1990). 'Distribution and Taxonomy of Birds of the World'. Yale University Press, New Haven.
Simon, P. (1965). Synthèse de l'avifaune du massif montagneux du Tibesti. *Gerfaut* **55**, 26–71.
Sinclair, J. C. (1984). 'Field Guide to the Birds of Southern Africa'. C. Struik, Cape Town.
Smith, K. D. (1957). An annotated check list of the birds of Eritrea. *Ibis* **99**, 307–337.
Smith, K. D. (1965). On the birds of Morocco. *Ibis* **107**, 493–526.
Smith, K. D. (1968). Spring migration through southeast Morocco. *Ibis* **110**, 452–492.
Smithers, R. H. N. (1964). 'A Check List of the Birds of the Bechuanaland Protectorate and the Caprivi Strip'. Trustees of the National Museum of Southern Rhodesia, Salisbury.
van Someren, V. G. L. (1922, 1932). Notes on the birds of East Africa. *Novit. Zool.* **29**, 1–246; **37**, 252–380.
van Someren, V. G. L. (1939). Reports on the Coryndon Museum expedition to the Chyulu Hills, Part. 2. The birds of the Chyulu Hills. *J. E. Afr. Nat. Hist. Soc.* **XIV**, 1–2, 15–129.
van Someren, V. G. L. (1956). Days with birds. *Fieldiana Zool.* **38**, 1–520.
van Someren, V. G. L. and van Someren, G. R. C. (1949). The birds of Bwamba. *Uganda J.*, Special Suppl. **13**.
Stresemann, E. and Stresemann, V. (1966). Die Mauser der Vögel. *J. Orn.* **107** (Suppl.), 1–445.
Stuart, S. N. (Ed.) (1986). 'Conservation of Cameroon Montane Forests'. Report of the ICBP Cameroon Montane Forest Survey. Int. Council for Bird Preserv., Cambridge.
Stuart, S. N. and Jensen, F. P. (1985). The avifauna of the Uluguru Mountains, Tanzania. *Gerfaut* **75**, 155–197.
Stuart, S. N., Jensen, F. P. and Brogger-Jensen, S. (1987). Altitudinal zonation of the avifauna in the Mwanihana and Magombera forests, eastern Tanzania. *Gerfaut* **77**, 165–186.
Tarboton, W. R., Kemp, M. I. and Kemp, A. C. (1987). 'Birds of the Transvaal'. Transvaal Museum, Pretoria.
Taylor, P. B. (1979). Palaearctic and intra-African migrant birds in Zambia: a report for the period May 1971 to December 1976. *Zambian Orn. Soc. Occ. Pap.* **1**, 1–169.
Thévenot, M. (1982). Contribution à l'étude écologique des passereaux forestiers du Plateau Central et de la corniche du Moyen-Atlas (Maroc). *Oiseau et R.F.O.* **52**, 21–85, 97–152.
Thévenot, M., Bergier, P., and Beaubrun, P. (1981). Compte-rendu d'ornithologie Marocaine, année 1980. *Documents de l'Institut Scientifique* **6**, Univ. Mohammed V, Rabat.
Thévenot, M., Beaubrun, P., Baouab, R. E. and Bergier, P. (1982). Compte-rendu d'ornithologie Marocaine, année 1981. *Documents de l'Institut Scientifique* **7**, Univ. Mohammed V, Rabat.
Thiollay, J. M. (1985). The birds of Ivory Coast. *Malimbus* **7**, 1–59.
Thomsen, P. and Jacobsen, P. (1979). 'The Birds of Tunisia'. Odense, Denmark.
Thonnerieux, Y. (1988). Etat des connaissances sur la reproduction de l'avifaune du Burkina Faso (ex Haute-Volta). *Oiseau et R.F.O.* **58**, 120–146.
Thonnerieux, Y., Walsh, J. F. and Bortoli, L. (1989). L'avifaune de la ville de Ouagadougou et ses environs. *Malimbus* **11**, 7–40.
Traylor, M. A. Jr. (1963). 'Check-list of Angolan Birds'. Comp. Diam. Angola, Museo do Dundo, Lisbon.
Urban, E. K. and Brown, L. H. (1971). 'A Checklist of the Birds of Ethiopia'. Haile Sellassie I University Press, Addis Ababa.
Van Tyne, J. and Berger, A. J. (1959). 'Fundamentals of Ornithology'. John Wiley & Sons, New York.
Vande weghe, J.-P. (1973). Les périodes de nidification des oiseaux du Parc National de l'Akagera au Rwanda. *Gerfaut* **63**, 235–255.
Vande weghe, J.-P. and Loiselle, B. A. (1987). The bird fauna of Bururi Forest, Burundi. *Gerfaut* **77**, 147–164.
Vaurie, C. (1959). 'The Birds of the Palearctic Fauna', Order Passeriformes. Witherby, London.
Verheyen, R. (1953). Exploration du Parc National de L'Upemba. Oiseaux. *Mission G.F. de Witte*, **19**, 1–687, Bruxelles.
Vieillard, J. (1971–1972). Données biogéographiques sur l'avifaune d'Afrique Centrale, II. *Alauda* **40**, 63–92.
Vincent, A. W. (1946, 1947, 1949). On the breeding habits of

some African birds. *Ibis* **88**, 462–477; **89**, 163–204; **91**, 111–139.
Vincent, J. (1935). The birds of northern Portuguese East Africa. *Ibis* **13**(5), 1–37, 355–397, 485–529, 707–762.
Voous, K. H. (1977). List of recent Holarctic bird species. Passerines, Part 1. *Ibis* **119**, 223–250.
Walsh, J. F., Cheke, R. A. and Sowah, S. A. (1990). Additional species and breeding records of birds in the Republic of Togo. *Malimbus* **12**, 2–18.
Welch, G. R. and Welch, H. J. (1984). 'Djibouti Expedition, March 1984'. Published by the authors (21a East Delph, Whittlesey, Cambridgeshire PE7 1RH, UK).
Welch, G. R. and Welch, H. J. (1986). 'Djibouti II, autumn 1985'. Published by the authors (21a East Delph, Whittlesey, Cambridgeshire PE7 1RH, UK).
White, C. M. N. (1961). 'A Revised Check List of African Broadbills, Pittas, Larks, Swallows, Wagtails and Pipits'. Government Printer, Lusaka.
White, C. M. N. (1962). 'A Revised Check List of African Shrikes, Orioles, Drongos, Starlings, Crows, Waxwings, Cuckoo-Shrikes, Bulbuls, Accentors, Thrushes and Babblers'. Government Printer, Lusaka.
Williams, J. G. and Arlott, N. (1980). 'A Field Guide to the Birds of East Africa'. Collins, London.
Winterbottom, J. M. (1966a). Results of the Percy Fitzpatrick Institute – Windhoek State Museum joint ornithological expeditions. 3. Report on the birds of the Okavango Valley. *Cimbebasia* No. 15, 1–78.
Winterbottom, J. M. (1966b). Results of the Percy Fitzpatrick Institute – Windhoek State Museum joint ornithological expeditions. 5. Report on the birds of the Kaokoveld and Kunene River. *Cimbebasia* No. 19, 1–71.
Winterbottom, J. M. (1968). A check list of the land and fresh water birds of the Western Cape Province. *Ann. S. Afr. Mus.* **53**.
Winterbottom, J. M. (1971). 'A Preliminary Check List of the Birds of South West Africa'. S.W. Afr. Scient. Soc., Windhoek.
Witherby, H. F., Jourdain, F. C. R., Ticehurst, N. F. and Tucker, B. W. (1938). 'The Handbook of British Birds'. H. F. & G. Witherby Ltd., London.
de Witte, G. F. (1938). Exploration du Parc National Albert. *Inst. Parcs Nat. Congo Belge*, Fasc. 9, Oiseaux.
Witwatersrand Bird Club, Johannesburg, South Africa (1975–). *Southern Birds*, Nos. 1 and following (ongoing series).
Zimmerman, D. A. (1972). The avifauna of the Kakamega Forest, western Kenya, including a bird population study. *Bull. Am. Mus. Nat. Hist.* **149**, Article 3.

2. References for Each Family

Order PASSERIFORMES

Family EURYLAIMIDAE: broadbills

Aspenlind, L. J. (1935). Huru jag fann Afrikas sällysntaste fågel, Grauers brednäbb. *Fauna och Flora* **30**, 173–179.
Chapin, R. T. (1978). Brief accounts of some central African birds, based on the journals of James Chapin. *Rev. Zool. Afr.* **92**, 806–836.
Clancey, P. A. (1963). Miscellaneous taxonomic notes on African Birds 20, 1. The South African races of the Broadbill *Smithornis capensis* (Smith). *Durban Mus. Novit.* **6**(19), 231–241.
Clancey, P. A. (1970). On *Smithornis capensis suahelicus* Grote, 1926. *Bull. Br. Orn. Club* **90**, 164–166.
Dean, W. R. J., Macdonald, I. A. W. and Vernon, C. J. (1974). Possible breeding record of *Cercococcyx montanus*. *Ostrich* **45**, 188.
Friedmann, H. (1970). The status and habits of Grauer's Broadbill in Uganda (Aves: Eurylaimidae). *Los Angeles County Mus. Contrib. Sci.* **176**.
Friedmann, H. and Williams, J. G. (1968). Notable records of rare or little-known birds from western Uganda. *Rev. Zool. Bot. Afr.* **57**, 11–36.
Grant, C. H. B. and Mackworth-Praed, C. W. (1939). Notes on some eastern African birds: (2) On the races of *Smitkornis capensis* (Smith) occurring in eastern Africa. *Bull. Br. Orn. Club* **59**, 113–115.
Lawson, W. J. (1961). Probable courtship behaviour of the Broadbill *Smithornis capensis*. *Ibis* **103**a, 289–290.
Lowe, P. R. (1924). On the presence of broadbills (Eurylaimidae) in Africa. *Proc. Zool. Soc. Lond.* 279–291.
Lowe, P. R. (1931). On the anatomy of *Pseudocalyptomena* and the occurrence of broadbills (Eurylaimidae) in Africa. *Proc. Zool. Soc. Lond.* 445–461.
Manson, A. J. (1983). On the presence of the African Broadbill in the Vumba. *Honeyguide* **113**, 33.
Olson, S. L. (1971). Taxonomic comments on the Eurylaimidae. *Ibis* **113**, 507–516.
Raikow, R. J. (1987). Hindlimb myology and evolution of the Old World suboscine passerine birds (Acanthisittidae, Pittidae, Philepittidae, Eurylaimidae). *Orn. Monogr.* 41.
Rockefeller, J. S. and Murphy, C. B. G. (1933). The rediscovery of *Pseudocalyptomena*. *Auk* **50**, 23–29.

Family PITTIDAE: pittas

Beakbane, A. J. and Boswell, E. M. (1984). Nocturnal Afrotropical migrants at Mufindi, southern Tanzania. *Scopus* **8**, 124–127.
Benson, C. W. and Irwin, M. P. S. (1964). The migrations of the pitta of eastern Africa. *N. Rhod. J.* **5**, 465–475.
Berry, P. S. M. and Ansell, P. D. H. (1974). African Pitta (*Pitta angolensis*) displaying in the upper Luangwa Valley. *Bull. Zamb. Orn. Soc.* **10**(1), 29.
Britton, P. L. and Rathbun, G. B. (1978). Two migratory thrushes and the African Pitta at the Kenya coast. *Scopus* **2**, 11–17.
Burke, V. E. M. (1969). The African Pitta, *Pitta angolensis* Vieillot. *J. E. Afr. Nat. Hist. Soc.* **27**, 233–234.
Clancey, P. A. and Maclean, G. L. (1980). Angola Pitta in Natal. *Ostrich* **51**, 191.
Harrison, G. (1966). Notes on some recent occurrences of the

Angola Pitta (*Pitta angolensis longipennis*) in Malaŵi. *J. Malaŵi Soc.* **19**, 26–28.

Harvey, W. O. (1938). The East African Pitta (*Pitta angolensis longipennis* Reichenow). *Ibis* **14**, 335–337.

Masterson, A. (1987). Pitta chatter. *Bokmakierie* **39**, 23–25.

Moreau, R. E. and Moreau, W. M. (1937). Biological and other notes on some East African birds, Part II. *Ibis* **14**(1), 321–345.

Petit, L. (1899). Ornithologie Congolais. *Mém. Soc. Zool. France* **12**, 59–106.

Raikow, R. J. (1987) Hindlimb myology and evolution of the Old World suboscine passerine birds (Acanthisittidae, Pittidae, Philepittidae, Eurylaimidae). *Orn. Monogr.* 41.

Rathbun, G. B. (1978). The African Pitta at Gedi Ruins, Kenya. *Scopus* **2**, 7–10.

Sibley, C. G. (1970). A comparative study of the egg-white proteins of passerine birds. *Peabody Mus. Nat. Hist. Bull., Yale Univ.*, **32**, 1–131.

Stuart, S. N. and Turner, D. A. (1980). Some range extensions and other notable records of forest birds from eastern and northeastern Tanzania. *Scopus* **4**, 36–41.

Urban, E. K. and Hakanson, T. (1971). An African Pitta, *Pitta angolensis longipennis*, from Ethiopia. *Bull. Br. Orn. Club* **91**, 9–10.

Family ALAUDIDAE: larks

Abs, M. (1963). Vergleichende Untersuchungen an Haubenlerche (*Galerida cristata* (L.)) und Theklalerche (*Galerida theklae* A. E. Brehm). *Bonn. zool. Beitr.* **14**, 1–128.

Acocks, J. P. H. (1975). Veld types of South Africa. 2nd ed. *Mem. Bot. Surv. S. Afr.*, **40**, 1–128.

Allan, D. G., Batchelor, G. R. and Tarboton, W. R. (1983). Breeding of Botha's Lark. *Ostrich* **54**, 55–57.

Ash, J. S. (1981). Field description of the Obbia Lark *Calandrella obbiensis*, its breeding and distribution. *Bull. Br. Orn. Club* **101**, 379–383.

Ash, J. S. and Olson, S. L. (1985). A second specimen of *Mirafra* (*Heteromirafra*) *sidamoensis* Erard. *Bull. Br. Orn. Club* **105**, 141–143.

Ash, J. S. and Gullick, T. M. (1989). The present situation regarding the endemic breeding birds of Ethiopia. *Scopus* **13**, 90–96.

Ash, J. S. and Gullick, T. M. (1990). Field observations on the Degodi Lark *Mirafra degodiensis*. *Bull. Br. Orn. Club* **110**, 90–93.

Aspinwall, D. R. (1979). Bird notes from Zambesi district, NW Province. *Zambian Orn. Soc. Occ. Pap.* **2**, 1–61.

Bates, G. L. (1934). Birds of the Southern Sahara and adjoining countries in French West Africa. *Ibis* Ser. **13**(4), 439–466.

Benson, C. W. (1946). Notes on the birds of Southern Abyssinia. *Ibis* **88**, 25–48.

Benson, C. W. (1956). New or unusual records from Northern Rhodesia. *Ibis* **98**, 595–605.

Benson, C. W. (1959). The Dusky Lark *Mirafra nigricans* (Sundevall). *Bull. Br. Orn. Club* **79**, 124–127.

Benson, C. W. (1963). The breeding seasons of birds in the Rhodesias and Nyasaland. *Proc. XIII Int. Orn. Cong.*, pp. 623–639.

Benson, C. W. and Irwin, M. P. S. (1965a). Some birds from the north-western province, Zambia. *Arnoldia* **1**(29).

Benson, C. W. and Irwin, M. P. S. (1965b). The Grey-backed Sparrow-Lark *Eremopterix verticalis* (Smith). *Arnoldia* **1**(36).

Beshir, E. S. A. (1978). The Black-breasted Lark (*Melanocorypha bimaculata*), a pest of sorghum in Butana region, Gezira Province, Sudan. *Proc. vert. Pest Conf.* **8**, 220–223.

Blondel, J. (1962). Migration prénuptiale dans les Monts des Ksours (Sahara septentrional). *Alauda* **30**, 1–29.

Borrett, R. P. and Wilson, K. J. (1971). Notes on the food of the Redcapped Lark. *Ostrich* **42**, 37–40.

Boyer, H. J. (1988). Breeding biology of the Dune Lark. *Ostrich* **59**, 30–37.

Brooke, R. K. (1988). Is the Bimaculated Lark *Melanocorypha bimaculata* a valid member of the southern African avifauna? *Ostrich* **59**, 76–77.

Cañadas, S., Castro, H., Manrique, J. and Miralles, J. M. (1988). Datos sobre la reproducción de la Alondra de Dupont (*Chersophilus duponti*) en Almería. *Ardeola* **35**, 158–162.

Cardwell, P. and Dunning, J. (1971). The mimicking ability of the Clapper Lark. *Wits. B. C. News* **76**, 1.

Chittenden, H. N. and Batchelor, G. R. (1977). Nesting habits of the Chestnutbacked Finchlark. *Ostrich* **48**, 112.

Clancey, P. A. (1957). On the range and status of *Certhilauda falcirostris* Reichenow, 1916: Port Nolloth, N. W. Cape. *Bull. Br. Orn. Club* **77**, 133–137.

Clancey, P. A. (1966). Subspeciation in the southern African populations of the Sabota Lark *Mirafra sabota* Smith. *Ostrich* **37**, 207–213.

Clancey, P. A. (1967). Comments on *Ammomanes burra* (Bangs). *Bull. Br. Orn. Club* **87**, 13–14.

Clancey, P. A. (1968a). Seasonal movement and variation in the southern populations of the Dusky Lark *Pinarocorys nigricans* (Sundevall). *Bull. Br. Orn. Club* **88**, 166–171.

Clancey, P. A. (1968b). Subspeciation in the southern populations of the Chestnut-backed Finchlark *Eremopterix leucotis* (Stanley). *Arnoldia* **4**, 1–5.

Clapham, C. S. (1964). The birds of the Dahlac Archipelago. *Ibis* **106**, 376–388.

Clarke, J. E. (1980). The avifauna of Shaumari Wildlife Reserve, Jordan. *Sandgrouse* **1**, 50–70.

Colebrook-Robjent, J. F. R. (1988). Nest and eggs of the Angola Lark *Mirafra angolensis*. *Bull. Br. Orn. Club* **108**, 27–28.

Colston, P. R. (1982). A new species of *Mirafra* (Alaudidae) and new races of the Somali Long-billed Lark *Mirafra somalica*, Thekla Lark *Galerida malabarica* and Malindi Pipit *Anthus melindae* from southern coastal Somalia. *Bull. Br. Orn. Club* **102**, 106–114.

Cox, G. W. (1983). Foraging behaviour of the Dune Lark. *Ostrich* **54**, 113–120.

Currie, P. W. E. (1965). Nest and young of Bifasciated Lark *Certhilauda alaudipes*. *Ibis* **107**, 253.

David, J. H. M. (1971). A short note on the frequency of feeding of Thick-billed Lark nestlings. *Ostrich* **42**, 152.

Dean, W. R. J. (1989a). A review of the genera *Calandrella*, *Spizocorys* and *Eremalauda* (Alaudidae). *Bull. Br. Orn. Club* **109**, 95–100.

Dean, W. R. J. (1989b). The nest, egg and nestlings of the Red Lark *Certhilauda burra*. *Ostrich* **60**, 158.

Dean, W. R. J. and Colston, P. R. (1988). The taxonomy of Sclater's Lark *Spizocorys sclateri*. *Bull. Br. Orn. Club* **108**, 173–174.

Dean, W. R. J. and Hockey, P. A. R. (1989). An ecological perspective of lark (Alaudidae) distribution and diversity in the southwest-arid zone of Africa. *Ostrich* **60**, 27–34.

Dean, W. R. J., Milton, S. J., Watkeys, M. K. and Hockey, P. A. R. (1991). Distribution, habitat preference and conservation status of the Red Lark *Certhilauda burra* in Cape Province, South Africa. *Biological Conservation* **58**, 257–274.

Delius, J. D. (1963). Das Verhalten der Feldlerche. *Z. Tierpsychol.* **20**, 297–348.

Delius, J. D. (1965). A population study of Skylarks. *Ibis* **107**, 466–492.

Dorst, J. and Roux, F. (1972). Esquisse écologique sur l'avifaune des Monts du Balé, Ethiopie. *Oiseau et R.F.O.* **42**, 203–240.

Dowsett, R. J. (1979). Recent additions to the Zambian list. *Bull. Br. Orn. Club* **99**, 94–98.

Dowsett, R. J. (1983). Fischer's Finch Lark (*Eremopterix leucopareia*) in Kasungu District. *Nyala* **9**, 57.

Dowsett-Lemaire, F. and Dowsett, R. J. (1978). Vocal mimicry in the lark *Mirafra hypermetra* as a possible species isolating mechanism. *Bull. Br. Orn. Club* **98**, 140–144.

Erard, C. (1975a). Une nouvelle alouette du sud de l'Ethiopie. *Alauda* **43**, 115–124.

Erard, C. (1975b). Variation géographique de *Mirafra gilletti* Sharpe: description d'une éspèce jumelle. *Oiseau et R.F.O.* **45**, 293–312.

Erard, C. and Jarry, G. (1973). A new race of Thekla Lark in Harrar, Ethiopia. *Bull. Br. Orn. Club* **93**, 139–140.

Erard, C. and Naurois, R. de (1973). A new race of Thekla Lark in Bale, Ethiopia. *Bull. Br. Orn. Club* **93**, 73–76.

Erlanger, C. F. von (1907). Beiträge zur Vogelfauna Nordostafrikas. *J. Orn.* **55**, 42–49.

Friedmann, H. (1930). A lark new to science from north-central Kenya Colony. *Auk* **47**, 418–419.

Fry, C. H. (1966). Crested Lark feeding on ant-lion larvae. *Bull. Niger. Orn. Soc.* **3**, 47.

Gallagher, M. D. and Rogers, T. D. (1978). On the breeding birds of Bahrein. *Bonn. zool. Beitr.* **29**, 5–17.

Gaston, A. J. (1970). Birds in the central Sahara in winter. *Bull. Br. Orn. Club* **90**, 53–66.

Goodman, S. M., Meininger, P. L. and Mullié, W. C. (1986). The birds of the Egyptian Western Desert. *Misc. Publ. Mus. zool. Mich.* **172**, 1–91.

Guichard, K. M. (1955). The birds of Fezzan and Tibesti. *Ibis* **97**, 393–424.

Haas, W. (1969). Observations ornithologiques dans le nord-ouest de l'Afrique. *Alauda* **37**, 28–36.

Hall, B. P. (1961). The status of *Mirafra pulpa* and *Mirafra candida*. *Bull. Br. Orn. Club* **81**, 108–111.

Harrison, C. J. O. (1966). The validity of some genera of larks (Alaudidae). *Ibis* **108**, 573–583.

Hazevoet, C. J. (1989). Wing-clapping display of Dupont's Lark *Chersophilus duponti*. *Bull. Br. Orn. Club* **109**, 181–183.

Hegazi, E. M. (1981). A study of the amount of some invertebrates that are eaten by wild birds in the Egyptian western desert. *J. Agric. Sci.* **96**, 497–501.

Hilgert, C. (1908). 'Katalog der Collection von Erlanger in Nieder-Ingelheim a. Rh.' Berlin, Friedlander und Sohn.

Hockey, P. A. R. and Sinclair, J. C. (1981). The nest and systematic position of Sclater's Lark. *Ostrich* **52**, 256–257.

Hockey, P. A. R., Allan, D. G., Rebelo, A. G. and Dean, W. R. J. (1988). The distribution, habitat requirements and conservation status of Rudd's Lark *Heteromirafra ruddi* in South Africa. *Biol. Conser.* **45**, 255–266.

Hollom, P. A. D., Porter, R. F., Christensen, S. and Willis, I. (1988). 'Birds of the Middle East and North Africa. A companion guide'. T. and A. D. Poyser, Calton, England.

Hustler, C. W. (1980). The song of the Short-clawed Lark *Mirafra chuana*. Unpublished B Sc Honours dissertation, University of Witwatersrand, Johannesburg.

Hustler, K. (1985). First breeding record of the Short-clawed Lark. *Honeyguide* **31**, 109–111.

Irwin, M. P. S. (1982). The status of the Chestnut-backed Finch Lark in the middle Zambezi and Luangwa valleys. *Honeyguide* **111/112**, 20–21.

Irwin, M. P. S. and Lorber, P. (1983). The breeding season of the Chestnut-backed Finch Lark in Zimbabwe. *Honeyguide* **114/115**, 22–24.

Jennings, M. C. (1980). Breeding birds of central Arabia. *Sandgrouse* **1**, 71–81.

Karcher, F. and Medland, R. D. (1989). Preferred habitat of Fischer's Finchlark. *Nyala* **14**, 45.

Keith, G. S. and Twomey, A. (1968). New distributional records of some East African birds. *Ibis* **110**, 537–548.

Lack, P. C. (1976). The first recorded nest of the Pink-breasted Lark *Mirafra poecilosterna* (Reichenow). *Bull. Br. Orn. Club* **96**, 111–112.

Lack, P. C. (1977). The status of Friedmann's Bush Lark *Mirafra pulpa*. *Scopus* **1**, 34–39.

Lack, P. C., Leuthold, W. and Smeenk, C. (1980). Check-list of the birds of Tsavo East National Park, Kenya. *J.E. Afr. Nat. Hist. Soc.* **170**, 1–25.

Lawson, W. J. (1961). The races of the Karoo Lark *Certhilauda albescens* (Lafresnaye). *Ostrich* **32**, 64–74.

Le Berre, M. and Rostan, J. C. (1976). Inventaire de l'avifaune d'une zone de mise en valeur agricole dans la Constantinois. *Bull. Soc. Hist. nat. Afr. Nord*, 243–270.

Lynes, H. (1912). Bird-notes in two Andalucian sierras. *Ibis* **6**(9), 454–489.

Lynes, H. (1920). Ornithology of the Moroccan 'Middle-Atlas'. *Ibis* **11**(2), 260–301.

MacDonald, J. D. (1953). Taxonomy of the Karroo and Redback Larks of western South Africa. *Bull. Brit. Mus. (Nat. Hist.), Zool. Ser.* **1**, 321–350.

Mackowicz, R. (1970). Biology of the woodlark *Lullula arborea* (Linnaeus, 1758) (Aves) in the Rzepin Forest (western Poland). *Acta zool. Cracov.* **15**, 61–160.

Maclean, G. L. (1970). Breeding behaviour of larks in the Kalahari Sandveld. *Ann. Natal Mus.* **20**, 381–401.

Meinertzhagen, R. (1951). Review of the Alaudidae. *Proc. Zool. Soc. Lond.* **121**, 81–132.

Miskell, J. E. and Ash, J. S. (1985). Gillett's Lark *Mirafra gilletti* new to Kenya. *Scopus* **9**, 53–54.

Morel, G. J. and Morel, M. Y. (1984). *Eremopterix nigriceps albifrons* et *Eremopterix leucotis melanocephala* (Alaudidés) au Sénégal. *Proc. V Pan-Afr. Orn. Congr.*, pp. 309–322.

Morgan, J. H. and Palfery, J. (1986). Some notes on the Black-crowned Finch Lark. *Sandgrouse* **8**, 58–73.

Mullié, W. C. and Keith, J. O. (in press). Notes on the breeding biology, food and weight of the Singing Bush-Lark *Mirafra javanica* in northern Senegal. *Malimbus*.

Mundy, P. J. and Cook, A. W. (1972). Birds of Sokoto. *Bull. Niger. Orn. Soc.* **9**, 26–47, 61–76.

Myburgh, N. and Steyn, P. (1989). Notes on a Red Lark's nest. *Birding in Southern Africa* **41**, 114–115.

de Naurois, R. (1974). Découverte de la reproduction d'*Eremalauda dunni* dans le Zemmour (Mauritanie septentrionale). *Alauda* **42**, 111–116.

Nelson, J. B. (1973). 'Azraq: Desert Oasis'. Allen Lane, London.

Niethammer, G. von (1954). Winterliche 'Männchenpaare' in der algerischen Sahara. *Vogelwarte* **17**, 194–196.

Niethammer, G. von (1969). Am neste der Namiblerche *Ammomanes grayi*. *J. Orn.* **110**, 503–504.

Norris, A. S. (1964). Nest and young of Bifasciated Lark *Certhilauda alaudipes*. *Ibis* **106**, 531–532.

North, M. E. W. and McChesney, D. S. (1964). 'More Voices of African Birds.' Houghton Mifflin Co., Boston.

Pasteur, G. (1958). Notes sur la biologie de la reproduction du Cochevis de Thekla. *Oiseau et R.F.O.* **28**, 73–76.

Payne, R. B. (1978). Local dialects in the wingflaps of Flappet Lark *Mirafra rufocinnamomea*. *Ibis* **120**, 204–207.

Reynolds, J. F. (1977). Thermo-regulatory problems of birds nesting in arid areas in East Africa: a review. *Scopus* **1**, 57–68.

Riley, S. (1989). Crested Lark using 'anvil'. *British Birds* **82**, 30–31.

Roberts, A. (1937). Some results of the Barlow-Transvaal Museum expedition to South-west Africa. *Ostrich* **8**, 84–111.

Safriel, U. N. (1990). Winter foraging behaviour of the Dune Lark in the Namib Desert, and the effect of prolonged drought on behaviour and population size. *Ostrich* **61**, 77–80.

Schmutterer, H. (1969). 'Pests of Crops in Northeast and Central Africa with particular reference to the Sudan.' Gustav Fischer Verlag, Stuttgart.

Serle, W. (1943). Notes on East African birds. *Ibis* **85**, 55–82.

Shannon, G. R. (1974). Studies of less familiar birds 174. Shore Lark and Temminck's Horned Lark. *British Birds* **67**, 502–511.

Shuel, R. (1938). Notes on the breeding habits of birds near Zaria, N. Nigeria, with descriptions of their nests and eggs. *Ibis* **14**, 2, 230–244.

Simmons, K. E. L. (1952). Social behaviour in the Desert Lark *Ammomanes deserti* (Licht.). *Ardea* **40**, 67–72.

Skead, D. M. (1975). Drinking habits of birds in the central Transvaal bushveld. *Ostrich* **46**, 139–146.

Steyn, P. (1988). Co-operative breeding in the Spikeheeled Lark. *Ostrich* **59**, 182.

Steyn, P. and Myburgh, N. (1989). Notes on Sclater's Lark. *Birding in Southern Africa* **41**, 67–69.

Suárez, F., Santos, T. and Tellería, J. L. (1982). The status of Dupont's Lark, *Chersophilus duponti*, in the Iberian Peninsula. *Gerfaut* **72**, 231–235.

Symmes, T. C. L. (1960). Notes on the Rufous-naped Lark (*Mirafra africana*) and the Red-capped Lark (*Calandrella cinerea*). *N. Rhod. J.* **4**, 377–381.

Tarboton, W. R. (1980). Avian populations in Transvaal savanna. *Proc. IV Pan-Afr. Orn. Congr.* 113–124.

Tellería, J. L. (1981). 'La migracion de las Aves en el Estrecho de Gibraltar', 2. 'Aves no Planeadoras'. Universidad Complutense, Madrid.

Thomson, W. R. (1983). On the Monotonous or White-tailed Bush Lark in Zimbabwe. *Honeyguide* **113**, 21–22.

Tomlinson, W. (1950). Some notes chiefly from the Northern Frontier District of Kenya. Part II. *J. E. Afr. Nat. Hist. Soc.* **19**, 225–250.

Took, J. M. E. (1972). Breeding records 1971. *Cyprus Orn. Soc. (1957) Ann. Rep.* **18**, 40–49.

Turner, D. A. (1985). On the claimed occurrence of the Spike-heeled Lark *Chersomanes albofasciata* in Kenya. *Scopus* **9**, 142.

Valverde, J. A. (1957). 'Aves del Sahara Español'. Madrid.

Vernon, C. J. (1973). Vocal imitation by southern African birds. *Ostrich* **44**, 23–30.

Vernon, C. J. (1983). Notes on the Monotonous or White-tailed Bush Lark in Zimbabwe. *Honeyguide* **113**, 19–20.

Vernon, C. J. and Dean, W. R. J. (1988). Further African bird-mammal feeding associations. *Ostrich* **59**, 38–39.

Vesey-FitzGerald, D. F. (1957). Nest of *Mirafra albicauda* White. *Bull. Br. Orn. Club* **77**, 23–24.

Walker, F. J. (1981). Notes on the birds of Dhofar, Oman. *Sandgrouse* **2**, 56–85.

Wallace, D. I. M. (1983). The breeding birds of the Azraq Oasis and its desert surround, Jordan, in the mid-1960s. *Sandgrouse* **5**, 1–18.

White, C. M. N. (1960a). The Ethiopian and allied forms of *Calandrella cinerea* (Gmelin). *Bull. Br. Orn. Club* **80**, 24–35.

White, C. M. N. (1960b). The Somali forms of *Calandrella rufescens*. *Bull. Br. Orn. Club* **80**, 132.

Willoughby, E. J. (1968). Water economy of the Stark's Lark and Grey-backed Finch-Lark from the Namib Desert of South West Africa. *Comp. Biochem. Physiol.* **27**, 723–745.

Willoughby, E. J. (1971). Biology of larks (Aves: Alaudidae) in the central Namib Desert. *Zool. Afr.* **6**, 133–176.

Winterbottom, J. M. (1967). Systematic notes on the birds of the Cape Province XXIX. The status of *Ammomanes burra* (Bangs). *Ostrich* **38**, 156–157.

Winterbottom, J. M. and Wilson, A. H. (1959). Notes on the breeding of the Red-capped Lark *Calandrella cinerea* (Gmel.) at Cape Town. *Ostrich* (Suppl.) **3**, 289–299.

Zedlitz, O. G. (1909). Ornithologische Beobachtungen aus Tunesien, speziell dem Chott-Gebiete. *J. Orn.* **57**, 121–211, 241–322.

Family HIRUNDINIDAE: swallows and martins

Acland, C. M. (1966). Grey-rumped Swallow *Pseudhirundo griseopyga*, seen at sea. *Ostrich* **37**, 56–57.

Allan, D. G. (1988). The Blue Swallow in with a chance. *Quagga* **22**, 5–7.

Allan, D. G., Gamble, K., Johnson, D. N., Parker, V., Tarboton, W. R. and Ward, D. M. (1987). Report on the Blue Swallow in South Africa and Swaziland. PFIAO, University of Cape Town, 41 pp.

Anon. (1967). European Swallows ringed with British rings recovered in the south west Cape Province. *Ostrich* **38**, 203.

Ash, J. S. (1969). Spring weights of trans-Saharan migrants in Morocco. *Ibis* **111**, 1–10.

Ash, J. S. (1983). Over fifty additions of birds to the Somalia list including two hybrids, together with notes from Ethiopia and Kenya. *Scopus* **7**, 54–79.

Ash, J. S. and Gullick, T. M. (1989). The present situation regarding the endemic breeding birds of Ethiopia. *Scopus* **13**, 90–96.

Ash, J. S., Ferguson-Lees, I. J. and Fry, C. H. (1967). B.O.U. expedition to Lake Chad, northern Nigeria, Mar–Apr. 1967: preliminary report. *Ibis* **109**, 478–486.

Ashford, R. W. (1968). Preuss's Cliff Swallow *Lecythoplastes preussi* breeding in western Nigeria. *Bull. Niger. Orn. Soc.* **5**, 42–44.

Aspinwall, D. R. (1977). Movement analysis charts (Lesser Striped Swallow). *Zambian Orn. Soc. Newsletter* **7**, 78–79.

Aspinwall, D. R. (1979a). Comments on White-throated Swallow (October 1976–September 1978). *Zambian Orn. Soc. Newsletter.* **9**, 72–74.

Aspinwall, D. R. (1979b). Comments on Pearl-breasted Swallow (October 1976–September 1978). *Zambian Orn. Soc. Newsletter* **9**, 168–169.

Aspinwall, D. R. (1980a). Comments on Grey-rumped Swallow (October 1976–September 1978). *Zambian Orn. Soc. Newsletter* **10**, 59–60.

Aspinwall, D. R. (1980b). Comments on Mosque Swallow (October 1976–September 1978). *Zambian Orn. Soc. Newsletter* **10**, 130–132.

Aspinwall, D. R. (1980c). Comments on Wire-tailed Swallow (October 1976–September 1978). *Zambian Orn. Soc. Newsletter* **10**, 147–148.

Aspinwall, D. R. (1980d). Comments on Red-breasted Swallow (October 1976–September 1978). *Zambian Orn. Soc. Newsletter* **10**, 166–167.

Aspinwall, D. R. (1981a). Comments on Lesser Striped Swallow (October 1976–September 1978, January 1979–January 1981). *Zambian Orn. Soc. Newsletter* **11**, 102–103.

Aspinwall, D. R. (1981b). Comments on African Saw-wing (October 1976–September 1978, January 1979–January 1981). *Zambian Orn. Soc. Newsletter* **11**, 149–150.

Aspinwall, D. R. (1982). Comments on Angola Swallow. *Zambian Orn. Soc. Newsletter* **12**, 118–120.

Aspinwall, D. R. (1983). Movement analysis charts. Comments on Banded Sand Martin. *Zambian Orn. Soc. Newsletter* **13**, 55–56.

Backhurst, G. (1974). Aerial roosting (*Delichon urbica*). *Bull. E. Afr. Nat. Hist. Soc.* **1974**, 134–135.

Backhurst, G. (1981). Eastern African ringing report 1977–1981. *J.E. Afr. Nat. Hist. Soc.* **174**, 1–19.

Backhurst, G. (1988). Eastern African ringing report 1981–1987. *Scopus* **12**, 1–52.

Batten, T. A. (1943). The Rock Martin (*Ptyonoprogne fuligula*). *Ostrich* **13**, 238–239.

Bennun, L. A., Gichuki, G., Darlington, J. and Ng'weno, F. (1986). The avifauna of Ol Doinyo Orok, a forest island: initial findings. *Scopus* **10**, 83–86.

Benson, C. W. (1942). A new species and ten new races from Southern Abyssinia. *Bull. Br. Orn. Club* **63**, 8–19.

Benson, C. W. (1946). Notes on the birds of Southern Abyssinia. *Ibis* **88**, 180–205.

Benson, C. W. (1949). The systematics and migrations of the Pearl-breasted Swallow. *Ostrich* **29**, 137–145.

Benson, C. W. (1956). New or unusual records from Northern Rhodesia. *Ibis* **98**, 595–605.

Benson, C. W. and Irwin, M. P. S. (1967). A contribution to the ornithology of Zambia. *Zambia Mus. Pap.* **1**, 139 pp.

Benson, C. W., Irwin, M. P. S. and White, C. M. N. (1962). The significance of valleys as avian geographical barriers. *Ann. Cap. Prov. Mus.* **II**, 155–189.

Benson, C. W., Irwin, M. P. S., Brooke, R. K., Dowsett, R. J. and Irwin, M. P. S. (1970). Notes on the birds of Zambia: Part V. *Arnoldia (Rhod.)* **4**(40), 1–59.

Best, J. R. (1977). A large hirundine roost in Uasin Gishu. *Bull. E. Afr. Nat. Hist. Soc.* **1977**, 39–40.

Bilby, H. A. (1957). Little recorded behaviour of the swallow family. *Bull. Br. Orn. Club* **77**, 5–7.

Bont, A. F. de (1957). Notes sur l'Hirondelle de cheminée, *Hirundo rustica* L., dans son quartier d'hiver. *Gerfaut* **47**, 127–133.

Bont, A. F. de (1960). Résultats du baguage d'oiseaux au Congo Belge et au Ruanda-Urundi, Exercise 1956–1959. *Gerfaut* **50**, 41–47.

Bont, A. F. de (1962). Composition des bandes d'hirondelles de cheminée *Hirundo rustica rustica* L. hivernant au Katanga, et analyse de la mue des rémiges primaires. *Gerfaut* **2**, 298–343.

Botha, M. C. (1988). Migrations of European Swallows at Hermanus. *Promerops* **185**, 15.

Bowen, P. St J. (1979a). The spread of the Red-throated Cliff Swallow (*Hirundo rufigula*) in North-Western Province. *Bull. Zamb. Orn. Soc.* **11**, 24–25.

Bowen, P. St J. (1979b). Tree-nesting by the Lesser Striped Swallow. *Bull. Zambian Orn. Soc.* **11**, 41–43.

Bowen, P. St J. (1983a). A second sight record of the South African Cliff Swallow *Hirundo spilodera* from Zambia. *Bull. Zambian Orn. Soc.* **13–15**, 122–123.

Bowen, P. St J. (1983b). Some observations on the Black-and-Rufous Swallow *Hirundo nigrorufa* in Zambia. *Bull. Zambian Orn. Soc.* **13–15**, 23–35.

Bowen, P. St J. and Colebrook-Robjent, J. F. R. (1984). The nest and eggs of the Black-and-Rufous Swallow *Hirundo nigrorufa*. *Bull. Br. Orn. Club* **104**, 146–147.

Broadbent, J. (1969). A note on the Preuss's Cliff-Swallows *Lecythoplastes preussi* at Oyo New Reservoir. *Bull. Niger. Orn. Soc.* **6**, 34.

Broekhuysen, G. J. (1952). *Hirundo rustica* feeding on Amphipoda. *Ostrich* **23**, 134–135.

Broekhuysen, G. J. (1953). A post mortem of the Hirundinidae which perished at Somerset West in April 1953. *Ostrich* **24**, 148–152.

Broekhuysen, G. J. (1961). Interesting roosting of European Swallow. *Bokmakierie* **13**, 5–6.

Broekhuysen, G. J. (1967). Bird migration in the most southern part of the African continent. *Vogelwarte* **24**, 6–15.

Broekhuysen, G. J. (1974). Third report on migration in south Africa. *Ostrich* **45**, 235–250.

Broekhuysen, G. J. and Brown, A. R. (1963). The moulting pattern of European Swallows, *Hirundo rustica*, wintering in the surroundings of Cape Town, South Africa. *Ardea* **51**, 25–43.

Broekhuysen, G. J. and Stanford, W. P. (1954). Display in S. A. Sand Martin (*Riparia paludicola paludicola*). *Ostrich* **25**, 99.

Brooke, R. K. (1956). Food of the European Swallow. *Ostrich* **27**, 88.

Brooke, R. K. (1966). Distribution and breeding notes on the birds of the central frontier of Rhodesia and Mozambique. *Ann. Natal Mus.* **18**(2), 429–453.

Brooke, R. K. (1971a). Breeding and breeding season notes on the birds of Mzimbiti and adjacent low-lying areas of Moçambique. *Ann. Natal Mus.* **21**, 55–69.

Brooke, R. K. (1971b). Field guide to the swallows of Rhodesia and adjacent areas. *Honeyguide* **66**, 19–26.

Brooke, R. K. (1972). Generic limits in old world Apodidae and Hirundinidae. *Bull. Br. Orn. Club* **92**, 53–57.

Brooke, R. K. (1974). Birds and bridges in Rhodesia. *Honeyguide* **80**, 42–45.

Brooke, R. K. (1975). *Cotyle paludibula* Rüppell, 1835. *Bull. Br. Orn. Club* **95**, 90.

Brooke, R. K. and Ryan, P. (1988). South African Cliff Swallow (528) colonies in the western Karroo. *Promerops* **185**, 15.

Brooke, R. K. and Vernon, C. J. (1961). Aspects of the breeding biology of the Rock Martin. *Ostrich* **32**, 51–52.

Brosset, A. and Erard, C. (1977). New faunistic records from Gabon. *Bull. Br. Orn. Club* **97**, 125–132.

van Bruggen, A. C. (1961). The Chirinda Forest, Mount Selinda, a montane rain forest in Southern Rhodesia. *Kurgl. Fysiografiska Sällskapets I Lund Förhandlingar* **31**(7), 61–75.

Bryant, D. M. (1975). Breeding biology of House Martins *Delichon urbica* in relation to insect abundance. *Ibis* **117**, 180–216.

Bub, H. and Klings, M. (1968). Ringfunde nord- und westdeutscher Uferschwalben *Riparia riparia*. *Auspicium* **3**, 69–95.

Bub, H., Eck, S. and Herroelen, P. (1981). 'Stelzen, Pieper und Würger'. Wittenberg Lutherstadt.

Burgerjon, J. J. (1964). Some census notes on a colony of South African Cliff Swallows (*Petrochelidon spilodera* (Sundevall)). *Ostrich* **35**, 77–85.

Butynski, T. (1987). Summary of nesting by Angola Swallows (*Hirundo angolensis*) at Ruhiza Forest Station (1986–87), Impenetrable (Bwindi) Forest Reserve (7,500 ft. a.s.l.). Unpub. MS, 2 pp.

Campbell, L. (1977). Interesting bird recovery – Red-rumped Swallow *Hirundo daurica* recovered after eight years. *E. Afr. Nat. Hist. Soc. Bull.* Nov./Dec. **1977**, 134.

Cannell, I. C. (1968). Notes from Angola. *Ostrich* **39**, 264–265.

Carr, B. A. (1984). Nest eviction of Rock Martins by Little Swifts. *Ostrich* 55, 223-224.
Chapin, J. P. (1925). A new swallow from Cameroon. *Ibis*, 148-151.
Chapin, J. P. (1948). Field notes on *Petrochelidon fuliginosa*. *Ibis* 90, 474-476.
Chapin, J. P. (1954). The African River Martin and its migration. *Annl. Mus. r. Congo belge* Sér. 4, 9-15.
Christy, P. (1984). L'Hirondelle des rochers d'Angola (*Petrochelidon rufigula*) au Gabon. *Oiseau et R.F.O.* 54, 362-363.
Claassen, J. (1991). Perelborsswaelnes met 6 eiers. *Promerops* 197, 11.
Clancey, P. A. (1965). Wintering European Swallows occurring in a city centre. *Ostrich* 36, 229-230.
Clancey, P. A. (1966). The Cliff Swallow in South West Africa. *Ostrich* 37, 197.
Clancey, P. A. (1969). Miscellaneous taxonomic notes on African birds, 27. *Durban Mus. Novit.* 8(15), 227-274.
Clancey, P. A. (1970). Miscellaneous taxonomic notes on African birds XXVIII. The races of the European Swallow wintering in southern and eastern Africa. *Durban Mus. Novit.* 8(17), 326-331.
Clancey, P. A. (1982). Miscellaneous taxonomic notes on African birds. *Durban Mus. Novit.* 13(6), 55-63.
Clancey, P. A. and Irwin, M. P. S. (1966). The South African races of the Banded Sand Martin *Riparia cincta* (Boddaert). *Durban Mus. Novit.* 8(3), 25-33.
Clancey, P. A., Lawson, W. J. and Irwin, M. P. S. (1969). The Mascarene Martin *Phedina borbonica* (Gmelin) in Mozambique: a new species to the South African list. *Ostrich* 40, 5-8.
Colebrook-Robjent, J. F. R. (1973). Some breeding records of birds in Zambia. *Zambia Mus. J.* 4, 7-18.
Colebrook-Robjent, J. F. R. (1976). Undescribed nests and eggs of birds breeding in Zambia. *Bull. Zambian Orn. Soc.* 8, 45-56.
Cramp, S. (1970). Studies of less familiar birds, 159. Crag Martin. *British Birds* 63, 239-243.
Crick, H. Q. P. and Marshall, P. J. (1981). The birds of Yankari Game Reserve, Nigeria: their abundance and seasonal occurrence. *Malimbus* 3, 103-114.
Critchley, R. A. (1975). Banded Sand Martin caught in sticky weed. *Bull. Zamb. Orn. Soc.* 7(2), 103.
Curry-Lindahl, K. (1963). Roosts of Swallows (*Hirundo rustica*) and House Martins (*Delichon urbica*) during the migration in tropical Africa. *Ostrich* 34, 99-101.
Davis, P. (1965). Recoveries of Swallows ringed in Britain and Ireland. *Bird Study* 12, 151-169.
Dean, W. R. J. (1988). Birds associating with fire at Nylsvlei Nature Reserve, Transvaal. *Ostrich* 58, 103-106.
Dean, W. R. J. (1989). Pearlbreasted Swallow apparently eating lime. *Birding in Southern Africa* 41(3), 88.
Delacour, J. (1932). Les oiseaux de la mission zoologique franco-anglo-américaine à Madagascar. *Oiseau et R.F.O.*, 2, 1-96.
Dial, K. P. and Vaughn, T. A. (1987). Opportunistic predation on alate termites in Kenya. *Biotropica* 19, 185-187.
Donnelly, B. G. (1966). Grey-rumped Swallows at sea. *Ostrich* 37, 227.
Donnelly, B. G. (1974). Vertical zonation of breeding swallows and swifts at Kariba, Rhodesia. *Ostrich* 45(4), 256-258.
Douaud, J. (1957). Les migrations au Togo (Afrique Occidentale). *Alauda* 25, 241-266.
Dowsett, R. J. (1966). The moulting pattern of European Swallows, *Hirundo rustica*, wintering in eastern Zambia. *Puku* 4, 91-100.
Dowsett, R. J. (1971). Suspended wing-moult of migrants. *Bird Study* 18, 53-54.
Dowsett, R. J. (1972). Geographical variation in *Pseudhirundo griseopyga*. *Bull. Br. Orn. Club* 92, 97-100.
Dowsett, R. J. (1978). A hybrid *Hirundo rustica* × *Delichon urbica* in Zambia. *Bull. Br. Orn. Club* 98, 113-114.
Dowsett, R. J. (1979). Sight record of a South African Cliff Swallow in Mwinilunga District. *Bull. Zambian Orn. Soc.* 11, 32.
Drost, R. and Schüz, E. (1952). Europäische Rauchschwalben (*Hirundo rustica*) in Afrika. *Vogelwarte* 16, 95-98.
Dunning, J. B. (1989). Gregariousness of Blue Swallows *Hirundo atrocaerulea* during nest building. *Ostrich* 60, 135-136.
Dyer, M. (1988). Nuptial flight display of the Blue Swallow (*Hirundo atrocaerulea*). *Nyala* 12(1-2), 27-30.
Dyson, W. G. (1976). Twice-yearly breeding of small birds at Muguga, Kenya. *Bull. E. Afr. Nat. Hist. Soc.* 1976, 132-134.
Earlé, R. A. (1984). 'Fishing' for Cliff Swallows. *Safring News* 13, 6-9.
Earlé, R. A. (1985a). Foraging behaviour and diet of the South African Cliff Swallow *Hirundo spilodera* (Aves: Hirundinidae). *Navors. nas. Mus., Bloemfontein* 5(4), 53-66.
Earlé, R. A. (1985b). Predators, parasites and symbionts of the South African Cliff Swallow *Hirundo spilodera* (Aves: Hirundinidae). *Navors. nas. Mus., Bloemfontein* 5, 1-18.
Earlé, R. A. (1985c). The nest of the South African cliff swallow *Hirundo spilodera* (Aves: Hirundinidae). *Navors. nas. Mus., Bloemfontein* 5(2), 21-36.
Earlé, R. A. (1985d). Ageing and sexing guide: South African Cliff Swallow *Hirundo spilodera*. *Safring News* 14, 48.
Earlé, R. A. (1985e). A description of the social, aggressive and maintenance behaviour of the South African Cliff Swallow *Hirundo spilodera* (Aves: Hirundinidae). *Navors. nas. Mus., Bloemfontein* 5(3), 37-50.
Earlé, R. A. (1985f). The Biology of the South African Cliff Swallow *Hirundo spilodera*. Unpubl. Ph.D. thesis, Rhodes University, pp. 154.
Earlé, R. A. (1986a). The breeding biology of the South African cliff swallow. *Ostrich* 57, 138-156.
Earlé, R. A. (1986b). Time budget of South African cliff swallows during breeding. *S. Afr. J. Zool.* 21, 57-59.
Earlé, R. A. (1986c). Dimensions and deformities of South African Cliff Swallows. *Ostrich* 57, 56-59.
Earlé, R. A. (1986d). Vocalizations of the South African Cliff Swallow (*Hirundo spilodera*). *S. Afr. J. Zool.* 21, 229-232.
Earlé, R. A. (1987a). Homing ability of the South African Cliff Swallow. *Safring News* 16, 3.
Earlé, R. A. (1987b). Distribution, migration and timing of moult in the South African Cliff Swallow. *Ostrich* 58, 118-121.
Earlé, R. A. (1987c). Moult and breeding seasons of the Greyrumped Swallow. *Ostrich* 58, 181-182.
Earlé, R. A. (1987d). A case of bigamy in the Redbreasted swallow *Hirundo semirufa*. *S. Afr. J. Zool.* 22, 325-325.
Earlé, R. A. (1987e). Measurements, moult and timing of breeding in the Blue Swallow. *Ostrich* 58, 182-185.
Earlé, R. A. (1987f). Nest eviction of Greater Striped Swallows by Anteating Chats. *Mirafra* 4, 35-36.
Earlé, R. A. (1987g). Ringing and recovery details of four southern African swallow species. *Safring News* 16, 67-72.
Earlé, R. A. (1987h). Notes on *Hirundo fuliginosa* and its status as a 'cliff swallow'. *Bull. Br. Orn. Club* 107, 59-63.
Earlé, R. A. (1988a). The timing of breeding and moult in the Lesser Striped Swallow *Hirundo abyssinica*. *Ibis* 130, 378-383.
Earlé, R. A. (1988b). Timing of breeding and moult in three African swallows. *J. Afr. Zool.* 102, 61-70.
Earlé, R. A. (1989). Breeding biology of the Redbreasted Swallow *Hirundo semirufa*. *Ostrich* 60, 13-21.
Earlé, R. A. and Brooke, R. K. (1988). South African Cliff Swallows overwintering near Bloemfontein. *Mirafra* 5, 53.

Earlé, R. and Brooke, R. K. (1989). Taxonomy, distribution, migration and moult of the Redbreasted Swallow *Hirundo semirufa*. *Ostrich* **60**, 151–158.
van Ee, C. A. (1988). Stomach contents. *Mirafra* **5**, 27–28.
Eisentraut, M. (1963). 'Die Wirbeltiere des Kamerungebirges'. Verlag Paul Parey, Hamburg.
Elgood, J. H. (1965). The birds of the Obudu Plateau, Eastern Region of Nigeria. *Nigr. Field* **30**, 60–69.
Elkins, N. and Etheridge, B. (1974). The Crag Martin in winter quarters at Gibraltar. *British Birds* **67**, 376–387.
Elkins, N. and Etheridge, B. (1977). Further studies of wintering Crag Martins. *Ringing and Migr.* **1**, 158–165.
Erard, C. (1981). Sur les migrations de *Pseudochelidon eurystomina* Hartlaub au Gabon. *Oiseau et R.F.O.* **51**, 244–246.
Every, B. (1988). Unusual feeding habits of swallows. *Bokmakierie* **40**, 127.
Farina, A. (1978). Breeding biology of the Crag Martin *Hirundo rupestris*. *Avocetta* **2**, 35–46.
Feare, C. J. (1977). *Phedina borbonica madagascariensis* in the Amirantes. *Bull. Br. Orn. Club.* **97**, 87–88.
Fellowes, E. C. (1971). House Martins apparently roosting in nests of Striped Swallows. *British Birds* **64**, 460.
Field, G. D. (1968). Utilization of mangroves by birds on the Freetown peninsula, Sierra Leone. *Ibis* **110**, 354–357.
Finch, B. W. (1989). Blue Swallows *Hirundo atrocaerulea* near Busia: the second record for Kenya. *Scopus* **13**, 125–126.
Francis, D. M. (1980). Moult of European Swallows in central Zambia. *Ringing Migr.* **3**, 4–8.
Fry, C. H. (1973). The juvenile plumage of *Pseudhirundo griseopyga* and identity of 'Vom Swallows'. *Bull. Br. Orn. Club* **93**, 138–139.
Fry, C. H. and Smith, D. A. (1985). A new swallow from the Red Sea. *Ibis* **127**, 1–6.
Fry, C. H., Ash, J. S. and Ferguson-Lees, I. J. (1970). Spring weights of some Palearctic birds at Lake Chad. *Ibis* **112**, 58–82.
Ginn, H. B. and Melville, D. S. (1983). 'Moult in Birds'. *Br. Trust Orn.* Guide 19, BTO, Tring.
Godfrey, R. (1943). The South African Cliff Swallow. *Ostrich* **13**, 219–226.
Grant, L. and Lewis, A. D. (1984). Breeding of the Ethiopian Swallow *Hirundo aethiopica* in interior Africa. *Scopus* **8**, 67–72.
Grobler, N. and Jacobs, J. (1985). Swallow predation by leguan. *Mirafra* **2**, 3.
Hallet, A. F. and Brown, A. R. (1964). A method of trapping European Swallows. *Ostrich* **35**, 293–296.
Hanmer, D. B. (1976). Birds of the lower Zambezi. *Southern Birds* **2**, 1–66.
Hanmer, D. B. (1977). European Sand Martin (*Riparia riparia*) and African Sand Martin (*R. paludicola*). *Nyala* **3**, 44.
Hanmer, D. B. (1980). A coastal form of the African Saw-wing (*Psalidoprocne pristoptera holomelaena*) in southern Malaŵi. *Nyala* **6**(2), 134–135.
Hanmer, D. B. (1981). Longevity from retraps. *Safring News* **10**, 12–22.
Hanmer, D. B. (1989). The Nchalo ringing station – bird longevity and migrant return. *Nyala* **14**, 21–27.
Hanmer, D. B. (1990). Saw-wings down and up. *Honeyguide* **36**, 92–93.
Harding, D. P. and Harding, R.S.O. (1982). A preliminary checklist of birds in the Kilimi area of northwest Sierra Leone. *Malimbus* **4**, 64–68.
Harper, J. and Harper, L. (1974). Where do all the House Martins go? *E. Afr. Nat. Hist. Soc. Bull.* 1974, 113–115.
Herholdt, J. J. (1989). South African Cliff Swallows in the Kalahari. *Mirafra* **6**, 19.
Herroelen, P. (1960). De rui van de boerenzwaluw, *Hirundo rustica* L. in Belgisch-Congo. *Gerfaut* **50**, 87–99.
Heu, R. (1961). Observations ornithologiques au Ténéré. *Oiseau et R.F.O.* **31**, 214–239.
Hofmeyr, J. (1989). European Swallows feeding on rooikrans. *Promerops* **187**, 15–17.
Holman, F. C. (1947). Birds of Gold Coast. *Ibis* **89**, 623–650.
Hornby, H. E. (1973). Roosting of migrant Swallows. *Honeyguide* **74**, 30–31, 37.
Hull, R. L. (1944). Nesting of the Red-throated Rock-Martin *Ptyonoprogne r. rufigula*. *J. East Afr. Uganda Nat. Hist. Soc.* **18**, 94–95.
Ingram, C. (1974). 'The Migration of the Swallow'. Witherby, London.
Irwin, M. P. S. (1977). Variation, geographical arcs and geneflow within the populations of the Rock Martin *Hirundo* (*Ptyonoprogne*) *fuligula* in eastern, southern and southwestern Africa. *Honeyguide* **91**, 11–19.
Jackson, H. D. (1970). Swimming ability of the Barn Swallow. *Auk* **87**, 577.
Jourdain, F. C. R. and Shuel, R. (1935). Notes on a collection of eggs and breeding-habits of birds near Lokoja, Nigeria. *Ibis* **13**(5), 623–663.
Kang, K.-W. (1971). Foreign recoveries of House Swallows banded in eastern Asian countries. *Ostrich* **42**, 179–189.
Kasparek, M. (1976). Über Populationsunterschiede im Mauserverhalten der Rauchschwalbe (*Hirundo rustica*). *Vogelwelt* **97**, 121–132.
Lack, P. C. and Quicke, D. L. J. (1978). Dietary notes on some Kenyan birds. *Scopus* **2**, 86–91.
Lewis, A. D. (1982). The breeding of the Rufous-chested Swallow *Hirundo semirufa* in East Africa. *Scopus* **6**, 103–105.
Lewis, A. D. (1989). Seasonality of Banded Martin *Riparia cincta* flocks in Kenya. *Scopus* **12**, 100–101.
Lind, E. A. (1960). Zur Ethologie und Ökologie der Mehlschwalbe, *Delichon urbica* (L.) *Ann. zool. Soc. Vanamo* **21**, 1–122.
Lockhart, P. S. (1970). House Martins nesting at Somerset West, Cape. *Ostrich* **41**, 254–255.
Loske, K.-H. (1986). The origins of European Swallows wintering in Namibia and Botswana. *Ringing and Migr.* **7**, 119–121.
Loske, K.-H. and Lederer, W. (1988). Moult, weight and biometrical data for some Palaearctic passerine migrants in Zambia. *Ostrich* **59**, 1–7.
Lowe, P. R. (1938). Some anatomical notes on the genus *Pseudochelidon* Hartlaub with reference to its taxonomic position. *Ibis* **2**(3), 429–437.
Madge, S. C. and Redman, N. J. (1989). The existence of a form of cliff swallow, *Hirundo* sp. in Ethiopia. *Scopus* **13**, 126–129.
Mayr, E. and Bond, J. (1943). Notes on the generic classification of the swallows, Hirundinidae. *Ibis* **85**, 334–341.
Mclean, S. (1988). Lesser Striped Swallows feeding on fruit. *Bokmakierie* **40**, 21.
Mead, C. J. (1970). The winter quarters of British Swallows. *Bird Study* **17**, 229–240.
Mead, C. J. (1979). Mortality and causes of death in British Sand Martins. *Bird Study* **26**, 107–112.
Mead, C. J. and Clark, J. A. (1987). Report on bird-ringing for 1987. *Ringing and Migr.* **8**, 135–200.
Mead, C. J. and Clark, J. A. (1988). Report on bird-ringing in Britain and Ireland for 1987. *Ringing and Migr.* **9**, 169–204.
Mead, C. J. and Harrison, J. D. (1979). Overseas movements of British and Irish Sand Martins. *Bird Study* **26**, 87–98.
Medland, R. D. (1985). South African Cliff Swallow, *Hirundo spilodera*, in Malaŵi. *Nyala* **11**, 26–27.
Medland, R. D. (1988). Mascarene Martin, *Phedina borbonica*, near Chiromo. *Nyala* **121**, 73.
Medland, R. D. (1989a). The White-throated Swallow in Malaŵi. *Nyala* **14**(2), 125–127.
Medland, R. D. (1989b). White-throated Swallow *Hirundo*

albigularis in Malaŵi – II. *Vocifer* **9**, 6–8.
Medland, R. D. (1989c). Movement of Pearl-breasted Swallow. *Vocifer* **12**, 10–11.
Mendelsohn, H. P. (1973). Races of the European Swallows, *Hirundo rustica*. *Safring News* **2**, 21–22.
Mendelsohn, J. M. (1973). Some observations on age ratio, weight and moult of the European Swallow, *Hirundo rustica* L. in the central Transvaal (Aves: Hirundinidae). *Ann. Trans. Mus.* **28**(6), 79–89.
van der Merwe, F. (1986). Banded Sand Martins in the south-western Cape. *Promerops* **175**, 9.
Milstein, P. le S. and Maclean, G. (1982). Flocks of migrant swallows and martins. *Bokmakierie* **34**, 69.
Møller, A. P. (1982). Clutch size in relation to nest size in the Swallow *Hirundo rustica*. *Ibis* **124**, 339–343.
Møller, A. P. (1984a). Parental defence of offspring in the Barn Swallow. *Bird Behav.* **5**, 110–117.
Møller, A. P. (1984b). Geographical trends in breeding parameters of Swallow *Hirundo rustica* and House Martins *Delichon urbica*. *Orn. Scand.* **15**, 43–54.
Moreau, R. E. (1939a). Parental care by some African swallows and swifts. *Bull. Br. Orn. Club* **59**, 145–149.
Moreau, R. E. (1939b). Numerical data on African birds' behaviour at the nest: *Hirundo s. smithii* Leach, the Wire-tailed Swallow. *Proc. Zool. Soc. Lond.*, Ser. A, **109**(2–3), 109–125.
Moreau, R. E. (1940). Numerical data on African birds' behaviour at the nest. II. *Psalidoprocne holomelaena massaica* Neum., the Rough-wing Bank-Martin. *Ibis* **14**(4), 234–248.
Moreau, R. E. (1947). Relations between number in brood, feeding-rates and nestling period in nine species of birds in Tanganyika Territory. *J. Anim. Ecol.* **16**, 205–209.
Moreau, R. E. (1961). Problems of Mediterranean-Saharan migration. *Ibis* **103a**, 373–427, 580–623.
Moreau, R. E. and Moreau, W. M. (1940). Incubation and fledging periods of African birds. *Auk* **57**, 313–325.
Niederfriniger, O. (1973). Crag Martins nesting on buildings. *British Birds* **66**, 121–123.
Nyandoro, R. (1987). Comments on territorial behaviour and weights of Wire-tailed Swallows. *Honeyguide* **33**, 62–63.
Oates, H. (1991). A swallow tale. *Natal Midlands Bird Club Newsletter* **4**, 4–5.
Oatley, T. B. (1983). Twenty-third ringing report for southern Africa. *Ostrich* **39**, 141–149.
Pearson, D. J. (1971). Weights of some Palearctic migrants in southern Uganda. *Ibis* **113**, 173–184.
Penry, E. H. (1979). Thousands of European Swallows on the ground. *Bull. Zambian Orn. Soc.* **11**, 33–44.
Penry, E. H. (1986). The distribution and status of the South African Cliff Swallow *Hirundo spilodera* in Botswana. *Babbler* **12**, 9–13.
Penry, E. H. (1987). A Cliff Swallow in the river. *Mirafra* **4**, 40.
Persson, C. (1973). The migration of Sand Martins *Riparia riparia* from Denmark and southern Scania. *Dansk Orn. Foren. Tidsskr.* **67**, 25–34.
Phillips, A. R. (1973). On the supposed genus *Petrochelidon*. *Bull. Br. Orn. Club* **93**, 20.
Piper, S. E. (1974). Moult of the European Swallow. *Safring News* **3**(3), 24–32.
Pitman, C. R. S. (1931). The breeding habits and eggs of *Hirundo senegalensis senegalensis* (Linn.). *Ool. Rec.* **11**, 14–16.
Pringle, V. L. (1989). Nest oddity. *Birding in Southern Africa* **41**(4), 120.
Prodon, R. (1982). Sur la nidification, le régime alimentaire et les vocalisations de l'Hirondelle rousseline en France (*Hirundo daurica rufula* Temm.). *Alauda* **50**, 176–192.
Rand, A. L. (1936). The distribution and habits of Madagascar birds. A summary of the field notes of the Mission Zoologique Franco-Anglo-Américaine à Madagascar. *Bull. Am. Mus. Nat. Hist.* **72**(V), 143–499.
Rebollo, F. de L. (1980). Biologie de la reproduction de l'Hirondelle rousseline *Hirundo daurica* en Espagne. *Alauda* **48**, 99–112.
Reynolds, J. F. (1971). A thousand Swallows. *Bull. E. Afr. Nat. Hist. Soc.* **1971**, 70–71.
de Ridder, M. (1978). Observations d'oiseaux au Basse-Casamance. II. *Biol. Jb. Dodonaea* **46**, 115–127.
Riols, C. (1978). Précisions sur le passage de l'Hirondelle rousseline *Hirundo daurica* en Tunisie. *Alauda* **46**, 183.
Roberts, L. (1989). Movements of European Swallows in the Hermanus area. *Promerops* **187**, 17.
Rolfe, J. G. and Pearson, D. J. (1973). Some recent records of Palaearctic migrants from eastern Uganda. *E. Afr. Nat. Hist. Soc. Bull.* 1973, 62.
Roos, L. and Roos, M. (1989). Contribution to the diet of Greater Striped Swallows. *Birding in Southern Africa* **41**(3), 88.
Rowan, M. K. (1963). Range of the Cliff Swallow. *Ostrich* **34**, 181–182.
Rowan, M. K. (1968). The origins of European Swallows 'wintering' in South Africa. *Ostrich* **39**, 76–84.
Rudebeck, G. (1955a). Some observations at a roost of European Swallows and other birds in the south-eastern Transvaal. *Ibis* **97**, 572–580.
Rudebeck, G. (1955b). Studies of some palaearctic and arctic birds in their winter quarters in South Africa: The European Swallow (*Hirundo rustica* L.) *S. Afr. Animal Life* **4**, 459–472.
Ruthke, P. (1971). Nächtlicher Gesang von Einfarbstaren (*Sturnus unicolor*) am Schlafplatz. *Vogelwelt* **92**, 191–192.
Sassi, M. and Zimmer, F. (1941). Beiträge zur Kenntnis der Vogelwelt des Songea-Distriktes mit besonderer Berücksichtigung des Matengo-Hochlandes (D.O.A.). *Ann. Naturhist. Mus. Wien* **51**, 236–346.
Saunders, C. R. (1981). Black Saw-wing Swallow in Selukwe. *Honeyguide* **106**, 35.
Schmidt, R. K. (1959). Notes on the Pearl-breasted Swallow *Hirundo dimidiata* in the South-western Cape. *Ostrich* **30**, 155–158.
Schmidt, R. K. (1962). Breeding of the Larger Striped Swallow *Cecropis cucullata* in the South West Cape. *Ostrich* **23**, 3–8.
Schmidt, R. K. (1964). Incubation period of Rock Martin (*Ptyonoprogne fuligula* (Lichtenstein). *Ostrich* **35**, 122.
Schouteden, H. (1922). Note sur la découverte du nid de *Pseudochelidon eurystomina* Hartl. *Rev. Zool. Afr.* **10**, 323–328.
Scott, A. J. (1986). Breeding data – Wire-tailed Swallow *Hirundo smithii*. *Zamb. Orn. Soc. Newsletter* **16**(3), c–d.
Sessions, P. H. B. (1966). Notes on the birds of Lengetia Farm, Mau Narok. *J. E. Afr. Nat. Hist. Soc.* **26**, 18–48.
Shaw, J. R. (1979). Some notes on the fledging of Wiretailed Swallow. *Honeyguide* **98**, 35–37.
Short, L. L. and Horne, J. F. M. (1985). Notes on some birds of Ol Ari Nyiro, Laikipia Plateau. *Scopus* **9**, 137–140.
Sibley, C. G. and Ahlquist, J. E. (1982). The relationship of the swallows (Hirundinidae). *J. Yamashina Inst. Ornith.* **14**, 122–130.
Siegfried, W. R. (1968). Ecological composition of the avifaunal community in a Stellenbosch suburb. *Ostrich* **2**, 105–129.
Skead, D. M. (1966). Birds frequenting the intertidal zone of the Cape Peninsula. *Ostrich* **37**, 10–16.
Skead, D. M. (1969). South African recoveries of birds ringed abroad. *Ostrich* **40**, 28.
Skead, D. M. (1979). Feeding associations of *Hirundo spilodera* with other animals. *Bokmakierie* **31**, 63.
Skead, D. M. and Skead, C. J. (1970). Hirundinid mortality during adverse weather, November 1968. *Ostrich* **41**, 247–251.

Snell, M. L. (1963). A study of the Blue Swallow (*Hirundo atrocaerulea*). *Bokmakierie* 15, 4–7.

Snell, M. L. (1969). Notes on the breeding of the Blue Swallow. *Ostrich* 40, 65–74.

Snell, M. L. (1970). Nesting behaviour of a pair of Blue Swallows. *Bokmakierie* 22, 27–29.

Snell, M. L. (1979). The vulnerable Blue Swallow. *Bokmakierie* 31, 74–78.

Snell, M. L. (1981). Blue Swallows at Inyanga. *Honeyguide* 106, 33.

Spearpoint, S. (1990). Observations on the Pearlbreasted Swallow at Tankatara. *Bee-eater* 41(1), 12.

Stark, A. C. and Sclater, W. L. (1900). 'The Fauna of South Africa: Birds'. R. H. Porter, London.

Steyn, P. (1968). European Swallows *Hirundo rustica* heavily infested with ticks *Hyalomma rufipes*. *Ostrich* 39, 35–36.

Steyn, P. (1988). Swallows feeding on fruit. *Bokmakierie* 40, 52.

Strahm, J. (1956). Nouvelles observations sur la reproduction de l'Hirondelle de rochers. *Nos Oiseaux* 23, 257–265.

Stresemann, E. and Stresemann, V. (1968). Im Sommer mausernde Populationen der Rauchschwalbe, *Hirundo rustica*. *J. Orn.* 109, 475–484.

de Swardt, D. H. (1988). Unusual mortalities of South African Cliff Swallows. *Mirafra* 5, 84.

Talbot, J. N. (1974). Feeding of European Swallows. *Honeyguide* 79, 43.

Taylor, J. S. (1942). Notes on the martins, swallows and swifts: Graaff-Reinet. *Ostrich* 13, 148–156.

Taylor, J. S. (1964). Feeding habits of Ruff (*Philomachus pugnax* (L.)) and European Swallow (*Hirundo rustica* L.). *Ostrich* 35, 66. (Also in *Ostrich* 34, 1963, 176–177).

Taylor, P. B. (1982). House Martins *Delichon urbica* associating with a breeding colony of Red-throated Cliff Swallows *Hirundo rufigula* in Zambia. *Scopus* 6, 43–45.

Thiollay, J.-M. (1977). Passage d'Hirondelles rousselines *Hirundo daurica* en Tunisie. *Alauda* 45, 343.

Thompson, A. L. (1966). The status of two swallow species in the Gambia. *Ibis* 108, 281–282.

Thonglongya, K. (1968). A new martin of the genus *Pseudochelidon* from Thailand. *Thai. Nat. Sci. Pap., Fauna Ser.* 1, 1–10.

Took, J. M. E. (1967). Grey-rumped Swallows seen at sea. *Ostrich* 38, 199.

Tree, A. J. (1964). The occurrence of the Cliff Swallow (*Hirundo spilodera* (Sundevall)) on the Copperbelt. *Ostrich* 35, 113–114.

Tree, A. J. (1970). European Swallow *Hirundo rustica*. *Ostrich* 41, 268.

Tree, A. J. (1973). Banded Sand Martin *Riparia cincta*. *Ostrich* 44(2), 130.

Tree, A. J. (1976). Movements of the Grey-rumped Swallow. *Honeyguide* 89, 35.

Tree, A. J. (1986a). The European Sand Martin in Zimbabwe. *Honeyguide* 32, 5–9.

Tree, A. J. (1986b). What is the status of the Pearl-breasted Swallow in Zimbabwe? *Honeyguide* 32, 65–67.

Tree, A. J. (1987). Recent reports. *Honeyguide* 33, 65–69.

Turner, A. K. (1982a). Timing of laying by Swallows (*Hirundo rustica*) and Sand Martins (*Riparia riparia*). *J. Anim. Ecol.* 51, 29–46.

Turner, A. K. (1982b). Optimal foraging by the Swallow (*Hirundo rustica*, L.): prey size selection. *Anim. Behav.* 30, 862–872.

Turner, A. K. and Rose, C. (1989). 'A Handbook to the Swallows and Martins of the World'. Christopher Helm, Bromley.

Tye, A. (1985). Preuss's Cliff Swallow *Hirundo preussi* breeding in Sierra Leone. *Malimbus* 7, 95–96.

Ussher, A. (1944). Nesting habits of Red-throated Rock-Martin (*Ptyonoprogne r. rufigula*). *J. E. Afr. Uganda Nat. Hist. Soc.* 17, 399.

Vaurie, C. (1950). Notes on some Asiatic swallows. *Am. Mus Novit.* 1529, 1–47.

Verheyen, R. (1952). Nos hirondelles (*Riparia riparia*, *Delichon urbica*, *Hirundo rustica*) dans leurs quartiers d'hiver. *Gerfaut* 42, 92–124.

Vernon, C. J. (1962). The occurrence of the Cliff Swallow, *Hirundo spilodera* in Southern Rhodesia. *Ostrich* 33, 53.

Vernon, C. J. and Dean, W. R. J. (1988). Further African bird-mammal feeding associations. *Ostrich* 59, 38–39.

Vietinghoff-Riesch, A. von. (1955). 'Die Rauchschwalbe'. Berlin.

Vincent, J. (1944). Nesting site of the Banded Sand Martin. *Ostrich* 15, 237.

Vincent, J. (1969). Mortality among Swallows *Hirundo rustica*. *Lammergeyer* 10, 97–98.

Vincent, J. (1973). Swallow ring recoveries. *Lammergeyer* 17, 33.

Voisin, J. C. (1958). Description du nid et des oeufs de *Lecythoplastes fuliginosa*. *Oiseau et R.F.O.* 28, 264–265.

Walsh, J. F. (1987). Records of birds seen in north-eastern Guinea in 1984–1985. *Malimbus* 9, 105–122.

Webber, J. (1975). Swallow roosts. *Honeyguide* 83, 46.

White, C. M. N. (1961). The African rough-winged swallows. *Bull. Br. Orn. Club* 81, 29–33.

Williams, J. G. (1966). A new species of swallow from Kenya. *Bull. Br. Orn. Club* 86, 40.

Winterbottom, J. M. (1966). Ringed Swallow recoveries. *Ostrich* 37, 62.

Woollard, E. and Woollard, J. (1989). Some nesting notes on the Greater Striped Swallow. *Babbler* 18, 44–46.

Zimmerman, D. A. (1978). Mascarene Martins in Kenya. *Scopus* 2, 74–75.

Zink, G. (1969). The migrations of European Swallows *Hirundo rustica* to Africa from data obtained through ringing in Europe. *Ostrich* (Suppl.) 8, 211–222.

Zino, P. A. (1978). Un cimetière d'hirondelles aux îles Salvages. *Oiseau et R.F.O.* 48, 73–74.

Zusi, R. L. (1978). Remarks on the generic allocation of *Pseudochelidon sirintarae*. *Bull. Br. Orn. Club* 98, 13–15.

Family MOTACILLIDAE
Genus *Motacilla*: wagtails

Altenburg, W., Engelmoer, M., Mes, R. and Piersma, T. (1982). Wintering waders on the Banc d'Arguin. *Rep. Netherlands Orn. Mauritanian Expedition* 1980, pp. 1–281. Groningen.

Ash, J. S. (1969). Spring weights of trans-Saharan migrants in Morocco. *Ibis* 111, 1–10.

Ash, J. S., Ferguson-Lees, I. J. and Fry, C. H. (1967). B.O.U. expedition to Lake Chad, northern Nigeria, March–April 1967: preliminary report. *Ibis* 109, 478–486.

Ashford, R. W. (1970). Yellow Wagtails *Motacilla flava* at a Nigerian winter roost: analysis of ringing data. *Bull. Niger. Orn. Soc.* 7(25/26), 24–26.

Aspinwall, D. R. (1981). Grey Wagtail sighting in Zambia. *Ostrich* 52, 128.

Backhurst, G. C. (1977). East African bird ringing report 1974–77. *J. E. Afr. Nat. Hist. Soc.* 31(163), 1–10.

Backhurst, G. C. (1988). Eastern African ringing report 1981–1987. *Scopus* 12, 1–52.

Begg, G. W. (1981). Cape Wagtail eating fiddler crabs. *Ostrich* **52**, 250.

Bergier, P. (1981). Mode de nidification inhabituel chez la Bergeronnette printanière *Motacilla flava* au Maroc. *Alauda* **49**, 309–310.

Berry, P. S. M. (1981). Cape and Long-tailed Wagtails in the Luangwa valley. *Bull. Zambian Orn. Soc.* **12**, 69–70.

Bowen, P. St J. (1979). A review of the genus *Motacilla* in the Northwestern Province with special reference to Mwinilunga District. *Bull. Zambian Orn. Soc.* **11**, 3–7.

Broadbent, J. (1969). Observations on the roosting of Yellow Wagtails *Budytes flavus* in Ibadan. *Bull. Niger. Orn. Soc.* **6**(21), 33.

Broekhuysen, G. J. (1969). A partial albino Cape Wagtail. *Ostrich* **40**, 62.

Casalis de Pury, R. J. (1979). Grey Wagtail *Motacilla cinerea* at Mwekera. *Bull. Zambian Orn. Soc.* **11**, 44–45.

Curry-Lindahl, K. (1958). Internal timers and spring migration in an equatorial migrant, the Yellow Wagtail (*Motacilla flava*). *Ark. Zool.* **11**, 541–557.

Curry-Lindahl, K. (1963). Moult, body weights, gonadal development, and migration in *Motacilla flava*. *Proc. XIII Int. Orn. Congr.* pp. 960–973.

Curry-Lindahl, K. (1964). Yellow Wagtails *Motacilla flava flavissima* found in great numbers on Mount Nimba, Liberia. *Ibis* **106**, 255–256.

Davies, N. B. (1977). Prey selection and social behaviour in wagtails (Aves: Motacillidae). *J. Anim. Ecol.* **46**, 37–57.

Davies, N. B. (1981). Calling as an ownership convention on Pied Wagtail territories. *Anim. Behav.* **39**, 529–534.

Davies, N. B. (1982). Territorial behaviour of Pied Wagtails in winter. *British Birds* **75**, 261–267.

Dent, A. C. and Benson, C. W. (1966). The Grey Wagtail in Malawi. *Nyasaland J.* **19**, 20.

Dittberner, H. and Dittberner, W. (1984). 'Die Schafstelze.' Wittenberg, Lutherstadt.

Donnelly, A. V. (1976). They are at it again! *Bull. E. Afr. Nat. Hist. Soc.* **1976**, 62–63.

Donnelly, A. V. (1977). More and more *Motacilla alba vidua*. *Bull. E. Afr. Nat. His. Soc.* **1977**, 4–5.

Donnelly, A. V. (1978). A year of wagtails, African Pied Wagtails – *Motacilla aguimp*. *Bull. E. Afr. Nat. His. Soc.* **1978**, 27–30.

Dowsett, R. J. (1965). The occurrence of the Yellow Wagtail *Motacilla flava flavissima* in Central Africa. *Ostrich* **36**, 32–33.

Dowsett, R. J. (1969). Migrants at Malamfatori, Lake Chad, autumn 1968. *Bull. Niger. Orn. Soc.* **6**(22), 39–45.

Earlé, R. A. (1986). Reproductive output of an urban pair of Cape Wagtails *Motacilla capensis*. *Mirafra* **3**, 44–46.

Edwards, S. J. (1988). African Pied Wagtail taking bream fry. *Honeyguide* **34**, 132.

Every, B. (1990). Grey Wagtail in the Eastern Cape. *Bee-eater* **41**, 17–20.

Farkas, T. (1962). Contribution to the bird fauna of Barberspan. *Ostrich* (Suppl.) **4**, 35.

Frost, P. G. H. and Cyrus, D. P. (1981). First records of the Grey Wagtail in South Africa. *Ostrich* **52**, 107.

Fry, C. H. (1961). Movements at sea between southwest Iberia and northwest Africa. *Ibis* **103a**, 291–293.

Fry, C. H., Ash, J. S. and Ferguson-Lees, I. J. (1970). Spring weights of some Palaearctic migrants at Lake Chad. *Ibis* **112**, 58–82.

Fry, C. H., Ferguson-Lees, I. J. and Dowsett, R. J. (1972). Flight muscle hypertrophy and ecophysiological variation of Yellow Wagtail *Motacilla flava* races at Lake Chad. *J. Zool. Lond.* **167**, 293–306.

Fry, C. H., Britton, P. L. and Horne, J. F. M. (1974). Lake Rudolf and the Palaearctic exodus from East Africa. *Ibis* **116**, 44–51.

Gatter, W. (1987a). Vogelzug in Westafrika: Beobachtungen und Hypothesen zu Zugstrategien und Wanderrouten Vogelzug in Liberia, Teil II. *Vogelwarte* **34**, 80–92.

Gatter, W. (1987b). Zugverhalten und Überwinterung von paläarktischen Vögeln in Liberia (Westafrika). *Verh. orn. Gesell. Bayern* **24**, 479–508.

Gatter, W. and Mattes, H. (1987). Anpassungen von Schafstelze *Motacilla flava* und afrikanischen Motacilliden an die Waldzerstörung in Liberia (Westafrika). *Verh. orn. Gesell. Bayern* **24**, 467–477.

Goodman, S. M. and Atta, G. A. M. (1987). The birds of southeastern Egypt. *Gerfaut* **77**, 3–31.

Goodwin, D. (1950). Behaviour, display and feeding-habits of White Wagtail in winter-quarters. *British Birds* **43**, 372.

Grant, C. H. B. and Mackworth-Praed, C. W. (1949). A new species and a new race of Yellow Wagtail from the Sudan and Turkestan. *Bull. Br. Orn. Club* **69**, 130–131.

Grant, C. H. B. and Mackworth-Praed, C. W. (1952). On the species and races of the yellow wagtails from western Europe to western North America. *Bull. Br. Mus (Nat. Hist.) Zool.* **1**(9), 255–268.

Greaves, R. H. (1941). Behaviour of White Wagtails wintering in Cairo district. *Ibis* **83**, 459–462.

Hammond, N. (1985). Israel. *Birds* (RSPB), **10**(8), 37–40.

Hartley, P. H. T. (1946). The song of the White Wagtail in winter quarters. *British Birds* **39**, 44–47.

Hazevoet, C. J. and Haafkens, L. B. (1988). Aves nuevas para España. *Garcilla* **73**, 27.

Herholdt, J. J. (1986). Unusual egg colouration in the Cape Wagtail. *Mirafra* **3**, 25.

Hustler, K. (1984). Long-tailed Wagtail at Victoria Falls. *Honeyguide* **30**, 35.

Irwin, M. P. S. (1960). Aspects of relationship between Palaearctic and Ethiopian Wagtails. *Bull. Br. Orn. Club* **80**, 61–64.

Irwin, M. P. S. (1984). Notes on the birds of the Matetsi Safari Area. *Honeyguide* **30**, 24–30.

Knox, A. (1989). Proposed changes to the Voous list. *British Birds* **82**, 119–120.

Larmuth, J. (1973). Migration of *Motacilla alba alba*. *Bull. Br. Orn. Club* **93**, 97–98.

Ledant, J.-P. and Jacobs, P. (1981). Observations sur l'écologie de la Bergeronnette printanière, de la Fauvette grisette et du Traquet motteux hivernant au nord-Cameroun. *Gerfaut* **71**, 433–442.

Liggitt, J. (1985). Wagtails 'washing' food. *Nyala* **11**, 28.

Marshall, A. J. and Williams, M. C. (1959). Pre-nuptial migration of Yellow Wagtails (*Motacilla flava*) from latitude 0.04°N. *Proc. Zool. Soc. Lond.* **132**, 313–320.

Masterson, A. N. B. (1976). On decline in numbers of *Motacilla capensis*. *Honeyguide* **85**, 48.

Mayr, E. (1956). The interpretation of variation among the Yellow Wagtails. *British Birds* **49**, 115–119.

Meininger, P. L., Sørensen, U. G. and Atta, G. A. M. (1986). Breeding birds of the lakes in the Nile delta, Egypt. *Sandgrouse* **7**, 1–20.

Moreau, R. E. (1949). The African Mountain Wagtail *Motacilla clara* at the nest. *Orn. Biol. Wissensch.* **60**, 183–191.

Moreau, R. E. (1961). Problems of Mediterranean-Saharan migration. *Ibis* **103a**, 373–427, 580–623.

Niven, C. (1981). Wagtails at Amanzi. *Bokmakierie* **33**, 19–20.

Nhlane, M. E. D. (1990). Breeding biology of the African Pied Wagtail *Motacilla aguimp* in Blantyre, Malaŵi. *Ostrich* **61**, 1–4.

Nhlane, M. E. D. (in press). The feeding behaviour of the African Pied Wagtail on a sewage farm in Blantyre City. *In* L. A. Bennun (ed.) *Abstr. Proc. VII Para. Orn. Congr.*

Oatley, T. (1988). Waggi tales. *Bokmakierie* **40**, 120–121.

Owen, D. F. (1969). The migration of the Yellow Wagtail from the equator. *Ardea* **57**, 77–85.

Pearson, D. J. (1971). Weights of some Palaearctic migrants in southern Uganda. *Ibis* **113**, 173–184.

Pearson, D. J. and Backhurst, G. C. (1973). The head plumage of Eastern Yellow-headed Yellow Wagtails wintering in Nairobi, Kenya. *Ibis* **115**, 589–591.

Piper, S. E. (1980). A ringing study of Long-tailed Wagtails in the Palmiet Nature Reserve. *Safring News* **9**, 10–13.

Piper, S. E. (1987). Blue over green, yellow over metal, is alive and well and ten. *Safring News* **16**, 79–81.

Piper, S. E. (1989). Breeding biology of the Longtailed Wagtail *Motacilla clara*. *Ostrich* (Suppl.) **14**, 7–15.

Piper, S. E. and Schultz, D. M. (1988). Monitoring territory, survival and breeding in the Longtailed Wagtail. *Safring News* **17**, 65–76.

Piper, S. E. and Schultz, D. M. (1989). Type, dimensionality and size of Longtailed Wagtail territories. *Ostrich* (Suppl.) **14**, 123–131.

Pitman, C. R. S. (1966). On movements, and a roost of the African Pied Wagtail, *Motacilla aguimp vidua* Sundevall. *Bull. Br. Orn. Club* **86**, 95.

Porter, O. (1932). Wagtails and conservatory pet chameleons, fish and frogs. *Ostrich* **3**, 46–48.

Reynolds, J. F. (1974). Palearctic birds in East Africa. *British Birds* **67**, 70–76.

Ruwet, J.-C. (1965). Les oiseaux des plaines et du lac-barrage de la Lufira supérieure (Katanga méridional). Liège Univ.

Sammalisto, L. (1961). An interpretation of variation in the dark-headed forms of the Yellow Wagtail. *British Birds* **54**, 54–69.

Schifferli, von L. (1972). Fütterungsfrequenz am Nest der Bergstelze *Motacilla cinerea* in verschiedenen Biotopen und Brutmonaten. *Orn. Beob.* **69**, 257–274.

Schouteden, H. (1940). Les Motacillidés du Congo Belge. *Rev. Zool. Bot. Afr.* **33**, 317–323.

Sharrock, J. T. R. and Dale, M. B. (1964). An interpretation of variation in the dark-headed forms of the Yellow Wagtail. *British Birds* **57**, 37–40.

Sibley, C. G. and Ahlquist, J. E. (1981). The relationships of the wagtails and pipits (Motacillidae) as indicated by DNA-DNA hybridization. *Oiseau et R.F.O.* **51**, 189–199.

Simmons, K. E. L. (1965). Pattern of dispersion of the White Wagtail and other birds outside the breeding season. *Bull. Br. Orn. Club* **85**, 161–168.

Skead, C. J. (1954). A study of the Cape Wagtail, *Motacilla capensis*. *Ibis* **96**, 91–103.

Smith, K. D. (1968). Some remarks on *Motacilla alba subpersonata*. *Ibis* **110**, 90–91.

Smith, K. D. (1969). Spring weights of trans-Saharan migrants in Morocco. *Ibis* **111**, 1–10.

Smith, S. (1950). 'The Yellow Wagtail'. Collins, London.

Smith, V. W. (1966). Autumn and spring weights of some Palaearctic migrants in central Nigeria. *Ibis* **108**, 492–512.

Smith, V. W. and Ebbutt, D. (1965). Notes on Yellow Wagtails *Motacilla flava* wintering in central Nigeria. *Ibis* **107**, 390–393.

Taylor, P. B. (1978). A melanistic Pied Wagtail (*Motacilla aguimp*) at Ndola. *Bull. Zamb. Orn. Soc.* **10**, 43.

Taylor, P. B. and Taylor, C. (1975). Races of the Yellow Wagtail (*Motacilla flava*) at Ndola. *Bull. Zambian Orn. Soc.* **7**, 34.

Tellería, J. L. (1981). 'La Migracion de las Aves en el Estrecho de Gibraltar'. Universidad Complutense, Madrid.

Thévenot, M., Bergier, P. and Beaubrun, P. (1980). Compte-rendu d'ornithologie marocaine. *Documents de l'Institut Scientifique* **5**, Univ. Mohammed V, Rabat.

Tree, A. J. (1963). Pied Wagtail *Motacilla aguimp* displaying to a Common Sandpiper *Tringa hypoleucos*. *Ostrich* **34**, 181.

Tyler, S. J. and Ormerod, S. J. (1986). Interactions between resident and migratory wagtails *Motacilla* spp. in Ethiopia – an ecological conundrum. *Scopus* **10**, 10–19.

Tyler, S. J. and Ormerod, S. J. (1987). Dietary overlap between Mountain Wagtails *Motacilla clara*, Grey Wagtails *M. cinerea* and Green Sandpipers *Tringa ochropus* in Ethiopia. *Scopus* **11**, 33–37.

Vader, W. (1982). Pied Wagtails catching young ghost crabs. *Ostrich* **53**, 205.

Vaurie, C. (1957). Systematic notes on Palearctic birds, 25. Motacillidae: the genus *Motacilla*. *Am. Mus. Novit.* **1832**, 1–16.

Wallace, D. I. M. (1955). The mixing of the races of the Yellow Wagtail in Kenya. *British Birds* **48**, 337–340.

Wallace, D. I. M. (1969). Palearctic migrants in west Lagos: November 1968 to May 1969. *Bull. Niger. Orn. Soc.* **6**(22), 45–49.

Ward, P. (1964). The fat reserves of Yellow Wagtails *Motacilla flava* wintering in southwest Nigeria. *Ibis* **106**, 370–375.

Welch, G. R. and Welch, H. J. (1990). Around the region. *Orn. Soc. Middle East Bull.* **24**, 37–38.

Williams, J. (1984). Grey Wagtail at Nyanga. *Honeyguide* **30**, 77–78.

Williamson, K. (1955). Migrational drift and the Yellow Wagtail complex. *British Birds* **48**, 382–403.

Winterbottom, J. M. (1959). Review of the races of the Cape Wagtail *Motacilla capensis* L. *Bull. Br. Orn. Club* **79**, 89–100.

Winterbottom, J. M. (1961). Note on the relations of the species of wagtails. *Bull. Br. Orn. Club* **81**, 46–47.

Winterbottom, J. M. (1964). Notes on the wagtails *Motacilla* of southern Africa. *Ostrich* **35**, 129–141.

Wood, B. (1975). The distribution of races of the Yellow Wagtail overwintering in Nigeria. *Bull. Niger. Orn. Soc.* **11**, 19–26.

Wood, B. (1976). The biology of Yellow Wagtails *Motacilla flava* L. overwintering in Nigeria. Ph.D. thesis, University of Aberdeen.

Wood, B. (1978). Weights of Yellow Wagtails wintering in Nigeria. *Ringing and Migr.* **2**, 20–26.

Wood, B. (1979). Changes in numbers of overwintering Yellow Wagtails *Motacilla flava* and their food supplies in a West African savanna. *Ibis* **121**, 228–231.

Wood, B. (1982). The trans-Saharan spring migration of Yellow Wagtails *Motacilla flava*. *J. Zool. Lond.* **197**, 267–284.

Zahavi, A. (1971). The social behaviour of the White Wagtail *Motacilla alba alba* wintering in Israel. *Ibis* **113**, 203–211.

Zink, G. (1975). 'Der Zug Europäischer Singvögel', 2. Vogelwarte Radolfzell am Max-Planck-Inst., Möggingen.

Family MOTACILLIDAE
Genera *Tmetothylacus*, *Anthus*: pipits

Ash, J. S. and Miskell, J. E. (1990). Presumed breeding of Tawny Pipits *Anthus campestris* in the Afrotropics. *Bull. Br. Orn. Club* **110**, 222–225.

Backhurst, G. C. and Pearson, D. J. (1977). Ethiopian region birds attracted to the lights of Ngulia Safari Lodge, Kenya. *Scopus* **1**, 98–103.

Benson, C. W. (1976). A breeding record of the Short-tailed Pipit (*Anthus brachyurus*). *Bull. Zambian Orn. Soc.* **8**, 67.

Benson, C. W., Irwin, M. P. S. and White, C. M. N. (1959). Some aspects of speciation in the birds of Rhodesia and Nyasaland. *Proc. I Pan-Afr. Orn. Congr.* pp. 397–414.

Borrett, R. P. and Wilson, K. J. (1971). Comparative feeding ecology of *Anthus novaeseelandiae* and *Anthus vaalensis* in Rhodesia. *Ostrich* (Suppl.) **8**, 333–341.

Britton, P. L. and Britton, H. A. (1978). The Malindi Pipit *Anthus melindae* in coastal Kenya. *Ibis* **120**, 215–219.

Brooke, R. K. and Irwin, M. P. S. (1972). A second southern record of the pipit *Tmetothylacus tenellus*. *Bull. Br. Orn. Club* **92**, 91.

Clancey, P. A. (1954). A revision of the South African races of Richard's Pipit *Anthus richardi* Vieillot. *Durban Mus. Novit.* **4**, 101–115.

Clancey, P. A. (1960). On the South African record of the Golden Pipit *Tmetothylacus tenellus* (Cabanis). *Ostrich* **31**, 176.

Clancey, P. A. (1964). On the South African races of the Long-billed Pipit *Anthus similis* Jerdon. *Durban Mus. Novit.* **7**, 177–182.

Clancey, P. A. (1977). On the southern limits of *Anthus novaeseelandiae lichenya* Vincent, 1933. *Durban Mus. Novit.* **11**, 263–264.

Clancey, P. A. (1978). On some enigmatic pipits associated with *Anthus novaeseelandiae* (Gmelin) from central and southern Africa (Aves, Motacillidae). *Bonn. zool. Beitr.* **29**, 148–164.

Clancey, P. A. (1984a). On the so-called Mountain Pipit of the Afrotropics. *Durban Mus. Novit.* **13**, 189–194.

Clancey, P. A. (1984b). The Long-billed Pipit *Anthus similis* Jerdon in equatorial West Africa. *Durban Mus. Novit.* **13**, 226–227.

Clancey, P. A. (1984c). Further on the status of *Anthus latistriatus* Jackson 1899. *Gerfaut* **74**, 375–382.

Clancey, P. A. (1984d). Subspeciation in the Striped Pipit *Anthus lineiventris* Sundevall. *Durban Mus. Novit.* **13**, 227–232.

Clancey, P. A. (1985a). Species limits in the Long-billed Pipits of the southern Afrotropics. *Ostrich* **56**, 157–169.

Clancey, P. A. (1985b). Subspeciation in *Anthus brachyurus* Sundevall 1850. *Bull. Br. Orn. Club* **105**, 133–135.

Clancey, P. A. (1986a). Subspeciation in the pipit *Anthus cinnamomeus* Rüppell of the Afrotropics. *Gerfaut* **76**, 187–211.

Clancey, P. A. (1986b). On the Mountain Pipit in Botswana. *Honeyguide* **32**, 44.

Clancey, P. A. (1986c). The eastern and northeast African subspecies of *Anthus similis* Jerdon. *Bull. Br. Orn. Club* **106**, 80–84.

Clancey, P. A. (1986d). On the status of *Anthus richardi bannermani* Bates, 1930. *Durban Mus. Novit.* **14**, 19–23.

Clancey, P. A. (1987a). Longbilled Pipit systematics. *Ostrich* **58**, 45–46.

Clancey, P. A. (1987b). The Tree Pipit *Anthus trivialis* (Linnaeus) in Southern Africa. *Durban Mus. Novit.* **14**(3), 29–42.

Clancey, P. A. (1989). The status of *Anthus caffer maimbaensis* Benson 1955. *Bull. Br. Orn. Club* **109**, 43–47.

Clancey, P. A. (1990). A review of the indigenous pipits (Genus *Anthus* Bechstein: Motacillidae) of the Afrotropics. *Durban Mus. Novit.* **15**, 42–72.

Colston, P. R. (1982). A new species of *Mirafra* (Alaudidae) and new races of the Somali Long-billed Lark *Mirafra somalica*, Thekla Lark *Galerida malabarica* and the Malindi Pipit *Anthus melindae* from southern coastal Somalia. *Bull. Br. Orn. Club* **102**, 106–114.

Cooper, M. R. (1985). A review of the genus *Macronyx* and its relationships to the Yellow-bellied Pipit. *Honeyguide* **31**, 81–92.

Donnelly, B. G. (1982). First record of the Short-tailed Pipit in Zimbabwe. *Honeyguide* **70**, 138–139.

Dowsett, R. J. and Dowsett-Lemaire, F. (1986). Long-billed Pipit systematics. *Ostrich* **57**, 115.

Dowsett-Lemaire, F. (1989). On the voice of the Mountain Pipit. *Ostrich* **60**, 85–87.

Farkas, T. (1988). Birds of Korannaberg. *Navors. nas. Mus. Bloemfontein* **6**.

Gatter, W. (1987). Zugverhalten und Überwinterung von paläarktischen Vögeln in Liberia (Westafrika). *Verh. orn. Gesell. Bayern* **24**, 479–508.

Grant, C. B. H. and Mackworth-Praed, C. W. (1939). Notes on eastern African birds. *Bull. Br. Orn. Club* **60**, 24–26.

Hall, B. P. (1959). The Plain-backed Pipits of Angola. *Bull. Br. Orn. Club* **79**, 113–116.

Hall, B. P. (1961). The taxonomy and identification of pipits (genus *Anthus*). *Bull. Br. Mus. (Nat. Hist.) Zool.* **7**, 243–289.

Hockey, P. A. R. and the Rarities Committee (1990). Rare birds in southern Africa, 1988: sixth report of the SAOS Rarities Committee. *Birding in Southern Africa* **42**, 34–38.

Irwin, M. P. S. (1978). On the pair-bond in the Tree Pipit *Anthus trivialis* in its winter quarters. *Honeyguide* **96**, 21–22.

Kelsey, M. G. and Langton, T. E. S. (1984). The conservation of the Arabuko-Sokoke Forest. Kenya. *ICBP Report* No. 4. Int. Council for Bird Preserv., Cambridge and University of East Anglia.

Knox, A. G. (1988). Taxonomy of the Rock/Water Pipit superspecies *Anthus petrosus*, *spinoletta* and *rubescens*. *British Birds* **81**, 206–211.

Ludlow, A. R. (1966). Body weight changes and moult in some Palaearctic migrants in southern Nigeria. *Ibis* **108**, 129–132.

Meinertzhagen, R. (1922). Notes on some birds from the Near East and from Tropical East Africa. *Ibis* **11**(3), 621–671.

McCleland, W. (1987). Nesting Plain-backed Pipit. *Bee-eater* **38**, 55.

Mendelsohn, J. (1984). The mountain pipit in the Drakensberg. *Bokmakerie* **36**, 40–44.

Patterson, M. L. (1959). Richard's Pipit *Anthus novaeseelandiae* in southern Rhodesia. *Proc. I Pan-Afr. Orn. Congr.* pp. 435–439.

Prigogine, A. (1981). The status of *Anthus latistriatus* Jackson, and the description of a new subspecies of *Anthus cinnamomeus* from Itombwe. *Gerfaut* **71**, 537–573.

Roberts, A. (1911). On birds from Wakkerstroom. *J. Sth. Afr. Orn. Union* **7**, 21–23.

Shelley, G. E. (1900). 'The Birds of Africa,' Vol. II, pp. 296–297. R. H. Porter, London.

Sinclair, J. C., Garland, I. and Carte, A. (1986). Pied Wheatear and Red-throated Pipit in southern Africa. *Bokmakierie* **38**, 45.

Smith, V. W. (1966). Autumn and spring weights of some Palaearctic migrants in central Nigeria. *Ibis* **107**, 390–393.

Stresemann, E. (1938). *Anthus hoeschi* species nova, ein neuer Peiper aus Sudwest-Afrika. *Orn. Monatsber.* **46**, 149–151.

Stronach, B. W. H. (1967). An unusual record of the Golden Pipit *Tmetothylacus tenellus* (Cabanis). *Bull. Br. Orn. Club* **87**, 164–165.

Taylor, I. R. and MacDonald, M. A. (1979). A population of *Anthus similis* on the Togo range in eastern Ghana. *Bull. Br. Orn. Club* **99**, 29–30.

Taylor, P. B. (1979). Red-throated Pipit *Anthus cervinus* at Ndola, Zambia. *Scopus* **3**, 80.

Taylor, P. B. (1980). Further occurrences of Red-throated Pipits *Anthus cervinus* at Ndola, Zambia. *Scopus* **4**, 72.

Traylor, M. A. (1962). A new pipit from Angola. *Bull. Br. Orn. Club* **82**, 76–77.

Vande weghe, J.-P. (1981). Additions à l'avifaune du Rwanda. *Gerfaut* **71**, 175–184.

Vernon, C. J. (1983). Glimpses of unfamiliar birds – Yellow-breasted Pipit. *Bee-eater* (Suppl.) **10**, 4–6.

Vincent, J. (1951). The description of a new race of Richard's

Pipit *Anthus richardi* Vieillot from Basutoland. *Ann. Natal Mus.* **12**, 135–136.
Vieillard, J. (1967). Le pipit de Richard passe-t-il au Sahara? *Oiseau et R.F.O.* **37**, 146–147.
White, C. M. N. (1946). Notes on pipits of the *Anthus richardi* group and a new race of waxbill for Northern Rhodesia. *Bull. Br. Orn. Club* **67**, 8–10.
White, C. M. N. (1948). The African Plain-backed Pipits – a case of sibling species. *Ibis* **90**, 547–553.
White, C. M. N. (1951). The status of *Anthus richardi rufuloides*. *Ibis* **93**, 627.
White, C. M. N. (1957). Taxonomic notes on African pipits, with description of a new race of *Anthus similis*. *Bull. Br. Orn. Club* **77**, 30–34.
Winterbottom, J. M. (1963). The South African subspecies of the Buffy Pipit *Anthus vaalensis* Shelley. *Ann. S. Afr. Mus.* **46**, 341–352.

Family MOTACILLIDAE
Genus *Macronyx*: longclaws

Ash, J. S. and Gullick, T. M. (1989). The present situation regarding the endemic breeding birds of Ethiopia. *Scopus* **13**, 90–96.
Bangs, O. and Loveridge, A. (1933). Reports on the scientific results of an expedition to the southwestern highlands of Tanganyika Territory, III, Birds. *Bull. Mus. Comp. Zool. Harvard* **75**(3), 143–221.
Belcher, C. F. (1930). 'The Birds of Nyasaland'. Crosby Lockwood, London.
Benson, C. W. (1940). Further notes on Nyasaland birds (with particular reference to those of the Northern Province), Part 3. *Ibis* **14**(4), 584–629.
Benson, C. W. (1946). Notes on the birds of southern Abyssinia, Part III. *Ibis* **88**, 25–48.
Benson, C. W. (1955). New forms of pipit, longclaw, robin-chat, grass-warbler, sunbird, quail-finch and canary from Central Africa. *Bull. Br. Orn. Club* **75**, 101–109.
Benson, C. W. (1958). Birds from the Mwinilunga district, Northern Rhodesia. *Ibis* **100**, 281–285.
Blanford, W. T. (1870). 'Observations on the Geology and Zoology of Abyssinia'. Macmillan & Co., London.
Bowland, A. E. (1984). Some habitat parameters of the Orangethroated Longclaw. *Ostrich* **55**, 32–34.
Britton, P. L. (1970). Birds of the Balovale District of Zambia. *Ostrich* **41**, 145–190.
Brosset, A. and Erard, C. (1977). New faunistic records from Gabon. *Bull. Br. Orn. Club* **97**, 125–132.
Cheeseman, R. E. and Sclater, W. L. (1935). On a collection of birds from North-Western Abyssinia. *Ibis* **77**, 594–622.
Clancey, P. A. (1952). Miscellaneous taxonomic notes on African birds (4). A new geographical race of the Orange-throated Longclaw *Macronyx capensis* (Linnaeus) from Southern Rhodesia. *Durban Mus. Novit.* **4**(3), 51–54.
Clancey, P. A. (1958). The South African races of the Yellow-throated Longclaw *Macronyx croceus* (Vieillot). *Ostrich* **29**, 75–78.
Clancey, P. A. (1962). On the geographical variation of the Yellow-throated Longclaw *Macronyx croceus* (Vieillot). *Bull. Br. Orn. Club* **82**, 5–9.
Clancey, P. A. (1963). Miscellaneous taxonomic notes on African birds, 20. *Durban Mus. Novit.* **6**(19), 244–264.
Clancey, P. A. (1967). Subspecific variation in *Macronyx ameliae* de Tarragon. *Bull. Br. Orn. Club* **87**, 10–13.
Clancey, P. A. (1968). Subspeciation in some birds from Rhodesia, Part III. *Durban Mus. Novit.* **8**(12), 153–182.
Clancey, P. A. (1984). Taxonomic notes on African birds. *Durban Mus. Novit.* **13**(18), 221–238.
Cooper, M. R. (1985). A review of the genus *Macronyx* and its relationships to the Yellow-bellied Pipit. *Honeyguide* **31**, 81–92.

von Erlanger, C. F. (1907). Beitrage zur Vögelfauna Nordostafrikas (Part V). *J. Orn.* **55**, 1–58.
Farkas, T. (1962). Contribution to the bird fauna of Barberspan. *Ostrich* **4** (Suppl.), 1–39.
Friedmann, H. (1946). Ecological counterparts in birds. *Sci. Mon.* **63**, 395–398.
Guichard, K. M. (1950). A summary of the birds of the Addis Ababa region, Ethiopia. *J. E. Afr. Nat. Hist. Soc.* **19**(5), 154–179.
Hamling, H. H. (1953). Observations on the behaviour of birds in Southern Rhodesia. *Ostrich* **24**, 9–16.
Jourdain, F. C. R. and Shuel, R. (1935). Notes on a collection of eggs and breeding-habits of birds near Lokoja, Nigeria. *Ibis* **77**, 623–663.
Marchant, S. (1942). Some birds of the Owerri Province, S. Nigeria. *Ibis* **84**, 137–196.
Markus, M. B. (1972). Notes on the natal plumage of South African passeriform birds. *Ostrich* **43**, 17–22.
Mees, G. F. (1970). Birds of the Inyanga National Park, Rhodesia. *Zool. Verhandelingen* **109**, 1–74.
Ripley, S. D. and Heinrich, G. H. (1960). Additions to the avifauna of Northern Angola (1). *Postilla* **47**, 1–7.
Roberts, A. (1922). Review of the nomenclature of South African birds. *Ann. Transv. Mus.* **8**, 187–272.
Sessions, P. H. B. (1966). Notes on the birds of Lengetia Farm, Mau Narok. *J. E. Afr. Nat. Hist. Soc.* **26**, 18–48.
Skead, D. M. (1966). Birds frequenting the intertidal zone of the Cape Peninsula. *Ostrich* **37**, 10–16.
Traylor, M. A. (1965). A collection of birds from Barotseland and Bechuanaland (continued). *Ibis* **107**, 357–384.
Urban, E. K. (1980). 'Ethiopia's Endemic Birds'. Ethiopian Tourism Commission, Addis Ababa.
Wedgewood Bowen, W. (1931). East African birds collected during the Gray African Expedition, 1929. *Proc. Acad. Nat. Sci. Philadelphia* **83**, 11–79.
White, C. M. N. (1946). The ornithology of the Kaonde-Lunda Province, Northern Rhodesia, Part IV. Upupidae-Fringillidae. *Ibis* **88**, 68–103.
White, C. M. N. (1961). Variation in *Macronyx croceus* Vieillot. *Bull. Br. Orn. Club* **81**, 34.
Winterbottom, J. M. (1966). Ecological distribution of birds in the indigenous vegetation of the south-western Cape. *Ostrich* **37**, 76–91.
Winterbottom, J. M. (1968a). The avifaunas of three agricultural habitats in the south-western Cape. *Ostrich* **39**, 51–60.
Winterbottom, J. M. (1968b). The avifaunas of three fresh water habitats in the south-western Cape. *Ostrich* **39**, 130–138.
Wyndham, C. (1948). Some nests and eggs. *Ostrich* **19**, 158–166.

Family CAMPEPHAGIDAE: cuckoo-shrikes

Backhurst, G. C. and Pearson, D. J. (1977). Ethiopian region birds attracted to the lights of Ngulia Safari Lodge, Kenya. *Scopus* **1**, 98–102.

Balchin, C. S. (1988). Recent observations of birds from the Ivory Coast. *Malimbus* **10**, 201–206.
Britton, P. L. (1973). Seasonal movements of the black cuckoo-

shrikes *Campephaga phoenicea* and *C. flava*, especially in eastern Africa. *Bull. Br. Orn. Club* **93**, 41–48.
Brosset, A. (1972). Etude de la reproduction de l'échenilleur pourpré *Campephaga quiscalina* Finsch. *Alauda* **40**, 145–153.
Collar, N. J. and Andrew, P. (1988). 'Birds to Watch'. *ICBP Tech. Pub.* No. 8. Int. Council for Bird Preserv., Cambridge.
Colston, P. R. (1972). African passerine bird weights. *Bull. Br. Orn. Club* **92**, 115–116.
Dean, W. R. J. (1974). Bird weights from Angola. *Bull. Br. Orn. Club* **94**, 170–172.
Gartshore, M. E. and Carson, P. (1989). Tai Forest news. *World Birdwatch* **11**(1), 31.
Green, A. A. (1984). Additional bird records from Bamingui-Bangoran National Park, Central African Republic. *Malimbus* **6**, 70–72.
Greig-Smith, P. W. and Davidson, N. C. (1977). Weights of West African savanna birds. *Bull. Br. Orn. Club* **97**, 96–99.
Hanmer, D. B. (1985). Malaŵi longevity records. *Safring News* **14**, 51–60.
Jackson, H. D. (1969). Notes on a collection of birds from the Khwe River in Botswana. *Arnoldia* **4**, 1–24.
Jackson, H. D. (1989). Weights of birds collected in the Mutare Municipal area, Zimbabwe. *Bull. Br. Orn. Club* **109**, 100–106.

Liversidge, R. (1986). Bird weights. *Ostrich* **39**, 223–227.
Louette, M. and Prévost, J. (1987). Passereaux collectés par J. Prévost au Cameroun. *Malimbus* **9**, 83–96.
Madge, S. G. (1972). The nest and eggs of the Purple-throated Cuckoo-Shrike *Campephaga quiscalina*. *Bull. Br. Orn. Club* **92**, 145–147.
Pierce, M. A. (1984). Weights of birds from Balmoral, Zambia. *Bull. Br. Orn. Club* **104**, 84–85.
Prigogine, A. (1972). The seasonal migrations of the common Black Cuckoo-Shrike *Campephaga flava*. *Bull. Br. Orn. Club* **92**, 83–90.
Skead, C. J. (1966). A study of the Black Cuckoo-Shrike *Campephaga phoenicea* (Latham). *Ostrich* **37**, 71–75.
Skorupa, J. P. (1982). First nest record for Petit's Cuckoo-Shrike *Campephaga petiti*. *Scopus* **6**, 72–73.
Vande weghe, J.-P. (1988). Problems of passerine speciation in Rwanda, Burundi and adjacent areas. *Proc. XIX Int. Orn. Congr.*, pp. 2547–2552.
Vernon, C. J. (1972). A list of the birds of Bisley Valley, Pietermaritzberg. *South Afr. Ser. Percy Fitzpatrick Inst. Afr. Orn.* **79**, 1–23.
Walsh, J. F. (1987). Records of birds seen in north-eastern Guinea in 1984–1985. *Malimbus* **9**, 105–122.
Whittingham, A. P. (1964). Notes on the nesting habits of the White-breasted Cuckoo-Shrike (*Coracina pectoralis*). *Ostrich* **35**, 63–64.

Family PYCNONOTIDAE: bulbuls

Ash, J. S. (1973). Six species of birds new to Ethiopia. *Bull. Br. Orn. Club* **93**, 3–6.
Ash, J. S. (1977). Four species of birds new to Ethiopia and other notes. *Bull. Br. Orn. Club* **97**, 4–9.
Aspinwall, D. R. (1973). Bird notes from five provinces. *Bull. Zambian Orn. Soc.* **5**(2), 43–63.
Aspinwall, D. R. (1975). Red-eyed Bulbuls in Sesheke District. *Bull. Zambian Orn. Soc.* **7**, 102–103.
Astley-Maberly, C. T. (1970). Albino Blackeyed Bulbul. *Ostrich* **41**, 261.
Baker, N. E. (1990). Three deletions from the avifauna of Tanzania. *Scopus* **14**, 34–35.
Baker, N. E. and Hirslund, P. (1987). Minziro National Forest Reserve: an ornithological note including seven additions to the Tanzania list. *Scopus* **11**, 9–12.
Balchin, C. S. (1988). Recent observations of birds from the Ivory Coast. *Malimbus* **10**, 201–206.
Bates, G. L. (1909). Field-notes on the birds of southern Kamerun. *Ibis*, 1–74.
Bates, G. L. (1911). Further notes on the birds of Southern Cameroon. *Ibis*, 479–545, 581–631.
Bates, G. L. (1927). Notes on some birds of Cameroon and the Lake Chad region: their status and breeding times. *Ibis* **12**(3), 1–64.
Beakbane, M. (1983). Bird bath surprises. *Bull. E. Afr. Nat. Hist. Soc.* Jan–Apr **1983**, 9–10.
Benson, C. W. (1937). Miscellaneous notes on Nyasaland birds. *Ibis* **14**(1), 551–582.
Benson, C. W. (1947). Notes on Nyasaland birds, 2. On the occurrence of three other birds in Nyasaland. (*Turdoides leucopygia hartlaubi* (Bocage), *Atimastillas flavicollis pallidigula* (Sharpe) and *Estrilda perreini perreini* (Vieillot)). *Bull. Br. Orn. Club* **67**, 37–38.
Benson, C. W. (1951). Breeding and other notes from Nyasaland and the Lundazi district of Northern Rhodesia. *Bull. Mus. Comp. Zool.* **106**(2).
Benson, C. W. (1956). The Joyful Bulbul *Chlorocichla laetissima* (Sharpe) and other new birds from Northern Rhodesia. *Rev. Zool. Bot. Afr.* **54**, 118–120.
Benson, C. W. and Benson, F. M. (1947). Some breeding and other notes from Nyasaland. *Ibis* **89**, 279–290.
Benson, C. W. and Irwin, M. P. S. (1965). Some birds from the North-western province, Zambia. *Arnoldia (Rhod.)* **1**(29).
Benson, C. W., Brooke, R. K., Dowsett, R. J. and Irwin, M. P. S. (1970). Notes on the birds of Zambia, Part V. *Arnoldia (Rhod.)* **4**(40).
Brass, D. (1987). Young Cuckoo being fed by Common Bulbul. *Bull. E. Afr. Nat. Hist. Soc.* **17**(4), 59.
Britton, P. L. (1972). Weights of African bulbuls (Pycnonotidae). *Ostrich* **43**, 23–42.
Brooke, R. K. (1965). An albino brood of *Pycnonotus barbatus* (Desfontaines). *Bull. Br. Orn. Club* **85**, 114–115.
Brooke, R. K. (1973). Aerial feeding by bulbuls (*Pycnonotus barbatus*). *Ibis* **115**, 606.
Brooke, R. K. (1981). Notes on the food and foraging of the Black-eyed Bulbul. *Honeyguide* **81**, 19–21.
Brosset, A. (1966). Recherches sur la composition qualititive et quantative des populations de vertébrés dans la forêt primaire du Gabon. *Biol. Gabon.* **2**, 163–177.
Brosset, A. (1971a). Recherches sur la biologie des Pycnonotidés au Gabon. *Biol. Gabon.* **7**(4), 423–460.
Brosset, A. (1971b). Premières observations sur la reproduction de six oiseaux africains. *Alauda* **39**, 112–136.
Brosset, A. (1974). La nidification des oiseaux en forêt gabonaise: architecture, situation des nids et prédation. *Terre Vie* **28**, 579–610.
Brosset, A. (1981a). Occupation du milieu et structure d'une population du bulbul forestier *Andropadus latirostris*. *Oiseau et R.F.O.* **51**, 115–126.
Brosset, A. (1981b). La périodicité de la reproduction chez un bulbul de forêt equitoriale africaine *Andropadus latirostris*. Ses incidences démographiques. *Rev. Ecol. (Terre Vie)* **35**, 109–129.

Brosset, A. (1981c). Evolution divergente des comportements chez deux bulbuls sympatriques (Pycnonotidae). *Alauda* **49**, 94–111.

Brosset, A. (1981d). The social life of the African forest Yellow-whiskered Greenbul *Andropadus latirostris*. *Z. Tierpsychol.* **60**, 239–255.

Brown, L. H. and Newman, K. B. (1974). 'Anting' in African passerine birds. *Ostrich* **45**, 194–195.

Bruce-Miller, W. F. and Bruce-Miller, M. (1975). Sombre Bulbul (*Andropadus importunus*) in Namwala District. *Bull. Zambian Orn. Soc.* **7**(1), 24.

Button, J. A. (1964a). The identification of *Andropadus* bulbuls. *Bull. Niger. Orn. Soc.* **1**, 8.

Button, J. A. (1964b). Notes on the status of the Pycnonotidae at Ilaro. *Bull. Niger. Orn. Soc.* **2**, 8–9.

Chalton, D. O. (1976). Another record of a Yellowbellied Bulbul perching on a mammal. *Ostrich* **47**, 68–69.

Chapin, J. P. (1944). *Phyllastrephus icterinus* (Bonaparte) and its larger counterpart. *Ibis* **86**, 543–545.

Clancey, P. A. (1958). Miscellaneous taxonomic notes on African birds X. 2. The South African races of the Sombre Bulbul *Andropadus importunus* (Vieillot). *Durban Mus. Novit.* **5**, 105–110.

Clancey, P. A. (1959). Geographical variation in the South African populations of the Red-eyed Bulbul *Pycnonotus nigricans* (Vieillot). *Bull. Br. Orn. Club* **79**, 166–170.

Clancey, P. A. (1960). On some interesting bulbuls *Pycnonotus* sp. from the Transvaal. *Bull. Br. Orn. Club* **80**, 100–101.

Clancey, P. A. (1975a). On the endemic birds of the montane evergreen forest biome of the Transvaal. *Durban Mus. Novit.* **10**(12), 151–180.

Clancey, P. A. (1975b). Miscellaneous taxonomic notes on African birds XLIII. An additional subspecies of the Redeyed Bulbul *Pycnonotus nigricans* (Vieillot). *Durban Mus. Novit.* **11**, 17–20.

Clancey, P. A. (1977). Miscellaneous taxonomic notes on African birds XLVII. On *Andropadus hypoxanthus* Sharpe, 1876: Tete, Mozambique, and allied taxa. *Durban Mus. Novit.* **11**(10), 187–192.

Clarke, G. (1985). Bird observations from northwest Somalia. *Scopus* **9**, 24–42.

Clarke, J. B. (1974). *Pycnonotus nigricans* attacking *Clamator jacobinus* near nest. *Honeyguide* **78**, 48.

Crowe, T. M., Rebelo, A. G., Lawson, W. J. and Manson, A. J. (1981). Patterns of variation in body-mass of the Black-eyed Bulbul *Pycnonotus barbatus*. *Ibis* **123**, 336–345.

Cunningham-van Someren, G. R. and Schifter, H. (1981). New races of montane birds from Kenya and southern Sudan. *Bull. Br. Orn. Club* **101**, 355–363.

Decoux, J. P. and Fotso, R. C. (1988). Composition et organisation spatiale d'une communauté d'oiseaux dans la région de Yaoundé. Conséquences biogéographiques de la dégradation forestière et de l'aridité croissante. *Alauda* **56**, 126–152.

Delacour, J. (1943). A revision of the genera and species of the Family Pycnonotidae (Bulbuls). *Zoologica, N.Y.* **28**, 17–28.

De Roo, A. (1967). A new species of *Chlorocichla* from northeastern Congo (Aves: Pycnonotidae). *Rev. Zool. Bot. Afr.* **75**, 392–395.

Diesselhorst, G. (1960). Zur geographischen Variabilität von *Phyllastrephus fischeri* (Reichenow). *Veroff. zool. Staatssamml. München* **6**, 81–100.

Dowsett, R. J. (1971). A new species of bulbul for Zambia, *Phyllastrephus placidus*. *Bull. Zamb. Orn. Soc.* **3**(2), 52.

Dowsett, R. J. (1972). Is the bulbul *Phyllastrephus placidus* a good species? *Bull. Br. Orn. Club* **92**, 132–138.

Dowsett, R. J. (1974). Geographical variation in iris colour in the bulbul *Andropadus milanjensis*. *Bull. Br. Orn. Club* **94**, 102–104.

Dowsett, R. J. (1983). Diurnal weight variation in some montane birds in south-central Africa. *Ostrich* **54**, 126–128.

Dowsett, R. J. and Dowsett-Lemaire, F. (1980). The systematic status of some Zambian birds. *Gerfaut* **70**, 151–199.

Dowsett-Lemaire, F. (1988). Fruit choice and seed dissemination by birds and mammals in the evergreen forests of upland Malaŵi. *Rev. Ecol. (Terre Vie)* **43**, 251–285.

Dowsett-Lemaire, F. and Dowsett, R. J. (1989). Zoogeography and taxonomic relationships of the forest birds of the Cameroon Afromontane region. *Tauraco Res. Rep.* **1**, 48–56.

Dowsett-Lemaire, F. and Stjernstedt, R. (1987). Stripe-cheeked Greenbul *Andropadus milanjensis* in Mbulu District, northern Tanzania. *Scopus* **11**, 46.

Dryden, C. (1981). Longevity of a captive Blackeyed Bulbul. *Ostrich* **52**, 188.

Eisentraut, M. (1968). Beitrag zur Vogelfauna von Fernando Poo und Westkamerun. *Bonn. zool. Beitr.* **19**, 49–68.

Erard, C. (1977). Découverte du nid de *Baeopogon clamans* (Sjöstedt). *Alauda* **45**, 271–277.

Erard, C. (1981). Le nid et la ponte d'*Andropadus gracilis*, Pycnonotidé. *Oiseau et R.F.O.* **51**, 246–247.

Erard, C. (1991). Variation géographique de *Bleda canicapilla* (Hartlaub) 1854 (Aves, Pycnonotidae). Description d'une sous-espèce nouvelle en Sénégambie. *Oiseau et R.F.O.* **61**, 66–67.

Everitt, C. (1964a). Breeding the Joyful Greenbul. *Avic. Mag.* **70**, 170–171.

Everitt, C. (1964b). Breeding the Red-eyed Bulbul. *Avic. Mag.* **70**, 214–216.

Field, G. D. (1979). The genus *Criniger* (Pycnonotidae) in Africa. *Bull. Br. Orn. Club* **99**, 57–59.

Friedmann, H. (1968). Range and variation of the Icterine Bulbul in Uganda. *Bull. Br. Orn. Club* **88**, 110–112.

Friedmann, H. and Williams, J. G. (1969). Birds of the Sango Bay forests, Buddu County, Masaka District, Uganda. *Los Angeles County Mus. Contrib. Sci.* No. 162.

Frost, S. (1979). Pre-copulatory display of Yellowbellied Bulbul. *Ostrich* **50**, 185.

Fry, C. H. (1970). Migration, moult and weights of birds in northern Guinea savanna in Nigeria and Ghana. *Ostrich* (Suppl.) **8**, 239–263.

Gatter, W. (1985). Ein neuer Bülbül aus Westafrika (Aves, Pycnonotidae). *J. Orn.* **126**, 155–161.

Gee, J. and Heigham, J. (1977). Birds of Lagos, Nigeria, Pt. II. *Bull. Niger. Orn. Soc.* **13**, 103–132.

Gichuki, C. M. and Schifter, H. (1990). Long life-span and sedentariness of birds in North Nandi Forest, Kenya. *Scopus* **14**, 24–25.

Grant, C. H. B. and Mackworth-Praed, C. W. (1940). A new race of Brown Babbler and a new race of Bulbul from eastern Africa. *Bull. Br. Orn. Club* **60**, 61–63.

Greig-Smith, P. W. and Davidson, N. C. (1977). Weight changes of Guinea savanna birds in Ghana. *Bull. Niger. Orn. Soc.* **13**, 94–97.

Hall, B. P. and Moreau, R. E. (1964). Notes on *Andropadus masukuensis* Shelley and the status of *Andropadus tephrolaemus kungwensis* Moreau. *Bull. Br. Orn. Club* **84**, 133–134.

Hanmer, D. B. (1978). Measurements and moult of five species of bulbul from Mozambique and Malaŵi. *Ostrich* **49**, 116–131.

Hanmer, D. B. (1984). Life expectancy and productivity estimates for 3 species of bulbul from southern Malaŵi. *Proc. V Pan-Afr. Orn. Congr.* pp. 149–161.

Hanmer, D. B. (1986). Aberrant Black-eyed Bulbuls. *Honeyguide* **32**, 161–162.

Hanmer, D. B. (1989). The Nchalo ringing station – bird longevity and migrant return. *Nyala* **14**, 21–27.

Happel, R. E. (1986). Observations of birds and other frugivores feeding at *Tetrorchidium didymostemon*. *Malimbus* 8, 77–78.

Harrison, C. J. O. and Parker, S. A. (1965). The eggs of the White-throated Greenbul *Phyllastrephus albigularis* (Sharpe). *Bull. Br. Orn. Club* 85, 95.

Harvey, W. G. (1979). Red-tailed Greenbul *Criniger calurus* in Tanzania. *Scopus* 3, 28–29.

van der Heiden, J. T. (1975). Overlap in the range of Black-eyed and Red-eyed Bulbuls. *Honeyguide* 81, 31.

Herroelen, P. (1955). Notes sur quelques nids et oeufs inconnus d'oiseaux africains observés au Congo Belge. *Rev. Zool. Bot. Afr.* 52, 185–192.

Heyman, R. and Morlion, M. L. (1980). The Pterylosis in the genera *Pycnonotus* and *Andropadus*. *Gerfaut* 70, 225–244.

Holliday, C. S. (1965). A new record of the Red-eyed Bulbul *Pycnonotus nigricans* Vieillot, in Zambia. *Ostrich* 36, 39.

Hollom, P. A. D., Porter, R. F., Christensen, S. and Willis, I. (1988). 'Birds of the Middle East and North Africa. A companion guide'. T. and A. D. Poyser, Calton, England.

Horne, J. F. M. and Short, L. L. (1990). Wax-eating by African Common Bulbuls. *Wilson Bull.* 102, 339–341.

Howe, S. and Merrie, T. D. H. (1984). 'Birds in Egypt 1983–1984'. B.P. Petroleum Development Ltd., Egypt Branch.

Hudson, J. R. (1945). Pale specimen of Yellow-vented Bulbul (*Pycnonotus tricolor fayi*). *J. E. Afr. Uganda Nat. Hist. Soc.* 18, 161–162.

Irwin, M. P. S. (1956). Notes on the drinking habits of birds in semi-desertic Bechuanaland. *Bull. Br. Orn. Club* 76, 99–101.

Irwin, M. P. S. (1958). The relationships of the bulbuls *Pycnonotus barbatus* and *Pycnonotus nigricans*. *Occ. Pap. Nat. Mus. S. Rhodesia* 3, 198–201.

Irwin, M. P. S. (1980). Food niche imprinting: a possible instance in the Terrestrial Bulbul *Phyllastrephus terrestris*. *Honeyguide* 100, 40–41.

Jensen, F. P. and Stuart, S. N. (1982). New subspecies of forest birds from Tanzania. *Bull. Br. Orn. Club* 102, 95–99.

Johnson, D. N. (1989). The feeding habits of some Afrotropical forest bulbuls. *Ostrich* (Suppl.) 14, 49–56.

Jordaan, H. (1975). Further records of the Red-eyed and Black-eyed Bulbuls in Wankie National Park. *Honeyguide* 82, 43.

Julliard, J.-P. (1986). Reproduction du bulbul *Pycnonotus barbatus* au Maroc. *Alauda* 54, 279–285.

Kunkel, I. and Kunkel, P. (1969). Contribution à la connaissance de l'avifaune de la forêt du Rugege (Rwanda). *Rev. Zool. Bot. Afr.* 79, 327–351.

Lambert, F. R. (1984). Birds at ant swarms in Kenya and southern Sudan. *Scopus* 8, 31–32.

Lawson, W. J. (1962). The genus *Pycnonotus* in southern Africa. *Durban Mus. Novit.* 6, 165–180.

Leck, C. F. (1977). Black-faced Vervet attacks and kills a bulbul. *Ostrich* 48, 111.

Lewis, A. D. (1983). A leucistic Common Bulbul *Pycnonotus barbatus*. *Scopus* 7, 94–95.

Liversidge, R. (1970). The ecological life history of the Cape Bulbul. Unpubl. Ph.D thesis, University of Cape Town.

Liversidge, R. (1983). Habitat degradation and hybridization in bulbuls. *Proc. Symp. Birds and Man*, Johannesburg 1983, pp. 99–106.

Long, R. C. (1961). The birds of Port Herald District, Pt. III. *Ostrich* 32, 147–173.

Longrigg, T. D. (1978). Pre-copulatory display of Sombre Bulbul. *Ostrich* 49, 202–203.

Lowe, P. R. (1937). Report on the Lowe-Waldron expeditions to the Ashanta forests and northern territories of the Gold Coast, Part II. *Ibis* 79, 635–690.

Macdonald, I. A. W. and Dean, W. R. J. (1978). A record of egg predation by the East African egg-eater *Dasypeltis medici*. *Ostrich* 13, 163.

Macdonald, M. A. (1979). Breeding data for birds in Ghana. *Malimbus* 1, 36–42.

MacInnes, J. E. (1934). Notes on birds of Turkana Province. *J. E. Afr. Uganda Nat. Hist. Soc.* 12, 24–50.

Mackworth-Praed, C. W. and Grant, C. H. B. (1947). Notes on eastern African birds: 2. On the status of the so-called type of *Andropadus oleaginus* Peters, J. f. O. 1868, p. 133: Lorenço Marques, Portuguese East Africa. *Bull. Br. Orn. Club* 68, 59.

Malbrant, R. and Maclatchy, A. (1949). 'Faune de l'équateur africain français. II. Oiseaux.' 2nd ed., Lechevalier, Paris.

Manson, A. J. (1985). Results of a ringing programme at Muruwati Farm, Mazowe. *Honeyguide* 31, 203–210.

Marchant, S. (1953). Notes on the birds of southeastern Nigeria. *Ibis* 95, 38–69.

Markus, M. B. (1963a). Bulbuls from the zone of contact between *Pycnonotus barbatus layardi* Gurney, 1879, and *Pycnonotus nigricans* (Vieillot) in the Transvaal. *Ostrich* 34, 110.

Markus, M. B. (1963b). An aberrant *Pycnonotus* from Johannesburg, South Africa. *Bull. Br. Orn. Club* 83, 117–118.

Markus, M. B. (1965). Eyelid of Black-eyed Bulbul *Pycnonotus barbatus* (Desfontaines). *Ostrich* 36, 41.

Markus, M. B. (1966). Systematic notes on *Pycnonotus* from the South-Western Transvaal. *Ostrich* 37, 234.

Markus, M. B. (1967). Secondary intergradation amongst bulbuls of the genus *Pycnonotus* in the Transvaal Province, South Africa. *Bull. Br. Orn. Club* 87, 17–23.

Markus, M. B. (1974). Aerial feeding by bulbuls. *Ibis* 116, 232.

Marshall, B. E. (1969). Unusual nesting site of Black-eyed Bulbul. *Ostrich* 40, 135.

Martin, R. (1990). Redeyed Bulbul on the Peninsula. *Promerops* 192, 9.

Masterson, A. N. B. (1975). Avocado pears as a source of food. *Honeyguide* 82, 43.

Mattes, H. and Gatter, W. (1989). Jahresperiodik und Biometrie von *Andropadus latirostris* Strickland (Aves: Pycnonotidae) in Liberia. *Stuttgarter Beitr. Naturk. Ser. A*, No. 429.

McCulloch, D. (1967). Status of the Sombre Bulbul *Andropadus importunus* on the Moçambique Channel Islands. *Ostrich* 38, 290.

Meinertzhagen, R. (1930). 'Nicoll's Birds of Egypt'. Hugh Rees, London.

van der Merwe, F. (1987). Incubation period of Cape Bulbul (566). *Promerops* 178, 15.

Mills, M. G. L. (1976). A revised check-list of birds in the Kalahari-Gemsbok National Park. *Koedoe* 19, 49–62.

Moreau, R. E. (1937). On *Phyllastrephus fischeri* and related forms. *Bull. Br. Orn. Club* 57, 125–128.

Moreau, R. E. (1941). A new race of Yellow-streaked Bulbul and a new race of Yellow-moustached Bulbul from Tanganyika Territory. *Bull. Br. Orn. Club* 62, 29–30.

Moreau, R. E. (1945). On the status of *Phyllastrephus flavostriatus kungwensis* Moreau. *Ibis* 87, 100–101.

Moreau, R. E. (1947). The relationship between *Phyllastrephus placidus grotei* Rchw., *P. f. fischeri* Rchw. and *P. p. munzneri*. *Bull. Br. Orn. Club* 67, 88–90.

Moreau, R. E. and Sclater, W. L. (1938). The avifauna along the mountains of the Rift Valley in north central Tanganyika Territory (Mbulu District), Part. II. *Ibis* 14(2), 1–32.

Moreau, R. E., Wilk, A. L. and Rowan, W. (1947). The moult and gonad cycles of three species of birds at five degrees south of the equator. *Proc. Zool. Soc. Lond.* 117, 345–364.

Oatley, T. B. (1986). Longevity of *Pycnonotus barbatus*. *Safring*

Okia, N. O. (1976). Birds of the understorey of lake-shore forests on the Entebbe Peninsula, Uganda. *Ibis* **118**, 1-13.
Payne, R. B. (1969). Overlap of breeding and molting schedules in a collection of African birds. *Condor* **71**, 140-145.
Peirce, M. A. (1984). Weights of birds from Balmoral, Zambia. *Bull. Br. Orn. Club* **104**, 84-85.
Pomeroy, D. E. and Tengecho, B. (1982). Studies of birds in a semi-arid area of Kenya. II. Bird parties in two woodland areas. *Scopus* **6**, 25-32.
Powett, T. (1963). Red-eyed Bulbul *Pycnonotus nigricans* and Long-tailed Widowbird *Diatropura procne* feeding on insects. *Ostrich* **34**, 47.
Preston, G. (1975). The omnivorous bulbul. *Bull. E. Afr. Nat. Hist. Soc.* Oct-Dec, 112-113.
Prigogine, A. (1954). Un nouveau Bulbul de l'Est du Congo belge. *Rev. Zool. Bot. Afr.* **49**, 347-349.
Prigogine, A. (1961). Nids et oeufs récoltés au Kivu (République du Congo). *Rev. Zool. Bot. Afr.* **66**, 248-259.
Prigogine, A. (1969). Trois nouveaux oiseaux du Katanga, République démocratique du Congo. *Rev. Zool. Bot. Afr.* **79**, 110-116.
Prigogine, A. (1972). Description of a new green bulbul from the Republic of Zaire (*Andropadus hallae*). *Bull. Br. Orn. Club* **92**, 138-141.
Prigogine, A. (1973). Etude taxonomique des populations de *Phyllastrephus flavostriatus* de l'Afrique Centrale et description de deux nouvelles races de la République du Zaïre. *Gerfaut* **63**, 219-234.
Rand, A. L. (1958). Notes on African bulbuls. Family Pycnonotidae: Class Aves. *Fieldiana: Zoology*, Vol. 35, No. 6.
Roberts, E. L. (1987). Incubation period of Cape Bulbul (566). *Promerops* **179**, 13.
Serle, W. (1964). The lower altitudinal limit of the montane forest birds of the Cameroon Mountain, West Africa. *Bull. Br. Orn. Club* **84**, 87-91.
Shapland, P. (1975). Black-eyed Bulbul catching lizard. *Honeyguide* **84**, 46.
Smithers, R. H. N. (1956). Aberrant *Pycnonotus xanthopygos layardi* from Southern Rhodesia. *Ostrich* **27**, 88.
van Someren, V. D. (1958). 'A Bird Watcher in Kenya'. Oliver and Boyd, Edinburgh.
Stanfield, J. P. (1973). Aerial feeding by bulbuls. *Ibis* **115**, 606.
Steyn, P. (1968). Some cases of allopreening. *Ostrich* **39**, 36-38.
Steyn, P. (1973). Red-eyed and Black-eyed Bulbuls occurring together. *P. nigricans* and *P. barbatus* in the magnificent fig tree at Shumba, Wankie Nat. Park, 6 July 1973. *Honeyguide* **75**, 28.
Steyn, P. (1975). Yellow-breasted Bulbul feeding on an impala. *Lammergeier* **22**, 51.
Storey, G. W. and Harrison, J. M. (1969). Comments on an intersexual bulbul (*Pycnonotus nigricans*). *Bull. Br. Orn. Club* **89**, 160-162.
Stuart, S. N. and Jensen, F. P. (1981). Further range extensions and other notable records of forest birds from Tanzania. *Scopus* **5**, 106-115.
Sugg, M. and Sugg, G. (1972). Notes on some birds recorded in South Nyanza, Kenya. *Bull. E. Afr. Nat. Hist. Soc.* (Mar), 38-40.
Swynnerton, C. F. M. (1907). On the birds of Gazaland, Southern Rhodesia. *Ibis* **9**(1), 30-74.
Traylor, M. A. Jr (1962). 'Notes on the birds of Angola, passeres'. Comp. Diam. Angola, Museo do Dundo, Lisbon.
Traylor, M. A. Jr and Archer, A. L. (1982). Some results of the Field Museum 1977 expedition to south Sudan. *Scopus* **6**, 5-12.
Tree, A. J. (1987). Recent Reports. *Honeyguide* **33**, 155-160.
Tucker, J. J. (1975). Six species of bulbul together at a locality near Mufulira. *Bull. Zamb. Orn. Soc.* **7**(1), 37.
Turner, D. A. (1979). Greenbuls of the Taita Hills, SE Kenya. *Scopus* **3**, 27-28.
Turner, D. A. and Zimmerman, D. A. (1979). Field identification of Kenya greenbuls. *Scopus* **3**, 33-47.
Vande weghe, J.-P. (1981). L'avifaune des papyraies au Rwanda et au Burundi. *Gerfaut* **71**, 489-536.
Vande weghe, J.-P. (1984). Further additions to the avifauna of Rwanda. *Scopus* **8**, 60-63.
Vernon, C. J. (1972). *Chlorocichla flaviventris* perching on *Sylvicapra grimmia*. *Ostrich* **43**, 137.
Vernon, C. J. (1975). More notes on the habits of the Black-eyed Bulbul. *Honeyguide* **82**, 42.
White, C. M. N. (1944). Note on two Northern Rhodesian birds. *Ibis* **86**, 553.
White, C. M. N. (1956). Notes on the systematics of African Bulbuls. *Bull. Br. Orn. Club* **76**, 155-158.
Willis, E. O. (1983). Jays, mimids, icterids and bulbuls (Corvidae, Mimidae, Icteridae and Pycnonotidae) as ant-followers. *Gerfaut* **73**, 379-392.
Wimpfheimer, D., Bruun, B., Baha el Din, S. M. and Jennings, M. C. (1983). 'The migration of birds of prey in the northern Red Sea area'. The Holy Land Conservation Fund, New York.
Wood, B. (1989). Biometrics, iris and bill colouration, and moult of Somali forest birds. *Bull. Br. Orn. Club* **109**, 11-22.
Woodall, P. F. (1971). Yellow-breasted Bulbul in a Salisbury garden. *Honeyguide* **66**, 38.
Young, C. G. (1946). Notes on some birds of the Cameroon Mountain District. *Ibis* **88**, 348-382.
Zimmerman, D. A. (1986a). A twenty-year-old greenbul from western Kenya. *Scopus* **10**, 111-112.
Zimmerman, D. A. (1986b). The Yellow-streaked Greenbul in Kenya. *Scopus* **10**, 112.

Family BOMBYCILLIDAE: waxwings and Hypocolius

Bunni, M. K. and Siman, H. Y. (1978). Vocalization of the Grey Hypocolius, *Hypocolius ampelinus* Bonaparte. *Bull. Nat. Hist. Res. Cent. Baghdad* **7**(2), 21-26.
Delacour, J. and Amadon, D. (1949). The relationships of *Hypocolius*. *Ibis* **91**, 427-429.
El Negumi, A. (1949). List of desert animals seen or collected during the periods shown. *Bull. Zool. Soc. Egypt* **8**, 20-21.
Goodman, S. M. (1984). Report on two small bird collections from the Gebel Elba region, southeastern Egypt. *Bonn. Zool. Beitr.* **35**, 39-56.

Family CINCLIDAE: dippers

Chaworth-Musters, J. L. (1939). Some notes on the birds of the High Atlas of Morocco. *Ibis* **14**(3:2), 269-281.
Shaw, G. (1978). The breeding biology of the Dipper. *Bird Study* **25**, 149-160.
Tyler, S. J. and Ormerod, S. J. (1991). Aspects of the biology of Dippers *Cinclus cinclus minor* in the Atlas Mountains of Morocco outside the breeding season. *Bonn. zool. Beitr.* **42**(1), 35-45.
Yoerg, S. I. and O'Halloran, J. (1991). Dipper nestlings fed by a Gray Wagtail. *Auk* **108**, 427-429.

Family TROGLODYTIDAE: wrens

Armstrong, E.A. (1955). 'The Wren'. Collins, London.
Dallmann, M. (1977). Beobachtungen Zür Brutbiologie des Zaunkönigs *Troglodytes troglodytes*. *Anz. Orn. Gesell. Bayern* **16**, 153–170.
Haynes, V. M. (1980). Communal roosting by Wrens. *British Birds* **73**, 104–105.

Family PRUNELLIDAE: accentors

Davies, N. B. (1983). Polyandry, cloaca-pecking and sperm competition in dunnocks. *Nature* **302**, 334–336.
Davies, N. B. (1986). Reproductive success of Dunnocks, *Prunella modularis*, in a variable mating system. II. Conflicts of interest among breeding adults. *J. Anim. Ecol.* **55**, 139–154.
Goodman, S. M. and Watson, G. E. (1983). Bird specimen records of some uncommon or previously unrecorded forms in Egypt. *Bull. Br. Orn. Club* **103**, 101–106.
Juana, E. and Santos, T. (1981). Observations sur l'hivernage des oiseaux dans le Haut-Atlas (Maroc). *Alauda* **49**, 1–12.
Sibley, C. G. and Ahlquist, J. E. (1981). The relationships of the accentors (*Prunella*) as indicated by DNA-DNA hybridization. *J. Orn.* **122**, 369–378.

Family TURDIDAE: thrushes
Genus *Oenanthe*: wheatears

Alexander, R. McN. (1985). Locomotion, terrestrial. In 'A Dictionary of Birds' (Campbell, B. and Lack, E. Eds), pp. 329–330. T. and A. D. Poyser, Calton, England.
Andersson, C. J. (1872). 'Notes on the birds of Damaraland and the adjacent countries of SW Africa.' J van Voorst, London.
Arnault, C. (1926). Le traquet à tête blanche. *Oiseau et R.F.O.* **7**, 156–160.
Arnault, C. (1934). Notes de Laghouat (Algérie). *Oiseau et R.F.O.* **4**, 740.
Arnould, M. (1961). Six mois d'observations ornithologiques à Hassi Messaoud (octobre 1959–avril 1960). *Oiseau et R.F.O.* **31**, 140–152.
Ash, J. S. (1980). Migrational status of Palaearctic birds in Ethiopia. *Proc. IV Pan-Afr. Orn. Congr.* pp. 199–208.
Ash, J. S. (1981). Desert Wheatears *Oenanthe deserti* in Ethiopia and Somalia. *Scopus* **5**, 35–36.
Aspinwall, D. R. (1975). A record of an overwintering *Oenanthe oenanthe* from Lusaka, Zambia. *Bull. Br. Orn. Club* **95**, 46–48.
Aspinwall, D. R. (1977). Isabelline Wheatears in eastern Zambia. *Bull. Zambian Orn. Soc.* **9**, 26–28.
Backhurst, D., Pearson, D. J. and Richards, D. K. (1984). A Kenya record of the Black-eared Wheatear *Oenanthe hispanica*. *Scopus* **8**, 50–51.
Backhurst, G. C. (1985). Desert Wheatear *Oenanthe deserti* in Kenya. *Scopus* **9**, 140–141.
Baha el Din, S. (1984). The Black Wheatear *Oenanthe leucura* in Egypt. *Courser* **1**, 5–8.
Bates, G. L. (1927). Notes on some birds of Cameroon and the Lake Chad region: their status and breeding times. *Ibis* **12**(3), 1–64.
Bates, G. L. (1935). On *Oenanthe xanthoprymna* and *O. chrysopygia*. *Ibis* **13**(5), 198–201.
Bates, G. L. (1936). Birds of Jidda and Central Arabia collected in 1934 and early in 1935, chiefly by Mr. Philby. *Ibis* **13**(6), 531–556, 674–712.
Beesley, J. H. (1956). Note on *Oenanthe pileata livingstonei* (Tristram). *Bull. Br. Orn. Club* **76**, 33.
Benson, C. W. (1940). Further notes on Nyasaland birds (with particular reference to those of the Northern Province). *Ibis* **14**(4), 257–298, 583–629.
Benson, C. W. (1946). Notes on the birds of southern Abyssinia. *Ibis* **88**, 180–205.
Benson, C. W. (1971). Some records of the Capped Wheatear. *Honeyguide* **68**, 25–27.
Benson, C. W. (1982). Migrants in the Afrotropical region south of the equator. *Ostrich* **53**, 31–49.
Benson, C. W., Brooke, R. K. and Vernon, C. J. (1964). Bird breeding data for the Rhodesias and Nyasaland. *Occ. Pap. Nat. Mus. Sth. Rhod.* **27B**, 30–105.
Blondel, J. (1962). Données écologiques sur l'avifaune des Monts des Ksours (Sahara septentrional). *Terre Vie* **109**, 209–251.
Borrett, R. P. and Jackson, H. D. (1970). The European Wheatear *Oenanthe oenanthe* (L.) in southern Africa. *Bull. Br. Orn. Club.* **90**, 124–129.
Bowen, W. W. (1926–1931). 'Catalogue of Sudan Birds'. Sudan Govt. Museum (Nat. Hist.), Khartoum.
Bulman, J. F. H. (1942). Notes on the birds of the Libyan desert. *Bull. Zool. Soc. Egypt* **4**, 5–12.
Bundy, G. and Morgan, J. H. (1969). Notes on Tripolitanian birds. *Bull. Br. Orn. Club.* **89**, 151–158.
Bundy, G. and Sharrock, J. T. R. (1986). Hooded Wheatear. *British Birds* **79**, 120–123.
Burnier, E. (1979). Notes sur l'ornithologie algérienne. *Alauda* **47**, 93–102.
Butler, A. L. (1905). A contribution to the ornithology of the Egyptian Soudan. *Ibis* **8**(5), 301–401.
Butler, A. L. (1908). A second contribution to the ornithology of the Egyptian Soudan. *Ibis* **9**(2), 205–263.
Butler, A. L. (1909). Contribution to the ornithology of the Sudan. No. III. On birds collected by Capt. E. P. Blencowe in the Bahr-el-Ghazal Province. *Ibis* **9**(3), 74–90, 389–405.
Butler, E. A., Feilden, H. W. and Reid, S. G. (1883). On the variations in plumage of *Saxicola monticola* as observed in Natal. *Ibis* **5**(1), 331–336.
Cheke, R. A. (1982). More bird records from the Republic of Togo. *Malimbus* **4**, 55–62.
Christensen, S. (1974). Notes on the plumage of the female Cyprus Pied Wheatear *Oenanthe pleschanka cypriaca*. *Ornis Scand.* **5**, 47–52.
Clancey, P. A. (1951). Notes on birds of the South African subcontinent. *Ann. Natal Mus.* **12**, 137–152.
Clarke, G. (1985). Bird observations from N. W. Somalia. *Scopus* **9**, 24–42.
Clement, P. (1987). Field identification of West Palaearctic

wheatears. *British Birds* **80**, 137–157, 187–238.
Cornwallis, L. (1975). The comparative ecology of eleven species of wheatear (genus *Oenanthe*) in S. W. Iran. D.Phil thesis, University of Oxford.
Cott, H. B. (1985). Palatability of birds and eggs. In 'A Dictionary of Birds' (Campbell, B. and Lack, E. Eds). T. and A. D. Poyser, Calton, England.
Curry, P. J. and Sayer, J. A. (1979). The inundation zone of the Niger as an environment for Palaearctic migrants. *Ibis* **121**, 20–40.
Davies, C. G. (1910). Notes on the plumage of the Mountain Chat (*Saxicola monticola*, Bechst.). *J. S. Afr. Orn. Un.* **6**, 33–37.
De Greling, C. (1972). New records from northern Cameroun. *Bull. Br. Orn. Club* **92**, 24–27.
Dekeyser, L. (1954). Contribution à l'étude du peuplement de la Mauritanie. Oiseaux (Note récapitulative). *Bull. Inst. fr. Afr. noire* Sér. A **16**, 1248–1292.
Dekeyser, L. and Villiers, A. (1950). Contribution à l'étude du peuplement de la Mauritanie. Oiseaux récoltés par A. Villiers. *Bull. Inst. fr. Afr. noire* **12**, 660–699.
Dementiev, G. P. and Gladkov, N. A. (1954). 'The Birds of the Soviet Union,' Vol. VI. Moscow.
Destre, R. (1984). Le traquet à tête blanche *Oenanthe leucopyga* (Brehm) dans le Tafilalt (Sud-est Marocain). *Bull. Inst. Sci. Rabat* **8**, 157–170.
Dittami, J. (1981). Observations on wintering Wheatears in burned grass areas at Lake Nakuru, Kenya. *Vogelwarte* **13**, 177–178.
Dowsett, R. J. (1965). Weights of some Zambian birds. *Bull. Br. Orn. Club* **85**, 150–152.
Dowsett, R. J. (1971). Identification of the European Wheatear *Oenanthe oenanthe*. *Ostrich* **42**, 296.
Duhart, F. and Deschamps, M. (1963). Notes sur l'avifaune du Delta Central Nigérien et régions avoisinantes. *Oiseau et R.F.O.* **33**. no. spécial. 1–106.
Ebbutt, D. P. (1967). Vocal mimicry in the Red-breasted Chat *Oenanthe heuglini*. *Bull. Niger. Orn. Soc.* **4**, 36–37.
Farrow, D. (1990). Eastern Pied Wheatear. *Orn. Soc. Middle East Bull.* **24**, 38.
Fischman, L. (1977). The White-crowned Black Wheatears at Saint Catherine. *Israel Land and Nature* **2**(3), 101–106.
Folse, L. J. (1982). An analysis of avifauna-resource relationships on the Serengeti plains. *Ecol. Monogr.* **52**, 111–127.
François, J. (1975). Contribution à la connaissance de l'avifaune d'Afrique du nord. *Alauda* **43**, 279–293.
Fry, C. H. (1965). The birds of Zaria. *Bull. Niger. Orn. Soc.* **2**, 9–17, 35–44.
Fry, C. H. (1970). Migration, moult and weights of birds in northern Guinea savanna in Nigeria and Ghana. *Ostrich* (Suppl.) **8**, 239–263.
Fry, C. H. and Erikson, J. (1989). The Eastern Pied Wheatear in Arabia. *Oman Bird News* **7**, 4–7.
Gaston, A. J. (1970). Birds in the Central Sahara in winter. *Bull. Br. Orn. Club* **90**, 53–66.
Gee, J. P. (1984). The birds of Mauritania. *Malimbus* **6**, 31–66.
George, U. (1978). 'In the Deserts of this Earth.' Hamish Hamilton, London.
Germain, M. (1965). Observations ornithologiques en Algérie Occidentale. *Oiseau et R.F.O.* **35**, 46–58, 117–134.
Gillet, H. (1960). Observations sur l'avifaune du Massif de l'Ennedi (Tchad). *Oiseau et R.F.O.* **30**, 45–82, 99–134.
Goodwin, D. (1957). Note on the immature plumages of *Oenanthe monacha* (Temminck). *Bull. Br. Orn. Club* **77**, 17–18.
Grimes, L. G. (1976). The occurrence of cooperative breeding behaviour in African birds. *Ostrich* **47**, 1–15.
Guichard, K. M. (1955). The birds of Fezzan and Tibesti. *Ibis* **97**, 393–424.
Guichard, K. M. (1957). The spring migration in Tripolitania – 1955. *Ibis* **99**, 106–114.

Haas, V. (1986). Social organization and role of helpers, in Anteater Chats (*Myrmecocichla aethiops*) and Mourning Wheatears (*Oenanthe lugens schalowi*). *Proc. XVIII Int. Orn. Congr.* pp. 1109–1110.
Hass, W. and Beck, P. (1979). Zum Frühjarhrszug paläarktische Vögel über die westliche Sahara. *J. Orn.* **120**, 237–246.
Haffer, J. (1977). Secondary contact zones of birds in northern Iran. *Bonn. zool. Monog.* **10**, 1–64.
Hartert, E. (1921). Captain Angus Buchanan's Aïr expedition. 4. The birds collected by Captain Angus Buchanan during his journey from Kano to Aïr or Asben. *Novit. Zool.* **28**, 78–141.
Hartley, P. H. T. (1949). The biology of the Mourning Chat in winter quarters. *Ibis* **91**, 393–413.
Heim de Balsac, H. (1926). Contributions à l'ornithologie du Sahara central et du Sud-Algérien. *Mém. Soc. Hist. Nat. Afr. Nord* **1**.
Heim de Balsac, H. (1975). Note sur l'extension éventuelle du milieu saharien. *Alauda* **43**, 293.
Heim de Balsac, H. and Blanchet, A. (1951–57). Oiseaux de Tunisie. *Mém. Soc. Sci. Nat. Tunisie* **1**.
Heim de Balsac, H. and Heim de Balsac, T. (1949–51). Les migrations des oiseaux dans l'ouest du continent Africain. *Alauda* **17–18**, 129–143, 206–221; **19**, 19–39, 97–112, 157–171, 193–210.
Heinzel, H., Fitter, R. and Parslow, J. (1974). 'The Birds of Britain and Europe.' Collins, London.
Heu, R. (1961). Observations ornithologiques au Ténéré. *Oiseau et R.F.O.* **31**, 214–239.
Irwin, M. P. S. (1971). Further notes on the movements of the Capped Wheatear. *Honeyguide* **68**, 27–28.
James, G. L. (1947). Unusual nesting site of the Capped Wheatear *Campicola pileata livingstonii*. *Ostrich* **18**, 188–189.
Jarry, G. (1969). Notes sur les oiseaux nicheurs de Tunisie. *Oiseau et R.F.O.* **39**, 112–120.
Jehl, H. (1974). Quelques migrateurs paléarctiques en République Centrafricaine. *Alauda* **42**, 397–406.
Knight, P. J., Marchant, J. and Pienkowski, M. W. (1973). Birds observed in Morocco by the expeditions. In 'Studies on coastal birds and wetlands in Morocco 1972' (Pienkowski, M. W., Ed.). UEA Expeditions to Morocco 1971–72, Norwich.
Lack, D. (1971). 'Ecological Isolation in Birds.' Blackwell, Oxford.
Lack, P. C., Leuthold, W. and Smeenk, C. (1980). Check-list of the birds of Tsavo East National Park, Kenya. *J. E. Afr. Nat. Hist. Soc.* **170**, 1–25.
Ledant, J.-P. and Jacobs, P. (1981). Observations sur l'écologie de la Bergeronette printanière, de la Fauvette grisette et du Traquet motteux hivernant au nord-Cameroun. *Gerfaut* **71**, 433–442.
Leisler, B., Heine, G. and Siebenrock, K. H. (1983). Einnischung und interspezifische Territorialität überwinternder Steinschmätzer (*Oenanthe isabellina*, *O. oenanthe*, *O. pleschanka*) in Kenia. *J. Orn.* **124**, 393–413.
Loskot, V. (1986). Morphism and hybridisation of wheatears *Oenanthe hispanica* and *Oenanthe pleschanka*. *Proc. XVIII Int. Orn. Congr.* p. 1031.
Lowden Stoole, E. W. (1968). Display behaviour in the Capped Wheatear. *Honeyguide* **56**, 23.
Lynes, H. (1934). Contribution to the ornithology of southern Tanganiyka Territory: birds of the Ubena-Uhehe highlands and Iringa upland. *J. Orn.* **82a**, 1–147.
MacDonald, J. D. (1952). Variation in the Capped Wheatear *Oenanthe pileata*. *Ostrich* **23**, 160–161.
MacGregor, D. E. (1950). Notes on the breeding of the Red-breasted Chat *Oenanthe heuglini*. *Ibis* **92**, 380–383.
Mackworth-Praed, C. W. and Grant, C. H. B. (1951). On the races of the Wheatear *Oenanthe oenanthe* (Linnaeus)

occurring in eastern Africa. *Ibis* **93**, 234–236.

Madden, J. F. (1935). Notes on the birds of southern Darfur, Part II – Passerine birds. *Sudan Notes Rec.* **18**, 103–118.

Malbrant, R. (1954). Contribution à l'étude des oiseaux du Borkou-Ennedi-Tibesti. *Oiseau et R.F.O.* **24**, 1–47.

Malbrant, R. and Maclatchy, A. (1949). 'Faune de l'équateur africain français. II. Oiseaux'. 2nd ed., Lechevalier, Paris.

Mayaud, N. (1951). Le plumage prénuptial d'*Oenanthe oenanthe seebohmi* (Dixon). *Alauda* **19**, 88–96.

Mayr, E. and Stresemann, E. (1950). Polymorphism in the chat genus *Oenanthe* (Aves). *Evolution* **4**, 291–300.

Meinerthagen, R. (1930). 'Nicoll's Birds of Egypt'. Hugh Rees Ltd, London.

Meinertzhagen, R. (1934). The biogeographical status of the Ahaggar Plateau in the central Sahara, with special reference to birds. *Ibis* **13**(4), 528–571.

Meinertzhagen, R. (1940). Autumn in Central Morocco. *Ibis* **14**(4), 106–136, 187.

Meinertzhagen, R. (1949). New races of a courser, woodpecker, swift, lark, wheatear and serin from Africa. *Bull. Br. Orn. Club* **69**, 104–108.

Meinertzhagen, R. (1954). 'Birds of Arabia.' Oliver and Boyd, Edinburgh.

Miskell, J. E. (in press). Vocalizations of the Somali Wheatear *Oenanthe phillipsi*. *Scopus*.

Monard, A. (1934). Ornithologie de l'Angola. *Arq. Mus. Bocage* **5**, 1–110.

Monk, J. F. and Johnson, E. D. H. (1975). Palaearctic bird migration in the northern Algerian Sahara, spring 1973. *Ardeola* **21**, 875–902.

Moreau, R. E. (1941). The ornithology of the Siwa Oasis, with particular reference to the results of the Armstrong College Expedition, 1935. *Bull. Inst. Egypte* **23**, 247–261.

Moreau, R. E. and Moreau, W. M. (1928). Some notes on the habits of Palaearctic migrants while in Egypt. *Ibis* **12**(4), 233–252.

Morel, G. (1968). Contribution à la synécologie des oiseaux du sahel sénégalais. *Mém. Off. Rech. Scient. Tech. Outre-Mer Paris* **29**, 1–179.

Morel, G. J. (1972). Liste commentée des oiseaux du Sénégal et de la Gambie. Off. Rech. Scient. Tech. Outre-Mer, Dakar.

Morrison, A. F. (1945). Notes on the occurrence of migrant wheatears at Nairobi. *J. E. Afr. Uganda Nat. Hist. Soc.* **18**, 116–121.

Palfery, J. (1988). Observations on the behaviour of the White-crowned Black Wheatear in eastern Arabia. *Sandgrouse* **10**, 1–25.

Pasteur, G. (1956). Premières observations sur le traquet, le bruant et l'ammomane du poste d'Aouïnet-Torkoz (bas Dra). *Bull. Soc. Sci. Nat. Phys. Maroc* **36**, 165–184.

Péris, S. J. (1981). Observations ornithologiques dans le sud-ouest du Maroc. *Bull. Inst. Sci. Rabat* **5**, 135–141.

Pineau, J. (1976). Première capture au Maroc du Traquet isabelle *Oenanthe isabellina*. *Oiseau et R.F.O.* **46**, 75–76.

Pineau, J. and Giraud-Audine, M. (1977). Notes sur les oiseaux nicheurs de l'extrême nord-ouest du Maroc: reproduction et mouvements. *Alauda* **45**, 75–103.

Plowes, D. C. H. (1948). The Mountain Chat at the nest. *Ostrich* **19**, 80–88.

Priest, C. D. (1948). 'Eggs of Birds Breeding in Southern Africa.' Glasgow University Press, Glasgow.

Raw, W., Sparrow, R. and Jourdain, F. C. R. (1921). Field notes on the birds of Lower Egypt. *Ibis* **11**(3), 238–264.

Richardson, F. (1965). Breeding and feeding habits of the Black Wheatear *Oenanthe leucura* in southern Spain. *Ibis* **107**, 1–16.

Ripley, S. D. (1963). Brief bird observations in Nubia. *Ibis* **105**, 108–109.

Rothschild, W. and Hartert, E. (1912). Ornithological explorations in Algeria. *Novit. Zool.* **18**, 456–550.

Salvan, J. (1964). Nouvelles observations de migrateurs du Paléarctique dans l'Est du Tchad (printemps 1963). *Oiseau et R.F.O.* **34**, 78–97.

Santos, T. and Suárez, F. (1985). The intersexual differentiation in the foraging behaviour of *Oenanthe hispanica* L. during the breeding season. *Doñana Acta Vert.* **12**, 93–103.

Sclater, W. L. and Mackworth-Praed, C. (1918). A list of the birds of the Anglo-Egyptian Sudan, based on the collections of Mr. A. L. Butler, Mr. A. Chapman and Capt. H. Lynes RN, and Major Cuthbert Christy, RAMC(TF). *Ibis* **10**(6), 416–476, 602–721.

Sharland, R. E. (1967). Wheatears in winter quarters. *Bull. Niger. Orn. Soc.* **16**, 23.

Short, L. L. and Horne, J. F. M. (1981). Bird observations along the Egyptian Nile. *Sandgrouse* **3**, 43–61.

Simmons, K. E. L. (1954). Field-notes on the behaviour of some Passerines migrating through Egypt. *Ardea* **42**, 140–151.

Sinclair, A. R. E. (1978). Factors affecting the food supply and breeding season of resident birds and movements of palaearctic migrants in a tropical savannah. *Ibis* **120**, 480–497.

Sinclair, J. C., Garland, I. and Carte, A. (1986). Pied Wheatear and Red-throated Pipit in southern Africa. *Bokmakierie* **38**, 45.

Sluys, R. and Van den Berg, M. (1982). On the specific status of the Cyprus Pied Wheatear *Oenanthe cypriaca*. *Ornis Scand.* **13**, 123–128.

Smith, K. D. (1971). Notes on *Oenanthe* species in winter in Africa. *Bird Study* **18**, 71–79.

Smith, V. W. (1966a). Birds seen on a trans-Saharan overland crossing in Spring 1966. *Bull. Niger. Orn. Soc.* **3**, 50–58.

Smith, V. W. (1966b). Breeding records for the Plateau Province over 3000 feet, 1957–1966. *Bull. Niger. Orn. Soc.* **3**(12), 78–91.

Spencer, R. and Hudson, R. (1978). Report on bird-ringing for 1976. *Ringing Migr.* **1**, 189–252.

Sutton, R. W. W. (1970). Bird records from Niger and Mali: July–August 1969. *Bull. Niger. Orn. Soc.* **7**, 56–59.

Taylor, J. S. (1946). Notes on the Mountain Chat (*Oenanthe monticola* Vieill.). *Ostrich* **17**, 248–253.

Thévenot, M., Bergier, P. and Beaubrun, P. (1980). Compte-rendu d'ornithologie marocaine, année 1979. *Documents de l'Institut Scientifique* **5**, Univ. Mohammed V, Rabat.

Ticehurst, C. B. (1922). Notes on some Indian wheatears. *Ibis* **11**(4), 151–158.

Ticehurst, C. B. (1927). On *Oenanthe hispanica*, *Oenanthe finschii* and *Oenanthe picata*. *Ibis* **12**(3), 65–74.

Traylor, M. A. (1961). Two new birds from Angola. *Bull. Br. Orn. Club.* **81**, 43–45.

Traylor, M. A. (1965). A collection of birds from Barotseland and Bechuanaland. *Ibis* **107**, 137–172.

Tristram, H. B. (1859). On the ornithology of northern Africa. *Ibis* **1**, 153–162, 277–301.

Tye, A. (1982). 'Social organization and feeding in the Wheatear and Fieldfare.' Ph.D. Thesis, University of Cambridge.

Tye, A. (1984). Attacks by shrikes *Lanius* spp. on wheatears *Oenanthe* spp.: competition, kleptoparasitism or predation? *Ibis* **126**, 95–102.

Tye, A. (1986). Plumage stages, moults, sexual dimorphism and systematic position of the Somali Wheatear *Oenanthe phillipsi*. *Bull. Br. Orn. Club* **106**, 104–111.

Tye, A. (1987). Clinal variation and subspeciation in the White-crowned Black Wheatear *Oenanthe leucopyga*. *Bull. Br. Orn. Club* **107**, 157–165.

Tye, A. (1988a). Foraging behaviour and selection of prey and perches by the Buffstreaked Chat *Oenanthe bifasciata*. *Ostrich* **59**, 105–115.

Tye, A. (1988b). Vocalizations and territorial behaviour by wheatears *Oenanthe* spp. in winter quarters. *Proc. VI Pan-Afr. Orn. Congr.* pp. 297–305.

Tye, A. (1989a). The systematic position of the Buff-streaked Chat (*Oenanthe/Saxicola bifasciata*). *Bull. Br. Orn. Club* 109, 53–58.

Tye, A. (1989b). Superspecies in the genus *Oenanthe* (Aves, Turdidae). *Bonn. zool. Beitr.* 40, 165–182.

Tye, A. and Tye, H. (1983). Field identification of Wheatear and Isabelline Wheatear. *British Birds* 76, 427–437.

Valverde, J. A. (1957). 'Aves del Sahara Español.' Inst. de Estudios Africanos, Madrid.

Vande weghe, J.-P. (1979). The wintering and migration of Palaearctic passerines in Rwanda. *Gerfaut* 69, 29–43.

Vaurie, C. (1949). Notes on the bird genus *Oenanthe* in Persia, Afghanistan and India. *Am. Mus. Novit.* 1425, 1–47.

Vaurie, C. (1950). Variation in *Oenanthe lugubris*. *Ibis* 92, 540–544.

Villiers, A. (1950). Contribution à l'étude de l'Aïr. Oiseaux. *Mém. Inst. Fr. afr. Noire* 10, 345–385.

Ward, P. (1963). Lipid levels in birds preparing to cross the Sahara. *Ibis* 105, 109–111.

Whitaker, J. I. S. (1905). 'The Birds of Tunisia'. Vol. I. R. H. Porter, London.

White, C. M. N. (1961). Notes on *Oenanthe pileata* (Gmelin). *Bull. Br. Orn. Club*. 81, 166–168.

Wilkinson, R. and Beecroft, R. (1985). Birds in Falgore Game Reserve, Nigeria. *Malimbus* 7, 63–72.

Zachai, G. (1984). [The reproductive biology of the Mourning Wheatear (*Oenanthe lugens*)]. (Hebrew, Eng. summary). *Sunbird* 2, 44–56.

Family TURDIDAE: thrushes
Other genera

Ames, P. L. (1975). The application of syringeal morphology to the classification of the Old World insect eaters (Muscicapidae). *Bonn. zool. Beitr.* 26, 107–134.

Ash, J. S. (1969). Spring weights of trans-Saharan migrants in Morocco. *Ibis* 111, 1–10.

Ash, J. S. (1973). *Luscinia megarhynchos* and *L. luscinia* in Ethiopia. *Ibis* 115, 267–269.

Ash, J. S. (1980). Migrational status of Palaearctic birds in Ethiopia. *Proc. IV Pan-Afr. Orn. Congr.* pp. 199–208.

Aspinwall, D. R. (1979). Bird notes from Zambezi District, North-western Province. *Zambian Orn. Soc. Occ. Pap.* 2, 1–60.

Backhurst, G. C., Britton, P. L. and Mann, C. F. (1973). The less common Palaearctic migrant birds of Kenya and Tanzania. *J. East Afr. Nat. Hist. Soc. Nat. Mus.* 140, 1–38.

Baker, N. E. (1983). Eastern Bearded Scrub Robin *Cercotrichas quadrivirgata* bathing in a tree hole. *Scopus* 7, 95–96.

Barbour, D. (1972). Some notes on Arnot's Chat. *Bokmakierie* 24, 16–17.

Bates, G. L. (1927). Notes on some birds of Cameroon and the Lake Chad Region: their status and breeding-times. *Ibis* 69, 1–64.

Bates, G. L. (1934). Birds of the southern Sahara and adjoining countries in French West Africa. Part 4. *Ibis* 76, 439–466.

Belcher, C. F. (1925). Birds of the Luchenya Plateau, Mlanje, Nyasaland. *Ibis* 12(1), 797–814.

Belcher, C. F. (1930). 'The Birds of Nyasaland'. Technical Press, London.

Benson, C. W. (1940). Further notes on Nyasaland birds (with particular reference to those of the Northern Province). Part III. *Ibis* 14(4), 583–629.

Benson, C. W. (1944). Notes from Nyasaland. *Ibis* 86, 445–480.

Benson, C. W. (1946a). Notes on the birds of southern Abyssinia. *Ibis* 88, 180–205.

Benson, C. W. (1946b). Notes on eastern and southern African birds. *Bull. Br. Orn. Club* 67, 28–33.

Benson, C. W. (1947). The birds of Mzimbiti, near Beira, Portuguese East Africa. *Ostrich* 28, 125–128.

Benson, C. W. (1951). Breeding and other notes from Nyasaland and the Lundazi District of Northern Rhodesia. *Bull. Mus. Comp. Zool.* 106, 69–114.

Benson, C. W. (1962). The type-locality of Sharpe's Akalat. *Nyasaland Mus. Ann. Rep. Bull. 1961–1962.*

Benson, C. W. (1982). Migrants in the Afrotropical region south of the Equator. *Ostrich* 53, 31–49.

Benson, C. W. and Benson, F. M. (1947). On some breeding and other records from Nyasaland. *Ibis* 89, 285.

Benson, C. W. and Benson, F. M. (1949). Notes on some birds of northern Nyasaland and adjacent Tanganyika Territory. *Ann. Transvaal Mus.* 21, 155–177.

Benson, C. W. and Irwin, M. P. S. (1966). The *Brachystegia* Avifauna. *Ostrich* (Suppl.) 6, 297–321.

Benson, C. W. and Irwin, M. P. S. (1967). A contribution to the ornithology of Zambia. *Zambia Mus. Pap.* 1, 1–139.

Benson, C. W. and Irwin, M. P. S. (1975). The systematic position of *Phyllastrephus orostruthus* and *Phyllastrephus xanthophrys*, two species incorrectly placed in Pycnonotidae (Aves). *Arnoldia (Rhod.)* 7(17), 1–10.

Benson, C. W. and Pitman, C. R. S. (1963). Further breeding records from Northern Rhodesia. *Bull. Br. Orn. Club* 83, 32–36.

Berlioz, J. and Roche, J. (1970). Note sur le *Cercomela spectatrix* Clarke et description d'une sous-espèce nouvelle. *Monit. Zool. Ital. (N.S.) (Suppl.)* 3, 267–271.

Best, J. R. (1977). Thrush Nightingale *Luscinia luscinia* new to Nigeria and West Africa. *Bull. Niger. Orn. Soc.* 14, 81.

Beven, G. (1970). Studies of less familiar birds. 160. Rufous Bush Chat. *British Birds* 63, 294–299.

Blondel, J. (1962). Données écologiques sur l'avifaune des Monts des Ksours (Sahara septentrional). *Terre Vie* 109, 209–251.

Boulton, R. and Rand, A. L. (1952). A collection of birds from Mount Cameroon. *Fieldiana Zool.* 34, 35–64.

Braun, R. H. (1956). Beitrag zur biologie von *Xenocopsychus ansorgei* Hartert. *J. Orn.* 97, 41–43.

Brelsford, W. V. (1943). Field observations, Northern Province, Northern Rhodesia. *Ostrich* 14, 170–178.

Britton, H. A. (1979). E.A.N.H.S. nest record scheme: 1979. *Scopus* 3, 121–131.

Britton, P. L. (1971). On the apparent movements of *Cossypha natalensis*. *Bull. Br. Orn. Club* 91, 137–144.

Britton, P. L. and Rathbun, G. B. (1978). Two migratory thrushes and the African Pitta at the Kenya coast. *Scopus* 2, 11–17.

Brooke, R. K. (1964). Avian observations on a journey across central Africa and additional information on some of the species seen. *Ostrich* 35, 277–292.

Brooks, D. J., Evans, M. I., Martins, P. R. and Porter, R. F. (1987). The status of birds in north Yemen and the records of the OSME Expedition in autumn 1985. *Sandgrouse* 9, 4–66.

Brosset, A. and Erard, C. (1976). Première description de la nidification de quatre espèces en forêt Gabonaise. *Alauda*

44, 205-235.
Brown, L. H. (1970). Recent new breeding records for Kenya. *Bull. Br. Orn. Club* **90**, 2-6.
Bundy, G. (1986). Blackstarts in southern Oman. *Sandgrouse* **7**, 43-46.
Calder, D. R. (1961). *Natal Bird Club News Letter* **75**.
Calder, D. R. (1962). *Natal Bird Club News Letter* **83**.
Campbell, N. A. (1988). Questions about Dabchicks and unfamiliar chat. *Honeyguide* **34**, 178.
Carter, C. (1978). First recorded nest site of *Alethe fuelleborni*. *Scopus* **2**, 25.
Chapin, R. T. (1978). Brief accounts of some central African birds, based on the journals of James Chapin. *Rev. Zool. Afr.* **92**, 805-836.
Cheesman, R. E. (1935). On a collection of birds from north-western Abyssinia. *Ibis* **77**, 151-191.
Cheeseman, R. E. and Sclater, W. L. (1935). On a collection of birds from north-western Abyssinia. Part III. *Ibis* **77**, 594-622.
Clancey, P. A. (1952). Birds collected on the Natal Museum Expedition. *Ann. Natal Mus.* **12**(2), 252-253.
Clancey, P. A. (1956). Miscellaneous taxonomic notes on African birds VII. *Durban Mus. Novit.* **4**(17), 273-291.
Clancey, P. A. (1958). Miscellaneous taxonomic notes on African birds X. *Durban Mus. Novit.* **5**(8), 99-110.
Clancey, P. A. (1961a). The South African races of the Stonechat *Saxicola torquata* (Linnaeus). *Durban Mus. Novit.* **6**, 87-96.
Clancey, P. A. (1961b). Miscellaneous taxonomic notes on African birds XVI. *Durban Mus. Novit.* **6**, 79-104.
Clancey, P. A. (1962). The South African races of the Familiar Chat *Cercomela familiaris* (Stephens.) *Ostrich* **33**, 24-28.
Clancey, P. A. (1962a). The South African races of the Mocking Chat *Thamnolaea cinnamomeiventris* (Lafresnaye). *Occ. Pap. Nat. Mus. Sth. Rhod.* **3**(26b), 739-745.
Clancey, P. A. (1962b). Miscellaneous taxonomic notes on African birds XIX. *Durban Mus. Novit.* **6**(15), 181-194.
Clancey, P. A. (1966). A catalogue of birds of the South African sub-region. *Durban Mus. Novit.* **7**(11), 389-464.
Clancey, P. A. (1981a). Variation in the Chorister Robin *Cossypha dichroa* (Gmelin), 1789. Miscellaneous taxonomic notes on African birds, 60. *Durban Mus. Novit.* **13**(1), 6-10.
Clancey, P. A. (1981b). Miscellaneous taxonomic notes on African birds, 61. *Durban Mus. Novit.* **13**(4), 41-47.
Clancey, P. A. (1982). On the robins *Cossypha dichroa* and *C. natalensis* (Aves: Turdidae) in southern Africa. *Bonn. zool. Beitr.* **33**, 293-302.
Clancey, P. A. and Lawson, W. J. (1969). A new race of White-breasted Alethe from Mozambique. *Bull. Br. Orn. Club* **89**, 4-6.
Clarke, G. (1985). Bird observations from northwest Somalia. *Scopus* **9**, 24-42.
Cornwallis, L. and Porter, R. F. (1982). Spring observations of the birds of North Yemen. *Sandgrouse* **4**, 1-36.
Courteney-Latimer, M. (1961). On the races of the Stonechat occurring in the Cape Province, South Africa, with the description of a new form. *Bull. Br. Orn. Club* **81**, 114-117.
Cunningham-van Someren, G. R. (1977). On the nest of the Brown-chested Alethe (*Alethe poliocephala*). *Bull. E. Afr. Nat. Hist. Soc.*, Jan-Feb, p. 6.
Cunningham-van Someren, G. R. and Schifter, H. (1981). New races of montane birds from Kenya and southern Sudan. *Bull. Br. Orn. Club* **101**, 355-363.
Curio, E. (1989). Is avian mortality preprogrammed? *Trends Ecol. Evol.* **4**, 81-82.
Cyrus, D. P. (1989). Seasonal and spatial distribution of the Cape Robin *Cossypha caffra* on the coastal plain of Zululand. *Ostrich* **60**, 22-26.
Davis, P. G. (1975). Probable bigamy in Nightingale. *British Birds* **68**, 77-78.

De Roo, A., De Vree, F. and Verheyen, W. (1969). Contribution à l'ornithologie de la Republique du Togo. *Rev. Zool. Bot. Afr.* **79**, 309-322.
Dean, W. R. J. (1976). Breeding records of *Crex egregia*, *Myrmecocichla nigra* and *Cichladusa ruficauda* from Angola. *Bull. Br. Orn. Club* **96**, 48-49.
Dean, W. R. J. and Dean, S. J. (1987). A hazard for birds. *Promerops* **179**, 10.
Dick, J. A. (1981). A comparison of foods eaten by Swynnerton's Robin and Starred Robin in the Chirinda Forest. *Ostrich* **52**, 251-253.
Dittami, J. P. and Gwinner, E. (1985). Annual cycles in the African stonechat *Saxicola torquata axillaris* and their relationship to environmental factors. *J. Zool. Lond. (A)* **207**, 357-370.
Donnelly, G. B. (1967). Some observations on behaviour and call in the Palm Thrush *Cichladusa arquata*. *Ostrich* **38**, 230-232.
Douaud, J. (1957). Les migrations au Togo (Afrique Occidentale). *Alauda* **25**, 241-266.
Dowsett, R. J. (1989). A preliminary natural history survey of Mambilla Plateau and some lowland forests of eastern Nigeria. *Tauraco Res. Rep.* **1**.
Dowsett, R. J. and Dowsett-Lemaire, F. (1986). Homing ability and territorial replacement in some forest birds in south-central Africa. *Ostrich* **57**, 25-31.
Dowsett, R. J. and Fry, C. H. (1971). Weight loss of trans-Saharan migrants. *Ibis* **113**, 531-533.
Dowsett-Lemaire, F. (1987). On the distribution, ecology and voice of two *Alethe* species in Malaŵi. *Scopus* **11**, 25-32.
Duhart, F. and Deschamps, M. (1963). Notes sur l'avifaune du delta Central Nigeria. *Oiseau et R.F.O.* (special no.), 106 pp.
Dymond, J. N., Fraser, P. A. and Gantlett, S. J. M. (1989). 'Rare Birds in Britain and Ireland.' T. and A. D. Poyser, Calton.
Eames, J. C. (1986). A record of the Black Bush Chat in southern Israel. *Sandgrouse* **7**, 60-61.
Earlé, R. A. (1981a). An unusual nest of the Rock Pigeon. *Bokmakierie* **83**, 6-7.
Earlé, R. A. (1981b). Factors governing avian breeding in *Acacia* savanna, Pietermaritzburg. Pt 3: Breeding success, recruitment and clutch size. *Ostrich* **52**, 235-243.
Earlé, R. A. (1987). Nest eviction of Greater Striped Swallows by Anteating Chat. *Mirafra* **4**, 35-36.
Earlé, R. A. and Herholdt, J. J. (1986). Cooperative breeding in the Anteating Chat. *Ostrich* **57**, 188-189.
Earlé, R. A. and Herholdt, J. J. (1987). Notes on a Greater Honeyguide *Indicator indicator* chick raised by Anteating Chats *Myrmecocichla formicivora*. *Bull. Br. Orn. Club* **107**, 70-73.
Earlé, R. A. and Herholdt, J. J. (1988). Breeding and moult of the anteating chat *Myrmecocichla formicivora*. *Ostrich* **59**, 155-161.
Earlé, R. A. and Louw, S. v. d. M. (1988). Diet of the anteating chat *Myrmecocichla formicivora* in relation to terrestrial arthropod abundance. *S. Afr. J. Zool.* **23**(3), 224-229.
Eisentraut, M. (1963). 'Die Wirbeltiere des Kamerungebirges'. Paul Parey, Hamburg and Berlin.
Elliott, H. F. I. and Fuggles-Couchman, N. R. (1948). An ecological survey of the birds of the Crater Highlands and Rift Valley lakes, northern Tanganyika territory. *Ibis* **90**, 394-425.
Erard, C. (1987). 'Ecologie et comportement des Gobemouches (Aves: Muscicapinae, Platysteirinae, Monarchinae) du Nord-Est du Gabon. Vol. 1. Morphologie des espèces et organisation du peuplement'. *Mém. Mus. natn. Hist. nat., A. Zool.* **138**.
Erard, C. (1990). 'Ecologie et comportement des Gobemouches (Aves: Muscicapinae, Platysteirinae, Monarchinae) du

Nord-Est du Gabon. Vol. 2. Organisation sociale, comportements, biologie de la reproduction'. *Mém Mus. natn. Hist. nat.*, A. Zool. **146**.

Erard, C. and Yeatman, L. (1967). Sur les migrations de *Phoenicurus ochruros gibraltariensis* (Gmelin) d'après les données du baguage. *Oiseau et R.F.O.* **37**, 20–47.

Everitt, C. (1964). Breeding the Natal Robin. *Foreign Birds* **30**(3), 111–112.

Farkas, T. (1961). Notes on the behaviour of the Mocking Chat *Thamnolaea cinnamomeiventris* (Lafr.) in western Transvaal. *Ostrich* **32**, 122–127.

Farkas, T. (1962). Contribution to the bird fauna of Barberspan. *Ostrich* (Suppl.) **4**, 1–39.

Farkas, T. (1966). Notes on the breeding activities and post-embryo developments in the Mocking Chat *Thamnolaea cinnamomeiventris cinnamomeiventris* (Lafresnaye). *Ostrich* (Suppl.) **6**, 95–107.

Farkas, T. (1969). Notes on the biology and ethology of the Natal Robin *Cossypha natalensis*. *Ibis* **111**, 281–292.

Farkas, T. (1973). Notes on the biology and ethology of Heuglin's Robin *Cossypha heuglini*. *Ostrich* **44**, 95–105.

Farkas, T. (1988). The Birds of Korranaberg, eastern Orange Free State, South Africa. *Navors. nas. Mus. Bloemfontein* **6**(3), 35–108.

Friedmann, H. (1966). A contribution to the ornithology of Uganda. *Bull. Los Angeles County Mus. Nat Hist. Sci.* **3**, 1–55.

Friedmann, H. and Williams, J. G. (1968). Notable records of rare and little-known forest birds from western Uganda. *Rev. Zool. Bot. Afr.* **77**, 11–36.

Fry, C. H. (1964). *Neocossyphus* and *Stizorhina* and the relationship of thrushes with flycatchers. *Bull. Niger. Orn. Soc.* **1**(2), 10–12.

Fry, C. H. (1966a). Preliminary notice of the Bambara Cliff-Chat in Nigeria. *Bull. Niger. Orn. Soc.* **3**, 48.

Fry, C. H. (1966b). Notes on status of Bluethroat in Nigeria. *Bull. Niger. Orn. Soc.* **3**, 98–99.

Fry, C. H. (1969). Migrations, moult and weights of birds in northern Guinea savanna in Nigeria and Ghana. *Ostrich* (Suppl.) **8**, 239.

Fuggles-Couchman, N. R. (1986). Breeding records of some Tanzanian birds. *Scopus* **10**, 20–26.

Fuggles-Couchman, N. R. and Elliott, H. F. I. (1946). Some records and field-notes from north-eastern Tanganyika territory. *Ibis* **88**, 327–347.

Gallagher, M. and Woodcock, M. (1980). 'Birds of Oman'. Quartet Books, London.

Gee, J. P. (1984). The birds of Mauritania. *Malimbus* **6**, 31–66.

Gichuki, C. M. and Schifter, H. (1990). Long life-span and sedentariness of birds in North Nandi Forest, Kenya. *Scopus* **14**, 24–25.

Goodman, S. M. and Atta, G. A. M. (1987). The birds of southeastern Egypt. *Gerfaut* **77**, 3–31.

Goodman, S. M. and Meininger, P. L. (1989). Discovery of the Arabian Warbler *Sylvia leucomelaena* in southeastern Egypt. *Courser* **2**, 24–26.

Grant, C. H. B. and Mackworth-Praed, C. W. (1940a). On the resident races of the Stonechat in eastern Africa. *Bull. Br. Orn. Club* **61**, 16–18.

Grant, C. H. B. and Mackworth-Praed, C. W. (1940b). On the races of the Black Redstart occurring in eastern Africa. *Bull. Br. Orn. Club* **61**, 22.

Grant, C. H. B. and Mackworth-Praed, C. W. (1940c). Notes on eastern African birds: (4) On the races of *Pinarochroa sordida* (Rüppell, N. Wirbelt, Vög. 1837, p. 75, pl. 26, fig. 2: Simen, northern Abyssinia) occurring in Abyssinia. *Bull. Br. Orn. Club* **61**, 8.

Grant, C. H. B. and Mackworth-Praed, C. W. (1946–47). On the migratory Stonechats of eastern Africa. *Bull. Br. Orn. Club* **67**, 47–48.

Green, A. A. (1980). Two populations of cliff-chats in the Arli-Pendjari region. *Malimbus* **2**, 99–101.

Grobler, J. H. and Steyn, P. (1980). Breeding habits of the Boulder Chat and its parasitism by the Red-chested Cuckoo. *Ostrich* **51**, 253–254.

Guichard, K. M. (1955). The birds of Fezzan and Tibesti. *Ibis* **97**, 393–424.

Haas, V. (1986). Social organisation and role of helpers in Anteater Chats (*Myrmecocichla aethiops*) and Mourning Wheatear (*Oenanthe lugens schalowi*). *Proc. XVIII Int. Orn. Congr.* pp. 1109–1110.

Hall, B. P. (1960). The faunistic importance of the scarp of Angola. *Ibis* **102**, 420–442.

Hall, B. P. (1961). Is *Muscicapa gabela* an Akalat? *Bull. Br. Orn. Club* **81**, 45–46.

Hall, B. P. and Moreau, R. E. (1962). A study of the rare birds of Africa. *Bull. Brit. Mus. (Nat. Hist.) Zool.* **8**, 313–378.

Hanmer, D. B. (1979). A trapping study of Palaearctic passerines at Nchalo, southern Malaŵi. *Scopus* **3**, 81–92.

Hanmer, D. B. (1989a). First record of the Rufous Bush Chat *Cercotrichas galactotes* from Malaŵi. *Scopus* **12**, 93–94.

Hanmer, D. B. (Ed.) (1989b). Flora and Fauna Records. *Nyala* **14**, 55–66.

Harcus, J. L. (1977). The functions of mimicry in the vocal behaviour of the Chorister Robin. *Z. Tierpsychol.* **44**, 178–193.

Harpum, J. (1978). Species-pair association of Stonechat and Black-lored Cisticola in southwest Tanzania. *Scopus* **2**, 99–101.

Harrison, I. D. (1977). Extension of the range of the White-headed Robin-chat *Cossypha heinrichi*. *Bull. Br. Orn. Club* **97**, 20–21.

Hartley, P. H. T. (1952). The biology of *Cercomela melanura* Temminck. *Bull. Br. Orn. Club* **72**, 71–73.

von Hecke, P. (1965). The migration of the West European Stonechat *Saxicola torquata* (L.), according to ringing-data. *Gerfaut* **55**, 146–194.

Heim de Balsac, H. and Mayaud, N. (1951). Sur la morphologie, la biologie et la systématique de *Cercotrichas podobe* (P. L. S. Müller). *Alauda* **19**, 137–151.

von Helversen, D. (1980). Structure and function of antiphonal duets. In: Nohring, R. (ed.) *Acta XVII Cong. Int. Orn.* pp. 682–688.

Herholdt, J. J. (1987). Some notes on the general behaviour of the Anteating Chat *Myrmecocichla formicivora*. *Mirafra* **4**, 63–66.

Herrera, C. M. (1977). Ecología alimenticia del Petirrojo (*Erithacus rubecula*) durante sur invernada en encinares del sur de España. *Doñana Acta Vert.* **4**, 35–59.

Herrera, C. M. (1984). A study of avian frugivores, bird-dispersed plants, and their interaction in Mediterranean scrublands. *Ecol. Monogr.* **54**, 1–23.

Hezekia, G. (1987). Fishing by Common Stonechat. *Honeyguide* **33**, 18.

Hockey, P. A. R. and the Rarities Committee (1990). Rare birds in southern Africa: 1988. *Birding in Southern Africa* **42**(2), 34–38.

Huff, J. N. and Auta, J. (1977). Nests of White-fronted Black Chats *Myrmecocichla albifrons*. *Bull. Niger. Orn. Soc.* **44**, 148.

Irwin, M. P. S. (1957). On the type locality of *Pinarornis plumosus* Sharpe. *Bull. Br. Orn. Club* **77**, 9–10.

Irwin, M. P. S. (1963). Systematic and distributional notes on southern African birds. *Durban Mus. Novit.* **7**, 1–26.

Irwin, M. P. S. (1983). The present status of the Collared Palm Thrush in the Middle Zambesi Valley. *Honeyguide* **114/115**, 59–60.

Irwin, M. P. S. and Clancey, P. A. (1974). A re-appraisal of the generic relationships of some African forest-dwelling robins (Aves: Turdidae). *Arnoldia (Rhod.)* **6**(34), 1–19.

Irwin, M. P. S. and Clancey, P. A. (1985). The systematic position of the genus *Modulatrix*. *Honeyguide* **31**, 162–165.

Irwin, M. P. S. and Clancey, P. A. (1986). A new generic status for the Dappled Mountain Robin. *Bull. Br. Orn. Club* **106**, 111–115.

James, H. W. (1929). Nesting of *Phoenicurus f. familiaris* (Vieill.) – Cape Familiar Chat. *Ool. Rec.* **9**, 19–21.

Jarry, G. and Larigauderie, F. (1971). *Saxicola torquata* (Linnaeus) breeding in Senegal. *Bull. Br. Orn. Club* **91**, 32.

Jehl, H. (1976). Les oiseaux de l'Ile de Kembe (R.C.A.). *Alauda* **44**, 153–167.

Jennings, M. C. (1981). 'The Birds of Saudi Arabia: a Checklist'. Published by the author (10, Mill Lane, Whittlesford, Cambridge, England).

Jensen, F. P. (1990). A review of some genera of African chats (Aves, Muscicapidae, Erithacini). *Steenstrupia* **15**, 161–175.

Jensen, F. P. and Stuart, S. N. (1982). New subspecies of forest birds from Tanzania. *Bull. Br. Orn. Club* **102**, 95–99.

Jensen, F. P., Brogger-Jensen, S. and Petersen, G. (1985). The White-chested Alethe *Alethe fuelleborni* in Tanzania. *Scopus* **9**, 127–132.

Jensen, R. A. C. and Jensen, M. K. (1971). First breeding records of the Herero Chat *Namibornis herero*, and taxonomic implications. *Ostrich* (Suppl.) **8**, 105–116.

Johnson, E. D. H. (1971a). Observations on a resident population of Stonechats in Jersey. *British Birds* **64**, 201–213, 267–279.

Johnson, E. D. H. (1971b). Wintering of *Saxicola torquata* in the Algerian Sahara. *Bull. Br. Orn. Club* **91**, 103–107.

Johnston-Stewart, N. G. B. (1984). Evergreen forest birds in the southern third of Malaŵi. *Nyala* **10**, 99–119.

Johnston-Stewart, N. G. B. (1989). Thyolo Alethe breeding in Mulanje at 720 m a.s.l. *Nyala* **13**, 80–81.

Kalina, J. and Baranga, J. (1991). Nesting association between the Red-throated Alethe *Alethe p. poliophrys* and the Equatorial Akalat *Sheppardia aequatoralis*. *Scopus* **15**, 61–62.

Kalina, J. and Butynski, T. M. (1989). First nest record of the Equatorial Akalat *Sheppardia aequatorialis*. *Scopus* **13**(2), 132–133.

Keith, S. (1968). Notes on birds of East Africa, including additions to the avifauna. *Amer. Mus. Novit.* **2321**, 1–15.

Keith, S. (1980). Birds observed during the winter 1980 Nile Cruise of the American Museum of Natural History. Report of American Museum of Natural History.

Kersten, M., Piersma, T., Smit, C. and Zegers, P. (1983). 'Rep. Nether. Morocco Exped. 1981'. Res. Inst. Nature Management, Texel.

King, B. (1978). April bird observations in Saudi Arabia. *J. Saudi Arab. Nat. Hist. Soc.* **1**(21), 3–24.

Ledant, J. P. (1986). L'habitat du Traquet tarier dans le centre de la Côte d'Ivoire. *Gerfaut* **71**, 433–442.

Louette, M. (1981). Sur quelques specimens nouveaux de *Cossypha heinrichi* du Zaire (Aves, Turdinae). *Rev. Zool. Afr.* **95**, 356–358.

Ludlow, A. (1966). Body weight changes and moult in some Palaearctic migrants in southern Nigeria. *Ibis* **108**, 129–132.

Lynes, H. (1938). Contribution to the ornithology of the southern Congo basin. *Rev. Zool. Bot. Afr.* **31**, 1–129.

Macdonald J. D. (1940). On the genus *Sheppardia*. *Ibis* **14**(4), 663–671.

Mann, C. F. (1974). Songs of two Kenya turdids. *Auk* **91**, 166–167.

Manson, A. J. (1990). The biology of Swynnerton's Robin. *Honeyguide* **36**, 5–13.

Manson, C. (1985). Swynnerton's Robin at the forest edge. *Honeyguide* **31**, 220.

Marchant, S. (1942). Some birds of the Owerri Province, S. Nigeria. *Ibis* **14**(6), 137–196.

Masterson, A. N. B. (1981). Notes from the Ruwenzori Mountains, including a description of the nest and eggs of Archer's Ground Robin *Dryocichloides archeri*. *Scopus* **5**, 33–34.

Masterson, B. A. (1916). Observations on the birds of the district of Humansdorp, Cape Province. *J. S. Afr. Orn. Un.* **11**, 119–142.

Mayaud, N. (1989). Les oiseaux du nord-ouest de l'Afrique. Notes complémentaires. *Alauda* **57**, 10–16.

Medland, R. D. (1985). First breeding record of Boulderchat *Pinarornis plumosus* in Malaŵi. *Nyala* **11**, 83.

Mees, G. F. (1988). The type locality of *Alethe poliocephala* (Bonaparte). *Bull. Br. Orn. Club* **108**, 125–127.

Meinertzhagen, R. (1930). 'Nicoll's Birds of Egypt'. Hugh Rees, London.

Meinertzhagen, R. (1950). On *Oenanthe tractrac* (Wilkes) and *Oenanthe albicans* (Wahlberg) and the development of the 'sickle wing'. *Bull. Br. Orn. Club* **70**, 9–10.

Monroe, B. L. Jr. (1964). Wing-flashing in Red-backed Scrub-robin, *Erythropygia zambesiana*. *Auk* **81**, 91–92.

Moreau, R. E. (1935). A contribution to the ornithology of Kilimanjaro and Mount Meru. *Proc. Zool. Soc. London* **1935**, 843–891.

Moreau, R. E. (1951). Geographical variation and plumage sequence in *Pogonocichla*. *Ibis* **93**, 383–401.

Moreau, R. E. (1961). Problems of Mediterranean-Sahara migration. *Ibis* **103a**, 373–427, 580–623.

Moreau, R. E. and Benson, C. W. (1956). *Cossypha insulana* Grote conspecific with *Cossypha bocagei* Finsch and Hartlaub. *Bull. Br. Orn. Club* **76**, 62–63.

Moreau, R. E. and Dolp, R. M. (1970). Fat, water, weights and wing lengths of autumn migrants in transit on the northwest coast of Egypt. *Ibis* **112**, 209–228.

Moreau, R. E. and Moreau, W. M. (1928). Some notes on the habits of Palaearctic migrants while in Egypt. *Ibis* **12**(4), 233–252.

Moreau, R. E. and Moreau, W. M. (1937). Biological and other notes on some East African birds. Part 2. *Ibis* **14**(1), 321–345.

Morel, G. J. (1973). The Sahel zone as an environment for Palaearctic migrants. *Ibis* **115**, 413–417.

Morel, G. J. (1985). Les oiseaux des milieux rocheux au Sénégal. *Malimbus* **7**, 115–119.

Mundy, P. J. and Cook, A. W. (1974). Birds of Sokoto, Part 3, *Bull. Niger. Orn. Soc.* **10**, 1–28.

Newby, J., Grettenberger, J. and Watkins, J. (1987). The birds of the northern Aïr, Niger. *Malimbus* **9**, 4–16.

Niethammer, G. (1955). Zur Vogelwelt des Ennedi-Gebirges (Französisch Äquatorial-Afrika). *Bonn. zool. Beitr.* **6**, 29–80.

Nikolaus, G. (1983). An important passerine ringing site near the Sudan Red Sea coast. *Scopus* **7**, 15–18.

Nikolaus, G. and Pearson, D. J. (1982). Autumn passage of Marsh Warbler *Acrocephalus palustris* and Sprosser *Luscinia luscinia* on the Sudan Red Sea coast. *Scopus* **6**, 17–19.

North, M. E. W. (1958). 'Voices of African birds'. Cornell University Press, Ithaca, NY.

North, M. E. W. and McChesney, D. S. (1964). 'More Voices of African Birds'. Houghton Mifflin Co., Boston.

Oatley, T. B. (1959). Notes on the genus *Cossypha*, with particular reference to *C. natalensis* Smith and *C. dichroa* (Gmelin). *Ostrich* (Suppl.) **3**, 426–434.

Oatley, T. B. (1961). Notes on *Sheppardia aequatorialis*. *Ibis* **103a**, 290–291.

Oatley, T. B. (1964). The probing of *Aloe* flowers by birds. *Lammergeyer* **3**, 2–8.

Oatley, T. B. (1966). Competition and local migration in some African Turdidae. *Ostrich* (Suppl.) **6**, 409–418.

Oatley, T. B. (1969). Observations on bird ecology in the evergreen forests of north-western Zambia. *Puku* **5**, 141–180.

Oatley, T. B. (1970a). Observations on the food and feeding habits of some African robins (Aves: Turdinae). *Ann. Natal Mus.* **20**(2), 293–327.

Oatley, T. B. (1970b). Robin hosts of the Red-chested Cuckoo in Natal. *Ostrich* **41**, 232–236.

Oatley, T. B. (1971). The function of vocal imitation by African Cossyphas. *Ostrich* (Suppl.) **8**, 85–89.

Oatley, T. B. (1980). Eggs of two cuckoo genera in one nest. *Ostrich* **51**, 126–127.

Oatley, T. B. (1982a). The Starred Robin in Natal, Part 1: Behaviour, territory and habitat. *Ostrich* **53**, 135–146.

Oatley, T. B. (1982b). The Starred Robin in Natal, Part 2: Annual cycles and feeding ecology. *Ostrich* **53**, 193–205.

Oatley, T. B. (1982c). The Starred Robin in Natal, Part 3: Breeding, populations and plumages. *Ostrich* **53**, 206–221.

Oatley, T. B. and Tinley, K. L. (1989). The forest avifauna of Gorongoza Mountain, Mozambique. *Ostrich* (Suppl.) **14**, 57–61.

Oatley, T. B. (1990). Identifying Robin-Chats. *Birding in Southern Africa* **42**(3), 79–81.

Ogilvie-Grant, W. R. (1900). On the birds collected during an expedition through Somaliland and southern Abyssinia to the Blue Nile. *Ibis* Ser. 7 (**6**), 115–178.

Olson, S. L. (1984). Syringeal morphology and relationships of *Chaetops* (Timaliidae) and certain South African Muscicapidae. *Ostrich* **55**, 30–32.

Pearson, D. J. (1972). The wintering and migration of Palaearctic passerines at Kampala, southern Uganda. *Ibis* **114**, 43–60.

Pearson, D. J. (1984). The Nightingale, *Luscinia megarhynchos*, the Sprosser *L. luscinia* and the Irania *Irania gutturalis* in Kenya. *Scopus* **8**, 18–23.

Pearson, D. J. (1989). Palaearctic migrants in the Middle and Lower Jubba valley, southern Somalia. *Scopus* **13**, 53–60.

Pearson, D. J. (1990). East African bird report 1988 and Kenya – general review. *Scopus* **12**, 105–126.

Pearson, D. J. and Backhurst, G. C. (1976). The southward migration of Palaearctic birds over Ngulia, Kenya. *Ibis* **118**, 78–105.

Pearson, D. J. and Turner, D. A. (1986). The less common Palaearctic migrant birds of Uganda. *Scopus* **10**, 61–82.

Pearson, D. J., Nikolaus, G. and Ash, J. S. (1988). The southward migration of Palaearctic passerines through northeast and east tropical Africa: a review. *Proc. VI Pan-Afr. Orn. Congr.* pp. 243–262.

Penry, E. H. (1976). Notes on Sooty Chat. *Bull. Zambian Orn. Soc.* **8**, 27–28.

Penry, E. H. and Talbot, J. N. (1975). Notes on the birds of the higher altitudes on Mulanje Mountain, Malawi. *Honeyguide* **82**, 14–25.

Pitman, C. R. S. (1930). The nest and eggs of the Speckled Babbler *Cichladusa guttata guttata* (Heuglin). *Ool. Rec.* **10**, 7–10.

Pitman, C. R. S. (1956). Remarks on the nidification of the Kenya Hill Chat *Pinarochroa sordida ernesti* Sharpe. *Bull. Br. Orn. Club* **76**, 107.

Plowes, D. C. H. (1943). Bird-life at the Orange River mouth. *Ostrich* **14**, 133.

Plowes, D. C. H. (1946). Data on eggs in my collection. *Ostrich* **17**, 111–121.

Pollard, J. R. T. (1946). Nest and eggs of the Kenya Hill-Chat, *Pinarochroa sordida ernesti* Sharpe. *Ibis* **88**, 520–521.

Prigogine, A. (1954). Notes sur les oiseaux du genre *Sheppardia* du Congo Belge. *Rev. Zool. Bot. Afr.* **50**, 10–12.

Prigogine, A. (1984). L'Alèthe à poitrine brune, *Alethe poliocephala*, au Rwanda et au Burundi. *Gerfaut* **74**, 181–184.

Prigogine, A. (1987). Non-conspecificity of *Cossypha insulana* Grote and *Cossypha bocagei* Finsch and Hartlaub, with the description of a new subspecies of *Cossypha bocagei* from western Tanzania. *Bull. Br. Orn. Club* **107**, 49–55.

Rand, A. L. (1955). A new species of thrush from Angola. *Fieldiana Zool.* **34**, 327–329.

Reynolds, J. F. (1977). Thermo-regulatory problems of birds nesting in arid areas in East Africa: a review. *Scopus* **1**, 57–68.

Ripley, S. D. (1952). A new genus of thrush from eastern Africa. *Postilla* **12**.

Ripley, S. D. and Heinrich, G. H. (1966). Additions to the avifauna of northern Angola, II. *Postilla* **95**.

Ripley, S. D. and Heinrich, G. H. (1969). Comments on the avifauna of Tanzania, II. *Postilla* **134**.

Robertson, I. S. (1977). Identification and European status of eastern Stonechats. *British Birds* **70**, 237–245.

Roddis, J. (1964). Breeding the Red-capped Robin-Chat or Natal Robin. *Foreign Birds* **30**(5), 197.

Rooke, K. B. (1947). Notes on Robins wintering in North Algeria. *Ibis* **89**, 204–210.

da Rosa Pinto, A. A. (1962). As observacoes de maior destaque das expedicoes ornitologicas do instituto de investigacao cientifica de Angola. *Bol. Inst. Invest. Cient. Ang.* **1**, 21–38.

Rowan, M. K. (1969). A study of the Cape Robin in southern Africa. *Living Bird* **8**, 5–32.

Rowan, M. K. (1983). 'The Doves, Parrots, Louries and Cuckoos of Southern Africa'. David Philip, Cape Town.

Ryan, P. G., Cooper, J., Stutterheim, C. J., and Loutit, R. (1984). An annotated list of the birds of the Skeleton Coast Park. *Madoqua* **14**, 79–90.

Schouteden, H. (1965a). La faune ornithologique du Rwanda. *Mus. Roy. Afr. cent. Tervuren. Doc. Zool.* **9**.

Schouteden, H. (1965b). La faune ornithologique des territoires de Dilolo et Kolwezi de la Province du Katanga. *Mus. Roy. Afr. cent. Tervuren. Doc. Zool.* **9**.

Schouteden, H. (1970). Le rossignol européen au Congo (*Luscinia megarhynchos* Brehm). *Rev. Zool. Bot. Afr.* **81**, 301–302.

Serle, W. (1955). The bird-life of the Angolan littoral. *Ibis* **97**, 425–431.

Sharland, R. E. and Wilkinson, R. (1981). The birds of Kano State, Nigeria. *Malimbus* **3**, 7–30.

Shaughnessy, P. D. and Shaughnessy, G. L. (1987). Birds at Wolf and Van Reenen Bays, Diamond Coast, SWA/Namibia. *Lanioturdus* **23**, 27–43.

Sherry, B. Y. (1985). Boulder Chat *Pinarornis plumosus* in Majete Game Reserve. *Nyala* **11**, 25.

Short, L. L. and Horne, J. F. M. (1985). Notes on some birds of the Arabuko-Sokoke forest. *Scopus* **9**, 117–126.

Sim, L. (1979). 'Birds of Wondo Genet'. Orgut-Swedforest Consortium, Stockholm.

Simon, P. (1965). Synthèse de l'avifaune du massif montagneux du Tibesti. *Gerfaut* **55**, 26–71.

Skead, D. M. (1966). Birds frequenting the intertidal zone of the Cape Peninsula. *Ostrich* **37**, 10–16.

Skead, D. M. (1971). New Didric Cuckoo host record. *Ostrich* **42**, 74.

Skead, D. M. (1974). Habitats and feeding preferences of birds in S. A. Lombard Nature Reserve, Transvaal. *Ostrich* **45**, 15–21.

Smith, K. D. (1951). The behaviour of some birds on the British list in their winter quarters or on migration in Southern Rhodesia. *British Birds* **44**, 113–117.

Smith, K. D. (1955). The winter breeding season of land-birds in eastern Eritrea. *Ibis* **97**, 480–507.

Smith, V. W. (1966). Spring weights of some Palaearctic migrants in central Nigeria. *Ibis* **108**, 492–512.

Smith, V. W., Woods, P. J. and Ebbutt, D. P. (1966). Breeding notes on the White-crowned Cliff-Chat *Thamnolaea coronata* in central Nigeria. *Ibis* **108**, 137–138.

Steyn, P. (1966). Observations on the breeding biology of the

Familiar Chat, *Cercomela familiaris* (Stephens). *Ostrich* **37**, 176–183.

Steyn, P. (1967). Mobile Familiar Chat's nest. *Bokmakierie* **19**, 67.

Steyn, P. (1968). Longevity of Familiar Chat. *Ostrich* **39**, 267.

Steyn, P. and Hosking, E. (1988). Familiar Chats associating with Klipspringers. *Ostrich* **59**, 182.

Stuart, S. N. (1981). An explanation for the disjunct distributions of *Modulatrix orostruthus* and *Apalis* (or *Orthotomus*) *moreaui*. *Scopus* **5**, 1–4.

Stuart, S. N. and Gartshore, M. E. (1986). The Red-capped Robin-chat *Cossypha natalensis* in West Africa. *Malimbus* **8**, 73–76.

Stuart, S. N. and Turner, D. A. (1980). Some range extensions and other notable records of forest birds from eastern and northeastern Tanzania. *Scopus* **4**, 36–41.

Swardt, D. de (1989). A possible range expansion of the Karoo Chat? *Mirafra* **6**, 42–43.

Swynnerton, C. F. M. (1907). On the birds of Gazaland, Southern Rhodesia. *Ibis* **9**(1), 30–74.

Swynnerton, C. F. M. (1908). Further notes on the birds of Gazaland. *Ibis* **9**(2), 1–107.

Taylor, J. S. (1936). Birds in the garden. *Ostrich* **7**, 45–48.

Thiollay, J. M. (1985). The West African forest avifauna: a review. *In* 'Conservation of Tropical Forest Birds' (Diamond A. W. and Lovejoy J. E. Eds). ICBP Tech. Pub. No. **4**, pp. 171–186. Int. Council for Bird Preserv., Cambridge.

Tomlinson, W. (1950). Some notes chiefly from the Northern Frontier District of Kenya. Part II. *J. E. Afr. Nat. Hist. Soc.* **19**, 225–250.

Traylor, M. A. Jr. (1962). 'Notes on the birds of Angola, passeres'. Comp. Diam. Angola, Museo do Dundo, Lisbon.

Tye, A. (1988). Foraging behaviour and selection of prey and perches by the Buffstreaked Chat *Oenanthe bifasciata*. *Ostrich* **59**, 105–115.

Tye, A. (1989). The systematic position of the Buff-streaked Chat (*Oenanthe/Saxicola bifasciata*). *Bull. Br. Orn. Club* **109**, 53–59.

Tye, A. (1991). A new subspecies of Forest Scrub-Robin *Cercotrichas leucosticta* from West Africa. *Malimbus* **13**, 74–77.

Urban, E. K. (1978). 'Ethiopia's Endemic Birds'. Ethiopian Tourist Organization, Addis Ababa. 30 pp.

Uys, C. J. (1974). Vignettes on the chats found in the Western Cape Province. *Bokmakierie* **26**, 61–65.

Valverde, J. A. (1957). 'Aves del Sahara Español'. Inst. Estudios Africanos, Madrid.

Vande weghe, J.-P. (1979). The wintering and migration of Palaearctic passerines in Rwanda. *Gerfaut* **69**, 29–43.

Vande weghe, J.-P. (1988). Problems in passerine speciation in Rwanda, Burundi, and adjacent areas. *Proc. XIX Int. Orn. Congr.* pp. 2545–2552.

Vernon, C. J. (1967). Some observations from the journals of K. W. Greenhow. *Ostrich* **38**, 48–49.

Voous, K. H. (1960). 'Atlas van de Europese Vogels'. Elsevier, Amsterdam.

Vuilleumier, F. (1979). Ornithological observations during the fall 1979 Nile Cruise of the American Museum of Natural History, Oct. 15 to Oct. 30. Report of American Museum of Natural History.

Walker, F. J. (1981). Notes on the birds of Dhofar, Oman. *Sandgrouse* **2**, 56–58.

Walker, G. R. (1939). Notes on the birds of Sierra Leone. *Ibis* **3**(14), 401–450.

Walker, H. (1966). Nests in banana plantations. *Natal Bird Club Newsletter* **139**.

Walsh, J. F. (1987). Notes on the birds of Ivory Coast. *Malimbus* **8**, 89–92.

Walsh, J. F. and Grimes, L. E. (1981). Observations on some Palaearctic land birds in Ghana. *Bull. Br. Orn. Club* **101**, 327–334.

Watson, G. E. (1971). 'A Serological and Ectoparasite Survey of Migratory Birds in Northeast Africa.' Smithsonian Institution, Washington, D. C.

Webb, D. (1976). An albinistic Familiar Chat. *Honeyguide* **87**, 34.

Welch, G. R. and Welch, H. J. (1983). PSPB Nile Cruise, 9–25 March 1983: Ornithological Report. Roy. Soc. Preserv. Birds.

Welch, G. R. and Welch, H. J. (1984). Birds seen on an expedition to Djibouti. *Sandgrouse* **6**, 1–23.

Whitaker, J. I. S. (1905). 'The Birds of Tunisia'. R. H. Porter, London.

White, C. M. N. (1945). The genus *Cossypha* in Northern Rhodesia. *Ostrich* **16**, 194–200.

Wilkinson, R. (1979). Palaearctic Rufous Scrub-Robin: new to Nigeria. *Malimbus* **1**, 65.

Van der Willigen, T. A. and Lovett, J. C. (Eds.) (1981). Report of the Oxford Expedition to Tanzania 1979. Unpublished.

Willis, E. O. (1985). East African Turdidae as safari ant followers. *Gerfaut* **75**, 140–153.

Willis, E. O. (1986). West African thrushes as safari ant followers. *Gerfaut* **76**, 95–108.

Wilson, J. D. (1989). Range extensions of some bird species of Cameroon. *Bull. Br. Orn. Club* **109**, 110–115.

Wilson, K. J. (1964). A note on the stomach contents of a Palm Thrush (*Cichladusa arquata* Peters). *Puku* **2**, 130–131.

Winterbottom, J. M. (1952). Some notes on Northern Rhodesian birds. Part II. *N. Rhod. J.* **1**, 32–40.

Winterbottom, J. M. (1966). The comparative ecology of the birds of some Karoo habitats in the Cape Province. *Ostrich* **37**, 109–127.

Winterbottom, J. M. (1968a). The avifaunas of three agricultural habitats in the south-western Cape. *Ostrich* **39**, 51–60.

Winterbottom, J. M. (1968b). The bird population of karroid broken veld in the Klaarstroom area, Prince Albert Division. *Ostrich* **39**, 85–90.

Wood, B. (1989). Biometric, iris and bill colouration, and moult of Somali forest birds. *Bull. Br. Orn. Club* **109**, 11–22.

Zimmerman, D. A. (1967). *Agapornis fischeri*, *Lybius guifsobalito* and *Stiphrornis erythrothorax* in Kenya. *Auk* **84**, 594–595.

Zink, G. (1973, 1975, 1981). 'Der Zug Europäischer Singvögel', 1–3 Lfg. Vogelzug-Verlag, Möggingen.

Zink, G. (1980). The winter distribution of European passerines in Africa. *Proc. IV Pan-Afr. Orn. Congr.* pp. 209–213.

3. Acoustic References

Section A: Discs and Cassettes

5. North, M. E. W. (1958). Voices of African Birds. Cornell University Press. 159 Sapsucker Woods Road, Ithaca, N.Y. 14850. One 12-inch, $33\frac{1}{3}$ r.p.m. disc. 42 species. The first African record concerned mainly with identification. Species are presented in systematic order, grouped on separate bands, and details given of circumstances, place and date of recording.

7. Haagner, C. H. (1961). Birds of the Kruger National Park. International Library of African Music, P.O. Box 138, Roodeport, near Johannesburg, South Africa. Two 7-inch, 45 r.p.m. discs, Nos. XTR 17044 and XTR 27045. 31 species in systematic order following Roberts (1957. 'Birds of South Africa', Trustees of the John Voelcker Bird Book Fund, Cape Town.) and with the Roberts number; each on a separate band.

10. North, M. E. W. and McChesney, D. S. (1964). More Voices of African Birds. Houghton Mifflin Co., Boston, U.S.A. One 12-inch, $33\frac{1}{3}$ r.p.m. disc. 90 species. Details of recordings are given in an accompanying booklet. These 2 discs (Nos. 5 and 10) together contain the voices of 132 species, and are the first major reference work for African bird voices.

12. Pooley, A. C. (1966). Wildlife Calls of Africa. Percy Fitzpatrick Institute, University of Cape Town, Rondebosch 7700, South Africa. One 12-inch, $33\frac{1}{3}$ r.p.m. disc. 24 species.

13. Hayes, C. and Hayes, J. (1966). East African Birdsong; No. 2 in *Heartbeat of Africa*, Series 1. Sapra Studios, Box 5882, Kimathi and York Streets, Nairobi, Kenya. One 7-inch, 45 r.p.m. disc. 25 species.

14. Stannard, J. and Niven, P. (1967). Bird Song of the Forest. Percy Fitzpatrick Institute (address under No. 12). One 12-inch $33\frac{1}{3}$ r.p.m. disc. GALP 1559. 32 species.

15. Walker, A. (1967). Bird Song of Southern Africa. African Music Society and International Library of African Music, Roodeport, South Africa. One 12-inch $33\frac{1}{3}$ r.p.m. disc, GALP 1501. 33 species.

17. Reucassel, R. and Pooley, A. C. (1967). Calls of the Bushveld. Published by the authors and obtainable from the Wildlife Society of South Africa. One 12-inch $33\frac{1}{3}$ r.p.m. disc, WL2; also available as a cassette. 28 species.

20. Henley, A. and Pooley, A. C. (1970). Birds of the Drakensberg. Published by the authors and obtainable from Wildlife Society of South Africa. One 12-inch $33\frac{1}{3}$ r.p.m. stereo disc, BD 100. 41 species.

21. Reucassel, R. and Adendroff, A. (1970). Nature's Melody. Published by the authors; obtainable from Wildlife Society of South Africa. One 12-inch $33\frac{1}{3}$ r.p.m. stereo disc, SWL 3. 53 species.

22. Walker, A. (1970). Garden Birds of Southern Africa. Gallo (Africa) Ltd., Johannesburg; obtainable from Wildlife Society of South Africa. One 12-inch $33\frac{1}{3}$ r.p.m. stereo disc, SGALP 1598. 40 species.

24. Roché, J.-C. (1970). *L'Oiseau Musicien*; The Bird as Musician, Nos. 10, 11 and 12. Châteaubois, F – 38350 La Mure, France, Three 7-inch, 45 r.p.m. discs. 2 species on each disc.

25. Dangerfield, G. (1970). *Sounds of the Serengeti*. Music for Pleasure Ltd., Astronaut House, Hounslow, Road, Feltham, England. One 12-inch $33\frac{1}{3}$ r.p.m. stereo disc, MFP 1371. 24 species.

27. Hayes, J. (c. 1970). Bird Song of Africa. *Heartbeat of Africa* series 2. Sapra Studios (address under No. 13). One 7-inch, 45 r.p.m. disc. 12 species.

30. Roché, J.-C. (1971). Birds of Kenya. Birds and Wild Beasts of Africa, No. 1. *L'Oiseau Musicien*, see No. 24. One 12-inch, $33\frac{1}{3}$ r.p.m. stereo disc, G. 07. 32 species.

32. Keith, G. S. and Gunn, W. W. H. (1971). Birds of the African Rain Forests. *Sounds of Nature* No. 9. Federation of Ontario Naturalists, 1262 Don Mills Road, Don Mills, Ontario M3B 2WB, Canada, and American Museum of Natural History, New York. Two 12-inch, $33\frac{1}{3}$ r.p.m. discs. 95 species. The most important reference work since the records of North (Nos. 5 and 10) and the first specializing in forest birds, many of which are here published for the first time. Most species are from East Africa, some from central Africa. Species are arranged in systematic order and grouped in bands; a simple announcement of the name accompanies each species, but a lot of information is provided on the jacket.

33. Stannard, J. (1971). Bird Sounds and Songs. FitzPatrick Institute (address under No. 12). Issued in conjunction with *Ostrich* Supplement 9. One 7-inch, 45 r.p.m. disc, NV1. 20 species.

35. Martin, R. B. (1971). Journey Across Africa. Parlophone PCSJ (D) 12.79. Obtainable from Wildlife Society of South Africa. One 12-inch, $33\frac{1}{3}$ r.p.m. disc. 34 species.

36. Ker, A. (1972). Safari 99. Equator Sound Studios Ltd., P.O. Box 30068, Nairobi, Kenya. One 12-inch $33\frac{1}{3}$ r.p.m. disc, ESS 1001. 63 species.

38. Keibel, W. D. (1972). Wildlife of South West Africa. Wildlife Society of South Africa, P.O. Box 3508, Windhoek, Namibia. 1 cassette, 48 species.

39. Roché, J.-C. (1973a). Birds of South Africa. Birds and Wild beasts of Africa. No. 2. *L'Oiseau Musicien*, see No. 24. One 12-inch, $33\frac{1}{3}$ r.p.m. stereo disc, G. 08. About 65 species. 7 environments are presented without commentary, created by 3 or 4 birds singing simultaneously.

40. Roché, J.-C. (1973b). Birds of West Africa – Senegal. Birds and Wild Beasts of Africa No. 3. *L'Oiseau Musicien*, see No. 24. One 12-inch, $33\frac{1}{3}$ r.p.m. disc, G. 09. 26 Species.

42. Worman, D. (1974). African Birds. Soundpics Enterprises (Pty) Ltd, P.O. Box 61055, Marshalltown 2107, South Africa. One 10-inch $33\frac{1}{3}$ r.p.m. disc, SP 002, and 16 colour slides. 16 species.

46. Anon. (1966). A Night at Treetops. Sapra Studios (address under No. 13). *Heartbeat of Africa*, Series 1, No. 3. One 17-cm 45 r.p.m. disc. About 10 species.

50. Hart, S. (1975). Listen to the Wild – in the Bush. EMI/Brigadiers (Pty) Ltd., South Africa. 30-cm, $33\frac{1}{3}$ r.p.m. stereo disc, Brigadiers Music LTW(W)1. 17 species.

51. Hart, S. (1975). Listen to the Wild – Among the Rocks. See No. 50. 16 species.

53. Chappuis, C. (1975). *Les Oiseaux de l'Ouest Africain*. Disc 5: Timaliidae, Pycnonotidae (first part), 32 species. Disc 6: Pycnonotidae (end), Turdidae (first part), 44 species. *Alauda*, Sound Supplement accompanying commentary in *Alauda* **43**, 450–474, M.N.H.N., Laboratoire d'Ecologie, 4 Avenue du Petit Château, 91.800 Brunoy, France. Two 12-inch, $33\frac{1}{3}$ r.p.m. discs. These records are a part of a series whose aim is to present all known recordings for species of a particular region, including different forms of songs and calls and geographical variation. Details of the recordings are provided in the accompanying article in *Alauda*, of which reprints may be requested when ordering the records.

These records represent a landmark in the history of African voice-recording. This is a lengthy series covering large numbers of species in great detail, and the accompanying commentaries in *Alauda* are of considerable scientific value.

56. Gunn, W. W. H. and Gulledge, J. L. (1977). *Beautiful Birds Songs of the World*. Cornell Laboratory of Ornithology, Sapsucker Woods, Ithaca, New York 14850. Two 12-inch, $33\frac{1}{3}$ r.p.m. discs, 13 species in Vol. IV.
57. Chappuis, C. (1978). *Les Oiseaux de l'Ouest Africain*, Sound Supplement to *Alauda*. Disc 8: Turdidae (end), Sylviidae 3; 35 species. One 12-inch, $33\frac{1}{3}$ r.p.m. mono disc, ALA 15 and 16; commentary in *Alauda* **46**, no. 4, 327–335. For details, see No. 53.
58. Natal Bird Club (*c.* 1978). Bird Calls, Vols 1 and 2. Natal Bird Club, P.O. Box 10909, Marine Parade, Durban 4056, South Africa. 2 cassettes. 136 species presented in random order. Lengthy and numerous cuts are provided for each species.
60. Roché, J.-C. (1968). *Guide sonore des oiseaux d'Europe*, Tome II: Maghreb. Edwards Records, 58, Rue du Docteur Calmette, 59320 Sequedin, France. Five 17-cm, $33\frac{1}{3}$ r.p.m. discs.
62. Palmer, S. and Boswall, J. (1969–1972). A Field Guide to the Bird Songs of Britain and Europe. SR Records, Swedish Broadcasting Corp., 105 10 Stockholm, Sweden. Twelve 12-inch, $33\frac{1}{3}$ r.p.m. discs, RFLP 5001–5012. 530 species, nesting or accidental in Europe, mostly wintering in Africa. Presented in systematic order, on separate bands, announced by scientific name. The most important reference work for Palearctic birds wintering in Africa.
63. Palmer, S. and Boswall, J. (1973). A sequel to No. 62. 2 discs, RFLP 5013 and 5014. Includes 23 African species.
66. Kabaya, T. (1978). Birds of the World. I: Africa. King Records Co., Japan. One 30-cm, $33\frac{1}{3}$ r.p.m. stereo disc, King Records SKS (H) 2007. 20 species.
69. Walker, A. (1980). Sounds of the Zimbabwe Bush. Available from the author at 1 Northmoor Road, Oxford OXZ 6UW, England, or Queen Victoria Museum, Harare, Zimbabwe. One stereo cassette. 27 species.
72. Audio Three (1981). Bird Calls. See No. 58. 3 cassettes, of which the first 2 are the same as those of No. 58; the third contains additional species.
73. Palmer, S. and Boswall, J. (1981). A Field Guide to the Bird Songs of Britain and Europe. 16 cassettes, RFLP 5021–5036. An updated edition of Nos. 62, 63 and 70 (see 'The Birds of Africa', Vol. III). 612 species, in boxes of 4 cassettes with commentary and list of species in each box. A first class reference collection.
75. Audio Three. Bird Calls: Bird Families, Vol IV. 2 cassettes, 171 species. Many of these species already appear on No. 72, but here all are in systematic order.
76. Chappuis, C. (1984). *Oiseaux de France: Migrateurs et Hivernants*, parts I and II. Obtainable from the author, 10 Vallon du Fer à Cheval, 76530 La Bouille, France. Two cassettes with booklets. Present mainly flight and contact calls, not full songs, of Palearctic birds; useful because these are the vocalizations typically made in Africa by migrants. 147 species.
78. Chappuis, C. (1985). *Les Oiseaux de l'Ouest Africain*, disc *Alauda* 13: Upupidae, Phoeniculidae, Apodidae, Picidae, Pittidae, Eurylaemidae, Alaudidae and Motacillidae; 66 species. One 12-inch, $33\frac{1}{3}$ r.p.m. mono disc; commentary in *Alauda* 53(2), 115–136. For details, see No. 53.
81. Gibbon, G. (1983). *Common Bird Calls of Southern Africa*. Obtainable from the author, P.O. Box 10123, Ashwood 3605, S.A. One mono cassette. 177 species.
86. Stjernstedt, R. (1986–1990). *Bird Songs of Zambia*. Obtainable from the author, Tan y Coed, Derwenlas, Machynlleth, Powys SY20 8PZ, UK. Two mono cassettes; No. 1: Non-passerines, Eurylaemidae, Pittidae 138 species. No. 2: Alaudidae to Sylviidae, 108 species. No. 3: cisticolas to sunbirds, 98 species. One of the major collections of African bird voices. The species are presented in systematic order, often with several types of vocalization per species.
88. Gillard, L. (1987). *Southern African Bird Calls*. Gillard Bird Cassettes, P.O. Box 72059, Parkview 2122, Johannesburg, South Africa. Revised and enlarged edition of Nos. 77 and 79 (see 'The Birds of Africa', Vol. III). Three cassettes of 90 min., 540 species presented in systematic order, often with several types of song and call per species. The large number of species makes this one of the most important and comprehensive collections of African bird voices so far published.
89. Chappuis, C. (1990). *Sounds of Migrant and Wintering Birds*, Western Europe. Obtainable from F. Franklin, 13 Carden Hill, Hollingbury, Brighton BN1 8AA, UK, or from C. Chappuis (see No. 76). Two mono cassettes; English (and revised) version of No. 76.
90. Veprintsev, B. N. (1986). *Birds of the Soviet Union, a sound guide: Buntings, Larks and Pipits*. Two 30-cm., $33\frac{1}{3}$ r.p.m. discs. Melodiya C 90, 24177003 and 24179008.

Section B: Most Important Discs and Cassettes by Region

East Africa: Nos 5, 10, 32, 36
West Africa: Nos 53, 57, 78

Southern Africa: Nos 58, 75, 81, 86, 88
Palearctic migrants: Nos 62, 63, 73, 76, 89

Section C: Institutions with Sound Libraries

A. Audio Three, 6, Larch Road, Durban, South Africa.
B. British Library of Wildlife Sounds (BLOWS). The British Library, National Sound Archive, 29 Exhibition Road, London SW7 2AS, UK.
C. Cornell University, Library of Natural Sounds, Laboratory of Ornithology, 159 Sapsucker Woods Road, Ithaca N.Y. 14850, USA.
E. Fonoteca Zoologica, Museo de Zoologia, Parc de la Ciutadella, 08003 Barcelona, Spain.

F. Fitzpatrick Bird Communication Library, Bird Department, Transvaal Museum, P.O. Box 413, Pretoria 0001, South Africa.
N. Natal Bird Club, P.O. Box 10909, Marine Parade, Durban 4056, South Africa.
S: South African Broadcasting Corporation. Library of Wildlife Sounds, P.O. Box 4559, Johannesburg 2000, South Africa.

Section D: Individual Recordists

(Recordists whose names are followed by an institution have deposited copies of their tapes in that institution).

ADE	Adendorff, A.	JOHN	Johnson, E. D. H.
ALDG	Allan, D. G.	KAB	Kabaya, T.
ANO	Anon.	KEIB	Keibel, W. D.
ASP	Aspinwall, D. R.	KEI	Keith, S., Cornell
BERG	Bergman, H. H.	KER	Ker, A.
BOY	Boyer, H.	LEM	Dowsett-Lemaire, F.
BROS	Brosset, A.	LEO	Leonovich, V. V.
BRU	Brunel, J.	LOW	Low, G. C., BLOWS
BUN	Bunni, M. K.	LUT	Lutgens, H., BLOWS
CA	Carter, A.	McVIC	McVicker, R. A.
CANE	Cane, P.	MAN	Manson, A.
CART	Carter, C.	MAR	Martin, R. B.
CHA	Chappuis, C.	MEES	Mees, V., BLOWS
CHAR	Charron, F.	MOR	Morel, G.
DAN	Dangerfield, G	MOY	Moyer, D.
DYE	Dyer, M.	NIV	Niven, P. N. F.
EAR	Earlé, R. A.	NOR	North, M. E. W.
ERA	Erard, C.	OAT	Oatley, T. B.
FAR	Farkas, T.	PAN	Panov, E. N.
FEE	Feeley, J. M.	PAR	Parker, T., Cornell
GI	Gill, F.	PAY	Payne, R. B.
GIB	Gibbon, G.	PEA	Pearson, D. J.
GIL	Gillard, L.	POO	Pooley, T.
GREG	Gregory, A. R.	REU	Reucassel, D.
GRI	Grimes, L., BLOWS	ROC	Roché, J.-C.
GUN	Gunn, W. W. H.	SEL	Sellar, P. J.
HA	Haagner, C. H.	STA	Stannard, J.
HAR	Hart, S.	STJ	Stjernstedt, R., BLOWS
HAYC	Hayes, C.	SVEN	Svensson, L.
HAYJ	Hayes, J., BLOWS	TYE	Tye, A.
HAZ	Hazevoet, C. J.	VEP	Veprinsef, B. N.
HEL	Helb, H. W.	VIEL	Vielliard, J.
HEN	Henley, A.	WALK	Walker, A.
HOL	Hollom, P. A. D.	WOOD	Woodcock, M. W.
HOR	Horne, J. F. M.	WOR	Worman, D.
HRS	Harris, A.	ZIM	Zimmerman, D. and M.
HUS	Hustler, C. W.		

ERRATA

The following errors and omissions have been brought to our attention by reviewers and others.

Volume I
See also Volume II, p. 538 and Volume III, p. 591.
White-faced Storm-Petrel. On p. 68, rt. column 1, after Bourne, simplify to 'Breeds on Cape Verde Is.'
Brown Booby. On p. 106, Range and Status, line 2 should read '...occurs Cape Verde Is., French Guinea...'.

Volume II
See also Volume III, p. 591.

Range and Status
White-throated Francolin, p. 31, line 8. Delete 'not seen in recent years.'
Wattled Crane, p. 134. Map should show it occurring in Transvaal (southeast highlands at Belfast and L. Chrissie) and Natal (Drakensberg escarpment) but not elsewhere in South Africa.
Striped Crake, p. 111, under Voice. The recording of Striped Crake by R. Stjernstedt is now believed to refer to African Crake (R. Stjernstedt, pers. comm.).
Bar-tailed Godwit, p. 310, line 13. Add 'Zimbabwe (sight record L. McIlwaine: Irwin 1981).'
Pomarine Skua, p. 333, map: hatching in the legend boxes should be reversed (main wintering range should be vertical hatching).
Grey-headed Gull, p. 356. Add, after 'Erard et al. 1984', 'Grey-headed and Hartlaub's Gull interbreed fairly regularly Namibian coast (Walvis and Swakopmund: Harrison 1983).'
Little Tern, p. 398, line 15. Insert, after 'in east,' '(Also rare South Africa, west and south coasts).'
Western Bronze-naped Pigeon, p. 459, line 9. Add, after 'Congo basin,' 'and uncommon Zambia (Lisombo Stream, N Mwinilunga).'

Other
Genus *Bugeranus*, p. 133. For 'Golger' read 'Gloger.'

Volume III

Range and Status
Grey Parrot, p. 2. Map should show it occurring in Togo and Benin north to about 8°N.
Great Blue Turaco, p. 27. Map should show it occurring along coastal Togo and Benin.
Pel's Fishing Owl. On p. 134, line 1, should read 'Endemic resident, Ivory Coast and Nigeria to Natal...'. On p. 135, map should show it occurring in Ivory Coast north to 9°30'N.
African Wood Owl, p. 148. Replace 'single record S Somalia' with 'Somalia (1 old and 5 recent records, c 2°N, 42°E; 1°N, 43°E; and 1°S, 42°E: Ash and Miskell 1983).'
Striped Kingfisher. On p. 276, end of Description, Clancey 1984a,b should read 1984c.

Field Characters
Tawny Owl. On p. 146, last three lines should read 'Sympatric with Long-eared Owl *Asio otus* which is more slender, prominently "eared", and has orange eyes.'

Other
p. 157, *Caprimulgus rufigena* superspecies map, add 1 to *C. europaeus*, 2 to *C. rufigena* and 3 to *C. fraenatus*.
p. 269, Grey-headed Kingfisher, General Habits paragraph 4, lines 11–12, change '2 Ethiopian adults recovered in Uganda and Kenya' to '2 Ethiopian adults recovered in Kenya, 1 near Uganda border (960 km away) and 1 at Mombasa (1750 km away).' Other than this, the statement in the review by G. Backhurst in *Scopus* 12 (3/4), 104 that 'Ringing recoveries of *Halcyon leucocephala* and *Ceryle rudis* have been muddled and distorted' is incorrect.
p. 285, White-bellied Kingfisher, change last 3 words of French name to 'à ventre blanc'.
p. 372, Figure labels A, B, C and D should read C, D, A and B respectively (text references to A, B, C and D do not need emending).
p. 562, add 'Oatley, T. B. (1964). The probing of aloe flowers by birds. *Lammergeyer* 3, 2–8.'
p. 569, Friedmann and Williams, item 5, date is 1971.
p. 577, add Clancey, P. A. (1984c). Miscellaneous taxonomic notes on African birds, 64. *Durban Mus. Novit.* 13(14), 169–187.
p. 589, add 71. Chappuis, C. (1981). Les Oiseaux de l'Ouest africain, disc *Alauda* 12: Caprimulgidae, Trogonidae, Coliidae, Capitonidae, Indicatoridae, 46 espèces. Société d'études ornithologiques, Laboratoire d'écologie, 4 Avenue du Petit Chateau, 91800 Brunoy, France. One 12-inch $33\frac{1}{3}$ r.p.m. disc, commentary in *Alauda* 49(1), 35–58.
Plate 16. Bill of Blue-breasted Kingfisher should show the lower mandible black and the upper mandible with a black mark by the gape. Its legs and feet should be red or red-brown (see text). Bills of both Woodland Kingfishers should have the same black marks by the gape as the bill of the Blue-breasted Kingfisher.

INDEXES

Bold page numbers indicate the main account of an entry in the index; italic, the relevant plate illustration.

Scientific Names

A

abyssinica, Alcippe 284
  Hirundo 135, 136, 146, 147, 150, 152, 155, *161*, 162, 164, 179, 186, 195, 555
  Hirundo a. **152**
Acanthisittides 1
Accipiter erythropus 295
  tachiro 297, 354, 355, 457, 458
acedis, Pyrrhurus s. 322
Achaetops pycnopygius 488
acholiensis, Sheppardia a. **398**
  Sheppardia c. 398
Acrocephalus 412, 485, 546
  scirpaceus 439
Actitis hypoleucos 213
actuosa, Cercomela f. **537**
acutirostris, Calandrella 77, **78**
adsimilis, Dicrurus 295, 526
adusta, Muscicapa 396
aegra, Oenanthe l. **502**
aequatorialis, Callene 397
  Sheppardia 353, 393, 394, 395, 396, **397**, 398, *401*, 402, 403, 415, 420
  Sheppardia a. 353, **397**, *401*
  Sheppardia c. 398
  Tachymarptis 162
aethiopica, Hirundo 113, 179, 183, **184**, 191
  Hirundo a. 113, **185**
aethiops, Myrmecocichla 465, 544, **545**, 547, 549, 551
  Myrmecocichla a. 465, **546**
affinis, Apus 164, 167, 171, 186
  Mirafra 35
africana, Luscinia m. **410**
  Mirafra 14, **22**, 24, 27, 28, *48*, 50, 96, 101
  Mirafra a. **23**
africanoides, Mirafra **35**, 41, *48*, 96
  Mirafra a. **35**, *96*
africanus, Bubo 167, 370
aguimp, Motacilla 176, 197, 204, 209, 211, **212**
  Motacilla a. *176*, **212**
aguirrei, Galerida m. **103**
airensis, Cercomela m. 464, **542**
akeleyi, Alethe p. **446**
  Ammomanes d. 74
Alaemon 14, **61**, 72
  alaudipes 33, **61**, 64, 102
  hamertoni 33, 61, **64**
Alauda 14, 67, 95, 105, **107**
  arvensis 16, *33*, 68, 70, 81, 92, 101, 104, 106, **107**, 385
  gulgula 107
  japonica 107
  razae 107
Alaudidae 13

alaudipes, Alaemon 33, **61**, 64, 102
  Alaemon a. 33, **62**, 64
  Upupa 61
alba, Motacilla 122, *176*, 197, 200, 203, 204, **209**, 212, 227, 414
  Motacilla a. *176*, **210**
albescens, Alauda 51
  Certhilauda 46, **51**, 52, 55, 97
  Certhilauda a. **51**, 97
albicans, Cercomela t. **536**
albicapilla, Cossypha 400, 418, 438, **439**
  Cossypha a. 400, **439**
albicapillus, Turdus 439
albicauda, Mirafra 16, **18**, *48*, 83
albiceps, Psalidoprocne 128, 131, **132**, 133, *160*
  Psalidoprocne a. **133**, *160*
albifrons, Eremopterix n. 33, **115**
  Myrmecocichla 465, 544, 546, 549, 550, **551**, 552
  Myrmecocichla a. 465, **551**
  Saxicola 551
albigula, Phyllastrephus d. **347**
albigularis, Hirundo 113, 175, 179, 184, **186**
  Phyllastrephus 320, 330, **342**
  Phyllastrephus a. 320, **342**
  Smithornis c. 8
  Xenocichla 342
albimentalis, Cossypha a. 420
albinotata, Oenanthe p. 526
albipileata, Oenanthe m. **507**
albiscapulata, Myrmecocichla c. 465, **554**
albofasciata, Certhilauda 59
  Chersomanes 43, 47, **59**, 97
  Chersomanes a. **59**, 97
  Saxicola t. 417, **495**
alboniger, Oenanthe 501, 502, 504, 506, 521
Alcippe 444
  abyssinica 284
Alethe 351, 387, 422, **441**, 463
  choloensis 368, *401*, 422, 441, 447, **451**
  diademata 10, 355, 368, 393, *401*, 441, **442**, 444, 446, 455
  fuelleborni 368, *401*, 404, 422, 441, **447**, 451
  poliocephala xi, 355, 368, *401*, 403, 441, 442, 444, **445**, 455
  poliophrys xi, 368, *401*, 441, **443**
alexanderi, Galerida c. 49, **101**
alfredi, Phyllastrephus 333
  Phyllastrephus f. 312, 320, **344**
algeriensis, Ammomanes d. 73
algida, Certhilauda c. 47
alopex, Mirafra a. **36**, 38, *48*
alpestris, Alauda 121
  Eremophila 33, 14, **121**, 123

altanus, Macronyx a. 260
altera, Alaemon h. **64**
altirostris, Galerida c. **101**
alticola, Chersomanes a. **59**
amadoni, Hirundo a. **185**
amani, Arcanator o. 400, **460**
ameliae, Macronyx 241, 246, 250, 255, **260**, 262
Ammomanes 55, 65, **71**, 77, 115, 519
  cincturus 33, **71**, 74, 89, 124
  deserti 33, 49, 71, **73**, 89
  grayi 71, **75**, 97
  phoenicurus 71
Ammoperdix heyi 517
ampelinus, Hypocolius 257, 377, **378**, 379
Ampelis 264
ampliformis, Hirundo a. **153**
Anas penelope 380
anderssoni, Calandrella c. **79**
  Hirundo f. **170**
andrewi, Pseudhirundo g. 112, **146**
Andropadus 279, 291, 444
  ansorgei 281, 288, 290, **291**, 293, 305
  curvirostris 288, 290, **292**, *305*, 332
  gracilirostris **294**, 301, 304, 332
  gracilis 288, **289**, 291, 293, *305*
  hallae 288
  importunus 279, **298**, *305*, 312, 327
  latirostris 279, 288, **295**, *305*, 307
  masukuensis xi, 279, **280**, 282, 286, *304*
  milanjensis 280, 283, **285**, *304*
  montanus xi, 279, 280, **281**, 283, *304*, 337
  tephrolaemus 280, **282**, 286, *304*, 337, 346, 450
  virens 279, **287**, 290, 293, 296, *305*, 306, 330, 332, 335, 355
angolensis, Cercomela f. **537**
  Dryoscopus 347
  Hirundo 113, 146, 147, 187, **188**, 189, 190
  Mirafra 27, 96
  Mirafra a. 27
  Pitta 9, 10, **11**, *32*
  Pitta a. **11**, *32*
annae, Anthus n. **218**
anomala, Callene 421
  Cossypha 369, 405, 418, **421**, 451
  Cossypha a. 369, **421**
anonyma, Cisticola 213
ansorgeanus, Criniger b. 321, 357
ansorgei, Andropadus 281, 288, 290, **291**, 293, *305*
  Andropadus a. **291**, *305*
  Anthus l. **226**
  Mirafra s. **41**
  Xenocopsychus 400, 418, 426, **440**
Anthreptes seimundi 295

Anthus 197, 215, **216**, 244
  brachyurus 216, **230**, 232, *240*
  caffer **231**, *240*
  campestris *177*, 219, **221**, 238, 531
  cervinus 216, 217, 234, **236**, 238, *240*
  chloris *177*, 216, **244**, 246
  cinnamomeus 218
  crenatus *177*, 242, **243**
  godlewskii 216
  hoeschi *177*, 216, 217, 219, **220**, 243
  leucophrys 65, *177*, 216, 217, 219, 224, **225**, 227, 228, 243
  lineiventris 240, **242**
  lutescens 216
  melindae **229**, *240*
  novaeseelandiae *177*, 216, **217**, 220, 222, 224, 226, 229, 234, 243, 244
  pallidiventris *177*, 216, **228**
  petrosus 216, 236, 238, **239**, *240*
  pratensis 216, 234, **235**, 237, 238, 239, *240*
  rubescens 238, 239
  similis 216, 217, 219, 221, 222, **223**, 226, 227, 230, *240*, 243, 500
  sokokensis **233**, *240*
  spinoletta 216, 236, **238**, 239, *240*
  trivialis 216, 217, 225, **233**, 236, 237, *240*, 242
  vaalensis *177*, 216, 217, 219, 226, **227**, 228, 243
antinorii, Psalidoprocne p. **131**, *160*
antonii, Mirafra a. **28**
Apalis binotata 3
apiaster, Merops 380, 426
apiata, Alauda 30
  Mirafra 15, 29, **30**, *96*
  Mirafra a. **30**
Apus affinis 164, 167, 171, 186
  caffer 154, 155, 164, 167, 186
aquaticus, Cinclus c. 381
Aquila chrysaetos 504
arabica, Hirundo f. **169**
arada, Cyphorhinus 355
arborea, Alauda 105
  Lullula x, 16, *33*, 101, **105**, 108, 441
  Lullula a. 106
Arcanator 387, 458, **460**
  orostruthus 400, 458, 459, **460**
archeri, Cossypha 369, 418, **420**
  Cossypha a. 369, **420**
  Heteromirafra 43, **44**, 45, *48*
Ardea melanocephala 167
Arenaria interpres 101
arenaria, Chersomanes a. **59**
arenicola, Galerida c. **100**
arenicolor, Ammomanes c. 33, **72**
argenta, Oenanthe o. 510
aridula, Certhilauda b. 55
armenica, Saxicola t. **495**
arnotti, Myrmecocichla 465, 507, 544, 549, 551, **552**
  Myrmecocichla a. **552**
  Saxicola 552
arorihensis, Mirafra g. **38**
arquata, Cichladusa 400, 463, **466**, 467, 469

Numenius 66
arsinoe, Pycnonotus b. 257, 364, **366**
arvensis, Alauda 16, *33*, 68, 70, 81, 92, 101, 104, 106, **107**, 385
  Alauda a. 108
asbenaicus, Anthus s. 224
ascensi, Macronyx f. **249**
ashi, Mirafra **26**, *48*
assabensis, Ammomanes d. **49**, **74**
ater, Molothrus 292
athensis, Calandrella s. **49**, **83**
athi, Mirafra a. **23**, *48*
Atimastillas 309
atlas, Eremophila a. 33, **121**
atmorii, Oenanthe m. **507**
atricapillus, Donacobius 382
atripennis, Dicrurus 456
atrocaerulea, Hirundo 155, *160*, **173**, 178, 187
atrogularis, Oenanthe d. 524
Atticora 128
aurantiigula, Macronyx 241, 246, 255, **258**
aurigaster, Pycnonotus 364, 365, 372, 374
aussae, Cercomela m. **542**
austin-robertsi, Mirafra a. **35**
australis, Andropadus l. 296
  Eremopterix 97, **109**, 111, 117
  Megalotis 109
australoabyssinicus, Anthus c. **232**
autochthones, Myrmecocichla c. **554**
axillaris, Saxicola t. 417, **495**
ayresii, Cisticola 247
azurea, Coracina 256, 273, **277**
azureus, Graucalus 277

B

baddeleyi, Chersomanes a. 59
Baeopogon **302**, 318
  clamans 303, **306**, 352
  indicator **302**, 306, 352
balsaci, Galerida c. **101**
bambarae, Myrmecocichla c. **554**
bamendae, Andropadus t. 283, *304*
  Mirafra a. **23**
bannermani, Anthus s. 224
  Hirundo a. **153**
bansoensis, Hirundo f. **169**
barbata, Cercotrichas 416, 470, **471**, 473, 474
  Cossypha 471
barbatus, Criniger 321, 342, 356, **357**, 358, 360, 363
  Criniger b. 321, **357**
  Pycnonotus 95, 213, 257, 291, 295, 297, 316, 327, 331, 351, 364, 365, 372, 374, 431
  Pycnonotus b. 257, **366**
  Trichophorus 357
  Turdus 365
barlowi, Cercomela t. **536**
  Chersomanes a. **60**
  Spizocorys c. **84**
bartteloti, Cossypha c. 369, **428**
batesi, Cossypha i. 369, **418**
  Mirafra a. **23**
Bathmocercus 444
bathoeni, Chersomanes a. 59
Batis 345

fratrum 312
molitor 312
baumanni, Phyllastrephus 305, 324, **330**, 331
  Phyllostrephus 330
beema, Motacilla f. **176**, **199**
beesleyi, Chersomanes a. **60**
benguelensis, Certhilauda c. **47**
benguellensis, Cercomela s. **535**
  Cercotrichas p. **483**
bensoni, Ammomanes d. 73
  Phyllastrephus t. 325
  Sheppardia g. **402**
Bessornornis 428, 442
biarmicus, Falco 76
bicolor, Ploceus 332, 347, 389
  Spreo 139, 144, 507
bifasciata, Saxicola 417, **494**, **499**, 532, 533
bilineatus, Pogoniulus 444
bilopha, Alauda 123
  Eremophila 33, 104, 121, **123**
bimaculata, Alauda 69
  Melanocorypha 33, 66, 67, 68, **69**
  Melanocorypha b. **69**
binotata, Apalis 3
blanfordi, Calandrella c. **49**, **79**
  Psalidoprocne p. **131**
blayneyi, Anthus c. **232**
Bleda 296, 301, **348**, 356, 443, 446
  canicapilla 321, 348, 351, **354**, 446
  exima 321, 339, 348, **350**, 355
  syndactyla 321, **348**, 351, 355, *455*
bocagei, Anthus n. **218**
  Cossypha 395, 420
  Sheppardia 353, 394, **395**, 396, 397, 398, *401*, 402, 403, 415, 419, 427
  Sheppardia b. 353, **395**, *401*
boehmi, Neafrapus 136
bohndorffi, Anthus l. **226**
bollei, Phoeniculus 3
Bombycilla **377**
  garrulus 257, **378**, 379
Bombycillidae **377**
borbonica, Phedina 134, 135, **136**, *161*
  Phedina b. 136
borosi, Ammomanes d. 73
boscaweni, Oenanthe l. 519
bottae, Campicola 528
  Oenanthe 448, 502, 510, 526, **528**, 529
  Oenanthe b. 528
boweni, Chersomanes a. **60**
brachydactila, Alauda 77
  Calandrella 33, 68, 70, 76, **77**, 78, 81, 101
  Calandrella b. 33, **77**
brachyrhynchus, Oriolus 273
brachyura, Galerida c. **100**
  Pitta 9
brachyurus, Anthus 216, **230**, 232, *240*
  Anthus b. **231**, *240*
bradfieldi, Mirafra s. **41**
  Motacilla c. 203
Bradornis 487
bradshawi, Certhilauda c. **47**
Bradyornis 467
brazzae, Phedina 134, **135**, 136, *161*

INDEXES: SCIENTIFIC NAMES

brevirostris, Certhilauda c. **47**
 Pericrocotus 263
brunneiceps, Cercotrichas l. **478**
Bubo africanus 167, 370
Buccanodon duchaillui 308
buchanani, Hirundo f. **169**
buckleyi, Mirafra r. **29**, 48
bucolica, Galerida m. **95**
budongoensis, Smithornis r. 5, *32*
bugoma, Ixonotus g. 307
bullockoides, Merops 538
burra, Ammomanes 54
 Certhilauda 52, **54**, *97*
bushmanensis, Chersomanes a. 59
Buteo buteo 157, 167
buteo, Buteo 157, 167

C

cabanisi, Criniger 334
 Phyllastrephus 320, 323, 324, 328, 330, 332, 333, **334**, 337, 343, 452, 459, 461
 Phyllastrephus c. *320*, **335**
caesia, Ceblepyris 274
 Coracina 256, 273, **274**, 277, 424
 Coracina c. **274**
cafer, Pycnonotus 364, 365, 372, 374
caffer, Anthus **231**, *240*
 Anthus c. **232**, *240*
 Apus 154, 155, 164, 167, 186
caffra, Cossypha 369, 391, *401*, 418, 420, **423**, 426, 430, 475
 Cossypha c. 369, *401*, **423**
 Motacilla 423
cailliautii, Campethera 361
calandra, Alauda 67
 Emberiza 66, 68
 Melanocorypha 33, 66, **67**, 69, 92
 Melanocorypha c. 33, **67**
Calandrella 14, **76**, 83
 acutirostris 77, 78
 brachydactyla 33, 68, 70, 76, **77**, 78, 81, 101
 cinerea 49, 76, 77, **78**, *97*
 raytal 80, 82
 rufescens 33, 70, 72, 76, 77, **80**, 82, 101, 104, 108
 somalica 22, 49, 76, 80, **82**
Calendula 89, 94
Callene 396, 397, 398, 415, 421
calurus, Criniger *321*, 343, 357, 358, **359**, 362, 363
 Criniger c. *321*, **360**
calva, Treron 429
calviniensis, Certhilauda a. 52
Calyptocichla **301**
 serina **301**, 310, 340, *352*
Calyptomena 2
camaroonensis, Anthus n. **218**
camerunensis, Smithornis c. 8, *32*
Campephaga 263, 271, 345
 flava 256, 263, 264, **266**, 268, 270
 petiti 256, 263, 264, 266, **268**, 270
 phoenicea 256, 263, **264**, 266, 268
 quiscalina 256, 267, **269**
Campephagidae **263**, 279
campestris, Alauda 221
 Anthus 177, 219, **221**, *238*, 531
 Anthus c. 177, **222**
Campethera 361

cailliautii 361
caroli 361
 nivosa 361
 taeniolaema 345, 347
Campicola 528
campicolina, Oenanthe b. 528
canaria, Serinus 213
candida, Mirafra 22
canicapilla, Bleda 321, 348, 351, **354**, 446
canicapillus, Trichophorus 354
cannabina, Carduelis 68, 92
canorus, Cuculus 494
cantarella, Alauda a. **108**
cantillans, Mirafra **15**, 20, 21, 22, 27, 44, 48, 83
capensis, Alauda 251
 Macronyx 241, 245, 246, 249, **251**, 255
 Macronyx c. **252**
 Motacilla 176, 197, 200, **203**, 208, 211, 212, 252
 Motacilla c. 176, **203**
 Platyrhynchus 7
 Pycnonotus 257, 299, 364, 365, 367, 372, **374**
 Smithornis 3, 4, 5, 7, 10, *32*
 Smithornis c. 7, *32*
 Spizocorys s. 86
 Turdus 374
caprata, Saxicola 532
Caprimulgus poliocephalus 554
caprius, Chrysococcyx 205, 214, 372, 480, 484, 485
Carduelis cannabina 68, 92
 carduelis 68
 chloris 92, 157
carduelis, Carduelis 68
caroli, Campethera 361
carolinae, Galerida m. **103**
carruthersi, Alethe p. 368, *401*, 444, **445**
carthaginis, Galerida c. **100**
Cassina 457
cassini, Malimbus 290
castanea, Alethe d. 368, **442**
cavei, Certhilauda a. **52**
cavernicola, Myrmecocichla c. **554**
Ceblepyris 271, 274
Cecropis 147
centralis, Chlorocichla f. **312**, *352*
 Psalidoprocne n. **129**
Centropus superciliosus 377
Cercococcyx mechowi 394
 montanus 8, 399
Cercomela 487, 499, 500, 507, **532**
 dubia 464, 532, **540**
 familiaris 426, 464, 488, 500, 508, 532, 533, 535, 536, **537**, 539
 fusca 532
 melanura 464, 532, 540, **541**
 schlegelii 464, 532, 533, **534**, 536
 scotocerca 464, 532, 537, **539**, 541
 sinuata 464, 532, **533**, 538
 sordida 464, 532, 540, **543**
 tractrac 75, 464, 532, **535**
Cercotrichas 452, **470**
 barbata *416*, 470, **471**, 473, 474
 coryphaeus *416*, 483, **484**
 galactotes *416*, 479, **480**, 486
 hartlaubi *416*, 471, **476**, 481

leucophrys x, 391, *416*, 471, 472, 473, **477**, 481, 483
 leucosticta 403, *416*, **470**, 471, 473, 474, 477
 paena *416*, 479, **483**, 485
 podobe *416*, 470, **486**
 quadrivirgata 402, *416*, 451, 470, 471, **473**, 474, 481
 signata 391, *416*, 470, 471, 473, **474**
Certhia 281
Certhilauda 25, **46**, 55
 albescens 46, **51**, 52, 55, *97*
 burra 52, **54**, *97*
 chuana 46, **50**, *96*
 curvirostris **46**, 50, 91, *96*, 99
 erythrochlamys 46, 51, **52**, 55, *97*
cerviniventris, Phyllastrephus 320, 324, 327, **328**, 329
 Phyllostrophus 328
cervinus, Anthus 216, 217, 234, **236**, *238*, *240*
 Motacilla 236
chadensis, Mirafra c. **16**, *48*
Chaetops 387
chagwensis, Andropadus g. **294**
chalybea, Nectarinia 389
 Psalidoprocne p. **131**
chapini, Mirafra a. **23**
 Motacilla c. **208**
 Sheppardia b. **395**
Chelidon 130, 194
cheniana, Mirafra 17, **19**, *96*
Chersomanes 59
 albofasciata 43, 47, **59**, *97*
Chersophilus **91**
 duponti 33, **91**, 104, 108
chicquera, Falco 60, 76, 91, 118, 511, 531
chirindensis, Pogonocichla s. 389
chlorigula, Andropadus t. 283, 286, **304**
chloris, Anthus 177, 216, **244**, 246
 Carduelis 92, 157
 Nicator 308, 341, *352*
chlorocephalus, Oriolus 434
Chlorocichla **309**, 318
 falkensteini 309, **313**, *321*
 flavicollis 309, 314, 315, **316**, *321*, 322, 359
 flaviventris 300, 309, **311**, 314, 315, 327, 330, *352*
 laetissima 301, 309, **311**, *352*
 prigoginei 309, **310**, *352*
 simplex 309, **314**, 317, *321*
chloronotus, Criniger *321*, 356, 357, **358**, 360
 Trichophorus 358
Chloropsis 279
chlorosaturatus, Baeopogon i. 302
chobiensis, Anthus v. **227**
choloensis, Alethe 368, *401*, 422, 441, 447, **451**
 Alethe c. 368, *401*, **451**
chorsophilus, Anthus s. 224
chrysaetos, Aquila 504
Chrysococcyx caprius 205, 214, 372, 480, 484, 485
 cupreus 298, 372, 436, 444
 klaas 372
chrysopygia, Oenanthe x. 522, **523**

chuana, Alauda 50
   Certhilauda 46, **50**, 96
chyulu, Andropadus m. 285
   Pycnonotus b. 366
chyuluensis, Anthus s. 224
cia, Emberiza 316
Cichladusa **463**
   arquata *400*, 463, **466**, 467, 469
   guttata *400*, **468**
   ruficauda *400*, 463, 466, **467**
Cinclidae **380**
Cinclus **380**
   cinclus 257, **380**
cinclus, Cinclus 257, **380**
   Sturnus 380
cincta, Hirundo 143
   Riparia 112, 138, 142, **143**, 146, 157, 170, 187
   Riparia c. 112, **143**
cinctura, Melanocorypha 71
cincturus, Ammomanes 33, **71**, 74, 89, 124
cinerea, Alauda 78
   Calandrella 49, 76, 77, **78**, 97
   Calandrella c. **79**, 97
   Motacilla 176, 197, 200, 204, **205**, 207, 210, 382
   Motacilla c. 176, **206**
cinereocapilla, Motacilla f. 176, **199**
cinereus, Cercotrichas c. 416, **484**
cinnamomeiventris, Myrmecocichla 154, 155, 462, **465**, 544, 550, 553, 556
   Myrmecocichla c. **553**
   Turdus 553
cinnamomeus, Anthus 218
   Anthus n. **218**
Circus macrourus 529
   ranivorus 204, 377
Cisticola 231, **500**
   anonyma 213
   ayresii 247
   nigriloris 496
Cisticolidae **279**
citrea, Protonotaria 433
citreola, Motacilla 197, 200, **202**
   Motacilla c. **202**
clamans, Baeopogon 303, **306**, 352
   Xenocichla 306
Clamatores 1
clanceyi, Saxicola t. **495**
clara, Hirundo l. **189**
   Motacilla 176, 197, 204, 206, **207**, 211, 212
   Motacilla c. **208**
clericalis, Myrmecocichla a. **551**
clot-bey, Melanocorypha 65
   Ramphocoris 33, **65**, 68, 72
codea, Certhilauda a. **51**
coelebs, Fringilla 104
collaris, Lanius 167, 500, 507, 513
   Mirafra 27, **34**, 48
   Prunella 257, **384**
   Prunella c. 257, **384**
   Sturnis 384
colletti, Macronyx c. 241, **251**
collsi, Cercotrichas l. 416, **471**
colstoni, Cercotrichas l. **471**
Columba guinea 529
compsonota, Alethe p. 368, **445**
concolor, Andropadus m. 281

Hirundo 147, 169, 171
congener, Andropadus l. **296**
congensis, Andropadus g. 294, *304*
congica, Cotile 140
   Riparia 112, 135, 137, **140**
conirostris, Alauda 83
   Spizocorys **83**, 85, 86, 90, 97
   Spizocorys c. **83**, 97
conjunctus, Smithornis c. **8**
Copsychus 440
Coracina **273**, 345
   azurea 256, 273, **277**
   caesia 256, 273, **274**, 277, 424
   graueri 256, 274, **277**, 278
   pectoralis 256, 273, 274, **275**, 277
Coraphaites 119
cordofanica, Mirafra 20, *33*, 37, 89
coronata, Myrmecocichla c. 465, **554**
coronatus, Stephanoaetus 252, 429
   Vanellus 526
corone, Corvus 68
Corvida 13, 279
Corvus 1
   corone 68
coryphaeus, Cercotrichas 416, 483, **484**
   Cercotrichas c. 416, **484**
   Sylvia 484
Cossypha 387, 414, 415, **418**, 440, 450, 463, 466, 544
   albicapilla *400*, 418, 438, **439**
   anomala 369, 405, 418, **421**, 451
   archeri 369, 418, **420**
   caffra 369, 391, *401*, 418, 420, **423**, 426, 430, 475
   cyanocampter 369, 418, **428**, 429, 431
   dichroa 369, *401*, 418, 433, **435**
   heinrichi *400*, 418, **437**
   heuglini 369, *401*, 418, 427, 428, 429, **431**, 434, 436
   humeralis *400*, 418, **425**, 440
   insulana 395
   isabellae 369, 396, *401*, **418**
   natalensis 369, *401*, 402, 418, 427, 429, **433**, 436, 439
   niveicapilla *400*, *401*, 418, **437**, 439
   polioptera 353, 395, *401*, 418, **427**
   semirufa 369, *401*, 418, 428, **429**, 431
Cossyphicula 415, 418
   roberti 369, 398, **415**, 419, 420
Cotile 140
Cotyle rufigula 169
coutellii, Anthus s. **238**
Crateropus 468
crenatus, Anthus 177, 242, **243**
crepuscula, Cossypha h. 425
Criniger 279, 349, **356**, 360, 363
   barbatus 321, 342, 356, **357**, 358, 360, 363
   calurus 321, 343, 357, 358, **359**, 362, 363
   chloronotus 321, 356, 357, **358**, 360
   ndussumensis 321, 360, **362**
   olivaceus 321, 342, 357, 360, **362**
cristata, Alauda 100
   Galerida 33, 49, 94, 95, **100**, 101, 103, 108
crocea, Alauda 253
croceus, Macronyx 241, 246, 250,

252, **253**, 259, 526
crypta, Spizocorys c. **84**
cryptoleuca, Myrmecocichla a. **546**
cryptoleucus, Smithornis c. **8**
cubla, Dryoscopus 3, 312
cucullata, Hirundo 147, 150, **154**, 161, 175, 548, 555
Cuculus canorus 494
   solitarius 205, 214, 317, 372, 390, 392, 402, 424, 427, 430, 433, 463, 474, 480
cummingi, Oenanthe x. 523
cupreus, Chrysococcyx 298, 372, 436, 444
curruca, Sylvia 542
Cursorius 63
   temminckii 526
curvirostris, Alauda 46
   Andropadus 288, 290, **292**, *305*, 332
   Andropadus c. **292**, *305*
   Certhilauda **46**, 50, 91, 96, 99
   Certhilauda c. **46**, 96
cyanecula, Luscinia s. 368, **412**
cyanocampter, Bessonornis 428
   Cossypha 369, 418, **428**, 429, 431
   Cossypha c. 369, **428**
cyanomelas, Trochocercus 312
cyornithopsis, Callene 396
   Shepppardia 353, 393, 394, 395, **396**, 397, 398, *401*, 402, 403
   Sheppardia c. **397**, *401*
Cyphorhinus arada 355
cypriaca, Oenanthe p. **513**
Cypsiurus parvus 136, 179

D

dacotiae, Saxicola 532
damarensis, Cercotrichas p. **483**
   Eremopterix v. **117**
   Spizocorys c. **84**
daroodensis, Calandrella c. **79**
Dasycephala 348
dasypus, Delichon 194
daurica, Hirundo 95, *113*, 146, 147, 149, 151, **156**
daviesi, Anthus v. 227
   Certhilauda c. 47
debilis, Phyllastrephus 320, 345, **347**
   Phyllastrephus d. **348**
   Xenocichla 347
degodiensis, Mirafra 15, 37, **38**, 48
deichleri, Galerida m. **103**
delacouri, Smithornis c. **8**
delamerei, Pseudalaemon f. 49, **94**
Delichon 125, **194**
   dasypus 194
   urbica 157, *161*, 162, 172, 184, **194**
Dendronanthus 197
   indicus 197
dendrophilus, Phyllastrephus f. 344
Dendropicos elliotii 347
deserti, Alauda 73
   Ammomanes 33, 49, 71, **73**, 89
   Ammomanes d. 33, **73**
   Mirafra a. **31**, 96
   Oenanthe 449, 502, 504, 510, 514, 519, 521, 522, 523, **524**, 526, 529, 530
   Oenanthe d. **524**
   Saxicola 524

desertorum, Alaemon a. **62**, 64
Dessonornis 425
Deutero-oscines 1
dewittei, Anthus s. **224**
diademata, Alethe 10, 355, 368, 393, *401*, 441, **442**, 444, 446, 455
 Alethe d. *368*, **442**
diadematus, Bessornis 442
 Turdus 442
dichroa, Cossypha 369, *401*, 418, 433, **435**
 Cossypha d. **435**
 Muscicapa 435
Dicruridae 279
Dicrurus 267, 345
 adsimilis 295, 526
 atripennis 456
 modestus 295
diluta, Riparia r. **142**
dimidiata, Hirundo *161*, 179, **183**, 187, 195
 Hirundo d. *161*, **183**
disjunctus, Andropadus m. 286
distans, Phyllastrephus f. 344
dodsoni, Pycnonotus b. *257*, **366**
dombrowskii, Motacilla f. **200**
domesticus, Passer 120, 167
domicella, Hirundo d. **157**
dominicus, Dulus 377
Donacobius atricapillus 382
donaldsoni, Cossypha s. **430**
dorsomaculatus, Ploceus 295
dryobates, Phyllastrephus f. 344
Dryoscopus 298, 345
 angolensis 347
 cubla 3, 312
dubia, Cercomela 464, 532, **540**
 Myrmecocichla 540
duchaillui, Buccanodon 308
ducis, Riparia p. *112*, **138**
dukhunensis, Motacilla a. **210**
Dulus 377
 dominicus 377
dunni, Calendula 89
 Eremalauda 20, *33*, 37, 88, **89**, 115
 Eremalauda d. *33*, **89**
duponti, Alauda 91
 Chersophilus *33*, **91**, 104, 108
 Chersophilus d. *33*, **92**

E

editus, Anthus r. 219, 221
elfriedae, Mirafra s. 41
elgonensis, Pogonocichla s. **388**
ellioti, Galerida m. *49*, **104**
elliotii, Dendropicos 347
eludens, Anthus b. 230
eluta, Cercotrichas l. **478**
Emberiza 72
 calandra 66, 68
 cia 316
 schoeniclus 92
Emberizidae 13
emini, Criniger c. **360**, 362
 Hirundo d. *113*, **157**
ensifera, Cercomela s. **533**
enunciator, Anthus l. **226**
Eremalauda 14, **88**
 dunni 20, *33*, 37, 88, **89**, 115
 starki 84, 86, 88, **90**, 97

Eremophila 72, **121**
 alpestris 14, *33*, **121**, 123
 bilopha *33*, 104, 121, **123**
Eremopterix 66, 89, 102, **109**, 114
 australis 97, **109**, 111, 117
 grisea 119
 leucopareia 49, 118, **119**
 leucotis 49, 97, **109**, 111, 115, 117, 119
 nigriceps *33*, 65, 72, 109, **115**, 119
 signata 49, 109, 111, 114, 116, **118**, 119
 verticalis 97, 109, **116**, 118
erikssoni, Chersomanes a. **59**
Erithacus **406**
 rubecula 368, *401*, **406**, 462, 481, 544
erlangeri, Calandrella c. *49*, **79**
 Galerida m. **103**
 Riparia c. **144**
ernesti, Cercomela s. **543**
 Oenanthe l. **503**
erongo, Mirafra s. 41
errolius, Andropadus i. 299
Erythacus 391
erythrochlamys, Alauda 52
 Certhilauda 46, 51, **52**, 55, 97
erythrochrous, Ammomanes d. **73**
erythroptera, Mirafra 35
erythropterus, Andropadus v. **287**, *305*
erythropus, Accipiter 295
Erythropygia leucophrys 469
erythropygia, Alauda 56
 Pinarocorys *48*, **56**, 57
erythrothorax, Stiphrornis 353, **392**, *401*, 415
 Stiphrornis e. *353*, **392**
esobe, Anthus p. **228**
Estrilda 3, 444
 nonnula 431
Estrildinae 197, 384
eugenius, Andropadus l. **296**, *305*
Eupodotis senegalensis 529
eurylaemus, Smithornis s. **4**
Eurylaimidae x, 1
Eurylaimides 1
Eurylaimoidea 1
eurystomina, Pseudochelidon **125**, 159, *161*
exasperatus, Anthus v. **227**
excubitor, Lanius 379
exima, Bleda *321*, 339, 348, **350**, 355
 Bleda e. *321*, 348, **351**
eximus, Trichophorus 350
extremus, Andropadus g. **290**

F

falcirostris, Certhilauda c. **47**
Falco biarmicus 76
 chicquera 60, 76, 91, 118, 511, 531
 tinnunculus 180
falkensteini, Cercomela f. 464, **537**
 Chlorocichla 309, **313**, *321*
 Criniger 313
familiaris, Cercomela 426, 464, 488, 500, 508, 532, 533, 535, 536, **537**, 539
 Cercomela f. **537**
 Cercotrichas g. **481**
 Saxicola 537

fayi, Pycnonotus b. 367
feldegg, Motacilla 198
 Motacilla f. *176*, **199**
fenestrarum Delichon u. 195
ferrea, Saxicola 532
ferruginea, Alauda 54
ferrugineus, Laniarius 377, 426, 436
festae, Galerida c. **100**
finschi, Cassina 457
 Neocossyphus 400, 452, 453, 455, 456, **457**
finschii, Oenanthe 449, 502, 511, 513, 514, 518, **520**, 525
 Oenanthe f. 449, **520**
 Saxicola 520
fischeri, Criniger 333
 Mirafra r. 29
 Phyllastrephus 312, 320, 324, 326, **333**, 334
flava, Campephaga 256, 263, 264, **266**, 268, 270
 Eremophila a. 122
 Motacilla 122, *176*, 197, **198**, 202, 204, 207, 211, 222, 230, 237, 513
 Motacilla f. *176*, **199**
flavicollis, Chlorocichla 309, 314, 315, **316**, *321*, 322, 359
 Chlorocichla f. **317**, *321*
 Haematornis 316
 Macronyx 241, 246, **248**
flavigula, Chlorocichla f. 314, **316**
flavissima, Motacilla 198
 Motacilla f. *176*, **199**
flaviventris, Chlorocichla 300, 309, **311**, 314, 315, 327, 330, 352
 Chlorocichla f. *312*, 352
 Motacilla 197, 203
 Trichophorus 311
flavostriatus, Andropadus 343
 Phyllastrephus 312, 320, 324, 336, **343**, 346, 348
 Phyliastrephus f. *320*, **344**
forbeswatsoni, Eremopterix n. **115**
formicivora, Myrmecocichla 155, 465, 507, 544, 545, **547**, 552
 Oenanthe 547
fradei, Mirafra s. 41
fraseri, Muscicapa 455
 Neocossyphus 400, 452, 453, **455**, 457
 Neocossyphus f. **456**
fratrum, Batis 312
fremantlii, Calendula 94
 Pseudalaemon *49*, **94**
 Pseudalaemon f. *49*, **94**
frenata, Oenanthe b. 448, **528**
fricki, Andropadus i. 299
friedmanni, Pogonocichla s. 388
Fringilla coelebs 104
fringillaris, Alauda 85
 Spizocorys 84, **85**, 86, 97
Fringillidae 13
frondicolus, Anthus s. **224**
frontalis, Myrmecocichla a. **465**, 551
fuelleborni, Alethe 368, *401*, 404, 422, 441, **447**, 451
 Macronyx 241, 246, **249**, 251, 255, 263
 Macronyx f. **250**
fuertisi, Calandrella c. 79

fuliginosa, Hirundo 138, 147, **158**, *160*
    Lecythoplastes 158
    Petrochelidon 147
    Psalidoprocne 128, 130, 131, 132, **133**, 159, *160*
fuligula, Hirundo 112, 138, 147, 157, 158, 163, 165, **169**, 171, 187, 555
    Hirundo f. 112, **170**
fulviventris, Phyllastrephus 320, 324, 328, **329**
    Phyllostrephus 329
furensis, Cercomela s. **540**
    Mirafra r. **29**
fusca, Cercomela 532
fusciceps, Andropadus t. **283**, *304*
fusciventris, Hirundo 147
    Hirundo f. 112, **169**
    Ptyonoprogne f. 169

G

gabar, Micronisus 36
gabela, Muscicapa 403
    Sheppardia 353, 394, 395, 396, 397, 398, 402, **403**
gabonensis, Pycnonotus b. 257, **366**
    Stiphrornis e. **393**, *401*
gabunensis, Neocossyphus r. **453**
galactotes, Cercotrichas 416, 479, **480**, 486
    Cercotrichas g. *416*, **481**
    Sylvia 480
Galerida 14, 72, **95**, 101, 105, 107
    cristata 33, 49, 94, 95, **100**, 101, 103, 108
    magnirostris 95, 97, **98**
    malabarica 33, 49, 94, 95, 101, **103**
    modesta 49, **95**
gallarum, Mirafra h. **24**
galtoni, Cercomela f. **537**
garguensis, Cossypha a. **433**
garrula, Chersomanes a. **59**
garrulus, Bombycilla 257, **378**, 379
    Bombycilla g. 257, **378**
    Lanius 378
geyri, Ammomanes d. **73**
ghansiensis, Mirafra a. **23**
gibraltariensis, Phoenicurus o. 417, **489**
giffardi, Cossyphus a. *400*, **439**
giffordi, Galerida m. **95**
gilberti, Kupeornis 347
gilletti, Mirafra 15, **37**, 38, *48*, 83
    Mirafra g. **37**
gilli, Certhilauda c. **47**
giloensis, Alethe p. **446**
githaginea, Rhodopechys 63, 66
glareola, Tringa 529
gobabisensis, Mirafra a. **35**
godlewskii, Anthus 216
    Anthus n. 216
gomesi, Mirafra a. **23**
goodi, Pycnonotus b. 366
goodsoni, Anthus l. *177*, **226**
gordoni, Hirundo s. **149**
gouldii, Anthus l. *177*, **226**
gracilirostris, Andropadus **294**, 301, 304, 332
    Andropadus g. **294**, *304*
gracilis, Andropadus 288, **289**, 291,
293, *305*
    Andropadus g. **290**, *305*
grandis, Motacilla 197, 209, 212
granti, Sheppardia b. **395**
Graucalus 275, 277
graueri, Coracina 256, 274, **277**, 278
    Phyllastrephus f. **320**, *344*
    Pseucocalyptomena **2**, *32*
grayi, Alauda 75
    Ammomanes 71, **75**, 97
    Ammomanes g. **75**, 97
greenwayi, Cercotrichas q. **473**
gregaria, Eremalauda s. **90**
grimwoodi, Cossypha p. **427**
    Macronyx 241, 246, 250, 261, **262**
grisea, Eremopterix 119
griseiceps, Oenanthe m. **506**
griseigularis, Myoparus 456
griseopyga, Hirundo 145
    Pseudhirundo 112, 140, **145**, *162*, 164, 179, 184
    Pseudhirundo g. 112, **146**
grisescens, Mirafra a. **23**
grisescentior, Pycnonotus n. **372**
griseus, Anthus c. **222**
grotei, Anthus n. **218**
    Cossypha a. 369, **421**
    Phyllastrephus f. 333
guinea, Columba 529
gularis, Nicator 352
gulgula, Alauda 107
gunningi, Sheppardia 353, 394, 395, 396, 397, 398, *401*, **402**, 403
    Sheppardia g. **402**
gurneyi, Zoothera 459, 461, 475
gurue, Cossypa a. 421
guttata, Certhilauda a. **51**
    Cichladusa *400*, **468**
    Cichladusa g. **468**
    Zoothera 434
guttatus, Crateropus 468
    Ixonotus *307*, 318, 341, *352*
guttifer, Pogonocichla s. **388**
gutturalis, Cossypha 413
    Hirundo r. **191**
    Irania 368, **413**

H

haagneri, Cossypha d. **435**
Haematornis 316
hafizi, Luscinia m. **410**
halfae, Galerida c. **101**
Haliaeetus vocifer 434
hallae, Alethe p. **446**
    Andropadus 288
    Anthus s. 224
halophila, Oenanthe l. 448, **518**
hamertoni, Alaemon 33, 61, **64**
    Alaemon h. 33, **64**
    Cercotrichas g. **481**
hararensis, Anthus s. **224**, *240*
harei, Certhilauda b. 55
    Galerida m. **99**
    Mirafra a. **35**
harrarensis, Galerida m. **104**
harrisoni, Eremopterix s. **119**
harterti, Alauda a. 33, **107**
    Mirafra a. **23**
    Myrmecocichla a. **552**

harti, Eremopterix v. **117**
    Spizocorys c. **84**
hartlaubi, Cercotrichas 416, 471, **476**, 481
    Erythropygia 476
heinei, Calandrella r. **81**
heinrichi, Cossypha 400, 418, **437**
helenae, Galerida c. **101**
helleri, Pogonocichla s. **389**
hellmayri, Cercomela f. **464**, **537**
Hemimacronyx 216, 246
henrici, Mirafra a. **23**
herero, Bradornis 487
    Mirafra s. **41**
    Namibornis 416, **487**
Heteromirafra **42**
    archeri 43, **44**, 45, *48*
    ruddi 43, 44, 45, *96*
    sidamoensis 43, 44, **45**, *48*
Heteronyx 43
Heterotrogon narina 444
heuglini, Cossypha 369, *401*, 418, 427, 428, 429, **431**, 434, 436
    Cossypha h. 369, *401*, **431**
    Oenanthe b. 448, 526, **528**
hewitti, Mirafra a. **31**, 96
heyi, Ammoperdix 517
Hippolais pallida 486
Hirundinidae **125**
Hirundininae **127**
Hirundo 126, 128, 134, 146, **147**
    abyssinica 135, 136, 146, 147, 150, **152**, 155, *161*, 162, 164, 179, 186, 195, 555
    aethiopica 113, 179, 183, **184**, 191
    albigularis 113, 175, 179, 184, **186**
    angolensis 113, 146, 147, 187, **188**, 189, 190
    atrocaerulea 155, *160*, **173**, 178, 187
    concolor 147, 169, 171
    cucullata 147, 150, **154**, *161*, 175, 548, 555
    daurica 95, *113*, 146, 147, 149, 151, **156**
    dimidiata *161*, 179, **183**, 187, 195
    fuliginosa 138, 147, **158**, *160*
    fuligula 112, 138, 147, 157, 158, 163, 165, **169**, 171, 187, 555
    fusciventris 147
    leucosoma 129, *161*, **181**
    lucida 113, 147, 185, 187, 188, **189**, 190
    megaensis *161*, **182**
    neoxena 188, 189, 190
    nigricans 147
    nigrita 129, *160*, **180**
    nigrorufa 113, **175**
    obsoleta 147
    perdita 147, *161*, 165, **168**
    preussi 146, 147, **159**, *161*, 163, 195
    rufigula 146, 147, 159, *161*, **163**, 165, 169, 178, 195
    rupestris 112, 138, 142, 147, 157, 169, **171**
    rustica 68, *113*, 139, 141, 144, 147, 157, 162, 168, 185, 187, 188, 189, **190**, 194, 513
    semirufa 113, 147, **149**, 151, 153, 155, 157, 182
    senegalensis 113, 147, 149, **150**, 157
    smithii 113, 153, **178**, 185

spilodera 147, *161*, 164, **165**, 168, 170
tahitica 188, 189, 190
hispanica, Motacilla 514
  Oenanthe 449, 510, 512, **514**, 519, 520, 526, 528, 530
  Oenanthe h. 449, 513, **514**
hoeschi, Ammomanes g. 75, 97
  Anthus 177, 216, 217, 219, **220**, 243
  Cercomela t. **536**
  Eremopterix l. **111**
  Mirafra s. 41
holochlorus, Andropadus v. 287
holomelaena, Psalidoprocne p. **131**, 160
homochroa, Oenanthe d. 449, **524**
houghtoni, Sheppardia c. **396**
huei, Galerida m. **103**
humeralis, Cossypha 400, 418, **425**, 440
  Dessonornis 425
huriensis, Galerida m. 49, **104**
hygrica, Pogonocichla s. **389**
hygricus, Macronyx c. 254
hypermetra, Mirafra 14, 22, **24**, 26, 27, 48
  Mirafra h. **24**, 48
hypermetrus, Spilocorydon 24
hypernephela, Cercomela s. **533**
hypochloris, Phyllastrephus 305, 324, 330, **331**
  Phyllastrephus b. 331
  Phyllastrephus c. 334
  Stelgidillas 331
Hypocolius **378**
  ampelinus 257, 377, 378, **379**
hypoleucos, Actitis 213
hypospodia, Cercomela s. 464, **543**
hypoxanthus, Andropadus i. 299, *305*

I

iberiae, Motacilla f. 176, **199**
Icteridae 245
icterinus, Phyllastrephus 301, **338**, 340, 342, 343, *352*
  Trichophorus 338
ilyai, Sheppardia b. **395**
immaculata, Prunella 386
importunus, Andropadus 279, **298**, *305*, 312, 327
  Andropadus i. **299**, *305*
  Turdus 298
Indicator indicator 547, 549
  minor 370
indicator, Baeopogon 302, 306, *352*
  Baeopogon i. **302**
  Criniger 302
  Indicator 547, 549
Indicatoridae 302
indicus, Dendronanthus 197
infelix, Certhilauda c. **47**, *96*
inornatus, Pycnonotus b. 257, **366**
inquieta, Scotocerca 517
insignis, Saxicola 532
insulana, Cossypha 395
  Sheppardia b. 353, **395**
insularis, Andropadus i. 299, *305*
integer, Oenanthe o. 510
intensa, Cossypha n. 369, *401*, **433**

Pogonocichla s. **388**
Spizocorys p. 49, **88**
intercalans, Cichladusa g. **469**
intercedens, Cossypha s. 369, **430**
  Mirafra a. 21, **35**, *48*
intermedia, Alauda a. 107
  Cossypha h. **431**
intermedius, Phyllastrephus t. 325
interpres, Arenaria 101
iolaema, Cossypha c. 369, **423**
Irania **413**
  gutturalis 368, **413**
isabellae, Cossypha 369, 396, *401*, **418**
  Cossypha i. 369, *401*, **419**
isabellina, Ammomanes d. **73**
  Galerida c. **101**
  Oenanthe 449, 502, 510, 513, 515, 519, 523, 525, 526, 528, **529**
  Saxicola 529
isolata, Mirafra a. 23
isseli, Mirafra a. 35
itoculo, Phyllastrephus a. 333
itombwensis, Anthus n. 218
  Phyllastrephus f. 344
Ixonotus **307**
  guttatus **307**, 318, 341, *352*
Ixus 364

J

jacobinus, Oxylophus 214, 300, 372, 374, 377
janetti, Ammomanes d. 73
japonica, Alauda 107
jappi, Mirafra a. **31**
jardinei, Turdoides 326
javanica, Mirafra 15
jebelmarrae, Anthus s. **224**
  Saxicola t. **495**
jerdoni, Saxicola 532
jocosus, Pycnonotus 364
jordansi, Galerida c. **101**
josensis, Anthus s. 224
juniperi, Troglodytes t. **383**

K

kaballii, Mirafra a. **23**, *96*
kaboboensis, Alethe p. 368, **444**
  Sheppardia b. **395**
kabylorum, Troglodytes t. 257, **382**
kakamegae, Andropadus m. 280, 283, 304
kakamegoes, Neocossyphus p. **454**
kalahariae, Chersomanes a. **59**, *97*
kaokensis, Certhilauda c. **47**, *96*
karruensis, Certhilauda a. **52**
katangae, Anthus n. 218
  Phyllastrephus t. 325
kathangorensis, Mirafra h. **25**
kavirondensis, Andropadus a. **291**
kawirondensis, Mirafra r. **29**, *48*
kenia, Cercotrichas h. 476
khama, Eremopterix v. **117**
kidepoensis, Mirafra h. **25**
kikuyuensis, Andropadus t. 280, **282**, 304
kimbutui, Cossypha a. **420**
kivuensis, Cossypha c. **423**
klaas, Chrysococcyx 372

kleinschmidti, Galerida c. **100**
kollmanspergeri, Ammomanes d. **73**
kordofanensis, Myrmecocichla c. **554**
kretschmeri, Macrosphenus 312
kumboensis, Hirundo d. **157**
kungwensis, Alethe p. **446**
  Andropadus m. 280, *283*
  Andropadus t. 280, *283*
  Phyllastrephus f. **344**
  Sheppardia b. **395**
Kupeornis gilberti 347
kurrae, Mirafra a. 23

L

lacuum, Anthus n. 177, **218**
laetissima, Chlorocichla 301, **309**, 311, *352*
  Chlorocichla l. **309**
laetissimus, Andropadus 309
Laniarius 432, 443
  ferrugineus 377, 426, 436
Lanius 501, 511, 531
  collaris 167, 500, 507, 513
  excubitor 379
  senator 104
larischi, Cossypha n. 369, **433**
larvatus, Oriolus 436
latimerae, Macronyx c. 251
latirostris, Andropadus 279, 288, **295**, 305, 307
  Andropadus l. **296**, *305*
latistriatus, Anthus n. 216, **218**, *224*
layardi, Pycnonotus b. 257, **366**
lebombo, Pogonocichla s. 389
Lecythoplastes 158, 159
leggei, Anthus b. **230**
leoninus, Andropadus c. **293**
leucocephala, Motacilla f. 176, **199**
leucocraspedon, Anthus s. 224, **240**
leucogenys, Pycnonotus 364, 365, 372, 374
leucolepis, Phyllastrephus 341, *352*
leucomelaena, Sylvia 542
leucomelaina, Tricholaema 167
leucomelas, Parus 551
leucopareia, Coraphaites 119
  Eremopterix 49, 118, **119**
leucophrys, Anthus 65, *177*, 216, 217, 219, 224, **225**, 227, 228, 243
  Anthus l. 177, **226**
  Cercotrichas x, *391*, *416*, 471, 472, 473, **477**, 481, 483
  Cercotrichas l. *416*, **478**
  Erythropygia 469
  Sylvia 477
leucopleura, Thescelocichla 318, 323, *352*
leucopleurus, Phyllostrophus 318
leucoptera, Cercotrichas l. *416*, **478**
  Melanocorypha 67
leucopyga, Oenanthe 448, 501, 502, 504, 506, 513, 516, 519
  Oenanthe l. 448, **503**
  Vitiflora 502
leucorrhoa, Oenanthe o. 510, 530
leucosoma, Hirundo 129, *161*, **181**
leucosticta, Cercotrichas 403, *416*, **470**, 471, 473, 474, 477
  Cercotrichas l. *416*, **470**
  Cossypha 470

598 INDEXES: SCIENTIFIC NAMES

leucotis, Eremopterix 49, 97, 109, **111**, 115, 117, 119
  Eremopterix l. **111**
  Loxia 111
  Pycnonotus 364, 365, 372, **374**
leucura, Oenanthe 448, 501, 502, **504**, 506
leucurus, Baeopogon i. **303**
  Turdus 504
libanotica, Oenanthe o. **510**
libonyanus, Turdus 426
lichenya, Anthus n. **218**
limbata, Myrmecocichla a. **551**
lineiventris, Anthus 240, **242**
Linurgus olivaceus 2
litsitsirupa, Psophocichla 58
  Turdus 526
littoralis, Anthus p. **239**
livingstonii, Oenanthe p. **526**
lobatus, Ceblepyris 271
  Lobotus 256, **271**, 272
Lobotus **271**
  lobatus 256, **271**, 272
  oriolinus 256, 271, **272**
longicauda, Motacilla 207
longipennis, Calandrella b. **77**
  Pitta a. **11**
lopezi, Sheppardia c. 353, **397**
loquax, Andropadus i. **299**
lorenzi, Phyllastrephus 305, 324, **332**
lowei, Alethe 405
  Sheppardia 353, 394, 399, 404, **405**, 422
  Loxia 111
lucida, Hirundo 113, 147, 185, 187, 188, **189**, 190
  Hirundo l. 113, **189**
lugens, Motacilla 197, 209, 212
  Oenanthe 448, 502, 503, 511, 513, 516, **518**, 522, 523, 525, 527
  Oenanthe l. **518**
  Saxicola 518
lugentoides, Oenanthe l. **519**
lugubris, Oenanthe l. 448, 502, **518**
Lullula 14, 95, **105**, 106, 107
  arborea x, 16, 33, 101, **105**, 108, 441
Luscinia 406, **407**, 414
  luscinia 368, **408**, 410, 468
  megarhynchos 368, 397, 408, **409**, 412, 414, 429, 439, 469
  svecica 368, **412**
luscinia, Luscinia 368, **408**, 410, 468
  Motacilla 408
lutea, Motacilla 198
  Motacilla f. 176, **199**
lutescens, Anthus 216
lwenarum, Anthus r. 221
  Mirafra r. **29**
lynesi, Anthus n. 177, **218**
  Mirafra r. **37**
lypura, Cercomela m. **542**

M

mababiensis, Mirafra r. **29**, 96
mabirae, Stiphrornis e. **393**
maccarthuri, Pogonocichla s. **388**
macclounii, Cossypha a. 369, **421**
macdonaldi, Chersomanes a. **59**
  Mirafra a. **36**

Macronyx **245**
  ameliae 241, 246, 250, 255, **260**, 262
  aurantiigula 241, 246, 255, **258**
  capensis 241, 245, 246, 249, **251**, 255
  croceus 241, 246, 250, 252, **253**, 259, 526
  flavicollis 241, 246, **248**
  fuelleborni 241, 246, **249**, 251, 255, 263
  grimwoodi 241, 246, 250, 261, **262**
  sharpei 177, 216, **246**, 249, 255
macrorhyncha, Galerida c. **100**
  Saxicola 532
Macrosphenus 345
  kretschmeri 312
macrourus, Circus 529
maculata, Galerida c. **101**
madagascariensis, Phedina 136
  Phedina b. **136**, *161*
madaraspatensis, Motacilla 197, 209, 212
madaraszi, Eremopterix l. **111**
magna, Luscinia s. **412**
magnirostris, Alauda 98
  Galerida 95, 97, **98**
  Galerida m. 97, **98**
major, Parus 104
makalaka, Cercotrichas l. **478**
makarikari, Mirafra a. **35**
makawai, Spizocorys c. **84**
malabarica, Alauda 103
  Galerida 33, 49, 94, 95, 101, **103**
Malaconotidae 279
Malaconotus 342
malbranti, Mirafra a. **23**
malimbicus, Merops 127
Malimbus 342
  cassini 290
mallablensis, Anthus m. **229**
  Galerida m. **104**
mangbettorum, Psalidoprocne p. **131**
margaritae, Chersophilus d. 33, **92**
margaritata, Pogonocichla s. 388
marginata, Mirafra c. **15**, 18, *48*
marjoriae, Mirafra a. **31**
martini, Campephaga q. 256, **270**
marungensis, Mirafra a. **28**
marwitzi, Andropadus v. 287
  Hirundo d. **183**
massaica, Psalidoprocne p. **131**
masukuensis, Andropadus xi, 279, 280, 282, 286, **304**
  Andropadus m. 280, **304**
mauritanica, Riparia p. **138**, 140
maxima, Hirundo a. **153**
  Melanocorypha 65
mbuluensis, Cossypha a. 369, **422**
mcchesneyi, Spizocorys p. **88**
mechowi, Cercococcyx 394
medianus, Smithornis c. **8**
megaensis, Calendrella s. **83**
  Hirundo 161, **182**
  Pseudalaemon f. **94**
Megalophoneus 28
Megalotis 109, 116
megarhynchos, Luscinia 368, 397, 408, **409**, 412, 414, 429, 439, 469
  Luscinia m. 368, **410**

meinertzhageni, Chersomanes a. **59**
  Smithornis c. **8**
melaena, Myrmecocichla 465, 544, 550, 551
  Saxicola 550
Melaenornis 444, 453
melanauchen, Eremopterix n. **115**
melanocephala, Ardea 167
  Eremopterix l. 49, **111**
  Sylvia 104, 542
Melanocorypha 14, 65, **67**
  bimaculata 33, 66, 67, 68, **69**
  calandra 33, 66, **67**, 69, 92
  leucoptera 67
  maxima 65, 67
  mongolica 67
  yeltoniensis 67
melanocrissa, Hirundo d. **157**
melanogrisea, Motacilla f. **200**
melanoleuca, Oenanthe h. 449, 510, 513, **514**, 520
  Saxicola 532
melanota, Cossypha n. 400, *401*, **438**
melanura, Cercomela 464, 532, 540, **541**
  melanura, Cercomela m. 464, **542**
  Saxicola 541
melanurus, Passer 167
melbina, Pseudhirundo g. **146**
meleagris, Numida 166
Melignomon zenkeri 295
melindae, Anthus **229**, 240
  Anthus m. **229**, *240*
mentor, Andropadus i. 299
meridionalis, Delichon u. 195
Merops 139, 142
  apiaster 380, 426
  bullockoides 538
  malimbicus 127
  pusillus 526
merula, Turdus 453, 548
Micronisus gabar 36
micrus, Pycnonotus b. 366
migrans, Milvus 529
milanjensis, Andropadus 280, 283, 285, *304*
  Andropadus m. **286**, *304*
  Xenocichla 285
millardi, Calandrella c. **79**
Milvus migrans 529
mimica, Cossypha d. **435**
minor, Calandrella r. 33, **81**
  Cercotrichas g. **481**
  Cinclus c. **380**
  Indicator 370
  Myrmecocichla f. **547**
  Pycnonotus b. 367
  Riparia p. 138
minyanyae, Mirafra a. 28
Mirafra **14**, 42, 55, 56, 76, 91, 101
  affinis 35
  africana 14, 22, 24, 27, 28, *48*, 50, 96, 101
  africanoides 35, 41, *48*, 96
  albicauda 16, **18**, *48*, 83
  angolensis 27, 96
  apiata 15, 29, **30**, 96
  ashi **26**, *48*
  candida 22
  cantillans 15, 20, 21, 22, 27, 44, *48*, 83

cheniana 17, **19**, 96
collaris 27, **34**, *48*
cordofanica 20, *33*, 37, 89
degodiensis 15, 37, **38**, *48*
erythroptera 35
gilletti 15, **37**, 38, **48**, 83
hypermetra 14, 22, **24**, 26, 27, *48*
javanica 15
passerina **17**, 19, 96
poecilosterna 27, 38, **39**, *48*
pulpa 21, *48*
rufa 33, **36**
rufocinnamomea 15, **28**, 30, 31, 37, *48*, 95, 96
sabota **40**, 96
somalica **25**, 27, *48*
williamsi 20, 22, *48*
moco, Anthus s. **224**
modesta, Galerida 49, **95**
Galerida m. 49, **95**
modestus, Dicrurus 295
modularis, Motacilla 386
Prunella 257, 385, **386**
Prunella m. 257, **386**
Modulatrix 387, 452, **458**, 461
orostruthus 458
stictigula **400**, 447, **458**, 461
moesta, Oenanthe **449**, 502, 516, 519, **521**, 523, 525
Oenanthe m. **449**, **521**
Saxicola 521
molitor, Batis 312
Molothrus ater 292
monacha, Oenanthe **448**, 501, 503, 510, 513, **516**, 519
Saxicola 516
mongolica, Melanocorypha 67
monodi, Ammomanes d. 73
montana, Alethe 404
Sheppardia 353, 394, 399, **404**, 405
montanus, Andropadus xi, 279, 280, **281**, 283, **304**, 337
Cercococcyx 8, 399
monteiri, Hirundo s. **113**, **151**
Monticola x, 387, 505
saxatilis 511, 513
solitarius 503
monticola, Oenanthe **448**, 500, **506**, 535, 552
Oenanthe m. **448**, **506**
montivaga, Galerida m. **99**
moptana, Saxicola t. **495**
mossambiquensis, Mirafra a. 35
Motacilla 124, 142, **197**, 215
aguimp **176**, 197, 204, 209, 211, **212**
alba 122, **176**, 197, 200, 203, 204, **209**, 212, 227, 414
capensis **176**, 197, 200, **203**, 208, 211, 212, 252
cinerea **176**, 197, 200, 204, **205**, 207, 210, 382
citreola 197, 200, **202**
clara **176**, 197, 204, 206, **207**, 211, 212
feldegg 198
flava 122, *176*, 197, **198**, 202, 204, 207, 211, 222, 230, 237, 513
flavissima 198
flaviventris 197, 203
grandis 197, 209, 212
longicauda 207

lugens 197, 209, 212
lutea 198
madaraspatensis 197, 209, 212
thunbergi 198
Motacillidae **197**, 384
Motacillinae x, 197, 384
motitensis, Passer 252
moussieri, Erithacus 493
Phoenicurus **417**, 491, **493**
multicolor, Bleda s. 349
munda, Cercotrichas l. **478**
munzneri, Campephaga q. **270**
Phyllastrephus f. 333
Muscicapa 8, 444
adusta 396
striata 511
Muscicapidae 387, 452
Muscicapinae 387
mya, Ammomanes d. **73**
Myadestes 452, 461
Myadestinae 387, 452
Myoparus griseigularis 456
plumbeus 456
Myrmecocichla 452, 500, **544**
aethiops 465, 544, **545**, 547, 549, 551
albifrons 465, 544, 546, 549, 550, **551**, 552
arnotti 465, 507, 544, 549, 551, **552**
cinnamomeiventris 154, 155, 462, 465, 544, 550, **553**, 556
formicivora 155, 465, 507, 544, 545, **547**, 552
melaena 465, 544, **550**, 551
nigra 465, 544, 545, 546, **549**, 551, 552
semirufa 465, 544, 550, **556**
tholloni 465, 544, **545**, 549, 552
mzimbaensis, Anthus c. **232**

N

naevia, Mirafra s. **41**, 96
namaquensis, Cercomela s. **535**
Cossypha c. **423**
namibicus, Anthus v. **227**
Namibornis **487**
herero **416**, **487**
namuli, Alethe c. **451**
nandensis, Alethe p. **446**
Bleda s. **349**
Phyllastrephus c. **335**
narina, Heterotrogon 444
nata, Mirafra a. **31**, 96
natalensis, Cossypha 369, *401*, 402, 418, 427, 429, **433**, 436, 439
Cossypha n. **433**
naumanni, Pycnonotus b. 366
ndussumensis, Criniger 321, 360, **362**
Criniger v. 362
Neafrapus boehmi 136
nebularia, Tringa 426, 529
nebularum, Saxicola t. **495**
nebulosa, Cercomela t. **536**
Nectarinia 342
chalybea 389
senegalensis 296
Neocossyphus 350, 351, 387, 443, 444, 446, **452**, 461
finschi **400**, 452, 453, 455, 456, **457**
fraseri **400**, 452, 453, **455**, 457

poensis 355, *400*, 452, 453, **454**, 456, 457
rufus 350, 351, *400*, 452, **453**, 455, 456, 457
Neolestes 279, 352
torquatus 352
neoxena, Hirundo 188, 189, 190
neseri, Oenanthe p. **526**
neumanni, Andropadus t. 283, *304*
Anthus v. **227**
Hirundo s. 149
newtoni, Riparia p. **138**
ngamii, Pycnonotus b. 367
ngurumanensis, Phyllastrephus c. **335**
Nicator 279, 352
chloris 308, 341, *352*
gularis 352
vireo 352
nicholsoni, Anthus s. **224**, 240
nicolli, Calandrella r. **81**
nigeriae, Pycnonotus b. 366
nigra, Myrmecocichla 465, 544, 545, 546, **549**, 551, 552
Oenanthe 549
nigrescens, Mirafra a. **23**, 96
nigricans, Alauda 57
Galerida c. 33, **101**
Hirundo 147
Petrochelidon 147
Pinarocorys 56, **57**, 97
Pinarocorys n. 57
Pycnonotus 257, 364, 365, 367, **372**, 374
Pycnonotus n. 257, **372**
Turdus 372
nigricauda, Oenanthe m. **507**
nigriceps, Andropadus t. 283, *304*
Cossypha p. *401*, **427**
Eremopterix 33, 65, 72, 109, **115**, 119
Eremopterix n. 115
Pyrrhulauda 115
nigridorsalis, Neocossyphus p. **455**
nigriloris, Cisticola 496
nigripennis, Oriolus 273
nigrita, Galerida m. **95**
Hirundo 129, *160*, **180**
nigriticola, Mirafra r. 37
nigrorufa, Hirundo 113, **175**
nitens, Atticora 128
Psalidoprocne **128**, 159, *160*
Psalidoprocne n. **128**
nivea, Oenanthe o. 510
niveicapilla, Cossypha 400, *401*, 418, **437**, 439
Cossypha n. **438**
niveicapillus, Turdus 437
nivescens, Anthus s. **224**
nivosa, Campethera 361
nonnula, Estrilda 431
noomei, Andropadus i. 299
notata, Bleda c. 351
Bleda e. 321, **351**
novae Seelandiae, Alauda 217
novaeseelandiae, Anthus 177, 216, **217**, 220, 222, 224, 226, 229, 234, 243, 244
Numenius arquata 66
Numida meleagris 166
nyassae, Anthus s. **224**
nyikae, Mirafra a. **23**

O

oatleyi, Cercotrichas s. 475
obbiensis, Spizocorys 49, **87**
obscura, Hirundo 129
   Prunella m. **386**
   Psalidoprocne 128, **129**, 130, 131, 134, *160*, *162*
obscurata, Chersomanes a. **60**
obsoleta, Hirundo 147
   Hirundo f. *112*, **169**, *172*
occidentalis, Chlorocichla f. **311**
   Mirafra a. **23**
occidentis, Pinarocorys n. **58**
ochropus, Tringa 529
Ochruros, Motacilla 489
ochruros, Phoenicurus *417*, **489**, 491, 538
odica, Myrmecocichla c. **554**
Oenanthe 494, **500**, 530, 532, 544
   alboniger 501, 502, 504, 506, 521
   bottae 448, 502, 510, 526, **528**, 529
   deserti 449, 502, 504, 510, 514, 519, 521, 522, 523, **524**, 526, 529, 530
   finschii 449, 502, 511, 513, 514, 518, **520**, 525
   hispanica 449, 510, 512, **514**, 519, 520, 526, 528, 530
   isabellina 449, 502, 510, 513, 515, 519, 523, 525, 526, 528, **529**
   leucopyga 448, 501, **502**, 504, 506, 513, 516, 519
   leucura 448, 501, 502, **504**, 506
   lugens 448, 502, 503, 511, 513, 516, **518**, 522, 523, 525, 527
   moesta 449, 502, 516, 519, **521**, 523, 525
   monacha 448, 501, 503, 510, 513, **516**, 519
   monticola 448, **500**, **506**, 535, 552
   oenanthe 449, 501, 504, 506, 508, **509**, 513, 515, 519, 521, 525, 526, 528, 530
   phillipsi 449, 501, 506, **508**, 509
   picata 502, 505, **517**, 518
   pileata 448, 502, 507, 511, 513, 519, **526**, 528, 529, 533
   pleschanka 449, 503, 510, **512**, 514, 519, 521, 525, 527, 530
   xanthoprymna 449, 502, 516, 519, 522, **523**
oenanthe, Motacilla 509
   Oenanthe 449, 501, 504, 506, 508, **509**, 513, 515, 519, 521, 525, 526, 528, 530
   Oenanthe o. 449, 509, **510**
oenanthoides, Oenanthe o. 510
oleaginea, Psalidoprocne p. 131
oleaginus, Andropadus i. **299**, *305*
Oligomyodi 1
olimotiensis, Cercomela s. **543**
olivaceiceps, Andropadus m. **285**, *304*
olivaceogriseus, Phyllastrephus f. *320*, **344**
olivaceus, Criniger 321, 342, 357, 360, **362**
   Linurgus 2
   Trichophorus 362
   Turdus 367
omaruru, Mirafra a. **35**
omoensis, Anthus l. **226**
   Cercomela f. **538**
   Cossypha a. **439**
   Mirafra r. **29**
ongumensis, Calandrella c. **79**
Onychognathus 444
   walleri 302, 347
opistholeuca, Oenanthe p. 505, 517
oreas, Picathartes 159
oreobates, Saxicola t. **495**
oreophila, Oenanthe d. **525**
oriens, Cercotrichas p. **483**
orientalis, Pogonocichla s. **389**
   Psalidoprocne p. **131**, *160*
   Pyrrhurus s. *320*, **322**
Oriolidae 279
oriolinus, Lobotos 256, 271, **272**
Oriolus brachyrhynchus 273
   chlorocephalus 434
   larvatus 436
   nigripennis 273
   percivali 444
orostruthus, Arcanator **400**, 458, 459, **460**
   Arcanator o. **460**
   Modulatrix 458
   Phyllastrephus 460
ortiva, Chlorocichla f. **311**
Oscines 1, 13
ovambensis, Mirafra a. **35**
ovamboensis, Cercotrichas l. **478**
Oxylophus jacobinus 214, 300, 372, 374, 377

P

pachyrhyncha, Myrmecocichla a. **551**
paena, Cercotrichas **416**, 479, **483**, 485
   Cercotrichas p. **416**, **483**
   Erythropygia 483
pallens, Ammomanes c. **72**
pallida, Galerida c. **100**
   Hippolais 486
   Hirundo f. 169
   Lullula a. *33*, **105**
   Mirafra a. **23**, **96**
pallidiflava, Pogonocichla s. **388**
pallidigula, Chlorocichla f. **316**
pallidigularis, Sheppardia a. **397**
palliditinctus, Anthus s. **224**
pallidiventris, Anthus 177, 216, **228**
   Anthus p. **228**
   Cossypha c. **428**
pallidus, Andropadus l. **296**
   Pycnonotus b. **366**
paludibula, Riparia p. *112*, **138**
paludicola, Hirundo 137
   Riparia *112*, 135, 136, **137**, 140, 146
   Riparia p. **138**
Parisoma subcaeruleum 485
Parus 23
   leucomelas 551
   major 104
parvula, Riparia c. **144**
parvus, Cypsiurus 136, 179
Passer domesticus 120, 167
   melanurus 167
   motitensis 252
Passeri 1, **13**
Passerida 13, 279

Passeridae x, 197, 384
Passeriformes **1**, 14, 127, 279
passerina, Mirafra **17**, 19, *96*
Passerinae 197, 384
patae, Certhilauda a. **51**
payni, Ammomanes d. **73**
peasi, Pycnonotus b. **366**
pectoralis, Cercotrichas l. **478**
   Coracina 256, 273, 274, **275**, 277
   Graucalus 275
peli, Scotopelia 453
pelios, Turdus 457
penelope, Anas 380
Pentholaea 544
percivali, Andropadus g. **294**, *304*
   Oriolus 444
perconfusa, Calandrella s. **82**
perconfusus, Motacilla f. **200**
perdita, Hirundo 147, *161*, 165, **168**
Pericrocotus 263
   brevirostris 263
persica, Oenanthe l. **518**
personata, Spizocorys 49, **88**
   Spizocorys p. **88**
petiti, Campephaga 256, 263, 264, 266, **268**, 270
   Psalidoprocne p. 130, **131**, 134
petricolus, Anthus s. **224**
Petrochelidon 147
Petrocinchla 429
petrosa, Alauda 239
petrosus, Anthus 216, 236, 238, **239**, *240*
   Anthus p. 239, *240*
Phaeornis 452
Phainopepla 377
Phainoptila 377
Phedina **134**, 138, 142
   borbonica 134, 135, **136**, *161*
   brazzae 134, **135**, 136, *161*
Philepittidae 1
Philetairus socius 538
phillipsi, Oenanthe 449, 501, 506, **508**, 509
   Saxicola 508
Philomachus pugnax 192
philomelos, Turdus 101, 272
phoenicea, Ampelis 264
   Campephaga 256, 263, **264**, 266, 268
Phoeniculus 327
   bollei 3
phoenicuroides, Phoenicurus o. *417*, **489**, 491
Phoenicurus **489**, 523
   moussieri *417*, 491, **493**
   ochruros *417*, **489**, 491, 538
   phoenicurus *417*, 490, **491**, 493, 504, 538
phoenicurus, Ammomanes 71
   Motacilla 491
   Phoenicurus *417*, 490, **491**, 493, 504, 538
   Phoenicurus p. *417*, **491**
Phyllastrephus 298, 322, **323**, 329, 363
   albigularis *320*, **330**, **342**
   baumanni 305, 324, **330**, 331
   cabanisi *320*, 323, 324, 328, 330, 332, 333, **334**, 337, 343, 452, 459, 461

cerviniventris 320, 324, 327, **328**, 329
debilis 320, 345, **347**
fischeri 312, *320*, 324, 326, **333**, 334
flavostriatus 312, *320*, 324, 336, **343**, 346, 348
fulviventris 320, 324, 328, **329**
hypochloris 305, 324, 330, **331**
icterinus 301, **338**, 340, 342, 343, 352
leucolepis **341**, 352
lorenzi 305, 324, **332**
poensis 282, *305*, 324, **337**
poliocephalus 301, *320*, 324, 337, 343, **346**
strepitans 320, 325, **327**, 328, 334
terrestris 320, *325*, 327, 328, 329, 334, 348, 434
xavieri 301, 338, **340**, 343, 349, *352*
Phylloscopus 348
sibilatrix 68
trochilus 426
Phyllostrephus 329, 330
Phyllostrophus 318, 328, 337
picata, Oenanthe 502, 505, **517**, 518
Oenanthe p. 517
Saxicola 517
Picathartes oreas 159
Picus viridis 68
pileata, Motacilla 526
Oenanthe 448, 502, 507, 511, 513, 519, **526**, 528, 529, 533
Oenanthe p. 448, **526**
Pinarochroa 532
Pinarocorys 56
erythopygia 48, **56**, 57
nigricans 56, **57**, 97
Pinarornis 387, 452, **461**
plumosus 416, **461**
pintoi, Mirafra r. **29**
Pitta 9
angolensis 9, 10, **11**, *32*
brachyura 9
reichenowi 9, 11, *32*
Pittidae x, 1, **9**
Pittoidea 9
placidus, Phyllastrephus c. *320*, **334**
Platyrhynchus 7
Platysteira 5, 8
plebeja, Mirafra s. **41**
pleschanka, Motacilla 512
Oenanthe 449, 503, 510, **512**, 514, 519, 521, 525, 527, 530
Oenanthe p. 449, **512**
Ploceinae 197, 384
Ploceus 126, 345, 367
bicolor 332, 347, 389
dorsomaculatus 295
plumbeus, Myoparus 456
plumosus, Pinarornis 416, **461**
podobe, Cercotrichas 416, 470, **486**
Cercotrichas p. 416, **486**
Turdus 486
poecilosterna, Alauda 39
Mirafra 27, 38, **39**, *48*
poensis, Cossypha 454
Neocossyphus 355, *400*, 452, 453, **454**, 456, 457
Neocossyphus p. *400*, **454**
Phyllastrephus 282, *305*, 324, **337**
Phyllostrophus 337

Sheppardia b. 353, **395**
Pogoniulus bilineatus 444
Pogonocichla 387, 390, 444, 450
stellata 353, **388**, 391, 393, 404, 405, 422, 452
poliocephala, Alethe xi, 355, **368**, *401*, 403, 441, 442, 444, **445**, 455
Alethe p. 368, **446**
Xenocichla 346
poliocephalus, Caprimulgus 554
Phyllastrephus 301, *320*, 324, 337, 343, **346**
Trichophorus 445
poliophrys, Alethe xi, 368, *401*, 441, **443**
Alethe p. 368, *401*, **444**
polioptera, Cossypha 353, 395, *401*, 418, **427**
Cossypha p. 353, **427**
pollux, Cercomela s. 464, **534**
praepectoralis, Neocossyphus p. **454**
praetermissa, Galerida m. **103**
pratensis, Alauda 235
Anthus 216, 234, **235**, 237, 238, 239, *240*
Anthus p. **235**, *240*
presaharica, Hirundo f. **169**
pressa, Modulatrix s. **459**
pretoriae, Hirundo f. **169**
preussi, Hirundo 146, 147, **159**, *161*, 163, 195
Lecythoplastes 159
prigoginei, Chlorocichla 309, **310**, 352
Prinia 444
Prionopidae 279
Prionops 279
pristoptera, Chelidon 130
Hirundo 130
Psalidoprocne 128, 129, **130**, 133, 136, 139, 159, *160*, 173, 187
Psalidoprocne p. **131**
Progne 127
subis 127
Promeropidae 387
promiscua, Saxicola t. **495**
Protonotaria citrea 433
Prunella 384
collaris 257, **384**
immaculata 386
modularis 257, 385, **386**
rubida 386
Prunellidea **384**
Prunellinae 197, 384
prunus, Anthus l. 226
Psalidoprocne 125, **128**, 129, 158
albiceps 128, 131, **132**, 133, *160*
fuliginosa 128, 130, 131, 132, **133**, 159, *160*
nitens **128**, 159, *160*
obscura 128, **129**, 130, 131, 134, *160*, 162
pristoptera 128, 129, **130**, 133, 136, 139, 159, *160*, 173, 187
psammochroa, Melanocorypha c. 68
Pseudalaemon 93
fremantlii 49, **94**
Pseudammomanes 55
Pseudhirundo 145
griseopyga *112*, 140, **145**, 162, 164, 179, 184

Pseudocalyptomena 2
graueri 2, 32
Pseudochelidon 125
eurystomina **125**, 159, *161*
sirintarae 125
Pseudochelidoninae **125**
Pseudocossyphus 453
Psophocichla x, 387
litsitsirupa 58
Ptilogonys 377
Ptyonoprogne 147
fuligula 169
puella, Hirundo a. **153**
pugnax, Philomachus 192
pulcher, Spreo 546
pulih, Pitta a. **11**
pulpa, Mirafra **21**, *48*
pura, Coracina c. 256, **274**
pusilla, Hirundo f. **169**
pusillus, Merops 526
Pycnonotidae x, **279**, 324, 331
Pycnonotus 279, 298, **363**, 374, 444
aurigaster 364, 365, 372, 374
barbatus 95, 213, 257, 291, 295, 297, 316, 327, 331, 351, 364, **365**, 372, 374, 431
cafer 364, 365, 372, 374
capensis 257, 299, 364, 365, 367, 372, **374**
jocosus 364
leucogenys 364, 365, 372, 374
leucotis 364, 365, 372, 374
nigricans 257, 364, 365, 367, **372**, 374
xanthopygos 257, **364**, 365, 372, 374
pycnopygius, Achaetops 488
pygmaea, Motacilla f. 176, **199**
Pyrrhula pyrrhula 92
pyrrhula, Pyrrhula 92
Pyrrhulauda 115, 118
Pyrrhurus 309, 322
scandens 319, *320*, **322**

Q

quadrivirgata, Cercotrichas 402, *416*, 451, 470, 471, **473**, 474, 481
Cercotrichas q. *416*, **473**
Thamnobia 473
quaesita, Mirafra a. **35**
quiscalina, Campephaga 256, 267, **269**
Campephaga q. 256, **269**

R

rabai, Phyllastrephus d. *320*, **347**
Ramphocoris 14, **65**
clot-bey 33, **65**, 68, 72
randonii, Galerida c. **100**
ranivorus, Circus 204, 377
raytal, Calandrella 80, 82
razae, Alauda 107
reclusa, Cercotrichas s. 475
Regulidae 279
reichenowi, Cercotrichas l. **471**
Pitta 9, 11, *32*
Psalidoprocne p. **131**
reynoldsi, Mirafra a. **31**
Rhodopechys githaginea 63, 66
richardi, Anthus n. 216, **218**

riggenbachi, Galerida c. **100**
   Oenanthe l. 505
Riparia 134, 135, 136, **137**, 146
   cincta 112, 138, 142, **143**, 146, 157, 170, 187
   congica 112, 135, 137, **140**
   paludicola 112, 135, 136, **137**, 140, 146
   riparia 112, 135, 136, 137, **140**, 144, 146, 157, 162, 164, 168, 170, 172, 187
riparia, Hirundo 140
   Riparia 112, 135, 136, 137, **140**, 144, 146, 157, 162, 164, 168, 170, 172, 187
   Riparia r. 112, **141**
roberti, Callene 415
   Cossyphicula 369, 398, **415**, 419, 420
   Cossyphicula r. 369, **415**
robertsi, Phyllastrephus t. 325
rochei, Mirafra s. 25
rodgersi, Swynnertonia s. **391**
roehli, Andropadus m. **280**, *304*
rostrata, Oenanthe o. 510
rothschildi, Hirundo l. **189**
rubecula, Erithacus 368, *401*, **406**, 462, 481, 544
   Erithacus r. **406**
   Motacilla 406
rubescens, Anthus 238, 239
rubetra, Motacilla 497
   Saxicola 417, 494, 496, **497**, 513, 529, 532
rubicola, Saxicola t. 417, **495**
rubicunda, Neocossyphus f. **400**, **455**
rubida, Prunella 386
rubidior, Mirafra a. 35
rubiginosa, Calandrella b. 33, **77**
ruddi, Heteromirafra **43**, 44, 45, 96
   Heteronyx 43
rufa, Mirafra 33, **36**
   Mirafra r. **36**
rufescens, Alauda 80
   Calandrella 33, 70, 72, 76, 77, **80**, 82, 101, 104, 108
   Melanocorypha b. 33, **70**
rufescentior, Cossyphicula r. **415**
ruficauda, Bradyornis 467
   Cercotrichas l. 478
   Cichladusa 400, 463, 466, **467**
ruficolor, Galerida m. **103**
rufigula, Cotyle 169
   Hirundo 146, 147, 159, *161*, **163**, 165, 169, 178, 195
   Hirundo f. 169
   Ptyonoprogne f. 169
rufipennis, Cichladusa g. **400**, **469**
   Trichastoma 459, 461
rufocinerea, Terpsiphone 290
rufocinnamomea, Megalophoneus 28
   Mirafra 15, **28**, 30, 31, 37, *48*, 95, 96
   Mirafra r. **29**
rufolateralis, Smithornis 3, 4, 5, 7, 8, *32*
   Smithornis r. 5
rufula, Hirundo d. 113, **156**
rufuloides, Anthus n. 177, 216, **218**, 221

rufus, Neocossyphus 350, 351, *400*, 452, **453**, 455, 456, 457
   Neocossyphus r. **400**, **453**
   Pseudocossyphus 453
rupestris, Hirundo 112, 138, 142, 147, 157, 169, **171**
rustica, Hirundo 68, *113*, 139, 141, 144, 147, 157, 162, 168, 185, 187, 188, 189, **190**, 194, 513
   Hirundo r. 113, **191**
ruwenzori, Psalidoprocne p. **131**, *160*
ruwenzoria, Mirafra a. 23
ruwenzorii, Phyllastrephus f. 344
   Pogonocichla s. **388**

S

sabota, Mirafra **40**, 96
   Mirafra s. **40**, 96
sabotoides, Mirafra s. **41**
salax, Saxicola t. **495**
saldanhae, Certhilauda a. 51
samamisicus, Phoenicurus p. **417**, 490, **491**
samharensis, Ammomanes d. **74**
sanjei, Arcanator o. **460**
saphiroi, Anthus l. **226**
sarwensis, Mirafra a. **35**, 96
saturatior, Calandrella c. **79**
   Hirundo s. **151**
saturatus, Andropadus l. **296**
savignii, Hirundo r. 113, **191**
saxatilis, Monticola 511, 513
Saxicola **494**, 500, 532
   bifasciata 417, 494, **499**, 532, 533
   caprata 532
   dacotiae 532
   ferrea 532
   insignis 532
   jerdoni 532
   macrorhyncha 532
   melanoleuca 532
   rubetra 417, 494, 496, **497**, 513, 529, 532
   torquata 398, *417*, **494**, 498, 499, 511, 532
Saxicolini 387
scandens, Phyllastrephus 322
   Pyrrhurus 319, *320*, **322**
   Pyrrhurus c. 320, **323**
schalowi, Oenanthe l. 448, 513, **519**, 527
schillingsi, Mirafra c. **15**
schioeleri, Oenanthe o. 510
schlegelii, Cercomela 464, 532, 533, **534**, 536
   Cercomela s. 464, 532, **535**
   Erithacus 534
schoana, Cercomela s. 543
schoanus, Pycnonotus b. 257, **366**
schoeniclus, Emberiza 92
schoensis, Riparia p. **138**
schoutedeni, Anthus s. **224**, *240*
   Chlorocichla l. **310**
   Mirafra r. **29**
   Phyllastrephus c. **328**
   Sheppardia b. **395**
scirpaceus, Acrocephalus 439
sclateri, Calandrella 86
   Cercotrichas l. **478**

Spizocorys 84, **86**, 90, 97
Scotocerca inquieta 517
scotocerca, Cercomela 464, 532, 537, **539**, 541
   Cercomela s. **539**
   Saxicola 539
Scotopelia peli 453
seclusa, Cossypha n. 433
seebohmi, Oenanthe o. *449*, 509, **510**
seimundi, Anthreptes 295
semirufa, Cossypha 369, *401*, 418, 428, **429**, 431
   Cossypha s. 369, *401*, **429**
   Hirundo 113, 147, **149**, 151, 153, 155, 157, 182
   Hirundo s. 113, **149**
   Myrmecocichla 465, 544, 550, **556**
   Petrocinchla 429
   Saxicola 556
semitorquata, Certhilauda 47
   Certhilauda c. **47**
senator, Lanius 104
senegalensis, Eupodotis 529
   Hirundo 113, 147, 149, **150**, 157
   Hirundo s. 113, **151**
   Nectarinia 296
senegallensis, Galerida c. 49, **101**
senegallus, Vanellus 526, 529
serina, Calyptocichla **301**, 310, 340, 352
Serinus canaria 213
serinus, Criniger 301
serlei, Mirafra r. **29**
   Phyllastrephus x. **340**
sethsmithi, Phyllastrephus i. 340
sharpei, Callene 398
   Macronyx 177, 216, **246**, 249, 255
   Mirafra a. **23**, 26, *48*
   Sheppardia 353, 394, 395, 396, 397, **398**, *401*, 402, 403, 404, 405
   Sheppardia s. 353, **399**, *401*
   Smithornis 4, 5, 8, *32*
   Smithornis s. **4**
shelleyi, Riparia r. **141**
Sheppardia 387, 392, **394**, 415, 418, 444, 450
   aequatorialis 353, 393, 394, 395, 396, **397**, 398, *401*, 402, 403, 415, 420
   bocagei 353, 394, **395**, 396, 397, 398, *401*, 402, 403, 415, 419, 420, 427
   cyornithopsis 353, 393, 394, 395, **396**, 397, 398, *401*, 402, 403
   gabela 353, 394, 395, 396, 397, 398, 402, **403**
   gunningi 353, 394, 395, 396, 397, 398, *401*, **402**, 403
   lowei 353, 394, 399, 404, **405**, 422
   montana 353, 394, 399, **404**, 405
   sharpei 353, 394, 395, 396, 397, **398**, *401*, 402, 403, 404, 405
shimba, Andropadus v. 287
shimbanus, Phyllastrephus d. 347
sibilatrix, Phylloscopus 68
sidamoensis, Heteromirafra 43, 44, **45**, *48*
   Mirafra 45
sierrae, Alauda a. **108**

signata, Cercotrichas 391, *416*, 470, 471, 473, **474**
  Cercotrichas s. **416, 475**
  Cossypha 474
  Eremopterix 49, 109, 111, 114, 116, **118**, 119
  Eremopterix s. **49, 118**
  Pyrrhulauda 118
similis, Anthus 216, 217, 219, 221, 222, **223**, 226, 227, 230, *240*, 243, 500
simplex, chlorocichla 309, **314**, 317, *321*
  Tricophorus 314
simplicicolor, Chlorocichla f. 317
simplicissima, Motacilla c. 176, **204**, 208
sinuata, Cercomela 464, 532, **533**, 538
  Cercomela s. 464, 532, **533**
  Luscinia 533
sirintarae, Pseudochelidon 125
smithersi, Mirafra r. **29**, 96
smithii, Eremopterix l. 97, **111**
  Hirundo 113, 153, **178**, 185
  Hirundo s. 113, **179**
Smithornis **3**, 29
  capensis **3**, 4, **5**, **7**, 10, *32*
  rufolateralis 3, 4, **5**, 7, **8**, *32*
  sharpei **4**, 5, **8**, *32*
sobatensis, Mirafra r. **29**
socius, Philetarius 538
sokokensis, Anthus **233**, *240*
  Phyllastrephus f. 333
  Sheppardia g. 353, 401, **402**
sokotrae, Anthus s. **224**
solitarius, Cuculus 205, 214, 317, 372, 390, 392, 402, 424, 427, 430, 433, 463, 474, 480
  Monticola 503
somalica, Alauda 82
  Calandrella 22, 49, 76, 80, **82**
  Calandrella s. **83**
  Certhilauda 25
  Mirafra 25, 27, **48**
  Mirafra s. **25, 48**
somaliensis, Andropadus i. 299
  Galerida c. 49, **101**
  Pycnonotus b. 257, **366**
sordida, Cercomela 464, 532, 540, **543**
  Cercomela s. 464, **543**
  Saxicola 543
soror, Chlorocichla f. **317**, 320
spatzi, Hirundo f. 112, **169**
spectratrix, Cercomela s. 464, **540**
Spilocorydon 24
spilodera, Hirundo 147, *161*, 164, **165**, 168, 170
spinoletta, Alauda 238
  Anthus 216, 236, **238**, 239, *240*
  Anthus s. **238**, *240*
Spizixos 279
Spizocorys 14, **83**
  conirostris **83**, 85, 96, **90**, 97
  fringillaris 84, **85**, 86, 97
  obbiensis 49, **87**
  personata 49, **88**
  sclateri 84, **86**, 90, 97
spleniata, Calandrella c. **79**
Spreo bicolor 139, 144, 507
  pulcher 546
spurium, Anthus n. **218**

spurius, Pycnonotus b. 257, **366**
stabilior, Macronyx c. 251
stabilis, Anthus n. **218**
starki, Calandrella 90
  Eremalauda 84, 86, 88, **90**, 97
Stelgidillas 331
Stelgidopteryx 125
stellata, Pogonocichla 353, **388**, 391, 393, 404, 405, 422, 452
  Pogonocichla s. 353, **388**
Stephanoaetus coronatus 252, 429
Sterna 126
stictigula, Modulatrix 400, **447, 458**, 461
  Modulatrix s. **400, 459**
  Turdinus 458
Stiphrornis 387, **392**, 415
  erythrothorax 353, **392**, *401*, 415
Stizorhina 452, 461
stonei, Saxicola t. **495**
strepitans, Criniger 327
  Phyllastrephus 320, 325, **327**, 328, 334
stresemanni, Mirafra a. **23**, 48
striata, Muscicapa 511
striifacies, Andropadus m. **285**, *304*
strumpelli, Galerida m. **95**
Sturnella 245
Sturnidae 279
Sturnus vulgaris 127, 294, 378
suahelica, Riparia c. **144**
suahelicus, Phyllastrephus t. 320, **325**
  Smithornis c. **8**
subalaris, Andropadus i. 299
  Hirundo l. **189**
subcaeruleum, Parisoma 485
subcoronata, Certhilauda c. **47**
subis, Progne 127
Suboscines 1
subpallida, Chersomanes a. **59**
subpersonata, Motacilla a. 176, **210**
subrufescens, Cossypha h. **431**
subrufipennis, Myrmecocichla c. **465**, 554
succosus, Phyllastrephus f. 334
sucosus, Phyllastrephus c. 330, 332, **335**, 343
sudanensis, Myrmecocichla a. **546**
suffusa, Psalidoprocne a. **133**
suffusca, Mirafra s. **41**
superciliaris, Motacilla f. **200**
superciliosus, Centropus 377
superflua, Galerida m. **33**, **103**
superior, Pycnonotus n. **373**
svecica, Luscinia 368, **412**
  Luscinia s. 368, **412**
  Motacilla 412
swynnertoni, Erythacus 391
  Swynnertonia 353, 389, **391**
  Swynnertonia s. 353, **391**
Swynnertonia 387, **390**
  swynnertoni 353, 389, **391**
syenitica, Oenanthe l. 448, **505**
Sylvia 104, 414
  curruca 542
  leucomelaena 542
  melanocephala 104, 542
Sylvioidea x, 125, 279
syndactyla, Bleda 321, **348**, 351, 355, 455
  Bleda s. **349**

Dasycephala 348
syriacus, Cercotrichas g. 416, **481**

T

tachiro, Accipiter 297, 354, 355, 457, 458
Tachymarptis aequatorialis 162
taeniolaema, Campethera 345, 347
tahitica, Hirundo 188, 189, 190
teitensis, Pycnonotus b. 366
temminckii, Cursorius 526
tenebrior, Pycnonotus b. 366
tenellus, Macronyx 214
  Tmetothylacus 177, **214**
tennenti, Cossypha n. 433
tenuirostris, Phyllastrephus f. 320, **344**
tephridorsus, Anthus l. **226**
tephrolaemus, Andropadus 280, **282**, 286, *304*, 337, 346, 450
  Andropadus t. **283**, *304*
  Trichophorus 282
Terpsiphone 444
  rufocinerea 290
terrestris, Phyllastrephus 320, **325**, 327, 328, 329, 334, 348, 434
  Phyllastrephus t. 320, **325**
tertia, Alaemon h. **64**
tertius, Macronyx c. 254
tessmanni, Cossypha p. **427**
Thamnobia 473
thamnodytes, Cercotrichas b. 472
Thamnolaea 500, 544
theklae, Galerida m. 103
theresae, Oenanthe m. **521**
  Spizocorys s. **86**
Thescelocichla 309, **318**, 322
  leucopleura **318**, 323, 352
tholloni, Myrmecocichla **465**, 544, **545**, 549, 552
  Saxicola 545
Thraupidae 13
thunbergi, Motacilla 198
  Motacilla f. 176, **199**
tigrina, Mirafra r. **29**
Timaliidae 387, 458, 460
tinnunculus, Falco 180
Tmetothylacus 197, **214**
  tenellus 177, **214**
togoensis, Baeopogon i. 303
  Cercotrichas s. **475**
torquata, Motacilla 494
  Saxicola 398, *417*, **494**, 498, 499, 511, 532
  Saxicola t. 417, **495**
torquatus, Neolestes 352
torrentium, Motacilla c. **208**
torrida, Mirafra r. **29**
tractrac, Cercomela 75, 464, 532, **535**
  Cercomela t. 464, 532, **535**
  Motacilla 535
transiens, Spizocorys c. **84**
transitiva, Hirundo r. **191**
transvaalensis, Certhilauda s. 47
  Mirafra africana 23, 96
  Mirafra africanoides 35
  Pogonocichla s. **389**
trapnelli, Mirafra a. 35
traylori, Anthus c. **232**
Treron calva 429

Trichastoma 345
　rufipennis 459, 461
Tricholaema leucomelaina 167
Trichophorus 282, 311, 314, 338, 350, 354, 357, 358, 359, 362, 445
tricolor, Phyllastrephus i. 338
　Pycnonotus b. 257, **367**
Tringa glareola 529
　nebularia 426, 529
　ochropus 529
trivialis, Alauda 233
　Anthus 216, 217, 225, **233**, 236, 237, *240*, 242
　Anthus t. **234**, *240*
trochilus, Phylloscopus 426
Trochocercus 345
　cyanomelas 312
Troglodytes **382**
　troglodytes 257, **382**
troglodytes, Motacilla 382
　Troglodytes 257, **382**
　Troglodytes t. 383
Troglodytidae **382**
tropicalis, Mirafra a. **23**
tsumebensis, Mirafra a. 35
Turdidae x, **387**, 452, 458, 460
Turdinae 387, 452
Turdinus 458
Turdoides 327
　jardinei 326
Turdus x, 104, 370, 381, 387, 455, 456, 481, 505, 552
　libonyanus 426
　litsitsirupa 526
　merula 453, 548
　olivaceus 367
　pelios 457
　philomelos 101, 272
turkana, Cercomela s. *464*, 538, **540**
Tyranni **1**, 13
Tyrannides 1
Tyranniformes 1

U

ufipae, Alethe p. **446**
ugandae, Andropadus g. **289**
　Bleda c. 351
　Bleda e. **351**
uis, Mirafra s. 41
ultima, Cercomela m. **542**
umbratica, Swynnertonia s. 391
unitatis, Hirundo a. 153, *161*

upembae, Pyrrhurus s. 322
urbica, Delichon 157, *161*, 162, 172, 184, **194**
　Delichon u. **195**
　Hirundo 194
usambarae, Alethe f. *401*, 447
　Andropadus t. **283**, *304*
　Sheppardia s. 399
uzungwensis, Phyllastrephus f. **344**

V

vaalensis, Anthus 177, 216, 217, 219, 226, **227**, 228, 243
　Anthus v. 177, **227**
validior, Cercomela s. **540**
vandeweghei, Alethe p. **446**
Vanellus coronatus 526
　senegallus 526, 529
variegata, Saxicola t. 417, **496**
vaughan-jonesi, Pycnonotus b. 367
vauriei, Oenanthe l. **518**
verreauxi, Criniger 362
　Criniger c. *321*, **360**
verticalis, Eremopterix 97, 109, **116**, 118
　Eremopterix v. 97, **116**
　Megalotis 116
vesey-fitzgeraldi, Mirafra s. 41
vidua, Motacilla a. 176, **212**
vincenti, Mirafra a. 35
　Phyllastrephus f. **344**
virago, Oenanthe o. 510
virens, Andropadus 279, **287**, 290, 293, 296, *305*, 306, 330, 332, 335, 355
　Andropadus v. **287**, *305*
vireo, Nicator 352
viridiceps, Phyllastrephus a. 320, **343**
viridis, Picus 68
Vitiflora 502
vittata, Oenanthe p. 512
vocifer, Haliaeetus 434
volgae, Luscinia s. **412**
vulgaris, Sturnus 127, 294, 378
vulpecula, Calandrella s. 83
vulpina, Cercotrichas l. **478**
　Neocossyphus f. **456**
vulturinus, Macronyx c. 254

W

waibeli, Mirafra s. **41**, 96

walleri, Onychognathus 302, 347
wellsi, Motacilla c. 176, **204**
whitakeri, Ammomanes d. 73
williamsi, Calandrella c. 79
　Mirafra **20**, 22, 48
winterbottomi, Anthus s. 224
wintoni, Macronyx a. 260
witherbyi, Erithacus r. 368, *401*, **406**
witputzi, Calandrella c. 79
woosnami, Alethe d. **442**
　Bleda s. 321, **349**

X

xanthogaster, Striphrornis e. 353, **393**
xanthoprymna, Oenanthe 449, 502, 516, 519, 522, **523**
　Oenanthe x. 449, 522, **523**
　Saxicola 523
xanthopygos, Ixus 364
　Pycnonotus 257, **364**, 365, 372, 374
xavieri, Phyllastrephus 301, 338, **340**, 343, 349, 352
　Phyllastrephus x. **340**, *352*
　Xenocichla 340
Xenocichla 285, 306, 340, 342, 346, 347
Xenocopsychus 387, **440**
　ansorgei 400, 418, 426, **440**
xerica, Riparia c. **144**
xuthura, Alethe f. 447

Y

yarrellii, Motacilla a. **210**
yavelloensis, Spizocorys p. 88
yeltoniensis Melanocorypha 67

Z

zambesiae, Chlorocichla f. 311
zambesiana, Cercotrichas l. 416, **478**
zanzibaricus, Andropadus v. **288**
zenkeri, Anthus l. 177, **226**
　Melignomon 295
　Smithornis s. **4**, *32*
zombensis, Andropadus v. **287**
Zoothera x, 387
　gurneyi 459, 461, 475
　guttata 434
Zosterops 345

English Names

A

Accentor, Alpine 257, **384**
　Hedge 257, 385, **386**
Akalat, Bocage's 353, **395**, *401*, 403, 415, 419, 420, 427
　East Coast 353, *401*, **402**
　Equatorial 353, 393, **397**, 399, *401*, 415, 420
　Gabela 353, **403**
　Iringa 353, 399, **405**, 422
　Lowland 353, 393, **396**, 398, *401*

　Sharpe's 353, **398**, *401*, 404, 405
　Usambara 353, 399, **404**
Alethe, Brown-chested 368, *401*, 403, 442, 444, **445**, 455
　Cholo 368, *401*, 422, **451**
　Fire-crested 10, 368, 393, *401*, **442**, 444, 446, 455
　Red-throated 368, *401*, **443**
　White-chested 368, *401*, 404, 422, **447**, 451
Anteater-Chat, Northern 465, **545**, 549, 551

Southern 155, *465*, 507, **547**
Ant-Thrush, Red-tailed 350, 351, *400*, **453**, 455, 456, 457
　White-tailed 355, *400*, 453, **454**, 456, 457
Apalis, Masked 3

B

Babbler, Abyssinian Hill 284
　Arrow-marked 326
　Gilbert's 347

INDEXES: ENGLISH NAMES

Tit- [see Tit-Babbler]
Barbet, Pied 167
  Yellow-spotted 308
Bee-eater, European 380, 426
  Little 526
  Rosy 127
  White-fronted 538
Blackbird, Eurasian 453, 548
Blackstart *464*, **540**, **541**
Bluethroat *368*, **412**
Boubou, Southern 377, 426, 436
Bristlebill, Green-tailed *321*, 339, **350**, 355
  Grey-headed *321*, 351, **354**, 446
  Red-tailed *321*, **348**, 351, 355, 455
Broadbill, African 4, 5, 7, 10, *32*
  African Green **2**, *32*
  Grauer's 2
  Grey-headed **4**, 5, 8, *32*
  Red-sided 5
  Rufous-sided 4, **5**, *8*, *32*
Brownbul, Northern *320*, 325, **327**, 328, 334
  Terrestrial *320*, **325**, 327, 328, 329, 334, 348, 434
Bulbul, African Red-eyed 257, 367, **372**, 374
  Black-eyed 365
  Black-fronted 372
  Cape 257, 299, 367, 373, **374**
  Common 95, 213, *257*, 291, 295, 297, 327, 331, 351, 364, **365**, 373, 374, 431
  Dark-capped 365
  Slender 347
  Swamp Palm **318**, 323, *352*
  White-eyed *257*, **364**
Bullfinch, Eurasian 92
Bunting, Corn 66, 68
  Reed 92
Bush-Lark, Red-winged **24**, 26, 27, *48*
  Rusty *33*, **36**
  Singing **15**, 18, 20, 21, 22, 27, 44, *48*, 83
  White-tailed 16, **18**, *48*, 83
Bustard, White-bellied 529
Buzzard, Common 157
  Steppe 167

C

Canary, Island 213
Cave-Chat, Angola *400*, 426, **440**
Chaffinch 104
Chat, Anteater- [see Anteater-Chat]
  Arnott's 552
  Boulder *416*, **461**
  Buff-streaked *417*, 494, **499**, 533
  Cave- [see Cave-Chat]
  Cliff- [see Cliff-Chat]
  Familiar 426, *464*, 488, 500, 508, 533, 535, 536, **537**, 540
  Herero *416*, **487**
  Hill 543
  Karoo *464*, 533, **534**, 536
  Mocking 553
  Moor- [see Moor-Chat]
  Moorland *464*, 540, **543**
  Mountain 506
  Red-tailed 537
  Robin- [see Robin-Chat]
  Rock- [see Rock-Chat]
  Rüppell's Black **465**, **550**, 551
  Sickle-winged *464*, **533**, 538
  Sooty *465*, 545, 546, **549**, 551, 552
  Tractrac 75, *464*, **535**
  White-fronted Black *465*, 546, 549, 550, **551**
  White-headed Black *465*, 507, 549, **552**
Cisticola, Ayres's 247
  Black-lored 496
  Chattering 213
Cliff-Chat, Mocking 154, 155, 462, *465*, 550, **553**, 556
  White-winged *465*, 550, **556**
Coucal, White-browed 377
Courser, Temminck's 526
Cowbird, Brown-headed 292
Crow, Carrion 68
Cuckoo, Barred Long-tailed 8, 399
  Black and White 214
  Diederik 205, 214, 372, 480, 484, 485
  Dusky Long-tailed 394
  Emerald 298, 372, 436, 444
  European 494
  Jacobin 300, 372, 374, 377
  Klaas's 372
  Red-chested 205, 214, 317, 372, 390, 392, 402, 424, 427, 430, 433, 463, 474, 480
Cuckoo-Shrike, Black 256, 265, **266**, 269, 270
  Blue *256*, **277**
  Eastern Wattled *256*, **272**
  Grauer's *256*, 274, **277**, 278
  Grey *256*, **274**, 277, 424
  Petit's *256*, 267, **268**, 270
  Purple-throated *256*, 267, **269**
  Red-shouldered *256*, **264**, 267, 269
  Western Wattled *256*, **271**
  White-breasted *256*, 274, **275**, 277
Curlew 66

D

Dipper, White-throated *257*, **380**
Drongo, Fork-tailed 526
  Shining 456
  Velvet-mantled 295
Dunnock 386

E

Eagle, Crowned 252, 429
  Fish- [see Fish-Eagle]
  Golden 504
Eagle-Owl, Spotted 167, 370

F

Falcon, Lanner 76
  Red-necked 60, 76, 91, 118, 511, 531
Finch, Oriole 2
  Trumpeter 63, 66
Fish-Eagle, African 434
Fishing-Owl, Pel's 453
Flycatcher, Dusky 396
  Rufous 455
  Spotted 511
  Tit- [see Tit-Flycatcher]
Flycatcher-Thrush, Finsch's *400*, 453, 455, 456, **457**
  Rufous *400*, 453, **455**, 457

G

Goldfinch, European 68
Goshawk, African 457, 458
  Gabar 36
Greenbul, Ansorge's 281, 288, 290, **291**, 293, *305*
  Baumann's *305*, **330**
  Cabanis's *320*, 323, 324, 328, 330, 332, **334**, 337, 343, 452, 459, 461
  Cameroon Montane **281**, 283, *304*, 337
  Cameroon Olive 282, *305*, **337**
  Cameroon Sombre 292, 332
  Eastern Bearded *321*, **358**, 360
  Falkenstein's 313
  Fischer's 312, *320*, 326, **333**, 334
  Golden **301**, 310, 340, *352*
  Grey-headed 301, *320*, 337, **346**
  Grey-olive *320*, 327, **328**
  Honeyguide **302**, 306, *352*
  Icterine 301, **338**, 340, 342, 343, *352*
  Joyful 301, **309**, 311, *352*
  Liberian **341**, *352*
  Little **287**, 290, 293, 296, *305*, 306, 330, 332, 335, 355
  Little Grey 288, **289**, 291, 293, *305*
  Mango 317
  Mountain 280, **282**, 286, *304*, 337, 346, 450
  Pale Olive *320*, **329**
  Plain 288, 290, **292**, *305*
  Prigogine's **310**, *352*
  Red-tailed *321*, 343, 357, 358, **359**, 362, 363
  Sassi's Olive *305*, **332**
  Shelley's **280**, 282, 283, 286, *304*
  Simple **314**, 317, *321*
  Sjöstedt's Honeyguide 303, **306**, *352*
  Slender-billed **294**, 301, *304*, 332
  Smaller Yellow-streaked 347
  Sombre **298**, *305*, 312, 327
  Spotted **307**, 318, 341, *352*
  Stripe-cheeked 280, 283, **285**, *304*
  Tiny *320*, 345, **347**
  Toro Olive *305*, **331**
  Western Bearded *321*, 342, **357**, 358, 360, 363
  White-bearded *321*, 360, **362**
  White-tailed 318
  White-throated *320*, 330, **342**
  Xavier's 301, 338, **340**, 343, 349, *352*
  Yellow-bearded *321*, 342, 357, 360, **362**
  Yellow-bellied 300, **311**, 314, 315, 327, 330, *352*
  Yellow-necked **313**, *321*
  Yellow-streaked 312, *320*, 336, **343**, 346, 348
  Yellow-whiskered 288, 295, *305*, 307
  Zanzibar Sombre 298
Greenfinch, European 92, 157
Greenshank, Common 426, 529

Ground-Thrush, Orange 459, 461, 475
  Spotted 434
Guineafowl, Helmeted 166

H

Harrier, African Marsh 204, 377
  Pallid 529
Heron, Black-headed 167
Honeyguide, Greater 547, 549
  Lesser 370
  Zenker's 295
Hoopoe-Lark *33*, **61**, 102
  Lesser *33*, 63, **64**
Hypocolius, Grey *257*, 377, **379**

I

Illadopsis, Pale-breasted 459, 461
Irania *368*, **413**

K

Kestrel, Common 180
Kite, Black 529

L

Lapwing, African Wattled 526, 529
  Crowned 526
Lark, Angola *27*, *96*
  Archer's **44**, *48*
  Ash's **26**, *48*
  Bar-tailed *33*, **71**, 74, 89, 124
  Bimaculated *33*, 66, 68, **69**
  Botha's 84, **85**, 86, *97*
  Bush- [see Bush-Lark]
  Calandra *33*, 66, **67**, 69, 92
  Clapper 29, **30**, *96*
  Collared *27*, **34**, *48*
  Crested *33*, *49*, *94*, **100**, 103, 108
  Degodi 38, *48*
  Desert *33*, *49*, **73**, 77, 89
  Dune 52, **55**, *97*
  Dunn's *20*, *33*, 37, **89**, 115
  Dupont's *33*, **91**, 104, 108
  Dusky **57**, *97*
  Fawn-coloured *21*, **35**, 38, 41, *48*, *96*
  Flappet 28, 31, **37**, *48*, 95, *96*
  Friedmann's **21**, *48*
  Gillett's 37, 38, *48*, **83**
  Gray's **75**, *97*
  Greater Short-toed *33*, 68, 70, **77**, 81, 101
  Hoopoe- [see Hoopoe-Lark]
  Horned *33*, **121**, 123
  Karoo **51**, 53, 55, *97*
  Kordofan *20*, *33*, 37, **89**
  Large-billed *97*, **98**
  Lesser Short-toed *33*, 70, 72, 77, **80**, 101, 104, 108
  Long-billed **46**, 50, *96*, *99*
  Masked *49*, **88**
  Melodious 17, **19**, *96*
  Monotonous 17, **19**, *96*
  Obbia *49*, **87**
  Pink-billed **83**, 85, 86, 90, *97*
  Pink-breasted *27*, 38, **39**, *48*
  Red 52, **54**, *97*
  Red-capped *49*, **78**, *97*
  Rudd's **43**, *96*
  Rufous-naped **22**, 25, 26, 27, 28, *48*, 50, *96*, 101
  Rufous-rumped *48*, **56**
  Sabota **40**, *96*
  Sclater's 84, **86**, 90, *97*
  Shore 121
  Short-clawed **50**, *96*
  Short-crested 103
  Short-tailed *49*, **94**
  Sidamo **45**, *48*
  Somali 25, 27, *48*
  Somali Short-toed **22**, *49*, **82**
  Sparrow- [see Sparrow-Lark]
  Spike-heeled 43, 47, **59**, *97*
  Stark's 84, 86, **90**, *97*
  Sun *49*, **95**
  Temminck's Horned *33*, 104, **123**
  Thekla *33*, *49*, *94*, 101, **103**
  Thick-billed *33*, **65**, 68, 72
  William's **20**, 22, *48*
Leaf-love 319, *320*, **322**
  Yellow-throated 314, 315, **316**, *321*, 359
Linnet, Brown 68, 92
Longbill, Kretschmer's 312
Longclaw, Abyssinian *241*, **248**
  Cape *241*, 245, 250, **251**, 255
  Fülleborn's *241*, **249**, 255, 263
  Grimwood's *241*, 250, 261, **262**
  Pangani *241*, 247, 255, **258**
  Pink-throated 260
  Rosy-breasted *241*, 247, 250, 255, **260**, 262
  Sharpe's *177*, **246**, 249, 255
  Yellow-throated *241*, 247, 250, 252, **253**, 259, 526

M

Malimbe, Cassin's 290
Martin, African River **125**, 158, *161*
  African Rock 169, 555
  African Sand 137
  Banded *112*, 138, 142, **143**, 146, 157, 170, 187
  Brazza's **135**, *161*
  Brown-throated Sand *112*, 136, **137**, 140, 142, 146
  Common House 157, *161*, 162, 172, 184, **194**
  Common Sand *112*, 136, 138, **140**, 144, 146, 157, 162, 164, 168, 170, 172, 187
  Congo Sand *112*, 138, **140**, 142
  Crag *112*, 138, 142, 157, 169, **171**
  Mascarene 135, **136**, *161*
  Pale Crag 169
  Purple 127
  Rock *112*, 138, 157, 158, 165, **169**, 172, 187
Moor-Chat, Congo 465, 545, 549, 552
Mountain-Robin, Dappled *400*, 459, **460**

N

Nicator, Western 308, 341
Nightingale *368*, 397, 408, **409**, 412, 414, 429, 439, 469
  Thrush 408
Nightjar, Mountain 554

O

Oriole, Black-headed 436
  Black-tailed 444
  Black-winged 273
  Green-headed 434
  Western Black-headed 273
Owl, Eagle- [see Eagle-Owl]
Owl, Fishing- [see Fishing-Owl]

P

Palmchat, West Indian 377
Palm-Thrush, Collared *400*, **466**, 468, 469
  Red-tailed *400*, **466**, **467**
  Spotted *400*, **468**
Paradise Flycatcher, Rufous-vented 290
Partridge, Sand 517
Pigeon, African Green 429
  Speckled 529
Pipit, African Rock *177*, 242, **243**
  Buffy *177*, 219, 225, 226, **227**, 228, 243
  Bush **231**, *240*
  Bushveld 231
  European Rock 236, 238, **239**, *240*
  Golden *177*, **214**
  Long-billed 219, 221, 222, **223**, 226, 227, 230, *240*, 243, 500
  Long-legged *177*, **228**
  Malindi **229**, *240*
  Meadow 234, **235**, 237, 238, 239, *240*
  Mountain *177*, **220**, 243
  Plain-backed 65, *177*, 219, 224, **225**, 227, 228, 243
  Red-throated 234, **236**, 238, *240*
  Richard's *177*, **217**, 221, 222, 224, 226, 229, 234, 243, 244
  Short-tailed **230**, 232, *240*
  Sokoke **233**, *240*
  Striped *240*, **242**
  Tawny *177*, 219, **221**, 238, 531
  Tree 225, **233**, 236, 237, *240*, 242
  Water 236, **238**, 239, *240*
  Yellow-breasted *177*, **244**
Pitta, African 10, **11**, *32*
  Green-breasted **9**, *32*
Puffback, Black-headed 3
  Pink-footed 347

R

Redstart, Black 417, **489**, 491, 538
  Common 417, 490, **491**, 493, 504, 538
  Moussier's 417, 491, **493**
River Martin [see Martin]
Robin, Black Bush 486
  European *368*, *401*, **406**, 462, 481, 544
  Forest *353*, **392**, *401*, 415
  Mountain- [see Mountain-Robin]
  Persian 413
  Rufous Bush 480

Scrub- [see Scrub-Robin]
Swynnerton's *353*, 389, **391**
White-starred *353*, **388**, 391, 393, 404, 405, 422, 452
Robin-Chat, Archer's *369*, **420**
Blue-shouldered *369*, **428**, 431
Cape *369*, 391, *401*, 420, **423**, 426, 430, 475
Chorister *369*, *401*, 433, **435**
Grey-winged *353*, 395, *401*, **427**
Mountain *369*, 396, *401*, **418**
Olive-flanked *369*, 405, **421**, 451
Red-capped *369*, *401*, 402, 427, 429, **433**, 436, 439
Rüppell's *369*, *401*, **429**, 431
Snowy-crowned *400*, *401*, **437**, 439
White-bellied *369*, 398, **415**, 419, 420
White-browed *369*, *401*, 427, 429, 430, **431**, 434, 436
White-crowned *400*, 438, **439**
White-headed *400*, **437**
White-throated *400*, **425**
Rock-Chat, Black-tailed 541
Brown-tailed *464*, 538, **539**, 541
Sombre *464*, **540**
Rockfowl, Grey-necked 159
Rockjumper, Damara 488
Rock-Thrush, Blue 503
Mountain 511, 513
Ruff 192

S

Sandpiper, Common 213
Green 529
Wood 529
Saw-wing, Black 129, **130**, 133, 134, 136, 139, 159, *160*, 173, 187
Fanti *129*, 131, 134, *160*, 162
Mountain 130, 131, **133**, 159, *160*
Square-tailed **128**, 159, *160*
White-headed 131, **132**, *160*
Scrub-Robin, Black *416*, **486**
Brown 391, *416*, 473, **474**
Brown-backed *416*, 471, **476**, 481
Eastern Bearded 402, *416*, 451, 472, **473**, 475, 481
Forest 403, *416*, **470**, 477
Kalahari *416*, 479, **483**, 485
Karoo *416*, 483, **484**
Miombo Bearded *416*, **471**, 473
Rufous *416*, 479, **480**, 486
White-browed x, 391, *416*, 469, 471, 472, **477**, 481, 483
White-winged 479
Shrike, Fiscal 167, 500, 507, 513
Great Grey 379
Woodchat 104
Skylark, Eurasian 16, *33*, 68, 70, 81, 92, 101, 104, 106, **107**, 385
Sparrow, Cape 167
House 120, 167
Rufous 252
Sparrowhawk, Red-thighed 295
Sparrow-Lark, Black-crowned *33*, 65, 72, **115**, 119
Black-eared *97*, **109**, 111, 117
Chestnut-backed *49*, *97*, 110, **111**, 115, 117, 119
Chestnut-headed *49*, 114, **118**, 119

Fischer's *49*, 118, **119**
Grey-backed *97*, **116**
Spinetail, Bat-like 136
Spot-throat *400*, 447, **458**, 461
Sprosser *368*, **408**, 410, 468
Starling, Chestnut-bellied 546
European 127, 294, 378
Pied 139, 144, 507
Waller's Chestnut-winged 302, 347
Stonechat, Common 398, *417*, **494**, 498, 499, 511
Sunbird, Lesser Double-collared 389
Little Green 295
Scarlet-chested 296
Swallow, Angola 113, 146, 187, **188**, 191
Bank 140
Barn 68, *113*, 139, 141, 144, 157, 162, 168, 185, 187, 188, 189, **190**, 194, 513
Black and Rufous *113*, **175**
Blue 155, *160*, **173**, 178, 187
Dusky Cliff 158
Ethiopian *113*, 179, 183, **184**, 191
Forest 129, 138, **158**, *160*
Greater Striped 150, **154**, *161*, 175, 548, 555
Grey-rumped *112*, 140, **145**, 162, 164, 179, 184
Lesser Striped 135, 136, 146, 150, **152**, 155, *161*, 162, 164, 179, 186, 196, 555
Mosque *113*, 149, **150**, 157
Pearl-breasted *161*, 179, **183**, 187, 195
Pied-winged 129, *161*, **181**
Preuss's Cliff 146, **159**, *161*, 164, 195
Red Sea Cliff *161*, **168**
Red-breasted *113*, **149**, 151, 153, 155, 157, 182
Red-chested *113*, 185, 187, 188, **189**, 191
Red-rumped 95, *113*, 146, 149, 151, **156**
Red-throated Cliff 146, *161*, **163**, 165, 169, 178, 195
South African Cliff *161*, 164, **165**, 168, 170
White-tailed *161*, **182**
White-throated *113*, 175, 179, 184, **186**
White-throated Blue 129, *160*, **180**
Wire-tailed *113*, 153, **178**, 185
Swift, African Palm 136, 179
Little 164, 167, 171, 186
Mottled 162
White-rumped 154, 155, 164, 167, 186

T

Thrush, African 457
Ant- [see Ant-Thrush]
Flycatcher- [see Flycatcher-Thrush]
Ground- [see Ground-Thrush]
Groundscraper 58, 526
Kurrichane 426
Olive 367
Palm- [see Palm-Thrush]

Rock- [see Rock-Thrush]
Song 101, 272
Tinkerbird, Lemon-rumped 444
Tit, Black 551
Great 104
Tit-Babbler 485
Tit-Flycatcher, Grey 456
Tit-Flycatcher, Grey-throated 456
Trogon, Narina's 444
Turnstone, Ruddy 101

W

Wagtail, African Pied *176*, 204, 211, **212**
Ashy-headed 199
Black-headed 199
Blue-headed 199
British Yellow 199
Cape *176*, 200, **203**, 208, 211, 212, 252
Citrine 200, **202**
Eastern Yellow-headed 199
Egyptian Yellow 199
Grey *176*, 200, 204, **205**, 208, 210, 382
Grey-headed 199
Long-tailed 207
Mountain *176*, 204, 206, **207**, 211, 212
Spanish Yellow 199
Sykes's 199
White 122, *176*, 200, 203, 204, **209**, 212, 227, 414
White-headed 199
Yellow 122, *176*, **198**, 202, 204, 207, 211, 222, 230, 237, 513
Warbler, Morning 466
Olivaceous 486
Prothonotary 433
Red-tailed Morning 467
Reed 439
Sardinian 104
Scrub 517
Spotted Morning 468
Willow 426
Wood 68
Waxbill, Black-cheeked 431
Waxwing, Bohemian 257, **378**, 379
Weaver, Dark-backed 332, 347
Forest 389
Sociable 538
Yellow-capped 295
Wheatear, Black *448*, 503, **504**
Black-eared *449*, 510, 513, **514**, 519, 520, 526, 528, 530
Capped *448*, 507, 511, 513, 519, **526**, 529, 533
Cyprus 513
Desert *449*, 504, 510, 514, 519, 521, 522, 523, **524**, 526, 529, 530
Eastern Pied 505, **517**
Finsch's *449*, 511, 513, 514, 519, **520**, 525
Hooded *448*, 503, 510, 513, **516**, 519
Hume's 521
Isabelline *449*, 510, 513, 515, 519, 523, 525, 526, 528, **529**
Mountain *448*, **506**, 535, 552

Mourning *448*, 503, 511, 513, 516, **518**, 522, 523, 525
Northern *449*, 504, 507, **509**, 513, 515, 519, 521, 525, 526, 528, 530
Pied *449*, 503, 510, **512**, 514, 519, 521, 525, 527, 530
Red-breasted *448*, 510, 526, **528**, 530
Red-rumped *449*, 516, 519, **521**, 523, 525

Red-tailed *449*, 516, 519, 522, **523**
Schalow's 513, 527
Somali *449*, **508**, 510
Spanish 514
Tristram's 521
White-crowned Black *448*, **502**, 505, 513, 516, 519
Whinchat *417*, 496, **497**, 513, 529
Wigeon, Eurasian 380

Wren 382
 Musician 355
 Winter *257*, **382**
Wood-Hoopoe, White-headed 3
Woodlark x, 16, *33*, 101, **105**, 108, 441
Woodpecker, Elliot's 347
 Green 68
 Tullberg's 345, 347

French Names

A

Accenteur alpin *257*, **384**
 moucher *257*, **386**
Agrobate du Kalahari *416*, **483**
 du Karoo *416*, **484**
 roux *416*, **480**
Akalat étoilé *353*, **388**
 de Gabela *353*, **403**
 de Gunning *353*, **401**, **402**
 à joues rousses *353*, **395**, *401*
 de Sharpe *353*, **398**, *401*
 de Swynnerton *353*, **391**
Alèthe à gorge rousse *368*, *401*, **443**
 à huppe rousse *368*, *401*, **442**
 du Mont Cholo *368*, *401*, **451**
 à poitrine blanche *368*, *401*, **447**
 à poitrine brune *368*, *401*, **445**
Alouette d'Angola 27, *96*
 d'Archer **44**, *48*
 d'Ash **26**, *48*
 bateleuse **30**, *96*
 à bec rose **83**, *97*
 bilophe *33*, **123**
 de Botha **85**, *97*
 bourdonnante **28**, *48*, *96*
 brune **57**, *97*
 calandra *33*, **67**
 calandrelle *33*, **77**
 cendrille *49*, **78**, *97*
 des champs *33*, **107**
 chanteuse **15**, *48*
 de Clot-Bey *33*, **65**
 à collier **34**, *48*
 de Degodi **38**, *48*
 à dos roux **52**, *97*
 de Dunn *33*, **89**
 éperonée **59**, *97*
 d'Erard **45**, *48*
 fauve **35**, *48*, *96*
 ferruginea **54**, *97*
 de Friedmann **21**, *48*
 du Gillett **37**, *48*
 hausse-col *33*, **121**
 du Karoo **51**, *97*
 du Kordofan **20**, *33*
 à long bec **46**, *96*
 lulu *33*, **105**
 masquée *49*, **88**
 melodieuse **19**, *96*
 monotone **17**, *96*
 monticole *33*, **69**
 à nuque rousse **22**, *48*, *96*
 d'Obbia *49*, **87**
 à ongles courts **50**, *96*

 pispolette *33*, **80**
 à poitrine rose **39**, *48*
 polyglotte **24**, *48*
 à queue blanche **18**, *48*
 à queue rousse *48*, **56**
 roussâtre *49*, **82**
 rousse *33*, **36**
 de Rudd **43**, *96*
 sabota **40**, *96*
 de Sclater **86**, *97*
 de Somalie **25**, *48*
 de Stark **90**, *97*
 de Williams **20**, *48*
Alouette-moineau à dos gris *97*, **116**
 de Fischer *49*, **119**
 à front blanc *33*, **115**
 à oreillons blancs *49*, *97*, **111**
 à oreillons noirs *97*, **109**
 d'Oustalet *49*, **118**
Ammomane élégante *33*, **71**
 de Gray **75**, *97*
 isabelline *33*, *49*, **73**

B

Bergeronnette du Cap *176*, **203**
 citrine **202**
 grise *176*, **209**
 à longue queue *176*, **207**
 pie *176*, **212**
 printanière *176*, **198**
 des ruisseaux *176*, **205**
Brève d'Angola **11**, *32*
 à poitrine verte **9**, *32*
Bulbul d'Ansorge **291**, *305*
 à barbe blanche *321*, **359**
 à barbe jaune *321*, **362**
 de Baumann *305*, **330**
 à bec grêle **294**, *304*
 brun *320*, **327**
 brunoir *257*, **372**
 bruyant **306**, *352*
 de Cabanis *320*, **334**
 du Cap *257*, **374**
 commun *257*, **365**
 concolore **281**, *304*
 crinon occidental *321*, **357**
 crinon oriental *321*, **358**
 curvirostre **292**, *305*
 doré **301**, *352*
 de Falkenstein **313**, *321*
 de Fischer *320*, **333**
 fourmilier *321*, **354**
 à gorge blanche *320*, **342**
 à gorge claire **316**, *321*

 à gorge grise **282**, *304*
 gracile **289**, *305*
 ictérin **338**, *352*
 ictérin tâcheté **341**, *352*
 importun **298**, *305*
 jaboteur *320*, **325**
 des jardins *257*, **364**
 joyeux **309**, *352*
 de Lorenz *305*, **332**
 des Masuku **280**, *304*
 minute *320*, **347**
 modeste **314**, *321*
 montagnard **285**, *304*
 moustac *321*, **348**
 à moustaches jaunes **295**, *305*
 olivâtre *305*, **337**
 à poitrine jaune **311**, *352*
 de Prigogine **310**, *352*
 à queue blanche **302**, *352*
 à queue rousse *320*, **322**
 à queue verte *321*, **350**
 des raphias **318**, *352*
 de Reichenow *321*, **362**
 à stries jaunes *320*, **343**
 tâcheté **307**, *352*
 du Toro *305*, **331**
 à ventre jaune *320*, **346**
 à ventre roux *320*, **329**
 verdâtre **287**, *305*
 vert-olive *320*, **328**
 de Xavier **340**, *352*

C

Cichladuse de Peters *400*, **466**
 à poitrine tachetée *400*, **468**
 à queue rousse *400*, **467**
Cincle plongeur *257*, **380**
Cochevis à gros bec *97*, **98**
 huppé *33*, *49*, **100**
 modeste *49*, **95**
 à queue courte *49*, **94**
 de Thékla *33*, *49*, **103**
Cossyphe à ailes bleues *369*, **428**
 d'Archer *369*, **420**
 à calotte rousse *369*, *401*, **433**
 du Cap *369*, *401*, **423**
 choriste *369*, *401*, **435**
 à flancs olives *369*, **421**
 à gorge blanche *400*, **425**
 grand à tête blanche *400*, **439**
 des grottes *400*, **440**
 d'Heuglin *369*, *401*, **431**
 d'Iringa *353*, **405**
 d'Isabelle *369*, *401*, **418**

petit à tête blanche *400*, *401*, **437**
de Rüppell *369*, *401*, **429**
à sourcils blancs *353*, *401*, **427**
à tête blanche d'Angola *400*, **437**
des Usambaras *353*, **404**
à ventre blanc *369*, **415**

E

Echenilleur à barbillons *256*, **271**
 bleu *256*, **277**
 à épaulettes rouges *256*, **264**
 à gorge blanche *256*, **275**
 de Grauer *256*, **277**
 gris *256*, **274**
 loriot *256*, **272**
 noir *256*, **266**
 de Petit *256*, **268**
 pourpré *256*, **269**
Eurylaime du Cap **7**, *32*
 à flancs roux **5**, *32*
 de Grauer **2**, *32*
 à tête grise **4**, *32*

G

Gorgebleue à miroir *368*, **412**
Grive de Finsch *400*, **457**
 fourmilière à queue blanche *400*, **454**
 fourmilière à queue rousse *400*, **453**
 fourmilière rousse *400*, **455**
 à gorge tachetée *400*, **458**
 tachetée *400*, **460**

H

Hirondelle à ailes tachetés *161*, **181**
 d'Angola *113*, **188**
 bleue *160*, **173**
 de Brazza **135**, *161*
 brune **133**, *160*
 de cheminée *113*, **190**
 à collier *112*, **143**
 à croupion gris *112*, **145**
 du désert **169**
 d'Ethiopie *113*, **184**
 fanti **129**, *160*
 de fenêtre *161*, **194**
 de forêt **158**, *160*
 à gorge blanche *113*, **186**
 à gorge fauve *161*, **163**
 à gorge perlée *161*, **183**
 à gorge rousse *113*, **189**
 hérissée **130**, *160*
 isabelline *112*, **169**
 à longs brins *113*, **178**
 des Mascareignes **136**, *161*
 de la Mer Rouge *161*, **168**
 des mosquées *113*, **150**
 noire *160*, **180**
 paludicole *112*, **137**
 de Preuss **159**, *161*

à queue blanche *161*, **182**
à queue courte **128**, *160*
de rivage *112*, **140**
de rivage du Congo *112*, **140**
de rivière **125**, *161*
de rochers *112*, **169**, **171**
rousse-et-noire *113*, **175**
rousseline *113*, **156**
striée **152**, *161*
sud-africaine *161*, **165**
à tête blanche **132**, *160*
à tête rousse **154**, *161*
à ventre roux *113*, **149**
Hypocolius gris *257*, **379**

I

Irania à gorge blanche *368*, **413**

J

Jaseur de Bohême *257*, **378**

M

Merle podobe *416*, **486**
 des rochers *416*, **461**
 rougegorge *353*, **396**, *401*
 rougegorge équatorial *353*, **397**, *401*

P

Pipit des arbres **233**, *240*
 cafre **231**, *240*
 doré *177*, **214**
 à dos roux *177*, **225**
 du Drakensberg *177*, **220**
 à gorge jaune *177*, **244**
 à gorge rousse **236**, *240*
 à long bec **223**, *240*
 à longes pattes *177*, **228**
 de Malindi **229**, *240*
 maritime **239**, *240*
 des prés **235**, *240*
 à queue courte **230**, *240*
 de Richard *177*, **217**
 des rochers *177*, **243**
 rousseline *177*, **221**
 de Sokoke **233**, *240*
 spioncelle **238**, *240*
 strié de Sundevall *240*, **242**
 du Vaal *177*, **227**

R

Robin-agrobate barbu brun *416*, **474**
 barbu du Miombo *416*, **471**
 barbu oriental *416*, **473**
 à dos brun *416*, **476**
 à dos roux *416*, **477**
 de Ghana *416*, **470**
Rossignol philomèle *368*, **409**

progné *368*, **408**
Rougegorge familier *368*, *401*, **406**
 de forêt *353*, **392**, *401*
Rougequeue à front blanc *417*, **491**
 noir *417*, **489**
Rubiette de Moussier *417*, **493**

S

Sentinelle d'Abyssinie *241*, **248**
 du Cap *241*, **251**
 dorée *241*, **258**
 de Fülleborn *241*, **249**
 à gorge jaune *241*, **253**
 à gorge rose *241*, **260**
 de Grimwood *241*, **262**
 de Sharpe *177*, **246**
Sirli du desert *33*, **61**
 du Dupont *33*, **91**
 de Witherby *33*, **64**

T

Traquet afroalpin *464*, **543**
 d'Arnott *465*, **552**
 bifascié *417*, **499**
 du Cap *448*, **526**
 à capuchon *448*, **516**
 du désert *449*, **524**
 deuil *448*, **518**
 de Finsch *449*, **520**
 du Herero *416*, **487**
 isabelle *449*, **529**
 du Karoo *464*, **534**
 montagnard *448*, **506**
 motteux *449*, **509**
 noir d'Abyssinie *465*, **550**
 noir à front blanc *465*, **551**
 oreillard *449*, **514**
 pâtre *417*, **494**
 pie *449*, **512**
 pie d'Orient **517**
 á poitrine rousse *448*, **528**
 à queue rousse *449*, **523**
 rieur *448*, **504**
 de roche d'Abyssinie *465*, **556**
 de roche brun *464*, **533**
 de roche pâle *464*, **535**
 de roche à queue brune *464*, **539**
 de roche à queue noire *464*, **541**
 de roche à queue rousse *464*, **537**
 de roche sombre *464*, **540**
 de roche à ventre roux *465*, **553**
 somali *449*, **508**
 tarier *417*, **497**
 à tête blanche *448*, **502**
 à tête grise *449*, **521**
Traquet-fourmilier brun du nord *465*, **545**
 brun du sud *465*, **547**
 du Congo *465*, **545**
 noir *465*, **549**
Troglodyte mignon *257*, **382**